W9-AEN-530

The History of Mathematics

The History of Mathematics
A Brief Course

Second Edition

Roger Cooke
University of Vermont

A JOHN WILEY & SONS, INC., PUBLICATION

Published by John Wiley & Sons, Inc., Hoboken, New Jersey.
Published simultaneously in Canada.

Library of Congress Cataloging-in-Publication Data:

Cooke, Roger, 1942–
The history of mathematics: a brief course/Roger Cooke – 2nd ed.
p.cm.
Includes bibliographical references and indexes.
ISBN 0-471-44459-6 (cloth: acid-free paper)
1. Mathematics–History. I. Title
QA 21.C649 2005
510′.9–dc22

2004042299

Printed in the United States of America

10 9 8 7 6 5 4 3 2

Contents

Preface

This second edition of *The History of Mathematics: A Brief Course* must begin with a few words of explanation to all users of the first edition. The present volume constitutes such an extensive rewriting of the original that it amounts to a considerable stretch in the meaning of the phrase *second edition.* Although parts of the first edition have been retained, I have completely changed the order of presentation of the material. A comparison of the two tables of contents will reveal the difference at a glance: In the first edition each chapter was devoted to a single culture or period within a single culture and subdivided by mathematical topics. In this second edition, after a general survey of mathematics and mathematical practice in Part 1, the primary division is by subject matter: numbers, geometry, algebra, analysis, mathematical inference.

For reasons that mathematics can illustrate very well, writing the history of mathematics is a nearly impossible task. To get a proper orientation for any particular event in mathematical history, it is necessary to take account of three independent "coordinates": the time, the mathematical subject, and the culture. To thread a narrative that is to be read linearly through this three-dimensional array of events is like drawing one of Peano's space-filling curves. Some points on the curve are infinitely distant from one another, and the curve must pass through some points many times. From the point of view of a reader whose time is valuable, these features constitute a glaring defect. The problem is an old one, well expressed eighty years ago by Felix Klein, in Chapter 6 of his *Lectures on the Development of Mathematics in the Nineteenth Century*:

> I have now mentioned a large number of more or less famous names, all closely connected with Riemann. They can become more than a mere list only if we look into the literature associated with the names, or rather, with those who bear the names. One must learn how to grasp the main lines of the many connections in our science out of the enormous available mass of printed matter without getting lost in the time-consuming discussion of every detail, but also without falling into superficiality and dilettantism.

Klein writes as if it were possible to achieve this laudable goal, but then his book was by intention only a collection of essays, not a complete history. Even so, he used more pages to tell the story of one century of European mathematics than a modern writer has available for the history of all of mathematics. For a writer who hates to leave any threads dangling the necessary sacrifices are very painful. My basic principle remains the same as in the first edition: not to give a mere list of names and results described in general terms, but to show the reader what important results were achieved and in what context. Even if unlimited pages were available, time is an important consideration for authors as well as readers. To switch metaphors, there were so many times during the writing when tempting digressions arose which I could not resist pursuing, that I suspected that I might be traversing the boundary of a fractal snowflake or creating the real-life example of Zeno's dichotomy. Corrections and supplementary material relating to this book can be found at my website at the University of Vermont. The url is:

`http://www.cem.uvm.edu/~cooke/history/seconded.html`

Fortunately, significant mathematical events are discrete, not continuous, so that a better analogy for a history of mathematics comes from thermodynamics. If the state of mathematics at any given time is a system, its atoms are mathematical problems and propositions, grouped into molecules of theory. As they evolve, these molecules sometimes collide and react chemically, as happened with geometry and algebra in the seventeenth century. The resulting development of the mathematical system resembles a Brownian motion; and while it is not trivial to describe a Brownian motion in detail, it is easier than drawing a space-filling curve.

Now let me speak more literally about what I have tried to do in the present book. As mentioned above, Part 1 is devoted to a broad survey of the world of mathematics. Each of the six subsequent parts, except Part 3, where the color plates are housed, concentrates on a particular aspect of mathematics (arithmetic, geometry, algebra, analysis, and mathematical inference) and discusses its development in different cultures over time. I had two reasons for reorganizing the material in this way.

First, I am convinced that students will remember better what they learn if they can focus on a single area of mathematics, comparing what was done in this area by different cultures, rather than studying the arithmetic, geometry, and algebra of each culture by turns. Second, although reviewers were for the most part kind, I was dissatisfied with the first edition, feeling that the organization of the book along cultural lines had caused me to omit many good topics, especially biographical material, and sources that really ought to have been included. The present edition aims to correct these omissions, along with a number of mistakes that I have noticed or others have pointed out. I hope that the new arrangement of material will make it possible to pursue the development of a single area of mathematics to whatever level the instructor wishes, then turn to another area and do the same. A one-semester course in mostly elementary mathematics from many cultures could be constructed from Chapters 1–7, 9–11, and 13–14. After that, one could use any remaining time to help the students write term papers (which I highly recommend) or go on to read other chapters in the book. I would also point out that, except for Chapters 8–12, and 15–19, the chapters, and even the sections within the chapters, can be read independently of one another. For a segment on traditional Chinese mathematics, for example, students could be assigned Section 3 of Chapter 2, Subsection 3.4 of Chapter 5, Section 2 of Chapter 6, Section 3 of Chapter 7, Section 3 of Chapter 9, Section 4 of Chapter 13, and Section 2 of Chapter 14.

Because of limitations of time and space, the present book will show the reader only a few of the major moments in the history of mathematics, omitting many talented mathematicians and important results. This restriction to the important moments makes it impossible to do full justice to what Grattan-Guinness has stated as the question the historian should answer: What happened in the past? We are reconstructing an evolutionary process, but the "fossil record" presented in any general history of mathematics will have many missing links. Unavoidably, history gets distorted in this process. New results appear more innovative than they actually are. To take just one example (not discussed elsewhere in this book), it was a very clever idea of Hermann Weyl to trivialize the proof of Kronecker's theorem that the fractional parts of the multiples of an irrational number are uniformly distributed in the unit interval; Weyl made this result a theorem about discrete and continuous averages of integrable periodic functions. One would expect that in an evolutionary process, there might be an intermediate step—someone who realized that these fractional parts are dense but not necessarily that they are uniformly distributed. And indeed there was: Nicole d'Oresme, 500 years before Kronecker. There are hundreds of results in mathematics with names on them, in many cases incorrectly attributed, and in many more cited in a much more polished form than the discoverer ever imagined. History ought to correct this misimpression, but a general history has only a limited ability to do so.

The other question mentioned by Grattan-Guinness—How did things come to be the way they are?—is often held up in history books as the main justification for requiring students to study political and social history.[1] That job is somewhat easier to do in a general textbook, and I hope the reader will be pleased to learn how some of the current parts of the curriculum arose.

I would like to note here three small technical points about the second edition.

Citations. In the first edition I placed a set of endnotes in each chapter telling the sources from which I had derived the material of that chapter. In the present edition I have adopted the more scholarly practice of including a bibliography organized by author and date. In the text itself, I include citations at the points where they are used. Thus, the first edition of this book would be cited as (Cooke, *1997*). Although I dislike the interruption of the narrative that this practice entails, I do find it convenient when reading the works of others to be able to note the source of a topic that I think merits further study without having to search for the citation. On balance, I think the advantage of citing a source on the spot outweighs the disadvantage of having to block out parenthetical material in order to read the narrative.

Translations. Unless another source is cited, all translations from foreign languages are my own. The reader may find smoother translations in most cases. To bring out significant concepts, especially in quotations from ancient Greek, I have made translations that are more literal than the standard ones. Since I don't know Sanskrit, Arabic, or Chinese, the translations from those languages are not mine; the source should be clear from the surrounding text.

Cover. Wiley has done me the great favor of producing a cover design in four colors rather than the usual two. That consideration made it possible to use a picture that I took at a quilt exposition at Norwich University (Northfield, Vermont) in 2003. The design bears the title "A Number Called Phi," and its creator, Mary Knapp of Watertown, New York, incorporated many interesting mathematical connections through the geometric and floral shapes it contains. I am grateful for her permission to use it as the cover of this second edition.

[1] In a lecture at the University of Vermont in September 2003 Grattan-Guinness gave the name *heritage* to the attempt to answer this question. Heritage is a perfectly respectable topic to write on, but the distinction between history and heritage is worth keeping in mind. See his article on this distinction (Grattan-Guinness, *2004*).

Acknowledgements. I am grateful to the editors at Wiley, Steve Quigley and especially Susanne Steitz, for keeping in touch throughout the long period of preparation for this book. I would also like to thank the copy editor, Barbara Zeiders, who made so many improvements to the text that I could not begin to list them all. I also ask Barbara to forgive my obstinacy on certain issues involving commas, my stubborn conviction that *independently of* is correct usage, and my constitutional inability to make all possessives end in *'s*. I cannot bring myself to write *Archimedes's* or *Descartes's*; and if we are going to allow some exceptions for words ending in a sibilant, as we must, I prefer—in defiance of the *Chicago Manual of Style*—to use an unadorned apostrophe for the possessive of *all* words ending in *s*, *z*, or *x*.

The diagram of Florence Nightingale's statistics on the Crimean War (Plate 5) is in the public domain; I wish to thank *Cabinet* magazine for providing the electronic file for this plate.

Many of the literature references in the chapters that follow were given to me by the wonderful group of mathematicians and historians on the Historia Mathematica e-mail list. It seemed that, no matter how obscure the topic on which I needed information, there was someone on the list who knew something about it. To Julio Gonzalez Cabillon, who maintains the list as a service to the community, I am deeply grateful.

Roger Cooke

January 2005

Part 1

The World of Mathematics and the Mathematics of the World

This first part of our history is concerned with the "front end" of mathematics (to use an image from computer algebra)—its relation to the physical world and human society. It contains some general considerations about mathematics, what it consists of, how it may have arisen, and how it has developed in various cultures around the world. Because of the large number of cultures that exist, a considerable paring down of the available material is necessary. We are forced to choose a few sample cultures to represent the whole, and we choose those that have the best-recorded mathematical history. The general topics studied in this part involve philosophical and social questions, which are themselves specialized subjects of study, to which a large amount of scholarly literature has been devoted. Our approach here is the naive commonsense approach of an author who is not a specialist in either philosophy or sociology. Since present-day governments have to formulate *policies* relating to mathematics and science, it is important that such questions not be left to specialists. The rest of us, as citizens of a republic, should read as much as time permits of what the specialists have to say and make up our own minds when it comes time to judge the effects of a policy.

This section consists of four chapters. In Chapter 1 we consider the nature and prehistory of mathematics. In this area we are dependent on archaeologists and anthropologists for the comparatively small amount of historical information available. We ask such questions as the following: "What is the subject matter of mathematics?" "Is new mathematics created to solve practical problems, or is it an expression of free human imagination, or some of each?" "How are mathematical concepts related to the physical world?"

Chapter 2 begins a broad survey of mathematics around the world. This chapter is subdivided according to a selection of cultures in which mathematics has arisen as an indigenous creation, in which borrowings from other cultures do not play a prominent role. For each culture we give a summary of the development of mathematics in that culture, naming the most prominent mathematicians and their works. Besides introducing the major works and their authors, an important goal of this chapter is to explore the question, "Why were these works written?" We quote the authors themselves as often as possible to bring out their motives. Chapters 2 and 3 are intended as background for the topic-based presentation that follows beginning with Chapter 5.

In Chapter 3 we continue the survey with a discussion of mathematical cultures that began on the basis of knowledge and techniques that had been created elsewhere. The contributions made by these cultures are found in the extensions, modifications, and innovations—some very ingenious—added to the inherited materials. In dividing the material over two chapters we run the risk of seeming to minimize the creations of these later cultures. Creativity is involved in mathematical innovations at every stage, from earliest to latest. The reason for having two chapters instead of one is simply that there is too much material for one chapter.

Chapter 4 is devoted to the special topic of women mathematicians. Although the *subject* of mathematics is gender neutral in the sense that no one could determine the gender of the author of a mathematical paper from an examination of the mathematical arguments given, the *profession* of mathematics has not been and is not yet gender neutral. There are obvious institutional and cultural explanations for this fact; but when an area of human endeavor has been polarized by gender, as mathematics has been, that feature is an important part of its history and deserves special attention.

CHAPTER 1

The Origin and Prehistory of Mathematics

In this chapter we have two purposes: first, to consider what mathematics is, and second, to examine some examples of *protomathematics*, the kinds of mathematical thinking that people naturally engage in while going about the practical business of daily life. This agenda assumes that there is a mode of thought called *mathematics* that is intrinsic to human nature and common to different cultures. The simplest assumption is that counting and common shapes such as squares and circles have the same meaning to everyone. To fit our subject into the space of a book of moderate length, we partition mathematical modes of thought into four categories:

Number. The concept of number is almost always the first thing that comes to mind when mathematics is mentioned. From the simplest finger counting by pre-school children to the recent sophisticated proof of Fermat's last theorem (a theorem at last!), numbers are a fundamental component of the world of mathematics.

Space. It can be argued that space is not so much a "thing" as a convenient way of organizing physical objects in the mind. Awareness of spatial relations appears to be innate in human beings and animals, which must have an instinctive understanding of space and time in order to move purposefully. When people began to intellectualize this intuitive knowledge, one of the first efforts to organize it involved reducing geometry to arithmetic. Units of length, area, volume, weight, and time were chosen, and *measurement* of these continuous quantities was reduced to *counting* these imaginatively constructed units. In all practical contexts measurement becomes counting in exactly this way. But in pure thought there is a distinction between what is *infinitely divisible* and what is *atomic* (from the Greek word meaning *indivisible*). Over the 2500 years that have elapsed since the time of Pythagoras this collision between the discrete modes of thought expressed in arithmetic and the intuitive concept of continuity expressed in geometry has led to puzzles, and the solution of those puzzles has influenced the development of geometry and analysis.

Symbols. Although early mathematics was discussed in ordinary prose, sometimes accompanied by sketches, its usefulness in science and society increased greatly when symbols were introduced to mimic the mental operations performed in solving problems. Symbols for numbers are almost the only *ideograms* that exist in languages written with a phonetic alphabet. In contrast to ordinary words, for example, the symbol 8 stands for an idea that is the same to a person in Japan, who reads it as *hachi*, a person in Italy who reads it as *otto*, and a person in Russia, who reads it as *vosem'*. The introduction of symbols such as $+$ and $=$ to stand for the common operations and relations of mathematics has led to both the clarity that mathematics has for its initiates and the obscurity it suffers from in the eyes of the nonmathematical. Although it is primarily in studying algebra that we become aware of the use of symbolism, symbols are used in other areas, and algebra, considered as the study of processes inverse to those of arithmetic, was originally studied without symbols.

Symbol-making has been a habit of human beings for thousands of years. The wall paintings on caves in France and Spain are an early example, even though one might be inclined to think of them as pictures rather than symbols. It is difficult to draw a line between a painting such as the *Mona Lisa*, an animé representation of a human being, and the ideogram for a person used in languages whose written form is derived from Chinese. The last certainly *is* a symbol, the first two usually are not thought of that way. Phonetic alphabets, which establish a symbolic, visual representation of sounds, are another early example of symbol-making. A similar spectrum presents itself in the many ways in which human beings convey instructions to one another, the purest being a computer program. Very often, people who think they are not mathematical are quite good at reading abstractly written instructions such as music, blueprints, road maps, assembly instructions for furniture, and clothing patterns. All these symbolic representations exploit a basic human ability to make correspondences and understand analogies.

Inference. Mathematical reasoning was at first numerical or geometric, involving either counting something or "seeing" certain relations in geometric figures. The finer points of logical reasoning, rhetoric, and the like, belonged to other areas of study. In particular, philosophers had charge of such notions as cause, implication, necessity, chance, and probability. But with the Pythagoreans, verbal reasoning came to permeate geometry and arithmetic, supplementing the visual and numerical arguments. There was eventually a countercurrent, as mathematics began to influence logic and probability arguments, eventually producing specialized mathematical subjects: mathematical logic, set theory, probability, and statistics. Much of this development took place in the nineteenth century and is due to mathematicians with a strong interest and background in philosophy. Philosophers continue to speculate on the meaning of all of these subjects, but the parts of them that belong to mathematics are as solidly grounded (apart from their applications) as any other mathematics.

We shall now elaborate on the origin of each of these components. Since these origins are in some cases far in the past, our knowledge of them is indirect, uncertain, and incomplete. A more detailed study of all these areas begins in Part 2. The present chapter is confined to generalities and conjectures as to the state of mathematical knowledge preceding these records.

1. Numbers

Counting objects that are distinct but similar in appearance, such as coins, goats, and full moons, is a universal human activity that must have begun to occur as soon as people had language to express numbers. In fact, it is impossible to imagine that numbers could have arisen without this kind of counting. Several closely related threads can be distinguished in the fabric of elementary arithmetic. First, there is a distinction that we now make between cardinal and ordinal numbers. We think of cardinal numbers as applying to *sets* of things—the word *sets* is meant here in its ordinary sense, not the specialized meaning it has in mathematical set theory—and ordinal numbers as applying to the individual elements of a set by virtue of an ordering imposed on the set. Thus, the cardinal number of the set $\{a, b, c, d, e, f, g\}$ is 7, and e is the fifth element of this set by virtue of the standard alphabetical ordering. These two notions are not so independent as they may appear in this illustration, however. Except for very small sets, whose cardinality can be perceived immediately, the cardinality of a set is usually determined by *counting*, that is, arranging its elements linearly as first, second, third, and so on, even though it may be the corresponding cardinal numbers—one, two, three, and so on—that one says aloud when doing the counting.

A second thread closely intertwined with counting involves the elementary operations of arithmetic. The commonest actions that are carried out with any collection of things are taking objects out of it and putting new objects into it. These actions, as everyone recognizes immediately, correspond to the elementary operations of subtraction and addition. The etymology of these words shows their origin, *subtraction* having the meaning of *pulling out* (literally pulling up or under) and *addition* meaning *giving to*. All of the earliest mathematical documents use addition and subtraction without explanation. The more complicated operations of multiplication and division may have arisen from comparison of two collections of different sizes (counting the number of times that one collection fits into another, or copying a collection a fixed number of times and counting the result), or perhaps as a shortened way of peforming addition or subtraction. It is impossible to know much for certain, since most of the early documents also assume that multiplication of small integers is understood without explanation. A notable exception occurs in certain ancient Egyptian documents, where computations that would now be performed using multiplication or division are reduced to repeated doubling, and the details of the computation are shown.

1.1. Animals' use of numbers. Counting is so useful that it has been observed not only in very young children, but also in animals and birds. It is not clear just how high animals and birds can count, but they certainly have the ability to distinguish not merely patterns, but actual numbers. The counting abilities of birds were studied in a series of experiments conducted in the 1930s and 1940s by O. Koehler (1889–1974) at the University of Freiburg. Koehler (*1937*) kept the trainer isolated from the bird. In the final tests, after the birds had been trained, the birds were filmed automatically, with no human beings present. Koehler found that parrots and ravens could learn to compare the number of dots, up to 6, on the lid of a hopper with a "key" pattern in order to determine which hopper contained food. They could make the comparison no matter how the dots were arranged, thereby demonstrating an ability to take account of the *number* of dots rather than the *pattern*.

1.2. Young children's use of numbers. Preschool children also learn to count and use small numbers. The results of many studies have been summarized by Karen Fuson (*1988*). A few of the results from observation of children at play and at lessons were as follows:

1. A group of nine children from 21 to 45 months was found to have used the word *two* 158 times, the word *three* 47 times, the word *four* 18 times, and the word *five* 4 times.
2. The children seldom had to count "one–two" in order to use the word *two* correctly; for the word *three* counting was necessary about half the time; for the word *four* it was necessary most of the time; for higher numbers it was necessary all the time.

One can thus observe in children the capacity to recognize groups of two or three without performing any conscious numerical process. This observation suggests that these numbers are primitive, while larger numbers are a conscious creation. It also illustrates what was said above about the need for arranging a collection in some linear order so as to be able to find its cardinal number.

1.3. Archaeological evidence of counting. Very ancient animal bones containing notches have been found in Africa and Europe, suggesting that some sort of counting procedure was being carried on at a very early date, although what exactly was being counted remains unknown. One such bone, the radius bone of a wolf, was discovered at Veronice (Czech Republic) in 1937. This bone was marked with two series of notches, grouped by fives, the first series containing five groups and the second six. Its discoverer, Karel Absolon (1887–1960), believed the bone to be about 30,000 years old, although other archaeologists thought it considerably younger. The people who produced this bone were clearly a step above mere survival, since a human portrait carved in ivory was found in the same settlement, along with a variety of sophisticated tools. Because of the grouping by fives, it seems likely that this bone was being used to count something. Even if the groupings are meant to be purely decorative, they point to a use of numbers and counting for a practical or artistic purpose.

Another bone, named after the fishing village of Ishango on the shore of Lake Edward in Zaire where it was discovered in 1960 by the Belgian archaeologist Jean de Heinzelin de Braucourt (1920–1998), is believed to be between 8500 and 11,000 years old. The Ishango Bone, which is now in the Musée d'Histoire Naturelle in Brussels, contains three columns of notches. One column consists of four series of notches containing 11, 21, 19, and 9 notches. Another consists of four series containing 11, 13, 17, and 19 notches. The third consists of eight series containing 3, 6, 4, 8, 10, 5, 5, and 7 notches, with larger gaps between the second and third series and between the fourth and fifth series. These columns present us with a mystery. Why were they put there? What activity was being engaged in by the person who carved them? Conjectures range from abstract experimentation with numbers to keeping score in a game. The bone could have been merely decorative, or it could have been a decorated tool. Whatever its original use, it comes down to the present generation as a reminder that human beings were engaging in abstract thought and creating mathematics a very, very long time ago.

2. Continuous magnitudes

In addition to the ability to count, a second important human faculty is the ability to perceive spatial and temporal relations. These perceptions differ from the discrete objects that elicit counting behavior in that the objects involved are perceived as being divisible into arbitrarily small parts. Given any length, one can always imagine cutting it in half, for example, to get still smaller lengths. In contrast, a penny cut in half does not produce two coins each having a value of one-half cent. Just as human beings are endowed with the ability to reason numerically and understand the concept of equal distribution of money or getting the correct change with a purchase, it appears that we also have an innate ability to reason spatially, for example, to understand that two areas are equal even when they have different shapes, provided that they can be dissected into congruent pieces, or that

The Veronice wolf bone, from the *Illustrated London News*, October 2, 1937.

two vessels of different shape have the same volume if one each holds exactly enough water to fill the other.

One important feature of counting as opposed to measuring—arithmetic as opposed to geometry—is its exactitude. Two sets having the same number of members are numerically *exactly equal.* In contrast, one cannot assert that two sticks, for example, are *exactly* the same length. This difference arises in countless contexts important to human society. Two people may have exactly the same amount of money in the bank, and one can make such an assertion with complete confidence after examining the balance of each of them. But it is only within some limit of error that one could assert that two people are of the same height. The word *exact* would be inappropriate in this context. The notion of absolute equality in relation to continuous objects means *infinite precision* and can be expressed only through the concept of a *real number*, which took centuries to distill. That process is one important thread in the tapestry of mathematical history.

Very often, a spatial perception is purely geometrical or topological, involving similarity (having the same shape), connectivity (having holes or being solid), boundedness or infinitude, and the like. We can see the origins of these concepts in many aspects of everyday life that do not involve what one would call formal geometry. The perception of continuous magnitudes such as lengths, areas, volumes, weights, and time is different from the perception of multiple copies of a discrete object. The two kinds of perception work both independently and together to help a human being or animal cope with the physical world. Getting these two "draft horses" harnessed together as parts of a common subject called mathematics has led to a number of interesting problems to be solved.

2.1. Perception of shape by animals. Obviously, the ability to perceive shape is of value to an animal in determining what is or is not food, what is a predator, and so forth; and in fact the ability of animals to perceive space has been very well documented. One of the most fascinating examples is the ability of certain species of bees to communicate the direction and distance of sources of plant nectar by performing a dance inside the beehive. The pioneer in this work was Karl von

Frisch (1886–1982), and his work has been continued by James L. Gould and Carol Grant Gould (*1995*). The experiments of von Frisch left many interpretations open and were challenged by other specialists. The Goulds performed more delicately designed experiments which confirmed the bee language by deliberately misleading the bees about the food source. The bee will traverse a circle alternately clockwise and counterclockwise if the source is nearby. If it is farther away, the alternate traversals will spread out, resulting in a figure 8, and the dance will incorporate sounds and waggling. By moving food sources, the Goulds were able to determine the precision with which this communication takes place (about 25%). Still more intriguing is the fact that the direction of the food source is indicated by the direction of the axis of the figure 8, oriented relative to the sun if there is light and relative to the vertical if there is no light.

As another example, in his famous experiments on conditioned reflexes using dogs as subjects the Russian scientist Pavlov (1849–1936) taught dogs to distinguish ellipses of very small eccentricity from circles. He began by projecting a circle of light on the wall each time he fed the dog. Eventually the dog came to expect food (as shown by salivation) every time it saw the circle. When the dog was conditioned, Pavlov began to show the dog an ellipse in which one axis was twice as long as the other. The dog soon learned not to expect food when shown the ellipse. At this point the malicious scientist began making the ellipse less eccentric, and found, with fiendish precision, that when the axes were nearly equal (in a ratio of 8 : 9, to be exact) the poor dog had a nervous breakdown (Pavlov, *1928*, p. 122).

2.2. Children's concepts of space. The most famous work on the development of mathematical concepts in children is due to Jean Piaget (1896–1980) of the University of Geneva, who wrote many books on the subject, some of which have been translated into English. Piaget divided the development of the child's ability to perceive space into three periods: a first period (up to about 4 months of age) consisting of pure reflexes and culminating in the development of primary habits, a second period (up to about one year) beginning with the manipulation of objects and culminating in purposeful manipulation, and a third period in which the child conducts experiments and becomes able to comprehend new situations. He categorized the primitive spatial properties of objects as proximity, separation, order, enclosure, and continuity. These elements are present in greater or less degree in any spatial perception. In the baby they come together at the age of about 2 months to provide recognition of faces. The human brain seems to have some special "wiring" for recognizing faces.

The interesting thing about these concepts is that mathematicians recognize them as belonging to the subject of topology, an advanced branch of geometry that developed in the late nineteenth and early twentieth centuries. It is an interesting paradox that the human ability to perceive shape depends on synthesizing various topological concepts; this progression reverses the pedagogical and historical ordering between geometry and topology. Piaget pointed out that children can make topological distinctions (often by running their hands over models) before they can make geometric distinctions. Discussing the perceptions of a group of 3-to-5-year-olds, Piaget and Inhelder (*1967*) stated that the children had no trouble distinguishing between open and closed figures, surfaces with and without holes, intertwined rings and separate rings, and so forth, whereas the seemingly simpler relationships of geometry—distinguishing a square from an ellipse, for example—were not mastered until later.

2.3. Geometry in arts and crafts. Weaving and knitting are two excellent examples of activities in which the spatial and numerical aspects of the world are combined. Even the sophisticated idea of a rectangular coordinate system is implicit in the placing of different-colored threads at intervals when weaving a carpet or blanket so that a pattern appears in the finished result. One might even go so far as to say that curvilinear coordinates occur in the case of sweaters.

Not only do arts and crafts *involve* the kind of abstract and algorithmic thinking needed in mathematics, their themes have often been inspired by mathematical topics. We shall give several examples of this inspiration in different parts of this book. At this point, we note just one example, which the author happened to see at a display of quilts in 2003. The quilt, shown on the cover of this book, embodies several interesting properties of the *Golden Ratio* $\Phi = (1+\sqrt{5})/2$, which is the ratio of the diagonal of a pentagon to its side. This ratio is known to be involved in the way many trees and flowers grow, in the spiral shell of the chambered nautilus, and other places. The quilt, titled "A Number Called Phi," was made by Mary Knapp of Watertown, New York. Observe how the quilter has incorporated the spiral connection in the sequence of nested circles and the rotation of each successive inscribed pentagon, as well as the phyllotaxic connection suggested by the vine.

Marcia Ascher (*1991*) has assembled many examples of rather sophisticated mathematics inspired by arts and crafts. The Bushoong people of Zaire make part of their living by supplying embroidered cloth, articles of clothing, and works of art to others in the economy of the Kuba chiefdom. As a consequence of this work, perhaps as preparation for it, Bushoong children amuse themselves by tracing figures on the ground. The rule of the game is that a figure must be traced without repeating any strokes and without lifting the finger from the sand. In graph theory this problem is known as the *unicursal tracing problem.* It was analyzed by the Swiss mathematician Leonhard Euler (1707–1783) in the eighteenth century in connection with the famous Königsberg bridge problem. According to Ascher, in 1905 some Bushoong children challenged the ethnologist Emil Torday (1875–1931) to trace a complicated figure without lifting his finger from the sand. Torday did not know how to do this, but he did collect several examples of such figures. The Bushoong children seem to learn intuitively what Euler proved mathematically: A unicursal tracing of a connected graph is possible if there are at most two vertices where an odd number of edges meet. The Bushoong children become very adept at finding such a tracing, even for figures as complicated as that shown in Fig. 1.

3. Symbols

We tend to think of symbolism as arising in algebra, since that is the subject in which we first become aware of it as a concept. The thing itself, however, is implanted in our minds much earlier, when we learn to talk. Human languages, in which sounds correspond to concepts and the temporal order or inflection of those sounds maps some relation between the concepts they signify, exemplify the process of abstraction and analogy, essential elements in mathematical reasoning. Language is, all by itself, ample proof that the symbolic ability of human beings is highly developed. That symbolic ability lies at the heart of mathematics.

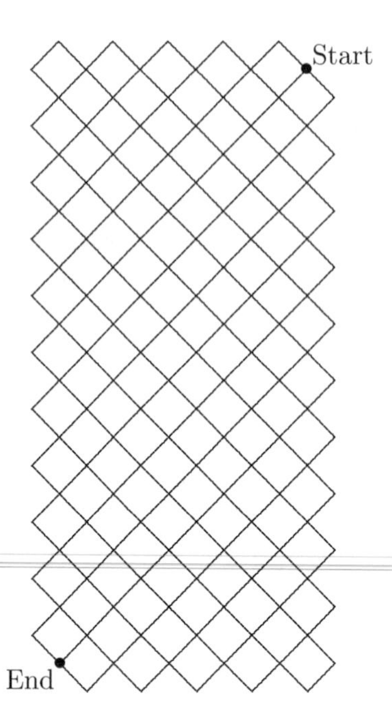

FIGURE 1. A graph for which a unicursal tracing is possible.

Once numbers have been represented symbolically, the next logical step would seem to be to introduce symbols for arithmetic operations or for combining the number symbols in other ways. However, this step may not be necessary for rapid computation, since mechanical devices such as counting rods, pebbles, counting boards, and the like can be used as analog computers. The operations performed using these methods can rise to a high level of sophistication without the need for any written computations. An example of the use of an automatic counting device is given by Ascher (*1997*) in a discussion of a system of divination used by the Malagasy of Madagascar, in which four piles of seeds are arranged in a column and the seeds removed from each pile two at a time until only one or two seeds remain. Each set of seeds in the resulting column can be interpreted as "odd" or "even." After this procedure is performed four times, the four columns and four rows that result are combined in different pairs using the ordinary rules for adding odds and evens to generate eight more columns of four numbers. The accuracy of the generation is checked by certain mathematical consequences of the method used. If the results are satisfactory, the 16 sets of four odds and evens are used as an oracle for making decisions and ascribing causes to such events as illnesses.

The Malagasy system of divination bears a resemblance to the procedures described in the Chinese classic *I Ching* (*Permutation Classic*). In the latter, a set of 50 yarrow sticks is used, the first stick being laid down to begin the ceremony. One stick is then placed between the ring and small fingers of the left hand to represent the human race. The remaining 48 sticks are then divided without counting into two piles, and one pile held in each hand. Those in the right hand are then discarded four at a time until four or fewer remain. These are then transferred to the left hand, and the same reduction is applied to the other pile, so that at the end, the left hand contains either five or nine sticks. After a sequence of such procedures, a final step begins with 32 or 36 or 40 sticks, and as a result the number of remaining sticks will be 24, 28, 32, or 36. This number is divided by four and the quotient determines the bottom row of the symbol to be used for divination. Six is called *lesser yang*, seven *greater ying*, eight *lesser ying*, and nine *greater yang*. The *ying* and *yang* are respectively female and male principles. The greater cases correspond to flux (tending to their opposites) and the lesser to stability. When this entire procedure has been carried out six times, the result is a stack of six symbols that can be interpreted according to the principles of divination. There are 64, that is, 2^6, different possible stackings of *ying* and *yang*, all discussed in the *I Ching*, and the duality between stability and flux makes for 4096 possible symbols. One must beware of attaching too much significance to numerical coincidences, but it is intriguing that both Malagasy and Chinese forms of divination are based on the number four.[1]

Divination seems to fulfill a nearly universal human desire to feel in control of the powerful forces that threaten human happiness and prosperity. It manifests itself in a variety of ways, as just shown by the examples of the Malagasy and the *I Ching*. We could also cite large parts of the Jewish *Kabbalah*, the mysticism of the Pythagoreans, and many others, down to the geometric logic of Ramon Lull (1232–1316), who was himself steeped in the *Kabbalah*. The variety of oracles that people have consulted for advice about the conduct of their lives—tarot cards, crystal balls, astrology, the entrails of animals and birds, palmistry, and the like—seems endless. For the purposes of this book, however, we shall be interested only in those aspects of divination that involve mathematics. Magic squares, for example, occur in both the *Kabbalah* and the *I Ching*. Although the author puts no stock whatsoever in the theories behind all this mysticism, it remains an important fact about human behavior over the centuries and deserves to be studied for that reason alone. But for now it is time to return to more prosaic matters.

Aids to computation, either tabular or mechanical, must be used to perform computations in some of the more cumbersome notational systems. Just imagine trying to multiply XLI by CCCIV! (However, Detlefsen and co-authors (*1975*) demonstrate that this task is not so difficult as it might seem.) Even to use the 28×19 table of dates of Easter discussed in Problem 6.26, the Slavic calculators had to introduce simplifications to accommodate the fact that dividing a four-digit number by a two-digit number was beyond the skill of many of the users of the table.

The earliest mathematical texts discuss arithmetical operations using everyday words that were probably emptied of their usual meaning. Students had to learn to generalize from a particular example to the abstract case, and many problems that refer to specific objects probably became archetypes for completely abstract reasoning, just as we use such expressions as "putting the cart before the horse" and "comparing apples and oranges" to refer to situations having no connection at all with horse-and-buggy travel or the harvesting of fruit. For example, problems of the

[1] Like all numbers, the number four is bound to occur in many contexts. One website devoted to spreading the lore found in the *I Ching* notes the coincidental fact that DNA code is written with four amino acids as its alphabet and rhapsodizes that "The sophistication of this method has not escaped modern interpretation, and the four-valued logic has been compared to the biochemistry of DNA amino acids. How a Neolithic shaman's divination technique presaged the basic logic of the human genome is one of the ageless mysteries."

type, "If 3 bananas cost 75 cents, how much do 7 bananas cost?" occur in the work of Brahmagupta from 1300 years ago. Brahmagupta named the three data numbers *argument* (3), *fruit* (75), and *requisition* (7). As another example, cuneiform tablets from Mesopotamia that are several thousand years old contain general problems that we would now solve using quadratic equations. These problems are stated as variants of the problem of finding the length and width of a rectangle whose area and perimeter are known. The mathematician and historian of mathematics B. L. van der Waerden (1903–1996) claimed that the words for *length* and *width* were being used in a completely abstract sense in these problems.

In algebra symbolism seems to have occurred for the first time in the work of the Greek mathematician Diophantus of Alexandria, who introduced the symbol ς for an unknown number. The Bakshali Manuscript, a document from India that may have been written within a century of the work of Diophantus, also introduces an abstract symbol for an unknown number. In modern algebra, beginning with the Muslim mathematicians more than a millennium ago, symbolism evolved gradually. Originally, the Arabic word for *thing* was used to represent the unknown in a problem. This word, and its Italian translation *cosa*, was eventually replaced by the familiar x most often used today. In this way an entire word was gradually pared down to a single letter that could be manipulated graphically.

4. Mathematical inference

Logic occurs throughout modern mathematics as one of its key elements. In the teaching of mathematics, however, the student generally learns all of arithmetic and the rules for manipulating algebraic expressions by rote. Any justification of these rules is purely experimental. Logic enters the curriculum, along with proof, in the study of Euclidean geometry. This sequence is not historical and may leave the impression that mathematics was an empirical science until the time of Euclid (ca. 300 BCE). Although one can imagine certain facts having been discovered by observation, such as the rule for comparing the area of a rectangle with the area of a square unit, there is good reason to believe that some facts were *deduced* from simpler considerations at a very early stage. The main reason for thinking so is that the conclusions reached by some ancient authors are not visually obvious.

4.1. Visual reasoning. As an example, it is immediately obvious that a diagonal divides a rectangle into two congruent triangles. If through any point on the diagonal we draw two lines parallel to the sides, these two lines will divide the rectangle into four rectangles. The diagonal divides two of these smaller rectangles into pairs of congruent triangles, just as it does the whole rectangle, thus yielding three pairs of congruent triangles, one large pair and two smaller pairs. It then follows (see Fig. 2) that the two remaining rectangles must have equal area, even though their shapes are different and *to the eye they do not appear to be equal.* For each of these rectangles is obtained by subtracting the two smaller triangles from the large triangle in which they are contained. When we find an ancient author mentioning that these two rectangles of different shape are equal, as if it were a well-known fact, we can be confident that this knowledge does not rest on an experimental or inductive foundation. Rather, it is the result of a combination of numerical and spatial reasoning.

Ancient authors often state *what* they know without saying *how* they know it. As the example just cited shows, we can be confident that the basis was not

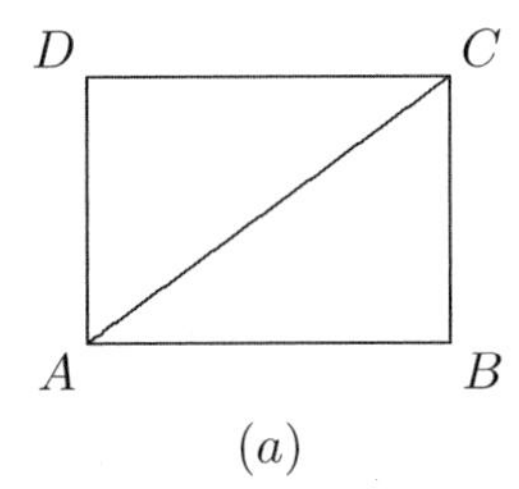

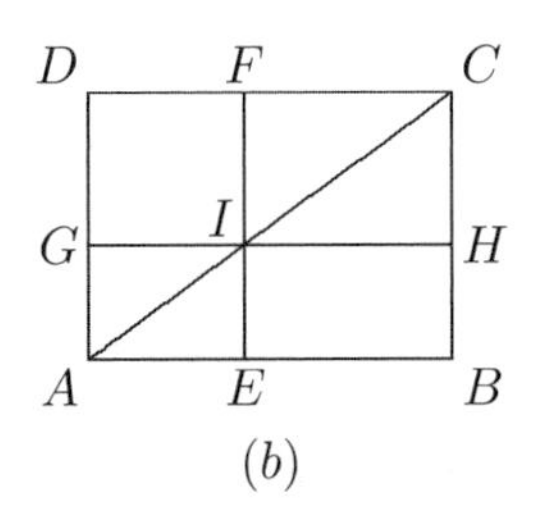

FIGURE 2. (*a*) The diagonal AC divides the rectangle $ABCD$ into congruent triangles ABC and CDA. (*b*) When the congruent pairs (AEI, IGA) and (IHC, CFI) are subtracted from the congruent pair (ABC, CDA), the remainders (rectangles $EBHI$ and $GIFD$) must be equal.

always induction or experiment. Perminov (*1997*) points out that solutions of complicated geometric problems which can be shown to be correct are stated without proof—but apparently with absolute confidence—by the writers of the very earliest mathematical documents, such as the Rhind Papyrus from Egypt and cuneiform tablets from Mesopotamia. The fact that an author presents not merely a solution but a sequence of steps leading to that solution, and the fact that this solution can now be reconstructed and verified by mathematical reasoning, justify the conclusion that the result was arrived at through mathematical deduction, even though the author does not write out the details.

4.2. Chance and probability. Logic is concerned with getting conclusions that are as reliable as the premises. From a behavioral point of view, the human tendency to make inferences based on logic is probably hardwired and expressed as the same mechanism by which habits are formed. This same mechanism probably accounts for the metaphysical notion of *cause*. If A implies B, one feels that in some sense A *causes* B to be true. The dogs in Pavlov's experiments, described above, were given *total* reinforcement as they learned geometry and came to make associations based on the constant conjunction of a given shape and a given reward or lack of reward. In the real world, however, we frequently encounter a weaker type of cause, where A is usually, but not always, followed by B. For example, lightning is always followed by thunder, but if the lightning is very distant, the thunder will not be heard. The analog of this weaker kind of cause in conditioning is *partial reinforcement.* A classical example is a famous experiment of Skinner (*1948*), who put hungry pigeons in a cage and attached a food hopper to the cage with an automatic timer to permit access to the food at regular intervals. The pigeons at first engaged in aimless activity when not being fed, but tended to repeat whatever activity they happened to be doing when the food arrived, as if they made an association between the activity and the arrival of food. Naturally, the more they repeated a given activity, the more likely that activity was to be reinforced by the arrival of food. Since they were always hungry, it was not long before they were engaged full-time in an activity that they apparently considered an infallible food producer. This activity varied from one bird to another. One pigeon thrust its head into an upper corner of the cage; another made long sweeping movements with its head; another tossed its head back; yet another made pecking motions toward the floor of the cage.

The difficulties that people, even mathematicians, have in understanding and applying probability can be seen in this example. For example, the human body has some capacity to heal itself. Like the automatic timer that eventually provided food to the pigeons, the human immune system often overcomes the disease. Yet sick people, like hungry pigeons, try various methods of alleviating their misery. The consequence is a wide variety of nostrums said to cure a cold or arthritis. One of the triumphs of modern mathematical statistics is the establishment of reliable systems of inference to replace the inferences that Skinner called "superstitious."

Modern logic has purged the concept of implication of all connection with the notion of cause. The statement "If Abraham Lincoln was the first President of the United States, then $2 + 2 = 4$" is considered a true implication, even though Lincoln was not the first President and in any case his being such would have no *causal* connection with the truth of the statement "$2 + 2 = 4$." In standard logic the statement "If A is true, then B is true" is equivalent to the statement "Either B is true, or A is false, or both." Absolute truth or falsehood is not available in relation to the observed world, however. As a result, science must deal with propositions of the form "If A is true, then B is *highly probable*." One cannot infer from this statement that "If B is false, then A is highly *im*probable." For example, an American citizen, taken at random, is probably not a U. S. Senator. It does not follow that if a person *is* a U.S. Senator, that person is probably not an American citizen.

Questions and problems

1.1. At what point do you find it necessary to count in order to say how large a collection is? Can you look at a word such as *tendentious* and see immediately how many letters it has? The American writer Henry Thoreau (1817–1863) was said to have the ability to pick up exactly one dozen pencils out of a pile. Try as an experiment to determine the largest number of pencils you can pick up out of a pile without counting. The point of this exercise is to see where direct perception needs to be replaced by counting.

1.2. In what practical contexts of everyday life are the fundamental operations of arithmetic—addition, subtraction, multiplication, and division—needed? Give at least two examples of the use of each. How do these operations apply to the problems for which the theory of proportion was invented?

1.3. What significance might there be in the fact that there are three columns of notches on the Ishango Bone? What might be the significance of the numbers of notches in the three series?

1.4. Is it possible that the Ishango Bone was used for divination? Can you think of a way in which it could be used for this purpose?

1.5. Is it significant that one of the yarrow sticks is isolated at the beginning of each step in the Chinese divination procedure described above? What difference does this step make in the outcome?

1.6. Measuring a continuous object involves finding its ratio to some standard unit. For example, when you measure out one-third of a cup of flour in a recipe, you are choosing a quantity of flour whose ratio to the standard cup is $1:3$. Suppose that you have a standard cup without calibrations, a second cup of unknown size, and a large bowl. How could you determine the volume of the second cup?

1.7. Units of time, such as a day, a month, and a year, have ratios. In fact you probably know that a year is about $365\frac{1}{4}$ days long. Imagine that you had never been taught that fact. How would you—how did people originally—determine how many days there are in a year?

1.8. Why is a calendar needed by an organized society? Would a very small society (consisting of, say, a few dozen families) require a calendar if it engaged mostly in hunting, fishing, and gathering vegetable food? What if the principal economic activity involved following a reindeer herd? What if it involved tending a herd of domestic animals? Finally, what if it involved planting and tending crops?

1.9. Describe three different ways of measuring time, based on different physical principles. Are all three ways equally applicable to all lengths of time?

1.10. In what sense is it possible to know the *exact* value of a number such as $\sqrt{2}$? Obviously, if a number is to be known only by its decimal expansion, nobody does know and nobody ever will know the exact value of this number. What immediate practical consequences, if any, does this fact have? Is there any other sense in which one could be said to know this number *exactly*? If there are no direct consequences of being ignorant of its exact value, is there any practical value in having the *concept* of an exact square root of 2? Why not simply replace it by a suitable approximation such as 1.41421? Consider also other "irrational" numbers, such as π, e, and $\Phi = (1+\sqrt{5})/2$. What is the value of having the *concept* of such numbers as opposed to approximate rational replacements for them?

1.11. Find a unicursal tracing of the graph shown in Fig. 1.

1.12. Does the development of personal knowledge of mathematics mirror the historical development of the subject? That is, do we learn mathematical concepts as individuals in the same order in which these concepts appeared historically?

1.13. Topology, which may be unfamiliar to you, studies (among other things) the mathematical properties of knots, which have been familiar to the human race at least as long as most of the subject matter of geometry. Why was such a familiar object not studied mathematically until the twentieth century?

1.14. One aspect of symbolism that has played a large role in human history is the mystical identification of things that exhibit analogous relations. The divination practiced by the Malagasy is one example, and there are hundreds of others: astrology, alchemy, numerology, tarot cards, palm reading, and the like, down to the many odd beliefs in the effects of different foods based on their color and shape. Even if we dismiss the validity of such divination (as the author does), is there any value for science in the development of these subjects?

1.15. What function does logic fulfill in mathematics? Is it needed to provide a psychological feeling of confidence in a mathematical rule or assertion? Consider, for example, any simple computer program that you may have written. What really gave you confidence that it worked? Was it your logical analysis of the operations involved, or was it empirical testing on an actual computer with a large variety of different input data?

1.16. Logic enters the mathematics curriculum in high-school geometry. The reason for introducing it at that stage is historical: Formal treatises with axioms, theorems, and proofs were a Greek innovation, and the Greeks were primarily geometers. There is no *logical* reason why logic is any more important in geometry than in algebra or arithmetic. Yet it seems that without the explicit statement of assumptions, the parallel postulate of Euclid (discussed in Chapter 10) would never have been questioned. Suppose things had happened that way. Does it follow that non-Euclidean geometry would never have been discovered? How important is non-Euclidean geometry, anyway? What other kinds of geometry do you know about? Is it necessary to be guided by axioms and postulates in order to discover or fully understand, say, the non-Euclidean geometry of a curved surface in Euclidean space? If it is not necessary, what is the value of an axiomatic development of such a geometry?

1.17. Perminov (*1997*, p. 183) presents the following example of tacit mathematical reasoning from an early cuneiform tablet. Given a right triangle ACB divided into a smaller triangle DEB and a trapezoid $ACED$ by the line DE parallel to the leg AC, such that EC has length 20, EB has length 30, and the trapezoid $ACED$ has area 320, what are the lengths AC and DE? (See Fig. 3.) The author of the tablet very confidently computes these lengths by the following sequence of operations: (1) $320 \div 20 = 16$; (2) $30 \cdot 2 = 60$; (3) $60 + 20 = 80$; (4) $320 \div 80 = 4$; (5) $16 + 4 = 20 = AC$; (6) $16 - 4 = 12 = DE$. As Perminov points out, to present this computation with any confidence, you would have to know exactly what you are doing. What *was* this anonymous author doing?

To find out, fill in the reasoning in the following sketch. The author's first computation shows that a rectangle of height 20 and base 16 would have exactly the same area as the trapezoid. Hence if we draw the vertical line FH through the midpoint G of AD, and complete the resulting rectangles as in Fig. 3, rectangle $FCEI$ will have area 320. Since $AF = MI = FJ = DI$, it now suffices to find this common length, which we will call x; for $AC = CF + FA = 16 + x$ and $DE = EI - DI = 16 - x$. By the principle demonstrated in Fig. 2, $JCED$ has the same area as $DKLM$, so that $DKLM + FJDI = DKLM + 20x$. Explain why $DKLM = 30 \cdot 2 \cdot x$, and hence why $320 = (30 \cdot 2 + 20) \cdot x$.

Could this procedure have been obtained experimentally?

1.18. A famous example of mathematical blunders committed by mathematicians (not statisticians, however) occurred some two decades ago. At the time, a very popular television show in the United States was called *Let's Make a Deal.* On that show, the contestant was often offered the chance to keep his or her current winnings, or to trade them for a chance at some other unknown prize. In the case in question the contestant had chosen one of three boxes, knowing that only one of them contained a prize of any value, but not knowing the contents of any of them. For ease of exposition, let us call the boxes A, B, and C, and assume that the contestant chose box A.

The emcee of the program was about to offer the contestant a chance to trade for another prize, but in order to make the program more interesting, he had box B opened, in order to show that it was empty. Keep in mind that the emcee *knew* where the prize was and would not have opened box B if the prize had been there. Just as the emcee was about to offer a new deal, the contestant asked to exchange the chosen box (A) for the unopened box (C) on stage. The problem posed to the

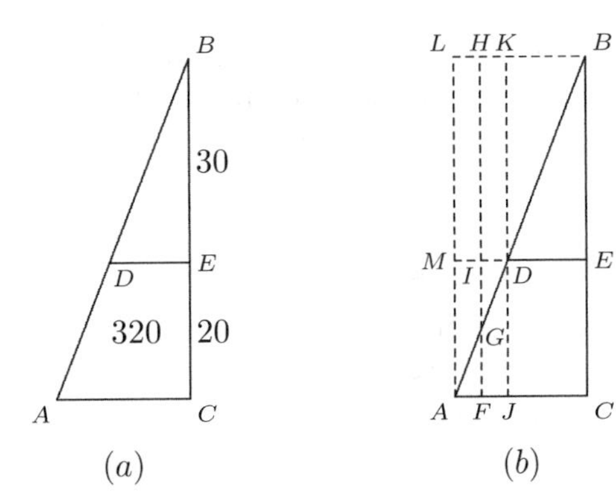

FIGURE 3. (*a*) Line DE divides triangle ABC into triangle DEB and trapezoid $ACED$. (*b*) Line $FGIH$ bisects line AD. Rectangle $FCEI$ has the same area as trapezoid $ACED$, and rectangle $JCED$ equals rectangle $MDKL$.

reader is: Was this a good strategy? To decide, analyze 300 hypothetical games, in which the prize is in box A in 100 cases, in box B in 100 cases (in these cases, of course, the emcee will open box C to show that it is empty), and in box C in the other 100 cases. First assume that in all 300 games the contestant retains box A. Then assume that in all 300 games the contestant exchanges box A for the unopened box on stage. By which strategy does the contestant win more games?

1.19. Explain why the following analysis of the game described in Problem 1.18 leads to an erroneous result. Consider all the situations in which the contestant has chosen box A and the emcee has shown box B to be empty. Imagine 100 games in which the prize is in box A and 100 games in which it is in box C. Suppose the contestant retains box A in all 200 games; then 100 will be won and 100 lost. Likewise, if the contestant switches to box C in all 200 games, then 100 will be won and 100 lost. Hence there is no advantage to switching boxes.

1.20. The fallacy discussed in Problem 1.19 is not in the mathematics, but rather in its application to the real world. The question involves what is known as *conditional probability*. Mathematically, the probability of event E, *given that event F has occurred*, is defined as the probability that E and F both occur, divided by the probability of F. The many mathematicians who analyzed the game erroneously proceeded by taking E as the event "The prize is in box C" and F as the event "Box B is empty." Given that box B has a 2/3 probability of being empty and the event "E and F" is the same as event E, which has a probability of 1/3, one can then compute that the probability of E given F is $(1/3)/(2/3) = 1/2$. Hence the contestant seems to have a 50% probability of winning as soon as the emcee opens Box B, revealing it to be empty.

Surely this conclusion cannot be correct, since the contestant's probability of having chosen the box with the prize is only 1/3 and the emcee can always open an empty box on stage. Replace event F with the more precise event "The emcee has *shown* that box B is empty" and redo the computation. Notice that the emcee is *going* to show that either box B or box C is empty, and that the two outcomes are equally likely. Hence the probability of this new event F is 1/2. Thus, even though the mathematics of conditional probability is quite simple, it can be a subtle problem to describe just what event has occurred. Conclusion: To reason correctly in cases of conditional probability, *one must be very clear in describing the event that has occurred.*

1.21. Reinforcing the conclusion of Problem 1.20, exhibit the fallacy in the following "proof" that *lotteries are all dishonest.*

Proof. The probability of winning a lottery is less than one chance in 1,000,000 ($=10^{-6}$). Since all lottery drawings are independent of one another, the probability of winning a lottery five times is less than $(10^{-6})^5 = 10^{-30}$. But this probability is far smaller than the probability of any conceivable event. Any scientist would disbelieve a report that such an event had actually been observed to happen. Since the lottery has been won five times in the past year, it must be that winning it is not a random event; that is, the lottery is fixed.

What is the event that has to occur here? Is it "Person A (specified in advance) wins the lottery," or is it "At least one person in this population (of 30 million people) wins the lottery"? What is the difference between those two probabilities? (The same fallacy occurs in the probabilistic arguments purporting to prove that evolution cannot occur, based on the rarity of mutations.)

1.22. The relation between mathematical creativity and musical creativity, and the mathematical aspects of music itself are a fascinating and well-studied topic. Consider just the following problem, based on the standard tuning of a piano keyboard. According to that tuning, the frequency of the major fifth in each scale should be 3/2 of the frequency of the base tone, while the frequency of the octave should be twice the base frequency. Since there are 12 half-tones in each octave, starting at the lowest A on the piano and ascending in steps of a major fifth, twelve steps will bring you to the highest A on the piano. If all these fifths are tuned properly, that highest A should have a frequency of $\left(\frac{3}{2}\right)^{12}$ times the frequency of the lowest A. On the other hand, that highest A is seven octaves above the lowest, so that, if all the octaves are tuned properly, the frequency should be 2^7 times as high. The difference between these two frequency ratios, $7153/4096 \approx 1.746$ is called the *Pythagorean comma.* (The Greek word *komma* means a break or cutoff.) What is the significance of this discrepancy for music? Could you hear the difference between a piano tuned so that all these fifths are exactly right and a piano tuned so that all the octaves are exactly right? (The ratio of the discrepancy between the two ratios to either ratio is about 0.01%.)

1.23. What meaning can you make of the statement attributed to the French poet Sully (René François Armand) Prudhomme (1839–1907), "Music is the pleasure the soul experiences from counting without realizing it is counting"?

CHAPTER 2

Mathematical Cultures I

In Chapter 1 we looked at the origin of mathematics in the everyday lives of people. The evidence for the conclusions presented there is indirect, coming from archaeology, anthropology, and other studies not directly mathematical. Wherever there are written documents to refer to, we can know much more about what was done and why. The present chapter is a broad survey of the development of mathematics that arose spontaneously, as far as is known, in a number of cultures around the world. We are particularly interested in highlighting the motives for creating mathematics.

1. The motives for creating mathematics

As we saw in Chapter 1, a certain amount of numerical and geometric knowledge is embedded in the daily lives of people and even animals. Human beings at various times have developed more intricate and sophisticated methods of dealing with numbers and space, leading to arithmetic, geometry, and beyond. That kind of knowledge must be taught systematically if it is to be passed on from generation to generation and become a useful part of a civilization. Some group of people must devote at least a part of their time to learning and perhaps improving the knowledge that has been acquired. These people are professional mathematicians, although their primary activity may be commercial, administrative, or religious, and sometimes a combination of the three, as in ancient Egypt.

1.1. Pure versus applied mathematics. How does the mathematics profession arise? Nowadays people choose to enter this profession for a variety of reasons. Undoubtedly, an important motive is that they find mathematical ideas interesting to contemplate and work with; but if there were no way of making a living from having some expertise in the subject, the number of its practitioners would be far smaller. The question thus becomes: "Why are some people paid for solving mathematical problems and creating new mathematical knowledge?" Industry and government find uses for considerable numbers of mathematicians and statisticians. For those of a purer, less applied bent, the universities and offices of scientific research subsidized by governments provide the opportunity to do research on questions of pure mathematical interest without requiring an immediate practical application. This kind of research has been pursued for thousands of years, and it has always had some difficulty justifying itself. Here, for example, is a passage from Book 7 of Plato's *Republic* in which, discussing the education of the leaders of an ideal state, Socrates and Glaucon decide on four subjects that they must learn, namely arithmetic, geometry, astronomy, and music. These four subjects were later to form the famous *quadrivium* (fourfold path) of education in medieval Europe. After listing arithmetic and geometry, they come to astronomy, and Socrates (the narrator of the dialogue) reports that Glaucon was in favor of including it:

> "I certainly am," he said, "to have a clearer perception of time periods, both months and eras is proper in agriculture and navigation, and no less so in military strategy."
>
> "You amuse me," I said, "in that you apparently fear the crowd, lest you seem to prescribe useless studies. But it is not an easy thing, it is difficult, to believe that some organ of the soul of each person is purified and refreshed in these studies, while it is lost and blinded by other pursuits; it is more to be preserved than ten thousand eyes, for by it alone is truth seen. Now some people will agree that what you are proposing is extremely good; but those who have never felt these things will regard you as having said nothing; for they see in them no profit worthy of the name."

Plato adopted the Pythagorean doctrine that there is a human faculty attuned to eternal truth and working through human reason. If there really is such a faculty, then of course mathematics is of high value for everyone. Plato, speaking here through Socrates, admits that some people seem to lack this faculty, so that it is difficult to believe in its universality. The difference in outlook between the two classes of people mentioned by Plato continues right down to the present time. Here, for example, is the view of the famous modern applied mathematician, R. W. Hamming (1915–1998), inventor of the Hamming codes, who if he believed at all in the "eye of the soul," at least did not believe it had a claim on public funds:

> [T]he computing expert needs to be wary of believing much that he learns in his mathematics courses... [M]uch of modern mathematics is not related to science but... to the famous scholastic arguing of the Middle Ages... I believe it is important to make these distinctions... [T]he failure to do so has... caused government money appropriated for numerical analysis to be diverted to the art form of pure mathematics.

On the opposite side of the question is the following point of view, expressed by the famous British mathematician G. H. Hardy (1877–1947), who in 1940 wrote *A Mathematician's Apology.* After quoting an earlier address, in which he had said, "after all, the scale of the universe is large and, if we are wasting our time, the waste of the lives of a few university dons is no such overwhelming catastrophe,"[1] he gave what he considered the justification of his life:

> I have never done anything "useful." No discovery of mine has made, or is likely to make, directly or indirectly, for good or ill, any difference to the amenity of the world... The case for my life, then,..., is this: that I have added something to knowledge, and have helped others to add more; and that these somethings have a value which differs in degree only, not in kind, from that of the creations of the great mathematicians.

[1] From a cosmic point of view, no doubt, this is a cogent argument. From the point of view of that second group of people mentioned by Plato, however, it ignores the main question: Why should a person expect to receive a salary for doing work that others regard as useless?

As Plato said, one class of people will find this argument a sufficient justification for being a mathematician; another class will not. By reading the motivations given by other mathematicians for their work, the reader may either find arguments to convince that second group of people, or else come to agree with them and Hamming.

2. India

From archaeological excavations at Mohenjo Daro and Harappa on the Indus River in Pakistan it is known that an early civilization existed in this region for about a millennium starting in 2500 BCE. This civilization may have been an amalgam of several different cultures, since anthropologists recognize five different physical types among the human remains. Many of the artifacts that were produced by this culture have been found in Mesopotamia, evidence of trade between the two civilizations.

The Aryan civilization. The early civilization of these five groups of people disappeared around 1500 BCE, and its existence was not known in the modern world until 1925. The cause of its extinction is believed to be an invasion from the northwest by a sixth group of people, who spoke a language closely akin to early Greek. Because of their language these people are referred to as Aryans. The Aryans gradually expanded and formed a civilization of small kingdoms, which lasted about a millennium.

Sanskrit literature. The language of the Aryans became a literary language known as Sanskrit, in which great classics of literature and science have been written. Sanskrit thus played a role in southern Asia analogous to that of Greek in the Mediterranean world and Chinese in much of eastern Asia. That is, it provided a means of communication among scholars whose native languages were not mutually comprehensible and a basis for a common literature in which cultural values could be preserved and transmitted. During the millennium of Aryan dominance the spoken language of the people gradually diverged from written Sanskrit. Modern descendants of Sanskrit are Hindi, Gujarati, Bengali, and others. Sanskrit is the language of the *Mahabharata* and the *Ramayana*, two epic poems whose themes bear some resemblance to the Homeric epics, and of the *Upanishads*, which contain much of the moral teaching of Hinduism.

Among the most ancient works of literature in the world are the Hindu *Vedas*. The word means *knowledge* and is related to the English word *wit*. The composition of the *Vedas* began around 900 BCE, and additions continued to be made to them for several centuries. Some of these *Vedas* contain information about mathematics, conveyed incidentally in the course of telling important myths.

Hindu religious reformers. Near the end of the Aryan civilization, in the second half of the sixth century BCE, two figures of historical importance arise. The first of these was Gautama Buddha, the heir to a kingdom near the Himalayas, whose spiritual journey through life led to the principles of Buddhism. The second leader, Mahavira, is less well known but has some importance for the history of mathematics. Like his contemporary Buddha, he began a reform movement within Hinduism. This movement, known as Jainism, still has several million adherents in India. It is based on a metaphysic that takes very seriously what is known in some Western ethical systems as the *chain of being*. Living creatures are ranked according to their awareness. Those having five senses are the highest, and those having only one sense are the lowest.

Islam in India. The amazingly rapid Muslim expansion from the Arabian desert in the seventh century brought Muslim invaders to India by the early eighth century. The southern valley of the Indus River became a province of the huge Umayyad Empire, but the rest of India preserved its independence, as it did 300 years later when another Muslim people, the Turks and Afghans, invaded. Still, the contact was enough to bring certain Hindu works, including the Hindu numerals, to the great center of Muslim culture in Baghdad. The complete and destructive conquest of India by the Muslims under Timur the Lame came at the end of the fourteenth century. Timur did not remain in India but sought new conquests; eventually he was defeated by the Ming dynasty in China. India was desolated by his attack and was conquered a century later by Akbar the Lion, a descendant of both Genghis Khan and Timur the Lame and the first of the Mogul emperors. The Mogul Empire lasted nearly three centuries and was a time of prosperity and cultural resurgence. One positive effect of this second Muslim expansion was a further exchange of knowledge between the Hindu and Muslim worlds. Interestingly, the official adminstrative language used for Muslim India was neither Arabic nor an Indian language; it was Persian.

British rule. During the seventeenth and eighteenth centuries British and French trading companies were in competition for the lucrative trade with the Mogul Empire. British victories during the Seven Years War left Britain in complete control of this trade. Coming at the time of Mogul decline due to internal strife among the Muslims and continued resistance on the part of the Hindus, this trade opened the door for the British to make India part of their empire. British colonial rule lasted nearly 200 years, coming to an end only after World War II. British rule made it possible for European scholars to become acquainted with Hindu classics of literature and science. Many Sanskrit works were translated into English in the early nineteenth century and became part of the world's science and literature.

We can distinguish three periods in the development of mathematics in the Indian subcontinent. The first period begins around 900 BCE with individual mathematical results forming part of the Vedas. The second begins with systematic treatises concerned mostly with astronomy but containing explanations of mathematical results, which appear in the second century CE. These treatises led to continuous progress for 1500 years, during which time much of algebra, trigonometry, and certain infinite series that now form part of calculus were discovered, a century or more before Europeans developed calculus. In the third stage, which began during the two centuries of British rule, this Hindu mathematics came to be known in the West, and Indian mathematicians began to work and write in the modern style of mathematics that is now universal.

2.1. The *Sulva Sutras*. In the period from 800 to 500 BCE a set of verses of geometric and arithmetic content were written and became part of the *Vedas*.[2]

---

[2] Most of this discussion of the *Sulva Sutras* is based on the work of Srinivasiengar (*1967*), which gives a clear exposition but contains statements that are rather alarming for one who is forced to rely on a secondary source. For example, on p. 6 we learn that the unit of length known as the *vyayam* was "about 96 inches," and "possibly this represented the height of the average man in those days." Indeed. Where, Mr. Srinivasiengar, have archaeologists discovered 8-foot-tall human skeletons?

These verses are known collectively as the *Sulva Sutras* or *Sulba Sutras*. The name means *Cord Rules* and probably reflects the use of a stretched rope or cord as a way of measuring length. The root *sulv* originally meant *to measure* or *to rule*, although it also has the meaning of a cord or rope; *sutra* means *thread* or *cord*, a common measuring instrument. In the case of the *Vedas* the objects being measured with the cords were altars. The maintenance of altar fires was a duty for pious Hindus, and because Hinduism is polytheistic, it was necessary to consider how elaborate and large the fire dedicated to each deity was to be. This religious problem led to some interesting problems in arithmetic and geometry.

Two scholars who studied primarily the Sanskrit language and literature made important contributions to mathematics. Pingala, who lived around 200 BCE, wrote a treatise known as the *Chandahsutra*, containing one very important mathematical result, which, however, was stated so cryptically that one must rely on a commentary written 1200 years later to know what it meant. Later, a fifth-century scholar named Panini standardized the Sanskrit language, burdening it with some 4000 grammatical rules that make it many times more difficult to learn than any other Indo-European language. In the course of doing so, he made extensive use of combinatorics and the kind of abstract reasoning that we associate with algebra. These subjects set the most ancient Hindu mathematics apart from that of other nations.

2.2. Buddhist and Jaina mathematics. As with any religion that encourages quiet contemplation and the renunciation of sensual pleasure, Jainism often leads its followers to study mathematics, which provides a different kind of pleasure, one appealing to the mind. There have always been some mathematicians among the followers of Jainism, right down to modern times, including one in the ninth century bearing the same name as the founder of Jainism. The early work of Jaina mathematicians is notable for algebra (the *Sthananga Sutra*, from the second century BCE), for its concentration on topics that are essentially unique to early Hindu mathematics, such as combinatorics (the *Bhagabati Sutra*, from around 300 BCE), and for speculation on infinite numbers (the *Anuyoga Dwara Sutra*, probably from the first century BCE). Buddhist monks were also very fond of large numbers, and their influence was felt when Buddhism spread to China in the sixth century CE.

2.3. The Bakshali Manuscript. A birchbark manuscript unearthed in 1881 in the village of Bakshali, near Peshawar, is believed by some scholars to date from the seventh century CE, although Sarkor (*1982*) believes it cannot be later than the end of the third century, since it refers to coins named *dīnāra* and *dramma*, which are undoubtedly references to the Greek coins known as the denarius and the drachma, introduced into India by Alexander the Great. These coins had disappeared from use in India by the end of the third century. The Bakshali Manuscript contains some interesting algebra, which is discussed in Chapter 14.

2.4. The *siddhantas*. During the second, third, and fourth centuries CE, Hindu scientists compiled treatises on astronomy known as *siddhantas*. The word *siddhanta* means a system.[3] One of these treatises, the *Surya Siddhanta* (System of

[3] A colleague of the author suggested that this word may be cognate with the Greek *idōn*, the aorist participle of the verb meaning *see*.

the Sun), from the late fourth century, has survived intact. Another from approximately the same time, the *Paulisha Siddhanta*, was frequently referred to by the Muslim scholar al-Biruni (973–1048). The name of this treatise seems to have been bestowed by al-Biruni, who says that the treatise was written by an Alexandrian astrologer named Paul.

2.5. Aryabhata I. With the writing of treatises on mathematics and astronomy, we at last come to some records of the motives that led people to create Hindu mathematics, or at least to write expositions of it. A mathematician named Aryabhata, the first of two mathematicians bearing that name, lived in the late fifth and early sixth centuries at Kusumapura (now Pataliputra, a village near the city of Patna) and wrote a book called the *Aryabhatiya.* This work had been lost for centuries when it was recovered by the Indian scholar Bhau Daji (1822–1874) in 1864. Scholars had known of its existence through the writings of commentators and had been looking for it. Writing in 1817, the English scholar Henry Thomas Colebrooke (1765–1837), who translated other Sanskrit mathematical works into English, reported, "A long and diligent research of various parts of India has, however, failed of recovering any part of the... *Algebra* and other works of Aryabhata." Ten years after its discovery the *Aryabhatiya* was published at Leyden and attracted the interest of European and American scholars. It consists of 123 stanzas of verse, divided into four sections, of which the first, third, and fourth are concerned with astronomy and the measurement of time.

Like all mathematicians, Aryabhata I was motivated by intellectual interest. This interest, however, was closely connected with his Hindu piety. He begins the *Aryabhatiya* with the following tribute to the Hindu deity:

> Having paid reverence to Brahman, who is one but many, the true deity, the Supreme Spirit, Aryabhata sets forth three things: mathematics, the reckoning of time, and the sphere. [Clark, *1930*, p. 1]

The translator adds phrases to explain that Brahman is one as the sole creator of the universe, but is many via a multitude of manifestations.

Aryabhata then continues his introduction with a list of the astronomical observations that he will be accounting for and concludes with a promise of the reward awaiting the one who learns what he has to teach:

> Whoever knows this *Dasagitika Sutra* which describes the movements of the Earth and the planets in the sphere of the asterisms passes through the paths of the planets and asterisms and goes to the higher Brahman. [Clark, *1930*, p. 20]

As one can see, students in Aryabhata's culture had an extra reason to study mathematics and astronomy, beyond the concerns of practical life and the pleasures of intellectual edification. Learning mathematics and astronomy helped to advance the soul through the cycle of births and rebirths that Hindus believed in.

After setting out his teaching on the three subjects, Aryabhata concludes with a final word of praise for the Hindu deity and invokes divine endorsement of his labors:

> By the grace of God the precious sunken jewel of true knowledge has been rescued by me, by means of the boat of my own knowledge, from the ocean which consists of true and false knowledge. He who disparages this universally true science of astronomy, which formerly was revealed by Svayambhu and is now described by me in this Aryabhatiya, loses his good deeds and his long life. [Clark, *1930*, p. 81]

2.6. Brahmagupta. The establishment of research centers for astronomy and mathematics at Kusumapura and Ujjain produced a succession of good mathematicians and mathematical works for many centuries after Aryabhata I. About a century after Aryabhata I another Hindu mathematician, Brahmagupta, was born in the city of Sind, now in Pakistan. He was primarily an astronomer, but his astronomical treatise, the *Brahmasphutasiddhanta* (literally *The* Corrected *Brahma Siddhanta*), contains several chapters on computation (*Ganita*). The Hindu interest in astronomy and mathematics continued unbroken for several centuries, producing important work on trigonometry in the tenth century.

2.7. Bhaskara II. Approximately 500 years after Brahmagupta, in the twelfth century, the mathematician Bhaskara, the second of that name, was born on the site of the modern city of Bijapur. He is the author of the *Siddhanta Siromani*, in four parts, a treatise on algebra and geometric astronomy. Only the first of these parts, known as the *Lilavati*, and the second, known as the *Vija Ganita*,[4] concern us here. Bhaskara says that his work is a compendium of knowledge, a sort of textbook of astronomy and mathematics. The name *Lilavati*, which was common among Hindu women, seems to have been a fancy of Bhaskara himself. Many of the problems are written in the form of puzzles addressed to this Lilavati.

Bhaskara II apparently wrote the *Lilavati* as a textbook to form part of what we would call a liberal education. His introduction reads as follows:

> Having bowed to the deity, whose head is like an elephant's, whose feet are adored by gods; who, when called to mind, relieves his votaries from embarrassment; and bestows happiness on his worshippers; I propound this easy process of computation, delightful by its elegance, perspicuous with words concise, soft and correct, and pleasing to the learned. [Colebrooke, *1817*, p. 1]

As a final advertisement at the end of his book, Bhaskara extols the pleasure to be derived from learning its contents:

> Joy and happiness is indeed ever increasing in this world for those who have *Lilavati* clasped to their throats, decorated as the members are with neat reduction of fractions, multiplication, and involution, pure and perfect as are the solutions, and tasteful as is the speech which is exemplified. [Colebrooke, *1817*, p. 127]

---

[4] This Sanskrit word means literally *source computation*. It is compounded from the Sanskrit root *vij*- or *bij*-, which means *seed*. As discussed in Chapter 13, the basic idea of algebra is to find one or more numbers (the "source") knowing the result of operating on them in various ways. The word is usually translated as *algebra*.

The *Vija Ganita* consists of nine chapters, in the last of which Bhaskara tells something about himself and his motivation for writing the book:

> On earth was one named Maheswara, who followed the eminent path of a holy teacher among the learned. His son Bhaskara, having from him derived the bud of knowledge, has composed this brief treatise of elemental computation. As the treatises of algebra [vija ganita] by Brahmagupta, Shidhara and Padmanabha are too diffusive, he has compressed the substance of them in a well-reasoned compendium for the gratification of learners... to augment wisdom and strengthen confidence. Read, do read, mathematician, this abridgement, elegant in style, easily understood by youth, comprising the whole essence of computation, and containing the demonstration of its principles, replete with excellence and void of defect. [Colebrooke, *1817*, pp. 275–276]

2.8. Muslim India. Indian mathematical culture reflects the religious division between the Muslim and Hindu communities to some extent. The Muslim conquest brought Arabic and Persian books on mathematics to India. Some of these works were translated from ancient Greek, and among them was Euclid's *Elements*. These translations of later editions of Euclid contained certain obscurities and became the subject of commentaries by Indian scholars. Akbar the Lion decreed a school curriculum for Muslims that included three-fourths of what was known in the West as the quadrivium. Akbar's curriculum included arithmetic, geometry, and astronomy, leaving out only music.[5] Details of this Indian Euclidean tradition are given in the paper by de Young (*1995*).

2.9. Indian mathematics in the colonial period and after. One of the first effects of British rule in India was to acquaint European scholars with the treasures of Hindu mathematics described above. It took a century before the British colonial rulers began to establish universities along European lines in India. According to Varadarajan (*1983*), these universities were aimed at producing government officials, not scholars. As a result, one of the greatest mathematical geniuses of all time, Srinivasa Ramanujan (1887–1920), was not appreciated and had to appeal to mathematicians in Britain to gain a position that would allow him to develop his talent. The necessary conditions for producing great mathematics were present in abundance, however, and the establishment of the Tata Institute in Bombay (now Mumbai) and the Indian Statistical Institute in Calcutta were important steps in this direction. After Indian independence was achieved, the first prime minister, Jawaharlal Nehru (1889–1964), made it a goal to achieve prominence in science. This effort has been successful in many areas, including mathematics. The names of Komaravolu Chandrasekharan (b. 1920), Harish-Chandra (1923–1983), and others have become celebrated the world over for their contributions to widely diverse areas of mathematics.

[5] The quadrivium is said to have been proposed by Archytas (ca. 428–350 BCE), who lived in southern Italy and apparently communicated it to Plato when the latter was there to consult with the ruler of Syracuse; Plato incorporated it in his writings on education, as discussed in Sect. 1.

Srinivasa Ramanujan. The topic of power series is one in which Indian mathematicians had anticipated some of the discoveries in seventeenth- and eighteenth-century Europe. It was a facility with this technique that distinguished Ramanujan, who taught himself mathematics after having been refused admission to universities in India. After publishing a few papers, starting in 1911, he was able to obtain a stipend to study at the University of Madras. In 1913 he took the bold step of communicating some of his results to G. H. Hardy. Hardy was so impressed by Ramanujan's ability that he arranged for Ramanujan to come to England. Thus began a collaboration that resulted in seven joint papers with Hardy, while Ramanujan alone was the author of some 30 others. He rediscovered many important formulas and made many conjectures about functions such as the hypergeometric function that are represented by power series.

Unfortunately, Ramanujan was in frail health, and the English climate did not agree with him. Nor was it easy for him to maintain his devout Hindu practices so far from his normal Indian diet. He returned to India in 1919, but succumbed to illness the following year. Ramanujan's notebooks have been a subject of continuing interest to mathematicians. Hardy passed them on to G. N. Watson (1886–1965), who published a number of "theorems stated by Ramanujan." The full set of notebooks was published in the mid-1980s (see Berndt, *1985*).

3. China

The name *China* refers to a region unified under a central government but whose exact geographic extent has varied considerably over the 4000 years of its history. To frame our discussion we shall sometimes refer to the following dynasties:[6]

The Shang Dynasty (sixteenth to eleventh centuries BCE). The Shang rulers controlled the northern part of what is now China and had an extensive commercial empire.

The Zhou Dynasty (eleventh to eighth centuries BCE). The Shang Dynasty was conquered by people from the northwest known as the Zhou. The great Chinese philosophers known in the West as Confucius, Mencius, and Lao-Tzu lived and taught during the period of disorder that came after the decay of this dynasty.

The Period of Warring States (403–221 BCE) and the *Qin Dynasty* (221–206 BCE). Warfare was nearly continuous in the fourth and third centuries BCE, but in the second half of the third century the northwestern border state of Qin gradually defeated all of its rivals and became the supreme power under the first Qin emperor. The name *China* is derived from the Qin.

The Han Dynasty (206 BCE–220 CE). The empire was conquered shortly after the death of the great emperor by people known as the Han, who expanded their control far to the south, into present-day Viet Nam, and established a colonial rule in the Korean peninsula. Contact with India during this dynasty brought Buddhism to China for the first time. According to Mikami (*1913*, pp. 57–58), mathematical and astronomical works from India were brought to China and studied. Certain topics, such as combinatorics, are common to both Indian and Chinese treatises, but "there

[6] Because of total ignorance of the Chinese language, the author is forced to rely on translations of all documents. We shall write Chinese words in the Latin alphabet but not strive for consistency among the various sources that use different systems. We shall also omit the accent marks used to indicate the pitch of the vowels, since these cannot be pronounced by foreigners without special training.

is nothing positive that serves as an evidence of any actual Indian influence upon the Chinese mathematics."

The Tang Dynasty (seventh and eighth centuries). The Tang Dynasty was a period of high scholarship, in which, for example, block printing was invented.

The Song Dynasty (960–1279). The period of disorder after the fall of the Tang Dynasty ended with the accession of the first Song emperor. Confucianism underwent a resurgence in this period, supplementing its moral teaching with metaphysical speculation. As a result, a large number of scientific treatises on chemistry, zoology, and botany were written, and the Chinese made great advances in algebra.

The Mongol conquest and the closing of China. The Song Dynasty was ended in the thirteenth century by the Mongol conquest under the descendants of Genghis Khan, whose grandson Kublai Khan was the first emperor of the dynasty known to the Chinese as the Yuan. As the Mongols were Muslims, this conquest brought China into contact with the intellectual achievements of the Muslim world. Knowledge flowed both ways, of course, and the sophisticated Chinese methods of root extraction seem to be reflected in the works of later Muslim scholars, such as the fifteenth-century mathematician al-Kashi. The vast Mongol Empire facilitated East–West contacts, and it was during this period that Marco Polo (1254–1324) made his famous voyage to the Orient.

The Ming Dynasty (fourteenth to seventeenth centuries). While the Mongol conquest of Russia lasted 240 years, the Mongols were driven out of China in less than a century by the first Ming emperor. During the Ming Dynasty, Chinese trade and scholarship recovered rapidly. The effect of the conquest, however, was to encourage Chinese isolationism, which became the official policy of the later Ming emperors during the period of European expansion. The first significant European contact came in the year 1582, when the Jesuit priest Matteo Ricci (1552–1610) arrived in China. The Jesuits were particularly interested in bringing Western science to China to aid in converting the Chinese to Christianity. They persisted in these efforts despite the opposition of the emperor. The Ming Dynasty ended in the mid-seventeenth century with conquest by the Manchus.

The Ching (Manchu) Dynasty (1644–1911). After two centuries of relative prosperity the Ching Dynasty suffered from the depredations of foreign powers eager to control its trade. Perhaps the worst example was the Opium War of 1839–1842, fought by the British in order to gain control of the opium trade. From that time on Manchu rule declined. In 1900 the Boxer Rebellion against the Western occupation was crushed and the Chinese were forced to pay heavy reparations. In 1911 the government disintegrated entirely, and a republic was declared.

The twentieth century. The establishment of a republic in China did not quell the social unrest, and there were serious uprisings for several decades. China suffered badly from World War II, which began with a Japanese invasion in the 1930s. Although China was declared one of the major powers when the United Nations was formed in 1946, the Communist revolution of 1949 drove the ruler Chiang Kai-Shek to the island of Taiwan. The United States recognized only the government in Taiwan until the early 1970s. Then, bowing to the inevitable, it recognized the Communist government and did not use its veto power when that government replaced the Taiwanese government on the Security Council of the United Nations. At present, the United States recognizes two Chinese governments, one in Beijing and one on Taiwan. This view is contradicted by the government in Beijing, which claims the authority to rule Taiwan. The American mathematician Walter Feit (1930–2004) visited China in May 1976 and reported that there was a heavy emphasis on combining mathematical theory with practice to solve social problems (Feit, *1977*). At first zealous in adhering to the Maoist version of Marxism, the Chinese Communist Party undertook major reforms during the 1990s and has been moving in the direction of a more market-driven economy since then, although democracy seems to be slow in arriving. China is now engaged in extensive cultural and commercial exchanges with countries all over the world and hosted the International Congress of Mathematicians in 2002. Its mathematicians have made outstanding contributions to the advancement of mathematics, and Chinese students are eagerly welcomed at universities in nearly every country.

3.1. Works and authors. Mathematics became a recognized and respected area of intellectual endeavor in China more than 2000 years ago. That its origins are at least that old is established by the existence of books on mathematics, at least one of which was probably written before the order of the Emperor Shih Huang Ti in 213 BCE that all books be burned.[7] A few books survived or were reconstituted after the brief reign of Shih Huang-Ti, among them the mathematical classic just alluded to. This work and three later ones now exist in English translation, with commentaries to provide the proper context for readers who are unfamiliar with the history and language of China. Under the Tang dynasty a standardized educational system came into place for the training of civil servants, based on literary and scientific classics, and the works listed below became part of a mathematical curriculum known as the *Suan Jing Shishu* (*Ten Canonical Mathematical Classics*—there are actually 12 of them). Throughout this long period mathematics was cultivated together with astronomy both as an art form and for their practical application in the problem of obtaining an accurate lunisolar calendar. In addition, many problems of commercial arithmetic appear in the classic works.

The Zhou Bi Suan Jing. The early treatise mentioned above, the *Zhou Bi Suan Jing*, has been known in English as the *Arithmetic Classic of the Gnomon and the Circular Paths of Heaven.* A recent, very thorough study and English translation has been carried out by Christopher Cullen of the University of London (*1996*). According to Cullen, the title *Zhou Bi* could be rendered as *The Gnomon of the Zhou.* The phrase *suan jing* occurs in the titles of several early mathematical works; it means *mathematical treatise* or *mathematical manual.* According to a tradition, the *Zhou Bi Suan Jing* was written during the Western Zhou dynasty, which overthrew the earlier Shang dynasty around 1025 BCE and lasted until 771 BCE. Experts now believe, however, that the present text was put together during the Western Han dynasty, during the first century BCE, and that the commentator Zhao Shuang, who wrote the version we now have, lived during the third century CE, after the fall of the Han dynasty. However, the astronomical information in the book could only have been obtained over many centuries of observation, and therefore must be much earlier than the writing of the treatise.

As the traditional title shows, the work is concerned with astronomy and surveying. The study of astronomy was probably regarded as socially useful in two ways: (1) It helped to regulate the calendar, a matter of great importance when rituals were to be performed; (2) it provided a method of divination (astrology),

[7] The Emperor was not hostile to learning, since he did not forbid the *writing* of books. Apparently, he just wanted to be remembered as the emperor in whose reign everything began.

also of importance both for the individual and for the state. Surveying is of use in any society where it is necessary to erect large structures such as dams and bridges, and where land is often flooded, requiring people to abandon their land holdings and reclaim them later. These considerations at least provide a reason for people to regard mathematics as useful in practice. However, the preface, written by the commentator Zhao Shuang, gives a different version of the motive for compiling this knowledge. Apparently a student of traditional Chinese philosophy, he had realized that it was impossible to understand fully all the mysteries of the changing universe. He reports that he had looked into this work while convalescing from an illness and had been so impressed by the acuity of the knowledge it contained that he decided to popularize it by writing commentaries to help the reader over the hard parts, saying, "Perhaps in time gentlemen with a taste for wide learning may turn their attention to this work" (Cullen, *1996*, p. 171).

Here we see mathematics being praised simply because it confers understanding where ignorance would otherwise be; it is regarded as a liberal art, to be studied by a leisured class of *gentlemen* scholars, people fortunate enough to be free of the daily grind of physical labor that was the lot of the majority of people in all countries until very recent times.

The Jiu Zhang Suanshu. Another ancient Chinese treatise, the *Jiu Zhang Suanshu*, meaning *Nine Chapters on the Mathematical Art*,[8] has been partly translated into English, with commentary, by Lam (*1994*). A corrected and commented edition was published in Chinese in 1992, assembled by Guo (*1992*). This work has claim to be called *the* classic Chinese mathematical treatise. It reflects the level of mathematics in China in the later Han dynasty, around the year 100 CE. The nine chapters that give this monograph its name contain 246 applied problems of a sort useful in teaching how to handle arithmetic and elementary algebra and how to apply them in commercial and administrative work. Unfortunately, these chapters have no prefaces in which the author explains their purpose, and so we must assume that the purpose was the obvious one of training people engaged in surveying, administration, and trade. Some of the problems have an immediately practical nature, explaining how to find areas, convert units of length and area, and deal with fractions and proportions. Yet when we analyze the algebraic parts of this work, we shall see that it contains impractical puzzle-type problems leading to systems of linear equations and resembling problems that have filled up algebra books for centuries. Such problems are apparently intended to train the mind in algebraic thinking.

The Sun Zi Suan Jing. The most elementary of the early treatises is the *Sun Zi Suan Jing*, or *Mathematical Classic of Sun Zi*, even though its date is several centuries later than the *Jiu Zhang Suanshu*. This work begins with a preface praising the universality of mathematics for its role in governing the lives of all creatures, and placing it in the context of Chinese philosophy and among the six fundamental arts (propriety, music, archery, charioteership, calligraphy, and mathematics).

The preface makes it clear that mathematics is appreciated as both a practical skill in life and as an intellectual endeavor. The practicality comes in the use of compasses and gnomons for surveying and in the use of arithmetic for computing weights and measures. The intellectual skill, however, is emphasized. Mathematics

---

[8] Chinese titles are apparently very difficult to render in English. Martzloff (*1994*) translates this title as *Computational Prescriptions in Nine Chapters.*

is valued because it trains the mind. "If one neglects its study, one will not be able to achieve excellence and thoroughness" (Lam and Ang, *1992*, p. 151).

As in the quotation from the commentary on the *Zhou Bi Suan Jing*, we find that an aura of mystery and "elitism" surrounds mathematics. It is to be pursued by a dedicated group of initiates, who expect to be respected for learning its mysteries. At the same time, it has a practical value that is also respected.

Liu Hui. The Hai Dao Suan Jing. The fall of the Han Dynasty in the early third century gave rise to three separate kingdoms in the area now known as China. The north-central kingdom is known as the Kingdom of Wei. There, in the late third century CE, a mathematician named Liu Hui (ca. 220–280) wrote a commentary on the final chapter of the *Jiu Zhang Suanshu.* This chapter is devoted to the theorem we know as the Pythagorean theorem, and Liu Hui's book, the *Hai Dao Suan Jing* (*Sea Island Mathematical Classic*), shows how to use pairs of similar right triangles to measure inacessible distances. The name of the work comes from the first problem in it, which is to find the height of a mountain on an offshore island and the distance to the base of the mountain. The work consists of nine problems in surveying that can be solved by the algebraic techniques practiced in China at the time. A translation of these problems, a history of the text itself, and commentary on the mathematical techniques can be found in the paper by Ang and Swetz (*1986*).

Liu Hui wrote a preface explaining that because of the burning of the books 400 years earlier, the few ancient texts still around had deteriorated, but that a minister of agriculture named Zhang Cang had produced a revised and corrected edition. However, most historians think that the *Jiu Zhang Suanshu* was written around 200 BC, after Shih Huang Ti ordered the burning of the books.

Zu Chongzhi and Zu Geng. According to Li and Du (*1987*, pp. 80–82), fifth-century China produced two outstanding mathematicians, father and son. Zu Chongzhi (429–501) and his son Zu Geng (ca. 450–520) were geometers who devised a method resembling what is now called *Cavalieri's principle* for calculating volumes bounded by curved surfaces. The elder Zu was also a numerical analyst, who wrote a book on approximation entitled *Zhui Shu* (*Method of Interpolation*), which became for a while part of the classical curriculum. However, this book was apparently regarded as too difficult for nonspecialists, and it was dropped from the curriculum and lost. Zu Geng continued working in the same area as his father and had a son who also became a mathematician.

Yang Hui. We now leave a considerable (700-year) gap in the story of Chinese mathematics. The next mathematician we wish to mention is Yang Hui (ca. 1238–1298), the author of a number of mathematical texts. According to Li and Du (*1987*, pp. 110, 115), one of these was *Xiangjie Jiuzhang Suan Fa* (*Detailed Analysis of the Mathematical Rules in the* Jiu Zhang Suanshu), a work of 12 chapters, one on each of the nine chapters of the *Jiu Zhang Suanshu*, plus three more containing other methods and more advanced analysis. In 1274 and 1275 he wrote two other works, which were later collected in a single work called the *Yang Hui Suan Fa* (*Yang Hui's Computational Methods*). In these works he discussed not only mathematics, but also its pedagogy, advocating real understanding over rote learning.

Zhu Shijie. Slightly later than Yang Hui, but still contemporary with him, was Zhu Shijie (ca. 1260–1320). He was still a young man in 1279, when China was united by the Mongol emperor Kublai Khan and its capital established at what is now Beijing. Unification of the country enabled Zhu Shijie to travel more widely than had previously been possible. As a result, his *Suan Shu Chimeng* (*Introduction to Mathematical Studies*), although based on the *Jiu Zhang Suanshu*, went beyond it, discussing the latest methods of Chinese algebra. The original of this book was lost in Chinese, but a Korean version was later exported to Japan, where it had considerable influence. Eventually, a translation back from Korean into Chinese was made in the nineteenth century. According to Zharov (*2001*), who analyzed four fragments from this work, it shows some influence of Hindu or Arabic mathematics in its classification of large numbers. Zharov also proposed that the title be translated as "Explanation of some obscurities in mathematics" but says that his Chinese colleagues argued that the symbols for *chi* and *meng* were written as one and should be considered a single concept.

The Suan Fa Tong Zong *of Cheng Dawei.* A later work, the *Suan Fa Tong Zong* (*Treatise on Arithmetic*) by Cheng Dawei (1533–1606), was published in 1592. This book is well described by its title. It contains a systematic treatment of the kinds of problems handled in traditional Chinese mathematics, and at the end has a bibliography of some 50 other works on mathematics. The author, according to one of his descendents, was fascinated by books discussing problems on fields and grain, and assembled this book of problems over a lifetime of purchasing such books. Like the book of Zhu Shijie, Cheng Dawei's book had a great influence on the development of mathematics in Korea and Japan. According to Li and Du (*1987*, p. 186), Cheng Dawei left a record of his mathematical studies, saying that he had been involved in travel and trade when young and had sought teachers everywhere he went. He retired from this profession while still young and spent 20 years consolidating and organizing his knowledge, so that "finally I rooted out the false and the nonsensical, put all in order, and made the text lucid."

3.2. China's encounter with Western mathematics. Jesuit missionaries who entered China during the late sixteenth century brought with them some mathematical works, in particular Euclid's *Elements*, the first six books of which the missionary Matteo Ricci and the Chinese scholar Xu Guangchi (1562–1633) translated into Chinese (Li and Du, *1987*, p. 193). The version of Euclid that they used, a Latin translation by the German Jesuit Christopher Clavius (1538–1612) bearing the title *Euclidis elementorum libri XV* (*The Fifteen Books of Euclid's* Elements), is still extant, preserved in the Beijing Library. This book aroused interest in China because it was the basis of Western astronomy and therefore offered a new approach to the calendar and to the prediction of eclipses. According to Mikami (*1913*, p. 114), the Western methods made a correct prediction of a solar eclipse in 1629, which traditional Chinese methods got wrong. It was this accurate prediction that attracted the attention of Chinese mathematicians to Euclid's book, rather than the elaborate logical structure which is its most prominent distinguishing characteristic. Martzloff (*1993*) has studied a commented (1700) edition of Euclid by the mathematician Du Zhigeng and has noted that it was considerably abridged, omitting many proofs of propositions that are visually or topologically obvious. As Martzloff says, although Du Zhigeng retained the logical form of Euclid, that is, the definitions, axioms, postulates, and propositions, he neglected proofs, either omitting them entirely or giving only a fraction of a proof, "a fraction not necessarily containing the part of the Euclidean argument relative to a given proposition and devoted to the mathematical proof in the proper sense of the term." Du Zhigeng also attempted to synthesize the traditional Chinese classics, such as the *Jiu Zhang Suanshu* and the *Suan Fa Tong Zong*, with works imported from Europe, such as Archimedes' treatise on the measurement of the circle. Thus in China, Western mathematics supplemented, but did not replace, the mathematics that already existed.

The first Manchu Emperor Kang Xi (1654–1722) was fascinated by science and insisted on being taught by two French Jesuits, Jean-François Gerbillon (1654–1707) and Joachim Bouvet (1656–1730), who were in China in the late 1680s. This was the time of the Sun King, Louis XIV, who was vying with Spain and Portugal for influence in the Orient. The two Jesuits were required to be at the palace from before dawn until long after sunset and to give lessons to the Emperor for four hours in the middle of each day (Li and Du, *1987*, pp. 217–218).

The encounter with the West came at a time when mathematics was undergoing an amazing efflorescence in Europe. The first books on the use of Hindu–Arabic numerals for computation had appeared some centuries before, and now trigonometry, logarithms, analytic geometry, and calculus were all being developed at a rapid pace. These new developments took a firmer hold in China than the ancient Greek mathematics of Euclid and Archimedes. Jami (*1988*) reports on an eighteenth-century work by Ming Antu (d. 1765) deriving power-series expansions for certain trigonometric functions. She notes that even though the proofs of these expansions would not be regarded as conclusive today, the greatest of the eighteenth-century Chinese mathematicians, Wang Lai (1768–1813), professed himself satisfied with them.[9] Thus, she concludes, there was a difference between the reception of Euclid in China and the reception of the more computational modern mathematics. The Chinese took Euclid's treatise on its own terms and attempted to fit it into their own conception of mathematics; but they reinterpreted contemporary mathematics completely, since it came to them in small pieces devoid of context (Jami, *1988*, p. 327).

Given the increasing contacts between East and West in the nineteenth century, some merging of ideas was inevitable. During the 1850s the mathematician Li Shanlan (1811–1882), described by Martzloff (*1982*) as "one of the last representatives of Chinese traditional mathematics," translated a number of contemporary works into Chinese, including an 1851 calculus textbook of the American astronomer–mathematician Elias Loomis (1811–1889) and an algebra text by Augustus de Morgan (1806–1871). Li Shanlan had a power over formulas that reminds one in many ways of the twentieth-century Indian genius Srinivasa Ramanujan. One of his combinatorial formulas, stated without proof in 1867, was finally proved through the ingenuity of the prominent Hungarian mathematician Paul Turán (1910–1976). By the early twentieth century Chinese mathematical schools had marked out their own territory, specializing in standard areas of mathematics such as analytic function theory. Despite the difficulties of war, revolution, and a period of isolation during the 1960s, transmission of mathematical literature between China and the West continued and greatly expanded through exchanges of students and faculty from the 1980s onward. Kazdan (*1986*) gives an interesting snapshot of the situation in China at the beginning of this period of expansion.

---

[9] European mathematicians of the time also used methods that would not be considered completely rigorous today, and their arguments have some resemblance to those reported by Jami.

4. Ancient Egypt

Although mathematics has been practiced in Egypt continuously starting at least 4000 years ago, it merged with Greek mathematics during the Hellenistic period that began at the end of the fourth century BCE, and it formed part of the larger Muslim culture centered in Baghdad starting about 1200 years ago. What we shall call Egyptian mathematics in this section had a beginning and an end. It began with hieroglyphic inscriptions containing numbers and dating to the third millennium BCE and ended in the time of Euclid, around 300 BCE. The city of Alexandria in the Nile delta was the main school of mathematics in the Hellenistic world, and many of the most prominent mathematicians who wrote in Greek studied there.

The great architectural monuments of ancient Egypt are covered with hieroglyphs, some of which contain numbers. In fact, the ceremonial mace of the founder of the first dynasty contains records that mention oxen, goats, and prisoners and contain hieroglyphic symbols for the numbers 10,000, 100,000, and 1,000,000. These hieroglyphs, although suitable for ceremonial recording of numbers, were not well adapted for writing on papyrus or leather. The language of the earliest written documents that have been preserved to the present time is a cursive known as *hieratic*.

The most detailed information about Egyptian mathematics comes from a single document written in the hieratic script on papyrus around 1650 BCE and preserved in the dry Egyptian climate. This document is known properly as the Ahmose Papyrus, after its writer, but also as the Rhind Papyrus after the British lawyer Alexander Rhind (1833–1863), who went to Egypt for his health and became an Egyptologist. Rhind purchased the papyrus in Luxor, Egypt, in 1857. Parts of the original document have been lost, but a section consisting of 14 sheets glued end to end to form a continuous roll $3\frac{1}{2}$ feet wide and 17 feet long remains. Part of it is on public display in the British Museum, where it has been since 1865 (see Plate 1). Some missing pieces of this document were discovered in 1922 in the Egyptian collection of the New York Historical Society; these are now housed at the Brooklyn Museum of Art. A slightly earlier mathematical papyrus, now in the Moscow Museum of Fine Arts, consists of sheets about one-fourth the size of the Ahmose Papyrus. This papyrus was purchased by V. S. Golenishchev (1856–1947) in 1893 and donated to the museum in 1912. A third document, a leather roll purchased along with the Ahmose Papyrus, was not unrolled for 60 years after it reached the British Museum because the curators feared it would disintegrate if unrolled. It was some time before suitable techniques were invented for softening the leather, and the document was unrolled in 1927. The contents turned out to be a collection of 26 sums of unit fractions, from which historians were able to gain insight into Egyptian methods of calculation. A fourth set of documents, known as the Reisner Papyri after the American archaeologist George Andrew Reisner (1867–1942), who purchased them in 1904, consists of four rolls of records from dockyard workshops, apparently from the reign of Senusret I (1971–1926 BCE). They are now in the Boston Museum of Fine Arts. Another document, the Akhmim Wooden Tablet, is housed in the Egyptian Museum in Cairo. These documents show the practical application of Egyptian mathematics in construction and commerce.

We are fortunate to be able to date the Ahmose Papyrus with such precision. The author himself gives us his name and tells us that he is writing in the fourth month of the flood season of the thirty-third year of the reign of Pharaoh A-user-re

(Apepi I). From this information Egyptologists arrived at a date of around 1650 BCE for this papyrus. Ahmose tells us, however, that he is merely copying work written down in the reign of Pharaoh Ny-maat-re, also known as Amenemhet III (1842–1797 BCE), the sixth pharaoh of the Twelfth Dynasty. From that information it follows that the mathematical knowledge contained in the papyrus is nearly 4000 years old.

What do these documents tell us about the practice of mathematics in ancient Egypt? Ahmose begins his work by describing it as a "correct method of reckoning, for grasping the meaning of things, and knowing everything that is, obscurities... and all secrets."[10] The author seems to value mathematics because of its explanatory power, but that explanatory power was essentially practical. The problems that are solved bear a very strong resemblance to those in other treatises such as the *Jiu Zhang Suanshu*.

The Akhmim Wooden Tablet contains several ways of expressing reciprocals of integers based on dividing unity (64/64) by these integers. According to Milo Gardner,[11] the significance of the number 64 is that it is the number of *ro* in a *hekat* of grain. This origin for the numbers makes sense and gives a solid practical origin for Egyptian arithmetic.

5. Mesopotamia

Some quite sophisticated mathematics was developed four millennia ago in the portion of the Middle East now known as Iraq and Turkey. Unfortunately, this knowledge was preserved on small clay tablets, and nothing like a systematic treatise contemporary with this early mathematics exists. Scholars have had to piece together a mosaic picture of this mathematics from a few hundred clay tablets that show how to solve particular problems. In contrast to Egypt, which had a fairly stable culture throughout many millennia, the region known as Mesopotamia (Greek for "between the rivers") was the home of many civilizations. The name of the region derives from the two rivers, the Euphrates and the Tigris, that flow from the mountainous regions around the Mediterranean, Black, and Caspian seas into the Persian Gulf. In ancient times this region was a very fertile floodplain, although it suffered from an unpredictable climate. It was invaded and conquered many times, and the successive dynasties spoke and wrote in many different languages. The convention of referring to all the mathematical texts that come from this area between 2500 and 300 BCE as "Babylonian" gives undue credit to a single one of the many dynasties that ruled over this region. The cuneiform script is used for writing several different languages. The tablets themselves date to the period from 2000 to about 300 BCE.

Of the many thousands of cuneiform texts scattered through museums around the world, a few hundred have been found to be mathematical in content. Deciphering them has not been an easy task, although the work was made simpler by mutilingual tablets that were created because the cuneiform writers themselves had need to know what had been written in earlier languages. It was not until 1854 that enough tablets had been deciphered to reveal the system of computation used, and not until the early twentieth century were significant numbers of mathematical

---

[10] This is the translation given by Robins and Shute (*1987*, p. 11). Chace (*1927*, p. 49) gives the translation as "the entrance into the knowledge of all existing things and all secrets."

[11] See `http://mathworld.wolfram.com/AkhmimWoodenTablet.html`.

texts deciphered and analyzed. The most complete analysis of these is the 1935 two-volume work by Otto Neugebauer (1899–1992), *Mathematische Keilschrifttexte*, recently republished by Springer-Verlag. A more up-to-date study has been published by the Oxford scholar Eleanor Robson (*1999*).

Cuneiform tablet BMP 15 285. © The British Museum.

Since our present concern is to introduce authors and their works and discuss motivation, there is little more to say about the cuneiform tablets at this point, except to speculate on the uses for these tablets. Some of the tablets that have been discussed by historians of mathematics appear to be "classroom materials," written by teachers as exercises for students. This conclusion is based on the fact that the answers so often "come out even." As Robson (*1995*, p. 11, quoted by Melville, *2002*, p. 2) states, "Problems were constructed from answers known beforehand." Melville provides an example of a different kind from tablet 4652 of the Yale Babylonian Collection in which the figures are not adjusted this way, but a certain technique is presumed. Thus, although there is an unavoidable lack of unity and continuity in the Mesopotamian texts compared with mathematics written on more compact and flexible media, the cuneiform tablets nevertheless contain many problems like those considered in India, China, and Egypt. The applications that these techniques had must be inferred, but we may confidently assume that they were the same everywhere: commerce, government administration, and religious rites, all of which call for counting and measuring objects on Earth and making mathematical observations of the sky in order to keep track of months and years.

6. The Maya

The Maya civilization of southern Mexico and Central America began around 2600 BCE. Its period of greatest material wealth lasted from the third to tenth centuries CE. Archaeologists have found evidence of economic decline from the tenth century onward. This civilization was conquered by Spanish explorers in the sixteenth century, with devastating effect on the ancient culture. Some authorities estimate that as much as 90% of the population may have perished of smallpox. Those who survived were dispersed into the countryside and forbidden to practice their ancient religion. In the 1550s the Franciscan friar Diego de Landa (1524–1579) undertook to destroy all Maya books.[12] Fortunately, three Maya books had already been sent to Europe by earlier colonizers. From these few precious remnants, something can be learned of Maya religion and astronomy, which are their main subjects. The authors remain unknown,[13] so that the books are named for their present locations: the Dresden Codex, the Madrid Codex, and the Paris Codex. A fourth work, consisting of parts of 11 pages of Venus tables, was recovered in Mexico in 1965. It was shown to the Maya scholar Michael Coe, who published it (Coe, *1973*). It is known as the Grolier Codex, after the New York publisher of Coe's book, and it now resides in Mexico City.

6.1. The Dresden Codex. The information on the Dresden Codex given here comes from the following website.

`http://www.tu-dresden.de/slub/proj/maya/maya.html`

An English summary of this information can be found at the following website.

`http://www.tu-dresden.de/slub/proj/maya/mayaeng.html`

The codex that is now in the Sächsische Landesbibliothek of Dresden was purchased in Vienna in 1739 by Johann Christian Goetze (1692–1749), who was at the time director of the royal library at the court of Saxony. It is conjectured that the codex was sent to the Hapsburg Emperor Charles V (1500–1558; he was also King Carlos I of Spain). The codex consists of 74 folios, folded like an accordion, with texts and illustrations in bright colors (see Plate 3). It suffered some damage from the British–American bombing of Dresden during World War II. Fortunately,

---

[12] Sources differ on the dates of Diego de Landa's life. Ironically, de Landa's own work helped in deciphering the Maya hieroglyphs, which he tried to summarize in a history of events in the Yucatan. He had no quarrel with the Maya language, only with the religious beliefs embodied in its books. According to Sharer (*1994*, p. 558), de Landa recognized and wrote about the vigesimal place system used by the Maya.

[13] One can infer something about Maya mathematics from the remains of Maya crafts and architecture. Closs (*1992*, p. 12) notes that one Maya vase contains a painting with a scribe figure who is apparently female and bears the name "Ah Ts'ib, The Scribe."

considerable work had been done earlier on the codex by another director of the Dresden library, a philologist named Ernst Förstemann (1822–1906), who had 200 copies of it made.

The work consists of eight separate treatises and, according to the experts, shows evidence of having been written by eight different people. Dates conjectured for it vary from the thirteenth to the fifteenth centuries, and it may have been a copy of an earlier document. The first 15 folios are devoted to almanacs and astronomy/astrology, while folios 16–23 are devoted to the Moon Goddess. Both of these sections are based on a 260-day calendar known as the *Tzolkin* (see Chapter 5). It is believed that these pages were consulted to determine whether the gods were favorably inclined toward proposed undertakings. Folio 24 and folios 46–50 are Venus tables, containing 312 years of records of the appearance of Venus as morning and evening star. Such records help to establish the chronology of Maya history as well as the date of the manuscript itself. The pictures accompanying the text (Plate 3) seem to indicate a belief that Venus exerted an influence on human life. These pages are followed by eclipse tables over the 33-year period from 755 to 788 CE. Folios 25–28 describe new-year ceremonies, and folios 29–45 give agricultural almanacs. Folios 61–73 give correlations of floods and storms with the 260-day calendar in order to predict the end of the next world cycle. Finally, folio 74 describes the coming end of the current world cycle. The Maya apparently believed there had been at least three such cycles before the current one.

The mathematics that can be gleaned from these codices and the steles that remain in Maya territory is restricted to applications to astronomy and the calendar. The arithmetic that is definitely attested by documents is rudimentary. For example, although there is no reason to doubt that the Maya performed multiplication and division, there is no record showing how they did so. Undoubtedly, there was a Maya arithmetic for commerce, but it is very difficult to reconstruct, since no treatises on the subject exist. Thus, our understanding of the achievements of Maya scientists and mathematicians is limited by the absence of sources. The Maya documents that have survived to the present are "all business" and contain no whimsical or pseudo-practical problems of an algebraic type such as can be found in ancient Chinese, Hindu, Mesopotamian, and Egyptian texts.

Questions and problems

2.1. Does mathematics realize Plato's program of understanding the world by contemplating eternal, unchanging forms that are perceived only by reason, not by the senses?

2.2. To what extent do the points of view expressed by Hamming and Hardy on the value of pure mathematics reflect the nationalities of their authors and the prevailing attitudes in their cultures? Consider that unlike the public radio and television networks in the United States, the CBC in Canada and the BBC in Britain do not spend four weeks a year pleading with their audience to send voluntary donations to keep them on the air. The BBC is publicly funded out of revenues collected by requiring everyone who owns a television set to pay a yearly license fee.

2.3. In an article in the *Review of Modern Physics*, **51**, No. 3 (July 1979), the physicist Norman David Mermin (b. 1935) wrote, "Bridges would not be safer if only people who knew the proper definition of a real number were allowed to design them" (quoted by Mackay, *1991*, p. 172). Granting that at the final point of contact between theory and the physical world, when a human design is to be executed in concrete and steel, every number is only an approximation, is there any value for science and engineering in the concept of an infinitely precise real number? Or is this concept only for idealistic, pure mathematicians? (The problems below may influence your answer.)

2.4. In 1837 and 1839 the crystallographer Auguste Bravais (1811–1863) and his brother Louis (1801–1843) published articles on the growth of plants.[14] In these articles they studied the spiral patterns in which new branches grow out of the limbs of certain trees and classified plants into several categories according to this pattern. For one of these categories they gave the amount of rotation around the limb between successive branches as $137^\circ\,30'\,28''$. Now, one could hardly measure the limb of a tree so precisely. To measure within 10° would require extraordinary precision. To refine such crude measurements by averaging to the claimed precision of $1''$, that is, $1/3600$ of a degree, would require thousands of individual measurements. In fact, the measurements were carried out in a more indirect way, by counting the total number of branches after each full turn of the spiral. Many observations convinced the brothers Bravais that normally there were slightly more than three branches in two turns, slightly less than five in three turns, slightly more than eight in five turns, and slightly less than thirteen in eight turns. For that reason they took the actual amount of revolution between successive branches to be the number we call $1/\Phi = (\sqrt{5}-1)/2 = \Phi - 1$ of a complete (360°) revolution, since

$$\frac{3}{2} < \frac{8}{5} < \Phi < \frac{13}{8} < \frac{5}{3}.$$

Observe that $360^\circ \div \Phi \approx 222.4922359^\circ \approx 222^\circ\,29'\,32'' = 360^\circ - \bigl(137^\circ\,30'\,28''\bigr)$. Was there scientific value in making use of this *real* (infinitely precise) number Φ even though no actual plant grows exactly according to this rule?

2.5. Plate 4 shows a branch of a flowering crab apple tree from the author's garden with the twigs cut off and the points from which they grew marked by pushpins. The "zeroth" pin at the left is white. After that, the sequence of colors is red, blue, yellow, green, pink, clear, so that the red pins correspond to 1, 7, and 13, the blue to 2 and 8, the yellow to 3 and 9, the green to 4 and 10, the pink to 5 and 11, and the clear to 6 and 12. (The green pin corresponding to 4 and part of the clear pin corresponding to 12 are underneath the branch and cannot be seen in the picture.) Observe that when these pins are joined by string, the string follows a helical path of nearly constant slope along the branch. Which pins fall nearest to the intersection of this helical path with the meridian line marked along the length of the branch? How many turns of the spiral correspond to these numbers of twigs? On that basis, what is a good approximation to the number of twigs per turn? Between which pin numbers do the intersections between the spiral and the meridian line fall? For example, the fourth intersection is between pins 6 and 7, indicating that the average number of pins per turn up to that point is between

[14] See the article by I. Adler, D. Barabe, and R. V. Jean, "A history of the study of phyllotaxis," *Annals of Botany*, **80** (1997), 231–244, especially p. 234. The articles by Auguste and Louis Bravais are "Essai sur la disposition générale des feuilles curvisériées," *Annales des sciences naturelles*, **7** (1837), 42–110, and "Essai sur la disposition générale des feuilles rectisériées," *Congrès scientifique de France*, **6** (1839), 278–330.

$\frac{6}{4} = 1.5$ and $\frac{7}{4} = 1.75$. Get upper and lower estimates in this way for all numbers of turns from 1 to 8. What are the narrowest upper and lower bounds you can place on the number of pins per turn in this way?

2.6. Suppose that the pins in Plate 4 had been joined by a curve winding in the opposite direction. How would the numbers of turns of the spiral and the number of pins joined compare? What change would occur in the slope of the spiral?

2.7. With which of the two groups of people mentioned by Plato do you find yourself more in sympathy: the "practical" people, who object to being taxed to support abstract speculation, or the "idealists," who regard abstract speculation as having value to society?

2.8. The division between the practical and the ideal in mathematics finds an interesting reflection in the interpretation of what is meant by solving an equation. Everybody agrees that the problem is to find a number satisfying the equation, but interpretations of "finding a number" differ. Inspired by Greek geometric methods, the Muslim and European algebraists looked for algorithms to invert the operations that defined the polynomial whose roots were to be found. Their object was to generate a sequence of arithmetic operations and root extractions that could be applied to the coefficients in order to exhibit the roots. The Chinese, in contrast, looked for numerical processes to approximate the roots with arbitrary accuracy. What advantages and disadvantages do you see in each of these approaches? What would be a good synthesis of the two methods?

2.9. When a mathematical document such as an early treatise or cuneiform tablet contains problems whose answers "come out even," should one suspect or conclude that it was a teaching device—either a set of problems with simplified data to build students' confidence or a manual for teachers showing how to construct such problems?

2.10. From what is known of the Maya codices, is it likely that they were textbooks intended for teaching purposes, like many of the cuneiform tablets and the early treatises from India, China, and Egypt?

2.11. Why was the Chinese encounter with the Jesuits so different from the Maya encounter with the Franciscans? What differences were there in the two situations, and what conditions account for these differences? Was it merely a matter of the degree of zeal that inspired Diego de Landa and Matteo Ricci, or were there institutional or national differences between the two as well? How much difference did the relative strength of the Chinese and the Maya make?

CHAPTER 3

Mathematical Cultures II

Many cultures borrow from others but add their own ideas to what they borrow and make it into something richer than the pure item would have been. The Greeks, for example, never concealed their admiration for the Egyptians or the debt that they owed to them, yet the mathematics that they passed on to the world was vastly different from what they learned in Egypt. To be sure, much of it was created in Egypt, even though it was written in Greek. The Muslim culture that flourished from 800 to 1500 CE learned something from the Greeks and Hindus, but also made many innovations in algebra, geometry, and number theory. The Western Europeans are still another example. Having learned about algebra and number theory from the Byzantine Empire and the Muslims, they went on to produce such a huge quantity of first-rate mathematics that for a long time European scholars were tempted to think of the rest of the world as merely a footnote to their own work. For example, in his history of western philosophy (*1945*), the British philosopher Bertrand Russell wrote (p. xvi), "In the Eastern Empire, Greek civilization, in a desiccated form, survived, as in a museum, till the fall of Constantinople in 1453, but nothing of importance to the world came out of Constantinople except an artistic tradition and Justinian's Codes of Roman law." He wrote further (p. 427), "Arabic philosophy is not important as original thought. Men like Avicenna and Averroes are essentially commentators." Yet Russell was not *consciously* a chauvinist. In the same book (p. 400), he wrote, "I think that, if we are to feel at home in the world after the present war [World War II], we shall have to admit Asia to equality in our thoughts, not only politically, but culturally."

Until very recently, many textbooks regarded as authoritative were written from this "it all started with the Greeks" point of view. As Chapter 2 has shown, however, there was mathematics before the Greeks, and the Greeks learned it before they began making their own remarkable innovations in it.

1. Greek and Roman mathematics

The Greeks of the Hellenic period (to the end of the fourth century BCE) traced the origins of their mathematical knowledge to Egypt and the Middle East. This knowledge probably came in "applied" form in connection with commerce and astronomy/astrology. The evidence of Mesopotamian numerical methods shows up most clearly in the later Hellenistic work on astronomy by Hipparchus (second century BCE) and Ptolemy (second century CE). Earlier astronomical models by Eudoxus (fourth century BCE) and Apollonius (third century BCE) were more geometrical. Jones (*1991*, p. 445) notes that "the astronomy that the Hellenistic Greeks received from the hands of the Babylonians was by then more a skill than a science: the quality of the predictions was proverbial, but in all likelihood the practitioners knew little or nothing of the origins of their schemes in theory and observations." Among the techniques transmitted to the Greeks and ultimately to the modern world was the convention of dividing a circle into 360 equal parts (degrees). Greek astronomers divided the radius into 60 equal parts so that the units of length on the radius and on the circle were very nearly equal.

The amount that the Greeks learned from Egypt is the subject of controversy. Many scholars who have read the surviving mathematical texts from papyri have concluded that Egyptian methods of computing were too cumbersome for application to the complicated measurements of astronomers. Yet both Plato and Aristotle speak approvingly of Egyptian computational methods and the ways in which they were taught. As for geometry, it is generally acknowledged that the Egyptian insight was extraordinary; the Egyptians knew how to find the volume of a pyramid, for example. They even found the area of a hemisphere, the only case known before Archimedes in which the area of a curved surface is found.[1] The case for advanced Egyptian mathematics is argued in some detail by Bernal (*1992*), who asserts that Ptolemy himself was an Egyptian. The argument is difficult to settle, since little is known of Ptolemy personally; for us, he is simply the author of certain works on physics and astronomy.

Because of their extensive commerce, with its need for counting, measuring, navigation, and an accurate calendar, the Ionian Greek colonies such as Miletus on the coast of Asia Minor and Samos in the Aegean Sea provided a very favorable environment for the development of mathematics, and it was there, with the philosophers Thales of Miletus (ca. 624–547 BCE) and Pythagoras of Samos (ca. 570–475 BCE), that Greek mathematics began.

1.1. Sources. Since the material on which the Greeks wrote was not durable, all the original manuscripts have been lost except for a few ostraca (shells) found in Egypt. We are dependent on copyists for preserving the information in early Greek works, since few manuscripts that still exist were written more than 1000 years ago. We are further indebted to the many commentators who wrote summary histories of philosophy, including mathematics, for the little that we know about the works that have not been preserved and their authors. The most prominent among these commentators are listed below. They will be mentioned many times in the chapters that follow.

Marcus Vitruvius (first century BCE) was a Roman architect who wrote an extremely influential treatise on architecture in 10 books. He is regarded as a rather unreliable source for information about mathematics, however.

Plutarch (45–120 CE) was a pagan author, apparently one of the best educated people of his time, who wrote on many subjects. He is best remembered as the author of the *Parallel Lives of the Greeks and Romans*, in which he compares famous Greeks with eminent Romans who engaged in the same occupation, such as the orators Demosthenes and Cicero.[2] Plutarch is important to the history of mathematics for what he reports on natural philosophers such as Thales.

---

[1] Some authors claim that the surface in question was actually half of the lateral surface of a cylinder, but the words used seem more consistent with a hemisphere. In either case it was a curved surface.

[2] Shakespeare relied on Plutarch's account of the life of Julius Caesar, even describing the miraculous omens that Plutarch reported as having occurred just before Caesar's death.

Theon of Smyrna (ca. 100 CE) was the author of an introduction to mathematics written as background for reading Plato, a copy of which still exists. It contains many quotations from earlier authors.

Diogenes Laertius (third century CE) wrote a comprehensive history of philosophy, *Lives of Eminent Philosophers*, which contains summaries of many earlier works and gives details of the lives and work of many of the pre-Socratic philosophers. He appears to be the source of the misnomer "Pythagorean theorem" that has come down to us (see Zhmud, *1989*, p. 257).

Iamblichus (285–330 CE) was the author of many treatises, including 10 books on the Pythagoreans, five of which have been preserved.

Pappus (ca. 300 CE) wrote many books on geometry, including a comprehensive treatise of eight mathematical books. He is immortalized in calculus books for his theorem on the volume of a solid of revolution. Besides being a first-rate geometer in his own right, he wrote commentaries on the *Almagest* of Ptolemy and the tenth book of Euclid's *Elements*.

Proclus (412–485 CE) is the author of a commentary on the first book of Euclid, in which he quoted a long passage from a history of mathematics, now lost, by Eudemus, a pupil of Aristotle.

Simplicius (500–549 CE) was a commentator on philosophy. His works contain many quotations from the pre-Socratic philosophers.

Eutocius (ca. 700 CE) was a mathematician who lived in the port city of Askelon in Palestine and wrote an extensive commentary on the works of Archimedes.

Most of these commentators wrote in Greek. Knowledge of Greek sank to a very low level in western Europe as a result of the upheavals of the fifth century. Although learning was preserved by the Church and all of the New Testament was written in Greek, a Latin translation (the Vulgate) was made by Jerome in the fifth century. From that time on, Greek documents were preserved mostly in the Eastern (Byzantine) Empire. After the Muslim conquest of North Africa and Spain in the eighth century, some Greek documents were translated into Arabic and circulated in Spain and the Middle East. From the eleventh century on, as secular learning began to revive in the West, scholars from northern Europe made journeys to these centers and to Constantinople, copied out manuscripts, translated them from Arabic and Greek into Latin, and tried to piece together some long-forgotten parts of ancient learning.

1.2. General features of Greek mathematics. Greek mathematics—that is, mathematics written in ancient Greek—is exceedingly rich in authors and works. Its most unusual feature, compared with what went before, is its formalism. Mathematics is developed systematically from definitions and axioms, general theorems are stated, and proofs are given. This formalism is the outcome of the entanglement of mathematics with Greek philosophy. It became a model to be imitated in many later scientific treatises, such as Newton's *Philosophiæ naturalis principia mathematica.* Of course, Greek mathematics did not arise in the finished form found in the treatises. Tradition credits Thales only with knowing four geometric propositions. By the time of Pythagoras, much more was known. The crucial formative period was the first half of the fourth century BCE, when Plato's Academy flourished. Plato himself was interested in mathematics because he hoped for a sort of "theory of everything," based on fundamental concepts perceived by the mind.

Plato is famous for his theory of ideas, which had both metaphysical and epistemological aspects. The metaphysical aspect was a response to two of his predecessors, Heraclitus of Ephesus (ca. 535–475 BCE), who asserted that everything is in constant flux, and Parmenides (born around 515 BCE), who asserted that knowledge is possible only in regard to things that do not change. One can see the obvious implication: Everything changes (Heraclitus). Knowledge is possible only about things that do not change (Parmenides). *Therefore*... . To avoid the implication that no knowledge is possible, Plato restricted the meaning of Heraclitus' "everything" to objects of sense and invented eternal, unchanging Forms that could be objects of knowledge.

The epistemological aspect of Plato's philosophy involves universal propositions, statements such as "Lions are carnivorous" (our example, not Plato's), meaning "*All* lions are carnivorous." This sentence is grammatically inconsistent with its meaning, in that the grammatical subject is the set of all lions, while the assertion is not about this set but about its individual members. It asserts that each of them is a carnivore, and therein lies the epistemological problem. What *is* the real subject of this sentence? It is not any particular lion. Plato tried to solve this problem by inventing the Form or Idea of a lion and saying that the sentence really asserts a relation perceived in the mind between the Form of a lion and the Form of a carnivore. Mathematics, because it dealt with objects and relations perceived by the mind, appeared to Plato to be the bridge between the world of sense and the world of Forms. Nevertheless, mathematical objects were not the same thing as the Forms. Each Form, Plato claimed, was unique. Otherwise, the interpretation of sentences by use of Forms would be ambiguous. But mathematical objects such as lines are not unique. There must be at least three lines, for example, in order for a triangle to exist. Hence, as a sort of hybrid of sense experience and pure mental creation, mathematical objects offered a way for the human soul to ascend to the height of understanding, by perceiving the Forms themselves. Incorporating mathematics into education so as to realize this program was Plato's goal, and his pupils studied mathematics in order to achieve it. Although the philosophical goal was not reached, the effort expended on mathematics was not wasted; certain geometric problems were solved by people associated with Plato, providing the foundation of Euclid's famous work, known as the *Elements.*

Within half a century of Plato's death, Euclid was writing that treatise, which is quite free of all the metaphysical accoutrements that Plato's pupils had experimented with. However, later neo-Platonic philosophers such as Proclus attempted to reintroduce philosophical ideas into their commentary on Euclid's work. The historian and mathematician Otto Neugebauer (*1975*, p. 572) described the philosophical aspects of Proclus' introduction as "gibberish," and expressed relief that scientific methodology survived despite the prevalent dogmatic philosophy.

According to Diels (*1951*, 44A5), Plato met the Pythagorean Philolaus in Sicily in 390. In any case, Plato must certainly have known the work of Philolaus, since in the *Phaedo* Socrates says that both Cebes and Simmias are familiar with the work of Philolaus and implies that he himself knows of it at second hand. It seems likely, then, that Plato's interest in mathematics began some time after the death of Socrates and continued for the rest of his life, that mathematics played an important role in the curriculum of his Academy and in the research conducted there, and that Plato himself played a leading role in directing that research. We do not, however, have any theorems that can with confidence be attributed to Plato himself. Lasserre (*1964*, p. 17) believes that the most important mathematical work at the Academy was done between 375 and 350 BCE.

Socrates explained that arithmetic was needed both to serve the eye of the soul and as a practical instrument in planning civic projects and military campaigns:

> The kind of knowledge we are seeking seems to be as follows. It is necessary for a military officer to learn (*matheîn*) these things for the purpose of proper troop deployment, and the philosopher must have risen above change, in order to grasp the essence of things, or else never become skilled in calculation (*logistikôͅ*).

Plato, through Socrates, complains of the lack of a government subsidy for geometry. In his day solid geometry was underdeveloped in comparison with plane geometry, and Socrates gave what he thought were the reasons for its backwardness:

> First, no government holds [the unsolved problems in solid geometry] in honor; and they are researched in a desultory way, being difficult. Second, those who are doing the research need a mentor, without which they will never discover anything. But in the first place, to become a mentor is difficult; and in the second place, after one became a mentor, as things are just now, the arrogant people doing this research would never listen to him. But if the entire state were to act in concert in conducting this research with respect, the researchers would pay heed, and by their combined intensive work the answers would become clear.

Plato himself was among that group of people mentioned in Chapter 2, for whom the "eye of the soul" was sufficient justification for intellectual activity. He seems to have had a rather dim view of the second group, the practical-minded people. In his long dialogue *The Laws*, one of the speakers, an Athenian, rants about the shameful Greek ignorance of incommensurables, surely a topic of limited application in the lives of most people.

1.3. Works and authors. Books on mathematics written in Greek begin appearing early in Hellenistic times (third century BCE) and continue in a steady stream for hundreds of years. We list here only a few of the most outstanding authors.

Euclid. This author lived and worked in Alexandria, having been invited by Ptolemy Soter (Ptolemy I) shortly after the city was founded. Essentially nothing is known of his life beyond that fact, but his famous treatise on the basics of geometry (the *Elements*) has become a classic known all over the world. Several of his minor works—the *Optics*, the *Data*, and the *Phænomena*—also have been preserved. Unlike Aryabhata and Bhaskara, Euclid did not provide any preface to tell us why he wrote his treatise. We do, however, know enough of the Pythagorean philosophy to understand why they developed geometry and number theory to the extent that they did, and it is safe to conclude that this kind of work was considered valuable because it appealed to the intellect of those who could understand it.

Archimedes. Much more is known of Archimedes (ca. 287–212 BCE). About 10 of his works have been preserved, including the prefaces that he wrote in the form of "cover letters" to the people who received the works. Here is one such letter, which accompanied a report of what may well be regarded as his most profound achievement—proving that the area of a sphere is four times the area of its equatorial circle.

> On a former occasion I sent you the investigations which I had up to that time completed, including the proofs, showing that any segment bounded by a straight line and a section of a right-angled cone [parabola] is four-thirds of the triangle which has the same base with the segment and equal height. Since then certain theorems not hitherto demonstrated have occurred to me, and I have worked out the proofs of them. They are these: first, that the surface of any sphere is four times its greatest circle... For, though these properties also were naturally inherent in the figures all along, yet they were in fact unknown to all the many able geometers who lived before Eudoxus, and had not been observed by anyone. Now, however, it will be open to those who possess the requisite ability to examine these discoveries of mine. [Heath, *1897*, Dover edition, pp. 1–2]

As this letter shows, mathematics was a "going concern" by Archimedes' time, and a community of mathematicians existed. Archimedes is known to have studied in Alexandria. He perished when his native city of Syracuse was taken by the Romans during the Second Punic War. Some of Archimedes' letters, like the one quoted above, give us a glimpse of mathematical life during his time. Despite being widely separated, the mathematicians of the time sent one another challenges and communicated their achievements.

Apollonius. Apollonius, about one generation younger than Archimedes, was a native of what is now Turkey. He studied in Alexandria somewhat after the time of Euclid and is also said to have taught there. He eventually settled in Pergamum (now Bergama in Turkey). He is the author of eight books on conic sections, four of which survive in Greek and three others in an Arabic translation. We know that there were originally eight books because commentators, especially Pappus, described the work and told how many propositions were in each book.

In his prefaces Apollonius implies that geometry was simply part of what an educated person would know, and that such people were as fascinated with it in his time as they are today about the latest scientific achievements. Among other things, he said the following.

> During the time I spent with you at Pergamum I observed your eagerness to become aquainted with my work in conics. [Book I]
>
> I undertook the investigation of this subject at the request of Naucrates the geometer, at the time when he came to Alexandria and stayed with me, and, when I had worked it out in eight books, I gave them to him at once, too hurriedly, because he was on the point of sailing; they had therefore not been thoroughly revised,

> indeed I had put down everything just as it occurred to me, postponing revision until the end. [Book II]

Ptolemy. Claudius Ptolemy was primarily an astronomer and physicist, although these subjects were hardly distinct from mathematics in his time. He lived in Alexandria during the second century, as is known from the astronomical observations that he made between 127 and 141 CE. He created an intricate and workable Earth-centered mathematical system of explaining the motion of the planets and systematized it in a treatise known as the *Syntaxis*, which, like Euclid's, consisted of 13 books. Also like Euclid's treatise, Ptolemy's *Syntaxis* became a classic reference and was used for well over a thousand years as the definitive work on mathematical astronomy. It soon became known as the "greatest" work (*megistos* in Greek) and when translated into Arabic became *al-megista* or the *Almagest*, as we know it today.

Diophantus. Little is known about this author of a remarkable treatise on what we now call algebra and number theory. He probably lived in the third century CE, although some experts believe he lived earlier than that. His treatise is of no practical value in science or commerce, but its problems inspired number theorists during the seventeenth century and led to the long-standing conjecture known as Fermat's last theorem. The 1968 discovery of what may be four books from this treatise that were long considered lost was the subject of a debate among the experts, some of whom believed the books might be commentaries, perhaps written by the late fourth-century commentator Hypatia. If so, they would be the only work by Hypatia still in existence.

Pappus. Pappus, who is known to have observed a solar eclipse in Alexandria in 320 CE, was the most original and creative of the later commentators on Greek geometry and arithmetic. His *Synagōgē* (*Collection*) consists of eight books of insightful theorems on arithmetic and geometry, as well as commentary on the works of other authors. In some cases where works of Euclid, Apollonius, and others have been lost, this commentary tells something about these works. Pappus usually assumes that the reader is interested in what he has to say, but sometimes he gives in addition a practical justification for his study, as in Book 8:

> The science of mechanics, my dear Hermodorus, has many important uses in practical life, and is held by philosophers to be worthy of the highest esteem, and is zealously studied by mathematicians, because it takes almost first place in dealing with the nature of the material elements of the universe. [Thomas, *1941*, p. 615]

As a commentator, Pappus was highly original, and the later commentators Theon of Alexandria (late fourth century) and his daughter Hypatia (ca. 370–415) produced respectable work, including a standard edition of Euclid's *Elements*. Several of Theon's commentaries still exist, but nothing authored by Hypatia has been preserved, unless the books of Diophantus mentioned above were written by her. Very little of value can be found in Greek mathematics after the fourth century. As Gow (*1884*, p. 308) says:

> The *Collection* of Pappus is not cited by any of his successors, and none of them attempted to make the slightest use of the proofs and *aperçus* in which the book abounds... His work is only the

> last convulsive effort of Greek geometry which was now nearly dead and was never effectually revived.

Greek mathematics held on longer in the Byzantine Empire than in Western Europe. Although Theon of Alexandria had found it necessary to water down the more difficult parts of Greek geometry for the sake of his weak students, the degeneration in Latin works was even greater. The philosopher Boethius (480–524) wrote Latin translations of many classical Greek works of mathematics and philosophy. His works on mathematics were translations based on Nicomachus and Euclid. Boethius' translation of Euclid has been lost. However, it is believed to be the basis of many other medieval manuscripts, some of which use his name. These are referred to as "Boethius" or pseudo-Boethius. The works of Boethius fit into the classical quadrivium of arithmetic, geometry, music, and astronomy.

Politically and militarily, the fifth century was full of disasters in Italy, and some of the best minds of the time turned from public affairs to theological questions. For many of these thinkers mathematics came to be valued only to the extent that it could inspire religious feelings. The pseudo-Boethius gives a good example of this point of view. He writes:[3]

> The utility of geometry is threefold: for work, for health, and for the soul. For work, as in the case of a mechanic or architect; for health, as in the case of the physician; for the soul, as in the case of the philosopher. If we pursue this art with a calm mind and diligence, it is clear in advance that it will illuminate our senses with great clarity and, more than that, will show what it means to subordinate the heavens to the soul, to make accessible all the supernal mechanism that cannot be investigated by reason in any other way and through the sublimity of the mind beholding it, also to integrate and recognize the Creator of the world, who veiled so many deep secrets.

In the tenth century, Gerbert of Aurillac (940–1003), who became Pope Sylvester II in 999, wrote a treatise on geometry based on Boethius. His reasons for studying geometry were similar:

> Indeed the utility of this discipline to all lovers of wisdom is the greatest possible. For it leads to vigorous exercises of the soul, and the most subtle demands on the intuition, and to many certain inquiries by true reasoning, in which wonderful and unexpected and joyful things are revealed to many along with the wonderful vigor of nature, and to contemplating, admiring, and praising the power and ineffable wisdom of the Creator who apportioned all things according to number and measure and weight; it is replete with subtle speculations.

These uses of geometry were expressed in the last Canto of Dante's *Divine Comedy*, which describes the poet's vision of heaven:

[3] This quotation and the next can be read online at `http://pld.chadwyck.com`, a commercial website. This passage is from Vol. 63; the next is from Vol. 139. Both can be reached by searching under "geometria" as title.

Like the geometer who applies all his powers
To measure the circle, but does not find
By thinking the principle he needs,

Such was I, in this new vista.
I wished to see how the image came together
With the circle and how it could be divined there.

But my own wings could not have made the flight
Had not my mind been struck
By a flash in which his will came to me.

In this lofty vision I could do nothing.
But now turning my desire and will,
Like a wheel that is uniformly moved,

Was the love that moves the sun and the other stars.

The quadrivium, from Boethius' *Arithmetic*. From left to right: Music holding an instrument, Arithmetic doing a finger computation, Geometry studying a set of figures, Astrology holding a set of charts for horoscopes.

The Byzantine Empire and modern Greece. Mathematics continued in the Byzantine Empire until the Turks conquered Constantinople in 1453. Of several figures who contributed to it, the one most worthy of mention is the monk Maximus Planudes (ca. 1260–1310), who is best known for the literature that he preserved (including Aesop's *Fables*). Planudes wrote commentaries on the work of Diophantus and gave an account of the Hindu numerals that was one of the sources from which these numerals eventually came down to us (Heath, *1921*, pp. 546–547).

The mainland of Greece was partitioned and disputed among various groups for centuries: the Latin West, the Byzantine Empire, the Ottoman Empire, the Venetians, and the Normans invaded or ruled over parts of it. In the fourteenth century it became a part of the Ottoman Empire, from which it gained independence only in the 1820s and 1830s. Even before independence, however, Greek scholars, inspired by the great progress in Europe, were laying the foundations of a modern mathematical school (see Phili, *1997*).

2. Japan

Both Korea and Japan adopted the Chinese system of writing their languages. The Chinese language was the source of a huge amount of technical vocabulary in Korea and Japan over many centuries, and even in recent times in Viet Nam (Koblitz, *1990*, p. 26). The establishment of Buddhism in Japan in the sixth century increased the rate of cultural importation from China and even from India.[4]

The influence of Chinese mathematics on both Korea and Japan was considerable. The courses of university instruction in this subject in both countries were based on reading (in the original Chinese language) the Chinese classics we discussed in Chapter 2. In relation to Japan the Koreans played a role as transmitters, passing on Chinese learning and inventions. This transmission process began in 553–554 when two Korean scholars, Wang Lian-tung and Wang Pu-son, journeyed to Japan. For many centuries both the Koreans and the Japanese worked within the system of Chinese mathematics. The earliest records of new and original work in these countries date from the seventeenth century. By that time mathematical activity was exploding in Europe, and Europeans had begun their long voyages of exploration and colonization. There was only a brief window of time during which indigenous mathematics independent of Western influence could grow up in these countries. The following synopsis is based mostly on the work of Mikami (*1913*), Smith and Mikami (*1914*), and Murata (*1994*). Following the usage of the first two of these sources, all Japanese names are given surname first. A word of caution is needed about the names, however. Most Chinese symbols (*kanji* in Japanese) have at least two readings in Japanese. For example, the symbol read as *chu* in the Japanese word for China (*Chugoku*), is also read as *naka* (meaning *middle*) in the surname Tanaka. These variant readings often cause trouble in names from the past, so that one cannot always be sure how a name was pronounced. As Mikami (*1913*, p. viii) says, "We read Seki Kōwa, although his personal name Kōwa should have been read Takakazu." Several examples of such alternate readings will be encountered below. A list of these names and their kanji rendering can be found in a paper of Martzloff (*1990*, p. 373).

[4] The Japanese word for China—*Chūgoku*—means literally *Midland*, that is, between Japan and India.

2.1. Chinese influence and calculating devices. All the surviving Japanese records date from the time after Japan had adopted the Chinese writing system. Japanese mathematicians were for a time content to read the Chinese classics. In 701 the emperor Monbu established a university system in which the mathematical part of the curriculum consisted of 10 Chinese classics. Some of these are no longer known, but the *Zhou Bi Suan Jing*, *Sun Zi Suan Jing*, *Jiu Zhang Suanshu*, and *Hai Dao Suan Jing* were among them. Japan was disunited for many centuries after this early encounter with Chinese culture, and the mathematics that later grew up was the result of a reintroduction in the sixteen and seventeenth centuries. In this reintroduction, the two most important works were the *Suan Fa Tong Zong* by Cheng Dawei, and the *Suan Shu Chimeng* of Zhu Shijie, both mentioned in Chapter 2. The latter became part of the curriculum in Korea very soon after it was written and was published in Japan in the mid-seventeenth century. The evidence of Chinese influence is unmistakable in the mechanical methods of calculation used for centuries—counting rods, counting boards, and the abacus, which played an especially important role in Japan.

The Koreans adopted the Chinese counting rods and counting boards, which the Japanese subsequently adopted from them. The abacus (*suan pan*) was invented in China, probably in the fourteenth century, when methods of computing with counting rods had become so efficient that the rods themselves were a hindrance to the performance of the computation. From China the invention passed to Korea, where it was known as the *sanbob*. Because it did not prove useful in Korean business, it did not become widespread there. It passed on to Japan, where it is known as the *soroban*, which may be related to the Japanese word for an orderly table (*soroiban*). The Japanese made two important technical improvements in the abacus: (1) they replaced the round beads by beads with sharp edges, which are easier to manipulate; and (2) they eliminated the superfluous second 5-bead on each string.

2.2. Japanese mathematicians and their works. A nineteenth-century Japanese historian reported that the emperor Hideyoshi sent the scholar Mōri Shigeyoshi (Mōri Kambei) to China to learn mathematics. According to the story, the Chinese ignored the emissary because he was not of noble birth. When he returned to Japan and reported this fact, the emperor conferred noble status on him and sent him back. Unfortunately, his second visit to China coincided with Hideyoshi's unsuccessful attempt to invade Korea, which made his emissary unwelcome in China. Mōri Shigeyoshi did not return to Japan until after the death of Hideyoshi, but when he did return (in the early seventeenth century), he brought the abacus with him. Whether this story is true or not, it is a fact that Mōri Shigeyoshi was one of the most influential early Japanese mathematicians. He wrote several treatises, all of which have been lost, but his work led to a great flowering of mathematical activity in seventeenth-century Japan, through the work of his students. This mathematics was known as *wasan*, and written using two Chinese characters. The first is *wa*, a word still used to denote Japanese-style work in arts and crafts, meaning literally *harmony*. The second is *san*, meaning calculation, the same Chinese symbol that respresents *suan* in the many Chinese classics mentioned above.[5] Murata (*1994*, p. 105) notes that the primary concern in *wasan* was to obtain elegant results, even when those results required very complicated calculations, and that "many

[5] The modern Japanese word for mathematics is *sūgaku*, meaning literally *number study*.

Wasanists were men of fine arts rather than men of mathematics in the European sense."

According to Murata (*1994*), the stimulus for the development of *wasan* came largely from the two Chinese classics mentioned above, the 1593 arithmetical treatise *Suan Fa Tong Zong* and the algebraic treatise *Suan Shu Chimeng*. The latter was particularly important, since it came with no explanatory notes and a rebellion in China had made communication with Chinese scholars difficult. By the time this treatise was understood, the Japanese mathematicians had progressed beyond its contents.

Sangaku. The shoguns of the Tokugawa family (1600–1868) concentrated their foreign policy on relations with China and held Western visitors at arms length, with the result that Japan was nearly closed to the Western world for 250 years. During this time a fascinating form of mathematics known as *sangaku* (mathematical study, the "study" being a physical plaque) arose, involving the posting of mathematical plaques at sacred shrines (see Plate 2). These problems are discussed in detail in the book of Fukagawa and Pedoe (*1989*).

Yoshida Koyu. Mōri Shigeyoshi trained three outstanding students during his lifetime, of whom we shall discuss only the first. This student was Yoshida Koyu (Yoshida Mitsuyoshi, 1598–1672). Being handicapped in his studies at first by his ignorance of Chinese, Yoshida Koyu devoted extra effort to this language in order to read the *Suan Fa Tong Zong*. Having read this book, Yoshida Koyu made rapid progress in mathematics and soon excelled even Mōri Shigeyoshi himself. Eventually, he was called to the court of a nobleman as a tutor in mathematics. In 1627 Yoshida Koyu wrote a textbook in Japanese, the *Jinkō-ki* (*Treatise on Large and Small Numbers*), based on the *Suan Fa Tong Zong*. This work helped to popularize the abacus (soroban) in Japan. It concluded with a list of challenge questions and thereby stimulated a great deal of further work. These problems were solved in a later treatise, which in turn posed new mathematical problems to be solved; this was the beginning of a tradition of posing and solving problems that lasted for 150 years.

Seki Kōwa and Takebe Kenkō. One figure in seventeenth-century Japanese mathematics stands out far above all others, a genius who is frequently compared with Archimedes, Newton, and Gauss.[6] His name was Seki Kōwa, and he was born around the year 1642, the year in which Isaac Newton was born in England. The stories told of him bear a great resemblance to similar stories told about other mathematical geniuses. For example, one of his biographers says that at the age of 5 Seki Kōwa pointed out errors in a computation that was being discussed by his elders. A very similar story is told about Gauss. Being the child of a samurai father and adopted by a noble family, Seki Kōwa had access to books. He was mostly self-educated in mathematics, having paid little attention to those who tried to instruct him; in this respect he resembles Newton. Like Newton, he served as an advisor on high finance to the government, becoming examiner of accounts to the lord of Koshu. Unlike Newton, however, he was a popular teacher and physically vigorous. He became a shogunate samurai and master of ceremonies in the household of the

[6] His biography suggests that the real comparison should be with Pythagoras, since he assembled a devoted following, and his followers were inclined to attribute results to him even when his direct influence could not be established. Newton and Gauss were not "people persons," and Gauss hated teaching.

Shogun. He died at the age of 66, leaving no direct heirs. His tomb in the Buddhist cemetery in Tokyo was rebuilt 80 years after his death by mathematicians of his school. His pedagogical activity earned him the title of *Sansei*, meaning *Arithmetical Sage*, a title that was carved on his tomb. Although he published very little during his lifetime, his work became known through his teaching activity, and he is said to have left copious notebooks.

Seki Kōwa made profound contributions to several areas of mathematics, in some cases anticipating results that were being obtained independently in Europe about this time. According to Mikami (*1913*, p. 160), he kept his technique a secret from the world at large; but apparently he confided it to his pupil Takebe Kenkō (Takebe Katahiro, 1664–1739). Some scholars say that Takebe Kenkō refused to divulge the secret, saying, "I fear that one whose knowledge is so limited as mine would tend to misrepresent its significance." However, other scholars claim that Takebe Kenkō did write an exposition of the latter method, and that it amounts to the principles of cancellation and transposition. These two scholars, together with Takebe Kenkō's brother, compiled a 20-volume encyclopedia, the *Taisei Sankyō* (*Great Mathematical Treatise*), containing all the mathematics known in their day.

Takebe Kenkō also wrote a book that is unique in its time and place, bearing the title *Tetsujutsu Sankyō* (roughly, *The Art of Doing Mathematics*, published in 1722), in which he speculated on the metaphysics of mathematical concepts and the kind of psychology needed to solve different types of mathematical problems (Murata, *1994*, pp. 107–108).

In Japan, knowledge of the achievements of Western mathematicians became widespread in the late nineteenth century, while the flow of knowledge in the opposite direction has taken longer. A book entitled *The Theory of Determinants in the Historical Order of Development*, which is a catalog of papers on the subject with commentaries, was written by the South African mathematician Thomas Muir (1844–1934) in 1905. Although this book consists of four volumes totaling some 2000 pages, it does not mention Seki Kōwa, the true discoverer of determinants!

Other treatises. The book *Sampō Ketsugi-shō* (*Combination Book*, but it contains many results on areas and volumes that we now compute using calculus) was published in 1661 by Isomura Yoshinori (Isomura Kittoku), a student of a student of Mōri Shigeyoshi. Although Isomura is known to have died in 1710, his birth date is uncertain The book was revised in 1684. Sawaguchi Kazuyuki, whose exact dates also are not known, wrote *Kokon Sampō-ki* (*Mathematics Ancient and Modern*) in 1671. This work is cited by Murata as the proof that *wasan* had developed beyond its Chinese origins.

The modern era in Japan. In the seventeenth century the Tokugawa shoguns had adopted a very strict policy vis-à-vis the West, one that could be enforced in an island kingdom such as Japan. Commercial contacts with the Dutch, however, resulted in some cultural penetration, and Western mathematical advances came to be known little by little in Japan. By the time Japan was opened to the West in the mid-nineteenth century, Japanese mathematicians were already aware of many European topics of investigation. In joining the community of nations for trade and politics, Japan also joined it intellectually. In the early nineteenth century, Japanese mathematicians were writing about such questions as the rectification of the ellipse, a subject of interest in Europe at the same period. By the end of the nineteenth century there were several Japanese mathematical journals publishing (in European languages) mathematical work comparable to what was being done in Europe at the same period, and a few European scholars were already reading these journals to see what advances were being made by the Japanese. In the twentieth century the number of Japanese works being read in the West multiplied, and Japanese mathematicians such as Gorō Shimura (b. 1930), Shōshichi Kobayashi (b. 1932), and many others have been represented among the leaders in nearly every field of mathematics.

3. The Muslims

From the end of the eighth century through the period referred to as Medieval in European history, the Umayyad and Abbasid Caliphates, centered in what is now Spain and Iraq respectively, produced an artistically and scientifically advanced culture, with works on mathematics, physics, chemistry, and medicine written in Arabic, the common language of scholars throughout the Muslim world. Persian, Hebrew, and other languages were also used by scholars working in this predominantly Muslim culture. Hence the label "Islamic mathematics" that we prefer to use is only a rough description of the material we shall be discussing. It is convenient, like the label "Greek mathematics" used above to refer to works written in the culture where scholars mostly wrote in Greek.

3.1. Islamic science in general. The religion of Islam calls for prayers facing Mecca at specified times of the day. That alone would be sufficient motive for studying astronomy and geography. Since the Muslim calendar is lunar rather than lunisolar, religious feasts and fasts are easy to keep track of. Since Islam forbids representation of the human form in paintings, mosques are always decorated with abstract geometric patterns (see Özdural, *2000*). The study of this *ornamental geometry* has interesting connections with the theory of transformation groups.

Hindu influences. According to Colebrooke (*1817*, pp. lxiv–lxv), in the year 773 CE, al-Mansur, the second caliph of the Abbasid dynasty, who ruled from 754 to 775, received at his court a Hindu scholar bearing a book on astronomy referred to in Arabic as *Sind-hind* (most likely, *Siddhanta*). Al-Mansur had this book translated into Arabic. No copies survive, but the book seems to have been the *Brahmasphutasiddhanta* mentioned above. This book was used for some decades, and an abridgement was made in the early ninth century, during the reign of al-Mamun (caliph from 813 to 833), by Muhammed ibn Musa al-Khwarizmi (ca. 780–850), who also wrote his own treatise on astronomy based on the Hindu work and the work of Ptolemy. Al-Mamun founded a "House of Wisdom" in Baghdad, the capital of his empire. This institution was much like the Library at Alexandria, a place of scholarship analogous to a modern research institute.

In the early days of this scientific culture, one of the main concerns of the scholars was to find and translate into Arabic as many scientific works as possible. The effort made by Islamic rulers, administrators, and merchants to acquire and translate Hindu and Hellenistic texts was prodigious. The works had first to be located, a job requiring much travel and expense. Next, they needed to be understood and adequately translated; that work required a great deal of labor and time, often involving many people. The world is much indebted to the scholars who undertook this work, for two reasons. First, some of the original works have been lost, and only their Arabic translations survive.[7] Second, the translators, inspired by the work they were translating, wrote original works of their own. The mechanism of this two-part process has been well described by Berggren (*1990*, p. 35):

> Muslim scientists and patrons were the main actors in the acquisition of Hellenistic science inasmuch as it was they who initiated the process, who bore the costs, whose scholarly interests dictated the choice of material to be translated and on whom fell the burden of finding an intellectual home for the newly acquired material within the Islamic *dār al-'ilm* ("abode of learning").

We shall describe the two parts of the process as "acquisition" and "development." The acquisitions were too many to be listed here. Some of the major ones were listed by Berggren (*2002*). They include Euclid's *Elements*, *Data*, and *Phænomena*, Ptolemy's *Syntaxis* (which became the *Almagest* as a result) and his *Geography*, many of Archimedes' works and commentaries on them, and Apollonius' *Conics*.

The development process as it affected the *Conics* of Apollonius was described by Berggren (*1990*, pp. 27–28). This work was used to analyze the astrolabe in the ninth century and to trisect the angle and construct a regular heptagon in the tenth century. It continued to be used down through the thirteenth century in the theory of optics, for solving cubic equations, and to study the rainbow. To the two categories that we have called acquisition and development Berggren adds the process of editing the texts to systematize them, and he emphasizes the very important role of mathematical philosophy or criticism engaged in by Muslim mathematicians. They speculated and debated Euclid's parallel postulate, for example, thereby continuing a discussion that began among the ancient Greeks and continued for 2000 years until it was finally settled in the nineteenth century.

The scale of the Muslim scientific schools is amazing when looked at in comparison with the populations and the general level of economic development of the time. Here is an excerpt from a letter of the Persian mathematician al-Kashi (d. 1429) to his father, describing the life of Samarkand, in Uzbekistan, where the great astronomer Ulugh Beg (1374–1449), grandson of the conqueror Timur the Lame, had established his observatory (Bagheri, *1997*, p. 243):

> His Royal Majesty had donated a charitable gift... amounting to thirty thousand... dinars, of which ten thousand had been ordered to be given to students. [The names of the recipients] were written down; [thus] ten thousand-odd students steadily engaged in learning and teaching, and qualifying for a financial aid, were listed... Among them there are five hundred persons who have begun [to study] mathematics. His Royal Majesty the World-Conqueror, may God perpetuate his reign, has been engaged in this art... for the last twelve years.

[7] Toomer (*1984*) points out that in the case of Ptolemy's *Optics* the Arabic translation has also been lost, and only a Latin translation from the Arabic survives. As Toomer notes, some of the most interesting works were not available in Spain and Sicily, where medieval scholars went to translate Arabic and Hebrew manuscripts into Latin.

3.2. Some Muslim mathematicians and their works. Continuing with our list of the major writers and their works, we now survey some of the more important ones who lived and worked under the rule of the caliphs.

Muhammed ibn Musa al-Khwarizmi. This scholar translated a number of Greek works into Arabic but is best remembered for his *Hisab al-Jabr w'al-Mugabalah* (*Book of the Calculation of Restoration and Reduction*). The word *restoration* here (*al-jabr*) is the source of the modern word *algebra.* It refers to the operation of keeping an equation in balance by transferring a term from one side to the opposite side with the opposite sign. The word *reduction* refers to the cancellation of like terms or factors from the two sides of an equation. The author came to be called simply al-Khwarizmi, which may be the name of his home town (although this is not certain); this name gave us another important term in modern mathematics, *algorithm.*

The integration of intellectual interests with religious piety that we saw in the case of the Hindus is a trait also possessed by the Muslims. Al-Khwarizmi introduces his algebra book with a hymn of praise of Allah, then dedicates his book to al-Mamun:

> That fondness for science, by which God has distinguished the Imam al-Mamun, the Commander of the Faithful..., that affability and condescension which he shows to the learned, that promptitude with which he protects and supports them in the elucidation of obscurities and in the removal of difficulties—has encouraged me to compose a short work on Calculating by (the rules of) Completion and Reduction, confining it to what is easiest and most useful in arithmetic, such as men constantly require in cases of inheritance, legacies, partition, law-suits, and trade, and in all their dealings with one another, or where the measuring of lands, the digging of canals, geometrical computation, and other objects of various sorts... My confidence rests with God, in this as in every thing, and in Him I put my trust... May His blessing descend upon all the prophets and heavenly messengers. [Rosen, *1831*, pp. 3–4]

Thabit ibn-Qurra. The Sabian (star-worshipping) sect centered in the town of Harran in what is now Turkey produced an outstanding mathematician/astronomer in the person of Thabit ibn-Qurra (826–901). Being trilingual (besides his native Syriac, he spoke Arabic and Greek), he was invited to Baghdad to study mathematics. His mathematical and linguistic skills procured him work translating Greek treatises into Arabic, including Euclid's *Elements.* He was a pioneer in the application of arithmetic operations to ratios of geometric quantities, which is the essence of the idea of a real number. The same idea occurred to René Descartes (1596–1650) and was published in his famous work on analytic geometry. It is likely that Descartes drew some inspiration from the works of the fourteenth-century Bishop of Lisieux Nicole d'Oresme (1323–1382); Oresme, in turn, is likely to have read translations from the Arabic. Hence it is possible that our modern concept of a real number can be traced back to the genius of Thabit ibn-Qurra. He also wrote on mechanics, geometry, and number theory.

Abu-Kamil. Although nothing is known of the life of Abu-Kamil (ca. 850–93), he is the author of certain books on algebra, geometry, and number theory that had a marked influence on both Islamic mathematics and the recovery of mathematics in Europe. Many of his problems were reproduced in the work of the Leonardo of Pisa (Fibonacci, 1170–1226).

Abu'l-Wafa. Mohammad Abu'l-Wafa (940–998) was born in Khorasan (now in Iran) and died in Baghdad. He was an astronomer–mathematician who translated Greek works and commented on them. In addition he wrote a number of works on practical arithmetic and geometry. According to R¯ashid (*1994*), his book of practical arithmetic for scribes and merchants begins with the claim that it "comprises all that an experienced or novice, subordinate or chief in arithmetic needs to know" in relation to taxes, business transactions, civil administration, measurements, and "all other practices... which are useful to them in their daily life."

Al-Biruni. Abu Arrayhan al-Biruni (973–1048), was an astronomer, geographer, and mathematician who as a young man worked out the mathematics of maps of Earth. Civil wars in the area where he lived (Uzbekistan and Afghanistan) made him into a wanderer, and he came into contact with astronomers in Persia and Iraq. He was a prolific writer. According to the *Dictionary of Scientific Biography*, he wrote what would now be well over 10,000 pages of texts during his lifetime, on geography, geometry, arithmetic, and astronomy.

Omar Khayyam. The Persian mathematician Omar Khayyam was born in 1044 and died in 1123. He is thought to be the same person who wrote the famous skeptical and hedonistic poem known as the *Rubaiyat* (*Quatrains*), but not all scholars agree that the two are the same. Since he lived in the turbulent time of the invasion of the Seljuk Turks, his life was not easy, and he could not devote himself wholeheartedly to scholarship. Even so, he advanced algebra beyond the elementary linear and quadratic equations that one can find in al-Khwarizmi's book and speculated on the foundations of geometry. He explained his motivation for doing mathematics in the preface to his *Algebra.* Like the Japanese *wasanists*, he was inspired by questions left open by his predecessors. Also, as with al-Khwarizmi, this intellectual curiosity is linked with piety and thanks to the patron who supported his work.

> In the name of God, gracious and merciful! Praise be to God, lord of all Worlds, a happy end to those who are pious, and ill-will to none but the merciless. May blessings repose upon the prophets, especially upon Mohammed and all his holy descendants.
>
> One of the branches of knowledge needed in that division of philosophy known as mathematics is the science of completion and reduction, which aims at the determination of numerical and geometrical unknowns. Parts of this science deal with certain very difficult introductory theorems, the solution of which has eluded most of those who have attempted it... I have always been very anxious to investigate all types of theorems and to distinguish those that can be solved in each species, giving proofs for my distinctions, because I know how urgently this is needed in the solution of difficult problems. However, I have not been able to find time to complete this work, or to concentrate my thoughts on it, hindered as I have been by troublesome obstacles. [Kasir, *1931*, pp. 43–44]

Al-Tusi. Nasir al-Din al-Tusi (1201–1274) had the misfortune to live during the time of the westward expansion of the Mongols, who subdued Russia during the 1240s, then went on to conquer Baghdad in 1258. Al-Tusi himself joined the Mongols and was able to continue his scholarly work under the new ruler Hulegu, grandson of Genghis Khan. Hulegu, who died in 1265, conquered and ruled Iraq and Persia over the last decade of his life, taking the title *Ilkhan* when he declared himself ruler of Persia. A generation later the Ilkhan rulers converted from Buddhism to Islam. Hulegu built al-Tusi an observatory at Maragheh, a city in the Azerbaijan region of Persia that Hulegu had made his seat of goverment. Here al-Tusi was able to improve on the earlier astronomical theory of Ptolemy, in connection with which he developed both plane and spherical trigonometry into much more sophisticated subjects than they had been previously. Because of his influence, the loss of Baghdad was less of a blow to Islamic science than it would otherwise have been. Nevertheless, the constant invasions had the effect of greatly reducing the vitality and the quantity of research. Al-Tusi played an important role in the flow of mathematical ideas back into India after the Muslim invasion of that country; it was his revised and commented edition of Euclid's *Elements* that was mainly studied (de Young, *1995*, p. 144).

4. Europe

As the western part of the world of Islam was growing politically and militarily weaker because of invasion and conquest, Europe was entering on a period of increasing power and vigor. One expression of that new vigor, the stream of European mathematical creativity that began as a small rivulet 1000 years ago, has been steadily increasing until now it is an enormous river and shows no sign of subsiding.

4.1. Monasteries, schools, and universities. From the sixth to the ninth centuries a considerable amount of classical learning was preserved in the monasteries in Ireland, which had been spared some of the tumult that accompanied the decline of Roman power in the rest of Europe. From this source came a few scholars to the court of Charlemagne to teach Greek and the quadrivium (arithmetic, geometry, music, and astronomy) during the early ninth century. Charlemagne's attempt to promote the liberal arts, however, encountered great obstacles, as his empire was divided among his three sons after his death. In addition, the ninth and tenth centuries saw the last waves of invaders from the north—the Vikings, who disrupted commerce and civilization both on the continent and in Britain and Ireland until they became Christians and adopted a settled way of life. Despite these obstacles, Charlemagne's directive to create cathedral and monastery schools had a permanent effect, contributing to the synthesis of observation and logic known as modern science.

Gerbert. In the chaos that accompanied the breakup of the Carolingian Empire and the Viking invasions the main source of stability was the Church. A career in public life for one not of noble birth was necessarily an ecclesiastical career, and church officials had to play both pastoral and diplomatic roles. That some of them also found time for scholarly activity is evidence of remarkable talent. Such a talent was Gerbert of Aurillac (ca. 940–1002). He was born to lower-class but free parents in south-central France. He benefited from Charlemagne's decree that monasteries and cathedrals must have schools and was educated in Latin grammar at the monastery of St. Gerald in Aurillac. Throughout a vigorous career in the Church that led to his coronation as Pope Sylvester II in the year 999 he worked for a revival of learning, both literary and scientific. (He was not a successful clergyman or pope. He got involved in the politics of his day, offended the Emperor, and was suspended from his duties as Archbishop of Reims by Pope Gregory V in 998. He was installed as pope by the 18-year-old Emperor Otto II in 999, but after only three years both he and Otto were driven from Rome by a rebellion. Otto died trying to reclaim Rome, and Sylvester II died shortly afterward.)

4.2. The high Middle Ages. By the midtwelfth century European civilization had absorbed much of the learning of the Islamic world and was nearly ready to embark on its own explorations. This was the zenith of papal power in Europe, exemplified by the ascendancy of the popes Gregory VII (1073–1085) and Innocent III (1198–1216) over the emperors and kings of the time. The Emperor Frederick I, known as Frederick Barbarossa because of his red beard, who ruled the empire from 1152 to 1190, tried to maintain the principle that his power was not dependent on the Pope, but was ultimately unsuccessful. His grandson Frederick II (1194–1250) was a cultured man who encouraged the arts and sciences. To his court in Sicily he invited distinguished scholars of many different religions, and he corresponded with many others. He himself wrote a treatise on the principles of falconry. He was in conflict with the Pope for much of his life and even tried to establish a new religion, based on the premise that "no man should believe aught but what may be proved by the power and reason of nature," as the papal document excommunicating him stated.

4.3. Authors and works. A short list of European mathematicians prominent in their time from the twelfth through sixteenth centuries begins in the empire of Frederick II.

Leonardo of Pisa. Leonardo says in the introduction to his major book that he accompanied his father on an extended commercial mission in Algeria with a group of Pisan merchants. There, he says, his father had him instructed in the Hindu–Arabic numerals and computation, which he enjoyed so much that he continued his studies while on business trips to Egypt, Syria, Greece, Sicily, and Provence. Upon his return to Pisa he wrote a treatise to introduce this new learning to Italy. The treatise, whose author is given as "Leonardus filius Bonaccij Pisani," that is, "Leonardo, son of Bonaccio of Pisa," bears the date 1202. In the nineteenth century Leonardo's works were edited by the Italian nobleman Baldassare Boncompagni (1821–1894), who also compiled a catalog of locations of the manuscripts (Boncompagni, *1854*). The name Fibonacci by which the author is now known seems to have become generally used only in the nineteenth century.

Jordanus Nemorarius. The works of Archimedes were translated into Latin in the thirteenth century, and his work on the principles of mechanics was extended. One of the authors involved in this work was Jordanus Nemorarius. Little is known about this author except certain books that he wrote on mathematics and statics for which manuscripts still exist dating to the actual time of composition.

Nicole d'Oresme. One of the most distinguished of the medieval philosophers was Nicole d'Oresme, whose clerical career brought him to the office of Bishop of Lisieux in 1377. D'Oresme had a wide-ranging intellect that covered economics, physics, and mathematics as well as theology and philosophy. He considered the motion of physical bodies from various points of view, formulated the Merton rule of uniformly accelerated motion (named for Merton College, Oxford), and for the first time in history explicitly used one line to represent time, a line perpendicular to it to represent velocity, and the area under the graph (as we would call it) to represent distance.

Regiomontanus. The work of translating the Greek and Arabic mathematical works went on for several centuries. One of the last to work on this project was Johann Müller of Königsberg (1436–1476), better known by his Latin name of Regiomontanus, a translation of Königsberg (King's Mountain). Although he died young, Regiomontanus made valuable contributions to astronomy, mathematics, and the construction of scientific measuring instruments. In all this he bears a strong resemblance to al-Tusi, mentioned above. He studied in Leipzig while a teenager, then spent a decade in Vienna and the decade following in Italy and Hungary. The last five years of his life were spent in Nürnberg. He is said to have died of an epidemic while in Rome as a consultant to the Pope on the reform of the calendar.

Regiomontanus checked the data in copies of Ptolemy's *Almagest* and made new observations with his own instruments. He laid down a challenge to astronomy, remarking that further improvement in theoretical astronomy, especially the theory of planetary motion, would require more accurate measuring instruments. He established his own printing press in Nürnberg so that he could publish his works. These works included several treatises on pure mathematics. He established trigonometry as an independent branch of mathematics rather than a tool in astronomy. The main results we now know as plane and spherical trigonometry are in his book *De triangulis omnimodis*, although not exactly in the language we now use.

Chuquet. The French Bibliothèque Nationale is in possession of the original manuscript of a comprehensive mathematical treatise written at Lyons in 1484 by one Nicolas Chuquet. Little is known about the author, except that he describes himself as a Parisian and a man possessing the degree of Bachelor of Medicine. The treatise consists of four parts: a treatise on arithmetic and algebra called *Triparty en la science des nombres*, a book of problems to illustrate and accompany the principles of the *Triparty*, a book on geometrical mensuration, and a book of commercial arithmetic. The last two are applications of the principles in the first book.

Luca Pacioli. Written at almost the same time as Chuquet's *Triparty* was a work called the *Summa de arithmetica, geometrica, proportioni et proportionalite* by Luca Pacioli (or Paciuolo) (1445–1517). Since Chuquet's work was not printed until the nineteenth century, Pacioli's work is believed to be the first Western printed work on algebra. In comparison with the *Triparty*, however, the *Summa* seems less original. Pacioli has only a few abbreviations, such as *co* for *cosa*, meaning *thing* (the unknown), *ce* for *censo* (the square of the unknown), and *æ* for *æquitur* (equals). Despite its inferiority to the *Triparty*, the *Summa* was much the more influential of the two books, because it was published. It is referred to by the Italian algebraists of the early sixteenth century as a basic source.

Leon Battista Alberti. In art the fifteenth century was a period of innovation. In an effort to give the illusion of depth in two-dimensional representations some artists looked at geometry from a new point of view, studying the projection of two- and three-dimensional shapes in two dimensions to see what properties were preserved and how others were changed. A description of such a procedure, based partly on the work of his predecessors, was given by Leon Battista Alberti (1404–1472) in a treatise entitled *Della pictura*, published posthumously in 1511.

Sixteenth-century Italy produced a group of sometimes quarrelsome but always brilliant algebraists, who worked to advance their science for the sheer pleasure of making new mathematical achievements. As happened in Japan a century later, each new advance brought a challenge for further progress.

Scipione del Ferro. A method of solving a cubic equations was discovered by a lector (reader, that is, a tutor) at the University of Bologna, Scipione del Ferro (1465–1525), around the year 1500. He communicated this discovery to another mathematician, Antonio Maria Fior (dates unknown), who then used the knowledge to win mathematical contests.

Niccolò Tartaglia. Fior met his match in 1535, when he challenged Niccolò Fontana (1500–1557) of Brescia, known as Tartaglia (the Stammerer) because a wound he received as a child when the French overran Brescia in 1512 left him with a speech impediment. Tartaglia had also discovered how to solve certain cubic equations and so won the contest.

Girolamo Cardano. A brilliant mathematician and gambler, who became rector of the University of Padua at the age of 25, Girolamo Cardano (1501–1576) was writing a book on mathematics in 1535 when he heard of Tartaglia's victory over Fior. He wrote to Tartaglia asking permission to include this technique in his work. Tartaglia at first refused, hoping to work out all the details of all cases of the cubic and write a treatise himself. According to his own account, Tartaglia confided the secret of one kind of cubic to Cardano in 1539, after Cardano swore a solemn oath not to publish it without permission and gave Tartaglia a letter of introduction to the Marchese of Vigevano. Tartaglia revealed a rhyme by which he had memorized the procedure.

Tartaglia did not claim to have given Cardano any proof that his procedure works. It was left to Cardano himself to find the demonstration. Cardano kept his promise not to publish this result until 1545. However, as Tartaglia delayed his own publication, and in the meantime Cardano had discovered the solution of other cases of the cubic himself and had also heard that del Ferro had priority anyway, he published the result in his *Ars magna* (*The Great Art*), giving credit to Tartaglia. Tartaglia was furious and started a bitter controversy over Cardano's alleged breach of faith.

Ludovico Ferrari. Cardano's student Ludovico Ferrari (1522–1565) worked with him in the solution of the cubic, and between them they had soon found a way of solving certain fourth-degree equations.

Rafael Bombelli. In addition to the mathematicians proper, we must also mention an engineer in the service of an Italian nobleman. Rafael Bombelli (1526–1572) is the author of a treatise on algebra that appeared in 1572. In the introduction to this treatise we find the first mention of Diophantus in the modern era. Bombelli said that, although all authorities are agreed that the Arabs invented algebra, he, having been shown the work of Diophantus, credits the invention to the latter. In making sense of what his predecessors did he was one of the first to consider the square root of a negative number and to formulate rules for operating with such numbers.

The work being done in Italy did not escape the notice of French and British scholars of the time, and important mathematical works were soon being produced in those two countries.

François Viète. A lawyer named François Viète (1540–1603), who worked as tutor in a wealthy family and later became an advisor to Henri de Navarre (who became the first Bourbon king, Henri IV, in 1598), found time to study Diophantus and to introduce his own ideas into algebra. His book *Artis analyticae praxis* (*The Practice of the Analytic Art*) contained some of the notational innovations that make modern algebra much less difficult than the algebra of the sixteenth century.

Girard Desargues. Alberti's ideas on projection were extended by the French architect and engineer Girard Desargues (1591–1661), who studied the projections of figures in general and the conic sections in particular.

John Napier. In the late sixteenth century the problem of simplifying laborious multiplications, divisions, root extractions, and the like, was attacked by the Scottish laird John Napier, Baron of Murchiston (1550–1617). His work consisted of two parts, a theoretical part, based on a continuous geometric model, and a computational part, involving a discrete (tabular) approximation of the continuous model. The computational part was published in 1614. However, Napier hesitated to publish his explanation of the theoretical foundation. Only in 1619, two years after his death, did his son publish an English translation of Napier's theoretical work under the title *Mirifici logarithmorum canonis descriptio* (*A Description of the Marvelous Law of Logarithms*). This subject, although aimed at a practical end, turned out to have enormous value in theoretical studies as well.

The European colonies. Wherever Europeans went during their great age of expansion, science and mathematics followed once the new lands were settled and acquired political stability and a certain level of economic prosperity. Like the mathematics of Europe proper, the story of this "colonial" mathematics is too large to fit into the present volume, and so we shall, with regret, omit South America and South Africa from the story and concentrate on the origins of mathematics in Mexico, the United States, Canada, Australia, and New Zealand.

5. North America

During the American colonial period and for nearly a century after the founding of the United States, mathematical research in North America was extremely limited. Educational institutions were in most cases directed toward history, literature, and classics, the major exception being the academy at West Point, which became the United States Military Academy in 1802. Modeling itself consciously on the École Polytechnique, the Academy taught engineering and applied mathematics.[8] For most of the period up to 1875 there were no professional journals devoted entirely to mathematics and no mathematical societies of any size. A period of rapid growth began in the 1870s, coinciding with the closing of the American frontier. By 1900 a respectable school of American mathematical researchers existed, although it was

[8] Rensselaer Polytechnic Institute was founded to teach engineering in 1824, and civil engineering was taught at the University of Vermont as early as 1829.

still puny compared with the schools in Germany, Britain, France, and Italy. Even as late as 1940, only about half a dozen mathematical journals were published in the United States. The United States vaulted to a position of world leadership in mathematics following World War II, and it has remained among the strongest nations in this area, thanks to its possession of a powerful university system and equally well-developed professional organizations such as the American Mathematical Society, the Mathematical Association of America, the Society for Industrial and Applied Mathematics, and the National Council of Teachers of Mathematics, together with over 100 professional journals devoted to mathematics in general or specific areas within it.

5.1. The United States and Canada before 1867. Until the late nineteenth century most of the mathematics done in North America was purely practical, and to find more than one or two examples of its practitioners we shall have to leave mathematics proper and delve into related areas. Nevertheless, one can find a few examples of Americans who practiced mathematics for its own sake, even in the eighteenth century.

David Rittenhouse. Like his younger brother Benjamin (1740–1825), David Rittenhouse (1732–1796) was primarily a manufacturer of compasses and clocks. He made two compasses for George Washington. He also got involved in surveying and in 1763 helped to settle a border dispute between William Penn and Lord Baltimore. He became the first director of the United States Mint by appointment of President Washington in 1792, and he became president of the American Philosophical Society in 1791, after the death of Benjamin Franklin. According to Homann (*1987*), he was self-taught in mathematics, but enjoyed calculation very much and so was able to read Newton's *Principia* on his own. He developed a continued-fraction method of approximating the logarithm of a positive number, described in detail by Homann. Like the Japanese tradition of challenge problems, some of Rittenhouse's papers asked for proofs of results the author himself had not been able to supply. In one case this challenge was taken up by Nathaniel Bowditch (discussed below).

Robert Adrain. An immigrant of great mathematical talent—he came to the United States from his native Ireland after being wounded by friendly fire in the rebellion of 1798—was Robert Adrain (1775–1843). He taught at Princeton until 1800, when he moved to York, Pennsylvania; in 1804 he moved again, to Reading, Pennsylvania. He contributed to, and in 1807 became editor of, the *Mathematical Correspondent*, the first mathematical research journal in the United States. Parshall (*2000*, p. 381) has noted that even as late as 1874 "[t]here were no journals in the United States devoted to mathematical research, and, in fact, up to that time all attempts to sustain such publication outlets had failed almost immediately." The *Mathematical Correspondent* appears to have ended with the first issue of Vol. 2, that is, the first one edited by Adrain. In an interesting article on the original editor of the *Mathematical Correspondent*, George Baron (b. 1769, date of death unknown), V. Fred Rickey notes that perhaps it may not have been merely the American ignorance of mathematics that led to an early demise for this journal. Rickey points out that the journal had 347 subscribers and published 487 copies of its first issue, but that an article in *The Analyst* in 1875 (**2**, No. 5, 131–138) by one David S. Hart contains the following interesting comment:

> The writer has a copy of No. 2. stitched in a blue cover, on which is an advertisement of a Lecture delivered in New York by G. Baron, which contains (as he says) "a complete refutation of the false and spurious principles, ignorantly imposed on the public, in the 'New American Practical Navigator,' written by N. Bowditch and published by E. M. Blunt." The sub-editors endorsing the above say, "We agree with the author that he has shown in the most incontrovertible manner, that the principles on which the 'New American Practical Navigator' is founded, are universally false, and gross impositions on the public."

Since Bowditch was, next to Adrain, the strongest mathematician in the country at the time, this sort of internecine feuding could only have been harmful to the development of a community of mathematicians. Rickey's article can be found by following links from the following website.

`http://www.dean.usma.edu/math/people/rickey/`

Adrain is best remembered for discovering, independently of Legendre and Gauss, the theory of least-squares and the normal (Gaussian) distribution. However, given the low state of science in general in the United States, it is not surprising that no one in Europe noticed Adrain's work. Kowalewski (*1950*, pp. 84–85) notes that the Göttingen astronomer Tobias Mayer (1723–1762) had used a similar method as early as 1748.

Commerce requires a certain amount of mathematics and astronomy to meet the needs of navigation, and all the early American universities taught dialing (theory of the sundial), astronomy, and navigation. These subjects were standard, long-known mathematics, a great contrast to the rapid pace of innovation in Europe at this period. Nevertheless, to write the textbooks of navigation and calculate the tides a year in advance required some ability. It is remarkable that this knowledge was acquired by two Americans who were not given even the limited formal education that could be obtained at an American university. Although neither was a mathematician in the strict sense, both of them understood and used the mathematics of astronomy.

Benjamin Banneker. In the fall of 1791 the Baltimore publishing house of William Goddard and James Angell published a book bearing the title *Banneker's Almanac and Ephemeris for the Year of our Lord 1792....* The author, Benjamin Banneker (1731–1806), was 60 years old at the time, the only child of parents of African descent[9] who had left him a small parcel of land as an inheritance. For most of his life Banneker lived near Baltimore, struggling as a poor farmer with a rudimentary formal education. Nevertheless, he acquired a reputation for cleverness due to his skill in arithmetic. In middle age he made the acquaintance of the Ellicotts, a prominent local family, who lent him a few books on astronomy. From these meager materials Banneker was able to construct an almanac for the year 1791. Encouraged by this success, he prepared a similar almanac for 1792. In that year the Ellicotts put him in contact with James McHenry (who had been Surgeon General of the American Army during the Revolutionary War). McHenry wrote to the editors:

---

[9] Banneker's grandmother was an Englishwoman who married one of her slaves. Their daughter Mary, Banneker's mother, also married a slave, who had the foresight to purchase a farm jointly in his own name and in the name of his son Benjamin.

> [H]e began and finished [this almanac] without the least information or assistance from any person, or other books than those I have mentioned; so that whatever merit is attached to his present performance is exclusively and peculiarly his own.

Banneker's *Almanac* was published and sold all over the United States in the decade from 1792 until 1802. The contents of the *Almanac* are comparable with those of other almanacs that have been published in the United States: On alternate pages one finds calendars for each week or month, giving the phases of the Moon, the locations of the planets and bright stars visible during the period in question, and the times of sunrise, high and low tides, and conjunctions and oppositions of planets. Recognition came late to Banneker. The money he earned from his *Almanac* gave him some leisure in his old age, and his name was praised by Pitt in Parliament and by Condorcet before the French Academy of Sciences.

African-American mathematicians. Although the antislavery movement had begun in Banneker's time, African Americans were to endure two more generations of slavery followed by three generations of institutionalized, legalized discrimination and disenfranchisement before the civil rights movement gained sufficient strength to open to them the opportunities that a white American of very modest means could expect. It is therefore no wonder that very few African Americans became noted scholars. Nevertheless, the scientific creativity of African Americans has been a significant factor in the economic life of the United States, as can be seen, for example, in the book of James (*1989*). The first African American to obtain a doctorate in mathematics was Elbert Cox (1895–1969), who became a professor at Howard University after obtaining the doctorate at Cornell in 1925, one of only 28 doctorates awarded to Americans (of any color) that year. The first African-American women to receive the doctorate in mathematics, both of them in 1949, were Marjorie Lee Brown (1914–1979) and Evelyn Boyd Granville (b. 1924). Brown was a differential topologist who received her degree at the University of Michigan and taught at North Carolina Central University. Granville received the Ph.D. from Yale University and worked in the space program during the 1960s. She later taught at California State University in Los Angeles.

The number of African Americans choosing to enter mathematics and science is still comparatively small. In fact, the author of an article entitled "Black Women Ph.D.'s in Mathematics" in the 1980s was able to interview *all* of the people described in the title who were still alive. A career in research, after all, requires a long apprenticeship, during which financial support must be provided either by family, by extra work, or by grants and loans. For people who do not come from wealthy families, other careers, promising earlier financial rewards, are likely to seem more attractive. Undoubtedly, if the average income of African Americans were higher, more of them would choose scientific careers. Lest these comments seem unduly pessimistic, it should be noted that a conference devoted to the research of African Americans in 1996 brought together 79 African-American mathematicians (Dean, *1996*).

Nathaniel Bowditch. Benjamin Banneker was about 40 years old and living in obscurity near Baltimore when Nathaniel Bowditch (1773–1838) was born in Salem, Massachusetts. His ancestors had been shipbuilders but had accumulated no substantial amount of money by this trade. His father abandoned it and became a cooper, a trade that barely provided for his family of seven children. Nathaniel received only a rudimentary public education before being apprenticed to a ship chandler at the age of 10. Twelve years later, when Banneker's *Almanac* had been published for only a year or two, he signed on board a ship and, like Banneker, used his few intervals of leisure to study mathematics and astronomy. Bowditch was a natural teacher who enthusiastically shared his knowledge of navigation with his shipmates. With his aptitude for mathematics, he managed to get through Newton's *Principia*, learning a considerable amount of Latin on the way. Later he taught himself French, which had displaced Latin as the language of science as a result of the pre-eminence of French mathematicians and scientists.

Bowditch first gained a scholarly reputation by pointing out errors in the standard navigational tables. His abilities immediately attracted interest, and his *Practical Navigator*, first published in 1800, gained him wide recognition[10] while he was still in his twenties. Bowditch became a member of the American Academy of Arts and Letters, and in 1818 was elected a member of the Royal Society. With recognition came leisure time to devote to purely scholarly pursuits, a luxury denied to Banneker in his most vigorous years. For the last quarter-century of his life Bowditch labored on his monumental translation and commentary of the *Mécanique céleste* by Pierre-Simon Laplace (1749–1827). This work amounts really to a complete rewriting of Laplace's treatise, which shows the effects of a pronounced stinginess with ink and paper. Bowditch filled in all the missing details of arguments that Laplace had merely waved his hand at, not having the patience to write down arguments that had sometimes taken him weeks to discover. These pursuits brought Bowditch international fame, and he died covered with honors. The *American Journal of Science* published his obituary with a portrait of him in a classical Roman tunic which it is unlikely he ever actually wore.

5.2. The Canadian Federation and post Civil War United States. The end of the American Civil War in 1865 was followed closely by the founding of the Canadian Federation in 1867. The Federation was the result of the North America Act, which reserved some constitutional controls for Britain. Full independence came in 1982. From that time on, both countries experienced a cultural flowering, which included advances in mathematics. Americans and Canadians began to go to Europe to learn advanced mathematics. This early generation of European-trained mathematicians generally found no incentive to continue research upon returning home. However, they at least made the curriculum more sophisticated and prepared the way for the next generation.

In Europe there were more Ph.D. mathematicians being produced than the universities could absorb. Most of these entered other professions, but a few emigrated across the Atlantic. A scholarly coup was scored by Johns Hopkins University, which opened in 1876 with a first-rate mathematician on board, James Joseph Sylvester. Despite being 62 years old, Sylvester was still a creative algebraist, whose presence in America attracted international attention. One of his first acts was to found the first mathematical research journal in the United States, the *American Journal of Mathematics*. The founding of this journal had been suggested by William Edward Story (1850–1930), one of many Americans who went abroad to get the Ph.D. degree but, atypically, continued to do mathematical research after returning to the United States. Before Johns Hopkins was founded, there had been a few graduate

---

[10] And apparently some detractors associated with the *Mathematical Correspondent* (see above).

programs in mathematics in places such as Harvard and the University of Michigan, but now such programs began to multiply. Bryn Mawr College opened in the mid-1880s with a graduate program in mathematics. The founding of Clark University in Worcester, Massachusetts and the University of Chicago in the late 1880s and early 1890s promised that the United States would soon begin to make respectable contributions to mathematical research. An account of this development giving the details of the mathematical areas studied in American universities can be found in the article by David Rowe (*1997*). A review of a number of professional "self-studies" made by American mathematicians can be found in the article by Karen Hunger Parshall (*2000*); both of these articles contain extensive bibliographies on the development of mathematics in the United States. We now continue our list of prominent mathematicians.

George William Hill. The mathematical side of astronomy, known as celestial mechanics, was pursued in the United States by the Canadian Simon Newcomb, who is discussed below, and by George William Hill (1838–1914). Hill worked for a time at the Nautical Almanac Office in Cambridge, Massachusetts, but was perfectly content to work in isolation at his home in Nyack, New York, most of his life. His work on the motion of the Moon was so profound that it received extravagant praise from Henri Poincaré (1854–1912), one of the greatest mathematicians of the late nineteenth and early twentieth centuries. In a paper on the motion of the lunar perigee published in the Swedish journal *Acta mathematica* in 1886, Hill derived a differential equation that bears his name and even today continues to generate new work.

Although the rise of the United States to a position of world leadership in mathematics after World War II was partly the result of the turbulence of the 1930s and 1940s, which drove many of the best European intellectuals to seek refuge far from the dangers that threatened them in their homelands, one should not think that the country was intellectually backward before that time. Americans had made significant contributions to algebra and logic in the nineteenth century, and in the early twentieth century a number of Americans achieved worldwide fame for their mathematical contributions. We mention only two here.

George David Birkhoff. Harvard professor George David Birkhoff (1884–1944) made contributions to differential equations, difference equations, ergodic theory, and mathematical physics (the kinetic theory of gases, in which the ergodic theorem plays a role, quantum mechanics, and relativity). He was held in such high esteem that a crater on the Moon now bears his name.

Norbert Wiener. An early prodigy who graduated from high school at age 11 and received the doctoral degree at age 18, despite having changed universities and majors more than once, Norbert Wiener (1894–1964) contributed to harmonic analysis, probability, quantum mechanics, and cybernetics, of which he was one of the founders. (The name comes from the Greek word *kybernetēs*, meaning a ship's captain or pilot.)

Like American schools of the same period, English-language Canadian institutions of higher learning tended to rely on British textbooks such as those of Charles Hutton (1737–1823, a professor at the Miltary School in Woolwich). In French Canada there was a long tradition of educational institutions, and a French calculus text written by Abbé Jean Langevin, who was to become Bishop of Rimouski in 1867, was published in 1848. For Canadians, as for Americans, the importance of research as an activity of the mathematics professor arose only after the founding of Johns Hopkins University in 1876. In fact, the early volumes of the *American Journal of Mathematics* contain articles by two Canadians, J. G. Glashan (1844–1932), superintendent of schools in Ottawa, and G. Paxton Young (1818–1889), a professor of philosophy at the University of Toronto.

Simon Newcomb. An outstanding nineteenth-century Canadian mathematician was Simon Newcomb (1835–1909), a native of Nova Scotia who taught school in a number of places in the United States before procuring a job at the Nautical Almanac Office in Cambridge, Massachusetts, where he attended Harvard. He eventually became director of the Naval Observatory in Washington, and after 1884 professor of mathematics at Johns Hopkins.

H. S. M. Coxeter. The geometer Harold Scott MacDonald Coxeter (1907–2003), a native of Britain, emigrated to Canada in 1936 and played a leading role in Canadian research in symmetry groups and symmetric geometric objects of all kinds. His work on tessellations inspired many famous paintings by the Dutch artist Maurits Escher (1898–1972).

John Synge. Although, strictly speaking, he counts as an Irish mathematician, who was born in Dublin and died there, John Synge (1897–1995) taught at the University of Toronto from 1920 to 1925 and again during the 1930s. From 1939 until 1948 he worked in the United States before returning to Ireland. He is listed here because of his daughter, Cathleen Synge Morawetz, who is discussed below.

John Charles Fields. One of the best-remembered Canadian mathematicians, John Charles Fields (1863–1932), was a native of Hamilton, Ontario. He received the Ph.D. from Johns Hopkins in 1887 and studied in Europe during the 1890s. In 1902 he became a professor at the University of Toronto. He wrote one book (on algebraic functions). Like many other mathematicians on the intellectual periphery of Europe, much of his activity was devoted to encouraging research in his native country. In the last few years of his life he established the Fields Medals, the highest international recognition for mathematicians, which are awarded at the quadrennial International Congress of Mathematicians. Beginning in 1936, when two awards were given, then resuming in 1950, the Fields Medals have by tradition been awarded to researchers early in their careers. As of 2002 about 40 mathematicians had been so honored, among them natives of China, Japan, New Zealand, the former Soviet Union, many European countries, and the United States.

Cecilia Krieger Dunaij. Canada has always taken in those fleeing oppression elsewhere, and some of these refugees have become prominent mathematicians. One example is Cypra Cecilia Krieger Dunaij (1894–1974), who studied mathematical physics at the University of Vienna before coming to Toronto in 1920, where she entered the university and took courses given by John Synge and John Fields. In 1930 she became the first woman to receive the doctoral degree in mathematics at a Canadian university (Toronto) and only the third woman to receive a doctoral degree in Canada.

Abraham Robinson. Among the mathematicians that the turbulent twentieth century condemned to wander the world was Abraham Robinson (1918–1974). He was born in what was then Germany and is now Poland, but emigrated with his family to Jerusalem in 1933, when the Nazis came to power in Germany. In 1940 he was studying in Paris, but evacuated to London when Paris fell to the Nazi invasion. After obtaining the Ph. D. and teaching in Britain for a few years, he spent six of his most productive years at the University of Toronto, beginning in 1951. While there he produced several Ph. D. students in mathematical logic. He left Toronto in 1957 to return to Jerusalem, but eventually moved to California and finally to Yale. The story of his Toronto years can be found in the article by Dauben (*1996*).

Cathleen Morawetz. Cathleen Synge Morawetz (b. 1923 in Toronto) is the daughter of John Synge. She attended the University of Toronto during World War II and then obtained the master's degree at the Massachusetts Institute of Technology in 1946. For a dissertation in mathematical physics she received the doctoral degree at New York University in 1951. Her subsequent career was very distinguished. She became associate director of the Courant Institute of Mathematical Sciences in 1978. In 1995–1996 she was president of the American Mathematical Society, the second woman to hold this post. In 1998 she was awarded the National Medal of Science, the highest scientific honor bestowed by the United States.

5.3. Mexico. The area that is now Mexico was the first part of the North American mainland to be colonized by Europeans and was the site of the first university in North America, the Universidad Real y Pontificia de México, founded in 1551. Unfortunately, the history of mathematics in modern Mexico has not been thoroughly studied, despite the fact that the mathematics of the earlier rulers of this part of the world, the Aztecs and Maya, has received quite a bit of scholarly attention. The present discussion amounts to a summary of the article by A. Garciadiego (*2002*), in which the author remarks that "the professionalization of the history of mathematics in Mexico is comparatively recent."

The Royal University opened its doors just two years after its founding, offering a curriculum that was essentially medieval, consisting of theology, law, and related subjects. The first technical and scientific studies came more than a century later, with the establishment of a chair of astrology and mathematics. Unfortunately, some important scientific works were on the Index Librorum Prohibitorum, and Catholics were forbidden to read them.[11] Among these books were the works of Galileo and Newton, so that no real progress in science was to be expected.

Mexico became an independent country in 1821. An attempt by the French Emperor Louis Napoleon to make Mexico part of a renewed French Empire by establishing the puppet emperor Maximilian in 1864 soon failed. The French army left, and Maximilian was executed in 1867. The new leaders of Mexico sought to establish intellectual freedom that would incorporate material progress based on science. Despite these intentions, the University was closed at various times for political reasons. The National University of Mexico opened in 1910 and was accompanied by preparatory schools and schools for advanced studies. The most prominent name in the advance of mathematics and science in Mexico was Sotero Prieto (1884–1935), who taught advanced mathematics and physics and advocated

---

[11] It is interesting that astrology, belief in which is listed as a sin in the Catholic catechism, was not only permitted, but actually encouraged. The Index was not officially abolished until 1966, long after it had ceased to be taken seriously by either the faithful or the clergy.

the use of history in teaching. He is quoted as saying, "The history of a science clarifies the origins of its fundamental concepts and exhibits the evolution of its methods" (Garciadiego, *2002*, p. 259).

Four years after the death of Prieto the Department of Mathematics was established as part of the Faculty of Sciences of the University, now known as the Universidad Nacional Autónoma de México. At this point, it could be said that the University had reached academic maturity. Seminars on current research opened, including one on scientific and philosophical problems. Foreign scholars came there to visit, and graduates from the University were able to find admission to first-rate universities in other countries. Upon his retirement from Princeton University in 1954, the distinguished topologist Solomon Lefschetz (1884–1972) accepted a position at the University of Mexico and began sending students to the graduate program at Princeton.[12] An amusing anecdote revealing the relations between the uninhibited Lefschetz and the Mexicans was reported by his student Gian-Carlo Rota (1932–1999). (See Rota's article *1989*.)

6. Australia and New Zealand

Because of their proximity to each other, we discuss Australia and New Zealand together, although they are not twins. Australia was settled by pioneers from Asia around 70,000 years ago, when the ocean levels were much lower than now. Even with the lower ocean levels, this settlement involved a long sea journey. When ocean levels rose after the last ice age, many of the original settlements were offshore and under water. New Zealand, in contrast, was settled by seafaring people only 1500 years ago. Europeans first arrived in this area in the sixteenth century, but actual settlement by Europeans did not begin until late in the eighteenth century. As in the United States, there were conflicts between the aboriginal inhabitants and the new settlers, very fierce in Australia but surprisingly mild in New Zealand. Britain proclaimed sovereignty over New Zealand in 1840 by including it in the Australian colony of New South Wales. This merger lasted only 10 months, at which time New Zealand became an independent colony. At this time a declaration of equal rights for settlers and Maoris was made; a constitution followed 12 years later. In 1850 the six Australian states gained self-government by act of Parliament, and in 1901 they united in the Commonwealth of Australia.

Comparatively little has been written about the development of mathematics in these countries, and the present account is based largely on an article (*1988*) by Garry J. Tee, a professor of computer science at the University of Auckland, who has written a great deal on the history of mathematics in general. Tee says that the indigenous peoples of this area had a well-developed system of numeration and makes the point that "the common assertions to the effect that 'Aborigines have only one, two, many' derive mostly from reports by nineteenth century Christian missionaries, who commonly understood less mathematics than did the people on whom they were reporting." At the same time, he notes that these missionaries did teach Western-style mathematics to indigenous people.

6.1. Colonial mathematics. As in other countries, European colonists were not long in establishing universities in these new lands. Australia acquired universities at Sydney (1850), Melbourne (1853), Adelaide (1874), and Hobart (University of

[12] During the present author's years at Princeton (1963–1966) several of the graduate students in mathematics were Mexican students of Lefschetz.

Tasmania, 1890). In New Zealand universities opened at Dunedin (University of Otago, 1869), Christchurch (University of Canterbury, 1873), Auckland (1883), and Wellington (Victoria University, 1897). The New Zealand universities were from the beginning co-educational. The Australian Mathematical Society was founded in 1956 and the New Zealand Mathematical Society in 1974. Long before that, however, good mathematicians were being born and working in these two countries. The following short list is far from complete, but it does show that world-class mathematics and science have been produced in this region almost from the beginning.

The Bragg family. In 1915 two Australians, father and son, were awarded the Nobel Prize for physics. Each of them served as director of the Royal Institution in London. William Henry Bragg (1862–1942) was a professor of mathematics at the University of Adelaide from 1885 to 1908. His son William Lawrence Bragg (1890–1971) became Cavendish Professor of physics at Cambridge and director of the National Physical Laboratory.

Horatio Scott Carslaw. One of the standard texts on Fourier series, which was reprinted many times and eventually became immortalized in a Dover edition, was written by H. S. Carslaw (1870–1954), the third professor of mathematics at the University of Sydney (1903–1935). Carslaw was born in Scotland but moved to Australia in 1903 to take up the position at the University of Sydney. Besides his book on Fourier series, he also collaborated on a standard textbook on the Laplace transform and had an interest in the history of logarithms.

Thomas Gerald Room. Carslaw's successor at the University of Sydney was Thomas Gerald Room (1902–1986), a native of London who, like Carslaw, moved to Australia to take up an academic position. He is well remembered by combinatoricists for the concept of Room squares, about which he published a paper in 1955.

Ernest Rutherford. Another physicist with mathematical gifts was the New Zealander Ernest Rutherford (1871–1937), who studied at Canterbury University, worked at McGill University in Montreal (1898–1907), and eventually, working at Manchester University, performed a famous experiment that helped to determine the structure of the atom (the positively charged nucleus surrounded by electrons that is still the popular picture of atoms).

V. F. R. Jones. One of the brightest stars in the mathematical firmament at the moment is Vaughan Frederick Randal Jones (b. 1952), who graduated from the University of Auckland in 1973. From there he went to Switzerland, where he received the doctoral degree for a prize-winning dissertation. Since 1980 he has worked in the United States. He won the Fields Medal in 1990 for his groundbreaking work in knot theory. (The Jones polynomial is named after him.) This discovery came about while he was working in a seemingly unrelated area (von Neumann algebras) and had links to areas of mathematical physics (topological quantum field theories) that were studied by mathematical physicists such as the American Edward Witten (b. 1951) and the British topologist Simon Donaldson (b. 1957), both of whom also won the Fields Medal, Donaldson in 1986 and Witten alongside Jones in 1990. The Jones polynomial was described by the French journal *La recherche* in its issue of July–August 1997 as one of the 300 most important discoveries of the last three centuries.

Refugee mathematicians. Like the United States and Canada, Australia took in some prominent European mathematicians who were fleeing persecution during the Nazi era. Among them were Kurt Mahler (1903–1988), Hans Schwerdtfeger (1902–1990), George Szekeres (b. 1911), Hanna Neumann (1914–1971), and her husband Bernhard Neumann (1909–2002). In honor of the mathematical achievements of these refugees the Australian Mathematical Society sponsors a Mahler Lectureship, a George Szekeres Medal, and a B. H. Neumann Prize.

Ties provided by the British Commonwealth seem to have facilitated the careers of many of these people. Rutherford and Schwerdtfeger, for example, both worked for a time at McGill University in Montreal, besides the time they spent in New Zealand, Australia, Britain, and elsewhere.

7. The modern era

The advanced work in number theory, geometry, algebra, and calculus that began in the seventeenth century will be incorporated into the discussion of the mathematics itself beginning in Chapter 5. There are two reasons for not discussing it here. First, many of the names from this time on, such as Pascal, Descartes, Leibniz, Newton, Cauchy, Riemann, Weierstrass are probably already familiar to the reader from mathematics courses. Second, the increasing unity of the world makes it less meaningful to talk of "European mathematics" or "Chinese mathematics" or "Indian mathematics," since in the modern era mathematicians the world over work on the same types of problems and use the same approaches to them. We shall now look at some general features of modern mathematics the world over.

Up to the nineteenth century mathematics for the most part grew as a wild plant. Although the academies of science of some of the European countries nourished mathematical talent once it was exhibited, there were no mathematical societies dedicated to producing mathematicians and promoting their work. This situation changed with the French Revolution and the founding of technical and normal schools to make education systematic. The effects of this change were momentous. The curriculum shifted its emphasis from classical learning to technology, and research and teaching became linked.

7.1. Educational institutions. At the time of the French Revolution the old universities began to be supplemented by a system of specialized institutions of higher learning. The most famous of these was the École Polytechnique, founded in 1795. A great deal of the content of modern textbooks of physics and mathematics was first worked out and set down in the lectures given at this institution. Admission to the École Polytechnique was a great honor, and only a few hundred of the brightest young scholars in France were accepted each year. This institution and several others founded during the time of the French Revolution, such as the École Normale Supérieure, produced a large number of brilliant mathematicians during the nineteenth century. Some of their research was devoted to questions of practical importance, such as cartography and canal building, but basic research into theoretical questions also flourished.

In Germany the unification of teaching and research proceeded from the other direction, as professors at reform-minded universities such as Göttingen (founded in 1737) began to undertake research along with their teaching. This model of development was present at the founding of the University of Berlin in 1809. This educational trend was duplicated elsewhere in the world. During his Italian campaign Napoleon founded the Scuola Normale Superiore in Pisa, which reopened in 1843 after a long hiatus. In Russia a university opened along with the Petersburg Academy of Sciences in 1726, and the University of Moscow was founded a generation later (1755) with the aim of producing qualified professionals. It was not until the nineteenth century, however, that the faculty in Moscow began to engage in research. The University of Stockholm opened in 1878 with aims similar to those of the institutions just named. In Japan an office of translations was opened in the Shogunate Observatory in 1811. It was renamed the Institute for the Study of Foreign Books in 1857 and became the home of a department of Western mathematics in 1863, taking on two Dutch faculty members in 1865. By 1869 only Western mathematics was being taught, and the teaching was being done by French and British teachers.

7.2. Mathematical societies. Another aspect of the professionalization of mathematics was the founding of professional societies to supplement the activities of the mathematical sections in academies of sciences. The oldest of these is the Moscow Mathematical Society (founded in 1864). The London Mathematical Society was founded in 1866, the Japanese Mathematical Society in 1877. The American Mathematical Society (originally the New York Mathematical Society) was founded in 1888 and the Canadian Mathematical Society in 1945.

7.3. Journals. These educational institutions and professional societies also published their own research journals, such as the *Journal de l'École Polytechnique* and the *Journal de l'École Normale Supérieure.* These journals contained some of the most profound research of the nineteenth century. Other nations soon emulated the French. The German *Journal für die reine und angewandte Mathematik* was founded by August Leopold Crelle (1780–1855) in 1826. Informally, it is still called *Crelle's Journal.* The Italian *Annali di scienze matematiche e fisiche* appeared in 1850; the Moscow Mathematical Society began publishing the *Matematicheskii Sbornik* (*Mathematical Collection*) in 1866; the Swedish *Acta mathematica* was founded in 1881. By the end of the nineteenth century there were mathematical research journals in every European country, in North America, and in Japan. The first American research journal, *The American Journal of Mathematics*, was founded at Johns Hopkins University in 1881 with the British mathematician J. J. Sylvester as its principal editor, assisted by the American William Edward Story. The first issue of *The Canadian Journal of Mathematics* was dated 1949.

Questions and problems

3.1. Compare the way in which mathematicians have been supported in various societies discussed in this chapter. If you were in charge of distributing the federal budget, how high a priority would you give to various forms of pure and applied research in mathematics? What justification would you give for your decision? Would it involve a practical "payoff" in economic terms, or do you believe that the government has a responsibility to support the creation of new mathematics, without regard to its economic value?

3.2. Why is Seki Kōwa the central figure in Japanese mathematics? Are comparisons between him and his contemporary Isaac Newton justified?

3.3. What is the justification for the statement by the historian of mathematics T. Murata that Japanese mathematics was not a science but an art?

3.4. Why might Seki Kōwa and other Japanese mathematicians have wanted to keep their methods secret, and why did their students, such as Takebe Kenkō, honor this secrecy?

3.5. For what purpose was algebra developed in Japan? Was it needed for science and/or government, or was it an "impractical" liberal-arts subject?

3.6. Dante's final stanza, quoted above, uses the problem of squaring the circle to express the sense of an intellect overwhelmed, which was inspired by his vision of heaven. What resolution does he find for the inability of his mind to grasp the vision rationally? Would such an attitude, if widely shared, affect mathematical and scientific activity in a society?

3.7. One frequently repeated story about Christopher Columbus is that he proved to a doubting public that the Earth was round. What grounds are there for believing that "the public" doubted this fact? Which people in the Middle Ages would have been likely to believe in a flat Earth? Consider also the frequently repeated story that people used to believe the stars were near the Earth. How is that story to be reconciled with Ptolemy's assertion that it was acceptable to regard Earth as having the dimensions of a point relative to the stars?

3.8. What are the possible advantages and disadvantages of eliminating or greatly reducing the volume of journals, placing all articles on electronic files that can be downloaded from various information systems?

3.9. Mathematical research is like any other commercial commodity in the sense that people have to be paid to do it. We have mentioned the debate over taxing the entire public to support such research and asked the student to consider whether there is a national interest that justifies this taxation. A similar taxation takes place in the form of tuition payments to American universities. Some of the money is spent to provide the salaries of professors who are required to do research. Is there an educational interest in such research that justifies its increased cost to the student?

CHAPTER 4

Women Mathematicians

The subject of women mathematicians has become a major area in the history of mathematics over the past generation, naturally connected with the women's movement in general. Any history of mathematics should include a discussion of the conditions under which mathematics flourishes and the reasons why some people and cultures develop mathematics to a high degree while others do not. To give as complete a picture of the history of mathematics as possible we need to examine these conditions, and the case of women in mathematics is a very instructive example.

The author, who began studying the history of mathematics by researching the career of Sof'ya Kovalevskaya (1850–1891), has heard it objected that women mathematicians are receiving attention out of proportion to their mathematical merit while many talented male mathematicians are being neglected by historians. Such an objection is beside the point. Male mathematicians did not have to overcome the energy- and time-consuming obstacles that women faced. The justification for devoting a full chapter to women mathematicians and for making women mathematicians a separate area of study is very simple: Until recently, all women mathematicians had one thing in common, a societal expectation that they would spend most of their time ministering to the needs of their families. A direct corollary of that expectation was that a mathematical career should not be a woman's first priority and that societal institutions need not support or even recognize any striving for such a career. In fact, Barnard College once had a policy of firing women who got married on the grounds that "the College cannot afford to have women on the staff to whom the college work is secondary; the College is not willing to stamp with approval a woman to whom self-elected home duties can be secondary."[1] In other words, if a woman chooses to marry, her duties as a wife should be first priority. If they aren't, she is a bad woman and hence unfit to be on the staff; if they are, her duties at the College must be secondary, and again, she is unfit to be on the staff.

The subject "women mathematicians" could be replaced by a category having no reference to gender, as "mathematics practiced under conditions of discrimination." In that way the subject would be enlarged so as to include minorities such as Jewish mathematicians in Europe and the United States from the Middle Ages until the twentieth century and African Americans up to very recent years. To keep this chapter of manageable size, however, we confine it to women.

[1] http://cwp.library.ucla.edu/Phase2/Maltby_Margaret_Eliza@901234567.html

1. Individual achievements and obstacles to achievement

A useful periodization of the progress—and it *is* a story of progress—of women in mathematics, is as follows: (1) before 1800, a time when only the most exceptional woman in the most exceptionally fortunate circumstances could hope to achieve anything in mathematics; (2) the nineteenth century, a period when the support of society for a woman to have a career in mathematics was missing, but a very determined, financially independent woman could at least break into the world of science and mathematics; (3) the twentieth century, when the dam restraining women from mathematical achievement developed cracks and finally burst completely, leading to a flood of women that continues to swell right up to the present. We first discuss in general terms the obstacles that needed to be overcome, and then give brief biographies describing the lives and achievements of a number of prominent women mathematicians.

1.1. Obstacles to mathematical careers for women. In the United States many of the best graduate schools were all-male until the 1960s. A classmate of the author at Northwestern University, a very bright and mathematically talented young woman, mentioned in 1962 that she had written to an Ivy League school to inquire about study for the doctoral degree and had received a reply saying, "We have no place to house you." A decade later, the women's movement began to focus attention on the small number of women in mathematics, and the resulting investigation into causes has helped to remove some of the obstacles to women's achievement in mathematics. Among the obstacles, the following have been identified:

Institutionalized discrimination. It required considerable time for society to realize that all-male institutions receiving government grants were discriminating against women. Indeed, the author's classmate mentioned above, whatever she may have thought, did not complain publicly of discrimination for being rejected by an Ivy League school. Ironically, the existence of women's colleges, which had arisen partly in response to this discrimination, was sometimes cited as proof that men's colleges were not discriminatory. If the opportunities and facilities at the women's colleges had been equal to those at the men's colleges, that argument would have had merit; but they were not.

Discrimination went beyond the student body; it was, if anything, even worse among the faculty. Until the 1970s most universities and many companies had "anti-nepotism" rules that forbade the hiring of both a husband and wife. Since women mathematicians often married men who were mathematicians, marriage became a serious impediment to a career, whether or not the husband was supportive of his wife's ambition. Karen Uhlenbeck (b. 1942) encountered this kind of discrimination and later wrote about it:

> I was told that there were nepotism rules and that they could not hire me for this reason, although when I called them on this issue years later, they did not remember saying these things.

In earlier times Ivy League universities were not the only places women were not allowed to be. In the eighteenth century, they were not allowed to attend meetings of the Academy of Sciences in Paris nor (by social convention) to enter cafés. These were the two places where the best scientific minds of the time assembled for conversation. The Marquise du Châtelet defied convention and went to cafés anyway, dressed as a man. In the nineteenth century women were not allowed into laboratories at some universities, so that Christine Ladd-Franklin (1847–1930) became a mathematics major even though she would have preferred physics. After writing a brilliant dissertation but being unable to obtain a degree, she turned to the new profession of psychology, but even there was shut out of professional life. In the twentieth century, when his colleagues were objecting to hiring Emmy Noether at Göttingen, Hilbert is reported (Dick, *1981*, p. 168; Mackay, *1991*, p. 117) to have ridiculed their objections, saying, "The Senate is not a locker room; why shouldn't a woman go there?" As our narrative proceeds, the same three institutions—the University of London, Bryn Mawr College, and the University of Göttingen—will appear repeatedly, showing how few opportunities there were for women to pursue advanced studies in mathematics until quite recently.

The situation in the early twentieth century was described by the mathematician Gerhard Kowalewski (1876–1950) in his memoirs:[2]

> At that time [1905] the first women students began to appear at the University of Bonn. They were still being met with harsh rejection on the part of distinguished professors at other universities, for example, Berlin, where Gustav Roethe, if he caught sight of women in the auditorium, simply refused to begin his lecture until they left the room.[3] People were not so narrow-minded at Bonn. The women students formed a Society and arranged balls to which they invited their professors. There was a whole series of talented women mathematicians. Many of them took the state examination under my supervision: [among them was] Maria Vaerting, who later became a famous novelist and whose first novel... was based on her student days... At the same time she was working on a very difficult topic for a doctoral dissertation under my direction. In the end, however, she didn't receive the doctorate as my student, since I was called to Prague. She then moved to Giessen, where her work was accepted by Professor Pasch. [Kowalewski, *1950*, pp. 206–207]

Discouragement from family, friends, and society in general. We do not know what attitudes were faced by the very earliest women mathematicians, but from the eighteenth century on there are many documented cases of family opposition to such a career; particularly good examples are Sophie Germain and Sof'ya Kovalevskaya, both of whom had to go to extraordinary lengths to participate in the mathematical community. (Kovalevskaya was fortunate in being able eventually to win her

[2] Kowalewski believed himself to be distantly related to Vladimir Kovalevskii, husband of Sof'ya Kovalevskaya, but this connection has never been verified.

[3] The kind of behavior exhibited by Roethe eventually disappeared, thanks in large part to the efforts of the Prussian Kultusminister Friedrich Althoff (1839–1908), who had asked Felix Klein (1854–1925) to be on the lookout for promising women students. In 1894, with Althoff's approval, Klein took Grace Chisholm Young as his student, and the doors of Göttingen University were thereafter open to women. One of Althoff's last acts as Kultusminister was to unify the education of boys and girls. Klein can be described as a liberal but not a radical, one who believed in equal opportunity for women and even affirmative action to recruit women; but he insisted that only women with demonstrated talent and background should be admitted to universities.

father's blessing on her career.) In addition, most women who have had both children and a career have had to invest more time in the children than men have done. This extra responsibility and a host of other societal expectations requiring time and effort on the part of women have made it more difficult for women to concentrate on their careers with the same single-mindedness that has characterized the most outstanding male mathematicians. In at least one case, that of Grace Chisholm Young (1868–1944), marriage meant a rather complete submersion of her talents for a time, with her husband (William H. Young, 1863–1942) getting all the credit for papers that were a joint effort. Such unequal partnerships, which seem terribly unfair a century later, were probably not common, but other such cases are known.[4]

Lack of role models. It cannot be a coincidence that many of the women "pioneers" in mathematics were the daughters of mathematicians or engineers. The absence of prominent women in these fields during the early days meant that many young girls thinking about their futures did not consider a career in technical areas. Most of the exceptions were in contact with mathematics and science from an early age because of the work their fathers did. The women who did choose such careers could get little advice from their male mentors as to how to deal with the special problems faced by a woman wishing a career in science. For example, Cathleen Morawetz, who was mentioned in Chapter 3, noticing how few job opportunities there were for women with doctorates in mathematics, nearly decided to choose a career in industry after getting her master's degree. It was her mentor Cecilia Krieger (1894–1974, later Cecilia Krieger Dunaij) who encouraged her to go to New York University. Such role models and encouragement were naturally present in greater degree at women's colleges.

Inappropriate teaching methods. The usefulness of women's colleges in helping women to develop their talents and ultimately overcome society's low expectations cannot be overemphasized. That girls, at least those being raised in traditional ways, needed to be taught differently from boys, is very clear from the following description of a geometry lesson given by Prince Bolkonskii to his daughter, Princess Mar'ya, in Leo Tolstoy's *War and Peace.*

> Leaning on the table, the prince pushed forward a notebook full of geometrical diagrams.
>
> "Now, young lady," the old man began, bending over the notebook close to his daughter and putting one hand on the arm of the chair in which the princess was sitting, so that she felt herself completely surrounded by her father's pungent old-man and tobacco scent, so long familiar to her. "Now, young lady, these triangles are similar. Notice the angle *abc*..."
>
> The princess looked nervously at her father's sparkling eyes close by; blushes rose to her cheeks, and it was apparent that she didn't understand anything and was so frightened that fear was preventing her from understanding any of her father's subsequent reasoning, no matter how clear it was. Whether it was the fault

[4] It is now well documented that Einstein's first wife made significant contributions to his 1905 paper on special relativity and deserved to be listed as a co-author. Although she never received the Nobel Prize in her own name, she did get Einstein's prize money under the terms of their divorce settlement.

> of the tutor or of the pupil, the same thing happened every day: everything swam in front of the princess' eyes, she saw and heard nothing, but only sensed her father's dry, stern face next to her, was aware of his breath and his scent, and thought only of getting out of the study as soon as possible so that she could understand the problem in the spacious freedom of her own room. The old man made extraordinary efforts: noisily moving the chair he was sitting in back and forth, he struggled not to lose his temper; but nearly always lost it, shouted at her, and sometimes threw the notebook.
>
> The princess had given an incorrect answer.
>
> "What a stupid thing to say!" shouted the prince, shoved the notebook aside, and quickly turned away. But then he immediately got up, walked around, touched the princess' hair, and sat down again.
>
> He came closer and continued his reasoning.
>
> "No, no, Princess," he said, when at last the princess had taken the notebook with the assignments in it and was preparing to leave. "Mathematics is a great thing, young lady. I don't want you to be like those silly debutantes. Perseverance brings pleasure." He stroked her cheek with his hand. "The frivolity will eventually jump out of your head." [*War and Peace*, Book 1, Part 1, Chapt. 22]

The vividness of this scene shows that Tolstoy must have drawn it from real life. Even an enlightened father, such as Tolstoy's Prince Bolkonskii, who loved his daughter and wanted more for her than the frivolous life offered to most women in the Russian aristocracy, did not know how to carry out his own good intentions.

Sexual harassment. This painful topic has apparently not been much talked about in relation to mathematics specifically. Keith and Keith (*2000*) report that at a 1988 conference on women in mathematics and the sciences every woman present had experienced discrimination, not only "gender harassment... but more brutal sexual harassment." The harm that can be done by sexual harassment includes creating anxiety that interferes with work, discouraging women from seeking help in a professor's office, and blocking professional advancement for women who protest harassment or reject unwanted advances.

To struggle against all of these obstacles was the task of heroic individual women for many centuries, and what they achieved seems in many ways miraculous. Who would have guessed, for example, that a journal named *The Woman Inventor* was published more than a century ago?[5] But real progress could be expected only when society as a whole undertook to provide support. To overcome these obstacles legislation was enacted at the federal level during the 1960s forbidding discrimination on the basis of gender. To overcome the more entrenched and subtle problems of societal discouragement and lack of role models a variety of measures have been introduced, including special workshops and institutes devoted to introducing women to mathematical research and the founding of the Association for Women in Mathematics in 1971. All major universities and corporations

[5] It was published by Charlotte Smith (1840–1917) and managed only two issues, in April and June of 1891 (Stanley, *1992*).

now have procedures for preventing and prosecuting sexual harassment. Although it cannot be said that all of these obstacles have been overcome, it is certainly the case that more and more women are choosing careers in mathematics. In many universities the number of undergraduate women majoring in mathematics is now larger than the number of men, and the number of women graduate students is approaching equality with the number of men. Equality of numbers, however, is not necessarily the goal. It may be that, given equal opportunity, more women than men would choose to be mathematicians; or the number of women freely choosing such a career might be less. What is (in the author's view) the ultimate goal—that each person should be aware of the opportunities for any career and accorded equal opportunity to pursue the career of her or his choice—has not quite been achieved, but it is fair to say that a woman can now pursue a career in mathematics and science with the same expectation of success, depending on her talent, as in any other major.

2. Ancient women mathematicians

Very few women mathematicians are known by name from early times. However, Closs (*1992*, p. 12) mentions a Maya ceramic with a picture of a female scribe/mathematician. From ancient Greece and the Hellenistic culture, at least two women are mentioned by name. Diogenes Laertius, in his work *Lives of Eminent Philosophers*, devotes a full chapter to the life of Pythagoras, and gives the names of his wife, daughter, and son. Since it is known that the Pythagoreans admitted women to their councils, it seems that Pythagoras' wife and daughter engaged in mathematical research at the highest levels of their day. However, nothing at all is known about any works they may have produced. All that we know about them is contained in the following paragraph from Diogenes Laertius:

> Pythagoras had a wife named Theano. She was the daughter of Brontinus of Croton, although some say that she was Brontinus' wife and Pythagoras' pupil. He also had a daughter named Damo, as Lysis mentions in a letter to Hipparchus. In this letter he speaks of Pythagoras as follows: "And many say that you [Hipparchus] give public lectures on philosophy, as Pythagoras once did. He entrusted his *Commentaries* to Damo, his daughter, and told her not divulge them to anyone not of their household. And she refused to part with them, even though she could have sold them for a considerable amount of money; for, despite being a woman,[6] she considered poverty and obedience to her father's instructions to be worth more than gold." He also had a son named Telauges, who succeeded him as head of the school, and who, according to some authors, was the teacher of Empedocles. Hippobotus, for one, reports that Empedocles described him as "Telauges, the noble youth, whom in due time, Theano bore to the sage Pythagoras." But no books by Telauges survive, although there are still some that are attributed to his mother Theano.

[6] It is hardly worth pointing out the slur on women's character implicit in this phrase.

Hypatia. There are two primary sources for information about the life of Hypatia. One is a passage in a seven-book history of the Christian Church written by Socrates Scholasticus, who was a contemporary of Hypatia but lived in Constantinople; the other is an article in the *Suda*, an encyclopedia compiled at the end of the tenth century, some five centuries after Hypatia.[7] In addition, several letters of Synesius, bishop of Ptolemais (in what is now Libya), who was a disciple of Hypatia, were written to her or mention her, always in terms of high respect. In one letter he requests her, being in the "big city," to procure him a scientific instrument (hygrometer) not available in the less urban area where he lived. In another he asks her judgment on whether to publish two books that he had written, saying

> If you decree that I ought to publish my book, I will dedicate it to orators and philosophers together. The first it will please, and to the other it will be useful, provided of course that it is not rejected by you, who are really able to pass judgment. If it does not seem to you worthy of Greek ears, if, like Aristotle, you prize truth more than friendship, a close and profound darkness will overshadow it, and mankind will never hear it mentioned. [Fitzgerald, *1926*]

The account of Hypatia's life written by Socrates Scholasticus occupies Chapter 15 of Book 7 of his *Ecclesastical History.* Socrates Scholasticus describes Hypatia as the pre-eminent philosopher of Alexandria in her own time and a pillar of Alexandrian society, who entertained the elite of the city in her home. Among that elite was the Roman procurator Orestes. There was considerable strife at the time among Christians, Jews, and pagans in Alexandria; Cyril, the bishop of Alexandria, was apparently in conflict with Orestes. According to Socrates, a rumor was spread that Hypatia prevented Orestes from being reconciled with Cyril. This rumor caused some of the more volatile members of the Christian community to seize Hypatia and murder her in March of 415.

The *Suda* devotes a long article to Hypatia, repeating in essence what was related by Socrates Scholasticus. It says, however, that Hypatia was the wife of the philosopher Isodoros, which is definitely not the case, since Isodoros lived at a later time. The *Suda* assigns the blame for her death to Cyril himself.

Yet another eight centuries passed, and Edward Gibbon came to write the story in his *Decline and Fall of the Roman Empire* (Chapter XLVII). In Gibbon's version Cyril's responsibility for the death of Hypatia is reported as fact, and the murder itself is described with certain gory details for which there is no factual basis. (The version given by Socrates Scholasticus is revolting enough and did not need the additional horror invented by Gibbon.)

A fictionalized version of Hypatia's life can be found in a nineteenth-century novel by Charles Kingsley, bearing the title *Hypatia, or New Foes with an Old Face.* What facts are known were organized into an article by Michael Deakin (*1994*) and a study of her life by Maria Dzielska (*1995*).

3. Modern European women

Women first began to break into the intellectual world of modern Europe in the eighteenth century, mingling with the educated society of their communities, but not allowed to attend the meetings of scientific societies. The eighteenth century

[7] This work bears the traditional name *Suidas*, erroneously thought to be the name of the person who compiled it.

produced three notable women mathematicians, whose biographies exhibit some noticeable similarities and some equally noticeable differences.

3.1. Continental mathematicians. The first two of three prominent eighteenth-century women mathematicians were the Marquise du Châtelet and Maria Gaetana Agnesi. Both were given strong classical educations at the insistence of their fathers, both took a strong interest in science, and both wrote expository works that incorporated their own original ideas. Apart from those similarities, however, there are a great many differences between the two women, beyond the obvious fact that the Marquise du Châtelet was French and Maria Gaetana Agnesi was Italian.

The Marquise du Châtelet. The Marquise du Châtelet was born Gabrielle-Émilie Tonnelier de Breteuil, the daughter of a court official of the "Sun King" Louis XIV, in 1706. She was presented at court at age 16, married to a nobleman at 19, and had a number of lovers throughout her life. She bore several children and died in 1749, apparently of complications from the birth of a child; she was 42 years old at the time.

In a preface to her translation and reworking of an English book entitled *The Fable of the Bees*, she wrote eloquently about the situation of women in general, and the difficulties she herself faced, saying

> I am convinced that many women are either unaware of their talents by reason of the fault in their education or that they bury them on account of prejudice for want of intellectual courage. My own experience confirms this. [Ehrman, *1986*, p. 61]

As a teenager Gabrielle-Émilie received encouragement to study mathematics from a family friend, M. de Mézières, but would have had contact with science in any case, just from being in a home where intellectual questions were taken seriously. Her scientific interests were in the area known as natural philosophy, which was the physics and chemistry of the time, but contained strong admixtures of philosophical doctrines that have since been purged. In 1740 she published *Institutions de physique*, in which she attempted a synthesis of the ideas of Newton, Descartes, and Leibniz. Five years later, she began the work for which she is best remembered, a French translation, with commentary, of Newton's *Philosophiæ naturalis principia mathematica.* This work was published in 1756, seven years after her death.

Maria Gaetana Agnesi. In contrast to the Marquise du Châtelet, Maria Gaetana Agnesi much preferred a simple, spartan life, even though her father was the heir to a fortune made in the silk trade. Born in 1718 in Bologna, which at the time was located in the Papal States, she wanted to be a nun, and only her father's pleading prevented her from going to a convent as a young woman. She never married and spent her time at home in activities that would be appropriate to a convent, reading religious books, praying, and studying mathematics.[8] She was encouraged in her interest in mathematics by a monk who was also a mathematician and who frequently visited her father. In the preface to her book *Istituzioni analitiche ad uso della gioventù italiana*, she expressed her gratitude for this support, saying

[8] If those three activities seem incongruous, one should keep in mind that a considerable portion of the women mathematicians in the United States during the 1930s were nuns.

that, despite her strong interest in mathematics, she would have gotten lost without his instruction. Using the famous mathematician Jacopo Riccati (1676–1754), for whom the Riccati equation is named, as an editor, she worked methodically on this textbook for many years. Riccati even gave her some of his own results on integration. The work was published in two volumes in 1748 and 1749 and immediately recognized as a masterpiece of organization and exposition, earning praise from the Paris Academy of Sciences. The Pope at the time, Benedict XIV, had an interest in mathematics, and he appointed her to a position as reader at the University of Bologna. Soon afterward, the Academy of Bologna offered her the chair of mathematics at the university, and the Pope confirmed this offer.

However, she does not seem to have accepted the offer. Her name remained on the rolls at the university, but she devoted herself to her charitable work, with ever more zeal after her father died in 1752. She gave away her fortune to the poor and died in poverty in 1799.

As is often the case with people who are kept away from full participation in scientific circles, the originality of Maria Agnesi's work is in the organization of the material. The small part of it that has immortalized her name is a curve that she called *la versiera*, meaning the *twisted curve*. It was translated into English by her contemporary John Colson, but the translation was not published until 1801. Colson apparently confused *la versiera* with *l'avversiera*, which means *wife of the devil*. Accordingly he gave this curve the name *witch of Agnesi*, a name that has unfortunately stuck to it and is both sad and ironic, considering the exemplary character of its author.[9]

Sophie Germain. Even though she was born much later than Maria Gaetana Agnesi and the Marquise du Châtelet, the third prominent woman mathematician of the eighteenth and early nineteenth centuries, Marie-Sophie Germain, was more isolated from the intellectual world than her two predecessors. She was born in Paris during the reign of Louis XVI, on April 1, 1776. Like Maria Gaetana Agnesi, her family had grown wealthy in the silk trade, and the family home was a center of intellectual activity. She, however, was strongly discouraged from scientific studies by her family and had to stay up late and study the works of Newton and Euler (1707–1783), teaching herself Latin in order to do so. Her persistence finally won acceptance, and she was allowed to remain unmarried and devoted to her studies. Even so, those studies were not easy to conduct. Even after the French Revolution, she was not allowed to attend school. She did venture to send some of her work to Joseph-Louis Lagrange (1736–1813) under the pseudonym "M. LeBlanc," work he found sufficiently impressive to seek her out. He was her only mentor, but the relationship between them was not nearly so close as that between Sof'ya Kovalevskaya and her adviser Weierstrass 80 years later. She conducted a famous correspondence with Adrien-Marie Legendre (1752–1833) on problems of number theory, some of which he included in the second edition of his treatise on the subject. Later she corresponded with Carl Friedrich Wilhelm Gauss (1777–1855), again disguised as "M. LeBlanc." Although they shared a love for number theory, the two never met face to face. Sophie Germain proved a special case of Fermat's last theorem, which asserts that there are no nonzero integer solutions of $a^n + b^n = c^n$ when $n > 2$. Her special case assumes that the prime number n does not divide a, b, or c and

[9] Despite the widely recognized name *witch of Agnesi*, Agnesi was not the first person to study this curve.

is less than 100.[10] Gauss also praised her work very highly. He did not learn her identity until 1806, when French troops occupied his homeland of Braunschweig. Remembering the death of Archimedes, Sophie Germain wrote to some friends, asking them to take care that Gauss came to no harm. Gauss' opinion of her, expressed in a letter to her the following year, is often quoted:

> But how to describe to you my admiration and astonishment at seeing my esteemed correspondent Monsieur LeBlanc metamorphose himself into this illustrious personage who gives such a brilliant example of what I would find it difficult to believe. The enchanting charms of this sublime science reveal themselves only to those who have the courage to go deeply into it. But when a woman, who because of her sex and our prejudices encounters infinitely more obstacles than a man in familiarizing herself with complicated problems, succeeds nevertheless in surmounting these obstacles and penetrating the most obscure parts of them, without doubt she must have the noblest courage, quite extraordinary talents and superior genius.

Remembering that Sophie Germain was completely self-taught in mathematics and had little time to learn physics, which was increasingly developing its own considerable body of theory, we can only marvel that she had the courage to enter a prize competition in 1811 for the best paper on the vibration of an elastic plate. She had to start from zero in this enterprise, and Lagrange had warned that the necessary mathematics simply did not yet exist. According to Dahan-Dalmédico (*1987*, p. 351), she had learned mechanics from Lagrange's treatise and from some papers of Euler "painfully translated" from Latin. It is not surprising that her paper contained errors and that she did not win the prize. Actually, no one did; she was the only one who ventured to enter. Even so, her paper contained valuable insights, in the form of modeling assumptions that allowed the aged Lagrange to derive the correct differential equations for the displacement of the middle plane of the plate. She then set to work on these equations and in 1816 was awarded a prize for her work. This work became fundamental in the development of the theory of elasticity during the nineteenth century.

Perhaps because of the inevitable deficiencies resulting from her inadequate education, but more likely because she was a woman, Sophie Germain never received the respect she obviously deserved from the French Academy of her time. Prominent academicians seem to have given her papers the minimum possible attention. In their defense, it should be said that they were a galaxy of brilliant stars—Cauchy, Poisson, Fourier, and others—and it is unfortunate that their occasional neglect of geniuses such as Sophie Germain and Niels Henrik Abel stand out so prominently.

Like the Marquise du Châtelet, Sophie Germain had a strong interest in philosophy and published her own philosophical works. She continued to work in mathematics right up to the end of her life, writing papers on number theory and

[10] The divisibility hypothesis makes for a nice theorem, since it is obviously impossible to satisfy if $n = 1$ or $n = 2$. It seems to explain why those cases are exceptions. However, we now know that it is not a necessary hypothesis.

the curvature of surfaces (another interest of Gauss, but also connected with elasticity through a principle Sophie Germain had derived from Euler's work) from the time she was stricken with breast cancer in 1829 until her death in 1831.

3.2. Nineteenth-century British women. In Britain, as on the Continent, the admission of women to universities began at a very slow pace in the late nineteenth century. Before that time women had to have some means of support for private study or otherwise blaze their own trails through the wilderness. As on the Continent, the earliest women were not specialists in mathematics but had general philosophical interests.

Mary Somerville. The work of Mary Somerville, coming about 75 years later than that of the Marquise du Châtelet, bears many resemblances to the latter, being largely expository and philosophical in nature. Mary Somerville was born in Jedburgh, Scotland, on December 26, 1780, to the family of William George Fairfax, a naval officer. Like Sophie Germain, she received no encouragement toward a scientific education. Indeed, although her mother taught her to read, she had to learn to write all by herself. Education was reserved for her brothers, although she did spend one year, which she hated, in a boarding school for girls. Like Sophie Germain, she decided to educate herself. With the encouragement of an uncle, she began learning Latin, so that alongside the education given to most girls in her day—piano, painting, needlework—she was undertaking technical subjects. A chance remark of her painting tutor, overheard by Mary, to the effect that Euclid was both the secret of perspective in art and the foundation of many other sciences, led her to study geometry with her younger brother's tutor. A brief first marriage, to a naval officer who had no appreciation of her ability or her desire to learn, led to the birth of two sons. When her husband died after only three years, she returned to Scotland with her sons, where she found a circle of sympathetic friends, including the geometer John Playfair (1748–1819), editor of a famous edition of Euclid and the man who formulated the now-common version of Euclid's fifth postulate, which is known as *Playfair's Axiom*. For one solution to a mathematical problem set in a popular journal, she received a silver medal in 1811 (which, it will be remembered, was the year in which Sophie Germain unsuccessfully sought a prize from the Paris Academy for work on a much more substantial problem).

The following year she married William Somerville, an inspector of hospitals. William proved to be much more supportive than her first husband, and together they studied geology. When he was appointed as inspector to the Army Medical Board in 1816 and elected to the Royal Society, they moved to London, where they made the acquaintance of the leading scientists of the day. In an 1826 treatise on electromagnetism by Harvard professor John Farrar (1779–1853), used widely throughout American universities in the 1830s, Mary Somerville is mentioned as having performed a vanguard experiment in electromagnetic theory. In Italy it had been discovered that when a beam of violet light was used to stroke a metal needle repeatedly in the same direction for a long time, the needle became magnetized. At the time physicists speculated that this effect might be due to the particular properties of sunlight in Italy. By verifying that the same effect could be obtained in Edinburgh, Mary Somerville showed that the explanation had to be in the physics of violet light itself. Her paper on this subject was reported in a paper bearing the title "The magnetic properties of the violet rays of the solar spectrum," published in the *Proceedings of the Royal Society* in 1826 and, as mentioned, quoted by John Farrar and thereby made famous throughout the United States.

In an interesting reciprocity with the French translation of Newton's *Principia* made by the Marquise du Châtelet, Mary Somerville made a translation of Laplace's *Mécanique céleste* (as, it will be recalled, Nathaniel Bowditch had also done). Like the Marquise and Bowditch, Mary Somerville went far beyond merely translating, supplementing Laplace's laconic style with extensive commentaries. This work was published in 1831 and was a great success. Her book *The Connection of the Physical Sciences* (1834) went through many editions, and its speculation on the existence of an eighth planet, eventually to be known as Neptune, beyond Uranus (which had been discovered in 1781), inspired one of the co-discoverers of that planet. According to Baker (*1948*), her next book, *Physical Geography* (1848), was less successful from her own point of view, although not from the point of view of the experts. She was disappointed that it went through only six editions, and blamed its lack of commercial success on the appearance of cheap imitations that were "just keeping within the letter of the law [on plagiarism]." She began it in 1839 but had to delay because of her husband's illness and the need to revise her earlier book. Then, just when the manuscript was ready to go to press, another book on geography, entitled *Cosmos* and written by the great German scholar Alexander von Humboldt (1769–1859) appeared, apparently discouraging her so greatly that she considered burning her manuscript and had to be persuaded by friends to allow it to be published. It finally appeared in 1848. The reviews from those capable of reading it were glowing. Humboldt himself wrote to her, "I do not know of any book on geography in any language that can be compared with yours. You have not missed any fact or any of the grand sights of nature," and he signed himself "the author of the imprudent *Cosmos*." The subject of geography was not yet established in the curriculum in Britain, and her two-volume work did much to gain it a secure place.

As a result of these and other works, Mary Somerville was elected to a number of professional societies, including the Société de Physique et d'Histoire Naturelle in Geneva (1834), the Royal Irish Academy (1834), the Royal Astronomical Society (1835), the American Geographical and Statistical Society (1857), and the Italian Geographical Society (1870). She also won a number of academic honors. Recognizing the need for women to be liberated from their traditional confinement to the home, Mary Somerville was the first to sign the petition to Parliament organized by the philosopher John Stuart Mill (1806–1873), asking that the right to vote be extended to women. (Together with his wife, Mill had written a book entitled *The Enfranchisement of Women* and had also published Mary Wollstonecraft's *The Subjugation of Women*.) As in the United States, this right was finally granted just after the end of World War I. In 1862 she petitioned the University of London on behalf of women seeking degrees. (Note that she was 82 years old at the time!) Although this petition was rejected at that time, the University was awarding degrees to women only a few years later. Her long life finally came to an end near the end of her ninety-second year, on November 29, 1872, in Naples, Italy, where much of her geographical research had been done.

Florence Nightingale. Occasionally, new mathematics is created when people who are not professional mathematicians exercise their mathematical imaginations to solve urgent practical problems. Most often today such work comes from physicists, who state mathematical conjectures based on their physical intuition; the conjectures are then either proved or modified by other mathematicians or mathematical physicists. The most useful mathematics from a social point of view is the mathematics used every day to settle important questions. Today, that generally means statistics. We are used to seeing histograms, line graphs, and pie charts in our newspapers, and most professional journals of even moderate technical pretensions will have articles referring to standard deviations, chi-square tests, p-numbers, and related concepts. The graphical representations of data that we are used to seeing in our newspapers owe something to the imagination of this remarkable woman.

She was born in Italy on May 12, 1820, the second daughter of a wealthy couple who were taking an extended trip. She was about one year younger than Victoria, heir to the British throne. As happened with many women of achievement, Florence's father took an interest in the education of his daughters and both encouraged and tutored them. Her decision to enter the health professions, taken in 1837, the year that Queen Victoria came to the throne, was made, she later said, as the result of a direct (though nonspecific) call from God. By the late 1840s she had persuaded her family to allow her to travel on the Continent and study the operation of hospitals. She had less technical training and inclination than did Maria Gaetana Agnesi, but she was able to integrate her technical competence with the charitable and public health activity that was her primary occupation.

The central episode in the life of Florence Nightingale was the Crimean War of 1854–1855, in which Britain and France compelled Russia to remove its fleet and fortifications from the Black Sea. Deaths from battle in this war, which was essentially a siege of the fortress of Sevastopol, were fewer than deaths from disease. Florence Nightingale was appointed to lead a party of 38 nurses to the front to treat wounded soldiers. Seeing the conditions that existed there, she was inspired to write, in collaboration with William Farr (1807–1883), a series of papers on public health, complete with statistics on the numbers and cause of deaths, which were presented in the form of a polar diagram, an early version of what we now recognize as a pie chart (Plate 5). In 1860, for this and other such innovations in data handling, she became the first woman elected a fellow of the Statistical Society. Because of her dedication to caring for the sick, comparisons with Maria Gaetana Agnesi naturally come to mind. One important difference between the two women appears to be Florence Nightingale's greater organizing skills and her belief in social rather than individual action. The explanation probably lies in the fact that the two women were born a century apart and that Florence Nightingale lived in a society where people felt themselves to have some influence over government. Maria Gaetana Agnesi, who grew up in the artistically fruitful but politically chaotic eighteenth-century Italy, probably did not have that sense of a duty to participate in political life.

Despite being an invalid for many years before her death at age 90 in 1910, Florence Nightingale worked constantly to improve health standards. To this end she published over 200 books and pamphlets, many of which are still read and still influential today. In 1907 she became the first woman awarded the Order of Merit. A museum in London is dedicated to her life and work, and links to information about her can be found at its website:

`http://www.florence-nightingale.co.uk`

3.3. Four modern pioneers. The struggle for a woman's right to be a scientist or mathematician was very much an obstacle course, similar to running the high hurdles. The first hurdle was to get the family to support a scientific education. That hurdle alone caused many to drop out at the very beginning, leaving only a few lucky or very determined women to go on to the second hurdle, gaining access to higher education. All of the women discussed above had only private tutoring in mathematics. The second hurdle began to be surmounted in the late nineteenth century. On the continent a few women were admitted to university lectures without being matriculated, as exceptional cases. These cases established a precedent, and the exceptions eventually became regularized. In Britain the University of London began admitting women in the 1870s, and in the United States there were women's colleges for undergraduate education. The opening of Bryn Mawr College in 1885 with a program of graduate studies in mathematics was an important milestone in this progress. Once a woman had surmounted the second hurdle, the third and highest of all had to be faced: getting hired and accepted as a scientist. The four pioneers we are about to discuss had to improvise their solutions to this problem. The fundamental societal changes needed to provide women with the same assured, routine access that men enjoyed when pursuing such a career required many decades to be recognized and partially implemented.

Charlotte Angas Scott. One of the first women to benefit from the relaxation of restrictions on women's education was Charlotte Angas Scott, who was born in Lincoln, England on June 8, 1858. Like many of the earlier women, she was fortunate in that her parents encouraged her to study mathematics with a tutor. She attended Girton College, Cambridge and took the comprehensive Tripos examination at Cambridge in 1880, being ranked as the eighth Wrangler (that is, she was eighth from the top of the class of mathematics majors). However, the Tripos alone was not enough to earn her a degree at Cambridge. She was very fortunate in being able to go on to graduate work in algebraic geometry under the direction of one of the greatest nineteenth-century mathematicians, Arthur Cayley (1821–1895). She earned a first (highest-rank) degree from the University of London in 1882 and, with Cayley's recommendation, the Ph. D. in 1885. Having now surmounted the second hurdle, she faced the third and highest one: finding an academic position.

Cayley, who had spent some time a few years earlier at Johns Hopkins University in Baltimore, knew that Bryn Mawr College was opening that year. On his recommendation, Scott was hired there as a professor of mathematics. There she was able to set rigorous standards for the mathematical curriculum. When the American Mathematical Society was founded a few years later, she was a member of its first Council. Another of the nine women among the original membership of the AMS was her first Ph. D. student. Her contributions to mathematical scholarship were impressive. She published one paper giving a new proof of an important theorem of Max Noether (1844–1921) in the *Mathematische Annalen*, a very prestigious German journal, and many papers in the *American Journal of Mathematics*, which had been founded by her countryman James Joseph Sylvester (1815–1897) when he was head of mathematics at Johns Hopkins. From 1899 to 1926 she was an editor of this journal, and in 1905 she became vice-president of the American Mathematical Society.

Near the end of her career the American Mathematical Society held a conference in her honor at Bryn Mawr, and one of the speakers was the great British philosopher–mathematician Alfred North Whitehead (1861–1947), who paid tribute to her work in promoting a community of scholars, saying, "A life's work such as that of Professor Charlotte Angas Scott is worth more to the world than many anxious efforts of diplomatists. She is a great example of the universal brotherhood of civilisations."

Charlotte Angas Scott retired from Bryn Mawr in 1924. The following year she returned to Cambridge, where she lived the rest of her life. She died in 1931.

Sof'ya Kovalevskaya. Most of the women discussed up to now came from a leisured class of people with independent incomes. Only such people can afford both to defy convention and to spend most of their time pursuing what interests them. However, merely having an independent income was not in itself sufficient to draw a young woman into a scientific career. In most cases, some contact with intellectual circles was present as well. Hypatia was the daughter of a distinguished scholar, and Maria Gaetana Agnesi's father encouraged her by hiring tutors to instruct her in classical languages. In the case of Sof'ya Kovalevskaya, the urge to study mathematics and science fused with her participation in the radical political and social movements of her time, which looked to science as the engine of material progress and aimed to establish a society in accordance with the ideals of democracy and socialism.

She was born Sof'ya Vasil'evna Kryukovskaya in Moscow, where her father was an officer in the army, on January 15, 1850 (January 3 on the Julian calendar in effect in the Russia of her day). As a child she looked with admiration on her older sister Anna (1843–1887) and followed Anna's lead into radical political and social activism. According to her Polish tutor, she showed talent for mathematics when still in her early teens. She also showed great sympathy for the cause of Polish independence during the rebellion of 1863, which was crushed by the Tsar's troops. When she was 15, one of her neighbors, a physicist, was impressed upon discovering that she had invented the rudiments of trigonometry all by herself in order to read a book on optics; he urged her father to allow her to study more science. She was allowed to study up through the beginnings of calculus with a private tutor in Saint Petersburg, but matriculation at a Russian university did not appear to be an option. Thinking that Western Europe was more enlightened in this regard, many young Russian women used a variety of methods to travel abroad. Some were able to persuade their parents to let them go. Others had to adopt more radical means, either running away or arranging a fictitious marriage, in Sof'ya's case to a young radical publisher named Vladimir Onufrevich Kovalevskii (1842–1883). They were married in 1868 and soon after left for Vienna and Heidelberg, where Kovalevskaya studied science and mathematics for a year without being allowed to enroll in the university, before moving on to Berlin with recommendations from her Heidelberg professors to meet the dominant influence on her professional life, Karl Weierstrass (1815–1897). At Berlin also, the university would not accept her as a regular student, but Weierstrass agreed to tutor her privately. (Comparisons with the relationship between Charlotte Angas Scott and Arthur Cayley inevitably come to mind here.)

Although the next four years were extremely stressful for a number of personal reasons, her regular meetings with Weierstrass brought her knowledge of mathematical analysis up to the level of the very best students in the world (those attending Weierstrass' lectures). By 1874, Weierstrass thought she had done more than enough work for a degree and proposed three of her papers as dissertations.

Since Berlin would not award the degree, he wrote to the more liberal University of Göttingen and requested that the degree be granted *in absentia*. It was, and one of the three papers became a classic work in differential equations, published the following year in the most distinguished German journal, the *Journal für die reine und angewandte Mathematik*.

The next eight years may well be described as Kovalevskaya's wandering in the intellectual wilderness. She and Vladimir, who had obtained a doctorate in geology from the University of Jena, returned to Russia; but neither found an academic position commensurate with their talents. They began to invest in real estate, in the hope of gaining the independent wealth they would need to pursue their scientific interests. In 1878 Kovalevskaya gave birth to a daughter, Sof'ya Vladimirovna Kovalevskaya (1878–1951). Soon afterward, their investments failed, and they were forced to declare bankruptcy. Vladimir's life began to unravel at this point, and Kovalevskaya, knowing that she would have to depend on herself, reopened her mathematical contacts and began to attend mathematical meetings. Recognizing the gap in her résumé since her dissertation, she asked Weierstrass for a problem to work on in order to re-establish her credentials. While she was in Paris in the spring of 1883, Vladimir (back in Russia) committed suicide, leading Sof'ya to an intense depression that nearly resulted in her own death. When she recovered, she resumed work on the problem Weierstrass that had given her. Meanwhile, Weierstrass and his student Gösta Mittag-Leffler (1846–1927) collaborated to find her a teaching position at the newly founded institution in Stockholm.[11] At first she was *Privatdozent*, meaning that she was paid a certain amount for each student she taught. After the first year, she received a regular salary. She was to spend the last eight years of her life teaching at this institution.

In the mid-1880s, Kovalevskaya made a second mathematical discovery of profound importance. Mathematical physics is made complicated by the fact that the differential equations used to describe even simple, idealized cases of physical laws are extremely difficult to solve. The obstacle consists of two parts. First, the equations must be reduced to a set of integrals to be evaluated; second, those integrals must be computed. In many important cases, such as the equations of the three-body problem, the first is impossible using only algebraic methods. When it is possible, the second is often impossible using only elementary functions. For example, the equation of pendulum motion can be reduced to an integral, but that integral involves the square root of a cubic or quartic polynomial; it is known as an *elliptic integral*. The six equations of motion for a rigid body in general cannot be reduced to integrals at all using only algebraic surfaces. In Kovalevskaya's day only two special cases were known in which such a reduction was possible, and the integrals in both cases were elliptic integrals. Only in the case of bodies satisfying the hypotheses of both of these cases simultaneously were the integrals elementary. With Weierstrass, however, Kovalevskaya had studied not merely elliptic integrals, but integrals of completely arbitrary algebraic functions. Such integrals were known as *Abelian integrals* after Niels Henrik Abel (1802–1829), the first person to make significant progress in studying them. She was not daunted by the prospect of working with such integrals, since she knew that the secret of taming them was to use the functions known as *theta functions*, which had been introduced earlier by Abel and his rival in the creation of elliptic function theory, Carl Gustav Jacobi

[11] It is now the University of Stockholm.

Cher Monsieur!

Je vous remercie pour Votre invitation pour demain et je viendrais avec plaisir.

Les equations dif. qu'il s'agit d'intégrer sont les suivantes

$$\begin{cases} A\,\frac{dp}{dt} = (B-C)qr + z_0\gamma' - y_0\gamma'' & \frac{d\gamma}{dt} = q\gamma'' - r\gamma' . \\ B\,\frac{dq}{dt} = (C-A)rp + x_0\gamma'' - z_0\gamma & \frac{d\gamma'}{dt} = r\gamma - p\gamma'' \\ C\,\frac{dr}{dt} = (A-B)pq + y_0\gamma' - x_0\gamma & \frac{d\gamma''}{dt} = p\gamma' - q\gamma \end{cases}$$

Jusqu'à présent il n'ont été intégré que dans 2 cas: 1) $x_0 = y_0 = z_0 = 0$ (le cas de Poisson et de Jacobi)

2) $A = B \quad x_0 = y_0 = 0$
les cas de Lagrange

Moi j'ai trouvé l'intégrale aussi dans le cas où $A = B = 2C \quad z_0 = 0$ et je puis montrer que ces 3 cas sont les seuls où l'intégrale

First page of an undated letter from Kovalevskaya. Probably the letter was written in June 1886 and meant for Charles Hermite. The reason it was not sent is probably that she saw Hermite in person before posting it. The letter communicates her discovery of a completely integrable case of the equations of motion of a rigid body about a fixed point under the influence of gravity. Courtesy of the Institut Mittag-Leffler.

(1804–1851). All she had to do was reduce the equations of motion to integrals; evaluating them was within her power, she knew. Unfortunately, it turns out that the completely general set of such equations cannot be reduced to integrals. But Kovalevskaya found a new case, much more general than the cases already known, in which this reduction was possible. The algebraic changes of variable by which she made this reduction are quite impressive, spread over some 16 pages of one of the papers she eventually published on this subject. Still more impressive is the 80-page argument that follows to evaluate these integrals, which turn out to be hyperelliptic, involving the square root of a fifth-degree polynomial. This work so impressed the leading mathematicians of Paris that they decided the time had come to propose a contest for work in this area. When the contest was held in 1888, Kovalevskaya submitted a paper and was awarded the prize. She had finally reached the top of her profession and was rewarded with a tenured position in Stockholm. Sadly, she was not to be in that lofty position for long. In January 1891 she contracted pneumonia while returning to Stockholm from a winter vacation in Italy and died on February 10.

Bronze bust of Sof'ya Kovalevskaya, placed outside the Institut Mittag-Leffler in Djursholm, Sweden on January 15, 2000, the 150th anniversary of her birth.

Resistance from conservatives. Lest it be thought that the presence of such powerful talents as Charlotte Angas Scott and Sof'ya Kovalevskaya removed all doubt as to women's ability to create mathematics, we must point out that minds did not simply change immediately. Confronted with the evidence that good women mathematicians had already existed, the geometer Gino Loria (1862–1954) rationalized his continuing opposition to the admission of women to universities as follows, in an article in *Revue scientifique* in 1904:

> As for Sophie Germain and Sonja Kowalevsky, the collaboration they obtained from first-rate mathematicians prevents us from fixing with precision their mathematical role. Nevertheless what we know allows us to put the finishing touches on a character portrait of any woman mathematician. She is always a child prodigy, who, because of her unusual aptitudes, is admired, encouraged, and strongly aided by her friends and teachers. In childhood she manages to surpass her male fellow-students; in her youth she succeeds only in equalling them; while at the end of her studies, when her comrades of the other sex are progressing vigorously and boldly, she always seeks the support of a teacher, friend, or relative; and after a few years, exhausted by efforts beyond her strength, she finally abandons a work which is bringing her no joy.

The analysis of the factual errors and statistical and logical fallacies in this farrago of nonsense is left to the reader (see Problem 4.9 below). Loria could have known better. Six years before Loria wrote these words Felix Klein was quoted by the journal *Le progrès de l'est* as saying that he found his women students to be in every respect the equals of their male colleagues.

Grace Chisholm Young. Klein began taking on women students in the 1890s. The first of these students was Grace Chisholm, who completed the doctorate under his supervision in 1895 with a dissertation on the algebraic groups of spherical trigonometry. Her life and career were documented by her daughter and written up in an article by I. Grattan-Guinness (*1972*), which forms the basis for the present essay.

She was born on March 15, 1868, near London, the fifth child of parents of modest but comfortable means and the third child to survive. As a child she was stricken with polio and never completely recovered the use of her right hand. Like Charlotte Angas Scott, she was tutored at home and passed the Cambridge Senior Examination in 1885. Also like Scott, she attended Girton College and met Cayley. Her impressions of him were not flattering. To her he seemed to be a lumbering intellectual dinosaur, preventing any new life from emerging to enjoy the mathematical sunshine. In a colorful phrase, she wrote, "Cayley, unconscious himself of the effect he was having on his entourage, sat, like a figure of Buddha on its pedestal, dead-weight on the mathematical school of Cambridge" (Grattan-Guinness, *1972*, p. 115).

In her first year at Cambridge she might have been tutored by William Young (1863–1942), who later became her husband, except that she heard that his teaching methods were ill suited to young women. She found that Newnham College, the other women's college at Cambridge, had a much more serious professional atmosphere than Girton. She made contacts there with two other young women who had the same tutor that she had. With the support of this tutor and her fellow women students, she began to move among the serious mathematicians at Cambridge. In particular, she made friends with a student named Isabel Maddison (1869–1950) of Newnham College, who was being tutored by William Young. It will be recalled that a decade earlier Charlotte Angas Scott had been eighth Wrangler in the Tripos. In 1890, after reading a few names of the top Wranglers, the moderator—W. W. Rouse Ball (1850–1925), the author of a best-selling popular history of mathematics—made a long pause to get the attention of the audience, then said in a loud, clear voice, "*Above* the Senior Wrangler: Fawcett, Newnham." The young woman, Philippa Fawcett[12] of Newnham College, had scored a major triumph for women's education, being the top mathematics student at Cambridge in her year. No better role model can be imagined for students such as Isabel Maddison and Grace Chisholm. They finished first and second respectively in the year-end examinations at Girton College the following year. That fall, due to the absence of her regular tutor, Chisholm was forced to take lessons from William Young. In 1892 she ranked between the 23rd and 24th men on the Tripos, and Isabel Maddison finished in a tie with the 27th. (The rankings went as far as 112.) As a result, each received a First in mathematics. That same year they became the first women to attempt the Final Honours examinations at Oxford, where Chisholm obtained a First and Maddison a Second. This achievement made Chisholm the first person—of either gender—to obtain a First in any subject from both Oxford and Cambridge.[13]

Unfortunately, Cambridge did not offer Grace Chisholm support for graduate study, and her application to Cornell University in the United States was rejected. As an interesting irony, then, she was forced to apply to a university with a higher standard of quality than Cornell at the time, and one that was the mathematical equal of Cambridge: the University of Göttingen. There, thanks to the liberal views of Felix Klein and Friedrich Althoff, she was accepted, along with two young American women, Mary Frances ("May") Winston (1869–1959) and Margaret Eliza Maltby (1860–1944). In 1895, Chisholm broached the subject of a Ph. D. with Klein, who agreed to use his influence in the faculty to obtain authorization for the degree. It turned out to be necessary to go all the way to the Ministry of Culture in Berlin and obtain permission from Althoff personally. Fortunately, Althoff continued to be an enthusiastic supporter, and her final oral examination took place on April 26 of that year. She passed it and was granted the Ph. D. *magna cum laude.* She herself could hardly take in the magnitude of her achievement. More than two decades had passed since the university had awarded the Ph. D. to Sof'ya Kovalevskaya *in absentia.* Grace Chisholm had become the first woman to obtain that degree in mathematics through regular channels anywhere in Germany. She and Mary Winston were left alone together for a few minutes, which they used "to execute a war dance of triumph." Her two companions Mary Winston and

[12] Philippa Garrett Fawcett (1868–1948) was the daughter of a professor of political economy at Cambridge. Her mother was a prominent advocate of women's rights, and her sister was the first woman to obtain a medical degree at St. Andrew's in Scotland. Philippa used her Cambridge education to go to the Transvaal in 1902 and help set up an educational system there. From 1905 to 1934 she was Director of Education of the London County Council.

[13] Isabel Maddison was awarded the Bachelor of Science degree at the University of London in 1892. She received the Ph. D. at Bryn Mawr in 1896 under the supervision of Charlotte Angas Scott. She taught at Bryn Mawr until her retirement in 1926.

Portraits of Felix Klein and David Hilbert in the Mathematisches Institut and streets in Göttingen named after them.

Margaret Maltby also received the Ph. D. degree at Göttingen, Maltby (in physics) in 1895 and Winston in 1896.[14]

Grace Chisholm sent a copy of her dissertation to her former tutor William Young, and in the fall of 1895 they began collaboration on a book on astronomy, a project that both soon forgot in the pleasant fog of courtship. They were married in June 1896. They planned a life in which Grace would do mathematical research and William would support the family by his teaching. Grace sent off her first research paper for publication, and William, who was then 33 years old, continued tutoring. Circumstances intervened, however, to change these plans. Cambridge began to reduce the importance of coaching, and the first of their four children was born in June 1897. Because of what they regarded as the intellectual dryness of Cambridge and the need for a more substantial career for William, they moved back to Germany in the autumn of 1897. With the help of Felix Klein, William sent off his first research paper to the London Mathematical Society. It was Klein's advice a few years later that caused both Youngs to begin working in set theory. William, once started in mathematics, proved to be a prolific writer. In the words of Grattan-Guinness (*1972*, p. 142), he "definitely belongs to the category of creative

[14] Margaret Maltby taught at Barnard College (now part of Columbia University in New York) for 31 years and was chair of physics for 20 of those years. Mary Winston had studied at Bryn Mawr with Charlotte Angas Scott. She had met Felix Klein at the World's Columbian Exposition in Chicago in 1893 and had moved to Göttingen at his invitation. After returning to the United States she taught at Kansas State Agricultural College, married Henry Newson, a professor of mathematics at the University of Kansas, bore three children, and went back to teaching after Henry's early death. From 1921 to 1942 she taught at Eureka College in Illinois.

men who published more than was good for him." Moreover, he received a great deal of collaboration from his wife that, apparently by mutual consent, was not publicly acknowledged. He himself admitted that much of his role was to lay out for Grace problems that he couldn't solve himself. To the modern eye he appears too eager to interpret this situation by saying that "we are rising *together* to new heights." As he explained in a letter to her:

> The fact is that our papers ought to be published under our joint names, but if this were done neither of us get the benefit of it. No. Mine the laurels now and the knowledge. Yours the knowledge only. Everything under my name now, and later when the loaves and fishes are no more procurable in that way, everything or much under your name. [Grattan-Guinness, *1972*, p. 141]

Perhaps the criticism Loria made of Sophie Germain and Sof'ya Kovalevskaya for obtaining help from first-rate mathematicians might more properly have been leveled against William Young. To the author, the rationalization in this quotation seems self-serving. Yet, the only person who could make that judgment, Grace Chisholm Young herself, never gave any hint that she felt exploited, and William was certainly a very talented mathematician in his own right, whose talent simply manifested itself very late in life.

In 1903 Cambridge University Press agreed to publish a work on set theory under both their names. That book appeared in 1906; a book on geometry appeared under both names in 1905. Grace was busy bearing children all this time (their last three children were born in 1903, 1904, and 1908) and studying medicine. She began to write mathematical papers under her own name in 1913, after William took a position in Calcutta, which of course required him to be away for long periods of time. These papers, especially her paper on the differentiability properties of completely arbitrary functions, added to her reputation and were cited in textbooks on measure theory for many decades.

Sadly, the fanaticism of World War I caused some strains between the Youngs and their old mentor Felix Klein. As a patriotic German, Klein had signed a declaration of support for the German position at the beginning of the war. Four years later, as the defeat of Germany drew near, Grace wrote to him, asking him to withdraw his signature. Of course, propaganda had been intense in all the belligerent countries during the war, and even the mildest-mannered people tended to believe what they were told and to hate the enemy. Klein replied diplomatically, saying that, "Everyone will hold to his own country in light and dark days, but we must free ourselves from passion if international cooperation such as we all desire is to assert itself again for the good of the whole" (Grattan-Guinness, *1972*, p. 160). If only other scholars had been as magnanimous as Klein, German scholars might have had less justification for complaining of exclusion in the bitter postwar period. At least there was no irreparable breach between the Youngs and Klein. When Klein died in 1925, his widow thanked the Youngs for sending their sympathy, saying, "From all over the world I received such lovely letters full of affection and gratitude, so many tell me that he showed them the way on which their life was built. I had him for fifty years, this wonderful man; how privileged I am above most women..." (Grattan-Guinness, *1972*, p. 171).

All four of their children eventually obtained doctoral degrees, and the pair had good grounds for being well-satisfied with their married life. When World War II began in September 1939 they were on holiday in Switzerland, and there was real fear that Switzerland would be invaded. Grace immediately returned to England, but William stayed behind. The fall of France in 1940 enforced a long separation on them. The health of William, who was by then in his late 70s, declined rapidly, and he died in a nursing home in June 1942. Grace survived for nearly two more years, dying in England in March 1944. Grattan-Guinness (*1972*, p. 181) has eloquently characterized this remarkable woman:

> She knew more than half a dozen languages herself, and in addition she was a good mathematician, a virtually qualified medical doctor, and in her spare time, pianist, poet, painter, author, Platonic and Elizabethan scholar—and a devoted mother to all her children. And in the blend of her rôles as scholar and as mother lay the fulfillment of her complicated personality.

Emmy Noether. Sof'ya Kovalevskaya and Grace Chisholm Young had had to improvise their careers, taking advantage of the opportunities that arose from time to time. One might have thought that Amalie Emmy Noether was better situated in regard to both the number of opportunities arising and the ability to take advantage of them. After all, she came a full generation later than Kovalevskaya, the University of Göttingen had been awarding degrees to women for five years when she enrolled, and she was the eldest child of the distinguished mathematician Max Noether.[15] According to Dick (*1981*), on whose biography of her the following account is based, she was born on March 23, 1882 in Erlangen, Germany, where her father was a professor of mathematics. She was to acquire three younger brothers in 1883, 1884, and 1889. Her childhood was quite a normal one for a girl of her day, and at the age of 18 she took the examinations for teachers of French and English, scoring very well. This achievement made her eligible to teach modern languages at women's educational institutions. However, despite the difficulties women were having at universities, as depicted by Gerhard Kowalewski, she decided to attend the University of Erlangen. There she was one of only two women in the student body of 986, and she was only an auditor, preparing simultaneously to take the graduation examinations in Nürnberg. After passing these examinations, she went to the University of Göttingen for one year, again not as a matriculated student. If it seems strange that Grace Chisholm was allowed to matriculate at Göttingen and Emmy Noether was not, the explanation seems to be precisely that Emmy Noether was a German.

In 1904 she was allowed to matriculate at Erlangen, where she wrote a dissertation under the direction of Paul Gordan (1837–1912). Gordan was a constructivist and disliked abstract proofs. According to Kowalewski (*1950*, p. 25) he is said to have remarked of one proof of the Hilbert basis theorem, "That is no longer mathematics; that is theology." In her dissertation Emmy Noether followed Gordan's constructivist methods; but she was later to become famous for work done from a much more abstract point of view. She received the doctorate *summa cum laude* in 1907. Thus, she surmounted the first two obstacles to a career in mathematics with only a small amount of difficulty, not much more than faced by her brother Fritz (1884–1941), who was also a mathematician. That third obstacle,

[15] It will be recalled that Charlotte Angas Scott had given a new proof of a theorem by Max Noether.

however, finding work at a university, was formidable. Emmy Noether spent many years working without salary at the Mathematical Institute in Erlangen. This position enabled her to look after her father, who had been frail since he contracted polio at the age of 14. It also allowed her to continue working on mathematical ideas. For nearly two decades she corresponded with Ernst Fischer (1875–1954, Gordan's successor in Erlangen), who is best remembered for having discovered the Riesz–Fischer theorem independently of F. Riesz (1880–1956). By staying in touch with the mathematical community and giving lectures on her discoveries, she kept her name before certain influential mathematicians, namely David Hilbert (1862–1943) and Felix Klein,[16] and in 1915 she was invited to work as a *Privatdozent* in Göttingen. (This was the same rank originally offered to Kovalevskaya at Stockholm in 1883.) Over the next four years Klein and Hilbert used all their influence to get her a regular appointment at Göttingen; during part of that time she lectured for Hilbert in mathematical physics. That work led her to a theorem in general relativity that was highly praised by both Hilbert and Einstein. Despite this brilliant work, however, she was not allowed to pass the *Habilitation* needed to acquire a professorship. Only after the German defeat in World War I, which was followed by the abdication of the Kaiser and a general spirit of reform in Germany, was she allowed to "habilitate." Between Sof'ya Kovalevskaya and Emmy Noether there was a curious kind of symmetry: Kovalevskaya was probably aided in her efforts to become a student in Berlin because many of the students were away at war at the time. Noether was aided in her efforts to become a professor by an influx of returning war veterans. She began lecturing in courses offered under the name Dr. Emmy Noether (without any mention of Hilbert) in the fall of 1919. Through the efforts of Richard Courant (1888–1972) she was eventually granted a small salary for her lectures.

In the 1920s she moved into the area of abstract algebra, and it is in this area that mathematicians know her work best. Noetherian rings became a basic area of study after her work, which became part of a standard textbook by her student Bartel Leendert van der Waerden (1903–1996). He later described her influence on this work (*1975*, p. 32):

> When I came to Göttingen in 1924, a new world opened up before me. I learned from Emmy Noether that the tools by which my questions could be handled had already been developed by Dedekind and Weber, by Hilbert, Lasker, and Macaulay, by Steinitz and by Emmy Noether herself.

Of all the women we have discussed Emmy Noether was unquestionably the most talented mathematically. Her work, both in quantity and quality, places her in the elite of twentieth-century mathematicians, and it was recognized as such during her lifetime. She became an editor of *Mathematische Annalen*, one of the two or three most prestigious journals in the world. She was invited to speak at the International Congress of Mathematicians in Bologna in 1928 and in Zürich in 1932, when she shared with Emil Artin (1898–1962) a prestigious prize for the advancement of mathematical knowledge. This recognition was clear and simple

[16] Klein wrote to Hilbert, "You know that Fräulein Noether is continually advising me in my projects and that it is really through her that I have become competent in the subject." (Dick, *1981*, p. 31)

Emmy Noether

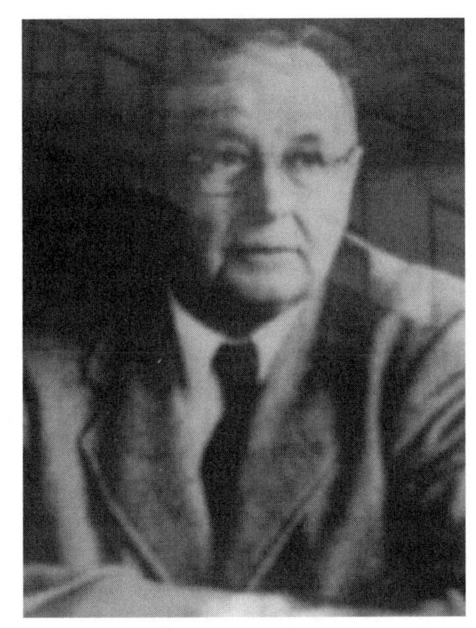

Hermann Weyl

proof of her ability, but it was still short of what she deserved. Hilbert's successor in Göttingen, Hermann Weyl (1885–1955), made this point when wrote her obituary:

> When I was called permanently to Göttingen in 1930, I earnestly tried to obtain from the Ministerium a better position for her, because I was ashamed to occupy such a preferred position beside her, whom I knew to be my superior as a mathematician in many respects. I did not succeed, nor did an attempt to push through her election as a member of the Göttinger Gesellschaft der Wissenschaften. Tradition, prejudice, external considerations, weighted the balance against her scientific merits and scientific greatness, by that time denied by no one. In my Göttingen years, 1930–1933, she was without doubt the strongest center of mathematical activity there. [Dick, *1981*, p. 169]

To have been recognized by one of the twentieth century's greatest mathematicians as "the strongest center of mathematical activity" at a university that was second to none in the quality of its research is high praise indeed. It is unfortunate that this recognition was beyond the capability of the Ministerium. The year 1932 was to be the summit of Noether's career. The following year, the advanced culture of Germany, which had enabled her to develop her talents to their fullest, turned its back on its own brilliant past and plunged into the nightmare of Nazism. Despite extraordinary efforts by the greatest scientists on her behalf, Noether was removed from the position that she had achieved through such a long struggle and the assistance of great mathematicians. Along with hundreds of other Jewish mathematicians, including her friends Richard Courant and Hermann Weyl (who was not Jewish, but whose wife was), she had to find a new life in a different land. She accepted a visiting professorship at Bryn Mawr, which allowed her also to lecture at the Institute for Advanced Study in Princeton.[17] Despite the gathering clouds in Germany, she returned there in 1934 to visit her brother Fritz, who was about to seek asylum in the Soviet Union. (Ironically, he was arrested in 1937, during one of the many purges conducted by Stalin, and executed as a German spy on the day the Germans occupied Smolensk in 1941.) She returned to Bryn Mawr in the spring of 1934.

Weyl, who went to Princeton in 1933, expressed his indignation at the Nazi policy of excluding "non-Aryans" from teaching. In a letter sent to Heinrich Brandt (1886–1954) in Halle he wrote:[18]

> What impresses me most about Emmy Noether is that her research has become more and more concrete and profound. Why should this Jewess not work in the area that has led to such great achievements in the hands of the "Aryan" Dedekind? I am happy to leave it to Herrn Spengler and Bieberbach to assign mathematical modes of thought according to cultures and races. [Jentsch, *1986*, p. 9]

At Bryn Mawr she was a great success and an inspiration to the women studying there. She taught several graduate and postdoctoral students who went on to successful careers, including her former assistant from Göttingen, Olga Taussky (1906–1995), who was forced to leave a tutoring position in Vienna in 1933. Her time, however, was to be very brief. She developed a tumor in 1935, but she does not seem to have been worried about its possible consequences. It was therefore a great shock to her colleagues in April 1935 when, after an operation at Bryn Mawr Hospital that seemed to offer a good prognosis, she developed complications and died within a few hours.

4. American women

In the United States higher education was open to women from the late nineteenth century on in the large, well-supported state universities. The elite eastern universities later known as the Ivy League remained mostly all-male for another century; but some of them were near women's colleges, and some of the women from those colleges were able to take courses at places like Harvard and the University of Pennsylvania. Although mathematics in general in the United States was not yet on a par with what was being done in Europe, American women began to participate in the profession in the late nineteenth century. Our summary of this story is very incomplete, and the reader is referred to the excellent article of Green and LaDuke (*1987*) for complete statistics on the women mathematicians and the institutions where they studied and worked.

---

[17] There was no chance of her lecturing at Princeton University itself, which was all-male at the time.

[18] Oswald Spengler (1880–1936) was a German philosopher of history, best known for having written *Der Untergang des Abendlandes* (*The Decline of the West*). His philosophy of history, which Weyl alludes to in this quote, suited the Nazis. Although at first sympathetic to them, he was repelled by their crudity and their antisemitism. By the time Weyl wrote this letter, the Nazis had banned all mention of Spengler on German radio. Ludwig Bieberbach (1886–1982) was a mathematician of some talent who worked at Berlin during the Nazi era and edited the Party-approved journal *Deutsche Mathematik*. At the time when Weyl wrote this letter, Bieberbach was wearing a Nazi uniform to the university and enthusiastically endorsing the persecution of non-Aryans.

Christine Ladd-Franklin. The first of the two American women we shall discuss was induced by prevailing prejudice to abandon mathematics for psychology, a field in which she also encountered firm exclusion. Christine Ladd was born in New York in 1847. Her mother and aunt were advocates of women's rights. Her mother died when she was 12, and she was sent to live with her father's mother. Education for girls had come to be seen as a necessity by the American middle class, and so she was enrolled at Wesleyan Academy along with boys her age who expected to be admitted to Harvard. She herself could have no such expectations, but she did dream of attending Vassar. Her father encouraged her in her studies at Wesleyan Academy, but the grandmother she was living with was opposed to Vassar. Nevertheless, she prevailed and her mother's sister supported her financially for the first year. At Vassar she was particularly encouraged by Maria Mitchell (1818–1889, the first American woman astronomer of note). After obtaining a bachelor's degree in 1869, she spent nine years as a teacher of science and mathematics, writing articles on mathematics education that were published in England. Burnout, that familiar phenomenon among those who teach adolescents, finally set in, and she began to cast about for other careers.

Such an opportunity came along at just the right time. In 1876 Johns Hopkins University opened in Baltimore, the first American university devoted exclusively to graduate studies. Moreover, it managed to hire one of the greatest European mathematicians, James Joseph Sylvester, who, being Jewish, could not obtain a position at Cambridge or Oxford.[19] By great good fortune, the name Christine Ladd was familiar to Sylvester from her articles on education. On his recommendation the university agreed to allow her to attend lectures, but only lectures by Sylvester. This restriction was lifted after the first year, and she was able to attend lectures by William Edward Story and by Charles Sanders Peirce (1839–1914, described by the British philosopher Bertrand Russell as "the greatest American thinker ever"). While working at Hopkins, she married Fabian Franklin (1853–1939), a young professor of mathematics who was born in Hungary but whose parents had moved to the United States when he was 2 years old. They were to have two children in rapid succession, one of whom died in infancy. After her marriage, she wrote her name with a hyphen as Ladd-Franklin. Under the influence of Peirce she wrote a dissertation bearing the title *The Algebra of Logic*, which was published in 1883 in the *American Journal of Mathematics*, the new journal founded by Sylvester at Story's suggestion. In fact, she published several papers in that journal, and was, by any objective standards, one of the best-qualified mathematicians in the United States. Nevertheless, Sylvester and Peirce together could not fulfill the mentoring role that Weierstrass performed for Kovalevskaya, Cayley for Charlotte Angas Scott and Klein for Grace Chisholm Young. She was unable to obtain either the Ph. D. degree or an academic position. Although she had overcome the first obstacle, getting her family's support for an education, the second and third stymied her for the rest of her life.

She had always been interested in areas of science other than mathematics, and her choice of mathematics as a major at Vassar had been partly the result of being excluded, as a woman, from the physics laboratories. In the mid-1880s she began to take an interest in psychology, especially the psychology of color perception.

[19] An earlier stay at the University of Virginia in 1841, when slavery still existed, had ended in disaster for Sylvester.

She wrote a study of this subject that was published in the first volume of the *American Journal of Psychology* in 1887. Vassar awarded her an honorary doctor of laws degree that year.

The laboratories she had not been allowed to enter at Vassar were finally opened to her in Germany, where her husband took a sabbatical (in Göttingen) in 1891–1892. She took advantage of the occasion to spend some time in Berlin with the great physicist Hermann von Helmholtz (1821–1894). She presented the results of her theory and experiments at a conference in London that year.

Upon returning to the United States, she began a long quest for a degree and an academic position suited to her talents. Hopkins, where her husband continued to teach, refused her applications year after year. She continued to work independently (what else could she do?) and for many years played an active role in administering fellowships to support postdoctoral work for women. Not until 1904 was she allowed to teach a course in psychology at Hopkins. The following year her husband gave up mathematics in favor of journalism. He found a position in New York in 1910, and they moved there. Remembering the "dean's rule" at Barnard College, no one will be surprised to learn that as a married woman, she had no hope of obtaining a position there. She was allowed, however, to lecture part-time at Columbia University during 1912–1913. In 1913 she lectured at Harvard and at Clark University, where her old professor from Johns Hopkins, William Edward Story, was chair of the Mathematics Department and the president, G. Stanley Hall (1844–1924), was a famous psychologist.[20] She also lectured at the University of Chicago in 1914. By this time, of course, she was no longer regarded as a mathematician; her lectures were on psychology.

Except for the position of editor for Baldwin's *Dictionary of Philosophy and Psychology*, which she occupied from 1901 to 1905, she was excluded rather completely from participation in the professional life of a psychologist. In particular, she was not allowed to attend meetings and deliver papers. Only in 1929 was she finally able to publish a lifetime of work in psychology in her treatise *Colour and Colour Theories*. This work was published simultaneously in London and New York (which explains the British spelling of the title). In a great anticlimax in 1926, Johns Hopkins finally awarded her the Ph. D. in mathematics that she had earned 43 years earlier. One hardly knows whether the old saying "better late than never" applies in such a case. She died in 1930.

Anna Johnson Pell Wheeler. A few decades of social change can make a great deal of difference to one's life. The social traditions and prejudice that deprived Christine Ladd-Franklin of what would have been a brilliant career were, with difficulty, overcome by one of the first American women to achieve recognition in mathematics, Anna Johnson (later Anna Johnson Pell, still later Anna Johnson Pell Wheeler). She was born in Iowa in 1883 and entered the University of South Dakota in 1899, graduating in 1903. The following year she earned a master's degree at the University of Iowa, and then in 1905 she earned a second master's degree at Radcliffe. She remained there another year in order to study with two of the first prominent American mathematicians, Harvard professors William Fogg Osgood

---

[20] G. Stanley Hall was the first American to obtain a doctoral degree in psychology; he had been a professor at Johns Hopkins during the early 1880s and had brought Sigmund Freud to lecture at Clark in 1910.

(1864–1943, a student of Max Noether at Erlangen) and Maxime Bôcher (1867–1918, a student of Felix Klein). Even though she was not a student at Wellesley, she was awarded a Wellesley fellowship for study abroad and went to Göttingen to attend lectures by Klein and Hilbert. Her recently widowed former professor at the University of South Dakota, Alexander Pell (1857–1921), had been corresponding with her for some time. In 1907 he came to Göttingen, and they were married. She returned with him to the United States but then went back to Göttingen to finish her doctorate. For reasons that are not clear, she did the work but did not receive the degree. Her family thought she had been pressured by her husband, who was suffering from the separation, to return to him. The only explanation she gave to the dean at Radcliffe for returning without the degree was that "in Göttingen I had some trouble with Professor Hilbert and came back to America without a degree" (Grinstein and Campbell, *1982*, p. 41). She emphasized that she had written her thesis without any help from Hilbert, and so was able to submit it to the University of Chicago, where, under the supervision of another early American mathematician of distinction, Eliakim Hastings Moore (1862–1932), she was awarded the degree *magna cum laude* in 1909.

Once again, a woman had obtained a Ph. D. in mathematics at the age of only 26. Her adviser Moore sought positions for her at many universities near Chicago, where her husband was a professor at the Armour Institute of Technology. But that third-stage hurdle that has been mentioned several times before once again proved nearly insurmountable. As she wrote,

> I had hoped for a position in one of the good univ. like Wisc., Ill. etc., but there is such an objection to women that they prefer a man even if he is inferior both in training and research. It seems that Professor Moore has also given up hope for he has inquired at some of the Eastern Girls' Colleges and Bryn Mawr is apparently the only one with a vacancy in Math. [Grinstein and Campbell, *1982*, p. 42]

As it happened, she did not go to Bryn Mawr immediately. Her husband had a stroke that year, and she took over his teaching duties at the Armour Institute of Technology while also lecturing at the University of Chicago. She proved extremely competent at both duties. Then, from 1911 to 1918 she taught at Mount Holyoke College in Massachusetts before moving on to Bryn Mawr. When Charlotte Angas Scott retired in 1924, Anna Pell became head of the mathematics department at Bryn Mawr. Four years after the death of Alexander Pell in 1921, she married another widower, Arthur Leslie Wheeler (1871–1932, a distinguished classics scholar whose books are still being reprinted 70 years after his death). Since Wheeler had just been appointed at Princeton at the time, Anna taught only part-time at Bryn Mawr for a few years. But when he died in 1932, she went back to full-time work at Bryn Mawr and presided over the invitation to Emmy Noether that brought that distinguished mathematician there. Her own work in linear algebra allowed her to supervise the theses of many students, and was so distinguished that in 1927 she was the first woman to be invited to give a Colloquium Lecture to the American Mathematical Society.[21] She remained at Bryn Mawr until her retirement in 1948, then continued to live in the area until her death in 1966.

---

[21] The second woman was Julia Bowman Robinson—in 1980!

5. The situation today

To bring the story of the progress of women in mathematics up to the present would require writing about people whose careers are still continuing. Even when people are willing to write about themselves, reporting on what they wrote is risky; there is a danger of putting the wrong emphasis on what they have said and giving an impression that they did not intend. For that reason we shall not discuss any more biographies, but consider only how the small window of opportunity available to the pioneering women mathematicians has been enlarged to a size comparable with that available to men, and ask what more needs to be done. For examples the reader is referred to the book of Henrion (*1997*), which contains interviews with a number of women and is aimed at overcoming persistent stereotypes about women in the profession of mathematics.

Although a mathematical education was *formally* available to women throughout the twentieth century, social conditioning discouraged young girls from aiming at such a career. As a result, few of them ever even realized that they might have the talent to be mathematicians or scientists. Colleges of engineering and medicine were full of young men; colleges of education and nursing were full of young women. There was very little "osmosis" between these two "cells" until the women's movement began in earnest in the 1970s. Only when significant numbers of women sought admission to scientific careers did the difficulties experienced by the few women already in those careers come to public attention. What was revealed was a wide variety of ways of discriminating—women not being admitted to some of the universities where the best work was being done, being ranked at a lower priority when applying for jobs, being asked personal questions about their families, offered lower salaries, being ignored in class, not being taken seriously in applications for graduate work, not being guided and mentored properly so as to encourage them to seek advanced degrees, being asked to do menial administrative work during probationary periods, and the like. Overcoming these problems required both antidiscrimination and affirmative-action legislation. It also involved patiently educating the public, men and women alike, in new ways of "conducting business." University administrators had to learn how to mentor young women faculty members to channel their work into areas likely to lead to tenure. The faculty members themselves had to learn to fight against the "good citizen" impulses that got them onto too many committees, into too much curriculum work, and the sacrifice of a great deal of time trying to be the best teacher possible, all at the expense of research.[22]

What now remains to be done? If we assume that the offices of affirmative action/equal opportunity at our major industries and universities are doing their job properly—and if they are not, legal redress is available for those with the courage to pursue it—the main work remaining is educational. Most of all, both boys and girls in their early teens need to be shown *how scientists actually spend their time, what their jobs consist of.* Without that kind of information, they are likely to judge a profession by the difficulty of the courses they are taking

[22] This sentence was written from the point of view of what is in the best interest of a faculty member seeking tenure. The best interest of the institution and its students and the greater good of society as a whole may very well be advanced through working on committees, developing curricula, and being the best possible teacher; but unless the prevailing attitudes change at major universities, a probationary faculty member is not advised to pursue tenure through those activities.

in school. And nearly *everyone*, even a very bright student, finds mathematics difficult. Students need to be shown that a career in mathematics does not require super intelligence. What students often imagine they must do—solve some difficult, long-open problem—is definitely optional, not a necessary part of a mathematical career. This task is being addressed by mathematical organizations such as the Mathematical Association of America and the American Mathematical Society, and by various programs supported by the Department of Education and the National Science Foundation.

A secondary task is to root out the remaining stereotypes from professional mathematicians themselves. The women interviewed by Henrion for her book (*1997*) pointed out a number of practices within the profession that "create a 'chilly climate' for women both in academia in general and in mathematics in particular." Henrion has quite astutely pointed out that the mathematical community, as a community, has its own set of expectations about how a person will work, and those expectations were set by men. How those expectations may change (or may not) as more and more women take on significant roles in the professional organizations is a development that will be interesting to observe in the future. And as the image of a set of successive obstacles that we have used above to interpret the lives of the early women mathematicians shows, the further a woman progresses, the higher the "hurdles" tend to be. Henrion (*1997*, p. xxxi) expressed the matter somewhat differently: "[W]omen are even further from equity the farther along in the pipeline we go." Making professional activities gender neutral is the primary challenge for the future.

Questions and problems

4.1. In the late fourth and early fifth centuries the city of Alexandria, where Hypatia lived, was divided into Christian, Jewish, and pagan cultures. Is it merely a random event that the only woman mathematician of the time in this city with a long history of scholarship happened to come from the pagan culture?

4.2. Compare the careers of Charlotte Angas Scott and Sof'ya Kovalevskaya. In what aspects were they similar? What significant differences were there? Were these differences due to the continental circles in which Kovalevskaya moved compared to the Anglo-American milieu of Scott's career? Or were they due to individual differences between the two women?

4.3. Choose two women mathematicians, either from among those discussed in this chapter or by going to a suitable website. Read brief biographical sketches of them. Then try to match each woman with a comparable male mathematician from the same era and country. Compare their motives for studying mathematics if any motives are given, the kind of education they received, the journals where they published their work, and the kind of academic positions they occupied.

4.4. How do you account for the fact that a considerable percentage (compared to their percentage of the general population) of the women studying higher mathematics in the United States during the 1930s were Roman Catholic nuns?[23]

[23] Some of these nuns produced mathematical research of high quality, for example, Sister Mary Celine Fasenmyer (1906–1996).

4.5. What were the advantages and disadvantages of marriage for a woman seeking an academic career before the twentieth century? How much of this depended on the particular choice of a husband at each stage of the career? The cases of Mary Somerville, Sof'ya Kovalevskaya, and Grace Chisholm Young will be illuminating, but it will be useful to seek more detailed sources than the narratives above.

4.6. How big a part did chance play in the careers of the early women mathematicians? (The word *chance* is used advisedly, rather than *luck*, since the opportunities that came for Sof'ya Kovalevskaya and Anna Johnson Pell Wheeler were the result of tragic misfortunes to their husbands.)

4.7. How important is (or was) encouragement from family and friends in the decision to study science? How important is it to have a mentor, an established professional in the same field, to help orient early career decisions? How important is it for a young woman to have an older woman as a role model? Try to answer these questions along a scale from "not at all important" through "somewhat important" and "very important" to "essential." Use the examples of the women whose careers are sketched above to support your rankings.

4.8. Why were most of the women who received the first doctoral degrees in mathematics at German universities foreigners? Why were there no Germans among them? In his lectures on the development of nineteenth-century mathematics (*1926*, Vol. 1, p. 284), Klein mentions that a 17-year-old woman named Dorothea Schlözer (apparently German, to judge by the name) had received a doctorate in economics at Göttingen a full century earlier.

4.9. How strong are the "facts" that Loria adduces in his argument against admitting women to universities? Were all the women discussed here encouraged by their families when they were young? Is it really true that it is impossible to "fix with precision" the original contributions of Sophie Germain and Sof'ya Kovalevskaya? You may wish to consult biographies of these women in which their correspondence is discussed. Would collaboration with other mathematicians make it impossible to "fix with precision" the work of any male mathematicians? Consider also the case of Charlotte Angas Scott and others. Is it true that they were exhausted after finishing their education?

Next, consider what we may call the "honor student" fallacy. Universities select the top students in high school classes for admission, so that a student who excelled the other students in high school might be able at best to equal the other students at a university. Further selections for graduate school, then for hiring at universities of various levels of prestige, then for academic honors, provide layer after layer of filtering. Except for an extremely tiny elite, those who were at the top at one stage find themselves in the middle at the next and eventually reach (what is ideally) a level commensurate with their talent. What conclusions could be justified in regard to any gender link in this universal process, based on a sample of fewer than five women? And how can Loria be sure he knows their proper level when all the women up to the time of writing were systematically locked out of the best opportunities for professional advancement? Look at the twentieth century and see what becomes of Loria's argument that women never reach the top.

Finally, examine Loria's logic in the light of the cold facts of society: A woman who wished to have a career in mathematics would naturally be well advised to find a mentor with a well-established reputation, as Charlotte Angas Scott and Sof'ya

Kovalevskaya did. A woman who did not do that would have no chance of being cited by Loria as an example, since she would never have been heard of. Is this argument not a classical example of catch-22?

4.10. Here is a policy question to consider. The primary undergraduate competition for mathematics majors is the Putnam Examination, administered the first weekend in December each year by the Mathematical Association of America. In addition to its rankings for the top teams and the top individuals, this examination also provides, for women who choose to enter, a prize for the highest-ranking woman. (The people grading the examinations do not know the identities of the entrants, and a woman can enter this competition without identifying herself to the graders.) Is this policy an important affirmative-action step to encourage talented young women in mathematical careers, or does it "send the wrong message," implying that women cannot compete with men on an equal basis in mathematics? If you consider it a good thing, how long should it be continued? Forever? If not, what criterion should be used to determine when to discontinue the separate category? Bear in mind that the number of women taking the Putnam Examination is still considerably smaller than the number of men.

4.11. Continuing the topic of the Question 4.10, what criterion should be used to determine when affirmative action policies designed to overcome the effects of past discrimination against women will have achieved their aim? For example, are these policies to be continued until 50% of all mathematics professors are women within the universities of each ranking? (The American Mathematical Society divides institutions into different rankings according to the degrees they grant; there is also a less formal but still effective ranking in terms of the prestige of institutions.) What goal is being pursued: that each man and each woman should have equal access to the profession and equal opportunity for advancement in it, or that equal numbers of men and women will choose the profession and achieve advancement? Or is the goal different from both of these? If the goal is the first of these, how will we know when it has been achieved?

Part 2

Numbers

Numbers are the first association in the minds of most people when they hear the word *mathematics.* The word *arithmetic* comes into English from the Greek word *arithmós*, meaning *number.* What is nowadays called arithmetic—that is, calculation—had a different name among the Greek writers of ancient times: *logistikê*, the source of our modern word *logistics.* In the comedy *The Acharnians* by Aristophanes, the hero Dicaeopolis reflects that, arriving early for meetings of the Athenian assembly, "aporō̂, gráphō, paratíllomai, logízomai" ("I don't know what to do with myself; I doodle, pull my hair, and calculate").

The different levels of sophistication in the use and study of numbers provide a convenient division into chapters for the present part of the book. We distinguish three different stages in the advancement of human knowledge about numbers: (1) the elementary stage, in which a limited set of integers and fractions is used for counting and measuring; (2) the stage of calculation, in which the common operations of addition, subtraction, multiplication, and division are introduced; and (3) the theoretical stage, in which numbers themselves become an object of interest, different kinds of numbers are distinguished, and new number systems are invented.

The first stage forms the subject matter of Chapter 5. Even when dealing with immediate problems of trade, mere counting is probably not quite sufficient numeracy for practical life; some way of comparing numbers in terms of size is needed. And when administering a more populous society, planning large public works projects, military campaigns, and the like, sophisticated ways of calculating are essential. In Chapter 6 we discuss the second stage, the methods of calculating used in different cultures.

In Chapter 7 we examine the third stage, number theory. This theory begins with the mathematicians of ancient Greece, India, and China. We look at the unique achievements of each civilization: prime, composite, triangular, square, and pentagonal numbers among the Pythagoreans, combinatorics and congruences among the Chinese and Hindus.

In Chapter 8 we discuss number systems and number theory in the modern world. Here we see how algebra led to the concept of irrational (algebraic) numbers, and the geometric representation of such numbers brought along still more (transcendental) numbers. When combined with geometry and calculus, this new algebraic view of numbers led to the theory of complex variables, which in turn made it possible to answer some very delicate questions on the relative density of prime numbers (the famous "prime number theorem"). The continuing development of these connections, as well as connections with the theory of trigonometric series, has made it possible to settle some famous conjectures: the Wiles–Taylor proof of Fermat's last theorem and a partial proof by I. M. Vinogradov of the famous Goldbach conjecture, for example.

CHAPTER 5

Counting

Counting could conceivably occur without number words. What is required is merely a matching of the objects in two sets. The legendary American gunslinger, putting a notch in the handle of his gun for every person he has killed, is an example of such counting. A vivid example is cited by Closs (*1986*, p. 16) as a folk tale of the Copper Eskimo. In this story, a hunter who has killed a wolf argues with another hunter who has killed a caribou as to which animal has more hair. To decide the question, they pull out the hairs one at a time and pair them off. This mode of thought has become very familiar to mathematics students over the past century because of the rise of set theory in the undergraduate curriculum.

1. Number words

Every human language that we know about has words for numbers. In the case of languages whose long history is known—English, for example—the number words seem to be of such ancient origin that they have no obvious relation to the non-numerical words in the language. Some attempts have been made to find clues as to the origin of number words and number concepts in the grammar of various languages, but very few reliable conclusions have been reached. The guesses involved are interesting, however, and we shall look at a few of them below, taken mostly from the books of Menninger (*1969*) and Gow (*1884*).

To begin with modern English, when a person says, "I know a number of ways to prove the Pythagorean theorem," the listener will interpret the phrase "a number of" to mean "three or more." The word *number* here is a synonym for the fancier word *multitude* and is used to refer to any set of things having three or more members. If the speaker knew only two ways to prove the Pythagorean theorem, the word *number* would almost certainly be replaced by *couple*. This last word is one of the few collective words in English with a definite numerical meaning. It is used as a synonym for *two* when the two objects mentioned are related to each other in some way, in phrases like "I have a couple of errands to run downtown." The connection with the number two is not quite exact here, since in very informal speech the word *couple* is often stretched to mean simply a small number. From such considerations we might form the hypothesis that one and two are instinctive concepts, and that numbers as a deliberate, conscious creation of the human mind begin with three. Support for this idea comes from the reflection that English has special words for ordinal (*second*) and partitive (*half*) concepts connected with the number two and a special word (*both*) to apply to the whole of a set of two objects, while the ordinal and partitive concepts merge for numbers three and higher (*third*, *one-third*, and so on) and the same word (*all*) is used to denote the whole of any set having more than two members. Of course, what is true of English does not always apply to other languages.

Ancient Greek and Sanskrit, as well as modern English and other European languages, share a great many root words and morphological features. In a book on Greek mathematics (*1884*) the British mathematician James Gow (1854–1923) of Trinity College, Cambridge speculated on the possibility of using comparative philology to discover the history of mathematical terms. He noted in particular that the words for one, two, three, and four are declinable in Greek, but not the words for five and above. That fact suggested to him that numbers above four are an artificial creation. (It also dovetails neatly with the observations of Karen Fuson, discussed in Chapter 1, on the counting abilities of children.) Gow noted that in Slavonic, which is a European language, *all* numerals are declined as feminine singular nouns (those ending in 5 or above still are, in modern Russian), but he regarded this usage as later and hence not relevant to his inquiry. He also noted that all numerals are declined in Sanskrit, but thought it an important difference that no gender could be assigned to them.

In a comprehensive study of numbers and counting (*1969*) the mathematician Karl Menninger (1898–1963) conjectured that the words for *one* and *two* may be connected with personal and demonstrative pronouns. In favor of Menninger's conjecture, we note that in formal writing in English, and sometimes also in formal speaking, the word *one* is used to mean an unspecified person. This English usage probably derives from similar usage of the special third-person pronoun *on* in French. Speaking of French, there *is* a suggestive similarity between this pronoun and the word (*un*) for one in that language. The Russian third-person pronouns (*on, ona, ono* in the three genders) are the short forms of the archaic demonstrative pronoun *onyi, onaya, onoe* (meaning *that one*), and the word for the number one is *odin, odna, odno.*

Menninger also suggested that the word *two*, or at least the word *dual*, may be related to the archaic second-person singular pronoun *thou* (*du*, still used in Menninger's native language, which is German). Menninger noted that the concept of two is closely related to the concept of "other." Consider, for example, the following sentences:

This is my favorite style of gloves. I have a second pair in my closet.

This is my favorite style of gloves. I have another pair in my closet.

Menninger suggested a connection between *three* and *through*, based on other languages, such as the Latin *tres* and *trans.* Despite these interesting connections, Menninger emphasized that the words for *cardinal* numbers have left no definite traces of their origin in the modern Indo-European languages. All the connections mentioned above could be merely coincidental. On the other hand, the *ordinal* number words *first* and *second* have a more obvious connection with non-mathematical language. The word *first* is an evolved form of *fore-est* (*foremost*), meaning the one farthest forward. The word *second* comes from the Latin word *sequor* (*I follow*).

In cultures where mathematics and counting are developed less elaborately, number words sometimes retain a direct relation with physical objects exemplifying the numbers. Nearly always, the words for numbers are also used for body parts in the corresponding number, especially fingers. In English also, we find the word *digit* used to describe both a finger or toe and the special kinds of numbers that occur in representations of the positive integers in terms of a base.

Are there languages in which body parts stand for *cardinal* numbers? Could the number two be the word for eyes, for example? Gow (*1884*) cited a number of examples to show that in many languages the word for five also means *hand*, and that words for eight, nine, and ten also designate specific fingers of the hand in some languages. A survey of ways of counting around the world provides some evidence for Gow's thesis. The Bororo of Mato Grosso, for example (Closs, *1986*, p. 23), use a phrase for the number five that translates literally "as many of them as my hand complete." In that same language the number seven is "my hand and another with a partner," 10 is "my fingers all together in front," and 15 is "now my foot is finished."

2. Bases for counting

Children have to be taught to count before they can talk about groups of more than four things. Beyond certain sizes, it becomes impossible for anyone to tell at a glance how many objects are present. Most people, for example, can say immediately how many letters are in a word of eight letters or fewer, but have to count for longer words. When the limit of immediate perception is reached, human ingenuity goes to work and always arrives at the idea of grouping the objects to be counted into *sets* of some definite size, then counting the number of *sets*. Thus arises the notion of a *base* for counting. It is well known and seems completely natural that in most cases this base is five or ten, the normal number of fingers on one or two hands.

2.1. Decimal systems. Decimal systems arose spontaneously in ancient Egypt, India, China, and elsewhere. The choice of 10 as a base is not itself a sign of superior wisdom. Only when combined with an efficient notation does the usefulness of a base make itself known. A place-value system greatly simplifies calculation, which is just as difficult in base 10 as in any other base when done without a place-value system.

The modern decimal system. Modern American counting—and increasingly also, British counting—has special words for thousand, million, billion,[1] then trillion (a thousand billions), quadrillion, quintillion, and so forth. Because these names change with every third decimal place, we are effectively using 1000 as a base for counting large sets. That fact shows through in the use of a comma (or period, in Europe) to separate each group of three digits from its predecessor. The largest of these units that anyone is likely to encounter in newspapers is the trillion, since it is the most convenient unit for discussing the national budget or the national debt in dollars. The Greek prefixes kilo- (thousand), mega- (million), giga- (billion), and tera- (trillion) are used to discuss the memory cells in computers, and the march of technology has made the first of these essentially negligible. The prefixes milli- (one-thousandth), micro- (one-millionth), and nano- (one-billionth) for reciprocals are used in discussing computing time. These are the units needed nowadays, and those that have names at present provide a comfortable margin around the objects to which they will be applied, so that no new units will need to be invented in the foreseeable future.

---

[1] A billion is a thousand millions in American and increasingly in British usage, where it originally meant a million millions.

Named powers of 10. The powers of 10 that have a name, such as hundred, thousand, and so on, vary from one society to another. Ancient Greek and modern Japanese contain special words for 10 thousand: *myrias* in Greek, *man* in Japanese. With this unit one million becomes "100 myriads" in Greek or *hyakuman* in Japanese (*hyaku* means 100). In these systems it would make more sense to insert commas every four places, rather than every three, to make reading easier. The ancient Hindus gave special names to numbers that one would think go beyond any practical use. One early poem, the *Valmiki Ramayana*, from about 500 BCE, explains the numeration system in the course of recounting the size of an army. The description uses special words for 10^7, 10^{12}, 10^{17}, and other denominations, all the way up to 10^{55}.

2.2. Nondecimal systems. The systems still used in the United States—the last bastion of resistance to the metric system—show abundant evidence that people once counted by twos, threes, fours, sixes, and eights. In the United States, eggs and pencils, for example, are sold by the *dozen* or the *gross*. In Europe, eggs are packed in cartons of 10. Until recently, stock averages were quoted in eighths rather than tenths. Measures of length, area, and weight show other groupings. Consider the following words: *fathom* (6 feet), *foot* (12 inches), *pound* (16 ounces), *yard* (3 feet), *league* (3 miles), *furlong* (1/8 of a mile), *dram* (1/8 or 1/16 of an ounce, depending on the context), *karat* (1/24, used as a pure number to indicate the proportion of gold in an alloy),[2] *peck* (1/4 of a bushel), *gallon* (1/2 peck), *pint* (1/8 of a gallon), and *teaspoon* (1/3 of a tablespoon).

Even in science there remain some vestiges of nondecimal systems of measurement, inherited from the ancient Middle East. In the measurement of both angles and time, minutes and seconds represent successive divisions by 60. A day is divided into 24 hours, each of which is divided into 60 minutes, each of which is divided into 60 seconds. At that point, our division of time becomes decimal; we measure races in tenths and hundredths of a second. A similar renunciation of consistency came in the measurement of angles as soon as hand-held calculators became available. Before these calculators came into use, students (including the present author) were forced to learn how to interpolate trigonometric tables in minutes (one-sixtieth of a degree) and seconds (one-sixtieth of a minute). In physical measurements, as opposed to mathematical theory, we still divide circles into 360 equal degrees. But our hand-held calculators have banished minutes and seconds. They divide degrees decimally and of course make interpolation an obsolete skill. Since π is irrational, it seems foolish to adhere to any rational fraction of a circle as a standard unit; hand-held calculators are perfectly content to use the natural (radian) measure, and we could eliminate a useless button by abandoning the use of degrees entirely. That reform, however, is likely to require even more time than the adoption of the metric system.[3]

---

[2] The word is a variant of *carat*, which also means 200 milligrams when applied to the size of a diamond.

[3] By abandoning another now-obsolete decimal system—the Briggsian logarithms—we could eliminate *two* buttons on the calculators. The base 10 was useful in logarithms only because it allowed the tables to omit the integer part of the logarithm. Since no one uses tables of logarithms any more, and the calculators don't care how messy a computation is, there is really no reason to do logarithms in any base except the natural one, the number e, or perhaps base 2 (in number theory). Again, don't expect this reform to be achieved in the near future.

Most peculiar of all in the English system is the common land measure, the *acre*, which is an area of 43,560 square feet.[4] That means that a square 1-acre plot of land is $\sqrt{43560} = 66\sqrt{10} \approx 208.710$ feet on a side. The unit turns out to be convenient, in that there are exactly 640 acres in a square mile (known as a *section*), which can thus be quartered into 160-acre (quarter-section) plots, a convenient size for a farm in the American Middle West during the nineteenth century. At that time a larger unit of 36 sections (an area 6 miles by 6 miles) was called a *township*. There would thus be 144 farms in a typical township.

These examples lead to an interesting inference about the origins of practical mathematics. It seems likely that numbers were not developed as an abstract tool and then applied in particular situations. If such were the case, we would expect the same base to be used in all forms of measurement. But the distillation of a preferred base, usually 10, to be applied in all measurements, took thousands of years to arrive. Even today it is resisted fiercely in the United States, which was ironically one of the earliest countries to use a decimal system of coinage. The grouping of numbers seems to have evolved in a manner specific to each particular application, just as the English language once had specific collective nouns to refer to different groups of things: a blush of boys, a bevy of girls, a herd of cattle, a flock of sheep, a gaggle of geese, a school of fish, and others.

Bases used in other cultures. A nondecimal system reported (*1937*) by the American mathematicians David Eugene Smith (1860–1944) and Jekuthiel Ginsburg (1889–1957) as having been used by the Andaman of Australia illustrates how one can count up to certain limits in a purely binary system. The counting up to 10, translated into English, goes as follows: "One two, another one two, another one two, another one two, another one two. That's all." In saying this last phrase, the speaker would bring the two hands together. This binary counting *appears* to be very inefficient from a human point of view, but it is the system that underlies the functioning of computers, since a switch has only two positions. The binary digits or *bits*, a term that seems to be due to the American mathematician Claude Shannon (1916–2001), are generally grouped into larger sets for processing.

Although bases smaller than 10 are used for various purposes, some societies have used larger bases. Even in English, the word *score* for 20 (known to most Americans only from the first sentence of Lincoln's Gettysburg Address) does occur. In French, counting between 60 and 100 is by 20s. Thus, 78 is *soixante dix-huit* (sixty-eighteen) and 97 is *quatre-vingt dix-sept* (four-twenty seventeen). Menninger describes a purely *vigesimal* (base 20) system used by the Ainu of Sakhalin. Underlying this system is a base 5 system and a base 10 system. Counting begins with *shi-ne* (*begin-to-be* = 1), and progresses through such numbers as *aschick-ne* (*hand* = 5), *shine-pesan* (*one away from* [10] = 9), *wan* (*both sides* = *both hands* = 10), to *hot-ne* (*whole-* [person]-*to-be* = 20). In this system 100 is *ashikne hotne* or 5 twenties; 1000, the largest number used, is *ashikne shine wan hotne* or 5 ten-twenties. There are no special words for 30, 50, 70, or 90, which are expressed in terms of the basic 20-unit. For example, 90 is *wan e ashikne hotne* (10 from 5 twenties). Counting by subtraction probably seems novel to most people, but it does occur in

[4] The word *acre* is related to *agriculture* and comes from the Latin *ager* or Greek *agrós*, both meaning *field*.

Roman numerals (IV $= 5 - 1$), and we use subtraction to tell time in expressions such as *ten minutes to four* and *quarter to five*.[5]

3. Counting around the world

We now examine the ways used to count in a selected set of cultures in which mathematics eventually developed to the point of being written down.

3.1. Egypt. In Egypt the numbers appearing in hieroglyphics (the oldest writing) are represented as vertical strokes (|) for each individual digit, up to 9; then 10 is written as ∩, 20 as ∩∩, and so on. To represent 100 the Egyptians used a symbol resembling a coil of rope. Such a system requires new symbols to be invented for higher and higher groupings, as larger and larger numbers become necessary. As the accompanying photograph shows, the Egyptians had hieroglyphic symbols for 1000 (a lotus blossom), 10,000 (a crooked thumb), 100,000 (a turbot fish), and 1,000,000 (said to be the god of the air). With this system of recording numbers, no symbol for zero was needed, nor was the order of digits of any importance, since, for example, ||| ∩ ∩ and ∩ ∩ ||| both mean 23. The disadvantage of the notation is that the symbol for each power of 10 must be written a number of times equal to the digit that we would put in its place. When hieroglyphics were invented, the Egyptians had apparently not realized that it would be useful to have names for the numbers 1 through 9, and then to name the powers of 10. Later on, in the hieratic and demotic scripts that replaced hieroglyphics, they had special symbols for 1 through 9, 10 through 90, 100 through 900, and so on, a system that was reproduced in the Greek numeration with Greek letters replacing the hieratic symbols.

3.2. Mesopotamia. As the examples of angle and time measurement show, the successive divisions or regroupings need not have the same number of elements at every stage. The sexagesimal system appears to have been superimposed on a decimal system. In the cuneiform tablets in which these numbers are written the numbers 1 through 9 are represented by a corresponding number of wedge-shaped vertical strokes, and 10 is represented by a new symbol, a hook-shaped mark that resembles a boomerang (Fig. 1). So far we seem to have a decimal system of representation, like the Egyptian hieroglyphics. However, the next grouping is not *ten* groups of 10, but rather *six* groups of 10. Even more strikingly, the symbol for the next higher group is again a vertical stroke. Logically, this system is equivalent to a base-60 place-value system with a floating "decimal" (sexagesimal) point that the reader or writer had to keep track of mentally. Within each unit (sexagesimal rank) of this system there is a truncated decimal system that is not place-value, since the ones and tens are distinguished by different symbols rather than physical location. The number that we write as 85.25, for example, could be transcribed into this notation as $1, 25; 15$, meaning $1 \cdot 60 + 25 \cdot 1 + 15 \cdot \frac{1}{60}$.

This place-value sexagesimal system goes back some 4000 years in the Middle East. However, in its original form it lacked one feature that we regard as essential today, a symbol for an empty place. The later Greek writers, such as Ptolemy in

[5] Technology, however, is rapidly removing this last vestige of the old way of counting from everyday life. Circular clock faces have been largely replaced by linear digital displays, and ten minutes to four has become 3:50. This process began long ago when railroads first imposed standard time in place of mean solar time and brought about the first 24-hour clocks.

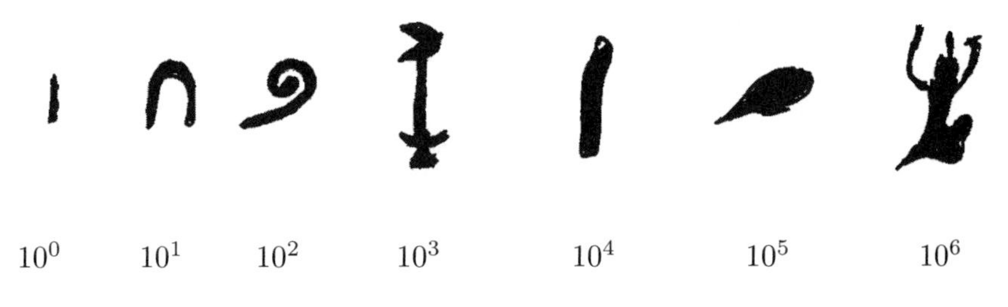

Powers of 10 from 10^0 to 10^6 in hieroglyphics.

Hieratic symbols, arranged as a multiplication table.

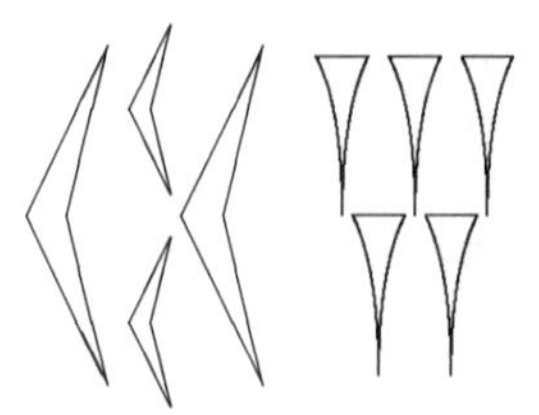

FIGURE 1. The cuneiform number 45.

the second century CE, used the sexagesimal notation with a circle to denote an empty place.

3.3. India. The modern system of numeration, in which 10 symbols are used and the value of a symbol depends only on its physical location relative to the other symbols in the representation of a number, came to the modern world from India by way of the medieval Muslim civilization. The changes that these symbols have undergone in their migration from ancient India to the modern world are shown in Fig. 2. The idea of using a symbol for an empty place was the final capstone on the creation of a system of counting and calculation that is in all essential aspects the one still in use. This step must have been taken well over 1500 years ago in India. There is some evidence, not conclusive, that symbols for an empty place were used earlier, but no such symbol occurs in the work of Arbyabhata. On the other hand, such a symbol, called in Sanskrit *sunya* (empty), occurs in the work of Brahmagupta a century after Arybhata.

3.4. China. The idea of having nine digits combined with names for the powers of 10 also occurred to the Chinese, who provided names for powers of 10 up to 100,000,000. The Chinese system of numbering is described in the *Sun Zi Suan Jing*. A certain redundancy is built into the Chinese system. To understand this redundancy, consider that in English we could write out the number 3875 in words as three thousand eight hundred seventy-five. Since Chinese uses symbols rather than letters for words, the distinction between a written number such as "seven" and the corresponding numeral 7 does not exist in Chinese. But in writing their numbers the Chinese did not use physical location as the *only* indication of the value of a digit. Rather, that value was written out in full, just as here. To convey the idea in English, we might write 3875 as 3 thousands 8 hundreds 7 tens and 5. Because of this way of writing, there is no need for a zero symbol to hold an empty place. For example, 1804 would simply be 1 thousand 8 hundreds and 4. Large numbers were handled very efficiently, with a special name for 10,000 (*wan*). Its square [*wan wan*, that is, 100,000,000 ($= 10^8$)] was called *yi*. Thereafter the Chinese had special names for each power of 10^8. Thus, 10^{16} was *zhao*, 10^{24} was *jing*, and so on, up to 10^{80} (*zai*), which was surely large enough to meet any needs of commerce or science until the twentieth century.[6] In that sense 10^8 amounted to a second base for arithmetic in Chinese usage.

[6] An estimate ascribed to Sir Arthur Eddington (1882–1944) of the number of protons in the universe put the number at $136 \cdot 2^{256}$, which is approximately $1.575 \cdot 10^{79}$.

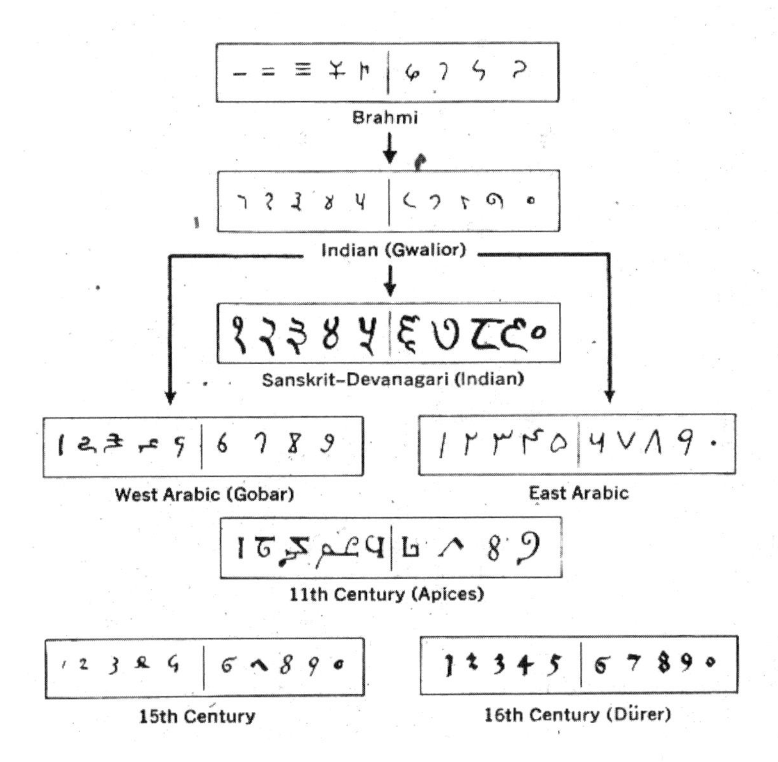

FIGURE 2. Evolution of the Hindu–Arabic numerals from India to modern Europe. ©Vandenhoeck & Ruprecht, from the book by Karl Menninger, *Zahlwort und Ziffer*, 3rd ed., Göttingen, 1979.

Some later Chinese numbering seems to reflect contact with India. Buddhism entered China during the Han dynasty (202 BCE–220 CE), and Buddhist monks had a fondness for large numbers. Unmistakable evidence of influence from India can be found in the *Suan Shu Chimeng* by Zhu Shijie, who introduced names for very large powers of 10, including the term "sand of the Ganges" for 10^{96}.

3.5. Greece and Rome. You are familiar with Roman numerals, since books still use them to number pages in the front matter, and some clock faces still show the hours in Roman numerals. Although these numerals were adequate for counting and recording, you can well imagine that they were rather inefficient for any kind of calculation. Adding, say MDCLIX to CCCIV, would take noticeably longer than adding 1659 to 304, and the idea of multiplying or dividing these two numbers seems almost too horrible to contemplate.[7] The Greek numeral system was hardly better as far as calculation is concerned. The 24-letter Greek alphabet used today, together with 3 older letters, provided symbols for 1,..., 9, 10,..., 90, 100,..., 900, essentially the system used by the Egyptians. These symbols are shown in Fig. 3. The 3 older letters were Ϝ (digamma) for 6 (now usually written as the letter sigma in the form ς that it assumes at the end of a word), Ϙ (qoppa) for 90, and ϡ (sampi) for 900. When letters were used as numbers, they were usually given a prime, so

[7] Nevertheless, the procedure for doing so can be learned in a fairly short time. Detlefsen and co-authors (*1975*) analyzed the procedure and compared it to a "paper-and-pencil abacus."

1	2	3	4	5	6	7	8	9
α'	β'	γ'	δ'	ε'	ς'	ζ'	η'	θ'
10	20	30	40	50	60	70	80	90
ι'	$\varkappa'$	λ'	μ'	ν'	ξ'	o'	π'	ϙ'
100	200	300	400	500	600	700	800	900
ρ'	σ'	τ'	υ'	φ'	χ'	ψ'	ω'	ϡ'

FIGURE 3. The ancient Greek numbering system

that the number 4 would be represented by the fourth letter of the Greek alphabet (δ) and written δ'. When they reached 1000 (*khiliás*), the Greeks continued with numerical prefixes such as *tetrakiskhílioi* for 4000 or by prefixing a subscripted prime to indicate that the letter stood for thousands. Thus, ͵δ' stood for 4000, and the number 5327 would be written ͵$\varepsilon'\tau'\varkappa'\zeta'$. The largest independently named number in the ancient Greek language was 10,000, called *myriás*. This word is the source of the English word *myriad* and is picturesquely derived from the word for an ant (*mýrmēx*). Just how large 10,000 seemed to the ancient Greeks can be seen from the related adjective *myríos*, meaning *countless*.

The Sand-reckoner *of Archimedes.* The Egyptian–Greek system has the disadvantage that it requires nine new names and symbols each time a higher power of 10 is needed, and the Roman system is even worse in this regard. One would expect that mathematicians having the talent that the Greeks obviously had would realize that a better system was needed. In fact, Archimedes produced a system of numbering that was capable of expressing arbitrarily large numbers. He wrote this method down in a work called the *Psammítēs* (*Sand-reckoner*, from *psámmos*, meaning *sand*).

The problem presented as the motivation for the *Psammítēs* was a childlike question: *How many grains of sand are there?* Archimedes noted that some people thought the number was infinite, while others thought it finite but did not believe there was a number large enough to express it. That the Greeks had difficulty imagining such a number is a reflection of the system of naming numbers that they used. To put the matter succinctly, they did not yet have an awareness of the immense potential that lies in the operation of exponentiation. The solution given by Archimedes for the sand problem is one way of remedying this deficiency.

Archimedes saw that solution of the problem required a way of "naming" arbitrarily large numbers. He naturally started with the largest available unit, the myriad (10,000), and proceeded from there by multiplication and a sort of induction. He defined the first *order* of numbers to be all the numbers up to a myriad of myriads (100,000,000), which was the largest number he could make by using the available counting categories to count themselves. The second order would then consist of the numbers from that point on up to a myriad of myriads of first-order numbers, that is, all numbers up to what we would call $\left(10^8\right)^2 = 10^{16}$. The third order would then consist of all numbers beyond the second order up to a myriad of myriads of second-order numbers (10^{24}). He saw that this process could be continued up to an *order* equal to a myriad of myriads, that is, to the number $(10^8)^{10^8}$. This is a gargantuan number, a 1 followed by 800 million zeros, surely larger than any number science has ever needed or will ever need. But Archimedes realized

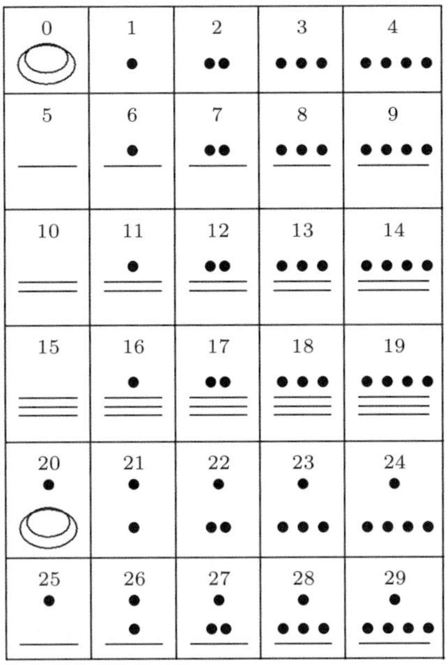

0	1	2	3	4
5	6	7	8	9
10	11	12	13	14
15	16	17	18	19
20	21	22	23	24
25	26	27	28	29

FIGURE 4. Maya numerals

the immensity of infinity. He saw that once the process just described had been completed, he could label the numbers that were named up to that point the first *period*. The mind feels carried to dizzying heights by such a process. Archimedes did not stop until he reached a number that we would represent as a 1 followed by *80 quadrillion* zeros! And of course, there is not the slightest reason to suspect that Archimedes thought the creation of integers needed to stop there. It must stop somewhere, of course.

By applying reasonable and generous estimates of the size of the universe as it was known to him, Archimedes showed that the number of grains of sand needed to fill it up could not go beyond the 1000 sixth-order units (1000 units into the seventh order, or 10^{51} in our terms). Allowing an assumption of an even larger universe, as imagined by the astronomer Aristarchus, he showed that it could not hold more than 10^{63} grains of sand.

3.6. The Maya. Although geographically far removed from Egypt, the Maya culture that existed in what is now southern Mexico and Central America from 300 BCE to 1500 CE shows some intriguing resemblances to that of ancient Egypt, especially in the building of pyramidal structures and in a hieroglyphic type of writing. On the other hand, the Maya system of counting resembles more the Mesopotamian sexagesimal system, except that it is vigesimal (base 20). As with the Mesopotamian system, only two symbols are needed to write all the numbers up to the base: a dot for ones and a horizontal line for fives. Thus the smaller base on which the vigesimal system is built is five in the case of the Maya, whereas it was 10 in the Mesopotamian system. The Maya numerals illustrate the principle that higher-level groupings need not always have the same number of members as the lower. As Fig. 4 shows, four groups of five are consolidated as a single unit of

20, and there is a cowrie-shell figure standing for zero. The second vigesimal digit in a Maya number should normally represent units of $20 \cdot 20 = 400$. However, in the *Haab* calendar, discussed below, it represents 360. The reason probably comes from the objects being counted, namely days. Even today a "business year" is 360 days (twelve 30-day months), and this cycle was also important to the Maya. Beyond this point the unit for each place value is 20 times the value of its predecessor.

4. What was counted?

People have counted an endless list of things since time immemorial. But if we were to name three items whose count was of most importance, these would be days, years, and new moons. One of the earliest uses of both arithmetic and geometry was in the construction of reliable calendars. Calendars have a practical economic value in organizing the activities of nomadic and agricultural peoples; this value is in addition to the social value associated with the scheduling of religious rites. For these reasons, calendars have been regarded as both sacred lore and applied science. At the base of any calendar must lie many years of record keeping, simply counting the days between full moons and solstices. Only after a sufficient data base has been collected can the computations needed to chart the days, weeks, months, and years be carried out. We know that such observations have been made for a long time, since the prominent lines of sight at many megalithic structures such as Stonehenge mark the summer and winter solstices. It does not require very acute observation to notice that the sun rises and sets at points farther and farther north for about 182 or 183 days, then begins to move south for the next 182 or 183 days. Once that observation was made, setting up posts to keep track of the exact location of sunrise and sunset would not have taken very long. This progression of the sun could also be correlated with the star patterns (constellations) that rise at sunset, marking the cycle we call a tropical or sidereal year. These two years actually differ by about 20 minutes, but obviously it would require a long time for that discrepancy to be noticed.

4.1. Calendars. The first broad division in calendars is between what we may call (for purposes of the present discussion only) *linear* and *cyclic* calendars. In a linear calendar the basic unit is the day, and days are simply numbered (positively or negatively) from some arbitrary day to which the number zero or 1 is assigned. Such calendars are highly artificial and used mostly for scientific purposes. For civil use, calendars attempt to repeat cycles after a month or a year or both. In traditional calendars, years were counted within the reign of a particular ruler and began with 1 as each new ruler came to power, but in the Gregorian calendar the year number does not cycle. Days and months, however, do cycle; they have their names and numbers repeated at fixed intervals. Cyclic calendars may be classified as *solar*, *lunar*, and *lunisolar*.

Ancient Egypt. The Egyptians observed the world about them with considerable accuracy, as the precise north–south orientation of some of the pyramids shows. Anyone who observes the sky for any extended period of time cannot help noticing the bright blue-green star Sirius, which is overhead at midnight during winter in the northern hemisphere. According to Montet (*1974*), it was recorded on the outside wall of the Temple of Ramesses III at Medinet Habu that the first day of the Egyptian year was to be the day on which Sirius and the Sun rose at the same time. To the Egyptians Sirius was the goddess Sôpdit, and they had a special reason for noticing it. Like all stars, Sirius gains about four minutes per day on the Sun, rising a little earlier each night until finally it rises just as the Sun is setting. Then for a while it cannot be seen when rising, since the Sun is still up, but it can be seen setting, since the Sun will have gone down before it sets. It goes on setting earlier and earlier until finally it sets just after the Sun. At that point it is too close to the Sun to be seen for about two months. Then it reappears in the sky, rising just before the Sun in the early dawn. It was during these days that the Nile began its annual flood in ancient times[8] Thus the heliacal rising of Sirius (simultaneous with the Sun) signaled the approach of the annual Nile flood. The Egyptians therefore had a very good basis for an accurate solar calendar, using the heliacal rising of Sirius as the marker of the year.

The Egyptians seem originally to have used a lunar calendar with 12 lunar cycles per year. However, such a calendar is seriously out of synchronicity with the Sun, by about 11 or 12 days per year, so that it was necessary to add an extra "intercalary" month every two or three years. All lunar calendars must do this, or else wander through the agricultural year. However, at an early date the Egyptians cut their months loose from the Moon and simply defined a month to consist of 30 days. Their calendar was thus a "civil" calendar, neither strictly lunar nor strictly solar. Each month was divided into three 10-day "weeks" and the entire system was kept from wandering from the Sun too quickly by adding five extra days at the end of the year, regarded as the birthdays of the gods Osiris, Horus, Seth, Isis, and Nephthys. This calendar is still short by about $\frac{1}{4}$ day per year, so that in 1456 years it would wander through an entire cycle of seasons. The discrepancy between the calendar and the Sun accumulated slowly enough to be adjusted for, so that no serious problems arose. In fact, this wandering has been convenient for historians, since the heliacal rising of Sirius was recorded. It was on the first day of the Egyptian year in 2773 BCE, 1317 BCE, and 139 CE. Hence a document that says the heliacal rising occurred on the sixteenth day of the fourth month of the second season of the seventh year of the reign of Senusret III makes it possible to state that Senusret III began his reign in 1878 BCE (Clayton, *1994*, pp. 12–13). On the other hand, some authorities claim that the calendar was adjusted by the addition of intercalary days from time to time to keep it from wandering too far. When the Greeks came to Egypt, they used the name Sothis to refer to Sôpdit. Consequently, the period of 1456 years is known as the *Sothic cycle.*

The Julian calendar. In a solar calendar, the primary period of time being tracked is the solar year, which we now know to be 365.2422 mean solar days long. Taking 365.25 days as an approximation to this period, the Julian calendar (a solar calendar) makes an ordinary year 365 days long and apportions it out among the months, with January, March, May, July, August, October, and December each getting 31 days, while April, June, September, and November each get 30 days and February gets 28 days. For historiographical purposes this calendar has been projected back to the time before it was actually created. In that context it is called the *proleptic* Julian calendar. In a solar calendar the month is not logically necessary, and the months have only an approximate relation to the phases of the Moon.

[8] The floods no longer occur since the Aswan Dam was built in the 1950s.

The Gregorian calendar. A modification of the Julian calendar was introduced in 1582 on the recommendation of Pope Gregory XIII. The Gregorian correction removed the extra day from years divisible by 100, but restored it for years divisible by 400. A 400-year period on the Gregorian calendar thus contains 303 years of 365 days and 97 leap years of 366 days, for a total of 146,097 days. It would be perfectly accurate if the year were of the following length:

$$365 + \frac{1}{4} - \frac{1}{100} + \frac{1}{400} = 365.2425 \text{ days.}$$

Since this figure is slightly too large, there are still too many leap years in the Gregorian calendar, and a discrepancy of one day accumulates in about 3300 years.

The Muslim calendar. The prophet Muhammed decreed that his followers should regulate their lives by a purely lunar calendar. In a lunar calendar the months are in close synchrony with the phases of the Moon, while the years need not have any close relationship to the seasons or the position of the Sun among the stars. The Muslim calendar, taking as its epoch the date of the Hijra (July 15, 622 CE), consists of 12-month years in which the odd-numbered months have 30 days and even-numbered months 29 days, except that the final month has 30 days in a leap year. Thus, the year is 354 or 355 days long, and as a result, the years wander through the tropical year.

The Hebrew calendar. More common than purely lunar calendars are lunisolar calendars, in which the months are kept in synchrony with the phases of the Moon and extra months are inserted from time to time to keep the years in synchrony with the Sun. These calendars lead to a need for *calculation* and therefore take us right up to the development of arithmetic. Several such calendars have been used since ancient times and continue to be used today in Israel, China, and elsewhere. It must have required many centuries of record keeping for the approximate equation "19 solar years = 235 lunar months" to be recognized. Since $235 = 12 \cdot 12 + 7 \cdot 13$, the addition of the extra month 7 times in 19 years will keep both years and months in balance, with an error of only about 2 hours in each 19-year cycle, or one day in 220 years.

The Julian day calendar. An example of what we have called a linear calendar is the Julian day calendar, which is to be distinguished from the Julian calendar. The Julian day calendar was invented by Joseph Justus Scaliger (1540–1609) and apparently named in honor of his father Julius Caesar Scaliger (1484–1558). It was advocated by the British astronomer John Frederick William Herschel (1792–1871). In this calendar each day is counted starting from what would be the date January 1, 4713 BCE on the Julian calendar. Thus the day on which the first draft of this paragraph was written (August 9, 2002, which is July 27, 2002 on the Julian calendar) was Julian day 2,452,496.

The Maya calendar. The most unusual calendar of all was kept by the Maya. The three Maya calendars account for a number of phenomena of astronomical and agricultural importance. As discussed above, numbers were written in a place-value system in which each unit is 20 times the next smaller unit, except that when days were being counted, the third-place unit, instead of being $20 \cdot 20 = 400$, was $20 \cdot 18 = 360$. This apparent inconsistency was probably because there are 360 days in the "regular" part of the 365-day Maya calendar known as the *Haab*, and the other 5 days were apparently regarded as unlucky (and so, best not included in the count). This *Haab* calendar is another point of resemblance between the Maya and Egyptian civilizations.

Counting by 20 also helps to explain the rather mysterious grouping of days in a second, 260-day calendar, known as the *Tzolkin*, which is said to be from the words *tzol*, meaning *to put in order*, and *kin*, meaning *day*. One conjecture to account for this calendar is that the Maya gave a 20-day period to each of 13 gods that they worshipped. A modification of this conjecture is that the Maya formed from two different groups, one naming its days after 13 gods, the other after 20 gods, and that the *Tzolkin* made mutual comprehension easier. A second conjecture is that 260 days is approximately the growing period for maize from the time of planting to harvesting. Still a third conjecture is based on the fact that 260 days is the length of time each year in which the Sun culminates south of the zenith at a latitude of 15° N, where the southern portion of the Maya territory was located. This last explanation, however, seems inconsistent with the obvious fact that the Sun then culminates north of the zenith for the next 105 days, yet the calendar begins another 260-day cycle immediately. The two cycles coincide after 52 *Haab* years, a period called the *Calendar Round*.

An important aspect of Maya astronomy was a close observation of Venus. The Maya established that the synodic period of Venus (the time between two successive conjunctions with the Sun when Venus is moving from the evening to the morning sky) is 584 days. By coincidence, 65 synodic periods of Venus equal two Calendar Rounds (37,960 days), so that the Maya calendar bears a particularly close relation to this planet.

The third Maya calendar, the *Long Count*, resembles the Julian day calendar, in that it counts the days since its epoch, believed to be August 12, 3113 BCE on the proleptic Julian calendar. This date is not certain; it is based on the dates given in Maya inscriptions, which are presumed to be historical. Most of these dates are five-digit numbers starting with 9. Since $9.0.0.0.0 = 9 \cdot 20^2 \cdot 18 \cdot 20 = 1,296,000$ days, that is, approximately 3548 years and 4 months, and this date is associated with events believed to have occurred in 436 CE, one arrives at the stated epoch. Even though the Long Count does not explicitly mention months or years and counts only days, the Mayan place-value notation for numbers makes it possible to convert any date immediately to years, months, and days since the beginning. The digits excluding the final two represent the vigesimal (base 20) notation for a multiple of 360, while the next-to-last represents a multiple of 20, and the final digit is the number of units. Thus the Long Count date of December 31, 2000, which is 1,867,664, would be written in Maya notation with the vigesimal digits separated by commas as 12, 19, 7, 17, 4. It therefore represents 5187 Haab years of 360 days each ($5187 = 12 \cdot 20^2 + 19 \cdot 20 + 7$), plus 17 Tzolkin months of 20 days each, plus 4 days ($1,867,664 = 5187 \cdot 360 + 17 \cdot 20 + 4$). The Long Count may not be a "perpetual" day calendar. That is, it may be cyclic rather than what we have called linear. Some scholars believe that it cycles back to zero when the first of the five vigesimal digits reaches 13. Since $13.0.0.0.0 = 1,872,000$ days, or about 5125 years, the Long Count should have recycled around 1992 on the Gregorian calendar.

4.2. Weeks. The seven-day week was laid down as a basic human labor cycle in the Book of Genesis. If we look for human origins of this time period, we might associate it with the waxing and waning of the Moon, since one week is the time required for the Moon to go halfway from full to new or vice versa. There is another, more plausible, astronomical connection, however, since there are exactly seven heavenly bodies visible to the unaided eye that move around among the fixed stars: the Sun, the Moon, and the five planets Mercury, Venus, Mars, Jupiter, and Saturn. These planets were also gods to many ancient peoples and gave their names to the weekdays in some of the Romance languages, such as French and Italian. In English the Norse gods serve the same purpose, with some identifications, such as Tiw (Tiu) with Mars, Odin (Wotan) with Mercury, Thor with Jupiter, and Frigga with Venus. Saturn was not translated; perhaps the Norse simply didn't have a god with his lugubrious reputation. This dual origin of the week, from Jewish law and from astrology, seems to have spread very widely throughout the world. It certainly reached India by the fifth century, and apparently even went as far as Japan. References to a seven-day week have been found in Japanese literature of 1000 years ago. When referring to the Gregorian calendar, the Japanese also give the days of the week the names of the planets, in exactly the same order as in the French and Italian calendars, that is, Moon, Mars, Mercury, Jupiter, Venus, Saturn, and Sun.

Colson (*1926*) gives a thorough discussion of what he calls the "planetary week" and explains the more plausible of two analyses for the particular order derived from a history of Rome written by Dion Cassius in the early third century. The natural ordering of the nonfixed heavenly bodies, from a geocentric point of view, is determined by the rapidity with which they move among the stars. The Moon is by far the fastest, moving about 13° per day, whereas the Sun moves only 1° per day. Mercury and Venus sometimes loop around the Sun, and hence move faster than the Sun. When these bodies are arranged from slowest to fastest in their movement across the sky, as seen from the earth, the order is: Saturn, Jupiter, Mars, Sun, Venus, Mercury, Moon. Taking every third one in cyclic order, starting with the Sun, we get the arrangement Sun, Moon, Mars, Mercury, Jupiter, Venus, Saturn, which is the cyclic order of the days of the week. Dion Cassius explains this order as follows: The planets take turns keeping one-hour tours of guard duty, so to speak. In any such cyclic arrangement, the planet that was on duty during the fourth hour of each day will be on duty during the first hour of the next day. The days are named after the planet that is on watch at sunrise. Since there are 24 hours in the day and seven planets, you can see that number of the first hour of successive days will be 1, 25, 49, 73, 97, 121, and 145. Up to multiples of seven, these numbers are equal to 1, 4, 7, 3, 6, 2, 5 respectively, and that is the cyclic order of our weekdays.

In his *Aryabhatiya* the Hindu writer Arybhata I (476–550) uses the planetary names for the days of the week and explains the correlation in a manner consistent with the hypothesis of Dion Cassius. He writes:

> [C]ounting successively the fourth in the order of their swiftness they become the Lords of the days from sunrise. [Clark, *1930*, p. 56]

The hypothesis of Dion Cassius is plausible and has been widely accepted for centuries. More than 600 years ago the poet Geoffrey Chaucer wrote a treatise on the astrolabe (Chaucer, *1391*) in which he said

> The firste houre inequal of every Saturday is to Saturne, and the seconde to Jupiter, the thirde to Mars, the fourthe to the sonne,

> the fifte to Venus, the sixte to Mercurius, the seventhe to the mone. And then ageyn the 8 houre is to Saturne, the 9 is to Jupiter, the 10 to Mars, the 11 to the sonne, the 12 to Venus... And in this manner succedith planete under planete fro Saturne unto the mone, and fro the mone up ageyn to Saturne. [Chaucer, *1391*, Robinson, p. 553]

Chaucer made references to these hours in the *Canterbury Tales.*[9] Thus, all this planetary lore and the seven-day week have their origin in the sexagesimal system of counting and the division of the day into 24 hours, which we know is characteristic of ancient Mesopotamia. But if the Mesopotamians had used a decimal system and divided their days into 10 hours, the days would still occur in the same order, since 10 and 24 are congruent modulo 7.

Questions and problems

5.1. Find an example, different from those given in the text, in which English grammar makes a distinction between a set of two and a set of more than two objects.

5.2. Consider the following three-column list of number names in English and Russian. The first column contains the cardinal numbers (those used for counting), the second column the ordinal numbers (those used for ordering), and the third the fractional parts. Study and compare the three columns. The ordinal numbers and fractions and the numbers 1 and 2 are grammatically adjectives in Russian. They are given in the feminine form, since the fractions are always given that way in Russian, the noun *dolya*, meaning *part* or *share*, always being understood. If you know another language, prepare a similar table for that language, then describe your observations and inferences. What does the table suggest about the origin of counting?

English			**Russian**		
one	first	whole	odna	pervaya	tselaya
two	second	half	dve	vtoraya	polovina
three	third	third	tri	tret'ya	tret'
four	fourth	fourth	chetyre	chetvyortaya	chetvert'
five	fifth	fifth	pyat'	pyataya	pyataya
six	sixth	sixth	shest'	shestaya	shestaya

5.3. How do you account for the fact that the ancient Greeks used a system of counting and calculating that mirrored the notation found in Egypt, whereas in their astronomical measurements they borrowed the sexigesimal system of Mesopotamia? Why were they apparently blind to the computational advantages of the place-value system used in Mesopotamia?

5.4. A tropical year is the time elapsed between successive south-to-north crossings of the celestial equator by the Sun. A sidereal year is the time elapsed between two successive conjunctions of the Sun with a given star; that is, it is the time required for the Sun to make a full circuit of the ecliptic path that it appears (from Earth) to follow among the stars each year. Because the celestial equator is rotating (one

[9] See the 1928 edition edited by John Matthews Manly, published by Henry Holt and Company, New York, especially the third part of the Knight's Tale, pp. 198–213.

revolution in 26,000 years) in the direction opposite to the Sun's motion along the ecliptic, a tropical year is about 20 minutes shorter than a sidereal year. Would you expect the flooding of the Nile to be synchronous with the tropical year or with the sidereal year? If the flooding is correlated with the tropical year, how long would it take for the heliacal rising of Sirius to be one day out of synchronicity with the Nile flood? If the two were synchronous 4000 years ago, how far apart would they be now, and would the flood occur later or earlier than the heliacal rising of Sirius?

5.5. How many *Tzolkin* cycles are there in a Calendar Round?

5.6. The pattern of leap-year days in the Gregorian calendar has a 400-year cycle. Do the days of the week also recycle after 400 years?

5.7. (*The revised Julian calendar*) The Gregorian calendar bears the name of the Pope who decreed that it should be used. It was therefore adopted early in many countries with a Catholic government, somewhat later in Anglican and Protestant countries. Countries that are largely Orthodox in faith resisted this reform until the year 1923, when a council suggested that century years should be leap years only when they leave a remainder of 2 or 6 when divided by 9. (This reform was not mandated, but was offered as a suggestion, pending universal agreement among all Christians on a date for Easter.) This modification would retain only two-ninths of the century years as leap years, instead of one-fourth, as in the Gregorian calendar. What is the average number of days in a year of this calendar? How does it compare with the actual length of a year? Is it more or less accurate than the Gregorian calendar?

5.8. In constructing a calendar, we encounter the problem of measuring time. Measuring *space* is a comparatively straightforward task, based on the notion of congruent lengths. One can use a stick or a knotted rope stretched taut as a standard length and compare lengths or areas using it. Two lengths are congruent if each bears the same ratio to the standard length. In many cases one can move the objects around and bring them into coincidence. But what is meant by congruent *time intervals*? In what sense is the interval of time from 10:15 to 10:23 congruent to the time interval from 2:41 to 2:49?

5.9. It seems clear that the decimal place-value system of writing integers is *potentially infinite*; that is there is no limit on the size of number that can be written in this system. But in practical terms, there is always a largest number for which a name exists. In ordinary language, we can talk about trillions, quadrillions, quintillions, sextillions, septillions, octillions, and so on. But somewhere before the number 10^{60} is reached, most people (except Latin scholars) will run out of names. Some decades ago, a nephew of the American mathematician Edward Kasner (1878–1955) coined the name *googol* for the number 10^{100}, and later the name *googolplex* for $10^{10^{100}}$. This seems to be the largest number for which a name exists in English. Does there exist a positive integer for which no name *could* possibly be found, not merely an integer larger than all the integers that have been or will have been named before the human race becomes extinct? Give a logical argument in support of your answer. (And, while you are at it, consider what is meant by saying that an integer "exists.")

CHAPTER 6

Calculation

In the present chapter we are going to look at processes that the modern calculator has rendered obsolescent, that is, the basic operations of arithmetic: addition, subtraction, multiplication, division, and the extraction of (square) roots. The word *obsolescent* is used instead of the more emphatic *obsolete* because these processes are still being taught to children in schools. But the skill that children are acquiring becomes weaker with every passing year. In fact, it has been at least 30 years since high-school students were actually taught to extract a square root. Even then, it was easier to consult a table of square roots than to carry out the error-prone, complicated operation of finding the root. Of course, what has caused these skills to fall out of use is the ready availability of hand-held calculators. This latest technological marvel is a direct continuation of earlier technology to ease the burden of concentration required in doing arithmetic, starting with counting rods and counting boards, then moving on through the abacus and the slide rule. The need to calculate has been a motivating force behind the development of mechanical methods of computation, and thus an important part of the history of mathematics. In this chapter we look first at the earliest methods developed for calculating, concentrating on multiplication and division (or, in the case of Egypt, processes equivalent to these) and the extraction of roots. We shall also look at three important motives for calculating: (1) commercial transactions involving labor, construction, and trade; (2) geometric problems of area and volume involving surveying and engineering; and (3) regulation of the calendar, especially finding important dates such as Easter.

1. Egypt

The richest source of information on Egyptian methods of calculation is the Ahmose (Rhind) Papyrus described in Chapter 2. After the descriptive title, the papyrus begins with the table of numbers shown in Fig. 1 below. In the modern world, we think of arithmetic as consisting of the four operations of addition, subtraction, multiplication, and division performed on whole numbers and fractions. We learn the rules for carrying out these operations in childhood and do them automatically, without attempting to prove that they are correct. The situation was different for the Egyptian. To the Egyptian, it seems, the fundamental operations were addition and *doubling*, and these operations were performed on whole numbers and *parts*. We need to discuss both the operations and the objects on which they were carried out.

Let us consider first the absence of multiplication and division as we know them. The tables you looked at in Problem 5.2 should have convinced you that there is something special about the number 2. We don't normally say "one-twoth" for the result of dividing something in two parts. This linguistic peculiarity suggests that *doubling* is psychologically different from applying the general concept of multiplying in the special case when the multiplier is 2.

Next consider the absence of what we would call fractions. The closest Egyptian equivalent to a fraction is what we called a *part*. For example, what we refer to nowadays as the fraction $\frac{1}{7}$ would be referred to as "the seventh part." This language conveys the image of a thing divided into seven equal parts arranged in a row and the seventh (and last) one being chosen. For that reason, according to van der Waerden (*1963*), there can be only *one* seventh part, namely the last one; there would be no way of expressing what we call the fraction $\frac{3}{7}$. An exception was the fraction that we call $\frac{2}{3}$, which occurs constantly in the Ahmose Papyrus. There was a special symbol meaning "the two parts" out of three. In general, however, the Egyptians used only *parts*, which in our way of thinking are *unit fractions*, that is, fractions whose numerator is 1. Our familiarity with fractions in general makes it difficult to see what the fuss is about when the author asks what must be added to the two parts and the fifteenth part in order to make a whole (Problem 21 of the papyrus). If this problem is stated in modern notation, it merely asks for the value of $1 - \left(\frac{1}{15} + \frac{2}{3}\right)$, and of course, we get the answer immediately, expressing it as $\frac{4}{15}$. Both this process and the answer would have been foreign to the Egyptian, whose solution is described below.

To understand the Egyptians, we shall try to imitate their way of writing down a problem. On the other hand, we would be at a great disadvantage if our desire for authenticity led us to try to solve the entire problem using their notation. The best compromise seems to be to use our symbols for the whole numbers and express a *part* by the corresponding whole number with a bar over it. Thus, *the fifth part* will be written $\overline{5}$, *the thirteenth part* by $\overline{13}$, and so on. For "the two parts" $\left(\frac{2}{3}\right)$ we shall use a double bar, that is, $\overline{\overline{3}}$.

1.1. Multiplication and division. Since the only operation other than addition and subtraction of integers (which are performed automatically without comment) is doubling, the problem that we would describe as "multiplying 11 by 19" would have been written out as follows:

	19	1	*
	38	2	*
	76	4	
	152	8	*
Result	209	11	

Inspection of this process shows its justification. The rows are kept strictly in proportion by doubling each time. The final result can be stated by comparing the first and last rows: 19 is to 1 as 209 is to 11. The rows in the right-hand column that must be added in order to obtain 11 are marked with an asterisk, and the corresponding entries in the left-hand column are then added to obtain 209. In this way any two positive integers can easily be multiplied. The only problem that arises is to decide how many rows to write down and which rows to mark with an asterisk. But that problem is easily solved. You stop creating rows when the next entry in the right-hand column would be bigger than the number you are multiplying by (in this case 11). You then mark your last row with an asterisk, subtract the entry in its right-hand column (8) from 11 (getting a remainder of 3), then move up and mark the next row whose right-hand column contains an entry not larger than this remainder (in this case the second row), subtract the entry in its right-hand column (2), from the previous remainder to get a smaller remainder (in this case 1), and so forth.

We shall refer to this general process of doubling and adding as *calculating.* What we call division is carried out in the same way, by reversing the roles of the two columns. For example, what we would call the problem of dividing 873 by 97 amounts to calculating with 97 so as to obtain 873. We can write it out as follows:

*	97	1	
	194	2	
	388	4	
*	776	8	
	873	9	Result.

The process, including the rules for creating the rows and deciding which ones to mark with an asterisk, is exactly the same as in the case of multiplication, except that now it is the left-hand column that is used rather than the right-hand column. We create rows until the next entry in the left-hand column would be larger than 873. We then mark the last row, subtract the entry in its left-hand column from 873 to obtain the remainder of 97, then look for the next row above whose left-hand entry contains a number not larger than 97, mark that row, and so on.

1.2. "Parts". Obviously, the second use of the two-column system can lead to complications. While in the first problem we can always express any positive integer as a sum of powers of 2, the second problem is a different matter. We were just lucky that we happened to find multiples of 97 that add up to 873. If we hadn't found them, we would have had to deal with those *parts* that have already been discussed. For example, if the problem were "calculate with 12 so as to obtain 28," it might have been handled as follows:

	12	1	
*	24	2	
	8	$\overline{\overline{3}}$	
*	4	$\overline{3}$	
	28	$2\ \overline{3}$	Result.

What is happening in this computation is the following. We stop creating rows after 24 because the next entry in the left-hand column (48) would be bigger than 28. Subtracting 24 from 28, we find that we still need 4, yet no 4 is to be found. We therefore go back to the first row and multiply by $\frac{2}{3}$, getting the row containing 8 and $\overline{\overline{3}}$. Dividing by 2 again gets a 4 in the left-hand column. We then have the numbers we need to get 28, and the answer is expressed as $2\ \overline{3}$. Quite often the first multiplication by a *part* involves the two-thirds part $\overline{\overline{3}}$. The scribes probably began with this part instead of one-half for the same reason that a carpenter uses a plane before sandpaper: the work goes faster if you take bigger "bites."

The parts that are negative powers of 2 play a special role. When applied to a hekat of grain, they are referred to as the *Horus-eye* parts.[1] Since $1/2 + 1/4 +$

[1] According to Egyptian legend, the god Horus lost an eye in a fight with his uncle, and the eye was restored by the god Thoth. Each of these fractions was associated with a particular part of Horus' eye.

$1/8 + 1/16 + 1/32 + 1/64 = 63/64$, the scribes apparently saw that unity could be restored (approximately), as Horus' eye was restored, by using these parts. The fact that (in our terms) 63 occurs as a numerator, shows that division by 3, 7, and 9 is facilitated by the use of the Horus-eye series. In particular, since $1/7 = (1/7) \cdot \big((63/64) + 1/64\big) = 9/64 + 1/448 = 8/64 + 1/64 + 1/448$, the seventh part could have been written as $\overline{8}\ \overline{64}\ \overline{448}$. In this way, the awkward seventh part gets replaced by the better-behaved Horus-eye fractions, plus a corrective term (in this case $\overline{448}$, which might well be negligible in practice. Five such replacements are implied, though not given in detail, in the Akhmim Wooden Tablet.[2] As another example, since $64 = 4 \cdot 13 + 8 + 4$, we find that $\overline{13}$ becomes $\overline{16}\ \overline{104}\ \overline{208}$.

There are two more complications that arise in doing arithmetic the Egyptian way. The first complication is obvious. Since the procedure is based on doubling, but the double of a *part* may not be expressible as a part, how does one "calculate" with parts? It is easy to double, say, the twenty-sixth part: The double of the twenty-sixth part is the thirteenth part. If we try to double again, however, we are faced with the problem of doubling a part involving an odd number. The table at the beginning of the papyrus gives the answer: The double of the thirteenth part is the eighth part plus the fifty-second part plus the one hundred fourth part. In our terms this tabular entry expresses the fact that

$$\tfrac{2}{13} = \tfrac{1}{8} + \tfrac{1}{52} + \tfrac{1}{104}.$$

Gillings (*1972*, p. 49) lists five precepts apparently followed by the compiler of this table in order to make it maximally efficient for use. The most important of these are the following three. One would like each double (1) to have as few terms as possible, (2) with each term as small as possible (that is, the "denominators" as small as possible), and (3) with even "denominators" rather than odd ones. These principles have to be balanced against one another, and the table in Fig. 1 represents the resulting compromise. However, Gillings' principles are purely negative ones, telling what *not* to do. The positive side of creating such a table is to find simple patterns in the numbers. One pattern that occurs frequently is illustrated by the double of $\overline{5}$, and amounts to the identity $2/p = 1/\big((p+1)/2\big) + 1/\big(p(p+1)/2\big)$. Another, illustrated by the double of $\overline{13}$, probably arises from the Horus-eye representation of the original part.

With this table, which gives the doubles of all parts involving an odd number up to 101, calculations involving parts become feasible. There remains, however, one final complication before one can set out to solve problems. The calculation process described above requires subtraction at each stage in order to find what is lacking in a given column. When the column already contains *parts*, this leads to the second complication: the problem of *subtracting parts*. (*Adding* parts is no problem. The author merely writes them one after another. The sum is condensed if, for example, the author knows that the sum of $\overline{3}$ and $\overline{6}$ is $\overline{2}$.) This technique, which is harder than the simple procedures discussed above, is explained in the papyrus itself in Problems 21 to 23. As mentioned above, Problem 21 asks for the parts that must be added to the sum of $\overline{\overline{3}}$ and $\overline{15}$ to obtain 1. The procedure used to solve this problem is as follows. Begin with the two parts in the first row:

---

[2] See `http://www.mathworld.com/AkhmimWoodenTablet.html`. In a post to the history of mathematics mailing list in December 2004 the author of that article, Milo Gardner, noted that recent analysis of this tablet has upset a long-held belief about the meaning of a certain term in these equations.

5	3 15	55	30 318 795
7	4 28	57	38 114
9	6 18	59	36 236 531
11	6 66	61	40 244 488 610
13	8 52 104	63	42 126
15	10 30	65	39 195
17	12 51 68	67	40 335 536
19	12 76 114	69	46 138
21	14 42	71	40 568 710
23	12 276	73	60 219 292 365
25	15 75	75	50 150
27	18 54	77	44 308
29	24 58 174 232	79	60 237 316 790
31	20 124 155	81	54 162
33	22 66	83	60 332 415 498
35	30 42	85	51 255
37	24 111 296	87	58 174
39	26 78	89	60 356 534 890
41	24 246 328	91	70 130
43	42 86 129 301	93	62 186
45	30 90	95	60 380 570
47	30 141 470	97	56 679 776
49	28 196	99	66 198
51	34 102	101	101 202 303 606

FIGURE 1. Doubles of unit fractions in the Ahmose Papyrus

$\bar{\bar{3}}$ $\overline{15}$ 1.

Now the problem is to see what must be added to the two terms on the left-hand side in order to obtain the right-hand side. Preserving proportions, the author multiplies the row by 15, getting

10 1 15

It is now clear that when the problem is "magnified" by a factor of 15, we need to add 4 units. Therefore, the only remaining problem is, as we would put it, to divide 4 by 15, or in language that may reflect better the thought process of the author, to "calculate with 15 so as to obtain 4." This operation is carried out in the usual way:

15	1	
1	$\overline{15}$	
2	$\overline{10}$ $\overline{30}$	[from the table]
4	$\overline{5}$ $\overline{15}$	Result.

Thus, the parts that must be added to the sum of $\bar{\bar{3}}$ and $\overline{15}$ in order to reach 1 are $\overline{5}$ and $\overline{15}$. This "subroutine," which is essential to make the system of computation work, was written in red ink in the manuscripts, as if the writers distinguished between computations made within the problem to find the answer and computations made in order to operate the system. Having learned how to complement (subtract) parts, what are called *hau* (or *aha*) computations by the author, one can confidently attack any arithmetic problem whatsoever. Although there is no single way of doing these problems, specialists in this area have detected systematic procedures by which the table of doubles was generated and patterns in the solution of problems that indicate, if not an algorithmic procedure, at least a certain habitual approach to such problems.

Let us now consider how these principles are used to solve a problem from the papyrus. The one we pick is Problem 35, which, translated literally and misleadingly, reads as follows:

> Go down I times 3. My third part is added to me. It is filled. What is the quantity saying this?

To clarify: This problem asks for a number that yields 1 when it is tripled and the result is then increased by the third part of the original number. In other words, "calculate with $3\ \overline{3}$ so as to obtain 1." The solution is as follows:

$3\ \overline{3}$	1	
10	3	[multiplied by 3]
5	$1\ \overline{2}$	
1	$\overline{5}\ \overline{10}$	Result.

1.3. Practical problems. One obvious application of calculation in everyday life is in surveying, where one needs some numerical way of comparing the sizes of areas of different shapes. This application is discussed in Chapter 9. The papyrus also contains several problems that involve proportion in the guise of the slope of pyramids and the strength of beer. Both of these concepts involve what we think of as a ratio, and the technique of finding the fourth element in a proportion by the procedure once commonly taught to grade-school students and known as the *Rule of Three*. It is best explained by a sample question. If three bananas cost 69 cents, what is the cost of five bananas? Here we have three numbers: 3, 69, and 5. We need a fourth number that has the same ratio to 69 that 5 has to 3, or, equivalently, the same ratio to 5 that 69 has to 3. The rule says that such a number is $69 \times 5 \div 3 = 105$. Since the Egyptian procedure for multiplication was based on an implicit notion of proportion, such problems yield easily to the Egyptian techniques. We shall reserve the discussion of pyramid slope problems until we examine Egyptian geometry in Chapter 9. Several units of weight are mentioned in these problems, but the measurement we shall pay particular attention to is a measure of the dilution of bread or beer. It is called a *pesu* and defined as the number of loaves of bread or jugs of beer obtained from one *hekat* of grain. A hekat was slightly larger than a gallon, 4.8 liters to be precise. Just how much beer or bread it would produce under various circumstances is a technical matter that need not concern us. The thing we need to remember is that the number of loaves of bread or jugs of beer produced by a given amount of grain equals the *pesu* times the number of hekats of grain. A large *pesu* indicates weak beer or bread. In the problems in the Ahmose Papyrus the *pesu* of beer varies from 1 to 4, while that for bread varies from 5 to 45.

Problem 71 tells of a jug of beer produced from half a hekat of grain (thus its *pesu* was 2). One-fourth of the beer is poured off and the jug is topped up with water. The problem asks for the new *pesu*. The author reasons that the eighth part of a hekat of grain was removed, leaving (in his terms) $\overline{4}\ \overline{8}$, that is, what we would

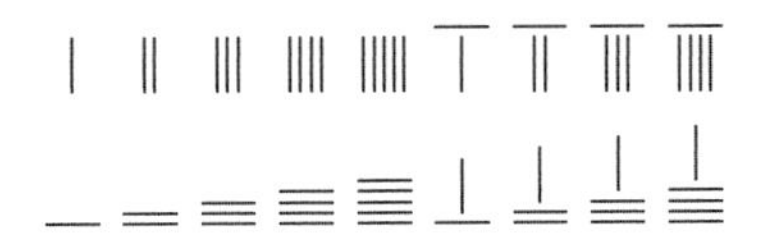

FIGURE 2. The Shang numerals.

call $\frac{3}{8}$ of a hekat of grain. Since this amount of grain goes into one jug, it follows that the *pesu* of that beer is what we call the *reciprocal* of that number, namely $2\ \overline{\overline{3}}$. The author gives this result immediately, apparently assuming that by now the reader will know how to "calculate with $\overline{4}\ \overline{8}$ until 1 is reached." The Rule of Three procedure is invoked in Problem 73, which asks how many loaves of 15-*pesu* bread are required to provide the same amount of grain as 100 loaves of 10-*pesu* bread. The answer is found by dividing 100 by 10, then multiplying by 15, which is precisely the Rule of Three.

2. China

In contrast to the Egyptians, who computed with ink on papyrus, the ancient Chinese, starting in the time of the Shang dynasty, used rods representing numerals to carry out computations. Chinese documents from the second century BCE mention the use of counting rods, and a set of such rods from the first century BCE was discovered in 1970. The rods can be arranged to form the Shang numerals (Fig. 2) and thereby represent decimal digits. They were used in conjunction with a counting board, which is a board ruled into squares so that each column (or row, depending on the direction of writing) represents a particular item. In pure computations, the successive rows in the board indexed powers of 10. These rods could be stacked to represent any digit from 1 to 9. Since they were placed on a board in rows and columns, the empty places are logically equivalent to a use of 0, but not psychologically equivalent. The use of a circle for zero in China is not found before the thirteenth century. On the other hand, according to Lam and Ang (*1987*, p. 102), the concept of negative numbers (*fu*), represented by black rods instead of the usual red ones for positive numbers (*cheng*), was also present as early as the fourth century BCE.

It is difficult to distinguish between, say, 22 (|| ||) and 4 (||||) if the rods are placed too close together. To avoid that difficulty, the Chinese rotated the rods in alternate rows through a right angle, in effect using a positional system based on 100 rather than 10. Since this book is being published in a language that is read from left to right, then from top to bottom, we shall alternate columns rather than rows. In our exposition of the system the number 22 becomes =|| and 4 remains ||||. The Shang numerals are shown in Fig. 2, the top row being used to represent digits multiplied by an even power of 10 and the bottom row digits multiplied by an odd power of 10.

Addition and subtraction with rods representing Shang numerals are obvious operations. Multiplication and division require somewhat more work, and those procedures are explained in the *Sun Zi Suan Jing*.

Except that multiplication was carried out starting with the largest denominations rather than the smallest, the procedure for multiplying digits and carrying resembles all other systems for multiplying. Using numerals in place of the rods, we can illustrate the multiplication $324 \cdot 29$ as follows:

$$
\begin{array}{cccc} & 3 & 2 & 4 \\ & & & \\ 2 & 9 & & \end{array}
\longrightarrow
\begin{array}{cccc} & 3 & 2 & 4 \\ 6 & & & \\ 2 & 9 & & \end{array}
\longrightarrow
\begin{array}{cccc} & 3 & 2 & 4 \\ 8 & 7 & & \\ 2 & 9 & & \end{array}
\longrightarrow
\begin{array}{cccc} & & 2 & 4 \\ 8 & 7 & & \\ 2 & 9 & & \end{array}
\longrightarrow
\begin{array}{cccc} & & 2 & 4 \\ 9 & 1 & & \\ & 2 & 9 & \end{array}
\longrightarrow
$$

$$
\longrightarrow
\begin{array}{cccc} & & 2 & 4 \\ 9 & 2 & 8 & \\ & 2 & 9 & \end{array}
\longrightarrow
\begin{array}{cccc} & & & 4 \\ 9 & 2 & 8 & \\ & & 2 & 9 \end{array}
\longrightarrow
\begin{array}{cccc} & & & 4 \\ 9 & 3 & 6 & \\ & & 2 & 9 \end{array}
\longrightarrow
\begin{array}{cccc} & & & 4 \\ 9 & 3 & 9 & 6 \\ & & 2 & 9 \end{array}
\longrightarrow
\begin{array}{cccc} 9 & 3 & 9 & 6 \end{array}
$$

The sequence of operations is very easy to understand from this illustration. First, the larger number is written on top (on the right in Chinese, of course, since the writing is vertical). The smaller number is written on the bottom (actually on the left), with its units digit opposite the largest digit of the larger number. Then, working always from larger denominations to smaller, we multiply the digits one at a time and enter the products between the two numbers. Once a digit of the larger number has been multiplied by all the digits of the smaller one, it is "erased" (the rods are picked up), and the rods representing the smaller number are moved one place to the right (actually downward). At that point, the process repeats until all the digits have been multiplied. When that happens, the last digit of the larger number and all the digits of the smaller number are picked up, leaving only the product.

Long division was carried out in a similar way. The partial quotients were kept in the top row, and the remainder at each stage occupied the center row (with the same caveat as above, that rows are actually columns in Chinese writing). For example, to get the quotient $438 \div 7$, one proceeds as follows.

$$
\begin{array}{ccc} & & \\ 4 & 3 & 8 \\ & & 7 \end{array}
\longrightarrow
\begin{array}{ccc} & & \\ 4 & 3 & 8 \\ 7 & & \end{array}
\longrightarrow
\begin{array}{ccc} & & \\ 4 & 3 & 8 \\ & 7 & \end{array}
\longrightarrow
\begin{array}{ccc} & 6 & \\ & 1 & 8 \\ & 7 & \end{array}
\longrightarrow
\begin{array}{ccc} & 6 & \\ & 1 & 8 \\ & & 7 \end{array}
\longrightarrow
\begin{array}{ccc} & 6 & 2 \\ & & 4 \\ & & 7 \end{array}
$$

The first step here is merely a statement of the problem. The procedure begins with the second step, where the divisor (7) is moved to the extreme left, then moved rightward until a division is possible. Thereafter, one does simple divisions, replacing the dividend by the remainder at each stage. The original dividend can be thought of as the remainder of a fictitious "zeroth" division. Except for the "erasures" when the rods are picked up, the process looks very much like the algorithm taught to school children in the United States. The final display allows the answer to be read off: $438 \div 7 = 62\frac{4}{7}$. It would be only a short step to replace this last common fraction by a decimal; all one would have to do is continue the algorithm as if there were zeros on the right of the dividend. However, no such procedure is described in the *Sun Zi Suan Jing*. Instead, the answer is expressed as an integer plus a proper fraction.

2.1. Fractions and roots. The early Chinese way of handling fractions is much closer to our own ideas than that of the Egyptians. The *Sun Zi Suan Jing* gives a procedure for reducing fractions that is equivalent to the familiar Euclidean algorithm for finding the greatest common divisor of two integers. The rule is to subtract the smaller number from the larger until the difference is smaller than the originally smaller number. Then begin subtracting the difference from the smaller number. Continue this procedure until two equal numbers are obtained. That number can then be divided out of both numerator and denominator.

With this procedure for reducing fractions to lowest terms, a complete and simple theory of computation with fractions is feasible. Such a theory is given in the *Sun Zi Suan Jing*, including the standard procedure for converting a mixed number to an improper fraction and the procedures for adding, subtracting, multiplying, and dividing fractions. Thus, the Chinese had complete control over the system of rational numbers, including, as we shall see below, the negative rational numbers.

At an early date the Chinese dealt with roots of integers, numbers like $\sqrt{355}$, which we now know to be irrational; and they found mixed numbers as approximations when the integer is not a perfect square. In the case of $\sqrt{355}$, the approximation would have been given as $18\frac{31}{36}$. (The denominator is always twice the integer part, as a result of the particular algorithm used.)

From arithmetic to algebra. Sooner or later, constantly solving problems of more and more complexity in order to find unknown quantities leads to the systematization of ways of imagining operations performed on a "generic" number (unknown). When the point arises at which a name or a symbol for an unknown number is invented, so that *expressions* can be written representing the result of operations on the unknown number, we may take it that algebra has arisen. There is a kind of twilight zone between arithmetic and algebra, in which certain problems are solved imaginatively without using symbols for unknowns, but later are seen to be easily solvable by the systematic methods of algebra. A good example is Problem 15 of Chapter 3 of the *Sun Zi Suan Jing*, which asks how many carts and how many people are involved, given that there are two empty carts (and all the others are full) when people are assigned three to a cart, but nine people have to walk if only two are placed in each cart. We would naturally make this a problem in two linear equations in two unknowns: If x is the number of people and y the number of carts, then

$$\begin{aligned} x &= 3(y-2)\,, \\ x &= 2y+9\,. \end{aligned}$$

However, that would be using algebra, and Sun Zi does not quite do that in this case. His solution is as follows:

> Put down 2 carts, multiply by 3 to give 6, add 9, which is the number of persons who have to walk, to obtain 15 carts. To find the number of persons, multiply the number of carts by 2 and add 9, which is the number of persons who have to walk.

Probably the reasoning in the first sentence here is pictorial. Imagine each cart filled with three people. When loaded in this way, the carts would accomodate all the "real" people in the problem, plus six "fictitious" people, since we are given that two carts would be empty if the others each carried three people. Let us imagine then, that six of the carts contain two real people and one fictitious person, while the others contain three real people. Now imagine one person removed from each cart, preferably a fictitious person if possible. The number of people removed would obviously be equal to the number of carts. The six fictitious people would then be removed, along with the nine real people who have to walk when there are only two people in each cart. It follows that there must be 15 carts. Finding the number of people is straightforward once the number of carts is known.

2.2. The *Jiu Zhang Suanshu*. This work is the most fundamental of the early Chinese mathematical classics. For the most part, it assumes that the methods of calculation explained in the *Sun Zi Suan Jing* are known and applies them to problems very similar to those discussed in the Ahmose Papyrus. In fact, Problems 5, 7, 10, and 15 from the first chapter reappear as the first four problems of Chapter 2 of the *Sun Zi Suan Jing*. As its title implies, the book is divided into nine chapters. These nine chapters contain a total of 246 problems. The first eight of these chapters discuss calculation and problems that we would now solve using linear algebra. The last chapter is a study of right triangles. The first chapter, whose title is "Rectangular Fields," discusses how to express the areas of fields given their sides. Problem 1, for example, asks for the area of a rectangular field that is 15 *bu* by 16 *bu*.[3] The answer, we see immediately, is 240 "*square bu*." However, the Chinese original does not distinguish between linear and square units. The answer is given as "1 *mu*." The *Sun Zi Suan Jing* explains that as a unit of *length*, 1 *mu* equals 240 *bu*. This ambiguity is puzzling, since a *mu* is both a length equal to 240 *bu* and the area of a rectangle whose dimensions are 1 *bu* by 240 *bu*. It would seem more natural for us if 1 *mu* of area were represented by a square of side 1 *mu*. If these units were described consistently, a square of side 1 linear *mu* would have an area equal to 240 "areal" *mu*. That there really is such a consistency appears in Problems 3 and 4, in which the sides are given in *li*. Since 1 *li* equals 300 *bu* (that is, 1.25 *mu*), to convert the area into *mu* one must multiply the lengths of the sides in *li*, then multiply by $1.25^2 \cdot 240 = 375$. In fact, the instructions say to multiply by precisely that number.

Rule of Three problems. Chapter 2 ("Millet and Rice") of the *Jiu Zhang Suanshu* contains problems very similar to the *pesu* problems from the Ahmose Papyrus. The proportions of millet and various kinds of rice and other grains are given as empirical data at the beginning of the chapter. If the Ahmose Papyrus were similarly organized into chapters, the chapter in it corresponding to this chapter would be called "Grain and Bread." Problems of the sort studied in this chapter occur frequently in all commercial transactions in all times. In the United States, for example, a concept analogous to *pesu* is the *unit price* (the number of dollars the merchant will obtain by selling 1 unit of the commodity in question). This number is frequently printed on the shelves of grocery stores to enable shoppers to compare the relative cost of purchasing different brands. Thus, the practicality of this kind of calculation is obvious. The 46 problems in Chapter 2, and also the 20 problems in Chapter 3 ("Proportional Distribution") of the *Jiu Zhang Suanshu* are of this type, including some extensions of the Rule of Three. For example, Problem 20 of Chapter 3 asks for the interest due on a loan of 750 *qian* repaid after 9 days if a loan of 1000 *qian* earns 30 *qian* interest each month (a month being 30 days). The result is obtained by forming the product 750 *qian* times 30 *qian* times 9 days, then dividing by the product 1000 *qian* times 30 days, yielding $6\frac{3}{4}$ *qian*. Here the product of the monthly interest on a loan of 1 *qian* and the number of days the loan is outstanding, divided by 30, forms the analog of the *pesu* for the loan, that is, the number of *qian* of interest produced by each *qian* loaned. Further illustrations are given in the problems at the end of the chapter.

[3] One *bu* is 600,000 *hu*, a *hu* being the diameter of a silk thread as it emerges from a silkworm. Estimates are that 1 *bu* is a little over 2 meters.

Chapter 6 ("Fair Transportation") is concerned with the very important problem of fair allocation of the burdens of citizenship. The Chinese idea of fairness, like that in many other places, including modern America, involves direct proportion. For example, Problem 1 considers a case of collecting taxes in a given location from four counties lying at different distances from the collection center and having different numbers of households. To solve this problem a constant of proportionality is assigned to each county equal to the number of its households divided by its distance from the collection center. The amount of tax (in millet) each county is to provide is its constant divided by the sum of all the constants of proportionality and multiplied by the total amount of tax to be collected. The number of carts (of a total prescribed number) to be provided by each county is determined the same way. The data in the problem are as follows.

County	Number of Households	Distance to Collection Center
A	10,000	8 days
B	9,500	10 days
C	12,350	13 days
D	12,200	20 days

A total of 250,000 *hu* of millet were to be collected as tax, using 10,000 carts. The proportional parts for the four counties were therefore 1250, 950, 950, and 610, which the author reduced to 125, 95, 95, and 61. These numbers total 376. It therefore followed that county A should provide $\frac{125}{376} \cdot 250,000$ *hu*, that is, approximately 83,111.7 *hu* of millet and $\frac{125}{376} \cdot 10,000$, or 3324 carts. The author rounded off the tax to three significant digits, giving it as 83,100 *hu*.

Along with these administrative problems, the 28 problems of Chapter 6 also contain some problems that have acquired an established place in algebra texts throughout the world and will be continue to be worked by students as long as there are teachers to require it. For example, Problem 26 considers a pond used for irrigation and fed by pipes from five different sources. Given that these five canals, each "working" alone, can fill the pond in $\frac{1}{3}$, 1, $2\frac{1}{2}$, 3, and 5 days, the problem asks how long all five "working" together will require to fill it. The author realized that the secret is to add the rates at which the pipes "work" (the reciprocals of the times they require individually to fill the pond), then take the reciprocal of this sum, and this instruction is given. The answer is $1/(3 + 1 + 2/5 + 1/3 + 1/5) = 15/74$.

3. India

We have noted a resemblance between the mathematics developed in ancient Egypt and that developed in ancient China. We should not be surprised at this resemblance, since these techniques arose in response to universal needs in commerce, industry, government, and society. They form a universal foundation for mathematics that remained at the core of any practical education until very recent times. Only the widespread use of computers and computer graphics has, over the past two decades, made these skills obsolete, just as word processors have made it unimportant to develop elegant handwriting.

To avoid repetition, we simply note that much of the Hindu method of computation is similar to what is now done or what is discussed in other sections of this chapter. A few unusual aspects can be noted, however. Brahmagupta gives the standard rules for handling common fractions. However, his arithmetic contains some original ways of looking at many things that we take for granted. For example, to do a long division with remainder, say, $\frac{750}{22}$, he would look for the next number after 22 that divides 750 evenly (25) and write

$$\tfrac{750}{22} = \tfrac{750}{25} + \left(\tfrac{750}{25}\right)\tfrac{3}{22},$$

that is,

$$\tfrac{750}{22} = 30\left(1 + \tfrac{3}{22}\right) = 30 + \tfrac{90}{22} = 34\tfrac{1}{11}.$$

Beyond these simple operations, he also codifies the methods of taking square and cube roots, and he states clearly the Rule of Three (Colebrooke, *1817*, p. 283). Brahmagupta names the three terms the "argument," the "fruit," and the "requisition," and points out that the argument and the requisition must be the same kind of thing. The unknown number he calls the "produce," and he gives the rule that the produce is the requisition multiplied by the fruit and divided by the argument.

4. Mesopotamia

Cuneiform tablets from the site of Senkereh (also known as Larsa), kept in the British Museum, contain tables of products, reciprocals, squares, cubes, square roots, and cube roots of integers. It appears that the people who worked with mathematics in Mesopotamia learned by heart, just as we do, the products of all the small integers. Of course, for them a theoretical multiplication table would have to go as far as 59×59, and the consequent strain on memory would be large. That fact may account for the existence of so many written tables. Just as most of us learn, without being required to do so, that $\frac{1}{3} = 0.3333\ldots$, the Mesopotamians wrote their fractions as sexagesimal fractions and came to recognize certain reciprocals, for example $\frac{1}{9} = 0;6,40$. With a system based on 30 or 60, all numbers less than 10 except 7 have terminating reciprocals. In order to get a terminating reciprocal for 7 one would have to go to a system based on 210, which would be far too complicated.

Even with base 60, multiplication can be quite cumbersome, and historians have conjectured that calculating devices such as an abacus might have been used, although none have been found. Høyrup (*2002*) has analyzed the situation by considering the errors in two problems on Old Babylonian cuneiform tablets and deduced that any such device would have had to be some kind of counting board, in which terms that were added could not be identified and subtracted again (like pebbles added to a pile).

Not only are sexagesimal fractions handled easily in all the tablets, the concept of a square root occurs explicitly, and actual square roots are approximated by sexagesimal fractions, showing that the mathematicians of the time realized that they hadn't been able to make these square roots come out even. Whether they realized that the square root would never come out even is not clear. For example, text AO 6484 (the AO stands for Antiquités Orientales) from the Louvre in Paris contains the following problem on lines 19 and 20:

> The diagonal of a square is 10 Ells. How long is the side? [To find the answer] multiply 10 by 0;42,30. [The result is] 7;5.

Now $0;42,30$ is $\frac{42}{60} + \frac{30}{3600} = \frac{17}{24} = 0.7083$, approximately. This is a very good approximation to $1/\sqrt{2} \approx 0.7071$, and the answer $7;5$ is, of course, $7\frac{1}{12} = 7.083 = 10 \cdot 0.7083$. It seems that the writer of this tablet knew that the ratio of the side of a square to its diagonal is approximately $\frac{17}{24}$. The approximation to $\sqrt{2}$ that arises from what is now called the *Newton–Raphson* method, starting from $\frac{3}{2}$ as the first approximation, turns up the number $\frac{17}{12}$ as the next approximation. Thus the fraction $\frac{17}{24}$ represents an approximation to $\frac{\sqrt{2}}{2} = \frac{1}{\sqrt{2}}$. The method of approximating square roots can be understood as an averaging procedure. In the present case, it works as follows. Since 1 is smaller than $\sqrt{2}$ and 2 is larger, let their average be the first approximation, that is, $\frac{3}{2}$. This number happens to be larger than $\sqrt{2}$, but it is not necessary to know that fact to improve the approximation. Whether it errs by being too large or two small, the result of dividing 2 by this number will err in the other direction. Thus, since $\frac{3}{2}$ is too large to be $\sqrt{2}$, the quotient $\frac{2}{3/2} = \frac{4}{3}$ is too small. The average of these two numbers will be closer to $\sqrt{2}$ than either number;[4] the second approximation to $\sqrt{2}$ is then $\frac{1}{2}\left(\frac{3}{2} + \frac{4}{3}\right) = \frac{17}{12}$. Again, whether this number is too large or two small, the number $\frac{2}{17/12} = \frac{24}{17}$ will err in the opposite direction, so that we can average the two numbers again and continue this process as long as we like. Of course, we cannot *know* that this procedure was used to get the approximate square root unless we find a tablet that says so.

The writers of these tablets realized that when numbers are combined by arithmetic operations, it may be of interest to know how to recover the original data from the result. This realization is the first step toward attacking the problem of inverting binary operations. Although we now solve such problems by solving quadratic equations, the Mesopotamian approach was more like the Chinese approach described above. That is, certain arithmetic processes that could be pictured were carried out, but what we call an equation was not written explicitly. With every pair of numbers, say 13 and 27, they associated two other numbers: their average $(13 + 27)/2 = 20$ and their *semidifference*[5] $(27 - 13)/2 = 7$. The average and semidifference can be calculated from the two numbers, and the original data can be calculated from the average and semidifference. The larger number (27) is the sum of the average and semidifference: $20 + 7 = 27$, and the smaller number (13) is their difference: $20 - 7 = 13$. The realization of this mutual connection makes it possible essentially to "change coordinates" from the number pair (a, b) to the pair $\bigl((a + b)/2, (a - b)/2\bigr)$.

At some point lost to history some Mesopotamian mathematician came to realize that the product of two numbers is the difference of the squares of the average and semidifference: $27 \cdot 13 = (20)^2 - 7^2 = 351$ (or 5, 51 in Mesopotamian notation). This principle made it possible to recover two numbers knowing their sum and product or knowing their difference and product. For example, given that the sum is 10 and the product is 21, we know that the average is 5 (half of the sum), hence that the square of the semidifference is $5^2 - 21 = 4$. Therefore, the semidifference is 2, and the two numbers are $5 + 2 = 7$ and $5 - 2 = 3$. Similarly, knowing that the difference is 9 and the product is 52, we conclude that the semidifference is 4.5 and the square of the average is $52 + (4.5)^2 = 72.25$. Hence the average is $\sqrt{72.25} = 8.5$.

---

[4] The error made by the average is half of the *difference* of the two errors.

[5] This word is coined because English contains no one-word description of this concept, which must otherwise be described as half of the difference of the two numbers. It is clear from the way in which the semidifference occurs constantly that the writers of these tablets automatically looked at this number along with the average when given two numbers as data. However, there seems to be no word in the Akkadian, Sumerian, and ideogram glossary given by Neugebauer to indicate that the writers of the clay tablets had a special word for these concepts. But at the very least, they were trained to calculate these numbers when dealing with this type of problem. In the translations given by Neugebauer the average and semidifference are obtained one step at a time, by first adding or subtracting the two numbers, then taking half of the result.

Therefore, the two numbers are $8.5+4.5 = 13$ and $8.5-4.5 = 4$. The two techniques just illustrated occur constantly in the cuneiform texts, and seem to be procedures familiar to everyone, requiring no explanation.

5. The ancient Greeks

It is fairly obvious how to do addition and subtraction within the Greek system of numbering. Doing multiplication in modern notation, as you know, involves memorizing all products up to $9 \cdot 9$, and learning how to keep columns straight, plus carrying digits where necessary. With our place-value notation, we have little difficulty multiplying, say $23 \cdot 42$. For the ancient Greeks the corresponding problem of multiplying $\kappa'\gamma'$ by $\mu'\beta'$ was more complicated. It would be necessary to find four products: $\kappa' \cdot \mu'$, $\kappa' \cdot \beta'$, $\gamma' \cdot \mu'$, and $\gamma' \cdot \beta'$. The first of three of these require the one doing the calculation to keep in mind that κ' is 10 times β' and μ' is 10 times δ'.

These operations are easier for us, since we use the same nine digits in different contexts, keeping track of the numbers they represent by keeping the columns straight while multiplying. They were more difficult for the ancient Greeks, since going from 30 to 3 was not merely a matter of ignoring a zero; it involved a shift forward by 10 letters in the alphabet, from λ' to γ'. In addition to the carrying that we must do, the Greek calculator had to commit to memory 20 such alphabet shifts (10 by 10 letters, and 10 shifts by 20 letters), and, while computing, remember how many such shifts of each kind were performed, so as to know how many factors of 10 were being "stored" during the calculation.

The procedure is explained in the *Synagōgē* of the third-century mathematician Pappus of Alexandria. The surviving portion of this work begins in the middle of Book 2, explaining how to multiply quickly numbers that are multiples of 10. Pappus illustrates the procedure by the following example, in which number *names* are translated into English but number *symbols* are transcribed directly from the original. (See Fig. 3 in Chapter 5 for the numerical values represented by these Greek letters.)

> Let the numbers be ν', ν', ν', μ', μ', and λ'. Then the basic numbers will be ε', ε', ε', δ', δ', and γ'. Their product will be $,\varsigma'$. Since there are ς' tens, and since ς' divided by four leaves a remainder of two, the product [of the six reduced numbers] will contain altogether one hundred myriads... and these ρ' myriads multiplied by the $,\varsigma'$ units will make ξ' twofold [that is, "square"] myriads.

When we translate the problem into our notation, it becomes trivial. We are trying to find the product $50 \cdot 50 \cdot 50 \cdot 40 \cdot 40 \cdot 30$. We have no trouble factoring out the six zeros and rewriting the problem as $5 \cdot 5 \cdot 5 \cdot 4 \cdot 4 \cdot 3 \cdot 1{,}000{,}000 = 6000 \cdot 10^6 = 60 \cdot (10{,}000)^2$. Converting from 40 to 4 is considerably easier than converting from μ' to δ'. Pappus divided the number of tens by 4, since he was counting in myriads (10^4), and he expressed the answer as "*myriadōn* ξ' *diplōn*," that is, "of myriads 60 twofold," or, in better English, "60 twofold myriads." To describe what we now call the square of a number, the ancient Greeks had to extend the normal meaning of the word *diploos* (double). A nonmathematical reader of Pappus' Greek might

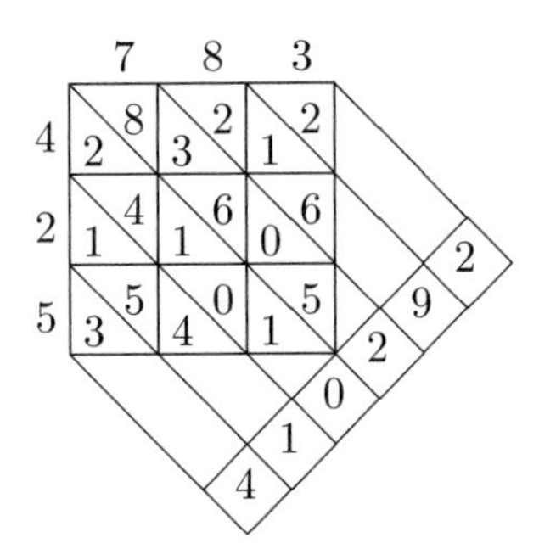

FIGURE 3. Multiplication with Hindu–Arabic numerals.

be inclined to think that the phrase *myrias diploos* meant $2 \cdot 10,000$ rather than $10,000^2$.

As you can see from this example, calculating (*logistikê*) was not quite so trivial for the Greeks as for us.

6. The Islamic world

It is well known that the numerals used all over the world today are an inheritance from both the Hindu and Arabic mathematicians of 1000 years ago. The Hindu idea of using nine symbols in a place-value system was known in what is now Iraq in the late seventh century, before that area became part of the Muslim Empire. In the late eighth century a scholar from India came to the court of Caliph al-Mansur with a work on Hindu astronomy using these numerals, and this work was translated into Arabic. An Arabic treatise on these numbers, containing the first known discussion of decimal fractions, was written by al-Uqlidisi (ca. 920–ca. 980).

Having inherited works from the time of Mesopotamia and also Greek and Hindu works that used the sexagesimal system in astronomy, the Muslim mathematicians of a thousand years ago also used that system. The sexagesimal system did not yield immediately to its decimal rival, and the technique of place-value computation developed in parallel in the two systems. Ifrah (*2000*, pp. 539–555) gives a detailed description of the long resistance to the new system. The sexagesimal system is mentioned in Arabic works of Abu'l-Wafa (940–988) and Kushar ben Laban (ca. 971–1029). It continued to appear in Arabic texts through the time of al-Kashi (1427), although the decimal system also occurs in the work of al-Kashi. In addition to the sexagesimal and decimal systems, the Muslim mathematicians used an elaborate system of finger reckoning. Some implementations of the decimal system require crossing out or erasing in the process of computation, and that was considered a disadvantage. Nevertheless, the superiority of decimal notation in computation was recognized early. For example, al-Daffa (*1973*, pp. 56–57) mentions that there there are manuscripts still extant dating to the twelfth century, in which multiplication is performed by the very efficient method illustrated in Fig. 3 for the multiplication $524 \cdot 783 = 410,292$.

7. Europe

The system of Roman numerals that now remains in countries settled by Europeans is confined to a few cases where numbers have only to be read, not computed with.

For computations these cumbersome numerals were supplanted centuries ago by the Hindu–Arabic place-value decimal system. Before that time, computations had been carried out using common fractions, although for geometric and astronomical computations, the sexagesimal system inherited from the Middle East was also used. It was through contacts with the Muslim culture that Europeans became familiar with the decimal place-value system, and such mathematicians as Gerbert of Aurillac encouraged the use of the new numbers in connection with the abacus. In the thirteenth century Leonardo of Pisa also helped to introduce this system of calculation into Europe, and in 1478 an arithmetic was published in Treviso, Italy, explaining the use of Hindu–Arabic numerals and containing computations in the form shown in Fig. 3. In the sixteenth century many scholars, including Robert Recorde (1510–1558) in Britain and Adam Ries (1492–1559) in Germany, advocated the use of the Hindu–Arabic system and established it as a universal standard.

The system was elegantly explained by the Flemish mathematician and engineer Simon Stevin (1548–1620) in his 1585 book *De Thiende (Decimals)*. Stevin took only a few pages to explain, in essentially modern terms, how to add, subtract, multiply, and divide decimal numbers. He then showed the application of this method of computing in finding land areas and the volumes of wine vats. He wrote concisely, as he said, "because here we are writing for teachers, not students." His notation appears slightly odd, however, since he put a circled 0 where we now have the decimal point, and thereafter indicated the rank of each digit by a similarly encircled number. For example, he would write 13.4832 as 13 ⓪ 4 ① 8 ② 3 ③ 2 ④ . Here is his explanation of the problem of expressing $0.07 \div 0.00004$:

> When the divisor is larger [has more digits] than the dividend, we adjoin to the dividend as many zeros as desired or necessary. For example, if 7 ② is to be divided by 4 ⑤ , I place some 0s next to the 7, namely 7000. This number is then divided as above, as follows:
>
> 3̸ 2̸
> 7̸ 0̸ 0̸ 0̸   (1 7 5 0 ⓪
> 4̸ 4̸ 4̸ 4̸
>
> Hence the quotient is 1750 ⓪ . [Gericke and Vogel, *1965*, p. 19]

Except for the location of the digits and the cross-out marks, this notation is essentially what is now done by school children in the United States. In other countries—Russia, for example—the divisor would be written just to the right of the dividend and the quotient just below the divisor.

Stevin also knew what to do if the division does not come out even. He pointed out that when 4 ① is divided by 3 ② , the result is an infinite succession of 3s and that the exact answer will never be reached. He commented, "In such a case, one may go as far as the particular case requires and neglect the excess. It is certainly true that 13 ⓪ 3 ① $3\frac{1}{3}$ ② , or 13 ⓪ 3 ① 3 ② $3\frac{1}{3}$ ③ , and so on are exactly equal to the required result, but our goal is to work only with whole numbers in this decimal computation, since we have in mind what occurs in human business, where [small parts of small measures] are ignored." Here we have a clear case in which the existence of infinite decimal expansions is admitted, without any hint of the possibility of irrational numbers. Stevin was an engineer, not a theoretical mathematician. His examples were confined to what is of practical value in business and engineering, and he made no attempt to show how to calculate with an actually infinite decimal expansion.

Stevin did, however, suggest a reform in trigonometry that was ignored until the advent of hand-held calculators, remarking that, "if we can trust our experience (with all due respect to Antiquity and thinking in terms of general usefulness), it is clear that the series of divisions by 10, not by 60, is the most efficient, at least among those that are by nature possible." On those grounds, Stevin suggested that degrees be divided into decimal fractions rather than minutes and seconds. Modern hand-held calculators now display angles in exactly this way, despite the scornful remark of a twentieth-century mathematician that "it required four millennia to produce a system of angle measurement that is completely absurd."

8. The value of calculation

One cannot help noticing, alongside a few characteristics that are unique to a given culture, a large core of commonality in all this elementary mathematics. All of the treatises we have looked at pose problems of closely similar structure. This commonality is so great that any textbook of arithmetic published in the modern world up to very recent times is almost certain to repeat problems from the *Jiu Zhang Suanshu* or the Ahmose Papyrus or the *Brahmasphutasiddhanta* almost word for word. Thus, where Brahmagupta instructs the reader to "multiply the fruit and the requisition and divide by the argument in order to obtain the produce," Greenleaf (*1876*, p. 233) tells the reader to "find the required term by dividing the product of the second and third terms by the first." As far as clarity of exposition is concerned, one would have to give the edge to Brahmagupta. The number of ways of solving a mathematical problem is, after all, quite small; it is not surprising if two people in widely different circumstances come to the same conclusion.

Of course, when looking at the history of mathematics in the late nineteenth century, we tend to focus on the new research occurring at that time and overlook mere expositions of long-known mathematics such as one finds in the book of Greenleaf. But what Greenleaf was expounding was a set of mathematical skills that had been useful for many centuries. Like the *Jiu Zhang Suanshu* and the *Sun Zi Suan Jing*, his book contains discussions of the relations of various units of measure to one another and a large number of examples, both realistic and fanciful, showing how to carry out all the elementary operations. His job, in fact, was harder than that of the earlier authors, since he had to explain the relation between common fractions and decimal fractions, exchange rates for different kinds of currency, and many principles of commercial and inheritance law. If we now tend to regard such books as being of secondary importance in the history of mathematics, that is only because such a high superstructure has been built on that foundation. When we read the classics from Egypt, India, China, and Mesopotamia, on the other hand, we are looking at the frontier of knowledge in their time. It is a tribute to the authors of the treatises discussed in this chapter that they worked out and explained in clear terms a set of useful mathematical skills and bequeathed it to the world. For many centuries it could be said that the standard mathematical curriculum had a permanent value. Only in very recent years have the computational skills needed in commerce and law been superseded by the higher-level skills needed for deciding *when* and *what* to compute and how to interpret the results.

9. Mechanical methods of computation

Any study of the history of calculation must take account of the variety of computing hardware that people have invented and the software algorithms that are developed from time to time. In ancient China the software (decimal place-value system) was so good that the hardware (counting rods, counting boards, and abacus) worked with it very smoothly. The Greek and Roman system of writing numbers, however, was not a good representation of the decimal system, and the abacus was probably an essential tool of computation. When the graphical methods associated with Hindu–Arabic numerals were introduced into Europe, they were thought to be superior to the abacus.

9.1. Software: prosthaphæresis and logarithms. The graphic arithmetic that had vanquished the counting board a few centuries earlier still had certain laborious aspects connected with multiplication and division, which mathematicians kept trying to simplify. Consider, for example, the two three-digit numbers 476 and 835. To add these numbers we must perform three simple additions, plus two more that result from "carrying," a total of eight simple additions. In general, at most $3n - 1$ simple additions with $n - 1$ carryings will be required to add two n-digit numbers. Similarly, subtracting these numbers will require at most two borrowings, with consequent modification of the digits borrowed from, and three simple subtractions. For an n-digit number that is at most n simple subtractions and $n - 1$ borrowings.

On the other hand, to multiply two three-digit numbers will require nine simple multiplications followed by addition of the partial products, which will involve up to 10 more simple additions if carrying is involved. Thus we are looking at considerably more labor, with a number of additions and multiplications on the order of $2n^2$ if the two numbers each have n digits. Not only is a greater amount of time and effort needed, the procedure is obviously more error-prone. On the other hand, in a practical application in which we are multiplying, say, two seven-digit numbers (which would involve more than 100 simple multiplications and additions), we seldom need all 14 or 15 digits of the result. If we could improve the speed of the operation at the expense of some precision, the trade-off would be worthwhile.

Prosthaphæresis. The increased accuracy of astronomical instruments, among other applications, led to a need to multiply numbers having a large number of digits. As just pointed out, the amount of labor involved in multiplying two numbers increases as the product of the numbers of digits, while the labor of adding increases according to the number of digits in the smaller number. Thus, multiplying two 15-digit numbers requires over 200 one-digit multiplications, while adding the two numbers requires only 15 such operations (not including carrying). It was the large number of digits in the table entries that caused the problem in the first place, but the key to the solution turned out to be in the structural properties of sines and cosines. The process was called *prosthaphæresis*, from two Greek prefixes *pros-*, meaning *toward*, and *apo-*, meaning *from*, together with the root verb *haírō*, meaning *I seize* or *I take*. Together these parts mean simply *addition and subtraction*.

There are hints of this process in several sixteenth-century works, but we shall quote just one example. In his *Trigonometria*, first published in Heidelberg in 1595, the theologian and mathematician Bartholomeus Pitiscus (1561–1613) posed the following problem: *to solve the proportion in which the first term is the radius, while the second and third terms are sines, avoiding multiplication and division.*

The problem here is to find the fourth proportional x, satisfying $r : a = b : x$, where r is the radius of the circle and a and b are two sines (half-chords) in the circle. We can see immediately that $x = ab/r$, but as Pitiscus says, the idea is to avoid the multiplication and division, since in the trigonometric tables of the time a and b might easily have seven or eight digits each.

The key to prosthaphæresis is the well-known formula

$$\sin\alpha\cos\beta = \frac{\sin(\alpha+\beta)+\sin(\alpha-\beta)}{2}.$$

This formula is applied as follows: If you have to multiply two large numbers, regard one of them as the sine of an angle, the other as the cosine of a second angle. (Since Pitiscus had only tables of sines, he had to use the complement of the angle having the second number as a sine.) Add the angles and take the sine of their sum to obtain the first term; then subtract the angles and take the sine of their difference to obtain a second term. Finally, divide the sum of the two terms by 2 to obtain the product. To take a very simple example, suppose that we wish to multiply 155 by 36. A table of trigonometric functions shows that $\sin 8^\circ\,55' = 0.15500$ and $\cos 68^\circ\,54' = 0.36000$. Hence

$$36 \cdot 155 = 10^5\frac{\sin 77^\circ\,49' + \sin(-59^\circ\,59')}{2} = \frac{97748 - 86588}{2} = 5580.$$

In general, some significant figures will be lost in this kind of multiplication. For large numbers this procedure saves labor, since multiplying even two seven-digit numbers would tax the patience of most modern people. A further advantage is that prosthaphæresis is less error-prone than multiplication. Its advantages were known to the Danish astronomer Tycho Brahe (1546–1601), who used it in the astronomical computations connected with the extremely precise observations he made at his observatory during the latter part of the sixteenth century.

Logarithms. The problem of simplifying laborious multiplications, divisions, root extractions, and the like, was being attacked at the same time in another part of the world and from another point of view. The connection between geometric and arithmetic proportion had been noticed earlier by Chuquet, but the practical application of this fact had never been worked out. The Scottish laird John Napier, Baron of Murchiston (1550–1617), tried to clarify this connection and apply it. His work consisted of two parts, a theoretical part based on a continuous geometric model, and a computational part, involving a discrete (tabular) approximation of the continuous model. The computational part was published in 1614. However, Napier hesitated to publish his explanation of the theoretical foundation. Only in 1619, two years after his death, did his son publish the theoretical work under the title *Mirifici logarithmorum canonis descriptio* (*A Description of the Marvelous Rule of Logarithms*). The word *logarithm* means *ratio number*, and it was from the concept of ratios (geometric progressions) that Napier proceeded.

To explain his ideas Napier used the concept of moving points. He imagined one point P moving along a straight line from a point T toward a point S with decreasing velocity such that the ratio of the distances from the point P to S at two different times depends only on the difference in the times. (Actually, he called the line ending at S a sine and imagined it shrinking from its initial size TS, which he called the radius.) A second point is imagined as moving along a second line at a constant velocity equal to that with which the first point began. These two motions can be clarified by considering Fig. 4.

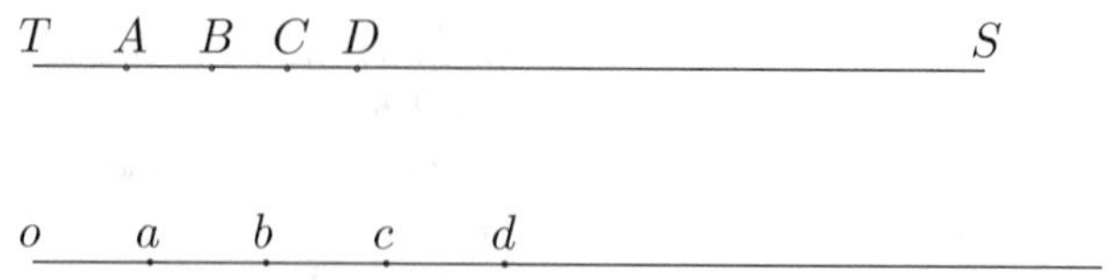

FIGURE 4. Geometric basis of logarithms.

The first point sets out from T at the same time and with the same speed with which the second point sets out from o. The first point, however slows down, while the second point continues to move at constant speed. The figure shows the locations reached at various times by the two points: When the first point is at A, the second is at a; when the first point is at B, the second is at b; and so on. The point moving with decreasing velocity requires a certain amount of time to move from T to A, the same amount of time to move from A to B, from B to C, and from C to D. Consequently, $TS : AS = AS : BS = BS : CS = CS : DS$.

The first point will never reach S, since it keeps slowing down, and its velocity at S would be zero. The second point will travel indefinitely far, given enough time. Because the points are in correspondence, the division relation that exists between two positions in the first case is mirrored by a subtractive relation in the corresponding positions in the second case. Thus, this diagram essentially changes division into subtraction and multiplication into addition. The top scale in Fig. 4 resembles a slide rule, and this resemblance is not accidental: a slide rule is merely an analog computer that incorporates a table of logarithms.

Napier's definition of the logarithm can be stated in the modern notation of functions by writing $\log(AS) = oa$, $\log(BS) = ob$, and so on; in other words, the logarithm increases as the "sine" decreases. These considerations contain the essential idea of logarithms. The quantity Napier defined is not the logarithm as we know it today. If points T, A, and P correspond to points o, a, and p, then

$$\overline{op} = \overline{oa} \log_k \left(\frac{\overline{PS}}{\overline{TS}} \right),$$

where $k = \overline{AS}/\overline{TS}$.

Arithmetical implementation of the geometric model. The geometric model just discussed is theoretically perfect, but of course one cannot put the points on a line into a table of numbers. It is necessary to construct the table from a finite set of points; and these points, when converted into numbers, must be rounded off. Napier was very careful to analyze the maximum errors that could arise in constructing such a table. In terms of Fig. 4, he showed that oa, which is the logarithm of AS, satisfies

$$TA < oa < TA\Big(1 + \frac{TA}{AS}\Big).$$

(These inequalities are simple to prove, since the point describing oa has a velocity larger than the velocity of the point describing TA but less than TS/AS times the velocity of that point.) Thus, the tabular value for the logarithm of AS can be taken as the average of the two extremes, that is, $TA\big[1 + (TA/2AS)\big]$, and the relative error will be very small when TA is small.

Napier's death at the age of 67 prevented him from making some improvements in his system, which are sketched in an appendix to his treatise. These improvements consist of scaling in such a way that the logarithm of 1 is 0 and the logarithm of 10 is 1, which is the basic idea of what we now call *common logarithms.* These further improvements to the theory of logarithms were made by Henry Briggs (1561–1630), who was in contact with Napier for the last two years of Napier's life and wrote a commentary on the appendix to Napier's treatise. As a consequence, logarithms to base 10 came to be known as *Briggsian logarithms.*

9.2. Hardware: slide rules and calculating machines. The fact that logarithms change multiplication into addition and that addition can be performed mechanically by sliding one ruler along another led to the development of rulers with the numbers arranged in proportion to their logarithms (slide rules). When one such scale is slid along a second, the numbers pair up in proportion to the distance slid, so that if 1 is opposite 5, then 3 will be opposite 15. Multiplication and division are then just as easy to do as addition and subtraction would be. The process is the same for both multiplication and division, as it was in the Egyptian graphical system, which was also based on proportion. Napier himself designed a system of rods for this purpose. A variant of this linear system was a system of sliding circles. Such a circular slide rule was described in a pamphlet entitled *Grammelogia* written in 1630 by Richard Delamain (1600–1644), a mathematics teacher living in London. Delamain urged the use of this device on the grounds that it made it easy to compute compound interest. Two years later the English clergyman William Oughtred (1574–1660) produced a similar description of a more complex device. Oughtred's *circles of proportion*, as he called them, gave sines and tangents of angles in various ranges on eight different circles. Because of their portability, slide rules remained the calculating machine of choice for engineers for 350 years, and improvements were still being made in them in the 1950s. Different types of slide rule even came to have different degrees of prestige, according to the number of different scales incorporated into them.

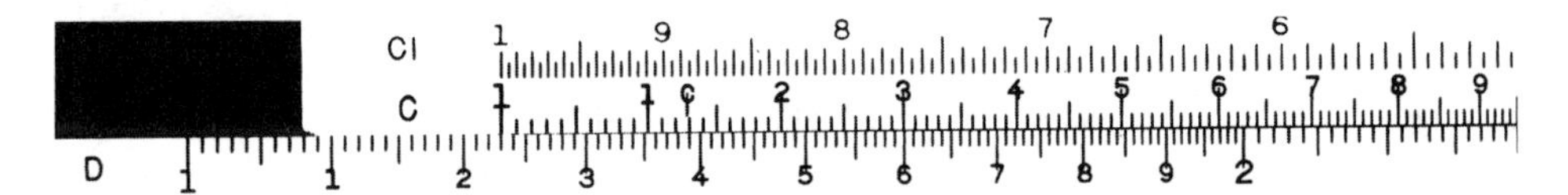

Portions of the C, D, and CI scales of a slide rule. Adjacent numbers on the C and D scales are in proportion, so that $1 : 1.23 :: 1.3 : 1.599 :: 1.9 : 2.337$. Thus, the position shown here illustrates the multiplication $1.23 \cdot 1.3 = 1.599$, the division $1.722 \div 1.4 = 1.23$, and many other computations. Some visual error is inevitable. The CI (inverted) scale gives the reciprocals of the numbers on the C scale, so that division can be performed as multiplication, only using the CI scale instead of the C scale. Decimal points have to be provided by the user.

Slide rule calculations are floating-point numbers with limited accuracy and necessary round-off error. When computing with integers, we often need an exact answer. To achieve that result, adding machines and other digital devices have been developed over the centuries. An early design for such a device with a series of interlocking wheels can be found in the notebooks of Leonardo da Vinci (1452–1519). Similar machines were designed by Blaise Pascal (1623–1662) and Gottfried Wilhelm Leibniz (1646–1716). Pascal's machine was a simple adding machine that depended on turning a crank a certain number of times in order to find a sum. Leibniz used a variant of this machine with a removable set of wheels that would multiply, provided that the user kept count of the number of times the crank was turned.

Machines designed to calculate for a specific purpose continued to be built for centuries, but all were doomed to be replaced by the general-purpose information processor that has spread like wildfire around the world in the past two decades. The first prefiguration of such a machine was Charles Babbage's difference engine, designed in the 1830s but built only partially many decades later. Although only one such machine seems to have been built, and only part of Babbage's more ambitious analytical engine was constructed, the idea of a general-purpose computer that could accept instructions and modify its operation in accordance with them was a brilliant innovation. Unfortunately, the full implementation of this idea could not be carried out by mechanical devices with moving parts. It needed the reliability of electronics, first thermionic valves (vacuum tubes) and then transistors, to produce the marvels of technology that we all use nowadays. That technology was developed in Britain and the United States, greatly stimulated by the needs of code breaking during World War II.

Meanwhile, fixed-purpose machines that could perform only a limited number of set arithmetic operations, continued to be built and improved upon. In the late days of World War II the "latest thing" in automated calculation was a machine with many fragile moving parts.

Hand-held calculators, at first very limited, began to supplant the slide rule during the 1970s. They were easier to use, infinitely faster, and much more accurate than slide rules, and they soon became much cheaper. The only disadvantage they had in comparison with slide rules was in durability.[6] A popular American textbook of college algebra published in 1980 weighed the advantages and disadvantages of slide rules, calculators, and tables of logarithms, and summed up for the jury in favor of using tables of logarithms, which had the cheapness of slide rules but more accuracy. Even that recently it was an extremely refined and expensive hand calculator that had more than a few dozen memory cells. It was not possible to foresee the explosion of computing power that was to result from the development of methods of producing huge quantities of memory on tiny chips at extremely low prices.

For several decades following World War II there were two types of calculating devices: Slide rules and cheap adding machines served the individual; more expensive calculators and the early gigantic computers such as ENIAC (Electronic Numerical Integrator And Computer) were used by large corporations. The overwhelming penetration of modern computers into nearly every human activity, especially their use for word processing and graphics, is due to the vision of people such as Charles Babbage, who realized that they must be able to use Boolean logic in addition to their calculating capacity.

Two mathematicians figure prominently in the development of this vision of the computer. One was Alan Turing (1912–1954), a British mathematician whose 1937 paper "On computable numbers" contained the idea of a universal computer

[6] One might think that there would be no further market for slide rules. To the contrary, an entire website is devoted to buying and selling them. The webmaster points out to the site visitor that the computer on which the purchase is being made will be in a landfill 50 years hence, whereas the slide rule will be only well broken in.

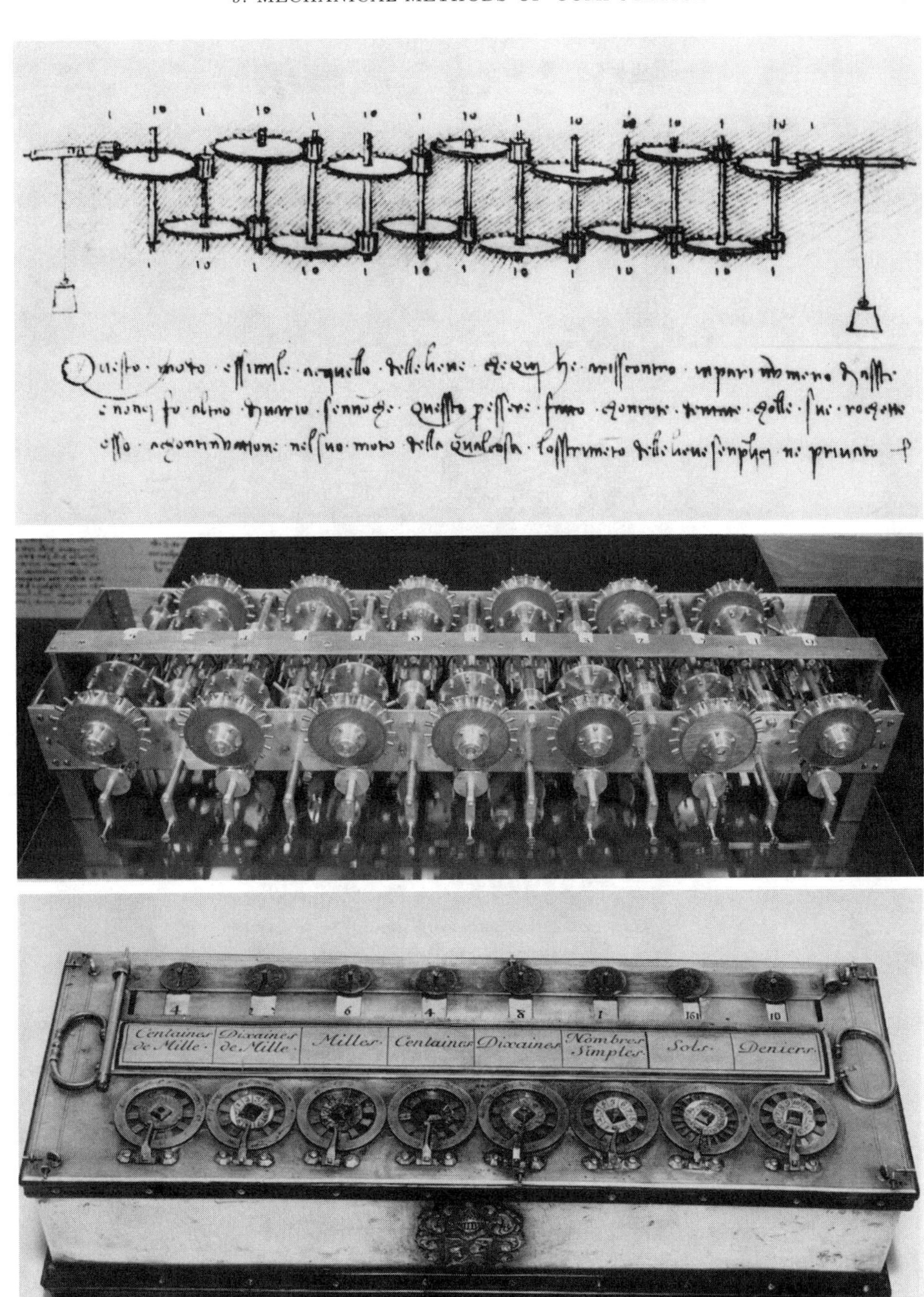

Top: Leonardo da Vinci's design for an adding machine. Center: model of the machine. Bottom: Pascal's adding machine. Courtesy of IBM Corporate Images.

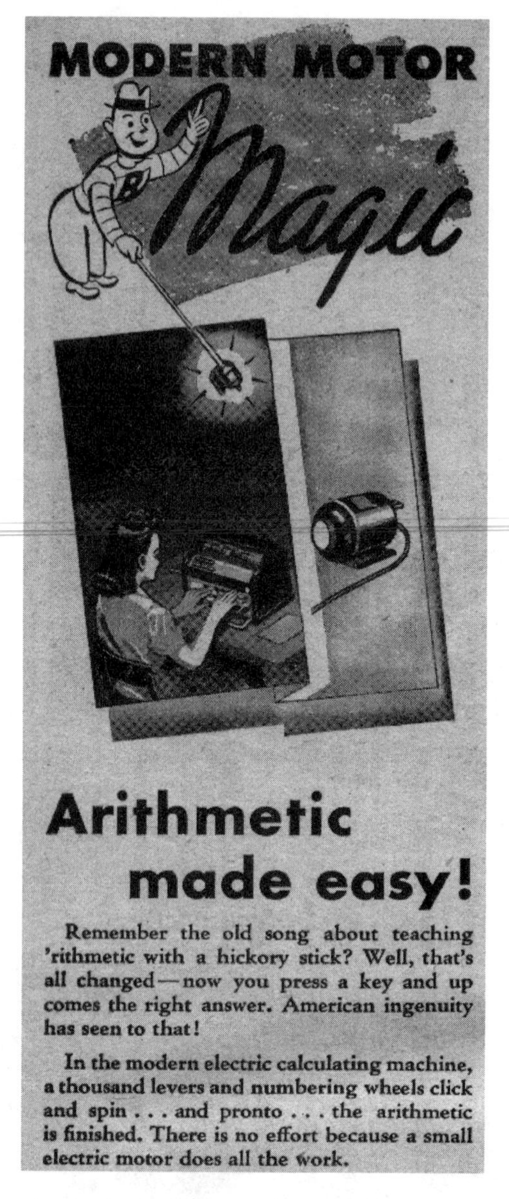

The future of calculation, as depicted in the April 17, 1944 issue of *Newsweek*. The potential of mechanical computing devices is limited by wear and tear on its moving parts. A slide rule suffers very little from such wear, but the more sophisticated adding machines of Pascal had to be designed with counterweights to delay the inevitable lopsidedness that results from wear. Courtesy of the Bodine Electric Company, Chicago.

now known as a *Turing machine*. The crucial element was programmability: Turing envisioned a tape filled with information that could be fed into the machine. As the machine read the information, it would modify its own internal state and then move ahead or back to an adjacent instruction. In this way, Turing went a step further than Babbage had envisioned. Babbage's assistant Augusta Ada Lovelace (1815–1852) had written that Babbage's analytical engine could do only what it was told to do, but Turing believed that the difference between human intelligence and a computer was not so stark as that. He considered it possible that what appeared as human creativity might be the result of some information delivered at an earlier time, and that computers might mimic this apparent creativity.

The other mathematician was John von Neumann (1903–1957), who was at Princeton in 1936 when Turing came there as a graduate student.[7] Von Neumann became involved in the development of the computer while working as a consultant at the Aberdeen Proving Grounds in Maryland. There, in August 1944, he met Herman H. Goldstine (1913–2004), who told him about ENIAC. From our perspective, two generations on, ENIAC looks like the brontosaurus of computing technology. Here is a description of it from a website devoted to its history:[8]

> When it was finished, the ENIAC filled an entire room, weighed thirty tons, and consumed two hundred kilowatts of power. It generated so much heat that it had to be placed in one of the few rooms at the University [of Pennsylvania] with a forced air cooling system. Vacuum tubes, over 19,000 of them, were the principal elements in the computer's circuitry. It also had fifteen hundred relays and hundreds of thousands of resistors, capacitors, and inductors. All of this electronics was held in forty-two panels nine feet tall, two feet wide, and one foot thick.

Despite its size, ENIAC had very little memory—only 1000 bits of RAM! Moreover the use of vacuum tubes (thermionic valves) meant frequent breakdowns—one every 8 minutes on the average, until the operators reduced the voltage and current to the minimum; after that, breakdowns occurred only once every two days on the average. Most seriously, it could not be programmed in the present-day sense of the word. It simply had to be set up for each particular computation. Von Neumann and the builders of ENIAC collaborated on the construction of EDVAC (Electronic Discrete Variable Automatic Computer), which had the ability to read stored programs. Of course, those programs had to be written in machine language, a serious drawback for human interaction, but von Neumann's basic idea was sound. We have engineers to thank that the vacuum tube has been replaced by the transistor, that transistors can now be etched onto tiny computer chips, and that production methods have made it possible to produce for a very modest price the machines that everyone now uses with such ease.

9.3. The effects of computing power. A crystal ball can be very cloudy, even in relation to the eternal truths of mathematics. A book of mathematical tables and formulas (Burington, *1958*) purchased by the author nearly half a century ago confidently assured its readers in a note from the publishers that

> the subject matter [of this book] is not ephemeral but everlasting—as true in the future as it has been in the past. By all means, retain

[7] According to Heppenheimer (*1990*), von Neumann offered Turing a position as his assistant, but Turing preferred to return to Cambridge.

[8] `http://ei.cs.vt.edu/~history/ENIAC.Richey.HTML`.

> this book for your own reference library. You will need it many times in years to come.

That book remains on the author's shelf, unopened since about 1985. The publishers in their confidence had overlooked the fact that the eternal truths of mathematics need not be reconstructed every time they are needed. Machines can store them and do the unimaginative computational work more efficiently and accurately than people can.

One result of all this magnificent computer engineering is that mathematics education faces a dilemma. On the one hand, the skills involved in doing elementary arithmetic, algebra, and calculus are now as obsolete as the skill of writing a letter in longhand. What is the point of teaching students how to solve quadratic equations, factor polynomials, carry out integration by parts, and solve differential equations when readily available programs such as *Mathematica*, *Maple*, *Matlab*, and others can produce the result in a split second with guaranteed accuracy? On the other hand, solving mathematical problems requires quantitative reasoning, and no one has yet found any way to teach quantitative reasoning without assuming a familiarity with these basic skills. How can you teach what multiplication *is* without making students learn the multiplication table? How can you explain the theory of equations without making students solve a few equations? If mathematics education is to be in any way relevant to the lives of the students who are its clients, it must be able to explain in cogent terms the reason for the skills it asks them to undergo so much boredom to learn, or else find other skills to teach them.

Questions and problems

6.1. Double the hieroglyphic number $\begin{smallmatrix} ||| & \cap \\ |||| & \cap\cap \end{smallmatrix}$.

6.2. Multiply 27 times 42 the Egyptian way.

6.3. (Stated in the Egyptian style.) Calculate with 13 so as to obtain 364.

6.4. Problem 23 of the Ahmose Papyrus asks what parts must be added to the sum of $\overline{4}$, $\overline{8}$, $\overline{10}$, $\overline{30}$, and $\overline{45}$ to obtain $\overline{\overline{3}}$. See if you can obtain the author's answer of $\overline{9}\ \overline{40}$, starting with his technique of magnifying the first row by a factor of 45. Remember that $\frac{5}{8}$ must be expressed as $\overline{2}\ \overline{8}$.

6.5. Problem 24 of the Ahmose Papyrus asks for a number that yields 19 when its seventh part is added to it, and concludes that one must perform on 7 the same operations that yield 19 when performed on 8. Now in Egyptian terms, 8 must be multiplied by $2\ \overline{4}\ \overline{8}$ in order to obtain 19. Multiply this number by 7 to obtain the scribe's answer, $16\ \overline{2}\ \overline{8}$. Then multiply that result by $\overline{7}$, add the product to the result itself, and verify that you do obtain 19, as required.

6.6. Problem 33 of the Ahmose Papyrus asks for a quantity that yields 37 when increased by its two parts (two-thirds), its half, and its seventh part. Try to get the author's answer: The quantity is $16\ \overline{56}\ \overline{679}\ \overline{776}$. [*Hint*: Look in the table of doubles of parts for the double of $\overline{97}$. The scribe first tried the number 16 and found that the result of these operations applied to 16 fell short of 37 by the double of $\overline{42}$, which, as it happens, is exactly $1\ \overline{\overline{3}}\ \overline{2}\ \overline{7}$ times the double of $\overline{97}$.]

6.7. Verify that the solution to Problem 71 ($2\ \overline{\overline{3}}$) is the correct *pesu* of the diluted beer discussed in the problem.

6.8. Compare the *pesu* problems in the Ahmose Papyrus with the following problem, which might have been taken from almost any algebra book written in the past century: *A radiator is filled with 16 quarts of a 10% alcohol solution. If it requires a 30% alcohol solution to protect the radiator from freezing when it is turned off, how much 95% solution must be added (after an equal amount of the 10% solution is drained off) to provide this protection?* Think of the alcohol as the grain in beer and the liquid in the radiator as the beer. The liquid has a *pesu* of 10. What is the *pesu* that it needs to have, and what is the *pesu* of the liquid that is to be used to achieve this result?

6.9. Verify that the solution $\overline{5}\ \overline{10}$ given above for Problem 35 is correct, that is, multiply this number by 3 and by $\overline{3}$ and verify that the sum of the two results is 1.

6.10. Why do you suppose that the author of the Ahmose Papyrus did not choose to say that the double of the thirteenth part is the seventh part plus the ninety-first part, that is,

$$\tfrac{2}{13} = \tfrac{1}{7} + \tfrac{1}{91}?$$

Why is the relation

$$\tfrac{2}{13} = \tfrac{1}{8} + \tfrac{1}{52} + \tfrac{1}{104}$$

made the basis for the tabular entry instead?

6.11. Generalizing Question 6.10, investigate the possibility of using the identity

$$\frac{2}{p} = \frac{1}{\left(\frac{p+1}{2}\right)} + \frac{1}{p\left(\frac{p+1}{2}\right)}$$

to express the double of the reciprocal of an odd number p as a sum of two reciprocals. Which of the entries in the table of Fig. 1 can be obtained from this pattern? Why was it not used to express $\frac{2}{15}$?

6.12. Why not simply write $\overline{13}\ \overline{13}$ to stand for what we call $\frac{2}{13}$? What is the reason for using two or three other "parts" instead of these two obvious parts?

6.13. Could the ability to solve a problem such as Problem 35, discussed in Subsection 1.2 of this chapter, have been of any practical use? Try to think of a situation in which such a problem might arise.

6.14. We would naturally solve many of the problems in the Ahmose Papyrus using an equation. Would it be appropriate to say that the Egyptians solved equations, or that they did algebra? What does the word *algebra* mean to you? How can you decide whether you are performing algebra or arithmetic?

6.15. Why did the Egyptians usually begin the process of division by multiplying by $\overline{\overline{3}}$ instead of the seemingly simpler $\overline{2}$?

6.16. Early mathematicians must have been adept at thinking in terms of expressions. But considering the solutions to the riders-and-carts problem and the colorful language of Brahmagupta in relation to the Rule of Three, one might look at the situation from a different point of view. Perhaps these early mathematicians were good "dramatists." In any algorithm the objects we now call variables amount to special "roles" played, with different numbers being assigned to "act" in those roles;

an algorithm amounts to the drama that results when these roles are acted. That is why it is so important that each part of the algorithm have its own name. The letters that we use for variables amount to names assigned to roles in the drama. A declaration of variables at the beginning of a program is analogous to the section that used to be titled "Dramatis Personæ" at the beginning of a play.

Explain long division from this point of view, using the roles of dividend, divisor, quotient, and remainder.

6.17. Imitate the reasoning used in solving the problem of riders and carts above to solve Problem 17 of the *Sun Zi Suan Jing.* The problem asks how many guests were at a banquet if every two persons shared a bowl of rice, every three persons a bowl of soup, and every four persons a bowl of meat, leading to a total of 65 bowls. Don't use algebra, but try to explain the rather cryptic solution given by Sun Zi: Put down 65 bowls, multiply by 12 to obtain 780, and divide by 13 to get the answer.

6.18. Compare the following loosely interpreted problems from the *Jiu Zhang Suanshu* and the Ahmose Papyrus. First, from the *Jiu Zhang Suanshu*: Five officials went hunting and killed five deer. Their ranks entitle them to shares in the proportion $1 : 2 : 3 : 4 : 5$. What part of a deer does each receive?

Second, from the Ahmose Papyrus (Problem 40): 100 loaves of bread are to be divided among five people (in arithmetic progression), in such a way that the amount received by the last two (together) is one-seventh of the amount received by the first three (together). How much bread does each person receive?

6.19. Compare the interest problem (Problem 20 of Chapter 3) from the *Jiu Zhang Suanshu* discussed above, with the following problem, taken from the American textbook *New Practical Arithmetic* by Benjamin Greenleaf (*1876*):

> The interest on \$200 for 4 months being \$4, what will be the interest on \$590 for 1 year and 3 months?

Are there any significant differences at all in the nature of the two problems, written nearly 2000 years apart?

6.20. Problem 4 in Chapter 6 of the *Jiu Zhang Suanshu* involves what is called *double false position.* The problem reads as follows: A number of families contribute equal amounts to purchase a herd of cattle. If the contribution (the same for each family) were such that seven families contribute a total of 190 [units of money], there would be a deficit of 330 [units of money]; but if the contribution were such that nine families contribute 270 [units of money], there would be a surplus of 30 [units of money]. Assuming that the families each contribute the correct amount, how much does the herd cost, and how many families are involved in the purchase? Explain the solution given by the author of the *Jiu Zhang Suanshu*, which goes as follows. Put down the proposed values (assessment to each family, that is, $\frac{190}{7}$ and $\frac{270}{9} = 30$), and below each put down the corresponding surplus or deficit (a positive number in each case). Cross-multiply and add the products to form the *shi* ($30 \cdot \frac{190}{7} + 330 \cdot 2709 = \frac{75000}{7}$). Add the surplus and deficit to form the *fa* ($330 + 30 = 360$). Subtract the smaller of the proposed values from the larger, to get the difference ($\frac{270}{9} - \frac{190}{7} = \frac{20}{7}$). Divide the *shi* by the difference to get the cost of the goods ($\frac{75000}{20} = 3750$); divide the *fa* by the difference to get the number of families ($\frac{360}{20/7} = 126$).

6.21. Compare the pond-filling problem (Problem 26 of Chapter 6) of the *Jiu Zhang Suanshu* (discussed above) with the following problem from Greenleaf (*1876*, p. 125): *A cistern has three pipes; the first will fill it in 10 hours, the second in 15 hours, and the third in 16 hours. What time will it take them all to fill it?* Is there any real difference between the two problems?

6.22. The fair taxation problem from the *Jiu Zhang Suanshu* considered above treats distances and population with equal weight. That is, if the population of one county is double that of another, but that county is twice as far from the collection center, the two counties will have exactly the same tax assessment in grain and carts. Will this impose an equal burden on the taxpayers of the two counties? Is there a direct proportionality between distance and population that makes them interchangeable from the point of view of the taxpayers involved? Is the growing of extra grain to pay the tax fairly compensated by a shorter journey?

6.23. Perform the division $\frac{980}{45}$ following the method used by Brahmagupta.

6.24. Convert the sexagesimal number 5; 35, 10 to decimal form and the number 314.7 to sexagesimal form.

6.25. As mentioned in connection with the lunisolar calendar, 19 solar years equal almost exactly 235 lunar months. (The difference is only about two hours.) In the Julian calendar, which has a leap year every fourth year, there is a natural 28-year cycle of calendars. The 28 years contain exactly seven leap-year days, giving a total of exactly 1461 weeks. These facts conjoin to provide a natural 532-year cycle ($532 = 28 \cdot 19$) of calendars incorporating the phases of the Moon. In particular, Easter, which is celebrated on the Sunday after the first full Moon of spring, has a 532-year cycle (spoiled only by the two-hour discrepancy between 19 years and 235 months). According to Simonov (*1999*), this 532-year cycle was known to Cyrus (Kirik) of Novgorod when he wrote his "Method by which one may determine the dates of all years" in the year 6644 from the creation of the world (1136 CE). Describe how you would create a table of dates of Easter that could, in principle, be used for all time, so that a user knowing the number of the current year could look in the table and determine the date of Easter for that year. How many rows and how many columns should such a table have, and how would it be used?

6.26. From 1901 through 2099 the Gregorian calendar behaves like the Julian calendar, with a leap year every four years. Hence the 19-year lunar cycle and 28-year cycle of days interact in the same way during these two centuries. As an example, we calculate the date of Easter in the year 2039. The procedure is first to compute the remainder when 2039 is divided by 19. The result is 6 ($2039 = 19 \times 107 + 6$). This number tells us where the year 2039 occurs in the 19-year lunar cycle. In particular, by consulting the table below for year 6, we find that the first full Moon of spring in 2039 will occur on April 8. (Before people became familiar with the use of the number 0, it was customary to add 1 to this remainder, getting what is still known in prayer books as the *golden number*. Thus the golden number for the year 2039 is 7.)

We next determine by consulting the appropriate calendar in the 28-year cycle which day of the week April 8 will be. In fact, it will be a Friday in 2039, so that Easter will fall on April 10 in that year. The dates of the first full Moon in spring for the years of the lunar cycle are as follows. The year numbers are computed as above, by taking the remainder when the Gregorian year number is divided by 19.

Year	0	1	2	3	4	5	6
Full Moon	Apr. 14	Apr. 3	Mar. 23	Apr. 11	Mar. 31	Apr. 18	Apr. 8
Year	7	8	9	10	11	12	13
Full Moon	Mar. 28	Apr. 16	Apr. 5	Mar. 25	Apr. 13	Apr. 2	Mar. 22
Year	14	15	16	17	18		
Full Moon	Apr. 10	Mar. 30	Apr. 17	Apr. 7	Mar. 27		

Using this table, calculate the date of Easter for the years from 2040 through 2045. You can easily compute the day of the week for each of these dates in a given year, starting from the fact that March 21 in the year 2000 was a Tuesday. [*Note*: If the first full Moon of spring falls on a Sunday, Easter is the following Sunday.]

6.27. Prosthaphæresis can be carried out using only a table of cosines by making use of the formula

$$\cos\alpha\,\cos\beta = \frac{\cos(\alpha+\beta)+\cos(\alpha-\beta)}{2}.$$

Multiply 3562 by 4713 using this formula and a table of cosines. (It is fair to use your calculator as a table of cosines; just don't use its arithmetical capabilities.)

6.28. Do the multiplication $742518 \cdot 635942$ with pencil and paper without using a hand calculator, and time yourself. Also count the number of simple multiplications you do. Then get a calculator that will display 12 digits and do the same problem on it to see what errors you made, if any. (The author carried out the 36 multiplications and 63 additions in just under 5 minutes, but had two digits wrong in the answer as a result of incorrect carrying.)

Next, do the same problem using prosthaphæresis. (Again, you may use your hand calculator as a trigonometric table.) How much accuracy can you obtain this way? With a five-place table of cosines, using interpolation, the author found the two angles to be 50.52° and 42.05°. The initial digits of the answer would thus be those of $\big(\cos(8.47°)+\cos(92.57°)\big)/2$, yielding 47213 as the initial digits of the 12-digit number. On the other hand, using a calculator that displays 14 digits, one finds the angles to be 50.510114088363° and 42.053645425939°. That same calculator then returns all 12 digits of the correct answer as the numerical value of $\big(\cos(8.45646866242°)+\cos(92.563759514302°)\big)/2$. Compared with the time to do the problem in full the time saved was not significant.

Finally, do the problem using logarithms. Again, you may use your calculator to look up the logarithms, since a table is probably not readily available.

CHAPTER 7

Ancient Number Theory

The impossibility of getting square roots to come out even, in connection with applications of the Pythagorean theorem, may have caused mathematicians to speculate on the difference between numbers that have (rational) square roots and those that do not. We shall take this problem as the starting point for our discussion of number theory, and we shall see two responses to this problem: first, in the present chapter, to find out when indeterminate quadratic equations have rational solutions; second, in Chapter 8, to create new numbers to play the role of square roots when no rational square root exists.

1. Plimpton 322

Rational numbers satisfying a quadratic equation are at the heart of a cuneiform tablet from the period 1900–1600 BCE, number 322 of the Plimpton collection at Columbia University. The numbers on this tablet have intrigued many mathematically oriented people, leading to a wide variety of speculation as to the original purpose of the tablet. We are not offering any new conjectures as to that purpose here, only a discussion of some earlier ones.

As you can see from the photograph on p. 160, there are a few chips missing, so that some of the cuneiform numbers in the tablet will need to be restored by plausible conjecture. Notice also that the column at the right-hand edge contains the cuneiform numbers in the sequence $1, 2, 3, 4, \ldots, \ldots, 7, 8, 9, 10, 11, 12, 13, \ldots,$ $\ldots.$ Obviously, this column merely numbers the rows. The column second from the right consists of identical symbols that we shall ignore entirely. Pretending that this column is not present, if we transcribe only what we can see into our version of sexagesimal notation, denoting the chipped-off places with brackets ([...]), we get the four-column table shown below.

Before analyzing the mathematics of this table, we make one preliminary observation: Row 13 is anomalous, in that the third entry is smaller than the second entry. For the time being we shall ignore this row and see if we can figure out how to correct it. Since the long numbers in the first column must be the result of computation—it is unlikely that measurements could be carried out with such precision—we make the reasonable conjecture that the shorter numbers in the second and third columns are data. As mentioned in Chapter 6, the Mesopotamian mathematicians routinely associated with any pair of numbers (a, b) two other numbers: their average $(a + b)/2$ and their semidifference $(b - a)/2$. Let us compute these numbers for all the rows except rows 13 and 15, to see how they would have appeared to a mathematician of the time. We get the following 13 pairs of numbers, which we write in decimal notation: $(144, 25)$, $(7444, 4077)$, $(5625, 1024)$, $(15625, 2916)$, $(81, 16)$, $(400, 81)$, $(2916, 625)$, $(1024, 225)$, $(655, 114)$, $(6561, 1600)$, $(60, 15)$, $(2304, 625)$, $(2500, 729)$.

Plimpton 322. © Rare Book and Manuscript Library, Columbia University.

	Width	Diagonal	
[...] 15	1,59	2,49	1
[...] 58,14,50,6,15	56,7	3,12,1	2
[...] 41,15,33,45	1,16,41	1,50,49	3
[...] 29,32,52,16	3,31,49	5,9,1	4
48,54,1,40	1,5	1,37	5
47,6,41,40	5,19	8,1	6
43,11,56,28,26,40	38,11	59,1	7
41,33,59,3,45	13,19	20,49	8
38,33,36,36	9,1	12,49	9
35,10,2,28,27,24,26	1,22,41	2,16,1	10
33,45	45	1,15	11
29,21,54,2,15	27,59	48,49	12
27,[...],3,45	7,12,1	4,49	13
25,48,51,35,6,40	29,31	53,49	14
23,13,46,40	[...]	[...]	[...]

You will probably recognize a large number of perfect squares here. Indeed, *all* of these numbers, except for those corresponding to rows 2, 9, and 11 are perfect squares: 10 pairs of perfect squares out of thirteen! That is too unusual to be a mere coincidence. A closer examination reveals that they are squares of numbers whose only prime factors are 2, 3, and 5. Now these are precisely the prime factors of the number 60, which the Mesopotamian mathematicians used as a base. That means that the reciprocals of these numbers will have terminating sexagesimal expansions.

We should therefore keep in mind that the reciprocals of these numbers may play a role in the construction of the table.

Notice also that these ten pairs are all *relatively prime* pairs. Let us now denote the square root of the average by p and the square root of the semidifference by q. Column 2 will then be $p^2 - q^2$, and column 3 will be $p^2 + q^2$. Having identified the pairs (p, q) as important clues, we now ask *which* pairs of integers occur here and how they are arranged. The values of q, being smaller, are easily handled. The smallest q that occurs is 5 and the largest is 54, which also is the largest number less than 60 whose only prime factors are 2, 3, and 5. Thus, we could try constructing such a table for all values of q less than 60 having only those prime factors. But what about the values of p? Again, ignoring the rows for which we do not have a pair (p, q), we observe that the rows occur in decreasing order of p/q, starting from $12/5 = 2.4$ and decreasing to $50/27 = 1.85185185\ldots$. Let us then impose the following conditions on the numbers p and q:

1. The integers p and q are relatively prime.
2. The only prime factors of p and q are 2, 3, and 5.
3. $q < 60$.
4. $1.8 \leq p/q \leq 2.4$

Now, following an idea of Price (*1964*), we ask which possible (p, q) satisfy these four conditions. We find that every possible pair occurs with only five exceptions: $(2, 1)$, $(9, 5)$, $(15, 8)$, $(25, 12)$, and $(64, 27)$. There are precisely five rows in the table—rows 2, 9, 11, 13, and 15—for which we did not find a pair of perfect squares. Convincing proof that we are on the right track appears when we arrange these pairs in decreasing order of the ratio p/q. We find that $(2, 1)$ belongs in row 11, $(9, 5)$ in row 15, $(15, 8)$ in row 13, $(25, 12)$ in row 9, and $(64, 27)$ in row 2, precisely the rows for which we did not previously have a pair p, q. The evidence is overwhelming that these rows were intended to be constructed using these pairs (p, q). When we replace the entries that we can read by the corresponding numbers $p^2 - q^2$ in column 2 and $p^2 + q^2$ in column 3, we find the following:

In row 2, the entry 3,12,1 has to be replaced by 1,20,25, that is, 11521 becomes 4825. The other entry in this row, 56,7, is correct.

In row 9, the entry 9,1 needs to be replaced by 8,1, so here the writer simply inserted an extra unit character.

In row 11, the entries 45 and 75 must be replaced by 3 and 5; that is, both are divided by 15. It has been remarked that if these numbers were interpreted as $45 \cdot 60$ and $75 \cdot 60$, then in fact, one would get $p = 60$, $q = 30$, so that this row was not actually "out of step" with the others. But of course when that interpretation is made, p and q are no longer relatively prime, in contrast to all the other rows.

In row 13 the entry 7,12,1 must be replaced by 2,41; that is, 25921 becomes 161. In other words, the table entry is the square of what it should be.

The illegible entries in row 15 now become 56 and 106. The first of these is consistent with what can be read on the tablet. The second appears to be 53, half of what it should be.

The final task in determining the mathematical meaning of the tablet is to explain the numbers in the first column and interpolate the missing pieces of that column. Notice that the second and third columns in the table are labeled "width" and "diagonal." Those labels tell us that we are dealing with dimensions of a rectangle here, and that we should be looking for its length. By the Pythagorean theorem, that length is $\sqrt{(p^2+q^2)^2 - (p^2-q^2)^2} = \sqrt{4p^2q^2} = 2pq$. Even with this auxiliary number, however, it requires some ingenuity to find a formula involving p and q that fits the entries in the first column that can be read. If the numbers in the first column are interpreted as the sexagesimal representations of numbers between 0 and 1, those in rows 5 through 14—the rows that can be read—all fit the formula[1]

$$\left(\frac{p/q - q/p}{2}\right)^2.$$

Assuming this interpretation, since it works for the 10 entries we can read, we can fill in the missing digits in the first four and last rows. This involves adding two digits to the beginning of the first four rows, and it appears that there is just the right amount of room in the chipped-off place to allow this to happen. The digits that occur in the bottom row are 23,13,46,40, and they are consistent with the parts that can be read from the tablet itself.

The purpose of Plimpton 322: some conjectures. The *structure* of the tablet is no longer a mystery, unless one counts the tiny mystery of explaining the misprint in row 2, column 3. Its *purpose*, however, is not clear. What information was the table intended to convey? Was it intended to be used as people once used tables of products, square roots, and logarithms, that is, to look up a number or pair of numbers? If so, which columns contained the input and which the output? One geometric problem that can be solved by use of this tablet is that of multiplying a square by a given number, that is, given a square of side a, it is possible to find the side b of a square whose ratio to the first square is given in the first column. To do so, take a rope whose length equals the side a and divide it into the number of equal parts given in the second column, then take a second rope with the same unit of length and total length equal to the number of units in the third column and use these two lengths to form a leg and the hypotenuse of a right triangle. The other leg will then be the side of a square having the given ratio to the given square. The problem of shrinking or enlarging squares was considered in other cultures, but such an interpretation of Plimpton 322 has only the merit that there is no way of proving the tablet *wasn't* used in this way. There is no proof that the tablet was ever put to this use.

Friberg (*1981*) suggested that the purpose of the tablet was trigonometrical, that it was a table of squares of tangents. Columns 2 and 3 give one leg and the hypotenuse of 15 triangles with angles intermediate between those of the standard 45-45-90 and 30-60-90 triangles. What is very intriguing is that the table contains all possible triangles whose shapes are between these two and whose legs have lengths that are multiples of a standard unit by numbers having only 2, 3, and 5

[1] In some discussions of Plimpton 322 the claim is made that a sexagesimal 1 should be placed before each of the numbers in the first column. Although the tablet is clearly broken off on the left, it does not appear from pictures of the tablet—the author has never seen it "live"—that there were any such digits there before. Neugebauer (*1952*, p. 37) claims that parts of the initial 1 remain from line 4 on "as is clearly seen from the photograph" and that the initial 1 in line 14 is completely preserved. When that assumption is made, however, the only change in the interpretation is a trivial one: The negative sign in the formula must be changed to a positive sign, and what we are interpreting as a column of squares of tangents becomes a column of squares of secants, since $\tan^2\theta + 1 = \sec^2\theta$.

as factors. Of all right triangles, the 45–45–90 and the 30–60–90 are the two that play the most important role in all kinds of geometric applications; plastic models of them were once used as templates in mechanical drawing, and such models are still sold. It is easy to imagine that a larger selection of triangle shapes might have been useful in the past, before modern drafting instruments and computer-aided design. Using this table, one could build 15 model triangles with angles varying in increments of approximately 1°. One can imagine such models being built and the engineer of 4000 years ago reaching for a "number 7 triangle" when a slope of $574/675 = .8504$ was needed. However, this scenario still lacks plausibility. Even if we assume that the engineer kept the tablet around as a reference when it was necessary to know the slope, the tablet stores the *square* of the slope in column 1. It is difficult to imagine any engineering application for that number.

Having failed to find a geometric explanation of the tablet, we now explore possible associations of the tablet with Diophantine equations, that is, equations whose solutions are to be rational numbers, in this case numbers whose numerators and denominators are products of only the first three prime numbers. The left-hand column contains numbers that are perfect squares and remain perfect squares when 1 is added to them. In other words, it gives u^2 for solutions to the Diophantine equation $u^2+1 = v^2$. This equation was much studied in other cultures, as we shall see below. If the purpose of the table were to generate solutions of this equation, there would of course be no reason to give v^2, since it could be obtained by placing a 1 before the entry in the first column. The use of the table would then be as follows: Square the entry in column 3, square the entry in column 2, then divide each by the difference of these squares. The results of these two divisions are v^2 and u^2 respectively. In particular, u^2 is in column 1. The numbers p and q that generate the two columns can be arbitrary, but in order to get a sexagesimally terminating entry in the first column, the difference $(p^2+q^2)^2 - (p^2-q^2)^2 = 4p^2q^2$ should have only 2, 3, and 5 as prime factors, and hence p and q also should have only these factors. Against this interpretation there lies the objection that p and q are concealed from the casual reader of the tablet. If the purpose of the tablet was to show how to generate u and v or u^2 and $v^2 = u^2 + 1$, some explanation should have been given as to how columns 2 and 3 were generated. But of course, the possibility exists that such an explanation was present originally. After all, it is apparent that the tablet is broken on the left-hand side. Perhaps it originally contained more columns of figures that might shed light on the entire tablet if we only had them. Here we enter upon immense possibilities, since the "vanished" portion of the tablet could have contained a huge variety of entries. To bring this open-ended discussion to a close, we look at what some experts in the area have to say.

In work that was apparently never published (see Buck, *1980*, p. 344), D. L. Voils pointed out that tablets amounting to "teacher's manuals" have been found in which the following problem is set: *Find a number that yields a given number when its reciprocal is subtracted.* In modern terms this problem requires solving the equation

$$x - \frac{1}{x} = d,$$

where d is the given number. Obviously, if you were a teacher setting such a problem for a student, you would want the solution x to be such that both x and $1/x$ have terminating sexagesimal digits. So, if the solution is to be $x = p/q$, we already see why we need both p and q to be products of 2, 3, and 5. This problem amounts to the quadratic equation $x^2 - dx - 1 = 0$, and its unique positive solution is $x = d/2 + \sqrt{1 + (d/2)^2}$. Column 1 of the tablet, which contains $(d/2)^2$ then appears as part of the solution process. It is necessary to take its square root and also the square root $\sqrt{1 + d^2/4}$ in order to find the solution $x = p/q$. This explanation seems to fit very well with the tablet. One could assume that the first column gives values of d that a teacher could use to set such a problem with the assurance that the pupil would get terminating sexagesimal expansions for both x and $1/x$. On the other hand, it does not fully explain why the tablet gives the numbers $p^2 - q^2$ and $p^2 + q^2$, rather than simply p and q, in subsequent columns. Doing our best for this theory, we note that columns 2 and 3 contain respectively the numerators of $x - 1/x$ and $x + 1/x$, and that their common denominator is the square root of the difference of the squares of these two numerators. Against that explanation is the fact that the Mesopotamians did not work with common fractions. The concepts of numerator and denominator to them would have been the concepts of dividend and divisor, and the final sexagesimal quotient would not display these numbers. The recipe for getting from columns 2 and 3 to column 1 would be first to square each of these columns, then find the reciprocal of the difference of the squares as a sexagesimal expansion, and finally, multiply the last result by the square in column 2.

In the course of a plea that historians look at Mesopotamian mathematics in its own terms rather than simply in relation to what came after, Robson (*2001*) examined several theories about the purpose of the tablet and gave some imaginative scenarios as to what may be in the lost portion of the tablet. Her conclusion, the only one justified by the present state of knowledge is that "the Mystery of the Cuneiform Tablet has not yet been fully solved."[2]

And we have not claimed to solve it here. Plimpton 322 is a fascinating object of contemplation and serves as a *possible* example of an early interest in what we now call quadratic Diophantine equations. Without assuming that there is some continuous history between Plimpton 322 and modern number theory, we can still take quadratic Diophantine equations as a convenient starting point for discussing the history of number theory.

2. Ancient Greek number theory

Our knowledge of Pythagorean number theory is based on several sources, of which two important ones are Books 7–9 of Euclid's *Elements* and a treatise on arithmetic by the neo-Pythagorean Nicomachus of Gerasa, who lived about 100 CE. Just as the *Sun Zi Suan Jing* preserves more of ancient Chinese arithmetic than the earlier *Jiu Zhang Suanshu*, it happens that the treatise of Nicomachus preserves more of Pythagorean lore than the earlier work of Euclid. For that reason, we discuss Nicomachus first.

The Pythagoreans knew how to find the greatest common divisor of two numbers. A very efficient procedure for doing so is described in Chapter 13 of Book 1 of Nicomachus' *Arithmetica* and in Proposition 2 of Book 7 of Euclid's *Elements*. This procedure, now known as the *Euclidean algorithm*, is what the Chinese called the *mutual-subtraction procedure*. Nicomachus applies it only to integers, any two

[2] In a posting at a mathematics history website, Robson noted that reciprocal pairs and cut-and-paste geometry seem to be the most plausible motives for the tablet.

of which naturally have 1 as a common divisor. Euclid, on the other hand, does not confine it to integers, but states the procedure for "magnitudes," which may lack a common measure. It is significant that the procedure terminates if and only if there is a common measure, and Euclid makes use of that fact in discussing incommensurables. The algorithm was certainly invented long before the time of Euclid, however. Zverkina (*2000*) believes that this procedure could not have arisen intuitively, but must have come about as the result of solving specific problems, most likely the problem of reducing ratios by canceling a common divisor. It is used for that purpose in Chinese mathematics. What follows is a description of the general procedure.

For definiteness, we shall imagine that the two quantities whose greatest common *measure* is to be found are two lengths, say a and b. Suppose that a is longer than b. (If the two are equal, their common value is also their greatest common divisor.) The general procedure is to keep subtracting the smaller quantity from the larger until the remainder is equal to the smaller quantity or smaller than it. It is not difficult to show that the smaller quantity and the remainder have the same common measures as the smaller quantity and the larger. Hence one can start over with the smaller quantity and the remainder, which is no more than half of the larger quantity. Either this process terminates with an equal pair, or it continues and the pairs become arbitrarily small.

An example will make the procedure clear. Let us find the greatest common measure (divisor) of 26173996849 and 180569389. A common measure does exist: the integer 1. Since the repeated subtraction process amounts to division with remainder, we do it this way: $26173996849 \div 180569389$ is 144 with a remainder of 172004833. We then divide 180569389 by 172004833, getting a quotient of 1 and a remainder of 8564556. Next we divide 172004833 by 8564556, getting a quotient of 20 and a remainder of 713713. We then divide 8564556 by 713713 and get a quotient of 12 with no remainder, so that the greatest common divisor is 713713.

This computation can be arranged as follows:

$$
\begin{array}{rrrrr}
 & 12 & 20 & 1 & 144\\
713713) & \overline{8564556}) & \overline{172004833}) & \overline{180569389}) & \overline{26173996849}\\
 & \underline{8564556} & \underline{171291120} & \underline{172004833} & \underline{26001992016}\\
 & 0 & 713713 & 8564556 & 172004833
\end{array}
$$

2.1. The *Arithmetica* of Nicomachus. In his first book Nicomachus makes the elementary distinction between odd and even numbers. Having made this distinction, he proceeds to refine it, distinguishing between even numbers divisible by 4 (evenly even) and those that are not (doubles of odd numbers). He goes on to classify odd numbers in a similar way, coming thereby to the concept of prime and composite numbers. Nicomachus also introduces what we now call pairs of *relatively prime numbers*. These are pairs of numbers that have no common prime divisor and hence no common divisor except 1. The notion of a relational property was difficult for Greek philosophers, and Nicomachus expresses the concept of relatively prime numbers in a confused manner, referring to three species of odd numbers: the prime and incomposite, the secondary and composite, and "the variety which, in itself is secondary and composite, but relatively is prime and incomposite." This way of writing seems to imply that there are three kinds of integers, prime and incomposite, secondary and composite, and a third kind midway between the other two. It also seems to imply that one can look at an individual integer and classify it into exactly one of these three classes. Such is not the case, however. The property of primeness is a property of a number alone. The property of being relatively prime is a property of a pair of numbers. On the other hand, the property of being relatively prime *to a given number* is a property of a number alone. Nicomachus explains the property in a rather wordy fashion in Chapter 13 of Book 1, where he gives a method of identifying prime numbers that has become famous as the *sieve of Eratosthenes.*

Nicomachus attributes this method to Eratosthenes. To use it, start with a list of all the odd numbers from 3 on, that is,

$$3,\ 5,\ 7,\ 9,\ 11,\ 13,\ 15,\ 17,\ 19,\ 21,\ 23,\ 25,\ 27,\ 29,\ 31,\ 33,\ 35,\ 37, \ldots .$$

From this list, remove every third number after 3, that is, remove 9, 15, 21, 27, 33,... . These numbers are multiples of 3 and hence not prime. The reduced list is then

$$3,\ 5,\ 7,\ 11,\ 13,\ 17,\ 19,\ 23,\ 25,\ 29,\ 31,\ 35,\ 37,\ 41,\ 43,\ 47,\ 49, \ldots .$$

From this list, remove all multiples of 5 larger than 5. The first non-prime in the new list is $49 = 7 \cdot 7$. In this way, you can generate in short order a complete list of primes up to the square of the first prime whose multiples were not removed. Thus, after removing the multiples of 7, we have the list

$$3,\ 5,\ 7,\ 11,\ 13,\ 17,\ 19,\ 23,\ 29,\ 31,\ 37,\ 41,\ 43,\ 47,\ 53,\ 59,\ 61 \ldots .$$

The first non-prime in this list would be $11 \cdot 11 = 121$.

Nicomachus' point of view on this sieve was different from ours. Where we think of the *factors* of, say 60, as being 2, 2, 3, and 5, Nicomachus thought of the quotients by these numbers and products of them as the *parts* of a number. Thus, in his language 60 has the parts 30 (half of 60), 20 (one-third of 60), 15 (one-fourth of 60), 12 (one-fifth of 60), 10 (one-sixth of 60), 6 (one-tenth) of 60, 5 (one-twelfth of 60), 4 (one-fifteenth of 60), 3 (one-twentieth of 60), 2 (one-thirtieth of 60) and 1 (one-sixtieth of 60). If these parts are added, the sum is 108, much larger than 60. Nicomachus called such a number *superabundant* and compared it to an animal having too many limbs. On the other hand, 14 is larger than the sum of its parts. Indeed, it has only the parts 7, 2, and 1, which total 10. Nicomachus called 14 a *deficient number* and compared it to an animal with missing limbs like the one-eyed Cyclops of the *Odyssey.* A number that is exactly equal to the sum of its parts, such as $6 = 1 + 2 + 3$, he called a *perfect number.* He gave a method of finding perfect numbers, which remains to this day the only way known to generate such numbers, although it has not been proved that there are no other such numbers. This procedure is also stated by Euclid: *If the sum of the numbers* $1, 2, 4, \ldots, 2^{n-1}$ *is prime, then this sum multiplied by the last term will be perfect.* The modern statement of this fact is given in the exercises below. To see the recipe at work, start with 1, then double and add: $1 + 2 = 3$. Since 3 is prime, multiply it by the last term, that is, 2. The result is 6, a perfect number. Continuing, $1 + 2 + 4 = 7$, which is prime. Multiplying 7 by 4 yields 28, the next perfect number. Then, $1 + 2 + 4 + 8 + 16 = 31$, which is prime. Hence $31 \cdot 16 = 496$ is a perfect number. The next such number is $8128 = 64(1 + 2 + 4 + 8 + 16 + 32 + 64)$. In this way, Nicomachus was able to generate the first four perfect numbers. He seems to hint at a conjecture, but draws back from stating it explicitly:

> When these have been discovered, 6 among the units and 28 in the tens, you must do the same to fashion the next...the result is 496, in the hundreds; and then comes 8,128 in the thousands, and so on, as far as it is convenient for one to follow [D'ooge, *1926*, p. 211].[3]

This quotation seems to imply that Nicomachus expected to find one perfect number N_k having k digits. Actually, the fifth perfect number is 33,550,336, so we have jumped from four digits to eight here. The sixth is 8,589,869,056 (10 digits) and the seventh is 137,438,691,328 (12 digits), so that there is no regularity about the distribution of perfect numbers. Thus, Nicomachus was wise to refrain from making conjectures too explicitly. According to Dickson (*1919*, p. 8), later mathematicians, including Cardano, were less restrained, and this incorrect conjecture has been stated more than once.

For a topic that is devoid of applications, perfect numbers have attracted a great deal of attention from mathematicians. Dickson (*1919*) lists well over 100 mathematical papers devoted to this topic over the past few centuries. From the point of view of pure number theory, the main questions about them are the following: (1) Is there an odd perfect number?[4] (2) Are all even perfect numbers given by the procedure described by Nicomachus?[5] (3) Which numbers of the form $2^n - 1$ are prime? These are called *Mersenne primes*, after Marin Mersenne (1588–1648), who, according to Dickson (*1919*, pp. 12–13), first noted their importance, precisely in connection with perfect numbers. Obviously, n must itself be prime if $2^n - 1$ is to be prime, but this condition is not sufficient, since $2^{11} - 1 = 23 \cdot 89$. The set of known prime numbers is surprisingly small, considering that there are infinitely many to choose from, and the new ones being found tend to be Mersenne primes, mostly because that is where people are looking for them. The largest currently known prime (as of June 2004) is $2^{24036583} - 1$, only the forty-first Mersenne prime known.[6] It was found on May 15, 2004 by the GIMPS (Great Internet Mersenne Prime Search) project, which links over 200,000 computers via the Internet and runs prime-searching software in the background of each while their owners are busy with their own work. This prime has 7,235,733 decimal digits. The fortieth Mersenne prime, $2^{20996011} - 1$, was found on November 17, 2003; it has 6,320,430 decimal digits. In contrast, the largest known non-Mersenne prime is $3 \cdot 2^{303093} + 1$, found by Jeff Young in 1998.[7] It is rather tiny in comparison with the last few Mersenne primes discovered, having "only" 91,241 decimal digits.

Beginning in Chapter 6 of Book 2, Nicomachus studies figurate numbers: polygonal numbers through heptagonal numbers, and then polyhedral numbers. These numbers are connected with geometry, with an identification of the number 1 with a geometric point. To motivate this discussion Nicomachus speculated that the

[3] D'ooge illustrates the procedure in a footnote, but states erroneously that 8191 is not a prime.

[4] The answer is unknown at present.

[5] The answer is yes. The result is amazingly easy to prove, but no one seems to have noticed it until a posthumous paper of Leonhard Euler gave a proof. Victor-Amédée Lebesgue (1791–1875) published a short proof in 1844.

[6] The reader will correctly infer from previous footnotes that exactly 41 perfect numbers are now known.

[7] See his article "Large primes and Fermat factors" in *Mathematics of Computation*, **67** (1998), 1735–1738, which gives a method of finding probable primes of the form $k \cdot 2^n + 1$.

simplest way to denote any integer would be repeating a symbol for 1 an appropriate number of times. Thus, he said, the number 5 could be denoted $\alpha\,\alpha\,\alpha\,\alpha\,\alpha$. This train of thought, if followed consistently, would lead back to a notation even more primitive than the hieroglyphic notation for numbers, since it would use only the symbol for units and discard the symbols for higher powers of 10. The Egyptians had gone beyond this principle in their hieratic notation, and the standard Greek notation was essentially a translation of the hieratic into the Greek alphabet. You can easily see where this speculation leads. The outcome is shown in Fig. 1, which illustrates triangular, square, pentagonal, and hexagonal numbers but using dots instead of the letter α. Observe that the figures are *not* associated with regular polygons except in the case of triangles and squares. The geometry alone makes it clear that a square number is the sum of the corresponding triangular number and its predecessor. Similarly, a pentagonal number is the sum of the corresponding square number and the preceding triangular number, a hexagonal number is the sum of the corresponding pentagonal number and the preceding triangular number, and so forth. This is the point at which modern mathematics parts company with Nicomachus, Proclus, and other philosophers who push analogies further than the facts will allow. As Nicomachus states at the beginning of Chapter 7:

> The point, then, is the beginning of dimension, but not itself a dimension, and likewise the beginning of a line, but not itself a line; the line is the beginning of surface, but not surface; and the beginning of the two-dimensional, but not itself extended in two dimensions... Exactly the same in numbers, unity is the beginning of all number that advances unit by unit in one direction; linear number is the beginning of plane number, which spreads out like a plane in one more dimension. [D'ooge, *1926*, p. 239]

This mystical mathematics was transmitted to Medieval Europe by Boethius. It is the same kind of analogical thinking found in Plato's *Timaeus*, where it is imagined that atoms of fire are tetrahedra, atoms of earth are cubes, and so forth. Since the Middle Ages, this topic has been of less interest to mathematicians. The phrase *of less interest*—rather than *of no interest*—is used advisedly here: There are a few theorems about figurate numbers in modern number theory, and they have some connections with analysis as well. For example, a formula of Euler asserts that

$$\prod_{k=1}^{\infty}(1-x^k) = \sum_{n=-\infty}^{\infty}(-1)^n x^{n(3n-1)/2}.$$

Here the exponents on the right-hand side range over the pentagonal numbers for n positive. By making this formula the definition of the nth pentagonal number for negative n, we thereby gain an interesting formula that can be stated in terms of figurate numbers. Carl Gustav Jacobi (1804–1851) was pleased to offer a proof of this theorem as evidence of the usefulness of elliptic function theory. Even today, these numbers crop up in occasional articles in graph theory and elsewhere.

2.2. Euclid's number theory. Euclid devotes his three books on number theory to divisibility theory, spending most of the time on proportions among integers and on prime and composite numbers. Only at the end of Book 9 does he prove a theorem of a different sort, giving the method of searching for perfect numbers described above. It is interesting that Euclid does not mention figurate numbers. Although

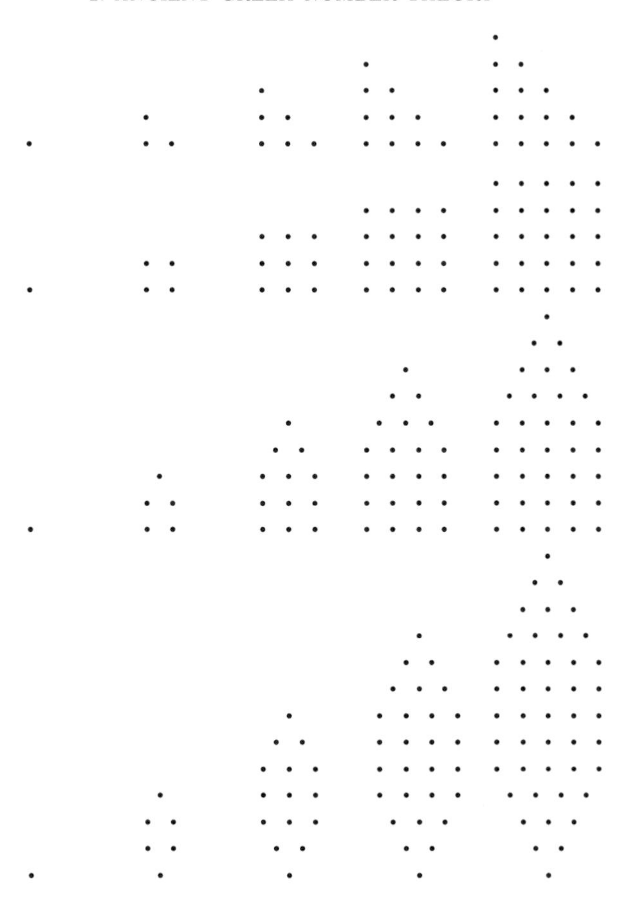

FIGURE 1. Figurate numbers. Top row: triangular numbers $T_n = n(n+1)/2$. Second row: square numbers $S_n = n^2$. Third row: pentagonal numbers $P_n = n(3n-1)/2$. Bottom row: hexagonal numbers $H_n = n(2n-1)$.

the Pythagorean and Platonic sources of Euclid's treatise are obvious, Euclid appears to the modern eye to be much more a mathematician than Pythagoras or Plato, much less addicted to flights of fanciful speculation on the nature of the universe. In fact, he never mentions the universe at all and suggests no practical applications of the theorems in his *Elements*.

Book 7 develops proportion for positive integers as part of a general discussion of how to reduce a ratio to lowest terms. The notion of relatively prime numbers is introduced, and the elementary theory of divisibility is developed as far as finding least common multiples and greatest common factors. Book 8 resumes the subject of proportion and extends it to squares and cubes of integers, including the interesting theorem that the mean proportional of two square integers is an integer (Proposition 11), and between any two cubes there are two such mean proportionals (Proposition 12): for example, $25 : 40 :: 40 : 64$, and $27 : 45 :: 45 : 75 :: 75 : 125$. Book 9 continues this topic; it also contains the famous theorem that there are infinitely many primes (Proposition 20) and ends by giving the only known method of constructing perfect numbers (Proposition 36), quoted above. No perfect number has yet been found that is *not* generated by this procedure, although no proof exists that all perfect numbers are of this form. Any exception would have to be an odd number, since it *is* known (see Problem 7.8) that all even perfect numbers are of this form.

From the modern point of view, Euclid's number theory is missing an explicit statement of the *fundamental theorem of arithmetic.* This theorem, which asserts that every positive integer can be written in only one way as a product of prime numbers, can easily be deduced from Book 7, Proposition 24: *If two numbers are relatively prime to a third, their product is also relatively prime to it.* However, modern historians (Knorr, *1976*) have pointed out that Euclid doesn't actually prove the fundamental theorem.

2.3. The *Arithmetica* of Diophantus. Two works of Diophantus have survived in part, a treatise on polygonal numbers and the work for which he is best known, the *Arithmetica.* Like many other ancient works, these two works of Diophantus survived because of the efforts of a ninth-century Byzantine mathematician named Leon, who organized a major effort to copy and preserve these works. There is little record of the influence the works of Diophantus may have exerted before this time.

According to the introduction to the *Arithmetica*, this work consisted originally of 13 books, but until recently only six were known to have survived; it was assumed that these were the first six books, on which Hypatia wrote a commentary. However, more books were recently found in an Arabic manuscript that the experts say is a translation made very early—probably in the ninth century. Sesiano (*1982*) stated that these books are in fact the books numbered 4 to 7, and that the books previously numbered 4 to 6 must come after them.

Diophantus begins with a small number of determinate problems that illustrate how to think algebraically, in terms of expressions involving a variable. Since these problems belong properly to algebra, they are discussed in Chapter 14. Indeterminate problems, which are number theory because the solutions are required to be rational numbers (the only kind recognized by Diophantus), begin in Book 2. A famous example of this type is Problem 8 of Book 2, *to separate a given square number into two squares.* Diophantus illustrates this problem using the number 16 as an example. His method of solving this problem is to express the two numbers in terms of a single unknown, which we shall denote ς, in such a way that one of the conditions is satisfied automatically. Thus, letting one of the two squares be ς^2, which Diophantus wrote as Δ^υ (as explained in Chapter 14), he noted that the other will automatically be $16-\varsigma^2$. To get a determinate equation for ς, he assumes that the other number to be squared is 4 less than an unspecified multiple of ς. The number 4 is chosen because it is the square root of 16. In our terms, it leads to a quadratic equation one of whose roots is zero, so that the other root can be found by solving a linear equation. As we would write it, assuming that $16-\varsigma^2 = (k\varsigma - 4)^2$, we find that $(k^2+1)\varsigma^2 = 8k\varsigma$, and—cancelling ς, since Diophantus does not operate with 0—we get $\varsigma = 8k/(k^2+1)$. This formula generates a whole infinite family of solutions of the equation that we would call $x^2 + y^2 = 16$ via the identity

$$\left(\frac{8k}{k^2+1}\right)^2 + \left(\frac{4(k^2-1)}{k^2+1}\right)^2 = 16.$$

You may be asking why it was necessary to use a square number (16) here. Why not separate any positive rational number, say 5, into a sum of two squares? If you look carefully at the solution, you will see that Diophantus had to make the constant term drop out of the quadratic equation, and that could only be done by introducing the square root of the given number.

Diophantus' procedure is slightly less general than what we have just shown, although his illustrations show that he knows the general procedure and could generate other solutions. In his illustration he assumes that the other square is $(2\varsigma - 4)^2$. Since this number must be $16 - \varsigma^2$, he finds that $4\varsigma^2 - 16\varsigma + 16 = 16 - \varsigma^2$, so that $\varsigma = \frac{16}{5}$. It is clear that this procedure can be applied very generally, showing an infinite number of ways of dividing a given square into two other squares.

At first sight it appears that number theory really is not involved in this problem, that it is a matter of pure algebra. However, the topic of the problem naturally leads to other questions that definitely do involve number theory, that is, the theory of divisibility of integers. The most obvious one is the problem of finding *all possible* representations of a positive rational number as the sum of the squares of two rational numbers. One could then generalize and ask how many ways a given rational number can be represented as the sum of the cubes or fourth powers, and so forth, of two rational numbers. Those of a more Pythagorean bent might ask how many ways a number can be represented as a sum of triangular, pentagonal, or hexagonal numbers. In fact, all of these questions have been asked, starting in the seventeenth century.

The problem just solved achieved lasting fame when Fermat, who was studying the *Arithmetica*, remarked that the analogous problem for cubes and higher powers had no solutions; that is, one cannot find positive integers x, y, and z satisfying $x^3 + y^3 = z^3$ or $x^4 + y^4 = z^4$, or, in general $x^n + y^n = z^n$ with $n > 2$. Fermat stated that he had found a proof of this fact, but unfortunately did not have room to write it in the margin of the book. Fermat never published any general proof of this fact, although the special case $n = 4$ is a consequence of a method of proof developed by Fermat, known as the method of infinite descent. The problem became generally known after 1670, when Fermat's son published an edition of Diophantus' work along with Fermat's notes. It was a tantalizing problem because of its comprehensibility. Anyone with a high-school education in mathematics can understand the statement of the problem, and probably the majority of mathematicians dreamed of solving it when they were young. Despite the efforts of hundreds of amateurs and prizes offered for the solution, no correct proof was found for more than 350 years. On June 23, 1993, the British mathematician Andrew Wiles announced at a conference held at Cambridge University that he had succeeded in proving a certain conjecture in algebraic geometry known as the Shimura–Taniyama conjecture, from which Fermat's conjecture is known to follow. This was the first claim of a proof by a reputable mathematician using a technique that is known to be feasible, and the result was tentatively endorsed by other mathematicians of high reputation. After several months of checking, some doubts arose. Wiles had claimed in his announcement that certain techniques involving what are called Euler systems could be extended in a particular way, and this extension proved to be doubtful. In collaboration with another British mathematician, Richard Taylor, Wiles eventually found an alternative approach that simplified the proof considerably, and there is now no doubt among the experts in number theory that the problem has been solved.

To give another illustration of the same method, we consider the problem following the one just discussed, that is, Problem 9 of Book II: *to separate a given number that is the sum of two squares into two other squares.* (That is, given one representation of a number as a sum of two squares, find a new representation of the same type.) Diophantus shows how to do this using the example $13 = 2^2 + 3^2$. He lets one of the two squares be $(\varsigma + 2)^2$ and the other $(2\varsigma - 3)^2$, resulting in the equation $5\varsigma^2 - 8\varsigma = 0$. Thus, $\varsigma = \frac{8}{5}$, and indeed $\left(\frac{18}{5}\right)^2 + \left(\frac{1}{5}\right)^2 = 13$. It is easy to see here that Diophantus is deliberately choosing a form for the solution that will cause the constant term to drop out. This amounts to a general method, used throughout the first two books, and based on the proportion

$$(a + Y) : X = X : (a - Y)$$

for solving the equation $X^2 + Y^2 = a^2$.

The method Diophantus used to solve such problems in his first two books was conjectured by Maximus Planudes (1255–1305) and has recently been explained in simple language by Christianidis (*1998*).

Some of Diophantus' indeterminate problems reach a high degree of complexity. For example, Problem 19 of Book 3 asks for four numbers such that if any of the numbers is added to or subtracted from the square of the sum of the numbers, the result is a square number. Diophantus gives the solutions as

$$\frac{17,136,600}{163,021,824},\ \frac{12,675,000}{163,021,824},\ \frac{15,615,600}{163,021,824},\ \frac{8,517,600}{163,021,824}.$$

3. China

Although figurate numbers were not a topic of interest to early Chinese mathematicians, there was always in China a great interest in the use of numbers for divination. According to Li and Du (*1987*, pp. 95–97), the magic square

4	9	2
3	5	7
8	1	6

appears in the treatise *Shushu Jiyi* (*Memoir on Some Traditions of the Mathematical Art*) by the sixth-century mathematician Zhen Luan. In this figure each row, column, and diagonal totals 15. In the early tenth century, during the Song Dynasty, a connection was made between this magic square and a figure called the *Luo-chu-shu* (*book that came out of the River Lo*) found in the famous classic work *I Ching*, which was mentioned in connection with divination in Chapter 1. The *I Ching* states that a tortoise crawled out of the River Lo and delivered to the Emperor Yu the diagram in Fig. 2. Because of this connection, the diagram came to be called the *Luo-shu* (*Luo book*). Notice that the purely numerical aspects of the magic square are enhanced by representing the even (female, ying) numbers as solid disks and the odd (male, yang) numbers as open circles. Like so much of number theory, the theory of magic squares has continued to attract attention from specialists, all the while remaining essentially devoid of any applications. In this particular case, the interest has come from specialists in combinatorics, for whom magic squares and Latin squares form a topic of continuing research.

Another example of the use of numbers for divination comes from the last problem (Problem 36 of Chapter 3) of the *Sun Zi Suan Jing*. The data for the problem are very simple. A woman, aged 29, is pregnant. The period of human

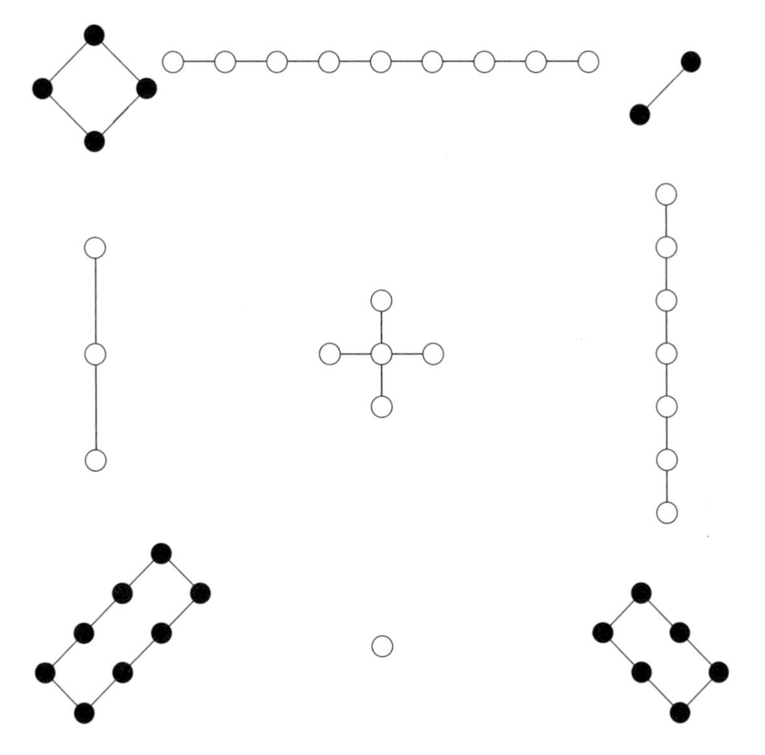

FIGURE 2. The *Luo-shu*.

gestation is nine months. The problem is to determine the gender of the unborn child. In what is apparently an echo of the *I Ching* method of divination, the author begins with 49 (the number of yarrow stalks remaining after the first one has been laid down to begin the divination process). He then says to add the number of months of gestation, then subtract the woman's age. From the remainder (difference?) one is then to subtract succesively 1 (heaven), 2 (Earth), 3 (man), 4 (seasons), 5 (phases), 6 (musical tones), 7 (stars in the Dipper), 8 (wind directions), and 9 (provinces of China under the Emperor Yu) and then use the final difference to determine the gender.[8]

The nature of divisibility for integers is also studied in Chinese treatises, in particular in the *Sun Zi Suan Jing*, which contains the essence of the result still known today as the Chinese remainder theorem. It was mentioned above that in general the *Sun Zi Suan Jing* is more elementary than the earlier *Jiu Zhang Suanshu*, but this bit of number theory is introduced for the first time in the *Sun Zi Suan Jing*. The problem asks for a number that leaves a remainder of 2 when divided by 3, a remainder of 3 when divided by 5, and a remainder of 2 when divided by 7. As in the case of Diophantus, the problem appears to be algebra, but it also involves the notion of divisibility with specified remainders. The assertion

[8] Although no explanation is given in the translation by Lam and Ang (*1992*, p. 182), and no value is given for the final difference in this problem, the child is said by the author to be male. Perhaps the subtracting of successive integers was meant to continue only until the number left was smaller than the next number to be subtracted. In the present case, that number would be 1, resulting after 7 was subtracted. This interpretation seems to make sense; otherwise, the result of the procedure would be determined entirely by the parity of the woman's age.

that any number of such congruences can be solved simultaneously if the divisors are all pairwise relatively prime is the content of what we know now as the Chinese remainder theorem. According to Dickson, (*1920*, p. 57) this name arose when the mathematically astute British missionary Alexander Wylie (1815–1887) wrote an article on it in the English-language newspaper *North China Herald* in 1852. By that time the result was already known in Europe, having been discovered by Gauss and published in his *Disquisitiones arithmeticæ* (Art. 36) in 1836.

Sun Zi's answer to this problem shows that he knew a general method of proceeding. He says, "Since the remainder on division by 3 is 2, take 140. The remainder on division by 5 is 3, so take 63. The remainder on division by 7 is 2, so take 30. Add these numbers, getting 233. From this subtract 210, getting the answer as 23."

It is just possible that the reader may not discern the underlying reasoning here, and so a bit of explanation may help. Sun Zi reasons that all multiples of $35 = 7 \cdot 5$ will leave a remainder of zero when divided by 5 or 7. He therefore took a multiple of 35, $140 = 35 \cdot 4$, that leaves a remainder of 2 when divided by 3. One may well ask why he didn't simply take 35 itself, since it also leaves a remainder of 2 when divided by 3. The seemingly clumsy use of 140 instead reveals still more of Sun Zi's thought processes. He must have looked first for a multiple of 35 that leaves a remainder of 1 when divided by 3, found it to be 70, then multiplied it by 2.[9] Similarly, 63 is the smallest multiple of $21 = 7 \cdot 3$ that leaves a remainder of 3 when divided by 5, and 30 is the smallest multiple of $15 = 3 \cdot 5$ that leaves a remainder of 2 when divided by 7. Adding all three numbers, we get the number 233, which leaves the desired remainders on all three divisions. Sun Zi also knew that any multiple of $105 = 3 \cdot 5 \cdot 7$ could be added or subtracted without affecting any of the remainders. Hence subtracting 210 produced the smallest possible solution. It is obvious from this explanation that Sun Zi's method is perfectly general and can be used to find all possible solutions to such problems. What is concealed in his exposition is the general hypothesis that the divisors must be pairwise relatively prime. Sun Zi does not discuss this concept, but obviously he must have encountered cases where such problems cannot be solved. Almost certainly he would have traced the difficulty back to the existence of common factors among the divisors.

The importance of this kind of problem to the Chinese was not merely theoretical. Given that the ratio of a month—the time between two successive full moons—to a year is 19:235, questions involving calendars lead very often to finding numbers that leave a given remainder when divided by 19 and another given remainder when divided by 235. For example, suppose we know that the moon was full on June 1, 1996. What is the next year on which it will be full on June 4? (See Problem 7.10.)

The secret of solving problems of this sort is the Euclidean algorithm. This algorithm was known in China from the first century CE (see Shen, *1988*) and used to solve a variety of problems, including the conversion of a long decimal expansion into a common fraction approximation with a small denominator (see Problem 7.12).

---

[9] We can assume that Sun Zi found 70 by trial and error. The appropriate multiple would not be so easy if the divisors involved had seven digits each. A method for handling such harder cases was discovered by the Hindus and is discussed below.

4. India

The *Sulva Sutras* contain rules for finding Pythagorean triples of integers, such as $(3, 4, 5)$, $(5, 12, 13)$, $(8, 15, 17)$, and $(12, 35, 37)$. It is not certain what practical use these arithmetic rules had. They may have been motivated by religious ritual. A Hindu home was required to have three fires burning at three different altars. The three altars were to be of different shapes, but all three were to have the same area. These conditions led to certain "Diophantine" problems, a particular case of which is the generation of Pythagorean triples, so as to make one square integer equal to the sum of two others.

One class of mathematical problems associated with altar building involves an altar of prescribed area in layers. In one problem from the *Bodhayana Sutra* the altar is to have five layers of bricks, each layer containing 21 bricks. Now one cannot simply divide a pile of 105 identical bricks into five layers and pile them up. Such a structure would not be stable. It is necessary to stagger the edges of the bricks. So that the outside of the altar will not be jagged, it is necessary to have at least two different sizes of bricks. The problem is to decide how many different sizes of bricks will be needed and how to arrange them. Assuming an area of one square unit (actually, the unit is 1 square *vyayam*, about 7 square meters), the author suggests using three kinds of square bricks, of areas $\frac{1}{36}$, $\frac{1}{16}$, and $\frac{1}{9}$ square unit. The first, third, and fifth layers are to have nine of the first kind and 12 of the second. The second and fourth layers get 16 of the first kind and five of the third.

4.1. Varahamihira's mystical square. According to Hayashi (*1987*), around the year 550 the mathematician Varahamihira wrote the *Brhatsamhita*, a large book devoted mainly to divination. However, Chapter 76 also discusses the mixing of perfumes from 16 substances, grouped in fours and mixed according to proportions given by the rows of the following square array:

2	3	5	8
5	8	2	3
4	1	7	6
7	6	4	1

Thus, the mysticism surrounding these squares penetrated even practical aspects of life. Hayashi notes that the Sanskrit word for the square itself, *kacchaputa*, means a box with compartments, but originally meant a tortoise shell. The resemblance to the *Luo Shu* is probably a coincidence.

4.2. Aryabhata I. In verses 32 and 33 of the *Aryabhatiya* we find a method of solving problems related to the problem of Sun Zi that leads to the Chinese remainder theorem. However, the context of the method and the description leave much to be desired in terms of clarity. It would have helped if Aryabhata had included specific examples. Such examples were provided by later commentators, and the process was described more clearly by Brahmagupta.

4.3. Brahmagupta. A century after Aryabhata, Brahmagupta called the method the *kuttaka* (pulverizer). We shall exclude certain complications in Brahmagupta's presentation and present the method as simply as possible. The *kuttaka* provides the following visual implementation of an algorithm for solving the equation $ax = by + c$, with $b > a > 0$ and a and b relatively prime. As an example, we shall find all solutions of the equation $4x = 23y + 5$. First, we carry out the Euclidean algorithm until 1 appears as a remainder:

$$\begin{aligned} 23 &= 5 \cdot 4 + 3\,, \\ 4 &= 1 \cdot 3 + 1\,. \end{aligned}$$

We then write the quotients (5 and 1 in this case) from the Euclidean algorithm in a column, and beneath them we write the additive term c if the number of quotients is even (in this case, two), otherwise $-c$. At the bottom of the column we write a 0. This zero is inserted so that the same transformation rule applies at the beginning as in all other steps of the algorithm. We then reduce the number of rows successively by operating on the bottom three rows at each stage. The second-from-last row is replaced by its product with the next-to-last row plus the last row; the next-to-last row is simply copied, and the last row is discarded. Thus to solve this system the *kuttaka* method amounts to the transformations

$$\begin{matrix} 5 \\ 1 \\ 5 \\ 0 \end{matrix} \longrightarrow \begin{matrix} 5 \\ 5 \\ 5 \end{matrix} \longrightarrow \begin{matrix} 30 \\ 5 \end{matrix}\,.$$

This column now gives x and y, and indeed, $4{\cdot}30 = 23{\cdot}5{+}5$. Diophantus showed how to find a particular solution of such a congruence. Brahmagupta, however, found *all* the solutions. He took the solutions x and y obtained by the *kuttaka* method, which were generally quite large numbers, divided x by b and y by a, replaced them by the remainders, and gave the general x and y as a pair of arithmetic sequences with differences b and a, respectively. In the present case, the general solution is $x = 30 + 23k$, $y = 5 + 4k$. The smallest positive solution $x = 7$, $y = 1$, is obtained by taking $k = -1$.

Brahmagupta's rule for finding the solutions is more complicated than the discussion just given, since he does not assume that the numbers a and b are relatively prime. However, the greater generality is only apparent. If the greatest common divisor of a and b is not a factor of c, the problem has no solution; if it is a factor of c, it can be divided out of the problem.

Brahmagupta also considers such equations with negative data and is not in the least troubled by this complication. It seems clear that the name *pulverizer* was applied because the original data are repeatedly broken down by the Euclidean algorithm (they are "pulverized").

Astronomical applications. It was mentioned above that this kind of remainder arithmetic, which we now call the *theory of linear congruences*, has applications to the calendar. Brahmagupta proposed the problem of finding the (integer) number of elapsed days when Jupiter is $22°$, $30'$ into the sign of Aries[10] (Colebrooke, *1817*, p. 334). Brahmagupta converted the $22°$, $30'$ into $1350'$. He had earlier taken the sidereal period of Saturn to be 30 years and to be a common multiple of all cycles.

[10] Obviously, Jupiter will pass this point once in each revolution, but it will reach *exactly* this point at the expiration of an *exact* number of days (no fractional hours or minutes) only once in a *yuga*, which is a common period for all the heavenly bodies. Brahmagupta took 30 years as a *yuga*, but his method is general and will yield better results if a more accurate *yuga* is provided by observation. He says that the value of 30 years is given for a *yuga* only to make the computation easier.

On that basis Jupiter was given (inaccurately) a sidereal period of 10 years. Again inaccurately, using a year of $365\frac{1}{3}$ days, we find that Jupiter undergoes three cycles in 10,960 days. Thus, in one day, Jupiter moves 3/10960 of a revolution. Since there are $360 \cdot 60 = 21600$ minutes in a revolution, we find that each day Jupiter moves $64800/10960 = 810/137$ minutes. The problem, then, is to solve $810x/137 = 21,600y + 1350$ or dividing out the common factor of 270, $3x = 137 \cdot (80y + 5)$; that is, $3x = 10,960y + 685$. Here x will be the number of days elapsed in the cycle and y the number of revolutions that Jupiter will have made. The Euclidean algorithm yields $10960 = 3 \cdot 3653 + 1$, so that we have

$$\begin{array}{c} 3653 \\ -685 \\ 0 \end{array} \longrightarrow \begin{array}{c} -2502305 \\ -685 \end{array}.$$

That is, $x = -2502305 + 10960t$ and $y = -685 + 3t$. The smallest positive solution occurs when $t = 229$: $x = 7535$, $y = 2$, which is Brahmagupta's answer.

Brahmagupta illustrated the formulas for right triangles by creating Pythagorean triples. In Chapter 12 of the *Brahmasphutasiddhanta* (Colebrooke, *1817*, p. 306) he gives the rule that "the sum of the squares of two unlike quantities are the sides of an isosceles triangle; twice the product of the same two quantities is the perpendicular; and twice the difference of their square is the base." This rule amounts to the formula $(a^2 + b^2)^2 = (2ab)^2 + (a^2 - b^2)^2$, but it is stated as if the right triangle has been doubled by gluing another copy to the side of length $a^2 - b^2$, thereby producing an isosceles triangle. The relation stated is a purely geometric relation, showing (in our terms) that the sides and altitude of an isosceles triangle of any shape can be generated by choosing the two lengths a and b suitably.

Brahmagupta also considered generalizations of the problem of Pythagorean triples to a more general equation called[11] *Pell's equation* and written $y^2 - Dx^2 = 1$. He gives a recipe for generating a new equation of this form and its solutions from a given solution. The recipe proceeds by starting with two rows of three entries, which we shall illustrate for the case $D = 8$, which has the solution $x = 1$, $y = 3$. We write

$$\begin{array}{ccc} 1 & 3 & 1 \\ 1 & 3 & 1 \end{array}.$$

The first column contains x, called the *lesser solution*, the second contains y, called the *greater solution*, and the third column contains the additive term 1. From these two rows a new row is created whose first entry is the sum of the cross-multiplied first two columns, that is $1{\cdot}3{+}3{\cdot}1 = 6$. The second entry is the product of the second entries plus 8 times the product of the first entries, that is $3 \cdot 3 + 8 \cdot 1 \cdot 1 = 17$, and the third row is the product of the third entries. Hence we get a new row 6 17 1, and indeed $8 \cdot 6^2 + 1 = 289 = 17^2$. In our terms, this says that if $8x^2 + 1 = y^2$ and $8u^2 + 1 = v^2$, then $8(xv + yu)^2 + 1 = (8xu + yv)^2$. More generally, Brahmagupta's rule says that if $ax^2 + d = y^2$ and $au^2 + c = v^2$, then

$$a(xv + yu)^2 + cd = (axu + yv)^2. \tag{1}$$

[11] Erroneously so-called, according to Dickson (*1920*, p. 341), who asserts that Fermat had studied the equation earlier than John Pell (1611–1685). However, the website at St. Andrew's University gives evidence that Euler's attribution of this equation to Pell was accurate. In any case, everybody agrees that the solutions of the equation were worked out by Lagrange, not Pell.

Although it is trivial to verify that this rule is correct using modern algebraic notation, one would like to know how it was discovered.[12] Although the route by which this discovery was made is not known, the *motivation* for studying the equation can be plausibly ascribed to a desire to approximate irrational square roots with rational numbers. Brahmagupta's rule with $c = d = 1$ gives a way of generating larger and larger solutions of the *same* Diophantine equation $ax^2 + 1 = y^2$. If you have two solutions (x, y) and (u, v) of this equation, which need not be different, then you have two approximations y/x and v/u for $\sqrt{a}$ whose squares are respectively $1/x^2$ and $1/u^2$ larger than a. The new solution generated will have a square that is only $1/(xv + yu)^2$ larger than a. This aspect of the problem of Pell's equation turns out to have a close connection with its complete solution in the eighteenth century.

4.4. Bhaskara II. In his treatise *Vija Ganita* (*Algebra*), Bhaskara states the rule for *kuttaka* more clearly than either Aryabhata or Brahmagupta had done and illustrates it with specific cases. For example, in Chapter 2 (Colebrooke, *1817*, p. 162) he asks, "What is that multiplier which, when it is multiplied by 221 and 65 is added to the product, yields a multiple of 195?" In other words, solve the equation $195x = 221y+65$. Dividing out 13, which is a common factor, reduces this equation to $15x = 17y+5$. The *kuttaka*, whose steps are shown explicitly, yields as a solution $x = 40$, $y = 35$.

In his writing on algebra, Bhaskara considered many Diophantine equations. For example, in Section 4 of Chapter 3 of the *Lilavati* (Colebrooke, *1817*, p. 27), he posed the problem of finding pairs of (rational) numbers such that the sum and difference of their squares are each 1 larger than a square. It would be interesting to know how he found the answer to this difficult problem. All he says is that the smaller number should be obtained by starting with any number, squaring, multiplying by 8, subtracting 1, then dividing by 2 and by the original number. The larger number is then obtained by squaring the smaller one, dividing by 2, and adding 1. In our terms, these recipes say that if u is any rational number, then

$$\left(8u^2 - 1 + \frac{1}{8u^2}\right)^2 \pm \left(4u - \frac{1}{2u}\right)^2 - 1$$

is the square of a rational number. The reader can easily verify that it is $\big(8u^2 - 1/(8u^2)\big)^2$ when the positive sign is chosen and $\big(8u^2 - 2 + \frac{1}{8u^2}\big)^2$ when the negative sign is taken.

Chapter 4 of the *Vija Ganita* contains many algebraic problems involved with solving triangles, interspersed with some pure Diophantine equations. One of the most remarkable (Colebrooke, *1817*, p. 200) is the problem of finding four unequal (rational) numbers whose sum equals the sum of their squares or the sum of the cubes of which equals the sum of their squares. In the first case he gives $\frac{1}{3}$, $\frac{2}{3}$, $\frac{3}{3}$, $\frac{4}{3}$. In the second case he gives $\frac{3}{10}$, $\frac{6}{10}$, $\frac{9}{10}$, $\frac{12}{10}$. In both cases the numbers are in the proportion $1 : 2 : 3 : 4$. These three extra conditions (three ratios of numbers) were deliberately added by Bhaskara so that the problem would become a determinate one.

The characteristic that makes problems like the preceding one easy is that the requirement imposed on the four numbers amounts to a single equation with

[12] Weil (*1984*, pp. 17, 83, 204) refers to Eq. 1 and the more general relation $(x^2+Ny^2)(z^2+Nt^2) = (xz \pm Nyt)^2 + N(xt \mp yz)^2$ as "Brahmagupta's identity" (his quotation marks).

more than one unknown. But Bhaskara also asks harder questions. For example (Colebrooke, *1817*, p. 202): *Find two (rational) numbers such that the sum of the cubes is a square and the sum of the squares is a cube.* Bhaskara manages to find a solution using the trick of assigning the ratio a of the two numbers. It is necessary for the technique that this ratio satisfy $1 + a^3 = b^2$. Bhaskara chooses $a = 2$, $b = 3$.[13] The smaller number is then chosen to be of the form $(1+a^2)w^3$ for some number w. The sum of the squares will then be $(1+a^2)^2w^6 + a^2(1+a^2)^2w^6 = (1+a^2)^3w^6 = \big((1+a^2)w^2\big)^3$, and the sum of their cubes will be $(1+a^2)^3w^9 + a^3(1+a^2)^3w^9 = (1+a^2)^3b^2w^9 = b^2\big((1+a^2)w^3\big)^3$. Hence, if w is chosen so that $(1+a^2)w$ is a square, this will be a perfect square. The simplest choice obviously is $w = 1 + a^2$. In Bhaskara's example, that choice gives the pair 625 and 1250.

5. The Muslims

The Muslims continued the work of Diophantus in number theory. Abu-Kamil (ca. 850–ca. 930) wrote a book on "indeterminate problems" in which he studied quadratic Diophantine equations and systems of such equations in two variables. The first 38 problems that he studied are arranged in a very strict ordering of coefficients, exponents, and signs, making it a very systematic exposition of these equations. Later scholars noted the astonishing fact that the first 25 of these equations are what are now known as algebraic curves of genus 0, while the last 13 are of genus 1, even though the concept of genus of an algebraic curve is a nineteenth-century invention (Baigozhina, *1995*).

Muslim mathematicians also went beyond what is in Euclid and Nicomachus, generalizing perfect numbers. In a number of articles, Rashed (see, for example, *1989*) points out that a large amount of theory of abundant, deficient, and perfect numbers was assembled in the ninth century by Thabit ibn-Qurra and others, and that ibn al-Haytham (965–1040) was the first to state and attempt to prove that Euclid's formula gives all the even perfect numbers. Thabit ibn Qurra made an interesting contribution to the theory of amicable numbers. A pair of numbers is said to be *amicable* if each is the sum of the proper divisors of the other. The smallest such pair of numbers is 220 and 284. Although these numbers are not discussed by Euclid or Nicomachus, the commentator Iamblichus (see Dickson, *1919*, p. 38) ascribed this notion to Pythagoras, who is reported as saying, "A friend is another self." This definition of a friend is given by Aristotle in his *Nicomachean Ethics* (Bekker 1170b, line 7).

We mentioned above the standard way of generating perfect numbers, namely the Euclidean formula $2^{n-1}(2^n - 1)$, whenever $2^n - 1$ is a prime. Thabit ibn-Qurra found a similar way of generating pairs of amicable numbers. His formula is

$$2^n(3 \cdot 2^n - 1)(3 \cdot 2^{n-1} - 1) \text{ and } 2^n(9 \cdot 2^{2n-1} - 1),$$

whenever $3 \cdot 2^n - 1$, $3 \cdot 2^{n-1} - 1$, and $9 \cdot 2^{2n-1} - 1$ are all prime. The case $n = 2$ gives the pair 220 and 284. Whatever one may think about the impracticality of amicable numbers, there is no denying that Thabit's discovery indicates very

[13] It was conjectured in 1844 by the Belgian mathematician Eugène Charles Catalan (1814–1894) that the only nonzero solutions to the Diophantine equation $p^m - q^n = 1$ are $q = 2 = n$, $p = 3 = n$. This conjecture was proved in 2002 by Predhu Mihailescu, a young mathematician at the Institute for Scientific Computing in Zürich. Lucky Bhaskara! He found the only possible solution.

profound insight into the divisibility properties of numbers. It is very difficult to imagine how he could have discovered this result. A conjecture, which cannot be summarized in a few lines, can be found in the article by Brentjes and Hogendijk (*1989*).

It is not clear how many new cases can be generated from this formula, but there definitely are some. For example, when $n = 4$, we obtain the amicable pair $17,296 = 16 \cdot 23 \cdot 47$ and $18,416 = 16 \cdot 1151$. Hogendijk (*1985*) gives Thabit ibn-Qurra's proof of his criterion for amicable numbers and points out that the case $n = 7$ generates the pair 9,363,584 and 9,437,056, which first appeared in Arabic texts of the fourteenth century.

Unlike some other number-theory problems such as the Chinese remainder theorem, which arose in a genuinely practical context, the theory of amicable numbers is an offshoot of the theory of perfect numbers, which was already a completely "useless" topic from the beginning. It did not seem useless to the people who developed it, however. According to M. Cantor (*1880*, p. 631) the tenth-century mystic al-Majriti recommended as a love potion writing the numbers on two sheets of paper and eating the number 284, while causing the beloved to eat the number 220. He claimed to have verified the effectiveness of this charm by personal experience! Dickson (*1919*, p. 39) mentions the Jewish scholar Abraham Azulai (1570–1643), who described a work purportedly by the ninth-century commentator Rau Nachshon, in which the gift of 220 sheep and 220 goats that Jacob sent to his brother Esau as a peace offering (Genesis 32:14) is connected with the concept of amicable numbers.[14] In any case, although their theory seems more complicated, amicable numbers are easier to find than perfect numbers. Euler alone found 62 pairs of them (see Erdős and Dudley, *1983*).

Another advance on the Greeks can be found in the work of Kamal al-Din al-Farisi, a Persian mathematician who died around 1320. According to Ağargün and Fletcher (*1994*), he wrote the treatise *Memorandum for Friends Explaining the Proof of Amicability*, whose purpose was to give a new proof of Thabit ibn-Qurra's theorem. Proposition 1 in this work asserts the existence (but not uniqueness) of a prime decomposition for every number. Propositions 4 and 5 assert that this decomposition is unique, that two distinct products of primes cannot be equal.

6. Japan

In 1627 Yoshida Koyu wrote a textbook of arithmetic called the *Jinkō-ki* (*Treatise on Large and Small Numbers*). This book contained a statement of what is known in modern mathematics as the *Josephus problem*. The Japanese version of the problem involves a family of 30 children choosing one of the children to inherit the parents' property. The children are arranged in a circle and count off by tens; the unlucky children who get the number 10 are eliminated; that is, numbers 10, 20, and 30 drop out. The remaining 27 children then count off again. The children originally numbered 11 and 22 will be eliminated in this round, and when the second round of numbering is complete, the child who was first will have the number 8. Hence the children originally numbered 3, 15, and 27 will be eliminated on the next

[14] The peace offering was necessary because Jacob had tricked Esau out of his inheritance. But if the gift was symbolic and associated with amicable numbers, the story seems to imply that Esau was obligated to give Jacob 284 sheep and 284 goats. Perhaps there was an ulterior motive in the gift!

round, and the first child will start the following round as number 3. The problem is to see which child will be the last one remaining. Obviously, solving this problem in advance could be very profitable, as the original Josephus story indicates.[15] The Japanese problem is made more interesting and more complicated by considering that half of the children belong to the couple and half are the husband's children by a former marriage. The wife naturally wishes one of her own children to inherit, and she persuades the husband to count in different ways on different rounds. The problem was reprinted by several later Japanese mathematicians.

The eighteenth-century mathematician Matsunaga Ryohitsu (1718–1749) discussed a variety of equations similar to the Pell equation and representations of numbers in general as sums and differences of powers. For example, his recipe for solving the equation $x^3 - y^3 = z^4$ was to take $z = m^3 - n^3$, then let $x = mz$, $y = nz$. But he also tackled some much more sophisticated problems, such as the problem of representing a given integer k as a sum of two squares and finding an integer that is simultaneously of the forms $y_1^2 + 69y_1 + 15$ and $y_2^2 + 72y^2 + 7$. Matsunaga gave the solution as 11,707, obtained by taking $y_1 = 79$, $y_2 = 78$.

7. Medieval Europe

In his *Liber quadratorum* (*Book of Squares*) Leonardo of Pisa (Fibonacci, 1170–1250) speculated on the difference between square and nonsquare numbers. In the prologue, addressed to the Emperor Frederick II, Leonardo says that he had been inspired to write the book because a certain John of Palermo, whom he had met at Frederick's court, had challenged him to find a square number such that if 5 is added to it or subtracted from it, the result is again a square. This question inspired him to reflect on the difference between square and nonsquare numbers. He then notes his pleasure on learning that Frederick had actually read one of his previous books and uses that fact as justification for writing on the challenge problem.

The *Liber quadratorum* is written in the spirit of Diophantus and shows a keen appreciation of the conditions under which a rational number is a square. Indeed, the ninth of its 24 propositions *is* a problem of Diophantus: *Given a nonsquare number that is the sum of two squares, find a second pair of squares having this number as their sum.* As mentioned above, this problem is Problem 9 of Book 2 of Diophantus.

The securest basis of Leonardo's fame is a single problem from his *Liber abaci*, written in 1202:

> *How many pairs of rabbits can be bred from one pair in one year given that each pair begins to breed in the second month after its birth, producing one new pair per month?*

[15] Josephus tells us that, faced with capture by the Romans after the fall of Jotapata, he and his Jewish comrades decided to commit mass suicide rather than surrender. Later commentators claimed that they stood in a circle and counted by threes, agreeing that every third soldier would be killed by the person on his left. The last one standing was duty bound to fall on his sword. According to this folk legend, Josephus immediately computed where he should position himself in order to be that last person, but decided to surrender instead of carrying out the bargain. Josephus himself, however, writes in *The Jewish Wars*, Book III, Chapter 8 that the order of execution was determined by drawing lots and that he and his best friend survived either by chance or by divine intervention in these lots. The mathematical problem we are discussing is also said to have been invented by Abraham ben Meir ibn Ezra (1092–1167), better known as Rabbi Ben Ezra, one of many Jewish scholars who flourished in the Caliphate of Cordoba.

By brute-force enumeration of cases, the author concludes that there will be 377 pairs, and "in this way you can do it for the case of infinite numbers of months."

The sequence generated here (1, 1, 2, 3, 5, 8,...), in which each term after the second is the sum of its two predecessors, has been known as the *Fibonacci sequence* ever since the *Liber abaci* was first printed in the nineteenth century. The Fibonacci sequence has been an inexhaustible source of identities. Many curious representations of its terms have been obtained, and there is a mathematical journal, the *Fibonacci Quarterly*, named in its honor and devoted to its lore. The Fibonacci sequence has been a rich source of interesting pure mathematics, but it has also had some illuminating practical applications, one of which is discussed in Problems 2.4–2.6.

Questions and problems

7.1. Compute the sexagesimal representation of the number

$$\left(\frac{p/q - q/p}{2}\right)^2$$

for the following pairs of integers (p, q): $(12, 5)$, $(64, 27)$, $(75, 32)$, $(125, 54)$, and $(9, 5)$. Then correct column 1 of Plimpton 322 accordingly.

7.2. On the surface the Euclidean algorithm looks easy to use, and indeed it *is* easy to use when applied to integers. The difficulty arises when it is applied to continuous objects (lengths, areas, volumes, weights). In order to execute a loop of this algorithm, you must be able to decide which element of the pair (a, b) is larger. But all judgments as to relative size run into the same difficulty that we encounter with calibrated measuring instruments: limited precision. There is a point at which one simply cannot say with certainty that the two quantities are either equal or unequal. Does this limitation have any practical significance? What is its theoretical significance? Show how it could give a wrong value for the greatest common measure even when the greatest common measure exists. How could it ever show that two quantities have *no* common measure?

7.3. The remainders in the Euclidean algorithm play an essential role in finding the greatest common divisor. The greatest common divisor of 488 and 24 is 8, so that the fraction 24/488 can be reduced to 3/61. The Euclidean algorithm generates two *quotients*, 20 and 3 (in order of generation). What is their relation to the two numbers? Observe the relation

$$\frac{1}{20 + \frac{1}{3}} = \frac{3}{61}$$

If you find the greatest common divisor of 23 and 56 (which is 1) this way, you will generate the quotients 2, 2, 3, 3. Verify that

$$\frac{23}{56} = \frac{1}{2 + \frac{1}{2 + \frac{1}{3 + \frac{1}{3}}}}.$$

This expression is called the *continued fraction representation* representation of 23/56. Formulate a general rule for finding the continued fraction representation of a proper fraction.

7.4. Draw dot figures for the first five heptagonal and octagonal numbers. What kind of figure would you need for nonagonal numbers?

7.5. Prove the formulas given in the caption of Fig. 1 for T_n, S_n, P_n, and H_n. Then prove that $S_n = T_n + T_{n-1}$, $P_n = S_n + T_{n-1} = T_n + 2T_{n-1}$, $H_n = P_n + T_{n-1} = T_n + 3T_{n-1}$. If $P_{k,n}$ is the nth k-gonal number, give a general formula for $P_{k,n}$ in terms of k and n.

7.6. Prove that the Pythagorean procedure always produces a perfect number. That is, if $p = 2^n - 1$ is prime, then $N = 2^{n-1}p$ is perfect. This theorem is not difficult to prove nowadays, since the "parts" (proper divisors) of N are easy to list and sum.

7.7. Let N_n be the nth perfect number, so that $N_1 = 6$, $N_2 = 28$, $N_3 = 496$, $N_4 = 8128$. Assuming that all perfect numbers are given by the Pythagorean formula, that is, they are of the form $2^{n-1}(2^n - 1)$ when $2^n - 1$ is a prime, prove that $N_{n+1} > 16N_n$ if $n > 1$. Conclude that there cannot be more than one k-digit perfect number for each k.

7.8. (*V. A. Lebesgue's proof of Euler's theorem on even perfect numbers*) Suppose that the perfect number N has the prime factorization $N = 2^\alpha p_1^{n_1} \cdots p_k^{n_k}$, where $p_1, \ldots, p_k$ are distinct odd primes and α, $n_1, \ldots, n_k$ are nonnegative integers. Since N is perfect, the sum of *all* its divisors is $2N$. This means that

$$\begin{aligned} 2^{\alpha+1}p_1^{n_1} \cdots p_k^{n_k} &= (1 + 2 + \cdots + 2^\alpha)(1 + p_1 + \cdots + p_1^{n_1}) \cdots (1 + p_k + \cdots + p_k^{n_k}) \\ &= (2^{\alpha+1} - 1)(1 + p_1 + \cdots + p_1^{n_1}) \cdots (1 + p_k + \cdots + p_k^{n_k}). \end{aligned}$$

Rewrite this equation as follows:

$$\begin{aligned} &(2^{\alpha+1} - 1)p_1^{n_1} \cdots p_k^{n_k} + p_1^{n_1} \cdots p_k^{n_k} = \\ &\qquad = (2^{\alpha+1} - 1)(1 + p_1 + \cdots + p_1^{n_1}) \cdots (1 + p_k + \cdots + p_k^{n_k}), \\ &p_1^{n_1} \cdots p_k^{n_k} + \frac{p_1^{n_1} \cdots p_k^{n_k}}{2^{\alpha+1} - 1} = (1 + p_1 + \cdots + p_1^{n_1}) \cdots (1 + p_k + \cdots + p_k^{n_k}). \end{aligned}$$

Since the second term on the left must be an integer, it follows that $2^{\alpha+1} - 1$ must divide $p_1^{n_1} \cdots p_k^{n_k}$. This is not a significant statement if $\alpha = 0$ (N is an odd number). But if N is even, so that $\alpha > 0$, it implies that $2^{\alpha+1} - 1 = p_1^{m_1} \cdots p_k^{m_k}$ for integers $m_1 \le n_1, \ldots, m_k \le n_k$, *not all zero*. Thus, the left-hand side consists of the two *distinct* terms $p_1^{n_1} \cdots p_k^{n_k} + p_1^{r_1} \cdots p_k^{r_k}$. It follows that the right-hand side must also be equal to this sum. Now it is obvious that the right-hand side *contains* these two terms. That means the sum of the remaining terms on the right-hand side must be zero. But since the coefficients of all these terms are positive, there *can be* only two terms on the right. Since the right-hand side obviously contains $(n_1 + 1)(n_2 + 1) \cdots (n_k + 1)$ terms, we get the equation

$$2 = (n_1 + 1)(n_2 + 1) \cdots (n_k + 1).$$

Deduce from this equation that N must be of the form $2^{n-1}(2^n - 1)$ and that $2^n - 1$ is prime.

7.9. Generalize Diophantus' solution to the problem of finding a second representation of a number as the sum of two squares, using his example of $13 = 2^2 + 3^2$ and letting one of the numbers be $(\varsigma + 3)^2$ and the other $(k\varsigma - 2)^2$.

7.10. Take as a unit of time $T = \frac{1}{235}$ of a year, about 37 hours, 18 minutes, say a day and a half in close approximation. Then one average lunar month is $M = 19T$, and one average solar year is $Y = 235T$. Given that the Moon was full on June 1, 1996, what is the next year in which it will be full on June 4? Observe that June 4 in whatever year that is will be 3 days ($2T$) plus an integer number of years. We are seeking integer numbers of months (x) and years (y), counting from June 1, 1996, such that $Mx = Yy + 2T$, that is (canceling T), $19x = 235y + 2$. Use the *kuttaka* to solve this problem and check your answer against an almanac. If you use this technique to answer this kind of question, you will get the correct answer most of the time. When the answer is wrong, it will be found that the full moon in the predicted year is a day earlier or a day later than the prescribed date. The occasional discrepancies occur because (1) the relation $M = 19T$ is not precise, (2) full moons occur at different times of day, and (3) the greatest-integer function is not continuous.

7.11. Use Bhaskara's method to find two integers such that the square of their sum plus the cube of their sum equals twice the sum of their cubes. (This is a problem from Chapter 7 of the *Vija Ganita.*)

7.12. The Chinese mutual-subtraction algorithm (the Euclidean algorithm) can be used to convert a decimal expansion to a common fraction and to provide approximations to it with small denominators. Consider, for example, the number $e \approx 2.71828$. By the Euclidean algorithm, we get

$$\begin{aligned}
271,828 &= 2 \cdot 100,000 + 71,828\\
100,000 &= 1 \cdot 71,828 + 28,172\\
71,828 &= 2 \cdot 28,172 + 15,484\\
28,172 &= 1 \cdot 15,484 + 12,688\\
15,484 &= 1 \cdot 12,688 + 2,796\\
12,688 &= 4 \cdot 2,796 + 1,504\\
2,796 &= 1 \cdot 1,504 + 1,292\\
1,504 &= 1 \cdot 1,292 + 212\\
1,292 &= 6 \cdot 212 + 20\\
212 &= 10 \cdot 20 + 12\\
20 &= 1 \cdot 12 + 8\\
12 &= 2 \cdot 8 + 4\\
8 &= 2 \cdot 4
\end{aligned}$$

Thus the greatest common divisor of 271,828 and 100,000 is 4, and if it is divided out of all of these equations, the quotients remain the same. We can thus write

$$2.71828 = \frac{271828}{100000} = \frac{67957}{25000} = 2 + \cfrac{1}{1 + \cfrac{1}{2 + \cfrac{1}{1 + \cfrac{1}{1 + \cfrac{1}{4 + \cdots}}}}}\,.$$

The first few partial fractions here give

$$2+\cfrac{1}{1} = 3\,,$$

$$2+\cfrac{1}{1+\cfrac{1}{2}} = 2\frac{2}{3} = \frac{8}{3} = 2.666\ldots\,,$$

$$2+\cfrac{1}{1+\cfrac{1}{2+\cfrac{1}{1}}} = \frac{11}{4} = 2.75\,,$$

$$2+\cfrac{1}{1+\cfrac{1}{2+\cfrac{1}{1+\cfrac{1}{1}}}} = 2\frac{5}{7} = \frac{19}{7} = 2.714285712485\ldots\,,$$

$$2+\cfrac{1}{1+\cfrac{1}{2+\cfrac{1}{1+\cfrac{1}{1+\cfrac{1}{4}}}}} = 2+\frac{23}{32} = \frac{87}{32} = 2.71875\,,$$

so that the approximations get better and better. Do the same with $\pi \approx 3.14159265$, and calculate the first five approximate fractions. Do you recognize any of these?

7.13. Can the pair of amicable numbers 1184 and 1210 be constructed from Thabit ibn-Qurra's formula?

7.14. Solve the generalized problem stated by Matsunaga of finding an integer N that is simultaneously of the form $x^2 + a_1x + b_1$ and $y^2 + a_2y + b_2$. To do this, show that it is always possible to factor the number $(a_2^2 + 4b_1) - (a_1^2 + 4b_2)$ as a product mn, where m and n are either both even or both odd, and that the solution is found by taking $x = \frac{1}{2}\left(\frac{m-n}{2} - a_1\right)$, $y = \frac{1}{2}\left(\frac{m+n}{2} - a_2\right)$.

7.15. Leonardo's solution to the problem of finding a second pair of squares having a given sum is explained in general terms, then illustrated with a special case. He considers the case $4^2 + 5^2 = 41$. He first finds two numbers (3 and 4) for which the sum of the squares *is* a square. He then forms the product of 41 and the sum of the squares of the latter pair, obtaining $25 \cdot 41 = 1025$. Then he finds two squares whose sum equals this number: 31^2 and 8^2 or 32^2 and 1^2. He thus obtains the results $\left(\frac{31}{5}\right)^2 + \left(\frac{8}{5}\right)^2 = 41$ and $\left(\frac{32}{5}\right)^2 + \left(\frac{1}{5}\right)^2 = 41$. Following this method, find another pair of rational numbers whose sum is 41. Why does the method work?

7.16. If the general term of the Fibonacci sequence is a_n, show that $a_n < a_{n+1} < 2a_n$, so that the ratio a_{n+1}/a_n always lies between 1 and 2. Assuming that this ratio has a limit, what is that limit?

7.17. Suppose that the pairs of rabbits begin to breed in the *first* month after they are born, but die after the second month (having produced two more pairs). What sequence of numbers results?

7.18. Prove that if x, y, and z are relatively prime integers such that $x^2 + y^2 = z^2$, with x and z odd and y even, there exist integers u and v such that $x = u^2 - v^2$,

$y = 2uv$, and $z = u^2 + v^2$. [*Hint*: Start from the fact that $x^2 = (z - y)(z + y)$, so that $z - y = a^2$ and $z + y = b^2$ for some a and b.]

CHAPTER 8

Numbers and Number Theory in Modern Mathematics

Beginning with the work of Fermat in the seventeenth century, number theory has become ever more esoteric and theoretical, developing connections with algebra and analysis that lie very deep and require many years of study to master. Obviously, we cannot explain in any satisfactory detail what has happened in this area in recent years. For that reason, we shall carry the story forward only as far as the beginning of the twentieth century.

A second topic that we must discuss before leaving the subject of numbers is the variety of invented number systems, starting with the natural positive integers. The number zero, negative numbers, and rational numbers do not require a long explanation, but we need to focus in more detail on real and complex numbers and the cardinal and ordinal arithmetic that came along with set theory.

Finally, mere counting turns out to be very difficult in some cases; for example, given twelve points on a circle, each pair of which is joined by a chord, into how many regions will these chords divide the circle if no three chords intersect in a common point? To solve such problems, sophisticated methods of counting have been developed, leading to the modern subject of combinatorics. A survey of its history concludes our study of numbers.

1. Modern number theory

We are forced to leave out many important results in our survey of modern number theory. Dickson's summary of the major results (*1919*, *1920*, *1923*) occupies 1600 pages, and an enormous amount of work has been added since it was published. Obviously, the present discussion is going to be confined to a few of the most significant authors and results.

1.1. Fermat. Pierre de Fermat (1603–1665) was a lawyer in Toulouse whose avid interest in mathematics led him to create, in his spare time, some analytic geometry, calculus, and modern number theory. According to one source book (Smith, *1929*, p. 214), he was "the first to assert that the equation $x^2 - Dy^2 = 1$ has infinitely many solutions in integers." As we have seen, given that it has *one* solution in integers, Brahmagupta knew 900 years before Fermat that it must have infinitely many, since he knew how to create new solutions from old ones. It was mentioned above that Fermat wrote in the margin of his copy of Diophantus that the sum of two positive rational cubes could not be a rational cube, and so on (Fermat's last theorem). Although Fermat never communicated his claimed proof of this fact, he was one of the first to make use of a method of proof—the method of infinite descent—by which many facts in number theory can be proved, including the case of fourth powers in Fermat's last theorem. A proof of the case of fourth powers was given by Euler in 1738. Actually, the proof shows that there can be no positive integers x, y, z such that $x^4 + y^4 = z^2$.

Another area of number theory pioneered by Fermat arises naturally from consideration of quadratic Diophantine equations. The question is: "In how many ways can an integer be represented as a sum of two squares?" A number of the form $4n + 3$ cannot be the sum of two squares. This is an easy result, since when a square is divided by 4 the remainder is either 0 or 1, and it is impossible to write 3 as the sum of two numbers, each of which is either 0 or 1. But Fermat proved the much more difficult result that a prime number of the form $4n + 1$ can be written as the sum of two squares in exactly one way. Thus, $73 = 8^2 + 3^2$, for example.

The work of Fermat in number theory was continued by many mathematicians in the eighteenth century. We shall discuss very briefly the lives and work of three of them.

1.2. Euler. The Swiss mathematician Leonhard Euler (1707–1783) was one of the most profound and prolific mathematical writers who ever lived, despite having lost the sight of one eye early in life and the other later on. His complete works have only recently been assembled in good order. In 1983, the two hundredth anniversary of his death, many memorial volumes were dedicated to him, including an entire issue (**45**, No. 5) of *Mathematics Magazine*. An even larger celebration is planned for 2007, the three hundredth anniversary of his birth.[1] He spent most of the years from 1726 to 1741 and from 1766 until his death in St. Petersburg, Russia, where he was one of the first members of the Russian Academy of Sciences, founded by Tsar Peter I (1672–1725) just before his death. From 1741 to 1766 he was in Berlin, at the Prussian Academy of Sciences of Frederick II (1712–1788). The exact date at which he made many of his great discoveries is sometimes difficult to establish, and different dates are sometimes given in the literature. Euler's contributions to the development of calculus, differential equations, algebra, geometry, and mathematical physics are enormous. The following paragraphs describe some of his better-known results in number theory.

Fermat primes. Fermat had conjectured that the number $F_n = 2^{(2^n)} + 1$ is always a prime. This statement is true for $n = 0, 1, 2, 3, 4$, as the reader can easily check. For $n = 5$ this number is $4,294,967,297$, and to prove that it is prime using the sieve of Eratosthenes, one must attempt to divide it by every prime less than $F_4 = 65,537$. In 1732 Euler found that this fifth Fermat number is divisible by 641. No Fermat number beyond F_4 has ever been shown to be prime, and well over 200 are now known to be composite, including $F_{2478782}$, discovered by John Cosgrave and others at St. Patrick's College, Dublin, on October 10, 2003. The smallest Fermat number not definitely known to be either prime or composite is F_{33}. The problem of Fermat primes is almost, but not quite, an idle question, that is, one without connections to anything else in mathematics. The connection in this case is that the regular polygons that have an odd number of sides and can be constructed with straightedge and compass are precisely those whose number of sides is a product of Fermat primes. Thus, until such time as another Fermat number is proved to be prime, the only Euclidean-constructible regular polygons

[1] See the website http://www.euler2007.com/.

with an odd number of sides will be those whose number of sides is a product of distinct numbers from the following short list: 3, 5, 17, 257, 65,537.[2]

Fermat's last theorem: the cases $n = 4$ and $n = 3$. We pointed out above that Fermat's method of infinite descent can be used to prove Fermat's last theorem for $n = 4$, as shown by Euler in 1738. In a textbook of algebra published in 1770 (see Struik, *1986*, pp. 36–40), Euler also gave a proof of the impossibility for the case $n = 3$, which is much more difficult. In 1772 Euler proved that every positive integer is the sum of at most four square integers and conjectured that no sum of fewer than n nth powers could be an nth power, a conjecture that was finally refuted for $n = 5$ in 1966.[3]

Fermat's little theorem. A second assertion of Fermat, which Euler proved in 1736 (see Struik, *1986*, p. 35), is known as *Fermat's little theorem.* It asserts that if p is a prime number that does not divide a number a, then p does divide $a^d - 1$ for some positive integer d; moreover, the smallest d for which this statement is true divides $p - 1$. In particular, p divides $a^{p-1} - 1$. Fermat's discovery of this theorem has an interesting history (Fletcher, *1989*). Through Mersenne Fermat had received a challenge in 1640 from Bernard Frénicle de Bessy (ca. 1612–1675) to find the first perfect number having at least 20 digits. It was known that $2^{30}(2^{31} - 1)$ was a perfect number having 19 digits.[4] Fermat did not succed in doing this; but after studying the problem, he noted that if n is composite, then $2^n - 1$ is also and that if n is a prime, then $2^n - 2$ is divisible by $2n$, that is, n divides $2^{n-1} - 1$. Moreover, he said, all other prime divisors of $2^n - 2$ leave a remainder of 1 when divided by $2n$. These theorems, in Fermat's view, were the secret of discovering perfect numbers, and he asserted that there were none having 20 or 21 digits. In a letter to Frénicle in October 1640, Fermat stated his "little" theorem—so called to distinguish it from his greater, "last" theorem. It is by no means a "little" result. Euler extended this result and showed that m divides $a(a^{\phi(m)} - 1)$, where $\phi(m)$ is the number of positive integers less than m and relatively prime to m (now called *Euler's ϕ-function*). That function provides the theoretical basis for constructing the RSA[5] codes that are an essential part of communications security. Thus, the completely "useless" topic of perfect numbers actually inspired a number-theoretic result of great practical value.

Residues modulo a prime. Euler defined an integer n to be a λ-power residue with respect to ("modulo," as we now say, using Euler's Latin term) a prime p if there is an integer a such that p divides $a^\lambda - n$. This concept has proved to be a rich source of investigation in number theory. In particular, the case $\lambda = 2$ (quadratic residues) has led to some deep theorems. In works published in 1751 and 1783,

[2] Heinrich Wefelscheid informs me that one Johann Hermes (1846–1912), a student in Königsberg from 1866 to 1870, actually attempted to work out the case 65,537 and published his method in the *Göttinger Nachrichten* in 1894. The Australian mathematician Joan Taylor has used a computer to complete Hermes' project and found that the algebraic expression for $\cos(2\pi/65537)$ occupies 12.5 megabytes and contains an integer of 19,717 digits.

[3] In connection with this result, we note that Fermat had stated a positive conjecture: *Every positive integer is the sum of at most n n-gonal numbers, that is, three triangular numbers, four squares, five pentagonal numbers, and so on.* This result was first proved for the general case by Augustin-Louis Cauchy (1789–1856) in 1813.

[4] Specifically, it is 2,305,843,008,139,952,128.

[5] Invented in 1977 and named from the initials of its three inventors, Ronald L. Rivest, Adi Shamir, and Leonard Adleman.

Euler conjectured what we now know as the law of *quadratic reciprocity*: *Given two primes p and q both of which equal 3 modulo 4, exactly one of them is a quadratic residue modulo the other. In all other cases, either each is a quadratic residue modulo the other or neither is.* For example, $11 \equiv 2^2 \equiv 5^2 \bmod 7$, but the quadratic residues modulo 11 are 0, 1, 4, 3 ($\equiv 5^2 \equiv 6^2 \bmod 11$), 5 ($\equiv 4^2 \equiv 7^2 \bmod 11$), and 9; 7 is not among them. That is because both 7 and 11 are equal to 3 modulo 4. On the other hand, since 5 equals 1 modulo 4, we find that $11 \equiv 1^2 \bmod 5$ and $5 \equiv 7^2 \bmod 11$; similarly, neither 5 nor 7 is a quadratic residue modulo the other. The fact that Euler did not succeed in proving the law of quadratic reciprocity shows how difficult a result it is.

The Goldbach conjecture. A problem of number theory whose fame is second only to the Fermat conjecture is a conjecture of Christian Goldbach (1690–1764), who wrote to Euler in 1742 that every integer seemed to be a sum of at most three prime integers (Struik, *1986*, pp. 47–49). Euler wrote back that he believed, but was unable to prove, the stronger proposition that every even integer larger than 4 is the sum of two odd primes—in other words, that one of the three primes conjectured by Goldbach can be chosen arbitrarily. Euler's statement is known as the *Goldbach conjecture.* In 1937 the Russian mathematician Ivan Matveevich Vinogradov (1891–1983) proved that every sufficiently large odd integer is the sum of at most three primes.

1.3. Lagrange. The generation after Euler produced the Italian–French mathematician Joseph-Louis Lagrange (1736–1813). His name gives the impression that he was French, and indeed his ancestry was French and he wrote in French; but then so did many others, as French was literally the "lingua franca," the common language of much scientific correspondence during the eighteenth and nineteenth centuries. Lagrange was born in Turin, however, and lived there for the first 30 years of his life, signing his first name as "Luigi" on his first mathematical paper in 1754. When the French Revolution came, he narrowly escaped arrest as a foreigner; and we have the word of Jean-Joseph Fourier, who heard him lecture, that he spoke French with a noticeable Italian accent. Thus it appears that the Italians are correct in claiming him as one of their own, even though his most prominent works were published in France and he was a member of the Paris Academy of Sciences for the latter part of his life.

Lagrange's early work impressed Euler, then in Berlin, very favorably, and attempts were made to bring him to Berlin. But the introverted Lagrange seems to have been intimidated by Euler's power as a mathematician and refused all such offers until Euler went back to St. Petersburg in 1766. He then came to Berlin and remained there until the death of Frederick II in 1788, at which point he accepted a position at the Paris Academy of Sciences, where he spent the last 15 years of his life. Lagrange did important work in algebra and mechanics that is discussed in later chapters. At this point we note only some of his number-theoretic results.

The Pell equation. Shortly after arriving in Berlin in 1766, Lagrange gave a definitive discussion of the solutions of the Pell equation $x^2 = Dy^2 \pm 1$, using the theory of continued fractions. In the course of this work he proved the important fact that any irrational number satisfying a quadratic equation with integer coefficients has a periodic continued fraction expansion. The converse of that statement is also true, and it turns out that the continued-fraction expansion of $\sqrt{D}$ can be used to construct *all* solutions of the Pell equation $Dx^2 \pm 1 = y^2$ (see Scharlau and Opolka, *1985*, pp. 45–56).

The four-squares theorem. In 1770 Lagrange gave a proof that every integer is the sum of at most four square integers (which Euler also proved a year or so later).

"Wilson's theorem". In 1771 Lagrange proved that an integer n is prime if and only if n divides $(n-1)! + 1$. Thus 5 is prime because $4! + 1 = 25$, but 6 is not prime because it does not divide $5! + 1 = 121$. This theorem was attributed to John Wilson (1741–1793) by the Cambridge professor Edward Waring (1736–1798), who was apparently unaware that it was first stated by al-Haytham (965–1040). No proof of it can be found in the work of Wilson, who left mathematics to become a lawyer.

Quadratic binary forms. The study of quadratic Diophantine equations involves expressions of the form $ax^2 + bxy + cy^2$. The integers that can be represented in this way for given values of a, b, and c were the subject of two memoirs by Lagrange, amounting to nearly 100 pages of work, during the years 1775–1777.

1.4. Legendre. The volume of work on number theory increased greatly in the last half of the eighteenth century, and the first treatises devoted specifically to that subject appeared. One of the prominent figures in this development was Adrien-Marie Legendre (1752–1833). Like all other mathematicians of the time, Legendre worked in many areas of mathematics, including calculus (elliptic functions) and mechanics. He also worked in number theory and produced several profound results there in an early textbook of the subject, which went through three editions before his death.

In 1785 he published the paper "Recherches d'analyse indéterminée," in which he proved the elegant result that there are integers x, y, z satisfying an equation $ax^2 + by^2 + cz^2 = 0$ with a, b, c not all of the same sign if and only if the products $-ab$, $-bc$, and $-ca$ are quadratic residues modulo $|c|$, $|a|$, and $|b|$ respectively. He also stated the law of quadratic reciprocity, which Euler had been unable to prove, and gave a flawed proof of it. He invented the still-used Legendre symbol $\left(\frac{p}{q}\right)$ whose value is 1 if p is a quadratic residue modulo q and -1 if not. The law of quadratic reciprocity can then be elegantly stated as $\left(\frac{p}{q}\right)\left(\frac{q}{p}\right) = (-1)^{\frac{(p-1)(q-1)}{4}}$. This proof was improved in his treatise *Théorie des nombres*, published in 1798, with a subsequent edition in 1808 and a third in 1830. He also conjectured, but did not prove, that any arithmetic sequence in which the constant difference is relatively prime to the first term will contain infinitely many primes. In fact, it was this unproved assumption that invalidated his proof of quadratic reciprocity (see Weil, *1984*, pp. 329–330). He quoted Fermat's conjecture that every number is the sum of at most n n-gonal numbers, noting with regret that either Fermat never completed the treatise he intended to write or that his executors never found the manuscript. Legendre gave a proof of this fact for all numbers larger than $50n - 79$. Further continuing the work of Fermat, Euler, and Lagrange, Legendre discovered some important facts in the theory of quadratic forms.

His most original contribution to number theory, however, lay in a different direction entirely. Since no general law had been found for describing the nth prime number or even producing a polynomial whose values are all prime numbers, Legendre's attempt to estimate the number of primes among the first n integers, published in the second (1808) edition of *Théorie des nombres*, was an important step forward. Legendre's estimate for this number, which is now denoted $\pi(n)$ (π for *prime*, of course), was

$$\pi(n) \approx \frac{n}{\log n - 1.08366} = \frac{n}{\log n}\left(1 + \frac{1.08366}{\log n} + \frac{1.08366^2}{\log^2 n} + \cdots\right).$$

Here the logarithm is understood as the natural logarithm, what calculus books usually denote $\ln n$. In particular, the ratio $\dfrac{\pi(n)}{n/\log n}$ tends to 1 as n tends to infinity. Legendre did not have a proof of this result, but merely conjecturing it was an important advance in the understanding of primes.

Legendre also worked on the classification of real numbers; his contributions to this area are described below.

Number theory blossomed in the nineteenth century due to the attention of many brilliant mathematicians. Again, we have space to discuss only a few of the major figures.

1.5. Gauss. Carl Friedrich Wilhelm Gauss (1777–1855), one of the giants of modern mathematics, lived his entire life in Germany. He studied at the University of Göttingen from 1795 to 1798 and received a doctoral degree in 1799 from the University of Helmstedt. Thereafter most of his life was spent in and around Göttingen, where he did profound work in several areas of both pure and applied mathematics. In particular, he worked in astronomy, geodesy, and electromagnetic theory, producing fundamental results on the use of observational data (least squares), mapping (Gaussian curvature), and applied electromagnetism (the telegraph). But his results in pure number theory are among the deepest ever produced. Here we look at just a few of them.

Street in Göttingen named after Gauss.

The 1801 work *Disquisitiones arithmeticae* became a classical work on the properties of integers. One the earliest discoveries that Gauss made, when he was still a teenager, was a proof of the law of quadratic reciprocity. This proof was published in the *Disquisitiones*, and over the next two decades he found seven more proofs of this fundamental fact. The *Disquisitiones* also contain a proof of the fundamental theorem of arithmetic and a construction of the regular 17-sided polygon, which is possible because 17 is a Fermat prime.

A considerable portion of the *Disquisitiones* is devoted to quadratic binary forms, in an elegant and sophisticated treatment that contemporaries found difficult to understand. As Weil says (*1984*, p. 354),

> No doubt the Gaussian theory... is far more elaborate [than Legendre's treatment of the subject]; so much so, indeed, that it remained a stumbling-block for all readers of the *Disquisitiones*

until Dirichlet restored its simplicity by going back very nearly to Legendre's original construction.

In attempting to extend the law of quadratic reciprocity to higher powers, Gauss was led to consider what are now called the *Gaussian integers*, that is, complex numbers of the form $m+n\sqrt{-1}$. Gauss showed that the concepts of prime and composite number make sense in this context just as in the ordinary integers and that every such number has a unique representation (up to multiplication by the units ± 1 and $\pm\sqrt{-1}$) as a product of irreducible factors. Notice that no prime integer of the form $4n + 1$ can be "prime" in this context, since it is a sum of two squares: $4n + 1 = p^2 + q^2 = (p + q\sqrt{-1})(p - q\sqrt{-1})$. The generalization of the notion of prime number to the Gaussian integers is an early example of the endless generalization and abstraction that characterizes modern mathematics.

Gauss also gave an estimate of the number of primes not larger than x, in the form of the integral

$$\pi(x) \approx \mathrm{Li}\,(x) = \int_2^x \frac{dt}{\log(t)}\,.$$

Here, as above, the logarithm means the natural logarithm. He did not, however, prove that this approximation is asymptotically good, that is, that $\pi(x)/\mathrm{Li}\,(x)$ tends to 1 as x tends to infinity. That is the content of the prime number theorem.

1.6. Dirichlet. The works of Gauss on number theory were read by another bright star of nineteenth-century mathematics, Johann Peter Gustav Lejeune-Dirichlet (1805–1859), who contributed several gems to this difficult area. He was of Belgian ancestry (hence his French-sounding name, even though he was a German). He was born in the city of Düren, which lies between Aachen (Aix) and Köln (Cologne), but went to Paris to study at the age of 16. At the age of 20 he proved the case $n = 5$ of Fermat's last theorem. (Legendre, who was the referee for Dirichlet's paper, contributed his own proof of one subcase of this case.) That same year he returned to Germany and took up a position at the University of Breslau. In 1828 he went to Berlin and was the first star in a bright galaxy of Berlin mathematicians. In 1831 he was elected to the Berlin Academy of Sciences. That year he married Rebekah Mendelssohn, sister of the composers Felix and Fanny Mendelssohn. In 1855, dissatisfied with the heavy teaching loads in Berlin, he moved to Göttingen as the successor of Gauss, who had died that year. In 1858 he suffered a heart attack and the death of his wife, and in 1859 he himself succumbed to heart disease.

Although Dirichlet also worked in the theory of Fourier series and analytic function theory, having given the first rigorous discussion of the convergence of a Fourier series in 1829 and the modern definition of a function in 1837, we are at the moment concerned with his contributions to number theory. One of these is his 1837 theorem, already mentioned, that each arithmetic sequence in which the first term and the common difference are relatively prime contains an infinite number of primes. To prove this result, he introduced what is now called the *Dirichlet character* $\chi(n) = (-1)^k$ if $n = 2k + 1$, $\chi(n) = 0$ if n is even, along with the *Dirichlet series*

$$\sum_{n=1}^{\infty} \frac{\chi(n)}{n^s} = 1 - \frac{1}{3^s} + \frac{1}{5^s} - \frac{1}{7^s} + \cdots\,.$$

This work brought number theory and analysis together in the subject now called analytic number theory. According to Weil (*1984*, pp. 252–256), the two subjects had been drawing closer together ever since Euler began his study of elliptic functions. Elliptic functions have played a prominent role in number theory since 1830. Carl Gustav Jacob Jacobi (1804–1851), whose work is discussed in Chapter 17, published a treatise on elliptic functions in 1829 in which he used these functions to derive a formula equivalent to

$$(1)\qquad \left(\sum_{n=-\infty}^{+\infty} q^{n^2}\right)^4 = 1 + 8\sum_{k=1}^{\infty}\sigma(2k-1)q^{2k-1} + 24\sum_{k=1}^{\infty}\sum_{j=1}^{\infty}\sigma(2k-1)q^{2^j(2k-1)},$$

where $\sigma(r)$ is the sum of the divisors of r. For $r > 0$, it is obvious that the coefficient of q^r on the left is 16 times the number of ways in which r can be represented as the sum of four ordered nonzero squares, plus 32 times the number of such representations as a sum of three squares, plus 24 times the number of representations as a sum of two squares, plus 8 if the number happens to be a square. The coefficient of q^r on the right is either eight or 24 times the sum of the divisors of the largest odd number that divides r. Since that sum is always positive, the four-square theorem is a consequence, but much additional information is added on the number of such representations.

The study of Dirichlet series, in particular the simplest one of all, which defines what is now called the *Riemann zeta function*

$$\zeta(z) = \sum_{n=1}^{\infty}\frac{1}{n^z}$$

(one of several zeta functions named after distinguished mathematicians), turned out to be important in both complex analysis and number theory. The zeta function was introduced, though not under that name, by Euler, who gave the formula

$$(2)\qquad \sum_{n=1}^{\infty}\frac{1}{n^z} = \prod_{p\ \text{prime}}\left(1-\frac{1}{p^z}\right).$$

The fact that the terms in the sum are indexed by all positive integers while the factors in the product are indexed by the prime numbers accounts for the deep connections of this function with number theory. Its values at the even integers can be computed in terms of the Bernoulli numbers.[6] In fact, the Bernoulli numbers were originally introduced this way. Nowadays, the nth Bernoulli number B_n is defined to be $n!$ times the coefficient of x^n in the Maclaurin series of $x/(e^x - 1)$.

1.7. Riemann. Another giant of nineteenth-century mathematics was Georg Bernhard Riemann (1826–1866), who despite his brief life managed to make major contributions to real and complex analysis, geometry, algebraic topology, and mathematical physics. He was inspired by Legendre's work on number theory and studied under Gauss at Göttingen, where he also became a professor, Dirichlet's successor after 1859. Because of his frail health (he succumbed to tuberculosis at the age of 40), he spent considerable time in Italy, where he made the acquaintance of the productive school of Italian geometers, including Enrico Betti (1823–1892). His greatest contribution to number theory was to attempt a rigorous estimate of $\pi(n)$. For this purpose he studied the zeta function introduced above and made the famous conjecture that except for its obvious zeros at the even negative integers, all

[6] The Bernoulli numbers were the object of the first computer program written for the Babbage analytical engine.

other zeros have real part equal to $\frac{1}{2}$. This *Riemann hypothesis* forms one of the still-outstanding unsolved problems of modern mathematics, standing alongside the Goldbach conjecture and a famous conjecture in topology due to Henri Poincaré (1854–1912).[7] The first two were mentioned by Hilbert in his address at the 1900 International Congress of Mathematicians in Paris. Hilbert gave the Riemann hypothesis as the eighth of his list of 23 unsolved problems and suggested that solving it would also solve the Goldbach conjecture. Despite a great many partial results, the complete problem remains open a century later. A summary of the work on this problem through the mid-twentieth century can be found in the book by Edwards (*1974*). The zeros of $\zeta(z)$ are now being computed at a furious rate by the ZetaGrid project, an Internet-based distributed program linking tens of thousands of computers, similar to the GIMPS mentioned above (see `www.zetagrid.net`).

Poster of Riemann at the Mathematisches Institut and street named after him in Göttingen.

1.8. Fermat's last theorem. Work on Fermat's most famous conjecture continued in the nineteenth century. In 1847 Gabriel Lamé (1795–1870), published a paper in which he claimed to have proved the result. Unfortunately, he assumed

[7] This conjecture may now have been solved (see Section 4 of Chapter 12).

that complex numbers of the form $a_0 + a_1\theta + \cdots + a_{n-1}\theta^{n-1}$, where $\theta^n = 1$ and $a_0, \ldots, a_{n-1}$ are integers, can be factored uniquely, just like ordinary integers. Ernst Eduard Kummer (1810–1893) had noticed some 10 years earlier that such is not the case. This was just one of the many ways in which the objects studied by mathematicians became increasingly abstract, and the old objects of numbers and space became merely special cases of the general objects about which theorems are proved. Kummer was the first to make general progress toward a proof of Fermat's last theorem. The conjecture that $x^p + y^p = z^p$ has no solutions in positive integers x, y, and z when p is an odd prime had been proved only for the cases $p = 3$, 5, and 7 until Kummer showed that it was true for a class of primes called *regular primes*, which included all the primes less than 100 except 37, 59, and 67. This step effectively closed off the possibility that Fermat might be proved wrong by calculating a counterexample.

1.9. The prime number theorem. A good estimate of the number of primes less than or equal to a given integer N is given by $N/(\log N)$. This estimate follows from the unproved estimate of Gauss given above. The estimate suggested by Legendre, $N/(A \log N + B)$ with $A = 1$, $B = -1.08366$, turns out to be correct only in its first term. This fact was realized by Dirichlet, but only after he had written approvingly of the estimate in print. (He corrected himself in a marginal note on a copy of his paper given to Gauss.) Dirichlet suggested $\sum_{k=2}^{N} \bigl(1/(\log k)\bigr)$ as a better approximation. This problem was also studied by the Russian mathematician Pafnutii L'vovich Chebyshëv (1821–1894).[8] In 1851 Chebyshëv proved that if $\alpha > 0$ is any positive number (no matter how small) and m is any positive number (no matter how large), the inequality

$$\pi(n) > \int_2^n \frac{dx}{\log x} - \frac{\alpha n}{\log^m n}$$

holds for infinitely many positive integers n, as does the inequality

$$\pi(n) < \int_2^n \frac{dx}{\log x} + \frac{\alpha n}{\log^m n}.$$

This result suggests that $\pi(n) \sim [n/(\ln n)]$, but it would be desirable to know if there is a constant A such that

$$\pi(n) = \frac{An}{\log n} + \varepsilon_n,$$

where ε_n is of smaller order than $\pi(n)$. It would also be good to know the rate at which $\varepsilon_n/\pi(n)$ tends to zero. Chebyshëv's estimates imply that if A exists, it must be equal to 1, and as a result, Legendre's approximation cannot be valid beyond the first term. Chebyshëv was able to show that

$$0.92129 < \frac{\pi(n)}{\log n} < 1.10555.$$

Chebyshëv mentions only Legendre in his memoir on this subject and shows that his estimates refute Legendre's conjecture. He makes no mention of Gauss, whose integral $\mathrm{Li}\,(x)$ appears in his argument. Similarly, Riemann makes no mention of

---

[8] In Russian this name is pronounced "Cheb-wee-SHAWF," approximately. However, because he wrote so often in French, where he signed his name as "Tchebycheff," it is usually prounced "CHEB-ee-shev" in the West.

Chebyshëv in his 1859 paper on $\pi(x)$, even though he was in close contact with Dirichlet, and Chebyshëv's paper had been published in a French journal.

The full proof of the prime number theorem turned out to involve the use of complex analysis. As mentioned above, Riemann had studied the zeros of the Riemann zeta function. This function was also studied by two long-lived twentieth-century mathematicians, the Belgian Charles de la Vallée Poussin (1866–1962) and the Frenchman Jacques Hadamard (1865–1963), who showed independently of each other (Hadamard, *1896*; Vallée Poussin, *1896*) that the Riemann zeta function has no zeros with real part equal to 1.[9] Vallée Poussin showed later (*1899*) that

$$\pi(n) = \int_2^n \frac{dx}{\log x} + \varepsilon_n,$$

where for some $\alpha > 0$ the error term ε_n is bounded by a multiple of $ne^{-\alpha\sqrt{\log n}}$.

Number theory did not slow down or stop after the proof of the prime number theorem. On the contrary, it exploded into a huge number of subfields, each producing a prodigious amount of new knowledge year by year. However, we must stop writing on this subject sometime and move on to other topics, and so we shall close our account of number theory at this point.

2. Number systems

To the ancient mathematicians in the Middle East and Europe, numbers meant positive integers or ratios of them, in other words, what we call rational numbers. In India and China negative numbers were recognized, and 0 was recognized as a number in its own right, as opposed to merely an absence of numbers, at a very early stage. Those numbers reached Europe only a brief while before algebra led to the consideration of imaginary numbers. In this section we explore the gradual expansion of the concept of a number to include not only negative and imaginary numbers, which at least had the merit of being understandable in finite terms, but also irrational roots of equations and transcendental real numbers such as π and e, and the infinite cardinal and ordinal numbers mathematicians routinely speak about today. It is a story of the gradual enlargement of the human imagination and the clarification of vague, intuitive ideas.

2.1. Negative numbers and zero. It was mentioned in Chapter 7 that place-value systems of writing numbers were invented in Mesopotamia, India, China, and Mesoamerica. What is known about the Maya system has already been described in Chapter 5. We do not know how or even if they performed multiplication or division or how they worked with fractions. Thus, for this case all we know is that they had a place-value system and that it included a zero to occupy empty places. The Mesopotamian system was sexagesimal and had no zero for at least the first 1000 years of its existence. The other three systems were decimal, and they too were rather late in acquiring the zero. Strange though it may seem to one who has a modern education, in India and China negative numbers seem to have been used before zero was invented.

---

[9] The fact that $\zeta(z)$ has no zeros with real part equal to 1 is an elementary theorem (see Ivić, *1985*, pp. 7–8). That does not make the prime number theorem trivial, however, since the equivalence between this result and the prime number theorem is very difficult to prove. A discussion of the reasons why the two are equivalent was given by Norbert Wiener; see his paper "Some prime-number consequences of the Ikehara theorem," in his *Collected Works*, Vol. 2, pp. 254–257.

China. Chinese counting rods were red or black according as the numbers represented were positive or negative, yet no zero occurs, since obviously it would be absurd to have a rod representing no rods at all. A Chinese work on astronomy and the calendar written in the late second century CE (Li and Du, *1987*, p. 49) gives rules for adding and subtracting "strong" (positive) and "weak" (negative): When adding, like signs add and opposite signs subtract; when subtracting, like signs subtract and opposite signs add. The same kinds of rules are given in Chapter 8 ("Rectangular Tabulation") of the *Jiu Zhang Suanshu.* Yet it was a full thousand years after that time when the rules for multiplying and dividing signed numbers first appeared in the *Suanshu Chimeng* of 1303. When one is using a counting board or an abacus, no symbol for zero is needed, since it is visually apparent that a given square has no numbers written in it or that the beads on a string are in their "empty" position. The first known occurrence of the symbol 0 for zero occurs in a work dating to the year 1247.

India. Around the year 500, Aryabhata I, used a place-value decimal system without zero. A century later Brahmagupta introduced zero in connection with the *kuttaka* method described above. Although he used the word *sunya* (empty) for this concept and it really does denote an empty place in that method, the idea that the algorithm $x \mapsto ax + b$ can be executed as $x \mapsto ax$ when no b is present suggests the use of a neutral element for addition, and that is what the zero is. Brahmagupta gave complete rules for addition, subtraction, multiplication, and division of both positive and negative quantities and zero. As we know, division by zero must be considered separately and either rejected or given some special meaning. Brahmagupta (Colebrooke, *1817*, pp. 339–340) showed some puzzlement about this, and he wrote:

> Cipher, divided by cipher, is nought. Positive, divided by negative, is negative. Negative, divided by affirmative, is negative. Positive or negative, divided by cipher, is a fraction with [cipher] for denominator, or cipher divided by negative or affirmative [is a fraction with the latter for denominator].

The word *cipher* here translates the Sanskrit *sunya* or *kha*, both meaning empty space. The last rule given here is not a happy effort at a definition; it is rather like saying that a jar contains its contents. Not much new information is conveyed by the sentence. But the obscurity is natural due to the complete absence of any human experience with situations corresponding to division by zero. Five hundred years later Bhaskara was still having trouble with this concept (Colebrooke, *1817*, p. 19):

> A definite quantity divided by cipher is the submultiple of nought [that is, a fraction with zero for its denominator, just as Brahmagupta had said]. The product of cipher is nought: but it must be retained as a multiple of cipher, if any further operation impend. Cipher having become a multiplier, should nought afterwards become a divisor, the definite quantity must be understood to be unchanged.

Although these principles might be more clearly stated, it seems that Bhaskara may have in mind here some operations similar to those that occur in limiting operations, for example, considering the appropriate value of a fraction such as $(5x^2 + 4x)/(3x^2 - 2x)$ when x becomes zero. One can formally cancel the x without thinking about whether or not it is zero. After cancellation, setting $x = 0$ gives the fraction the value -2. Bhaskara is explicit in saying that zero added to any number leaves that number unchanged. Hence for him it is more than a mere placeholder; arithmetic operations can be performed with it. The use of an empty circle or a circle with its center marked as a symbol for zero seems to be culturally invariant, since it appears in inscriptions in India from the ninth century, in Greek documents from the second century, and in Chinese documents from the thirteenth century.

Islamic number systems. The transmission of Hindu treatises to Baghdad led ultimately to the triumph of the numerals used today. According to al-Daffa (*1973*, p. 51) the Sanskrit words for an empty place were translated as the Arabic word *sifr*, which became the English words *cipher* and *zero* and their cognates in other European languages. Al-Daffa also points out that the earliest record of the symbol for zero in India comes from an inscription at Gwalior dating to 876, and that there is a document in Arabic dating from 873 in which this symbol occurs.

2.2. Irrational and imaginary numbers. In a peculiar way, the absence of a place-value system of writing numbers may have stimulated the creation of mathematics in ancient Greece in the case of irrational numbers. Place-value notation provides approximate square roots in practical form, even when the expansion does not terminate.[10] A cuneiform tablet from Iraq (Yale Babylonian Collection 7289) shows a square with its diagonals drawn and the sexagesimal number 1;24,51,10, which gives the length of the diagonal of a square of side 1 to great precision. But in all the Chinese, Mesopotamian, Egyptian,[11] and Hindu texts there is nothing that can be considered a theoretical discussion of "numbers" whose expansions do not terminate.

The word *numbers* is placed in inverted commas here because the meaning of the square root of 2 is not easy to define. It is very easy to go around in circles making the definition. The difficulty came in a clash of geometry and arithmetic, the two fundamental modes of mathematical thinking. From the arithmetical point of view the problem is minimal. If numbers must be what we now call positive rational numbers, then some of them are squares and some are not, just as some integers are triangular, square, pentagonal, and so forth, while others are not. No one would be disturbed by this fact; and since the Greeks had no place-value system to suggest an infinite process leading to an exact square root, they might not have speculated deeply on the implications of this arithmetical distinction in geometry. But in fact, they did speculate on both the numerical and geometric aspects of the problem, as we shall now see. We begin with the arithmetical problem.

The arithmetical origin of irrationals: nonsquare rational numbers. In Plato's dialogue *Theatetus*, the title character reports that a certain Theodorus proved that the integers 2, 3, 5, and so on, up to 17 have no (rational) square roots, except of course the obvious integers 1, 4, and 9; and he says that for some reason, Theodorus stopped at that point. On that basis the students decided to classify numbers as

---

[10] In the case of Chinese mathematics the end of a nonterminating square root was given as a common fraction, and Simon Stevin likewise terminated infinite decimals with common fractions.

[11] Square roots, called *corners*, are rarely encountered in the Egyptian papyri, and Gillings (*1972*, p. 214) suggests that they were found from tables of squares.

equilateral and *oblong*. The former class consists of the squares of rational numbers, for example $\frac{25}{9}$, and the latter are all other positive rational numbers, such as $\frac{3}{2}$.

One cannot help wondering why Theodorus stopped at 17 after proving that the numbers 3, 5, 6, 7, 8, 10, 11, 12, 13, 14, and 15 have no square roots. The implication is that Theodorus "got stuck" trying to prove this fact for a square of area 17. What might have caused him to get stuck? Most assuredly the square root of 17 is irrational, and the proof commonly given nowadays to show the irrationality of $\sqrt{3}$, for example, based on the unique prime factorization of integers, works just as well for 17 as for any other number. If Theodorus had our proof, he wouldn't have been stuck doing 17, and he wouldn't have bothered to do so many special cases, since the proofs are all the same. Therefore, we must assume that he was using some other method.

An ingenious conjecture as to Theodorus' method was provided by Knorr (1945–1997) (*1975*). Knorr suggests that the proof was based on the elementary distinction between even and odd. To see how such a proof works, suppose that 7 is an equilateral number in the sense mentioned by Theatetus. Then there must exist two integers such that the square of the first is seven times the square of the second. We can assume that both integers are odd, since if both are even, we can divide them both by 2, and it is impossible for one of them to be odd and the other even. For the fact that the square of one of them equals seven times the square of the other would imply that an odd integer equals an even integer if this were the case. But it is well known that the square of an odd integer is always 1 larger than a multiple of 8. The supposition that the one square is seven times the other then implies that an integer 1 larger than a multiple of 8 equals an integer 7 larger than a multiple of 8, which is clearly impossible.

This same argument shows that none of the odd numbers 3, 5, 7, 11, 13, and 15 can be the square of a rational number. With a slight modification it can also be made to show that none of the numbers 2, 6, 8, 10, 12, and 14 is the square of a rational number, although no argument is needed in the case of 8 and 12, since it is already known that $\sqrt{2}$ and $\sqrt{3}$ are irrational. Notice that the argument fails, as it must, for 9: A number 9 larger than a multiple of 8 is also 1 larger than a multiple of 8. However, it also breaks down for 17 and for the same reason: A number 17 larger than a multiple of 8 is also 1 larger than a multiple of 8. Thus, even though it is *true* that 17 is not the square of a rational number, the argument just given, based on what we would call arithmetic modulo 8, cannot be used to *prove* this fact. In this way the conjectured method of proof would explain why Theodorus got stuck at 17.

The Greeks thus found not only that there was no integer whose square is, say, 11 (which is a simple matter of ruling out the few possible candidates), but also that there was not even any rational number having this property; that is, 11 is not the square of anything they recognized as a number.

The geometric origin of irrationals: incommensurable magnitudes. A second, "geometric" theory of the origin of irrational numbers comes from geometry and seems less plausible. If we apply the Euclidean algorithm to the side and diagonal of the regular pentagon in Fig. 1, we find that the pair AD and CD get replaced by lines equal to CD and CF, which are the diagonal and side of a smaller pentagon. Thus, no matter how many times we apply the procedure of the Euclidean algorithm, the result will always be a pair consisting of the side and diagonal of a

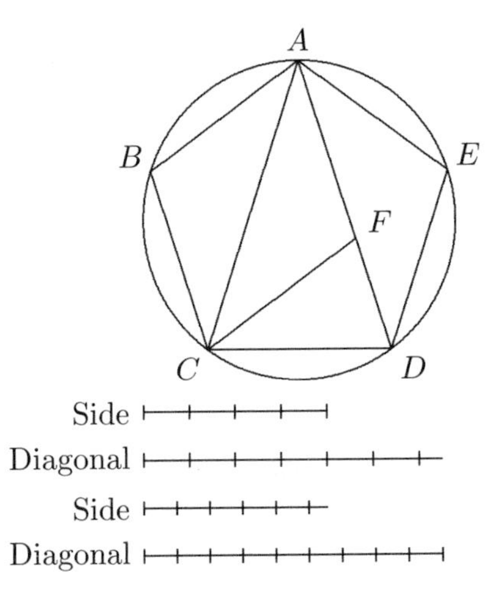

FIGURE 1. Diagonal and side of a regular pentagon. If a unit is chosen that divides the side into equal parts, it cannot divide the diagonal into equal parts, and vice versa.

pentagon. Therefore, in this case the Euclidean algorithm will *never* produce an equal pair of lines. We know, however, that it *must* produce an equal pair if a common measure exists. We conclude that *no common measure can exist for the side and diagonal of a pentagon.*

The argument just presented was originally given by von Fritz (*1945*). Knorr (*1975*, pp. 22–36) argues against this approach, however, pointing out that the simple arithmetic relation $d^2 = 2s^2$ satisfied by the diagonal and side of a square can be used in several ways to show that d and s could not both be integers, no matter what length is chosen as unit. Knorr prefers a reconstruction closer to the argument given in Plato's *Meno*, in which the problem of doubling a square is discussed.[12] Knorr points out that when discussing irrationals, Plato and Aristotle always invoke the side and diagonal of a square, never the pentagon or the related problem of dividing a line in mean and extreme ratio, which they certainly knew about.

Whatever the argument used, the Greeks discovered the existence of incommensurable pairs of line segments before the time of Plato. For Pythagorean metaphysics this discovery was disturbing: Number, it seems, is *not* adequate to explain all of nature. A legend arose that the Pythagoreans attempted to keep secret the discovery of this paradox.[13] However, scholars believe that the discovery of incommensurables came near the end of the fifth century BCE, when the original Pythagorean group was already defunct.

---

[12] In Chapter 9 we invoke the same passage to speculate on the origin of the Pythagorean theorem.
[13] The legend probably arose from a passage in Chapter 18, Section 88 of the *Life of Pythagoras* by Iamblichus. Iamblichus says that a certain Hippasus perished at sea, a punishment for his impiety because he published "the sphere of the 12 pentagons" (probably the radius of the sphere circumscribed about a dodecahedron), taking credit as if he had discovered it, when actually everything was a discovery of That Man (Pythagoras, who was too august a personage to be called by name). Apparently, new knowledge was to be kept in-house as a secret of the initiated and attributed in a mystical sense to Pythagoras.

The existence of incommensurables throws doubt on certain oversimplified proofs of geometric proportion, and this question is discussed in detail in Chapter 10. At present we are concerned with its effect on the concept of a *number*. At the beginning, one would have to say that the effect was almost nil. Geometry and arithmetic were separate subjects in the Greek tradition. But when algebra arose and the Persian mathematician Omar Khayyam (1050–1130) discovered that some cubic equations that could not be solved arithmetically had geometric solutions, the idea of a *real number* as a *ratio of lines* began to take shape.

The idea of using a line to stand for a number, the numbers being regarded as the length of the line, is very familiar to us and has its origin in the work of the ancient Greeks and medieval Muslim mathematicians. In Europe this idea received some development in the work of the fourteenth-century Bishop of Lisieux, Nicole d'Oresme, whose graphical representation of relationships was a forerunner of our modern analytic geometry. Oresme was familiar with the concept of incommensurable lines, a subject that was missing from earlier medieval work in geometry, and he was careful to keep the distinction between commensurable and incommensurable clear. Indeed, Oresme was even more advanced than the average twentieth-century person, in that he recognized a logical difficulty in talking about a power of, say, $\frac{1}{2}$ that equals $\frac{1}{3}$, whereas modern students are taught how to use the *rules* of exponents, but not encouraged to ask what is meant by expressions such as $\sqrt{2}^{\sqrt{3}}$.

A great advance came in the seventeenth century, when analytic geometry as we know it today was invented by Descartes and Fermat. Fermat's work seems somewhat closer to what we know, in the sense that he used a pair of mutually perpendicular axes; on the other hand, he believed that only dimensionally equivalent expressions could be added. This is the restriction that led Omar Khayyam to write a cubic equation in the form equivalent to $x^2 + ax^2 + b^2x = b^2c$, in which each term is of degree 3. In his *Géométrie*, Descartes showed how to avoid this complication. The difficulty lay in the geometric representation of the operation of multiplication. Because ratios of lines were not always numbers, Euclid did not make the association of a line with a number called its length. The product of two numbers is a number, but Euclid did not speak of the product of two lines. He spoke instead of the rectangle on the two lines. That was the tradition Omar Khayyam was following. Stimulated by algebra, however, and the application of geometry to it, Descartes looked at the product of two lengths in a different way. As pure numbers, the product ab is simply the fourth proportional to $1 : a : b$. That is, $ab : b :: a : 1$. He therefore fixed an arbitrary line that he called I to represent the number 1 and represented ab as the line that satisfied the proportion $ab : b :: a : I$, when a and b were lines representing two given numbers.

The notion of a *real number* had at last arisen, not as most people think of it today—an infinite decimal expansion—but as a ratio of line segments. Only a few decades later Newton defined a (real) number to be "the ratio of one magnitude to another magnitude of the same kind, arbitrarily taken as a unit." Newton classified numbers as integers, fractions, and surds (Whiteside, *1967*, Vol. 2, p. 7). Even with this amount of clarity introduced, however, mathematicians were inclined to gloss over certain difficulties. For example, there is an arithmetic rule according to which $\sqrt{ab} = \sqrt{a}\sqrt{b}$. Even with Descartes' geometric interpretation of these results, it is not obvious how this rule is to be proved. The use of the decimal system, with its easy approximations to irrational numbers, soothed the consciences of mathematicians and gave them the confidence to proceed with their development of the calculus. No one even seemed very concerned about the absence of any good geometric construction of cube roots and higher roots of real numbers. The real line answered the needs of algebra in that it gave a representation of any real root there might be of any algebraic equation with real numbers as coefficients. It was some time before anyone realized that geometry still had resources that even algebra did not encompass and would lead to numbers for which pure algebra had no use.

Those resources included the continuity of the geometric line, which turned out to be exactly what was needed for the limiting processes of calculus. It was this property that made it sensible for Euler to talk about the number that we now call e, that is,

$$e = \lim_{n\to\infty}\left(1+\frac{1}{n}\right)^n = \sum_{n=0}^{\infty}\frac{1}{n!} = 2.7182818284590\ldots,$$

and the other Euler constant

$$\gamma = \lim_{n\to\infty}\left[\left(\sum_{k=1}^{n}\frac{1}{k}\right) - \log n\right] = 0.5772156649\ldots.$$

The intuitive notion of continuity assured mathematicians that there were points on the line, and hence infinite decimal expansions, that must represent these numbers, even though no one would ever know the full expansions. The geometry of the line provided a geometric representation of real numbers and made it possible to reason about them without having to worry about the decimal expansion.

The continuity of the line brought the realization that the real numbers had more to offer than merely convenient representations of the solutions of equations. They could even represent some numbers such as e and γ that had not been found to be solutions of any equations. The line was richer than it needed to be for algebra alone. The concept of a real number had allowed arithmetic to penetrate into parts of geometry where even algebra could not go. The sides and diagonals of regular figures such as squares, cubes, pentagons, pyramids, and the like all had ratios that could be represented as the solutions of equations, and hence are algebraic. For example, the diagonal D and side S of a pentagon satisfy $D^2 = S(D+S)$. For a square the relationship is $D^2 = 2S^2$, and for a cube it is $D^2 = 3S^2$. But what about the number we now call π, the ratio of the circumference C of a circle to its diameter D? In the seventeenth century Leibniz noted that any line that could be constructed using Euclidean methods (straightedge and compass) would have a length that satisfied some equation with rational coefficients. In a number of letters and papers written during the 1670s, Leibniz was the first to contrast what is algebraic (involving polynomials with rational coefficients) with objects that he called *analytic* or *transcendental* and the first to suggest that π might be transcendental. In the preface to his pamphlet *De quadratura arithmetica circuli* (*On the Arithmetical Quadrature of the Circle*), he wrote:

> A complete quadrature would be one that is both analytic and linear; that is, it would be constructed by the use of curves whose equations are of [finite] degrees. The brilliant Gregory [James Gregory, 1638–1675], in his book *On the Exact Quadrature of the Circle*, has claimed that this is impossible, but, unless I am mistaken,

> has given no proof. I still do not see what prevents the circumference itself, or some particular part of it, from being measured [that is, being commensurable with the radius], a part whose arc has a ratio to its sine [half-chord] that can be expressed by an equation of finite degree. But to express the ratio of the arc to the sine *in general* by an equation of finite degree is impossible, as I shall prove in this little work. [Gerhardt, *1971*, Vol. 5, p. 97]

No representation of π as the root of a polynomial with rational coefficients was ever found. This ratio had a long history of numerical approximations from all over the world, but no one ever found any nonidentical equation satisfied by C and D. The fact that π is transcendental was first proved in 1881 by Ferdinand Lindemann (1852–1939). The complete set of real numbers thus consists of the positive and negative rational numbers, all real roots of equations with integer coefficients (the *algebraic* numbers), and the transcendental numbers. All transcendental numbers and some algebraic numbers are irrational. Examples of transcendental numbers are rather difficult to produce. The first number to be proved transcendental was the base of natural logarithms e, and this proof was achieved only in 1873, by the French mathematician Charles Hermite (1822–1901). It is still not known whether the Euler constant $\gamma \approx 0.57712$ is even irrational.

The arithmetization of the real numbers. Not until the nineteenth century, when mathematicians took a retrospective look at the magnificent edifice of calculus that they had created and tried to give it the same degree of logical rigor possessed by algebra and Euclidean geometry, were attempts made to define real numbers arithmetically, without mentioning ratios of lines. One such definition by Richard Dedekind (1831–1916), a professor at the Zürich Polytechnikum, was inspired by a desire for rigor when he began lecturing to students in 1858. He found the rigor he sought without much difficulty, but did not bother to publish what he regarded as mere common sense until 1872, when he wished to publish something in honor of his father. In his book *Stetigkeit und irrationale Zahlen* (*Continuity and Irrational Numbers*) he referred to Newton's definition of a real number:

> ... the way in which the irrational numbers are usually introduced is based directly upon the conception of extensive magnitudes—which itself is nowhere carefully defined—and explains number as the result of measuring such a magnitude by another of the same kind. Instead of this I demand that arithmetic shall be developed out of itself.

As Dedekind saw the matter, it was really the *totality* of rational numbers that defined a ratio of continuous magnitudes. Although one might not be able to say that two continuous quantities a and b had a ratio *equal* to, or defined by, a ratio $m : n$ of two integers, an inequality such as $ma < nb$ could be interpreted as saying that the real number $a : b$ (whatever it was) was *less than* the rational number n/m. Thus a positive real number could be defined as a way of dividing the positive rational numbers into two classes, those that were larger than the number and those that were equal to it or smaller, and every member of the first class was larger than every member of the second class. But, so reasoned Dedekind, once the positive rational numbers have been partitioned in this way, the two classes themselves can be regarded as the number. They are a well-defined object, and one can define arithmetic operations on such classes so that the resulting system has all the properties we want the real numbers to have, especially the essential one for calculus: continuity. Dedekind claimed that in this way he was able to prove rigorously for the first time that $\sqrt{2}\sqrt{3} = \sqrt{6}$.[14]

The practical-minded reader who is content to use approximations will probably be getting somewhat impatient with the discussion at this point and asking if it was really necessary to go to so much trouble to satisfy a pedantic desire for rigor. Such a reader will be in good company. Many prominent mathematicians of the time asked precisely that question. One of them was Rudolf Lipschitz (1832–1903). Lipschitz didn't see what the fuss was about, and he objected to Dedekind's claims of originality (Scharlau, *1986*, p. 58). In 1876 he wrote to Dedekind:

> I do not deny the validity of your definition, but I am nevertheless of the opinion that it differs only in form, not in substance, from what was done by the ancients. I can only say that I consider the definition given by Euclid... to be just as satisfactory as your definition. For that reason, I wish you would drop the claim that such propositions as $\sqrt{2}\sqrt{3} = \sqrt{6}$ have never been proved. I think the French readers especially will share my conviction that Euclid's book provided necessary and sufficient grounds for proving these things.

Dedekind refused to back down. He replied (Scharlau, *1986*, pp. 64–65):

> I have never imagined that my concept of the irrational numbers has any particular merit; otherwise I should not have kept it to myself for nearly fourteen years. Quite the reverse, I have always been convinced that any well-educated mathematician who seriously set himself the task of developing this subject rigorously would be bound to succeed... Do you really believe that such a proof can be found in any book? I have searched through a large collection of works from many countries on this point, and what does one find? Nothing but the crudest circular reasoning, to the effect that $\sqrt{a}\sqrt{b} = \sqrt{ab}$ because $\left(\sqrt{a}\sqrt{b}\right)^2 = \left(\sqrt{a}\right)^2\left(\sqrt{b}\right)^2 = ab$; not the slightest explanation of how to multiply two irrational numbers. The proposition $(mn)^2 = m^2n^2$, which is proved for rational numbers, is used unthinkingly for irrational numbers. Is it not scandalous that the teaching of mathematics in schools is regarded as a particularly good means to develop the power of reasoning, while no other discipline (for example, grammar) would tolerate such gross offenses against logic for a minute? If one is to proceed scientifically, or cannot do so for lack of time, one should at least honestly tell the pupil to believe a proposition on the word of the teacher, which the students are willing to do anyway. That is better than destroying the pure, noble instinct for correct proofs by giving spurious ones.

[14] In his paper (*1992*) David Fowler (1937–2004) investigated a number of approaches to the arithmetization of the real numbers and showed how the specific equation $\sqrt{2}\sqrt{3} = \sqrt{6}$ could have been proved geometrically, and also how difficult this proof would have been using many other natural approaches.

Mathematicians have accepted the need for Dedekind's rigor in the teaching of mathematics majors, although the idea of defining real numbers as partitions of the rational numbers (Dedekind cuts) is no longer the most popular approach to that rigor. More often, students are now given a set of axioms for the real numbers and asked to accept on faith that those axioms are consistent and that they characterize a set that has the properties of a geometric line. Only a few books attempt to start with the rational numbers and construct the real numbers. Those that do tend to follow an alternative approach, defining a real number to be a sequence of rational numbers (more precisely, an equivalence class of such sequences, one of which is the sequence of successive decimal approximations to the number).

2.3. Imaginary and complex numbers. Although imaginary numbers seem more abstract to moderns than irrational numbers, that is because their physical interpretation is more remote from everyday experience. One interpretation of $i = \sqrt{-1}$, for example, is as a rotation through a right angle (the effect of multiplying by i in the complex plane). We have an intuitive concept of the length of a line segment and decimal approximations to describe that length as a number; that is what gives us confidence that irrational numbers really are numbers. But it is difficult to think of a rotation as a number. On the other hand, the rules for multiplying complex numbers—at least those whose real and imaginary parts are rational—are much simpler and easier to understand than the definition just given for irrationals. In fact, complex numbers were understood before real numbers were properly defined; mathematicians began trying to make sense of them as soon as there was a clear need to do so. That need came not, as one might expect, from trying to solve quadratic equations such as $x^2 - 2x + 2 = 0$, where the quadratic formula produces $x = -1 \pm \sqrt{-1}$. It was possible in this case simply to say that the equation had no solution. On the other hand, as discussed in Chapter 14, the sixteenth-century Italian mathematicians succeeded in giving an arithmetic solution of the general cubic equation. However, the algorithm for finding the solution had the peculiar property that it involved taking the square root of a negative number precisely when there were three real solutions. Looking at their algorithm as a formula, one would find that the solution of the equation $x^3 - 7x + 6 = 0$ is

$$x = \sqrt[3]{3 - \sqrt{-\frac{100}{27}}} - \sqrt[3]{3 + \sqrt{-\frac{100}{27}}}\,.$$

We cannot say that the equation has no roots, since it obviously has 1, 2, and -3 as roots. Thus the challenge arose: Make sense of this formula. Make it say "1, 2, and -3."

This challenge was taken up by Rafael Bombelli (1526–1572), an engineer in the service of an Italian nobleman. Bombelli was the author of a treatise on algebra which he wrote in 1560, but which was not published until 1572. In that treatise he invented the name "plus of minus" to denote a square root of -1 and "minus of minus" for its negative. He did not think of these two concepts as different numbers, but rather as the *same* number being added in the first case and subtracted in the second. What is most important is that he realized what rules must apply to them in computation: plus of minus times plus of minus makes minus and minus of minus times minus of minus makes minus, while plus of minus times minus of minus makes plus. Bombelli had no systematic way of taking the cube root of a complex number. In considering the equation $x^3 = 15x + 4$, he found by applying the formula that

$x = \sqrt[3]{2+\sqrt{-121}}+\sqrt[3]{2-\sqrt{-121}}$. In this case, however, Bombelli was able to work backward, since he knew in advance that one root is 4; the problem was to make the formula *say* "4." Bombelli had the idea that the two cube roots must consist of real numbers together with his "plus of minus" or "minus of minus." Since the imaginary parts in the sum of the two cube roots must cancel out and the real parts must add up to 4, it seems obvious that the real parts of the cube roots must be 2. In our terms, the cube roots must be $2 \pm t\sqrt{-1}$ for some t. Then since the cube of the cube roots must be $2 \pm 11\sqrt{-1}$ (what Bombelli called 2 plus 11 times "plus of minus"), it is clear that the cube roots must be 2 plus "plus of minus" and 2 minus "plus of minus," that is, $2 \pm \sqrt{-1}$. As a way of solving the equation, this reasoning is circular, but it does allow the formula for solving the cubic equation to make sense.

In an attempt to make these numbers more familiar, the English mathematician John Wallis (1616–1703) pointed out that while no positive or negative number could have a negative square, nevertheless it is also true that no physical quantity can be negative, that is, less than nothing. Yet negative numbers were accepted and interpreted as retreats when the numbers measure advances along a line. Wallis thought that what was allowed in lines might also apply to planes, pointing out that if 30 acres are reclaimed from the sea, and 40 acres are flooded, the net amount "gained" from the sea would be -10 acres. Although he did not say so, it appears that he regarded this real loss of 10 acres as an imaginary gain of a square of land $\sqrt{-435600} = 660\sqrt{-1}$ feet on a side.

What he did say in his 1673 treatise on algebra was that one could represent $\sqrt{-bc}$ as the mean proportional between $-b$ and c. The mean proportional is easily found for two positive line segments b and c. Simply lay them end to end, use the union as the diameter of a circle, and draw the half-chord perpendicular to that diameter at the point where the two segments meet. That half-chord is the mean proportional. When one of the numbers $(-b)$ was regarded as negative, Wallis regarded the negative quantity as an oppositely directed line segment. He then modified the construction of the mean proportional between the two segments. When two oppositely directed line segments are joined end to end, one end of the shorter segment lies between the point where the two segments meet and the other end of the longer segment, so that the point where the segments meet lies *outside* the circle passing through the other two endpoints. Wallis interpreted the mean proportional as the tangent to the circle from the point where the two segments meet. Thus, whereas the mean proportional between two positive quantities is represented as a sine, that between a positive and negative quantity is represented as a tangent.

Wallis applied this procedure in an "imaginary" construction problem. First he stated the following "real" problem. Given a triangle having side AP of length 20, side PB of length 15, and altitude PC of length 12, find the length of side AB, taken as base in Fig. 2. Wallis pointed out that two solutions were possible. Using the foot of the altitude as the reference point C and applying the Pythagorean theorem twice, he found that the possible lengths of AB were 16 ± 9, that is, 7 and 25. This construction is a well-known method of solving quadratic equations geometrically, given earlier by Descartes. It always works when the roots are real, whether positive or negative. He then proposed reversing the data, in effect considering an impossible triangle having side AP of length 20, side PB of length 12, and altitude PC of length 15. Although the algebraic problem has no real solution, a fact verified by

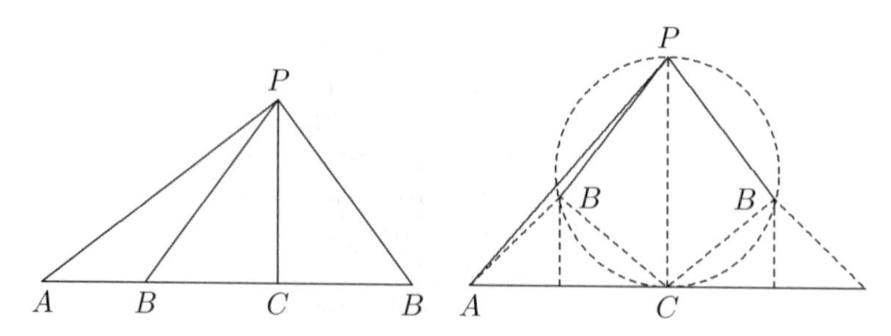

FIGURE 2. Wallis' geometric solution of quadratic equation with real roots (left) and complex roots (right).

the geometric figure (Fig. 2), one could certainly draw the two line segments AB. These line segments could therefore be interpreted as the numerical solutions of the equation, representing a triangle with one side having imaginary length.

The rules given by Bombelli made imaginary and complex numbers accessible, and they turned out to be very convenient in many formulas. Euler made free use of them, studying power series in which the variables were allowed to be complex numbers and deriving a famous formula

$$e^{v\sqrt{-1}} = \cos v + \sqrt{-1}\sin v\,.$$

Euler derived this result in a paper on ballistics written around 1727 (see Smith, *1929*, pp. 95–98), just after he moved to Russia. But he had no thought of representing $\sqrt{-1}$ as we now do, on a line perpendicular to the real axis.

Wallis' work had given the first indication that complex numbers would have to be interpreted as line segments in a plane, a discovery made again a century later by the Norwegian surveyor Caspar Wessel (1745–1818). The only mathematical paper he ever wrote was delivered to the Royal Academy in Copenhagen, Denmark in 1797, but he had been in possession of the results for about a decade at that time. In that paper (Smith, *1929*, pp. 55–66), he explained how to multiply lines in a plane by multiplying their lengths and adding the angles they make with a given reference line, on which a length is chosen to represent $+1$:

> Let $+1$ designate the positive rectilinear unit and $+\epsilon$ a certain other unit perpendicular to the positive unit and having the same origin; the direction angle of $+1$ will be equal to $0°$, that of -1 to $180°$, that of $+\epsilon$ to $90°$, and that of $-\epsilon$ to $-90°$ or $270°$. By the rule that the direction angle of the product shall equal the sum of the angles of the factors, we have: $(+1)(+1) = +1$; $(+1)(-1) = -1$; $(-1)(-1) = +1$; $(+1)(+\epsilon) = +\epsilon$; $(+1)(-\epsilon) = -\epsilon$; $(-1)(+\epsilon) = -\epsilon$; $(-1)(-\epsilon) = +\epsilon$; $(+\epsilon)(+\epsilon) = -1$; $(+\epsilon)(-\epsilon) = +1$; $(-\epsilon)(-\epsilon) = -1$. From this it is seen that ϵ is equal to $\sqrt{-1}$. [Smith, *1929*, p. 60]

Wessel noticed the connection of these rules with the addition and subtraction formulas for sign and cosine and gave the formula $(\cos x + \epsilon \sin x)(\cos y + \epsilon \sin y) = \cos(x + y) + \epsilon \sin(x + y)$. On that basis he was able to reduce the extraction of the nth root of a complex number to extracting the same root for a positive real number and dividing the polar angle by n.

The reaction of the mathematical community to this simple but profound idea was less than overwhelming. Wessel's work was forgotten for a full century. In the meantime another mathematician by avocation, the French accountant Jean Argand (1768–1822), published the small book *Essai sur une manière de représenter les quantités imaginaires dans les constructions géométriques* at his own expense in 1806, modestly omitting to name himself as its author, in which he advocated essentially the same idea, thinking, as Wallis had done, of an imaginary number as the mean proportional between a positive number and a negative number. Through a complicated series of events this book and its author gradually became known in the mathematical community. There was, however, resistance to the idea of interpreting complex numbers geometrically, since they had arisen in algebra. But geometry was essential to the algebra of complex numbers, as shown by the fact that a proof of the fundamental theorem of algebra by Gauss in 1799 is based on the idea of intersecting curves in a plane. The lemmas that Gauss used for the proof had been proved earlier by Euler using the algebra of imaginary numbers, but Gauss gave a new proof using only real numbers, precisely to avoid invoking any properties of imaginary numbers.

Even though he avoided the algebra of imaginary numbers, Gauss still needed the continuity properties of real numbers, which, as we just saw, were not fully arithmetized until many years later.[15] Continuity was a geometric property not explicitly found in Euclid, but Gauss expressed the opinion that continuity could be arithmetized. In giving a fifth proof of this theorem half a century later, he made full use of complex numbers. In fact, the complex plane is sometimes called the *Gaussian plane.*

2.4. Infinite numbers. The problem of infinity has occupied mathematicians for a very long time. Neither arithmetic nor geometry can place any preassigned limit on the sizes of objects. An integer can be as large as we like, and a line can be bisected as many times as we like. These are *potential infinities* and *potential infinitesimals.* Geometry can lead to the concept of an *actual infinity* and an *actual infinitesimal.* A line, plane, or solid is an infinite set of points; and in a sense a point is an infinitesimal (infinitely short) line, a line is an infinitesimal (infinitely narrow) plane, and a plane is an infinitesimal (infinitely thin) solid. These notions of the infinite and the infinitesimal present a logical problem for beings whose experience extends over only a finite amount of space, whose senses cannot resolve impressions below a certain threshold, and whose reasoning is presented using a finite set of words. The difficulties of dividing by zero and the problem of incommensurables, mentioned above, are two manifestations of this difficulty. We shall see others in later chapters.

The infinite in Hindu mathematics. Early Hindu mathematics had a prominent metaphysical component that manifested itself in the handling of the infinite. The Hindus accepted an actual infinity and classified different kinds of infinities. This part of Hindu mathematics is particularly noticeable with the Jainas. They classified numbers as enumerable, unenumerable, and infinite, and space as one-dimensional, two-dimensional, three-dimensional, and infinitely infinite. Further, they seem to have given a classification of infinite numbers remarkably similar to

[15] The Czech scholar Bernard Bolzano (1781–1848) showed how to approach the idea of continuity analytically in an 1817 paper. One could argue that his work anticipated Dedekind's arithmetization of real numbers.

the modern theory of infinite ordinals. The idea is to progress through the finite numbers $2, 3, 4, \ldots$ until the "first unenumerable" number is reached. This number corresponds to what is now called ω, the first infinite ordinal number. Then, exactly as in modern set theory, one can consider the unenumerable numbers $\omega + 1$, $\omega + 2, \ldots, \omega^2$, and so on. We do not have enough specifics to say any more, but there is a very strong temptation to say that the Jaina classification of enumerable, unenumerable, infinite corresponds to our modern classification of finite, countably infinite, and uncountably infinite, but of course it is only a coincidental prefiguration.

Infinite ordinals and cardinals. A fuller account of the creation of the theory of cardinal and ordinal numbers in connection with set theory is given in Chapter 19. At this point, we merely note that these theories were created along with set theory in the late nineteenth century through the work of several mathematicians, most prominently Georg Cantor (1854–1918). The relation between cardinal and ordinal numbers is an important one that has led to a large amount of research. Briefly, ordinal numbers arise from continuing the ordinary series of natural numbers "past infinity." Cardinal numbers arise from comparing two sets by matching their elements in a one-to-one manner.

3. Combinatorics

From earliest times mathematicians have been concerned with counting things and with space, that is, with the arrangement of objects of interest. Counting arrangements of things became a separate area of study within mathematics. We now call this area *combinatorics*, and it has ramified to include a number of distinct areas of interest, such as formulas for summation of powers, graph theory, magic squares, Latin squares, Room squares, and others. We have seen already that magic squares were used in divination, and there is a very prominent connection between this area and some varieties of mystical thinking. It may be coincidental that the elementary parts of probability theory, the parts that students find most frustrating, involve these sophisticated methods of counting. Probability is the mathematization of possible outcomes of events, exactly the matters that are of interest to people who consult oracles. These hypothetical happenings are usually too many to list, and some systematic way of counting them is needed.

3.1. Summation rules. The earliest example of a summation problem comes from the Ahmose Papyrus. Problem 79 describes seven houses in which there are seven cats, each of which had eaten seven mice, each of which had eaten seven seeds, each of which would have produced seven *hekats* of grain if sown. The author asks for the total, that is, for the sum $7 + 7^2 + 7^3 + 7^4 + 7^5$, and gives the answer correctly as 19,607. The same summation with a different illustration is found in Fibonacci's *Liber abaci* of 1202. In this example we encounter the summation of a finite geometric progression.

A similar example is Problem 34 of Chapter 3 of the *Sun Zi Suan Jing*, which tells of 9 hillsides, on each of which 9 trees are growing, with 9 branches on each tree, 9 bird's nests on each branch, and 9 birds in each nest. Each bird has 9 young, each young bird has 9 feathers, and each feather has 9 colors. The problem asks for the total number of each kind of object and gives the answer: 81 trees, 729 branches, 6561 nests, 59,049 birds, 531,441 young birds, 4,782,969 feathers, and

43,046,721 colors. It does *not* ask for the sum of this series, which indeed is an absurd operation, given that the objects are of different kinds.

Hindu mathematicians gave rules for summing geometric progressions and also the terms of arithmetic progressions and their squares. In Section 3 of Chapter 12 of the *Brahmasphutasiddhanta* (Colebrooke, *1817*, pp. 290–295), Brahmagupta gives four rules for dealing with arithmetic progressions. The first rule gives the sum of an arithmetic progression as its average value (half the sum of its first and last terms) times its period, which is the number of terms in the progression. We would write this rule as the formula

$$\sum_{k=0}^{n}(a+kd) = (n+1)\frac{a+(a+nd)}{2} = (n+1)a + d\frac{n(n+1)}{2}.$$

For the case $a = 0$ and $d = 1$, this formula gives the familiar rule

$$\sum_{k=1}^{n} k = \frac{n(n+1)}{2},$$

and Brahmagupta then says that the sum of the squares will be this number multiplied by twice the period added to 1 and divided by 3; in other words

$$\sum_{k=1}^{n} k^2 = \frac{n(n+1)(2n+1)}{6}.$$

For visual proofs of these results Brahmagupta recommended using piles of balls or cubes.

These same rules were given in Chapter 5 of Bhaskara's *Lilavati* (Colebrooke, *1817*, pp. 51–57). Bhaskara goes a step further, saying, "The sum of the cubes of the numbers one, and so forth, is pronounced by the ancients equal to the square of the addition." This also is the correct rule that we write as

$$\sum_{k=1}^{n} k^3 = \left(\sum_{k=1}^{n} k\right)^2 = \left(\frac{n(n+1)}{2}\right)^2.$$

Bhaskara also gives the general rule for the sum of a geometric progression of $n+1$ terms $(a, ar, ar^2, \ldots, ar^n)$ in terms that amount to $a(r^{n+1}-1)/(r-1)$. He illustrates this rule with several examples, finding that $2+6+18+54+162+486+1456 = 2(3^7-1)/2 = 3^7-1 = 2186$.

About a century later than Bhaskara, Fibonacci's *Liber quadratorum* gives the same rule for the sum of the squares (Proposition 10): "If beginning with the unity, a number of consecutive numbers, both even and odd numbers, are taken in order, then the triple product of the last number and the number following it and the sum of the two is equal to six times the sum of the squares of all the numbers, namely from the unity to the last."

Proposition 11 gives a more elaborate summation rule, which we can express simply as

$$(2n+1)(2n+3)(4n+4) = 12\bigl(1^2+3^2+5^2+\cdots+(2n+1)^2\bigr).$$

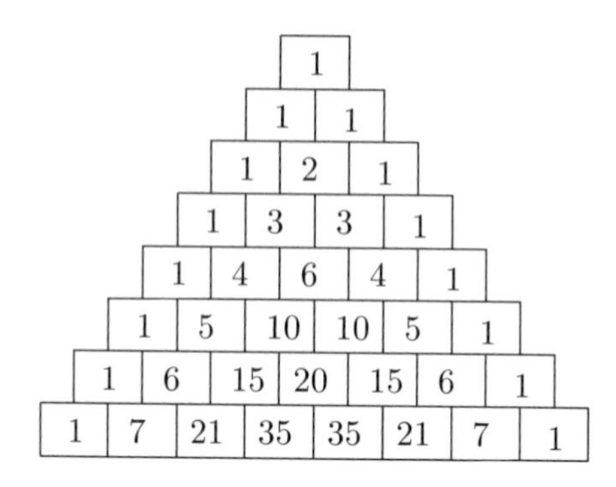

FIGURE 3. The *Meru Prastara*.

Permutations and combinations. The metaphysics of the Jainas, based on a classification of sentient beings according to the number of senses possessed, led them to a mathematical topic related to number theory. They called it *vikalpa*, and we know it as the basic part of combinatorics.

A typical question might be: "How many groups of three can be formed from a collection of five objects?" We know the answer, as did the early Jaina mathematicians. In the *Bhagabati Sutra*, written about 300 BCE, the author asks how many philosophical systems can be formed by taking a certain number of doctrines from a given list of basic doctrines. After giving the answers for 2, 3, 4, and so on, the author says that enumerable, unenumerable, and infinite numbers of things can be discussed, and "as the number of combinations are formed, all of them must be worked out."

The general process for computing combinatorial coefficients was known to the Hindus at an early date. Combinatorial questions seemed to arise everywhere for the Hindus, not only in the examples just given but also in a work on medicine dating from the sixth century BCE (Biggs, *1979*, p. 114) that poses the problem of the number of different flavors that can be made by choosing subsets of six basic flavors (bitter, sour, salty, astringent, sweet, hot). The author gives the answer as $6+15+20+15+6+1$, that is, 63. We recognize here the combinatorial coefficients that give the subsets of various sizes that can be formed from six elements. The author did not count the possibility of no flavor at all.

Combinatorics also arose in the study of Sanskrit in the third century BCE when the writer Pingala gave a rule for finding the number of different words that could be formed from a given number of letters. This rule was written very obscurely, but a commentator named Halayudha, who is believed to have lived in the tenth century CE (Needham, *1959*, p. 37), explained it as follows. First draw a square. Below it and starting from the middle of the lower side, draw two squares. Then draw three squares below these, and so on. Write the number 1 in the middle of the top square and inside the first and last squares of each row. Inside every other square the number to be written is the sum of the numbers in the two squares above it and overlapping it.

This array of numbers, which is known as *Pascal's triangle* because of a treatise on it written by Pascal in the seventeenth century, was studied four centuries before Pascal by Jordanus Nemorarius, who developed many of its properties (Hughes, *1989*). Pascal's triangle also occurs in Chinese manuscripts some four centuries before Pascal's treatise. In China the inspiration for the study of this diagram arose in connection with the extraction of cube roots and higher roots. The diagram appears in the *Xiangjie Jiuzhang Suan Fa* (*Detailed Analysis of the Mathematical Methods in the Nine Chapters*) of Yang Hui, written in 1261 (Li and Du, *1987*, p. 122). But in India we find it 300 years before it was published in China and 700 years before Pascal. Moreover it purports to be only a clarification of a rule invented 1200 years earlier![16] Its Sanskrit name is *Meru Prastara* (see Fig. 3), which means the *staircase of Mount Meru*.[17]

According to Singh (*1985*), Pingala's work on poetry also leads to another interesting combinatorial topic, recognized as such by the Hindu mathematicians. We treated this topic above as number theory, but it will bear repeating as combinatorics. To simplify the explanation as much as possible, suppose that a line of poetry is to be written using short beats and long ones, a long one being equivalent to two short ones. If a line contains n beats, how many arrangements are possible. Just to get started, we see that there is obviously one line of one beat (short), two lines of two beats (two short or one long), three lines of three beats (short-long, long-short, short-short-short), and five lines of four beats (long-long, short-short-long, short-long-short, long-short-short, short-short-short-short). Since a line with $n + 1$ beats must begin with either a short or a long beat, we observe that those beginning with a short beat are in one-to-one correspondence with the lines of n beats, all of which can be obtained by removing the initial short beat, while those beginning with a long beat are in a similar correspondence with lines of $n-1$ beats. It follows that the number of lines with $n+1$ beats is the sum of the numbers with $n - 1$ and n. Once again we generate the Fibonacci sequence.

Bhaskara II knew the rules for combinatorial coefficients very well. In Chapter 4 of the *Lilavati* (Colebrooke, *1817*, pp. 49–50), he gives an example of hexameter and asks how many possible combinations of long and short syllables are possible. He prescribes setting the numbers from 1 to 6 down "in direct and inverse order," that is, setting down the 2×6 matrix

$$\begin{matrix} 6 & 5 & 4 & 3 & 2 & 1 \\ 1 & 2 & 3 & 4 & 5 & 6 \end{matrix}.$$

From this array, by forming the products from the left and dividing, he finds the number of verses with different numbers of short syllables from 1 to 5 as

$$\frac{6}{1} = 6\,, \quad \frac{6 \cdot 5}{1 \cdot 2} = 15\,, \quad \frac{6 \cdot 5 \cdot 4}{1 \cdot 2 \cdot 3} = 20\,, \quad \frac{6 \cdot 5 \cdot 4 \cdot 3}{1 \cdot 2 \cdot 3 \cdot 4} = 15\,, \quad \frac{6 \cdot 5 \cdot 4 \cdot 3 \cdot 2}{1 \cdot 2 \cdot 3 \cdot 4 \cdot 5} = 6\,.$$

---

[16] That claim cannot be verified, however. Evidence indicates that knowledge of the combinatorial coefficients arose in India around the time of Aryabhata I, in the sixth century (Biggs, *1979*, p. 115).

[17] In Hindu mythology Mount Meru plays a role similar to that of Mount Olympus in Greek mythology. One Sanskrit dictionary gives this mathematical meaning of *Meru Prastara* as a separate entry. The word *prastara* apparently has some relation to the notion of expansion as used in connection with the binomial theorem.

Bhaskara recognized that there was one more possibility, six short syllables, but did not mention the possibility of no short syllables. The convention that an empty product equals 1 was not part of his experience.

Magic squares and Latin squares. In 1274 (see Li and Du, *1987*, p. 166) Yang Hui wrote *Xugu Zhaiqi Suanfa* (*Continuation of Ancient Mathematical Methods for Elucidating the Strange* [Properties of Numbers]), in which he listed magic squares of order up to 10. According to Biggs (*1979*, p. 121), there is evidence that the topic of magic squares had reached a high degree of development in China before this time and that Yang Hui was merely listing ancient results that were not a topic of current research, since he seems to have no concept of a general rule for constructing magic squares. Magic-square-type figures were a source of fascination in Korea, and they also seem to have spread west from China. Whether from China or as an indigenous product, magic squares of order up to 6 appear in the Muslim world around the year 990 (Biggs, *1979*, p. 119) and squares of order up to 9 are mentioned. Rules for constructing such squares were given by the Muslim scholar al-Buni (d. 1225). From the Muslim world, they entered Europe around the year 1315 in the works of the Greek scholar named Manuel Moschopoulos, who was claimed as a student by Maximus Planudes, who was mentioned in Chapter 2. They exerted a fascination on European scholars also, and the artist Albrecht Dürer incorporated a 4×4 magic square in his famous engraving *Melencolia*, with the year of its composition (1514) in the bottom row. The most fascinating thing about them is their sheer number. Difficult as they are to construct, there are nevertheless 880 distinct 4×4 magic squares.

Magic squares occur profusely throughout Indian, Chinese, and Japanese mathematics, alongside more elaborate numerical figures such as magic circles and magic hexagons. A variant of the idea of a magic square is that of a Latin square, an $n \times n$ array in which each of n letters appears once in each row and once in each column. It is easy to construct such a square by writing the letters down in order along the first row and then cyclically permuting them by one step in each subsequent row. To make the problem harder, mathematicians beginning with Euler in 1781 have sought pairs of *orthogonal* Latin squares: that is, two Latin squares that can be superimposed in such a way that each of the n^2 possible ordered pairs of letters occurs exactly once. An example, given by Biggs (*1979*, p. 123), with one of the squares using Greek letters for additional clarity, is

$$\begin{array}{cccc} A\alpha & B\beta & C\gamma & D\delta \\ B\gamma & A\delta & D\alpha & C\beta \\ C\delta & D\gamma & A\beta & B\alpha \\ D\beta & C\alpha & B\delta & A\gamma \end{array}.$$

Modern combinatorics. The usefulness of combinatorics in elementary probability has already been noted. It is an interesting exercise to compute, for example, the probability of a particular poker hand, say a full house, and see why the rules of the game make three of a kind a better hand (because less likely) than two pairs.

The strongest impetus to combinatorial studies in Europe, however, came from Gottfried Wilhelm Leibniz (1646–1716), who is best remembered for his brilliant discoveries in the calculus. He was also a profound philosopher and a diplomat with a deep interest in Oriental cultures. Many fundamental results are found in his *De arte combinatoria*, published in 1666. In this work Leibniz gave tables of

Albrecht Dürer's *Melencolia*, containing a magic square showing the date of composition as 1514. © Corbis Images (No. BE005826).

the number of permutations of n objects. There are many very curious aspects of this work. Although written mostly in Latin, it is rather polyglot. Leibniz occasionally breaks into Greek or German, and the tables are labeled with Hebrew letters. For permutations Leibniz used the word *numerus* to denote the size of the set from which objects are chosen, and *exponent* (literally, *placing out*) for the number of objects chosen. The total number of permutations of a number of objects he called its *variationes*, and for the number of combinations of a set of objects taken, say, four at a time, he wrote *con4natio*, an abbreviation for *conquattuornatio.* The case of two objects taken at a time provides the modern word *combination.* These combinations, now called *binomial coefficients*, were referred to generically as *complexiones.* The first problem posed by Leibniz was: *Given the* numerus *and the* exponent, *find the* complexiones. In other words, given n and k, find the number of combinations of n things taken k at a time.

Like the Hindu mathematicians, Leibniz applied combinatorics to poetry and music. He considered the hexameter lines possible with the Guido scale *ut, re, mi, fa, sol, la,* finding a total of 187,920.[18]

De arte combinatoria contains 12 sophisticated counting problems and a number of exotic applications of the counting techniques. It appears that Leibniz intended these techniques to be a source by which all possible propositions about the world could be generated. Then, combined with a good logic checker, this technique would provide the key to all knowledge. His intent was philosophical as well as mathematical, as evidenced by his claimed mathematical proof of the existence of God at the beginning of the work. Thus once again, this particular area of mathematics seems to be linked, more than other kinds of mathematics, with mysticism. The frontispiece of *De arte combinatoria* shows a mystical arrangement of the opposite pairs wet/dry, cold/hot, with the four elements of earth, air, fire, and water as cardinal points. This figure resembles an elaborate version of the famous ying/yang symbol from Chinese philosophy and it also recalls the proposition generator of the mystic theologian Ramon Lull (1232–1316), which consisted of a series of nested circles with words inscribed on them. When rotated independently, they would generate sentences. Leibniz was familiar with Lull's work, but was not a proponent of it.

The seeds planted by Leibniz in *De arte combinatoria* sprouted and grew during the nineteenth century, as problems from algebra, probability, and topology required sophisticated techniques of counting. One of the pioneers was the British clergyman Thomas Kirkman (1806–1895). The first combinatorial problem he worked on was posed in the *Lady's and Gentleman's Diary* in 1844: *Determine the maximum number of distinct sets of p symbols that can be formed from a set of n symbols subject to the restriction that no combination of q symbols can be repeated in different sets.* Kirkman himself posed a related problem in the same journal five years later: *Fifteen young ladies in a school walk out three abreast for 7 days in succession; it is required to arrange them daily so that no two shall walk twice abreast.* This problem is an early example of a problem in combinatorial design. The problem of constructing a Latin square is another example. This kind of combinatorial design has a practical application in the scheduling of athletic tournaments, and in fact colleagues of the author specializing in combinatorial design procured a contract to design the schedule for the short-lived XFL Football League in 2001.

We shall terminate our discussion of combinatorics with these nineteenth-century results. We note in parting that it remains an area with a plenitude of unsolved problems whose statement can be understood without long preparation.

---

[18] The first five of these tones are the first syllables of a medieval Latin chant on ascending tones. The replacement of *ut* by the modern *do* came later.

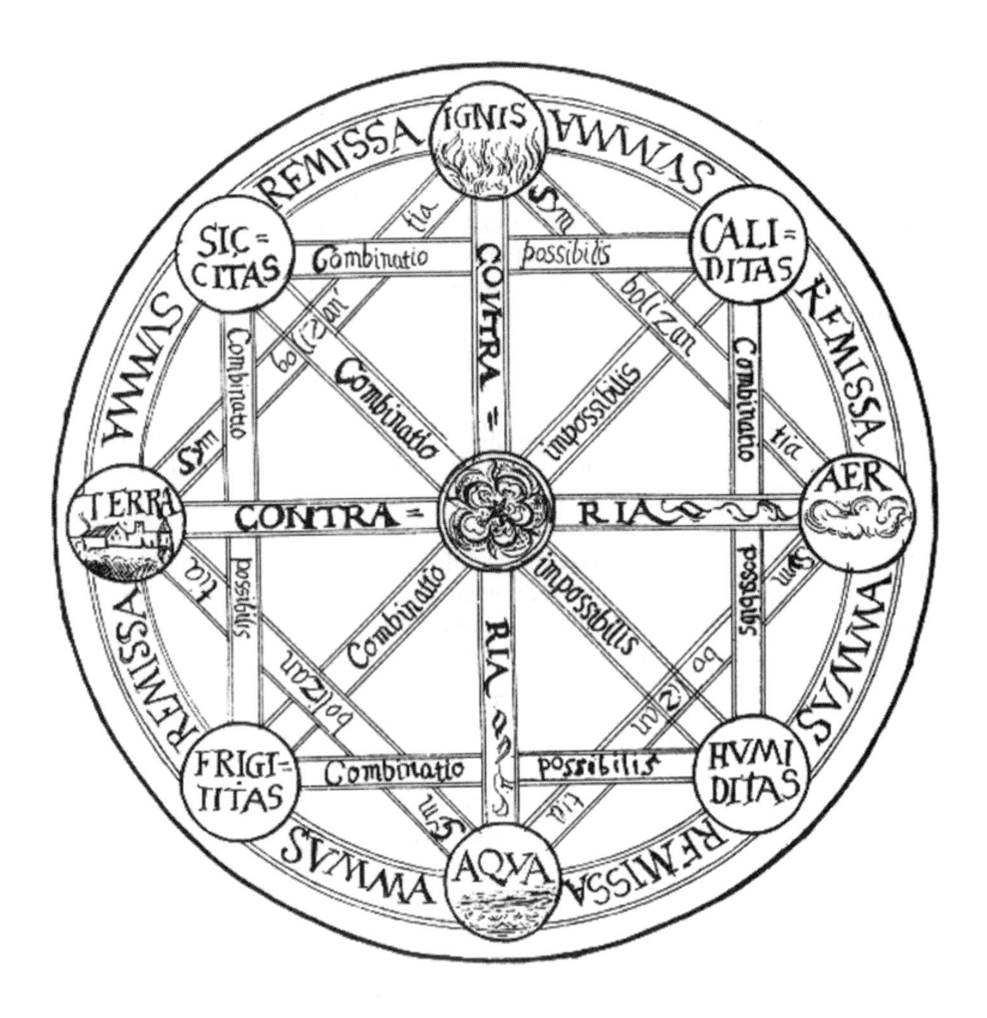

Frontispiece of *De arte combinatoria*, from Vol. 5, p. 7 of the collected works of Leibniz. © Georg Olms Verlag, Hildesheim.

It was for that reason that combinatorial problems were used as the mathematical background of the film *Good Will Hunting*.

Questions and problems

8.1. We know a mathematical algorithm for computing as many decimal digits of $\sqrt{2}$ as we have time for, and $\sqrt{2}$ has a precise representation in Euclidean geometry as the ratio of the diagonal of a square to its side. It is a provable theorem of Euclidean geometry that that ratio is the same for all squares, so that two observers using different squares should get the same result. To the extent that physical space really is Euclidean, this definition makes it possible to determine $\sqrt{2}$ empirically by measuring the sides and diagonals of physical squares. In that sense, we could theoretically determine $\sqrt{2}$ with arbitrarily prescribed precision by physical measurements. In particular, it makes perfectly good sense to ask what the 50th decimal digit of $\sqrt{2}$ is—it happens to be 4, but rounds up to 5—and we could try to get instruments precise enough to yield this result from measurement.

Consider, in contrast, the case of a physical constant, say the universal gravitational constant, usually denoted G_0, which occurs in Newton's law of gravitation:

$$F = G_0 \frac{Mm}{r^2}\,.$$

Here F is the force each of two bodies exerts on the other, M and m are the masses of the two bodies, and r is the distance between their centers of gravity. The accepted value of G_0, given as upper and lower assured limits, is 6.674215 ± 0.000092 N $\cdot$ m^2/kg^2, although some recent measurements have cast doubt on this value. From a mathematical point of view, G_0 is determined by the equation

$$G_0 = \frac{Fr^2}{Mm}\,,$$

and its value is found—as Cavendish actually did—by putting two known masses M and m at a known distance r from each other and measuring the force each exerts on the other. The assertion that the ratio Fr^2/Mm is the same for all masses and all distances is precisely the content of *Newton's law of gravity*, so that two experimenters using different masses and different distances should get the same result. But Newton's law of gravity is not deducible from axioms; it is, rather, an empirical hypothesis, to be judged by its explanatory power and its consistency with observation. What should we conclude if two experimenters do *not* get the same result for the value of G_0? Did one of them do something wrong, or is Newton's law not applicable in all cases? Does it even *make sense* to ask what the 50th decimal digit of G_0 is?

8.2. You can represent $\sqrt{ab}$ geometrically by putting a line of length b end-to-end with a line of length a, drawing a circle having this new line as diameter, and then drawing the perpendicular to the circle from the point where the two lines meet. To get $\sqrt{a}$ and $\sqrt{b}$, you would have to use Descartes' unit length I as one of the factors. Is it possible to prove by use of this construction that $\sqrt{abI} = \sqrt{a}\sqrt{b}$? Was Dedekind justified in claiming that this identity had never been proved?

8.3. Try to give a definition of real numbers—perhaps using decimal expansions—that will enable you to say what the numbers $\sqrt{2}$, $\sqrt{3}$, and $\sqrt{6}$ are, and how they can be added and multiplied. Does your definition enable you to prove that $\sqrt{2}\sqrt{3} = \sqrt{6}$?

8.4. Use the method of infinite descent to prove that $\sqrt{3}$ is irrational. [*Hint*: Assuming that $m^2 = 3n^2$, where m and n are positive integers having no common factor, that is, they are as small as possible, verify that $(m - 3n)^2 = 3(m - n)^2$. Note that $m < 2n$ and hence $m-n < n$, contradicting the minimality of the original m and n.]

8.5. Show that $\sqrt[3]{3}$ is irrational by assuming that $m^3 = 3n^3$ with m and n positive integers having no common factor. [*Hint*: Show that $(m-n)(m^2+mn+n^2) = 2n^3$. Hence, if p is a prime factor of n, then p divides either $m - n$ or $m^2 + mn + n^2$. In either case p must divide m. Since m and n have no common factor, it follows that $n = 1$.]

8.6. Suppose that x, y, and z are positive integers, no two of which have a common factor, none of which is divisible by 3, and such that $x^3 + y^3 = z^3$. Show that there exist integers p, q, and r such that $z - x = p^3$, $z - y = q^3$, and $x + y = r^3$. Then, letting $m = r^3 - (p^3 + q^3)$ and $n = 2pqr$, verify from the original equation that $m^3 = 3n^3$, which by Problem 8.5 is impossible if m and n are nonzero. Hence $n = 0$, which means that $p = 0$ or $q = 0$ or $r = 0$, that is, at least one of x and y equals 0. Conclude that no such positive integers x, y, and z can exist.

8.7. Verify that

$$27^5 + 84^5 + 110^5 + 133^5 = 144^5.$$

[See L. J. Lander and T. R. Parkin, "Counterexample to Euler's conjecture on sums of like powers," *Bulletin of the American Mathematical Society*, **72** (1966), p. 1079. Smaller counterexamples to this conjecture have been discovered more recently.]

8.8. Prove Fermat's little theorem by induction on a. [*Hint*: The theorem can be restated as the assertion that p divides $a^p - a$ for every positive integer a. Use the binomial theorem to show that $(a+1)^p - (a+1) = mp + a^p - a$ for some integer m.]

8.9. Verify the law of quadratic reciprocity for the primes 17 and 23 and for 67 and 71.

8.10. Show that the factorization of numbers of the form $m + n\sqrt{-3}$ is *not* unique by finding two different factorizations of 4. Is factorization unique for numbers of the form $m + n\sqrt{-2}$?

8.11. Prove that the number of primes less than or equal to N is at least $\log_2(N/3)$, by proceeding as follows. Let $p_1, \ldots, p_n$ be the prime numbers among $1, \ldots, N$, and let $\theta(N)$ be the number of square-free integers among $1, \ldots, N$, that is, the integers not divisible by any square number. We then have the following relation, since it is known that $\sum_{k=1}^{\infty}(1/k^2) = \pi^2/6$.

$$\begin{aligned}
\theta(N) &> N - \sum_{k=1}^{n}\left[\frac{N}{p_k^2}\right] \\
&> N\left(1 - \sum_{k=1}^{n}\frac{1}{p_k^2}\right) \\
&> N\left(1 - \sum_{k=2}^{\infty}\frac{1}{k^2}\right) \\
&= N\left(2 - \frac{\pi^2}{6}\right) > \frac{N}{3}.
\end{aligned}$$

(Here the square brackets denote the greatest-integer function.) Now a square-free integer k between 1 and N is of the form $k = p_1^{e_1} \cdots p_n^{e_n}$, where each e_j is either 0 or 1. Hence $\theta(N) \le 2^n$, and so $n > \log_2(N/3)$. This interesting bit of mathematical trivia is due to the Russian–American mathematician Joseph Perott (1854–1924).

8.12. Assuming that $\lim_{n\to\infty} \dfrac{\log(n)\pi(n)}{n}$ exists, use Chebyshëv's estimates to show that this limit must be 1 and hence that Legendre's estimate cannot be valid beyond the first term.

Part 3

Color Plates

Plate 1. Top to bottom: Problems 49–54 of the Ahmose Papyrus. © The British Museum.

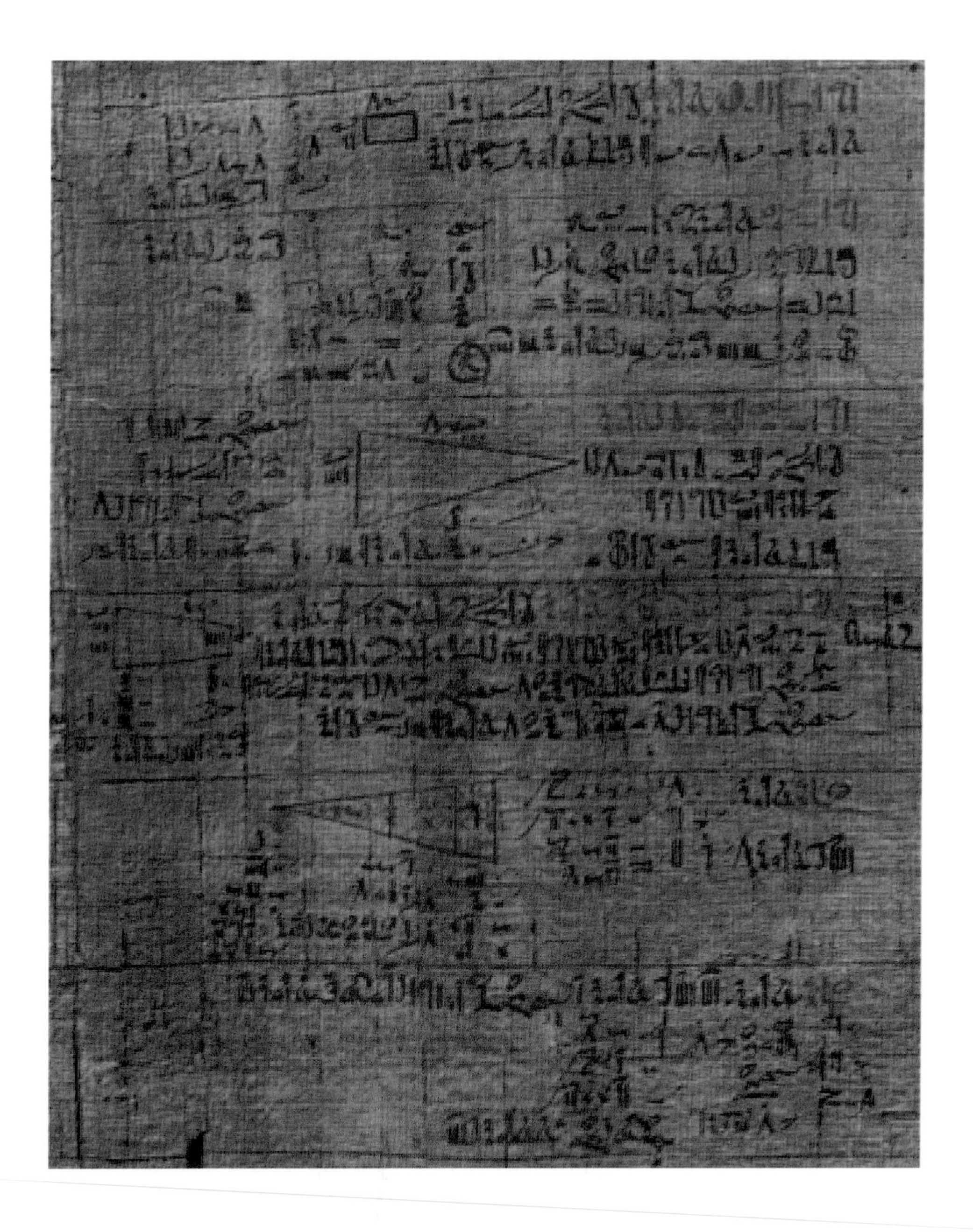

Plate 2. (Overleaf): *Sangaku* on display at the Suguwara Shrine in Mie Prefecture, 1854. Courtesy of Mr. Hidetoshi Fukagawa.

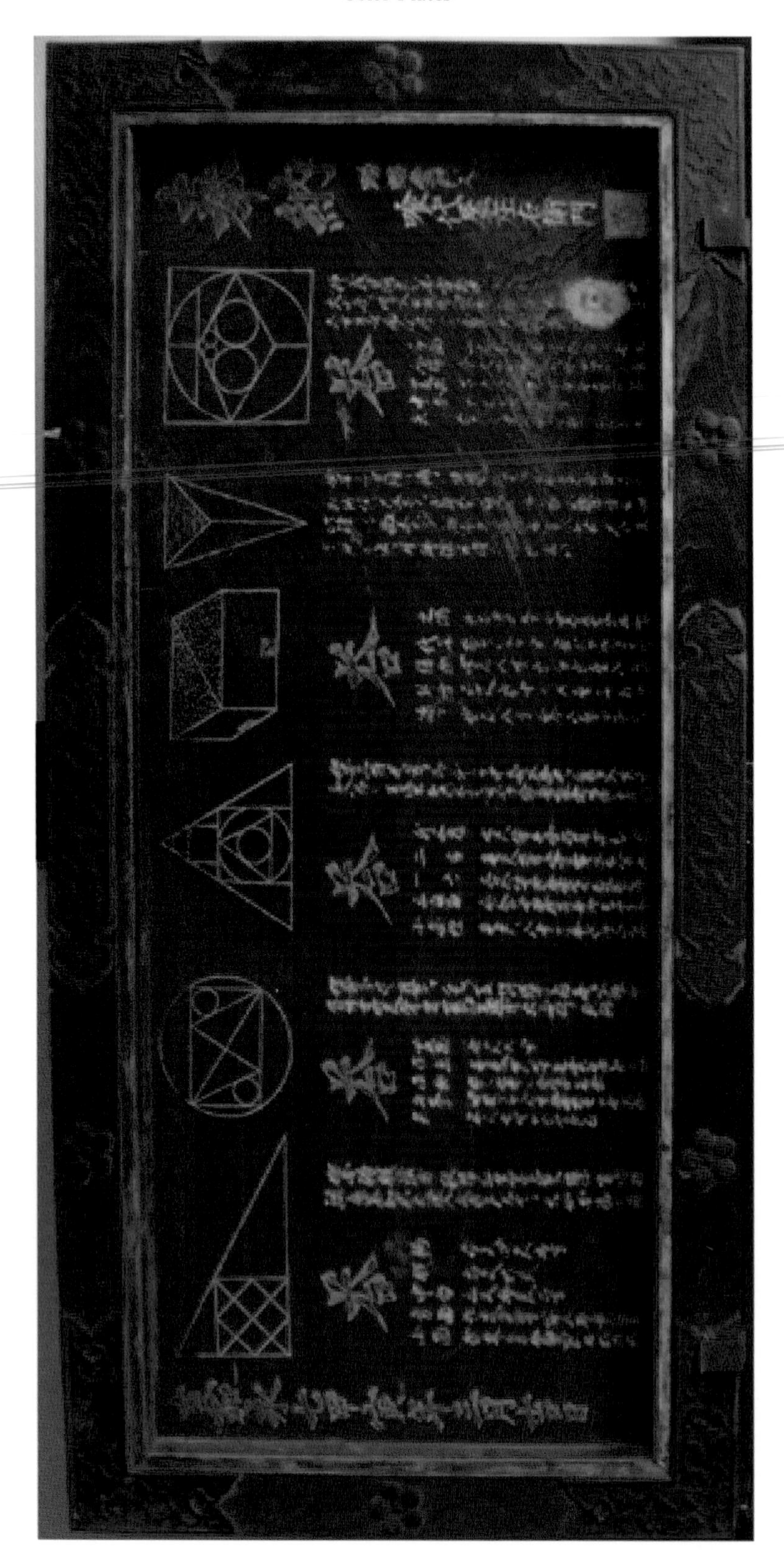

Plate 3. Folio 47 of the Dresden Codex. © Sächsische Landesbibliothek, Dresden.

Plate 4. A branch of a flowering crabapple tree, illustrating the Fibonacci/golden ratio pattern of twig growth.

Plate 5. Florence Nightingale's "batwing" or "coxcomb" diagram, forerunner of the modern pie chart. Unlike the modern pie chart, the sectors here do not represent percentages. Rather, they give the monthly death tolls from wounds and disease during the Crimean War of 1854–1855. Nowadays, such data would be presented as a line graph.

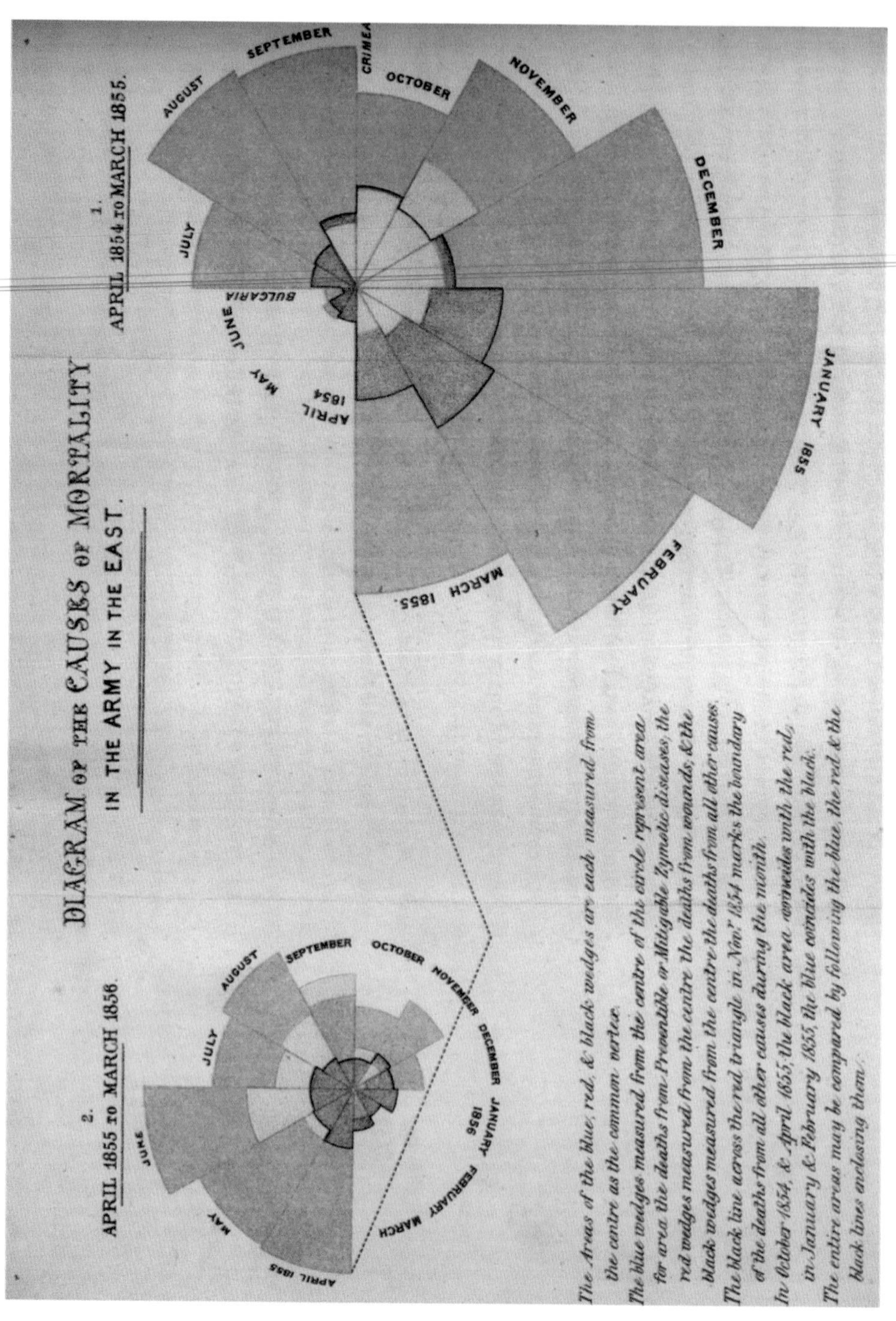

Plate 6. A page of the Archimedes Palimpsest. © Christie's Images Incorporated 2005.

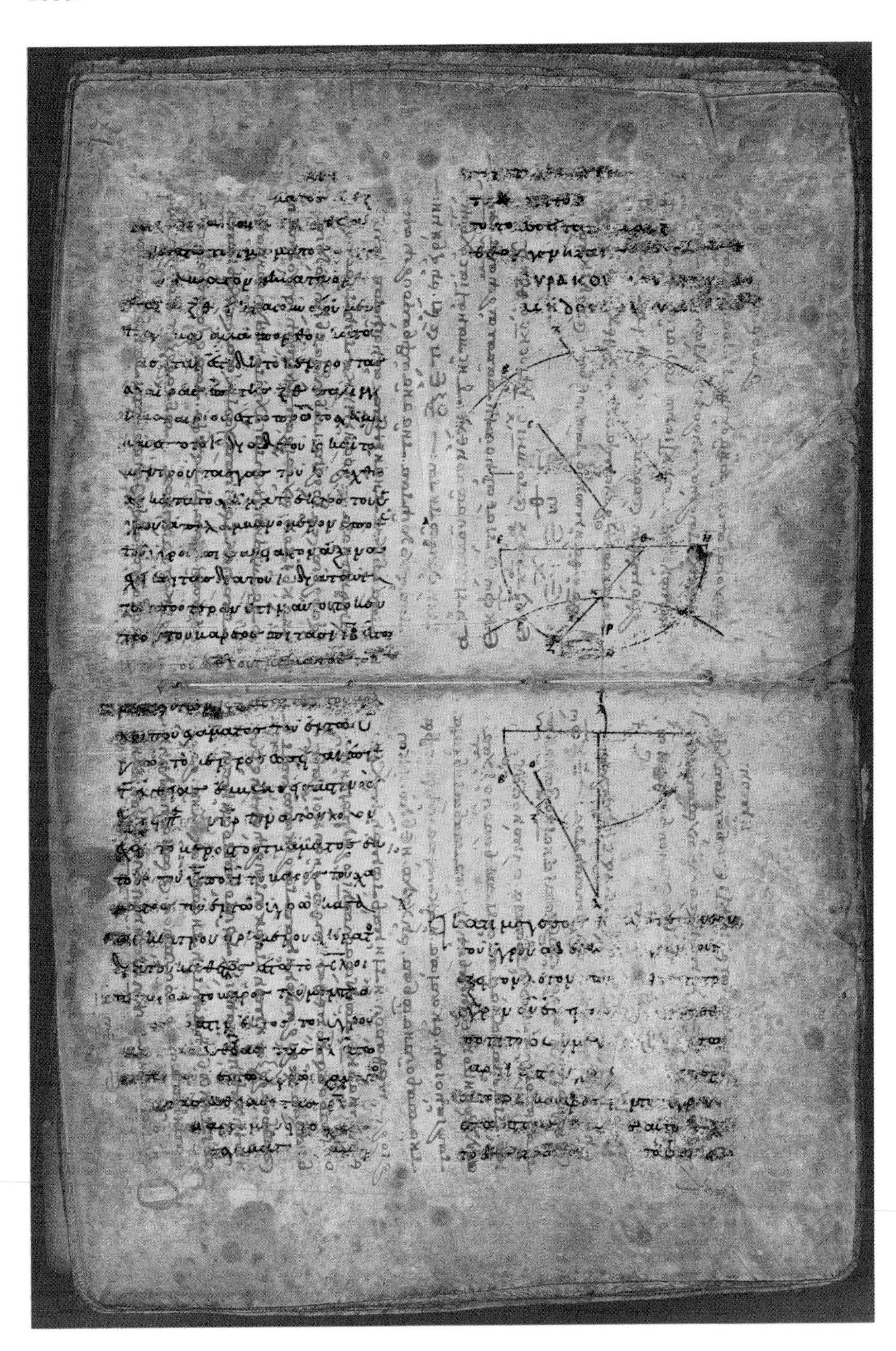

Plate 7. Egyptian field measurers: a wall painting from the tomb of Menna, Scribe of the Fields, around 1400 BCE. © Corbis Images (No. WF004135).

Plate 8. Reconstruction of one of Ptolemy's maps of the world. © The Vatican Library (Urb. gr. 82, ff. 60v–61r).

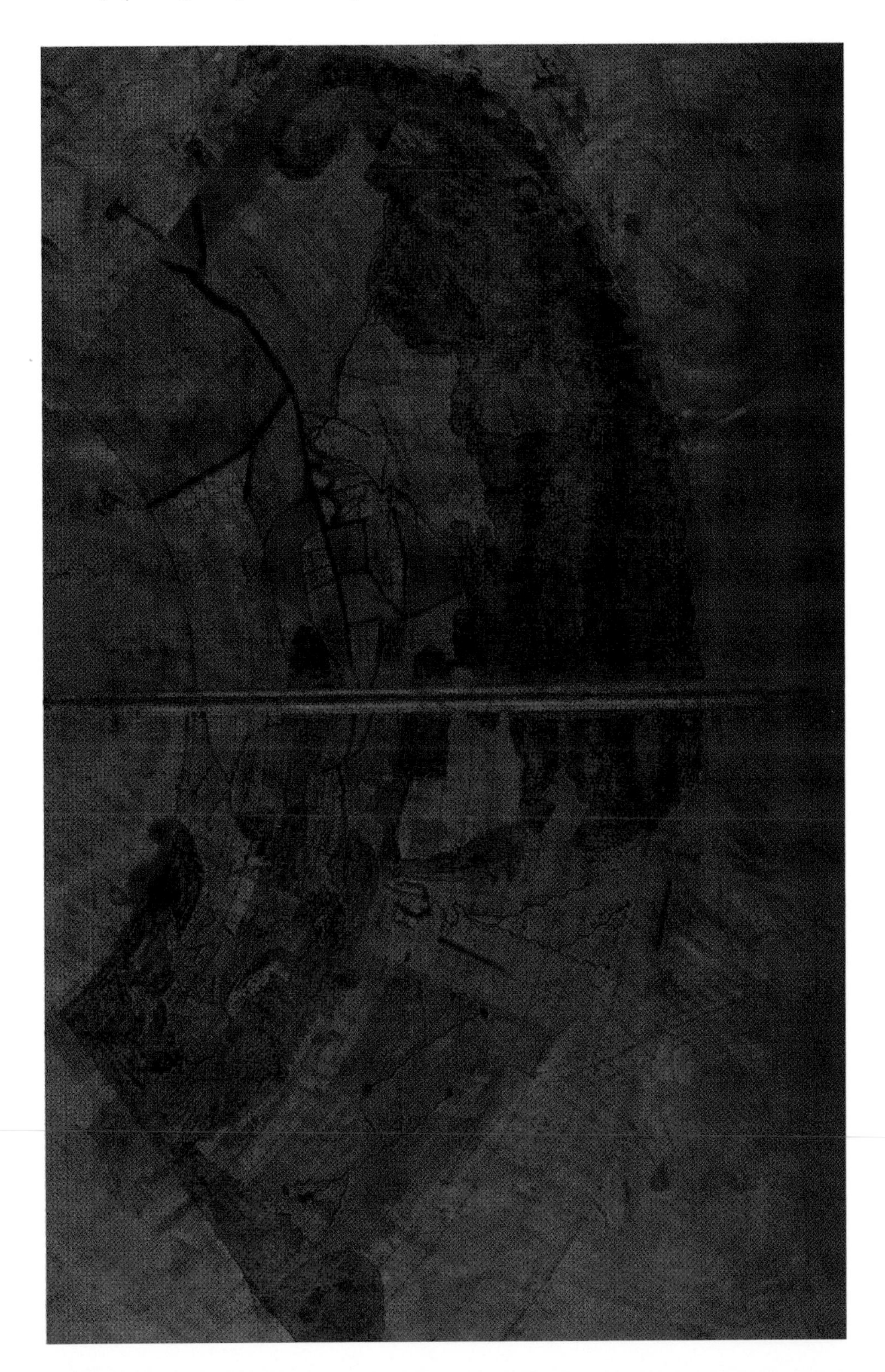

Plate 9. Portrait of Luca Pacioli in 1495 working with a text of Euclid's *Elements*, by Jacopo de Barbari (1440–1515). It is not certain who the young man behind him is, but it may be Albrecht Dürer.

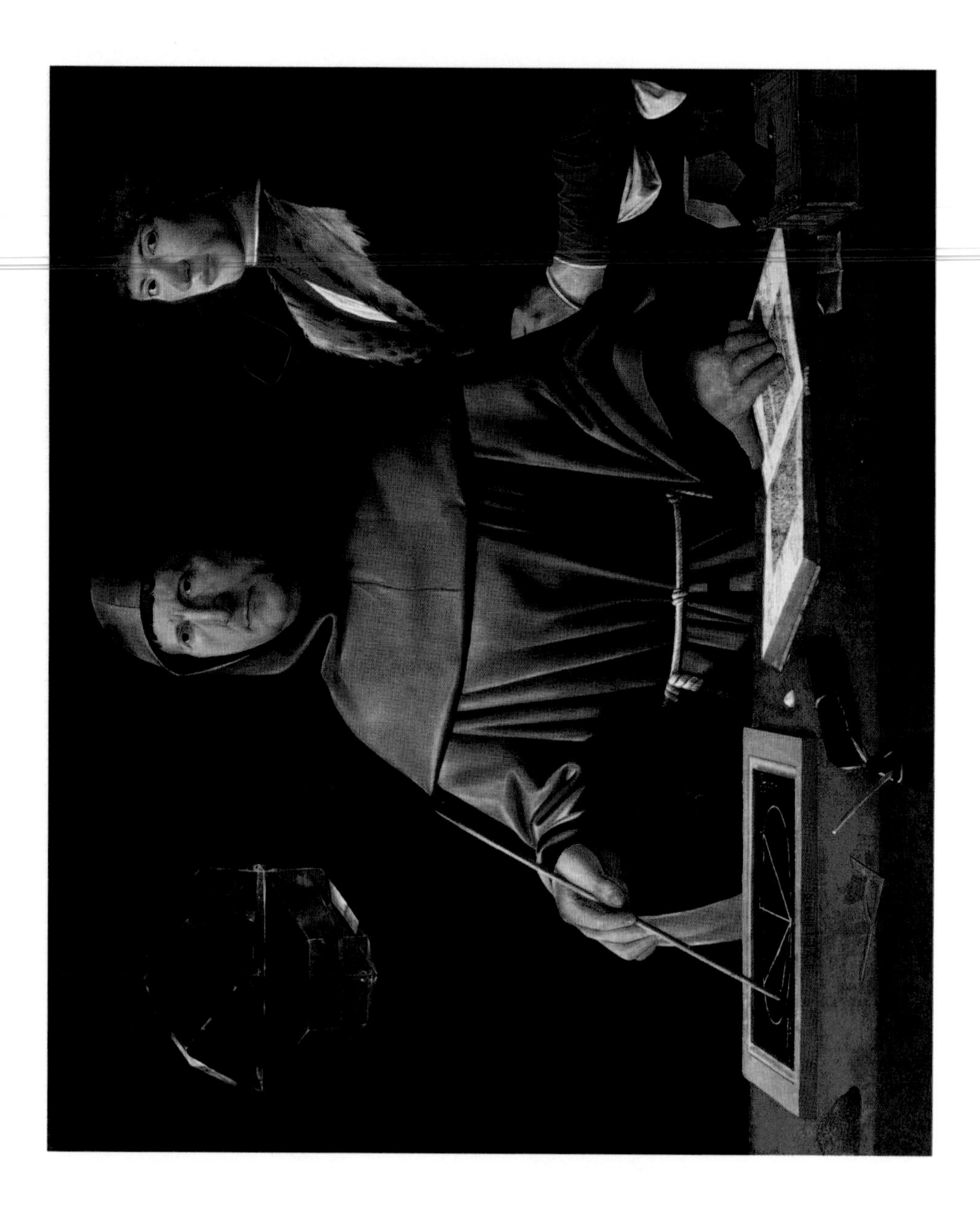

Part 4

Space

Geometry is a way of organizing our perception of shape. As in arithmetic, we can distinguish levels of sophistication in the development of geometry. The first level is that of measurement: comparing the sizes of objects having different shapes. Measurement is *arithmetic applied to space*, beginning with figures having flat sides or faces. The second level is the study of the proportions among the parts of geometric figures, such as triangles, squares, rectangles, and circles. A good marker for the beginning of this stage of development is the Pythagorean relation for right triangles.

Although regular polygons and polyhedra can be measured using simple dissection techniques, algebra is needed to measure more complicated figures, such as the portion of a disk remaining after three other disks, all tangent to one another and to the original disk and whose radii are in given ratios, are removed from it. One of the uses of geometry in many ancient treatises is as a source of interesting equations to be solved. Although the problems are posed as problems in measurement, the shapes being measured are so unusual that it is hard to think of them as a motive. The suspicion begins to arise that the author's real purpose was to exhibit some algebra.

The Pythagoreans, Plato, and Aristotle gave geometry a unique philosophical grounding that turned it out of the path it would probably have followed otherwise. Their insistence on a logical development based on a system of axioms made explicit many assumptions—especially the parallel postulate—that otherwise might not have been noticed. As a result, there is a marked difference between the mathematical practitioners who learned geometry from Euclid (the medieval Muslims and renaissance Europeans) and those who learned it from a different tradition.

At its highest level, elementary geometry employs algebra and the infinite processes of calculus in order to find the areas and volumes of ever more complicated curvilinear figures. Like arithmetic, geometry has given rise to many specialties, such as projective, analytic, and differential geometry. As more and more general properties of space became mathematized, geometry generated the subject known as *topology*, from Greek roots meaning *theory of position*.

In the four chapters that constitute the present part of our history, we shall look at all these aspects of geometry. In Chapter 9 we study the way space was measured in a number of civilizations. Chapters 10 and 11 form a unit devoted to the most influential form of elementary geometry, the Euclidean geometry that arose in the Hellenistic civilization. Chapter 12 contains a survey of the variety of forms of geometry that have arisen over the past three centuries.

CHAPTER 9

Measurement

The word *geometry* comes from the Greek words *gê*, meaning *earth*, and *metreîn*, meaning *measure*. It seems that all human societies have had to measure fields for agriculture or compute the amount of work involved in excavating a building site. That there is a basic similarity in approaches to these problems is attested by the presence of words for circles, rectangles, squares, and triangles in every language. Geometric intuition seems to be innate to human beings. Many different societies independently discovered the Pythagorean theorem, for example. These external similarities conceal certain differences in outlook, however. For example, we are taught to think of a line as having no thickness. But did the Hindus, Egyptians, and others think of it that way? The word *line* comes from the Greek *línon* (Latin *linea*) meaning *string*; the physical object (a stretched string) on which the abstraction is based is clear. Early Hindu work on geometry also uses Sanskrit words for ropes and cords. It is very likely that ancient engineers thought of a line as a physical object: a rope, stretched taut. The quantity of rope was given as a number (length). Geometry at this stage of development was a matter of relating lengths to other quantities of geometric interest, such as areas, slopes, and volumes. It was an application of arithmetic and useful in planning public works projects, for example, since it provided an estimate of the size of a job and hence the number of workers and the amount of materials and time required to excavate and build a structure. It also proved useful in surveying, since geometric relationships could be used to compute inacessible distances from accessible ones. The fact that similar triangles are the basic tool in this science has caused it to be named *trigonometry*. The elementary rules for measuring regular geometric figures persist in treatises for many centuries. Nearly always the author begins by describing the standard units of length, area, volume, and weight, then presents a variety of procedures that have a great resemblance to the procedures described in all other treatises of the same kind. This geometry, though elementary, should not be thought of as primitive. Textbooks of "practical mathematics" containing exactly this material are still being written and published today.

Although geometry looks very much the same across cultures, there is one place where we must exercise a little care in order to understand it from the point of view of the people we are studying. Textbooks often give approximate values of the number π allegedly used in different cultures without being clear about which constant they mean. When calculating the circumference C of a circle in terms of its diameter, we use the formula $C = \pi d$. When calculating the area of a circle (disk) in terms of its radius, we use the familiar formula $A = \pi r^2$. When calculating the area of a sphere in terms of its radius, we use $A = 4\pi r^2$, and for the volume, the formula $V = \frac{4}{3}\pi r^3$. There are really four different values of π here, depending on the dimension and flatness or curvature. The first formula reflects the fact that the circumference of a circle is proportional to its diameter. That is, given two circles with circumferences C_i and diameters D_i, $i = 1, 2$, we have $C_1/C_2 = D_1/D_2$. We shall call the ratio C/D the *one-dimensional* π. The second formula implies that if Δ_1 and Δ_2 are the areas of disks of radius r_1 and r_2 and S_1 and S_2 are the areas of squares of sides r_1 and r_2, then $\Delta_1/\Delta_2 = S_1/S_2$. We shall call the ratio Δ/S the *two-dimensional* π, and similarly for volumes. We won't need the π that occurs in the formula for the area of a sphere, since everybody seemed to relate that one to one of the others. Thus, we are dealing with several direct proportions with different constants of proportionality. It is not obvious that these constants for different dimensions have any simple relationship to one another. That fact requires some digging in geometry to discover. Without the abstract concept of a *constant of proportionality*, when a mathematician is seeking only numerical approximations that accord with observation, there is no reason to suspect any connection between these constants in different dimensions. To be sure, only a small amount of intuition is required to establish the connection, as shown in Problems 9.13 and 9.20, but in any discussion of supposed approximations to π used in different cultures, we need to keep in mind the dimension of the object being studied: Was it a circle, a disk, a cone, a cylinder, a sphere, or a ball?

1. Egypt

Foreigners have been interested in the geometry of the Egyptians for a very long time. In Section 109 of Book 2 of his *History*, the Greek historian Herodotus writes that King Sesostris[1] dug a multitude of canals to carry water to the arid parts of Egypt. He goes on to connect this Egyptian engineering with Greek geometry:

> It was also said that this king distributed the land to all the Egyptians, giving an equal quadrilateral farm to each, and that he got his revenue from this, establishing a tax to be paid for it. If the river carried off part of someone's farm, that person would come and let him know what had happened. He would send surveyors to remeasure and determine the amount by which the land had decreased, so that the person would pay less tax in proportion to the loss. It seems likely to me that it was from this source that geometry was found to have come into Greece. For the Greeks learned of the sundial and the twelve parts of the day from the Babylonians.

The main work of Egyptian surveyors was measuring fields. That job corresponds well to the Latin word *agrimensor*, which means *surveyor*. Our word *surveyor* comes through French, but has its origin in the Latin *supervideo*, meaning *I oversee*. The equivalent word in Greek was used by Herodotus in the passage above. He said that the king would send *episkepsoménous kaì anametrĕsontas*, literally *overseeing and remeasuring men*. The process of measuring a field is shown in a painting from the tomb of an Egyptian noble named Menna at Sheikh Abd el-Qurna in Thebes (Plate 7). Menna bore the title Scribe of the Fields of the Lord of the Two Lands during the eighteenth dynasty, probably in the reign of Amenhotep III or Thutmose IV, around 1400 BCE. His job was probably that of

[1] There were several pharaohs with this name. Some authorities believe that the one mentioned by Herodotus was actually Ramses II, who ruled from 1279 to 1212 BCE.

a steward, to oversee planting and harvest. As the painting shows, the instrument used to measure distance was a rope that could be pulled taut. That measuring instrument has given rise to another name often used to refer to these surveyors: *harpedonáptai*, from the words *harpedónē*, meaning *rope*, and *háptein*, meaning *attach*. The philosopher Democritus (d. 357 BCE) boasted, "In demonstration no one ever surpassed me, not even those of the Egyptians called *harpedonáptai*."[2]

The geometric problems considered in the Egyptian papyri all involve measurement. These problems show considerable insight into the properties of simple geometric figures such as the circle, the triangle, the rectangle, and the pyramid; and they rise to a rather high level of sophistication in computing the area of a hemisphere. Those involving flat boundaries (polygons and pyramids) are correct from the point of view of Euclidean geometry, while those involving curved boundaries (disks and spheres) are correct up to the constant of proportionality chosen.

1.1. Areas. Since the areas of rectangles and triangles are easy to compute, it is understandable that very little attention is given to these problems. Only four problems in the Ahmose Papyrus touch on these questions: Problems 6, 49, 51, and 52 (see Plate 1).

Rectangles, triangles, and trapezoids. Problem 49 involves computing the area of a rectangle that has dimensions 1 *khet* by 10 *khets*. This in itself would be a trivial problem, except that areas are to be expressed in square cubits rather than square *khets*. Since a *khet* is 100 cubits, the answer is given correctly as 100,000 square cubits. Problem 51 is a matter of finding the area of a triangle, and it is illustrated by a figure (see Plate 1) showing the triangle. The area is found by multiplying half of the base by the height. In Problem 52, this technique is generalized to a trapezoid, and half of the sum of the upper and lower bases is multiplied by the height.

Of all these problems, the most interesting is Problem 6, which involves a twist that makes it equivalent to a quadratic equation. A rectangle is given having area 12 *cubit strips*; that is, it is equal to an area 1 cubit by 12 cubits, though not of the same shape. The problem is to find its dimensions given that the width is three-fourths of the length ($\overline{2}\ \overline{4}$ in the notation of the papyrus). The first problem is to "calculate with $\overline{2}\ \overline{4}$, until 1 is reached," that is, in our language, dividing 1 by $\overline{2}\ \overline{4}$. The result is $1\ \overline{3}$. Then 12 is multiplied by $1\ \overline{3}$, yielding 16, after which the scribe takes the *corner* (square root) of 16—unfortunately, without saying how—getting 4 as the length. This is a very nice example of thinking in terms of expressions. The scribe seems to have in mind a picture of the length being multiplied by three-fourths of the length, and the result being 12. Then the length squared has to be found by multiplying by what we would call the reciprocal of its coefficient, after which the length is found by taking the square root.

Slopes. The beginnings of trigonometry can be seen in Problems 56–60 of the papyrus, which involve the slope of the sides of pyramids and other figures. There is a unit of slope analogous to the *pesu* that we saw in Chapter 5 in the problems involving strength of bread and beer. The unit of slope is the *seked*, defined as the number of palms of horizontal displacement associated with a vertical displacement of 1 royal cubit. One royal cubit was 7 palms. Because of the relative sizes of

[2] Quoted by the second-century theologian Clement of Alexandria, in his *Miscellanies*, Book 1, Chapter 15.

horizontal and vertical displacements, it makes sense to use a larger unit of length for vertical distances than is used for horizontal distances, even at the expense of introducing an extra factor into computations of slope. In our terms the *seked* is seven times the tangent of the angle that the sloping side makes with the vertical. In some of the problems the *seked* is given in such a way that the factor of 7 drops out. Notice that if you were ordering a stone from the quarry, the *seked* would tell the stonecutter immediately where to cut. One would mark a point one cubit (distance from fingertip to elbow) from the corner in one direction and a point at a number of palms equal to the *seked* in the perpendicular direction, and then simply cut between the two points marked.

In Problem 57 a pyramid with a *seked* of $5\ \overline{4}$ and a base of 140 cubits is given. The problem is to find its height. The *seked* given here ($\frac{3}{4}$ of 7) is exactly that of one of the actual pyramids, the pyramid of Khafre, who reigned from 2558 to 2532 BCE. It appears that stones were mass-produced in several standard shapes with a *seked* that could be increased in intervals of one-fourth. Pyramid builders and designers could thereby refer to a standard brick shape, just as architects and contractors since the time of ancient Rome have been able to specify a standard diameter for a water pipe. Problem 58 gives the dimensions of the same pyramid and asks for its *seked*, apparently just to reinforce the reader's grasp of the relation between *seked* and dimension.

The circle. Five of the problems in the Ahmose Papyrus (41–43, 48, and 50) involve calculating the area of a circle. The answers given are approximations, but would be precise if the value $64/81$ used in the papyrus where we would use $\pi/4$ were exact. The author makes no distinction between the two. When physical objects such as grain silos are built, the parts used to build them have to be measured. In addition, the structures and their contents have a commercial, monetary value. Some number has to be used to express that value. It would therefore *not* be absurd—although it would probably be unnecessary—for a legislature to pass a bill prescribing a numerical value to be used for π.[3] Similarly, the claim often made that the "biblical" value of π is 3, based on the description of a vat 10 cubits from brim to brim girdled by a line of 30 cubits (1 Kings 7:23) is pure pedantry. It assumes more precision than is necessary in the context. The author may have been giving measurements only to the nearest 10 cubits, not an unreasonable thing to do in a literary description.[4]

---

[3] However, in the most notorious case where such a bill was nearly passed—House Bill 246 of the 1897 Indiana legislature—it *was* absurd. The bill was written by a physician and amateur mathematician named Edwin J. Goodwin. Goodwin had copyrighted what he thought was a quadrature of the circle. He offered to allow textbooks sold in Indiana to use his proof royalty-free provided that the Indiana House would pass this bill, whose text mostly glorified his own genius. Some of the mathematical statements the legislature was requested to enact were pure gibberish. For example, "a circular area is to the square on a line equal to the quadrant of the circumference, as the area of an equilateral rectangle is to the square on one side." The one clear statement is that "the ratio of the chord and arc of ninety degrees... is as seven to eight." That statement implies that $\pi = 16\sqrt{2}/7 \approx 3.232488\ldots$. The square root in this expression did not trouble Dr. Goodwin, who declared that $\sqrt{2} = 10/7$. At this point, one might have taken his value of π to be $160/49 = 3.265306122\ldots$. But, in a rare and uncalled-for manifestation of consistency, since he "knew" that $100/49 = (10/7)^2 = 2$, Goodwin declared this fraction equal to $16/5 = 3.2$. The bill was stopped at the last minute by lobbying from a member of the Indiana Academy of Sciences and was tabled without action.

[4] However, like everything in the Bible, this passage has been subject to exhaustive and repeated analysis. For a summary of the conclusions reached in the Talmud, see Tsaban and Garber (*1998*).

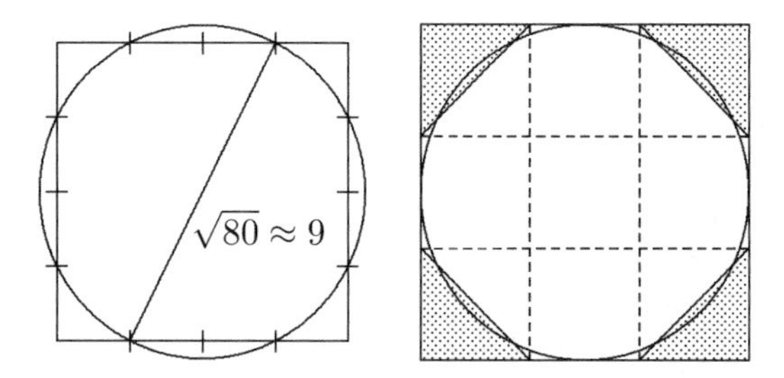

FIGURE 1. Conjectured explanations of the Egyptian squaring of the circle.

Ahmose takes the area of a circle to be the area of the square whose side is obtained by removing the ninth part of the diameter. In our language the area is the square on eight-ninths of the diameter, that is, it is the square on $\frac{16}{9}$ of the radius. In our language, not that of Egypt, this gives a value of π for area problems equal to $\frac{256}{81}$. Please remember, however, that the Egyptians had no concept of the number π. The constant of proportionality that they always worked with represents what we would call $\pi/4$. There have been various conjectures as to how the Egyptians might have arrived at this result. One such conjecture given by Robins and Shute (*1987*, p. 45) involves a square of side 8. If a circle is drawn through the points 2 units from each corner, it is visually clear that the four fillets at the corners, at which the square is outside the circle, are nearly the same size as the four segments of the circle outside the square; hence this circle and this square may be considered equal in area. Now the diameter of this circle can be obtained by connecting one of the points of intersection to the opposite point, as shown on the left-hand diagram in Fig. 1, and measurement will show that this line is very nearly 9 units in length (it is actually $\sqrt{80}$ in length). A second theory due to K. Vogel (see Gillings, *1972*, pp. 143–144) is based on the fact that the circle inscribed in a square of side nine is roughly equal to the unshaded region in the right-hand diagram in Fig. 1. This area is $\frac{7}{9}$ of 81, that is, 63. A square of equal size would therefore have side $\sqrt{63} \approx 7.937 \approx 8$. In favor of Vogel's conjecture is the fact that a figure very similar to this diagram accompanies Problem 48 of the papyrus. A detailed discussion of various conjectures, giving connections with traditional African crafts, was given by Gerdes (*1985*).

The Pythagorean theorem. Inevitably in the discussion of ancient cultures, the question of the role played by the Pythagorean theorem is of interest. Did the ancient Egyptians know this theorem? It has been reported in numerous textbooks, popular articles, and educational videos that the Egyptians laid out right angles by stretching a rope with 12 equal intervals knotted on it so as to form a 3–4–5 right triangle. What is the evidence for this assertion? First, the Egyptians *did* lay out very accurate right angles. Also, as mentioned above, it is known that their surveyors used ropes as measuring instruments and were referred to as *rope-fixers* (see Plate 7). That is the evidence that was cited by the person who originally made the conjecture, the historian Moritz Cantor (1829–1920) in the first volume of his history of mathematics, published in 1882. The case can be made stronger, however. In his essay *Isis and Osiris* Plutarch says the following.

> It has been imagined that the Egyptians regarded one triangle above all others, likening it to the nature of the universe. And in his *Republic* Plato seems to have used it in arranging marriages. This triangle has 3 on the vertical side, 4 on the base, and a hypotenuse of 5, equal in square to the other two sides. It is to be imagined then that it was constituted of the masculine on the vertical side, and the feminine on the base; also, Osiris as the progenitor, Isis as the receptacle, and Horus as the offspring. For 3 is the first odd number and is a perfect number; the 4 is a square formed from an even number of dyads; and the 5 is regarded as derived in one way from the father and another way from the mother, being made up of the triad and the dyad.

Still further, Berlin Papyrus 6619 contains a problem in which one square equals the sum of two others. It is hard to imagine anyone being interested in such conditions without knowing the Pythagorean theorem. Against the conjecture, we could note that the earliest Egyptian text that mentions a right triangle and finds the length of all its sides using the Pythagorean theorem dates from about 300 BCE, and by that time the presence of Greek mathematics in Alexandria was already established. None of the older papyri mention or use by implication the Pythagorean theorem.

On balance, one would guess that the Egyptians *did* know the Pythagorean theorem. However, there is no evidence that they used it to construct right angles, as Cantor conjectured. There are much simpler ways of doing that (even involving the stretching of ropes), which the Egyptians must have known. Given that the evidence for this conjecture is so meager, why is it so often reported as fact? Simply because it has been repeated frequently since it was originally made. We know precisely the source of the conjecture, but that knowledge does not seem to reach the many people who report it as fact.[5]

Spheres or cylinders? Problem 10 of the Moscow Papyrus has been subject to various interpretations. It asks for the area of a curved surface that is either half of a cylinder or a hemisphere. In either case it is worth noting that the area is obtained by multiplying the length of a semicircle by another length in order to obtain the area. Finding the area of a hemisphere is an extremely difficult problem. Intuitive techniques that work on flat or ruled surfaces break down, as shown in Problem 9.20. If the Egyptians did compute this area, no one has given any reasonable conjecture as to how they did so. The difficulty of this problem was given as one reason for interpreting the figure as half of a cylinder. Yet the plain language of the problem implies that the surface is a hemisphere. The problem was translated into German by the Russian scholar V. V. Struve (1889–1965); the following is a translation from the German:

> The way of calculating a basket, if you are given a basket with an opening of 4 $\overline{2}$. O, tell me its surface!

---

[5] This point was made very forcefully by van der Waerden (*1963*, p. 6). In his later book, *Geometry and Algebra in Ancient Civilizations*, van der Waerden claimed that integer-sided right triangles, which seem to imply knowledge of the Pythagorean theorem, are ubiquitous in the oldest megalithic structures. Thus, he seems to imply that the Egyptians knew the theorem, but didn't use it as Cantor suggested.

> Calculate $\overline{9}$ of 9, since the basket is half of an egg. The result is 1. Calculate what is left as 8. Calculate $\overline{9}$ of 8. The result is $\overline{\overline{3}}\ \overline{6}\ \overline{18}$. Calculate what is left of this 8 after this $\overline{\overline{3}}\ \overline{6}\ \overline{18}$ is taken away. The result is $7\ \overline{9}$. Calculate $4\ \overline{2}$ times with $7\ \overline{9}$. The result is 32. Behold, this is the surface. You have found it correctly.

If we interpret the basket as being a hemisphere, the scribe has first doubled the diameter of the opening from $4\ \overline{2}$ to 9 "because the basket is half of an egg." (If it had been the *whole* egg, the diameter would have been quadrupled.) The procedure used for finding the area here is equivalent to the formula $2d \cdot \frac{8}{9} \cdot \frac{8}{9} \cdot d$. Taking $\left(\frac{8}{9}\right)^2$ as representing $\pi/4$, we find it equal to $(\pi d^2)/2$, or $2\pi r^2$, which is indeed the area of a hemisphere of radius r.

This value is also the lateral area of half of a cylinder of height d and base diameter d. If the basket is interpreted as half of a cylinder, the opening would be square and the number $4\ \overline{2}$ would be the side of the square. That would mean also that the "Egyptian π" ($\pi/4 = 64/81$), used for area problems was also being applied to the ratio of the circumference to the diameter. The numerical answer is consistent with this interpretation, but it does seem strange that only the lateral surface of the cylinder was given. That would indicate that the basket was open at the sides. It would be strange to describe such a basket as "half of an egg." The main reason given by van der Waerden (*1963*, pp. 33–34) for preferring this interpretation is an apparent inaccuracy in Struve's statement of the problem. Van der Waerden quotes T. E. Peet, who says that the number $4\ \overline{2}$ occurs twice in the statement of the problem, as the opening of the top of the basket and also as its depth. This interpretation, however, leads to further difficulties. If the surface is indeed half of a cylinder of base diameter $4\ \overline{2}$, its depth is not $4\ \overline{2}$; it is $2\ \overline{4}$. Van der Waerden also mentions a conjecture of Neugebauer, that this surface was intended to be a domelike structure of a sort seen in some Egyptian paintings, resembling very much the small end of an egg. That interpretation restores the idea that this problem was the computation of the area of a nonruled surface, and the approximation just happens to be the area of a hemisphere.

1.2. Volumes. One of the most remarkable achievements of the Egyptians is the discovery of accurate ways of computing volumes. As in the case of surface areas, the most remarkable result is found in the Moscow, not the Ahmose, Papyrus. In Problem 41 of the Ahmose Papyrus we find the correct procedure used for finding the volume of a cylindrical silo, that is, the area of the circular base is multiplied by the height. To make the numbers easy, the diameter of the base is given as 9 cubits, as in Problems 48 and 50, so that the area is 64 square cubits. The height is 10 cubits, giving a volume of 640 cubic cubits. However, the standard unit of grain volume was a *khar*, which is two-thirds of a cubic cubit, resulting in a volume of 960 *khar*. In a further twist, to get a smaller answer, the scribe divides this number by 20, getting 48 "hundreds of quadruple *hekats*." (A *khar* was 20 *hekats*.) Problem 42 is the same problem, only with a base of diameter 10 cubits. Apparently, once the reader has the rule well in hand, it is time to test the limits by making the data more cumbersome. The answer is computed to be $1185\ \overline{6}\ \overline{54}$ *khar*, again expressed in hundreds of quadruple *hekats*. Problems 44–46 calculate the volume of prisms on a rectangular base by the same procedure.

Given that pyramids are so common in Egypt, it is surprising that the Ahmose Papyrus does not discuss the volume of a pyramid. However, Problem 14 from the Moscow Papyrus asks for the volume of the frustum of a square pyramid given that the side of the lower base is 4, the side of the upper base is 2, and the height is 6. The author gives the correct recipe: Add the areas of the two bases to the area of the rectangle whose sides are the sides of the bases, that is, $2 \cdot 2 + 4 \cdot 4 + 2 \cdot 4$, then multiply by one-third of the height, getting the correct answer, 56. This technique could not have been arrived at through experience. Some geometric principle must be involved, since the writer knew that the sides of the bases, which are *parallel* lines, need to be multiplied. Normally, the lengths of two lines are multiplied only when they are perpendicular to each other, so that the product represents the area of a rectangle. Gillings (*1972*, pp. 190–193) suggests a possible route. Robins and Shute (*1987*, pp. 48–49) suggest that the result may have been obtained by completing the frustum to a full pyramid, and then subtracting the volume of the smaller pyramid from the larger. In either case, the power of visualization involved in seeing that the relation is the correct one is remarkable.

Like the surface area problem from the Moscow Papyrus just discussed, this problem reflects a level of geometric insight that must have required some accumulation of observations built up over time. It is very easy to see that if a right pyramid with a square base is sliced in half by a plane through its vertex and a pair of diagonally opposite vertices of the base, the base is bisected along with the pyramid. Thus, a tetrahedron whose base is half of a square has volume exactly half that of the pyramid of the same height having the whole square as a base.

It is also easy to visualize how a cube can be cut into two wedges, as in the top row of Fig. 2. Each of these wedges can then be cut into a pyramid on a face of the cube plus an extra tetrahedron, as in the bottom row. The tetrahedron $P'Q'R'S'$ has a base that is half of the square base of the pyramid $PQRST$, and hence has half of its volume. It follows that the volume of the tetrahedron is one-sixth that of the cube, and so the pyramid $PQRST$ is one-third of the volume. A "mixed" strategy is also possible, involving weighing of the parts. The two tetrahedra would, in theory, balance one of the square pyramids. This model could be sawn out of stone or wood. From that special case one might generalize the vital clue that the volume of a pyramid is one-third the area of the base times the altitude.

Once the principle is established that a pyramid equals a prism on the same base with one-third the height, it is not difficult to chop a frustum of a pyramid into the three pieces described in the Moscow Papyrus. Referring to Fig. 3, which shows a frustum with bottom base a square of side a and upper base a square of side b with $b < a$, we can cut off the four corners and replace them by four rectangular solids with square base of side $(a - b)/2$ and height $h/3$. These four fit together to make a single solid with square base of side $a - b$ and height $h/3$. One opposite pair of the four sloping faces that remain after the corners are removed can be cut off, turned upside down, and laid against the other two sloping faces so as to make a single slab with a rectangular base that is $a \times b$ and has height h. The top one-third of this slab can then be cut off and laid aside. It has volume $(h/3)ab$. The top half of what remains can then be cut off, and a square prism of side b and height $h/3$ cut off from it. If that square prism is laid aside (it has volume $(h/3)b^2$), the remaining piece, which is $(a - b) \times b \times (h/3)$, will fill out the other corner of the bottom layer, resulting in a square prism of volume $(h/3)a^2$. Thus, we obtain the

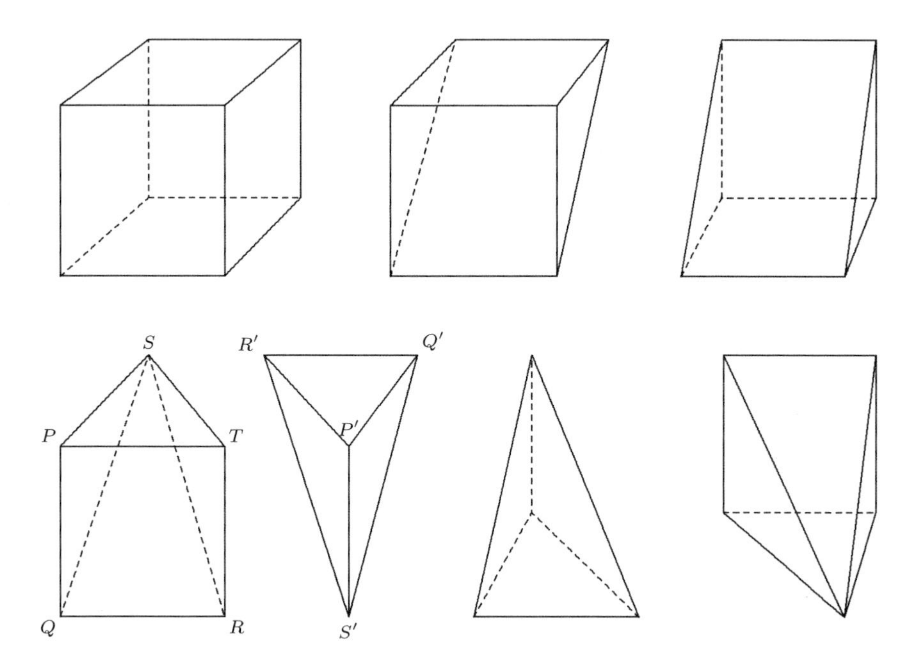

FIGURE 2. Dissection of a cube into two square pyramids and two tetrahdra.

three pieces that the scribe added to get the volume of the frustum in a way that is not terribly implausible.

It goes without saying that the last few paragraphs and Figs. 2 and 3 are conjectures, not facts of history. We do not know how the Egyptians discovered that the volume of a pyramid is one-third the volume of a prism of the same base and height or how they found the volume of a frustum. The little story just presented is merely one possible scenario.

2. Mesopotamia

Mesopotamian geometry, like its Egyptian counterpart, was regarded more as an application of mathematics than as mathematics proper. The primary emphasis was on areas and volumes. However, the Mesopotamian tablets suggest a very strong algebraic component. Many of the problems that are posed in geometric garb have no apparent practical application but are very good exercises in algebra. For example, British Museum tablet 13901 contains the following problem: *Given two squares such that the side of one is two-thirds that of the other plus 5 GAR and whose total area is 25,25 square GAR, what are the sides of the squares?* Where in real life would one encounter such a problem? The tablet itself gives no practical context, and we conclude that this apparently geometric problem is really a problem in algebraic manipulation of expressions. As Neugebauer states (*1952*, p. 41), "It is easy to show that geometrical concepts play a very secondary part in Babylonian algebra, however extensively a geometrical terminology may be used." Both Neugebauer and van der Waerden (*1963*, p. 72) point out that the cuneiform tablets contain operations that are geometrically absurd, such as adding

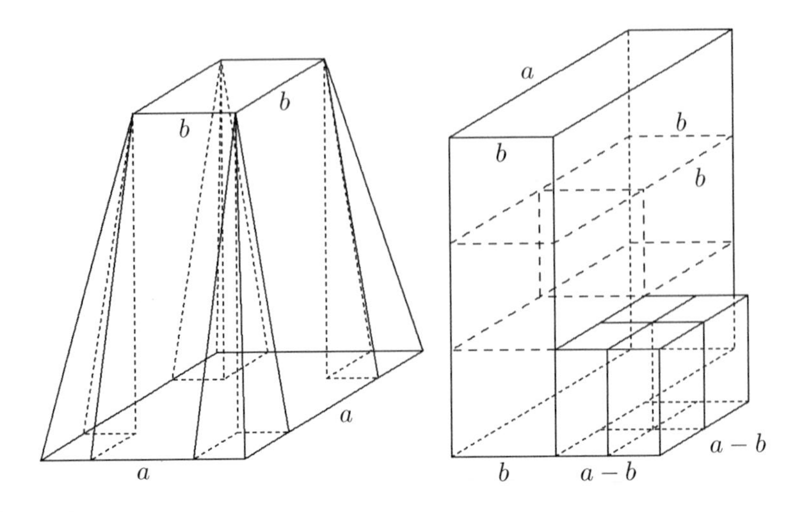

FIGURE 3. Dissection of a frustum of a pyramid.

a length to an area or multiplying two areas. For that reason, our discussion of such problems is postponed to Chapter 13.

2.1. The Pythagorean theorem. In contrast to the case of Egypt, there is clear proof that the Mesopotamians knew the Pythagorean theorem in full generality at least 1000 years before Pythagoras. They were thus already on the road to finding more abstract properties of geometric figures than mere size. Of course, this theorem was known at an early date in India and China, so that one cannot say certainly where the earliest discovery was and whether the appearance of this theorem in different localities was the result of independent discovery or transmission. But as far as present knowledge goes, the earliest examples of the use of the "Pythagorean" principle that the square on the hypotenuse of a right triangle equals the sum of the squares on the other two legs occur in the cuneiform tablets. Specifically, the old Babylonian text known as BM 85 196 contains a problem that has appeared in algebra books for centuries. We give it below as Problem 9.4. In this problem we are dealing with a right triangle of hypotenuse 30 with one leg equal to $30 - 6 = 24$. Obviously, this is the famous 3–4–5 right triangle with all sides multiplied by 6. Obviously also, the interest in this theorem was more numerical than geometric. How often, after all, are we called upon to solve problems of this type in everyday life?

How might the Pythagorean theorem have been discovered? The following hypothesis was presented by Allman (*1889*, pp. 35–37), who cited a work (*1870*) by Carl Anton Bretschneider (1808–1878). Allman thought this dissection was due to the Egyptians, since, he said, it was done in their style. If he was right, the Egyptians did indeed discover the theorem.

Suppose that you find it necessary to construct a square twice as large as a given square. How would you go about doing so? (This is a problem the Platonic Socrates poses in the dialogue *Meno*.) You might double the side of the square, but you would soon realize that doing so actually quadruples the size of the square. If you drew out the quadrupled square and contemplated it for a while, you might be

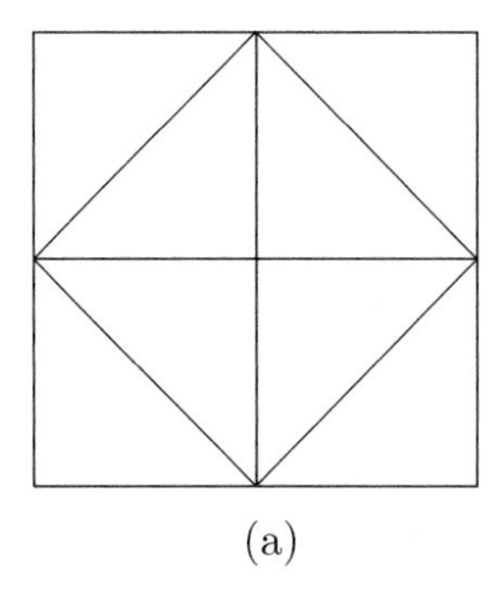
(a)

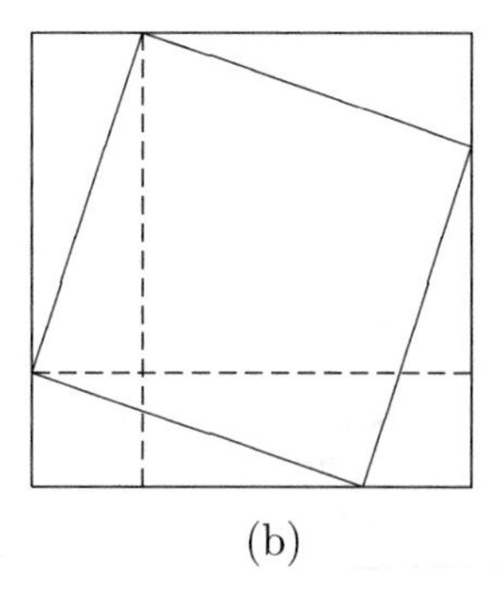
(b)

FIGURE 4. (a) Doubling a square; (b) the Pythagorean theorem.

led to join the midpoints of its sides in order, that is, to draw the diagonals of the four copies of the original square. Since these diagonals cut the four squares in half, they will enclose a square twice as big as the original one (Fig. 4). It is quite likely that someone, either for practical purposes or just for fun, discovered this way of doubling a square. If so, someone playing with the figure might have considered the result of joining in order the points at a given distance from the corners of a square instead of joining the midpoints of the sides. Doing so creates a square in the center of the larger square surrounded by four copies of a right triangle whose hypotenuse equals the side of the center square (Fig. 4); it also creates the two squares on the legs of that right triangle and two rectangles that together are equal in area to four copies of the triangle. (In Fig. 4 one of these rectangles is divided into two equal parts by its diagonal, which is the hypotenuse of the right triangle.) Hence the larger square consists of four copies of the right triangle plus the center square. It also consists of four copies of the right triangle plus the squares on the two legs of the right triangle. The inevitable conclusion is that *the square on the hypotenuse of any right triangle equals the sum of the squares on the legs.* This is the Pythagorean theorem, and it is used in many places in the cuneiform texts.

2.2. Plane figures. Some cuneiform tablets give the area of a circle of unit radius, which we have called the two-dimensional π, as 3. On the other hand (Neugebauer, *1952*, p. 46), the one-dimensional π was known to slightly more accuracy. On a tablet excavated at Susa in 1936, it was stated that the perimeter of a regular hexagon, which is three times its diameter, is 0;57,36 times the circumference of the circumscribed circle. That makes the circumference of a circle of unit diameter equal to

$$\frac{3}{0;57,36} = \frac{25}{8} = 3.125\,.$$

That the Mesopotamian mathematicians recognized the relation between the area and the circumference of a circle is shown by two tablets from the Yale Babylonian Collection (YBC 7302 and YBC 11120, see Robson, *2001*, p. 180). The first contains a circle with the numbers 3 and 9 on the outside and 45 on the inside. These numbers fit perfectly the formula $A = C^2/(4\pi)$, given that the scribe was using $\pi = 3$. Assuming that the 3 represents the circumference, 9 its square and 45 the quotient, we find $9/(4 \cdot 3) = 3/4 = 0;45$. Confirmation of this hypothesis comes from the other tablet, which contains $1;30$ outside and $11;15$ inside, since $(1;30^2)/(4 \cdot 3) = (2;15)/12 = 135/12 = 11.25 = 11;15$.

The strongest area of Mesopotamian science that has been preserved is astronomy, and it is here that geometry becomes most useful. The measurement of angles—arcs of circles—is essential to observation of the Sun, Moon, stars, and planets, since to the human eye they all appear to be attached to a large sphere rotating overhead. The division of the circle into 360 degrees is one convention that came from Mesopotamia, was embraced by the Greeks, and became an essential part of applied geometry down to the present day. The reason for the number 360 is the base-60 computational system used in Mesopotamia. The astronomers divided all circles into 360 or 720 equal parts and the radius into 60 equal parts. In that way, a unit of length along the radius was approximately equal to a unit of length on the circle.

2.3. Volumes. The cuneiform tablets contain computations of some of the same volumes as the Egyptian papyri. For example, the volume of a frustum of a square pyramid is computed in an old Babylonian tablet (British Museum 85 194). This volume is computed correctly in the Moscow Papyrus, but the Mesopotamian scribe seems to have generalized incorrectly from the case of a trapezoid and reasoned that the volume is the height times the average area of the upper and lower faces. This rule overestimates the volume by twice the volume of the four corners cut out in Fig. 3. There is, however, some disagreement as to the correct translation of the tablet in question. Neugebauer (*1935*, Vol. 1, p. 187) claimed that the computation was based on an algebraic formula that is geometrically correct. The square bases are given as having sides 10 and 7 respectively, and the height is given as 18. The incorrect rule we are assuming would give a volume of 1341, which is 22,21 in sexagesimal notation; but the actual text reads 22,30. The discrepancy could be a simple misprint, with three ten-symbols carelessly written for two ten-symbols and a one-symbol. The computation used is not entirely clear. The scribe first took the average base side $(10 + 7)/2$ and squared it to get 1,12;15 in sexagesimal notation (72.25). At this point there is apparently some obscurity in the tablet itself. Neugebauer interpreted the next number as 0;45, which he assumed was calculated as one-third of the square of $(10 - 7)/2$. The sum of these two numbers is 1,13, which, multiplied by 18, yields 21,54 (that is, 1314), which is the correct result. But it is difficult to see how this number could have been recorded incorrectly as 22,30. If the number that Neugebauer interprets as 0;45 is actually 2;15 (which is a stretch—three ten-symbols would have to become two one-symbols), it would be exactly the square of $(10 - 7)/2$, and it would yield the same incorrect formula as the assumption that the average of the areas of the two bases was being taken. In any case, the same procedure is used to compute the volume of the frustum of a cone (Neugebauer, *1935*, p. 176), and in that case it definitely is the incorrect rule stated here, taking the average of the two bases and multiplying by the height.

3. China

Three early Chinese documents contain a considerable amount of geometry, always connected with the computation of areas and volumes. We shall discuss the geometry in them in chronological order, omitting the parts that repeat procedures we have already discussed in connection with Egyptian geometry.

3.1. The *Zhou Bi Suan Jing*. As mentioned in Chapter 2, the earliest Chinese mathematical document still in existence, the *Zhou Bi Suan Jing*, is concerned

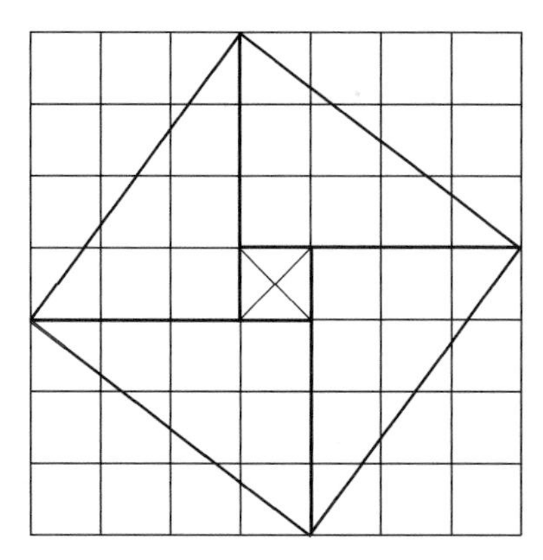

FIGURE 5. Chinese illustration of the Pythagorean theorem.

with astronomy and the applications of mathematics to the study of the heavens. The title refers to the use of the sundial or gnomon in astronomy. This is the physical model that led the Chinese to discover the Pythagorean theorem. Here is a paraphrase of the discussion:

> Cut a rectangle whose width is 3 units and whose length is 4 (units) along its diagonal. After drawing a square on this diagonal, cover it with half-rectangles identical to the piece of the original rectangle that lies outside the square, so as to form a square of side 7. [See Fig. 5.] Then the four outer half-rectangles, each of width 3 and length 4 equal two of the original rectangle, and hence have area 24. When this amount is subtracted from the square of area 49, the difference, which is the area of the square on the diagonal, is seen to be 25. The length of the diagonal is therefore 25.

Although the proof is given only for the easily computable case of the 3–4–5 right triangle, it is obvious that the geometric method is perfectly general, lacking only abstract symbols for unspecified numbers. In our terms, the author has proved that the length of the diagonal of a rectangle whose width is a and whose length is b is the square root of $(a+b)^2 - 2ab$. Note that this form of the theorem is not the "$a^2 + b^2 = c^2$" that we are familiar with.

The *Zhou Bi Suan Jing* contains three diagrams accompanying the discussion of the Pythagorean theorem. According to Cullen (*1996*, p. 69), one of these diagrams was apparently added in the third century by the commentator Zhao Shuang. This diagram is shown in Fig. 5 for the special case of a 3–4–5 triangle. The other two were probably added by later commentators in an attempt to elucidate Zhao Shuang's commentary.

According to Li and Du (*1987*, p. 29), the vertical bar on a sundial was called *gu* in Chinese, and its shadow on the sundial was called *gou*; for that reason the Pythagorean theorem was known as the *gougu* theorem. Cullen (*1996*, p. 77) says that *gu* means *thigh* and *gou* means *hook*. All authorities agree that the hypotenuse was called *xian* (bowstring). The *Zhou Bi Suan Jing* says that the Emperor Yu was able to bring order into the realm because he knew how to use this theorem to compute distances. Zhao Shuang credited the Emperor Yu with saving his people from floods and other great calamities, saying that in order to do so he had to survey the shapes of mountains and rivers. Apparently the Emperor had drainage canals dug to channel floods out of the valleys and into the Yangtze and Yellow Rivers.

The third-century commentary on the *Zhou Bi Suan Jing* by Zhao Shuang explains a method of surveying that was common in China, India, and the Muslim world for centuries. The method is illustrated in Fig. 6, which assumes that the height H of an inacessible object is to be determined. To determine H, it is necessary to put two poles of a known height h vertically into the ground in line with the object at a known distance D apart. The height h and the distance D are theoretically arbitrary, but the larger they are, the more accurate the results will be. After the poles are set up, the lengths of the shadows they would cast if the Sun were at the inacessible object are measured as s_1 and s_2. Thus the lengths s_1, s_2, h, and D are all known. A little trigonometry and algebra will show that

$$H = h + \frac{Dh}{s_2 - s_1}.$$

We have given the result as a formula, but as a set of instructions it is very easy to state in words: *The required height is found by multiplying the height of the poles by the distance between them, dividing by the difference of the shadow lengths, and adding the height of the poles.*

This method was expounded in more detail in a commentary on the *Jiu Zhang Suanshu* written by Liu Hui in 263 CE. This commentary, along with the rest of the material on right triangles in the *Jiu Zhang Suanshu* eventually became a separate treatise, the *Hai Dao Suan Jing* (*Sea Island Mathematical Manual*, see Ang and Swetz, *1986*). Liu Hui mentioned that this method of surveying could be found in the *Zhou Bi Suan Jing* and called it the *double difference method* (*chong cha*). The name apparently arises because the difference $H - h$ is obtained by dividing Dh by the difference $s_2 - s_1$.

We have described the lengths s_1 and s_2 as shadow lengths here because that is the problem used by Zhao Shuang to illustrate the method of surveying. He attempts to calculate the height of the Sun, given that at the summer solstice a stake 8 *chi* high casts a shadow 6 *chi* long and that the shadow length decreases by 1 *fen* for every 100 *li* that the stake is moved south, casting no shadow at all when moved 60,000 *li* to the south. This model assumes a flat Earth, under which the shadow length is proportional to the distance from the pole to the foot of the perpendicular from the Sun to the plane of the Earth. Even granting this assumption, as we know, the Sun is so distant from the Earth that no lengthening or shortening of shadows would be observed. To any observable precision the Sun's rays are parallel at all points on the Earth's surface. The small change in shadow length that we observe is due entirely to the curvature of the Earth. But let us continue, accepting Zhao Shuang's assumptions.

The data here are $D = 1000$ *li*, $s_2 - s_1 = 1$ *fen*, $h = 8$ *chi*. One *chi* is about 25 centimeters, one *fen* is about 2.5 cm, and one *li* is 1800 *chi*, that is, about 450 meters. Because the pole height h is obviously insignificant in comparison with the height of the Sun, we can neglect the first term in the formula we gave above, and write

$$H = \frac{Dh}{s_2 - s_1}.$$

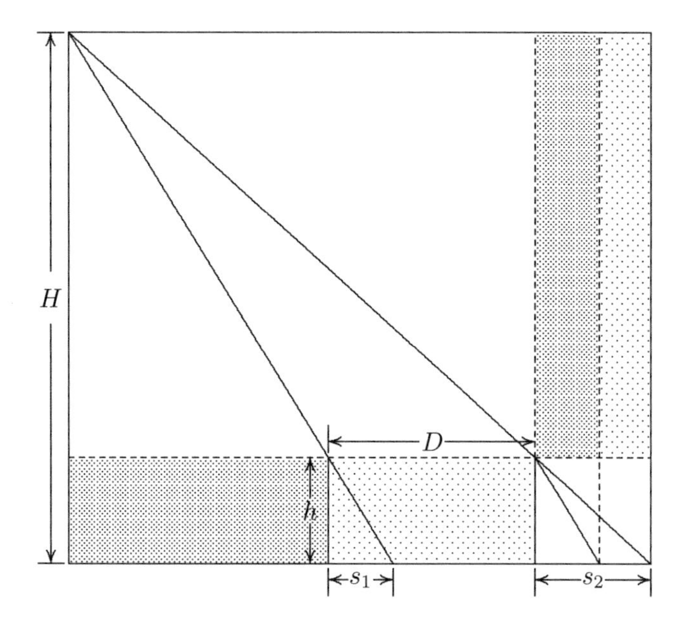

FIGURE 6. The double-difference method of surveying.

When we insert the appropriate values, we find, as did Zhao Shuang, that the Sun is 80,000 *li* high, about 36,000 kilometers. Later Chinese commentators recognized that this figure was inaccurate, and in the eighth century an expedition to survey accurately a north–south line found the actual lengthening of the shadow to be 4 *fen* per thousand *li*. Notice that there seem to be two methods of computing the height here. In the method just discussed, the fact that the Sun is directly overhead at a distance of 60,000 *li* to the south is irrelevant to the computation. If it is taken into account, one can immediately use the similar triangles to infer the height of 80,000 *li*. This fact suggests that the original text was modified by later commentators, but that not all the parts that became irrelevant as a result of the modifications were removed.

3.2. The *Jiu Zhang Suanshu*. The *Jiu Zhang Suanshu* contains all the standard formulas for the areas of squares, rectangles, triangles, and trapezoids, and also the recognition of a relation between the circumference and the area of a circle, which we could interpret as a connection between the one-dimensional π and the two-dimensional π. The geometric formulas given in this treatise are more extensive than those of the Ahmose Papyrus; for example, there are approximate formulas for the volume of segment of a sphere and the area of a segment of a circle. It is perhaps not fair to compare the two documents, since the Ahmose Papyrus was written nearly two millennia earlier, and the *Jiu Zhang Suan Shu* was intended to cover all the mathematics known at the time. The implied value of one-dimensional π, however, is $\pi = 3$. It is surprising to find this value so late, since it is known that the value 3.15147 had been obtained in China by the first century. According to Li and Du (*1987*, p. 68), Liu Hui refined it to $3.14 + 64/62500 = 3.141024$ by approximating the area of a 192-sided polygon.[6] That is, he started with a hexagon and doubled the number of sides five times. However, since the area of the polygon with twice the number of sides is the radius of the circumscribed circle times the perimeter of the original polygon, it was only necessary to find the perimeter of a 96-sided polygon and multiply by the radius.

Problems 31 and 32 ask for the area of a circular field of a given diameter and circumference.[7] The method is to multiply half of the circumference by half of the diameter, which is exactly right in terms of Euclidean geometry; equivalently, the reader is told that one may multiply the two quantities and divide by 4. However, in the actual data for problems the diameter given is exactly one-third of the given circumference; in other words, the value assumed for one-dimensional π is 3. The assumption of that value leads to two other procedures for calculating the area: squaring the diameter, then multiplying by 3 and dividing by 4, or squaring the circumference and dividing by 12. An elaboration of this problem occurs in Problems 37 and 38, in which the area of an annulus (the region outside the smaller of two concentric circles and inside the larger) is given in terms of its width and the circumferences of the two circles.

The authors knew also how to find the volume of a pyramid. Problem 15 of Chapter 5 asks for the volume of a pyramid whose base is a rectangle 5 *chi* by 7 *chi* and whose height is 8 *chi*. The answer is given as $93\frac{1}{3}$ (cubic) *chi*. For a frustum of a pyramid having rectangular bases the recipe is to add twice the length of the upper base to the lower base and multiply by the width of the upper base to get one term. A second term is obtained symmetrically as twice the length of the lower base plus the length of the upper base, multiplied by the width of the lower base. These two terms are then added and multiplied by the height, after which one divides by 6. If the bases are $a \times b$ and $c \times d$ (the sides of length a and c being parallel) and the height is h, this yields what we would write (correctly) as

$$V = \frac{h}{6}\big[(2a+c)b + (2c+a)d\big] .$$

Notice that this result is more general than the formula in the Moscow Papyrus, which is given for a frustum with square bases.

The Pythagorean theorem. The last of the nine chapters of the *Jiu Zhang Suanshu* contains 24 problems on the *gougu* theorem. After a few "warm-up" problems in which two of the three sides of a right triangle are given and the third is to be computed, the problems become more complicated. Problem 11, for example, gives a rectangular door whose height exceeds its width by 6 *chi*, 8 *cun* and has a diagonal of 1 *zhang*. One *zhang* is 10 *chi* and 1 *chi* is 10 *cun* (apparently a variant rendering of *fen*). The recipe given is correct: Take half the difference of the height and width, square it, double, subtract from the square of the diagonal, then take the square root of half of the result. That process yields the average of the height and width, and given their semidifference of 3 *chi*, 4 *cun*, one can easily get both the width and the height.

3.3. The *Sun Zi Suan Jing*.

The *Sun Zi Suan Jing* contains a few problems in measurement that are unusual enough to merit some discussion. An inverse area

---

[6] Lam and Ang (*1986*) give the value as $3.14 + 169/625 = 3.142704$.

[7] All references to problem numbers and nomenclature in this section are based on the article of Lam (*1994*).

FIGURE 7. The double square umbrella.

problem occurs in Problem 20, in which a circle is said to have area 35,000 square *bu*, and its circumference is required. Since the area is taken as one-twelfth of the square of the circumference, the author multiplies by 12, then takes the square root, getting $648\frac{96}{1296}$ *bu*.

3.4. Liu Hui. Chinese mathematics was greatly enriched from the third through the sixth centuries by a series of brilliant geometers, whose achievements deserve to be remembered alongside those of Euclid, Archimedes, and Apollonius. We have space to discuss only three of these, beginning with the third-century mathematician Liu Hui (ca. 220–ca. 280). Liu Hui had a remarkable ability to visualize figures in three dimensions. In his commentary on the *Jiu Zhang Suanshu* he asserted that the circumference of a circle of diameter 100 is 314. In solid geometry he provided dissections of many geometric figures into pieces that could be reassembled to demonstrate their relative sizes beyond any doubt. As a result, real confidence could be placed in the measurement formulas that he provided. He gave correct procedures, based on such dissections, for finding the volumes enclosed by many different kinds of polyhedra. But his greatest achievement is his work on the volume of the sphere.

The *Jiu Zhang Suanshu* made what appears to be a very reasonable claim: that the ratio of the volume enclosed by a sphere to the volume enclosed by the circumscribed cylinder can be obtained by slicing the sphere and cylinder along the axis of the cylinder and taking the ratio of the area enclosed by the circular cross section of the sphere to the area enclosed by the square cross section of the cylinder. In other words, it would seem that the ratio is $\pi : 4$. This conjecture seems plausible, since every such section produces exactly the same figure. It fails, however because of what is called *Pappus' principle*: The volume of a solid of revolution equals the area revolved about the axis times the distance traveled by the centroid of the area. The half of the square that is being revolved to generate the cylinder has a centroid that is farther away from the axis than the centroid of the semicircle inside it whose revolution produces the sphere; hence when the two areas are multiplied by the two distances, their ratios get changed. When a circle inscribed in a square is rotated, the ratio of the volumes generated is 2 : 3, while that of the original areas is $\pi : 4$. Liu Hui noticed that the sections of the figure parallel to the base of the cylinder do not all have the same ratios. The sections of the cylinder are all disks of the same size, while the sections of the sphere shrink as the section moves from the equator to the poles. He also formed a solid by intersecting two cylinders circumscribed about the sphere whose axes are at right angles to each other, thus producing a figure he called a *double square umbrella*, which is now known as a *bicylinder* or *Steinmetz solid* (see Hogendijk, *2002*). A representation of the double square umbrella, generated using *Mathematica* graphics, is shown in Fig. 7. Its volume *does* have the same ratio to the sphere that the square has to its inscribed circle, that is, $4 : \pi$. This proportionality between the double square umbrella and the sphere is easy to see intuitively, since every horizontal slice of this figure by a plane parallel to the plane of the axes of the two circumscribed cylinders intersects the double square umbrella in a square and intersects the sphere in the circle inscribed in that square. Liu Hui inferred that the volume enclosed by the double umbrella would have this ratio to the volume enclosed by the sphere. This inference is correct and is an example of what is called *Cavalieri's principle*: *Two solids such that the section of one by each horizontal plane bears a fixed ratio to the section of the other by the same plane have volumes in that same ratio.* This principle had been used by Archimedes five centuries earlier, and in the introduction to his *Method*, Archimedes uses *this very example*, and asserts that the volume of the intersection of the two cylinders is two-thirds of the volume of the cube in which they are inscribed.[8] But Liu Hui's use of it (see Lam and Shen, *1985*) was obviously independent of Archimedes. It amounts to a limiting case of the dissections he did so well. The solid is cut into *infinitely thin* slices, each of which is then dissected and reassembled as the corresponding section of a different solid. This realization was a major step toward an accurate measurement of the volume of a sphere. Unfortunately, it was not granted to Liu Hui to complete the journey. He maintained a consistent agnosticism on the problem of computing the volume of a sphere, saying, "Not daring to guess, I wait for a capable man to solve it."

3.5. Zu Chongzhi. That "capable man" required a few centuries to appear, and he turned out to be two men. "He" was Zu Chongzhi (429–500) and his son Zu Geng (450–520). Zu Chongzhi was a very capable geometer and astronomer who said that if the diameter of a circle is 1, then the circumference lies between 3.1415926 and 3.1415927. From these bounds, probably using the Chinese version of the Euclidean algorithm, the method of mutual subtraction (see Problem 7.12), he stated that the circumference of a circle of diameter 7 is (approximately) 22 and that of a circle of diameter 113 is (approximately) 355.[9] These estimates are very good, far too good to be the result of any inspired or hopeful guess. Of course, we don't have to imagine that Zu Chongzhi actually *drew* the polygons. It suffices to know how to compute the perimeter, and that is a simple recursive process: If s_n is the length of the side of a polygon of n sides inscribed in a circle of unit radius, then

$$s_{2n}^2 = 2 - \sqrt{4 - s_n^2}\,.$$

---

[8] Hogendijk (*2002*) argues that Archimedes also knew the surface area of the bicylinder.

[9] The approximation $\pi \approx \frac{22}{7}$ was given earlier by He Chengtian (370–447), and of course much earlier by Archimedes. A more sophisticated approach by Zhao Youqin (b. 1271) that gives $\frac{355}{113}$ was discussed by Volkov (*1997*).

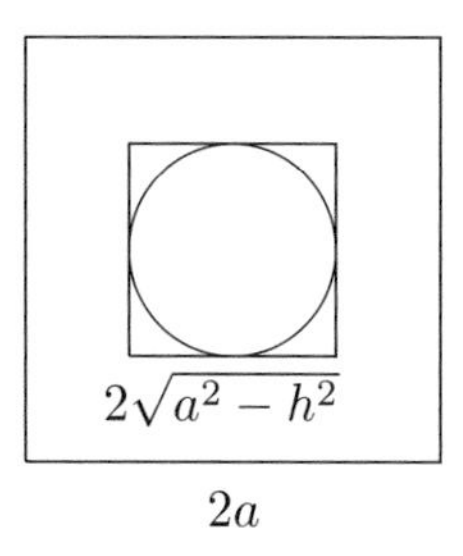

FIGURE 8. Sections of the cube, double square umbrella, and sphere at height h.

Hence each doubling of the number of sides makes it necessary to compute a square root, and the approximation of these square roots must be carried out to many decimal places in order to get enough guard digits to keep the errors from accumulating when you multiply this length by the number of sides. In principle, however, given enough patience, one could compute any number of digits of π this way.

One of Zu Chongzhi's outstanding achievements, in collaboration with his son, was finding the volume enclosed by Liu Hui's double square umbrella. As Fu (*1991*) points out, this volume was not approachable by the direct method of dissection and recombination that Liu Hui had used so successfully.[10] An indirect approach was needed. The trick turned out to be to enclose the double square umbrella in a cube and look at the volume inside the cube and outside the double square umbrella. Suppose that the sphere has radius a. The double square umbrella can then be enclosed in a cube of side $2a$. Consider a horizontal section of the enclosing cube at height h above the middle plane of that cube. In the double umbrella this section is a square of side $2\sqrt{a^2 - h^2}$ and area $4(a^2 - h^2)$, as shown in Fig. 8. Therefore the area outside the double umbrella and inside the cube is $4h^2$.

It was no small achievement to look at the region in question. It was an even keener insight on the part of the family Zu to realize that this cross-sectional area is equal to the area of the cross section of an upside-down pyramid with a square base of side $2a$ and height a. Hence *the volume of the portion of the cube outside the double umbrella in the upper half of the cube equals the volume of a pyramid with square base* $2a$ *and height* a. But thanks to earlier work contained in Liu Hui's commentaries on the *Jiu Zhang Suanshu*, Zu Chongzhi knew that this volume was $(4a^3)/3$. It therefore follows, after doubling to include the portion below the middle plane, that the region inside the cube but outside the double umbrella has volume $(8a^3)/3$, and hence that the double umbrella itself has volume $8a^3 - (8a^3)/3 = (16a^3)/3$.

Since, as Liu Hui had shown, the volume of the sphere is $\pi/4$ times the volume of the double square umbrella, it follows that the sphere has volume $(\pi/4)\cdot(16a^3)/3$, or $(4\pi a^3)/3$.

[10] Lam and Shen (*1985*, p. 223), however, say that Liu Hui *did* consider the idea of setting the double umbrella inside the cube and trying to find the volume between the two. Of course, that volume also is not accessible through direct, finite dissection.

4. Japan

The *Wasanists* mentioned in Chapter 3, whose work extended from 1600 to 1850, inherited a foundation of mathematics established by the great Chinese mathematicians, such as Liu Hui, Zu Chongzhi, and Yang Hui. They had no need to work out procedures for computing the areas and volumes of simple figures. The only problems in elementary measurement of figures that had not been solved were those involving circles and spheres, connected, as we know, with the value of π in various dimensions. Nevertheless, during this time there was a strong tradition of geometric challenge problems. It has already been mentioned that religious shrines in Japan were frequently decorated with the solutions of such problems (see Plate 2). The geometric problems that were solved usually involved combinations of simple figures whose areas or volumes were known but which were arranged in such a way that finding their parts became an intricate problem in algebra. The word *algebra* needs to be emphasized here. The challenge in these problems was only incidentally geometric; it was largely algebraic, as the book of Fukagawa and Pedoe (*1989*) shows very convincingly. New geometry arose in Japan near the end of the seventeenth century, with better approximations to π and the solution of problems involving the rectification of arcs and the computation of the volume and area of a sphere by methods using infinite series and sums that approximate integrals.

We begin by mentioning a few of the challenge problems without giving their solutions, since they are really problems in algebra. Afterward we shall briefly discuss the infinitesimal methods used to solve the problems of measuring arcs, areas, and volumes in spheres.

4.1. The challenge problems.

In 1627 Yoshida Koyu wrote the *Jinkō-ki* (*Treatise on Large and Small Numbers*), concluding it with a list of challenge questions, and thereby stimulated a great deal of further work. Here are some of the questions:

1. There is a log of precious wood 18 feet long whose bases are 5 feet and $2\frac{1}{2}$ feet in circumference. Into what lengths should it be cut to trisect the volume?
2. There have been excavated 560 measures of earth, which are to be used for the base of a building. The base is to be 3 measures square and 9 measures high. Required, the size of the upper base.
3. There is a mound of earth in the shape of a frustum of a circular cone. The circumferences of the bases are 40 measures and 120 measures and the mound is 6 measures high. If 1200 measures of earth are taken evenly off the top, what will be the height?
4. A circular piece of land 100 [linear] measures in diameter is to be divided among three persons so that they shall receive 2900, 2500, and 2500 [square] measures respectively. Required, the lengths of the chords and the altitudes of the segments.

These problems were solved in a later treatise, which in turn posed new mathematical problems to be solved; this was the beginning of a tradition of posing and solving problems that lasted for 150 years. Seki Kōwa solved a geometric problem that would challenge even the best algebraist today. It was the fourteenth in a list of challenge problems posed by Sawaguchi Kazuyuki: *There is a quadrilateral whose sides and diagonals are u, v, w, x, y, and z* [as shown in Fig. 9].

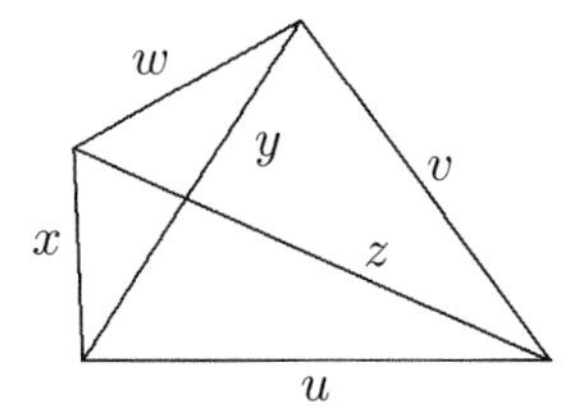

FIGURE 9. Sawaguchi Kazuyuki's quadrilateral problem.

It is given that

$$\begin{aligned} z^3 - u^3 &= 271 \\ u^3 - v^3 &= 217 \\ v^3 - y^3 &= 60.8 \\ y^3 - w^3 &= 326.2 \\ w^3 - x^3 &= 61\,. \end{aligned}$$

Required, to find the values of u, v, w, x, y, z.

The fact that the six quantities are the sides and diagonals of a quadrilateral provides one equation that they must satisfy, namely:

$$u^4w^2 + x^2\bigl(v^4 + w^2y^2 - v^2(w^2 - x^2 + y^2)\bigr) - \bigl(y^2(w^2 + x^2 - y^2) + v^2(-w^2 + x^2 + y^2)\bigr)z^2 \\ + y^2z^4 - u^2\bigl(v^2(w^2 + x^2 - y^2) + w^2(-w^2 + x^2 + y^2) + (w^2 - x^2 + y^2)z^2\bigr) = 0\,.$$

This equation, together with the five given conditions, provides a complete set of equations for the six quantities. However, Seki Kōwa's explanation, which is only a sketch, does not mention this sixth equation, so it may be that what he solved was the indeterminate problem given by the other five equations. That, however, would be rather strange, since then the quadrilateral would play no role whatsoever in the problem. His solution is discussed in Sect. 3 of Chapter 14. Whatever the case, it is known that such equations were solved numerically by the Chinese using a counting board. Here once again it is very clear that the motive for the problem is algebraic, even though it does amount to a nontrivial investigation of the relations among the parts of a quadrilateral.

4.2. Beginnings of the calculus in Japan. By the end of the seventeenth century the *wasanists* were beginning to use techniques that resemble the infinitesimal methods being used in Europe about this time. Of course, in one sense Zu Chongzhi had used some principles of calculus 1000 years earlier in his application of Cavalieri's principle to find the volume of a sphere. The intuitive basis of the principle is that equals added to equals yield equal sums, and a solid can be thought of as the sum of its horizontal sections. It isn't really, of course. No finite sum of areas and no limit of such a sum can ever have positive volume. Students in calculus courses learn to compute volumes using approximating sums that are very thin prisms, but not infinitely thin.

In Japan these techniques were first applied in the area called *yenri* (circle theory),[11] a topic that had been studied extensively in China. The idea of approximating by shells or disks can be seen in the 1684 edition of the *Ketsugi-shō* (*Combination Book*), first published in 1660 by Isomura Kittoku.

Isomura Kittoku explained the method as follows (Mikami, *1913*, p. 204):

> If we cut a sphere of diameter 1 foot into 10,000 slices, the thickness of each slice is 0.001 feet, which will be something like that of a very thin paper. Finding in this way the volume of each of them, we sum up the results, 10,000 in number, when we get 532.6 measures [that is, a volume of 0.5326 cubic foot]. Besides, it is true, there are small incommensurable parts, which are neglected.

The technique of obtaining extraordinary precision and using it to perform numerical experiments which provide the basis for general assertions also appears in some remarkable infinite series attributed to Takebe Kenkō, as we shall see below. Takebe Kenkō's method of squaring the circle was based on a relation, which he apparently discovered in 1722, between the square of half of an arc, the height h of the arc,[12] and the diameter d of the circle. Here is his own description of this discovery, as explained by Smith and Mikami (*1914*, pp. 147–149). He began with height $h = 0.000001 = 10^{-6}$ and $d = 10$, finding the square of the arc geometrically with accuracy to 53 decimal places.

The value of the square of this arc is

$$0.00001\ 00000\ 00333\ 33335\ 11111\ 12253\ 96833\ 52381\ 01394\ 90188\ 203+.$$

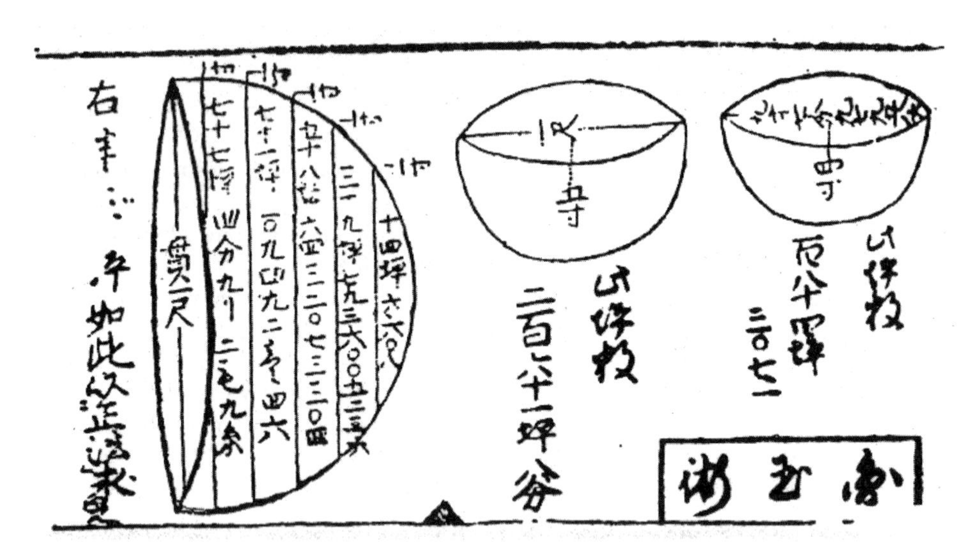

Isomura Kittoku's method of computing the volume of a sphere.

According to Smith and Mikami (*1914*, p. 148), the value given by Takebe Kenkō was

[11] The symbol for circle here (*yen*) is also the symbol for the Japanese unit of currency; it is actually pronounced "en."

[12] This height is called the *sagitta* (arrow) by lens grinders, a name first bestowed on it in India. It is now called the *versed sine* in mathematics.

$$0.00000\ 00000\ 33333\ 35111\ 11225\ 39690\ 66667\ 28234\ 77694\ 79595\ 875+.$$

But this value does not fit with the procedure followed by Takebe Kenkō; it does not even yield the correct first approximation. The figure given by Smith and Mikami appears to represent the value obtained by Takebe Kenkō *after* the first approximation was subtracted, but with the result multiplied by the square of the diameter.[13] In appreciating Takebe Kenkō's method, the first problem to be solved is the source of this extremely accurate measurement of the circle. According to Smith and Mikami (*1914*, p. 148), Takebe Kenkō said that the computation was given in two other works, both of which are now lost, leaving us to make our own conjectures. The first clue that strikes us in this connection is the seemingly strange choice of the *square* of the arc rather than the arc itself. Why would it be easier to compute the square of the arc than the arc itself? An answer readily comes to mind: The arc is approximated by its chord, and the chord is one side of a convenient right triangle. In fact, the chord is the mean proportional between the diameter of the circle and the height of the arc, so that in this case it is $\sqrt{dh} = \sqrt{10^{-5}}$. When we square it, we get just $dh = 10^{-5}$, which acts as Takebe Kenkō's first approximation. That result suggests that the length of the arc was reached by repeatedly bisecting the arc, taking the chord as an approximation. This hypothesis gains plausibility, since it is known that this technique had been used earlier to approximate π. Since $a^2 = 4(a/2)^2$, it was only necessary to find the square of half the arc, then multiply by 4. The ratio of the chord to the diameter is even easier to handle, especially since Takebe Kenkō has taken the diameter to be 10. If x is the square of this ratio for a given chord, the square of ratio for the chord of half of the arc is $\bigl(1-\sqrt{1-x}\bigr)/2$. In other words, the iterative process $x \mapsto \bigl(1-\sqrt{1-x}\bigr)/2$ makes the bisection easy. If we were dealing with the arc instead of its square, each step in that process would involve two square roots instead of one. Even as it is, Takebe Kenkō must have been a calculating genius to iterate this process enough times to get 53 decimal places of accuracy without making any errors. The result of 50 applications yields a ratio which, multiplied by $100 \cdot 4^{50}$, is

$$0.00001\ 00000\ 00333\ 33335\ 11111\ 12253\ 96833\ 52381\ 01131\ 94822\ 94294\ 362+.$$

This number of iterations gives 38 decimal places of accuracy. Even with this plausible method of procedure, it still strains credibility that Takebe Kenkō achieved the claimed precision. However, let us pass on to the rest of his method.

After the first approximation hd is subtracted, the new error is 10^{-12} times $0.3333333\ldots$, which suggests that the next correction should be $10^{-12}/3$. But this is exactly $h^2/3$, in other words $h/(3d)$ times the first term. When it is subtracted from the previously corrected value, the new error is

$$10^{-19} \cdot 0.17777\ 77892\ 06350\ 01904\ 76806\ 15685\ 4870+\ .$$

The long string of 7's here suggests that this number is 10^{-19} times $\frac{1}{10} + \frac{7}{90} = \frac{16}{90} = \frac{8}{45}$, which is $(8h)/(15d)$ times the previous correction. By continuing for a few more terms, Takebe Kenkō was able to observe a pattern: The corrections are obtained by multiplying successively by $h/(3d)$, $(8h)/(15d)$, $(9h)/(14d)$, $(32h)/(45d)$, $(25h)/(33d), \ldots$. Some sensitivity to the factorization of integers is necessary to

[13] Even so, there is one 3 missing at the beginning and, after it is restored, the accuracy is "only" 33 decimal places. That precision, however, would have been all that Takebe Kenkō needed to compute the four corrections he claimed to have computed.

see the recursive operation: multiplication by $(h/d)[2n^2/(n+1)(2n+1)]$. Putting these corrections together as an infinite series leads to the expression

$$\frac{a^2}{4} = dh\left[1 + \sum_{n=1}^{\infty} \frac{2^{2n+1}(n!)^2}{(2n+2)!} \cdot \left(\frac{h}{d}\right)^n\right]$$

when the full arc has length a.

In using this numerical approach, Takebe Kenkō had reached his conclusion inductively. This induction was based on a faith (which turns out to be justified) that the coefficients of the power series are rational numbers that satisfy a fairly simple recursive formula. As you know, the power series for the sine, cosine, exponential, and logarithm have this happy property, but the series for the tangent, for example, does not.

This series solves the problem of rectification of the circle and hence all problems that depend on knowing the value of π. In modern terms the series given by Takebe Kenkō represents the function

$$\left(d \arcsin\left(\sqrt{\frac{h}{d}}\right)\right)^2.$$

Takebe Kenkō's discovery of this result in 1722 falls between the discovery of the power series for the arcsine function by Newton in 1676 and its publication by Euler in 1737.

Was European calculus transmitted to Japan in the seventeenth century? The methods used by Isomura Kittoku to compute the volume and surface area of a sphere and by Takebe Kenkō to compute the square of a half-arc in terms of the versed sine of the arc are at the heart of calculus. Smith and Mikami (*1914*, pp. 148–155) argue that some transmission from Europe at this time is plausible in the case of Takebe Kenkō. They note that there was some contact, although very limited, between Japanese and European scholars, even during the period of "closure," and that a Jesuit missionary in China named Pierre Jartoux (1668–1721) communicated some of the latest European discoveries to his Chinese hosts. After noting that "there is no evidence that Seki or his school borrowed their methods from the West" (*1914*, p. 142), they argue as follows (*1914*, p. 155):

> Here then is a scholar, Jartoux, in correspondence with Leibnitz [sic], giving a series not difficult of deduction by the calculus, which series Takebe uses and which is the essence of the *yenri*, but which Takebe has difficulty in explaining... [I]t seems a reasonable conjecture that Western learning was responsible for [Jartoux'] work, that he was responsible for Takebe's series, and that Takebe explained the series as best he could.

Probably the question of Western influence on Japanese mathematics cannot be decided. However, in allowing for the possibility of communication from West to East, we must not neglect the possibility of some transmission in the opposite direction, in addition to what was transmitted from the Muslims and Hindus earlier. Leibniz, in particular, was fascinated with oriental cultures, and at least two of his results, one of them a simple observation on determinants and the other a more extensive development of combinatorics, were known earlier in India and Japan. It should also be noted that in contrast to the Chinese mathematicians,

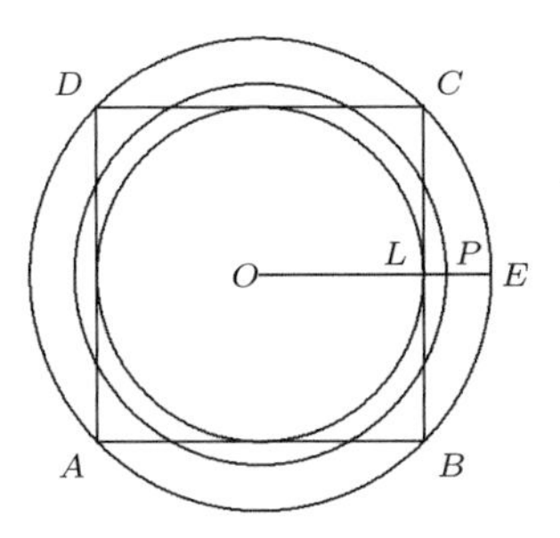

FIGURE 10. Rounding a square.

the practitioners of *wasan* were not immediately attracted to Euclid when his work arrived in Japan in the nineteenth century. According to Murata (*1994*, p. 109), having seen Chinese translations of Euclid, they were repelled by the great amount of fuss required to derive elementary facts. They may have taken just the ideas that appealed to them out of the information reaching them through contacts with the Chinese.

5. India

The *Sulva Sutras* contain many transformation-of-area constructions such as are later found in Euclid. In particular, the Pythagorean theorem, and constructions for finding the side of a square equal to a rectangle, or the sum or difference of two other squares are given. This construction resembles the one found in Proposition 5 of Book 2 of Euclid rather than Euclid's construction of the mean proportional in Book 6, both of which are discussed in Chapter 10. The Pythagorean theorem is not given a name, but is stated as the fact that "the diagonal of a rectangle produces both [areas] which its length and breadth produce separately." Among other transformation of area problems the Hindus considered in particular the problem of squaring the circle. The *Bodhayana Sutra* states the converse problem of constructing a circle equal to a given square. The construction is shown in Fig. 10, where $LP = \frac{1}{3}LE$.

In terms that we can appreciate, this construction gives a value for two-dimensional π of $18(3 - 2\sqrt{2})$, which is about 3.088.

5.1. Aryabhata I. Chapter 2 of Aryabhata's *Aryabhatiya* (Clark, *1930*, pp. 21–50) is called *Ganitapada* (*Mathematics*). In Stanza 6 of this chapter Aryabhata gives the correct rule for area of a triangle, but declares that the volume of a tetrahedron is half the product of the height and the area of the base. He says in Stanza 7 that the area of a circle is half the diameter times half the circumference, which is correct, and shows that he knew that one- and two-dimensional π were the same number. But he goes on to say that the volume of a sphere is the area of a great circle times its own square root. This would be correct only if three-dimensional π equaled $\frac{16}{9}$, very far from the truth! Yet Aryabhata knew a very good approximation to one-dimensional π. In Stanza 10 he writes:

> Add 4 to 100, multiply by 8, and add 62,000. The result is approximately the circumference of a circle of which the diameter is 20,000.

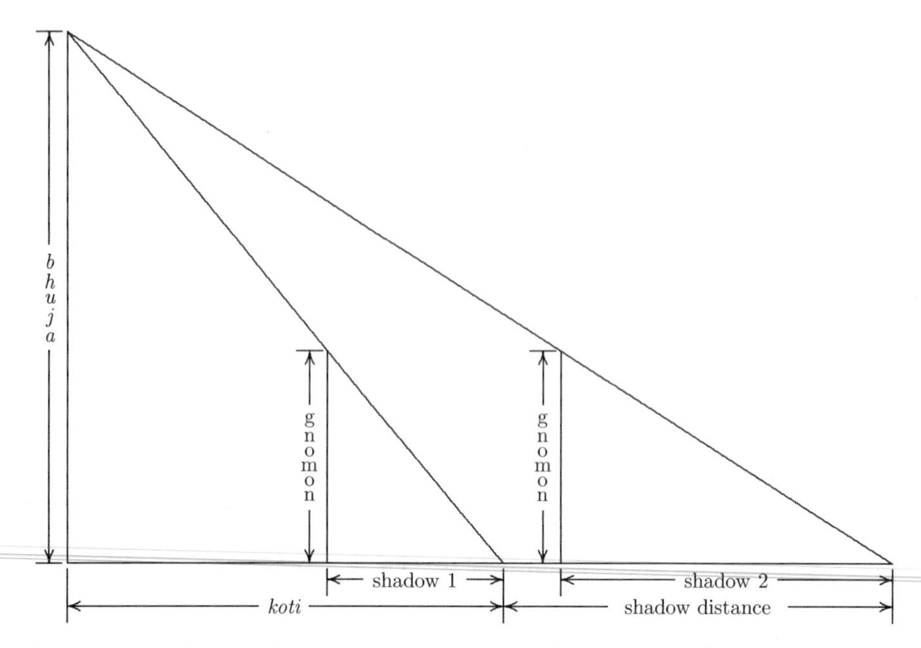

FIGURE 11. The Hindu variant of the double-difference method of surveying.

This procedure gives a value of one-dimensional π equal to 3.1416, which is quite accurate indeed. It exceeds the true value by less than 0.01%.

Aryabhata also knows about the double-difference method of surveying that we discussed above. Whether this knowledge is a case of transmission or independent discovery is not clear. The rule given is slightly different from the discussion that accompanies Fig. 6 and is illustrated by Fig. 11.

> The distance between the ends of the two shadows multiplied by the length of the shadow and divided by the difference in length of the two shadows give the *koti*. The *koti* multiplied by the length of the gnomon and divided by the length of the shadow gives the length of the *bhuja*. [Clark, *1930*, p. 32]

Trigonometry. The inclusion of this variant of the double-difference method of surveying in the *Aryabhatiya* presents us with a small puzzle. As a method of surveying, it is not efficient. It would seem to make more sense to measure angles rather than using only right angles and measuring many more lines. But angles are really not involved here. It is possible to have a clear picture of two mutually perpendicular lines without thinking "right angle." The notion of angles in general as a species of mathematical objects—the figures formed by intersecting lines, which can be measured, added, and subtracted—appears to be a Greek innovation in the sixth and fifth centuries BCE, and it seems to occur only in plane geometry. Its origins may be in stonemasonry and carpentry, where regular polygons have to be fitted together. Astronomy probably also made some contribution.

The earliest form of trigonometry that we can recognize was a table of correspondences between arcs and their chords. We know exactly how such a table was originally constructed, since an explanation can be found in Ptolemy's treatise on astronomy, written around 150 CE.

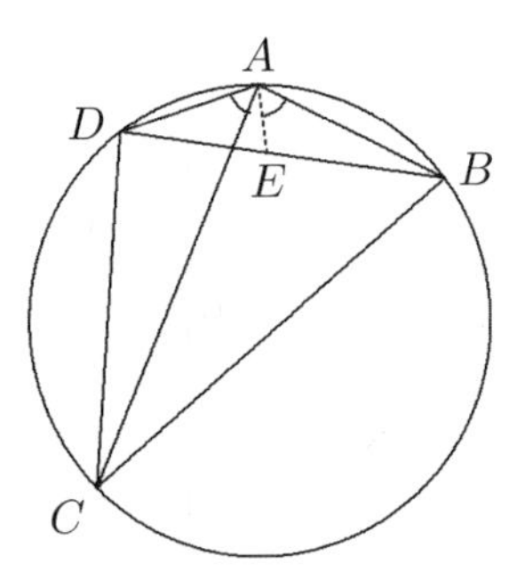

FIGURE 12. For a quadrilateral inscribed in a circle, the product of the diagonals equals the sum of the products of the two pairs of opposite sides.

To construct his table of chords, Ptolemy had to make use of some subtle geometry developed earlier: in particular, the fact that for a quadrilateral inscribed in a circle the product of the diagonals is the sum of the products of the two pairs of opposite sides.[14] Ptolemy proved this result by drawing AE (see Fig. 12) so that $\angle BAE = \angle DAC$, thus obtaining two pairs of similar triangles: $\triangle BAE \sim \triangle CAD$ and $\triangle ADE \sim \triangle ACB$. (Angles ABD and ACD are equal, both being inscribed in the same arc $\overset{\frown}{AD}$; and similarly $\angle ACB = \angle ADB$.) Ptolemy used this relation to compute the chord of the difference of two arcs and the chord of half an arc.

Since Ptolemy knew the construction of the regular dodecagon and the regular decagon, he was easily able to compute the chords of 36° and 30°, expressed in units of one-sixtieth of the radius. His difference theorem then gave the chord of 6°. Then by repeated bisection he got the chord of 3°, then 1° 30′, and finally, 45′. Using these two values and certain inequalities, he was able to set upper and lower bounds on the chord of 1° with sufficient precision for his purposes. He then set out a table with 360 entries, giving the chords of arcs at half-degree increments up to 180°.

Although this table fulfilled its purpose in astronomy, the chord is a cumbersome tool to use in studying plane geometry. For example, it was well known that in any triangle, the angle opposite the larger of two sides will be larger than the angle opposite the smaller side. But what is the exact, quantitative relation between the two sides and the two angles? The ratio of the sides has no simple relationship to the ratio of the angles or to the chords of those angles. There is, however, a very simple relation between the sides and the chords of *twice* the opposite angles, that is, the chords these angles cut off on the circumscribed circle. One might have thought that the constant comparison of a chord with the diameter would have inspired someone to associate the arc with the angle inscribed in it rather than the central angle it subtends. After all, a side of any triangle is the chord of a central angle in the circumscribed circle equal to the double of the opposite angle. Hindu astronomers discovered that trigonometry is simpler if you express the relations between circular arcs and chords in terms of half-chords, what are now called *sines*.

[14] When the quadrilateral is a rectangle, this fact is the form of the Pythagorean theorem given in the *Sulva Sutras*. Gow (*1884*, p. 194) describes this result as "now appended to Euclid VI."

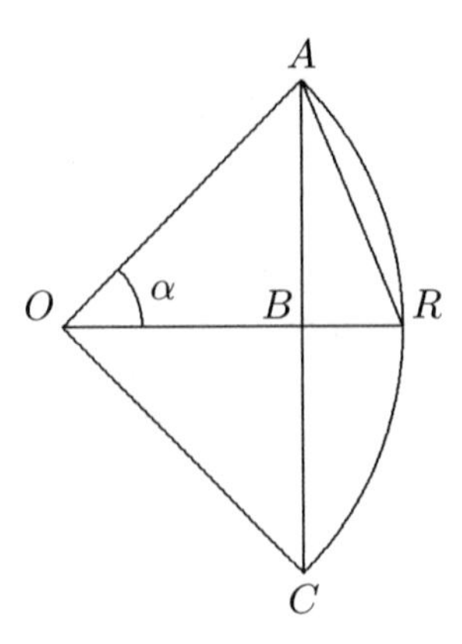

FIGURE 13. The "bowstring" diagram. The sine of the arc $\widehat{AR}$ is the line AB.

In Fig. 13 the arc $\widehat{AR}$ can be measured by either line AB or AR. Ptolemy chose AR and was led to the complications already mentioned. The Hindus preferred AB, which is succinctly described as half the chord of twice the arc. We mentioned above that the Chinese word (*xian*) for the hypotenuse of a right triangle means *bowstring*. The Hindus used the Sanskrit term for a bowstring (*jya* or *jiva*) to mean the sine. The reason for the colorful language is obvious from the figure.

To all appearances, then, trigonometry began to assume its modern form among the Hindus some 1500 years ago. A few reservations are needed, however. First, for the Hindu mathematicians the *sine* was not, as it is to us, a *ratio*. It was a *length*, and that physical dimension had to be taken into account in all computations. Second, the only Hindu concept corresponding approximately to our trigonometric functions was the sine. The tangent, secant, cosine, cotangent, and cosecant were not used. Third, the use of trigonometry was restricted to astronomy. Surveying, which is the other natural place to use trigonometry, did not depend on angle measurement.

Aryabhata used the sine function developed in the *Surya Siddhanta*, giving a table for computing its values at intervals of $225'$ ($3^\circ\ 45'$) of arc from 0° to 90° degrees and expressing these values in units of $1'$ of arc, rounded to the nearest integer, so that the sine of 90° is $3438 = 360 \cdot 60 \cdot \pi$. This interval suggests that the tables were computed independently of Ptolemy's work. If the Hindu astronomers had read Ptolemy, their tables of sines could easily have been constructed from his table of chords, and with more precision than is actually found. Almost certainly, this interval was reached by starting with an angle of 30°, whose sine was known to be half of the radius, then applying the formula for a half-angle to get successively 15°, $7^\circ\ 30'$, and finally $3^\circ\ 45'$. Arybhata's table is actually a list of the *differences* of 24 successive sines at intervals of 225 minutes. Since one minute of arc is a very small quantity relative to the radius, the 24 values provide sufficient precision for the observational technology available at the time. Notice, however, that to calculate the sine of half of an angle θ one would have to apply the cumbersome formula

$$\sin\frac{\theta}{2} = \sqrt{\frac{3438 - \sqrt{3438^2 - \sin^2\theta}}{2}}\,.$$

We can therefore well understand why Aryabhata did not refine his table further. Aryabhata's list of sine differences is the following:

225, 224, 222, 219, 215, 210, 205, 199, 191, 183, 174,
164, 154, 143, 131, 119, 106, 93, 79, 65, 51, 37, 22, 7.

A comparison with a computer-generated table for the same differences reveals that Aryabhata's table is accurate except that his sixth entry should be 211 and the eighth should be 198. But surely an error of less than half of 1% is not a practical matter, and Aryabhata definitely took the practical approach. He explained that his table of sine differences was computed by a recursive procedure, which can be described in our terms as follows (Clark, *1930*, p. 29). Starting with $d_1 = 225$,

$$d_{n+1} = d_n - \frac{d_1 + \cdots + d_n}{d_1}\,,$$

where each term is rounded to the nearest integer after being calculated from this formula.

Aryabhata applied the sine function to determine the elevation of the Sun at a given hour of the day. The procedure is illustrated in Fig. 14 for an observer located at O in the northern hemisphere on a day in spring or summer. This figure shows a portion of the celestial sphere. The arc $RETSWV$ is the portion of the great circle in which the observer's horizontal plane intersects the sphere. The Sun will rise for this observer at the point R and set at the point V. The arc is slightly larger than a semicircle, since we are assuming a day in spring or summer. The chord RV runs from east to west. The Sun will move along the small circle RHV at a uniform rate, and the plane of this circle is parallel to the equatorial circle EMW. (At the equinox, the "day-circle" RV coincides with the equatorial circle EW.) Aryabhata gave the correct formula for finding the radius of this day-circle in terms of the elevation of the Sun above the celestial equator and the radius of the celestial sphere. That radius is the sine of the co-declination of the Sun. Although Aryabhata had the concept of co-latitude, which served him in places where we would use the cosine function, for some reason he did not use the analogous concept of co-declination. As a result, he had to subtract the square of the sine of the declination from the square of the radius of the celestial sphere, then take the square root.

The point Z is the observer's zenith, M is the point on the celestial equator that is due south to the observer, and S is the point due south on the horizon, so that the arc $\overset{\frown}{ZM}$ is the observer's latitude, and the two arcs $\overset{\frown}{ZN}$ and $\overset{\frown}{MS}$ are both equal to the observer's co-latitude. The point H is the location of the Sun at a given time, MF and HG are the projections of M and H respectively on the horizontal plane, and HK is the projection of H on the chord RV. Finally, the great-circle arc HT, which runs through Z, is the elevation of the Sun. The problem is to determine its sine HG in terms of lengths that can be measured.

Because their sides are parallel lines, the triangles MOF and HKG are similar, so that $MO : HK = MF : HG$. Hence we get

$$HG = \frac{HK \cdot MF}{MO}\,.$$

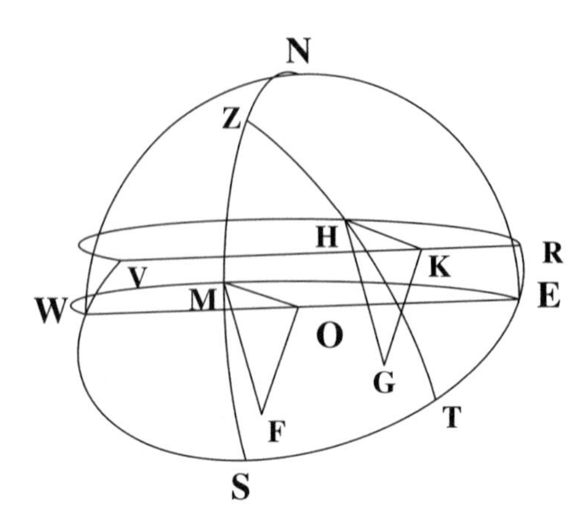

FIGURE 14. Finding the Sun's elevation at a given hour.

In this relation, MF is the sine of MS, that is, the sine of the observer's co-latitude, and MO is the radius of the celestial sphere. The line HK is, in a loose sense, the sine of the arc $\overset{\frown}{RH}$, which is proportional to the time elapsed since sunrise. It is perpendicular to the chord RV and would be a genuine sine if RV were the diameter of its circle. As it is, that relation holds only at the equinoxes. It is not certain whether Aryabhata meant his formula to apply only on the equinox, or whether he intended to use the word *sine* in this slightly inaccurate sense. Because the radius of the Sun's small circle is never less than 90% of the radius of the celestial sphere, probably no observable inaccuracy results from taking HK to be a sine. In any case, that is the way Aryabhata phrased the matter:

> The sine of the Sun at any given point from the horizon on its day-circle multiplied by the sine of the co-latitude and divided by the radius is the [sine of the altitude of the Sun] when any given part of the day has elapsed or remains. [Clark, *1930*, p. 72]

Notice that it is necessary to divide by the radius, because for Aryabhata the sine of an arc is a length, not a ratio.

5.2. Brahmagupta. Brahmagupta devotes five sections of Chapter 12 of the *Brahmasphutasiddhanta* to geometric results (Colebrooke, *1817*, pp. 295–318). Like Aryabhata, he has a practical bent. In giving the common area formulas for triangles and quadrilaterals, he first gives a way of getting a rough estimate of the area: Take the product of the averages of the two pairs of opposite sides. For this purpose a triangle counts as a quadrilateral having one side equal to zero. In the days when calculation had to be done by hand, this was a quick approximation that worked well for quadrilaterals and triangles that are nearly rectangular (that is, tall, thin isosceles triangles). He also gave a formula that he says is exact, and this formula is a theorem commonly known as *Brahmagupta's theorem*: *Half the sum of the sides set down four times and severally lessened by the sides, being multiplied together,* *the square root of the product is the area.* In our terms this rule says that the area of a quadrilateral of sides a, b, c, and d is $\sqrt{(s-a)(s-b)(s-c)(s-d)}$, where s is half of the sum of the lengths of the sides. The case when $d = 0$, which is a triangle, is known as *Heron's formula.* Brahmagupta did not mention the restriction that the quadrilateral must be a cyclic quadrilateral, that is, it must be inscribed in a circle.

Like Aryabhata, Brahmagupta knew that what we are calling one- and two-dimensional π were the same number. In Stanza 40 he says that when the diameter and the square of the radius respectively are multiplied by 3, the results are the "practical" circumference and area. In other words, $\pi = 3$ is a "practical" value. He also gives the "neat" ("exact") value as $\sqrt{10}$. Since $\sqrt{10} = 3.1623$, this value is not an improvement on Aryabhata's 3.1416 in terms of accuracy. If one had to work with π^2, however, it might be more convenient. But π^2 occurs in very few contexts in mathematics, and none at all in elementary mathematics.

Section 5 of Chapter 12 of the *Brahmasphutasiddhanta* gives a rule for finding the volume of a frustum of a rectangular pyramid. In keeping with his approach of giving approximate rules, Brahmagupta says to take the product of the averages of the sides of the top and bottom in the two directions, then multiply by the depth. He calls this result the "practical measure" of the volume, and he knew that this simple rule gave a volume that was too small.

For his second approximation, which he called the"rough" volume, he took the average of the areas of the top and bottom and multiplied by the depth.[15] He also knew that this procedure gave a volume that was too large. The actual volume lies between the "practical" volume and the "rough" volume, but where? We know that the actual volume is obtained as a mixture of two parts "practical" and one part "rough", and so did Brahmagupta. His corrective procedure to give the "neat" (exact) volume was: Subtract the practical from the rough, divide the difference by three, then add the quotient to the practical value.

The phrasing of this result cries out for speculation on its origin. Why use the "practical" volume twice? Why not simply say, "The exact volume is two-thirds of the practical volume plus one-third of the rough volume"? Surely Brahmagupta could do this computation as well as we can and could have used this simpler language. Perhaps his roundabout way of expressing the result reveals the analysis by which he discovered it. Let us investigate what happens when we subtract the "practical" volume from the "rough" volume. First of all, since each is merely an area times the height of the frustum, we are really just subtracting the average area of two rectangles from the area of the rectangle formed by the averages of their parallel sides. Let us simplify by taking the case of two squares of sides a and b. What we are getting, then, is the average of the squares minus the square of the average:

$$\frac{a^2+b^2}{2} - \left(\frac{a+b}{2}\right)^2.$$

Figure 15 shows immediately that this difference is just the square on side $(a-b)/2$. In that figure, half of the squares of sides a and b are set down with their diagonals in a straight line. The two isosceles right triangles below and to the right of the dashed lines fit together to form a square of side $(a+b)/2$. If the

[15] This is the same procedure followed in the cuneiform tablet BM 85 194, discussed above in Subsection 2.3.

rectangle that is shaded dark, which lies inside these two isosceles triangles but outside the squares of sides b and a, is moved inside the square of side a so as to cover the rectangle that is shaded light, we see that the two isosceles triangles cover all of the two half-squares except for a square of side $(a-b)/2$. Since this figure is a very simple one, it seems likely that Brahmagupta would have known that the difference between his two estimates of the volume of a square frustum amounted to the volume of a prism of square base $(a-b)/2$ and height h.

But how did he know that he needed to take one-third of this prism, that is, the volume of a pyramid of the same base and height, and add it to the practical volume? To answer that question, consider a slight variant of the dissection shown in Fig. 3. First remove the four pyramids in the corners, each of which has volume $(h/3)((a-b)/2)^2$, which is exactly one-third of the difference between the gross and practical volumes. Doing so leaves a square platform with four "ramps" running down its sides. In our previous dissection we sliced off two of these ramps on opposite sides and glued them upside down on the other two ramps to make a "slab" of dimensions $a \times b \times h$. This time we slice off the outer half of all four ramps and bend them up to cover their upper halves. The result, shown in Fig. 16, is the cross-shaped prism of height h whose base is a square of side $(a+b)/2$ having a square indentation of side $\frac{a-b}{4}$ at each corner. Filling in these square prisms produces the volume that Brahmagupta called the practical measure. The volume needed to do so is $4h\big((a-b)/4\big)^2 = h\big((a-b)/2\big)^2$. Now three of the four pyramids removed from the corners, taken together, have exactly this much volume. If we use these three to fill in the practical volume, we have one pyramid left over, and its volume is one-third of the difference between the rough and practical volumes. A person who followed the dissection outlined above would then very naturally describe the volume of the pyramid as the practical volume plus one-third of the difference between the gross and practical volumes. That would be natural, but it would be rash to infer that Brahmagupta *did* imagine this dissection; all we have shown is that he *might have done* that.

Questions and problems

9.1. Show how it is possible to square the circle using ruler and compass given the assumption that $\pi = (16\sqrt{2})/7$.

9.2. Prove that the implied Egyptian formula for the volume of a frustum of a square pyramid is correct. If the sides of the upper and lower squares are a and b and the height is h, the implied formula is:

$$V = \frac{h}{3}(a^2 + ab + b^2).$$

9.3. Looking at the Egyptian pyramids, with their layers of brick revealed, now that most of the marble facing that was originally present has been removed, one can see that the total number of bricks must be $1+4+9+\cdots+n^2$ if the slope (*seked*) is constant. Assuming that the Egyptian engineers had the kind of numerical knowledge that would enable them to find this sum as $\frac{1}{6}n(n+1)(2n+1)$, can you conjecture how they may have arrived at the formula for the volume of a frustum? Is it significant that in the only example we have for this computation, the height is 6 units?

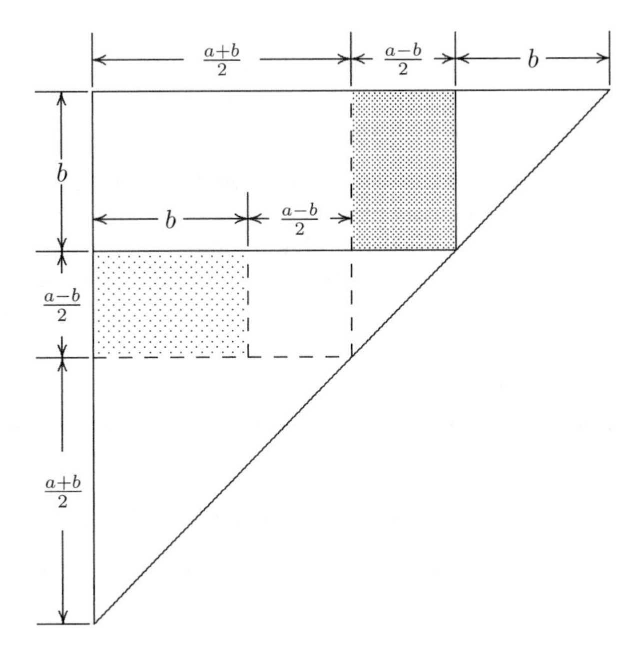

FIGURE 15. The average of the squares minus the square of the average is the square of the semi-difference.

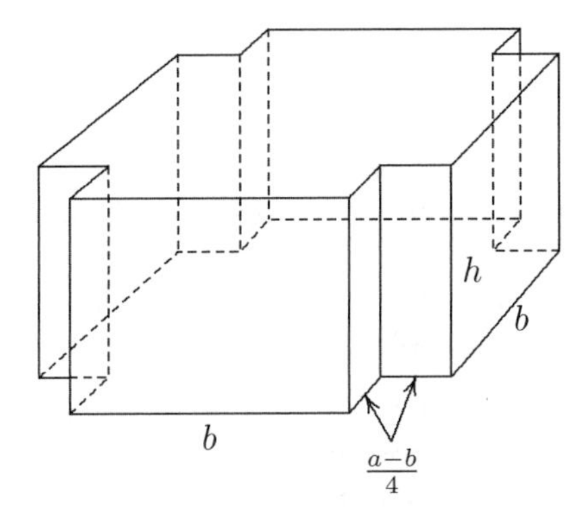

FIGURE 16. Frustum of a pyramid with its corners removed and side "ramps" folded up to form a cross-shaped prism.

9.4. Explain the author's solution of the following problem from the cuneiform tablet BM 85 196. Here the numbers in square brackets were worn off the tablet and have been reconstructed.

> A beam of length 0;30 GAR is leaning against a wall. Its upper end is 0;6 GAR lower than it would be if it were perfectly upright. How far is its lower end from the wall?
>
> Do the following: Square 0;30, obtaining 0;15. Subtracting 0;6 from 0;30 leaves 0;24. Square 0;24, obtaining 0;9,36. Subtract

> 0;9,36 from [0;15], leaving 0;5,24. What is the square root of 0;5,24? The lower end of the beam is [0;18] from the wall.
>
> When the lower end is 0;18 from the wall, how far has the top slid down? Square 0;18, obtaining 0;5;24... .

9.5. Show that the average of the areas of the two bases of a frustum of a square pyramid is the sum of the squares of the average and semidifference of the sides of the bases. Could this fact have led the Mesopotamian mathematicians astray in their computation of the volume of the frustum? Could the analogy with the area of a trapezoid have been another piece of misleading evidence pointing toward the wrong conclusion?

9.6. The author of the *Zhou Bi Suan Jing* had a numerical method of finding the length of the diagonal of a rectangle of width a and length b, which can be described as follows. Square the sum of width and length, subtract twice the area, then take the square root. Should one conclude from this that the author knew that the square on the hypotenuse was the sum of the squares on the legs?

9.7. What happens to the estimate of the Sun's altitude (36,000 km) given by Zhao Shuang if the "corrected" figure for shadow lengthening (4 *fen* per 1000 *li*) is used in place of the figure of 1 *fen* per 1000 *li*?

9.8. The *gougu* section of the *Jiu Zhang Suanshu* contains the following problem:

> Under a tree 20 feet high and 3 in circumference there grows a vine, which winds seven times the stem of the tree and just reaches its top. How long is the vine?

Solve this problem.

9.9. Another right-triangle problem from the *Jiu Zhang Suanshu* is the following. "There is a string hanging down from the top of a pole, and the last 3 feet of string are lying flat on the ground. When the string is stretched, it reaches a point 8 feet from the pole. How long is the string?" Solve this problem. You can also, of course, figure out how high the pole is from this information.

9.10. A frequently reprinted problem from the *Jiu Zhang Suanshu* is the "broken bamboo" problem: A bamboo 10 feet high is broken and the top touches the ground at a point 3 feet from the stem. What is the height of the break? Solve this problem, which reappeared several centuries later in the writings of the Hindu mathematician Brahmagupta.

9.11. The *Jiu Zhang Suanshu* implies that the diameter of a sphere is proportional to the cube root of its volume. Since this fact is equivalent to saying that the volume is proportional to the cube of the diameter, should we infer that the author knew both proportions? More generally, if an author knows (or has proved) "fact A," and fact A is logically equivalent to fact B, is it accurate to say that the author knew or proved fact B? (See also Problem 9.6 above.)

9.12. Show that the solution to the quadrilateral problem of Sawaguchi Kazuyuki is $u = 9$, $v = 8$, $w = 5$, $x = 4$, $y = \sqrt{(1213 + 69\sqrt{273})/40}$, $z = 10$. (The approximate value of y is 7.6698551.) From this result, explain how Sawaguchi Kazuyuki must have invented the problem and what the two values 60.8 and 326.2

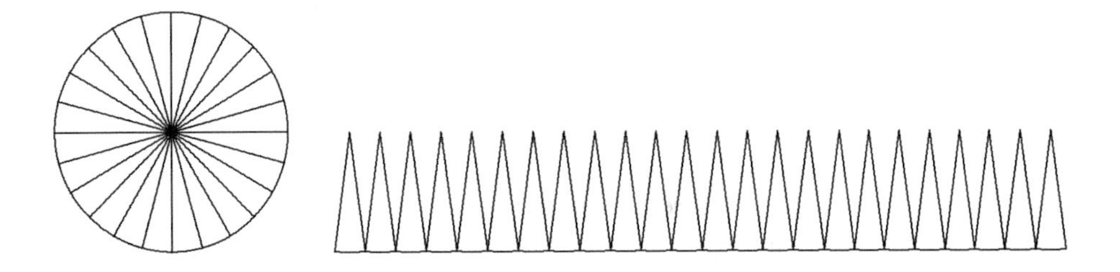

FIGURE 17. A disk cut into sectors and opened up.

are approximations for. How does this problem illustrate the claim that these challenge problems were algebraic rather than geometric?

9.13. How is it possible that some Japanese mathematicians believed the area of the sphere to be one-fourth the square of the circumference, that is, $\pi^2 r^2$ rather than the true value $4\pi r^2$? Smith and Mikami (*1914*, p. 75) suggest a way in which this belief might have appeared plausible. To explain it, we first need to see an example in which the same line of reasoning really does work.

By imagining a circle sliced like a pie into a very large number of very thin pieces, one can imagine it cut open and all the pieces laid out next to one another, as shown in Fig. 17. Because these pieces are very thin, their bases are such short segments of the circle that each base resembles a straight line. Neglecting a very tiny error, we can say that if there are n pieces, the base of each piece is a straight line of length $2\pi r/n$. The segments are then essentially triangles of height r (because of their thinness), and hence area $(1/2)\cdot(2\pi r^2)/n$. Since there are n of them, the total area is πr^2. This heuristic argument gives the correct result. In fact, this very figure appears in a Japanese work from 1698 (Smith and Mikami, *1914*, p. 131).

Now imagine a hemispherical bowl covering the pie. If the slices are extended upward so as to slice the bowl into equally thin segments, and those segments are then straightened out and arranged like the segments of the pie, they also will have bases equal to $\frac{2\pi r}{n}$, but their height will be one-fourth of the circumference, in other words, $\pi r/2$, giving a total area for the hemisphere of $(1/2)\cdot\pi^2 r^2$. Since the area is $2\pi r^2$, this would imply that $\pi = 4$. What is wrong with the argument? How much error would there be in taking $\pi = 4$?

9.14. What is the justification for the statement by the historian of mathematics T. Murata that Japanese mathematics (*wasan*) was not a science but an art?

9.15. Show that Aryabhata's list of sine differences can be interpreted in our language as the table whose nth entry is

$$3438\left[\sin\left(\frac{n\pi}{48}\right) - \sin\left(\frac{(n-1)\pi}{48}\right)\right].$$

Use a computer to generate this table for $n = 1, \ldots, 24$, and compare the result with Aryabhata's table.

9.16. If the recursive procedure described by Aryabhata is followed faithfully (as a computer can do), the result is the following sequence.

225, 224, 222, 219, 215, 210, 204, 197, 189, 181, 172,
162, 151, 140, 128, 115, 102, 88, 74, 60, 45, 30, 15, 0

Compare this list with Aryabhata's list, and note the systematic divergence. These differences should be approximately 225 times the cosine of the appropriate angle. That is, $d_n \approx 225 \cdot \cos\bigl(225(n + 0.5)\,\text{minutes}\bigr)$. What does that fact suggest about the source of the systematic errors in the recursive procedure described by Aryabhata?

9.17. Use Aryabhata's procedure to compute the altitude of the Sun above the horizon in London (latitude $51^\circ\,32'$) at 10:00 AM on the vernal equinox. Assume that the sun rises at 6:00 AM on that day and sets at 6:00 PM.

9.18. Why is it necessary that a quadrilateral be inscribed in a circle in order to compute its diagonals knowing the lengths of its sides? Why is it not possible to do so in general?

9.19. Show that the formula given by Brahmagupta for the area of a quadrilateral is correct if and only if the quadrilateral can be inscribed in a circle.

9.20. Imagine a sphere as a polyhedron having a large number of very small faces. Deduce the relation between the volume of a sphere and its area by considering the pyramids obtained by joining the points of each face to the center of the sphere.

CHAPTER 10

Euclidean Geometry

We shall divide the history of Greek mathematics into four periods. The first period, from about 600 to 400 BCE, was the time when the Greeks acquired geometry from Egypt and Mesopotamia and turned it in the direction of logical argument. The second period came in the fourth century, when the logical aspects of the subject were debated in Plato's Academy[1] and Aristotle's Lyceum,[2] proofs were improved, and basic principles and assumptions were isolated. The third period began in the third century, when the mature subject was expounded in Euclid's *Elements*, and further research continued on more complicated curves and surfaces. The fourth and final period was a long decline in originality, in which no revolutionary changes occurred and commentaries were the main literary form.

1. The earliest Greek geometry

The history of Greek mathematics up to the time of Euclid (300 BCE) was written by Eudemus, a pupil of Aristotle. This history was lost, but it is believed to be the basis of the first paragraph of a survey given by Proclus in the fifth century CE in the course of his commentary on the first book of Euclid. In this passage Proclus mentions 25 men who were considered to have made significant contributions to mathematics. Of these 25, five are well known as philosophers (Thales, Pythagoras, Anaxagoras, Plato, and Aristotle); three are famous primarily as mathematicians and astronomers (Euclid, Eratosthenes, and Archimedes). The other 17 have enjoyed much less posthumous fame. Some of them are so obscure that no mention of them can be found anywhere except in Proclus' summary. Some others (Theodorus, Archytas, Menaechmus, Theaetetus, and Eudoxus) are mentioned by other commentators and by Plato. The 13 just named are the main figures we shall use to sketch the history of Greek geometry. It is clear from what Proclus writes that something important happened to mathematics during the century of Plato and Aristotle, and the result was a unique book, Euclid's *Elements*.

Missing from the survey of Proclus is any reference to Mesopotamian influence on Greek geometry. This influence is shown clearly in Greek astronomy, in the

[1] This word has become so common in English that its original, legendary meaning is mostly forgotten. In his biography of the Athenian king Theseus, who had slain the Minotaur on Crete as a youth, Plutarch says that at the age of 50 the widowed king abducted the beautiful 12-year-old Helen of Sparta and hid her away. (This was before she married Menelaus and ran off with Paris, becoming the cause of the Trojan War.) Her twin brothers Castor and Polydeukes (Pollux) threatened to destroy Athens in revenge. Akademos, however, averted the calamity by telling them where she was hidden. For this deed he was venerated as the savior of the city, and a grove of trees on its northwest side, supposedly his burial place, was dedicated to his memory. Plato gave his lectures in that grove, and hence arose the phrase "the groves of Academe."

[2] Here is another word whose origins are lost in common usage. The Lyceum was so named because it was near the temple to Apollo Lykeios ("Apollo of the Wolves").

use of the sexagesimal system of measuring angles and in Ptolemy's explicit use of Mesopotamian astronomical observations. It *may* also appear in Book 2 of Euclid's *Elements*, which contains geometric constructions equivalent to certain algebraic relations that are frequently encountered in the cuneiform tablets. This relation, however, is controversial. Leaving aside the question of Mesopotamian influence, we do see a recognition of their debt to Egypt, which the Greeks never concealed. And how could they? Euclid actually lived in Egypt, and the other two of the "big three" Greek geometers, Archimedes and Apollonius, both studied there, in the Hellenistic city of Alexandria at the mouth of the Nile.

1.1. Thales. The philosopher Thales, who lived in the early sixth century BCE, was a citizen of Miletus, a Greek colony on the coast of Asia Minor. The ruins of Miletus are now administered by Turkey. Herodotus mentions Thales in several places. Discussing the war between the Medes and the Lydian king Croesus, which had taken place in the previous century, he says that an eclipse of the Sun frightened the combatants into making peace. Thales, according to Herodotus, had predicted that an eclipse would occur no later than the year in which it actually occurred. Herodotus goes on to say that Thales had helped Croesus to divert the river Halys so that his army could cross it.

These anecdotes show that Thales had both scientific and practical interests. His prediction of a solar eclipse, which, according to the astronomers, occurred in 585 BCE, seems quite remarkable, even if, as Herodotus says, he gave only a period of several years in which the eclipse was to occur. Although solar eclipses occur regularly, they are visible only over small portions of the Earth, so that their regularity is difficult to discover and verify. Lunar eclipses exhibit the same period as solar eclipses and are easier to observe. Eclipses recur in cycles of about 19 solar years, a period that seems to have been known to many ancient peoples. Among the cuneiform tablets from Mesopotamia there are many that discuss astronomy, and Ptolemy uses Mesopotamian observations in his system of astronomy. Thales could have acquired this knowledge, along with certain simple facts about geometry, such as the fact that the base angles of an isosceles triangle are equal. Bychkov (*2001*) argues that the recognition that the base angles of an isosceles triangle are equal probably did come from Egypt. In construction, for example, putting a roof on a house, it is not crucial that the cross section be exactly an isosceles triangle, since it is the ridge of the roof that must fit precisely, not the edges. However, when building a symmetric square pyramid, errors in the base angles of the faces would make it impossible for the faces to fit together tightly. Therefore, he believes, Thales must have derived this theorem from his travels in Egypt.

In his *Discourses on the Seven Wise Men*, Plutarch reports that Thales traveled to Egypt and was able to calculate the height of the Great Pyramid by driving a pole into the ground and observing that the ratio of the height of the pyramid to that of the pole was the same as the ratio of their shadow lengths. In his *Lives of Eminent Philosophers*, Diogenes Laertius cites the historian Hieronymus (fourth or third century BCE) in saying that Thales calculated the height of the pyramid by waiting until his shadow was exactly as long as he was tall, then measuring the length of the shadow of the Great Pyramid.[3] There are practical difficulties in executing this

[3] A very interesting mystery/historical novel by Denis Guedj, called *Le théorème du perroquet*, uses this history to connect its story line. An English translation of this novel now exists, *The Parrot's Theorem*, St. Martin's Press, New York, 2002.

plan, since one could not get into the Pyramid to measure the distance from the center to the tip of the shadow directly. One might use the Pythagorean theorem, which Thales could well have known, to measure the distance from the center of the pyramid to the point where its outer wall intersects the vertical plane through the top of the pyramid and the tip of its shadow. A simpler way of computing the distance, however, is to reflect a triangle about one of its vertices. This technique is known to have been used by Roman surveyors to measure the distance across a river without leaving shore.

According to Diogenes Laertius, a Roman historian named Pamphila, who lived in the time of Nero, credits Thales with being the first to inscribe a right triangle in a circle. To achieve this construction, one would have to know that the hypotenuse of the inscribed triangle is a diameter. Diogenes Laertius goes on to say that others attribute this construction to Pythagoras.

1.2. Pythagoras and the Pythagoreans. Half a century later than Thales the philosopher Pythagoras was born on the island of Samos, another of the Greek colonies in Ionia. No books of Pythagoras survive, but many later writers mention him, including Aristotle. Diogenes Laertius devotes a full chapter to the life of Pythagoras. He acquired even more legends than Thales. According to Diogenes Laertius, who cites the logicist Apollodorus, Pythagoras sacrificed 100 oxen when he discovered the theorem that now bears his name. If the stories about Pythagoras can be believed, he, like Thales, traveled widely, to Egypt and Mesopotamia. He gathered about him a large school of followers, who observed a mystical discipline and devoted themselves to contemplation. They lived in at least two places in Italy, first at Croton, then, after being driven out,[4] at Metapontion, where he died sometime around 500 BCE.

According to Book I, Chapter 9 of *Attic Nights*, by the Roman writer Aulus Gellius (ca. 130–180), the Pythagoreans first looked over potential recruits for physical signs of being educable. Those they accepted were first classified as *akoustikoí* (auditors) and were compelled to listen without speaking. After making sufficient progress, they were promoted to *mathēmatikoí* (learners).[5] Finally, after passing through that state they became *physikoí* (natural philosophers). In his *Life of Pythagoras* Iamblichus uses these terms to denote the successors of Pythagoras, who split into two groups, the *akoustikoí* and the *mathēmatikoí*. According to Iamblichus, the *mathēmatikoí* recognized the *akoustikoí* as genuine Pythagoreans, but the sentiment was not reciprocated. The *akoustikoí* kept the pure Pythagorean doctrine and regarded the *mathēmatikoí* as followers of the disgraced Hippasus mentioned in Chapter 8.

Diogenes Laertius quotes the philosopher Alexander Polyhistor (ca. 105–35 BCE) as saying that the Pythagoreans generated the world from *monads* (units). By adding a single monad to itself, they generated the natural numbers. By allowing the monad to move, they generated a line, then by further motion the line generated plane figures (polygons), the plane figures then moved to generate solids (polyhedra). From the regular polyhedra they generated the four elements of earth, air, fire, and water.

---

[4] Like modern cults, the Pythagoreans attracted young people, to the despair of their parents. Accepting new members from among the local youth probably aroused the wrath of the citizenry.
[5] Gellius remarks at this point that the word *mathēmatikoí* was being inappropriately used in popular speech to denote a "Chaldean" (astrologer).

1.3. Pythagorean geometry. Euclid's geometry is an elaboration and systematization of the geometry that came from the Pythagoreans via Plato and Aristotle. From Proclus and other later authors we have a glimpse of a fairly sophisticated Pythagorean geometry, intertwined with mysticism. For example, Proclus reports that the Pythagoreans regarded the right angle as ethically and aesthetically superior to acute and obtuse angles, since it was "upright, uninclined to evil, and inflexible." Right angles, he says, were referred to the "immaculate essences," while the obtuse and acute angles were assigned to divinities responsible for changes in things. The Pythagoreans had a bias in favor of the eternal over the changeable, and they placed the right angle among the eternal things, since unlike acute and obtuse angles, it cannot change without losing its character. In taking this view, Proclus is being a strict Platonist; for Plato's ideal Forms were defined precisely by their absoluteness; they were incapable of undergoing any change without losing their identity.

Proclus mentions two topics of geometry as being Pythagorean in origin. One is the theorem that the sum of the angles of a triangle is two right angles (Book 1, Proposition 32). Since this statement is equivalent to Euclid's parallel postulate, it is not clear what the discovery amounted to or how it was made.

The other topic mentioned by Proclus is a portion of Euclid's Book 6 that is not generally taught any more, called application of areas. However, that topic had to be preceded by the simpler topic of transformation of areas. In his *Nine Symposium Books*[6] Plutarch called the transformation of areas "one of the most geometrical" problems. He thought solving it was a greater achievement than discovering the Pythagorean theorem and said that Pythagoras was led to make a sacrifice when he solved the problem. The basic idea is to convert a figure having one shape to another shape while preserving its area, as in Fig. 1. To describe the problem in a different way: Given two geometric figures A and B, construct a third figure C the same size as A and the same shape as B. One can imagine many reasons why this problem would be attractive. If one could find, for example, a square equal to any given figure, then comparing sizes would be simple, merely a matter of converting all areas into squares and comparing the lengths of their sides. But why stop at that point? Why not do as the Pythagoreans apparently did, and consider the general problem of converting any shape into any other? For polygons this problem was solved very early, and the solution appears in very elegant form as Proposition 25 of Euclid's Book 6.

Related to the transformation of areas is the problem of application of areas. There are two such problems, both involving a given straight line segment AB and a planar polygon Γ. The first problem is to construct a parallelogram equal to Γ on part of the line segment AB in such a way that the parallelogram needed to fill up a parallelogram on the entire base, called the *defect*, will have a prescribed shape. This is the problem of *application with defect*, and the solution is given in Proposition 28 of Book 6. The second application problem is to construct a parallelogram equal to Γ on a base containing the line AB and such that the portion of the parallelogram extending beyond AB (the *excess*) will have a prescribed shape. This is the problem of *application with excess*, and the solution is Proposition 29 of Book 6. The construction for application with defect is shown in Fig. 2. This

[6] The book is commonly known as *Convivial Questions.* The Greek word *sympósion* means literally *drinking together.*

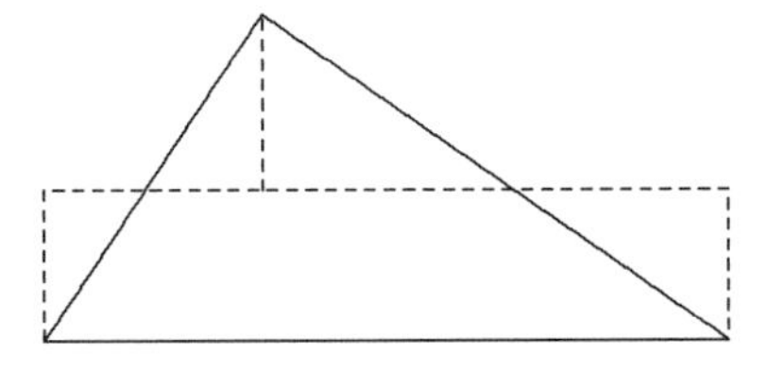

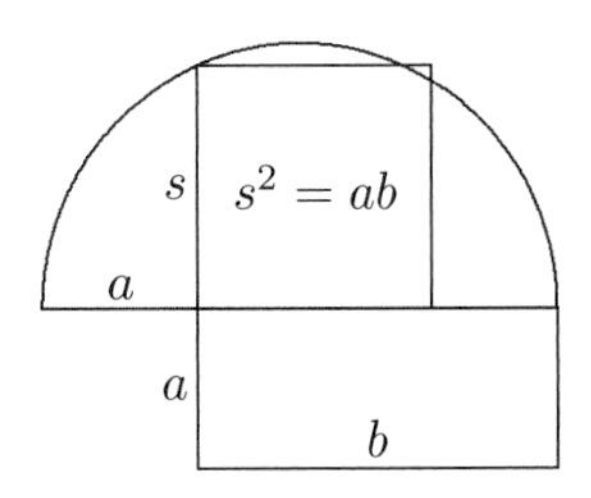

FIGURE 1. Left: turning a triangle into a rectangle. Right: turning a rectangle into a square ($s^2 = ab$).

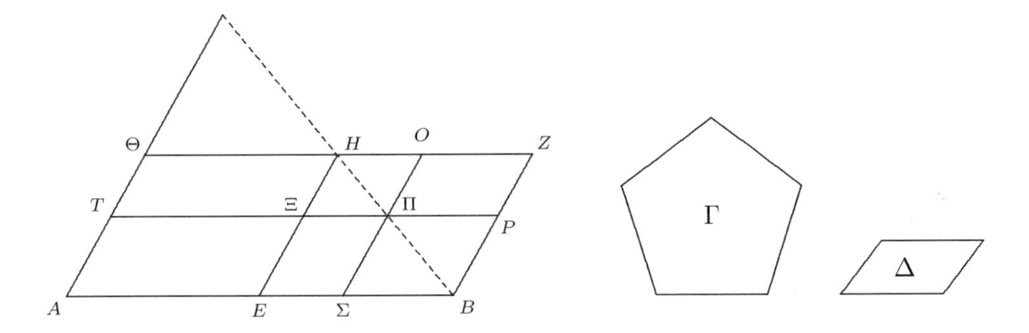

FIGURE 2. Application with defect. Euclid, Book 6, Proposition 28.

problem does not have a solution for all given lines and areas, since the largest parallelogram that can be formed under these conditions is the one whose base is half of the given line (Book 6, Proposition 26). Assuming that condition, let AB be the given line, Γ the given polygonal region, and Δ the given parallelogram shape. The dashed line from B makes the same angle with AB that the diagonal of the parallelogram Δ makes with its base. The line $A\Theta$ is drawn to make the same angle as the corresponding sides of Δ. Then any parallelogram having its sides along AB and $A\Theta$ and opposite corner on the dashed line will automatically generate a "defect" that is similar to Δ. The remaining problem is to find the one that has the same area as Γ. That is achieved by constructing the parallelogram $H\Xi\Pi O$ similar to Δ and equal to the difference between $AEH\Theta$ and Γ.

The Greek word for application is *parabolê*. Proclus cites Eudemus in asserting that the solution of the application problems was an ancient discovery of the Pythagoreans, and that they gave them the names *ellipse* and *hyperbola*, names that were later transferred to the conic curves by Apollonius. This version of events is also confirmed by Pappus. We shall see the reason for the transfer below.

Although most of Euclid's theorems have obvious interest from the point of view of anyone curious about the world, the application problems raise a small mystery. Why were the Pythagoreans interested in them? Were they merely a refinement of the transformation problems? Why would anyone be interested in applying an area so as to have a defect or excess of a certain shape? Without restriction on the shape of the defect or excess, the application problem does not have a unique solution. Were the additional conditions imposed simply to make the problem determinate? Some historians have speculated that there was a further motive.

In the particular case when the excess or defect is a square, these problems amount to finding two unknown lengths given their sum and product (application with deficiency) or given their difference and product (application with excess). In modern terms these two problems amount to quadratic equations. Some authors have argued that this "geometric algebra" was a natural response to the discovery of incommensurable magnitudes, described in Chapter 7, indeed a logically necessary response. On this point, however, many others disagree. Gray, for example, says that, while the discovery of incommensurables did point out a contradiction in a naive approach to ratios, "it did not provoke a foundational crisis." Nor did it force the Pythagoreans to recast algebra as geometry:

> There is no logical necessity about it. It would be quite possible to persevere with an arithmetic of natural numbers to which was adjoined such new quantities as, say, arose in the solution of equations. There is nothing more intelligible about a geometric segment than a root of an equation, unless you have already acquired a geometric habit of thought. Rather than turning from algebra to geometry, I suggest that the Greeks were already committed to geometry. [Gray, *1989*, p. 16]

1.4. Challenges to Pythagoreanism: unsolved problems. Supposing that this much was known to the early Pythagoreans, we can easily guess what problems they would have been trying to solve. Having learned how to convert any polygon to a square of equal area, they would naturally want to do the same with circles and sectors and segments of circles. This problem was known as *quadrature (squaring) of the circle.* Also, having solved the transformation problems for a plane, they would want to solve the analogous problems for solid figures, in other words, to convert a polyhedron to a cube of equal volume. Finding the cube would be interpreted as finding the length of its side. Now, the secret of solving the planar problem was to triangulate a polygon, construct a square equal to each triangle, then add the squares to get bigger squares using the Pythagorean theorem. By analogy, the three-dimensional program would be to cut a polyhedron into tetrahedra, convert any tetrahedron into a cube of equal volume, then find a way of adding cubes analogous to the Pythagorean theorem for adding squares. The natural first step of this program (as we imagine it to have been) was to construct a cube equal to the double of a given cube, the problem of *doubling the cube,* just as we imagined that doubling a square may have led to the Pythagorean theorem.

Yet another example of such a problem is that of dividing an arc (or angle) into equal parts. If we suppose that the Pythagoreans knew how to bisect arcs (Proposition 9 of Book 1 of the *Elements*) and how to divide a line into any number of equal parts (Proposition 9 of Book 6), this asymmetry between their two basic figures—lines and circles—would very likely have been regarded as a challenge. The first step in this problem would have been to divide an arc into three equal parts, the problem of *trisection of the angle.*

The three problems just listed were mentioned by later commentators as an important challenge to all geometers. To solve them, geometers had to enlarge their set of basic objects beyond lines and planes. They were rather conservative in

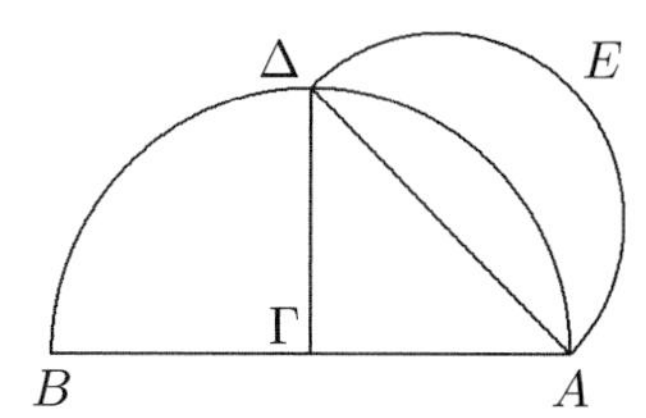

FIGURE 3. Hippocrates' quadrature of a lune, acording to Simplicius.

doing so, first invoking familiar surfaces such as cones and cylinders, which could be generated by moving lines on circles, and intersecting them with planes so as to get the conic sections that we know as the ellipse, parabola, and hyperbola. These curves made it possible to solve two of the three problems (trisecting the angle and doubling the cube). Later, a number of more sophisticated curves were invented, among them spirals, the cissoid, and the quadratrix. This last curve got its name from its use in squaring the circle. Although it is not certain that the Pythagoreans had a program like the one described here, it is known that all three of these problems were worked on in antiquity.

Squaring the circle. Proclus mentions Hippocrates of Chios as having discovered the quadratures of lunes. In fact this mathematician (ca. 470–ca. 410 BCE), who lived in Athens at the time of the Peloponnesian War (430–404), is said to have worked on all three of the classical problems. A lune is a figure resembling a crescent moon: the region inside one of two intersecting circles and outside the other. In the ninth volume of his commentary on Aristotle's books on physics, the sixth-century commentator Simplicius discusses several lunes that Hippocrates squared, including the one we are about to discuss. After detailing the criticism by Eudemus of earlier attempts by Antiphon (480–411) to square the circle by the kind of polygonal approximation we discussed in Chapter 9, Simplicius reports one of Hippocrates' quadratures (Fig. 3), based on Book 12 of Euclid's *Elements*. The result needed is that semicircles are proportional to the squares on their diameters.

Simplicius' reference to Book 12 of Euclid is anachronistic, since Hippocrates lived before Euclid; but it was probably well known that similar segments are proportional to the squares on their bases. Even that theorem is not needed here, except in the case of semicircles, and that special case is easy to derive from the theorem for whole circles. The method of Hippocrates does not achieve the quadrature of a whole circle; we can see that his procedure works because the "irrationalities" of the two circles cancel each other when the segment of the larger circle is removed from the smaller semicircle.

In his essay *On Exile*, Plutarch reports that the philosopher Anaxagoras worked on the quadrature of the circle while imprisoned in Athens. (He was brought there by Pericles, who was eventually compelled to send him away.) Other attempts are reported, one by Dinostratus (ca. 390–ca. 320 BCE), who is said to have used the curve called (later, no doubt) the *quadratrix* (*squarer*), invented by Hippias of Elis (ca. 460–ca. 410 BCE) for the purpose of trisecting the angle. It is discussed below in that connection.

Doubling the cube. Although the problem of doubling the cube fits very naturally into what we have imagined as the Pythagorean program, some ancient authors

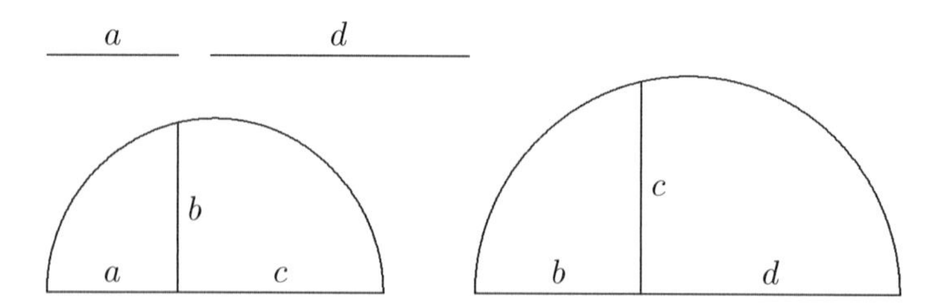

FIGURE 4. The problem of two mean proportionals: Given a and d, find b and c.

gave it a more exotic origin. In *The Utility of Mathematics*, the commentator Theon of Smyrna, who lived around the year 100 CE, discusses a work called *Platonicus* that he ascribes to Eratosthenes, to the effect that the citizens of Delos (the island that was the headquarters of the Athenian Empire) consulted an oracle in order to be relieved of a plague, and the oracle told them to double the size of an altar (probably to Apollo). Plagues were common in ancient Greece; one is described in Sophocles' *Oedipus the King*, and another decimated Athens early in the Peloponnesian War, claiming Pericles as one of its victims. According to Theon, Eratosthenes depicted the Delians as having turned for technical advice to Plato, who told them that the altar was not the point: The gods really wanted the Delians to learn geometry better. In his commentary on Archimedes' work on the sphere and cylinder, Eutocius gives another story, also citing Eratosthenes, but he says that Eratosthenes told King Ptolemy in a letter that the problem arose on the island of Crete when King Minos ordered that a tomb built for his son be doubled in size.

Whatever the origin of the problem, both Proclus and Eutocius agree that Hippocrates was the first to reduce it to the problem of two mean proportionals. The Pythagoreans knew that the mean proportional between any two square integers is an integer, for example, $\sqrt{16 \cdot 49} = 28$ and that between any two cubes such as 8 and 216 there are two mean proportionals (Euclid, Book 8, Propositions 11 and 12); for example, $8 : 24 :: 24 : 72 :: 72 : 216$. If two mean proportionals could be found between two cubes—as seems possible, since every volume can be regarded as the cube on some line—the problem would be solved. It would therefore be natural for Hippocrates to think along these lines when comparing two cubes. Eutocius, however, was somewhat scornful of this reduction, saying that the new problem was just as difficult as the original one. That claim, however, is not true: One can easily draw a figure containing two lines and their mean proportional (Fig. 1): the two parts of the diameter on opposite sides of the endpoint of the half-chord of a circle and the half-chord itself. The only problem is to get two such figures with the half-chord and one part of the diameter reversing roles between the two figures and the other parts of the diameters equal to the two given lines, as shown in Fig. 4. It is natural to think of using two semicircles for this purpose and moving the chords to meet these conditions.

In his commentary on the treatise of Archimedes on the sphere and cylinder, Eutocius gives a number of solutions to this problem, ascribed to various authors, including Plato. The earliest one that he reports is due to Archytas (ca. 428–350 BCE). This solution requires intersecting a cylinder with a torus and a cone.

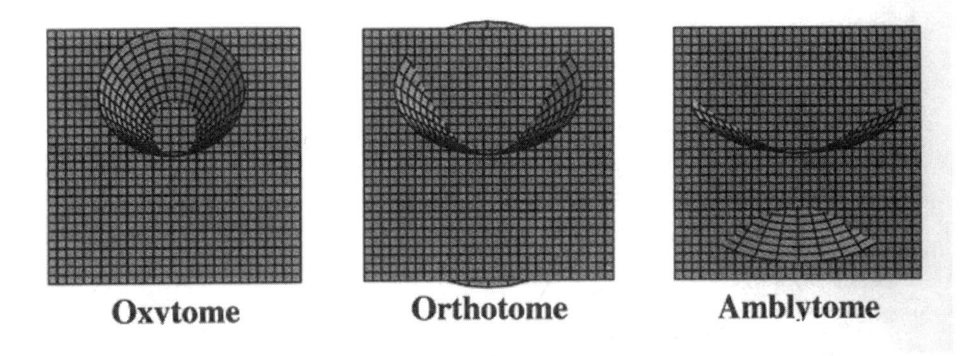

FIGURE 5. The three conic sections, according to Menaechmus.

The three surfaces intersect in a point from which the two mean proportionals can easily be determined. A later solution by Menaechmus may have arisen as a simplification of Archytas' rather complicated construction. It requires intersecting two cones, each having a generator parallel to a generator of the other, with a plane perpendicular to both generators. These intersections form two conic sections, a parabola and a rectangular hyperbola; where they intersect, they produce the two mean proportionals.

If Eutocius is correct, the conic sections first appeared, but not with the names they now bear, in the late fourth century BCE. Menaechmus created these sections by cutting a cone with a plane perpendicular to one of its generators. When that is done, the shape of the section depends on the vertex angle of the cone. If that angle is acute, the section will be an ellipse; if it is a right angle, the section will be a parabola; if it is obtuse, the section will be a hyperbola. In his commentary on Archimedes' treatise on the sphere and cylinder, Eutocius tells how he happened to find a work written in the Doric dialect which seemed to be a work of Archimedes. He mentions in particular that instead of the word *parabola*, used since the time of Apollonius, the author used the phrase *section of a right-angled cone*, and instead of *hyperbola*, the phrase *section of an obtuse-angled cone*. Since Proclus refers to "the conic section triads of Menaechmus," it is inferred that the original names of the conic sections were *oxytomē̆* (sharp cut), *orthotomē̆* (right cut), and *amblytomē̆* (blunt cut), as shown in Fig. 5. However, Menaechmus undoubtedly thought of the cone as the portion of the figure from the vertex to some particular circular base. In particular, he wouldn't have thought of the hyperbola as having two nappes, as we now do.

How Apollonius came to give them their modern names a century later is described below. Right now we shall look at the consequences of Menaechmus' approach and see how it enabled him to solve the problem of two mean proportionals. It is very difficult for a modern mathematician to describe this work without breaking into modern algebraic notation, essentially using analytic geometry. It is very natural to do so; for Menaechmus, if Eutocius reports correctly,[7] comes very close to stating his theorem in algebraic language.

[7] That is a big "if." Eutocius clearly had read Apollonius; Menaechmus, just as clearly, could not have done so.

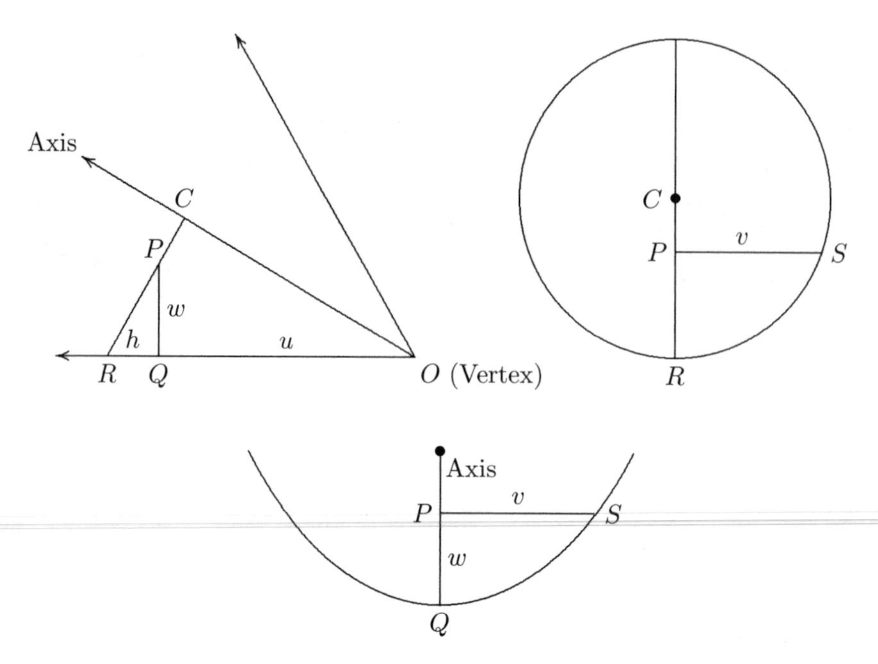

FIGURE 6. Sections of a cone. Top left: through the axis. Top right: perpendicular to the axis. Bottom: perpendicular to the generator OR at a point Q lying at distance u from the vertex O. The fundamental relation is $v^2 = h^2 + 2uh - w^2$. The length h has a fixed ratio to w, depending only on the shape of the triangle OCR.

We begin by looking at a general conic section, shown in Fig. 6. When a cone is cut by a plane through its axis, the resulting figure is simply a triangle. The end that we have left open by indicating with arrows the direction of the axis and two generators in this plane would probably have been closed off by Menaechmus. If it is cut by a plane perpendicular to the axis, the result is a circle. The conic section itself is obtained as the intersection with a plane perpendicular to one of its generators at a given distance (marked u in the figure) from the vertex. The important relation needed is that between the length of a horizontal chord (double the length marked v) in the conic section and its height (marked w) above the generator that has been cut. Using only similar triangles and the fact that a half chord in a circle is the mean proportional between the segments of the diameter through its endpoint, Menaechmus would easily have derived the fundamental relation

$$v^2 = h^2 + 2uh - w^2 . \tag{1}$$

Although we have written this relation as an equation with letters in it, Menaechmus would have been able to describe what it says in terms of the lines v, u, h, and w, and squares and rectangles on them. He would have known the value of the ratio h/w, which is determined by the shape of the triangle ROC. In our terms $h = w\tan(\varphi/2)$, where φ is the vertex angle of the cone.

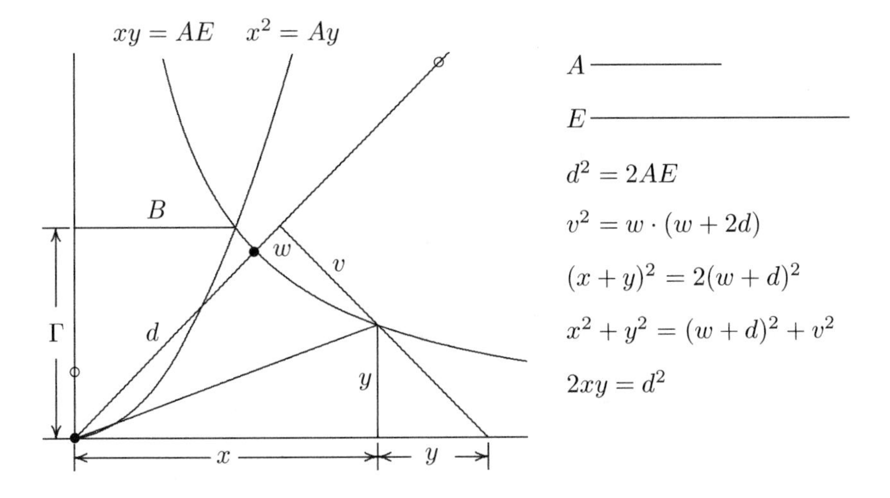

FIGURE 7. One of Menaechmus' solutions to the problem of two mean proportionals, as reported by Eutocius.

The simplest case is that of the parabola, where the vertex angle is 90° and $h = w$. In that case the relation between v and w is

$$v^2 = 2uw\,.$$

In the problem of putting two mean proportionals B and Γ between two lines A and E, Menaechmus took this u to be $\frac{1}{2}A$, so that $v^2 = Aw$.

The hyperbola Menaechmus needed for this problem was a rectangular hyperbola, which results when the triangle ROC is chosen so that $\overline{RC}^2 = 2\overline{OC}^2$, and therefore $\overline{OR}^2 = 3\overline{OC}^2$. Such a triangle is easily constructed by extending one side of a square to the same length as the diagonal and joining the endpoint to the opposite corner of the square. In any triangle of this shape the legs are the side and diagonal of a square. For that case Menaechmus would have been able to show that the relation

$$v^2 = w(w + 2d)$$

holds, where d is the diagonal of a square whose side is u. To solve the problem of two mean proportionals, Menaechmus took $u = \sqrt{AE}$, that is, the mean proportional between A and E. Menaechmus' solution is shown in Fig. 7.

The solution fits perfectly within the framework of Pythagorean–Euclidean geometry, yet people were not satisfied with it. The objection to it was that the data and the resulting figure all lie within a plane, but the construction requires the use of cones, which cannot be contained in the plane.

Trisecting the angle. The practicality of trisecting an angle is immediately evident: It is the vital first step on the way to dividing a circular arc into any number of equal pieces. If a right angle can be divided into n equal pieces, a circle also can be divided into n equal pieces, and hence the regular n-gon can be constructed. The success of the Pythagoreans in constructing the regular pentagon must have encouraged them to pursue this program. It is possible to construct the regular n-gon for $n = 3, 4, 5, 6, 8, 10$, but not 7 or 9. The number 7 is awkward, being the only prime between 5 and 10, and one could expect to have difficulty constructing

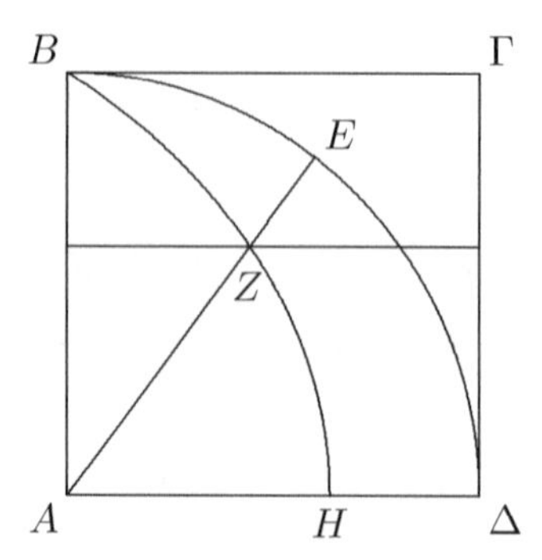

FIGURE 8. The quadratrix of Hippias.

the regular heptagon. Surprisingly, however, the regular 17-sided polygon *can* be constructed using only compass and straightedge. Since $9 = 3 \cdot 3$, it would seem natural to begin by trying to construct this figure, that is, to construct an angle of $40°$. That would be equivalent to constructing an angle of $20°$, hence trisecting the angles of an equilateral triangle.

Despite the seeming importance of this problem, less has been written about the ancient attempts to solve it than about the other two problems. For most of the history we are indebted to two authors. In his commentary on Euclid's *Elements*, Proclus mentions the problem and says that it was solved by Nicomedes using his conchoid and by others using the quadratrices of Hippias and Nicomedes. In Book 4 of his *Synagōgḗ* (*Collection*), Pappus says that the circle was squared using the curve of Dinostratus and Nicomedes. He then proceeds to describe that curve, which is the one now referred to as the quadratrix of Hippias.[8]

The quadratrix is described physically as follows. The radius of a circle rotates at a uniform rate from the vertical position AB in Fig. 8 to the horizontal position $A\Delta$, while in exactly the same time a horizontal line moves downward at a constant speed from the position $B\Gamma$ to the position $A\Delta$. The point of intersection Z traces the curve BZH, which is the quadratrix. The diameter of the circle is the mean proportional between its circumference and the line AH. Unfortunately, H is the one point on the quadratrix that is not determined, since the two intersecting lines coincide when they both reach $A\Delta$. This point was noted by Pappus, citing an earlier author named Sporos. In order to draw the curve, which is mechanical, you first have to know the ratio of the circumference of a circle to its diameter. But if you knew that, you would already be able to square the circle. One can easily see, however, that since the angle $ZA\Delta$ is proportional to the height of Z, this curve makes it possible to divide an angle into any number of equal parts.

Pappus also attributed a trisection to Menelaus of Alexandria. Pappus gave a classification of geometric construction problems in terms of three categories: planar, solid, and [curvi]linear. The first category consisted of constructions that used only straight lines and circles, the second those that used conic sections. The last,

[8] Hippias should be thankful for Proclus, without whom he would apparently be completely forgotten, as none of the other standard commentators discuss him, except for a mention in passing by Diogenes Laertius in his discussion of Thales. Allman (*1889*, pp. 94–95) argues that the Hippias mentioned in connection with the quadratrix is not the Hippias of Elis mentioned in the Eudemian summary, and other historians have agreed with him, but most do not.

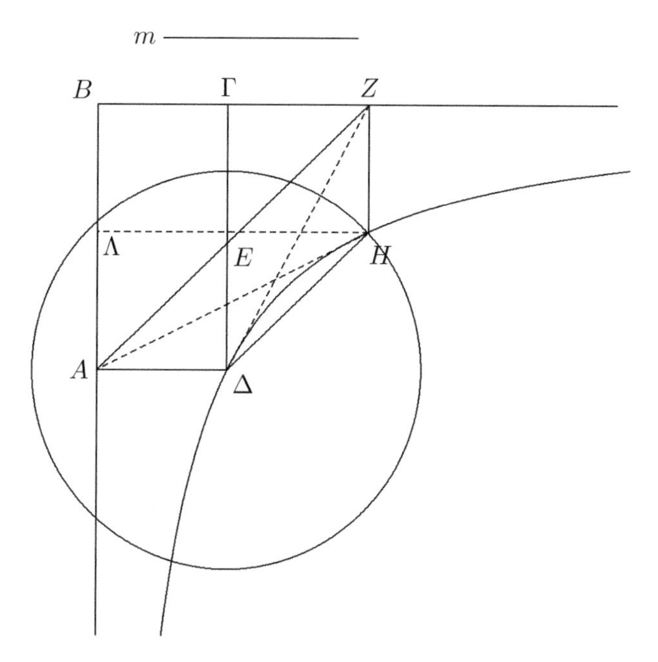

FIGURE 9. Pappus' construction of a *neûsis* using a rectangular hyperbola.

catch-all category consisted of problems requiring all manner of more elaborate and less regular curves, which were harder to visualize than the first two and presumably required some mechanical device to draw them. Pappus says that some of these curves come from locus problems, and lists the inventors of some others, among them a curve that Menelaus called the *paradox*. Other spirals of the same type, he says, are the quadratrices, the conchoids, and the cissoids. He goes on to say that geometers regard it as a major defect when a planar problem is solved using conics and other curves.

> Based on this classification of problems, the first geometers were unable to solve the abovementioned problem of [trisecting] the angle, which is by nature a solid problem, through planar methods. For they were not yet familiar with the conic sections; and for that reason they were at a loss. But later they trisected the angle through conics, using the convergence described below.

The word *convergence* (*neûsis*) comes from the verb *neúein*, one of whose meanings is *to incline toward*. In this particular case, it refers to the following construction. We are given a rectangle $AB\Gamma\Delta$ and a prescribed length m. It is required to find a point E on $\Gamma\Delta$ such that when AE is drawn and extended to meet the extension of $B\Gamma$ at a point Z, the line EZ will have length m. The construction is shown in Fig. 9, where the circle drawn has radius m and the hyperbola is rectangular, so that $A\Delta \cdot \Gamma\Delta = \Lambda H \cdot ZH$.

Given the *neûsis*, it becomes a simple matter to trisect an angle, as Pappus pointed out. Given any acute angle, label its vertex A, choose an arbitrary point Γ on one of its sides, and let Δ be the foot of the perpendicular from Γ to the other side of the angle. Complete the rectangle $AB\Gamma\Delta$, and carry out the *neûsis* with $m = 2A\Gamma$. Then let H be the midpoint of ZE, and join ΓH, as shown in Fig. 10.

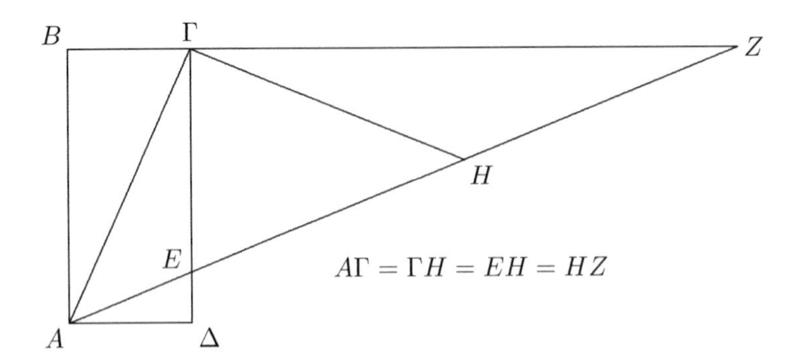

FIGURE 10. Trisection of an arbitrary angle by *neûsis* construction.

A mechanical (curvilinear) solution of the neûsis problem. Finding the point E in the *neûsis* problem is equivalent to finding the point Z. Either point allows the line AEZ to be drawn. Now one line that each of these points lies on is known. If some other curve that Z must lie on could be drawn, the intersection of that curve with the line $B\Gamma$ would determine Z and hence solve the *neûsis* problem. If we use the condition that the line ZE must be of constant length, we have a locus-type condition for Z, and it is easy to build a device that will actually draw this locus. What is needed is the T-shaped frame shown in Fig. 11, consisting of two pieces of wood or other material meeting at right angles. The horizontal part of the T has a groove along which a peg (shown as a hollow circle in the figure) can slide. The vertical piece has a fixed peg (shown as a solid circle) at distance $A\Delta$ from its top. Onto this frame a third piece is fitted with a fixed peg (the hollow circle) at distance m from its end and a groove between the peg and the other end that fits over the peg on the vertical bar. The frame is then laid down with its horizontal groove over the line $\Gamma\Delta$ and its fixed peg over A. When the moving piece is fitted over the frame so that its peg slides along the horizontal groove over $\Gamma\Delta$ and its groove slides over the peg at A, its endpoint (where a stylus is located to draw the curve) traces the locus on which Z must lie. The point Z lies where that locus meets the extension of $B\Gamma$. In practical terms, such a device can be built, but the rigid pegs must be located at exactly the distance from the ends determined by the rectangle and the fixed distance given in the *neûsis* problem. Thus the device must be modified by moving the pegs to the correct locations for each particular problem. If oxymoron is permitted, we might say that the practical value of this device is mostly theoretical. The locus it draws is the *conchoid of Nicomedes*, mentioned by Pappus and Proclus.

Because of the objections reported by Pappus to the use of methods that were more elaborate than the problems they were intended to solve, the search for planar (ruler-and-compass) solutions to these problems continued for many centuries. It was not until the 1830s that it was proved that no ruler-and-compass solution exists. The proof had no effect on the cranks of the world, of course. The problems continue to be of interest since that time, and not only to cranks who imagine they have solved them. Felix Klein, a leading German mathematician and educator in

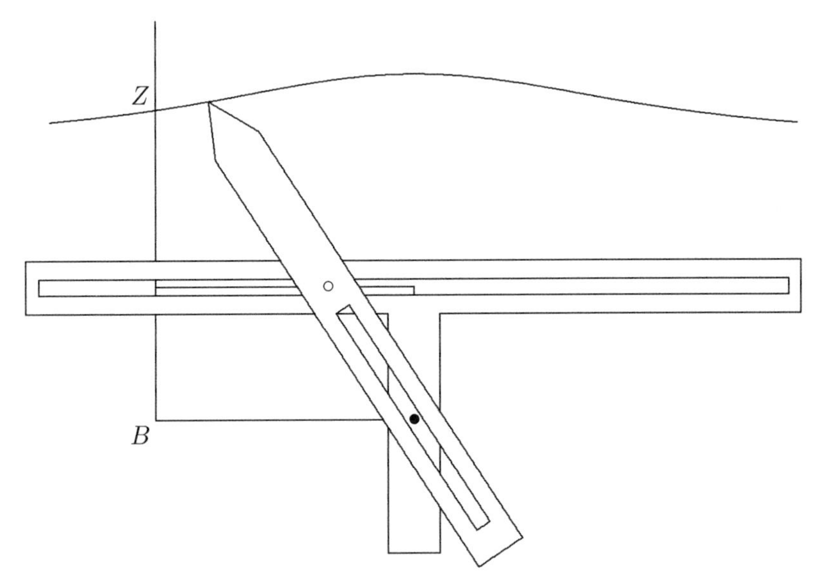

FIGURE 11. A mechanical device for drawing the conchoid of Nicomedes.

the late nineteenth and early twentieth centuries, urged that they be studied as a regular part of the curriculum (Beman and Smith, *1930*).

1.5. Challenges to Pythagoreanism: the paradoxes of Zeno of Elea. Although we have some idea of the geometric results proved by the Pythagoreans, our knowledge of their interpretation of these results is murkier. How did they conceive of geometric entities such as points, lines, planes, and solids? Were these objects physically real or merely ideas? What properties did they have? Some light is shed on this question by the philosophical critics of Pythagoreanism, one of whom has become famous for the paradoxes he was able to spin out of Pythagorean principles.

In the Pythagorean philosophy the universe was generated by numbers and motion. That these concepts needed to be sharpened up became clear from critics of the Pythagorean school. It turned out that the Pythagorean view of geometry and number contained paradoxes within itself, which were starkly pointed out by the philosopher Zeno of Elea. Zeno died around 430 BCE, and we do not have any of his works to rely on, only expositions of them by other writers. Aristotle, in particular, says that Zeno gave four puzzles about motion, which he called the Dichotomy (division), the Achilles, the Arrow, and the Stadium. Here is a summary of these arguments in modern language, based on Book 6 of Aristotle's *Physics*.

The Dichotomy. Motion is impossible because before an object can arrive at its destination it must first arrive at the middle of its route. But before it can arrive at the middle, it must travel one-fourth of the way, and so forth. Thus we see that the object must do infinitely many things in a finite time in order to move.

The Achilles. (This paradox is apparently so named because in Homer's *Iliad* the legendary warrior Achilles chased the Trojan hero Hector around the walls of Troy, overtook him, and killed him.) If given a head start, the slower runner will never be overtaken by the faster runner. Before the two runners can be at the same point at the same instant, the faster runner must first reach the point from which the slower runner started. But at that instant the slower runner will have reached another point ahead of the faster. Hence the race can be thought of as beginning again at that instant, with the slower runner still having a head start. The race will "begin again" in this sense infinitely many times, with the slower runner always having a head start. Thus, as in the dichotomy, infinitely many things must be accomplished in a finite time in order for the faster runner to overtake the slower.

The Arrow. An arrow in flight is at rest at each instant of time. That is, it does not move from one place to another during that instant. But then it follows that it cannot traverse any positive distance because successive additions of zero will never result in anything but zero.

The Stadium. (In athletic stadiums in Greece the athletes ran from the goal, around a halfway post and then back. This paradox seems to have been inspired by imagining two lines of athletes running in opposite directions and meeting each other.) Consider two parallel line segments of equal length moving toward each other with equal speeds. The speed of each line is measured by the number of points of space it passes over in a given time. But each point of one line passes *twice* that many points of the other line in the same time as the two lines move past each other. Hence the velocity of the line must equal its double, which is absurd.

The Pythagoreans had built their system on lines "made of" points, and now Zeno was showing them that space cannot be "made of" points in the same way that a building can be made of bricks. For assuredly the number of points in a line segment cannot be finite. If it were, the line would not be infinitely divisible as the dichotomy and Achilles paradoxes showed that it must be; moreover, the stadium paradox would show that the number of points in a line segment equals its double. There must therefore be an infinity of points in a line. But then each of these points must take up no space; for if each point occupied some space, an infinite number of them would occupy an infinite amount of space. But if points occupy no space, how can the arrow, whose tip is at a single point at each instant of time, move through a *positive* quantity of space? A continuum whose elements are points was needed for geometry, yet it could not be thought of as being made up of points in the way that discrete collections are made up of individuals.

1.6. Challenges to Pythagoreanism: incommensurables. The difficulties pointed out by Zeno affected the intuitive side of geometry. The challenge they posed may have been an impetus to the kind of logical rigor that we know as Euclidean. There is, however, an even stronger impetus to that rigor, one that was generated from within Pythagorean geometry. To the modern mathematician, this second challenge to Pythagorean principles is much more relevant and interesting than the paradoxes of Zeno. That challenge is the problem of incommensurables, which led ultimately to the concept of a real number.

The existence of incommensurables throws doubt on certain oversimplified proofs of geometric proportion. When two lines or areas are commensurable, one can describe their ratio as, say, $5 : 7$, meaning that there is a common measure such that the first object is five times this measure and the second is seven times it. A proportion such as $a : b \; :: \; c : d$, then, is the statement that ratios $a : b$ and $c : d$ are both represented by the same pair of numbers.

This theory of proportion is extremely important in geometry if we are to have such theorems as Proposition 1 of Book 6 of Euclid's *Elements*, which says that the areas of two triangles or two parallelograms having the same height are proportional to their bases, or the theorem (Book 12, Proposition 2) that the areas of two circles are proportional to the squares on their diameters. Even the simplest constructions, such as the construction of a square equal in area to a given rectangle or the application problems mentioned above, may require the concept of proportionality of lines. Because of the importance of the theory of proportion for geometry, the discovery of incommensurables made it imperative to give a definition of proportion without relying on a common measure to define a ratio.

Fowler (*1998*) argues for the existence of a Pythagorean theory of proportion based on *anthyphaíresis*, the mutual subtraction procedure we have now described many times.[9] He makes a very telling point (p. 18) in citing a passage from Aristotle's *Topics* where the assertion is made that having the same *antanaíresis* is tantamount to having the same ratio. Fowler takes *antanaíresis* to be a synonym of *anthyphaíresis*. Like Gray (quoted above), Knorr (*1975*) argues that the discovery of irrationals was not a major "scandal," and that it was not responsible for the "geometric algebra" in Book 2 of Euclid. While arguing that incommensurability forced some modifications in the way the Pythagoreans thought about physical magnitudes, he says (p. 41):

> It is thus thoroughly obvious that, far from being in a state of paralysis, fifth- and fourth-century geometers proceeded with their studies of similar figures as if they were still unaware of the foundational consequences of the existence of incommensurable lines.

1.7. The influence of Plato. Plato is still held in high esteem by philosophers, and it is well recognized that his philosophy contains a strong mathematical element. But since Plato was a follower of Socrates, who was almost entirely concerned with questions of ethics and the right conduct of life,[10] his interest in mathematical questions needs to be explained. Born in 427 BCE, Plato served in the Athenian army during the Peloponnesian War. He was also a devoted follower of Socrates. Socrates enjoyed disputation so much and was so adept at showing up the weakness in other people's arguments that he made himself very unpopular. When Athens was defeated in 404 BCE, Plato sided with the party of oligarchs who ruled the city temporarily. When the democratic rule was restored, the citizens took revenge on their enemies, among whom they counted Socrates. Plato was devastated by the trial and execution of Socrates in 399 BCE. He left Athens and traveled to Italy, where he became acquainted with the Pythagorean philosophy. He seems to have met the Pythagorean Philolaus in Sicily in 390. He also met the Pythagorean Archytas at Tarentum (where some Pythagoreans had fled to escape danger at Croton). Plato returned to Athens and founded the Academy in 387 BCE. There he hoped to train the young men[11] for public service and establish good government. At the behest of Archytas and a Syracusan politician named Dion, brother-in-law of the ruler Dionysus I, Plato made several trips to Syracuse, in Sicily, between 367 and 361 BCE, to act as advisor to Dionysus II. However, there was virtual civil war between Dion and Dionysus, and Plato was arrested and nearly executed. Diogenes

---

[9] Fowler avoids as far as possible using the phrase *Euclidean algorithm*.

[10] The Socrates depicted by Plato is partly a literary device through which Plato articulated his own thoughts on many subjects that the historical Socrates probably took little notice of.

[11] In his writing, especially *The Republic*, Plato argues for equal participation by women in government. There is no record of any female student at his Academy, however. His principles were far in advance of what the Athenians would tolerate in practice.

Laertius quotes a letter allegedly from Archytas to Dionysus urging that Plato be released. Plato returned to the Academy in 360 and remained there for the last 13 years of his life. He died in 347.

Archytas. Archytas, although a contemporary of Plato, is counted paradoxically among the "pre-Socratics" in philosophy; but that is because he worked outside Athens and continued the earlier Pythagorean tradition. Archytas' solution of the problem of two mean proportionals using two half-cylinders intersecting at right angles was mentioned above. In his *Symposium Discourses*, Plutarch claimed that

> for that reason Plato also lamented that the disciples of Eudoxus, Archytas, and Menaechmus attacked the duplication of a solid by building tools and machinery hoping to get two ratios through the irrational, by which it might be possible to succeed, [saying that by doing so they] immediately ruined and destroyed the good of geometry by turning it back toward the physical and not directing it upward or striving for the eternal and incorporeal images, in which the god is ever a god.

Although the sentiment Plutarch ascribes to Plato is consistent with the ideals expressed in the *Republic*, Eutocius reports one such mechanical construction as being due to Plato himself. From his upbringing as a member of the Athenian elite and from the influence of Socrates, Plato had a strong practical streak, concerned with life as it is actually lived.[12] Platonic idealism in the purely philosophical sense does not involve idealism in the sense of unrealistic striving for perfection.

Archytas and Philolaus provided the connection between Pythagoras and Plato, whose interest in mathematics began some time after the death of Socrates and continued for the rest of his life. Mathematics played an important role in the curriculum of his Academy and in the research conducted there, and Plato himself played a leading role in directing that research. Lasserre (*1964*, p. 17) believes that the most important mathematical work at the Academy began with the arrival of Theaetetus in Athens around 375 and ended with Eudoxus' departure for Cnidus around 350.

The principle that knowledge can involve only eternal, unchanging entities led Plato to some statements that sound paradoxical. For example, in Book 7 of the *Republic* he writes:

> Thus we must make use of techniques such as geometry when we take up astronomy and let go of the things in the heavens if we really intend to create something intrinsically useful and practical in the soul by mastering astronomy.

If Plato's mathematical concerns seem to be largely geometrical, that is probably because he encountered Pythagoreanism at the time when the challenges discussed above were still current topics. (Recall the quotation from the *Republic* in

---

[12] In the famous allegory of the cave in Book 7 of the *Republic*, Plato depicts the nonphilosophical person as living in a cave with feet in chains, seeing only flickering shadows on the wall of the cave, while the philosopher is the person who has stepped out of the cave into the bright sunshine and wishes to communicate that reality to the people back in the cave. While he encouraged his followers to "think outside the cave," his trips to Syracuse show that he understood the need to make philosophy work inside the cave, where everyday life was going on.

Chapter 3, where he laments the lack of public support for research into solid geometry.) There is a long-standing legend that Plato's Academy bore the following sign above its entrance:[13]

ΑΓΕΩΜΕΤΡΗΤΟΣ ΜΗΔΕΙΣ ΕΙΣΙΤΩ

("Let no ungeometrical person enter.") If Plato was indeed more concerned with geometry than with arithmetic, there is an obvious explanation for his preference: The imperfections of the real world relate entirely to geometry, not at all to arithmetic. For example, it is sometimes asserted that there are no examples of exact equality in the real world. But in fact, there are many. Those who make the assertion always have in mind continuous magnitudes, such as lengths or weights, in other words, geometrical concepts. Where arithmetic is concerned, exact equality is easy to achieve. If I have $11,328.75 in the bank, and my neighbor has $11,328.75 in the bank, the two of us have *exactly* the same amount of money. Our bank accounts are interchangeable for all monetary purposes. But Plato's love for geometry should not be overemphasized. In his ideal curriculum, described in the *Republic*, arithmetic is still regarded as the primary subject.

1.8. Eudoxan geometry. To see why the discovery of incommensurables created a problem for the Pythagoreans, consider the following conjectured early proof of a fundamental result in the theory of proportion: the proposition that two triangles having equal altitudes have areas proportional to their bases. This assertion is half of Proposition 1 of Book 6 of Euclid's *Elements*. Let ABC and ACD in Fig. 12 be two triangles having the same altitude. Euclid draws them as having a common side, but that is only for convenience. This positioning causes no loss in generality because of the proposition that any two triangles of equal altitude and equal base have equal areas, proved as Proposition 38 of Book 1.

Suppose that the ratio of the bases $BC : CD$ is $2 : 3$, that is, $3BC = 2CD$. Extend BD leftward to H so that $BC = BG = GH$, producing triangle AHC, which is three times triangle ABC. Then extend CD rightward to K so that $CD = DK$, yielding triangle ACK equal to two times triangle ACD. But then, since $GC = 3BC = 2CD = CK$, triangles AGC and ACK are equal. Since $AGC = 3ABC$ and $ACK = 2ACD$, it follows that $ABC : ACD = 2 : 3$. We, like Euclid, have no way of actually *drawing* an unspecified number of copies of a line, and so we are forced to *illustrate* the argument using specific numbers (2 and 3 in the present case), while expecting the reader to understand that the argument is completely general.

An alternative proof could be achieved by finding a common measure of BC and CD, namely $\frac{1}{2}BC = \frac{1}{3}CD$. Then, dividing the two bases into parts of this length, one would have divided ABC into two triangles, ACD into three triangles, and all five of the smaller triangles would be equal. But both of these arguments fail if no integers m and n can be found such that $mBC = nCD$, or (equivalently) no common measure of BC and CD exists. This proof needs to be shored up, but how is that to be done?

---

[13] These words are the earliest version of the legend, which Fowler (*1998*, pp. 200–201) found could not be traced back earlier than a scholium attributed to the fourth-century orator Sopatros. The commonest source cited for this legend is the twelfth-century Byzantine Johannes Tzetzes, in whose *Chiliades*, VIII, 975, one finds Μηδεὶς ἀγεωμέτρητος εἰσίτω μου τὴν στέγην. "Let no ungeometrical person enter my house."

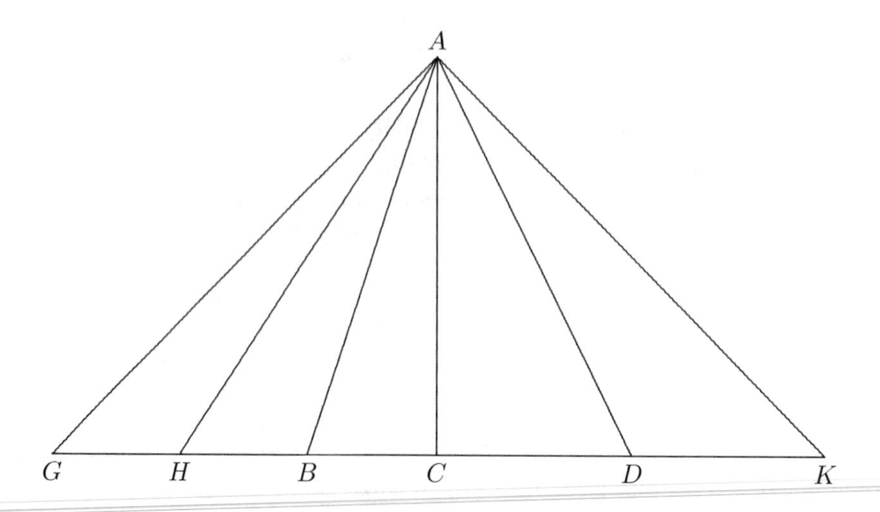

FIGURE 12. A fundamental theorem in the theory of proportion. Proposition 1 of Book 6 of the *Elements*.

The solution to the difficulty was provided by Eudoxus of Cnidus (ca. 407–354 BCE), whom Diogenes Laertius describes as "astronomer, geometer, physician, and lawgiver." He learned geometry from Archytas. Diogenes Laertius cites another commentator, named Sotion, who said that Eudoxus spent two months in Athens and attended lectures by Plato. Because of his poverty, he could not afford to live in Athens proper. He lived at the waterfront, known as the Piraeus, supported by a physician named Theomedus, and walked 11 km from there into Athens. Then, with a subsidy from friends, he went to Egypt and other places, and finally returned, "crammed full of knowledge," to Athens, "some say, just to annoy Plato for snubbing him earlier." Plato was not in Eudoxus' league as a mathematician; and if Eudoxus felt that Plato had patronized him in his earlier visit, perhaps because Plato and his other students were wealthy and Eudoxus was poor, his desire to return and get his own back from Plato is quite understandable. He must have made an impression on Plato on his second visit. In his essay *On Socrates' Daemon*, Plutarch reports that when the Delians consulted Plato about doubling the cube, in addition to advising them to study geometry, he told them that the problem had already been solved by Eudoxus of Cnidus and Helicon of Cyzicus. If true, this story suggests that the Delians appealed to Plato after Eudoxus had left for Cnidus, around 350. By that time Plato was a very old man, and perhaps mellower than he had been a quarter-century earlier during Eudoxus' first stay in Athens. In Cnidus, Eudoxus made many astronomical observations that were cited by the astronomer Hipparchus, and one set of his astronomical observations has been preserved. Although the evidence is not conclusive, it seems that while he was in Athens, he contributed two vital pieces to the mosaic that is Euclid's *Elements*.

The Eudoxan definition of proportion. The first piece of the *Elements* contributed by Eudoxus was the solution of the problem of incommensurables. This solution is attributed to him on the basis of two facts: (1) Proclus' comment that Euclid "arranged many of the theorems of Eudoxus"; (2) an anonymous scholium (commentary) on Euclid's Book 5, which asserts that the book is the creation "of a certain Eudoxus, [the student] of the teacher Plato" (Allman, *1889*, p. 132).

His central observation is a very simple one: Suppose that D and S are respectively the diagonal and side of a square (or pentagon). Even though there are no integers m and n such that $mD = nS$, so that the ratio $D : S$ cannot be defined as $n : m$ for any integers, it remains true that for *every* pair of integers m and n there is a trichotomy: Either $mD < nS$ or $mD = nS$ or $mD > nS$. That fact makes it possible at least to define what is meant by saying that the ratio of D to S is the same for all squares. We simply define the proportion $D_1 : S_1 :: D_2 : S_2$ for two different squares to mean that whatever relation holds between mD_1 and nS_1 for a given pair of integers m and n, that same relation holds between mD_2 and nS_2. Accordingly, as defined by Euclid at the beginning of Book 5, "A relation that two magnitudes of the same kind have due to their sizes is a *ratio*." As a definition, this statement is somewhat lacking, but we may paraphrase it as follows: "the relative size of one magnitude in terms of a second magnitude of the same kind is the *ratio* of the first to the second." We think of size as resulting from measurement and relative size as the result of *dividing* one measurement by another, but Euclid keeps silent on both of these points. Then, "Two magnitudes are said to have a ratio to each other if they are capable of exceeding each other when multiplied." That is, some multiple of each is larger than the other. Thus, the periphery of a circle and its diameter can have a ratio, but the periphery of a circle and its center cannot. Although the definition of ratio would be hard to use, fortunately there is no need to use it. What is needed is equality of ratios, that is, proportion. That definition follows from the trichotomy just mentioned. Here is the definition given in Book 5 of Euclid, with the material in brackets added from the discussion just given to clarify the meaning:

> Magnitudes are said to be in the same ratio, the first to the second $[D_1 : S_1]$ and the third to the fourth $[D_2 : S_2]$, when, if any equimultiples whatever be taken of the first and third $[mD_1$ and $mD_2]$ and any equimultiples whatever of the second and fourth $[nS_1$ and $nS_2]$, the former equimultiples alike exceed, are alike equal to, or are alike less than the latter equimultiples taken in corresponding order [that is, $mD_1 > nS_1$ and $mD_2 > nS_2$, or $mD_1 = nS_1$ and $mD_2 = nS_2$, or $mD_1 < nS_1$ and $mD_2 < nS_2$].

Let us now look again at our conjectured early Pythagorean proof of Euclid's Proposition 1 of Book 6 of the *Elements*. How much change is required to make this proof rigorous? Very little. Where we have assumed that $3BC = 2CD$, it is only necessary to consider the cases $3BC > 2CD$ and $3BC < 2CD$ and show with the same figure that $3ABC > 2ACD$ and $3ABC < 2ACD$ respectively, and that is done by using the trivial corollary of Proposition 38 of Book 1: *If two triangles have equal altitudes and unequal bases, the one with the larger base is larger.* Eudoxus has not only shown how proportion can be defined so as to apply to incommensurables, he has done so in a way that fits together seamlessly with earlier proofs that apply only in the commensurable case. If only the fixes for bugs in modern computer programs were so simple and effective!

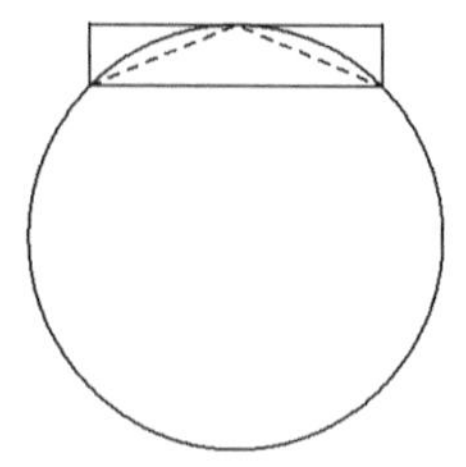

FIGURE 13. The basis of the method of exhaustion.

The method of exhaustion. Eudoxus' second contribution is of equal importance with the first; it is the proof technique known as the *method of exhaustion.* This method is used by both Euclid and Archimedes to establish theorems about areas and solids bounded by curved lines and surfaces. As in the case of the definition of proportion for incommensurable magnitudes, the evidence that Eudoxus deserves the credit for this technique is not conclusive. In his commentary on Aristotle's *Physics*, Simplicius credits the Sophist Antiphon (480–411) with inscribing a polygon in a circle, then repeatedly doubling the number of sides in order to square the circle. However, the perfected method seems to belong to Eudoxus. Archimedes says in the cover letter accompanying his treatise on the sphere and cylinder that it was Eudoxus who proved that a pyramid is one-third of a prism on the same base with the same altitude and that a cone is one-third of the cylinder on the same base with the same altitude. What Archimedes meant by proof we know: He meant proof that meets Euclidean standards, and that can be achieved for the cone only by the method of exhaustion. Like the definition of proportion, the basis of the method of exhaustion is a simple observation: When the number of sides in an inscribed polygon is doubled, the excess of the circle over the polygon is reduced by more than half, as one can easily see from Fig. 13. This observation works together with the theorem that *if two magnitudes have a ratio* and more than half of the larger is removed, then more than half of what remains is removed, and this process continues, then at some point what remains will be less than the smaller of the original two magnitudes (*Elements*, Book 10, Proposition 1). This principle is usually called *Archimedes' principle* because of the frequent use he made of it. The phrase *if two magnitudes have a ratio* is critical, because Euclid's proof of the principle depends on converting the problem to a problem about integers. Since some multiple (n) of the smaller magnitude exceeds the larger, it is only a matter of showing that a finite sequence $a_1, a_2, \ldots$ in which each term is less than half of the preceding will eventually reach a point where the ratio $a_k : a_1$ is less than $1/n$.

The definition of ratio and proportion allowed Eudoxus/Euclid to establish all the standard facts about the theory of proportion, including the important fact that similar polygons are proportional to the squares on their sides (*Elements*, Book 6, Propositions 19 and 20). Once that result is achieved, the method of exhaustion makes it possible to establish rigorously what the Pythagoreans had long believed: that similar curvilinear regions are proportional to the squares on similarly situated chords. In particular, it made it possible to prove the fundamental fact that was being used by Hippocrates much earlier: Circles are proportional to the squares on their diameters. This fact is now stated as Proposition 2 of Book 12 of the *Elements*, and the proof given by Euclid is illustrated in Fig. 14.

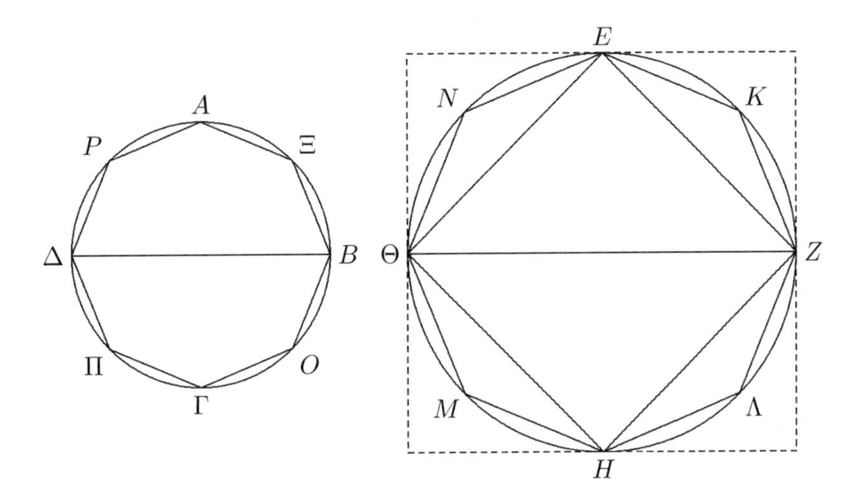

FIGURE 14. Proof that circles are proportional to the squares on their diameters.

Let $AB\Gamma\Delta$ and $EZH\Theta$ be two circles with diameters $B\Delta$ and ΘZ, and suppose that the circles are not proportional to the squares on their diameters. Let the ratio $B\Delta^2 : \Theta Z^2$ be the same as $AB\Gamma\Delta : \Sigma$, where Σ is an area larger or smaller than $EZH\Theta$. Suppose first that Σ is smaller than the circle $EZH\Theta$. Draw the square $EZH\Theta$ inscribed in the circle $EZH\Theta$. Since this square is half of the circumscribed square with sides perpendicular and parallel to the diameter ΘZ, and the circle is smaller than the circumscribed square, the inscribed square is more than half of the circle. Now bisect each of the arcs EZ, ZH, $H\Theta$, and ΘE at points K, Λ, M, and N, and join the polygon $EKZ\Lambda H\Theta NE$. As shown above, doing so produces a larger polygon, and the excess of the circle over this polygon is less than half of its excess over the inscribed square. If this process is continued enough times, the excess of the circle over the polygon will eventually be less than its excess over Σ, and therefore the polygon will be larger than Σ. For definiteness, Euclid assumes that this polygon is the one reached at the first doubling: $EKZ\Lambda H\Theta NE$. In the first circle $AB\Gamma\Delta$, inscribe a polygon $A\Xi BO\Gamma\Pi\Delta P$ similar to $EKZ\Lambda H\Theta NE$. Now the square on $B\Delta$ is to the square on $Z\Theta$ as $A\Xi BO\Gamma\Pi\Delta P$ is to $EKZ\Lambda H\Theta NE$. But also the square on $B\Delta$ is to the square on $Z\Theta$ as the circle $AB\Gamma\Delta$ is to Σ. It follows that $A\Xi BO\Gamma\Pi\Delta P$ is to $EKZ\Lambda H\Theta NE$ as the circle $AB\Gamma\Delta$ is to Σ. Since the circle $AB\Gamma\Delta$ is larger than $A\Xi BO\Gamma\Pi\Delta P$, it follows that Σ must be larger than $EKZ\Lambda H\Theta NE$, But by construction, it is smaller, which is impossible. A similar argument shows that it is impossible for Σ to be larger than $EZH\Theta$.

A look ahead. Ratios as defined by Euclid are always between two magnitudes of the same type. He never considered what we call density, for example, which is the ratio of a mass to a volume. Being always between two magnitudes of the same type, ratios are "dimensionless" in our terms, and could be used as numbers, if only they could be added and multiplied. However, the Greeks obviously did not think of operations on ratios as being the same thing they could do with numbers. In terms of adding, Euclid does say (Book 6, Proposition 24) that if two proportions have the same second and fourth terms, then their first terms and third terms can be added (first to first and third to third), that is, if $a : b :: c : d$ and $e : b :: f : d$, then $(a+e) : b :: (c+f) : d$. But he did not think of the second and fourth terms in a proportion as denominators or try to get a common denominator. For multiplication of ratios, Euclid gives three separate definitions. In Book 5, Definition 9, he defines the duplicate (which we would call the square) of the ratio $a : b$ to be the ratio $a : c$ if b is the mean proportional between a and c, that is, $a : b :: b : c$. Similarly, when there are four terms in proportion, as in the problem of two mean proportionals, so that $a : b :: b : c :: c : d$, he calls the ratio $a : d$ the triplicate of $a : b$. We would call it the cube of this ratio. Not until Book 6, Definition 5 is there any kind of general definition of the product of two ratios. Even that definition is not in all manuscripts and is believed to be a later interpolation. It goes as follows: *A ratio is said to be the* composite *of two ratios when the sizes in the two ratios produce something when multiplied by themselves.*[14] This rather vague definition is made worse by the fact that the word for *composed* (*sygkeímena*) is simply a general word for *combined.* It means literally *lying together* and is the same word used when two lines are placed end to end to form a longer line. In that context it corresponds to addition, whereas in the present one it corresponds (but only very loosely) to multiplication. It can be understood only by seeing the way that Euclid operates with it. Given four lines a, b, c, and d, to form the compound ratio $a : b.c : d$, Euclid first takes any two lines m and n such that $a : b :: m : n$. He then finds a line p such that $n : p :: c : d$ and *defines* the compound ratio $a : b.c : d$ to be $m : p$.

There is some arbitrariness in this procedure, since m could be any line. A modern mathematician looking at this proof would note that Euclid could have shortened the labor by taking $m = a$ and $n = b$. The same mathematician would add that Euclid ought to have shown that the final ratio is the same independently of the choice of m, which he did not do. But one must remember that the scholarly community around Euclid was much more intimate than in today's world; he did not have to write a "self-contained" book. In the present instance a glance at Euclid's *Data* shows that he knew what he was doing. The first proposition in that book says that "if two magnitudes A and B are given, then their ratio is given." In modern language, any quantity can be replaced by an equal quantity in a ratio without changing the ratio. The proof is that if $A = \Gamma$ and $B = \Delta$, then $A : \Gamma :: B : \Delta$, and hence by Proposition 16 of Book 5 of the *Elements*, $A : B :: \Gamma : \Delta$. The second proposition of the *Data* draws the corollary that if a given magnitude has a given ratio to a second magnitude, then the second magnitude is also given. That is, if two quantities have the same ratio to a given quantity, then they are equal. From these principles, Euclid could see that the final ratio $m : p$ is what mathematicians now call "well-defined," that is, independent of the initial choice of m.[15] The first use made of this process is in Proposition 23 of Book 6, which asserts that equiangular parallelograms are in the compound ratio of their (corresponding) sides.

With the departure of Eudoxus for Cnidus, we can bring to a close our discussion of Plato's influence on mathematics. If relations between Plato and Eudoxus were less than intimate, as Diogenes Laertius implies, Eudoxus may have drawn off some of Plato's students whose interests were more scientific (in modern terms)

---

[14] I am aware that the word "in" here is not a literal translation, since the Greek has the genitive case—the sizes *of* the two ratios. But I take *of* here to mean *belonging to*, which is one of the meanings of the genitive case.

[15] A good exposition of the purpose of Euclid's *Data* and its relation to the *Elements* was given by Il'ina (*2002*), elaborating a thesis of I. G. Bashmakova.

and less philosophical. It is likely that even Plato realized that his attempt to understand the universe through his Forms was not going to work. His late dialogue *Parmenides* gives evidence of a serious rethinking of this doctrine. In any case, it is clear that Plato could not completely dominate the intellectual life of his day.

1.9. Aristotle. Plato died in 347 BCE, and his place as the pre-eminent scholar of Athens was taken a decade after his death by his former pupil Aristotle (384–322 BCE). Aristotle became a student at the Academy at the age of 18 and remained there for 20 years. After the death of Plato he left Athens, traveled, got married, and in 343 became tutor to the future Macedonian King Alexander (the Great), who was 13 years old when Aristotle began to teach him and 16 when he became king on the death of his father. In 335 Aristotle set up his own school, located in the Lyceum, over the hill from the Academy. For the next 12 years he lived and wrote there, producing an enormous volume of speculation on a wide variety of subjects, scientific, literary, and philosophical. In 322 Alexander died, and the Athenians he had conquered turned against his friends. Unlike Socrates, Aristotle felt no obligation to be a martyr to the laws of the polis. He fled to escape the persecution, but died the following year. Aristotle's writing style resembles very much that of a modern scholar, except for the absence of footnotes. Like Plato, in mathematics he seems more like a well-informed generalist than a specialist.

The drive toward the logical organization of science reached its full extent in the treatises of Aristotle. He analyzed reason itself and gave a very thorough and rigorous discussion of formal inference and the validity of various kinds of arguments in his treatise *Prior Analytics*, which was written near the end of his time at the Academy, around 350 BCE. It is easy to picture debates at the Academy, with the mathematicians providing examples of their reasoning, which the logician Aristotle examined and criticized in order to distill his rules for making inferences. In this treatise Aristotle discusses subjects, predicates, and syllogisms connecting the two, occasionally giving a glimpse of some mathematics that may indicate what the mathematicians were doing at the time.

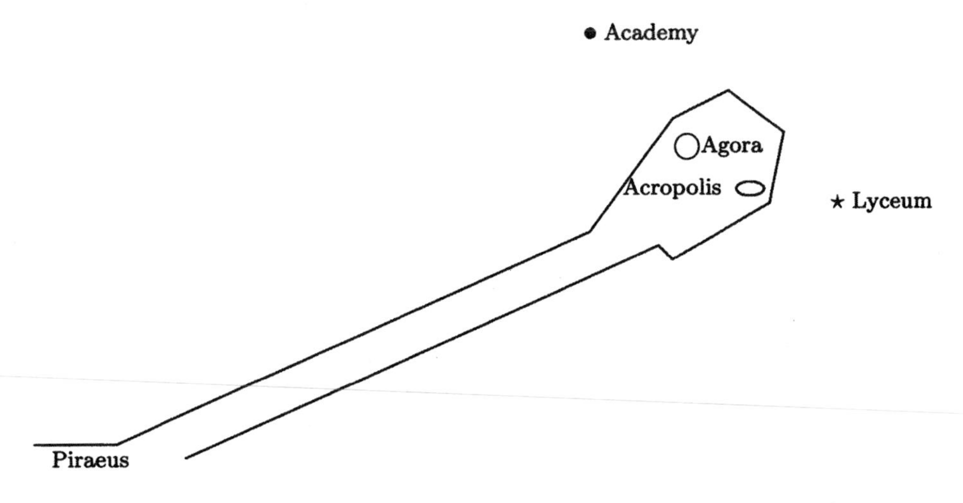

Athens in the fourth century BCE: the waterfront (Piraeus), Academy, and Lyceum.

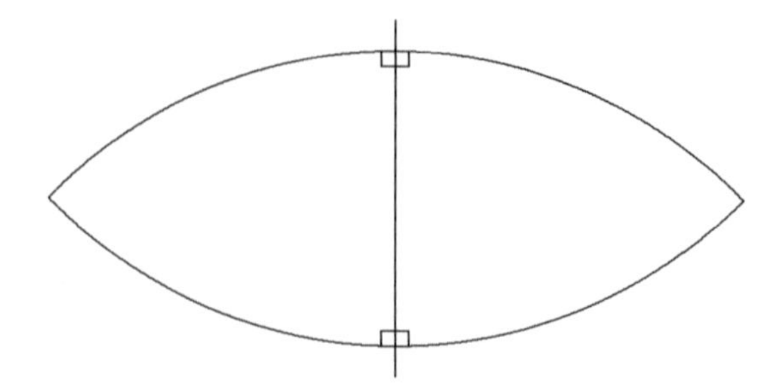

FIGURE 15. How do we exclude the possibility that two lines perpendicular to the same line may intersect each other?

In Book 1 Aristotle describes how to organize the study of a subject, looking for all the attributes and subjects of both of the terms that are to appear in a syllogism. The subject–attribute relation is mirrored in modern thought by the notion of elements belonging to a set. The element is the subject, and the set it belongs to is defined by attributes that can be predicated of all of its elements and no others. Just as sets can be elements of other sets, Aristotle said that the same object can be both a subject and a predicate. He thought, however, that there were some absolute subjects (individual people, for example) that were not predicates of anything and some absolute predicates (what we call abstractions, such as beauty) that were never the subject of any proposition.[16] Aristotle says that the postulates appropriate to each subject must come from experience. If we are thorough enough in stating all the attributes of the fundamental terms in a subject, it will be possible to prove certain things and state clearly what must be assumed.

In Book 2 he discusses ways in which reasoning can go wrong, including the familiar fallacy of "begging the question" by assuming what is to be proved. In this context he offers as an example the people who claim to construct parallel lines. According to him, they are begging the question, starting from premises that cannot be proved without the assumption that parallel lines exist. We may infer that there were around him people who did claim to show how to construct parallel lines, but that he was not convinced. It seems obvious that two lines perpendicular to the same line are parallel, but surely that fact, so obvious to us, would also be obvious to Aristotle. Therefore, he must have looked beyond the obvious and realized that the existence of parallel lines does *not* follow from the immediate properties of lines, circles, and angles. Only when this realization dawns is it possible to see the fallacy in what appears to be common sense. Common sense—that is, human intuition—suggests what can be proved: If two perpendiculars to the same line meet on one side of the line, then they must meet on the other side also, as in Fig. 15. Indeed, Ptolemy did prove this, according to Proclus. But Ptolemy then concluded that two lines perpendicular to the same line cannot meet at all. "But," Aristotle would have objected, "you have not proved that two lines cannot meet in two different points." And he would have been right: the assumptions that two lines can meet in only one point and that the two sides of a line are different regions (not connected to each other) are equivalent to assuming that parallel lines exist.

---

[16] In modern set theory it is necessary to assume that one cannot form an infinite chain of sets a, b, c,... such that $b \in a$, $c \in b$,.... That is, at the bottom of any particular element of a set, there is an "atom" that has no elements.

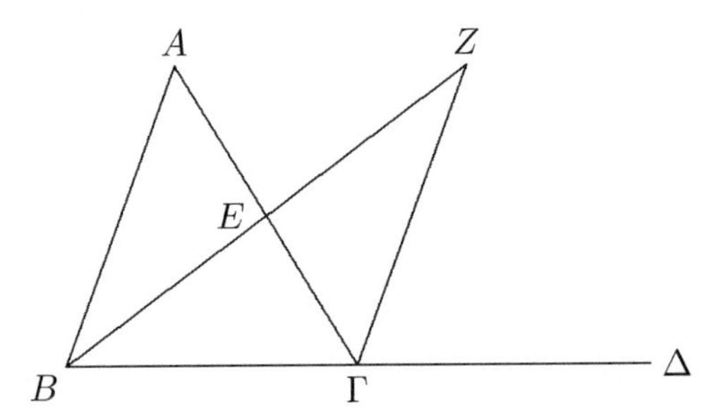

FIGURE 16. The exterior angle theorem.

Euclid deals with this issue in the *Elements* by stating as the last of his assumptions that "two straight lines do not enclose an area." Oddly, however, he seems unaware of the need for this assumption when proving the main lemma (Book 1, Proposition 16) needed to prove the existence of parallel lines.[17] This proposition asserts that an exterior angle of a triangle is larger than either of the opposite interior angles. Euclid's proof is based on Fig. 16, in which a triangle $AB\Gamma$ is given with side $B\Gamma$ extended to Δ, forming the exterior angle $A\Gamma\Delta$. He wishes to prove that this angle is larger than the angle at A. To do so, he bisects $A\Gamma$ at E, draws AE, and extends it to Z so that $EZ = AE$. When $Z\Gamma$ is joined, it is seen that the triangles ABE and ΓZE are congruent by the side–angle–side criterion. It follows that the angle at A equals $\angle E\Gamma Z$, which is smaller than $\angle E\Gamma\Delta$, being only a part of it.

In the proof Euclid assumes that the points E and Z are on the same side of line $B\Gamma$. But that is obvious only for triangles small enough to see. It needs to be proved. To be sure, Euclid could have proved it by arguing that if E and Z were on opposite sides of $B\Gamma$, then EZ would have to intersect either $B\Gamma$ or its extension in some point H, and then the line BH passing through Γ and the line BEH would enclose an area. But he did not do that. In fact, the only place where Euclid invokes the assumption that two lines cannot enclose an area is in the proof of the side–angle–side criterion for congruence (Book 1, Proposition 4).[18]

Granting that Aristotle was right about this point, we still must wonder why he considered the existence of parallel lines to be in need of proof. Why would he have doubts about something that is so clear on an intuitive level? One possible reason is that parallelism involves the infinite: Parallelism involves the concept that two finite line segments will *never* meet, no matter how far they are extended. If geometry is interpreted physically (say, by regarding a straight line as the path of a light ray), we really have no assurance whatever that parallel lines exist—how could anyone assert with confidence what will happen if two apparently parallel lines are extended to a length of hundreds of light years?

---

[17] In standard editions of Euclid, there are 14 assumptions, but three of them, concerned with adding equals to equals, doubling equals, and halving equals, are not found in some manuscripts. Gray (*1989*, p. 46) notes that the fourteenth assumption may be an interpolation by the Muslim mathematician al-Nayrizi, the result of speculation on the foundations of geometry. That would explain its absence from the proof of Proposition 16.

[18] This proof also uses some terms and some hidden assumptions that are visually obvious but which mathematicians nowadays do not allow.

As Aristotle's discussion of begging the question continues, further evidence comes to light that this matter of parallel lines was being debated around 350, and proofs of the existence of parallel lines (Book 1, Proposition 27 of the *Elements*) were being proposed, based on the exterior-angle principle. In pointing out that different false assumptions may lead to the same wrong conclusion, Aristotle notes in particular that the nonexistence of parallel lines would follow if an internal angle of a triangle could be greater than an external angle (not adjacent to it), and also if the angles of a triangle added to more than two right angles.[19] One is almost tempted to say that the mathematicians who analyzed the matter in this way foresaw the non-Euclidean geometry of Riemann, but of course that could not be. Those mathematicians were examining what must be assumed in order to get parallel lines into their geometry. They were not exploring a geometry without parallel lines.

2. Euclid

In retrospect the third century BCE looks like the high-water mark of Greek geometry. Beginning with the *Elements* of Euclid around 300 BCE, this century saw the creation of sublime mathematics in the treatises of Archimedes and Apollonius. It is very tempting to regard Greek geometry as essentially finished after Apollonius, to see everything that came before as leading up to these creations and everything that came after as "polishing up." And indeed, although there were some bright spots afterward and some interesting innovations, none had the scope or the profundity of the work done by these three geometers.

The first of the three major figures from this period is Euclid, who is world famous for his *Elements*, which we have in essence already discussed. This work is so famous, and dominated all teaching in geometry throughout much of the world for so long, that the man and his work have essentially merged. For centuries people said not that they were studying geometry, but that they were studying Euclid. This one work has eclipsed both Euclid's other books and his biography. He did write other books, and two of them—the *Data* and *Optics*—still exist. Others—the *Phænomena*, *Loci*, *Conics*, and *Porisms*—are mentioned by Pappus, who quotes theorems from them.

Euclid is defined for us as the author of the *Elements*. Apart from his writings, we know only that he worked at Alexandria in Egypt just after the death of Alexander the Great. In a possibly spurious passage in Book 7 of his *Synagōgḗ*, Pappus gives a brief description of Euclid as the most modest of men, a man who was precise but not boastful, like (he implies) Apollonius.

2.1. The *Elements*.

As for the *Elements* themselves, the editions that we now have came to us through many hands, and some passages seem to have been added by hands other than Euclid's, especially Theon of Alexandria. We should remember, of course, that Theon was not interested in preserving an ancient literary artifact unchanged; he was trying to produce a good, usable treatise on geometry. Some manuscripts have 15 books, but the last two have since been declared spurious by the experts, so that the currently standard edition has 13 books, the last of which looks suspiciously less formal than the first 12, leading some to doubt that Euclid wrote it. Leaving aside the thorny question of which parts were actually written

[19] Field and Gray (*1987*, p. 64) note that this point has been made by many authors since Aristotle.

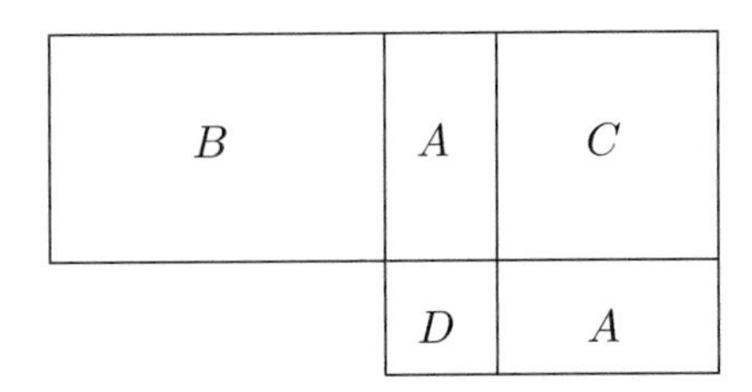

FIGURE 17. Expression of a rectangle as the difference of two squares.

by Euclid, we give just a summary description of the contents, since we have seen them coming together in the work of the Pythagoreans, Plato, and Aristotle.

The contents of the first book of the *Elements* are covered in the standard geometry courses given in high schools. This material involves the elementary geometric constructions of copying angles and line segments, drawing squares, and the like and the basic properties of parallelograms, culminating in the Pythagorean theorem (Proposition 47). In addition, these properties are applied to the problem of transformation of areas, leading to the construction of a parallelogram with a given base angle, and equal in area to any given polygon (Proposition 45). There the matter rests until the end of Book 2, where it is shown (Proposition 14) how to construct a square equal to any given polygon.

Book 2 contains geometric constructions needed to solve problems that may involve quadratic incommensurables without resorting to the Eudoxan theory of proportion. For example, a fundamental result is Proposition 5: *If a straight line is cut into equal and unequal segments, the rectangle contained by the unequal segments of the whole together with the square on the straight line between the points of the section is equal to the square on the half.* This proposition is easily seen using Fig. 17, in terms of which it asserts that $(A+B)+D = 2A+C+D$; that is, $B = A+C$.

This proposition, in arithmetic form, appeared as a fundamental tool in the cuneiform tablets. For if the unequal segments of the line are regarded as two unknown quantities, then half of the segment is precisely their average, and the straight line between the points (that is, the segment between the midpoint of the whole segment and the point dividing the whole segment into unequal parts) is precisely what we called earlier the semidifference. Thus, this proposition says that the square of the average equals the product plus the square of the semidifference; and that result was fundamental for solving the important problems of finding two numbers given their sum and product or their difference and product. However, those geometric constructions do not appear until Book 6. These application problems could have been solved in Book 2 in the case when the excess or defect is a square. Instead, these special cases were passed over and the more general results, which depend on the theory of proportion, were included in Book 6.

Book 2 also contains the construction of what came to be known as the *Section*, that is, the division of a line in mean and extreme ratio so that the whole is to one part as that part is to the other. But Euclid is not ready to prove that version yet, since he doesn't have the theory of proportion. Instead, he gives what must have been the original form of this proposition (Proposition 11): *to cut a line so that the rectangle on the whole and one of the parts equals the square on the other part.* After it is established that four lines are proportional when the rectangle on the means equals the rectangle on the extremes (Proposition 16, Book 6), it becomes possible to convert this construction into the construction of the Section (Proposition 30, Book 6).

Books 3 and 4 take up topics familiar from high-school geometry: circles, tangents and secants, and inscribed and circumscribed polygons. In particular, Book 4 shows how to inscribe a regular pentagon in a circle (Proposition 11) and how to circumscribe a regular pentagon about a circle (Proposition 12), then reverses the figures and shows how to get the circles given the pentagon (Propositions 13 and 14). After the easy construction of a regular hexagon (Proposition 15), Euclid finishes off Book 4 with the construction of a regular pentakaidecagon (15-sided polygon, Proposition 16).

Book 5 contains the Eudoxan theory of geometric proportion, in particular the construction of the mean proportional between two lines (Proposition 13). In Book 6 this theory is applied to solve the problems of application with defect and excess. A special case of the latter, in which it is required to construct a rectangle on a given line having area equal to the square on the line and with a square excess is the very famous Section (Proposition 30). Euclid phrases the problem as follows: *to divide a line into mean and extreme ratio.* This means to find a point on the line so that the whole line is to one part as that part is to the second part. The Pythagorean theorem is then generalized to cover not merely the squares on the sides of a right triangle, but any similar polygons on those sides (Proposition 31). The book finishes with the well-known statement that central and inscribed angles in a circle are proportional to the arcs they subtend.

Books 7–9 were discussed in Chapter 7. They are devoted to Pythagorean number theory. Here, since irrationals cannot occur, the notion of proportion is redefined to eliminate the need for the Eudoxan technique.

Book 10 occupies fully one-fourth of the entire length of the *Elements.* For its sheer bulk, one would be inclined to consider it the most important of all the 13 books, yet its 115 propositions are among the least studied of all, principally because of their technical nature. The irrationals constructed in this book by taking square roots are needed in the theory developed in Book 13 for inscribing regular solids in a sphere (that is, finding the lengths of their sides knowing the radius of the sphere). The book begins with the operating principle of the method of exhaustion, also known as the principle of Archimedes. The way to demonstrate incommensurability through the Euclidean algorithm then follows as Proposition 2: *If, when the smaller of two given quantities is continually subtracted from the larger, that which is left never divides evenly the one before it, the quantities are incommensurable.* We used this method of showing that the side and diagonal of a regular pentagon are incommensurable in Chapter 8.

Book 11 contains the basic parts of the solid geometry of planes, parallelepipeds, and pyramids. The theory of proportion for these solid figures is developed in Book 12, where one finds neatly tucked away the theorem that circles are proportional to the squares on their diameters (Proposition 2), which we quoted above.

Book 12 continues the development of solid geometry by establishing the usual proportions and volume relations for solid figures; for example, a triangular prism can be divided by planes into three pyramids, all having the same volume (Proposition 7), a cone has one-third the volume of a cylinder on the same base, similar cones and cylinders are proportional to the cubes of their linear dimensions, ending with the proof that spheres are proportional to the cubes on their diameters (Proposition 18). As we noted above, Archimedes (or someone who edited his works) credited these theorems to Eudoxus.

Book 13, the last book of the *Elements*, is devoted to the construction of the regular solids and the relation between their dimensions and the dimensions of the sphere in which they are inscribed. The last proposition (Proposition 18) sets out the sides of these regular solids and their ratios to one another. An informal discussion following this proposition concludes that there can be only five regular solids.

2.2. The *Data*. Euclid's *Elements* assume a certain familiarity with the principles of geometric reasoning, principles that are explained in more detail in the *Data*. The Greek name of this work (*Dedómena*) means [Things That Are] *Given*, just as *Data* does in Latin. The propositions in this book can be interpreted in various ways. Some can be looked at as *uniqueness* theorems. For example (Proposition 53), if the shapes—that is, the angles and ratios of the sides—are given for two polygons, and the ratio of the areas of the polygons is given, then the ratio of any side of one to any side of the other is given. Here, being given means being uniquely determined. Uniqueness is needed in proofs and constructions so that one can be sure that the result will be the same no matter what choices are made. It is an issue that arises frequently in modern mathematics, where operations on sets are defined by choosing representatives of the sets; when that is done, it is necessary to verify that the operation is *well defined*, that is, independent of the choice made. In geometry we frequently say, "Let ABC be a triangle having the given properties *and having such-and-such a property*," such as being located in a particular position. In such cases we need to be sure that the additional condition does not restrict the generality of the argument. In another sense, this same proposition reassures the reader that an explicit construction is *possible*, and removes the necessity of including it in the exposition of a theorem.

Other propositions assert that certain properties are *invariant*. For example (Proposition 81), when four lines A, B, Γ, and Δ are given, and the line H is such that $\Delta : E = A : H$, where E is the fourth proportional to A, B, and Γ, then $\Delta : \Gamma = B : H$. This last proposition is a lemma that can be useful in working out locus problems, which require finding a curve on which a point must lie if it satisfies certain prescribed conditions. Finally, a modern mathematician might interpret the assertion that an object is "given" as saying that the object "exists" and can be meaningfully talked about. To Euclid, that existence would mean that the object was explicitly constructible.

3. Archimedes

Archimedes is one of a small number of mathematicians of antiquity of whose works we know more than a few fragments and of whose life we know more than the approximate time and place. The man indirectly responsible for his death, the Roman general Marcellus, is also indirectly responsible for the preservation of some of what we know about him. Archimedes lived in the Greek city of Syracuse on the island of Sicily during the third century BCE and is said by Plutarch to have been "a relative and a friend" of King Hieron II. Since Sicily lies nearly on a direct line between Carthage and Rome, it became embroiled in the Second Punic War.

Marcellus took the city of Syracuse after a long siege, and Archimedes was killed by a Roman soldier in the chaos of the final fall of the city. In the course of writing a biography of Marcellus, the polymath Plutarch included some information on mathematics and philosophy in general.

According to Plutarch's biography of Marcellus, the general was very upset that Archimedes had been killed and had his body buried in a suitably imposing tomb. According to Eutocius, a biography of Archimedes was written by a certain Heracleides, who is mentioned in some of Archimedes' letters. However, no copy of this biography is known to exist today.

There are many legends connected with Archimedes, scattered among the various sources. Plutarch, for instance, says that Archimedes made many mechanical contrivances but generally despised such work in comparison with pure mathematical thought. Plutarch also reports three different stories of the death of Archimedes and tells us that Archimedes wished to have a sphere inscribed in a cylinder carved on his tombstone. The famous story that Archimedes ran naked through the streets shouting "Eureka!" ("I've got it!") when he discovered the principle of specific gravity in the baths is reported by the Roman architect Vitruvius. Proclus gives another well-known anecdote: that Archimedes built a system of pulleys that enabled him (or King Hieron) single-handedly to pull a ship through the water. Finally, Plutarch and Pappus both quote Archimedes as saying in connection with his discovery of the principle of the lever that if there were another Earth, he could move this one by standing on it.

With Archimedes we encounter the first author of a considerable body of original mathematical research that has been preserved to the present day. He was one of the most versatile, profound, creative, imaginative, rigorous, and influential mathematicians who ever lived. Ten of Archimedes' treatises have come down to the present, along with a *Book of Lemmas* that seems to be Archimedean. Some of these works are prefaced by a "cover letter" intended to explain their contents to the person to whom Archimedes sent them. These correspondents of Archimedes were: Gelon, son of Hieron II and one of the kings of Syracuse during Archimedes' life; Dositheus, a student of Archimedes' student and close friend Conon; and Eratosthenes, an astronomer who worked in Alexandria. Like the manuscripts of Euclid, all of the Archimedean manuscripts date from the ninth century or later. These manuscripts have been translated into English and published by various authors. A complete set of Medieval manuscripts of Archimedes' work has been published by Marshall Clagett in the University of Wisconsin series on Medieval Science.

The 10 treatises referred to above are the following.

1. *On the Equilibrium of Planes*, Part I
2. *Quadrature of the Parabola*
3. *On the Equilibrium of Planes*, Part II
4. *On the Sphere and the Cylinder*, Parts I and II
5. *On Spirals*
6. *On Conoids and Spheroids*
7. *On Floating Bodies*
8. *Measurement of a Circle*
9. *The Sand-reckoner*
10. *The Method*

References by Archimedes himself and other mathematicians tell of the existence of other works by Archimedes of which no manuscripts are now known to exist. These include works on the theory of balances and levers, optics, the regular polyhedra, the calendar, and the construction of mechanical representations of the motion of heavenly bodies. In 1998 a palimpsest of Archimedes' work was sold at auction for $2 million (see Plate 6).

From this list we can see the versatility of Archimedes. His treatises on the equilibrium of planes and floating bodies contain principles that are now fundamental in mechanics and hydrostatics. The works on the quadrature of the parabola, conoids and spheroids, the measurement of the circle, and the sphere and cylinder extend the theory of proportion, area, and volume found in Euclid for polyhedra and polygons to the more complicated figures bounded by curved lines and surfaces. The work on spirals introduces a new class of curves, and develops the theory of length, area, and proportion for them.

Since we do not have space to discuss all of Archimedes' geometry, we shall confine our discussion to what may be his greatest achievement: finding the surface area of a sphere. In addition, because of its impact on the issues involving proof that we have been discussing, we shall discuss his *Method.*

3.1. The area of a sphere. Archimedes' two works on the sphere and cylinder were sent to Dositheus. In the letter accompanying the first of these he gives some of the history of the problem. Archimedes considered his results on the sphere to be rigorously established, but he did have one regret:

> It would have been beneficial to publish these results when Conon was alive, for he is the one we regard as most capable of understanding and rendering a proper judgment on them. But, as we think it well to communicate them to the initiates of mathematics, we send them to you, having rewritten the proofs, which those versed in the sciences may scrutinize.

The fact that a pyramid is one-third of a prism on the same base and altitude is Proposition 7 of Book 12 of Euclid's *Elements*. Thus Archimedes could say confidently that this theorem was well established. Archimedes sought the surface area of a sphere by finding the lateral surface area of a frustum of a cone and the lateral area of a right cylinder. In our terms the area of a frustum of a cone with upper radius r, lower radius R, and side of slant height h is $\pi h(R+r)$. Archimedes phrased this fact by saying that the area is that of a circle whose radius is the mean proportional between the slant height and the sum of the two radii; that is, the radius is $\sqrt{h(R+r)}$. Likewise, our formula for the lateral surface area of a cylinder of radius r and height h is $2\pi rh$. Archimedes said it was the area of a circle whose radius is the mean proportional between the diameter and height of the cylinder.

These results can be applied to the figures generated by revolving a circle about a diameter with certain chords drawn. Archimedes showed (Proposition 22) that

$$(BB' + CC' + \cdots + KK' + LM) : AM = A'B : BA$$

in Fig. 18.

This result is easily derived by connecting B' to C, C' to K, and K' to L and considering the ratios of the legs of the resulting similar triangles. These ratios can be added. All that then remains is to cross-multiply this proportion and use the expressions already derived for the area of a frustum of a cone. One finds easily

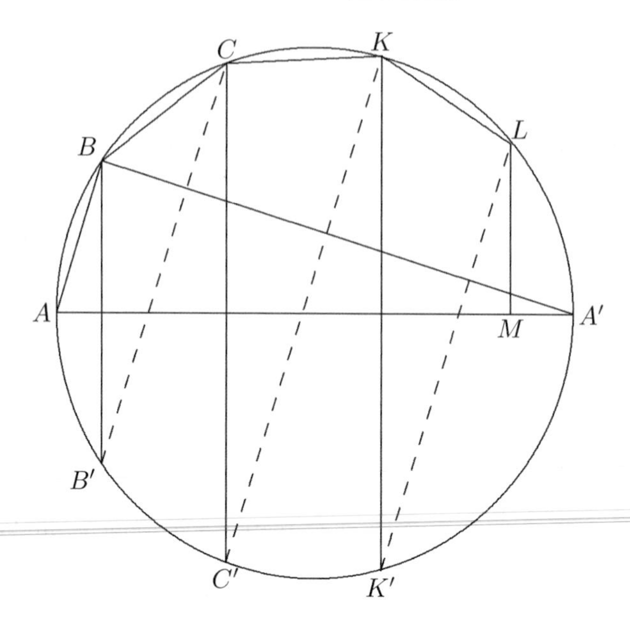

FIGURE 18. Finding the surface area of a sphere.

that the area of the surface obtained by revolving the broken line $ABCKL$ about the axis AA' is $\pi AM \cdot A'B$. The method of exhaustion then shows that the product $AM \cdot A'B$ can be made arbitrarily close to the square of AA'; it therefore gives the following result (Proposition 33): *The surface of any sphere is equal to four times the greatest circle in it.*

By the same method, using the inscribed right circular cone with the equatorial circle of the sphere as a base, Archimedes shows that the volume of the sphere is four times the volume of this cone. He then obtains the relations between the areas and volumes of the sphere and circumscribed closed cylinder. He finishes this first treatise with results on the area and volume of a segment of a sphere, that is, the portion of a sphere cut off by a plane. This argument is the only ancient proof of the area and volume of a sphere that meets Euclidean standards of rigor.

Three remarks should be made on this proof. First, in view of the failure of efforts to square the circle, it seems that the later Greek mathematicians had two standard areas, the circle and the square. Archimedes expressed the area of a sphere in terms of the area of a circle. Second, as we have seen, the volume of a sphere was found in China several centuries after Archimedes' time, but the justification for it involved intuitive principles such as Cavalieri's principle that do not meet Euclidean standards. Third, Archimedes did not *discover* this theorem by Euclidean methods. He told how he came to discover it in his *Method.*

3.2. The *Method*. Early in the twentieth century the historian of mathematics J. L. Heiberg, reading in a bibliographical journal of 1899 the account of the discovery of a tenth-century manuscript with mathematical content, deduced from a few quotations that the manuscript was a copy[20] of a work of Archimedes. In

[20] A copy, not Archimedes' own words, since it was written in the Attic dialect, while Archimedes wrote in Doric. It is interesting that in the statement of his first theorem Archimedes refers to a "section of a right-angled cone $AB\Gamma$," and then immediately in the proof says, "since $AB\Gamma$ is a

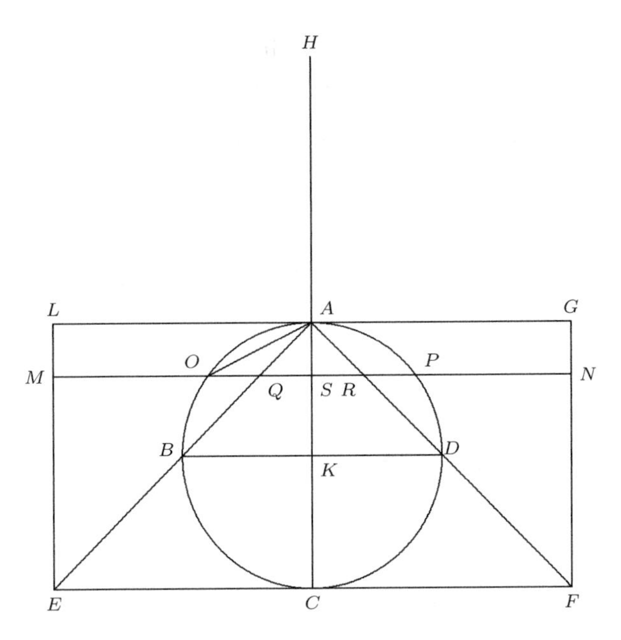

FIGURE 19. Volumes of sphere, cone, and cylinder.

1906 and 1908 he journeyed to Constantinople and established the text, as far as was possible. Attempts had been made to wash off the mathematical text during the Middle Ages so that the parchment could be used to write a book of prayers. The 177 pages of this manuscript contain parts of the works just discussed and a work called *Method.* The existence of such a work had been known because of the writings of commentators on Archimedes.

There are quotations from the *Method* in a work of the mathematician Heron called the *Metrica*, which was discovered in 1903. The *Method* had been sent to the astronomer Eratosthenes as a follow-up to a previous letter that had contained the statements of two theorems without proofs and a challenge to discover the proofs. Both of the theorems involve the volume and surface of solids of revolution. In contrast to his other work on this subject, however, Archimedes here makes free use of the principle now commonly known as *Cavalieri's principle*, which we mentioned in connection with the Chinese computation of the volume of a sphere. Archimedes' *Method* is a refinement of this principle, obtained by imagining the sections of a region balanced about a fulcrum. The reasoning is that if each pair of corresponding sections balance at distances a and b, then the bodies themselves will balance at these distances, and therefore, by Archimedes' principle of the lever, the area or volume of the two bodies must be have the ratio $b : a$.

The volume of a sphere is four times the volume of the cone with base equal to a great circle of the sphere and height equal to its radius, and the cylinder with base equal to a great circle of the sphere and height equal to the diameter is half again as large as the sphere.

Archimedes' proof is based on Fig. 19. If this figure is revolved about the line CAH, the circle with center at K generates a sphere, the triangle AEF generates

parabola...". If these were the original words, it appears that the nomenclature for conic sections was changing in Archimedes' time.

a cone, the rectangle $LGFE$ generates a cylinder, and each horizontal line such as MN generates a disk. The point A is the midpoint of CH. Archimedes shows that the area of the disk generated by revolving QR plus the area of the disk generated by revolving OP has the same ratio to the area of the disk generated by revolving MN that AS has to AH. It follows from his work on the equilibrium of planes that if the first two of these disks are hung at H, they will balance the third disk about A as a fulcrum. Archimedes concluded that the sphere and cone together placed with their centers of gravity at H would balance (about the point A) the cylinder, whose center of gravity is at K.

Therefore,

$$HA : AK = (\text{cylinder}) : (\text{sphere} + \text{cone}).$$

But $HA = 2AK$. Therefore, the cylinder equals twice the sum of the sphere and the cone AEF. And since it is known that the cylinder is three times the cone AEF, it follows that the cone AEF is twice the sphere. But since $EF = 2BD$, cone AEF is eight times cone ABD, and the sphere is four times the cone ABD.

From this fact Archimedes easily deduces the famous result allegedly depicted on his tombstone: *The cylinder circumscribed about a sphere equals the volume of the sphere plus the volume of a right circular cone inscribed in the cylinder.*

Having concluded the demonstration, Archimedes reveals that this method enabled him to discover the area of a sphere. He writes

> For I realized that just as every circle equals a triangle having as its base the circumference of the circle and altitude equal to the [distance] from the center to the circle [that is, the radius], in the same way every sphere is equal to a cone having as its base the surface [area] of the sphere and altitude equal to the [distance] from the center to the sphere.

The *Method* gives an inside view of the route by which Archimedes discovered his results. The method of exhaustion is convincing as a method of proving a theorem, but useless as a way of discovering it. The *Method* shows us Archimedes' route to discovery.

4. Apollonius

From what we have already seen of Greek geometry we can understand how the study of the conic sections came to seem important. From commentators like Pappus we know of treatises on the subject by Aristaeus, a contemporary of Euclid who is said to have written a book on *Solid Loci*, and by Euclid himself. We have also just seen that Archimedes devoted a great deal of attention to the conic sections. The only treatise on the subject that has survived, however, is that of Apollonius, and even for this work, unfortunately, no faithful translation into English exists. The version most accessible is that of Heath, who says in his preface that writing his translation involved "the substitution of a new and uniform notation, the condensation of some propositions, the combination of two or more into one, some slight re-arrangements of order for the purpose of bringing together kindred propositions in cases where their separation was rather a matter of accident than indicative of design, and so on." He might also have mentioned that he supplemented Apollonius' purely synthetic methods with analytic arguments, based on the algebraic notation we are familiar with. All this labor has no doubt made Apollonius more readable. On the other hand, Apollonius' work is no longer current research, and from the historian's point of view this kind of tinkering with the text only makes it harder to place the work in proper perspective.

In contrast to his older contemporary Archimedes, Apollonius remains a rather obscure figure. His dates can be determined from the commentary written on the *Conics* by Eutocius. Eutocius says that Apollonius lived in the time of the king Ptolemy Euergetes and defends him against a charge by Archimedes' biographer Heracleides that Apollonius plagiarized results of Archimedes. Eutocius' information places Apollonius reliably in the second half of the third century BCE, perhaps a generation or so younger than Archimedes.

Pappus says that as a young man Apollonius studied at Alexandria, where he made the acquaintance of a certain Eudemus. It is probably this Eudemus to whom Apollonius addresses himself in the preface to Book 1 of his treatise. From Apollonius' own words we know that he had been in Alexandria and in Perga, which had a library that rivaled the one in Alexandria. Eutocius reports an earlier writer, Geminus by name, as saying that Apollonius was called "the great geometer" by his contemporaries. He was highly esteemed as a mathematician by later mathematicians, as the quotations from his works by Ptolemy and Pappus attest. In Book 12 of the *Almagest*, Ptolemy attributes to Apollonius a geometric construction for locating the point at which a planet begins to undergo retrograde motion. From these later mathematicians we know the names of several works by Apollonius and have some idea of their contents. However, only two of his works survive to this day, and for them we are indebted to the Islamic mathematicians who continued to work on the problems that Apollonius considered important. Our present knowledge of Apollonius' *Cutting Off of a Ratio*, which contains geometric problems solvable by the methods of application with defect and excess, is based on an Arabic manuscript, no Greek manuscripts having survived. Of the eight books of Apollonius' *Conics*, only seven have survived in Arabic and only four in Greek.

4.1. History of the *Conics*. The evolution of the *Conics* was reported by Pappus five centuries after they were written in Book 7 of his *Collection*.

> By filling out Euclid's four books on the conics and adding four others Apollonius produced eight books on the conics. Aristaeus... and all those before Apollonius, called the three conic curves sections of acute-angled, right-angled, and obtuse-angled cones. But since all three curves can be produced by cutting any of these three cones, as Apollonius seems to have objected, [noting] that some others before him had discovered that what was called a section of an acute-angled cone could also be [a section of] a right- or obtuse-angled cone... changing the nomenclature, he named the so-called acute section an ellipse, the right section a parabola, and the obtuse section a hyperbola.

As already mentioned, the first four books of Apollonius' *Conics* survived in Greek, and seven of the eight books have survived in Arabic; the astronomer Edmund Halley (1656–1743) published a Latin edition of them in 1710.

4.2. Contents of the *Conics*. In a preface addressed to the aforementioned Eudemus, Apollonius lists the important results of his work: the description of the sections, the properties of the figures relating to their diameters, axes, and asymptotes, things necessary for analyzing problems to see what data permit a solution, and the three- and four-line locus. He continues:

> The third book contains many remarkable theorems of use for the construction of solid loci and for distinguishing when problems have a solution, of which the greatest part and the most beautiful are new. And when we had grasped these, we knew that the three-line and four-line locus had not been constructed by Euclid, but only a chance part of it and that not very happily. For it was not possible for this construction to be completed without the additional things found by us.

We have space to discuss only the definition and construction of the conic sections and the four-line locus problem, which Apollonius mentions in the passage just quoted.

4.3. Apollonius' definition of the conic sections. The earlier use of conic sections had been restricted to cutting cones with a plane perpendicular to a generator. As we saw in our earlier discussion, this kind of section is easy to analyze and convenient in the applications for which it was intended. In fact, only one kind of hyperbola, the rectangular, is needed for duplicating the cube and trisecting the angle. The properties of a general section of a general cone were not discussed. Also, it was considered a demerit that the properties of these planar curves had to be derived from three-dimensional figures. Apollonius set out to remove these gaps in the theory.

First it was necessary to define a cone as the figure generated by moving a line around a circle while one of its points, called the *apex* and lying outside the plane of the circle, remains fixed. Next, it was necessary to classify all the sections of a cone that happen to be circles. Obviously, those sections include all sections by planes parallel to the plane of the generating circle (Book 1, Proposition 4). Surprisingly, there is another class of sections that are also circles, called *subcontrary* sections. Once the circles are excluded, the remaining sections must be parabolas, hyperbolas, and ellipses. We have space only to consider Apollonius' construction of the ellipse. His construction of the other conics is very similar. Consider the planar section of a cone in Fig. 20, which cuts all the generators of the cone on the same side of its apex. This condition is equivalent to saying that the cutting intersects both sides of the axial triangle. Apollonius proved that there is a certain line, which he called *the* [up]*right side*, now known by its Latin name *latus rectum*, such that the square on the ordinate from any point of the section to its axis equals the rectangle applied to the portion of the axis cut off by this ordinate (the abscissa) and whose defect on the axis is similar to the rectangle formed by the axis and the latus rectum. He gave a rule, too complicated to go into here, for constructing the latus rectum. This line characterized the shape of the curve. Because of its connection with the problem of application with defect, he called the resulting conic section an *ellipse*. Similar connections with the problems of application and application with excess respectively arise in Apollonius' construction of the parabola and hyperbola. These connections motivated the names he gave to these curves.

In Fig. 20 the latus rectum is the line EH, and the locus condition is that the square on LM equal the rectangle on EO and EM; that is, $LM^2 = EO \cdot EM$.

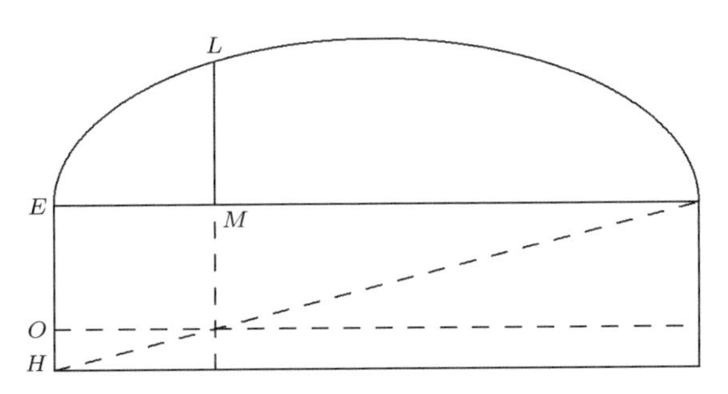

FIGURE 20. Apollonius' construction of the ellipse.

In one sense, this locus definition for an ellipse is not far removed from what we now think of as the equation of the ellipse, but that small gap was unbridgeable in Apollonius' time. If we write $LM = y$ and $EM = x$ in Fig. 20 (so that we are essentially taking rectangular coordinates with origin at E), we see that Apollonius is claiming that $y^2 = x \cdot EO$. Now, however, $EO = EH - OH$, and EH is constant, while OH is directly proportional to EM, that is, to x. Specifically, the ratio of OH to EM is the same as the ratio of EH to the axis. Thus, if we write $OH = kx$—the one, crucial step that Apollonius could not take, since he did not have the concept of a dimensionless constant of proportionality—and denote the latus rectum EH by C, we find that Apollonius' locus condition can be stated as the equation $y^2 = Cx - kx^2$. By completing the square on x, transposing terms, and dividing by the constant term, we can bring this equation into what we now call the standard form for an ellipse with center at $(a, 0)$:

$$(2) \qquad \frac{(x-a)^2}{a^2} + \frac{y^2}{b^2} = 1 ,$$

where $a = C/(2k)$ and $b = C\sqrt{k}$. In these terms the latus rectum C is $2b^2/a$. Apollonius, however, did *not* have the concept of an equation nor the symbolic algebraic notation we now use, and if he did have, he would still have needed the letter k used above as a constant of proportionality. These "missing" pieces gave his work on conics a ponderous character with which most mathematicians today have little patience.

Apollonius' constructions of the parabola and hyperbola also depend on the latus rectum. He was the first to take account of the fact that a plane that produces a hyperbola must cut both nappes of the cone. He regarded the two branches as two hyperbolas, referring to them as "opposites" and reserving the term *hyperbola* for either branch. For the hyperbola Apollonius proves the existence of *asymptotes*, that is, a pair of lines through the center that never meet the hyperbola but such that any line through the center passing into the region containing the hyperbola does meet the hyperbola. The word *asymptote* means literally *not falling together*, that is, not intersecting.

Books 1 and 2 of the *Conics* are occupied with finding the proportions among line segments cut off by chords and tangents on conic sections, the analogs of results on circles in Books 3 and 4 of Euclid. These constructions involve finding the tangents to the curves satisfying various supplementary conditions such as being parallel to a given line.

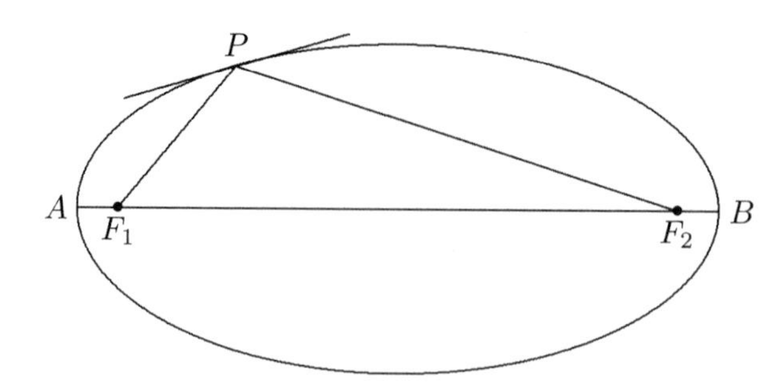

FIGURE 21. Focal properties of an ellipse.

4.4. Foci and the three- and four-line locus. We are nowadays accustomed to constructing the conic sections using the focus–directrix property, so that it comes as a surprise that the original expert on the subject does not seem to recognize the importance of the foci. He never mentions the focus of a parabola, and for the ellipse and hyperbola he refers to these points only as "the points arising out of the application." The "application" he has in mind is explained in Book 3. Propositions 48 and 52 read as follows:

(Proposition 48) *If in an ellipse a rectangle equal to the fourth part of the figure is applied from both sides to the major axis and deficient by a square figure, and from the points resulting from the application straight lines are drawn to the ellipse, the lines will make equal angles with the tangent at that point.*

(Proposition 52) *If in an ellipse a rectangle equal to the fourth part of the figure is applied from both sides to the major axis and deficient by a square figure, and from the points resulting from the application straight lines are drawn to the ellipse, the two lines will be equal to the axis.*

The "figure" referred to is the rectangle whose sides are the major axis of the ellipse and the latus rectum. In Fig. 21 the points F_1 and F_2 must be chosen on the major axis AB so that $AF_1 \cdot F_1B$ and $AF_2 \cdot BF_2$ both equal one-fourth of the area of the rectangle formed by the axis AB and the latus rectum. Proposition 48 expresses the focal property of these two points: Any ray of light emanating from one will be reflected to the other. Proposition 52 is the *string property* that characterizes the ellipse as the locus of points such that the sum of the distances to the foci is constant. These are just two of the theorems Apollonius called "strange and beautiful." Apollonius makes little use of these properties, however, and does not discuss the use of the string property to draw an ellipse.

A very influential part of the *Conics* consists of Propositions 54–56 of Book 3, which contain the theorems that Apollonius claimed (in his cover letter) would provide a solution to the three- and four-line locus problems. Both in their own time and because of their subsequent influence, the three- and four-line locus problems have been of great importance for the development of mathematics. These propositions involve the proportions in pieces of chords inscribed in a conic section. Three propositions are needed because the hyperbola requires two separate statements to cover the cases when the points involved lie on different branches of the hyperbola.

Proposition 54 asserts that given a chord $A\Gamma$ such that the tangents at the endpoints meet at Δ, and the line from Δ to the midpoint E of the chord meets the conic at B, any point Θ on the conic has the following property (Fig. 22). *The*

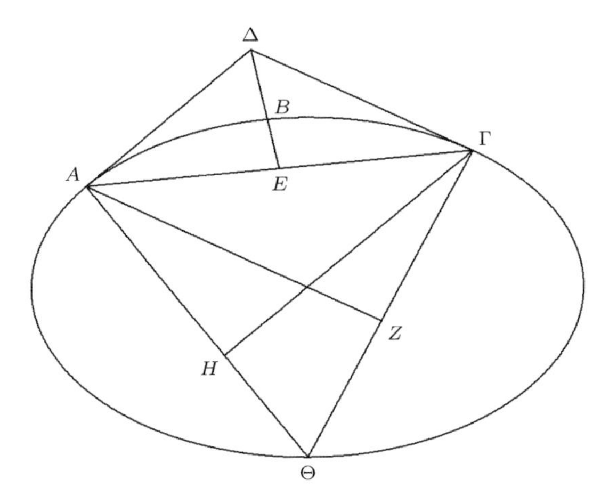

FIGURE 22. The basis for solving the four-line locus problem.

lines from Θ to A and Γ meet the lines through Γ and A respectively, each parallel to the tangent through the other endpoint, in points H and Z respectively such that the following proportion holds: The ratio of the rectangle on AZ and ΓH to the square on $A\Gamma$ is the composite of the ratio of the square on EB to the square on $B\Delta$ and the ratio of the rectangle on $A\Delta$ and $\Delta\Gamma$ to the rectangle on AE and $E\Gamma$. As we would write this relation,

$$AZ \cdot \Gamma H : A\Gamma^2 :: EB^2 : B\Delta^2 . A\Delta \cdot \Delta\Gamma : AE \cdot E\Gamma\,.$$

It is noteworthy that this theorem involves an expression that seldom occurs in Greek mathematics: a composition of ratios involving squares. If we thought of it in our terms, we would say that Apollonius was working in four dimensions. But he wasn't. The composition of two ratios is possible only when both ratios are between quantities of the same type, in this case areas. The proof given by Apollonius has little in common with the way we would proceed nowadays, by multiplying and dividing. Apollonius had to find an area C such that $AZ \cdot \Gamma H : C :: EB^2 : B\Delta^2$ and $C : A\Gamma^2 :: A\Delta \cdot \Delta\Gamma : ZE \cdot E\Gamma$. He could then "cancel" C in accordance with the definition of compound ratio.

It is not at all obvious how this proposition makes it possible to solve the four-line locus, and Apollonius does not fill in the details. We shall not attempt to do so, either. To avoid excessive complexity, we merely state the four-line locus problem and illustrate it. The data for the problem are four lines, which for definiteness we suppose to intersect two at a time, and four given angles, one corresponding to each line. The problem requires the locus of points P such that if lines are drawn from P to the four lines, each making the corresponding angle with the given line (for simplicity all shown as right angles in Fig. 23), the rectangle on two of the lines will have a constant ratio to the rectangle on the other two. The solution is in general a pair of conics.

The origin of this kind of problem may lie in the problem of two mean proportionals, which was solved by drawing fixed reference lines and finding the loci of points satisfying two conditions resembling this. The square on the line drawn perpendicular to one reference line equals the rectangle on a fixed line and the line drawn to the other reference line. The commentary on this problem by Pappus, who mentioned that Apollonius had left a great deal unfinished in this area, inspired Fermat and Descartes to take up the implied challenge and solve the problem completely. Descartes offered his success in solving the locus problem to any number of lines as proof of the value of his geometric methods.

Questions and problems

10.1. Show how it would be possible to compute the distance from the center of a square pyramid to the tip of its shadow without entering the pyramid, after first driving a stake into the ground at the point where the shadow tip was located at the moment when vertical poles cast shadows equal to their length.

10.2. Describe a mechanical device to draw the quadratrix of Hippias. You need a smaller circle of radius $2/\pi$ times the radius that is rotating, so that you can use it to wind up a string attached to the moving line; or conversely, you need the rotating radius to be $\pi/2$ times the radius of the circle pulling the line. How could you get such a pair of circles?

10.3. Prove that the problem of constructing a rectangle of prescribed area on part of a given base a in such a way that the defect is a square is precisely the problem of finding two numbers given their sum and product (the two numbers are the lengths of the sides of the rectangle). Similarly, prove that the problem of application with square excess is precisely the problem of finding two numbers (lengths) given their difference and product.

10.4. Show that the problem of application with square excess has a solution for any given area and any given base. What restrictions are needed on the area and base in order for the problem of application with square defect to have a solution?

10.5. Use an argument similar to the argument in Chapter 8 showing that the side and diagonal of a pentagon are incommensurable to show that the side and diagonal of a square are incommensurable. That is, show that the Euclidean algorithm, when applied to the diagonal and side of a square, requires only two steps to produce the side and diagonal of a smaller square, and hence can never produce an equal pair. To do so, refer to Fig. 24.

In this figure $AB = BC$, angle ABC is a right angle, AD is the bisector of angle CAB, and DE is drawn perpendicular to AC. Prove that $BD = DE$, $DE = EC$, and $AB = AE$. Then show that the Euclidean algorithm starting with the pair (AC, AB) leads first to the pair $(AB, EC) = (BC, BD)$, and then to the pair $(CD, BD) = (CD, DE)$, and these last two are the diagonal and side of a square.

10.6. It was stated above that Thales might have used the Pythagorean theorem in order to calculate the distance from the center of the Great Pyramid to the tip of its shadow. How could this distance be computed without the Pythagorean theorem?

10.7. State the paradoxes of Zeno in your own words and tell how you would have advised the Pythagoreans to modify their system in order to avoid these paradoxes.

10.8. Do we share any of the Pythagorean mysticism about geometric shapes that Proclus mentioned? Think of the way in which we refer to an honorable person as *upright*, or speak of getting a *square deal*, while a person who cheats is said to be *crooked*. Are there other geometric images in our speech that have ethical connotations?

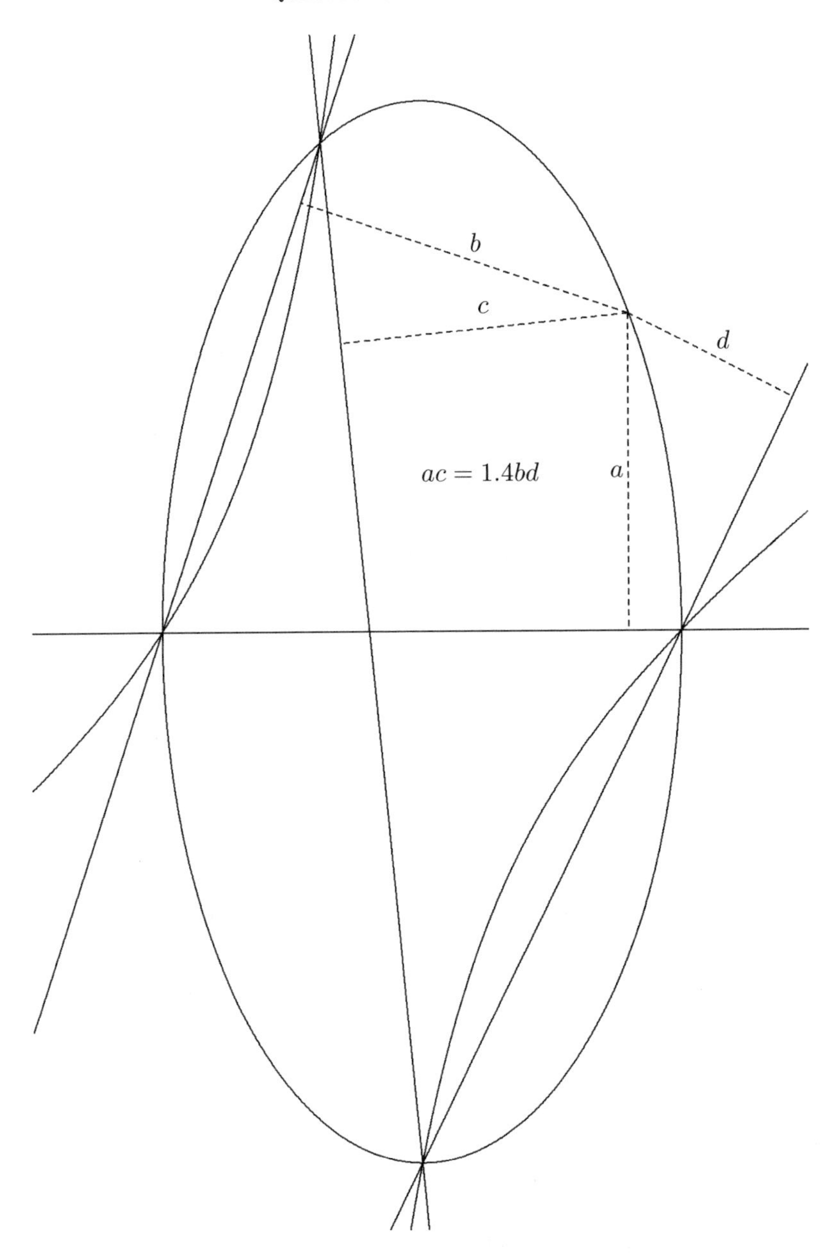

FIGURE 23. The four-line locus. If a point moves so that the product of its distances to two lines bears a constant ratio to the product of its distances to two other lines, it must move in a conic. In this illustration, two conics satisfy the condition: one an ellipse, the other a hyperbola.

10.9. In the Pythagorean tradition there were two kinds of mathematical activity. One kind, represented by the attempt to extend the theory of the transformation of polygons to circles and solid figures, is an attempt to discover new facts and enlarge the sphere of mathematics—to generalize. The other, represented by the discovery

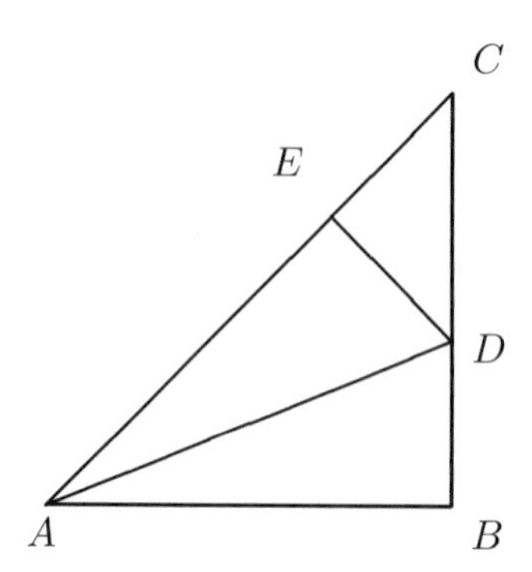

FIGURE 24. Diagonal and side of a square.

of incommensurables, is an attempt to bring into sharper focus the theorems already proved and to test the underlying assumptions of a theory—to rigorize. Are these kinds of activity complementary, opposed, or simply unrelated to each other?

10.10. Hippocrates' quadrature of a lune used the fact that the areas of circles are proportional to the squares on their radii. Could Hippocrates have known this fact? Could he have proved it?

10.11. Plato apparently refers to the famous 3–4–5 right triangle in the *Republic*, 546*c*. Proclus alludes to this passage in a discussion of right triangles with commensurable sides. We can formulate the recipes that Proclus attributes to Pythagoras and Plato respectively as

$$(2n+1)^2 + (2n^2+2n)^2 = (2n^2+2n+1)^2$$

and

$$(2n)^2 + (n^2-1)^2 = (n^2+1)^2.$$

Considering that Euclid's treatise is regarded as a compendium of Pythagorean mathematics, why is this topic not discussed? In which book of the *Elements* would it belong?

10.12. Proposition 14 of Book 2 of Euclid shows how to construct a square equal in area to a rectangle. Since this construction is logically equivalent to constructing the mean proportional between two line segments, why does Euclid wait until Book 6, Proposition 13 to give the construction of the mean proportional?

10.13. Show that the problem of squaring the circle is equivalent to the problem of squaring one segment of a circle when the central angle subtended by the segment is known. (Knowing a central angle means having two line segments whose ratio is the same as the ratio of the angle to a full revolution.)

10.14. Referring to Fig. 18, show that all the right triangles in the figure formed by connecting B' with C, C' with K, and K' with L are similar. Write down a string of equal ratios (of their legs). Then add all the numerators and denominators to deduce the equation

$$(BB' + CC' + \cdots + KK' + LM) : AM = A'B : BA.$$

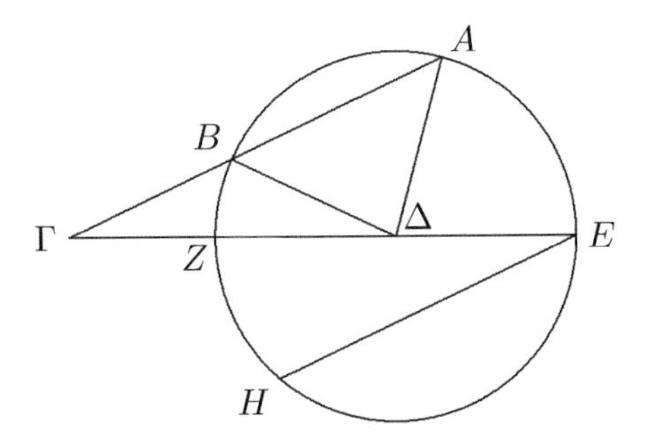

FIGURE 25. Archimedes' trisection of an angle: $\angle A\Gamma\Delta = \frac{1}{3}\angle A\Delta E$.

10.15. Show that Archimedes' result on the relative volumes of the sphere, cylinder, and cone can be obtained by considering the cylinder, sphere and double-napped cone formed by revolving a circle inscribed in a square about a midline of the square, the cone being generated by the diagonals of the square. In this case the area of a circular section of the cone plus the area of the same section of the sphere equals the area of the section of the cylinder since the three radii form the sides of a right triangle. The radius of a section of the sphere cuts off a segment of the axis of rotation from the center equal to the radius of the section of the cone, since the vertex angle of the cone is a right angle. These two segments form the legs of a right triangle whose hypotenuse is a radius of the sphere, which is equal to the radius of the section of the cylinder.

10.16. A minor work attributed to Archimedes called the *Book of Lemmas* contains an angle trisection. In Fig. 25 we are given an acute angle $\angle A\Delta E$, whose trisection is required. We draw a circle of any radius r about Δ, the vertex of the angle. Then, using a straightedge, we mark off on it two points P and Q separated by the distance r. Setting the straightedge down so that P is at point Γ on the extension of the diameter $E\Delta Z$, Q is at point B on the circle, and the point A is also on the edge of the straightedge, we draw the line $A\Gamma$. By drawing EH parallel to $A\Gamma$, we get $\angle A\Gamma E = \angle\Gamma EH$. By joining ΔB, we obtain the isosceles triangle $\Gamma B\Delta$. Now since $\angle B\Delta Z$ is a central angle on the arc $\overset{\frown}{BZ}$ and is equal to $\angle B\Gamma\Delta$, which is equal to $\angle ZEH$, which is inscribed in the arc $\overset{\frown}{ZH}$, it follows that $\overset{\frown}{ZH} = 2\,\overset{\frown}{BZ}$. Since the arcs $\overset{\frown}{AE}$ and $\overset{\frown}{BH}$ are equal (being cut off by parallel chords), we now get $\overset{\frown}{AE} = \overset{\frown}{BH} = 3\,\overset{\frown}{BZ}$. Therefore, $\angle A\Gamma E = \angle B\Delta Z = \frac{1}{3}\angle A\Delta E$.

Why is this construction *not* a straightedge-and-compass trisection of the angle, which is known to be impossible? How does it compare with the *neûsis* trisection shown above? Show how to obtain this same result more simply by erasing everything in the figure below the diameter of the circle.

10.17. Show that the problem of increasing the size of a sphere by half is equivalent to the problem of two mean proportionals (doubling the cube).

10.18. A circle can be regarded as a special case of an ellipse. What is the *latus rectum* of a circle?

10.19. When the equation $y^2 = Cx - kx^2$ is converted to the standard form

$$\frac{(x-h)^2}{a^2} + \frac{y^2}{b^2} = 1,$$

what are the quantities h, a, and b in terms of C and k?

10.20. Show from Apollonius' definition of the foci that the product of the distances from each focus to the ends of the major axis of an ellipse equals the square on half of the minor axis.

10.21. We have seen that the three- and four-line locus problems have conic sections as their solutions. State and solve the two-line locus problem. You may use modern analytic geometry and assume that the two lines are the x axis and the line $y = ax$. The locus is the set of points whose distances to these two lines have a given ratio. What curve is this?

10.22. Show that the apparent generality of Apollonius' statement of the three-line locus problem, in which arbitrary angles can be prescribed at which lines are drawn from the locus to the fixed lines, is illusory. (To do this, show that the ratio of a line from a point P to line l making a fixed angle θ with the line l bears a constant ratio to the line segment from P perpendicular to l. Hence if the problem is solved for all ratios in the special case when lines are drawn from the locus perpendicular to the given lines, then it is solved for all ratios in any case.)

10.23. Show that the line segment from a point $P = (x, y)$ to a line $ax + by = c$ making angle θ with the line has length

$$\frac{|ax + by - c|}{\sqrt{a^2 + b^2}\sin\theta}.$$

Use this expression and three given lines $l_i : a_i x + b_i y = c_i$, $i = 1, 2, 3$, to formulate the three-line locus problem analytically as a quadratic equation in two variables by setting the square of the distance from (x, y) to line l_1 equal to a constant multiple of the product of the distances to l_2 and l_3. Show that the locus passes through the intersection of the line l_1 with l_2 and l_3, but not through the intersection of l_2 with l_3. Also show that its tangent line where it intersects l_i is l_i itself, $i = 2, 3$.

10.24. One reason for doubting Cavalieri's principle is that it breaks down in one dimension. Consider, for instance, that every section of a right triangle parallel to one of its legs meets the other leg and the hypotenuse in congruent figures (a single point in each case). Yet the other leg and the hypotenuse are obviously of different lengths. Is there a way of redefining "sections" for one-dimensional figures so that Cavalieri's principle can be retained? If you could do this, would your confidence in the validity of the principle be restored?

10.25. We know that interest in conic sections *arose* because of their application to the problem of two mean proportionals (doubling the cube). Why do you think interest in them was *sustained* to the extent that caused Euclid, Aristaeus, and Apollonius to write treatises developing their properties in such detail?

10.26. Pappus' history of the conics implies that people knew that the ellipse, for example, could be obtained by cutting a right-angled cone with a plane. Can *every* ellipse be obtained by cutting a right-angled cone with a plane? Prove that it can, by showing that any a and b whatsoever in Eq. 2 can be obtained as the section of the right-angled cone whose equation is $y^2 = zx$ by the plane $x = 2a - (a^2z/b^2)$. Then show that by taking $a = eu/(1 - e^2)$, $b = a\sqrt{1 - e^2}$, $x = w$, $y = v$, where $e = h/w$, you get Eq. 1. [*Hint*: Recall that e is constant in a given conic section. Also, observe that $0 < e < 1$ for a section of an acute-angle cone, since $h = w\tan(\theta/2)$, where θ is the vertex angle of the cone.]

10.27. As we have seen, Apollonius was aware of the string property of ellipses, yet he did not mention that this property could be used to draw an ellipse. Do you think that he did not *notice* this fact, or did he omit to mention it because he considered it unimportant?

10.28. Prove Proposition 54 of Book 3 of Apollonius' *Conics* in the special case in which the conic is a circle and the point Θ is at the opposite end of the diameter from B (Fig. 22).

CHAPTER 11

Post-Euclidean Geometry

A certain dullness came over Greek geometry from the beginning of the second century BCE. The preceding century had seen the beginning of Roman expansion, whose early stage took the life of the aged Archimedes. Julius Caesar (100–44 BCE), who did more than anyone before him to turn Rome from a republic into an empire, took an army to Egypt to fight his rival Pompey and incidentally help Cleopatra, the last of the heirs of Ptolemy Soter, defeat her brother in a civil war. In pursuing his aim he sent fire ships into the harbor of Alexandria to set it ablaze. Although he himself naturally says nothing about any destruction of the city, later writers, such as Plutarch in his *Life of Caesar* and Gellius in his *Attic Nights*, say that the fire damaged the Library. Gellius claims that 700,000 books were destroyed. After Caesar's heir Octavian defeated Mark Antony and Cleopatra in 31 BCE, Egypt became a province of the Roman Empire. Whether because of this disruption or from limitations inherent in the Pythagorean philosophy, the level of brilliant achievements of Euclid, Archimedes, and Apollonius was not sustained. Nevertheless, geometry did not die out entirely, and some of the later commentators are well worth reading. Very little new geometry was written in Greek after the sixth century, however. From the ninth century to the fifteenth the Euclidean tradition in geometry was pursued by Muslim mathematicians. Since these mathematicians were also interested in the philosophy of Aristotle, in their work mathematics once again began to be mixed with philosophy, as it was in the time just before Euclid.

When the Roman Empire was vigorous, all upper-class Romans understood Greek, and many seemed to prefer it to Latin. The Emperor Marcus Aurelius, for example, who ruled from 161 to 180, wrote his meditations in Greek. After the Emperor Diocletian (284–305) split the empire into eastern and western halves to make it governable and the eastern Emperor Constantine (307–337), who proclaimed Christianity the official religion of the Empire, moved his capital to Constantinople, knowledge of Greek began to decline in the western part of the empire. Many books were translated into Latin, or replacements for them were written in Latin.

The repeated ravaging of Italy by invaders from the north caused an irreversible decline in scholarship there. In the east, which fared somewhat better, scholarship continued for another thousand years, until the Turkish conquest of Constantinople in 1453. The eastern Emperor Justinian (525–565) managed to reassert his rule over part of Italy, but this project proved too expensive to sustain, and Italy was soon once again beyond the control of the Emperor. For several centuries before the reign of Justinian an entirely new civilization based on Christianity had been replacing the ancient Greco-Roman world, symbolically marked by the Justinian's closing of the pagan Academy at Athens in 529.

1. Hellenistic geometry

Although the Euclidean restrictions set limits to the growth of geometry, there remained people who attempted to push the limits beyond the achievements of Archimedes and Apollonius, and they produced some good work over the next few centuries. We shall look at just a few of them.

1.1. Zenodorus.

The astronomer Zenodorus lived in Athens in the century following Apollonius. Although his exact dates are not known, he is mentioned by Diocles in his book *On Burning Mirrors* and by Theon of Smyrna. According to Theon, Zenodorus wrote *On Isoperimetric Figures*, in which he proved four theorems: (1) If two regular polygons have the same perimeter, the one with the larger number of sides encloses the larger area; (2) a circle encloses a larger area than any regular polygon whose perimeter equals its circumference; (3) of all polygons with a given number of sides and perimeter, the regular polygon is the largest; (4) of all closed surfaces with a given area, the sphere encloses the largest volume. With the machinery inherited from Euclidean geometry, Zenodorus could not have hoped for any result more general than these. Let us examine his proof of the first two, as reported by Theon.

Referring to Fig. 1, let $AB\Gamma$ and ΔEZ be two regular polygons having the same perimeter, with $AB\Gamma$ having more sides than ΔEZ. Let H and Θ be the centers of these polygons, and draw the lines from the centers to two adjacent vertices and their midpoints, getting triangles $B\Gamma H$ and $EZ\Theta$ and the perpendicular bisectors of their bases HK and $\Theta\Lambda$. Then, since the two polygons have the same perimeter but $AB\Gamma$ has more sides, BK is shorter than $E\Lambda$. Mark off M on $E\Lambda$ so that $M\Lambda = BK$. Then if P is the common perimeter, we have $E\Lambda : P :: \angle E\Theta\Lambda : 4$ right angles and $P : BK :: 4$ right angles $: \angle BHK$. By composition, then $E\Lambda : BK :: \angle E\Theta\Lambda : \angle BHK$, and therefore $E\Lambda : M\Lambda :: \angle E\Theta\Lambda : \angle BHK$. But, Zenodorus claimed, the ratio $E\Lambda : M\Lambda$ is *larger* than the ratio $\angle E\Theta\Lambda : \angle M\Theta\Lambda$, asking to postpone the proof until later. Granting that lemma, he said, the ratio $\angle E\Theta\Lambda : \angle BHK$ will be larger than the ratio $\angle E\Theta\Lambda : \angle M\Theta\Lambda$, and therefore $\angle BHK$ is smaller than $\angle M\Theta\Lambda$. It then follows that the complementary angles $\angle HBK$ and $\angle\Theta M\Lambda$ satisfy the reverse inequality. Hence, copying $\angle HBK$ at M so that one side is along $M\Lambda$, we find that the other side intersects the extension of $\Lambda\Theta$ at a point N beyond Θ. Then, since triangles BHK and $MN\Lambda$ are congruent by angle–side–angle, it follows that $HK = N\Lambda > \Theta\Lambda$. But it is obvious that the areas of the two polygons are $\frac{1}{2}HK \cdot P$ and $\frac{1}{2}\Theta\Lambda \cdot P$, and therefore $AB\Gamma$ is the larger of the two.

The proof that the ratio $E\Lambda : M\Lambda$ is larger than the ratio $\angle E\Theta\Lambda : \angle M\Theta\Lambda$ was given by Euclid in his *Optics*, Proposition 8. But Theon does not cite Euclid in his quotation of Zenodorus. He gives the proof himself, implying that Zenodorus did likewise. The proof is shown on the top right in Fig. 1, where the circular arc ΞMN has been drawn through M with Θ as center. Since the ratio $\triangle E\Theta M$: sector $N\Theta M$ is larger than the ratio $\triangle M\Theta\Lambda$: sector $M\Theta\Xi$ (the first triangle is larger than its sector, the second is smaller), it follows, interchanging means, that $E\Theta M : M\Theta\Lambda >$ sector $N\Theta M$: sector $M\Theta\Xi$. But $E\Theta M : M\Theta\Lambda :: EM : M\Lambda$, since the two triangles have the same altitude measured from the base line $EM\Lambda$. And sector $N\Theta M$: sector $M\Theta\Xi :: \angle E\Theta M : \angle M\Theta\Lambda$. Therefore, $EM : M\Lambda$ is larger than the ratio $\angle E\Theta M : \angle M\Theta\Lambda$, and it then follows that $E\Lambda : M\Lambda$ is larger than $\angle E\Theta\Lambda : \angle M\Theta\Lambda$.

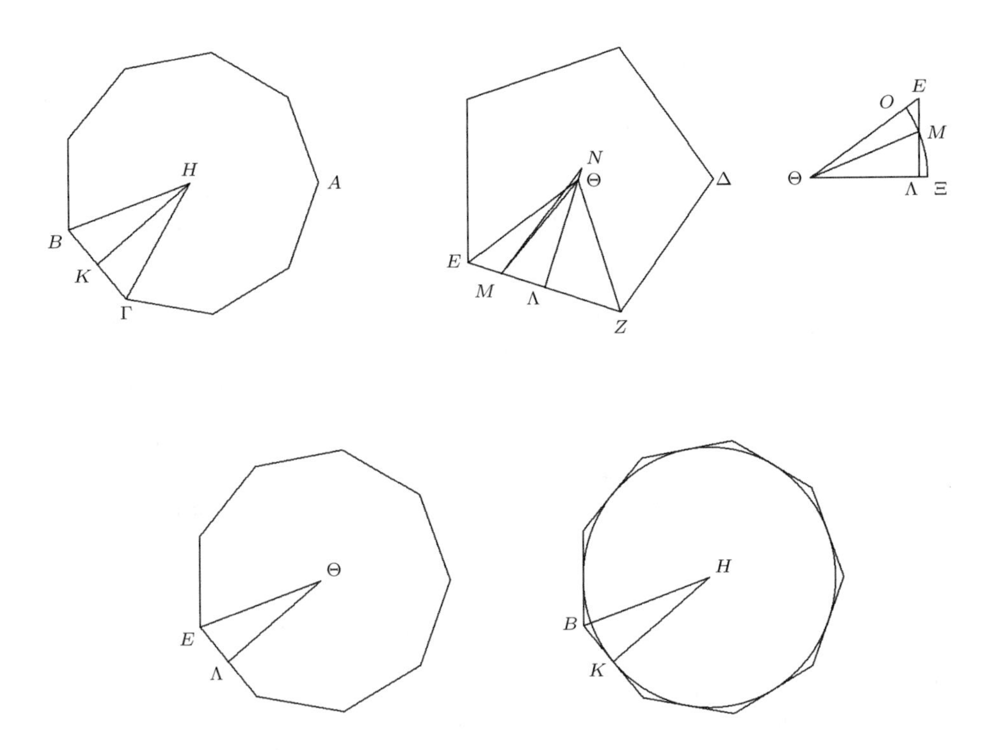

FIGURE 1. Two theorems of Zenodorus. Top: When two regular polygons have the same perimeter, the one with the larger number of sides is larger. Bottom: A circle is larger than a regular polygon whose perimeter equals the circumference of the circle.

Zenodorus' proof that a circle is larger than a regular polygon whose perimeter equals the circumference of the circle is shown at the bottom of Fig. 1. Given such a polygon and circle, circumscribe a similar polygon around the circle. Since this polygon is "convex on the outside," as Archimedes said in his treatise on the sphere and cylinder, it can be assumed longer than the circumference. (Both Archimedes and Zenodorus recognized that this was an assumption that they could not prove; Zenodorus cited Archimedes as having assumed this result.) That means the circumscribed polygon is larger than the original polygon since it has a larger perimeter. But then by similarity, HK is larger than $\Theta\Lambda$. Since the area of the circle equals half of the rectangle whose sides are its circumference and HK, while the area of the polygon is half of the rectangle whose sides are its perimeter and $\Theta\Lambda$, it follows that the circle is larger.

1.2. The parallel postulate. We saw in the Chapter 10 that there was a debate about the theory of parallel lines in Plato's Academy, as we infer from the writing of Aristotle. This debate was not ended by Euclid's decision to include a parallel postulate explicitly in the *Elements*. This foundational issue was discussed at length by the Stoic philosopher Geminus, whose dates are a subject of disagreement among experts, but who probably lived sometime between 50 BCE and 50 CE. Geminus wrote an encyclopedic work on mathematics, which has been entirely lost, except for certain passages quoted by Proclus, Eutocius, and others. Proclus said that the parallel postulate should be completely written out of the list of postulates, since it is really a theorem. The asymptotes of hyperbolas provided the model on which he reasoned that converging is not the same thing as intersecting. But still he thought that such behavior was impossible for straight lines. He claimed that a line that intersected one of two parallel lines must intersect the other,[1] and he reports a proof of Geminus that assumes in many places that certain lines drawn will intersect, not realizing that by doing so he was already assuming the parallel postulate.

Proclus also reports an attempt by Ptolemy to prove the postulate by arguing that a pair of lines could not be parallel on one side of a transversal "rather than" on the other side. (Proclus did not approve of this argument.) But of course the assumption that parallelism is two-sided is one of the properties of Euclidean geometry that does not extend to hyperbolic geometry. These early attempts to prove the parallel postulate began the process of unearthing more and more plausible alternatives to the postulate, but of course did not lead to a proof of it.

1.3. Heron. We have noted already the limitations of the Euclidean approach to geometry, the chief one being that lengths are simply represented as lines, not numbers. After Apollonius, however, the metric aspects of geometry began to resurface in the work of later writers. One of these writers was Heron (ca. 10–ca. 75), who wrote on mechanics; he probably lived in Alexandria. Pappus discusses his work at some length in Book 8 of his *Synagōgē*. Heron's geometry is much more concerned with measurement than was the geometry of Euclid. The change of interest in the direction of measurement and numerical procedures signaled by his *Metrica* is shown vividly by his repeated use (130 times, to be exact) of the word *area* (*embadón*), a word never once used by Euclid, Archimedes, or Apollonius.[2] There is a difference in point of view between saying that two plane figures are equal and saying that they have the same area. The first statement is geometrical and is the stronger of the two. The second is purely numerical and does not necessarily imply the first. Heron discusses ways of finding the areas of triangles from their sides. After giving several examples of triangles that are either integer-sided right triangles or can be decomposed into such triangles by an altitude, such as the triangle with sides of length 13, 14, 15, which is divided into a 5–12–13 triangle and a 9–12–15 triangle by the altitude to the side of length 14, he gives "a direct method by which the area of a triangle can be found without first finding its altitude." He

[1] This assertion is an *assumption* equivalent to the parallel postulate and obviously equivalent to the form of the postulate commonly used nowadays, known as Playfair's axiom: *Through a given point not on a line, only one parallel can be drawn to the line.*

[2] Reporting (in his commentary on Ptolemy's *Almagest*) on Archimedes' measurement of the circle, however, Theon of Alexandria did use this word to describe what Archimedes did; but that usage was anachronistic. In his work on the sphere, for example, Archimedes referred to its *surface* (*epipháneia*), not its *area*. On the other hand, Dijksterhuis (*1956*, pp. 412–413) reports the Arabic mathematician al-Biruni as having said that "Heron's formula" is really due to Archimedes. Considering the contrast in style between the proof and the applications, it does appear plausible that Heron learned the proof from Archimedes. Heath (*1921*, Vol. 2, p. 322) endorses this assertion unequivocally.

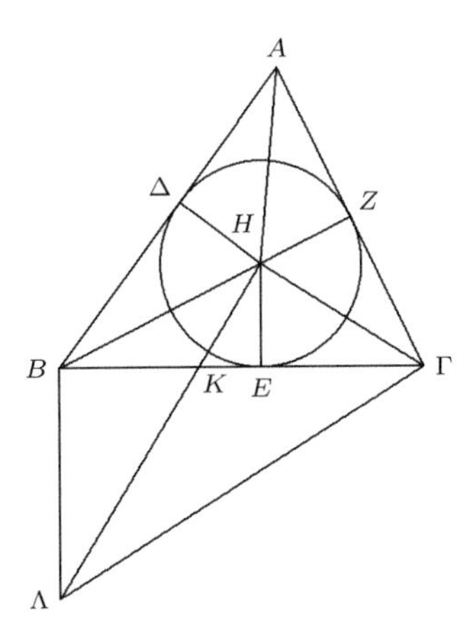

FIGURE 2. Heron's proof of his direct method of computing the area of a triangle.

gave as an example a triangle whose sides were 7, 8, and 9 units. His prescription was: Add 9 and 8 and 7, getting 24. Take half of this, getting 12. Subtract 7 units from this, leaving 5. Then subtract 8 from 12, leaving 4. Finally, subtract 9, leaving 3. Multiply 12 by 5, getting 60. Multiply this by 4, getting 240. Multiply this by 3, getting 720. Take the square root of this, and that will be the area of the triangle. He went on to explain that since 720 is not a square, it will be necessary to approximate, starting from the nearest square number, 729.

This result seems anomalous in Greek geometry, since Heron is talking about multiplying an area by an area. That is probably why he emphasizes that his results are numerical rather than geometric. An examination of his proof of the formula shows that he need not have multiplied two areas together. He must have made a deliberate choice to express himself this way. His proof is based on Fig. 2, in which one superfluous line has been omitted to streamline it. In the following proof, some rewording has been done to accommodate this minor modification of the figure.

The lines ΛB and ΛH are perpendicular respectively to $B\Gamma$ and $H\Gamma$. The proof follows easily once it is shown that the quadrilateral $\Lambda BH\Gamma$ is cyclic, that is, can be inscribed in a circle. In fact, if Σ denotes the semiperimeter, then

$$\Sigma^2 : \Sigma \cdot (\Sigma - B\Gamma) :: \Sigma : (\Sigma - B\Gamma) :: (\Sigma - A\Gamma) : KE :: (\Sigma - A\Gamma) \cdot \Gamma E : KE \cdot \Gamma E$$
$$= (\Sigma - A\Gamma) \cdot (\Sigma - AB) : EH^2 .$$

Here $KE \cdot \Gamma E = EH^2$ because EH is the altitude to the base of the right triangle $HK\Gamma$.

Heron *could have* stated the result in Euclidean language if he had wanted to. If he were to regard each term in the proportion

$$\Sigma^2 : \Sigma \cdot (\Sigma - B\Gamma) :: (\Sigma - A\Gamma) \cdot (\Sigma - AB) : EH^2$$

as an area and take the sides of squares equal to them, he would have four squares in proportion, of sides Σ, α, β, EH, where α is the mean proportional between Σ and $\Sigma - B\Gamma$ and β the mean proportional between $\Sigma - A\Gamma$ and $\Sigma - AB$. It would need to be proved that if four squares are in proportion, then their sides are also in proportion; however, that fact follows immediately from the Eudoxan theory of

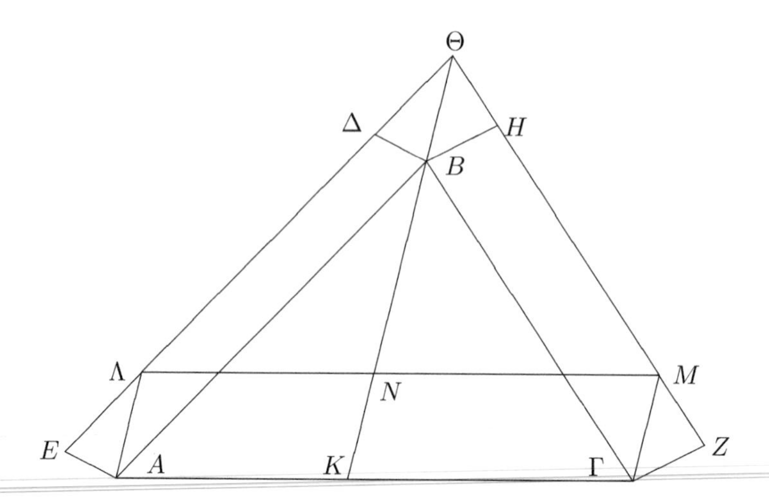

FIGURE 3. Pappus' generalization of the Pythagorean theorem.

proportion (see Problem 11.3). Working with the sides of the squares, it would then be legitimate to multiply means and extremes—that is, to form rectangles on the sides—since the appropriate theorems were proved in Book 6 of Euclid. He could have said that the triangle $AB\Gamma$ equals the rectangle on Σ and EH, which in turn equals the rectangle on α and β. The assertion that the triangle $AB\Gamma$ is the rectangle on α and β is precisely Heron's theorem. What he has done up to this point would not have offended a logical Euclidean purist. Why did he not finish the proof in this way?

The most likely explanation is that the proof came from Archimedes, as many scholars believe, and that Heron was aiming at numerical results. Another possible explanation is that our reconstruction of what Heron *could have* done lacks the symmetry of the process described by Heron, since α and β do not contain the sides in symmetric form. Whatever the reason, his summing up of the argument leaves no doubt that he was willing to accept the product of two areas as a product of numbers.

1.4. Pappus. Book 4 of Pappus' *Synagōgē* contains a famous generalization of the Pythagorean theorem: Given a triangle $AB\Gamma$ and any parallelograms $B\Gamma ZH$ and $AB\Delta E$ constructed on two sides, it is possible to construct (with straightedge and compass) a parallelogram $A\Gamma M\Lambda$ on the third side equal in area to the sum of the other two (see Fig. 3).

The isoperimetric problem. In Book 5 Pappus states almost verbatim the argument that Theon of Alexandria, quoting Zenodorus, gave for the proof of the isoperimetric inequality. Pappus embroiders the theorem with a beautiful literary device, however. He speaks poetically of the divine mission of the bees to bring from heaven the wonderful nectar known as honey and says that in keeping with this mission they must make their honeycombs without any cracks through which honey could be lost. Being endowed with a divine sense of symmetry as well, the bees had to choose among the regular shapes that could fulfill this condition, that is, triangles, squares, and hexagons. They chose the hexagon because a hexagonal prism required the least material to enclose a given volume, out of all the possible prisms whose base would tile the plane.[3]

Analysis, locus problems, and Pappus' theorem. Book 7 of the *Synagōgē* is a treasure trove of fascinating information about Greek geometry for several reasons. First, Pappus describes the kinds of techniques used to carry on the research that was current at the time. He lists a number of books of this *analysis* and tells who wrote them and what their contents were, in general terms, thereby providing valuable historical information. What he means by analysis, as opposed to synthesis, is a kind of algebraic reasoning in geometry. As he puts it, when a construction is to be made or a relation is to be proved, one imagines the problem to have been solved and then deduces consequences connecting the result with known principles, after which the process is reversed and a proof can be synthesized. This process amounts to thinking about objects not yet determined in terms of properties that they must have; when applied to numbers, that process is algebra.

Second, Book 7 also contains a general discussion of locus problems, such as we have already encountered in Apollonius' *Conics*. This discussion exerted a strong influence on the development of geometry in seventeenth-century France.

Proposition 81 of Euclid's *Data*, discussed above, inspired Pappus to create a very general proposition about plane loci. Referring to the points of intersection of a set of lines, he writes:

> To combine these discoveries in a single proposition, we have written the following. *If three points are fixed on one line... and all the others except one are confined to given lines, then that last one is also confined to a given line.* This is asserted only for four lines, no more than two of which intersect in the same point. It is not known whether this assertion is true for every number.

Pappus could not have known that he had provided the essential principle by which a famous theorem of projective geometry known as Desargues' theorem (see Section 2 of Chapter 12) was to be proved 1400 years later. Desargues certainly knew the work of Pappus, but may not have made the connection with this theorem. The connection was pointed out by van der Waerden (*1963*, p. 287).

Pappus discusses the three- and four-line locus for which the mathematical machinery is found in Book 3 of Apollonius' *Conics*. For these cases the locus is always one of the three conic sections. Pappus mentions that the two-line locus is a planar problem; that is, the solution is a line or circle. He says that a point satisfying the conditions of the locus to five or six lines is confined to a definite curve (a curve "given in position" as the Greeks said), but that this curve is "not yet familiar, and is merely called a curve." The curve is defined by the condition that the rectangular parallelepiped spanned by the lines drawn from a point to three fixed lines bears a fixed ratio to the corresponding parallelepiped spanned by the lines drawn to three other fixed lines. In our terms, this locus is a cubic curve.

[3] If one is looking for mathematical explanations of this shape, it would be simpler to start with the assumption that the body of a bee is approximately a cylinder, so that the cells should be approximately cylinders. Now one cylinder can be tightly packed with six adjacent cylinders of the same size. If the cylinders are flexible and there is pressure on them, they will flatten into hexagonal prisms.

Third, in connection with the extension of these locus problems, Pappus considers the locus to more than six lines and says that a point satisfying the corresponding conditions is confined to a definite curve. This step was important, since it proposed the possibility that a curve could be determined by certain conditions without being explicitly constructible. Moreover, it forced Pappus to go beyond the usual geometric interpretation of products of lines as rectangles, thus pushing the same boundary that Heron had gone through. Noting that "nothing is subtended by more than three dimensions," he continues:

> It is true that some of our recent predecessors have agreed among themselves to interpret such things, but they have not made a meaningful clear definition in saying that what is subtended by certain things is multiplied by the square on one line or the rectangle on others. But these things can be stated and proved using the composition of ratios.

It appears that Pappus was on the very threshold of the creation of the modern concept of a real number as a ratio of lines. Why did he not cross that threshold? The main reason was probably the cumbersome Euclidean definition of a composite ratio, discussed in Chapter 10. But there was a further reason: he wasn't interested in foundational questions. He made no attempt to prove or justify the parallel postulate, for example. And that brings us to the fourth attraction of Book 7. In that book Pappus investigated some very interesting problems, which he preferred to foundational questions. After concluding his discussion of the locus problems, he implies that he is merely reporting what other people, who are interested in them, have claimed. "But," he says,

> after proving results that are much stronger and promise many applications,... to show that I do not come boasting and empty-handed... I offer my readers the following: *The ratio of rotated bodies is the composite of the ratio of the areas rotated and the ratio of straight lines drawn similarly* [at the same angle] *from their centers of gravity to the axes of rotation. And the ratio of incompletely rotated bodies is the composite of the ratio of the areas rotated and the ratio of the arcs described by their centers of gravity.*

Pappus does not say how he discovered these results, nor does he give the proof. The proof would have been fairly easy, given that he had read Archimedes' *Quadrature of the Parabola*, in which the method of exhaustion is used. For the first theorem it would have been sufficient to compute the volume generated by revolving a right triangle with one leg parallel to the axis of rotation, and in that case the volume could be computed by subtracting the volume of a cylinder from the volume of a frustum of a cone. If the theorem is true for two nonoverlapping areas, it is easily seen to be true for the union of those areas. Pappus could then have applied the method of exhaustion to get the general result. The second result is an immediate application of the Eudoxan theory of proportion, since the volume generated is obviously in direct proportion to the angle of rotation, as are the arcs traversed by individual points. The modern theorem that is called Pappus' theorem asserts that the volume of a solid of revolution is equal to the product of the area rotated and the distance traversed by its center of gravity (which is 2π times the length of the line from the center of gravity to the axis of rotation). In the modern form this theorem was first stated in 1609 by the Swiss astronomer–mathematician Paul Guldin (1577–1643), a Jesuit priest, and published between 1635 and 1640 in the second volume of his four-volume work *Centrobaryca seu de centro gravitatis trium specierum quantitatis continuae* (*The Barycenter, or on the Center of Gravity of the Three Kinds of Continuous Magnitude*). Guldin had apparently not read Pappus and made the discovery independently. He did not prove the result, and the first proof is due to Bonaventura Cavalieri (1598–1647).

2. Roman geometry

In the Roman Empire geometry found applications in mapmaking. The way back from the abstractness of Euclidean geometry was led by Heron, Ptolemy, and other geometers who lived during the early Empire. We have already mentioned Ptolemy's *Almagest*, which was an elegant arithmetization of some basic Euclidean geometry applied to astronomy. In it, concrete computations using the table of chords are combined with rigorous geometric demonstration of the relations involved. But Ptolemy studied the Earth as well as the sky, and his contribution to geography is also a large one, and also very geometric.

Ptolemy was one of the first scholars to look at the problem of representing large portions of the Earth's surface on a flat map. His data, understandably very inaccurate from the modern point of view, came from his predecessors, including the astronomers Eratosthenes (276–194) and Hipparchus (190–120) and the geographers Strabo (ca. 64 BCE–24 CE) and Marinus of Tyre (70–130), whom he followed in using the now-familiar lines of latitude and longitude. These lines have the advantage of being perpendicular to one another, but the disadvantage that the parallels of latitude are of different sizes. Hence a degree of longitude stands for different distances at different latitudes.

Ptolemy assigned latitudes to the inhabited spots that he knew about by computing the length of sunlight on the longest day of the year. This computational procedure is described in Book 2, Chapter 6 of the *Almagest*, where Ptolemy describes the latitudes at which the longest day lasts $12\frac{1}{4}$ hours, $12\frac{1}{2}$ hours, and so on up to 18 hours, then at half-hour intervals up to 20 hours, and finally at 1-hour intervals up to 24. Although he knew theoretically what the Arctic Circle is, he didn't know of anyone living north of it, and took the northernmost location on the maps in his *Geography* to be Thoúlē, described by the historian Polybius around 150 BCE as an island six days sail north of Britain that had been discovered by the merchant–explorer Pytheas (380–310) of Masillia (Marseille) some two centuries earlier.[4] It has been suggested that Thoúlē is the Shetland Islands (part of Scotland since 1471), located between 60° and 61° north; that is just a few degrees south of the Arctic Circle, which is at $66^\circ\,30'$. It is also sometimes said to be Iceland, which is on the Arctic Circle, but west of Britain rather than north. Whatever it was, Ptolemy assigned it a latitude of 63°, although he said in the *Almagest* that some "Scythians" (Scandinavians and Slavs) lived still farther north at $64\frac{1}{2}^\circ$. Ptolemy did know of people living south of the equator and took account of places as far south as Agisymba (Ethiopia) and the promontory of Prasum (perhaps Cabo Delgado in Mozambique, which is 14° south). Ptolemy placed it $12^\circ\,30'$ south of

[4] The Latin idiom *ultima Thule* means roughly *the last extremity.*

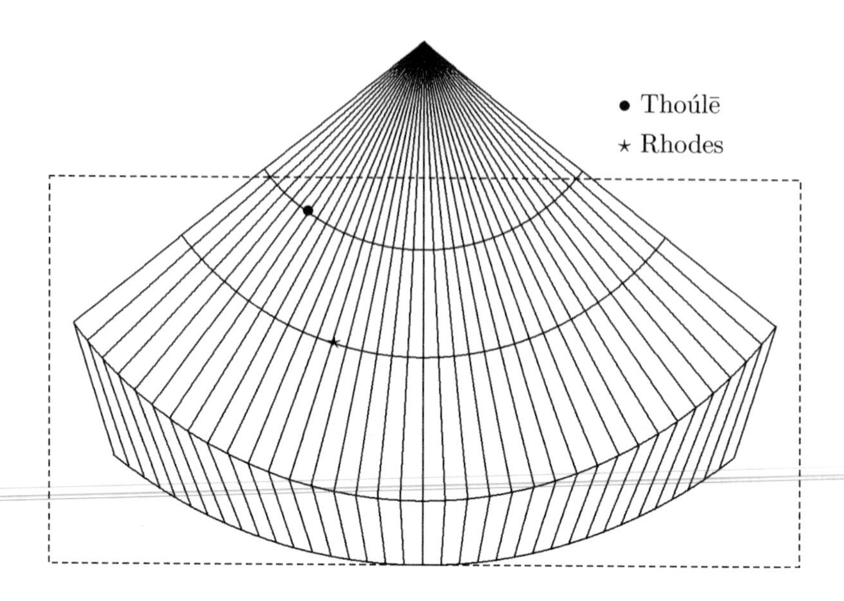

FIGURE 4. Ptolemy's first method of mapping.

the Equator. The extreme southern limit of his map was the circle $16^\circ\,25'$ south of the equator, which he called "anti-Meróē," since Meróē was $16^\circ\,25'$ north.

Since he knew only the geography of what is now Europe, Africa, and Asia, he did not need 360° of longitude. He took his westernmost point to be the Blessed Islands (possibly the Canary Islands, at 17° west). That was his prime meridian, and he measured longitude out to 180° eastward from there, to the Sêres[5] and the Chinese (Sínai) and "Kattígara." According to Dilke (*1985*, p. 81), "Kattígara" may refer to Hanoi. Actually, the east–west span from the Canary Islands to Shanghai (about 123° east) is only 140° of longitude. Ptolemy's inaccuracy is due partly to unreliable reports of distances over trade routes and partly to his decision to accept 500 stades as the length of a degree of latitude when the true distance is about 600 stades.[6] We are not concerned with geography, however, only with its mathematical aspects.

The problem Ptolemy faced was to draw a flat map of the Earth's surface spanning 180° of longitude and about 80° degrees of latitude, from $16^\circ\,25'$ south to 63° north. Ptolemy described three methods of doing this, the first of which we shall now discuss. The latitude and longitude coordinates of the inhabited world (*oikuménē*) known to Ptolemy represent a rectangle whose width is $\frac{5}{9}$ of its length. But Ptolemy did not like to represent parallels of latitude as straight lines; he preferred to draw them as arcs of concentric circles while keeping the meridians

[5] The Sêres were a Hindu people known to the Greeks from the silk trade.

[6] It has become a commonplace that Christopher Columbus, relying on Ptolemy's geography, expected to reach the Orient at a distance that would have placed him in the middle of the Pacific Ocean had North America not been in the way. If he believed Ptolemy, he would have thought it about 180° of longitude, which at a latitude of 40° would have been about 138 great-circle degrees. But he thought a degree was 500 stades (92 km), and hence that the distance to Japan was about 12,700 km. Since a degree is actually 600 stades (110 km), the journey would have been more than 15,000 km. But the latitude of Japan is slightly south of the latitude of Spain.

of longitude as straight lines emanating from the common center, representing the north pole. Thus, his plan is to map this portion of the Earth into the portion of a sector of a disk bounded by two radii and two concentric circles. In terms of Fig. 4, his first problem is to decide which radii and which circles are to form these boundaries. Ptolemy recognized that it would be impossible in such a map to place all the parallels of latitude at the correct distances from one another and still get their lengths in proportion. He decided to keep his northernmost parallel, through Thoúlē, in proportion to the parallel through the equator. That meant these arcs should be in the proportion of about 9:20—to be precise, $\cos(63°)$ in our terms. Since there would be 63 equal divisions between that parallel and the equator, he needed the upper radius x to satisfy $x : (x + 63) :: 9 : 20$. Solving this proportion is not hard, and one finds that $x = 52$, to the nearest integer. The next task was to decide on the angular opening. For this principle he decided, like his predecessor Marinus, to get the parallel of latitude through Rhodes in the correct proportion. Since Rhodes is at 36° latitude, half of the parallel through it amounts to about $\frac{4}{5}$ of the 180° arc of a great circle, which is about 145°. Since the radius of Rhodes must be 79 (27 great-circle degrees more than the radius of Thoúlē), he needed the opening angle of the sectors θ to satisfy $\theta : 180° :: 146 : 79\pi$, so that $\theta \approx 106°$. After that, he inserted meridians of longitude every one-third of an hour of longitude (5°) fanning out from the north pole to the equator.

Ptolemy recognized that continuing to draw the parallels of latitude in the same way for points south of the equator would lead to serious distortion, since the circles in the sector continue to increase as the distance south of the north pole increases, while the actual parallels on the Earth begin to decrease at that point. The simplest solution to that problem, he decided, was to let his southernmost parallel at $16° 25'$ south have its actual length, then join the meridians through that parallel by straight lines to the points where they intersect the equator. Once that decision was made, he was ready to draw the map on a rectangular sheet of paper. He gave instructions for how to do that: Begin with a rectangle that is approximately twice as long as it is wide, draw the perpendicular bisector of the horizontal (long) sides, and extend it above the upper edge so that the portion above that edge and the whole bisector are in the ratio $34° : 131°, 25'$. In that way, the 106° arc through Thoúlē will begin and end just slightly above the upper edge of the rectangle, while the lowest point of the map will be at the foot of the bisector, being about 80 units below the lowest point on the parallel of Thoúlē, as indicated by the dashed line in Fig. 4.

This way of mapping is *not* a conical projection, as it might appear to be, since it preserves north–south distances. It does a tolerably good job of mapping the parts of the world for which Ptolemy had reliable data. One can recognize Europe and the Middle East in the map of Plate 8, constructed around the year 1300 CE to accompany an edition of the *Geography*.

2.1. Roman civil engineering. Dilke (*1985*, pp. 88–90) describes the use of geometry in Roman civil engineering as follows. The center of a Roman village would be at the intersection of two perpendicular roads, a (usually) north–south road called the *kardo maximus* (literally, the *main hinge*) and an east–west road called *decumanus maximus*, the *main tenth.* Lots were laid out in blocks (*insulæ*) called *hundredths* (*centuriæ*), each block being assigned a pair of numbers, telling how many units it was *dextra decumani* (on the right decumanus) or *sinistra decumani*

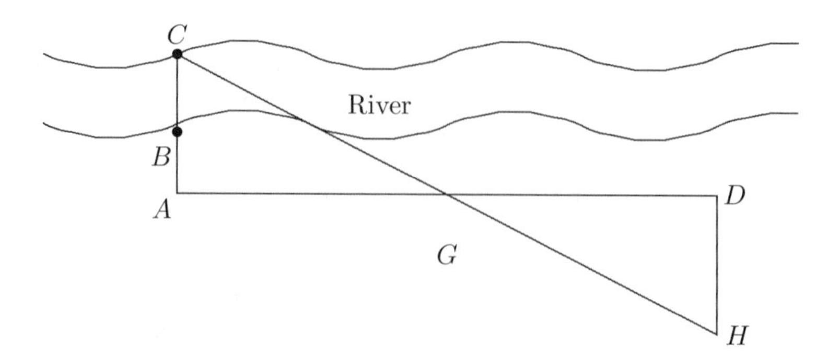

FIGURE 5. Nipsus' method of computing the width of a river.

(on the left decumanus) and how many units it was *ultra kardinem* (on the far kardo) or *citra kardinem* (on the near kardo).[7]

A collection of Roman writings on surveying was collected, translated into German, and published in Berlin in the middle of the nineteenth century. This two-volume work bears the title *Corpus Agrimensorum Romanorum*, the word *agrimensor* (field measurer) being the Latin name for a surveyor. Among the agrimensores was one named M. Iunius Nipsus, a second-century surveyor, who, according to Dilke (*1985*, p. 99), gives the following directions for measuring the width of a river (Fig. 5).

You mark the point C on the opposite bank from B (a part of the procedure Nipsus neglects to mention until later), continue the straight line CB to some convenient point A, lay down the crossroads sign at A, then move along the direction perpendicular to AC until you reach a point G, where you erect a pole, then continue on to D so that $GD = AG$. You then move away from D along the direction perpendicular to AD until you see G and C in a straight line from the point H. Since the triangles AGC and DGH are congruent (by angle–side–angle), it follows that $CB = CA - AB = HD - AB$.

For this procedure to work in practice, it is necessary to have an accessible and level piece of land covering the lines shown as AD and DH. If the river is large, such a stretch of land may not exist, since the river banks are likely to be hilly. In its neglect of similar triangles, this method seems a large step backward in applied geometry.

3. Medieval geometry

Among the translations of Greek works into Latin mentioned above was a translation of Euclid's *Elements* written by Boethius (ca. 480–524). This work has been lost, although references to it survive.[8] A "pseudo-Boethius" text of geometry, written some centuries later, has survived. It may have been a standard text during the Middle Ages. There were anonymous treatises on geometry during this time, some attributed to Boethius, usually containing Latin translations of a few of the

[7] In modern terms these would be *First Avenue East*, *First Avenue West*, *North Main Street*, and *South Main Street*.

[8] For example, in his *Encyclopedia of Liberal and Literary Studies* the early sixth-century writer Magnus Aurelius Cassiodorus refers to the great Greek mathematicians "of whom Euclid was given to us translated into the Latin language by the same great man Boethius."

early books of Euclid, often drastically edited. The tradition of attributing these works to Boethius continued even in the twelfth century, when translations from Arabic began to appear, as one can see in the booklets of Folkerts (*1970, 1971*), the second of which compares an anonymous Latin version ascribed to Boethius with the translation (from Arabic) of Athelhard of Bath. In a series of papers (*1999*, *2000*) Zaitsev has argued that the pseudo-Boethius was enlarging the commentaries on Euclid by including material from surveying and geometric astronomy. As a result,

> [I]n the writing process geometric concepts were systematically translated into the language of surveying, and the resulting melange of surveying and geometry was used as the basis for discussing the theological–cosmological significance of the discipline. [Zaitsev, *2000*, p. 222]

Thus, in the West also, mathematics became once again mixed with philosophy, but this time with the philosophy of Christianity.[9] Zaitsev also notes (*2000*, p. 223) that the idea of multiple layers of meaning was dear to the authors of medieval texts; but in contrast to biblical commentaries, which were strictly separated from the texts used as a source, commentaries and sources were routinely intermixed in the geometric work.

Mathematics sank to a rather low level in Europe after 500 CE, recovering only slightly if at all in the Carolingian Renaissance of the ninth century. One of the better-informed scholars of the tenth century was Gerbert of Aurillac (ca. 940–1002), who reigned as Pope Sylvester II during the last three years of his life. Even though Gerbert was one of the leading scholars of his day, who advocated the use of Hindu–Arabic numerals, one of his letters to a certain Adalbold of Liège is occupied with a discussion of the rule for finding the area of a triangle given its base and altitude! The general level of geometry, however, was not so bad as the correspondence between Gerbert and Adalbold seems to imply. In fact, Gerbert wrote, but did not finish, *Geometria*, a practical manual of surveying based on what was probably in Boethius' textbook. This work, which can be read online,[10] consists of 89 brief chapters devoted to triangles, circles, spheres, and regular polygons. It gives the names of standard units of length and finds the areas of such simple figures as a trapezoid (Chapter XVLIII) and a semicircle (Chapter LXXIX, where the rule is given to multiply the square of the diameter by 11 and divide the product by 28). A specimen that may be typical of the level of geometric knowledge used in civil engineering, architecture, surveying, and geometric astronomy in the twelfth century, just as translations from Arabic works began to circulate in Europe, is provided by Hugh of St. Victor's *Practical Geometry* (Homann, *1991*), in which one can find a description of the construction and use of an astrolabe (a fundamental tool used by

[9] Compare with the quotations from the pseudo-Boethius and Gerbert in Chapter 3.
[10] `http:\\pld.chadwyck.com`, a commercial site that provides the *Patrologia Latina* of Jacques-Paul Migne (1800–1875). Search for the title "geometria."

navigators and explorers for many centuries)[11] and a discussion of different ways of using similar triangles to determine distances to inacessible objects.

3.1. Late Medieval and Renaissance geometry. In the late Middle Ages Europeans from many countries eagerly sought Latin translations of Arabic treatises, just as some centuries earlier Muslim scholars had sought translations of Hindu and Greek treatises. In both cases, those treatises were made the foundation for ever more elaborate and beautiful mathematical theories. An interesting story arises in connection with these translations, showing the unreliability of transmission. The Sanskrit word "bowstring" (*jya*) used for a half-chord of a circle was simply borrowed by the Arab translators and written as *jb*, apparently pronounced *jiba*, since Arabic was written without vowels. Over time, this word came to be interpreted as *jaib*, meaning a pocket or fold in a garment. When the Arabic works on trigonometry were translated into Latin in the twelfth century, this word was translated as *sinus*, which also means a pocket or cavity. The word caught on very quickly, apparently because of the influence of Leonardo of Pisa, and is now well established in all European languages. That is the reason we now have three trigonometric functions, the secant (Latin for cutting), the tangent (Latin for touching), and the sine (Latin for a concept having nothing at all to do with geometry!).

We think of analytic geometry as the application of algebra to geometry. Its origins in Europe, however, antedate the high period of European algebra by a century or more. The first adjustment in the way mathematicians think about physical dimensions, an essential step on the way to analytic geometry, occurred in the fourteenth century.

Nicole d'Oresme. The first prefiguration of analytic geometry occurs in the work of Nicole d'Oresme (1323–1382). The *Tractatus de latitudinibus formarum*, published in Paris in 1482 and ascribed to Oresme but probably written by one of his students, contains descriptions of the graphical representation of intensities. The crucial realization that he came to was that since the area of a rectangle is computed by multiplying length and width and the distance traveled at constant speed is computed by multiplying velocity and time, it follows that if one line is taken proportional to time and a line perpendicular to it is proportional to a (constant) velocity, the area of the resulting rectangle is proportional to the distance traveled.

Oresme considered three forms of qualities, which he labeled *uniform*, *uniformly difform*, and *difformly difform*. We would call these classifications constant, linear, and nonlinear. Examples are provided in Fig. 6, although Oresme realized that the "difformly difform" constituted a large class of qualities and mentioned specifically that a semicircle could be the representation of such a quality.

The advantage of representing a *distance* by an *area* rather than a line appeared in the case when the velocity changed during a motion. In the simplest nontrivial case the velocity was uniformly difform. In that case, the distance traversed is what it would have been had the body moved the whole time with the velocity it

[11] The French explorer Samuel de Champlain (1567–1635) apparently lost his astrolabe while exploring the Ottawa River in 1613. Miraculously, that astrolabe was found 254 years later, in 1867, and the errors in Champlain's diaries were used by an author named Alex Jamieson Russell (1807–1887) to establish the fact and date of the loss (Russell, *1879*). The discovery of this astrolabe only a month after the founding of the Canadian Federation was of metaphorical significance to Canadian poets. See, for example, *The Buried Astrolabe*, by Craig Stewart Walker, McGill–Queens University Press, Montreal, 2001, a collection of essays on Canadian dramatists.

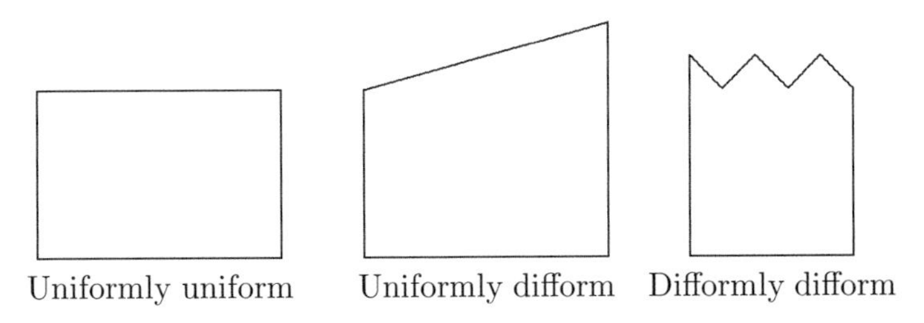

FIGURE 6. Nicole Oresme's classification of motions.

had at the midpoint of the time of travel. This is the case now called *uniformly accelerated* motion. According to Clagett (*1968*, p. 617), this rule was first stated by William Heytesbury (ca. 1313–ca. 1372) of Merton College, Oxford around 1335 and was well known during the Middle Ages.[12] It is called the *Merton Rule.* In his book *De configurationibus qualitatum et motuum*, Oresme applied these principles to the analysis of such motion and gave a simple geometric proof of the Merton Rule. He illustrated the three kinds of motion by drawing a figure similar to Fig. 6. He went on to say that if a difformly difform quality was composed of uniform or uniformly difform parts, as in the example in Fig. 6, its quantity could be measured by (adding) its parts. He then pushed this principle to the limit, saying that if the quality was difform but not made up of uniformly difform parts, say being represented by a curve, then "it is necessary to have recourse to the mutual measurement of curved figures" (Clagett, *1968*, p. 410). This statement must mean that the distance traveled is the "area under the velocity curve" in all three cases. Oresme unfortunately did not give any examples of the more general case, but he could hardly have done so, since the measurement of figures bounded by curves was still very primitive in his day.

Trigonometry. Analytic geometry would be unthinkable without plane trigonometry. Latin translations of Arabic texts of trigonometry, such as those of al-Tusi and al-Jayyani, which will be discussed below, began to circulate in Europe in the late Middle Ages. These works provided the foundation for such books as *De triangulis omnimodis* by Regiomontanus, published in 1533, after his death, which contained trigonometry almost in the form still taught. Book 2, for example, contains as its first theorem the law of sines for plane triangles, which asserts that the sides of triangles are proportional to the sines of the angles opposite them. The main difference between this trigonometry and ours is that a sine remains a *length* rather than a *ratio*. It is referred to an *arc* rather than to an *angle*. It was once believed that Regiomontanus discovered the law of sines for spherical triangles (Proposition 16 of Book 4) as well; but we now know that this theorem was known at least 500 years earlier to Muslim mathematicians whose work Regiomontanus must have read. A more advanced book on the subject, which reworked the reasoning of Heron on the area of a triangle given its sides, was *Trigonometriæ sive de dimensione triangulorum libri quinque* (*Five Books of Trigonometry, or, On the Size of Triangles*), first published in 1595, written by the Calvinist theologian Bartholomeus Pitiscus

[12] Boyer (*1949*, p. 83) says that the rule was stated around this time by another fourteenth-century Oxford scholar named Richard Suiseth, known as Calculator for his book *Liber calculatorum.* Suiseth shares with Oresme the credit for having proved that the harmonic series diverges.

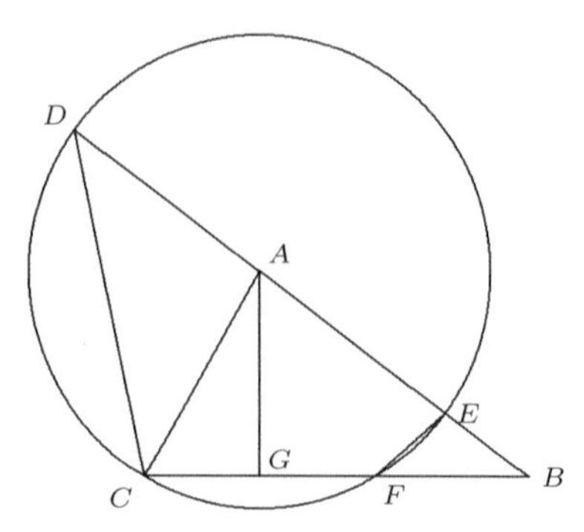

FIGURE 7. Pitiscus' derivation of the proportions in which an altitude divides a side of a triangle.

(1561–1613). This was, incidentally, the book that established the name *trigonometry* for this subject. Pitiscus showed how to determine the parts into which a side of a triangle is divided by the altitude, given the lengths of the three sides. To guarantee that the angles adjacent to the side were acute, he stated the theorem only for the altitude from the vertex of the largest angle.

Pitiscus' way of deriving this proportion was as follows. If the shortest side of the triangle ABC is AC and the longest is BC, let the altitude to BC be AG, as in Fig. 7. Draw the circle through C with center at A, so that B lies outside the circle, and let the intersections of the circle with AB and BC be E and F respectively. Then extend BA to meet the circle at D, and connect CD. Then $\angle BFE$ is the supplement of $\angle CFE$, which subtends the arc EDC, which in turn is an arc of $180°$ plus the arc $\overset{\frown}{CD}$. Hence $\angle BFE$ is the complement of the angle subtended by the arc $\overset{\frown}{CD}$. That in turn is the angle subtended by the supplementary arc $\overset{\frown}{CE}$; thus $\angle BFE = \angle CDB$, and so the triangles BCD and BEF are similar. It follows that

$$AB^2 = AC^2 + BC^2 - 2AC.BC\cos(\angle ACB)\,,$$

which is what we now know as the *law of cosines.*

Pitiscus also gave an algebraic solution of the trisection problem discovered by an earlier mathematician, Jobst Bürgi (1552–1632). The solution had been based on the fact that the chord of triple an angle is three times the chord of the angle minus the cube of the chord of the angle. This relation makes no sense in terms of geometric dimension; it is a purely numerical relation. It is interesting that it is stated in terms of chords, since Pitiscus surely knew about sines.

4. Geometry in the Muslim world

In the Western world most of the advancement of geometry in the millennium from the fall of the Western Roman Empire to the fall of the Eastern Empire occurred among the Muslim and Jewish mathematicians of Baghdad, Samarkand, Cordoba, and other places. This work had some features of Euclid's style and some of Heron's. Matvievskaya (*1999*) has studied the extensive commentaries on the tenth book of Euclid's *Elements* written by Muslim scholars from the ninth through twelfth centuries and concluded that while formally preserving a Euclidean distinction between magnitude and number, they actually operated with quadratic and quartic irrationals as if they were numbers.

4.1. The parallel postulate. The Islamic mathematicians continued the later Hellenistic speculation on Euclid's parallel postulate. According to Sabra (*1969*), this topic came into Islamic mathematics through a commentary by Simplicius on Book 1 of the *Elements*, whose Greek original is lost, although an Arabic translation exists. In fact, Sabra found a manuscript that contains Simplicius' attempted proof. The reworking of this topic by Islamic mathematicians consisted of a criticism of Simplicius' argument followed by attempts to repair its defects. Gray (*1989*, pp. 42–54) presents a number of these arguments, beginning with the ninth-century mathematician al-Gauhari. Al-Gauhari attempted to show that two lines constructed so as to be parallel, as in Proposition 27 of Book 1 of the *Elements* must also be equidistant at all points. If he had succeeded, he would indeed have proved the parallel postulate.

4.2. Thabit ibn-Qurra. Thabit ibn-Qurra (826–901), whose revision of the Arabic translation of Euclid became a standard in the Muslim world, also joined the debate over the parallel postulate. According to Gray (*1989*, pp. 43–44), he considered a solid body moving without rotating so that one of its points P traverses a straight line. He claimed that the other points in the body would also move along straight lines, and obviously, they would remain equidistant from the line generated by the point P. By regarding these lines as completed loci, he avoided a certain objection that could be made to a later argument of ibn al-Haytham, discussed below. Thabit ibn-Qurra's work on this problem was ground-breaking in a number of ways, anticipating much that is usually credited to the eighteenth-century mathematicians Lambert and Saccheri. He proved, for example, that if a quadrilateral has two equal adjacent angles, and the sides not common to these two angles are equal, then the other two angles are also equal to each other. In the case when the equal angles are right angles, such a figure is called—unjustly, we may say—a *Saccheri quadrilateral*, after Giovanni Saccheri (1667–1733), who like Thabit ibn-Qurra, developed it in an attempt to prove the parallel postulate. Gray prefers to call it a *Thabit quadrilateral*, and we shall use this name. Thabit ibn-Qurra's proof amounted to the claim that a perpendicular drawn from one leg of such a quadrilateral to the opposite leg would also be perpendicular to the leg from which it was drawn. Such a figure, a quadrilateral having three right angles, or half of a Thabit quadrilateral, is now called—again, unjustly—a *Lambert quadrilateral*, after Johann Heinrich Lambert (1728–1777), who used it for the same purpose. We should probably call it a *semi-Thabit quadrilateral*. Thabit's claim is that either type of Thabit quadrilateral is in fact a rectangle. If this conclusion is granted, it follows by consideration of the diagonals of a rectangle that the sum of the acute angles in a right triangle is a right angle, and this fact makes Thabit's proof of the parallel postulate work.

The argument of Thabit ibn-Qurra, according to Gray, is illustrated in Fig. 8.[13] Given three lines l, m, and n such that l is perpendicular to n at E and m intersects it at A, making an acute angle, let W be any point on m above n and draw a perpendicular WZ from W to n. If E is between A and Z, then l must intersect m by virtue of what is now called *Pasch's theorem*. That much of the argument would

[13] We are supplementing the figure and adding steps to the argument for the sake of clarity.

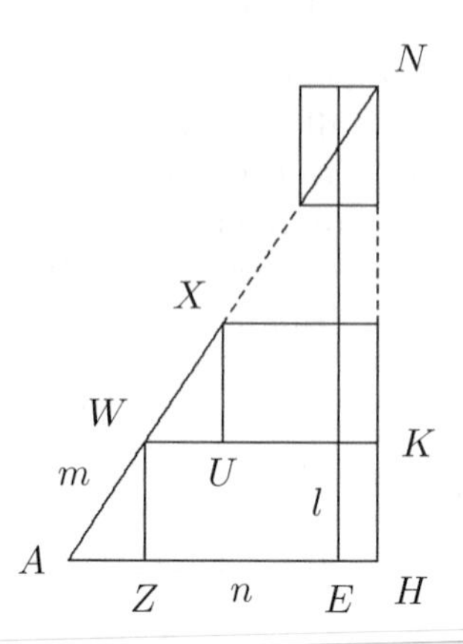

FIGURE 8. Thabit ibn-Qurra's attempted proof of the parallel postulate.

be uncontroversial. The difficult part occurs when Z is between A and E. Thabit ibn-Qurra argued as follows. By Archimedes' principle, some multiple of AZ, say AH, exceeds AE, so that E lies between A and H. Now by drawing a perpendicular to n at H and making HK equal to ZW, we get a Thabit quadrilateral, so that if WK is joined, we have a rectangle $WZHK$. Then, if X is chosen so that $AW = WX$ and a perpendicular XU is drawn to WK, the triangles AWZ and WXU will be congruent because the sum of the acute angles of a right triangle is a right angle. Then WU will equal AZ. We can then start over, since WK will be less than AH by a length equal to AZ. In this way, in a finite number of steps, we will reach a point N on line m that is also on the extension of HK. Hence m contains points on both sides of l and therefore intersects l.

Gray has called Thabit ibn-Qurra's mistake "an interesting and deep one," since it makes use of motion in geometry in a way that seems to be implied by Euclid's own arguments involving coinciding figures; that is, that they can be moved without changing their size or shape. Euclid makes this assumption in Proposition 4 of Book 1 where he "proves" the side–angle–side criterion for congruence by superimposing one triangle on another. He does not speak explicitly of moving a triangle, but how else is one to imagine this superimposition taking place?

Thabit ibn-Qurra also created a generalization of the Pythagorean theorem. His theorem is easily derived from similar triangles. Consider a triangle ABC whose longest side is BC. Copy angle B with A as vertex and AC as one side, extending the other side to meet BC in point C', then copy angle C with A as vertex and BA as one side, extending the other side to meet BC in point B', so that angle $AB'B$ and angle $AC'C$ both equal angle A. It then follows that the triangles $B'AB$ and CAC' are similar to the original triangle ABC, and so $\overline{AB}^2 = \overline{BC} \cdot \overline{BB'}$ and $\overline{AC}^2 = \overline{BC} \cdot \overline{CC'}$, hence

$$\overline{AB}^2 + \overline{AC}^2 = \overline{BC}(\overline{BB'} + \overline{CC'}).$$

The case when angle A is acute is shown in Fig. 9.

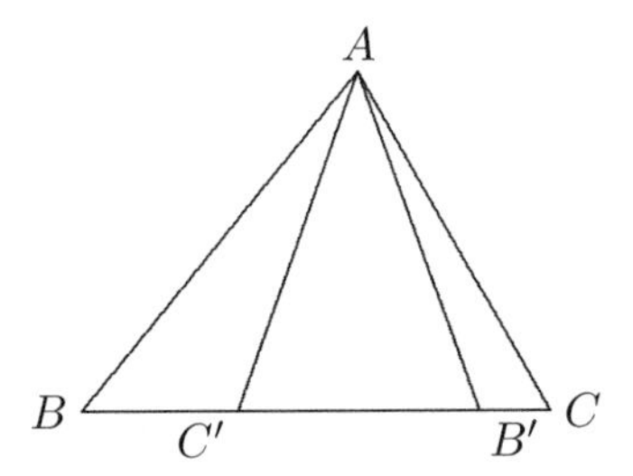

FIGURE 9. Thabit ibn-Qurra's Pythagorean theorem.

4.3. Al-Kuhi. A mathematician who devoted himself almost entirely to geometry was Abu Sahl al-Kuhi (ca. 940–ca. 1000), the author of many works, of which some 30 survive today. Berggren (*1989*), who has edited these manuscripts, notes that 14 of them deal with problems inspired by the reading of Euclid, Archimedes, and Apollonius, while 11 others are devoted to problems involving the compass, spherical trigonometry, and the theory of the astrolabe. Berggren presents as an example of al-Kuhi's work the angle trisection shown in Fig. 10. In that figure the angle φ to be trisected is ABG, with the base BG horizontal. The idea of the trisection is to extend side AB any convenient distance to D. Then, at the midpoint of BD, draw a new set of mutually perpendicular lines making an angle with the horizontal equal to $\varphi/2$, and draw the rectangular hyperbola through B having those lines as asymptotes. Apollonius had shown (*Conics*, Book 1, Propositions 29 and 30) that D lies on the other branch of the hyperbola. Then BE is drawn equal to BD, that is a circle through D with center at B is drawn, and its intersection with the hyperbola is labeled E. Finally, EZ is drawn parallel to BG. It then follows that $\varphi = \angle AZE = \angle ZBE + \angle ZEB = 3\theta$, as required.

4.4. Al-Haytham. One of the most prolific and profound of the Muslim mathematician–scientists was Abu Ali ibn al-Haytham (965–1040), known in the West as Alhazen. He was the author of more than 90 books, 55 of which survive.[14] A significant indication of his mathematical prowess is that he attempted to reconstruct the lost Book 8 of Apollonius' *Conics*. His most famous book is his *Treatise on Optics* (*Kitab al-Manazir*) in seven volumes. The fifth volume contains the problem known as Alhazen's problem: *Given the location of a surface, an object, and an observer, find the point on the surface at which a light ray from the object will be reflected to the observer.* Rashed (*1990*) points out that burning-mirror problems of this sort had been studied extensively by Muslim scholars, especially by Abu Saad ibn Sahl some decades before al-Haytham. More recently (see Guizal and Dudley, *2002*) Rashed has discovered a manuscript in Teheran written by ibn Sahl containing precisely the law of refraction known in Europe as *Snell's law*, after Willebrod Snell (1591–1626) or *Descartes' law*.[15] The law of refraction as given by Ptolemy in the form of a table of values of the angle of refraction and the angle of incidence implied that the angle of refraction was a quadratic function of the angle of incidence. The actual relation is that the ratio of the sines of the two angles is a constant for refraction at the interface between two different media.

---

[14] Rashed (*1989*) suggested that these works and the biographical information about al-Haytham may actually refer to two different people. The opposite view was maintained by Sabra (*1998*).
[15] According to Guizal and Dudley, this law was stated by Thomas Harriot (1560–1621) in 1602.

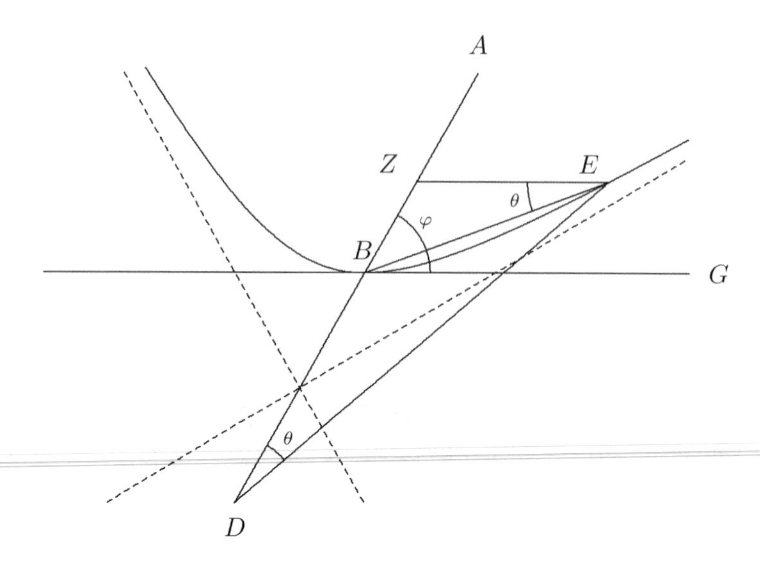

FIGURE 10. Al-Kuhi's angle trisection.

Al-Haytham also attempted to prove the parallel postulate. According to Gray (*1989*, p. 45), the argument given by al-Haytham in his *Commentary on the Premises to Euclid's Book* The Elements, and later in his *Book on the Resolution of Doubts* was based on the idea of translating a line perpendicular to a given line in such a way that it always remains perpendicular. The idea is that the endpoint of the line must trace a straight line parallel to the directing line.

4.5. Omar Khayyam. In his paper "Discussion of difficulties in Euclid" (Amir-Moez, *1959*), the Persian mathematician Omar Khayyam (1048–1131) raised a number of questions about al-Haytham's argument. He asked how a line could move while remaining perpendicular to a given line, and more generally, how geometry and motion could be connected. Even admitting that Euclid allowed a line to be generated by a moving point and a surface by a moving line, he pointed out that al-Haytham was requiring something more in demanding that one line remain perpendicular to another at each instant during its motion.[16]

Having refuted al-Haytham's proof, Omar Khayyam himself attempted a proof (Amir-Moez, *1959*) based on a proposition that he claimed Aristotle had proved: *If two lines converge, they will (eventually) intersect.* This claim raises an interesting question, since as we have seen, Aristotle did not accept the arguments given by scholars in Plato's Academy to prove that parallel lines exist. Given his disbelief in a completed infinity, he probably would have liked an argument proving that

[16] Omar Khayyam's objection is right on target from the point of view of modern physics. If the special theory of relativity is correct, no sense can be attached to the statement that two events occurring at different places are simultaneous. One observer may find them so, while another does not agree. The same objection applies to Thabit ibn-Qurra's argument, which assumes a rigid body. In special relativity rigid bodies do not exist. What al-Haytham did was to ignore all points from the moving solid except those lying along a certain line. The relation between motion and geometry lies at the heart of relativity theory.

converging lines must intersect. Although none of the writings now attributed to Aristotle contain such an argument, Gray (*1989*, p. 47) points out that Omar Khayyam may have had access to Aristotelian treatises that no longer exist. In any case, he concluded on the basis of Aristotle's argument that two lines that converge on one side of a transversal must diverge on the other side. With that, having proved correctly that the perpendicular bisector of the base of a Thabit quadrilateral is also the perpendicular bisector of the summit, Omar Khayyam concluded that the base and summit could not diverge on either side, and hence must be equidistant. Like Thabit ibn-Qurra's proof, his proof depended on building one Thabit quadrilateral on top of another by doubling the common bisector of the base and summit, then crossing its endpoint with a perpendicular which (he said) would intersect the extensions of the lateral sides. Unfortunately, if that procedure is repeated often enough in hyperbolic geometry, those intersections will not occur.

All of these mathematicians were well versed in the Euclidean tradition of geometry. In the preface to his book on algebra, Omar Khayyam says that no one should attempt to read it who has not already read Euclid's *Elements* and *Data* and the first two books of Apollonius' *Conics*. His reason for requiring this background was that he intended to use conic sections to solve cubic and quartic equations geometrically. This book contains Euclidean rigor attached to algebra in a way that fits equally well into the history of both algebra and geometry. However, in other places it seems clear that Omar Khayyam was posing geometric problems for the sake of getting interesting equations to solve. For example (Amir-Moez, *1963*), he posed the problem of finding a point on a circle such that the perpendicular from the point to a radius has the same ratio to the radius that the two segments into which it divides the radius have to each other. If the radius is r and the length of the longer segment cut off on the radius is the unknown x, the equation to be satisfied is $x^3 + rx^2 + r^2x = r^3$. Without actually writing out this equation, Omar Khayyam showed that the geometric problem amounted to using the stated condition to find the second asymptote of a rectangular hyperbola, knowing one of its asymptotes and one point on the hyperbola. However, he regarded that analysis as merely an introduction to his real purpose, which was a discussion of the kinds of cubic equations that require conic sections for their solution. After a digression to classify these equations, he returned to the original problem, and finally, showed how to solve it using a rectangular hyperbola. He found the arc to be about 57°, so that $x \approx r\cos(57°) = 0.544r$. Omar Khayyam described x as being about $30\frac{2}{3}$ *pieces*, that is, sixtieths of the radius. We reserve the discussion of this combination of algebra and geometry for Chapter 14.

As his work on the parallel postulate shows, Omar Khayyam was very interested in logical niceties. In the preface to his *Algebra* and elsewhere (for example, Amir-Moez, *1963*, p. 328) he shows his adherence to Euclidean standards, denying the reality of a fourth dimension:

> If the algebraist were to use the square of the square in measuring areas, his result would be figurative [theoretical] and not real, because it is impossible to consider the square of the square as a magnitude of a measurable nature... This is even more true in the case of higher powers. [Kasir, *1931*, p. 48]

4.6. Nasir al-Din al-Tusi. The thirteenth century was as disruptive to the Islamic world as the fifth century had been to the Roman world. This was the time of the Mongol expansion, which brought the conquest of China in the early part of the century, then the conquest of Kievan Rus in 1243, and finally, the sack of Baghdad in 1254. Despite the horrendous times, the astronomer–mathematician Nasir al-Din al-Tusi (1201–1274) managed to produce some of the best mathematics of the era. Al-Tusi was treated with respect by the Mongol conqueror of Baghdad, who even built for him an astronomical observatory, at which he made years of accurate observations and improved the models in Ptolemy's *Almagest*.[17] Al-Tusi continued the Muslim work on the problem of the parallel postulate. According to Gray (*1989*, pp. 50–51), al-Tusi's proof followed the route of proving that the summit angles of a Thabit quadrilateral are right angles. He showed by arguments that Euclid would have accepted that they cannot be obtuse angles, since if they were, the summit would diverge from the base as a point moves from either summit vertex toward the other. Similarly, he claimed, they could not be acute, since in that case the summit would converge toward the base as a point moves from either summit vertex toward the other. Having thus argued that a Thabit quadrilateral must be a rectangle, he could give a proof similar to that of Thabit ibn-Qurra to establish the parallel postulate.

In a treatise on quadrilaterals written in 1260, al-Tusi also reworked the trigonometry inherited from the Greeks and Hindus and developed by his predecessors in the Muslim world, including all six triangle ratios that we know today as the trigonometric functions. In particular, he gave the law of sines for spherical triangles, which states that the sines of great-circle arcs forming a spherical triangle are proportional to the sines of their opposite angles. According to Hairetdinova (*1986*) trigonometry had been developing in the Muslim world for some centuries before this time, and in fact the mathematician Abu Abdullah al-Jayyani (989–1079), who lived in the Caliphate of Cordoba, wrote *The Book on Unknown Arcs of a Sphere*, a treatise on plane and spherical trigonometry. Significantly, he treated ratios of lines as numbers, in accordance with the evolution of thought on this subject in the Muslim world. Like other Muslim mathematicians, though, he does not use negative numbers. As Hairetdinova mentions, there is clear evidence of the Muslim influence in the first trigonometry treatise written by Europeans, the book *De triangulis omnimodis* by Regiomontanus, whose exposition of plane trigonometry closely follows that of al-Jayyani.

Among these and many other discoveries, al-Tusi discovered the interesting theorem that if a circle rolls without slipping inside a circle twice as large, each point on the smaller circle moves back and forth along a diameter of the larger circle. This fact is easy to prove and an interesting exercise in geometry. It has obvious applications in geometric astronomy, and was rediscovered three centuries later by Copernicus and used in Book 3, Chapter 4 of his *De revolutionibus*.

5. Non-Euclidean geometry

The centuries of effort by Hellenistic and Islamic mathematicians to establish the parallel postulate as a fact of nature began to be repeated in early modern Europe with the efforts of a number of mathematicians to replace the postulate with some other assumption that seemed indubitable. Then, around the year 1800, a change

[17] The world's debt to Muslim astronomers is shown in the large number of stars bearing Arabic names, such as Aldebaran (the Follower), Altair (the Flyer), Algol (the Ghoul), Betelgeuse (either the Giant's Hand or the Giant's Armpit), and Deneb (the Tail).

in attitude took place, as a few mathematicians began to explore non-Euclidean geometries as if they might have some meaning after all. Within a few decades the full light of day dawned on this topic, and by the late nineteenth century, models of the non-Euclidean geometries inside Euclidean and projective geometry removed all doubt as to their consistency. This history exhibits a sort of parallelism with the history of the classical construction problems and with the problem of solving higher-degree equations in radicals, all of which were shown in the early nineteenth century to be impossible tasks. In all cases, the result was a deeper insight into the original questions. In all three cases, group theory came to play a role, although a much smaller one in the case of non-Euclidean geometry than in the other two.

5.1. Girolamo Saccheri. The Jesuit priest Girolamo Saccheri (1667–1733), a professor of mathematics at the University of Pavia, published in the last year of his life the treatise *Euclides ab omni nævo vindicatus* (*Euclid Acquitted of Every Blemish*), a good example of the creativity a very intelligent person will exhibit when trying to retain a strongly held belief. Some of his treatise duplicates what had already been done by the Islamic mathematicians, including the study of Thabit quadrilaterals, that is, quadrilaterals having a pair of equal opposite sides and equal base angles and also quadrilaterals having three right angles. Saccheri deduced with strict rigor all the basic properties of Thabit quadrilaterals with right angles at the base.[18] He realized that the fundamental question involved the summit angles of these quadrilaterals—Saccheri quadrilaterals, as they are now called. Since these angles were equal, the only question was whether they were obtuse, right, or acute angles. He showed in Propositions 5 and 6 that if one such quadrilateral had obtuse summit angles, then all of them did likewise, and that if one had right angles, then all of them did likewise. It followed by elimination and without further proof (Proposition 7, which Saccheri proved anyway) that if one of them had acute angles, then all of them did likewise. Not being concerned to eliminate the possibility of the right angle, which he believed was the true one, he worked to eliminate the other two hypotheses.

He showed that the postulate as Euclid stated it is true under the hypothesis of the obtuse angle. That is, two lines cut by a transversal in such a way that the interior angles on one side are less than two right angles will meet on that side of the transversal. As we now know, that is because they will meet on *both* sides of the transversal, assuming it makes sense to talk of opposite sides. Saccheri remarked that the intersection must occur at a finite distance. This remark seems redundant, since all distances in geometry were finite until projective geometers introduced points at infinity. But Saccheri, in the end, would be reasoning about points at infinity as if something were known about them, even though he had no careful definition of them.

It is true, as many have pointed out, that his proof of this fact uses the exterior angle theorem (Proposition 16 of Book 1 of Euclid) and hence assumes that lines are infinite.[19] But Euclid himself, at least as later edited, states explicitly that

---

[18] It is unlikely that Saccheri knew of the earlier work by Thabit ibn-Qurra and others. Although Arabic manuscripts stimulated a revival of mathematics in Europe, they were apparently soon forgotten as Europeans began writing their own treatises. Coolidge (*1940*) gives the history of the parallel postulate jumping directly from Proclus and Ptolemy to Saccheri, never mentioning any of the Muslim mathematicians.

[19] Actually, the use of that proposition is confined to elaborations by the modern reader. The proof stated by Saccheri uses only the fact that lines are *unbounded*, that is, can be extended to

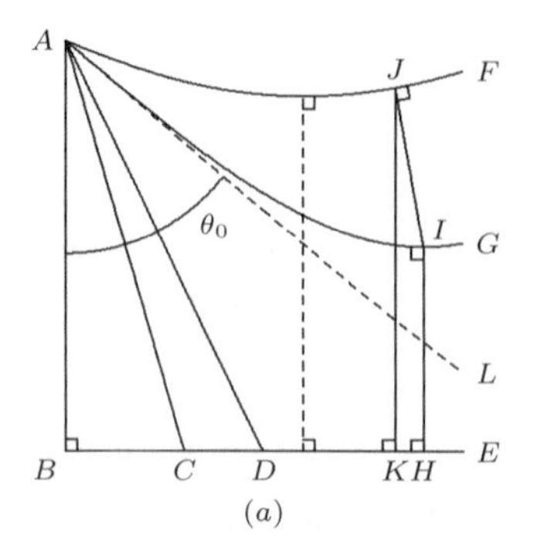

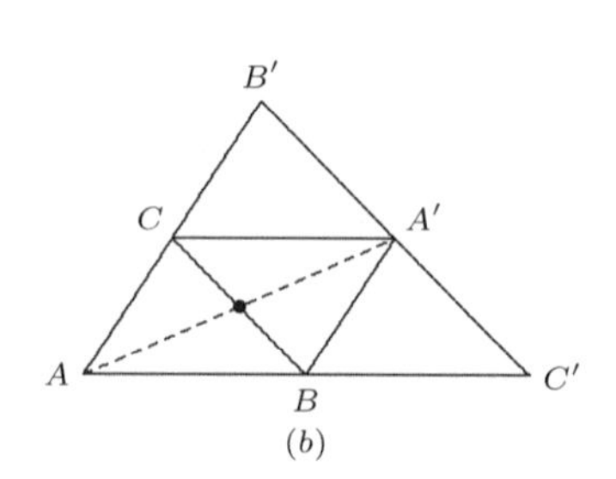

FIGURE 11. (a) Lines through A that intersect BE and those that share a common perpendicular with BE are separated by a line (AL) that is asymptotic to BE. (b) The angle defect of $\triangle AB'C'$ is more than twice the defect of $\triangle ABC$.

two lines cannot enclose an area, so that Saccheri can hardly be faulted for dealing with only one Euclidean postulate at a time. Since the parallel postulate implies that the summit and base of a Saccheri quadrilateral must meet on *both* sides of the quadrilateral under the hypothesis of the obtuse angle, even a severe critic should be inclined to give Saccheri a passing grade when he rejects this hypothesis.

Having disposed of the hypothesis of the obtuse angle, Saccheri then joined battle (his phrase) with the hypothesis of the acute angle. Here again, he proved some basic facts about what we now call hyperbolic geometry. Given any quadrilateral having right angles at the base and acute angles at the summit, it follows from continuity considerations that the length of a perpendicular dropped from the summit to the base must reach a minimum at some point, and at that point it must also be perpendicular to the summit. Saccheri analyzed this situation in detail, describing in the process a great deal of what must occur in what is now called hyperbolic geometry. In terms of Fig. 11(a),[20] he considered all the lines like AF through the point A such that angle BAF is acute. He wished to show that they all intersected the line BE.

Saccheri proved that there must be at least one angle θ_0 for which the line AL making that angle neither intersects BE nor has a common perpendicular with it. This line, as Saccheri showed in Proposition 23, must approach BE asymptotically as we would say. At that point he made the small slip that had been warned against even in ancient times, assuming that "approaching" implies "meeting." His intuition for hyperbolic geometry was very good, as he imagined a line perpendicular to BE moving away from AB and the lines from A perpendicular to it rotating

---

any length. It is not necessary to require that the extension never overlap the portion already present.

[20] Since the flat page is not measurably non-Euclidean, and wouldn't be even if spread out to cover the entire solar system, the kinds of lines that occur in hyperbolic geometry cannot be drawn accurately on paper. Our convention is the usual one: When asymptotic properties are not involved, draw the lines straight. When asymptotic properties need to be shown, draw them as hyperbolas. Actually, if the radius of curvature of the plane were comparable to the width of the page, two lines with a common perpendicular would diverge from each other like the graphs of $\cosh x$ and $-\cosh x$, very rapidly indeed.

clockwise about A to make angles that decreased to θ_0. He then—too hastily, as we now know—drew the conclusion that θ_0 would have the properties of *both* of the sets of angles that it separated, that is, the line making this angle would intersect BE and would also have a common perpendicular with it. In fact, it has neither property. But Saccheri was determined to have both. As he described the situation, the hypothesis of the acute angle implied the existence of two straight lines that have a common perpendicular *at the same point*. In other words, there could be two distinct lines perpendicular to the same line at a point, which is indeed a contradiction. Unfortunately, the point involved was not a point of the plane, but is infinitely distant, as Saccheri himself realized. But he apparently believed that points and lines at infinity must obey the same axioms as those in the finite plane.

Once again, as in the case of Ptolemy, Thabit ibn-Qurra, ibn al-Haytham, and others, Saccheri had developed a new kind of geometry, but resorted to procrustean methods to reconcile it with the geometry he believed in.

5.2. Lambert and Legendre. The writings of the Swiss mathematician Johann Heinrich Lambert (1728–1777) seem modern in many ways. For example, he proved that π is irrational (specifically, that $\tan x$ and x cannot both be rational numbers), studied the problem of constructions with straightedge and a fixed compass, and introduced the hyperbolic functions and their identities as they are known today, including the notation $\sinh x$ and $\cosh x$. He wrote, but did not publish, a treatise on parallel lines, in which he pointed out that the hypothesis of the obtuse angle holds for great circles on a sphere and that the area of a spherical triangle is the excess of its angle sum over π times the square of the radius. He concluded that in a sphere of imaginary radius ir, whose area would be negative, the area of a triangle might be proportional to the excess of π over the angle sum. What a sphere of imaginary radius looks like took some time to discern, a full century, to be exact.

By coincidence, the hyperbolic functions that he studied turned out to be the key to trigonometry in this imaginary world. Just as on the sphere there is a natural unit of length (the radius of the sphere, for example), the same would be true, as Lambert realized, on his imaginary sphere. Such a unit could be selected in a number of ways. The angle θ_0 mentioned above, for example, decreases steadily as the length AB increases. Hence every length is associated with an acute angle, and a natural unit of length might be the one associated with half of a right angle. Or, it might be the length of the side of an equilateral triangle having a specified angle. In any case, Lambert at least recognized that he had not proved the parallel postulate. As he said, it was always possible to develop a proof of the postulate to the point that only some small, seemingly obvious point remained unproved, but that last point nearly always concealed an assumption equivalent to what was being proved.

Some of Lambert's reasoning was recast in more precise form by Legendre, who wrote a textbook of geometry used in many places during the nineteenth century, including (in English translation) the United States. Legendre, like Lambert and Saccheri, refuted the possibility that the angle sum of a triangle could be more than two right angles and attempted to show that it could not be less. Since the defect of a triangle—the difference between two right angles and its angle sum—is additive, in the sense that if a triangle is cut into two smaller triangles, the defect of the larger triangle is the sum of the defects of the two smaller ones, he saw correctly that if one could repeatedly double a triangle, eventually the angle sum would have to become negative, which was surely impossible. Unfortunately, the possibility of repeated doubling that he had in mind was just one of those small points mentioned by Lambert that turn out to be equivalent to the parallel postulate. In fact, it is rather easy to see that such is the case, since (Fig. 11(b)) the possibility of drawing a line $B'C'$ through a point A' inside the angle CAB that intersects both AB and AC is simply another way of saying that the lines AB and AC must both intersect *some* line through A', that is, AC cannot be parallel to *every* line through A' that intersects AB.

5.3. Gauss. The true situation in regard to the parallel postulate was beginning to be understood by the end of the eighteenth century. Gauss, who read Lambert's work on parallels (which had been published posthumously), began to explore this subject as a teenager, although he kept his thoughts to himself except for letters to colleagues and never published anything on the subject. His work in this area was published in Vol. 8 of the later edition of his collected works. It is nicely summarized by Klein (*1926*, pp. 58–59). In 1799 he wrote to Farkas Bólyai (1775–1856), his classmate from Göttingen, that he could prove the parallel postulate provided that triangles of arbitrarily large area were admitted. Such a confident statement can only mean that he had developed the metric theory of hyperbolic geometry to a considerable extent. Five years later he wrote again to explain the error in a proof of the parallel postulate proposed by Bólyai. Gauss, like Lambert, realized that a non-Euclidean space would have a natural unit of length, and mentioned this fact in a letter of 1816 to his student Christian Ludwig Gerling (1788–1864), proposing as unit the side of an equilateral triangle whose angles were $59^\circ\,59'\,59.99999\ldots''$.[21] To Gauss' surprise, in 1818 he received from Gerling a paper written by Ferdinand Karl Schweikart (1780–1859), a lawyer then in Marburg, who had developed what he called *astral geometry*. It was actually hyperbolic geometry, and Schweikart had gone far into it, since he knew that there was an upper bound to the area of a triangle in this geometry, that its metric properties depended on an undetermined constant C (its radius of curvature), and that it contained a natural unit of length, which he described picturesquely by saying that if that length were the radius of the earth, then the line joining two stars would be tangent to the earth. Gauss wrote back to correct some minor points of bad drafting on Schweikart's part (for example, Schweikart neglected to say that the stars were assumed infinitely distant), but generally praising the work. In fact, he communicated his formula for the limiting area of a triangle:

$$\frac{\pi C^2}{\bigl(\ln(1+\sqrt{2})\bigr)^2}\,.$$

By coincidence, Schweikart's nephew Franz Adolph Taurinus (1794–1874), also a lawyer, who surely must have known of his uncle's work in non-Euclidean geometry, sent Gauss his attempt at a proof of the parallel postulate in 1824. Gauss explained the true situation to Taurinus under strict orders to keep the matter secret. The following year, Taurinus published a treatise *Geometriæ prima elementa* (*First Elements of Geometry*) in which he accepted the possibility of other geometries. Gauss wrote to the astronomer–mathematician Friedrich Wilhelm Bessel

[21] In comparison with the radius of curvature of space, this would be an extremely small unit of length; however, if space is curved negatively at all, its radius of curvature is so enormous that in fact this unit might be very large.

(1784–1846) in 1829 that he had been thinking about the foundations of geometry off and on for nearly 40 years (in other words, from the age of 13 on), saying that his investigations were "very extensive," but probably wouldn't be published, since he feared the controversy that would result. Some time during the mid-1820s, the time when he was writing and publishing his fundamental work on differential geometry, Gauss wrote a note—which, typically, he never published—in which he mentioned that revolving a tractrix about its asymptote produced a surface that is the opposite of a sphere. This surface turns out to be a perfect local model of the non-Euclidean geometry in which the angle sum of a triangle is less than two right angles. It is now called a pseudosphere. This same surface was discussed a decade later by Ferdinand Minding (1806–1885), who pointed out that some pairs of points on this surface can be joined by more than one minimal path, just like antipodal points on a sphere.

5.4. Lobachevskii and János Bólyai. From what has been said so far, it is clear that the full light of day was finally dawning on the subject of the parallel postulate. As more and more mathematicians worked over the problem and came to the same conclusion, from which others gained insight little by little, all that remained was a slight push to tip the balance from attempts to prove the parallel postulate to the exploration of alternative hypotheses. The fact that this extra step was taken by several people nearly simultaneously can be expressed poetically, as it was by Felix Klein (*1926*, p. 57), who referred to "one of the remarkable laws of human history, namely that the times themselves seem to hold the great thoughts and problems and offer them to heads gifted with genius when they are ripe." But we need not be quite so lyrical about a phenomenon that is entirely to be expected: When many intelligent people who have received similar educations work on a problem, it is quite likely that more than one of them will make the same discovery.

The credit for first putting forward hyperbolic geometry for serious consideration must belong to Schweikart, since Gauss was too reticent to do so. However, credit for the first full development of it, including its trigonometry, is due to the Russian mathematician Nikolai Ivanovich Lobachevskii (1792–1856) and the Hungarian János Bólyai (1802–1860), son of Farkas Bólyai. Their approaches to the subject are very similar. Both developed the geometry of the hyperbolic plane and then extended it to three-dimensional space. In three-dimensional space they considered the entire set of directed lines parallel to a given directed line in a given direction. Then they showed that a surface (now called a *horosphere*) that cuts all of these lines at right angles has all the properties of a Euclidean plane. By studying sections of this surface they were able to deduce the trigonometry of their new geometry. In modern terms the triangle formulas fully justify Lambert's assertion that this kind of geometry is that of a sphere of imaginary radius. Here, for example, is the Pythagorean theorem for a right triangle of sides a, b, c in spherical and hyperbolic geometry, derived by both Lobachevskii and Bólyai, but not in the notation of hyperbolic functions. Since $\cos(ix) = \cosh(x)$ the hyperbolic formula can be obtained from the spherical formula by replacing the radius r with ir, just as Lambert stated.

Spherical geometry	Hyperbolic geometry
$\cos\left(\frac{a}{r}\right)\cos\left(\frac{b}{r}\right) = \cos\left(\frac{c}{r}\right)$	$\cosh\left(\frac{a}{r}\right)\cosh\left(\frac{b}{r}\right) = \cosh\left(\frac{c}{r}\right).$

Lobachevskii's geometry. Lobachevskii connected the parts of a hyperbolic triangle through his formula for the angle of parallelism, which is the angle θ_0 referred to above, as a function of the length AB. He gave this formula as

$$\tan\left(\frac{1}{2}F(\alpha)\right) = e^{\alpha}\,,$$

where α denotes the length AB and $F(\alpha)$ the angle θ_0. Here e could be any positive constant, since the radius of curvature of the hyperbolic plane could not be determined. However, Lobachevskii found it convenient to take this constant to be $e = 2.71828\ldots$. In effect, he took the radius of curvature of the plane as the unit of length. Lobachevskii gave the Pythagorean theorem, for example, as

$$\sin F(a)\sin F(b) = \sin F(c)\,.$$

Of the two nearly simultaneous creators of hyperbolic geometry and trigonometry, Lobachevskii was the first to publish, unfortunately in a journal of limited circulation. He was a professor at the provincial University of Kazan' in Russia and published his work in 1826 in the proceedings of the Kazan' Physico-Mathematical Society. He reiterated this idea over the next ten years or so, developing its implications. Like Gauss, he drew the conclusion that only observation could determine if actual space was Euclidean or not. As luck would have it, the astronomers were just beginning to attempt measurements on the interstellar scale. In particular, by measuring the angles formed by the lines of sight from the Earth to a given fixed star at intervals of six months, one could get the base angles of a gigantic triangle and thereby (since the angle sum could not be larger than two right angles, as everyone agreed) place an upper bound on the size of the parallax of the star (the angle subtended by the Earth's orbit from that star). Many encyclopedias claim that the first measurement of stellar parallax was carried out in Königsberg by Bessel in 1838, and that he determined the parallax of 61 Cygni to be 0.3 seconds. Russian historians credit another Friedrich Wilhelm, namely Friedrich Wilhelm Struve (1793–1864), who emigrated to Russia and is known there as Vasilii Yakovlevich Struve. He founded the Pulkovo Observatory in 1839. Struve determined the parallax of the star Vega in 1837. Attempts to determine stellar parallax must have been made earlier, since Lobachevskii cited such measurements in an 1829 work and claimed that the measured parallax was less than $0.000372''$, which is much smaller than any observational error.[22] As he said (see his collected works, Vol. 1, p. 207, quoted by S. N. Kiro, *1967*, Vol. 2, p. 159):

> At the very least, astronomical observations prove that all the lines amenable to our measurements, even the distances between celestial bodies, are so small in comparison with the length taken as a unit in our theory that the equations of (Euclidean) plane trigonometry, which have been used up to now must be true without any sensible error.

[22] The vast distances between stars make terrestrial units of length inadequate. The light-year (about $9.5 \cdot 10^{12}$ km) is the most familiar unit now used, particularly good, since it tells us "what time it was" when the star emitted the light we are now seeing. Stellar parallax provides another unit, the parsec, which is the distance at which the radius of the Earth's orbit subtends an angle of $1''$. A parsec is about 3.258 light-years.

Thus, ironically, the acceptance of the logical consistency of hyperbolic geometry was accompanied by a nearly immediate rejection of any practical application of it in astronomy or physics. That situation was to change only much later, with the advent of relativity.

Lobachevskii was unaware of the work of Gauss, since Gauss kept it to himself and urged others to do likewise. Had Gauss been more talkative, Lobachevskii would easily have found out about his work, since his teacher Johann Martin Christian Bartels (1769–1836) had been many years earlier a teacher of the 8-year-old Gauss and had remained a friend of Gauss. As it was, however, although he continued to perfect his "imaginary geometry," as he called it, and wrote other mathematical papers, he made his career in administration, as rector of Kazan' University. He at least won some recognition for his achievement during his lifetime, and his writings were translated into French and German after his death and highly regarded.

Even though his imaginary geometry was not used directly to describe the world, Lobachevskii found some uses for it in providing geometric interpretations of formulas in analysis. In particular, his paper "Application of imaginary geometry to certain integrals," which he published in 1836, was translated into German in 1904, with its misprints corrected (Liebmann, *1904*). Just as we can compute the seemingly complicated integral

$$\int_0^r \sqrt{r^2 - x^2}\, dx = \frac{\pi}{4} r^2$$

immediately by recognizing that it represents the area of a quadrant of a circle of radius r, he could use the differential form for the element of area in rectangular coordinates in the hyperbolic plane given by $dS = (1/\sin y')\, dx\, dy$, where y' is the angle of parallelism for the distance y (in our terms $\sin y' = \operatorname{sech} y$) to express certain integrals as the non-Euclidean areas of simple figures. In polar coordinates the corresponding element of area is $dS = \cot r'\, dr\, d\theta = \sinh r\, dr\, d\theta$. Lobachevskii also gave the elements of volume in rectangular and spherical coordinates and computed 49 integrals representing hyperbolic areas and volumes, including the volumes of pyramids. These volumes turn out to involve some very complicated integrals indeed. He proved, for example, that

$$\int_0^\pi \int_0^\infty (e^x - e^{-x}) F'\bigl[a(e^x + e^x) + b \cos\omega (e^x - e^{-x})\bigr]\, dx\, d\omega = \frac{-\pi}{\sqrt{a^2 - b^2}} F\Bigl[2\sqrt{a^2 - b^2}\Bigr].$$

Bólyai's fate. János Bólyai's career turned out less pleasantly than Lobachevskii's. Even though he had the formula for the angle of parallelism in 1823, a time when Lobachevskii was still hoping to vindicate the parallel postulate, he did not publish it until 1831, five years after Lobachevskii's first publication. Even then, he had only the limited space of an appendix to his father's textbook to explain himself. His father sent the appendix to Gauss for comments, and for once Gauss became quite loquacious, explaining that he had had the same ideas many years earlier, and that none of these discoveries were new to him. He praised the genius of the young Lobachevskii for discovering it, nevertheless. Bólyai the younger was not overjoyed at this response. He suspected Gauss of trying to steal his ideas. According to Paul Stäckel (1862–1919), who wrote the story of the Bólyais, father and son (quoted in Coolidge, *1940*, p. 73), when Lobachevskii's work began to be known, Bólyai immediately thought that Gauss was stealing his work and publishing it under the pseudonym Lobachevskii, since "it is hardly likely that two or even three people knowing nothing of one another would produce almost the same result by different routes."

5.5. The reception of non-Euclidean geometry. Some time was required for the new world revealed by Lobachevskii and Bólyai to attract the interest of the mathematical community. Because it seemed possible—even easy—to prove that parallel lines exist, or equivalently, that the sum of the angles of a triangle could not be *more* than two right angles, one can easily understand why a sense of symmetry would lead to a certain stubbornness in attempts to refute the opposite hypothesis as well. Although Gauss had shown the way to a more general understanding with the concept of curvature of a surface, which could be either negative or positive, in the 1825 paper on differential geometry (published in 1827, to be discussed in detail in Sect. 3 of Chapter 12), it took Riemann's inaugural lecture in 1854 (published in 1867, also discussed in detail in Sect. 3 of Chapter 12), which made the crucial distinction between the unbounded and the infinite, to give the proper perspective. After that, acceptance of non-Euclidean geometry was quite rapid. In 1868, the year after the publication of Riemann's lecture, Eugenio Beltrami (1835–1900) realized that Lobachevskii's theorems provide a model of the Lobachevskii–Bólyai plane in a Euclidean disk. This model is described by Gray (*1989*, p. 112), as follows.

Imagine a directed line perpendicular to the Lobachevskii–Bólyai plane in Lobachevskii–Bólyai three-dimensional space. The entire set of directed lines that are parallel (asymptotic) to this line on the same side of the plane generates a unique horosphere tangent to the plane at its point of intersection with the line. Some of the lines parallel to the given perpendicular in the given direction intersect the original plane, and others do not. Those that do intersect it pass through the portion of the horosphere denoted Ω in Fig. 12. Shortest paths on the horosphere are obtained as its intersections with planes passing through the point at infinity that serves as its "center." These paths are called *horocycles.* But there is only one horocycle through a given point in Ω that does not intersect a given horocycle, so that the geometry of Ω is Euclidean. As a result, we have a faithful mapping of the Lobachevskii–Bólyai plane onto the interior of a disk Ω in a Euclidean plane, under which lines in the plane correspond to chords on the disk. This model provides an excellent picture of points at infinity: they correspond to the boundary of the disk Ω. Lines in the plane are parallel if and only if the chords corresponding to them have a common endpoint. Lines that have a common perpendicular in the Lobachevskii–Bólyai plane correspond to chords whose extensions meet outside the circle. It is somewhat complicated to compute the length of a line segment in the Lobachevskii–Bólyai plane from the length of its corresponding chordal segment in Ω or vice versa, and the angle between two intersecting chords is not simply related to the angle between the lines they correspond to.[23] Nevertheless these computations can be carried out from the trigonometric rules given by Lobachevskii. The result is a perfect model of the Lobachevskii–Bólyai plane *within the Euclidean plane*, obtained by formally reinterpreting the words *line*, *plane*, and *angle*. If

[23] It can be shown that perpendicular lines correspond to chords having the property that the extension of each passes through the point of intersection of the tangents at the endpoints of the other. But it is far from obvious that this property is symmetric in the two chords, as perpendicularity is for lines.

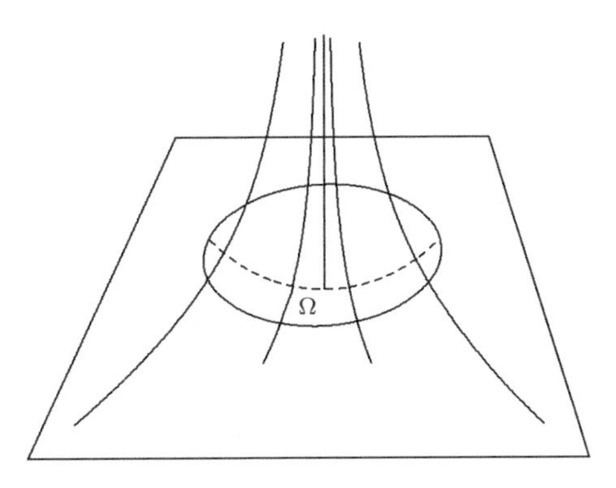

FIGURE 12. Projection of the Lobachevskii–Bólyai plane onto the interior of a Euclidean disk.

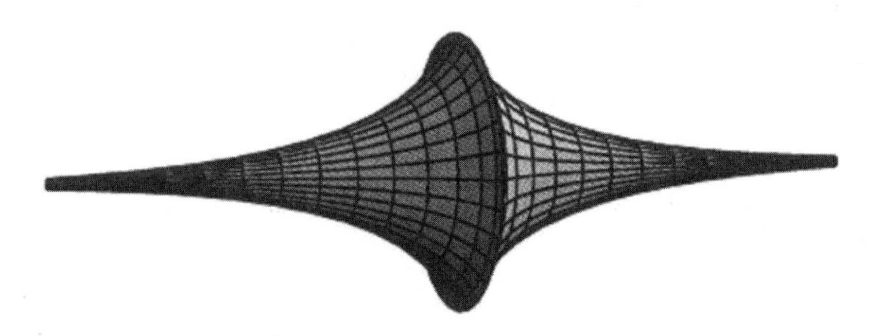

FIGURE 13. The pseudosphere. Observe that it has no definable curvature at its cusp. Elsewhere its curvature is constant and negative.

there were any contradiction in the new geometry, there would be a corresponding contradiction in Euclidean geometry itself.

A variant of this model was later provided by Henri Poincaré (1854–1912), who showed that the diameters and the circular arcs in a disk that meet the boundary in a right angle can be interpreted as lines, and in that case angles can be measured in the ordinary way.

Beltrami also provided a model of a portion of the Lobachevskii–Bólyai plane that could be embedded in three-dimensional Euclidean space: the pseudosphere obtained by revolving a tractrix about its asymptote, as shown in Fig. 13.

In 1871 Felix Klein gave a discussion of the three kinds of plane geometry in his article "Über die sogennante nicht-Euklidische Geometrie" ("On the so-called non-Euclidean geometry"), published in the *Mathematische Annalen.* In that article he gave the classification of them that now stands, saying that the points at infinity on a line were distinct in hyperbolic geometry, imaginary in spherical geometry, and coincident in parabolic (Euclidean) geometry.

The pseudosphere is not a model of the entire Lobachevskii–Bólyai plane, since its curvature has a very prominent discontinuity. The problem of finding a surface in three-dimensional Euclidean space that was a perfect model for the Lobachevskii–Bólyai plane, in the sense that its geodesics corresponded to straight lines and lengths and angles were measured in the ordinary way, remained open until Hilbert, in an article "Über Flächen von konstanter Gaußscher Krümmung" ("On surfaces of constant Gaussian curvature"), published in the *Transactions of the American Mathematical Society* in 1901, showed that no such surface exists.

5.6. Foundations of geometry. The problem of the parallel postulate was only one feature of a general effort on the part of mathematicians to improve on the rigor of their predecessors. This problem was particularly acute in the calculus, but the parts of calculus that raised the most doubts were those that were geometric in nature. Euclid, it began to be realized, had taken for granted not only the infinitude of the plane, but also its continuity, and had not specified in many cases what ordering of points was needed on the line for a particular theorem to be true. If one attempts to prove these theorems without drawing any figures, it becomes obvious what is being assumed. It seemed obvious, for example, that a line joining a point inside a circle to a point outside the circle must intersect the circle in a point, but that fact could not be deduced from Euclid's axioms. A complete reworking of Euclid was the result, expounded in detail in Hilbert's *Grundlagen der Geometrie* (*Foundations of Geometry*), published in 1903. This book went through many editions and has been translated into English (Bernays, *1971*). In Hilbert's exposition the axioms of geometry are divided into axioms of incidence, order, congruence, parallelism, and continuity, and examples are given to show what cannot be proved when some of the axioms are omitted.

One thing is clear: No new comprehensive geometries are to be expected by pursuing the axiomatic approach of Hilbert. In a way, the geometry of Lobachevskii and Bólyai was a throwback even in its own time. The development of projective and differential geometry would have provided—indeed, *did* provide—non-Euclidean geometry by a natural expansion of the study of surfaces. It was Riemann, not Lobachevskii and Bólyai, who showed the future of geometry. The real "action" in geometry since the early nineteenth century has been in differential and projective geometry. That is not to say that no new theorems can be produced in Euclidean geometry, only that their scope is very limited. There are certainly many such theorems. Coolidge, who undertook the herculean task of writing his *History of Geometric Methods* in 1940, stated in his preface that the subject was too vast to be covered in a single treatise and that "the only way to make any progress is by a rigorous system of exclusion." In his third chapter, on "later elementary geometry," he wrote that "the temptation to run away from the difficulty by not considering elementary geometry after the Greek period at all is almost irresistible." But to attempt to build an entire theory as Apollonius did, on the synthetic methods and limited techniques in the Euclidean tool kit, would be futile. Even Lobachevskii and Bólyai at least used analytic geometry and trigonometry to produce their results. Modern geometries are much more algebraic, as we shall see in Chapter 12.

6. Questions and problems

11.1. The figure used by Zenodorus at the main step in his proof of the isoperimetric inequality had been used earlier by Euclid to show that the apparent size of objects is not inversely proportional to their distance. Prove this result by referring

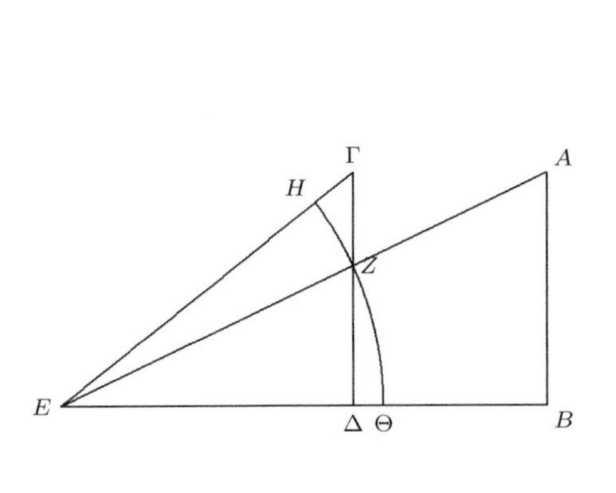

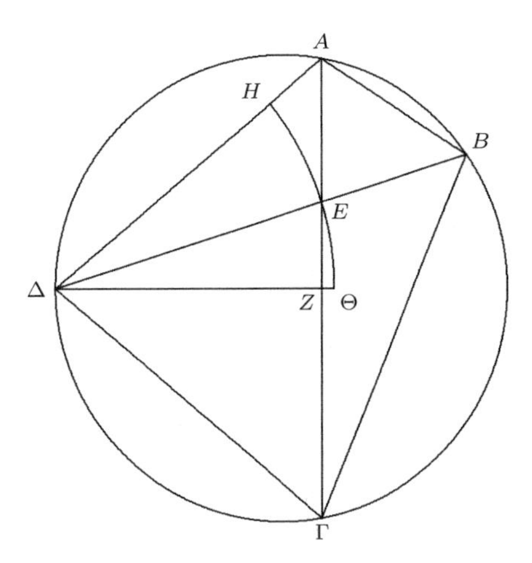

FIGURE 14. Greek use of a fundamental inequality. Left: from Euclid's *Optics*. Right: from Ptolemy's *Almagest*.

to the diagram on the left in Fig. 14. Show that $BE : E\Delta :: AB : Z\Delta :: \Gamma\Delta : Z\Delta$ and that this last ratio is larger than $\overset{\frown}{H\Theta}{:}\overset{\frown}{Z\Theta}$.

11.2. Use the diagram on the right in Fig. 14 to show that the ratio of a larger chord to a smaller is less than the ratio of the arcs they subtend, that is, show that $B\Gamma : AB$ is less than $\overset{\frown}{B\Gamma}{:}\overset{\frown}{AB}$, where $A\Gamma$ and ΔZ are perpendicular to each other. (Hint: $B\Delta$ bisects angle $AB\Gamma$.) Ptolemy said, paradoxically, that the chord of $1°$ had been proved "both larger and smaller than the same number" so that it must be *approximately* $1;2,50$.) Carry out the analysis carefully and get accurate upper and lower bounds for the chord of $1°$. Convert this result to decimal notation, and compare with the actual chord of $1°$ which you can find from a calculator. (It is $120\sin\left(\frac{1}{2}°\right)$.)

11.3. Let A, B, C, and D be squares such that $A : B :: C : D$, and let r, s, t, and u be their respective sides. Show that $r : s :: t : u$ by strict Eudoxan reasoning, giving the reason for each of the following implications. Let m and n be any positive integers. Then

$$mr > ns \Rightarrow m^2A > n^2B \Rightarrow m^2C > n^2D \Rightarrow mt > nu\,.$$

11.4. Sketch a proof of Pappus' theorem on solids of revolution by beginning with right triangles having a leg parallel to the axis of rotation, then progressing to unions of areas for which the theorem holds, and finally to general areas that can be approximated by unions of triangles.

11.5. Explain how Thabit ibn-Qurra's generalization of the Pythagorean theorem reduces to that theorem when angle A is a right angle. What does the figure look like if angle A is obtuse? Is there an analogous theorem if BC is not the longest side of the triangle?

11.6. One form of non-Euclidean geometry, known as doubly elliptic geometry, is formed by replacing the plane with a sphere and straight lines with great circles, that is, the intersections of the sphere with planes passing through its center. Let one "line" (great circle) be the equator of the sphere. Describe the equidistant curve generated by the endpoint of a "line segment" (arc of a great circle) of fixed length and perpendicular to the equator when the other endpoint moves along the equator. Why is this curve not a "line"?

11.7. Al-Haytham's attempted proof of the parallel postulate is fallacious because in non-Euclidean geometry two straight lines cannot be equidistant at all points. Thus in a non-Euclidean space the two rails of a railroad cannot both be straight lines. Assuming Newton's laws of motion (an object that does not move in a straight line must be subject to some force), show that in a non-Euclidean universe one of the wheels in a pair of opposite wheels on a train must be subject to some unbalanced force at all times. [Note: The spherical earth that we live on happens to be non-Euclidean. Therefore the pairs of opposite wheels on a train cannot both be moving in a great circle on the earth's surface.]

11.8. Prove that in any geometry, if a line passes through the midpoint of side AB of triangle ABC and is perpendicular to the perpendicular bisector of the side BC, then it also passes through the midpoint of AC. (This is easier than it looks: Consider the line that *does* pass through both midpoints, and show that it is perpendicular to the perpendicular bisector of BC; then argue that there is only one line passing through the midpoint of BC that is perpendicular to the perpendicular bisector of BC.)

11.9. Use the previous result to prove, independently of the parallel postulate, that the line joining the midpoints of the lateral sides of a Thabit (Saccheri) quadrilateral bisects the diagonals.

CHAPTER 12

Modern Geometries

In geometry, as in number theory, the seventeenth century represents a break with the past. The two main reasons for the sudden surge of mathematical activity are the same in both cases: first, the availability of translations from the Arabic, which stimulated European mathematicians to try to recover and extend the fascinating results achieved by the ancient Greeks and medieval Muslims; second, the development of algebra and its evolution into a symbolic form in the Italian city-states during the sixteenth century. This development suggested new ways of thinking about old problems. The result was a variety of new forms of geometry that came about as a result of the calculus: analytic geometry, algebraic geometry, projective geometry, descriptive geometry, differential geometry, and topology.

1. Analytic and algebraic geometry

The creation of what we now know as analytic geometry had to wait for algebraic thinking about geometry (the type of thinking Pappus called *analytic*) to become a standard mode of thinking. No small contribution to this process was the creation of the modern notational conventions, many of which were due to François Viète (1540–1603) and Descartes. It was Descartes who started the very useful convention of using letters near the beginning of the alphabet for constants and data and those near the end of the alphabet for variables and unknowns. Viète's convention, which was followed by Fermat, had been to use consonants and vowels respectively for these purposes.

1.1. Fermat. Besides working in number theory, Fermat studied the works of Apollonius, including references by Pappus to lost works. This study inspired him to write a work on plane and solid loci, first published with his collected works in 1679. He used these terms in the sense of Pappus: A plane locus is one that can be constructed using straight lines and circles, and a solid locus is one that requires conic sections for its construction. He says in the introduction that he hopes to systematize what the ancients, known to him from Book 7 of Pappus' *Synagōgē*, had left haphazard. Pappus had written that the locus to more than six lines had hardly been touched. Thus, locus problems were the context in which Fermat invented analytic geometry.

Apart from his adherence to a dimensional uniformity that Descartes (finally!) eliminated, Fermat's analytic geometry looks much like what we are now familiar with. He stated its basic principle very clearly, asserting that the lines representing two unknown magnitudes should form an angle that would usually be assumed a right angle. He began with the equation of a straight line:[1] $Z^2 - DA = BE$. This equation looks strange to us because we automatically (following Descartes) tend

[1] Fermat actually wrote "*Z pl.* − *D* in *A* æquetur *B* in *E*."

to look at the Z as a variable and the A and E as constants, exactly the reverse of what Fermat intended. If we make the replacements $Z \mapsto c$, $D \mapsto a$, $A \mapsto x$, $E \mapsto y$, this equation becomes $c^2 - ax = by$, and now only the exponent looks strange, the result of Fermat's adherence to the Euclidean niceties of dimension.

Fermat illustrated the claim of Apollonius that a locus was determined by the condition that the sum of the pairwise products of lines from a variable point to given lines is given. His example was the case of two lines, where it is the familiar rectangular hyperbola that we have now seen used many times for various purposes. Fermat wrote its equation as $ae = z^2$. He showed that the graph of any quadratic equation in two variables is a conic section.

1.2. Descartes. Fermat's work on analytic geometry was not published in his lifetime, and therefore was less influential than it might have been. As a result, his contemporary René Descartes is remembered as the creator of analytic geometry, and we speak of "Cartesian" coordinates, even though Fermat was more explicit about their use.

René Descartes is remembered not only as one of the most original and creative modern mathematicians, but also as one of the leading voices in modern philosophy and science. Both his scientific work on optics and mechanics and his geometry formed part of his philosophy. Like Plato, he formed a grand project of integrating all of human knowledge into a single system. Also like Plato, he recognized the special place of mathematics in such a system. In his *Discourse on Method*, published at Leyden in 1637, he explained that logic, while it enabled a person to make correct judgments about inferences drawn through syllogisms, did not provide any actual knowledge about the world, what we would call empirical knowledge. In what was either a deadpan piece of sarcasm or a sincere tribute to Ramon Lull (mentioned above in Chapter 8), he said that in the art of "Lully" it enabled a person to speak fluently about matters on which he is entirely ignorant. He seems to have agreed with Plato that mathematical concepts are real objects, not mere logical relations among words, and that they are perceived directly by the mind. In his famous attempt at doubting everything, he had brought himself back from utter skepticism by deducing the principle that whatever he could clearly and distinctly perceive with his mind must be correct.

As Davis and Hersh (*1986*) have written, the *Discourse on Method* was the fruit of a decade and a half of hard work and thinking on Descartes' part, following a series of three vivid dreams on the night of November 10, 1619, when he was a 23-year-old soldier of fortune. The link between Descartes' philosophy and his mathematics lies precisely in the matter of "clear and distinct perception." For there seems to be no other area of thought in which human ideas are so clear and distinct. As Grabiner (*1995*, p. 84) says, when Descartes attacked, for example, a locus problem, the answer had to be "it is this curve, it has this equation, and it can be constructed in this way." Descartes' *Géométrie*, which contains his ideas on analytic geometry, was published as the last of three appendices to the *Discourse*.

What Descartes meant by "clear and distinct" ideas in mathematics is shown in a method of generating curves given in his *Géométrie* that appears mechanical, but can be stated in pure geometric language. A pair of lines intersecting at a fixed point Y coincide initially (Fig. 1). The point A remains fixed on the horizontal line. As the oblique line rotates about Y, the point B, which remains fixed on it, describes a circle. The tangent at B intersects the horizontal line at C, and

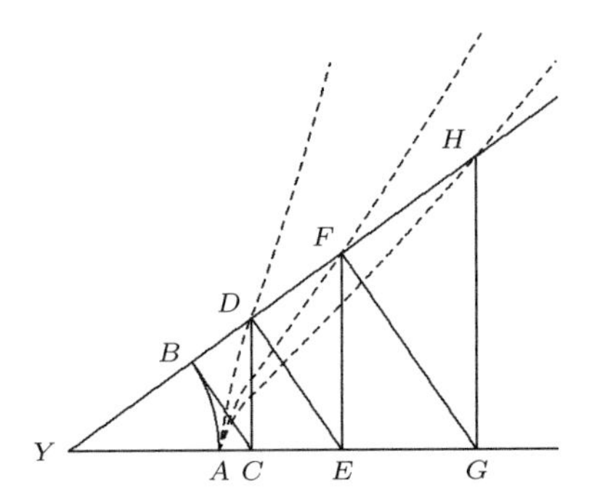

FIGURE 1. Descartes' linkage for generating curves. The curve $x^{4n} = a^2(x^2 + y^2)^{2n-1}$ is shown for $n = 0, 1, 2, 3$.

the point on the oblique line directly above C is D. The line perpendicular to the oblique line at D intersects the horizontal line at E, from which a vertical line intersects the oblique line at F, and so forth in a zigzag pattern. Descartes imagined a mechanical linkage that could actually draw these curves.

Descartes regarded determinate curves of this sort, depending on one parameter, as we would say, as legitimate to use in geometry. He offered the opinion that the opposition to "mechanical" curves by ancient Greek mathematicians arose because the curves they knew about—he mentioned the spiral of Archimedes and the quadratix—were indeterminate. In the case of the spiral of Archimedes, which is generated by a point moving at constant linear velocity along a line that is rotating with constant angular velocity, the indeterminacy arises because the two velocities need to be coordinated with infinite precision. For the quadratrix, the same problem arises, as the ratio of the velocity of a rotating line and that of a translating line needs to be known with infinite precision.

Descartes' *Géométrie* resembles a modern textbook of analytic geometry less than does Fermat's *Introduction to Plane and Solid Loci.* He does not routinely use a system of "Cartesian" coordinates, as one might expect from the name. But he does remove the dimensional difficulties that had complicated geometric arguments since Euclid's cumbersome definition of a composite ratio.

> [U]nity can always be understood, even when there are too many or too few dimensions; thus, if it be required to extract the cube root of $a^2b^2 - b$, we must consider the quantity a^2b^2 divided once by unity, and the quantity b multiplied twice by unity. [Smith and Latham, *1954*, p. 6]

Here Descartes is explaining that all four arithmetic operations can be performed on *lines* and yield *lines* as a result. He illustrated the product and square root by the diagrams in Fig. 2, where $AB = 1$ on the left and $FG = 1$ on the right.

Descartes went a step further than Oresme in eliminating dimensional considerations, and he went a step further than Pappus in his classification of locus problems. Having translated these problems into the language of algebra, he realized that the three- and four-line locus problems always led to polynomial equations of degree at most 2 in x and y, and conversely, any equation of degree 2 or less represented a three- or four-line locus. He asserted with confidence that he had solved

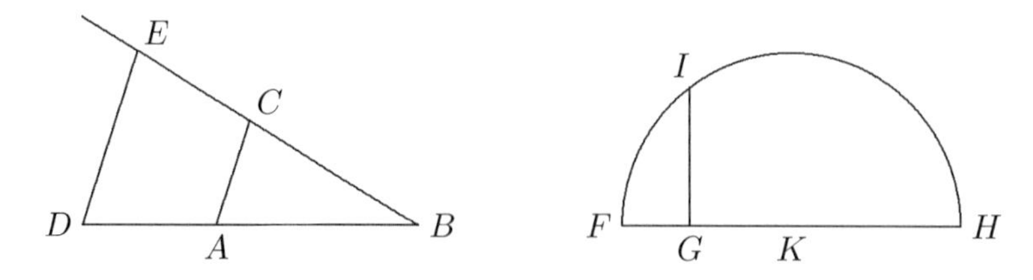

FIGURE 2. Left: $AB = 1$, so that $BE = BC \cdot BD$. Right: $FG = 1$, so that $GI = \sqrt{GH}$.

the problem that Pappus reported unsolved in his day. It was in this context that he formulated the idea of using two intersecting lines as a frame of reference, saying that

> since so many lines are confusing, I may simplify matters by considering one of the given lines and one of those to be drawn... as the principal lines, to which I shall try to refer all the others. [Smith and Latham, *1954*, p. 29]

The idea of using two coordinate lines is psychologically very close to the linkages illustrated in Fig. 1. In terms of Fig. 3, Descartes took one of the fixed lines as a horizontal axis AB, since a line was to be drawn from point C on the locus making a fixed angle θ with AB. He thought of this line as sliding along AB and intersecting it at point B, and he denoted the variable length AB by x. Then since C needed to slide along this moving line so as to keep the proportions demanded by the conditions of the locus problem, he denoted the distance CB by y. All the lines were fixed except CB, which moved parallel to itself, causing x to vary, while on it y adjusted to the conditions of the problem. For each of the other fixed lines, say AR, the angles ψ, θ, and φ will all be given, ψ by the position of the fixed lines AB and AR, and the other two by the conditions prescribed in the problem. Since these three angles determine the shape of the triangles ADR and BCD, they determine the ratios of any pair of sides in these triangles through the law of sines, and hence all sides can be expressed in terms of constants and the two lengths x and y. If the set of $2n$ lines is divided into two sets of n as the $2n$-line locus problem requires, the conditions of the problem can be stated as an equation of the form

$$p(x, y) = q(x, y)\,,$$

where p and q are of degree at most n in each variable. The analysis was mostly "clear and distinct."

Descartes argued that the locus could be considered known if one could locate as many points on it as desired.[2] He next pointed out that in order to locate points on the locus one could assign values to either variable x and y, then compute the value of the other by solving the equation.[3]

Everyone who has studied analytic geometry in school must have been struck at the beginning by how much clearer and easier it was to use than the synthetic geometry of Euclid. That aspect of the subject is nicely captured in the words the poet Paul Valéry (1871–1945) applied to Descartes' philosophical method in

[2] The validity of this claim is somewhat less than "clear and distinct."

[3] This claim also involves a great deal of hope, since equations of degree higher than 4 were unknown territory in his day.

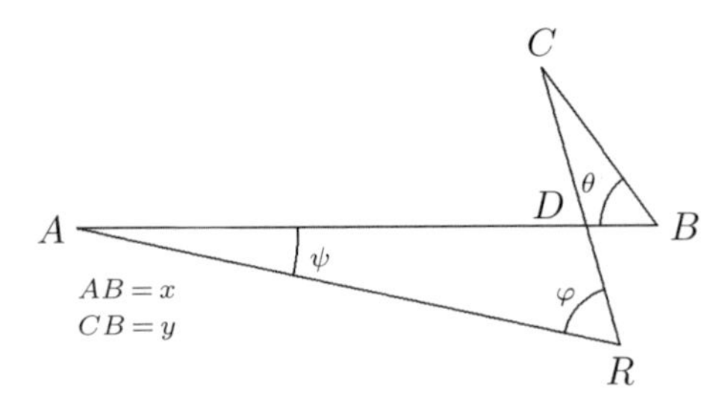

FIGURE 3. Descartes' analysis of the n-line locus problem.

general: "the most brilliant victory ever achieved by a man whose genius was applied to reducing the need for genius" (quoted by Davis and Hersh, *1986*, p. 7).

This point was ignored by Newton in a rather ungenerous exhibition of his own remarkable mathematical talent (Whiteside, *1967*, Vol. IV, pp. 275–283). Newton said that Descartes "makes a great show" about his solution of the three- and four-line locus problems, "as if he had achieved something so earnestly sought after by the ancients." He also expressed a distaste for Descartes' use of symbolic algebra to solve this problem (a distaste that would be echoed by other mathematicians), saying that if this algebra were written out in words, it "would prove to be so tedious and entangled as to provoke nausea." One is inclined to say, on Descartes' behalf, "Precisely! That's why it's better to use algebraic symbolism and avoid the tedium, confusion, and nausea."

1.3. Newton's classification of curves. Like Descartes, Newton made a classification of curves according to the degree of the equations that represent them, or rather, according to the maximal number of points in which they could intersect a straight line. As Descartes had argued for the use of any curves that could be generated by one parameter, excluding spirals and the quadratrix because they required two independent motions to be coordinated, Newton likewise argued that geometers should either confine themselves to conic sections or else allow any curve having a clear description. In his *Universal Arithmetick*, he mentioned in particular the trochoid,[4] which makes it possible to divide an angle into any number of equal parts, as a useful curve that is simple to describe.

1.4. Algebraic geometry. As we have just seen, Descartes began the subject of algebraic geometry with his classification of algebraic curves into genera, and Newton gave an alternative classification of curves also, based on algebra, although he included some curves that we would call transcendental, curves that could intersect a line in infinitely many points. The general study of algebraic curves $p(x, y) = 0$, where $p(x, y)$ is a polynomial in two variables, began with Colin Maclaurin (1698–1746), who in his *Geometria organica* of 1720 remarked that a cubic curve was not uniquely determined by nine points, even though nine points apparently suffice to determine the coefficients of any polynomial $p(x, y)$ of degree 3, up to proportionality and hence determine a unique curve $p(x, y) = 0$. Actually, however, two *distinct*

[4] A trochoid is the locus of a point rigidly attached to a rolling wheel. If the point lies between the rim and the center, the trochoid is called a *curtate cycloid.* If the point lies outside the rim, the trochoid is a *prolate cycloid.* If the point is right on the rim, the trochoid is called a *cycloid.* The names come from the Greek words *trokhós* (*wheel*) and *kýklos* (*circle*).

cubic curves generally intersect in nine points, so that *some* sets of nine points do not determine the curve uniquely (see Problem 12.7). This fact was later (1748) noted by Euler as well, and finally, by Gabriel Cramér (1704–1752), who also noted Maclaurin's priority in the discovery that a curve of degree m and a curve of degree n meet generally in mn points. This interesting fact is called *Cramér's paradox* after Cramér published it in a 1750 textbook on algebraic curves. Although he correctly explained why more than one curve of degree n can sometimes be made to pass through $n(n+3)/2$ points—because the equations for determining the coefficients from the coordinates of the points might not be independent—he noted that in that case there were actually infinitely many such curves. That, he said, was a real paradox. Incidentally, it was in connection with the determination of the coefficients of an algebraic curve through given points that Cramér stated Cramér's rule for solving a system of linear equations by determinants.[5]

2. Projective and descriptive geometry

It is said that Euclid's geometry is tactile rather than visual, since the theorems tell you what you can measure and feel with your hands, not what your eye sees. It is a commonplace that a circle seen from any position except a point on the line through its center perpendicular to its plane appears to be an ellipse. If figures did not distort in this way when seen in perspective, we would have a very difficult time navigating through the world. We are so accustomed to adjusting our judgments of what we see that we usually recognize a circle automatically when we see it, even from an angle. The distortion is an essential element of our perception of depth. Artists, especially those of the Italian renaissance, used these principles to create paintings that were astoundingly realistic. As Leonardo da Vinci (1452–1519) said, "the primary task of a painter is to make a flat plane look like a body seen in relief projecting out of it." Many records of the principles by which this effect was achieved have survived, including treatises of Leonardo himself and a very famous painter's manual of Albrecht Dürer (1471–1528), first published in 1525. Over a period of several centuries these principles gave rise to the subject now known as projective geometry.

2.1. Projective properties.

Projective geometry studies the mathematical relations among figures that remain constant in perspective. Among these things are points and lines, the number of intersections of lines and circles, and consequently also such things as parallelism and tangency, but not things that depend on shape, such as angles or circles.

A less obvious property that is preserved is what is now called the *cross-ratio* of four points on a line.[6] If A, B, C, and D are four points on a line, with B and

---

[5] As mentioned in Chapter 8, the solution of linear equations by determinants had been known to Seki Kōwa and Leibniz. Thus, Cramér has two mathematical concepts named after him, and in both cases he was the third person to make the discovery.

[6] Although this ratio has been used for centuries, the name it now bears in English seems to go back only to an 1869 treatise on dynamics by William Kingdon Clifford (1845–1879). Before that it was called the *anharmonic ratio*, a phrase translated from an 1837 French treatise by Michel Chasles (1809–1880). This information came from the website on the history of mathematical terms maintained by Jeff Miller of Gulf High School in New Port Richey, Florida. The url of the website is `http://members.aol.com/jeff570/mathword.html`.

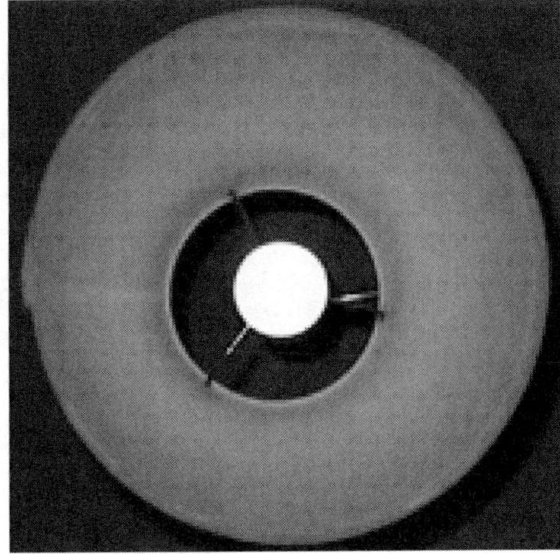

A circle seen in perspective is an ellipse.

C both between A and D and C between B and D, their cross-ratio is

$$(A, B, C, D) = \frac{AC \cdot BD}{AD \cdot BC}.$$

It is not difficult to show, for example, that if the rays PA, PB, PC, and PD from a point P intersect a second line in points A', B', C', and D', the cross-ratio of these new points is the same as that of the original four points. Coolidge (*1940*, p. 88) speculated that Euclid may have known about the cross-ratio, and he asserted that the early second-century mathematician Menelaus did know about it.

Some theorems that might appear difficult to prove from the standard Euclidean techniques of proportion and congruence can be quite easy when looked at "in perspective," so to speak. For example, it is easy to prove that if two tangents to a circle from points A and C meet at a point P, then the line from P to the midpoint of the chord AC meets the circle in a point (namely the midpoint of the arc $\overset{\frown}{AC}$) at which the tangent to the circle is parallel to the chord AC. To prove that same theorem for an ellipse using analytic geometry is a very tedious computation. However, remembering that the ellipse was obtained as the intersection of a cone with a plane oblique to its base, one has only to note that projection preserves tangency, intersections, parallelism (usually), and midpoints. Then, projecting the cone and all the lines into the base plane yields the result immediately, as shown in Fig. 4.[7] Similarly, it could be shown that the bisector of a chord from the point of intersection of the tangents at the endpoints of the chord passes through the center of the ellipse.

2.2. The Renaissance artists. The revival of interest in ancient culture in general during the Renaissance naturally carried with it an interest in geometry. The famous artist Piero della Francesca (1410?–1492) was inspired by the writings of Leonardo of Pisa and others to write treatises on arithmetic and the five regular solids. The scholar Luca Pacioli (1445–1517), who was influenced by Piero della Francesca and was a friend of Leonardo da Vinci, published a comprehensive

[7] Of course, in the figure the "circle" in the base is really an ellipse because it has been projected onto the page.

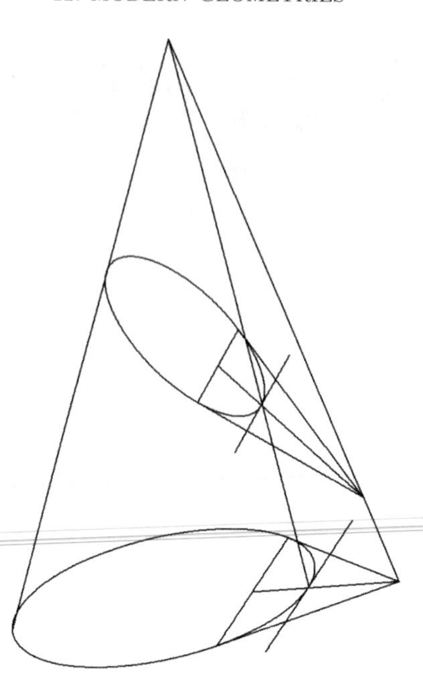

FIGURE 4. Central projections preserve tangency, midpoints, and (usually) parallelism.

treatise on arithmetic and geometry in 1494, and a second book, *De divina proportione*, in 1509. He gave the name *divine proportion* to what is now called the *Golden Section*, the division of a line into mean and extreme ratios. Interest in the five regular solids branched out into an interest in semiregular solids. Leonardo da Vinci designed wooden models of these, which were depicted in Pacioli's treatise.

The regular and semiregular solids formed an important part of Dürer's manual for painters, published in 1525. He showed how to cut out a paper model of a truncated icosahedron, which consists of 12 pentagons and 20 hexagons (Fig. 5). The solid, although not the name, has become very familiar to modern people through its application in athletics and organic chemistry.

A geometric description of perspective was given by Leon Battista Alberti (1404–1472) in a treatise entitled *Della pictura*, published posthumously in 1511. If the eye is at fixed height above a horizontal plane, parallel horizontal lines in that plane receding from the imagined point where the eye is located can be drawn as rays emanating from a point (the vanishing point) at the same height above the plane, giving the illusion that the vanishing point is infinitely distant. The application to art is obvious: Since the canvas can be thought of as a window through which the scene is viewed, if you want to draw parallel horizontal lines as they would appear through a window, you must draw them as if they all converged on the vanishing point. Thus, a family of lines having a common property (passing through the vanishing point) projects to a family having a different common property (being parallel to one another). Obviously, lines remain lines under such a projection. However, perpendicular lines will not remain perpendicular, nor will circles remain circles.

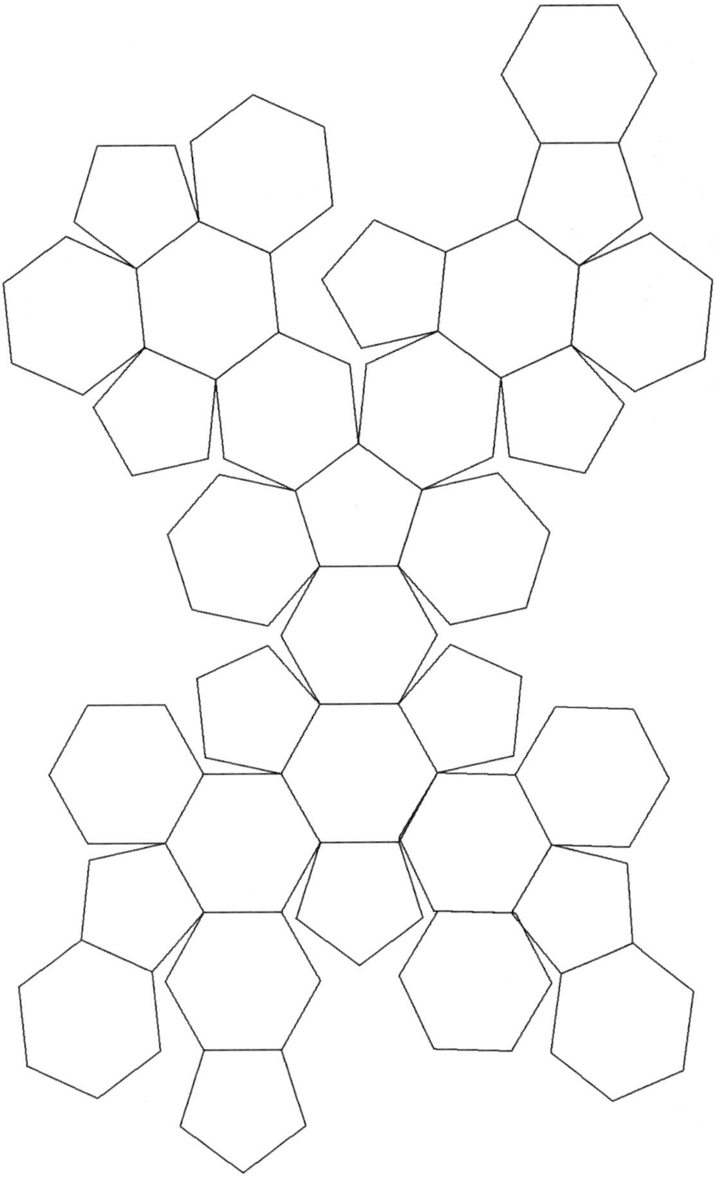

FIGURE 5. Dürer's paper model of a truncated icosahedron.

In those days before photography and computers, the mechanical aspects of drawing according to Alberti's rules apparently did not disturb artists. Dürer, in particular, seemed to enjoy thinking up mechanical ways of producing technical perfection. One of his devices is shown below. Although the device seems a very

Two modern applications of the truncated icosahedron: a molecule of buckminsterfullerene ("buckyball"); a soccer ball.

strange and inefficient way of painting, it does illustrate the use of projection very vividly, even if it was only a "thought experiment."[8]

2.3. Girard Desargues. The mathematical development of the theory of projection began with the work of Girard Desargues (1593–1662). In 1636, one year before the publication of Descartes' *Géométrie*, Desargues published a pamphlet with the ponderous title *An Example of One of the General Methods of S.G.D.L.*[9] *Applied to the Practice of Perspective Without the Use of Any Third Point, Whether of Distance or Any Other Kind, Lying Outside the Work Area.* The reference to a "third point" was aimed at the primary disadvantage of Alberti's rules, the need to use a point not on the canvas in order to get the perspective correct. Three years later he produced a *Rough Draft of an Essay on the Consequences of Intersecting a Cone with a Plane.* In both works, written in French rather than the more customary Latin, he took advantage of the vernacular to invent new names, not only for the conic sections,[10] as Dürer had done, but also for a large number of concepts that called attention to particular aspects of the distribution and proportions of points and lines. He was particularly fond of botanical names,[11] and included *tree*, *trunk*, *branch*, *shoot*, and *stem*, among many other neologisms. Although the new language might seem distracting, using standard terms for what he had in mind would have been misleading, since the theory he was constructing unified concepts that had been distinct before. For example, he realized that a cylinder could be regarded as a limiting case of a cone, and so he gave the name *scroll* to the class consisting of both surfaces. Desargues had very little need to refer to any specific conic section; his theorems applied to all of them equally. As he said (Field and Gray, *1987*, p. 102—I have changed their *roll* to *scroll*):

[8] According to Strauss (*1977*, p. 31), painters of Dürer's time who actually tried to build such devices found them quite impracticable.

[9] Sieur Girard Desargues Lyonnois.

[10] He gave the standard names, but suggested *deficit*, *equalation*, and *exceedence* as alternatives.

[11] Ivins (*1947*, cited by Field and Gray, *1987*, p. 62) suggested that these names were inspired by similar names in Alberti's treatise.

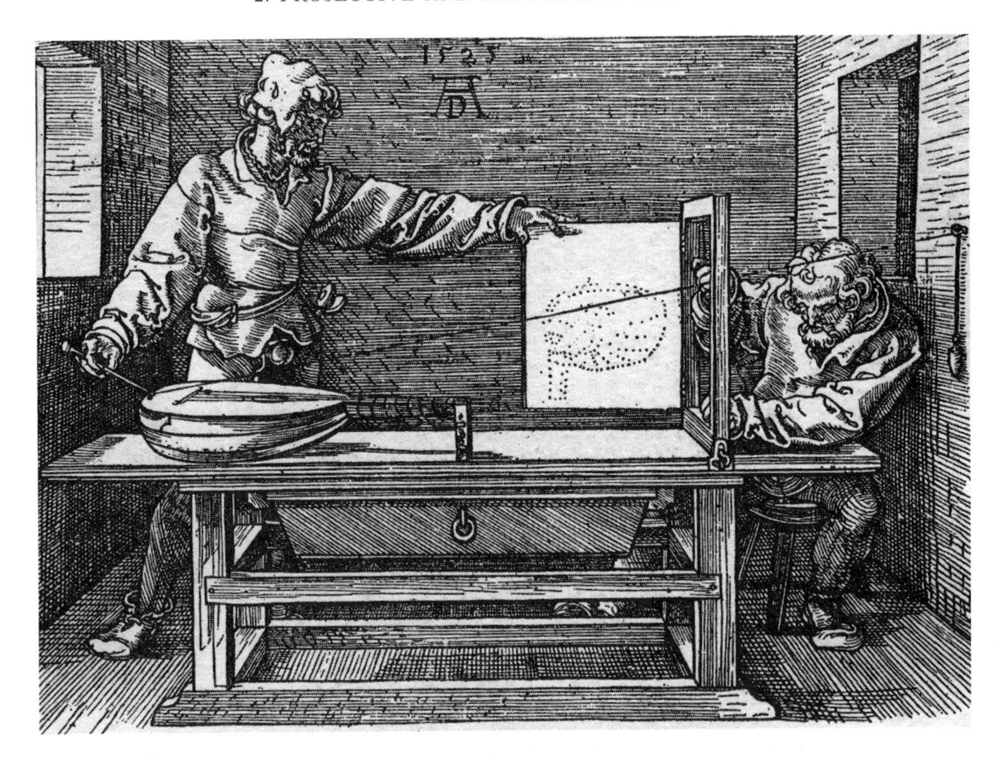

One of Dürer's devices for producing an accurate painting. The artist's assistant at the left holds a needle at a particular point on the lute being painted, while the artist sticks a pair of crosshairs on the frame to mark the exact point where the thread passes through the window. The needle and thread are then to be removed, the door holding the canvas closed, and the spot where the crosshairs meet marked on the canvas.

> The most remarkable properties of the sections of a scroll are common to all types, and the names *Ellipse*, *Parabola*, and *Hyperbola* have been given them only on account of matters extraneous to them and to their nature.

Desargues was among the first to regard lines as infinitely long, in the modern way. In fact, he opens his treatise by saying that he will consider both the infinitely large and the infinitely small in his work, and he says firmly that "in this work every straight line is, if necessary, taken to be produced to infinity in both directions." He also had the important insight that a family of parallel lines and a family of lines with a common point of intersection have similar properties. He said that lines belonged to the same *order*[12] if either they all intersected at a common point or were all mutually parallel. This term was introduced "[to] indicate that in the one case as well as in the other, it is *as if* they all converged to the same place" [emphasis added].

[12] Now called a *pencil* or *sheaf*.

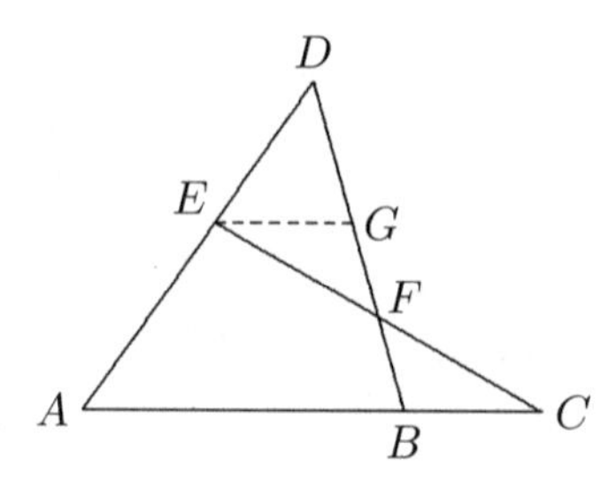

FIGURE 6. Menelaus' theorem for a plane triangle.

Although Desargues' terminology is very difficult to follow, his *Rough Draft* contained some elegant theorems about points on conics. Two significant results are the following:[13]

First: *If four lines in a plane intersect two at a time, and the points of intersection on the first line are A, B, and C, with B between A and C, and the lines through A and B intersect in the point D, those through A and C in E and those through B and C in F, then*

$$\frac{BD}{BF} = \frac{AD}{AE} \cdot \frac{CE}{CF}. \tag{1}$$

The situation here was described by Pappus, and the result is also known as Menelaus' theorem. The proof is easily achieved by drawing the line through E parallel to AB, meeting BD in a point G, then using the similarity of triangles EGF and CBF and of triangles DEG and DAB, as in Fig. 6. From Eq. 1 it is easy to deduce that $BD \cdot AE \cdot CF = BF \cdot AD \cdot CE$. Klein (*1926*, p. 80) attributes this form of the theorem to Lazare Carnot (1753–1823).

Second: *The converse of this statement is also true, and can be interpreted as stating that three points lie on a line.* That is, if ADB is a triangle, and E and F are points on AD and BD respectively such that $AD : AE < BD : BF$, then the line through E and F meets the extension of AB on the side of B in a point C, which is characterized as the only point on the line EF satisfying Eq. 1.

In 1648 the engraver Abraham Bosse (1602–1676), who was an enthusiastic supporter of Desargues' new ideas, published *La Perspective de Mr Desargues*, in which he reworked these ideas in detail. Near the end of the book he published the theorem that is now known as Desargues' theorem. Like Desargues' work, Bosse's statement of the theorem is a tangled mess involving ten points denoted by four uppercase letters and six lowercase letters. The points lie on nine different lines. When suitably clarified, the theorem states that if the lines joining the three pairs of vertices from two different triangles intersect in a common point, the pairs of lines containing the corresponding sides of these triangles meet in three points all on the same line. This result is easy to establish if the triangles lie in different planes, since the three points must lie on the line of intersection of the two planes containing the triangles, as shown in Fig. 7.

For two triangles in the same plane, the theorem, illustrated in Fig. 8, was proved by Bosse by applying Menelaus' theorem to the three sets of collinear points

[13] To keep the reader's eye from getting *too* tangled up, we shall use standard letters in the statement and figure rather than Desargues' weird mixture of uppercase and lowercase letters and numbers, which almost seems to anticipate the finest principles of computer password selection.

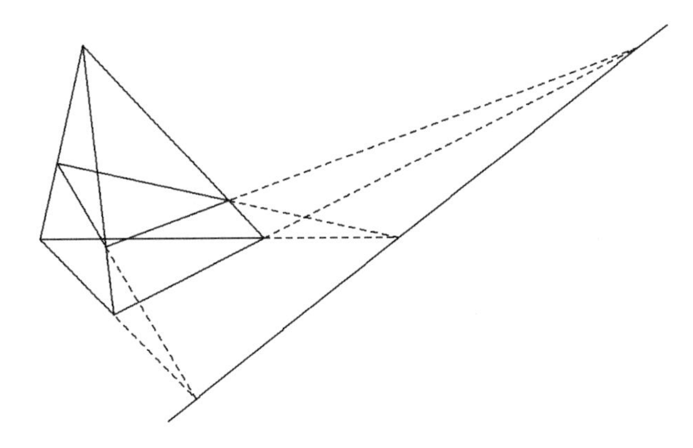

FIGURE 7. Desargues' theorem for triangles lying in different planes.

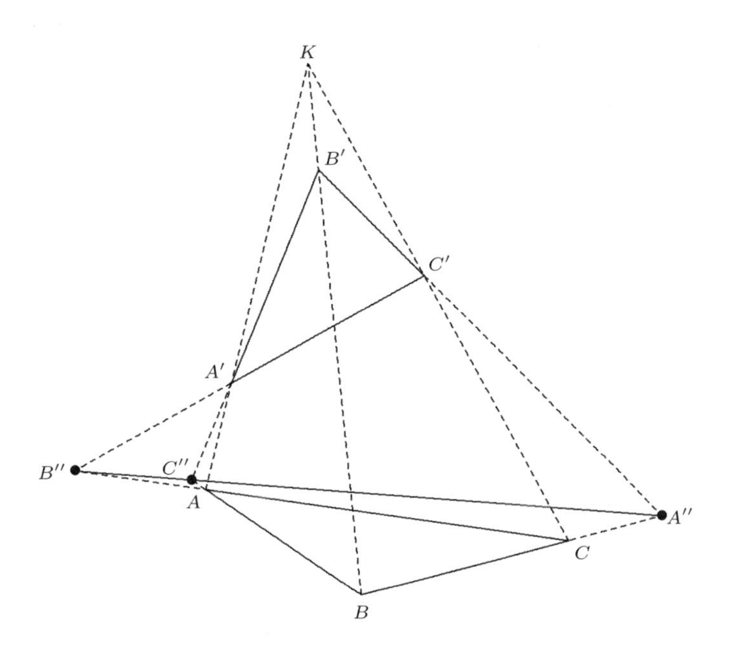

FIGURE 8. Desargues' theorem for two triangles in the same plane.

$\{A'', C, B\}$, $\{B'', A, C\}$, and $\{C'', A, B\}$, with K as the third vertex of the triangle whose base ends in the second and third points in all three cases. (There is no other conceivable way to proceed, so that in a sense the proof is a mere computation.) When the ratios $AK : AA'$, $BK : BB'$, and $CK : CC'$ are eliminated from the three resulting equations, the result can be written as the equation

$$\frac{C''B'}{C''A'} = \frac{A''B'}{A''C'} \cdot \frac{B''C'}{B''A'} .$$

Having received a copy of this work from Marin Mersenne, Descartes took the word *draft* literally and regarded it as a proposal to write a treatise—which it may have been—such as a modern author would address to a publisher, and a publisher would send to an expert for review. He wrote to Desargues to express his opinion of "what I can conjecture of the *Treatise on Conic Sections*, of which [Mersenne] sent me the *Draft*." Descartes' "review" of the work contained the kind of advice reviewers still give: that the author should decide more definitely who the intended audience was. As he said, if Desargues was aiming to present new ideas to scholars, there was no need to invent new terms for familiar concepts. On the other hand, if the book was aimed at the general public, it would need to be very thick, since everything would have to be explained in great detail (Field and Gray, *1987*, p. 176).

2.4. Blaise Pascal. Desargues' work was read by a teenage boy named Blaise Pascal (1623–1662), who was to become famous for his mathematical work and renowned for his *Pensées* (*Meditations*), which are still read by many people today for inspiration. He began working on the project of writing his own treatise on conics. Being very young, he was humble and merely sketched what he planned to do, saying that his mistrust of his own abilities inclined him to submit the proposal to experts, and "if someone thinks the subject worth pursuing, we shall try to carry it out to the extent that God gives us the strength." Pascal admired Desargues' work very much, saying that he owed "what little I have discovered to his writings" and would imitate Desargues' methods, which he considered especially important because they treated conic sections without introducing the extraneous axial section of the cone. He did indeed use much of Desargues' notation for points and lines, including the word *order* for a family of concurrent lines. His work, like that of Desargues, remained only a draft, although Struik (*1986*, p. 165) reports that Pascal did work on this project and that Leibniz saw a manuscript of it—not the rough draft, apparently—in 1676. All that has been preserved, however, is the rough draft. That draft contains several results in the spirit of Desargues, one of which, called by Pascal a "third lemma," is still known as Pascal's theorem. Referring to Fig. 9, in which four lines MK, MV, SK, and SV are drawn and then a conic is passed through K and V meeting these four lines in four other points P, O, N, and Q respectively, Pascal asserted that the lines PQ, NO, and MS would be concurrent (belong to the same *order*).

2.5. Newton's degree-preserving mappings. Newton also made contributions to projective geometry, in a way that related it to Descartes' analytic geometry and to algebraic geometry. He described the mapping shown in Fig. 10 (Whiteside, *1967*, Vol. VI, p. 269). In that figure the parallel lines BL and AO and the points A, B, and O are fixed from the outset, and the angle θ is specified in advance. Thus the distances h and Δ and the angles φ and θ are given before the mapping is defined. Then, to map the figure GHI to its image ghi, first project each point G parallel to BL so as to meet the extension of AB at a point D. Next, draw the line OD meeting BL in point d. Finally, from d along the line making angle θ with BL, choose the image point g so that $gd : Od :: GD : OD$. The original point, according to Newton, had coordinates (BD, DG) and its image the coordinates (Bd, dg). Thus, if we let $x = BD$ and $y = DG$, the coordinate transformation in

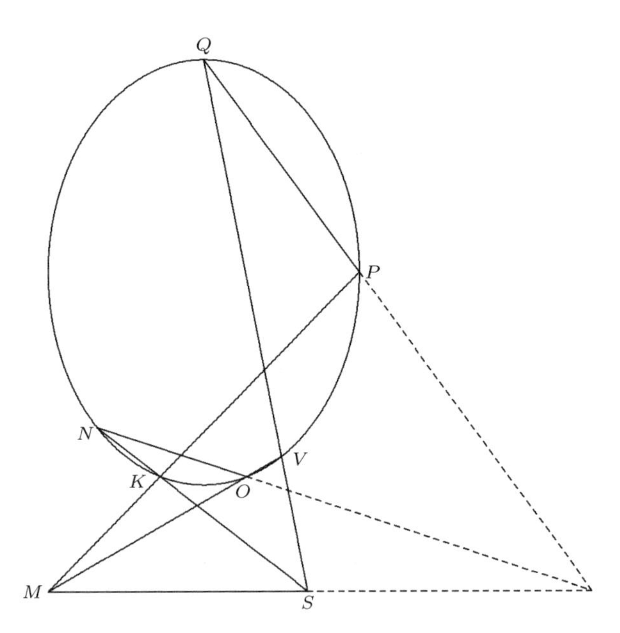

FIGURE 9. Pascal's theorem (third lemma).

the two directions can be described as

$$\begin{aligned}(x, y) &\mapsto (\xi, \eta) = \Bigl(\frac{\Delta x \sin\varphi}{h + x\sin\varphi}, \frac{hy}{h + x \sin\varphi}\Bigr),\\ (\xi, \eta) &\mapsto \Bigl(\frac{h\xi}{(\Delta - \xi)\sin\varphi}, \frac{\Delta\eta}{\Delta - \xi}\Bigr).\end{aligned}$$

Newton noted that this kind of projection preserves the degree of an equation. Hence a conic section will remain a conic section, a cubic curve will remain a cubic curve, and so on, under such a mapping. In fact, if a polynomial equation $p(x, y) = 0$ is given whose highest-degree term is $x^m y^n$, then every term $x^p y^q$, when expressed in terms of ξ and η, will be a multiple of $\xi^p\eta^q/(\Delta - \xi)^{p+q}$, so that if the entire equation is converted to the new coordinates and then multiplied by $(\Delta - \xi)^{m+n}$, this term will become $\xi^p\eta^q(\Delta - \xi)^{m+n-p-q}$, which will be of degree $m+n$. Thus the degree of an equation does not change under Newton's mapping. These mappings are special cases of the transformations known as *fractional-linear* or *Möbius* transformations, after August Ferdinand Möbius (1790–1868), who developed them more fully. They play a vital role in algebraic geometry and complex analysis, being the only one-to-one analytic mappings of the extended complex plane onto itself. According to Coolidge (*1940*, p. 269), it was Edward Waring (1736–1798) who first remarked, in 1762, that fractional-linear transformations were the most general degree-preserving transformations.

2.6. Charles Brianchon. Pascal's work on the projective properties of conics was extended by Charles Julien Brianchon (1785–1864), who was also only a teenager when he proved what is now recognized as the dual of Pascal's theorem: *The pairs*

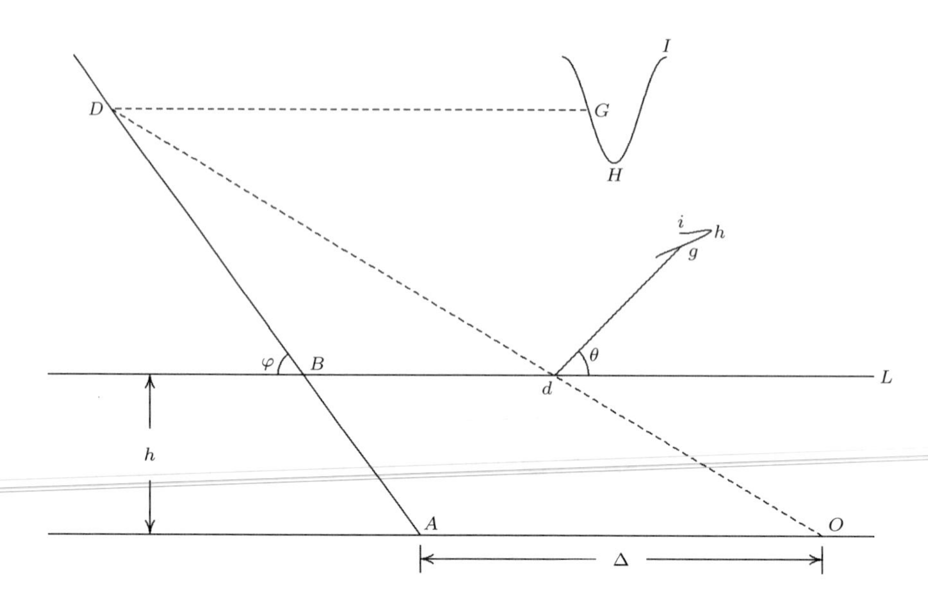

FIGURE 10. Newton's degree-preserving projection.

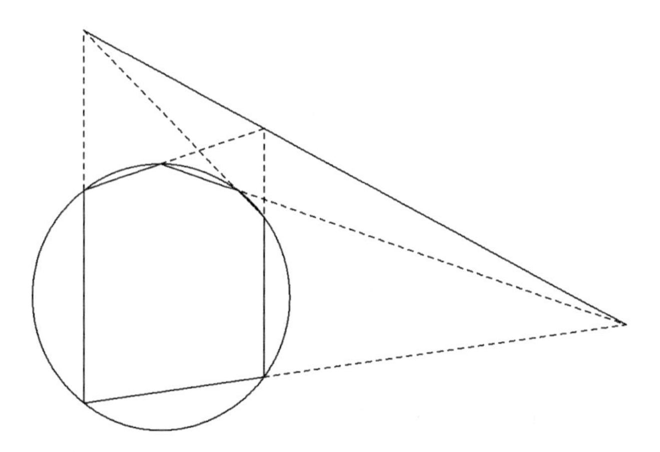

FIGURE 11. Brianchon's theorem for a circle.

of opposite sides of a hexagon inscribed in a conic meet in three collinear points. The case of a circle is illustrated in Fig. 11.

2.7. Monge and his school. After a century of relative neglect, projective geometry revived at the École Polytechnique under the students of Gaspard Monge (1746–1818), who was a master of the application of calculus to geometry. Klein (*1926*, pp. 77–78) described his school as distinguished by "the liveliest spatial intuition combined in the most natural way possible with analytic operations." Klein went on to say that he taught his students to make physical models, "not to make up for the deficiencies of their intuition but to develop an already clear and lively intuition." As a military engineer, Monge had used his knowledge of geometry to design fortifications. His work in this area was highly esteemed by his superiors and declared a military secret. He wrote a book on descriptive geometry and one on the applications of analysis to geometry, whose influence appeared in the work of his students. Klein says of the second book that it "reads like a novel." In this book, Monge analyzed quadric surfaces with extreme thoroughness.

Monge is regarded as the founder of descriptive geometry, which is based on the same principles of perspective as projective geometry but more concerned with the mechanics of representing three-dimensional objects properly in two dimensions and the principles of interpreting such representations. Monge himself described the subject as the science of giving a complete description in two dimensions of those three-dimensional objects that can be defined geometrically. As such, it continues to be taught today under other names, such as mechanical drawing; it is the most useful form of geometry for engineers.

Monge's greatest student (according to Klein) was Jean-Victor Poncelet (1788–1867). He participated as a military engineer in Napoleon's invasion of Russia in 1812, was wounded, and spent a year in a Russian prison, where he busied himself with what he had learned from Monge. Returning to France, he published his *Treatise on the Projective Properties of Figures* in 1822, the founding document of modern projective geometry. Its connection with its historical roots in the work of Desargues shows in the first chapter, where Poncelet says he will be using the word *projective* in the same sense as the word *perspective*. In Chapter 3 he introduces the idea that all points at infinity in a plane can be regarded as belonging to a single line at infinity.[14] These concepts brought out fully the duality between points and lines in a plane and between points and planes in three-dimensional space, so that interchanging these words in a theorem of projective geometry results in another theorem. The theorems of Pascal and Brianchon, for example, are dual to each other.

2.8. Jacob Steiner. The increasing algebraization of geometry was opposed by the Swiss mathematician Jacob Steiner (1796–1863), described by Klein (*1926*, pp. 126–127) as "the only example known to me... of the development of mathematical abilities after maturity." Steiner had been a farmer up to the age of 17, when he entered the school of the Swiss educational reformer Johann Heinrich Pestalozzi (1746–1827), whose influence was widespread, extending through the philosopher–psychologist Johann Friedrich Herbart (1776–1841) down to Riemann, as will be explained in the next section.[15] Steiner was a peculiar character in the history of mathematics, who when his own originality was in decline, adopted the ideas of others as his own without acknowledgement (see Klein, *1926*, p. 128). But in his best years, around 1830, he had the brilliant idea of building space using higher-dimensional objects such as lines and planes instead of points, recognizing that these objects were projectively invariant. He sought to restore the ancient Greek "synthetic" approach to geometry, which was independent of numbers and the concept of length. To this end, in his 1832 work on geometric figures he considered a family of mappings of one plane on another that resembles somewhat Newton's projection. Klein (*1926*, p. 129) found nothing materially new in this work, but admired the systematization that it contained. The Steiner principle of successively

[14] Field and Gray (*1987*, p. 185) point out that Johannes Kepler (1571–1630) had introduced points at infinity in a 1604 work on conic sections, so that a parabola would have two foci.

[15] Klein (*1926*, pp. 127–128), has nothing good to say about the more extreme recommendations of these men, calling these recommendations "pedagogical monstrosities."

building more and more intricate figures by allowing simpler ones to combine geometrically was novel and had its uses, but according to Klein, encompassed only one part of geometry.

2.9. August Ferdinand Möbius. Projective geometry was enhanced through the barycentric calculus invented by August Ferdinand Möbius (1790–1868) and expounded in a long treatise in 1827. This work contained a number of very useful innovations. Möbius' use of barycentric coordinates to specify the location of a point anticipated vector methods by some 20 years, and proved its value in many parts of geometry. He used his barycentric coordinates to classify plane figures in new ways. As he explained in Chapter 3 of the second section of his barycentric calculus (Baltzer, *1885*, pp. 177–194), if the vertices of a triangle were specified as A, B, C, and one considered all the points that could be written as $aA + bB + cC$, with the lengths of the sides and the proportions of the coefficients $a : b : c$ given, all such figures would be congruent (he used the phrase "equal and similar"). If one specified only the proportions of the sides instead of their lengths, all such figures would be similar. If one specified only the proportions of the coefficients, the figures would be in an *affine* relationship, a word still used to denote a linear transformation followed by a translation in a vector space. Finally, he introduced the relation of equality (in area).

Cauchy, then at the height of his powers, reviewed Möbius' work[16] on the barycentric calculus. In his review, as reported by Baltzer (*1885*, pp. xi–xii), he was cautious at first, saying that the work was "a different method of analytic geometry whose foundation is certainly not so simple; only a deeper study can enable us to determine whether the advantages of this method will repay the difficulties." After reporting on the new classification of figures in Part 2, he commented:

> One must be very confident of taking a large step forward in science to burden it with so much new terminology and to demand that your readers follow you in investigations presented to them in such a strange manner.

Finally, after reporting some of the results from Part 3, he concluded that, "It seems that the author of the barycentric calculus is not familiar with the general theory of duality between the properties of systems of points and lines established by M. Gergonne." This comment is difficult to explain on the assumption that Cauchy had actually read Chapters 4 and 5 of Part 3, since this duality (*gegenseitiges Entsprechen*) was part of the title of both chapters; but perhaps Cauchy was alluding to ideas in Gergonne's papers not found in the work of Möbius. Chapters 4 and 5 contain some of the most interesting results in the work. Chapter 4, for example, discusses conic sections and uses the barycentric calculus to prove that two distinct parabolas can be drawn through four coplanar points, provided none of them lies inside the triangle formed by the other three.

---

[16] It might appear that Cauchy was able to read German, not a common accomplishment for French mathematicians in the 1820s, when the vast majority of mathematical papers of significance were written in French. But perhaps he read a French or Latin version of the work.

Möbius is best remembered for two concepts, the Möbius transformation, and the Möbius band. A Möbius transformation, by which we now understand a mapping of the complex plane into itself, $z \mapsto w$, of the form

$$w = \frac{az + b}{cz + d}, \quad ad - bc \neq 0,$$

can be found in his 1829 paper on metric relations in line geometry. He gave such transformations with real coefficients in terms of the two coordinates (x, y), the real and imaginary parts of what we now write as the complex number z, and showed that they were the most general one-to-one transformations that preserve collinearity. The Möbius band is discussed in Section 4 below.

2.10. Julius Plücker. A number of excellent German, Swiss, and Italian geometers arose in the nineteenth century. Their work cannot be classified as purely projective geometry, since it also relates to algebraic geometry. As an example, we take Julius Plücker (1801–1868), who was a professor at the University of Bonn for the last 30 years of his life. Plücker himself remembered (Coolidge, *1940*, p. 144) that when young he had discovered a theorem in Euclidean geometry: The three lines containing the common chords of pairs of three intersecting circles are all concurrent. Plücker's proof of this theorem is simplicity itself. Suppose that the equations of the three circles are $A = 0$, $B = 0$, $C = 0$, where each equation contains $x^2 + y^2$ plus linear terms. By subtracting these equations in pairs, we get the quadratic terms to drop out, leaving the equations of the three lines containing the three common chords: $A - B = 0$, $A - C = 0$, $B - C = 0$. But it is manifest that any two of these equations imply the third, so that the point of intersection of any two also lies on the third line.

Plücker's student Felix Klein (*1926*, p. 122) described a more sophisticated specimen of this same kind of reasoning by Plücker to prove Brianchon's theorem[17] that the opposite sides of a hexagon inscribed in a conic, when extended, intersect in three collinear points. The proof goes as follows: The problem involves two sets, each containing three lines, six of whose nine pairwise intersections lie on a conic section. The conic section has an equation of the form $q(x, y) = 0$, where $q(x, y)$ is quadratic in both x and y. Represent each line by a linear polynomial of the form $a_j x + b_j y + c_j$, the jth line being the set of (x, y) where this polynomial equals zero, and assume that the lines are numbered in clockwise order around the hexagon. Form the polynomial

$$\begin{aligned} s(x, y) = (a_1 x + b_1 y + c_1)(a_3 x + b_3 y + c_3)(a_5 x + b_5 y + c_5) \\ - \mu(a_2 x + b_2 y + c_2)(a_4 x + b_4 y + c_4)(a_6 x + b_6 y + c_6) \end{aligned}$$

with the parameter μ to be chosen later. This polynomial vanishes at all nine intersections of the lines. Line 1, for example, meets lines 2 and 6 inside the conic and line 4 outside it.[18]

Now, when y is eliminated from the equations $q(x, y) = 0$ and $s(x, y) = 0$, the result is an equation $t(x) = 0$, where $t(x)$ is a polynomial of degree at most 6 in x. This polynomial must vanish at all of the simultaneous zeros of $q(x, y)$ and $s(x, y)$. We know that there are six such zeros for every μ. However, it is very easy

[17] Klein called it Pascal's theorem.

[18] This polynomial is the difference of two completely factored cubics, by coincidence exactly the kind of polynomial that arises in the six-line locus problem, even though we are not dealing with the distances to any lines here.

to choose μ so that there will be a seventh common zero. With that choice of μ, the polynomial $t(x)$ must have seven zeros, and hence must vanish identically. But since $t(x)$ was the result of eliminating y between the two equations $q(x, y) = 0$ and $s(x, y) = 0$, it now follows that $q(x, y)$ divides $s(x, y)$ (see Problem 12.5 below). That is, the equation $s(x, y) = 0$ can be written as $(ax + by + c)q(x, y) = 0$. Hence its solution set consists of the conic and the line $ax + by + c = 0$, and this line must contain the other three points of intersection.

Conic sections and quadratic functions in general continued to be a source of new ideas for geometers during the early nineteenth century. Plücker liked to use homogeneous coordinates to give a symmetric description of a quadric surface. To take the simplest example, consider the sphere of radius 2 in three-dimensional space with center at $(2, 3, 1)$, whose equation is

$$(x - 2)^2 + (y - 3)^2 + (z - 1)^2 = 4\,.$$

If x, y, and z, are replaced by ξ/τ, η/τ, and ζ/τ and each term is multiplied by τ^2, this equation becomes a homogeneous quadratic relation in the four variables (ξ, η, ζ, τ):

$$(\xi - 2\tau)^2 + (\eta - 3\tau)^2 + (\zeta - \tau)^2 = 4\tau^2\,.$$

The sphere of unit radius centered at the origin then has the simple equation $\tau^2 - \xi^2 - \eta^2 - \zeta^2 = 0$. Plücker introduced homogeneous coordinates in 1830. One of their advantages is that if $\tau = 0$, but the other three coordinates are not all zero, the point (ξ, η, ζ, τ) can be considered to be located on a sphere of infinite radius. The point $(0, 0, 0, 0)$ is excluded, since it seems to correspond to all points at once.

Homogeneous coordinates correspond very well to the ideas of projective geometry, in which a point in a plane is identified with all the points in three-dimensional space that project to that point from a point outside the plane. If, for example, we take the center of projection as $(0, 0, 0)$ and identify the plane with the plane $z = 1$, that is, each point (x, y) is identified with the point $(x, y, 1)$, the points that project to (x, y) are all points (tx, ty, t), where $t \neq 0$. Since the equation of a line in the (x, y)-plane has the form $ax + by + c = 0$, one can think of the coordinates (a, b, c) as the coordinates of the line. Here again, multiplication by a nonzero constant does not affect the equation, so that these coordinates can be identified with (ta, tb, tc) for any $t \neq 0$. Notice that the condition for the point (x, y) to lie on the line (a, b, c) is that $\langle (a, b, c), (x, y, 1) \rangle = a \cdot x + b \cdot y + c \cdot 1 = 0$, and this condition is unaffected by multiplication by a constant. The duality between points and lines in a plane is then clear. Any triple of numbers, not all zero, can represent either a point or a line, and the incidence relation between a point and a line is symmetric in the two. We might as well say that the line lies on the point as that the point lies on the line.

Equations can be written in either line coordinates or point coordinates. For example, the equation of an ellipse can be written in homogeneous point coordinates (ξ, η, ζ) as

$$b^2c^2\xi^2 + a^2c^2\eta^2 = a^2b^2\zeta^2\,,$$

or in line coordinates (λ, μ, ν) as

$$a^2\lambda^2 + b^2\mu^2 = c^2\nu^2\,,$$

where the geometric meaning of this last expression is that the line (λ, μ, ν) is tangent to the ellipse.

2.11. Arthur Cayley. Homogeneous coordinates provided important invariants and covariants[19] in projective geometry. One such invariant under orthogonal transformations (those that leave the sphere fixed) is the angle between two planes $Ax + By + Cz = D$ and $A'x + B'y + C'z = D'$, given by

$$\arccos\left(\frac{AA' + BB' + CC'}{\sqrt{A^2 + B^2 + C^2}\sqrt{(A')^2 + (B')^2 + (C')^2}}\right). \tag{2}$$

In his "Sixth memoir on quantics," published in the *Transactions of the London Philosophical Society* in 1858, Cayley fixed a "quantic" (quadratic form) $\sum \alpha_{ij}u_iu_j$, whose zero set was a quadric surface that he called the *absolute*, and defined angles by analogy with Eq. (2) and other metric concepts by a similar analogy. In this way he obtained the *general projective metric*, commonly called the *Cayley metric.* It allowed metric geometry to be included in descriptive–projective geometry. As Cayley said, "Metrical geometry is thus a part of descriptive geometry and descriptive geometry is all geometry." By suitable choices of the absolute, one could obtain the geometry of all kinds of quadric curves and surfaces, including the non-Euclidean geometries studied by Gauss, Lobachevskii, Bólyai, and Riemann. Klein (*1926*, p. 150) remarked that Cayley's models were the most convincing proof that these geometries were consistent.

3. Differential geometry

Differential geometry is the study of curves and surfaces (from 1852 onwards, manifolds) using the methods of differential calculus, such as derivatives and local series expansions. This history falls into natural periods defined by the primary subject matter: first, the tangents and curvatures of plane curves; second, the same properties for curves in three-dimensional space; third, the analogous properties for surfaces, geodesics on surfaces, and minimal surfaces; fourth, the application (conformal mapping) of surfaces on one another; fifth, very broad expansions of all these topics, to embrace n-dimensional manifolds and global properties instead of local.

3.1. Huygens. Struik (*1933*) and Coolidge (*1940*, p. 319) agree that credit for the first exploration of secondary curves generated by a plane curve—the involute and evolute—occurred in Christiaan Huygens' work *Horologium oscillatorium* (*Of Oscillating Clocks*) in 1673, even though calculus had not yet been developed. The involute of a curve is the path followed by the endpoint of a taut string being wound onto the curve or unwound from it. Huygens did not give it a name; he simply called it the "line [curve] described by evolution." There are as many involutes as there are points on the curve to begin or end the winding process.

Huygens was seeking a truly synchronous pendulum clock, and he needed a pendulum that would have the same period of oscillation no matter how great the amplitude of the oscillation was.[20] Huygens found the mathematically ideal solution of the problem in two properties of the cycloid. First, a frictionless particle

[19] According to Klein (*1926*, p. 148), the distinction between an invariant and a covariant is not essential. Any algebraic expression that remains unchanged under a family of changes of coordinates is a covariant if it contains variables, and is an invariant if it contains only constants.

[20] Despite the legend that Galileo observed a chandelier swinging and noticed that all its swings, whether wide or short, required the same amount of time to complete, for circular arcs that observation is only true approximately for small amplitudes, as anyone who has done the experiment in high-school physics will have learned.

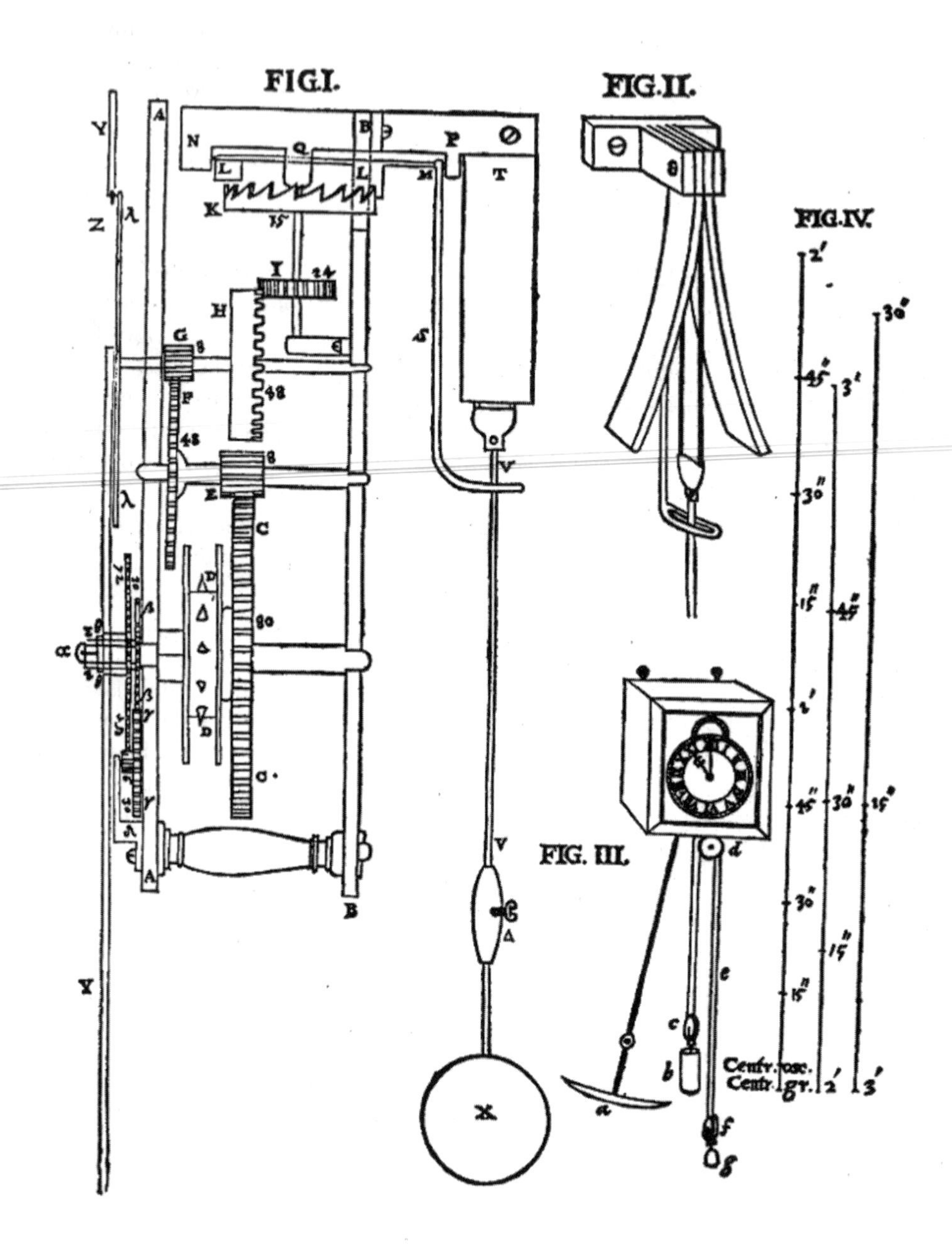

Huygens' cycloidal pendulum, from his *Horologium oscillatorium*.

requires the same time to slide to the bottom of a cycloid no matter where it begins; second, the involute of a cycloid is another cycloid. He therefore designed a pendulum clock in which the pendulum bob was attached to a flexible leather strap that is confined between two inverted cycloidal arcs. The pendulum is thereby forced to fall along the involute of a cycloid and hence to be truly tautochronous. Reality being more complicated than our dreams, however, this apparatus—like

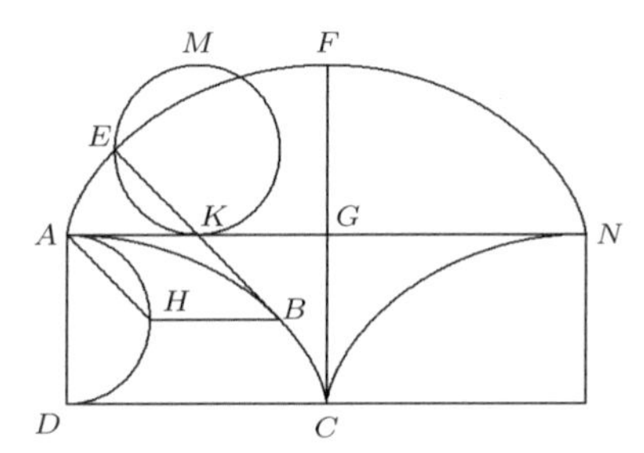

FIGURE 12. Huygens' construction of the "curve formed by evolution" from a cycloid—an identical cycloid.

Dürer's mechanical drawing methods—does not really work any better than the standard methods.[21]

Referring to Fig. 12, in which a line is drawn from one peak of a cycloid to the next and an identical cycloid then drawn atop that line, he showed that BE is perpendicular to the cycloid. But the curve that cuts all the tangents to another curve at right angles is precisely the "curve generated by evolution."

3.2. Newton. In his *Fluxions*, which was first published in 1736, after his death, even though it appears to have been written in 1671, Newton found the circle that best fits a curve. Struik (*1933*, **19**, p. 99) doubted that this material was really in the 1671 manuscript. Be that as it may, the topic occurs as Problem 5 in the *Fluxions*: *At any given Point of a given Curve, to find the Quantity of Curvature.* Newton needed to find a circle tangent to the curve at a given point, which meant finding its center. However, Newton wanted not just any tangent circle. He assumed that if a circle was tangent to a curve at a point and "no other circle can be interscribed in the angles of contact near that point,... that circle will be of the same curvature as the curve is of, in that point of contact." In this connection he introduced terms *center of curvature* and *radius of curvature* still used today. His construction is shown in Fig. 13, in which one unnecessary letter has been removed and the figure has been rotated through a right angle to make it fit the page. The weak point of Newton's argument was his claim that, "If CD be conceived to move, while it insists [remains] perpendicularly on the Curve, that point of it C (if you except the motion of approaching to or receding from the Point of Insistence C,) will be least moved, but will be as it were the Center of Motion." Huygens had had this same problem with clarity. Where Huygens had referred to points that *can be treated as* coincident, Newton used the phrase *will be as it were.*

Newton also treated the problem of the cycloidal pendulum in his *Principia Mathematica*, published in 1687. Huygens had found the evolute of a complete arch of a cyloid. That is, the complete arch is the involute of the portion of two half-arches starting at the halfway point on the arch. In Proposition 50, Problem 33 of Book 1, Newton found the evolute for an arbitrary piece of the arch, which was

[21] The master's thesis of Robert W. Katsma at California State University at Sacramento in the year 2000 was entitled "An analysis of the failure of Huygens' cycloidal pendulum and the design and testing of a new cycloidal pendulum." Katsma was granted patent 1992-08-18 in Walla Walla County for a cycloidal pendulum. However, the theoretical consensus is that "in every case, such devices would introduce greater errors into the going of a good clock than the errors they are supposed to eliminate." (See the website `http://www.ubr.com/clocks/nawec/hsc/hsn95a.html`.)

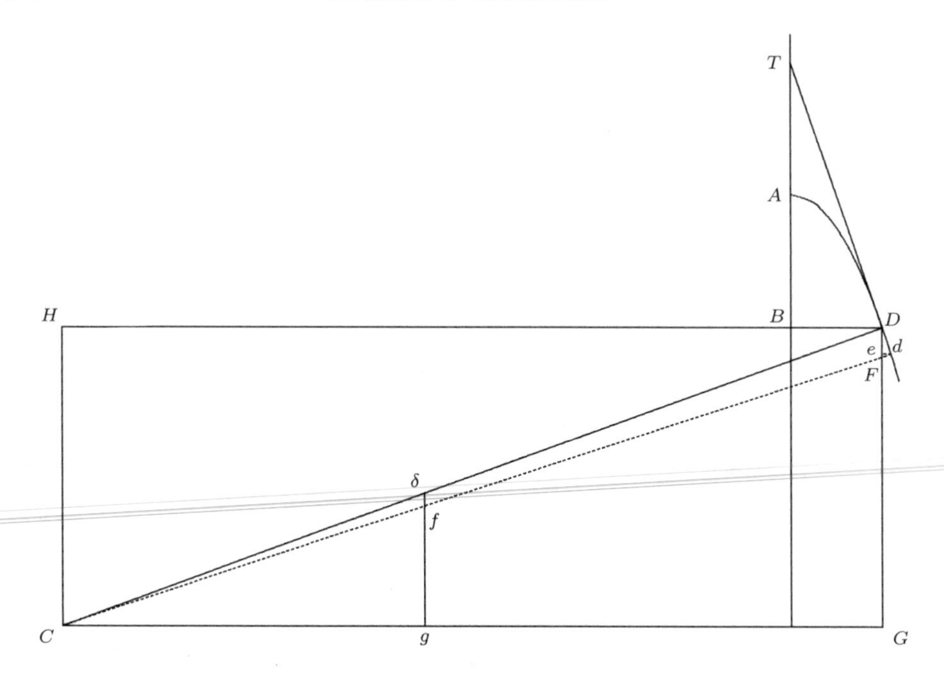

FIGURE 13. Newton's construction of the radius of curvature, from his posthumously published *Fluxions*.

a much more complicated problem. It was, however, once again a cycloid. This evolute made it possible to limit the oscillations of a cycloidal pendulum by putting a complete cycloidal frame in place to stop the pendulum when the thread was completely wound around the evolute.

3.3. Leibniz. Leibniz' contributions to differential geometry began in 1684, when he gave the rules for handling what we now call differentials. His notation is essentially the one we use today. He regarded x and $x + dx$ as infinitely near values of x and v and dv as the corresponding infinitely near values of v on a curve defined by an equation relating x and v. At a maximum or minimum point he noted that $dv = 0$, so that the equation defining the curve had a double root (v and $v + dv$) at that point. He noted that the two cases could be distinguished by the concavity of the curve, defining the curve to be concave if the difference of the increments ddv (which we would now write as $(d^2v/dx^2)\,dx^2$) was positive, so that the increments dv themselves increased with increasing v. He defined a point where the increments changed from decreasing to increasing to be a *point of opposite turning* (*punctum flexus contrarii*), and remarked that at such a point (if it was a point where $dv = 0$ also), the equation had a triple root. What he said is easily translated into the language of today, by looking at the equation $0 = f(x + h) - f(x)$. Obviously, $h = 0$ is a root. At a maximum or minimum, it is a double root. If the point x yields $dv = 0$ (that is, $f'(x) = 0$) but is not a maximum or minimum, then $h = 0$ is a triple root.

In 1686 he was the first to use the phrase *osculating circle*. He explained the matter thus:

> In the infinitely small parts of a curve it is possible to consider not only the direction or inclination or declination, as has been done up to now, but also the change in direction or curvature (*flexura*), and as the measures of the direction of curves are the simplest lines of geometry having the same direction at the same point, that is, the tangent lines, likewise the measure of curvature is the simplest curve having at the same point not only the same direction but also the same curvature, that is a circle not only tangent to the given curve but, what is more, osculating.[22]

Leibniz recognized the problem of finding the evolute as that of constructing "not merely an arbitrary tangent to a single curve at an arbitrary point, but a unique common tangent[23] of infinitely many curves belonging to the same order." That meant differentiating with respect to the parameter and eliminating it between the equation of the family and the differentiated equation. In short, Leibniz was the first to discuss what is now called the envelope of a family of curves defined by an equation containing a parameter.

3.4. The eighteenth century. Compared to calculus, differential equations, and analysis in general, differential geometry was not the subject of a large number of papers in the eighteenth century. Nevertheless, there were important advances.

Euler. According to Coolidge (*1940*, p. 325), Euler's most important contribution to differential geometry came in a 1760 paper on the curvature of surfaces. In that paper he observed that different planes cutting a surface at a point would generally intersect it in curves having different curvatures, but that the two planes for which this curvature was maximal or minimal would be at right angles to each other. For any other plane, making angle α with one of these planes, the radius of curvature would be

$$r = \frac{2fg}{f + g + (g - f)\cos 2\alpha},$$

where f and g are the minimum and maximum radii of curvature at the point. Nowadays, because of an 1813 treatise of Monge's student Pierre Dupin (1784–1873), this formula is written in terms of the curvature $1/r$ as

$$\frac{1}{r} = \frac{\cos^2\alpha}{g} + \frac{\sin^2\alpha}{f},$$

where α is the angle between the given cutting plane and the plane in which the curvature is minimal ($1/g$). The equation obviously implies that in a plane perpendicular to the given plane the curvature would be the same expression with the cosine and sine reversed, or, what is the same, with f and g reversed.

Another fundamental innovation due to Euler was the introduction of the now-familiar idea of a parameterized surface, in a 1770 paper on surfaces that can be mapped into a plane. The canvas on which an artist paints and the paper on which an engineer or architect draws plans are not only two-dimensional but also *flat*, having curvature zero. Parameters allow the mathematician or engineer to represent information about any curved surface, as Euler remarked, in the form of

---

[22] Literally, *kissing*.

[23] The tangent was not necessarily to be a straight line.

functions $(t, u) \mapsto \bigl(x(t, u), y(t, u), z(t, u)\bigr)$. Quantities such as curvature and area are then expressed as functions of the parameters (t, u).

Lagrange. Another study of surfaces, actually a paper in the calculus of variations, was Lagrange's 1762 work on extremal values of integrals.[24] The connection with differential geometry is in the problem of minimal surfaces and isoperimetric problems, although he began with the brachystochrone problem (finding the curve of most rapid descent for a falling body). Lagrange found a necessary condition for a surface $z = f(x, y)$ to be minimal.

The French geometers. After these "preliminaries" we finally arrive at the traditional beginning of differential geometry, a 1771 paper of Monge on curves in space and his 1780 paper on curved surfaces. Monge elaborated Leibniz' idea for finding the envelope of a family of lines, considering a family of planes parametrized by their intersections with the z-axis, and obtained the equation of the surface that is the envelope of the family of planes and can be locally mapped into a plane without stretching or shrinking.

3.5. Gauss. With the nineteenth century, differential geometry entered on a period of growth and has continued to reach new heights for two full centuries. The first mathematician to be mentioned is Gauss, who during the 1820s was involved in mapping the region of Hannover in Lower Saxony, where Göttingen is located. This mapping had been ordered by King George IV of England, who was also Elector of Hannover by inheritance from his great grandfather George I. Gauss had been interested in geodesy for many years (Reich, *1977*, pp. 29–34) and had written a paper in response to a problem posed by the Danish Academy of Sciences. This paper, which was published in 1825, discussed conformal mapping, that is, mappings that are a pure magnification at each point, so that directions are preserved and the limiting ratio of the actual distance between two points to the map distance between them as one of them approaches the other is the same for approach from any direction.

Involvement with the mapping project inspired Gauss to reflect on the mathematical aspects of developing a curved surface on a flat page and eventually, the more general problem of developing one curved surface on another, that is, mapping the surfaces so that the ratio that the distance from a given point P to a nearby point Q has to the distance between their images P' and Q' tends to 1 as Q tends to P. Gauss apparently planned a full-scale treatise on geodesy but never completed it. Two versions of his major work *Disquisitiones generales circa superficies curvas* (*General Investigations of Curved Surfaces*) were written in the years 1825 and 1827. In the preface to the latter Gauss explained the problem he had set: "to find all representations of a given surface upon another in which the smallest elements remain unchanged." He admitted that some of what he was doing needed to be made more precise through a more careful statement of hypotheses, but wished to show certain results of fundamental importance in the general problem of mapping.

A simple and fruitful technique that Gauss used was to represent any line in space by a point on a fixed sphere of unit radius: the endpoint of the radius parallel to the line.[25] This idea, he said, was inspired by the use of the celestial sphere

---

[24] *Œuvres de Lagrange*, T. 1, pp. 335–362.

[25] An oriented line is meant here, since there are obviously two opposite radii parallel to the line. Gauss surely knew that the order of the parameters could be used to fix this orientation.

in geometric astronomy. This unit sphere is used in mapping a curved surface by taking the normal line at each point of the surface and mapping it to a point on the sphere, as described, so that the sphere and the surface have parallel normal lines at corresponding points. Obviously a plane maps to a single point under this procedure, since all of its normal lines are parallel to one another. Gauss proposed to use the area of the portion of the sphere covered by this map as a measure of curvature of the surface in question. He called this area the *total curvature* of the surface. He refined this total curvature by specifying that it was to be positive if the surface was convex in both of two mutually perpendicular directions and negative if it was convex in one direction and concave in the other (like a saddle). Gauss gave an informal discussion of this question in terms of the side of the surface on which the normals were to be erected. When the quality of convexity varied in different parts of a surface, Gauss said, a still more refined definition was necessary, which he found it necessary to omit. Along with the total curvature he defined what we would call its density function and he called the *measure of curvature*, namely the ratio of the total curvature of an element of surface to the area of the same element of surface, which he denoted k. The simplest example is provided by a sphere of radius R, any region of which projects to the similar region on the unit sphere. The ratio of the areas is $k = 1/R^2$, which is therefore the measure of curvature of a sphere at every point.

Gauss used two mappings from the parameter space (p, q) into three-dimensional space. The first was the mapping onto the surface itself:

$$(p, q) \mapsto \bigl(x(p, q), y(p, q), z(p, q)\bigr) .$$

The second was the mapping

$$(p, q) \mapsto \bigl(X(p, q), Y(p, q), Z(p, q)\bigr)$$

to the unit sphere, which takes (p, q) to the three direction cosines of the normal to the surface at the point $\bigl(x(p, q), y(p, q), z(p, q)\bigr)$.

From these preliminaries, Gauss was able to derive very simply what he himself described as "almost everything that the illustrious Euler was the first to prove about the curvature of curved surfaces." In particular, he showed that his measure of curvature k was the reciprocal of the product of the two principal radii of curvature that Euler called f and g. He then went on to consider more general parameterized surfaces. Here he introduced the now-standard quantities E, F, and G, given by

$$\begin{aligned} E &= \Bigl(\frac{\partial x}{\partial p}\Bigr)^2 + \Bigl(\frac{\partial y}{\partial p}\Bigr)^2 + \Bigl(\frac{\partial z}{\partial p}\Bigr)^2 , \\ F &= \frac{\partial x}{\partial p}\frac{\partial x}{\partial q} + \frac{\partial y}{\partial p}\frac{\partial y}{\partial q} + \frac{\partial z}{\partial p}\frac{\partial z}{\partial q} , \\ G &= \Bigl(\frac{\partial x}{\partial q}\Bigr)^2 + \Bigl(\frac{\partial y}{\partial q}\Bigr)^2 + \Bigl(\frac{\partial z}{\partial q}\Bigr)^2 , \end{aligned}$$

and what is now called the first fundamental form for the square of the element of arc length:

$$ds^2 = E\,dp^2 + 2F\,dp\,dq + G\,dq^2 .$$

It is easy to compute that the element of area—the area of an infinitesimal parallelogram whose sides are $\bigl(\frac{\partial x}{\partial p}\,dp, \frac{\partial y}{\partial p}\,dp, \frac{\partial z}{\partial p}\,dp\bigr)$ and $\bigl(\frac{\partial x}{\partial q}\,dq, \frac{\partial y}{\partial q}\,dq, \frac{\partial z}{\partial q}\,dq\bigr)$—is just

$\Delta\, dp\, dq$, where $\Delta = \sqrt{EG - F^2}$. Gauss denoted the analogous expression for the mapping $(p,q) \mapsto \bigl(X(p,q), Y(p,q), Z(p,q)\bigr)$, by

$$D\, dp^2 + 2D'\, dp\, dq + D''\, dq^2\,. \tag{3}$$

It turns out that D is just Δ times the cosine of the angle between the normal line to the surface and the line through the origin passing through the point

$$\Bigl(\frac{\partial^2 x}{\partial p^2}, \frac{\partial^2 x}{\partial p\, \partial q}, \frac{\partial^2 x}{\partial q^2}\Bigr),$$

and similarly for D' and D'' with x replaced by y and z respectively. This coincidence is particular to three-dimensional space, since there just happen to be three different second-order partial derivatives.

The expression in formula (3) is now divided by Δ and the quotient, called the *second fundamental form*, is written $e\, dp^2 + 2f\, dp\, dq + g\, dq^2$. The element of area on the sphere is $\bigl(DD'' - (D')^2\bigr)\, dp\, dq$. Hence the measure of curvature—what is now called the *Gaussian curvature* and denoted k—is

$$\frac{DD'' - (D')^2}{(EG - F^2)^2}\,,$$

or as it is now written,

$$\frac{eg - f^2}{EG - F^2}\,.$$

Gauss found another expression for k involving only the quantities E, F, and G and their first and second partial derivatives with respect to the parameters p and q. The expression was complicated, but it was needed for theoretical purposes, not computation.

In a very prescient remark that was later to be developed by Riemann, Gauss noted that "for finding the measure of curvature, there is no need of finite formulæ, which express the coordinates x, y, z as functions of the indeterminates p, q; but that the general expression for the magnitude of any linear element is sufficient." The idea is that the geometry of a surface is to be built up from the infinitesimal level using the parameters, not derived from the metric imposed on it by its position in Euclidean space. That is the essential idea of what is now called a *differentiable manifold.*

It is also clear from Gauss' correspondence (Klein, *1926*, p. 16) that Gauss already realized that non-Euclidean geometry was consistent. In fact, the question of consistency did not trouble him; he was more interested in measuring large triangles to see if the sum of their angles could be demonstrably less than two right angles. If so, what we now call hyperbolic geometry would be more convenient for physics than Euclidean geometry.

Gauss considered the possibility of developing one surface on another, that is, mapping it in such a way that lengths are preserved on the infinitesimal level. If the mapping is $(x, y, z) \mapsto (u, v, w)$, then by composition, u, v, and w are all functions of the same parameters that determine x, y, and z, and they generate functions E', F', and G' for the second surface that must be equal to E, F, and G at the corresponding points, since that is what is meant by developing one surface on another. But since he had just derived an expression for the measure of curvature that depended only on E, F, G and their partial derivatives, he was able to state the profound result that has come to be called his *theorema egregium* (*outstanding theorem*):

If a curved surface is mapped on any other surface, the measure of curvature at each point remains unchanged.

Among other consequences, this meant that surfaces that can be developed on a plane, such as a cone or cylinder, must have Gaussian curvature 0 at each point.

With the first fundamental form Gauss was able to derive a pair of differential equations that must be satisfied by geodesic lines, which he called *shortest lines*,[26] and prove that the endpoints of a geodesic circle—the set of geodesics originating at a given point and having a given length—form a curve that intersects each geodesic at a right angle. This result was the foundation for a generalized theory of polar coordinates on a surface, using p as the distance along a geodesic from a variable point to a pole of reference and q as the angle between that geodesic and a fixed geodesic through the pole. This topic very naturally led to the subject of geodesic triangles, formed by joining three points to one another along geodesics. Since he had shown earlier that the element of surface area was

$$d\sigma = \sqrt{EG - F^2}\, dp\, dq\,,$$

and that this expression was particularly simple when one of the sets of coordinate lines consisted of geodesics (as in the case of a sphere, where the lines of longitude are geodesics), the total curvature of such a triangle was easily found for a geodesic triangle and turned out to be

$$A + B + C - \pi\,,$$

where A, B, and C are the angles of the triangle, expressed in radians. For a plane triangle this expression is zero. For a spherical triangle it is, not surprisingly, the area of the triangle divided by the square of the radius of the sphere. In this way, area, curvature, and the sum of the angles of a triangle were shown to be linked on curved surfaces. This result was the earliest theorem on global differential geometry, since it applies to any surface that can be triangulated. In its modern, developed version, it relates curvature to the topological property of the surface as a whole known as the Euler characteristic. It is called the *Gauss–Bonnet theorem* after Pierre Ossian Bonnet (1819–1892), who introduced the notion of the geodesic curvature of a curve on a surface (that is, the tangential component of the acceleration of a point moving along the curve with unit speed)[27] and generalized the formula to include this concept.

3.6. The French and British geometers. In France differential geometry was of interest for a number of reasons connected with physics. In particular, it seemed applicable to the problem of heat conduction, the theory of which had been pioneered by such outstanding mathematicians as Jean-Baptiste Joseph Fourier (1768–1830), Siméon-Denis Poisson (1781–1840), and Gabriel Lamé (1795–1870), since isothermal surfaces and curves in a body were a topic of primary interest. It also applied to the theory of elasticity, studied by Lamé and Sophie Germain, among others. Lamé developed a theory of elastic waves that he hoped would explain light propagation in an elastic medium called ether. Sophie Germain noted that the average

[26] According to Klein (*1926*, Vol. 2, p. 148), the term *geodesic* was first used by Joseph Liouville (1809–1882) in 1850. Klein cites an 1893 history of the term by Paul Stäckel (1862–1919) as source.

[27] According to Struik (*1933*, **20**, pp. 163, 165), even this concept was anticipated by Gauss in an unpublished paper of 1825 and followed up on by Ferdinand Minding (1806–1885) in a paper in Crelle's *Journal* in 1830.

of the two principal curvatures derived by Euler would be the same for any two mutually perpendicular planes cutting a surface. She therefore recommended this average curvature as the best measure of curvature. Her approach does indeed make sense in elasticity theory,[28] but turns out not to be so useful for pure geometry.[29] Joseph Liouville (1809–1882), who founded the *Journal de mathématiques pures et appliquées* in 1836 and edited it until 1874, proved that conformal maps of three-dimensional regions are far less varied than those in two dimensions, being necessarily either inversions or similarities or rigid motions. He published this result in the fifth edition of Monge's book on the applications of analysis to geometry. In contrast, a mapping $(x, y) \mapsto (u, v)$ is conformal if and only if one of the functions $u(x, y) \pm iv(x, y)$ is analytic. As a consequence, there is a rich supply of conformal mappings of the plane.

After Newton differential geometry languished in Britain until the nineteenth century, when William Rowan Hamilton (1805–1865) published papers on systems of rays, building the foundation for the application of differential geometry to differential equations. Another British mathematician, George Salmon (1819–1904), made the entire subject more accessible with his famous textbooks *Higher Plane Curves* (1852) and *Analytic Geometry of Three Dimensions* (1862).

3.7. Riemann. Once the idea of using parameters to describe a surface has been grasped, the development of geometry can proceed algebraically, without reference to what is possible in three-dimensional Euclidean space. This idea was understood by Hermann Grassmann (1809–1877), a secondary-school teacher, who wrote a philosophically inclined mathematical work published in 1844 under the title *Die lineale Ausdehnungslehre, ein neuer Zweig der Mathematik* (*The Theory of Lineal Extensions, a New Branch of Mathematics*). This work, which developed ideas Grassmann had conceived earlier in a work on the ebb and flow of tides, contained much of what is now regarded as multilinear algebra. What we call the coefficients in a linear combination of vectors Grassmann called the numbers by means of which the quantity was derived from the other quantities. He introduced what we now call the tensor product and the wedge product for what he called extensive quantities. He referred to the tensor product simply as the *product* and the wedge product as the *combinatory product.* The tensor product of two extensive quantities $\sum \alpha_r e_r$ and $\sum \beta_s e_s$ was

$$\Bigl[\sum_r \alpha_r e_r, \sum_s \beta_s e_s\Bigr] = \sum_{r,s} \alpha_r \beta_s [e_r, e_s] \, .$$

The combinatory product was obtained by applying to this product the rule that $[e_r, e_s] = -[e_s, e_r]$ (antisymmetrizing). The determinant is a special case of the combinatory product. Grassmann remarked that when the factors are "numerically related" (which we call linearly dependent), the combinatory product would be zero. When the basic units e_r and e_s were entirely distinct, Grassmann called the combinatory product the *outer product* to distinguish it from the *inner product*, which is still called by that name today and amounts to the ordinary dot product

[28] In particular, her concept of the average curvature plays a role in the Navier–Stokes equations (`http://www.navier-stokes.net/nsbcst.htm`).

[29] However, the average curvature must be zero on a minimal surface.

when applied to vectors in physics. Grassmann remarked that parentheses have no effect on the outer product—in our terms, it is an associative operation.[30]

Working with these concepts, Grassmann defined the *numerical value* of an extended quantity as the positive square root of its inner square, exactly what we now call the absolute value of a vector in n-dimensional space. He proved that "the quantities of an orthogonal system are not related numerically," that is, an orthogonal set of nonzero vectors is linearly independent.

Historians of mathematics seem to agree that, because of its philosophical tone and unusual nomenclature, *Ausdehnungslehre* did not attract a great deal of notice until Grassmann revised it and published a more systematic exposition in 1862. If that verdict is correct, there is a small coincidence in Riemann's use of the term "extended," which appears to mimic Grassmann's use of the word, and in his focus on a general number of dimensions in his inaugural lecture at the University of Göttingen. Riemann's most authoritative biographer Laugwitz (*1999*, p. 223) says that Grassmann's work would have been of little use to Riemann, since for him linear algebra was a trivial subject.[31] This lecture was read in 1854, with the aged Gauss in the audience.[32] Although Riemann's lecture "Über die Hypothesen die der Geometrie zu Grunde liegen" ("On the hypotheses that form the basis of geometry") occupies only 14 printed pages and contains almost no mathematical symbolism—it was aimed at a largely nonmathematical audience—it set forth ideas that had profound consequences for the future of both mathematics and physics. As Hermann Weyl said:

> The same step was taken here that was taken by Faraday and Maxwell in physics, the theory of electricity in particular, ... by passing from the theory of action at a distance to the theory of local action: the principle of understanding the world from its behavior on the infinitesimal level. [Narasimhan, *1990*, p. 740]

In the first section Riemann began by developing the concept of an n-fold extended quantity, asking the indulgence of his audience for delving into philosophy, where he had limited experience. He cited only some philosophical work of Gauss and of Johann Friedrich Herbart (1776–1841), a mathematically inclined philosopher whose attempts to quantify sense impressions was an early form of mathematical psychology.[33] He began with the concept of quantity in general, which arises when some general concept can be defined (measured or counted) in different ways. Then, according as there is or is not a continuous transformation from one of the

---

[30] To avoid confusing the reader who knows that the cross product is not an associative product, we note that the outer product applies only when each of the factors is orthogonal to the others. In three dimensional space the cross product of three such vectors, however they are grouped, is always zero.

[31] One can't help wondering about the *multi*linear algebra that Grassmann was developing. The recognition of this theory as an essential part of geometry is explicit in Felix Klein's 1908 work on elementary geometry from a higher viewpoint, but Riemann apparently did not make the connection.

[32] At the time of the lecture Gauss had less than a year of life remaining. Yet his mind was still active, and he was very favorably impressed by Riemann's performance.

[33] Herbart's 1824 book *Psychologie als Wissenschaft, neu gegründet auf Erfahrung, Metaphysik, und Mathematik* (*Psychology as Science on a New Foundation of Experiment, Metaphysics, and Mathematics*) is full of mathematical formulas involving the strength of sense impressions, manipulated by the rules of algebra and calculus.

ways into another, the various determinations of it form a continuous or discrete manifold. He noted that discrete manifolds (sets of things that can be counted, as we would say) are very common in everyday life, but continuous manifolds are rare, the spatial location of objects of sense and colors being almost the only examples.

The main part of his lecture was the second part, in which he investigated the kinds of metric relations that could exist in a manifold if the length of a curve was to be independent of its position. Assuming that the point was located by a set of n coordinates $x_1, \ldots, x_n$ (almost the only mathematical symbols that appear in the paper), he considered the kinds of properties needed to define an infinitesimal element of arc length ds along a curve. The simplest function that met this requirements was

$$ds = \sqrt{\sum a_{ij}(x_1, \ldots, x_n)\, dx_i\, dx_j}\,,$$

where the coefficients a_{ij} were continuous functions of position and the expression under the square root is always nonnegative. The next simplest case, which he chose not to develop, occurred when the Maclaurin series began with fourth-degree terms. As Riemann said,

> The investigation of this more general type, to be sure, would not require any essentially different principles, but it would be rather time-consuming and cast relatively little new light on the theory of space; and moreover the results could not be expressed geometrically.

For the case in which coordinates could be chosen so that $a_{ii} = 1$ and $a_{ij} = 0$ when $i \neq j$, Riemann called the manifold *flat.*

Having listed the kinds of properties space was assumed to have, Riemann asked to what extent these properties could be verified by experiment, especially in the case of continuous manifolds. What he said at this point has become famous. He made a distinction between the infinite and the unbounded, pointing out that while space is always assumed to be unbounded, it might very well not be infinite. Then, as he said, assuming that solid bodies exist independently of their position, it followed that the curvature of space would have to be constant, and all astronomical observation confirmed that it could only be zero. But, if the volume occupied by a body varied as the body moved, no conclusion about the infinitesimal nature of space could be drawn from observations of the metric relations that hold on the finite level. "It is therefore quite conceivable that the metric relations of space are not in agreement with the assumptions of geometry, and one must indeed assume this if phenomena can be explained more simply thereby." Riemann evidently intended to follow up on these ideas, but his mind produced ideas much faster than his frail body would allow him to develop them. He died before his 40th birthday with this project one of many left unfinished. He did, however, send an essay to the Paris Academy in response to a prize question proposed (and later withdrawn): *Determine the thermal state of a body necessary in order for a system of initially isothermal lines to remain isothermal at all times, so that its thermal state can be expressed as a function of time and two other variables.* Riemann's essay was not awarded the prize because its results were not developed with sufficient rigor. It was not published during his lifetime.[34]

---

[34] Klein (*1926*, Vol. 2, p. 165) notes that very valuable results were often submitted for prizes at that time, since professors were so poorly paid.

Differential geometry and physics. The work of Grassmann and Riemann was to have a powerful impact on the development of both geometry and physics. One has only to read Einstein's accounts of the development of general relativity to understand the extent to which he was imbued with Riemann's outlook. The idea of geometrizing physics seems an attractive one. The Aristotelian idea of force, which had continued to serve through Newton's time, began to be replaced by subtler ideas developed by the Continental mathematical physicists of the nineteenth century, with the introduction of such principles as conservation of energy and minimal action. In his 1736 treatise on mechanics, Euler had shown that a particle constrained to move along a surface by forces normal to the surface, but on which no forces tangential to the surface act, would move along a shortest curve on the surface. And when he discovered the variational principles that enabled him to solve the isoperimetric problem (see Chapter 17), he applied them to the theory of elasticity and vibrating membranes. As he said,

> Since the material of the universe is the most perfect and proceeds from a supremely wise Creator, nothing at all is found in the world that does not illustrate some maximal or minimal principle. For that reason, there is absolutely no doubt that everything in the universe, being the result of an ultimate purpose, is amenable to determination with equal success from these efficient causes using the method of maxima and minima. [Euler, *1744*, p. 245]

It is known that Riemann was searching for a connection between light, electricity, magnetism, and gravitation at this time.[35] In 1846, Gauss' collaborator Wilhelm Weber (1804–1891) had incorporated the velocity of light in a formula for the force between two moving charged particles. According to Hermann Weyl (Narasimhan, *1990*, p. 741), Riemann did not make any connection between that search and the content of his inaugural lecture. Laugwitz (*1999*, p. 222), however, cites letters from Riemann to his brother which show that he did make precisely that connection. In any case, four years later Riemann sent a paper[36] to the Royal Society in Göttingen in which he made the following remarkable statement:

> I venture to communicate to the Royal Society a remark that brings the theory of electricity and magnetism into a close connection with the theory of light and heat radiation. I have found that the electrodynamic effects of galvanic currents can be understood by assuming that the effect of one quantity of electricity on others is not instantaneous but propagates to them with a velocity that is constant (equal to that of light within observational error).

3.8. The Italian geometers. The unification of Italy in the mid-nineteenth century was accompanied by a surge of mathematical activity even greater than the sixteenth-century work in algebra (discussed in Chapter 14). Gauss had analyzed a general surface by using two parameters and introducing six functions: the coefficients of the first and second fundamental forms. The question naturally arises

---

[35] His lecture was given nearly a decade before Maxwell discovered his famous equations connecting the speed of light with the propagation of electromagnetic waves.

[36] This paper was later withdrawn, but was published after his death (Narasimhan, *1990*, pp. 288–293).

whether a surface can be synthesized from any six functions regarded as the coefficients of these forms. Do they determine the surface, up to the usual Euclidean motions of translation, rotation, and reflection that can be used to move any prescribed point to a prescribed position and orientation? Such a theorem does hold for curves, as was established by two French mathematicians, Jean Frenet (1816–1900) and Joseph Serret (1818–1885), who gave a set of equations—the Frenet–Serret[37] equations—determining the curvature and torsion of a curve in three-dimensional spaces. A curve can be reconstructed from its curvature and torsion up to translation, rotation, and reflection. A natural related question is: Which sets of six functions, regarded as the components of the two fundamental forms, can be used to construct a surface? After all, one needs generally only three functions of two parameters to determine the surface, so that the six given by Gauss cannot be independent of one another.

In an 1856 paper, Gaspare Mainardi (1800–1879) provided consistency conditions in the form of four differential equations, now known as the Mainardi–Codazzi equations,[38] that must be satisfied by the six functions E, F, G, D, D', and D'' if they are to be the components of the first and second fundamental forms introduced by Gauss. Mainardi had learned of Gauss' work through a French translation, which had appeared in 1852. These same equations were discovered by Delfino Codazzi (1824–1875) two years later, using an entirely different approach, and helped him to win a prize from the Paris Academy of Sciences. Codazzi published these equations only in 1883.

When Riemann's lecture was published in 1867, the year after his death, it became the point of departure for a great deal of research in Italy.[39] One who worked to develop these ideas was Riemann's friend Enrico Betti (1823–1892), who tried to get Riemann a chair of mathematics in Palermo. These ideas led Betti to the notion of the connectivity of a surface. On the simplest surfaces, such as a sphere, every closed curve is the boundary of a region. On a torus, however, the circles of latitude and longitude are not boundaries. These ideas belong properly to topology, discussed in the next section. In his fundamental work on this subject, Henri Poincaré named the maximum number of independent non-boundary cycles in a surface the *Betti number* of the surface, a concept that is now generalized to n dimensions. The nth Betti number is the rank of the nth homology group.

Another Italian mathematician who extended Riemann's ideas was Eugenio Beltrami (1835–1900), whose 1868 paper on spaces of constant curvature contained a model of a three-dimensional space of constant negative curvature. Beltrami had previously given the model of a pseudosphere, as explained in Chapter 11, to represent the hyperbolic plane. It was not obvious before his work that three-dimensional hyperbolic geometry and a three-dimensional manifold of constant negative curvature were basically the same thing. Beltrami also worked out the appropriate n-dimensional analogue of the Laplacian $\frac{\partial^2 u}{\partial x^2} + \frac{\partial^2 u}{\partial y^2} + \frac{\partial^2 u}{\partial z^2}$, which plays a fundamental role in mathematical physics. By working with an integral considered earlier by

[37] Frenet gave six equations for the direction cosines of the tangent and principal normal to the curve and its radius of curvature. Serret gave the full set of nine now called by this name, which are more symmetric but contain no more information than the six of Frenet.

[38] The Latvian mathematician Karl Mikhailovich Peterson (1828–1881) published an equivalent set of equations in Moscow in 1853, but they went unnoticed for a full century.

[39] Riemann went to Italy for his health and died of tuberculosis in Selasca. He was in close contact with Italian mathematicians and even published a paper in Italian.

Jacobi (see Klein, *1926*, Vol. 2, p. 190), Beltrami arrived at the operator

$$\Delta u = \frac{1}{\sqrt{a}} \sum_{i=1}^{n} \frac{\partial}{\partial x^i} \left(\sqrt{a} \sum_{j=1}^{n} a_{ij} \frac{\partial u}{\partial x^j} \right),$$

where, with the notation slightly modernized, the Riemannian metric is given by the usual $ds^2 = \sum_{i,j=1}^{n} a_{ij}\, dx^i\, dx^j$, and a denotes the determinant $\det(a_{ij})$. The generalized operator is now referred to as the Laplace–Beltrami operator on a Riemannian manifold.

The algebra of Grassmann and its connection with Riemann's general metric on an n-dimensional manifold was not fully codified until 1901, in "Méthodes de calcul différentiel absolu et leurs applications" ("Methods of absolute differential calculus and their applications"), published in *Mathematische Annalen* in 1901, written by Gregorio Ricci-Curbastro (1853–1925) and Tullio Levi-Civita (1873–1941). This article contained the critical ideas of tensor analysis as it is now taught. The absoluteness of the calculus consisted in the great generality of the transformations that it permitted, showing how differential forms changed when coordinates were changed. Although Ricci-Curbastro competed in a prize contest sponsored that year by the Accademia dei Lincei, he was not successful, as some of the judges regarded his absolute differential calculus as "useful but not essential"[40] to the development of mathematics—the same sort of criticism leveled by Weierstrass against the work of Hamilton in quaternions (see Section 2 of Chapter 15).

The following year Luigi Bianchi (1873–1928) published "Sui simboli a quattro indice e sulla curvatura di Riemann" ("On the quadruply-indexed symbols and Riemannian curvature"), in which he gave the relations among the covariant derivatives of the Riemann curvature tensor, which, however, he derived by a direct method for manifolds of constant curvature, not following the route of Ricci-Curbastro and Levi-Civita. The Bianchi identity was later to play a crucial role in general relativity, assuring local conservation of energy when Einstein's gravitational equation is assumed.

4. Topology

Projections distort the shape of geometric objects, so that some metric properties are lost. Some properties, such as parallelism, however, remain simply because the number of intersections of two curves does not change. The study of space focusing on such very general properties as connections and intersections has been known by various names over the centuries. Latin has two words, *locus* and *situs*, meaning roughly *place* and *position*. The word *locus* is one that we still use today to denote the path followed by a point moving subject to stated constraints. It was the translation of the Greek word *tópos* used by Pappus for the same concept. Since *locus* was already in use, Leibniz fastened on *situs* and mentioned the need for a geometry or analysis of *situs* in a 1679 letter to Huygens.[41] The meaning of *geometria situs* and *analysis situs* evolved gradually. It seems to have been Johann Benedict Listing (1808–1882) who, some time during the 1830s, realized that the Greek root

---

[40] See the article on Ricci-Curbastro's paper at `http://www.math.unifi.it/matematicaitaliana/`.

[41] This letter was published in Huygens' *Œuvres complètes*, M. Nijhoff, La Haye, 1888, Vol. 8, p. 216. From the context it appears that Leibniz was calling for some simple way of expressing position "as algebra expresses magnitude." If so, perhaps we now have what he wanted in the form of vector analysis.

was available. The word *topology* first appeared in the title of his 1848 book *Vorstudien zur Topologie* (*Prolegomena to Topology*). Like geometry itself, topology has bifurcated several times, so that one can now distinguish combinatorial, algebraic, differential, and point-set topology.

4.1. Early combinatorial topology. The earliest result that deals with the combinatorial properties of figures is now known as the *Euler characteristic*, although Descartes is entitled to some of the credit.[42] In a work on polyhedra that he never published, Descartes defined the solid angle at a vertex of a closed polyhedron to be the difference between 2π and the sum of the angles at that vertex. He asserted that the sum of the solid angles in any closed polyhedron was exactly eight right angles. (In our terms, that number is 4π, the area of a sphere of unit radius.) Descartes' work was found among his effects after he died. By chance Leibniz saw it a few decades later and made a copy of it. When it was found among Leibniz' papers, it was finally published. In the eighteenth century, Euler discovered this same theorem in the form that the sum of the angles at the vertices of a closed polyhedron was $4n - 8$ right angles, where n is the number of vertices. Euler noted the equivalent fact that the number of faces and vertices exceeded the number of edges by 2. That is the formula now generally called *Euler's formula*:

$$V - E + F = 2\,.$$

Somewhat peripheral to the general subject of topology was Euler's analysis of the famous problem of the seven bridges of Königsberg in 1736. In Euler's day there were two islands in the middle of the River Pregel, which flows through Köngisberg (now Kaliningrad, Russia). These islands were connected to each other by a bridge, and one of them was connected by two bridges to each shore, the other by one bridge to each shore. The problem was to go for a walk and cross each bridge exactly once, returning, if possible to the starting point. In fact, as one can easily see, it is impossible even to cross each bridge exactly once without boating or swimming across the river. Returning to the starting point merely adds another condition to a condition that is already impossible to fulfill. Euler proved this fact by labeling the two shores and the two islands A, B, C, and D, and representing a stroll as a "word," such as $ABCBD$, in which the bridges are "between" the letters. He showed that any such path as required would have to be represented by an 8-letter word containing three of the letters twice and the other letter three times, which is obviously impossible. This topic belongs to what is now called graph theory; it is an example of the problem of unicursal tracing.

4.2. Riemann. The study of analytic functions of a complex variable turned out to require some concepts from topology. These issues were touched on in Riemann's 1851 doctoral dissertation at Göttingen, "Grundlagen für eine allgemeine Theorie der Functionen einer veränderlichen complexen Grösse" ("Foundations for a general theory of functions of a complex variable"). Although all analytic functions of a complex variable, both algebraic and transcendental, were encompassed in Riemann's ideas, he was particularly interested in algebraic functions, that is functions $w = f(z)$ that satisfy a nontrivial polynomial equation $p(z, w) = 0$. Algebraic functions are essentially and unavoidably multivalued. To take the simplest example,

---

[42] Much of the information in this paragraph is based on the following website: `http://www.math.sunysb.edu/~tony/whatsnew/column/descartes-0899/descartes2.html`.

in which $z - w^2 = 0$, every complex number $z = a + bi$ has two distinct complex square roots:

$$w = \pm(u + iv)\,, \quad \text{where } u = \sqrt{\frac{\sqrt{a^2 + b^2} + a}{2}} \quad \text{and } v = \operatorname{sgn}(b)\sqrt{\frac{\sqrt{a^2 + b^2} - a}{2}}\,.$$

The square roots of the positive real numbers that occur here are assumed positive. There is no way of choosing just one of the two values at each point that will result in a continuous function $w = \sqrt{z}$. In particular, it is easy to show that any such choice must have a discontinuity at some point of the circle $|z| = 1$.

One way to handle this multivaluedness was to take two copies of the z-plane, labeled with subscripts as z_1 and z_2 and place one of the square roots in one plane and the other in the other. This technique was used by Cauchy and had been developed into a useful way of looking at complex functions by Victor Puiseux (1820–1883) in 1850. Indeed, Puiseux seems to have had the essential insights that can be found in Riemann's work, although differently expressed. Riemann is known to have seen the work of Puiseux, although he did not cite it in his own work. He generally preferred to work out his own way of doing things and tended to ignore earlier work by other people. In any case, the essential problem with choosing one square root and sticking to it is that a single choice cannot be continuous on a closed path that encloses the origin without going through it. At some point on such a path, there will be nearby points at which the function assumes two values that are close to being negatives of each other.

Riemann had the idea of cutting the two copies of the z-plane along a line running from zero to infinity (both being places where there is only one square root, assuming a bit about complex infinity). Then if the lower edge of each plane is imagined as being glued to the upper edge of the other,[43] the result is a single connected surface in which the origin belongs to both planes. On this new surface a continuous square-root function can be defined. It was the gluing that was really new here. Cauchy and Puiseux both had the idea of cutting the plane to keep a path from winding around a branch point and of using different copies of the plane to map different branches of the function.

Riemann introduced the idea of a *simply connected surface*, one that is disconnected by any cut from one boundary point to another that passes through its interior without intersecting itself. He stated as a theorem that the result of such a cut would be two simply connected surfaces. In general, when a connected surface is cut by a succession of such *crosscuts*, as he called them, the difference between the number of crosscuts and the number of connected components that they produce is a constant, called the *order of connectivity* of the surface. A sphere, for example, can be thought of as a square with adjacent edges glued together, as in Fig. 14. It is simply connected because a diagonal cut disconnects it. The torus, on the other hand, can be thought of as a square with opposite edges identified (see Fig. 14). To disconnect this surface, it is necessary to cut it at least twice, for example, either by drawing both diagonals or by cutting it through its midpoint with two lines parallel to the sides. No single cut will do. The torus is thus doubly connected.

---

[43] You can visualize this operation being performed if you imagine one copy of the plane picked up and turned upside down above the other so that the upper edge is glued to the upper edge and the lower to the lower.

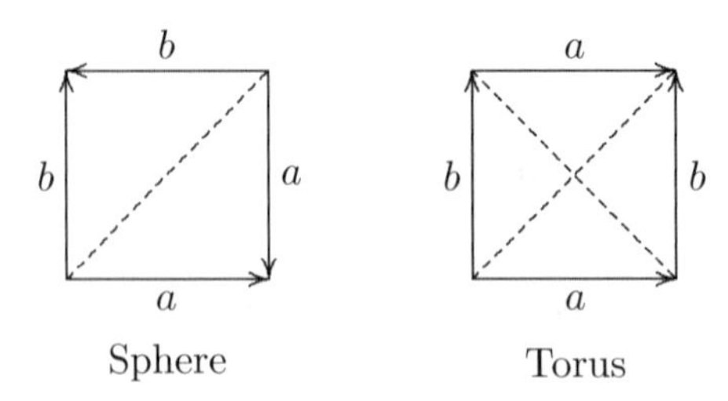

FIGURE 14. Left: The sphere, regarded as a square with edges identified, is disconnected by a diagonal. Right: The torus requires two cuts to disconnect.

4.3. Möbius. One fact that had been thought well established about polyhedra was that in any polyhedron it was possible to direct the edges in such a way that one could trace around the boundary of each face by following the prescribed direction of its edges. Each face would be always to the left or always to the right as one followed the edges around it while looking at it from outside the polyhedron. This fact was referred to as the *edge law* (*Gesetz der Kanten*). The first discovery of a closed polyhedron that violated this condition[44] was due to Möbius, sometime during the late 1850s. Möbius did not publish this work, although he did submit some of it to the Paris Academy as his entry to a prize competition in 1858. This work was edited and introduced by Curt Reinhardt (dates unknown) and published in Vol. 2 of Möbius' collected works. There in the first section, under the heading "one-sided polyhedra," is a description of the Möbius band as we now know it (Fig. 14). After describing it, Möbius went on to say that although a triangulated polyhedron whose surface was two-sided will apparently contain only two-sided bands, *nevertheless a triangulated polyhedron with a one-sided surface can contain both one- and two-sided bands.*

Möbius explored polyhedra and made a classification of them according to the number of boundary curves they possessed. He showed how more complicated polyhedra could be produced by gluing together a certain set of basic figures. He found an example of a triangulated polyhedron consisting of 10 triangles, six vertices, and 15 edges, rather than 14, as would be expected from Euler's formula for a closed polyhedron: $V - E + F = 2$. This figure is the projective plane, and cannot be embedded in three-dimensional space. If one of the triangles is removed, the resulting figure is the Möbius band, which can be embedded in three-dimensional space.

4.4. Poincaré's *Analysis situs*. Poincaré seemed to be dealing constantly with topological considerations in his work in both complex function theory and differential equations. To set everything that he discovered down in good order, he wrote a treatise on topology called *Analysis situs* in 1895, published in the *Journal de l'École Polytechnique*, that has been regarded as the founding document of modern algebraic topology.[45] He introduced the notion of homologous curves—curves that (taken together) form the boundary of a surface. This notion could be formalized, so that one could consider formal linear combinations (now called

[44] In fact, a closed nonorientable polyhedron cannot be embedded in three-dimensional space, so that the edge law is actually true for *closed* polyhedra in three-dimensional space.
[45] Poincaré followed this paper with a number of supplements over the next decade.

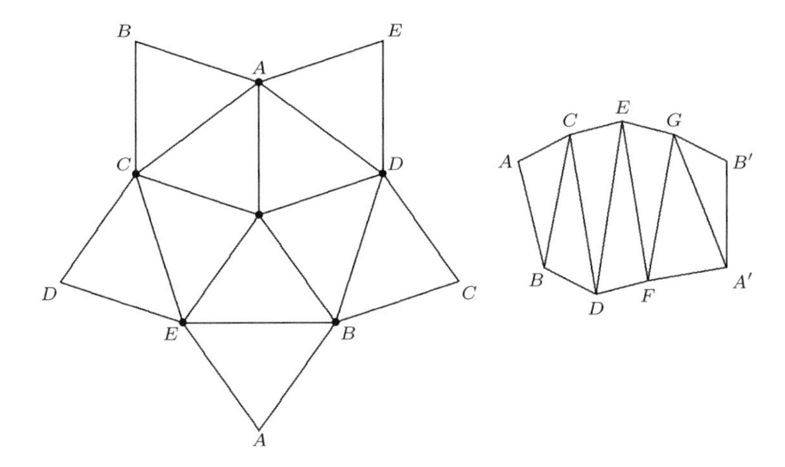

FIGURE 15. Left: the projective plane triangulated and cut open. If two opposite edges with corresponding endpoints are glued together, the figure becomes a Möbius band. In three-dimensional space it is not possible to glue all the edges together as indicated. Right: the Möbius band as originally described by Möbius.

chains) $C = n_1C_1 + \cdots + n_rC_r$ of oriented curves C_i with integer coefficients n_i. The interpretation of such a combination came from analysis: A line integral over C was interpreted as the number $I = n_1I_1 + \cdots + n_rI_r$, where I_j was the line integral over C_j. When generalized to k-dimensional manifolds (called *varieties* by Poincaré) and combined with the concept of the boundary of an oriented manifold as a cycle, this idea was the foundation of homology theory: The k-cycles (k-chains whose boundaries are the zero $(k-1)$-chain—Poincaré called them *closed varieties*) form a group, of which the k-cycles that are the boundary of a $(k+1)$-cycle form a subgroup. When two homologous cycles (cycles whose difference is a boundary) are identified, the resulting classes of cycles form the kth *homology group*. For example, in the sphere shown in Fig. 14, the diagonal that is drawn forms a cycle. This cycle is the complete boundary of the upper and lower triangles in the figure, and it turns out that any cycle on the sphere is a boundary. The first homology group of the sphere is therefore trivial (consists of only one element). For the torus depicted in Fig. 14, a and b are each cycles, but neither is a boundary, nor is any cycle $ma + nb$. On the other hand, the cycle formed by adding either diagonal to $a+b$ is the boundary of the two triangles with these edges. Thus, the first homology group of the torus can be identified with the set of cycles $ma+nb$. Any other cycle will be homologous to one of these.

Poincaré also introduced the notion of the *fundamental group* of a manifold. He had been led to algebraic topology partly by his work in differential equations. He discovered the fundamental group by imagining functions satisfying a set of differential equations and being permuted as a point moved around a closed loop. He was thus led to consider formal sums of loops starting and ending at a given point, two loops being equivalent if tracing them successively left the functions invariant. The resulting set of permutations was what he called the *fundamental group*. He cautioned that, despite appearances, the fundamental group was not the same thing as the first homology group, since there was no base point involved in the homology group. Moreover, he noted, while the order in which the cycles in a chain were traversed was irrelevant, the fundamental group was not necessarily commutative. He suggested redefining the term *simply connected* to mean having a trivial fundamental group. He gave examples to show that the homology groups do not determine the topological nature of a manifold, exhibiting three three-dimensional manifolds all having the same homology groups, but different fundamental groups and therefore not topologically the same (homeomorphic). He then asked a number of questions about fundamental groups, one of which has become famous. *Given two manifolds of the same number of dimensions having the same fundamental group, are they homeomorphic?* Like Fermat's last theorem, this question has been attacked by many talented mathematicians, and proofs have been proposed for a positive answer to the question, but—at least until recently—all such proofs have been found wanting.[46]

4.5. Point-set topology. Topology is sometimes popularly defined as "rubber-sheet geometry," in the sense that the concepts it introduces are invariant under moving and stretching, provided that no tearing takes place. In the kinds of combinatorial topology just discussed, those concepts usually involve numbers in some form or other—the number of independent cycles on a manifold, the Euler characteristic, and so forth. But there are also topological concepts not directly related to number.

Continuity and connectedness. The most important of these is the notion of *connectedness* or *continuity.* This word denotes a deep intuitive idea that was the source of many paradoxes in ancient times, such as the paradoxes of Zeno. As we shall see in Chapter 15, it is impossible to prove the fundamental theorem of algebra without this concept.[47] For analysts, it was crucial to know that if a certain function was negative at one point on a line and positive at another, it must assume the value zero at some point between the two points. That property eventually supplanted earlier definitions of continuity, and the property now taken as the definition of continuity is designed to make this proposition true. The clarification of the ideas surrounding continuity occurred in the early part of the nineteenth century and is discussed in more detail in Chapter 17. Once serious analysis of this concept was undertaken, it became clear that many intuitive assumptions about the connectedness of curves and surfaces had been made from the beginning of deductive geometry. These continuity considerations complicated the theory of functions of a real variable for some decades until adequate explanations of it were found. A good example of such problems is provided by Dedekind's construction of the real numbers, discussed in Chapter 8, which he presented as a solution to the problem of defining what is meant by a continuum.

[46] As of this writing, evidence begins to accumulate that the Russian mathematician Grigorii Perlman of the Steklov Institute in St. Petersburg has settled the Poincaré conjecture (Associated Press, January 7, 2004). As a graduate student at Princeton in 1964, when a mathematician came to town claiming to have proved this elusive result, I discussed it with Norman Steenrod (1910–1971), one of the twentieth century's greatest topologists. He told me that proving the conjecture, although difficult, would be a rather uninteresting thing to do, since it would only confirm what people already thought was true. It would have been much more exciting to disprove it.

[47] Even the second of the four proofs that Gauss gave, which is generally regarded as a purely algebraic proof, required the assumption that an equation of odd degree with real coefficients has a real solution—a fact that relies on connectedness.

Compactness. Another basic concept of point-set topology is that of *compactness.* This concept is needed to make the distinction between being bounded and having a minimum or maximum. The concepts of compactness, connectedness, and continuity are used together nowadays to prove such theorems as Rolle's theorem in calculus.

At least three lines of thought led to the notion of compactness. The first was the search for maxima and minima of functions, that is, points at which the function assumed the largest or smallest possible value. It was clear that a sequence of points x_n could always be found such that $f(x_n)$ tended to a maximum value; that was what a maximum value meant. But did the sequence x_n itself, or some subsequence of it, also converge to a point x? If so, it was clear from the definition of continuity that x must be a maximum or minimum. This property was studied by the Czech mathematician Bernard Bolzano (1781–1848), who was looking for a proof of the continuity property discussed above. He showed as early as 1817 (see Manheim, *1964*, p. 67) that the continuity property could be made to follow from the property that a set of numbers that is bounded above has a least upper bound. He phrased this statement differently, of course, saying that if there is a property possessed by a function at some points, but not all, and that property holds for all points less than some a, there is a smallest number U such that the property holds for all numbers less than U. Bolzano proved this fact by repeated bisection of an interval such that the property holds at the lower endpoint but not the upper. Some 50 years later, after defining real numbers as sequences of rational numbers (with a suitable notion of equivalent sequences), Weierstrass used arguments of this type to deduce that a bounded sequence of real numbers has a convergent subsequence. This theorem, in several closely equivalent forms, is now known as the *Bolzano–Weierstrass theorem.*

The second line of thought leading to compactness was the now-familiar distinction between pointwise continuity and uniform continuity. This distinction was brought to the fore in the mid-1850s, and Dirichlet proved that on an interval $[a, b]$ (including the endpoints) a continuous function was uniformly continuous. He was really the first person to use the idea of replacing a covering by open sets with a finite subcovering. The same theorem was proved by Eduard Heine (1821–1881) in 1872; as a result, Heine found his name attached to one form of the basic theorem.

The third line was certain work in complex analysis by Émile Borel (1871–1956) in the 1890s. Borel was studying analytic continuation, whereby a complex-valued function is expanded as a power series about some point:

$$f(z) = f(z_0) + \sum_{n=1}^{\infty} \frac{f^{(n)}(z_0)}{n!}(z - z_0)^n .$$

If the series has a finite radius of convergence, it represents $f(z)$ only inside a disk. However, it enables all the derivatives of $f(z)$ to be computed at all points of the disk, so that if one forms the analogous series at some point z_1 in the disk different from z_0, it is possible that the new series will converge at some points outside the original disk. In this way, one can continue a function uniquely along a path γ from one point a to another point b, provided there is such a series with a positive radius of convergence at each point of γ. What is needed is some way of proving that only a finite number of such disks will be required to cover the whole curve γ. The resulting covering theorem was further refined by a number of mathematicians, including Henri Lebesgue (1875–1941), and is now generally known as the *Heine–Borel theorem*. The word *compact* was first used in 1906 in two equivalent senses in two different papers, by Maurice Fréchet (1878–1973).

Closed and open sets. The word *set*, without which modern mathematicians would not be able to talk at all, was not introduced formally until the 1870s. The history of set theory is discussed in more detail in Chapter 19. At present we merely mention that the idea of a closed set arose from consideration of the set of limit points of a given set (its derived set). In an 1884 paper, Georg Cantor (1854–1918) called a set *closed* if it contained all of its limit points. Since it was easy to show that a limit point of limit points of a set P is itself a limit point of P, it followed that the derived set P' is always a closed set.

Although the phrase *closed set* appears in 1884, its dual—the phrase *open set*—did not appear for nearly two more decades. Weierstrass had used the *concept* of an open set in discussing analytic functions, since he used power series, which required that the function be defined a small disk called a *neighborhood* about each point in its domain of definition. Weierstrass used the German term *Gebiet* (*region*) for such a domain of definition. The phrase *open set* seems to have been used for the first time by W. H. Young in 1902.[48] In a 1905 paper in descriptive function theory (that is, discussing what it means for a function to be "analytic" in a very general sense), Henri Lebesgue referred specifically to *ensembles ouverts* (*open sets*) and defined them to be the complements of closed sets.

Metric spaces. The frequent repetition of certain basic patterns of reasoning, and perhaps just a normal human penchant for order, led to the creation of very abstract structures around the beginning of the twentieth century. The kind of continuity argument we now associate with δ's and ε's was generalized in 1905 by Maurice Fréchet, who considered abstract sets on which there was a sort of distance between two points A and B, denoted (A, B). This distance had the properties normally associated with distance, that is, $(A, B) = (B, A)$ (the distance from A to B is the same as the distance from B to A), $(A, B) > 0$ if $A \neq B$, and $(A, A) = 0$. Further, he assumed that there was a real-valued function $f(t)$ tending to 0 as t tends to 0, and such that $(A, C) < f(\varepsilon)$ if $(A, B) < \varepsilon$ and $(B, C) < \varepsilon$. Such a structure is now called a *metric space*, although the definition is streamlined somewhat, the third property being replaced by the *triangle inequality*. It can be shown that for each distance function introduced by Fréchet there is an equivalent metric in the modern sense.

In 1906 Fréchet also gave two definitions of the term *compact* (for metric spaces) in the modern sense. In one paper he defined a space to be compact if every infinite subset of it had at least one limit point. In the other he defined compactness to mean that every decreasing sequence of nonempty closed sets had a nonempty intersection. Thus he used both the Bolzano–Weierstrass property and the Heine–Borel property (which are equivalent for metric spaces).

General topology. The notion of a topological space in the modern sense arose in 1914 in the work of the Youngs and in the work of Felix Hausdorff (1868–1942), who was at the time a professor at Bonn. Hausdorff's influential book *Grundzüge der Mengenlehre* (*Elements of Set Theory*) was translated into many languages.

[48] The early papers of W. H. Young and his wife G. C. Young were published under his name alone, as mentioned in Chapter 4.

The first part of the book is an exposition of abstract set theory as it existed at the time, including cardinal and ordinal numbers, and the early stages of what is now called *descriptive set theory*, that is, the classification of sets according to their complexity, starting with a ground class consisting of closed sets and open sets, and then proceeding up a hierarchy by passing to countable unions and intersections. He invented the term *ring* for a class of sets that was closed under finite unions and intersections and *field* for a class that was closed under set differences and finite unions, but warned in a footnote that "the expressions *ring* and *field* are taken from the theory of algebraic numbers based on an approximate analogy that it will not do to push too far."[49]

Hausdorff introduced metric spaces, being the first to use that name for them, via the axioms now used, then gave a set of "neighborhood axioms" for a more general type of space:

1. To each point x there corresponds at least one neighborhood U_x; every neighborhood U_x contains the point x.
2. If U_x and V_x are two neighborhoods of the point x, there is another neighborhood W_x of x contained in both of them.
3. If the point y lies in U_x, there is a neighborhood U_y contained in U_x.
4. For any distinct points x and y, there are two neighborhoods U_x and U_y whose intersection is empty.

These were Hausdorff's axioms for topology, and they were well designed for discussing the local behavior of functions on a highly abstract level. A quarter-century later, the group of French authors known collectively as Nicolas Bourbaki introduced a global point of view, defining a topological space axiomatically as we know it today, in terms of open sets. The open sets of a space can be any collection that has the empty set and the whole set as members and is closed under arbitrary unions and finite intersections. In those terms, one of Hausdorff's neighborhoods U_x is any open set O with $x \in O$. Conversely, given a set on which the first three of Hausdorff's axioms hold, it is easy to show that the sets that are neighborhoods of all of their points form a topology in the sense of Bourbaki. Bourbaki omitted the last property specified by Hausdorff. Spaces having this extra property are now called (appropriately enough) Hausdorff spaces.

Questions and problems

12.1. Judging from Descartes' remarks on mechanically drawn curves, should he have admitted the conchoid of Nicomedes among the legitimate curves of geometry?

12.2. Prove Menelaus' theorem and its converse. What happens if the points E and F are such that $AD : AE :: BD : BE$? (Euclid gave the answer to this question.)

12.3. Use Menelaus' theorem to prove that two medians of a triangle intersect in a point that divides each in the ratio of 1:2.

---

[49] The word *ring* in the abstract algebraic sense was also introduced in 1914, in a paper of A. Fraenkel (see Section 2 of Chapter 15). A very influential work in measure theory, written in 1950 by Paul Halmos (b. 1916), caused Hausdorff's *ring* to fall into disuse and appropriated the term *ring* to mean what Hausdorff called a *field*. Halmos reserved the term *algebra* for a ring, one of whose elements was the entire space. Probabilists, however, use the term *field* for what Halmos called an algebra.

12.4. Deduce Brianchon's theorem for a general conic from the special case of a circle. How do you interpret the case of a regular hexagon inscribed in a circle?

12.5. Fill in the details of Plücker's proof of Brianchon's theorem, as follows: Suppose that the equation of the conic is $q(x,y) = y^2 + r_1(x)y + r_2(x) = 0$, where $r_1(x)$ is a linear polynomial and $r_2(x)$ is quadratic. Choose coordinate axes not parallel to any of the sides of the inscribed hexagon and such that the x-coordinates of all of its vertices will be different, and also choose the seventh point to have x-coordinate different from those of the six vertices. Then suppose that the polynomial generated by the three lines is $s(x,y) = y^3 + t_1(x)y^2 + t_2(x)y + t_3(x) = 0$, where $t_j(x)$ is of degree j, $j = 1,2,3$. Then there are polynomials $u_j(x)$ of degree j, $j = 1,2,3$, such that

$$s(x,y) = q(x,y)\bigl(y - u_1(x)\bigr) + \bigl(u_2(x)y + u_3(x)\bigr).$$

We need to show that $u_2 \equiv 0$ and $u_3 \equiv 0$. At the seven points on the conic where both $q(x,y)$ and $s(x,y)$ vanish it must also be true that $u_2(x)y + u_3(x) = 0$. Rewrite the equation $q(x,y) = 0$ at these seven points as

$$(u_2 y)^2 + r_1 u_2 (u_2 y) + u_2^2 r_2 = 0$$

observe that at these seven points $u_2 y = -u_3$, so that the polynomial $u_3^2 - r_1 u_2 u_3 + u_2^2 r_2$, which is of degree 6, has seven distinct zeros. It must therefore vanish identically, and that means that

$$(2u_3 - r_1 u_2)^2 = u_2^2 (r_1^2 - 4r_2).$$

This means that either u_2 is identically zero, which implies that u_3 also vanishes identically, or else u_2 divides u_3. Prove that in the second case the conic must be a pair of lines, and give a separate argument in that case.

12.6. Consider the two equations

$$\begin{aligned} xy &= 0, \\ x(y-1) &= 0. \end{aligned}$$

Show that these two equations are independent, yet have infinitely many common solutions. What kind of conic sections do these equations represent?

12.7. Consider the general cubic equation

$$Ax^3 + Bx^2y + Cxy^2 + Dy^3 + Ex^2 + Fxy + Gy^2 + Hx + Iy + J = 0,$$

which has 10 coefficients. Show that if this equation is to hold for the 10 points $(1,0)$, $(2,0)$, $(3,0)$, $(4,0)$, $(0,1)$, $(0,2)$, $(0,3)$, $(1,1)$, $(2,2)$, $(1,-1)$, all 10 coefficients $A, \ldots, J$ must be zero. In general, then, it is not possible to pass a curve of degree 3 through any 10 points in the plane. Use linear algebra to show that it is always possible to pass a curve of degree 3 through any nine points, and that the curve is generally unique.

On the other hand, two *different* curves of degree 3 generally intersect in 9 points, a result known as Bézout's theorem after Étienne Bézout (1730–1783), who stated it around 1758, although Maclaurin had stated it earlier. How does it happen that while nine points generally determine a *unique* cubic curve, yet *two distinct* cubic curves generally intersect in nine points? [*Hint*: Suppose that a set of eight points $\{(x_j, y_j) : j = 1, \ldots, 8\}$ is given for which the system of equations for $A, \ldots, J$ has rank 8. Although the system of linear equations for the coefficients is generally of rank 9 if another point is adjoined to this set, there generally is a point

(x_9, y_9), the ninth point of intersection of two cubic curves through the other eight points, for which the rank will remain at 8.]

12.8. Find the Gaussian curvature of the hyperbolic paraboloid $z = (x^2 - y^2)/a$ at each point using x and y as parameters.

12.9. Find the Gaussian curvature of the pseudosphere obtained by revolving a tractrix about the x-axis. Its parameterization can be taken as

$$\mathbf{r}(u, v) = \left(u - a \tanh\left(\frac{u}{a}\right), a \operatorname{sech}\left(\frac{u}{a}\right) \cos(v), a \operatorname{sech}\left(\frac{u}{a}\right) \sin(v) \right).$$

Observe that the elements of area on both the pseudosphere and its map to the sphere vanish when $u = 0$. (In terms of the first and second fundamental forms, $E = 0 = g$ when $u = 0$.) Hence curvature is undefined along the circle that is the image of that portion of the parameter space. Explain why the pseudosphere can be thought of as "a sphere of imaginary radius." Notice that it has a cusp along the circle in which it intersects the plane $x = 0$.

12.10. Prove that the Euler relation $V - E + F = 2$ for a closed polyhedron is equivalent to the statement that the sum of the angles at all the vertices is $(2V-4)\pi$, where V is the number of vertices. [*Hint:* Assume that the polygon has F faces, and that the numbers of edges on the faces are $e_1, \ldots, e_F$. Then the number of edges in the polyhedron is $E = (e_1 + \cdots + e_F)/2$, since each edge belongs to two faces. Observe that a point traversing a polygon changes direction by an amount equal to the exterior angle at each vertex. Since the point returns to its starting point after making a complete circuit, the sum of the *exterior* angles of a polygon is 2π. Since the interior angles are the supplements of the exterior angles, we see that their sum is $e_i\pi - 2\pi = (e_i - 2)\pi$. The sum of all the interior angles of the polyhedron is therefore $(2E - 2F)\pi$.]

12.11. Give an informal proof of the Euler relation $V - E + F = 2$ for closed polyhedra, assuming that every vertex is joined by a sequence of edges to every other vertex. [*Hint:* Imagine the polyedron inflated to become a sphere. That stretching will not change V, E, or F. Start drawing the edges on a sphere with a single vertex, so that $V = 1 = F$ and $E = 0$. Show that adding a new vertex by distinguishing an interior point of an edge as a new vertex, or by distinguishing an interior point of a face as a new vertex and joining it to an existing vertex, increases both V and E by 1 and leaves F unchanged, while drawing a diagonal of a face increases E and F by 1 and leaves V unchanged. Show that the entire polyhedron can be constructed by a sequence of such operations.]

Part 5

Algebra

Occasionally, a practical problem arises in which it is necessary to invert a sequence of arithmetic operations. That is, we know the result of the operations but not the data. The best examples of this kind of problem come from geometry, and a typical specimen can be seen in the *sangaku* plaque shown in Color Plate 2. This type of problem is the seed of the area we call algebra, whose development can be conveniently divided into three stages. In the first stage, knowing the procedure followed and the result, one is forced to think in the terms that Pappus referred to as analysis, that is, deducing consequences of the formula until one arrives at the data. The main tool in this analysis is the equation, but equations occur explicitly only after a stock of examples has been accumulated. At the second stage, equations are identified as an object of independent interest, and techniques for solving them are developed. In the third stage, a higher-level analysis of the algorithms for solution leads to the subject we now know as modern algebra. We shall devote one chapter to each of these stages.

CHAPTER 13

Problems Leading to Algebra

Algebra suffers from a motivational problem. Examples of the useless artificiality of most algebraic problems abound in every textbook ever written on the subject. Here, for example, is a problem from Girolamo Cardano's book *Ars magna* (1545):

> Two men go into business together and have an unknown capital. Their gain is equal to the cube of the tenth part of their capital. If they had made three ducats less, they would have gained an amount exactly equal to their capital. What was the capital and their profit? [Quoted by Pesic, (*2003*), pp. 30–31]

If reading this problem makes you want to suggest, "Let's just ask them what their capital and profit were," you are to be congratulated on your astuteness. The second statement of the problem, in particular, marks the entire scenario as an airy flight of fancy. Where in the world would anyone get this kind of information? What data banks is it kept in? How could anyone know this relationship between capital and profit without knowing what the capital and profit were? One of the hardest questions to answer in teaching either algebra or its history is "What is it *for*?" Although some interesting *algebra problems* can be generated from geometric figures, it is not clear that these problems are interesting *as geometry*. Reading the famous treatises on algebra, we might conclude that it is pursued for amusement by people who like puzzles.[1] That answer is not very satisfying, however, and we shall be on the alert for better motivations as we study the relevant documents.

1. Egypt

Although arithmetic and geometry fill up most of the Egyptian papyri, there are some problems in them that can be considered algebra. These problems tend to be what we now classify as linear problems, since they involve the implicit use of direct proportion. The concept of proportion is the key to the problems based on the "rule of false position." Problem 24 of the Ahmose Papyrus, for example, asks for the quantity that yields 19 when its seventh part is added to it. The author notes that if the quantity were 7 (the "false [sup]position"), it would yield 8 when its seventh part is added to it. Therefore, the correct quantity will be obtained by performing the same operations on the number 7 that yield 19 when performed on the number 8. The Egyptian format for such computations is well adapted for handling problems of this sort. The key to the solution seems to be, implicitly, the

[1] In one episode of a popular American situation comedy series during the 1980s, a young policewoman was working undercover, pretending to be a high-school student. While studying algebra with a classmate, she encountered a problem akin to the following. "Johnny is one-third as old as his father; in 15 years he will be half as old. How old are Johnny and his father?" Her response—a triumph of common sense over a mindless educational system—was: "Do we know these people?"

notion that multiplication is distributive over addition (another way of saying that proportions are preserved). But of course, since multiplication was thought of in a peculiar way in Egyptian culture, the algebraic reasoning was very likely as follows: Such-and-such operations applied to 8 will yield 19. If I first add the seventh part of 7 to 7, I will get 8 as a result. If I then perform those operations on 8, I will get 19. *Therefore*, if I first perform those operations on 7, and then add the seventh part of the result to itself, I will also get 19.

The computation is carried out by the standard Egyptian method. First find the operations that must be performed on 8 in order to yield 19:

$$\begin{array}{rl}
1 & 8 \\
2 & 16\,* \\
\overline{2} & 4 \\
\overline{4} & 2\,* \\
\overline{8} & 1\,* \\
2\ \overline{4}\ \overline{8} & 19 \text{ Result}.
\end{array}$$

Next, perform these operations on 7:

$$\begin{array}{rl}
1 & 7 \\
2 & 14\,* \\
\overline{2} & 3\ \overline{2} \\
\overline{4} & 1\ \overline{2}\ \overline{4}\,* \\
\overline{8} & \overline{2}\ \overline{4}\ \overline{8}\,* \\
2\ \overline{4}\ \overline{8} & 16\ \overline{2}\ \overline{8} \text{ Result}.
\end{array}$$

This is the answer. The scribe seems quite confident of the answer and does not carry out the computation needed to verify that it works.

The Egyptian scribes were capable of performing operations more complicated than mere proportion. They could take the square root of a number, which they called a *corner*. The Berlin Papyrus 6619, contains the following problem (Gillings, *1972*, p. 161):

> The area of a square of 100 is equal to that of two smaller squares. The side of one is $\overline{2}\ \overline{4}$ the side of the other. Let me know the sides of the two unknown squares.

Here we are asking for two quantities given their ratio $(\frac{3}{4})$ and the sum of their squares (100). The scribe assumes that one of the squares has side 1 and the other has side $\overline{2}\ \overline{4}$. Since the resulting total area is $1\ \overline{2}\ \overline{16}$, the square root of this quantity is taken $(1\ \overline{4})$, yielding the side of a square equal to the sum of these two given squares. This side is then multiplied by the correct proportionality factor so as to yield 10 (the square root of 100). That is, the number 10 is divided by $1\ \overline{4}$, giving 8 as the side of the larger square and hence 6 as the side of the smaller square. This example, incidentally, was cited by van der Waerden as evidence of early knowledge of the Pythagorean theorem in Egypt.

2. Mesopotamia

If we interpret Mesopotamian algebra in our own terms, we can credit the mathematicians of that culture with knowing how to solve some systems of two linear equations in two unknowns, any quadratic equation having at least one real positive root, some systems of two equations where one of the equations is linear and the other quadratic, and a potentially complete set of cubic equations. Of course, it must be remembered that these people were solving *problems*, not *equations*. They did not have any classification of equations in which some forms were solvable and others not. What they knew was that they could find certain numbers from certain data.

2.1. Linear and quadratic problems. As mentioned in Section 4 of Chapter 6, the Mesopotamian approach to algebraic problems was to associate with every pair of numbers another pair: their average and their *semidifference*. These linear problems arise frequently as a subroutine in the solution of more complex problems involving squares and products of unknowns. In Mesopotamia, quadratic equations occur most often as problems in two unknown quantities, usually the length and width of a rectangle. The Mesopotamian mathematicians were able to reduce a large number of problems to the form in which the sum and product or the difference and product of two unknown numbers are given. We shall consider an example that has been written about by many authors. It occurs on a tablet from the Louvre in Paris, known as AO 8862.[2]

A loose translation of the text of this tablet, made from Neugebauer's German translation, reads as follows:

> I have multiplied the length and width so as to make the area. Then I added to the area the amount by which the length exceeds the width, obtaining 3,3. Then I added the length and width together, obtaining 27. What are the length, width, and area?
>
> 27 3,3 the sums
> 15 length
> 3,0 area
> 12 width
>
> You proceed as follows:
>
> Add the sum (27) of the length and width to 3,3. You thereby obtain 3,30. Next add 2 to 27, getting 29. You then divide 29 in half, getting 14;30. The square of 14;30 is 3,30;15. You subtract 3,30 from 3,30;15, leaving the difference of 0;15. The square root of 0;15 is 0;30. Adding 0;30 to the original 14;30 gives 15, which is the length. Subtracting 0;30 from 14;30 gives 14 as width. You then subtract 2, which was added to the 27, from 14, giving 12 as the final width.

The author continues, verifying that these numbers do indeed solve the problem. This text requires some commentary, since it is baffling at first. Knowing the general approach of the Mesopotamian mathematicians to problems of this sort, one can understand the reason for dividing 29 in half (so as to get the average of two numbers) and the reason for subtracting 3,30 from the square of 14;30 (the

[2] AO stands for Antiquités Orientales.

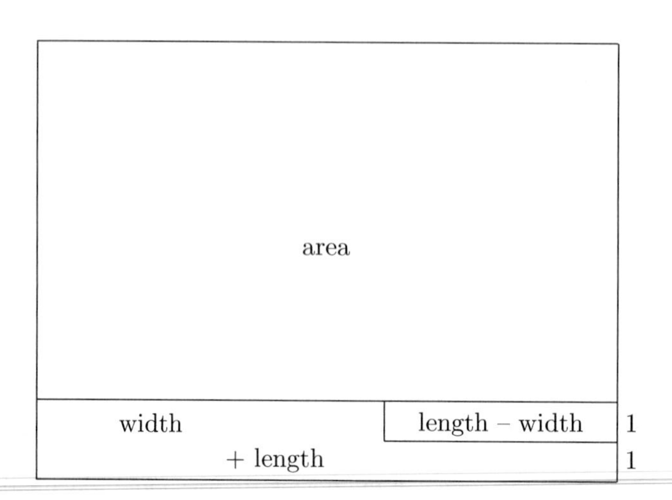

FIGURE 1. Reduction of a problem to standard form.

difference between the square of the average and the product will be the square of the semidifference of the two numbers whose sum is 29 and whose product is 3,30, that is, 210). What is not clear is the following: Why add 27 to the number 3,3 in the first place, and why add 2 to 27? Possibly the answer is contained in Fig. 1, which shows that adding the difference between length and width to the area amounts to gluing a smaller rectangle of unit width onto a larger rectangle. Then adding the sum of length and width amounts to gluing a gnomon onto the resulting figure in order to complete a rectangle two units wider than the original. Finding the dimensions of that rectangle from its perimeter and area is the standard technique of solving a quadratic equation, and that is what the author does.

The tablet AO 6670, discussed by van der Waerden (*1963*, pp. 73–74) contains a rare explanation of the procedure for solving a problem that involves two unknowns and two conditions, given in abstract terms without specific numbers. Unfortunately, the explanation is very difficult to understand. The statement of the problem is taken directly from Neugebauer's translation: *Length and width as much as area; let them be equal.* Thereafter, the translation given by van der Waerden, due to François Thureau-Dangin (1872–1944), goes as follows:

> The product you take twice. From this you subtract 1. You form the reciprocal. With the product that you have taken you multiply, and the width it gives you.

Van der Waerden asserts that the formula $y = (1/(x-1)) \cdot x$ is "stated in the text" of Thureau-Dangin's translation. If so, it must have been stated in a place not quoted by van der Waerden, since x is not a "product" here, nor is it taken twice. Van der Waerden also notes that according to Evert Marie Bruins (1909–1990), the phrase "length and width" does not mean the *sum* of length and width. Van der Waerden says that "the meaning of the words has to be determined in relation to the mathematical content." The last two sentences in the description tell how to determine the width once the length has been found. That is, you take

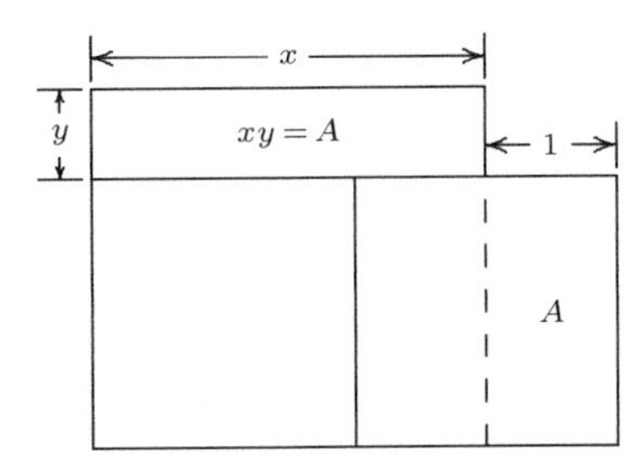

FIGURE 2. A scenario that may "fit" a text from cuneiform tablet AO 6670.

reciprocal of the length and multiply it by the product of length and width, which must be given in the problem as the area. The mystery is then pushed into the first two instructions. What product is being "taken twice"? Does taking a product twice mean multiplying by 2, or does it mean cubing? Why is the number 1 being subtracted? Perhaps we should go back to the original statement and ask whether "as much as area" implies an equation, or whether it simply means that length and width *form* an area. What does the word *them* refer to in the statement, "Let them be equal"? Is it the length and the width, or some combination of them and the area? Without knowing the original language and seeing the original text, we cannot do anything except suggest possible meanings, based on what is mathematically correct, to those who do know the language.

We can get a geometric problem that fits this description by considering Fig. 2, where two equal squares have been placed side by side and a rectangle of unit length, shown by the dashed line, has been removed from the end. If the problem is to construct a rectangle on the remaining base equal to the part that was cut off, we have conditions that satisfy the instructions in the problem. That is, the length x of the base of the new rectangle is obtained numerically by subtracting 1 from twice the given area. This scenario is fanciful, however, and is not seriously proposed as an explanation of the text. Another scenario that "explains" the text can be found in Problem 13.4.

2.2. Higher-degree problems. Cuneiform tablets have been found that give the sum of the square and the cube of an integer for many values of the integer. These tablets may have been used for finding the numbers to which this operation was applied in order to obtain a given number. In our terms these tablets make it possible to solve the equation $x^3 + x^2 = a$, a very difficult problem indeed. In fact, given a complete table of $x^3 + x^2$, one can solve every cubic equation $ay^3 + by^2 + cy = d$, where b and c are nonnegative numbers and a and d are positive. (See Problem 13.5.)

Neugebauer (*1935*, p. 99; *1952*, p. 43) reports that the Mesopotamian mathematicians moved beyond algebra proper and investigated the laws of exponents, compiling tables of successive powers of numbers and determining the power to which one number must be raised in order to yield another. Such problems occur in a commercial context, involving compound interest. For example, the tablet AO 6484 gives the sum of the powers of 2 from 0 to 9 as the last term plus one less than the last term, and the sum of the squares of the first segment of integers as the sum of the same integers multiplied by the sum of $\frac{1}{3}$ and $\frac{2}{3}$ of the last term.

This recipe is equivalent to the modern formula for the sum of the squares of the first n integers.

3. India

Problems leading to algebra can be found in the *Sulva Sutras* and the Bakshali Manuscript, mentioned in Section 2 of Chapter 2. Since we have already discussed the Diophantine-type equations that result from the altar-construction problems in the *Sulva Sutras*, we confine ourselves here to a few problems that lead to linear Diophantine equations, determinate quadratic equations, and the summation of progressions.

3.1. Jaina algebra. According to Srinivasiengar (*1967*, p. 25), by the year 300 BCE Jaina mathematicians understood certain cases of the laws of exponents. They could make sense of an expression like $a^{m/2^n}$, interpreting it as extracting the square root n times and then raising the result to the power m. The notation used was of course not ours. The power $\frac{3}{4}$, for example, was described as "the cube of the second square root." That the laws of exponents were understood for these special values is attested by such statements as "the second square root multiplied by the third square root, or the cube of the third square root," indicating an understanding of the equality

$$a^{1/4}a^{1/8} = a^{3/8}\,.$$

3.2. The Bakshali Manuscript. The birchbark manuscript discovered in the village of Bakshali, near Peshawar, in 1881 uses the symbol $\smile$ to denote an unknown quantity. One of the problems in the manuscript is written as follows, using modern number symbols and a transliteration of the Sanskrit into the Latin alphabet:

$$\begin{array}{cccccccccc} \smile & 5 & & & \smile & & \smile & 7+ & m\bar{u} & \smile \\ 1 & 1 & yu & m\bar{u} & 1 & sa & 1 & 1 & & 1 \end{array}.$$

This symbolism can be translated as, "a certain thing is increased by 5 and the square root is taken, giving [another] thing; and the thing is decreased by 7 and the square root is taken, giving [yet another] thing." In other words, we are looking for a number x such that $x + 5$ and $x - 7$ are both perfect squares. This problem is remarkably like certain problems in Diophantus. For example, Problem 11 of Book 2 of Diophantus is to add the same number to two given numbers so as to make each of them a square. If the two given numbers are 5 and -7, this is *exactly* the problem stated here; Diophantus, however, did not use negative numbers.

The Bakshali Manuscript also contains problems in linear equations, of the sort that have had a long history in elementary mathematics texts. For example, three persons possess seven thoroughbred horses, nine draft horses, and 10 camels respectively. Each gives one animal to each of the others. The three are then equally wealthy. Find the (relative) prices of the three animals. Before leaping blindly into the set of two linear equations in three unknowns that this problem prescribes, we should take time to note that the problem can be solved by imagining the experiment actually performed. Suppose that these donations have been made and the three people are now equally wealthy. They will remain equally wealthy if each gives away one thoroughbred horse, one draft horse, and one camel. It follows that four thoroughbred horses, six draft horses, and seven camels are all of equal value. The problem has thereby been solved, and no actual algebra has been performed. Srinivasiengar (*1967*, p. 39) gives the solution using symbols for the unknown values of the animals, but does not assert that the solution is given this way in the manuscript itself.

4. China

With the exception of the the *Zhou Bi Suan Jing*, which is mostly about geometry and astronomy, algebra forms a major part of early Chinese mathematical works. The difficulty with finding early examples of problems leading to algebra is that the earliest document after the *Zhou Bi Suan Jing*, the *Jiu Zhang Suanshu*, contains not only many problems leading to systems of linear equations but also a sophisticated method of solving these equations, fully equivalent to what we now call Gaussian elimination (row reduction) of matrices and known as *fang cheng* or *the rectangular algorithm*. Li and Du (*1987*, pp. 46–47) discuss one example involving the yield of three different kinds of grain, in which a matrix is triangularized so that the solution can be obtained by working from bottom to top. Our discussion of this technique, like the discussion of quadratic equations, is reserved for Chapter 14.

4.1. The *Jiu Zhang Suanshu*. In Chapter 6 of the *Jiu Zhang Suanshu* we find some typical problems leading to one linear equation in one unknown. This type of problem can be solved using algebra, but does not necessarily *require* algebraic reasoning to solve, since the answer lies very close to the surface. For example (Mikami, *1913*, p. 16), if a fast walker goes 100 paces in the time required for a slow walker to go 60 paces, and the slower walker has a head start of 100 paces, how many paces will be required for the fast walker to overtake the slow one? The instruction given is to multiply the head start by the faster speed and divide by the difference in speeds. That will obviously give the number of paces taken by the faster runner. The author says nothing about the number of paces that will be taken by the slower runner, but he probably noticed that that number could be obtained in two ways: by subtracting 100 (the head start) or by multiplying by the slower speed instead of the faster. This equivalence, if noticed, would give some insight into manipulating expressions for numbers.

Chapter 7 contains the kind of excess–deficiency problems discussed in Section 2 of Chapter 6. The solutions are described in some detail, so that we can judge the extent to which they are to be considered algebra. For example, an unknown number of people are buying hens. If each gives nine (units of money), there will be a surplus of 11 units. If each gives six, there will be a deficit of 16. The instructions for solution are to arrange the data in a rectangle, cross-multiply, and add the products. In other words, form the number $9 \cdot 16 + 6 \cdot 11 = 210$. If this number is divided by the difference $9 - 6$, the result, 70, represents the total price to be paid. Adding the surplus and deficit gives 27, and when this is divided by $9 - 6$, we get 9, the number of people buying. This solution is far too sophisticated and general to be an early method aimed at one specific problem. It is algebra proper.

4.2. The *Suanshu Shu*. Li and Du (*1987*, pp. 56–57) describe a set of bamboo strips discovered in 1983–1984 in three tombs from the western Han Dynasty containing a *Suanshu Shu* (*Arithmetical Book*) and dated no later than the first half of the second century BCE. This work contains instructions on the performance of arithmetical operations and some applications that border on algebra. For example, one problem is to find the width of a field whose area is 1 *mu* (240 square *bu*) and whose length is $1\frac{1}{2}$ *bu*. This problem amounts to one linear equation in one unknown. Dividing the area by the length yields 160 *bu* as the answer. The whole difficulty of the problem lies in the complicated rules for dividing by a fraction.

4.3. The *Sun Zi Suan Jing*. A large number of problems leading to algebra are considered in the *Sun Zi Suan Jing*. Some of these are the kind of excess and deficit problems already discussed. Others involve arithmetic and geometric progressions and are solved by clever numerical reasoning. As an example of an arithmetic progression, Problem 25 of Chapter 2 of the *Sun Zi Suan Jing* discusses the distribution of 60 tangerines among five noblemen of different ranks in such a way that each will receive three more than the one below him. Sun Zi says first to give the lowest-ranking nobleman three, then six to the next-higher rank, and so on, until the fifth person gets 15. That accounts for $3 + 6 + 9 + 12 + 15 = 45$ tangerines and leaves fifteen more to be divided equally among the five. Thus the numbers given out are 6, 9, 12, 15, and 18.

4.4. Zhang Qiujian. To the fifth-century mathematician Zhang Qiujian (ca. 430–490) we owe one of the most famous and long-lasting problems in the history of algebra. It goes by the name of the Hundred Fowls Problem, and reads as follows: Roosters cost 5 qian each, hens 3 qian each, and three baby chicks cost 1 qian. If 100 fowls are bought for 100 qian, how many roosters, hens, and chicks were bought? The answer is not unique, but Zhang gives all the physically possible solutions: $(4, 18, 78)$, $(8, 11, 81)$, and $(12, 4, 84)$. Probably this answer was obtained by enumeration. Given that one is to buy at least one of each type of chicken, at most 19 roosters can be bought. Zhang observed that the number of roosters must increase in increments of 4, the number of hens must decrease in increments of 7, and the number of baby chicks must increase in increments of 3. That is because $4 - 7 + 3 = 0$ and $4 \cdot 5 - 7 \cdot 3 + 3 \cdot \frac{1}{3} = 0$.

According to Mikami (*1913*, p. 41), three other "hardy perennials" of algebra can be traced to Zhan Qiujian's treatise. One involves arithmetic progression. A weaver produces 5 feet of fabric on the first day, and the output diminishes (by the same amount) each day, until only 1 foot is produced on the thirtieth day. What was the total production? The recipe for the answer is to add the amounts on the first and last days, divide by 2, and multiply by the number of days.

The second is a rate problem of the type found in the *Jiu Zhang Suanshu*. A horse thief rode 37 miles before his theft was discovered. The owner then pursued him for 145 miles and narrowed the distance between them to 23 miles, but gave up at that point and returned home. If he had continued the pursuit, how many more miles would he have had to ride to catch the thief? Here we have the case of one person traveling 145 miles in the same time required for the other to travel 131 miles, and the other person having a 23-mile head start. Following the formula given in the *Jiu Zhang Suanshu*, Zhang Qiujian gives the answer as $145 \times 23 \div 14$.

Finally, we have another rate problem: If seven men construct $12\frac{1}{2}$ bows in nine days, how many days will be required for 17 men to construct 15 bows?

All these problems can be solved by *reasoning about numbers* without necessarily writing down any equations. But they are definitely proto-algebra in that they require thinking about performing operations on abstract, unspecified numbers.

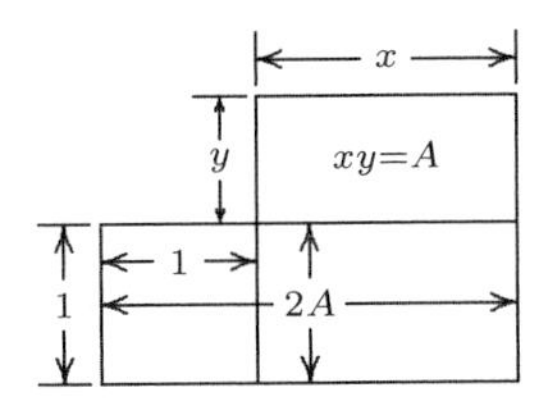

FIGURE 3. Another scenario to "fit" a text on cuneiform tablet AO 6670.

Questions and problems

13.1. What do the two problems of recovering two numbers from their sum and product or from their difference and product have to do with quadratic equations as we understand them today? Can we conclude that the Mesopotamians "did algebra"?

13.2. You can verify that the solution of the problem from tablet AO 8862 (15 and 12) given by the author is not the only possible one. The numbers 14 and 13 will also satisfy the conditions of the problem. Why didn't the author give this solution?

13.3. Of what practical value are the problems we have called "algebra"? Taking just the quadratic equation as an example, the data can be construed as the area and the semiperimeter of a rectangle and the solutions as the sides of the rectangle. What need, if any, could there be for solving such a problem? Where are you ever given the perimeter and area of a rectangle and asked to find its shape?

13.4. Figure 3 gives a scenario that can be fit to the data in AO 6670. Given a square 1 unit on a side, in the right angle opposite one of its corners construct a rectangle of prescribed area A that will be one-third of the completed gnomon. Explain how the figure fits the statement of the problem. (As in Section 2, this scenario is *not* being proposed as a serious explanation of the text.)

13.5. Given a cubic equation

$$ax^3 + bx^2 + cx = d\,,$$

where all coefficients are assumed positive, let $A = d + bc/(3a) - 2b^3/(9a^2)$, $B = b^2/(3a) - c$, and $t = 3aA/(3aBx - bB)$, that is, $x = A/(Bt) + b/(3a)$. Show that in terms of these new parameters, this equation is

$$t^3 + t^2 = \frac{aA^2}{B^3}\,.$$

It could therefore be solved numerically by consulting a table of values of $t^3 + t^2$. [*Again a caution*: The fact that such a table exists and could be used this way does not imply that it *was* used this way, any more than the fact that a saucer can be used to hold paper clips implies that it was designed for that purpose.]

13.6. Considering the origin of algebra in the mathematical traditions we have studied, do you find a point in their development at which mathematics ceases to be a disjointed collection of techniques and becomes systematic? What criteria would you use for defining such a point, and where would you place it in the mathematics of Egypt, Mesopotamia, Greece, China, and India?

CHAPTER 14

Equations and Algorithms

In this chapter we take up the history of algebra from the point at which equations appear explicitly and carry it forward along two parallel lines. In one line of development the emphasis is on finding numerical approximations to the roots of an equation. In the second line the emphasis is on finding an algorithm involving only the four operations of arithmetic and the extraction of roots that will yield the solution. The second line of development reached its highest point of achievement in sixteenth-century Italy, with the arithmetical solution of equations of degree 4. That is the point at which the present chapter ends. Standing somewhat to one side of both lines of evolution was the work of Diophantus, which contains a mixture of topics that now form part of number theory and algebra.

1. The *Arithmetica* of Diophantus

The work of Diophantus of Alexandria occupies a special place in the history of algebra. To judge it, one should know something of its predecessors and its influence. Unfortunately, information about either of these is difficult to come by. The Greek versions of the treatise, of which there are 28 manuscripts, according to Sesiano (*1982*, p. 14), all date to the thirteenth century. Among the predecessors of Diophantus, we can count Heron of Alexandria and one very obscure Thymaridas, who showed how to solve a particular set of linear equations, the *epanthēma* (blossom) of Thymaridas. Because the work of Diophantus is so different from the Pythagorean style found in Euclid and his immediate successors, the origins of his work have been traced to other cultures, notably Egypt and Mesopotamia. The historian of mathematics Paul Tannery (1843–1904) printed an edition of Diophantus' work and included a fragment supposedly written by the eleventh-century writer Michael Psellus (1018–ca. 1078), which stated that "As for this *Egyptian* method, while Diophantus developed it in more detail,... ." It was on this basis, identifying Anatolius with a third-century Bishop of Laodicea originally from Alexandria, that Tannery assigned Diophantus to the third century. Neugebauer (*1952*, p. 80) distinguishes two threads in Hellenistic mathematics, one in the logical tradition of Euclid, the other having roots in the Babylonian and Egyptian procedures and says that, "the writings of Heron and Diophantus... form part of this oriental tradition which can be followed into the Middle Ages both in the Arabic and in the western world." Neugebauer sees Diophantus as reflecting an earlier type of mathematics practiced in Greece alongside the Pythagorean mathematics and temporarily eclipsed by the Euclidean school. As he says (*1952*, p. 142):

> It seems to me characteristic, however, that Archytas of Tarentum could make the statement that not geometry but arithmetic alone

could provide satisfactory proofs. If this was the opinion of a leading mathematician of the generation just preceding the birth of the axiomatic method, then it is rather obvious that early Greek mathematics cannot have been very different from the Heronic Diophantine type.

1.1. Diophantine equations. An equation containing two or more unknowns for which only rational (or more often, integer) solutions are sought is nowadays called a *Diophantine equation.* Diophantus wrote a treatise commonly known under the somewhat misleading name *Arithmetica.* As mentioned previously, the six books of this treatise that have been known for some centuries may now be supplemented by parts of four other books, discovered in 1968, but that is not certain. The title itself is of some interest. Its suffix *-tica* has come into English from Greek in a large number of words such as *logistics*, *mathematics*, and *gymnastics*. It has a sense of how-to, that is, the techniques involved in using numbers (*arithmoí*) or reasoning (*lógoi*) or learning (*mathêmata*) or physical training (*gýmnasis*).[1] The plural *-s* on these English words, even though they are now regarded as singular, reflects the fact that these words were originally intended to be plural—the neuter plural form of the corresponding adjectives *arithmētikós* (adept with number), *logistikós* (skilled in calculating), *mathēmatikós* (disposed to learn), and *gymnastikós* (skilled in bodily exercise), but which evolved into a feminine singular form. The Greek title of the work is *Diophántou Alexándreōs Arithmētikôn*, meaning [The Books] *of Arithmetics of Diophantus of Alexandria.*

1.2. General characteristics of the *Arithmetica*. In contrast to other ancient works containing problems that lead to algebra, the problems that require algebraic techniques in the *Arithmetica* all involve purely numerical relations. They are not problems about *things* that have been counted or measured. They are about *counting itself.* The work begins with a note to one Dionysius, whom the author characterizes as "eager to learn" how to solve problems in arithmetic.[2] In a number of ways Diophantus seems to be doing something that resembles the algebra taught nowadays. In particular, he has a symbol for an unknown or abstract number that is to be found in a problem, and he appears to know what an equation is, although he doesn't exactly use the word *equation.*

Diophantus began by introducing a symbol for a constant unit $\overset{\circ}{M}$, from *monás* ($\mu o \nu \acute{\alpha}\varsigma$), along with a symbol for an unknown number ς, conjectured to be an abbreviation of the first two letters of the Greek word for number: *arithmós* ($\grave{\alpha}\rho\iota\theta\mu\acute{o}\varsigma$). For the square of an unknown he used Δ^{υ}, the first two letters of *dýnamis* ($\Delta\acute{\upsilon}\nu\alpha\mu\iota\varsigma$), meaning *power.* For its cube he used K^{υ}, the first two letters of *kýbos* ($K\acute{\upsilon}\beta o\varsigma$), meaning *cube.* He then combined these letters to get fourth ($\Delta^{\upsilon}\Delta$), fifth (ΔK^{υ}), and sixth ($K^{\upsilon}K$) powers. For the reciprocals of these powers of the unknown he invented names by adjoining the suffix *-ton* ($-\tau o\nu$) to the names of the corresponding powers. These various powers of the unknown were called *eída* ($\grave{\varepsilon}\acute{\iota}\delta\alpha$), meaning *species.* Diophantus' system for writing down the equivalent of a polynomial in the unknown consisted of writing down these symbols in order

[1] From the root *gymnós*, meaning *naked.*

[2] One of the reasons that Tannery assigned Dionysius to the third century was that this date made it easy to imagine that Dionysius was the man appointed Bishop of Alexandria in 247—not that Dionysius (Dennis) was exactly a rare name in those days.

to indicate addition, each term followed by the corresponding number symbol (for which the Greeks used their alphabet). Terms to be added were placed first, separated by a pitchfork (⋔) from those to be subtracted. Heath conjectured that this pitchfork symbol is a condensation of the letters lambda and iota, the first two letters of a Greek root meaning *less* or *leave*. Thus what we would call the expression $2x^4 - x^3 - 3x^2 + 4x + 2$ would be written $\Delta^{\upsilon}\Delta\bar{\beta}\varsigma\bar{\delta}\ \overset{\circ}{M}\ \bar{\beta}$ ⋔ $K^{\upsilon}\bar{\alpha}\Delta^{\upsilon}\bar{\gamma}$.

Diophantus' use of symbolism is rather sparing by modern standards; he often uses words where we would use symbolic manipulation. For this reason his algebra was described by the nineteenth-century German historian of mathematics Nesselmann as a transitional "syncopated" phase between the earliest "rhetorical" algebra, in which everything is written out in words, and the modern "symbolic" algebra.[3]

1.3. Determinate problems. The determinate problems in the *Arithmetica* require that one or more unknown numbers be found from conditions that we would nowadays write as systems of linear or quadratic equations. The 39 problems of Book 1 and the first ten problems of Book 2 are of these types. Some of these problems have a unique solution. For example, Problem 7 of Book 1 is: *From a given unknown number subtract two given numbers so that the remainders have a given ratio.* In our terms, this condition says

$$x - a = m(x - b),$$

where x is unknown, a and b are the given numbers, and m is the given ratio. Since it is obvious that $m \geq 1$ if all quantities are positive and $a \leq b$, Diophantus has no need to state this restriction.

Similarly, Problem 15 of Book 1 asks for two numbers (x and y, we would say) such that for given numbers a and b the ratios $x + a : y - a$ and $y + b : x - b$ are equal to two given ratios r and s.

The symbolic notation of Diophantus extended only as far as the unknown and representations of sums, products, and differences. He had no way of forming mathematical expressions containing the phrases "a given number" (a and b above) and "a given ratio" (r and s above). As a result, he could explain his methods of solution only by using a particular example, in the present case taking $a = 30$, $r = 2$, $b = 50$, $s = 3$. He then assumed that $y = \varsigma + 30$ and $x = 2\varsigma - 30$, so that the first equation was satisfied automatically and the second became $\varsigma + 80 = 3(2\varsigma - 80)$. Here it is very easy to recognize the explicit manipulation of formal expressions, leading to the discovery of the unknown number. This manipulation of expressions is characteristic of algebraic technique.

Some of the problems that are determinate from our point of view may have no positive rational solutions for certain data, and in such cases Diophantus requires a restriction on the data so that positive rational solutions will exist. For example, Problem 8 of Book 1 is *to add the same (unknown) number to two given numbers so that the sums have a given ratio.* This problem amounts to the equation

$$x + a = m(x + b).$$

---

[3] Nesselmann is quoted by Jacob Klein (*1934–36*, p. 146). In the author's opinion, there is not much for Diophantus to be transitional between, since little is known of his algebraic predecessors, and later algebraists wrote everything out in words. Jacob Klein seems to share these reservations.

If $x > 0$ and $a > b$, then $1 < m = (x + a)/(x + b) < a/b$. That is, the given ratio must be larger than 1 and less than the ratio of the larger number to the smaller.

1.4. The significance of the *Arithmetica*. The existence of Arabic manuscripts of Diophantus' treatise shows that his work was known to the Muslim mathematicians of the Middle Ages. Sesiano (*1982*, pp. 9–20) discusses the extent to which a number of Islamic and Byzantine mathematicians were influenced by his work or commented on it. He comments (p. 9) that, "There is nothing to suggest that the Egyptian Abu-Kamil had any direct (or even indirect) knowledge of Diophantus' *Arithmetica*, although the problems in his *Algebra* dealing with indeterminate analysis are perfectly Diophantine in form and the basic methods are attested to in the *Arithmetica*." In contrast, the Diophantine connection is clear in the case of the eleventh-century mathematician al-Karkhi, (also known as al-Karaji, 953–1029), whose *Fakhri* has many points of contact with Diophantus. Tracing the influence of Diophantus, however, is more difficult. Jacob Klein (*1934–36*, p. 5), citing nineteenth-century work of Tannery and others, says that "the special influence of the *Arithmetic* of Diophantus on the content, but even more so on the form, of this Arabic science is unmistakable—if not in the *Liber Algorismi* of Al-Khowarizmi himself, at any rate from the tenth century on." In a treatise on algebra published in the late sixteenth century, the engineer–mathematician Rafael Bombelli stated that, although it had been agreed up to his time that algebra was an invention of the Muslims, he was convinced, after reading the work of Diophantus, that the invention should be ascribed to the latter.

At the very least, Diophantus used equations and developed a symbolism for handling algebraic expressions, and that, in the long run, was an important innovation. As two prominent Russian historians of science say:

> Diophantus was the first to deduce that it was possible to formulate the conditions of a problem as equations or systems of equations; as a matter of fact, before Diophantus, there were no equations at all, either determinate or indeterminate. Problems were studied that we can now reduce to equations, but nothing more than that. [Bashmakova and Smirnova *1997*, p. 132]

1.5. The view of Jacob Klein. In several places in Part 2 and in the present part we have used without comment the "standard view" among historians of a contrast between *logistikḗ* and *arithmētikḗ* in the science and philosophy of ancient Greece, *logistikḗ* being counting or computation and *arithmētikḗ* being the study of the theoretical properties of numbers. A different point of view is contained in the extended essay by Jacob Klein (*1934–36*). Klein maintains that even the word *arithmós* itself has been misinterpreted, that Euclid and Diophantus did not have in mind cardinal numbers in the abstract, but used the word *arithmós* to mean a set or collection. As he says (p. 7), "*arithmós never* means anything other than 'a definite number of definite objects.'" He goes on to say (p. 19) that for Plato "'arithmetic' is, accordingly, not 'number theory,' but first and foremost the art of correct counting."[4] In particular, Klein denies that Euclid was thinking about

[4] Such may well be the case. If so, that is unfortunate for Plato's reputation. Neugebauer (*1952*, p. 146) offers the opinion that "Plato's role has been widely exaggerated. His own direct contributions to mathematical knowledge were obviously nil... The often adopted notion that Plato 'directed' research fortunately is not borne out by the facts."

numbers in the abstract and illustrating them geometrically with lines. It does seem strange that Euclid clings to what appears to be a completely unnecessary geometric representation of a number. According to Klein, the mystery is solved if we recognize that specific numbers were always intended, even though an abstract symbol (a letter or two letters) was used for them. Klein (p. 124) cites Tannery in arguing that these letters did not represent general, unspecified numbers, because they were not amenable to being operated on.

2. China

The development of algebra in China began early and continued for many centuries. The aim was to find numerical approximate solutions to equations, and the Chinese mathematicians were not intimidated by equations of high degree.

2.1. Linear equations. We have already mentioned the Chinese technique of solving simultaneous linear equations and pointed out its similarity to modern matrix techniques. Examples of this method are found in the *Jiu Zhang Suanshu* (Mikami, *1913*, pp. 18–22; Li and Du, *1987*, pp. 46–49). Here is one example of the technique.

> There are three kinds of [wheat]. The grains contained in two, three and four bundles, respectively, of these three classes of [wheat], are not sufficient to make a whole measure. If however we add to them one bundle of the second, third, and first classes, respectively, then the grains would become one full measure in each case. How many measures of grain does then each one bundle of the different classes contain?

The following counting-board arrangement is given for this problem.

1		2	1st class
	3	1	2nd class
4	1		3rd class
1	1	1	measures

Here the columns from right to left represent the three samples of wheat. Thus the right-hand column represents 2 bundles of the first class of wheat, to which one bundle of the second class has been added. The bottom row gives the result in each case: 1 measure of wheat. The word problem might be clearer if the final result is thought of as the result of threshing the raw wheat to produce pure grain. We can easily, and without much distortion in the procedure followed by the author, write down this counting board as a matrix and solve the resulting system of three equations in three unknowns. The author gives the solution: A bundle of the first type of wheat contains $\frac{9}{25}$ measure, a bundle of the second $\frac{7}{25}$ measure, and a bundle of the third $\frac{4}{25}$ measure.

2.2. Quadratic equations. The last chapter of the *Jiu Zhang Suanshu*, which involves right triangles, contains problems that lead to linear and quadratic equations. For example (Mikami, *1913*, p. 24), there are several problems involving a town enclosed by a square wall with a gate in the center of each side. In some cases the problem asks at what distance from the south gate a tree a given distance east of the east gate will first be visible. The data are the side s of the square and the distance d of the tree from the gate. For that kind of data, the problem is the linear equation $(x + s/2)/(s/2 + d) = s/(2d)$. When the side of the town is the unknown, a quadratic equation results, as in one case, in which it is asserted that the tree is 20 paces north of the north gate and is just visible to a person who walks 14 paces south of the south gate, then 1775 paces west. This problem proposes a quadratic equation as a problem to be solved for a single unknown number, in contrast to the occurrence of quadratic equations in Mesopotamia, where they amount to finding two numbers given their sum and product. Since the Chinese technique of solving equations numerically is practically independent of degree, we shall not bother to discuss the techniques of solving quadratic equations separately.

2.3. Cubic equations. Cubic equations first appear in Chinese mathematics (Li and Du, *1987*, p. 100; Mikami, *1913*, p. 53) in the seventh-century work *Xugu Suanjing* (*Continuation of Ancient Mathematics*) by Wang Xiaotong. This work contains some intricate problems associated with right triangles. For example, compute the length of a leg of a right triangle given that the product of the other leg and the hypotenuse is $1337\frac{1}{20}$ and the difference between the hypotenuse and the leg is $1\frac{1}{10}$.[5] Obviously, the data are perfectly general for a product P and a difference D. Wang Xiaotong gives a general description of the result of eliminating the hypotenuse and the other leg that amounts to the equation

$$x^3 + \frac{5D}{2}x^2 + 2D^2x = \frac{p^2}{2D} - \frac{D^3}{2}.$$

In this particular case the equation is

$$x^3 + \frac{1}{4}x^2 + \frac{1}{50}x - 8938513\frac{64}{125} = 0.$$

He then says to compute the root (which he gives as $92\frac{2}{5}$) "according to the rule of the cubic root extraction." Li and Du (*1987*, pp. 118–119) report that the eleventh-century mathematician Jia Xian developed the following method for extracting the cube root. This method generalizes from the case $x^3 = N$ to the general cubic equation quite easily, as we shall see.

The computation is arranged in rows (or columns) of five elements. We shall use columns for convenience. The top entry is always the current approximation a to the cube root, the bottom entry is always 1. The entries in the next-to-bottom and middle rows are obtained successively by multiplying the entry that was just below at the preceding stage by the adjustment and adding to the entry that was in the same row at the preceding step. The entry next to the top is obtained the same way, except that the adjustment is subtracted instead of being added. This row always contains the current or adjusted error. The adjustment procedure works first from the bottom to the second row, then from the bottom to the third row, and finally, from the bottom to the fourth row. For example, the first four steps go as follows, assuming a "zeroth" approximation of 0, which is to be improved by an initial guess a:

---

[5] Mikami gives $\frac{1}{10}$ as the difference, which is incompatible with the answer given by Wang Xiaotong. I do not know if the mistake is due to Mikami or is in the original.

$$\begin{matrix} 0 & & a & & a & & a \\ N & & N-a^3 & & N-a^3 & & N-a^3 \\ 0 & \longrightarrow & a^2 & \longrightarrow & 3a^2 & \longrightarrow & 3a^2 \\ 0 & & a & & 2a & & 3a \\ 1 & & 1 & & 1 & & 1 \end{matrix} .$$

Next, given any approximation a, the approximation is improved by adding an adjustment b, and the rows are then recomputed, again, first working from the bottom to the second row, then from the bottom to the third row, and finally, from the bottom to the fourth row:

$$\begin{matrix} a & & a+b & & a+b & & a+b \\ N-a^3 & & N-(a+b)^3 & & N-(a+b)^3 & & N-(a+b)^3 \\ 3a^2 & \longrightarrow & 3a^2+3ab+b^2 & \longrightarrow & 3(a+b)^2 & \longrightarrow & 3(a+b)^2 \\ 3a & & 3a+b & & 3a+2b & & 3(a+b) \\ 1 & & 1 & & 1 & & 1 \end{matrix} .$$

By introduction of a counting board ruled into squares analogous to the registers in a calculator, the procedure could be made completely mechanical. Using an analogous procedure, one can take fifth roots, seventh roots, and so on, with increasingly messy computations, of course. Composite roots can be reduced to prime roots, but since the generalization of this method works so well, there is really no need to do so. The sixth root, for example, can be taken by extracting the square root of the cube root, or it could be extracted directly following this method.

2.4. The numerical solution of equations. The Chinese mathematicians of 800 years ago invented a method of finding numerical approximations of a root of an equation, similar to a method that was rediscovered independently in the nineteenth century in Europe and is commonly called *Horner's method*, in honor of the British school teacher William Horner (1786–1837).[6] The first appearance of the method is in the work of the thirteenth-century mathematician Qin Jiushao, who applied it in his 1247 treatise *Sushu Jiu Zhang* (*Arithmetic in Nine Chapters*, not to be confused with the *Jiu Zhang Suanshu*).

The connection of this method with the cube root algorithm will be obvious. We illustrate with the case of the cubic equation. Suppose in attempting to solve the cubic equation $px^3 + qx^2 + rx + s = 0$ we have found the first digit (or any approximation) a of the root. We then "reduce" the equation by setting $x = y + a$ and rewriting it. What will the coefficients be when the equation is written in terms of y? The answer is immediate; the new equation is

$$\begin{matrix} py^3 & + & 3pay^2 & + & 3pa^2y & + & pa^3 & & \\ & + & qy^2 & + & 2qay & + & qa^2 & & \\ & & & + & ry & + & ra & & \\ & & & & & + & s & = & 0 . \end{matrix}$$

[6] Besides being known to the Chinese mathematicians 600 years before Horner, this procedure was used by Sharaf al-Tusi (ca. 1135–1213), as discussed in Section 5 below, and was discovered by the Italian mathematician Paolo Ruffini (1765–1822) a few years before Horner published it. In fairness to Horner, it must be said that he applied the method not only to polynomials, but to infinite series representations. To him it was a theorem in calculus, not algebra.

We see that we need to make the following conversion of the coefficients (reading from bottom to top):

$$\begin{array}{c} s \\ r \\ q \\ p \end{array} \longrightarrow \begin{array}{c} pa^3 + qa^2 + ra + s \\ 3pa^2 + 2qa + r \\ 3pa + q \\ p \end{array}.$$

The procedure followed in the cube root algorithm works perfectly. That is, start at the bottom and at each stage, multiply the element below by a and add it to the element in the same row at the preceding stage. Going from bottom all the way to the top gets the top row correct. Then going from bottom to the second row gets the second row correct; and finally going from the bottom to the third row completes the transition:

$$\begin{array}{c} s \\ r \\ q \\ p \end{array} \longrightarrow \begin{array}{c} pa^3 + qa^2 + ra + s \\ pa^2 + qa + r \\ pa + q \\ p \end{array} \longrightarrow \begin{array}{c} pa^3 + qa^2 + ra + s \\ 3pa^2 + 2qa + r \\ 2pa + q \\ p \end{array} \longrightarrow \begin{array}{c} pa^3 + qa^2 + ra + s \\ 3pa^2 + 2qa + r \\ 3pa + q \\ p \end{array}.$$

In this context the cube root algorithm itself becomes merely the case $p = 1$, $q = 0 = r$, $s = -N$, with the top row omitted and the subtraction in the second row (now the top row) replaced by addition, since N has been replaced by $-N$. Not only is this algorithm simple to use; it also provides the most efficient and accurate way of computing a polynomial numerically. Before the advent of computer algebra programs, numerical analysis books instructed the student to compute the polynomial $px^3 + qx^2 + rx + s$ at different values of x by the sequence of operations

$$p \to px \to px + q \to x(px + q) \to x(px + q) + r \to$$
$$\to x\bigl(x(px + q) + r\bigr) \to x\bigl(x(px + q) + r\bigr) + s\,.$$

This sequence of operations avoids the error that tends to accumulate when large numbers of opposite sign are added.[7]

Wang Xiaotong's reference to the use of cube root extraction for solving his equation seems to suggest that this method was known as early as the seventh century. However, as we have just noted, the earliest explicit record of it seems to be in the treatise of Qin Jiushao, who illustrated it by solving the quartic equation

$$-x^4 + 763200x^2 - 40642560000 = 0\,.$$

The method of solution gives proof that the Chinese did not think in terms of a quadratic formula. If they had, this equation would have been solved for x^2 using that formula and then x could have been found by taking the square root of any positive root. But Qin Jiushao applied the method described above to get the solution $x = 840$. (He missed the smaller solution $x = 240$.)

The efficiency of this method in finding approximate roots allowed the Chinese to attack equations involving large coefficients and high degrees. Qin Jiushao (Libbrecht, *1973*, pp. 134–136) considered the following problem: *Three* li *north of the wall of a circular town there is a tree. A traveler walking east from the southern gate of the town first sees the tree after walking 9* li. *What are the diameter and circumference of the town?*

[7] In addition, a very simple hand calculator with no memory cells can carry out this sequence of operations without the need to stop entering and write down a partial result.

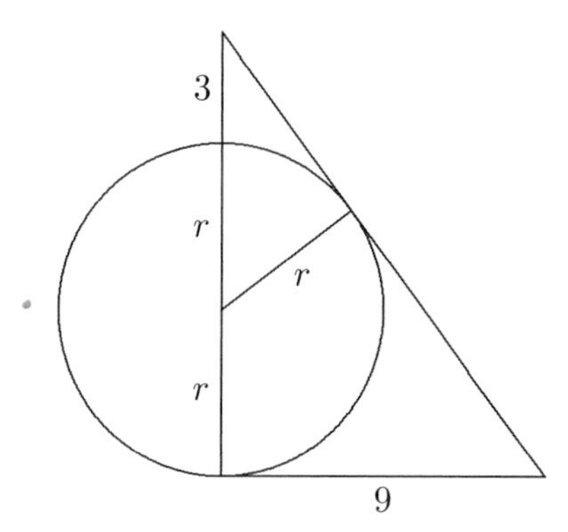

FIGURE 1. A quartic equation problem.

This problem is obviously concocted so as to lead to an equation of higher degree. (The diameter of the town could surely be measured directly from inside, so that it is highly unlikely that anyone would ever need to solve such a problem for a practical purpose.) Representing the diameter of the town as x^2, Qin Jiushao obtained the equation[8]

$$x^{10} + 15x^8 + 72x^6 - 864x^4 - 11664x^2 - 34992 = 0\,.$$

The reasoning behind such a complicated equation is difficult to understand. Perhaps the approach to the problem was to equate two expressions for the area of the triangle formed by the center of the town, the tree, and the traveler. In that case, if the line from the traveler to the tree is represented as $b + \sqrt{a(x^2 + a)}$, the formula for the area of a triangle in terms of its sides is used, and the resulting area is equated to $\frac{1}{2}(a + (x^2)/2)b$, the result, after all radicals are cleared, will be an equation of degree 10 in x, but not the one mentioned by Qin Jiushao. It will be

$$ax^{10} + (a^2 - 4b^2)x^8 - 8abx^6 - 8a^2b^2x^4 + 16ab^4x^2 + 16a^2b^4 = 0\,.$$

One has to be very unlucky to get such a high-degree equation. Even a very simplistic approach leads only to a quartic equation. It is easy to see (Fig. 1) that if the diameter of the town rather than its square root is taken as the unknown, and the radius is drawn to the point of tangency, trigonometry will yield a quartic equation. If the radius is taken as the unknown, the similar right triangles in Fig. 1 lead to the cubic equation $2r^3 + 3r^2 = 243$. But, of course, the object of this game was probably to practice the art of algebra, not to get the simplest possible equation, no matter how virtuous it may seem to do so in other contexts. In any case, the historian's job is not that of a commentator trying to improve a text. It is to try to understand what the original author was thinking.

3. Japan

The Japanese mathematicians showed themselves to be superb algebraists from the beginning. We have already mentioned (Section 4 of Chapter 9) the quadrilateral problem of Sawaguchi Kazuyuki, which led to an equation of degree 1458, solved by Seki Kōwa. This problem, like many of the problems in the *sangaku* plaques,

[8] Even mathematicians working within the Chinese tradition seem to have been puzzled by the needless elevation of the degree of the equation (Libbrecht, *1973*, p. 136).

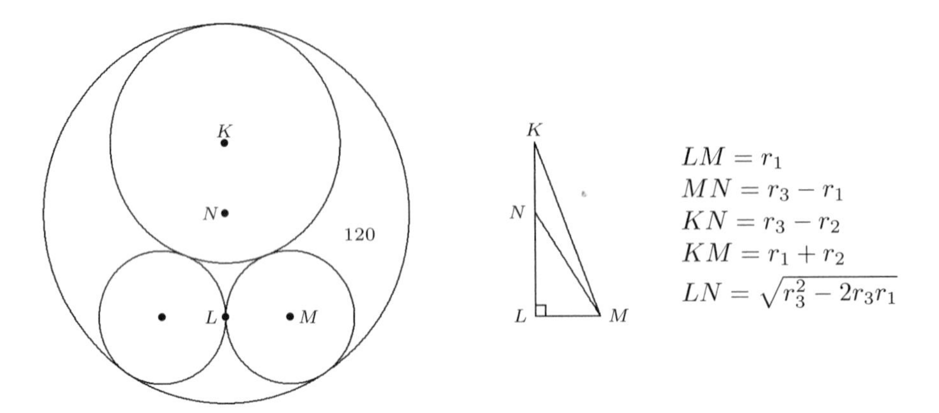

FIGURE 2. Sawaguchi Kazuyuki's first problem.

seems to be inspired by the desire to do some complicated algebra rather than by any pressing geometric need.

One impetus to the development of mathematics in Japan came with the arrival of the Chinese "method of the celestial element" (*tian yuan shu*), used in China. This name was given to the unknown in an equation by Li Ye in his 1248 treatise *Ceyuan Haijing* (*Sea Mirror of Circle Measurements*, see Mikami, *1913*, p. 81).[9] This term spread to Korea as *ch'onwonsul* and thence to Japan as *tengen jutsu*. This Chinese algebra became part of the standard Japanese curriculum before the seventeenth century.

Fifteen problems were published by Sawaguchi Kazuyuki in his 1670 work *Kokon Sampō-ki* (*Ancient and Modern Mathematics*). As an example of the great difficulty of these problems, consider the first of them. In this problem there are three circles each externally tangent to the other two and internally tangent to a fourth circle, as in Fig. 2. The diameters of two of the enclosed circles are equal and the third enclosed circle has a diameter five units larger. The area inside the enclosing circle and outside the three smaller circles is 120 square units. The problem is to compute the diameters of all four circles. This problem, although it yields to modern algebra, is complicated. In fact, Fig. 2 shows that the problem leads to the simultaneous equations

$$\begin{aligned} r_1 + \frac{5}{2} &= r_2\,, \\ 2\pi r_1^2 + \pi r_2^2 + 120 &= \pi r_3^2\,, \\ 4r_2^2 r_3 + 2r_1 r_2 r_3 + r_1 r_2^3 + r_1 r_3^2 &= 4r_2 r_3^2\,, \end{aligned}$$

where r_1, r_2, and r_3 are the radii of the circles. The last of these relations results from applying the Pythagorean theorem first to the triangle LMN to get LM, then to KLM.

3.1. Seki Kōwa. This problem was solved by Seki Kōwa (Smith and Mikami, *1914*, pp. 96–97). In case Seki Kōwa's prowess in setting up and solving equations was not clear from his solution of Sawaguchi Kazuyuki's first problem, remember

[9] The same word was used in a rather different and obscure sense by Qin Jiushao a year earlier in his *Sushu Jiu Zhang* (Libbrecht, *1973*, pp. 345–346).

that he also solved the fourteenth problem (see Section 4 of Chapter 9), which involved an equation of degree 1458. Although the procedure was a mechanical one using counting boards, prodigious concentration must have been required to execute it. What a chess player Seki Kōwa could have been! As Mikami (*1913*, p. 160) remarks, "Perseverance and hard study were a part of the spirit that characterized Japanese mathematics of the old times."

Seki Kōwa was primarily an algebraist who converted the celestial element method into two sophisticated techniques for handling equations, known as *the method of explanation* and the *method of clarifying things of obscure origin.* He kept the latter method a secret. According to some scholars, his pupil Takebe Kenkō (1664–1739) refused to divulge the secret, saying, "I fear that one whose knowledge is so limited as mine would tend to misrepresent its significance." However, other scholars claim that Takebe Kenkō did write an exposition of the latter method, and that it amounts to the principles of cancellation and transposition. (See Section 2 of Chapter 3.)

Determinants. Seki Kōwa is given the credit for inventing one of the central ideas of modern mathematics: determinants. He introduced this subject in 1683 in *Kai Fukudai no Hō* (*Method of Solving* Fukudai *Problems*).[10] Nowadays determinants are usually introduced in connection with linear equations, but Seki Kōwa developed them in relation to equations of higher degree as well. The method is explained as follows. Suppose that we are trying to solve two simultaneous quadratic equations

$$\begin{aligned} ax^2 + bx + c &= 0 \\ a'x^2 + b'x + c' &= 0. \end{aligned}$$

When we eliminate x^2, we find the linear equation

$$(a'b - ab')x + (a'c - ac') = 0.$$

Similarly, if we eliminate the constant term from the original equations and then divide by x, we find

$$(ac' - a'c)x + (bc' - b'c) = 0.$$

Thus from two quadratic equations we have derived two linear equations. Seki Kōwa called this process *tatamu* (*folding*).

We have written out expressions for the simple 2×2 determinants here. For example,

$$\begin{vmatrix} a & c \\ a' & c' \end{vmatrix} = ac' - a'c;$$

but, as everyone knows, the full expanded expressions for determinants are very cumbersome even for the 3×3 case. It is therefore important to know ways of simplifying such determinants, using the structural properties we now call the *multilinear property* and the *alternating property.* Seki Kōwa knew how to make use of the multilinear property to take out a common factor from a given row. He not only formulated the concept of a determinant but also knew many of their properties, including how to determine which terms are positive and which are negative in the expansion of a determinant. It is interesting that determinants were introduced in

[10] The word *fukudai* seems to be related to *fukugen suru,* meaning *reconstruct* or *restore.* According to Smith and Mikami (*1914*, p. 124), Seki Kōwa's school offered five levels of diploma, the third of which was called the *fukudai menkyo* (*fukudai* license) because it involved knowledge of determinants.

Europe around the same time (1693, by Leibniz), but in a comparatively limited context. As Smith and Mikami say (*1914*, p. 125),

> It is evident that Seki was not only the discoverer but that he had a much broader idea than that of his great German contemporary.

4. Hindu algebra

The promising symbolic notation of the Bakshali Manuscript was not adopted immediately throughout the world of Hindu mathematics. In particular, Aryabhata I tended to work in prose sentences. He considered the problem of finding two numbers given their product and their difference and gave the standard recipe for solving it.

4.1. Brahmagupta. The techniques involved with the *kuttaka* (pulverizer) belong to algebra, but since they are applied in number theory, we discussed them in that connection in Section 4 of Chapter 7. Brahmagupta also considered many problems that require finding the lengths of lines partitioning a polygon into triangles and quadrilaterals.

Brahmagupta's algebra is done entirely in words; for example (Colebrooke, *1817*, p. 279), his recipe for the cube of a binomial is:

> The cube of the last term is to be set down; and, at the first remove from it, thrice the square of the last multiplied by the preceding; then thrice the square of the preceding term taken into that last one; and finally the cube of the preceding term. The sum is the cube.

In short, $(a+b)^3 = a^3+3a^2b+3ab^2+b^3$. This rule is used for finding successive approximations to the cube root, just as in China and Japan. Similarly, in Section 4 (Colebrooke, *1817*, p. 346), he tells how to solve a quadratic equation:

> Take the absolute number from the side opposite to that from which the square and simple unknown are to be subtracted. To the absolute number multiplied by four times the [coefficient of the] square, add the square of the [coefficient of the] middle term; the square root of the same, less the [coefficient of the] middle term, being divided by twice the [coefficient of the] square is the [value of the] middle term.

Here the "middle term" is the unknown, and this statement is a very involved description of what we write as the quadratic formula:

$$x = \frac{\sqrt{4ac+b^2}-b}{2a} \quad \text{when} \quad ax^2+bx=c\,.$$

Brahmagupta does not consider equations of degree higher than 2.

4.2. Bhaskara II. In the five centuries between Brahmagupta and Bhaskara II (who will henceforth be referred to simply as Bhaskara), the idea of using symbols for the unknown in an equation seems to have taken hold in Hindu mathematics. In Section 4 of his *Vija Ganita* (*Algebra*) Bhaskara reports that the initial syllables of the names for colors "have been selected by venerable teachers for names of values of unknown quantities, for the purpose of reckoning therewith" (Colebrooke, *1817*, p. 139). He proceeds to give the rules for manipulating expressions involving such quantities; for example, the rule that we would write as $(-x-1)+(2x-8) = x-9$ is written

$$ya\,\dot{1}\;\;ru\,\dot{1}$$
$$ya\,2\;\;ru\,\dot{8}$$
$$\text{Sum}\;\;ya\,1\;\;ru\,\dot{9},$$

where the dots indicate negative quantities. The syllable *ya* is the first syllable of the word for *black*, and *ru* is the first syllable of the word for *species*.[11]

By the time of Bhaskara, the distinction between a rational and an irrational square root was well known. The Sanskrit word is *carani*, according to the commentator Krishna (Colebrooke, *1817*, p. 145), who defines it as a number "the root of which is required but cannot be found without residue." Bhaskara gives rules such as $\sqrt{8}+\sqrt{2}=\sqrt{18}$ and $\sqrt{8}-\sqrt{2}=\sqrt{2}$.

Bhaskara's algebraic rules go beyond what is taught even today as standard algebra. He says that a nonzero number divided by zero gives an infinite quotient.

> This fraction [3/0], of which the denominator is cipher, is termed an infinite quantity.
>
> In this quantity consisting of that which has cipher for its divisor, there is no alteration, though many be inserted or extracted; as no change takes place in the infinite and immutable GOD, at the period of the destruction or creation of worlds, though numerous orders of beings are absorbed or put forth. [Colebrooke, *1817*, pp. 137–138]

Both the *Vija Ganita* and the *Lilavati* contain problems on simple interest in which an unknown principal is to be found given the rate of simple interest and the amount to which it accrues after a given time. These equations are linear equations in one unknown.

The *Lilavati* contains a collection of problems in algebra, which are sometimes stated as though they were intended purely for amusement. For example, the rule for solving quadratic equations is applied in the *Vija Ganita* (Colebrooke, *1871*, p. 212) to find the number of arrows x that Arjuna (hero of the *Bhagavad Gita*) had in his quiver, given that he shot them all, using $\frac{1}{2}x$ to deflect the arrows of his antagonist, $4\sqrt{x}$ to kill his antagonist's horse, six to kill the antagonist himself, three to demolish his antagonist's weapons and shield, and one to decapitate him. In other words, $x = \frac{1}{2}x + 4\sqrt{x} + 10$.

Bhaskara gives a criterion for a quadratic equation to have two (positive) roots. He also says that "if the solution cannot be found in this way, as in the case of cubic

[11] There is no evidence that Bhaskara knew of Diophantus; the fact that both describe a power of the unknown using a word whose meaning is approximated by the English word *species* is simply a coincidence.

or quartic equations, it must be found by the solver's own ingenuity" (Colebrooke, *1817*, pp. 207–208). That ingenuity includes some work that would nowadays be regarded as highly inventive, not to say suspect; for example (Colebrooke, *1817*, p. 214), how to solve the equation

$$\frac{\left(0(x+\frac{1}{2}x)\right)^2+2\left(0(x+\frac{1}{2}x)\right)}{0}=15\,.$$

Bhaskara warns that multiplying by zero does not make the product zero, since further operations are to be performed. Then he simply cancels the zeros, saying that, since the multiplier and divisor are both zero, the expression is unaltered. The result is the equation we would write as $\frac{9}{4}x^2+3x=15$. Bhaskara clears the denominator and writes the equivalent of $9x^2+12x=60$. Even if the multiplication by zero is interpreted as multiplication by an expression that is tending to zero, as a modern mathematician would like to do, this cancellation is not allowed, since the first term in the numerator is a higher-order infinitesimal than the second. Bhaskara is handling 0 here as if it were 1. Granting that operation, he does correctly deduce, by completing the square (adding 4 to each side), that $x=2$.

5. The Muslims

It has always been recognized that Europe received algebra from the Muslims; the very word *algebra* (*al-jabr*) is an Arabic word meaning *transposition* or *restoration*. Its origins in the Muslim world date from the ninth century, in the work of Muhammed ibn Musa al-Khwarizmi (780–850), as is well established.[12] What is less certain is how much of al-Khwarizmi's algebra was original with him and how much he learned from Hindu sources. According to Colebrooke (*1817*, pp. lxiv–lxxx), he was well versed in Sanskrit and translated a treatise on Hindu computation[13] into Arabic at the request of Caliph al-Mamun, who ruled from 813 to 833. Colebrooke cites the Italian writer Pietro Cossali[14] who presented the alternatives that al-Khwarizmi learned algebra either from the Greeks or the Hindus and opted for the Hindus. These alternatives are a false dichotomy. We need not conclude that al-Khwarizmi took everything from the Hindus or that he invented everything himself. It is very likely that he expounded some material that he read in Sanskrit and added his own ideas to it. Rosen (*1831*, p. x) explains the difference in the preface to his edition of al-Khwarizmi's algebra text, saying that "at least the method which he follows in expounding his rules, as well as in showing their application, differs considerably from that of the Hindu mathematical writers."

---

[12] Colebrooke (*1817*, p. lxxiii) noted that a manuscript of this work dated 1342 was in the Bodleian Library at Oxford. Obviously, this manuscript could not be checked out, and Colebrooke complained that the library's restrictions "preclude the study of any book which it contains, by a person not enured to the temperature of apartments unvisited by artificial warmth." If he worked in the library in 1816, his complaint would be understandable: Due to volcanic ash in the atmosphere, there was no summer that year. This manuscript is the source that Rosen (*1831*) translated and reproduced.

[13] It is apparently this work that brought al-Khwarizmi's name into European languages in the form *algorism*, now *algorithm*. A Latin manuscript of this work in the Cambridge University Library, dating to the thirteenth century, has recently been translated into English (Crossley and Henry, *1990*).

[14] His dates are 1748–1813. He was Bishop of Parma and author of *Origine, trasporto in Italia, primi progressi in essa dell' algebra* (*The Origins of Algebra and Its Transmission to Italy and Early Progress There*), published in Parma in 1797.

Colebrooke also notes (p. lxxi) that Mohammed Abu'l-Wafa al-Buzjani (940–998) wrote a translation or commentary on the *Arithmetica* of Diophantus. This work, however, is now lost. Apart from these possible influences of Greek and Hindu algebra, whose effect is difficult to measure, it appears that the progress of algebra in the Islamic world was an indigenous growth. We shall trace that growth through several of its most prominent representatives, starting with the man recognized as its originator, Muhammed ibn Musa al-Khwarizmi.

5.1. Al-Khwarizmi. Besides the words *algebra* and *algorithm*, there is a common English word whose use is traceable to Arabic influence (although it is not an Arabic word), namely *root* in the sense of a square or cube root or a root of an equation. The Greek picture of the square root was the side of a square, and the word *side* (*pleura*) was used accordingly. The Muslim mathematicians apparently thought of the root as the part from which the equation was generated and used the word *jadhr* accordingly. According to al-Daffa (*1973*, p. 80), translations into Latin from Greek use the word *latus* while those from Arabic use *radix*. In English the word *side* lost out completely in the competition.

Al-Khwarizmi's numbers correspond to what we call positive real numbers. Theoretically, such a number could be defined by any convergent sequence of rational numbers, but in practice some rule is needed to generate the terms of the sequence. For that reason, it is more accurate to describe al-Khwarizmi's numbers as positive *algebraic numbers*, since all of his numbers are generated by equations with rational coefficients. The absence of negative numbers prevented al-Khwarizmi from writing all quadratic equations in the single form "squares plus roots plus numbers equal zero" ($ax^2 + bx + c = 0$). Instead, he had to consider three basic cases and three others, in which either the square or linear term is missing. He described the solution of "squares plus roots equal numbers" by the example of "a square plus 10 roots equal 39 dirhems." (A dirhem is a unit of money.) Al-Khwarizmi's solution of this problem is to draw a square of unspecified size (the side of the square is the desired unknown) to represent the square (Fig. 3). To add 10 roots, he then attaches to each side a rectangle of length equal to the side of the square and width $2\frac{1}{2}$ (since $4 \cdot 2\frac{1}{2} = 10$). The resulting cross-shaped figure has, by the condition of the problem, area equal to 39. He then fills in the four corners of the figure (literally "completing the square"). The total area of these four squares is $4 \cdot \left(2\frac{1}{2}\right)^2 = 25$. Since $39 + 25 = 64$, the completed square has side 8. Since this square was obtained by adding rectangles of side $2\frac{1}{2}$ to each side of the original square, it follows that the original square had side 3.

This case is the one al-Khwarizmi considers first and is the simplest to understand. His figures for the other two cases of quadratic equations are more complicated, but all are based on Euclid's geometric illustration of the identity $\left((a+b)/2\right)^2 + \left((a-b)/2\right)^2 = ab$ (Fig. 17 of Chapter 10).

Al-Khwarizmi did not consider any cubic equations. Roughly the first third of the book is devoted to various examples of pure mathematical problems leading to quadratic equations, causing the reader to be somewhat skeptical of his claim to be presenting the material needed in commerce and law. In fact, there are no genuine applications of quadratic equations in the book. But if quadratic equations have no practical applications (outside of technology, of course), there are occasions when a practical problem requires solving linear equations. Al-Khwarizmi found many

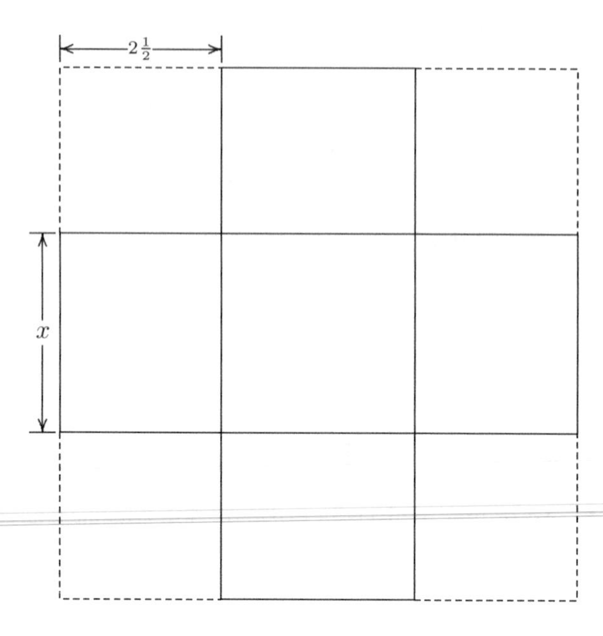

FIGURE 3. Al-Khwarizmi's solution of "square plus 10 roots equals 39 dirhems."

such cases in problems of inheritance, which occupy more than half of his *Algebra*. Here is a sample:

> A man dies, leaving two sons behind him, and bequeathing one-fifth of his property and one dirhem to a friend. He leaves 10 dirhems in property and one of the sons owes him 10 dirhems. How much does each legatee receive?

Although mathematics is cross-cultural, its applications are very specific to the culture in which they are used. The difference between the modern solution of this legal problem and al-Khwarizmi's solution is considerable. Under modern law the man's estate would be considered to consist of 20 dirhems, the 10 dirhems cash on hand, and the 10 dirhems owed by one of the sons. The friend would be entitled to 5 dirhems (one-fifth plus one dirhem), and the indebted son would owe the estate 10 dirhems. His share of the estate would be one-half of the 15 dirhems left after the friend's share is taken out, or $7\frac{1}{2}$ dirhems. He would therefore have to pay $2\frac{1}{2}$ dirhems to the estate, providing it with cash on hand equal to $12\frac{1}{2}$ dirhems. His brother would receive $7\frac{1}{2}$ dirhems.

Now the notion of an estate as a legal entity that can owe and be owed money is a modern European one, alien to the world of al-Khwarizmi. Apparently in al-Khwarizmi's time, money could be owed only to a *person*. What principles are to be used for settling accounts in this case? Judging from the solution given by al-Khwarizmi, the estate is to consist of the 10 dirhems cash on hand, plus a *certain portion* (not all) of the debt the second son owed to his deceased father. This "certain portion" is the unknown in a linear equation and is the reason for invoking algebra in the solution. It is to be chosen so that *when the estate is divided up, the indebted son neither receives any more money nor owes any to the other heirs.* This condition leads to an equation that can be solved by algebra. Al-Khwarizmi explains the solution as follows (we put the legal principle that provides the equation in capital letters):

> Call the amount taken out of the debt *thing*. Add this to the property; the sum is 10 dirhems plus *thing*. Subtract one-fifth of this, since he has bequeathed one-fifth of his property to the friend. The remainder is 8 dirhems plus $\frac{4}{5}$ of *thing*. Then subtract the 1 dirhem extra that is bequeathed to the friend. There remain 7 dirhems and $\frac{4}{5}$ of *thing*. Divide this between the two sons. The portion of each of them is $3\frac{1}{2}$ dirhems plus $\frac{2}{5}$ of *thing*. THIS MUST BE EQUAL TO *THING*. Reduce it by subtracting $\frac{2}{5}$ of *thing* from *thing*. Then you have $\frac{3}{5}$ of *thing* equal to $3\frac{1}{2}$ dirhems. Form a complete *thing* by adding to this quantity $\frac{2}{3}$ of itself. Now $\frac{2}{3}$ of $3\frac{1}{2}$ dirhems is $2\frac{1}{3}$ dirhems, so that *thing* is $5\frac{5}{6}$ dirhems.

Rosen (*1831*, p. 133) suggested that the many arbitrary principles used in these problems were introduced by lawyers to protect the interests of next-of-kin against those of other legatees.

5.2. Abu Kamil. A commentary on al-Khwarizmi's *Algebra* was written by the mathematician Abu Kamil (ca. 850–930). His exposition of the subject contained none of the legacy problems found in al-Khwarizmi's treatise, but after giving the basic rules of algebra, it listed 69 problems of considerable intricacy to be solved. For example, a paraphrase of Problem 10 is as follows:

> The number 50 is divided by a certain number. If the divisor is increased by 3, the quotient decreases by $3\frac{3}{4}$. What is the divisor?

Abu Kamil is also noteworthy because many of his problems were copied by Leonardo of Pisa, one of the first to introduce the mathematics of the Muslims into Europe.

5.3. Omar Khayyam. Although al-Khwarizmi did not consider any equations of degree higher than 2, such equations were soon to be considered by Muslim mathematicians. As we saw in Section 1 of Chapter 10, a link between geometry and algebra appeared in the use of the rectangular hyperbola by Pappus to carry out the *neûsis* construction for trisecting an angle (Fig. 9 of Chapter 10). The mathematician Omar Khayyam, of the late eleventh and early twelfth centuries (see Amir-Moez, *1963*), realized that a large class of geometric problems of this type led to cubic equations that could be solved using conic sections. His treatise on algebra[15] was largely occupied with the classification and solution of cubic equations by this method.

Omar Khayyam did not have modern algebraic symbolism. He lived within the confines of the universe constructed by the Greeks. His classification of equations, like al-Khwarizmi's, is conditioned by the use of only positive numbers as data. For that reason his classification is even more complicated than al-Khwarizmi's, since he is considering cubic equations as well as quadratics. He lists 25 types of equations (Kasir, *1931*, pp. 51–52), six of which do not involve any cubic terms. The particular cubic we shall consider is *cubes plus squares plus sides equal number*, or, as

[15] This treatise was little noticed in Europe until a French translation by Franz Woepcke (1827–1864) appeared in 1851 (Kasir, *1931*, p. 7).

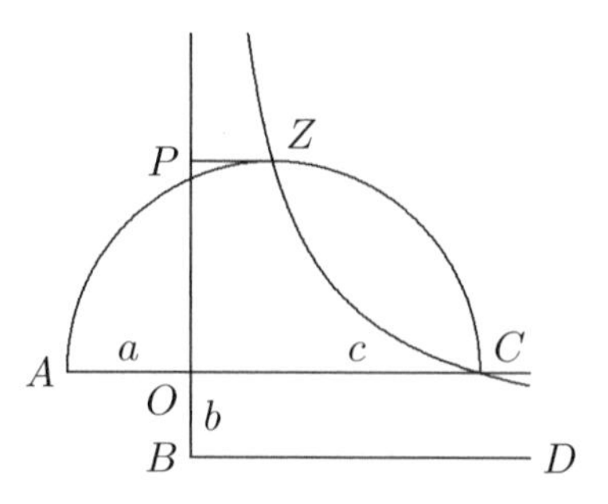

FIGURE 4. Omar Khayyam's solution of $x^3 + ax^2 + b^2x = b^2c$.

we would phrase it, $x^3 + ax^2 + bx = c$. In keeping with his geometric interpretation of magnitudes as line segments, Omar Khayyam had to regard the coefficient b as a square, so that we shall write b^2 rather than b. Similarly, he regarded the constant term as a solid, which without any loss of generality he considered to be a rectangular prism whose base was an area equal to the coefficient of the unknown. In keeping with this reduction we shall write b^2c instead of c. Thus Omar Khayyam actually considered the equation $x^3 + ax^2 + b^2x = b^2c$, where a, b, and c are data for the problem, to be represented as lines. His solution is illustrated in Fig. 4. He drew a pair of perpendicular lines intersecting at a point O and marked off $OA = a$ and $OC = c$ in opposite directions on one of the lines and $OB = b$ on the other line. He then drew a semicircle having AC as diameter, the line DB through B perpendicular to OB (parallel to AC), and the rectangular hyperbola through C having DB and the extension of OB as asymptotes. This hyperbola intersects the semicircle in the point C and in a second point Z. From Z he drew ZP perpendicular to the extension of OB, and ZP represented the solution of the cubic.

When it comes to actually producing a root by numerical procedures, Omar Khayyam's solution is circular, a mere restatement of the problem. He has broken the cubic equation into two quadratic equations in two unknowns, but any attempt to eliminate one of the two unknowns merely leads back to the original problem. In fact, no method of solution exists or can exist that reduces the solution of every cubic equation with real roots to the extraction of real square and cube roots of real numbers. What Omar Khayyam had created was an *analysis* of cubic equations using conic sections. He said that no matter how hard you look, you will never find a numerical solution "because whatever is obtained by conic sections cannot be obtained by arithmetic" (Amir-Moez, *1963*, p. 336).

5.4. Sharaf al-Din al-Muzaffar al-Tusi. A generation after the death of Omar Khayyam, another Muslim mathematician, Sharaf al-Tusi (ca. 1135–1213, not to be confused with Nasir-Eddin al-Tusi, whose work was discussed in Section 4 of Chapter 11), wrote a treatise on equations in which he analyzed the cubic equation using methods that are surprisingly modern in appearance. This work has been analyzed by Hogendijk (*1989*). Omar Khayyam had distinguished between the eight types of cubic equations that always have a solution and the five that could fail to have a solution. Al-Tusi provided a numerical method of solution for the first eight types that was essentially the Chinese method of solving cubic equations.

He then turned to the five types that could have no (positive) solutions for some values of the data. As an example, one of these forms is

$$x^3 + ax^2 + c = bx\,.$$

For each of these cases, al-Tusi considered a particular value of x, which for this example is the value m satisfying

$$3m^2 + 2am = b\,.$$

Let us denote the positive root of this equation (the larger root, if there are two) by m. The reader will undoubtedly have noticed that the equation can be obtained by differentiating the original equation and setting x equal to m. The point m is thus in all cases a relative minimum of the difference of the left- and right-hand sides of the equation. That, of course, is precisely the property that al-Tusi wanted. Hogendijk comments that it is unlikely that al-Tusi had any concept of a derivative. In fact, the equation for m can be derived without calculus, by taking m as the value at which the minimum occurs, subtracting the values at x from the value at m, and dividing by $x - m$. The result is the inequality $m^2 + mx + x^2 + a(m+x) > b$ for $x > m$ and the opposite inequality for $x < m$. Therefore equality must hold when $x = m$, that is, $3m^2 + 2am = b$, which is the condition given by al-Tusi. After finding the point m, al-Tusi concluded that there will be no solutions if the left-hand side of the equation is larger than the right-hand side when $x = m$. There will be one unique solution, namely $x = m$ if equality holds there. That left only the case when the left-hand side was smaller than the right-hand side when $x = m$. For that case he considered the auxiliary cubic equation

$$y^3 + py^2 = d\,,$$

where p and d were determined by the type of equation. The quantity d was simply the difference between the right- and left-hand sides of the equation at $x = m$, that is, $bm - m^3 - am^2 - c$ in the present case, with p equal to $3m + a$. Al-Tusi was replacing x with $y = x - m$ here. The procedure was precisely the method we know as Horner's method, and the linear term drops out because the condition by which m was chosen ordains that it be so (see Problem 14.9.) The equation in y was known to have a root because it was one of the other 13 types, which always have solutions. Thus, it followed that the original equation must also have a solution, $x = m + y$, where y was the root of the new equation. The added bonus was that a lower bound of m was obtained for the solution.

6. Europe

As soon as translations from Arabic into Latin became generally available in the twelfth and thirteenth centuries, Western Europeans began to learn about algebra. The first of these was a Latin translation of al-Khwarizmi's *Algebra*, made in 1145 by Robert of Chester (dates unknown). Several talented mathematicians appeared early on who were able to make original contributions to its development. In some cases the books that they wrote were not destined to be published for many centuries, but at least one of them formed part of an Italian tradition of algebra that continued for several centuries.

6.1. Leonardo of Pisa (Fibonacci). Many of the problems in the *Liber abaci* reflect the routine computations that must be performed when converting currencies. These are applications of the Rule of Three that we have found in Brahmagupta and Bhaskara. Many of the other problems are purely fanciful. Leonardo's indebtedness to Arabic sources was detailed by Levey (*1966*, pp. 217–220), who listed 29 problems in the *Liber abaci* that are identical to problems in the *Algebra* of Abu Kamil. In particular, the problem of separating the number 10 into two parts satisfying an extra condition occurs many times. For example, one problem is to find x such that $10/x + 10/(10 - x) = 6\frac{1}{4}$.

The Liber quadratorum. The *Liber quadratorum* is written in the spirit of Diophantus. The resemblance in some points is so strong that it would be very strange if Leonardo had not seen a copy of Diophantus. This question is discussed by the translator of the *Liber quadratorum* (Sigler, *1987*, pp. xi–xii), who notes that strong resemblances have been pointed out between the *Liber quadratorum* and al-Karaji's *Fakhri*, parts of which were copied from the *Arithmetica*, but that there are also parts of the *Liber quadratorum* that are original. The resemblance to Diophantus is shown in such statements as the ninth of its 24 propositions: *Given a nonsquare number that is the sum of two squares, find a second pair of squares having this number as their sum.* Leonardo's solution of this problem, like that of Diophantus, involves a great deal of arbitrariness, since the problem does not have a unique solution.

One advance in the *Liber quadratorum* is the use of general letters in an argument. Although in some proofs Leonardo argues much as Diophantus does, using specific numbers, he becomes more abstract in others. For example, Proposition 5 requires finding two numbers the sum of whose squares is a square that is also the sum of the squares of two given numbers (Problem 9 of Book 2 of Diophantus). He says to proceed as follows. Let the two given numbers be $.a.$ and $.b.$ and the sum of their squares $.g.$. Now take any other two numbers $.de.$ and $.ez.$ [not proportional to the given numbers] the sum of whose squares is a square. These two numbers are arranged as the legs of a right triangle. If the square on the hypotenuse of this triangle is $.g.$, the problem is solved. If the square on the hypotenuse is larger than $.g.$, mark off the square root of $.g.$ on the hypotenuse. The projections (as we would call them) of this portion of the hypotenuse on each of the legs are known, since their ratios to the square root of $.g.$ are known. Moreover, that ratio is rational, since they are the same as the ratios of $.a.$ and $.b.$ to the hypotenuse of the original triangle. These two projections therefore provide the new pair of numbers. Being proportional to $.a.$ and $.b.$, which are not proportional to the two numbers given originally, they must be different from those numbers. This argument is more convincing, because more abstract, than proofs by example, but the geometric picture plays an important role in making the proof comprehensible.

The Flos. Leonardo's approach to algebra begins to look modern in other ways as well. In one of his works, called the *Flos super solutionibus quarumdam questionum ad numerum et ad geometriam vel ad utrumque pertinentum* (*The Full Development*[16] *of the Solutions of Certain Questions Pertaining to Number or Geometry or Both*, see Boncompagni, *1854*, p. 4), he mentions a challenge from John of Palermo to find a number satisfying $x^3 + 2x^2 + 10x = 20$ using the methods

[16] The word *flos* means *bloom*, and is used in the figurative sense of "the bloom of youth." That appears to be its meaning here.

given by Euclid in Book 10 of the *Elements*, that is, to construct a line of this length using ruler and compass. In working on this question, Leonardo made two important contributions to algebra, one numerical and one theoretical. The numerical contribution was to give the unique positive root in sexagesimal notation correct to six places. The theoretical contribution was to show by using divisibility properties of numbers that there cannot be a rational solution or a solution obtained using only rational numbers and square roots of rational numbers.

6.2. Jordanus Nemorarius. The translator and editor of the book *De numeris datis* (*On Given Numbers*), written by Jordanus Nemorarius, says (Hughes, *1981*, p. 11) "It is reasonable to assume... that Jordanus was influenced by al-Khwarizmi's work." This conclusion was reached on the basis of Jordanus' classification of quadratic equations and his order of expounding the three types, among other resemblances between the two works.

De numeris datis is the algebraic equivalent of Euclid's *Data*. Where Euclid says that a line is given (determined) if its ratio to a given line is given, Jordanus Nemorarius says that a number is given if its ratio to a given number is given. The well-known elementary fact that two numbers can be found if their sum and difference are known is generalized to the theorem that any set of numbers can be found if the differences of the successive numbers and the sum of all the numbers is known.[17] In general, this book contains a large variety of data sets that determine numbers. For example, *if the sum of the squares of two numbers is known, and the square of the difference of the numbers is known, the numbers can be found.* The four books of *De numeris datis* contain about 100 such results. These results admit a purely algebraic interpretation. For example, in Book 4 Jordanus Nemorarius writes:

> If a square with the addition of its root multiplied by a given number makes a given number, then the square itself will be given. [Hughes, *1981*, p. 100][18]

Where earlier mathematicians would have proved this proposition with examples, Jordanus Nemorarius uses letters representing abstract numbers. The assertion is that there is only one (positive) number x such that $x^2 + \alpha x = \beta$, and that x can be found if α and β are given.

6.3. The fourteenth and fifteenth centuries. The century in which Nicole d'Oresme made such remarkable advances in geometry, coming close to the creation of analytic geometry, was also a time of rapid advance in algebra, epitomized by Antonio de' Mazzinghi (ca. 1353–1383). His *Trattato d'algebra* contains some complicated systems of linear and quadratic equations in as many as three unknowns (Franci, *1988*). He was one of the earliest algebraists to move the subject toward the numerical and away from the geometric interpretation of problems.

In the following century Luca Pacioli wrote *Summa de arithmetica, geometrica, proportioni et proportionalita* (*Treatise on Arithmetic, Geometry, Proportion, and Proportionality*), which was closer to the elementary work of al-Khwarizmi and more geometrical in its approach to algebra than the work of Mazzinghi. Actually

[17] This statement is a variant of the *epanthēma* (blossom) of Thymaridas.

[18] This translation is my own and is intended to be literal; Hughes gives a smoother, more idiomatic translation on p. 168.

(see Parshall, *1988*), the work was largely a compilation of the works of Leonardo of Pisa, but it did bring the art of abbreviation closer to true symbolic notation. For example, what we now write as $x - \sqrt{x^2 - 36}$ was written by Pacioli as

$$1.\text{co}.\tilde{\text{m}}\text{Rv}.1.\text{ce}\ \tilde{\text{m}}36\,.$$

Here *co* means *cosa* (*thing*), the unknown; *ce* means *censo* (*power*), and *Rv* is probably a printed version of *Rx*, from the Latin *radix*, meaning *root*.[19] Pacioli's work was both an indication of how widespread knowledge of algebra had become by this time and an important element in propagating it. The sixteenth-century Italian algebraists who moved to the forefront of the subject and advanced it far beyond where it had been up to that time had all read Pacioli's treatise thoroughly.

6.4. Chuquet. The *Triparty en la science des nombres* by Nicolas Chuquet is accompanied by a book of problems to illustrate its principles, a book on geometrical mensuration, and a book of commercial arithmetic. The last two are applications of the principles in the first book. Thus the subject matter is similar to that of al-Khwarizmi's *Algebra* or Leonardo's *Liber abaci.*

There are several new things in the *Triparty.* One is a superscript notation similar to the modern notation for the powers of the unknown in an equation. The unknown itself is called the *premier* or "first." Algebra in general is called the *rigle des premiers* or "rule of firsts." Chuquet listed the first 20 powers of 2 and pointed out that when two such numbers are multiplied, their indices are added. Thus, he had a clear idea of the laws of integer exponents. A second innovation in the *Triparty* is the free use of negative numbers as coefficients, solutions, and exponents. Still another innovation is the use of some symbolic abbreviations. For example, the square root is denoted R^2 (R for the Latin *radix*, or perhaps the French *racine*). The equation we would write as $3x^2 + 12 = 9x$ was written $.3.^2\ \bar{p}.12.$ egaulx a $.9.^1.$ Chuquet called this equation impossible, since its solution would involve taking the square root of -63.

His instructions are given in words. For example (Struik, *1986*, p. 62), consider the equation

$$R^2\underline{4^2_{.}\tilde{p}.4^1}\tilde{p}.2^1_{.}\tilde{p}.1\ \text{egaulx a}\ .100\,,$$

which we would write

$$\sqrt{4x^2 + 4x} + 2x + 1 = 100\,.$$

Chuquet says to subtract $.2^1_{.}\tilde{p}.1$ from both sides, so that the equation becomes

$$R^2\underline{4^2_{.}\tilde{p}.4^1}\ \text{egaulx a}\ .99\tilde{m}.2^1_{.}\,.$$

Next he says to square, getting

$$4^2_{.}\tilde{p}.4^1\ \text{egaulx a}\ 9801.\tilde{m}.396^1_{.}\tilde{p}.4^2_{.}\,.$$

Subtracting 4^2 from both sides and adding $396_{.}1$ to both sides then yields

$$400^1_{.}\ \text{egaulx a}\ .9801.\,.$$

Thus $x = 9801/400$.

---

[19] The symbol *Rx* should not be confused with the same symbol in pharmacy, which comes from the Latin *recipe*, meaning *take*.

6.5. Solution of cubic and quartic equations. In Europe algebra was confined to linear and quadratic equations for many centuries, whereas the Chinese and Japanese had not hesitated to attack equations of any degree. The difference in the two approaches is a result of different ideas of what constitutes a solution. This distinction is easy to make nowadays: The European mathematicians were seeking an exact solution using only arithmetic operations and root extractions, what is called *solution by radicals.* However, it will not do to press the distinction too far: It is impossible to do good numerical work without a sound theoretical basis. As we saw in the work of Sharaf al-Tusi, the coefficients that appear in the course of his numerical solution have theoretical significance.

The Italian algebraists of the early sixteenth century made advances in the search for a general algorithm for solving higher-degree equations. We discussed the interesting personal aspects of the solution of cubic equations in Section 4 of Chapter 3. Here we concentrate on the technical aspects of the solution.

The verses Tartaglia had memorized say, in modern language, that to solve the problem $x^3 + px = q$, one should look for two numbers u and v satisfying $u - v = q$, $uv = (p/3)^3$. The problem of finding u and v is that of finding two numbers given their difference and their product, and of course, that is merely a matter of solving a *quadratic* equation, a problem that had already been completely solved. Once this quadratic has been solved, the solution of the original cubic is $x = \sqrt[3]{u} - \sqrt[3]{v}$. The solution of the cubic has thus been reduced to solving a quadratic equation, taking the cube roots of its two roots, and subtracting. Cardano illustrated with the case of "a cube and six times the side equal to 20." Using his complicated rule (complicated because he stated it in words), he gave the solution as

$$\sqrt[3]{\sqrt{108} + 10} - \sqrt[3]{\sqrt{108} - 10}\,.$$

He did not add that this number equals 2.

Ludovico Ferrari. Cardano's student Ludovico Ferrari worked with him in the solution of the cubic, and between them they had soon found a way of solving certain fourth-degree equations. Ferrari's solution of the quartic was included near the end of Cardano's *Ars magna.* Counting cases as for the cubic, one finds a total of 20 possibilities. The principle in most cases is the same, however. The idea is to make a perfect square in x^2 equal to a perfect square in x by adding the same expression to both sides. Cardano gives the example

$$60x = x^4 + 6x^2 + 36.$$

It is necessary to add to both sides an expression $rx^2 + s$ to make them squares, that is, so that both sides of

$$rx^2 + 60x + s = x^4 + (6 + r)x^2 + (36 + s)$$

are perfect squares. Now the condition for this to happen is well known: ax^2+bx+c is a perfect square if and only if $b^2 - 4ac = 0$. Hence we need to have simultaneously

$$3600 - 4sr = 0, \quad (6 + r)^2 - 4(36 + s) = 0.$$

Solving the second of these equations for s in terms of r and substituting in the first leads to the equation

$$r^3 + 12r^2 = 108r + 3600.$$

This is a cubic equation called the *resolvent* cubic. Once it is solved, the original quartic breaks into two quadratic equations upon taking square roots and adding an ambiguous sign.

A few aspects of the solution of cubic and quartic equations should be noted. First, the problem is not a practical one. Second, the Cardano recipe for solving an equation sometimes gives the solution in a rather strange form. For example, Cardano says that the solution of $x^3+6x=20$ is $\sqrt[3]{\sqrt{108}+10}-\sqrt[3]{\sqrt{108}-10}$. The expression is correct, but can you tell at a glance that it represents the number 2?

Third, the procedure does not always work. For example, the equation $x^3+6=7x$ has to be solved by guessing a number that can be added to both sides so as to produce a common factor that can be canceled out. The number in this case is 21, but there is no *algorithm* for finding such a number. For equations of this type the algebraic procedures for finding x involve square roots of negative numbers. The search for an algebraic procedure using only real numbers to solve this case of the cubic continued for 300 years, until finally it was shown that no such procedure can exist.

6.6. Consolidation. There were two natural ways to build on what had been achieved in algebra by the end of the sixteenth century. One was to find a notation that could unify equations so that it would not be necessary to consider so many different cases and so many different possible numbers of roots. The other was to solve equations of degree five and higher. We shall discuss the first of these here and devote Chapter 15 to the quest for the second and its consequences.

All original algebra treatises written up to and including the treatise of Bombelli are very tiresome for the modern student, who is familiar with symbolic notation. For that reason we have sometimes allowed ourselves the convenience of modern notation when doing so will not distort the thought process involved. In the years between 1575 and 1650 several innovations in notation were introduced that make treatises written since that time appear essentially modern. The symbols + and − were originally used in bookkeeping in warehouses to indicate excess and deficiencies; they first appeared in a German treatise on commercial arithmetic in 1489 but were not widely used in the rest of Europe for another century. The sign for equality was introduced by a Welsh medical doctor, physician to the short-lived Edward VI, named Robert Recorde (1510–1558). His symbol was a very long pair of parallel lines, because, as he said, "noe 2. thynges, can be moare equalle." The use of abbreviations for the various powers of the unknown in an equation was eventually achieved, but there were two other needs to be met before algebra could become a mathematical subject on a par with geometry: a unified way of writing equations and a concept of number in which every equation would have a solution. The use of exponential notation and grouping according to powers was discussed by Simon Stevin (see Section 7 of Chapter 6). Stevin used the abbreviation M for the first unknown in a problem, *sec* for the second, and *ter* for the third. Thus (see Zeuthen, *1903*, p. 95), what we would write as the equation

$$\frac{6x^3}{y} \div 2xz^2 = \frac{3x^2}{yz^2}$$

was expressed as follows: If we divide

$$6\ M\ ③\ D\ sec\ ①\ \text{by}\ 2\ M\ ①\ ter\ ②\ ,$$

we obtain

$$3\ M\ ②\ D\ sec\ ①\ D\ ter\ ②\ .$$

Although notation still had far to go, from the modern point of view, at least it was no longer necessary to use a different letter to represent each power of the unknown in a problem.

François Viète. The French lawyer François Viète (1540–1603), who worked as tutor in a wealthy family and later became an advisor to Henri de Navarre (the future king Henri IV), found time to study Diophantus and to introduce his own ideas into algebra. Viète is credited with several crucial advances in the subject. In his book *Artis analyticae praxis* (*The Practice of the Analytic Art*) he begins by giving the rules for powers of binomials (in words). For example, he describes the fifth power of a binomial as "the fifth power of the first [term], plus the product of the fourth power of the first and five times the second,... ." Viète's notation was slightly different from ours, but is more recognizable to us than that of Stevin. He would write the equation $A^3 + 3BA = D$, where the vowel A represented the unknown and the consonants B and D were taken as known, as follows (Zeuthen, *1903*, p. 98):

$$A\ cubus + B\ planum\ in\ A3\ aequatur\ D\ solido.$$

As this quotation shows, Viète appears to be following the tedious route of writing everything out in words, and to be adhering to the requirement that all the terms in an equation be geometrically homogeneous.

This introduction is followed by five books of *zetetics* (research, from the Greek word *zētein*, meaning *seek*). The mention of "roots" in connection with the binomial expansions was not accidental. Viète studied the relation between roots and coefficients in general equations. By using vowels to represent unknowns and consonants to represent data for a problem, Viète finally achieved what was lacking in earlier treatises: a convenient way of talking about general data without having to give specific examples. His consonants could be thought of as representing numbers that would be known in any particular application of a process, but were left unspecified for purposes of describing the process itself. His first example was the equation $A^2 + AB = Z^2$, in other words, a standard quadratic equation. According to Viète these three letters are associated with three numbers in direct proportion, Z being the middle, B the difference between the extremes, and A the smallest number. In our terms, $Z = Ar$ and $B = Ar^2 - A$. Thus, the general problem reduces to finding the smallest of three numbers A, Ar, Ar^2 given the middle value and the difference of the largest and smallest. Viète had already shown how to do that in his books of zetetics.

This analysis showed Viète the true relation between the coefficients and the roots. For example, he knew that in the equation $x^3 - 6x^2 + 11x = 6$, the sum and product of the roots must be 6 and the sum of the products taken two at a time must be 11. This observation still did not enable him to solve the general cubic equation, but he did study the problem geometrically and show that any cubic could be solved provided that one could solve two of the classical problems of antiquity: constructing two mean proportionals between two given lines and trisecting any angle. As he concluded at the end of his geometric chapter: "It is very worthwhile to note this."

Questions and problems

14.1. Problem 6 of Book 1 of the *Arithmetica* is *to separate a given number into two numbers such that a given fraction of the first exceeds a given fraction of the other by a given number.* In our terms this is a problem in two unknowns x and y, and there are four bits of data: the sum of the two numbers, which we denote by a, the two proper fractions r and s, and the amount b by which rx exceeds sy. Write down and solve the two equations that this problem involves. Under what conditions will the solutions be positive rational numbers (assuming that a, b, r, and s are positive rational numbers)? Compare your statement of this condition with Diophantus' condition, stated in very complicated language: *The last given number must be less than that which arises when that fraction of the first number is taken which exceeds the other fraction.*

14.2. Carry out the solution of the bundles of wheat problem from the *Jiu Zhang Suanshu.* Is it possible to solve this problem without the use of negative numbers?

14.3. Solve the equation for the diameter of a town considered by Li Rui. [*Hint:* Since $x = -3$ is an obvious solution, this equation can actually be written as $x^3 + 3x^2 = 972$.]

14.4. Solve the following legacy problem from al-Khwarizmi's *Algebra*: *A woman dies and leaves her daughter, her mother, and her husband, and bequeaths to some person as much as the share of her mother and to another as much as one-ninth of her entire capital. Find the share of each person.* It was understood from legal principles that the mother's share would be $\frac{2}{13}$ and the husband's $\frac{3}{13}$.

14.5. Solve the problem of Abu Kamil in the text.

14.6. If you know some modern algebra, explain, by filling in the details of the following argument, why it is not surprising that Omar Khayyam's geometric solution of the cubic cannot be turned into an algebraic procedure. Consider a cubic equation with rational coefficients but no rational roots,[20] such as $x^3 + x^2 + x = 2$. By Omar Khayyam's method, this equation is replaced with the system $y(z + 1) = 2$, $z^2 = (y+1)(2-y)$, one obvious solution of which is $y = 2$, $z = 0$. The desired value of x is the y-coordinate of the other solution. The procedure for eliminating one variable between the two quadratic equations representing the hyperbola and circle is a rational one, involving only multiplication and addition. Since the coefficients of the two equations are rational, the result of the elimination will be a polynomial equation with rational coefficients. If the root is irrational, that polynomial will be divisible by the minimal polynomial for the root over the rational numbers. However, a cubic polynomial with rational coefficients but no rational roots is itself the minimal polynomial for all of its roots. Hence the elimination will only return the original problem.

14.7. Why did al-Khwarizmi include a complete discussion of the solution of quadratic equations in his treatise when he had no applications for them at all?

14.8. Contrast the modern Western solution of the Islamic legacy problem discussed in the text with the solution of al-Khwarizmi. Is one solution "fairer" than the other? Can mathematics make any contribution to deciding what is fair?

---

[20] If the coefficients are rational, their denominators can be cleared. Then all rational roots will be found among the finite set of fractions whose numerators divide the constant term and whose denominators divide the leading coefficient. There is an obvious algorithm for finding these roots.

14.9. Consider the cubic equation of Sharaf al-Tusi's third type, which we write as $-x^3 - ax^2 + bx - c = 0$. Using Horner's method, as described in Section 2, show that if the first approximation is $x = m$, where m satisfies $3m^2 + 2am - b = 0$, then the equation to be satisfied at the second approximation is $y^2 - (3m+a)y^2 - (m^3 + am^2 - bm + c) = 0$. That is, carry out the algorithm for reduction and show that the process is

$$\begin{matrix} -c & & -m^3 - am^2 + bm - c \\ b & \longrightarrow & -3m^2 - 2am + b\ (=0) \\ -a & & -3m - a \\ -1 & & -1 \end{matrix}\ .$$

14.10. Consider Problem 27 of Book 1 of *De numeris datis*: *Two numbers are given whose sum is 10. If one is divided by 4 and the other by 2, the product of the quotients is 2. What are the two numbers?* Solve this problem in your own way, then solve it following Jordanus' recipe, which we paraphrase as follows. Let the two numbers be x and y, and let the quotients be e and f when x and y are divided by c and d respectively; let the product of the quotients be $ef = b$. Let $bc = h$, which is the same as fce or fx. Then multiply d by h to produce j, which is the same as xdf or xy. Since we now know both $x + y$ and xy, we can find x and y.

14.11. Solve the equation $x^3 + 60x = 992$ using the recipe given by Tartaglia.

14.12. How can you *prove* that $\sqrt[3]{\sqrt{108} + 10} - \sqrt[3]{\sqrt{108} - 10} = 2$?

14.13. If you know the polar form of complex numbers $z = r\cos\theta + ir\sin\theta$, show that the problem of taking the cube root of a complex number is equivalent to solving two of the classical problems of antiquity simultaneously, just as Viète claimed: the problem of two mean proportionals and the problem of trisecting the angle.

14.14. Consider Viète's problem of finding three numbers in direct proportion given the middle number and the difference between the largest and smallest. Show that this problem amounts to finding x and y given $\sqrt{xy}$ and $y - x$. How do you solve such a problem?

14.15. Show that the equation $x^3 = px + q$, where $p > 0$ and $q > 0$, has the solution $x = \sqrt{4p/3}\cos\theta$, where $\theta = \frac{1}{3}\arccos\big((q\sqrt{27})/(2\sqrt{p^3})\big)$. In order for this inverse cosine to exist it is necessary and sufficient that $q^2/4 - p^3/27 \le 0$, which is precisely the condition under which the Cardano formula requires the cube root of a complex number. [*Hint:*Use the formula $4\cos^3\theta - 3\cos\theta = \cos(3\theta)$.]

Observe that

$$\theta = \frac{1}{3}\int_a^1 \frac{1}{\sqrt{1-t^2}}\,dt\,,$$

where $a = (q\sqrt{27})/(2\sqrt{p^3})$. Thus, the solution of the cubic equation has a connection with the integral of an algebraic function $1/y$, where y satisfies the quadratic equation $y^2 = 1 - x^2$. This kind of connection turned out to be the key to the solution of higher-degree algebraic equations. As remarked in the text, Viète's solution of the cubic uses a transcendental method, even though an algebraic method exists.

CHAPTER 15

Modern Algebra

By the mid-seventeenth century, the relation between the coefficients and roots of a general equation was understood, and it was conjectured that if you counted roots according to multiplicity and allowed complex roots, an equation of degree n would have n roots. Algebra had been consolidated to the point that the main unsolved problem, the solution of equations of degree higher than 4, could be stated simply and analyzed.

The solution of this problem took nearly two centuries, and it was not until the late eighteenth and early nineteenth centuries that enough insight was gained into the process of determining the roots of an equation from its coefficients to prove that arithmetic operations and root extractions were not sufficient for this purpose. Although the solution was a negative result, it led to the important concepts of modern algebra that we know as groups, rings, and fields; and these, especially groups, turned out to be applicable in many areas not directly connected with algebra. Also on the positive side, nonalgebraic methods of solving higher-degree equations were also sought and found, and a theoretically perfect way of deciding whether a given equation can be solved in radicals was produced.

1. Theory of equations

Viète understood something of the relation between the roots and the coefficients of some equations. His understanding was not complete, because he was not able to find all the roots. Before the connection could be made completely, there had to be a domain in which an equation of degree n would have n roots. Then the general connection could be made for quadratic, cubic, and quartic equations and generalized from there. The missing theorem was eventually to be called the *fundamental theorem of algebra.*[1]

1.1. Albert Girard.

This fundamental theorem was first stated by Albert Girard (1595–1632), the editor of the works of Simon Stevin. In 1629 he wrote *L'invention nouvelle en l'algèbre* (*New Discovery (Invention) in Algebra*). This work contained some of the unifying concepts that make modern algebra the compact, efficient system that it now is. One of these, for example, is regarding the constant term as the coefficient of the zeroth power of the unknown. He introduced the notion of *factions* of a finite set of numbers. The first faction is the sum of the numbers, the second the sum of all products of two distinct numbers from the set, and so on. The last faction is the product of all the numbers, so that "there are as many

[1] In his textbook on analytic function theory (*Analytic Function Theory*, Ginn & Co., Boston, 1960, Vol. 1, p. 24), Einar Hille (1894–1980) wrote that "modern algebraists are inclined to deny both its algebraic and its fundamental character." Hille does not name the modern algebraists, but he was a careful writer who must have had someone in mind. In the context of its time, the theorem was both algebraic and fundamental.

factions as there are numbers given." He noted that the number of terms in each faction could be found by using Pascal's triangle.

Girard always regarded the leading coefficient as 1. Putting the equation into this form, he stated as a theorem (see, for example, Struik, *1986*, p. 85) that "all equations of algebra receive as many solutions as the denomination [degree] of the highest form shows, except the incomplete, and the first faction of the solutions is equal to the number of the first mixed [that is, the cofficient of the power one less than the degree of the equation], their second faction is equal to the number of the second mixed, their third to the third mixed, and so on, so that the last faction is equal to the closure [product], and this according to the signs that can be observed in the alternate order." This recognition that the coefficients of a polynomial are elementary symmetric polynomials in its zeros was the first ray of light at the dawn of modern algebra.

By "incomplete," Girard meant equations with some terms missing. In some cases, he said, these may not have a full set of solutions. He gave the example of the equation $x^4 = 4x - 3$, whose solutions he gave as 1, 1, $-1 + \sqrt{-2}$, and $-1 - \sqrt{-2}$, showing that he realized the need to count both complex roots and multiple real roots for the sake of the general rule. He invoked the simplicity of the general rule as justification for introducing the multiple and complex roots, along with the fact that complex numbers provide solutions where otherwise none would exist.

1.2. Tschirnhaus transformations. Every complex number has nth roots—exactly n of them except in the case of 0—that are also complex numbers. As a consequence, any formula for solving equations that involves only the application of rational operations and root extractions starting with the coefficients will remain within the domain of complex numbers. This elementary fact led to the proposition stated by Girard, which we know as the fundamental theorem of algebra. Finding such a formula for equations of degree five and higher was to become a preoccupation of algebraists for the next two centuries.

Analysis of the cubic. By the year 1600 equations of degrees 2, 3, and 4 could all be solved, assuming that one could extract the cube root of a complex number. The methods used suggest an inductive process in which the solution of an equation of degree n, say

$$x^n - a_1 x^{n-1} + \cdots \mp a_{n-1}x \pm a_n = 0\,,$$

would be found by a substitution $y = x^{n-1} - b_1 x^{n-2} + \cdots \pm b_{n-2}x \mp b_{n-1}$ with the coefficients $b_1, \ldots, b_{n-1}$ chosen so that the original equation becomes $y^n = C$. Observe that there are $n-1$ coefficients b_k at our disposal and $n-1$ coefficients $a_1, \ldots, a_{n-1}$ to be removed from the original equation. The program looks feasible. Something of the kind must have been the reasoning that led Ehrenfried Walther von Tschirnhaus (1652–1708) to the belief that he had discovered a general solution to all polynomial equations. In 1677 he wrote to Leibniz:

> In Paris I received some letters from Mr. Oldenburg, but from lack of time have not yet been able to write back that I have found a new way of determining the irrational roots of all equations... The entire problem reduces to the following: We must be able to remove all the middle terms from any equation. When that is done, and as a result only a single power and a single known quantity remain, one need only extract the root.

Tschirnhaus claimed that the the middle terms (the a_k above) would be eliminated by a polynomial of the sort just discussed, provided that the b_k are suitably chosen. Such a change of variable is now called a *Tschirnhaus transformation.* If a Tschirnhaus transformation could be found for the *general* equation of degree n, and a formula existed for solving the *general* equation of degree $n-1$, the two could be combined to generate a formula for solving the general equation of degree n. At the time, there was not even a Tschirnhaus transformation for the cubic equation. Tschirnhaus was to provide one.

He illustrated his transformation using the example $x^3 - qx - r = 0$. Taking $y = x^2 - ax - b$, he noted that y satisfied the equation

$$y^3 + (3b - 2q)y^2 + (3b^2 + 3ar - 4qb + q^2 - a^2q)y \\ + (b^2 - 2qb^2 + 3bar + q^2b - aqr - a^2qb + a^3r - r^2) = 0\,.$$

He eliminated the square term by choosing $b = 2q/3$, then removed the linear term by solving for a in the quadratic equation

$$qa^2 - 3ra + 4q^2/3 = 0\,.$$

In this way, he had found at the very least a second solution of the general cubic equation, independent of the solution given by Cardano. And, what is more important, he had indicated a plausible way by which any equation might be solved. If it worked, it would prove that every polynomial equation could be solved using rational operations and root extractions, thereby proving at the same time that the complex numbers are algebraically closed. Unfortunately, detailed examination of the problem revealed difficulties that Tschirnhaus had apparently not noticed at the time of his letter to Leibniz.

The main difficulty is that when the variable x is eliminated between two polynomial equations $p_n(x) = 0$ and $y = p_{n-1}(x)$, where p_n is of degree n and p_{n-1} of degree $n-1$, the degrees of the equations needed to eliminate the successive coefficients in the equation for y increase to $(n-1)!$, not $n-1$.[2] It is only in the case of a cubic, where $(n-1)! = n-1$, that the program can be made to work in general. It may, however, work for a *particular* equation of higher degree. Leibniz, at any rate, was not convinced. He wrote to Tschirnhaus,

> I do not believe that [your method] will be successful for equations of higher degree, except in special cases. I believe that I have a proof for this. [Kracht and Kreyszig, *1990*, p. 27]

Tschirnhaus' method had intuitive plausibility: If there existed an algorithm for solving all equations, that algorithm should be a procedure like the Tschirnhaus transformation. Because the method does *not* work, the thought suggests itself that there may be equations that cannot be solved algebraically. The work of Tschirnhaus and Girard had produced two important insights into the general problem of polynomial equations: (1) the coefficients are symmetric functions of the roots; (2) solving the equation should be a matter of finding a sequence of operations that would eliminate coefficients until a pure equation $y^n = C$ was obtained. Since the problem was still unresolved, still more new insights were needed.

---

[2] Seki Kōwa knew the rational procedures (what he called *folding*, as discussed in Section 3 of Chapter 14) for eliminating x. It does seem a pity that the contemporaries Tschirnhaus and Seki Kōwa lived so far apart. They would have had much to talk about if they could have met.

To explain the most important of these new insights, let us consider what Girard's result means when applied to Cardano's solution of the cubic $y^3 + py = q$. If the roots of this equation are r, s, and t, then $p = st + tr + rs$, $q = rst$, $t = -r - s$, since the coefficient of y^2 is zero. The sequence of operations implied by Cardano's formula is

$$\begin{aligned} u &= \frac{p}{3}; \quad v = \frac{q}{2}; \\ a &= \sqrt{u^3 + v^2}; \\ y &= \sqrt[3]{v + a} + \sqrt[3]{v - a}. \end{aligned}$$

Girard's work implies that the quantity a, which is an *irrational* function of the coefficients p and q, is a *rational* function of the roots r, s, and t:

$$a = \pm \frac{i}{\sqrt{108}}(r - s)(s - t)(t - r);$$

that is, it does not involve taking the square root of any expression containing a root.

1.3. Newton, Leibniz, and the Bernoullis. In the 1670s Newton wrote a textbook of algebra called *Arithmetica universalis*, which was published in 1707, in which he stated more clearly and generally than Girard had done the relation between the coefficients and roots of a polynomial. Moreover, he showed that other symmetric polynomials of the roots could be expressed as polynomials in the coefficients by giving a set of rules that are still known by his name, although Edward Waring also proved that such an expression is possible.

Another impetus toward the fundamental theorem of algebra came from calculus. The well-known method known as partial fractions for integrating a quotient of two polynomials reduces all such problems to the purely algebraic problem of factoring the denominator. It is not immediately obvious that the denominator can be factored into linear and quadratic real factors; that is the content of the fundamental theorem of algebra. Johann Bernoulli (1667–1748, the first of three mathematicians named Johann in the Bernoulli family) asserted in a paper in the *Acta eruditorum* in 1702 that such a factoring was always possible, and therefore all rational functions could be integrated. Leibniz did not agree, arguing that the polynomial $x^4 + a^2$, for example, could not be factored into quadratic factors over the reals. Here we see a great mathematician being misled by following a method. He recognized that the factorization had to be $(x^2 + a^2\sqrt{-1})(x^2 - a^2\sqrt{-1})$ and that the first factor should therefore be factored as $(x + a\sqrt{-\sqrt{-1}})(x - a\sqrt{-\sqrt{-1}})$ and the second factor as $(x + a\sqrt{\sqrt{-1}})(x - a\sqrt{\sqrt{-1}})$, but he did not realize that these factors could be combined to yield $x^4 + a^2 = (x^2 - \sqrt{2}ax + a^2)(x^2 + \sqrt{2}ax + a^2)$. It was pointed out by Niklaus Bernoulli (1687–1759, known as Niklaus I) in the *Acta eruditorum* of 1719 (three years after the death of Leibniz) that this last factorization was a consequence of the identity $x^4 + a^4 = (x^2 + a^2)^2 - 2a^2x^2$.

1.4. Euler, d'Alembert, and Lagrange. The eighteenth century saw considerable progress in the understanding of equations in general and the procedures needed to solve them. Much of this new understanding came from the two men who dominated mathematical life in that century, Euler and Lagrange.

Euler. In his 1749 paper "Recherches sur les racines imaginaires des équations" ("Investigations into the imaginary roots of equations"), devoted to equations whose degree is a power of 2 and published in the memoirs of the Berlin Academy, Euler showed that when the coefficients of a polynomial are real, its roots occur in conjugate pairs, and therefore produce irreducible real quadratic factors of the form $(x-a)^2+b^2$. In this paper Euler argued that every polynomial of degree 2^n with real coefficients can be factored as a product of two polynomials of degree 2^{n-1} with real coefficients. In the course of the proof Euler presented the germ of an idea that was to have profound consequences. In showing that a polynomial of degree 8 could be written as a product of two polynomials of degree 4, he assumed that the coefficient of x^7 was made equal to zero by means of a linear substitution. The remaining polynomial $x^8-ax^6+bx^5-cx^4-dx^2+ex-f$ was then to be written as a product

$$(x^4-ux^3+\alpha x^2+\beta x+\gamma)(x^4+ux^3+\delta x^2+\varepsilon x+\zeta)\,.$$

Euler noted that since u was the sum of four roots of the equation, it could assume (potentially) 70 values (the number of combinations of eight things taken four at a time), and its square would satisfy an equation of degree 35.

In this paper, Euler also conjectured that the roots of an equation of degree higher than 4 cannot be constructed by applying a finite number of algebraic operations to the coefficients. This was the first explicit statement of such a conjecture.

In his 1762 paper "De resolutione aequationum cuiusque gradus" ("On the solution of equations of any degree"), published in the proceedings of the Petersburg Academy, Euler tried a different approach,[3] assuming a solution of the form

$$x=w+A\sqrt[n]{v}+B\sqrt[n]{v^2}+\cdots+Q\sqrt[n]{v^{n-1}}\,,$$

where w is a real number and v and the coefficients $A,\ldots,Q$ are to be found by a procedure resembling a Tschirnhaus transformation. This approach was useful for equations of degree 2^n, but fell short of being a general solution of the problem.

D'Alembert. Euler's contemporary and correspondent Jean le Rond d'Alembert (1717–1783) tried to prove that all polynomials could be factored into linear and quadratic factors in order to prove that all rational functions could be integrated by partial fractions. In the course of his argument he assumed that any algebraic function could be expanded in a series of fractional powers of the independent variable. While Euler was convinced by this proof, he also wrote to d'Alembert to say that this assumption would be questioned (Bottazzini, *1986*, pp. 15–18).

Lagrange. In 1770 Lagrange made a survey of the methods known up to his time for solving general equations. He devoted a great deal of space to a preliminary analysis of the cubic and quartic equations. In particular, he was intrigued by the fact that the resolvent equation, which he called the *reduced* equation (*équation reduite*), for the cubic was actually an equation of degree 6 that just happened to be quadratic in the third power of the unknown. He showed that if the roots of the cubic equation $x^3+px=q$ being solved were a, b, and c, then a root of the resolvent would be

$$y=\frac{a+\alpha b+\alpha^2 c}{3}\,,$$

[3] This approach was discovered independently by Étienne Bézout (1730–1783).

where $\alpha^3 = 1$, $\alpha \neq 1$. He argued that since the original equation was symmetric in a, b, and c, the resolvent would have to admit this y as a root, no matter how the letters a, b, and c were permuted. It therefore followed that the resolvent would in general have six different roots.

For the quartic equation with roots a, b, c, and d, he showed that the resolvent cubic equation would have a root

$$t = \frac{ab + cd}{2}.$$

Since this expression could assume only three different values when the roots were permuted—namely, half of $ab + cd$, $ac + bd$, or $ad + bc$—it would have to satisfy an equation of degree three.

Proceeding to equations of fifth degree, Lagrange noted the only methods proposed up to that time, by Tschirnhaus and Euler–Bézout, and showed that the resolvent to be expected in all cases would be of degree 24. Pointing out that even Tschirnhaus, Euler, and Bézout themselves had not seriously attacked equations of degree five or higher, nor had anyone else tried to extend their methods, he said, "It is therefore greatly to be desired that one could estimate *a priori* the success that is to be expected in applying these methods to degrees higher than the fourth." He then set out to provide proof that, in general, one could not expect the resolvent equation to reduce to lower degree than the original equation in such cases, at least using the methods mentioned.

To prove his point, Lagrange analyzed the method of Tschirnhaus from a more general point of view. For cubic and quartic equations, in which only two coefficients needed to be eliminated (the linear and quadratic terms in the cubic, the linear and cubic terms in the quartic) the substitution $y = x^2 + ax + b$ would always work, since the elimination procedure resulted in linear and quadratic expressions in a and b in the coefficients that needed to be eliminated. Still, as Lagrange remarked, that meant two pairs of possible values (a, b) and hence really two cubic resolvents to be solved. The resolvent was therefore once again an equation of degree 6, which happened to factor into the product of two cubics. He noted what must be an ominous sign for those hoping to solve all algebraic equations by algebraic methods: The construction of the coefficients in the resolvent for an equation of degree n appeared to require solving $n - 1$ equations in $n - 1$ unknowns, of degrees $1, 2, \ldots, n - 1$, so that eliminating the variable x in these equations therefore led to an expression for x that was of degree $(n - 1)!$ in y, and hence to a resolvent equation of degree $n!$ in y.

Lagrange summed up his analysis as follows:

> To apply, for example, the method of Tschirnhaus to the equation of degree 5, one must solve four equations in four unknowns, the first being of degree 1, the second of degree 2, and so on. Thus the final equation resulting from the elimination of three of these unknowns will in general be of degree 24. But apart from the immense amount of labor needed to derive this equation, it is clear that after finding it, one will be hardly better off than before, unless one can reduce it to an equation of degree less than 5; and if such a reduction is possible, it can only be by dint of further labor, even more extensive than before.

The technique of counting the number of different values the root of the resolvent will have when the roots of the original equation are permuted among themselves was an important clue in solving the problem of the quintic.

1.5. Gauss and the fundamental theorem of algebra. The question of the theoretical existence of roots was settled on an intuitive level in the 1799 dissertation of Gauss. Gauss distinguished between the abstract *existence* of a root, which he proved, and an algebraic *algorithm* for finding it, the existence of which he doubted. He pointed out that attempts to prove the existence of a root and any possible algorithm for finding it must assume the possibility of extracting the nth root of a complex number. He also noted the opinion, first stated by Euler, that no algebraic algorithm existed for solving the general quintic.

The reason we say that the existence of roots was settled only on the intuitive level is that the proof of the fundamental theorem of algebra is as much topological as algebraic. The existence of real roots of an equation of odd degree with real coefficients seems obvious since a real polynomial of odd degree tends to oppositely signed infinities as the independent variable ranges from one infinity to the other. It thus follows by connectivity that it must assume a zero at some point. Gauss' proof of the existence of complex roots was similar. Much of what he was doing was new at the time, and he had to explain it in considerable detail. For that reason, he preferred to use only real-variable methods, so as not to raise any additional doubts with the use of complex numbers. In fact, he stated his purpose in that way: to prove that every equation with real coefficients has a complete factorization into linear and quadratic real polynomials.

The complex-variable background of the proof is obvious nowadays, and Gauss admitted that his lemmas were normally proved using complex numbers. The steps were as follows. First, considering the equation $z^m + Az^{m-1} + Bz^{m-2} + \cdots + Kz^2 + Lz + M = 0$, where all coefficients $A, \ldots, M$ were real numbers,[4] taking $z = r(\cos\varphi + i\sin\varphi)$ and using the relation $z^m = r^m(\cos m\varphi + i\sin m\varphi)$, one can see that finding a root amounts to setting the real and imaginary parts equal to zero simultaneously, that is, finding r and φ such that

$$\begin{aligned} r^m\cos m\varphi + Ar^{m-1}\cos(m-1)\varphi + \cdots + Kr^2\cos 2\varphi + Lr\cos\varphi + M &= 0\,, \\ r^m\sin m\varphi + Ar^{m-1}\sin(m-1)\varphi + \cdots + Kr^2\sin 2\varphi + Lr\sin\varphi &= 0\,. \end{aligned}$$

What remained was to show that there actually were points where the two curves intersected. For that purpose, Gauss divided both equations by r^m and argued that for large values of r the two curves must have zeros near the zeros of $\cos m\varphi = 0$ and $\sin m\varphi = 0$. That would mean that on a sufficiently large circle, each would have $2m$ zeros, and moreover the zeros of one curve, being near the points with polar angles $(k+1/2)\pi/m$ must separate those of the other, which are near the points with polar angles $k\pi/m$. Then, arguing that the portion of each curve inside the disk of radius r was connected, he said that it was obvious that one could not join all the pairs from one set and all the pairs from the other set using two curves that do not intersect.

Gauss was uneasy about the intuitive aspect of the proof. During his lifetime he gave several other proofs of the theorem that he regarded as more rigorous.

---

[4] This restriction involves no loss of generality (see Problem 15.1).

1.6. Ruffini. As it turned out, Gauss had no need to publish his own research on the quintic equation. In the very year in which he wrote his dissertation, the first claim of a proof that it is impossible to find a formula for solving all quintic equations by algebraic operations was made by the Italian physician Paolo Ruffini (1765–1822). Ruffini's proof was based on Lagrange's count of the number of values a function can assume when its variables are permuted.[5] The principles of such a proof were gradually coming into focus. Waring's proof that every symmetric function of the roots of a polynomial is a function of its coefficients was an important step, as was the idea of counting the number of different values a rational function of the roots can assume. To get the general proof, it was necessary to show that the root extractions performed in the course of a hypothetical solution would also be rational functions of the roots. That this is the case for quadratic and cubic equations is not difficult to see, since the quadratic formula for solving $x^2 - (r_1 + r_2)x + r_1r_2 = 0$ involves taking only one square root:

$$\sqrt{(r_1 + r_2)^2 - 4r_1r_2} = \sqrt{(r_1 - r_2)^2}\,.$$

Similarly, the Cardano formula for solving $y^3 + (r_1r_2 + r_2r_3 + r_3r_1)y = r_1r_2r_3$, where $r_1 + r_2 + r_3 = 0$, involves taking

$$\sqrt{\frac{(r_1r_2 + r_2r_3 + r_3r_1)^3}{27} + \frac{(r_1r_2r_3)^2}{4}} = \sqrt{\frac{-1}{108}}\Big((r_1 - r_2)(2r_1^2 + 5r_1r_2 + 2r_2^2)\Big)\,,$$

followed by extraction of the cube roots of the two numbers

$$\frac{i}{3\sqrt{3}}(r_1 + \omega r_2)^2 \text{ and } \frac{i}{3\sqrt{3}}(r_1 + \omega^2 r_2)^2\,,$$

where $\omega = -1/2 + i\sqrt{3}/2$ is a complex cube root of 1. These radicals are consequently rational (but not symmetric) functions of the roots.

1.7. Cauchy. Although Ruffini's proof was not generally accepted by his contemporaries, it was endorsed many years later by Augustin-Louis Cauchy (1789–1856). In 1812 Cauchy wrote a paper "Essai sur les fonctions symétriques" in which he proved the crucial fact that a function of n variables that assumes fewer values than the largest prime number less than n when the variables are permuted, actually assumes at most two values. In 1815 he published this result.

Cauchy gave credit to Lagrange, Alexandre Théophile Vandermonde (1735–1796), and Ruffini for earlier work in this area. Vandermonde, in particular, exhibited the Vandermonde determinant

$$\det\begin{bmatrix} 1 & x_1 & x_1^2 & \cdots & x_1^{n-1} \\ 1 & x_2 & x_2^2 & \cdots & x_2^{n-1} \\ \vdots & \vdots & \cdots & \vdots & \\ 1 & x_n & x_n^2 & \cdots & x_n^{n-1} \end{bmatrix} =$$

$$= -(x_1 - x_2)(x_1 - x_3)\cdots(x_1 - x_n)(x_2 - x_3)\cdots(x_2 - x_n)\cdots(x_{n-1} - x_n)\,,$$

which assumes only two values, since interchanging two variables permutes the rows of the determinant and hence reverses the sign of the determinant.

[5] An exposition of Ruffini's proof, clothed in modern terminology that Ruffini would not have recognized, can be found in the paper of Ayoub (*1980*).

Ruffini had shown that it was not possible to exhibit a function of five variables that could be changed into three different functions or four different functions by permuting the variables. It was this work that Cauchy proposed to generalize.

Cauchy's theorem was an elegant piece of work in the theory of finite permutation groups. To prove it, he had to invent a good deal of that theory. He pointed out that the number of permutations N equals $n!$, and that the number of those permutations that leave the function unchanged is a divisor of N, which he denoted M. In a manner now familiar, he showed that the number of different values (that is, different functions of the variables) that can be obtained by permuting the variables is $R = N/M$, and that if S is a permutation that leaves the function unchanged and T changes its value from K to K', then ST also changes its value from K to K'. He then introduced cyclic permutations and what we now call the order of a cyclic permutation. The matrix notation now sometimes used for permutations and the notation $(\alpha\beta\gamma)$ for a permutation that maps α to β, β to γ, and γ to α, leaving all other elements fixed, was introduced in this paper.

Cauchy showed that if a permutation U is of order m, the complete set of permutations breaks up into N/m pairwise disjoint subsets (now called cosets) of m elements each. If $m > R$, which means $M > N/m$, some coset must contain two distinct elements S and T that leave the function invariant. When m is a prime p, this fact implies that some power U^s with s between 1 and $p-1$ leaves the function invariant, and since every power U^{sk} then leaves the function invariant, it follows that all powers of U leave the function invariant. If p is 2, this is not a strong statement, since $R = 1$ in that case, and all permutations whatsoever leave the function invariant. For $p > 2$, it implies that the set of permutations that leave the function invariant contains all permutations of order p.

Cauchy then showed that this set must contain all permutations of order 3, by explicitly writing any permutation of order 3 as the composition of two permutations of order p.[6] It then followed that the permutation group can produce at most two different functions. For this case Cauchy showed that the function must be of the form $K + SV$, where K and S are symmetric and V is the Vandermonde determinant mentioned above, which switches sign when any two of its arguments are interchanged.

Besides the notation for permutations and cycles, Cauchy also invented some of the terminology of group theory, including the word *index* (*indice*) still used for the number of cosets of a subgroup of a finite group. For the number of elements M in the subgroup, he used the term *indicial* (or *indicative*) *divisor* (*diviseur indicatif*). He proposed the name *substitution* (of one permutation into another) for the composition of two permutations, and he called two permutations *equivalent* if they produce the same function, that is, they are equal modulo the subgroup of permutations that leave the function invariant. To picture cyclic permutations of finite order, he suggested arranging the distinct powers as the vertices of a regular polygon and thinking of the composition of two of them as a clockwise rotation (he said "a rotation from east to west") of the polygon. Such an arrangement suggests studying the symmetries of these polygons. However, although he frequently referred to "groups of indices" in this paper, he did not define the notion of a group in its modern sense.

---

[6] The number $N = n!$ has no prime factors larger than n, so that $p \leq n$ in any case.

1.8. Abel. Cauchy's work had a profound influence on two young geniuses whose lives were destined to be very short. The first of these, the Norwegian mathematician Niels Henrik Abel (1802–1829), believed in 1821 that he had succeeded in solving the quintic equation. He sent his solution to the Danish mathematician Ferdinand Degen (1766–1825), who asked him to provide a worked-out example of a quintic equation that could be solved by Abel's method. While working through the details of an example, Abel realized his mistake. In 1824 he constructed an argument to show that such a solution was impossible and had the proof published privately. A formal version was published in the *Journal für die reine und angewandte Mathematik* in 1826. Abel was aware of Ruffini's work, and mentioned it in his argument. He attempted to fill in the gap in Ruffini's work with a proof that the intermediate radicals in any supposed solution by formula can be expressed as rational functions of the roots.

Abel's idea was that if some finite sequence of rational operations and root extractions applied to the coefficients produces a root of the equation

$$x^5 - ax^4 + bx^3 - cx^2 + dx - e = 0\,,$$

the final result must be expressible in the form

$$x = p + R^{\frac{1}{m}} + p_2 R^{\frac{2}{m}} + \cdots + p_{m-1} R^{\frac{m-1}{m}}\,,$$

where p, $p_2, \ldots, p_{m-1}$, and R are also formed by rational operations and root extractions applied to the coefficients, m is a prime number,[7] and $R^{1/m}$ is not expressible as a rational function of the coefficients a, b, c, d, e, p, $p_2, \ldots, p_{m-1}$.[8] By straightforward reasoning on a system of linear equations for the coefficients p_j, he was able to show that R is a symmetric function of the roots, and hence that $R^{1/m}$ must assume exactly m different values as the roots are permuted. Moreover, since there are 5! permutations of the roots and m is a prime, it followed that $m = 2$ or $m = 5$, the case $m = 3$ having been ruled out by Cauchy. The hypothesis that $m = 5$ led to an equation in which the left-hand side assumed only five values while the right-hand side assumed 120 values as the roots were permuted. Then the hypothesis $m = 2$ led to a similar equation in which one side assumed 120 values and the other only 10. Abel concluded that the hypothesis that there exists an algorithm for solving the equation was incorrect.

The standard version of the history of mathematics credits Abel with being "the" person who proved the impossibility of solving the quintic equation. But according to Ayoub (*1980*, p. 274), in 1832 the Prague Scientific Society declared the proofs of Ruffini and Abel unsatisfactory and offered a prize for a correct proof. The question was investigated by William Rowan Hamilton in a report to the Royal Society in 1836 and published in the *Transactions of the Royal Irish Academy* in 1839. Hamilton's report was so heavily laden with subscripts and superscripts bearing primes that only the most dedicated reader would attempt to understand it, although Felix Klein was later (*1884*) to describe it as being "as lucid as it is voluminous." The proof was described by the American number theorist and historian

[7] Extracting any root is tantamount to the sequential extraction of prime roots. Hence every root extraction in the hypothetical process of solving the equation can be assumed to be the extraction of a prime root.

[8] Abel incorporated the apparently missing coefficient p_1 into R here, since he saw no loss of generality in doing so. A decade later, Hamilton pointed out that doing so might increase the index of the root that needed to be extracted, since p_1 might itself require the extraction of an mth root.

of mathematics Leonard Eugene Dickson as "a very complicated reconstruction of Abel's proof." Hamilton regarded the problem of the solvability of the quintic as still open. He wrote:

> [T]he opinions of mathematicians appear to be not yet entirely agreed respecting the possibility or impossibility of expressing a root as a function of the coefficients by any finite combination of radicals and rational functions.

The verdict of history has been that Abel's proof, suitably worded, is correct. Ruffini also had a sound method (see Ayoub, *1980*), but needed to make certain subtle distinctions that were noticed only after the problem was better understood. By the end of the nineteenth century, Klein (see *1884*) referred to "the proofs of *Ruffini* and *Abel*, by which it is established that a solution of the general equation of the fifth degree by extracting a finite number of roots is impossible."

Besides his impossibility proof, Abel made positive contributions to the solution of equations. He generalized the work of Gauss on the cyclotomic (circle-splitting) equation $x^n + x^{n-1} + \cdots + x + 1 = 0$, which had led Gauss to the construction of the regular 17-sided polygon. Abel showed that if every root of an equation could be generated by applying a given rational function successively to a single (primitive) root, the equation could be solved by radicals. Any two permutations that leave this function invariant necessarily commute with each other. As a result, nowadays any group whose elements commute is called an *Abelian* group.

1.9. Galois. More light was shed on the solution of equations by the work of Abel's contemporary Évariste Galois (1811–1832), a volatile young man who did not live to become even mature. As is well known, he died at the age of 20 in a duel fought with one of his fellow Republicans.[9]

The neatly systematized concepts of group, ring, and field that now make modern algebra the beautiful subject that it is grew out of the work of Abel and Galois, but neither of these two short-lived geniuses had a full picture of any of them. The absence of the notion of a field seems to be the most noticeable lacuna in the theorems they were proving. Where we now talk easily about *algebraic and transcendental field extensions* and regard the general equation of degree n over a field F as $x^n + a_1x^{n-1} + \cdots + a_{n-1}x + a_n = 0$, where a_j is transcendental over F, Galois had to explain that the concept of a rational function was relative to what was given. For an equation with numerical coefficients, a rational function was simply a quotient of two polynomials with numerical coefficients, while if the equation had letters as coefficients, a rational function meant a quotient of two polynomials whose coefficients were rational functions of the coefficients of the equation. Even the concept of a group, which is associated with Galois, is not stated formally in any of his work. He does use the word *group* frequently in referring to a set of permutations of the roots of an equation, and he uses the properties that

---

[9] The word *Republican* (*republicain*) is being used in its French sense, of course, not the American sense. It is approximately the opposite of *royaliste*. There are murky details about the duel, but it appears that the gun Galois used was not loaded, probably because he did not wish to kill a comrade-in-arms. It is also possible that the combatants had jointly decided to let fate determine the outcome and each picked up a weapon not knowing which of the two guns was loaded. The cause of the duel is also not entirely clear. The notes that Galois left behind seem to imply that he felt it necessary to warn his friends about what he considered to be the wiles of a certain young woman by whom he felt betrayed, and they felt obliged to defend her honor against his remarks.

we associate with a group: the composition of permutations. However, it is clear from his language that what makes a set of permutations a group is that *all of them have the same effect on certain rational functions of the roots*. In particular, when what we now call a group is decomposed into cosets over a subgroup, Galois refers to the cosets as groups, since any two elements of a given coset have the same effect on the rational functions. He says that a group, in this sense, may begin with any permutation at all, since there is no need to specify any natural initial order of the roots.

Besides the shortness of their lives, Abel and Galois had another thing in common: neglect of their achievements by the Paris Academy of Sciences. We shall see some details of Abel's case in Chapter 17. As for Galois, he had been expelled from the École Normale because of his Republican activities and had been in prison. He left a second paper on the subject among his effects, which was finally published in 1846.[10] It had been written in January 1831, 17 months before his death, and it contained the following plaintive preface:

> The attached paper is excerpted from a work that I had the honor to present to the Academy a year ago. Since this work was not understood, and doubt was cast on the propositions that it contains, I have had to settle for giving the general principles and *only one* application of my theorie in systematic order. I beg the referees at least to read these few pages with attention. [Picard, *1897*, p. 33]

The language and notation used by Galois are very close to those of Lagrange. He considers an equation of degree n and claims that there exists a function (polynomial) $\varphi(a, b, c, d, \ldots)$ that takes on $n!$ different values when the roots are permuted. Such a polynomial, he says, can be $\varphi(a, b, c, d, \ldots) = Aa + Bb + Cc + Dd + \cdots$, where A, B, C, D, and so on, are positive integers. He then fixes one root a and forms a function of two variables

$$f(V, a) = \prod \left(V - \varphi(a, b, c, d, \ldots)\right),$$

(the Galois resolvent), in which the product extends over all permutations that leave a fixed. Since the function on the right is symmetric in b, c, d,..., all of these variables can be replaced by suitable combinations of a and the coefficients $p_1, \ldots, p_n$ (see Problem 15.4). The equation $f\bigl(\varphi(a, b, c, d, \ldots), x\bigr) = 0$ then has the solution $x = a$, but has no other roots in common with the equation $p(x) = 0$. Finding the greatest common divisor of these two polynomials then makes it possible to express a as a rational function of $\varphi(a, b, c, d, \ldots)$. Galois cited one of Abel's memoirs (on elliptic functions) as having stated this theorem without proof.

The main theorem of the memoir was the following: *For any equation, there is a group of permutations of the roots such that every function of the roots that is invariant under the group can be expressed rationally in terms of the coefficients of the equation, and conversely, every such function is invariant under the group.* We would nowadays say that the elements of this group generate automorphisms of the splitting field of the equation that leaves the field of coefficients invariant. As his formulation shows, Galois had only the skeleton of that result. He called the group of permutations in question the *group of the equation*. His groups are

[10] Abel's great work on integrals of algebraic functions, submitted in 1827, was finally published, at the insistence of Jacobi, in 1841.

all concrete objects—permutations of the roots of equations. He developed Galois theory to the extent of analyzing what happened to the group of an equation when, in modern terms, a new element is adjoined to the base field. Galois could not be so clear. He said, "When we agree thus to regard certain quantities as known, we shall say that we are *adjoining* them to the equation being solved and that these quantities are *adjoined* to the equation." He thought of the new element as a root of an auxiliary polynomial (the minimal polynomial of the new element, in our terms), since that is where he got the elements that he adjoined. Instead of saying that the original group might be decomposed into the cosets of the group of the new equation when all the roots of the auxiliary equation are adjoined, he said it might split into p groups, each belonging to the equation. He noted that "these groups have the remarkable property that one can pass from one to the other by operating on all the permutations with the same letter substitution."

In a letter to a friend written the night before the duel in which he died, Galois showed that he had gone still further into this subject, making the distinction between proper and improper decompositions of the group of an equation, that is, the distinction we now make between normal and nonnormal subgroups.

Galois theory. The ideas of Galois and his predecessors were developed further by Laurent Wantzel (1814–1848) and Enrico Betti. In 1837 Wantzel used Galois' ideas to prove that it is impossible to double the cube or trisect the angle using ruler and compass; in 1845 he proved that it is impossible to solve all equations in radicals. In 1852 Betti published a series of theorems elucidating the theory of solvability by radicals. In this way, group theory proved to be the key not only to the solvability of equations but to the full understanding of classical problems. When Ferdinand Lindemann (1852–1939) proved in 1881 that π is a transcendental number, it followed that no ruler-and-compass quadrature of the circle was possible.

The proof that the general quintic equation of degree 5 was not solvable by radicals naturally raised two questions: (1) How *can* the general quintic equation of degree 5 be solved? (2) Which particular quintic equations *can* be solved by radicals? These questions required some time to answer.

Solution of the general quintic by elliptic integrals. A partial answer to the first question came from the young mathematician Ferdinand Eisenstein (1823–1852), who showed in 1844 that the general quintic equation could be solved in terms of a function $\chi(\lambda)$ that satisfies the special quintic equation

$$\big(\chi(\lambda)\big)^5 + \chi(\lambda) = \lambda\,,$$

This function is in a sense an analog of root extraction, since the square root function φ and the cube root function ψ satisfy the equations

$$\begin{aligned}\big(\varphi(\lambda)\big)^2 &= \lambda\,,\\ \big(\psi(\lambda)\big)^3 &= \lambda\,.\end{aligned}$$

Eisenstein's solution stands somewhat apart from the main line of development, but in modern times it begins to look more reasonable. To solve all quadratic equations in a field of characteristic 2, for example, it is necessary to assume, in addition to the possibility of extracting a square root, that one has solutions to the equation $x^2 + x + 1 = 0$; these roots must be created by fiat. For a full discussion of Eisenstein's paper, see the article of Patterson (*1990*).

As elliptic integrals—integrals containing the square root of a cubic or quartic polynomial—became better understood, both computational and theoretical considerations brought about a focus on transformations of one elliptic integral to another. In 1828 Jacobi studied rational changes of variable $y = U(x)/V(x)$, where U and V are polynomials of degree at most n, and found an algebraic equation that U and V must satisfy in order for this transformation to convert an elliptic integral containing one parameter (modulus) into another.

The transformation

$$\int \frac{dx}{\sqrt{(1-x^2)(1-\kappa^2x^2)}} \mapsto \int \frac{dy}{\sqrt{(1-y^2)(1-\lambda^2y^2)}}$$

corresponds to an equation

$$u^6 - v^6 + 5u^2v^2(u^2 - v^2) + 4uv(1 - u^4v^4) = 0\,,$$

where $u = \sqrt[4]{\kappa}$ and $v = \sqrt[4]{\lambda}$ (see Klein, *1884*, Part II, Chapter 1, Section 3). Galois had recognized this connection and noted that the general modular equation of degree 6 could be reduced to an equation of degree 5 of which it was a resolvent. The parameter u can be expressed as a quotient of two infinite series (theta functions) in the number $q = e^{-\pi}K'/K$, where K and K' are the complete elliptic integrals of first kind with moduli κ and $\sqrt{1-\kappa^2}$. Thus, a family of equations of degree 6 containing a parameter could be solved using the elliptic modular function. It was finally Charles Hermite (1822–1901) who, in 1858, made all these facts fit together in a solution of the general quintic equation using elliptic functions.

Solution of particular quintics by radicals. The study of particular quintics that are solvable by radicals has occupied considerably more time. It is not difficult to reduce the general problem to the study of equations of the form $x^5 + px + q = 0$ via a Tschirnhaus transformation. This topic was studied by Carl Runge (1856–1927) in an 1886 paper in the *Acta mathematica*. There are only five groups of permutations of five letters that leave no letter fixed and hence could be the group of an irreducible quintic equation. They contain respectively 5, 10, 20, 60, and 120 permutations. A quintic equation having one of the first three as its group will be solvable by radicals, whereas an equation having either of the other two groups will not be. The actual construction of the solution, however, is by no means trivial. The situation is similar to that involved in the construction of regular polygons with ruler and compass. Thanks to Galois theory, we now know that it is possible for a person with sufficient patience to construct a 17-sided regular polygon—that is, partition a circle into 17 equal arcs—using ruler and compass, and Gauss actually did so.[11] The details of the construction, however, are quite complicated. The same theory assures us that it is similarly possible to divide the circle into 65,537 congruent arcs, a task attempted by Johann Hermes (see p. 189). In contrast, algorithms have been produced for solving quintics by radicals where it is possible to do so.[12] An early summary of results in this direction was the famous book by Felix Klein on the icosahedron (*1884*). An up-to-date study of the theory of solvability of equations of all degrees, with historical documentation, is the book of R. Bruce King (*1996*).

[11] Abel, using elliptic functions, partitioned the lemniscate into 17 arcs of equal length.

[12] See the paper by D.S. Dummit "Solving solvable quintics," in *Mathematics of Computation*, **57** (1991), No. 195, 387–401.

2. Algebraic structures

The concept of a group was the first of the many abstractions that make up the world of modern algebra. We have seen how it arises through the study of the permutations of the roots of an equation. In the work of Lagrange, Ruffini, Cauchy, and Abel, only the number of different forms that a function of the roots could take was studied. Then Galois focused attention on the structure of the permutations themselves, and the result was the first abstract structure, a permutation group. Another two decades passed before the idea of a group was made abstract by Arthur Cayley (1821–1895) in 1849. Cayley defined a group as a set of symbols that could be combined in a way that was associative (he used the word) but not necessarily commutative, and such that the elements must repeat themselves if all are operated on by the same element. (From Cayley's language it is not clear whether he intended this last property as an axiom or believed that it followed from the other properties of a group.) An important example given by Cayley was a group of matrices.[13] The complete set of axioms for an abstract group was stated by Walther von Dyck (1856–1934), a student of Felix Klein, in 1883.

2.1. Fields, rings, and algebras. The concept of a group arose in the study of the procedures used to solve equations, but that study involved other concepts that were also somewhat vague and in need of clarification. What exactly did Galois mean when he said "if we agree to regard certain objects as known" and spoke of adjoining roots to an equation? Rational *expressions* in variables representing unspecified numbers had long been part of the discourse in the solution of algebraic equations. Both Abel and Galois made frequent use of this concept. Over the course of the nineteenth century this domain of rationality evolved into what Dedekind in 1858 called a *Zahlkörper* (*number body*). Leopold Kronecker (1823–1891) preferred the term *Rationalitäts-Bereich* (*domain of rationality*). The abstract object that grew out of this concept eventually came to be known in French as a *corps*, in German as a *Körper*, and in English as a *field*.[14] Dedekind considered only fields built on top of the rational numbers. Finite fields were first introduced, along with the word *field* itself, by E. H. Moore (1862–1932) in a paper published in the *Bulletin of the New York Mathematical Society* in 1893.

Other algebraic concepts arise as generalizations of number systems. In particular, the integers, the complex numbers of the form $m+ni$ (the Gaussian integers), and the integers modulo a fixed integer m led to the general concept that Hilbert, in his exhaustive 1897 report to the German Mathematical Union "Die Theorie des allgemeinen Zahlkörpers" ("The theory of the general number field") , called a *Zahlring* (*Number ring*). He gave as an equivalent term *Integritätsbereich* (*integral domain*). Both, however, were names for sets of complex numbers. Nowadays, an integral domain is defined abstractly as a commutative ring with identity in which the product of nonzero elements is nonzero; that is, there do not exist *zero divisors*. An element that is not a zero divisor is said to be *regular*. These structures were consciously abstracted and developed into the concept of an abstract *ring* in the paper "Über die Teiler der Null und die Zerlegung von Ringen" ("On zero divisors and the decomposition of rings") by Adolf Fraenkel (1891–1965), which appeared

---

[13] The word *matrix* was Cayley's invention; the word is Latin for *womb* and is used figuratively in mining to denote an ore-bearing rock. Cayley's "wombs" bore numbers rather than ore or babies.

[14] A *corps*, however, is not necessarily commutative. Strictly speaking, it corresponds to what is called in English a *division ring*.

in the *Journal für die reine und angewandte Mathematik* in 1914. In his introduction Fraenkel cited the large number of particular examples as a reason for defining the abstract object. He required that the ring have at least one right identity, an element ε such that $a\varepsilon = a$ for all a, and that for at least one of the right identities every regular element should have an inverse. The English term was introduced by Eric Temple Bell (pseudonym of John Taine, 1883–1960) in a paper in the *Bulletin of the American Mathematical Society* in 1930.

Although mathematical communication was very extensive from the nineteenth century on, there was still enough difficulty due to language and transportation that British and Continental mathematicians sometimes took very different approaches to the same subject. Such was the case in algebra, where the solution of equations and abstract number theory led Continental mathematicians in one direction, at the time when British mathematicians were pursuing an abstract approach to algebra having connections with an outstanding British school of symbolic logic. For linguistic and academic reasons, the British approach also caught on in the United States, the first American foray into mathematical research. This Anglo-American algebra has been studied by Parshall (*1985*).

One of the first examples of this British algebra was the algebra of quaternions, invented by William Rowan Hamilton in 1843. Hamilton had been intrigued by the complex numbers since his teenage years. He questioned the meaningfulness of writing, for example $3 + \sqrt{-5}$, since this notation made it appear that two objects of different kinds—real and imaginary numbers—were being added. To rationalize this process, he took the step that seems obvious now, regarding the two numbers as ordered pairs, so that 3 is merely an abbreviation for $(3, 0)$ and $\sqrt{-5}$ an abbreviation for $(0, \sqrt{5})$, thereby algebraizing the plane. Hamilton was very much a physicist, and he saw complex multiplication, when the numbers were put in polar form $r(\cos\theta + i\sin\theta)$, as representing rotations and dilations. For him, complex addition represented all the possible translations of the plane, and complex multiplication sufficed to describe all its rotations and dilations.

Influenced by the mysticism of the poet Coleridge (1772–1834),[15] whom he knew personally, he felt that great insight would be obtained if he could similarly algebraize three-dimensional space, that is, find a way to multiply triples of numbers (x, y, z) similar to the complex multiplication of pairs (x, y). In particular, he wished to find algebraic operations corresponding to all translations and rotations of three-dimensional space. Translations were not a problem, since ordinary addition took care of them. After much reflection, during a walk in Dublin on October 10, 1843 that has become one of the most famous events in the history of mathematics, he realized that he needed a fourth quantity, since if he used one coordinate to provide a unit x, having the property that $xy = y$ and $xz = z$, the product yz would have to be expressible symmetrically in x, y, and z. When the formulas we now write as $i^2 = j^2 = k^2 = -1$, $ij = -ji = k$, $ki = -ik = j$, $jk = -kj = i$ occurred to him during this walk, he scratched them in the stone on Brougham Bridge.[16] In his 1845 paper in the *Quarterly Journal* he referred to $ix + jy + kz$ as "the vector from 0 to the point x, y, z." The word *vector* (Latin for *carrier*) occurs in this context for the first time. A quaternion thus consists of a number and a vector. Very

[15] Coleridge's most famous poem, *The Rime of the Ancient Mariner*, is full of mystical uses of the numbers 3, 7, and 9.

[16] In the 160 years since then, they have been effaced.

likely it was his physical intuition that led him to make this discovery. A rotation in two-dimensional space requires only one parameter for its determination: the angle of rotation. In three-dimensional space it is necessary to specify the axis of rotation by a point on the unit sphere, and then the angle of rotation, a total of three parameters. Dilations then require a fourth parameter. Thus, although quaternions were invented to describe transformations of three-dimensional space, they require four parameters to do so.

Hamilton, an excellent physicist and astronomer, worried about simply making up symbols out of his head and manipulating them. He soon found applications of them, however; and a school of his followers grew up, dedicated to spreading the lore of quaternions. By throwing away one of Hamilton's dimensions (the one that contained the unit) and using only the three symbols i, j, and k, the American mathematician Josiah Willard Gibbs (1839–1903) developed the vector calculus as we know it today, in essentially the same language that is used now. In the language of vectors quaternions can be explained easily. A quaternion is simply the formal sum of a number and a point in three-dimensional space, such as $A = a + \alpha$ or $B = b + \beta$. As Hamilton had done with complex numbers, it is possible to rationalize this seeming absurdity by regarding the number, the point, and the quaternion itself as quadruples of numbers: $a = (a, 0, 0, 0)$, $\alpha = (0, a_1, a_2, a_3)$, $A = (a, a_1, a_2, a_3)$. The familiar cross product developed by Gibbs is obtained by regarding two vectors as quaternions, multiplying them, then setting the numerical part equal to zero (projecting from four-dimensional space to three-dimensional space). Conversely, quaternion multiplication can be defined in terms of the vector (dot and cross) products: $AB = (ab - \alpha \cdot \beta) + (a\beta + b\alpha + \alpha \times \beta)$. The quaternion $\bar{A} = a - \alpha$ is the conjugate, analogous to the complex conjugate, and has the analogous property $A\bar{A} = a^2 + \alpha \cdot \alpha = a^2 + |\alpha|^2 = a^2 + a_1^2 + a_2^2 + a_3^2$. Thus $A\bar{A}$ represents the square of Euclidean distance from A to $(0, 0, 0, 0)$ and can be denoted $|A|^2$. This equation in turn shows how to divide quaternions, multiplying by the reciprocal: $1/A = (1/|A|^2)\bar{A}$. The absolute value of a quaternion has the pleasant property that $|AB| = |A|\,|B|$.

The Harvard professor Benjamin Peirce (1808–1880) became an enthusiast of quaternions and was already lecturing on them in 1848, only a few years after their invention. Like many mathematicians before and after, he was philosophically attracted to algebra and believed it encapsulated pure thought in a way that was unique to itself. His treatise *Linear Associative Algebra* was one of the earliest treatises in this surprisingly late-arising subject.[17]

On the Continent algebra developed from other roots, more geometric in nature, exemplified by Grassmann's *Ausdehnungslehre*, which was described in Section 3 of Chapter 12. To some Continental mathematicians, what the British were doing did not seem sufficiently substantial. On New Year's Day 1875, Weierstrass wrote to his pupil Sof'ya Kovalevskaya that she had much more important things to learn than Hamilton's quaternions, whose algebraic foundations, he said, were of a very trivial nature. In his discussion of quaternions Klein (*1926*, p. 182) remarked, "It is hardly necessary to mention that the Grassmannians and the quaternionists were bitter rivals, while each of the two schools in turn split into fiercely warring subspecies." Weierstrass himself, in 1884, gave a discussion of an algebra, including

[17] I say "surprisingly" because, as anyone would agree, its basic subject matter—linear equations—is much simpler than many parts of algebra that developed earlier.

the structure constants that constitute the multiplication table for the elements of a basis. Even so, the subject seems to have caught on in only a few places in Germany. At Göttingen Emmmy Noether revolutionized the subject of algebras and representations of finite groups, and the concept of a Noetherian ring is now one of the basic parts of ring theory. Yet Salomon Bochner (1899–1982), who was educated in Germany and spent the first 15 years of his professional career there before coming to Princeton, recalled that the concept of an algebra was completely new to him when he first heard a young American woman lecture on it at Oxford in 1925.[18]

> She came from Chicago and she gave at this seminar a lecture on algebras, which left all of us totally uncomprehending what it was all about. She spoke in a well-articulated, self-confident manner, but none of us had remotely heard before the terms she used, and we were lost. [Bochner, *1974*, p. 832]

Bochner went on to say that a decade later he was taken aback to find a German book with the title *Algebren* (*Algebras*—it was use of the plural that Bochner found jarring). Bochner was an extremely creative and productive analyst and differential geometer. Not until the mid-1960s did he have time to ferret out Peirce's book and sit down to read it.

2.2. Abstract groups. The general theory of groups of permutations was developed in great detail in an 1869 treatise of Camille Jordan (1838–1922). This work made the importance of groups widely known. But despite Cayley's 1849 paper in group theory, the word *group* was still being used in an imprecise sense as late as 1871, in the work of two of the founders of group theory, Felix Klein and Sophus Lie (see Hawkins, *1989*, p. 286). All groups were pictured concretely, as one-to-one mappings of sets. That assumption made a cancellation law $ab = ac \Longrightarrow b = c$ valid automatically. For permutations of finite sets, the cancellation law implied that every element had an inverse. The corresponding inference for mappings of infinite sets is not valid, but Klein and Lie did not notice the difference at first. Lie even thought this inference could be proved. Groups as an abstract concept, characterized by the three, four, or five axioms one finds in modern textbooks, did not arrive until the twentieth century. On the abstract concept of a group Klein (*1926*, pp. 335–336) commented:

> This abstract formulation is helpful in the construction of proofs, but not at all adapted to the discovery of new ideas and methods; on the contrary, it rather represents the culmination of an earlier development.

Klein was quick to recognize the potential of groups of transformations as a useful tool in the study of many areas of geometry and analysis. The double periodicity of elliptic functions, for example, meant that these functions were invariant

[18] The woman's name was Echo Dolores Pepper, and the Mathematics Genealogy website lists her as having received the Ph. D. at the University of Chicago in 1925. Her dissertation, *Theory of Algebras over a Quasi-field* is on record, and she published at least one paper the year after receiving the degree, "Asymptotic expression for the probability of trials connected in a chain," *Annals of Mathematics*, **2** **(28)** (1926–27), 318–326. I have been unable to find out any more about her.

under an infinite group of translations of the plane. Klein introduced the concept of an *automorphic function*, an analytic function $f(z)$ that is invariant under a group of fractional-linear (Möbius) transformations

$$z \mapsto \frac{az+b}{cz+d}, \quad ad - bc \neq 0 .$$

Both Klein and Lie made use of groups to unify many aspects of projective geometry, and Klein suggested that various kinds of geometry could be classified in relation to the groups of transformations that leave their basic objects invariant.

Lie groups. One of the most fundamental and far-reaching applications of the group concept is due to Lie, Klein's companion from 1869, when both young geometers felt like outsiders in the intensely analytic and algebraic world of Berlin mathematics. In studying surfaces in three-dimensional space, Lie and Klein naturally encountered the problem of solving the differential equations that lead to such surfaces. Lie had the idea of solving these equations using continuous transformations that leave the differential equation invariant, in analogy with what Galois had done for algebraic equations. Klein noticed an analogy between this early work of Lie and Abel's work on the solution of equations and wrote to Lie about it. Lie was very pleased at Klein's suggestion. He believed as a matter of faith in the basic validity of this analogy and developed it into the theory of Lie groups. Lie himself did not present a whole Lie group, only a portion of it near its identity element. He considered a set of one-to-one transformations indexed by n-tuples of sufficiently small real numbers in a neighborhood of $(0, 0, \ldots, 0)$ in such a way that the composition of the transformations corresponded to addition of the points that indexed them. This subject was developed by Lie, Wilhelm Killing (1847–1923), Élie Cartan (1869–1951), Hermann Weyl (1885–1955), Claude Chevalley (1909–1984), Harish-Chandra (1923–1983), and others into one of the most imposing edifices of modern mathematics. A *Lie group* is a manifold in Riemann's sense that also happens to be a group, in which the group operations (multiplication and inversion) are analytic functions of the coordinates.

Lie's work is far too complicated to summarize, but we can explain his basic ideas with a simple example. The sphere S^3 in four-dimensional space can be regarded as the set of quaternions of unit norm, that is, $A = a + \alpha$ such that $|A|^2 = a^2 + |\alpha|^2 = 1$. Because $|AB| = |A|\,|B|$, this set is closed under quaternion multiplication and inverses. But this sphere is also a three-dimensional manifold and can be parameterized by, say, the stereographic projection from $(-1, 0, 0, 0)$ through the equatorial hyperplane consisting of points $(0, x, y, z)$. This projection maps $(0, x, y, z)$ to

$$\Bigl(\frac{1 - x^2 - y^2 - z^2}{1 + x^2 + y^2 + z^2}, \frac{2x}{1 + x^2 + y^2 + z^2}, \frac{2y}{1 + x^2 + y^2 + z^2}, \frac{2z}{1 + x^2 + y^2 + z^2}\Bigr).$$

This parameterization covers the entire group except for the point $(-1, 0, 0, 0)$. To parameterize a portion of the sphere containing this point requires a second parametrization, which can be projection from the opposite pole $(1, 0, 0, 0)$. When the points in the group with coordinates (u, v, w) and (x, y, z) are multiplied, the result is the point whose first coordinate is

$$\frac{-\bigl(u^2x + (-1 + v^2 + w^2)x + 2wy - 2vz + u(-1 + x^2 + y^2 + z^2)\bigr)}{1 - 2(ux + vy + wz) + (u^2 + v^2 + w^2)(x^2 + y^2 + z^2)}.$$

Since we have no need to do any computations, we omit the other two coordinates. The point to be noticed is that this function is *differentiable*, so that the group operation, when interpreted in terms of the parameters, is a differentiable operation.

To study a Lie group, one passes to the tangent spaces it has as a manifold: in particular, the tangent space at the group identity. This tangent space is determined by the directions in the parameter space around the point corresponding to the identity. Each direction gives a directional derivative that operates on differentiable functions. For that reason, the tangent space is defined to be the set of differential operators of the form $X = \sum a_j(x)\frac{\partial}{\partial x_j}$. The composition of two such operators involves second derivatives, so that XY is not in general an element of the tangent space. However, the second partial derivatives cancel in the expression $XY - YX$ (the Lie bracket). This multiplication operation makes the tangent space into a *Lie algebra*. This algebra, being a finite-dimensional vector space, is determined as an algebra once the multiplication table for the elements of a basis is given. To take the simplest nontrivial example, the Lie group of rotations of three-dimensional space (represented as 3×3 rotation matrices) has the vector algebra developed by Gibbs (with the cross product as multiplication) as its Lie algebra.

Elements of the group can be generated from the Lie algebra by applying a mapping called the exponential mapping from the Lie algebra into the Lie group. Finding this mapping amounts to solving a differential equation. The resulting combination of algebra, geometry, and analysis is both profound and beautiful.

It would have been pleasant for mathematics in general and for Lie in particular if this beauty and profundity had been recognized immediately. Unfortunately, Lie's work was not well understood at first. In January 1884 he wrote to his friend, the German mathematician Adolf Mayer (1839–1908):

> I am so certain, so absolutely certain, that these theories will be acknowledged as fundamental one day in the future. If I wish to procure such an opinion *soon*, it is... because I could produce ten times more. [Engel, *1899*, quoted by Parshall, *1985*, p. 265]

Lie's vision was soon vindicated. By the end of his life, the potential of the theory was being recognized, and its development has never slowed in the century that has elapsed since that time.

Group representations. The road to abstraction is a two-way street. Once an axiomatic characterization of an object is stated, a classification program starts automatically, aimed at answering two important questions: First, how abstract *is* the abstract object, really? Second, how many abstractly different objects fit the axioms?

The first question leads to the search for concrete representations of abstract groups given only by a multiplication table. We know, for example, that every group G can be thought of as a group of one-to-one mappings by associating with each $a \in G$ the mapping $L_a : G \to G$ given by $z \mapsto az$. This fact was noted by Cayley when he introduced the abstract concept. It is also easy to show that any finite group can be represented in a trivial way as a group of invertible matrices whose entries are all zeros and ones. First regard the group as a group of permutations of a set of k objects. Then make the k objects into the basis of a vector space and associate with each permutation the matrix of the linear transformation it defines. Essentially, this representation was introduced by the American logician–philosopher Charles Sanders Peirce in 1879.

An early prefiguration of an important concept in the representation of groups occurred in an 1837 paper of Dirichlet proving that an arithmetic progression whose first term and difference are relatively prime contains infinitely many primes. As discussed in Section 1 of Chapter 8, that paper contains the *Dirichlet character* $\chi(n)$, defined as $(-1)^{(n-1)/2}$ if n is odd and 0 if n is even. This character has the property that $\chi(mn) = \chi(m)\chi(n)$. The definition of a character as a homomorphism into the multiplicative group of nonzero complex numbers was given by Dedekind in an 1879 supplement to Dirichlet's lectures. About the same time, Sylvester was showing how matrices could be used to represent the quaternions.[19]

The theory of representations of finite groups was developed by the German mathematician Ferdinand Georg Frobenius (1849–1917), responding to a question posed by Dedekind. The original question was simply to factor the determinant of a certain matrix associated with a finite group. In trying to solve this problem, Frobenius introduced the idea of a representation and a character of a representation. Although the subject is too technical for details to be given here, the characters, being computable, reveal certain facts about the structure of a finite group, and in some cases determine it completely.

This theory was extended to Lie groups in 1927 by Hermann Weyl and his student F. Peter. In that context, it turned out, representation theory subsumes and unifies the theories of Fourier series and Fourier integrals, both of which are ways of analyzing functions defined on a Lie group (the circle or line) by transferring them to functions defined on a separate (dual) group. The subject is now called *abstract harmonic analysis.*

Finite groups. The subject of finite groups grew up in connection with the solution of equations, as we have already seen. For that purpose, one of the most important questions was to decide which groups corresponded to equations that are solvable. Such a group G has a chain of normal subgroups $G \supset G_1 \supset G_2 \supset \cdots \supset \{1\}$ in which each factor group G_i/G_{i+1} is commutative. Because of the connection with equations, a group having such a chain of subgroups is said to be *solvable.* A solvable group can be built up from the simplest type of group, the group of integers modulo a prime, and so its structure may be regarded as known. It would be desirable to have a classification that can be used to break down any finite group into its simplest elements in a similar way. The general problem is so difficult that it is nowhere near solution. However, a significant piece of the program has been achieved: the classification of simple groups. A simple group is one whose only normal subgroups are itself and $\{1\}$.

The project of classifying these groups was referred to by one of its leaders, Daniel Gorenstein (1923–1992), as the Thirty Years' War, since a strategy for the classification was suggested by Richard Brauer (1901–1977) at the International Congress of Mathematicians in 1954 and the classification was completed in the

[19] Cayley, Peirce, and Sylvester were well acquainted personally and professionally with one another. Cayley and Peirce were at Johns Hopkins University during part of the time that Sylvester was chair of mathematics there. They formed the strong core of the Anglo-American school of abstract algebra described by Parshall (*1985*). At the heart of this abstraction, at least in the case of Peirce, was a philosophical program of creating a universal symbolic algebra that could be applied in any situation. Such a program required that the symbols be mere symbols until applied to some specific situation.

1980s with the discovery of the last "sporadic" group. The project consists of so many complicated parts that the process of streamlining and clarifying it, with all the new projects that process will no doubt spawn, is likely to continue for many decades to come.

An important part of the project was the 1963 proof by Walter Feit and John Thompson (b. 1932) that all finite simple groups have an even number of elements. The proof was 250 pages long and occupied an entire issue of the *Pacific Journal of Mathematics*. As it turned out, this project was destined to generate large numbers of long papers. In fact, the mathematician who contributed the last step in the classification later wrote, "At least 3,000 pages of mathematically dense preprints appeared in the years 1976–1980 and simply overwhelmed the digestive system of the group theory community" (Solomon, *1995*, p. 236). The outcome of the project is intensely satisfying from an aesthetic point of view. It turns out that there are three (or four) infinite classes of finite simple groups: (1) groups of prime order; (2) the group of even permutations on n letters ($n \geq 5$);[20] (3) certain finite linear groups, a class that can be subdivided into classical matrix groups and twisted groups of Lie type, whose exact definition is not important for present purposes. Outside those classes are the "sporadic" groups.

If this classification seems to resemble the old classification of constructions as planar, solid, and curvilinear, in which the final class is merely a catchall term for anything that doesn't fit into the other classes, that impression is misleading. The class of sporadic groups turns out to contain precisely 26 groups, whose properties have been tabulated. The smallest of them is M_{11} with 7920 elements, one of five sporadic simple groups discovered by Émil Mathieu (1835–1890). The largest, officially denoted F_1, is informally known as the Monster, since it consists of

$$2^{46} \cdot 3^{20} \cdot 5^9 \cdot 7^6 \cdot 11^2 \cdot 13^3 \cdot 17 \cdot 19 \cdot 23 \cdot 29 \cdot 31 \cdot 41 \cdot 47 \cdot 59 \cdot 71$$

elements. It was constructed in 1980.

2.3. Number systems. Rings and fields can be regarded as generalized number systems, since they admit addition, subtraction, and multiplication, and sometimes division as well. As noted above, such general systems have been used since Gauss began the study of arithmetic modulo an integer and proved that the factorization of a Gaussian integer $m + ni$ into irreducible Gaussian integers is unique up to a power of i. Gauss was particularly interested in these numbers, since when they are introduced, 2 is not a prime number ($2 = (1 + i)(1 - i)$), nor is any number of the form $4n + 1$ prime. For example, $5 = (2 + i)(2 - i)$. The number 3, however, remains prime. If we pass to numbers of the form $m + n\sqrt{-2}$, factorization is still unique, but this uniqueness is lost for numbers of the form $m + n\sqrt{-3}$, since $4 = 2 \cdot 2 = (1 + \sqrt{-3})(1 - \sqrt{-3})$. The Gaussian integers were the first of an increasingly abstract class of structures on which multiplication is defined and obeys a cancellation law (that is $ac = bc$ and $c \neq 0$ implies that $a = b$), but division is not necessarily always possible. If the factorization of each element is essentially unique, such a structure is called a *Gaussian domain.*

[20] The fact that this group is simple and noncommutative implies that the symmetric group consisting of all permutations on n letters cannot be solvable for $n \geq 5$, and hence that the general equation of degree n is not solvable by radicals.

Ideal complex numbers. Numbers of the form $m + n\omega$, where $\omega = -1/2 + \sqrt{-3}/2$, a primitive cube root of unity, satisfies the equation $\omega^2 = -1 - \omega$, have properties similar to the Gaussian integers. For this system the number 3 is not a prime, since $3 = (2+\omega)(1-\omega)$. Numbers of the form $m+n\omega$ do have a unique factorization into primes, but Ernst Eduard Kummer (1810–1893) discovered in 1844 that "complex integers" of the form $m+n_1\alpha+n_2\alpha^2+\cdots+n_{\lambda-2}\alpha^{\lambda-2}$, where $1+\alpha+\cdots+\alpha^{\lambda-1} = 0$ and λ is prime, do not necessarily have the property of unique factorization.[21] This fact seems to have a connection with Fermat's last theorem, which can be stated by saying that the number

$$x^\lambda + y^\lambda = (x + y)(x + \alpha y)(x + \alpha^2 y)\cdots(x + \alpha^{\lambda-1} y)$$

is never equal to z^λ for any nonzero integers x, y, z. For that reason, Klein (*1926*, p. 321) asserted that it was precisely in this context that Kummer made the discovery. But Edwards (*1977*, pp. 79–81) argues convincingly that the discovery was connected with the search for higher reciprocity laws, analogs of the quadratic reciprocity discussed in Section 1 of Chapter 8.

However that may be, what Kummer made of the discovery is quite interesting. He introduced "ideal" numbers that would divide some of the otherwise irreducible numbers, just as the imaginary number $2 + i$ divides 5, and (he said) just as one introduces ideal chords to be held in common by two circles that do not intersect in the ordinary sense. By his definition, a prime number p that equals 1 modulo λ has $\lambda - 1$ factors.[22] These factors may be actual complex integers of the form stated. For example, when $\lambda = 3$, $p = 13$, we have $13 = (4+\omega)(3-\omega)$. If they were not actual complex numbers, he assigned an ideal factor of p to correspond to each root ξ of the congruence $\xi^\lambda \equiv 1 \bmod p$, thereby obtaining $\lambda - 1$ nonunit factors. His rationale was that if $f(\alpha)$ divided p in the ordinary sense, then $f(\xi)$ would be divisible by p in the ordinary sense. More generally, a complex integer $\Phi(\alpha)$ was to have the ideal factor corresponding to ξ if $\Phi(\xi) \equiv 0 \bmod p$. Then if $\Phi(\alpha)$ was divisible by p, it would be divisible by all of the factors of p, whether actual complex numbers or ideal complex numbers. When these ideal complex numbers were introduced, unique factorization was restored.

Questions and problems

15.1. Prove that if every polynomial with real coefficients has a zero in the complex numbers, then the same is true of every polynomial with complex coefficients. To get started, let $p(z) = z^n + a_1 z^{n-1} + \cdots + a_{n-1}z + a_n$ be a polynomial with complex coefficients $a_1, \ldots, a_n$. Consider the polynomial $q(z)$ of degree $2n$ given by $q(z) = p(z)\overline{p(\bar{z})}$, where the overline indicates complex conjugation. This polynomial has real coefficients, and so by hypothesis has a complex zero z_0.

15.2. Formulate Cauchy's 1812 result as the following theorem and prove it: *Let p be a prime number, $3 \le p \le n$. If a subgroup of the symmetric group on n letters contains all permutations of order p, it is either the entire symmetric group or the alternating group.*

---

[21] The first prime for which unique factorization fails is $p = 23$, just in case the reader was hoping to see an example.

[22] This relation between the primes p and λ speaks in favor of Edwards' argument that Kummer's goal had been a higher reciprocity law.

15.3. Cauchy's theorem that every cycle of order 3 can be written as the composition of two cycles of order m if $m > 3$ looks as if it ought to apply to cycles of order 2 also. What goes wrong when you try to prove this "theorem"?

15.4. Let $S_j(a, b, c, d)$ be the jth elementary symmetric polynomial, that is, the sum of all products of j distinct factors chosen from $\{a, b, c, d\}$. Prove that $S_j(a, b, c, d) = S_j(b, c, d) + aS_{j-1}(b, c, d)$. Derive as a corollary that given a polynomial equation $x^4 - S_1(a, b, c, d)x^3 + S_2(a, b, c, d)x^2 + S_3(a, b, c, d)x + S_r(a, b, c, d) = 0 = x^4 - p_1x^3 + p_2x^2 - p_3x + p_4$ having a, b, c, d as roots, each elementary symmetric function in b, c, d can be expressed in terms of a and the coefficients p_j: $S_1(b, c, d) = p_1 - a$, $S_2(b, c, d) = p_2 - aS_1(b, c, d) = p_2 - ap_1 + a^2$, $S_3(b, c, d) = p_3 - ap_2 + a^2p_1 - a^3$.

15.5. Prove that if z is a prime in the ring obtained by adjoining the pth roots of unity to the integers (where p is a prime), the equation

$$z^p = x^p + y^p$$

can hold only if $x = 0$ or $y = 0$.

15.6. Consider the complex numbers of the form $z = m + n\omega$, where $\omega = -1/2 + \sqrt{-3}/2$ is a cube root of unity. Show that $N(z) = m^2 - mn + n^2$ has the property $N(zw) = N(z)N(w)$ and that $N(z + w) \leq 2\big(N(z) + N(w)\big)$. Then show that a Euclidean algorithm exists for such complex numbers: Given z and $w \neq 0$, there exist q and r. Such that $z = qw + r$ where $N(r) < N(w)$. Thus, a Euclidean algorithm exists for these numbers, and so they must exhibit unique factorization. [*Hint*: $N(z) = |z|^2$. Show that for every complex number u there exists a number q of this form such that $|q - u| < 1$. Apply this fact with $u = z/w$ and *define* r to be $z - qw$.]

15.7. Show that in quaternions the equation $X^2 + r^2 = 0$, where r is a positive real number (scalar), is satisfied precisely by the quaternions $X = x + \xi$ such that $x = 0$, $|\xi| = r$, that is, by all the points on the sphere of radius r. In other words, in quaternions the square roots of negative numbers are simply the nonzero vectors in three-dimensional space. Thus, even though quaternions act "almost" like the complex numbers, the absence of a commutative law makes a great difference when polynomial algebra is considered. A linear equation can have only one solution, but a quadratic equation can have an uncountable infinity of solutions.

Part 6

Analysis

The great watershed in the history of mathematics is the invention of the calculus. It synthesized nearly all the algebra and geometry that had come before and generated problems that led to much of the mathematics studied today. Although calculus is an amalgam of algebra and geometry, it soon developed results that were indispensible in other areas of mathematics. Even theories whose origins seem to be independent of all forms of geometry—combinatorics, for example—turn out to involve concepts such as generating functions, for which the calculus is essential.

Elements of the calculus had existed from the earliest times in the form of infinitesimal methods in geometry, and such techniques were refined in the early seventeenth century. Although strict boundaries in history tend to be artificial constructions, there is such a boundary between analytic geometry and calculus. That boundary is the introduction of infinitesimal methods. As approximate, easy-to-remember formulas, we can write: algebra + geometry = analytic geometry, and analytic geometry + infinitesimals = calculus.

The introduction of infinitesimals into a geometry that had only recently struggled back to the level of rigor achieved by Archimedes raised alarms in certain quarters, but the new methods led to spectacular advances in theoretical physics and geometry that have continued to the present time. Like the Pythagoreans, modern mathematicians faced the twin challenges of extending the range of applicability of their mathematics while making it more rigorous. The responses to these challenges led to the modern subject of analysis. Starting from a base of real numbers, represented as ratios of line segments and written in the symbolic language of modern algebra, mathematicians extended their formulas to complex numbers, opening up a host of new applications and creating the beautiful subject of analytic function theory (complex analysis). At the same time, they were examining the hidden assumptions in their methods and making their limiting processes more rigorous by introducing the appropriate definitions of integrals, derivatives, and series, leading to the subject of functions of a real variable (real analysis).

Part 6 consists of two chapters. The creation of calculus and its immediate outgrowths, differential equations and the calculus of variations, is described in Chapter 16, while the further development of these subjects into modern analysis is the theme of Chapter 17.

CHAPTER 16

The Calculus

The infinite occurs in three forms in calculus: the derivative, the integral, and the power series. Integration, in the form of finding areas and volumes, was developed as a particular theory before the other two subjects came into general use. As we have seen, infinitesimal methods were used in geometry by the Chinese and Japanese, and the latter also used infinite series to solve geometric problems. In India also, mathematicians used infinite series to solve geometric problems via trigonometry. According to Rajagopal (*1993*), the mathematician Nilakanta, who lived in South India and whose dates are given as 1444–1543, gave a general proof of the formula for the sum of a geometric series. The most advanced of these results is attributed to Madhava (1340–1425), but is definitively stated in the work of Jyeshtadeva (1530–ca. 1608):

> The product of the given Sine and the radius divided by the Cosine is the first result. From the first,... etc., results obtain... a sequence of results by taking repeatedly the square of the Sine as the multiplier and the square of the Cosine as the divisor. Divide ... in order by the odd numbers one, three, etc... From the sum of the odd terms, subtract the sum of the even terms. [The result] becomes the arc. [Rajagopal, *1993*, p. 98]

These instructions give in words an algorithm that we would write as the following formula, remembering that the Sine and Cosine used in earlier times correspond to our $r \sin\theta$ and $r\cos\theta$, where r is the radius of the circle:

$$r\theta = \frac{r^2 \sin\theta}{r\cos\theta} - \frac{r^4\sin^3\theta}{3r^3\cos^3\theta} + \frac{r^6\sin^5\theta}{5r^5\cos^5\theta} - \cdots.$$

The bulk of calculus was developed in Europe during the seventeenth century, and it is on that development that the rest of this chapter is focused.

Since analytic geometry was discussed in Section 1 of Chapter 12, we take up the story at the point where infinitesimal methods begin to be used in finding tangents and areas. The crucial step is the realization of the mutually inverse nature of these two processes and their consolidation as a set of algebraic and limit operations that can be applied to any function. At the center of the entire process lies the very concept of a function, which was a seventeenth-century innovation.

1. Prelude to the calculus

In his comprehensive history of the calculus (*1949*), Boyer described "a century of anticipation" during which the application of algebra to geometric problems began to incorporate some of the less systematic parts of ancient geometry, especially the infinitesimal ideas contained in what was called the method of indivisibles. Let us

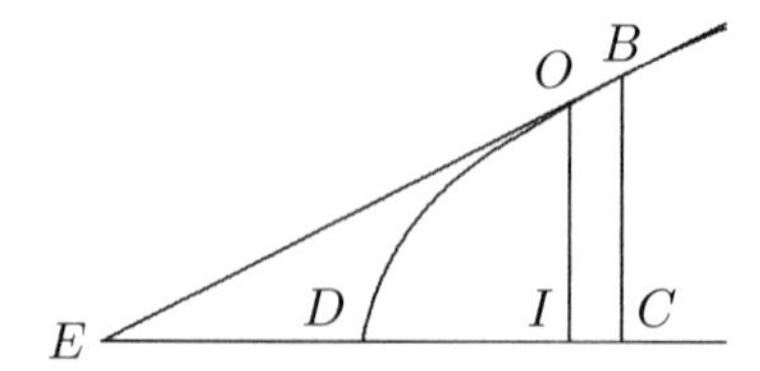

FIGURE 1. Fermat's method of finding the subtangent.

take up the story of calculus at the point where algebra enters the picture, beginning with some elementary problems of finding tangents and areas.

1.1. Tangent and maximum problems. The main problem in finding a tangent to a curve at a given point is to find some second condition, in addition to passing through the point, that this line must satisfy so as to determine it uniquely. It suffices to know either a second point that it must pass through or the angle that it must make with a given line. Fermat had attacked the problem of finding maxima and minima of variables even before the publication of Descartes' *Géométrie*. As his works were not published during his lifetime but only circulated among those who were in a rather select group of correspondents, his work in this area was not recognized for some time. His method is very close to what is still taught in calculus books. The difference is that whereas we now use the derivative to find the slope of the tangent line, that is, the tangent of the angle it makes with a reference axis, Fermat looked for the point where the tangent intercepted that axis. If the two lines did not intersect, obviously the tangent was easily determined as the unique parallel through the given point to the given axis. In all other cases Fermat needed to determine the length of the projection of the tangent on the axis from the point of intersection to the point below the point of tangency, a length known as the *subtangent*. In a letter sent to Mersenne and forwarded to Descartes in 1638 Fermat explained his method of finding the subtangent.

In Fig. 1 the curve DB is a parabola with axis CE, and the tangent at B meets the axis at E. Since the parabola is convex, a point O between B and E on the tangent lies outside the parabola. That location provided Fermat with two inequalities, one of which was $\overline{CD} : \overline{DI} > \overline{BC}^2 : \overline{OI}^2$. (Equality would hold here if $\overline{OI}$ were replaced by the portion of it cut off by the parabola.) Since $\overline{BC} : \overline{OI} = \overline{CE} : \overline{EI}$, it follows that $\overline{CD} : \overline{DI} > \overline{CE}^2 : \overline{EI}^2$. Then abbreviating by setting $\overline{CD} = g$, $\overline{CE} = x$, and $\overline{CI} = y$, we have $g : g - y > x^2 : x^2 + y^2 - 2xy$, and cross-multiplying,

$$gx^2 + gy^2 - 2gxy > gx^2 - x^2y.$$

Canceling the term gx^2 and dividing by y, we obtain $gy - 2gx > -x^2$. Since this inequality must hold for all y (no matter how small), it follows that $x^2 \geq 2gx$, that is, $x \geq 2g$ if $x > 0$. Choosing a point O beyond B on the tangent and reasoning in the same way would give $x \leq 2g$, so that $x = 2g$. Since x was the quantity to be determined, the problem is solved.

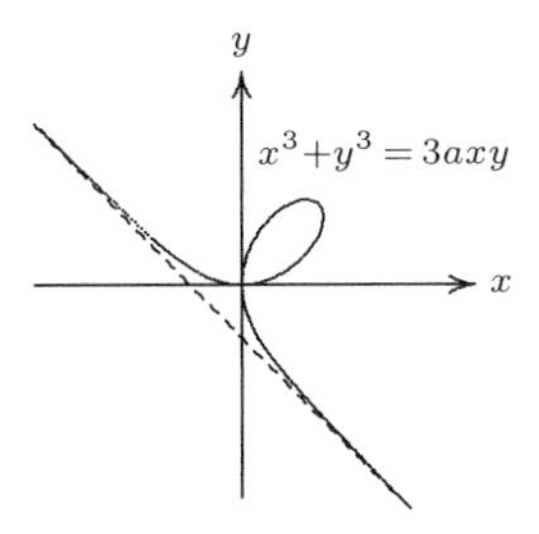

FIGURE 2. The folium of Descartes. Descartes and Fermat considered only the loop in this curve.

In this paper Fermat asserted, "And this method never fails...." This assertion provoked an objection from Descartes,[1] who challenged Fermat with the curve of Fig. 2, now known as the folium of Descartes, having equation $x^3 + y^3 = 3axy$.

Descartes did not regard curves such as the spiral and the quadratrix as admissible in argument, since they are generated by two motions whose relationship to each other cannot be determined exactly. A few such curves, however, were to prove a very fruitful source of new constructions and applications. One of them, which had first been noticed in the early sixteenth century by an obscure mathematician named Charles Bouvelles (ca. 1470–ca. 1553), is the cycloid, the curve generated by a point on a circle (called the generating circle) that rolls without slipping along a straight line. We have already mentioned this curve, assuming that the reader will have heard of it, in Section 3 of Chapter 12. It is easily pictured by imagining a painted spot on the rim of a wheel as the wheel rolls along the ground. Since the linear velocity of the rim relative to its center is exactly equal to the linear velocity of the center, it follows that the point is at any instant moving along the bisector of the angle formed by a horizontal line and the tangent to the generating circle. In this way, given the generating circle, it is an easy matter to construct the tangent to the cycloid. This result was obtained independently around 1638 by Descartes, Fermat, and Gilles Personne de Roberval (1602–1675), and slightly later by Evangelista Torricelli (1608–1647), a pupil of Galileo Galilei (1564–1642).

1.2. Lengths, areas, and volumes. Seventeenth-century mathematicians had inherited two conceptually different ways of applying infinitesimal ideas to find areas and volumes. One was to regard an area as a "sum of lines." The other was to approximate the area by a sum of regular figures and try to show that the approximation got better as the individual regular figures got smaller. The rigorous version of the latter argument, the method of exhaustion, was tedious and of limited application.

Cavalieri's principle. In the "sum of lines" approach, a figure whose area or volume was required was sliced into parallel sections, and these sections were shown to be equal or proportional to corresponding sections of a second figure whose area or

[1] There was little love lost between Descartes and Fermat, since Fermat had dismissed Descartes' derivation of the law of refraction. (Descartes assumed that light traveled faster in denser media; Fermat assumed that it traveled slower. Yet they both arrived at the same law! For details, see Indorato and Nastasi, *1989*.) Descartes longed for revenge, and even though he eventually ended the controversy over Fermat's methods with the equivalent of, "You should have said so in the first place, and we would never have argued...," he continued to attack Fermat's construction of the tangent to a cycloid.

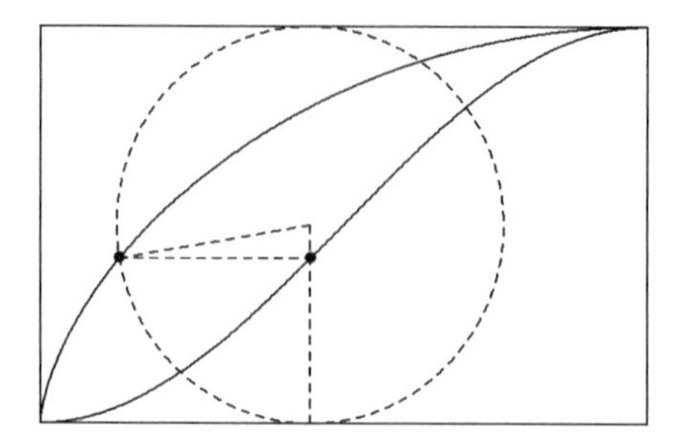

FIGURE 3. Roberval's quadrature of the cycloid.

volume was known. The first figure was then asserted to be equal or proportional to the second. The principle was stated in 1635 by Bonaventura Cavalieri (1598–1647), a Jesuit priest and a student of Galileo. At the time it was customary for professors to prove their worthiness for a chair of mathematics by a learned dissertation. Cavalieri proved certain figures equal by pairing off congruent sections of them, in a manner similar to that of Archimedes' *Method* and the method by which Zu Chongzhi and Zu Geng found the volume of a sphere. This method implied that figures in a plane lying between two parallel lines and such that all sections parallel to those lines have the same length must have equal area. This principle is now called *Cavalieri's principle.* The idea of regarding a two-dimensional figure as a sum of lines or a three-dimensional figure as a sum of plane figures was extended by Cavalieri to consideration of the squares on the lines in a plane figure, then to the cubes on the lines in a figure, and so on.

The cycloid. Cavalieri's principle was soon applied to find the area of the cycloid. Roberval, who found the tangent to the cycloid, also found the area beneath it by a clever use of the method of indivisibles. He considered along with half an arch of the cycloid itself a curve he called the *companion* to the cycloid. This companion curve is generated by a point that is always directly below or above the center of the generating circle as it rolls along and at the same height as the point on the rim that is generating the cycloid. As the circle makes half a revolution (see Fig. 3), the cycloid and its companion first diverge from the ground level, then meet again at the top. Symmetry considerations show that the area under the companion curve is exactly one-half of the rectangle whose vertical sides are the initial and final positions of the diameter of the generating circle through the point generating the cycloid. But by definition of the two curves their generating points are always at the same height, and the horizontal distance between them at any instant is half of the corresponding horizontal section of the generating circle. Hence by Cavalieri's principle the area between the two curves is exactly half the area of the circle.

Rectangular approximations and the method of exhaustion. Besides the method of indivisibles (Cavalieri's principle), mathematicians of the time also applied the method of polygonal approximation to find areas. In 1640 Fermat wrote a paper on quadratures in which he found the areas under certain figures by a method that he saw could easily be generalized. He considered a "general hyperbola," as in Fig. 4, a curve referred to asymptotes AR and AC and defined by the property

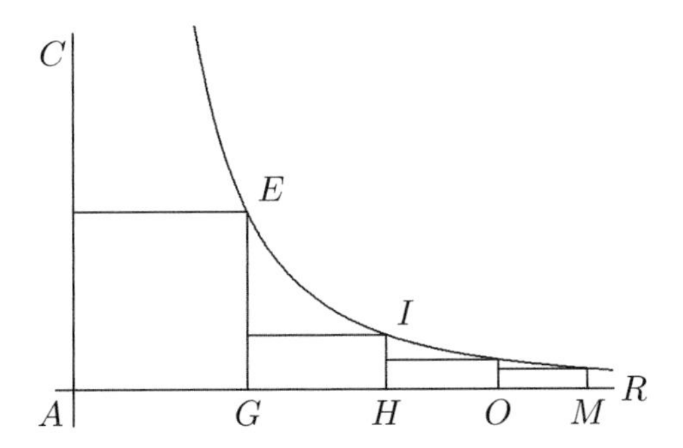

FIGURE 4. Fermat's quadrature of a generalized hyperbola.

that the ratio $AH^m : AG^m = EG^n : HI^n$ is the same for any two points E and I on the curve; we would describe this property by saying that $x^m y^n = \text{const}$.

Powers of sines. Cavalieri found the "sums of the powers of the lines" inside a triangle. In 1659 Pascal did the same for the "sums of the powers of the lines inside a quadrant of a circle." Now a line inside a quadrant of a circle is what up to now has been called a sine. Thus, Pascal found the sum of the powers of the sines of a quadrant of a circle. In modern terms, where Cavalieri found $\int_0^a x^n\,dx = a^{n+1}/(n+1)$, Pascal found $\int_\alpha^\beta (R\sin\varphi)\,R\,d\varphi = R(R\cos\alpha - R\cos\beta)$.

1.3. The relation between tangents and areas. The first statement of a relation between tangents and areas appears in 1670 in a book entitled *Lectiones geometricae* by Isaac Barrow (1630–1677), a professor of mathematics at Cambridge and later chaplain to Charles II. Barrow gave the credit for this theorem to "that most learned man, Gregory of Aberdeen" (James Gregory, 1638–1675). Barrow states several theorems resembling the fundamental theorem of calculus. The first theorem (Section 11 of Lecture 10) is the easiest to understand. Given a curve referred to an axis, Barrow constructs a second curve such that the ordinate at each point is proportional to the area under the original curve up to that point. We would express this relation as $F(x) = (1/R)\int_a^x f(t)\,dt$, where $y = f(x)$ is the first curve, $y = F(x)$ is the second, and $1/R$ is the constant of proportionality. If the point $T = (t, 0)$ is chosen on the axis so that $(x - t)\cdot f(x) = RF(x)$, then, said Barrow, T is the foot of the subtangent to the curve $y = F(x)$; that is, $x - t$ is the length of the subtangent. In modern language the length of the subtangent to the curve $y = F(x)$ is $|F(x)/F'(x)|$. This expression would replace $(x - t)$ in the equation given by Barrow. If both $F(x)$ and $F'(x)$ are positive, this relation really does say that $f(x) = RF'(x) = (d/dx)\int_a^x f(t)\,dt$.

Later, in Section 19 of Lecture 11, Barrow shows the other version of the fundamental theorem, that is, that if a curve is chosen so that the ratio of its ordinate to its subtangent (this ratio is precisely what we now call the derivative) is proportional to the ordinate of a second curve, the area under the second curve is proportional to the ordinate of the first.

1.4. Infinite series and products. The methods of integration requiring the summing of infinitesimal rectangles or all the lines inside a plane figure led naturally to the consideration of infinite series. Several special series were known by the mid-seventeenth century. For example, the Scottish mathematician James Gregory published a work on geometry in 1668 in which he stated the equivalent of the formula given earlier (unbeknown to Gregory, of course) by Jyeshtadeva:

$$\arctan x = x - \frac{x^3}{3} + \frac{x^5}{5} - \frac{x^7}{7} + \cdots .$$

Similarly, infinite product expansions were known by this time for the number π. One, due to Wallis, is

$$\frac{2}{\pi} = \frac{1 \cdot 3 \cdot 3 \cdot 5 \cdot 5 \cdot 7 \cdots}{2 \cdot 2 \cdot 4 \cdot 4 \cdot 6 \cdot 6 \cdots} .$$

The binomial series. It was the binomial series that really established the use of infinite series in analysis. The expansion of a power of a binomial leads to finite series when the exponent is a nonnegative integer and to an infinite series otherwise. This series, which we now write in the form

$$(1+x)^r = 1 + \sum_{k=1}^{\infty} \frac{r(r-1)\cdots(r-k+1)}{1\cdots k} x^k ,$$

was discovered by Isaac Newton (1642–1727) around 1665, although, of course, he expressed it in a different language, as a recursive procedure for finding the terms. In a 1676 letter to Henry Oldenburg (1615–1677), the Secretary of the Royal Society, Newton wrote this expansion as

$$\overline{P+PQ}|\frac{m}{n} = P|\frac{m}{n} + \frac{m}{n}AQ + \frac{m-n}{2n}BQ + \frac{m-2n}{3n}CQ + \frac{m-3n}{4n}DQ + \text{etc.}$$

"where $P+PQ$ stands for a quantity whose root or power or whose root of a power is to be found, P being the first term of that quantity, Q being the remaining terms divided by the first term and m/n the numerical index of the powers of $P+PQ$... A stands for the first term $P|\frac{m}{n}$, B for the second term $\frac{m}{n}AQ$, and so on... ."

Newton's explanation of the meaning of the terms A, B, C,..., means that the kth term is obtained from its predecessor via multiplication by $\{[(m/n) - k]/(k+1)\}Q$. He said that m/n could be any fraction, positive or negative.

2. Newton and Leibniz

The results we have just examined show that parts of the calculus were recognized by the mid-seventeenth century, like the pieces of a jigsaw puzzle lying loose on a table. What was needed was someone to see the pattern and fit all the pieces together. The unifying principle was the concept of a derivative, and that concept came to Newton and Leibniz independently and in slightly differing forms.

2.1. Isaac Newton. Isaac Newton discovered the binomial theorem, the general use of infinite series, and what he called the *method of fluxions* during the mid-1660s. His early notes on the subject were not published until after his death, but a revised version of the method was expounded in his *Principia*.

Newton's first version of the calculus. Newton first developed the calculus in what we would call parametric form. Time was the universal independent variable, and the relative rates of change of other variables were computed as the ratios of their absolute rates of change with respect to time. Newton thought of variables as moving quantities and focused attention on their velocities. He used the letter o to represent a small time interval and p for the velocity of the variable x, so that the change in x over the time interval o was op. Similarly, using q for the velocity of y, if y and x are related by $y^n = x^m$, then $(y + oq)^n = (x + op)^m$. Both sides can be expanded by the binomial theorem. Then if the equal terms y^n and x^m are subtracted, all the remaining terms are divisible by o. When o is divided out, one side is $nqy^{n-1} + oA$ and the other is $mpx^{m-1} + oB$. Ignoring the terms containing o, since o is small, one finds that the *relative* rate of change of the two variables, q/p is given by $q/p = (mx^{m-1})/(ny^{n-1})$; and since $y = x^{m/n}$, it follows that $q/p = (m/n)x^{(m/n)-1}$. Here at last was the concept of a derivative, expressed as a relative rate of change.

Newton recognized that reversing the process of finding the relative rate of change provides a solution of the area problem. He was able to find the area under the curve $y = ax^{m/n}$ by working backward.

Fluxions and fluents. Newton's "second draft" of the calculus was the concept of fluents and fluxions. A *fluent* is a moving or flowing quantity; its *fluxion* is its rate of flow, which we now call its velocity or derivative. In his *Fluxions*, written in Latin in 1671 and published in 1742 (an English translation appeared in 1736), he replaced the notation p for velocity by $\dot{x}$, a notation still used in mechanics and in the calculus of variations. Newton's notation for the opposite operation, finding a fluent from the fluxion has been abandoned: Where we write $\int x(t)\,dt$, he wrote $\acute{x}$.

The first problem in the *Fluxions* is: *The relation of the flowing quantities to one another being given, to determine the relation of their fluxions.* The rule given for solving this problem is to arrange the equation that expresses the given relation (assumed algebraic) in powers of one of the variables, say x, multiply its terms by any arithmetic progression (that is, the first power is multiplied by c, the square by $2c$, the cube by $3c$, etc.), and then multiply by $\dot{x}/x$. After this operation has been performed for each of the variables, the sum of all the resulting terms is set equal to zero.

Newton illustrated this operation with the relation $x^3 - ax^2 + axy - y^2 = 0$, for which the corresponding fluxion relation is $3x^2\dot{x} - 2ax\dot{x} + a\dot{x}y + ax\dot{y} - 2y\dot{y} = 0$, and by numerous examples of finding tangents to well-known curves such as the spiral and the cycloid. Newton also found their curvatures and areas. The combination of these techniques with infinite series was important, since fluents often could not be found in finite terms. For example, Newton found that the area under the curve $\dot{z} = 1/(1 + x^2)$ was given by the Jyeshtadeva–Gregory series $z = x - \frac{1}{3}x^3 + \frac{1}{5}x^5 - \frac{1}{7}x^7 + \cdots$.

Later exposition of the calculus. Newton made an attempt to explain fluxions in terms that would be more acceptable logically, calling it the "method of first and last ratios," in his treatise on mechanics, the *Philosophiae naturalis principia mathematica* (*Mathematical Principles of Natural Philosophy*), where he said,

> Quantities, and the ratios of quantities, which in any finite time converge continually toward equality, and before the end of that

> time approach nearer to each other than by any given difference, become ultimately equal.
>
> If you deny it, suppose them to be ultimately unequal, and let D be their ultimate difference. Therefore they cannot approach nearer to equality than by that given difference D; which is contrary to the supposition.

If only the phrase *become ultimately equal* had some clear meaning, as Newton seemed to assume, this argument might have been convincing. As it is, it comes close to being a definition of *ultimately equal*, or, as we would say, equal in the limit. Newton came close to stating the modern concept of a limit, when he described the "ultimate ratios" (derivatives) as "limits towards which the ratios of quantities decreasing without limits do always converge, and to which they approach nearer than by any given difference." Here one can almost see the "arbitrarily small ε" that plays the central role in the concept of a limit.

2.2. Gottfried Wilhelm von Leibniz. Leibniz believed in the reality of infinitesimals, quantities so small that any finite sum of them is still less than any assignable positive number, but which are nevertheless not zero, so that one is allowed to divide by them. The three kinds of numbers (finite, infinite, and infinitesimal) could, in Leibniz' view, be multiplied by one another, and the result of multiplying an infinite number by an infinitesimal might be any one of the three kinds. This position was rejected in the nineteenth century but was resurrected in the twentieth century and made logically sound. It lies at the heart of what is called *nonstandard analysis*, a subject that has not penetrated the undergraduate curriculum. The radical step that must be taken in order to believe in infinitesimals is a rejection of the Archimedean axiom that for any two positive quantities of the same kind a sufficient number of bisections of one will lead to a quantity smaller than the second. This principle was essential to the use of the method of exhaustion, which was one of the crowning glories of Euclidean geometry. It is no wonder that mathematicians were reluctant to give it up.

Leibniz invented the expression dx to indicate the difference of two infinitely close values of x, dy to indicate the difference of two infinitely close values of y, and dy/dx to indicate the ratio of these two values. This notation was beautifully intuitive and is still the preferred notation for thinking about calculus. Its logical basis at the time was questionable, since it avoided the objections listed above by claiming that the two quantities have not vanished at all but have yet become less than any assigned positive number. However, at the time, consistency would have been counterproductive in mathematics and science.

The integral calculus and the fundamental theorem of calculus flowed very naturally from Leibniz' approach. Leibniz could argue that the ordinates to the points on a curve represent infinitesimal rectangles of height y and width dx, and hence finding the area under the curve—"summing all the lines in the figure"—amounted to summing infinitesimal differences in area dA, which collapsed to give the total area. Since it was obvious that on the infinitesimal level $dA = y\,dx$, the fundamental theorem of calculus was an immediate consequence. Leibniz first set it out in geometric form in a paper on quadratures in the 1693 *Acta eruditorum*. There he considered two curves: one, which we denote $y = f(x)$ with its graph above a horizontal axis, the other, which we denote $z = F(x)$, with its graph below

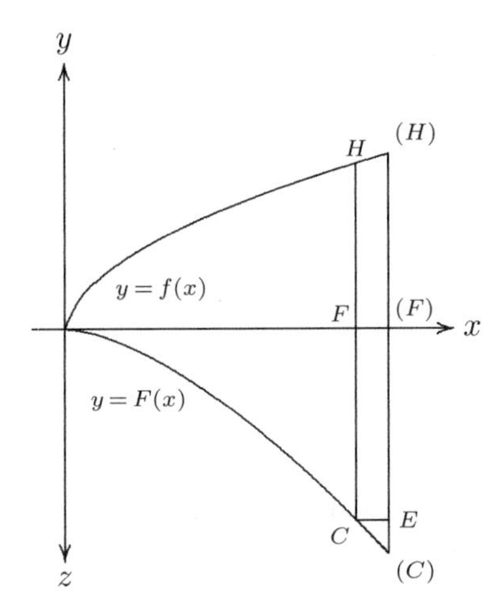

FIGURE 5. Leibniz' proof of the fundamental theorem of calculus.

the horizontal axis.[2] The second curve has an ordinate proportional to the area under the first curve. That is, for a positive constant a, having the dimension of length, $aF(x)$ is the area under the curve $y = f(x)$ from the origin up to the point with abscissa x. As we would write the relation now,

$$aF(x) = \int_0^x f(t)\,dt\,.$$

In this form the relation is dimensionally consistent. What Leibniz proved was that the curve $z = F(x)$, which he called the *quadratrix* (squarer), could be constructed from its infinitesimal elements. In Fig. 5 the parentheses around letters denote points at an infinitesimal distance from the points denoted by the same letters without parentheses. In the infinitesimal triangle $CE(C)$ the line $E(C)$ represents dF, while the infinitesimal quadrilateral $HF(F)(H)$ represents dA, the element of area under the curve. The lines $F(F)$ and CE both represent dx. Leibniz argued that by construction, $a\,dF = f(x)\,dx$, and so $dF : dx = f(x) : a$. That meant that the quadratrix could be constructed by antidifferentiation.

Leibniz eventually abbreviated the sum of all the increments in the area (that is, the total area) using an elongated S, so that $A = \int dA = \int y\,dx$. Nearly all the basic rules of calculus for finding the derivatives of the elementary functions and the derivatives of products, quotients, and so on, were contained in Leibniz' 1684 paper on his method of finding tangents. However, he had certainly obtained these results much earlier. His collected works contain a paper written in Latin with the title *Compendium quadraturae arithmeticae*, to which the editor assigns a date of 1678 or 1679. This paper shows Leibniz' approach through infinitesimal differences and their sums and suggests that it was primarily the problem of squaring the circle and other conic sections that inspired this work. The work consists of 49 propositions and two problems. Most of the propositions are stated without

[2] The vertical axis is to be assumed positive in both directions from the origin. We are preserving in Fig. 5 only the lines needed to explain Leibniz' argument. He himself merely labeled points on the two curves with letters and referred to those letters.

proof; they contain the basic results on differentiation and integration of elementary functions, including the Taylor series expansions of logarithms, exponentials, and trigonometric functions. Although the language seems slightly archaic, one can easily recognize a core of standard calculus here.

Later reflections on the calculus. Like Newton, Leibniz was forced to answer objections to the new methods of the calculus. In the *Acta eruditorum* of 1695 Leibniz published (in Latin) a "Response to certain objections raised by Herr Bernardo Niewentiit regarding differential or infinitesimal methods." These objections were three: (1) that certain infinitely small quantities were discarded as if they were zero (this principle was set forth as fundamental in the following year in the textbook of calculus by the Marquis de l'Hospital); (2) the method could not be applied when the exponent is a variable; and (3) the higher-order differentials were inconsistent with Leibniz' claim that only geometry could provide a foundation. In answer to the first objection Leibniz attempted to explain different orders of infinitesimals, pointing out that one could neglect all but the lowest orders in a given equation. To answer the second, he used the binomial theorem to demonstrate how to handle the differentials dx, dy, dz when $y^x = z$. To answer the third, Leibniz said that one should not think of $d(dx)$ as a quantity that fails to yield a (finite) quantity even when multiplied by an infinite number. He pointed out that if x varies geometrically when y varies arithmetically, then $dx = (x\,dy)/a$ and $ddx = (dx\,dy)/a$, which makes perfectly good sense.

2.3. The disciples of Newton and Leibniz. Newton and Leibniz had disciples who carried on their work. Among Newton's followers was Roger Cotes (1682–1716), who oversaw the publication of a later edition of Newton's *Principia* and defended Newton's inverse square law of gravitation in a preface to that work. He also fleshed out the calculus with some particular results on plane loci and considered the extension of functions defined by power series to complex values, deriving the important formula $i\phi = \log(\cos\phi + i\sin\phi)$, where $i = \sqrt{-1}$. Another of Newton's followers was Brook Taylor (1685–1731), who developed a calculus of finite differences that mirrors in many ways the "continuous" calculus of Newton and Leibniz and is of both theoretical and practical use today. Taylor is famous for the infinite power series representation of functions that now bears his name. It appeared in his 1715 treatise on finite differences. We have already seen, however, that many particular "Taylor series" were known to Newton and Leibniz; Taylor's merit is to have recognized a general way of producing such a series in terms of the derivatives of the generating function. This step, however, was also taken by Johann Bernoulli.

Leibniz also had a group of active and intelligent followers who continued to develop his ideas. The most prominent of these were the Bernoulli brothers Jakob (1654–1705) and Johann (1667–1748), citizens of Switzerland, between whom relations were not always cordial. They investigated problems that arose in connection with calculus and helped to systematize, extend, and popularize the subject. In addition, they pioneered new mathematical subjects such as the calculus of variations, differential equations, and the mathematical theory of probability. A French nobleman, the Marquis de l'Hospital (1661–1704), took lessons from Johann Bernoulli and paid him a salary in return for the right to Bernoulli's mathematical discoveries. As a result, Bernoulli's discovery of a way of assigning values to what are now called indeterminate forms appeared in L'Hospital's 1696 textbook *Analyse des infiniment petits* (*Infinitesimal Analysis*) and has ever since been known as L'Hospital's rule. Like the followers of Newton, who had to answer the objections of Bishop Berkeley (see Section 3 below) Leibniz' followers encountered objections from Michel Rolle (1652–1719), which were answered by Johann Bernoulli with the claim that Rolle didn't understand the subject.

The priority dispute. One of the better-known and less edifying incidents in the history of mathematics is the dispute between the disciples of Newton and those of Leibniz over the credit for the invention of the calculus. Although Newton had discovered the calculus by the early 1670s and had described it in a paper sent to James Collins, the librarian of the Royal Society, he did not publish his discoveries until 1687. Leibniz made his discoveries a few years later than Newton but published some of them earlier, in 1684. Newton's vanity was wounded in 1695 when he learned that Leibniz was regarded on the Continent as the discoverer of the calculus, even though Leibniz himself made no claim to this honor. In 1699 a Swiss immigrant to England, Nicolas Fatio de Duillier (1664–1753), suggested that Leibniz had seen Newton's paper when he had visited London and talked with Collins in 1673. (Collins died in 1683, before his testimony in the matter was needed.) This unfortunate rumor poisoned relations between Newton and Leibniz and their followers. In 1711–1712 a committee of the Royal Society (of which Newton was President) investigated the matter and reported that it believed Leibniz had seen certain documents that in fact he had not seen. Relations between British and Continental mathematicians reached such a low ebb that Newton deleted certain laudatory references to Leibniz from the third edition of his *Principia*. This dispute confirmed the British in the use of the clumsy Newtonian notation for more than a century, a notation far inferior to Leibniz's elegant and intuitive symbolism. But in the early nineteenth century the impressive advances made by Continental scholars such as Euler, Lagrange, and Laplace won over the British mathematicians, and scholars such as William Wallace (1768–1843) rewrote the theory of fluxions in terms of the theory of limits. Wallace asserted that there was never any need to introduce motion and velocity into this theory, except as illustrations, and that indeed Newton himself used motion only for illustration, recasting his arguments in terms of limits when rigor was needed (see Panteki, *1987*, and Craik, *1999*). Eventually, even the British began using the term *integral* instead of *fluent* and *derivative* instead of *fluxion*, and these Newtonian terms became mathematically part of a dead language.

Certain relevant facts were concealed by the terms in which the priority dispute was cast. One of these is the extent to which Fermat, Descartes, Cavalieri, Pascal, Roberval, and others had developed the techniques in isolated cases that were to be unified by the calculus as we know it now. In any case, Newton's teacher Isaac Barrow had the insight into the connection between subtangents and area before either Newton or Leibniz thought of it. Barrow's contributions were shunted aside in the heat of the dispute; their significance has been pointed out by Feingold (*1993*).

Early textbooks on calculus. The secure place of calculus in the mathematical curriculum was established by the publication of a number of excellent textbooks. One of the earliest was the *Analyse des infiniment petits*, mentioned above, which was published by the Marquis de l'Hospital in 1696.

Most students of calculus know the Maclaurin series as a special case of the Taylor series. Its discoverer was a Scottish contemporary of Taylor, Colin Maclaurin (1698–1746), whose treatise on fluxions (1742) contained a thorough and rigorous exposition of calculus. It was written partly as a response to the philosophical attacks on the foundations of calculus by the philosopher George Berkeley.

The Italian textbook *Istituzioni analitiche ad uso della gioventù italiana* (*Analytic Principles for the Use of Italian Youth*) became a standard treatise on analytic geometry and calculus and was translated into English in 1801. Its author was Maria Gaetana Agnesi, who was mentioned in Chapter 4 as one of the first women to achieve prominence in mathematics.

The definitive textbooks of calculus were written by the greatest mathematician of the eighteenth century, the Swiss scholar Leonhard Euler. In his 1748 *Introductio in analysin infinitorum*, a two-volume work, Euler gave a thorough discussion of analytic geometry in two and three dimensions, infinite series (including the use of complex variables in such series), and the foundations of a systematic theory of algebraic functions. The modern presentation of trigonometry was established in this work. The *Introductio* was followed in 1755 by *Institutiones calculi differentialis* and a three-volume *Institutiones calculi integralis* (1768–1774), which included the entire theory of calculus and the elements of differential equations, richly illustrated with challenging examples. It was from Euler's textbooks that many prominent nineteenth-century mathematicians such as the Norwegian genius Niels Henrik Abel first encountered higher mathematics, and the influence of Euler's books can be traced in their work.

The state of the calculus around 1700. Most of what we now know as calculus—rules for differentiating and integrating elementary functions, solving simple differential equations, and expanding functions in power series—was known by the early eighteenth century and was included in the standard textbooks just mentioned. Nevertheless, there was much unfinished work. We list here a few of the open questions:

Nonelementary integrals. Differentiation of elementary functions is an algorithmic procedure, and the derivative of any elementary function whatsoever, no matter how complicated, can be found if the investigator has sufficient patience. Such is not the case for the inverse operation of integration. Many important elementary functions, such as $(\sin x)/x$ and e^{-x^2}, are not the derivatives of elementary functions. Since such integrals turned up in the analysis of some fairly simple motions, such as that of a pendulum, the problem of these integrals became pressing.

Differential equations. Although integration had originally been associated with problems of area and volume, because of the importance of differential equations in mechanical problems the solution of differential equations soon became the major application of integration. The general procedure was to convert an equation to a form in which the derivatives could be eliminated by integrating both sides (reduction to quadratures). As these applications became more extensive, more and more cases began to arise in which the natural physical model led to equations that could not be reduced to quadratures. The subject of differential equations began to take on a life of its own, independent of the calculus.

Foundational difficulties. The philosophical difficulties connected with the use of infinitesimal methods were paralleled by mathematical difficulties connected with the application of the algebra of finite polynomials to infinite series. These difficulties were hidden for some time, and for a blissful century mathematicians and physicists operated formally on power series as if they were finite polynomials. They did so even though it had been known since the time of Oresme that the partial sums of the harmonic series $1 + \frac{1}{2} + \frac{1}{3} + \cdots$ grow arbitrarily large.

3. Branches and roots of the calculus

The calculus grew organically, sending forth branches while simultaneously putting down firm roots. The roots were the subject of philosophical speculation that eventually led to new mathematics as well, but the branches were natural outgrowths of pure mathematics that appeared very early in the history of the subject. We begin this section with the branches and will end it with the roots.

3.1. Ordinary differential equations. Ordinary differential equations arose almost as soon as there was a language (differential calculus) in which they could be expressed.[3] These equations were used to formulate problems from geometry and physics in the late seventeenth century, and the natural approach to solving them was to apply the integral calculus, that is, to reduce a given equation to quadratures. Leibniz, in particular, developed the technique now known as separation of variables as early as 1690 (Grosholz, *1987*). In the simplest case, that of an ordinary differential equation of first order and first degree, one is seeking an equation $f(x, y) = c$, which may be interpreted as a conservation law if x and y are functions of time having physical significance. The conservation law is expressed as the differential equation

$$\frac{\partial f}{\partial x}\,dx + \frac{\partial f}{\partial y}\,dy = 0\,.$$

The resulting equation is known as an exact differential equation. To solve this equation, one has only to integrate the first differential with respect to x, adding an arbitrary function $g(y)$ to the solution, then differentiate with respect to y and compare the result with $\frac{\partial f}{\partial y}$ in order to get an equation for $g'(y)$, which can then be integrated.

If all equations were this simple, differential equations would be a very trivial subject. Unfortunately, it seems that nature tries to confuse us, multiplying these equations by arbitrary functions $\mu(x, y)$. That is, when an equation is written down as a particular case of a physical law, it often looks like

$$M(x, y)\,dx + N(x, y)\,dy = 0\,,$$

where $M(x, y) = \mu(x, y)\frac{\partial f}{\partial x}$ and $N(x, y) = \mu(x, y)\frac{\partial f}{\partial y}$, and no one can tell from looking at M just which factors in it constitute μ and which constitute $\frac{\partial f}{\partial x}$. To take the simplest possible example, the mass y of a radioactive substance that remains undecayed in a sample after time t satisfies the equation

$$dy - ky\,dx = 0\,,$$

where k is a constant. The mathematician's job is to get rid of $\mu(x, y)$ by looking for an "integrating factor" that will make the equation exact. One integrating factor for this equation is $1/y$; another is e^{-kx}.

[3] This subsection is a summary of an unpublished paper that can be found in full at the following website: `http://www.emba.uvm.edu/~cooke/ckthm.pdf`

It appeared at a very early stage that finding an integrating factor is not in general possible, and both Newton and Leibniz were led to the use of infinite series with undetermined coefficients to solve such equations. Later, Maclaurin, was to warn against too hasty recourse to infinite series, saying that certain integrals could be better expressed geometrically as the arc lengths of various curves. But the idea of replacing a differential equation by a system of algebraic equations was very attractive. The earliest examples of series solutions were cited by Feigenbaum (*1994*). In his *Fluxions*, which was written in 1671 and left unpublished during his lifetime (see Whiteside, *1967*, Vol. 3, p. 99), Newton considered the linear differential equation that we would now write as

$$\frac{dy}{dx} = 1 - 3x + x^2 + (1+x)y\,.$$

Newton wrote it as $n/m = 1 - 3x + y + xx + xy$ and found that

$$y = x - x^2 + \frac{1}{3}x^3 - \frac{1}{6}x^4 + \frac{1}{30}x^5 - \frac{1}{45}x^6 - \cdots\,.$$

Similarly, in a paper published in the *Acta eruditorum* in 1693 (Gerhardt, *1971*, Vol. 5, p. 287), Leibniz studied the differential equations for the logarithm and the arcsine in order to obtain what we now call the Maclaurin series of the logarithm, exponential, and sine functions. For example, he considered the equation $a^2\,dy^2 = a^2\,dx^2 + x^2\,dy^2$ and assumed that $x = by + cy^3 + ey^5 + fy^7 + \cdots$, thereby obtaining the series that represents the function $x = a\sin(y/a)$. Neither Newton nor Leibniz mentioned that the coefficients in these series were the derivatives of the functions represented by the series divided by the corresponding factorials. However, that realization came to Johann Bernoulli very soon after the publication of Leibniz' work. In a letter to Leibniz dated September 2, 1694 (Gerhardt, *1971*, Vol. 3/1, p. 350), Bernoulli described essentially what we now call the Taylor series of a function. In the course of this description, he gave in passing what became a standard definition of a function, saying, "I take n to be a quantity formed in an arbitrary manner from variables and constants." Leibniz had used this word as early as 1673, and in an article in the 1694 *Acta eruditorum* had defined a function to be "the portion of a line cut off by lines drawn using only a fixed point and a given point lying on a curved line." As Leibniz said, a given curve defines a number of functions: its abscissas, its ordinates, its subtangents, and so on. The problem that differential equations solve is to reconstruct the curve given the ratio between two of these functions.

In classical terms, the solution of a differential equation is a function or family of functions. Given that fact, the ways in which a function can be presented become an important issue. With the modern definition of a function and the familiar notation, one might easily forget that in order to apply the theory of functions it is necessary to deal with particular functions, and these must be *presented* somehow. Bernoulli's description addresses that issue, although it leaves open the question of what methods of combining variables and constants are legal.

A digression on time. The Taylor series of a given function can be generated knowing the values of the function over any interval of the independent variable, no matter how short. Thus, a quantity represented by such a series is determined for all values of the independent variable when the values are given on any interval at all. Given that the independent variable is usually time, that property corresponds to physical determinacy: Knowing the full state of a physical quantity for some interval of time determines its values for all time. Lagrange, in particular, was a proponent of power series, for which he invented the term *analytic function*. However, as we now know, the natural domain of analytic function theory is the complex numbers. Now in mechanics the independent variable represents time, and that fact raises an interesting question: Why should time be a complex variable? How do complex numbers turn out to be relevant to a problem where only real values of the variables have any physical meaning? To this question the eighteenth- and nineteenth-century mathematicians gave no answer. Indeed, it does not appear that they even asked the question very often. Extensive searches of the nineteenth-century literature by the present author have produced only the following comments on this interesting question, made by Weierstrass in 1885 (see his *Werke*, Bd. 3, S. 24):

> It is very remarkable that in a problem of mathematical physics where one seeks an unknown function of two variables that, in terms of their physical meaning, can have only real values and is such that for a particular value of one of the variables the function must equal a prescribed function of the other, an expression often results that is an analytic function of the variable and hence also has a meaning for complex values of the latter.

It is indeed very remarkable, but neither Weierstrass nor anyone since seems to have explained the mystery. But, just as complex numbers were needed to produce the three real roots of a cubic equation, it may not have seemed strange that the complex-variable properties of solutions of differential equations are relevant, even in the study of problems generated by physical considerations involving only real variables. Time is, however, sometimes represented as a two-dimensional quantity, in connection with what are known as Gibbs random fields.

3.2. Partial differential equations. In the middle of the eighteenth century mathematical physicists began to consider problems involving more than one independent variable. The most famous of these is the vibrating string problem discussed by Euler, d'Alembert, and Daniel Bernoulli (1700–1782, son of Johann Bernoulli) during the 1740s and 1750s. This problem led to the one-dimensional wave equation

$$\frac{\partial^2 u}{\partial t^2} = c^2 \frac{\partial^2 u}{\partial^2 x},$$

with the initial condition $u(x,0) = f(x)$, $\frac{\partial u}{\partial t}(x,0) = 0$, which Bernoulli solved in the form of an infinite trigonometric series

$$\sum_{n=1}^{\infty} a_n \sin nx \cos nct,$$

the a_n being chosen so that $\sum\limits_{n=1}^{\infty} a_n \sin nx = f(x)$.[4]

With this problem, partial differential equations arose, leading to new methods of solution. The developments that grew out of trigonometric-series techniques are

[4] This solution was criticized by Euler, leading to a debate over the allowable methods of defining functions and the proper definition of a function.

discussed in Chapter 17, along with the development of real analysis in general. For the rest of the present section, we confine our discussion to power-series techniques.

The heat equation

$$\frac{\partial u}{\partial t} = a\frac{\partial^2 u}{\partial x^2}$$

was the first partial differential equation to which the power-series method was applied. Fourier used this method to produce the solution

$$u(x,t) = \sum_{r=0}^{\infty} \frac{\varphi^{(2r)}(x)}{r!} t^r$$

when $a = 1$, without realizing that this solution "usually" diverges.

It was not until the nineteenth century that mathematicians began to worry about the convergence of their series solutions. Then Cauchy and Weierstrass produced proofs that the series do converge for ordinary differential equations, provided that the coefficients have convergent series representations. For partial differential equations, it turned out that the form of the equation had some influence. Weierstrass' student Sof'ya Kovalevskaya showed that in general the power series solution for the heat equation diverges if the intial temperature distribution is prescribed, even when that temperature is an analytic function of position. She showed, however, that the series converges if the temperature and temperature gradient at one point are prescribed as analytic functions of time. More generally, she showed that the power-series solution of any equation in "normal form" (solvable for a pure derivative of order equal to the order of the equation) would converge.

3.3. Calculus of variations. The notion of function lies at the heart of calculus. The usual picture of a function is of one *point* being mapped to another *point*. However, the independent variable in a function can be a curve or surface as well as a point. For example, given a curve γ that is the graph of a function $y = f(x)$ between $x = a$ and $x = b$, we can define its length as

$$\Lambda(\gamma) = \int_a^b \sqrt{1 + \bigl(f'(x)\bigr)^2}\, dx\,.$$

One of the important problems in the history of geometry has been to pick out the curve γ that minimizes $\Lambda(\gamma)$ and satisfies certain extra conditions, such as joining two fixed points P and Q on a surface or enclosing a fixed area A. The calculus technique of "setting the derivative equal to zero" needs to be generalized for such problems, and the techniques for doing so constitute the calculus of variations. The history of this outgrowth of the calculus has been studied very thoroughly in a number of classic works, such as Woodhouse (*1810*),[5] Todhunter (*1861*), and Goldstine (*1980*), as well as many articles, such as Kreyszig (*1993*).

[5] The treatise of Woodhouse is a textbook as much as a history, and its last chapter is a set of 29 examples posed as exercises for the reader with solutions provided. The book also marks an important transition in British mathematics. Woodhouse says in the preface that, "In a former Work, I adopted the foreign notation...". The foreign notation was the Leibniz notation for differentials, in preference to the dot above the letter that Newton used to denote his fluxions. He says that he found this notation even more necessary in calculus of variations, since he would otherwise have had to adopt some new symbol for Lagrange's variation. But he then goes on to marvel that Lagrange had taken the reverse step of introducing Newton's fluxion notation into the calculus of variations.

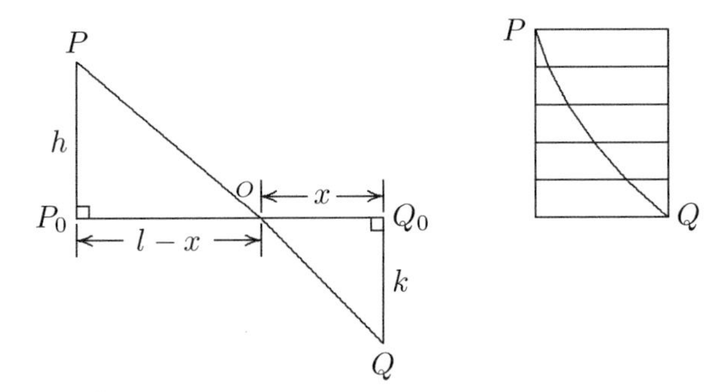

FIGURE 6. Left: Fermat's principle. Choosing the point O so that the time of travel from P to Q through O is a minimum. Right: the principle applied layer by layer when the speed increases proportionally to the square root of the distance of descent.

As with the ordinary calculus, the development of calculus of variations proceeded from particular problems solved by special devices to general techniques and algorithms based on theoretical analysis and rigorous proof. In the seventeenth century there were three such special problems that had important consequences. The first was the brachystochrone (shortest-time) problem for an object crossing an interface between two media while moving from one point to another. In the simplest case (Fig. 6), the interface is a straight line, and the point O is to be chosen so that the time required to travel from P to O at speed v, then from O to Q at speed w, is minimized. If the two speeds are not the same, it is clear that the path of minimum time will not be a straight line, since time can be saved by traveling a slightly longer distance in the medium in which the speed is greater.

The second problem, that of finding the cross-sectional shape of the optimally streamlined body moving through a resisting medium, is discussed in the scholium to Proposition 34 (Theorem 28) of Book 2 of Newton's *Principia*.

Fermat's principle, which asserts that the path of a light ray is the one that requires least time, came into play in a challenge problem stated by Johann Bernoulli in 1696: To find the curve down which a frictionless particle will slide from point P to point Q under the influence of gravity in minimal time. Since the speed of a falling body is proportional to the square root of the distance fallen, Bernoulli reasoned that the sine of the angle between the tangent and the vertical would be proportional to the square root of the vertical coordinate (assuming the vertical axis directed downward). (Recall that ibn Sahl, al-Haytham, Harriot, Snell, and Descartes had all derived the law of refraction which asserts that the ratio of the sines of the angles of incidence and refraction at an interface are proportional to the velocities in the two media.) In that way he arrived at a differential equation for the curve:

$$\frac{dy}{dx} = \sqrt{\frac{y}{a+y}}\,.$$

(We have taken y as the vertical coordinate. Bernoulli apparently took x.) He recognized this equation as the differential equation of a cycloid and thus came to the fascinating conclusion that this curve, which Huygens had studied because it enabled a clock to keep theoretically perfect time (the tautochrone property), also had the brachystochrone property. The challenge problem was solved by Bernoulli himself, his brother Jakob, and by both Newton and Leibniz.[6] According to Woodhouse (*1810*, p. 150), Newton's anonymously submitted solution was so concise and elegant that Johann Bernoulli knew immediately who it must be from. He wrote, "Even though the author, from excessive modesty, does not give his name, we can nevertheless tell certainly by a number of signs that it is the famous Newton; and even if these signs were not present, seeing a small sample would suffice to recognize him, as *ex ungue Leonem.*"[7]

Euler. Variational problems were categorized and systematized by Euler in a large treatise in 1744 named *Methodus inveniendi lineas curvas* (*A Method of Finding Curves*). In this treatise Euler set forth a series of problems of increasing complexity, each involving the finding of a curve having certain extremal properties, such as minimal length among all curves joining two points on a given surface.[8] Proposition 3 in Chapter 2, for example, asks for the minimum value of an integral $\int Z\,dx$, where Z is a function of variables, x, y, and $p = y' = \frac{dy}{dx}$. Based on his previous examples, Euler derived the differential equation

$$0 = N\,dx - dP\,,$$

where $dZ = M\,dx + N\,dy + P\,dp$ is the differential of the integrand Z. Since $N = \frac{\partial Z}{\partial y}$ and $P = \frac{\partial Z}{\partial p}$, this equation could be written in the form that is now the basic equation of the calculus of variations, and is known as Euler's equation:

$$\frac{\partial Z}{\partial y} = \frac{d}{dx}\left(\frac{\partial Z}{\partial y'}\right).$$

In Chapter 3, Euler generalized this result by allowing Z to depend on additional parameters and applied his result to find minimal surfaces. In an appendix he studied elastic curves and surfaces, including the problem of the vibrating membrane. This work was being done at the very time when Euler's friend Daniel Bernoulli was studying the simpler problem of the vibrating string. In a second appendix he showed how to derive the equations of mechanics from variational principles, thus providing a unifying mathematical principle that applied to both optics (Fermat's principle) and mechanics.

Lagrange. The calculus of variations acquired "variations" and its name as the result of a letter written by Lagrange to Euler in 1755. In that letter, Lagrange generalized Leibniz' differentials from points to curves, using the Greek δ instead of the Latin d to denote them. Thus, if $y = f(x)$ was a curve, its *variation* δy was a small perturbation of it. Just as dy was a small change in the value of y at a point, δy was a small change in all the values of y at all points. The variation operator δ can be manipulated quite easily, since it commutes with differentiation and integration: $\delta y' = (\delta y)'$ and $\delta \int Z\,dx = \int \delta Z\,dx$. With this operator, Euler's equation and its many applications, were easy to derive. Euler immediately recognized the usefulness of what Lagrange had done and gave the new theory the name it has borne ever since: calculus of variations.

Lagrange also considered extremal problems with constraint and introduced the famous Lagrange multipliers as a way of turning these relative (constrained)

[6] Newton apparently recognized structural similarities between this problem and his own optimal-streamlining problem (see Goldstine, *1980*, pp. 7–35).

[7] A Latin proverb much in vogue at the time. It means literally "from [just] the claw [one can recognize] the Lion."

[8] This problem was Example 4 in Chapter 4 of the treatise.

extrema into absolute (unconstrained) extrema. Euler had given an explanation of this process earlier. Woodhouse (*1810*, p. 79) thought that Lagrange's systematization actually deprived Euler's ideas of their simplicity.

Second-variation tests for maxima and minima. Like the equation $f'(x) = 0$ in calculus, the Euler equation is only a necessary condition for an extremal, not sufficient, and it does not distinguish between maximum, minimum, and neither. In general, however, if Euler's equation has only one solution, and there is good reason to believe that a maximum or minimum exists, the solution of the Euler equation provides a basis to proceed in practice. Still, mathematicians were bound to explore the question of distinguishing maxima from minima. Such investigations were undertaken by Lagrange and Legendre in the late eighteenth century.

In 1786 Legendre was able to show that a sufficient condition for a minimum of the integral

$$I(y) = \int_a^b f(x, y, y')\, dx\,,$$

at a function satisfying Euler's necessary condition, was $\frac{\partial^2 f}{\partial^2 y'} > 0$ for all x and that a sufficient condition for a maximum was $\frac{\partial^2 f}{\partial y'^2} < 0$.

In 1797 Lagrange published a comprehensive treatise on the calculus, in which he objected to some of Legendre's reasoning, noting that it assumed that certain functions remained finite on the interval of integration (Dorofeeva, *1998*, p. 209).[9]

Jacobi: sufficiency criteria. The second-variation test is strong enough to show that a solution of the Euler equation really is an extremal among the smooth functions that are "nearby" in the sense that their values are close to those of the solution and their derivatives also take values close to those of the derivative of the solution. Such an extremal was called a *weak extremal* by Adolf Kneser (1862–1930). Jacobi had the idea of replacing the curve $y(x)$ that satisfied Euler's equation with a family of such curves depending on parameters (two in the case we have been considering) $y(x, \alpha_1, \alpha_2)$ and replacing the nearby curves $y + \delta y$ and $y' + \delta y'$ with values corresponding to different parameters. In 1837—see Dorofeeva (*1998*) or Fraser (*1993*)—he finally solved the problem of finding sufficient conditions for an extremal. He included his solution in the lectures on dynamics that he gave in 1842, which were published in 1866, after his death. The complication that had held up Jacobi and others was the fact that sometimes the extremals with given endpoints are not unique. The most obvious example is the case of great circles on the sphere, which satisfy the Euler equations for the integral that gives arc length subject to fixed endpoints. If the endpoints happen to be antipodal points, all great circles passing through the two points have the same length. Weierstrass was later to call such pairs of points *conjugate points*. Jacobi gave a differential equation whose solutions had zeros at these points and showed that Legendre's criterion was correct, provided that the interval $(a, b]$ contained no points conjugate to a.

Weierstrass and his school. A number of important advances in the calculus of variations were due to Karl Weierstrass, such as the elimination of some of the more

---

[9] More than that was wrong, however, since great circles on a sphere satisfy Legendre's criteria, but do not give a minimum distance between their endpoints if they are more than 180° long.

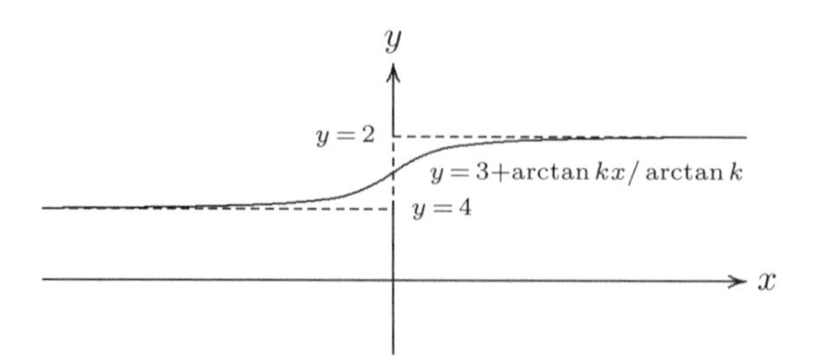

FIGURE 7. The functional $\Phi(y) = \int_{-1}^{+1} \left(xy'(x)\right)^2 dx$ does not assume its minimum value for continuously differentiable functions $y(x)$ satisfying $y(-1) = 2$, $y(+1) = 4$. The limiting position of a minimizing sequence is the dashed line.

restrictive assumptions about differentiability and taking account of the distinction between a lower bound and a minimum.[10]

An important example in this connection was Riemann's use of *Dirichlet's principle* to prove the Riemann mapping theorem, which asserts that any simply connected region in the plane except the plane itself can be mapped conformally onto the unit disk $\Delta = \{(x, y) : x^2 + y^2 < 1\}$. That principle required the existence of a real-valued function $u(x, y)$ that minimizes the integral

$$\iint\limits_{\Delta} \left(\frac{\partial u}{\partial x}\right)^2 + \left(\frac{\partial u}{\partial y}\right)^2 dx\, dy$$

among all functions $u(x, y)$ taking prescribed values on the boundary of the disk. That function is the unique harmonic function in Δ with the given boundary values. In 1870 Weierstrass called attention to the integral

$$\Phi(\varphi) = \int_{-1}^{+1} x^2 \left(\varphi'(x)\right)^2 dx\,,$$

which when combined with the boundary condition $\varphi(-1) = a$, $\varphi(+1) = b$, can be made arbitrarily small by the function

$$\varphi(x) = \frac{a+b}{2} + \frac{b-a}{2} \frac{\arctan(kx)}{\arctan(k)}\,,$$

yet (if $a \neq b$) cannot be zero for any function φ satisfying the boundary conditions and such that φ' exists at every point.

Weierstrass' example was a case where it was necessary to look outside the class of smooth functions for a minimum of the functional. The limiting position of the graphs of the functions for which the integral approximates its minimum value consists of the two horizontal lines from $(-1, a)$ to $(0, a)$, from $(0, b)$ to $(+1, b)$, and the section of the y-axis joining them (see Fig. 7).

Weierstrass thought of the smoothness assumptions as necessary evils. He recognized that they limited the generality of the results, yet he saw that without them no application of the calculus was possible. The result is a certain vagueness about the formulation of minimal principles in physics. A certain functional must be

[10] This distinction was pointed out by Gauss as early as 1799, in his criticism of d'Alembert's 1746 proof of the fundamental theorem of algebra.

a minimum *assuming* that all the relevant quantities are differentiable a sufficient number of times. Obviously, if a functional can be extended to a wider class of functions in a natural way, the minimum reached may be smaller, or the maximum larger. To make the restrictions as weak as possible, Weierstrass imposed the condition that the partial derivatives of the integrand should be continuous at corners. An extremal among all functions satisfying these less restrictive hypotheses was called a *strong* extremal. The corner condition was also found by G. Erdmann, a teacher at the Gymnasium in Königsberg, who proved that Jacobi's sufficient condition for a weak extremal was also necessary.

3.4. Foundations of the calculus. The British and Continental mathematicians both found the power of the calculus so attractive that they applied and developed it (sending forth new branches), all the while struggling to be clear about the principles they were using (extending its roots). The branches grew more or less continuously from the beginning. The development of the roots was slower and more sporadic. A satisfactory consensus was achieved only late in the nineteenth century, with the full development of real analysis, which is discussed in the Chapter 17.

The source of all the difficulty was the introduction of the infinite into analysis, in the form of infinitesimal reasoning. Leibniz believed in actual infinitesimals, levels of magnitude that were real, not zero, but so small that no accumulation of them could ever exceed any finite quantity. His dx was such an infinitesimal, and a product of two, such as $dx\,dy$ or dx^2, was a higher-order infinitesimal, so small that no accumulation of such could ever exceed any infinitesimal of the first order. On this view, even though theorems established using calculus were not absolutely accurate, the errors were below the threshold of human perception and therefore could not matter in practice. Newton was probably alluding to Leibniz when in his discussion of the quadrature of curves, he wrote, "Errores quam minimi in rebus mathematicis non sunt contemnendi" ("Errors, no matter how small, are not to be considered in mathematics"). Newton knew that his arguments could have been phrased using the Eudoxan method of exhaustion. In his *Principia* he wrote that he used his method of first and last ratios "to avoid the tediousness of deducing involved demonstrations *ad absurdum*, according to the method of the ancient geometers."

There seemed to be three approaches that would allow the operation that we now know as integration to be performed by antidifferentiation of tangents. One is the infinitesimal approach of Leibniz, characterized by Mancosu (*1989*) as "static." That is, a tangent is a state or position of a line, namely that of passing through two infinitely near points. The second is Newton's "dynamic" approach, in which a fluxion is the velocity of a moving object. The third is the ancient method of exhaustion. In principle, a reduction of calculus to the Eudoxan theory of proportion is possible. Psychologically, it would involve not only a great deal of tedium, as Newton noted, but also a great deal of unnecessary confusion, which he did not point out. If mathematicians had been shackled by the requirements of this kind of rigor, the amount of geometry and analysis created would have been incomparably less than it was. Still, Newton felt the objection and tried to phrase his exposition of the method of first and last ratios in such a way as not to outrage anyone's logical scruples. He said:

> Perhaps it may be objected, that there is no ultimate proportion of evanescent quantities; because the proportion, before the quantities have vanished, is not the ultimate; and when they are vanished, is [not defined]. But by the same argument it may be alleged that a body arriving at a certain place, and there stopping, has no ultimate velocity, because the velocity before the body comes to the place, is not its ultimate velocity; when it has arrived, there is none. But the answer is easy; for by the ultimate velocity is meant that with which the body is moved, neither before it arrives at its last place and the motion ceases, nor after, but at the very instant it arrives.

Was this explanation adequate? Do human beings in fact have any conception of what is meant by an instant of time? Do we have a clear idea of the velocity of a body *at the very instant* when it stops moving? Or do some people only imagine that we do? We are here very close to the arrow paradox of Zeno. At any given instant, the arrow does not move; therefore it is at rest. How can there be a motion (a traversal of a positive distance) as a result of an accumulation of states of rest, in each of which no distance is traveled? Newton's "by the same argument" practically invited the further objection that his attempted explanation merely stated the same fallacy in a new way.

That objection was raised in 1734 by the philosopher George Berkeley[11] (1685–1753, Anglican Bishop of Cloyne, Ireland), for whom the city of Berkeley[12] in California is named. Berkeley first took on Newton's fluxions:

> It is said that the minutest errors are not to be neglected in mathematics: that the fluxions are celerities [speeds], not proportional to the finite increments, though ever so small; but only to the moments or nascent increments, whereof the proportion alone, and not the magnitude, is considered. And of the aforesaid fluxions there be other fluxions, which fluxions of fluxions are called second fluxions. And the fluxions of the second fluxions are called third fluxions: and so on, fourth, fifth, sixth, &c. *ad infinitum*. Now, as our sense is strained and puzzled with the perception of objects extremely minute, even so the imagination, which faculty derives from sense, is very much strained and puzzled to frame clear ideas of the least particles of time... and much more so to comprehend... those increments of the flowing quantities... in their very first origin, or beginning to exist, before they become finite particles... The incipient celerity of an incipient celerity, the nascent augment of a nascent augment, *i.e.*, of a thing which hath no magnitude: take it in what light you please, the clear conception of it will, if I mistake not, be found impossible.

He then proceeded to attack the views of Leibniz:

---

[11] Pronounced "Barkley."

[12] Pronounced "Birkley".

> The foreign mathematicians are supposed by some, even of our own, to proceed in a manner less accurate, perhaps, and geometrical, yet more intelligible... Now to conceive a quantity infinitely small, that is, infinitely less than any sensible or imaginable quantity or than any the least finite magnitude is, I confess, above my capacity. But to conceive a part of such infinitely small quantity that shall be still infinitely less than it, and consequently though multiplied infinitely shall never equal the minutest finite quantity, is, I suspect, an infinite difficulty to any man whatsoever.

Berkeley analyzed a curve whose area up to x was x^3 (he wrote xxx). If $z - x$ was the increment of the abscissa and $z^3 - x^3$ the increment of area, the quotient would be $z^2 + zx + x^2$. He said that, if $z = x$, of course this last expression is $3x^2$, and that must be the ordinate of the curve in question. That is, its equation must be $y = 3x^2$. But, he pointed out,

> [H]erein is a direct fallacy: for, in the first place, it is supposed that the abscisses z and x are unequal, without which supposition no one step could have been made [that is, the division by $z - x$ would have been undefined]; which is a manifest inconsistency, and amounts to the same thing that hath been before considered... The great author of the method of fluxions felt this difficulty, and therefore he gave in to those nice abstractions and geometrical metaphysics without which he saw nothing could be done on the received principles... It must, indeed, be acknowledged that he used fluxions, like the scaffold of a building, as things to be laid aside or got rid of as soon as finite lines were found proportional to them... And what are these fluxions? The velocities of evanescent increments? And what are these same evanescent increments? They are neither finite quantities, nor quantities infinitely small, nor yet nothing. May we not call them the ghosts of departed quantities?

The debate on the Continent. Calculus disturbed the metaphysical assumptions of philosophers and mathematicians on the Continent as well as in Britain. L'Hospital's textbook had made two explicit assumptions: first, that if a quantity is increased or diminished by a quantity that is infinitesimal in comparison with itself, it may be regarded as remaining unchanged. Second, that a curve may be regarded as an infinite succession of straight lines. L'Hospital's justification for these claims was not commensurate with the strength of the assumptions. He merely said:

> [T]hey seem so obvious to me that I do not believe they could leave any doubt in the mind of attentive readers. And I could even have proved them easily after the manner of the Ancients, if I had not resolved to treat only briefly things that are already known, concentrating on those that are new. [Quoted by Mancosu, *1989*, p. 228]

The idea that $x + dx = x$, implicit in l'Hospital's first assumption, leads algebraically to the equation $dx = 0$ if equations are to retain their previous meaning. Rolle raised this objection and was answered by the claim that dx represents the distance traveled in an instant of time by an object moving with finite velocity. This debate was carried on in private in the Paris Academy during the first decade of the eighteenth century, and members were at first instructed not to discuss it in public, as if it were a criminal case! Rolle's criticism could be answered, but it was *not* answered at the time. According to Mancosu (*1989*), the matter was settled in a most unacademic manner, by making l'Hospital into an icon after his death in 1704. His eulogy by Bernard Lebouyer de Fontenelle (1657–1757) simply declared the anti-infinitesimalists wrong, as if the Academy could decide metaphysical questions by fiat, just as it can define what is proper usage in French:

> [T]hose who knew nothing of the mysteries of this new infinitesimal geometry were shocked to hear that there are infinities of infinities, and some infinities larger or smaller than others; for they saw only the top of the building without knowing its foundation. [Quoted by Mancosu, *1989*, 241]

In the eighteenth century, however, better expositions of the calculus were produced by d'Alembert. In his article on the differential for the famous *Encyclopédie* he wrote that 0/0 could be equal to anything, and that the derivative $\frac{dy}{dx}$ was not actually 0 divided by 0, but the limit of finite quotients as numerator and denominator tended to zero.

Lagrange's algebraic analysis. The attempt to be clear about infinitesimals or to banish them entirely took many forms during the eighteenth and nineteenth centuries. One of the most prominent (see Fraser, *1987*) was Lagrange's exposition of analytic functions. Lagrange understood the term *function* to mean a formula composed of symbols representing variables and arithmetic operations. He argued that "in general" (with certain obvious exceptions) every function $f(x)$ could be expanded as a power series, based on Taylor's theorem, for which he provided his own form of the remainder term. Using an argument that resembles the one given by Ruffini and Abel to prove the insolvability of the quintic, he claimed that the hypothetical expansion

$$\sqrt{x+h} = \sqrt{x} + ph + qh^2 + \cdots + h^{m/n}$$

could not occur, since the left-hand side has only two values, while the right-hand side has n values. In this way, he ruled out fractional exponents. Negative exponents were ruled out by the mere fact that the function was defined at $h = 0$. The determinacy property of analytic functions was used implicitly by Lagrange when he assumed that any zero of a function must have finite order, as we would say (Fraser, *1987*, p. 42).

The advantage of confining attention to functions defined by power series is that the derivative and integral of such a function have a perfectly definite meaning. Lagrange advocated it on the grounds that it pointed up the qualitative difference between the new functions produced by infinitesimal analysis: dx was a completely different function from x.

Cauchy's calculus. The modern presentation of calculus owes a great deal to the textbooks of Cauchy, written for his lectures at the École Polytechnique during the 1820s.[13] Cauchy recognized that calculus could not get by without something

[13] Although we have mentioned particular results of Cauchy in connection with the solution of algebraic and differential equations, his treatises on analysis are the contributions for which he is

equivalent to infinitesimals. He defined a function $f(x)$ to be continuous if the absolute value of the difference $f(x+\alpha)-f(x)$ "decreases without limit along with that of α." He continues:

> In other words, *the function $f(x)$ remains continuous with respect to x in a given interval, if an infinitesimal increase in the variable within this interval always produces an infinitesimal increase in the function itself.*

Certain distinctions that we now make to clarify whether x is a fixed point or the increase is thought of as occurring at all points simultaneously are not stated here. In particular, uniform convergence and continuity are assumed but not stated. Cauchy defined a limit in terms of the "successive values attributed to a variable," approaching a fixed value and ultimately differing from it by an arbitrarily small amount. This definition can be regarded as an informal version of what we now state precisely with deltas and epsilons, and Cauchy is generally regarded, along with Weierstrass, as being one of the people who finally made the foundations of calculus secure. Yet Cauchy's language clearly presumes that infinitesimals are real. As Laugwitz (*1987*, p. 272) says:

> All attempts to understand Cauchy from a 'rigorous' theory of real numbers and functions including uniformity concepts have failed... One advantage of modern theories like the Nonstandard Analysis of Robinson... [which includes infinitesimals] is that they provide consistent reconstructions of Cauchy's concepts and results in a language which sounds very much like Cauchy's.

The secure foundation of modern analysis owes much to Cauchy's treatises. As Grabiner (*1981*) says, he applied ancient Greek rigor and modern algebraic techniques to derive results from analysis. The contributions of other nineteenth-century mathematicians to this rigor are discussed in Chapter 17.

Questions and problems

16.1. Show that the Madhava–Jyeshtadeva formula given at the beginning of the chapter is equivalent to

$$\theta = \sum_{k=0}^{\infty}(-1)^k \frac{\tan^{2k+1}\theta}{2k+1},$$

or, letting $x = \tan\theta$,

$$\arctan x = \sum_{k=0}^{\infty}(-1)^k \frac{x^{2k+1}}{2k+1}.$$

16.2. Consider an ellipse with semiaxes a and b and a circle of radius b, both circle and ellipse lying between a pair of parallel lines a distance $2b$ apart. For every line between the two lines and parallel to them, show that the portion inside the ellipse will be a/b times the portion inside the circle. Use this fact and Cavalieri's principle to compute the area of the ellipse. This result was given by Kepler.

---

best remembered. Incidentally, he became a mathematician only after practicing as an engineer for several years.

16.3. Show that the point at which the tangent to the curve $y = f(x)$ intersects the y axis is $y = f(x) - xf'(x)$, and verify that the area under this curve—more precisely, the integral of $f(x) - xf'(x)$ from $x = 0$ to $x = a$—is twice the area between the curve $y = f(x)$ and the line $ay = f(a)x$ between the points $(0, 0)$ and $(a, f(a))$. This result was used by Leibniz to illustrate the power of his infinitesimal methods.

16.4. Recall that Eudoxus solved the problem of incommensurables by changing the definition of proportion, or rather, *making* a definition to cover cases where no definition existed before. Newton's "theorem" asserting that quantities that approach each other continually (we would say monotonically) and become arbitrarily close to each other in a finite time must become equal in an infinite time assumes that one has a definition of equality at infinity. What is the definition of equality at infinity? Since we cannot *actually* reach infinity, the definition will have to be stated as a potential infinity, that is, a statement about all possible finite times. Formulate the definition and compare Newton's solution of this difficulty with Eudoxus' solution of the problem of incommensurables.

16.5. Draw a square and one of its diagonals. Then draw a very fine "staircase" by connecting short horizontal and vertical line segments in alternation, each segment crossing the diagonal. The total length of the horizontal segments is the same as the side of the square, and the same is true of the vertical segments. Now in a certain intuitive sense these segments approximate the diagonal of the square, seeming to imply that the diagonal of a square equals twice its side, which is absurd. Does this argument show that the method of indivisibles is wrong?

16.6. In the passage quoted from the *Analyst*, Berkeley asserts that the experience of the senses provides the only foundation for our imagination. From that premise he concludes that we can have no understanding of infinitesimals. Analyze whether the premise is true, and if so, whether it implies the conclusion. Assuming that our thinking processes have been shaped by the evolution of the brain, for example, is it possible that some of our spatial and counting intuition is "hard-wired" and not the result of any previous sense impressions? The philosopher Immanuel Kant (1724–1804) thought so. Do we have the power to make correct judgments about spaces and times on scales that we have not experienced? What would Berkeley have said if he had heard Riemann's argument that space may be finite, yet unbounded? How would he have explained the modern computer chip, on which unimaginable amounts of data can be recorded in space far too small for the senses to perceive? Go a step further and consider how quantum mechanics is understood and interpreted.

CHAPTER 17

Real and Complex Analysis

In the mid-1960s Walter Rudin (b. 1921), the author of a number of standard graduate textbooks in mathematics, wrote a textbook with the title *Real and Complex Analysis*, aimed at showing the considerable unity and overlap between the two subjects. It was necessary to write such a book because the two subjects, while sharing common roots in the calculus, had developed quite differently. The contrasts between the two are considerable. Complex analysis considers the smoothest, most orderly possible functions, those that are analytic, while real analysis allows the most chaotic imaginable functions. Complex analysis was, to pursue our botanical analogy, fully a "branch" of calculus, and foundational questions hardly entered into it. Real analysis had a share in both roots and branches, and it was intimately involved in the debate over the foundations of calculus.

What caused the two varieties of analysis to become so different? Both are dealing with functions, and both evolved under the stimulus of the differential equations of mathematical physics. The central point is the concept of a function. We have already seen the early definitions of this concept by Leibniz and Johann Bernoulli. All mathematicians from the seventeenth and eighteenth centuries had an intuitive picture of a function as a formula or expression in which variables are connected by rules derived from algebra or geometry. A function was regarded as continuous if it was given by a single formula throughout its range. If the formula changed, the function was called "mechanical" by Euler. Although "mechanical" functions may be continuous in the modern sense, they are not usually analytic. All the "continuous" functions in the older sense are analytic. They have power-series expansions, and those power-series expansions are often sufficient to solve differential equations. As a general signpost indicating where the paths diverge, the path of power-series expansions and the path of trigonometric-series expansions is a very good guide. A consequence of the development was that real-variable theory had to deal with very irregular and "badly behaved" functions. It was therefore in real analysis that the delicate foundational questions arose.

1. Complex analysis

Calculus began with a limited stock of geometry: a few curves and surfaces, all of which could be described analytically in terms of rational, trigonometric, exponential, and logarithmic functions of real variables. Soon, however, calculus was used to formulate problems in mathematical physics as differential equations. To solve those equations, the preferred technique was integration, but where integration failed, power series were the technique of first resort. These series automatically brought with them the potential of allowing the variables to assume complex values. But then integration and differentiation had to be suitably defined for complex functions of a complex variable. The result was a theory of analytic functions of a complex variable whose range was much vaster than the materials that led to its creation.

In his 1748 *Introductio*, Euler emended the definition of a function, saying that a function is an *analytic expression* formed from a variable and constants. The rules for manipulating symbols were agreed on as long as only finite expressions were involved. But what did the symbols *represent*? Euler stated that variables were allowed to take on negative and imaginary values. Thus, even though the physical quantities the variables represented were measured as *positive rational* numbers, the algebraic and geometric properties of negative, irrational, and complex numbers could be invoked in the analysis. The extension from finite to infinite expressions was not long in coming. The extension of the calculus to complex numbers turned out to have monumental importance.

Lagrange undertook to reformulate the calculus in his treatises *Théorie des fonctions analytiques* (1797) and *Leçons sur le calcul des fonctions* (1801), basing it entirely on algebraic principles and stating as a fundamental premise that the functions to be considered are those that can be expanded in power series (having no negative or fractional powers of the variable). With this approach the derivatives of a function need not be defined as ratios of infinitesimals, since they can be defined in terms of the coefficients of the series that represents the function. Functions having a power series representation are known nowadays as *analytic functions* from the title of Lagrange's work.

1.1. Algebraic integrals. Early steps toward complexification were taken only on a basis of immediate necessity. As we have already seen, the applications of calculus in solving differential equations made the computation of integrals extremely important. Where computing the derivative never leads outside the class of elementary functions and leaves algebraic functions algebraic, trigonometric functions trigonometric, and exponential functions exponential, integrals are a very different matter. Algebraic functions often have nonalgebraic integrals, as Leibniz realized very early. The relation we now write as

$$\arccos(1-x) = \int_0^x \frac{1}{\sqrt{2t-t^2}}\,dt$$

was written by him as

$$a = \int dx : \sqrt{2x - x^2}\,,$$

where $x = 1 - \cos a$.[1] Eighteenth-century mathematicians were greatly helped in handling integrals like this by the use of trigonometric functions. It was therefore natural that they would see the analogy when more complicated integrals came to be considered. Such problems arose from the study of pendulum motion and the rotation of solid bodies in physics, but we shall illustrate it with examples from pure geometry: the rectification of the ellipse and the division of the lemniscate into equal arcs. For the circle, we know that the corresponding problems lead to the integral

$$\int_0^x \frac{1}{\sqrt{1-t^2}}\,dt$$

[1] The limits of integration that we now use were introduced by Joseph Fourier in the nineteenth century.

for rectification and an equation

$$\int_0^y \frac{1}{\sqrt{1-t^2}}\,dt = \frac{1}{n}\int_0^x \frac{1}{\sqrt{1-t^2}}\,dt\,,$$

which can be written in differential form as

$$\frac{dx}{\sqrt{1-x^2}} = \frac{n\,dy}{\sqrt{1-y^2}}\,,$$

for the division of an arc.

Trigonometry helps to solve this last equation. Instead of the arccosine function that the integral actually represents, it makes more sense to look at the inverse of it, the cosine function. This function provides an algebraic equation through its addition formula,

$$a_0 y^n - a_2 y^{n-2} + a_3 y^{n-4} - \cdots = x\,,$$

relating the abscissas of the end of the given arc (x) and the end of the nth part of it (y). The algebraic nature of this equation determines whether the division problem can be solved with ruler and compass. In particular, for $n = 3$ and a 60-degree arc ($x = 1/2$), for which the equation is $4y^3 - 3y = 1/2$, such a solution does not exist. Thus the problems of computing arc length on a circle and equal division of its arcs lead to an interesting combination of algebra, geometry, and calculus. Moreover, the periodicity of the inverse function makes this equation easy to solve (see Problem 17.1).

The division problem was fated to play an important role in study of integrals of algebraic functions. The Italian nobleman Fagnano (1682–1766) studied the problem of rectifying the lemniscate, whose polar equation is $r^2 = 2\cos(2\theta)$. Its element of arc is $\sqrt{2}(1 - 2\sin^2\theta)^{-1/2}\,d\theta$, and the substitution $u = \tan\theta$ turns this expression into $\sqrt{2}(1-u^4)^{-1/2}\,du$. Thus, the rectification problem involves evaluating the integral

$$\int_0^x \frac{\sqrt{2}}{\sqrt{1-u^4}}\,du\,,$$

while the division problem involves solving the differential equation

$$\frac{dz}{\sqrt{1-z^4}} = \frac{n\,du}{\sqrt{1-y^4}}\,.$$

Fagnano gave the solution for $n = 2$ as the algebraic relation

$$\frac{u\sqrt{2}}{\sqrt{1-u^4}} = \frac{1}{z}\sqrt{1-\sqrt{1-z^4}}\,.$$

Euler observed the analogy between these integrals and the circular integrals just discussed, and suggested that it would be reasonable to study the inverse function. But Euler lived at a time when the familiar functions were still the elementary ones. He found a large number of integrals that could be expressed in terms of algebraic, logarithmic, and trigonometric functions and showed that there were others that could not be so expressed.

Legendre, Jacobi, and Abel. The foundation for further work in integration was laid by Legendre, who invented the term *elliptic integral.* Off and on for some 40 years between 1788 and 1828, he thought about integrals like those of Fagnano and Euler, classified them, computed their values, and studied their properties. He found their algebraic addition formulas and thereby reduced the division problem for these integrals to the solution of algebraic equations. Interestingly, he found that whereas the division problem requires solving an equation of degree n for the circle, it requires solving an equation of degree n^2 for the ellipse. After publishing his results as exercises in integral calculus in 1811, he wrote a comprehensive treatise in the 1820s. As he was finishing the third volume of this treatise he discovered a new set of transformations of elliptic integrals that made their computation easier. (He already knew one set of such transformations.) Just after the treatise appeared in 1827, he found to his astonishment that Jacobi had discovered the same transformations, along with others, and had connected them with the division problem. Jacobi's results in turn were partially duplicated by those of Abel.

Abel, who admired Gauss, was proud of having achieved the division of the lemniscate into 17 equal parts,[2] just as Gauss had done for the circle. The secret for the circle was to use the algebraic addition formula for trigonometric functions. For the lemniscate, as Legendre had shown, the equation was of higher degree. Abel was able to solve it by using complex variables, and in the process, he discovered that the inverse functions of the elliptic integrals, when regarded as functions of a complex variable, were *doubly periodic.* The double period accounted for the fact that the division equation was of degree n^2 rather than n. Without complex variables, the theory of elliptic integrals would have been a disconnected collection of particular results. With them, a great simplicity and unity was achieved. Abel went on to study algebraic addition formulas for very general integrals of the type

$$\int R\bigl(x, y(x)\bigr)\, dx\,,$$

where $R(x, y)$ is a rational function of x and y and $y(x)$ satisfies a polynomial equation $P\bigl(x, y(x)\bigr) = 0$. Such integrals are now called *Abelian integrals* in his honor. In particular, he established that for each polynomial $P(x, y)$ there was a number p, now called the *genus* of $P(x, y)$, such that a sum of any number of integrals $R(x, y)$ with different limits could be expressed in terms of just p integrals, whose limits of integration were rational functions of those in the given sum. For elliptic integrals, $p = 1$, and that is the content of the algebraic addition formulas discovered by Legendre. For a more complicated integral, say

$$\int \frac{1}{\sqrt{q(x)}}\, dx\,,$$

where $q(x)$ is a polynomial of degree 5 or higher, the genus may be higher. If $P(x, y) = y^2 - q(x)$, where the polynomial q is of degree $2p + 1$ or $2p + 2$, the genus is p.

After Abel's premature death, Jacobi continued to develop algebraic function theory. In 1832, he realized that for algebraic integrals of higher genus, the inverse functions could not be well defined, since there were p integrals and only one equation connecting them to the variable in terms of which they were to be expressed. He therefore had the idea of adjoining extra equations in order to get well-behaved

[2] Or, more generally, a Fermat prime number of parts.

inverses. For example, if $q(x)$ is of degree 5, he posed the problem of solving for x and y in terms of u and v in the equations

$$\begin{aligned} u &= \int_0^x \frac{1}{\sqrt{q(t)}}\,dt + \int_0^y \frac{1}{\sqrt{q(t)}}\,dt \\ v &= \int_0^x \frac{t}{\sqrt{q(t)}}\,dt + \int_0^y \frac{t}{\sqrt{q(t)}}\,dt\,. \end{aligned}$$

This problem became known as the *Jacobi inversion problem*. Solving it took a quarter of a century and led to progress in both complex analysis and algebra.

Jacobi himself gave it a start in connection with elliptic integrals. Although a nonconstant function that is analytic in the whole plane cannot be doubly periodic (because its absolute value cannot attain a maximum), a quotient of such functions can be, and Jacobi found the ideal numerators and denominators to use for expressing the doubly periodic elliptic functions as quotients: theta functions. The secret of solving the Jacobi inversion problem was to use theta functions in more than one complex variable, but working out the proper definition of those functions and the mechanics of applying them to the problem required the genius of Riemann and Weierstrass. These two giants of nineteenth-century mathematics solved the problem independently and simultaneously in 1856, but considerable preparatory work had been done in the meantime by other mathematicians. The importance of algebraic functions as the basic core of analytic function theory cannot be overemphasized. Klein (*1926*, p. 280) goes so far as to say that Weierstrass' purpose in life was

> to conquer the inversion problem, even for hyperelliptic integrals of arbitrarily high order, as Jacobi had foresightedly posed it, perhaps even the problem for general Abelian integrals, using rigorous, methodical work with power series (including series in several variables).
>
> It was in this way that the topic called the Weierstrass theory of analytic functions arose as a by-product.

1.2. Cauchy. Cauchy's name is associated most especially with one particular approach to the study of analytic functions of a complex variable, that based on complex integration. A complex variable is really two variables, as Cauchy was saying even as late as 1821. But a function is to be given by the same symbols, whether they denote real or complex numbers. When we integrate and differentiate a given function, which variable should we use? Cauchy discovered the answer, as early as 1814, when he first discussed such questions in print. The value of the function is also a pair of real numbers $u + iv$, and if the derivative is to be independent of the variable on which it is taken, these must satisfy the equations we now call the Cauchy–Riemann equations:

$$\frac{\partial u}{\partial x} = \frac{\partial v}{\partial y}; \quad \frac{\partial u}{\partial y} = -\frac{\partial v}{\partial x}\,.$$

In that case, as Cauchy saw, if we are integrating $u + iv$ in a purely formal way, separating real and imaginary parts, over a path from the lower left corner of a rectangle (x_0, y_0) to its upper right corner (x_1, y_1), the same result is obtained whether the integration proceeds first vertically, then horizontally or first horizontally, then vertically. As Gauss had noted as early as 1811, Cauchy observed that the function

$1/(x+iy)$ did not have this property if the rectangle contained the point $(0,0)$. The difference between the two paths was $2\pi i$, which Cauchy called the *residue.* Over the period from 1825 to 1840, Cauchy developed from this theorem what is now known as the Cauchy integral theorem, the Cauchy integral formula, Taylor's theorem, and the calculus of residues. The Cauchy integral theorem states that if γ is a curve enclosing a region in which $f(z)$ has a derivative then

$$\int_\gamma f(z)\,dz = 0\,.$$

If the real and imaginary parts of this integral are written out and compared with the Cauchy–Riemann equations, this formula becomes a simple consequence of what is known as Green's theorem (the two-variable version of the divergence theorem), published in 1828 by George Green (1793–1841) and simultaneously in Russia by Mikhail Vasilevich Ostrogradskii (1801–1862). When combined with the fact that the integral of $1/z$ around a curve that winds once around 0 is $2\pi i$, this theorem immediately yields as a consequence the Cauchy integral formula

$$f(z_0) = \frac{1}{2\pi i}\int_\gamma \frac{f(z)}{z-z_0}\,dz\,.$$

When generalized, this formula becomes the residue theorem. Also from it, one can obtain estimates for the size of the derivatives. Finally, by expanding the denominator as a geometric series in powers of $z - z_1$, where z_1 lies inside the curve γ, one can obtain the Taylor series expansion of $f(z)$. These theorems form the essential core of modern first courses in complex analysis. This work was supplemented by a paper of Pierre Laurent (1813–1854), submitted to the Paris Academy in 1843, in which power series expansions about isolated singularities (Laurent series) were studied.

Cauchy was aware of the difficulties that arise in the case of multivalued functions and introduced the idea of a boundary curve (*ligne d'arrêt*) to prevent a function from assuming more than one value at a given point. As mentioned in Chapter 12, his student Puiseux studied the behavior of algebraic functions in the neighborhood of what we now call branch points, which are points c such that the function assumes many different values at each point of every neighborhood of c. Puiseux showed that at a branch point c near which there are n values of the function each of the n values of the function could be expanded in its own series of powers of a variable u such that $u^n = x-c$. The work of Cauchy, Laurent, Puiseux, and others thus brought complex analysis into existence as a well-articulated theory containing important principles and theorems.

1.3. Riemann. The work of Puiseux on algebraic functions of a complex variable was to be subsumed in two major papers of Riemann. The first of these, his doctoral dissertation, contained the concept now known as a Riemann surface. It was aimed especially at simplifying the study of an algebraic function $w(z)$ satisfying a polynomial equation $P\bigl(z, w(z)\bigr) \equiv 0$. In a sense, the Riemann surface revealed that all the significant information about the function was contained precisely in its singularities—the way it branched at its branch points. Information about the surface was contained in its *genus*, defined as half the total number of branch points, counted according to order, less the number of sheets in the surface, plus 1.[3] The Riemann surface of $w = \sqrt{z}$, for example, has two branch points (0 and ∞), each of order 1, and two sheets, resulting in genus 0. Riemann's geometric approach to the subject brought out the duality between surfaces and mappings of them, encapsulated in a formula known as the Riemann–Roch theorem (after Gustav Roch, 1839–1866). This formula connects the dimension of the space of functions on a Riemann surface having prescribed zeros and poles with the genus of the surface. Although it is a simple formula to write down, explaining the meaning of the terms in it requires considerable space, and so we omit the details.

In 1856 Riemann used his theory to give a very elegant solution of the Jacobi inversion problem. Since an analytic function must be constant if it has no poles on a Riemann surface, it was possible to use the periods of the integrals that occur in the problem to determine the function up to a constant multiple and then to find quotients of theta functions having the same periods, thereby solving the problem.

1.4. Weierstrass. Of the three founders of analytic function theory, Weierstrass was the most methodical. He had found his own solution to the Jacobi inversion problem and submitted it simultaneously with Riemann. When he saw Riemann's work, he withdrew his own paper and spent many years working out in detail how the two approaches related to each other. Where Riemann had allowed his geometric intuition to create castles in the air, so to speak, Weierstrass was determined to have a firm algebraic foundation. Instead of picturing kinematically a point wandering from one sheet of a Riemann surface to another, Weierstrass preferred a static object that he called a *Gebilde* (*structure*). His *Gebilde* was based on the set of pairs of complex numbers (z, w) satisfying a polynomial equation $p(z, w) = 0$, where $p(z, w)$ was an irreducible polynomial in the two variables. These pairs were supplemented by certain ideal points of the form (z, ∞), (∞, w), or (∞, ∞) when one or both of w or z tended to infinity as the other approached a finite or infinite value. Around all but a finite set of points, it was possible to expand w in an ordinary Taylor series in nonnegative integer powers of $z - z_0$. For each of the exceptional points, there would be one or more expansions in fractional or negative powers of $z - z_0$, as Puiseux and Laurent had found. These power series were Weierstrass' basic tool in analytic function theory.

Comparison of the three approaches. At first sight, it appears that Cauchy's approach, which is simultaneously analytic and geometric, subsumes the work of both Riemann and Weierstrass. Riemann, to be sure, had a more elegant way of overcoming the difficulty presented by multivalued functions, but Cauchy and Puiseux between them came very close to doing something logically equivalent. Weierstrass begins with the power series and considers only functions that have a power-series development, whereas Cauchy assumes only that the function is continuously differentiable.[4] On the other hand, before you can verify Cauchy's basic assumption that a function is differentiable, you have to know what the function is. How is that information to be communicated, if not through some formula like a power series? Weierstrass saw this point clearly; in 1884 he said, "No matter how you twist and turn, you cannot avoid using some sort of analytic expressions" (quoted by Siegmund-Schultze, *1988*, p. 253).

[3] Klein (*1926*, p. 258) ascribes this definition to Alfred Clebsch (1833–1872).

[4] It was shown by Edouard Goursat (1858–1936) in 1900 that differentiability implies continuous differentiability on open subsets of the plane.

2. Real analysis

In complex analysis attention is restricted from the outset to functions that have a complex derivative. That very strong assumption automatically ensures that the functions studied will have convergent Taylor series. If only mathematical physics could manage with just such smooth functions, the abstruse concepts that fill up courses in real analysis would not be needed. But the physical world is full of boundaries, where the density of matter is discontinuous, temperatures undergo abrupt changes, light rays reflect and refract, and vibrating membranes are clamped. For these situations the imaginary part of the variable, which often has no physical interpretation anyway, might as well be dropped, since its only mathematical role was to complete the analytic function. From that point on, analysis proceeds on the basis of real variables only. Real analysis, which represents another extension of calculus, has to deal with much more general and "rough" functions. All of the logical difficulties about calculus poured into this area of analysis, including the important questions of convergence of series, existence of maxima and minima, allowable ways of defining functions, continuity, and the meaning of integration. As a result, real analysis is so much less unified than complex analysis that it hardly appears to be a single subject. Its basic theorems do not follow from one another in any canonical order, and their proofs tend to be a bag of special tricks, rarely remembered for long except by professors who lecture on the subject constantly.

The free range of intuition suffered only minor checks in complex analysis. In that subject, what one wanted to believe very often turned out to be true. But real analysis almost seemed to be trapped in a hall of mirrors at times, as it struggled to gain the freedom to operate while avoiding paradoxes and contradictions. The generality of operations allowed in real analysis has fluctuated considerably over the centuries. While Descartes had imposed rather strict criteria for allowable curves (functions), Daniel Bernoulli attempted to represent very arbitrary functions as trigonometric series, and the mathematical physicist André-Marie Ampère (1775–1836) attempted to prove that a continuous function (in the modern sense, but influenced by preconceptions based on the earlier sense) would have a derivative at most points. The critique of this proof was followed by several decades of backtracking, as more and more exceptions were found for operations with series and integrals that appeared to be formally all right. Eventually, when a level of rigor was reached that eradicated the known paradoxes, the time came to reach for more generality. Georg Cantor's set theory played a large role in this increasing generality, while developing paradoxes of its own. In the twentieth century, the theories of generalized functions and distributions restored some of the earlier freedom by inventing a new object to represent the derivative of functions that have no derivative in the ordinary sense.

2.1. Fourier series, functions, and integrals. There is a symmetry in the development of real and complex analysis. Broadly speaking, both arose from differential equations, and complex analysis grew out of power series, while real analysis grew out of trigonometric series. These two techniques, closely connected with each other through the relation $z^n = r^n(\cos n\theta + i \sin n\theta)$, led down divergent paths that nevertheless crossed frequently in their meanderings. The real and complex viewpoints in analysis began to diverge with the study of the vibrating string problem in the 1740s by d'Alembert, Euler, and Daniel Bernoulli.

For a string fastened at two points, say $(0, 0)$ and $(L, 0)$ and vibrating so that its displacement above or below the point $(x, 0)$ at time t is $y(x, t)$, mathematicians agreed that the best compromise between realism and comprehensibility to describe this motion was the *one-dimensional wave equation*, which d'Alembert studied in 1747,[5] publishing the results in 1749:

$$\frac{\partial^2 y}{\partial t^2} = c^2 \frac{\partial^2 y}{\partial x^2}.$$

D'Alembert pointed out that the solution must be of the form

$$y(x, t) = \Psi(t + x) + \Gamma(t - x),$$

where for simplicity he assumed that $c = 1$. The equation alone does not determine the function, of course, since the vibrations depend on the initial position and velocity of the string. Accordingly, d'Alembert followed up with a prescribed initial position $f(x) = y(x, 0)$ and velocity $v(x) = \frac{\partial y}{\partial t}\big|_{t=0}$. He considered first the case when the initial position is identically zero, for which the function Ψ must be an even function of period $2L$, then the more general case.

The following year Euler took up this problem and commented on d'Alembert's solution. He observed that the initial position could be any shape at all, "either regular or irregular and mechanical." D'Alembert found that claim hard to accept. After all, the functions Ψ and Γ had to have periodicity and parity properties. How else could they be defined except as power series containing only odd or only even powers? Euler and d'Alembert were not interpreting the word "function" in the same way. Euler was even willing to consider initial positions $f(x)$ with corners (a "plucked" string), whereas d'Alembert insisted that $f(x)$ must have two derivatives simply to satisfy the equation.

Three years later Daniel Bernoulli tried to straighten this matter out, giving a solution in the form

$$y(x, t) = \sum_{n=1}^{\infty} a_n \sin\left(\frac{n\pi x}{L}\right) \cos\left(\frac{n\pi ct}{L}\right),$$

which he did not actually write out. Here the coefficients a_n were to be chosen so that the initial condition was satisfied, that is,

$$f(x) = y(x, 0) = \sum_{n=1}^{\infty} a_n \sin\left(\frac{n\pi x}{L}\right).$$

Observing that he had an infinite set of coefficients at his disposal for "fitting" the function, Bernoulli claimed that "any" function $f(x)$ had such a representation. Bernoulli's solution was the first of many instances in which the classical partial differential equations of mathematical physics—the wave, heat, and potential equations—were studied by separating variables and superposing the resulting solutions. The technique was ultimately to lead to what are called Sturm–Liouville problems, which we shall mention again below.

Before leaving the wave equation, we must mention one more important crossing between real and complex analysis in connection with it. In studying the action

[5] Thirty years earlier Brook Taylor (1685–1731) had analyzed the problem geometrically and concluded that the normal acceleration at each point would be proportional to the normal curvature at that point. That statement is effectively the same as this equation, and was quoted by d'Alembert.

of gravity, Pierre-Simon Laplace (1749–1827) was led to what is now known as Laplace's equation in three variables. The two-variable version of this equation is

$$\frac{\partial^2 u}{\partial x^2} + \frac{\partial^2 u}{\partial y^2} = 0.$$

The operator on the left-hand side of this equation is known as the *Laplacian.* Since Laplace's equation can be thought of as the wave equation with velocity $c = \sqrt{-1}$, complex numbers again enter into a physical problem. Recalling d'Alembert's solution of the wave equation, Laplace suggested that the solutions of his equation might be sought in the form $f(x + y\sqrt{-1}) + g(x - y\sqrt{-1})$. Once again a problem that started out as a real-variable problem led inexorably to the need to study functions of a complex variable.

The definition of a function. Daniel Bernoulli accepted his father's definition of a function as "an expression formed in some manner from variables and constants," as did Euler and d'Alembert. But those words seemed to have different meanings for each of them. Daniel Bernoulli thought that his solution met the criterion of being "an expression formed from variables and constants." His former colleague in the Russian Academy of Sciences,[6] Euler, saw the matter differently. This time it was Euler who argued that the concept of function was being used too loosely. According to him, since the right-hand side of Bernoulli's formula consisted of odd functions of period $2L$, it could represent only an odd function of period $2L$. Therefore, he said, it did not have the generality of the solution he and d'Alembert had given. Bottazzini (*1986*, p. 29) expresses the situation very well, saying, "We are here facing a misunderstanding that reveals one aspect of the contradictions between the old and new theory of functions, even though they are both present in the same man, Euler, the protagonist of this transformation." The difference between the old and new concepts is seen in the simplest example, the function $|x|$, which equals x when $x \geq 0$ and $-x$ for $x \leq 0$. We have no difficulty thinking of this function as one function. It appeared otherwise to nineteenth-century mathematicians. Fourier described what he called a "discontinuous function represented by a definite integral" in 1822: the function

$$\frac{2}{\pi}\int_0^{\infty} \frac{\cos qx}{1+q^2}\,dq = \begin{cases} e^{-x} & \text{if } x \geq 0\,, \\ e^{x} & \text{if } x \leq 0\,. \end{cases}$$

Fifty years later Gaston Darboux (1844–1918) gave the modern point of view, that this function is not truly discontinuous but merely a function expressed by two different analytic expressions in different parts of its domain.

The change in point of view came about gradually, but an important step was Cauchy's refinement of the definition in the first chapter of his 1821 *Cours d'analyse*:

> When variable quantities are related so that, given the value of one of them, one can infer those of the others, we normally consider that the quantities are all expressed in terms of one of them, which is called the *independent* variable, while the others are called *dependent variables.*

[6] Bernoulli had left St. Petersburg in 1733, Euler in 1741.

Cauchy's definition still does not specify what *ways* of expressing one variable in terms of another are legitimate, but this definition was a step toward the basic idea that the value of the independent variable determines (uniquely) the value of the dependent variable or variables.

Fourier series. Daniel Bernoulli's work introduced trigonometric series as an alternative to power series. In his classic work of 1811, a revised version of which was published in 1821,[7] *Théorie analytique de chaleur* (*Analytic Theory of Heat*), Fourier established the standard formulas for the Fourier coefficients of a function. For an even function of period 2π, these formulas are

$$f(x) = \frac{1}{2}a_0 + \sum_{n=1}^{\infty} a_n \cos nx\,; \qquad a_n = \frac{1}{\pi}\int_0^{2\pi} f(x)\cos nx\,dx\,, \qquad n = 0, 1, \ldots\,.$$

A trigonometric series whose coefficients are obtained from an integrable function $f(x)$ in this way is called a *Fourier series.*

After trigonometric series had become a familiar technique, mathematicians were encouraged to look for other simple functions in terms of which solutions of more general differential equations than Laplace's equation could be expressed. Between 1836 and 1838 this problem was attacked by Charles Sturm (1803–1855) and Joseph Liouville, who considered general second-order differential equations of the form

$$[p(x)y'(x)]' + [\lambda r(x) + q(x)]y(x) = 0\,.$$

When a solution of Laplace's equation is sought in the form of a product of functions of one variable, the result is often an equation of this type for the one-variable functions. It often happens that only isolated values of λ yield solutions satisfying given boundary conditions. Sturm and Liouville found that in general there will be an infinite set of values $\lambda = \lambda_n$, $n = 1, 2, \ldots$, satisfying the equation and a pair of conditions at the endpoints of an interval $[a, b]$, and that these values increase to infinity. The values can be arranged so that the corresponding solutions $y_n(x)$ have exactly n zeros in $[a, b]$, and any solution of the differential equation can be expressed as a series

$$y(x) = \sum_{n=1}^{\infty} c_n y_n(x)\,.$$

The sense in which such series converge was still not clear, but it continued to be studied by other mathematicians. It required some decades for all these ideas to be sorted out clearly.

Proving that a Fourier series actually did converge to the function that generated it was one of the first places where real analysis encountered greater difficulties than complex analysis. In 1829 Dirichlet proved that the Fourier series of $f(x)$ converged to $f(x)$ for a bounded periodic function $f(x)$ having only a finite number of discontinuities and a finite number of maxima and minima in each period.[8] Dirichlet tried to get necessary and sufficient conditions for convergence, but that is a problem that has never been solved. He showed that some kind of continuity would be required by giving the famous example of the function whose value at x is one of two different values according as x is rational or irrational. This function

---

[7] The original version remained unpublished until 1972, when Grattan-Guinness published an annotated version of it.

[8] We would call such a function *piecewise monotonic.*

is called the *Dirichlet function.* For such a function, he thought, no integral could be defined, and therefore no Fourier series could be defined.[9]

Fourier integrals. The convergence of the Fourier series of $f(x)$ can be expressed as the equation

$$f(x) = \frac{1}{\pi}\int_0^\pi f(y)\,dy + \frac{2}{\pi}\sum_{n=1}^\infty \int_0^\pi f(y)\cos(ny)\cos(nx)\,dy\,.$$

That equation may have led to an analogous formula for Fourier integrals, which appeared during the early nineteenth century in papers on the wave and heat equations written by Poisson, Laplace, Fourier, and Cauchy. The central discovery in this area was the Fourier inversion formula, which we now write as

$$f(x) = \frac{2}{\pi}\int_0^\infty \int_0^\infty f(y)\cos(zy)\cos(zx)\,dy\,dz\,.$$

The analogy with the formula for series is clear: The continuous variable z replaces the discrete index n, and the integral on z replaces the sum over n. Once again, the validity of the representation is much more questionable than the validity of the formulas of complex analysis, such as the Cauchy integral formula for an analytic function. The Fourier inversion formula has to be interpreted very carefully, since the order of integration cannot be reversed. If the integrals make sense in the order indicated, that happy outcome can only be the result of some special properties of the function $f(x)$. But what are those properties?

The difficulty was that the integral extended over an infinite interval so that convergence required the function to have two properties: It needed to be continuous, and it needed to decrease sufficiently rapidly at infinity to make the integral converge. These properties turned out to be, in a sense, dual to each other. Considering just the inner integral as a function of z:

$$\widehat{f}(z) = \int_0^\infty f(y)\cos(zy)\,dy\,,$$

it turns out that the more rapidly $f(y)$ decreases at infinity, the more derivatives $\widehat{f}(z)$ has, and the more derivatives $f(y)$ has, the more rapidly $\widehat{f}(z)$ decreases at infinity. The converses are also, broadly speaking, true. Could one insist on having both conditions, so that the representation would be valid? Would these assumptions impair the usefulness of these techniques in mathematical physics? Alfred Pringsheim (1850-1941, father-in-law of the great writer Thomas Mann) studied the Fourier integral formula (*1910*), noting especially the two kinds of conditions that $f(x)$ needed to satisfy, which he called "conditions in the finite region" ("im Endlichen") and "conditions at infinity" ("im Unendlichen"). Nowadays, they are called local and global conditions. Pringsheim noted that the local conditions could be traced all the way back to Dirichlet's work of 1829, but said that "a rather obvious backwardness reveals itself" in regard to the global conditions.

---

[9] The increasing latitude allowed in analysis, mentioned above, is illustrated very well by this example. When the Lebesgue integral is used, this function is regarded as identical with the constant value it assumes on the irrational numbers.

> [They] seem in general to be limited to a relatively narrow condition, one which is insufficient for even the simplest type of application, namely that of absolute integrability of $f(x)$ over an infinite interval. There are, as far as I know, only a few exceptions.

Thus, to the question as to whether physics could get by with sufficiently smooth functions $f(x)$ that decay sufficiently rapidly, the answer turned out to be, in general, no. Physics needs to deal with discontinuous integrable functions $f(y)$, and for these $\widehat{f}(z)$ cannot decay rapidly enough at infinity to make its integral converge, at least not absolutely. What was to be done?

One solution involved the introduction of *convergence factors*, leading to a more general sense of convergence, called Abel–Poisson convergence. In a paper on wave motion published in 1818 Poisson used the representation

$$f(x) = \frac{1}{\pi}\int_0^{\infty}\int_{-\infty}^{+\infty} f(\alpha)\cos a(x-\alpha)e^{-ka}\,d\alpha\,da\,.$$

The exponential factor provided enough decrease at infinity to make the integral converge. Poisson claimed that the resulting integral tended toward $f(x)$ as k decreased to 0.

Abel used a similar technique to justify the natural value assigned to nonabsolutely convergent series such as

$$\ln(2) = 1 - \frac{1}{2} + \frac{1}{3} - \frac{1}{4} + \cdots \quad \text{and} \quad \frac{\pi}{4} = 1 - \frac{1}{3} + \frac{1}{5} - \frac{1}{7} + \cdots\,.$$

which can be obtained by expanding the integrands of the following integrals as geometric series and integrating termwise:

$$\int_0^1 \frac{1}{1+r}\,dr\,; \quad \int_0^1 \frac{1}{1+r^2}\,dr\,.$$

In Abel's case, the motive for making a careful study of continuity was his having noticed that a trigonometric series could represent a discontinuous function. From Paris in 1826 he wrote to a friend that the expansion

$$\frac{x}{2} = \sin x - \frac{1}{2}\sin 2x + \frac{1}{3}\sin 3x - \frac{1}{4}\sin 4x + \cdots$$

was provable for $0 \le x < \pi$, although obviously it could not hold at $x = \pi$. Thus, while the representation might be a good thing, it meant, on the other hand, that the sum of a series of continuous functions could be discontinuous. Abel also believed that many of the difficulties mathematicians were encountering were traceable to the use of divergent series. He gave, accordingly, a thorough discussion of the convergence of the binomial series, the most difficult of the elementary Taylor series to analyze.[10]

For the two conditionally convergent series shown above and the general Fourier integral, continuity of the sum was needed. In both cases, what appeared to be a necessary evil—the introduction of the convergence factor e^{-ka} or r—turned out to have positive value. For the functions $r^n \cos n\theta$ and $r^n \sin n\theta$ are harmonic functions

[10] Unknown to Abel, Bolzano had discussed the binomial series in 1816, considering integer, rational, and irrational (real) exponents, admitting that he could not cover all possible cases, due to the incomplete state of the theory of complex numbers at the time (Bottazzini, *1986*, pp. 96–97). He performed a further analysis of series in general in 1817, with a view to proving the intermediate value property (see Section 4 of Chapter 12).

if r and θ are regarded as polar coordinates, while $e^{-ay}\cos(ax)$ and $e^{-ay}\sin(ax)$ are harmonic if x and y are regarded as rectangular coordinates. The factor used to ensure convergence was providing harmonic functions, at no extra cost.

General trigonometric series. The study of trigonometric functions advanced real analysis once again in 1854, when Riemann was required to give a lecture to qualify for the position of *Privatdocent* (roughly what would be an assistant professor nowadays). As the rules required, he was to propose three topics and the faculty would choose the one he lectured on. One of the three, based on conversations he had had with Dirichlet over the preceding year, was the representation of functions by trigonometric series.[11] Dirichlet was no doubt hoping for more progress toward necessary and sufficient conditions for convergence of a Fourier series, the topic he had begun so promisingly a quarter-century earlier. Riemann concentrated on one question in particular: *Can a function be represented by more than one trigonometric series?* That is, *can two trigonometric series with different coefficients have the same sum at every point?* In the course of his study, Riemann was driven to examine the fundamental concept of integration. Cauchy had defined the integral

$$\int_a^b f(x)\,dx$$

as the number approximated by the sums

$$\sum_{n=1}^{N} f(x_n)(x_n - x_{n-1})$$

as N becomes large, where $a = x_0 < x_1 < \cdots < x_{n-1} < x_n = b$. Riemann refined the definition slightly, allowing $f(x_n)$ to be replaced by $f(x_n^*)$ for any x_n^* between x_{n-1} and x_n. The resulting integral is known as the Riemann integral today. Riemann sought necessary and sufficient conditions for such an integral to exist. The condition that he formulated led ultimately to the concept of a set of measure zero,[12] half a century later: *For each $\sigma > 0$ the total length of the intervals on which the function $f(x)$ oscillates by more than σ must become arbitrarily small if the partition is sufficiently fine.*

2.2. Completeness of the real numbers. The concept now known as completeness of the real numbers is associated with the *Cauchy convergence criterion*, which asserts that a sequence of real numbers $\{a_n\}_{n=1}^{\infty}$ converges to some real number a if it is a *Cauchy sequence*; that is, for every $\varepsilon > 0$ there is an index n such that $|a_n - a_k| < \varepsilon$ for all $k \geq n$. This condition was stated somewhat loosely by Cauchy in his *Cours d'analyse*, published in the mid-1820s, and the proof given there was also somewhat loose. The same criterion had been stated, and for sequences of functions rather than sequences of numbers, a decade earlier by Bolzano.

---

[11] As the reader will recall from Chapter 12, this topic was *not* the one Riemann did lecture on. Gauss preferred the topic of foundations of geometry, and so Riemann's paper on trigonometric series was not published until 1867, after his death.

[12] A set of points on the line has measure zero if for every $\varepsilon > 0$ it can be covered by a sequence of intervals (a_k, b_k) whose total length is less than ε.

2.3. Uniform convergence and continuity. Cauchy was not aware at first of any need to make the distinction between pointwise and uniform convergence, and he even claimed that the sum of a series of continuous functions would be continuous, a claim contradicted by Abel, as we have seen. The distinction is a subtle one. It is all too easy not to notice whether choosing n large enough to get a good approximation when $f_n(x)$ converges to $f(x)$ requires one to take account of which x is under consideration. That point was rather difficult to state precisely. The first clear statement of it is due to Philipp Ludwig von Seidel (1821–1896), a professor at Munich, who in 1847 studied the examples of Dirichlet and Abel, coming to the following conclusion:

> When one begins from the certainty thus obtained that the proposition cannot be generally valid, then its proof must basically lie in some still hidden supposition. When this is subject to a precise analysis, then it is not difficult to discover the hidden hypothesis. One can then reason backwards that this [hypothesis] cannot occur [be fulfilled] with series that represent discontinuous functions. [Quoted in Bottazzini, *1986*, p. 202]

In order to reason confidently about continuity, derivatives, and integrals, mathematicians began restricting themselves to cases where the series converged uniformly. Weierstrass, in particular, provided a famous theorem known as the M-test for uniform convergence of a series. But, although the M-test is certainly valuable in dealing with power series, uniform convergence in general is too severe a restriction. The important trigonometric series studied by Abel, for example, represented a discontinuous function as the sum of a series of continuous functions and therefore did not converge uniformly. Yet it could be integrated term by term. One could provide many examples of series of continuous functions that converged to a continuous function but not uniformly. Weaker conditions were needed that would justify the operations rigorously without restricting their applicability too strongly.

2.4. General integrals and discontinuous functions. The search for less restrictive hypotheses and the consideration of more general figures on a line than just points and intervals led to more general notions of length, area, and integral, allowing more general functions to be integrated. Analysts began generalizing the integral beyond the refinements introduced by Riemann. Foundational problems also added urgency to this search. For example, in 1881, Vito Volterra (1860–1940) gave an example of a continuous function having a derivative at every point, but whose derivative was not Riemann integrable. What could the fundamental theorem of calculus mean for such a function, which had an antiderivative but no integral, as integrals were then understood?

New integrals were created by the Latvian mathematician Axel Harnack (1851–1888), by the French mathematicians Émile Borel (1871–1956), Henri Lebesgue (1875–1941), and Arnaud Denjoy (1884–1974), and by the German mathematician Oskar Perron (1880–1975). By far the most influential of these was the Lebesgue integral, which was developed between 1899 and 1902. This integral was to have profound influence in the area of probability, due to its use by Borel, and in trigonometric series representations, an application that Lebesgue developed, perhaps as proof of the usefulness of his highly abstruse integral, which, as a former colleague of the author was fond of saying, "did not change any tables of integrals." Lebesgue justified his more general integral in the following words, from the preface to his 1904 monograph.

> [I]f we wished to limit ourselves always to these good [that is, smooth] functions, we would have to give up on the solution of a number of easily stated problems that have been open for a long time. It was the solution of these problems, rather than a love of complications, that caused me to introduce in this book a definition of the integral that is more general than that of Riemann and contains the latter as a special case.

Despite its complexity—to develop it with proofs takes four or five times as long as developing the Riemann integral—the Lebesgue integral was included in textbooks as early as 1907: for example, *Theory of Functions of a Real Variable*, by E. W. Hobson (1856–1933). Its chief attraction was the greater generalilty of the conditions under which it allowed termwise integration. Following the typical pattern of development in real analysis, the Lebesgue integral soon generated new questions. The Hungarian mathematician Frigyes Riesz (1880–1956) introduced the classes now known as L_p-spaces, the spaces of measurable functions[13] f for which $|f|^p$ is Lebesgue integrable, $p > 0$. (The space L_∞ consists of functions that are bounded on a set whose complement has measure zero.) How the Fourier series and integrals of functions in these spaces behave became a matter of great interest, and a number of questions were raised. For example, in his 1915 dissertation at the University of Moscow, Nikolai Nikolaevich Luzin (1883–1950) posed the conjecture that the Fourier series of a square-integrable function converges except on a set of measure zero. Fifty years elapsed before this conjecture was proved by the Swedish mathematician Lennart Carleson (b. 1928).

2.5. The abstract and the concrete. The increasing generality allowed by the notation $y = f(x)$ threatened to carry mathematics off into stratospheric heights of abstraction. Although Ampère had tried to show that a continuous function is differentiable at most points, the attempt was doomed to failure. Bolzano constructed a "sawtooth" function in 1817 that was continuous, yet had no derivative at any point. Weierstrass later used an absolutely convergent trigonometric series to achieve the same result,[14] and a young Italian mathematician Salvatore Pincherle (1853–1936), who took Weierstrass' course in 1877–1888, wrote a treatise in 1880 in which he gave a very simple example of such a function (Bottazzini, *1986*, p. 286):

$$f(x) = \sum_{n=1}^{\infty} \frac{\sin(n!x)}{n!}.$$

Volterra's example of a continuous function whose derivative was not integrable, together with the examples of continuous functions having no derivative at any point naturally cast some doubt on the applicability of the abstract concept of continuity and even the abstract concept of a function. Besides the construction of more general integrals and the consequent ability to "measure" more complicated

[13] See below for the definition.

[14] This example was communicated by his student Paul du Bois-Reymond (1831–1889) in 1875. The following year du Bois-Reymond constructed a continuous periodic function whose Fourier series failed to converge at a set of points that came arbitrarily close to every point.

geometric figures, it was necessary to investigate differentiation in more detail as well.

The secret of that quest turned out to be not continuity, but monotonicity. A continuous function may fail to have a derivative, but in order to fail, it must oscillate very wildly, as the examples of Bolzano and Weierstrass did. A function that did not oscillate or oscillated only a finite total amount, necessarily had a derivative except on a set of measure zero. The ultimate result in this direction was achieved by Lebesgue, who showed that a monotonic function has a derivative on a set whose complement has measure zero. Such a function might or might not be the integral of its derivative, as the fundamental theorem of calculus states. In 1902 Lebesgue gave necessary and sufficient conditions for the fundamental theorem of calculus to hold; a function that satisfies these conditions, and is consequently the integral of its derivative, is called *absolutely continuous.*

To return to the problem of abstractness, we note that it had been known at least since the time of Lagrange that any finite set of n data points (x_k, y_k), $k = 1, \ldots, n$, with x_k all different, could be fitted perfectly with a polynomial of degree at most $n - 1$. Such a polynomial might—indeed, probably would—oscillate wildly in the intervals between the data points. Weierstrass showed in 1884 that any continuous function, no matter how abstract, could be uniformly approximated by a polynomial over any bounded interval $[a, b]$. Since there is always some observational error in any set of data, this result meant that polynomials could be used in both practical and theoretical ways, to fit data, and to establish general theorems about continuous functions. Weierstrass also proved a second version of the theorem, for periodic functions, in which he showed that for these functions the polynomial could be replaced by a finite sum of sines and cosines. This connection to the classical functions freed mathematicians to use the new abstract functions, confident that in applications they could be replaced by computable functions.

Weierstrass lived before the invention of the new abstract integrals mentioned above arose, although he did encourage the development of the abstract set theory of Georg Cantor on which these integrals were based. With the development of the Lebesgue integral a new category of functions arose, the *measurable functions.* These are functions $f(x)$ such that the set of x for which $f(x) > c$ always has a meaningful measure, although it need not be a geometrically simple set, as it is in the case of continuous functions. It appeared that Weierstrass' work needed to be repeated, since his approximation theorem did not apply to measurable functions. In his 1915 dissertation Luzin produced two beautiful theorems in this direction. The first was what is commonly called by his name nowadays, the theorem that for every measurable function $f(x)$ and every $\varepsilon > 0$ there is a continuous function $g(x)$ such that $g(x) \neq f(x)$ only on a set of measure less than ε. As a consequence of this result and Weierstrass' approximation theorem, it followed that every measurable function is the limit of a sequence of polynomials on a set whose complement has measure zero. Luzin's second theorem was that every finite-valued measurable function is the *derivative* of a continuous function at the points of a set whose complement has measure zero. He was able to use this result to show that any prescribed set of measurable boundary values on the disk could be the boundary values of a harmonic function.

With the Weierstrass approximation theorem and theorems like those of Luzin, modern analysis found some anchor in the concrete analysis of the "classical" period that ran from 1700 to 1900. But that striving for generality and freedom of operation still led to the invocation of some strong principles of inference in the context of set theory. By mid-twentieth century mathematicians were accustomed to proving concrete facts using abstract techniques. To take just one example, it can be proved that some differential equations have a solution because a contraction mapping of a complete metric space must have a fixed point. Classical mathematicians would have found this proof difficult to accept, and many twentieth-century mathematicians have preferred to write in "constructivist" ways that avoid invoking the abstract "existence" of a mathematical object that cannot be displayed explicitly. But most mathematicians are now comfortable with such reasoning.

2.6. Discontinuity as a positive property. The Weierstrass approximation theorems imply that the property of being the limit of a sequence of continuous functions is no more general than the property of being the limit of a sequence of polynomials or the sum of a trigonometric series. That fact raises an obvious question: What kind of function *is* the limit of a sequence of continuous functions? As noted above, du Bois-Reymond had shown that it can be discontinuous on a set that is, as we now say, dense. But can it, for example, be discontinuous at *every* point? That was one of the questions that interested René-Louis Baire (1874–1932). If one thinks of discontinuity as simply the absence of continuity, classifying mathematical functions as continuous or discontinuous seems to make no more sense than classifying mammals as cats or noncats. Baire, however, looked at the matter differently. In his 1905 *Leçons sur les fonctions discontinues* (*Lectures on Discontinuous Functions*) he wrote

> Is it not the duty of the mathematician to begin by studying in the abstract the relations between these two concepts of continuity and discontinuity, which, while mutually opposite, are intimately connected?

Strange as this view may seem at first, we may come to have some sympathy for it if we think of the dichotomy between the continuous and the discrete, that is, between geometry and arithmetic. At any rate, to a large number of mathematicians at the turn of the twentieth century, it did not seem strange. The Moscow mathematician Nikolai Vasilevich Bugaev (1837–1903, father of the writer Andrei Belyi) was a philosophically inclined scholar who thought it possible to establish two parallel theories, one for continuous functions, the other for discontinuous functions. He called the latter theory *arithmology* to emphasize its arithmetic character. There is at least enough of a superficial parallel between integrals and infinite series and between continuous and discrete probability distributions (another area in which Russia has produced some of the world's leaders) to make such a program plausible. It is partly Bugaev's influence that caused works on set theory to be translated into Russian during the first decade of the twentieth century and brought the Moscow mathematicians Luzin and Dmitrii Fyodorovich Egorov (1869–1931) and their students to prominence in the area of measure theory, integration, and real analysis.

Baire's monograph was single-mindedly dedicated to the pursuit of one goal: to give a necessary and sufficient condition for a function to be the pointwise limit of a sequence of continuous functions. He found the condition, building on earlier ideas introduced by Hermann Hankel (1839–1873): The necessary and sufficient condition is that the discontinuities of the function form a set of *first category.*

A set is of first category if it is the union of a sequence of sets A_k such that every interval (a, b) contains an interval (c, d) disjoint from A_k. All other sets are of second category.[15] Although interest in the specific problems that inspired Baire has waned, the importance of his work has not. The whole edifice of what is now functional analysis rests on three main theorems, two of which are direct consequences of what is called the Baire category theorem (that a complete metric space is of second category as a subset of itself) and cannot be proved without it. Here we have an example of an unintended and fortuitous consequence of one bit of research turning out to be useful in an area not considered by its originator.

Questions and problems

17.1. The familiar formula $\cos\theta = 4\cos^3(\theta/3) - 3\cos(\theta/3)$, can be rewritten as $p(\cos\theta/3, \cos\theta) = 0$, where $p(x, y) = 4x^3 - 3x - y$. Observe that $\cos(\theta + 2m\pi) = \cos\theta$ for all integers m, so that

$$p\biggl(\cos\Bigl(\frac{\theta + 2m\pi}{3}\Bigr), \cos\theta\biggr) \equiv 0\,,$$

for all integers m. That makes it very easy to construct the roots of the equation $p(x, \cos\theta) = 0$. They must be $\cos\bigl((\theta + 2m\pi)/3\bigr)$ for $m = 0, 1, 2$. What is the analogous equation for dividing a circular arc into five equal pieces?

Suppose (as is the case for elliptic integrals) that the inverse function of an integral is doubly periodic, so that $f(x + m\omega_1 + n\omega_2) = f(x)$ for all m and n. Suppose also that there is a polynomial $p(x)$ of degree n^2 such that $p\bigl(f(\theta/n)\bigr) = f(\theta)$. Show that the roots of the equation $p(x) = f(\theta)$ must be $f(\theta/n + (k/n)\omega_1 + (l/n)\omega_2)$, where k and l range independently from 0 to $n - 1$.

17.2. Show that if $y(x, t) = \bigl(f(x + ct) + f(x - ct)\bigr)/2$ is a solution of the one-dimensional wave equation that is valid for all x and t, and $y(0, t) = 0 = y(L, t)$ for all t, then $f(x)$ must be an odd function of period $2L$.

17.3. Show that the problem $X''(x) - \lambda X(x) = 0$, $Y''(y) + \lambda Y(y) = 0$, with boundary conditions $Y(0) = Y(2\pi)$, $Y'(0) = Y'(2\pi)$, implies that $\lambda = n^2$, where n is an integer, and that the function $X(x)Y(y)$ must be of the form $\bigl(c_n e^{nx} + d_n e^{-nx}\bigr)\bigl(a_n\cos(ny) + b_n\sin(ny)\bigr)$ if $n \neq 0$.

17.4. Show that the differential equation

$$\frac{dx}{\sqrt{1 - x^4}} + \frac{dy}{\sqrt{1 - y^4}} = 0$$

has the solution $y = [(1 - x^2)/(1 + x^2)]^{1/2}$. Find another obvious solution of this equation.

17.5. Show that Fourier series can be obtained as the solutions to a Sturm–Liouville problem on $[0, 2\pi]$ with $p(x) = r(x) \equiv 1$, $q(x) = 0$, with the boundary conditions $y(0) = y(2\pi)$, $y'(0) = y'(2\pi)$. What are the possible values of λ?

[15] In his work on set theory, discussed in Section 4 of Chapter 12, Hausdorff criticized this terminology as "colorless."

Part 7

Mathematical Inferences

At various points in this survey of mathematics we have found mathematicians debating the meaning of what they were doing and the legitimacy of their procedures. In this last part of our study we examine the ways in which mathematics enters into the process of *drawing conclusions.* Human beings draw conclusions with differing levels of confidence, based on experience. At the one extreme are opinions about the most complicated phenomena around us, other human beings and human society as a whole. These matters are so complex that very few statements about them can be simultaneously free of significant doubt and of great importance. At the other end are matters that are so simple and obvious that we do not hesitate to label as insane anyone who doubts them. At the very least, we do not bother refuting a person who says that the word *yes* requires 10 letters to spell or that the city of Tuscaloosa is located in Siberia. Mathematics itself lies on the "confident" end of the spectrum, but its applications to practical life do not always share that certainty. Chapter 18 surveys the uncertain, in the form of the history of probability and statistics. Chapter 19 covers the more certain, in the form of the history of logic.

CHAPTER 18

Probability and Statistics

The need to make decisions on the basis of incomplete data is very widespread in human life. We need to decide how warmly to dress and whether to carry an umbrella when we leave home in the morning. We may have to decide whether to risk a dangerous but potentially life-saving medical procedure. Such decisions rely on statistical reasoning. Statistics is a science that is not exactly mathematics. It *uses* mathematics, in the form of probability, but its procedures are the inverse ones of fitting probability distributions to real-world data. Probability theory, on the other hand, is a form of pure mathematics, with theorems that are just as certain as those in algebra and analysis. We begin this chapter with the pure mathematics and end with its application.

1. Probability

The word *probability* is related to the English words *probe*, *probation*, *prove*, and *approve*. All of these words originally had a sense of *testing* or *experimenting*,[1] reflecting their descent from the Latin *probo*, which has these meanings. In other languages the word used in this mathematical sense has a meaning more like *plausibility*,[2] as in the German *Wahrscheinlichkeit* (literally, *truth resemblance*) or the Russian *veroyatnost'* (literally, *credibility*, from the root *ver-*, meaning *faith*). The concept is very difficult to define in declarative sentences, precisely because it refers to phenomena that are normally described in the subjunctive mood. This mood has nearly disappeared in modern English; it clings to a precarious existence in the past tense, "If it were true that..." having replaced the older "If it be true that...". The language of Aristotle and Plato, however, who were among the first people to discuss chance philosophically, had two such moods, the subjunctive and the optative, sometimes used interchangeably. As a result, they could express more easily than we the intuitive concepts involved in discussing events that are imagined rather than observed.

Intuitively, probability attempts to express the relative strength of the feeling of confidence we have that an event will occur. How surprised would we be if the event happened? How surprised would we be if it did not happen? Because we do have different degrees of confidence in certain future events, quantitative concepts become applicable to the study of probability. Generally speaking, if an event occurs essentially all the time under specified conditions, such as an eclipse

[1] The common phrase "the exception that proves the rule" is nowadays misunderstood and misused because of this shift in the meaning of the word *prove*. Exceptions *test* rules, they do not *prove* them in the current sense of that word. In fact, quite to the contrary, exceptions *disprove* rules.

[2] Here is another interesting word etymology. The root is *plaudo*, meaning *strike*, but specifically meaning to clap one's hands together, to applaud. Once again, *approval* is involved in the notion of probability.

of the Sun, we use a deterministic model (geometric astronomy, in this case) to study and predict it. If it occurs sometimes under conditions frequently associated with it, we rely on probabilistic models. Some earlier scientists and philosophers regarded probability as a measure of our ignorance. Kepler, for example, believed that the supernova of 1604 in the constellation Serpent may have been caused by a random collision of particles; but in general he was a determinist who thought that our uncertainty about a roll of dice was merely a matter of insufficient data being available. He admitted, however, that he could find no law to explain the apparently random pattern of eccentricities in the elliptical orbits of the six planets known to him.

Once the mathematical subject got started, however, it developed a life of its own, in which theorems could be proved with the same rigor as in any other part of mathematics. Only the application of those theorems to the physical world remained and remains clouded by doubt. We use probability informally every day, as the weather forecast informs us that the chance of rain is 30% or 80% or 100%,[3] or when we are told that one person in 30 will be afflicted with Alzheimer's disease between the ages of 65 and 74. Much of the public use of such probabilistic notions is, although not meaningless, at least irrelevant. For example, we are told that the life expectancy of an average American is now 77 years. Leaving aside the many questionable assumptions of environmental and political stability used in the model that produced this fascinating number, we should at least ask one question: Can the number be related to the life of any person in any meaningful way? What plans can one base on it, since anyone may die on any given day, yet very few people can confidently rule out the possibility of living past age 90?[4]

The many uncertainties of everyday life, such as the weather and our health, occur mixed with so many possibly relevant variables that it would be difficult to distill a theory of probability from those intensely practical matters. What is needed is a simpler and more abstract model from which principles can be extracted and gradually made more sophisticated. The most obvious and accessible such models are games of chance. On them probability can be given a quantitative and empirical formulation, based on the frequency of wins and losses. At the same time, the imagination can arrange the possible outcomes symmetrically and in many cases assign equal probabilities to different events. Finally, since money generally changes hands at the outcome of a game, the notion of a random variable (payoff to a given player, in this case) as a quantity assuming different values with different probabilities can be modeled.

1.1. Cardano. The systematic mathematization of probabiliy began in sixteenth-century Italy with Cardano. Cardano gambled frequently with dice and attempted to count the favorable cases for a throw of three dice. His table of values, as reported by Todhunter (*1865*, p. 3) is as follows.

---

[3] These numbers are generated by computer models of weather patterns for squares in a grid representing a geographical area. The modeling of their accuracy also uses probabilistic notions (see Problem 18.1).

[4] The Russian mathematician Yu. V. Chaikovskii (*2001*) believes that some of this cloudiness is about to be removed with the creation of a new science he calls *aleatics* (from the Latin word *alea*, meaning *dice-play* or *gambling*). We must wait and see. A century ago, other Russian mathematicians confidently predicted a bright future for "arithmology." Prophecy is the riskiest of all games of chance.

1	2	3	4	5	6	7	8	9	10	11	12
108	111	115	120	126	133	33	36	37	36	33	26

Readers who enjoy playing with numbers may find some amusement here. Since it is impossible to roll a 1 with three dice, the table value should perhaps be interpreted as the number of ways in which 1 may appear on *at least* one of the three dice. If so, then Cardan has got it wrong. One can imagine him thinking that if a 1 appears on one of the dice, the other two may show 36 different numbers, and since there are three dice on which the 1 may appear, the total number of ways of rolling a 1 must be $3 \cdot 36$ or 108. That way of counting ignores the fact that in some of these cases 1 appears on two of the dice or all three. By what is now known as the inclusion-exclusion principle, the total should be $3 \cdot 36 - 3 \cdot 6 + 1 = 91$. But it is difficult to say what Cardano had in mind. The number 111 given for 2 may be the result of the same count, increased by the three ways of choosing two of the dice to show a 1. Todhunter worked out a simple formula giving these numbers, but could not imagine any gaming rules that would correspond to them. If indeed Cardano made mistakes in his computations, he was not the only great mathematician to do so.

Cardano's *Liber de ludo* (*Book on Gambling*) was published about a century after his death. In this book Cardano introduces the idea of assigning a probability p between 0 and 1 to an event whose outcome is not certain. The principal applications of this notion were in games of chance, where one might bet, for example, that a player could roll a 6 with one die given three chances. The subject is not developed in detail in Cardano's book, much of which is occupied by descriptions of the actual games played. However, Cardano does state the multiplicative rule for a run of successes in independent trials. Thus the probability of getting a six on each of three successive rolls with one die is $\left(\frac{1}{6}\right)^3$. Most important, he recognized the real-world application of what we call the law of large numbers, saying that when the probability for an event is p, then after a large number n of repetitions, the number of times it will occur does not lie far from the value np. This law says that it is not certain that the number of occurrences will be near np, but "that is where the smart money bets."

After a bet has been made and before it is settled, a player cannot unilaterally withdraw from the bet and recover her or his stake. On the other hand, an accountant computing the net worth of one of the players ought to count part of the stake as an asset owned by that player; and perhaps the player would like the right to sell out and leave the game. What would be a fair price to charge someone for taking over the player's position? More generally, what happens if the game is interrupted? How are the stakes to be divided? The principle that seemed fair was that, *regardless of the relative amount of the stake each player had bet, at each moment in the game a player should be considered as owning the portion of the stakes equal to that player's probability of winning at that moment.* Thus, the net worth of each player is constantly changing as the game progresses, in accordance with what we now call *conditional probability.* Computing these probabilities in games of chance usually involves the combinatorial counting techniques the reader has no doubt encountered.

1.2. Fermat and Pascal. A French nobleman, the Chevalier de Méré, who was fond of gambling, proposed to Pascal the problem of dividing the stakes in a game where one player has bet that a six will appear in eight rolls of a single die, but the game is terminated after three unsuccessful tries. Pascal wrote to Fermat that the player should be allowed to sell the throws one at a time. If the first throw is foregone, the player should take one-sixth of the stake, leaving five-sixths. Then if the second throw is also foregone, the player should take one-sixth of the remaining five-sixths or $\frac{5}{36}$, and so on. In this way, Pascal argued that the fourth through eighth throws were worth $\frac{1}{6}\left[\left(\frac{5}{6}\right)^3 + \left(\frac{5}{6}\right)^4 + \left(\frac{5}{6}\right)^5 + \left(\frac{5}{6}\right)^6 + \left(\frac{5}{6}\right)^7\right]$.

This expression *is* the value of those throws *before* any throws have been made. If, after the bets are made but before any throws of the die have been made, the bet is changed and the players agree that only three throws shall be made, then the player holding the die should take this amount as compensation for sacrificing the last five throws. Remember, however, that the net worth of a player is constantly changing as the game progresses and the probability of winning changes. The value of the fourth throw, for example, is smaller to begin with, since there is some chance that the player will win before it arrives, in which case it will not arrive. At the beginning of the game, the chance of winning on the fourth roll is $\left(\frac{5}{6}\right)^3\frac{1}{6}$, the factor $\left(\frac{5}{6}\right)^3$ representing the probability that the player will *not* have won before then. After three unsuccesful throws, however, the probability that the player "will not have" won (because he *did not* win) on the first three throws is 1, and so the probability of winning on the fourth throw becomes $\frac{1}{6}$.

Fermat expressed the matter as follows:

> [T]he three first throws having gained nothing for the player who holds the die, the total sum thus remaining at stake, he who holds the die and who agrees not to play his fourth throw should take $\frac{1}{6}$ as his reward. And if he has played four throws without finding the desired point and if they agree that he shall not play the fifth time, he will, nevertheless, have $\frac{1}{6}$ of the total for his share. Since the whole sum stays in play it not only follows from the theory, but it is indeed common sense that each throw should be of equal value.

Pascal wrote back to Fermat, proclaiming himself satisfied with Fermat's analysis and overjoyed to find that "the truth is the same at Toulouse and at Paris."

1.3. Huygens. Huygens wrote a treatise on probability in 1657. His *De ratiociniis in ludo aleæ* (*On Reasoning in a Dice Game*) consisted of 14 propositions and contained some of the results of Fermat and Pascal. In addition, Huygens was able to consider multinomial problems, involving three or more players. Cardano's idea of an *estimate of the expectation* was elaborated by Huygens. He asserted, for example, that if there are p (equally likely) ways for a player to gain a and q ways to gain b, then the player's expectation is $(pa + qb)/(p + q)$.

Even simple problems involving these notions can be subtle. For example, Huygens considered two players A and B taking turns rolling the dice, with A going first. Any time A rolls a 6, A wins; any time B rolls a 7, B wins. What are the relative chances of winning? (The answer to that question would determine the fair proportions of the stakes to be borne by the two players.) Huygens concluded that the odds were 31:30 in favor of B, that is, A's probability of of winning was $\frac{30}{61}$ and B's probability was $\frac{31}{61}$.

1.4. Leibniz. Although Leibniz wrote a full treatise on combinatorics, which provides the mathematical apparatus for computing many probabilities in games of chance, he did not himself gamble. But he did analyze many games of chance and suggest modifications of them that would make them fair (zero-sum) games. Some of his manuscripts on this topic have been analyzed by de Mora-Charles (*1992*). One of the games he analyzed is known as quinquenove. This game is played between two players using a pair of dice. One of the players, called the banker, rolls the dice, winning if the result is either a double or a total number of spots showing equal to 3 or 11. There are thus 10 equally likely ways for the banker to win with this roll, out of 36 equally likely outcomes. If the banker rolls a 5 or 9 (hence the name "quinquenove"), the other player wins. The other player has eight ways of winning of the equally likely 36 outcomes, leaving 18 ways for the game to end in a draw. The reader will be fascinated and perhaps relieved to learn that the great Leibniz, author of *De arte combinatoria*, confused permutations and combinations in his calculations for this game and got the probabilities wrong.

1.5. The *Ars Conjectandi* of Jakob Bernoulli. One of the classic founding documents of probability theory was published in 1713, eight years after the death of its author, Leibniz' disciple Jakob Bernoulli. This work, *Ars conjectandi* (*The Art of Prediction*), moved probability theory beyond the limitations of analyzing games of chance. It was intended by its author to apply mathematical methods to the uncertainties of life. As he said in a letter to Leibniz, "I have now finished the major part of the book, but it still lacks the particular examples, the principles of the art of prediction that I teach how to apply to society, morals, and economics. . . ." That was an ambitious undertaking, and Bernoulli had not quite finished the work when he died in 1705.

Bernoulli gave a very stark picture of the gap between theory and application, saying that only in simple games such as dice could one apply the equal-likelihood approach of Fermat and Pascal, whereas in the cases of interest, such as human health and longevity, no one had the power to construct a suitable model. He recommended statistical studies as the remedy to our ignorance, saying that if 200 people out of 300 of a given age and constitution were known to have died within 10 years, it was a 2-to-1 bet that any other person of that age and constitution would die within a decade.

In this treatise Bernoulli reproduced the problems solved by Huygens and gave his own solution of them. He considered what are now called *Bernoulli trials* in his honor. These are repeated experiments in which a particular outcome either happens (success) with probability b/a or does not happen (failure) with probability c/a, the same probability each time the experiment is performed, each outcome being independent of all others. (A simple nontrivial example is rolling a single die, counting success as rolling a 5. Then the probabilities are $\frac{1}{6}$ and $\frac{5}{6}$.) Since $b/a + c/a = 1$, Bernoulli saw correctly that the binomial expansion, and hence Pascal's triangle, would be useful in computing the probability of getting at least m successes in n trials. He gave that probability as

$$\sum_{k=m}^{n} \binom{n}{k} \left(\frac{b}{a}\right)^k \left(\frac{c}{a}\right)^{n-k}.$$

It was, incidentally, in this treatise, when computing the sum of the cth powers of the first n integers, that Bernoulli introduced what are now called the *Bernoulli numbers*, defined by the formula

$$\sum_{k=1}^{n} k^c = \frac{n^{c+1}}{c+1} + \frac{n}{2} + \frac{c}{2}An^{c-1} + \frac{c(c-1)(c-2)}{2 \cdot 3 \cdot 4}Bn^{c-3} + \cdots .$$

Nowadays we define these numbers as $B_0 = 1$, $B_1 = -\frac{1}{2}$, and thence $B_2 = A$, $B_3 = -B$, and so forth. He illustrated his formula by finding

$$\sum_{k=1}^{1000} k^{10} = 91409924241424243424241924242500 .$$

The law of large numbers. Bernoulli imagined an urn containing numbers of black and white pebbles, whose ratio is to be determined by sampling with replacement. Here it is possible that you will always get a white pebble, no matter how many times you sample. However, if black pebbles constitute a significant proportion of the contents of the urn, this outcome is very unlikely. After discussing the degree of certainty that would suffice for practical purposes (he called it *virtual certainty*),[5] he noted that this degree of certainty could be attained empirically by taking a sufficiently large sample. The probability that the empirically determined ratio would be close to the true ratio increases as the sample size increases, but the result would be accurate only within certain limits of error. More precisely, given certain limits of tolerance, by a sufficient number of trials,

> [W]e can attain any desired degree of probability that the ratio found by our many repeated observations will lie between these limits.

This last assertion is an informal statement of the law of large numbers for what are now called *Bernoulli trials*, that is, repeated independent trials with the same probability of a given outcome at each trial. If the probability of the outcome is p and the number of trials is n, this law can be phrased precisely by saying that for any $\varepsilon > 0$ there exists a number n_0 such that if m is the number of times the outcome occurs in n trials and $n > n_0$, the probability that the inequality $|(m/n) - p| > \varepsilon$ will hold is less than ε.[6] Bernoulli stated this principle in terms of the segment of the binomial series of $(r + s)^{n(r+s)}$ consisting of the n terms on each side of the largest term (the term containing $r^{nr}s^{ns}$), and he proved it by giving an estimate on n sufficient to make the ratio of this sum to the sum of the remaining terms at least c, where c is specified in advance. This problem is the earliest in which probability and statistics were combined to solve a problem of practical application.

---

[5] This phrase is often translated more literally as *moral* certainty, which has the wrong connotation.

[6] Probabilists say that the frequency of successes converges "in probability" to the probability of success at each trial. Analysts say it converges "in measure." There is also a strong law of large numbers, more easily stated in terms of independent random variables, which asserts that (under suitable hypotheses) there is a set of probability 1 on which the convergence to the mean occurs. That is, the convergence is "almost surely," as probabilists say and "almost everywhere," as analysts phrase the matter. On a finite measure space such as a probability space, almost everywhere convergence implies convergence in measure, but the converse is not true.

1.6. De Moivre. In 1711, even before the appearance of Jakob Bernoulli's treatise, another groundbreaking book on probability appeared, the *Doctrine of Chances*, written by Abraham de Moivre (1667–1754), a French Huguenot who took refuge in England after 1685, when Louis XIV revoked the Edict of Nantes, which had guaranteed civil rights for Huguenots when Henri IV took the French throne in 1598.[7] De Moivre's book went through several editions. Its second edition, which appeared in 1738, introduced a significant piece of numerical analysis, useful for approximating sums of terms of a binomial expansion $(a+b)^n$ for large n. De Moivre had published the work earlier in a paper written in 1733. Having no notation for the base e, which was introduced by Euler a few years later, de Moivre simply referred to the hyperbolic (natural) logarithm and "the number whose logarithm is 1." De Moivre first considered only the middle term of the expansion. That is, for an even power $n = 2m$, he estimated the term

$$\binom{2m}{m} = \frac{(2m)!}{(m!)^2}$$

and found it equal to $\frac{2}{B\sqrt{n}}$, where B was a constant for which he knew only an infinite series. At that point, he got stuck, as he admitted, until his friend James Stirling (1692–1770) showed him that "the Quantity B did denote the Square-root of the Circumference of a Circle whose Radius is Unity." In our terms, $B = \sqrt{2\pi}$, but de Moivre simply wrote c for B. Without having to know the exact value of B de Moivre was able to show that "the Logarithm of the Ratio, which a Term distant from the middle by the Interval l, has the the middle Term, is [approximately, for large n] $-\frac{2ll}{n}$." In modern language,

$$\binom{2n}{n+l}\bigg/\binom{2n}{n} \approx e^{-2l^2/n}\,.$$

De Moivre went on to say, "The Number, which answers to the Hyperbolic Logarithm $-2ll/n$, [is]

$$1 - \frac{2ll}{n} + \frac{4l^2}{2nn} - \frac{8l^6}{6n^3} + \frac{16l^8}{24n^4} - \frac{32l^{10}}{120n^5} + \frac{64l^{12}}{720n^7}, \text{ \&c."}$$

By scaling, de Moivre was able to estimate segments of the binomial distribution. In particular, the fact that the numerator was l^2 and the denominator n allowed him to estimate the probability that the number of successes in Bernoulli trials would be between fixed limits. He came close to noticing that the natural unit of probability for n trials was a multiple of $\sqrt{n}$. In 1893 this natural unit of measure for probability was named the *standard deviation* by the British mathematician Karl Pearson (1857–1936). For Bernoulli trials with probability of success p at each trial the standard deviation is $\sigma = \sqrt{np(1-p)}$.

For what we would call a coin-tossing experiment in which $p = \frac{1}{2}$—he imagined tossing a metal disk painted white on one side and black on the other—de Moivre observed that with 3600 coin tosses, the odds would be more than 2 to 1 against a deviation of more than 30 "heads" from the expected number of 1800. The standard deviation for this experiment is exactly 30, and 68 percent of the area under a normal curve lies within one standard deviation of the mean. De Moivre

[7] The spirit of sectarianism has infected historians to the extent that Catholic and Protestant biographers of de Moivre do not agree on how long he was imprisoned in France for being a Protestant. They do agree that he was imprisoned, however. To be fair to the French, they did elect him a member of the Academy of Sciences a few months before his death.

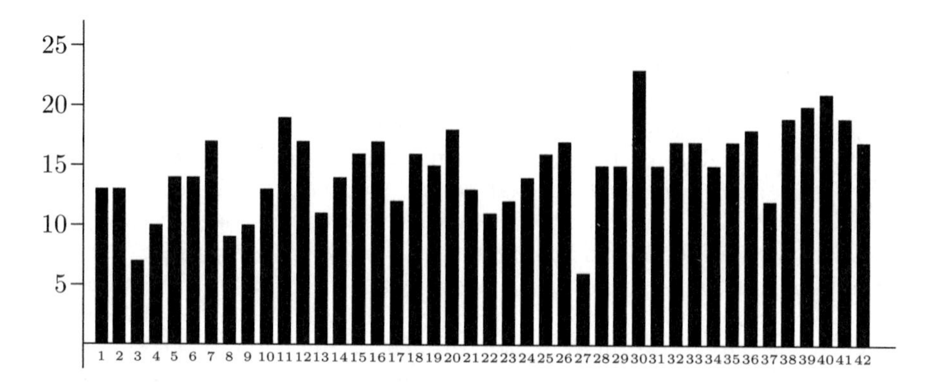

FIGURE 1. Frequencies of numbers in a state lottery over a one-year period.

could imagine the bell-shaped normal curve that we are familiar with, but he could not give it an equation. Instead he described it as the curve whose ordinates were numbers having certain logarithms. What seems most advanced in his analysis is that he recognized the area under the curve as a probability and computed it by a mechanical quadrature method that he credited jointly to Newton, Roger Cotes, James Stirling, and himself. This tendency of the average of many independent trials to look like the bell-shaped curve is called the *central limit theorem.*

It is difficult to appreciate the work of Bernoulli and de Moivre in applications without seeing it applied in a real-world illustration. To take a very simple example, consider Fig. 1, which is a histogram of the frequencies with which the numbers from 1 to 42 were drawn in a state lottery over a period of one year[8] Six numbers are drawn twice a week, for a total of 624 numbers each year. At each drawing a given number has a probability of $\frac{1}{7}$ of being drawn. Thus, focusing attention only on the occurrence of a fixed integer k, we can think of the lottery as a series of 104 independent trials with a probability of success (drawing the number k) equal to $\frac{1}{7}$ at each trial.

Although the individual data do not reveal the binomial distribution or show any bell-shaped curve, we can think of the frequencies with which the 42 numbers are drawn as the data for a second probabilistic model. By the binomial distribution, for each frequency r from 0 to 104, The probability that a given number will be drawn r times should theoretically be

$$\binom{104}{r}\left(\frac{1}{7}\right)^r\left(\frac{6}{7}\right)^{104-r}.$$

If the probability of an event is proportional to the number of times that the event occurs in a large number of trials, then the number of numbers drawn r times should be 42 times this expression. The resulting theoretical frequencies are negligibly small for $r < 6$ or $r > 23$. The values predicted by this theoretical model for r between 6 and 23, rounded to the nearest integer, are given in the second row of the following table, while the experimentally observed numbers are given in the bottom row.

[8] The Tri-state Megabucks of Maine, New Hampshire, and Vermont, from mid-December 2000 to mid-December 2001.

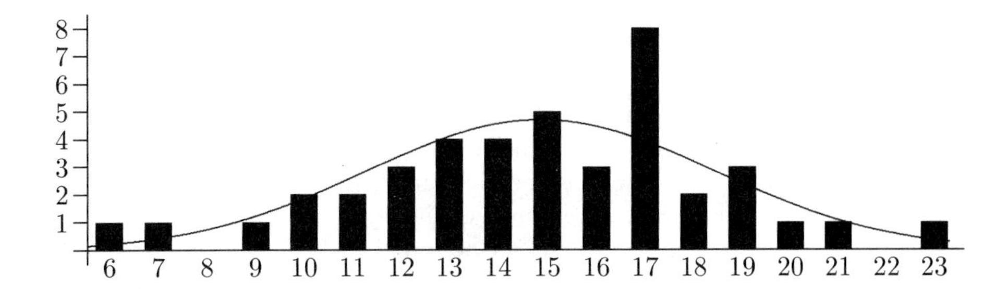

FIGURE 2. Histogram of the frequencies of the frequencies in Fig. 1, compared with a normal distribution.

Freq.	6	7	8	9	10	11	12	13	14	15	16	17	18	19	20	21	22	23
Pred.	0	0	1	1	2	3	4	4	5	5	4	4	3	2	2	1	1	0
Obs.	1	1	0	1	2	2	3	4	4	5	3	8	2	3	1	1	0	1

The agreement is not perfect, nor would we expect it to be. But it is remarkably close, except for the one "outlier" at a frequency of 17, attained by eight numbers instead of the theoretically predicted four. The mean for this model is $104/7 \approx 14.85$, and the standard deviation is $\sqrt{624}/7 \approx 3.569$. The histogram for the frequencies of the frequencies, compared with the graph of the standard bell-shaped curve with this mean and standard deviation are shown in Fig. 2. The fact that mere numerical reasoning *compels* even the most chaotic phenomena to exhibit some kind of order is one of the most awe-inspiring aspects of applied probability theory. It is the phenomenon that led the British mathematician Francis Galton (1822–1911) to describe the normal distribution as "the supreme law of unreason."

The Petersburg paradox. Soon after its introduction by Huygens and Jakob Bernoulli the concept of mathematical expectation came in for some critical appraisal. While working in the Russian Academy of Sciences, Daniel Bernoulli discussed the problem now known as the *Petersburg paradox* with his brother Niklaus (1695–1726, known as Niklaus II). We can describe this paradox informally as follows. Suppose that you toss a coin until heads appears. If it appears on the first toss, you win \$2, if it first appears on the second toss, you win \$4, and so on; if heads first appears on the nth toss, you win 2^n dollars. How much money would you be willing to pay to play this game? Now by "rational" computations the expected winning is infinite, being $2 \cdot \frac{1}{2} + 4 \cdot \frac{1}{4} + 8 \cdot \frac{1}{8} + \cdots$, so that you should be willing to pay, say, \$10,000 to play each time. On the other hand, who would bet \$10,000 knowing that there was an even chance of winning back only \$2, and that the odds are 7 to 1 against winning more than \$10? Something more than mere expectation was involved here. Daniel Bernoulli discussed the matter at length in an article in the *Comentarii* of the Petersburg Academy for 1730–1731 (published in 1738). He argued for the importance of something that we now call *utility*. If you already possess an amount of money x and you receive a small additional amount of money dx, how much *utility* does the additional money have for you, subjectively? Bernoulli assumed that the increment of utility dy was directly proportional to dx and inversely proportional to x, so that

$$dy = \frac{k\,dx}{x},$$

and as a result, the total utility of personal wealth is a logarithmic function of total wealth. One consequence of this assumption is a law of diminishing returns: The additional satisfaction from additional wealth decreases as wealth increases. Bernoulli used this idea to explain why any rational person would refuse to play the game. Obviously, the expected gain in utility from each of these wins, being proportional to the logarithm of the money gained, has a finite total, and so one should be willing to pay only an amount of money that has an equal utility to the gambler. A different explanation can be found in Problem 18.4 below. This explanation seems to have been given first by the mathematician John Venn (1834–1923) of Caius[9] College, Cambridge in 1866.

The utility y, which Bernoulli called the *emolumentum* (*gain*), is an important tool in economic analysis, since it provides a dynamic model of economic behavior: Buyers exchange money for goods or services of higher personal utility; sellers exchange goods and services for money of higher personal utility. If money, goods, and services did not have different utility for different people, no market could exist at all.[10] That idea is valid independently of the actual formula for utility given by Bernoulli, although, as far as measurements of pyschological phenomena can be made, Bernoulli's assumption was extremely good. The physiologist Ernst Heinrich Weber (1795–1878) asked blindfolded subjects to hold weights, which he gradually increased, and to say when they noticed an increase in the weight. He found that the threshold for a noticeable difference was indeed inversely proportional to the weight. That is, if S is the perceived weight and W the actual weight, then $dS = k\,dW/W$, where dW is the smallest increment that can be noticed and dS the corresponding perceived increment. Thus he found exactly the law *assumed* by Bernoulli for perceived increases in wealth.[11] Utility is of vital importance to the insurance industry, which makes its profit by having a large enough stake to play "games" that resemble the Petersburg paradox.

Mathematically, there was an important concept missing from the explanation of the Petersburg paradox. Granted that one should expect the "expected" value of a quantity depending on chance, how *confidently* should one expect it? The question of *dispersion* or *variance* of a random quantity lies beneath the surface here and needed to be brought out. It turns out that when the expected value is infinite, or even when the variance is infinite, no rational projections can be made. However, since we live in a world of finite duration and finite resources, each "game" will be played only a finite number of times. It follows that every actual game has a finite expectation and variance and is subject to rational analysis using them.

1.7. Laplace. Although Laplace is known primarily as an astronomer, he developed a great deal of theoretical physics. (The differential equation satisfied by harmonic functions is named after him.) He also understood the importance of probabilistic methods for processing the results of measurements. In his *Théorie analytique des probabilités*, he proved that the distribution of the average of random observational errors that are uniformly distributed in an interval symmetric about zero tends to the normal distribution as the number of observations increases.

[9] Pronounced "Keys."

[10] One feels the lack of this concept very strongly in the writing on economics by Aristotle and his followers, especially in their condemnation of the practice of lending money at interest.

[11] Weber's result was publicized by Gustave Theodor Fechner (1801–1887) and is now known as the Weber–Fechner law.

Except for using the letter c where we now use e to denote the base of natural logarithms, he had what we now call the central limit theorem for independent uniformly distributed random variables.

1.8. Legendre. In a treatise on ways of determining the orbits of comets, published in 1805, Legendre dealt with the problem that frequently results when observation meets theory. Theory prescribes a certain number of equations of a certain form to be satisfied by the observed quantities. These equations involve certain theoretical parameters that are not observed, but are to be determined by fitting observations to the theoretical model. Observation provides a large number of empirical, approximate solutions to these equations, and thus normally provides a number of equations far in excess of the number of parameters to be chosen. If the law is supposed to be represented by a straight line, for example, only two constants are to be chosen. But the observed data will normally not lie on the line; instead, they may cluster around a line. How is the observer to choose canonical values for the parameters from the observed values of each of the quantities?

Legendre's solution to this problem is now a familiar technique. If the theoretical equation is $y = f(x)$, where $f(x)$ involves parameters $\alpha, \beta, \ldots$, and one has data points (x_k, y_k), $k = 1, \ldots, n$, sum the squares of the "errors" $f(x_k) - y_k$ to get an expression in the parameters

$$E(\alpha, \beta, \ldots) = \sum_{k=1}^{n} \bigl(f(x_k) - y_k\bigr)^2 ,$$

and then choose the parameters so as to minimize E. For fitting with a straight line $y = ax + b$, for example, one needs to choose $E(a, b)$ given by

$$E(a, b) = \sum_{k=1}^{n} (ax_k + b - y_k)^2$$

so that

$$\frac{\partial E}{\partial a} = 0 = \frac{\partial E}{\partial b} .$$

1.9. Gauss. Legendre was not the first to tackle the problem of determining the most likely value of a quantity x using the results of repeated measurements of it, say x_k, $k = 1, \ldots, n$. In 1799 Laplace had tried the technique of taking the value x that minimizes the sum of the absolute errors[12] $|x - x_k|$. But still earlier, in 1794 as shown by his diary and correspondence, the teenager Gauss had hit on the least-squares technique for the same purpose. However, as Reich (*1977*, p. 56) points out, Gauss did not consider this discovery very important and did not publish it until 1809. In 1816 Gauss published a paper on observational errors, in which he discussed the most probable value of a variable based on a number of observations of it. His discussion was much more modern in its notation than those that had gone before, and also much more rigorous. He found the likelihood of an error of size x to be

$$\frac{h}{\sqrt{\pi}} e^{-h^2 x^2} ,$$

where h was what he called the *measure of precision*. He showed how to estimate this parameter by inverse-probability methods. In modern terms, $1/\sqrt{2}h$ is the

[12] This method has the disadvantage that one large error and many small errors count equally. The least-squares technique avoids that problem.

standard deviation. This work brought the normal distribution into a more or less standard form, and it is now often referred to as the *Gaussian distribution*.

1.10. Philosophical issues. The notions of chance and necessity have always played a large role in philosophical speculation; in fact, most books on logic are kept in the philosophy sections of libraries. Many of the mathematicians who have worked in this area have had a strong interest in philosophy and have speculated on what probability means. In so doing, they have come up against the same difficulties that confront natural philosophers when trying to explain how induction works. Granted that like Pavlov's dogs and Skinner's pigeons (see Chapter 1), human beings tend to form expectations based on frequent, but not necessarily invariable conjunctions of events and seem to find it very difficult to suspend judgment and live with no belief where there is no evidence,[13] can philosophy offer us any assurance that proceeding by induction based on probability and statistics is any better than, say, divination such as one finds in the *I Ching*? Are insurance companies acting on *pure faith* when they offer to bet us that we will survive long enough to pay them more money in premiums than they will pay out when we die? If probability is a subjective matter, is subjectivity the same as arbitrariness?

What, then, *is* probability, when applied to the physical world? Is it merely a matter of frequency of observation, and consequently objective? Or do human beings have some innate faculty for assigning probabilities? For example, when we toss a coin twice, there are four distinguishable outcomes: HH, HT, TH, TT. Are these four equally likely? If one does not know the order of the tosses, only three possibilities can be distinguished: two heads, two tails, and one of each. Should those be regarded as equally likely, or should we imagine that we do know the order and distinguish all four possibilities?[14] Philosophers still argue over such matters. Siméon-Denis Poisson (1781–1840) seemed to be having it both ways in his *Recherches sur la probabilité des jugemens* (*Investigations into the Plausibility of Inferences*) when he wrote that

> The probability of an event is the reason we have to believe that it has taken place, or that it will take place.

and then immediately followed up with

> The measure of the probability of an event is the ratio of the number of cases favorable to that event, to the total number of cases favorable or contrary.

In the first statement, he appeared to be defining probability as a subjective event, one's own *personal* reason, but then proceeded to make that reason an objective thing by assuming equal likelihood of all outcomes. Without some restriction on the universe of discourse, these definitions are not very useful. We do not know, for example, whether our automobile will start tomorrow morning or not, but if

[13] In his *Formal Logic*, Augustus de Morgan imagined asking a person selected at random for an opinion whether the volcanoes—he meant craters—on the unseen side of the moon were larger than those on the side we can see. He concluded, "The odds are, that though he has never thought of the question, he has a pretty stiff opinion in three seconds."

[14] If the answer to that question seems intuitively obvious, please note that in more exotic applications of statistics, such as in quantum mechanics, either possibility can occur. Fermions have wave functions that are antisymmetric, and they distinguish between HT and TH; bosons have symmetric wave functions and do not distinguish them.

the probability of its doing so were really only 50% because there are precisely two possible outcomes, most of us would not bother to buy an automobile. Surely Poisson was assuming some kind of symmetry that would allow the imagination to assign equal likelihoods to the outcomes, and intending the theory to be applied only in those cases. Still, in the presence of ignorance of causes, equal probabilities seem to be a reasonable starting point. The law of entropy in thermodynamics, for example, can be deduced as a tendency for an isolated system to evolve to a state of maximum probability, and maximum probability means the maximum number of equally likely states for each particle.

1.11. Large numbers and limit theorems. The idea of the law of large numbers was stated imprecisely by Cardano and with more precision by Jakob Bernoulli. To better carry out the computations involved in using it, de Moivre was led to approximate the binomial distribution with what we now realize was the normal distribution. He, Laplace, and Gauss all grasped with different degrees of clarity the principle (central limit theorem) that when independent measurements are averaged, they tend to shape themselves into the bell-shaped curve.

The law of large numbers was given its name in the 1837 work of Poisson just mentioned. Poisson discovered an approximation to the probability of getting at most k successes in n trials, valid when n is large and the probability p is small. He thereby introduced what is now known as the *Poisson distribution*, in which the probability of k successes is given by

$$p_k = e^{-\lambda} \frac{\lambda^k}{k!} .$$

The Russian mathematician Chebyshëv introduced the concept of a random variable and its mathematical expectation. He is best known for his 1846 proof of the weak law of large numbers for repeated independent trials. That is, he showed that the probability that the actual proportion of successes will differ from the expected proportion by less than any specified $\varepsilon > 0$ tends to 1 as the number of trials increases. In 1867 he proved what is now called *Chebyshëv's inequality*: *The probability that a random variable will assume a value more than [what is now called] k standard deviations from its mean is at most* $1/k^2$. This inequality was published by Chebyshëv's friend and translator Irénée-Jules Bienaymé (1796–1878) and is sometimes called the *Chebyshëv–Bienaymé inequality* (see Heyde and Seneta, *1977*). This inequality implies the weak law of large numbers. In 1887 Chebyshëv also gave an explicit statement of the central limit theorem for independent random variables.

The extension of the law of large numbers to dependent trials was achieved by Chebyshëv's student Andrei Andreevich Markov (1856–1922). The subject of dependent trials—known as *Markov chains*—remains an object of current research. In its simplest form it applies to a system in one of a number of states $\{S_1, \ldots, S_n\}$ which at specified times may change from one state to another. If the probability of a transition from S_i to S_j is p_{ij}, the matrix

$$P = \begin{pmatrix} p_{11} & \cdots & p_{1n} \\ \vdots & \ddots & \vdots \\ p_{n1} & \cdots & p_{nn} \end{pmatrix}$$

is called the *transition matrix*. If successive transitions are all independent of one another, one can easily verify that the matrix power P^k gives the probabilities of the transitions in k steps.

2. Statistics

The subject of probability formed the theoretical background for the empirical science known as statistics. Some theoretical analysis of the application of probability to hypothesis testing and modification is due to Thomas Bayes (1702–1761), a British clergyman. Bayes' articles were published in 1764–1765 (after his death) by Rev. Richard Price (1723–1791). Bayes considered the problem opposite to that considered by Jakob Bernoulli. Where Bernoulli assigned probabilities to the event of getting k successes in n independent trials, assuming the probability of success in each trial was p, Bayes analyzed the problem of finding the probability p based on an observation that k successes and $n-k$ failures have occurred. In other words, he tried to estimate the parameter in a distribution from observed data. His claim was that p would lie between a and b with a probability proportional to the area under the curve $y = x^k(1-x)^{n-k}$ between those limits. He then analyzed a more elaborate example. Suppose we know that the probability of event B is p, given that event A has occurred. Suppose also that, after a number of trials, without reference to whether A has occurred or not, we find that event B has occurred m times and has not occurred n times. What probability should be assigned to A? Bayes' example of event A was a line drawn across a billiard table parallel to one of its sides at an unknown distance x from the left-hand edge. A billiard ball is rolled at random across the table, coming to rest on the left of the line m times and on the right of it n times. Assuming that the width of the table is a, the probability of the ball resting left of the line is x/a, and the probability that it rests on the right is $1 - x/a$. How can we determine x from the actual observed frequencies m and n? Bayes' answer was that the probability that x lies between b and c is proportional to the area under the curve $y = x^m(a-x)^n$ between those two values. This first example of statistical estimation is also the first *maximum-likelihood* estimation, since the "density" function $x^m(a-x)^n$ has its maximum value where $m(a-x) = nx$, that is, $x = a\frac{m}{m+n}$, so that the proportion $m : n = x : a-x$ holds. It seems intuitively reasonable that the most likely value of x is the value that makes this proportion correct, and that the likelihood decreases as x moves away from this value. This is the kind of reasoning used by Gauss in his 1816 paper on the estimation of observational errors to find the parameter (measure of precision) in the normal distribution. To derive this result, Bayes had to introduce the concept of conditional probability. The probability of A, *given that* B has occurred, is equal to the probability that both events happen divided by the probability of B. (If B has occurred, it must have positive probability, and therefore the division is legitimate.) Although Bayes stated this much with reasonable clarity (see Todhunter, *1865*, p. 298), the full statement of what is now called Bayes' theorem (see below) is difficult to discern in his analysis.

The word *statistics* comes from the state records of births, deaths, and other economic facts that governments have always found it necessary to keep for administrative purposes. The raw data form far too large a set of numbers to be analyzed individually in most cases, and that is where probabilistic models and inverse-probability reasoning, such as that used by Bayes and Gauss become most useful. An early example was an argument intended to prove that the world was designed for human habitation based on the ratio of male to female births. In 1710 Queen Anne's physician John Arbuthnott (1667–1735) published in the *Philosophical Transactions of the Royal Society* a paper with the title, "An argument for Divine Providence, taken from the constant regularity observ'd in the births of both sexes." In that paper Arbuthnott presented baptismal records from the years 1629 through 1710 giving the number of boys and girls baptized during those years. In each of the 82 years, without exception, the number of boys exceeded the number of girls by amounts varying from less than 3% in 1644 (4107 boys, 3997 girls) to more than 15% in 1659 (3209 boys, 2781 girls). Arbuthnott inferred correctly that the hypothesis that births of boys and girls were equally likely was not plausible, since it implied that an event with probability 2^{-82} had occurred. He even consulted a table of logarithms to write this number out in decimal form, so as to impress his readers:

$$\frac{1}{4\,8360\,0000\,0000\,0000\,0000\,0000}.$$

Exhibiting the usual haste to reach conclusions in such matters, Arbuthnott concluded that this constant imbalance must be the result of a divine plan to offset the higher mortality of males due to violence and accidents.[15] He did not, for example, consider the possibility that more girls than boys were simply abandoned by mothers and fathers unable to support them. His final conclusion was that polygamy was against nature.

2.1. Quetelet. The first work on statistics proper was a treatise of 1835 entitled *Physique social*, written by the Belgian scientist Lambert Quetelet (1796–1874). Quetelet had been trained in both mathematics and astronomy, and he was familiar with the normal curve. He was the first to use it to describe variables other than those representing observational errors. He noticed certain analogies between probabilistic concepts and physical concepts, and he introduced them into social analysis. The most famous of these concepts was the "average person" (*l'homme moyen*), which he hoped could play a mathematical role similar to its physical analog, the center of gravity of a physical body.

2.2. Statistics in physics. One of the places in which individual phenomena are too numerous and too chaotic for analysis is in physics at the molecular level and below. Statistics has become an important tool in analyzing such systems. A very good example is thermodynamics, in which thermal energy is considered to be stored in a hypothetical (unobserved) translational and/or rotational motion of molecules against resisting forces that are equally hypothetical. In the simplest case, that of an ideal gas, there are no resisting forces and there is no rotational motion of molecules. All the thermal energy is stored as the translational kinetic energy of the molecules, which determine its temperature. At room temperature helium, which is monatomic, is the best approximation to an ideal gas.

In a way, thermodynamics, and in particular its famous second law, is only common sense, but physics needs to explain that common sense. Why can heat flow only from higher temperature to lower, just as water can flow only downhill? If temperature is determined by the translational kinetic energy of molecules, objects

[15] In his day, death from contagious disease was at least as common in women as in men, perhaps even more common, due to the dangers of childbirth. Death from the debilities associated with old age was relatively uncommon.

at a higher temperature have molecules with higher kinetic energy—they are either more massive or moving faster. When two bodies are in contact, their molecules collide along the interface. Thermal energy then diffuses just as a gas diffuses when the boundaries confining it are removed. James Clerk Maxwell (1831–1879) created a theory of gases in which, he thought, the second law of thermodynamics could be violated. In an 1867 letter to Peter Guthrie Tait (1831–1901), he imagined a person or other agency, later dubbed "Maxwell's demon" by Willam Thomson (Lord Kelvin, 1824–1907). The demon's job was to stand guard at a small interface between two objects at different temperatures and allow only those molecules to pass through that would cause the temperature difference to increase.[16] Statistically, it was *possible* that thermal energy might flow "uphill," so to speak. The question was a quantitative one: How *likely* was that to happen?

The workings of this process can most easily be seen in the case of a sample of an ideal monatomic gas, whose pressure (P), volume (V), and absolute Kelvin temperature T satisfy the equation of state $PV = nRT = kT$, where n is the number of moles of gas present and R is a universal constant of proportionality. The quantity $S = k\ln(T^{3V/2})$ is called the *entropy* of the sample.[17]

The evolution of a thermally isolated system can be thought of as the effect of bringing many small samples of gas at different temperatures into contact. If a sample of n_1 moles of the ideal monatomic gas occupying volume V_1 at temperature T_1 is placed in thermally isolated contact with a sample of n_2 moles occupying volume V_2 at temperature T_2, the total internal energy will be $\frac{3}{2}(k_1 + k_2)T = \frac{3}{2}k_1T_1 + \frac{3}{2}k_2T_2$, where T is the temperature after equilibrium is reached. Thus $T = (k_1T_1)/(k_1 + k_2) + (k_2T_2)/(k_1 + k_2) = cT_1 + (1 - c)T_2$. The ultimate entropy of the combined system will then be

$$(k_1 + k_2)\ln\Big(\big(cT_1 + (1 - c)T_2\big)^{\frac{3}{2}}(V_1 + V_2)\Big).$$

This quantity is larger than the combined initial entropy of the two parts,

$$k_1\ln\left(T_1^{3/2}V_1\right) + k_2\ln\left(T_2^{3/2}V_2\right),$$

as one can see easily since $c = (k_1)/(k_1 + k_2)$.[18] Thus, *entropy increases* for this system of two samples, and by extension in any thermally isolated system.

Maxwell began to urge a statistical view of thermodynamics in 1868, comparing the velocities of gas molecules with the white and black balls in the urn models that had been used for 150 years. In particular, he noted the tendency of these velocities to assume the normal distribution, as a consequence of the central limit theorem. When he became head of the Cavendish Laboratory at Cambridge in 1871, he said in his inaugural lecture that the statistical method

---

[16] There is more to Maxwell's demon than is implied here. Consider, for example, what energy is required for the demon to acquire the information about each molecule, decide whether to allow it to pass, and enforce the decision.

[17] Strictly speaking, it is not possible to take the logarithm of a quantity having a physical dimension. The expression $\int(1/V)\,dV = \ln(V)$ should be interpreted in dimensionless terms. That is, V is really V/V_u, where V_u is a unit volume, and likewise $dV = 1/V_u\,dV$ and $\ln(V)$ is $\ln(V/V_u)$.

[18] The function $3\ln(x)/2$ is concave, so that a point on an arc of its graph lies above the chord of that arc, and (assuming without loss of generality that $V_1 \geq V_2$) $\ln\big(1 + (V_1)/(V_2)\big) \geq \ln\big((V_1)/(V_2)\big) \geq c\ln\big((V_1)/(V_2)\big)$.

> involves an abandonment of strict dynamical principles and an adoption of the mathematical methods belonging to the theory of probability... if the scientific doctrines most familiar to us had been those which must be expressed in this way, it is possible that we might have considered the existence of a certain kind of contingency a self-evident truth, and treated the doctrine of philosophical necessity as a mere sophism. [Quoted by Porter, (*1986*), pp. 201–202]

2.3. The metaphysics of probability and statistics. The statistical point of view required an adjustment in thinking. Maxwell appeared to like the indeterminacy that it introduced; Einstein was temperamentally opposed to it. Although deterministic and probabilistic models might both produce the same predictions because of the law of large numbers, there was a theoretical difference that could be seen clearly in thermodynamics. If the laws of Newtonian mechanics applied to the point-particles that theoretically made up, say, an ideal gas, the state of the gas should evolve equally well in either direction, since those mechanical laws are time-symmetric. Imagine, then, two identical containers containing the same number of molecules of an ideal gas that, at a given instant, are in exactly the same positions relative to the boundaries of the containers and such that each particle is moving with equal speed but in exactly opposite direction, to the particle in the corresponding place in the other container. By the laws of mechanics, the past states of each container must be the future states of the other. But then one of the two must be evolving in a direction that decreases entropy, in contradiction to the second law of thermodynamics.

The explanation of that "must be" is statistical. It is not *absolutely* impossible for the mechanical system to be in a state that would cause it to evolve, following the deterministic laws of mechanics, in a direction of decreasing entropy. But the initial conditions that lead to this evolution are *extremely unlikely*, so unlikely that no one ever expects to observe such a system. As an illustration, Newtonian mechanics can perfectly well explain water flowing uphill given that the initial velocities of all the water molecules are uphill. But no one ever expects these initial conditions to be satisfied in practice, in the absence of a tsunami. An additional consideration, which Maxwell regarded as relevant, was that in some cases initial-value problems do not have a unique solution. For example, the equation $\frac{3}{4} \cdot \frac{dy}{dt} - y^{1/4} = 0$ with initial condition $y = 0$ when $t = 0$ is satisfied for $t \geq 0$ by both relations $y \equiv 0$ and $y = t^{\frac{4}{3}}$. Shortly before his death, Maxwell wrote to Francis Galton:

> There are certain cases in which a material system, when it comes to a phase in which the particular path which it is describing coincides with the envelope of all such paths may either continue in the particular parth or take to the envelope (which in these cases is also a possible path) and which course it takes is not determined by the forces of the system (which are the same for both cases) but when the bifurcation of the path occurs, the system, ipso facto, invokes some determining principle which is extra physical (but not extra natural) to determine which of the two paths it is to follow. [Quoted in Porter, *1986*, p. 206]

Statistics has been the focus of metaphysical debate, just like mathematics. For some early thinkers, such as Augustus de Morgan, the applications of probability were simply a matter of human ignorance: If we knew any reason for a system to be in one state rather than another, we would posit that reason as a physical law. In the absence of such a reason, all possible states are equally likely. A principle very close to this one is the basis of the second law of thermodynamics as now deduced in statistical physics. Yet other thinkers took a different point of view, positing some resemblance of the future to the past. This principle is the basis of the "frequentist" philosophy, which asserts that the probability of a future event is to be hypothesized from its occurrence in the past. The standard example is the question "What is the probability that the sun will rise tomorrow?" Assuming that we have adequate records that would have noted any exceptions to this very regular event over the past 5,000 years, a purely frequentist statistician would offer odds of $1,800,000$ to 1 in favor of the event happening tomorrow. However, as William Feller (1906–1970) pointed out in his classic textbook of probability, our records really do not guarantee that there have been no exceptions. Our confidence that the sun rose in the remote past is based on the same considerations that give us confidence that it will rise tomorrow.

Opposed to the frequentists are the Bayesians, who believe it is possible to assign a probability to an event before a similar event has occurred. Classical probabilists, with their urn models, drawings from a deck of cards, and throws of dice, were in effect Bayesians who believed that symmetry considerations and intuition enabled people to assign probabilities to hypothetical events. The results of experiment, where available, helped to revise those assignments through Bayes' theorem: *If the events $A_1, \ldots, A_n$ are mutually exclusive and exhaustive, and B is any event, then*

$$P(A_k|B) = \frac{P(A_k \wedge B)}{P(B)} = \frac{P(A_k)P(B|A_k)}{\sum_j P(B|A_j)P(A_j)}.$$

The use of this formula is as follows. From some basic principles of symmetry, or purely subjectively, the events A_k are assigned hypothetical probabilities. Then the conditional probability of event B is computed assuming each of these events. After an experiment in which event B occurs, the probability of A_k is "updated" to the value of $P(A_K|B)$ computed from this formula. The simplest illustration is the case of a chest containing two drawers. Drawer 1 contains two gold coins, and drawer 2 contains a silver coin and a gold coin. The two events are A_1: drawer 1 is chosen; A_2: drawer 2 is chosen. Each is assigned a preliminary probability of $\frac{1}{2}$. Event B is "a gold coin is drawn from the drawer." The conditional probabilities are easily seen to be $P(B|A_1) = 1$, $P(B|A_2) = \frac{1}{2}$. If in fact a gold coin is drawn, then

$$P(A_1|B) = \frac{1 \cdot \frac{1}{2}}{1 \cdot \frac{1}{2} + \frac{1}{2}\frac{1}{2}} = \frac{\frac{1}{2}}{\frac{3}{4}} = \frac{2}{3}$$

and

$$P(A_2|B) = \frac{\frac{1}{2} \cdot \frac{1}{2}}{1 \cdot \frac{1}{2} + \frac{1}{2}\frac{1}{2}} = \frac{\frac{1}{4}}{\frac{3}{4}} = \frac{1}{3}.$$

2.4. Correlations and statistical inference. The many intricate techniques that statisticians have developed today for analyzing data to determine if two variables are correlated are far too complex to be discussed in full here. But we cannot leave this topic without at least mentioning one of the giants in this area, Karl Pearson (1856–1936), a British polymath who studied philosophy and law and was even called to the bar, although he never practiced. As professor at University College, London, he became interested in Darwin's theory of evolution and wrote a number of papers between 1893 and 1912 on the mathematics of evolution. In an 1893 paper he coined the term *standard deviation* to denote the natural unit of probability, and in 1900 he introduced the chi-square test of significance, a mainstay of applied statistics nowadays.[19] Mathematically, the chi-square distribution with n degrees of freedom is the distribution of the sum of the squares of n independent standard normal distributions. What that means is that if the probability that X_k lies between a and b is given by the normal density,:

$$P(a \le X_k \le b) = \frac{1}{\sqrt{2\pi}} \int_a^b e^{-\frac{1}{2}t^2}\,dt\,,$$

and each of these probabilities is independent of all the others, the probability that $X_1^2 + \cdots + X_n^2$ lies between 0 and c is given by the chi-square density with n degrees of freedom:

$$P(X_1^2 + \cdots + X_n^2 \le c) = \frac{1}{2^{n/2}\Gamma\left(\frac{n}{2}\right)} \int_0^c x^{(n/2)-1} e^{-x/2}\,dx\,.$$

The chi-square distribution is useful because, if $X_1, \ldots, X_n$ are independent random variables with expected positive values $\mu_1, \ldots, \mu_n$, the random variable

$$\chi^2 = \frac{(X_1 - \mu_1)^2}{\mu_1} + \cdots + \frac{(X_n - \mu_n)^2}{\mu_n}$$

has the chi-square distribution with $n - 1$ degrees of freedom. One can then determine whether actual deviations of the variables X_k from their expected values are likely to be random (hence whether *bias* is present) by computing the value of χ^2 and comparing it with a table of chi-square values.

To illustrate the connection between the chi-square and the standard normal distribution, Fig. 3 shows the frequency histogram for a computer experiment in which 1000 random values were computed for the sum of the squares of 10 standard normal random variables. This histogram is superimposed on the graph of the chi-square density function with 10 degrees of freedom.

The word *bias* in the preceding paragraph has a purely statistical meaning of "not random." The rather pejorative meaning it has in everyday life is an indication of the connection that people tend to make between fairness and *equal outcomes.* If we find that some identifiable group of people is underrepresented or overrepresented in some other population—prisons or universities or other institutions—we proceed on the assumption that some cause is operating. In doing so, one must beware of jumping to conclusions, as Arbuthnott did. He was quite correct in his conclusion that the sexual imbalance was not a random deviation from a general rule of equality, but there are all kinds of possible explanations for the *bias.* An even larger sexual imbalance exists in China today, for example, as a result of the one-child policy of the Chinese government, combined with a traditional social pressure to produce male heirs. Evolutionary theory produces an explanation very similar to Arbuthnott's, but based on adaptation rather than intelligent, human-centered design.

---

[19] The symbol χ^2 was Pearson's abbreviation for $x^2 + y^2$.

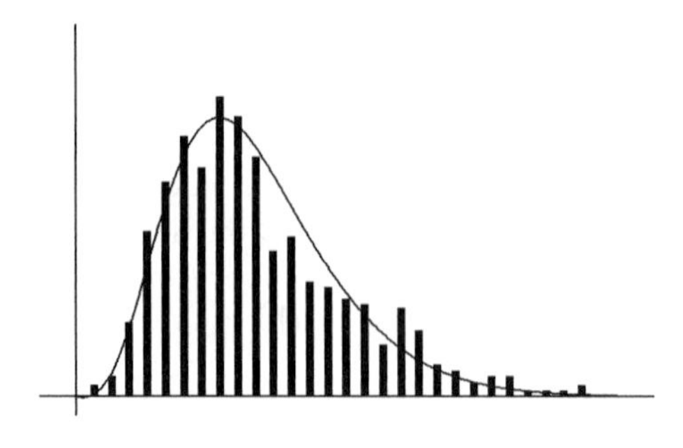

FIGURE 3. Comparison of the chi-square distribution and the frequencies for the sum of the squares of ten independent standard normal random variables when the experiment is performed 1000 times.

One of the many pitfalls of statistical inference was pointed out by Pearson's colleague George Udny Yule (1871–1951) in 1903. Following up on Pearson's 1899 paper "On the spurious correlation produced by forming a mixture of heterogeneous but uncorrelated material," Yule produced a set of two 2×2 tables, each of which had no correlation, but produced a correlation when combined (see David and Edwards, *2001*, p. 137). Yule's result was, for some reason, not given his name; but because it was publicized by Edward Hugh Simpson in 1951,[20] it came to be known as *Simpson's paradox*.[21]

Simpson's paradox is a counterintuitive oddity, not a contradiction. It arises frequently in practice. An example of it occurred in the admissions data from the graduate school of the University of California at Berkeley in 1973. These data raised some warning flags. Of the 12,763 applicants, 5232 were admitted, giving an admission rate of 41%. However, investigation revealed that 44% of the male applicants had been admitted and only 35% of the female applicants. There were 8442 male applicants, 3738 of whom were accepted, and 4321 female applicants, 1494 of whom were accepted. Simple chi-square testing showed that the hypothesis that these numbers represent a random deviation from a sex-independent acceptance rate of 41% was not plausible. There was unquestionably *bias*. The question was: Did this bias amount to discrimination? If so, who was doing the discriminating?

For more information on this case study and a very surprising conclusion, see "Sex bias in graduate admissions: data from Berkeley," *Science*, **187**, 7 February 1975, 398–404. In that paper, the authors analyzed the very evident *bias* in admissions to look for evidence of *discrimination*. Since admission decisions are made by the individual departments, it seemed logical to determine which departments had a noticeably higher admission rate for men than for women. Surprisingly, the authors found only four such departments (out of 101), and the imbalance resulting from those four departments was more than offset by six other departments that had a higher admission rate for women. It appears that the source of the bias was

[20] See "The interpretation of interaction in contingency tables," *Journal of the Royal Statistical Society*, Series B, **13**, 238–241.

[21] The name *Simpson's paradox* goes back at least to the article of C. R. Blyth, "On Simpson's paradox and the sure-thing principle," in the *Journal of the American Statistical Association*, **67** (1972), 364–366.

hiding itself very well. The curious reader is referred to Problem 18.7, where the paradox is explained with an even more extreme example.

Questions and problems

18.1. Weather forecasters are evaluated for accuracy using the *Briers score.* The *a posteriori* probability of rain on a given day, judged from the observation of that day, is 0 if rain did not fall and 1 if rain did fall. A weather forecaster who said (the day before) that the chance of rain was 30% gets a Briers score of $30^2 = 900$ if no rain fell and $70^2 = 4900$ if rain fell. Imagine a very good forecaster, who over many years of observation learns that a certain weather pattern will bring rain 30% of the time. Also assume that for the sake of negotiating a contract that forecaster wishes to optimize (minimize) his or her Briers score. Should that forecaster state truthfully that the probability of rain is 30%? If we assume that the prediction and the outcome are independent events, we find that, for the days on which the true probability of rain is 30% the forecaster who makes a prediction of a 30% probability would in the long run average a Briers score of $0.3 \cdot 70^2 + 0.7 \cdot 30^2 = 2100$. This score is better (in the sense of a golf score—it is lower) than would result from randomly predicting a 100% probability of rain 30% of the time and a 0% probability 70% of the time. That strategy will be correct an expected 58% of the time ($.58 = .3^2 + .7^2$) and incorrect 42% of the time, resulting in a Briers score of $.42 \cdot 100^2 = 4200$. Let p be the actual probability of rain and x the forecast probability. Assuming that the event and the forecast are independent, show that the expected Briers score $10^4\bigl(p(1-x)^2 + (1-p)x^2\bigr)$ is minimized when $x = p$. [*Note*: If this result did not hold, a meteorologist who prized his/her reputation as a forecaster, based on the Briers measure, would be well advised to predict an incorrect probability, so as to get a better score for accuracy!]

18.2. We saw above that Cardano (probably) and Pascal and Leibniz (certainly) miscalculated some elementary probabilities. As an illustration of the counter-intuitive nature of many simple probabilities, consider the following hypothetical games. (A casino could probably be persuaded to open such games if there was enough public interest in them.) In game 1 the dealer lays down two randomly-chosen cards from a deck on the table and turns one face up. If that card is not an ace, no game is played. The cards are replaced in the deck, the deck is shuffled, and the game begins again. If the card is an ace, players are invited to bet against a fixed winning amount offered by the house that the other card is also an ace. What winning should the house offer (in order to break even in the long run) if players pay one dollar per bet?

In game 2 the rules are the same, except that the game is played only when the card turned up is the ace of hearts. What winning should the house offer in order to break even charging one dollar to bet? Why is this amount not the same as for game 1?

18.3. Use the Maclaurin series for $e^{-(1/2)t^2}$ to verify that the series given by de Moivre, which was

$$\sqrt{\frac{2}{\pi}}\Bigl(\frac{1}{0!\cdot 1\cdot 2} - \frac{1}{1!\cdot 3\cdot 4} + \frac{1}{2!\cdot 5\cdot 8} - \frac{1}{3!\cdot 7\cdot 16} + \cdots\Bigr),$$

represents the integral

$$\frac{1}{\sqrt{2\pi}} \int_0^1 e^{-\frac{1}{2}t^2}\, dt,$$

which is the area under a standard normal (bell-shaped) curve above the mean, but by at most one standard deviation, as given in many tables.

18.4. Use Daniel Bernoulli's concept of utility to explain why only a person with astronomical amounts of money should play a Petersburg paradox-type game. In your explanation, take account of what the utility of the stakes must be for a gambler versus the utility of the gain. Make an analogy between risk and work in this regard. A laborer exchanges time and effort for money; a gambler exchanges risk for potential gain. Remembering that all economic decisions are made "at the margin," at what point does additional work (or risk) not bring enough additional utility to be worth the exchange?

18.5. Radium-228 is an unstable isotope. Each atom of Ra-228 has a probability of 0.1145 (about 1 chance in 9, or about the probability of rolling a 5 with two dice) of decaying to form an atom of actinium within any given year. This means that the probability that the atom will survive the year as an atom of Ra-228 is $1-0.1145 = 0.8855$. Denote this "one-year survival" probability by p. Because any sample of reasonable size contains a huge number of atoms, that survival probability (0.8855) is the proportion of the weight of Ra-228 that we would expect to survive a year.

If you had one gram of Ra-228 to begin with, after one year you would expect to have $p = 0.8855$ grams. Each succeeding year, the weight of the Ra-228 left would be multiplied by p, so that after two years you would expect to have $p^2 = (0.8855)^2 = 0.7841$ grams. In general, after t years, if you started with W_0 grams, you would expect to have $W = W_0 p^t$ grams. Now push these considerations a little further and determine *how strongly* you can rely on this expectation. Recall Chebyshëv's inequality, which says that the probability of being more than k standard deviations from the expected value is never larger than $(1/k)^2$. What we need to know to answer the question in this case is the standard deviation σ.

Our assumption is that each atom decays at random, independently of what happens to any other atom. This independence allows us to think that observing our sample for a year amounts to a large number of "independent trials," one for each atom. We test each atom to see if it survived as an Ra-228 atom or decayed into actinium. Let N_0 be the number of atoms that we started with. Assuming that we started with 1 gram of Ra-228, there will be $N_0 = 2.642 \cdot 10^{21}$ atoms of Ra-228 in the original sample.[22] That is a very large number of atoms. The survival probability is $p = 0.8855$. For this kind of independent trial, as mentioned the standard deviation with N_0 trials is

$$\sqrt{N_0 p(1-p)} = \sqrt{\frac{p(1-p)}{N_0}}\, N_0 \, .$$

We write the standard deviation in this odd-looking way so that we can express it as a fraction of the number N_0 that we started with. Since weights are proportional to the number of atoms, that same fraction will apply to the weights as well.

---

[22] According to chemistry, the number of atoms in one gram of Ra-228 is the *Avogadro number* $6.023 \cdot 10^{23}$ divided by 228.

Put in the given values of p and N_0 to compute the fraction of the initial sample that constitutes one standard deviation. Since the original sample was assumed to be one gram, you can regard the answer as being expressed in grams. The use Chebyshëv's inequality to estimate the probability that the amount of the sample remaining will differ from the theoretically predicted amount by 1 millionth of a gram (1 microgram, that is, 10^{-6} grams)? [*Hint*: How many standard deviations is one millionth of a gram?]

18.6. Analyze the revised probabilities in the problem of two drawers, one containing two gold coins, the other a gold and a silver coin, given an experiment in which event B occurs, if B is the event, "a silver coin is drawn."

18.7. Consider the case of 200 men and 200 women applying to a university consisting of only two different departments, and assume that the acceptance rates are given by the following table.

	Men	**Women**
Department A	120/160	32/40
Department B	8/40	40/160

Observe that the admission rate for men in department A is $\frac{3}{4}$, while that for women is $\frac{4}{5}$. In department B the admission rate for men is $\frac{1}{5}$ and for women it is $\frac{1}{4}$. In both cases, the people actually making the decisions are admitting a higher proportion of women than of men. Now explain the source of the bias, in our example and at Berkeley in simple, nonmathematical language.

CHAPTER 19

Logic and Set Theory

Logic has been an important part of western mathematics since the time of Plato. It also has a long history in other cultures, such as the Hindu and Buddhist culture (see Vidyabhusana, *1971*). Logic became mathematized in the nineteenth century, in the work of mostly British mathematicians such as George Peacock (1791–1858), George Boole (1815–1864), William Stanley Jevons (1835–1882), and Augustus de Morgan, and a few Americans, notably Charles Sanders Peirce.

Set theory was the creation of nineteenth-century analysts and geometers, prominent among them Georg Cantor (1845–1918), whose inspiration came from geometry and analysis, mostly the latter. It resonated with the new abstraction that was entering mathematics from algebra and geometry, and its use by the French mathematicians Borel, Lebesgue, and Baire as the framework for their theories of integration and continuity helped to establish it as the foundation of all mathematics.

1. Logic

The mathematization of logic has a prehistory that goes back to Leibniz (not published in his lifetime), but we shall focus on mostly the nineteenth-century work. After a brief discussion of the preceding period, we examine the period from 1847 to 1930. This period opens with the treatises of Boole and de Morgan and closes with Gödel's famous incompleteness theorem. Our discussion is not purely about logic in the earlier parts, since the earlier writers considered both logical and probabilistic reasoning.

1.1. From algebra to logic. Leibniz was one of the first to conceive the idea of creating an artificial language in which to express propositions. He compared formal logic to the lines drawn in geometry as guides to thought. If the language encoded thought accurately, thought could be analyzed in a purely mechanical manner:

> Then, in case of a difference of opinion, no discussion between two philosophers will be any longer necessary, as (it is not) between two calculators. It will rather be enough for them to take pen in hand, set themselves to the abacus and (if it so pleases, at the invitation of a friend) say to one another: *Let us calculate!* [Quoted by Bochenski, *1961*, p. 275]

In another place he wrote:

> Ordinary languages, though mostly helpful for the inferences of thought, are yet subject to countless ambiguities and cannot do the task of a calculus, which is to expose mistakes in inference... This remarkable advantage is afforded up to date only by the symbols of

> arithmeticians and algebraists, for whom inference consists only in the use of characters, and a mistake in thought and in the calculus is identical. [Quoted by Bochenski, *1961*, p. 275]

The ideal enunciated by Leibniz remains largely unfulfilled when it comes to settling philosophical disagreements. It reflects an oversimplified and optimistic view of human beings as basically rational creatures. This sort of optimism continued into the early nineteenth century, as exemplified by the *Handbook of Political Fallacies* by the philosopher Jeremy Bentham (1748–1832). But if the complex questions of the world of nature and society could not be mastered through logic alone, mathematics proved more amenable to the influences of logic. The influence, however, was bidirectional. In fact, there is a paradox, if one thinks of logic as being the rudder that steers mathematical arguments and keeps them from going astray. As Charles Sanders Peirce wrote in 1896, reviewing a book on logic:

> It is a remarkable historical fact that there is a branch of science in which there has never been a prolonged dispute concerning the proper objects of that science. It is mathematics... Hence, we homely thinkers believe that, considering the immense amount of disputation there has always been concerning the doctrines of logic, and especially concerning those which would otherwise be applicable to settle disputes concerning the accuracy of reasonings in metaphysics, the safest way is to appeal for our logical principles to the science of mathematics. [Quoted in Bochenski, *1961*, pp. 279–280]

Peirce seemed to believe that far from sorting out the mathematicians, logicians should turn to them for guidance. But we may dispute his assertion that there has never been a prolonged dispute about the proper objects of mathematics. Zeno confronted the Pythagoreans over that very question. In Peirce's own day, Kronecker and Cantor were at opposite ends of a dispute about what is and is not proper mathematics, and that discussion continues, politely, down to the present day. (See, for example, Hersh, *1997*.)

Leibniz noted in the passage quoted above that algebra had the advantage of a precise symbolic language, which he held up as an ideal for clarity of communication. Algebra, in fact, was one of the mathematical sources of mathematical logic. When de Morgan translated a French algebra textbook into English in 1828, he defined algebra as "the part of mathematics in which symbols are employed to abridge and generalize the reasonings which occur in questions relating to numbers." Thus, for de Morgan at the time, the symbols represented numbers, but *unspecified* numbers, so that reasoning about them applied to any particular numbers. Algebra was a ship anchored in numbers, but it was about to slip its anchor. In fact, only two years later (in 1830) George Peacock wrote a treatise on algebra in which he proposed that algebra be a purely symbolic science independent of any arithmetical interpretation. This step was a radical innovation at the time, considering that abstract groups, for example, were not to appear for several more decades. The assertion that the formula $(a - b)(a + b) = a^2 - b^2$ holds independently of any numerical values that replace a and b, for example, almost amounts to an axiomatic approach to mathematics. De Morgan's ideas on this subject matured during the 1830s, and at the end of the decade he wrote:

> When we wish to give the idea of symbolical algebra... we ask, firstly, what symbols shall be used (without any reference to meaning); next, what shall be the laws under which such symbols are to be operated upon; the deduction of all subsequent consequences is again an application of common logic. Lastly, we explain the meanings which must be attached to the symbols, in order that they may have prototypes of which the assigned laws of operation are true. [Quoted by Richards, *1987*, pp. 15–16]

This set of procedures is still the way in which mathematical logic operates, although the laws under which the symbols are to be operated on are now more abstract than de Morgan probably had in mind. To build a formal language, you first specify which sequences of symbols are to be considered "well-formed formulas," that is, formulas capable of being true or false. The criterion for being well-formed must be purely formal, capable of being decided by a machine. Next, the sequences of well-formed formulas that are to be considered deductions are specified, again purely formally. The *syntax* of the language is specified by these two sets of rules, and the final piece of the construction, as de Morgan notes, is to specify its *semantics*, that is, the interpretation of its symbols and formulas. Here again, the modern world takes a more formal and abstract view of "interpretation" than de Morgan probably intended. For example, the semantics of propositional calculus consists of truth tables. After specifying the semantics, one can ask such questions as whether the language is consistent (incapable of proving a false proposition), complete (capable of proving all true propositions), or categorical (allowing only one interpretation, up to isomorphism).

In his 1847 treatise *Formal Logic*, de Morgan went further, arguing that "we have power to invent new meanings for all the forms of inference, in every way in which we have power to make new meanings of *is* and *is not*... ." This focus on the meaning of *is* was very much to the point. One of the disputes that Peirce overlooked in the quotation just given is the question of what principles allow us to infer that an object "exists" in mathematics. We have seen this question in the eighteenth-century disagreement over what principles are allowed to define a function. In the case of symbolic algebra, where the symbols originally represented numbers, the existence question was still not settled to everyone's liking in the early nineteenth century. That is why Gauss stated the fundamental theorem of algebra in terms of real factorizations alone. Here de Morgan was declaring the right to create mathematical entities by *fiat*, subject to certain restrictions. That enigmatic "exists" is indispensible in first-order logic, where the negation of "For every x, P" is "For some x, not-P." But what can "some" mean unless there actually *exist* objects x? This defect was to be remedied by de Morgan's friend George Boole.

In de Morgan's formal logic, this "exists" remains hidden: When he talks about a class X, it necessarily has members. Without this assumption, even the very first example he gives is not a valid inference. He gives the following table by way of introduction to the symbolic logic that he is about to introduce:

Instead of:	*Write*:
All men will die	Every Y is X
All men are rational beings	Every Y is Z
Therefore some rational beings will die	Therefore some Zs are X's.

De Morgan's notation in this work was not the best, and very little of it has caught on. He used a parenthesis in roughly the same way as the modern notation for implication. For example, $X\,)\,Y$ denoted the proposition "Every X is a Y." Nowadays we would write $X \supset Y$ (read "X horseshoe Y") for "X implies Y." The rest of his notation—$X : Y$ for "Some X's are not Ys," $X.Y$ for "No X's are Ys" and XY for "Some X's are Ys"—is no longer used. For the negation of these properties he used lowercase letters, so that x denoted not-X. De Morgan introduced the useful "necessary" and "sufficient" language into implications: $X\,)\,Y$ meant that Y was *necessary* for X and X was *sufficient* for Y. He gave a table of the relations between X or x and Y or y for the relations $X\,)\,Y$, $X.Y$, $Y\,)\,X$, and $x\,.\,y$. For example, given that X implies Y, he noted that this relation made Y necessary for X, y an impossible condition for X, y a sufficient condition for x, and Y a contingent (not necessary, not sufficient, not impossible) condition for x.

For compound propositions, he wrote PQ for conjunction (his word), meaning both P and Q are asserted, and $P\,,\,Q$ for disjunction (again, his word), meaning either P or Q. He then noted what are still known as *De Morgan's laws*:

> The contrary of PQ is $p\,,\,q$. *Not both* is either not one or not the other, or not either. *Not either P nor Q* (which we might denote by $: P\,,\,Q$ or $.P\,,\,Q$) is logically '*not P and not Q*' or pq: and this is then the contrary of $P\,,\,Q$.

De Morgan's theory of probability. De Morgan devoted three chapters (Chapters 9 through 11) to probability and induction, starting off with a very Cartesian principle:

> That which we know, of which we are certain, of which we are well assured nothing could persuade us to the contrary, is the existence of our own minds, thoughts, and perceptions.

He then took the classical example of a certain proposition, namely that $2+2 = 4$ and showed by analyzing the meaning of 2, 4, and + that "It is true, no doubt, that 'two and two' is four, in amount, value, &c. but not in form, construction, definition, &c."[1] He continued:

> There is no further use in drawing distinction between the knowlege which we have of our own existence, and that of two and two amounting to four. This absolute and inassailable feeling we shall call *certainty*. We have lower grades of knowledge, which we usually call *degrees of belief*, but they are really *degrees of knowledge*... It may seem a strange thing to treat *knowledge* as a magnitude, in the same manner as length, or weight, or surface. This is what all writers do who treat of probability... But it is not customary to make the statement so openly as I now do.

As this passage shows, for de Morgan probability was a subjective entity. He said that *degree of probability* meant *degree of belief*. In this way he placed himself firmly against the frequentist position, saying, "I throw away objective *probability*

[1] There is some intellectual sleight-of-hand here. The effectiveness of this argument depends on the reader's not knowing—and de Morgan's not stating—what is meant by addition of integers and by equality of integers, so that $2 + 2$ cannot be broken down into any terms simpler than itself. It really can be proved that $2 + 2 = 4$.

altogether, and consider the word as meaning the state of the mind with respect to an assertion, a coming event, or any other matter on which absolute knowledge does not exist." But subjectivity is not the same thing as arbitrariness. De Morgan, like us, would have labeled insane a person who asserted that the probability of rolling double sixes with a pair of dice is 50%. In fact, he gave the usual rules for dealing with probabilites of disjoint and independent events, and even stated Bayes' rule of inverse probability. He considered two urns, one containing six white balls and one black ball, the other containing two white balls and nine black ones. Given that one has drawn a white ball, he asked what the probability is that it came from the first urn. Noting that the probability of a white ball was $\frac{6}{7}$ in the first case and $\frac{2}{11}$ in the second, he concluded that the odds that it came from either of the two urns must be in the same proportion, $\frac{6}{7} : \frac{2}{11}$ or $33 : 7$. He thus gave $\frac{33}{40}$ as the probability that the ball came from the first urn. This answer is in numerical agreement with the answer that would be obtained by Bayes' rule, but de Morgan did not think of it as revising a preliminary estimate of $\frac{1}{2}$ for the probability that the first urn was chosen.

1.2. Symbolic calculus. An example of the new freedom in the interpretation of symbols actually occurred somewhat earlier than the time of de Morgan, in Lagrange's algebraic approach to analysis. Thinking of Taylor's theorem as

$$\Delta_h f(x) = f(x+h) - f(x) = hDf(x)h + \frac{1}{2!}h^2D^2f(x) + \frac{1}{3!}h^3D^3f(x) + \ldots,$$

where $Df(x) = f'(x)$, and comparing with the Taylor's series of the exponential function,

$$e^t = 1 + t + \frac{1}{2!}t^2 + \frac{1}{3!}t^3 + \cdots,$$

Lagrange arrived at the formal equation

$$\Delta_h = e^{hD} - 1\,.$$

Although the equation is purely formal and should perhaps be thought of only as a convenient way of remembering Taylor's theorem, it does suggest a converse relation

$$Df(x) = \frac{1}{h}\big(\ln(1+\Delta_h)\big)f(x) = \frac{1}{h}\Big(\Delta_h f(x) + \frac{1}{2}\Delta_h^2 f(x) + \cdots\Big),$$

and this relation is literally true for polynomials $f(x)$. The formal use of this symbolic calculus may have been merely suggestive, but as Grattan-Guinness remarks (*2000*, p. 19), "some people regarded these methods as legitimate in themselves, not requiring foundations from elsewhere."

1.3. Boole's *Mathematical Analysis of Logic*. One such person was George Boole. In a frequently quoted passage from the introduction to his brief 1847 treatise *The Mathematical Analysis of Logic*, Boole wrote

> [T]he validity of the processes of analysis does not depend upon the interpretation of the symbols which are employed but solely upon the laws of their combination. Every system of interpretation which does not affect the truth of the relations supposed is equally admissible, and it is thus that the same process may under one scheme of interpretation represent the solution of a question or the

> properties of number, under another that of a geometrical problem, and under a third that of optics.

Here Boole, like de Morgan, was arguing for the freedom to create abstract systems and attach an interpretation to them later. This step was still something of an innovation at the time. It was generally accepted, for example, that irrational and imaginary numbers had a meaning in geometry but not in arithmetic. One could not, or should not, simply conjure them into existence. Cayley raised this objection shortly after the appearance of Boole's treatise (see Grattan-Guinness, *2000*, p. 41), asking whether it made any sense to write $\frac{1}{2}x$. Boole replied by comparing the question to the existence of $\sqrt{-1}$, which he said was "a symbol (i) which satisfies particular laws, and especially this: $i^2 = -1$." In other words, when we are inventing a formal system, we are nearly omnipotent. Whatever we prescribe will hold for the system we define. If we want a square root of -1 to exist, it will exist (whatever "exist" may mean).

Logic and classes. Although set theory had different roots on the Continent, we can see its basic concept—membership in a class—in Boole's work. Departing from de Morgan's notation, he denoted a generic member of a class by an uppercase X, and used the lowercase x "operating on any subject," as he said, to denote the class itself. Then xy was to denote the class "whose members are both X's and Ys." This language rather blurs the distinction between a set, its members, and the properties that determine what the members are; but we should expect that clarity would take some time to achieve. The connection between logic and set theory is an intimate one and one that is easy to explain. But the kind of set theory that logic alone would have generated was different from the geometric set theory of Georg Cantor, which is discussed in the next section.

The influence of the mathematical theory of probability on logic is both extensive and interesting. The subtitle of de Morgan's *Formal Logic* is *The Calculation of Inference, Necessary and Probable*, and, as noted above, three chapters (some 50 pages) of *Formal Logic* are devoted to probability and induction. Probability deals with events, whereas logic deals with propositions. The connnection between the two was stated by Boole in his later treatise, *An Investigation of the Laws of Thought*, as follows:

> [T]here is another form under which all questions in the theory of probabilities may be viewed; and this form consists in substituting for *events* the propositions which assert that those events have occurred, or will occur; and viewing the element of numerical probability as having reference to the *truth* of those *propositions*, not to the *occurrence* of the *events*.

Two events can combine in different ways: exactly one of E and F may occur, or E and F may both occur. If the events E and F are independent, the probability that both E and F occur is the *product* of their individual probabilities. If the two events cannot both occur, the probability that at least one occurs is the *sum* of their individual probabilities. More generally,

$$P(E \,\text{or}\, F) + P(E \,\text{and}\, F) = P(E) + P(F)\,.$$

When these combinations of events are translated into logical terms, the result is a *logical calculus*.

The idea of probability 0 as indicating impossibility and probability 1 as indicating certainty must have had some influence on Boole's use of these symbols to denote "nothing" and "the universe." He expressed the proposition "all X's are Y's," for example, as $xy = x$ or $x(1-y) = 0$. Notice that $1-y$ appears, not $y-1$, which would have made no sense. Here $1-y$ corresponds to the things that are not-y. From there, it is not far to thinking of 0 as false and 1 as true. The difference between probability and logic here is that the probability of an event may be any number between 0 and 1, while propositions are either true or false.[2] These analogies were brought out fully in Boole's major work, to which we now turn.

1.4. Boole's *Laws of Thought*. Six years later, after much reflection on the symbolic logic that he and others had developed, Boole wrote an extended treatise, *An Investigation of the Laws of Thought*, which began by recapping what he had done earlier. The *Laws of Thought* began with a very general proposition that laid out the universe of symbols to be used. These were:

> 1st. Literal symbols, as x, y, &c., representing things as subjects of our conceptions.
> 2nd. Signs of operation, as $+$, $-$, $\times$, standing for those operations of the mind by which the conceptions of things are combined or resolved so as to form new conceptions involving the same elements.
> 3rd. The sign of identity, $=$.
> And these symbols of Logic are in their use subject to definite laws, partly agreeing with and partly differing from the laws of the corresponding symbols in the science of Algebra.

Boole used $+$ to represent disjunction (or) and juxtaposition, used in algebra for multiplication, to represent conjunction (and). The sign $-$ was used to stand for "and not." In his examples, he used $+$ only when the properties were, as we would say, disjoint and $-$ only when the property subtracted was, as we would say, a subset of the property from which it was subtracted. He illustrated the equivalence of "European men and women" (where the adjective *European* is intended to apply to both nouns) with "European men and European women" as the equation $z(x+y) = zx + zy$. Similarly, to express the idea that the class of men who are non-Asiatic and white is the same as the class of white men who are not white Asiatic men, he wrote $z(x-y) = zx - zy$. He attached considerable importance to what he was later to call the *index law*, which expresses the fact that affirming a property twice conveys no more information than affirming it once. That is to say, $xx = x$, and he adopted the algebraic notation x^2 for xx. This piece of algebraization led him, by analogy with the rules $x0 = 0$ and $x1 = x$, to conclude that "the respective interpretations of the symbols 0 and 1 in the system of Logic are *Nothing* and *Universe*." From these considerations he deduced the principle of contradiction:

---

[2] Classical set theory deals with propositions of the form $x \in E$, which are either true or false: Either x belongs to E, or it does not, and there is no other possibility. The recently created *fuzzy set theory* restores the analogy with probability, allowing an element to belong partially to a given class and expressing the degree of membership by a function $\varphi(x)$ whose values are between 0 and 1. Thus, for example, whether a woman is pregnant or not is a classical set-theory question; whether she is tall or not is a fuzzy set-theory question. Fuzzy-set theorists point out that their subject is not subsumed by probability, since it deals with the properties of individuals, not those of large sets.

$x^2 = x \Rightarrow x(1-x) = 0$, that is, no object can have a property and simultaneously not have that property.[3]

Boole was carried away by his algebraic analogies. Although he remained within the confines of his initial principles for a considerable distance, when he got to Chapter 5 he introduced the concept of *developing* a function. That is, for each algebraic expression $f(x)$, no matter how complicated, to find an equivalent linear expression $ax + b(1-x)$, one that would have the same values as $f(x)$ for $x = 0$ and $x = 1$. That expression would obviously be $f(1)x + f(0)(1-x)$. Boole gave a convoluted footnote to explain this simple fact by deriving it from Taylor's theorem and the idempotence property.

Like De Morgan, after discussing his 0–1 logic, Boole then turned to philosophy, metaphysics, and probability, placing himself in the philosophical camp of Poisson and de Morgan. He gave the now-familiar rule for the conditional probability of A given B as the probability of both A and B divided by the probability of B. He also gave a formal definition of independence, saying that two events are independent if "the probability of the happening of either of them is unaffected by our expectation of the occurrence or failure of the other." All of this was done in words, but could have been done symbolically, as he surely realized.

Application to jurisprudence. Nearly all of the early writers on probability, statistics, and logic had certain applications in mind, to insurance in the case of statistics, especially to the decisions of courts. The question of the believability of witnesses and the probability that a jury has been deceived interested Laplace, Quetelet, Poisson, de Morgan, and Boole, among others. Boole, for example, gave as an example, the following problem:

> The probability that a witness A speaks the truth is p, the probability that another witness B speaks the truth is q, and the probability that they disagree in a statement is r. What is the probability that if they agree, their statement is true?

Boole gave the answer as $(p + q - r)/\bigl(2(1-r)\bigr)$. He claimed to prove as a theorem the following proposition:

> From the records of the decisions of a court or deliberative assembly, it is not possible to deduce any definite conclusion respecting the correctness of the individual judgments of its members.

1.5. Venn. It is interesting to compare the mathematization of logic with the mathematization of probability. Both have ultimately been successful, but both were resisted to some extent as an intrusion of mathematics into areas of philosophy where it had no legitimate business. The case for removing mathematics from philosophy was made by John Venn, whose name is associated with a common tool of set theory: *Venn diagrams*, so-called, although the idea really goes back to Euler. In his book *The Logic of Chance*, which was first published in 1867, then revised a decade later and revised once again after another decade, Venn

---

[3] Nowadays, a ring in which every element is idempotent, that is, the law $x^2 = x$ holds, is called a *Boolean ring*. It is an interesting exercise to show that such a ring is always commutative and of characteristic 2, that is, $x + x = 0$ for all x. The subsets of a given set form a Boolean ring when addition is interpreted as symmetric difference, that is, $A + B$ means "either A or B but not both."

proudly proclaimed that "Not only... will 'no knowledge of mathematics beyond the simple rules of Arithmetic' be required to understand these pages, but it is not intended that any such knowledge should be acquired by the process of reading them." Venn was particularly exercised about the attempts to apply probability theory in jurisprudence. Referring to Laplace, Quetelet, and the others, he wrote:

> When they have searched for illustrations drawn from the practical business of life, they have very generally, but unfortunately, hit upon just the sort of instances which, as I shall endeavour to show hereafter, are among the very worst that could be chosen for the purpose. It is scarcely possible for any unprejudiced person to read what has been written about the credibility of witnesses by eminent writers, without his experiencing an invincible distrust of the principles which they adopt.

He went on to say that, although probability may require considerable mathematical knowledge, "the discussion of the fundamental principles on which the rules are based does not necessarily require any such qualification." Moreover,

> The opinion that Probability, instead of being a branch of the general science of evidence which happens to make much use of mathematics, *is* a portion of mathematics, erroneous as it is, has yet been very disadvantageous to the science in several ways.

As one might expect, he took a dim view of the writings of de Morgan and Boole, saying that de Morgan had "given an investigation into the foundations of Probability as conceived by him, and nothing can be more complete and precise than his statement of principles and his deductions from them. If I could at all agree with these principles there would have been no necessity for the following essay." As for Boole, "Owing to his peculiar treatment of the subject, I have scarely anywhere come into contact with any of his expressed opinions," a subtle, but acerbic way of saying that Boole had failed to convince anyone.

In Venn's view, expressed at the beginning of his fourth chapter, the practical application of probability in such matters as insurance was simply one more aspect of induction, the extrapolation of past experience into the future:

> We cannot tell how many persons will be born or die in a year, or how many houses will be burnt or ships wrecked, without actually counting them. When we thus speak of "experience," we mean to employ the term in its widest signification; we mean experience supplemented by all the aids which inductive or deductive logic can afford. When, for instance, we have found the series which comprises the numbers of persons of any assigned class who die in successive years, we have no hesitation in extending it some way into the future as well as into the past. The justification of such a procedure must be sought in the ordinary canons of Induction.

Venn thus proclaimed himself a frequentist. The justification for applied probability and statistics was to be induction. But how firm a foundation was induction? The skeptical Scot David Hume (1711–1776) had leveled a devastating criticism against the principles of induction and cause in the preceding century. Venn's reduction risked exposing probability theory to the same demolition.

1.6. Jevons. Both de Morgan and Boole used the syllogism or *modus ponens* (*inferring method*) as the basis of logical inference, although de Morgan did warn against an overemphasis on it. Their successor William Stanley Jevons, formulated this law algebraically and adjoined to it a principle of indirect inference, which amounted to inference by exhaustive enumeration of cases. The possibility of doing the latter by sorting through slips of paper led him to the conclusion that this sorting could be done by machine. Since he had removed much of the mathematical notation used by Boole, he speculated that the mathematics could be entirely removed from it. He also took the additional step of suggesting, rather hesitantly, that mathematics was itself a branch of logic. According to Grattan-Guinness (*2000*, p. 59), this speculation apparently had no influence on the mathematical philosophers who ultimately developed its implications, Russell and Frege.

2. Set theory

Set theory is all-pervasive in modern mathematics. It is the common language used to express concepts in all areas of mathematics. Because it is the language everyone writes in, it is difficult to imagine a time when mathematicians did not use the word *set* or think of sets of points. Yet that time was not long ago, less than 150 years. Before that time, mathematicians spoke of geometric figures. Or they spoke of points and numbers having certain properties, without thinking of those points and numbers as being assembled in a set. We have seen how concepts similar to those of set theory arose, in the notion of classes of objects having certain properties, in the British school of logicians. On the Continent, geometry and analysis provided the grounds for a development that resulted in a sort of "convergent evolution" with mathematical logic.

2.1. Technical background. Although the founder of set theory, Georg Cantor, was motivated by both geometry and analysis, for reasons of space we shall discuss only the analytic connection, which was the more immediate one. It is necessary to be slightly technical to explain how a problem in analysis leads to the general notion of a set and an ordinal number. We begin with the topic that Riemann developed for his 1854 lecture but did not use because Gauss preferred his geometric lecture. That topic was uniqueness of trigonometric series, and it was published in 1867, the year after Riemann's death. Riemann aimed at proving that if a trigonometric series converged to zero at every point, all of its coefficients were zero. That is,

$$\frac{1}{2}a_0 + \sum_{n=1}^{\infty}(a_n \cos nx + b_n \sin nx) \equiv 0 \Longrightarrow a_n = 0 = b_n \,.$$

Riemann assumed that the coefficients a_n and b_n tend to zero, saying that it was clear to him that without that assumption, the series could converge only at isolated points.[4] In order to prove this theorem, Riemann integrated twice to form the continuous function

$$F(x) = Ax + B + \frac{1}{4}a_0x^2 - \sum_{n=1}^{\infty}\frac{(a_n \cos nx + b_n \sin nx)}{n^2} \,.$$

[4] Kronecker pointed out later that this assumption was dispensible; Cantor showed that it was deducible from the mere convergence of the series.

His object was to show that $F(x)$ must be a linear function, so that $G(x) = F(x) - Ax - B - \frac{1}{4}a_0x^2$ would be a quadratic polynomial that was also the sum of a uniformly convergent trigonometric series, and hence itself a constant, from which it would follow, first that $a_0 = 0$, and then that all the other a_n and all the b_n are zero. To that end, he showed that its generalized second derivative

$$F_g''(x) = \lim_{h \to 0} \frac{F(x+h) + F(x-h) - 2F(x)}{h^2}$$

was zero wherever the original series converged to zero.[5] A weaker theorem, similar to the theorem that a differentiable function must be continuous, implied that

$$\lim_{h \to 0} \frac{\big(F(x+h) - F(x)\big) + \big(F(x-h) - F(x)\big)}{h} = 0\,.$$

The important implication of this last result is that *the function $F(x)$ cannot have a corner.* If it has a right-hand derivative at a point, it also has a left-hand derivative at the point, and the two one-sided derivatives are equal. This fact, which at first sight appears to have nothing to do with set theory, was a key step in Cantor's work.

2.2. Cantor's work on trigonometric series. In 1872 Cantor published his first paper on uniqueness of trigonometric series, finishing the proof that Riemann had set out to give: that a trigonometric series that converges to zero at every point must have all its coefficients equal to zero. In following the program of proving that $F(x)$ is linear and hence constant, he observed that it was not necessary to assume that the series converged to zero at every point. A finite number of exceptions could be allowed, at which the series either diverged or converged to a nonzero value. For $F(x)$ is certainly continuous, and if it is linear on $[a, b]$ and also on $[b, c]$, the fact that it has no corners implies that it must be linear on $[a, c]$. Hence any isolated exceptional point b could be discounted.

The question therefore naturally arose: Can one allow an infinite number of exceptional points? Here one comes up against the Bolzano–Weierstrass theorem, which asserts that the exceptional points cannot all be isolated. They must have at least one point of accumulation. But exceptional points isolated from other exceptional points could be discounted, just as before. That left only their points of accumulation. If these were isolated—in particular, if there were only finitely many of them—the no-corners principle would once again imply uniqueness of the series.

Ordinal numbers. Cantor saw the obvious induction immediately. Denoting the set of points of accumulation of a set P (what we now call the derived set) by P', he knew that $P' \supseteq P'' \supseteq P''' \supseteq \cdots$. Thus, if at some finite level of accumulation points of accumulation points a finite set was obtained, the uniqueness theorem would remain valid. But the study of these sets of points of accumulation turned out to be even more interesting than trigonometric series themselves. No longer dealing with geometrically regular sets, Cantor was delving into point-set topology, as we now call it. No properties of a geometric nature were posited for the exceptional points he was considering, beyond the assumption that the sequence of derived sets must terminate at some finite level. Although the points of any particular set (as

[5] Hermann Amandus Schwarz later showed that if $F_g''(x) \equiv 0$ on an open interval (a, b), then $F(x)$ is linear on the closed interval $[a, b]$.

we now call it) might be easily describable, Cantor needed to discuss the general case. He needed the abstract concept of "sethood." Cantor felt compelled to dig to the bottom of this matter and soon abandoned trigonometric series to write a series of papers on "infinite linear point-manifolds."

Early on, he noticed the possibility of the transfinite. If the nth-level derived set is $P^{(n)}$, the nesting of these sets allows the natural definition of the derived set of infinite order $P^{(\infty)}$ as the intersection of all sets of finite order. But then one could consider derived sets even at the transfinite level: the derived set of $P^{(\infty)}$ could be defined as $P^{(\infty+1)} = \big(P^{(\infty)}\big)'$. Cantor had discovered the infinite ordinal numbers. However, he did not at first recognize them as numbers, but rather regarded them as "symbols of infinity" (see Ferreirós, *1995*).

Cardinal numbers. Cantor was not only an analyst, however. He had written his dissertation under Kronecker and Kummer on number-theoretic questions. Only two years after he wrote his first paper in trigonometric series, he noticed that his set-theoretic principles led to another interesting conclusion. The set of algebraic real or complex numbers is a countable set (as we would now say in the familiar language that we owe to Cantor), but the set of real numbers is not. Cantor had proved this point to his satisfaction in a series of exchanges of letters with Dedekind.[6] Hence there must exist transcendental numbers. This second hierarchy of sets led to the concept of a cardinal number, two sets being of the same cardinality if they could be placed in one-to-one correspondence. To establish such correspondences, Cantor allowed himself certain powers of defining sets and functions that went beyond what mathematicians had been used to seeing. The result was a controversy that lasted some two decades.

Grattan-Guinness (*2000*, p. 125) has pointed out that Cantor emphasized five different aspects of point sets: their topology, dimension, measure, cardinality, and ordering. In the end, point-set topology was to become its own subject, and dimension theory became part of both algebraic and point-set topology. Measure theory became an important part of modern integration theory and had equally important applications to the theory of probability and random variables. Cardinality and ordering remained as an essential core of set theory, and the study of sets in relation to their complexity rather than their size became known as *descriptive set theory.*

Although descriptive set theory produces its own questions, it had at first a close relation to measure theory, since it was necessary to specify which sets could be measured. Borel was very careful about this procedure, allowing that the kinds of sets one could clearly define would have to be obtained by a finite sequence of operations, each of which was either a countable union or a countable intersection or a complementation, starting from ordinary open and closed sets. Ultimately those of a less constructive disposition than Borel honored him with the creation of

---

[6] There are two versions of this proof, one due to Cantor and one due to Dedekind, but both involve getting nested sequences of closed intervals that exclude, one at a time, the elements of any given sequence $\{a_n\}$ of numbers. The intersection of the intervals must then contain a number not in the sequence. In his private speculations, Luzin noted that Cantor was actually assuming more than the mere *existence* of the countable set $\{a_n\}$. In order to construct a point not in it, one had to know something about each of its elements, enough to find a subinterval of the previous closed interval that would exclude the next element. On that basis, he concluded that Cantor had proved that there was no effective enumeration of the reals, not that the reals were uncountable. Luzin thus raised the question of what it could mean for an enumeration to "exist" if it was not effective. He too delved into philosophy to find out the meaning of "existence."

the *Borel sets*, which is the smallest class that contains all closed subsets and also contains the complement of any of its sets and the union of any countable collection of its sets. This class can be "constructed" only by a transfinite induction.

Set theory, although it was an attempt to provide a foundation of clear and simple principles for all of mathematics, soon threw up its own unanswered mathematical questions. The most important was the continuum question. Cantor had shown that the set of all real numbers could be placed in one-to-one correspondence with the set of all subsets of the integers. Since he denoted the cardinality of the integers as $\aleph_0$ and the cardinality of the real numbers as $\mathfrak{c}$ (where $\mathfrak{c}$ stands for "continuum"), the question naturally arose whether there was any subset of the real numbers that had a cardinality between these two. Cantor struggled for a long time to settle this issue. One major theorem of set theory, known as the *Cantor–Bendixson theorem*,[7] after Ivar Bendixson (1861–1935), asserts that every closed set is the union of a countable set and a perfect set, one equal to its derived set. Since it is easily proved that a nonempty perfect subset of the real numbers has cardinality $\mathfrak{c}$, it follows that every uncountable closed set contains a subset of cardinality $\mathfrak{c}$. Thus a set of real numbers having cardinality between $\aleph_0$ and $\mathfrak{c}$ cannot be a closed set. Many mathematicians, especially the Moscow mathematicians after the arrival of Luzin as professor in 1915, worked on this problem. Luzin's students Pavel Sergeevich Aleksandrov (1896–1982) and Mikhail Yakovlevich Suslin (1894–1919) proved that any uncountable Borel set must contain a nonempty perfect subset, and so must have cardinality $\mathfrak{c}$. Indeed, they proved this fact for a slightly larger class of sets called *analytic sets*. Luzin then proved that a set was a Borel set if and only if the set and its complement were both analytic sets.

The problem of the continuum remained open until 1938, when Kurt Gödel (1906–1978) partially closed it by showing that set theory is consistent with the continuum hypothesis and the axiom of choice,[8] provided that it is consistent without them. Closure came to this question in 1963, when Paul Cohen (b. 1934—like Cantor, he began his career by studying uniqueness of trigonometric series representations) showed that the continuum hypothesis and the axiom of choice are independent of the other axioms of set theory.

2.3. The reception of set theory. If Venn believed that probability was an unwarranted intrusion of mathematics into philosophy, there were many mathematicians who believed that set theory was an equally unwarranted intrusion of philosophy into mathematics. One of those was Cantor's teacher Leopold Kronecker. Although Cantor was willing to consider the existence of a transcendental number proved just because the real numbers were "too numerous" to be exhausted by the algebraic numbers, Kronecker preferred a more constructivist approach. His most famous utterance,[9] and one of the most famous in all the history of mathematics, is: "The good Lord made the integers; everything else is a human creation." (*"Die ganzen Zahlen hat der liebe Gott gemacht; alles andere ist Menschenwerk."*) That is, the only infinity he admitted was the series of positive integers $1, 2, \ldots$.

[7] Ferreirós (*1995*) points out that it was the desire to prove this theorem adequately, in 1882, that really led Cantor to treat transfinite ordinal numbers as numbers. He was helped toward this discovery by Dedekind's pointing out to him the need to use finite ordinal numbers to define finite cardinal numbers.

[8] Gödel actually included four additional assumptions in his consistency proof, one of the other two being that there exists a set that is analytic but is not a Borel set.

[9] He made this statement at a meeting in Berlin in 1886 (see Grattan-Guiness, *2000*, p. 122).

Beyond that point, everything was human-made and therefore had to be finite. If you spoke of a number or function, you had an obligation to say how it was defined. His 1845 dissertation, which he was unable to polish to his satisfaction until 1881, when he published it as "Foundations of an arithmetical theory of algebraic quantities" in honor of his teacher Kummer, shows how careful he was in his definitions. Instead of an arbitrary *field* defined axiomatically as we would now do, he wrote:

> A domain of rationality is in general an *arbitrarily* bounded domain of magnitudes, but only to the extent that the concept permits. To be specific, since a domain of rationality can be enlarged only by the adjoining of arbitrarily chosen elements $\mathfrak{R}$, each arbitrary extension of its boundary requires the simultaneous inclusion of *all* quantities rationally expressible in terms of the new Element.

In this way, while one could enlarge a field to make an equation solvable, the individual elements of the larger field could still be described constructively. Kronecker's concept of a general field can be described as "finitistic." It is the minimal object that contains the necessary elements. Borel took this point of view in regard to measurable sets, and Hilbert was later to take a similar point of view in describing formal languages, saying that a meaningful formula must be obtained from a specified list of elements by a finite number of applications of certain rules of combination. This approach was safer and more explicit than, for example, Bernoulli's original definition of a function as an expression formed "in some manner" from variables and constants. The "manner" was limited in a very definite way.

Cantor believed that Kronecker had delayed the publication of his first paper on infinite cardinal numbers. Whether that is the case or not, it is clear that Kronecker would not have approved of some of his principles of inference. As Grattan-Guinness points out, much of what is believed about the animosity between Cantor and Kronecker is based on Cantor's own reports, which may be unreliable. Cantor was subject to periodic bouts of depression, probably caused by metabolic imbalances having nothing to do with his external circumstances. In fact, he had little to complain of in terms of the acceptance of his theories. It is true that there was some resistance to it, notably from Kronecker (until his death in 1891) and then from Poincaré. But there was also a great deal of support, from Weierstrass, Klein, Hilbert, and many others. In fact, as early as 1892, the journal *Bibliotheca mathematica* published a "Notice historique" on set theory by Giulio Vivanti (1859–1949), who noted that there had already been several expositions of the theory, and that it was still being developed by mathematicians, applied to the theory of functions of a real variable, and studied from a philosophical point of view.

2.4. Existence and the axiom of choice. In the early days Cantor's set theory seemed to allow a remarkable amount of freedom in the "construction" or, rather, the calling into existence, of new sets. Cantor seems to have been influenced in his introduction of the term *set* by an essay that Dedekind began in 1872, but did not publish until 1887 (see Grattan-Guinness, *2000*, p. 104), in which he referred to a "system" as "various things a, b, c... comprehended from any cause under one point of view." Dedekind defined a "thing" to be "any object of our thought." Just as Descartes was able to conceive many things clearly and distinctly, mathematicians seemed to be able to form many "things" into "systems." For example, given *any* set A, one could conceive of another set whose members were the subsets of A. This set is nowadays denoted 2^A and called the *power set* of A. If A has a finite number n of elements, then 2^A has 2^n elements, counting the improper subsets $\varnothing$ and A.

It was not long, however, before the indiscriminate use of this freedom to form sets led to paradoxes. The most famous of these is Russell's paradox, discussed in the next section. The source of the difficulty is that "existence" has a specialized mathematical meaning. The abstraction that comes with set theory has the consequence that much of the action in a proof takes place "offstage." That is, certain objects needed in a proof are called into existence by saying, "Let there be...," but no procedure for constructing them is given. Proofs relying on the abstract existence of such objects, when it is not possible to choose a particular object and examine it, became more and more common in the twentieth century. Indeed, much of measure theory, topology, and functional analysis would be impossible without such proofs. The principle behind these proofs later came to be known as *Zermelo's axiom*, after Ernst Zermelo (1871–1953), who first formulated it in 1904 to prove that every set could be well ordered.[10] It was also known as the principle of free choice (in German, *Auswahlprinzip*) or, more commonly in English, the axiom of choice. In its broadest form this axiom states that *there exists a function f defined on the class of all nonempty sets such that $f(A) \in A$ for every nonempty set.* Intuitively, if A is nonempty, there exist elements of A, and $f(A)$ chooses one such element from every nonempty set.

This axiom is used in very many proofs, but probably the earliest (see Moore, *1982*, p. 9) is Cantor's proof that a countable union of countable sets is countable. The proof goes as follows. Assume that A_1, A_2,... are countable sets, and let $A = A_1 \cup A_2 \cup \cdots$. Then A is countable. For, let the sets A_j be enumerated, as follows

$$\begin{aligned} A_1 &= a_{11}, a_{12}, \ldots, \\ A_2 &= a_{21}, a_{22}, \ldots, \\ &\vdots \\ A_n &= a_{n1}, a_{n2}, \ldots, \\ &\vdots \quad . \end{aligned}$$

Then the elements of A can be enumerated as follows: $a_{11}, a_{12}, a_{21}, a_{13}, a_{22}, a_{31}, \ldots$, where the elements whose ranks are larger than the triangular number $T_n = n(n+1)/2$ but not larger than $T_{n+1} = (n+1)(n+2)/2$ are those for which the sum of the subscripts is $n+2$. There are $n+1$ such elements and $n+1$ such ranks. It is a very subtle point to notice that this proof assumes more than the mere existence of an enumeration of *each* of the sets, which is given in the hypothesis. It assumes the *simultaneous* existence of infinitely many enumerations, one for each set. The reasoning appears to be so natural that one would hardly question it. If a real choice exists at each stage of the proof, why can we not assume that infinitely many such choices have been made? As Moore notes, without the axiom of choice,

---

[10] A set is *well ordered* if any two elements can be compared and every nonempty subset has a smallest element. The positive integers are well ordered by the usual ordering. The positive real numbers are not.

it is consistent to assume that the real numbers can be expressed as a countable union of countable sets.[11]

Zermelo made this axiom explicit and showed its connection with ordinal numbers. The problem then was either to justify the axiom of choice, or to find a more intuitively acceptable substitute for it, or to find ways of doing without such "noneffective" concepts. A debate about this axiom took place in 1905 in the pages of the *Comptes rendus* of the French Academy of Sciences, which published a number of letters exchanged among Hadamard, Borel, Lebesgue, and Baire.[12] Borel had raised objections to Zermelo's proof that every set could be well-ordered on the grounds that it assumed an infinite number of enumerations. Hadamard thought it an important distinction that in some cases the enumerations were all independent, as in Cantor's proof above, but in others each depended for its definition on other enumerations having been made in correspondence with a smaller ordinal number. He agreed that the latter should not be used transfinitely. Borel had objected to using the axiom of choice nondenumeratively, but Hadamard thought that this usage brought no further damage, once a denumerable infinity of choices was allowed. He also mentioned the distinction due to Jules Tannery (1848–1910) between *describing* an object and *defining* it. To Hadamard, describing an object was a stronger requirement than defining it. To supply an example for him, we might mention a well-ordering of the real numbers, which is *defined* by the phrase itself, but effectively *indescribable*. Hadamard noted Borel's own work on analytic continuation and pointed out how it would change if the only power series admitted were those that could be effectively described. The difference, he said, belongs to psychology, not mathematics.

Hadamard received a response from Baire, who took an even more conservative position than Borel. He said that once an infinite set was spoken of, "the comparison, *conscious or unconscious*, with a bag of marbles passed from hand to hand must disappear completely."[13] The heart of Baire's objection was Zermelo's *supposition* that to each (nonempty) subset of a set M there corresponds one of its elements." As Baire said, "all that it proves, as far as I am concerned, is that we do not perceive a contradiction" in imagining any set well-ordered.

Responding to Borel's request for his opinion, Lebesgue gave it. As far as he was concerned, Zermelo had very ingeniously shown how to solve problem A (to well-order any set) provided one could solve problem B (to choose an element from every nonempty subset of a given set). He remarked, probably with some irony, that, "Unfortunately, problem B is not easy to resolve, it seems, except for the sets that we know how to well-order." Lebesgue mentioned a concept that was to play a large role in debates over set theory, that of "effectiveness," roughly what we would call constructibility. He interpreted Zermelo's claim as the assertion that a well-ordering *exists* (that word again!) and asked a question, which he said was "hardly new": *Can one prove the existence of a mathematical object without defining it?* One would think not, although Zermelo had apparently proved the existence of a well-ordering (and Cantor had proved the existence of a transcendental number) without *describing* it. Lebesgue and Borel preferred the verb *to name* (*nommer*)

---

[11] Not *every* countable union of countable sets is uncountable, however; the rational numbers remain countable, because an explicit counting function can be constructed.

[12] These letters were translated into English and published by Moore (*1982*, pp. 311–320).

[13] Luzin said essentially the same in his journal: What makes the axiom of choice seem reasonable is the picture of reaching into a set and helping yourself to an element of it.

when referring to an object that was defined effectively, through a finite number of uses of well-defined operations on a given set of primitive objects.

After reading Lebesgue's opinion, Hadamard was sure that the essential distinction was between what is determined and what is described. He compared the situation with the earlier debate over the allowable definitions of a function. But, he said, uniqueness was not an issue. If one could say "For each x, there exists a number satisfying.... Let y be this number," surely one could also say "For each x, there exists an infinity of numbers satisfying.... Let y be one of these numbers." But he put his finger squarely on one of the paradoxes of set theory (the Burali-Forti paradox, discussed in the next section). "It is the very existence of the set W that leads to a contradiction... the general definition of the word *set* is incorrectly applied." (Question to ponder: What *is* the definition of the word *set*?)

The validity and value of the axiom of choice remained a puzzle for some time. It leads to short proofs of many theorems whose statements are constructive. For example, it proves the existence of a nonzero translation-invariant Borel measure on any locally compact Abelian group. Since such a measure is provably unique (up to a constant multiple), there ought to be effective proofs of its existence that do not use the axiom of choice (and indeed there are). One benefit of the 1905 debate was a clarification of equivalent forms of the axiom of choice and an increased awareness of the many places where it was being used. A list of important theorems whose proof used the axiom was compiled for Luzin's seminar in Moscow in 1918. The list showed, as Luzin wrote in his journal, that "almost nothing is proved without it." Luzin was horrified, and spent some restless nights pondering the situation.

The axiom of choice is ubiquitous in modern analysis; almost none of functional analysis or point-set topology would remain if it were omitted entirely (although weaker assumptions might suffice). It is fortunate, therefore, that its consistency with, and independence of, the other axioms of set theory has been proved. However, the consequences of this axiom are suspiciously strong. In 1924 Alfred Tarski (1901–1983) and Stefan Banach (1892–1945) deduced from it that any two sets A and B in ordinary three-dimensional Euclidean space, each of which contains some ball, can be decomposed into pairwise congruent subsets. This means, for example, that a cube the size of a grain of salt (set A) and a ball the size of the Sun (set B) can be written as disjoint unions of sets $A_1, \ldots, A_n$ and $B_1, \ldots, B_n$ respectively such that A_i is congruent to B_i for each i. This result (the Banach–Tarski paradox) is very difficult to accept. It can be rationalized only by realizing that the notion of existence in mathematics has no metaphysical content. To say that the subsets A_i, B_i "exist" means only that a certain formal statement beginning $\exists \ldots$ is deducible from the axioms of set theory.

2.5. Doubts about set theory. The powerful and counterintuitive results obtained from the axiom of choice naturally led to doubts about the consistency of set theory. Since it was being inserted under the rest of mathematics as a foundation, the consistency question became an important one. A related question was that of completeness. Could one provide a foundation for mathematics, that is, a set of basic objects and rules of proof, that would allow any meaningful proposition to be proved true or false? The two desirable qualities are in the abstract opposed to each other, just as avoiding disasters and avoiding false alarms are opposing goals.

The most influential figure in mathematical logic during the twentieth century was Gödel. The problems connected with consistency and completeness of arithmetic, the axiom of choice, and many others all received a fully satisfying treatment at his hands that settled many old questions and opened up new areas of investigation. In 1931, he astounded the mathematical world by producing a proof that any consistent formal language in which arithmetic can be encoded is necessarily incomplete, that is, contains statements that are true according to its metalanguage but not deducible within the language itself. The intuitive idea behind the proof is a simple one, based on the statement that follows:

This statement cannot be proved.

Assuming that this statement has a meaning—that is, its context is properly restricted so that "proved" has a definite meaning—we can ask whether it is *true.* The answer must be positive if the system in which it is made is consistent. For if this statement is false, by its own content, it *can* be proved; and in a consistent deductive system, a false statement cannot be proved. Hence we agree that the statement is true, but, again by its own content, it cannot be proved.

The example just given is really nonsensical, since we have not carefully delineated the universe of axioms and rules of inference in which the statement is made. The word "proved" that it contains is not really defined. Gödel, however, took an accepted formalization of the axioms and rules of inference for arithmetic and showed that the metalanguage of arithmetic could be encoded within arithmetic. In particular each formula can be numbered uniquely, and the statement that formula n is (or is not) deducible from those rules can itself be coded as a well-formed formula of arithmetic. Then, when n is chosen so that the statement, "Formula number n cannot be proved" happens to *be* formula n, we have exactly the situation just described. Gödel showed how to construct such an n. Thus, if Gödel's version of arithmetic is consistent, it contains statements that are formally undecidable; that is, they are true (based on the metalanguage) but not deducible. This is Gödel's first incompleteness theorem. His second incompleteness theorem is even more interesting: *The assertion that arithmetic is consistent is one of the formally undecidable statements.*[14] If the formalized version of arithmetic that Gödel considered is consistent, it is incapable of proving itself so. It is doubtful, however, that one could truly formalize every kind of argument that a rational person might produce. For that reason, care should be exercised in drawing inferences from Gödel's work to the actual practice of mathematics.

3. Philosophies of mathematics

Besides Cantor, other mathematicians were also considering ways of deriving mathematics logically from simplest principles. Gottlob Frege (1848–1925), a professor in Jena, who occasionally lectured on logic, attempted to establish logic on the basis of "concepts" and "relations" to which were attached the labels *true* or *false.* He was the first to establish a complete predicate calculus, and in 1884 wrote a treatise called *Grundgesetze der Arithmetik* (*Principles of Arithmetic*). Meanwhile in Italy, Giuseppe Peano (1858–1939) was axiomatizing the natural numbers. Peano took the successor relation as fundamental and based his construction of the natural

[14] Detlefsen (*2001*) has analyzed the meaning of proving consistency in great detail and concluded that the generally held view of this theorem—that the consistency of a "sufficiently rich" theory cannot be proved by a "finitary" theory—is incorrect.

numbers on this one relation and nine axioms, together with a symbolic logic that he had developed. The work of Cantor, Frege, and Peano attracted the notice of a young student at Cambridge, Bertrand Russell, who had written his thesis on the philosophy of Leibniz. Russell saw in this work confirmation that mathematics is merely a prolongation of formal logic. This view, that mathematics can be deduced from logic without any new axioms or rules of inference, is now called *logicism.* Gödel's work was partly inspired by it, and can be interpreted as a counterargument to its basic thesis—that mathematics can be axiomatized. Logicism had encountered difficulties still earlier, however. Even the seemingly primitive notion of membership in a set turned out to require certain caveats.

3.1. Paradoxes. In 1897 Peano's assistant Cesare Burali-Forti (1861–1931), apparently unintentionally, revealed a flaw in the ordinal numbers.[15] To state the problem in the clear light of hindsight, if two ordinal numbers satisfy $x < y$, then $x \in y$, but $y \notin x$. In that case, what are we to make of the set of all ordinal numbers? Call this set A. Like any other ordinal number, it has a successor $A+1$ and $A \in A+1$. But since $A+1$ is an ordinal number, we must also have $A+1 \in A$, and hence $A < A+1$ and $A+1 < A$. This was the first paradox of uncritical set theory, but others were to follow.

The most famous paradox of set theory arose in connection with cardinal numbers rather than ordinal numbers. Cantor had defined equality between cardinal numbers as the existence of a one-to-one correspondence between sets representing the cardinal numbers. Set B has larger cardinality than set A if there is no function $f : A \to B$ that is "onto," that is, such that every element of B is $f(x)$ for some $x \in A$. Cantor showed that the set of all subsets of A, which we denote 2^A, is always of larger cardinality than A, so that there can be no largest cardinal number. If $f : A \to 2^A$, the set $C = \{t \in A : t \notin f(t)\}$ is a subset of A, hence an element of 2^A, and it cannot be $f(x)$ for any $x \in A$. For if $C = f(x)$, we ask whether $x \in C$ or not. If $x \in C$, then $x \in f(x)$ and so by definition of C, $x \notin C$. On the other hand, if $x \notin C$, then $x \notin f(x)$, and again by definition of C, $x \in C$. Since the whole paradox results from the assumption that $C = f(x)$ for some x, it follows that no such x exists, that is, the mapping f is not "onto." This argument was at first disputed by Russell, who wrote in an essay entitled "Recent work in the philosophy of mathematics" (1901) that "the master has been guilty of a very subtle fallacy." Russell thought that there was a largest set, the set of *all* sets. In a later reprint of the article he added a footnote explaining that Cantor was right.[16] Russell's first attempt at a systematic exposition of mathematics as he thought it ought to be was his 1903 work *Principles of Mathematics.* According to Grattan-Guinness (*2000*, p. 311), Russell removed his objection to Cantor's proof and published his paradox in this work, but kept the manuscript of an earlier version, made before he was able to work out where the difficulty lay.

To explain Russell's paradox, consider the set of all sets. We must, by its definition, believe it to be *equal* to the set of all its subsets. Therefore the mapping $f(E) = E$ should have the property that Cantor says no mapping can have. Now

[15] Moore (*1982*, p. 59) notes that Burali-Forti himself did not see any paradox and (p. 53) that the difficulty was known earlier to Cantor.

[16] Moore (*1982*, p. 89) points out that Zermelo had discovered Russell's paradox two years before Russell discovered it and had written to Hilbert about it. Zermelo, however, did not consider it a very troubling paradox. To him it meant only that no set should contain all of its subsets as elements.

if we apply Cantor's argument to this mapping, we are led to consider $S = \{E : E \notin E\}$. By definition of the mapping f we should have $f(S) = S$, and so, just as in the case of Cantor's argument, we ask if $S \in S$. Either way, we are led to a contradiction. This result is known as *Russell's paradox.*

After Russell had straightened out the paradox with a theory of types, he collaborated with Alfred North Whitehead on a monumental derivation of mathematics from logic, published in 1910 as *Principia mathematica.*

3.2. Formalism. A different view of the foundations of mathematics was advanced by Hilbert, who was interested in the problem of axiomatization (the axiomatization of probability theory was the sixth of his famous 23 problems) and particularly interested in preserving as much as possible of the freedom to reason that Cantor had provided while avoiding the uncomfortable paradoxes of logicism. The essence of this position, now known as formalism, is the idea stated by de Morgan and Boole that the legal manipulation of the symbols of mathematics and their interpretation are separate issues. Hilbert is famously quoted as having claimed that the words *point*, *line*, and *plane* should be replaceable by *table*, *chair*, and *beer mug* when a theorem is stated. Grattan-Guinness (*2000*, p. 208) notes that Hilbert may not have intended this statement in quite the way it is generally perceived and may not have thought the matter through at the time. He also notes (p. 471) that Hilbert never used the name *formalism.* Characteristic of the formalist view is the assumption that any mathematical object whatever may be defined, provided only that the definition does not lead to a contradiction. Cantor was a formalist in this sense (Grattan-Guinness, *2000*, p. 119). In the formalist view mathematics is the study of formal systems, but the rules governing those systems must be stated with some care. In that respect, formalism shares some of the caution of the earlier constructivist approach. It involves a strict separation between the symbols and formulas of mathematics and the meaning attached to them, that is, a distinction between syntax and semantics. Hilbert had been interested in logical questions in the 1890s and early 1900s, but his work on formal languages such as propositional calculus dates from 1917. In 1922, when the intuitionists (discussed below) were publishing their criticism of mathematical methodologies, he formulated his own version of mathematical logic. In it he introduced the concept of metamathematics, the study whose subject matter is the structure of a mathematical system.[17] A formal language consists of certain rules for recognizing legitimate formulas, certain formulas called axioms, and certain rules of inference (such as syllogism, generalization over unspecified variables, and the rules for manipulating equations). These elements make up the syntax of the language. One can therefore always tell by following clearly prescribed rules whether a formula is meaningful (well formed) and whether a sequence of formulas constitutes a valid deduction. To avoid infinity in this system while preserving sufficient generality, Hilbert resorted to a "finitistic" device called a *schema.* Certain basic formulas are declared to be legitimate by fiat. Then a few rules are adopted, such as the rule that if A and B are legitimate formulas, so is $[A \Rightarrow B]$. This way of defining legitimate (well-formed) formulas makes it possible to determine in a finite number of steps whether or not a formula is well formed. It replaces the synthetic constructivist approach with an analytic

[17] This distinction had been introduced by L. E. J. Brouwer in his 1907 thesis, but not given a name and never developed (see Grattan-Guinness, *2000*, p. 481).

approach (which can be reversed, once the analysis is finished, to synthesize a given well-formed formula from primitive elements).

The formalist approach makes a distinction between statements *of* arithmetic and statements *about* arithmetic. For example, the assertion that there are no positive integers x, y, z such that $x^3 + y^3 = z^3$ is a statement *of* arithmetic. The assertion that this statement can be proved from the axioms of arithmetic is a statement *about* arithmetic. The *metalanguage*, in which statements are made about arithmetic, contains all the meaning to be assigned to the propositions of arithmetic. In particular, it becomes possible to distinguish between what is true (that is, what can be known to be true from the metalanguage) and what is provable (what can be deduced within the object language). Two questions thus arise in the metalanguage: (1) *Is every deducible proposition true?* (the problem of consistency); (2) *Is every true proposition deducible?* (the problem of completeness). As we saw in Section 2, Gödel showed that the answer, for first-order recursive arithmetic and more generally for systems of that type, is very pessimistic. This language is not complete and is incapable of proving its own consistency.

3.3. Intuitionism. The most cautious approach to the foundations of mathematics, known as *intuitionism*, was championed by the Dutch mathematician Luitzen Egbertus Jan Brouwer (1881–1966). Brouwer was one of the most mystical of mathematicians, and his mysticism crept into his early work. He even published a pamphlet in 1905, claiming that true happiness came from the inner world, and that contact with the outer world brought pain (Franchella, *1995*, p. 305). In his dissertation at the University of Amsterdam in 1907, he criticized the logicism of Russell and Zermelo's axiom of choice. Although he was willing to grant the validity of constructing each particular denumerable ordinal number, he questioned whether one could meaningfully form a set of all denumerable ordinals.[18] In a series of articles published form 1918 to 1928, Brouwer laid down the principles of intuitionism. These principles include the rejection not only of the axiom of choice beyond the countable case, but also of proof by contradiction. That is, the implication "A implies not-(not-A)" is accepted, but not its converse, "Not-(not-A) implies A." Intuitionists reject any proof whose implementation leaves choices to be made by the reader. Thus it is not enough in an intuitionist proof to say that objects of a certain kind exist. One must choose such an object and use it for the remainder of the proof. This extreme caution has rather drastic consequences. For example, the function $f(x)$ defined in ordinary language as

$$f(x) = \begin{cases} 1, & x \geq 0, \\ 0, & x < 0, \end{cases}$$

is not considered to be defined by the intuitionists, since there are ways of defining numbers x that do not make it possible to determine whether the number is negative or positive. For example, is the number $(-1)^n$, where n is the trillionth decimal digit of π, positive or negative? This restrictedness has certain advantages, however. The objects that are acceptable to the intuitionists tend to have pleasant properties. For example, every rational-valued function of a rational variable is continuous.

The intuitionist rejection of proof by contradiction needs to be looked at in more detail. Proof by contradiction was always used somewhat reluctantly by

[18] This objection seems strange at first, but the question of whether an effectively defined set must have effectively defined members is not at all trivial.

mathematicians, since such proofs seldom give insight into the structures being studied. For example, Euclid's proof that there are infinitely many primes proceeds by assuming that the set of prime numbers is a finite set $P = \{p_1, p_2, \ldots, p_n\}$ and showing that in this case the number $1 + p_1 \cdots p_n$ must either itself be a prime number or be divisible by a prime different from $p_1, \ldots, p_n$, which contradicts the original assumption that $p_1, \ldots, p_n$ formed the entire set of prime numbers.

The appearance of starting with a false assumption and deriving a contradiction can be avoided here by stating the theorem as follows: If there exists a set of n primes $p_1, \ldots, p_n$, there exists a set of $n + 1$ primes. The proof is exactly as before. Nevertheless, the proof is still not intuitionistically valid, since there is no way of saying whether or not $1 + p_1 \cdots p_n$ is prime.

In 1928 and 1929, a quarter-century after the debate over Zermelo's axiom of choice, there was debate about intuitionism in the bulletin of the Belgian Royal Academy of Sciences. Two Belgian mathematicians, M. Barzin and A. Errera, had argued that Brouwer's logic amounted to a three-valued logic, since a statement could be true, false, or undecidable. The opposite point of view was defended by two distinguished Russian mathematicians, Aleksandr Yakovlevich Khinchin (1894–1959) and Valerii Ivanovich Glivenko (1897–1940). Barzin and Errera had suggested that to avoid three-valued logic, intuitionists ought to adopt as an axiom that if p implies "q or r", then either p implies q or p implies r,[19] and also that if "p or q" implies r, then p implies r and q implies r. Starting from these principles of Barzin and Errera and the trivial axiom "p or not-p" implies "p or not-p", Khinchin deduced that p implies not-p and not-p implies p, thus reducing the suggestions of Barzin and Errera to nonsense. Glivenko took only a little longer to show that, in fact, Brouwer's logic was not three-valued. He proved that the statement "p or not-p is false" is false in Brouwer's logic, and ultimately derived the theorem that the statement "p is neither true nor false" is false (see Novosyolov, *2000*).

A more "intuitive" objection to intuitionism is that intuition by its nature cannot be codified as a set of rules. In adopting such rules, the intuitionists were not being intuitionistic in the ordinary sense of the word. In any case, intuitionist mathematics is obviously going to be somewhat sparser in results than mathematics constructed on more liberal principles. That may be why it has attracted only a limited group of adherents.

3.4. Mathematical practice. The paradoxes of naive set theory (such as Russell's paradox) were found to be avoidable if the word *class* is used loosely, as Cantor had previously used the word *set*, but the word *set* is restricted to mean only a class that is a member of some other class. (Classes that are not sets are called *proper classes.*) Then to belong to a class A, a class B must not only fulfill the requirements of the definition of the class A but must also be known in advance to belong to some (possibly different) class.

This approach avoids Russell's paradox. The class $C = \{x : x \notin x\}$ is a class; its elements are those classes that *belong to some class and* are not elements of themselves. If we now ask the question that led to Russell's paradox—Is C a member of itself?—we do not reach a contradiction. If we assume $C \in C$, then we conclude that $C \notin C$, so that this assumption is not tenable. However, the opposite assumption, that $C \notin C$, *is* acceptable. It no longer leads to the conclusion

[19] In the currently accepted semantics (metalanguage) of intuitionistic propositional calculus, if "q or r" is a theorem, then either q is a theorem or r is a theorem.

that $C \in C$. For an object x to belong to C, it no longer suffices that $x \notin x$; it must also be true that $x \in A$ for some class A, an assumption not made for the case when x is C. A complete set of axioms for set theory avoiding all known paradoxes was worked out by Paul Bernays (1888–1977) and Adolf Fraenkel (1891–1965). It forms part of the basic education of mathematicians today. It is generally accepted because mathematics can be deduced from it. However, it is very far from what Cantor had hoped to create: a clear, concise, and therefore *obviously* consistent foundation for mathematics. The axioms of set theory are extremely complicated and nonintuitive, and far less obvious than many things deduced from them. Moreover, their consistency is not only not obvious, it is even unprovable. In fact, one textbook of set theory, *Introduction to Set Theory*, by J. Donald Monk (McGraw-Hill, New York, 1969), p. 22, asserts of these axioms: "Naturally no inconsistency has been found, and *we have faith* that the axioms are, in fact, consistent"! (Emphasis added.)

Questions and problems

19.1. Bertrand Russell pointed out that some applications of the axiom of choice are easier to avoid than others. For instance, given an infinite collection of pairs of shoes, describe a way of choosing one shoe from each pair. Could you do the same for an infinite set of pairs of socks?

19.2. Prove that $C = \{x : x \notin x\}$ is a proper class, not a set, that is, it is not an element of any class.

19.3. Suppose that the only allowable way of forming new formulas from old ones is to connect them by an implication sign; that is, given that A and B are well formed, $[A \Rightarrow B]$ is well formed, and conversely, if A and B are not both well formed, then neither is $[A \Rightarrow B]$. Suppose also that the only basic well-formed formulas are p, q, and r. Show that

$$\Big[[p \Rightarrow r] \Rightarrow \big[[p \Rightarrow q] \Rightarrow r\big]\Big]$$

is well formed but

$$[[p \Rightarrow r] \Rightarrow [r \Rightarrow]]$$

is not. Describe a general algorithm for determining whether a finite sequence of symbols is well formed.

19.4. Consider the following theorem. There exists an irrational number that becomes rational when raised to an irrational power. *Proof:* Consider the number $\theta = \sqrt{3}^{\sqrt{2}}$. If this number is rational, we have an example of such a number. If it is irrational, the equation $\theta^{\sqrt{2}} = \sqrt{3}^{2} = 3$ provides an example of such a number. Is this proof intuitionistically valid?

19.5. Show that any two distinct *Fermat numbers* $2^{2^m} + 1$ and $2^{2^n} + 1$, $m < n$, are relatively prime. (Use mathematical induction on n.) Apply this result to deduce that there are infinitely many primes. Would this proof of the infinitude of the primes be considered valid by an intuitionist?

19.6. Suppose that you prove a theorem by assuming that it is false and deriving a contradiction. What you have then proved is that either the axioms you started with are inconsistent or the assumption that the theorem is false is itself false.

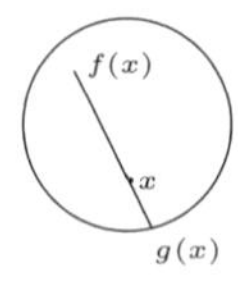

FIGURE 1. The Brouwer fixed-point theorem.

Why should you conclude the latter rather than the former? Is this why some mathematicians have claimed that the practice of mathematics requires faith?

19.7. What are the advantages, if any, of building a theory by starting with abstract definitions, then later proving a structure theorem showing that the abstract objects so defined are actually familiar objects?

19.8. Brouwer, the leader of the intuitionist school of mathematicians, is also known for major theorems in topology, including the invariance of geometric dimension under homeomorphisms and the *Brouwer fixed-point theorem*, which asserts that for any continuous mapping f of a closed disk into itself there is a point x such that $x = f(x)$. To prove this theorem, suppose there is a continuous mapping f for which $f(x) \neq x$ at every point x. Construct a continuous mapping g by drawing a line from $f(x)$ to x and extending it to the point $g(x)$ at which it meets the boundary circle (see Fig. 1). Then $g(x)$ maps the disk continuously onto its boundary circle and leaves each point of the boundary circle fixed. Such a continuous mapping is intuitively impossible (imagine stretching the entire head of a drum onto the rim without moving any point already on the rim and without tearing the head) and can be shown rigorously to be impossible (the disk and the circle have different homotopy groups). How can you explain the fact that the champion of intuitionism produced theorems that are not intuitionistically valid?

19.9. A naive use of the formula for the sum of the geometric series $1/(1+x) = 1 - x + x^2 - x^3 + \cdots$ seems to imply that $1 - 5 + 25 - 125 + \cdots = 1/(1+5) = 1/6$. Nineteenth-century analysts rejected this use of infinite series and confined themselves to series that converge in the ordinary sense. However, Kurt Hensel (1861–1941) showed in 1905 that it is possible to define a notion of distance (the p-adic metric) by saying that an integer is close to zero if it is divisible by a large power of the prime number p (in the present case, $p = 5$). Specifically, the distance from m to 0 is given by $d(m, 0) = 5^{-k}$, where 5^k divides m but 5^{k+1} does not divide m. The distance between 0 and the rational number $r = m/n$ is then by definition $d(m, 0)/d(n, 0)$. Show that $d(1, 0) = 1$. If the distance between two rational numbers r and s is defined to be $d(r - s, 0)$, then in fact the series just mentioned does converge to $\frac{1}{6}$ in the sense that $d(S_n, \frac{1}{6}) \to 0$, where S_n is the nth partial sum.

What does this historical experience tell you about the truth or falsity of mathematical statements? Is there an "understood context" for every mathematical statement that can never be fully exhibited, so that certain assertions will be *verbally* true in some contexts and verbally false in others, depending on the meaning attached to the terms?

19.10. Are there true but unknowable propositions in everyday life? Suppose that your class meets on Monday, Wednesday, and Friday. Suppose also that your instructor announces one Friday afternoon that you will be given a surprise exam at one of the regular class meetings the following week. One of the brighter students then reasons as follows. The exam will not be given on Friday, since if it were, having been told that it would be one of the three days, and not having had it on Monday or Wednesday, we would know on Thursday that it was to be given on Friday, and so it wouldn't be a surprise. Therefore it will be given on Monday or Wednesday. But then, since we *know* that it can't be given on Friday, it also can't be given on Wednesday. For if it were, we would know on Tuesday that it was to be given on Wednesday, and again it wouldn't be a surprise. Therefore it must be given on Monday, we know that now, and therefore it isn't a surprise. Hence it is impossible to give a surprise examination next week.

Obviously something is wrong with the student's reasoning, since the instructor can certainly give a surprise exam. Most students, when trying to explain what is wrong with the reasoning, are willing to accept the first step. That is, they grant that it is impossible to give a *surprise* exam on the *last* day of an assigned window of days. Yet they balk at drawing the conclusion that this argument implies that the originally next-to-last day must thereby become the last day. Notice that, if the professor had said nothing to the students, it would be possible to give a surprise exam on the last day of the window, since the students would have no way of knowing that there was any such window. The conclusion that the exam cannot be given on Friday therefore does not follow from assuming a surprise exam within a limited window alone, but rather from these assumptions supplemented by the following proposition: *The students* know *that the exam is to be a surprise and they* know *the window in which it is to be given.*

This fact is apparent if you examine the student's reasoning, which is full of statements about what the students *would know*. Can they truly *know* a statement (even a true statement) if it leads them to a contradiction?

Explain the paradox in your own words, deciding whether the exam would be a surprise if given on Friday. Can the paradox be avoided by saying that the conditions under which the exam is promised are true but the students cannot *know* that they are true?

How does this puzzle relate to Gödel's incompleteness result?

Literature

Note: CSHPM/SCHPM = Canadian Society for the History and Philosophy of Mathematics/Société Canadienne d'Histoire et Philosophie des Mathématiques

Adler, Ada, 1971. *Suidae Lexicon*, Teubner-Verlag, Stuttgart.

Ağargün, Ahmet G.; Fletcher, Colin R., 1994. "Al-Farisi and the fundamental theorem of arithmetic," *Historia Mathematica*, **21**, No. 2, 162–173.

Allman, George Johnston, 1889. *Greek Geometry from Thales to Euclid*, Longmans, Green & Co., London.

Amir-Moez, Ali R., 1959. "Discussion of difficulties in Euclid by Omar ibn Abrahim al-Khayyami," *Scripta Mathematica*, **XXIV**, 275–303.

Amir-Moez, Ali R., 1963. "A paper of Omar Khayyam," *Scripta Mathematica*, **XXVI**, No. 4, 323–337.

Andrews, George E., 1979. "An introduction to Ramanujan's 'lost' notebook," *American Mathematical Monthly*, **86**, No. 2, 89–108.

Ang Tian-Se; Swetz, Frank J., 1986. "A Chinese mathematical classic of the third century: *The Sea Island Mathematical Manual* of Liu Hui, *Historia Mathematica*, **13**, No. 2, 99–117.

Aschbacher, Michael, 1981. "The classification of the finite simple groups," *The Mathematical Intelligencer*, **3**, No. 2, 59–65.

Ascher, Marcia, 1991. *Ethnomathematics*, Brooks/Cole, New York.

Ascher, Marcia, 1992. "Before the conquest," *Mathematics Magazine*, **65**, No. 4, 211–218.

Ascher, Marcia, 1995. "Models and maps from the Marshall Islands: a case in ethnomathematics," *Historia Mathematica*, **22**, 347–370.

Ascher, Marcia, 1997. "Malagasy *Sikidy*: a case in ethnomathematics," *Historia Mathematica*, **24**, No. 4, 376–395.

Ascher, Marcia; Ascher, Robert, 1997. *Mathematics of the Incas: Code of the Quipu*, Dover, Mineola, NY.

Ayoub, R. 1980. "Paolo Ruffini's contributions to the quintic," *Archive for History of Exact Sciences*, **23**, No. 3, 253–277.

Baatz, Simon, 1991. "'Squinting at Silliman': scientific periodicals in the early American Republic, 1810–1833," *Isis*, **82**, 223–244.

Bag, Amulya Kumar, 1966. "Binomial theorem in ancient India," *Indian Journal of History of Science*, **1**, No. 1, 68–74.

Bagheri, Mohammad, 1997. "A newly found letter of al-Kashi on scientific life in Samarkand," *Historia Mathematica*, **24**, No. 3, 241–256.

Baigozhina, G. O., 1995. "On the classification principle of the problems in Abu-Kamil's *Book of Indeterminate Problems*," *Istoriko-Matematicheskie Issledovaniya*, **1 (36)**, 61–66 (Russian).

Baker, J. N. L., 1948. "Mary Somerville and geography in England," *The Geographical Journal*, **111**, No. 4/6, 207–222.

Baltzer, R., ed., 1885. *August Ferdinand Möbius, Gesammelte Werke*, S. Hirzel, Leipzig.

Bashmakova, I. G; Smirnova, G. S., 1997. "The origin and development of algebra," in: B. V. Gnedenko, ed., *Essays on the History of Mathematics*, Moscow University Press, pp. 94–246 (Russian). English translation published separately as *The Beginnings and Evolution of Algebra*, A. Shenitzer (transl.), Mathematical Association of America, Oberlin, OH, 2000.

Beckers, Danny J., 1999. "Lagrange in the Netherlands: Dutch attempts to obtain rigor in calculus, 1797–1840," *Historia Mathematica*, **26**, No. 3, 234–238.

Bedini, Silvio A., 1972. *The Life of Benjamin Banneker*, Charles Scribner's Sons, New York.

Beman, Wooster Woodruff; Smith, David Eugene, 1930. *Famous Problems of Elementary Geometry*, G. E. Stechert & Co., New York.

Berggren, J. L., 1989. "Abu Sahl al-Kuhi: what the manuscripts say," *Proceedings of the 15th annual meeting of the CSHPM/SCHPM*, Université Laval, Montréal, Québec, pp. 31–48.

Berggren, J. L., 1990. "Greek and Islamic elements in Arabic mathematics," *Proceedings of the 16th annual Meeting of the CSHPM/SCHPM*, University of Victoria, Victoria, British Columbia, pp. 25–38.

Berggren, J. L., 2002. "The transmission of Greek geometry to medieval Islam," *CUBO*, **4**, No. 2, 1–13.

Bernal, Martin, 1992. "Animadversions on the origins of western science," *Isis*, **83**, No. 4, 596–607.

Bernays, Paul, ed., 1971. *Foundations of Geometry*, by David Hilbert, translated by Leo Unger, Open Court, La Salle, IL.

Berndt, Bruce C., ed., 1985. *Ramanujan's Notebooks*, 5 vols., Springer-Verlag, New York.

Betti, E. 1852. "Sulla risoluzione delle equazioni algebriche," *Tortolini Annali*, **III**, 49–51.

Biggs, N. L., 1979. "The roots of combinatorics," *Historia Mathematica*, **6**, No. 2, 109–136.

Biggs, N. L., 1981. "T. P. Kirkman, mathematician," *Bulletin of the London Mathematical Society*, **13**, 97–120.

Billard, Lynne, 1991. "The past, present, and future of academic women in the mathematical sciences," *Notices of the American Mathematical Society*, **38**, No. 7, 707–714.

Blackwell, Richard, transl., 1986. *Christiaan Huygens'* The Pendulum Clock, Iowa State University Press, Ames, IA.

Blum, Lenore, 1991. "A brief history of the Association for Women in Mathematics: the presidents' perspectives," *Notices of the American Mathematical Society*, **38**, No. 7, 738–754.

Bochenski, I. M., 1961. *A History of Formal Logic*, University of Notre Dame Press.

Bochner, Salomon, 1974. "Mathematical reflections," *American Mathematical Monthly*, **81**, No. 8, 827–840.

Boncompagni, Baldassare, 1854. *Intorno ad alcune opere matematica notizie di Leonardo, matematico del secolo decimoterzo*, Tipografia Delle Belle Arti, Rome.

Bottazzini, Umberto, 1986. *The* Higher Calculus: *A History of Real and Complex Analysis from Euler to Weierstrass*, Springer-Verlag, New York.

Boyer, Carl B., 1949. *The History of the Calculus and its Conceptual Development*, Hafner, New York. Reprint: Dover, New York, 1959.

Brentjes, Sonja; Hogendijk, Jan P., 1989. "Notes on Thabit ibn Qurra and his rule for amicable numbers," *Historia Mathematica*, **16**, No. 4, 373–378.

Bretschneider, Carl Anton, 1870. *Die Geometrie und die Geometer vor Euklides. Ein historischer Versuch*, Teubner, Leipzig. Reprint: M. Sändig, Wiesbaden, 1968.

Brown, James Robert, 1999. *Philosophy of Mathematics: An Introduction to the World of Proofs and Pictures*, Routledge, New York.

Buck, R. C., 1980. "Sherlock Holmes in Babylon," *The American Mathematical Monthly*, **87**, No. 5, 335–345.

Burington, Richard Stevens, 1958. *Handbook of Mathematical Tables and Formulas*, Handbook Publishers, Sandusky, OH.

Butzmann, Hans, 1970. *Codex Agrimensorum Romanorum: Codex Arcerianusa der Herzog-August-Bibliothek zu Wolfenbüttel*, A. W. Sijthoff, Lugduni Batavorum.

Bychkov, S. N., 2001. "Egyptian geometry and Greek science," *Istoriko-Matematicheskie Issledovaniya*, **6 (41)**, 277–284 (Russian).

Cantor, Moritz, 1880. *Vorlesungen über Geschichte der Mathematik*, Vol. 1, Teubner, Leipzig.

Cauchy, A.-L., 1815. "Mémoire sur le nombre des valeurs qu'une fonction peut acquérir," *Journal de l'École Polytechnique, XVIIe* Cahier, Tome X = *Œuvres Complètes*, Tome 13, pp. 64–90. Gauthier-Villars, Paris.

Caveing, Maurice, 1985. "La tablette babylonienne AO 17264 du Musée du Louvre et le problème des six frères," *Historia Mathematica*, **12**, No. 1, 6–24.

Chace, A. B., Bull, L., Manning, H. P., and Archibald, R. C., 1927. *The Rhind Mathematical Papyrus*, Mathematical Association of America, Oberlin, OH, Vol. 1.

Chaikovskii, Yu. V., 2001. "What is probability? The evolution of the concept (from antiquity to Poisson)," *Istoriko-Matematicheskie Issledovaniya*, **6 (41)**, 34–56 (Russian).

Chaucer, Geoffrey, 1391. *A Treatise on the Astrolabe*, in: F. N. Robinson, ed., *The Works of Geoffrey Chaucer*, 2nd ed., Houghton Mifflin, Boston, 1957, pp. 545–563.

Chemla, Karine, 1991. "Theoretical aspects of the Chinese algorithmic tradition (first to third century)," *Historia Scientiarum*, No. 42, 75–98.

Christianidis, Jean, 1998. "Une interprétation byzantine de Diophante," *Historia Mathematica*, **25**, No. 1, 22–28.

Clagett, Marshall, 1968. *Nicole Oresme and the Medieval Geometry of Qualities and Motions*, University of Wisconsin Press, Madison, WI.

Clark, Walter Eugene, ed., 1930. *The* Aryabhatiya *of Aryabhata*, University of Chicago Press, Chicago.

Clayton, Peter A., 1994. *Chronicle of the Pharaohs*, Thames & Hudson, London.

Closs, Michael P., ed., 1986. *Native American Mathematics*, University of Texas Press, Austin, TX.

Closs, Michael P., 1992. "Ancient Maya mathematics and mathematicians," *Proceedings of the 18th Annual Meeting of the CSHPM/SCHPM*, University of Prince Edward Island, Charlottetown, P.E.I., pp. 1–13.

Coe, Michael D., 1973. *The Maya Scribe and His World*, The Grolier Club, New York.

Colebrooke, Henry Thomas, 1817. *Algebra with Arithmetic and Mensuration from the Sanscrit of Brahmegupta and Bhascara*, J. Murray, London.

Colson, F. H., 1926. *The Week: An Essay on the Origin and Development of the Seven-Day Cycle*, Greenwood Press, Westport, CT.

Coolidge, Julian Lowell, 1940. *A History of Geometrical Methods*, Clarendon Press, Oxford.

Craik, Alex D. D., 1999. "Calculus and analysis in early nineteenth-century Britain: the work of William Wallace," *Historia Mathematica*, **26**, No. 3, 239–267.

Crossley, John N.; Henry, Alan S., 1990. "Thus spake al-Khwārizmī: A translation of the text of Cambridge University Library Ms. ii.vi.5," *Historia Mathematica*, **17**, No. 2, 103–131.

Cullen, Christopher, 1996. *Astronomy and Mathematics in Ancient China: The* Zhou Bi Suan Jing, Cambridge University Press.

al-Daffa, Ali Abdullah, 1977. *The Muslim Contribution to Mathematics*, Humanities Press, Atlantic Highlands, NJ.

Dahan, Amy, 1980. "Les travaux de Cauchy sur les substitutions; Étude de son approche du concept de groupe," *Archive for History of Exact Sciences*, **23**, No. 4, 279–319.

Dahan-Dalmédico, Amy, 1987. "Mécanique et théorie des surfaces: les travaux de Sophie Germain," *Historia Mathematica*, **14**, No. 4, 347–365.

Dauben, Joseph W., 1996. "Mathematics at the University of Toronto: Abraham Robinson in Canada (1951–1957)," in: Dauben, Folkerts, Knobloch, and Wussing, *History of Mathematics: States of the Art*, Academic Press, New York.

Dauben, Joseph W.; Scriba, Christoph J., eds., 2002. *Writing the History of Mathematics: Its Historical Development*, Birkhäuser, Boston.

David, H. A.; Edwards, A. W. F., 2001. *Annotated Readings in the History of Statistics*, Springer-Verlag, New York.

Davis, Margaret Daly, 1977. *Piero Della Francesca's Mathematical Treatises: the* Tratto d'abaco *and* Libellus de quinque corporibus regularibus, Longe Editore, Ravenna.

Davis, Philip J.; Hersh Reuben, 1986. *Descartes' Dream: The World According to Mathematics*, Harcourt Brace Jovanovich, New York.

Deakin, Michael, 1994. "Hypatia and her mathematics," *American Mathematical Monthly*, **101**, No. 3, 234–243.

Dean, Nathaniel, ed., 1996. *African Americans in Mathematics. DIMACS Workshop, June 26–28, 1996*, American Mathematical Society, Providence, RI.

Detlefsen, Michael, 2001. "What does Gödel's second theorem say?" *Philosophia Mathematica* (3), **9**, No. 1, 37–71.

Detlefsen, Michael; Erlandson, Douglas K.; Heston, J. Clark; Young, Charles M., 1975. "Computation with Roman numerals," *Archive for History of Exact Science*, **15**, No. 2, 141–148.

Dick, Auguste, 1981. *Emmy Noether, 1882–1935*, translated by H. I. Blocher. Birkhäuser, Boston.

Dickson, Leonard Eugene, 1919. *History of the Theory of Numbers I: Divisibility and Primality*, Carnegie Institute, Washington, DC. Reprint: Chelsea, New York, 1966.

Dickson, Leonard Eugene, 1920. *History of the Theory of Numbers II: Diophantine Analysis*, Carnegie Institute, Washington, DC. Reprint: Chelsea, New York, 1966.

Dickson, Leonard Eugene, 1923. *History of the Theory of Numbers III: Quadratic and Higher Forms*, Carnegie Institute, Washington, DC. Reprint: Chelsea, New York, 1966.

Diels, Hermann, 1951. *Die Fragmente der Vorsokratiker*, 6th corrected edition, (Walther Kranz, ed.), Weidmann, Berlin.

Dijksterhuis, E. J., 1956. *Archimedes*, Munksgård, Copenhagen.

Dilke, O. A. W., 1985. *Greek and Roman Maps*, Cornell University Press, Ithaca, NY.

D'ooge, Martin Luther, 1926, translator. *Introduction to Arithmetic* (Nicomachus of Gerasa), Macmillan, New York.

Dorofeeva, A. V., 1998. "The calculus of variations," in: *Mathematics of the 19th Century*, Birkhäuser, Basel, pp. 197–260.

Duren, Peter, 1989. *A Century of Mathematics in America* (3 Vols.), American Mathematical Society, Providence, RI.

Dutka, Jacques, 1988. "On the Gregorian revision of the Julian calendar," *Mathematical Intelligencer*, **10**, No. 1, 56–64.

Dzielska, Maria, 1995. *Hypatia of Alexandria*, Harvard University Press, Cambridge, MA.

Edwards, H. M., 1974. *Riemann's Zeta Function*, Academic Press, New York.

Edwards, H. M., 1977. *Fermat's Last Theorem*, Springer-Verlag, New York.

Ehrman, Esther, 1986. *Mme du Châtelet*, Berg Publishers, Leamington Spa, Warkwickshire.

Engel, Friedrich; Heegaard, Poul, eds., 1960. *Sophus Lie, Gesammelte Abhandlungen*, Teubner, Leipzig.

Erdős, P.; Dudley, U., 1983. "Some remarks and problems in number theory related to the work of Euler," *Mathematics Magazine*, **56**, No. 5, 292–298.

Euler, L., 1732. "De formes radicum aequationum cuiusque ordinis coniectatio," *Commentarii Academiae Petropolitanae*, **6** (1738), p. 216.

Euler, L., 1744. *Methodus inveniendi lineas curvas maximi minimive proprietate gaudentes*, Bousquet & Co., Lausanne.

Euler, L., 1749. "Recherches sur les racines imaginaires des équations," *Histoire de l'Académie des Sciences de Berlin*, p. 222.

Euler, L., 1762. "De resolutione aequationum cuiusque gradus," *Novi Commentarii Academiae Petropolitanae* **9**, p. 70.

Farrar, John, 1826. *Elements of Electricity, Magnetism, and Electro-magnetism*, Hilliard and Metcalf, Cambridge, MA.

Feigenbaum, L., 1994. "Infinite series and solutions of ordinary differential equations, 1670–1770," in: *Companion Encyclopedia of the History and Philosophy of the Mathematical Sciences*, Vol. 1, Routledge, London and New York, pp. 504-519.

Feingold, Mordechai, 1993. "Newton, Leibniz, and Barrow too: an attempt at a reinterpretation," *Isis*, **84**, No. 2, 310–338.

Feit, Walter, 1977. "A mathematical visit to China, May 1976," *Notices of the American Mathematical Society*, **24**, No. 2, 110–113.

Fennema, Elizabeth; Leder, Gilah C., eds., 1990. *Mathematics and Gender*, Teachers College Press, New York.

Ferreirós, José, 1995. "'What fermented in me for years': Cantor's discovery of transfinite numbers," *Historia Mathematica*, **22**, 33–42.

Field, J. V.; Gray, J. J., 1987. *The Geometrical Work of Girard Desargues*, Springer-Verlag, Berlin.

Fitzgerald, Augustine, transl., 1926. *The Letters of Synesius of Cyrene*, Oxford University Press.

Fletcher, Colin R., 1989. "Fermat's theorem," *Historia Mathematica*, **16**, No. 2, 149–153.

Folkerts, Menso, 1970. *Ein neuer Text des Euclides latinus*, H. A. Gerstenberg, Hildesheim.

Folkerts, Menso, 1971. *Anonyme lateinische Euklidbearbeitungen aus dem 12. Jahrhundert*, Österreichische Akademie der Wissenschaften, Mathematisch-Naturwissenschaftliche Klasse, Denkschriften, **116**, erste Abhandlung.

Fowler, David, 1992. "Dedekind's theorem: $\sqrt{2} \times \sqrt{3} = \sqrt{6}$," *The American Mathematical Monthly*, **99**, No. 8, 725–733.

Fowler, David, 1998. *The Mathematics of Plato's Academy*, 2nd ed., Clarendon Press, Oxford.

Franchella, Miriam, 1995. "L. E. J. Brouwer: toward intuitionistic logic," *Historia Mathematica*, **22**, No. 3, 304–322.

Franci, Rafaella, 1988. "Antonio de' Mazzinghi, an algebraist of the fourteenth century," *Historia Mathematica*, **15**, No. 3, 240–249.

Fraser, Craig, 1987. "Joseph-Louis Lagrange's algebraic vision of the calculus," *Historia Mathematica*, **14**, No. 1, 38–53.

Fraser, Craig, 1993. "A history of Jacobi's theorem in the calculus of variations," *Proceedings of the 19th Annual Meeting of the CSPHM/SCHPM*, Carleton University, Ottawa, Ontario, pp. 168–185.

Freidel, David; Schele, Linda; Parker, Joy, 1993. *Maya Cosmos: Three Thousand Years on the Shaman's Path*, William Morrow, New York.

Friberg, Jöran, 1981. "Methods and traditions of Babylonian mathematics: Plimpton 322, Pythagorean triples and the Babylonian triangle parameter equations," *Historia Mathematica*, **8**, No. 3, 277–318.

von Fritz, Kurt, 1945. "The discovery of incommensurability by Hippasus of Metapontum," *Annals of Mathematics*, **46**, 242–264.

Fu Daiwie, 1991. "Why did Lui Hui fail to derive the volume of a sphere?", *Historia Mathematica*, **18**, No. 3, 212–238.

Fukagawa Hidetoshi; Pedoe, D., 1989. *Japanese Temple Geometry Problems: San Gaku*, Winnipeg, Manitoba.

Fuson, Karen, 1988. *Children's Counting and Concepts of Number*, Springer-Verlag, New York.

Garciadiego, Alejandro, 2002. "Mexico," in: Dauben and Scriba, (*2002*, pp. 256–263).

Gauss, Carl Friedrich, 1799. "Demonstratio nova theorematis omnem functionem algebraicam rationalem integralam unius variabilis in factores reales primi vel secundi gradus resolvi posse," in: *Werke*, Vol. 3, Königlichen Gesellschaft der Wissenschaften, Göttingen, 1866, pp. 3–31.

Gauss, Carl Friedrich, 1965. *General Investigations of Curved Surfaces*, translated from the Latin and German by Adam Hiltebeitel and James Morehead, Raven Press, Hewlett, NY.

Geijsbeek, John B., ed. and transl., 1914. *Ancient Double-Entry Bookkeeping. Lucas Pacioli's Treatise (A.D. 1494—The Earliest Known Writer on Bookkeeping)*, Denver, CO.

Gerdes, Paulus, 1985. "Three alternate methods of obtaining the ancient Egyptian formula for the area of a circle," *Historia Mathematica*, **12**, No. 3, 261–268.

Gerhardt, C. I., ed., 1971. *Leibniz: Mathematische Schriften*, G. Olms Verlag, Hildesheim.

Gericke, Helmut, 1996. "Zur Geschichte der negativen Zahlen," in: Dauben, Folkerts, Knobloch, and Wussing, eds., *History of Mathematics: States of the Art*, Academic Press, New York, pp. 279–306.

Gericke, Helmut; Vogel, Kurt, 1965. *De Thiende von Simon Stevin*, Akademische Verlagsgesellschaft, Frankfurt am Main.

Gillings, Richard J., 1972. *Mathematics in the Time of the Pharaohs*, MIT Press, Cambridge, MA. Reprint: Dover, New York, 1982.

Gold, David; Pingree, David, 1991. "A hitherto unknown Sanskrit work concerning Mādhava's derivation of the power series for sine and cosine," *Historia Scientiarum*, No. 42, 49–65.

Goldstine, Herman H., 1980. *A History of the Calculus of Variations from the 17th Through the 19th Century*, Springer-Verlag, New York.

Gottwald, Siegfried; Ilgauds, Hans-Joachim; Schlote, Karl-Heinz, eds., 1990. *Lexikon Bedeutender Mathematiker*, Bibliographisches Institut, Leipzig.

Gould, James L.; Gould, Carol Grant, 1995. *The Honey Bee*, Scientific American Library, New York.

Gow, James, 1884. *A Short History of Greek Mathematics*. Reprint: Chelsea, New York, 1968.

Grabiner, Judith, 1995. "Descartes and problem-solving," *Mathematics Magazine*, **68**, No. 2, 83–97.

Grattan-Guinness, Ivor, 1972. "A mathematical union: William Henry and Grace Chisholm Young," *Annals of Science*, **29**, No. 2, 105–186.

Grattan-Guinness, Ivor, 1975. "Mathematical bibliography for W. H. and G. C. Young," *Historia Mathematica*, **2**, 43–58.

Grattan-Guinness, Ivor, 1990. *Convolutions in French Mathematics, 1800-1840*, Birkhäuser, Basel.

Grattan-Guinness, Ivor, 2000. *The Search for Mathematical Roots, 1870–1940: Logics, Set Theories, and the Foundations of Mathematics from Cantor through Russell to Gödel*, Princeton University Press, Princeton, NJ.

Grattan-Guinness, Ivor, 2004. "History or heritage? An important distinction in mathematics and for mathematics education," *American Mathematical Monthly*, **111**, 1–12.

Gray, J. J., 1989. *Ideas of Space: Euclidean, Non-Euclidean, and Relativistic*, 2nd ed., Clarendon Press, Oxford.

Gray, Robert, 1994. "Georg Cantor and transcendental numbers," *The American Mathematical Monthly*, **101**, No. 9, 819–832.

Green, Judy; LaDuke, Jeanne, 1987. "Women in the American mathematical community: the pre-1940 Ph.D.'s," *The Mathematical Intelligencer*, **9**, No. 1, 11–23.

Greenleaf, Benjamin, 1876. *New Practical Arithmetic; in Which the Science and Its Applications Are Simplified by Induction and Analysis*, Leach, Shewell, and Sanborn, Boston and New York.

Grinstein, Louise S.; Campbell, Paul J., 1982. "Anna Johnson Pell Wheeler: Her life and work," *Historia Mathematica*, **9**, No. 1, 37–53.

Grosholz, Emily, 1987. "Two Leibnizian manuscripts of 1690 concerning differential equations," *Historia Mathematica*, **14**, No. 1, 1–37.

Guizal, Brahim; Dudley, John, 2002. "Ibn Sahl: Inventeur de la loi de la réfraction," *Revue pour la science*, No. 301, November 2002.

Guo Shuchung, 1992. "Guo Shuchung's edition of the *Jiu Zhang Suan Shu*," *Historia Mathematica*, **19**, No. 2, 200–202.

Gupta, R. C., 1989. "Sino-Indian interaction and the great Chinese Buddhist astronomer-mathematician I-Hsing (A.D. 683–727)," *Bulletin of the Indian Society for History of Mathematics*, **11**, Nos. 1–4, 38–49.

Gupta, R. C., 1991. "On the volume of a sphere in ancient India," *Historia Scientiarum*, No. 42, 33–44.

Gupta, R. C., 1994a. "Six types of Vedic mathematics," *Bulletin of the Indian Society for History of Mathematics*, **16**, Nos. 1–4, 5–15.

Gupta, R. C., 1994b. "A circulature rule from the *Agni Purāṇa*," *Bulletin of the Indian Society for History of Mathematics*, **16**, Nos. 1–4, 53–56.

Gustafson, W. H.; Halmos, P. R.; Moolgavkar, S. H.; Wheeler, W. H.; Ziemer, W. P., 1976. "American mathematics from 1940 to the day before yesterday," *The American Mathematical Monthly*, **83**, No. 7, 503–516.

Hadamard, J., 1896. "Sur la distribution des zéros de la fonction $\zeta(s)$ et ses consequences arithmétiques," *Bulletin de la Société Mathématique de France*, **24**, 199–220.

Hairetdinova, N. G., 1986. "On spherical trigonometry in the medieval Near East and in Europe," *Historia Mathematica*, **13**, No. 2, 136–146.

Hamilton, W. R., 1837. "On the argument of Abel, respecting the impossibility of expressing a root of any general equation above the fourth degree, by any finite combination of radicals and rational functions," *Transactions of the Royal Irish Academy*, **XVIII** (1839), 171–259; H. Halberstam and R. Ingram, eds., *The Mathematical Papers of Sir William Rowan Hamilton*, Vol. III, pp. 517–569.

Hari, K. Chandra, 2002. "Genesis and antecedents of Āryabhaṭīya," *Indian Journal of History of Science*, **37**, No. 2, 101–113.

Harrison, Jenny, 1991. "The Escher staircase," *Notices of the American Mathematical Society*, **38**, No. 7, 730–734.

al-Hassan, Ahmad Y.; Hill, Donald R., 1986. *Islamic Technology: An Illustrated History*, Cambridge University Press.

Hawkins, Thomas, 1989. "Line geometry, differential equations, and the birth of Lie's theory of groups," in: David Rowe and John McCleary, eds, *The History of Modern Mathematics*, Vol. 1, Academic Press, New York.

Hayashi Takao, 1987. "Varahamihira's pandiagonal magic square of the order four," *Historia Mathematica*, **14**, No. 2, 159–166.

Hayashi Takao, 1991. "A note on Bhāskara I's rational approximation to sine," *Historia Scientiarum*, No. 42, pp. 45–48.

Heath, T. L., 1910. *Diophantus of Alexandria: A Study in the History of Greek Algebra*, 2nd ed., Cambridge University Press, 1910.

Heath, T. L., 1897–1912. *The Works of Archimedes Edited in Modern Notation with Introductory Chapters*, Reprint: Dover, New York, 1953.

Heath, T. L., 1921. *A History of Greek Mathematics*, Clarendon Press, Oxford.

Henrion, Claudia, 1991. "Merging and emerging lives: women in mathematics," *Notices of the American Mathematical Society*, **38**, No. 7, 724–729.

Henrion, Claudia, 1997. *Women in Mathematics: The Addition of Difference*, Indiana University Press, Bloomington, IN.

Heppenheimer, T. A., 1990. "How von Neumann showed the way," *American Heritage of Invention and Technology*, **6**, No. 2, 8–16.

Hersh, Reuben, 1997. *What Is Mathematics, Really?* Oxford University Press, New York.

Heyde, C. C.; Seneta, E., 1977. *I. J. Bienaymé: Statistical Theory Anticipated*, Springer-Verlag, New York.

Hilbert, David, 1971. *Foundations of Geometry*, translated by Leo Unger, Open Court, LaSalle, IL.

Hogendijk, Jan P., 1985. "Thabit ibn Qurra and the pair of amicable numbers 17296, 18416," *Historia Mathematica*, **12**, No. 3, 269–273.

Hogendijk, Jan P., 1989. "Sharaf al-Din al-Tusi on the number of positive roots of cubic equations," *Historia Mathematica*, **16**, No. 1, 69–85.

Hogendijk, Jan P., 1991. "Al-Khwarizmi's table of the 'sine of the hours' and underlying sine table," *Historia Scientiarum*, No. 42, pp. 1–12.

Hogendijk, Jan P., 2002. "The surface area of the bicylinder and Archimedes' *Method*," *Historia Mathematica*, **29**, No. 1, 199–203.

Homann, Frederick A., 1987. "David Rittenhouse: logarithms and leisure," *Mathematics Magazine*, **60**, No. 1, 15–20.

Homann, Frederick A., 1991. *Practical Geometry: Practica Geometriae, attributed to Hugh of St. Victor*, Marquette University Press, Milwaukee, WI.

Høyrup, Jens, 2002. "A note on Old Babylonian computational techniques," *Historia Mathematica*, **29**, No. 2, 193–198.

Hughes, Barnabas, 1981. *De numeris datis* (Jordan de Nemore), University of California Press, Berkeley, CA.

Hughes, Barnabas, 1989. "The arithmetical triangle of Jordanus de Nemore," *Historia Mathematica*, **16**, No. 3, 213–223.

Hultsch, F., ed., 1965. *Pappi Alexandrini Collectionis*, Vol. 1, Verlag Adolf M. Hakkert, Amsterdam.

Ifrah, Georges, 2000. *The Universal History of Numbers: From Prehistory to the Invention of the Computer*, translated from the French by David Bellos, Wiley, New York.

Il'ina, E. A., 2002. "On Euclid's *Data*," *Istoriko-Matematicheskie Issledovaniya*, **7 (42)**, 201–208 (Russian).

Indorato, Luigi; Nastasi, Pietro, 1989. "The 1740 resolution of the Fermat–Descartes controversy," *Historia Mathematica*, **16**, No. 2, 137–148.

Ivić, Aleksandar, 1985. *The Riemann Zeta-Function: Theory and Applications*, Dover, New York.

Ivins, W. M., 1947. "A note on Desargues' theorem," *Scripta Mathematica*, **13**, 203–210.

Jackson, Allyn, 1991. "Top producers of women mathematics doctorates," *Notices of the American Mathematical Society*, **18**, No. 7, 715–720.

Jacobs, Konrad; Utz, Heinrich, 1984. "Erlangen programs," *The Mathematical Intelligencer*, **6**, No. 1.

James, Portia P., 1989. *The Real McCoy: African-American Invention and Innovation, 1619–1930*, Smithsonian Institution Press, Washington, DC.

Jami, Catherine, 1988. "Western influence and Chinese tradition in an eighteenth century Chinese mathematical work," *Historia Mathematica*, **15**, No. 4, 311–331.

Jami, Catherine, 1991. "Scholars and mathematical knowledge during the late Ming and early Qing," *Historia Scientiarum*, No. 42, 95–110.

Jentsch, Werner, 1986. "Auszüge aus einer unveröffentlichten Korrespondenz von Emmy Noether und Hermann Weyl mit Heinrich Brandt," *Historia Mathematica*, **13**, No. 1, 5–12.

Jha, V. N., 1994. "Indeterminate analysis in the context of the Mahāsiddhānta of Āryabhaṭa II," *Indian Journal of History of Science*, **29**, No. 4, 565–578.

Jones, Alexander, 1991. "The adaptation of Babylonian methods in Greek numerical astronomy," *Isis*, **82**, No. 313, 441–453.

Kasir, Daoud, 1931. *The Algebra of Omar Khayyam*, Teachers College, Columbia University Contributions to Education, No. 385, New York.

Kawahara Hideki, 1991. "World-View of the *Santong-Li*," *Historia Scientiarum*, No. 42, 67–73.

Kazdan, Jerry L., 1986. "A visit to China," *The Mathematical Intelligencer*, **8**, No. 4, 22–32.

Keith, Natasha; Keith, Sandra Z., 2000. Review of Claudia Henrion's *Women in Mathematics*, *Humanistic Mathematics Network Journal*, Issue 22, 26–30.

Kenschaft, Patricia, 1981. "Black women in mathematics in the United States," *The American Mathematical Monthly*, **88**, No. 8, 592–604.

Kimberling, C. H., 1972a. "Emmy Noether," *The American Mathematical Monthly*, **79**, No. 2, 136–149.

Kimberling, C. H., 1972b. "Addendum to 'Emmy Noether'," *The American Mathematical Monthly*, **79**, No. 7, 755.

King, R. Bruce, 1996. *Beyond the Quartic Equation*, Birkhäuser, Boston.

Kiro, S. N., 1967, "N. I. Lobachevskii and mathematics at Kazan' University," in: *History of Russian and Soviet Mathematics* (*Istoriya Otechestvennoi Matematiki*), Vol. 2, Naukova Dumka, Kiev (Russian).

Klein, Felix, 1884. *Lectures on the Icosahedron and the Solution of Equations of the Fifth Degree*, translated by George Gavin Morrice. Reprint: Dover, New York, 1956.

Klein, Felix, 1926. *Vorlesungen über die Entwicklung der Mathematik im 19. Jahrhundert*, Springer-Verlag, Berlin, 2 vols. Reprint: American Mathematical Society (Chelsea Publishing Company), Providence, RI, 1967.

Klein, Jacob, 1933. *Plato's Trilogy: Theaetetus, the Sophist, and the Statesman*, University of Chicago Press, Chicago, 1977.

Klein, Jacob, 1934–1936. *Greek Mathematical Thought and the Origin of Algebra*, translated by Eva Brann, MIT Press, Cambridge, MA, 1968.

Klein, Jacob, 1965. *A Commentary on Plato's* Meno, University of North Carolina Press, Chapel Hill, NC.

Kleiner, Israel, 1991. "Emmy Noether: Highlights of her life and work," *Proceedings of the 17th Annual Meeting of the CSHPM/SCHPM*, Queen's University, Kingston, Ontario, pp. 19–42.

Knorr, Wilbur, 1975. *The Evolution of the Euclidean* Elements, Reidel, Boston.

Knorr, Wilbur, 1976. "Problems in the interpretation of Greek number theory: Euclid and the 'fundamental theorem of arithmetic'," *Studies in the Historical and Philosophical Sciences*, **7**, 353–368.

Knorr, Wilbur, 1982. "Techniques of fractions in ancient Egypt," *Historia Mathematica*, **9**, No. 2, 133–171.

Koblitz, Ann Hibner, 1983. *A Convergence of Lives. Sophia Kovalevskaia: Scientist, Writer, Revolutionary.* Birkhäuser, Boston.

Koblitz, Ann Hibner, 1984. "Sofia Kovalevskaia and the mathematical community," *The Mathematical Intelligencer*, **6**, No. 1, 20–29.

Koblitz, Ann Hibner, 1991. "Historical and cross-cultural perspectives on women in mathematics," *Proceedings of the 17th Annual Meeting of the CSHPM/SCHPM*, Queen's University, Kingston, Ontario, pp. 1–18.

Koblitz, Neal, 1990. "Recollections of mathematics in a country under siege," *The Mathematical Intelligencer*, **12**, No. 3, 16–34.

Koehler, Otto, 1937. *Bulletin of Animal Behavior*, No. 9. English translation in James R. Newman, ed., *The World of Mathematics*, vol. 1, Simon and Schuster, New York, 1956, pp. 491–492.

Kowalewski, Gerhard, 1950. *Bestand und Wandel*, Oldenbourg, München.

Kox, A. J.; Klein, Martin J.; Schulmann, Robert, eds., 1996. *The Collected Papers of Albert Einstein*, Princeton University Press, Princeton, NJ.

Kracht, Manfred; Kreyszig, Erwin, 1990. "E. W. von Tschirnhaus: his role in early calculus and his work and impact on algebra," *Historia Mathematica*, **17**, No. 1, 16–35.

Kreyszig, Erwin, 1993. "On the calculus of variations and its major influences on the mathematics of the first half of our century," *Proceedings of the 19th Annual Meeting of the CSHPM/SCHPM*, Carleton University, Ottawa, Ontario, pp. 119–149.

Krupp, E. C., 1983. *Echoes of the Ancient Skies*, Harper & Row, New York.

Kunoff, Sharon, 1990. "A curious counting/summation formula from the ancient Hindus," in: *Proceedings of the 16th Annual Meeting of the CSHPM/SCHPM*, University of Victoria, Victoria, British Columbia, pp. 101–107.

Kunoff, Sharon, 1991. "Women in mathematics: Is history being rewritten?" *Proceedings of the 17th Annual Meeting of the CSHPM/SCHPM*, Queen's University, Kingston, Ontario, pp. 43–52.

Kunoff, Sharon, 1992. "Some inheritance problems in ancient Hebrew literature," in: *Proceedings of the 18th Annual Meeting of the CSHPM/SCHPM*, University of Prince Edward Island, Charlottetown, P.E.I., pp. 14–20.

Lagrange, J.-L., 1771. "Réflexions sur la résolution algébrique des équations," *Nouveaux mémoires de l'Académie Royale des Sciences et Belles-lettres de Berlin*; *Œuvres* (1869), Vol. 3, pp. 205–421.

Lagrange. J.-L., 1795. *Lectures on Elementary Mathematics*, translated by Thomas J. McCormack, Open Court, Chicago, 1898.

Lam Lay-Yong, 1994. "*Jiu Zhang Suanshu* (*Nine Chapters on the Mathematical Art*): An Overview," *Archive for History of Exact Sciences*, **47**, No. 1, 1–51.

Lam Lay-Yong; Ang Tian-Se, 1986. "Circle measurements in ancient China," *Historia Mathematica*, **13**, No. 4, 325–340.

Lam Lay-Yong; Ang Tian-Se, 1987. "The earliest negative numbers: how they emerged from a solution of simultaneous linear equations," *Archive for History of Exact Sciences*, **37**, 222–267.

Lam Lay-Yong; Ang Tian-Se, 1992. *Fleeting Footsteps. Tracing the Conception of Arithmetic and Algebra in Ancient China*, World Scientific, River Edge, NJ.

Lam Lay-Yong; Shen Kangsheng, 1985. "The Chinese concept of Cavalieri's principle and its applications," *Historia Mathematica*, **12**, No. 3, 219–228.

Lasserre, François, 1964. *The Birth of Mathematics in the Age of Plato*, Hutchinson, London.

Laugwitz, Detlef, 1987. "Infinitely small quantities in Cauchy's textbooks," *Historia Mathematica*, **14**, No. 3, 258–274.

Laugwitz, Detlef, 1999. *Bernhard Riemann, 1826–1866: Turning Points in the Conception of Mathematics*, Birkhäuser, Boston.

Levey, Martin, 1966. *The Algebra of Abū Kāmil*, University of Wisconsin Press, Madison, WI.

Levey, Martin; Petruck, Marvin, 1965. *Kūshyār ibn Labbān: Principles of Hindu Reckoning*, University of Wisconsin Press, Madison, WI.

Lewis, D. J., 1991. "Mathematics and women: the undergraduate school and the pipeline," *Notices of the American Mathematical Society*, **38**, No. 7, 721–723.

Libbrecht, Ulrich, 1973. *Chinese Mathematics in the Thirteenth Century*, MIT Press, Cambridge, MA.

Liebmann, Heinrich, 1904. *N. J. Lobatschefskijs imaginäre Geometrie und Anwendung der imaginären Geometrie auf einige Integrale*, Teubner, Leipzig.

Li Yan; Du Shiran, 1987. *Chinese Mathematics: A Concise History*, translated by John N. Crossley and Anthony W.-C. Lun, Clarendon Press, Oxford.

Mack, John, 1990. *Emil Torday and the Art of the Congo. 1900–1909*, University of Washington Press, Seattle, WA.

Mackay, Alan L., 1991. *A Dictionary of Scientific Quotations*, Institute of Physics Publishing, Bristol and Philadelphia.

Mallayya, V. Madhukar, 1997. "Arithmetic operation of division with special reference to Bhāskara II's Līlāvatī and its commentaries," *Indian Journal of History of Science*, **32**, No. 4, 315–324.

Mancosu, Paolo, 1989. "The metaphysics of the calculus: a foundational debate in the Paris Academy of Sciences, 1700–1706," *Historia Mathematica*, **16**, No. 3, 224–248.

Manheim, Jerome H., 1964. *The Genesis of Point Set Topology*, Macmillan, New York.

Martzloff, Jean-Claude, 1982. "Li Shanlan (1811–1882) and Chinese traditional mathematics," *The Mathematical Intelligencer*, **14**, No. 4, 32–37.

Martzloff, Jean-Claude, 1990. "A survey of Japanese publications on the history of Japanese traditional mathematics (*Wasan*) from the last 30 years," *Historia Mathematica*, **17**, No. 4, 366–373.

Martzloff, Jean-Claude, 1993. "Eléments de réflexion sur les réactions chinoises à la géométrie euclidienne à la fin du XVIIième siècle—Le *Jihe lunyue* {a} de Du Zhigeng {b} vue principalement à partir de la préface de l'auteur et deux notices bibliographiques rédigées par des lettrés illustres," *Historia Mathematica*, **20**, 160–179.

Martzloff, Jean-Claude, 1994. "Chinese mathematics," in: I. Grattan-Guinness, ed., *Companion Encyclopedia of the History and Philosophy of the Mathematical Sciences*, Vol. 1, Routledge, London, pp. 93–103.

Matvievskaya, G. P., 1999. "On the Arabic commentaries to the tenth book of Euclid's *Elements*," *Istoriko-Matematicheskie Issledovaniya*, **4 (39)**, 12–25 (Russian).

Melville, Duncan, 2002. "Weighing stones in ancient Mesopotamia," *Historia Mathematica*, **29**, No. 1, 1–12.

Menninger, Karl, 1969. *Number Words and Number Symbols: A Cultural History of Numbers*, translated from the revised German edition by Paul Broneer. MIT Press, Cambridge, MA.

Mikami Yoshio, 1913. *The Development of Mathematics in China and Japan.* Reprint: Chelsea, New York, 1961.

Mikolás, M., 1975. "Some historical aspects of the development of mathematical analysis in Hungary," *Historia Mathematica*, **2**, No. 2, 304–308.

Milman, Dean; Guizot, M.; Smith, William, 1845. *The History of the Decline and Fall of the Roman Empire by Edward Gibbon*, John D. Morris, Philadelphia.

Montet, Pierre, 1974. *Everyday Life in Egypt in the Days of Ramesses The Great*, translated from the French by A. R. Maxwell-Hyslop and Margaret S. Drower, Greenwood Press, Westport, CT.

Moore, Gregory H., 1982. *Zermelo's Axiom of Choice*, Springer-Verlag, New York.

de Mora-Charles, S., 1992. "Quelques jeux de hazard selon Leibniz," *Historia Mathematica*, **19**, No. 2, 125–157.

Murata Tamotsu, 1994. "Indigenous Japanese mathematics, *wasan*," in: I. Grattan-Guinness, ed., *Companion Encyclopedia of the History and Philosophy of Mathematical Science*, Vol. 1, Routledge, London, pp. 104–110.

Narasimhan, Raghavan, 1990. *Bernhard Riemann: Gesammelte Mathematische Werke, Wissenschaftlicher Nachlaß und Nachträge*, Springer-Verlag, Berlin.

Needham, J., 1959. *Science and Civilisation in China*, Vol. 3: *Mathematics and the Sciences of the Heavens and the Earth*, Cambridge University Press, London.

Neeley, Kathryn A., 2001. *Mary Somerville: Science, Illumination, and the Female Mind*, Cambridge University Press, New York.

Neugebauer, O., 1935. *Mathematische Keilschrifttexte*, Springer-Verlag, Berlin.

Neugebauer, O., 1952. *The Exact Sciences in Antiquity*, Princeton University Press, Princeton, NJ.

Neugebauer, O., 1975. *A History of Ancient Mathematical Astronomy* (three vols.), Springer-Verlag Berlin.

Novosyolov, M. M., 2000. "On the history of the debate over intuitionistic logic," *Istoriko-Matematicheskie Issledovaniya*, **5 (40)**, 272–280 (Russian).

Ore, Oystein 1957. *Niels Henrik Abel, Mathematician Extraordinary.* Reprint: Chelsea, New York, 1974.

Özdural, Alpay, 2000. "Mathematics and arts: connections between theory and practice in the Medieval Islamic world," *Historia Mathematica*, **27**, 171–200.

Panteki, M., 1987. "William Wallace and the introduction of Continental calculus to Britain: a letter to George Peacock," *Historia Mathematica*, **14**, No. 2, 119–132.

Parshall, Karen Hunger, 1985. "Joseph H. M. Wedderburn and the structure theory of algebras," *Archive for History of Exact Sciences*, **32**, No. 3/4, 223–349.

Parshall, Karen Hunger, 1988. "The art of algebra from al-Khwarizmi to Viète: a study in the natural selection of ideas," *History of Science*, **26**, 129–164.

Parshall, Karen Hunger, 2000. "Perspectives on American mathematics," *Bulletin of the American Mathematical Society*, **37**, No. 4, 381–405.

Patterson, S. J., 1990. "Eisenstein and the quintic equation," *Historia Mathematica*, **17**, 132–140.

Pavlov, Ivan, 1928. *Conditioned Reflexes.* Reprint: Dover, New York, 1960.

Pavlov, Ivan, 1955. *Selected Works*, Foreign Languages Publishing House, Moscow.

Perminov, V. Ya., 1997. "On the nature of deductive reasoning in the pre-Greek era of the development of mathematics," *Istoriko-Matematicheskie Issledovaniya* **2 (37)**, 180–200 (Russian).

Pesic, Peter, 2003. *Abel's Proof*, MIT Press, Cambridge, MA.

Phili, Ch., 1997. "Sur le développement des mathématiques en Grèce durant la période 1850–1950. Les fondateurs," *Istoriko-Matematicheskie Issledovaniya*, 2nd ser., special issue on mathematical schools.

Piaget, Jean, 1952. *The Child's Conception of Number*, Humanities Press, New York.

Piaget, Jean; Inhelder, Bärbel, 1967. *The Child's Conception of Space*, Routledge & Kegan Paul, London.

Picard, É., ed., 1897. *Œuvres Mathématiques d'Évariste Galois*, Gauthier-Villars, Paris.

Pitcher, Everett, 1988. "The growth of the American Mathematical Society," *Notices of the American Mathematical Society*, **35**, No. 6, 781–782.

Poisson, S.-D., 1818. "Remarques sur les rapports qui existent entre la propagation des ondes à la surface de l'eau, et leur propagation dans une plaque élastique," *Bulletin des sciences, par la Société Philomatique de Paris*, 9799.

Porter, Theodore M., 1986. *The Rise of Statistical Thinking*, Princeton University Press, Princeton, NJ.

Price, D. J., 1964. "The Babylonian 'Pythagorean triangle'," *Centaurus*, **10**, 210–231.

Pringsheim, A., 1910. "Über neue Gültigkeitsbedingungen für die Fouriersche Integralformel," *Mathematische Annalen*, **68**, 367408.

Rajagopal, P., 1993. "Infinite series in south Indian mathematics, 1400–1600," *Proceedings of the CSHPM/SCHPM 19th Annual Meeting*, Carleton University, Ottawa, Ontario, pp. 86–118.

Rashed, Roshdi, 1989. "Ibn al-Haytham et les nombres parfaits," *Historia Mathematica*, **16**, No. 4, 343–352.

Rashed, Roshdi, 1990. "A pioneer in anaclastics: Ibn Sahl on burning mirrors and lenses," *Isis*, **81**, No. 308, 464–491.

Rashed, Roshdi, 1993. *Les mathématiques infinitésimales du IXe au XIe siècle*, Vol. II, Al-Furqan Islam Heritage Foundation, London.

R¯ashid, Rushd¯i, 1994. *The Development of Arabic Mathematics: between Arithmetic and Algebra*, translated by Angela Armstrong, Kluwer Academic, Dordrecht and Boston.

Reich, Karin, 1977. *Carl Friedrich Gauss: 1777/1977*, Inter Nationes, Bonn–Bad Godesberg.

van Renteln, M., 1987. *Aspekte zur Geschichte der Analysis im 20. Jahrhundert, von Hilbert bis J. v. Neumann*, Lecture notes, University of Karlsruhe.

van Renteln, M., 1989. *Geschichte der Analysis im 19. Jahrhundert, von Cauchy bis Cantor*, Lecture notes, University of Karlsruhe.

van Renteln, M., 1991. *Geschichte der Analysis im 18. Jahrhundert, von Euler bis Laplace*, Lecture notes, University of Karlsruhe.

Richards, Joan, 1987. "Augustus de Morgan and the history of mathematics," *Isis*, **78**, No. 291, 7–30.

Robins, Gay; Shute, Charles, 1987. *The Rhind Mathematical Papyrus: An Ancient Egyptian Text*, British Museum Publications, London.

Robson, Eleanor, 1995. *Old Babylonian coefficient lists and the wider context of mathematics in ancient Mesopotamia 2100–1600 BC*. Dissertation, Oxford University.

Robson, Eleanor, 1999. *Mesopotamian Mathematics, 2100–1600 BC: Technical Constants in Bureaucracy and Education*, Clarenden Press, Oxford and Oxford University Press, New York.

Robson, Eleanor, 2001. "Neither Sherlock Holmes nor Babylon: a reassessment of Plimpton 322," *Historia Mathematica*, **28**, No. 3, 167–206.

Rosen, Frederic, 1831. *The Algebra of Mohammed ben Musa*, Oriental Translation Fund, London.

Rota, Gian-Carlo, 1989. "Fine Hall in its golden age: remembrances of Princeton in the early fifties," in: Duren, *1989*, Vol. 3, pp. 223–236.

Rowe, David, 1997. "Research schools in the United States," *Istoriko-Matematicheskie Issledovaniya*, special issue, 103–127 (Russian).

Russell, Alex Jamieson, 1879. *On Champlain's Astrolabe, Lost on the 7th June, 1613, and Found in August, 1867*, Burland-Desbarats, Montreal.

Russell, Bertrand, 1945. *A History of Western Philosophy*, Simon and Schuster, New York.

Sabra, A. I., 1969. "Simplicius's proof of Euclid's parallel postulate," *Journal of the Warburg and Courtauld Institute*, **32**, 1–24.

Sabra, A. I., 1998. "One ibn al-Haytham or two? An exercise in reading the bio-bibliographical sources," *Zeitschrift für Geschichte der Arabisch-Islamischen Wissenschaft*, **12**, 1–50.

Sanderson, Marie, 1974. "Mary Somerville: her work in physical geography," *Geographical Review*, **64**, No. 3, 410–420.

Sarkor, Ramatosh, 1982. "The Bakhshali Manuscript," *Ganita-Bharati* (Bulletin of the Indian Society for the History of Mathematics), **4**, No. 1–2, 50–55.

Schafer, Alice T., 1991. "Mathematics and women: perspectives and progress," *Notices of the American Mathematical Society*, **38**, No. 7, 735–737.

Scharlau, W., 1986. *Rudolf Lipschitz, Briefwechsel mit Cantor, Dedekind, Helmholtz, Kronecker, Weierstraß*, Vieweg, Deutsche Mathematiker-Vereinigung, Braunschweig–Wiesbaden.

Scharlau, Winfried; Opolka, Hans, 1985. *From Fermat to Minkowski: Lectures on the Theory of Numbers and Its Historical Development*, Springer-Verlag, New York.

Servos, John W., 1986. "Mathematics and the physical sciences in America, 1880–1930," *Isis*, **77**, 611–629.

Sesiano, Jacques, 1982. *Books IV to VII of Diophantus'* Arithmetica *in the Arabic Translation Attributed to Qusta ibn Luqa*, Springer-Verlag, New York.

Shanker, Stuart, 1993. "Turing and the origins of AI," *Proceedings of the 19th Annual Meeting of the CSHPM/SCHPM*, Carleton University, Ottawa, Ontario, pp. 1–36.

Sharer, Robert J., 1994. *The Ancient Maya*, Stanford University Press, Pasadena, CA.

Shen Kangshen, 1988. "Mutual-subtraction algorithm and its applications in ancient China," *Historia Mathematica*, **15**, 135–147.

Siegmund-Schultze, Reinhard, 1988. *Ausgewählte Kapitel aus der Funktionenlehre*, Teubner, Leipzig.

Siegmund-Schultze, Reinhard, 1997. "The emancipation of mathematical research publication in the United States from German dominance (1878–1945)," *Historia Mathematica*, **24**, 135–166.

Sigler, L. E., transl., 1987. *Leonardo Pisano Fibonacci: The Book of Squares*, Academic Press, New York.

Simonov, R. A., 1999. "Recent research on methods of rationalizing the computation of the Slavic Easter calculators (from manuscripts of the 14th through 17th centuries)," *Istoriko-Matematicheskie Issledovaniya*, **3 (38)**, 11–31 (Russian).

Singh, Parmanand, 1985. "The so-called Fibonacci numbers in ancient and medieval India," *Historia Mathematica*, **12**, No. 3, 229–244.

Skinner, B. F., 1948. "'Superstition' in the pigeon," *Journal of Experimental Psychology*, **38**, No. 1 (February), 168–172.

Smith, David Eugene, 1929. *A Source Book in Mathematics*, 2 vols. Reprint: Dover, New York, 1959.

Smith, David Eugene; Ginsburg, Jekuthiel, 1934. *A History of Mathematics in America before 1900*, Mathematical Association of America/Open Court, Chicago (Carus Mathematical Monograph 5).

Smith, David Eugene; Ginsburg, Jekuthiel, 1937. *Numbers and Numerals*, National Council of Teachers of Mathematics, Washington, DC.

Smith, David Eugene; Latham, Marcia L., transl., 1954. *The Geometry of René Descartes*. Reprint: Dover, New York.

Smith, David Eugene; Mikami Yoshio, 1914. *A History of Japanese Mathematics*, Open Court, Chicago.

Solomon, Ron, 1995. "On finite simple groups and their classification," *Notices of the American Mathematical Society*, **42**, No. 2, 231–239.

Spicci, Joan, 2002. *Beyond the Limit: The Dream of Sofya Kovalevskaya*, Tom Doherty Associates, New York.

Srinivasiengar, C. N., 1967. *The History of Ancient Indian Mathematics*, World Press Private, Calcutta.

Stanley, Autumn, 1992. "The champion of women inventors," *American Heritage of Invention and Technology*, **8**, No. 1, 22–26.

Stevin, Simon, 1585. *De Thiende*. German translation by Helmuth Gericke and Kurt Vogel. Akademische Verlag, Frankfurt am Main, 1965.

Strauss, Walter, 1977. *Albrecht Dürer: The Painter's Manual*, Abaris Books, New York.

Struik, D. J., 1933. "Outline of a history of differential geometry," *Isis*, **19**, 92–121, **20**, 161–192.

Struik, D. J., ed., 1986. *A Source Book in Mathematics, 1200–1800*, Princeton University Press, Princeton, NJ.

Stubhaug, Arild, 2000. *Niels Henrik Abel and his Times: Called Too Soon by Flames Afar*, Springer-Verlag, New York.

Stubhaug, Arild, 2002. *The Mathematician Sophus Lie: It Was the Audacity of my Thinking*, translated from the Norwegian by Richard Daly, Springer-Verlag, Berlin.

Swetz, Frank, 1977. *Was Pythagoras Chinese? An Examination of Right Triangle Theory in Ancient China*, The Pennsylvania State University Studies, No. 40. The Pennsylvania State University Press, University Park, PA, and National Council of Teachers of Mathematics, Reston, VA.

Tattersall, J. J., 1991. "Women and mathematics at Cambridge in the late nineteenth century," *Proceedings of the 17th Annual Meeting of the CSHPM/SCHPM*, Queen's University, Kingston, Ontario, pp. 53–66.

Tee, Garry J. 1988. "Mathematics in the Pacific basin," *British Journal of the History of Science*, **21**, 401–417.

Tee, Garry J. 1999. "The first 25 years of the New Zealand Mathematical Society," *New Zealand Mathematical Society Newsletter*, No. 76, 30–35.

Thomas, Ivor, 1939. *Selections Illustrating the History of Greek Mathematics*, Vol. 1, *Thales to Euclid*, Harvard University Press, Cambridge, MA.

Thomas, Ivor, 1941. *Selections Illustrating the History of Greek Mathematics*, Vol. 2, *Aristarchus to Pappus*, Harvard University Press, Cambridge, MA.

Todhunter, Isaac, 1861. *A History of the Calculus of Variations in the Nineteenth Century*. Reprint: Chelsea, New York, 1962.

Todhunter, Isaac, 1865. *A History of the Mathematical Theory of Probability from the Time of Pascal to that of Laplace*. Reprint: Chelsea, New York, 1949.

Toomer, G. J., 1984. "Lost Greek mathematical works in Arabic translation," *The Mathematical Intelligencer*, **4**, No. 2, 32–38.

Tropfke, 1902. *Geschichte der Elementarmatematik*, 4. Auflage, completely revised by Kurt Vogel, Karin Reich, and Helmuth Gericke. Band 1: *Arithmetik und Algebra*. Walter de Gruyter, Berlin and New York, 1980.

Tsaban, Boaz; Garber, David, 1998. "On the rabbinical approximation of π," *Historia Mathematica*, **25**, No. 1, 75–84.

Urton, Gary, 1997. *The Social Life of Numbers: A Quechua Ontology of Numbers and Philosophy of Arithmetic*, University of Texas Press, Austin, TX.

de la Vallée Poussin, Charles, 1896. "Recherches analytiques sur la théorie des nombres (première partie)," *Annales de la Société des Sciences de Bruxelles*, **I**, 20_2, 183–256.

de la Vallée Poussin, Charles, 1899. "Sur la fonction $\zeta(s)$ de Riemann et le nombre des nombres premiers inférieurs à une limite donnée," *Mémoires couronnés et autres mémoires publiés par l'Académie Royale des Sciences, des Lettres, et des Beaux Arts de la Belgique*, **59**.

Varadarajan, V. S., 1983. "Mathematics in and out of Indian universities," *The Mathematical Intelligencer*, **5**, No. 1, 38–42.

Vidyabhusana, Satis Chandra, 1971. *A History of Indian Logic (Ancient, Mediaeval, and Modern Schools)*, Motilal Banarsidass, Delhi.

Volkov, Alexei, 1997. "Zhao Youqin and his calculation of π," *Historia Mathematica*, **24**, No. 3, 301–331.

van der Waerden, B. L., 1963. *Science Awakening*, Wiley, New York.

van der Waerden, B. L., 1975. "On the sources of my book *Moderne Algebra*," *Historia Mathematica*, **2**, 31–40.

van der Waerden, B. L. 1985. *A History of Algebra from al-Khwarizmi to Emmy Noether*, Springer-Verlag, Berlin.

Wald, Robert M., 1984. *General Relativity*, University of Chicago Press.

Walford, E., transl., 1853. *The Ecclesiastical History of Socrates, Surnamed Scholasticus. A History of the Church in Seven Books*, H. Bohn, London.

Wantzel, Laurent, 1845. "Démonstration de l'impossibilité de résoudre toutes les équations algébriques avec des radicaux," *Bulletin des sciences, par la Société Philomathique de Paris*, 5–7.

Waring, E. 1762. *Miscellanea analytica*, Oxford.

Waterhouse, William C., 1994. "A counterexample for Germain," *The American Mathematical Monthly*, **101**, No. 2, 140–150.

Weil, André, 1984. *Number Theory: An Approach Through History from Hammurapi to Legendre*, Birkhäuser, Boston.

Whiteside, T. L., 1967. *The Mathematical Papers of Isaac Newton*, Johnson Reprint Corporation, London.

Witmer, T. Richard, transl., 1968. *The Great Art, or The Rules of Algebra, by Girolamo Cardano*, MIT Press, Cambridge, MA.

Woepcke, Franz, 1852. "Notice sur une théorie ajoutée par Thábit ben Korrah à l'arithmétique spéculative des Grècs," *Journal asiatique*, **4**, No. 20, 420–429.

Woodhouse, Robert, 1810. *A History of the Calculus of Variations in the Eighteenth Century*. Reprint: Chelsea, New York, 1964.

Wright, F. A., 1933. *Select Letters of St. Jerome*, G. P. Putnam's Sons, New York.

Yoshida Kôsaku, 1980. "Mathematical works of Takakazu Seki," *The Mathematical Intelligencer*, **3**, No. 3: Reprint of material written for the centennial of the Japan Academy.

de Young, Gregg, 1995. "Euclidean geometry in the tradition of Islamic India," *Historia Mathematica*, **22**, No. 2, 138–153.

Zaitsev, E. A., 1999. "The meaning of early medieval geometry: from Euclid and surveyors' manuals to Christian philosophy," *Isis*, **90**, 522–553.

Zaitsev, E. A., 2000. "The Latin versions of Euclid's *Elements* and the hermeneutics of the twelfth century," *Istoriko-Matematicheskie Issledovaniya*, **5 (40)**, 222–232 (Russian).

Zaslavsky, Claudia, 1999. *Africa Counts: Number and Pattern in African Cultures*, Lawrence Hill Books, Chicago.

Zeuthen, H. G., 1903. *Geschichte der Mathematik im 16. und 17. Jahrhundert*, Teubner, Stuttgart. Johnson Reprint Corporation, New York, 1966.

Zharov, V. K., 2001. "On the 'Introduction' to the treatise *Suan Shu Chimeng* of Zhu Shijie," *Istoriko-Matematicheskie Issledovaniya*, 2nd Ser., No. 6 (41), 347–353 (Russian).

Zhmud, Leonid, 1989. "Pythagoras as a mathematician," *Historia Mathematica*, **16**, No. 3, 249–268.

Zverkina, G. A., 2000. "The Euclidean algorithm as a computational procedure in ancient mathematics," *Istoriko-Matematicheskie Issledovaniya*, **5 (40)**, 232–243 (Russian).

Subject Index

Name Index

Differential Diagnosis *for* Physical Therapists

SCREENING FOR REFERRAL

Differential Diagnosis *for* Physical Therapists

SCREENING FOR REFERRAL

Catherine Cavallaro Goodman, MBA, PT, CBP
Medical Multimedia Group
Faculty Affiliate
University of Montana
Missoula, Montana

Teresa E. Kelly Snyder, MN, RN, OCN
Oncology Treatment Area
Montana Cancer Specialists
Missoula, Montana

3251 Riverport Lane
St. Louis, Missouri 63043

DIFFERENTIAL DIAGNOSIS FOR PHYSICAL THERAPISTS:
Screening for Referral ISBN: 978-1-4377-2543-8

Notices

Knowledge and best practice in this field are constantly changing. As new research and experience broaden our understanding, changes in research methods, professional practices, or medical treatment may become necessary.

Practitioners and researchers must always rely on their own experience and knowledge in evaluating and using any information, methods, compounds, or experiments described herein. In using such information or methods they should be mindful of their own safety and the safety of others, including parties for whom they have a professional responsibility.

With respect to any drug or pharmaceutical products identified, readers are advised to check the most current information provided (i) on procedures featured or (ii) by the manufacturer of each product to be administered, to verify the recommended dose or formula, the method and duration of administration, and contraindications. It is the responsibility of practitioners, relying on their own experience and knowledge of their patients, to make diagnoses, to determine dosages and the best treatment for each individual patient, and to take all appropriate safety precautions.

To the fullest extent of the law, neither the Publisher nor the authors, contributors, or editors, assume any liability for any injury and/or damage to persons or property as a matter of products liability, negligence or otherwise, or from any use or operation of any methods, products, instructions, or ideas contained in the material herein.

International Standard Book Number: 978-1-4377-2543-8

Vice President: Linda Duncan
Executive Content Strategist: Kathy Falk
Senior Content Development Specialist: Christie M. Hart
Publishing Services Manager: Catherine Jackson
Senior Project Manager: David Stein
Design Direction: Jessica Williams

Printed in the United States

Last digit is the print number: 9 8 7 6 5 4 3 2 1

To Teresa...this edition is dedicated to you for all the years you gave to the instruction of physicians, nurses, and physical therapists. Thanks for taking all those late night calls, searching for references, writing and rewriting text...adjusting from the yellow legal pad to the electronic age...and offering your expertise and wise counsel through four editions of this text. This one's for you.

C.C.G.

FOREWORD TO THE FOURTH EDITION

Catherine Goodman and Teresa Snyder are to be commended for making several important contributions to the role of physical therapists as diagnosticians with this revision of their classic text. The first step in the diagnostic process is to determine if the patient's condition necessitates a referral to a medical doctor. Therefore this book is an invaluable guide because the authors have provided a model that is focused and complete. Although the focus of the text is on identifying the most common conditions that mimic musculoskeletal problems, Goodman and Snyder also note that this is just the first step in the diagnostic process and have made suggestions for future directions. Thus the authors are providing a timely guide to practice and professional development by addressing the issue of terminology associated with diagnosis.

As physical therapy seeks to clarify its professional responsibilities by providing education at the clinical doctoral level, emphasizing diagnostic skills, and providing direct access care, a necessary component is accuracy in communicating these responsibilities. For many years, the issue of appropriate terminology and/or the context in which it is used with regard to diagnosis in physical therapy has been one of confusion. The scope of the confusion is reflected in a variety of editorials,[1-8] textbooks, and advertisements that are inconsistent in their use of differential diagnosis.

Goodman and Snyder have provided a model for approaching this confusion. Appropriately, this book's title, *Differential Diagnosis for Physical Therapists: Screening for Referral*, clarifies that a primary responsibility of the physical therapist is to recognize the possible presence of a medical condition that supersedes or mimics a condition requiring physical therapy treatment. Clarification that differential diagnosis does not mean identifying the specific disease is important in our relationship with physicians and in maintaining our legal scope of practice, as physical therapists assume a larger role in direct access and primary care.[1]

As stated in this text, the first step in the diagnostic process is for the physical therapist to be able to identify medical conditions that are to be referred to the appropriate practitioner. Clearly this is a skill that any physical therapist must be able to demonstrate. Not only does this book provide the necessary information, but also the manner in which the material is presented should enable every reader to achieve a high level of skill. This book is intended to augment both the reader's skill in screening for medical conditions and also his or her skill in navigating the entire diagnostic process. The highly consumer-friendly and engaging format of this book is among the many reasons every student and clinician should include the book in their personal library.

But as Catherine Goodman and Teresa Snyder have so wisely stated in the preface, the primary focus of this book is just the first step in an evaluation that must ultimately lead to a diagnosis that directs physical therapy intervention. To their credit they have also provided an introduction to the next steps in the complete diagnostic process. In keeping with the *Guide to Physical Therapist Practice*, Goodman and Snyder have addressed the importance of the concept of the movement system to physical therapy and thus to another level of differential diagnosis. They have directed our attention to a developing system of diagnoses of movement system impairments. This system requires differentiating among movement system impairment conditions at both the tissue and the movement level and then using this information to establish a diagnosis that directs physical therapy treatment.

In addition to providing information for physical therapists, Goodman and Snyder have also attempted to assist other health professionals in identifying which conditions should be referred to a physical therapist. This effort is another reflection of their prescient recognition of the direction of practice. The examination and diagnostic skills of the physical therapist, whether for ruling out or identifying a medical condition or cogently labeling a movement impairment syndrome, must become the most highly visible aspects of the profession's role in health care.

Historically the profession has mainly been considered one in which the practitioner provided treatment based on the physician's diagnosis. Evaluation, examination, diagnosis, and program planning whether sought by a client, a physician, or another health professional is the necessary direction for the profession if we are to assume our role in health promotion, maintenance, and/or remediation. Exercise, which is the prevailing form of physical therapy treatment, continues to receive increased attention as the most effective form of preventive and restorative care for life-style–induced diseases. Yet physical therapists are not readily consulted for their expertise in developing programs that cannot only address life-style–induced diseases but that can also prevent inducing musculoskeletal problems.

An important goal of the profession is to promote recognition that we are the health profession with the expertise to appropriately screen, diagnose, and then develop treatment programs that are safe and effective for individuals with all levels of movement system dysfunction. We are indebted to Catherine Goodman and Teresa Snyder for their contributions to enabling us to achieve this goal.

Shirley Sahrmann, PT, PhD, FAPTA
Professor Physical Therapy, Neurology, Cell Biology & Physiology
Washington University School of Medicine—St. Louis, MO

REFERENCES

1. Boissonnault W, Goodman C: Physical therapists as diagnosticians: drawing the line on diagnosing pathology. *J Orthop Sports Phys Ther* 36(6):351–353, 2006.
2. Davenport TE, Kulig K, Resnick C: Diagnosing pathology to decide the appropriateness of physical therapy: what's our role? *J Orthop Sports Phys Ther* 36(1):1–2, 2006.
3. Guccione AA: Physical therapy diagnosis and the relationship between impairments and function. *Phys Ther* 71(7):499–503, 1991.
4. Jette AM: Diagnosis and classification by physical therapists: a special communication. *Phys Ther* 69(11):967–969, 1989.
5. Rose SJ: Physical therapy diagnosis: role and function. *Phys Ther* 69(7):535–537, 1989.
6. Sahrmann SA: Diagnosis by the physical therapist—a prerequisite for treatment: a special communication. *Phys Ther* 68(11): 1703–1706, 1988.
7. Sahrmann SA: Are physical therapists fulfilling their responsibilities as diagnosticians? *J Orthop Sports Phys Ther* 35(9):556–558, 2005.
8. Zimny NJ: Diagnostic classification and orthopaedic physical therapy practice: what can we learn from medicine? *J Orthop Sports Phys Ther* 34(3):105–111, 2004.

PREFACE

If you have ever looked in this book hoping for a way to figure out just what is wrong with your client's back or neck or shoulder but did not find the answer, then you understand the need for a title to clarify just what is in here.

The updated name for the 4th edition, *Differential Diagnosis for Physical Therapists: Screening for Referral,* did not reflect a change in the content of the text so much as it reflected a better understanding of the screening process as the first step in making a diagnosis. Before implementing a plan of care the therapist must confirm (or rule out) the need for physical therapy intervention. We must ask and answer these questions:

- Is this an appropriate physical therapy referral?
- Is there a problem that does not fall into one of the four categories of conditions outlined by the *Guide*?
- Are there any red flag histories, red flag risk factors, or cluster of red flag signs and/or symptoms?

This text provides students, physical therapist assistants, and physical therapy clinicians with a step-by-step approach to client evaluation that follows the standards for competency established by the American Physical Therapy Association (APTA) related to conducting a screening examination.

In fact, we present a screening model that can be used with each client. By following these steps—Past Medical History, Risk Factor Assessment, Clinical Presentation, Associated Signs and Symptoms, and Review of Systems—the therapist will avoid omitting any critical part of the screening process. With the physical therapy screening interview as a foundation for subjectively evaluating patients and clients, each organ system is reviewed with regard to the most common disorders encountered, particularly those that may mimic primary musculoskeletal or neuromuscular problems.

A cognitive processing-reasoning orientation is used throughout the text to encourage students to gather and analyze data, pose and solve problems, infer, hypothesize, and make clinical judgments. Many new case examples have been added. Case examples and case studies are used to integrate screening information and help the therapist make decisions about how and when to treat, refer, or treat AND refer.

The text is divided into three sections: Section I introduces the screening interview along with a new chapter on physical assessment for screening with many helpful photographs and illustrations. Another new chapter presents pain types and viscerogenic pain patterns. How and why the organs can refer pain to the musculoskeletal system is explained.

Section II presents a systems approach, looking at each organ system and the various diseases, illnesses, and conditions that can refer symptoms to the neuromuscular or musculoskeletal system. Red flag histories, risk factors, clinical presentation, and signs and symptoms are reviewed for each system. As in previous editions, helpful screening clues and guidelines for referral are included in each chapter.

In the third and final section, the last chapter in the previous editions has been expanded into five separate chapters. An individual screening focus is presented based on the various body parts from head to toe.

As always, while screening for medical disease, side effects of medications, or other unrecognized comorbidities, the therapist must still conduct a movement exam to identify the true cause of the pain or symptom(s) should there be a primary neuromuscular or primary musculoskeletal problem. And there are times when therapists are treating patients and/or clients with a movement system impairment who also report signs and symptoms associated with a systemic disease or illness. For many conditions, early detection and referral can reduce morbidity and mortality.

The goal of this text is to provide the therapist (both students and clinicians) with a consistent way to screen for systemic diseases and medical conditions that can mimic neuromusculoskeletal problems. It is not our intent to teach physical therapists how to diagnose pathology or medical conditions. However, we recognize the need to consider possible pathologic conditions in order to screen effectively and evaluate the need for referral or consult.

Catherine Cavallaro Goodman, MBA, PT, CBP
Teresa E. Kelly Snyder, MN, RN, OCN

ACKNOWLEDGMENTS

We never imagined our little book would ever go beyond a first edition. The first edition was a direct result of our experience in the military as nurse (Teresa) and physical therapist (Catherine), although we did not know each other at that time. So to the many men and women of the United States Armed Forces who have worked as independent practitioners and fine-tuned this material, we say thank you.

In addition, special thanks go to the many fine folks (past and present) at Elsevier Science:

Andrew Allen
Louise Beirig
Julie Burchett
Amy Buxton
Linda Duncan
Kathy Falk
Christie M. Hart
Sue Hontscharik
Kathy Macciocca
Jacqui Merrill
R.F. Schneider, Permissions Dept.
David Stein
Marion Waldman

Unnamed but appreciated copy editors, production staff, marketing personnel, sales representatives, editorial assistants, and many more we don't even know about! Please consider yourselves appreciated and thanked.

To all the others as well:

M.D. Bang
Maj. Richard E. Baxter
Theresa Bernsen (in memoriam)
Nancy Bloom
Bill Boissonnault
Chuck Ciccone
Nancy Ciesla
Carla Cleary
Jeff Damaschke
Gail Deyle
Brent Dodge
Jacquie Drouin
Ryan L. Elliott
Kenda Fuller
J. Gabbard
Brant Goode
Cliff Goodman
K. Grenne
Janet Hulme
Airelle Hunter-Giordan
Chelsea Jordan
Michael Keith, APTA Governance
Bonnie Lasinski
Allan Chong Lee
Leanne Lenker
Pam Little
Renee Mabey
Charles L. McGarvey, III
Brian Murphy
Barbara Norton
Dennis O'Connell
Lee Ann Odom
Phillip B. Palmer
Celeste Peterson
Cindy Pfalzer
Sue Queen
Daniel Rhon
M. Ross
Shirley Sahrmann
Saint Patrick's Hospital and Health Sciences Center, Center for Health Information (Dana Kopp, Ginny Bolten, and Lisa Autio)
Ken Saladin
Donald K. Shaw
MaryJane Strauhal
Jason Taitch (in memoriam)
Steve Tepper
Jody Tomasic
Peg and Doug Waltner
Valerie Wang
Mark Weber
Karen Wilson
University of Montana Physical Therapy: Reed Humphrey, Steve Fehrer, Dave Levison, Beth Ikeda, Alex Santos
University of Montana College of Health Professional Biomedical Sciences Drug Information Service: Tanner Higginbotham, Kimberly Swanson, Sherrill Brown, Nicole M. Marcellus

And to any other family member, friend, or colleague whose name should have been on this list but was inadvertently missed … a special hug of thanks.

Catherine Cavallaro Goodman, MBA, PT, CBP
Teresa E. Kelly Snyder, MN, RN, OCN

ENHANCE YOUR LEARNING and PRACTICE EXPERIENCE

The images below are QR (Quick Response) codes. Each code corresponds to one of the appendices or reference lists at the end of each chapter. Appendices can be accessed on your mobile device for quick reference in a lab or clinical setting. References are linked to the Medline abstract where available!

For fast and easy access, right from your mobile device, follow these instructions. You can also find these documents at:
www.DifferentialDiagnosisforPT.com

What you need:

- A mobile device, such as a smartphone or tablet, equipped with a camera and Internet access
- A QR code reader application (if you do not already have a reader installed on your mobile device, look for free versions in your app store.)

How it works:

- Open the QR code reader application on your mobile device.
- Point the device's camera at the code and scan.
- Each code opens an individual URL for instant viewing of the appendices and the references where you can further access the Medline links—no log-on required.

APPENDIX A: SCREENING SUMMARY

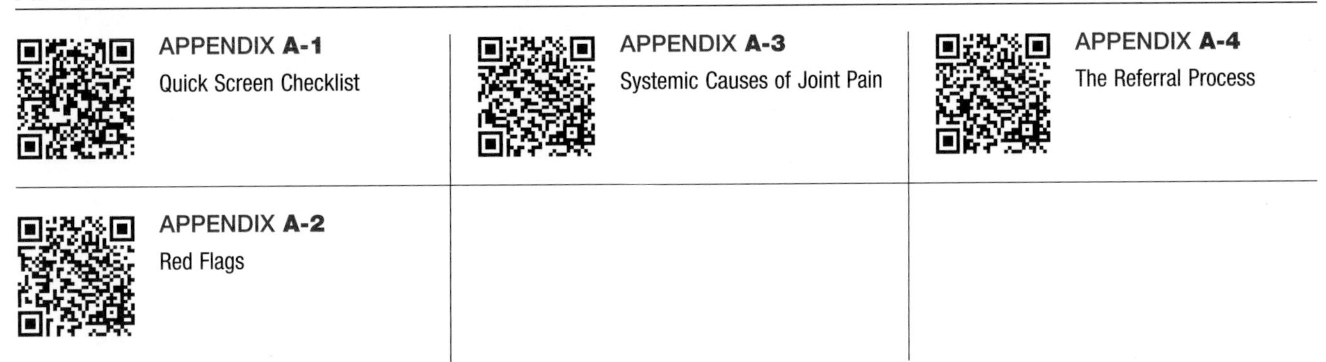

APPENDIX **A-1**
Quick Screen Checklist

APPENDIX **A-2**
Red Flags

APPENDIX **A-3**
Systemic Causes of Joint Pain

APPENDIX **A-4**
The Referral Process

APPENDIX B: SPECIAL QUESTIONS TO ASK (Screening for)

APPENDIX **B-1**
Alcohol Abuse: AUDIT Questionnaire

APPENDIX **B-2**
Alcohol Abuse: CAGE Questionnaire

APPENDIX **B-3**
Assault, Intimate Partner Abuse, or Domestic Violence

APPENDIX **B-4**
Bilateral Carpal Tunnel Syndrome

APPENDIX **B-5**
Bladder Function

APPENDIX **B-6**
Bowel Function

APPENDIX **B-7**
Breast

APPENDIX **B-8**
Chest/Thorax

APPENDIX **B-9**
Depression/Anxiety (see also Appendix B-10)

APPENDIX **B-10**
Depression in Older Adults

APPENDIX **B-11**
Dizziness

APPENDIX **B-12**
Dyspnea (Shortness of Breath [SOB]; Dyspnea on Exertion [DOE])

(continued next page)

APPENDIX B: SPECIAL QUESTIONS TO ASK (Screening for) (cont.)

APPENDIX **B-13A**
Eating Disorders

APPENDIX **B-13B**
Resources for Screening for Eating Disorders

APPENDIX **B-14**
Environmental and Work History

APPENDIX **B-15**
Fibromyalgia Syndrome (FMS)

APPENDIX **B-16**
Gastrointestinal (GI) Problems

APPENDIX **B-17**
Headaches

APPENDIX **B-18**
Joint Pain

APPENDIX **B-19**
Kidney and Urinary Tract Impairment

APPENDIX **B-20**
Liver (Hepatic) Impairment

APPENDIX **B-21**
Lumps (Soft Tissue) or Skin Lesions

APPENDIX **B-22**
Lymph Nodes

APPENDIX **B-23**
Medications

APPENDIX **B-24**
Men Experiencing Back, Hip, Pelvic, Groin, or Sacroiliac Pain

APPENDIX **B-25**
Night Pain

APPENDIX **B-26**
Nonsteroidal Antiinflammatories (NSAIDs) (Side Effects of)

APPENDIX **B-27**
Odors (Unusual)

APPENDIX **B-28**
Pain

APPENDIX **B-29**
Palpitations (Chest or Heart)

APPENDIX **B-30**
Prostate Problems

APPENDIX **B-31**
Psychogenic Source of Symptoms

APPENDIX **B-32A**
Taking a Sexual History

APPENDIX **B-32B**
Taking a Sexual History

APPENDIX **B-33**
Sexually Transmitted Diseases

APPENDIX **B-34**
Shoulder and Upper Extremity

APPENDIX **B-35**
Sleep Patterns

APPENDIX **B-36**
Substance Use/Abuse

See APPENDIX **B-19**
Urinary Tract Impairment (see Appendix B-19)

APPENDIX **B-37**
Women Experiencing Back, Hip, Pelvic, Groin, Sacroiliac (SI), or Sacral Pain

APPENDIX C: SPECIAL FORMS TO USE

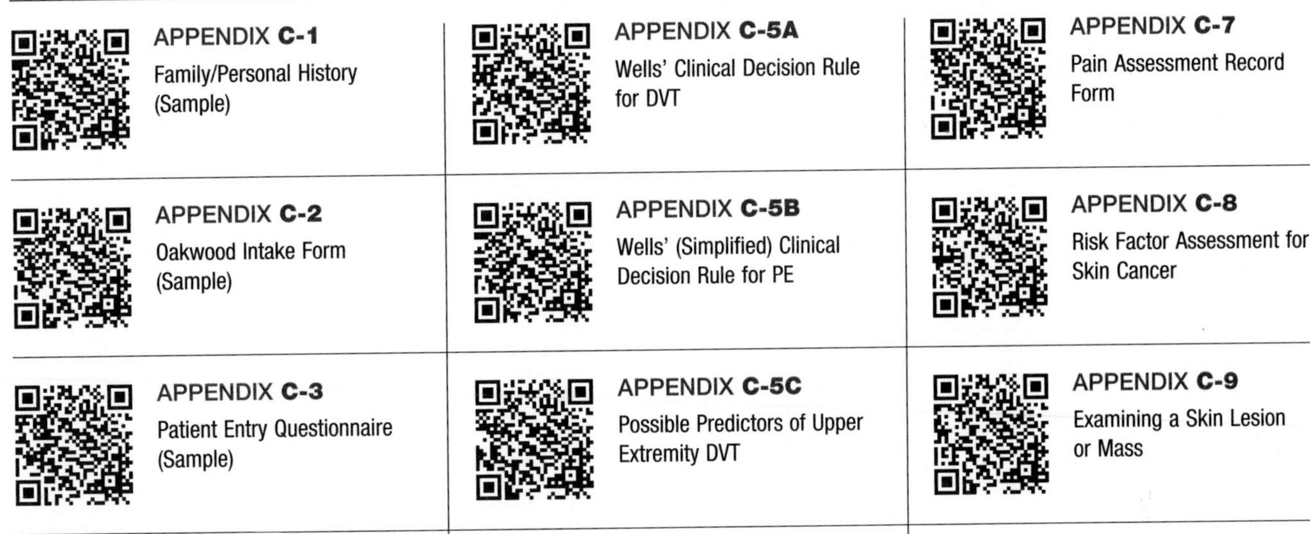

APPENDIX **C-1**
Family/Personal History (Sample)

APPENDIX **C-2**
Oakwood Intake Form (Sample)

APPENDIX **C-3**
Patient Entry Questionnaire (Sample)

APPENDIX **C-4**
Bogduk's Checklist for Red/Yellow Flags (Low Back Pain)

APPENDIX **C-5A**
Wells' Clinical Decision Rule for DVT

APPENDIX **C-5B**
Wells' (Simplified) Clinical Decision Rule for PE

APPENDIX **C-5C**
Possible Predictors of Upper Extremity DVT

APPENDIX **C-6**
Osteoporosis Screening Evaluation

APPENDIX **C-7**
Pain Assessment Record Form

APPENDIX **C-8**
Risk Factor Assessment for Skin Cancer

APPENDIX **C-9**
Examining a Skin Lesion or Mass

APPENDIX **C-10**
Breast and Lymph Node Examination

APPENDIX D: SPECIAL TESTS TO PERFORM

APPENDIX **D-1**
Guide to Physical Assessment in a Screening Examination

APPENDIX **D-2**
Extremity Examination Checklist

APPENDIX **D-3**
Hand and Nail Bed Assessment

APPENDIX **D-4**
Peripheral Vascular Assessment

APPENDIX **D-5**
Review of Systems

APPENDIX **D-6**
Breast Self-Examination (BSE)

APPENDIX **D-7**
Clinical Breast Examination: Recommended Procedures

APPENDIX **D-8**
Testicular Self-Examination

(continued next page)

REFERENCES

CHAPTER **1**	CHAPTER **7**	CHAPTER **13**
CHAPTER **2**	CHAPTER **8**	CHAPTER **14**
CHAPTER **3**	CHAPTER **9**	CHAPTER **15**
CHAPTER **4**	CHAPTER **10**	CHAPTER **16**
CHAPTER **5**	CHAPTER **11**	CHAPTER **17**
CHAPTER **6**	CHAPTER **12**	CHAPTER **18**

CONTENTS

SECTION I INTRODUCTION TO THE SCREENING PROCESS

SECTION II VISCEROGENIC CAUSES OF NEUROMUSCULOSKELETAL PAIN AND DYSFUNCTION

SECTION III SYSTEMIC ORIGINS OF NEUROMUSCULAR OR MUSCULOSKELETAL PAIN AND DYSFUNCTION

Differential Diagnosis *for* Physical Therapists

SCREENING FOR REFERRAL

CHAPTER 1

Introduction to Screening for Referral in Physical Therapy

It is the therapist's responsibility to make sure that each patient/client is an appropriate candidate for physical therapy. In order to be as cost-effective as possible, we must determine what biomechanical or neuromusculoskeletal problem is present and then treat the problem as specifically as possible.

As part of this process, the therapist may need to screen for medical disease. Physical therapists must be able to identify signs and symptoms of systemic disease that can mimic neuromuscular or musculoskeletal (herein referred to as neuromusculoskeletal or NMS) dysfunction. Peptic ulcers, gallbladder disease, liver disease, and myocardial ischemia are only a few examples of systemic diseases that can cause shoulder or back pain. Other diseases can present as primary neck, upper back, hip, sacroiliac, or low back pain and/or symptoms.

Cancer screening is a major part of the overall screening process. Cancer can present as primary neck, shoulder, chest, upper back, hip, groin, pelvic, sacroiliac, or low back pain/symptoms. Whether there is a primary cancer or cancer that has recurred or metastasized, clinical manifestations can mimic NMS dysfunction. The therapist must know how and what to look for to screen for cancer.

The purpose and the scope of this text are not to teach therapists to be medical diagnosticians. The purpose of this text is twofold. The first is to help therapists recognize areas that are beyond the scope of a physical therapist's practice or expertise. The second is to provide a step-by-step method for therapists to identify clients who need a medical (or other) referral or consultation.

As more states move toward direct access and advanced scope of practice, physical therapists are increasingly becoming the practitioner of choice and thereby the first contact that patient/clients seek,* particularly for care of musculoskeletal dysfunction. This makes it critical for physical therapists to be well versed in determining when and how referral to a physician (or other appropriate health care professional) is necessary. Each individual case must be reviewed carefully.

Even without direct access, screening is an essential skill because any client can present with red flags requiring reevaluation by a medical specialist. The methods and clinical decision-making model for screening presented in this text remain the same with or without direct access and in all practice settings.

EVIDENCE-BASED PRACTICE

Clinical decisions must be based on the best evidence available. The clinical basis for diagnosis, prognosis, and intervention must come from a valid and reliable body of evidence referred to as evidence-based practice. Each therapist must develop the skills necessary to assimilate, evaluate, and make the best use of evidence when screening patient/clients for medical disease.

*The *Guide to Physical Therapist Practice*[1] defines *patients* as "individuals who are the recipients of physical therapy care and direct intervention" and *clients* as "individuals who are not necessarily sick or injured but who can benefit from a physical therapist's consultation, professional advice, or prevention services." In this introductory chapter, the term *patient/client* is used in accordance with the patient/client management model as presented in the *Guide*. In all other chapters, the term "client" is used except when referring to hospital inpatient/clients or outpatient/clients.

Every effort has been made to sift through all the pertinent literature, but it remains up to the reader to keep up with peer-reviewed literature reporting on the likelihood ratios, predictive values, reliability, sensitivity, specificity, and validity of yellow (cautionary) and red (warning) flags and the confidence level/predictive value behind screening questions and tests. Each therapist will want to build his or her own screening tools based on the type of practice he or she is engaged in by using best evidence screening strategies available. These strategies are rapidly changing and will require careful attention to current patient-centered peer-reviewed research/literature.

Evidence-based clinical decision making consistent with the patient/client management model as presented in the *Guide to Physical Therapist Practice*[1] will be the foundation upon which a physical therapist's differential diagnosis is made. Screening for systemic disease or viscerogenic causes of NMS symptoms begins with a well-developed client history and interview.

The foundation for these skills is presented in Chapter 2. In addition, the therapist will rely heavily on clinical presentation and the presence of any associated signs and symptoms to alert him or her to the need for more specific screening questions and tests.

Under evidence-based medicine, relying on a red-flag checklist based on the history has proved to be a very safe way to avoid missing the presence of serious disorders. Efforts are being made to validate red flags currently in use (see further discussion in Chapter 2). When serious conditions have been missed, it is not for lack of special investigations but for lack of adequate and thorough attention to clues in the history.[2,3]

Some conditions will be missed even with screening because the condition is early in its presentation and has not progressed enough to be recognizable. In some cases, early recognition makes no difference to the outcome, either because nothing can be done to prevent progression of the condition or there is no adequate treatment available.[2]

STATISTICS

How often does it happen that a systemic or viscerogenic problem masquerades as a neuromuscular or musculoskeletal problem? There are very limited statistics to quantify how often organic disease masquerades or presents as NMS problems. Osteopathic physicians suggest this happens in approximately 1% of cases seen by physical therapists, but little data exist to confirm this estimate.[4,5] At the present time, the screening concept remains a consensus-based approach patterned after the traditional medical model and research derived from military medicine (primarily case studies).

Efforts are underway to develop a physical therapists' national database to collect patient/client data that can assist us in this effort. Again, until reliable data are available, it is up to each of us to look for evidence in peer-reviewed journals to guide us in this process.

Personal experience suggests the 1% figure would be higher if therapists were screening routinely. In support of this hypothesis, a systematic review of 64 cases involving physical therapist referral to physicians with subsequent diagnosis of a medical condition showed that 20% of referrals were for other concerns.[6] Physical therapists involved in the cases were routinely performing screening examinations, regardless of whether or not the client was initially referred to the physical therapist by a physician.

These results demonstrate the importance of therapists screening beyond the chief presenting complaint (i.e., for this group the red flags were not related to the reason physical therapy was started). For example, one client came with diagnosis of cervical stenosis. She did have neck problems, but the therapist also observed an atypical skin lesion during the postural exam and subsequently made the referral.[6]

KEY FACTORS TO CONSIDER

Three key factors that create a need for screening are:

- Side effects of medications
- Comorbidities
- Visceral pain mechanisms

If the medical diagnosis is delayed, then the correct diagnosis is eventually made when

1. The patient/client does not get better with physical therapy intervention.
2. The patient/client gets better then worse.
3. Other associated signs and symptoms eventually develop.

There are times when a patient/client with NMS complaints is really experiencing the side effects of medications. In fact, this is probably the most common source of associated signs and symptoms observed in the clinic. Side effects of medication as a cause of associated signs and symptoms, including joint and muscle pain, will be discussed more completely in Chapter 2. Visceral pain mechanisms are the entire subject of Chapter 3.

As for comorbidities, many patient/clients are affected by other conditions such as depression, diabetes, incontinence, obesity, chemical dependency, hypertension, osteoporosis, and deconditioning, to name just a few. These conditions can contribute to significant morbidity (and mortality) and must be documented as part of the problem list. Physical therapy intervention is often appropriate in affecting outcomes, and/or referral to a more appropriate health care or other professional may be needed.

Finally, consider the fact that some clients with a systemic or viscerogenic origin of NMS symptoms get better with physical therapy intervention. Perhaps there is a placebo effect. Perhaps there is a physiologic effect of movement on the diseased state. The therapist's intervention may exert an influence on the neuroendocrine-immune axis as the body tries to regain homeostasis. You may have experienced this phenomenon yourself when coming down with a cold or symptoms of a virus. You felt much better and even symptom-free after exercising.

Movement, physical activity, and moderate exercise aid the body and boost the immune system,[7-9] but sometimes such measures are unable to prevail, especially if other factors are

BOX 1-1 REASONS FOR SCREENING

- Direct access: Therapist has primary responsibility or first contact.
- Quicker and sicker patient/client base.
- Signed prescription: Clients may obtain a signed prescription for physical/occupational therapy based on similar past complaints of musculoskeletal symptoms without direct physician contact.
- Medical specialization: Medical specialists may fail to recognize underlying systemic disease.
- Disease progression: Early signs and symptoms are difficult to recognize, or symptoms may not be present at the time of medical examination.
- Patient/client disclosure: Client discloses information previously unknown or undisclosed to the physician.
- Client does not report symptoms or concerns to the physician because of forgetfulness, fear, or embarrassment.
- Presence of one or more yellow (caution) or red (warning) flags.

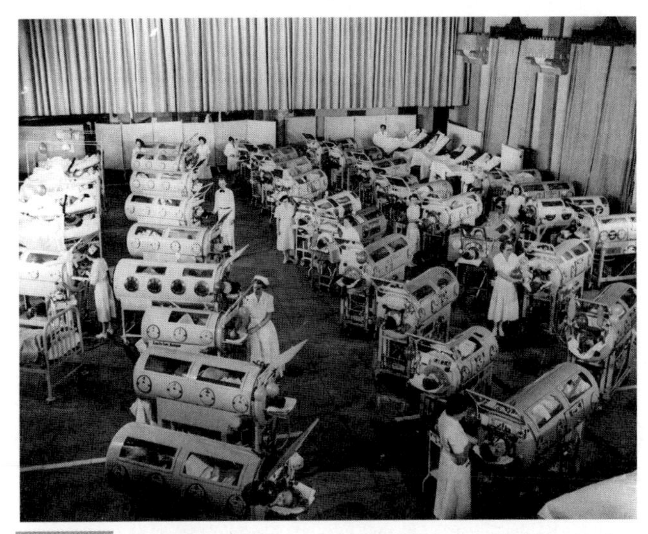

Fig. 1-1 Patients in iron lungs receive treatment at Rancho Los Amigos during the polio epidemic of the 1940s and 1950s. (Courtesy Rancho Los Amigos, 2005.)

present such as inadequate hydration, poor nutrition, fatigue, depression, immunosuppression, and stress. In such cases the condition will progress to the point that warning signs and symptoms will be observed or reported and/or the patient/client's condition will deteriorate. The need for medical referral or consultation will become much more evident.

REASONS TO SCREEN

There are many reasons why the therapist may need to screen for medical disease. Direct access (see definition and discussion later in this chapter) is only one of those reasons (Box 1-1).

Early detection and referral is the key to prevention of further significant comorbidities or complications. In all practice settings, therapists must know how to recognize systemic disease masquerading as NMS dysfunction. This includes practice by physician referral, practitioner of choice via the direct access model, or as a primary practitioner.

The practice of physical therapy has changed many times since it was first started with the Reconstruction Aides. Clinical practice, as it was shaped by World War I and then World War II, was eclipsed by the polio epidemic in the 1940s and 1950s. With the widespread use of the live, oral polio vaccine in 1963, polio was eradicated in the United States and clinical practice changed again (Fig. 1-1).

Today, most clients seen by therapists have impairments and disabilities that are clearly NMS-related (Fig. 1-2). Most of the time, the client history and mechanism of injury point to a known cause of movement dysfunction.

However, therapists practicing in all settings must be able to evaluate a patient/client's complaint knowledgeably and

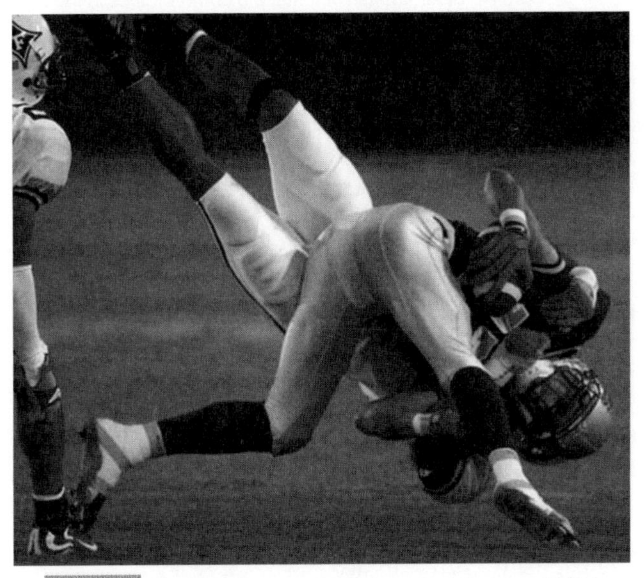

Fig. 1-2 (Courtesy Jim Baker, Missoula, Montana, 2005.)

determine whether there are signs and symptoms of a systemic disease or a medical condition that should be evaluated by a more appropriate health care provider. This text endeavors to provide the necessary information that will assist the therapist in making these decisions.

Quicker and Sicker

The aging of America has impacted general health in significant ways. "Quicker and sicker" is a term used to describe patient/clients in the current health care arena (Fig. 1-3).[10] "Quicker" refers to how health care delivery has changed in the last 10 years to combat the rising costs of health care. In the acute care setting, the focus is on rapid recovery protocols. As a result, earlier mobility and mobility with more complex

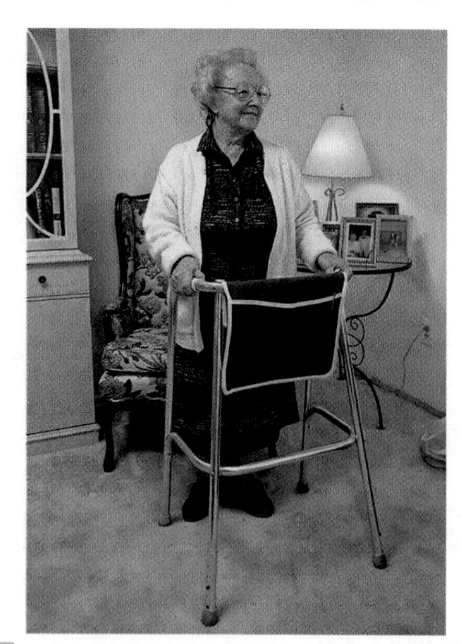

Fig. 1-3 The aging of America from the "traditionalists" (born before 1946) and the Baby Boom generation ("boomer" born 1946-1964) will result in older adults with multiple comorbidities in the care of the physical therapist. Even with a known orthopedic and/or neurologic impairment, these clients will require a careful screening for the possibility of other problems, side effects from medications, and primary/secondary prevention programs. (From Sorrentino SA: *Mosby's textbook for nursing assistants,* ed 7, St. Louis, 2008, Mosby.)

patients are allowed.[11] Better pharmacologic management of agitation has allowed earlier and safer mobility. Hospital inpatient/clients are discharged much faster today than they were even 10 years ago. Patients are discharged from the intensive care unit (ICU) to rehab or even home. Outpatient/client surgery is much more common, with same-day discharge for procedures that would have required a much longer hospitalization in the past. Patient/clients on the medical-surgical wards of most hospitals today would have been in the ICU 20 years ago.

Today's health care environment is complex and highly demanding. The therapist must be alert to red flags of systemic disease at all times but especially in those clients who have been given early release from the hospital or transition unit. Warning flags may come in the form of reported symptoms or observed signs. It may be a clinical presentation that does not match the recent history. Red warning and yellow caution flags will be discussed in greater detail later in this chapter.

"Sicker" refers to the fact that patient/clients in acute care, rehabilitation, or outpatient/client setting with any orthopedic or neurologic problems may have a past medical history of cancer or a current personal history of diabetes, liver disease, thyroid condition, peptic ulcer, and/or other conditions or diseases.

The number of people with at least one chronic disease or disability is reaching epidemic proportions. According to the National Institute on Aging,[12] 79% of adults over 70 have at least one of seven potentially disabling chronic conditions (arthritis, hypertension, heart disease, diabetes, respiratory diseases, stroke, and cancer).[13] The presence of multiple comorbidities emphasizes the need to view the whole patient/client and not just the body part in question.

In addition, the number of people who do not have health insurance and who wait longer to seek medical attention are sicker when they access care. This factor, combined with the American lifestyle that leads to chronic conditions such as obesity, hypertension, and diabetes, results in a sicker population base.[14]

Natural History

Improvements in treatment for neurologic and other conditions previously considered fatal (e.g., cancer, cystic fibrosis) are now extending the life expectancy for many individuals. Improved interventions bring new areas of focus such as quality-of-life issues. With some conditions (e.g., muscular dystrophy, cerebral palsy), the artificial dichotomy of pediatric versus adult care is gradually being replaced by a lifestyle approach that takes into consideration what is known about the natural history of the condition.

Many individuals with childhood-onset diseases now live well into adulthood. For them, their original pathology or disease process has given way to secondary impairments. These secondary impairments create further limitation and issues as the person ages. For example, a 30-year-old with cerebral palsy may experience chronic pain, changes or limitations in ambulation and endurance, and increased fatigue.

These symptoms result from the atypical movement patterns and musculoskeletal strains caused by chronic increase in tone and muscle imbalances that were originally caused by cerebral palsy. In this case, the screening process may be identifying signs and symptoms that have developed as a natural result of the primary condition (e.g., cerebral palsy) or long-term effects of treatment (e.g., chemotherapy, biotherapy, or radiotherapy for cancer).

Signed Prescription

Under direct access, the physical therapist may have primary responsibility or become the first contact for some clients in the health care delivery system. On the other hand, clients may obtain a signed prescription for physical therapy from their primary care physician or other health care provider, based on similar past complaints of musculoskeletal symptoms, without actually seeing the physician or being examined by the physician (Case Example 1-1).

? FOLLOW-UP QUESTIONS

Always ask a client who provides a signed prescription:

- Did you actually see the physician (chiropractor, dentist, nurse practitioner, physician assistant)?
- Did the doctor (dentist) examine you?

CASE EXAMPLE 1-1

Physician Visit Without Examination

A 60-year-old man retired from his job as the president of a large vocational technical school and called his physician the next day for a long-put-off referral to physical therapy. He arrived at an outpatient orthopedic physical therapy clinic with a signed physician's prescription that said, "Evaluate and Treat."

His primary complaint was left anterior hip and groin pain. This client had a history of three previous total hip replacements (anterior approach, lateral approach, posterior approach) on the right side, performed over the last 10 years.

Based on previous rehabilitation experience, he felt certain that his current symptoms of hip and groin pain could be alleviated by physical therapy.

- Social History: Recently retired as the director of a large vocational rehabilitation agency, married, three grown children
- Past Medical History (PMHx): Three total hip replacements (THRs) to the left hip (anterior, posterior, and lateral approaches) over the last 7 years
  - Open heart surgery 10 years ago
  - Congestive heart failure (CHF) 3 years ago
  - Medications: Lotensin daily, 1 baby aspirin per day, Zocor (20 mg) once a day
- Clinical presentation:
  - Extensive scar tissue around the left hip area with centralized core of round, hard tissue (4 × 6 cm) over the greater trochanter on the left
  - Bilateral pitting edema of the feet and ankles (right greater than left)
  - Positive Thomas (30-degree hip flexion contracture) test for left hip
  - Neurologic screen: Negative but general deconditioning and global decline observed in lower extremity strength
  - Vital signs*:

    Blood pressure (sitting, right arm) 92/58 mm Hg
    Heart rate 86 bpm
    Respirations 22/min
    Pulse oximeter (at rest) 89%
    Body temperature 97.8° F

The client arrived at the physical therapy clinic with a signed prescription in hand, but when asked if he had actually seen the physician, he explained that he received this prescription after a telephone conversation with his physician.

How Do You Communicate Your Findings and Concerns to the Physician? It is always a good idea to call and ask for a copy of the physician's dictation or notes. It may be that the doctor is well aware of the client's clinical presentation. Health Insurance Portability and Accountability Act (HIPAA) regulations require the client to sign a disclosure statement before the therapist can gain access to the medical records. To facilitate this process, it is best to have the paperwork requirements completed on the first appointment before the therapist sees the client.

Sometimes a conversation with the physician's office staff is all that is needed. They may be able to look at the client's chart and advise you accordingly. At the same time, in our litigious culture, outlining your concerns or questions almost always obligates the medical office to make a follow-up appointment with the client.

It may be best to provide the client with your written report that he or she can hand carry to the physician's office. Sending a fax, email, or mailed written report may place the information in the chart but not in the physician's hands at the appropriate time. It is always advised to do both (fax or mail along with a hand-carried copy).

Make your documentation complete, but your communication brief. Thank the physician for the referral. Outline the problem areas (human movement system diagnosis, impairment classification, and planned intervention). Be brief! The physician is only going to have time to scan what you sent.

Any associated signs and symptoms or red flags can be pointed out as follows:

During my examination, I noted the following:

Bilateral pitting edema of lower extremities

Vital signs:

Blood pressure (sitting, right arm) 92/58 mm Hg
Heart rate 86 bpm
Respirations 22/min
Pulse oximeter (at rest) 89%
Body temperature 97.8° F

Some of these findings seem outside the expected range. Please advise.

Note to the Reader: If possible, highlight this last statement in order to draw the physician's eye to your primary concern.

It is outside the scope of our practice to suggest possible reasons for the client's symptoms (e.g., congestive failure, side effect of medication). Just make note of the findings and let the physician make the medical diagnosis. An open-ended comment such as "Please advise" or question such as "What do you think?" may be all that is required.

Of course, in any collaborative relationship you may find that some physicians ask for your opinion. It is quite permissible to offer the evidence and draw some possible conclusions.

Result: An appropriate physical therapy program of soft tissue mobilization, stretching, and home exercise was initiated. However, the client was returned to his physician for an immediate follow-up appointment. A brief report from the therapist stated the key objective findings and outlined the proposed physical therapy plan. The letter included a short paragraph with the following remarks:

Given the client's sedentary lifestyle, previous history of heart disease, and blood pressure reading today, I would like to recommend a physical conditioning program. Would you please let me know if he is medically stable? Based on your findings, we will begin him in a pre-aerobic training program here and progress him to a home-based or fitness center program.

*The blood pressure and pulse measurements are difficult to evaluate given the fact that this client is taking antihypertensive medications. Ace inhibitors and beta-blockers, for example, reduce the heart rate so that the body's normal compensatory mechanisms (e.g., increased stroke volume and therefore increased heart rate) are unable to function in response to the onset of congestive heart failure. Low blood pressure and high pulse rate with higher respiratory rate and mildly diminished oxygen saturation (especially on exertion) must be considered red flags. Auscultation would be in order here. Light crackles in the lung bases might be heard in this case.

CASE EXAMPLE 1-2

Medical Specialization

A 45-year-old long-haul truck driver with bilateral carpal tunnel syndrome was referred for physical therapy by an orthopedic surgeon specializing in hand injuries. During the course of treatment the client mentioned that he was also seeing an acupuncturist for wrist and hand pain. The acupuncturist told the client that, based on his assessment, acupuncture treatment was indicated for liver disease.

Comment: Protein (from food sources or from a gastrointestinal bleed) is normally taken up and detoxified by the liver. Ammonia is produced as a by-product of protein breakdown and then transformed by the liver to urea, glutamine, and asparagine before being excreted by the renal system. When liver dysfunction results in increased serum ammonia and urea levels, peripheral nerve function can be impaired. (See detailed explanation on neurologic symptoms in Chapter 9.)

Result: The therapist continued to treat this client, but knowing that the referring specialist did not routinely screen for systemic causes of carpal tunnel syndrome (or even screen for cervical involvement) combined with the acupuncturist's information, raised a red flag for possible systemic origin of symptoms. A phone call was made to the physician with the following approach:

> Say, Mr. Y was in for therapy today. He happened to mention that he is seeing an acupuncturist who told him that his wrist and hand pain is from a liver problem. I recalled seeing some information here at the office about the effect of liver disease on the peripheral nervous system. Since Mr. Y has not improved with our carpal tunnel protocol, would you like to have him come back in for a reevaluation?

Comment: How to respond to each situation will require a certain amount of diplomacy, with consideration given to the individual therapist's relationship with the physician and the physician's openness to direct communication.

It is the physical therapist's responsibility to recognize when a client's presentation falls outside the parameters of a true neuromusculoskeletal condition. Unless prompted by the physician, it is not the therapist's role to suggest a specific medical diagnosis or medical testing procedures.

Medical Specialization

Additionally, with the increasing specialization of medicine, clients may be evaluated by a medical specialist who does not immediately recognize the underlying systemic disease, or the specialist may assume that the referring primary care physician has ruled out other causes (Case Example 1-2).

Progression of Time and Disease

In some cases, early signs and symptoms of systemic disease may be difficult or impossible to recognize until the disease has progressed enough to create distressing or noticeable symptoms (Case Example 1-3). In some cases, the patient/client's clinical presentation in the physician's office may be very different from what the therapist observes when days or

CASE EXAMPLE 1-3

Progression of Disease

A 44-year-old woman was referred to the physical therapist with a complaint of right paraspinal/low thoracic back pain. There was no reported history of trauma or assault and no history of repetitive movement. The past medical history was significant for a kidney infection treated 3 weeks ago with antibiotics. The client stated that her follow-up urinalysis was "clear" and the infection resolved.

The physical therapy examination revealed true paraspinal muscle spasm with an acute presentation of limited movement and exquisite pain in the posterior right middle to low back. Spinal accessory motions were tested following application of a cold modality and were found to be mildly restricted in right sidebending and left rotation of the T8-T12 segments. It was the therapist's assessment that this joint motion deficit was still the result of muscle spasm and guarding and not true joint involvement.

Result: After three sessions with the physical therapist in which modalities were used for the acute symptoms, the client was not making observable, reportable, or measurable improvement. Her fourth scheduled appointment was cancelled because of the "flu."

Given the recent history of kidney infection, the lack of expected improvement, and the onset of constitutional symptoms (see Box 1-3), the therapist contacted the client by telephone and suggested that she make a follow-up appointment with her doctor as soon as possible.

As it turned out, this woman's kidney infection had recurred. She recovered from her back sequelae within 24 hours of initiating a second antibiotic treatment. This is not the typical medical picture for a urologically compromised person. Sometimes it is not until the disease progresses that the systemic disorder (masquerading as a musculoskeletal problem) can be clearly differentiated.

Last, sometimes clients do not relay all the necessary or pertinent medical information to their physicians but will confide in the physical therapist. They may feel intimidated, forget, become unwilling or embarrassed, or fail to recognize the significance of the symptoms and neglect to mention important medical details (see Box 1-1).

Knowing that systemic diseases can mimic neuromusculoskeletal dysfunction, the therapist is responsible for identifying as closely as possible what neuromusculoskeletal pathologic condition is present.

The final result should be to treat as specifically as possible. This is done by closely identifying the underlying neuromusculoskeletal pathologic condition and the accompanying movement dysfunction, while at the same time investigating the possibility of systemic disease.

This text will help the clinician quickly recognize problems that are beyond the expertise of the physical therapist. The therapist who recognizes hallmark signs and symptoms of systemic disease will know when to refer clients to the appropriate health care practitioner.

weeks separate the two appointments. Holidays, vacations, finances, scheduling conflicts, and so on can put delays between medical examination and diagnosis and that first appointment with the therapist.

Given enough time, a disease process will eventually progress and get worse. Symptoms may become more readily apparent or more easily clustered. In such cases, the alert therapist may be the first to ask the patient/client pertinent questions to determine the presence of underlying symptoms requiring medical referral.

The therapist must know what questions to ask clients in order to identify the need for medical referral. Knowing what medical conditions can cause shoulder, back, thorax, pelvic, hip, sacroiliac, and groin pain is essential. Familiarity with risk factors for various diseases, illnesses, and conditions is an important tool for early recognition in the screening process.

Patient/Client Disclosure

Finally, sometimes patient/clients tell the therapist things about their current health and social history unknown or unreported to the physician. The content of these conversations can hold important screening clues to point out a systemic illness or viscerogenic cause of musculoskeletal or neuromuscular impairment.

Yellow or Red Flags

A large part of the screening process is identifying yellow (caution) or red (warning) flag histories and signs and symptoms (Box 1-2). A yellow flag is a cautionary or warning symptom that signals "slow down" and think about the need for screening. Red flags are features of the individual's medical

BOX 1-2 RED FLAGS

The presence of any one of these symptoms is not usually cause for extreme concern but should raise a red flag for the alert therapist. The therapist is looking for a pattern that suggests a viscerogenic or systemic origin of pain and/or symptoms. The therapist will proceed with the screening process, depending on which symptoms are grouped together. Often the next step is to conduct a risk factor assessment and look for associated signs and symptoms.

Past Medical History (Personal or Family)

- Personal or family history of cancer
- Recent (last 6 weeks) infection (e.g., mononucleosis, upper respiratory infection (URI), urinary tract infection (UTI), bacterial such as streptococcal or staphylococcal; viral such as measles, hepatitis), especially when followed by neurologic symptoms 1 to 3 weeks later (Guillain-Barré syndrome), joint pain, or back pain
- Recurrent colds or flu with a cyclical pattern (i.e., the client reports that he or she just cannot shake this cold or the flu—it keeps coming back over and over)
- Recent history of trauma, such as motor vehicle accident or fall (fracture, any age), or minor trauma in older adult with osteopenia/osteoporosis
- History of immunosuppression (e.g., steroids, organ transplant, human immunodeficiency virus [HIV])
- History of injection drug use (infection)

Risk Factors

Risk factors vary, depending on family history, previous personal history, and disease, illness, or condition present. For example, risk factors for heart disease will be different from risk factors for osteoporosis or vestibular or balance problems. As with all decision-making variables, a single risk factor may or may not be significant and must be viewed in context of the whole patient/client presentation. This represents only a partial list of all the possible health risk factors.

Substance use/abuse	Alcohol use/abuse
Tobacco use	Sedentary lifestyle
Age	Race/ethnicity
Gender	Domestic violence
Body mass index (BMI)	Hysterectomy/oophorectomy
Exposure to radiation	Occupation

Clinical Presentation

No known cause, unknown etiology, insidious onset

Symptoms that are not improved or relieved by physical therapy intervention are a red flag.

Physical therapy intervention does not change the clinical picture; client may get worse!

Symptoms that get better after physical therapy, but then get worse again is also a red flag identifying the need to screen further

Significant weight loss or gain without effort (more than 10% of the client's body weight in 10 to 21 days)

Gradual, progressive, or cyclical presentation of symptoms (worse/better/worse)

Unrelieved by rest or change in position; no position is comfortable

If relieved by rest, positional change, or application of heat, in time, these relieving factors no longer reduce symptoms

Symptoms seem out of proportion to the injury

Symptoms persist beyond the expected time for that condition

Unable to alter (provoke, reproduce, alleviate, eliminate, aggravate) the symptoms during exam

Does not fit the expected mechanical or neuromusculoskeletal pattern

No discernible pattern of symptoms

A growing mass (painless or painful) is a tumor until proved otherwise; a hematoma should decrease (not increase) in size with time

Continued

BOX 1-2 RED FLAGS—cont'd

Postmenopausal vaginal bleeding (bleeding that occurs a year or more after the last period [significance depends on whether the woman is on hormone replacement therapy and which regimen is used])

Bilateral symptoms:

Edema	Clubbing
Numbness, tingling	Nail-bed changes
Skin-pigmentation changes	Skin rash

Change in muscle tone or range of motion (ROM) for individuals with neurologic conditions (e.g., cerebral palsy, spinal-cord injured, traumatic-brain injured, multiple sclerosis)

Pain Pattern

Back or shoulder pain (most common location of referred pain; other areas can be affected as well, but these two areas signal a particular need to take a second look)

Pain accompanied by full and painless range of motion (see Table 3-1)

Pain that is not consistent with emotional or psychologic overlay (e.g., Waddell's test is negative or insignificant; ways to measure this are discussed in Chapter 3); screening tests for emotional overlay are negative

Night pain (constant and intense; see complete description in Chapter 3)

Symptoms (especially pain) are constant and intense (Remember to ask anyone with "constant" pain: Are you having this pain right now?)

Pain made worse by activity and relieved by rest (e.g., intermittent claudication; cardiac: upper quadrant pain with the use of the lower extremities while upper extremities are inactive)

Pain described as throbbing (vascular) knifelike, boring, or deep aching

Pain that is poorly localized

Pattern of coming and going like spasms, colicky

Pain accompanied by signs and symptoms associated with a specific viscera or system (e.g., GI, GU, GYN, cardiac, pulmonary, endocrine)

Change in musculoskeletal symptoms with food intake or medication use (immediately or up to several hours later)

Associated Signs and Symptoms

Recent report of confusion (or increased confusion); this could be a neurologic sign; it could be drug-induced (e.g., NSAIDs) or a sign of infection; usually it is a family member who takes the therapist aside to report this concern

Presence of constitutional symptoms (see Box 1-3) or unusual vital signs (see Discussion, Chapter 4); body temperature of 100° F (37.8° C) usually indicates a serious illness

Proximal muscle weakness, especially if accompanied by change in DTRs (see Fig. 13-3)

Joint pain with skin rashes, nodules (see discussion of systemic causes of joint pain, Chapter 3; see Table 3-6)

Any cluster of signs and symptoms observed during the Review of Systems that are characteristic of a particular organ system (see Box 4-19; Table 13-5)

Unusual menstrual cycle/symptoms; association between menses and symptoms

It is imperative at the end of each interview that the therapist ask the client a question like the following:

- Are there any other symptoms or problems anywhere else in your body that may not seem related to your current problem?

history and clinical examination thought to be associated with a high risk of serious disorders such as infection, inflammation, cancer, or fracture.[15] A red-flag symptom requires immediate attention, either to pursue further screening questions and/or tests or to make an appropriate referral.

The presence of a single yellow or red flag is not usually cause for immediate medical attention. Each cautionary or warning flag must be viewed in the context of the whole person given the age, gender, past medical history, known risk factors, medication use, and current clinical presentation of that patient/client.

Clusters of yellow and/or red flags do not always warrant medical referral. Each case is evaluated on its own. It is time to take a closer look when risk factors for specific diseases are present or both risk factors and red flags are present at the same time. Even as we say this, the heavy emphasis on red flags in screening has been called into question.[16,17]

It has been reported that in the primary care (medical) setting, some red flags have high false-positive rates and have very little diagnostic value when used by themselves.[5] Efforts are being made to identify reliable red flags that are valid based on patient-centered clinical research. Whenever possible, those yellow/red flags are reported in this text.[5,18,19]

The patient/client's history, presenting pain pattern, and possible associated signs and symptoms must be reviewed along with results from the objective evaluation in making a treatment-versus-referral decision.

Medical conditions can cause pain, dysfunction, and impairment of the

- Back/neck
- Shoulder
- Chest/breast/rib
- Hip/groin
- Sacroiliac (SI)/sacrum/pelvis

For the most part, the organs are located in the central portion of the body and refer symptoms to the nearby major muscles and joints. In general, the back and shoulder represent the primary areas of referred viscerogenic pain patterns. Cases of isolated symptoms will be presented in this text as they occur in clinical practice. Symptoms of any kind that present bilaterally always raise a red flag for concern and further investigation (Case Example 1-4).

Monitoring vital signs is a quick and easy way to screen for medical conditions. Vital signs are discussed more completely in Chapter 4. Asking about the presence of constitutional symptoms is important, especially when there is no known cause. Constitutional symptoms refer to a constellation of signs and symptoms present whenever the patient/client is experiencing a systemic illness. No matter what system is involved, these core signs and symptoms are often present (Box 1-3).

MEDICAL SCREENING VERSUS SCREENING FOR REFERRAL

Therapists can have an active role in both primary and secondary prevention through screening and education. Primary prevention involves stopping the process(es) that lead to the development of diseases such as diabetes, coronary artery disease, or cancer in the first place (Box 1-4).

According to the *Guide*,[1] physical therapists are involved in primary prevention by "preventing a target condition in a susceptible or potentially susceptible population through such specific measures as general health promotion efforts" [p. 33]. Risk factor assessment and risk reduction fall under this category.

Secondary prevention involves the regular screening for early detection of disease or other health-threatening conditions such as hypertension, osteoporosis, incontinence,

CASE EXAMPLE 1-4

Bilateral Hand Pain

A 69-year-old man presented with pain in both hands that was worse on the left. He described the pain as "deep aching" and reported it interfered with his ability to write. The pain got worse as the day went on.

There was no report of fever, chills, previous infection, new medications, or cancer. The client was unaware that joint pain could be caused by sexually transmitted infections but said that he was widowed after 50 years of marriage to the same woman and did not think this was a problem.

There was no history of occupational or accidental trauma. The client viewed himself as being in "excellent health." He was not taking any medications or herbal supplements.

Wrist range of motion was limited by stiffness at end ranges in flexion and extension. There was no obvious soft tissue swelling, warmth, or tenderness over or around the joint. A neurologic screening examination was negative for sensory, motor, or reflex changes.

There were no other significant findings from various tests and measures performed. There were no other joints involved. There were no reported signs and symptoms of any kind anywhere else in the muscles, limbs, or general body.

What Are the Red-Flag Signs and Symptoms Here? Should a Medical Referral Be Made? Why or Why Not?

Red Flags

Age

Bilateral symptoms

Lack of other definitive findings

It is difficult to treat as specifically as possible without a clear differential diagnosis. You can treat the symptoms and assess the results before making a medical referral. Improvement in symptoms and motion should be seen within one to three sessions.

However, in light of the red flags, best practice suggests a medical referral to rule out a systemic disorder before initiating treatment. This could be rheumatoid arthritis, osteoarthritis, osteoporosis, the result of a thyroid dysfunction, gout, or other arthritic condition.

How Do You Make this Suggestion to the Client, Especially if He Was Coming to You to Avoid a Doctor's Visit/Fee?

Perhaps something like this would be appropriate:

> Mr. J,
>
> You have very few symptoms to base treatment on. When pain or other symptoms are present on both sides, it can be a sign that something more systemic is going on. For anyone over 40 with bilateral symptoms and a lack of other findings, we recommend a medical exam.
>
> Do you have a regular family doctor or primary care physician? It may be helpful to have some x-rays and lab work done before we begin treatment here. Who can I call or send my report to?

Result: X-rays showed significant joint space loss in the radiocarpal joint, as well as sclerosis and cystic changes in the carpal bones. Calcium deposits in the wrist fibrocartilage pointed to a diagnosis of calcium pyrophosphate dihydrate (CPPD) crystal deposition disease (pseudogout).

There was no osteoporosis and no bone erosion present.

Treatment was with oral nonsteroidal antiinflammatory drugs for symptomatic pain relief. There is no evidence that physical therapy intervention can change the course of this disease or even effectively treat the symptoms.

The client opted to return to physical therapy for short-term palliative care during the acute phase.

To read more about this condition, consult the *Primer on the Rheumatic Diseases,* 13th edition. Arthritis Foundation (www.arthritis.org), Atlanta, 2008.

Data from Raman S, Resnick D: Chronic and increasing bilateral hand pain, *J Musculoskeletal Med* 13(6):58-61, 1996.

BOX 1-3 CONSTITUTIONAL SYMPTOMS

Fever
Diaphoresis (unexplained perspiration)
Sweats (can occur anytime night or day)
Nausea
Vomiting
Diarrhea
Pallor
Dizziness/syncope (fainting)
Fatigue
Weight loss

BOX 1-4 PHYSICAL THERAPIST ROLE IN DISEASE PREVENTION

Primary Prevention: Stopping the process(es) that lead to the development of disease(s), illness(es), and other pathologic health conditions through education, risk factor reduction, and general health promotion.

Secondary Prevention: Early detection of disease(es), illness(es), and other pathologic health conditions through regular screening; this does not prevent the condition but may decrease duration and/or severity of disease and thereby improve the outcome, including improved quality of life.

Tertiary Prevention: Providing ways to limit the degree of disability while improving function in patients/clients with chronic and/or irreversible diseases.

Health Promotion and Wellness: Providing education and support to help patients/clients make choices that will promote health or improved health. The goal of wellness is to give people greater awareness and control in making choices about their own health.

diabetes, or cancer. This does not prevent any of these problems but improves the outcome. The *Guide* outlines the physical therapist's role in secondary prevention as "decreasing duration of illness, severity of disease, and number of sequelae through early diagnosis and prompt intervention" [p. 33].

Although the terms *screening for medical referral* and *medical screening* are often used interchangeably, these are really two separate activities. Medical screening is a method for detecting disease or body dysfunction before an individual would normally seek medical care. Medical screening tests are usually administered to individuals who do not have current symptoms, but who may be at high risk for certain adverse health outcomes (e.g., colonoscopy, fasting blood glucose, blood pressure monitoring, assessing body mass index, thyroid screening panel, cholesterol screening panel, prostate-specific antigen, mammography).

In the context of a human movement system diagnosis, the term *medical screening* has come to refer to the process of screening for referral. The process involves determining whether the individual has a condition that can be addressed by the physical therapist's intervention and if not, then whether the condition requires evaluation by a medical doctor or other medical specialist.

Both terms (*medical screening* and *screening for referral*) will probably continue to be used interchangeably to describe the screening process. It may be important to keep the distinction in mind, especially when conversing/consulting with physicians whose concept of medical screening differs from the physical therapist's use of the term to describe screening for referral.

DIAGNOSIS BY THE PHYSICAL THERAPIST

The term "diagnosis by the physical therapist" is language used by the American Physical Therapy Association (APTA). It is the policy of the APTA that physical therapists shall establish a diagnosis for each patient/client. Prior to making a patient/client management decision, physical therapists shall utilize the diagnostic process in order to establish a diagnosis for the specific conditions in need of the physical therapist's attention.[20]

In keeping with advancing physical therapist practice, the current education strategic plan and Vision 2020, *Diagnosis by Physical Therapists* (HOD P06-97-06-19), has been updated to include ordering of tests that are performed and interpreted by other health professionals (e.g., radiographic imaging, laboratory blood work). The position now states that it is the physical therapist's responsibility in the diagnostic process to organize and interpret all relevant data.[21]

The diagnostic process requires evaluation of information obtained from the patient/client examination, including the history, systems review, administration of tests, and interpretation of data. Physical therapists use diagnostic labels that identify the impact of a condition on function at the level of the system (especially the human movement system) and the level of the whole person.[22]

The physical therapist is qualified to make a diagnosis regarding primary NMS conditions, though we must do so in accordance with the state practice act. The profession must continue to develop the concept of human movement as a physiologic system and work to get physical therapists recognized as experts in that system.[23]

Further Defining Diagnosis

Whenever diagnosis is discussed, we hear this familiar refrain: diagnosis is both the process and the end result of evaluating examination data, which the therapist organizes into defined clusters, syndromes, or categories to help determine the prognosis and the most appropriate intervention strategies.[1]

It has been described as the decision reached as a result of the diagnostic process, which is the evaluation of information obtained from the patient/client examination.[20] Whereas the physician makes a medical diagnosis based on the pathologic or pathophysiologic state at the cellular level, in a diagnosis-based physical therapist's practice, the therapist places an emphasis on the identification of specific human movement impairments that best establish effective interventions and reliable prognoses.[24]

Others have supported a revised definition of the physical therapy diagnosis as: a process centered on the evaluation of multiple levels of movement dysfunction whose purpose is to inform treatment decisions related to functional restoration.[25] According to the *Guide,* the diagnostic-based practice requires the physical therapist to integrate five elements of patient/client management (Box 1-5) in a manner designed to maximize outcomes (Fig. 1-4).

BOX 1-5 ELEMENTS OF PATIENT/CLIENT MANAGEMENT

Examination: History, systems review, and tests and measures
Evaluation: Assessment or judgment of the data
Diagnosis: Determined within the scope of practice
Prognosis: Projected outcome
Intervention: Coordination, communication, and documentation of an appropriate treatment plan for the diagnosis based on the previous four elements

Data from *Guide to physical therapist practice,* ed 2 (Revised), Alexandria, VA, 2003, American Physical Therapy Association (APTA).

One of those proposed modifications is in the Elements of Patient/Client Management offered by the APTA in the *Guide.* Fig. 1-4 does not illustrate all decisions possible.

Boissonnault proposed a fork in the clinical decision-making pathway to show three alternative decisions[6,26] (Fig. 1-5), including

1. Referral/consultation (no treatment; referral may be a nonurgent consult or an immediate/urgent referral)
2. Diagnose and treat
3. Both (treat and refer)

The decision to refer or consult with the physician can also apply to referral to other appropriate health care professionals and/or practitioners (e.g., dentist, chiropractor, nurse practitioner, psychologist).

In summary, there has been considerable discussion that evaluation is a process with diagnosis as the end result.[27] The concepts around the "diagnostic process" remain part of an evolving definition that will continue to be discussed and clarified by physical therapists. We will present some additional pieces to the discussion as we go along in this chapter.

APTA Vision Sentence for Physical Therapy 2020

By 2020, physical therapy will be provided by physical therapists who are doctors of physical therapy, recognized by consumers and other health care professionals as the practitioners

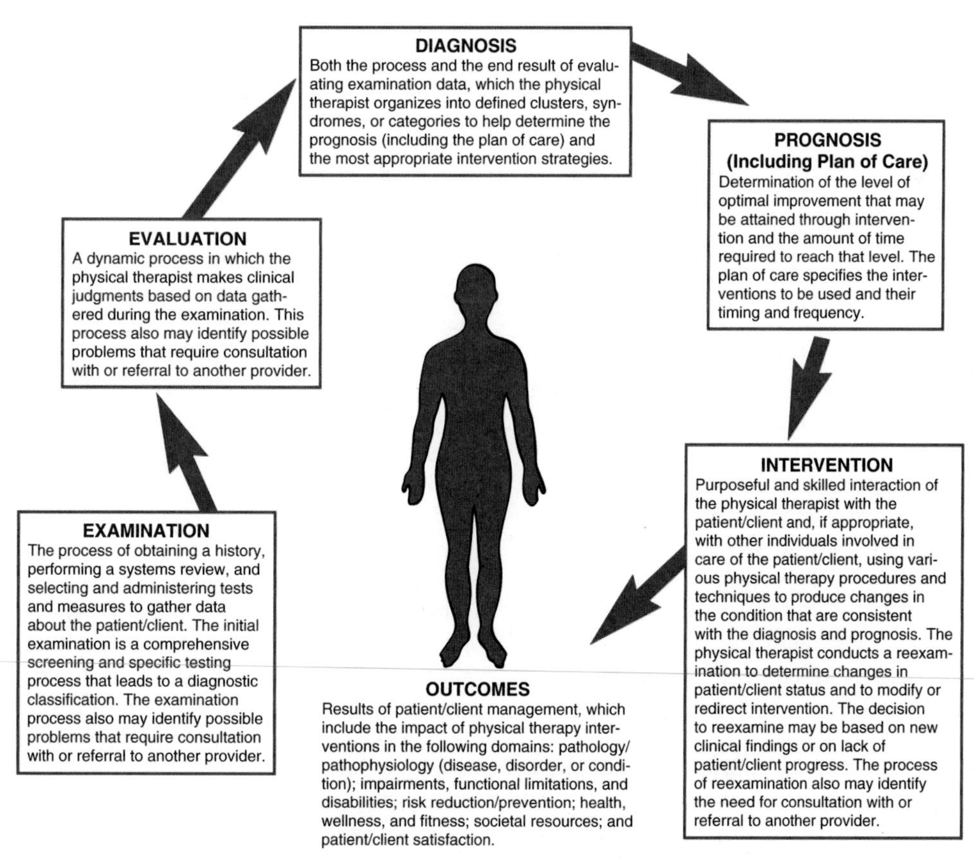

Fig. 1-4 The elements of patient/client management leading to optimal outcomes. Screening takes place anywhere along this pathway. (Reprinted with permission from *Guide to physical therapist practice,* ed 2 [Revised], 2003, Fig. 1-4, p. 35.)

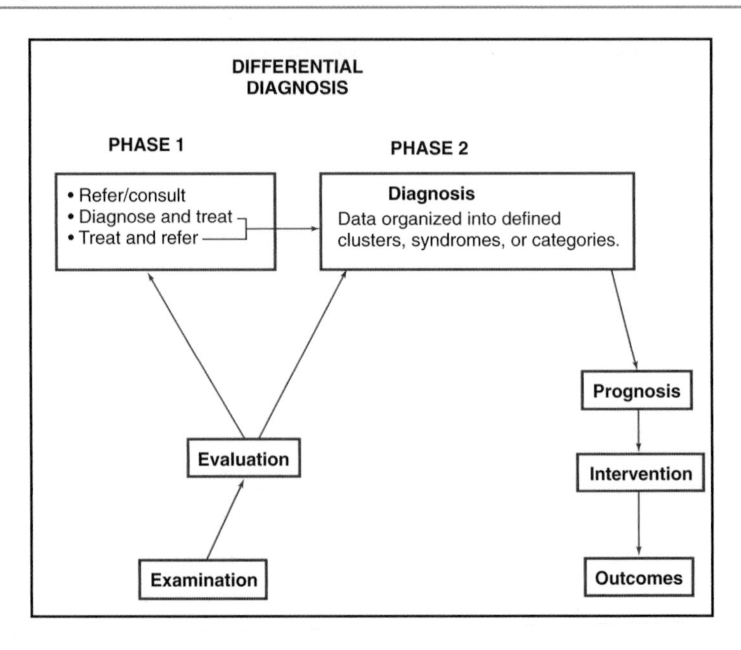

Fig. 1-5 Modification to the patient/client management model. On the left side of this figure, the therapist starts by collecting data during the examination. Based on the data collected, the evaluation leads to clinical judgments. The current model in the *Guide* gives only one decision-making option and that is the diagnosis. In this adapted model, a fork in the decision-making pathway allows the therapist the opportunity to make one of three alternative decisions as described in the text. This model is more in keeping with recommended clinical practice. (From Boissonnault WG, Umphred DA: Differential diagnosis phase I. In Umphred DA, editors: *Neurological rehabilitation,* ed 6, St. Louis, 2012, Mosby.)

of choice to whom consumers have direct access for the diagnosis of, interventions for, and prevention of impairments, functional limitations, and disabilities related to movement, function, and health.[28]

The vision sentence points out that physical therapists are capable of making a diagnosis and determining whether the patient/client can be helped by physical therapy intervention. In an autonomous profession the therapist can decide if physical therapy should be a part of the plan, the entire plan, or not needed at all.

When communicating with physicians, it is helpful to understand the definition of a medical diagnosis and how it differs from a physical therapist's diagnosis. The medical diagnosis is traditionally defined as the recognition of disease.

It is the determination of the cause and nature of pathologic conditions. Medical differential diagnosis is the comparison of symptoms of similar diseases and medical diagnostics (laboratory and test procedures performed) so that a correct assessment of the patient/client's actual problem can be made.

A differential diagnosis by the physical therapist is the comparison of NMS signs and symptoms to identify the underlying human movement dysfunction so that treatment can be planned as specifically as possible. If there is evidence of a pathologic condition, referral is made to the appropriate health care (or other) professional. This step requires the therapist to at least consider the possible pathologic conditions, even if unable to verify the presence or absence of said condition.[29]

One of the APTA goals is that physical therapists will be universally recognized and promoted as the practitioners of choice for persons with conditions that affect human movement, function, health, and wellness.[30]

Purpose of the Diagnosis

In the context of screening for referral, the purpose of the diagnosis is to:

- Treat as specifically as possible by determining the most appropriate plan of care and intervention strategy for each patient/client
- Recognize the need for a medical referral

More broadly stated the purpose of the human movement system diagnosis is to guide the physical therapist in determining the most appropriate intervention strategy for each patient/client with a goal of decreasing disability and increasing function. In the event the diagnostic process does not yield an identifiable cluster, disorder, syndrome, or category, intervention may be directed toward the alleviation of symptoms and remediation of impairment, functional limitation, or disability.[20]

Sometimes the patient/client is too acute to examine fully on the first visit. At other times, we evaluate nonspecific referral diagnoses such as problems medically diagnosed as "shoulder pain" or "back pain." When the patient/client is referred with a previously established diagnosis, the physical therapist determines that the clinical findings are consistent with that diagnosis[20] (Case Example 1-5).

Sometimes the screening and diagnostic process identifies a systemic problem as the underlying cause of NMS symptoms. At other times, it confirms that the patient/client has a human movement system syndrome or problem after all (see Case Examples 1-5[31] and 1-7).

CASE EXAMPLE 1-5

Verify Medical Diagnosis

A 31-year-old man was referred to physical therapy by an orthopedic physician. The diagnosis was "shoulder-hand syndrome." This client had been evaluated for this same problem by three other physicians and two physical therapists before arriving at our clinic. Treatment to date had been unsuccessful in alleviating symptoms.

The medical diagnosis itself provided some useful information about the referring physician. "Shoulder-hand syndrome" is an outdated nomenclature previously used to describe reflex sympathetic dystrophy syndrome (RSDS or RSD), now known more accurately as complex regional pain syndrome (CRPS).[31]

Shoulder-hand syndrome was a condition that occurred following a myocardial infarct, or MI (heart attack), usually after prolonged bedrest. This condition has been significantly reduced in incidence by more up-to-date and aggressive cardiac rehabilitation programs. Today CRPS, primarily affecting the limbs, develops after injury or surgery, but it can still occur as a result of a cerebrovascular accident (CVA) or heart attack.

This client's clinical presentation included none of the typical signs and symptoms expected with CRPS such as skin changes (smooth, shiny, red skin), hair growth pattern (increased dark hair patches or loss of hair), temperature changes (increased or decreased), hyperhidrosis (excessive perspiration), restricted joint motion, and severe pain. The clinical picture appeared consistent with a trigger point of the latissimus dorsi muscle, and in fact, treatment of the trigger point completely eliminated all symptoms.

Conducting a thorough physical therapy examination to identify the specific underlying cause of symptomatic presentation was essential to the treatment of this case. Treatment approaches for a trigger point differ greatly from intervention protocols for CRPS.

Accepting the medical diagnosis without performing a physical therapy diagnostic evaluation would have resulted in wasted time and unnecessary charges for this client.

The International Association for the Study of Pain replaced the term *RSDS* with *CRPS I* in 1995.[31] Other names given to RSD included neurovascular dystrophy, sympathetic neurovascular dystrophy, algodystrophy, "red-hand disease," Sudeck's atrophy, and causalgia.

Historical Perspective

The idea of "physical therapy diagnosis" is not a new one. In fact, from its earliest beginnings until now, it has officially been around for at least 20 years. It was first described in the literature by Shirley Sahrmann[32] as the name given to a collection of relevant signs and symptoms associated with the primary dysfunction toward which the physical therapist directs treatment. The dysfunction is identified by the physical therapist based on the information obtained from the history, signs, symptoms, examination, and tests the therapist performs or requests.

In 1984, the APTA House of Delegates (HOD) made a motion that *the physical therapist may establish a diagnosis within the scope of their knowledge, experience, and expertise.* This was further qualified in 1990 when the Education Standards for Accreditation described "Diagnosis" for the first time.

In 1990, teaching and learning content and the skills necessary to determine a diagnosis became a required part of the curriculum standards established then by the Standards for Accreditation for Physical Therapist Educational Program. At that time the therapist's role in developing a diagnosis was described as:

- Engage in the diagnostic process in an efficient manner consistent with the policies and procedures of the practice setting.
- Engage in the diagnostic process to establish differential diagnoses for patient/clients across the lifespan based on evaluation of results of examinations and medical and psychosocial information.
- Take responsibility for communication or discussion of diagnoses or clinical impressions with other practitioners.

In 1995, the HOD amended the 1984 policy to make the definition of diagnosis consistent with the then upcoming *Guide to Physical Therapist Practice.* The first edition of the *Guide* was published in 1997. It was revised and published as a second edition in 2001; the second edition was revised in 2003.

Classification System

According to Rothstein,[33] in many fields of medicine when a medical diagnosis is made, the pathologic condition is determined and stages and classifications that guide treatment are also named. Although we recognize that the term diagnosis relates to a pathologic process, we know that pathologic evidence alone is inadequate to guide the physical therapist.

Physical therapists do not diagnose disease in the sense of identifying a specific organic or visceral pathologic condition. However, identified clusters of signs, symptoms, symptom-related behavior, and other data from the patient/client history and other testing can be used to confirm or rule out the presence of a problem within the scope of the physical therapist's practice. These diagnostic clusters can be labeled as *impairment classifications* or *human movement dysfunctions* by physical therapists and can guide efficient and effective management of the client.[34]

Although not diagnostic labels, the *Guide* groups the preferred practice patterns into four categories of conditions that can be used to guide the examination, evaluation, and intervention. These include musculoskeletal, neuromuscular, cardiovascular/pulmonary, and integumentary categories. An individual may belong to one or more of these groups or patterns.

Diagnostic classification systems that direct treatment interventions are being developed based on client prognosis and definable outcomes demonstrated in the literature.[1,35] At the same time, efforts are being made and ongoing discussions are taking place to define diagnostic categories or diagnostic descriptors for the physical therapist.[36-40] There is also a trend toward identification of subgroups within a particular group of individuals (e.g., low back pain, shoulder dysfunction) and predictive factors (positive and negative) for treatment and prognosis.

Diagnosis Dialog

Since 2006, a group of physical therapists across the United States have been meeting to define diagnosis, the purpose of diagnoses, and developing a template for universal use for all physical therapists to use in making a diagnosis. In keeping with our expertise in the human movement system, it has been suggested that the primary focus of the physical therapist's diagnostic expertise should be on diagnosing syndromes of the human movement system.[41] To see more about this group and the work being done, go to http://dxdialog.wusm.wustl.edu.

Earlier in this text discussion, we attempted to summarize various opinions and thoughts presented in our literature defining diagnosis. Here is an added component to that discussion. The "working" definition of diagnosis put forth by the Diagnosis Dialog group is:

> Diagnosis is both a ***process*** and a ***descriptor.*** The diagnostic process includes integrating and evaluating the data that are obtained during the examination for the purpose of guiding the prognosis, the plan of care, and intervention strategies. Physical therapists assign diagnostic descriptors that identify a condition or syndrome at the level of the system, especially the human movement system, and at the level of the whole person.[41]

In keeping with the APTA's Vision 2020 establishing our professional identity with the movement system, the human movement system has become the focus of the physical therapist's "diagnosis." The suggested template for this diagnosis under discussion and development is currently as follows[41]:

- Use recognized anatomic, physiologic, or movement-related terms to describe the condition or syndrome of the human movement system.
- Include, if deemed necessary for clarity, the name of the pathology, disease, disorder, or symptom that is associated with the diagnosis.
- Be as short as possible to improve clinical usefulness.

DIFFERENTIAL DIAGNOSIS VERSUS SCREENING

If you are already familiar with the term *differential diagnosis,* you may be wondering about the change in title for this text. Previous editions were entitled *Differential Diagnosis in Physical Therapy.*

The new name *Differential Diagnosis for Physical Therapists: Screening for Referral,* first established for the fourth edition of this text, does not reflect a change in the content of the text as much as it reflects a better understanding of the screening process and a more appropriate use of the term "differential diagnosis" to identify and describe the specific movement impairment present (if there is one).

When the first edition of this text was published, the term "physical therapy diagnosis" was not yet commonly used nomenclature. Diagnostic labels were primarily within the domain of the physician. Over the years, as our profession has changed and progressed, the concept of diagnosis has evolved.

A *diagnosis by the physical therapist* as outlined in the *Guide* describes the patient/client's primary dysfunction(s). This can be done through the classification of a patient/client within a specific practice pattern. The diagnostic process begins with the collection of data (examination), proceeds through the organization and interpretation of data (evaluation), and ends in the application of a label (i.e., the diagnosis).[1]

As part of the examination process, the therapist may conduct a screening examination. This is especially true if the diagnostic process does not yield an identifiable movement dysfunction. Throughout the evaluation process, the therapist must ask himself or herself:

- Is this an appropriate physical referral?
- Is there a problem that falls into one of the four categories of conditions described in the *Guide?*
- Is there a history or cluster of signs and/or symptoms that raises a yellow (cautionary) or red (warning) flag?

The presence of risk factors and yellow or red flags alerts the therapist to the need for a screening examination. Once the screening process is complete and the therapist has confirmed the client is appropriate for physical therapy intervention, then the objective examination continues.

Sometimes in the early presentation, there are no red flags or associated signs and symptoms to suggest an underlying systemic or viscerogenic cause of the client's NMS symptoms or movement dysfunction.

It is not until the disease progresses that the clinical picture changes enough to raise a red flag. This is why the screening process is not necessarily a one-time evaluation. Screening can take place anywhere along the circle represented in Fig. 1-4.

The most likely place screening occurs is during the examination when the therapist obtains the history, performs a systems review, and carries out specific tests and measures. It is here that the client reports constant pain, skin lesions, gastrointestinal problems associated with back pain, digital clubbing, palmar erythema, shoulder pain with stair climbing, or any of the many indicators of systemic disease.

The therapist may hear the client relate new onset of symptoms that were not present during the examination. Such new information may come forth anytime during the episode of care. If the patient/client does not progress in physical therapy or presents with new onset of symptoms unreported before, the screening process may have to be repeated.

Red-flag signs and symptoms may appear for the first time or develop more fully during the course of physical therapy intervention. In some cases, exercise stresses the client's

CASE EXAMPLE 1-6

Scope of Practice

A licensed physical therapist volunteered at a high school athletic event and screened an ankle injury. After performing a heel strike test (negative), the physical therapist recommended RICE. (Rest, Ice, Compression, and Elevation) and follow-up with a medical doctor if the pain persisted.

A complaint was filed 2 years later claiming that the physical therapist violated the state practice act by "... engaging in the practice of physical therapy in excess of the scope of physical therapy practice by undertaking to diagnose and prescribe appropriate treatment for an acute athletic injury."

The therapist was placed on probation for 2 years. The case was appealed and amended as it was clearly shown that the therapist was practicing within the legal bounds of the state's practice act. Imagine the impact this had on the individual in the community and as a private practitioner.

Know your state practice act and make sure it allows physical therapists to draw conclusions and make statements about findings of evaluations (i.e., diagnosis).

physiology enough to tip the scales. Previously unnoticed, unrecognized, or silent symptoms suddenly present more clearly.

As mentioned, a lack of progress signals the need to conduct a reexamination or to modify/redirect intervention. The process of reexamination may identify the need for consultation with or referral to another health care provider (*Guide,*[1] Figure 1: Intervention, p. 43). The medical doctor is the most likely referral recommendation, but referral to a nurse practitioner, physician assistant, chiropractor, dentist, psychologist, counselor, or other appropriate health care professional may be more appropriate at times.

Scope of Practice

A key phrase in the APTA standards of practice is "within the scope of physical therapist practice." Establishing a diagnosis is a professional standard within the scope of a physical therapist practice but may not be permitted according to the therapist's state practice act (Case Example 1-6).

As we have pointed out repeatedly, an organic problem can masquerade as a mechanical or movement dysfunction. Identification of causative factors or etiology by the physical therapist is important in the screening process. By remaining within the scope of our practice the diagnosis is limited primarily to those pathokinesiologic problems associated with faulty biomechanical or neuromuscular action.

When no apparent movement dysfunction, causative factors, or syndrome can be identified, the therapist may treat symptoms as part of an ongoing diagnostic process. Sometimes even physicians use physical therapy as a diagnostic tool, observing the client's response during the episode of care to confirm or rule out medical suspicions.

If, however, the findings remain inconsistent with what is expected for the human movement system and/or the patient/client does not improve with intervention,[16,42] then referral to

CASE EXAMPLE 1-7

Identify Mechanical Problems: Cervical Spine Arthrosis Presenting as Chest Pain

A 42-year-old woman presented with primary chest pain of unknown cause. She was employed as an independent pediatric occupational therapist. She has been seen by numerous medical doctors who have ruled out cardiac, pulmonary, esophageal, upper gastrointestinal (GI), and breast pathology as underlying etiologies.

Since her symptoms continued to persist, she was sent to physical therapy for an evaluation.

She reported symptoms of chest pain/discomfort across the upper chest rated as a 5 or 6 and sometimes an 8 on a scale from 0 to 10. The pain does not radiate down her arms or up her neck. She cannot bring the symptoms on or make them go away. She cannot point to the pain but reports it as being more diffuse than localized.

She denies any shortness of breath but admits to being "out of shape" and hasn't been able to exercise due to a failed bladder neck suspension surgery 2 years ago. She reports fatigue but states this is not unusual for her with her busy work schedule and home responsibilities.

She has not had any recent infections, no history of cancer or heart disease, and her mammogram and clinical breast exam are up-to-date and normal. She does not smoke or drink but by her own admission has a "poor diet" due to time pressure, stress, and fatigue.

Final Result: After completing the evaluation with appropriate questions, tests, and measures, a Review of Systems pointed to the cervical spine as the most likely source of this client's symptoms. The jaw and shoulder joint were cleared, although there were signs of shoulder movement dysfunction.

After relaying these findings to the client's primary care physician, radiographs of the cervical spine were ordered. Interestingly, despite the thousands of dollars spent on repeated diagnostic work-ups for this client, a simple x-ray had never been taken.

Results showed significant spurring and lipping throughout the cervical spine from early osteoarthritic changes of unknown cause. Cervical spine fusion was recommended and performed for instability in the midcervical region.

The client's chest pain was eliminated and did not return even up to 2 years after the cervical spine fusion. The physical therapist's contribution in pinpointing the location of referred symptoms brought this case to a successful conclusion.

an appropriate medical professional may be required. Always keep in mind that the screening process may, in fact, confirm the presence of a musculoskeletal or neuromuscular problem.

The flip side of this concept is that client complaints that cannot be associated with a medical problem should be referred to a physical therapist to identify mechanical problems (Case Example 1-7). Physical therapists have a responsibility to educate the medical community as to the scope of our practice and our role in identifying mechanical problems and movement disorders.

Staying within the scope of physical therapist practice, the therapist communicates with physicians and other health care practitioners to request or recommend further medical evaluation. Whether in a private practice, school or home health setting, acute care hospital, or rehabilitation setting, physical therapists may observe and report important findings outside the realm of NMS disorders that require additional medical evaluation and treatment.

DIRECT ACCESS AND SELF-REFERRAL

Direct access and self-referral is the legal right of the public to obtain examination, evaluation, and intervention from a licensed physical therapist without previous examination by, or referral from, a physician, gatekeeper, or other practitioner. In the civilian sector, the need to screen for medical disease was first raised as an issue in response to direct-access legislation. Until direct access, the only therapists screening for referral were the military physical therapists.

Before 1957 a physician referral was necessary in all 50 states for a client to be treated by a physical therapist. Direct access was first obtained in Nebraska in 1957, when that state passed a licensure and scope-of-practice law that did not mandate a physician referral for a physical therapist to initiate care.[43]

One of the goals of the APTA as outlined in the APTA 2020 vision statement is to achieve direct access to physical therapy services for citizens of all 50 states by the year 2020. At the present time, all but a handful of states in the United States permit some form of direct access and self-referral to allow patient/clients to consult a physical therapist without first being referred by a physician, dentist, or chiropractor. Direct access is relevant in all practice settings and is not limited just to private practice or outpatient services.

It is possible to have a state direct-access law but a state practice act that forbids therapists from seeing Medicare clients without a referral. A therapist in that state can see privately insured clients without a referral, but not Medicare clients. Passage of the Medicare Patient/Client Access to Physical Therapists Act (PAPTA) will extend direct access nationwide to all Medicare Part B beneficiaries who require outpatient physical therapy services, in states where direct access is authorized without a physician's referral or certification of the plan of care.

Full, unrestricted direct access is not available in all states with a direct-access law. Various forms of direct access are available on a state-by-state basis. Many direct-access laws are permissive, as opposed to mandatory. This means that consumers are permitted to see therapists without a physician's referral; however, a payer can still require a referral before providing reimbursement for services. Each therapist must be familiar with the practice act and direct-access legislation for the state in which he or she is practicing.

Sometimes states enact a two- or three-tiered restricted or provisional direct-access system. For example, some states' direct-access law only allows evaluation and treatment for therapists who have practiced for 3 years. Some direct-access laws only allow physical therapists to provide services for up to 14 days without physician referral. Other states list up to 30 days as the standard.

There may be additional criteria in place, such as the patient/client must have been referred to physical therapy by a physician within the past 2 years or the therapist must notify the patient/client's identified primary care practitioner no later than 3 days after intervention begins.

Some states require a minimum level of liability insurance coverage by each therapist. In a three tiered–direct access state, three or more requirements must be met before practicing without a physician referral. For example, licensed physical therapists must practice for a specified number of years, complete continuing education courses, and obtain references from two or more physicians before treating clients without a physician referral.

There are other factors that prevent therapists from practicing under full direct-access rights even when granted by state law. For example, Boissonnault[44] presents regulatory barriers and internal institutional policies that interfere with the direct access practice model.

In the private sector, some therapists think that the way to avoid malpractice lawsuits is to continue operating under a system of physician referral. Therapists in a private practice driven by physician referral may not want to be placed in a position as competitors of the physicians who serve as a referral source.

Internationally, direct access has become a reality in some, but not all, countries. It has been established in Australia, New Zealand, Canada, the United Kingdom, and the Netherlands. Direct access is not uniformly defined, implemented, or reimbursed from country to country.[45]

Primary Care

Primary care is the coordinated, comprehensive, and personal care provided on a first-contact and continuous basis. It incorporates primary and secondary prevention of chronic disease states, wellness, personal support, education (including providing information about illness, prevention, and health maintenance), and addresses the personal health care needs of patient/clients within the context of family and community.[25] Primary care is not defined by who provides it but rather it is a set of functions as described. It is person- (not disease- or diagnosis-) focused care over time.[46]

In the primary care delivery model, the therapist is responsible as a patient/client advocate to see that the patient/client's NMS and other health care needs are identified and prioritized, and a plan of care is established. The primary care model provides the consumer with first point-of-entry access to the physical therapist as the most skilled practitioner for human movement system dysfunction. The physical therapist may also serve as a key member of a multidisciplinary primary care team that works together to assist the patient/client in maintaining his or her overall health and fitness.

Through a process of screening, triage, examination, evaluation, referral, intervention, coordination of care, education, and prevention, the therapist prevents, reduces, slows, or remediates impairments, functional limitations, and disabilities while achieving cost-effective clinical outcomes.[1,47]

Expanded privileges beyond the traditional scope of the physical therapist practice may become part of the standard future physical therapist primary care practice. In addition to the usual privileges included in the scope of the physical therapist practice, the primary care therapist may eventually refer patient/clients to radiology for diagnostic imaging and other diagnostic evaluations. In some settings (e.g., U.S. military), the therapist is already doing this and is credentialed to prescribe analgesic and nonsteroidal antiinflammatory medications.[48]

Direct Access Versus Primary Care

Direct access is the vehicle by which the patient/client comes directly to the physical therapist without first seeing a physician, dentist, chiropractor, or other health care professional. Direct access does not describe the type of practice the therapist is engaging in.

Primary care physical therapy is not a setting but rather describes a philosophy of whole-person care. The therapist is the first point-of-entry into the health care system. After screening and triage, patient/clients who do not have NMS conditions are referred to the appropriate health care specialist for further evaluation.

The primary care therapist is not expected to diagnose conditions that are not neuromuscular or musculoskeletal. However, risk factor assessment and screening for a broad range of medical conditions (e.g., high blood pressure, incontinence, diabetes, vestibular dysfunction, peripheral vascular disease) is possible and an important part of primary and secondary prevention.

Autonomous Practice

Autonomous physical therapist practice is the centerpiece of the APTA Vision 2020 statement.[49] It is defined as "self-governing;" "not controlled (or owned) by others."[50] Autonomous practice is described as independent, self-determining professional judgment and action.[51] Autonomous practice for the physical therapist does not mean practice independent of collaborative and collegial communication with other health care team members (Box 1-6) but rather, interdependent evidence-based practice that is patient- (client-) centered care. Professional autonomy meets the health needs of people who are experiencing disablement by providing a service that supports the autonomy of that individual.[49]

Five key objectives set forth by the APTA in achieving an autonomous physical therapist practice include (1) demonstrating professionalism, (2) achieving direct access to physical therapist services, (3) basing practice on the most up-to-date evidence, (4) providing an entry-level education at the level of Doctor of Physical Therapy, and (5) becoming the practitioner of choice.[51]

BOX 1-6 ATTRIBUTES OF AUTONOMOUS PRACTICE

Direct and unrestricted access: The physical therapist has the professional capacity and ability to provide to all individuals the physical therapy services they choose without legal, regulatory, or payer restrictions

Professional ability to refer to other health care providers: The physical therapist has the professional capability and ability to refer to others in the health care system for identified or possible medical needs beyond the scope of physical therapist practice

Professional ability to refer to other professionals: The physical therapist has the professional capability and ability to refer to other professionals for identified or patient/client needs beyond the scope of physical therapy services

Professional ability to refer for diagnostic tests: The physical therapist has the professional capability and ability to refer for diagnostic tests that would clarify the patient/client situation and enhance the provision of physical therapy services

Data from American Physical Therapy Association. Board of Directors minutes (Program 32, Competencies of the Autonomous Physical Therapist Practitioner, BOD 11/01). Available at www.apta.org [governance Board of Directors policies Section 1–Professional and Societal page 41 of the PDF/page 36 of the actual document]. Accessed Sept. 1, 2010.

APTA Vision Statement for Physical Therapy 2020

Physical therapy, by 2020, will be provided by physical therapists who are doctors of physical therapy and who may be board-certified specialists. Consumers will have direct access to physical therapists in all environments for patient/client management, prevention, and wellness services.

Physical therapists will be practitioners of choice in patient/clients' health networks and will hold all privileges of autonomous practice. Physical therapists may be assisted by physical therapist assistants who are educated and licensed to provide physical therapist–directed and supervised components of interventions.[28]

Self-determination means the privilege of making one's own decisions, but only after key information has been obtained through examination, history, and consultation. The autonomous practitioner independently makes professional decisions based on a distinct or unique body of knowledge. For the physical therapist, that professional expertise is confined to the examination, evaluation, diagnosis, prognosis, and intervention of human movement system impairments.

Physical therapists have the capability, ability, and responsibility to exercise professional judgment within their scope of practice. In this context, the therapist must conduct a thorough examination, determine a diagnosis, and recognize when physical therapy is inappropriate, or when physical therapy is appropriate, but the client's condition is beyond the therapist's training, experience, or expertise. In such a case, referral is required, but referral may be to a qualified physical therapist who specializes in treating such disorders or conditions.[52,53]

Reimbursement Trends

Despite research findings that episodes of care for patient/clients who received physical therapy via direct access were shorter, included fewer numbers of services, and were less costly than episodes of care initiated through physician referral,[54] many payers, hospitals, and other institutions still require physician referral.[44,55]

Direct-access laws give consumers the legal right to seek physical therapy services without a medical referral. These laws do not always make it mandatory that insurance companies, third-party payers (including Medicare/Medicaid), self-insured, or other insurers reimburse the physical therapist without a physician's prescription.

Some state home-health agency license laws require referral for all client care regardless of the payer source. In the future, we hope to see all insurance companies reimburse for direct access without further restriction. Further legislation and regulation are needed in many states to amend the insurance statutes and state agency policies to assure statutory compliance.

This policy, along with large deductibles, poor reimbursement, and failure to authorize needed services has resulted in a trend toward a cash-based, private-pay business. This trend in reimbursement is also referred to as direct contracting, first-party payment, direct consumer services, or direct fee-for-service.[56] In such an environment, decisions can be made based on the good of the clients rather than on cost or volume.

In such circumstances, consumers are willing to pay out-of-pocket for physical therapy services, bypassing the need for a medical evaluation unless requested by the physical therapist. A therapist can use a cash-based practice only where direct access has been passed and within the legal parameters of the state practice act.

In any situation where authorization for further intervention by a therapist is not obtained despite the therapist's assessment that further skilled services are needed, the therapist can notify the client and/or the family of their right to an appeal with the agency providing health care coverage.

The client has the right to make informed decisions regarding pursuit of insurance coverage or to make private-pay arrangements. Too many times the insurance coverage ends, but the client's needs have not been met. Creative planning and alternate financial arrangements should remain an option discussed and made available.

DECISION-MAKING PROCESS

This text is designed to help students, physical therapist assistants, and physical therapy clinicians screen for medical disease when it is appropriate to do so. But just exactly how is this done? The proposed Goodman screening model can be used in conducting a screening evaluation for any client (Box 1-7).

BOX 1-7 GOODMAN SCREENING FOR REFERRAL MODEL

- Past medical history
- Personal and family history
- Risk factor assessment
- Clinical presentation
- Associated signs and symptoms of systemic diseases
- Review of systems

By using these decision-making tools, the therapist will be able to identify chief and secondary problems, identify information that is inconsistent with the presenting complaint, identify noncontributory information, generate a working hypothesis regarding possible causes of complaints, and determine whether referral or consultation is indicated.

The screening process is carried out through the client interview and verified during the physical examination. Therapists compare the subjective information (what the client tells us) with the objective findings (what we find during the examination) to identify movement impairment or other neuromuscular or musculoskeletal dysfunction (that which is within the scope of our practice) and to rule out systemic involvement (requiring medical referral). This is the basis for the evaluation process.

Given today's time constraints in the clinic, a fast and efficient method of screening is essential. Checklists (see Appendix A-1), special questions to ask (see companion website; see also Appendix B), and the screening model outlined in Box 1-7 can guide and streamline the screening process. Once the clinician is familiar with the use of this model, it is possible to conduct the initial screening exam in 3 to 5 minutes when necessary. This can include (but is not limited to) the following:

- Take vital signs
- Use the word "symptom(s)" rather than "pain" during the screening interview
- Watch for red flag histories, signs, and symptoms
- Review medications; observe for signs and symptoms that could be a result of drug combinations (polypharmacy), dual drug dosage; consult with the pharmacist
- Ask a final open-ended question such as:
  1. Are you having any other symptoms of any kind anywhere else in your body we haven't talked about yet?
  2. Is there anything else you think is important about your condition that we haven't discussed yet?

If a young, healthy athlete comes in with a sprained ankle and no other associated signs and symptoms, there may be no need to screen further. But if that same athlete has an eating disorder, uses anabolic steroids illegally, or is on antidepressants, the clinical picture (and possibly the intervention) changes. Risk factor assessment and a screening physical examination are the most likely ways to screen more thoroughly.

Or take, for example, an older adult who presents with hip pain of unknown cause. There are two red flags already present (age and insidious onset). As clients age, the past medical history and risk factor assessment become more important assessment tools. After investigating the clinical presentation, screening would focus on these two elements next.

Or, if after ending the interview by asking, "Are there any symptoms of any kind anywhere else in your body that we have not talked about yet?" the client responds with a list of additional symptoms, it may be best to step back and conduct a Review of Systems.

Past Medical History

Most of history taking is accomplished through the client interview and includes both family and personal history. The client/patient interview is very important because it helps the physical therapist distinguish between problems that he or she can treat and problems that should be referred to a physician (or other appropriate health care professional) for medical diagnosis and intervention.

In fact, the importance of history taking cannot be emphasized enough. Physicians cite a shortage of time as the most common reason to skip the client history, yet history taking is the essential key to a correct diagnosis by the physician (or physical therapist).[57,58] At least one source recommends performing a *history and differential diagnosis* followed by *relevant examination.*[58]

In Chapter 2, an interviewing process is described that includes concrete and structured tools and techniques for conducting a thorough and informative interview. The use of follow-up questions (FUPs) helps complete the interview. This information establishes a solid basis for the therapist's objective evaluation, assessment, and therefore intervention.

During the screening interview it is always a good idea to use a standard form to complete the personal/family history (see Fig. 2-2). Any form of checklist assures a thorough and consistent approach and spares the therapist from relying on his or her memory.

The types of data generated from a client history are presented in Fig. 2-1. Most often, age, race/ethnicity, gender, and occupation (general demographics) are noted. Information about social history, living environment, health status, functional status, and activity level is often important to the patient/client's clinical presentation and outcomes. Details about the current condition, medical (or other) intervention for the condition, and use of medications is also gathered and considered in the overall evaluation process.

The presence of any yellow or red flags elicited during the screening interview or observed during the physical examination should prompt the therapist to consider the need for further tests and questions. Many of these signs and symptoms are listed in Appendix A-2.

Psychosocial history may provide insight into the client's clinical presentation and overall needs. Age, gender, race/ethnicity, education, occupation, family system, health habits, living environment, medication use, and medical/surgical history are all part of the client history evaluated in the screening process.

Risk Factor Assessment

Greater emphasis has been placed on risk factor assessment in the health care industry recently. Risk factor assessment is an important part of disease prevention. Knowing the various risk factors for different kinds of diseases, illnesses, and conditions is an important part of the screening process.

Therapists can have an active role in both primary and secondary prevention through screening and education. According to the *Guide,*[1] physical therapists are involved in primary prevention by preventing a target condition in a susceptible or potentially susceptible population through such specific measures as general health promotion efforts.

Educating clients about their risk factors is a key element in risk factor reduction. Identifying risk factors may guide the therapist in making a medical referral sooner than would otherwise seem necessary.

In primary care, the therapist assesses risk factors, performs screening exams, and establishes interventions to prevent impairment, dysfunction, and disability. For example, does the client have risk factors for osteoporosis, urinary incontinence, cancer, vestibular or balance problems, obesity, cardiovascular disease, and so on? The physical therapist practice can include routine screening for any of these, as well as other problems.

More and more evidence-based clinical prediction rules for specific conditions (e.g., deep venous thrombosis) are available and included in this text; research is needed to catch up in the area of clinical prediction rules and identification of specificity and sensitivity of specific red flags and screening tests currently being presented in this text and used in clinical practice. Prediction models based on risk that would improve outcomes may eventually be developed for many diseases, illnesses, and conditions currently screened by red flags and clinical findings.[59,60]

Eventually, genetic screening may augment or even replace risk factor assessment. Virtually every human illness is believed to have a hereditary component. The most common problems seen in a physical therapist practice (outside of traumatic injuries) are now thought to have a genetic component, even though the specific gene may not yet be discovered for all conditions, diseases, or illnesses.[61]

Exercise as a successful intervention for many diseases, illness, and conditions will become prescriptive as research shows how much and what kind of exercise can prevent or mediate each problem. There is already a great deal of information on this topic published, and an accompanying need to change the way people think about exercise.[62]

Convincing people to establish lifelong patterns of exercise and physical activity will continue to be a major focus of the health care industry. Therapists can advocate disease prevention, wellness, and promotion of healthy lifestyles by delivering health care services intended to prevent health problems or maintain health and by offering annual wellness screening as part of primary prevention.

Clinical Presentation

Clinical presentation, including pain patterns and pain types, is the next part of the decision-making process. To assist the physical therapist in making a treatment-versus-referral decision, specific pain patterns corresponding to systemic diseases are provided in Chapter 3. Drawings of primary and referred pain patterns are provided in each chapter for quick reference. A summary of key findings associated with systemic illness is listed in Box 1-2.

The presence of any one of these variables is not cause for extreme concern but should raise a yellow or red flag for the alert therapist. The therapist is looking for a pattern that suggests a viscerogenic or systemic origin of pain and/or symptoms. This pattern will not be consistent with what we might expect to see with the neuromuscular or musculoskeletal systems.

The therapist will proceed with the screening process, depending on all findings. Often the next step is to look for associated signs and symptoms. Special follow-up questions (FUPs) are listed in the subjective examination to help the physical therapist determine when these pain patterns are accompanied by associated signs and symptoms that indicate visceral involvement.

Associated Signs and Symptoms of Systemic Diseases

The major focus of this text is the recognition of yellow- or red-flag signs and symptoms either reported by the client subjectively or observed objectively by the physical therapist.

Signs are observable findings detected by the therapist in an objective examination (e.g., unusual skin color, clubbing of the fingers [swelling of the terminal phalanges of the fingers or toes], hematoma [local collection of blood], effusion [fluid]). Signs can be seen, heard, smelled, measured, photographed, shown to someone else, or documented in some other way.

Symptoms are reported indications of disease that are perceived by the client but cannot be observed by someone else. Pain, discomfort, or other complaints, such as numbness, tingling, or "creeping" sensations, are symptoms that are difficult to quantify but are most often reported as the chief complaint.

Because physical therapists spend a considerable amount of time investigating pain, it is easy to remain focused exclusively on this symptom when clients might otherwise bring to the forefront other important problems.

Thus the physical therapist is encouraged to become accustomed to using the word symptoms instead of pain when interviewing the client. It is likewise prudent for the physical therapist to refer to symptoms when talking to clients with chronic pain in order to move the focus away from pain.

Instead of asking the client, "How are you today?" try asking:

FOLLOW-UP QUESTIONS

- Are you better, same, or worse today?
- What can you do today that you couldn't do yesterday? (Or last week/last month?)

This approach to questioning progress (or lack of progress) may help you see a systemic pattern sooner than later.

The therapist can identify the presence of associated signs and symptoms by asking the client:

FOLLOW-UP QUESTIONS

- Are there any symptoms of any kind anywhere else in your body that we have not talked about yet?
- *Alternately:* Are there any symptoms or problems anywhere else in your body that may not be related to your current problem?

The patient/client may not see a connection between shoulder pain and blood in the urine from kidney impairment or blood in the stools from chronic nonsteroidal anti-inflammatory drug (NSAID) use. Likewise the patient/client may not think the diarrhea present is associated with the back pain (gastrointestinal [GI] dysfunction).

The client with temporomandibular joint (TMJ) pain from a cardiac source usually has some other associated symptoms, and in most cases, the client does not see the link. If the therapist does not ask, the client does not offer the extra information.

Each visceral system has a typical set of core signs and symptoms associated with impairment of that system (see Box 4-19). Systemic signs and symptoms that are listed for each condition should serve as a warning to alert the informed physical therapist of the need for further questioning and possible medical referral.

For example, the most common symptoms present with pulmonary pathology are cough, shortness of breath, and pleural pain. Liver impairment is marked by abdominal ascites, right upper quadrant tenderness, jaundice, and skin and nailbed changes. Signs and symptoms associated with *endocrine* pathology may include changes in body or skin temperature, dry mouth, dizziness, weight change, or excessive sweating.

Being aware of signs and symptoms associated with each individual system may help the therapist make an early connection between viscerogenic and/or systemic presentation of NMS problems. The presence of constitutional symptoms is always a red flag that must be evaluated carefully (see Box 1-3).

Systems Review Versus Review of Systems

The Systems Review is defined in the *Guide* as a brief or limited exam of the anatomic and physiologic status of the cardiovascular/pulmonary, integumentary, musculoskeletal, and neuromuscular systems. The Systems Review also includes assessment of the client's communication ability, affect, cognition, language, and learning style.

The specific tests and measures for this type of Systems Review are outlined in the *Guide*[1] (Appendix 5, Guidelines for Physical Therapy Documentation, pp. 695-696). As part of this Systems Review, the client's ability to communicate, process information, and any barriers to learning are identified.

The Systems Review looks beyond the primary problem that brought the client to the therapist in the first place. It gives an overview of the "whole person," and guides the therapist in choosing appropriate tests and measures. The Systems Review helps the therapist answer the questions, "What should I do next?" and "What do I need to examine in depth?" It also answers the question, "What don't I need to do?"[63]

In the screening process, a slightly different approach may be needed, perhaps best referred to as a Review of Systems. After conducting an interview, performing an assessment of the pain type and/or pain patterns, and reviewing the clinical presentation, the therapist looks for any characteristics of systemic disease. Any identified clusters of associated signs and symptoms are reviewed to search for a potential pattern that will identify the underlying system involved.

The Review of Systems as part of the screening process (see discussion, Chapter 4) is a useful tool in recognizing clusters of associated signs and symptoms and the possible need for medical referral. Using this final tool, the therapist steps back and looks at the big picture, taking into consideration all of the presenting factors, and looking for any indication that the client's problem is outside the scope of a physical therapist's practice.

The therapist conducts a Review of Systems in the screening process by categorizing all of the complaints and associated signs and symptoms. Once these are listed, compare this list to Box 4-19. Are the signs and symptoms all genitourinary (GU) related? GI in nature?

Perhaps the therapist observes dry skin, brittle nails, cold or heat intolerance, or excessive hair loss and realizes these signs could be pointing to an endocrine problem. At the very least the therapist recognizes that the clinical presentation is not something within the musculoskeletal or neuromuscular systems.

If, for example, the client's signs and symptoms fall primarily within the GU group, turn to Chapter 10 and use the additional, pertinent screening questions at the end of the chapter. The client's answers to these questions will guide the therapist in making a decision about referral to a physician or other health care professional.

The physical therapist is not responsible for identifying the specific systemic or visceral disease underlying the clinical signs and symptoms present. However, the alert therapist who classifies groups of signs and symptoms in a Review of Systems will be more likely to recognize a problem outside the scope of physical therapy practice and make a timely referral.

As a final note in this discussion of Systems Review versus Review of Systems, there is some consideration being given to possibly changing the terminology in the *Guide* to reflect the full measure of these concepts, but no definitive decision had been made by the time this text went to press. The concept will be discussed, and any decision made will go through both an expert and wide review process. Results will be reflected in future editions of this text.

CASE EXAMPLES AND CASE STUDIES

Case examples and case studies are provided with each chapter to give the therapist a working understanding of how to recognize the need for additional questions. In addition, information is given concerning the type of questions to ask and how to correlate the results with the objective findings.

Cases will be used to integrate screening information in making a physical therapy differential diagnosis and deciding when and how to refer to the physician or other health care professional. Whenever possible, information about when and how to refer a client to the physician is presented.

Each case study is based on actual clinical experiences in a variety of inpatient/client and outpatient/client physical therapy practices to provide reasonable examples of what to expect when the physical therapist is functioning under any of the circumstances listed in Box 1-1.

PHYSICIAN REFERRAL

As previously mentioned, the therapist may treat symptoms as part of an ongoing medical diagnostic process. In other words, sometimes the physician sends a patient/client to physical therapy "to see if it will help." This may be part of the medical differential diagnosis. Medical consultation or referral is required when no apparent movement dysfunction, causative factors, or syndrome can be identified and/or the findings are not consistent with a NMS dysfunction.

Communication with the physician is a key component in the referral process. Phone, email, and fax make this process faster and easier than ever before. Persistence may be required in obtaining enough information to glean what the doctor knows or thinks to avoid sending the very same problem back for his/her consideration. This is especially important when the physician is using physical therapy intervention as part of the medical differential diagnostic process.

The hallmark of professionalism in any health care practitioner is the ability to understand the limits of his or her professional knowledge. The physical therapist, either on reaching the limit of his or her knowledge or on reaching the limits prescribed by the client's condition, should refer the patient/client to the appropriate personnel. In this way, the physical therapist will work within the scope of his or her level of skill, knowledge, and practical experience.

Knowing when and how to refer a client to another health care professional is just as important as the initial screening process. Once the therapist recognizes red flag histories, risk factors, signs and symptoms, and/or a clinical presentation that do not fit the expected picture for NMS dysfunction, then this information must be communicated effectively to the appropriate referral source.

Knowing how to refer the client or how to notify the physician of important findings is not always clear. In a direct access or primary care setting, the client may not have a personal or family physician. In an orthopedic setting, the client in rehab for a total hip or total knee replacement may be reporting signs and symptoms of a nonorthopedic condition. Do you send the client back to the referring (orthopedic) physician or refer him or her to the primary care physician?

Suggested Guidelines

When the client has come to physical therapy without a medical referral (i.e., self-referred) and the physical therapist recommends medical follow-up, the patient/client should be referred to the primary care physician if the patient/client has one.

Occasionally, the patient/client indicates that he or she has not contacted a physician or was treated by a physician (whose name cannot be recalled) a long time ago or that he or she has just moved to the area and does not have a physician.

In these situations the client can be provided with a list of recommended physicians. It is not necessary to list every physician in the area, but the physical therapist can provide several appropriate choices. Whether the client makes or does not make an appointment with a medical practitioner, the physical therapist is urged to document subjective and objective findings carefully, as well as the recommendation made for medical follow-up. The therapist should make every effort to get the physical therapy records to the consulting physician.

Before sending a client back to his or her doctor, have someone else (e.g., case manager, physical therapy colleague or mentor, nursing staff if available) double check your findings and discuss your reasons for referral. Recheck your own findings at a second appointment. Are they consistent?

Consider checking with the medical doctor by telephone. Perhaps the physician is aware of the problem, but the therapist does not have the patient/client records and is unaware of this fact. As mentioned it is not uncommon for physicians to send a client to physical therapy as part of their own differential diagnostic process. For example, they may have tried medications without success and the client does not want surgery or more drugs. The doctor may say, "Let's try physical therapy. If that doesn't change the picture, the next step is ..."

As a general rule, try to send the client back to the referring physician. If this does not seem appropriate, call and ask the physician how he or she wants to handle the situation. Describe the problem and ask:

FOLLOW-UP QUESTIONS

- Do you want Mr. X/Mrs. Y to check with his/her family doctor ... or do you prefer to see him/her yourself?

Perhaps an orthopedic client is demonstrating signs and symptoms of depression. This may be a side effect from medications prescribed by another physician (e.g., gynecologist, gastroenterologist). Provide the physician with a list of the observed cluster of signs and symptoms and an open-ended question such as:

FOLLOW-UP QUESTIONS

- How do you want to handle this? or How do you want me to handle this?

Do not suggest a medical diagnosis. When providing written documentation, a short paragraph of physical therapy findings and intervention is followed by a list of concerns, perhaps with the following remarks, "These do not seem consistent with a neuromuscular or musculoskeletal problem (choose the most appropriate description of the human movement system syndrome/problem or name the medical diagnosis [e.g., S/P THR])." Then follow-up with one of two questions/comments:

FOLLOW-UP QUESTIONS

- What do you think? or Please advise.

Special Considerations

What if the physician refuses to see the client or finds nothing wrong? We recommend being patiently persistent. Sometimes it is necessary to wait until the disease progresses to a point that medical testing can provide a diagnosis. This is unfortunate for the client but a reality in some cases.

Sometimes it may seem like a good idea to suggest a second opinion. You may want to ask your client:

FOLLOW-UP QUESTIONS

- Have you ever thought about getting a second opinion?

It is best not to tell the client what to do. If the client asks you what he or she should do, consider asking this question:

FOLLOW-UP QUESTIONS

- What do you think your options are? or What are your options?

It is perfectly acceptable to provide a list of names (more than one) where the client can get a second opinion. If the client asks which one to see, suggest whoever is closest geographically or with whom he or she can get an appointment as soon as possible.

What do you do if the client's follow-up appointment is scheduled 2 weeks away and you think immediate medical attention is needed? Call the physician's office and see what is advised: does the physician want to see the client in the office or send him/her to the emergency department?

For example, what if a patient/client with a recent total hip replacement develops chest pain and shortness of breath during exercise? The client also reports a skin rash around the surgical site. This will not wait for 2 weeks. Take the client's vital signs (especially body temperature in case of infection) and report these to the physician. In some cases the need for medical care will be obvious such as in the case of acute myocardial infarct or if the client collapses.

Documentation and Liability

Documentation is any entry into the patient/client record. Documentation may include consultation reports, initial examination reports, progress notes, recap of discussions with physicians or other health care professionals, flow sheets, checklists, reexamination reports, discharge summaries, and so on.[1] Various forms are available for use in the *Guide* to aid in collecting data in a standardized fashion. Remember, in all circumstances, in a court of law, if you did not document it, you did not do it (a common catch phrase is "not documented, not done").

The U.S. Department of Health and Human Services (HHS) is taking steps in building a national electronic health care system that will allow patient/clients and health care providers access to their complete medical records anytime and anywhere they are needed, leading to reduced medical errors, improved care, and reduced health care costs. The goal is to have digital health records for most Americans by the year 2014.[64]

Documentation is required at the onset of each episode of physical therapy care and includes the elements described in Box 1-5. Documentation of the initial episode of physical therapy care includes examination, comprehensive screening, and specific testing leading to a diagnostic classification and/or referral to another practitioner (*Guide,* p. 695).[1]

Clients with complex medical histories and multiple comorbidities are increasingly common in a physical therapist's practice. Risk management has become an important consideration for many clients. Documentation and communication must reflect this practice.

Sometimes the therapist will have to be more proactive and assertive in communicating with the client's physician. It may not be enough to suggest or advise the client to make a follow-up appointment with his or her doctor. Leaving the decision up to the client is a passive and indirect approach. It does encourage client/consumer responsibility but may not be in his/her best interest.

In the APTA *Standards of Practice and the Criteria* (HOD 06-00-11-22), it states, "The physical therapy service collaborates with all disciplines as appropriate [Administration of the Physical Therapy Service, Section II, Item J]. In HOD 06-90-15-28 *(Referral Relationships),* it states, "The physical therapist must refer patients/clients to the referring practitioner or other health care practitioners if symptoms are present for which physical therapy is contraindicated or are indicative of conditions for which treatment is outside the scope of his/her knowledge."[65]

In cases where the seriousness of the condition can affect the client's outcome, the therapist may need to contact the physician directly and describe the problem. If the therapist's assessment is that the client needs medical attention, advising the client to see a medical doctor as soon as possible may not be enough.

Good risk management is a proactive process that includes taking action to minimize negative outcomes. If a client is advised to contact his or her physician and fails to do so, the therapist should call the doctor.[66]

Failure on the part of the therapist to properly report on a client's condition or important changes in condition reflects a lack of professional judgment in the management of the client's case. A number of positions and standards of the APTA Board of Directors emphasize the importance of physical therapist communication and collaboration with other health care providers. This is a key to providing the best possible client care (Case Example 1-8).[67]

HOD 06-97-06-19 (Policy on *Diagnosis by Physical Therapists*) states that, "as the diagnostic process continues, physical therapists may identify findings that should be shared with other health professionals, including referral sources, to ensure optimal patient/client care." Part of this process may require "appropriate follow-up or referral."

Failure to share findings and concerns with the physician or other appropriate health care provider is a failure to enter into a collaborative team approach. Best-practice standards of optimal patient/client care support and encourage interactive exchange.

Prior negative experiences with difficult medical personnel do not exempt the therapist from best practice, which means making every attempt to communicate and document clinical findings and concerns.

The therapist must describe his or her concerns. Using the key phrase "scope of practice" may be helpful. It may be necessary to explain that the symptoms do not match the expected pattern for a musculoskeletal or neuromuscular problem. The problem appears to be outside the scope of a physical therapist's practice ... , or the problem requires a greater collaborative effort between health care disciplines.

It may be appropriate to make a summary statement regarding key objective findings with a follow-up question for the physician. This may be filed in the client's chart or electronic medical record in the hospital or sent in a letter to the outpatient/client's physician (or other health care provider).

For example, after treatment of a person who has not responded to physical therapy, a report to the physician may include additional information: "Miss Jones reported a skin rash over the backs of her knees 2 weeks before the onset of joint pain and experiences recurrent bouts of sore throat and fever when her knees flare up. These features are not consistent with an athletic injury. Would you please take a look?" (For an additional sample letter, see Fig. 1-6.)

Other useful wording may include "Please advise" or "What do you think?" The therapist does not suggest a

CASE EXAMPLE 1-8

Failure to Collaborate and Communicate with the Physician

A 43-year-old woman was riding a bicycle when she was struck from behind and thrown to the ground. She was seen at the local walk-in clinic and released with a prescription for painkillers and muscle relaxants. X-rays of her head and neck were unremarkable for obvious injury.

She came to the physical therapy clinic 3 days later with complaints of left shoulder, rib, and wrist pain. There was obvious bruising along the left chest wall and upper abdomen. In fact, the ecchymosis was quite extensive and black in color indicating a large area of blood extravasation into the subcutaneous tissues.

She had no other complaints or problems. Shoulder range of motion was full in all planes, although painful and stiff. Ribs 9, 10, and 11 were painful to palpation but without obvious deformity or derangement.

A neurologic screening exam was negative. The therapist scheduled her for 3 visits over the next 4 days and started her on a program of Codman's exercises, progressing to active shoulder motion. The client experienced progress over the next 5 days and then reported severe back muscle spasms.

The client called the therapist and cancelled her next appointment because she had the flu with fever and vomiting. When she returned, the therapist continued to treat her with active exercise progressing to resistive strengthening. The client's painful shoulder and back symptoms remained the same, but the client reported that she was "less stiff."

Three weeks after the initial accident, the client collapsed at work and had to be transported to the hospital for emergency surgery. Her spleen had been damaged by the initial trauma with a slow bleed that eventually ruptured.

The client filed a lawsuit in which the therapist was named. The complaint against the therapist was that she failed to properly assess the client's condition and failed to refer her to a medical doctor for a condition outside the scope of physical therapy practice.

Did the Physical Therapist Show Questionable Professional Judgment in the Evaluation and Management of this Case? There are some obvious red-flag signs and symptoms in this case that went unreported to a medical doctor. There was no contact with the physician at any time throughout this client's physical therapy episode of care. The physician on-call at the walk-in clinic did not refer the client to physical therapy—she referred herself.

However, the physical therapist did not send the physician any information about the client's self-referral, physical therapy evaluation, or planned treatment.

Subcutaneous blood extravasation is not uncommon after a significant accident or traumatic impact such as this client experienced. The fact that the physician did not know about this and the physical therapist did not report it demonstrates questionable judgment. Left shoulder pain after trauma may be Kehr's sign, indicating blood in the peritoneum (see the discussion in Chapter 18).

The new onset of muscle spasm and unchanging pain levels with treatment are potential red-flag symptoms. Concomitant constitutional symptoms of fever and vomiting are also red flags, even if the client thought it was the flu.

The therapist left herself open to legal action by failing to report symptoms unknown to the physician and failing to report the client's changing condition. At no time did the therapist suggest the client go back to the clinic or see a primary care physician. She did not share her findings with the physician either by phone or in writing.

The therapist exercised questionable professional judgment by failing to communicate and collaborate with the attending physician. She did not screen the client for systemic involvement, based on the erroneous thinking that this was a traumatic event with a clear etiology.

She assumed in a case like this where the client was a self-referral and the physician was a "doc-in-a-box" that she was "on her own." She failed to properly report on the client's condition, failed to follow the APTA's policies governing a physical therapist's interaction with other health care providers, and was legally liable for mismanagement in this case.

medical cause or attempt to diagnose the findings medically. Providing a report and stating that the clinical presentation does not follow a typical neuromuscular or musculoskeletal pattern may be all that is needed.

Guidelines for Immediate Medical Attention

After each chapter in this text, there is a section on Guidelines for Physician Referral. Guidelines for immediate medical attention are provided whenever possible. An overall summary is provided here, but specifics for each viscerogenic system and NMS situation should be reviewed in each chapter as well.

Keep in mind that prompt referral is based on the physical therapist's overall evaluation of client history and clinical presentation, including red/yellow flag findings and associated signs and symptoms. The recent focus on validity, reliability, specificity, and sensitivity of individual red flags has shown that there is little evidence on the diagnostic accuracy of red flags in the primary care medical (physician) practice.[68]

Experts agree that red flags are important and ignoring them can result in morbidity and even mortality for some individuals. On the other hand, accepting them uncritically can result in unnecessary referrals.[69] Until the evidence supporting or refuting red flags is complete, the therapist is advised to consider all findings in context of the total picture.

For now, immediate medical attention is still advised when:

- Client with anginal pain not relieved in 20 minutes with reduced activity and/or administration of nitroglycerin; angina at rest
- Client with angina has nausea, vomiting, profuse sweating

Referral. A 32-year-old female university student was referred for physical therapy through the student health service 2 weeks ago. The physician's referral reads: "Possible right oblique abdominis tear/possible right iliopsoas tear." A faculty member screened this woman initially, and the diagnosis was confirmed as being a right oblique abdominal strain.

History. Two months ago, while the client was running her third mile, she felt "severe pain" in the right side of her stomach, which caused her to double over. She felt immediate nausea and had abdominal distention. She cannot relieve the pain by changing the position of her leg. Currently, she still cannot run without pain.

Presenting Symptoms. Pain increases during sit-ups, walking fast, reaching, turning, and bending. Pain is eased by heat and is reduced by activity. Pain in the morning versus in the evening depends on body position. Once the pain starts, it is intermittent and aches. The client describes the pain as being severe, depending on her body position. She is currently taking aspirin when necessary.

SAMPLE LETTER

Date

John Smith, M.D.
University of Montana Health Service
Eddy Street
Missoula, MT 59812

Re: Jane Doe

Dear Dr. Smith,

Your client, Jane Doe, was evaluated in our clinic on 5/2/11 with the following pertinent findings:

She has severe pain in the right lower abdominal quadrant associated with nausea and abdominal distention. Although the onset of symptoms started while the client was running, she denies any precipitating trauma. She describes the course of symptoms as having begun 2 months ago with temporary resolution and now with exacerbation of earlier symptoms. Additionally, she reports chronic fatigue and frequent night sweats.

Presenting pain is reproduced by resisted hip or trunk flexion with accompanying tenderness/tightness on palpation of the right iliopsoas muscle (compared with the left iliopsoas muscle). There are no implicating neurologic signs or symptoms.

Evaluation. A musculoskeletal screening examination is consistent with the proposed medical diagnosis of a possible iliopsoas or abdominal oblique tear. Jane does appear to have a combination of musculoskeletal and systemic symptoms, such as those outlined earlier. Of particular concern are the symptoms of fatigue, night sweats, abdominal distention, nausea, repeated episodes of exacerbation and remission, and severe quality of pain and location (right lower abdominal quadrant). These symptoms appear to be of a systemic nature rather than caused by a primary musculoskeletal lesion.

Recommendations. The client has been advised to return to you for further medical follow-up to rule out any systemic involvement before the initiation of physical therapy services. I am concerned that my proposed plan of care, including soft tissue mobilization and stretching may aggravate an underlying infectious or disease process.

I will contact you directly by telephone by the end of the week to discuss these findings and to answer any questions that you may have. Thank you for this interesting referral.

Sincerely,

Catherine C. Goodman, M.B.A., P.T.

Result. This client returned to the physician, who then ordered laboratory tests. After an acute recurrence of the symptoms described earlier, she had exploratory surgery. A diagnosis of a ruptured appendix and peritonitis was determined at surgery. In retrospect, the proposed plan of care would have been contraindicated in this situation.

Fig. 1-6 Sample letter of the physical therapist's findings that is sent to the referring physician.

- Client presents with bowel/bladder incontinence and/or saddle anesthesia secondary to cauda equina lesion or cervical spine pain concomitant with urinary incontinence
- Client is in anaphylactic shock (see Chapter 12)
- Client has symptoms of inadequate ventilation or CO_2 retention (see the section on Respiratory Acidosis in Chapter 7)
- Client with diabetes appears confused or lethargic or exhibits changes in mental function (perform fingerstick glucose testing and report findings)
- Client has positive McBurney's point (appendicitis) or rebound tenderness (inflamed peritoneum) (see Chapter 8)
- Sudden worsening of intermittent claudication may be due to thromboembolism and must be reported to the physician immediately
- Throbbing chest, back, or abdominal pain that increases with exertion accompanied by a sensation of a heartbeat when lying down and palpable pulsating abdominal mass may indicate an aneurysm
- Changes in size, shape, tenderness, and consistency of lymph nodes; detection of palpable, fixed, irregular mass in the breast, axilla, or elsewhere, especially in the presence of a previous history of cancer

Guidelines for Physician Referral

Medical attention must be considered when any of the following are present. This list represents a general overview of warning flags or conditions presented throughout this text. More specific recommendations are made in each chapter based on impairment of each individual visceral system.

General Systemic

- Unknown cause
- Lack of significant objective NMS signs and symptoms
- Lack of expected progress with physical therapy intervention
- Development of constitutional symptoms or associated signs and symptoms any time during the episode of care
- Discovery of significant past medical history unknown to physician
- Changes in health status that persist 7 to 10 days beyond expected time period
- Client who is jaundiced and has not been diagnosed or treated

For Women

- Low back, hip, pelvic, groin, or sacroiliac symptoms without known etiologic basis and in the presence of constitutional symptoms
- Symptoms correlated with menses
- Any spontaneous uterine bleeding after menopause
- For pregnant women:
  - Vaginal bleeding
  - Elevated blood pressure
  - Increased Braxton-Hicks (uterine) contractions in a pregnant woman during exercise

Vital Signs (Report These Findings)

- Persistent rise or fall of blood pressure
- Blood pressure elevation in any woman taking birth control pills (should be closely monitored by her physician)
- Pulse amplitude that fades with inspiration and strengthens with expiration
- Pulse increase over 20 bpm lasting more than 3 minutes after rest or changing position
- Difference in pulse pressure (between systolic and diastolic measurements) of more than 40 mm Hg
- Persistent low-grade (or higher) fever, especially associated with constitutional symptoms, most commonly sweats
- Any unexplained fever without other systemic symptoms, especially in the person taking corticosteroids
- See also yellow cautionary signs presented in Box 4-7 and the section on Physician Referral: Vital Signs in Chapter 4

Cardiac

- More than three sublingual nitroglycerin tablets required to gain relief from angina
- Angina continues to increase in intensity after stimulus (e.g., cold, stress, exertion) has been eliminated
- Changes in pattern of angina
- Abnormally severe chest pain
- Anginal pain radiates to jaw/left arm
- Upper back feels abnormally cool, sweaty, or moist to touch
- Client has any doubts about his or her condition
- Palpitation in any person with a history of unexplained sudden death in the family requires medical evaluation; more than six episodes of palpitation in 1 minute or palpitations lasting for hours or occurring in association with pain, shortness of breath, fainting, or severe lightheadedness requires medical evaluation
- Clients who are neurologically unstable as a result of a recent cerebrovascular accident (CVA), head trauma, spinal cord injury, or other central nervous system insult often exhibit new arrhythmias during the period of instability; when the client's pulse is monitored, any new arrhythmias noted should be reported to the nursing staff or physician
- Anyone who cannot climb a single flight of stairs without feeling moderately to severely winded or who awakens at night or experiences shortness of breath when lying down should be evaluated by a physician
- Anyone with known cardiac involvement who develops progressively worse dyspnea should notify the physician of these findings
- Fainting (syncope) without any warning period of lightheadedness, dizziness, or nausea may be a sign of heart valve or arrhythmia problems; unexplained syncope in the presence of heart or circulatory problems (or risk factors

for heart attack or stroke) should be evaluated by a physician

Cancer

Early warning sign(s) of cancer:

- The CAUTIONS mnemonic for early warning signs is pertinent to the physical therapy examination (see Box 13-1)
- All soft tissue lumps that persist or grow, whether painful or painless
- Any woman presenting with chest, breast, axillary, or shoulder pain of unknown etiologic basis, especially in the presence of a positive medical history (self or family) of cancer
- Any man with pelvic, groin, sacroiliac, or low back pain accompanied by sciatica and a history of prostate cancer
- New onset of acute back pain in anyone with a previous history of cancer
- Bone pain, especially on weight-bearing, that persists more than 1 week and is worse at night
- Any unexplained bleeding from any area

Pulmonary

- Shoulder pain aggravated by respiratory movements; have the client hold his or her breath and reassess symptoms; any reduction or elimination of symptoms with breath holding or the Valsalva maneuver suggests pulmonary or cardiac source of symptoms
- Shoulder pain that is aggravated by supine positioning; pain that is worse when lying down and improves when sitting up or leaning forward is often pleuritic in origin (abdominal contents push up against diaphragm and in turn against parietal pleura; see Figs. 3-4 and 3-5)
- Shoulder or chest (thorax) pain that subsides with autosplinting (lying on painful side)
- For the client with asthma: Signs of asthma or abnormal bronchial activity during exercise
- Weak and rapid pulse accompanied by fall in blood pressure (pneumothorax)
- Presence of associated signs and symptoms, such as persistent cough, dyspnea (rest or exertional), or constitutional symptoms (see Box 1-3)

Genitourinary

- Abnormal urinary constituents, for example, change in color, odor, amount, flow of urine
- Any amount of blood in urine
- Cervical spine pain accompanied by urinary incontinence (unless cervical disk protrusion already has been medically diagnosed)

Gastrointestinal

- Back pain and abdominal pain at the same level, especially when accompanied by constitutional symptoms
- Back pain of unknown cause in a person with a history of cancer
- Back pain or shoulder pain in a person taking NSAIDs, especially when accompanied by GI upset or blood in the stools
- Back or shoulder pain associated with meals or back pain relieved by a bowel movement

Musculoskeletal

- Symptoms that seem out of proportion to the injury or symptoms persisting beyond the expected time for the nature of the injury
- Severe or progressive back pain accompanied by constitutional symptoms, especially fever
- New onset of joint pain following surgery with inflammatory signs (warmth, redness, tenderness, swelling)

Precautions/Contraindications to Therapy

- Uncontrolled chronic heart failure or pulmonary edema
- Active myocarditis
- Resting heart rate 120 or 130 bpm*
- Resting systolic rate 180 to 200 mm Hg*
- Resting diastolic rate 105 to 110 mm Hg*
- Moderate dizziness, near-syncope
- Marked dyspnea
- Unusual fatigue
- Unsteadiness
- Irregular pulse with symptoms of dizziness, nausea, or shortness of breath or loss of palpable pulse
- Postoperative posterior calf pain
- For the client with diabetes: Chronically unstable blood sugar levels must be stabilized (fasting target glucose range: 60 to 110 mg/dL; precaution: <70 or >250 mg/dL)

Clues to Screening for Medical Disease

Some therapists suggest a lack of time as an adequate reason to skip the screening process. A few minutes early in the evaluation process may save the client's life. Less dramatically, it may prevent delays in choosing the most appropriate intervention.

Listening for yellow- or red-flag symptoms and observing for red-flag signs can be easily incorporated into everyday practice. It is a matter of listening and looking intentionally. If you do not routinely screen clients for systemic or viscerogenic causes of NMS impairment or dysfunction, then at least pay attention to this red flag:

RED FLAG

- Client does not improve with physical therapy intervention or gets worse with treatment.[16]
- Client is not making progress consistent with the prognosis.

If someone fails to improve with physical therapy intervention, gets better and then worse, or just gets worse, the

*Unexplained or poorly tolerated by client.

treatment protocol may not be in error. Certainly, the first steps are to confirm your understanding of the clinical presentation, repeat appropriate exams, and review selected intervention(s), but also consider the possibility of a systemic or viscerogenic origin of symptoms. Use the screening tools outlined in this chapter to evaluate each individual client (see Box 1-7).

Key Points to Remember

- Systemic diseases can mimic NMS dysfunction.
- It is the therapist's responsibility to identify what NMS impairment is present.
- There are many reasons for screening of the physical therapy client (see Box 1-1).
- Screening for medical disease is an ongoing process and does not occur just during the initial evaluation.
- The therapist uses several parameters in making the screening decision: client history, risk factors, clinical presentation including pain patterns/pain types, associated signs and symptoms, and Review of Systems. Any red flags in the first three parameters will alert the therapist to the need for a screening examination. In the screening process, a Review of Systems includes identifying clusters of signs and symptoms that may be characteristic of a particular organ system.
- The two body parts most commonly affected by visceral pain patterns are the back and the shoulder, although the thorax, pelvis, hip, sacroiliac, and groin can be involved.
- The physical therapist is qualified to make a diagnosis regarding primary NMS conditions referred to as human movement system syndromes.
- The purpose of the diagnosis, established through the subjective and objective examinations, is to identify as closely as possible the underlying NMS condition involving the human movement system. In this way the therapist is screening for medical disease, ruling out the need for medical referral, and treating the physical therapy problem as specifically as possible.
- Sometimes in the diagnostic process the symptoms are treated because the client's condition is too acute to evaluate thoroughly. Usually, even medically diagnosed problems (e.g., "shoulder pain" or "back pain") are evaluated.
- Careful, objective, detailed evaluation of the client with pain is critical for accurate identification of the sources and types of pain (underlying impairment process) and for accurate assessment of treatment effectiveness.[67]
- Painful symptoms that are out of proportion to the injury or that are not consistent with objective findings may be a red flag indicating systemic disease. The therapist must be aware of and screen for other possibilities such as physical assault (see the section on Domestic Violence in Chapter 2) and emotional overlay (see Chapter 3).
- If the client or the therapist is in doubt, communication with the physician, dentist, family member, or referral source is indicated.
- The therapist must be familiar with the practice act for the state in which he or she is practicing. These can be accessed on the APTA website at: http://www.apta.org (search window type in: State Practice Acts).

PRACTICE QUESTIONS

1. In the context of screening for referral, primary purpose of a diagnosis is:
   a. To obtain reimbursement
   b. To guide the plan of care and intervention strategies
   c. To practice within the scope of physical therapy
   d. To meet the established standards for accreditation
2. Direct access is the only reason physical therapists must screen for systemic disease.
   a. True
   b. False
3. A patient/client gives you a written prescription from a physician, chiropractor, or dentist. The first screening question to ask is:
   a. What did the physician (dentist, chiropractor) say is the problem?
   b. Did the physician (dentist, chiropractor) examine you?
   c. When do you go back to see the doctor (dentist, chiropractor)?
   d. How many times per week did the doctor (dentist, chiropractor) suggest you come to therapy?
4. Screening for medical disease takes place:
   a. Only during the first interview
   b. Just before the client returns to the physician for his/her next appointment
   c. Throughout the episode of care
   d. None of the above
5. Physical therapists are qualified to make a human movement system diagnosis regarding primary neuromusculoskeletal conditions, but we must do so in accordance with:
   a. The *Guide to Physical Therapist Practice*
   b. The State Practice Act
   c. The screening process
   d. The SOAP method

PRACTICE QUESTIONS—*cont'd*

6. Medical referral for a problem outside the scope of the physical therapy practice occurs when:
   a. No apparent movement dysfunction exists
   b. No causative factors can be identified
   c. Findings are not consistent with neuromuscular or musculoskeletal dysfunction
   d. Client presents with suspicious red-flag symptoms
   e. Any of the above
   f. None of the above
7. Physical therapy evaluation and intervention may be part of the physician's differential diagnosis.
   a. True
   b. False
8. What is the difference between a yellow- and a red-flag symptom?
9. What are the major decision-making tools used in the screening process?
10. See if you can quickly name 6 to 10 red flags that suggest the need for further screening.

REFERENCES

1. *Guide to Physical Therapist Practice*, ed 2 (Revised), Alexandria, VA, 2003, American Physical Therapy Association.
2. Bogduk N: *Evidence-based clinical guidelines for the management of acute low back pain,* The National Musculoskeletal Medicine Initiative, National Health and Medical Research Council, Nov. 1999.
3. McGuirk B, King W, Govind J, et al: Safety, efficacy, and cost effectiveness of evidence-based guidelines for the management of acute low back pain in primary care. *Spine* 26(23):2615–2622, 2001.
4. Kuchera ML: *Foundations for integrative musculoskeletal medicine*, Philadelphia, Pennsylvania, 2005, Philadelphia College of Osteopathic Medicine.
5. Henschke N: Prevalence of and screening for serious spinal pathology in patients presenting to primary care settings with acute low back pain. *Arthritis & Rheum* 60(10):3072–3080, 2009.
6. Boissonnault WG, Ross M: Clinical factors leading to physical therapists referring patients to physicians. A systematic review paper. *J Orthop Sports Phys Ther* 41(1):A23, 2011.
7. Goodman CC, Kapasi Z: The effect of exercise on the immune system. *Rehab Oncol* 20(1):13–26, 2002.
8. Malm C, Celsing F, Friman G: Immune defense is both stimulated and inhibited by physical activity. *Lakartidningen* 102(11):867–873, 2005.
9. Kohut ML, Senchina DS: Reversing age-associated immunosenescence via exercise. *Exerc Immunol Rev* 10:6–41, 2004.
10. Sinnott M: Challenges 2000: Acute care/hospital clinical practice. *PT Magazine* 8(1):43–46, 2000.
11. Morris PE: Moving our critically ill patients: mobility barriers and benefits. *Crit Care Clin* 23:1–20, 2007.
12. National Institute on Aging. Fiscal Year 2011 Justification. Available online at www.nia.nih.gov. Accessed Sept. 1, 2010.
13. National Center for Health Statistics (NCHS), Health, United States, 2010 Health and aging. Available online at www.cdc.gov/nchs. Accessed Sept. 1, 2010.
14. National Center for Health Statistics (NCHS). Health, United States, 2009: Trends in the health of Americans. Available online at http://www.cdc.gov/nchs/hus.htm. Accessed December 12, 2010.
15. Waddell G: *The back pain revolution*, ed 2, Edinburgh, 2004, Churchill Livingstone.
16. Ross MD, Boissonnault WG: Red flags: to screen or not to screen? *J Orthop Sports Phys Ther* 40(11):682–684, 2010.
17. Underwood M: Diagnosing acute nonspecific low back pain: time to lower the red flags? *Arthritis Rheum* 60:2855–2857, 2009.
18. Henschke N: Screening for malignancy in low back pain patients: a systematic review. *Eur Spine J* 16:1673–1679, 2007.
19. Henschke N: A systematic review identifies five "red flags" to screen for vertebral fractures in patients with low back pain. *J Clin Epidemiol* 61:110–118, 2008.
20. American Physical Therapy Association (APTA) House of Delegates (HOD): Diagnosis by physical therapists HOD 06-97-06-19 (Program 32) [Amended HOD 06-95-12-07; HOD 06-94-22-35, Initial HOD 06-84-19-78]. APTA Governance.
21. *PT Bulletin OnLine* 9(26), June 17, 2008. Available at http://www.apta.org/AM/Template.cfm?Section=Archives2&Template=/Customsource/TaggedPage/PTIssue.cfm&Issue=06/17/2008#article49315; accessed August 8, 2009
22. *A normative model of physical therapist professional education: version 2004,* APTA, 2004, Alexandria, Va.
23. Ellis J: Paving the path to a brighter future. Sahrmann challenges colleagues to move precisely during 29th McMillan lecture at PT '98. *PT Bulletin* 13(29):4–10, 1998.
24. Sahrmann S: A challenge to diagnosis in physical therapy: tradition, APTA. CSM Opening lecture, 1997.
25. Spoto MM, Collins J: Physiotherapy diagnosis in clinical practice: a survey of orthopaedic certified specialists in the USA. *Physiother Res Int* 13(1):31–41, 2008.
26. Boissonnault WG: Differential diagnosis: taking a step back before stepping forward. *PT Magazine* 8(11):45–53, 2000.
27. Fosnaught M: A critical look at diagnosis. *PT Magazine* 4(6):48–54, 1996.
28. Vision sentence and vision statement for physical therapy 2020 [Hod 06-00-24-35 (Program 01)].
29. Quinn L, Gordon J: *Documentation for rehabilitation: a guide to clinical decision making,* ed 2, Philadelphia, 2009, Saunders.
30. American Physical Therapy Association: Vision 2020, Annual Report 2010. APTA.
31. Raj PP: *Pain medicine: a comprehensive review,* St Louis, 1996, Mosby.
32. Sahrmann S: Diagnosis by the physical therapist—a prerequisite for treatment. A special communication. *Phys Ther* 68:1703–1706, 1988.
33. Rothstein JM: Patient classification. *Phys Ther* 73(4):214–215, 1993.
34. Delitto A, Snyder-Mackler L: The diagnostic process: examples in orthopedic physical therapy. *Phys Ther* 75(3):203–211, 1995.

35. Guccione A: *Diagnosis and diagnosticians: the future in physical therapy,* Combined sections meeting, Dallas, February 13-16, 1997.
36. Diagnosis Dialog I: *Defining the 'x' in DxPT,* Washington University in St. Louis, July 19-21, 2006.
37. Zimny JN: Diagnostic classification and orthopaedic physical therapy practice: what we can learn from medicine. *J Orthop Sports Phys Ther,* 34:105–115, 2004.
38. Zadai C: Disabling our diagnostic dilemmas. *Phys Ther* 87:641–653, 2007.
39. Sullivan KJ: Role of the physical therapist in neurologic differential diagnosis: a reality in neurologic physical therapist practice. *JNPT* 31:236–237, 2007.
40. Norton B: Diagnosis dialog: progress report. *Phys Ther* 87(10):1270–1273, 2007.
41. Norton B: *Diagnosis dialog: defining the 'x' in DxPT,* Combined Sections Meeting, New Orleans, 2011.
42. Desai MJ, Padmanabhan G: Spinal schwannoma in a young adult. *J Orthop Sports Phys Ther* 40(11):762, 2010.
43. Moore J: Direct access under Medicare Part B: the time is now! *PT Magazine* 10(2):30–32, 2002.
44. Boissonnault WG: Pursuit and implementation of hospital-based outpatient direct access to physical therapy services: an administrative case report. *Phys Ther* 90(1):100–109, 2010.
45. Leemrijse CJ, Swinkels Ilse CS, Veenhof C: Direct access to physical therapy in the Netherlands: Results from the first year in community-based physical therapy. *Phys Ther* 88(8):936–946, 2008.
46. Roland M: The future of primary care: lessons from the UK. *N Engl J Med* 359(20):2087–2092, 2008.
47. APTA primary care and the role of the physical therapist, *HOD* 06-02-23-46 (Program 32) [Initial HOD 06-95-26-16] [RC 1-06].
48. Ryan GG, Greathouse D, Matsui I, et al: Introduction to primary care medicine. In Boissonnault WG: *Primary care for the physical therapist,* ed 2, Philadelphia, 2010, WB Saunders.
49. Sandstrom RW: The meanings of autonomy for physical therapy. *Phys Ther* 87(1):98–110, 2007.
50. Merriam-Webster On-line Dictionary.
51. Autonomous physical therapist practice: definitions and privileges, BOD 03-03-12-28 [RC 13-06]. Available online at www.apta.org. Type in search box: Autonomous Physical Therapist Practice. Accessed August 26, 2010.
52. Schunk C, Thut C: Autonomous practice: issues of risk. *PT Magazine* 11(5):34–40, May 2003.
53. Cooperman J, Lewis DK: A physical therapist's road to referral, HSPO Risk Advisor, *Physical Therapist Edition* 4(2):Summer, 2001.
54. Mitchell IM, de Lissovoy G: A comparison of resource use and cost in direct access versus physician referral episodes of physical therapy. *Phys Ther* 77(1):10–18, 1997.
55. Fosnaught M: Direct access: exploring new opportunities. *PT Magazine* 10(2):58–62, 2002.
56. Johnson LH: Is cash-only reimbursement for you? *PT Magazine* 11(1):35–39, 2003.
57. Gonzalez-Urzelai V, Palacio-Elua L, Lopez-de-Munain J: Routine primary care management of acute low back pain: adherence to clinical guidelines. *Eur Spine J* 12(6):589–594, 2003.
58. Sandler G: The importance of the history in the medical clinic and cost of unnecessary tests. *Amer Heart J* 100:928–931, 1980.
59. Vickers AJ: Against diagnosis. *Ann Intern Med* 149(3):200–203, 2008.
60. Steyerberg EW: Assessing the performance of prediction models. *Epidemiology* 21(1):128–138, 2010.
61. Poirot L: Genetic disorders and engineering: implications for physical therapists. *PT Magazine* 13(2):54–60, 2005.
62. Goodman C, Helgeson K: *Exercise prescription for medical conditions,* Philadelphia, 2011, FA Davis.
63. Giallonardo L: Guide in action. *PT Magazine* 8(9):76–88, 2000.
64. US Department of Health and Human Services (HHS): Use of Electronic Medical Record. Available online at www.hhs.gov. Accessed Sept. 1, 2010.
65. American Physical Therapy Association (APTA): Referral relationships HOD 06-90-15-28.
66. Arriaga R: Stories from the front, Part II: Complex medical history and communication. *PT Magazine* 11(7):23–25, July 2003.
67. Management of the individual with pain. I. Physiology and evaluation. *PT Magazine* 4(11):54–63, 1996.
68. Moffett J, McLean S: The role of physiotherapy in the management of non-specific back pain and neck pain. *Rheumatology* 45(4):371–378, 2006.
69. Moffett J, McLean S: Red flags need more evaluation: reply. *Rheumatology* 45(7):921, 2006.

CHAPTER

2

Interviewing as a Screening Tool

The client interview, including the personal and family history, is the single most important tool in screening for medical disease. The client interview as it is presented here is the first step in the screening process.

Interviewing is an important skill for the clinician to learn. It is generally agreed that 80% of the information needed to clarify the cause of symptoms is given by the client during the interview. This chapter is designed to provide the physical therapist with interviewing guidelines and important questions to ask the client.

Medical practitioners (including nurses, physicians, and therapists) begin the interview by determining the client's chief complaint. The **chief complaint** is usually a symptomatic description by the client (i.e., symptoms reported for which the person is seeking care or advice). The **present illness,** including the chief complaint and other current symptoms, gives a broad, clear account of the symptoms—how they developed and events related to them.

Questioning the client may also assist the therapist in determining whether an injury is in the acute, subacute, or chronic stage. This information guides the clinician in addressing the underlying pathology while providing symptomatic relief for the acute injury, more aggressive intervention for the chronic problem, and a combination of both methods of treatment for the subacute lesion.

The interviewing techniques, interviewing tools, Core Interview, and review of the inpatient hospital record in this chapter will help the therapist determine the location and potential significance of any symptom (including pain).

The interview format provides detailed information regarding the frequency, duration, intensity, length, breadth, depth, and anatomic location as these relate to the client's chief complaint. The physical therapist will later correlate this information with objective findings from the examination to rule out possible systemic origin of symptoms.

The subjective examination may also reveal any contraindications to physical therapy intervention or indications for the kind of intervention that is most likely to be effective. The information obtained from the interview guides the therapist in either referring the client to a physician or planning the physical therapy intervention.

CONCEPTS IN COMMUNICATION

Interviewing is a skill that requires careful nurturing and refinement over time. Even the most experienced health care professional should self-assess and work toward improvement. Taking an accurate medical history can be a challenge. Clients' recollections of their past symptoms, illnesses, and episodes of care are often inconsistent from one inquiry to the next.[1]

Clients may forget, underreport, or combine separate health events into a single memory, a process called *telescoping.* They may even (intentionally or unintentionally) fabricate or falsely recall medical events and symptoms that never occurred. The individual's personality and mental state at the time of the illness or injury may influence their recall abilities.[1]

Adopting a compassionate and caring attitude, monitoring your communication style, and being aware of cultural differences will help ensure a successful interview. Using the tools and techniques presented in this chapter will get you started or help you improve your screening abilities throughout the subjective examination.

Compassion and Caring

Compassion is the desire to identify with or sense something of another's experience and is a precursor of caring. Caring is the concern, empathy, and consideration for the needs and values of others. Interviewing clients and communicating effectively, both verbally and nonverbally, with compassionate caring takes into consideration individual differences and the client's emotional and psychologic needs.[2,3]

Establishing a trusting relationship with the client is essential when conducting a screening interview and examination. The therapist may be asking questions no one else has asked before about body functions, assault, sexual dysfunction, and so on. A client who is comfortable physically and emotionally is more likely to offer complete information regarding personal and family history.

Be aware of your own body language and how it may affect the client. Sit down when obtaining the history and keep an appropriate social distance from the client. Take notes while maintaining adequate eye contact. Lean forward, nod, or encourage the individual occasionally by saying, "Yes, go ahead. I understand."

Silence is also a key feature in the communication and interviewing process. Silent attentiveness gives the client time to think or organize his or her thoughts. The health care professional is often tempted to interrupt during this time, potentially disrupting the client's train of thought. Silence can give the therapist time to observe the client and plan the next question or step.

Communication Styles

Everyone has a slightly different interviewing and communication style. The interviewer may need to adjust his or her personal interviewing style to communicate effectively.

Relying on one interviewing style may not be adequate for all situations.

There are gender-based styles and temperament/personality-based styles of communication for both the therapist and the client. There is a wide range of ethnic identifications, religions, socioeconomic differences, beliefs, and behaviors for both the therapist and the client.

There are cultural differences based on family of origin or country of origin, again for both the therapist and the client. In addition to spoken communication, different cultural groups may have nonverbal, observable differences in communication style. Body language, tone of voice, eye contact, personal space, sense of time, and facial expression are only a few key components of differences in interactive style.[4]

Illiteracy

Throughout the interviewing process and even throughout the episode of care, the therapist must keep in mind that an estimated 44 million American adults are illiterate and an additional 35 million read only at a functional level for social survival. According to the National Center for Education Statistics, illiteracy is on the rise in the United States.

Nearly 24 million people in the United States do not speak or understand English. More than one third of English-speaking patients and half of Spanish-speaking patients at U.S. hospitals have low health literacy.[5] According to the findings of the Joint Commission, health literacy skills are not evident during most health care encounters. Clear communication and plain language should become a goal and the standard for all health care professionals.[6]

Low health literacy means that adults with below basic skills have no more than the most simple reading skills. They cannot read a physician's (or physical therapist's) instructions or food or pharmacy labels.[7]

It is likely that the rates of health illiteracy defined as the inability to read, understand, and respond to health information are much higher. It is a problem that has gone largely unrecognized and unaddressed. Health illiteracy is more than just the inability to read. People who can read may still have great difficulty understanding what they read.

The Institute of Medicine (IOM) estimates nearly half of all American adults (90 million people) demonstrate a low health literacy. They have trouble obtaining, processing, and understanding the basic information and services they need to make appropriate and timely health decisions.

Low health literacy translates into more severe, chronic illnesses and lower quality of care when care is accessed. There is also a higher rate of health service utilization (e.g., hospitalization, emergency services) among people with limited health literacy. People with reading problems may avoid outpatient offices and clinics and utilize emergency departments for their care because somebody else asks the questions and fills out the form.[7]

It is not just the lower socioeconomic and less-educated population that is affected. Interpreting medical jargon and diagnostic test results and understanding pharmaceuticals are challenges even for many highly educated individuals.

We are living at a time when the amount of health information available to us is almost overwhelming, and yet most Americans would be shocked at the number of their friends and neighbors that (sic) can't understand the instructions on their prescription medications or how to prepare for a simple medical procedure.[8,9]

English as a Second Language

The therapist must keep in mind that many people in the United States speak English as a second language (ESL) or are limited English proficient (LEP), and many of those people do not read, or write English.[10] More than 14 million people age 5 and older in the United States speak English poorly or not at all. Up to 86% of non–English speakers who are illiterate in English are also illiterate in their native language.

In addition, millions of immigrants (and illegal or unregistered citizens) enter U.S. communities every year. Of these people, 1.7 million who are age 25 and older have less than a fifth-grade education. There is a heavy concentration of persons with low literacy skills among the poor and those who are dependent on public financial support.

Although the percentages of illiterate African-American and Hispanic adults are much higher than those of white adults, the actual number of white nonreaders is twice that of African-American and Hispanic nonreaders, a fact that dispels the myth that literacy is not a problem among Caucasians.[11]

People who are illiterate cannot read instructions on bottles of prescription medicine or over-the-counter medications. They may not know when a medicine is past the date of safe consumption nor can they read about allergic risks, warnings to diabetics, or the potential sedative effect of medications.

They cannot read about "the warning signs" of cancer or which fasting glucose levels signal a red flag for diabetes. They cannot take online surveys to assess their risk for breast cancer, colon cancer, heart disease, or any other life-threatening condition.

The Physical Therapist's Role

The therapist should be aware of the possibility of any form of illiteracy and watch for risk factors such as age (over 55 years old), education (0 to 8 years or 9 to 12 years but without a high school diploma), lower paying jobs, living below the poverty level and/or receiving government assistance, and ethnic or racial minority groups or history of immigration to the United States.

Health illiteracy can present itself in different ways. In the screening process, the therapist must be careful when having the client fill out medical history forms. The illiterate or functionally illiterate adult may not be able to understand the written details on a health insurance form, accurately complete a Family/Personal History form, or read the details of exercise programs provided by the therapist. The same is true for individuals with learning disabilities and mental impairments.

When given a choice between "yes" and "no" answers to questions, functionally illiterate adults often circle "no" to everything. The therapist should briefly review with each client to verify the accuracy of answers given on any questionnaire or health form.

For example, you may say, "I see you circled 'no' to any health problems in the past. Has anyone in your immediate family (or have you) ever had cancer, diabetes, hypertension . . ." and continue to name some or all of the choices provided. Sometimes, just naming the most common conditions is enough to know the answer is really "no"—or that there may be a problem with literacy.

Watch for behavioral red flags such as misspelling words, not completing intake forms, leaving the clinic before completing the form, outbursts of anger when asked to complete paperwork, asking no questions, missing appointments, or identifying pills by looking at the pill rather than naming the medication or reading the label.[12]

The IOM has called upon health care providers to take responsibility for providing clear communication and adequate support to facilitate health-promoting actions based on understanding. Their goal is to educate society so that people have the skills they need to obtain, interpret, and use health information appropriately and in meaningful ways.[13,14]

Therapists should minimize the use of medical terminology. Use simple but not demeaning language to communicate concepts and instructions. Encourage clients to ask questions and confirm knowledge or tactfully correct misunderstandings.[13]

Consider including the following questions:

FOLLOW-UP QUESTIONS

- What questions do you have?
- What would you like me to go over?

Resources

There is a text available specifically for physical therapists to help us identify our own culture and recognize the importance of understanding and communicating with clients of different cultural backgrounds. Widely accepted cultural practices of various ethnic groups are included along with descriptions of cultural and language nuances of subcultures within each ethnic group.[15] A text on this same topic for health care professionals is also available.[16]

Identifying individual personality style may be helpful for each therapist as a means of improving communication. Resource materials are available to help with this.[17,18] The Myers-Briggs Type Indicator, a widely used questionnaire designed to identify one's personality type, is also available on the Internet at www.myersbriggs.org.

For the experienced clinician, it may be helpful to reevaluate individual interviewing practices. Making an audio or videotape during a client interview can help the therapist recognize interviewing patterns that may need to improve. Watch and/or listen for any of the guidelines listed in Box 2-1.

Texts are available with the complete medical interviewing process described. These resources are helpful not only to give the therapist an understanding of the training physicians receive and methods they use when interviewing clients, but also to provide helpful guidelines when conducting a physical therapy screening or examination interview.[19,20]

The therapist should be aware that under federal civil rights laws and the Medicaid Act, any client with LEP has the right to an interpreter free of charge if the health care provider receives federal funding. But keep in mind that quality of care for individuals who are LEP is compromised when qualified interpreters are not used (or available). Errors of omission, false fluency, substitution, editorializing, and addition are common and can have important clinical consequences.[10] Standards for medical interpreting professionals in the United States have been published and are available online.[21]

The American Physical Therapy Association (APTA) makes available a distance-learning course that provides listening and speaking skills needed to communicate effectively with Spanish-speaking clients and their families. Contact Member Services for information at 800-999-2782 and ask for *Spanish for Physical Therapists: Tools for Effective Patient Communication.*

The Joint Commission's 2007 report, *What did the doctor say? Improving health literacy to protect patient safety,* is a must read. It is available online at www.jointcommission.org.

CULTURAL COMPETENCE

Interviewing and communication require a certain level of cultural competence as well. *Culture* refers to integrated patterns of human behavior that include the language, thoughts, communications, actions, customs, beliefs, values,

BOX 2-1 INTERVIEWING DO's AND DON'Ts

DO's

Do extend small courtesies (e.g., shaking hands if appropriate, acknowledging others in the room)

Do use a sequence of questions that begins with open-ended questions.

Do leave closed-ended questions for the end as clarifying questions.

Do select a private location where confidentiality can be maintained.

Do give your undivided attention; listen attentively and show it both in your body language and by occasionally making reassuring verbal prompts, such as "I see" or "Go on." Make appropriate eye contact.

Do ask one question at a time and allow the client to answer the question completely before continuing with the next question.

Do encourage the client to ask questions throughout the interview.

Do listen with the intention of assessing the client's current level of understanding and knowledge of his or her current medical condition.

Do eliminate unnecessary information and speak to the client at his or her level of understanding.

Do correlate signs and symptoms with medical history and objective findings to rule out systemic disease.

Do provide several choices or selections to questions that require a descriptive response.

DON'Ts

Don't jump to premature conclusions based on the answers to one or two questions. (Correlate all subjective and objective information before consulting with a physician.)

Don't interrupt or take over the conversation when the client is speaking.

Don't destroy helpful open-ended questions with closed-ended follow-up questions before the person has a chance to respond (e.g., How do you feel this morning? Has your pain gone?).

Don't use professional or medical jargon when it is possible to use common language (e.g., don't use the term myocardial infarct instead of heart attack).

Don't overreact to information presented. Common overreactions include raised eyebrows, puzzled facial expressions, gasps, or other verbal exclamations such as "Oh, really?" or "Wow!" Less dramatic reactions may include facial expressions or gestures that indicate approval or disapproval, surprise, or sudden interest. These responses may influence what the client does or does not tell you.

Don't use leading questions. Pain is difficult to describe, and it may be easier for the client to agree with a partially correct statement than to attempt to clarify points of discrepancy between your statement and his or her pain experience.

Leading Questions	Better Presentation of Same Questions
Where is your pain?	Do you have any pain associated with your injury? If yes, tell me about it.
Does it hurt when you first get out of bed?	When does your back hurt?
Does the pain radiate down your leg?	Do you have this pain anywhere else?
Do you have pain in your lower back?	Point to the exact location of your pain.

and institutions of racial, ethnic, religious, or social groups.[2,22] *Multiculturalism* is a term that takes into account that every member of a group or country does not have the same ideals, beliefs, and views.

Cultural competence can be defined as the ability to understand, honor, and respect the beliefs, lifestyles, attitudes, and behaviors of others.[23] Cultural competency goes beyond being "politically correct." As health care professionals, we must develop a deeper sense of understanding of how ethnicity, language, cultural beliefs, and lifestyles affect the interviewing, screening, and healing process.

Minority Groups

The need for culturally competent physical therapy care has come about, in part, because of the rising number of groups in the United States. Groups other than "white" or "Caucasian" counted as race/ethnicity by the U.S. Census are listed in Box 2-2. Previously these groups were referred to as "minorities," but social scientists are looking for a different term to describe these groups. Terms such as "dominant" and "nondominant" have been suggested when discussing race and ethnicity.

This has come about because some minority groups are no longer a "minority" in the United States due to changing demographics. According to the U.S. Census Bureau, 31% of the U.S. population belongs to a racial/ethnic minority group. By the year 2042, Caucasians will represent less than 50% of the population (currently at approximately 75%).[24] Hispanic Americans will comprise nearly a quarter of the American population (currently 12.5% and expected to reach 30% by 2042). African Americans make up 12.5% of the population (as of 1990). This will increase to approximately 15% so that Hispanic Americans will outnumber African Americans by 2:1. Asian/Pacific Island Americans will make up almost 10% in 2050.[24]

BOX 2-2 RACIAL/ETHNIC DESIGNATIONS

Some individuals may consider themselves "multiracial" based on the combination of their father and mother's racial background. The categories below are used by the U.S. government for census-taking but do not recognize multiple racial combinations. This grouping was adopted for use by the APTA in the *Guide to Physical Therapist Practice,* ed 2 (Revised), 2003.

- American Indian/Alaska Native
- Asian
- Black/African American
- Hispanic or Latino (of any race)
- Native Hawaiian/Pacific Islander
- White/Caucasian

CASE EXAMPLE 2-1

Cultural Competency

A 25-year-old African-American woman who is also a physical therapist came to a physical therapy clinic with severe right knee joint pain. She could not recall any traumatic injury but reported hiking 3 days ago in the Rocky Mountains with her brother. She lives in New York City and just returned yesterday.

A general screening examination revealed the following information:

- Frequent urination for the last 2 days
- Stomach pain (related to stress of visiting family and traveling)
- Fatigue (attributed to busy clinic schedule and social activities)
- Past medical history: Acute pneumonia, age 11
- Nonsmoker, social drinker (1-3 drinks/week)

What Are the Red-Flag Signs/Symptoms?
How Do You Handle a Case Like This?

- Young age
- African American

With the combination of red flags (change in altitude, increased fatigue, increased urination, and stomach pain), there could be a possible systemic cause, not just life's stressors as attributed by the client. The physical therapist treated the symptoms locally but not aggressively and referred the client immediately to a medical doctor.

Result: The client was subsequently diagnosed with sickle cell anemia. Medical treatment was instituted along with client education and a rehab program for local control of symptoms and a preventive strengthening program.

BOX 2-3 CULTURAL COMPETENCY IN A SCREENING INTERVIEW

- Wait until the client has finished speaking before interrupting or asking questions.
- Allow "wait time" (time gaps) for some cultures (e.g., Native Americans, English as a second language [ESL]).
- Be aware that eye contact, body-space boundaries, even handshaking may differ from culture to culture.

When Working with an Interpreter

- Choosing an interpreter is important. A competent medical interpreter is familiar with medical terminology, cultural customs, and the policies of the health care facility in which the client is receiving care.
- There may be problems if the interpreter is younger than the client; in some cultures it is considered rude for a younger person to give instructions to an elder.
- In some cultures (e.g., Muslim), information about the client's diagnosis and condition are relayed to the head of the household who then makes the decision to share the news with the client or other family members.
- Listen to the interpreter but direct your gaze and eye contact to the client (as appropriate; sustained direct eye contact may be considered aggressive behavior in some cultures).
- Watch the client's body language while listening to him or her speak.
- Head nodding and smiling do not necessarily mean understanding or agreement; when in doubt, always ask the interpreter to clarify any communication.
- Keep comments, instructions, and questions simple and short. Do not expect the interpreter to remember everything you said and relay it exactly as you said it to the client if you do not keep it short and simple.
- Avoid using medical terms or professional jargon.

Cultural Competence in the Screening Process

Clients from a racial/ethnic background may have unique health care concerns and risk factors. It is important to learn as much as possible about each group served (Case Example 2-1). Clients who are members of a cultural minority are more likely to be geographically isolated and/or underserved in the area of health services. Risk-factor assessment is very important, especially if there is no primary care physician involved.

Communication style may be unique from group to group; be aware of groups in your area or community and learn about their distinctive health features. For example, Native Americans may not volunteer information, requiring additional questions in the interview or screening process. Courtesy is very important in Asian cultures. Clients may act polite, smiling and nodding, but not really understand the clinician's questions. ESL may be a factor; the client may need an interpreter. The client may not understand the therapist's questions but will not show his or her confusion and will not ask the therapist to repeat the question.

Cultural factors can affect the way a person follows through on instructions, interprets questions, and participates in his or her own care. In addition to the guidelines in Box 2-1, Box 2-3 offers some "Dos" in a cultural context for the physical therapy or screening interview.

Resources

Learning about cultural preferences helps therapists become familiar with factors that could impact the screening process. More information on cultural competency is available to help therapists develop a deeper understanding of culture and cultural differences, especially in health and health care.[4,25,26]

The Health Policy and Administration Section of the APTA has a Cross-Cultural & International Special Interest Group (CCISIG) with information available regarding international physical therapy, international health-related issues, and physical therapists working in third world countries or with ethnic groups.[27] The APTA also has a department dedicated to Minority and International Affairs with additional information available online regarding cultural competence.[23,36]

Information on laws and legal issues affecting minority health care are also available. Best practices in culturally competent health services are provided, including summary recommendations for medical interpreters, written materials, and cultural competency of health professionals.[36]

The APTA's *Tips to Increase Cultural Competency* offers information on values and principles integral to culturally competent education and delivery systems, a Publications Corner that includes articles on cultural competence, links to resources, resources for treating patients/clients from diverse background, and more.[28] Also, there is a *Blueprint for Teaching Cultural Competence in Physical Therapy Education* now available that was created by the Committee on Cultural Competence.[29] This program is a guide to help physical therapists develop core knowledge, attitudes, and skills specific to developing cultural competence as we meet the needs of diverse consumers and strive to reduce or eliminate health disparities.[29]

The U.S. Department of Health and Human Services' Office of Minority Health has published national standards for culturally and linguistically appropriate services (CLAS) in health care. These are available on the Office of Minority Health's Web site (www.omhrc.gov/clas).[30]

Resources on the language and cultural needs of minorities, immigrants, refugees, and other diverse populations seeking health care are available, including strategies for overcoming language and cultural barriers to health care.[31]

The American Academy of Orthopaedic Surgeons offers a free online mini-test of cultural competence for residents and medical students that physical therapist may find helpful and informative.[32] For more specific information about the Muslim culture, visit The Council on American-Islamic Relations[33] or the Muslim American Society.[34,35]

The Gay and Lesbian Medical Association (GLMA) offers publications on professional competencies in providing a safe clinical environment for Lesbian-Gay-Bisexual-Transgender-Intersex (LGBTI) health.[37]

THE SCREENING INTERVIEW

The therapist will use two main interviewing tools during the screening process. The first is the Family/Personal History form (see Fig. 2-2). With the client's responses on this form and/or the client's chief complaint in hand, the interview begins.

The overall client interview is referred to in this text as the Core Interview (see Fig. 2-3). The Core Interview as presented in this chapter gives the therapist a guideline for asking questions about the present illness and chief complaint. Screening questions may be interspersed throughout the Core Interview as seems appropriate, based on each client's answers to questions.

There may be times when additional screening questions are asked at the end of the Core Interview or even on a subsequent date at a follow-up appointment. Specific series of questions related to a single symptom (e.g., dizziness, heart palpitations, night pain) or event (e.g., assault, work history, breast examination) are included throughout the text and compiled in the Appendix for the clinician to use easily.

Interviewing Techniques

An organized interview format assists the therapist in obtaining a complete and accurate database. Using the same outline with each client ensures that all pertinent information related to previous medical history and current medical problem(s) is included. This information is especially important when correlating the subjective data with objective findings from the physical examination.

The most basic skills required for a physical therapy interview include:

- Open-ended questions
- Closed-ended questions
- Funnel sequence or technique
- Paraphrasing technique

Open-Ended and Closed-Ended Questions

Beginning an interview with an *open-ended question* (i.e., questions that elicit more than a one-word response) is advised, even though this gives the client the opportunity to control and direct the interview.[38]

People are the best source of information about their own condition. Initiating an interview with the open-ended directive, "Tell me why you are here" can potentially elicit more information in a relatively short (5- to 15-minute) period than a steady stream of *closed-ended questions* requiring a "yes" or "no" type of answer (Table 2-1).[39,40] This type of interviewing style demonstrates to the client that what he or she has to say is important. Moving from the open-ended line of questions to the closed-ended questions is referred to as the *funnel technique* or *funnel sequence.*

Each question format has advantages and limitations. The use of open-ended questions to initiate the interview may allow the client to control the interview (Case Example 2-2), but it can also prevent a false-positive or false-negative response that would otherwise be elicited by starting with closed-ended (yes or no) questions.

False responses elicited by closed-ended questions may develop from the client's attempt to please the health care provider or to comply with what the client believes is the correct response or expectation.

TABLE 2-1 Interviewing Techniques

Open-Ended Questions	Closed-Ended Questions
1. How does bed rest affect your back pain?	1. Do you have any pain after lying in bed all night?
2. Tell me how you cope with stress and what kinds of stressors you encounter on a daily basis.	2. Are you under any stress?
3. What makes the pain (better) worse?	3. Is the pain relieved by food?
4. How did you sleep last night?	4. Did you sleep well last night?

CASE EXAMPLE 2-2

Monologue

You are interviewing a client for the first time, and she tells you, "The pain in my hip started 12 years ago, when I was a waitress standing on my feet 10 hours a day. It seems to bother me most when I am having premenstrual symptoms.

"My left leg is longer than my right leg, and my hip hurts when the scars from my bunionectomy ache. This pain occurs with any changes in the weather. I have a bleeding ulcer that bothers me, and the pain keeps me awake at night. I dislocated my shoulder 2 years ago, but I can lift weights now without any problems." She continues her monologue, and you feel out of control and unsure how to proceed.

This scenario was taken directly from a clinical experience and represents what we call "an organ recital." In this situation the client provides detailed information regarding all previously experienced illnesses and symptoms, which may or may not be related to the current problem.

How Do You Redirect This Interview? A client who takes control of the interview by telling the therapist about every ache and pain of every friend and neighbor can be rechanneled effectively by interrupting the client with a polite statement such as:

Follow-Up Questions

- I'm beginning to get an idea of the nature of your problem. Let me ask you some other questions.

At this point the interviewer may begin to use closed-ended questions (i.e., questions requiring the answer to be "yes" or "no") in order to characterize the symptoms more clearly.

Closed-ended questions tend to be more impersonal and may set an impersonal tone for the relationship between the client and the therapist. These questions are limited by the restrictive nature of the information received so that the client may respond only to the category in question and may omit vital, but seemingly unrelated, information.

Use of the funnel sequence to obtain as much information as possible through the open-ended format first (before moving on to the more restrictive but clarifying "yes" or "no" type of questions at the end) can establish an effective forum for trust between the client and the therapist.

Follow-Up Questions. The funnel sequence is aided by the use of *follow-up* questions, referred to as *FUPs* in the text. Beginning with one or two open-ended questions in each section, the interviewer may follow up with a series of closed-ended questions, which are listed in the Core Interview presented later in this chapter.

For example, after an open-ended question such as: "How does rest affect the pain or symptoms?" the therapist can follow up with clarifying questions such as:

FOLLOW-UP QUESTIONS

- Are your symptoms aggravated or relieved by any activities? If yes, what?
- How has this problem affected your daily life at work or at home?
- How has it affected your ability to care for yourself without assistance (e.g., dress, bathe, cook, drive)?

Paraphrasing Technique. A useful interviewing skill that can assist in synthesizing and integrating the information obtained during questioning is the *paraphrasing technique.* When using this technique, the interviewer repeats information presented by the client.

This technique can assist in fostering effective, accurate communication between the health care recipient and the health care provider. For example, once a client has responded to the question, "What makes you feel better?" the therapist can paraphrase the reply by saying, "You've told me that the pain is relieved by such and such, is that right? What other activities or treatment brings you relief from your pain or symptoms?"

If the therapist cannot paraphrase what the client has said, or if the meaning of the client's response is unclear, then the therapist can ask for clarification by requesting an example of what the person is saying.

Interviewing Tools

With the emergence of evidence-based practice, therapists are required to identify problems, to quantify symptoms (e.g., pain), and to demonstrate the effectiveness of intervention.

Documenting the effectiveness of intervention is called *outcomes management.* Using standardized tests, functional tools, or questionnaires to relate pain, strength, or range of motion to a quantifiable scale is defined as *outcome measures.* The information obtained from such measures is then compared with the functional outcomes of treatment to assess the effectiveness of those interventions.

In this way, therapists are gathering information about the most appropriate treatment progression for a specific diagnosis. Such a database shows the efficacy of physical therapy intervention and provides data for use with insurance companies in requesting reimbursement for service.

Along with impairment-based measures therapists must use reliable and valid functional outcome measures. No single instrument or method of assessment can be considered the best under all circumstances.

Pain assessment is often a central focus of the therapist's interview, so for the clinician interested in quantifying pain, some way to quantify and describe pain is necessary. There are numerous pain assessment scales designed to determine the quality and location of pain or the percentage of impairment or functional levels associated with pain (see further discussion in Chapter 3).

There are a wide variety of anatomic region, function, or disease-specific assessment tools available. Each test has a specific focus—whether to assess pain levels, level of balance, risk for falls, functional status, disability, quality of life, and so on.

Some tools focus on a particular kind of problem such as activity limitations or disability in people with low back pain (e.g., Oswestry Disability Questionnaire,[41] Quebec Back Pain Disability Scale,[42] Duffy-Rath Questionnaire).[43] The Simple Shoulder Test[44] and the Disabilities of the Arm, Shoulder, and Hand Questionnaire (DASH)[45] may be used to assess physical function of the shoulder. Nurses often use the PQRST mnemonic to help identify underlying pathology or pain (see Box 3-3).

Other examples of specific tests include the

- Visual Analogue Scale (VAS; see Figure 3-6)
- Verbal Descriptor Scale (see Box 3-1)
- McGill Pain Questionnaire (see Fig. 3-11)
- Pain Impairment Rating Scale (PAIRS)
- Likert Scale
- Alzheimer's Discomfort Rating Scale

A more complete evaluation of client function can be obtained by pairing disease- or region-specific instruments with the Short-Form Health Survey (SF-36 Version 2).[46,47] The SF-36 is a well-established questionnaire used to measure the client's perception of his or her health status. It is a generic measure, as opposed to one that targets a specific age, disease, or treatment group. It includes eight different subscales of functional status that are scored in two general components: physical and mental.

An even shorter survey form (the SF-12 Version 2) contains only 1 page and takes about 2 minutes to complete. There is a Low-Back SF-36 Physical Functioning survey[48] and also a similar general health survey designed for use with children (SF-10 for children). All of these tools are available at www.sf-36.org. To see a sample of the SF-36 v.2 go to www.sf-36.org/demos/SF-36v2.html.

The initial Family/Personal History form (see Fig. 2-2) gives the therapist some idea of the client's previous medical history (personal and family), medical testing, and current general health status. Make a special note of the box inside the form labeled "Therapists." This is for liability purposes. Anyone who has ever completed a deposition for a legal case will agree it is often difficult to remember the details of a case brought to trial years later.

A client may insist that a condition was (or was not) present on the first day of the examination. Without a baseline to document initial findings, this is often difficult, if not impossible to dispute. The client must sign or initial the form once it is complete. The therapist is advised to sign and date it to verify that the information was discussed with the client.

Resources. The Family/Personal History form presented in this chapter is just one example of a basic intake form. See the companion website for other useful examples with a different approach. If a client has any kind of literacy or writing problem, the therapist completes it with him or her. If not, the therapist goes over the form with the client at the beginning of the evaluation. The *Guide to Physical Therapist Practice*[49] provides an excellent template for both inpatient and outpatient histories (see the *Guide*, Appendix 6). Other commercially available forms have been developed for a wide range of prescreening assessments.[50]

Therapists may modify the information collected from these examples depending on individual differences in client base and specialty areas served. For example, hospital or institution accreditation agencies such as Commission on Accreditation of Rehabilitation Facilities (CARF) and the Joint Commission on Accreditation of Health Care Organizations (JCAHO) may require the use of their own forms.

An orthopedic-based facility or a sports-medicine center may want to include questions on the intake form concerning current level of fitness and the use of orthopedic devices used, such as orthotics, splints, or braces. Therapists working with the geriatric population may want more information regarding current medications prescribed or levels of independence in activities of daily living.

The Review of Systems (see Box 4-19; see also Appendix D-5), which provides a helpful chart of signs and symptoms characteristic of each visceral system, can be used along with the Family/Personal History form. The *Guide* also provides both an outpatient and an inpatient documentation template for similar purposes (see the *Guide*, Appendix 6).

A teaching tool with practice worksheets is available to help students and clinicians learn how to document findings from the history, systems review, tests and measures, problems statements, and subjective and objective information using both the SOAP note format and the Patient/Client Management model shown in Fig. 1-4.[51]

SUBJECTIVE EXAMINATION

The subjective examination is usually thought of as the "client interview." It is intended to provide a database of information that is important in determining the need for medical referral or the direction for physical therapy intervention. Risk-factor assessment is conducted throughout the subjective and objective examinations.

Key Components of the Subjective Examination

The subjective examination must be conducted in a complete and organized manner. It includes several components, all gathered through the interview process. The order of flow may vary from therapist to therapist and clinic to clinic (Fig. 2-1).

The traditional medical interview begins with family/personal history and then addresses the chief complaint. Therapists may find it works better to conduct the Core Interview and then ask additional questions after looking over the client's responses on the Family/Personal History form.

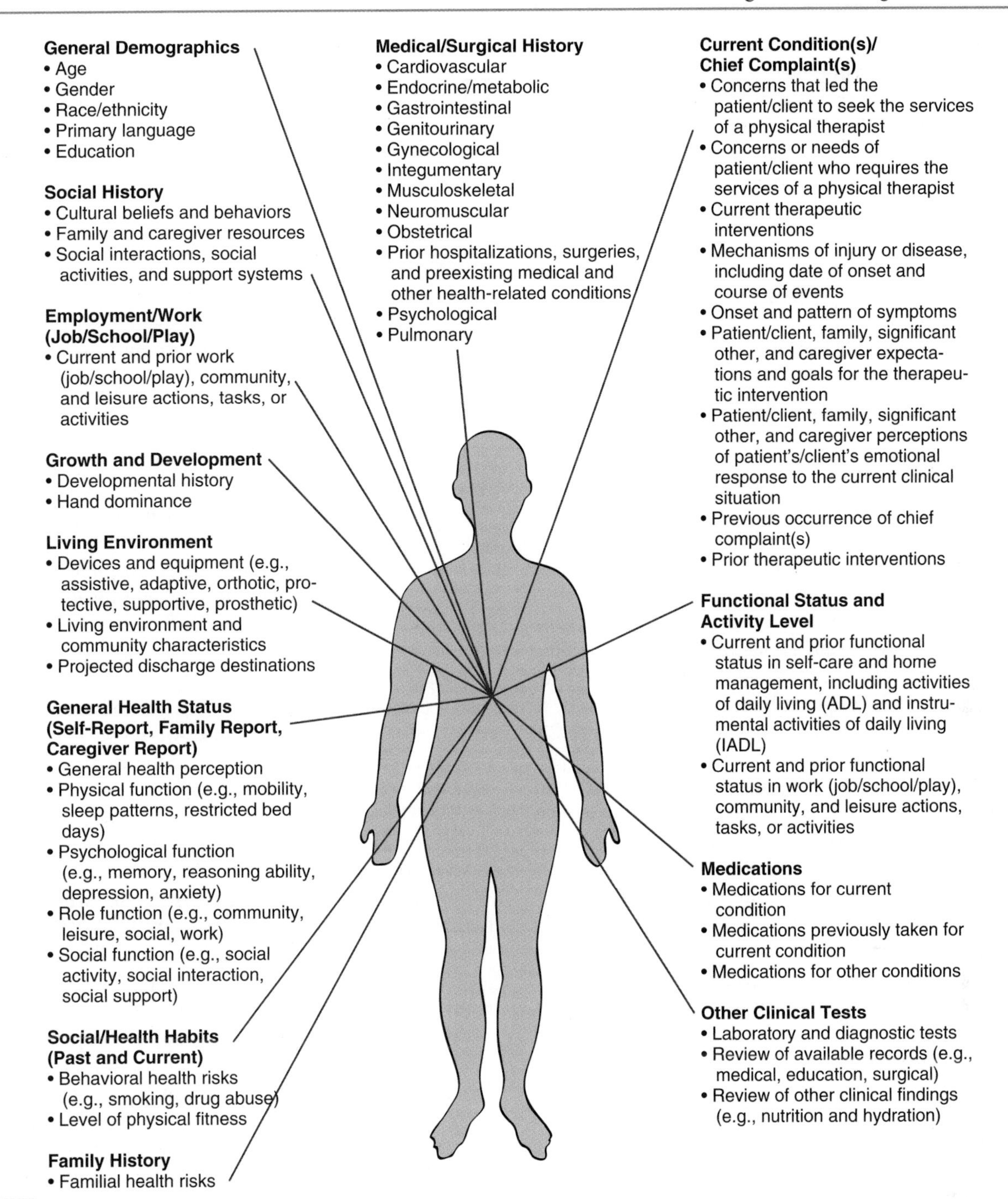

Fig. 2-1 Types of data that may be generated from a client history. In this model, data about the visceral systems is reflected in the Medical/Surgical history. The data collected in this portion of the patient/client history is not the same as information collected during the Review of Systems (ROS). It has been recommended that the ROS component be added to this figure.[52] (From *Guide to physical therapist practice,* ed 2 [Revised], Alexandria, VA, 2003, American Physical Therapy Association.)

In a screening model, the therapist is advised to have the client complete the Family/Personal History form before the client-therapist interview. The therapist then quickly reviews the history form, making mental note of any red-flag histories. This information may be helpful during the subjective and objective portions of the examination. Information gathered will include:

- Family/Personal History (see Fig. 2-2)
  Age
  Sex
  Race and Ethnicity
  Past Medical History
  General Health
  Past Medical and Surgical History

Clinical Tests
Work and Living Environment
- The Core Interview (see Fig. 2-3)
  History of Present Illness
    Chief Complaint
    Pain and Symptom Assessment
    Medical Treatment and Medications
    Current Level of Fitness
    Sleep-related History
    Stress (Emotional/Psychologic screen)
    Final Questions
  Associated Signs and Symptoms
  Special Questions
- Review of Systems

Family/Personal History

It is unnecessary and probably impossible to complete the entire subjective examination on the first day. Many clinics or health care facilities use some type of initial intake form before the client's first visit with the therapist.

The Family/Personal History form presented here (Fig. 2-2) is one example of an initial intake form. Throughout the rest of this chapter, the text discussion will follow the order of items on the Family/Personal History form. The reader is encouraged to follow along in the text while referring to the form.

As mentioned, the *Guide* also offers a form for use in an outpatient setting and a separate form for use in an inpatient setting. This component of the subjective examination can elicit valuable data regarding the client's family history of disease and personal lifestyle, including working environment and health habits.

The therapist must keep the client's family history in perspective. Very few people have a clean and unencumbered family history. It would be unusual for a person to say that nobody in the family ever had heart disease, cancer, or some other major health issue.

A check mark in multiple boxes on the history form does not necessarily mean the person will have the same problems. Onset of disease at an early age in a first-generation family member (sibling, child, parent) can be a sign of genetic disorders and is usually considered a red flag. But an aunt who died of colon cancer at age 75 is not as predictive.

A family history brings to light not only shared genetic traits but also shared environment, shared values, shared behavior, and shared culture. Factors such as nutrition, attitudes toward exercise and physical activity, and other modifiable risk factors are usually the focus of primary and secondary prevention.

Resources

The U.S. Department of Health and Human Services (HHS) has developed a computerized tool to help people learn more about their family health history. "My Family Health Portrait" is available online at: www.hhs.gov/familyhistory/download.html.

The download is free and helps identify common diseases that may run in the family. The therapist can encourage each client to use this tool to create and print out a graphic representation of his or her family's generational health disorders. This information should be shared with the primary health care provider for further screening and evaluation.

Follow-Up Questions (FUPs)

Once the client has completed the Family/Personal History intake form, the clinician can then follow-up with appropriate questions based on any "yes" selections made by the client. Beware of the client who circles one column of either all "Yeses" or all "Nos." Take the time to carefully review this section with the client. The therapist may want to ask some individual questions whenever illiteracy is suspected or observed.

Each clinical situation requires slight adaptations or alterations to the interview. These modifications, in turn, affect the depth and range of questioning. For example, a client who has pain associated with a traumatic anterior shoulder dislocation and who has no history of other disease is unlikely to require in-depth questioning to rule out systemic origins of pain.

Conversely, a woman with no history of trauma but with a previous history of breast cancer who is self-referred to the therapist without a previous medical examination and who complains of shoulder pain should be interviewed more thoroughly. The simple question "How will the answers to the questions I am asking permit me to help the client?" can serve as your guide.[53]

Continued questioning may occur both during the objective examination and during treatment. In fact, the therapist is encouraged to carry on a continuous dialogue during the objective examination, both as an educational tool (i.e., reporting findings and mentioning possible treatment alternatives) and as a method of reducing any apprehension on the part of the client. This open communication may bring to light other important information.

The client may wonder about the extensiveness of the interview, thinking, for example, "Why is the therapist asking questions about bowel function when my primary concern relates to back pain?"

The therapist may need to make a qualifying statement to the client regarding the need for such detailed information. For example, questions about bowel function to rule out stomach or intestinal involvement (which can refer pain to the back) may seem to be unrelated to the client but make sense when the therapist explains the possible connection between back pain and systemic disease.

Throughout the questioning, record both positive and negative findings in the subjective and objective reports in order to correlate information when making an initial assessment of the client's problem. Efforts should be made to quantify all information by frequency, intensity, duration, and exact location (including length, breadth, depth, and anatomic location).

Family/Personal History

Date: ____________________

Client's name: ____________________ DOB: __________ Age: __________

Race/ethnicity:
- ☐ American Indian/Alaska Native
- ☐ Black/African American
- ☐ Hispanic/Latino
- ☐ Multiracial
- ☐ Asian
- ☐ Caucasian/white
- ☐ Native Hawaiian/Pacific Islander
- ☐ Other/unknown

Language: ☐ English understood ☐ Interpreter needed ☐ Primary language:

Medical diagnosis: ____________________ Date of onset: ____________________

Physician: __________ Date of surgery (if any): __________ Therapist: __________

Past Medical History

Have you or any immediate family member (parent, sibling, child) ever been told you have:

Circle one:

• Allergies	Yes	No
• Angina or chest pain	Yes	No
• Anxiety/panic attacks	Yes	No
• Arthritis	Yes	No
• Asthma, hay fever, or other breathing problems	Yes	No
• Cancer	Yes	No
• Chemical dependency (alcohol/drugs)	Yes	No
• Cirrhosis/liver disease	Yes	No
• Depression	Yes	No
• Diabetes	Yes	No
• Eating disorder (bulimia, anorexia)	Yes	No
• Headaches	Yes	No
• Heart attack	Yes	No
• Hemophilia/slow healing	Yes	No
• High cholesterol	Yes	No
• Hypertension or high blood pressure	Yes	No
• Kidney disease/stones	Yes	No
• Multiple sclerosis	Yes	No
• Osteoporosis	Yes	No
• Stroke	Yes	No
• Tuberculosis	Yes	No
• Other (please describe)	Yes	No

(Do **NOT** complete) **For the therapist:**

Relation to client	Date of onset	Current status

Therapists: Use this space to record baseline information. This is important in case something changes in the client's status. You are advised to record the date and sign or initial this form for documentation and liability purposes, indicating that you have reviewed this form with the client. You may want to have the client sign and date it as well.

Fig. 2-2 Sample of a Family/Personal History Form.

Continued

Personal History

Have you ever had:

• Anemia	Yes	No	• Chronic bronchitis	Yes	No
• Epilepsy/seizures	Yes	No	• Emphysema	Yes	No
• Fibromyalgia/myofascial pain syndrome	Yes	No	• GERD	Yes	No
• Hepatitis/jaundice	Yes	No	• Gout	Yes	No
• Joint replacement	Yes	No	• Guillain-Barré syndrome	Yes	No
• Parkinson's disease	Yes	No	• Hypoglycemia	Yes	No
• Polio/postpolio	Yes	No	• Peripheral vascular disease	Yes	No
• Shortness of breath	Yes	No	• Pneumonia	Yes	No
• Skin problems	Yes	No	• Prostate problems	Yes	No
• Urinary incontinence (dribbling, leaking)	Yes	No	• Rheumatic/scarlet fever	Yes	No
• Urinary tract infection	Yes	No	• Thyroid problems	Yes	No
			• Ulcer/stomach problems	Yes	No
			• Varicose veins	Yes	No

For Women:

History of endometriosis Yes No

History of pelvic inflammatory disease Yes No

Are you/could you be pregnant? Yes No

Any trouble with leaking or dribbling urine? Yes No

Number of pregnancies ________ Number of live births ________

Have you ever had a miscarriage/abortion? Yes No

General Health

1. I would rate my health as (circle one): Excellent Good Fair Poor
2. Are you taking any prescription or over-the-counter medications? If yes, please list: ________ Yes No
3. Are you taking any nutritional supplements (any kind, including vitamins) Yes No
4. Have you had any illnesses within the last 3 weeks (e.g., colds, influenza, bladder or kidney infection)? Yes No
   If yes, have you had this before in the last 3 months? Yes No
5. Have you noticed any lumps or thickening of skin or muscle anywhere on your body? Yes No
6. Do you have any sores that have not healed or any changes in size, shape, or color of a wart or mole? Yes No
7. Have you had any unexplained weight gain or loss in the last month? Yes No
8. Do you smoke or chew tobacco? Yes No
   If yes, how many packs/pipes/pouches/sticks a day? ________ How many months or years? ________
9. I used to smoke/chew but I quit Yes No
   If yes: pack or amount/day ________ Year quit ________
10. I would like to quit smoking/using tobacco Yes No
11. How much alcohol do you drink in the course of a week? (One drink is equal to 1 beer, 1 glass of wine, or 1 shot of hard liquor) ________
12. Do you use recreational or street drugs (marijuana, cocaine, crack, meth, amphetamines, or others)? Yes No
    If yes, what, how much, how often? ________
13. How much caffeine do you consume daily (including soft drinks, coffee, tea, or chocolate)? ________
14. Are you on any special diet? Yes No

Fig. 2-2, cont'd

15. Do you have (or have you recently had) any of these problems:
- ☐ Blood in urine, stool, vomit, mucus
- ☐ Dizziness, fainting, blackouts
- ☐ Fever, chills, sweats (day or night)
- ☐ Nausea, vomiting, loss of appetite
- ☐ Changes in bowel or bladder
- ☐ Throbbing sensation/pain in belly or anywhere else
- ☐ Skin rash or other skin changes
- ☐ Cough, dyspnea
- ☐ Dribbling or leaking urine
- ☐ Heart palpitations or fluttering
- ☐ Numbness or tingling
- ☐ Swelling or lumps anywhere
- ☐ Problems seeing or hearing
- ☐ Unusual fatigue, drowsiness
- ☐ Difficulty swallowing/speaking
- ☐ Memory loss
- ☐ Confusion
- ☐ Sudden weakness
- ☐ Trouble sleeping
- ☐ Other: ____________
- ☐ None of these

Medical/Surgical History

1. Have you ever been treated with chemotherapy, radiation therapy, biotherapy, or brachytherapy (radiation implants)? Yes No
   If yes, please describe: ____________
2. Have you had any x-rays, sonograms, computed tomography (CT) scans, or magnetic resonance imaging (MRI) or other imaging done recently? Yes No
   If yes, what? ____________ When? ____________ Results? ____________
3. Have you had any laboratory work done recently (urinalysis or blood tests)? Yes No
   If yes, what? ____________ When? ____________ Results (if known)? ____________
4. Any other clinical tests? Yes No
   Please describe: ____________
5. Please list any operations that you have ever had and the date(s):
   Operation ____________ Date ____________
6. Do you have a pacemaker, transplanted organ, joint replacement, breast implants, or any other implants? Yes No
   If yes, please describe: ____________

Work/Living Environment

1. What is your job or occupation? ____________
2. Military service: (When and where): ____________
3. Does your work involve:
   - ☐ Prolonged sitting (e.g., desk, computer, driving)
   - ☐ Prolonged standing (e.g., equipment operator, sales clerk)
   - ☐ Prolonged walking (e.g., mill worker, delivery service)
   - ☐ Use of large or small equipment (e.g., telephone, forklift, computer, drill press, cash register)
   - ☐ Lifting, bending, twisting, climbing, turning
   - ☐ Exposure to chemicals, pesticides, toxins, or gases
   - ☐ Other: please describe
   - ☐ Not applicable; none of these
4. Do you use any special supports:
   - ☐ Back cushion, neck cushion
   - ☐ Back brace, corset
   - ☐ Other kind of brace or support for any body part
   - ☐ None; not applicable

History of falls:
- ☐ In the past year, I have had no falls
- ☐ I have just started to lose my balance/fall
- ☐ I fall occasionally
- ☐ I fall frequently (more than two times during the past 6 months)
- ☐ Certain factors make me cautious (e.g., curbs, ice, stairs, getting in and out of the tub)

I live:
- ☐ Alone ☐ With family, spouse, partner
- ☐ Nursing home ☐ Assisted living ☐ Other ____________

For the physical therapist:

Exercise history: determine level of activity, exercise, fitness (type, frequency, intensity, duration)

Vital signs (also complete Pain Assessment Record Form, Appendix C-7)

Resting pulse rate: ________ Body temperature: ________ Respirations: ________ Oxygen saturation: ________

Blood pressure: 1st reading ____________ 2nd reading ____________
Position: Sitting Standing Extremity: Right Left

Fig. 2-2, cont'd

TABLE 2-2 Some Age- and Sex-Related Medical Conditions

Diagnosis	Sex	Age (in Years)
NEUROMUSCULOSKELETAL		
Guillain-Barré syndrome	Men > women	Any age; history of infection/alcoholism
Multiple sclerosis	Women > men	15-35 (peak)
Rotator cuff degeneration		30+
Spinal stenosis	Men > women	60+
Tietze's syndrome		Before 40, including children
Costochondritis	Women > men	40+
Neurogenic claudication		40-60+
SYSTEMIC		
AIDS/HIV	Men > women	20-49
Ankylosing spondylitis	Men > women	15-30
Abdominal aortic aneurysm	(hypertensive) Men > women	40-70
Buerger's disease	Men > women	20-40 (smokers)
Cancer	Men > women	Any age; incidence rises after age 50
Breast cancer	Women > men	45-70 (peak incidence)
Hodgkin's disease	Men > women	20-40, 50-60
Osteoid osteoma (benign)	Men > women	10-20
Pancreatic carcinoma	Men > women	50-70
Rheumatoid arthritis	Women > men	20-50
Skin cancer	Men = women	Rarely before puberty; increasing incidence with increasing age
Gallstones	Women > men	40+
Gout	Men > women	40-59
Gynecologic conditions	Women	20-45 (peak incidence)
Paget's disease of bone	Men > women	60+
Prostatitis	Men	40+
Primary biliary cirrhosis	Women > men	40-60
Reiter's syndrome	Men > women	20-40
Renal tuberculosis	Men > women	20-40
Rheumatic fever	Girls > boys	4-9; 18-30
Shingles		60+; increasing incidence with increasing age
Spontaneous pneumothorax	Men > women	20-40
Systemic backache		45+
Thyroiditis	Women > Men	30-50
Vascular claudication		40-60+

Age and Aging

Age is the most common primary risk factor for disease, illness, and comorbidities. It is the number one risk factor for cancer. The age of a client is an important variable to consider when evaluating the underlying neuromusculoskeletal (NMS) pathologic condition and when screening for medical disease.

Age-related changes in metabolism increase the risk for drug accumulation in older adults. Older adults are more sensitive to both the therapeutic and toxic effects of many drugs, especially analgesics.

Functional liver tissue diminishes and hepatic blood flow decreases with aging, thus impairing the liver's capacity to break down and convert drugs. Therefore aging is a risk factor for a wide range of signs and symptoms associated with drug-induced toxicities.

It is helpful to be aware of NMS and systemic conditions that tend to occur during particular decades of life. Signs and symptoms associated with that condition take on greater significance when age is considered. For example, prostate problems usually occur in men after the fourth decade (age 40+). A past medical history of prostate cancer in a 55-year-old man with sciatica of unknown cause should raise the suspicions of the therapist. Table 2-2 provides some of the age-related systemic and NMS pathologic conditions.

Epidemiologists report that the U.S. population is beginning to age at a rapid pace, with the first baby boomers turning 65 in 2011. Between now and the year 2020, the number of individuals age 65 and older (referred to by some as the "Big Gray Wave") will double, reaching 70.3 million and making up a larger proportion of the entire population (increasing from 13% in 2000 to 20% in 2030).[54]

Of particular interest is the explosive growth expected among adults age 85 and older. This group is at increased risk for disease and disability. Their numbers are expected to grow from 4.3 million in the year 2000 to at least 19.4 million in 2050. As mentioned previously, the racial and ethnic makeup of the older population is expected to continue changing, creating a more diverse population of older Americans.

Human aging is best characterized as the progressive constriction of each organ system's homeostatic reserve. This decline, often referred to as "homeostenosis," begins in the third decade and is gradual, linear, and variable among individuals. Each organ system's decline is independent of changes in other organ systems and is influenced by diet, environment, and personal habits.

Dementia increases the risk of falls and fracture. Delirium is a common complication of hip fracture that increases the length of hospital stay and mortality. Older clients take a disproportionate number of medications, predisposing them to adverse drug events, drug-drug interactions, poor adherence to medication regimens, and changes in pharmacokinetics and pharmacodynamics related to aging.[55,56]

An abrupt change or sudden decline in any system or function is always due to disease and not to "normal aging." In the absence of disease the decline in homeostatic reserve should cause no symptoms and impose no restrictions on activities of daily living regardless of age. In short, "old people are sick because they are sick, not because they are old."

The onset of a new disease in older people generally affects the most vulnerable organ system, which often is different from the newly diseased organ system and explains why disease presentation is so atypical in this population. For example, at presentation, less than one fourth of older clients with hyperthyroidism have the classic triad of goiter, tremor, and exophthalmos; more likely symptoms are atrial fibrillation, confusion, depression, syncope, and weakness.

Because the "weakest links" with aging are so often the brain, lower urinary tract, or cardiovascular or musculoskeletal system, a limited number of presenting symptoms predominate no matter what the underlying disease. These include:

- Acute confusion
- Depression
- Falling
- Incontinence
- Syncope

The corollary is equally important: The organ system usually associated with a particular symptom is less likely to be the cause of that symptom in older individuals than in younger ones. For example, acute confusion in older adults is less often due to a new brain lesion, incontinence is less often due to a bladder disorder; falling, to a neuropathy; or syncope, to heart disease.

Sex and Gender

In the screening process, sex (male versus female) and gender (social and cultural roles and expectations based on sex) may be important issues (Case Example 2-3). To some extent, men and women experience some diseases that are different from each other. When they have the same disease, the age at onset, clinical presentation, and response to treatment is often different.

Men. It may be appropriate to ask some specific screening questions just for men. A list of these questions is provided in Chapter 14 (see also Appendices B-24 and B-37). Taking a sexual history (see Appendix B-32, A and B) may be appropriate at some point during the episode of care.

For example, the presentation of joint pain accompanied by (or a recent history of) skin lesions in an otherwise healthy, young adult raises the suspicion of a sexually transmitted infection (STI). Being able to recognize STIs is helpful in the clinic. The therapist who recognized the client presenting with joint pain of "unknown cause" and also demonstrating signs of an STI may help bring the correct diagnosis to light sooner than later. Chronic pelvic or low back pain of unknown cause may be linked to incest or sexual assault.

The therapist may need to ask men about prostate health (e.g., history of prostatitis, benign prostatic hypertrophy, prostate cancer) or about a history of testicular cancer. In some cases, a sexual history (see Appendix B-32, A and B) may be helpful. Many men with a history of prostate problems are incontinent. Routinely screening for this condition may bring to light the need for intervention.

Men and Osteoporosis. In an awkward twist of reverse bias, many men are not receiving intervention for osteoporosis. In fact the overall prevalence of osteoporosis among men of all ages remains unknown, with ranges from 20% to 36% reported in the literature.[57,58] Osteoporosis is prevalent

CASE EXAMPLE 2-3

Sex as a Risk Factor

Clinical Presentation: A 45-year-old woman presents with midthoracic pain that radiates to the interscapular area on the right. There are two red flags recognizable immediately: age and back pain. Female sex can be a red flag and should be considered during the evaluation.

Referred pain from the gallbladder is represented in Fig. 9-10 as the light pink areas. If the client had a primary pain pattern with gastrointestinal symptoms, she would have gone to see a medical doctor first.

Physical therapists see clients with referred pain patterns, often before the disease has progressed enough to be accompanied by visceral signs and symptoms. They may come to us from a physician or directly.

Risk-Factor Assessment: Watch for specific risk factors. In this case, look for the five Fs associated with gallstones: fat, fair, forty (or older), female, and flatulent.

Clients with gallbladder disease do not always present this way, but the risk increases with each additional risk factor. Other risk factors for gallbladder disease include:

- Age: increasing incidence with increasing age
- Obesity
- Diabetes mellitus
- Multiparity (multiple pregnancies and births)

Women are at increased risk of gallstones because of their exposure to estrogen. Estrogen increases the hepatic secretion of cholesterol and decreases the secretion of bile acids. Additionally, during pregnancy, the gallbladder empties more slowly, causing stasis and increasing the chances for cholesterol crystals to precipitate.

For any woman over 40 presenting with midthoracic, scapular, or right shoulder pain, consider gallbladder disease as a possible underlying etiology. To screen for systemic disease, look for known risk factors and ask about:

Associated Signs and Symptoms: When the disease advances, gastrointestinal distress may be reported. This is why it is always important to ask clients if they are having any symptoms of any kind anywhere else in the body. The report of recurrent nausea, flatulence, and food intolerances points to the gastrointestinal system and a need for medical attention.

but poorly documented in men in long-term care facilities.[59]

Men have a higher mortality rate after fracture compared with women.[60] Thirty percent of older men who suffer a hip fracture will die within a year of that fracture—double the rate for older adult women. Only 1.1% of the men brought to the hospital for a serious fracture ever receive a bone density test to evaluate their overall risk. Only 1% to 5% of men discharged from the hospital following hip fracture are treated for osteoporosis. This is compared to 27% or more for women.[61,62]

Keeping this information in mind and watching for risk factors of osteoporosis (see Fig. 11-9) can guide the therapist in recognizing the need to screen for osteoporosis in men and women.

Women. According to the Health Resources and Services Administration (HRSA), women today are more likely than men to die of heart disease, and women between the ages of 26 and 49 are nearly twice as likely to experience serious mental illness as men in the same age group.[63]

Women have a unique susceptibility to the neurotoxic effects of alcohol. Fewer drinks with less alcohol content have a greater physiologic impact on women compared to men. Women may be at greater risk of alcohol-induced brain injury than men, suggesting medical management of alcoholism in women may require a different approach from that for men.[64]

Sixty-two percent of American women are overweight and 33% are obese. Lung cancer caused an estimated 27% of cancer deaths among women in 2004, followed by breast cancer (15%) and cancer of the colon and rectum (10%).[65]

These are just a few of the many ways that being female represents a unique risk factor requiring special consideration when assessing the overall individual and when screening for medical disease.

Questions about past pregnancies, births and deliveries, past surgical procedures (including abortions), incontinence, endometriosis, history of sexually transmitted or pelvic inflammatory disease(s), and history of osteoporosis and/or compression fractures are important in the assessment of some female clients (see Appendix B-37). The therapist must use common sense and professional judgment in deciding what questions to ask and which follow-up questions are essential.

Life Cycles. For women, it may be pertinent to find out where each woman is in the life cycle (Box 2-4) and correlate this information with age, personal and family history, current health, and the presence of any known risk factors. It may be necessary to ask if the current symptoms occur at the same time each month in relation to the menstrual cycle (e.g., day 10 to 14 during ovulation or at the end of the cycle during the woman's period).

Each phase in the life cycle is really a process that occurs over a number of years. There are no clear distinctions most of the time as one phase blends gradually into the next one.

Perimenopause is a term that was first coined in the 1990s. It refers to the transitional period from physiologic ovulatory menstrual cycles to eventual ovarian shut down. During the perimenopausal time before cessation of menses, signs and symptoms of hormonal changes may become evident. These can include fatigue, memory problems, weight gain, irritability, sleep disruptions, enteric dysfunction, painful intercourse, and change in libido.

BOX 2-4 LIFE CYCLES OF A WOMAN

- Premenses (before the start of the monthly menstrual cycle; may include early puberty)
- Reproductive years (including birth, delivery, miscarriage and/or abortion history; this time period may include puberty)
- Perimenopause (usually begins without obvious symptoms in the mid-30s and continues until symptoms of menopause occur)
- Menopausal (may be natural or surgical menopause [i.e., hysterectomy])
- Postmenopausal (cessation of blood flow associated with menstrual cycle)

Early stages of physiologic perimenopause can occur when a woman is in her mid-30s. Symptoms may not be as obvious in this group of women; infertility may be the most obvious sign in women who have delayed childbirth.[66]

Menopause is an important developmental event in a woman's life. Menopause means pause or cessation of the monthly, referring to the menstrual, cycle. The term has been expanded to include approximately 1½ to 2 years before and after cessation of the menstrual cycle.

Menopause is not a disease but rather a complex sequence of biologic aging events, during which the body makes the transition from fertility to a nonreproductive status. The usual age of menopause is between 48 and 54 years. The average age for menopause is still around 51 years of age, although many women stop their periods much earlier.[67-69]

The pattern of menstrual cessation varies. It may be abrupt, but more often it occurs over 1 to 2 years. Periodic menstrual flow gradually occurs less frequently, becoming irregular and less in amount. Occasional episodes of profuse bleeding may be interspersed with episodes of scant bleeding.

Menopause is said to have occurred when there have been no menstrual periods for 12 consecutive months. Postmenopause describes the remaining years of a woman's life when the reproductive and menstrual cycles have ended. *Any spontaneous uterine bleeding after this time is abnormal and is considered a red flag.*

The significance of postmenopausal bleeding depends on whether or not the woman is taking hormone replacement therapy (HRT) and which regimen she is using. Women who are on continuous-combined HRT (estrogen in combination with progestin taken without a break) are likely to have irregular spotting until the endometrium atrophies, which takes about 6 months. Medical referral is advised if bleeding persists or suddenly appears after 6 months without bleeding.

Women on sequential HRT (estrogen taken daily or for 25 days each month with progestin taken for 10 days) normally bleed lightly each time the progestin is stopped. Postmenopausal bleeding in women who are not on HRT always requires a medical evaluation.

Within the past decade, removal of the uterus (hysterectomy) has become a common major surgery in the United States. In fact, more than one third of the women in the United States have hysterectomies. The majority of these women have this operation between the ages of 25 and 44 years.

Removal of the uterus and cervix, even without removal of the ovaries, usually brings on an early menopause (surgical menopause), within 2 years of the operation. Oophorectomy (removal of the ovaries) brings on menopause immediately, regardless of the age of the woman, and early surgical removal of the ovaries (before age 30) doubles the risk of osteoporosis.

CLINICAL SIGNS AND SYMPTOMS

Menopause

- Fatigue and malaise
- Depression, mood swings
- Difficulty concentrating; "brain fog"
- Headache
- Altered sleep pattern (insomnia/sleep disturbance)
- Hot flashes
- Irregular menses, cessation of menses
- Vaginal dryness, pain during intercourse
- Atrophy of breasts and vaginal tissue
- Pelvic floor relaxation (cystocele/rectocele)
- Urge incontinence

Women and Hormone Replacement Therapy. For a time, it was enough to find out which women in their menopausal years were taking HRT. It was thought these women were protected against cardiac events, osteoporosis, and hip fractures.

Women who were not on HRT were targeted with information about the increased risk of osteoporosis and hip fractures. Anyone with cardiac risk factors was encouraged to begin taking HRT. Research from the landmark Women's Health Initiative study[70] has shown that HRT is not cardioprotective as was once thought. In fact there is an increase in myocardial infarction (MI) and stroke in healthy women taking HRT along with an increase in breast cancer and blood clots. HRT is associated with a decrease in colorectal cancer and hip fracture.[70]

The next wave of research reported that these findings applied to long-term use, not short-term use to alleviate symptoms. Doctors started prescribing HRT as a short-term intervention to manage symptoms rather than with the intention of replacing naturally diminishing hormones. However, a newer study[71] reported there are only 1- to 2-point differences (scale 0-100) for a large study comparing women taking versus not taking HRT for symptomatic relief. After 3 years, even those slight differences disappeared.

Women and Heart Disease. When a 55-year-old woman with a significant family history of heart disease comes to the therapist with shoulder, upper back, or jaw pain, it will be necessary to take the time and screen for possible cardiovascular involvement.

For women, sex-linked protection against coronary artery disease ends with menopause. At age 45 years, one in nine women develops heart disease. By age 65 years, this statistic changes to one in three women.[72]

Ten times as many women die of heart disease and stroke as they do of breast cancer (about one half million every year in the United States for heart disease compared to about 41,000 from breast cancer).[65] More women die of heart disease each year in the United States than the combined deaths from the next seven causes of death in women. In fact, more women than men die of heart disease every year.[72,73]

Women under 50 are more than twice as likely to die of heart attacks compared to men in the same age group. Two thirds of women who die suddenly have no previously recognized symptoms. Prodromal symptoms as much as 1 month before a myocardial infarction go unrecognized (see Table 6-4).

Therapists who recognize age combined with the female sex as a risk factor for heart disease will look for other risk factors and participate in heart disease prevention. See Chapter 6 for further discussion of this topic.

Women and Osteoporosis. As health care specialists, therapists have a unique opportunity and responsibility to provide screening and prevention for a variety of diseases and conditions. Osteoporosis is one of those conditions.

To put it into perspective, a woman's risk of developing a hip fracture is equal to her combined risk of developing breast, uterine, and ovarian cancer. Women have higher fracture rates than men of the same ethnicity. Caucasian women have higher rates than black women.

Assessment of osteoporosis and associated risk factors along with further discussion of osteoporosis as a condition are discussed in Chapter 11.

Race and Ethnicity

The distinction between the terms "race" and "ethnicity" is not always clear, and the terms are used interchangeably or combined and discussed as "racial/ethnic minorities." Social scientists make a distinction in that race describes membership in a group based on physical differences (e.g., color of skin, shape of eyes). Ethnicity refers to being part of a group with shared social, cultural, language, and geographic factors (e.g., Hispanic, Italian).[74]

An individual's ethnicity is defined by a unique sociocultural heritage that is passed down from generation to generation but can change as the person changes geographic locations or joins a family with different cultural practices. A child born in Korea but adopted by a Caucasian-American family will grow up speaking English, eating American food, and studying U.S. history. Ethnically, the child is American but will be viewed racially by others as Asian.

The Genome Project dispelled previous ideas of biologic differences based on race. It is now recognized that humans are the same biologically regardless of race or ethnic background.[75,76] In light of these new findings, the focus of research is centered now on cultural differences, including religious, social, and economic factors and how these might explain health differences among ethnic groups.

Ethnicity is a risk factor for health outcomes. Despite tremendous advances and improved public health in America, the non-Caucasian racial/ethnic groups listed in Box 2-2 are medically underserved and suffer higher levels of illness, premature death, and disability. These include cancer, heart disease and stroke, diabetes, infant mortality, and HIV and AIDS.[77]

Racial/ethnic minorities living in rural areas may be at greater risk when health care access is limited.[78] For example, American Indians (also referred to as Native Americans) living on reservations may benefit from many services for free that might not be available in other areas, while city-dwelling (urban) American Indians are more likely than the general population to die from diabetes, alcohol-related causes, lung cancer, liver disease, pneumonia, and influenza.[79] The therapist must remember to look for these risk factors when conducting a risk-factor assessment.

Black men have a higher risk factor for hypertension and heart disease than white men. Black women have 250% higher incidence with twice the mortality of white women for cervical cancer. Black women are more likely to die of pneumonia, influenza, diabetes, and liver disease. Scientists and epidemiologists ask if this could be the result of socioeconomic factors such as later detection. Perhaps the lack of health insurance prevents adequate screening and surveillance.

Epidemiologists tracking cancer statistics point out that African Americans still have the highest mortality and worst survival of any population and the statistics have not improved significantly over the last 20 years.[80] Studies have shown that equal treatment yields equal outcomes among individuals with equal disease.[81] Conversely, minority status can be translated into disparities in health care with worse outcomes in many cases for a variety of illnesses.[80,82]

African-American teenagers and young adults are three to four times more likely to be infected with hepatitis B than whites. Asian Americans and Pacific Islanders are twice as likely as whites to be infected with hepatitis B. Of all the cases of tuberculosis reported in the United States over the last 10 years, almost 80% were in racial/ethnic minorities.[77]

Mexican Americans, who make up two thirds of Hispanics, are also the largest minority group in the United States. Stroke prevention and early intervention are important in this group because their risk for stroke is much higher than for non-Hispanic or white adults.

Mexican Americans ages 45 to 59 are twice as likely to suffer a stroke, and those in their 60s and early 70s are 60% more likely to have a stroke. Family history of stroke or transient ischemic attack (TIA) is a warning flag in this population.[83,84]

Other studies are underway to compare ethnic differences among different groups for different diseases (Case Example 2-4).

Resources. Definitions and descriptions for race and ethnicity are available through the Centers for Disease Control and Prevention (CDC).[85] For a report on racial and ethnic disparities, see the IOM's *Unequal Treatment, Confronting Racial and Ethnic Disparities in Health Care.*[82]

The U.S. National Library of Medicine and the National Institutes of Health offer the latest news on health care issues and other topics related to African Americans.[86] Baylor College of Medicine's Intercultural Cancer Council provides information about cancer and various racial/ethnic groups.[87] See also the Kagawa-Singer article for an excellent discussion of cancer, culture, and health disparities.[80]

Past Medical and Personal History

It is important to take time with these questions and to ensure that the client understands what is being asked. A "yes" response to any question in this section would require further questioning, correlation to objective findings, and consideration of referral to the client's physician.

CASE EXAMPLE 2-4

Risk Factors Based on Ethnicity

A 25-year-old African-American woman who is also a physical therapist went to a physical therapy clinic with severe right knee joint pain. She could not recall any traumatic injury but reported hiking 3 days ago in the Rocky Mountains with her brother. She lives in New York City and just returned yesterday.

- A general screening examination revealed the following information:
- Frequent urination for the last 2 days
- Stomach pain (attributed by the client to the stress of visiting family and traveling)
- Self-reported increased fatigue (attributed to busy clinic schedule and social activities)
- PMHx: Acute pneumonia, age 11
- Nonsmoker, social drinker (1-3 drinks/week)

What Are the Red-Flag Signs/Symptoms?
How Do You Handle a Case Like This?

Red-Flag Signs and Symptoms:

- Young age
- African American
- Combination: change in altitude, increased fatigue, increased urination, and stomach pain: possible systemic cause, not just life's stressors

Intervention: Treat locally but not aggressively, refer immediately

Medical Diagnosis: Sickle cell anemia. The therapist applied the correct intervention using the Rest Ice Compression Elevation (RICE) formula to treat the knee joint. Local treatment is not enough in such cases given the underlying pathology. Early referral and medical intervention reduced morbidity in this case.

For example, a "yes" response to questions on this form directed toward *allergies, asthma,* and *hay fever* should be followed up by asking the client to list the allergies and to list the symptoms that may indicate a manifestation of allergies, asthma, or hay fever. The therapist can then be alert for any signs of respiratory distress or allergic reactions during exercise or with the use of topical agents.

Likewise, clients may indicate the presence of *shortness of breath* with only mild exertion or without exertion, possibly even after waking at night. This condition of breathlessness can be associated with one of many conditions, including heart disease, bronchitis, asthma, obesity, emphysema, dietary deficiencies, pneumonia, and lung cancer.

Some "no" responses may warrant further follow-up. The therapist can screen for diabetes, depression, liver impairment, eating disorders, osteoporosis, hypertension, substance use, incontinence, bladder or prostate problems, and so on. Special questions to ask for many of these conditions are listed in the appendices.

Many of the screening tools for these conditions are self-report questionnaires, which are inexpensive, require little or no formal training, and are less time consuming than formal testing. Knowing the risk factors for various illnesses, diseases, and conditions will help guide the therapist in knowing when to screen for specific problems. Recognizing the signs and symptoms will also alert the therapist to the need for screening.

Eating Disorders and Disordered Eating. Eating disorders, such as bulimia nervosa, binge eating disorder, and anorexia nervosa, are good examples of past or current conditions that can impact the client's health and recovery. The therapist must consider the potential for a negative impact of anorexia on bone mineral density, while also keeping in mind the psychologic risks of exercise (a common intervention for osteopenia) in anyone with an eating disorder.

The first step in screening for eating disorders is to look for risk factors for eating disorders. Common risk factors associated with eating disorders include being female, Caucasian/white, having perfectionist personality traits, a personal or family history of obesity and/or eating disorders, sports or athletic involvement, and history of sexual abuse or other trauma.

Distorted body image and disordered eating are probably underreported, especially in male athletes. Athletes participating in sports that use weight classifications, such as wrestling and weightlifting, are at greater risk for anorexic behaviors such as fasting, fluid restriction, and vomiting.[88]

Researchers have recently described a form of body image disturbance in male bodybuilders and weightlifters referred to as *muscle dysmorphia.* Previously referred to as "reverse anorexia" this disorder is characterized by an intense and excessive preoccupation or dissatisfaction with a perceived defect in appearance, even though the men are usually large and muscular. The goal in disordered eating for this group of men is to increase body weight and size. The use of performance-enhancing drugs and dietary supplements is common in this group of athletes.[89,90]

Gay men tend to be more dissatisfied with their body image and may be at greater risk for symptoms of eating disorders compared to heterosexual men.[91] Screening is advised for anyone with risk factors and/or signs and symptoms of eating disorders. Questions to ask may include:

? FOLLOW-UP QUESTIONS

- Are you satisfied with your eating patterns?
- Do you force yourself to exercise, even when you don't feel well?
- Do you exercise more when you eat more?
- Do you think you will gain weight if you stop exercising for a day or two?
- Do you exercise more than once a day?
- Do you take laxatives, diuretics (water pills), or any other pills as a way to control your weight or shape?
- Do you ever eat in secret? (Secret eating refers to individuals who do not want others to see them eat or see what they eat; they may eat alone or go into the bathroom or closet to conceal their eating.)
- Are there days when you don't eat anything?
- Do you ever make yourself throw up after eating as a way to control your weight?

CLINICAL SIGNS AND SYMPTOMS

Eating Disorders

Physical

- Weight loss or gain
- Skeletal myopathy and weakness
- Chronic fatigue
- Dehydration or rebound water retention; pitting edema
- Discoloration or staining of the teeth from contact with stomach acid
- Broken blood vessels in the eyes from induced vomiting
- Enlarged parotid (salivary) glands (facial swelling) from repeated contact with vomit
- Tooth marks, scratches, scars, or calluses on the backs of hands from inducing vomiting (Russell's sign)
- Irregular or absent menstrual periods; delay of menses onset in young adolescent girls
- Inability to tolerate cold
- Dry skin and hair; brittle nails; hair loss and growth of downy hair (lanugo) all over the body, including the face
- Reports of heartburn, abdominal bloating or gas, constipation, or diarrhea
- Vital signs: Slow heart rate (bradycardia); low blood pressure
- In women/girls: Irregular or absent menstrual cycles

Behavioral

- Preoccupation with weight, food, calories, fat grams, dieting, clothing size, body shape
- Mood swings, irritability
- Binging and purging (bulimia) or food restriction (anorexia); frequent visits to the bathroom after eating
- Frequent comments about being "fat" or overweight despite looking very thin
- Excessive exercise to burn off calories
- Use of diuretics, laxatives, enemas, or other drugs to induce urination, bowel movements, and vomiting (purging)

General Health

Self-assessed health is a strong and independent predictor of mortality and morbidity. People who rate their health as "poor" are four to five times more likely to die than those who rate their health as "excellent."[92,93] Self-assessed health is also a strong predictor of functional limitation.[94]

At least one study has shown similar results between self-assessed health and outcomes after total knee replacement (TKR).[95] The therapist should consider it a red flag anytime a client chooses "poor" to describe his or her overall health.

Medications. Although the Family/Personal History form includes a question about prescription or over-the-counter (OTC) medications, specific follow-up questions come later in the Core Interview under Medical Treatment and Medications. Further discussion about this topic can be found in that section of this chapter.

It may be helpful to ask the client to bring in any prescribed medications he or she may be taking. In the older adult with multiple comorbidities, it is not uncommon for the client to bring a gallon-sized Ziploc bag full of pill bottles. Taking the time to sort through the many prescriptions can be time consuming.

Start by asking the client to make sure each one is a drug that is being taken as prescribed on a regular basis. Many people take "drug holidays" (skip their medications intentionally) or routinely take fewer doses than prescribed.[96] Make a list for future investigation if the clinical presentation or presence of possible side effects suggests the need for consultation with the pharmacy.

Recent Infections. Recent infections, such as mononucleosis, hepatitis, or upper respiratory infections, may precede the onset of Guillain-Barré syndrome. Recent colds, influenza, or upper respiratory infections may also be an extension of a chronic health pattern of systemic illness.

Further questioning may reveal recurrent influenza-like symptoms associated with headaches and musculoskeletal complaints. These complaints could originate with medical problems such as endocarditis (a bacterial infection of the heart), bowel obstruction, or pleuropulmonary disorders, which should be ruled out by a physician.

Knowing that the client has had a recent bladder, vaginal, uterine, or kidney infection, or that the client is likely to have such infections, may help explain back pain in the absence of any musculoskeletal findings.

The client may or may not confirm previous back pain associated with previous infections. If there is any doubt, a medical referral is recommended. On the other hand, repeated coughing after a recent upper respiratory infection may cause chest, rib, back, or sacroiliac pain.

Screening for Cancer. Any "yes" responses to early screening questions for cancer (General Health questions 5, 6, and 7) must be followed up by a physician. An in-depth discussion of screening for cancer is presented in Chapter 13.

Changes in appetite and unexplained weight loss can be associated with cancer, onset of diabetes, hyperthyroidism, depression, or pathologic anorexia (loss of appetite). Weight loss significant for neoplasm would be a 10% loss of total body weight over a 4-week period unrelated to any intentional diet or fasting.

A significant, unexplained weight gain can be caused by congestive heart failure, hypothyroidism, or cancer. The person with back pain who, despite reduced work levels and decreased activity, experiences unexplained weight loss demonstrates a key "red flag" symptom.

Weight gain/loss does not always correlate with appetite. For example, weight gain associated with neoplasm may be accompanied by appetite loss, whereas weight loss associated with hyperthyroidism may be accompanied by increased appetite.

Substance Abuse. Substances refer to any agents taken nonmedically that can alter mood or behavior. Addiction refers to the daily need for the substance in order to function, an inability to stop, and recurrent use when it is harmful physically, socially, and/or psychologically. *Addiction* is based on physiologic changes associated with drug use but also has psychologic and behavioral components. Individuals who are addicted will use the substance to relieve psychologic symptoms even after physical pain or discomfort is gone.

Dependence is the physiologic dependence on the substance so that withdrawal symptoms emerge when there is a rapid dose reduction or the drug is stopped abruptly. Once a medication is no longer needed, the dosage will have to be tapered down for the client to avoid withdrawal symptoms.

Tolerance refers to the individual's need for increased amounts of the substance to produce the same effect. Tolerance develops in many people who receive long-term opioid therapy for chronic pain problems. If undermedicated, drug-seeking behaviors or unauthorized increases in dosage may occur. These may seem like addictive behaviors and are sometimes referred to as "pseudoaddiction," but the behaviors disappear when adequate pain control is achieved. Referral to the prescribing physician is advised if you suspect a problem with opioid analgesics (misuse or abuse).[97,98]

Among the substances most commonly used that cause physiologic responses but are not usually thought of as drugs are alcohol, tobacco, coffee, black tea, and caffeinated carbonated beverages.

Other substances commonly abused include *depressants,* such as alcohol, barbiturates (barbs, downers, pink ladies, rainbows, reds, yellows, sleeping pills); *stimulants,* such as amphetamines and cocaine (crack, crank, coke, snow, white, lady, blow, rock); *opiates* (heroin); *cannabis derivatives* (marijuana, hashish); and *hallucinogens* (LSD or acid, mescaline, magic mushroom, PCP, angel dust).

Methylenedioxymethamphetamine (MDMA; also called Ecstasy, hug, beans, and love drug), a synthetic, psychoactive drug chemically similar to the stimulant methamphetamine and the hallucinogen mescaline, has been reported to be sold in clubs around the country. It is often given to individuals without their knowledge and used in combination with alcohol and other drugs.

BOX 2-5 POPULATION GROUPS AT RISK FOR SUBSTANCE ABUSE

- Teens and adults with attention deficit disorder or attention deficit disorder with hyperactivity (ADD/ADHD)
- History of posttraumatic stress disorder (PTSD)
- Baby boomers with a history of substance use
- Individuals with sleep disorders
- Individuals with depression and/or anxiety disorders

Public health officials tell us that alcohol and other drug use/abuse is the number one health problem in the United States. Addictions (especially alcohol) have reached epidemic proportions in this country. Yet, it is largely ignored and often goes untreated.[99,100] A well-known social scientist in the area of drug studies published a new report showing that overall, alcohol is the most harmful drug (to the individual and to others) with heroin and crack cocaine ranked second and third.[101]

Up to one third of workers use these illegal, psychoactive substances to face up to job strain.[102] Alcohol and other drugs are commonly used to self-medicate mental illness, pain, and the effects of posttraumatic stress disorder (PTSD).

Risk Factors. Many teens and adults are at risk for using and abusing various substances (Box 2-5). Often, they are self-medicating the symptoms of a variety of mental illnesses, learning disabilities, and personality disorders. The use of alcohol to self-medicate depression is very common, especially after a traumatic injury or event in one's life.

Baby boomers (born between 1946 and 1964) with a history of substance use, aging adults (or others) with sleep disturbances or sleep disorders, and anyone with an anxiety or mood disorder is at increased risk for use and abuse of substances.

Think about this in terms of risk-factor assessment. According to the CDC, at least two thirds of boomers who enter drug treatment programs have been drinking, taking drugs, or both during the bulk of their adult lives.

It is estimated that 50% of all traumatic brain-injured (TBI) cases are alcohol- or drug-related—either by the clients themselves or by the perpetrator of the accident. Some centers estimate this figure to be much higher, around 80%.[103]

Risk factors for opioid misuse in people with chronic pain have been published.[98] These include family or personal history of substance abuse or previous alcohol or other drug rehabilitation), young age, history of criminal activity or legal problems (including driving under the influence [DUIs]), risk-taking or thrill-seeking behaviors, heavy tobacco use, and history of severe depression or anxiety). Physicians and clinical psychologists may use one of several tools (e.g., Current Opioid Misuse Measure, Screener and Opioid Assessment for Patients in Pain) to screen for risk of opioid misuse.

Signs and Symptoms of Substance Use/Abuse. Behavioral and physiologic responses to any of these substances depend on the characteristics of the chemical itself, the route of administration, and the adequacy of the client's circulatory system (Table 2-3).

Behavioral red flags indicating a need to screen can include consistently missed appointments (or being chronically late to scheduled sessions), noncompliance with the home program or poor attention to self-care, shifting mood patterns (especially the presence of depression), excessive daytime sleepiness or unusually excessive energy, and/or deterioration of physical appearance and personal hygiene.

The physiologic effects and adverse reactions have the additional ability to delay wound healing or the repair of soft tissue injuries. Soft tissue infections such as abscess and cellulitis are common complications of injection drug use (IDU). Affected individuals may present with swelling and tenderness in a muscular area from intramuscular injections. Low-grade fever may be found when taking vital signs.[104]

Substance abuse in older adults often mimics many of the signs of aging: memory loss, cognitive problems, tremors, and falls. Even family members may not recognize when their loved one is an addict. Late-stage abuse (age 60 and older) contributes to weight loss, muscle wasting, and among those who abuse alcohol elevated rates of breast cancer (especially among women).[105]

Screening for Substance Use/Abuse. Questions designed to screen for the presence of chemical substance abuse need to become part of the physical therapy assessment. Clients who depend on alcohol and/or other substances require lifestyle intervention. However, direct questions may be offensive to some people, and identifying a person as a substance abuser (i.e., alcohol or other drugs) often results in referral to professionals who treat alcoholics or drug addicts, a label that is not accepted in the early stage of this condition.

Because of the controversial nature of interviewing the alcohol- or drug-dependent client, the questions in this section of the Family/Personal History form are suggested as a guideline for interviewing.

After (or possibly instead of) asking questions about use of alcohol, tobacco, caffeine, and other chemical substances, the therapist may want to use the Trauma Scale Questionnaire[106] that makes no mention of substances but asks about previous trauma. Questions include[106]:

FOLLOW-UP QUESTIONS

- Have you had any fractures or dislocations to your bones or joints?
- Have you been injured in a road traffic accident?
- Have you injured your head?
- Have you been in a fight or assault?

These questions are based on the established correlation between trauma and alcohol or other substance use for individuals 18 years old and older. "Yes" answers to two or more of these questions should be discussed with the physician or

TABLE 2-3 Physiologic Effects and Adverse Reactions to Substances

Caffeine	Cannabis	Depressants	Narcotics	Stimulants	Tobacco
EXAMPLES					
Coffee, espresso Chocolate, some over-the-counter "alert aids" used to stay awake, black tea and other beverages with caffeine (e.g., Red Bull, caffeinated water)	Marijuana, hashish	Alcohol, sedatives/ sleeping pills, barbiturates, tranquilizers	Heroin, opium, morphine, codeine	Cocaine and its derivatives, amphetamines, methamphetamine, MDMA (ecstasy)	Cigarettes, cigars, pipe smoking, smokeless tobacco products (chew, snuff)
EFFECTS					
Vasoconstriction Irritability Enhances pain perception Intestinal disorders Headaches Muscle tension Fatigue Sleep disturbances Urinary frequency Tachypnea Sensory disturbances Agitation Nervousness Heart palpitation	Short-term memory loss Sedation Tachycardia Euphoria Increased appetite Relaxed inhibitions Fatigue Paranoia Psychosis Ataxia, tremor	Agitation; mood swings; anxiety; depression Vasodilation; red eyes Fatigue Altered pain perception Excessive sleepiness or insomnia Coma (overdose) Altered behavior Slow, shallow breathing Clammy skin Slurred speech	Euphoria Drowsiness Respiratory depression	Increased alertness Excitation Euphoria Loss of appetite Increased blood pressure Insomnia Increased pulse rate Agitation, increased body temperature, hallucinations, convulsions, death (overdose)	Increased heart rate Vasoconstriction Decreased oxygen to heart Increased risk of thrombosis Loss of appetite Poor wound healing Poor bone grafting Increased risk of pneumonia Increased risk of cataracts Disk degeneration

Adapted from Goodman CC, Fuller KS: *Pathology: implications for the physical therapist,* ed 3, Philadelphia, 2009, WB Saunders.

used to generate a referral for further evaluation of alcohol use. It may be best to record the client's answers with a simple + for "yes" or a – for "no" to avoid taking notes during the discussion of sensitive issues.

The RAFFT Questionnaire[107] (Relax, Alone, Friends, Family, Trouble) poses five questions that appear to tap into common themes related to adolescent substance use such as peer pressure, self-esteem, anxiety, and exposure to friends and family members who are using drugs or alcohol. Similar dynamics may still be present in adult substance users, although their use of drugs and alcohol may become independent from these psychosocial variables.

- **R:** Relax—Do you drink or take drugs to relax, feel better about yourself, or fit in?
- **A:** Alone—Do you ever drink or take drugs while you are alone?
- **F:** Friends—Do any of your closest friends drink or use drugs?
- **F:** Family—Does a close family member have a problem with alcohol or drugs?
- **T:** Trouble—Have you ever gotten into trouble from drinking or taking drugs?

Depending on how the interview has proceeded thus far, the therapist may want to conclude with one final question: "Are there any drugs or substances you take that you haven't mentioned?" Other screening tools for assessing alcohol abuse are available, as are more complete guidelines for interviewing this population.[20,108]

Resources. Several guides on substance abuse for health care professionals are available.[109,110] These resources may help the therapist learn more about identifying, referring, and preventing substance abuse in their clients.

The University of Washington provides a Substance Abuse Screening and Assessments Instruments database to help health care providers find instruments appropriate for their work setting.[111] The database contains information on more than 225 questionnaires and interviews; many have proven clinical utility and research validity, while others are newer instruments that have not yet been thoroughly evaluated.

Many are in the public domain and can be freely downloaded from the Web; others are under copyright and can only be obtained from the copyright holder. The Partnership for a Drug-Free America also provides information on the effects of drugs, alcohol, and other illicit substances available at www.drugfree.org.

Alcohol. Other than tobacco, alcohol is the most dominant addictive agent in the United States. Alcohol use disorder rates are higher among males than females and highest in the youngest age cohort (18 to 29 years).[112]

According to the National Institute on Alcohol Abuse and Alcoholism (NIAAA), alcoholism, also known as alcohol abuse or alcohol dependence is characterized by four symptoms: *craving* (the strong need or urge to drink), *loss of control* (unable to stop drinking once started), *physical dependence* (withdrawal symptoms develop after stopping drinking), and *tolerance* (a need for greater amounts of alcohol to get the same effects).[112]

ICD-10 includes the concept of "harmful use." This category was created so that health problems related to alcohol and other drug use would not be underreported. Harmful use implies alcohol use that causes either physical or mental damage in the absence of dependence.[113]

As the graying of America continues, the number of adults affected by alcoholism is expected to increase, especially as baby boomers, having grown up in an age of alcohol and substance abuse, carry that practice into old age.

Older adults are not the only ones affected. Alcohol consumption is a major contributor to risky behaviors and adverse health outcomes in adolescents and young adults. Motor vehicle accidents, homicides, suicides, and accidental injuries are the four leading causes of death in individuals aged 15 to 20 years, and alcohol plays a substantial role in many of these events.[114] In addition, the use of alcohol is associated with risky sexual behavior, teen pregnancy, and sexually transmitted diseases (STDs).

Binge drinking, defined as consuming five or more alcoholic drinks within a couple of hours, is a serious problem among adults and high-school aged youths. Binge drinking contributes to more than half of the 79,000 deaths caused by excessive drinking annually in the United States.[115]

Effects of Alcohol Use. Excessive alcohol use can cause or contribute to many medical conditions. Alcohol is a toxic drug that is harmful to all body tissues. Certain social and behavioral changes, such as heavy regular consumption, binge drinking, frequent intoxication, concern expressed by others about one's drinking, and alcohol-related accidents, may be early signs of problem drinking and unambiguous signs of dependence risk.[116]

Alcohol has both vasodilatory and depressant effects that may produce fatigue and mental depression or alter the client's perception of pain or symptoms. Alcohol has deleterious effects on the gastrointestinal (GI), hepatic, cardiovascular, hematopoietic, genitourinary (GU), and neuromuscular systems.

Efforts to look at the relationship between alcohol and cancer have revealed that even moderate drinking (defined as no more than one drink per day for women and two for men) is associated with an increased risk of mouth, pharynx, larynx, esophageal, breast, and colon cancers.[117]

Many conditions are made worse by drinking alcohol: hypertension; gout; diabetes; depression, anxiety, or other mental disorders; cirrhosis or other liver problems; and GI bleeding, ulcers, or gastroesophageal reflux disease.[118] Signs and symptoms of alcohol abuse in older adults may not be as obvious as in younger individuals.

CLINICAL SIGNS AND SYMPTOMS

Alcohol Use Disorders in Older Adults[118]

- Memory loss or cognitive impairment (new onset or worsening of previous condition)
- Depression or anxiety
- Neglect of hygiene and appearance
- Poor appetite and nutritional deficits
- Sleep disruption
- Refractory (resistant) hypertension
- Blood glucose control problems
- Refractory seizures
- Impaired gait, impaired balance, and falls
- Recurrent gastritis and esophagitis
- Difficulty managing warfarin dosing

Prolonged use of excessive alcohol may affect bone metabolism, resulting in reduced bone formation, disruption of the balance between bone formation and resorption, and incomplete mineralization.[119] Alcoholics are often malnourished, which exacerbates the direct effects of alcohol on bones. Alcohol-induced osteoporosis (the predominant bone condition in most people with cirrhosis) may progress for years without any obvious symptoms.

Regular consumption of alcohol may indirectly perpetuate trigger points through reduced serum and tissue folate levels and because of poor nutrition from eating habits. Ingestion of alcohol reduces the absorption of folic acid, while increasing the body's need for it.[120]

Therapists may also see alcoholic polyneuropathy, alcoholic myopathy, nontraumatic hip osteonecrosis, injuries from falls, and stroke[121] from heavy alcohol use. In fact, alcohol-related problems often mimic signs and symptoms associated with aging such as falls or memory loss.

Alcohol may interact with prescribed medications to produce various effects, including death. Prolonged drinking changes the way the body processes some common prescription drugs, potentially increasing the adverse effects of medications or impairing or enhancing their effects.

Binge drinking commonly seen on weekends and around holidays can cause atrial fibrillation, a condition referred to as "holiday heart." The affected individual may report dyspnea, palpitations, chest pain, dizziness, fainting or near-fainting, and signs of alcohol intoxication. Strenuous physical activity is contraindicated until the cardiac rhythm converts to normal sinus rhythm. Medical evaluation is required in cases of suspected holiday heart syndrome.[122]

Of additional interest to the therapist is the fact that alcohol diminishes the accumulation of neutrophils necessary for "clean-up" of all foreign material present in inflamed areas. This phenomenon results in delayed wound healing and recovery from inflammatory processes involving tendons, bursae, and joint structures.

Signs and Symptoms of Alcohol Withdrawal. The therapist must be alert to any signs or symptoms of alcohol withdrawal, a potentially life-threatening condition. This is especially true in the acute care setting,[123] especially for individuals who are recently hospitalized for a motor vehicle accident or other trauma or the postoperative orthopedic patient (e.g., total hip or total knee patient).[124] Alcohol withdrawal may be a factor in recovery for any orthopedic or neurologic patient (e.g., stroke, total joint, fracture), especially trauma patients.

Early recognition can bring about medical treatment that can reduce the symptoms of withdrawal as well as identify the need for long-term intervention. Withdrawal begins 3 to 36 hours after discontinuation of heavy alcohol consumption. Symptoms of autonomic hyperactivity may include diaphoresis (excessive perspiration), insomnia, general restlessness, agitation, and loss of appetite. Mental confusion, disorientation, and acute fear and anxiety can occur.

Tremors of the hands, feet, and legs may be visible. Symptoms may progress to hyperthermia, delusions, and paranoia called *alcohol hallucinosis* lasting 1 to 5 or more days. Seizures occur in up to one third of affected individuals, usually 12 to 48 hours after the last drink or potentially sooner when the blood alcohol level returns to zero. Five percent have delirium tremens (DTs) following cessation of alcohol consumption.[125] This is an acute and sometimes fatal psychotic reaction caused by cessation of excessive intake of alcohol.[126] It consists of autonomic hyperactivity (tachycardia, agitation), confusion, and disorientation with an increased potential for alcohol-withdrawal seizures.

CLINICAL SIGNS AND SYMPTOMS

Alcohol Withdrawal

- Agitation, irritability
- Headache
- Insomnia
- Hallucinations
- Anorexia, nausea, vomiting, diarrhea
- Loss of balance, incoordination (apraxia)
- Seizures (occurs 12 to 48 hours after the last drink)
- Delirium tremens (occurs 2 to 3 days after the last drink)
- Motor hyperactivity, tachycardia
- Elevated blood pressure

The Clinical Institute Withdrawal of Alcohol Scale (CIWA)[127] is an assessment tool used to monitor alcohol withdrawal symptoms. Although it is used primarily to determine the need for medication, it can provide the therapist with an indication of stability level when determining patient safety before initiating physical therapy. The assessment requires about 5 minutes to administer and is available online with no copyright restrictions.

Screening for Alcohol Abuse. In the United States, alcohol use/abuse is often considered a moral problem and may pose an embarrassment for the therapist and/or client when asking questions about alcohol use. Keep in mind the goal is to obtain a complete health history of factors that can affect healing and recovery as well as pose risk factors for future health risk.

Based on the definition of alcohol abuse defined earlier in this section, four broad categories of drinking patterns exist[128]:

1. Abstaining or infrequent drinking (fewer than 12 drinks/year)
2. Drinking within the screening limits
3. Exceeding daily limits, occasionally and frequently
4. Exceeding weekly limits

There is little to no risk of developing an alcohol disorder in categories 1 and 2. Individuals in category 3 have a 7% chance of becoming alcohol dependent whether in the occasional or frequent group. Group 4 have a 1 in 4 or 25% chance of developing alcohol dependence.[128]

There are several tools used to assess a client's history of alcohol use, including the Short Michigan Alcoholism Screening Test (SMAST),[129] the CAGE questionnaire, and a separate list of alcohol-related screening questions (Box 2-6). The SMAST has a geriatric version (MAST-G) available online (search for MAST-G or SMAST-G).

The CAGE questionnaire helps clients unwilling or unable to recognize a problem with alcohol, although it is possible for a person to answer "no" to all of the CAGE questions and still be drinking heavily and at risk for alcohol dependence. The specificity of this test is high for assessing alcohol abuse pretraumatic and posttraumatic brain injury.[130] After 25 years of use, the CAGE questionnaire is still widely used and

BOX 2-6 SCREENING FOR EXCESSIVE ALCOHOL

CAGE Questionnaire

C: Have you ever thought you should cut down on your drinking?

A: Have you ever been annoyed by criticism of your drinking?

G: Have you ever felt guilty about your drinking?

E: Do you ever have an eye-opener (a drink or two) in the morning?

Key

- One "yes" answer suggests a need for discussion and follow-up; taking the survey may help some people in denial to accept that a problem exists
- Two or more "yes" answers indicates a problem with alcohol; intervention likely needed

Alcohol-Related Screening Questions

- Have you had any fractures or dislocations to your bones or joints since your eighteenth birthday?
- Have you been injured in a road traffic accident?
- Have you ever injured your head?
- Have you been in a fight or been hit or punched in the last 6 months?

Key

- "Yes" to two or more questions is a red flag

considered one of the most efficient and effective screening tools for the detection of alcohol abuse.[131]

The AUDIT (Alcohol Use Disorders Identification Test) developed by the World Health Organization to identify persons whose alcohol consumption has become hazardous or harmful to their health is another popular, valid,[132,133] and easy to administer screening tool (Box 2-7).[116]

The AUDIT is designed as a brief, structured interview or self-report survey that can easily be incorporated into a general health interview, lifestyle questionnaire, or medical history. It is a 10-item screening questionnaire with questions on the amount and frequency of drinking, alcohol dependence, and problems caused by alcohol.

When presented in this context by a concerned and interested interviewer, few clients will be offended by the questions. Results are most accurate when given in a nonthreatening, friendly environment to a client who is not intoxicated and who has not been drinking.[116]

The experience of the WHO collaborating investigators indicated that AUDIT questions were answered accurately regardless of cultural background, age, or sex. In fact, many individuals who drank heavily were pleased to find that a health worker was interested in their use of alcohol and the problems associated with it.

The best way to administer the test is to give the client a copy and have him or her fill it out (see Appendix B-1). This is suggested for clients who seem reliable and literate. Alternately, the therapist can interview clients by asking them the questions. Some health care workers use just two questions (one based on research in this area and one from the AUDIT) to quickly screen.

FOLLOW-UP QUESTIONS

- How often do you have six or more drinks on one occasion?
  - 0 = Never
  - 1 = Less than monthly
  - 2 = Monthly
  - 3 = Weekly
  - 4 = Daily or almost daily
- How many drinks containing alcohol do you have each week?
  - More than 14/week for men constitutes a problem
  - More than 7/week for women constitutes a problem

When administered during the screening interview, it may be best to use a transition statement such as:

Now I am going to ask you some questions about your use of alcoholic beverages during the past year. Because alcohol use can

BOX 2-7 ALCOHOL USE DISORDERS IDENTIFICATION TEST (AUDIT)

Therapists: This form is available in Appendix B-1 for clinical use. It is also available from the National Institute on Alcohol Abuse and Alcoholism (NIAAA) online: www.niaaa.nih.gov. Type AUDIT in search window.

1) How often do you have a drink containing alcohol?
   (0) NEVER (1) MONTHLY OR LESS (2) TWO TO FOUR TIMES A MONTH
   (3) TWO TO THREE TIMES A WEEK (4) FOUR OR MORE TIMES A WEEK
2) How many drinks containing alcohol do you have on a typical day when you are drinking?
   (0) 1 OR 2 (1) 3 OR 4 (2) 5 OR 6 (3) 7 OR 8 (4) 10 OR MORE
3) How often do you have six or more drinks on one occasion?
   (0) NEVER (1) LESS THAN MONTHLY (2) MONTHLY (3) WEEKLY (4) DAILY OR ALMOST DAILY
4) How often during the last year have you found that you were unable to stop drinking once you had started?
   (0) NEVER (1) LESS THAN MONTHLY (2) MONTHLY (3) WEEKLY (4) DAILY OR ALMOST DAILY
5) How often during the last year have you failed to do what was normally expected from you because of drinking?
   (0) NEVER (1) LESS THAN MONTHLY (2) MONTHLY (3) WEEKLY (4) DAILY OR ALMOST DAILY
6) How often during the last year have you needed a first drink in the morning to get going after a heavy drinking session?
   (0) NEVER (1) LESS THAN MONTHLY (2) MONTHLY (3) WEEKLY (4) DAILY OR ALMOST DAILY
7) How often during the last year have you had a feeling of guilt or remorse after drinking?
   (0) NEVER (1) LESS THAN MONTHLY (2) MONTHLY (3) WEEKLY (4) DAILY OR ALMOST DAILY
8) How often during the last year have you been unable to remember the night before because you had been drinking?
   (0) NEVER (1) LESS THAN MONTHLY (2) MONTHLY (3) WEEKLY (4) DAILY OR ALMOST DAILY
9) Have you or someone else been injured as the result of your drinking?
   (0) NO (2) YES, BUT NOT IN THE LAST YEAR (4) YES, DURING THE LAST YEAR
10) Has a relative, friend, or health professional been concerned about your drinking or suggested you cut down?
   (0) NO (2) YES, BUT NOT IN THE LAST YEAR (4) YES, DURING THE LAST YEAR
   TOTAL SCORE: _______

Key:
The numbers for each response are added up to give a composite score. Scores above 8 warrant an in-depth assessment and may be indicative of an alcohol problem. See options presented to clients in Appendix B-1: AUDIT Questionnaire.

Data from World Health Organization, 1992. Available for clinical use without permission.

affect many areas of health (and may interfere with healing and certain medications), it is important for us to know how much you usually drink and whether you have experienced any problems with your drinking. Please try to be as honest and as accurate as you can be.

Alternately, if the client's breath smells of alcohol, the therapist may want to say more directly:

FOLLOW-UP QUESTIONS

- I can smell alcohol on your breath right now. How many drinks have you had today?
  As a follow-up to such direct questions, you may want to say:
- Alcohol, tobacco, and caffeine often increase our perception of pain, mask or even increase other symptoms, and delay healing. I would like to ask you to limit as much as possible your use of any of these stimulants. At the very least, it would be better if you didn't drink alcohol before our therapy sessions, so I can see more clearly just what your symptoms are. You may progress and move along more quickly through our plan of care if these substances aren't present in your body.

A helpful final question to ask at the end of this part of the interview may be:

FOLLOW-UP QUESTIONS

- Are there any other drugs or substances you take that you haven't mentioned?

Physical Therapist's Role. Incorporating screening questions into conversation during the interview may help to engage individual clients. Honest answers are important to guiding treatment. Reassure clients that all information will remain confidential and will be used only to ensure the safety and effectiveness of the plan of care. Specific interviewing techniques, such as normalization, symptom assumption, and transitioning, may be helpful.[126,134]

Normalization involves asking a question in a way that lets the person know you find a behavior normal or at least understandable under the circumstances. The therapist might say, "Given the stress you're under, I wonder if you've been drinking more lately?"

Symptom assumption involves phrasing a question that assumes a certain behavior already occurs and the therapist will not be shocked by it. For example, "What kinds of drugs do you use when you're drinking?" or "How much are you drinking?"

Transitioning is a way of using the client's previous answer to start a question such as, "You mentioned your family is upset by your drinking. Have your coworkers expressed similar concern?"[126]

What is the best way to approach alcohol and/or substance use/abuse? Unless the client has a chemical dependency on alcohol, appropriate education may be sufficient for the client experiencing negative effects of alcohol use during the episode of care.

It is important to recognize the distinct and negative physiologic effects each substance or addictive agent can have on the client's physical body, personality, and behavior. Some physicians advocate screening for and treating suspected or known excessive alcohol consumption no differently than diabetes, high blood pressure, or poor vision. The first step may be to ask all clients: Do you drink alcohol, including beer, wine, or other forms of liquor? If yes, ask about consumption (e.g., days per week/number of drinks). Then proceed to the CAGE questions before advising appropriate action.[135]

If the client's health is impaired by the use and abuse of substances, then physical therapy intervention may not be effective as long as the person is under the influence of chemicals.

Encourage the client to seek medical attention or let the individual know you would like to discuss this as a medical problem with the physician (Case Example 2-5).

Research shows that the longer people spend in treatment, the more likely they are to recover. A California study by the Rand Corporation showed that for every $1.00 spent on treatment for addictions, $7.00 is saved in social costs.[136] Often the general sentiment in the medical community is that alcoholism and addictions are not treatable. Yet, the national statistics show that one third stay sober after one year. Of the two thirds that relapse, 50% will get well if they go back to treatment.[136]

Alcohol-related trauma patients have a high reinjury rate. Even a brief intervention can reduce this by up to half. A single question or single suggestion from a health care worker can make a difference.[137,138]

Physical therapists are not chemical dependency counselors or experts in substance abuse, but armed with a few questions, the therapist can still make a significant difference. Hospitalization or physical therapy intervention for an injury is potentially a teachable moment. Clients with substance abuse problems have worse rehabilitation outcomes, are at increased risk for reinjury or new injuries, and additional comorbidities.

Therapists can actively look for and address substance use/abuse problems in their clients. At the very minimum, therapists can participate in the National Institute on Alcohol Abuse and Alcoholism's National Alcohol Screening Day with a program that includes the CAGE questionnaire, educational materials, and an opportunity to talk with a health care professional about alcohol.

For those who want to participate in an anonymous self-screening process that includes alcohol, anxiety, eating disorders, and depression, go to www.mentalhealthscreening.org/screening/alcohol.asp. Click on your state and the nearest program and follow the directions. When screening in any setting or circumstance, if a red flag is raised after completing any of the screening questions, the therapist may want to follow-up with:

CASE EXAMPLE 2-5

Substance Abuse

A 44-year-old man previously seen in the physical therapy clinic for a fractured calcaneus returns to the same therapist 3 years later because of new onset of midthoracic back pain. There was no known cause or injury associated with the presenting pain. This man had been in the construction business for 30 years and attributed his symptoms to "general wear and tear."

Although there were objective findings to support a musculoskeletal cause of pain, the client also mentioned symptoms of fatigue, stomach upset, insomnia, hand tremors, and headaches. From the previous episode of care, the therapist recalled a history of substantial use of alcohol, tobacco, and caffeine (three six-packs of beer after work each evening, 2 pack/day cigarette habit, 18+ cups of caffeinated coffee during work hours).

The therapist pointed out the potential connection between the client's symptoms and the level of substance use, and the client agreed to "pay more attention to cutting back." After 3 weeks the client returned to work with a reduction of back pain from a level of 8 to a level of 0-3 (intermittent symptoms), depending on the work assignment.

Six weeks later this client returned again with the same symptomatic and clinical presentation. At that time, given the client's age, the insidious onset, the cyclic nature of the symptoms, and significant substance abuse, the therapist recommended a complete physical with a primary care physician.

Medical treatment began with nonsteroidal antiinflammatory drugs (NSAIDs), which caused considerable gastrointestinal (GI) upset. The GI symptoms persisted even after the client stopped taking the NSAIDs. Further medical diagnostic testing determined the presence of pancreatic carcinoma. The prognosis was poor, and the client died 6 months later, after extensive medical intervention.

In this case, it could be argued that the therapist should have referred the client to a physician immediately because of the history of substance abuse and the presence of additional symptoms. A more thorough screening examination during the first treatment for back pain may have elicited additional red-flag GI symptoms (e.g., melena or bloody diarrhea in addition to the stomach upset).

FOLLOW-UP QUESTIONS

- How do you feel about the role of alcohol in your life?
- Is there something you want or need to change?

Earlier referral for a physical examination may have resulted in earlier diagnosis and treatment for the cancer. Unfortunately, these clinical situations occur often and are very complex, requiring ongoing screening (as happened here).

Finally, the APTA recognizes that physical therapists and physical therapist assistants can be adversely affected by alcoholism and other drug addictions. Impaired therapists or assistants should be encouraged to enter into the recovery process. Reentry into the work force should occur when the well-being of the physical therapy practitioner and patient/client are assured.[139]

Recreational Drug Use. As with tobacco and alcohol use, recreational or street drug use can lead to or compound already present health problems. Although the question "Do you use recreational or street drugs?" is asked on the Family/Personal History form (see Fig. 2-2), it is questionable whether the client will answer "yes" to this question.

At some point in the interview, the therapist may need to ask these questions directly:

FOLLOW-UP QUESTIONS

- Have you ever used "street" drugs such as cocaine, crack, crank, "downers," amphetamines ("uppers"), methamphetamine, or other drugs?
- Have you ever injected drugs?
  - If yes, have you been tested for HIV or hepatitis?

Cocaine and amphetamines affect the cardiovascular system in the same manner as does stress. The drugs stimulate the sympathetic nervous system to increase its production of adrenaline. Surging adrenaline causes severe constriction of the arteries, a sharp rise in blood pressure, rapid and irregular heartbeats, and seizures.[140]

Heart rate can accelerate by as much as 60 to 70 beats per minute (bpm). In otherwise healthy and fit people, this overload can cause death in minutes, even in first-time cocaine users. In addition, cocaine can cause the aorta to rupture, the lungs to fill with fluid, the heart muscle and its lining to become inflamed, blood clots to form in the veins, and strokes to occur as a result of cerebral hemorrhage.

Tobacco. Nearly half a million tobacco-related deaths are reported in the United States every year with an estimated 200 billion dollars spent each year treating tobacco-related diseases (e.g., stroke, cancer, chronic obstructive pulmonary disease [COPD], heart disease).[141] Tobacco and tobacco products are known carcinogens. This includes secondhand smoke, pipes, cigars, cigarettes, and chewing (smokeless) tobacco.

Tobacco is well documented in its ability to cause vasoconstriction and delay wound healing. More people die from tobacco use than alcohol and all the other addictive agents combined. Cigarettes sold in the United States reportedly contain 600 chemicals and additives, ranging from chocolate to counteract tobacco's bitterness to ammonia, added to increase nicotine absorption. Cigarette smoke contains approximately 4000 chemicals (250 are known to be harmful, 50 are carcinogenic).[142]

As health care providers, the therapist has an important obligation to screen for tobacco use and incorporate smoking cessation education into the physical therapy plan of care whenever possible.[143] The American Cancer Society publishes a chart (and pamphlet for distribution) of the benefits of smoking cessation starting from 20 minutes since the last cigarette until 15 years later.[144] Therapists can encourage the clients to decrease (or eliminate) tobacco use while in treatment.

Client education includes a review of the physiologic effects of tobacco (see Table 2-3). Nicotine in tobacco, whether in the form of chewing tobacco or from cigar, pipe, or cigarette smoking, acts directly on the heart, blood vessels, digestive tract, kidneys, and nervous system.[145]

For the client with respiratory or cardiac problems, nicotine stimulates the already compensated heart to beat faster, narrows the blood vessels, increases airflow obstruction,[145] reduces the supply of oxygen to the heart and other organs, and increases the chances of developing blood clots. Narrowing of the blood vessels is also detrimental for anyone with peripheral vascular disease, diabetes, or delayed wound healing.

Smoking markedly increases the need for vitamin C, which is poorly stored in the body. One cigarette can consume 25 mg of vitamin C (one pack would consume 500 mg/day). The capillary fragility associated with low ascorbic acid levels greatly increases the tendency for tissue bleeding, especially at injection sites.[146]

Smoking has been linked with disc degeneration[147,148] and acute lumbar and cervical intervertebral disc herniation.[149,150] Nicotine interacts with cholinergic nicotinic receptors, which leads to increased blood pressure, vasoconstriction, and vascular resistance. These systemic effects of nicotine may cause a disturbance in the normal nutrition of the disc.[147]

The combination of coffee ingestion and smoking raises the blood pressure of hypertensive clients about 15/30 mm Hg for as long as 2 hours. All these effects have a direct impact on the client's ability to exercise and must be considered when the client is starting an exercise program. Careful monitoring of vital signs during exercise is advised.

The commonly used formula to estimate cigarette smoking history is done by taking the number of packs smoked per day multiplied by the number of years smoked. If a person smoked 2 packs per day for 30 years, this would be a 60-pack year history (2 packs per day × 20 years = 60-pack years). A 60-pack year history could also be achieved by smoking 3 packs of cigarettes per day for 20 years, and so on (Case Example 2-6).

A significant smoking history is considered 20-pack years and is a risk factor for lung disease, cancer, heart disease, and other medical comorbidities. Less significant smoking habits must still be assessed in light of other risk factors present, personal/family history, and other risky lifestyle behaviors.

If the client indicates a desire to quit smoking or using tobacco (see Fig. 2-2, General Health: Question 10), the therapist must be prepared to help him or her explore options for smoking cessation. Many hospitals, clinics, and community organizations, such as the local chapter of the American Lung Association, sponsor annual (or ongoing) smoking cessation programs. Pamphlets and other reading material should be available for any client interested in tobacco cessation. Referral to medical doctors who specialize in smoking cessation may be appropriate for some clients.

CASE EXAMPLE 2-6

Recognizing Red Flags

A 60-year-old man was referred to physical therapy for weakness in the lower extremities. The client also reports dysesthesia (pain with touch).

Social/Work History: Single, factory worker, history of alcohol abuse, 60-pack year* history of tobacco use.

Clinically, the client presented with mild weakness in distal muscle groups (left more than right). Over the next 2 weeks, the weakness increased and a left footdrop developed. Now the client presents with weakness of right wrist and finger flexors and extensors.

What Are the Red Flags Presented in This Case? Is Medical Referral Required?

- Age
- Smoking history
- Alcohol use
- Bilateral symptoms
- Progressive neurologic symptoms

Consultation with the physician is certainly advised given the number and type of red flags present, especially the progressive nature of the neurologic symptoms in combination with other key red flags.

*Pack years = # packs/day × number of years. A 60-pack year history could mean 2 packs/day for 30 years or 3 packs/day for 20 years.

Caffeine. Caffeine is a substance with specific physiologic (stimulant) effects. Caffeine ingested in toxic amounts has many effects, including nervousness, irritability, agitation, sensory disturbances, tachypnea (rapid breathing), heart palpitations (strong, fast, or irregular heartbeat), nausea, urinary frequency, diarrhea, and fatigue.

The average cup of coffee or tea in the United States is reported to contain between 40 and 150 mg of caffeine; specialty coffees (e.g., espresso) may contain much higher doses. OTC supplements used to combat fatigue typically contain 100 to 200 mg caffeine per tablet. Many prescription drugs and OTC analgesics contain between 32 and 200 mg of caffeine.

People who drink 8 to 15 cups of caffeinated beverages per day have been known to have problems with sleep, dizziness, restlessness, headaches, muscle tension, and intestinal disorders. Caffeine may enhance the client's perception of pain. Pain levels can be reduced dramatically by reducing the daily intake of caffeine.

In large doses, caffeine is a stressor, but abrupt withdrawal from caffeine can be equally stressful. Withdrawal from caffeine induces a syndrome of headaches, lethargy, fatigue, poor concentration, nausea, impaired psychomotor performance, and emotional instability, which begins within 12 to 24 hours and lasts about 1 week.[151,152] Anyone seeking to break free from caffeine dependence should do so gradually over a week's time or more.

Fatal caffeine overdoses in adults are relatively rare; physiologically toxic doses are measured as more than 250 mg/day or 3 average cups of caffeinated coffee.[153]

New research suggests that habitual, moderate caffeine intake from coffee and other caffeinated beverages may not represent a health hazard after all and may even be associated with beneficial effects on cardiovascular health.[154]

Other sources of caffeine are tea (black and green), cocoa, chocolate, caffeinated-carbonated beverages, and some drugs, including many OTC medications. Some people also take caffeine in pill form (e.g., Stay Awake, Vivarin). There are even off-label uses of drugs such as Provigil, normally used as an approved therapy for narcolepsy. This unauthorized use is for increasing alertness and cutting short the number of hours required for sleep.

Decaffeinated coffee may not have caffeine in it, but coffee contains several hundred different substances. It has been shown to have specific cardiovascular effects.[155] Drinking decaf also increases the risk of rheumatoid arthritis among older women.[156]

Artificial Sweeteners. According to the American Dietetic Association (ADA), artificial sweeteners are safe when used in amounts specified by the Food and Drug Administration (FDA).[157] Other experts still question the potential toxic effects of these substances.[158-160]

From the author's own clinical experience, it appears that some individuals may react to artificial sweeteners and can experience generalized joint pain, myalgias, fatigue, headaches, and other nonspecific symptoms.

For anyone with these symptoms, connective tissue disorders, fibromyalgia, multiple sclerosis, or other autoimmune disorders such as systemic lupus erythematosus or Hashimoto thyroid disease, it may be helpful to ask about the use of products containing artificial sweeteners.

FOLLOW-UP QUESTIONS

- Do you drink diet soda or diet pop or use aspartame, Equal, saccharin, NutraSweet, Splenda, or other artificial sweeteners?

If the client uses these products in any amount, the therapist can suggest eliminating them on a trial basis for 30 days. Artificial sweetener-induced symptoms may disappear in some people. Symptoms will develop again in susceptible people if use of artificial sweeteners is resumed.

Client Checklist. Screening for medical conditions can be aided by the use of a client checklist of associated signs and symptoms. Any items checked will alert the therapist to the possible need for further questions or tests.

A brief list here of the most common systemic signs and symptoms is one option for screening. It may be preferable to use the Review of Systems checklist (see Box 4-19; see also Appendix D-5).

Medical and Surgical History. Tests contributing information to the physical therapy assessment may include radiography (x-rays, sonograms), computed tomography (CT) scans, magnetic resonance imaging (MRI), bone scans and other imaging, lumbar puncture analysis, urinalysis, and blood tests. The client's medical records may contain information regarding which tests have been performed and the results of the test. It may be helpful to question the client directly by asking:

FOLLOW-UP QUESTIONS

- What medical test have you had for this condition?
- After giving the client time to respond, the therapist may need to probe further by asking:
  - Have you had any x-ray films, sonograms, CT scans, MRIs, or other imaging studies done in the last 2 years?
  - Do you recall having any blood tests or urinalyses done?

If the response is affirmative, the therapist will want to know when and where these tests were performed and the results (if known to the client). Knowledge of where the test took place provides the therapist with access to the results (with the client's written permission for disclosure).

Surgical History. Previous surgery or surgery related to the client's current symptoms may be indicated on the Family/Personal History form (see Fig. 2-2). Whenever treating a client postoperatively, the therapist should read the surgical report. Look for notes on complications, blood transfusions, and the position of the client during the surgery and the length of time in that position.

Clients in an early postoperative stage (within 3 weeks of surgery) may have stiffness, aching, and musculoskeletal pain unrelated to the diagnosis, which may be attributed to position during the surgery. Postoperative infections can lie dormant for months. Accompanying constitutional symptoms may be minimal with no sweats, fever, or chills until the infection progresses with worsening of symptoms or significant change in symptoms.

Specific follow-up questions differ from one client to another, depending on the type of surgery, age of client, accompanying medical history, and so forth, but it is always helpful to assess how quickly the client recovered from surgery to determine an appropriate pace for physical activity and exercise prescribed during an episode of care.

Clinical Tests. The therapist will want to examine the available test results as often as possible. Familiarity with the results of these tests, combined with an understanding of the clinical presentation. Knowledge of testing and test results also provides the therapist with some guidelines for suggesting or recommending additional testing for clients who have not had a radiologic workup or other potentially appropriate medical testing.

Laboratory values of interest to therapists are displayed on the inside covers of this book.

Work/Living Environment. Questions related to the client's daily work activities and work environments are included in the Family/Personal History form to assist the therapist in planning a program of client education that is consistent with the objective findings and proposed plan of care.

For example, the therapist is alerted to the need for follow-up with a client complaining of back pain who sits for prolonged periods without a back support or cushion. Likewise, a worker involved in bending and twisting who complains of lateral thoracic pain may be describing a muscular strain from repetitive overuse. These work-related questions may help the client report significant data contributing to symptoms that may otherwise have gone undetected.

Questions related to occupation and exposure to toxins such as chemicals or gases are included because well-defined physical (e.g., cumulative trauma disorder) and health problems occur in people engaging in specific occupations.[161] For example pesticide exposure is common among agricultural workers. Asthma and sick building syndrome are reported among office workers. Lung disease is associated with underground mining and silicosis is found in those who must work near silica. There is a higher prevalence of tuberculosis in health care workers compared to the general population.

Each geographic area has its own specific environmental/occupational concerns but overall, the chronic exposure to chemically based products and pesticides has escalated the incidence of environmental allergies and cases of multiple chemical sensitivity. Frequently, these conditions present in a physical therapy setting with nonspecific NMS manifestations.[162]

Exposure to cleaning products can be an unseen source of problems. Headaches, fatigue, skin lesions, joint arthralgias, myalgias, and connective tissue disorders may be the first signs of a problem. The therapist may be the first person to put the pieces of the puzzle together. Clients who have seen every kind of specialist end up with a diagnosis of fibromyalgia, rheumatoid arthritis, or some other autoimmune disorder and find their way to the physical therapy clinic (Case Example 2-7).

Military service at various periods and associated with specific countries or geographic areas has potential association with known diseases. For example, survivors of the Vietnam War who have been exposed to the defoliant mixtures, including Agent Orange, are at risk for developing soft tissue sarcoma, non-Hodgkin's lymphoma, Hodgkin's disease, and a skin-blistering disease called chloracne.[163]

About 30,000 U.S. soldiers who served in the Gulf War have reported symptoms linked to Gulf War syndrome, including chronic fatigue, headaches, chemical sensitivity, memory loss, joint pain and inflammation, and other fibromyalgia-like musculoskeletal disorders.[164]

Survivors of the Gulf War are nearly twice as likely to develop amyotrophic lateral sclerosis (ALS; Lou Gehrig's disease) than other military personnel. Classic early symptoms include irregular gait and decreased muscular coordination. Other occupational-related illnesses and diseases have been reported (Table 2-4).

When to Screen. Taking an environmental, occupational, or military history may be appropriate when a client has a history of asthma, allergies, fibromyalgia, chronic fatigue syndrome, or connective tissue or autoimmune disease or in the presence of other nonspecific disorders.

Conducting a quick survey may be helpful when a client presents with puzzling, nonspecific symptoms, including myalgias, arthralgias, headaches, back pain, sleep disturbance, loss of appetite, loss of sexual interest, or recurrent upper respiratory symptoms.

CASE EXAMPLE 2-7
Cleaning Products

A 33-year-old dental hygienist came to physical therapy for joint pain in her hands and wrists. In the course of taking a symptom inventory, the therapist discovered that the client had noticed multiple arthralgias and myalgias over the last 6 months.

She reported being allergic to many molds, dusts, foods, and other allergens. She was on a special diet but had obtained no relief from her symptoms. The doctor, thinking the client was experiencing painful symptoms from repetitive motion, sent her to physical therapy.

A quick occupational survey will include the following questions[161]:

- What kind of work do you do?
- Do you think your health problems are related to your work?
- Are your symptoms better or worse when you're at home or at work?
- Do others at work have similar problems?

The client answered "No" to all work-related questions but later came back and reported that other dental hygienists and dental assistants had noticed some of the same symptoms, although in a much milder form.

None of the other support staff (receptionist, bookkeeper, secretary) had noticed any health problems. The two dentists in the office were not affected either. The strongest red flag came when the client took a 10-day vacation and returned to work symptom-free. Within 24-hours of her return to work, her symptoms had flared up worse than ever.

This is not a case of emotional stress and work avoidance. The women working in the dental cubicles were using a cleaning spray after each dental client to clean and disinfect the area. The support staff was not exposed to it and the dentists only came in after the spray had dissipated. When this was replaced with an effective cleaning agent with only natural ingredients, everyone's symptoms were relieved completely.

TABLE 2-4 Common Occupational Exposures

Occupation	Exposure
Agriculture	Pesticides, herbicides, insecticides, fertilizers
Industrial	Chemical agents or irritants, fumes, dusts, radiation, loud noises, asbestos, vibration
Health care workers	Tuberculosis, hepatitis
Office workers	Sick building syndrome
Military service	Gulf War syndrome, connective tissue disorders, amyotrophic lateral sclerosis (ALS), non-Hodgkin's lymphoma, soft tissue sarcoma, chloracne (skin blistering)

After determining the client's occupation and depending on the client's chief complaint and accompanying associated signs and symptoms, the therapist may want to ask[161]:

FOLLOW-UP QUESTIONS

- Do you think your health problems are related to your work?
- Do you wear a mask at work?
- Are your symptoms better or worse when you are at home or at work?
  - Follow-up if worse at work: Do others at work have similar problems?
  - Follow-up if worse at home: Have you done any remodeling at home in the last 6 months?
- Are you now or have you previously been exposed to dusts, fumes, chemicals, radiation, loud noise, tools that vibrate, or a new building/office space?
- Have you ever been exposed to chemical agents or irritants such as asbestos, asphalt, aniline dyes, benzene, herbicides, fertilizers, wood dust, or others?
- Do others at work have similar problems?
- Have you ever served in any branch of the military?
  - If yes, were you ever exposed to dusts, fumes, chemicals, radiation, or other substances?

The idea in conducting a work/environmental screening is to look for patterns in the past medical history that might link the current clinical presentation with the reported or observed associated signs and symptoms. Further follow-up questions are listed in Appendix B-14.

The mnemonic CH^2OPD^2 (Community, Home, Hobbies, Occupation, Personal habits, Diet, and Drugs) can be used as a tool to identify a client's history of exposure to potentially toxic environmental contaminants[165]:

- **Community** — Live near a hazardous waste site or industrial site
- **Home** — Home is more than 40 years old; recent renovations; pesticide use in home, garden, or on pets
- **Hobbies** — Work with stained glass, oil-based paints, varnishes
- **Occupation** — Air quality at work; exposure to chemicals
- **Personal habits** — Tobacco use, exposure to secondhand smoke
- **Diet** — Contaminants in food and water
- **Drugs** — Prescription, over-the-counter drugs, home remedies, illicit drug use

Resources. Further suggestions and tools to help health care professionals incorporate environmental history questions can be found online. The Children's Environmental Health Network (www.cehn.org) has an online training manual, Pediatric Environmental Health: Putting It into Practice. Download and review the chapter on environmental history taking.

The Agency for Toxic Substances and Disease Registry (ATSDR) website, (www.atsdr.cdc.gov) offers information on specific chemical exposures.

History of Falls. In the United States, falls are the second leading cause of TBI among persons aged 65 or older.[166] Older adults who fall often sustain more severe head injuries than their younger counterparts. Falls are a major cause of intracranial lesion among older persons because of their greater susceptibility to subdural hematoma.[166]

It is reported that approximately one in four Americans in this age category who are living at home will fall during the next year. There is a possibility that older adults are falling even more often than is generally reported.[167]

By assessing risk factors (prediction) and offering preventive and protective strategies, the therapist can make a significant difference in the number of fall-related injuries and fractures. There are many ways to look at falls assessment. For the screening process, there are four main categories:

- Well-adult (no falling pattern)
- Just starting to fall
- Falls frequently (more than once every 6 months)
- Fear of falling

Healthy older adults who have no falling patterns may have a fear of falling in specific instances (e.g., getting out of the bath or shower; walking on ice, curbs, or uneven terrain). Fear of falling can be considered a mobility impairment or functional limitation. It restricts the client's ability to perform specific actions, thereby preventing the client from doing the things he or she wants to do. Functionally, this may appear as an inability to take a tub bath, walk on grass unassisted, or even attempt household tasks such as getting up on a sturdy step stool to change a light bulb (Case Example 2-8).

Risk Factors for Falls. If all other senses and reflexes are intact and muscular strength and coordination are normal, the affected individual can regain balance without falling. Many times, this does not happen. The therapist is a key health care professional to make early identification of adults at increased risk for falls.

With careful questioning, any potential problems with balance may come to light. Such information will alert the therapist to the need for testing static and dynamic balance and to look for potential risk factors and systemic or medical causes of falls (Table 2-5).

All of the variables and risk factors listed in Table 2-5 for falls are important. Older adults may have impaired balance, slower reaction times, and decreased strength, leading to more frequent falls. Sleep deprivation can lead to slowing in motor reaction time, thus increasing the risk of falls.[168] Medications, especially polypharmacy or hyperpharmacotherapy (see definition and discussion of Medications in this chapter), can contribute to falls.[169] There are five key areas that are the most common factors in falls among the aging adult population:

- Vision/hearing
- Balance
- Blood pressure regulation
- Medications/substances
- Elder assault

As we age, cervical spinal motion declines, as does peripheral vision. These two factors alone contribute to changes in

CASE EXAMPLE 2-8

Fracture After a Fall

Case Description: A 67-year-old woman fell and sustained a complete transverse fracture of the left fibula and an incomplete fracture of the tibia. The client reported she lost her footing while walking down four steps at the entrance of her home.

She was immobilized in a plaster cast for 9 weeks. Extended immobilization was required after the fracture because of slow rate of healing secondary to osteopenia/osteoporosis. She was non-weight bearing and ambulated with crutches while her foot was immobilized. Initially this client was referred to physical therapy for range of motion (ROM), strengthening, and gait training.

Client is married and lives with her husband in a single-story home. Her goals were to ambulate independently with a normal gait.

Past Medical History: Type II diabetes, hypertension, osteopenia, and history of alcohol use. Client used tobacco (1½ packs a day for 35 years) but has not smoked for the past 20 years. Client described herself as a "weekend alcoholic," meaning she did not drink during the week but drank six or more beers a day on weekends.

Current medications include tolbutamide, enalapril, hydrochlorothiazide, Fosamax and supplemental calcium, and multivitamin.

Intervention: The client was seen six times before a scheduled surgery interrupted the plan of care. Progress was noted as increased ROM and increased strength through the left lower extremity, except dorsiflexion.

Seven weeks later, the client returned to physical therapy for strengthening and gait training secondary to a "limp" on the left side. She reported that she noticed the limping had increased since she had both her big toenails removed. She also noted increased toe dragging, stumbling, and leg cramps (especially at night). She reported she had decreased her use of alcohol since she fractured her leg because of the pain medications and recently because of fear of falling.

Minimal progress was noted in improving balance or improving strength in the lower extremity. The client felt that her loss of strength could be attributed to inactivity following the foot surgery, even though she reported doing her home exercise program.

Neurologic screening exam was repeated with hyperreflexia observed in the lower extremities, bilaterally. There was a positive Babinski reflex on the left. The findings were reported to the primary care physician who requested that physical therapy continue.

During the next week and a half, the client reported that she fell twice. She also reported that she was "having some twitching in her [left] leg muscles." The client also reported "coughing a lot while [she] was eating; food going down the wrong pipe."

Outcome: The client presented with a referral for weakness and gait abnormality thought to be related to the left fibular fracture and fall that did not respond as expected and, in fact, resulted in further loss of function.

The physician was notified of the client's need for a cane, no improvement in strength, fasciculations in the left lower extremity, and the changes in her neurologic status. The client returned to her primary care provider who then referred her to a neurologist.

Results: Upon examination by the neurologist, the client was diagnosed with amyotrophic lateral sclerosis (ALS). A new physical therapy plan of care was developed based on the new diagnosis.

From Chanoski C: Adapted from case report presented in partial fulfillment of DPT 910, Principles of Differential Diagnosis, Institute for Physical Therapy Education, Widener University, Chester, Pennsylvania, 2005. Used with permission.

our vestibular system and the balance mechanism. Macular degeneration, glaucoma, cataracts, or any other visual problems can result in loss of depth perception and even greater loss of visual acuity.

The autonomic nervous system's (ANS) ability to regulate blood pressure is also affected by age. A sudden drop in blood pressure can precipitate a fall. Coronary heart disease, peripheral vascular disease, diabetes mellitus, and blood pressure medications are just a few of the factors that can put additional stress on the regulating function of the ANS.

Lower standing balance, even within normotensive ranges, is an independent predictor of falls in community-dwelling older adults. Older women (65 years old or older) with a history of falls and with lower systolic blood pressure should have more attention paid to the prevention of falls and related accidents.[170]

The subject of balance impairment and falls as it relates to medical conditions and medications is very important in the diagnostic and screening process. Chronic diseases and multiple pathologies are more important predictors of falling than even polypharmacy (use of four or more medications during the same period).[171] The presence of chronic musculoskeletal pain is associated with a 1.5-fold increased risk of falling for adults ages 70 and up.[172]

Multiple comorbidities often mean the use of multiple drugs (polypharmacy/hyperpharmacotherapy). These two variables together increase the risk of falls in older adults. Some medications (especially psychotropics such as tranquilizers and antidepressants, including amitriptyline, doxepin, Zoloft, Prozac, Paxil, Remeron, Celexa, Wellbutrin) are red flag–risk factors for loss of balance and injuries from falls.

The therapist should watch for clients with chronic conditions who are taking any of these drugs. Anyone with fibromyalgia, depression, cluster migraine headaches, chronic pain, obsessive-compulsive disorders (OCD), panic disorder, and anxiety who is on a psychotropic medication must be monitored carefully for dizziness, drowsiness, and postural orthostatic hypotension (a sudden drop in blood pressure with an increase in pulse rate). In addition, alcohol can interact with many medications, increasing the risk of falling.

It is not uncommon for clients on hypertensive medication (diuretics) to become dehydrated, dizzy, and lose their

TABLE 2-5 Risk Factors for Falls

Age Changes	Environmental/ Living Conditions	Pathologic Conditions	Medications	Other
Muscle weakness; loss of joint motion (especially lower extremities) Abnormal gait Impaired or abnormal balance Impaired proprioception or sensation Delayed muscle response/ increased reaction time ↓Systolic blood pressure (<140 mm Hg in adults age over 65 years old) Stooped or forward bent posture	Poor lighting Throw rugs, loose carpet complex carpet designs Cluster of electric wires or cords Stairs without handrails Bathroom without grab bars Slippery floors (water, urine, floor surface, ice); icy sidewalks, stairs, or streets Restraints Use of alcohol or other drugs Footwear, especially slippers	Vestibular disorders; episodes of dizziness or vertigo from any cause Orthostatic hypotension (especially before breakfast) Chronic pain condition Neuropathies Cervical myelopathy Osteoarthritis; rheumatoid arthritis Visual or hearing impairment; multifocal eyeglasses; change in perception of color; loss of depth perception; decreased contrast sensitivity Cardiovascular disease Urinary incontinence Central nervous system disorders (e.g., stroke, Parkinson's disease, multiple sclerosis) Motor disturbance Osteopenia, osteoporosis Pathologic fractures Any mobility impairments (e.g., amputation, neuropathy, deformity) Cognitive impairment; dementia; depression	Antianxiety; benzodiazepines Anticonvulsants Antidepressants Antihypertensives Antipsychotics Diuretics Narcotics Sedative-hypnotics Phenothiazines Use of more than four medications (polypharmacy/ hyperpharmacotherapy)	History of falls Female sex; postmenopausal status Living alone Elder abuse/assault Nonambulatory status (requiring transfers) Gait changes (decreased stride length or speed) Postural instability; reduced postural control Fear of falling; history of falls Dehydration from any cause Recent surgery (general anesthesia, epidural) Sleep disorder/disturbance; sleep deprivation; daytime drowsiness; brief disorientation after waking up from a nap[249]

balance. Postural orthostatic hypotension can (and often does) occur in the aging adult—even in someone taking blood pressure–regulating medications.

Orthostatic hypotension as a risk factor for falls may occur as a result of volume depletion (e.g., diabetes mellitus, sodium or potassium depletion), venous pooling (e.g., pregnancy, varicosities of the legs, immobility following a motor vehicle or cerebrovascular accident), side effects of medications such as antihypertensives, starvation associated with anorexia or cachexia, and sluggish normal regulatory mechanisms associated with anatomic variations or secondary to other conditions such as metabolic disorders or diseases of the central nervous system (CNS).

Remember too that falling is a primary symptom of Parkinson's disease. Any time a client reports episodes of dizziness, loss of balance, or a history of falls, further screening and possible medical referral is needed. This is especially true in the presence of other neurologic signs and symptoms such as headache, confusion, depression, irritability, visual changes, weakness, memory loss, and drowsiness or lethargy.

Screening for Risk of Falls. Aging adults who have just started to fall or who fall frequently may be fearful of losing their independence by revealing this information even to a therapist. If the client indicates no difficulty with falling, the therapist is encouraged to review this part of the form (see Fig. 2-2) carefully with each older client.

Some potential screening questions may include (see Appendix B-11 for full series of questions):

FOLLOW-UP QUESTIONS

- Do you have any episodes of dizziness?
  If yes, does turning over in bed cause (or increase) dizziness?
- Do you have trouble getting in or out of bed without losing your balance?
- Can you/do you get in and out of your tub or shower?
- Do you avoid walking on grass or curbs to avoid falling?
- Have you started taking any new medications, drugs, or pills of any kind?
- Has there been any change in the dosage of your regular medications?

During the Core Interview, the therapist will have an opportunity to ask further questions about the client's Current Level of Fitness (see discussion later in this chapter).

Performance-based tests such as the Functional Reach Test,[173,174] One-Legged Stance Test,[175] Berg Balance Scale (BBS),[176,177] and the Timed "Up and Go" Test (TUGT)[178-180] can help identify functional limitations, though not necessarily the causes, for balance impairment.

Fear of falling can be measured using the Falls Efficacy Scale (FES)[181,182] and the Survey of Activities and Fear of Falling in the Elderly (SAFE) assessment. The Activities-Specific Balance Confidence Scale (ABC) can measure balance confidence.[183,184]

No one balance scale best predicts falls risk in older adults. A simple clinical scale to stratify risk of recurrent falls in community-dwelling older adults as low, moderate, or high risk using four easy-to-obtain items has been proposed.[185] The ABC and FES are highly correlated with each other, meaning they measure similar constructs. These two tests are moderately correlated with the SAFE, indicating they predict differently. It is likely that using more than one scale will help identify individuals who may be at risk and are candidates for an intervention program.[186]

Measuring vital signs and screening for postural orthostatic hypotension is another important tool in predicting falls. Positive test results for any of the tests mentioned require further evaluation, especially in the presence of risk factors predictive of falls.

Resources. As the population of older people in the United States continues to grow, the number of TBIs, fractures, and other injuries secondary to falls also is likely to grow.[166] Therapists are in a unique position to educate people on using strength, flexibility, and endurance activities to help maintain proper posture, improve balance, and prevent falls. The APTA has a Balance and Falls Kit (Item number PR-294) available to assist the therapist in this area.[187]

National Committee on Aging (NCOA) has partnered with the APTA to provide a Falls Free plan that can help reduce fall dangers for older adults. More information on the plan is accessible at www.healthyagingprograms.org. The American Geriatric Society (AGS) also provides excellent evidence-based guidelines for the screening and prevention of falls, including clinical algorithms, assessment materials, and intervention strategies (available on line at http://www.americangeriatrics.org/education/summ_of_rec.shtml).

Vital Signs. Taking a client's vital signs remains the single easiest, most economic, and fastest way to screen for many systemic illnesses. Dr. James Cyriax, a renowned orthopedic physician, admonishes therapists to always take the body temperature in any client with back pain of unknown cause.[188]

A place to record vital signs is provided at the end of the Family/Personal History form (see Fig. 2-2). The clinician must be proficient in taking vital signs, an important part of the screening process. All vital signs are important, but the client's temperature and blood pressure have the greatest utility as early screening tools. An in-depth discussion of vital signs as part of the screening physical assessment is presented in Chapter 4.

CORE INTERVIEW

Once the therapist reviews the results of the Family/Personal History form and reviews any available medical records for the client, the client interview (referred to as the Core Interview in this text) begins (Fig. 2-3).

Screening questions may be interspersed throughout the Core Interview and/or presented at the end. When to screen depends on the information provided by the client during the interview.

Special questions related to sensitive topics such as sexual history, assault or domestic violence, and substance or alcohol use are often left to the end or even on a separate day after the therapist has established sufficient rapport to broach these topics.

History of Present Illness

Chief Complaint

The history of present illness (often referred to as the chief complaint and other current symptoms) may best be obtained through the use of open-ended questions. This section of the interview is designed to gather information related to the client's reason(s) for seeking clinical treatment.

The following open-ended statements may be appropriate to start an interview:

FOLLOW-UP QUESTIONS

- Tell me how I can help you.
- Tell me why you are here today.
- Tell me about your injury.
- (Alternate) What do you think is causing your problem or pain?

During this initial phase of the interview, allow the client to carefully describe his or her current situation. Follow-up questions and paraphrasing as shown in Fig. 2-3 can be used in conjunction with the primary, open-ended questions.

Pain and Symptom Assessment

The interview naturally begins with an assessment of the chief complaint, usually (but not always) pain. Chapter 3 of this text presents an in-depth discussion of viscerogenic sources of NMS pain and pain assessment, including questions to ask to identify specific characteristics of pain.

For the reader's convenience, a brief summary of these questions is included in the Core Interview (see Fig. 2-3). In addition, the list of questions is included in Appendices B-28 and C-7 for use in the clinic.

Beyond a pain and symptom assessment, the therapist may conduct a screening physical examination as part of the objective assessment (see Chapter 4). Table 4-13 and Boxes 4-15 and 4-16 are helpful tools for this portion of the examination and evaluation.

THE CORE INTERVIEW

HISTORY OF PRESENT ILLNESS

Chief Complaint (Onset)

- Tell me why you are here today.
- Tell me about your injury.

  Alternate question: What do you think is causing your problem/pain?

  FUPs: How did this injury or illness begin?

  - Was your injury or illness associated with a fall, trauma, assault, or repetitive activity (e.g., painting, cleaning, gardening, filing papers, driving)?
  - Have you been hit, kicked, or pushed? (For the therapist: See text [Assault] before asking this question.)
  - When did the present problem arise and did it occur gradually or suddenly?

  Systemic disease: Gradual onset without known cause.

  - Have you ever had anything like this before? If yes, when did it occur?
  - Describe the situation and the circumstances.
  - How many times has this illness occurred? Tell me about each occasion.
  - Is there any difference this time from the last episode?
  - How much time elapses between episodes?
  - Do these episodes occur more or less often than at first?

  Systemic disease: May present in a gradual, progressive, cyclical onset: worse, better, worse.

PAIN AND SYMPTOM ASSESSMENT

- Do you have any pain associated with your injury or illness? if yes, tell me about it.

Location

- Show me exactly where your pain is located.

  FUPs: Do you have this same pain anywhere else?

  - Do you have any other pain or symptoms anywhere else?
  - If yes, what causes the pain or symptoms to occur in this other area?

Description

- What does it feel like?

  FUPS: Has the pain changed in quality, intensity, frequency, or duration (how long it lasts) since it first began?

Pattern

- Tell me about the pattern of your pain or symptoms.

  Alternate question: When does your back/shoulder (name the body part) hurt?

  Alternate question: Describe your pain/symptoms from first waking up in the morning to going to bed at night. (See special sleep-related questions that follow.)

  FUPs: Have you ever experienced anything like this before?

  - If yes, do these episodes occur more or less often than at first?
  - How does your pain/symptom(s) change with time?
  - Are your symptoms worse in the morning or in the evening?

Frequency

- How often does the pain/symptom(s) occur?

  FUPs: Is your pain constant, or does it come and go (intermittent)?

  - Are you having this pain now?
  - Did you notice these symptoms this morning immediately after awakening?

Duration

- How long does the pain/symptom(s) last?

  Systemic disease: Constant.

Fig. 2-3 Core Interview.

Continued

Intensity

- On a scale from 0 to 10, with 0 being no pain and 10 being the worst pain you have experienced with this condition, what level of pain do you have right now?

  Alternate question: How strong is your pain?

  1 = Mild

  2 = Moderate

  3 = Severe

  FUPs: Which word describes your pain right now?
  - Which word describes the pain at its worst?
  - Which word describes the least amount of pain?

  Systemic disease: Pain tends to be intense.

Associated Symptoms

- What other symptoms have you had that you can associate with this problem?

  FUPs: Have you experienced any of the following?

☐ Blood in urine, stool, vomit, mucus
☐ Dizziness, fainting, blackouts
☐ Fever, chills, sweats (day or night)
☐ Nausea, vomiting, loss of appetite
☐ Changes in bowel or bladder
☐ Throbbing sensation/pain in belly or anywhere else
☐ Skin rash or other skin changes
☐ Headaches
☐ Cough, dyspnea
☐ Dribbling or leaking urine
☐ Heart palpitations or fluttering
☐ Numbness or tingling
☐ Swelling or lumps anywhere
☐ Problems seeing or hearing
☐ Unusual fatigue, drowsiness
☐ Joint pain
☐ Difficulty swallowing/speaking
☐ Memory loss
☐ Confusion
☐ Sudden weakness
☐ Trouble sleeping

Systemic disease: Presence of symptoms bilaterally (e.g., edema, nail bed changes, bilateral weakness, paresthesia, tingling, burning). Determine the frequency, duration, intensity, and pattern of symptoms. Blurred vision, double vision, scotomas (black spots before the eyes), or temporary blindness may indicate early symptoms of multiple sclerosis (MS), cerebral vascular accident (CVA), or other neurologic disorders.

Aggravating Factors

- What kinds of things affect the pain?

  FUPs: What makes your pain/symptoms worse (e.g., eating, exercise, rest, specific positions, excitement, stress)?

Relieving Factors

- What makes it better?

  Systemic disease: Unrelieved by change in position or by rest.
- How does rest affect the pain/symptoms?

  FUPs: Are your symptoms aggravated or relieved by any activities? If yes, what?
  - How has this problem affected your daily life at work or at home?
  - How has it affected your ability to care for yourself without assistance (e.g., dress, bathe, cook, drive)?

MEDICAL TREATMENT AND MEDICATIONS

Medical Treatment

- What medical treatment have you had for this condition?

  FUPs: Have you been treated by a physical therapist for this condition before? If yes:
  - When?
  - Where?
  - How long?
  - What helped?
  - What didn't help?
  - Was there any treatment that made your symptoms worse? If yes, please elaborate.

Medications

- Are you taking any prescription or over-the-counter medications?

  FUPs: If no, you may have to probe further regarding use of laxatives, aspirin, acetaminophen (Tylenol), and so forth. If yes:
  - What medication do you take?
  - How often?

Fig. 2-3, cont'd

- What dose do you take?
- Why are you taking these medications?
- When was the last time that you took these medications? Have you taken these drugs today?
- Do the medications relieve your pain or symptoms?
- If yes, how soon after you take the medications do you notice an improvement?
- Do you notice any increase in symptoms or perhaps the start of symptoms after taking your medication(s)? (This may occur 30 minutes to 2 hours after ingestion.)
- If prescription drugs, who prescribed them for you?
- How long have you been taking these medications?
- When did your physician last review these medications?
- Are you taking any medications that weren't prescribed for you?

If no, follow-up with: Are you taking any pills given to you by someone else besides your doctor?

CURRENT LEVEL OF FITNESS

- What is your present exercise level?

  FUPs: What type of exercise or sports do you participate in?
  - How many times do you participate each week (frequency)?
  - When did you start this exercise program (duration)?
  - How many minutes do you exercise during each session (intensity)?
  - Are there any activities that you could do before your injury or illness that you cannot do now? If yes, please describe.

  Dyspnea: Do you ever experience any shortness of breath (SOB) or lack of air during any activities (e.g., walking, climbing stairs)?

  FUPs: Are you ever short of breath without exercising?
  - If yes, how often?
  - When does this occur?
  - Do you ever wake up at night and feel breathless? If yes, how often?
  - When does this occur?

SLEEP-RELATED HISTORY

- Can you get to sleep at night? If no, try to determine whether the reason is due to the sudden decrease in activity and quiet, which causes you to focus on your symptoms.
- Are you able to lie or sleep on the painful side? If yes, the condition may be considered to be chronic, and treatment would be more vigorous than if no, indicating a more acute condition that requires more conservative treatment.
- Are you ever wakened from a deep sleep by pain?

  FUPs: If yes, do you awaken because you have rolled onto that side? Yes may indicate a subacute condition requiring a combination of treatment approaches, depending on objective findings.
  - Can you get back to sleep?

  FUPs: If yes, what do you have to do (if anything) to get back to sleep? (The answer may provide clues for treatment.)
- Have you had any unexplained fevers, night sweats, or unexplained perspiration?

  Systemic disease: Fevers and night sweats are characteristic signs of systemic disease.

STRESS

- What major life changes or stresses have you encountered that you would associate with your injury/illness?

  Alternate question: What situations in your life are "stressors" for you?
- On a scale from 0 to 10, with 0 being no stress and 10 being the most extreme stress you have ever experienced, in general, what number rating would you give to your stress at this time in your life?
- What number would you assign to your level of stress today?
- Do you ever get short of breath or dizzy or lose coordination with fatigue (anxiety-produced hyperventilation)?

FINAL QUESTION

- Do you wish to tell me anything else about your injury, your health, or your present symptoms that we have not discussed yet?

  Alternate question: Is there anything else you think is important about your condition that we haven't discussed yet?

FUPs, Follow-up Questions

Fig. 2-3, cont'd

Insidious Onset

When the client describes an insidious onset or unknown cause, it is important to ask further questions. Did the symptoms develop after a fall, trauma (including assault), or some repetitive activity (such as painting, cleaning, gardening, filing, or driving long distances)?

The client may wrongly attribute the onset of symptoms to a particular activity that is really unrelated to the current symptoms. The alert therapist may recognize a true causative factor. Whenever the client presents with an unknown etiology of injury or impairment or with an apparent cause, always ask yourself these questions:

FOLLOW-UP QUESTIONS

- Is it really insidious?
- Is it really caused by such and such (whatever the client told you)?

Trauma

When the symptoms seem out of proportion to the injury or when the symptoms persist beyond the expected time for that condition, a red flag should be raised in the therapist's mind. Emotional overlay is often the most suspected underlying cause of this clinical presentation. But trauma from assault and undiagnosed cancer can also present with these symptoms.

Even if the client has a known (or perceived) cause for his or her condition, the therapist must be alert for trauma as an etiologic factor. Trauma may be intrinsic (occurring within the body) or extrinsic (external accident or injury, especially assault or domestic violence).

Twenty-five percent of clients with primary malignant tumors of the musculoskeletal system report a prior traumatic episode. Often the trauma or injury brings attention to a preexisting malignant or benign tumor. Whenever a fracture occurs with minimal trauma or involves a transverse fracture line, the physician considers the possibility of a tumor.

Intrinsic Trauma. An example of intrinsic trauma is the unguarded movement that can occur during normal motion. For example, the client who describes reaching to the back of a cupboard while turning his or her head away from the extended arm to reach that last inch or two. He or she may feel a sudden "pop" or twinge in the neck with immediate pain and describe this as the cause of the injury.

Intrinsic trauma can also occur secondary to extrinsic (external) trauma. A motor vehicle accident, assault, fall, or known accident or injury may result in intrinsic trauma to another part of the musculoskeletal system or other organ system. Such intrinsic trauma may be masked by the more critical injury and may become more symptomatic as the primary injury resolves.

Take, for example, the client who experiences a cervical flexion/extension (whiplash) injury. The initial trauma causes painful head and neck symptoms. When these resolve (with treatment or on their own), the client may notice midthoracic spine pain or rib pain.

The midthoracic pain can occur when the spine fulcrums over the T4-6 area as the head moves forcefully into the extended position during the whiplash injury. In cases like this, the primary injury to the neck is accompanied by a secondary intrinsic injury to the midthoracic spine. The symptoms may go unnoticed until the more painful cervical lesion is treated or healed.

Likewise, if an undisplaced rib fracture occurs during a motor vehicle accident, it may be asymptomatic until the client gets up the first time. Movement or additional trauma may cause the rib to displace, possibly puncturing a lung. These are all examples of intrinsic trauma.

Extrinsic Trauma. Extrinsic trauma occurs when a force or load external to the body is exerted against the body. Whenever a client presents with NMS dysfunction, the therapist must consider whether this was caused by an accident, injury, or assault.

The therapist must remain aware that some motor vehicle "accidents" may be reported as accidents but are, in fact, the result of domestic violence in which the victim is pushed, shoved, or kicked out of the car or deliberately hit by a vehicle.

Assault. Domestic violence is a serious public health concern that often goes undetected by clinicians. Women (especially those who are pregnant or disabled), children, and older adults are at greatest risk, regardless of race, religion, or socioeconomic status. Early intervention may reduce the risk of future abuse.

It is imperative that physical therapists and physical therapist assistants remain alert to the prevalence of violence in all sectors of society. Therapists are encouraged to participate in education programs on screening, recognition, and treatment of violence and to advocate for people who may be abused or at risk for abuse.[189]

Addressing the possibility of sexual or physical assault/abuse during the interview may not take place until the therapist has established a working relationship with the client. Each question must be presented in a sensitive, respectful manner with observation for nonverbal cues.

Although some interviewing guidelines are presented here, questioning clients about abuse is a complex issue with important effects on the outcome of rehabilitation. All therapists are encouraged to familiarize themselves with the information available for screening and intervening in this important area of clinical practice.

Generally, the term *abuse* encompasses the terms physical abuse, mental abuse, sexual abuse, neglect, self-neglect, and exploitation (Box 2-8). *Assault* is by definition any physical, sexual, or psychologic attack. This includes verbal, emotional, and economic abuse. *Domestic violence (DV)* or *intimate partner violence* (IPV) is a pattern of coercive behaviors perpetrated by a current or former intimate partner that may include physical, sexual, and/or psychologic assaults.[190,191]

Violence against women is more prevalent and dangerous than violence against men,[192] but men can be in an abusive

BOX 2-8 DEFINITIONS OF ABUSE

Abuse—Infliction of physical or mental injury, or the deprivation of food, shelter, clothing, or services needed to maintain physical or mental health

Sexual abuse—Sexual assault, sexual intercourse without consent, indecent exposure, deviate sexual conduct, or incest; adult using a child for sexual gratification without physical contact is considered sexual abuse

Neglect—Failure to provide food, shelter, clothing, or help with daily activities needed to maintain physical or mental well-being; client often displays signs of poor hygiene, hunger, or inappropriate clothing

Material exploitation—Unreasonable use of a person, power of attorney, guardianship, or personal trust to obtain control of the ownership, use, benefit, or possession of the person's money, assets, or property by means of deception, duress, menace, fraud, undue influence, or intimidation

Mental abuse—Impairment of a person's intellectual or psychologic functioning or well-being

Emotional abuse—Anguish inflicted through threats, intimidation, humiliation, and/or isolation; belittling, embarrassing, blaming, rejecting behaviors from adult toward child; withholding love, affection, approval

Physical abuse—Physical injury resulting in pain, impairment, or bodily injury of any bodily organ or function, permanent or temporary disfigurement, or death

Self-neglect—Individual is not physically or mentally able to obtain and perform the daily activities of life to avoid physical or mental injury

Data from Smith L, Putnam DB: The abuse of vulnerable adults. Montana State Bar. *The Montana Lawyer* magazine, June/July 2001. Available at http://www.montanabar.org/montanalawyer/junejuly2001/elderabuse.html. Accessed July 5, 2005.

relationship with a parent or partner (male or female).[193,194] For the sake of simplicity, the terms "she" and "her" are used in this section, but this could also be "he" and "his."

Intimate partner assault may be more prevalent against gay men than against heterosexual men.[195] Many men have been the victims of sexual abuse as children or teenagers.

Child abuse includes neglect and maltreatment that includes physical, sexual, and emotional abuse. Failure to provide for the child's basic physical, emotional, or educational needs is considered neglect even if it is not a willful act on the part of the parent, guardian, or caretaker.[187]

Screening for Assault or Domestic Violence. The American Medical Association (AMA) and other professional groups recommend routine screening for domestic violence. At least one study has shown that screening does not put victims at increased risk for more violence later. Many victims who participated in the study contacted community resources for victims of domestic violence soon after completing the study survey.[196]

As health care providers, therapists have an important role in helping to identify cases of domestic violence and abuse. Routinely incorporating screening questions about domestic violence into history taking only takes a few minutes and is advised in all settings. When interviewing the client it is often best to use some other word besides "assault."

Many people who have been physically struck, pushed, or kicked do not consider the action an assault, especially if someone they know inflicts it. The therapist may want to preface any general screening questions with one of the following lead-ins:

FOLLOW-UP QUESTIONS

- Abuse in the home is so common today we now ask all our clients:
  - Are you threatened or hurt at home or in a relationship with anyone?
  - Do you feel safe at home?
- Many people are in abusive relationships but are afraid to say so. We ask everyone about this now.
  - FUP: Has this ever happened to you?
- We are required to ask everyone we see about domestic violence. Many of the people I treat tell me they are in difficult, hurtful, sometimes even violent relationships. Is this your situation?

Several screening tools are available with varying levels of sensitivity and specificity. The Woman Abuse Screening Tool (WAST) has direct questions that are easy to understand (e.g., Have you been abused physically, emotionally, or sexually by an intimate partner?) but have not been independently validated.[197] There is also the Composite Abuse Scale (CAS),[198] the Partner Violence Screen (PVS),[199] and Index of Spouse Abuse.

The PVS, a quick three-question screening tool, may be easiest to use as it is positive for partner violence if even one question is answered "yes"[199] (see FUPs just below). When compared with other screening tools, the PVS has 64.5% to 71.4% sensitivity in detecting partner abuse and 80.3% to 84.4% specificity.[199]

FOLLOW-UP QUESTIONS

- Have you been kicked, hit, pushed, choked, punched or otherwise hurt by someone in the last year?
- Do you feel safe in your current relationship?
- Is anyone from a previous relationship making you feel unsafe now?
- Alternate: Are your symptoms today caused by someone kicking, hitting, pushing, choking, throwing, or punching you?
- Alternate: I'm concerned someone hurting you may have caused your symptoms. Has anyone been hurting you in any way?
- FUP: Is there anything else you would like to tell me about your situation?

Indirect Questions[187]

- I see you have a bruise here. It looks like it's healing well. How did it happen?
- Are you having problems with your partner?
- Have you ever gotten hurt in a fight?

- You seem concerned about your partner. Can you tell me more about that?
- Does your partner keep you from coming to therapy or seeing family and friends?

Follow-up questions will depend on the client's initial response.[187] The timing of these personal questions can be very delicate. A private area for interviewing is best at a time when the client is alone (including no children, friends, or other family members). The following may be helpful:

FOLLOW-UP QUESTIONS

- May I ask you a few more questions?
- If yes, has anyone ever touched you against your will?
- How old were you when it started? When it stopped?
- Have you ever told anyone about this?
- *Client denies abuse*

Response: I know sometimes people are afraid or embarrassed to say they've been hit. If you are ever hurt by anyone, it's safe to tell me about it.

- *Client is offended*

Response: I'm sorry to offend you. Many people need help but are afraid to ask.

- *Client says "Yes"*

Response: Listen, believe, document if possible. Take photographs if the client will allow it. If the client does not want to get help at this time, offer to give her/him the photos for future use or to keep them on file should the victim change his/her mind. See documentation guidelines. Provide information about local resources.

During the interview (and subsequent episode of care), watch out for any of the risk factors and red flags for violence (Box 2-9) or any of the clinical signs and symptoms listed in this section. The physical therapist should not turn away from signs of physical or sexual abuse.

In attempting to address such a sensitive issue, the therapist must make sure that the client will not be endangered by intervention. Physical therapists who are not trained to be counselors should be careful about offering advice to those believed to have sustained abuse (or even those who have admitted abuse).

The best course of action may be to document all observations and, when necessary or appropriate, to communicate those documented observations to the referring or family physician. When an abused individual asks for help or direction, the therapist must always be prepared to provide information about available community resources.

In considering the possibility of assault as the underlying cause of any trauma, the therapist should be aware of cultural differences and how these compare with behaviors that suggest excessive partner control. For example:

- Abusive partner rarely lets the client come to the appointment alone (partner control).
- Collectivist cultures (group-oriented) often come to the clinic with several family members; such behavior is a cultural norm.
- Noncompliance/missed appointments (could be either one).

BOX 2-9 RISK FACTORS AND RED FLAGS FOR DOMESTIC VIOLENCE

- Women with disabilities
- Cognitively impaired adult
- Chronically ill and dependent adult (especially adults over age 75)
- Chronic pain clients
- Physical and/or sexual abuse history (men and women)
- Daily headache
- Previous history of many injuries and accidents (including multiple motor vehicle accidents)
- Somatic disorders
  - Injury seems inconsistent with client's explanation; injury in a child that is not consistent with the child's developmental level
  - Injury takes much longer to heal than expected
- Pelvic floor problems
  - Incontinence
  - Infertility
  - Pain
- Recurrent unwanted pregnancies
- History of alcohol abuse in male partner

Elder Abuse. Health care professionals are becoming more aware of elder abuse as a problem. Last year, more than 5 million cases of elder abuse were reported. It is estimated that 84% of elder abuse and neglect is never reported. The International Network for the Prevention of Elder Abuse has more information (www.inpea.net).

The therapist must be alert at all times for elder abuse. Skin tears, bruises, and pressure ulcers are not always predictable signs of aging or immobility. During the screening process, watch for warning signs of elder abuse (Box 2-10).

Clinical Signs and Symptoms. Physical injuries caused by battering are most likely to occur in a central pattern (i.e., head, neck, chest/breast, abdomen). Clothes, hats, and hair easily hide injuries to these areas, but they are frequently

BOX 2-10 WARNING SIGNS OF ELDER ABUSE

- Multiple trips to the emergency department
- Depression
- "Falls"/fractures
- Bruising/suspicious sores
- Malnutrition/weight loss
- Pressure ulcers
- Changing physicians/therapists often
- Confusion attributed to dementia

observable by the therapist in a clinical setting that requires changing into a gown or similar treatment attire.

Assessment of cutaneous manifestations of abuse is discussed in greater detail in this text in Chapter 4. The therapist should follow guidelines provided when documenting the nature (e.g., cut, puncture, burn, bruise, bite), location, and detailed description of any injuries. The therapist must be aware of Mongolian spots, which can be mistaken for bruising from child abuse in certain population groups (see Fig. 4-25).

In the pediatric population, fractures of the ribs, tibia/fibula, radius/ulna, and clavicle are more likely to be associated with abuse than with accidental trauma, especially in children less than 18 months old. In the group older than 18 months, a rib fracture is highly suspicious of abuse.[200]

A link between a history of sexual or physical abuse and multiple somatic and other medical disorders in adults (e.g., cardiovascular,[201] GI, endocrine,[202] respiratory, gynecologic, headache and other neurologic problems) has been confirmed.[203]

CLINICAL SIGNS AND SYMPTOMS

Domestic Violence

Physical Cues[204]

- Bruises, black eyes, malnutrition
- Sprains, dislocations, foot injuries, fractures in various stages of healing
- Skin problems (e.g., eczema, sores that do not heal, burns); see Chapter 4
- Chronic or migraine headaches
- Diffuse pain, vague or nonspecific symptoms
- Chronic or multiple injuries in various stages of healing
- Vision and hearing loss
- Chronic low back, sacral, or pelvic pain
- Temporomandibular joint (TMJ) pain
- Dysphagia (difficulty swallowing) and easy gagging
- Gastrointestinal disorders
- Patchy hair loss, redness, or swelling over the scalp from violent hair pulling
- Easily startled, flinching when approached

Social Cues

- Continually missing appointments; does not return phone calls; unable to talk on the phone when you call
- Bringing all the children to a clinic appointment
- Spouse, companion, or partner always accompanying client
- Changing physicians often
- Multiple trips to the emergency department
- Multiple car accidents

Psychologic Cues

- Anorexia/bulimia
- Panic attacks, nightmares, phobias
- Hypervigilance, tendency to startle easily or be very guarded
- Substance abuse
- Depression, anxiety, insomnia
- Self-mutilation or suicide attempts
- Multiple personality disorders
- Mistrust of authority figures
- Demanding, angry, distrustful of health care provider

Workplace Violence. Workers in the health care profession are at risk for workplace violence in the form of physical assault and aggressive acts. Threats or gestures used to intimidate or threaten are considered assault. Aggressive acts include verbal or physical actions aimed at creating fear in another person. Any unwelcome physical contact from another person is battery. Any form of workplace violence can be perpetrated by a co-worker, member of a co-worker's family, by a client, or a member of the client's family.[205]

Predicting violence is very difficult, making this occupational hazard one that must be approached through preventative measures rather than relying on individual staff responses or behavior. Institutional policies must be implemented to protect health care workers and provide a safe working environment.[206]

Therapists must be alert for risk factors (e.g., dependence on drugs or alcohol, depression, signs of paranoia) and behavioral patterns that may lead to violence (e.g., aggression toward others, blaming others, threats of harm toward others) and immediately report any suspicious incidents or individuals.[205]

Clients with a mental disorder and history of substance abuse have the highest probability of violent behavior. Adverse drug events can lead to violent behavior, as well as conditions that impair judgment or cause confusion, such as alcohol- or HIV-induced encephalopathy, trauma (especially head trauma), seizure disorders, and senility.[205]

The Physical Therapist's Role. Providing referral to community agencies is perhaps the most important step a health care provider can offer any client who is the victim of abuse, assault, or domestic violence of any kind. Experts report that the best approach to addressing abuse is a combined law enforcement and public health effort.

Any health care professional who asks these kinds of screening questions must be prepared to respond. Having information and phone numbers available is imperative for the client who is interested. Each therapist must know what reporting requirements are in place in the state in which he or she is practicing (Case Example 2-9).

The therapist should avoid assuming the role of "rescuer" but rather recognize domestic violence, offer a plan of care and intervention for injuries, assess the client's safety, and offer information regarding support services. The therapist should provide help at the pace the client can handle. Reporting a situation of domestic violence can put the victim at risk.

The client usually knows how to stay safe and when to leave. Whether leaving or staying, it is a complex process of decision making influenced by shame, guilt, finances, religious beliefs, children, depression, perceptions, and realities. The therapist does not have to be an expert to help someone who is a victim of domestic violence. Identifying the problem for the first time and listening is an important first step.

During intervention procedures, the therapist must be aware that hands-on techniques, such as pushing, pulling, stretching, compressing, touching, and rubbing, may impact a client with a history of abuse in a negative way. Behaviors, such as persistence in cajoling, cheerleading, and demanding

CASE EXAMPLE 2-9

Elder Abuse

An 80-year-old female (Mrs. Smith) was referred to home health by her family doctor for an assessment following a mild cerebrovascular accident (CVA). She was living with her 53-year-old divorced daughter (Susan). The daughter works full-time to support herself, her mother, and three teenage children.

The CVA occurred 3 weeks ago. She was hospitalized for 10 days during which time she had daily physical and occupational therapy. She has residual left-sided weakness.

Home health nursing staff notes that she has been having short-term memory problems in the last week. When the therapist arrived at the home, the doors were open, the stove was on with the stove door open, and Mrs. Smith was in front of the television set. She was wearing a nightgown with urine and feces on it. She was not wearing her hearing aid, glasses, or false teeth.

Mrs. Smith did not respond to the therapist or seem surprised that someone was there. While helping her change into clean clothes, the therapist noticed a large bruise on her left thigh and another one on the opposite upper arm. She did not answer any of the therapist's questions but talked about her daughter constantly. She repeatedly said, "Susan is mean to me."

How Should the Therapist Respond in this Situation? Physical therapists do have a role in prevention, assessment, and intervention in cases of abuse and neglect. Keeping a nonjudgmental attitude is helpful.

Assessment: Examination and Evaluation

1. Attempt to obtain a detailed history.
2. Conduct a thorough physical exam. Look for warning signs of pressure ulcers, burns, bruises, or other signs suggesting force. Include a cognitive and neurologic assessment. Document findings with careful notes, drawings, and photographs whenever possible.

Intervention: Focus on Providing the Client with Safety and the Family with Support and Resources

1. Contact the case manager or nurse assigned to Mrs. Smith.
2. Contact the daughter before calling the county's Adult Protective Services (APS).
3. Team up with the nurse if possible to assess the situation and help the daughter obtain help.
4. When meeting with the daughter, acknowledge the stress the family has been under. Offer the family reassurance that the home health staff's role is to help Mrs. Smith get the best care possible.
5. Let the daughter know what her options are but acknowledge the need to call APS (if required by law).
6. Educate the family and prevent abuse by counseling them to avoid isolation at home. Stay involved in other outside activities (e.g., church/synagogue, school, hobbies, friends).
7. Encourage the family to recognize their limits and seek help when and where it is available.

Result: APS referred Mrs. Smith to an adult day health care program covered by Medicaid. She receives her medications, two meals, and programming with other adults during the day while her daughter works.

The daughter received counseling to help cope with her mother's declining health and loss of mental faculties. She also joined an Alzheimer's "36-hour/day" support group. Respite care was arranged through the adult day care program once every 6 weeks.

compliance, meant as encouragement on the part of the therapist may further victimize the individual.[207]

Reporting Abuse. The law is clear in all U.S. states regarding abuse of a minor (under age 18 years) (Box 2-11):

> *When a professional has reasonable cause to suspect, as a result of information received in a professional capacity, that a child is abused or neglected, the matter is to be reported promptly to the department of public health and human services or its local affiliate.*[208]

Guidelines for reporting abuse in adults are not always so clear. Some states require health care professionals to notify law enforcement officials when they have treated any individual for an injury that resulted from a domestic assault. There is much debate over such laws as many domestic violence advocate agencies fear mandated police involvement will discourage injured clients from seeking help. Fear of retaliation may prevent abused persons from seeking needed health care because of required law enforcement involvement.

The therapist should be familiar with state laws or statutes regarding domestic violence for the geographic area in which he or she is practicing. The Elder Justice Act of 2003 requires reporting of neglect or assault in long-term care facilities in all 50 U.S. states. The Elder Justice Act and the Patient Safety Abuse Prevention Act of 2010 provides funding ($3.9 billion) to establish advisory departments, justice resource centers, ombudsman training, nursing home training, and support for Adult Protective Services (APS). The National Center

BOX 2-11 REPORTING CHILD ABUSE

- The law requires professionals to report suspected child abuse and neglect.
- The therapist must know the reporting guidelines for the state in which he or she is practicing.
- Know who to contact in your local child protective service agency and police department.
- The duty to report findings only requires a reasonable suspicion that abuse has occurred, not certainty.[208]
- A professional who delays reporting until doubt is eliminated is in violation of the reporting law.
- The decision about maltreatment is left up to investigating officials, not the reporting professional.

Data from Mudd SS, Findlay JS: The cutaneous manifestations and common mimickers of physical child abuse, *J Pediatr Health Care* 18(3):123-129, 2004.

on Elder Abuse (NCEA) has more information (www.ncea.aoa.gov).

Documentation. Most state laws also provide for the taking of photographs of visible trauma on a child without parental consent. Written permission must be obtained to photograph adults. Always offer to document the evidence of injury. The APTA publications on domestic violence, child abuse, and elder abuse provide reproducible documentation forms and patient resources.[187,209,210]

Even if the client does not want a record of the injury on file, he or she may be persuaded to keep a personal copy for future use if a decision is made to file charges or prosecute at a later time. Polaroid and digital cameras make this easy to accomplish with certainty that the photographs clearly show the extent of the injury or injuries.

The therapist must remember to date and sign the photograph. Record the client's name and injury location on the photograph. Include the client's face in at least one photograph for positive identification. Include a detailed description (type, size, location, depth) and how the injury/injuries occurred.

Record the client's own words regarding the assault and the assailant. For example, "Ms. Jones states, 'My partner Doug struck me in the head and knocked me down.'" Identifying the presumed assailant in the medical record may help the client pursue legal help.[211]

Resources. Consult your local directory for information about adult and child protection services, state elder abuse hotlines, shelters for the battered, or other community services available in your area. For national information, contact:

- National Domestic Violence Hotline. Available 24 hours/day with information on shelters, legal advocacy and assistance, and social service programs. Available at www.ndvh.org or 1-800-799-SAFE (1-800-799-7233).
- Family Violence Prevention Fund. Updates on legislation related to family violence, information on the latest tools and research on prevention of violence against women and children. Posters, displays, safety cards, and educational pamphlets for use in a health care setting are also available at http://endabuse.org/ or 1-415-252-8900.
- U.S. Department of Justice. Office on Violence Against Women provides lists of state hotlines, coalitions against domestic violence, and advocacy groups (www.ovw.usdoj.gov/).
- Elder Care Locator. Information on senior services. The service links those who need assistance with state and local area agencies on aging and community-based organizations that serve older adults and their caregivers www.eldercare.gov/ or 1-800-677-1116.
- U.S. Department of HHS Administration for Children and Families. Provides fact sheets, laws and policies regarding minors, and phone numbers for reporting abuse. Available at www.acf.hhs.gov/ or 1-800-4-A-CHILD (1-800-422-4453).

Specific websites devoted to just men, just women, or any other specific group are available. Anyone interested can go to www.google.com and type in key words of interest.

The APTA offers three publications related to domestic violence, available online at www.apta.org (click on Areas of Interest>Publications):

- Guidelines for Recognizing and Providing Care for Victims of Child Abuse (2005)
- Guidelines for Recognizing and Providing Care for Victims of Domestic Abuse (2005)
- Guidelines for Recognizing and Providing Care for Victims of Elder Abuse (2007)

Medical Treatment and Medications

Medical Treatment

Medical treatment includes any intervention performed by a physician (family practitioner or specialist), dentist, physician's assistant, nurse, nurse practitioner, physical therapist, or occupational therapist. The client may also include chiropractic treatment when answering the question:

FOLLOW-UP QUESTIONS

- What medical treatment have you had for this condition?
- Alternate: What treatment have you had for this condition? (allows the client to report any and all modes of treatment including complementary and alternative medicine)

In addition to eliciting information regarding specific treatment performed by the medical community, follow-up questions relate to previous physical therapy treatment:

FOLLOW-UP QUESTIONS

- Have you been treated by a physical therapist for this condition before?
- If yes, when, where, and for how long?
- What helped and what didn't help?
- Was there any treatment that made your symptoms worse? If yes, please describe.

Knowing the client's response to previous types of treatment techniques may assist the therapist in determining an appropriate treatment protocol for the current chief complaint. For example, previously successful treatment intervention described may provide a basis for initial treatment until the therapist can fully assess the objective data and consider all potential types of treatments.

Medications

Medication use, especially with polypharmacy, is important information. Side effects of medications can present as an impairment of the integumentary, musculoskeletal, cardiovascular/pulmonary, or neuromuscular system. Medications may be the most common or most likely cause of systemically induced NMS signs and symptoms.

Please note the use of a new term: hyperpharmacotherapy. Whereas polypharmacy is often defined as the use of multiple medications to treat health problems, the term has also been expanded to describe the use of multiple pharmacies to fill the same (or other) prescriptions, high-frequency medications, or multiple-dose medications. Hyperpharmacotherapy is the current term used to describe the excessive use of drugs to treat disease, including the use of more medications than are clinically indicated or the unnecessary use of medications.

Medications (either prescription, shared, or OTC) may or may not be listed on the Family/Personal History form at all facilities. Even when a medical history form is used, it may be necessary to probe further regarding the use of over-the-counter preparations such as aspirin, acetaminophen (Tylenol), ibuprofen (e.g., Advil, Motrin), laxatives, antihistamines, antacids, and decongestants or other drugs that can alter the client's symptoms.

It is not uncommon for adolescents and seniors to share, borrow, or lend medications to friends, family members, and acquaintances. In fact, medication borrowing and sharing is a behavior that has been identified in patients of all ages.[212]

Most of the sharing and borrowing is done without consulting a pharmacist or medical doctor. The risk of allergic reactions or adverse drug events is much higher under these circumstances than when medications are prescribed and taken as directed by the person for whom they were intended.[213]

Risk Factors for Adverse Drug Events. Pharmacokinetics (the processes that affect drug movement in the body) represents the biggest risk factor for adverse drug events (ADEs). An ADE is any unexpected, unwanted, abnormal, dangerous, or harmful reaction or response to a medication. Most ADEs are medication reactions or side effects.

A *drug-drug* interaction occurs when medications interact unfavorably, possibly adding to the pharmacologic effects. A *drug-disease* interaction occurs when a medication causes an existing disease to worsen. Absorption, distribution, metabolism, and excretion are the main components of pharmacokinetics affected by age,[56] size, polypharmacy or hyperpharmacotherapy, and other risk factors listed in Box 2-12.

Once again, ethnic background is a risk factor to consider. Herbal and home remedies may be used by clients based on their ethnic, spiritual, or cultural orientation. Alternative healers may be consulted for all kinds of conditions from diabetes to depression to cancer. Home remedies can be harmful or interact with some medications.

Some racial groups respond differently to medications. Effectiveness and toxicity can vary among racial and ethnic groups. Differences in metabolic rate, clinical drug responses, and side effects of many medications, such as antihistamines, analgesics, cardiovascular agents, psychotropic drugs, and CNS agents, have been documented. Genetic factors also play a significant role.[214,215]

Women metabolize drugs differently throughout the month as influenced by hormonal changes associated with menses. Researchers are investigating the differences in drug metabolism in women who are premenopausal versus postmenopausal.[216]

BOX 2-12 RISK FACTORS FOR ADVERSE DRUG EVENTS (ADEs)

- Age (over 65, but especially over 75)
- Small physical size or stature (decrease in lean body mass)
- Sex (men and women respond differently to different drugs)
- Polypharmacy (taking several drugs at once; duplicate or dual medications) or hyperpharmacotherapy (excessive use of drugs to treat disease)
- Prescribing cascade (failure to recognize signs and symptoms as an ADE and treating it as the onset of a new illness; taking medications to counteract side effects of another medication)
- Taking medications prescribed for someone else
- Organ impairment and dysfunction (e.g., renal or hepatic insufficiency)
- Concomitant alcohol consumption
- Concomitant use of certain nutraceuticals
- Previous history of ADEs
- Mental deterioration or dementia (unintentional repeated dosage; failure to take medications as prescribed)
- Difficulty opening medication bottles, difficulty swallowing, unable to read or understand directions
- Racial/ethnic variations

Clients receiving home health care are at increased risk for medication errors such as uncontrolled hypertension despite medication, confusion or falls while on psychotropic medications, or improper use of medications deemed dangerous to the older adult such as muscle relaxants. Nearly one third of home health clients are misusing their medications as well.[217]

Potential Drug Side Effects. Side effects are usually defined as predictable pharmacologic effects that occur within therapeutic dose ranges and are undesirable in the given therapeutic situation. Doctors are well aware that drugs have side effects. They may even fully expect their patients to experience some of these side effects. The goal is to obtain maximum benefit from the drug's actions with the minimum amount of side effects. These are referred to as "tolerable" side effects.

The most common side effects of medications are constipation or diarrhea, nausea, abdominal pain, and sedation. More severe reactions include confusion, drowsiness, weakness, and loss of coordination. Adverse events, such as falls, anorexia, fatigue, cognitive impairment, urinary incontinence, and constipation, can occur.[55]

Medications can mask signs and symptoms or produce signs and symptoms that are seemingly unrelated to the client's current medical problem. For example, long-term use of steroids resulting in side effects, such as proximal muscle weakness, tissue edema, and increased pain threshold, may alter objective findings during the examination of the client.

A detailed description of GI disturbances and other side effects caused by nonsteroidal antiinflammatory drugs (NSAIDs) resulting in back, shoulder, or scapular pain is presented in Chapter 8. Every therapist should be very familiar with these.

Physiologic or biologic differences can result in different responses and side effects to drugs. Race, age, weight, metabolism, and for women, the menstrual cycle can impact drug metabolism and effects. In the aging population, drug side effects can occur even with low doses that usually produce no side effects in younger populations. Older people, especially those who are taking multiple drugs, are two or three times more likely than young to middle-aged adults to have adverse drug events.

Seventy-five percent of all older clients take OTC medications that may cause confusion, cause or contribute to additional symptoms, and interact with other medications. Sometimes the client is receiving the same drug under different brand names, increasing the likelihood of drug-induced confusion. Watch for the four *D*s associated with OTC drug use:

- Dizziness
- Drowsiness
- Depression
- Visual disturbance

Because many older people do not consider these "drugs" worth mentioning (i.e., OTC drugs "don't count"), it is important to ask specifically about OTC drug use. Additionally, alcoholism and other drug abuse are more common in older people than is generally recognized, especially in depressed clients. Screening for substance use in conjunction with medication use and/or prescription drug abuse may be important for some clients.

Common medications in the clinic that produce other signs and symptoms include:

- Skin reactions, noninflammatory joint pain (antibiotics; see Fig. 4-12)
- Muscle weakness/cramping (diuretics)
- Muscle hyperactivity (caffeine and medications with caffeine)
- Back and/or shoulder pain (NSAIDs; retroperitoneal bleeding)
- Hip pain from femoral head necrosis (corticosteroids)
- Gait disturbances (Thorazine/tranquilizers)
- Movement disorders (anticholinergics, antipsychotics, antidepressants)
- Hormonal contraceptives (elevated blood pressure)
- Gastrointestinal symptoms (nausea, indigestion, abdominal pain, melena)

This is just a partial listing, but it gives an idea why paying attention to medications and potential side effects is important in the screening process. Not all, but some, medications (e.g., antibiotics, antihypertensives, antidepressants) must be taken as prescribed in order to obtain pharmacologic efficacy.

Nonsteroidal Antiinflammatory Drugs (NSAIDs). NSAIDs are a group of drugs that are useful in the symptomatic treatment of inflammation; some appear to be more useful as analgesics. OTC NSAIDs are listed in Table 8-3. NSAIDs are commonly used postoperatively for discomfort; for painful musculoskeletal conditions, especially among the older adult population; and in the treatment of inflammatory rheumatic diseases.

The incidence of adverse reactions to NSAIDs is low—complications develop in about 2% to 4% of NSAID users each year.[218] However, 30 to 40 million Americans are regular users of NSAIDs. The widespread use of readily available OTC NSAIDs results in a large number of people being affected. It is estimated that approximately 80% of outpatient orthopedic clients are taking NSAIDs. Many are taking dual NSAIDs (combination of NSAIDs and aspirin) or duplicate NSAIDs (two or more agents from the same class).[219]

Side Effects of NSAIDs. NSAIDs have a tendency to produce adverse effects on multiple-organ systems, with the greatest damage to the GI tract.[220] GI impairment can be seen as subclinical erosions of the mucosa or more seriously, as ulceration with life-threatening bleeding and perforation. People with NSAID-induced GI impairment can be asymptomatic until the condition is advanced. NSAID-related gastropathy causes thousands of hospitalizations and deaths annually.

For those who are symptomatic, the most common side effects of NSAIDs are stomach upset and pain, possibly leading to ulceration. GI ulceration has been reported in up to 30% of adults using NSAIDS.[221] With the use of cyclooxygenase-2 (COX-2) inhibitors, serious GI side effects have modified this figure down to 4% of chronic NSAID users.[222] Physical therapists are seeing a large percentage of people taking NSAIDs routinely and are likely to be the first to identify a problem. NSAID use among surgical patients can cause postoperative complications such as wound hematoma, upper GI tract bleeding, hypotension, and impaired bone or tendon healing.[223]

NSAIDs are also potent renal vasoconstrictors and may cause increased blood pressure and peripheral edema. Clients with hypertension or congestive heart failure are at risk for renal complications, especially those using diuretics or angiotensin-converting enzyme (ACE) inhibitors.[224] NSAID use may be associated with confusion and memory loss in the older adult.

People with coronary artery disease taking NSAIDs may also be at a slightly increased risk for a myocardial event during times of increased myocardial oxygen demand (e.g., exercise, fever).

Older adults taking NSAIDs and antihypertensive agents must be monitored carefully. Regardless of the NSAID chosen, it is important to check blood pressure when exercise is initiated and periodically afterwards.

CLINICAL SIGNS AND SYMPTOMS

NSAID Complications

- May be asymptomatic
- May cause confusion and memory loss in the older adult

Gastrointestinal

- Indigestion, heartburn, epigastric or abdominal pain
- Esophagitis, dysphagia, odynophagia
- Nausea
- Unexplained fatigue lasting more than 1 or 2 weeks
- Ulcers (gastric, duodenal), perforations, bleeding
- Melena

Renal

- Polyuria, nocturia
- Nausea, pallor
- Edema, dehydration
- Muscle weakness, restless legs syndrome

Integumentary

- Pruritus (symptom of renal impairment)
- Delayed wound healing
- Skin reaction to light (photodermatitis)

Cardiovascular/Pulmonary

- Elevated blood pressure
- Peripheral edema
- Asthma attacks in individuals with asthma

Musculoskeletal

- Increased symptoms after taking the medication
- Symptoms linked with ingestion of food (increased or decreased depending on location of GI ulcer)
- Midthoracic back, shoulder, or scapular pain
- Neuromuscular
- Muscle weakness (sign of renal impairment)
- Restless legs syndrome (sign of renal impairment)
- Paresthesias (sign of renal impairment)

Screening for Risk Factors and Effects of NSAIDs. Screening for risk factors is as important as looking for clinical manifestations of NSAID-induced complications. High-risk individuals are older with a history of ulcers and any coexisting diseases that increase the potential for GI bleeding. Anyone receiving treatment with multiple NSAIDs is at increased risk, especially if the dosage is high and/or includes aspirin.

As with any risk-factor assessment, we must know what to look for before we can recognize signs of impending trouble. In the case of NSAID use, back and/or shoulder pain can be the first symptom of impairment in its clinical presentation.

Any client with this presentation in the presence of the risk factors listed in Box 2-13 raises a red flag of suspicion.[225] Look for the presence of associated GI distress such as indigestion, heartburn, nausea, unexplained chronic fatigue, and/or melena (tarry, sticky, black or dark stools from oxidized blood in the GI tract) (Case Example 2-10). A scoring system to estimate the risk of GI problems in clients with rheumatoid arthritis who are also taking NSAIDs is presented in Table 2-6 (Case Example 2-11).

BOX 2-13 RISK FACTORS FOR NSAID GASTROPATHY

Back, shoulder, neck, or scapular pain in any client taking NSAIDs in the presence of the following risk factors for NSAID-induced gastropathy raises a red flag of suspicion:

- Age (65 years and older)
- History of peptic ulcer disease, GI disease, or rheumatoid arthritis
- Tobacco or alcohol use
- NSAIDs combined with oral corticosteroid use
- NSAIDs combined with anticoagulants (blood thinners; even when used for cardioprevention at a lower dose [e.g., 81 to 325 mg aspirin/day, especially for those already at increased risk])[225]
- NSAIDs combined with selective serotonin reuptake inhibitors (SSRIs; antidepressants such as Prozac, Zoloft, Celexa, Paxil)
- Chronic use of NSAIDs (duration: 3 months or more)
- Higher doses of NSAIDs, including the use of more than one NSAID (dual or duplicate use)
- Concomitant infection with *Helicobacter pylori* (under investigation)
- Use of acid suppressants (e.g., H_2-receptor antagonists, antacids); these agents may mask the warning symptoms of more serious GI complications, leaving the client unaware of ongoing damage

Correlate increased musculoskeletal symptoms after taking medications. Expect to see a decrease (not an increase) in painful symptoms after taking analgesics or NSAIDs. Ask about any change in pain or symptoms (increase or decrease) after eating (anywhere from 30 minutes to 2 hours later).

Ingestion of food should have no effect on the musculoskeletal tissues, so any change in symptoms that can be consistently linked with food raises a red flag, especially for the client with known GI problems or taking NSAIDs.

The peak effect for NSAIDs when used as an *analgesic* varies from product to product. For example, peak analgesic effect of aspirin is 2 hours, whereas the peak for naproxen sodium (Aleve) is 2 to 4 hours (compared to acetaminophen, which peaks in 30 to 60 minutes). Therefore the symptoms may occur at varying lengths of time after ingestion of food or drink. It is best to find out the peak time for each antiinflammatory taken by the client and note if maximal relief of symptoms occurs in association with that time.

The time to impact *underlying tissue impairment* also varies by individual and severity of impairment. There is a big difference between 220 mg (OTC) and 500 mg (by prescription) of naproxen sodium. For example, 220 mg may appear to "do nothing" in the client's subjective assessment (opinion) after a week's dosing.

What most adults do not know is that it takes more than 24 to 48 hours to build up a high enough level in the body to impact inflammatory symptoms. The person may start

CASE EXAMPLE 2-10

Assessing for NSAID Complications

A 72-year-old orthopedic outpatient presented 4 weeks status post (S/P) left total knee replacement (TKR). She did not attain 90 degrees of knee flexion and continued to walk with a stiff leg. Her orthopedic surgeon sent her to physical therapy for additional rehabilitation.

Past Medical History: The client reports generalized osteoarthritis. She had a left shoulder replacement 18 months ago with very slow recovery and still does not have full shoulder ROM. She has a long-standing hearing impairment of 60 years and lost her left eye to macular degeneration 2 years ago.

Reported Drug Use: Darvocet for pain 3x/day. Vioxx daily for arthritis (this drug was later removed from the market). She also took Feldene when her shoulder bothered her and daily ibuprofen.

The client walks with a bilateral Trendelenburg gait and drags her left leg using a wheeled walker. Her current symptoms include left knee and shoulder pain, intermittent dizziness, sleep disturbance, finger/hand swelling in the afternoons, and early morning nausea.

How Do You Assess for NSAID Complications?

1. First, review Box 2-14 for any risk factors:
   Shoulder pain
   - Age: 65 years old or older (72 years old)
   - Ask about tobacco and alcohol use
   - Nausea: ask about the presence of other gastrointestinal (GI) symptoms and previous history of peptic ulcer disease
   - Ask about use of corticosteroids, anticoagulants, antidepressants, and acid suppressants
2. Ask about the timing of symptoms in relation to taking her Vioxx, Feldene, and ibuprofen (i.e., see if her shoulder pain is worse 30 minutes to 2 hours after taking the NSAIDs)
3. Take blood pressure
4. Observe for peripheral edema

adding more drugs before an effective level has been reached in the body. Five hundred milligrams (500 mg) can impact tissue in a shorter time, especially with an acute event or flare-up; this is one reason why doctors sometimes dispense prescription NSAIDs instead of just using the lower dosage OTC drugs.

Older adults taking NSAIDs and antihypertensive agents must be monitored carefully. Regardless of the NSAID chosen, it is important to check blood pressure when exercise is initiated and periodically afterwards.

Ask about muscle weakness, unusual fatigue, restless legs syndrome, polyuria, nocturia, or pruritus (signs and symptoms of renal failure). Watch for increased blood pressure and peripheral edema (perform a visual inspection of the feet and ankles). Document and report any significant findings.

Women who take nonaspirin NSAIDs or acetaminophen (Tylenol) are twice as likely to develop high blood pressure. This refers to chronic use (more than 22 days/month). There is not a proven cause-effect relationship, but a statistical link exists between the two.[226]

Acetaminophen. Acetaminophen, the active ingredient in Tylenol and other OTC and prescription pain relievers and cold medicines, is an analgesic (pain reliever) and antipyretic (fever reducer) but not an antiinflammatory agent. Acetaminophen is effective in the treatment of mild-to-moderate pain and is generally well tolerated by all age groups.

It is the analgesic least likely to cause GI bleeding, but taken in large doses over time, it can cause liver toxicity, especially when used with vitamin C or alcohol. Women are more quickly affected than men at lower levels of alcohol consumption.

Individuals at increased risk for problems associated with using acetaminophen are those with a history of alcohol use/

TABLE 2-6 Is Your Client at Risk for NSAID-Induced Gastropathy?

This scoring system allows clinicians to estimate the risk of gastrointestinal (GI) problems in clients with rheumatoid arthritis who are also taking NSAIDs. Risk is equal to the sum of:

Age in years	× 2 =
History of NSAID symptoms, e.g., upper abdominal pain, bloating, nausea, heartburn, loss of appetite, vomiting	+ 50 points
ARA class*	add 0, 10, 20 or 30 based on class 1-4
NSAID dose† (fraction of maximum recommended)	× 15
If currently using prednisone	add 40 points
TOTAL SCORE	

*American Rheumatism Association (ARA) FUNCTIONAL CLASS
+0 points for class 1 (normal)
+10 points for ARA class 2 (adequate)
+20 points for ARA class 3 (limited)
+30 points for class 4 (unable)

ARA Criteria for Classification of Functional Status in Rheumatoid Arthritis:

Class 1 Completely able to perform usual ADLs (self-care, vocational, avocational)
Class 2 Able to perform usual self-care and vocational activities, but limited in avocational activities
Class 3 Able to perform usual self-care activities, but limited in vocational and avocational activities
Class 4 Limited in ability to perform usual self-care, vocational, and avocational activities

†NSAID dose used in this formulation is the fraction of the manufacturer's highest recommended dose. The manufacturer's highest recommended dose on the package insert is given a value of 1.00. The dose of each individual is then normalized to this dose.

For example, the value 1.03 indicates the client is taking 103% of the manufacturer's highest recommended dose. Most often, clients are taking the highest dose recommended. They receive a 1.0. Anyone taking less will have a fraction percentage less than 1.0. Anyone taking more than the highest dose recommended will have a fraction percentage greater than 1.0. See Case Example 2-11.

Risk Calculation: To determine the risk (%) of hospitalization or death caused by GI complications over the next 12 months, use the TOTAL SCORE in the following formula:

Risk %/year = [TOTAL SCORE − 100] ÷ 40

Risk percentage is the likelihood of a GI event leading to hospitalization or death over the next 12 months for the person with rheumatoid arthritis on NSAIDs. Higher Total Scores yield greater predictive risk. The risk ranges from 0.0 (low risk) to 5.0 (high risk).

Data from Fries JF, Williams CA, Bloch DA, et al: Nonsteroidal anti-inflammatory drug-associated gastropathy: incidence and risk factor models, *Amer J Med* 91(3):213-222, 1991.

CASE EXAMPLE 2-11

Risk Calculation for NSAID-Induced Gastropathy

A 66-year-old woman with a history of rheumatoid arthritis (class 3) has been referred to physical therapy after three metacarpal-phalangeal (MCP) joint replacements.

Although her doctor has recommended maximum dosage of ibuprofen (800 mg tid; 2400 mg), she is really only taking 1600/day. She says this is all she needs to control her symptoms. She was taking prednisone before the surgery, but tapered herself off and has not resumed its use.

She has been hospitalized 3 times in the past 6 years for gastrointestinal (GI) problems related to NSAID use but does not have any apparent GI symptoms at this time.

Use the following model to calculate her risk for serious problems with NSAID use:

Age in years	66 × 2 =	132
History of NSAID symptoms, e.g., abdominal pain, bloating, nausea	+50 points	50
ARA class	Add 0, 10, 20, or 30 based on class 1-4	20
NSAID dose (fraction of maximum recommended)	1600/2400 × 15 (0.67 × 15)	10
If currently using prednisone	Add 40 points	0
TOTAL Score		212

Risk/year = [Total score − 100] ÷ 40
Risk/year = [212 − 100] ÷ 40
Risk/year = 112 ÷ 40 = 2.80

The scores range from 0.0 (very low risk) to 5.0 (very high risk). A predictive risk of 2.8 is moderately high. This client should be reminded to report GI distress to her doctor immediately. Periodic screening for GI gastropathy is indicated with early referral if warranted.

abuse, anyone with a history of liver disease (e.g., cirrhosis, hepatitis), and anyone who has attempted suicide using an overdose of this medication.[227]

Some medications (e.g., phenytoin, isoniazid) taken in conjunction with acetaminophen can trigger liver toxicity. The effects of oral anticoagulants may be potentiated by chronic ingestion of large doses of acetaminophen.[228]

Clients with acetaminophen toxicity may be asymptomatic or have anorexia, mild nausea, and vomiting. The therapist may ask about right upper abdominal quadrant tenderness, jaundice, and other signs and symptoms of liver impairment (e.g., liver palms, asterixis, carpal tunnel syndrome, spider angiomas); see discussion in Chapter 9.

Corticosteroids. Corticosteroids are often confused with the singular word "steroids." There are three types or classes of steroids:

1. Anabolic-androgenic steroids such as testosterone, estrogen, and progesterone
2. Mineralocorticoids responsible for maintaining body electrolytes
3. Glucocorticoids, which suppress inflammatory processes within the body

All three types are naturally occurring hormones produced by the adrenal cortex; synthetic equivalents can be prescribed as medication. Illegal use of a synthetic derivative of testosterone is a concern with athletes and millions of men and women who use these drugs to gain muscle and lose body fat.[229]

Corticosteroids used to control pain and reduce inflammation are associated with significant side effects even when given for a short time. Administration may be by local injection (e.g., into a joint), transdermal (skin patch), or systemic (inhalers or pill form).

Side effects of *local injection* (catabolic glucocorticoids) may include soft tissue atrophy, changes in skin pigmentation, accelerated joint destruction, and tendon rupture, but it poses no problem with liver, kidney, or cardiovascular function. *Transdermal* corticosteroids have similar side effects. The incidence of skin-related changes is slightly higher than with local injection, whereas the incidence of joint problems is slightly lower.

Systemic corticosteroids are associated with GI problems, psychologic problems, and hip avascular necrosis. Physician referral is required for marked loss of hip motion and referred pain to the groin in a client on long-term systemic corticosteroids.

Long-term use can lead to immunosuppression, osteoporosis, and other endocrine-metabolic abnormalities. Therapists working with athletes may need to screen for nonmedical (illegal) use of anabolic steroids. Visually observe for signs and symptoms associated with anabolic steroid use. Monitor behavior and blood pressure.

CLINICAL SIGNS AND SYMPTOMS

Anabolic Steroid Use

- Rapid weight gain
- Elevated blood pressure (BP)
- Peripheral edema associated with increased BP
- Acne on face and upper body
- Muscular hypertrophy
- Stretch marks around trunk
- Abdominal pain, diarrhea
- Needle marks in large muscle groups
- Personality changes (aggression, mood swings, "roid" rages)
- Bladder irritation, urinary frequency, urinary tract infections
- Sleep apnea, insomnia
- Altered ejection fraction (lower end of normal: under 55%)[229]

Opioids. Opioids, such as codeine, morphine, tramadol, hydrocodone, or oxycodone, are safe when used as directed. They do not cause kidney, liver, or stomach impairments and have few drug interactions. Side effects can include nausea, constipation, and dry mouth. The client may also experience impaired balance and drowsiness or dizziness, which can increase the risk of falls.

Addiction (physical or psychologic dependence) is often a concern raised by clients and family members alike. Addiction to opioids is uncommon in individuals with no history of substance abuse. Adults over age 60 are often good candidates for use of opioid medications. They obtain greater pain control with lower doses and develop less tolerance than younger adults.[230]

Prescription Drug Abuse. The U.S. Drug Enforcement Administration has reported that more than 7 million Americans abuse prescription medications.[231] The CDC reports drug overdose of opioids are now the second leading cause of accidental death in the United States (second only to motor vehicle accidents). Opioid misuse and dependence among prescription opioid patients in the United States is likely higher than currently documented.[232] Medical and nonmedical prescription drug abuse has become an increasing problem, especially among young adolescents and teenagers.[233]

Oxycodone, hydrocodone, methadone, benzodiazepines, and muscle relaxants used to treat pain and anxiety and stimulants used to treat learning disorders are listed as the most common medications involved in nonmedical use.[234,235] Prescription opioids are monitored carefully and withdrawn or stopped gradually to avoid withdrawal symptoms. Psychologic dependence tends to occur when opioids are used in excessive amounts and often does not develop until after the expected time for pain relief has passed.

Risk factors for prescription drug abuse and nonmedical use of prescription drugs include age under 65, previous history of opioid abuse, major depression, and psychotropic medication use.[232] Teen users raiding the family medicine cabinet for prescription medications (a practice referred to as "pharming") often find a wide range of mood stabilizers, painkillers, muscle relaxants, sedatives, and tranquilizers right within their own homes. Combining medications and/or combining prescription medicines with alcohol can lead to serious drug-drug interactions.[231]

Hormonal Contraceptives. Some women use birth control pills to prevent pregnancy while others take them to control their menstrual cycle and/or manage premenstrual and menstrual symptoms, including excessive and painful bleeding.

Originally, birth control pills contained as much as 20% more estrogen than the amount present in the low-dose, third-generation oral contraceptives available today. Women taking the newer hormonal contraceptives (whether in pill, injectable, or patch form) have a slightly increased risk of high blood pressure, which returns to normal shortly after the hormone is discontinued.

Age over 35, smoking, hypertension, obesity, bleeding disorders, major surgery with prolonged immobilization, and diabetes are risk factors for blood clots (venous thromboembolism, not arterial), heart attacks, and strokes in women taking hormonal contraceptives.[236] Adolescents using the injectable contraceptive Depo-Provera (DMPA) are at risk for bone loss.[237]

Anyone taking hormonal contraception of any kind, but especially premenopausal cardiac clients, must be monitored by taking vital signs, especially blood pressure, during physical activity and exercise. Assessing for risk factors is an important part of the plan of care for this group of individuals.

Any woman on combined oral contraceptives (estrogen and progesterone) reporting break-through bleeding should be advised to see her doctor.

Antibiotics. Skin reactions (see Fig. 4-12) and noninflammatory joint pain (see Box 3-4) are two of the most common side effects of antibiotics seen in a therapist's practice. Often these symptoms are delayed and occur up to 6 weeks after the client has finished taking the drug.

Fluoroquinolones, a class of antibiotics used to treat bacterial infections (e.g., urinary tract; upper respiratory tract; infectious diarrhea; gynecologic infections; and skin, soft tissue, bone and joint infections) are known to cause tendinopathies ranging from tendinitis to tendon rupture.

Commonly prescribed fluoroquinolones include ciprofloxacin (Cipro), ciprofloxacin extended release (Cipro ER, Proquin XR), gemifloxacin (Factive), levofloxacin (Levaquin), norfloxacin (Noroxin), ofloxacin (Floxin), and moxifloxacin (Avelox). Although tendon injury has been reported with most fluoroquinolones, most of the fluoroquinolone-induced tendinopathies of the Achilles tendon are due to ciprofloxacin.

The incidence of this adverse event has been enough that in 2008, the U.S. FDA required makers of fluoroquinolone antimicrobial drugs for systemic use to add a boxed warning to the prescribing information about the increased risk of developing tendinitis and tendon rupture. At the same time, the FDA issued a notice to health care professionals about this risk, the known risk factors, and what to advise anyone taking these medications who report tendon pain, swelling, or inflammation (i.e., stop taking the fluoroquinolone, avoid exercise and use of the affected area, promptly contact the prescribing physician).[238]

The concomitant use of corticosteroids and fluoroquinolones in older adults (over age 60) are the major risk factors for developing musculoskeletal toxicities.[239,240] Other risk factors include organ transplant (kidney, heart, and lung) recipients and previous history of tendon ruptures or other tendon problems.

Other common side effects include depression, headache, convulsions, fatigue, GI disturbance (nausea, vomiting, diarrhea), arthralgia (joint pain, inflammation, and stiffness), and neck, back, or chest pain. Symptoms may occur as early as 2 hours after the first dose and as late as 6 months after treatment has ended[241] (Case Example 2-12).

Nutraceuticals. Nutraceuticals are natural products (usually made from plant substances) that do not require a prescription to purchase. They are often sold at health food stores, nutrition or vitamin stores, through private distributors, or on the Internet. Nutraceuticals consist of herbs, vitamins, minerals, antioxidants, and other natural supplements.

The use of herbal and other supplements has increased dramatically in the last decade. Exposure to individual herbal ingredients may continue to rise as more of them are added

CASE EXAMPLE 2-12

Fluoroquinolone-Induced Tendinopathy

A 57-year-old retired army colonel (male) presented to an outpatient physical therapy clinic with a report of swelling and pain in both ankles.

Symptoms started in the left ankle 4 days ago. Then the right ankle and foot became swollen. Ankle dorsiflexion and weight bearing made it worse. Staying off the foot made it better.

Past Medical History:

- Prostatitis diagnosed and treated 2 months ago with antibiotics; placed on levofloxacin
- 11 days ago when urinary symptoms recurred
- Chronic benign prostatic hypertrophy
- Gastroesophageal reflux (GERD)
- Hypertension

Current Medications:

- Omeprazole (Prilosec)
- Lisinopril (Prinivil, Zestril)
- Enteric-coated aspirin
- Tamsulosin (Flomax)
- Levofloxacin (Levaquin)

Clinical Presentation:

- Moderate swelling of both ankles; malleoli diminished visually by 50%
- No lymphadenopathy (cervical, axillary, inguinal)
- Fullness of both Achilles tendons with pitting edema of the feet extending to just above the ankles, bilaterally
- No nodularity behind either Achilles tendon
- Ankle joint tender to minimal palpation; reproduced when Achilles tendons are palpated
- Range of motion (ROM): normal subtalar and plantar flexion of the ankle; dorsiflexion to neutral (limited by pain); inversion and eversion within normal limits (WNL) and pain-free; unable to squat due to painfully limited ROM
- Neuro screen: negative
- Knee screen: no apparent problems in either knee

Associated Signs and Symptoms: The client reports fever and chills the day before the ankle started swelling, but this has gone away now. Urinary symptoms have resolved. Reports no other signs or symptoms anywhere else in his body.

Vital Signs:

- Blood pressure 128/74 mm Hg taken seated in the left arm
- Heart rate 78 bpm
- Respiratory rate 14 breaths per minute
- Temperature 99.0° F (client states "normal" for him is 98.6° F)

What Are the Red-Flag Signs and Symptoms Here? Should a Medical Referral Be Made? Why or Why Not?

Red Flags:

- Age
- Bilateral swelling
- Recent history of new medication (levofloxacin) known to cause tendon problems in some cases
- Constitutional symptoms × 1 day; presence of low-grade fever at the time of the initial evaluation

A cluster of red flags like this suggests medical referral would be a good idea before initiating intervention. If there is an inflammatory process going on, early diagnosis and medical treatment can minimize damage to the joint.

If there is a medical problem, it is not likely to be life-threatening, so theoretically the therapist could treat symptomatically for three to five sessions and then evaluate the results. Medical referral could be made at that time if symptoms remain unchanged by treatment. If this option is chosen, the client's vital signs must be monitored closely.

Decision: The client was referred to his primary care physician with the following request:

Date

Dr. Smith,

This client came to our clinic with a report of bilateral ankle swelling. I observed the following findings:

Moderate swelling of both ankles; malleoli diminished visually by 50%

No lymphadenopathy (cervical, axillary, inguinal)

Fullness of both Achilles tendons with pitting edema of the feet extending to just above the ankles, bilaterally

No nodularity behind either Achilles tendon

Ankle joint tender to minimal palpation; reproduced when Achilles tendons are palpated

ROM: normal subtalar and plantar flexion of the ankle; dorsiflexion to neutral (limited by pain); inversion and eversion WNL and pain-free; unable to squat due to painfully limited ROM

Neuro screen: negative

Knee screen: no apparent problems in either knee

Associated Signs and Symptoms:

The client reports fever and chills the day before the ankle started swelling, but this went away by the time he came to physical therapy. Urinary symptoms also had resolved. The client reported no other signs or symptoms anywhere else in his body.

Vital signs:

Blood pressure	128/74 mm Hg taken seated in the left arm
Heart rate	78 bpm
Respiratory rate	14 breaths per minute
Temperature	99.0° F (client states "normal" for him is 98.6° F)

I'm concerned by the following cluster of red flags:

Age

Bilateral swelling

Recent history of new medication (levofloxacin)

Constitutional symptoms × 1 day; presence of low-grade fever at the time of the initial evaluation

I would like to request a medical evaluation before beginning any physical therapy intervention. I would appreciate a copy of your report and any recommendations you may have if physical therapy is appropriate.

Thank you. Best regards,

Result: The client was diagnosed (x-rays and diagnostic lab work) with levofloxacin-induced bilateral Achilles tendonitis. Medical treatment included NSAIDs, rest, and discontinuation of the levofloxacin.

Symptoms resolved completely within 7 days with full motion and function of both ankles and feet. There was no need for physical therapy intervention. Client was discharged from any further PT involvement for this episode of care.

Recommended Reading: Greene BL: Physical therapist management of fluoroquinolone-induced Achilles tendinopathy, *Phys Ther* 82(12):1224-1231, 2002.

Data from McKinley BT, Oglesby RJ: A 57-year-old male retired colonel with acute ankle swelling, *Mil Med* 169(3):254-256, 2004.

to mainstream multivitamin products and advertised as cancer and chronic disease preventatives.[242]

These products may be produced with all natural ingredients, but this does not mean they do not cause problems, complications, and side effects. When combined with certain food items or taken with some prescription drugs, nutraceuticals can have potentially serious complications.

Herbal and home remedies may be used by clients based on their ethnic, spiritual, or cultural orientation. Alternative healers may be consulted for all kinds of conditions from diabetes to depression to cancer. Home remedies and nutraceuticals can be harmful when combined with some medications.

The therapist should ask clients about and document their use of nutraceuticals and dietary supplements. In a survey of surgical patients, more than one in three adults had taken an herb that had effects on coagulation, blood pressure, cardiovascular function, sedation, and electrolytes or diuresis within 2 weeks of the scheduled surgery. As many as 70% of these individuals failed to disclose this use during the preoperative assessment.[243]

FOLLOW-UP QUESTIONS

- Are you taking any remedies from a naturopathic physician or homeopathic healer?
- Are you taking any other vitamins, herbs, or supplements?
- If yes, does your physician have a list of these products?
- Are you seeing anyone else for this condition (e.g., alternative practitioner, such as an acupuncturist, massage therapist, or chiropractor, or Reiki, BodyTalk, Touch for Healing, or Ayurveda practitioner)?

A pharmacist can help in comparing signs and symptoms present with possible side effects and drug-drug or drug-nutraceutical interactions. Mayo Clinic offers a list of herbal supplements that should not be taken in combination with certain types of medications (www.mayoclinic.com; type the words: herbal supplements in the search window). Other resources are available as well.[244]

The Physical Therapist's Role. For every client the therapist is strongly encouraged to take the time to look up indications for use and possible side effects of prescribed medications. Drug reference guidebooks that are updated and published every year are available in hospital and clinic libraries or pharmacies. Pharmacists are also invaluable sources of drug information. Websites with useful drug information are included in the next section (see Resources).

Distinguishing drug-related signs and symptoms from disease-related symptoms may require careful observation and consultation with family members or other health professionals to see whether these signs tend to increase following each dose.[245] This information may come to light by asking the question:

FOLLOW-UP QUESTIONS

- Do you notice any increase in symptoms, or perhaps the start of symptoms, after taking your medications? (This may occur 30 minutes to 2 hours after taking the drug.)

Because clients are more likely now than ever before to change physicians or practitioners during an episode of care, the therapist has an important role in education and screening. The therapist can alert individuals to watch for any red flags in their drug regimen. Clients with both hypertension and a condition requiring NSAID therapy should be closely monitored and advised to make sure the prescribing practitioner is aware of both conditions.

The therapist may find it necessary to reeducate the client regarding the importance of taking medications as prescribed, whether on a daily or other regular basis. In the case of antihypertensive medication, the therapist should ask whether the client has taken the medication today as prescribed.

It is not unusual to hear a client report, "I take my blood pressure pills when I feel my heart starting to pound." The same situation may occur with clients taking antiinflammatory drugs, antibiotics, or any other medications that must be taken consistently for a specified period to be effective. Always ask the client if he or she is taking the prescription every day or just as needed. Make sure this is with the physician's knowledge and approval.

Clients may be taking medications that were not prescribed for them, taking medications inappropriately, or not taking prescribed medications without notifying the doctor.

Appropriate FUPs include the following:

FOLLOW-UP QUESTIONS

- Why are you taking these medications?
- When was the last time that you took these medications?
- Have you taken these drugs today?
- Do the medications relieve your pain or symptoms?
  - If yes, how soon after you take the medications do you notice an improvement?
- If prescription drugs, who prescribed this medication for you?
- How long have you been taking these medications?
- When did your physician last review these medications?
- Are you taking any medications that weren't prescribed for you?
  - If no, follow-up with: Are you taking any pills given to you by someone else besides your doctor?

Many people who take prescribed medications cannot recall the name of the drug or tell you why they are taking it. It is essential to know whether the client has taken OTC or prescription medication before the physical therapy examination or intervention because the symptomatic relief or possible side effects may alter the objective findings.

Similarly, when appropriate, treatment can be scheduled to correspond with the time of day when clients obtain maximal relief from their medications. Finally, the therapist may be the first one to recognize a problem with medication or dosage. Bringing this to the attention of the doctor is a valuable service to the client.

Resources. Many resources are available to help the therapist identify potential side effects of medications, especially in the presence of polypharmacy or hyperpharmacotherapy with the possibility of drug interactions.

Find a local pharmacist willing to answer questions about medications. The pharmacist can let the therapist know when associated signs and symptoms may be drug-related. Always bring this to the physician's attention. It may be that the "burden of tolerable side effects" is worth the benefit, but often, the dosage can be adjusted or an alternative drug can be tried.

Several favorite resources include *Mosby's Nursing Drug Handbook,* published each year by Elsevier Science (Mosby, St. Louis), *The People's Pharmacy Guide to Home and Herbal Remedies,*[246] *PDR for Herbal Medicines ed 4,*[247] and *Pharmacology in Rehabilitation.*[245]

A helpful general guide regarding potentially inappropriate medications for older adults called the Beers' list has been published and revised. This list along with detailed information about each class of drug is available online at: www.dcri.duke.edu/ccge/curtis/beers.html.

Easy-to-use websites for helpful pharmacologic information include:

- MedicineNet (www.medicinenet.com)
- University of Montana Drug Information Service (DIS) (www.umt.edu/druginfo or by phone: 1-800-501-5491) [our personal favorite—an excellent resource]
- RxList: The Internet Drug Index (www.rxlist.com)
- DrugDigest (www.drugdigest.com)
- National Council on Patient Information and Education: BeMedWise. Advice on use of OTC medications. Available at www.bemedwise.org.

Current Level of Fitness

An assessment of current physical activity and level of fitness (or level just before the onset of the current problem) can provide additional necessary information relating to the origin of the client's symptom complex.

The level of fitness can be a valuable indicator of potential response to treatment based on the client's motivation (i.e., those who are more physically active and healthy seem to be more motivated to return to that level of fitness through disciplined self-rehabilitation).

It is important to know what type of exercise or sports activity the client participates in, the number of times per week (frequency) that this activity is performed, the length (duration) of each exercise or sports session, as well as how long the client has been exercising (weeks, months, years), and the level of difficulty of each exercise session (intensity). It is very important to ask:

FOLLOW-UP QUESTIONS

- Since the onset of symptoms, are there any activities that you can no longer accomplish?

The client should give a description of these activities, including how physical activities have been affected by the symptoms. Follow-up questions include:

FOLLOW-UP QUESTIONS

- Do you ever experience shortness of breath or lack of air during any activities (e.g., walking, climbing stairs)?
- Are you ever short of breath without exercising?
- Are you ever awakened at night breathless?
  - If yes, how often and when does this occur?

If the Family/Personal History form is not used, it may be helpful to ask some of the questions shown in Fig. 2-2: Work/Living Environment or History of Falls. For example, assessing the history of falls with older people is essential. One third of community-dwelling older adults and a higher proportion of institutionalized older people fall annually. Aside from the serious injuries that may result, a debilitating "fear of falling" may cause many older adults to reduce their activity level and restrict their social life. This is one area that is often treatable and even preventable with physical therapy.

Older persons who are in bed for prolonged periods are at risk for secondary complications, including pressure ulcers, urinary tract infections, pulmonary infections and/or infarcts, congestive heart failure, osteoporosis, and compression fractures. See previous discussion in this chapter on History of Falls for more information.

Sleep-Related History

Sleep patterns are valuable indicators of underlying physiologic and psychologic disease processes. The primary function of sleep is believed to be the restoration of body function. When the quality of this restorative sleep is decreased, the body and mind cannot perform at optimal levels.

Physical problems that result in pain, increased urination, shortness of breath, changes in body temperature, perspiration, or side effects of medications are just a few causes of sleep disruption. Any factor precipitating sleep deprivation can contribute to an increase in the frequency, intensity, or duration of a client's symptoms.

For example, fevers and sweats are characteristic signs of systemic disease. Sweats occur as a result of a gradual increase in body temperature followed by a sudden drop in temperature; although they are most noticeable at night, sweats can occur anytime of the day or night. This change in body temperature can be related to pathologic changes in immunologic, neurologic, or endocrine function.

Be aware that many people, especially women, experience sweats associated with menopause, poor room ventilation, or too many clothes and covers used at night. Sweats can also occur in the neutropenic client after chemotherapy or as a side effect of other medications such as some antidepressants, sedatives or tranquilizers, and some analgesics.

Anyone reporting sweats of a systemic origin must be asked if the same phenomenon occurs during the waking hours. Sweats (present day and/or night) can be associated with medical problems such as tuberculosis, autoimmune diseases, and malignancies.[248]

An isolated experience of sweats is not as significant as intermittent but consistent sweats in the presence of risk factors for any of these conditions or in the presence of other constitutional symptoms (see Box 1-3). Assess vital signs in the client reporting sweats, especially when other symptoms are present and/or the client reports back or shoulder pain of unknown cause.

Certain neurologic lesions may produce local changes in sweating associated with nerve distribution. For example, a client with a spinal cord tumor may report changes in skin temperature above and below the level of the tumor. At presentation, any client with a history of either sweats or fevers should be referred to the primary physician. This is especially true for clients with back pain or multiple joint pain without traumatic origin.

Pain at night is usually perceived as being more intense because of the lack of outside distraction when the person lies quietly without activity. The sudden quiet surroundings and lack of external activity create an increased perception of pain that is a major disrupter of sleep.

It is very important to ask the client about pain during the night. Is the person able to get to sleep? If not, the pain may be a primary focus and may become continuously intense so that falling asleep is a problem.

FOLLOW-UP QUESTIONS

- Does a change in body position affect the level of pain?

If a change in position can increase or decrease the level of pain, it is likely to be a musculoskeletal problem. If, however, the client is awakened from a deep sleep by pain in any location that is unrelated to physical trauma and is unaffected by a change in position, this may be an ominous sign of serious systemic disease, particularly cancer. FUPs include:

FOLLOW-UP QUESTIONS

- If you wake up because of pain, is it because you rolled onto that side?
- Can you get back to sleep?
- If yes, what do you have to do (if anything) to get back to sleep? (This answer may provide clues for treatment.)

Many other factors (primarily environmental and psychologic) are associated with sleep disturbance, but a good, basic assessment of the main characteristics of physically related disturbances in sleep pattern can provide valuable information related to treatment or referral decisions. The McGill Home Recording Card (see Fig. 3-7) is a helpful tool for evaluating sleep patterns.

Stress (see also Chapter 3)

By using the interviewing tools and techniques described in this chapter, the therapist can communicate a willingness to consider all aspects of illness, whether biologic or psychologic. Client self-disclosure is unlikely if there is no trust in the health professional, if there is fear of a lack of confidentiality, or if a sense of disinterest is noted.

Most symptoms (pain included) are aggravated by unresolved emotional or psychologic stress. Prolonged stress may gradually lead to physiologic changes. Stress may result in depression, anxiety disorders, and behavioral consequences (e.g., smoking, alcohol and substance abuse, accident proneness).

The effects of emotional stress may be increased by physiologic changes brought on by the use of medications or poor diet and health habits (e.g., cigarette smoking or ingestion of caffeine in any form). As part of the Core Interview, the therapist may assess the client's subjective report of stress by asking:

FOLLOW-UP QUESTIONS

- What major life changes or stresses have you encountered that you would associate with your injury/illness?
- Alternate: What situations in your life are "stressors" for you?

It may be helpful to quantify the stress by asking the client:

- On a scale from 0 to 10, with 0 being no stress and 10 being the most extreme stress you have ever experienced, what number rating would you give your stress in general at this time in your life?
- What number would you give your stress level today?

Emotions, such as fear and anxiety, are common reactions to illness and treatment intervention and may increase the client's awareness of pain and symptoms. These emotions may cause autonomic (branch of nervous system not subject to voluntary control) distress manifested in such symptoms as pallor, restlessness, muscular tension, perspiration, stomach pain, diarrhea or constipation, or headaches.

It may be helpful to screen for anxiety-provoked hyperventilation by asking:

FOLLOW-UP QUESTIONS

- Do you ever get short of breath or dizzy or lose coordination when you are fatigued?

After the objective evaluation has been completed, the therapist can often provide some relief of emotionally amplified symptoms by explaining the cause of pain, outlining a plan of care, and providing a realistic prognosis for improvement. This may not be possible if the client demonstrates signs of hysterical symptoms or conversion symptoms (see discussion in Chapter 3).

Whether the client's symptoms are systemic or caused by an emotional/psychologic overlay, if the client does not respond to treatment, it may be necessary to notify the physician that there is not a satisfactory explanation for the client's complaints. Further medical evaluation may be indicated at that time.

Final Questions

It is always a good idea to finalize the interview by reviewing the findings and paraphrasing what the client has reported. Use the answers from the Core Interview to recall specifics about the location, frequency, intensity, and duration of the symptoms. Mention what makes it better or worse.

Recap the medical and surgical history including current illnesses, diseases, or other medical conditions; recent or past surgeries; recent or current medications; recent infections; and anything else of importance brought out by the interview process.

It is always appropriate to end the interview with a few final questions such as:

FOLLOW-UP QUESTIONS

- Are there any other symptoms of any kind anywhere else in your body that we haven't discussed yet?
- Is there anything else you think is important about your condition that we have not discussed yet?
- Is there anything else you think I should know?

If you have not asked any questions about assault or partner abuse, this may be the appropriate time to screen for domestic violence.

Special Questions for Women

Gynecologic disorders can refer pain to the low back, hip, pelvis, groin, or sacroiliac joint Any woman having pain or symptoms in any one or more of these areas should be screened for possible systemic diseases. The need to screen for systemic disease is essential when there is no known cause of the pain or symptoms.

Any woman with a positive family/personal history of cancer should be screened for medical disease even if the current symptoms can be attributed to a known NMS cause.

Chapter 14 has a list of special questions to ask women (see also Appendix B-37). The therapist will not need to ask every woman each question listed but should take into consideration the data from the Family/Personal History form, Core Interview, and clinical presentation when choosing appropriate FUPs.

Special Questions for Men

Men describing symptoms related to the groin, low back, hip, or sacroiliac joint may have prostate or urologic involvement. A positive response to any or all of the questions in Appendix B-24 must be evaluated further. Answers to these questions correlated with family history, the presence of risk factors, clinical presentation, and any red flags will guide the therapist in making a decision regarding treatment versus referral.

HOSPITAL INPATIENT INFORMATION

Medical Record

Treatment of hospital inpatients or residents in other facilities (e.g., step down units, transition units, extended care facilities) requires a slightly different interview (or information-gathering) format. A careful review of the medical record for information will assist the therapist in developing a safe and effective plan of care. Watch for conflicting reports (e.g., emergency department, history and physical, consult reports). Important information to look for might include:

- Age
- Medical diagnosis
- Surgery report
- Physician's/nursing notes
- Associated or additional problems relevant to physical therapy
- Medications
- Restrictions
- Laboratory results
- Vital signs

An evaluation of the patient's medical status in conjunction with age and diagnosis can provide valuable guidelines for the plan of care.

If the patient has had recent surgery, the physician's report should be scanned for preoperative and postoperative orders (in some cases there is a separate physician's orders book or link to click on if the medical records are in an electronic format). Read the operative report whenever available. Look for any of the following information:

- Was the patient treated preoperatively with physical therapy for gait, strength, range of motion, or other objective assessments?
- Were there any unrelated preoperative conditions?
- Was the surgery invasive, a closed procedure via arthroscopy, fluoroscopy, or other means of imaging, or virtual by means of computerized technology?
- How long was the operative procedure?
- How much fluid and/or blood products were given?
- What position was the patient placed in during the procedure?

Fluid received during surgery may affect arterial oxygenation, leaving the person breathless with minimal exertion and experiencing early muscle fatigue. Prolonged time in any one position can result in residual musculoskeletal complaints.

The surgical position for men and for women during laparoscopy (examination of the peritoneal cavity) may place patients at increased risk for thrombophlebitis because of the decreased blood flow to the legs during surgery.

Other valuable information that may be contained in the physician's report may include:

- What are the current short-term and long-term medical treatment plans?
- Are there any known or listed contraindications to physical therapy intervention?
- Does the patient have any weight-bearing limitations?

Associated or additional problems to the primary diagnosis may be found within the record (e.g., diabetes, heart disease, peripheral vascular disease, respiratory involvement). The physical therapist should look for any of these conditions in order to modify exercise accordingly and to watch for any related signs and symptoms that might affect the exercise program:

- Are there complaints of any kind that may affect exercise (e.g., shortness of breath [dyspnea], heart palpitations, rapid heart rate [tachycardia], fatigue, fever, or anemia)?

If the patient has diabetes, the therapist should ask:

- What are the current blood glucose levels and recent A1C levels?
- When is insulin administered?

Avoiding peak insulin levels in planning exercise schedules is discussed more completely in Chapter 11. Other questions related to medications can follow the Core Interview outline with appropriate follow-up questions:

- Is the patient receiving oxygen or receiving fluids/medications through an intravenous line?
- If the patient is receiving oxygen, will he or she need increased oxygen levels before, during, or following physical therapy? What level(s)? Does the patient have chronic obstructive pulmonary disease (COPD) with restrictions on oxygen use?
- Are there any dietary or fluid restrictions?
- If so, check with the nursing staff to determine the full limitations. For example:
- Are ice chips or wet washcloths permissible?
- How many ounces or milliliters of fluid are allowed during therapy?
- Where should this amount be recorded?

Laboratory values and vital signs should be reviewed. For example:

- Is the patient anemic?
- Is the patient's blood pressure stable?

Anemic patients may demonstrate an increased normal resting pulse rate that should be monitored during exercise. Patients with unstable blood pressure may require initial standing with a tilt table or monitoring of the blood pressure before, during, and after treatment. Check the nursing record for pulse rate at rest and blood pressure to use as a guide when taking vital signs in the clinic or at the patient's bedside.

Nursing Assessment

After reading the patient's chart, check with the nursing staff to determine the nursing assessment of the individual patient. The essential components of the nursing assessment that are of value to the therapist may include:

- Medical status
- Pain
- Physical status
- Patient orientation
- Discharge plans

The nursing staff are usually intimately aware of the patient's current medical and physical status. If pain is a factor:

- What is the nursing assessment of this patient's pain level and pain tolerance?

Pain tolerance is relative to the medications received by the patient, the number of days after surgery or after injury, fatigue, previous history of substance abuse or chemical addiction, and the patient's personality.

To assess the patient's physical status, ask the nursing staff or check the medical record to find out:

- Has the patient been up at all yet?
- If yes, how long has the patient been sitting, standing, or walking?
- How far has the patient walked?
- How much assistance does the patient require?

Ask about the patient's orientation:

- Is the patient oriented to time, place, and person?

In other words, does the patient know the date and the approximate time, where he or she is, and who he or she is? Treatment plans may be altered by the patient's awareness; for example, a home program may be impossible without family compliance.

- Are there any known or expected discharge plans?
- If yes, what are these plans and when is the target date for discharge?

Cooperation between nurses and therapists is an important part of the multidisciplinary approach in planning the patient's plan of care. The questions to ask and factors to consider provide the therapist with the basic information needed to carry out an objective examination and to plan the intervention. Each individual patient's situation may require that the therapist obtain additional pertinent information (Box 2-14).

PHYSICIAN REFERRAL

The therapist will be using the questions presented in this chapter to identify symptoms of possible systemic origin. The therapist can screen for medical disease and decide if referral to the physician (or other appropriate health care professional) is indicated by correlating the client's answers with family/personal history, vital signs, and objective findings from the physical examination.

For example, consider the client with a chief complaint of back pain who circles "yes" on the Family/Personal History form, indicating a history of ulcers or stomach problems.

BOX 2-14 HOSPITAL INPATIENT INFORMATION

Medical Record
- **Patient age**
- **Medical diagnosis**
- **Surgery:** Did the patient have surgery? What was the surgery for?

FUPs:
- Was the patient seen by a physical therapist preoperatively?
- Were there any unrelated preoperative conditions?
- Was the surgery invasive, a closed procedure via arthroscopy, fluoroscopy, or other means of imaging, or virtual by means of computerized technology?
- How long was the procedure? Were there any surgical complications?
- How much fluid and/or blood products were given?
- What position was the patient placed in and for how long?
- **Physician's report**
  - What are the short-term and long-term medical treatment plans?
  - Are there precautions or contraindications for treatment?
  - Are there weight-bearing limitations?
- **Associated or additional problems** such as diabetes, heart disease, peripheral vascular disease, respiratory involvement

FUPs:
- Are there precautions or contraindications of any kind that may affect exercise?
- If diabetic, what are the current blood glucose levels (normal range: 70 to 100 mg/dL)?
- When is insulin administered? (Use this to avoid the peak insulin levels in planning an exercise schedule.)
- **Medications** (what, when received, what for, potential side effects)

FUPs:
- Is the patient receiving oxygen or receiving fluids/medications through an intravenous line?
- **Restrictions:** Are there any dietary or fluid restrictions?

FUPs:
- If yes, check with the nursing staff to determine the patient's full limitation.
- Are ice chips or a wet washcloth permissible?
- How many ounces or milliliters of fluid are allowed during therapy?
- **Laboratory values:** Hematocrit/hemoglobin level (see inside cover for normal values and significance of these tests); exercise tolerance test results if available for cardiac patient; pulmonary function test (PFT) to determine severity of pulmonary problem; arterial blood gas (ABG) levels to determine the need for supplemental oxygen during exercise
- **Vital signs:** Is the blood pressure stable?

FUPs:
- If no, consider initiating standing with a tilt table or monitoring the blood pressure before, during, and after treatment.

Nursing Assessment
- **Medical status:** What is the patient's current medical status?
- **Pain:** What is the nursing assessment of this patient's pain level and pain tolerance?
- **Physical status:** Has the patient been up at all yet?

FUPs:
- If yes, is the patient sitting, standing, or walking? How long and (if walking) what distance, and how much assistance is required?
- **Patient orientation:** Is the patient oriented to time, place, and person? (Does the patient know the date and the approximate time, where he or she is, and who he or she is?)
- **Discharge plans:** Are there any known or expected discharge plans?

FUPs:
- If yes, what are these plans and when will the patient be discharged?
- **Final question:** Is there anything else that I should know before exercising the patient?

Obtaining further information at the first appointment by using Special Questions to Ask is necessary so that a decision regarding treatment or referral can be made immediately.

This treatment-versus-referral decision is further clarified as the interview, and other objective evaluation procedures continue. Thus, if further questioning fails to show any association of back pain with GI symptoms and the objective findings from the back evaluation point to a true musculoskeletal lesion, medical referral is unnecessary and the physical therapy intervention can begin.

This information is not designed to make a medical diagnosis but rather to perform an accurate assessment of pain and systemic symptoms that can mimic or occur simultaneously with a musculoskeletal problem.

Guidelines for Physician Referral

As part of the Review of Systems, correlate *history* with *patterns of pain* and any *unusual findings* that may indicate systemic disease. The therapist can use the decision-making tools discussed in Chapter 1 (see Box 1-7) to make a decision regarding treatment versus referral.

Some of the specific indications for physician referral mentioned in this chapter include the following:

- Spontaneous postmenopausal bleeding
- A growing mass, whether painful or painless
- Persistent rise or fall in blood pressure
- Hip, sacroiliac, pelvic, groin, or low back pain in a woman without traumatic etiologic complex who reports fever, sweats, or an association between menses and symptoms
- Marked loss of hip motion and referred pain to the groin in a client on long-term systemic corticosteroids
- A positive family/personal history of breast cancer in a woman with chest, back, or shoulder pain of unknown cause
- Elevated blood pressure in any woman taking birth control pills; this should be closely monitored by her physician

Key Points to Remember

- The process of screening for medical disease before establishing a diagnosis by the physical therapist and plan of care requires a broad range of knowledge.
- Throughout the screening process, a medical diagnosis is not the goal. The therapist is screening to make sure that the client does indeed have a primary NMS problem within the scope of a physical therapist practice.
- The screening steps begin with the client interview, but screening does not end there. Screening questions may be needed throughout the episode of care. This is especially true when progression of disease results in a changing clinical presentation, perhaps with the onset of new symptoms or new red flags after the treatment intervention has been initiated.
- The client history is the first and most basic skill needed for screening. Most of the information needed to determine the cause of symptoms is contained within the subjective assessment (interview process).
- The Family/Personal History form can be used as the first tool to screen clients for medical disease. Any "yes" responses should be followed up with appropriate questions. The therapist is strongly encouraged to review the form with the client, entering the date and his or her own initials. This form can be used as a document of baseline information.
- Screening examinations (interview and vital signs) should be completed for any person experiencing back, shoulder, scapular, hip, groin, or sacroiliac symptoms of unknown cause. The presence of constitutional symptoms will almost always warrant a physician's referral but definitely requires further follow-up questions in making that determination.
- It may be necessary to explain the need to ask such detailed questions about organ systems seemingly unrelated to the musculoskeletal symptoms.
- Not every question provided in the lists offered in this text needs to be asked; the therapist can scan the list and ask the appropriate questions based on the individual circumstances.
- When screening for domestic violence, sexual dysfunction, incontinence, or other conditions, it is important to explain that a standard set of questions is asked and that some may not apply.
- With the older client, a limited number of presenting symptoms often predominate—no matter what the underlying disease is—including acute confusion, depression, falling, incontinence, and syncope.
- A recent history of any infection (bladder, uterine, kidney, vaginal, upper respiratory), mononucleosis, influenza, or colds may be an extension of a chronic health pattern or systemic illness.
- The use of fluoroquinolones (antibiotic) has been linked with tendinopathies, especially in older adults who are also taking corticosteroids.
- Reports of dizziness, loss of balance, or a history of falls require further screening, especially in the presence of other neurologic signs and symptoms such as headache, confusion, depression, irritability, visual changes, weakness, memory loss, and drowsiness or lethargy.
- Special Questions for Women and Special Questions for Men are available to screen for gynecologic or urologic involvement for any woman or man with back, shoulder, hip, groin, or sacroiliac symptoms of unknown origin at presentation.
- Consider the possibility of physical/sexual assault or abuse in anyone with an unknown cause of symptoms, clients who take much longer to heal than expected, or any combination of physical, social, or psychologic cues listed.
- In screening for systemic origin of symptoms, review the subjective information in light of the objective findings. Compare the client's history with clinical presentation and look for any associated signs and symptoms.

CASE STUDY*

REFERRAL

A 28-year-old white man was referred to physical therapy with a medical diagnosis of progressive idiopathic Raynaud's syndrome of the bilateral upper extremities. He had this condition for the last 4 years.

The client was examined by numerous physicians, including an orthopedic specialist. The client had complete numbness and cyanosis of the right second, third, fourth, and fifth digits on contact with even a mild decrease in temperature.

He reported that his symptoms had progressed to the extent that they appear within seconds if he picks up a glass of cold water. This man works almost entirely outside, often in cold weather, and uses saws and other power equipment. The numbness has created a very unsafe job situation.

The client received a gunshot wound in a hunting accident 6 years ago. The bullet entered the posterior left thoracic region, lateral to the lateral border of the scapula, and came out through the anterior lateral superior chest wall. He says that he feels as if his shoulders are constantly rolled forward. He reports no cervical, shoulder, or elbow pain or injury.

PHYSICAL THERAPY INTERVIEW

Note that not all of these questions would necessarily be presented to the client because his answers may determine the next question and may eliminate some questions.

Tell me why you are here today. (Open-ended question)

PAIN

- Do you have any pain associated with your past gunshot wound? If yes, describe your pain.
  FUPs: Give the client a chance to answer and prompt only if necessary with suggested adjectives such as "Is your pain sharp, dull, boring, or burning?" or "Show me on your body where you have pain."
  To pursue this line of questioning, if appropriate:
  FUPs: What makes your pain better or worse?
- What is your pain like when you first get up in the morning, during the day, and in the evening?
- Is your pain constant or does it come and go?
- On a scale from 0 to 10, with zero being no pain and 10 being the worst pain you have ever experienced with this problem, what level of pain would you say that you have right now?
- Do you have any other pain or symptoms that are not related to your old injury?
- If yes, pursue as above to find out about the onset of pain, etc.
- You indicated that you have numbness in your right hand. How long does this last?
  FUPs: Besides picking up a glass of cold water, what else brings it on?
  How long have you had this problem?
- You told me that this numbness has progressed over time. How fast has this happened?
- Do you ever have similar symptoms in your left hand?

ASSOCIATED SYMPTOMS

Even though this client has been seen by numerous physicians, it is important to ask appropriate questions to rule out a systemic origin of current symptoms, especially if there has been a recent change in the symptoms or presentation of symptoms bilaterally. For example:

- What other symptoms have you had that you can associate with this problem?
- In addition to the numbness, have you had any of the following?
  - Tingling
  - Burning
  - Weakness
  - Vomiting
  - Hoarseness
  - Difficulty with breathing
  - Nausea
  - Dizziness
  - Difficulty with swallowing
  - Heart palpitations or fluttering
  - Unexplained sweating or night sweats
  - Problems with your vision
- How well do you sleep at night? (Open-ended question)
- Do you have trouble sleeping at night? (Closed-ended question)
- Does the pain awaken you out of a sound sleep? Can you sleep on either side comfortably?

MEDICATIONS

- Are you taking any medications? If yes, and the person does not volunteer the information, probe further:
  What medications?
  Why are you taking this medication?
  When did you last take the medication?
  Do you think the medication is easing the symptoms or helping in any way?
  Have you noticed any side effects? If yes, what are these effects?

PREVIOUS MEDICAL TREATMENT

- Have you had any recent medical tests, such as x-ray examination, MRI, or CT scan? If yes, find out the results.
- Tell me about your gunshot wound. Were you treated immediately?
- Did you have any surgery at that time or since then? If yes, pursue details with regard to what type of surgery and where and when it occurred.

CASE STUDY—*cont'd*

- Did you have physical therapy at any time after your accident? If yes, relate when, for how long, with whom, what was done, did it help?
- Have you had any other kind of treatment for this injury (e.g., acupuncture, chiropractic, osteopathic, naturopathic, and so on)?

ACTIVITIES OF DAILY LIVING (ADLs)

- Are you right-handed?
- How do your symptoms affect your ability to do your job or work around the house?
- How do your symptoms affect caring for yourself (e.g., showering, shaving, other ADLs such as eating or writing)?

FINAL QUESTION

- Is there anything else you feel that I should know concerning your injury, your health, or your present situation that I have not asked about?

Note: If this client had been a woman, the interview would have included questions about breast pain and the date when she was last screened for cancer (cervical and breast) by a physician.

*Adapted from Bailey W, Northwestern Physical Therapy Services, Inc., Titusville, Pennsylvania.

PRACTICE QUESTIONS

1. What is the effect of NSAIDs (e.g., Naprosyn, Motrin, Anaprox, ibuprofen) on blood pressure?
   a. No effect
   b. Increases blood pressure
   c. Decreases blood pressure
2. Most of the information needed to determine the cause of symptoms is contained in the:
   a. Subjective examination
   b. Family/Personal History Form
   c. Objective information
   d. All of the above
   e. a and c
3. With what final question should you always end your interview?
4. A risk factor for NSAID-related gastropathy is the use of:
   a. Antibiotics
   b. Antidepressants
   c. Antihypertensives
   d. Antihistamines
5. After interviewing a new client, you summarize what she has told you by saying, "You told me you are here because of right neck and shoulder pain that began 5 years ago as a result of a car accident. You also have a 'pins and needles' sensation in your third and fourth fingers but no other symptoms at this time. You have noticed a considerable decrease in your grip strength, and you would like to be able to pick up a pot of coffee without fear of spilling it."
   This is an example of:
   a. An open-ended question
   b. A funnel technique
   c. A paraphrasing technique
   d. None of the above
6. Screening for alcohol use would be appropriate when the client reports a history of accidents.
   a. True
   b. False
7. What is the significance of sweats?
   a. A sign of systemic disease
   b. Side effect of chemotherapy or other medications
   c. Poor ventilation while sleeping
   d. All of the above
   e. None of the above
8. Spontaneous uterine bleeding after 12 consecutive months without menstrual bleeding requires medical referral.
   a. True
   b. False
9. Which of the following are red flags to consider when screening for systemic or viscerogenic causes of neuromuscular and musculoskeletal signs and symptoms:
   a. Fever, (night) sweats, dizziness
   b. Symptoms are out of proportion to the injury
   c. Insidious onset
   d. No position is comfortable
   e. All of the above
10. A 52-year-old man with low back pain and sciatica on the left side has been referred to you by his family physician. He has had a discectomy and laminectomy on two separate occasions about 5 to 7 years ago. No imaging studies have been performed (e.g., x-ray examination or MRI) since that time. What follow-up questions should you ask to screen for medical disease?

Continued

PRACTICE QUESTIONS—*cont'd*

11. You should assess clients who are receiving NSAIDs for which physiologic effect associated with increased risk of hypertension?
    a. Decreased heart rate
    b. Increased diuresis
    c. Slowed peristalsis
    d. Water retention
12. Instruct clients with a history of hypertension and arthritis to:
    a. Limit physical activity and exercise
    b. Avoid over-the-counter medications
    c. Inform their primary care provider of both conditions
    d. Drink plenty of fluids to avoid edema
13. Alcohol screening tools should be:
    a. Used with every client sometime during the episode of care
    b. Brief, easy to administer, and nonthreatening
    c. Deferred when the client has been drinking or has the smell of alcohol on the breath
    d. Conducted with one other family member present as a witness

REFERENCES

1. Barsky AJ: Forgetting, fabricating, and telescoping: the instability of the medical history. *Arch Intern Med* 162(9):981–984, 2002.
2. A Normative Model of Physical Therapist Professional Education: Version 2004, Alexandria, VA, 2004, American Physical Therapy Association.
3. Davis CM: *Patient practitioner interaction: an experiential manual for developing the art of healthcare*, ed 5, Thorofare, NJ, 2011, Slack.
4. Leavitt RL: Developing cultural competence in a multicultural world, Part II. *PT Magazine* 11(1):56–70, 2003.
5. National Center for Education Statistics (NCES): Facts and Figures. Available online at http://nces.ed.gov. Accessed August 27, 2010.
6. The Joint Commission: What did the doctor say? Improving health literacy to protect patient safety. 2007. Available online at www.jointcommission.org. Accessed Dec. 7, 2010.
7. Marcus EN: The silent epidemic- the health effects of illiteracy. *N Engl J Med* 355(4):339–341, 2006.
8. William H, Mahood MD, president of the American Medical Association Foundation, the philanthropic arm of the AMA: www.amaassn.org/scipubs/amnews/pick_00/hlse0320.htm
9. National Adult Literacy Agency (NALA): Resource Room. Available at: http://www.nala.ie/. Accessed July 19, 2010.
10. Nordrum J, Bonk E: Serving patients with limited English proficiency. *PT Magazine* 17(1):58–60, 2009.
11. National Center for Education Statistics. U.S. Department of Education Institute of Education Sciences: Available at: http://nces.ed.gov. Accessed on June 26, 2010.
12. Weiss BD: Assessing health literacy in clinical practice. Available online at www.medscape.com/viewprogram/8203, 2007. Accessed March 4, 2010.
13. Nielsen-Bohlman L, Panzer AM, Kindig DA, editors: *Health literacy: A prescription to end confusion*, Washington, D.C., 2004, National Academies Press.
14. Magasi S: Rehabilitation consumers' use and understanding of quality information: a health literacy perspective. *Arch Phys Med Rehabil* 90(2):206–212, 2009.
15. Lattanzi JB, Purnell LD, editors: *Developing cultural competence in physical therapy practice*, Philadelphia, 2005, FA Davis.
16. Spector RE: *Cultural diversity in health and illness*, ed 7, Upper Saddle River, NJ, 2008, Prentice-Hall.
17. Berens LV, et al: *Quick guide to the 16 personality types in organizations: understanding personality differences in the workplace*, Huntington Beach, CA, 2002, Telos Publications.
18. Nardi D: *Multiple intelligences and personality type*, Huntington Beach, CA, 2001, Telos Publications.
19. Cole SA, Bird J: *The medical interview: The three-function approach*, ed 2, St. Louis, 2000, Mosby-Year Book.
20. Coulehan JL, Block MR: *The medical interview: Mastering skills for clinical practice*, ed 5, Philadelphia, 2006, FA Davis.
21. National Council on Interpreting in Health Care: National Standards of Practice for Interpreters in Health Care. September 2005. Available online at www.ncihc.org/mc/page.do?sitePageId=57768&orgId=ncihc. Accessed Aug. 31, 2010.
22. Assuring cultural competence in health care: Recommendations for national standards and an outcomes-focused research agenda, 1999, Office of Minority Health, Public Health Service, U.S. Department of Health and Human Services.
23. APTA: Advocacy: Minority and international affairs. Available at: http://www.apta.org/. Click on left sided Advocacy menu. Type in research box: Minority Affairs or Cultural Competence. Accessed Oct 29, 2010.
24. U.S. Census Bureau: United States Census 2010, American Fact Finder. U.S. Census Projections. Available: www.census.gov/. Accessed Oct. 1, 2010.
25. Bonder B, Martin L, Miracle A: *Culture in clinical care*, Clifton Park, 2001, Delmar Learning.
26. Leavitt RL: Developing cultural competence in a multicultural world, Part I. *PT Magazine* 10(12):36–48, 2002.
27. American Physical Therapy Association. Health and Policy Administration Section. Cross-Cultural & International Special Interest Group (CCISIG): Available at: http://www.aptasoa.org/. Accessed Oct. 21, 2010.
28. APTA: Tips to increase cultural competence. Available at http://www.apta.org; Accessed August 31, 2010. [Membership and Leadership section].
29. APTA: *Committee on Cultural Competence. Blueprint for Teaching Cultural Competence in Physical Therapy Education*, Alexandria, VA, 2008, American Physical Therapy Association.
30. Office of Minority Health (OMH): Assuring cultural competence in health care: recommendations for national standards and an outcomes-focused research agenda. Available at: www.omhrc.gov/clas. Accessed Oct 30, 2010.
31. Diversity Rx: Multicultural best practices overview. Available at: www.diversityrx.org. Accessed Oct 26, 2010.
32. The American Academy of Orthopaedic Surgeons (AAOS): Test your cultural competency. Available online at www.orthoinfo.org/diversity. Accessed Oct. 1, 2010.
33. The Council on American-Islamic Relations (CAIR): Available at: www.cair-net.org. Accessed Oct 1, 2010.

34. The Muslim American Society (MAS): Available at: www.masnet.org. Accessed Oct 1, 2010.
35. Siddiqui H: Healthcare barriers for Muslim Americans. *Hemaware* 9(1):18–20, 2004.
36. American Physical Therapy Association: Tips on how to increase cultural competency. Available at www.apta.org. Type in Search box: Cultural Competency. Accessed Oct 30, 2010.
37. Gay & Lesbian Medical Association: Available at www.glma.org. Type in search window: professional competency >see Bibliography of Professional Competency in LGBT health>click on DOC (document). Accessed Oct. 5, 2010.
38. Platt FW: The patient-centered interview. *Ann Intern Med* 134:1079–1085, 2001.
39. Baker LH: What else? Setting the agenda for the clinical interview. *Ann Intern Med* 143:766–770, 2005.
40. Churchill LR: Improving patient care. *Arch Int Med* 149:720–724, 2008.
41. Fairbank JC, Couper J, Davies JB, et al: The Oswestry Low Back Pain Disability Questionnaire. *Physiotherapy* 66:271–273, 1980.
42. Kopec JA, Esdaile JM, Abrahamowicz M, et al: The Quebec Back Pain Disability Scale. Measurement properties. *Spine* 20:341–352, 1995.
43. Ventre J, Schenk RJ: Validity of the Duffy-Rath Questionnaire. *Orthopaedic Practice* 17(1):22–28, 2005.
44. Lippitt SB, Harryman DT, Matsen FA: A practical tool for evaluating function: The Simple Shoulder Test. In Matsen FA, Hawkins RJ, Fu FH, editors: *The shoulder: A balance of mobility and stability*, Rosemont, 1993, American Academy of Orthopaedic Surgeons.
45. Hudak PL, Amadio PC, Bombardier C: Development of an upper extremity outcome measure: The DASH (disabilities of the arm, shoulder, and hand), The Upper Extremity Collaborative Group (UECG). *Am J Ind Med* 29:602–608, 1996. Erratum 30:372, 1996.
46. Ware JE, Sherbourne CD: The MOS 36-Item Short Form Health Survey (SF-36): I. Conceptual framework and item selection. *Med Care* 30:473–489, 1992.
47. Ware JE: SF-36 Health Survey update. *Spine* 25(24):3130–3139, 2000.
48. Wright BD, Linacre JM, editors: *Rasch measurement transactions. Part 2. Reasonable Mean-Square Fit Values*, Chicago, 1996, MESA Press.
49. Guide to physical therapist practice, ed 2 (Revised), Alexandria, 2003, American Physical Therapy Association.
50. Performance Physio Ltd: Pre-assessment therapy questionnaires. Available at: http://www.mystudiosoft.com/. Accessed June 23, 2005. Mention of these products does not constitute commercial endorsement. No financial benefit was gained by providing this reference.
51. Kettenbach G: *Writing patient/client notes: ensuring accuracy in documentation*, ed 4, Philadelphia, 2009, FA Davis.
52. Boissonnault WG: Differential diagnosis: Taking a step back before stepping forward. *PT Magazine* 8(11):45–53, 2000.
53. Wolf GA, Jr: *Collecting data from patients*, Baltimore, 1977, University Park Press.
54. Federal Interagency Forum on Aging Related Statistics: Older Americans 2010: Key indicators of well-being. Available online at www.agingstats.gov. Accessed Oct. 1, 2010.
55. Potter JF: The older orthopaedic patient. General considerations. *Clin Orthop Rel Res* 425:44–49, 2004.
56. Budnitz DS: Medication use leading to emergency department visits for adverse drug events in older adults. *Ann Intern Med* 148(8):628–629, 2007.
57. Richy F, Gourlay ML, Garrett J, et al: Osteoporosis prevalence in men varies by the normative reference. *J Clin Densitom* 7(2):127–133, 2004.
58. Ebeling PR: Clinical practice: osteoporosis in men. *NEJM* 358(14):1474–1482, 2008.
59. Elliott ME, Drinka PJ, Krause P, et al: Osteoporosis assessment strategies for male nursing home residents. *Maturitas* 48(3): 225–233, 2004.
60. Ringe JD, Faber H, Farahmand P, et al: Efficacy of risedronate in men with primary and secondary osteoporosis: Results of a 1-year study. *Rheumatol Int* 26(5):427–431, 2006.
61. Kiebzak GM, Beinart GA, Perser K, et al: Undertreatment of osteoporosis in men with hip fracture. *Arch Intern Med* 162:2217–2222, 2002.
62. Feldstein AC, Nichols G, Orwoll E, et al: The near absence of osteoporosis treatment in older men with fractures. *Osteoporos Int* June 1, 2005.
63. Health Resources and Services Administration (HRSA): *Women's health USA data book [Annual report on women's health in the U.S.]*, ed 9, Rockville, 2010, United States Department of Health and Human Services. Available at www.hrsa.gov/ Accessed Oct. 28, 2010.
64. Prendergast MA: Do women possess a unique susceptibility to the neurotoxic effects of alcohol? *J Am Med Womens Assoc* 59(3):225–227, 2004.
65. American Cancer Society (ACS) Cancer statistics: 2010. Available online at www.cancer.org. Click on "Cancer Facts and Figures." Accessed Oct. 01, 2010.
66. Baird DT, Collins J, Egozcue J, et al: Fertility and ageing. *Hum Reprod Update*11(3):261–276, 2005.
67. Moore M: *The only menopause guide you'll need*, ed 2, Baltimore, Johns Hopkins Press Health Book, 2004, Johns Hopkins University Press.
68. Northrup C: *The Wisdom of menopause*, ed 2, New York, 2006, Bantam Books (Random House).
69. National Institute on Aging: Menopause. Available online at www.nia.nih.gov. Accessed Oct. 1, 2010.
70. Rossouw JE, Anderson GL, Prentice RL, et al: Risks and benefits of estrogen plus progestin in healthy postmenopausal women: Principal results from the Women's Health Initiative randomized controlled trial. *JAMA* 288(3):321–333, 2002.
71. Grady D: Postmenopausal hormones—therapy for symptoms only. *N Engl J Med* 348(19):1835–1837, 2003.
72. U.S. Department of Health and Human Services, Health Resources and Services Administration: New Statistical Guide to Women's Health. Available at: www.hrsa.gov/. Accessed Oct 23, 2010.
73. Mosca L, Appel LJ, Benjamin EJ, et al: Evidence-based guidelines for cardiovascular disease prevention in women, 2007 update. American Heart Association Guidelines. *Circulation* 115:1481–1501, 2007.
74. LaViest T: *Minority populations and health*, San Francisco, 2005, Jossey-Boss.
75. National Human Genome Research Institute: Educational Resources. Available at: www.nhgri.nih.gov. Accessed Sept. 09, 2010.
76. Nature, International Weekly Journal of Science: The Human Genome. Available at: www.nature.com/genomics. Accessed Sept. 09, 2010.
77. Fowler K: PTs confront minority health and health disparities. *PT Magazine* 12(5):42–47, 2004.
78. Probst JC, Moore CG, Glover SH, et al: Person and place: The compounding effects of race/ethnicity and rurality on health. *Am J Public Health* 94(10):1695–1703, 2004.
79. Urban Indian Health Institute: Health status of urban American Indians. 2004. Available at: http://www.uihi.org/. Accessed Sept. 09, 2010.
80. Kawaga-Singer M: Cancer, culture, and health disparities. *Ca J Clin* 60(1):12–39, 2010.
81. Bach PB, Schrag D, Brawley OW: Survival of blacks and whites after a cancer diagnosis. *JAMA* 287:2106–2113, 2002.

82. Smedley B, Stith A, Nelson A, editors: *Unequal treatment—confronting racial and ethnic disparities in health care*, Washington, DC, 2002, National Academy Press.
83. Morgenstern LB, Smith MA, Lisabeth LD, et al: Excess stroke in Mexican Americans compared with non-Hispanic whites. *Am J Epidemiol* 160(4):376–383, 2004.
84. Lisabeth LD, Kardia SL, Smith MA, et al: Family history of stroke among Mexican-American and non-Hispanic white patients with stroke and TIA: Implications for the feasibility and design of stroke genetics research. *Neuroepidemiology* 24(1-2):96–102, 2005.
85. Centers for Disease Control and Prevention (CDC): Race and Ethnicity Code Set. Available at http://www.cdc.gov/nedss/DataModels/. Accessed June 1, 2010.
86. U.S. National Library of Medicine and the National Institutes of Health. African-American Health: Available at: www.nlm.nih.gov/medlineplus/africanamericanhealth.html May 2005. Accessed Sept 09, 2010.
87. Intercultural Cancer Council (ICC): Cancer fact sheets. Available at: http://iccnetwork.org/cancerfacts/. Accessed Sept 1, 2010.
88. Beals KA: Disordered eating and body-image disturbances in male athletes, *Health & Fitness*, ACSM, March/April 2003.
89. Beals KA: *Disordered eating among athletes. A comprehensive guide for health professionals*, Champaign, IL, 2004, Human Kinetics.
90. National Athletic Trainers Association (NATA): NATA Position Statement on Disordered Eating Among Athletes. 2008. Available online at www.nata.org. Accessed Sept. 09, 2010.
91. Kaminski PL, Chapman BP, Haynes SD, et al: Body image, eating behaviors, and attitudes toward exercise among gay and straight men. *Eat Behav* 6(3):179–187, 2005.
92. Long MJ, Marshall BS: The relationship between self-assessed health status, mortality, service use, and cost in a managed care setting. *Health Care Manage Rev* 4:20–27, 1999.
93. Gold DT, Burchett BM, Shipp KM, et al: Factors associated with self-rated health in patients with Paget's disease of bone. *J Bone Miner Res* 14 (Suppl 2):99–102, 1999.
94. Idler EL, Russell LB, Davis D: Survival, functional limitations, and self-rated health in the NHANES I Epidemiologic Follow-up Study 1992: First national health and nutrition examination survey. *Am J Epidemiol* 9:874–883, 2000.
95. Long MJ, McQueen DA, Bangalore VG, et al: Using self-assessed health to predict patient outcomes after total knee replacement. *Clin Ortho Rel Res* 434:189–192, 2005.
96. Osterberg L, Blaschke T: Drug therapy: adherence to medication. *NEJM* 353(5):487–497, 2005.
97. Weissman DE, Haddox JD: Opioid pseudoaddiction: an iatrogenic syndrome. *Pain* 36:363–366, 1989.
98. Ross EL, Holcomb C, Jamison RN: Chronic pain update: addressing abuse and misuse of opioid analgesics. *J Musculoskel Med* 25(6):268–277, 302, 2008.
99. Storr CL, Trinkoff AM, Anthony JC: Job strain and non-medical drug use. *Drug Alcohol Depend* 55(1-2): 45–51, 1999.
100. National Institute on Drug Abuse (NIDA): NIDA InfoFacts: Nationwide trends. Available at: http://www.nida.nih.gov/infofacts/nationtrends.html. Accessed Sept 09, 2010.
101. Nutt D: Drug harms in the UK: a multicriteria decision analysis. *Lancet* 6736(10):61462–61466, 2010.
102. Lapeyre-Mestre M, Sulem P, Niezborala M, et al: Taking drugs in the working environment: A study in a sample of 2106 workers in the Toulouse metropolitan area. *Therapie* 59(6): 615–623, 2004.
103. Kolakowsky-Hayner SA: Pre-injury substance abuse among persons with brain injury and persons with spinal cord injury. *Brain Inj* 13(8):571–581, 1999.
104. Soft-tissue infections among injection drug users. *MMWR* 50(19):381–384, 2001.
105. Colleran C, Jay D: *Aging and addiction: helping older adults overcome alcohol or medication dependence*, Center City, MN, 2002, Hazelden Guidebooks.
106. Skinner HA: Identification of alcohol abuse using laboratory tests and a history of trauma. *Ann Intern Med* 101(6):847–851, 1984.
107. Bastiaens L, Francis G, Lewis K: The RAFFT as a screening tool for adolescent substance use disorders. *Am J Addict* 9:10–16, 2000.
108. Goodman CC, Fuller K: *Pathology: implications for the physical therapist*, ed 3, Philadelphia, 2009, WB Saunders.
109. Center for Advanced Health Studies: *Substance abuse. A guide for health professionals*, ed 2, Glendive, 2001, American Academy of Pediatrics.
110. Rassool GH: *Alcohol and drug misuse: a handbook for students & health professionals*, United Kingdom, 2009, T & F Books.
111. University of Washington Alcohol and Drug Abuse Institute (ADAI): Substance Use and Screening Assessment Instruments database. Available at: http://lib.adai.washington.edu/instruments/. Accessed Sept. 09, 2010.
112. National Institute on Alcohol Abuse and Alcoholism (NIAAA), National Institutes of Health: Frequently Asked Questions about Alcoholism (2010). Available online at: www.niaaa.nih.gov. Accessed June 23, 2010.
113. World Health Organization: *International Classification of Diseases (ICD), Tenth Revision, Second edition, Vol. 2*, Geneva, 1994, World Health Organization. Available online at: www.who.int/classifications/icd/en. Accessed June 23, 2010.
114. Cook RL, Chung T, Kelly TM, et al: Alcohol screening in young persons attending a sexually transmitted disease clinic. *J Gen Intern Med* 20(1):1–6, 2005.
115. Kanny D: Vital signs: binge drinking among high school students and adults in the United States. *MMWR* 59(Early release):1–6, October 5, 2010.
116. Babor TF, de la Fuente JR, Saunders J, et al: *AUDIT (The Alcohol Use Disorders Identification Test): Guidelines for use in primary health care*, 1992, World Health Organization. Available: http://whqlibdoc.who.int/hq/1992/WHO_PSA_92.4.pdf. Accessed Sept. 09, 2010.
117. World Cancer Research Fund (WCRF): Alcohol and cancer prevention. Available online at http://www.wcrf-uk.org/preventing_cancer/recommendations/alcohol_and_cancer.php. Accessed December 2, 2010.
118. The American Geriatrics Society. Clinical Guidelines for Alcohol Use Disorders in Older Adults. http://www.americangeriatrics.org/products/positionpapers/alcoholPF.shtml. Updated November 2003. Accessed 9/9/10.
119. Shapira D: Alcohol abuse and osteoporosis. *Semin Arthritis Rheum* 19(6): 371–376, 1990.
120. Simons DG, Travell JG, Simons LS: *Myofascial pain and dysfunction. The trigger point manual. Volume 1. Upper half of body*, Baltimore, 1999, Williams & Wilkins.
121. Mukamal KJ, Ascherio A, Mittleman MA, et al: Alcohol and risk for ischemic stroke in men: The role of drinking patterns and usual beverage. *Ann Intern Med* 142(1):11–19, 2005.
122. Pittman HJ: Recognizing "holiday heart" syndrome. *Nursing 2004* 34(12):32cc6–32cc7, 2004.
123. Cranston AR: Physical therapy management of a patient experiencing alcohol withdrawal. *J Acute Care Phys Ther* 1(2):56–63, Winter 2010.
124. Vincent WR: Review of alcohol withdrawal in the hospitalized patient: diagnosis and assessment. *Orthopedics* 30(5):358–361, 2007.
125. Victor M: The effect of alcohol on the nervous system. Research Publications Association for Research in Nervous and Mental Diseases, 1953, p 526–573.
126. Henderson-Martin B: No more surprises: Screening patients for alcohol abuse. *Nursing 2000* 100(9):26–32, 2000.

127. Sullivan JT, Sykora K, Schneiderman J, et al: Assessment of alcohol withdrawal: The revised Clinical Institute Withdrawal Assessment for Alcohol scale (CIWA-Ar). *Br J Addiction* 84:1353–1357, 1989.
128. National Institute on Alcohol Abuse and Alcoholism (NIAAA): Alcohol Alert. Publications. Available at http://www.niaaa.nih.gov/. Accessed Sept. 09, 2010.
129. Selzer ML: A self-administered Short Michigan Alcoholism Screening Test (SMAST). *J Stud Alcohol* 36(1):117–126, 1975.
130. Ashman TA, Schwartz ME, Cantor JB, et al: Screening for substance abuse in individuals with traumatic brain injury. *Brain Inj* 18(2):191–202, 2004.
131. O'Brien CP: The CAGE Questionnaire for detection of alcoholism: a remarkably useful but simple tool. *JAMA* 300(17): 2054–2056, 2008.
132. Bush K: The AUDIT alcohol consumption questions (AUDIT-C); an effective brief screening test for problem drinking. Ambulatory Care Quality Improvement Project (ACQUIP). Alcohol Use Disorders Identification Test. *Arch Intern Med* 158:1789–1795, 1998.
133. Reinert DF, Allen JP: The Alcohol Use Disorders Identification Test (AUDIT): a review of recent research. *Alcohol Clin Exp Res* 26:272–279, 2002.
134. Carlat DJ: The psychiatric review of symptoms: A screening tool for family physicians. *Am Fam Physician* 58(7):1617–1624, 1998.
135. Vinson DC: Comfortably engaging: which approach to alcohol screening should we use? *Ann Fam Med* 2(5):398–404, 2004.
136. Sturm R, Stein B, Zhang W, et al: Alcoholism treatment in managed private sector plans. How are carve-out arrangements affecting costs and utilization? *Recent Dev Alcohol* 15:271–284, 2001.
137. Dunn C: Hazardous drinking by trauma patients during the year after injury. *J Trauma* 54(4):707–712, 2003.
138. Gentilello LM: Alcohol interventions in a trauma center as a means of reducing the risk of injury recurrence. *Ann Surg* 230(4):473–483, 1999.
139. American Physical Therapy Association (APTA): Substance Abuse HOD 06-93-25-49 (Program 32, Practice Department), June 2003.
140. Majid PA, Cheirif JB, Rokey R, et al: Does cocaine cause coronary vasospasm in chronic cocaine abusers? A study of coronary and systemic hemodynamics. *Clin Cardiol* 15(4): 253–258, 1992.
141. Dube SR: Current Cigarette Smoking among adults aged ≥18 years—United States, 2009. *MMWR* 59(35):1135–1140, 2010. Available on line at http://www.cdc.gov/mmwr/preview/mmwrhtml/mm5935a3.htm?s_cid=mm5935a3_w. Accessed December 20, 2010.
142. Centers for Disease Control and Prevention (CDC): CDC's Tobacco Laboratory. Available online at http://www.cdc.gov/biomonitoring/pdf/tobacco_brochure.pdf. Accessed December 20, 2010.
143. Frownfelter D, Ohtake PJ: Quitters are winners—the role of physical therapists in smoking cessation. Combined Sections Meeting. San Diego, California, February 20, 2010.
144. American Cancer Society (ACS): Health benefits over time. Available at: www.cancer.org [In the search box, type in: When Smokers Quit]. Accessed Sept 09, 2010.
145. Rodriguez J: The association of pipe and cigar use with cotinine levels, lung function, and airflow obstruction. The Multi-Ethnic Study of Atherosclerosis (MESA). *Ann Intern Med* 152(4);201–210, 2010.
146. Travell JG, Simons DG: *Myofascial pain and dysfunction: The trigger point manual: The lower extremities, vol 2*, Baltimore, 1992, Williams & Wilkins.
147. Kim KS, Yoon ST, Park JS, et al: Inhibition of proteoglycan and type II collagen synthesis of disc nucleus cells by nicotine. *J Neurosurg: Spine* 99(3 Suppl):291–297, 2003.
148. Akmal M, Kesani A, Anand B, et al: Effect of nicotine on spinal disc cells: A cellular mechanism for disc degeneration. *Spine* 29(5):568–575, 2004.
149. Frymoyer JW, Pope MH, Clements JH, et al: Risk factors in low back pain. *J Bone Joint Surg* 65-A: 213–218, 1983.
150. Holm S, Nachemson A: Nutrition of the intervertebral disc: acute effects of cigarette smoking. An experimental animal study. *Upsala J Med Sci* 83: 91–98, 1998.
151. Hughes JR, Oliveto AH, Helzer JE, et al: Should caffeine abuse, dependence, or withdrawal be added to DSM-IV and ICD-10? *Am J Psychiatry* 149(1): 33–40, 1992.
152. Hughes JR, Oliveto AH, Liguori A, et al: Endorsement of DSM-IV dependence criteria among caffeine users. *Drug Alcohol Depend* 52(2): 99–107, 1998.
153. Kerrigan S, Lindsey T: Fatal caffeine overdose. *Forensic Sci Int* 153(1):67–69, 2005.
154. Sudano I, Binggeli C, Spieker L: Cardiovascular effects of coffee: Is it a risk factor? *Prog Cardiovasc Nurs* 20(2):65–69, 2005.
155. Corti R, et al: Coffee acutely increases sympathetic nerve activity and blood pressure independently of caffeine content: Role of habitual versus nonhabitual drinking. *Circulation* 106(23):2935–2940, 2002.
156. Mikuls TR, Cerhan JR, Criswell LA, et al: Coffee, tea, and caffeine consumption and risk of rheumatoid arthritis: Results from the Iowa Women's Health Study. *Arthritis Rheum* 46(1):83–91, 2002.
157. Position of the American Dietetic Association (ADA): Use of nutritive and nonnutritive sweeteners. *J Amer Dietetic Assoc* 104(2):255–275, 2004. Available at: http://www.eatright.org/. Accessed Sept. 09, 2010.
158. Blaylock R: *Excitotoxins: the taste that kills*, Albuquerque, 1996, Health Press.
159. Roberts HJ: *Aspartame disease: the ignored epidemic*, West Palm Beach, 2001, Sunshine Sentinel Press.
160. Roberts HJ: *Defense against Alzheimer's disease*, Palm Beach, 2001, Sunshine Sentinel Press.
161. Newman LS: Occupational illness. *N Engl J Med* 333:1128–1134, 1995.
162. Radetsky P: *Allergic to the twentieth century: the explosion in environmental allergies*, Boston, 1997, Little, Brown.
163. Frumkin H: Agent orange and cancer: An overview for clinicians. *CA Cancer J Clin* 53(4):245–255, 2003.
164. Veterans Health Administration (VHA): Gulf War Illnesses. Available at: www.va.gov/gulfwar/. Accessed July Sept. 09, 2010.
165. Marshall L, Weir E, Abelsohn A, et al: Identifying and managing adverse environmental health effects: Taking an exposure history. *Can Med Assoc J* 166(8):1049–1054, 2002.
166. Cross J, Trent R: Public health and aging: Nonfatal fall-related traumatic brain injury among older adults. *MMWR* 52(13): 276–278, 2003.
167. Boulgarides LK, McGinty SM, Willett JA, et al: Use of clinical and impairment-based tests to predict falls by community-dwelling older adults. *Phys Ther* 83(4):328–339, 2003.
168. Stone KL: Actigraphy-measured sleep characteristics and risk of falls in older women. *Arch Intern Med* 168(16):1768–1775, 2008.
169. Woolcott JC: Meta-analysis of the impact of 9 medication classes on falls in elderly persons. *Arch Intern Med* 169:1952–1960, 2009.
170. Kario K, Tobin JN, Wolfson LI, et al: Lower standing systolic blood pressure as a predictor of falls in the elderly: A community-based prospective study. *J Am Coll Cardiol* 38(1):246–252, 2001.

171. Lawlor DA, Patel R, Ebrahim S: Association between falls in elderly women and chronic diseases and drug use: Cross sectional study. *BMJ* 327(7417):712–717, 2003.
172. Leveille S: Chronic musculoskeletal pain and the occurrence of falls in an older population. *JAMA* 302(20):2214–2221, 2009.
173. Weiner D, Duncan P, Chandler J, et al: Functional reach: A marker of physical frailty. *J Am Geriatr Soc* 40(3):203–207, 1992.
174. Newton R: Validity of the multi-directional reach test: A practical measure for limits of stability in older adults. *J Gerontol Biol Sci Med* 56(4):M248-M252, 2001.
175. Vellas BJ, Wayne S, Romero L, et al: One-leg balance is an important predictor of injurious falls in older persons. *J Am Geriatr Soc* 45:735–738, 1997.
176. Berg K, Wood-Dauphinee S, Williams JI, Gayton D: Measuring balance in the elderly: Preliminary development of an instrument. *Physiotherapy Canada* 41:304–311, 1989.
177. Berg K, Wood-Dauphinee S, Williams JI, Maki, B: Measuring balance in the elderly: Validation of an instrument. *Can J Pub Health* 83(Suppl 2):S7–11, 1992.
178. Mathias S, Nayak U, Isaacs B: Balance in elderly patients: The "Get-Up and Go Test," *Archives of Physical & Medical Rehabilitation* 67:387–389, 1986.
179. Podsiadlo D, Richardson S: The timed "Up & Go": A test of basic functional mobility for frail elderly persons. *J Am Geriatr Soc* 39:142–148, 1991.
180. Thompson M, Medley A: Performance of community dwelling elderly on the timed up and go test. *Physical and Occupational Therapy in Geriatrics* 13(3):17–30, 1995.
181. Tinetti ME, Mendes de Leon CF, Doucette JT, et al: Fear of falling and fall-related efficacy in relationship to functioning among community-living elders. *J Gerontol* 49:M140–M147, 1984.
182. Tinetti ME, Richman D, Powell LE: Falls efficacy as a measure of fear of falling. *J Gerontol* 45:P239–P243, 1990.
183. Powell LE, Myers AM: The Activities-Specific Balance Confidence (ABC) Scale. *J Gerontol A Biol Sci Med Sci* 50:M28-M34, 1995.
184. Myers AM, Fletcher PC, Myers AH: Discriminative and evaluative properties of the Activities-specific Balance Confidence Scale. *J Gerontol* 53:M287–M294, 1998.
185. Buatois S: A simple clinical scale to stratify risk of recurrent falls in community-dwelling adults aged 65 years and older. *Physical Therapy* 90(4):550–560, 2010.
186. Hotchkiss A, Fisher A, Robertson R, et al: Convergent and predictive validity of three scales to falls in the elderly. *Am J Occup Ther* 58(1):100–103, Jan-Feb 2004.
187. American Physical Therapy Association (APTA): Guidelines for recognizing and providing care for victims of domestic abuse. 2005. Available at: www.apta.org [1-800-999-2782, ext. 3395. Accessed Sept 6, 2011.
188. Cyriax JH: *Textbook of orthopedic medicine*, Philadelphia, 1998, W.B. Saunders.
189. American Physical Therapy Association (APTA): *New position on family violence outlines physical therapy role*, Alexandria, 2005, APTA.
190. Ketter P: Physical therapists need to know how to deal with domestic violence issues. *PT Bulletin* 12(31): 6–7, 1997.
191. McCloskey LA: Abused women disclose partner interference with health care: an unrecognized form of battering. *J Gen Intern Med* 22:1067–1072, 2007.
192. Janssen PA, Nicholls TL, Kumar RA, et al: Of mice and men: Will the intersection of social science and genetics create new approaches for intimate partner violence? *J Interpers Violence* 20(1):61–71, 2005.
193. Goldberg WG, Tomlanovich MC: Domestic violence victims in the emergency department. *JAMA* 251:3259–3264, 1984.
194. George MJ: A victimization survey of female perpetrated assaults in the United Kingdom. *Aggressive Behavior* 25:67–79, 1999.
195. Owen SS, Burke TW: An exploration of prevalence of domestic violence in same sex relationships. *Psychol Rep* 95(1):129–132, 2004.
196. Houry D: Does screening in the emergency department hurt or help victims of intimate partner violence? *Annals of Emergency Med* 51(4):433–442, 2008.
197. Brown JB, Lent B, Brett PJ, et al: Development of the Woman Abuse Screening Tool for use in family practice. *Fam Med* 28:422–428, 1996.
198. MacMillan HL: Approaches to screening for intimate partner violence in health care settings: a randomized trial. *JAMA* 296:530–536, 2006.
199. Feldhaus K: Accuracy of 3 brief screening questions for detecting partner violence in the emergency department. *JAMA* 277(17):1357–1361, 1997.
200. Pandya NK: Child abuse and othopaedic injury patterns. Paper #587. Presented at the American Academy of Orthopaedic Surgeons 76th Annual Meeting. Feb. 25–28, 2009. Las Vegas, Nevada.
201. Dong M, Giles WH, Felitti VJ, et al: Insights into causal pathways for ischemic heart disease: Adverse childhood experiences study. *Circulation* 110(13):1761–1766, 2004.
202. Friedman MJ, Wang, S, Jalowiec JE, et al: Thyroid hormone alterations among women with posttraumatic stress disorder due to childhood sexual abuse. *Biol Psychiatry* 57(10):1186–1192, 2005.
203. Paras ML: Sexual abuse and lifetime diagnosis of somatic disorders: a systematic review and meta-analysis. *JAMA* 302:550–561, 2009.
204. Bhandari M: Violence Against Women Health Research Collaborative: musculoskeletal manifestations of physical abuse after intimate partner violence. *J Trauma* 61:1473–1479, 2006.
205. Keely BR: Could your patient—or colleague—become violent? *Nursing 2002* 32(12):32cc1–32cc5, 2002.
206. Doody L: Defusing workplace violence. *Nursing 2003* 33(8): 32hn1–32hn3, 2003.
207. Kimmel D: Association of physical abuse and chronic pain explored. *ADVANCE for Physical Therapists,* February 17, 1997.
208. Myers JE, Berliner L, Briere J, et al: *The APSAC handbook on child maltreatment*, ed 3, Thousand Oaks, 2010, Sage Publications.
209. American Physical Therapy Association (APTA): Guidelines for recognizing and providing care for victims of child abuse. 2005. Available at: www.apta.org [1-800-999-2782, ext. 3395. Accessed Sept 6, 2010.
210. American Physical Therapy Association (APTA): Guidelines for recognizing and providing care for victims of elder abuse. 2007. Available at: www.apta.org [1-800-999-2782, ext. 3395. Accessed Sept 6, 2010.
211. Feldhaus KM: Fighting domestic violence: An intervention plan. *J Musculoskel Med* 18(4):197–204, 2001.
212. Ellis J: Prescription medication borrowing and sharing—risk factors and management. *Aust Fam Phys* 38(10):816–819, 2009.
213. Goldsworthy RC: Prescription medication sharing among adolescents: prevalence, risk, and outcomes. *J Adolesc Health* (online). August 3, 2009.
214. Burroughs VJ, Maxey RW, Levy RA: Racial and ethnic differences in response to medicines. *J Natl Med Assoc* 94(10 Suppl):1–26, 2002.
215. Morrison A, Levy R: Toward individualized pharmaceutical care of East Asians: The value of genetic testing for polymorphisms in drug-metabolizing genes. *Pharmacogenomics* 5(6): 673–689, 2004.

216. Gandhi M, Aweeka F, Greenblatt RM, et al: Sex differences in pharmacokinetics and pharmacodynamics. *Annu Rev Pharmacol Toxicol* 44:499–523, 2004.
217. Meredith S, Feldman PH, Frey D, et al: Possible medication errors in home healthcare patients. *J Am Geriatr Soc* 49(6):719–724, 2001.
218. Food and Drug Administration (FDA): Center for Drug Education and Research: Information for healthcare professionals: NSAIDs. Available at: www.fda.gov. Posted June 15, 2005. Accessed Sept 09, 2010.
219. Boissonnault WG, Meek PD: Risk factors for anti-inflammatory drug or aspirin induced GI complications in individuals receiving outpatient physical therapy services. *J Orthop Sports Phys Ther* 32(10):510–517, 2002.
220. Biederman RE: Pharmacology in rehabilitation: Nonsteroidal anti-inflammatory agents. *J Orthop Sports Phys Ther* 35(6):356–367, 2005.
221. Lefkowith JB: Cyclooxygenase-2 specificity and its clinical implications. *Am J Med* 106:43S-50S, 1999.
222. Van der Linden MW: The balance between severe cardiovascular and gastrointestinal events among users of selective and non-selective nonsteroidal antiinflammatory drugs. *Ann Rheum Dis* 68(5):668–673, 2009.
223. Reuben SS: Issues in perioperative use of NSAIDs. *J Musculoskel Med* 22(6):281–282, 2005.
224. Huerta C: Nonsteroidal antiinflammatory drugs and risk of acute renal failure in the general population. *Am J Kidney Dis* 45(3):531–539, 2005.
225. Cryer B: Gastrointestinal safety of low-dose aspirin. *Am J Manag Care* 8(22 Suppl):S701–S708, 2002.
226. Curhan GC, Willett WC, Rosner B: Frequency of analgesic use and risk of hypertension in younger women. *Arch Intern Med* 162(19):2204–2208, 2002.
227. Schiodt FV, Rochling FA: Acetaminophen toxicity in an urban country hospital. *N Engl J Med* 337(16):1112–1117, 1997.
228. Acello B: Administering acetaminophen safely. *Nursing 2003* 33(11):18, 2003.
229. Baggish AA: Long-term anabolic-androgenic steroid use is associated with left ventricular dysfunction. *Circulation: Heart Failure* 3:472–476, 2010 (epub). Available online (subscription) http://circheartfailure.ahajournals.org/cgi/content/full/circhf;3/4/470?hits=10&FIRSTINDEX=0&FULLTEXT=steroids&SEARCHID=1&gca=circhf%3B3%2F4%2F472&gca=circhf%3B3%2F4%2F470&. Accessed August 27, 2010.
230. Buntin-Mushock C, Phillip L, Moriyama K: Age-dependent opioid escalation in chronic pain patients. *Anesth Analg* 100(6):1740–1745, 2005.
231. Haller C, James L: Out of the medicine closet: time to talk straight about prescription drug abuse. *Clin Pharm Ther* 88(3):279–282, 2010.
232. Boscarino JA: Risk factors for drug dependence among outpatients on opioid therapy in a large US health-care system. *Addiction* 105(10):1776–1782, 2010.
233. Community Anti-Drug Coalitions of America (CADCA): Teen prescription drug abuse: an emerging threat. Available at http://www.theantidrug.com. Accessed September 22, 2010.
234. Cai R: Emergency department visits involving nonmedical use of selected prescription drugs in the United States. *J Pain Palliat Care Pharmacother* 24(3):293–297, 2010.
235. Hernandez SH: Prescription drug abuse: insight into epidemic. *Clin Pharmacol Ther* 88(3):307–317, 2010.
236. Burkman R, Schlesselman JJ, Zieman M: Safety concerns and health benefits associated with oral contraception. *Am J Obstet Gynecol* 190(4 Suppl):S5–22, 2004.
237. Barclay L, Lie D: Bone loss from Depot Medroxyprogesterone acetate may be reversible. *Arch Pediatr Adolesc Med* 159:139–144, 2005.
238. U.S. Food and Drug Administration: Information for Healthcare Professionals. Fluoroquinolone Antimicrobial Drugs. 2008. Available at http://www.fda.gov/Drugs/DrugSafety/PostmarketDrugSafetyInformationforPatientsandProviders/ucm126085.htm. Accessed on Nov. 15, 2010.
239. Filippucci E, Farina A, Bartolucci F, et al: Levofloxacin-induced bilateral rupture of the Achilles tendon: Clinical and sonographic findings. *Rheumatiso* 55(4):267–269, 2003.
240. Melhus A: Fluoroquinolones and tendon disorders. *Expert Opin Saf* 4(2):299–309, 2005.
241. Khaliq Y, Zhanel GG: Fluoroquinolone-associated tendinopathy: A critical review of the literature. *Clin Infect Dis* 36(11):1404–1410, 2003.
242. Kelly JP, Kaufman DW, Kelley K, et al: Recent trends in use of herbal and other natural products. *Arch Intern Med* 165(3):281–286, 2005.
243. Trapskin P, Smith KM: Herbal medications in the perioperative orthopedic surgery patient. *Orthopedics* 27(8):819–822, 2004.
244. *Herbal medicine handbook*, ed 3, 2005, Lippincott Williams & Wilkins.
245. Ciccone CD: *Pharmacology in rehabilitation, 4e*, Philadelphia, 2007, FA Davis.
246. Graedon J, Graedon T: *The people's pharmacy guide to home and herbal remedies*, New York, 2002, St. Martin's Press.
247. Gruenwald G: *PDR for herbal medicines*, ed 4, Stamford, 2007, Thomson Healthcare.
248. Mold JW, Roberts M, Aboshady HM: Prevalence and predictors of night sweats, day sweats, and hot flashes in older primary care patients. *Ann Fam Med* 2(5):391–397, 2004.
249. Stone KL: Self-reported sleep and nap habits and risk of mortality in a large cohort of older women. *J Am Geriatr Soc* 57(4):604–611, 2009.

CHAPTER

3

Pain Types and Viscerogenic Pain Patterns

Pain is often the primary symptom in many physical therapy practices. Pain assessment is a key feature in the physical therapy interview. Pain is now recognized as the "fifth vital sign,"[1] along with blood pressure, temperature, pulse, and respiration.

Recognizing pain patterns that are characteristic of systemic disease is a necessary step in the screening process. Understanding how and when diseased organs can refer pain to the neuromusculoskeletal (NMS) system helps the therapist identify suspicious pain patterns.

This chapter includes a detailed overview of pain patterns that can be used as a foundation for all the organ systems presented. Information will include a discussion of pain types in general and viscerogenic pain patterns specifically. Additional resources for understanding the mechanisms of pain are available.[2]

Each section discusses specific pain patterns characteristic of disease entities that can mimic pain from musculoskeletal or neuromuscular disorders. In the clinical decision-making process the therapist will evaluate information regarding the location, referral pattern, description, frequency, intensity, and duration of systemic pain in combination with knowledge of associated symptoms and relieving and aggravating factors.

This information is then compared with presenting features of primary musculoskeletal disorders that have similar patterns of presentation. Pain patterns of the chest, back, shoulder, scapula, pelvis, hip, groin, and sacroiliac (SI) joint are the most common sites of referred pain from a systemic disease process. These patterns are discussed in greater detail later in this text (see Chapters 14 to 18).

A large component in the screening process is being able to recognize the client demonstrating a significant emotional overlay. Pain patterns from cancer can be very similar to what we have traditionally identified as psychogenic or emotional sources of pain. It is important to know how to differentiate between these two sources of painful symptoms. To help identify psychogenic sources of pain, discussions of conversion symptoms, symptom magnification, and illness behavior are also included in this chapter.

MECHANISMS OF REFERRED VISCERAL PAIN

The neurology of visceral pain is not well understood at this time. Proposed models are based on what is known about the somatic (nonvisceral) sensory system. Scientists have not found actual nerve fibers and specific nociceptors in organs. Peripheral mechanisms are suspected.[3] We do know the afferent supply to internal organs is in close proximity to blood vessels along a path similar to the sympathetic nervous system.[4,5]

Research is ongoing to identify the sites and mechanisms of visceral nociception. During inflammation, increased nociceptive input from an inflamed organ can sensitize neurons that receive convergent input from an unaffected organ, but the site of visceral cross-sensitivity is unknown.[6]

Viscerosensory fibers ascend the anterolateral system to the thalamus with fibers projecting to several regions of the brain. These regions encode the site of origin of visceral pain, although they do it poorly because of low receptor density, large overlapping receptive fields, and extensive convergence in the ascending pathway. Thus the cortex cannot distinguish where the pain messages originate from.[7,8]

Studies show there may be multiple mechanisms operating at different sites to produce the sensation we refer to as "pain." The same symptom can be produced by different mechanisms and a single mechanism may cause different symptoms.[9]

In the case of referred pain patterns of viscera, there are three separate phenomena to consider from a traditional Western medicine approach. These are:

- Embryologic development
- Multisegmental innervation
- Direct pressure and shared pathways

Embryologic Development

Each system has a bit of its own uniqueness in how pain is referred. For example, the viscera in the abdomen comprise a large percentage of all the organs we have to consider. When a person gives a history of abdominal pain, the location

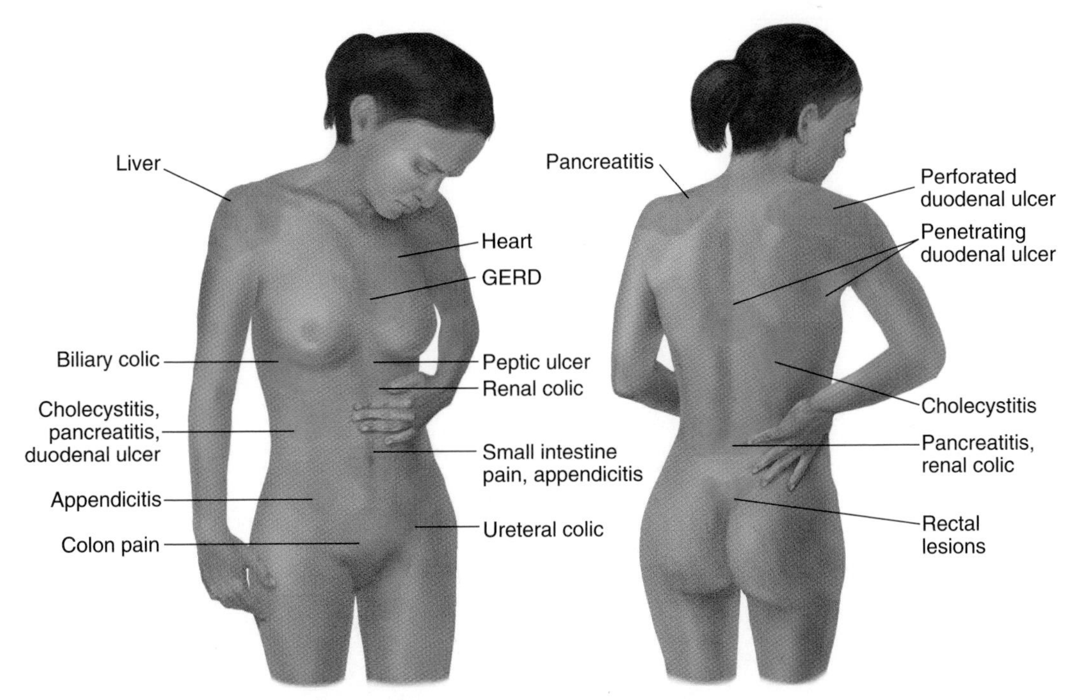

Fig. 3-1 Common sites of referred pain from the abdominal viscera. When a client gives a history of referred pain from the viscera, the pain's location may not be directly over the impaired organ. Visceral embryologic development is the mechanism of the referred pain pattern. Pain is referred to the site where the organ was located in fetal development. (From Jarvis C: *Physical examination and health assessment*, ed 5, Philadelphia, 2008, WB Saunders.)

of the pain may not be directly over the involved organ (Fig. 3-1).

Functional magnetic resonance imaging (fMRI) and other neuroimaging methods have shown activation of the inferolateral postcentral gyrus by visceral pain so the brain has a role in visceral pain patterns.[10,11] However, it is likely that embryologic development has the primary role in referred pain patterns for the viscera.

Pain is referred to a site where the organ was located in fetal development. Although the organ migrates during fetal development, its nerves persist in referring sensations from the former location.

Organs, such as the kidneys, liver, and intestines, begin forming by 3 weeks when the fetus is still less than the size of a raisin. By day 19, the notochord forming the spinal column has closed and by day 21, the heart begins to beat.

Embryologically, the chest is part of the gut. In other words, they are formed from the same tissue in utero. This explains symptoms of intrathoracic organ pathology frequently being referred to the abdomen as a viscero-viscero reflex. For example, it is not unusual for disorders of thoracic viscera, such as pneumonia or pleuritis, to refer pain that is perceived in the abdomen instead of the chest.[4]

Although the heart muscle starts out embryologically as a cranial structure, the pericardium around the heart is formed from gut tissue. This explains why myocardial infarction or pericarditis can also refer pain to the abdomen.[4]

Another example of how embryologic development impacts the viscera and the soma, consider the ear and the kidney. These two structures have the same shape since they come from the same embryologic tissue (otorenal axis of the mesenchyme) and are formed at the same time (Fig. 3-2).

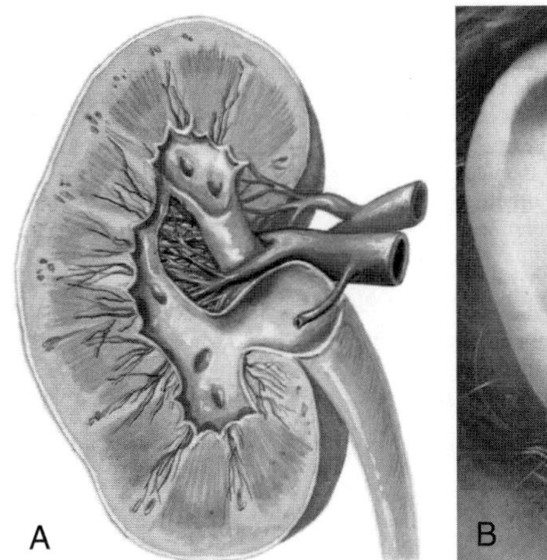

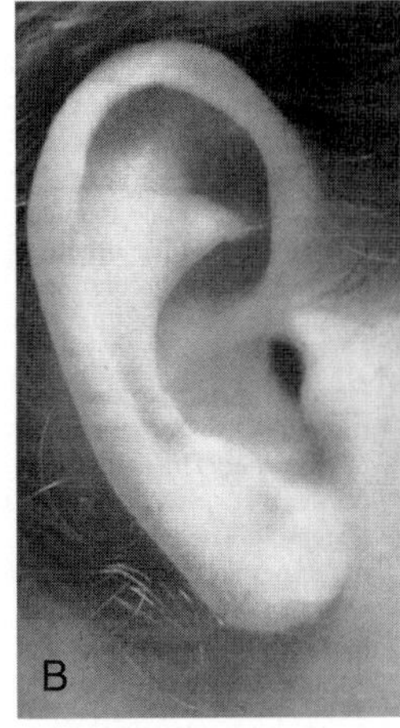

Fig. 3-2 The ear and the kidney have the same shape since they are formed at the same time and from the same embryologic tissue (otorenal axis of the mesenchyme). This is just one example of how fetal development influences form and function. When a child is born with a deformed or missing ear, the medical staff looks for a similarly deformed or missing kidney on the same side. (From Anderson KN: *Mosby's medical, nursing & allied health dictionary,* ed 5, St. Louis, 1988, Mosby; A-39; and from Seidel HM, Ball JW, Dains JE, et al: *Mosby's physical examination handbook,* St. Louis, 2003, Mosby.)

When a child is born with any anomaly of the ear(s) or even a missing ear, the medical staff knows to look for possible similar changes or absence of the kidney on the same side.

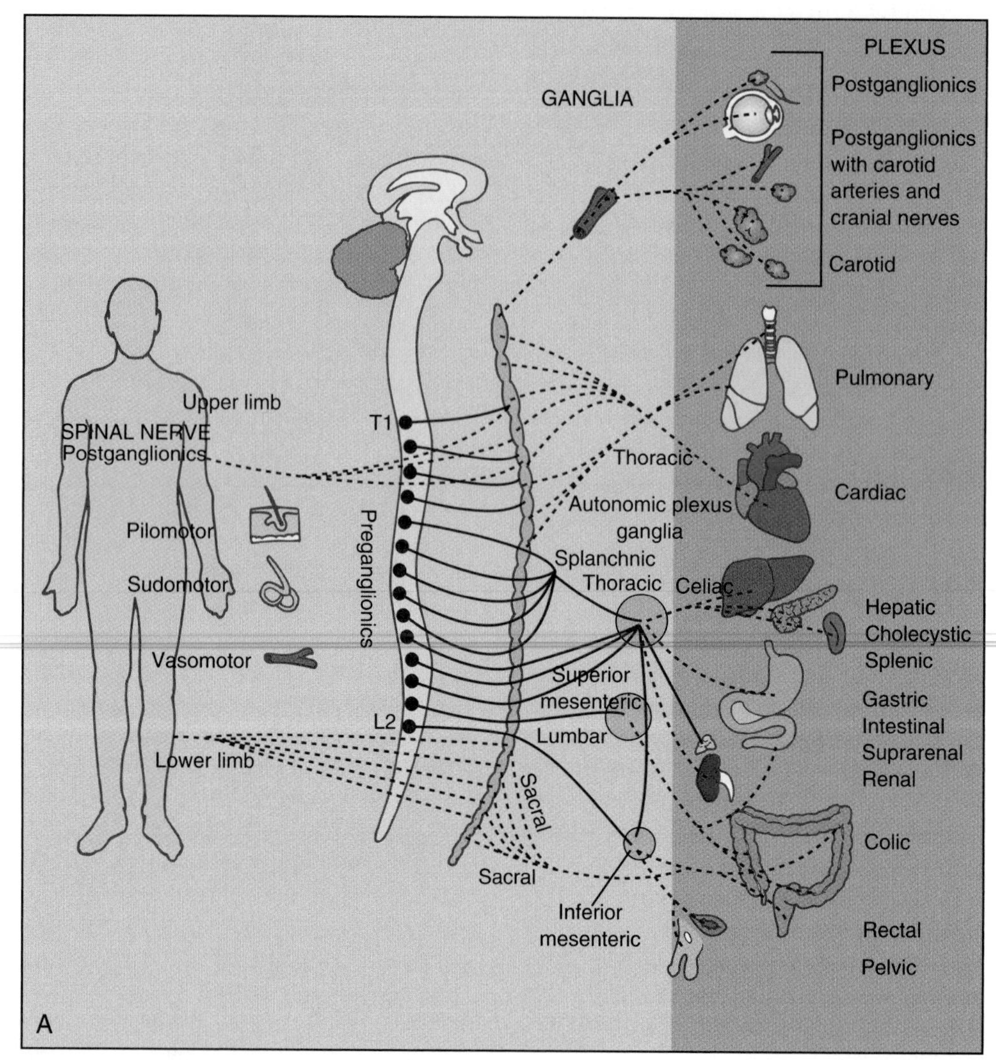

Fig. 3-3 Sympathetic **(A)** and parasympathetic **(B)** divisions of the autonomic nervous system. The visceral afferent fibers mediating pain travel with the sympathetic nerves, except for those from the pelvic organs, which follow the parasympathetics of the pelvic nerve. Major visceral organs have multisegmental innervations overlapping innervations of somatic structures. Visceral pain can be referred to the corresponding somatic area because sensory fibers for the viscera and somatic structures enter the spinal cord at the same levels converging on the same neurons. (From Levy MN, Koeppen BM: *Berne and Levy principles of physiology,* ed 4, St. Louis, 2006, Mosby.)

A thorough understanding of fetal embryology is not really necessary in order to recognize red flag signs and symptoms of visceral origin. Knowing that it is one of several mechanisms by which the visceral referred pain patterns occur is a helpful start.

However, the more you know about embryologic development of the viscera, the faster you will recognize somatic pain patterns caused by visceral dysfunction. Likewise, the more you know about anatomy, the origins of anatomy, its innervations, and the underlying neurophysiology, the better able you will be to identify the potential structures involved.

This will lead you more quickly to specific screening questions to ask. The manual therapist will especially benefit from a keen understanding of embryologic tissue derivations. An appreciation of embryology will help the therapist localize the problem vertically.

Multisegmental Innervation

Multisegmental innervation is the second mechanism used to explain pain patterns of a viscerogenic source (Fig. 3-3). The autonomic nervous system (ANS) is part of the peripheral nervous system. As shown in this diagram, the viscera have multisegmental innervations. The multiple levels of innervation of the heart, bronchi, stomach, kidneys, intestines, and bladder are demonstrated clearly.

There is new evidence to support referred visceral pain to somatic tissues based on overlapping or same segmental projections of spinal afferent neurons to the spinal dorsal horn. This concept is referred to as *visceral-organ cross-sensitization.* The mechanism is likely to be sensitization of viscera-somatic convergent neurons.[12]

For the first time ever, scientists showed that individuals diagnosed with multiple visceral problems obtained relief

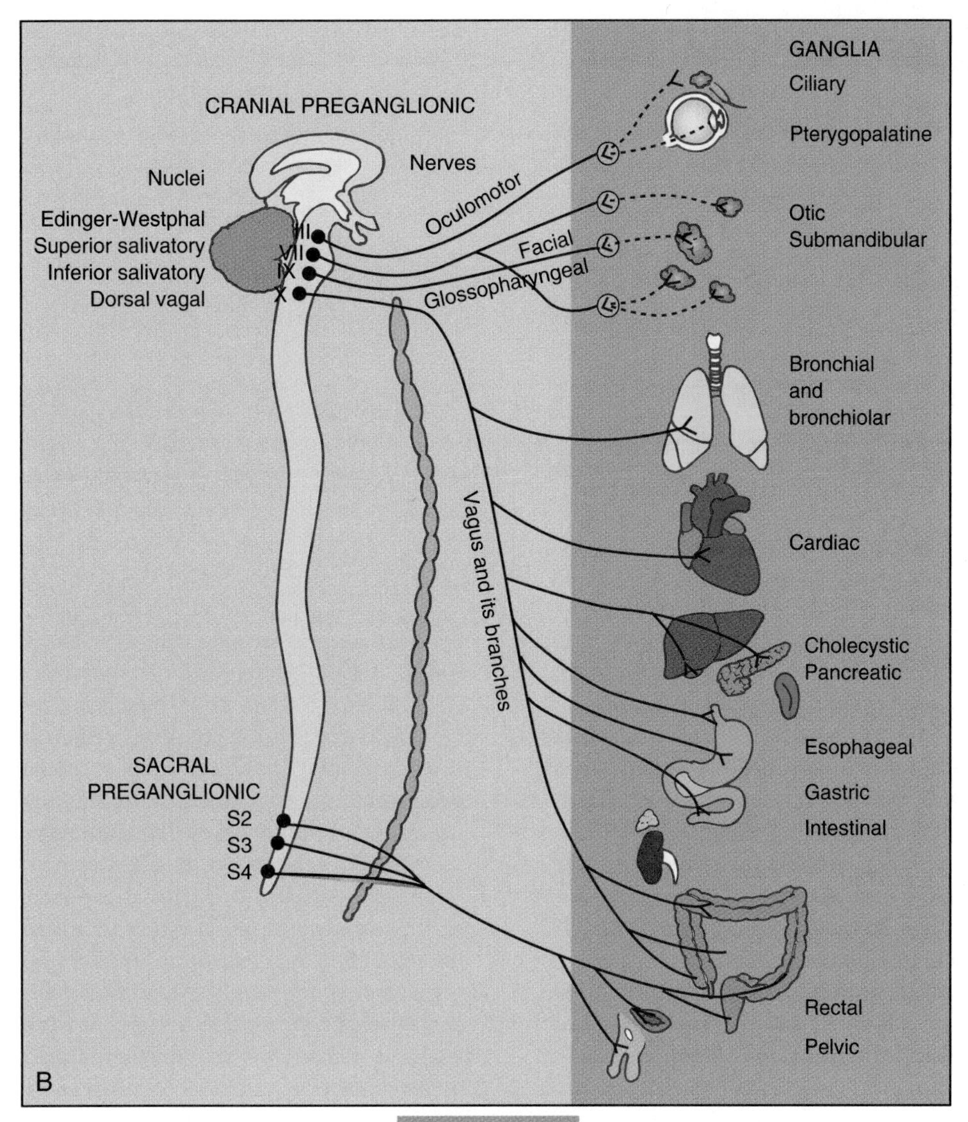

Fig. 3-3, cont'd

from pain in all organ systems with overlapping segmental projections when only one visceral area was treated. In other words, nontreated visceral disease significantly decreased when one viscera of the overlapping segments was addressed. For groups of people with no overlapping segments, spontaneous relief of referred pain was not obtained until and unless all involved visceral systems were treated.[12]

Pain of a visceral origin can be referred to the corresponding somatic areas. The example of cardiac pain is a good one. Cardiac pain is not felt in the heart, but is referred to areas supplied by the corresponding spinal nerves.

Instead of actual physical heart pain, cardiac pain can occur in any structure innervated by C3 to T4 such as the jaw, neck, upper trapezius, shoulder, and arm. Pain of cardiac and diaphragmatic origin is often experienced in the shoulder, in particular, because the C5 spinal segment supplies the heart, respiratory diaphragm, and shoulder.

Direct Pressure and Shared Pathways

A third and final mechanism by which the viscera refer pain to the soma is the concept of direct pressure and shared pathways (Fig. 3-4). As shown in this illustration, many of the viscera are near the respiratory diaphragm. Any pathologic process that can inflame, infect, or obstruct the organs can bring them in contact with the respiratory diaphragm.

Anything that impinges the *central diaphragm* can refer pain to the *shoulder* and anything that impinges the *peripheral diaphragm* can refer pain to the *ipsilateral costal margins* and/or *lumbar region* (Fig. 3-5).

This mechanism of referred pain through shared pathways occurs as a result of ganglions from each neural system gathering and sharing information through the cord to the plexuses. The visceral organs are innervated through the ANS.

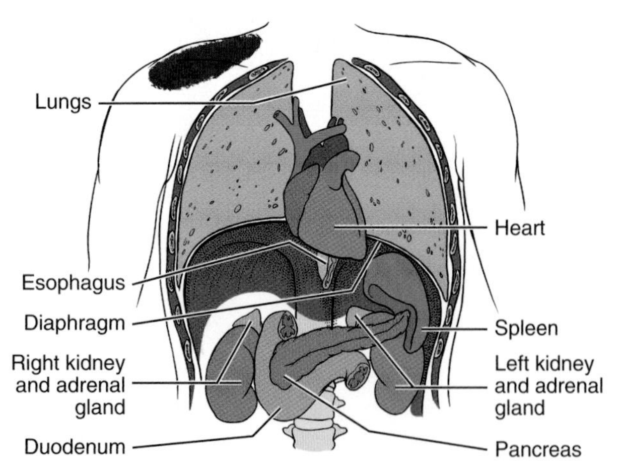

Fig. 3-4 Direct pressure from any inflamed, infected, or obstructed organ in contact with the respiratory diaphragm can refer pain to the ipsilateral shoulder. Note the location of each of the viscera. The spleen is tucked up under the diaphragm on the left side so any impairment of the spleen can cause left shoulder pain. The tail of the pancreas can come in contact with the diaphragm on the left side potentially causing referred pain to the left shoulder. The head of the pancreas can impinge the right side of the diaphragm causing referred pain to the right side. The gallbladder (not shown) is located up under the liver on the right side with corresponding right referred shoulder pain possible. Other organs that can come in contact with the diaphragm in this way include the heart and the kidneys.

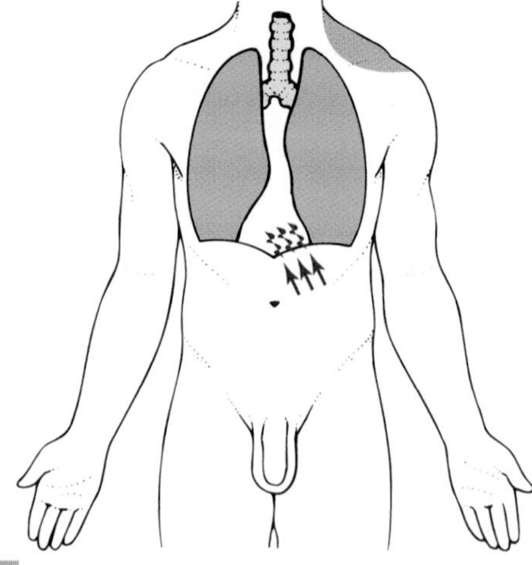

Fig. 3-5 Irritation of the peritoneal (outside) or pleural (inside) surface of the central area of the respiratory diaphragm can refer sharp pain to the upper trapezius muscle, neck, and supraclavicular fossa. The pain pattern is ipsilateral to the area of irritation. Irritation of the peripheral portion of the diaphragm can refer sharp pain to the costal margins and lumbar region (not shown).

The ganglions bring in information from around the body. The nerve plexuses decide how to respond to this information (what to do) and give the body finely-tuned, local control over responses.

Plexuses originate in the neck, thorax, diaphragm, and abdomen, terminating in the pelvis. The brachial plexus supplies the upper neck and shoulder while the phrenic nerve innervates the respiratory diaphragm. More distally, the celiac plexus supplies the stomach and intestines. The neurologic supply of the plexuses is from parasympathetic fibers from the vagus and pelvic splanchnic nerves.[4]

The plexuses work independently of each other but not independently of the ganglia. The ganglia collect information derived from both the parasympathetic and the sympathetic fibers. The ganglia deliver this information to the plexuses; it is the plexuses that provide fine, local control in each of the organ systems.[4]

For example, the lower portion of the heart is in contact with the center of the diaphragm. The spleen on the left side of the body is tucked up under the dome of the diaphragm. The kidneys (on either side) and the pancreas in the center are in easy reach of some portion of the diaphragm.

The body of the pancreas is in the center of the human body. The tail rests on the left side of the body. If an infection, inflammation, or tumor or other obstruction distends the pancreas, it can put pressure on the central part of the diaphragm.

Since the phrenic nerve (C3-5) innervates the central zone of the diaphragm, as well as part of the pericardium, the gallbladder, and the pancreas, the client with impairment of these viscera can present with signs and symptoms in any of the somatic areas supplied by C3-5 (e.g., shoulder).

In other words, the person can experience symptoms in the areas innervated by the same nerve pathways. So a problem affecting the pancreas can look like a heart problem, a gallbladder problem, or a mid-back/scapular or shoulder problem.

Most often, clients with pancreatic disease present with the primary pain pattern associated with the pancreas (i.e., left epigastric pain or pain just below the xiphoid process). The somatic presentation of referred pancreatic pain to the shoulder or back is uncommon, but it is the unexpected, referred pain patterns that we see in a physical or occupational therapy practice.

Another example of this same phenomenon occurs with peritonitis or gallbladder inflammation. These conditions can irritate the phrenic endings in the central part of the diaphragmatic peritoneum. The client can experience referred shoulder pain due to the root origin shared in common by the phrenic and supraclavicular nerves.

Not only is it true that any structure that touches the diaphragm can refer pain to the shoulder, but even structures adjacent to or in contact with the diaphragm in utero can do the same. Keep in mind there has to be some impairment of that structure (e.g., obstruction, distention, inflammation) for this to occur (Case Example 3-1).

ASSESSMENT OF PAIN AND SYMPTOMS

The interviewing techniques and specific questions for pain assessment are outlined in this section. The information gathered during the interview and examination provides a description of the client that is clear, accurate, and comprehensive. The therapist should keep in mind cultural rules and differences in pain perception, intensity, and responses to pain found among various ethnic groups.[13]

Measuring pain and assessing pain are two separate issues. A measurement assigns a number or value to give dimension to pain intensity.[14] A comprehensive pain assessment includes a detailed health history, physical exam, medication history (including nonprescription drug use and complementary and alternative therapies), assessment of functional status, and consideration of psychosocial-spiritual factors.[15]

The portion of the core interview regarding a client's perception of pain is a critical factor in the evaluation of signs and symptoms. Questions about pain must be understood by the client and should be presented in a nonjudgmental manner. A record form may be helpful to standardize pain assessment with each client (Fig. 3-6).

To elicit a complete description of symptoms from the client, the physical therapist may wish to use a term other than *pain*. For example, referring to the client's *symptoms* or using descriptors such as *hurt* or *sore* may be more helpful with some individuals. Burning, tightness, heaviness, discomfort, and aching are just a few examples of other possible word choices. The use of alternative words to describe a client's symptoms may also aid in refocusing attention away from pain and toward improvement of functional abilities.

If the client has completed the McGill Pain Questionnaire (see discussion of McGill Pain Questionnaire in this chapter),[16] the physical therapist may choose the most appropriate alternative word selected by the client from the list to refer to the symptoms (Table 3-1).

Pain Assessment in the Older Adult

Pain is an accepted part of the aging process, but we must be careful to take the reports of pain from older persons as serious and very real and not discount the symptoms as part of aging. Well over half of the older adults in the United States report chronic joint symptoms.[17] We are likely to see pain more often as a key feature among older adults as our population continues to age.

The American Geriatrics Society (AGS) reports the use of over-the-counter (OTC) analgesic medications for pain, aching, and discomfort is common in older adults along with routine use of prescription drugs. Many older adults have taken these medications for 6 months or more.[18]

CASE EXAMPLE 3-1

Mechanism of Referred Pain

A 72-year-old woman has come to physical therapy for rehabilitation after cutting her hand and having a flexor tendon repair. She uses a walker to ambulate, reports being short of breath "her whole life," and takes the following prescription and over-the-counter (OTC) medications:

- Feldene
- Vioxx*
- Ativan
- Glucosamine
- Ibuprofen "on bad days"
- Furosemide
- And one other big pill once a week on Sunday "for my bones"

During the course of evaluating and treating her hand, she reports constant, aching pain in her right shoulder and a sharp, tingling, burning sensation behind her armpit (also on the right side). She does not have any associated bowel or bladder signs and symptoms, but reports excessive fatigue "since the day I was born."

You suspect the combination of Feldene and Ibuprofen along with long-term use of Vioxx may be a problem.

What Is the Most Likely Mechanism of Pain: Embryologic Development, Multisegmental Innervation of the Stomach and Duodenum, or Direct Pressure on the Diaphragm? Even though Vioxx is a cyclooxygenase-2 (COX-2) inhibitor and less likely to cause problems, gastritis and gastrointestinal (GI) bleeding are still possible, especially with chronic long-term use of multiple nonsteroidal antiinflammatory drugs (NSAIDs).

Retroperitoneal bleeding from peptic ulcer can cause referred pain to the back at the level of the lesion (T6-10) or right shoulder and/or upper trapezius pain. Shoulder pain may be accompanied by sudomotor changes such as burning, gnawing, or cramping pain along the lateral border of the scapula. The scapular pain can occur alone as the only symptom.

Side effects of NSAIDs can also include fatigue, anxiety, depression, paresthesia, fluid retention, tinnitus, nausea, vomiting, dry mouth, and bleeding from the nose, mouth, or under the skin. If peritoneal bleeding is the cause of her symptoms, *the mechanism of pain* is blood in the posterior abdominal cavity irritating the diaphragm through direct pressure.

Be sure to take the client's vital signs and observe for significant changes in blood pressure and pulse. Poor wound healing and edema (sacral, pedal, hands) may be present. Ask if the same doctor prescribed each medication and if her physician (or physicians) knows which medications she is taking. It is possible that her medications have not been checked or coordinated from before her hospitalization to the present time.

*Removed from the market in 2004 by Merck & Co., Inc., due to reports of increased risk of cardiovascular events.

Pain Assessment Record Form

Client's name: **Date:**

Physician's diagnosis: **Physical therapist's diagnosis:**

Medications:________________________________

Onset of pain (circle one): Was there an:

Accident Injury Trauma (violence) Specific activity

If yes, describe:

Characteristics of pain/symptoms:

Location (Show me exactly where your pain/symptom is located):

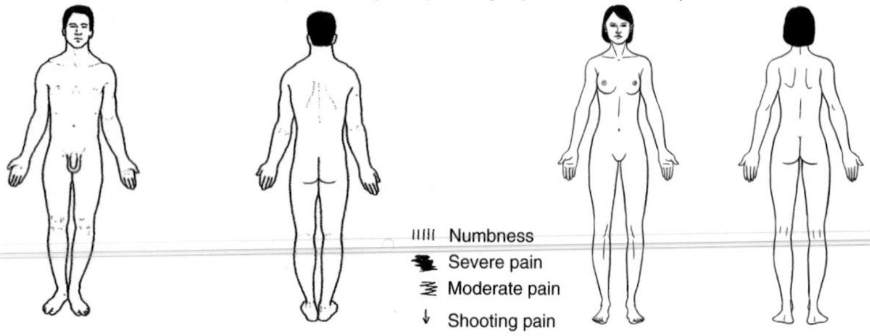

IIII Numbness
Severe pain
Moderate pain
Shooting pain

Do you have any pain or symptoms anywhere else? Yes No

Description (If yes, what does it feel like):

Circle any other words that describe the client's symptoms:

Knifelike	Dull	Aching	Other (describe):
Boring	Burning	Throbbing	
Heaviness	Discomfort	Sharp	
Stinging	Tingling	Stabbing	

Frequency (circle one): Constant Intermittent (comes and goes)

If constant: Do you have this pain right now? Yes No

If intermittent: How often is the pain present (circle all that apply):

Hourly Once/daily Twice/daily Unpredictable Other (please describe):________________

Intensity: *Numeric Rating Scale* and the *Faces Pain Scale*

Instructions: On a scale from 0 to 10 with zero meaning 'No pain' and 10 for 'Unbearable pain,' how would you rate your pain right now?

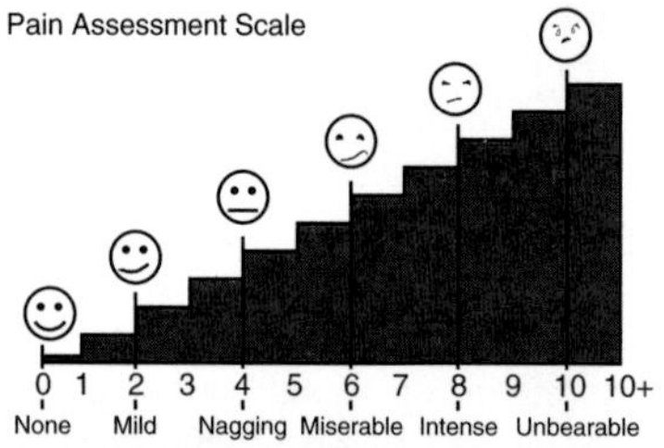

Pain Assessment Scale

0 1 2 3 4 5 6 7 8 9 10 10+

None Mild Nagging Miserable Intense Unbearable

Alternately: Point to the face that best shows how much pain you are having right now.

Intensity: *Visual Analog Scale*

Instructions: On the line below, put a mark (or point to) the place on the line between 'Pain free' and 'Worst possible pain' that best describes/shows how much pain you are having right now.

A **Pain Free**____**Worst Possible Pain**

Fig. 3-6 Pain Assessment Record Form. Use this form to complete the pain history and obtain a description of the pain pattern. The form is printed in the Appendix for your use. This form may be copied and used without permission. (From Carlsson AM: Assessment of chronic pain. I. Aspects of the reliability and validity of the visual analogue scale, *Pain* 16(1):87–101, 1983. Used with permission.)

Duration:
How long does your pain (name the symptom) last?

Aggravating factors (What makes it worse?)	**Relieving factors** (What makes it better?)

Pattern
Has the pain changed since it first began? Yes No
If yes, please explain:

What is your pain/symptom like from morning (am) to evening (pm)?

Circle one:	Worse in the morning	Worse midday/afternoon	Worse at night
Circle one:	Gradually getting better	Gradually getting worse	Staying the same

Circle all that apply:
Present upon waking up   Keeps me from falling asleep   Wakes me up at night

Therapist: Record any details or description about night pain. See also Appendix for *Screening Questions for Night Pain* when appropriate:

Associated symptoms (What other symptoms have you had with this problem?)

Circle any words the client uses to describe his/her symptoms. If the client says there are no other symptoms ask about the presence of any of the following:

Burning	Difficulty breathing	Shortness of breath	Cough
Skin rash (or other lesions)	Change in bowel/bladder	Difficulty swallowing	Painful swallowing
Dizziness	Heart palpitations	Hoarseness	Nausea/vomiting
Diarrhea	Constipation	Bleeding of any kind	Sweats
Numbness	Problems with vision	Tingling	Weakness
Joint pain	Weight loss/gain	Other:______	

Final question: Are there any other pain or symptoms of any kind anywhere else in your body that we have not talked about yet?

For the therapist:
Follow up questions can include:
Are there any positions that make it feel better? Worse?
How does rest affect the pain/symptoms?
How does activity affect the pain/symptoms?
How has this problem affected your daily life at work or at home?
Has this problem affected your ability to care for yourself without assistance (e.g. dress, bathe, cook, drive)?
Has this problem affected your sexual function or activity?

Therapist's evaluation:
Can you reproduce the pain by squeezing or palpating the symptomatic area?
Does resisted motion reproduce the pain/symptoms?
Is the client taking NSAIDs? Experiencing increased symptoms after taking NSAIDs?
If taking NSAIDs, is the client at risk for peptic ulcer? Check all that apply:
- ☐ Age>65 years
- ☐ History of peptic ulcer disease or GI disease
- ☐ Smoking, alcohol use
- ☐ Oral corticosteroid use
- ☐ Anticoagulation or use of other anticoagulants (even when used for heart patients at a lower dose, e.g., 81 to 325 mg aspirin/day)
- ☐ Renal complications in clients with hypertension or congestive heart failure (CHF) or who use diuretics or ACE inhibitors
- ☐ NSAIDs combined with selective serotonin reuptake inhibitors (SSRIs; antidepressants such as Prozac, Zoloft, Celexa, Paxil)
- ☐ Use of acid suppressants (e.g., H_2-receptor antagonists, antacids)

Other areas to consider:
- Sleep quality
- Bowel/bladder habits
- Correlation of symptoms with peak effect of medications (dosage, time of day)
- Evaluation of joint pain (see Appendix: Screening Questions for Joint Pain)
- Depression or anxiety screening score
- For women: correlation of symptoms with menstrual cycle

B

Fig. 3-6, cont'd

TABLE 3-1 Recognizing Pain Patterns

Vascular	Neurogenic	Musculoskeletal	Emotional
Throbbing	Sharp	Aching	Tiring
Pounding	Crushing	Sore	Miserable
Pulsing	Pinching	Heavy	Vicious
Beating	Burning	Hurting	Agonizing
	Hot	Deep	Nauseating
	Searing	Cramping	Frightful
	Itchy	Dull	Piercing
	Stinging		Dreadful
	Pulling		Punishing
	Jumping		Exhausting
	Shooting		Killing
	Electrical		Unbearable
	Gnawing		Annoying
	Pricking		Cruel
			Sickening
			Torturing

From Melzack R: The McGill Pain Questionnaire: major properties and scoring methods, *Pain* 1:277, 1975.

BOX 3-1 VERBAL DESCRIPTOR SCALE (VDS)

Directions: Show the scale to your client. Read the descriptors and ask the client to point to the one that best matches his or her pain (achiness, soreness, or discomfort) today. Give the client at least 30 seconds to respond. A verbal reply is acceptable. It is best if the client is sitting upright facing the interviewer. Provide the client with good lighting, his or her eyeglasses, and/or hearing aid(s) if appropriate.

Today I Have:
0 = NO PAIN
1 = SLIGHT PAIN
2 = MILD PAIN
3 = MODERATE PAIN
4 = SEVERE PAIN
5 = EXTREME PAIN
6 = PAIN AS BAD AS IT CAN BE

Older adults may avoid giving an accurate assessment of their pain. Some may expect pain with aging or fear that talking about pain will lead to expensive tests or medications with unwanted side effects. Fear of losing one's independence may lead others to underreport pain symptoms.[19]

Sensory and cognitive impairment in older, frail adults makes communication and pain assessment more difficult.[18] The client may still be able to report pain levels reliably using the visual analogue scales in the early stages of dementia. Improving an older adult's ability to report pain may be as simple as making sure the client has his or her glasses and hearing aid.

The Verbal Descriptor Scale (VDS) (Box 3-1) may be the most sensitive and reliable among older adults, including those with mild-to-moderate cognitive impairment.[20] But these and other pain scales rely on the client's ability to understand the scale and communicate a response. As dementia progresses, these abilities are lost as well.

A client with Alzheimer's-type dementia loses short-term memory and cannot always identify the source of recent painful stimuli.[21,22] The Alzheimer's Discomfort Rating Scale may be more helpful for older adults who are unable to communicate their pain.[23] The therapist records the frequency, intensity, and duration of the client's discomfort based on the presence of noisy breathing, facial expressions, and overall body language.

Another tool under investigation is the Pain Assessment in Advanced Dementia (PAINAD) scale. The PAINAD scale is a simple, valid, and reliable instrument for measurement of pain in noncommunicative clients developed by the same author as the Alzheimer's Discomfort Rating Scale.[24] A disadvantage of this pain scale is that the pain is inferred by the examiner or caregiver rather than self-reported directly by the individual experiencing the pain.[25]

Facial grimacing; nonverbal vocalization such as moans, sighs, or gasps; and verbal comments (e.g., ouch, stop) are the most frequent behaviors among cognitively impaired older adults during painful movement (Box 3-2). Bracing, holding onto furniture, or clutching the painful area are other behavioral indicators of pain. Alternately, the client may resist care by others or stay very still to guard against pain caused by movement.[26]

Untreated pain in an older adult with advanced dementia can lead to secondary problems such as sleep disturbances, weight loss, dehydration, and depression. Pain may be manifested as agitation and increased confusion.[21]

Older adults are more likely than younger adults to have what is referred to as atypical acute pain. For example, silent acute myocardial infarction (MI) occurs more often in the older adult than in the middle-aged to early senior adult. Likewise, the older adult is more likely to experience appendicitis without any abdominal or pelvic pain.[27]

Pain Assessment in the Young Child

Many infants and children are unable to report pain. Even so the therapist should not underestimate or prematurely conclude that a young client is unable to answer any questions about pain. Even some clients (both children and adults) with substantial cognitive impairment may be able to use pain-rating scales when explained carefully.[28]

The Faces Pain Scale (FACES or FPS) for children (see Fig. 3-6) was first presented in the 1980s.[29] It has since been revised (FPS-R)[30] and presented concurrently by other researchers with similar assessment measures.[31]

Most of the pilot work for the FPS was done informally with children from preschool through young school age. Researchers have used the FPS scale with adults, especially the elderly, and have had successful results. Advantages of the

BOX 3-2 SYMPTOMS OF PAIN IN CLIENTS WITH COGNITIVE IMPAIRMENT

- Verbal comments such as ouch or stop
- Nonverbal vocalizations (e.g., moans, sighs, gasps)
- Facial grimacing or frowning
- Audible breathing independent of vocalization (labored, short or long periods of hyperventilation)
- Agitation or increased confusion
- Unable to be consoled or distracted
- Bracing or holding onto furniture
- Decreased mobility
- Lying very still, refusing to move
- Clutching the painful area
- Resisting care provided by others, striking out, pushing others away
- Sleep disturbance
- Weight loss
- Depression

cartoon-type FPS scale are that it avoids gender, age, and racial biases.[32]

Research shows that use of the word "hurt" rather than pain is understood by children as young as 3 years old.[33,34] Use of a word such as "owie" or "ouchie" by a child to describe pain is an acceptable substitute.[32] Assessing pain intensity with the FPS scale is fast and easy. The child looks at the faces, the therapist or parent uses the simple words to describe the expression, and the corresponding number is used to record the score.

A review of multiple other measures of self-report is also available,[14] as well as a review of pain measures used in children by age, including neonates.[35]

When using a rating scale is not possible, the therapist may have to rely on the parent or caregiver's report and/or other measures of pain in children with cognitive or communication impairments and physical disabilities. Look for telltale behavior such as lack of cooperation, withdrawal, acting out, distractibility, or seeking comfort. Altered sleep patterns, vocalizations, and eating patterns provide additional clues.

In very young children and infants, the Child Facial Coding System (CFCS) and the Neonatal Facial Coding System (NFCS) can be used as behavioral measures of pain intensity.[36,37]

Facial actions and movements, such as brow bulge, eye squeeze, mouth position, and chin quiver, are coded and scored as pain responses. This tool has been revised and tested as valid and reliable for use postoperatively in children ages 0 to 18 months following major abdominal or thoracic surgery.[38]

Vital signs should be documented but not relied upon as the sole determinant of pain (or absence of pain) in infants or young children. The pediatric therapist may want to investigate other pain measures available for neonates and infants.[39,40]

Characteristics of Pain

It is very important to identify how the client's description of pain as a symptom relates to sources and types of pain discussed in this chapter. Many characteristics of pain can be elicited from the client during the Core Interview to help define the source or type of pain in question. These characteristics include:

- Location
- Description of sensation
- Intensity
- Duration
- Frequency and duration
- Pattern

Other additional components are related to factors that aggravate the pain, factors that relieve the pain, and other symptoms that may occur in association with the pain. Specific questions are included in this section for each descriptive component. Keep in mind that an increase in frequency, intensity, or duration of symptoms over time can indicate systemic disease.

Location of Pain

Questions related to the location of pain focus the client's description as precisely as possible. An opening statement might be as follows:

FOLLOW-UP QUESTIONS

- Show me exactly where your pain is located.

  Follow-up questions may include:
- Do you have any other pain or symptoms anywhere else?
- If yes, what causes the pain or symptoms to occur in this other area?

If the client points to a small, localized area and the pain does not spread, the cause is likely to be a superficial lesion and is probably not severe. If the client points to a small, localized area but the pain does spread, this is more likely to be a diffuse, segmental, referred pain that may originate in the viscera or deep somatic structure.

The character and location of pain can change and the client may have several pains at once, so repeated pain assessment may be needed.

Description of Pain

To assist the physical therapist in obtaining a clear description of pain sensation, pose the question:

FOLLOW-UP QUESTIONS

- What does it feel like?

  After giving the client time to reply, offer some additional choices in potential descriptors. You may want to ask: Is your pain/Are your symptoms:

- Knifelike
- Boring
- Throbbing
- Deep aching
- Dull
- Burning
- Prickly
- Sharp

Follow-up questions may include:

- Has the pain changed in quality since it first began?
- Changed in intensity?
- Changed in duration (how long it lasts)?

When a client describes the pain as knifelike, boring, colicky, coming in waves, or a deep aching feeling, this description should be a signal to the physical therapist to consider the possibility of a systemic origin of symptoms. Dull, somatic pain of an aching nature can be differentiated from the aching pain of a muscular lesion by squeezing or by pressing the muscle overlying the area of pain. Resisting motion of the limb may also reproduce aching of muscular origin that has no connection to deep somatic aching.

Intensity of Pain

The level or intensity of the pain is an extremely important but difficult component to assess in the overall pain profile. Psychologic factors may play a role in the different ratings of pain intensity measured between African Americans and Caucasians. African Americans tend to rate pain as more unpleasant and more intense than whites, possibly indicating a stronger link between emotions and pain behavior for African Americans compared with Caucasians.[41]

The same difference is observed between women and men.[42,43] Likewise, pain intensity is reported as less when the affected individual has some means of social or emotional support.[44]

Assist the client with this evaluation by providing a rating scale. You may use one or more of these scales, depending on the clinical presentation of each client (see Fig. 3-6). Show the pain scale to your client. Ask the client to choose a number and/or a face that best describes his or her current pain level. You can use this scale to quantify symptoms other than pain such as stiffness, pressure, soreness, discomfort, cramping, aching, numbness, tingling, and so on. Always use the same scale for each follow-up assessment.

The Visual Analog Scale (VAS)[45,46] allows the client to choose a point along a 10-cm (100 mm) horizontal line (see Fig. 3-6). The left end represents "No pain" and the right end represents "Pain as bad as it could possibly be" or "Worst Possible Pain." This same scale can be presented in a vertical orientation for the client who must remain supine and cannot sit up for the assessment. "No pain" is placed at the bottom, and "Worst pain" is put at the top.

The VAS scale is easily combined with the numeric rating scale with possible values ranging from 0 (no pain) to 10 (worst imaginable pain). It can be used to assess current pain, worst pain in the preceding 24 hours, least pain in the past 24 hours, or any combination the clinician finds useful.

The Numeric Rating Scale (NRS; see Fig. 3-6) allows the client to rate the pain intensity on a scale from 0 (no pain) to 10 (the worst pain imaginable). This is probably the most commonly used pain rating scale in both the inpatient and outpatient settings. It is a simple and valid method of measuring pain.

Although the scale was tested and standardized using 0 to 10, the plus is used for clients who indicate the pain is "off the scale" or "higher than a 10." Some health care professionals prefer to describe 10 as "worst pain experienced with this condition" to avoid needing a higher number than 10.

This scale is especially helpful for children or cognitively impaired clients. In general, even adults without cognitive impairments may prefer to use this scale.

An alternative method provides a scale of 1 to 5 with word descriptions for each number[16] and asks:

? FOLLOW-UP QUESTIONS

- How strong is your pain?
  1 = Mild
  2 = Discomforting
  3 = Distressing
  4 = Horrible
  5 = Excruciating

This scale for measuring the intensity of pain can be used to establish a baseline measure of pain for future reference. A client who describes the pain as "excruciating" (or a 5 on the scale) during the initial interview may question the value of therapy when several weeks later there is no subjective report of improvement.

A quick check of intensity by using this scale often reveals a decrease in the number assigned to pain levels. This can be compared with the initial rating, thus providing the client with assurance and encouragement in the rehabilitation process. A quick assessment using this method can be made by asking:

FOLLOW-UP QUESTIONS

- How strong is your pain?
  1 = Mild
  2 = Moderate
  3 = Severe

The description of intensity is highly subjective. What might be described as "mild" for one person could be "horrible" for another person. Careful assessment of the person's nonverbal behavior (e.g., ease of movement, facial grimacing, guarding movements) and correlation of the person's personality with his or her perception of the pain may help to clarify the description of the intensity of the pain. Pain of an intense, unrelenting (constant) nature is often associated with systemic disease.

BOX 3-3 NURSING ASSESSMENT OF PAIN (PQRST)

Provocation and palliation. What causes the pain and what makes it better or worse?
Quality of pain. What type of pain is present (aching, burning, sharp)?
Region and radiation. Where is the pain located? Does it radiate to other parts of the body?
Severity on a scale from 0 to 10. Does the pain interfere with daily activities, mood, function?
Timing. Did the pain come on suddenly or gradually? Is it constant or does it come and go (intermittent)? How often does it occur? How long does it last? Does it come on at the same time of the day or night?

The 36-Item Short-Form Health Survey discussed in Chapter 2 includes an assessment of bodily pain along with a general measure of health-related quality of life. Nurses often use the PQRST mnemonic to help identify underlying pathology or pain (Box 3-3).

Frequency and Duration of Pain

The frequency of occurrence is related closely to the pattern of the pain, and the client should be asked how often the symptoms occur and whether the pain is constant or intermittent. Duration of pain is a part of this description.

FOLLOW-UP QUESTIONS

- How long do the symptoms last?
  For example, pain related to systemic disease has been shown to be a *constant* rather than an *intermittent* type of pain experience. Clients who indicate that the pain is constant should be asked:
- Do you have this pain right now?
- Did you notice these symptoms this morning immediately when you woke up?

Further responses may reveal that the pain is perceived as being constant but in fact is not actually present consistently and/or can be reduced with rest or change in position, which are characteristics more common with pain of musculoskeletal origin. Symptoms that truly do not change throughout the course of the day warrant further attention.

Pattern of Pain

After listening to the client describe all the characteristics of his or her pain or symptoms, the therapist may recognize a vascular, neurogenic, musculoskeletal (including spondylogenic), emotional, or visceral pattern (see Table 3-1).

The following sequence of questions may be helpful in further assessing the pattern of pain, especially how the symptoms may change with time.

FOLLOW-UP QUESTIONS

- Tell me about the pattern of your pain/symptoms.
- *Alternate question:* When does your back/shoulder (name the involved body part) hurt?
- *Alternate question:* Describe your pain/symptoms from first waking up in the morning to going to bed at night. (See special sleep-related questions that follow.)
  Follow-up questions may include:
- Have you ever experienced anything like this before?
- If yes, do these episodes occur more or less often than at first?
- How does your pain/symptom(s) change with time?
- Are your symptoms worse in the morning or evening?

The pattern of pain associated with systemic disease is often a progressive pattern with a cyclical onset (i.e., the client describes symptoms as being alternately worse, better, and worse over a period of months). When there is back pain, this pattern differs from the sudden sequestration of a discogenic lesion that appears with a pattern of increasingly worse symptoms followed by a sudden cessation of all symptoms. Such involvement of the disk occurs without the cyclical return of symptoms weeks or months later, which is more typical of a systemic disorder.

If the client appears to be unsure of the pattern of symptoms or has "avoided paying any attention" to this component of pain description, it may be useful to keep a record at home assisting the client to take note of the symptoms for 24 hours. A chart such as the McGill Home Recording Card[16] (Fig. 3-7) may help the client outline the existing pattern of the pain and can be used later in the episode of care to assist the therapist in detecting any change in symptoms or function.

There is also a Short-Form McGill Pain Questionnaire that has been validated for use to assess treatment response. It is designed to measure all kinds of pain—both neuropathic and nonneuropathic—using a numeric rating scale to assess 22 pain descriptors from zero (none) to 10 (worst possible).[47]

Medications can alter the pain pattern or characteristics of painful symptoms. Find out how well the client's current medications reduce, control, or relieve pain. Ask how often medications are needed for breakthrough pain.

When using any of the pain rating scales, record the use of any medications that can alter or reduce pain or symptoms such as antiinflammatories or analgesics. At the same time remember to look for side effects or adverse reactions to any drugs or drug combinations.

Watch for clients taking nonsteroidal antiinflammatory drugs (NSAIDs) who experience an increase in shoulder, neck, or back pain several hours after taking the medication. Normally, one would expect symptom relief from NSAIDs so any increase in symptoms is a red flag for possible peptic ulcer.

McGill Home Recording Card

Name:______________________ Date started: ______________

	Morning	Noon	Dinner	Bedtime
M				
Tu				
W				
Th				
F				
Sa				
Su				

Please record:

1. Pain intensity #:
   0 = No pain
   1 = Mild
   2 = Discomforting
   3 = Distressing
   4 = Horrible
   5 = Excruciating
2. # analgesics taken
3. Note any unusual pain, symptoms, or activities on back of card.
4. Record # hours slept in morning column.

Please note: If the client previously rated the pain on a scale from 0–10, substitute the 0–10 scale in place of the 0–5 scale used to describe pain intensity.

Fig. 3-7 McGill Home Recording Card. When assessing constant pain, have the client complete this form for 24 to 48 hours. Pay attention to the client who describes a loss of sleep but who is not awake enough to record missed or interrupted sleep. This may help the physician in differentiating between a sleep disorder and sleep disturbance. You may want to ask the client to record sexual activity as a measure of function and pain levels. It is not necessary to record details, just when the client perceived him or herself as being sexually active. (From Melzack R: The McGill Pain Questionnaire: major properties and scoring methods, *Pain* 1:298, 1975.)

A client frequently will comment that the pain or symptoms have not changed despite 2 or 3 weeks of physical therapy intervention. This information can be discouraging to both client and therapist; however, when the symptoms are reviewed, a decrease in pain, increase in function, reduced need for medications, or other significant improvement in the pattern of symptoms may be seen.

The improvement is usually gradual and is best documented through the use of a baseline of pain activity established at an early stage in the episode of care by using a record such as the Home Recording Card (or other pain rating scale).

However, if no improvement in symptoms or function can be demonstrated, the therapist must again consider a systemic origin of symptoms. Repeating screening questions for medical disease is encouraged throughout the episode of care even if such questions were included in the intake interview.

Because of the progressive nature of systemic involvement, the client may not have noticed any constitutional symptoms at the start of the physical therapy intervention that may now be present. Constitutional symptoms (see Box 1-3) affect the whole body and are characteristic of systemic disease or illness.

Aggravating and Relieving Factors

A series of questions addressing aggravating and relieving factors must be included such as:

FOLLOW-UP QUESTIONS

- What brings your pain (symptoms) on?
- What kinds of things make your pain (symptoms) worse (e.g., eating, exercise, rest, specific positions, excitement, stress)?

  To assess relieving factors, ask:
- What makes the pain better?

  Follow-up questions include:
- How does rest affect the pain/symptoms?
- Are your symptoms aggravated or relieved by any activities?
- If yes, what?
- How has this problem affected your daily life at work or at home?
- How has this problem affected your ability to care for yourself without assistance (e.g., dress, bathe, cook, drive)?

The McGill Pain Questionnaire also provides a chart (Fig. 3-8) that may be useful in determining the presence of relieving or aggravating factors.

	Indicate a plus (+) for aggravating factors or a minus (−) for relieving factors.		
	Liquor		Sleep/rest
	Stimulants (e.g., caffeine)		Lying down
	Eating		Distraction (e.g., television)
	Heat		Urination/defecation
	Cold		Tension/stress
	Weather changes		Loud noises
	Massage		Going to work
	Pressure		Intercourse
	No movement		Mild exercise
	Movement		Fatigue
	Sitting		Standing

Fig. 3-8 Factors aggravating and relieving pain. (From Melzack R: The McGill Pain Questionnaire: major properties and scoring methods, *Pain* 1:277, 1975.)

Systemic pain tends to be relieved minimally, relieved only temporarily, or unrelieved by change in position or by rest. However, musculoskeletal pain is *often* relieved both by a change of position and by rest.

Associated Symptoms

These symptoms may occur alone or in conjunction with the pain of systemic disease. The client may or may not associate these additional symptoms with the chief complaint. The physical therapist may ask:

FOLLOW-UP QUESTIONS

- What other symptoms have you had that you can associate with this problem?

If the client denies any additional symptoms, follow-up this question with a series of possibilities such as:

Burning	Heart palpitations	Numbness/Tingling
Difficulty in breathing	Hoarseness	Problems with vision
Difficulty in swallowing	Nausea	Vomiting
Dizziness	Night sweats	Weakness

Whenever the client says "yes" to such associated symptoms, check for the presence of these symptoms bilaterally. Additionally, bilateral weakness, either proximally or distally, should serve as a red flag possibly indicative of more than a musculoskeletal lesion.

Blurred vision, double vision, scotomas (black spots before the eyes), or temporary blindness may indicate early symptoms of multiple sclerosis or may possibly be warning signs of an impending cerebrovascular accident. The presence of any associated symptoms, such as those mentioned here, would require contact with the physician to confirm the physician's knowledge of these symptoms.

In summary, careful, sensitive, and thorough questioning regarding the multifaceted experience of pain can elicit essential information necessary when making a decision regarding treatment or referral. The use of pain assessment tools such as Fig. 3-6 and Table 3-2 may facilitate clear and accurate descriptions of this critical symptom.

SOURCES OF PAIN

Between the twentieth and twenty-first centuries, the science of clinical pain assessment and management made a significant paradigm shift from an empiric approach to one that is based on identifying and understanding the actual mechanisms involved in the pathogenesis of pain.

The implications of this are immense as we move from classifying pain on the basis of disease, duration, and body part or anatomy to a mechanism-based classification. In this approach the major goal of assessment is to identify the pathophysiologic mechanism of the pain and use this information to plan appropriate intervention.[9,48]

Physical therapists frequently see clients whose primary complaint is pain, which often leads to a loss of function. However, focusing on sources of pain does not always help us to identify the causes of tissue irritation.

The most effective physical therapy diagnosis will define the syndrome and address the causes of pain rather than just identifying the sources of pain.[49] Usually, a careful assessment of pain behavior is invaluable in determining the nature and extent of the underlying pathology.

The clinical evaluation of pain usually involves identification of the primary disease/etiological factor(s) considered responsible for producing or initiating the pain. The client is placed within a broad pain category usually labeled as nociceptive, inflammatory, or neuropathic pain (see Table 3-4). Pain and sensory disturbances associated with central changes (sensitization) may be present with chronic pain.[9,50] It can be difficult in clinical practice to specify which of these, alone or in combination, may be present.[19]

We further classify the pain by identifying the anatomic distribution, quality, and intensity of the pain. Such an approach allows for physical therapy interventions for each identified mechanism involved.

From a screening perspective, we look at the possible *sources* of pain and *types* of pain. When listening to the client's description of pain, consider these possible sources of pain (Table 3-3):

- Cutaneous
- Somatic
- Visceral
- Neuropathic
- Referred

TABLE 3-2 Comparison of Systemic Versus Musculoskeletal Pain Patterns

	Systemic Pain	Musculoskeletal Pain
Onset	• Recent, sudden • Does not present as observed for years without progression of symptoms	May be sudden or gradual, depending on the history • **Sudden:** Usually associated with acute overload stress, traumatic event, repetitive motion; can occur as a side effect of some medications (e.g., statins) • **Gradual:** Secondary to chronic overload of the affected part; may be present off and on for years
Description	• Knifelike quality of stabbing from the inside out, boring, deep aching • Cutting, gnawing • Throbbing • Bone pain • Unilateral or bilateral	• Usually unilateral • May be stiff after prolonged rest, but pain level decreases • Achy, cramping pain • Local tenderness to pressure is present
Intensity	• Related to the degree of noxious stimuli; usually unrelated to presence of anxiety • Mild to severe • Dull to severe	• May be mild to severe • May depend on the person's anxiety level—the level of pain may increase in a client fearful of a "serious" condition
Duration	• Constant, no change, awakens the person at night	• Duration can be modified by rest or change in position • May be constant but is more likely to be intermittent, depending on the activity or the position

Pattern	• Although constant, may come in waves
	• Gradually progressive, cyclical
	• Night pain
	  • Location: chest/shoulder
	  • Accompanied by shortness of breath, wheezing
	  • Eating alters symptoms
	  • Sitting up relieves symptoms (decreases venous return to the heart: possible pulmonary or cardiovascular etiology)
	• Symptoms unrelieved by rest or change in position
	• Migratory arthralgias: Pain/symptoms last for 1 week in one joint, then resolve and appear in another joint
		• Restriction of active/passive/accessory movement(s) observed
		• One or more particular movements "catch" the client and aggravate the pain

Aggravating Factors	• Cannot alter, provoke, alleviate, eliminate, aggravate the symptoms
	• Organ dependent (examples):
	  • Esophagus—eating or swallowing affects symptoms
	  • Heart—cold, exertion, stress, heavy meal (especially when combined) bring on symptoms
	  • Gastrointestinal (GI)—peristalsis (eating) affects symptoms
		• Altered by movement; pain may become worse with movement or some myalgia decreases with movement

Relieving Factors	• Organ dependent (examples):
	  • Gallbladder—leaning forward may reduce symptoms
	  • Kidney—leaning to the affected side may reduce symptoms
	  • Pancreas—sitting upright or leaning forward may reduce symptoms
		• Symptoms reduced or relieved by rest or change in position
		• Muscle pain is relieved by short periods of rest without resulting stiffness, except in the case of fibromyalgia; stiffness may be present in older adults
		• Stretching
		• Heat, cold

TABLE 3-2 Comparison of Systemic Versus Musculoskeletal Pain Patterns—cont'd

	Systemic Pain	Musculoskeletal Pain

Associated Signs and Symptoms	• Fever, chills
	• Sweats (at any time day or night)
	• Unusual vital signs
	• Warning signs of cancer (see Chapter 13)
	• GI symptoms: Nausea, vomiting, anorexia, unexplained weight loss, diarrhea, constipation
	• Early satiety (feeling full after eating)
	• Bilateral symptoms (e.g., paresthesias, weakness, edema, nailbed changes, skin rash)
	• Painless weakness of muscles: more often proximal but may occur distally
	• Dyspnea (breathlessness at rest or after mild exertion)
	• Diaphoresis (excessive perspiration)
	• Headaches, dizziness, fainting
	• Visual disturbances
	• Skin lesions, rashes, or itching that the client may not associate with the musculoskeletal symptoms
	• Bowel/bladder symptoms
	  • Hematuria (blood in the urine)
	  • Nocturia
	  • Urgency (sudden need to urinate)
	  • Frequency
	  • Melena (blood in feces)
	  • Fecal or urinary incontinence
	  • Bowel smears
		• Usually none, although stimulation of trigger points (TrPs) may cause sweating, nausea, blanching

TABLE 3-3 Sources of Pain, Pain Types, and Pain Patterns

Sources	Types	Characteristics/Patterns

Cutaneous
Deep somatic
Visceral
Neuropathic
Referred
	Tension
	Inflammatory
	Ischemic
	Myofascial pain
	  • Muscle tension
	  • Muscle spasm
	  • Trigger points (TrPs)
	  • Muscle deficiency (weakness and stiffness)
	  • Muscle trauma
	Joint pain
	  • Drug-induced
	  • Chemical exposure
	  • Inflammatory bowel disease
	  • Septic arthritis
	  • Reactive arthritis
	Radicular pain
	Arterial, pleural, tracheal
	Gastrointestinal pain
	Pain at rest
	Night pain
	Pain with activity
	Diffuse pain
	Chronic pain
		Client describes:
		  • Location/onset
		  • Description
		  • Frequency
		  • Duration
		  • Intensity
		Therapist recognizes the pattern:
		  • Vascular
		  • Neurogenic
		  • Musculoskeletal/spondylotic
		  • Visceral
		  • Emotional

Cutaneous Sources of Pain

Cutaneous pain (related to the skin) includes superficial somatic structures located in the skin and subcutaneous tissue. The pain is well localized as the client can point directly to the area that "hurts." Pain from a cutaneous source can usually be localized with one finger. Skin pain or tenderness can be associated with referred pain from the viscera or referred from deep somatic structures.

Impairment of any organ can result in sudomotor changes that present as trophic changes such as itching, dysesthesia, skin temperature changes, or dry skin. The difficulty is that biomechanical dysfunction can also result in these same changes, which is why a careful evaluation of soft tissue structures along with a screening exam for systemic disease is required.

Cutaneous pain perception varies from person to person and is not always a reliable indicator of pathologic etiology. These differences in pain perception may be associated with different pain mechanisms. For example, differences in cutaneous pain perception exist based on gender and ethnicity. There may be differences in opioid activity and baroreceptor-regulated pain systems between the sexes to account for these variations.[42]

Somatic Sources of Pain

Somatic pain can be superficial or deep. Somatic pain is labeled according to its source as deep somatic, somatovisceral, somatoemotional (also referred to as *psychosomatic*), or viscerosomatic.

Most of what the therapist treats is part of the somatic system whether we call that the neuromuscular system, the musculoskeletal system, or the NMS system. When psychologic disorders present as somatic dysfunction, we refer to these conditions as psychophysiologic disorders.

Psychophysiologic disorders, including somatoform disorders, are discussed in detail elsewhere.[51-53]

Superficial somatic structures involve the skin, superficial fasciae, tendon sheaths, and periosteum. *Deep somatic pain* comes from pathologic conditions of the periosteum and cancellous (spongy) bone, nerves, muscles, tendons, ligaments, and blood vessels. Deep somatic structures also include deep fasciae and joint capsules. Somatic referred pain does not involve stimulation of nerve roots. It is produced by stimulation of nerve endings within the superficial and deep somatic structures just mentioned.

Somatic referred pain is usually reported as dull, aching, or gnawing or described as an expanding pressure too diffuse to localize. There are no neurologic signs associated with somatic referred pain since this type of pain is considered nociceptive and is not caused by compression of spinal nerves or nerve root. It is possible to have combinations of pain and neurologic findings when more than one pathway is disturbed.[54-56]

Deep somatic pain is poorly localized and may be referred to the body surface, becoming cutaneous pain. It can be associated with an autonomic phenomenon, such as sweating, pallor, or changes in pulse and blood pressure, and is commonly accompanied by a subjective feeling of nausea and faintness.

Pain associated with deep somatic lesions follows patterns that relate to the embryologic development of the musculoskeletal system. This explains why such pain may not be perceived directly over the involved organ (see Fig. 3-1).

Parietal pain (related to the wall of the chest or abdominal cavity) is also considered deep somatic. The visceral pleura (the membrane enveloping the organs) is insensitive to pain, but the parietal pleura is well supplied with pain nerve endings. For this reason, it is possible for a client to have extensive visceral disease (e.g., heart, lungs) without pain until the disease progresses enough to involve the parietal pleura.

When we talk about the "psycho-*somatic*" response, we refer to the mind-*body* connection.

Somatoemotional or *psychosomatic* sources of pain occur when emotional or psychologic distress produces physical symptoms either for a relatively brief period or with recurrent and multiple physical manifestations spanning many months or years. The person affected by the latter may be referred to as a *somatizer*, and the condition is called a *somatization disorder*.

Two different approaches to somatization have been proposed. One method treats somatization as a phenomenon that is secondary to psychologic distress. This is called *presenting somatization*. The second defines somatization as a primary event characterized by the presence of medically unexplained symptoms. This model is called *functional somatization*.[57]

Alternately, there are *viscerosomatic* sources of pain when visceral structures affect the somatic musculature, such as the reflex spasm and rigidity of the abdominal muscles in response to the inflammation of acute appendicitis or the pectoral trigger point associated with an acute myocardial infarction. These visible and palpable changes in the tension of skin and subcutaneous and other connective tissues that are segmentally related to visceral pathologic processes are referred to as connective tissue zones or reflex zones.[58]

Somatovisceral pain occurs when a myalgic condition causes functional disturbance of the underlying viscera, such as the trigger points (TrPs) of the abdominal muscles, causing diarrhea, vomiting, or excessive burping (Case Example 3-2).

Visceral Sources of Pain

Visceral sources of pain include the internal organs and the heart muscle. This source of pain includes all body organs located in the trunk or abdomen, such as those of the respiratory, digestive, urogenital, and endocrine systems, as well as the spleen, the heart, and the great vessels.

Visceral pain is not well localized for two reasons:

1. Innervation of the viscera is multisegmental.
2. There are few nerve receptors in these structures (see Fig. 3-3).

The pain tends to be poorly localized and diffuse.

Visceral pain is well known for its ability to produce referred pain (i.e., pain perceived in an area other than the site of the stimuli). Referred pain occurs because visceral fibers synapse at the level of the spinal cord close to fibers supplying specific somatic structures. In other words, visceral pain corresponds to dermatomes from which the organ receives its innervations, which may be the same innervations for somatic structures.

For example, the heart is innervated by the C3-T4 spinal nerves. Pain of a cardiac source can affect any part of the soma (body) also innervated by these levels. This is one

CASE EXAMPLE 3-2

Somatic Disorder Mimicking Visceral Disease

A 61-year-old woman reported left shoulder pain for the last 3 weeks. The pain radiates down the arm in the pattern of an ulnar nerve distribution. She had no known injury, trauma, or repetitive motion to account for the new onset of symptoms. She denied any constitutional symptoms (nausea, vomiting, unexplained sweating, or sweats). There was no reported shortness of breath.

Pain was described as "gripping" and occurred most often at night, sometimes waking her up from sleep. Physical activity, motion, and exertion did not bring on, reproduce, or make her symptoms worse.

After Completing the Interview and Screening Examination, What Final Question Should Always Be Asked Every Client?

- Do you have any other pain or symptoms of any kind anywhere else in your body?

Result: In response to this question, the client reported left-sided chest pain that radiated to her nipple and then into her left shoulder and down the arm. Palpation of the chest wall musculature revealed a trigger point (TrP) of the pectoralis major muscle. This trigger point was responsible for the chest and breast pain.

Further palpation reproduced a TrP of the left subclavius muscle, which was causing the woman's left arm pain. Releasing the TrPs eliminated all of the woman's symptoms.

Should You Make a Medical Referral for This Client? Yes, referral should be made to rule out a viscerosomatic reflex causing the TrPs. A clinical breast exam (CBE) and mammography may be appropriate, depending on client's history and when she had her last CBE and mammogram.

The client saw a cardiologist. Her echocardiogram and stress tests were negative. She was diagnosed with pseudocardiac disease secondary to a myofascial pain disorder.

From Murphy DR: *Myofascial pain and pseudocardiac disease,* Posted on-line April 22, 2004 [www.chiroweb.com].

reason why someone having a heart attack can experience jaw, neck, shoulder, mid-back, arm, or chest pain and accounts for the many and varied clinical pictures of MI (see Fig. 6-9).

More specifically, the pericardium (sac around the entire heart) is adjacent to the diaphragm. Pain of cardiac and diaphragmatic origin is often experienced in the shoulder because the C5-6 spinal segment (innervation for the shoulder) also supplies the heart and the diaphragm.

Other examples of organ innervations and their corresponding sensory overlap are as follows[4]:

- Sensory fibers to the heart and lungs enter the spinal cord from T1-4 (this may extend to T6).
- Sensory fibers to the gallbladder, bile ducts, and stomach enter the spinal cord at the level of the T7-8 dorsal roots (i.e., the greater splanchnic nerve).
- The peritoneal covering of the gallbladder and/or the central zone of the diaphragm are innervated by the phrenic nerve originating from the C3-5 (phrenic nerve) levels of the spinal cord.
- The phrenic nerve (C3-5) also innervates portions of the pericardium.
- Sensory fibers to the duodenum enter the cord at the T9-10 levels.
- Sensory fibers to the appendix enter the cord at the T10 level (i.e., the lesser splanchnic nerve).
- Sensory fibers to the renal/ureter system enter the cord at the L1-2 level (i.e., the splanchnic nerve).

As mentioned earlier, diseases of internal organs can be accompanied by cutaneous hypersensitivity to touch, pressure, and temperature. This viscerocutaneous reflex occurs during the acute phase of the disease and disappears with its recovery.

The skin areas affected are innervated by the same cord segments as for the involved viscera; they are referred to as *Head's zones.*[58] Anytime a client presents with somatic symptoms also innervated by any of these levels, we must consider the possibility of a visceral origin.

Keep in mind that when it comes to visceral pain, the viscera have few nerve endings. The visceral pleura are insensitive to pain. It is not until the organ capsule (deep somatic structure) is stretched (e.g., by a tumor or inflammation) that pain is perceived and possibly localized. This is why changes can occur within the organs without painful symptoms to warn the person. It is not until the organ is inflamed or distended enough from infection or obstruction to impinge nearby structures or the lining of the chest or abdominal cavity that pain is felt.

The neurology of visceral pain is not well understood. There is not a known central processing system unique to visceral pain. Primary afferent fibers innervating the viscera consist entirely of Aδ and C fibers. Nociceptors of the organs are polymodal, responding to heat, chemical stimuli, and mechanical stimuli (e.g., compression, distention).[5,59] It is known that the afferent supply to internal organs follows a path similar to that of the sympathetic nervous system, often in close proximity to blood vessels.[4] The origins of embryology explain far more of the visceral pain patterns than anything else (see discussion in this chapter).

In the early stage of visceral disease, sympathetic reflexes arising from afferent impulses of the internal viscera may be expressed first as sensory, motor, and/or trophic changes in the skin, subcutaneous tissues, and/or muscles. As mentioned earlier, this can present as itching, dysesthesia, skin temperature changes, or dry skin. The viscera do not perceive pain, but the sensory side is trying to get the message out that something is wrong by creating sympathetic sudomotor changes.

It appears that there is not one specific group of spinal neurons that respond only to visceral inputs. Since messages from the soma and viscera come into the cord at the same level (and sometimes visceral afferents converge over several segments of the spinal cord), the nervous system has trouble deciding: Is it somatic or visceral? It sends efferent information back out to the plexus for change or reaction, but the input results in an unclear impulse at the cord level.

The body may get skin or somatic responses such as muscle pain or aching periosteum or it may tell a viscus innervated at the same level to do something it can do (e.g., the stomach increases its acid content). This also explains how sympathetic signals from the liver to the spinal cord can result in itching or other sudomotor responses in the area embryologically related to the liver.[4]

This somatization of visceral pain is why we must know the visceral pain patterns and the spinal versus visceral innervations. We examine one (somatic) while screening for the other (viscera).

Because the somatic and visceral afferent messages enter at the same level, it is possible to get **somatic-somatic** reflex responses (e.g., a bruise on the leg causes knee pain), **somato-visceral** reflex responses (e.g., a biomechanical dysfunction of the tenth rib can cause gallbladder changes), or **viscero-somatic** reflex responses (e.g., gallbladder impairment can result in a sore 10th rib; pelvic floor dysfunction can lead to incontinence; heart attack causes arm or jaw pain). These are actually all referred pain patterns originating in the soma or the viscera.

A more in-depth discussion of the visceral-somatic response is available.[60] A visceral-somatic response can occur when biochemical changes associated with visceral disease affect somatic structures innervated by the same spinal nerves.

Prior to her death, Dr. Janet Travell[60] was researching how often people with anginal pain are really experiencing residual pectoralis major TrPs caused by previous episodes of angina or MI. This is another example of the viscero-somatic response mentioned.

A **viscero-viscero** reflex (also referred to as *cross-organ sensitization*) occurs when pain or dysfunction in one organ causes symptoms in another organ.[3] For example, the client presents with chest pain and has an extensive cardiac workup with normal findings. The client may be told "it's not in your heart, so don't worry about it."

The problem may really be the gallbladder. Because the gallbladder originates from the same tissue embryologically as the heart, gallbladder impairment can cause cardiac changes in addition to shoulder pain from its contact with the diaphragm. This presentation is then confused with cardiac pathology.[4]

On the other hand, the doctor may do a gallbladder workup and find nothing. The chest pain could be coming from arthritic changes in the cervical spine. This occurs because the cervical spine and heart share common sensory pathways from C3 to the spinal cord.

Information from the cardiac plexus and brachial plexus enter the cord at the same level. The nervous system is not able to identify who sent the message, just what level it came from. It responds as best it can, based on the information present, sometimes resulting in the wrong symptoms for the problem at hand.

Pain and symptoms of a visceral source are usually accompanied by an ANS response such as a change in vital signs, unexplained perspiration (diaphoresis), and/or skin pallor. Signs and symptoms associated with the involved organ system may also be present. We call these *associated signs and symptoms*. They are red flags in the screening process.

Neuropathic Pain

Neuropathic or neurogenic pain results from damage to or pathophysiologic changes of the peripheral or central nervous system (CNS).[61] Neuropathic pain can occur as a result of injury or destruction to the peripheral nerves, pathways in the spinal cord, or neurons located in the brain. Neuropathic pain can be acute or chronic depending on the timeframe.

This type of pain is not elicited by the stimulation of nociceptors or kinesthetic pathways as a result of tissue damage but rather by malfunction of the nervous system itself.[58] Disruptions in the transmission of afferent and efferent impulses in the periphery, spinal cord, and brain can give rise to alterations in sensory modalities (e.g., touch, pressure, temperature), and sometimes motor dysfunction.

It can be drug-induced, metabolic-based, or brought on by trauma to the sensory neurons or pathways in either the peripheral nervous system or CNS. It appears to be idiosyncratic: not all individuals with the same lesion will have pain.[62] Some examples are listed in Table 3-4.

TABLE 3-4 Causes of Neuropathic Pain

Central Neuropathic Pain	Peripheral Neuropathic Pain
Multiple sclerosis	Trigeminal neuralgia (Tic douloureux)
Headache (migraine)	Poorly controlled diabetes mellitus (metabolic-induced)
Stroke	Vincristine (Oncovin) (drug-induced, used in cancer treatment)
Traumatic brain injury (TBI)	Isoniazid (INH) (drug-induced, used to treat tuberculosis)
Parkinson's disease	Amputation (trauma)
Spinal cord injury (incomplete)	Crush injury/brachial avulsion (trauma)
	Herpes zoster (shingles, postherpetic neuralgia)
	Complex regional pain syndrome (CRPS2, causalgia)
	Nerve compression syndromes (e.g., carpel tunnel syndrome, thoracic outlet syndrome)
	Paraneoplastic neuropathy (cancer-induced)
	Cancer (tumor infiltration/ compression of the nerve)
	Liver or biliary impairment (e.g., liver cancer, cirrhosis, primary biliary cirrhosis)
	Leprosy
	Congenital neuropathy (e.g., porphyria)
	Guillain-Barré syndrome

It is usually described as sharp, shooting, burning, tingling, or producing an electric shock sensation. The pain is steady or evoked by some stimulus that is not normally considered noxious (e.g., light touch, cold). Some affected individuals report aching pain. There is no muscle spasm in neurogenic pain.[58] Acute nerve root irritation tends to be severe, described as burning, shooting, and constant. Chronic nerve root pain is more often described as annoying or nagging.

Neuropathic pain is not alleviated by opiates or narcotics, although local anesthesia can provide temporary relief. Medications used to treat neuropathic pain include antidepressants, anticonvulsants, antispasmodics, adrenergics, and anesthetics. Many clients have a combination of neuropathic and somatic pain, making it more difficult to identify the underlying pathology.

Referred Pain

By definition, referred pain is felt in an area far from the site of the lesion but supplied by the same or adjacent neural segments. Referred pain occurs by way of shared central pathways for afferent neurons and can originate from any somatic or visceral source (primary cutaneous pain is not usually referred to other parts of the body).

Referred pain can occur alone or with accompanying deep somatic or visceral pain. When caused by an underlying visceral or systemic disease, visceral pain usually precedes the development of referred musculoskeletal pain. However, the client may not remember or mention this previous pain pattern … and the therapist has not asked about the presence of any other symptoms.

Referred pain is usually well localized (i.e., the person can point directly to the area that hurts), but it does not have sharply defined borders. It can spread or radiate from its point of origin. Local tenderness is present in the tissue of the referred pain area, but there is no objective sensory deficit. Referred pain is often accompanied by muscle hypertonus over the referred area of pain.

Visceral disorders can refer pain to somatic tissue (see Table 3-8). On the other hand, as mentioned in the last topic on visceral sources of pain, some somatic impairments can refer pain to visceral locations or mimic known visceral pain patterns. Finding the original source of referred pain can be quite a challenge (Case Example 3-3).

Always ask one or both of these two questions in your pain interview as part of the screening process:

FOLLOW-UP QUESTIONS

- Are you having any pain anywhere else in your body?
- Are you having symptoms of any other kind that may or may not be related to your main problem?

CASE EXAMPLE 3-3

Type of Pain and Possible Cause

A 44-year-old woman has come to physical therapy with reports of neck, jaw, and chest pain when using her arms overhead. She describes the pain as sharp and "hurting." It is not always consistent. Sometimes she has it, sometimes she does not. Her job as the owner of a window coverings business requires frequent, long periods of time with her arms overhead.

A. Would you classify this as cutaneous, somatic, visceral, neuropathic, or referred pain?

B. What are some possible causes and how can you differentiate neuromusculoskeletal from systemic?

A. The client has not mentioned the skin hurting or pointed to a specific area to suggest a cutaneous source of pain. It could be referred pain, but we do not know yet if it is referred from the neuromusculoskeletal system (neck, ribs, shoulder) or from the viscera (given the description, most likely cardiac).

Without further information, we can say it is somatic or referred visceral pain. We can describe it as radiating since it starts in the neck and affects a wide area above and below that. No defined dermatomes have been identified to suggest a neuropathic cause, so this must be evaluated more carefully.

B. This could be a pain pattern associated with *thoracic outlet syndrome* (TOS) because the lower cervical plexus can innervate as far down as the nipple line. This can be differentiated when performing tests and measures for TOS.

Since TOS can impact the neuro- (brachial plexus) or vascular bundle, it is important to measure blood pressure in both arms and compare them for a possible vascular component.

Onset of *anginal pain* occurs in some people with the use of arms overhead. To discern if this may be a cardiac problem, have the client use the lower extremities to exercise without using the arms (e.g., stairs, stationary bike).

Onset of symptoms from a cardiac origin usually has a lag effect. In other words, symptoms do not start until 5 to 10 minutes after the activity has started. It is not immediate as it might be when using impaired muscles. If the symptoms are reproduced 3 to 5 or 10 minutes after the lower extremity activity, consider a cardiac cause. Look for signs and symptoms associated with cardiac impairment. Ask about a personal/family history of heart disease.

At age 44, she may be perimenopausal (unless she has had a hysterectomy, which brings on surgical menopause) and still on the young side for cardiac cause of upper quadrant symptoms. Still, it is possible and would have to be ruled out by a physician if you are unable to find a NMS cause of symptoms.

Chest pain can have a *wide range of causes,* including trigger points, anabolic steroid or cocaine use, breast disease, premenstrual symptoms, assault or trauma, lactation problems, scar tissue from breast augmentation or reduction, and so on. See further discussion, Chapter 17.

Differentiating Sources of Pain[4]

How do we differentiate somatic sources of pain from visceral sources? It can be very difficult to make this distinction. That is one reason why clients end up in physical therapy even though there is a viscerogenic source of the pain and/or symptomatic presentation.

The superficial and deep somatic structures are innervated unilaterally via the spinal nerves, whereas the viscera are innervated bilaterally through the ANS via visceral afferents. The quality of superficial somatic pain tends to be sharp and more localized. It is mediated by large myelinated fibers, which have a low threshold for stimulation and a fast conduction time. This is designed to protect the structures by signaling a problem right away.

Deep somatic pain is more likely to be a dull or deep aching that responds to rest or a non–weight-bearing position. Deep somatic pain is often poorly localized (transmission via small unmyelinated fibers) and can be referred from some other site.

Pain of a deep somatic nature increases after movement. Sometimes the client can find a comfortable spot, but after moving the extremity or joint, cannot find that comfortable spot again. This is in contrast to visceral pain, which usually is not reproduced with movement, but rather, tends to hurt all the time or with all movements.[4]

Pain from a visceral source can also be dull and aching, but usually does not feel better after rest or recumbency. Keep in mind pathologic processes occurring within somatic structures (e.g., metastasis, primary tumor, infection) may produce localized pain that can be mechanically irritated. This is why movement in general (rather than specific motions) can make it worse. Back pain from metastasis to the spine can become quite severe before any radiologic changes are seen.[4]

Visceral diseases of the abdomen and pelvis are more likely to refer pain to the back, whereas intrathoracic disease refers pain to the shoulder(s). Visceral pain rarely occurs without associated signs and symptoms, although the client may not recognize the correlation. Careful questioning will usually elicit a systemic pattern of symptoms.

Back or shoulder range of motion (ROM) is usually full and painless in the presence of visceral pain, especially in the early stages of disease. When the painful stimulus increases or persists over time, pain-modifying behaviors, such as muscle splinting and guarding, can result in subsequent changes in biomechanical patterns and pain-related disability[63] and may make it more difficult to recognize the systemic origin of musculoskeletal dysfunction.

TYPES OF PAIN

Although there are five sources of most physiologic pain (from a medical screening perspective), many types of pain exist within these categories (see Table 3-3).

When orienting to pain from these main sources, it may be helpful to consider some specific types of pain patterns. Not all pain types can be discussed here, but some of the most commonly encountered are included.

Tension Pain

Organ distention, such as occurs with bowel obstruction, constipation, or the passing of a kidney stone, can cause tension pain. Tension pain can also be caused by blood pooling from trauma and pus or fluid accumulation from infection or other underlying causes. In the bowel, tension pain may be described as "colicky" with waves of pain and tension occurring intermittently as peristaltic contractile force moves irritating substances through the gastrointestinal (GI) system. Tension pain makes it difficult to find a comfortable position.

Inflammatory Pain

Inflammation of the viscera or parietal peritoneum (e.g., acute appendicitis) may cause pain that is described as deep or boring. If the visceral peritoneum is involved, then the pain is usually poorly localized. If the parietal peritoneum is the primary area affected, the pain pattern may become more localized (i.e., the affected individual can point to it with one or two fingers). Pain arising from inflammation causes people to seek positions of quiet with little movement.

Ischemic Pain

Ischemia denotes a loss of blood supply. Any area without adequate perfusion will quickly die. Ischemic pain of the viscera is sudden, intense, constant, and progressive in severity or intensity. It is not typically relieved by analgesics, and no position is comfortable. The person usually avoids movement or change in positions.

Myofascial Pain

Myalgia, or muscle pain, can be a symptom of an underlying systemic disorder. Cancer, renal failure, hepatic disease, and endocrine disorders are only a few possible systemic sources of muscle involvement.

For example, muscle weakness, atrophy, myalgia, and fatigue that persist despite rest may be early manifestations of thyroid or parathyroid disease, acromegaly, diabetes, Cushing's syndrome, or osteomalacia.

Myalgia can be present in anxiety and depressive disorders. Muscle weakness and myalgia can occur as a side effect of drugs. Prolonged use of systemic corticosteroids and immunosuppressive drugs has known adverse effects on the musculoskeletal system, including degenerative myopathy with muscle wasting and tendon rupture.

Infective endocarditis caused by acute bacterial infection can present with myalgias and no other manifestation of endocarditis. The early onset of joint pain and myalgia as the first sign of endocarditis is more likely if the person is older and has had a previously diagnosed heart murmur. Joint pain (arthralgia) often accompanies myalgia, and the client is diagnosed with rheumatoid arthritis.

Polymyalgia rheumatica (PR), which literally means "pain in many muscles," is a disorder marked by diffuse pain and stiffness that primarily affects muscles of the shoulder and pelvic girdles.

With PR, symptoms are vague and difficult to diagnose resulting in delay in medical treatment. The person may wake up one morning with muscle pain and stiffness for no apparent reason or the symptoms may come on gradually over several days or weeks. Adults over age 50 are affected most often (white women have the highest incidence); most cases occur after age 70.[64]

Temporal arteritis occurs in 25% of all cases of PR. Watch for headache, visual changes (blurred or double vision), intermittent jaw pain (claudication), and cranial nerve involvement. The temporal artery may be prominent and painful to touch, and the temporal pulse absent.

From a screening point of view, there are many types of muscle-related pain such as tension, spasm, weakness, trauma, inflammation, infection, neurologic impairment, and trigger points (see Table 3-3).[65] The clinical presentation most common with systemic disease is presented here.

Muscle Tension

Muscle tension, or sustained muscle tone, occurs when prolonged muscular contraction or co-contraction results in local ischemia, increased cellular metabolites, and subsequent pain. Ischemia as a factor in muscle pain remains controversial. Interruption of blood flow in a resting extremity does not cause pain unless the muscle contracts during the ischemic condition.[66]

Muscle tension also can occur with physical stress and fatigue. Muscle tension and the subsequent ischemia may occur as a result of faulty ergonomics, prolonged work positions (e.g., as with computer or telephone operators), or repetitive motion.

Take for example the person sitting at a keyboard for hours each day. Constant typing with muscle co-contraction does not allow for the normal contract-relax sequence. Muscle ischemia results in greater release of substance P, a pain neurotransmitter (neuropeptide).

Increased substance P levels increase pain sensitivity. Increased pain perception results in more muscle spasm as a splinting or protective guarding mechanism, and thus the pain-spasm cycle is perpetuated. This is a somatic-somatic response.

Muscle tension from a visceral-somatic response can occur when pain from a visceral source results in increased muscle tension and even muscle spasm. For example, the pain from any inflammatory or infectious process affecting the abdomen (e.g., appendicitis, diverticulitis, pelvic inflammatory disease) can cause increased tension in the abdominal muscles.

Given enough time and combined with overuse and repetitive use or infectious or inflammatory disease, muscle tension can turn into muscle spasm. When opposing muscles such as the flexors and extensors contract together for long periods of time (called co-contraction), muscle tension and then muscle spasm can occur.

Muscle Spasm

Muscle spasm is a sudden involuntary contraction of a muscle or group of muscles, usually occurring as a result of overuse or injury of the adjoining NMS or musculotendinous attachments. A person with a painful musculoskeletal problem may also have a varying degree of reflex muscle spasm to protect the joint(s) involved (a somatic-somatic response). A client with painful visceral disease can have muscle spasm of the overlying musculature (a viscero-somatic response).

Spasm pain cannot be attributed to transient increased muscle tension because the intramuscular pressure is insufficiently elevated. Pain with muscle spasm may occur from prolonged contraction under an ischemic situation. An increase in the partial pressure of oxygen has been documented inside the muscle in spasm under these circumstances.[67]

Muscle Trauma

Muscle trauma can occur with acute trauma, burns, crush injuries, or unaccustomed intensity or duration of muscle contraction, especially eccentric contractions. Muscle pain occurs as broken fibers leak potassium into the interstitial fluid. Blood extravasation results from damaged blood vessels, setting off a cascade of chemical reactions within the muscle.[66]

When disintegration of muscle tissue occurs with release of their contents (e.g., oxygen-transporting pigment myoglobin) into the bloodstream, a potentially fatal muscle toxicity called *rhabdomyolysis* can occur. Risk factors and clinical signs and symptoms are listed in Table 3-5. Immediate medical attention is required (Case Example 3-4).

Muscle Deficiency

Muscle deficiency (weakness and stiffness) is a common problem as we age and even among younger adults who are deconditioned. Connective tissue changes may occur as small amounts of fibrinogen (produced in the liver and normally converted to fibrin to serve as a clotting factor) leak from the vasculature into the intracellular spaces, adhering to cellular structures.

The resulting microfibrinous adhesions among the cells of muscle and fascia cause increased muscular stiffness. Activity and movement normally break these adhesions; however, with the aging process, production of fewer and less efficient macrophages combined with immobility for any reason result in reduced lysis of these adhesions.[68]

Other possible causes of aggravated stiffness include increased collagen fibers from reduced collagen turnover, increased cross-links of aged collagen fibers, changes in the mechanical properties of connective tissues, and structural and functional changes in the collagen protein. Tendons and ligaments also have less water content, resulting in increased stiffness.[69]

When muscular stiffness occurs as a result of aging, increased physical activity and movement can reduce

TABLE 3-5 Risk Factors for Rhabdomyolysis

Risk Factors	Examples	Signs and Symptoms
Trauma	Crush injury Electric shock Severe burns Extended mobility	Profound muscle weakness Pain Swelling Stiffness and cramping Associated signs and symptoms • Reddish-brown urine (myoglobin) • Decreased urine output • Malaise • Fever • Sinus tachycardia • Nausea, vomiting • Agitation, confusion
Extreme Muscular Activity	Strenuous exercise Status epilepticus Severe dystonia	
Toxic Effects	Ethanol Ethylene glycol Isopropanol Methanol Heroin Barbiturates Methadone Cocaine Tetanus Ecstasy (street drug) Carbon monoxide Snake venom Amphetamines	
Metabolic Abnormalities	Hypothyroidism Hyperthyroidism Diabetic ketoacidosis	
Medication-Induced	Inadvertent intravenous (IV) infiltration (e.g., amphotericin B, azathioprine, cyclosporine) Cholesterol-lowering statins (e.g., Zocor, Lipitor, Crestor)	

Data from Fort CW: How to combat 3 deadly trauma complications, *Nursing2003* 33(5):58–64, 2003.

associated muscular pain. As part of the diagnostic evaluation, consider a general conditioning program for the older adult reporting generalized muscle pain. Even 10 minutes a day on a stationary bike, on a treadmill, or in an aquatics program can bring dramatic and fast relief of painful symptoms when caused by muscle deficiency.

Proximal muscle weakness accompanied by change in one or more deep tendon reflexes is a red flag sign of cancer or neurologic impairment. In the presence of a past medical history of cancer, further screening is advised with possible medical referral required, depending on the outcome of the examination/evaluation.

Trigger Points

TrPs, sometimes referred to as myofascial TrPs (MTrPs), are hyperirritable spots within a taut band of skeletal muscle or in the fascia. Taut bands are ropelike indurations palpated in the muscle fiber. These areas are very tender to palpation and are referred to as local tenderness.[70] There is often a history of immobility (e.g., cast immobilization after fracture or injury), prolonged or vigorous activity such as bending or lifting, or forceful abdominal breathing such as occurs with marathon running.

TrPs are reproduced with palpation or resisted motions. When pressing on the TrP, you may elicit a "jump sign." Some people say the jump sign is a local twitch response of muscle fibers to trigger point stimulation, but this is an erroneous use of the term.[60]

The *jump sign* is a general pain response as the client physically withdraws from the pressure on the point and may even cry out or wince in pain. The *local twitch response* is the visible contraction of tense muscle fibers in response to stimulation.

When TrPs are compressed, local tenderness with possible referred pain results. In other words, pain that arises from the trigger point is felt at a distance, often remote from its source.

The referred pain pattern is characteristic and specific for every muscle. Knowing the TrP locations and their referred pain patterns is helpful. By knowing the pain patterns, you can go to the site of origin and confirm (or rule out) the presence of the TrP. The distribution of referred TrP pain

CASE EXAMPLE 3-4

Military Rhabdomyolysis

A 20-year-old soldier reported to the military physical therapy clinic with bilateral shoulder pain and weakness. He was unable to perform his regular duties due to these symptoms. He attributed this to doing many push-ups during physical training 2 days ago.

When asked if there were any other symptoms of any kind to report, the client said that he noticed his urine was a dark color yesterday (the day after the push-up exercises).

The soldier had shoulder active range of motion (ROM) to 90 degrees accompanied by an abnormal scapulohumeral rhythm with excessive scapular elevation on both sides. Passive shoulder ROM was full but painful. Elbow active and passive ROM were also restricted to 90 degrees of flexion secondary to pain in the triceps muscles.

The client was unable to handle manual muscle testing with pain on palpation to the pectoral, triceps, and infraspinatus muscles, bilaterally. The rotator cuff tendon appeared to be intact.

What Are the Red Flags in This Case?

- Bilateral symptoms (pain and weakness)
- Age (for cancer, too young [under 25 years old] or too old [over 50] is a red-flag sign)
- Change in urine color

Result: The soldier had actually done hundreds of different types of push-ups, including regular, wide-arm, and diamond push-ups. Although the soldier was not in any apparent distress, laboratory studies were ordered. Serum CK level was measured as 9600 U/L (normal range: 55-170 U/L).

The results were consistent with acute exertional rhabdomyolysis (AER), and the soldier was hospitalized. Early recognition of a potentially serious problem may have prevented serious complications possible with this condition.

Physical therapy intervention for muscle soreness without adequate hydration could have led to acute renal failure. He returned to physical therapy for a recovery program following hospitalization.

Data from Baxter RE, Moore JH: Diagnosis and treatment of acute exertional rhabdomyolysis, *J Orthop Sports Phys Ther* 33(3):104–108, 2003.

rarely coincides entirely with the distribution of a peripheral nerve or dermatomal segment.[60]

TrPs can be categorized as active, latent, key, or satellite. *Active* TrPs refer pain locally or to another location and can cause pain at rest. *Latent* TrPs do not cause spontaneous pain but generate referred pain when the affected muscle(s) are put under pressure, palpated, or strained. *Key* TrPs have a pain-referral pattern along nerve pathways, and *satellite* TrPs are set off by key trigger points.

In the screening process, TrPs must be eliminated to rule out systemic pathology as a cause of muscle pain. Beware when your client fails to respond to TrP therapy. Consider this situation a yellow flag. It is *not* necessarily a red flag suggesting the need for screening for systemic or other causes of muscle pain. Muscle recovery from TrPs is not always so simple.

Muscles with active TrPs fatigue faster and recover more slowly. They show more abnormal neural circuit dysfunction. The pain and spasm of TrPs may not be relieved until the aberrant circuits are corrected.[71]

Any compromise of muscle energy metabolism, such as occurs with endocrine or cancer-related disorders, can aggravate and perpetuate TrPs making successful intervention a more challenging and lengthy process.

Remember, too, that visceral disease can create tender points. For those who understand the Jones' Strain/Counterstrain concept, some of the Jones' points might happen to fall in the same area as viscerogenic tender point, but the two are not the same points. A careful evaluation is required to differentiate between Jones' points and viscerogenic tender points.

Travell's TrPs can also produce visceral symptoms without actual organ impairment or disease. This is an example of a somato-visceral response. For example, the client may have an abdominal muscle TrP, but the history is one of upset stomach or chest (cardiac) pain. It is possible to have both tender points and TrPs when the underlying cause is visceral disease.

Pain and dysfunction of myofascial tissues is the subject of several texts to which the reader is referred for more information.[60,72,73]

Joint Pain

Noninflammatory joint pain (no redness, no warmth, no swelling) of unknown etiology can be caused by a wide range of pathologic conditions (Box 3-4). Fibromyalgia, leukemia, sexually transmitted infections, artificial sweeteners,[74-76] Crohn's disease (also known as regional enteritis), postmenopausal status or low estrogen levels, and infectious arthritis are all possible causes of joint pain.

Joint pain in the presence of fatigue may be a red flag for anxiety, depression, or cancer. The client history and screening interview may help the therapist find the true cause of joint pain. Look for risk factors for any of the listed conditions and review the client's recent activities.

When comparing joint pain associated with systemic versus musculoskeletal causes, one of the major differences is in the area of associated signs and symptoms (Table 3-6). Joint pain of a systemic or visceral origin usually has additional signs or symptoms present. The client may not realize there is a connection, or the condition may not have progressed enough for associated signs and symptoms to develop.

The therapist also evaluates joint pain over a 24-hour period. Joint pain from a systemic cause is more likely to be constant and present with all movements. Rest may help at first but over time even this relieving factor will not alter the symptoms. This is in comparison to the client with osteoarthritis (OA), who often feels better after rest (though stiffness may remain). Morning joint pain associated with OA is less than joint pain at the end of the day after using the joint(s) all day.

BOX 3-4 SYSTEMIC CAUSES OF JOINT PAIN

Infectious and noninfectious systemic causes of joint pain can include but are not limited to

- Allergic reactions (e.g., medications such as antibiotics)
- Side effect of other medications such as statins, prolonged use of corticosteroids, aromatase inhibitors
- Delayed reaction to chemicals or environmental factors
- Sexually transmitted infections (STIs) (e.g., human immunodeficiency virus [HIV], syphilis, chlamydia, gonorrhea)
- Infectious arthritis
- Infective endocarditis
- Recent dental surgery
- Lyme disease
- Rheumatoid arthritis
- Other autoimmune disorders (e.g., systemic lupus erythematosus, mixed connective tissue disease, scleroderma, polymyositis)
- Leukemia
- Tuberculosis
- Acute rheumatic fever
- Chronic liver disease (hepatic osteodystrophy affecting wrists and ankles; hepatitis causing arthralgias)
- Inflammatory bowel disease (e.g., Crohn's disease or regional enteritis)
- Anxiety or depression (major depressive disorder)
- Fibromyalgia
- Artificial sweeteners

On the other hand, muscle pain may be worse in the morning and gradually improves as the client stretches and moves about during the day. The Pain Assessment Record Form (see Fig. 3-6) includes an assessment of these differences across a 24-hour span as part of the "Pattern."

The therapist can use the specific screening questions for joint pain to assess any joint pain of unknown cause or with an unusual presentation or history. Joint pain and symptoms that do not fit the expected pattern for injury, overuse, or aging can be screened using a few important questions (Box 3-5).

Drug-Induced

Joint pain as an allergic response, sometimes referred to as "serum sickness" can occur up to 6 weeks after taking a prescription drug (especially antibiotics). Joint pain is also a potential side effect of statins (e.g., Lipitor, Zocor). These are cholesterol-lowering agents.

Musculoskeletal symptoms (e.g., morning stiffness, bone pain, arthralgia, arthritis) are a well-known side effect of chemotherapy and aromatase inhibitors used in the treatment of breast cancer. Low estrogen concentrations and postmenopausal status are linked with these symptoms. Risk factors for developing joint symptoms may include previous hormone replacement therapy, hormone-receptor positivity, previous chemotherapy, obesity, and treatment with anastrozole (Arimidex—aromatase inhibitor).[77]

Noninflammatory joint pain is typical of a delayed allergic reaction. The client may report fever, skin rash, and fatigue that go away when the drug is stopped.

Chemical Exposure

Likewise, delayed reactions can occur as a result of occupational or environmental chemical exposure. A work and/or military history may be required for anyone presenting with joint or muscle pain or symptoms of unknown cause. These clients can be mislabeled with a diagnosis of autoimmune disease or fibromyalgia. The alert therapist may recognize and report clues to help the client obtain a more accurate diagnosis.

Inflammatory Bowel Disease

Ulcerative colitis (UC) and regional enteritis (Crohn's disease [CD]) are accompanied by an arthritic component and skin rash in about 25% of all people affected by this inflammatory bowel condition.

The person may have a known diagnosis of inflammatory bowel disease (IBD) but may not know that new onset of joint symptoms can be part of this condition. The client interview should have brought out the personal history of either UC or CD. See the discussion of IBD in Chapter 8.

Peripheral joint disease associated with IBD involves the large joints, most often a single hip or knee. Joint symptoms often occur simultaneously with UC but less often at the same time as CD. Ankylosing spondylitis (AS) is also possible with either form of IBD.

As with typical AS, symptoms affect the low back, sacrum, or SI joint first. The most common symptoms are intermittent low back pain with decreased low back motion. The course of AS associated with IBD is the same as without the bowel component.

Joint problems usually respond to medical treatment of the underlying bowel disease but in some cases require separate management. Interventions for the musculoskeletal involvement follow the usual protocols for each area affected.

Arthritis

Joint pain (either inflammatory or noninflammatory) can be associated with a wide range of systemic causes, including bacterial or viral infection, trauma, and sexually transmitted diseases. There is usually a positive history or other associated signs and symptoms to help the therapist identify the need for medical referral.

Infectious Arthritis. Joint pain can be a *local* response to an infection. This is called infectious, septic, or bacterial arthritis. Invading microorganisms cause inflammation of the synovial membrane with release of cytokines (e.g., tumor necrosis factor [TNF], interleukin-1 [IL-1]) and proteases.

TABLE 3-6 Joint Pain: Systemic or Musculoskeletal?

	Systemic	Musculoskeletal
Clinical Presentation	Awakens at night Deep aching, throbbing Reduced by pressure* Constant or waves/spasm Cyclical, progressive symptoms	Decreases with rest Sharp Reduced by change in position Reduced or eliminated when stressful action is stopped Restriction of A/PROM Restriction of accessory motions 1 or more movements "catch," reproducing or aggravating pain/symptoms
Past Medical History	Recent history of infection: Hepatitis, bacterial infection from staphylococcus or streptococcus (e.g., cellulitis), mononucleosis, measles, URI, UTI, gonorrhea, osteomyelitis, cellulitis History of bone fracture, joint replacement or arthroscopy History of human bite Sore throat, headache with fever in the last 3 weeks or family/household member with recently diagnosed strep throat Skin rash (infection, medications) Recent medications (last 6 weeks); any drug but especially statins (cholesterol lowering), antibiotics, aromatase inhibitors, chemotherapy Hormone associated (postmenopausal status, low estrogen levels) History of injection drug use/abuse History of allergic reactions History of GI symptoms Recent history of enteric or venereal infection or new sexual contact (e.g., Reiter's) Presence of extensor surface nodules	Repetitive motion Arthritis Static postures (prolonged) Trauma (including domestic violence)
Associated Signs and Symptoms	Jaundice Migratory arthralgias Skin rash/lesions Nodules (extensor surfaces) Fatigue Weight loss Low grade fever Suspicious or aberrant lymph nodes Presence of GI symptoms Cyclical, progressive symptoms Proximal muscle weakness	Usually none Check for trigger points TrPs may be accompanied by some minimal ANS phenomenon (e.g., nausea, sweating)

A/PROM, Active/passive range of motion; *URI,* upper respiratory infection; *UTI,* urinary tract infection; *GI,* gastrointestinal; *TrPs,* trigger points; *ANS,* autonomic nervous system.
*This is actually a cutaneous or somatic response because the pressure provides a counterirritant; it does not really affect the viscera directly.

The end result can be cartilage destruction even after eradicating the offending organism.[78]

Bacteria can find its way to the joint via the bloodstream (most common) by:

- Direct inoculation (e.g., surgery, arthroscopy, intraarticular corticosteroid injection, central line placement, total joint replacement)
- Penetrating wound (e.g., human bite or fracture)
- Direct extension (e.g., osteomyelitis, cellulitis, diverticulitis, abscess)

Staphylococcus aureus, streptococci, and gonococci are the most common infectious causes. A connection between infection and arthritis has been established in Lyme disease. Arthritis can be the first sign of infective endocarditis.[79] Viruses, mycobacteria, fungal agents, and Lyme disease are other causes.[78]

Viral infections such as hepatitis B, rubella (after vaccination), and Fifth's (viral) disease can be accompanied by arthralgias and arthritis sometimes called *viral arthritis*. Joint symptoms appear during the prodromal state of hepatitis (prior to the clinical onset of jaundice).

Sexually transmitted (infectious) diseases (STIs/STDs) are often accompanied by joint pain and symptoms called *gonococcal arthritis*. Joint pain accompanied by skin lesions at the joint or elsewhere may be a sign of sexually transmitted infections.

BOX 3-5 SCREENING QUESTIONS FOR JOINT PAIN

- Please describe the pattern of pain/symptoms from when you wake up in the morning to when you go to sleep at night.
- Do you have any symptoms of any kind anywhere else in your body? (You may have to explain these symptoms don't have to relate to the joint pain; if the client has no other symptoms, offer a short list including constitutional symptoms, heart palpitations, unusual fatigue, nail or skin changes, sores or lesions anywhere but especially in the mouth or on the genitals, and so forth.)
- Have you ever had:
  - Cancer of any kind
  - Leukemia
  - Crohn's disease (regional enteritis)
  - Sexually transmitted infection (you may have to prompt with specific diseases such as chlamydia, genital herpes, genital warts, gonorrhea or "the clap," syphilis, Reiter's syndrome, human immunodeficiency virus [HIV])
  - Fibromyalgia
  - Joint replacement or arthroscopic surgery of any kind
- Have you recently (last 6 weeks) had any:
  - Fractures
  - Bites (human, animal)
  - Antibiotics or other medications
  - Infections [you may have to prompt with specific infections such as strep throat, mononucleosis, urinary tract, upper respiratory (cold or flu), gastrointestinal, hepatitis]
  - Skin rashes or other skin changes
- Do you drink diet soda/pop or use aspartame, Equal, or NutraSweet? (If the client uses these products in any amount, suggest eliminating them on a trial basis for 30 days; artificial sweetener–induced symptoms may disappear in some people; neurotoxic effects from use of newer products (e.g., Stevia, Splenda) have not been fully investigated.)

To the therapist: You may have to conduct an environmental or work history (occupation, military, exposure to chemicals) to identify a delayed reaction.

Quick Survey

- What kind of work do you do?
- Do you think your health problems are related to your work?
- Are your symptoms better or worse when you're at home or at work?
- Follow-up if worse at work: Do others at work have similar problems?
- Have you been exposed to dusts, fumes, chemicals, radiation, or loud noise?
- Follow-up: It may be necessary to ask additional questions based on past history, symptoms, and risk factors present.
- Do you live near a hazardous waste site or any industrial facilities that give off chemical odors or fumes?
- Do you live in a home built more than 40 years ago? Have you done renovations or remodeling?
- Do you use pesticides in your home, on your garden, or on your pets?
- What is your source of drinking water?
- Chronology of jobs (type of industry, type of job, years worked)
- How new is the building you are working in?
- Exposure survey (protective equipment used, exposure to dust, radiation, chemicals, biologic hazards, physical hazards)

In the case of STI/STDs with joint involvement, skin lesions over or near a joint have a typical appearance with a central black eschar or scab-like appearance surrounded by an area of erythema (Fig. 3-9). Alternately, the skin lesion may have a hemorrhagic base with a pustule in the center. Fever and arthritic-like symptoms are usually present (Fig. 3-10).

Anyone with human immunodeficiency virus (HIV) may develop unusual rheumatologic disorders. Diffuse body aches and pain without joint arthritis are common among clients with HIV. (See further discussion on HIV in Chapter 12.)

Other forms of arthritis, such as systemic lupus erythematosus (SLE), scleroderma, polymyositis, and mixed connective tissue disease, may have an infectious-based link, but the connection has never been proven definitively.

Infectious (septic) arthritis should be suspected in an individual with persistent joint pain and inflammation occurring in the course of an illness of unclear origin or in the course of a well-documented infection such as pneumococcal pneumonia, staphylococcal sepsis, or urosepsis.

Major risk factors include age (older than 80 years), diabetes mellitus, intravenous drug use, indwelling catheters, immunocompromised condition, rheumatoid arthritis, or osteoarthritis.[78] Look for a history of preexisting joint damage due to bone trauma (e.g., fracture) or degenerative joint disease.

Other predisposing factors are listed in Box 3-6. Infectious arthritis is a rare complication of anterior cruciate ligament (ACL) reconstruction using contaminated bone-tendon-bone allografts.[80,81] Infections in prosthetic joints can occur years after the implant is inserted. Indwelling catheters and urinary tract infections are major risk factors for seeding to prosthetic joints.[69]

Watch for joint symptoms in the presence of skin rash, low-grade fever, and lymphadenopathy. The rash may appear

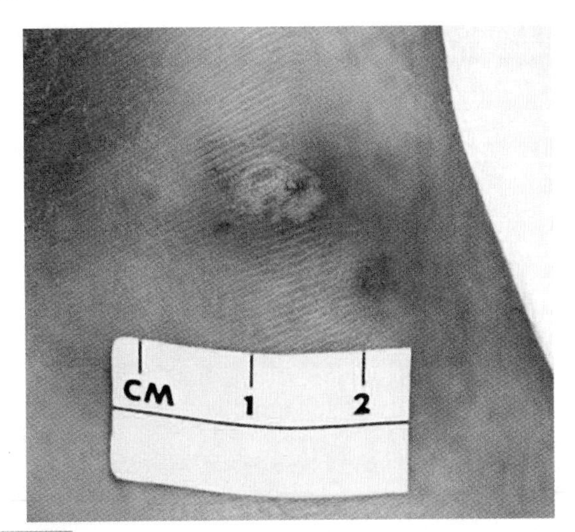

Fig. 3-9 Cutaneous gonococcal lesion secondary to disseminated *Neisseria gonorrhoeae* bacterial infection. Though a sexually transmitted disease, if gonorrhea is allowed to go untreated, the *N. gonorrhoeae* bacteria responsible for the infection can become disseminated throughout the body and form lesions in extragenital locations. This type of lesion can present as (gonococcal) arthritis in any joint; the ankle joint is the target here. (From Goldman L, Schafer AI, et al: *Goldman's Cecil medicine*, ed 24, Philadelphia, 2012, Saunders.)

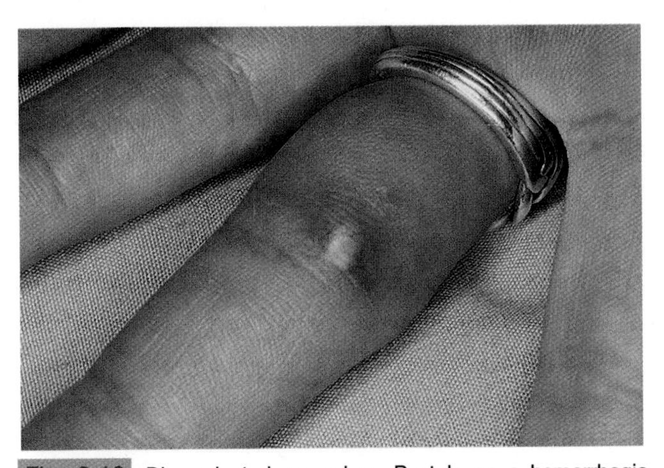

Fig. 3-10 Disseminated gonorrhea. Pustule on a hemorrhagic base. The typical client presents with fever, arthritis, and scattered lesions as show. Cultures from the lesions are often negative. The therapist should always use standard precautions. Medical referral is required. (From Callen JP, Paller AS, Greer KE, et al: *Color atlas of dermatology,* ed 2, Philadelphia, 2000, WB Saunders; Fig. 6-5, p 148.)

and disappear before the joint symptoms. Joints may be mildly to severely involved. Fingers, knees, shoulders, and ankles are affected most often (bilaterally). Inflammation is nonerosive, suggestive of rheumatoid arthritis.

Often one joint is involved (knee or hip), but sometimes two or more are also symptomatic, depending on the underlying pathologic mechanism.[64] Symptoms can range from mild to severe. Joint destruction can be rapid so immediate medical referral is required. Once treated (antibiotics, joint aspiration), the postinfectious inflammation may last for weeks.[82]

With infectious arthritis, the client may be unable to bear weight on the joint. Usually, there is an acute arthritic presentation and the client has a fever (often low grade in older adults or in anyone who is immunosuppressed).

Medical referral is important for the client with joint pain with no known cause and a recent history of infection of any kind. Ask about recent (last 6 weeks) skin lesions or rashes of any kind anywhere on the body, urinary tract infection, or respiratory infection.

Take the client's temperature and ask about recent episodes of fever, sweats, or other constitutional symptoms. Palpate for residual lymphadenopathy. Early diagnosis and intervention are essential to limit joint destruction and preserve function. Diagnosis can be difficult. The physician must differentiate infectious/septic arthritis from reactive arthritis (Case Example 3-5).

BOX 3-6 RISK FACTORS FOR INFECTIOUS ARTHRITIS

- History of:
  - Previous surgery, especially arthroscopy for joint repair or replacement
  - Human bite, tick bite (Lyme disease), fracture, central line placement
  - Direct, penetrating trauma
  - Infection of any kind (e.g., osteomyelitis, cellulitis, diverticulitis, abscess (located anywhere), hepatitis A or B, *Staphylococcus aureus, Streptococcus pneumoniae, gonococci,* or urinary tract or respiratory tract infection)
  - Rheumatoid arthritis, systemic lupus erythematosus, scleroderma, or mixed connective tissue disease
  - Diabetes mellitus
  - Sarcoidosis (inflammatory pulmonary condition can affect knees, proximal interphalangeal [PIP] joints, wrists, elbows)
- Sexually active, young adult
- Injection drug user
- Chronic joint damage (e.g., rheumatoid arthritis, gout)
- Previous infection of joint prosthesis
- Recent immunization
- Increasing age
- Indwelling catheter (especially in the client with a prosthetic joint)
- Malnutrition, skin breakdown
- Immunosuppression or immunocompromise (e.g., renal failure, steroid treatment, organ transplantation, chemotherapy)

CASE EXAMPLE 3-5

Septic Arthritis

A 62-year-old man presented in physical therapy with left wrist pain. There was no redness, warmth, or swelling. Active motion was mildly limited by pain. Passive motion could not be tested because of pain.

All other clinical tests were negative. Neuro screen was negative. Past medical history includes hypertension and non–insulin-dependent diabetes mellitus controlled by diet and exercise.

The client denied any history of fever, skin rashes, or other lesions. He reported a recent trip to Haiti (his native country) 3 weeks ago.

How Do You Screen This Client for Systemic-Induced Joint Pain?

- Review Box 3-6 (Risk Factors for Infectious Arthritis). Besides diabetes, what other risk factors are present? Ask the client about any that apply. Compile a list to review during the Review of Systems.
- Ask the client: Are there any other symptoms of any kind anywhere else in your body?
- Use the client's answer while reviewing clinical signs and symptoms of infectious arthritis for any signs and symptoms of infectious arthritis.
- Review Box 3-5 (Screening Questions for Joint Pain). Are there any further questions from this list appropriate for the screening process?
- Assess the joints above and below (e.g., elbow, shoulder, neck). Assess for trigger points.

Using the information obtained from these steps, look at past medical history, clinical presentation, and associated signs and symptoms. What are the red flags? Review the *Clues to Screening for Viscerogenic Sources of Pain* and *Guidelines for Physician Referral Required* in this chapter.

Based on your findings, decide whether to treat and reevaluate or make a medical referral now.

Result: In this case the therapist did not find enough red flags or suspicious findings to warrant immediate referral. Treatment intervention was initiated. The client missed three appointments because of the "flu." When he returned, his wrist pain was completely gone, but he was reporting left knee pain. There was mild effusion and warmth on both sides of the knee joint. The client stated that he still had some occasional diarrhea from his bout with the flu.

The therapist recognized some additional red flags, including ongoing gastrointestinal (GI) symptoms attributed by the client to the flu and new onset of inflammatory joint pain. The therapist decided to take the client's vital signs and found he was febrile (100° F).

Given his recent travel history, migratory noninflammatory and inflammatory arthralgias, and ongoing constitutional symptoms, the client was referred to his medical doctor. Lab tests resulted in a physician's diagnosis of joint sepsis with hematogenous seeding to the wrist and knee; possible osteomyelitis. Probable cause: Exposure to pathogens in contaminated water or soil during his stay in Haiti.

CLINICAL SIGNS AND SYMPTOMS

Infectious Arthritis

- Fever (low-grade or high), chills, malaise
- Recurrent sore throat
- Lymphadenopathy
- Persistent joint pain
- Single painful swollen joint (knee, hip, ankle, elbow, shoulder)*
- Multiple joint involvement (often migratory)*
- Pain on weight bearing
- Back pain (infective endocarditis)
- Skin lesions (characteristic of the specific underlying infection)
- Conjunctivitis, uveitis
- Other musculoskeletal symptoms depending on the specific underlying infection
  - Myalgias
  - Tenosynovitis (especially wrist and ankle extensor tendon sheaths)
- Elevated C-reactive protein and sedimentation rate

*The particular joint or joints involved and associated signs and symptoms will vary from client to client and are dependent upon the underlying infectious cause. For example, joint involvement with Lyme disease presents differently from Reiter's syndrome or Hepatitis B.

Reactive Arthritis. Reactive arthritis is sometimes used synonymously with Reiter's syndrome, a triad of nongonococcal urethritis, conjunctivitis, and multiple joint involvement of inflammatory arthritis (oligoarthropathy). However, joint symptoms can occur 1 to 4 weeks after infection (e.g., GI or genitourinary (GU) infection) or virus (e.g., Fifth's disease in adults).

The most common GI infections associated with reactive arthritis include *Salmonella*, *Shigella*, and *Campylobacter*, which occur in men and women equally. Reactive arthritis from sexually acquired urethritis is caused by *Chlamydia* or *Ureaplasma* and affects only men.

The joint is not septic (infected) but rather aseptic (without infection). Affected joints are often at a site remote from the primary infection. Often, only one joint is involved (knee, ankle, foot, distal interphalangeal joint), but two or more can be affected.

Reactive arthritis often causes inflammation along tendons or where tendons attach to the bone resulting in persistent pain from plantar fasciitis and sacroiliitis. Nail bed changes can include onycholysis (fingers or toes).

Anyone with joint pain of unknown cause who presents with a skin rash, lesions on the genitals, or recent history of infection (especially GI or GU, usually within the last 1 to 3 weeks) must be referred to a health care clinic or medical doctor for further evaluation.

Radicular Pain

Radicular pain results from direct irritation of axons of a spinal nerve or neurons in the dorsal root ganglion and is experienced in the musculoskeletal system in a dermatome, sclerotome, or myotome.

Radicular, radiating, and referred pain are not the same, although a client can have radicular pain that radiates. Radiating means the pain spreads or fans out from the originating point of pain.

Whereas radicular pain is caused by nerve root compression, referred pain results from activation of nociceptive free nerve endings (nociceptors) of the nervous system in somatic or visceral tissue. The physiologic basis for referred pain is convergence of afferent neurons onto common neurons within the CNS.

The term *sciatica* is outdated and reflects our previous (limited) understanding of referred pain. Regional pain anywhere near, around, or along the pathway of the sciatic nerve was automatically attributed to irritation of the sciatic nerve and labeled "sciatica." The International Association for the Study of Pain recommends replacing the term *sciatica* with radicular pain.[54]

Radiculopathy is another symptom that is separate from radicular pain. Radiculopathy describes a neurologic state in which conduction along a spinal nerve or its roots is blocked. Instead of pain, numbness is the primary symptom (when sensory fibers are blocked) or weakness (when there is a motor block). The numbness will be in a dermatomal pattern, whereas the weakness will present in a myotomal distribution. Radiculopathy is determined by these objective neurologic signs and symptoms rather than by pain. It is possible to have radiculopathy and radicular symptoms at the same time. Radiculopathy can occur alone (no pain) and radicular pain can occur without radiculopathy.[54]

Differentiating between radicular (pain from the peripheral nervous system) and referred pain from the ANS can be difficult. Both can start at one point and radiate outwards. Both can cause pain distal to the site of pathology.

As mentioned previously, the CNS may not be able to distinguish which part of the body is responsible for the input into these common neurons so, for example, ischemia of the heart results in shoulder pain, one of several somatic areas innervated by the same neural segments as the heart.[83]

Referred pain occurs most often far away from the site of pathologic origin of symptoms, whereas radicular pain does not skip myotomes, dermatomes, or sclerotomes associated with the affected peripheral nerves.

For example, cardiac pain may be described as beginning retrosternally (behind the sternum) and radiating to the left shoulder and down the inner side of the left arm. This radiating referred pain is generated via the pathways of the ANS but follows the somatic pattern of ulnar nerve distribution. It is not radicular pain from direct irritation of a spinal nerve of the peripheral nervous system but rather referred pain from shared pathways in the spinal cord.

Ischemic cardiac pain does not cause arm pain, hand pain, or pain in somatic areas other than those innervated at the C3 to T4 spinal levels of the ANS. Similarly, gallbladder pain may be felt to originate in the right upper abdomen and to radiate to the angle of the scapula. These are the somatic areas innervated by the same level of the ANS as the involved viscera mentioned.

Physical disease can localize pain in dermatomal or myotomal patterns. More often the therapist sees a client who describes pain that does not match a dermatomal or myotomal pattern. This is neither referred visceral pain from ANS involvement nor irritation of a spinal nerve. For example, the client who describes whole leg pain or whole leg numbness may be experiencing *inappropriate illness behavior.*

Inappropriate illness behavior is recognized clinically as illness behavior that is out of proportion to the underlying physical disease and is related more to associated psychologic disturbances than to actual physical disease.[84] This behavioral component to pain is discussed in the section on Screening for Systemic Versus Psychogenic Symptoms.

Arterial, Pleural, and Tracheal Pain

Pain arising from arteries, as with arteritis (inflammation of an artery), migraine, and vascular headaches, increases with systolic impulse so that any process associated with increased systolic pressure, such as exercise, fever, alcohol consumption, or bending over, may intensify the already throbbing pain.

Pain from the pleura, as well as from the trachea, correlates with respiratory movements. Look for associated signs and symptoms of the cardiac or pulmonary systems. Listen for a description of pain that is "throbbing" (vascular) or sharp and increased with respiratory movements such as breathing, laughing, or coughing.

Palpation and resisted movements will not reproduce the symptoms, which may get worse with recumbency, especially at night or while sleeping.

Gastrointestinal Pain

Pain arising from the GI tract tends to increase with peristaltic activity, particularly if there is any obstruction to forward progress of the food bolus. The pain increases with ingestion and may lessen with fasting or after emptying the involved segment (vomiting or bowel movement).

On the other hand, pain may occur secondary to the effect of gastric acid on the esophagus, stomach, or duodenum. This pain is relieved by the presence of food or by other neutralizing material in the stomach, and the pain is intensified when the stomach is empty and secreting acid. In these cases it is important to ask the client about the effect of eating on musculoskeletal pain. Does the pain increase, decrease, or stay the same immediately after eating and 1 to 3 hours later?

When hollow viscera, such as the liver, kidneys, spleen, and pancreas, are distended, body positions or movements that increase intraabdominal pressure may intensify the pain, whereas positions that reduce pressure or support the structure may ease the pain.

For example, the client with an acutely distended gallbladder may slightly flex the trunk. With pain arising from a tense, swollen kidney (or distended renal pelvis), the client flexes the trunk and tilts toward the involved side; with pancreatic pain, the client may sit up and lean forward or lie down with the knees drawn up to the chest.

Pain at Rest

Pain at rest may arise from ischemia in a wide variety of tissue (e.g., vascular disease or tumor growth). The acute onset of severe unilateral extremity involvement accompanied by the "five Ps"—pain, pallor, pulselessness, paresthesia, and paralysis—signifies acute arterial occlusion (peripheral vascular disease [PVD]). Pain in this situation is usually described by the client as burning or shooting and may be accompanied by paresthesia.

Pain related to ischemia of the skin and subcutaneous tissues is characterized by the client as burning and boring. All these occlusive causes of pain are usually worse at night and are relieved to some degree by dangling the affected leg over the side of the bed and by frequent massaging of the extremity.

Pain at rest secondary to neoplasm occurs usually at night. Although neoplasms are highly vascularized (a process called *angiogenesis*), the host organ's vascular supply and nutrients may be compromised simultaneously, causing ischemia of the local tissue. The pain awakens the client from sleep and prevents the person from going back to sleep, despite all efforts to do so. See the next section on Night Pain.

The client may describe pain noted on weight-bearing or bone pain that may be mild and intermittent in the initial stages, becoming progressively more severe and more constant. A series of questions to identify the underlying cause of night pain is presented later in this chapter.

Night Pain

Whenever you take a pain history, an evaluation of night pain is important (Box 3-7). As therapists, we are always gauging pain responses to identify where the client might be on the continuum from acute to subacute to chronic. This information helps guide our treatment plan and intervention.

For example, the client who cannot even lie on the involved side is probably fairly acute. Pain modulation is the first order of business. Modalities and cryotherapy may be most effective here. On the other hand, the client who can roll onto the involved side and stay there for 30 minutes to an hour may be more in the subacute phase. A combination of modalities, hands-on treatment, and exercise may be warranted.

The client who can lie on the involved side for up to 2 hours is more likely in the chronic phase of the musculoskeletal condition. Tissue ischemia brings on painful symptoms after prolonged static positioning. A more aggressive approach can usually be taken in these cases. These comments all apply to pain of an NMS origin.

BOX 3-7 SCREENING QUESTIONS FOR NIGHT PAIN

When screening someone with night pain for the possibility of a systemic or cancerous condition, some possible questions are:

- Tell me about the pattern of your symptoms at night (open-ended question).
- Can you lie on that side? For how long?
- (Alternate question): Does it wake you up when you roll onto that side?
- How are you feeling in general when you wake up?
- Follow-up question: Do you have any other symptoms when the pain wakes you up? Give the client time to answer before prompting with choices such as coughing, wheezing, shortness of breath, nausea, need to go to the bathroom, night sweats.

Always ask the client reporting night pain of any kind (not just bone pain) the following screening questions:

- What makes it better/worse?
- What happens to your pain when you sit up? (Upright posture reduces venous return to the heart; decreased pain when sitting up may indicate a cardiopulmonary cause.)
- How does taking aspirin affect your pain/symptoms? (Disproportionate pain relief can occur using aspirin in the presence of bone cancer.)
- How does eating or drinking affect your pain/symptoms (for shoulder, neck, back, hip, pelvic pain/symptoms; gastrointestinal [GI] system)?
- Does taking an antacid such as Tums change your pain/symptoms? (Some women with pain of a cardiac nature experience pain relief much like men do with nitroglycerin; remember, this would be a woman who is postmenopausal, possibly with a personal and/or family history of heart disease—check vital signs!)

Night Pain and Cancer

Pain at night is a classic red flag symptom of cancer, but it does not mean that all pain at night is caused by cancer or that all people with cancer will have night pain.[85] For example, the person who lies down at night and has not even fallen asleep who reports increased pain may just be experiencing the first moment in the day without distractions. Suddenly, his or her focus is on nothing but the pain, so the client may report the pain is much worse at night.

Bone pain at night is the most highly suspicious symptom, especially in the presence of a previous history of cancer. Neoplasms are highly vascularized at the expense of the host. This produces local ischemia and pain.

In the case of bone pain (deep pain; pain on weight bearing), perform a heel strike test. This is done by applying a percussive vertical force with the heel of your hand through the heel of the client's foot in a non–weight-bearing (supine) position. Reproduction of painful symptoms is positive and highly suspicious of a bone fracture or stress reaction.[86]

Keep in mind for the older adult that pain on weight bearing may be a symptom of a hip fracture. It is not uncommon for an older adult to fall and have hip pain and the x-rays are initially negative. If the pain persists, new x-rays or additional imaging may be needed. MRIs are extremely sensitive for a femoral neck fracture very early after the fracture. MRI may miss a pubic ramus fracture, requiring single-photon emission computed tomography (SPECT) bone scan to rule out an occult fracture in a client who has fallen and is still having hip pain.

In a physically capable client, clear the hip, knee, and ankle by asking the client to assume a full squat position. You may also ask him or her to hop on the involved side. These tests are used to screen for pubic ramus or hip stress fractures (reactions). Stress reactions or stress fractures are discussed in Chapter 16.

Pain with Activity

Pain with activity is common with NMS pathology. Mechanical and postural factors are common. Pain with activity from a systemic or disease process is most often caused by vascular compromise. In this context, activity pain of the upper quadrant is known as *angina* when the heart muscle is compromised and *intermittent vascular claudication* in the case of peripheral vascular compromise (lower quadrant).

Pain from an ischemic muscle (including heart muscle) builds up with the use of the muscle and subsides with rest. Thus there is a direct relationship between the degree of circulatory insufficiency and muscle work.

In other words, the interval between the beginning of muscle contraction and the onset of pain depends on how long it takes for hypoxic products of muscle metabolism to accumulate and exceed the threshold of receptor response. This means with vascular-induced pain there is usually a delay or lag time between the beginning of activity and the onset of symptoms.

The client complains that a certain distance walked, a certain level of increased physical activity, or a fixed amount of usage of the extremity brings on the pain. When a vascular pathologic condition causes ischemic muscular pain, the location of the pain depends on the location of the vascular pathologic source. This is discussed in greater detail later in this text (see the section on Arterial Disease in Chapter 6).

The timing of symptom onset offers the therapist valuable screening clues when determining when symptoms are caused by musculoskeletal impairment or by vascular compromise.

Look for immediate pain or symptoms (especially when these can be reproduced with palpation, resistance to movement, and/or a change in position) versus symptoms 5 to 10 minutes after activity begins. Further investigate for the presence of other signs and symptoms associated with cardiac impairment, appropriate risk factors, and positive personal and/or family history.

Diffuse Pain

Diffuse pain that characterizes some diseases of the nervous system and viscera may be difficult to distinguish from the equally diffuse pain so often caused by lesions of the moving parts.

Most clients in this category are those with obscure pain in the trunk, especially when the symptoms are felt only anteriorly.[87] The distinction between visceral pain and pain caused by lesions of the vertebral column may be difficult to make and will require a medical diagnosis.

Chronic Pain

Chronic pain persists past the expected physiologic time of healing. This may be less than 1 month or more often, longer than 6 months. An underlying pathology is no longer identifiable and may never have been present.[88] The International Association for the Study of Pain has fixed 3 months as the most convenient point of division between acute and chronic pain.[89]

There are some who suggest 6 weeks is a better cut-off point in terms of clinical progress. Any longer than that and the client is at increased risk for chronic pain and behavioral consequences of that pain.[90,91] Repeated pain stimuli changes how the body processes pain. Pain signals become faster and more intense, depleting the body's own pain blocking substances (e.g., norepinephrine, serotonin).

Chronic pain syndrome is characterized by a constellation of life changes that produce altered behavior in the individual and persist even after the cause of the pain has been eradicated. This syndrome is a complex multidimensional phenomenon that requires a focus toward maximizing functional abilities rather than treatment of pain.

With chronic pain, the approach is to assess how the pain has affected the person. Physical therapy intervention can be directed toward decreasing the client's emotional response to pain or developing skills to cope with stress and other changes that impair quality of life.

In acute pain, the pain is proportional and appropriate to the problem and is treated as a symptom. In the chronic pain syndrome, uncontrolled and prolonged pain alters both the peripheral nervous system and CNS through processes of neural plasticity and central sensitization and thus pain becomes a disease itself.[40,92]

Each person may have a unique response to pain called a neuromatrix or neurosignature. The *neuromatrix* is initially determined through genetics and early sensory development. Later, life experiences related to pain and coping shape the neural patterns. Each person develops individual perceptual and behavioral responses to pain that are unique to that person.[93]

The person's description of chronic pain often is not well defined and is poorly localized; objective findings are not identified. The person's verbal description of the pain may contain words associated with emotional overlay (see Table 3-1). This is in contrast to the predominance of sensory descriptors associated with acute pain.[88] It may be helpful to ask the client or caregiver to maintain a pain log (see Figs. 3-7 and 3-8).

This should include entries for pain intensity and its relationship to activities or intervention. Clients can be reevaluated regularly for improvement, deterioration, or complications, using the same scales that were used for the initial evaluation.

Always keep in mind that painful symptoms out of proportion to the injury or that are not consistent with the objective findings may be a red flag indicating systemic disease. Pain can be triggered by bodily malfunction or severe illness.

In some cases of chronic pain, a diagnosis is finally made (e.g., spinal stenosis or thyroiditis) and the intervention is specific, not merely pain management. More often, identifying the cause of chronic pain is unsuccessful.

Risk Factors

Research evidence has implicated biologic, psychologic, and social variables as key risk factors in chronic pain. These factors do not operate in isolation but often interact with each other.[94] Cognitive processes, such as thoughts, beliefs, and expectations, are important in understanding chronic pain, adaptation to chronic pain, response to intervention, and disability.[95]

Catastrophizing and/or inflammatory reactivity, which are linked to the tendency to express negative thoughts and emotions, exaggerate the impact of painful experiences, and view the situation as hopeless (and the person in the situation as helpless) are additional risk factors for the development of chronic pain.[96,97]

Risk factors for persistent postsurgical pain include pain in other areas of the body before the operation, high levels of psychosocial distress (e.g., anxiety, depression, panic disorder), tobacco use, sleep disturbance (e.g., insomnia, sleep disruption), and chronic use of opioids. It may be necessary to screen for risk of opioid misuse.

Some of the risk factors for the misuse of opioid analgesics include personal/family history of substance abuse, history of criminal activity and/or legal problems (including driving under the influence [DUI]), heavy tobacco use, history of severe depression or anxiety, and history of rehabilitation for alcohol or other drugs.[98] Special screening tools are available including the 5-item Opioid Risk Tool, Screener and Opioid Assessment for Patients in Pain (SOAPP tool) or the Current Opioid Misuse Measure (COMM). These tools have been validated and provide predictive measures of drug-related behaviors (see articles, discussions, and questionnaires at www.painedu.org).

The therapist should be aware that chronic pain can be associated with physical and/or sexual abuse in both men and women (see discussion of Assault in Chapter 2). The abuse may be part of the childhood history and/or a continuing part of the adult experience.

Fear-Avoidance Behavior

Fear-avoidance behavior can also be a part of disability from chronic pain. The Fear-Avoidance Model of Exaggerated Pain Perception (FAMEPP) or Fear Avoidance Model (FAM) was first introduced in the early 1980s.[99,100] The concept is based on studies that show a person's fear of pain (not physical impairments) is the most important factor in how he or she responds to musculoskeletal pain.

Anxiety, fear of pain, and pain catastrophizing can lead to avoiding physical or social activities. Screening for fear-avoidance behavior to determine whether an individual will resume normal activities (low psychologic distress) or will avoid normal activities due to anticipation of increased pain and/or reinjury (high psychologic distress) can be done using the Fear-Avoidance Beliefs Questionnaire (Table 3-7).[101,102] The therapist should not rely on his or her own perception of patient/client's fear-avoidance behaviors. In addition to the Fear-Avoidance Beliefs Questionnaire (FABQ), the Tampa Scale of Kinesophobia (TSK-11)[103] and Pain Catastrophizing Scale (PSC)[104] are available to identify psychologic beliefs linked with pain.[105]

Elevated fear-avoidance beliefs are not indicative of a red flag for serious medical pathology. They are indicative of someone who has a poorer prognosis for rehabilitation (e.g., poor clinical outcomes, elevated pain symptoms, development of depressive symptoms, greater physical impairments, continued disability).[106] They are more accurately labeled a "yellow flag" indicating psychosocial involvement and provide insight into the prognosis. Such a yellow flag signals the need to modify intervention and consider the need for referral to a psychologist or behavioral counselor.

When the client shows signs of fear-avoidance beliefs, then the therapist's management approach should include education that addresses the client's fear and avoidance behavior and should consider a graded approach to therapeutic exercise.[107]

The therapist can teach clients about the difference between pain and tissue injury. Chronic ongoing pain does not mean continued tissue injury is taking place. This common misconception can result in movement avoidance behaviors.

There are no known "cut-off" scores for referral to a specialist.[107,108] Some researchers categorize FABQ scores into "high" and "low" based on the *physical activity scale* (score range 0 to 24). Less than 15 is a "low" score (low risk for elevated fear-avoidance beliefs) and more than 15 is "high."

Higher numbers indicate increased levels of fear-avoidance beliefs. The distinction between these two categories is minor and arbitrary. It may be best to consider the scores as a continuum rather than dividing them into low or high.[107,108] A cut-off score for the *work scale* indicative of having a decreased chance of returning to work has been proposed. The work subscale of the FABQ is the strongest predictor of work status. There is a greater likelihood of return-to-work for scores less than 30 and less likelihood of return-to-work or increased risk of prolonged work restrictions for scores greater than 34.[109]

Examination of fear-avoidance beliefs may serve as a useful screening tool for identifying clients who are at risk for prolonged work restrictions. Caution is advised when interpreting and applying the results of the FABQ work subscale to individual clients. This screening tool may be a better predictor of low risk for prolonged work restrictions. The work subscale may be less effective in identifying clients at high risk for prolonged work restrictions.[109]

Efforts are underway to develop a single-item screening tool that could be used to identify people with elevated levels

TABLE 3-7 Fear-Avoidance Beliefs Questionnaire (FABQ)

Here are some of the things other patients have told us about their pain. For each statement, please circle any number from 0 to 6 to say how much physical activities, such as bending, walking, or driving, affect or would affect your back pain.

	Completely Disagree			Unsure			Completely Agree
1. My pain was caused by physical activity	0	1	2	3	4	5	6
2. Physical activity makes my pain worse	0	1	2	3	4	5	6
3. Physical activity might harm my back	0	1	2	3	4	5	6
4. I should not do physical activities which (might) make my pain worse	0	1	2	3	4	5	6
5. I cannot do physical activities which (might) make my pain worse	0	1	2	3	4	5	6

The following statements are about how your normal work affects or would affect your back pain.

	Completely Disagree			Unsure			Completely Agree
6. My pain was caused by my work or by an accident at work	0	1	2	3	4	5	6
7. My work aggravated my pain	0	1	2	3	4	5	6
8. I have a claim for compensation for my pain	0	1	2	3	4	5	6
9. My work is too heavy for me	0	1	2	3	4	5	6
10. My work makes or would make my pain worse	0	1	2	3	4	5	6
11. My work might harm my back	0	1	2	3	4	5	6
12. I should not do my normal work with my present pain	0	1	2	3	4	5	6
13. I cannot do my normal work with my present pain	0	1	2	3	4	5	6
14. I cannot do my normal work until my pain is treated	0	1	2	3	4	5	6
15. I do not think I will be back to my normal work	0	1	2	3	4	5	6
16. I do not think that I will ever be able to go back to that work	0	1	2	3	4	5	6

The FABQ is used to quantify the level of fear of pain and beliefs clients with low back pain have about the need to avoid movements or activities that might cause pain. The FABQ has 16 items, each scored from 0 to 6, with higher numbers indicating increased levels of fear-avoidance beliefs. There are 2 subscales: a 7-item work subscale (Sum of items 6, 7, 9, 10, 11, 12, and 15; score range = 0-42) and a 4-item physical activity subscale (Sum of items 2, 3, 4, and 5; score range = 0-24). The FABQ work subscale is associated with current and future disability and work loss in patients with acute and chronic low back pain.

From Waddell G, Somerville D, Henderson I, et al: Fear-avoidance beliefs questionnaire (FABQ) and the role of fear avoidance beliefs in chronic low back pain and disability, *Pain* 52:157–158, 1993.

of fear across a wide variety of impairments such as hip, knee, cervical, shoulder, and neck (not just for back pain).[110]

Differentiating Chronic Pain from Systemic Disease

Sometimes a chronic pain syndrome can be differentiated from a systemic disease by the nature and description of the pain. Chronic pain is usually dull and persistent. The chronic pain syndrome is characterized by multiple complaints, excessive preoccupation with pain, and, frequently, excessive drug use. With chronic pain, there is usually a history of some precipitating injury or event.

Systemic disease is more acute with a recent onset. It is often described as sharp, colicky, knifelike, and/or deep. Look for concomitant constitutional symptoms, any red flags in the personal or family history, and/or any known risk factors. Ask about the presence of associated signs and symptoms characteristic of a particular organ or body system (e.g., GI, GU, respiratory, gynecologic).

Because pain has an affective component, chronic pain can cause anxiety, depression, and anger. The amount of pain behaviors and the intensity of pain perceived can change with alterations in environmental reinforcers (e.g., increasing as the time to return to work draws near, decreasing when no one is watching). For more information and assessment tools, see the discussions related to anxiety and depression in this chapter.

Secondary gain may be a factor in perpetuating the problem. This may be primarily financial, but social and family benefits, such as increased attention or avoidance of unpleasant activities or work situations, may be factors (see later discussion of behavior responses to injury/illness).

Aging and Chronic Pain

Chronic pain in older adults is very common. One in five older Americans is taking analgesic medications regularly. Many take prescription pain medications for more than 6 months.[111]

Older adults are more likely to suffer from arthritis, bone and joint disorders, back problems, and other chronic conditions. Pain is the single most common problem for which aging adults seek medical care.

At the same time, older adults have been observed to present with unusually painless manifestations of common illnesses such as MI, acute abdomen, and infections.[112-114]

To address the special needs of older adults, the AGS has developed specific recommendations for assessment and management of chronic pain (Box 3-8).[115]

BOX 3-8 AGS RECOMMENDATIONS FOR CHRONIC PAIN ASSESSMENT IN THE GERIATRIC POPULATION

- All older clients should be assessed for signs of chronic pain.
- Use alternate words for pain when screening older clients (e.g., burning, discomfort, aching, sore, heavy, tight)
- Contact caregiver for pain assessment in adults with cognitive or language impairments
- Clients with cognitive or language impairments should be observed for nonverbal pain behaviors, recent changes in function, and vocalizations to suggest pain (e.g., irritability, agitation, withdrawal, gait changes, tone changes, nonverbal but vocal utterances such as groaning, crying, or moaning)
- Follow AGS guidelines for comprehensive pain assessment including
  - Medical history
  - Medication history, including current and previously used prescription and over-the-counter (OTC) drugs, as well as any nutraceuticals (natural products, "remedies")
  - Physical examination
  - Review pertinent laboratory results and diagnostic tests (look for clues to the sequence of events leading to present pain complaint)
  - Assess characteristics of pain (frequency, intensity, duration, pattern, description, aggravating and relieving factors); use a standard pain scale such as the visual analogue scale (see Fig. 3-6)
- Observe neuromusculoskeletal (NMS) system for:
  - Neurologic impairments
  - Weakness
  - Hyperalgesia; hyperpathia (exaggerated response to pain stimulus)
  - Allodynia (skin pain to nonnoxious stimulus)
  - Numbness, paresthesia
  - Tenderness, trigger points
  - Inflammation
  - Deformity
- Pain that affects function or quality of life should be included in the medical problem list

Data from American Geriatrics Society (AGS) Panel on Chronic Pain in Older Persons. Clinical practice guidelines, *JAGS* 46:635–651, 1998.

COMPARISON OF SYSTEMIC VERSUS MUSCULOSKELETAL PAIN PATTERNS

Table 3-2 provides a comparison of the clinical signs and symptoms of systemic pain versus musculoskeletal pain using the typical categories described earlier. The therapist must be very familiar with the information contained within this table. Even with these guidelines to follow, the therapist's job is a challenging one.

In the orthopedic setting, physical therapists are very aware that pain can be referred above and below a joint. So, for example, when examining a shoulder problem, the therapist always considers the neck and elbow as potential NMS sources of shoulder pain and dysfunction.

Table 3-8 reflects what is known about referred pain patterns for the musculoskeletal system. Sites for referred pain from a visceral pain mechanism are listed. Lower cervical and upper thoracic impairment can refer pain to the interscapular and posterior shoulder areas.

Likewise, shoulder impairment can refer pain to the neck and upper back, while any condition affecting the upper lumbar spine can refer pain and symptoms to the SI joint and hip. When examining the hip region, the therapist always considers the possibility of an underlying SI or knee joint impairment and so on.

If the client presents with the typical or primary referred pain pattern, he or she will likely end up in a physician's office. A secondary or referred pain pattern can be very deceiving. The therapist may not be able to identify the underlying pathology (in fact, it is not required), but it is imperative to recognize when the clinical presentation does not fit the expected pattern for NMS impairment.

A few additional comments about systemic versus musculoskeletal pain patterns are important. First, it is unlikely that the client with back, hip, SI, or shoulder pain that has been present for the last 5 to 10 years is demonstrating a viscerogenic cause of symptoms. In such a case, systemic origins are suspected only if there is a sudden or recent change in the clinical presentation and/or the client develops constitutional symptoms or signs and symptoms commonly associated with an organ system.

Secondly, note the word descriptors used with pain of a systemic nature: knifelike, boring, deep, throbbing. Pay attention any time someone uses these particular words to describe the symptoms.

Third, observe the client's reaction to the information you provide. Often, someone with a NMS problem gains immediate and intense pain relief just from the examination provided and evaluation offered. The reason? A reduction in the anxiety level.

Many people have a need for high control. Pain throws us in a state of fear and anxiety and a perceived loss of control.

TABLE 3-8 Common Patterns of Pain Referral

Pain Mechanism	Lesion Site	Referral Site
Somatic	C7, T1-5 vertebrae	Interscapular area, posterior
	Shoulder	Neck, upper back
	L1, L2 vertebrae	Sacroiliac (SI) joint and hip
	Hip joint	SI and knee
	Pharynx	Ipsilateral ear
	Temporomandibular joint (TMJ)	Head, neck, heart
Visceral	Diaphragmatic irritation	Shoulder, lumbar spine
	Heart	Shoulder, neck, upper back, TMJ
	Urothelial tract	Back, inguinal region, anterior thigh, and genitalia
	Pancreas, liver, spleen, gallbladder	Shoulder, midthoracic or low back
	Peritoneal or abdominal cavity	Hip pain from abscess of psoas or obturator muscle
Neuropathic	Nerve or plexus	Anywhere in distribution of a peripheral nerve
	Nerve root	Anywhere in corresponding dermatome
	Central nervous system	Anywhere in region of body innervated by damaged structure

Knowing what the problem is and having a plan of action can reduce the amplification of symptoms for someone with soft tissue involvement when there is an underlying psychologic component such as anxiety.

On the other hand, someone with cancer pain, viscerogenic origin of symptoms, or systemic illness of some kind will not obtain relief from or reduction of pain with reassurance. Signs and symptoms of anxiety are presented later in this chapter.

Fourth, aggravating and relieving factors associated with NMS impairment often have to do with change in position or a change (increased or decreased) in activity levels. There is usually some way the therapist can alter, provoke, alleviate, eliminate, or aggravate symptoms of a NMS origin.

Pain with activity is immediate when there is involvement of the NMS system. There may be a delayed increase in symptoms after the initiation of activity with a systemic (vascular) cause.

For the orthopedic or manual therapist, be aware that an upslip of the innominate that does not reduce may be a viscero-somatic reflex. It could be a visceral ligamentous problem. If the problem can be corrected with muscle energy techniques or other manual therapy intervention, but by the end of the treatment session or by the next day, the correction is gone and the upslip is back, then look for a possible visceral source as the cause.[4]

If you can reduce the upslip, but it does not hold during the treatment session, then look for the source of the problem at a lower level. It can even be a crossover pattern from the pelvis on the other side.[4]

Aggravating and relieving factors associated with systemic pain are organ dependent and based on visceral function. For example, chest pain, neck pain, or upper back pain from a problem with the esophagus will likely get worse when the client is swallowing or eating.

Back, shoulder, pelvic, or sacral pain that is made better or worse by eating, passing gas, or having a bowel movement is a red flag. Painful symptoms that start 3 to 5 minutes after initiating an activity and go away when the client stops the activity suggest pain of a vascular nature. This is especially true when the client uses the word "throbbing," which is a descriptor of a vascular origin.

Clients presenting with vascular-induced musculoskeletal complaints are not likely to come to the therapist with a report of cardiac-related chest pain. Rather, the therapist must be alert for the man over age 50 or postmenopausal woman with a significant family history of heart disease, who is borderline hypertensive. New onset or reproduction of back, neck, temporomandibular joint (TMJ), shoulder, or arm pain brought on by exertion with arms raised overhead or by starting a new exercise program is a red flag.

Leaning forward or assuming a hands and knees position sometimes lessens gallbladder pain. This position moves the distended or inflamed gallbladder out away from its position under the liver. Leaning or side bending toward the painful side sometimes ameliorates kidney pain. Again, for some people, this may move the kidney enough to take the pressure off during early onset of an infectious or inflammatory process.

Finally, notice the long list of potential signs and symptoms associated with systemic conditions (see Table 3-2). At the same time, note the *lack* of associated signs and symptoms listed on the musculoskeletal side of the table. Except for the possibility of some ANS responses with the stimulation of trigger points, there are no comparable constitutional or systemic signs and symptoms associated with the NMS system.

CHARACTERISTICS OF VISCEROGENIC PAIN

There are some characteristics of viscerogenic pain that can occur regardless of which organ system is involved. Any of these by itself is cause for suspicion and careful listening and watching. They often occur together in clusters of two or three. Watch for any of the following components of the pain pattern.

Gradual, Progressive, and Cyclical Pain Patterns

Gradual, progressive, and cyclical pain patterns are characteristic of viscerogenic disease. The one time this pain pattern occurs in an orthopedic situation is with the client who has low back pain of a discogenic origin. The client is given the appropriate intervention and begins to do his/her exercise program. The symptoms improve, and the client completes a full weekend of gardening, 18 holes of golf, or other excessive activity.

The activity aggravates the condition, and the symptoms return worse than before. The client returns to the clinic, gets firm reminders by the therapist regarding guidelines for physical activity, and is sent out once again with the appropriate exercise program. The "cooperate—get better—then overdo" cycle may recur until the client completes the rehabilitation process and obtains relief from symptoms and return of function.

This pattern can mimic the gradual, progressive, and cyclical pain pattern normally associated with underlying organic pathology. The difference between a NMS pattern of pain and symptoms and a visceral pattern is the NMS problem gradually improves over time, whereas the systemic condition gets worse.

Of course, beware of the client with discogenic back and leg pain who suddenly returns to the clinic completely symptom free. There is always the risk of disc herniation and sequestration when the nucleus detaches and becomes a loose body that may enter the spinal canal. In the case of a "miraculous cure" from disc herniation, be sure to ask about the onset of any new symptoms, especially changes in bowel and bladder function.

Constant Pain

Pain that is constant and intense should raise a red flag. There is a logical and important first question to ask anyone who says the pain is "constant." Can you think what this question might be?

 **FOLLOW-UP QUESTIONS**

- Do you have that pain right now?

It is surprising how often the client will answer "No" to this question. While it is true that pain of a NMS origin can be constant, it is also true there is usually some way to modulate it up or down. The client often has one or two positions that make it better (or worse).

Constant, intense pain in a client with a previous personal history of cancer and/or in the presence of other associated signs and symptoms raises a red flag. You may want to use the McGill Home Recording Card to assess the presence of true constant pain (see Fig. 3-7).

It is not necessary to have the client complete an entire week's pain log to assess constant pain. A 24- to 48-hour time period is sufficient. Use the recording scale on the right indicating pain intensity and medications taken (prescription and OTC).

Under item number 3, include sexual activity. The particulars are not necessary, just some indication that the client was sexually active. The client defines "sexually active" for him or herself, whether this is just touching and holding or complete coitus. This is another useful indicator of pain levels and functional activity.

Remember to offer clients a clear explanation for any questions asked concerning sexual activity, sexual function, or sexual history. There is no way to know when someone will be offended or claim sexual harassment. It is in your own interest to behave in the most professional manner possible.

There should be no hint of sexual innuendo or humor injected into any of your conversations with clients at any time. The line of sexual impropriety lies where the complainant draws it and includes appearances of misbehavior. This perception differs broadly from client to client.[4]

Finally, the number of hours slept is helpful information. Someone who reports sleepless nights may not actually be awake, but rather, may be experiencing a sleep disturbance. Cancer pain wakes the client up from a sound sleep. An actual record of being awake and up for hours at night or awakened repeatedly is significant (Case Example 3-6). See the discussion on Night Pain earlier in this chapter.

Physical Therapy Intervention "Fails"

If a client does not get better with physical therapy intervention, do not immediately doubt yourself. The lack of progression in treatment could very well be a red flag symptom. If the client reports improvement in the early intervention phase but later takes a turn for the worse, it may be a red flag. Take the time to step back, reevaluate the client and your intervention, and screen if you have not already done so (or screen again if you have).

If painful, tender, or sore points (e.g., TrPs, Jones' points, acupuncture/acupressure points/Shiatsu) are eliminated with intervention then return quickly (by the end of the individual session), suspect visceral pathology. If a tender point comes back later (several days or weeks), you may not be holding it long enough.[4]

Bone Pain and Aspirin

There is one odd clinical situation you should be familiar with, not because you are likely to see it, but because the physicians may use this scenario to test your screening knowledge. Before the advent of nonaspirin pain relievers, a major red flag was always the disproportionate relief of bone pain from cancer with a simple aspirin.

The client who reported such a phenomenon was suspected of having osteoid osteoma and a medical workup would be ordered. The mechanism behind this is explained by the fact that salicylates in the aspirin inhibit the pain-inducing prostaglandins produced by the bone tumor.

CASE EXAMPLE 3-6

Constant Night Pain

A 33-year-old man with left shoulder pain reports "constant pain at night." After asking all the appropriate screening questions related to night pain and constant pain, you see the following pattern:

Shoulder pain that is made worse by lying down whether it is at night or during the day. There are no increased pulmonary or breathing problems at night when lying down. Pain is described as a "deep aching." The client cannot find a comfortable position and moves from bed to couch to chair to bed all night long.

He injured his arm 6 months ago in a basketball game when he fell and landed on that shoulder. Symptoms have been gradually getting worse and nothing he does makes it go away. He reports a small amount of relief if he puts a rolled towel under his armpit.

He is not taking any medication; has no significant personal or family history for cancer, kidney, heart, or stomach disease; and has no other symptoms of any kind.

Do You Need to Screen any Further for Systemic Origin of Symptoms?

Probably not, even though there are what look like red flags:

- Constant pain
- Deep aching
- Symptoms beyond the expected time for physiologic healing
- No position is comfortable

Once you complete the objective tests and measures, you will have a better idea if further questions are needed. Although his pain is "constant" and occurs at night, it looks like it may be positional.

An injury 6 months ago with continued symptoms falls into the category of "symptoms persist beyond the expected time for physiologic healing." His description of not being able to find a position of comfort is a possible example of "no position is comfortable."

Given the mechanism of injury and position of mild improvement (towel roll under the arm), it may be more likely that a soft tissue tear is present and physiologic healing has not been possible.

Referral to a physician (or returning the client to the referring physician) may not be necessary just yet. Some clients do not want surgery and opt for a rehabilitation approach. Make sure you have all the information from the primary care physician if there is one involved. Your rehabilitation protocol will depend on a specific diagnosis (e.g., torn rotator cuff, labral tear, impingement syndrome).

If the client does not respond to physical therapy intervention, reevaluation (possibly including a screening component) is warranted with physician referral considered at that time.

When conversing with a physician, it is not necessary for the therapist to identify the specific underlying pathology as a bone tumor. Such a conclusion is outside the scope of a physical therapist's practice.

However, recognizing a sign of something that does not fit the expected mechanical or NMS pattern *is* within the scope of our practice and that is what the therapist can emphasize when communicating with medical doctors. Understanding this concept and being able to explain it in medical terms can enhance communication with the physician.

BOX 3-9 RANGE OF MOTION CHANGES WITH SYSTEMIC DISEASE

- **Early screening:** Full and pain-free range of motion (ROM)
- **Late screening:** Biomechanical response to pain results in changes associated with splinting and guarding

Pain Does Not Fit the Expected Pattern

In a primary care practice or under direct access, the therapist may see a client reporting back, hip, or SI pain of systemic or visceral origin early on in its development. In these cases, during early screening, the client often presents with full and pain-free ROM. Only after pain has been present long enough to cause splinting and guarding, does the client exhibit biomechanical changes (Box 3-9).

SCREENING FOR EMOTIONAL AND PSYCHOLOGIC OVERLAY

Pain, emotions, and pain behavior are all integral parts of the pain experience. There is no disease, illness, or state of pain without an accompanying psychologic component.[4] This does not mean the client's pain is not real or does not exist on a physical level. In fact, clients with behavioral changes may also have significant underlying injury.[116] Physical pain and emotional changes are two sides of the same coin.[117]

Pain is not just a physical sensation that passes up to consciousness and then produces secondary emotional effects. Rather, the neurophysiology of pain and emotions are closely linked throughout the higher levels of the CNS. Sensory and emotional changes occur simultaneously and influence each other.[90]

The sensory discriminative component of pain is primarily physiologic in nature and occurs as a result of nociceptive stimulation in the presence of organic pathology. The motivational-affective dimension of pain is psychologic in nature subject to the underlying principles of emotional behavior.[99]

The therapist's practice often includes clients with personality disorders, malingering, or other psychophysiologic disorder. Psychophysiologic disorders (also known as *psychosomatic* disorders) are any conditions in which the physical symptoms may be caused or made worse by psychologic factors.

Recognizing somatic signs of any psychophysiologic disorder is part of the screening process. Behavioral, psychologic, or medical treatment may be indicated. Psychophysiologic disorders are generally characterized by subjective complaints that exceed objective findings, symptom development in the presence of psychosocial stresses, and physical symptoms involving one or more organ systems. It is the last variable that can confuse the therapist when trying to screen for medical disease.

It is impossible to discuss the broad range of psychophysiologic disorders that comprise a large portion of the physical therapy caseload in a screening text of this kind. The therapist is strongly encouraged to become familiar with the *Diagnostic and Statistical Manual of Mental Disorders (DSM-IV-TR)*[51] to understand the psychologic factors affecting the successful outcome of rehabilitation.

However, recognizing clusters of signs and symptoms characteristic of the psychologic component of illness is very important in the screening process. Likewise, the therapist will want to become familiar with nonorganic signs indicative of psychologic factors.[118-120]

Three key psychologic components have important significance in the pain response of many people:

- Anxiety
- Depression
- Panic Disorder

Anxiety, Depression, and Panic Disorder

Psychologic factors, such as emotional stress and conflicts leading to anxiety, depression, and panic disorder, play an important role in the client's experience of physical symptoms. In the past, physical symptoms caused or exacerbated by psychologic variables were labeled psychosomatic.

Today the interconnections between the mind, the immune system, the hormonal system, the nervous system, and the physical body have led us to view psychosomatic disorders as psychophysiologic disorders.

There is considerable overlap, shared symptoms, and interaction between these emotions. They are all part of the normal human response to pain and stress[90] and occur often in clients with serious or chronic health conditions. Intervention is not always needed. However, strong emotions experienced over a long period of time can become harmful if excessive.

Depression and anxiety often present with somatic symptoms that may resolve with effective treatment of these disorders. Diagnosis of these conditions is made by a medical doctor or trained mental health professional. The therapist can describe the symptoms and relay that information to the appropriate agency or individual when making a referral.

Anxiety

Anyone who feels excessive anxiety may have a generalized anxiety disorder with excessive and unrealistic worry about day-to-day issues that can last months and even longer.

Anxiety amplifies physical symptoms. It is like the amplifier ("amp") on a sound system. It does not change the sound; it just increases the power to make it louder. The tendency to amplify a broad range of bodily sensations may be an important factor in experiencing, reporting, and functioning with an acute and relatively mild medical illness.[121]

Keep in mind the known effect of anxiety on the *intensity* of pain of a musculoskeletal versus systemic origin. Defining the problem, offering reassurance, and outlining a plan of action with expected outcomes can reduce painful symptoms amplified by anxiety. It does not ameliorate pain of a systemic nature.[122]

Musculoskeletal complaints, such as sore muscles, back pain, headache, or fatigue, can result from anxiety-caused tension or heightened sensitivity to pain. Anxiety increases muscle tension, thereby reducing blood flow and oxygen to the tissues, resulting in a buildup of cellular metabolites.

Somatic symptoms are diagnostic for several anxiety disorders, including panic disorder, agoraphobia (fear of open places, especially fear of being alone or of being in public places) and other phobias (irrational fears), obsessive-compulsive disorder (OCD), posttraumatic stress disorder (PTSD), and generalized anxiety disorders.

Anxious persons have a reduced ability to tolerate painful stimulation, noticing it more or interpreting it as more significant than do nonanxious persons. This leads to further complaining about pain and to more disability and pain behavior such as limping, grimacing, or medication seeking.

To complicate matters more, persons with an organic illness sometimes develop anxiety known as *adjustment disorder with anxious mood*. Additionally, the advent of a known organic condition, such as a pulmonary embolus or chronic obstructive pulmonary disease (COPD), can cause an agoraphobia-like syndrome in older persons, especially if the client views the condition as unpredictable, variable, and disabling.

According to C. Everett Koop, the former U.S. Surgeon General, 80% to 90% of all people seen in a family practice clinic are suffering from illnesses caused by anxiety and stress. Emotional problems amplify physical symptoms such as ulcerative colitis, peptic ulcers, or allergies. Although allergies may be inherited, anxiety amplifies or exaggerates the symptoms. Symptoms may appear as physical, behavioral, cognitive, or psychologic (Table 3-9).

The Beck Anxiety Inventory (BAI) quickly assesses the presence and severity of client anxiety in adolescents and adults ages 17 and older. It was designed to reduce the overlap between depression and anxiety scales by measuring anxiety symptoms shared minimally with those of depression.

The BAI consists of 21 items, each scored on a 4-point scale between 0 and 3, for a total score ranging from 0 to 63. Higher scores indicate higher levels of anxiety. The BAI is reported to have good reliability for clients with various psychiatric diagnoses.[123,124]

Both physiologic and cognitive components of anxiety are addressed in the 21 items describing subjective, somatic, or panic-related symptoms. The BAI differentiates between anxious and nonanxious groups in a variety of clinical settings and is appropriate for all adult mental health populations.

TABLE 3-9 Symptoms of Anxiety and Panic

Physical	Behavioral	Cognitive	Psychologic
Increased sighing respirations	Hyperalertness	Fear of losing mind	Phobias
Increased blood pressure	Irritability	Fear of losing control	Obsessive-compulsive behavior
Tachycardia	Uncertainty		
Muscle tension	Apprehension		
Dizziness	Difficulty with memory or concentration		
Lump in throat	Sleep disturbance		
Shortness of breath			
Clammy hands			
Dry mouth			
Diarrhea			
Nausea			
Muscle tension			
Profuse sweating			
Restlessness, pacing, irritability, difficulty concentrating			
Chest pain*			
Headache			
Low back pain			
Myalgia (muscle pain, tension, or tenderness)			
Arthralgia (joint pain)			
Abdominal (stomach) distress			
Irritable bowel syndrome (IBS)			

*Chest pain associated with anxiety accounts for more than half of all emergency department admissions for chest pain. The pain is substernal, a dull ache that does not radiate, and is not aggravated by respiratory movements but is associated with hyperventilation and claustrophobia. See Chapter 17 for further discussion of chest pain triggered by anxiety.

Depression

Once defined as a deep and unrelenting sadness lasting 2 weeks or more, depression is no longer viewed in such simplistic terms. As an understanding of this condition has evolved, scientists have come to speak of the *depressive illnesses*. This term gives a better idea of the breadth of the disorder, encompassing several conditions, including depression, dysthymia, bipolar disorder, and seasonal affective disorder (SAD).

Although these conditions can differ from individual to individual, each includes some of the symptoms listed. Often the classic signs of depression are not as easy to recognize in people older than 65, and many people attribute such symptoms simply to "getting older" and ignore them.

Anyone can be affected by depression at any time. There are, in fact, many underlying physical and medical causes of depression (Box 3-10), including medications used for Parkinson's disease, arthritis, cancer, hypertension, and heart disease (Box 3-11). The therapist should be familiar with these.

For example, anxiety and depressive disorders occur at a higher rate in clients with COPD, obesity, diabetes, asthma, arthritis, cancer, and cardiovascular disease.[125,126] Other risk factors for depression include lifestyle choices such as tobacco use, physical inactivity and sedentary lifestyle, and binge drinking.[127] There is also a link between depression and heart risks in women. Depressed but otherwise healthy postmenopausal women face a 50% higher risk of dying from heart disease than women who are not depressed.[128]

People with chronic pain have three times the average risk of developing depression or anxiety, and clients who are depressed have three times the average risk of developing chronic pain.[129]

Almost 500 million people are suffering from mental disorders today. One in four families has at least one member with a mental disorder at any point in time, and these numbers are on the increase. Depressive disorders are the fourth leading cause of disease and disability. Public health prognosticators predict that by 2020, clinical depression will be the leading cause of medical disability on earth. Adolescents are increasingly affected by depression.[130]

The reasons for the increased incidence are speculative at best. Rapid cultural change around the world, worldwide poverty, and the aging of the world's population (the incidence of depression and dementia increases with age) have been put forth by researchers as possibilities.[131-133]

Others suggest better treatment of the symptoms has resulted in fewer suicides.[134] Researchers think that genes may play a role in a person's risk of developing depression.[135-137] In earlier times, adults who had this genetic link may have committed suicide before bearing children and passing the gene on. Today, with better treatment and greater longevity, people with major depressive disorders may unwittingly pass the disease on to their children.[138]

New insights on depression have led scientists to see clinical depression as a biologic disease possibly originating in the brain with multiple visceral involvements (Table 3-10). One error in medical treatment has been to recognize and treat the client's esophagitis, palpitations, irritable bowel, heart

BOX 3-10 PHYSICAL CONDITIONS COMMONLY ASSOCIATED WITH DEPRESSION

Cardiovascular
Atherosclerosis
Hypertension
Myocardial infarction
Angioplasty or bypass surgery

Central Nervous System
Parkinson's disease
Huntington's disease
Cerebral arteriosclerosis
Stroke
Alzheimer's disease
Temporal lobe epilepsy
Postconcussion injury
Multiple sclerosis
Miscellaneous focal lesions

Endocrine, Metabolic
Hyperthyroidism
Hypothyroidism
Addison's disease
Cushing's disease
Hypoglycemia
Hyperglycemia
Hyperparathyroidism
Hyponatremia
Diabetes mellitus
Pregnancy (postpartum)

Viral
Acquired immunodeficiency syndrome (AIDS)
Hepatitis
Pneumonia
Influenza

Nutritional
Folic acid deficiency
Vitamin B_6 deficiency
Vitamin B_{12} deficiency

Immune
Fibromyalgia
Chronic fatigue syndrome
Systemic lupus erythematosus
Sjögren's syndrome
Rheumatoid arthritis
Immunosuppression (e.g., corticosteroid treatment)

Cancer
Pancreatic
Bronchogenic
Renal
Ovarian

Miscellaneous
Pancreatitis
Sarcoidosis
Syphilis
Porphyria
Corticosteroid treatment

From Goodman CC. Biopsychosocial-spiritual concepts related to health care. In Goodman CC, Fuller K: *Pathology: implications for the physical therapist,* ed 3, Philadelphia, 2009, WB Saunders.

disease, asthma, or chronic low back pain without seeing the real underlying impairment of the CNS (CNS dysregulation: depression) leading to these dysfunctions.[134,139,140]

A medical diagnosis is necessary because several known physical causes of depression are reversible if treated (e.g., thyroid disorders, vitamin B_{12} deficiency, medications [especially sedatives], some hypertensives, and H_2 blockers for stomach problems). About half of clients with panic disorder will have an episode of clinical depression during their lives.

Depression is not a normal part of the aging process, but it is a normal response to pain or disability and may influence the client's ability to cope. Whereas anxiety is more apparent in acute pain episodes, depression occurs more often in clients with chronic pain.

The therapist may want to screen for psychosocial factors, such as depression, that influence physical rehabilitation outcomes, especially when a client demonstrates acute pain that persists for more than 6 to 8 weeks. Screening is also important because depression is an indicator of poor prognosis.[141]

In the primary care setting, the physical therapist has a key role in identifying comorbidities that may have an impact on physical therapy intervention. Depression has been clearly identified as a factor that delays recovery for clients with low back pain. The longer depression is undetected, the greater the likelihood of prolonged physical therapy intervention and increased disability.[141,142]

Tests such as the Beck Depression Inventory (BDI) second edition (BDI-II),[143-145] the Zung Depression Scale,[146] or the Geriatric Depression Scale (short form) (Table 3-11) can be administered by a physical therapist to obtain baseline information that may be useful in determining the need for a medical referral. These tests do not require interpretation that is out of the scope of physical therapist practice.

The short form of the BDI, the most widely used instrument for measuring depression, takes five minutes to complete, and is also used to monitor therapeutic progress. The BDI consists of questions that are noninvasive and straightforward in presentation.

BOX 3-11 DRUGS COMMONLY ASSOCIATED WITH DEPRESSION

- Antianxiety medications (e.g., Valium, Xanax)
- Illegal drugs (e.g., cocaine, crack)
- Antihypertensive drugs (e.g., beta blockers, antiadrenergics)
- Cardiovascular medications (e.g., digitoxin, digoxin)
- Antineoplastic agents (e.g., vinblastine)
- Opiate analgesics (e.g., morphine, Demerol, Darvon)
- Anticonvulsants (e.g., Dilantin, phenobarbital)
- Corticosteroids (e.g., prednisone, cortisone, dexamethasone)
- Nonsteroidal antiinflammatory drugs (NSAIDs) (e.g., indomethacin)
- Alcohol
- Hormone replacement therapy and oral contraceptives

For a complete list of drugs that can cause depression, see Wolfe S: List of drugs that cause depression, Public Citizen's Health Research Group, Washington, DC, 2004 http://www.worstpills.org/

The BDI-II is a 21-item self-report instrument intended to assess the existence and severity of symptoms of depression in adults and adolescents 13 years of age and older as listed in the American Psychiatric Association's *DSM-IV-TR*.[51]

When presented with the BDI-II, a client is asked to consider each statement as it relates to the way they have felt for the past 2 weeks, to more accurately correspond to the DSM-IV criteria. The authors warn against the use of this instrument as a sole diagnostic measure because depressive symptoms may be part of other primary diagnostic disorders (see Box 3-10).

In the acute care setting, the therapist may see results of the BDI-II for Medical Patients in the medical record. This seven-item self-report measure of depression in adolescents and adults reflects the cognitive and affective symptoms of depression, while excluding somatic and performance symptoms that might be attributable to other conditions. It is a quick and effective way to assess depression in populations with biological, medical, alcohol, and/or substance abuse problems.

The Beck Scales for anxiety, depression, or suicide can help identify clients from ages 13 to 80 with depressive, anxious,

TABLE 3-10 Systemic Effects of Depression

System	Sign or Symptom
General (multiple system cross over)	Persistent fatigue Insomnia, sleep disturbance See clinical signs and symptoms of depression in the text
Cardiovascular	Chest pain • Associated with myocardial infarction • Can be atypical chest pain that is not associated with coronary artery disease
Gastrointestinal	Irritable bowel syndrome (IBS) Esophageal dysmotility Nonulcer dyspepsia Functional abdominal pain (heartburn)
Neurologic (often symmetric and nonanatomic)	Paresthesia Dizziness Difficulty concentrating and making decisions; problems with memory
Musculoskeletal	Weakness Fibromyalgia (or other unexplained rheumatic pain) Myofascial pain syndrome Chronic back pain
Immune	Multiple allergies Chemical hypersensitivity Autoimmune disorders Recurrent or resistant infections
Dysregulation	Autonomic instability • Temperature intolerance • Blood pressure changes Hormonal dysregulation (e.g., amenorrhea)
Other	Migraine and tension headaches Shortness of breath associated with asthma or not clearly explained Anxiety or panic disorder

Data from Smith NL: *The effects of depression and anxiety on medical illness,* Sandy, Utah, 2002, Stress Medicine Clinic, School of Medicine, University of Utah.

TABLE 3-11 Geriatric Depression Scale (Short Form)

For each question, choose the answer that best describes how you felt over the past week.

1. Are you basically satisfied with your life?	Yes/NO
2. Have you dropped many of your activities and interests?	YES/No
3. Do you feel that your life is empty?	YES/No
4. Do you often get bored?	YES/No
5. Are you in good spirits most of the time?	Yes/NO
6. Are you afraid that something bad is going to happen to you?	YES/No
7. Do you feel happy most of the time?	Yes/NO
8. Do you often feel helpless?	YES/No
9. Do you prefer to stay at home, rather than going out and doing new things?	YES/No
10. Do you feel you have more problems with memory than most people?	YES/No
11. Do you think it is wonderful to be alive now?	Yes/NO
12. Do you feel pretty worthless the way you are now?	YES/No
13. Do you feel full of energy?	Yes/NO
14. Do you feel that your situation is hopeless?	YES/No
15. Do you think that most people are better off than you are?	YES/No

NOTE: The scale is scored as follows: 1 point for each response in capital letters. A score of 0 to 5 is normal; a score above 5 suggests depression and warrants a follow-up interview; a score above 10 almost always indicates depression.

Used with permission from Sheikh JI, Yesavage JA: Geriatric Depression Scale (GDS): recent evidence and development of a shorter version, *Clin Gerontol* 5:165–173, 1986.

or suicidal tendencies even in populations with overlapping physical and/or medical problems.

The Beck Scales have been developed and validated to assist health care professionals in making focused and reliable client evaluations. Test results can be the first step in recognizing and appropriately treating an affective disorder. These are copyrighted materials and can be obtained directly from The Psychological Corporation now under the new name of Harcourt Assessment.[147]

If the resultant scores for any of these assessment tools suggest clinical depression, psychologic referral is not always necessary. Intervention outcome can be monitored closely, and if progress is not made, the therapist may want to review this outcome with the client and discuss the need to communicate this information to the physician. Depression can be treated effectively with a combination of therapies, including exercise, proper nutrition, antidepressants, and psychotherapy.

Symptoms of Depression. About one-third of the clinically depressed clients treated do not feel sad or blue. Instead, they report somatic symptoms such as fatigue, joint pain, headaches, or chronic back pain (or any chronic, recurrent pain present in multiple places).

Eighty per cent to 90% of the most common GI disorders (e.g., esophageal motility disorder, nonulcer dyspepsia, irritable bowel syndrome) are associated with depressive or anxiety disorders.[140-148]

Some scientists think the problem is overresponse of the enteric system to stimuli. The gut senses stimuli too early, receives too much of a signal, and responds with too much of a reaction. Serotonin levels are low and substance P levels are too high when, in fact, these two neurotransmitters are supposed to work together to modulate the GI response.[149,150]

Other researchers propose that one of the mechanisms underlying chronic disorders associated with depression such as irritable bowel syndrome and fibromyalgia is an increased activation of brain regions concerned with the processing and modulation of visceral and somatic afferent information, particularly in the subregions of the anterior cingulate cortex (ACC).[151]

Another red flag for depression is any condition associated with smooth muscle spasm such as asthma, irritable or overactive bladder, Raynaud's disease, and hypertension. Neurologic symptoms with no apparent cause such as paresthesias, dizziness, and weakness may actually be symptoms of depression. This is particularly true if the neurologic symptoms are symmetric or not anatomic.[134]

CLINICAL SIGNS AND SYMPTOMS

***Depression** (See Also Table 3-10)*

- Persistent sadness, low mood, or feelings of emptiness
- Frequent or unexplained crying spells
- A sense of hopelessness
- Feelings of guilt or worthlessness
- Problems in sleeping
- Loss of interest or pleasure in ordinary activities or loss of libido
- Fatigue or decreased energy
- Appetite loss (or overeating)
- Difficulty in concentrating, remembering, and making decisions
- Irritability
- Persistent joint pain
- Headache
- Chronic back pain
- Bilateral neurologic symptoms of unknown cause (e.g., numbness, dizziness, weakness)
- Thoughts of death or suicide
- Pacing and fidgeting
- Chest pain and palpitations

Drugs, Depression, Dementia, or Delirium? The older adult often presents with such a mixed clinical presentation, it is difficult to know what is a primary musculoskeletal problem and what could be caused by drugs or depression (Case Example 3-7). Family members confuse signs and

Selecting more than 10 is a red flag for emotional or psychologic overlay, especially when the word selections come from groups 10 through 20.

Illness Behavior Syndrome and Symptom Magnification

Pain in the absence of an identified source of disease or pathologic condition may elicit a behavioral response from the client that is now labeled *illness behavior syndrome.* Illness behavior is what people say and do to show they are ill or perceive themselves as sick or in pain. It does not mean there is nothing wrong with the person. Illness behavior expresses and communicates the severity of pain and physical impairment.[90]

This syndrome has been identified most often in people with chronic pain. Its expression depends on what and how the client thinks about his or her symptoms/illness. Components of this syndrome include:

- Dramatization of complaints, leading to overtreatment and overmedication
- Progressive dysfunction, leading to decreased physical activity and often compounding preexisting musculoskeletal or circulatory dysfunction
- Drug misuse
- Progressive dependency on others, including health care professionals, leading to overuse of the health care system
- Income disability, in which the person's illness behavior is perpetuated by financial gain[88]

Symptom magnification syndrome (SMS) is another term used to describe the phenomenon of illness behavior; conscious symptom magnification is referred to as *malingering,* whereas unconscious symptom magnification is labeled *illness behavior.* Conscious malingering may be described as exaggeration or faking symptoms for external gain. Some experts differentiate symptom amplification from malingering or factitious disorder (i.e., fakery or self-induced symptoms that enable the sick role).[176]

The term *symptom magnification* was first coined by Leonard N. Matheson, PhD,* in 1977 to describe clients whose symptoms have reinforced their behavior, that is, the symptoms have become the predominant force in the client's function rather than the physiologic phenomenon of the injury determining the outcome.

By definition, SMS is a self-destructive, socially reinforced behavioral response pattern consisting of reports or displays of symptoms that function to control the life of the sufferer.[177-179] The amplified symptoms rather than the physiologic phenomenon of the injury determine the outcome/function.

The affected person acts as if the future cannot be controlled because of the presence of symptoms. All present limitations are blamed on the symptoms: "My (back) pain won't let me … ." The client may exaggerate limitations beyond those that seem reasonable in relation to the injury, apply minimal effort on maximal performance tasks, and overreact to physical loading during objective examination.

It is important for physical therapists to recognize that we often contribute to SMS by focusing on the relief of symptoms, especially pain, as the goal of therapy. Reducing pain is an acceptable goal for some types of clients, but for those who experience pain after the injuries have healed, the focus should be restoration, or at least improvement, of function.

In these situations, instead of asking whether the client's symptoms are "better, the same, or worse," it may be more appropriate to inquire about functional outcomes, for example, what can the client accomplish at home that she or he was unable to attempt at the beginning of treatment, last week, or even yesterday.

Conscious or unconscious? Can a physical therapist determine when a client is consciously or unconsciously symptom magnifying? Is it within the scope of the physical therapist's practice to use the label "malingerer" without a psychologist or psychiatrist's diagnosis of such first?

Physical exam techniques available include McBride's, Mankopf's, Waddell's, Hoover's, Abductor, Arm Drop, and Midline Split. The evidence supporting strength of recommendation (SOR) for these tests to detect malingering is ranked as B (systematic review of low-quality studies) or C (expert opinion, small case studies). For a review of these tests and a summary of the evidence for each one, see Greer et al, 2005.[180] The American Psychiatric Association and the American Medical Association agree confirmation of malingering is extremely difficult and depends on direct observation. It is safest to assume a person is not malingering unless direct evidence is available.[51,181]

Keep in mind the goal is to screen for a psychologic or emotional component to the client's clinical presentation. The key to achieving this goal is to use objective test measures whenever possible. In this way, the therapist obtains the guidance needed for referral versus modification of the physical therapy intervention.

Compiling a list of nonorganic or behavioral signs and identifying how the client is reacting to pain may be all that is needed. Signs of illness behavior may point the therapist in the direction of more careful "management" of the psychosocial and behavioral aspects of the client's illness.[116]

Waddell's Nonorganic Signs

Waddell et al[182] identified five nonorganic signs and seven nonanatomic or behavioral descriptions of symptoms (Table 3-12) to help differentiate between physical and behavioral causes of back pain. Each of the nonorganic signs is determined by using one or two of the tests listed. These tests are used to assess a client's pain behavior and detect abnormal illness behavior. The literature supports that these signs may be present in 10% of clients with acute low back pain, but are found most often in people with chronic low back pain.

A score of three or more positive signs places the client in the category of *nonmovement dysfunction.* This person is said to have a clinical pattern of nonmechanical, pain-focused

*Director, ERIC Human Performance Laboratory, Washington University School of Medicine, St. Louis, Missouri.

TABLE 3-12 Waddell's Nonorganic Signs and Behavioral Symptoms

Test	Signs	Nonanatomic or Behavioral Description of Symptoms
Tenderness	*Superficial*—the client's skin is tender to light pinch over a wide area of lumbar skin; unable to localize to one structure. *Nonanatomic*—deep tenderness felt over a wide area, not localized to one structure; crosses multiple somatic boundaries.	1. Pain at the tip of the tailbone 2. Whole leg pain from the groin down to below the knee in a stocking pattern (not dermatomal or sclerotomal, intermittent) 3. Whole leg numbness or whole leg "going dead" (intermittent)
Simulation tests	*Axial loading*—light vertical loading over client's skull in the standing position reproduces lumbar (not cervical) spine pain. *Acetabular rotation*—lumbosacral pain from upper trunk rotation, back pain reported when the pelvis and shoulders are passively rotated in the same plane as the client stands, considered a positive test if pain is reported within the first 30 degrees.	4. Whole leg giving way or collapsing (intermittent, client maintains upright position) 5. Constant pain for years on end without relief 6. Unable to tolerate any treatment, reaction or side effects to every intervention
Distraction tests	*Straight-leg-raise* (SLR) discrepancy—marked improvement of SLR when client is distracted as compared with formal testing; different response to SLR in supine (worse) compared to sitting (better) when both tests should have the same result in the presence of organic pathology. *Double leg raise*—when both legs are raised after straight leg raising, the organic response would be a greater degree of double leg raising; clients with a nonorganic component demonstrate less double leg raise as compared with the single leg raise.	7. Emergency admission to hospital for back pain without precipitating traumatic event
Regional disturbances	*Weakness*—cogwheeling or giving way of many muscle groups that cannot be explained on a neurologic basis. *Sensory disturbance*—diminished sensation fitting a "stocking" rather than a dermatomal pattern.	
Overreaction	Disproportionate verbalization, facial expression, muscle tension, and tremor, collapsing, or sweating. Client may exhibit any of the following behaviors during the physical examination: guarding, bracing, rubbing, sighing, clenching teeth, or grimacing.	

Adapted from Karas R, McIntosh G, Hall H, et al: The relationship between nonorganic signs and centralization of symptoms in the prediction of return to work for patients with low back pain, *Phys Ther* 77(4):354–360, 1997.

behavior. This type of score is predictive of poor outcome and associated with delayed return-to-work or not working.

One or two positive signs is a low Waddell's score and does not classify the client with a nonmovement dysfunction. The value of these nonorganic signs as predictors for return-to-work for clients with low back pain has been investigated.[183] Less than two is a good prognosticator of return-to-work. The results of how this study might affect practice are available.[184]

A positive finding for nonorganic signs does not suggest an absence of pain but rather a behavioral response to pain (see discussion of symptom magnification syndrome). It does not confirm malingering or illness behavior. Neither do these signs imply the nonexistence of physical pathology.

Waddell and associates[117,182] have given us a tool that can help us identify early in the rehabilitation process those who need more than just mechanical or physical treatment intervention. Other evaluation tools are available (e.g., Oswestry Back Pain Disability Questionnaire, Roland-Morris Disability Questionnaire). A psychologic evaluation and possibly behavioral therapy or psychologic counseling may be needed as an adjunct to physical therapy.[185]

Conversion Symptoms

Whereas SMS is a behavioral, learned, inappropriate *behavior,* conversion is a psychodynamic phenomenon and quite rare in the chronically disabled population.

Conversion is a physical expression of an unconscious psychologic conflict such as an event (e.g., loss of a loved one) or a problem in the person's work or personal life. The conversion may provide a solution to the conflict or a way to express "forbidden" feelings. It may be a means of enacting the sick role to avoid responsibilities, or it may be a reflection of behaviors learned in childhood.[16]

Diagnosis of a conversion syndrome is difficult and often requires the diagnostic and evaluative input of the physical therapist. Presentation always includes a motor and/or sensory component that cannot be explained by a known medical or neuromusculoskeletal condition.

The clinical presentation is often mistaken for an organic disorder such as multiple sclerosis, systemic lupus erythematosus, myasthenia gravis, or idiopathic dystonias. At presentation, when a client has an unusual limp or bizarre gait pattern that cannot be explained by functional anatomy, family

members may be interviewed to assess changes in the client's gait and whether this alteration in movement pattern is present consistently.

The physical therapist can look for a change in the wear pattern of the client's shoes to decide if this alteration in gait has been long-standing. During manual muscle testing, true weakness results in smooth "giving way" of a muscle group; in hysterical weakness the muscle "breaks" in a series of jerks.

Often the results of muscle testing are not consistent with functional abilities observed. For example, the person cannot raise the arm overhead during testing but has no difficulty dressing, or the lower extremity appears flaccid during recumbency but the person can walk on the heels and toes when standing.

The physical therapist should carefully evaluate and document all sensory and motor changes. Conversion symptoms are less likely to follow any dermatome, myotome, or sclerotome patterns.

CLINICAL SIGNS AND SYMPTOMS

Conversion

- Sudden, acute onset
- Lack of concern about the symptoms
- Unexplainable motor or sensory function impairment

Motor

- Impaired coordination or balance and/or bizarre gait pattern
- Paralysis or localized weakness
- Loss of voice, difficulty swallowing, or sensation of a lump in the throat
- Urinary retention

Sensory

- Altered touch or pain sensation (paresthesia or dysesthesia)
- Visual changes (double vision, blindness, black spots in visual field)
- Hearing loss (mild-to-profound deafness)
- Hallucinations
- Seizures or convulsions
- Absence of significant laboratory findings
- Electrodiagnostic testing within normal limits
- Deep tendon reflexes within normal limits

Screening Questions for Psychogenic Source of Symptoms

Besides observing for signs and symptoms of psychophysiologic disorders, the therapist can ask a few screening questions (Box 3-13). The client may be aware of the symptoms but does not know that these problems can be caused by depression, anxiety, or panic disorder.

Medical treatment for physiopsychologic disorders can and should be augmented with exercise. Physical activity and exercise has a known benefit in the management of mild-to-moderate psychologic disorders, especially depression and anxiety. Aerobic exercise or strength training have both been shown effective in moderating the symptoms of these conditions.[186-189]

Patience is a vital tool for therapists when working with clients who are having difficulty adjusting to the stress of illness and disability or the client who has a psychologic disorder. The therapist must develop personal coping mechanisms when working with clients who have chronic illnesses or psychologic disturbances.

Recognizing clients whose symptoms are the direct result of organic dysfunction helps us in coping with clients who are hostile, ungrateful, noncompliant, negative, or adversarial. Whenever possible, involve a psychiatrist, psychologist, or counselor as part of the management team. This approach will benefit the client as well as the health care staff.

PHYSICIAN REFERRAL

Guidelines for Immediate Physician Referral

- Immediate medical attention is required for anyone with risk factors for and clinical signs and symptoms of rhabdomyolysis (see Table 3-5).
- Clients reporting a disproportionate relief of bone pain with a simple aspirin may have bone cancer. This red flag requires immediate medical referral in the presence of a personal history of cancer of any kind.
- Joint pain with no known cause and a recent history of infection of any kind. Ask about recent (last 6 weeks) skin lesions or rashes of any kind anywhere on the body, urinary tract infection, or respiratory infection. Take the client's temperature and ask about recent episodes of fever, sweats, or other constitutional symptoms. Palpate for residual lymphadenopathy. Early diagnosis and treatment are essential to limit joint destruction and preserve function.[78]

BOX 3-13 SCREENING QUESTIONS FOR PSYCHOGENIC SOURCE OF SYMPTOMS

- Do you have trouble sleeping at night?
- Do you have trouble focusing during the day?
- Do you worry about finances, work, or life in general?
- Do you feel a sense of dread or worry without cause?
- Do you ever feel happy?
- Do you have a fear of being in groups of people? Fear of flying? Public speaking?
- Do you have a racing heart, unexplained dizziness, or unexpected tingling in your face or fingers?
- Do you wake up in the morning with your jaw clenched or feeling sore muscles and joints?
- Are you irritable or jumpy most of the time?

Data from Davidson J, Dreher H: *The anxiety book: developing strength in the face of fear,* New York, 2003, Penguin Putnam.

Guidelines for Physician Referral Required

- Proximal muscle weakness accompanied by change in one or more deep tendon reflexes in the presence of a previous history of cancer.
- The physician should be notified of anyone with joint pain of unknown cause who presents with recent or current skin rash or recent history of infection (hepatitis, mononucleosis, urinary tract infection, upper respiratory infection, STI, streptococcus).
- A team approach to fibromyalgia requires medical evaluation and management as part of the intervention strategy. Therapists should refer clients suspected with fibromyalgia for further medical follow up.
- Diffuse pain that characterizes some diseases of the nervous system and viscera may be difficult to distinguish from the equally diffuse pain so often caused by lesions of the moving parts. The distinction between visceral pain and pain caused by lesions of the vertebral column may be difficult to make and may require a medical diagnosis.
- The therapist may screen for signs and symptoms of anxiety, depression, and panic disorder. These conditions are often present with somatic symptoms that may resolve with effective intervention. The therapist can describe the symptoms and relay that information to the appropriate agency or individual when making a referral. Diagnosis is made by a medical doctor or trained mental health professional.
- Clients with new onset of back, neck, TMJ, shoulder, or arm pain brought on by a new exercise program or by exertion with the arms raised overhead should be screened for signs and symptoms of cardiovascular impairment. This is especially important if the symptoms are described as "throbbing" and start after a brief time of exercise (3 to 5 up to 10 minutes) and diminish or go away quickly with rest. Look for significant risk factors for cardiovascular involvement. Check vital signs. Refer for medical evaluation if indicated.
- Persistent pain on weight bearing or bone pain at night, especially in the older adult with risk factors such as osteoporosis, postural hypotension leading to falls, or previous history of cancer.

Clues to Screening for Viscerogenic Sources of Pain

We know systemic illness and pathologic conditions affecting the viscera can mimic NMS dysfunction. The therapist who knows pain patterns and types of viscerogenic pain can sort through the client's description of pain and recognize when something does not fit the expected pattern for NMS problems.

We must keep in mind that pain from a disease process or viscerogenic source is often a late symptom rather than a reliable danger signal. For this reason the therapist must remain alert to other signs and symptoms that may be present but unaccounted for.

In this chapter, pain types possible with viscerogenic conditions have been presented along with three mechanisms by which viscera refer pain to the body (soma). Characteristics of systemic pain compared to musculoskeletal pain are presented, including a closer look at joint pain.

Pain with the following features raises a red flag to alert the therapist of the need to take a closer look:

- Pain of unknown cause.
- Pain that persists beyond the expected time for physiologic healing.
- Pain that is out of proportion to the injury.
- Pain that is unrelieved by rest or change in position.
- Pain pattern does not fit the expected clinical presentation for a neuromuscular or musculoskeletal impairment.
- Pain that cannot be altered, aggravated, provoked, reduced, eliminated, or alleviated.
- There are some positions of comfort for various organs (e.g., leaning forward for the gallbladder or side bending for the kidney), but with progression of disease the client will obtain less and less relief of symptoms over time.
- Pain, symptoms, or dysfunction are not improved or altered by physical therapy intervention.
- Pain that is poorly localized.
- Pain accompanied by signs and symptoms associated with a specific viscera (e.g., GI, GU, gynecologic [GYN], cardiac, pulmonary, endocrine).
- Pain that is constant and intense no matter what position is tried and despite rest, eating, or abstaining from food; a previous history of cancer in this client is an even greater red flag necessitating further evaluation.
- Pain (especially intense bone pain) that is disproportionately relieved by aspirin.
- Listen to the client's choice of words to describe pain. Systemic or viscerogenic pain can be described as deep, sharp, boring, knifelike, stabbing, throbbing, colicky, or intermittent (comes and goes in waves).
- Pain accompanied by full and normal ROM.
- Pain that is made worse 3 to 5 minutes after initiating an activity and relieved by rest (possible symptom of vascular impairment) versus pain that goes away with activity (symptom of musculoskeletal involvement); listen for the word descriptor "throbbing" to describe pain of a vascular nature.
- Pain is a relatively new phenomenon and not a pattern that has been present over several years' time.
- Constitutional symptoms in the presence of pain.
- Pain that is not consistent with emotional or psychologic overlay.
- When in doubt, conduct a screening exam for emotional overlay. Observe the client for signs and symptoms of anxiety, depression, and/or panic disorder. In the absence of systemic illness or disease and/or in the presence of suspicious psychologic symptoms, psychologic evaluation may be needed.
- Pain in the absence of any positive Waddell's signs (i.e., Waddell's test is negative or insignificant).

- Manual therapy to correct an upslip is not successful, and the problem has returned by the end of the session or by the next day; consider a somato-visceral problem or visceral ligamentous problem.
- If painful, tender or sore points (e.g., TrPs, Jones' points, acupuncture/acupressure points/Shiatsu) are eliminated with intervention then return quickly (by the end of the treatment session), suspect visceral pathology. If a tender point comes back later (several days or weeks), the clinician may not be holding it long enough.[4]
- Back, neck, TMJ, shoulder, or arm pain brought on by exertion with the arms raised overhead may be suggestive of a cardiac problem. This is especially true in the postmenopausal woman or man over age 50 with a significant family history of heart disease and/or in the presence of hypertension.
- Back, shoulder, pelvic, or sacral pain that is made better or worse by eating, passing gas, or having a bowel movement.
- Night pain (especially bone pain) that awakens the client from a sound sleep several hours after falling asleep; this is even more serious if the client is unable to get back to sleep after changing position, taking pain relievers, or eating or drinking something.
- Joint pain preceded or accompanied by skin lesions (e.g., rash or nodules), following antibiotics or statins, or recent infection of any kind (e.g., GI, pulmonary, GU); check for signs and symptoms associated with any of these systems based on recent client history.
- Clients can have more than one problem or pathology present at one time; it is possible for a client to have both a visceral AND a mechanical problem.[4]
- Remember Osler's Rule of Age*: Under age 60, most clients' symptoms are related to one problem, but over 60, it is rarely just one problem.[4]
- A careful general history and physical examination is still the most important screening tool; never assume this was done by the referring physician or other staff from the referring agency.[4]
- Visceral problems are unlikely to cause muscle weakness, reflex changes, or objective sensory deficits (exceptions include endocrine disease and paraneoplastic syndromes associated with cancer). If pain is referred from the viscera to the soma, challenging the somatic structure by stretching, contracting, or palpating will not reproduce the symptoms. For example, if a muscle is not sore when squeezed or contracted, the muscle is not the source of the pain.[4]

*Physicians often rely on ad hoc rules of thumb, or "heuristics," to guide them. These are often referred to as Osler's Rules. Sir William Osler, MD (1849-1919) promoted the idea that good medical science follows from gathering evidence by directly observing patients.

Key Points to Remember

- Pain of a visceral origin can be referred to the corresponding somatic areas. The mechanisms of referred visceral pain patterns are not fully known. Information in this chapter is based on proposed models from what is known about the somatic sensory system.
- Recognizing pain patterns that are characteristic of systemic disease is a necessary step in the screening process. Understanding how and when diseased organs can refer pain to the NMS system helps the therapist identify suspicious pain patterns.
- At least three mechanisms contribute to referred pain patterns of the viscera (embryologic development, multisegmental innervation, and direct pressure and shared pathways). Being familiar with each one may help the therapist quickly identify pain patterns of a visceral source.
- The therapist should keep in mind cultural rules and differences in pain perception, intensity, and responses to pain found among various ethnic groups.
- Pain patterns of the chest, back, shoulder, scapula, pelvis, hip, groin, and SI joint are the most common sites of referred pain from a systemic disease process.
- Visceral diseases of the abdomen and pelvis are more likely to refer pain to the back, whereas intrathoracic disease refers pain to the shoulder(s). Visceral pain rarely occurs without associated signs and symptoms, although the client may not recognize the correlation. Careful questioning will usually elicit a systemic pattern of symptoms.
- A comprehensive pain assessment includes a detailed health history, physical exam, medication history (including nonprescription drug use and complementary and alternative therapies), assessment of functional status, and consideration of psychosocial-spiritual factors. Assessment tools vary from the very young to the very old.
- Careful, sensitive, and thorough questioning regarding the multifaceted experience of pain can elicit essential information necessary when making a decision regarding treatment or referral. The use of pain assessment tools, such as Fig. 3-6 and Table 3-2, may facilitate clear and accurate descriptions of this critical symptom.
- The client describes the characteristics of pain (location, frequency, intensity, duration, description). It is up

Continued

Key Points to Remember—cont'd

to the therapist to recognize sources and types of pain and to know the pain patterns of a viscerogenic origin.

- Choose alternative words to "pain" when discussing the client's symptoms in order to get a complete understanding of the clinical presentation.
- Specific screening questions for joint pain are used to assess any joint pain of unknown cause, joint pain with an unusual presentation or history, or joint pain which does not fit the expected pattern for injury, overuse, or aging (see Box 3-5).
- It is important to know how to differentiate psychogenic and psychosomatic origins of painful symptoms from systemic origins, including signs and symptoms of cancer.
- Pain described as constant or present at night, awakening the client from sleep must be evaluated thoroughly. When assessing constant and/or night pain, the therapist must know how to differentiate the characteristics of acute versus chronic pain associated with a neuromusculoskeletal problem from a viscerogenic or systemic presentation.

SUBJECTIVE EXAMINATION

Special Questions to Ask

Pain Assessment

Location of Pain

Show me exactly where your pain is located.

Follow up questions may include:

- Do you have any other pain or symptoms anywhere else?
- *If yes,* what causes the pain or symptoms to occur in this other area?

Description of Pain

What does it feel like?

After giving the client time to reply, offer some additional choices in potential descriptors. You may want to ask: Is your pain/Are your symptoms

Knifelike	Dull
Boring	Burning
Throbbing	Prickly
Deep aching	Sharp

Follow up questions may include:

- Has the pain changed in quality since it first began?
- Changed in intensity?
- Changed in duration (how long it lasts)?

Frequency and Duration of Pain

How long do the symptoms last?

Clients who indicate that the pain is constant should be asked:

- Do you have this pain right now?
- Did you notice these symptoms this morning immediately when you woke up?

Pattern of Pain

Tell me about the pattern of your pain/symptoms.

- *Alternate question:* When does your back/shoulder (name the involved body part) hurt?
- *Alternate question:* Describe your pain/symptoms from first waking up in the morning to going to bed at night. (See special sleep-related questions that follow.)

  Follow up questions may include:
- Have you ever experienced anything like this before?

  If yes, do these episodes occur more or less often than at first?
- How does your pain/symptom(s) change with time?
- Are your symptoms worse in the morning or evening?

Aggravating and Relieving Factors

- What brings your pain (symptoms) on?
- What kinds of things make your pain (symptoms) worse (e.g., eating, exercise, rest, specific positions, excitement, stress)?

  To assess relieving factors, ask:
- What makes the pain better?

  Follow up questions include:
- How does rest affect the pain/symptoms?
- Are your symptoms aggravated or relieved by any activities?

  If yes, what?
- How has this problem affected your daily life at work or at home?
- How has this problem affected your ability to care for yourself without assistance (e.g., dress, bathe, cook, drive)?

SUBJECTIVE EXAMINATION—*cont'd*

Associated Symptoms

- What other symptoms have you had that you can associate with this problem?

If the client denies any additional symptoms, follow up this question with a series of possibilities such as:

Burning	Heart palpitations	Numbness/Tingling
Difficulty in breathing	Hoarseness	Problems with vision
Difficulty in swallowing	Nausea	Vomiting
Dizziness	Night sweats	Weakness

- Are you having any pain anywhere else in your body?

Alternately: Are you having symptoms of any other kind that may or may not be related to your main problem?

Anxiety/Depression (See Table 3-11)

- Have you been under a lot of stress lately?
- Are you having some trouble coping with life in general and/or life's tensions?
- Do you feel exhausted or overwhelmed mentally or physically?
- Does your mind go blank or do you have trouble concentrating?
- Do you have trouble sleeping at night (e.g., difficulty getting to sleep, staying asleep, restless sleep, feel exhausted upon awakening)? Focusing during the day?
- Do you worry about finances, work, or life in general?
- Do you get any enjoyment in life?
- Do you feel keyed up or restless? Irritable and jumpy? On edge most of the time?
- Do you have a general sense of dread or unknown fears?
- Do you have any of these symptoms: a racing heart, dizziness, tingling, muscle or joint pains?

For the Asian client:

- Do you feel you are having any imbalance of yin and yang?
- Is your Qi (or chi, pronounced "chee") (internal energy) low?
- Do you believe it is your destiny to have this condition or your destiny not to have this condition (fatalism versus well-being approach to illness)?

Joint Pain (See Box 3-5)

Night Pain (See Box 3-7)

Psychogenic Source of Symptoms (See Box 3-13)

CASE STUDY*

REFERRAL

A 44-year-old male was referred to physical therapy with a report of right-sided thoracic pain.

Past Medical History: The client reported a 20-pack year smoking history (one pack per day for 20 years) and denied the use of alcohol or drugs. There was no other significant past medical history reported. He had a sedentary job.

The client's symptoms began following chiropractic intervention to relieve left-sided lower extremity radiating pain. Within 6 to 8 hours after the chiropractor manipulated the client's thoracic spine, he reported sharp shooting pain on the right side of the upper thoracic spine at T4. The pain radiated laterally under the right axilla into the anterior chest. He also reported tension and tightness along the same thoracic level and moderate discomfort during inspiration. There was no history of thoracic pain prior to the upper thoracic manipulation by the chiropractor.

The client saw his primary care physician who referred him to physical therapy for treatment. No imaging studies were done prior to physical therapy referral. The client rated the pain as a constant 10/10 on the Numeric Rating Scale (NRS) during sitting activities at work. He also reported pain waking him at night.

The client was unable to complete a full day at work without onset of thoracic discomfort; pain was aggravated by prolonged sitting.

EVALUATION

The client was described as slender in build (ectomorph body type) with forward head and shoulders and kyphotic posturing as observed in the upright and sitting positions. There were no significant signs of inflammation or superficial tissue changes observed or palpated in the thoracic spine region. There was palpable tenderness at approximately the T4 costotransverse joint and along the corresponding rib.

A full orthopedic evaluation was conducted to determine the biomechanical and soft tissue dysfunction that produced the client's signs and symptoms. Active and passive motion and intersegmental mobility were tested. Findings were consistent with a physical therapy diagnosis of hypomobile costotransverse joint at level T4.

Continued

CASE STUDY—*cont'd*

This was further evidenced by pain at the posterior costovertebral joint with radiating pain laterally into the chest wall. Pain was increased on inspiration. Patient had a smoker's cough, but reported no other associated signs or symptoms of any kind. See the Pain Assessment Record Form on the companion website.

RESULT

The client obtained gradual relief from painful symptoms after 8 treatment sessions of stretches and costotransverse joint mobilization (grade 4, nonthrust progressive oscillations at the end of the available range). Pain was reduced from 10/10 to 3/10 and instances of night pain had decreased. The client could sit at work with only mild discomfort, which he could correct with stretching.

The client's thoracic pain returned on the 10th and 11th treatment sessions. He attributed this to increased stressors at work and long work hours. Night pain and pain with respiratory movements (inhalation) increased again.

Red flags in this case included:

- Age over 40
- History of smoking (20-pack years)
- Symptoms persisting beyond the expected time for physiologic healing
- Pain out of proportion to the injury
- Recurring symptoms (failure to respond to physical therapy intervention)
- Pain is constant and intense; night pain

The client was returned to his primary care physician for further diagnostic studies and later diagnosed with metastatic lung cancer.

SUMMARY

Working with clients several times a week allows the therapist to monitor their symptoms and the effectiveness of interventions. This case study shows the importance of reassessment and awareness of red flags that would lead a practitioner to suspect the symptoms may be pathologic.

*Leanne Lenker, DPT. This case was part of an internship experience at St. Luke's Outpatient Clinic, Allentown, PA under the supervision of Jeff Bays, MSPT (Clinical Instructor). Dr. Lenker is a graduate of the University of St. Augustine for Health Sciences program in St. Augustine, Florida. Used with permission, 2005.

PRACTICE QUESTIONS

1. What is the best follow-up question for someone who tells you that the pain is constant?
   a. Can you use one finger to point to the pain location?
   b. Do you have that pain right now?
   c. Does the pain wake you up at night after you have fallen asleep?
   d. Is there anything that makes the pain better or worse?
2. A 52-year-old woman with shoulder pain tells you that she has pain at night that awakens her. After asking a series of follow-up questions, you are able to determine that she had trouble falling asleep because her pain increases when she goes to bed. Once she falls asleep, she wakes up as soon as she rolls onto that side. What is the most likely explanation for this pain behavior?
   a. Minimal distractions heighten a person's awareness of musculoskeletal discomfort.
   b. This is a systemic pattern that is associated with a neoplasm.
   c. It is impossible to tell.
   d. This represents a chronic clinical presentation of a musculoskeletal problem.
3. Referred pain patterns associated with impairment of the spleen can produce musculoskeletal symptoms in:
   a. The left shoulder
   b. The right shoulder
   c. The mid-back or upper back, scapular, and right shoulder areas
   d. The thorax, scapulae, right shoulder, or left shoulder
4. Associated signs and symptoms are a major red flag for pain of a systemic or visceral origin compared to musculoskeletal pain.
   a. True
   b. False
5. Words used to describe neurogenic pain often include:
   a. Throbbing, pounding, beating
   b. Crushing, shooting, pricking
   c. Aching, heavy, sore
   d. Agonizing, piercing, unbearable
6. Pain (especially intense bone pain) that is disproportionately relieved by aspirin can be a symptom of:
   a. Neoplasm
   b. Assault or trauma
   c. Drug dependence
   d. Fracture
7. Joint pain can be a reactive, delayed, or allergic response to:
   a. Medications
   b. Chemicals
   c. Infections
   d. Artificial sweeteners
   e. All of the above
8. Bone pain associated with neoplasm is characterized by:
   a. Increases with weight bearing
   b. Negative heel strike
   c. Relieved by Tums or other antacid in women
   d. Goes away after eating

PRACTICE QUESTIONS—*cont'd*

9. Pain of a viscerogenic nature is not relieved by a change in position.
   a. True
   b. False
10. Referred pain from the viscera can occur alone but is usually preceded by visceral pain when an organ is involved.
   a. True
   b. False
11. A 48-year old man presented with low back pain of unknown cause. He works as a carpenter and says he is very active, has work-related mishaps (accidents and falls), and engages in repetitive motions of all kinds using his arms, back, and legs. The pain is intense when he has it, but it seems to come and go. He is not sure if eating makes the pain better or worse. He has lost his appetite because of the pain. After conducting an examination including a screening exam, the clinical presentation does not match the expected pattern for a musculoskeletal or neuromuscular problem. You refer him to a physician for medical testing. You find out later he had pancreatitis. What is the most likely explanation for this pain pattern?
   a. Toxic waste products from the pancreas are released into the intestines causing irritation of the retroperitoneal space.
   b. Rupture of the pancreas causes internal bleeding and referred pain called Kehr's sign.
   c. The pancreas and low back structures are formed from the same embryologic tissue in the mesoderm.
   d. Obstruction, irritation, or inflammation of the body of the pancreas distends the pancreas, thus applying pressure on the central respiratory diaphragm.

REFERENCES

1. Flaherty JH: Who's taking your fifth vital sign? *J Gerontol A Biol Sci Med Sci* 56:M397–M399, 2001.
2. Sluka KA: *Mechanisms and management of pain for the physical therapist*, Seattle, 2009, IASP Press.
3. Brumovsky PR, Gebhart GF: Visceral organ cross-sensitization—an integrated perspective. *Auton Neurosci* 153(1-2):106–115, 2010.
4. Rex L: *Evaluation and treatment of somatovisceral dysfunction of the gastrointestinal system*, Edmonds, WA, 2004, URSA Foundation.
5. Christianson JA: Development, plasticity, and modulation of visceral afferents. *Brain Res Rev* 60(1):171–178, 2009.
6. Chaban W: Peripheral sensitization of sensory neurons. *Ethn Dis* 20(1 Suppl 1):S1–S6, 2010.
7. Squire LR: *Fundamental neuroscience*, ed 3, Burlington, MA, 2008, Academic Press.
8. Saladin KS: Personal communication, Distinguished Professor of Biology, Milledgeville, Georgia, 2004, Georgia College and State University.
9. Woolf CJ, Decosterd I: Implications of recent advances in the understanding of pain pathophysiology for the assessment of pain in patients. *Pain Suppl* 6:S141–S147, 1999.
10. Strigio I: Differentiation of visceral and cutaneous pain in the human brain. *J Neurophysiol* 89:3294–3303, 2003. Available at http://jn.physiology.org/cgi/content/abstract/89/6/3294
11. Aziz Q: Functional neuroimaging of visceral sensation. *J Clin Neurophysiol* 17(6):604–612, 2000.
12. Giamberardino M: Viscero-visceral hyperalgesia: characterization in different clinical models. *Pain* 151(2):307–322, 2010.
13. Leavitt RL: Developing cultural competence in a multicultural world. Part II. *PT Magazine* 11(1):56–70, 2003.
14. O'Rourke D: The measurement of pain in infants, children, and adolescents: from policy to practice. *Phys Ther* 84(6):560–570, 2004.
15. Wentz JD: Assessing pain at the end of life, *Nursing2003* 33(8):22, 2003.
16. Melzack R: The McGill Pain Questionnaire: major properties and scoring methods. *Pain* 1:277, 1975.
17. American Geriatrics Society (AGS): Chronic pain in older persons. *J Am Geriatr Soc* 57(8):1331–1346, 2009.
18. American Geriatrics Society Panel on Persistent Pain in Older Persons (revised guideline). *J Am Geriatr Soc* 50(Suppl 6):205–224, 2002.
19. Argoff CE, Ferrell B: Pharmacologic therapy for persistent pain in older adults: the updated American Geriatrics Society Guidelines and their clinical implications. *Pain Medicine News* 8(5):1–8, 2010.
20. Herr KA, Spratt K, Mobily PR, et al: Pain intensity assessment in older adults: use of experimental pain to compare psychometric properties and usability of selected pain scales with younger adults. *Clin J Pain* 20(4):207–219, 2004.
21. Lane P: Assessing pain in patients with advanced dementia. *Nursing2004* 34(8):17, 2004.
22. D'Arcy Y: Assessing pain in patients who can't communicate. *Nursing2004* 34(10):27, 2004.
23. Hurley AC: Assessment of discomfort in advanced Alzheimer patients. *Res Nurs Health* 15(5):369–377, 1992.
24. Warden V, Hurley AC, Volicer L: Development and psychometric evaluation of the Pain Assessment in Advanced Dementia (PAINAD) scale. *J Am Med Dir Assoc* 4(1):9–15, 2003.
25. Herr K: Tools for assessment of pain in nonverbal older adults with dementia: a state of the science review. *J Pain Symptom Manage* 31(2):170–192, 2006.
26. Feldt K: The checklist of nonverbal pain indicators (CNPI). *Pain Manag Nurs* 1(1):13–21, 2000.
27. Pasero C, Reed BA, McCaffery M: Pain in the elderly. In McCaffery M, *Pain: clinical manual*, ed 2, St. Louis, 1999, Mosby, pp 674–710.
28. Ferrell BA: Pain in cognitively impaired nursing home patients. *J Pain Symptom Manage* 10(8):591–598, 1995.

29. Wong D, Baker C: Pain in children: comparison of assessment scales. *Pediatr Nurs* 14(1):9017, 1988.
30. Hicks CL, von Baeyer CL, Spafford PA, et al: The Faces Pain Scale-Revised: toward a common metric in pediatric pain measurement. *Pain* 93:173–183, 2001.
31. Bieri D: The Faces Pain Scale for the self-assessment of the severity of pain experienced by children: development, initial validation, and preliminary investigation for the ratio scale properties. *Pain* 41(2):139–150, 1990.
32. Wong on Web: FACES *Pain Rating Scale,* Elsevier Health Science Information, 2004.
33. Baker-Lefkowicz A, Keller V, Wong DL, et al: Young children's pain rating using the FACES Pain Rating Scale with original vs abbreviated word instructions, unpublished, 1996.
34. von Baeyer CL, Hicks CL: Support for a common metric for pediatric pain intensity scales. *Pain Res Manage* 4(2):157–160, 2000.
35. Ramelet AS, Abu-Saad HH, Rees N, et al: The challenges of pain measurement in critically ill young children: a comprehensive review. *Aust Crit Care* 17(1):33–45, 2004.
36. Grunau RE, Craig KD: Pain expression in neonates: facial action and cry. *Pain* 28:395–410, 1987.
37. Grunau RE, Oberlander T, Holsti L, et al: Bedside application of the Neonatal Facial Coding System in pain assessment of premature neonates. *Pain* 76:277–286, 1998.
38. Peters JW, Koot HM, Grunau RE, et al: Neonatal Facial Coding System for assessing postoperative pain in infants: item reduction is valid and feasible. *Clin J Pain* 19(6):353–363, 2003.
39. Stevens B: Pain in infants. In McCaffery M, Pasero C, editors: *Pain: clinical manual,* ed 2, St. Louis, 1999, Mosby, pp 626–673.
40. Turk DC, Melzack R, editors: *Handbook of pain assessment,* ed 3, New York, 2010, Guilford Press.
41. Riley JL, 3rd, Wade JB, Myers CD, et al: Racial/ethnic differences in the experience of chronic pain. *Pain* 100(3):291–298, 2002.
42. Sheffield D, Biles PL, Orom H, et al: Race and sex differences in cutaneous pain perception. *Psychosom Med* 62(4):517–523, 2000.
43. Unruh AM, Ritchie J, Merskey H: Does gender affect appraisal of pain and pain coping strategies? *Clin J Pain* 15(1):31–40, 1999.
44. Brown JL, Sheffield D, Leary MR, et al: Social support and experimental pain. *Psychosom Med* 65(2):276–283, 2003.
45. Huskinson EC: Measurement of pain. *Lancet* 2:1127–1131, 1974.
46. Carlsson AM: Assessment of chronic pain: aspects of the reliability and validity of the visual analog scale. *Pain* 16:87–101, 1983.
47. Dworkin RH: Development and initial validation of an expanded and revised version of the Short-form McGill Pain Questionnaire (SF-MPQ-2). *Pain* 144(1-2):35–42, 2009.
48. Turk DC, Melzack R, editors: *Handbook of pain assessment,* ed 3, New York, 2010, Guilford.
49. Sahrmann S: Diagnosis and diagnosticians: the future in physical therapy, Combined Sections Meeting, Dallas, February 13–16, 1997. Available at www.apta.org.
50. Courtney CA: Interpreting joint pain: quantitative sensory testing in musculoskeletal management. *J Orthop Sports Phys Ther* 40(12):818–825, 2010.
51. American Psychiatric Association (APA): *Diagnostic and statistical manual of mental disorders (DSM-IV-TR),* Washington, DC, 2000, APA.
52. Goodman CC, Fuller K: *Pathology: implications for the physical therapist,* ed 3, Philadelphia, 2009, WB Saunders.
53. Morrison J: *DSM-IV made easy: the clinician's guide to diagnosis,* New York, 2002, Guilford.
54. Bogduk N: On the definitions and physiology of back pain, referred pain, and radicular pain. *Pain* 147(1-3):17–19, 2009.
55. International Association for the Study of Pain: IASP pain terminology. Available at http://www.iasp-pain.org/AM/Template.cfm?Section=Pain_Defi...isplay.cfm&ContentID=1728. Accessed December 18, 2010.
56. Schaible HG: Joint pain. *Exp Brain Res* 196:153–162, 2009.
57. De Gucht V, Fischler B: Somatization: a critical review of conceptual and methodological issues. *Psychosomatics* 43:1–9, 2002.
58. Wells PE, Frampton V, Bowsher D: *Pain management in physical therapy,* ed 2, Oxford, 1994, Butterworth-Heinemann.
59. Gebhart GF: Visceral polymodal receptors. *Prog Brain Res* 113:101–112, 1996.
60. Simons D, Travell J: *Myofascial pain and dysfunction: the trigger point manual,* ed 2, Vol 1 and 2, Baltimore, 1999, Williams and Wilkins.
61. McMahon S, Koltzenburg M, editors: *Wall and Melzack's textbook of pain,* ed 5, New York, 2005, Churchill Livingstone.
62. Tasker RR: Spinal cord injury and central pain. In Aronoff GM, editor: *Evaluation and treatment of chronic pain,* ed 3, Philadelphia, 1999, Lippincott, Williams & Wilkins, pp 131–146.
63. Prkachin KM: Pain behavior and the development of pain-related disability: the importance of guarding. *Clin J Pain* 23(3):270–277, 2007.
64. Pachas WN: Joint pains and associated disorders. In Aronoff GM, editor: *Evaluation and treatment of chronic pain,* ed 3, Philadelphia, 1999, Lippincott Williams and Wilkins, pp 201–215.
65. Kraus H: Muscle deficiency. In Rachlin ES, editor: *Myofascial pain and fibromyalgia,* ed 2, St. Louis, 2002, Mosby.
66. Cailliet R: *Low back pain syndrome,* ed 5, Philadelphia, 1995, FA Davis.
67. Emre M, Mathies H: *Muscle spasms and pain,* Park Ridge, Illinois, 1988, Parthenon.
68. Sinnott M: *Assessing musculoskeletal changes in the geriatric population,* American Physical Therapy Association Combined Sections Meeting, February 3–7, 1993.
69. Potter JF: The older orthopaedic patient. General considerations. *Clin Orthop Rel Res* 425:44–49, 2004.
70. Myburgh C, Larsen AH, Hartvigsen J: A systematic, critical review of manual palpation for identifying myofascial trigger points: evidence and clinical significance. *Arch Phys Med Rehabil* 89(6):1169–1176, June 2008.
71. Headley BJ: When movement hurts: a self-help manual for treating trigger points, 1997, Innovative Systems. (Barbara Headley, MS, PT, EMS, director, and CEO of Headley Systems, Colorado, http://www.barbaraheadley.com/).
72. Kostopoulos D, Rizopoulos K: *The manual of trigger point and myofascial therapy,* Thorofare, NJ, 2001, Slack.
73. Rachlin ES, Rachlin IS, editors: *Myofascial pain and fibromyalgia: trigger point management,* ed 2, St. Louis, 2002, Mosby.
74. Blaylock RL: *Excitotoxins: the taste that kills,* New Mexico, 1996, Health Press. Available on-line: http://www.russellblaylockmd.com/.
75. Roberts HJ: *Aspartame disease: the ignored epidemic,* West Palm Beach, Fl, 2001, Sunshine Sentinel Press.
76. Roberts HJ: *Defense against Alzheimer's disease,* West Palm Beach, Fl, 2001, Sunshine Sentinel Press.
77. Sestak I, Cuzick J, Sapunar F: Risk factors for joint symptoms in patients enrolled in the ATAC trial: a retrospective, exploratory analysis. *Lancet Oncol* 9(9):866–872, Sept. 2008.
78. Issa NC, Thompson RL: Diagnosing and managing septic arthritis: a practical approach. *J Musculoskel Med* 20(2):70–75, 2003.

79. Sapico FL, Liquete JA, Sarma RJ: Bone and joint infections in patients with infective endocarditis: review of a 4-year experience. *Clin Infect Dis* 22:783–787, 1996.
80. Lutz B: Septic arthritis following anterior cruciate ligament reconstruction using tendon allografts–Florida and Louisiana. 2000, *MMWR* 50(48):1081–1083, 2001.
81. Pola E: Onset of Berger disease after *Staphylococcus aureus* infection: septic arthritis after anterior cruciate ligament reconstruction. *Arthroscopy* 19(4):E29, 2003.
82. Kumar S, Cowdery JS: Managing acute monarthritis in primary care practice. *J Musculoskel Med* 21(9):465–472, 2004.
83. Fishman SM, editor: *Bonica's management of pain*, ed 4, vol 1, Philadelphia, 2009, Lippincott, Williams & Wilkins.
84. Waddell G, Bircher, M, Finlayson D, et al: Symptoms and signs: physical disease or illness behaviour? *BMJ* 289:739–741, 1984.
85. Slipman CW: Epidemiology of spine tumors presenting to musculoskeletal physiatrists. *Arch Phys Med Rehabil* 84:492–495, 2003.
86. Ozburn MS, Nichols JW: Pubic ramus and adductor insertion stress fractures in female basic trainees. *Mil Med* 146(5):332–334, 1981.
87. Cyriax J: *Textbook of orthopaedic medicine*, ed 8, Vol 1, London, 1982, Baillière.
88. Management of the individual with pain: Part 1–physiology and evaluation. *PT Magazine* 4(11):54–63, 1996.
89. Merskey H, Bogduk N: *Classification of chronic pain*, ed 2, Seattle, 1994, International Association for the Study of Pain.
90. Waddell G: *The back pain revolution*, ed 2, Philadelphia, 2004, Churchill Livingstone.
91. Hellsing AL, Linton SJ, Kälvemark M: A prospective study of patients with acute back and neck pain in Sweden. *Phys Ther* 74(2):116–128, 1994.
92. Simmonds MJ: Pain, mind, and movement—an expanded, updated, and integrated conceptualization. *Clin J Pain* 24(4):279–280, 2008.
93. Melzack R: From the gate to the neuromatrix. *Pain* 6(suppl 6):S121–S126, 1999.
94. Smith BH: Epidemiology of chronic pain, from the laboratory to the bus stop: time to add understanding of biological mechanisms to the study of risk factors in population-based research? *Pain* 127:5–10, 2007.
95. Turk DC: Understanding pain sufferers: the role of cognitive processes. *Spine J* 4(1):1–7, 2004.
96. Berna C: Induction of depressed mood disrupts emotion regulation neurocircuitry and enhances pain unpleasantness. *Biol Psychiatry* 67(11)1038–1090, 2010.
97. Celestin J: Pretreatment psychosocial variables as predictors of outcomes following lumbar surgery and spinal cord stimulation: a systematic review and literature synthesis. *Pain Med* 10(4):639–653, 2009.
98. Ross EL: Chronic pain update: addressing abuse and misuse of opioid analgesics. *JOMM* 25(6):268–277, 2008.
99. Lethem J, Slade PD, Troup JDG, et al: Outline of a fear-avoidance model of exaggerated pain perception. I. *Behav Res Ther* 21(4):401–408, 1983.
100. Slade PD, Troup JDG, Lethem J, et al: The fear-avoidance model of exaggerated pain perception. II. *Behav Res Ther* 21(4):409–416, 1983.
101. Waddell G, Somerville D, Henderson I, et al: A fear avoidance beliefs questionnaire (FABQ) and the role of fear avoidance beliefs in chronic low back pain and disability. *Pain* 52:157–168, 1993.
102. George SZ: A psychometric investigation of fear-avoidance model measures in patients with chronic low back pain. *J Orthop Sports Phys Ther* 40(4):197–205, 2010.
103. Swinkels-Meewisse EJ: Psychometric properties of the Tampa Scale for kinesiophobia and the fear-avoidance beliefs questionnaire in acute low back pain. *Man Ther* 8:29–36, 2003.
104. Keefe FJ: An objective approach to quantifying pain behavior and gait patterns in low back pain patients. *Pain* 21:153–161, 1985.
105. Calley D: Identifying patient fear-avoidance beliefs by physical therapists managing patients with low back pain. *J Orthop Sports Phys Ther* 40(12):774–783, 2010.
106. Leeuw M: The fear-avoidance model of musculoskeletal pain: current state of scientific evidence. *J Behav Med* 30:77–94, 2000.
107. George SZ, Bialosky JE, Fritz JM: Physical therapist management of a patient with acute low back pain and elevated fear-avoidance beliefs. *Phys Ther* 84(6):538–549, 2004.
108. George SZ: Personal communication. May 2004.
109. Fritz JM, George SZ: Identifying psychosocial variables in patients with acute work-related low back pain. The importance of fear-avoidance beliefs. *Phys Ther* 82(10):973–983, 2002.
110. Hart DL: Screening for elevated levels of fear-avoidance beliefs regarding work or physical activities in people receiving outpatient therapy. *Phys Ther* 89(8):770–785, 2009.
111. Cooner E, Amorosi S: *The study of pain and older Americans*, New York, 1997, Louis Harris and Associates (Harris Opinion Poll).
112. Barsky AJ, Hochstrasser B, Coles NA, et al: Silent myocardial ischemia: is the person or the event silent? *JAMA* 364:1132–1135, 1990.
113. Kauvar DR: The geriatric acute abdomen. *Clin Geriatr Med* 9:547–558, 1993.
114. Norman DC, Toledo SD: Infections in elderly persons: an altered clinical presentation. *Clin Geriatr Med* 8:713–719, 1992.
115. AGS Panel on Chronic Pain in Older Persons: The management of chronic pain in older persons. *J Am Geriatr Soc* 46:635–651, 1998.
116. Connelly C: Managing low back pain and psychosocial overlie. *J Musculoskel Med* 21(8):409–419, 2004.
117. Main CJ, Waddell G: Behavioral responses to examination: a reappraisal of the interpretation of "nonorganic signs." *Spine* 23(21): 2367–2371, 1998.
118. Scalzitti DA: Screening for psychological factors in patients with low back problems: Waddell's nonorganic signs. *Phys Ther* 77(3):306–312, 1997.
119. Teasell RW, Shapiro AP: Strategic-behavioral intervention in the treatment of chronic nonorganic motor disorders. *Am J Phys Med Rehab* 73(1):44–50, 1994.
120. Waddell G: Symptoms and signs: physical disease or illness behavior? *BMJ* 289:739–741, 1984.
121. Barsky AJ, Goodson JD, Lane RS, et al: The amplification of somatic symptoms. *Psychosom Med* 50(5):510–519, 1988.
122. Turk DC: Understanding pain sufferers: the role of cognitive processes. *Spine J* 4:1–7, 2004.
123. Beck AT, Epstein N, Brown G, et al: An inventory for measuring clinical anxiety: psychometric properties. *J Consult Clin Psych* 56:893–897, 1988.
124. Steer RA, Beck AT: Beck anxiety inventory. In Zalaquett CP, Wood RJ, editors: *Evaluating stress: a book of resources*, Lanham, MD, 1997, Scarecrow Press.
125. Brenes GA: Anxiety and chronic obstructive pulmonary disease: prevalence, impact, and treatment. *Psychosom Med* 65(6):963–970, 2003.
126. Gonzalez O: Current depression among adults in the United States. *MMWR* 59(38):1229–1235, 2010.
127. Strine TW: Depression and anxiety in the United States: findings from the 2006 Behavioral Risk Factor Surveillance System. *Psychiatr Serv* 59:1383–1390, 2008.
128. Wassertheil-Smoller S, Shumaker S, Ockene J, et al: Depression and cardiovascular sequelae in postmenopausal women. The

Women's Health Initiative (WHI). *Arch Intern Med* 164(3):289–298, 2004.
129. Miller MC: Depression and pain. *Harvard Mental Health* 21(3):4, 2004.
130. World Health Organization (WHO): 2004. www.who.int [type in Depression in search window].
131. Andrade L, Caraveo-Anduaga JJ, Berglund P, et al: The epidemiology of major depressive episodes: results from the International Consortium of Psychiatric Epidemiology (ICPE) surveys. *Int J Methods Psychiatr Res* 12(3):165, 2003.
132. Abe T: Increased incidence of depression and its socio-cultural background in Japan. *Seishin Shinkeigaku Zasshi* 105(1):36–42, 2003.
133. Kessler RC: Epidemiology of women and depression. *J Affect Disord* 74(1):5–13, 2003.
134. Smith NL: *The effects of depression and anxiety on medical illness*, Sandy, Utah, 2002, University of Utah, Stress Medicine Clinic.
135. Corsico A, McGuffin P: Psychiatric genetics: recent advances and clinical implications. *Epidemiol Psychiatr Soc* 10(4):253–259, 2001.
136. Lotrich FE, Pollock BG: Meta-analysis of serotonin transporter polymorphisms and affective disorders. *Psychiatr Genet* 14(3):121–129, 2004.
137. Lee MS, Lee HY, Lee HJ, et al: Serotonin transporter promoter gene polymorphism and long-term outcome of antidepressant treatment. *Psychiatr Genet* 14(2):111–115, 2004.
138. McGuffin P, Marusic A, Farmer A: What can psychiatric genetics offer suicidology? *Crisis* 22(2):61–65, 2001.
139. Lespérance F, Jaffe AS: Beyond the blues: understanding the link between coronary artery disease and depression. Retrieved June 15, 2006, from *http://www.medscape.com/viewarticle/423461*
140. Lydiard RB: Irritable bowel syndrome, anxiety, and depression: what are the links? *J Clin Psychiatry* 62(Suppl 8):38–45, 2001.
141. Haggman S, Maher CG, Refshauge KM: Screening for symptoms of depression by physical therapists managing low back pain. *Phys Ther* 84(12):1157–1166, 2004.
142. Sartorius N, Ustun T, Lecrubier Y, et al: Depression comorbid with anxiety: results from the WHO study on psychological disorders in primary health care. *Br J Psychiatry* 168:38–40, 1996.
143. Beck AT, Ward CH, Mendelson M, et al: An inventory for measuring depression. *Arch Gen Psychiatry* 4:561–571, 1961.
144. C de C Williams A, Richardson PH: What does the BDI measure in chronic pain? *Pain* 55:259–266, 1993.
145. Yesavage JA: The geriatric depression scale, *J Psychiatr Res* 17(1):37–49, 1983.
146. Zung WWK: A self-rating depression scale, *Arch Gen Psychiatry* 12:63–70, 1965.
147. Harcourt Assessment (formerly The Psychological Corporation): *The Beck scales,* San Antonia, 2004.
148. Garakani A, Win T, Virk S, et al: Comorbidity of irritable bowel syndrome in psychiatric patients: a review, *Am J Ther* 10(1):61–67, 2003.
149. Campo JV, Dahl RE, Williamson DE, et al: Gastrointestinal distress to serotonergic challenge: a risk marker for emotional disorder? *J Am Acad Child Adolesc Psychiatry* 42(10):1221–1226, 2003.
150. Salt WB: *Irritable bowel syndrome and the mind-body/brain-gut connection*, ed 2, Columbus, Ohio, 2002, Parkview.
151. Chang L, Berman S, Mayer EA, et al: Brain responses to visceral and somatic stimuli in patients with irritable bowel syndrome with and without fibromyalgia. *Am J Gastroenterol* 98(6):1354–1361, 2003.
152. Fletcher JC, White, L: Use of the confusion assessment method (CAM) to screen for delirium in the acute care setting: a case report. *J Acute Care Phys Ther* 1(2):71–72, Winter 2010.
153. Miller MC: Understanding depression: a special health report from Harvard Medical School. Boston, 2003.
154. Inouye S: Clarifying confusion: the confusion assessment method. *Ann Intern Med* 113(12):941–948, 1990.
155. Hendrix ML: *Understanding panic disorder*, Washington, DC, 1993, National Institutes of Health.
156. Melzack R, Dennis SG: Neurophysiologic foundations of pain. In Sternbach RA, editor: *The psychology of pain*, New York, 1978, Raven Press, pp 1–26.
157. Wieseler-Frank J, Maier SF, Watkins LR: Glial activation and pathological pain. *Neurochem Int* 45(2-3):389–395, 2004.
158. Pert C: *Molecules of emotion: the science behind mind-body medicine*, New York, 1998, Simon and Schuster.
159. Knaster M: Remembering through the body. *Massage Therapy Journal* 33(1):46–59, 1994.
160. Pearsall P: *The heart's code: new findings about cellular memories and their role in the mind/body/spirit connection*, New York, 1998, Broadway Books (Random House).
161. van der Kolk BA: The body keeps the score: memory and the evolving psychobiology of posttraumatic stress. *Harvard Review of Psychiatry* 1(5):253–265, 1994.
162. Van Meeteren NLU, et al: Psychoneuroendocrinology and its relevance for physical therapy [Abstract]. *Phys Ther* 81(5):A66, 2001.
163. Yang J: UniSci International Science News, posted July 30, 2001 [http://unisci.com/], source: University of Rochester Medical Center, Rochester, NY, 2001.
164. Watkins LR, Milligan ED, Maier SF: Glial proinflammatory cytokines mediate exaggerated pain states: implications for clinical pain. *Adv Exp Med Biol* 521:1–21, 2003.
165. Wu CM, Lin MW, Cheng JT, et al: Regulated, electroporation-mediated delivery of pro-opiomelanocortin gene suppresses chronic constriction injury-induced neuropathic pain in rats. *Gene Ther* 11(11):933–940, 2004.
166. Maier SF, Watkins LR: Immune-to-central nervous system communication and its role in modulating pain and cognition: implications for cancer and cancer treatment. *Brain Behav Immun* 17(Suppl 1):S125–S131, 2003.
167. Watkins LR, Maier SF: The pain of being sick: implications of immune-to-brain communication for understanding pain. *Annu Rev Psychol* 51:29–57, 2000.
168. Watkins LR, Maier SF: Beyond neurons: evidence that immune and glial cells contribute to pathological pain states. *Physiol Rev* 82(4):981–1011, 2002.
169. Holguin A, O'Connor KA, Biedenkapp J, et al: HIV-1 gp120 stimulates proinflammatory cytokine-mediated pain facilitation via activation of nitric oxide synthase-I (nNOS). *Pain* 110(3):517–530, 2004.
170. Osman A: Factor structure, reliability, and validity of the pain catastrophizing scale. *J Behavioural Med* 20(6):589–605, 1997.
171. Osman A: The Pain Catastrophizing Scale: further psychometric evaluation with adult samples. *J Behav Med* 23(4):351–365, 2000.
172. Swinkels-Meewisse IEJ: Acute low back pain: pain-related fear and pain catastrophizing influence physical performance and perceived disability. *Pain* 120(1-2):36–43, 2006.
173. Sullivan M: The pain catastrophizing scale: development and validation. *Psych Assess* 7:524–532, 1995.
174. George SZ: A psychometric investigation of fear-avoidance model measures in patients with chronic low back pain. *J Orthop Sports Phys Ther* 40(4):197–205, 2010.
175. Melzack R: The short-form McGill Pain Questionnaire. *Pain* 30:191–197, 1987.

176. Dohrenwend A, Skillings JL: Diagnosis-specific management of somatoform disorders: moving beyond "vague complaints of pain." *J Pain* 10(11):1128–1137, 2009.
177. Matheson LN: *Work capacity evaluation: systematic approach to industrial rehabilitation*, Anaheim, CA, 1986, Employment and Rehabilitation Institute of California.
178. Matheson LN: *Symptom magnification casebook*, Anaheim, CA, 1987, Employment and Rehabilitation Institute of California.
179. Matheson LN: Symptom magnification syndrome structured interview: rationale and procedure. *J Occup Rehab* 1(1):43–56, 1991.
180. Greer S, Chambliss L, Mackler L: What physical exam techniques are useful to detect malingering? *J Fam Prac* 54(8):719–722, 2005.
181. Cocchiarella L, Anderson G: *Guides to the evaluation of permanent impairment*, ed 5, Chicago, 2001, AMA.
182. Waddell G, McCulloch JA, Kummer E, et al: Nonorganic physical signs in low back pain. *Spine* 5(2):117–125, 1980.
183. Karas R, McIntosh G, Hall H, et al: The relationship between nonorganic signs and centralization of symptoms in the prediction of return to work for patients with low back pain. *Phys Ther* 77(4):354–360, 1997.
184. Rothstein JM, Erhard RE, Nicholson GG, et al: Conference. *Phys Ther* 77(4):361–369, 1997.
185. Rothstein JM: Unnecessary adversaries (editorial). *Phys Ther* 77(4):352, 1997.
186. Goodwin RD: Association between physical activity and mental disorders among adults in the United States. *Prev Med* 36:698–703, 2003.
187. Lawlor DA, Hopker SW: The effectiveness of exercise as an intervention in the management of depression: systematic review and meta-regression analysis of randomized controlled trials. *BMJ* 322:1–8, 2001.
188. Dunn AL, Trivedi MH, Kampert JB, et al: The DOSE study: a clinical trial to examine efficacy and dose response of exercise as treatment for depression. *Control Clin Trials* 23:584–603, 2002.
189. Dowd SM, Vickers KS, Krahn D: Exercise for depression: physical activity boosts the power of medications and psychotherapy. *Psychiatry Online* 3(6), June 2004.

CHAPTER

4

Physical Assessment as a Screening Tool

In the medical model, clients are often assessed from head to toe. The doctor, physician assistant, nurse, or nurse practitioner starts with inspection, followed by percussion and palpation, and finally by auscultation.

In a screening assessment, the therapist may not need to perform a complete head-to-toe physical assessment. If the initial observations, client history, screening questions, and screening tests are negative, move on to the next step. A thorough examination may not be necessary.

In most situations, it is advised to assess one system above and below the area of complaint based on evidence supporting a regional-interdependence model of musculoskeletal impairments (i.e., symptoms present may be caused by musculoskeletal impairments proximal or distal to the site of presenting symptoms distinct from the phenomenon of referred pain).[1]

When screening for systemic origins of clinical signs and symptoms, the therapist first scans the area(s) that directly relate to the client's history and clinical presentation. For example, a shoulder problem can be caused by a problem in the stomach, heart, liver/biliary, lungs, spleen, kidneys, and ovaries (ectopic pregnancy). Only the physical assessment tests related to these areas would be assessed. These often can be narrowed down by the client's history, gender, age, presence of risk factors, and associated signs and symptoms linked to a specific system.

More specifically, consider the postmenopausal woman with a primary family history of heart disease who presents with shoulder pain that occurs 3 to 4 minutes after starting an activity and is accompanied by unexplained perspiration. This individual should be assessed for cardiac involvement. Or think about the 45-year-old mother of five children who presents with scapular pain that is worse after she eats. A cardiac assessment may not be as important as a scan for signs and symptoms associated with the gallbladder or biliary system.

Documentation of physical findings is important. From a legal standpoint, if you did not document it, you did not assess it. Look for changes from the expected norm, as well as changes for the client's baseline measurements. Use simple and clear documentation that can be understood and used by others. As much as possible, record both normal and abnormal findings for each client.[2] Keep in mind that the client's cultural and educational background, beliefs, values, and previous experiences can influence his or her response to questions.

Finally, screening and ongoing physical assessment is often a part of an exercise evaluation, especially for the client with one or more serious health concerns. Listening to the heart and lung sounds before initiating an exercise program may bring to light any contraindications to exercise. A compromised cardiopulmonary system may make it impossible and even dangerous for the client to sustain prescribed exercise levels.

The use of quick and easy screening tools such as the Physical Therapist Community Screening Form for Aging Adults can help therapists identify limitations to optimal heath, wellness, and fitness in any of seven areas (e.g., posture, flexibility, strength, balance, cardiovascular fitness) for adults aged 65 and older. With the 2007 House of Delegates position statement recommending that all individuals visit a physical therapist at least once a year to promote optimal health and wellness, evidence-based tests of this type will become increasingly available.[3]

GENERAL SURVEY

Physical assessment begins the moment you meet the client as you observe body size and type, facial expressions, evaluate self-care, and note anything unusual in appearance or presentation. Keep in mind (as discussed in Chapter 2) that cultural factors may dictate how the client presents himself

BOX 4-1 CONTENTS OF A SCREENING EXAMINATION KIT

- Stethoscope
- Sphygmomanometer
- Thermometer
- Pulse oximeter
- Reflex hammer
- Penlight
- Safety pin or sharp object (tongue depressor broken in half gives sharp and dull sides)
- Cotton-tipped swab or cotton ball
- Two test tubes
- Familiar objects (e.g., paper clip, coin, marble)
- Tuning fork (128 Hz)
- Watch with ability to count seconds
- Gloves for palpation of skin lesions
- Ruler or plastic tape measure to measure wound dimensions, skin lesions, leg length
- Goniometer

(e.g., avoiding eye contact when answering questions, hiding or exaggerating signs of pain).

A few pieces of equipment in a small kit within easy reach can make the screening exam faster and easier (Box 4-1). Using the same pattern in screening each time will help the therapist avoid missing important screening clues.

As the therapist makes a general survey of each client, it is also possible to evaluate posture, movement patterns and gait, balance, and coordination. For more involved clients the first impression may be based on level of consciousness, respiratory and vascular function, or nutritional status.

In an acute care or trauma setting the therapist may be using vital signs and the ABCDE (airway, breathing, circulation, disability, exposure) method of quick assessment. A common strategy for history taking in the trauma unit is the mnemonic AMPLE: **A**llergies, **M**edications, **P**ast medical history, **L**ast meal, and **E**vents of injury.

In any setting, knowing the client's personal health history will also help guide and direct which components of the physical examination to include. We are not just screening for medical disease masquerading as neuromusculoskeletal (NMS) problems. Many physical illnesses, diseases, and medical conditions directly impact the NMS system and must be taken into account. For example inspection of the integument, limb inspection, and screening of the peripheral vascular system is important for someone at risk for lymphedema.

Neurologic function, balance, reflexes, and peripheral circulation become important when screening a client with diabetes mellitus. Peripheral neuropathy is common in this population group, often making walking more difficult and increasing risk of other problems developing.

Therapists in all settings, especially primary care therapists, can use a screening physical assessment to provide education toward primary prevention, as well as intervention and management of current dysfunctions and disabilities.

Mental Status

Level of consciousness, orientation, and ability to communicate are all part of the assessment of a client's mental status. Orientation refers to the client's ability to answer correctly questions about time, place, and person. A healthy individual with normal mental status will be alert, speak coherently, and be aware of the date, day, and time of day.

The therapist must be aware of any factor that can affect a client's current mental status. Shock, head injury, stroke, hospitalization, surgery (use of anesthesia), medications, age, and the use of substances and/or alcohol (see discussion, Chapter 2) can cause impaired consciousness.

Other factors affecting mental status may include malnutrition, exposure to chemicals, and hypothermia or hyperthermia. Depression and anxiety (see discussion, Chapter 3) also can affect a client's functioning, mood, memory, ability to concentrate, judgment, and thought processes. Educational and socioeconomic background along with communication skills (e.g., English as a second language, aphasia) can affect mental status and function.

In a hospital, transition unit, or extended care facility, mental status is often evaluated and documented by the social worker or nursing service. It is always a good idea to review the client's chart or electronic record regarding this information before beginning a physical therapy evaluation.

Risk Factors for Delirium

It is not uncommon for older adults to experience a change in mental status or go through a stage of *confusion* about 24 hours after hospitalization for a serious illness or trauma, including surgery under a general anesthetic. Physicians may refer to this as iatrogenic delirium, anesthesia-induced dementia, or postoperative delirium. It is usually temporary but can last several hours to several weeks.

The cause of deterioration in mental ability is unknown. In some cases, delirium/dementia appears to be triggered by the shock to the body from anesthesia and surgery.[4] It may be a passing phase with complete recovery by the client, although this can take weeks to months. The likelihood of delirium associated with hospitalization is much higher with hip fractures and hip and knee joint replacements,[5,6] possibly attributed to older age, slower metabolism, and polypharmacy (more than four prescribed drugs at admission).[7]

The therapist should pay attention to risk factors (Box 4-2) and watch out for any of the signs or symptoms of delirium. Physical exam should include vital signs with oxygen concentration measured, neurologic screening exam, and surveillance for signs of infection. A medical diagnosis is needed to make the distinction between postoperative delirium, baseline dementia, depression, and withdrawal from drugs and alcohol.[5]

CLINICAL SIGNS AND SYMPTOMS

Iatrogenic Delirium

Cognitive Impairment

- Unable to concentrate during conversations
- Easily distracted or inattentive
- Switches topics often
- Unable to complete simple math or spell simple words backwards

Impaired Orientation

- Unable to remember familiar concepts (e.g., say the days of the week, unable to tell time)
- Does not know who he is or where he is
- Unable to recognize family or close friends without help

Impaired Speech

- Speech is difficult to understand
- Unable to speak in full sentences; sentences do not make sense

Psychologic Impairment

- Anxious and afraid; requires frequent reassurance
- Suspicious of others, paranoid
- Irritable, jumpy, or in constant motion
- Experiencing delusions and hallucinations (e.g., sees objects or people who are not there; smells scents that are not present)

Several scales are used to assess level of consciousness, performance, and disability. The Confusion Assessment Method (CAM) is a bedside rating scale physical therapists can use to assess hospitalized or institutionalized individuals for delirium. This tool has been adapted for use with patients who are ventilated and in an intensive care unit (CAM-ICU).[8]

There are two parts to the assessment instrument: part one screens for overall cognitive impairment. Part two includes four features that have the greatest ability to distinguish delirium or reversible confusion from other types of cognitive impairment. The tool identifies the presence of delirium but does not assess the severity of the condition.[9]

As a screening tool, the CAM has been validated for use by physicians and nurses in palliative care and intensive care settings (sensitivity of 94% to 100% and specificity of 90% to 95%). Values for positive predictive accuracy were 91% to 94%, and values for negative predictive accuracy were 100% and 90% for the two populations assessed (general medicine, outpatient geriatric center).[9]

The Glasgow Outcome Scale[10,11] describes patients/clients on a 5-point scale from good recovery (1) to death (5). Vegetative state, severe disability, and moderate disability are included in the continuum. This and other scales and clinical assessment tools are not part of the screening assessment but are available online for use by health care professionals.[12]

The Karnofsky Performance Scale (KPS) in Table 4-1 is used widely to quantify functional status in a wide variety of individuals, but especially among those with cancer. It can be used to compare effectiveness of intervention and to assess

BOX 4-2 RISK FACTORS FOR IATROGENIC OR POSTOPERATIVE DELIRIUM

- Stress, trauma, pain, infection
- Hospitalization (for hip fracture, serious illness or trauma including surgery) or change in residence
- Older age (65 years old or older)
- Anesthesia
- Hip or knee joint replacement
- Poor cognitive function, underlying dementia, previous cognitive impairment
- Vision or hearing deficits
- Decreased physical function
- History of alcohol abuse
- Medications (e.g., benzodiazepine, narcotics, NSAIDs, anticholinergics prescribed for sleep, psychoactive drugs/antidepressants/antipsychotics, dopamine agents, analgesics, sedative agents for pain and anxiety after surgery)*
- Dehydration
- Urinary retention, fecal impaction, diarrhea
- Sleep deprivation
- Postoperative low hemoglobin, abnormal fluid and/or electrolytes, low oxygen saturation
- Malnutrition, vitamin B_{12}/folate deficiency, low albumin

Data from Alfonso DT: Nonsurgical complications after total hip and total knee arthroplasty, *Am J Orthop* 35(11):503-510, 2006; Short M, Winstead PS: Delirium dilemma: pharmacology update, *Orthopedics* 30(4):273-277, 2007.

NSAIDs, Nonsteroidal antiinflammatory drugs.

*Higher risk medications commonly associated with delirium; lower risk medications associated with delirium include some cardiovascular agents (e.g., antiarrhythmics, beta blockers, clonidine, digoxin), antimicrobials (e.g., fluoroquinolones, penicillins, sulfonamides, acyclovir), anticonvulsants, and medications for gastroesophageal reflux or nausea.

TABLE 4-1 Karnofsky Performance Scale (Rating in %)

Score	Description
100	Normal, no complaints; no evidence of disease
90	Able to carry on normal activities; minor signs or symptoms of disease
80	Normal activity with effort; some signs or symptoms of disease
70	Cares for self; unable to carry on normal activity or to do active work
60	Requires occasional assistance but able to care for most of own personal needs
50	Requires considerable assistance and frequent medical care
40	Disabled; requires special care and assistance
30	Severely disabled; hospitalization indicated though death not imminent
20	Very ill; hospitalization required; active supportive treatment necessary
10	Failing rapidly; moribund
0	Dead

TABLE 4-2 ECOG Performance Status Scale

Grade	Level of Activity
0	Fully active, able to carry on all pre-disease performance without restriction (Karnofsky 90%-100%)
1	Restricted in physically strenuous activity but ambulatory and able to carry out work of a light or sedentary nature (e.g., light house work, office work) (Karnofsky 70%-80%)
2	Ambulatory and capable of all self-care but unable to carry out any work activities. Up and about more than 50% of waking hours (Karnofsky 50%-60%)
3	Capable of only limited self-care, confined to bed or chair more than 50% of waking hours (Karnofsky 30%-40%)
4	Completely disabled. Cannot carry on any self-care. Totally confined to bed or chair (Karnofsky 10%-20%)
5	Dead (Karnofsky 0%)

The Karnofsky Performance Scale allows individuals to be classified according to functional impairment. The lower the score, the worse the prognosis for survival for most serious illnesses.

ECOG, Eastern Cooperative Oncology Group.

From Oken MM, Creech RH, Tormey DC, et al: Toxicity and response criteria of the Eastern Cooperative Oncology Group, *Am J Clin Oncol* 5:649-655, 1982. Available at www.ecog.org/general/perf_stat.html.

individual prognosis. The lower the Karnofsky score, the worse the prognosis for survival.

The most practical performance scale for use in any rehabilitation setting for most clients is the Eastern Cooperative Oncology Group (ECOG) Performance Status Scale (Table 4-2). Researchers and health care professionals use these scales and criteria to assess how an individual's disease is progressing, to assess how the disease affects the daily living abilities of the client, and to determine appropriate treatment and prognosis.

Any observed change in level of consciousness, orientation, judgment, communication or speech pattern, or memory should be documented regardless of which scale is used. The therapist may be the first to notice increased lethargy, slowed motor responses, or disorientation or confusion.

Confusion is not a normal change with aging and must be reported and documented. Confusion is often associated with various systemic conditions (Table 4-3). Increased confusion in a client with any form of dementia can be a symptom of infection (e.g., pneumonia, urinary tract infection), electrolyte imbalance, or delirium. Likewise, a sudden change in muscle tone (usually increased tone) in the client with a neurologic disorder (adult or child) can signal an infectious process.

Nutritional Status

Nutrition is an important part of growth and development and recovery from infection, illness, wounds, and surgery. Clients can exhibit signs of malnutrition or overnutrition (obesity).

TABLE 4-3 Systemic Conditions Associated with Confusional States

System	Impairment/Condition
Endocrine	Hypothyroidism, hyperthyroidism
	Perimenopause, menopause
Metabolic	Severe anemia
	Fluid and/or electrolyte imbalances; dehydration
	Wilson's disease (copper disorder)
	Porphyria (inherited disorder)
Immune/Infectious	AIDS
	Cerebral amebiasis, toxoplasmosis, or malaria
	Fungal or tubercular meningitis
	Lyme disease
	Neurosyphilis
Cardiovascular	CHF
Cerebrovascular	Cerebral insufficiency (TIA, CVA)
	Postanoxic encephalopathy
Pulmonary	COPD
	Hypercapnia (increased CO_2)
	Hypoxemia (decreased arterial O_2)
Renal	Renal failure, uremia
	Urinary tract infection
Neurologic	Encephalopathy (hepatic, hypertensive)
	Head trauma
	Cancer
	CVA; stroke
Other	Chronic drug and/or alcohol use
	Medication (e.g., anticonvulsants, antidepressants, antiemetics, antihistamines, antipsychotics, benzodiazepines, narcotics, sedative-hypnotics, Zantac, Tagamet)
	Postoperative
	Severe anemia
	Cancer metastasized to the brain
	Sarcoidosis
	Sleep apnea
	Vasculitis (e.g., SLE)
	Vitamin deficiencies (B_{12}, folate, niacin, thiamine)
	Whipple's disease (severe intestinal disorder)

Modified from Dains JE, Baumann LC, Scheibel P: *Advanced health assessment & clinical diagnosis in primary care,* ed 2, St. Louis, 2003, Mosby; p 425.

AIDS, Acquired immunodeficiency syndrome; *CHF,* congestive heart failure; *TIA,* transient ischemic attack; *CVA,* cerebrovascular accident; *COPD,* chronic obstructive pulmonary disease; *SLE,* systemic lupus erythematosus.

CLINICAL SIGNS AND SYMPTOMS

Undernutrition or Malnutrition

- Muscle wasting
- Alopecia (hair loss)
- Dermatitis; dry, flaking skin
- Chapped lips, lesions at corners of mouth
- Brittle nails
- Abdominal distention
- Decreased physical activity/energy level; fatigue, lethargy
- Peripheral edema
- Bruising

BOX 4-3 RISK FACTORS FOR NUTRITIONAL DEFICIENCY

- Economic status
- Living alone
- Older age (metabolic rate slows in older adults; altered sense of taste and smell affects appetite)
- Depression, anxiety
- Eating disorders
- Lactose intolerance (common in Mexican Americans, African Americans, Asians, Native Americans)
- Alcohol/drug addiction
- Chronic diarrhea
- Nausea
- Gastrointestinal impairment (e.g., bowel resection, gastric bypass, pancreatitis, Crohn's disease, pernicious anemia)
- Chronic endocrine or metabolic disorders (e.g., diabetes mellitus, celiac sprue)
- Liver disease
- Dialysis
- Medications (e.g., captopril, chemotherapy, steroids, insulin, lithium) including over-the-counter drugs (e.g., laxatives)
- Chronic disability affecting activities of daily living (ADLs; e.g., problems with balance, mobility, food preparation)
- Burns
- Difficulty chewing or swallowing (dental problems, stroke or other neurologic impairment)

Be aware in the health history of any risk factors for nutritional deficiencies (Box 4-3). Remember that some medications can cause appetite changes and that psychosocial factors such as depression, eating disorders, drug or alcohol addictions, and economic variables can affect nutritional status.

It may be necessary to determine the client's ideal body weight by calculating the body mass index (BMI).[13,14] Several websites are available to help anyone make this calculation. There is a separate website for children and teens sponsored by the National Center for Chronic Disease Prevention and Health Promotion.[15]

Whenever nutritional deficiencies are suspected, notify the physician and/or request a referral to a registered dietitian.

Body and Breath Odors

Odors may provide some significant clues to overall health status. For example, a fruity (sweet) breath odor (detectable by some but not all health care professionals) may be a symptom of diabetic ketoacidosis. Bad breath (halitosis) can be a symptom of dental decay, lung abscess, throat or sinus infection, or gastrointestinal (GI) disturbances from food intolerances, *Helicobacter pylori* bacteria, or bowel obstruction. Keep in mind that ethnic foods and alcohol can affect breath and body odor.

Clients who are incontinent (bowel or bladder) may smell of urine, ammonia, or feces. It is important to ask the client about any unusual odors. It may be best to offer an introductory explanation with some follow-up questions:

FOLLOW-UP QUESTIONS

Mrs. Smith, as part of a physical therapy exam, we always look at our client's overall health and general physical condition. Do you have any other health concerns besides your shoulder/back (Therapist: name the involved body part)?

Are you being treated by anyone for any other problems? (Wait for a response but add prompts as needed: chiropractor? acupuncturist? naturopath?)

[If you suspect urinary incontinence]: Are you having any trouble with leaking urine or making it to the bathroom on time? (Ask appropriate follow-up questions about cause, frequency, severity, triggers, and so on; see Appendix B-5).

[If you suspect fecal incontinence]: Do you have trouble getting to the toilet on time for a bowel movement?

Do you have trouble wiping yourself clean after a bowel movement? (Ask appropriate follow-up questions about cause, frequency, severity, triggers, and so on).

[If you detect breath odor]: I notice an unusual smell on your breath. Do you know what might be causing this? (Ask appropriate follow-up questions depending on the type of smell you perceive; you may have to conduct an alcohol screening survey [see Chapter 2 or Appendices B-1 and B-2].)

Vital Signs

The need for therapists to assess vital signs, especially pulse and blood pressure is increasing.[16] Without the benefit of laboratory values, physical assessment becomes much more important. Vital signs, observations, and reported associated signs and symptoms are among the best screening tools available to the therapist.

Vital sign assessment is an important tool because high blood pressure is a serious concern in the United States. Many people are unaware they have high blood pressure. Often primary orthopedic clients have secondary cardiovascular disease.[17]

Physical therapists practicing in a primary care setting will especially need to know when and how to assess vital signs. The *Guide to Physical Therapist Practice*[18] recommends that heart rate (pulse) and blood pressure measurements be included in the examination of new clients. Exercise professionals are strongly encouraged to measure blood pressure during each visit.[19]

Taking a client's vital signs remains the single easiest, most economic, and fastest way to screen for many systemic illnesses. All the vital signs are important (Box 4-4); temperature and blood pressure have the greatest utility as early screening tools for systemic illness or disease, while pulse, blood pressure, and oxygen (O_2) saturation level offer

BOX 4-4 VITAL SIGNS

- Pulse (beats per minute [bpm])
- Blood pressure (BP)
- Core body temperature (oral or ear)
- Respirations
- Pulse oximetry (oxygen [O_2] saturation)
- Skin temperature—digits (thermister)*
- Pain (now called the 5th vital sign; see Chapter 3 for assessment)
- Walking speed (the 6th vital sign)[141]

*One thermistor, the PhysioQ, is a handheld device used to measure skin temperature (fingertips, hand). Similar tools are available as part of some biofeedback equipment. Using skin temperature is an excellent tool for teaching clients how to modulate the autonomic nervous system, a technique called *physiologic quieting.* This tool is commercially available with a guided relaxation tape [www.phoenixpub.com; 1-800-549-8371]. It can be a useful intervention with clients who have chronic fatigue syndrome, fibromyalgia, Raynaud's phenomenon or disease, sleep disturbance, and peripheral vascular disease. Results can be measured using all the vital signs, but especially by measuring and recording changes in skin temperature.

valuable information about the cardiovascular/pulmonary systems.

As an aside comment: using vital signs is an easy, yet effective way to document outcomes. In today's evidence-based practice, the therapist can use something as simple as pulse or blood pressure to document changes that occur with intervention.

For example, if ambulating with a client morning and afternoon results in no change in ease of ambulation, speed, or distance, consider taking blood pressure, pulse, and O_2 saturation levels before and after each session. Improvement in O_2 saturation levels or faster return to normal of heart rate after exercise are just two examples of how vital signs can become an important part of outcomes documentation.

Assessment of baseline vital signs should be a part of the initial data collected so that correlations and comparisons with future values are available when necessary. The therapist compares measurements taken against normal values and also compares future measurements to the baseline units to identify significant changes (normalizing values or moving toward abnormal findings) for each client.

Normal ranges of values for the vital signs are provided for the therapist's convenience. However, these ranges can be exceeded by a client and still represent normal for that person. Keep in mind that many factors can affect vital signs, especially pulse and blood pressure (Table 4-4). Substances such as alcohol, caffeine, nicotine, and cocaine/cocaine derivatives as well as pain and stress/anxiety can cause fluctuations in blood pressure. Adults who monitor their own blood pressure may report wide fluctuations without making the association between these and other factors listed. It is the unusual vital sign in combination with other signs and symptoms,

TABLE 4-4 Factors Affecting Pulse and Blood Pressure

Pulse	Blood Pressure*
Age	Age
Anemia	Alcohol
Autonomic dysfunction (diabetes, spinal cord injury)	Anxiety
Caffeine	Blood vessel size
Cardiac muscle dysfunction	Blood viscosity
Conditioned/deconditioned state	Caffeine
Dehydration (decreased blood volume increases heart rate)	Cocaine and cocaine derivatives
Exercise	Diet
Fear	Distended urinary bladder
Fever, heat	Force of heart contraction
Hyperthyroidism	Living at higher altitudes
Infection	Medications
Medications	• ACE inhibitors (lowers pressure)
• Antidysrhythmic (slows rate)	• Adrenergic inhibitors (lowers pressure)
• Atropine (increases rate)	• Beta-blockers (lowers pressure)
• Beta-blocker (slows rate)	• Diuretics (lowers pressure)
• Digitalis (slows rate)	• Narcotic analgesics (lowers pressure)
Sleep disorders or sleep deprivation	Nicotine
Stress (emotional or psychologic)	Pain
	Time of recent meal (increases SBP)

SBP, Systolic blood pressure; *ACE,* angiotensin-converting enzyme.
*Conditions, such as chronic kidney disease, renovascular disorders, primary aldosteronism, and coarctation of the aorta, are identifiable causes of elevated blood pressure. Chronic overtraining in athletes, use of steroids and/or nonsteroidal antiinflammatory drugs (NSAIDs), and large increases in muscle mass can also contribute to hypertension.[38] Treatment for hypertension, dehydration, heart failure, heart attack, arrhythmias, anaphylaxis, shock (from severe infection, stroke, anaphylaxis, major trauma), and advanced diabetes can cause low blood pressure.
From Goodman CC, Fuller K: *Pathology: implications for the physical therapist,* ed 3, Philadelphia, 2009, WB Saunders.

medications, and medical status that gives clinical meaning to the pulse rate, blood pressure, and temperature.

Pulse Rate

The pulse reveals important information about the client's heart rate and heart rhythm. A resting pulse rate (normal range: 60 to 100 beats per minute [bpm]) taken at the carotid artery or radial artery (preferred sites) pulse point should be available for comparison with the pulse rate taken during treatment or after exercise. A pulse rate above 100 bpm indicates tachycardia; below 60 bpm indicates bradycardia.

Do not rely on pulse oximeter devices for pulse rate because these units often take a sample pulse rate that reflects a mean average and may not reveal dysrhythmias (e.g., a regular irregular pulse rate associated with atrial fibrillation). It is recommended that the pulse always be checked in two places in older adults and in anyone with diabetes (Fig. 4-1). Pulse strength (amplitude) can be graded as

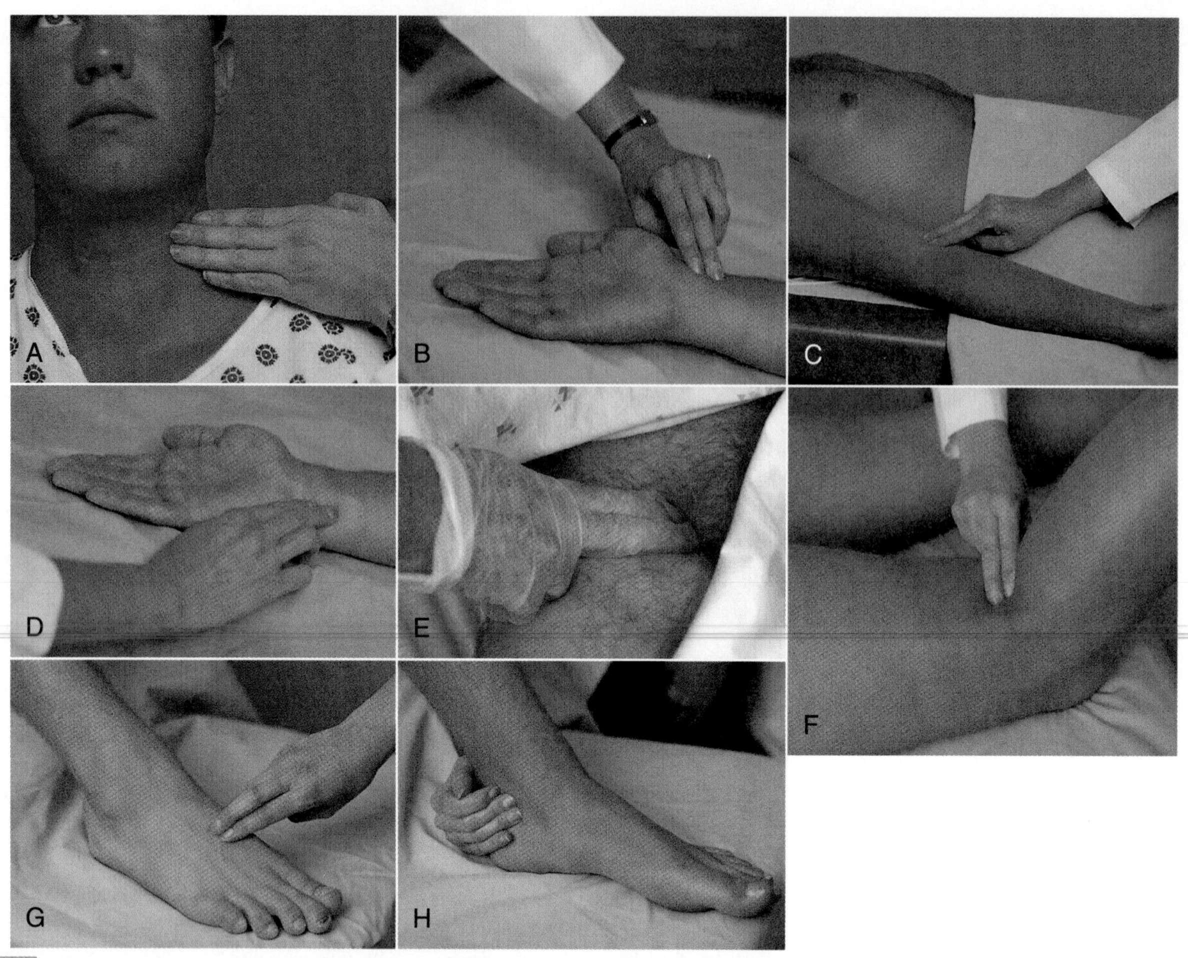

Fig. 4-1 Pulse points. The easiest and most commonly palpated pulses are the **(A)** carotid pulse and **(B)** radial pulse. Other pulse points include brachial pulse **(C),** ulnar pulse **(D),** femoral pulse **(E),** popliteal pulse (knee slightly flexed) **(F),** dorsalis pedis **(G),** and posterior tibial **(H).** The anterior tibial pulse becomes the dorsalis pedis and is palpable where the artery lies close to the skin on the dorsum of the foot. Peripheral pulses are more difficult to palpate in older adults and anyone with peripheral vascular disease. (From Potter PA: *Fundamentals of nursing,* ed 7, St. Louis, 2009, Mosby.)

0	Absent, not palpable
1+	Pulse diminished, barely palpable
2+	Easily palpable, normal
3+	Full pulse, increased strength
4+	Bounding, too strong to obliterate

Keep in mind that taking the pulse measures the peripheral arterial wave propagation generated by the heart's contraction—it is not the same as measuring the true heart rate (and should not be recorded as heart rate when measured by palpation). A true measure of heart rate requires auscultation or electrocardiographic recording of the electrical impulses of the heart. The distinction between pulse rate and heart rate becomes a matter of concern in documentation liability and even greater importance for individuals with dysrhythmias. In such cases, the output of blood by some beats may be insufficient to produce a detectable pulse wave that would be discernible with an electrocardiogram.[20]

Pulse amplitude (weak or bounding quality of the pulse) gives an indication of the circulating blood volume and the strength of left ventricle ejection. Normally, the pulse increases slightly with inspiration and decreases with expiration. This slight change is not considered significant.

Pulse amplitude that fades with inspiration instead of strengthening and strengthens with expiration instead of fading is *paradoxic* and should be reported to the physician. Paradoxic pulse occurs most commonly in clients with chronic obstructive pulmonary disease (COPD) but is also observed in clients with constrictive pericarditis.[21]

Constriction or compression around the heart from pericardial effusion, tension pneumothorax, pericarditis with fluid, or pericardial tamponade may be associated with paradoxical pulse. When the person breathes in, the increased mechanical pressure of inspiration added to the physiologic compression from the underlying disease prevents the heart from contracting fully and results in a reduced pulse. When the person breathes out, the pressure from chest expansion is reduced and the pulse increases.

A pulse increase with activity of more than 20 bpm lasting for more than 3 minutes after rest or changing position should also be reported. Other pulse abnormalities are listed in Box 4-5.

BOX 4-5 PULSE ABNORMALITIES

- Weak pulse beats alternating with strong beats
- Weak, thready pulse
- Bounding pulse (throbbing pulse followed by sudden collapse or decrease in the force of the pulse)
- Two quick beats followed by a pause (no pulse)
- Irregular rhythm (interval between beats is not equal)
- Pulse amplitude decreases with inspiration/increases with expiration
- Pulse rate too fast (greater than 100 bpm; tachycardia)
- Pulse rate too slow (less than 60 bpm; bradycardia)

The resting pulse may be higher than normal with fever, anemia, infections, some medications, hyperthyroidism, anxiety, or pain. A low pulse rate (below 60 bpm) is not uncommon among trained athletes. Medications, such as beta-blockers and calcium channel blockers, can also prevent the normal rise in pulse rate that usually occurs during exercise. In such cases the therapist must monitor rates of perceived exertion (RPE) instead of pulse rate.

When taking the resting pulse or pulse during exercise, some clinicians measure the pulse for 15 seconds and multiply by 4 to get the rate per minute. For a quick assessment, measure for 6 seconds and add a zero. A 6-second pulse count can result in an error of 10 bpm if a 1-beat error is made in counting. For screening purposes, it is always best to palpate the pulse for a full minute. Longer pulse counts give greater accuracy and provide more time for detection of some dysrhythmias (Box 4-6).[19]

Pulse assessment following vascular injuries (especially dislocation of the knee) should not be relied upon as the only diagnostic testing procedure as occult arterial injuries can be present even when pulses are normal. A meta-analysis of 284 dislocated knees concluded that *abnormal* pulse examinations have a sensitivity of .79 and specificity of .91 for detection of arterial injuries.[22,23] On the other hand, there are reports of normal pulse examinations at the time of the initial knee injury in people who later developed ischemia leading to amputation.[24,25]

Respirations

Try to assess the client's breathing without drawing attention to what is being done. This measure can be taken right after counting the pulse while still holding the client's wrist.

Count respirations for 1 minute unless respirations are unlabored and regular, in which case the count can be taken for 30 seconds and multiplied by 2. The rise and fall of the chest equals 1 cycle.

The normal rate is between 12 and 20 breaths per minute. Observe rate, excursion, effort, and pattern. Note any use of accessory muscles and whether breathing is silent or noisy. Watch for puffed cheeks, pursed lips, nasal flaring, or asymmetric chest expansion. Changes in the rate, depth, effort, or pattern of a client's respirations can be early signs of neurologic, pulmonary, or cardiovascular impairment.

BOX 4-6 TIPS ON PALPATING PULSES

- Assess each pulse for strength and equality for one full minute; pulse rate should NOT be taken for part of a minute and then multiplied by a factor (e.g., 15 seconds × 4, 30 seconds × 2, 6 seconds × 10).
- Expect to palpate 60 to 90 pulses per minute at all pulse sites. Begin the pulse count with zero, not "one."
- Normal pulse is 2+ and equal bilaterally (see scale in text).
- Apply gentle pressure; pulses are easily obliterated in some people.
- Popliteal pulse requires deeper palpation.
- Normal veins are flat; pulsations are not visible.
- Flat veins in supine that become distended in sitting may indicate heart disease.
- Pulses should be the same from side to side and should not change significantly with inspiration, expiration, or change in position.
- Pulses tend to diminish with age; distal pulses are not palpable in many older adults.
- If pulses are diminished or absent, listen for a bruit to detect arterial narrowing.
- Pedal pulses can be congenitally absent; the client may or may not know if absent pulse at this pulse site is normal or a change in pulse pressure.
- In the case of diminished or absent pulses observe the client for other changes (e.g., skin temperature, texture, color, hair loss, change in toenails); ask about pain in calf or leg with walking that goes away with rest (intermittent claudication, peripheral vascular disease [PVD]).
- Carotid pulse: Assess in the seated position; have client turn the head slightly toward the side being palpated. Gently and carefully palpate along the medial edge of the sternocleidomastoid muscle (see Fig. 4-1). Palpate one carotid artery at a time; apply light pressure; deep palpation can stimulate carotid sinus with a sudden drop in heart rate and blood pressure. Do not poke or mash around to find the pulse; palpation must not provide a massage to the artery due to the risk of liberating a thrombus or plaque, especially in older adults.
- Femoral pulse: Femoral artery is palpable below the inguinal ligament midway between the anterior superior iliac spine (ASIS) and the symphysis pubis. It can be difficult to assess in the obese client; place fingertips of both hands on either side of the pulse site; femoral pulse should be as strong (if not stronger) than radial pulse.
- Posterior tibial pulse: Foot must be relaxed with ankle in slight plantar flexion (see Fig. 4-1).

Pulse Oximetry

O_2 saturation on hemoglobin (SaO_2) and pulse rate can be measured simultaneously using pulse oximetry. This is a noninvasive, photoelectric device with a sensor that can be attached to a well-perfused finger, the bridge of the nose, toe, forehead, or ear lobe. Digital readings are less accurate with clients who are anemic, undergoing chemotherapy, or who use fingernail polish or nail acrylics. In such cases, attach the sensor to one of the other accessible body parts.

The sensor probe emits red and infrared light, which is transmitted to the capillaries. When in contact with the skin, the probe measures transmitted light passing through the vascular bed and detects the relative amount of color absorbed by the arterial blood. The SaO_2 level is calculated from this information.

The normal SaO_2 range at rest and during exercise is 95% to 100%. Referral for medical evaluation is advised when resting saturation levels fall below 90%. The exception to the normal range listed here is for clients with a history of tobacco use and/or COPD. Some individuals with COPD tend to retain carbon dioxide and can become apneic if the oxygen levels are too high. For this reason, SaO_2 levels are normally kept lower for this population.

The drive to breathe in a healthy person results from an increase in the arterial carbon dioxide level ($PaCO_2$). In the normal adult, increased CO_2 levels stimulate chemoreceptors in the brainstem to increase the respiratory rate. With some chronic lung disorders these central chemoreceptors may become desensitized to $PaCO_2$ changes resulting in a dependence on the peripheral chemoreceptors to detect a fall in arterial oxygen levels (PaO_2) to stimulate the respiratory drive.

Too much oxygen delivered as a treatment can depress the respiratory drive in those individuals with COPD who have a dampening of the CO_2 drive. Monitoring respiratory rate, level of oxygen administered by nasal canula, and SaO_2 levels is very important in this client population.

Some pulmonologists agree that supplemental oxygen levels can be increased during activity without compromising the individual because they will "blow it (carbon dioxide) off" anyway. To our knowledge, there is no evidence yet to support this clinical practice.

Any condition that restricts blood flow (including cold hands) can result in inaccurate SaO_2 readings. Relaxation and physiologic quieting techniques can be used to help restore more normal temperatures in the distal extremities. A handheld device such as the PhysioQ[26] can be used by the client to improve peripheral circulation. Do not apply a pulse oximetry sensor to an extremity with an automatic blood pressure cuff.[27]

SaO_2 levels can be affected also by positioning because positioning can impact a person's ability to breathe. Upright sitting in individuals with low muscle tone or kyphosis can cause forward flexion of the thoracic spine compromising oxygen intake. Tilting the person back slightly can open the trunk, ease ventilation, and improve SaO_2 levels.[28] Using SaO_2 levels may be a good way to document outcomes of positioning programs for clients with impaired ventilation.

Other factors affecting pulse oximeter readings can include nail polish and nail coverings, irregular heart rhythms, hyperemia (increased blood flow to the area), motion artifact, pressure on the sensor, electrical interference, and venous congestion.[20]

In addition to SaO_2 levels, assess other vital signs, skin and nail bed color and tissue perfusion, mental status, breath sounds, and respiratory pattern for all clients using pulse oximetry. If the client cannot talk easily whether at rest or while exercising, SaO_2 levels are likely to be inadequate.

Blood Pressure

Blood pressure (BP) is the measurement of pressure in an artery at the peak of systole (contraction of the left ventricle) and during diastole (when the heart is at rest after closure of the aortic valve, which prevents blood from flowing back to the heart chambers). The measurement (in mm Hg) is listed as:

$$\frac{\text{Systolic (contraction phase)}}{\text{Diastolic (relaxation phase)}}$$

BP depends on many factors; the normal range differs slightly with age and varies greatly among individuals (see Table 4-4). Normal systolic BP (SBP) ranges from 100 to 120 mm Hg, and diastolic BP (DBP) ranges from 60 to 80 mm Hg. Highly trained athletes may have much lower values. Target ranges for BP are listed in Table 4-5 and Box 4-7.

Assessing Blood Pressure. BP should be taken in the same arm and in the same position (supine or sitting) each time it is measured. The baseline BP values can be recorded on the Family/Personal History form (see Fig. 2-2).

Cuff size is important and requires the bladder width-to-length be at least 1:2. The cuff bladder should encircle at least 80% of the arm. BP measurements are overestimated with a cuff that is too small; if a cuff is too small, go to the next size up. Keep in mind that if the cuff is too large, falsely lower BPs may be recorded.[29]

Do not apply the blood pressure cuff above an intravenous (IV) line where fluids are infusing or an arteriovenous (AV) shunt, on the same side where breast or axillary surgery has been performed, or when the arm or hand have been traumatized or diseased. Until research data supports a change, it is recommended that clients who have undergone axillary node dissection (ALND) avoid having BP measurements taken on the affected side.

Although it is often recommended that anyone who has had bilateral axillary node dissection should have BP measurements taken in the leg, this is not standard clinical practice across the United States.[30] Leg pressures can be difficult to assess and inaccurate.

Some oncology staff advise taking BP in the arm with the least amount of nodal dissection. Technique in measuring BP is a key factor in all clients, especially those with ALND (Box 4-8).

TABLE 4-5 Classification of Blood Pressure

	Systolic Blood Pressure	Diastolic Blood Pressure
FOR ADULTS*		
Normal	<120 mm Hg	<80 mm Hg
Prehypertension	120-139	80-89
Stage 1 Hypertension	140-159	90-99
Stage 2 Hypertension	≥160	≥100
FOR CHILDREN AND ADOLESCENTS†		
Normal	<90th percentile; 50th percentile is the midpoint of the normal range	
Prehypertension	90th-95th percentile or if BP is greater than 120/80 (even if this figure is <90th percentile)	
Stage 1 Hypertension	95th-99th percentile + 5 mm Hg	
Stage 2 Hypertension	>99th percentile + 5 mm Hg	

The relationship between blood pressure (BP) and risk of coronary vascular disease (CVD) events is continuous, consistent, and independent of other risk factors. The higher the BP, the greater the chance of heart attack, heart failure, stroke, and kidney disease.

For individuals 40 to 70 years of age, each 20 mm Hg incremental increase in systolic BP (SBP) or 10 mm Hg in diastolic BP (DBP) doubles the risk of CVD across the entire BP range from 115/75 to 185/115 mm Hg.

*From The Seventh Report of the Joint National Committee on Prevention, Detection, Evaluation, and Treatment of High Blood Pressure, NIH Publication No. 03-5233, May 2003. National Heart, Lung, and Blood Institute (NHLBI) www.nhlbi.nih.gov/.

†From National Heart, Lung, and Blood Institute (NHLBI): Fourth Report on the diagnosis, evaluation, and treatment of high blood pressure in children and adolescents, *Pediatrics* 114(2):555-576, August 2004.

A common mistake is to pump the BP cuff up until the systolic measurement is 200 mm Hg and then take too long to lower the pressure or to repeat the measurement a second time without waiting. Repeating the BP without a 1-minute wait time may damage the blood vessel and set up an inflammatory response.[31] This poor technique is to be avoided, especially in clients at risk for lymphedema or who already have lymphedema.

Take the BP twice at least a minute apart in both arms. If both measurements are within 5 mm Hg of each other, record this as the resting (baseline) measurement. If not, wait 1 minute and take the BP a third time. Monitor the BP in the arm with the highest measurements.[21] Record measurements exactly; do not round numbers up or down as this can result in inaccuracies.[32]

For clients who have had a mastectomy without ALND (i.e., prophylactic mastectomy), BP can be measured in either arm. These recommendations are to be followed for life.[33]

Until automated BP devices are improved enough to ensure valid and reliable measurements, the BP response to exercise in all clients should be taken manually with a BP cuff (sphygmomanometer) and a stethoscope.[33]

BOX 4-7 GUIDELINES FOR BLOOD PRESSURE IN A PHYSICAL THERAPIST'S PRACTICE

Consider the following as yellow (caution) flags that require closer monitoring and possible medical referral:

- SBP greater than 120 mm Hg and/or DBP greater than 80 mm Hg, especially in the presence of significant risk factors (age, medications, personal or family history)
- Decrease in DBP below 70 mm Hg in adults age 75 or older (risk factor for Alzheimer's)
- Persistent rise or fall in BP over time (at least 3 consecutive readings over 2 weeks), especially in a client taking NSAIDs (check for edema) or any woman taking birth control pills (should be closely monitored by physician)
- Steady fall in BP over several years in adult over 75 (risk factor for Alzheimer's)
- Lower standing SBP (less than 140 mm Hg) in adults over age 65 with a history of falls (increased risk for falls)
- A difference in pulse pressure greater than 40 mm Hg
- More than 10 mm Hg difference (SBP or DBP) from side to side (upper extremities)
- Approaching or more than 40 mm Hg difference (SBP or DBP) from side to side (lower extremities)
- BP in lower extremities is lower than in the upper extremities
- DBP increases more than 10 mm Hg during activity or exercise
- SBP does not rise as workload increases; SBP falls as workload increases
- SBP exceeds 200 mm Hg during exercise or physical activity; DBP exceeds 100 mm Hg during exercise or physical activity; these values represent the upper limits and may be too high for the client's age, general health, and overall condition.
- BP changes in the presence of other warning signs such as first-time onset or unstable angina, dizziness, nausea, pallor, extreme diaphoresis
- Sudden fall in BP (more than 10 to 15 mm Hg SBP) or more than 10 mm Hg DBP with concomitant rise (10% to 20% increase) in pulse (orthostatic hypotension); watch for postural hypotension in hypertensive clients, especially anyone taking diuretics (decreased fluid volume/dehydration)
- Use a manual sphygmomanometer to measure BP during exercise; most standard automatic units are not designed for this purpose

BP, Blood pressure; *SBP*, systolic blood pressure; *DBP*, diastolic blood pressure; *NSAIDs*, nonsteroidal antiinflammatory drugs.

BOX 4-8 ASSESSING BLOOD PRESSURE

- Client should avoid tobacco for 30 minutes and caffeine for 60 minutes before BP reading[142]; let the client sit quietly for a few minutes; this can help offset the physical exertion of moving to the exam room or the emotional stress of being with a health care professional (white-coat hypertension). The client should be seated comfortably in a chair with the back and arm supported, legs uncrossed, feet on the floor, and not talking.
- Assess for factors that can affect BP (see Table 4-4).
- Position the arm extended in a forward direction (sitting) at the heart's level or parallel to the body (supine); avoid using an arm with a fistula, IV or arterial line, or with a previous history of lymph node biopsy or breast or axillary surgery.
- Wrap the cuff around the client's upper arm (place over bare skin) 1 inch (2.5 cm) above the antecubital fossa (inside of the elbow); cuff size is critical to accurate measurement.
- The length of the bladder cuff should encircle at least 80% of the upper arm. The width of the cuff should be about 40% of the upper arm circumference. If a cuff is too short or too narrow, the BP reading will be erroneously high; if the cuff is too long or too wide, the BP reading will be erroneously low. BP measurement errors are usually worse in cuffs that are too small compared to those that are too big.
- If you are measuring BP at the ankle, an arm cuff is usually appropriate, but BPs taken at the thigh require a thigh cuff unless the person is very thin.
- Slide your finger under the cuff to make sure it is not too tight.
- Close the valve on the rubber bulb.
- Place the stethoscope (diaphragm or bell side) lightly over the brachial artery at the elbow (about 2 to 3 cm above the antecubital fossa)[142]; you may hear the low-pitched Korotkoff sounds more clearly using the bell.
- Inflate the cuff until you no longer hear a pulse sound and then inflate 30 mm more; this is the point of reference for auscultation.
- Slowly release the valve on the bulb (deflate at a rate of 2 to 3 mm Hg/second) as you listen for the first Korotkoff sound (2 consecutive beats signals the systolic reading) and the last Korotkoff sound (diastolic reading).
- If you are new to BP assessment or if the BP is elevated, check the BP twice. Wait 1 minute and retest. Some sources say to wait at least 2 minutes before retaking BP on the same arm, but this is not always done in the typical clinical setting.
- Record date, time of day, client position, extremity measured (arm or leg, left or right), and results for each reading. Record any factors that might affect BP readings (e.g., recent tobacco use, caffeine intake).
- As soon as the blood begins to flow through the artery again, Korotkoff sounds are heard. The first sounds are tapping sounds that gradually increase in intensity. The initial tapping sound that is heard for at least 2 consecutive beats is recorded as SBP.
- The first phase of sound may be followed by a momentary disappearance of sounds that can last from 30 to 40 mm Hg as the needle descends. Following this temporary absence of sound, there are murmuring or swishing sounds (second Korotkoff sound). As deflation of the cuff continues, the sounds become sharper and louder. These sounds represent phase 3. During phase 4, the sounds become muffled rather abruptly and then are followed by silence, which represents phase 5. Phase 5 (the fifth Korotkoff sound or K5), the point at which sounds disappear, is most often used as the DBP.

From American Heart Association Updates Recommendations for Blood Pressure Measurements, Available at: http://www.americanheart.org. Accessed Jan. 25, 2011.
BP, Blood pressure; *SBP*, systolic blood pressure; *DBP*, diastolic blood pressure.

It is advised to invest in the purchase of a well-made, reliable stethoscope. Older models with tubing long enough to put the earpieces in your ears and still place the bell in a lab coat pocket should be replaced. Tubing should be no more than 50 to 60 cm (12 to 15 inches) and 4 mm in diameter. Longer and wider tubing can distort transmitted sounds.[34]

For the student or clinician learning to take vital signs, it may be easier to hear the BP (tapping, Korotkoff) sounds in adults using the left arm because of the closer proximity to the left ventricle. Arm position does make a difference in BP readings. BP measurements are up to 10% higher when the elbow is at a right angle to the body with the elbow flexed at heart level. The preferred position is seated with the arms parallel and extended in a forward direction (if supine, then parallel to the body).[35]

It is more accurate to evaluate consecutive BP readings over time rather than using an isolated measurement for reporting BP abnormalities. BP also should be correlated with any related diet or medication.

Before reporting abnormal BP readings, measure both sides for comparison, re-measure both sides, and have another health professional check the readings. Correlate BP measurements with other vital signs, and screen for associated signs and symptoms such as pallor, fatigue, perspiration, and/or palpitations. A persistent rise or fall in BP requires medical attention and possible intervention.

Pulse Pressure. The difference between the systolic and diastolic pressure readings (SBP − DBP) is called *pulse pressure* normally around 40 mm Hg. Pulse pressure is an index of vascular aging (i.e., loss of arterial compliance and indication of how stiff the arteries are). A widened resting pulse pressure often results from stiffening of the aorta secondary to atherosclerosis. *Resting* pulse pressure consistently greater than 60 to 80 mm Hg is a yellow (caution) flag and is a risk factor for new onset of atrial fibrillation.[36]

Widening of the pulse pressure is linked to a significantly higher risk of stroke and heart failure after the sixth decade. Some BP medications increase resting pulse pressure width by lowering diastolic pressure more than systolic while others (e.g., angiotensin-converting enzyme [ACE] inhibitors) can lower pulse pressure.[37]

Narrowing of the resting pulse pressure (usually by a drop in SBP as the DBP rises) can suggest congestive heart failure (CHF) or a significant blood loss such as occurs in hypovolemic shock. A high pulse pressure accompanied by bradycardia is a sign of increased intracranial pressure and requires immediate medical evaluation.

In a normal, healthy adult, the pulse pressure generally increases in direct proportion to the intensity of exercise as the SBP increases and DBP stays about the same.[38] A difference of *more than* 80 to 100 mm Hg taken during or right after exercise should be evaluated carefully. In a healthy adult, pulse pressure will return to normal within 3 to 10 minutes following moderate exercise.

The key is to watch for pulse pressures that are not accommodating during exercise. Expect to see the systolic rise slightly while diastolic stays the same. If diastolic drops while systolic rises or if the pulse width exceeds 100 mm Hg, further assessment and evaluation is needed. Depending on all other parameters (e.g., general health of the client, past medical history, medications, concomitant associated signs and symptoms), the therapist may monitor pulse pressures over a few sessions and look for a pattern (or lack of pattern) to report if/when generating a medical consult.[39]

Variations in Blood Pressure. There can be some normal variation in SBP from side to side (right extremity compared to left extremity). This is usually no more than 5 to 10 mm Hg DBP or SBP (arms) and 10 to 40 mm Hg SBP (legs). A difference of 10 mm Hg or more in either systolic or diastolic measurements from one extremity to the other may be an indication of vascular problems (look for associated symptoms; in the upper extremity test for thoracic outlet syndrome).

Normally the SBP in the legs is 10% to 20% higher than the brachial artery pressure in the arms. BP readings that are lower in the legs as compared with the arms are considered abnormal and should prompt a medical referral for assessment of peripheral vascular disease.[33]

With a change in position (supine to sitting), the normal fluctuation of BP and heart rate increases slightly (about 5 mm Hg for systolic and diastolic pressures and 5 to 10 bpm in heart rate).

Systolic pressure increases with age and with exertion in a linear progression. If systolic pressure does not rise as workload increases, or if this pressure falls, it may be an indication that the functional reserve capacity of the heart has been exceeded.

The deconditioned, menopausal woman with coronary heart disease (CHD) requires careful monitoring, especially in the presence of a personal or family history of heart disease and myocardial infarct (personal or family) or sudden death in a family member.

On the other hand, women of reproductive age taking birth control pills may be at increased risk for hypertension, heart attack, or stroke. The risk of a cardiovascular event is very low with today's low-dose oral contraceptives. However, smoking, hypertension, obesity, undiagnosed cardiac anomalies, and diabetes are factors that increase a woman's risk for cardiovascular events. Any woman using oral contraceptives who presents with consistently elevated BP values must be advised to see her physician for close monitoring and follow-up.[40,41]

The left ventricle becomes less elastic and more noncompliant as we age. The same amount of blood still fills the ventricle, but the pumping mechanism is less effective. The body compensates to maintain homeostasis by increasing the blood pressure. BP values greater than 120 mm Hg (systolic) and more than 80 mm Hg (diastolic) are treated with lifestyle modifications first then medication.

Blood Pressure Changes with Exercise. As mentioned, the SBP increases with increasing levels of activity and exercise in a linear fashion. In a healthy adult under conditions of minimal to moderate exercise, look for normal change (increase) in SBP of 20 mm Hg or more.

The American College of Sports Medicine (ACSM) suggests the normal SBP response to incremental exercise is a progressive rise, typically 10 mm + 2 mm Hg for each metabolic equivalent (MET) where 1 MET = 3.5 mL O_2/kg/min. Expect to see a 40 to 50 mm change in SBP with intense exercise (again, this is in the healthy adult). These values are less likely with individuals taking BP medications, anyone with a significant history of heart disease, and well-conditioned athletes.

Diastolic should be the same side to side with less than 10 mm Hg difference observed. DBP generally remains the same or decreases slightly during progressive exercise.[38]

In an exercise-testing situation, the ACSM recommends stopping the test if the SBP exceeds 260 mm Hg.[38] In a clinical setting without the benefit of cardiac monitoring, exercise or activity should be reduced or stopped if the systolic pressure exceeds 200 mm Hg.

This is a general guideline that can be changed according to the client's age, general health, use of cardiac medications, and other risk factors. DBP increases during upper extremity exercise or isometric exercise involving any muscle group. Activity or exercise should be monitored closely, decreased, or halted if the diastolic pressure exceeds 100 mm Hg.

This is a general (conservative) guideline when exercising a client without the benefit of cardiac testing (e.g., electrocardiogram [ECG]). This stop-point is based on the ACSM guideline to stop exercise testing at 115 mm Hg DBP. Other sources suggest activity should be decreased or stopped if the DBP exceeds 130 mm Hg.[42]

Other warning signs to moderate or stop exercising include the onset of angina, dyspnea, and heart palpitations. Monitor the client for other signs and symptoms such as fever, dizziness, nausea/vomiting, pallor, extreme diaphoresis, muscular cramping or weakness, and incoordination. Always honor the client's desire to slow down or stop.

Hypertension (See Further Discussion on Hypertension in Chapter 6). In recent years, an unexpected increase in illness and death caused by hypertension has prompted the National Institutes of Health (NIH) to issue new guidelines for more effective BP control. More than one in four Americans has high blood pressure, increasing their risk for heart and kidney disease and stroke.[43]

In adults hypertension is a systolic pressure above 140 mm Hg or a diastolic pressure above 90 mm Hg. Consistent BP measurements between 120 and 139 (systolic) and between 80 and 89 diastolic is classified as pre-hypertensive. The overall goal of treating clients with hypertension is to prevent morbidity and mortality associated with high blood pressure. The specific objective is to achieve and maintain arterial blood pressure below 120/80 mm Hg, if possible (Box 4-9).[34]

The older adult taking nonsteroidal antiinflammatory drugs (NSAIDs) is at risk for increased BP because these drugs are potent renal vasoconstrictors. Monitor BP carefully in these clients and look for sacral and lower extremity edema. Document and report these findings to the physician. Use the risk factor analysis for NSAIDs presented in Chapter 2 (see Box 2-13 and Table 2-6).

Always beware of *masked hypertension* (normal in the clinic but periodically high at home) and *white-coat hypertension*, a clinical condition in which the client has elevated BP levels when measured in a clinic setting by a health care professional. In such cases, BP measurements are consistently normal outside of a clinical setting.

Masked hypertension may affect up to 10% of adults; white-coat hypertension occurs in 15 to 20% of adults with stage I hypertension.[44] These types of hypertension are more common in older adults. Antihypertensive treatment for white coat hypertension may reduce office BP but may not affect ambulatory BP. The number of adults who develop sustained high BPs is much higher among those who have masked or white-coat hypertension.[44]

At-home BP measurements can help identify adults with masked hypertension, white-coat hypertension, ambulatory hypertension, and individuals who do not experience the usual nocturnal drop in BP (decrease of 15 mm Hg), which is a risk factor for cardiovascular events.[45] Excessive morning BP surge is a predictor of stroke in older adults with known hypertension and is also a red-flag sign.[46] Medical referral is indicated in any of these situations.

Hypertension in African Americans. Nearly 40% of African Americans suffer from heart disease and 13% have diabetes. Hypertension contributes to these conditions or makes them worse. African Americans are significantly more likely to die of high BP than the general public because current treatment strategies have been unsuccessful.[47,48]

BOX 4-9 GUIDELINES FOR HYPERTENSION AND MANAGEMENT

The Seventh Report of the Joint National Committee on Prevention, Detection, Evaluation, and Treatment of High Blood Pressure provides a new guideline for hypertension prevention and management. The following are the report's key messages:

- In persons older than 50 years, systolic blood pressure greater than 140 mm Hg is a much more important cardiovascular disease (CVD) risk factor than DBP. Elevated systolic pressure raises the risk of heart attacks, congestive heart failure (CHF), dementia, end-stage kidney disease, and cardiovascular mortality.
- The risk of CVD beginning at 115/75 mm Hg doubles with each increment of 20/10 mm Hg; individuals who are normotensive at age 55 have a 90% lifetime risk for developing hypertension.
- Individuals with a SBP of 120-139 mm Hg or a DBP of 80-89 mm Hg should be considered as prehypertensive and require health-promoting lifestyle modifications to prevent CVD.
- Thiazide-type diuretics should be used in drug treatment for most clients with uncomplicated hypertension, either alone or combined with drugs from other classes. Certain high-risk conditions are compelling indications for the initial use of other antihypertensive drug classes (angiotensin-converting enzyme inhibitors, angiotensin receptor blockers, beta-blockers, calcium channel blockers).
- Most clients with hypertension will require two or more antihypertensive medications to achieve goal BP (less than 140/90 mm Hg, or less than 130/80 mm Hg for clients with diabetes or chronic kidney disease).
- If BP is more than 20/10 mm Hg above goal BP, consideration should be given to initiating therapy with two agents, one of which usually should be a thiazide-type diuretic.
- The most effective therapy prescribed by the most careful clinician will control hypertension only if clients are motivated. Motivation improves when clients have positive experiences with and trust in the clinician.

From The Seventh Report of the Joint National Committee on Prevention, Detection, Evaluation, and Treatment of High Blood Pressure. NIH Publication No. 03-5233, May 2003. National Heart, Lung, and Blood Institute (NHLBI) www.nhlbi.nih.gov/.
BP, Blood pressure; *SBP*, systolic blood pressure; *DBP*, diastolic blood pressure.

Guidelines for treating high blood pressure in African Americans have been issued by the International Society on Hypertension in Blacks (ISHIB).[49] The ISHIB recommends a blood pressure target of less than 130/80 mm Hg for African Americans with BP screening for all African-American adults and early prevention for anyone in the pre-hypertensive range. Aggressive treatment for hypertension is advised using drug combinations.[48]

The therapist can incorporate blood pressure screening into any evaluation for clients with ethnic risk factors. Any client of any ethnic background with risk factors for hypertension should also be screened (see Table 6-6).

Hypertension in Hispanics. The Hispanic population in the United States is expected to be reported by the 2010 census as the largest minority group in the nation. Research on hypertension among Hispanics has shown that their incidence of high BP is greater than that of whites and Asians and less than that of blacks. More Hispanics than whites have undiagnosed hypertension and are generally less knowledgeable in heart disease prevention. Factors that contribute to elevated BP in Hispanics include high rates of obesity and diabetes, as well as a genetic predisposition and low socioeconomic status.[50] With equal access to medical care and medication, Hispanic men and women have as good or greater chance as non-Hispanics of controlling their high BP.[51]

Hypertension in Children and Adolescents[52]. Up to 3% of children under age 18 also have hypertension. New guidelines for children have been published by the National Heart, Lung, and Blood Institute (NHLBI) (see Table 4-5). Tables with BP levels for boys and girls by age and height percentile are available.[52]

The updated BP tables for children and adolescents are based on *recently revised child height percentiles.* Any child with readings above the 95th percentile for gender, age, and height on three separate occasions is considered to have hypertension. The 50th percentile has been added to the tables to provide the clinician with the BP level at the midpoint of the normal range.

Under the new guidelines, children whose readings fall between the 90th and 95th percentile are now considered to have pre-hypertension. Earlier guidelines called this category "high normal."

The long-term health risks for hypertensive children and adolescents can be substantial; therefore it is important that elevated BP is recognized early and measures taken to reduce risks and optimize health outcomes.[52]

Children ages 3 to 18 seen in any medical setting should have the BP measured at least once during each health care episode. The preferred method is auscultation with a BP cuff and stethoscope. Correct measurement requires a cuff that is appropriate to the size of the child's upper arm.

The right arm is preferred with children for comparison with standard tables and in the possible event there is a coarctation of the aorta (see Fig. 6-6), which can lead to a false low reading in the left arm.[53]

Preparation of the child can affect the BP level as much as technique. The child should be seated with feet and back supported. The right arm should be supported parallel to the floor with the cubital fossa at heart level.[54,55] Children can be affected by white-coat hypertension as much as adults. Follow the same guidelines for adults as presented in Box 4-8.

Hypotension. Hypotension is a systolic pressure below 90 mm Hg or a diastolic pressure below 60 mm Hg. A BP level that is borderline low for one person may be normal for another. When the BP is too low, there is inadequate blood flow to the heart, brain, and other vital organs.

The most important factor in hypotension is how the BP changes from the normal condition. Most normal BPs are in the range of 90/60 mm Hg to 120/80 mm Hg, but a significant change, even as little as 20 mm Hg, can cause problems for some people.

Lower standing SBP (less than 140 mm Hg) even within the normotensive range is an independent predictor of loss of balance and falls in adults over age 65.[56] DBP does not appear to be related to falls. Older adult women with lower standing SBP and a history of falls are at greatest risk. The therapist has an important role in educating clients with these risk factors in preventing falls and related accidents. See discussion in Chapter 2 related to taking a history of falls.

In older adults a decrease in BP may be an early warning sign of Alzheimer's disease. DBP below 70 or declines in systolic pressure equal to or greater than 15 mm Hg over a period of 3 years raises the risk of dementia in adults 75 or older. For each 10-point drop in pressure, the risk of dementia increases by 20%.[57,58]

It is unclear if the steady drop in BP during the 3 years before a dementia diagnosis is a cause or effect of dementia as reduced blood flow to the brain accelerates the development of dementia. Perhaps brain cell degeneration characteristic of dementia damages parts of the brain that regulate BP.[57]

Postural Orthostatic Hypotension. A common cause of low BP is orthostatic hypotension (OH), defined as a sudden drop in BP when changing positions, usually moving from supine to an upright position.

Physiologic responses of the sympathetic nervous system decline with aging putting them at greater risk for OH. Older adults are prone to falls from a combination of OH and antihypertensive medications. Volume depletion and autonomic dysfunction are the most common causes of OH (see Table 2-5).

Postural OH is more accurately defined as a decrease in SBP of at least 20 mm Hg *or* decrease in diastolic pressure of at least 10 mm Hg *and* a 10% to 20% increase in pulse rate. Changes must be noted in *both* the BP and the pulse rate with change in position (supine to sitting, sitting to standing) (Box 4-10 and Case Example 4-1).[45]

The client should lie supine 2 to 3 minutes prior to BP and pulse check. At least a 1-minute wait is recommended after each subsequent position change before taking the BP and pulse. Standing postural orthostatic hypotension is measured after 3 to 5 minutes of quiet standing. Food ingestion, time of day, age, and hydration can impact this form of hypotension, as can a history of Parkinsonism, diabetes, or multiple myeloma.[43]

Throughout the procedure assess the client for signs and symptoms of hypotension, including dizziness, lightheadedness, pallor, diaphoresis, or syncope (or arrhythmias if using a cardiac monitor). Assist the client to a seated or supine position if any of these symptoms develop and report the results. Do not test the client in the standing position if signs and symptoms of hypotension occur while sitting.

Gravitational effects on the circulatory system can cause a 10 mm Hg drop in SBP when a person changes position from

BOX 4-10 POSTURAL ORTHOSTATIC HYPOTENSION

For a diagnosis of postural orthostatic hypotension, the client must have:

- Decrease of 10 to 15 mm Hg of systolic pressure SBP AND/OR
- Increase of 10 mm Hg (or more) DBP AND
- 10% to 20% increase in pulse rate

These changes occur with change in position (supine to upright sitting, sit to stand). Another measurement after 1 to 5 minutes of standing may identify orthostatic hypotension missed by earlier readings.

Monitor the client carefully since fainting is a possible risk with low BP, especially when combined with the dehydrating effects of diuretics. Oncology patients receiving chemotherapy who are hypotensive are also at risk for dizziness and loss of balance during the repeated BP measurement in the standing position.

This repetition is useful in the older adult (65 years old or older). Waiting to repeat the BP measurements reveals a client's inability to regulate BP after a change in position. The presence of low BP after a prolonged time is a red-flag finding. Be sure to check pulse rate.[21]

SBP, Systolic blood pressure; *DBP*, diastolic blood pressure; *BP*, blood pressure.

supine to sitting to standing. This drop usually occurs without symptoms as the body quickly compensates to ensure there is no reduction in cardiac output.

In clients on prolonged bed rest or on antihypertensive drug therapy, there may be either no reflexive increase in heart rate or a sluggish vasomotor response. These clients may experience larger drops in BP and often experience lightheadedness.

Other clients at risk for postural OH include those who have just donated blood, anyone with autonomic nervous system disease or dysfunction, and postoperative patients. Other risk factors for OH in aging adults include hypovolemia associated with dehydration and the overuse of diuretics, anticholinergic medications, antiemetics, and various over-the-counter (OTC) cough/cold preparations.

Core Body Temperature

Normal body temperature is not a specific number but a range of values that depends on factors such as the time of day, age, medical status, medication use, activity level, or presence of infection. Oral body temperature ranges from 36° to 37.5° C (96.8° to 99.5° F), with an average of 37° C (98.6° F) (Table 4-6). Hypothermic core temperature is defined as less than 35° C (95° F). Hyperthermia is defined as a temperature greater than 38° C (100.4° F).

Older adults (over age 65) are less likely to have a fever even in the presence of severe infection, so the predictive value of taking the body temperature is less. Due to

CASE EXAMPLE 4-1

Vital Signs

A 74-year-old retired homemaker had a total hip replacement (THR) 2 days ago. She remains as an inpatient with complications related to congestive heart failure (CHF). She has a previous medical history of gallbladder removal 20 years ago, total hysterectomy 30 years ago, and surgically induced menopause with subsequent onset of hypertension.

Her medications include intravenous (IV) furosemide (Lasix), digoxin, and potassium replacement.

During the initial physical therapy intervention, the client complained of muscle cramping and headache but was able to complete the entire exercise protocol. Blood pressure was 100/76 mm Hg. Systolic measurement dropped to 90 mm Hg when the client moved from supine to standing. Pulse rate was 56 bpm with a pattern of irregular beats. Pulse rate did not change with postural change. Platelet count was 98,000 cells/mm^3 when it was measured yesterday.

What is the significance of her vital signs? How would you use vital sign monitoring in a patient like this?

Nurses will be monitoring the patient's signs and symptoms closely. Read the chart to stay up with what everyone else knows about Mrs. S. and/or has observed. Read the physician's notes to see what, if any, medical intervention has been ordered based on laboratory values (e.g., platelet levels) or vital signs (e.g., changes in medication).

Do not hesitate to discuss concerns and observations with the nursing staff. This helps them know you are aware of the medical side of care, but also gives you some perspective from the nursing side. What do they see as significant? What requires immediate medical attention?

Be sure to report anything observed but not already recorded in the chart such as muscle cramping, headache, irregular heartbeat with bradycardia, low pulse, and orthostatic hypotension.

Bradycardia is one of the first signs of digitalis toxicity. In some hospitals, a pulse less than 60 bpm in an adult would mean withholding the next dose of digoxin and necessitate physician contact. The protocol may be different from institution to institution.

In this case, report and document:

1. Irregular heart beat with bradycardia (a possible sign of digoxin/digitalis toxicity)
2. Muscle cramping (possible side effect of Lasix) and headache (possible side effect of digoxin)
3. Always chart vital signs; her blood pressure was not too unusual and pulse rate did not change with position change (probably because of medications) so she does not have medically defined orthostatic hypotension.

The response of vital signs to exercise must be monitored carefully and charted; monitor vital signs throughout intervention. Record the time it takes for the client's vital signs to return to normal after exercise or treatment. This can be used as a means of documenting measurable outcomes. Mrs. S. may not ambulate any further or faster in the afternoon compared with the morning, but her vital signs may reflect closer to normal values and a faster return to homeostasis as a measurable outcome.

TABLE 4-6 Core Body Temperature

Oral	96.8° to 99.5° F (36° to 37.5° C)
Rectal	97.3° to 100.2° F (36.3° to 37.9° C)
Tympanic membrane	97.2° to 100° F (36.2° to 37.8° C)

Body temperature below 95° F (35° C) is a sign of hypothermia.
Body temperature varies throughout the day (lowest in early morning, highest in late afternoon).
Body temperature varies over the lifespan (decreases with age due to lower metabolic rate, decreased subcutaneous fat, decreased activity levels, inadequate diet). No matter what the regular temperature is, a temperature of 100° F/37.8° C or higher usually indicates a systemic illness.

age-related changes in the thermoregulatory system, they are also more likely to develop hypothermia than young adults. There is a tendency among the aging population to develop an increase in temperature on hospital admission or in response to any change in homeostasis. However, some persons with infectious disease remain afebrile, especially the immunocompromised and those with chronic renal disease, alcoholics, and older adults. A low-grade fever can be an early sign of life-threatening infections (most commonly pneumonia, urinary tract infection). Unexplained fever in adolescents may be a manifestation of drug abuse or endocarditis.

Postoperative fever is common and may be from an infectious or noninfectious cause. Medical evaluation is needed to make this determination. In the home health setting, wound infection, abscess formation, or peritonitis may appear as a hectic fever pattern 3 to 4 days postoperatively with increases and declines of body temperature but no return to baseline (normal). Such a situation would warrant telephone consultation with the physician's office nurse.

Any client who has back, shoulder, hip, sacroiliac, or groin pain of unknown cause must have a temperature reading taken. Temperature should also be assessed for any client who has constitutional symptoms (see Box 1-3), especially "sweats" (gradual increase followed by a sudden drop in body temperature), pain, or symptoms of unknown etiologic basis and for clients who have not been medically screened by a physician. Ask about the presence of other signs and symptoms of infection.

When measuring body temperature, the therapist should ask if the person's normal temperature differs from 37° C (98.6° F). A persistent elevation of temperature over time is a red-flag sign; a single measurement may not be sufficient to cause concern. Any measurement outside of normal for that individual should be rechecked.

It is also important to ask whether the client has taken aspirin (or other NSAIDs) or acetaminophen (Tylenol) to reduce the fever, which might mask an underlying problem. Clients taking dopamine blockers, such as Thorazine, Mellaril, or the less commonly used Navane for schizophrenia, have a lowered "normal" temperature (around 96° F/35.6° C). Anyone who is chronically immunosuppressed (such as an organ transplant recipient, a person being treated with chemotherapy, and any older adult) may have an infection without elevation of temperature.

When using a tympanic membrane (ear) thermometer, perform a gentle ear tug to straighten the ear canal. In a child, pull the ear straight back; in an adult, pull it slightly upward and backward. While holding the ear in this position, use a small rotation movement to insert the probe gently and slowly.[59]

The probe must penetrate at least one-third of the external ear canal to prevent air temperature from affecting the reading. Aim the probe toward the tympanic membrane where it will indirectly measure core body temperature by taking infrared temperature readings of the tympanic membrane (eardrum).[59]

Temperatures can vary from side to side, so record which ear was used and try to use the same ear each time the temperature is recorded. For the client with hearing aid(s), take the temperature in the ear without an aid. Or, if hearing aids are present in both ears, remove one hearing aid and wait 20 minutes before measuring that side. The presence of excessive earwax will prevent an accurate reading.

There are some additional concerns reported in the literature about the accuracy of tympanic thermometers with evidence of significant variability possibly related to the condition of the censor, presence of ear wax, placement in the ear canal, operator error, and maintenance of the equipment.[60-62] Such concerns may be more important in critical care ICUs compared with outpatient screening, but no studies comparing these two populations have been published.

Newer handheld digital forehead thermometers are noninvasive and are quick and easy to use. The forehead plastic temperature strip (forehead thermometer, fever strip) and the pacifier thermometer for children are not the most reliable methods to take a temperature.

The therapist should use discretionary caution with any client who has a fever. Exercise with a fever stresses the cardiopulmonary system, which may be further complicated by dehydration. Severe dehydration can occur from vomiting, diarrhea, medications (e.g., diuretics), or heat exhaustion.

CLINICAL SIGNS AND SYMPTOMS

Dehydration

Mild
- Thirst
- Dry mouth, dry lips

Moderate
- Very dry mouth, cracked lips
- Sunken eyes, sunken fontanel (infants)
- Poor skin turgor (see Fig. 4-4)
- Postural hypotension
- Headache

Severe
- All signs of moderate dehydration
- Rapid, weak pulse (more than 100 bpm at rest)
- Rapid breathing
- Confusion, lethargy, irritability
- Cold hands and feet
- Unable to cry or urinate

Clients at greatest risk of dehydration include postoperative patients, aging adults, and athletes. Severe fluid volume deficit can cause vascular collapse and shock. Clients at risk of shock include burn or trauma patients, clients in anaphylactic shock or diabetic ketoacidosis, and individuals experiencing severe blood loss.

CLINICAL SIGNS AND SYMPTOMS

Shock

Stage 1 (Early Stage)
- Restlessness, anxiety, hyperalert
- Listless, lack of interest in play (children)
- Tachycardia
- Increased respiratory rate, shallow breathing, frequent sighs
- Rapid, bounding pulse (not weak)
- Distended neck veins
- Skin warm and flushed
- Thirst, nausea, vomiting

Stage 2
- Confusion, lack of focused eye contact (vacant look)
- Abrupt changes in affect or behavior
- No crying or excessive, unexplained crying in infant
- Cold, clammy skin, profuse sweating, chills
- Weak pulses (not bounding)
- Hypotension (low BP), dizziness, fainting
- Collapsed neck veins
- Weak or absent peripheral pulses
- Muscle tension

Stage 3 (late stage)
- Cyanosis (blue lips, gray skin)
- Dull eyes, dilated pupils
- Loss of bowel or bladder control
- Change in level of consciousness

Walking Speed: The Sixth Vital Sign

Walking speed is used by some as a general indicator of function[63] and as such, a reflection of many variables such as health status, motor control, muscle strength, and endurance to name only a few. It is a reliable, valid, and sensitive measure of functional ability with additional predictive value in assessing future health status, functional decline, potential for hospitalization, and even mortality.[64]

The test is conducted using a timed 10-meter walk test on a 20-meter long straight path. Complete descriptions of the test and expected results are available.[63-66] As a screening tool, walking speed may not indicate the presence of systemic pathology, but as specialists in human movement and function, the therapist can use it as a practical and predictive "vital sign" of general health that can be used to monitor change (improvement or decline) in health and function.

TECHNIQUES OF PHYSICAL EXAMINATION

There are four simple techniques used in the medical physical examination: inspection, palpation, percussion, and auscultation. Percussion and some auscultation techniques require advanced clinical skill and are beyond the scope of a screening examination.

Throughout any screening examination the therapist also assesses function of the integument, musculoskeletal, neuromuscular, and cardiopulmonary systems. Assessment techniques are relatively simple; it is using the finding that is more difficult. The saying, "What one knows, one sees" underscores the idea that knowledge of physical assessment techniques and experience in performing these are extremely important and come from practice.

Inspection

Good lighting and good exposure are essential. Always compare one side to the other. Assess for abnormalities in all of the following:

Texture	Tenderness
Size	Shape, contour, symmetry
Position, alignment	Mobility or movement
Color	Location

The therapist should try to follow the same pattern every time to decrease the chances of missing an assessment parameter and to increase accuracy and thoroughness.

Palpation

Palpation is used to discriminate between textures, dimensions, consistencies, and temperature. It is used to define things that are inspected and to reveal things that cannot be inspected. Textures are best detected using the fingertips, whereas dimension or contours are detected using several fingers, the entire hand, or both hands, depending on the area being examined.

Inspection and palpation are often performed at the same time; be sure and look at the client and not at your hands. Muscle tension interferes with palpation so the client must be positioned and draped appropriately in a room with adequate lighting and temperature.

Assess skin temperature with both hands at the same time. The back of the therapist's hands sense temperature best because of the thin layer of skin. Use the palm or heel of the hand to assess for vibration. The finger pads are best to assess texture, size, shape, position, pulsation, consistency, and turgor. Heavy or continued pressure dulls the examiner's palpatory skill and sensation.

Light palpation is used first, looking for areas of tenderness followed by deep palpation to examine organs or look for masses and elicit deep pain. Light palpation (skin is depressed up to ¼ to ½ inch) is also used to assess texture, temperature, moisture, pulsations, vibrations, and superficial lesions. Deep palpation is used for assessing abdominal structures. During deep palpation, enough pressure is used to depress the skin up to 1 inch; applying too much pressure decreases sensation.

Tender or painful areas are assessed last while carefully observing the client's face for signs of discomfort.

Percussion

Percussion (tapping) is used to determine the size, shape, and density of tissue using sound created by vibration. Percussion can also detect the presence of fluid or air in a body cavity such as the abdominal cavity. Most percussive techniques are beyond the scope of a screening examination and are not discussed in detail.

Percussion can be done directly over the client's skin using the fingertip of the examiner's index finger. Indirect percussion is performed by placing the middle finger of the examiner's nondominant hand firmly against the client's skin then striking above or below the interphalangeal joint with the pad of the middle finger of the dominant hand. The palm and fingers stay off the skin during indirect percussion. Blunt percussion using the ulnar surface of the hand or fist to strike the body surface (directly or indirectly) detects pain from infection or inflammation (see Fig. 4-54).

The examiner must be careful not to dampen the sound by dull percussing (sharp percussion is needed), holding a finger too loosely on the body surface, or resting the hand on the body surface. Percussive sounds lie on a continuum from tympany to flat based on density of tissue.

Auscultation

Some sounds of the body can be heard with the unaided ear; others must be heard by auscultation using a stethoscope. The bell side of the stethoscope is used to listen to low-pitched sounds such as heart murmurs and BP (although the diaphragm can also be used for BP).

Pressing too hard on the skin can obliterate sounds. The diaphragm side of the stethoscope is used to listen to high-pitched sounds such as normal heart sounds, bowel sounds, and friction rubs. Avoid holding either side of the stethoscope with the thumb to avoid hearing your own pulse.

Auscultation usually follows inspection, palpation, and percussion (when percussion is performed). The one exception is during examination of the abdomen, which should be assessed in this order: inspection, auscultation, percussion, then palpation as percussion and palpation can affect findings on auscultation.

Besides measuring BP, auscultation can be used to listen for breath sounds, heart sounds, bowel sounds, and abnormal sounds in the blood vessels called bruits. Bruits are abnormal blowing or swishing sounds heard on auscultation as blood travels through narrowed or obstructed arteries such as the aorta or renal, iliac, or femoral arteries. Bruits with both systolic and diastolic components suggest the turbulent blood flow of partial arterial occlusion possible with aneurysm or vessel constriction. All large arteries in the neck, abdomen, and limbs can be examined for bruits.

A medical assessment (e.g., physician, nurse, physician assistant) may routinely include auscultation of the temporal and carotid arteries and jugular vein in the head and neck, as well as vascular sounds in the abdomen (e.g., aorta, iliac, femoral, and renal arteries). The therapist is more likely to assess for bruits when the client's history (e.g., age over 65, history of coronary artery disease), clinical presentation (e.g., neck, back, abdominal, or flank pain), and associated signs and symptoms (e.g., syncopal episodes, signs and symptoms of peripheral vascular disease) warrant additional physical assessment.

The results from inspection, percussion (when appropriate), and palpation should always be correlated with the client's history, risk factors, clinical presentation, and any associated signs and symptoms before making the decision regarding medical referral.

INTEGUMENTARY SCREENING EXAMINATION

When screening for systemic disease, the therapist must increase attention to what is observable on the outside, primarily the skin and nail beds. Changes in the skin and nail beds may be the first sign of inflammatory, infectious, and immunologic disorders and can occur with involvement of a variety of organs.

For example, dermatitis can occur 6 to 8 weeks before primary signs and symptoms of pulmonary malignancy develop. Clubbing of the fingers can occur quickly in various acute illnesses and conditions. Skin, hair, and nail bed changes are common with endocrine disorders. Renal disease, rheumatic disorders, and autoimmune diseases are all accompanied by skin and nail bed changes in many physical therapy clients.

When assessing skin conditions of any kind, even benign lesions such as psoriasis (Fig. 4-2) or eczema, the therapist always should use standard precautions because any disruption of the skin increases the risk of infection. Chronic skin conditions of this type may have new, more effective treatment available. In such cases, the therapist may be able to guide the uninformed client to obtain updated medical treatment.

Consider all findings in relation to the client's age, ethnicity, occupation, and general health. The presence of skin lesions may point to a problem with the integumentary system or may be an integumentary response to a systemic problem.

For example, pruritus is the most common manifestation of dermatologic disease but is also a symptom of underlying systemic disease in up to 50% of individuals with generalized itching.[67] In both situations, skin rash is a common accompanying sign. The most common visceral system causing pruritus is the hepatic system. Look for other associated signs and symptoms of liver or gallbladder impairment such as

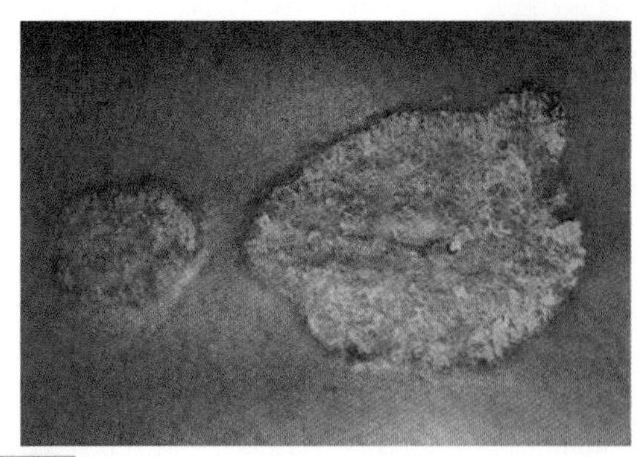

Fig. 4-2 Psoriasis. A common chronic skin disorder characterized by red patches covered by thick, dry silvery scales that are the result of excessive buildup of epithelial cells. Lesions often come and go and can be anywhere on the body but are most common on extensor surfaces, bony prominences, scalp, ears, and genitals. Arthritis of the small joints of the hands often accompanies the skin disease (psoriatic arthritis). (From Lookingbill DP, Marks JG: *Principles of dermatology,* ed 3, Philadelphia, 2000, WB Saunders.)

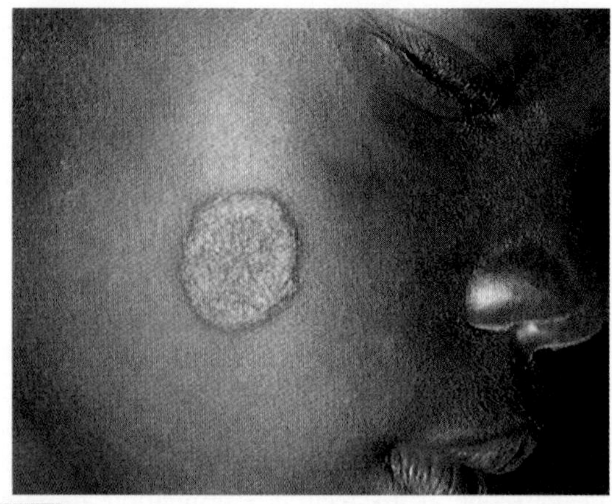

Fig. 4-3 Tinea corporis or ringworm of the body presents anywhere on the body in adult or children but more commonly on the chest, abdomen, back of arms, face, and dorsum of the feet. The circular lesions with clear centers can form singly or in clusters and represent a fungal infection that is both contagious and treatable. Tinea pedis (not shown), also known as ringworm of the feet or "athlete's foot" occurs most often between the toes, but also along the sides of the feet and the soles (easily spread and treatable). (From Hurwitz S: *Clinical pediatric dermatology: a textbook of skin disorders of childhood and adolescence,* ed 2, Philadelphia, 1993, WB Saunders.)

liver flap (asterixis), carpal tunnel syndrome, liver palms (palmar erythema), and spider angiomas (see Fig. 9-3).

At the same time, be aware that pruritus, or itch, is very common among aging adults. The natural attrition of glands that moisturize the skin combined with the effects of sun exposure, medications, excessive bathing, and harsh soaps can result in dry, irritable skin.[68]

Some clients may describe formication, also referred to as a tactile hallucination, the sensation of ants crawling on the skin, sometimes described as an itching, prickling, or crawling feeling. The most common cause is menopause, but chronic drug or alcohol use can also cause formication. Some schizophrenics also experience formication. As one of the many side effects of crystal methamphetamine addiction, formication is also referred to as speed bumps, meth sores, and crank bugs.

The therapist may see scratch marks or even broken skin where the sufferer has scratched violently. Open, red (often bleeding) sores appear most commonly on the face and arms but can be anywhere on the body. These lesions can become inflamed, swollen, and pus-filled in the presence of a *Staphylococcus* infection. Left untreated, pathogens can enter the bloodstream, causing dangerous sepsis or deeper abscess. There is no cure, but medical evaluation is needed; topical treatment and cryotherapy can help, and antibiotic treatment is needed when there is infection.

New onset of skin lesions, especially in children, should be medically evaluated (Fig. 4-3). Many conditions in adults and children can be treated effectively; some, but not all, can be cured.

Skin Assessment

With the possible exception of a dermatologist, the therapist sees more skin than anyone else in the health care system. Clients are more likely to point out skin lesions or ask the therapist about lumps and bumps. It is important to have a working knowledge of benign versus pathologic skin lesions and know when to refer appropriately.

The hands, arms, feet, and legs can be assessed throughout the physical therapy examination for changes in texture, color, temperature, clubbing, circulation including capillary filling, and edema (Box 4-11). Abnormal texture changes include shiny, stiff, coarse, dry, or scaly skin.

Skin mobility and turgor are affected by the fluid status of the client. Dehydration and aging reduce skin turgor (Fig. 4-4), and edema decreases skin mobility. The therapist should be aware of medications that cause skin to become sensitive to sunlight. The most commonly prescribed medications linked with photosensitivity are listed in Box 4-12.

Chronically ill or hospitalized patients should be examined frequently for signs of skin breakdown. Check all pressure points, including the ears, sacrum, scapulae, shoulders, area over the greater trochanters, heels, malleoli, and the back of the head. Document staging of any pressure ulcers (Table 4-7).

The staging system developed by National Pressure Ulcer Advisory Panel (NPUAP) is an anatomic description of tissue destruction or wound depth designed for use only with pressure ulcers or wounds created by pressure. While it is essential to have this information, it is also very important to document other wound characteristics, such as size, drainage, and granulation tissue, to make the wound assessment complete.

Coordinate with nursing staff to remove prostheses, restraints, and dressings to look beneath them. Anyone with an IV line, catheter, or other insertion sites must be examined for signs of infiltration (e.g., pus, erythema), phlebitis, and tape burns.

BOX 4-11 EXAMINING A SKIN LESION OR MASS

Record observations about any skin lesion or mass using the mnemonic:

5 **S**tudents and 5 **T**eachers around the **CAMPFIRE:**

- **S**ite (location, single versus multiple)
- **S**ize
- **S**hape
- **S**pider angiomas (pregnancy, alcoholism; see Figs. 9-3 and 9-4)
- **S**urface (smooth, rough, indurated, scratches, scarring; see Fig. 4-8), hair growth/loss, bruising [violence, hemophilia, liver damage, thrombocytopenia])
- **T**enderness or pain
- **T**exture
- **T**urgor (hydration)
- **T**emperature
- **T**ransillumination (shine flashlight through it from the side and from the top)
- **C**onsistency (soft, spongy, hard), **C**olor, **C**irculation
- **A**ppearance of the client
- **M**obility (move the lump in 2 directions: side-to-side and up-down; contract muscle and repeat test)
  - *Bone:* lump is immobile
  - *Muscle:* contraction decreases mobility of the lump
  - *Subcutaneous:* skin moves over lump
  - *Skin:* lump moves with skin
- **P**ulsation (place 2 fingers on mass: are fingers pushed in the same direction or apart from each other?)
- **F**luctuation (does the mass contain fluid: place 2 fingers in V-shape on either side of lump, tap center of lump with index finger of the opposite hand; fingers move if lump is fluid-filled)
- **I**rreducibility
  - *Compressible:* mass goes away or decreases with pressure but comes back when pressure is released
  - *Reducible:* mass goes away and only comes back with cough or change in position
- **R**egional lymph nodes (examine nearest lymph nodes); **R**ash (e.g., dermatitis, shingles, drug reaction)
- **E**dge (clearly defined, poorly defined, symmetric, asymmetric), **E**dema

If a lesion is present, assess for:

- Associated signs and symptoms (e.g., bleeding, pruritus, fever, joint pain)
- When did the lesion(s) first appear?
- Is it changing over time? How (increasing, decreasing)?
- Were there any known or suspected triggers? (e.g., perfumes, soaps, or cosmetics; medications; environmental/sunlight exposure (includes vectors such as ticks, spiders, scabies, fleas); diet; psychologic or emotional factors)
- A military history may be important.

From http://www.clinicalexam.com/pda/g_ref_mass_examination.htm.

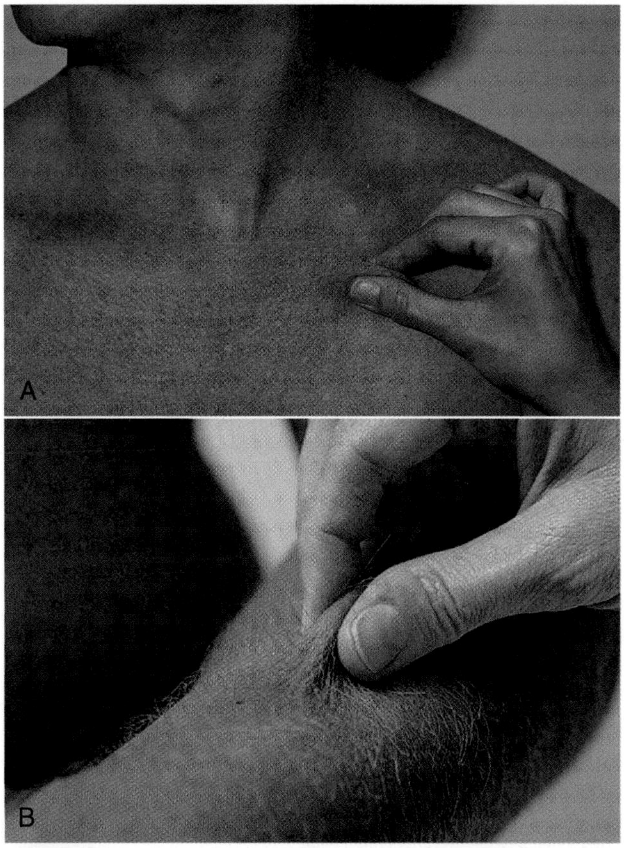

A

B

Fig. 4-4 To check skin turgor (elasticity or resiliency), gently pinch the skin between your thumb and forefinger, lifting it up slightly, then release. Skin turgor can be tested on the forehead or sternum, beneath the clavicle **(A),** and over the extensor surface of the arm **(B)** or hand. Expect to see the skin lift up easily and return to place quickly. The test is positive for decreased turgor (often caused by dehydration) when the pinched skin remains lifted 5 or more seconds after release and returns to normal very slowly. (**A** from Seidel HM: *Mosby's guide to physical examination,* ed 7, St. Louis, 2011, Mosby. **B** from Potter P, Perry A: *Basic nursing: essentials for practice,* ed 6, St. Louis, 2007, Mosby.)

BOX 4-12 MOST COMMON MEDICATIONS CAUSING PHOTOSENSITIVITY

- Ciprofloxacin (antibiotic)
- Doxycycline (antibiotic)
- Furosemide (diuretic)
- Glipizide (hypoglycemic)
- Glyburide (hypoglycemic)
- Ibuprofen (NSAID)
- Ketoprofen (NSAID)
- Naproxen (NSAID)
- Sulfonamides (wide range of antibiotics)
- Tetracycline (antibiotic)
- 5-Fluorouracil (cytotoxic drug)

NSAID, Nonsteroidal antiinflammatory drug.

From Bergamo BM, Elmets CA: Drug-induced photosensitivity. eMedicine available at: http://www.emedicine.com/DERM/topic108.htm. Posted October 27, 2004. Accessed January 2011.

TABLE 4-7 Staging of Pressure Ulcers*

Pressure ulcers have been defined by the National Pressure Ulcer Advisory Panel (NPUAP) in conjunction with the European Pressure Ulcer Advisory Panel (EPUAP) as localized injury to the skin and/or underlying tissue usually over a bony prominence, as a result of pressure or pressure in combination with shear.

Stage	Description
Stage I	Skin changes observable (increased or decreased temperature, tissue consistency), sensation (pain, itching).
Stage II	Epidermis and dermis layers are damaged (partial-thickness); ulcer is superficial and presents as an abrasion, blister, or shallow crater.
Stage III	Damage through to subcutaneous tissue (full-thickness skin loss); does not extend through fascia; appears as a deep crater; this is a "never event" (i.e., should never happen).
Stage IV	Involvement of muscle, bone, tendon, joint capsule or other supporting structures (full-thickness tissue loss); this is also a "never event" (i.e., should never happen).

*Staging does not indicate the process of wound healing. The NPUAP also provides the Pressure Ulcer Scale for Healing (PUSH Tool) as a quick, reliable tool to monitor the change in pressure ulcer status over time. Updated staging system available on-line at: http://www.npuap.org/push3-0.html.

From U.S. Department of Health and Human Services: *Pressure ulcers in adults: prediction and prevention.* Clinical practice guideline no. 3. AHCPR publication no. 92-0047, Rockville, Maryland, 1992, DHHS; European Pressure Ulcer Advisory Panel and National Pressure Ulcer Advisory Panel. Prevention and Treatment of Pressure Ulcers: Clinical Practice Guidelines. Washington, DC, 2009, NPUAP.

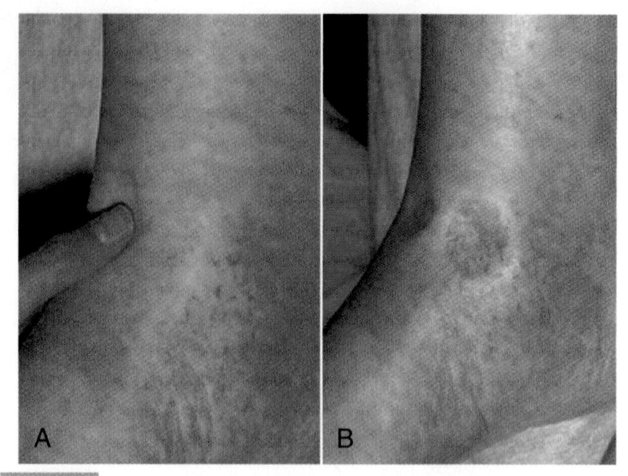

Fig. 4-5 Pitting edema in a patient with cardiac failure. A depression ("pit") remains in the edema for some minutes after firm fingertip pressure is applied. (From Forbes CD, Jackson WD. *Color atlas and text of clinical medicine*, ed 3, London, 2003, Mosby.)

Observe for signs of edema. Edema is an accumulation of fluid in the interstitial spaces. Pitting edema in which pressing a finger into the skin leaves an indentation often indicates a chronic condition (e.g., chronic kidney failure, liver failure, CHF) but can occur acutely as well (e.g., face) (Fig. 4-5). The location of edema helps identify the potential cause. Bilateral edema of the legs may be seen in clients with heart failure or with chronic venous insufficiency.

Abdominal and leg edema can be seen in clients with heart disease, cirrhosis of the liver (or other liver impairment), and protein malnutrition. Edema may also be noted in dependent areas, such as the sacrum, when a person is confined to bed. Localized edema in one extremity may be the result of venous obstruction (thrombosis) or lymphatic blockage of the extremity (lymphedema).

Change in Skin Temperature

Skin temperature can be an indication of vascular supply. A handheld, noninvasive, infrared thermometer can be used to measure skin surface temperature. The most common use of this device is for temperature observation and comparison of both feet in individuals with diabetes for the purpose of identifying increased skin temperatures, intended as an early warning of inflammation, impending infection, and possible foot ulceration. Temperature differences of four or more degrees between the right and left foot is a predictive risk factor for foot ulcers; self-monitoring has been shown to reduce the risk of ulceration in high-risk individuals.[69-72]

Other signs and symptoms of vascular changes of an affected extremity may include paresthesia, muscle fatigue and discomfort, or cyanosis with numbness, pain, and loss of hair from a reduced blood supply (Box 4-13).

Change in Skin Color

Capillary filling of the fingers and toes is an indicator of peripheral circulation. Perform a capillary refill test by pressing down on the nail bed and releasing. Observe first for blanching (whitening) followed by return of color within 3 seconds after release of pressure (normal response).

Skin color changes can occur with a variety of illnesses and systemic conditions. Clients may notice a change in their skin color before anyone else does, so be sure and ask about it. Look for pallor; increased or decreased pigmentation; yellow, green, or red skin color; and cyanosis.

Color changes are often observed first in the fingernails, lips, mucous membranes, conjunctiva of the eye, and palms and soles of dark-skinned people.

Skin changes associated with impairment of the hepatic system include jaundice, pallor, and orange or green skin.

In some situations, jaundice may be the first and only manifestation of disease. It is first noticeable in the sclera of the eye as a yellow hue when bilirubin level reaches 2 to 3 mg/dL. Dark-skinned persons may have a normal yellow color to the outer sclera. Jaundice involves the whole sclera up to the iris.

When the bilirubin level reaches 5 to 6 mg/dL, the skin becomes yellow. Other skin and nail bed changes associated with liver disease include palmar erythema (see Fig. 9-5), spider angiomas (see Figs. 9-3 and 9-4), and nails of Terry (see Fig. 9-6; see further discussion in Chapter 9).

BOX 4-13 PERIPHERAL VASCULAR ASSESSMENT

Inspection

Compare extremities side to side:

- Size
- Symmetry
- Skin
- Nail beds
- Color
- Hair growth
- Sensation

Palpation

Pulses *(see Fig. 4-1)*

Upper Quadrant

- Carotid
- Brachial
- Radial
- Ulnar

Lower Quadrant

- Femoral
- Popliteal
- Dorsalis pedis
- Posterior tibial

Characteristics of Pulses

Rate

Rhythm

Strength (Amplitude)

- +4 = bounding
- +3 = full, increased
- +2 = normal
- +1 = diminished, weak
- 0 = absent

Check for symmetry (compare right to left)

Compare upper extremity to lower extremity

Arterial Insufficiency of Extremities

Pulses	Decreased or absent
Color	Pale on elevation Dusky rubor on dependency
Temperature	Cool/cold
Edema	None
Skin	Shiny, thin pale skin; thick nails; hair loss Ulcers on toes
Sensation	Pain: increased with exercise (claudication) or leg elevation; relieved by dependent dangling position Paresthesias

Venous Insufficiency of Extremities

Pulses	Normal arterial pulses
Color	Pink to cyanotic Brown pigment at ankles
Temperature	Warm
Edema	Present
Skin	Discolored, scaly (eczema or stasis dermatitis) Ulcers on ankles, toes, fingers Varicose veins
Sensation	Pain: increased with standing or sitting; relieved with elevation or support hose

Special (Quick Screening) Tests

Capillary refill time (fingers and toes)

Arterial-brachial index (ABI)

Rubor on dependency

Allen test

A bluish cast to skin color can occur with cyanosis when oxygen levels are reduced in the arterial blood (central cyanosis) or when blood is oxygenated normally but blood flow is decreased and slow (peripheral cyanosis). Cyanosis is first observed in the hands and feet, lips, and nose as a pale blue change in color. The client may report numbness or tingling in these areas.

Central cyanosis is caused by advanced lung disease, congestive heart disease, and abnormal hemoglobin. Peripheral cyanosis occurs with CHF (decreased blood flow), venous obstruction, anxiety, and cold environment.

Rubor (dusky redness) is a common finding in peripheral vascular disease as a result of arterial insufficiency. When the legs are raised above the level of the heart, pallor of the feet and lower legs develops quickly (usually within 1 minute). When the same client sits up and dangles the feet down, the skin returns to a pink color quickly (usually in about 10 to 15 seconds). A minute later the pallor is replaced by rubor, usually accompanied by pain and diminished pulses. Skin is cool to the touch and trophic changes may be seen (e.g., hair loss over the foot and toes, thick nails, thin skin).

Diffuse hyperpigmentation can occur with Addison's disease, sarcoidosis, pregnancy, leukemia, hemochromatosis, celiac sprue (malabsorption syndrome), scleroderma, and chronic renal failure.

This presents as patchy tan to brown spots most often but may occur as yellow-brown or yellow to tan with scleroderma and renal failure. Any area of the body can be affected, although pigmentation changes in pregnancy tend to affect just the face (melasma or the mask of pregnancy).

Assessing Dark Skin

Clients with dark skin may require a slightly different approach to skin assessment than the Caucasian population. Observe for any obvious changes in the palms of the hands and soles of the feet; tongue, lips, and gums in the mouth; and in the sclera and conjunctiva of the eyes.

Pallor may present as a yellow or ashen-gray due to an absence of the normally present underlying red tones in the skin. The palms and the soles show changes more clearly than the skin. Skin rashes may present as a change in skin texture so palpating for changes is important. Edema can be palpated as "tightness" and darker skin may appear lighter. Inflammation may be perceived as a change in skin temperature instead of redness or erythema of the skin.

Jaundice may appear first in the sclera but can be confused for the normal yellow pigmentation of dark-skinned clients. Be aware that the normal oral mucosa (gums, borders of the tongue, and lining of the cheeks) of dark-skinned individuals may appear freckled.

Petechiae are easier to see when present over areas of skin with lighter pigmentation such as the abdomen, gluteal area, and volar aspect of the forearm. Petechiae and ecchymosis (bruising) can be differentiated from erythema by applying pressure over the involved area. Pressure will cause erythema to blanch, whereas the skin will not change in the presence of petechiae or ecchymosis.

Examining a Mass or Skin Lesion

When examining a skin lesion or mass of any kind, follow the guidelines provided earlier in Box 4-11. In addition, the American Cancer Society (ACS) and the Skin Cancer Foundation advocate using the following ABCDEs to assess skin lesions for cancer detection (Fig. 4-6):

A—Asymmetry
B—Border
C—Color
D—Diameter
E—Evolving

The ABCD (now including E) criteria have been verified in multiple studies, documenting the effectiveness and diagnostic accuracy of this screening technique. Their efficacy has been confirmed with digital image analysis; sensitivity ranges from 57% to 90% and specificity from 59% to 90%.[73]

Round, symmetric skin lesions such as common moles, freckles, and birthmarks are considered "normal." If an existing mole or other skin lesion starts to change and a line drawn down the middle shows two different halves, medical evaluation is needed.

Common moles and other "normal" skin changes usually have smooth, even borders or edges. Malignant melanomas, the most deadly form of skin cancer, have uneven, notched borders.

Benign moles, freckles, "liver spots," and other benign skin changes are usually a single color (most often a single shade of brown or tan) (Fig. 4-7). A single lesion with more than one shade of black, brown, or blue may be a sign of malignant melanoma.

Even though some of us have moles we think are embarrassingly large, the average mole is really less than ¼ of an inch (about the size of a pencil eraser). Anything larger than this should be inspected carefully.

The Skin Cancer Foundation (www.skincancer.org) has many public education materials available to help the therapist identify suspicious skin lesions. In addition to their website, they have posters, brochures, videos, and other materials available for use in the clinic. It is highly recommended that these types of education materials be available in waiting rooms as part of a nationwide primary prevention program.

Other websites (www.skincheck.com [Melanoma Education Foundation]; http://www.melanomafoundation.com.au/ [The Melanoma Foundation of the University of Sydney Australia]) provide additional photos of suspicious lesions with more screening guidelines. The therapist must become

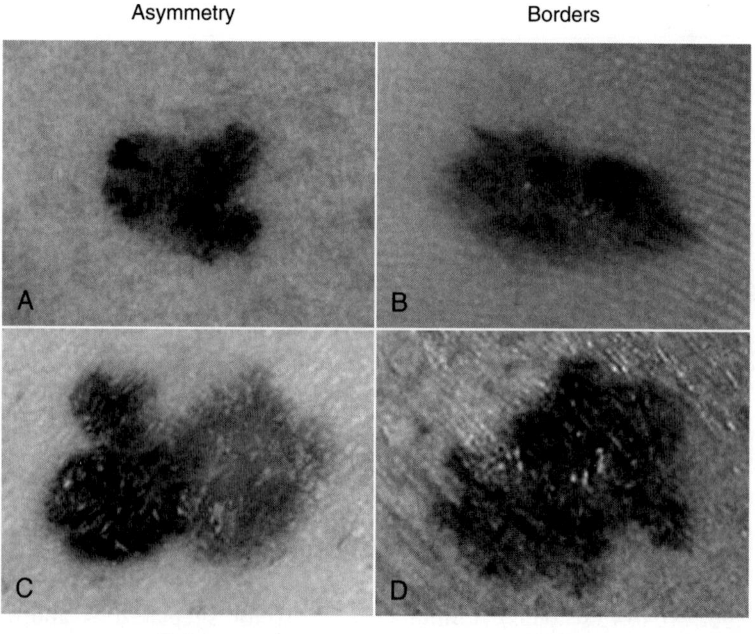

Fig. 4-6 Common characteristics associated with early melanomas are described and shown in this photo. **A, A**symmetry: a line drawn through the middle does not produce matching halves. **B, B**orders are uneven, fuzzy, or have notched or scalloped edges. **C, C**olor changes occur with shades of brown, black, tan, or other colors present at the same time. **D, D**iameter is greater than the width of a pencil eraser. Not shown: **E**volving (change), in size, shape, color, elevation, or another trait, or any new symptom, such as bleeding, itching, or crusting, requires evaluation. (From Dermik Laboratories [www.dermnet.com], 2005. Used with permission.)

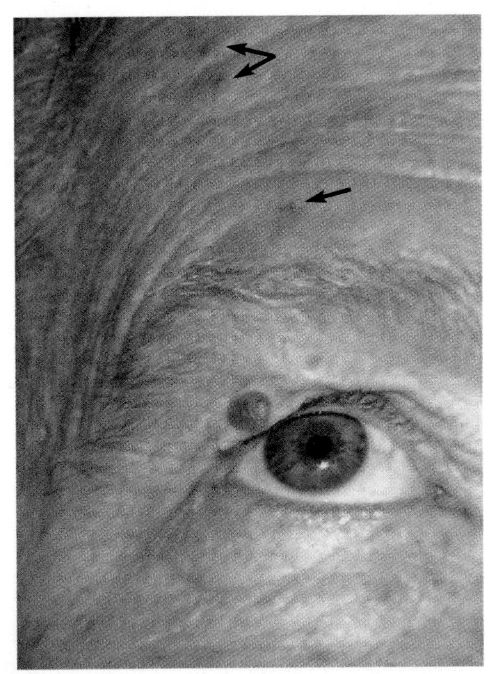

Fig. 4-7 Seborrheic keratosis, a benign well-circumscribed, raised, tan-to-black lesion often presents on the face, neck, chest, or upper back. This lesion represents a buildup of keratin, which is the primary component of the epidermis. There is a family tendency to develop these lesions. The more serious lesions are the red patches located on this client's forehead *(arrows),* precancerous lesions called *actinic keratosis,* the result of chronic sun exposure. These lesions have a "sandpaper" feel when palpated. Medical treatment is needed for this premalignant lesion. (Courtesy Catherine C. Goodman, 2005. Used with permission.)

as familiar as possible with what suspicious skin aberrations may look like in order to refer as early as possible.

Remember to evaluate risk factors when screening for skin cancer. The average lifetime risk of developing melanoma (Caucasians) is 1 in 58.[73] This has increased from 1 in 90 just in the last two decades. Your risk is much higher if you have any of the risk factors listed in Box 13-2.

For all lesions, masses, or aberrant tissue, observe or palpate for heat, induration, scarring, or discharge. Use the mnemonic in Box 4-11. Make note of how long the client has had the lesion, if it has changed in the last 6 weeks to 6 months, and whether it has been medically evaluated. Always ask appropriate follow-up questions with this assessment:

FOLLOW-UP QUESTIONS

- How long have you had this?
- Has it changed in the last 6 weeks to 6 months?
- Has your doctor seen it?
- Does it itch, hurt, feel sore, or burn?
- Does anyone else in your household have anything like this?
- Have you taken any new medications (prescribed or OTC) in the last 6 weeks?
- Have you traveled somewhere new in the last month?
- Have you been exposed to anything in the last month that could cause this? (Consider exposure due to occupational, environmental, and hobby interests.)
- Do you have any other skin changes anywhere else on your body?
- Have you had a fever or sweats in the last 2 weeks?
- Are you having any trouble breathing or swallowing?
- Have you had any other symptoms of any kind anywhere else in your body?

How you ask is just as important as *what* you say. Do not frighten people by first telling them you always screen for skin cancer. It may be better to introduce the subject by saying that as health care professionals, therapists are trained to observe many body parts, including the skin, joints, posture, and so on. You notice the client has an unusual mole (or rash … or whatever you have observed) and you wonder if this is something that has been there for years. Has it changed in the last 6 weeks to 6 months? Has the client ever shown it to the doctor?

Assess Surgical Scars

It is always a good idea to look at surgical scars (Fig. 4-8), especially sites of local cancer removal. Any suspicious scab or tissue granulation, redness, or discoloration must be noted (photographed if possible).

Start by asking the client if he or she has noticed any changes in the scar. Continue by asking:

FOLLOW-UP QUESTIONS

- Would you have any objections if I looked at (or examined) the scar tissue?

If the client declines or refuses, be sure to follow-up with counsel to perform self-inspection and report any changes to the physician.

In Fig. 4-9, the small scab and granular tissue forming above the scar represent red flags of suspicious local recurrence. Even if the client suggests this is from "picking" at the scar, a medical evaluation is well advised.

The therapist has a responsibility to report these findings to the appropriate health care professional and make every effort to ensure client/patient compliance with follow-up.

Common Skin Lesions

Vitiligo

A lack of pigmentation from melanocyte destruction (vitiligo) (Fig. 4-10) can be hereditary and have no significance or it can be caused by conditions such as hyperthyroidism, stomach cancer, pernicious anemia, diabetes mellitus, or autoimmune diseases.

Lesions can occur anywhere on the body but tend to develop in sun-exposed areas, body folds, and around body

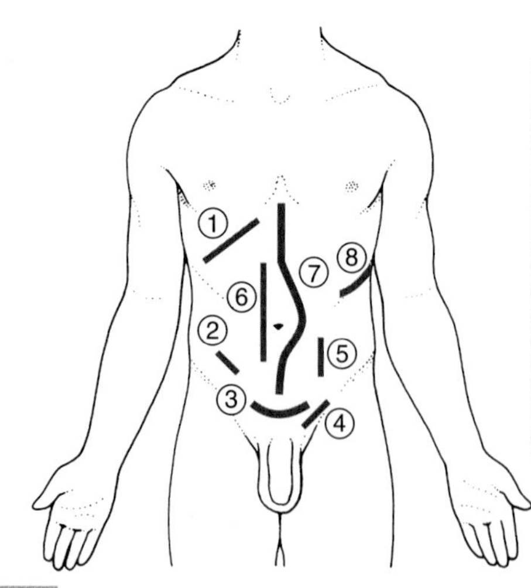

1. Cholecystectomy; a subcostal incision, when made on the right, provides exposure of the gallbladder and common bile duct; used on the left for splenectomy
2. Appendectomy, sometimes called McBurney's or Gridiron incision
3. Transverse suprapubic incision for hysterectomy and other pelvic surgeries
4. Inguinal hernia repair (herniorrhaphy or hernioplasty)
5. Anterior rectal resection (left paramedian incision)
6. Incision through the right flank (right paramedian); called laparotomy or celiotomy; sometimes used to biopsy the liver
7. Midline laparotomy
8. Nephrectomy (removal of the kidney) or other renal surgery

Fig. 4-8 Abdominal surgical scars. Not shown: puncture sites for laparoscopy, usually close to the umbilicus and one or two other sites.

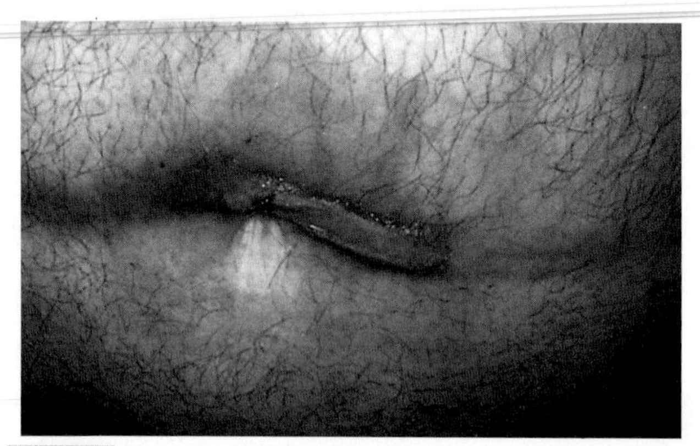

Fig. 4-9 Squamous cell carcinoma in scar. Always ask to see and examine scars from previous surgeries, especially when there has been a history of any kind of cancer, including skin cancer. Even in this black and white photo, you can see many skin changes to suggest the need for medical evaluation. Look to the far right of the raised scar tissue. You will see a normal, smooth scar. This is what the entire scar should look like. In this photo there is a horizontal line of granulation along the upper edge of the scar, as well as a scabbed over area in the middle of the raised scar. There is also a change in skin color on either side of the scar. (From Swartz MH: *Textbook of physical diagnosis,* ed 4, Philadelphia, 2001, WB Saunders.)

openings. Intraarticular steroid injections can cause temporary loss of pigmentation at the injection site. Anyone with any kind of skin type and skin color can be affected by vitiligo.

Café-au-lait

Café-au-lait (coffee with milk) spots describe the light-brown macules (flat lesion, different in color) on the skin as shown in Fig. 4-11. This benign skin condition may be associated with Albright's syndrome or a hereditary disorder called neurofibromatosis. The diagnosis is considered when a child presents with five or more of these skin lesions or if any single patch is greater than 1.5 cm in diameter.

Skin Rash

There are many possible causes of skin rash, including viruses (e.g., chicken pox, measles, Fifth disease, shingles), systemic conditions (e.g., meningitis, lupus, hives), parasites (e.g., lice, scabies), and reactions to chemicals.

A common cause of skin rash seen in a physical therapy practice is medications, especially antibiotics (Fig. 4-12). The reaction may occur immediately or there may be a delayed reaction of hours up to 6 to 8 weeks after the drug is stopped.

Skin rash can also occur before visceral malignancy of many kinds. Watch for skin rash or hives in someone who has never had hives before, especially if there has been no contact with medications, new foods, new detergents, or new perfumes or travel.

Hemorrhagic Rash

Hemorrhagic rash requires medical evaluation. A hemorrhagic rash occurs when small capillaries under the skin start to bleed forming tiny blood spots under the skin (petechiae). The petechiae increase over time as bleeding continues.

This type of rash does not fade under pressure with continued bleeding. Press a clear see-through drinking glass against the skin. Rashes from allergies or viral infections are more likely to fade and the skin will become white or pale. During later stages of hemorrhagic bleeding the rash does not fade or become pale with the pressure test; this test is not as reliable during early onset of hemorrhage. Left untreated, hemorrhagic spots may become bruises and then large red-purple areas of blood. Pressure on a bruise will not cause it to blanch.

Dermatitis

Dermatitis (sometimes referred to as eczema) is characterized by skin that is red, brown, or gray; sore; itchy; and sometimes swollen. The skin can develop blisters and weeping sores. Skin changes, especially in the presence of open lesions, puts the client at increased risk of infection. In chronic dermatitis, the skin can become thick and leathery.

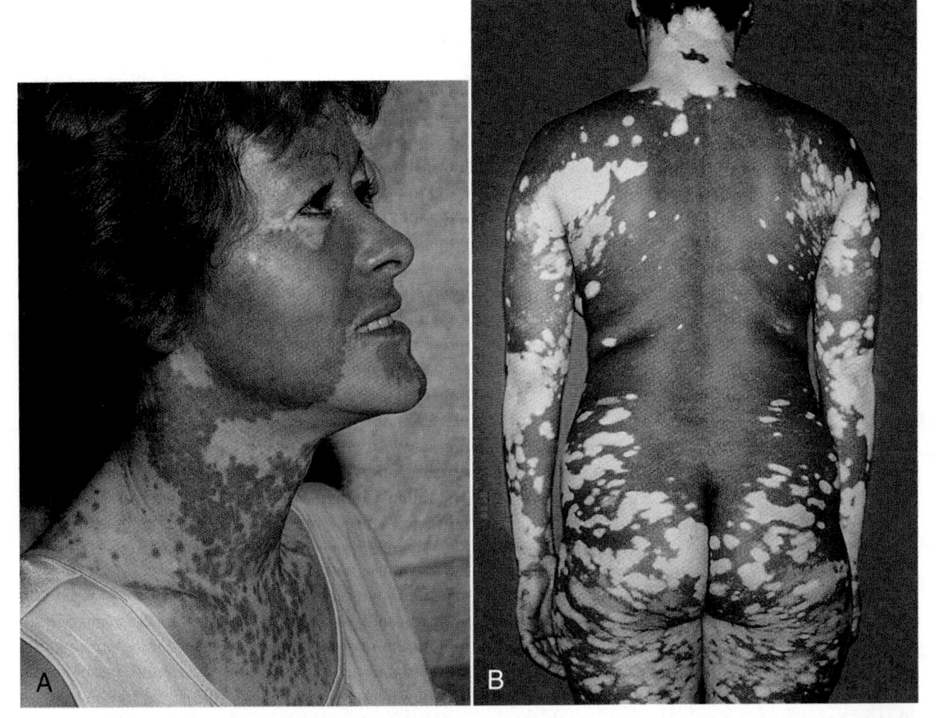

Fig. 4-10 *Vitiligo* is a term derived from the Greek word for "calf" used to describe patches of light skin caused by loss of epidermal melanocytes. Note the patchy loss of pigment on the face, trunk, and axilla. This condition can affect any part of the face, hands, or body and can be very disfiguring, especially in dark-skinned individuals. This skin change may be a sign of hyperthyroidism. (From Swartz MH: *Textbook of physical diagnosis,* ed 5, Philadelphia, 2006, WB Saunders.)

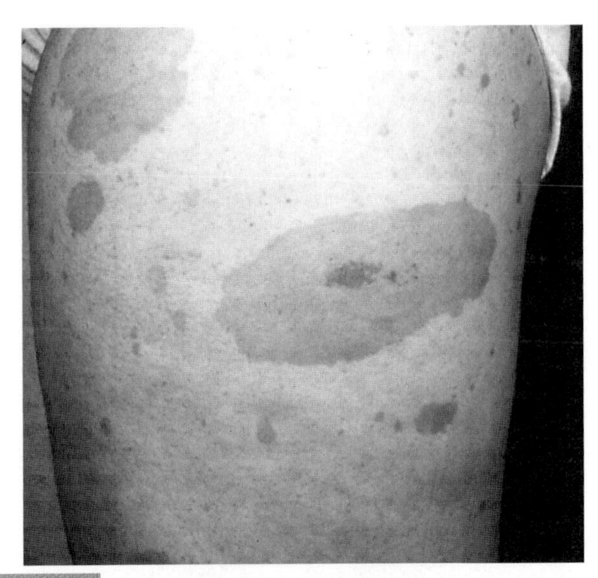

Fig. 4-11 Café-au-lait patches of varying sizes in a client with neurofibromatosis. Occasional (less than 5) tan macules are not significant and can occur normally. Patches 1.5 cm in diameter or larger raise the suspicion of underlying pathology even if there is only one present. (From Epstein O, Perkin GD, deBono DP, et al: *Clinical examination,* London, 1992, Gower Medical Publishing. Used with permission, Elsevier Science.)

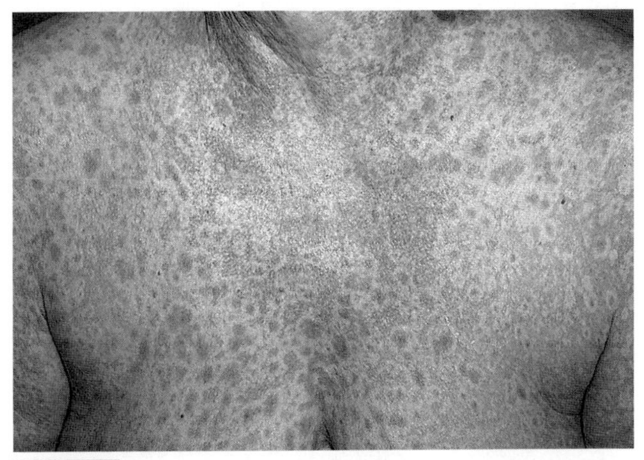

Fig. 4-12 Skin rash (reactive erythema) caused by a drug reaction to phenobarbital. Hypersensitivity reactions to drugs are most common with antibiotics (especially penicillin), sulfonamides ("sulfa drugs," antiinfectives), and phenobarbital as shown here. (From Callen JP, Paller AS, Greer KE, et al: *Color atlas of dermatology,* ed 2, Philadelphia, 2000, WB Saunders.)

There are different types of dermatitis diagnosed on the basis of medical history, etiology (if known), and presenting signs and symptoms. Contributing factors include stress, allergies, genetics, infection, and environmental irritants. For example, contact dermatitis occurs when the skin reacts to something it has come into contact with such as soap, perfume, metals in jewelry, and plants (e.g., poison ivy or oak).

Dyshidrotic dermatitis can affect skin that gets wet frequently. It presents as small, itchy bumps on the sides of the fingers or toes and progresses to a rash. Atopic dermatitis often accompanies asthma or hay fever. It appears to affect genetically predisposed clients who are hypersensitive to environmental allergens. This type of dermatitis can affect any part of the body, but often involves the skin inside the elbow and on the back of the knees.

Rosacea

Rosacea is a chronic facial skin disorder seen most often in adults between the ages of 30 and 60 years. It can cause a facial rash easily mistaken for the butterfly rash associated with lupus. Features include erythema, flushing, telangiectasia, papules, and pustules affecting the cheeks and nose of the face. An enlarged nose is often present, and the condition progressively gets worse (Fig. 4-13).

Rosacea can be controlled with dermatologic or other medical treatment in some cases. Recent studies suggest rosacea may be linked to GI disease caused by the *H. pylori* bacteria.[74-76] Such cases may respond favorably to antibiotics. Medical referral is needed for an accurate diagnosis.

Thrombocytopenia

Decrease in platelet levels can result in thrombocytopenia, a bleeding disorder characterized by petechiae (tiny purple or red spots), multiple bruises, and hemorrhage into the tissues (Fig. 4-14). Joint bleeds, nose and gum bleeds, excessive menstruation, and melena (dark, tarry, sticky stools from oxidized blood in the GI tract) can occur with thrombocytopenia.

There are many causes of thrombocytopenia. In a physical therapy practice, the most common causes seen are bone marrow failure from radiation treatment, leukemia, or metastatic cancer; cytotoxic agents used in chemotherapy; and drug-induced platelet reduction, especially among adults with rheumatoid arthritis treated with gold or inflammatory conditions treated with aspirin or other NSAIDs.

Postoperative thrombocytopenia can be heparin-induced for patients receiving IV heparin. Watch for limb ischemia, cyanosis of fingers or toes, signs and symptoms of a stroke, heart attack, or pulmonary embolus. (See further discussion on Thrombocytopenia in Chapter 5.)

Xanthomas

Xanthomas are benign fatty fibrous yellow plaques, nodules, or tumors that develop in the subcutaneous layer of skin (Fig. 4-15), often around tendons. The lesion is characterized by the intracellular accumulation of cholesterol and cholesterol esters.

These are seen most often associated with disorders of lipid metabolism, primary biliary cirrhosis, and uncontrolled diabetes (Fig. 4-16). They may have no pathologic significance but can occur in association with malignancy such as leukemia, lymphoma, or myeloma. Xanthomas require a medical referral if they have not been evaluated by a physician. When associated with diabetes, these nodules will resolve with adequate glucose control.

The therapist has an important role in education and prescriptive exercise for the client with xanthomas from poorly controlled diabetes. Gaining control of glucose levels using the three keys of intervention (diet, exercise, and insulin or oral hypoglycemic medication) is essential and requires a team management approach.

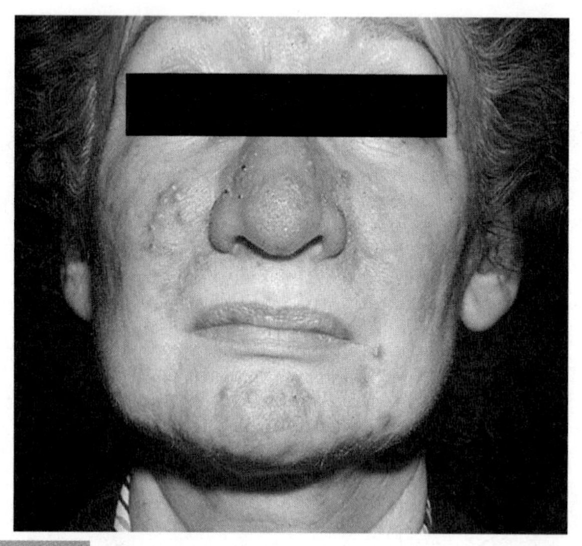

Fig. 4-13 Rosacea, a form of adult acne, may be associated with *Helicobacter pylori;* medical evaluation and treatment is needed to rule out this possibility. (Courtesy University of Iowa Virtual Hospital. Copyright protected material used with permission of the authors and the University of Iowa's Virtual Hospital; www.vh.org/adult/provider/dermatology/PietteDermatology/BlackTray/33Rosacea.html.)

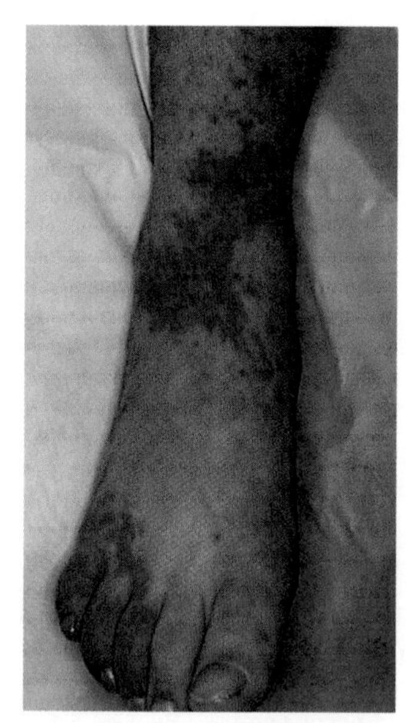

Fig. 4-14 Purpura. Petechiae and ecchymoses are seen in this flat macular hemorrhage from thrombocytopenia (platelet level less than 100,000/mm^3). This condition also occurs in older adults as blood leaks from capillaries in response to minor trauma. It can occur in fair-skinned people with skin damage from a lifetime of exposure to ultraviolet (UV) radiation. Exposure to UVB and UVA rays can cause permanent damage to the structural collagen that supports the walls of the skin's blood vessels. Combined with thinning of the skin that occurs with aging, radiation-impaired blood vessels are more likely to rupture with minor trauma. (From Hurwitz S: *Clinical pediatric dermatology: a textbook of skin disorders of childhood and adolescence,* ed 2, Philadelphia, 1993, WB Saunders.)

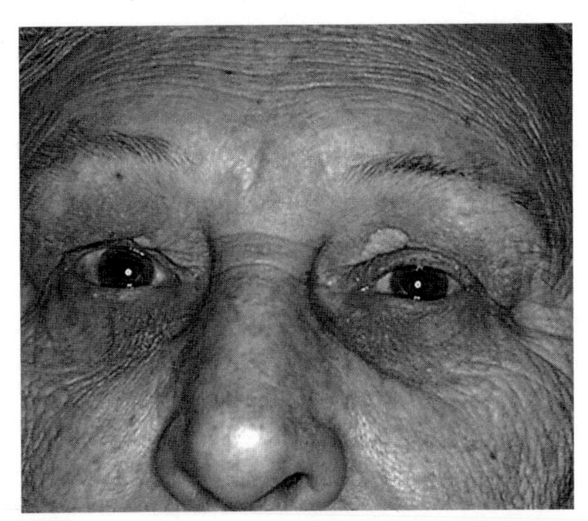

Fig. 4-15 Xanthelasma. Soft, raised yellow plaques, also known as *xanthomas,* commonly occur with aging and may be a sign of high cholesterol levels. Shown here on the eyelid, these benign lesions also occur on the extensor surfaces of tendons, especially in the hands, elbows, and knees. They may have no pathologic significance, but they often appear in association with disorders of lipid metabolism. (From Albert DM, Jakobeic FA: *Principles and practice of ophthalmology,* vol 3, Philadelphia, 1994, WB Saunders.)

Rheumatologic Diseases

Skin lesions are often the first sign of an underlying rheumatic disease (Box 4-14). In fact, the skin has been called a "map" to rheumatic diseases. The butterfly rash over the nose and cheeks associated with lupus erythematosus can be seen in the acute (systemic) phase, whereas discoid lesions are more common with chronic integumentary form of lupus (Fig. 4-17).[77]

Individuals with dermatomyositis often have a heliotrope rash and/or Gottron papules. Scleroderma is accompanied by many skin changes; pitting of the nails is common with psoriatic arthritis. Skin and nail bed changes are common with some sexually transmitted diseases that also have a rheumatologic component (see Figs. 3-10 and 4-22).[78]

BOX 4-14 RHEUMATIC DISEASES ACCOMPANIED BY SKIN LESIONS

- Acute rheumatic fever
- Discoid lupus erythematosus
- Dermatomyositis
- Gonococcal arthritis
- Lyme disease
- Psoriatic arthritis
- Reactive arthritis
- Rubella
- Scleroderma
- Systemic lupus erythematosus (SLE)
- Vasculitis

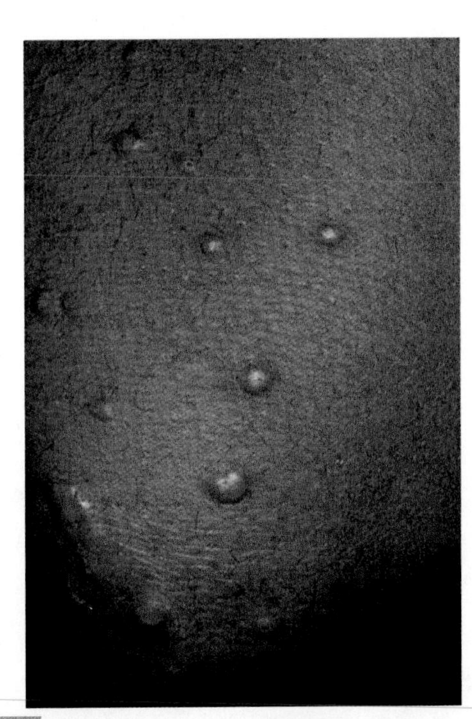

Fig. 4-16 A slightly different presentation of xanthomas, this time associated with poorly controlled diabetes mellitus. Although the lesion is considered "benign," the presence of these skin lesions in anyone with diabetes signals the need for immediate medical attention. The therapist also plays a key role in client education and the development of an appropriate exercise program to bring blood glucose levels under adequate control. (From Callen JP, Jorizzo J, Greer KE, et al: *Dermatological signs of internal disease,* Philadelphia, 1988, WB Saunders.)

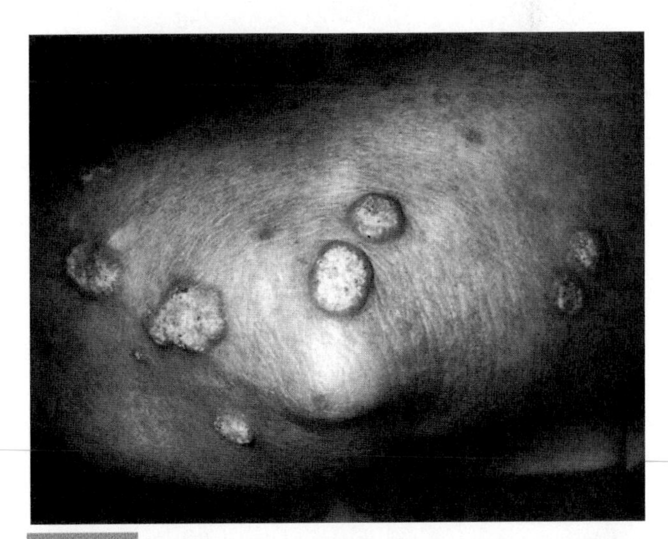

Fig. 4-17 Skin lesions associated with discoid lupus erythematosus. These disk-shaped lesions look like warts or squamous cell carcinoma. A medical examination is needed to make the definitive differential diagnosis. (From Callen JP, Jorizzo J, Greer KE, et al: *Dermatological signs of internal disease,* Philadelphia, 1988, WB Saunders.)

Steroid Skin and Steroid Rosacea

Steroid skin is the name given when bruising or ecchymosis occurs as a result of chronic use of topical or systemic corticosteroids (Fig. 4-18). In the case of topical steroid creams, this is a red flag that pain is not under control and medical attention for an underlying (probably inflammatory) condition is needed.

The use of topical corticosteroids for more than 2 weeks to treat chronic skin conditions affecting the face can cause a condition characterized by rosacea-like eruptions known as *steroid rosacea.* Attempts to stop using the medication may result in severe redness and burning called *steroid addiction syndrome* (Fig. 4-19).

Whenever signs and symptoms of chronic corticosteroids are seen, a medical evaluation may be needed to review medical management of the problem. In the case of steroid skin from chronic systemic corticosteroid use, ask if the physician has seen (or knows about) the signs and symptoms and how long it has been since medications have been reviewed. The multiple side effects of chronic corticosteroid use are discussed in association with Cushing's syndrome (see Chapter 11).

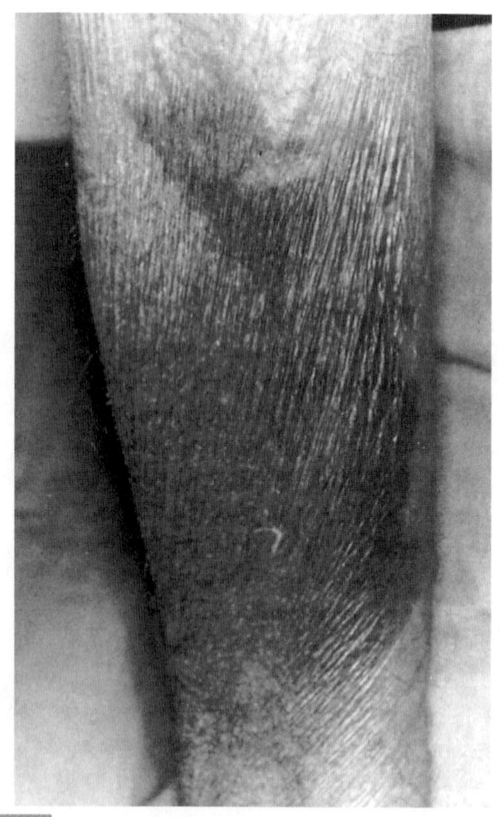

Fig. 4-18 Ecchymosis as a result of steroid application. Also note the cutaneous atrophy produced by topical steroids. This skin condition is referred to as "steroid skin" when associated with chronic oral or topical steroid use. Medical referral may be needed for better pain control. (From Callen JP, Jorizzo J, Greer KE, et al: *Dermatological signs of internal disease,* Philadelphia, 1988, WB Saunders.)

Erythema Chronicum Migrans

One or more erythema migrans rash may occur with Lyme disease. There is no one prominent rash. The rash varies in size and shape and may have purple, red, or bruised-looking rings. The rash may appear as a solid red expanding rash or blotch or as a central red spot surrounded by clear skin that is ringed by an expanding red rash. It may be smooth or bumpy to the touch, and it may itch or ooze. The Lyme Disease Foundation provides a photo gallery of possible rashes associated with Lyme disease.[78]

This rash, which develops in most people with Lyme disease, appears most often 1 to 2 weeks after the disease is transmitted (via tick bite) and may persist for 3 to 5 weeks. It usually is not painful or itchy but may be warm to the touch. The bull's-eye rash may be more difficult to see on darker-skinned people. A dark, bruise-like appearance is more common in those cases. Other symptoms are listed in Chapter 12.

Effects of Radiation

Radiation for the treatment of some cancers has some specific effects on the skin. Pigment producing cells can be affected by either low dose radiation causing hyperpigmentation or by high dose radiation resulting in depigmentation (vitiligo). Pigmentation changes can be localized or generalized.[79]

Radiation recall reaction can occur months later as a post-irradiation effect. The physiologic response is much like overexposure to the sun with erythema of the skin in the same pattern as the radiation exposure without evidence of disease progression at that site. It is usually precipitated by some external stimuli or event such as exposure to the sun, infection, or stress.

Radiation recall is also more likely to occur when an individual receives certain chemotherapies (e.g., cyclophosphamide, paclitaxel, doxorubicin, gemcitabine) after radiation.[80,81] The chemotherapy causes the previously radiated area to become inflamed and irritated.

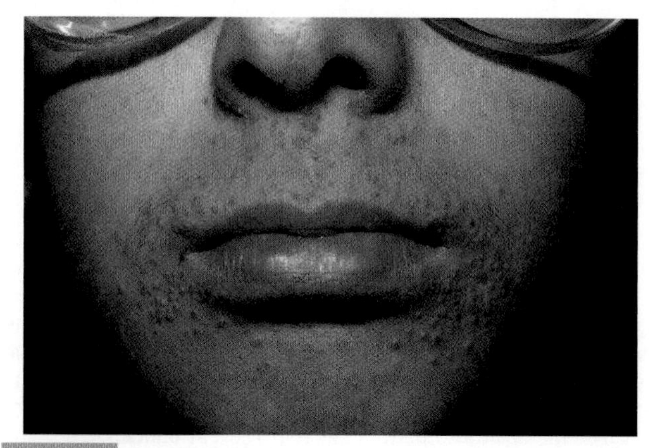

Fig. 4-19 Steroid addiction appearing to be acne. This condition was caused by long-term application of a moderate-potency steroid and Mycostatin combined. (From Weston WL: *Weston color textbook of pediatric dermatology*, ed 4, St. Louis, 2007, Mosby.)

Radiation dermatitis and *x-ray keratosis,* separate from radiation recall, are terms used to describe acute (expected) skin irritation caused by radiation at the time of radiation.

Skin changes can also occur as a long-term effect of radiation exposure. Radiation levels administered to oncology patients even 10 years ago were much higher than today's current treatment regimes. Always look at previous radiation sites for evidence of long-term effects.

Sexually Transmitted Diseases/Infections

Sexually transmitted diseases (STDs) are a variety of clinical syndromes caused by pathogens that can be acquired and transmitted through sexual activity.[82] STDs, also known as sexually transmitted infections (STIs), are often accompanied by skin and/or nail bed lesions and joint pain. Being able to recognize STIs is helpful in the clinic. Someone presenting with joint pain of "unknown cause" and also demonstrating signs of a STI (see Fig. 3-10) may help bring the correct diagnosis to light sooner than later.

Around the world, STIs pose a major health problem. In 1970, there were two major STIs; today, there are 25. The prevalence of STIs is rapidly increasing to epidemic proportions in the United States. Two-thirds of all STIs occur in people 25 years of age or younger.[83,84]

STIs have been positively identified as a risk factor for cancer. Not all STIs are linked with cancer, but studies have confirmed that human papillomavirus (HPV) is the primary cause of cervical cancer (Fig. 4-20).[85] HPV is the leading viral STI in the United States. More than 70 types of HPV have been identified: 23 infect the cervix and 13 types are associated with cancer in men and women. Infection with one of these viruses does not predict cancer, but the risk of cancer is increased.[82]

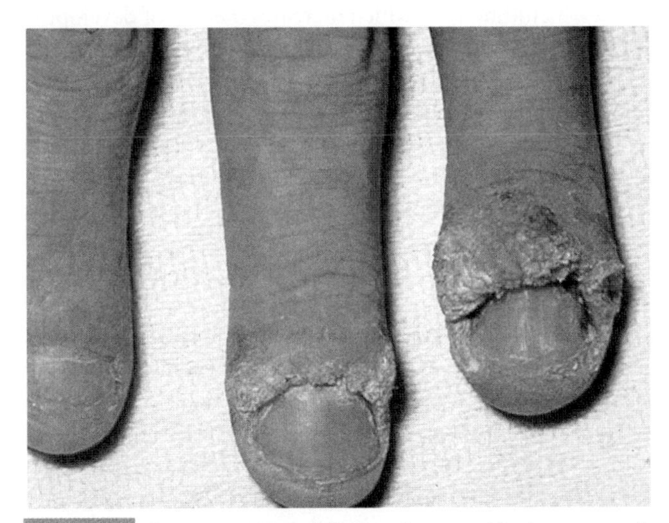

Fig. 4-20 Common warts of the hands caused by human papillomavirus (HPV) via nonsexual transmission. The virus can be transmitted through sexual contact and is a precursor to cancer of the cervix. Genital warts do not typically occur by autoinoculation from the hands. In other words, warts on the fingers caused by HPV are probably NOT transmitted from finger to genitals. They occur with contact of someone else's genital warts. Warts on the fingers caused by sexually transmitted HPV do not transmit the sexually transmitted infection (STI) to the therapist if the therapist shakes hands with the client or touches the warts. However, standard precautions are always recommended whenever skin lesions of any kind are present. For a summary of Standard Precautions, see Goodman et al., 2003.[85] (From Parkin JM, Peters BS: *Differential diagnosis in AIDS,* London, 1991, Mosby-Wolfe.)

Syphilis is on the rise again with the number of cases doubled in the last few years among gay and bisexual men, suggesting an erosion of safe sex practices.[86,87] It is highly contagious, spread from person to person by direct contact with a syphilis sore on the body of an infected person. Sores occur at the site of infection, mainly on the external genitals, vagina, anus, or rectum. Sores can also occur on the lips and in the mouth.

Transmission occurs during vaginal, anal, or oral sex. An infected pregnant woman can also pass the disease to her unborn child. Syphilis cannot be spread by contact with toilet seats, doorknobs, swimming pools, hot tubs, bathtubs, shared clothing, or eating utensils.

In the first stage of syphilis, a syphilis chancre may appear (Fig. 4-21) at the site of inoculation (usually the genitals, anus, or mouth). The chancre occurs 4 weeks after initial infection and is often not noticed in women when present in the genitalia. The chancre is often accompanied by lymphadenopathy.

Without treatment, the spread of the bacteria through the blood causes the second stage (secondary syphilis). Therapists may see lesions associated with secondary syphilis (Fig. 4-22). Neurologic (untreated) infection may present as cranial nerve dysfunction, meningitis, stroke, acute or chronic altered mental status, loss of vibration sense, and auditory or ophthalmic abnormalities.[82]

The client may report or present with a characteristic rash that can appear all over the body, most often on the palms and soles. The appearance of these skin lesions occurs after the primary chancre disappears. During this time the risk of human immunodeficiency virus (HIV) transmission from unsafe sexual practices is increased twofold to fivefold.

Syphilis can be tested for with a blood test and treated successfully with antibiotics. Left untreated, tertiary (late

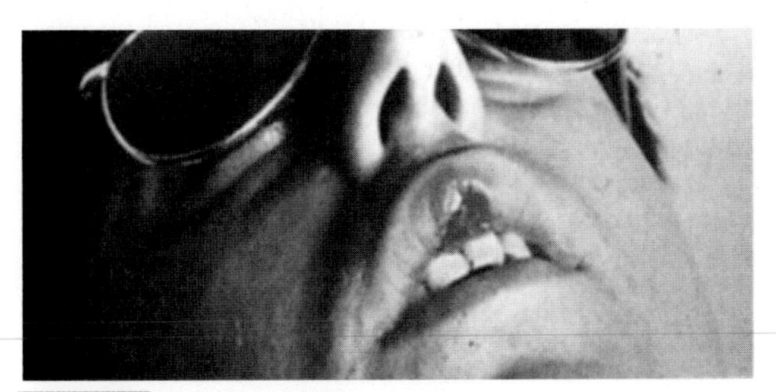

Fig. 4-21 The first stage (primary syphilis) is marked by a very infectious sore called a *chancre.* The chancre is usually small, firm, round with well-demarcated edges, and painless. It appears at the spot where the bacteria entered the body. Chancres last 1 to 5 weeks and heal on their own. (From *A close look at venereal disease,* Public service slide presentation. Courtesy Pfizer Laboratories, Pfizer Inc, New York. Permission granted 2004.)

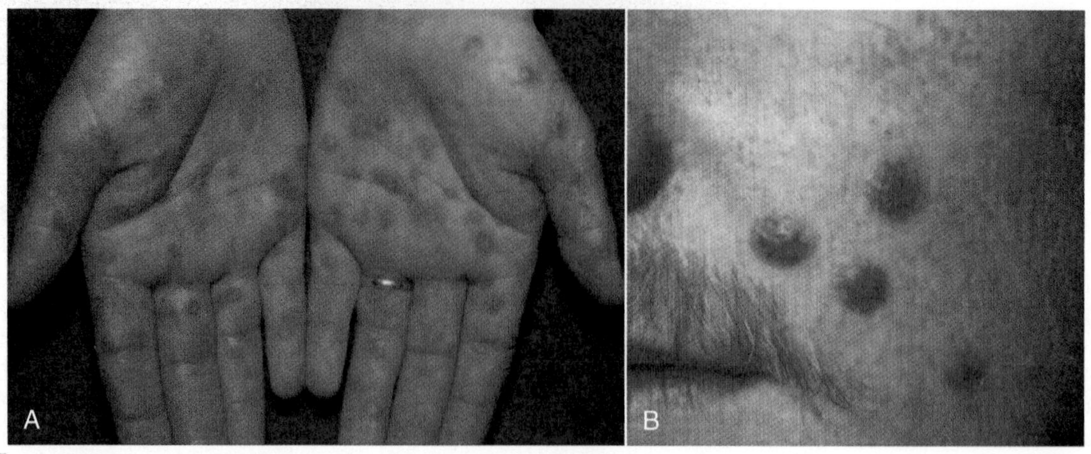

Fig. 4-22 Maculopapular rash associated with secondary syphilis appears as a pink, dusky, brownish-red or coppery, indurated, oval or round lesion with a raised border. These are referred to as "copper penny" spots. The lesions do not bleed and are usually painless. They usually appear scattered on the palms **(A)** or the bottom of the feet (not shown) but may also present on the face **(B).** The second stage begins 2 weeks to 6 months after the initial chancre disappears. The client may report joint pain with general flu-like symptoms (e.g., headache, sore throat, swollen glands, muscle aches, fatigue). Patchy hair loss may be described or observed. (From Mir MA: *Atlas of clinical diagnosis,* London, 1995, WB Saunders; p 198.)

stage) syphilis can cause paralysis, blindness, personality changes or dementia, and damage to internal organs and joints. The therapist can facilitate early detection and treatment through immediate medical referral. See further discussion on Infectious Causes of Pelvic Pain in Chapter 15, including special questions to ask concerning sexual activity and STIs. (See also Appendix B-33.)

Herpes Virus. Several herpes viruses are accompanied by characteristic skin lesions. Herpes simplex virus (HSV)-1 and -2 are the most common. Most people have been exposed at an early age and already have immunity. In fact, four out of five Americans harbor HSV-1. Due to the universal distribution of these viruses, most individuals have developed immunity by the ages of 1 to 2 years.

The HSV-1 and -2 viruses are virtually identical, sharing approximately 50% of their DNA. Both types infect the body's mucosal surfaces, usually the mouth or genitals, and then establish latency in the nervous system.[88] Both can cause skin and nail bed changes.

Cold sores caused by HSV-1 (also known as recurrent herpes labialis; "fever blister") are found on the lip or the skin near the mouth. HSV-1 usually establishes latency in the trigeminal ganglion, a collection of nerve cells near the ear. HSV-1 generally only infects areas above the waistline and occurs when oral secretions or mucous membranes infected with HSV come in contact with a break in the skin (e.g., torn cuticle, skin abrasion).

HSV-1 can be transmitted to the genital area during oral sex. In fact, HSV-1 can be transmitted oral-to-oral, oral-to-genital, anal-to-genital, and oral-to-anal. HSV-1 actually predominates for oral transmission, while a second herpesvirus (genital herpes; HSV-2) is more often transmitted sexually.

HSV 2, also known as "genital herpes" can cause cold sores but usually does not; rather, it is more likely to infect body tissues below the waistline as it resides in the sacral ganglion at the base of the spine.

Most HSV-1 and -2 infections are not a major health threat in most people but slowing the spread of genital herpes is important. The virus is more of a social problem than a medical one. The exception is that genital lesions from herpes can make it easier for a person to become infected with other viruses, including HIV, which increases the risk of developing acquired immunodeficiency syndrome (AIDS).

Nonmedical treatment with OTC products is now available for cold sores. Outbreaks of genital herpes can be effectively treated with medications, but these do not "cure" the virus. HSV-1 is also the cause of herpes whitlow, an infection of the finger and "wrestler's herpes," a herpes infection on the chest or face.

Herpetic Whitlow. Herpetic whitlow, an intense painful infection of the terminal phalanx of the fingers is caused by HSV-1 (60%) and HSV-2 (40%). The thumb and index fingers are most commonly involved. There may be a history of fever or malaise several days before symptoms occur in the fingers.

Common initial symptoms of infection include tingling pain or tenderness of the affected digit, followed by throbbing pain, swelling, and redness. Fluid-filled vesicles form and eventually crust over, ending the contagious period. The client with red streaks down the arm and lymphadenopathy may have a secondary infection. Take the client's vital signs (especially body temperature) and report all findings to the physician.[89]

As in other herpes infections, viral inoculation of the host occurs through exposure to infected body fluids via a break in the skin such as a paper cut or a torn cuticle. Autoinoculation can occur in anyone with other herpes infections such as genital herpes. It is an occupational risk among health care workers exposed to infected oropharyngeal secretions of clients, easily prevented by using standard precautions.[90]

Herpes Zoster. Varicella-zoster virus (VZV), or herpes zoster or "shingles," is another herpes virus with skin lesions characteristic of the condition. VZV is caused by the same virus that causes chickenpox. After an attack of chickenpox, the virus lies dormant in the nerve tissue, usually the dorsal root ganglion. If the virus is reactivated, the virus can reappear in the form of shingles.

Shingles is an outbreak of a rash or blisters (vesicles with an erythematous base) on the skin that may be associated with severe pain (Fig. 4-23). The pain is associated with the involved nerve root and associated dermatome and generally presents on one side of the body or face in a pattern characteristic for the involved site (Fig. 4-24). Early signs of shingles include burning or shooting pain and tingling or itching. The rash or blisters are present anywhere from 1 to 14 days.

Complications of shingles involving cranial nerves include hearing and vision loss. Postherpetic neuralgia (PHN), a condition in which the pain from shingles persists for months, sometimes years, after the shingles rash has healed, can also occur. PHN can be very debilitating. Early intervention within the first 72 hours of onset with anti-retroviral medications may diminish or eliminate PHN. Early identification and intervention is very important to outcomes.

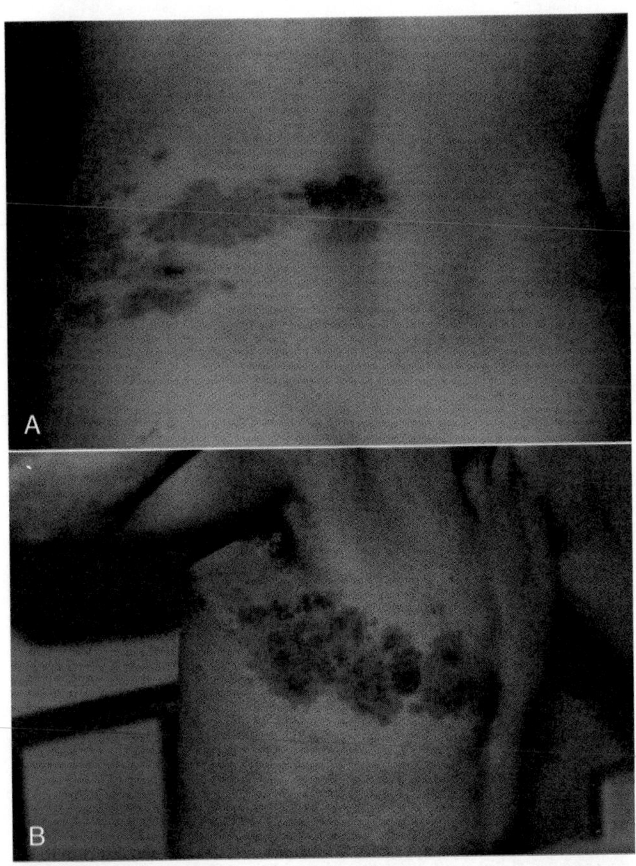

Fig. 4-23 Herpes zoster (shingles). **A,** Lesions appear unilaterally along the path of a spinal nerve. **B,** Eruptions involving the T4 dermatome. (**A** from Callen J, Greer K, Hood H, et al: *Color atlas of dermatology,* Philadelphia, 1993, WB Saunders; **B** from Marx J, Hockberger R, Walls R: *Rosen's emergency medicine: concepts and clinical practice,* ed 6, St. Louis, 2006, Mosby.)

Adults with shingles are infectious to anyone who has not had chickenpox. Anyone who has had chickenpox can develop shingles when immunocompromised. Other risk factors for VZV include age (young or old) and immunocompromise from HIV infection, chemotherapy or radiation treatment, transplants, aging, and stress. It is highly recommended that health care professionals with no immunity to VZV receive the varicella vaccine. Therapists who have never had chickenpox (and especially women of childbearing age who have not had the chickenpox) should be tested for immune status.

Cutaneous Manifestations of Abuse

Signs of child abuse or domestic violence in adults may be seen as skin lesions. Cigarette burns leave a punched out ulceration with dry, purple crusts (see www.dermatlas.org). Splash marks or scald lines from thermal (hot water) burns occur most often on the buttocks and distal extremities.[91] Bruising from squeezing and shaking involving the midportion of the upper arms is a suspicious sign.

Accidental bruising in young children is common; the therapist should watch for nonaccidental bruising found in atypical areas, such as the buttocks, hands, and trunk, or in a child who is not yet biped (up on two feet) and cruising (walking along furniture or holding an object while taking steps). To make an accurate assessment, it is important to differentiate between inflicted cutaneous injuries and mimickers of physical abuse.

For example, infants with bruising may be demonstrating early signs of bleeding disorders.[91] Mongolian spots can also be mistaken for bruising from child abuse (see next section). The therapist is advised to take digital or Polaroid photos of

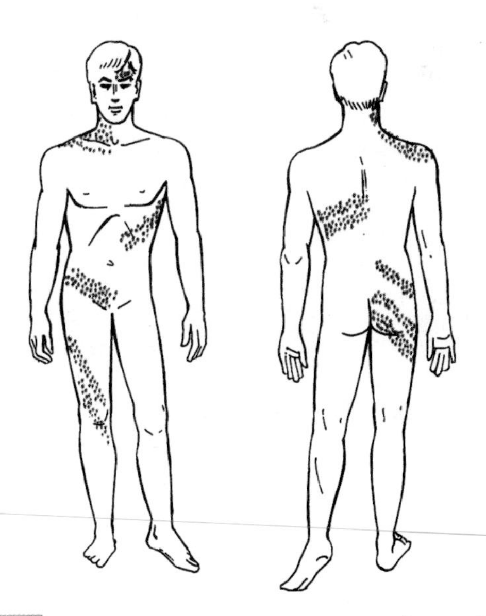

Fig. 4-24 Symptoms of shingles appear on only one side of the body, usually on the torso or face. Most often, the lesions are visible externally. In unusual cases, clients report the same symptoms internally along the dermatome but without a corresponding external skin lesion. (From Malasanos L, Barkauskas V, Stoltenberg-Allen K: *Health assessment,* ed 4, St. Louis, 1990, Mosby.)

any suspicious lesions in children under the age of 18. Document the date and provide a detailed description.

The law requires that professionals report suspected child abuse and neglect to the appropriate authorities. It is not up to the health care professional to determine child abuse has occurred; this is left up to investigating officials. See other guidelines regarding child abuse and domestic violence in Chapter 2. Understanding the reporting guidelines helps direct practitioners in their decision making.[91]

Mongolian Spots. Discoloration of the skin in newborn infants called Mongolian spots (Fig. 4-25) can be mistaken for signs of child abuse. The Mongolian spot is a congenital, developmental condition exclusively involving the skin and is very common in children of Asian, African, Indian, Native American, Eskimo, Polynesian, or Hispanic origins.

These benign pigmentation changes appear as flat dark blue or black areas and come in a variety of sizes, shapes, and colors. The skin changes result from entrapment of melanocytes (skin cells containing melanin, the normal pigment of the skin) during their migration from the neural crest into the epidermis.

Cancer-Related Skin Lesions

When screening for primary skin cancer, keep in mind there are other cancer-related skin lesions to watch out for as well. For example, skin rash can present as an early sign of a paraneoplastic syndrome before other manifestations of cancer or cancer recurrence (Fig. 4-26). See further discussion of paraneoplastic syndromes in Chapter 13.

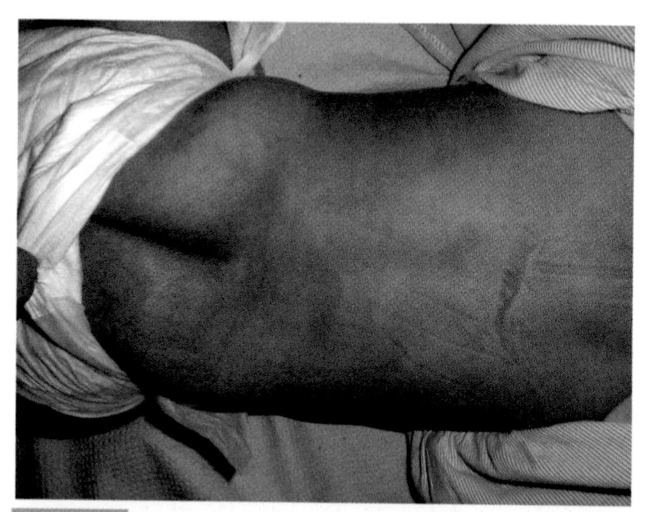

Fig. 4-25 Mongolian spots (congenital dermal melanocytosis). Mongolian spots are common among people of Asian, East Indian, Native American, Inuit, African, and Latino or Hispanic heritage. They are also present in about one in ten fair-skinned infants. Bluish gray to deep brown to black skin markings, they often appear on the base of the spine, on the buttocks and back, and even sometimes on the shoulders, ankles, or wrists. Mongolian spots may cover a large area of the back. When the melanocytes are close to the surface, they look deep brown. The deeper they are in the skin, the more bluish they look, often mistaken for signs of child abuse. These spots "fade" with age as the child grows and usually disappear by age 5. (Courtesy Dr. Dubin Pavel, 2004.)

Pinch purpura, a purplish, brown, or red discoloration of the skin can be mistaken by the therapist for a birthmark or port wine stain (Fig. 4-27). Using the question "How long have you had this?" can help differentiate between something the person has had his or her entire life and a suspicious skin lesion or recent change in the integument.

When purpura causes a raised and palpable skin lesion, it is called *palpable purpura.* The palpable hemorrhages are caused by red blood cells extravasated (escaped) from damaged vessels into the dermis. This type of purpura can be associated with cutaneous vasculitis, pulmonary-renal

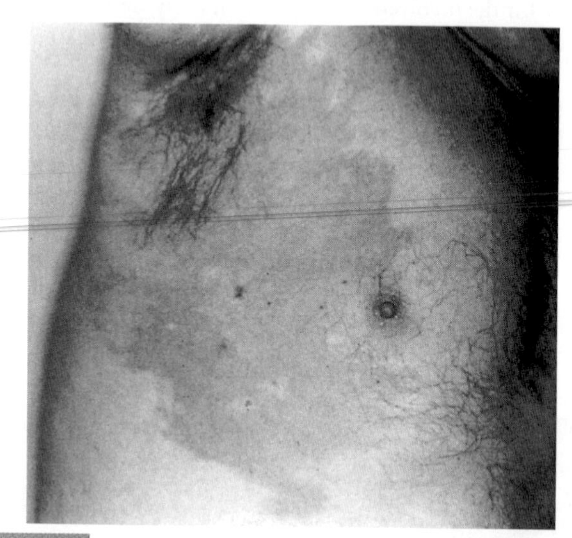

Fig. 4-26 Metastatic carcinoma presenting as a cellulitic skin rash on the anterior chest wall as a result of carcinoma of the lung. This rash can be red, tan, or brown with a flat or raised appearance. When associated with a paraneoplastic syndrome, it may appear far from the site of the primary cancer. (From Callen JP, Jorizzo J, Greer KE, et al: *Dermatological signs of internal disease,* Philadelphia, 1988, WB Saunders.)

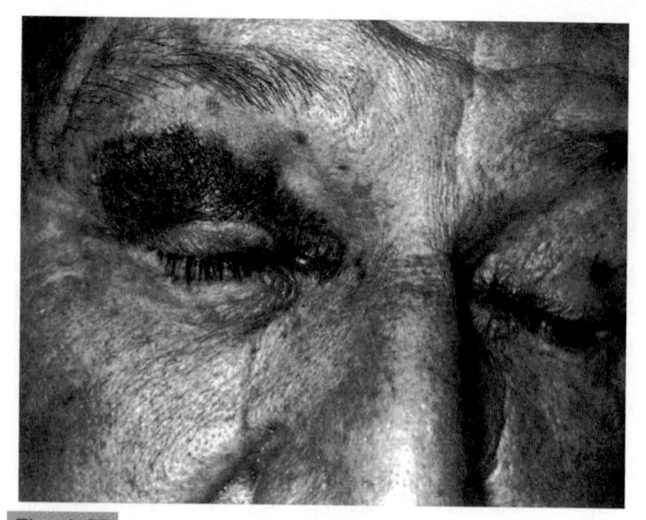

Fig. 4-27 Pinch purpura in an individual with multiple myeloma caused by amyloidosis of the skin. The purpura shown here is a recent skin change for this client. (From Callen JP, Jorizzo J, Greer KE, et al: *Dermatological signs of internal disease,* ed 2, Philadelphia, 1995, WB Saunders.)

syndrome, or drug reaction. The lower extremities are affected most often.

Many older adults assume this is a "normal" sign of aging (and in fact, purpura does occur more often in aging adults; see Fig. 4-14); they do not see a physician when it first appears. Early detection and referral is always the key to a better prognosis. In asking the three important questions, the therapist plays an instrumental part in the cancer screening process.

A client with a past medical history of cancer now presenting with a suspicious skin lesion (Fig. 4-28) that has not been evaluated by the physician must be advised to have this evaluated as soon as possible. We must be aware of how to present this recommendation to the client. There is a need to avoid frightening the client while conveying the importance of early diagnosis of any unusual skin lesions.

Kaposi's Sarcoma

Kaposi's sarcoma is a form of skin cancer common in older Jewish men of Mediterranean descent that presents with a wide range of appearance. A gallery of photos can be seen by doing a Google search of "Kaposi's sarcoma." (Go to www.Google.com and type in the words: Kaposi's sarcoma, then click on the word "Images" on the Google page.) It is not contagious to touch and does not usually cause death or disfigurement.

More recently, Kaposi's sarcoma has presented as an opportunistic disease in adults with HIV/AIDS. With the more successful treatment of AIDS with antiretroviral agents, opportunistic diseases, such as Kaposi's sarcoma, are on the decline.

Even though this skin lesion will not transmit skin cancer or HIV, the therapist is always advised to use standard precautions with anyone who has skin lesions of any type.

Lymphomas

Round patches of reddish-brown skin with hair loss over the area are lymphomas, a type of neoplasm of lymphoid tissue (Fig. 4-29). The most common forms of lymphoma are Hodgkin's disease and non-Hodgkin's lymphoma (NHL).

Typically, the appearance of a painless, enlarged lymph node or skin lesion of this type is followed by weakness, fever, and weight loss. A history of chronic immunosuppression (e.g., antirejection drugs for organ transplants, chronic use of immunosuppressant drugs for inflammatory or autoimmune diseases, cancer treatment) in the presence of this clinical presentation is a major red flag.

NAIL BED ASSESSMENT

As with assessment of the skin, nail beds (fingers and toes) should be evaluated for color, shape, thickness, texture, and the presence of lesions (Box 4-15). Systemic changes affect both fingernails and toenails, but the signs are typically more prominent in the faster-growing fingernails.[92]

The normal nail consists of three parts: the nail bed, the nail plate, and the cuticle (Fig. 4-30). The nail bed is highly vascularized and gives the nail its pink color. The hard nail is formed at the proximal end (the matrix). About one-fourth of the nail is covered by skin known as the proximal nail fold. The cuticle seals and protects the space between the proximal fold and the nail plate.[92]

Many individual variations in color, texture, and grooming of the nails are influenced by factors unrelated to disease, such as occupation, chronic use of nail polish or acrylics, or exposure to chemical dyes and detergents. Longitudinal lines of darker color (pigment) may be seen in the normal nails of clients with darker skin.

In assessing the older adult, minor variations associated with the aging process may be observed (e.g., gradual thickening of the nail plate, appearance of longitudinal ridges, yellowish-gray discoloration).

In the normal individual, pressing or blanching the nail bed of a finger or toe produces a whitening effect; when pressure is released, a return of color should occur within 3 seconds. If the capillary refill time exceeds 3 seconds, the lack of circulation may be due to arterial insufficiency from atherosclerosis or spasm.

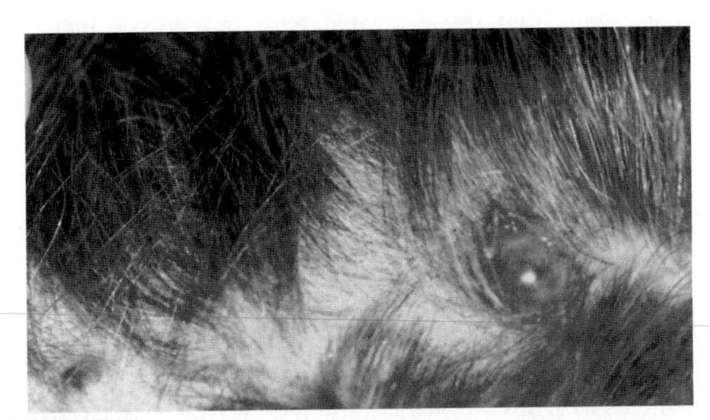

Fig. 4-28 Metastatic renal carcinoma presenting as a nodule in the scalp. Observing any skin lesions no matter what part of the body the therapist is examining must be followed by the three assessment questions listed in the text. (From Callen JP, Jorizzo J, Greer KE, et al: *Dermatological signs of internal disease,* Philadelphia, 1988, WB Saunders.)

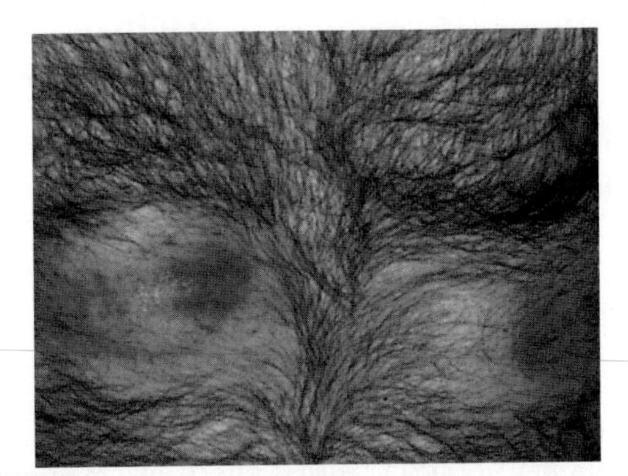

Fig. 4-29 Lymphomas seen here just below the nipples on the chest of an adult male arise in individuals who are chronically immunosuppressed for any reason. (From Conant MA: The link between HIV and skin malignancies, *The Skin Cancer Foundation Journal,* Vol XII, 1994.)

BOX 4-15 HAND AND NAIL BED ASSESSMENT

Observe the Hands for:
- Palmar erythema (see Fig. 9-5)
- Tremor (e.g., liver flap or asterixis; see Fig. 9-7)
- Pallor of palmar creases (anemia, gastrointestinal [GI] malabsorption)
- Palmar xanthomas (lipid deposits on palms of hands; hyperlipidemia, diabetes)
- Turgor (lift skin on back of hands; hydration status; see Fig 4-4)
- Edema

Observe The Fingers and Toenails For:
- Color (capillary refill time, nails of Terry: see Fig. 9-6)
- Shape and curvature
- Clubbing:
  - **C**rohn's or **C**ardiac/cyanosis
  - **L**ung (cancer, hypoxia, cystic fibrosis)
  - **U**lcerative colitis
  - **B**iliary cirrhosis
  - Present at **b**irth (harmless)
  - **N**eoplasm
  - **G**I involvement
- Nicotine stains
- Splinter hemorrhages (see Fig. 4-34)
- Leukonychia (whitening of nail plate with bands, lines, or white spots; inherited or acquired from malnutrition from eating disorders, alcoholism, or cancer treatment; myocardial infarction [MI], renal failure, poison, anxiety)
- Koilonychia ("spoon nails"; see Fig. 4-32); congenital or hereditary, iron-deficiency anemia, thyroid problem, syphilis, rheumatic fever)
- Beau's lines (see Fig. 4-33); decreased production of the nail by the matrix caused by acute illness or systemic insult such as chemotherapy for cancer; recent MI, chronic alcohol abuse, or eating disorders. This can also occur in isolated nail beds from local trauma
- Adhesion to the nail bed. Look for onycholysis (loosening of nail plate from distal edge inward; Graves' disease, psoriasis, reactive arthritis, obsessive compulsive behavior: "nail pickers")
- Pitting (psoriasis, eczema, alopecia areata)
- Thinning/thickening

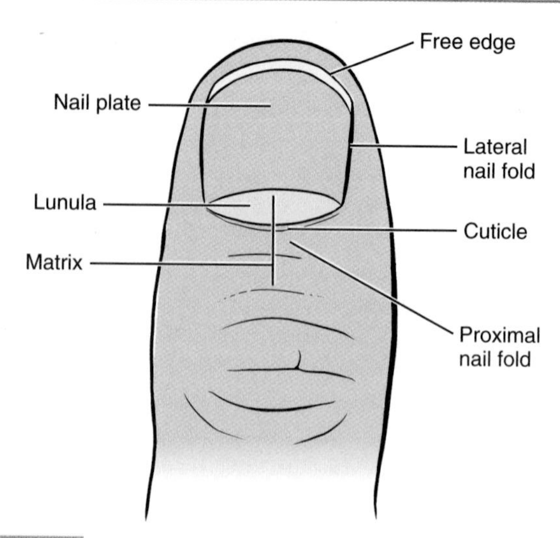

Fig. 4-30 Normal nail structure. The nail matrix forms the nail plate and begins about 5 mm to 8 mm beneath the proximal nail fold and extends distally to the edge of the lunula, where the nail bed begins. The lunula (half-moon) is the exposed part of the nail matrix, distal to the proximal nail fold; it is not always visible.

Nail Bed Changes

Some of the more common nail bed changes seen in a physical therapy practice are included in this chapter. With any nail or skin condition, ask if the nails have always been like this or if any changes have occurred in the last 6 weeks to 6 months. Referral may not be needed if the physician is aware of the new onset of nail bed changes. Ask about the presence of other signs and symptoms consistent with any of the conditions listed here that can cause any of these nail bed changes.

Again, as with visual inspection of the skin, this section of the text is only a cursory look at the most common nail bed changes. Many more are not included here. A well-rounded library should include at least one text with color plates and photos of various nail bed changes.[93-97] This is not to help the therapist diagnose a medical problem but rather to provide background information, which can be used in the referral decision-making process.

Onycholysis

Onycholysis, a painless loosening of the nail plate occurs from the distal edge inward (Fig. 4-31). Fingers and toes may both be affected as a consequence of dermatologic conditions such as dermatitis, fungal disease, lichen planus, and psoriasis. Systemic diseases associated with onycholysis include myeloma, neoplasia, Graves' disease, anemia, and reactive arthritis.[98]

Medications, such as tetracycline, fluoroquinolones, anticancer drugs, nonsteroidal antiinflammatories, psoralens, retinoids, zidovudine, and quinine, can cause photo-onycholysis (toes must be exposed to the sun for the condition to occur).[98]

Local causes from chemical, physical, cosmetic, or traumatic sources can bring on this condition. In the case of trauma, a limited number of nails are affected. For example, in clients with onycholysis as a result of nervous or obsessive-compulsive behaviors, only one or two nails are targeted. The individual picks around the edges until the nail is raised and separated from the nail bed. When there is an underlying systemic disorder, it is more common to see all the nail plates affected.

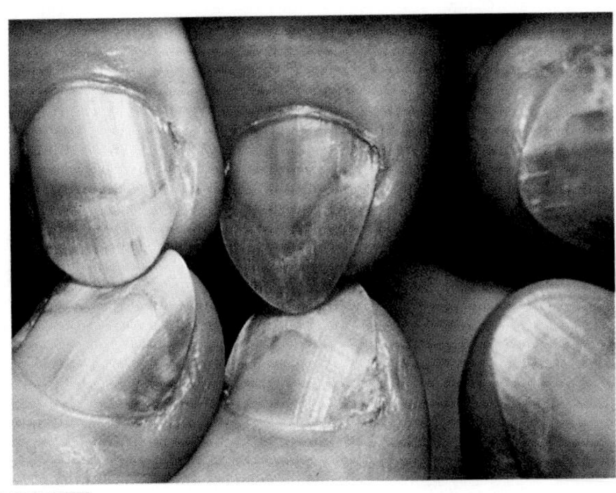

Fig. 4-31 Onycholysis. Loosening of the nail plate, usually from the tip of the nail, progressing inward and from the edge of the nail moving inward. Possible causes include Graves' disease, psoriasis, reactive arthritis, and obsessive-compulsive behaviors (nail pickers). (From Arndt KA, Wintroub BU, Robinson JK, et al: *Primary care dermatology,* Philadelphia, 1997, WB Saunders.)

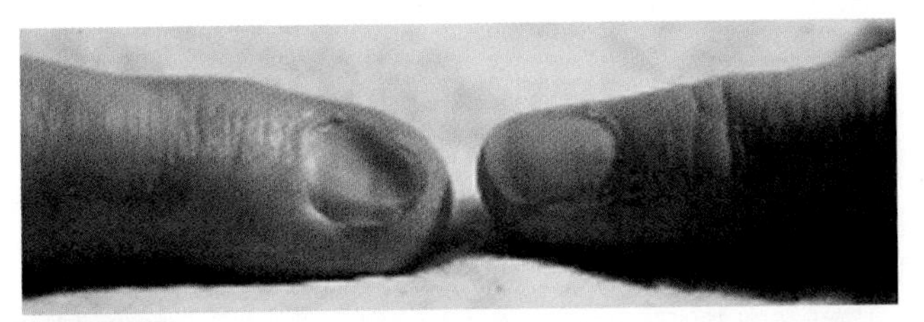

Fig. 4-32 Koilonychia (spoon nails). In this side-by-side view, the affected nail bed is on the left and the normal nail on the right. With a spoon nail, the rounded indentation would hold a drop or several drops of water, hence the name. (From Swartz MH: *Textbook of physical diagnosis,* ed 5, Philadelphia, 2006, WB Saunders.)

Koilonychia

Koilonychia or "spoon nails" may be a congenital or hereditary trait and as such is considered "normal" for that individual. These are thin, depressed nails with lateral edges tilted upward, forming a concave profile (Fig. 4-32).

Koilonychia can occur as a result of hypochromic anemia, iron deficiency (with or without anemia), poorly controlled diabetes of more than 15 years duration, chemical irritants, local injury, developmental abnormality, or psoriasis. It can also be an outward sign of thyroid problems, syphilis, and rheumatic fever.

Beau's lines

Beau's lines are transverse grooves or ridges across the nail plate as a result of a decreased or interrupted production of the nail by the matrix (Fig. 4-33). The cause is usually an acute illness or systemic insult such as chemotherapy for cancer. Other common conditions associated with Beau's lines are poor peripheral circulation, eating disorders, cirrhosis associated with chronic alcohol use, and recent myocardial infarction (MI).

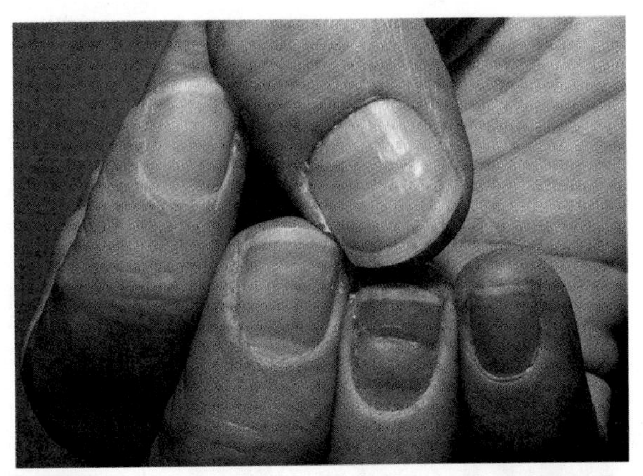

Fig. 4-33 Beau's lines or grooves across the nail plate. A depression across the nail extends down to the nail bed. This occurs with shock, illness, malnutrition, or trauma severe enough to impair nail formation such as acute illness, prolonged fever, or chemotherapy. A dent appears first at the cuticle and moves forward as the nail grows. All nails can be involved, but with local trauma, only the involved nail will be affected. This photo shows a client postinsult after full recovery. At the time of the illness, nail loss is obvious often with change in nail bed color such as occurs with chemotherapy. (From Callen JP, Greer KE, Hood AF, et al: *Color atlas of dermatology,* Philadelphia, 1994, WB Saunders.)

Since the nails grow at an approximate rate of 3 mm/month, the date of the initial onset of illness or disease can be estimated by the location of the line. The dent appears first at the cuticle and moves forward as the nail grows. Measure the distance (in millimeters) from the dent to the cuticle and add 3 to account for the distance from the cuticle to the matrix. This is the number of weeks ago the person first had the problem.

Beau's lines are temporary until the impaired nail formation is corrected (if and when the individual returns to normal health). These lines can also occur as a result of local trauma to the hand or fingers.

In the case of an injury, the dent may be permanent. Hand therapists see this condition most often. If it is not the result of a recent injury, the client may be able to remember sustaining an injury years ago.

Splinter Hemorrhages

Splinter hemorrhages may be the sign of a silent MI or the client may have a known history of MI. These red-brown linear streaks (Fig. 4-34) can also signal other systemic conditions such as bacterial endocarditis, vasculitis, or renal failure.

In a hospital setting, they are not uncommon in the cardiac care unit (CCU) or other ICU. In such a case, the therapist may just take note of the nail bed changes and correlate it with the pathologic insult probably already a part of the medical record.

When present in only one or two nail beds, local trauma may be linked to the nail bed changes. Asking the client about recent trauma or injury to the hand or fingers may bring this to light.

Whenever splinter hemorrhages are observed in the nails, visually inspect both hands and the toenails as well. If the

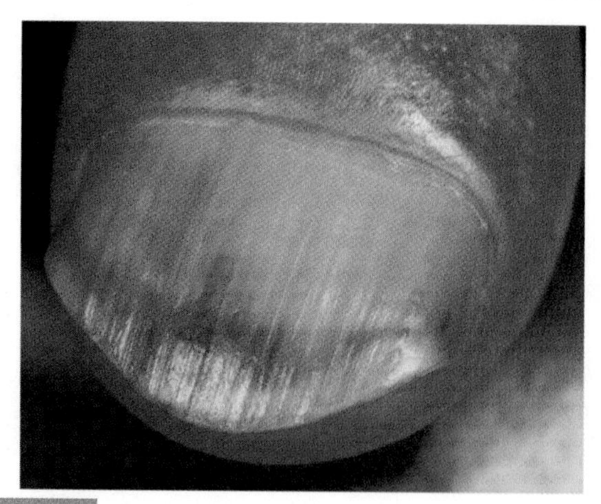

Fig. 4-34 Splinter hemorrhages. These red-brown streaks, embolic lesions, occur with subacute bacterial endocarditis, sepsis, rheumatoid arthritis, vitamin C deficiency, or hematologic neoplasm. They can occur from local trauma in which case only the injured nail beds will have the telltale streak. Splinter hemorrhages also may be a nonspecific sign. (From Hordinsky MK, Sawaya ME, Scher RK: *Atlas of hair and nails,* Philadelphia, 2000, Churchill Livingstone.)

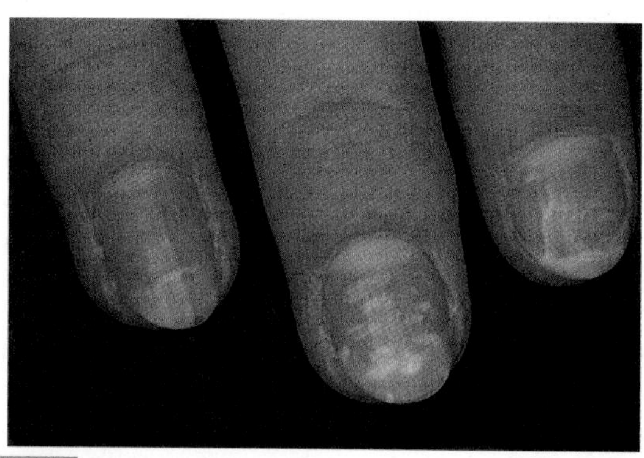

Fig. 4-35 Leukonychia, acquired or inherited white discoloration in the nail. There is a wide range of possibilities in the clinical presentation of leukonychia. Spots, vertical lines, horizontal lines, and even full nail bed changes can occur on individual nail beds of the fingers and/or toes. One or more nails may be affected. (From Jarvis C: *Physical examination and health assessment,* ed 5, Philadelphia, 2008, WB Saunders.)

client cannot recall any recent illness, look for a possible cardiac history or cardiac risk factors. In the case of cardiac risk factors with no known cardiac history, proper medical follow-up and diagnosis is essential in the event the client has had a silent MI.

Leukonychia

Leukonychia, or white nail syndrome, is characterized by dots or lines of white that progress to the free edge of the nail as the nail grows (Fig. 4-35). White nails can be congenital, but more often, they are acquired in association with hypocalcemia, severe hypochromic anemia, Hodgkin's disease, renal failure, malnutrition from eating disorders, MI, leprosy, hepatic cirrhosis, and arsenic poisoning.

Acquired leukonychia is caused by a disturbance to the nail matrix. Repeated trauma, such as keyboard punching, is a more recently described acquired cause of this condition.[99] When the entire nail plate is white, the condition is called *leukonychia totalis* (Case Example 4-2).

Paronychia

Paronychia (not shown) is an infection of the fold of skin at the margin of a nail. There is an obvious red, swollen site of inflammation that is tender or painful. This may be acute as with a bacterial infection or chronic in association with an occupationally induced fungal infection referred to as "wet work" from having the hands submerged in water for long periods of time.

The client may also give a history of finger exposure to chemical irritants, acrylic nails or nail glue, or sculpted nails.

CASE EXAMPLE 4-2

Leukonychia

A 24-year-old male (Caucasian) was seen in physical therapy for a work-related back injury. He was asked the final interview questions:

- Are there any other symptoms of any kind anywhere else in your body?
- Is there anything else about your condition that we have not discussed yet?

The client showed the therapist his nails and asked what could be the cause of the white discoloration in all the nail beds.

He reported the nails seem to grow out from time to time. There was tenderness along the sides of the nails and at the distal edge of the nails. It was obvious the nails were bitten and there were several nails that were red and swollen. The client admitted to picking at his nails when he was nervous. He was observed tapping his nails on the table repeatedly during the exam. No changes of any kind were observed in the feet.

Past medical history was negative for any significant health problems. He was not taking any medications, over-the-counter drugs, or using recreational drugs. He did not smoke and denied the use of alcohol. His job as a supervisor in a machine shop did not require the mechanical use of his hands. He was not exposed to any unusual chemicals or solvents at work.

Result: When asked, "How long have you had this?" the client reported for 2 years. When asked, "Have your nails changed in the last 6 weeks to 6 months?" the answer was, "Yes, the condition seems to come and go." When asked, "Has your doctor seen these changes?" the client did not think so.

The therapist did not know what was causing the nail bed changes and suggested the client ask his physician about the condition at his next appointment. The physician also observed the client repeatedly tapping his nails and performed a screening exam for anxiety.

The nail bed condition was diagnosed as leukonychia from repeated microtrauma to the nail matrix. The patient was referred to psychiatry to manage the observed anxiety symptoms. The nails returned to normal in about 3 months (90 to 100 days) after the client stopped tapping, restoring normal growth to the nail matrix.

From Maino K, Stashower ME: Traumatic transverse leukonychia, *SKINmed* 3(1):53-55, 2004. Accessed on-line http://www.medscape.com/viewarticle/467074 (posted 01/20/2004).

Paronychia of one or more fingers is not uncommon in people who pick, bite, or suck their nails. Health care professionals with these nervous habits working in a clinical setting (especially hospitals) are at increased risk for paronychia from infection with bacteria such as *Streptococcus* or *Staphylococcus.* Green coloration of the nail may indicate *Pseudomonas* infection.

Paronychia infections may spread to the pulp space of the finger, developing a painful felon (an infection with localized abscess). Untreated infection can spread to the deep spaces of the hand and beyond.

It is especially important to recognize any nail bed irregularity because it may be a clue to malignancy. Likewise, anyone with diabetes mellitus, immunocompromise, or history of steroids and retroviral use are at increased risk for paronychia formation. Early identification and medical referral are imperative to avoid more serious consequences.

Clubbing

Clubbing of the fingers (Fig. 4-36) and toes usually results from chronic oxygen deprivation in these tissue beds. It is most often observed in clients with advanced chronic obstructive pulmonary disease, congenital heart defects, and cor pulmonale but can occur within 10 days in someone with an acute systemic condition such as a pulmonary abscess, malignancy, or polycythemia. Clubbing may be the first sign of a paraneoplastic syndrome associated with cancer. Clubbing can be assessed by the Schamroth method (Fig. 4-37).

Any positive findings in the nail beds should be viewed in light of the entire clinical presentation. For example, a positive Schamroth test without observable clinical changes in skin color, capillary refill time, or shape of the fingertips may not signify systemic disease but rather a normal anatomic variation of nail curvature.

Nail Patella Syndrome

Nail patella syndrome (NPS), also called Fong's disease, hereditary onycho-osteodysplasia (HOOD), or Turner-Kieser syndrome, is a genetic disorder characterized by an absence or underdevelopment of nail bed changes as shown here (Fig. 4-38, *A*). Lack of skin creases is also a telltale sign (Fig. 4-39).

Nail abnormalities vary and range from a sliver on each corner of the nail bed to a full nail that is very thick with splits. Some people have brittle, underdeveloped, cracked, or ridged nails while others are absent entirely. They are often concave, causing them to split and flip up, catching on clothing and bedding. Often, the lunula (the light crescent "half-moons" of the nail near the cuticle) are pointed or triangular-shaped (Fig. 4-38, *B*).

The therapist may be the first to see this condition because skeletal and joint problems are a common feature with this condition. The elbows, hips, and knees are affected most often. Absence or hypoplasia (underdevelopment) of the patella and deformities of the knee joint itself often give them a square shape. Knee instability with patellar dislocation is not uncommon due to malformations in the bones, muscles, and ligaments; there is often much instability in the knee joint.

The client may also develop scoliosis, glaucoma, and kidney disease. Medical referral to establish a diagnosis is important, as clients with NPS need annual screening for renal disease, biannual screening for glaucoma in adulthood, and magnetic resonance imaging (MRI) for orthopedic abnormalities before physical therapy is considered.[100]

When a client presents with skin or nail bed changes of any kind, taking a personal medical history and reviewing

Fig. 4-36 Rapid development of digital clubbing (fingers as shown on the left or toes [not shown]) over the course of a 10-day to 2-week period requires immediate medical evaluation. Clubbing can be assessed using the Schamroth method shown in Fig. 4-37. (From Swartz MH: *Textbook of physical diagnosis,* ed 6, Philadelphia, 2009, WB Saunders.)

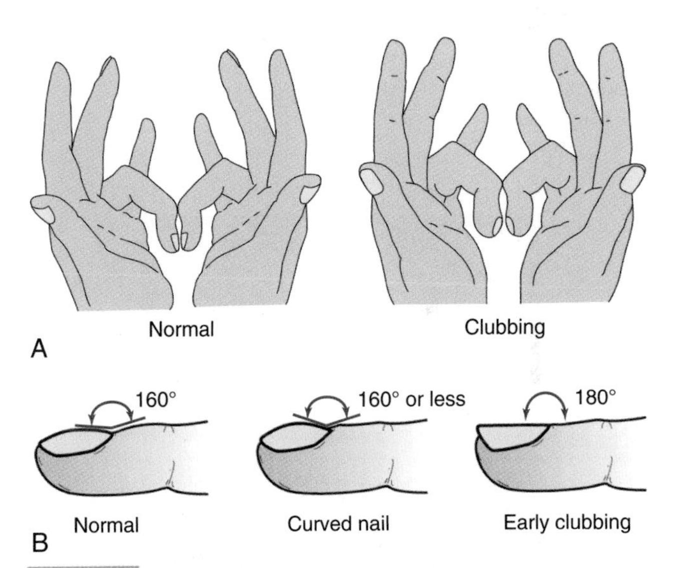

Fig. 4-37 Schamroth method. **A,** Assessment of clubbing by the Schamroth method. The client places the fingernails of opposite fingers together and holds them up to a light. If the examiner can see a diamond shape between the nails, there is no clubbing. Clubbing is identified by the absence of the diamond shape. It occurs first in the thumb and index finger. **B,** The index finger is viewed at its profile, and the angle of the nail base is noted (it should be about 160 degrees). The nail base is firm to palpation. Curved nails are a variation of normal with a convex profile. They may look like clubbed nails, but the angle between the nail base and the nail is normal (i.e., 160 degrees or less). Clubbing of nails occurs with congenital chronic cyanotic heart disease, emphysema, cystic fibrosis, and chronic bronchitis. In early clubbing the angle straightens out to 180 degrees, and the nail base feels spongy to palpation. (**A** from Ignatavicius DD, Bayne MV: Assessment of the cardiovascular system. In Ignatavicius DD, Bayne MV, editors: *Medical-surgical nursing,* Philadelphia, 1993, WB Saunders. **B** from Jarvis C: *Physical examination and health assessment,* Philadelphia, 2004, WB Saunders.)

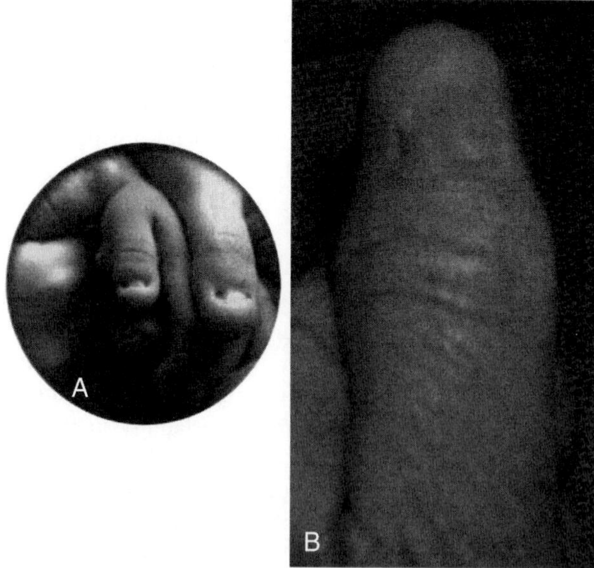

Fig. 4-38 Nail patella syndrome (NPS). Nail bed changes associated with NPS are presented here. Effects vary greatly between individuals but usually involve absence of part or all of the nail bed. For more photos of the skeletal changes associated with NPS, see http://members.aol.com/pacali/NPSpage.html (Accessed October 25, 2005). **A,** The thumbnails are affected the most with involvement decreasing toward the little finger. Nail changes occur more severely on the ulnar side of the affected nail. Toenails can be affected; it is often the little toenail that is involved.[100] **B,** Note the altered nail bed (more on the ulnar side) and especially the triangular-shaped lunula, a very distinctive sign of NPS. (Courtesy Nail Patella Syndrome Worldwide, 2005.)

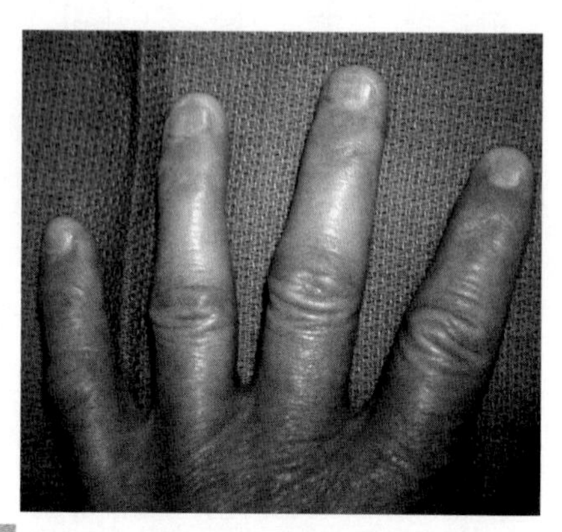

Fig. 4-39 Nail changes are the most constant feature of NPS and may be absent, underdeveloped (hypoplastic), or abnormal in size or shape (dysplastic). Alternately, there may be longitudinal ridges, pitting, or discoloration. Note the lack of creases in the distal interphalangeal (DIP) joints of the fingers in this photo. The condition may be present at birth and usually presents symmetrically and bilaterally. (Courtesy Gary Ross, RN, BSN, Board Member, Nail Patella Syndrome Worldwide, 2005.)

recent (last 6 weeks) and current medications can provide the therapist with important clues in knowing whether to make an immediate medical referral. Table 4-8 provides a summary of the most common skin changes encountered in a physical therapy practice and possible causes for each one.

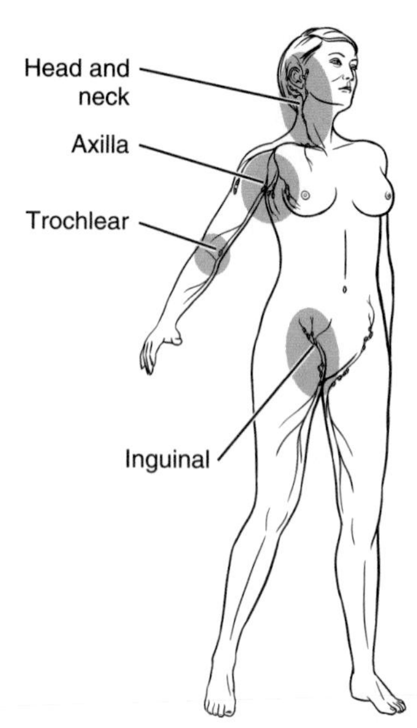

Fig. 4-40 Locations of most easily palpable lymph nodes. Epitrochlear nodes are located on the medial side of the arm above the elbow in the depression above and posterior to the medial condyle of the humerus. Horizontal and vertical chains of inguinal nodes may be palpated in the same areas as the femoral pulse. Popliteal lymph nodes (not shown) are deep but may be palpated in some clients with the knee slightly flexed.

LYMPH NODE PALPATION

Part of the screening process for the therapist may involve visual inspection of the skin overlying lymph nodes and palpation of the lymph nodes. Look for any obvious areas of swelling or redness (erythema) along with any changes in skin color or pigmentation. Ask about the presence or recent history of sores or lesions anywhere on the body.

Keep in mind the therapist cannot know what the underlying pathology may be when lymph nodes are palpable and questionable. Performing a baseline assessment and reporting the findings is the important outcome of the assessment.

Whenever examining a lump or lesion, use the mnemonic in Box 4-11 to document and report findings on location, size, shape, consistency, mobility or fixation, and signs of tenderness (see also Box 4-18). Review Special Questions to Ask: Lymph Nodes at the end of this chapter for appropriate follow-up questions.

There are several sites where lymph nodes are potentially observable and palpable (Fig. 4-40). Palpation must be done lightly. Excessive pressure can press a node into a muscle or between two muscles. "Normal" lymph nodes usually are not visible or easily palpable. Not all visible or palpable lymph nodes are a sign of cancer. Infections, viruses, bacteria, allergies, thyroid conditions, and food intolerances can cause changes in the lymph nodes.

People with seasonal allergies or allergic rhinitis often have enlarged, tender, and easily palpable lymph nodes in the submandibular and supraclavicular areas. This is a sign that the

TABLE 4-8 Common Causes of Skin and Nail Bed Changes

Skin/Nail Bed Changes	Possible Cause
SKIN CHANGES	
Dermatitis	Pulmonary malignancy, allergic reaction
Loss of turgor or elasticity	Dehydration
Rash (see Fig. 4-12)	Viruses (chicken pox, measles, Fifth disease)
	Systemic conditions: meningitis, lupus, hives, rosacea)
	Sexually transmitted diseases
	Lyme disease
	Parasites (e.g., lice, scabies)
	Reaction to chemicals, medications, food
	Malignancy, neoplastic syndromes
Hemorrhage (petechiae, ecchymosis, purpura) (see Fig. 4-14)	NSAIDs
	Anticoagulants (heparin, Coumadin/warfarin, aspirin)
	Hemophilia
	Thrombocytopenia (low platelet level) and anything that can cause thrombocytopenia
	Neoplasm; paraneoplastic syndrome
	Domestic violence
	Aging
Skin color	Jaundice (yellow, green, orange): hepatitis
	Chronic renal failure (yellow-brown)
	Cyanosis (pale, blue): anxiety, hypothermia, lung disease, congestive heart disease, venous obstruction
	Rubor (dusky red): arterial insufficiency
	Sunburn (red): radiation recall or radiation dermatitis
	Tan, black, blue: skin cancer
Hyperpigmentation (see Fig. 4-11)	Addison's disease, ACTH-producing tumors
	Sarcoidosis
	Pregnancy
	Leukemia
	Hemochromatosis
	Celiac sprue (malabsorption)
	Scleroderma
	Chronic renal failure
	Hereditary (nonpathognomonic)
	Low dose radiation
Café-au-lait (hyperpigmentation; see Fig. 4-11)	Neurofibromatosis (more than 5 lesions)
	Albright's syndrome
	Urticaria pigmentosa (less than 5 lesions)
Hypopigmentation (vitiligo; see Fig. 4-10)	Albinism
	Sun exposure
	Steroid injections
	Hyperthyroidism
	Stomach cancer
	Pernicious anemia
	Diabetes mellitus
	Autoimmune diseases
	High dose radiation
Xanthomas (see Fig. 4-16)	Disorders of lipid metabolism
	Primary biliary cirrhosis
	Diabetes mellitus (uncontrolled)
Mongolian spots (see Fig. 4-25)	Blue-black discoloration: normal in certain groups
NAIL BED CHANGES	
Onycholysis (see Fig. 4-31)	Graves' disease, psoriasis, reactive arthritis, persistent or chronic nail picking
Beau's lines (see Fig. 4-33)	Acute systemic illness
	Chemotherapy
	PVD
	Eating disorders
	Cirrhosis (chronic alcohol use)
	Recent heart attack
	Local trauma

Continued

TABLE 4-8 Common Causes of Skin and Nail Bed Changes—cont'd

Skin/Nail Bed Changes	Possible Cause
Koilonychia (see Fig. 4-32)	Congenital or hereditary Hypochromic anemia Iron deficiency Diabetes mellitus (chronic, uncontrolled) Psoriasis Syphilis Rheumatic fever Thyroid dysfunction
Splinter hemorrhages (see Fig. 4-34)	Heart attack Bacterial endocarditis Vasculitis Renal failure Any systemic insult
Leukonychia (see Fig. 4-35)	Acquired or congenital Acquired: hypocalcemia, hypochromic anemia, Hodgkin's disease, renal failure, malnutrition, heart attack, hepatic cirrhosis, arsenic poisoning
Paronychia	Fungal infection Bacterial infection
Digital clubbing (see Fig. 4-36)	*Acute*: pulmonary abscess, malignancy, polycythemia, paraneoplastic syndrome *Chronic*: COPD, cystic fibrosis, congenital heart defects, cor pulmonale
Absent or underdeveloped nail bed (s) (see Figs. 4-38 and 4-39)	NPS Congenital
Pitting	Psoriasis

NSAIDs, Nonsteroidal antiinflammatory drugs; *ACTH,* Adrenocorticotropic hormone; *PVD,* peripheral vascular disease; *COPD,* chronic obstructive pulmonary disease; *NPS,* nail patella syndrome.

immune system is working hard to stop as many perceived pathogens as possible.

Children often have easily palpable and tender lymph nodes because their developing immune system is continuously filtering out pathogens. Anyone with food intolerances or celiac sprue can have the same lymph node response in the inguinal area.

Studies show lymph node changes occur after total hip arthroplasty. The presence of polyethylene or metal debris in lymph nodes of the ipsilateral side have been demonstrated.[101]

The therapist is most likely to palpate enlarged lymph nodes in the neck, supraclavicular, and axillary areas during an upper quadrant examination (Fig. 4-41). Virchow's node, a palpable enlargement of one of the supraclavicular lymph nodes may be palpated in the supraclavicular area in the presence of primary carcinoma of thoracic or abdominal organs. Virchow's node is more often found on the left side.

Posterior cervical lymph node enlargement can occur during the icteric stage of hepatitis (see Table 9-3). Swelling of the regional lymph nodes often accompanies the first stage of syphilis that is usually painless. These glands feel rubbery, freely movable, and are not tender on palpation. This may be followed by a general lymphadenopathy palpable in the posterior cervical or epitrochlear nodes (located in the inner condyle of the humerus).

The axillary lymph nodes are divided into three zones based on regional orientation[102] (Fig. 4-42). Only zones I and II are palpable. Zone I nodes are superficial and palpable with the client sitting (preferred) or supine with the client's arm supported by the examiner's hand and forearm. Gently palpate the entire axilla for any lymph nodes.

Zone II lymph nodes are palpated with the client in sitting position. Zone II lymph nodes are below the clavicle in the area of the pectoralis muscle. The examiner must reach deep into the axilla to palpate for these lymph nodes.

To examine the right axilla, the examiner supports the client's right arm with his or her own right forearm and hand (reverse for palpation of the left axillary lymph nodes) (Fig. 4-43). This will help ensure relaxation of the chest wall musculature. Lift the client's upper arm up from under the elbow while reaching the fingertips of the left hand up as high as possible into the axilla.

The examiner's fingertips will be against the chest wall and should be able to feel the rib cage (and any palpable lymph nodes). As the client's arm is lowered slowly, allow the fingertips to move down over the rib cage. Feel for the central nodes by compressing them against the chest wall and muscles of the axilla. The examiner may want to repeat this motion a second or third time until becoming more proficient with this examination technique.

Whenever the therapist encounters enlarged or palpable lymph nodes, ask about a past medical history of cancer (Case Example 4-3), implants, mononucleosis, chronic fatigue, allergic rhinitis, and food intolerances. Ask about a recent cut or infection in the hand or arm. Any tender, moveable nodes present may be associated with these conditions, pharyngeal or dental infections, and other infectious diseases.

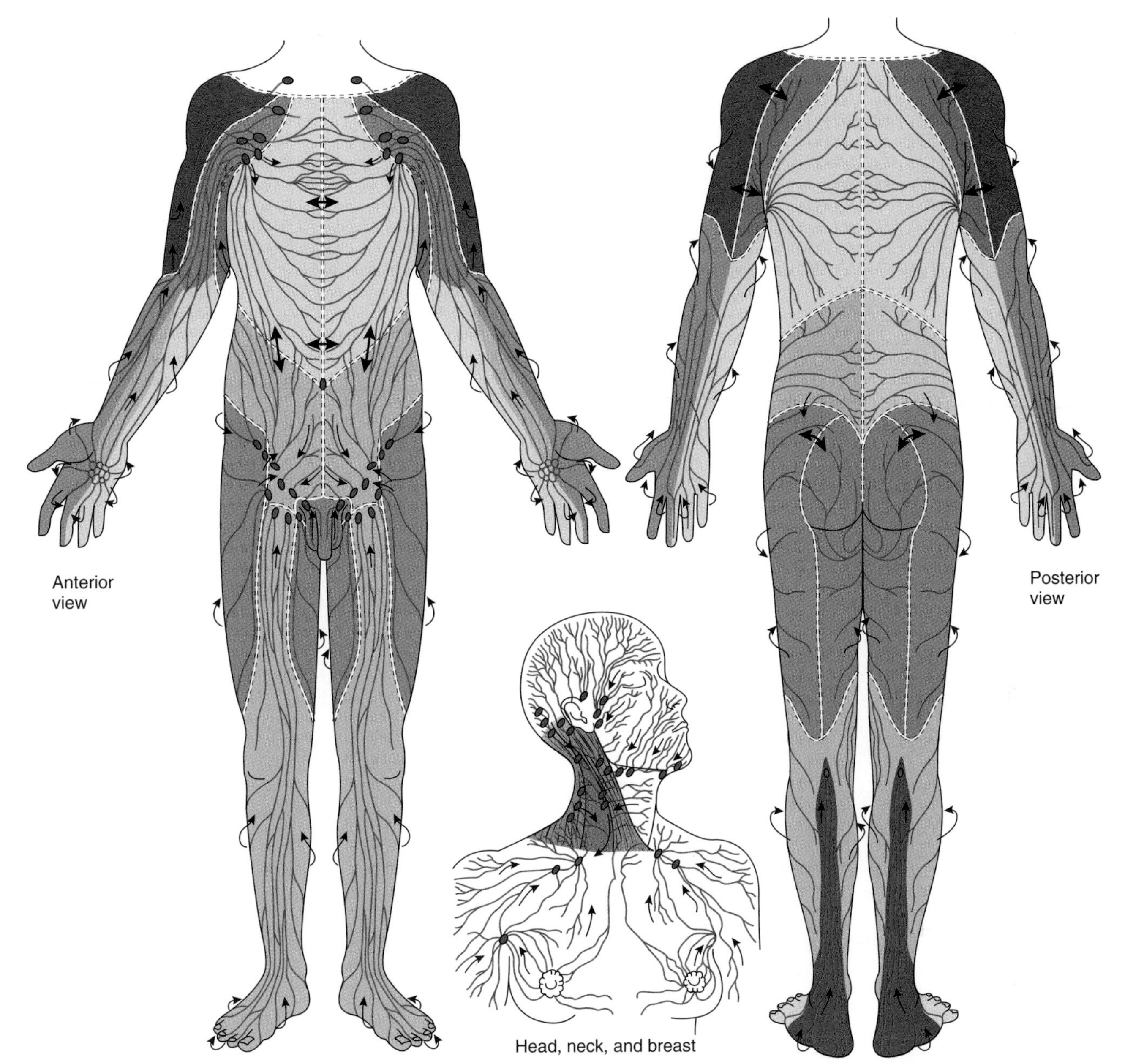

Fig. 4-41 Regional lymphatic system. **A,** Superficial and deep collecting channels and their lymph node chains. Lymph territories are indicated by different shadings. Territories are separated by watersheds (marked by = = =). Normal drainage is away from the watershed. **B,** Lymph nodes of the head, neck, and chest and the direction of their drainage. The head and neck areas are divided into the anterior and posterior triangles divided by the sternocleidomastoid muscle. There are an estimated 75 lymph nodes on each side of the neck. The deep cervical chain is located in the anterior triangle, largely obstructed by the overlying sternocleidomastoid muscle. (From Casley-Smith JR: *Modern treatment for lymphoedema,* ed 5, Adelaide, Australia, 1997, Lymphoedema Association of Australia. Modified from Földi M, Kubik S: *Lehrbuch der lymphologie fur mediziner und physiotherapeuter mit anhang: praktische linweise fur die physiotherape,* Stuttgart, Germany, 1989, Gustav Fischer Verlag.)

Lymph nodes that are hard, immovable, and nontender raise the suspicion of cancer, especially in the presence of a previous history of cancer. Previous editions of this textbook mentioned that any change in lymph nodes present for more than one month in more than one location was a red flag. This has changed with the increased understanding of cancer metastases via the lymphatic system and the potential for cancer recurrence. The most up-to-date recommendation is for physician evaluation of all suspicious lymph nodes.

MUSCULOSKELETAL SCREENING EXAMINATION

Muscle pain, weakness, poor coordination, and joint pain can be caused by many systemic disorders such as hypokalemia, hypothyroidism, dehydration, alcohol or drug use, vascular disorders, GI disorders, liver impairment, malnutrition, vitamin deficiencies, and psychologic factors.

In a screening examination of the musculoskeletal system, the client is observed for any obvious deformities,

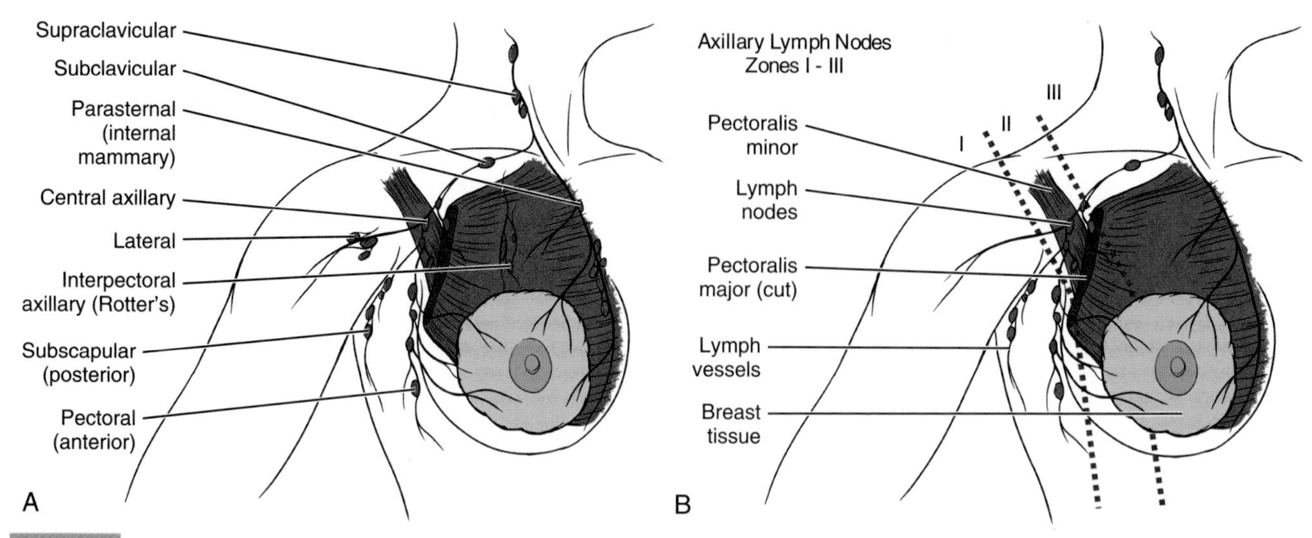

Fig. 4-42 **A,** The breast has an extensive network of venous and lymphatic drainage. Most of the lymphatic drainage empties into axillary nodes. The main lymph node chains and lymphatic drainage are labeled as shown here. **B,** Axillary lymph nodes are divided into three zones based on anatomic sites. Level I axillary lymph nodes are defined as those nodes lying lateral to the lateral border of the pectoralis minor muscle. Level II nodes are under the pectoralis minor muscle, between its lateral and medial borders. Level III nodes lie medial to the clavipectoral fascia, which invests the pectoralis minor muscle and covers the axillary nodes. Only Zones I and II are palpable. Level III nodes are accessed only through penetration of the investing layer, usually during axillary surgery. (From Townsend CM: *Sabiston textbook of surgery*, ed 18, Philadelphia, 2008, Saunders.)

Fig. 4-43 Palpation of Zone II lymph nodes in the sitting position. See description in text. (From Seidel HM, Ball JW, Dains JE: *Mosby's physical examination handbook,* ed 3, St. Louis, 2003, Mosby.)

abnormalities, disabilities, and asymmetries. Inspection and palpation of the skin, muscles, soft tissues, and joints often takes place simultaneously.

Assess each client from the front, back, and each side. Some general examination principles include[103]:

- Let the client know what to expect; offer simple but clear instructions and feedback.
- When comparing sides, test the "normal" side first.
- Examine the joint above and below the "involved" joint.
- Perform active, passive, and accessory or physiologic movements in that order unless circumstances direct otherwise.
- Resisted isometric movements (break test) follow accessory or physiologic motion.
- Resisted isometric motion is done in a physiologic neutral position (open pack position); the joint should not move.
- Painful joint motion or painful empty end feel of a joint should not be forced.
- Inspect and palpate the skin and surrounding tissue for erythema, swelling, masses, tenderness, temperature changes, and crepitus.
- The forward bend and full squat positions as well as walking on toes and heels and hopping on one leg are useful general screening tests.
- Perform specific special tests last based on client history, and results of the screening interview and clinical findings so far.

The *Guide* suggests key tests and measures to include in a comprehensive screening and specific testing process of the musculoskeletal system, including:

- Patient/client history (demographics, social and employment history, family and personal history, results of other clinical tests)
- Aerobic capacity and endurance
- Anthropometric characteristics
- Arousal, attention, and cognition
- Environmental, home, and work barriers
- Ergonomics and body mechanics
- Gait, locomotion, and balance
- Motor function (motor control and motor learning)
- Muscle performance (strength, power, and endurance)
- Posture
- Range of motion (ROM)
- Self-care and home management

CASE EXAMPLE 4-3

Lymphadenopathy

A 73-year-old woman was referred to a physical therapy clinic by her oncologist with a diagnosis of cervical radiculopathy. She had a history of uterine cancer 20 years ago and a history of breast cancer 10 years ago.

Treatment included a hysterectomy, left radical mastectomy, radiation therapy, and chemotherapy. She has been cancer free for almost 10 years. Her family physician, oncologist, and neurologist actually all evaluated her before she was referred to the physical therapist.

Examination by a physical therapist revealed obvious lymphadenopathy of the left cervical and axillary lymph nodes. When asked if the referring physician (or other physicians) saw the "swelling," she told the therapist that she had not disrobed during her medical evaluation and consultation.

The question for us as physical therapists in a situation like this one is how to proceed?

Several steps must be taken. First, the therapist must document all findings. If possible, photographs of the chest, neck, and axilla should be obtained.

Second, the therapist must ascertain whether the physician is already aware of the problem and has requested physical therapy as a palliative measure. Requesting the physician's dictation or notes from the examination is essential.

Contact with the physician will be important soon after obtaining the records, either to confirm the request as palliative therapy or to report your findings and confirm the need for medical reevaluation.

If it turns out that the physician is, indeed, unaware of these physical findings, it is best to make a problem list identified as "outside the scope of a physical therapist" when returning the client to the physician. Be careful to avoid making any statements that could be misconstrued as a medical diagnosis.

It is highly advised that the therapist offer to make the appointment for the client and do so immediately. We recommend writing a brief letter with the pertinent findings and ending with one of two one-liners:

What do you think?

Please advise.

Documentation of findings and recommendations must be complete even if the client declines. Every effort must be made to get the client to a physician. This may require follow-up phone calls and some persistence on the part of the therapist.

- Work, community, and leisure integration or reintegration

These steps lead to a diagnostic classification or when appropriate, to a referral to another practitioner.[18] Assessing joint or muscle pain is discussed in greater depth in Chapter 3 (see also Appendix B-18).

NEUROLOGIC SCREENING EXAMINATION

Much of the neurologic examination is actually completed in conjunction with other parts of the physical assessment. Acute insult or injury to the neurologic system may cause changes in neurologic status requiring frequent reassessment. Systemic disease can produce nerve damage; careful assessment can help pinpoint the area of pathology. A family history of neurologic disorders or personal history of diabetes, hypertension, high cholesterol, cancer, seizures, or heart disease may be significant.

There are six major areas to assess:

1. Mental and emotional status
2. Cranial nerves
3. Motor function (gross motor and fine motor; coordination, gait, balance)
4. Sensory function (light touch, vibration, pain, pressure, and temperature)
5. Reflexes
6. Neural tension

As always start with the client's history and note any previous trauma to the head or spine along with reports of headache, confusion (increased confusion), dizziness (see Appendix B-11), seizures, or other neurologic signs and symptoms. Note the presence of any incoordination, tremors, weakness, or abnormal speech patterns.

As mentioned previously, neurologic symptoms with no apparent cause such as paresthesias, dizziness, and weakness may actually be symptoms of depression. This is particularly true if the neurologic symptoms are symmetric or not anatomic.[104]

FOLLOW-UP QUESTIONS

- Have you ever been in a car accident?
  If yes, did you lose consciousness, have a concussion, or a fractured skull?
- Have you ever been knocked out, unconscious, or have a concussion at any time?
- Have you ever had a seizure?
- Have you ever been paralyzed in your arms or legs?
- Have you ever broken your neck or back?

CLINICAL SIGNS AND SYMPTOMS

Neurologic Impairment

- Confusion/increased confusion
- Depression
- Irritability
- Drowsiness/lethargy
- Dizziness/lightheadedness
- Loss of consciousness
- Blurred vision or other change in vision
- Slurred speech or change in speech pattern
- Headache
- Balance/coordination problems
- Weakness
- Change in memory
- Change in muscle tone for individual with previously diagnosed neurologic condition
- Seizure activity
- Nerve palsy; transient paralysis

If there are no positive findings upon gross examination of the nervous system (e.g., reflexes, muscle tone assessment, gross manual muscle testing, sensation), further testing may not be required. For example, sensory function is not assessed if motor function is intact and there are no client reports of specific sensory problems or changes.

Keep in mind that fatigue and side effects of medications can affect the results of a neurologic exam. Give instructions clearly and take the time needed to map out any areas of deficit observed during the initial screening examination.

Mental Status

See the previous discussion on General Survey in this chapter.

Cranial Nerves

A neurologic screening exam may not involve a survey of the cranial nerves unless the therapist finds reason to perform a more focused neurologic exam. Many of the cranial nerves can be tested during other portions of the physical assessment. When conducting a cranial nerve assessment, follow the information in Table 4-9.

Motor Function

Motor and cerebellar function can be screened most easily by observation of gait, posture, balance, strength, coordination, muscle tone, and motion. Most of these tests are performed as part of the musculoskeletal exam. See previous discussion on Musculoskeletal Screening Examination in this chapter.

Specific tests, such as tandem walking, Romberg's test, diadochokinesia (rapid, alternating movements of the hands or fingers such as repeatedly alternating forearm pronation to supination, the finger-to-nose/finger-to-finger, and thumb-to-finger opposition tests), can be added for more in-depth screening. Demonstrate all test maneuvers in order to prevent poor performance from a lack of understanding rather than from neurologic impairment.

Sensory Function

Screening for sensory function can begin with superficial pain (pinprick) and light touch (cotton ball) on the extremities. Show the client the items you will be using and how they will be used. For light touch, dab the skin lightly; do not stroke.

Tests are done with the client's eyes closed. Ask the client to tell you where the sensation is felt or to identify "sharp" or "dull." Apply the stimulus randomly and bilaterally over the face, neck, upper arms, hands, thighs, lower legs, and feet. Allow at least 2 seconds between the time the stimulus is applied and the next one is given. This avoids the summation effect.

Follow-up with temperature using test tubes filled with hot and cold water and vibration using a low-pitch tuning fork over peripheral joints. Temperature can be omitted if pain sensation is normal. Apply the stem of the vibrating tuning fork to the distal interphalangeal joint of fingers and interphalangeal joint of the great toe, elbow, and wrist. Ask the client to tell you when vibration is first felt and when it stops. Remember, aging adults often lose vibratory sense in the great toe and ankle on both sides.

Other tests include proprioception (joint position sense), kinesthesia (movement sense), stereognosis (identification of common object placed in the hand), graphesthesia (identifying number or letter when drawn on the palm of the hand), and two-point discrimination. Again, all tests should be performed on both sides.

Reflexes

Deep tendon reflexes (DTRs) are tested in a screening examination at the

- Jaw (cranial nerve V)
- Biceps (C5-6)
- Brachioradialis (C5-6)
- Triceps (C7-8)
- Patella (L3-4)
- Achilles (S1-2)

These reflexes are assessed for symmetry and briskness using the following scale:

0	No response, absent
+1	Low normal, decreased; slight muscle contraction
+2	Normal, visible muscle twitch producing movement of arm/leg
+3	More brisk than normal, increased or exaggerated; may not indicate disease
+4	Hyperactive; very brisk, clonus; spinal cord disorder suspected

A change in 1 or more DTRs is a yellow (caution) flag. Some individuals have very brisk reflexes normally, while others are much more hyporeflexive. Whenever encountering increased (hyper-) or decreased (hypo-) reflexes, the therapist routinely follows several guidelines:

FOLLOW-UP QUESTIONS

- Test reflexes above and below and from side to side in order to gauge overall reflexive response. A "normal" hyperreflexive response will be present in most, if not all, reflexes. The same is true for generalized hyporeflexive responses.
- Offer the client distraction while testing through conversation or by asking such silly questions as, "What color is your toothbrush?" "What day of the week were you born?" or "Count out loud backwards by three starting at 89."
- Retest unusual reflexes later in the day or on another day.
- Have another clinician test your client.

The isolated DTR that does not fit the client's physiologic pattern must be considered a red flag. As discussed in Chapter 2, one red flag by itself does not require immediate medical

TABLE 4-9 Cranial Nerve Function and Assessment

Cranial Nerve (CN)	Type	Function	Assessment
I Olfactory	Sensory	Sense of smell	Able to identify common odors (e.g., coffee, vanilla, orange or peppermint) with eyes closed. Close one nostril and test one nostril at a time.
II Optic	Sensory	Visual acuity	Visual acuity; test each eye separately with Snellen eye chart. If literate, able to read printed material.
III Oculomotor	Motor	Extraocular eye movement Pupil constriction and dilation	Assess CNs III, IV, and VI together. Look for equal pupil size and shape; equal response to light and accommodation; inspect eyelids for drooping (ptosis). Ask about blurry or double vision. Follow finger with eyes without moving head (six points in an H pattern; gaze test). Convergence (move finger towards client's nose).
IV Trochlear	Motor	Upward and downward movement of eyeball	See CN III; assess directions of gaze (eye can move out when intact/eye remains focused up and out when impaired); visual tracking. Ask about double vision.
V Trigeminal	Mixed	Sensory nerve to skin of face	Corneal reflex: client looks up and away, examiner lightly touches opposite cornea with wisp of cotton (look for blink in both eyes or report by client of blinking sensation). Facial sensation: apply sterile, sharp item to forehead, cheek, jaw; repeat with dull object; client reports "sharp" or "dull"; if abnormal, test for temperature, vibration, and light touch.
		Motor nerve to muscles of jaw (mastication)	Ask client to clench teeth together as you palpate muscles over temples (temporal muscle) and jaw (masseter) on each side. Look for symmetric tone (normal) or muscle atrophy, deviation of jaw to one side, or fasciculations (abnormal).
VI Abducens	Motor	Lateral movement of eyeballs	See CN III; assess directions of gaze (able to move eyes out laterally when intact/medially deviated when impaired).
VII Facial	Mixed	Facial expression	Look for symmetry with facial expressions (e.g., frown, smile, raise and lower eyebrows, puff cheeks out, close eyes tightly).
VIII Acoustic (auditory, vestibulocochlear)	Sensory	Hearing	Assess ability to hear spoken word and whisper. Examiner stands behind client with hands on either side of client's head/ears; rub candy wrapper or fingers together to make noise on one side, ask client to identify which side noise is coming from. Examiner stands 18 inches behind client and whispers 3 numbers. Assess for dizziness and imbalance.
IX Glossopharyngeal	Mixed	Taste Gag, swallow	Client identifies sour or sweet taste on back of tongue. Gag reflex (sensory IX and motor X) and ability to swallow.
X Vagus	Mixed	Pharyngeal sensation, voice, swallow	Client says, "Ah." Observe for normal palate and pharynx movement. Listen for hoarseness or nasal quality in voice. Observe for difficulty swallowing.
XI Spinal Accessory	Motor	Movement of head and shoulders	Client is able to shrug shoulders and turn head against resistance. Observe shoulders from behind for trapezius atrophy and/or asymmetry (abnormal finding). Assess for neck weakness.
XII Hypoglossal	Motor	Position and movement of tongue	Client can stick out tongue to midline and move it from side to side. Client can move tongue toward nose and chin. Clear articulation in speech pattern.

follow-up. But a hyporesponsive patellar tendon reflex that is reproducible and is accompanied by back, hip, or thigh pain in the presence of a past history of prostate cancer offers a different picture altogether.

A diminished reflex may be interpreted as the sign of a possible "space-occupying lesion"—most often, a disc protruding from the disc space and either pressing on a spinal nerve root or irritating the spinal nerve root (i.e., chemicals released by the herniation in contact with the nerve root can cause nerve root irritation).

Tumors (whether benign or malignant) can also press on the spinal nerve root, mimicking a disc problem. A small lesion can put just enough pressure to irritate the nerve root, resulting in a hyperreflexive DTR. A large tumor can obliterate the reflex arc resulting in diminished or absent reflexes. Either way, changes in DTRs must be considered a yellow (caution) or red (warning) flag sign to be documented, reported, and further investigated.

Superficial (cutaneous) reflexes (e.g., abdominal, cremasteric, plantar) can also be tested using the handle of the reflex hammer. The abdominal reflex is elicited by applying a stroking motion with a cotton-tipped applicator (or handle of the reflex hammer) toward the umbilicus. A positive sign of neurologic impairment is observed if the umbilicus moves toward the stroke. The test can be repeated in each abdominal quadrant (upper abdominal T7-9; lower abdominal T11-12).

The cremasteric reflex is elicited by stroking the thigh downward with a cotton-tipped applicator (or handle of the reflex hammer). A normal response in males is an upward movement of the testicle (scrotum) on the same side. The absence of a cremasteric reflex is indication of disruption at the T12-L1 level. Testing the cremasteric reflex may help the therapist identify neurologic impairment in any male with suspicious back, pelvic, groin (including testicular), and/or anterior thigh pain.

The plantar reflex occurs when the sole of the foot is stroked and the toes plantarflex downward.

Neural Tension

Excessive nerve tightness or adhesion can cause adverse neural tension in the peripheral nervous system. When the nerve cannot slide or glide in its protective sheath, neural extensibility and mobility are impaired. The clinical result can be numbness, tingling, and pain. This could be caused by disc protrusion, scar tissue, or space-occupying lesions including cysts, bone spurs, tumors, and cancer metastases.[105,106]

A positive neural tension test does not tell the therapist what is the underlying etiology—only that the peripheral nerve is involved. History and physical examination are still very important in assessing the clinical presentation.

Someone with full ROM accompanied by negative articular signs but with impaired neural extensibility and mobility raises a yellow (caution) flag. A second look at the history and a more thorough neurologic exam may be warranted.

Reducing symptoms with neural mobilization does not rule out the possibility of cancer. A red flag is raised with any client who responds well to neural mobilization but experiences recurrence of symptoms. This could be a sign that the tumor has grown larger or cancer metastases have progressed, once again interfering with neural mobility.

REGIONAL SCREENING EXAMINATION

Head and Neck

Screening of the head and neck areas takes place when client history and report of symptoms or clinical presentation warrant this type of examination. The head and neck assessment provides information about oral health and the general health of multiple systems including integumentary, neurologic, respiratory, endocrine, hepatic, and gastrointestinal.

The head, hair and scalp, and face are observed for size, shape, symmetry, cleanliness, and presence of infection. Position of the head over the spine and in relation to midline and range-of-motion testing of the cervical spine and temporomandibular joints can be part of the screening and posture assessment.

Because the head and neck have a large blood supply, infection from the mouth can quickly spread throughout the body increasing the risk of osteomyelitis, pneumonia, and septicemia in critically ill patients. Evidence of gum disease (e.g., bright red, enlarged, spongy, or bleeding) should be medically evaluated. Ulcerations on the tongue, lips, or gums also require further medical/dental evaluation.

The eyes can be examined for changes in shape, motor function, and color (conjunctiva and sclera). Conducting an assessment of cranial nerves II, III, and IV also will help screen for visual problems. The therapist should be aware that there are changes in the way older adults perceive color. This kind of change can affect function and safety; for example, some older adults are unable to tell when floor tiles end and the bathtub begins in a bathroom. Stumbling and loss of balance can occur at boundary changes.

Assessment of cranial nerves (see Table 4-9), regional lymph nodes (see discussion, this chapter), carotid artery pulses (see Fig. 4-1), and jugular vein patency (Fig. 4-44) are part of the head and neck screening examination. Therapists in a primary care setting may also examine the position of the trachea and thyroid for obvious deviations or palpable lesions.

Headaches are common and often the result of specific foods, stress, muscle tension, hormonal fluctuations, nerve compression, or cervical spine or temporomandibular joint

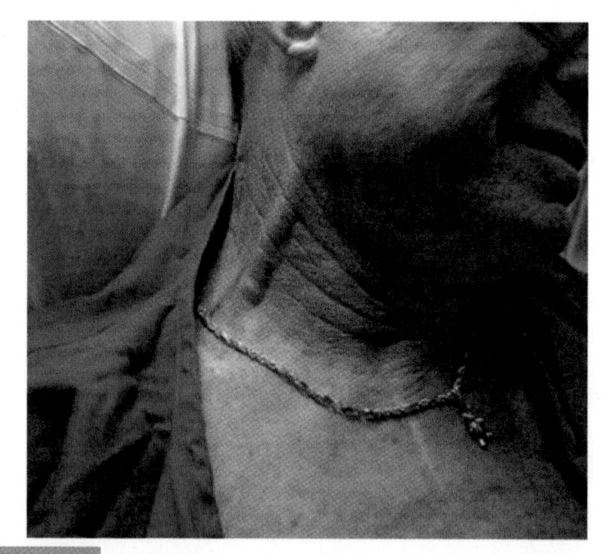

Fig. 4-44 Jugular venous distention is a sign of increased venous return (volume overload) or heart failure, especially congestive heart failure and requires immediate medical attention if not previously reported. Inspect the jugular veins with the client sitting up at a 90-degree angle and again with the head at a 30- to 45-degree angle. The cervical spine should be in a neutral position. Correlate findings of this exam with vital signs, breath and heart sounds, peripheral edema, and auscultation of the carotid arteries for bruits. (From http://courses.cvcc.vccs.edu/WisemanD/index.htm, and Goldman L: *Cecil medicine,* ed 23, Philadelphia, 2008, WB Saunders.)

dysfunction. Most headaches are acute and self-limited. Headaches can be a symptom of a serious medical condition and should be assessed carefully (see Appendix B-17; see also the discussion of viscerogenic causes of head and neck pain in Chapter 14).

Upper and Lower Extremities

The extremities are examined through a systematic assessment of various aspects of the musculoskeletal, neurologic, vascular, and integumentary systems. Inspection and palpation are two techniques used most often during the examination. A checklist can be very helpful (Box 4-16).

Peripheral Vascular Disease

Peripheral vascular disease (PVD), both arterial and venous conditions, is a common problem observed in the extremities of older adults, especially those with a history of heart disease. Knowing the risk factors for any condition, but especially problems like PVD, helps the therapist know when to screen. These conditions, including risk factors, are discussed in greater depth in Chapter 6.

BOX 4-16 EXTREMITY EXAMINATION CHECKLIST

- Inspect skin for color, scratch marks, inflammation, track marks, bruises, heat, or other obvious changes (see Box 4-11)
- Observe for hair loss or hair growth
- Observe for asymmetry, contour changes, edema, obvious atrophy, fractures or deformities; measure circumference if indicated
- Assess palpable lesions (see Box 4-11)
- Palpate for temperature, moisture, and tenderness
- Palpate pulses
- Palpate lymph nodes
- Check nail bed refill (normal: capillary refill time under 2-3 seconds for fingers and 3-4 seconds for toes)
- Observe for clubbing, signs of cyanosis, other nail bed changes
- Observe for PVD (see Box 4-13); listen for femoral bruits if indicated; test for thrombophlebitis
- Assess joint ROM and muscle tone
- Perform gross MMT (gross strength test); grip and pinch strength
- Sensory testing: light touch, vibration, proprioception, temperature, pinprick
- Assess coordination (UEs: dysmetria, diadochokinesia; LEs: gait, heel-to-shin test)
- Test deep tendon reflexes

PVD, Peripheral vascular disease, *ROM*, range of motion; *MMT*, manual muscle testing; *UEs*, upper extremities; *LEs*, lower extremities.

The first signs of vascular occlusive disease are often skin changes (see Box 4-13). The therapist must watch out for common risk factors, including bedrest or prolonged immobility, use of IV catheters, obesity, MI, heart failure, pregnancy, postoperative patients, and any problems with coagulation.

Screening assessment of peripheral arterial disease (PAD) can be done using the ankle-brachial index (ABI) (Fig. 4-45). In fact, the American Diabetes Association recommends screening for anyone with diabetes who is 50 years old or older. Others who can benefit from screening include clients with risk factors for PAD such as smoking, advancing age, hypertension, hyperlipidemia, and symptoms of claudication.

Baseline ABI should be taken on both sides for anyone who has (or may have) PAD. Clients with diabetes who have normal ABI levels should be retested periodically.[105,107] The ABI is the ratio of the SBP in the ankle divided by the SBP at the arm:

$$SBP_{ankle}/SBP_{arm}$$

Assess ankle SBP using both the dorsalis pedis pulse and the posterior tibial pulse. Divide the higher SBP from each leg by the higher brachial systolic pressure (i.e., if the SBP is higher in the right arm use that figure for the calculations in both legs).

Normal ABI values lie in the range of 0.91 to 1.3. A general guideline is provided in Table 4-10. Recall what was said earlier in this chapter about normal BPs in the legs versus the arms: SBP in the legs is normally 10% to 20% higher than the brachial artery pressure in the arms, resulting in an ABI greater than 1.0. PAD obstruction is indicated when the ABI values fall to less than 1.0.

TABLE 4-10 Ankle Brachial Index Reading*

INDICATORS OF PERIPHERAL ARTERIAL DISEASE	
1.0-1.3	Normal (blood pressure at the ankle and arm are the same; no significant narrowing or obstruction of blood flow)
0.8-0.9	Mild peripheral arterial occlusive disease
0.5-< 0.8	Moderate peripheral arterial occlusive disease
Less than 0.5	Severe peripheral arterial occlusive disease; critical limb ischemia
Less than 0.2	Ischemic or gangrenous extremity

*Different sources offer slightly different ABI values for normal to severe PAD. Some sources use values between 0.90 and 0.97 as the lower end of normal. Values greater than 1.3 are not considered reliable because calcified vessels show falsely elevated pressures. The therapist should follow guidelines provided by the physician or facility.

Data from Sacks D, et al.: Position statement on the use of the ankle brachial index in the evaluation of patients with peripheral vascular disease. A consensus statement developed by the Standards Division of the Society of Interventional Radiology, *J Vasc Interv Radiol* 13:353, 2002.

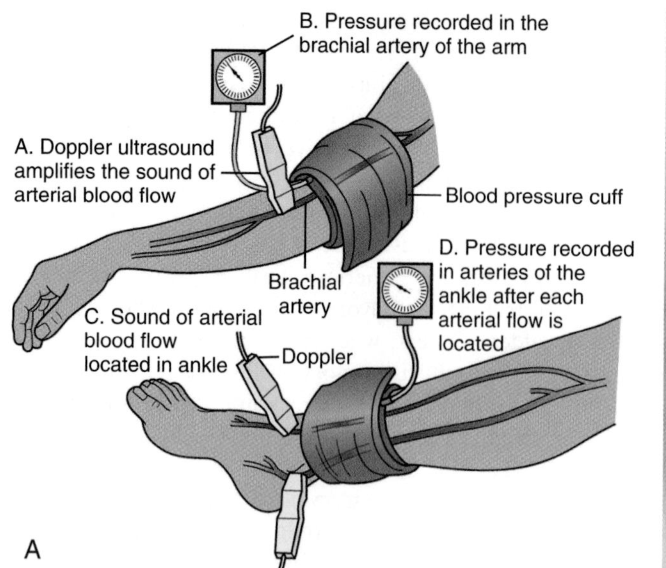

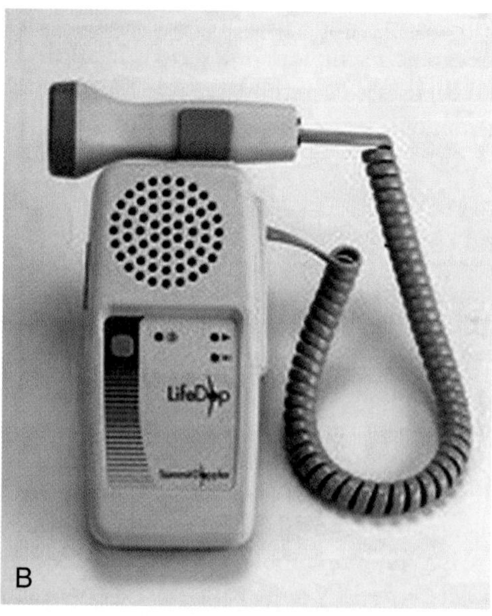

Fig. 4-45 **A,** Peripheral vascular testing is usually performed in a vascular laboratory, but an approximation of the integrity of the peripheral arterial circulation can be found by calculating the ankle/brachial index (ABI) by using a Doppler device to compare systolic blood pressures in the foot and arm. **B,** The ABI is measured using a simple, noninvasive tool. Example of a handheld Doppler device with speaker is shown. Devices with an attached stethoscope are also used to measure the ABI. The index correlates well with blood vessel disease severity and symptoms. In the normal individual, the systolic blood pressure in the legs is slightly greater than or equal to the brachial (arm) systolic blood pressure (ABI = 1.0 or above). Where there is arterial obstruction and narrowing, systolic blood pressure is reduced below the area of obstruction. When the ankle systolic blood pressure falls below the brachial systolic blood pressure, the ABI is less than 1.0 (a sign of peripheral vascular obstruction). (From Roberts JR: *Clinical procedures in emergency medicine*, ed 5, Philadelphia, 2010, WB Saunders.)

Rubor on dependency is another test used to observe the adequacy of arterial circulation. Place the client in the supine position and observe the color of the soles of the feet. Normal feet should be pink or flesh colored in Caucasians and tan or brown in clients with dark skin tones. The feet of clients with impaired circulation are often chalky white (Caucasians) or gray or white in clients with darker skin.

Elevate the legs to 45 degrees (above the heart level). For clients with compromise of the arterial blood supply, any color present will quickly disappear in this position; in other words, the elevated foot develops increased pallor. No change (or little change) is observed in the normal individual. Bring the individual to a sitting position with the legs dangling. Venous filling is delayed following foot elevation. Color change in the lower leg and foot may take 30 seconds or more and will be a very bright red (dependent rubor).[27,106]

Venous Thromboembolism

Another condition affecting the extremities is venous thrombophlebitis or venous thromboembolism (VTE), a common complication seen in clients with cancer or following abdominal or pelvic surgery (especially orthopedic surgery such as total hip and total knee replacements), major trauma, or prolonged immobilization. Other risk factors are discussed in Chapter 6.

VTE is an inflammation of a vein associated with thrombus formation affecting superficial or deep veins of the extremities. Superficial vein thrombophlebitis typically involves the veins of the upper extremities (usually unilateral) and is most commonly associated with trauma (e.g., insertion of IV lines and catheters in the subclavian vein). In fact, the presence of an indwelling central venous catheter (often present in cancer patients) is the strongest predictor of upper extremity deep vein thrombosis (DVT).[108] DVT usually involves the deep veins of the legs, primarily the calf.

DVTs can become dislodged and travel as an embolus where it can become lodged in the pulmonary artery, leading to comorbidity or even death from acute coronary syndrome and stroke. DVTs are often asymptomatic, making screening a key component in the prevention of this potentially life-threatening problem.[109]

Many clinicians use *Homans' sign* as physical evidence of DVTs. The test uses slow dorsiflexion of the foot or gentle squeezing of the affected calf to elicit deep calf pain. Homans' sign is not specific for DVT because a positive Homans' sign is possible with Achilles tendinitis and muscle injury of the gastrocnemius and plantar muscle.

Autar DVT Risk Assessment Scale (Table 4-11) is a much more sensitive and specific test developed as a predictive index of DVT. Seven risk categories are included: increasing age, build and BMI, immobility, special DVT risk, trauma, surgery, and high-risk disease.[110,111]

A similar (less comprehensive) tool, the Wells' Clinical Decision Rule (CDR),[112-114] frequently discussed in the general medical literature, can also be used by therapists (Table 4-12).[115] The CDR incorporates signs, symptoms, and risk factors for DVT and should be used in all cases of suspected DVT. A simple model to predict upper extremity DVT has also been proposed and remains under investigation.[108]

TABLE 4-11 Autar Deep Vein Thrombosis Scale

Category (Points)	Possible Score (Points)	Score
Age (years)	10-30	0
	31-40	1
	41-50	2
	51-60	3
	61+	4
BMI (Wt kg)/(Hgt m)[144]	16-19	0
	20-25	1
	26-30	2
	31-40	3
	41+	4
Mobility	Ambulant	0
	Limited (walks with aids)	1
	Very limited (needs help)	2
	Chair bound	3
	Bed rest	4
Special DVT risk	Birth control pills (less than 35 years of age)	1
	Birth control pills (35 years of age/older)	2
	Pregnant/up to 6 weeks postpartum	3
Trauma risk factors (score only preoperatively; score only one item)	Head or chest	1
	Head and chest	2
	Spinal	2
	Pelvic	3
	Lower limb	4
Surgical risk factors (score only one item)	Minor surgery (<30 minutes)	1
	Major surgery	2
	Emergency major	3
	Pelvic, abdominal, or thoracic	3
	Orthopedic (below waist)	4
	Spinal	4
High-risk diseases	Ulcerative colitis	1
	Anemia (sickle cell, polycythemic, hemolytic)	2
	Chronic heart disease	3
	MI	4
	Malignancy	5
	Varicose veins	6
	Previous DVT or CVA	7
Results based on total score	≤6	No risk
	7-10	Low risk (<10%)
	11-14	Moderate risk (11%-40%)
	≥15	High risk (>40%)

BMI, Body mass index; see text for web links for easy calculation; *MI*, myocardial infarction; *DVT*, deep venous thrombosis; *CVA*, cerebrovascular accident.

From Autar R: Nursing assessment of clients at risk of deep vein thrombosis (DVT): the Autar DVT scale, *J Advanced Nursing* 23(4): 763-770, 1996. Reprinted with permission of Blackwell Scientific.

TABLE 4-12 Wells' Clinical Decision Rule for Deep Vein Thrombosis

Clinical Presentation	Score
Previously diagnosed deep vein thrombosis (DVT)	1
Active cancer (within 6 months of diagnosis or receiving palliative care)	1
Paralysis, paresis, or recent immobilization of lower extremity	1
Bedridden for more than 3 days or major surgery in the last 4 weeks	1
Localized tenderness in the center of the posterior calf, the popliteal space, or along the femoral vein in the anterior thigh/groin	1
Entire lower extremity swelling	1
Unilateral calf swelling (more than 3 mm larger than uninvolved side)	1
Unilateral pitting edema	1
Collateral superficial veins (non-varicose)	1
An alternative diagnosis is as likely (or more likely) than DVT (e.g., cellulitis, postoperative swelling, calf strain)	–2
Total Points	
KEY	
–2 to 0	Low probability of DVT (3%)
1 to 2	Moderate probability of DVT (17%)
3 or more	High probability of DVT (75%)

Medical consultation is advised in the presence of low probability; medical referral is required with moderate or high score.

From Wells PS, Anderson DR, Bormanis J, et al: Value of assessment of pretest probability of deep-vein thrombosis in clinical management, *Lancet* 350:1795-1798, 1997. Used with permission.

The CDR is used more widely in the clinic, but the Autar scale is more comprehensive and incorporates BMI, postpartum status, and the use of oral contraceptives as potential risk factors.

Research shows some physical therapists may not be referring clients to a physician for additional workup when the individual's risk for developing DVT warrants referral.[116] Medical consultation is advised for anyone with a low probability of DVT; medical referral is required for any individual suspected of having DVT.

The Chest and Back (Thorax)

A screening examination of the thorax requires the same basic techniques of inspection, palpation, and auscultation. Once again, keep in mind this is a screening examination. Being familiar with normal findings of the chest and thorax will help the therapist identify abnormal results requiring further evaluation or referral. Only basic screening tools are included. Specialized training may be required for some acute care or primary care settings.

BOX 4-17 CLINICAL INSPECTION OF THE RESPIRATORY SYSTEM

- Respiratory rate, depth, and effort of breathing
  - Tachypnea
  - Dyspnea
  - Gasping respirations
- Breathing pattern or sounds (see also Box 7-1)
  - Cheyne-Stokes respiration
  - Hyperventilation or hypoventilation
  - Kussmaul's respiration
  - Lateral-costal breathing
  - Paradoxic breathing
  - Prolonged expiration
  - Pursed-lip breathing
  - Wheezing
  - Rhonchi
  - Crackles (formerly called rales)
- Cyanosis
- Pallor or redness of skin during activity
- Clubbing (toes, fingers)
- Nicotine stains on fingers and hands
- Retraction of intercostal, supraclavicular, or suprasternal spaces
- Use of accessory muscles
- Nasal flaring
- Tracheal tug
- Chest wall shape and deformity
  - Barrel chest
  - Pectus excavatum
  - Pectus carinatum
  - Kyphosis
  - Scoliosis
- Cough
- Sputum: clear or white (normal); frothy; red-tinged, green, or yellow (pathologic)

Adapted from Goodman CC, Fuller K: *Pathology: implications for the physical therapist,* ed 3, Philadelphia, 2009, WB Saunders.

Chest and Back: Inspection[35]

Clinical inspection of the chest and back encompasses both the cardiac and pulmonary systems, but some of the most obvious changes are observed in relation to the respiratory system (Box 4-17). Inspect the client while he or she is sitting upright without support if possible. Observe the client's thorax from the front, back, side, and over the shoulder (looking down over the anterior chest). Note any skin changes, signs of skin breakdown, and signs of cyanosis or pallor, scars, wounds, bruises, lesions, nodules, or superficial venous patterns.

Note the shape and symmetry of the thorax from the front and back. Note any obvious anatomic changes or deformities (e.g., pectus excavatum, pectus carinatum, barrel chest). Observe posture and spinal alignment. Record the presence of any deviations from normal (e.g., forward head, rounded shoulders, upward angle of clavicles, rib alignment) or deformities (e.g., kyphosis, scoliosis, rib hump) that can compromise chest wall excursion. Estimate the anterior-posterior diameter compared to the transverse diameter.

A normal ratio of 1:2 may be replaced by an equal diameter (1:1 ratio) typical of the barrel chest that develops as a result of hyperinflation. Look at the angle of the costal margins at the xiphoid process; for example, anything less than a 90-degree alignment may be indicative of a barrel chest. Observe (and palpate) for equal intercostal spaces and compare sides (right to left). Observe the client for muscular development and nutritional status by noting the presence of underlying adipose tissue and the visibility of the ribs.

Assess respiratory rate, depth, and rhythm or pattern of breathing while the client is breathing normally. A description of altered breathing patterns can be found in Chapter 7. Watch for symmetry of chest wall movement, costal versus abdominal breathing, the use of accessory muscles, and bulging or retraction of the intercostal spaces. Remember, the normal ratio of inspiration to expiration is 1:2.

Chest and Back: Palpation

Palpation of the thorax is usually combined with inspection to save time. Breast examination and lymph node assessment may be part of the screening exam of the chest in males and females (see discussion later in this section).

Palpation can reveal skin changes and alert the therapist to conditions that relate to the client's respiratory status. Look for crepitus, a crackly, crinkly sensation in the subcutaneous tissue. Feel for vibrations during inspiration as described next.

Palpate the entire thorax (anterior and posterior) for tactile fremitus by placing the palms of both hands (examiner's) over the client's upper anterior chest at the second intercostal space (Fig. 4-46). The normal response is a feeling of vibrations of equal intensity during vocalizations on either side of the midline, front to back.

Tactile fremitus is found most commonly in the upper chest near the bronchi around the second intercostal space. Stronger vibrations are felt whenever air is present; the absence of air such as occurs with atelectasis is marked by an absence of vibrations. Fluid outside the lung pressing on the lung increases the force of vibration.

Fremitus can be increased or decreased—increased over areas of compression because the dense tissue improves transmission of the vibrational wave while being decreased over areas of effusion (decreased density, decreased vibration).

Increased tactile fremitus often accompanies inflammation, infection, congestion, or consolidation of a lung or part of a lung. Diminished tactile fremitus may indicate the presence of pleural effusion or pneumothorax.[35] Record the location and beginning and ending points, and note whether the tactile fremitus is increased or decreased. Confirm all findings with auscultation; a medical evaluation with chest x-ray provides the differential diagnosis.

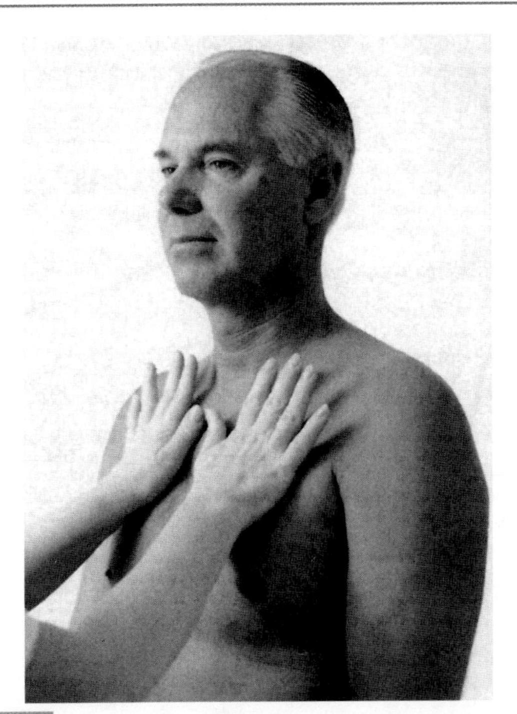

Fig. 4-46 Palpating for tactile fremitus. Place the palmar surfaces of both hands on the client's chest at the second intercostal space. Ask the client to say "99" repeatedly as you gradually move your hands over the chest, systematically comparing the lung fields. Start at the center and move out toward the arms. Drop the hands down one level and move back toward the center. Continue until the complete lung fields have been assessed. Repeat the exam on the client's back. The normal response is a feeling of vibrations of equal intensity during vocalizations on either side of the midline, front to back. (From *Expert 10-minute physical examinations,* St. Louis, 1997, Mosby.)

Assess respiratory excursion and symmetry with the client in the sitting position (preferred) or if unable to assume or maintain an upright position, then in the supine position. The upright position makes it easier to assess all areas of respiratory excursion: upper, middle, and lower. Stand in front of your client and place your thumbs along the client's costal margins wrapping your fingers around the rib cage. Observe how far the thumbs move apart (range and symmetry) during chest expansion (normal breathing and deep breathing). Normal respiratory excursion will separate the thumbs by 1¼ to 2 inches. Assess respiratory excursion at more than one level, front to back.

The presence of costovertebral tenderness should be followed up with Murphy's percussion test (see Fig. 4-54). Bone tenderness over the lumbar spinous processes is a red-flag symptom for osseous disorders, such as fracture, infection, or neoplasm, and requires a more complete evaluation.

Chest and Back: Percussion

Chest and back percussion is an important part of the screening exam. The clinician can use percussion of the chest and back to identify the left ventricular border of the heart and the depth of diaphragmatic excursion in the upper abdomen during breathing. Percussion can also help identify disorders that impair lung ventilation such as stomach distention, hemothorax, lung consolidation, and pneumothorax.[35]

Dullness over the lungs during percussion may indicate a mass or consolidation (e.g., pneumonia). Hyperresonance over the lungs may indicate hyperinflated or emphysemic lungs. Decreased diaphragmatic excursion on one side occurs with pleural effusion, diaphragmatic (phrenic nerve) paralysis, tension pneumothorax, stomach distention (left side), hepatomegaly (right side), or atelectasis. Clients with COPD often have decreased excursion bilaterally as a result of a hyperinflated chest depressing the diaphragm.

Chest and Back: Lung Auscultation

In a screening exam, the therapist should listen for normal breath sounds and air movement through the lungs during inspiration and expiration. Taking a deep breath or coughing clears some sounds. At times the examiner instructs the client to take a deep breath or cough and listens again for any changes in sound. With practice and training, the therapist can identify the most common abnormal sounds heard in clients with pulmonary involvement: crackles, wheezing, and pleural friction rub.

Crackles (formerly called rales) are the sound of air moving through an airway filled with fluid. It is heard most often on inspiration and sounds like strands of hair being rubbed together under a stethoscope. Crackles/rales are normal sounds in the morning as the alveolar spaces open up. Have the client take a deep breath to complete this process first before listening.

Abnormal crackles can be heard during exhalation in clients with pneumonia or CHF and during inhalation with the re-expansion of atelectatic areas. Crackles are described as fine (soft and high pitched), medium (louder and low pitched), or coarse (moist and more explosive).

Wheezing is the sound of air passing through a narrowed airway blocked by mucus secretions and usually occurs during expiration (wheezing on inspiration is a sign of a more serious problem). It is frequently described as a high-pitched, musical whistling sound but there are different tones to wheezing that can be identified in making the medical differential diagnosis.

There is some debate about the difference between *rhonchi* and wheezing. Some experts consider the low-pitched, rattling sound of rhonchi (similar to snoring) as a form of wheezing. Whether called wheezing or rhonchi, this sound is most often associated with asthma or emphysema. Take careful note when the wheezing goes away in any client who presents with wheezing. Disappearance of wheezing occurs when the person is not breathing.

Rhonchi is defined as air moving through thick secretions partially obstructing airflow and is almost always associated with bronchitis. These loud, low-pitched rumbling sounds can be heard on inspiration or expiration and may be cleared by coughing.

Pleural friction rub makes a high-pitched scratchy sound heard when a hand is cupped over the ear and scratched along the outside of the hand. It is caused by inflamed pleural surfaces and can be heard on inspiration and/or expiration. The pleural linings should move over each other smoothly and easily. In the presence of inflammation (pleuritis, pleurisy, pneumonia, tumors), the inflamed tissue (parietal pleura) is rubbing against other inflamed, irritated tissue (visceral pleura) causing friction.

Assess for *egophony* by asking the client to say and repeat the "ee" sound during auscultation of the lung fields. The "aa" sound heard as the client says the "ee" sound indicates pleural effusion or lung consolidation. *Bronchophony,* a clear and audible "99" sound suggests the sound is traveling through fluid or a mass; the sound should be muffled in the healthy adult. Finally, assess for consolidation using *whispered pectoriloquy.* Ask the client to whisper "1-2-3" as you listen to the chest and back. The examiner should hear a muffled noise (normal) instead of a clear and audible "1-2-3" (consolidation).

Describe sounds heard, the location of the sounds on the thorax, and when the sounds are heard during the respiratory cycle (inspiration versus expiration). Have the client gently cough (bronchial secretions causing a "gurgle" will often clear with a cough) and listen again. Compare results with the first exam (and with the baseline if available). The decision to treat, treat and consult/refer, or consult/refer a client for further evaluation is based on history, clinical findings, client distress, vital signs, and any associated signs and symptoms observed or reported.

Giving the physician details of your physical exam findings is important in order to give a clear idea of the reason for the consult and urgency (if there is one). Sometimes, abnormal findings are "normal" for people with chronic pulmonary problems. For example, always evaluate apparent abnormal breath (and heart) sounds in order to recognize change that is significant for the individual. The physician is more likely to respond immediately when told that the therapist identified crackles in the right lower lung and the client is producing yellow sputum than if the report is "client has altered sounds on auscultation."

Chest and Back: Heart Auscultation

The same general principles for auscultation of lung sounds apply to auscultation of heart sounds. The therapist's primary responsibility in a screening examination is to know what "normal" heart sounds are like and report any changes (absence of normal sounds or presence of additional sounds).

The normal cardiac cycle correlates with the direction of blood flow and consists of two phases: systole (ventricles contract and eject blood) and diastole (ventricles relax and atria contract to move blood into the ventricles and fill the coronary arteries).

Normal heart sounds (S1 and S2) occur in relation to the cardiac cycle. Just before S1, the mitral and tricuspid (AV) valves are open, and blood from the atria is filling the relaxed ventricles (see Fig. 6-1). The ventricles contract and raise pressure, beginning the period called systole.

Pressure in the ventricles increases rapidly, forcing the mitral and tricuspid valves to close causing the first heart sound (S1). The S1 sound produced by the closing of the AV valves is *lubb;* it can be heard at the same time the radial or carotid pulse is felt.

As the pressure inside the ventricles increases, the aortic and pulmonic valves open and blood is pumped out of the heart into the lungs and aorta. The ventricle ejects most of its blood and pressure begins to fall causing the aortic and pulmonic valves to snap shut. The closing of these valves produces the *dubb* (S2) sound. This marks the beginning of the diastole phase.

S3 (third heart sound), S4 (fourth heart sound), heart murmurs, and pericardial friction rub are the most common extra sounds heard. S3, also known as a ventricular gallop, is a faint, low-pitched *lubb-dup-ah* sound heard directly after S2. It occurs at the beginning of diastole and may be heard in healthy children and young adults as a result of a large volume of blood pumping through a small heart. This sound is also considered normal in the last trimester of pregnancy. It is not normal when it occurs as a result of anemia, decreased myocardial contractility, or volume overload associated with CHF.

S4, also known as an atrial gallop, occurs in late diastole (just before S1) if there is a vibration of the valves, papillary muscles, or ventricular walls from resistance to filling (stiffness). It is usually considered an abnormal sound *(ta-lup-dubb)* but may be heard in athletes with well-developed heart muscles.

Heart murmurs are swishing sounds made as blood flows across a stiff or incompetent (leaky) valve or through an abnormal opening in the heart wall. Most murmurs are associated with valve disease (stenosis, insufficiency), but they can occur with a wide variety of other cardiac conditions. They may be normal in children and during the third trimester of pregnancy.

A pericardial friction rub associated with pericarditis is a scratchy, scraping sound that is heard louder during exhalation and forward bending. The sound occurs when inflamed layers of the heart viscera (see Fig. 6-5) rub against each other causing friction.

Auscultate the heart over each of the six anatomic landmarks (Fig. 4-47), first with the diaphragm (firm pressure) and then with the bell (light pressure) of the stethoscope. Use a Z-path to include all landmarks while covering the entire surface area of the heart.

Practice at each site until you can hear the rate and rhythm, S1 and S2, extra heart sounds, including murmurs. The high-pitched sounds of S1 and S2 are heard best with the stethoscope diaphragm. Murmurs are not heard with a bell (listen for murmurs with the diaphragm); low-pitched sounds of S3 and S4 are heard with the bell.

As with lung sounds, describe heart sounds heard, rate and rhythm of the sounds, the location of the sounds on the thorax, and when the sounds are heard during the cardiac cycle. The decision to refer a client for further evaluation is based on history, age, risk factors, presence of pregnancy, clinical findings, client distress, vital signs, and any associated signs and symptoms observed or reported.

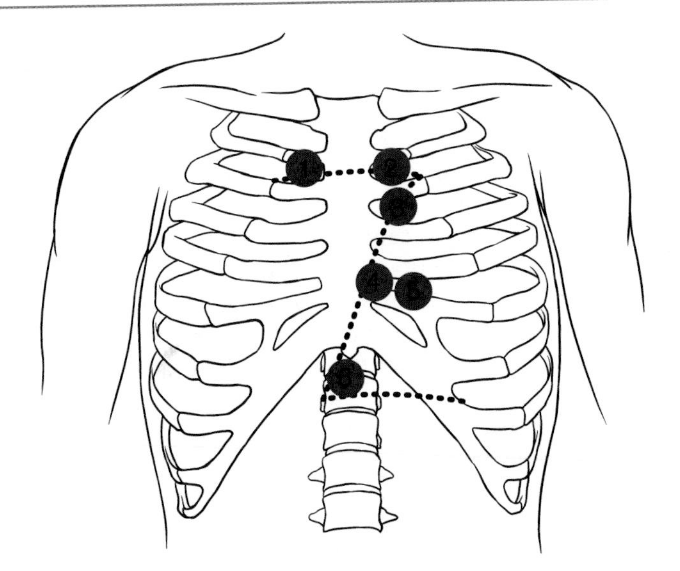

1. Aortic area (right 2nd intercostal space, ICS)
2. Pulmonic area (P1) (left 2nd ICS)
3. Erb's point (P2) (left 3rd ICS); S2 best auscultated here, murmurs are best heard at this point using the stethoscope bell
4. Tricuspid (left 5th ICS at sternum)
5. Mitral or apical point of maximal impulse (PMI), (left 5th ICS medial to the midclavicular line)
6. Epigastric area (just below the tip of the sternum)

Fig. 4-47 Auscultation of cardiac areas using cardiac anatomic landmarks. Inspect and palpate the anterior chest with the client first sitting up then lying down supine. Ask the individual to breathe quietly while you auscultate all six anatomic landmarks shown.

Chest: Clinical Breast Examination

Although the third edition of this text specifically said, "breast examination is not within the scope of a physical therapist's practice," this practice is changing.

As the number of cancer survivors increases in the United States, therapists treating postmastectomy women and clients of both genders with lymphedema are on the rise. Cancer recurrence after mastectomy with or without reconstruction is possible. Recurrences after reconstructive surgery have been detected first on physical examination.[117-119]

With direct and unrestricted access of consumers to physical therapists in many states and with the expanded role of physical therapists in primary care, advanced skills have become necessary. For some clients, clinical breast exam (CBE) is an appropriate assessment tool in the screening process.[120]

CBE and the *Guide*

The *Guide* includes a concise description of the physical therapist in primary care and our respective roles in primary and secondary prevention.[18] Currently, the *Guide* does not specifically include nor exclude examination of breast tissue but does include examination of the integument.

The *Guide* does not discriminate what parts of the body may, or may not, be examined as a part of professional practice. The section on examination of the integumentary system identifies the need to perform visual and palpatory inspection of the skin for abnormalities.

It is recognized that the *Guide* is a guide to practice that is intended to be revised periodically to reflect current physical therapy practice. Determining whether CBE should be included or excluded from the scope of a physical therapist's practice is the responsibility of each individual state.

If the therapist receives proper training to perform CBEs,* he or she must make sure this examination is allowed according to the state practice act for the state(s) in which the therapist is conducting CBEs. In some states, it is allowed by exclusion, meaning it is not mentioned and therefore included. When moving to a new state, the therapist is advised to take the time to review and understand that state's practice act.

Screening for Early Detection of Breast Cancer. The goal of screening examinations is early detection of breast cancer. Breast cancers that are detected because they are causing symptoms tend to be relatively larger and are more likely to have spread beyond the breast. In contrast, breast cancers found during screening examinations are more likely to be small and still confined to the breast.

The size of a breast neoplasm and how far it has spread are the most important factors in predicting the prognosis for anyone with this disease. According to the American Cancer Society (ACS) early detection tests for breast cancer saves many thousands of lives each year; many more lives could be saved if health care providers took advantage of these tests. Following the ACS's guidelines for the early detection of breast cancer improves the chances that breast cancer can be diagnosed at an early stage and treated successfully.[121,122]

CBE detects a small portion of cancers (4.6% to 5.9%) not found by mammography.[123] It may be important for women who do not receive regular mammograms, either because mammography is not recommended (e.g., women aged 40 and younger) or because some women do not receive screening mammography as recommended.[124]

ACS guidelines for CBE recommend CBE every 3 years for women ages 20 to 39 (more often if there are risk factors; see

*The American Cancer Society and National Cancer Institute support the provision of cancer screening procedures by qualified health specialists. With additional training, physical therapists can qualify.

An introductory course to breast cancer and CBE for the physical therapist is available. Charlie McGarvey, PT, MS, and Catherine Goodman, MBA, PT, present the course in various sites around the United States and upon request.

A certified training program is also available through *MammaCare Specialist.* The program is offered to health care professionals at training centers in the United States. The course teaches proficient breast examination skills. For more information, contact: http://www.mammacare.com/professional_training.htm.

Table 17-3). An annual CBE is advised every year for asymptomatic women ages 40 and older. Men can develop breast cancer, but this disease is about 100 times more common among women than men.[122]

Breast self-examination (BSE) is an option for women starting at age 20. Women should be educated about the benefits and limitations of the BSE. Regular BSE is one way for women to know how their breasts normally feel and to notice any changes. Advise all women during the client education portion of the physical therapy intervention to report any breast changes right away. The primary care provider should also review each client's method of performing a BSE (see Appendix D-6).

It has been generally accepted that mammography screening reduces the risk of death from breast cancer (except, perhaps, among younger women, for whom the benefit of mammography is controversial), and that this risk reduction is achieved through the detection of malignancy at an earlier stage, when treatment is more efficacious.[125] More recently, the role of mammography in lowering breast cancer–related mortality has come under closer scrutiny.[126,127]

The ideal interval between screening mammography remains under investigation. In the United States, the U.S. Preventive Services Task Force recommends screening every 1 to 2 years while the ACS recommends annual screening from age 40 years. In Europe, most countries focus on women aged 50 years and older and recommend screening every 2 years.[128]

Mammograms do not identify all breast cancers; for this reason, a *thorough* CBE remains an essential part of breast screening to complement mammograms and reduce false-negative results.[129,130]

"Thorough" is defined as palpation of the area from the collarbone to the bottom of the rib cage, one dime-size area at a time, at three levels of pressure from just below the skin, down to the mid-breast, and up against the chest wall. A thorough CBE by a specially trained practitioner should take at least five minutes per breast.[121]

It has been shown both in the laboratory and in the clinic that properly trained fingers can detect breast lesions as small as 3 mm in diameter.[131,132] As experts in the assessment and palpation of the integumentary and musculoskeletal systems, physical therapists are highly qualified professionals to receive this type of training.

Physical Therapist's Role in Screening for Breast Cancer. Historically the upper quadrant screening examination conducted by physical therapists has not included CBE. Procedures have been confined to questions posed to clients during the interview regarding past medical history (e.g., cancer, lactation, abscess, mastitis) and questions to identify the possibility of breast pathology as the underlying cause for back or shoulder pain and/or other symptoms.[120]

As health care specialists with advanced observational and palpatory skills, physical therapists can play an important role in the identification of aberrant soft tissue lesions requiring further medical evaluation. Physical therapists are professionally trained experts in tissue integrity and possess highly developed skills in the detection of various types of abnormalities. Properly trained physical therapists should be considered "qualified health care specialists" as defined by the ACS in the provision of cancer screening when the history, clinical presentation, and associated signs and symptoms point to the need for CBE.[120]

A physical therapist conducting a CBE could miss a lump (false negative finding) and not send the client for medical evaluation. However, this will most certainly occur if the therapist is not conducting a CBE to assess the integrity of the skin and soft tissues of the breast or axilla. Conversely, a physical therapist finding a lump or abnormality on CBE would refer the client to a physician for examination.[120]

A blueprint for the screening assessment is provided in Box 4-18 and Fig. 4-48. Standardized CBE and reporting of results has not been established. Best practice for CBE using inspection and palpation based on current recommendations[124,133] is presented in Appendix D-7. For a discussion of benign breast diseases and breast conditions that can refer pain to the shoulder, neck, or chest, see Chapter 17.

When a suspicious mass is found during examination, it must be medically evaluated, even if the client reports a recent mammography was "normal."[124] Clinical signs and symptoms of breast disease are listed in Chapter 17. More in-depth discussion of the role of the physical therapist in primary care and cancer screening as it relates to integrating CBE into an upper quarter examination is available.[120]

Abdomen

Anyone presenting with primary pain patterns from pathology of the abdominal organs will likely see a physician rather than a physical therapist. For this reason, abdominal and visceral assessment is not generally part of the physical therapy evaluation. When the therapist suspects referred pain from the viscera to the musculoskeletal system, this type of assessment can be helpful in the screening examination.

Abdomen: Inspection

From a screening or assessment point of view, the abdomen is divided into four quadrants centered on the umbilicus (as shown in Figs. 4-49 and 4-50). During the inspection, any abdominal scars (and associated history) should be identified (see Fig. 4-8).

Note the color of the skin and the presence and location of any scars, striae from pregnancy or weight gain/loss, petechiae, or spider angiomas (see Fig. 9-3). A bluish discoloration around the umbilicus (Cullen's sign) or along the lower abdomen and flanks (Grey Turner's sign) may be the sign of a retroperitoneal bleed (e.g., pancreatitis, ruptured ectopic pregnancy, posterior perforated ulcer). The color may be a shade of blue-red, blue-purple, or green-brown, depending on the stage of hemoglobin breakdown.[35]

From a seated position next to the client, note the contour of the abdomen and look for any asymmetry. Repeat the same visual inspection while standing behind the client's head. Generalized distention can accompany gas, whereas

BOX 4-18 SCREENING FOR BREAST DISEASE

Upper Quadrant Examination

Many conditions can cause breast pain or refer pain to the breast (see Table 17-2). For anyone with neck, upper back, chest, scapular, or shoulder symptoms, a screening examination of the upper quadrant may be necessary. The physical therapy differential diagnosis requires determination of the underlying soft tissue involvement and assesses the need for medical referral. The screening examination includes past and current medical history, clinical examination, review of systems, and review of any associated signs and symptoms.

Client Interview

Past Medical History

Look for a history of trauma, recent birth, overuse, increased abdominal breathing (e.g., prolonged running, lifting), cardiovascular disease, cancer, GI involvement, or other systemic conditions that can refer pain/symptoms to the upper quadrant.

Ask about a previous history of breast disease of any kind (e.g., chronic mastitis, cancer, Paget's disease).

Assess Risk Factors

See Table 17-3.

Special Questions to Ask: Breast

- Have you ever had any breast surgery (implants, lumpectomy, mastectomy, reconstructive surgery, or augmentation)?
- Have you ever been treated for cancer of any kind? If yes, when? What?
- Do you have any discharge from your breasts or nipples? If yes, do you know what is causing this discharge? Have you received medical treatment for this problem?
- Have you noticed any other changes in your breast(s)? For example, are there any noticeable, bulging, or distended veins, puckering, swelling, or any other skin changes?
- Are you nursing an infant (lactating)?
- Have you examined yourself for any lumps or nodules and found any thickening or lump? If yes, has your physician examined/treated this? If no, do you examine your own breasts? (Follow-up questions [e.g., last breast examination by self or health care professional?])
- Have you been involved in any activities of a repetitive nature that could cause sore muscles (e.g., painting, washing walls, push-ups or other calisthenics, heavy lifting or pushing, overhead movements, prolonged running, or fast walking)?
- Have you recently been coughing excessively?
- Have you ever had angina (chest pain) or a heart attack (residual trigger points)?
- Have you been in a fight or hit, punched, or pushed against any object that injured your chest or breast (assault)?
- See also Special Questions to Ask: Lymph Nodes (this chapter and Appendix B-22)

Clinical Presentation

Visual Inspection (Posture, Skin, Breast)

Look for asymmetry, nipple retraction, ulceration, erythema, peau d'orange, or other skin changes. Have the client raise arms overhead. Assess the lower half of the breast and the inframammary folds. Have the client place hands on hips and press down to contract the pectoralis major muscles. Observe for any undetected asymmetries or changes. Have the client bend forward with arms relaxed at his/her sides. Again, look for any undetected asymmetries or changes.

Screening Assessment/Evaluation

- MMT, neurologic screen (reflexes, sensation, proprioception), trigger point assessment
- Neck, shoulder, clavicle, sternum ROM (AROM, PROM, accessory motions)

Palpation

Palpate appropriate soft tissue structures according to history and evaluation results. This may include breast, axilla, and lymph nodes, especially the supraclavicular nodes, since these are the typical nodes assessed during CBE procedures in order to stage the disease. Palpation of the breast is performed with the client in the supine position. The shoulder/scapular area may be supported with a small pillow or foam wedge.

Palpate from the midaxillary line to the sternum and from the clavicle to inframammary fold. Assess the entire thickness of the breast parenchyma. Patterns of palpation include: horizontal strip, radial, circular, and vertical strip. There is some evidence to suggest that the vertical pattern (up and down pattern) is the most effective pattern for covering the entire breast, without missing any breast tissue.[143]

Other Clues

- Resisted movements reproduce pain/symptoms (look for musculoskeletal cause)
- Response to stretch (reduces or eliminates pain/symptoms [look for musculoskeletal cause])
- Eliminated by treatment (stretching, trigger point therapy [look for musculoskeletal cause])
- Assess effect of lower extremity exertion only (screen for cardiovascular disease)
- Assess for 3 P's: pain on **p**alpation (myalgia), change in symptoms with change in **p**osition, symptoms increase with respiratory movements (**p**leuritic pain)

Associated Signs and Symptoms of Breast Disease

Palpable breast nodules or lumps
Skin surface red, warm, edematous, firm, painful
Firm, painful site under the skin surface
Skin dimpling over the lesion with attachment of mass to surrounding tissues
Unusual nipple discharge
Pain aggravated by jarring or movement of the breasts
Pain that is not aggravated by resistance to isometric movements of the upper extremities

GI, Gastrointestinal; *MMT,* manual muscle testing; *ROM,* range of motion; *AROM,* active ROM; *PROM,* passive ROM; *CBE,* clinical breast examination.

Breast and Lymph Node Examination – Palpable Findings

Please refer to Appendix: Clinical Breast Examination (CBE) Recommended Procedures when conducting the assessment and completing this form.

Date:________________ First day (date) of last (menstrual) period:________________ Client's name:________________

Description of lump:
None (circle if exam results are normal); skip to lymph node assessment below

Location	**Size** (cm x cm)	**Contour**	**Shape**
Side: Right		Irregular Smooth	Describe:
Record o'clock position			
Side: Left		Irregular Smooth	Describe:
Record o'clock position			

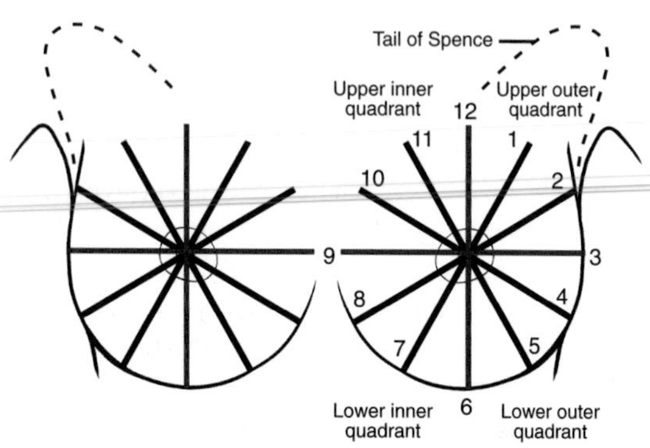

Note location of mass by quadrant, place on the clock and distance from the nipple

Density of lump/lesion	**Tenderness**	**Mobility**
Soft	Yes	Mobile
Rubbery	No	Fixed
Hard		

Description of lymph nodes:
None (exam results are normal)

Location	**Zone** (see Fig. 4-40)			**Density**	**Mobility**	**Tenderness**
Side: Right	I	II	III	Soft Rubbery Hard	Mobile Fixed	Yes No
Side: Left	I	II	III	Soft Rubbery Hard	Mobile Fixed	Yes No

Note: Keep in mind the following general guidelines for breast lumps.
Refer all suspicious lesions or aberrant tissue for medical evaluation and diagnosis.

Benign	**Malignant**
Soft	Firm
Smooth	Irregular
Moveable	Fixed

Notes:

Fig. 4-48 Breast and lymph node examination.

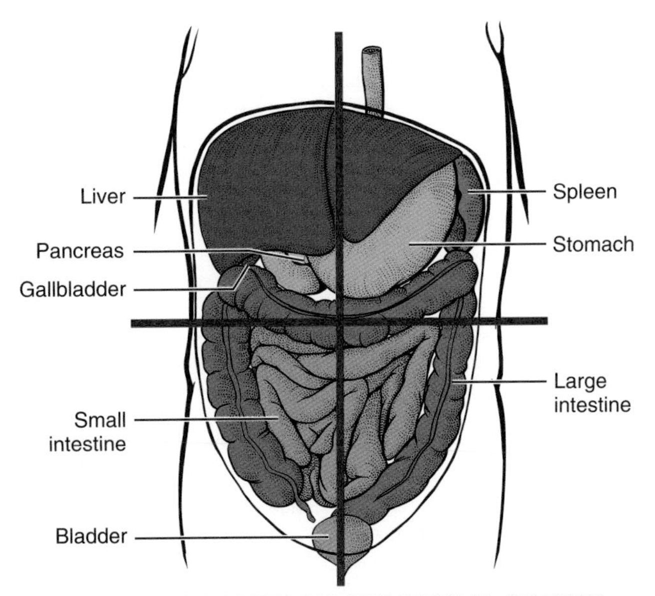

Fig. 4-49 Four abdominal (anterior) quadrants formed by two imaginary perpendicular lines running through the umbilicus. As a general rule, viscera in the right upper quadrant (RUQ) can refer pain to the right shoulder; viscera in the left upper quadrant (LUQ) can refer pain to the left shoulder; viscera in the lower quadrants are less specific and can refer pain to the pelvis, pelvic floor, groin, low back, hip, and sacroiliac/sacral areas (see also Figs. 3-4 and 3-5).

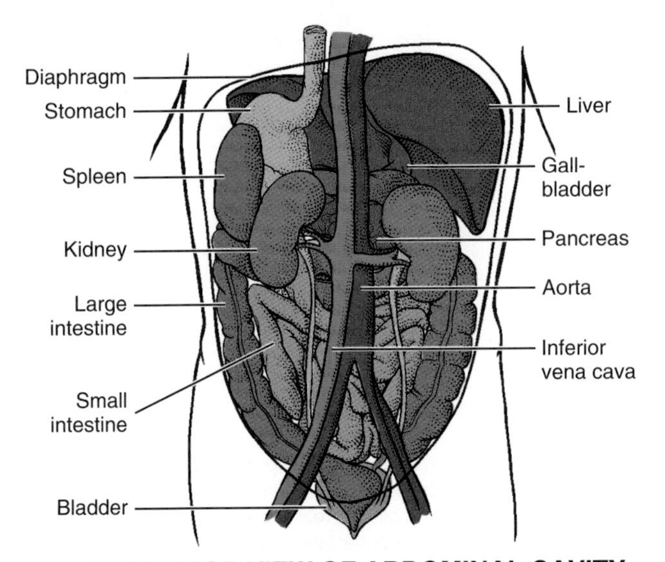

Fig. 4-50 Posterior view of the abdomen. The abdominal aorta passes from the diaphragm through the abdominal cavity, just to the left of the midline. It branches into the left and right common iliac arteries at the level of the umbilicus. Retroperitoneal bleeding from any of the posterior elements (e.g., stomach, spleen, aorta, kidneys) can cause back, hip, and/or shoulder pain.

local bulges may occur with a distended bladder or hernia. Make note if the umbilicus is displaced in any direction or if there are any masses, pulsations, or movements of the abdomen. Visible peristaltic waves are not normal and may signal a GI problem. Document the presence of ascites (see Fig. 9-8).

Clients with an organic cause for abdominal pain usually are not hungry. Ask the client to point to the location of the pain. Pain corresponding to the epigastric, periumbilical, and lower midabdominal regions is shown in Fig. 8-2. If the finger points to the navel, but the client seems well and is not in any distress, there may be a psychogenic source of symptoms.[134]

Abdomen: Auscultation

In a screening examination, the therapist may auscultate the four abdominal quadrants for the presence of abdominal sounds. Expect to hear clicks, rumblings, and gurgling sounds every few seconds throughout the abdomen. Auscultation should occur before palpation and/or percussion to avoid altering bowel sounds.

The absence of sounds or very few sounds in any or all of the quadrants is a red flag and is most common in the older adult with multiple risk factors such as recent abdominal, back, or pelvic surgery and the use of narcotics or other medications.

As previously mentioned the therapist may auscultate the abdomen for vascular sounds (e.g., bruits) when the history (e.g., age over 65, history of coronary artery disease), clinical presentation (e.g., neck, back, abdominal, or flank pain), and associated signs and symptoms (e.g., syncopal episodes, signs and symptoms of PVD) warrant this type of assessment.

Listen for pulsations/bruits over the aorta, renal arteries, iliac arteries, and femoral arteries first and remember to do so before palpation because palpation may stir up the bowel contents and increase peristalsis, making auscultation more difficult.

Abdomen: Percussion and Palpation

Remember to auscultate first before percussion and palpation in order to avoid altering the frequency of bowel sounds. Percussion over normal, healthy abdominal organs is an advanced skill even among doctors and nurses and is not usually an integral part of the physical therapy screening examination.

Palpation (light and deep) of all four abdominal quadrants is a separate skill used to assess for temperature changes, tenderness, and large masses. Keep in mind that even a skilled clinician will not be able to palpate abdominal organs in an obese person.

Most viscera in the normal adult are not palpable unless enlarged. Anatomical structures can be mistaken for an abdominal mass. Palpation is contraindicated in anyone with a suspected abdominal aortic aneurysm, appendicitis, or a known kidney disease, or who has had an abdominal organ transplantation.

When palpation is carried out, always explain to your client what test you are going to perform and why. Make sure the person being examined has an empty bladder. Examine any painful areas last. Use proper draping and warm your hands. Have the client bend the knees with the feet flat on the exam table to put the abdominal muscles in a relaxed position. During palpation, if the person is ticklish or tense, place his or her hand on top of your palpating hand. Ask him or her to breathe in and out slowly and regularly. The tickle response disappears in the presence of a truly acute abdomen.[135] To help distinguish between voluntary abdominal muscle guarding and involuntary abdominal rigidity, have the client exhale or breathe through the mouth. Voluntary guarding usually decreases using these techniques.

Start with a light touch, moving the fingers in a circular motion, slowly and gently. If the abdominal muscles are contracted, observe the client as he or she breathes in and out. The contraction is voluntary if the muscles are more strongly contracted during inspiration and less strong during expiration. Muscles firmly contracted throughout the respiratory cycle (inspiration and expiration) are more likely to be involuntary, possibly indicating an underlying abdominal problem. To check for rebound tenderness, the "pinch-an-inch" test is preferred over the Rebound Tenderness (Blumberg's sign) (see Figs. 8-10 to 8-12).

A variation on the Blumberg sign (for peritonitis) is the Rovsing sign (suggesting appendicitis). Rovsing sign is elicited by pushing on the abdomen in the left lower quadrant (away from the appendix as in most people the appendix is in the right lower quadrant). While this maneuver stretches the entire peritoneal lining, it only causes pain in any location where the peritoneum is irritating the muscle. In the case of appendicitis, the pain is felt in the right lower quadrant despite pressure being placed elsewhere.

If left lower quadrant pressure by the examiner leads only to left-sided pain or pain on both the left and right sides, then there may be some other pathologic etiology (e.g., bladder, uterus, ascending [right] colon, fallopian tubes, ovaries, or other structures).

Liver. Liver percussion to determine its size and identify its edges is a skill beyond the scope of a physical therapist for a screening examination. Therapists involved in visceral manipulation will be most likely to develop this advanced skill.

To palpate the liver (Fig. 4-51), have the client take a deep breath as you feel deeply beneath the costal margin. During inspiration, the liver will move down with the diaphragm so that the lower edge may be felt below the right costal margin.

A normal adult liver is not usually palpable and palpation is not painful. Cirrhosis, metastatic cancer, infiltrative leukemia, right-sided CHF, and third-stage (tertiary) syphilis can cause an enlarged liver. The liver in clients with COPD is more readily palpable as the diaphragm moves down and pushes the liver below the ribs. The liver is often palpable 2 to 3 cm below the costal margin in infants and young children.

If you come in contact with the bottom edge tucked up under the rib cage, the normal liver will feel firm, smooth, even, and rubbery. A palpable hard or lumpy edge warrants further investigation. Some clinicians prefer to stand next to the client near his head, facing his feet. As the client breathes in, curl the fingers over the costal margin and up under the ribs to feel the liver (Fig. 4-52).

Spleen. As with other organs, the spleen is difficult to percuss, even more so than the liver, and is not part of a screening examination.

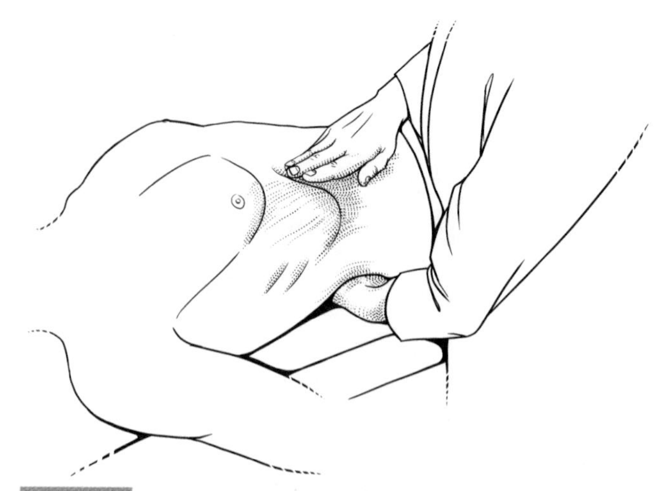

Fig. 4-51 Palpating the liver. Place your left hand under the client's right posterior thorax (as shown) parallel to and at the level of the last two ribs. Place your right hand on the client's RUQ over the midclavicular line. The fingers should be pointing toward the client's head and positioned below the lower edge of liver dullness (previously mapped out by percussion). Ask the client to take a deep breath while pressing inward and upward with the fingers of the right hand. Attempt to feel the inferior edge of the liver with your right hand as it descends below the last rib anteriorly.

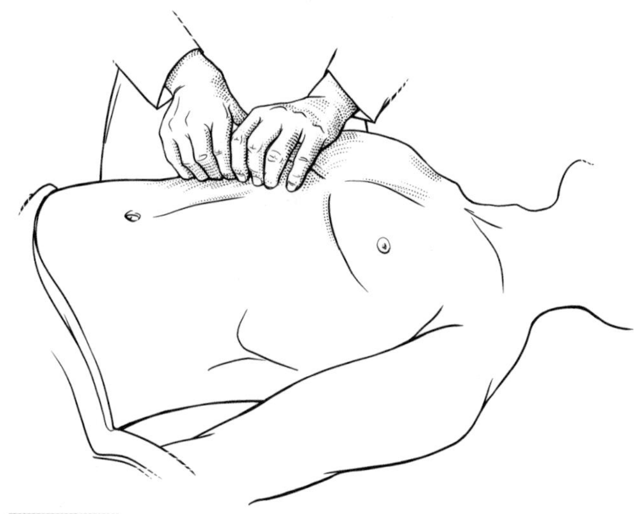

Fig. 4-52 An alternate way to palpate the liver. Hook the fingers of one or both hands (depending on hand size) up and under the right costal border. As the client breathes in, the liver descends and the therapist may come in contact with the lower border of the liver. If this procedure elicits exquisite tenderness, it may be a positive *Murphy's sign* for acute cholecystitis (not the same as Murphy's percussion test [costovertebral tenderness] of the kidney depicted in Fig. 4-54).

Palpation of the spleen is not possible unless it is distended and bulging below the left costal margin (Fig. 4-53). The spleen enlarges with mononucleosis and trauma. Do not continue to palpate an enlarged spleen because it can rupture easily. Report to the physician immediately how far it extends below the left costal margin and request medical evaluation.

Gallbladder and Pancreas. Likewise, the gallbladder tucked up under the liver (see Figs. 9-1 and 9-2) is not palpable unless grossly distended. To palpate the gallbladder, ask the person to take a deep breath as you palpate deep below

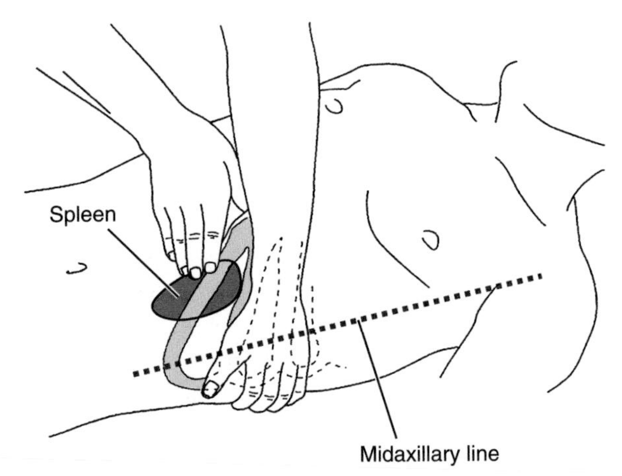

Fig. 4-53 Palpation of the spleen. The spleen is not usually palpable unless it is distended and bulging below the left costal margin. Left shoulder pain can occur with referred pain from the spleen. Stand on the person's right side and reach across the client with the left hand, placing it beneath the client over the left costovertebral angle. Lift the spleen anteriorly toward the abdominal wall. Place the right hand on the abdomen below left costal margin. Using findings from percussion, gently press fingertips inward toward the spleen while asking the client to take a deep breath. Feel for spleen as it moves downward toward the fingers. (From Leasia MS, Monahan FD: *A practical guide to health assessment,* ed 2, Philadelphia, 2002, WB Saunders.)

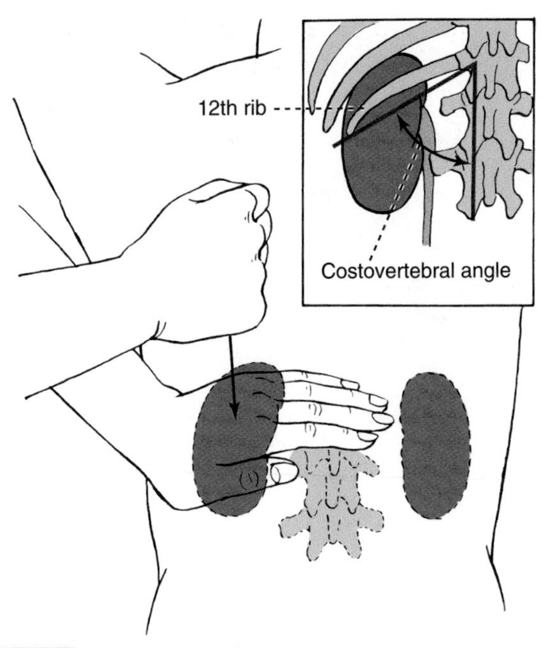

Fig. 4-54 Murphy's percussion also known as the test for *costovertebral tenderness.* Murphy's percussion is used to rule out involvement of the kidney and assess for pseudorenal pain (see discussion, Chapter 10). Indirect fist percussion causes the kidney to vibrate. To assess the kidney, position the client prone or sitting and place one hand over the rib at the costovertebral angle on the back. Give the hand a percussive thump with the ulnar edge of your other fist. The person normally feels a thud but no pain. Reproduction of back and/or flank pain with this test is a red-flag sign for renal involvement (e.g., kidney infection or inflammation). (From Black JM, Matassarin-Jacobs E, editors: *Luckmann and Sorensen's medical-surgical nursing,* ed 4, Philadelphia, 1993, WB Saunders.)

the liver margin. Only an abnormally enlarged gallbladder can be palpated this way.

The pancreas is also inaccessible; it lies behind and beneath the stomach with its head along the curve of the duodenum and its tip almost touching the spleen (see Fig. 9-1). A round, fixed swelling above the umbilicus that does not move with inspiration may be a sign of acute pancreatitis or cancer in a thin person.

Kidneys. The kidneys are located deep in the retroperitoneal space in both upper quadrants of the abdomen. Each kidney extends from approximately T12 to L3. The right kidney is usually slightly lower than the left.

Percussion of the kidney is accomplished using Murphy's percussion test (Fig. 4-54). Although this test is commonly performed, its diagnostic value has never been validated. Results of at least one Finnish study[136] suggested that in acute renal colic loin tenderness and hematuria (blood in the urine) are more significant signs than renal tenderness.[137] A diagnostic score incorporating independent variables, including results of urinalysis, presence of costovertebral angle tenderness and renal tenderness, as well as duration of pain, appetite level, and sex (male versus female), reached a sensitivity of 0.89 in detecting acute renal colic, with a specificity of 0.99 and an efficiency of 0.99.[136]

To palpate the kidney, stand on the right side of the supine client. Place your left hand beneath the client's right flank. Flex the left metacarpophalangeal joints (MCPs) in the renal angle while pressing downward with the right hand against the right outer edge of the abdomen. This method compresses the kidney between your hands. The left kidney is usually not palpable because of its position beneath the bowel.

Kidney transplants are often located in the abdomen. The therapist should not percuss or palpate the kidneys of anyone with chronic renal disease or organ transplantation.

Bladder. The bladder lies below the symphysis pubis and is not palpable unless it becomes distended and rises above the pubic bone. Primary pain patterns for the bladder are shown in Fig. 10-9. Sharp pain over the bladder or just above the symphysis pubis can also be caused by abdominal gas. The presence of associated GI signs and symptoms and lack of urinary tract signs and symptoms may be helpful in identifying this pain pattern.

Aortic Bifurcation. It may be necessary to assess for an abdominal aneurysm, especially in the older client with back pain and/or who reports a pulsing or pounding sensation in the abdomen during increased activity or while in the supine position.

The ease with which the aortic pulsations can be felt varies greatly with the thickness of the abdominal wall and the anteroposterior diameter of the abdomen. To palpate the aortic pulse, the therapist should press firmly deep in the upper abdomen (slightly to the left of the midline) to find the aortic pulsations (Fig. 4-55, *A*).

The therapist can assess the width of the aorta by using both hands (one on each side of the aorta) and pressing deeply. The examiner's fingers along the outer margins of the aorta should remain the same distance apart until the aortic

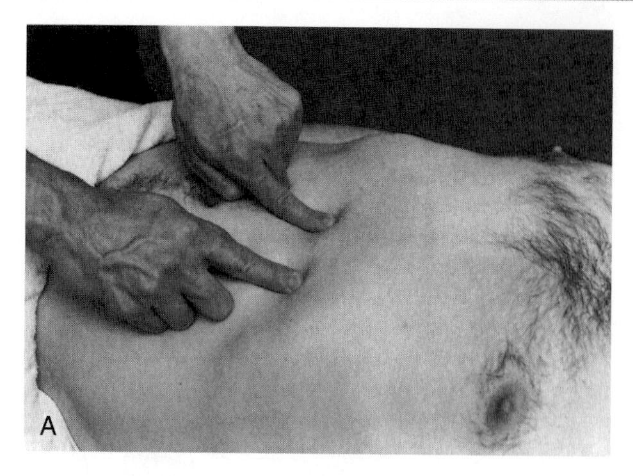

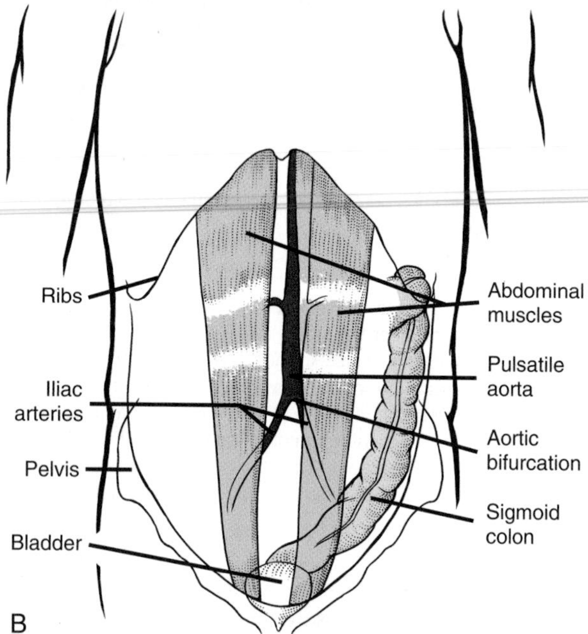

Fig. 4-55 **A,** Place one hand or one finger on either side of the aorta as shown here. Press firmly deep in the upper abdomen just to the left of the midline. You may feel aortic pulsations. These pulsations are easier to appreciate in a thin person and more difficult to feel in someone with a thick abdominal wall or large anteroposterior diameter of the abdomen. Obesity and abdominal ascites or distention makes this more difficult. Use the stethoscope (bell) to listen for bruits. Bruits are abnormal blowing or swishing sounds heard on auscultation of the arteries. Bruits with both systolic and diastolic components suggest the turbulent blood flow of partial arterial occlusion. If the renal artery is occluded as well, the client will be hypertensive. **B,** Visualize the location of the aorta slightly to the left of midline and its bifurcation just below the umbilicus. See text (this chapter and Chapter 6) for discussion of normal and average pulse widths. The pulse width expands at the aortic bifurcation (again, usually just below the umbilicus). An aneurysm can occur anywhere along the aorta; 95% of all abdominal aortic aneurysms occur just below the renal arteries. (**A** from Boissonnault WG, editor: *Examination in physical therapy practice,* ed 2, New York, 1995, Churchill Livingstone.)

bifurcation. Where the aorta bifurcates (usually near the umbilicus), the width of the pulse should expand (Fig. 4-55, *B*). The normal aortic pulse width is between 2.5 and 4.0 cm (some sources say the width must be no more than 3.0 cm; others list 4.0 cm). Average pulse width is 2.5 cm or about 1.2 inches wide.[138,139] See Chapter 6 for a more in-depth discussion of aortic pulse width norms.

Throbbing pain that increases with exertion and is accompanied by a sensation of a heart beat when lying down and of a palpable pulsating abdominal mass requires immediate medical attention. Remember, the presence of an abdominal bruit accompanied by risk factors for an aortic aneurysm may be a contraindication to abdominal palpation.

SYSTEMS REVIEW ... OR ... REVIEW OF SYSTEMS?

When a physician conducts a systems review, the examination is a routine physical assessment of each system, starting with ear, nose, and throat (ENT), followed by chest auscultation for pulmonary and cardiac function, palpation of lymph nodes, and so on.

The therapist does not conduct a systems review such as the medical doctor performs. The *Guide* uses the terminology "Systems Review" to describe a brief or limited exam of the anatomic and physiologic status of the cardiovascular/pulmonary, integumentary, musculoskeletal, and neuromuscular systems.

As part of this Systems Review, the client's ability to communicate and process information is identified as well as any learning barriers such as hearing or vision impairment, illiteracy, or English as a second language (see the *Guide*, pp. 713/s705).

A more appropriate term for what the therapist does in the screening process may be a "Review of Systems." In the screening examination as presented in this text, after conducting an interview and assessment of pain type/pain patterns, the therapist may conduct a review of systems (ROS) first by looking for any characteristics of systemic disease in the history or clinical presentation.

Then, the identified cluster(s) of associated signs and symptoms are reviewed to search for a potential pattern that will identify the underlying system involved (Box 4-19). The therapist may find cause to examine just the upper quadrant or just the lower quadrant more closely. A guide to physical assessment in a screening examination is provided in Table 4-13.

Therapists perform the ROS by categorizing all of the complaints and reported or observed associated signs and symptoms. This type of review helps bring to the therapist's attention any signs or symptoms the client has not recognized, has forgotten, or thought unimportant. After compiling a list of the client's signs and symptoms, compare those to the list in Box 4-19. Are there any identifying clusters to direct the decision-making process? See Case Example 4-4.

For example, cutaneous (skin) manifestations and joint pain may occur secondary to systemic diseases, such as

BOX 4-19 REVIEW OF SYSTEMS

When conducting a general review of systems, ask the client about the presence of any other problems anywhere else in the body. Depending on the client's answer you may want to prompt him or her about any of the following common signs and symptoms* associated with each system:

General Questions

___Fever, chills, sweating (constitutional symptoms)
___Appetite loss, nausea, vomiting (constitutional symptoms)
___Fatigue, malaise, weakness (constitutional symptoms)
___Excessive, unexplained weight gain or loss
___Vital signs: blood pressure, temperature, pulse, respirations, pain, walking speed
___Insomnia
___Irritability
___Hoarseness or change in voice, frequent or prolonged sore throat
___Dizziness, falls

Integumentary (Include Skin, Hair, and Nails)

___Recent rashes, nodules, or other skin changes
___Unusual hair loss or breakage
___Increased hair growth (hirsutism)
___Nail bed changes
___Itching (pruritus)

Musculoskeletal/Neurologic

___Joint pain, redness, warmth, swelling, stiffness, deformity
___Frequent or severe headaches
___Vision or hearing changes
___Vertigo
___Paresthesias (numbness, tingling, "pins and needles" sensation)
___Change in muscle tone
___Weakness; atrophy
___Abnormal deep tendon (or other) reflexes
___Problems with coordination or balance; falling
___Involuntary movements; tremors
___Radicular pain
___Seizures or loss of consciousness
___Memory loss
___Paralysis
___Mood swings; hallucinations

Rheumatologic

___Presence/location of joint swelling
___Muscle pain, weakness
___Skin rashes
___Reaction to sunlight
___Raynaud's phenomenon
___Nail bed changes

Cardiovascular

___Chest pain or sense of heaviness or discomfort
___Palpitations
___Limb pain during activity (claudication; cramps, limping)
___Discolored or painful feet; swelling of hands and feet
___Pulsating or throbbing pain anywhere, but especially in the back or abdomen
___Peripheral edema; nocturia
___Sudden weight gain; unable to fasten waistband or belt, unable to wear regular shoes
___Persistent cough
___Fatigue, dyspnea, orthopnea, syncope
___High or low blood pressure, unusual pulses
___Differences in blood pressure from side to side with position change (10 mm Hg or more; increase or decrease/diastolic or systolic; associated symptoms: dizziness, headache, nausea, vomiting, diaphoresis, heart palpitations, increased primary pain or symptoms)
___Positive findings on auscultation

Pulmonary

___Cough, hoarseness
___Sputum, hemoptysis
___Shortness of breath (dyspnea, orthopnea); altered breathing (e.g., wheezing, pursed-lip breathing)
___Night sweats; sweats anytime
___Pleural pain
___Cyanosis, clubbing
___Positive findings on auscultation (e.g., friction rub, unexpected breath sounds)

Psychologic

___Sleep disturbance
___Stress levels
___Fatigue, psychomotor agitation
___Changes in personal habits, appetite
___Depression, confusion, anxiety
___Irritability, mood changes

Gastrointestinal

___Abdominal pain
___Indigestion; heartburn
___Difficulty in swallowing
___Nausea/vomiting; loss of appetite
___Diarrhea or constipation
___Change in stools; change in bowel habits
___Fecal incontinence
___Rectal bleeding; blood in stool; blood in vomit
___Skin rash followed by joint pain (Crohn's disease)

*Cluster of three to four or more lasting longer than 1 month.

Continued

BOX 4-19 REVIEW OF SYSTEMS—cont'd

Hepatic/Biliary
___Change in taste/smell
___Anorexia
___Feeling of abdominal fullness, ascites
___Asterixis (muscle tremors)
___Change in urine color (dark, cola-colored)
___Light-colored stools
___Change in skin color (yellow, green)
___Skin changes (rash, itching, purpura, spider angiomas, palmar erythema)

Hematologic
___Skin color or nail bed changes
___Bleeding: nose, gums, easy bruising, melena
___Hemarthrosis, muscle hemorrhage, hematoma
___Fatigue, dyspnea, weakness
___Rapid pulse, palpitations
___Confusion, irritability
___Headache

Genitourinary
___Reduced stream, decreased output
___Burning or bleeding during urination; change in urine color
___Urinary incontinence, dribbling
___Impotence, pain with intercourse
___Hesitation, urgency
___Nocturia, frequency
___Dysuria (painful or difficult urination)
___Testicular pain or swelling
___Genital lesions
___Penile or vaginal discharge
___Impotence (males) or other sexual difficulty (males or females)
___Infertility (males or females)
___Flank pain

Gynecologic
___Irregular menses, amenorrhea, menopause
___Pain with menses or intercourse
___Vaginal discharge, vaginal itching
___Surgical procedures
___Pregnancy, birth, miscarriage, and abortion histories
___Spotting, bleeding, especially for the postmenopausal woman 12 months after last period (without hormone replacement therapy)

Endocrine
___Hair and nail changes
___Change in appetite, unexplained weight change
___Fruity breath odor
___Temperature intolerance, hot flashes, diaphoresis (unexplained perspiration)
___Heart palpitations, tachycardia
___Headaches
___Low urine output, absence of perspiration
___Cramps
___Edema, polyuria, polydipsia, polyphagia
___Unexplained weakness, fatigue, paresthesia
___Carpal/tarsal tunnel syndrome
___Periarthritis, adhesive capsulitis
___Joint or muscle pain (arthralgia, myalgia), trigger points
___Prolonged deep tendon reflexes
___Sleep disturbance

Cancer
___Constant, intense pain, especially bone pain at night
___Unexplained weight loss (10% of body weight in 10-14 days); most clients in pain are inactive and gain weight
___Loss of appetite
___Excessive fatigue
___Unusual lump(s), thickening, change in a lump or mole, sore that does not heal; other unusual skin lesions or rash
___Unusual or prolonged bleeding or discharge anywhere
___Change in bowel or bladder habits
___Chronic cough or hoarseness, change in voice
___Rapid onset of digital clubbing (10-14 days)
___Proximal muscle weakness, especially when accompanied by change in one or more deep tendon reflexes

Immunologic
___Skin or nail bed changes
___Fever or other constitutional symptoms (especially recurrent or cyclical symptoms)
___Lymph node changes (tenderness, enlargement)
___Anaphylactic reaction
___Symptoms of muscle or joint involvement (pain, swelling, stiffness, weakness)
___Sleep disturbance

Crohn's disease (regional enteritis) or psoriatic arthritis, or as a delayed reaction to medications. Likewise, hair and nail changes, temperature intolerance, and unexplained excessive fatigue are cluster signs and symptoms associated with the endocrine system.

Changes in urinary frequency, flow of urine, or color of urine point to urologic involvement. Other groupings of signs and symptoms associated with each system are listed as mentioned in Box 4-19. If, for example, the client's signs and symptoms fall primarily within the genitourinary group, then turn to Chapter 10 for additional, pertinent screening questions listed at the end of the chapter. The client's answers to these questions will guide the therapist in making a final decision regarding physician referral (see the Case Study at the end of the chapter).

The therapist is not responsible for identifying the specific pathologic disease underlying the clinical signs and symptoms present. However, the alert therapist who recognizes

TABLE 4-13 Guide to Physical Assessment in a Screening Examination

General Survey	Upper Quadrant Exam	Lower Quadrant Exam
Level of consciousness Mental and emotional status Vision and hearing Speech General appearance Nutritional status Level of self-care Body size and type (body mass index [BMI]) Obvious deformities Muscle atrophy Posture Body and breath odors Posture Movement patterns and gait Use of assistive devices or mobility aids Balance and coordination Inspect skin, hair, and nails Vital signs	Lymph node palpation Head and neck Cranial nerves Upper limbs • Muscle tone and strength • Trigger points • Joint range of motion (ROM) • Reflexes • Coordination • Vascular assessment • Motor and sensory function • Vascular assessment Chest and back (heart and lungs) • Inspection • Palpation • Auscultation Clinical breast examination (CBE)	Lymph node palpation Lower limbs • Muscle tone and strength • Trigger points • Joint ROM • Reflexes • Coordination • Motor and sensory function • Vascular assessment Abdomen • Inspection • Auscultation • Palpation

clusters of signs and symptoms of a systemic nature will be more likely to identify a problem outside the scope of physical therapy practice and make the appropriate referral. Early identification and intervention for many medical conditions can result in improved outcomes, including decreased morbidity and mortality.

As mentioned in Chapter 2, there is some consideration being given to possibly changing the terminology in the *Guide* to reflect the full measure of these concepts, but no definitive decision had been made by the time this text went to press. The idea of Systems Review versus Review of Systems will be discussed and any decision made will go through both an expert and wide review process.

PHYSICIAN REFERRAL

All skin and nail bed lesions must be examined and evaluated carefully because any tissue irregularity may be a clue to malignancy. It is better to err on the side of being too quick to refer for medical evaluation than to delay and risk progression of underlying disease.

Medical evaluation is advised when the therapist is able to palpate a distended liver, gallbladder, or spleen. This is especially true in the presence of any cluster of signs and symptoms observed during the ROS that are characteristic of that particular organ system.

CASE EXAMPLE 4-4

Steps in the Screening Process

A 47-year-old man with low back pain of unknown cause has come to you for exercises. After gathering information from the client's history and conducting the interview, you ask him:

- Are there any other symptoms of any kind anywhere else in your body?

The client tells you he does break out into an unexpected sweat from time to time but does not think he has a temperature when this happens. He has increased back pain when he passes gas or has a bowel movement, but then the pain goes back to the "regular" pain level (reported as 5 on a scale from 0 to 10).

Other reported symptoms include:

- Heartburn and indigestion
- Abdominal bloating after meals
- Chronic bronchitis from smoking (3 packs/day)
- Alternating diarrhea and constipation

Do these symptoms fall into any one category? See Box 4-19.

What is the next step?

It appears that many of the symptoms are gastrointestinal in nature. Since the client has mentioned unexplained sweating, but no known fevers, take the time to measure all vital signs, especially body temperature.

Turn to the Special Questions to Ask at the end of Chapter 8 and scan the list of questions for any that might be appropriate with this client.

For example, find out about the use of nonsteroidal antiinflammatories (prescription and over-the-counter [OTC]; be sure to include aspirin). Follow up with:

- Have you ever been treated for an ulcer or internal bleeding while taking any of these pain relievers?
- Have you experienced any unexpected weight loss in the last few weeks?
- Have you traveled outside the United States in the last year?
- What is the effect of eating or drinking on your abdominal pain? Back pain?
- Have the client pay attention to his symptoms over the next 24 to 48 hours:
  - Immediately after eating
  - Within 30 minutes of eating
  - One to two hours later
- Do you have a sense of urgency so that you have to find a bathroom for a bowel movement or diarrhea right away without waiting?

Your decision to refer this client to a physician depends on your findings from the clinical examination and the client's responses to these questions. This does not appear to be an emergency since the client is not in acute distress. An elevated temperature or other unusual vital signs might speed along the referral process. Documentation of the screening process is important, and the physician should be notified appropriately (by phone, fax, and/or report).

Headaches that cannot be linked to a musculoskeletal cause (e.g., dysfunction of the cervical spine, thoracic spine, or temporomandibular joints; muscle tension; poor posture) may need further medical referral and evaluation. Conducting an upper quadrant physical screening assessment, including a neurologic screening may help in making this decision.

Vital Signs

Vital sign assessment is a very important and valuable screening tool. If the therapist does not conduct any other screening physical assessment, vital signs should be assessed for a baseline value and then monitored. The following findings should always be documented and reported:

- Any of the yellow caution signs presented in Box 4-7.
- Individuals who report consistent ambulatory hypertension using out-of-office or at-home readings, adults who do not show a drop in BP at night (15 mm Hg lower than daytime measures), and older adults with diagnosed hypertension who have an excessive surge in morning BP measurements.
- African Americans with elevated BP should be evaluated by a medical doctor.
- Pulse amplitude that fades with inspiration and strengthens with expiration.
- Irregular pulse and/or irregular pulse combined with symptoms of dizziness or shortness of breath (SOB); tachycardia or bradycardia.
- Pulse increase over 20 bpm lasting more than 3 minutes after rest or changing position.
- Persistent low-grade (or higher) fever, especially associated with constitutional symptoms, most commonly sweats but also unintended weight loss, malaise, nausea, vomiting.
- Any unexplained fever without other systemic symptoms, especially in the person taking corticosteroids or otherwise immunosuppressed.
- Weak and rapid pulse accompanied by fall in BP (pneumothorax).
- Clients who are neurologically unstable as a result of a recent CVA, head trauma, spinal cord injury, or other central nervous system insult often exhibit new arrhythmias during the period of instability; when the client's pulse is monitored, any new arrhythmias noted should be reported to the nursing staff or physician.
- Always take BP in any client with neck pain, upper quadrant symptoms, or thoracic outlet syndrome (TOS).

Precautions/Contraindications to Therapy

The following parameters are listed as precautions/contraindications rather than one or the other because these signs and symptoms may have different significance depending on the client's overall health, age, medications taken, and other factors. What may be a precaution for one client may be a clear contraindication for another and vice versa.

- Resting heart rate 120 to 130 bpm*
- Resting systolic pressure 180 to 200 mm Hg*
- Resting diastolic pressure 105 to 110 mm Hg*
- Marked dyspnea
- Loss of palpable pulse or irregular pulse with symptoms of dizziness, nausea, or SOB

Anemic individuals may demonstrate an increased normal resting pulse rate that should be monitored during exercise. Anyone with unstable BP may require initial standing with a tilt table or monitoring of the BP before, during, and after treatment. Check the nursing record for pulse rate at rest and BP to use as a guide when taking vital signs in the clinic or at the patient's bedside.

*Unexplained or poorly tolerated by client.

Guidelines for Immediate Physician Referral

- Anyone with diabetes mellitus, who is immunocompromised, or who has a history of steroid and retroviral use now presenting with red, inflamed, swollen nail bed(s) or any skin lesion involving the feet must be referred for medical evaluation immediately.
- Any suspicious breast changes (e.g., unexplained nipple discharge, erythema, contour changes) must be reported to the physician immediately. Nipple discharge can occur during and after normal lactation or the use of oral contraceptives. Bloody discharge can occur during the last trimester of pregnancy or first 3 months of lactation.[124] The physician must decide when this finding is considered "normal" physiologically. Unilateral discharge is highly suspicious. Some medications can also produce nipple discharge (e.g., digitalis, tricyclic antidepressants, benzodiazepines, antipsychotics, isoniazid) as well as some drugs (e.g., marijuana, heroin).[140]
- Detection of palpable, fixed, irregular mass in the breast, axilla, or elsewhere requires medical referral or a recommendation to the client to contact a physician for evaluation of the mass. Suspicious lymph node enlargement or lymph node changes; generalized lymphadenopathy.[120]
- Unusual or suspicious findings during inspection, palpation, or auscultation of the chest or abdomen including positive Murphy's percussion test, Murphy's sign, the presence of rebound tenderness, or palpable distention of the spleen, liver, or gallbladder.
- Recurrent cancer can appear as a single lump, a pale or red nodule just below the skin surface, a swelling, a dimpling of the skin, or a red rash. Report any of these changes to a physician immediately.
- Immediate medical referral is advised for any client reporting new-onset of SOB who is tachypneic, diaphoretic, or cyanotic; any suspicion of anaphylaxis is also an emergency situation.
- Cough with sputum production that is yellow, green, or rust colored.
- Abrupt change in mental status, confusion or increasing confusion, and new onset of delirium requires medical attention.
- Outbreak of vesicular rash associated with herpes zoster. Medical referral within 72 hours of the initial appearance of skin lesions is needed; client will begin a course of antiretroviral medication to manage symptoms and help prevent postherpetic neuropathy.

Key Points to Remember

- A head-to-toe complete physical assessment is an advanced clinical skill and a challenge even to the most skilled physician, physician assistant, or nurse practitioner. The therapist conducts a screening assessment using appropriate portions of the physical assessment.
- The therapist carries out certain portions of the physical assessment with every client by observing general health and nutrition, mental status, mood or affect, skin and body contours, mobility, and function.
- The therapist conducts a formal screening examination using the subjective and objective portions of the evaluation whenever the client history, age, gender, or other risk factors or the clinical presentation raise yellow (caution) or red (warning) flags.
- Measuring vital signs is a key component of the screening assessment. Vital signs, observations, and reported associated signs and symptoms are among the best screening tools available to the therapist. These same parameters can be used to plan and progress safe and effective exercise programs for clients who have true neuromuscular or musculoskeletal problems and also have other health concerns or comorbidities.
- Documentation of physical findings is important. From a legal point of view, if it is not documented, it was not assessed. Record important normal and abnormal findings.
- The therapist must be able to recognize normal and abnormal results when conducting inspection, palpation, percussion, and auscultation of the chest, thorax, and abdomen. The order of these tests is important and differs from chest and thorax to abdomen.
- Auscultation usually follows inspection and palpation of the chest and thorax. Examination of the abdomen should be performed in this order: inspection, auscultation, and then palpation, because palpation can affect findings on auscultation.
- The therapist should try to follow the same pattern every time to decrease the chances of missing an assessment parameter and to increase accuracy and thoroughness.
- Skin and nail bed assessment should be a part of every patient/client assessment.
- Changes in the skin and nail beds may be the first sign of inflammatory, infectious, and immunologic disorders and can occur with involvement of a variety of organs.
- Consider all integumentary and nail bed findings in relation to the client's age, ethnicity, occupation, and general health. When analyzing any signs and symptoms present, assess if this is a problem with the integumentary system versus an integumentary response to a systemic problem.
- The therapist may encounter enlarged or palpable lymph nodes. Keep in mind the therapist cannot know what the underlying pathology may be when lymph nodes are palpable and questionable. Performing a baseline assessment and reporting the findings is the important outcome of the assessment.
- With direct and unrestricted access of consumers to physical therapists in many states and the role of physical therapists in primary care, advanced skills have become necessary. For some clients, CBE is an appropriate assessment tool in the screening process.
- Properly trained physical therapists should be considered "qualified health care specialists" as defined by the ACS in the provision of cancer screening when the history, clinical presentation, and associated signs and symptoms point to the need for CBE.
- When a suspicious mass is found during examination, it must be medically evaluated, even if the client reports a recent mammography was "normal."
- For therapists without adequate training in conducting a CBE, the screening process is confined to asking questions during the interview regarding past medical history (e.g., cancer, lactation, abscess, mastitis) and questions to identify the possibility of breast pathology as the underlying cause for back or shoulder pain and/or other symptoms.
- Medical referral is advised for any individual suspected of having DVT; medical consultation is advised for those with a low probability.

SUBJECTIVE EXAMINATION

Special Questions to Ask

Breath and Body Odors

[Introductory remarks]: Mrs. Smith, as part of a physical therapy exam we always look at our client's overall health and general physical condition. Do you have any other health concerns besides your shoulder/back (name the involved body part)?

Are you being treated by anyone for any other problems? [Wait for a response but add prompts as needed: chiropractor? acupuncturist? naturopath?]

[If you suspect urinary incontinence]: Are you having any trouble with leaking urine or making it to the bathroom on time?

(Ask appropriate follow-up questions about cause, frequency, severity, triggers, and so on; see Appendix B-5).

[If you suspect fecal incontinence]: Do you have trouble getting to the toilet on time for a bowel movement? Do you have trouble wiping yourself clean after a bowel movement?

(Ask appropriate follow-up questions about cause, frequency, severity, triggers, and so on.)

[If you detect breath odor]: I notice an unusual smell on your breath. Do you know what might be causing this? (Ask appropriate follow-up questions depending on the type of smell you perceive; you may have to conduct an alcohol screening survey—see Chapter 2 or Appendices B-1 and B-2.

Skin

- Does the skin itch?
- Yellow skin (jaundice). Look for associated risk factors for hepatitis (see Box 9-2).
- Have you recently had a serious blood loss (possibly requiring transfusion)? (**Anemia;** also consider **jaundice/hepatitis posttransfusion**)
- Soft tissue lumps or skin lesions:
  How long have you had this?
  Has it changed in the last 6 weeks to 6 months?
  Has your doctor seen it?
  Does it itch, hurt, feel sore, or burn?
  Does anyone else in your household have anything like this?
  Have you taken any new medications (prescribed or over-the-counter) in the last 6 weeks?
  Have you traveled somewhere new in the last month?
  Have you been exposed to anything in the last month that could cause this?
  (Consider exposure due to occupational, environmental, and hobby interests.)
  Do you have any other skin changes anywhere else on your body?
  Have you had a fever or sweats in the last 2 weeks?
  Are you having any trouble breathing or swallowing?
  Have you had any other symptoms of any kind anywhere else in your body?
- Surgical scars: Would you have any objections if I looked at (or examined) the scar tissue?

Lymph Nodes

Use the lymph node assessment form (see Fig. 4-48) to record and report baseline findings.

- [General screening question:] Have you examined yourself for any lumps or nodules and found any thickening or lump? *If yes,* has your physician examined/treated this?
- Do you have (now or recently) any sores, rashes, or lesions anywhere on your body?
- If any suspicious or aberrant lymph nodes are observed during palpation, ask the following question.

Have you (ever) had

- Cancer of any kind?
  If no, have you ever been treated with radiation or chemotherapy for any reason?
- Breast implants
- Mastectomy or prostatectomy
- Mononucleosis
- Chronic fatigue syndrome
- Allergic rhinitis
- Food intolerances, food allergies, or celiac sprue
- Recent dental work
- Infection of any kind
- Recent cut, insect bite, or infection in the hand or arm
- A sexually transmitted disease of any kind
- Sores or lesions of any kind anywhere on the body (including genitals)
  Breast: See Chapter 17 and Appendix B-7
  Headache: See Appendix B-17

Neurologic

- Have you ever been in a car accident?
  If yes, did you lose consciousness, have a concussion, or a fractured skull?
- Have you ever been knocked out, unconscious, or have a concussion at any time?
- Have you ever had a seizure?
- Have you ever been paralyzed in your arms or legs?
- Have you ever broken your neck or back?

CASE STUDY

REVIEW OF SYSTEMS

A 48-year-old male was seen at the hospital emergency department following a multiple car motor vehicle accident (MVA). X-rays were negative, and the patient was diagnosed with an uncomplicated sprain/strain of the left hip. He was referred to the rehab clinic associated with the hospital for therapy within 1 week of the MVA.

Carry out a review of systems and group the following findings in appropriate clusters. Look for any patterns suggesting the need for further screening and/or impairment or dysfunction in any one particular organ system.

- Unable to palpate lymph nodes; obesity may have prevented an accurate assessment
- Symptoms worse after one month of soft tissue therapy
- Unable to ambulate more than 5 feet without assistance due to severe pain (self-rated as 8.5/10)
- Lumbar range of motion (ROM) and left hip ROM limited due to pain
- Client reports night pain from time to time that wakes him up at night; sometimes he can get back to sleep easily, at other times, not so well
- Valsalva's maneuver: negative
- Initial complaint of low back and left hip pain described as "burning"
- Client was sweating profusely throughout initial exam; described intermittent "sweats" over the course of month long therapy
- Initial complaint of pain that radiates down the left leg into the left foot occasionally; never present during sessions with the therapist
- Patellar deep tendon reflexes 2/3 (right: average/left: slightly brisk)
- Achilles deep tendon reflexes 0 (absent) due to edema
- 8 cm mass on client's left buttock developed 1 month after MVA; firm and cool to the touch; no bruising or skin color changes; gluteal musculature atrophied around the mass
- Alert but agitated from time to time
- BP 120/80 bilateral upper extremities
- Heart rate 72 bpm, steady
- Height: 6′3″ Weight: 300 pounds
- No saddle anesthesia reported; no bowel or bladder dysfunction reported; no sexual dysfunction reported
- Any movement of the left hip increases low back and left hip pain
- Bilateral straight leg raise to 80 degrees without reproducing symptoms
- Signs of venous stasis in the feet and lower legs; client reports these have been present since before the accident
- Marked pitting edema of both ankles
- Gross Manual Muscle Test (MMT):
  - Right hip flexion 5/5
  - Left hip flexion 2/5
  - Ankle (right or left) 5/5
- Lower extremity sensation: within normal limits (WNL)
- Lab Values from emergency department intake exam:
  - Hemoglobin 116 g/dL (mildly decreased)
  - Serum lactate, dehydrogenase levels (Normal)

There is no "right" or "wrong" way to approach this, just different methods. Using the client history, clinical presentation, and associated signs and symptoms presented throughout this text, we suggest:

HISTORY

- Significant history of trauma

CLINICAL PRESENTATION

- Pain pattern: progressive, unremitting, radiating pain; night pain; pain described as "burning"; aggravated by any movement of the left hip
- Left hip flexor weakness
- Alert, but agitated
- Mass on left buttock with gluteal muscle atrophy at that site
- Signs of chronic venous stasis in lower extremities; bilateral pitting edema

ASSOCIATED SIGNS AND SYMPTOMS

- Constitutional symptoms: unexplained perspiration

REVIEW OF SYSTEMS

- The most significant cluster of findings seems to be neurologic: "burning" pain that radiates, muscle weakness, agitation, and progressive presentation of pain pattern. The new onset of a soft tissue mass is of great concern.
- There is no mention in the client's chart of changes (worsening/improving) of muscle strength over the course of treatment.
- Chronic pitting edema with signs of venous stasis suggests a peripheral vascular disorder.

ADDITIONAL CLUES

- Client is not better after physical therapy intervention; in fact, client is worse.

Continued

CASE STUDY—*cont'd*

There are enough red flags here to suggest further screening is needed if not immediate referral.

Age (over 40)

Left buttock mass with muscle atrophy

Chronic venous stasis present before the accident

Screening may be nothing more than asking some of the same questions as presented in the intake interview, as well as conducting some of the same tests again. For example:

Past Medical History:

Previous health history (especially including previous/current history of cancer, vascular or cardiac disease)

Review information collected using the Documentation Template for Physical Therapist Client Management, *Guide,* Appendix. Ask yourself:

Have I left anything out that is important?

Any history of drug or alcohol use/abuse?

Recent history of medications, over-the-counter drugs, nutraceuticals?

Recent history of infections?

Previous history of any surgeries? Remember back and/or hip pain can occur even years later after orthopedic surgery, for example from infection, implant loosening, fracture, hemorrhage, or other complication.

CLINICAL PRESENTATION:

Any signs of liver impairment (liver flap, nail bed changes, symptoms of carpal tunnel syndrome, angiomas, palmar erythema)?

Changes in muscle strength or deep tendon reflexes?

Left hip pain: conduct tests for psoas abscess (iliopsoas and obturator tests) or other infectious/inflammatory processes such as appendicitis (McBurney's Point), peritonitis (pinch-an-inch test).

ASSOCIATED SIGNS AND SYMPTOMS:

Effect of food on symptoms (especially pain at night)

Any significant weight loss?

Take the client's temperature

Changes in vital signs

From Busse JW: Delayed diagnosis of non-Hodgkin's lymphoma: a case report, *J Neuromusculoskeletal System* 9(2):60-64, Summer 2001.

PRACTICE QUESTIONS

1. When assessing the abdomen, what sequence of physical assessment is best?
   a. Auscultation, inspection, palpation, percussion
   b. Inspection, percussion, auscultation, palpation
   c. Inspection, auscultation, percussion, palpation
   d. Auscultation, inspection, percussion, palpation
2. A line drawn down the middle of a lesion with two different halves suggests:
   a. A malignant lesion
   b. A benign lesion
   c. A normal presentation
   d. A skin reaction to medications
3. Pulse strength graded as 1 means:
   a. Easily palpable, normal
   b. Present occasionally
   c. Pulse diminished, barely palpable
   d. Within normal limits
4. During auscultation of an adult client with rheumatoid arthritis, the heart rate gets stronger as she breathes in and decreases as she breathes out. This sign is:
   a. Characteristic of lung disease
   b. Typical in coronary artery disease
   c. A normal finding
   d. Common in anyone with pain
5. How do you plan or modify an exercise program for a client with cancer without the benefit of blood values?
6. Body temperature should be taken as part of vital sign assessment:
   a. Only for clients who have not been seen by a physician
   b. For any client who has musculoskeletal pain of unknown origin
   c. For any client reporting the presence of constitutional symptoms, especially fever or sweats
   d. b and c
   e. All of the above
7. When would you consider listening for femoral bruits?
8. A 23-year-old female presents with new onset of skin rash and joint pain followed 2 weeks later by GI symptoms of abdominal pain, nausea, and diarrhea. She has a previous history of Crohn's disease, but this condition has been stable for several years. She does not think her current symptoms are related to her Crohn's disease. What kind of screening assessment is needed in this case?
   a. Vital signs only.
   b. Vital signs and abdominal auscultation.
   c. Vital signs, neurologic screening examination, and abdominal auscultation.
   d. No further assessment is needed; there are enough red flags to advise this client to seek medical attention.

PRACTICE QUESTIONS—*cont'd*

9. A 76-year-old man was referred to physical therapy after a total hip replacement (THR). The goal is to increase his functional mobility. Is a health assessment needed since he was examined just before the surgery 2 weeks ago? The physician conducted a systems review and summarized the medical record by saying the client was in excellent health and a good candidate for THR.
10. You notice a new client has an unusual (strong) breath odor. How do you assess this?
11. Why does postural orthostatic hypotension occur upon standing for the first time in a young adult who has been supine in skeletal traction for 3 weeks?

REFERENCES

1. Wainner RS: Regional interdependence: a musculoskeletal examination model whose time has come. *J Orthop Sports Phys Ther* 37(11):658–660, 2007.
2. Danielson K, Solheim K: *Essential physical assessment skills*, Eau Claire, Wisconsin, 2003, PESI HealthCare.
3. Physical Therapists Community Screening Form for Aging Adults: *GeriNotes* 15(1):13–22, January 2008.
4. Sieber FE: Sedation depth during spinal anesthesia and the development of postoperative delirium in elderly patients undergoing hip fracture repair. *Mayo Clin Proc* 85(1):18–26, 2010.
5. Alfonso DT: Nonsurgical complications after total hip and total knee arthroplasty. *Am J Orthop* 35(11):503–510, 2006.
6. Erikson A, Wilmarth MA: The role of the physical therapist in the assessment and management of the elderly patient with postoperative delirium following hip fracture surgery: an evidence-based case report. *Orthopaedic Practice* 18(2):12–17, 2006.
7. Bjorklund KB: Reducing delirium in elderly patients with hip fracture: a multi-factorial intervention study. *Acta Anaesthesiol Scand* March 15, 2010. Epub ahead of print.
8. Sieber FE: Postoperative delirium in the elderly surgical patient. *Anesthesiolog Clin* 27(3):451–464, 2009.
9. Inouye S: Clarifying confusion: the confusion assessment method. *Ann Intern Med* 113(12):941–948, 1990.
10. Jennett B, Bond M: Assessment of outcome after severe brain damage. *Lancet* 1(7905):480–484, 1975.
11. Rowlett R: *Glasgow coma scale*, University of Carolina at Chapel Hill. Available at: www.unc.edu/~rowlett/units/scales/glasgow.htm. Accessed January 14, 2011.
12. The Internet Stroke Center: *Stroke scales and clinical assessment tools*. Available at: http://www.strokecenter.org/trials/scales/index.htm. Accessed January 14, 2011.
13. National Institutes of Health: *Calculate your body mass index*. Available at: http://nhlbisupport.com/bmi/bmicalc.htm. Accessed January 11, 2011.
14. Steven B: *Hall's Body Mass Index seeker*. Available at: http://www.halls.md/body-mass-index/bmi.htm. Accessed January 11, 2011.
15. National Center for Chronic Disease Prevention and Health Promotion: *Body mass index for children and teens*. Available at: http://www.cdc.gov/nccdphp/dnpa/bmi/bmi-for-age.htm. Accessed January 11, 2011.
16. Frese EM, Richter RR, Burlis TV: Self-reported measurement of heart rate and blood pressure in patients by physical therapy clinical instructors. *Phys Ther* 82(12):1192–1200, 2002.
17. Billek-Sawhney B, Sawhney R: Cardiovascular considerations in outpatient physical therapy. *J Orthop Sports Phys Ther* 27:57, 1998.
18. Guide to the Physical Therapist Practice, ed 2 (revised). *Phys Ther* 81(1):2003, Alexandria Virginia, American Physical Therapy Association.
19. American College of Sports Medicine (ACSM): *Resource manual for guidelines for exercise testing and prescription*, ed 5, Philadelphia, 2005, Lippincott, Williams, and Wilkins.
20. Shaw DK: What's so vital about vital signs? *Q Rev* 44(4):1–5, 2009.
21. Bates B, Bickley LS, Hoekelman RA: *A guide to physical examination and history taking*, ed 10, Philadelphia, 2008, J.B. Lippincott.
22. Barnes CJ: Does the pulse examination in patients with traumatic knee dislocation predict a surgical arterial injury? *J Trama* 53(6):1109–1114, 2002.
23. Johnson ME: Neurologic and vascular injuries associated with knee ligament injuries. *Am J Sports Med* 36(12):2448–2462, 2008.
24. Lohmann M: Arterial lesions in major knee trauma: pedal pulse a false sign of security? *Arch Orthop Trauma Surg* 109:238–239, 1990.
25. McCutchan JDS: Injury to the popliteal artery associated with dislocation of the knee: palpable pulses do not negate the requirement for arteriography. *Injury* 20(5):307–310, 1989.
26. Phoenix Publishing Product No. 4009-01: *PhysioQ*. Available at: http://www.phoenixpub.com/store/ [no financial or other benefit is derived by the authors of this textbook from posting this weblink].
27. Pullen RL: Using an ear thermometer. *Nursing2003* 33(5):24, 2003.
28. Hardwick KD: Insightful options. *Rehab Magazine* 15(7):30–33, 2002.
29. Ostchega Y, Prineas RJ, Paulose-Ram R, et al: National health and nutrition examination survey 1999-2000: effect of observer training and protocol standardization on reducing blood pressure measurement error. *J Clin Epidemiol* 56:768–774, 2003.
30. Personal communication with oncology staff (nurses and physical therapists) across the United States, 2005.
31. Pfalzer C: Personal communication, 2005.
32. Nelson MR: Cluster-randomized controlled trial of oscillometric vs. manual sphygmomanometer for blood pressure management in primary care (CRAB). *Am J Hypertens* 22(6):598–603, 2009.
33. Levin DK: Measuring blood pressure in legs. *Medscape Internal Medicine* 6(1), 2004. Available on-line at http://www.medscape.com/viewarticle/471829. Accessed January 25, 2011.
34. The Seventh Report of the Joint National Committee on Prevention, Detection, Evaluation, and Treatment of High Blood Pressure (JNC-7), NIH Publication No. 03-5233, May 2003. National Heart, Lung, and Blood Institute (NHLBI).
35. *Expert 10-minute physical examinations*, ed 2, St. Louis, 2004, Mosby.
36. Mitchell GF: Pulse pressure and risk of new onset atrial fibrillation. *JAMA* 297(7):709–715, 2007.
37. Franklin SS: Pulse pressure as a risk factor. *Clin Exp Hypertens* 26(7-8):645–652, 2004.

38. American College of Sports Medicine: *ACSM guidelines for exercise testing and prescription*, ed 8, Philadelphia, 2009, Lippincott, Williams, and Wilkins.
39. Lim MA, Townsend RR: Arterial compliance in the elderly: its effect on blood pressure measurement and cardiovascular outcomes. *Clin Geriatr Med* 25(2):191–205, 2009.
40. Hussain SF: Progestogen-only pills and high blood pressure: is there an association? A literature review. *Contraception* 69(2):89–97, 2004.
41. Burkman R, Schlesselman JJ, Zieman M: Safety concerns and health benefits associated with oral contraception. *Am J Obstet Gynecol* 190(4 Suppl):S5-S22, 2004.
42. Hillegass EA: *Essentials of cardiopulmonary physical therapy*, ed 3, Philadelphia, 2010, Saunders.
43. American Heart Association: *Postural orthostatic hypotension*. Available at: http://www.phoenixpub.com/store/ [no financial or other benefit is derived by the authors of this textbook from posting this weblink].
44. Mancia G: Long-term risk of sustained hypertension in white-coat or masked hypertension. *Hypertension* 54(2):226–232, 2009.
45. American Heart Association Updates Recommendations for Blood Pressure Measurements: Available at: http://www.americanheart.org. Accessed December 20, 2010.
46. Kario K: Morning surge and variability in blood pressure. A new therapeutic target? *Hypertension* 45(4):485–486, Feb. 21, 2005.
47. Saunders E: Managing hypertension in African-American patients. *J Clin Hypertens (Greenwich)* 6(4 Suppl 1):19–25, 2004.
48. International Society on Hypertension in Blacks (ISHIB): Available at: http://www.ishib.org/. Accessed January 25, 2011.
49. Douglas JG, Bakris GL, Epstein M, et al: Management of high blood pressure in African Americans: consensus statement of the Hypertension in African Americans Working Group of the International Society on Hypertension in Blacks. *Arch Intern Med* 163(5):521–522, 2003.
50. National Alliance for Hispanic Health: *Hypertension in Hispanics*. Available on-line at www.hispanichealth.org. Accessed March 19, 2010.
51. Margolis KL, Piller LB: Blood pressure control in Hispanics in the antihypertensive and lipid-lowering treatment to prevent heart attack. *Hypertension* 50(5):854–861, 2007.
52. National Heart, Lung, and Blood Institute (NHLBI): Fourth report on the diagnosis, evaluation, and treatment of high blood pressure in children and adolescents. *Pediatrics* 114(2): 555–576, 2004.
53. Rocchini AP: Coarctation of the aorta and interrupted aortic arch, In Moller JH, Hoffmann U, editors: *Pediatric cardiovascular medicine*, New York, 2000, Churchill Livingstone, p. 570.
54. Mourad A, Carney S, Gillies A, et al: Arm position and blood pressure: a risk factor for hypertension? *J Hum Hypertens* 17:389–395, 2003.
55. Netea RT, Lenders JW, Smits P, et al: Both body and arm position significantly influence blood pressure measurement. *J Hum Hypertens* 17:459–462, 2003.
56. Kario K, Tobin JN, Wolfson LI, et al: Lower standing systolic blood pressure as a predictor of falls in the elderly: a community-based prospective study. *J Am Coll Cardiol* 38(1):246–252, 2001.
57. Qui C, von Strauss E, Winblad B, et al: Decline in blood pressure over time and risk of dementia: a longitudinal study from the Kungsholmen project. *Stroke* 35(8):1810–1815. Epub July 01, 2004.
58. Verghese J, Lipton RB, Hall CB, et al: Low blood pressure and the risk of dementia in very old individuals. *Neurology* 61(12):1667–1672, 2003.
59. Pullen RL: Caring for a patient on pulse oximetry. *Nursing 2003* 33(9):30, 2003.
60. Bridges E, Thomas K: Noninvasive measurement of body temperature in critically ill patients. *Crit Care Nurse* 29(3):94–97, 2009.
61. Mcilvoy L: Comparison of brain temperature to core temperature: a review of the literature. *J Neurosci Nurs* 36(1):23–31, 2004.
62. Kirk D: Infra-red thermometry: the reliability of tympanic and temporal artery readings in predicting brain temperature after severe traumatic brain injury. *Crit Care* 13(3):R81, 2009. Available on-line at http://ccforum.com/content/13/3/R81. Accessed Dec. 12, 2010.
63. Montero-Odasso M: Gait velocity as a single predictor of adverse events in healthy seniors aged 75 years and older. *J Gerontol A Biol Sci Med Sci* 60:1304–1309, 2005.
64. Fritz S, Lusardi M: Walking speed: the sixth vital sign. *J Geriatric Phys Ther* 32(2):2–5, 2009.
65. Bohannon RW: Comfortable and maximum walking speed of adults aged 20–79 years: reference values and determinants. *Age and Ageing* 26:15–19, 1997.
66. Bohannon R: Population representative gait speeds and its determinants. *J Geriatr Phys Ther* 31:49–52, 2008.
67. Rupp JF, Kaplan DL: Pruritus: causes—cures, part 1. *Consultant* 39(11):3157–3160, 1999.
68. Webster GF: Common skin disorders in the elderly. *Clinical Cornerstone* 4(1):39–44, 2001.
69. Armstrong DG: Skin temperature monitoring reduces the risk for diabetic foot ulceration in high-risk patients. *Am J Med* 120(12):1042–1046, 2007.
70. Lavery LA, Higgins KR, Lanctot DR, et al: Preventing diabetic foot ulcer recurrence in high-risk patients: use of temperature monitoring as a self-assessment tool. *Diabetes Care* 30(1):14–20, 2007.
71. Sun PC: Assessing foot temperature using infrared thermography. *Foot Ankle Int* 26(10):847–853, 2005.
72. Armstrong DG: Does dermal thermometry predict clinical outcome in diabetic foot infection? Analysis of data from the SIDESTEP* trial. *Wound J.* 3(4):302–307, 2006.
73. Rigel DS: The evolution of melanoma diagnosis: 25 years beyond the ABCDs. *Ca J Clin* 60(5):301–316, 2010.
74. Candelli M, Carloni E, Nista EC, et al: *Helicobacter pylori* eradication and acne rosacea resolution: cause—effect or coincidence? *Dig Liver Dis* 36(2):163, 2004.
75. Diaz C, O'Callaghan CJ, Khan A, et al: Rosacea: a cutaneous marker of *Helicobacter pylori* infection? Results of a pilot study. *Acta Derm Venereol* 83(4):282–286, 2003.
76. Rebora A: The management of rosacea. *Am J Clin Dermatol* 3(7):489–496, 2002.
77. Yazici Y, Erkan D, Scott R, et al: The skin: a map to rheumatic diseases. *J Musculoskel Med* 18(1):43–53, 2001.
78. Lyme Disease Foundation: *Community Education*. Available at: www.lyme.org/gallery/rashes.html. Accessed January 25, 2011.
79. Guillot B, Bessis D, Dereure O: Mucocutaneous side effects of antineoplastic chemotherapy. *Expert Opin Drug Saf* 3(6):579–587, 2004.
80. Borroni G, Vassallo C, Brazzelli V, et al: Radiation recall dermatitis, panniculitis, and myositis following cyclophosphamide therapy: histopathologic findings of a patient affected by multiple myeloma. *Am J Dermatopathol* 26(3):213–216, June 2004.
81. Friedlander PA, Bansal R, Schwartz L, et al: Gemcitabine-related radiation recall preferentially involves internal tissues and organs. *Cancer* 100(9):1793–1799, 2004.
82. Workowski KA, Berman S: Sexually transmitted diseases treatment guidelines, 2010. *MMWR* 59(RR-12: 1-110), 2010. Available on-line at http://www.cdc.gov/mmwr/preview/

mmwrhtml/rr5912a1.htm?s_cid=rr5912a1_e. Accessed Dec. 17, 2010.
83. Centers for Disease Control and Prevention: *Trends in STDs in the United States*. 2009. Available at: http://www.cdc.gov/std/stats09/trends.htm. Accessed January 25, 2011.
84. Centers for Disease Control and Prevention: *Sexually transmitted disease surveillance, 2009*, Atlanta, 2001, U.S. Department of Health and Human Services. Available on-line at: http://www.cdc.gov/std/stats09/default.htm. Accessed January 25, 2011.
85. Goodman CC, Fuller K: *Pathology: implications for the physical therapist*, ed 3, 2009.
86. Primary and secondary syphilis—United States, 2002. *MMWR* 52(46):1117–1120, 2003. Available at: http://www.cdc.gov/mmwr/preview/mmwrhtml/mm5246a1.htm. Accessed January 25, 2011.
87. Heffelfinger J: *Syphilis trends in the U.S.* Presented at the 2004 National STD Convention, Philadelphia, PA, March 8–11, 2004.
88. American Social Health Association: *The truth about HSV-1 and HSV-2*. Available at: http://www.herpes.com/hsv1-2.html Accessed on January 25, 2011.
89. Walker BW: Getting the lowdown on herpetic whitlow. *Nursing 2004* 34(7):17, 2004.
90. Omori M: Herpetic whitlow. *eMedicine* 2004.
91. Mudd SS, Findlay JS: The cutaneous manifestations and common mimickers of physical child abuse. *J Pediatr Health Care* 18(3):123–129, 2004.
92. Stanley WJ: Nailing a key assessment: learn the significance of certain nail anomalies. *Nursing 2003* 33(8):50–51, August 2003.
93. Callen J, Jorizzo J, Bolognia J, et al: *Dermatological signs of internal disease*, ed 3, Philadelphia, 2003, Saunders.
94. Bolognia JL: *Bolognia: dermatology*, ed 2, Philadelphia, 2008, Saunders.
95. Hordinsky MK, Sawaya ME, Scher RK, et al: *Atlas of hair and nails*, Philadelphia, 2000, Churchill Livingstone.
96. Mir MA: *Atlas of clinical diagnosis*, ed 2, Philadelphia, 2003, Saunders.
97. Schalock PC: *Lippincott's primary care dermatology*, Baltimore, 2010, Lippincott, Williams & Wilkins.
98. Rabar D, Combemale P, Peyron P: Doxycycline-induced photo-onycholysis. *J Travel Med* 11(6):386–387, 2004.
99. Maino K, Stashower ME: Traumatic transverse leukonychia. *SKINmed* 3(1):53–55, 2004. Accessed on-line: http://www.medscape.com/viewarticle/467074 (posted 01/25/2004). Accessed January 25, 2011.
100. Sweeney E, Fryer A, Mountford R, et al: Nail patella syndrome: a review of the phenotype aided by developmental biology. *J Med Genet* 40:153–162, 2003.
101. Hicks DL, Judkins AR, Sickel JZ, et al: Granular histiocytosis of pelvic lymph nodes following total hip arthroplasty. *J Bone Joint Surg* 78A:482–496, 1996.
102. Harris J, Lippman M, Osborne K, et al, editors: *Diseases of the breast*, ed 4, Philadelphia, 2009, Lippincott, Williams & Wilkins.
103. Hosford D: On-Line University: *The Hosford's differential diagnosis tables*. Available at: http://www.ptcentral.com/university/diagnose_pdf.html. Accessed January 25, 2011.
104. Smith NL: *The effects of depression and anxiety on medical illness*, University of Utah, 2002, Stress Medicine Clinic.
105. Mohler ER: Peripheral arterial disease. *Arch Intern Med* 163(19):2306–2314, 2003.
106. Merli GJ, Weitz HH, Carabasi A: *Peripheral vascular disorders*, Philadelphia, 2004, Elsevier Science.
107. Sheehan P: Diabetes and PAD: consensus statement urges screening. *Medscape Cardiology* 8(1), 2004. Available at: http://www.medscape.com/viewarticle/467520. Accessed January 25, 2011.
108. Constans J: A clinical prediction score for upper extremity deep venous thrombosis. *Thromb Haemost* 99:202–207, 2008.
109. Tepper S, McKeough M: Deep venous thrombosis: risks, diagnosis, treatment interventions, and prevention. *Acute Care Perspectives* 9(1):1–7, 2000.
110. Autar R: Nursing assessment of clients at risk of deep vein thrombosis (DVT): the Autar DVT scale. *J Adv Nurs* 23(4):763–770, 1996.
111. Autar R: Calculating patients' risk of deep vein thrombosis. *Br J Nurs* 7(1):7–12, 1998.
112. Wells PS, Hirsch J, Anderson DR, et al: Accuracy of clinical assessment of deep-vein thrombosis. *Lancet* 345:1326–1330, 1995.
113. Wells PS, Anderson DR, Bormanis J, et al: Value of assessment of pretest probability of deep-vein thrombosis in clinical management. *Lancet* 350:1795–1798, 1997.
114. Wells PS, Hirsch J, Anderson DR, et al: A simple clinical model for the diagnosis of deep vein thrombosis combined with impedance plethysmography: potential for an improvement in the diagnostic process. *J Intern Med* 243:15–23, 1998.
115. Riddle DL, Wells PS: Diagnosis of lower-extremity deep vein thrombosis in outpatients. *Phys Ther* 84(8):729–735, 2004.
116. Riddle DL, Hillner BE, Wells PS, et al: Diagnosis of lower-extremity deep vein thrombosis in outpatients with musculoskeletal disorders: a national survey study of physical therapists. *Phys Ther* 84(8):717–728, 2004.
117. Shaikh N, LaTrenta G, Swistel A, et al: Detection of recurrent breast cancer after TRAM flap reconstruction. *Ann Plast Surg* 47(6):602–607, 2001.
118. Devon RK, Rosen MA, Mies C, et al: Breast reconstruction with a transverse rectus abdominis myocutaneous flap: spectrum of normal abnormal MR imaging findings. *Radiographics* 24(5):1287–1299, 2004.
119. Mustonen P, Lepisto J, Papp A, et al: The surgical and oncological safety of immediate breast reconstruction. *Eur J Surg Oncol* 30(8):817–823, 2004.
120. Goodman CC, McGarvey CL: The role of the physical therapist in primary care and cancer screening: integrating clinical breast examination (CBE) in the upper quarter examination. *Rehabilitation Oncology* 21(2):4–11, 2003.
121. American Cancer Society (ACS): Cancer reference information. Available at: http://www.cancer.org/docroot/CRI/content/CRI_2_4_3X_Can_breast_cancer_be_found_early_5.asp/ Accessed January 13, 2011.
122. Smith RA, Saslow D, Sawyer KA, et al: American cancer society guidelines for breast cancer screening: update 2003. *Cancer J Clin* 53(3):141–169, 2003.
123. Bancej C, Decker K, Chiarelli A, et al: Contribution of clinical breast examination to mammography screening in the early detection of breast cancer. *J Med Screen* 10:16–21, 2003.
124. Saslow D, Hannan J, Osuch J, et al: Clinical breast examination: practical recommendations for optimizing performance and reporting. *Cancer J Clin* 54:327–344, 2004.
125. White E, Miglioretti DL, Yankaskas BC, et al: Biennial versus annual mammography and the risk of late-stage breast cancer. *J Natl Cancer Inst* 96(24):1832–1839, 2004.
126. Kaunitz AM: Mammography revisited: does screening lower breast cancer-related mortality? *Journal Watch* 30(19), 2010. Medscape Today. Available on-line at http://www.medscape.com/viewarticle/731345. Accessed January 25, 2011.
127. Kalager M et al: Effect of screening mammography on breast-cancer mortality in Norway. *N Engl J Med* 363:1203, 2010.
128. Perry N: *European guidelines for quality assurance in breast cancer screening and diagnosis*, ed 4, Luxembourg, 2006, Office for Official Publications of the European Communities. Available on-line at: http://ec.europa.eu/health/ph_projects/2002/

cancer/fp_cancer_2002_ext_guid_01.pdf. Accessed January 25, 2011.

129. Park BW, Kim SI, Kim MH, et al: Clinical breast examination for screening of asymptomatic women: the importance of clinical breast examination for breast cancer detection. *Yonsei Med J* 41(3):312–318, 2000.
130. Shen Y, Zelen M: Screening sensitivity and sojourn time from breast cancer early detection clinical trials: mammograms and physical examinations. *J Clin Oncol* 19(15):3490–3499, 2001.
131. Pennypacker HS, Naylor L, Sander AA, et al: Why can't we do better breast examinations? *Nurse Practitioner Forum* 10(3): 122–128, 1999.
132. Bloom H, Criswell E, Pennypacker H, et al: Major stimulus dimensions determining detection of simulated breast lesions. *Percep Psychophys* 20:163–167, 1982.
133. McDonald S, Saslow D, Alciati MH: Performance and reporting of clinical breast examination: a review of the literature. *Cancer J Clin* 54:345–361, 2004.
134. Potter PA, Weilitz PB: *Pocket guide to health assessment*, ed 6, St. Louis, 2006, Mosby.
135. Schnur W: Tickle me not. *Postgrad Med* 96(6):35, 1994.
136. Eskelinen M: Usefulness of history-taking, physical examination and diagnostic scoring in acute renal colic. *Eur Urol* 34(6):467–473, 1998.
137. Houppermans RP, Brueren MM: Physical diagnosis—pain elicited by percussion in the kidney area. *Ned Tijdschr Geneeskd* 145(5):208–210, 2001.
138. O'Gara PT: Aortic aneurysm. *Circulation* 107:e43, 2003. Available on-line at http://circ.ahajournals.org/cgi/content/full/107/6/e43. Accessed January 6, 2010.
139. Moore KL: *Clinically oriented anatomy*, ed 6, Philadelphia, 2010, Wolters Kluwer/Lippincott Williams & Wilkins.
140. Dains JE, Baumann LC, Scheibel P: *Advanced health assessment and clinical diagnosis in primary care*, ed 3, St. Louis, 2007, Mosby.
141. Fritz S, Lusardi M: White paper: "Walking speed: the sixth vital sign." *J Geriatr Phys Ther* 32(2):2–5, 2009.
142. Kaplan NM: Technique of blood pressure measurement in the diagnosis of hypertension, *UpToDate*. Version 18.3. 2010. Available on-line at http://www.uptodate.com/online/content/topic.do?topicKey=hyperten/9469.
143. Barton M, Harris R, Fletcher S: Does this patient have breast cancer? The Screening Clinical Breast Examination. *JAMA* 282(13):1270–1280, 1999.
144. Holden U: Dementia in acute units: confusion, *Nurs Stand* 9(17):37–39, 1995.

CHAPTER 5

Screening for Hematologic Disease

The blood consists of two major components: plasma, a pale yellow or gray-yellow fluid; and formed elements, erythrocytes (red blood cells [RBCs]), leukocytes (white blood cells [WBCs]), and platelets (thrombocytes). Blood is the circulating tissue of the body; the fluid and its formed elements circulate through the heart, arteries, capillaries, and veins.

The *erythrocytes* carry oxygen to tissues and remove carbon dioxide from them. *Leukocytes* act in inflammatory and immune responses. The *plasma* carries antibodies and nutrients to tissues and removes wastes from tissues. *Platelets,* together with *coagulation factors* in plasma, control the clotting of blood.

Primary hematologic diseases are uncommon, but hematologic manifestations secondary to other diseases are common. Cancers of the blood are discussed in Chapter 13.

In the physical therapist's practice, symptoms of blood disorders are most common in relation to the use of nonsteroidal antiinflammatory drugs (NSAIDs) for inflammatory conditions, neurologic complications associated with pernicious anemia, and complications of chemotherapy or radiation.

Hematologic considerations in the orthopedic population fall into two main categories: bleeding and clotting. People with known abnormalities of hemostasis (either hypocoagulation or hypercoagulation problems) will require close observation.[1]

All surgical patients, neurologically compromised, or immobilized individuals must also be observed carefully for any signs or symptoms of venous thromboembolism.

SIGNS AND SYMPTOMS OF HEMATOLOGIC DISORDERS

There are many signs and symptoms that can be associated with hematologic disorders. Some of the most important indicators of dysfunction in this system include problems associated with exertion (often minimal exertion) such as dyspnea, chest pain, palpitations, severe weakness, and fatigue. Neurologic symptoms, such as headache, drowsiness, dizziness, syncope, or polyneuropathy, can also indicate a variety of possible problems in this system.[2,3]

Significant skin and fingernail bed changes that can occur with hematologic problems might include pallor of the face, hands, nail beds, and lips; cyanosis or clubbing of the fingernail beds; and wounds or easy bruising or bleeding in skin, gums, or mucous membranes, often with no reported trauma to the area. The presence of blood in the stool or emesis or severe pain and swelling in joints and muscles should also alert the physical therapist to the possibility of a hematologic-based systemic disorder and can sometimes be a critical indicator of bleeding disorders that can be life threatening.[2,3]

Many hematologic-induced signs and symptoms seen in the physical therapy practice occur as a result of medications. For example, chronic or long-term use of steroids and NSAIDs can lead to gastritis and peptic ulcer with gastrointestinal (GI) bleeding and subsequent iron deficiency anemia.[4] Leukopenia, a common problem occurring during chemotherapy, or as a symptom of certain types of cancer, can produce symptoms of infections such as fever, chills, tissue inflammation; severe mouth, throat and esophageal pain; and mucous membrane ulcerations.[5]

Thrombocytopenia (decreased platelets) associated with easy bruising and spontaneous bleeding is a result of the pharmacologic treatment of common conditions seen in a physical therapy practice such as rheumatoid arthritis and cancer. More about this condition will be included later in this chapter.

CLASSIFICATION OF BLOOD DISORDERS

Erythrocyte Disorders

Erythrocytes (red blood cells) consist mainly of hemoglobin and a supporting framework. Erythrocytes transport oxygen and carbon dioxide; they are important in maintaining normal acid-base balance. There are many more erythrocytes than leukocytes (600 to 1). The total number is determined by gender (women have fewer erythrocytes than men), altitude (less oxygen in the air requires more erythrocytes to carry sufficient amounts of oxygen to the tissues), and physical activity (sedentary people have fewer erythrocytes, athletes have more).

Disorders of erythrocytes are classified as follows (not all of these conditions are discussed in this text):

- Anemia (too few erythrocytes)
- Polycythemia (too many erythrocytes)
- Poikilocytosis (abnormally shaped erythrocytes)
- Anisocytosis (abnormal variations in size of erythrocytes)
- Hypochromia (erythrocytes deficient in hemoglobin)

TABLE 5-1 Changes Associated with Hematologic Disorders

Changes	Causes
SKIN	
Light, lemon-yellow tint	Untreated pernicious anemia
White, waxy appearance	Severe anemia resulting from acute hemorrhage
Gray-green yellow	Chronic blood loss
Gray tint	Leukemia
Pale hands or palmar creases	Anemia
NAIL BED	
Brittle	Long-standing iron deficiency anemia
Concave (rather than convex)	Long-standing iron deficiency anemia
ORAL MUCOSA/CONJUNCTIVA	
Pale or yellow color	Anemia

Anemia

Anemia is a reduction in the oxygen-carrying capacity of the blood as a result of an abnormality in the quantity or quality of erythrocytes. Anemia is not a disease but is a symptom of any number of different blood disorders. Excessive blood loss, increased destruction of erythrocytes, and decreased production of erythrocytes are the most common causes of anemia.[6]

In the physical therapy practice, anemia-related disorders usually occur in one of four broad categories:

1. Iron deficiency associated with chronic GI blood loss secondary to NSAID use
2. Chronic diseases (e.g., cancer, kidney disease, liver disease) or inflammatory diseases (e.g., rheumatoid arthritis or systemic lupus erythematosus)
3. Neurologic conditions (pernicious anemia)
4. Infectious diseases, such as tuberculosis or acquired immunodeficiency syndrome (AIDS), and neoplastic disease or cancer (bone marrow failure).

Anemia with neoplasia may be a common complication of chemotherapy or develop as a consequence of bone marrow metastasis.[7] Anemia can also occur as a symptom of leukemia. Adults with pernicious anemia have significantly higher risks for hip fracture even with vitamin B_{12} therapy. The hypothesized underlying factor is a lack of gastric acid (achlorhydria).[8]

Clinical Signs and Symptoms. Deficiency in the oxygen-carrying capacity of blood may result in disturbances in the function of many organs and tissues leading to various symptoms that differ from one person to another. Slowly developing anemia in young, otherwise healthy individuals is well tolerated, and there may be no symptoms until hemoglobin concentration and hematocrit fall below one half of normal (see values inside book cover).

However, rapid onset may result in symptoms of dyspnea, weakness and fatigue, and palpitations, reflecting the lack of oxygen transport to the lungs and muscles. Many people can have moderate-to-severe anemia without these symptoms. Although there is no difference in normal blood volume associated with severe anemia, there is a redistribution of blood so that organs most sensitive to oxygen deprivation (e.g., brain, heart, muscles) receive more blood than, for example, the hands and kidneys.

Changes in the hands and fingernail beds (Table 5-1) may be observed during the inspection/observation portion of the physical therapy evaluation (see Table 4-8 and Boxes 4-13 and 4-15). The physical therapist should look for pale palms with normal-colored creases (severe anemia causes pale creases as well). Observation of the hands should be done at the level of the client's heart. In addition, the anemic client's hands should be warm; if they are cold, the paleness is due to vasoconstriction.

Pallor in dark-skinned people may be observed by the absence of the underlying red tones that normally give brown or black skin its luster. The brown-skinned individual demonstrates pallor with a more yellowish-brown color, and the black-skinned person will appear ashen or gray.

Systolic blood pressure may not be affected, but diastolic pressure may be lower than normal, with an associated increase in the resting pulse rate. Resting cardiac output is usually normal in people with anemia, but cardiac output increases with exercise more than it does in people without anemia.[9] As the anemia becomes more severe, resting cardiac output increases and exercise tolerance progressively decreases until dyspnea, tachycardia, and palpitations occur at rest.

Diminished exercise tolerance is expected in the client with anemia. Exercise testing and prescribed exercise(s) in clients with anemia must be instituted with extreme caution and should proceed very gradually to tolerance and/or perceived exertion levels.[10,11] In addition, exercise for any anemic client should be first approved by his or her physician (Case Example 5-1).

CLINICAL SIGNS AND SYMPTOMS

Anemia

- Skin pallor (palms, nail beds) or yellow-tinged skin (mucosa, conjunctiva)
- Fatigue and listlessness
- Dyspnea on exertion accompanied by heart palpitations and rapid pulse (more severe anemia)
- Chest pain with minimal exertion
- Decreased diastolic blood pressure
- CNS manifestations (pernicious anemia):
  - Headache
  - Drowsiness
  - Dizziness, syncope
  - Slow thought processes
  - Apathy, depression
  - Polyneuropathy

CASE EXAMPLE 5-1

Anemia

A 72-year-old woman, status post hip fracture, was treated surgically with nails (used for the fixation of the ends of fractured bones) and was referred to physical therapy for follow-up treatment before hospital discharge. The physician's preoperative examination and surgical report were unremarkable for physical therapy precautions or contraindications.

When the therapist met with the client for the first time, she had already been ambulating alone in her room from the bed to the bathroom and back using a hospital wheeled walker. She was wearing thigh length support hose, hospital gown, and open-heeled slippers from home. Although the nursing report indicated she was oriented to time and place, she seemed confused and required multiple verbal cues to follow the physical therapist's directions.

After ambulating a distance of approximately 50 feet using her wheeled walker and standby assistance from the therapist, the client reported that she could not "catch her breath" and asked to sit down. She placed her hand over her heart and commented that her heart was "fluttering." Blood pressure and pulse measurements were taken and recorded as 145/72 mm Hg (blood pressure) and 90 bpm (pulse rate).

The physical therapist consulted with nursing staff immediately regarding this episode and was given the "go ahead" to complete the therapy session. The physical therapist documented the episode in the medical record and left a note for the physician, briefly describing the incident and ending with the question: Are there any medical contraindications to continuing progressive therapy?

Result: A significant fall in hemoglobin (Hb) often occurs after hip fracture and surgical intervention secondary to the blood loss caused by the fracture and surgery. Other contributing factors may include blood transfusion and alcoholic liver cirrhosis.[12-14]

In this case, although the physician did not offer a direct reply to the physical therapist, the physician's notes indicated a suspected diagnosis of anemia. Follow-up blood work was ordered, and the diagnosis was confirmed. Nursing staff conferred with the physician, and the therapist was advised to work within the patient's tolerance using perceived exertion as a guide while monitoring pulse and blood pressure.

Polycythemia

Polycythemia (also known as erythrocytosis) is characterized by increases in both the number of red blood cells and the concentration of hemoglobin. People with polycythemia have increased whole blood viscosity and increased blood volume.

The increased erythrocyte production results in this thickening of the blood and an increased tendency toward clotting. The viscosity of the blood limits its ability to flow easily, diminishing the supply of blood to the brain and to other vital tissues. Increased platelets in combination with the increased blood viscosity may contribute to the formation of intravascular thrombi.

There are two distinct forms of polycythemia: primary polycythemia (also known as *polycythemia vera*) and secondary polycythemia. *Primary polycythemia* is a relatively uncommon neoplastic disease of the bone marrow of unknown etiology. *Secondary polycythemia* is a physiologic condition resulting from a decreased oxygen supply to the tissues. It is associated with high altitudes, heavy tobacco smoking, radiation exposure, and chronic heart and lung disorders, especially congenital heart defects.

Clinical Signs and Symptoms. The symptoms of this disease are often insidious in onset with vague complaints. The most common first symptoms are shortness of breath and fatigue. The affected individual may be diagnosed only secondary to a sudden complication (e.g., stroke or thrombosis). Increased skin coloration and elevated blood pressure may develop as a result of the increased concentration of erythrocytes and increased blood viscosity.

Gout is sometimes a complication of primary polycythemia, and a typical attack of acute gout may be the first symptom of polycythemia. Gout is a metabolic disease marked by increased serum urate levels (hyperuricemia), which cause painfully arthritic joints. Uric acid level is an end product of purine metabolism. Purine metabolism is altered by excessive cellular proliferation and breakdown associated with increased red cells, granulocytes, and platelets. Hyperuricemia is uncommon in secondary polycythemia because the cellular proliferation is not as extensive as in primary polycythemia.

Blockage of the capillaries supplying the digits of either the hands or the feet may cause a peripheral vascular neuropathy with decreased sensation, burning, numbness, or tingling. This small blood vessel occlusion can also contribute to the development of cyanosis and clubbing. If the underlying disorder is not recognized and treated, the person may develop gangrene and have subsequent loss of tissue.

Watch for increase in blood pressure and elevated hematocrit levels.

CLINICAL SIGNS AND SYMPTOMS

Polycythemia

Clinical signs and symptoms of polycythemia (whether primary or secondary) are directly related to the increase in blood viscosity described earlier and may include

- General malaise and fatigue
- Shortness of breath
- Intolerable pruritus (skin itching; polycythemia vera)[15]*
- Headache
- Dizziness
- Irritability
- Blurred vision
- Fainting
- Decreased mental acuity
- Feeling of fullness in the head
- Disturbances of sensation in the hands and feet
- Weight loss
- Easy bruising
- Cyanosis (blue hue to the skin)
- Clubbing of the fingers
- Splenomegaly (enlargement of spleen)
- Gout
- Hypertension

*This condition of skin itching is particularly related to warm conditions, such as being in bed at night or in a bath, and is called the "hot bath sign."

Sickle Cell Anemia

Sickle cell disease is a generic term for a group of inherited, autosomal recessive disorders characterized by the presence of an abnormal form of hemoglobin, the oxygen-carrying constituent of erythrocytes. A genetic mutation resulting in a single amino acid substitution in hemoglobin causes the hemoglobin to aggregate into long chains, altering the shape of the cell. This sickled or curved shape causes the cell to lose its ability to deform and squeeze through tiny blood vessels, thereby depriving tissue of an adequate blood supply.[3]

The two features of sickle cell disorders, chronic hemolytic anemia and vasoocclusion, occur as a result of obstruction of blood flow to the tissues and early destruction of the abnormal cells. Anemia associated with this condition is merely a symptom of the disease and not the disease itself, despite the term *sickle cell anemia.*

Clinical Signs and Symptoms. A series of "crises," or acute manifestations of symptoms, characterize sickle cell disease. Some people with this disease have only a few symptoms, whereas others are affected severely and have a short lifespan. Recurrent episodes of vasoocclusion and inflammation result in progressive damage to most organs, including the brain, kidneys, lungs, bones, and cardiovascular system, which becomes apparent with increasing age. Cerebrovascular accidents (CVAs) and cognitive impairment are a frequent and severe manifestation.[16]

Stress from viral or bacterial infection, hypoxia, dehydration, emotional disturbance, extreme temperatures, fever, strenuous physical exertion, or fatigue may precipitate a crisis. Pain caused by the blockage of sickled RBCs forming sickle cell clots is the most common symptom; it may be in any organ, bone, or joint of the body. Painful episodes of ischemic tissue damage may last 5 or 6 days and manifest in many different ways, depending on the location of the blood clot (Case Example 5-2).

CLINICAL SIGNS AND SYMPTOMS

Sickle Cell Anemia

- Pain
  - Abdominal
  - Chest
  - Headaches
- Bone and joint episodes from the ischemic tissue, lasting for hours to days and subsiding gradually
  - Low-grade fever
  - Extremity pain
  - Back pain
  - Periosteal pain
  - Joint pain, especially in the shoulder and hip
- Vascular complications
  - Cerebrovascular accidents (affects children and young adults most often)
  - Chronic leg ulcers
  - Avascular necrosis of the femoral head
  - Bone infarcts
- Pulmonary episodes
  - Chest pain
  - Dyspnea
  - Tachypnea
- Neurologic manifestations
  - Seizures
  - Dizziness
  - Drowsiness
  - Stiff neck
  - Paresthesias
  - Cranial nerve palsies
  - Blindness
  - Nystagmus
  - Coma
- Hand-foot syndrome
  - Fever
  - Pain
  - Dactylitis (painful swelling of the dorsum of hands and feet)
- Splenic sequestration episode (occurs before adolescence)
  - Liver and spleen enlargement/tenderness due to trapped erythrocytes
  - Subsequent spleen atrophy due to repeated blood vessel obstruction
- Renal complications
  - Enuresis (bed-wetting)
  - Nocturia (excessive urination at night)
  - Hematuria (blood in the urine)
  - Pyelonephritis
  - Renal papillary necrosis
  - End-stage renal failure (older adult population)

CASE EXAMPLE 5-2

Sickle Cell Anemia

A 20-year-old African-American woman came to physical therapy with severe right knee joint pain. She could recall no traumatic injury but reported hiking 2 days previously in the Rocky Mountains with her brother, whom she was visiting (she was from New York City).

A general screen for systemic illness revealed frequent urination over the past 2 days. She also complained of stomach pain, but she thought this was related to the stress of visiting her family. Past medical history included one other similar episode when she had acute pneumonia at the age of 11 years. She stated that she usually felt fatigued but thought it was because of her active social life and busy professional career. She is a nonsmoker and a social drinker (1 to 3 drinks per week).

On examination, the right knee was enlarged and inflamed, with joint range of motion (ROM) limited by the local swelling. In fact, pain, swelling, and guarded motion in the joint prevented a complete evaluation. Given that restraint, there were no other physical findings, but not all special tests were completed. The neurologic screen was negative.

This woman was treated for local joint inflammation, but the combination of change in altitude, fatigue, increased urination, and stomach pains alerted the therapist to the possibility of a systemic process despite the client's explanation for the fatigue and stomach upset. Because the client was from out of town and did not have a local physician, the therapist telephoned the hospital emergency department for a telephone consultation. It was suggested that a blood sample be obtained for preliminary screening while the client continued to receive physical therapy. Laboratory results included the following:

- Hematocrit (Hct) 30% (normal 35% to 47%)
- Hemoglobin (Hb) 10 g/dL (normal 12 to 15 g/dL)
- White blood cells (WBC) 20,000/mm^3 (normal 4500 to 11,000/mm^3)

Based on these findings, the client was admitted to the hospital and diagnosed as having sickle cell anemia. It is likely that the change in altitude, the emotional stress of visiting family, and the physical exertion precipitated a "crisis" (now referred to as "episodes"). She received continued physical therapy treatment during her hospital stay and was discharged with further follow-up planned in her home city.

Adapted from Jennings B: Nursing role in management: hematological problems. In Lewis S, Collier I, editors: *Medical-surgical nursing: assessment and management of clinical problems,* St Louis, 1992, Mosby, pp 664-714. Used with permission.

Leukocyte Disorders

The blood contains three major groups of leukocytes, including:

1. Lymphoid cells (lymphocytes, plasma cells)
2. Monocytes
3. Granulocytes (neutrophils, eosinophils, and basophils)

Lymphocytes produce antibodies and react with antigens, thus initiating the immune response to fight infection. *Monocytes* are the largest circulating blood cells and represent an immature cell until they leave the blood and travel to the tissues where they form macrophages in response to foreign substances such as bacteria. *Granulocytes* contain lysing agents capable of digesting various foreign materials and defend the body against infectious agents by phagocytosing bacteria and other infectious substances.[17]

Disorders of leukocytes are recognized as the body's reaction to disease processes and noxious agents. The therapist will encounter many clients who demonstrate alterations in the blood leukocyte (WBC) concentration as a result of acute infections or chronic systemic conditions. The leukocyte count also may be elevated (leukocytosis) in women who are pregnant; in clients with bacterial infections, appendicitis, leukemia, uremia, or ulcers; in newborns with hemolytic disease; and normally at birth. The leukocyte count may drop below normal values *(leukopenia)* in clients with viral diseases (e.g., measles), infectious hepatitis, rheumatoid arthritis, cirrhosis of the liver, and lupus erythematosus and also after treatment with radiation or chemotherapy.

Leukocytosis

Leukocytosis characterizes many infectious diseases and is recognized by a count of more than 10,000 leukocytes/mm^3. It can be associated with an increase in circulating neutrophils (neutrophilia), which are recruited in large numbers early in the course of most bacterial infections.

Leukocytosis is a common finding and is helpful in aiding the body's response to any of the following:

- Bacterial infections
- Inflammation or tissue necrosis (e.g., infarction, myositis, vasculitis)
- Metabolic intoxications (e.g., uremia, eclampsia, acidosis, gout)
- Neoplasms (especially bronchogenic carcinoma, lymphoma, melanoma)
- Acute hemorrhage
- Splenectomy
- Acute appendicitis
- Pneumonia
- Intoxication by chemicals
- Acute rheumatic fever

CLINICAL SIGNS AND SYMPTOMS

Leukocytosis

These clinical signs and symptoms are usually associated with symptoms of the conditions listed earlier and may include

- Fever
- Symptoms of localized or systemic infection
- Symptoms of inflammation or trauma to tissue

Leukopenia

Leukopenia, or reduction of the number of leukocytes in the blood below 5000/mL, can be caused by a variety of factors. Unlike leukocytosis, leukopenia is never beneficial.

Leukopenia can occur in many forms of bone marrow failure such as that following antineoplastic chemotherapy or radiation therapy, in overwhelming infections, in dietary deficiencies, and in autoimmune diseases.[2]

It is important for the physical therapist to be aware of the client's most recent WBC count prior to and during the course of physical therapy. If the client is immunosuppressed, infection is a major problem. Constitutional symptoms, such as fever, chills, or sweats, warrant immediate medical referral.

Nadir, or the lowest point the WBC count reaches, usually occurs 7 to 14 days after chemotherapy or radiation therapy. At this time, the client is extremely susceptible to opportunistic infections and severe complications. The importance of good handwashing and hygiene practices cannot be overemphasized when treating any of these clients.[7]

CLINICAL SIGNS AND SYMPTOMS

Leukopenia

- Sore throat, cough
- High fever, chills, sweating
- Ulcerations of mucous membranes (mouth, rectum, vagina)
- Frequent or painful urination
- Persistent infections

Leukemia

Leukemia is a disease arising from the bone marrow and involves the uncontrolled growth of immature or dysfunctional WBCs; a complete discussion of this cancer is found in Chapter 13.

Platelet Disorders

Platelets (thrombocytes) function primarily in hemostasis (stopping bleeding) and in maintaining capillary integrity (see normal values listed inside book cover). They function in the coagulation (blood clotting) mechanism by forming hemostatic plugs in small ruptured blood vessels or by adhering to any injured lining of larger blood vessels.

A number of substances derived from the platelets that function in blood coagulation have been labeled "platelet factors." Platelets survive approximately 8 to 10 days in circulation and are then removed by the reticuloendothelial cells. *Thrombocytosis* refers to a condition in which the number of platelets is abnormally high, whereas *thrombocytopenia* refers to a condition in which the number of platelets is abnormally low.

Platelets are affected most often by anticoagulant drugs, including aspirin, heparin, warfarin (Coumadin), and other newer antithrombotic drugs now appearing on the market (e.g., Arixtra). Platelet levels can also be affected by diet (presence of lecithin preventing coagulation or vitamin K from promoting coagulation), by exercise that boosts the production of chemical activators that destroy unwanted clots, and by liver disease that affects the supply of vitamin K. Platelets are also easily suppressed by radiation and chemotherapy.[2,3]

Thrombocytosis

Thrombocytosis is an increase in platelet count that is usually temporary. It may be primary caused by unregulated production of platelets or secondary (reactive thrombocytosis) as a compensatory mechanism (exaggerated physiologic response) after severe hemorrhage, surgery, and splenectomy; in iron deficiency and polycythemia vera; and as a manifestation of an occult (hidden) neoplasm (e.g., lung cancer).[18,19]

It is associated with a tendency to clot because blood viscosity is increased by the very high platelet count, resulting in intravascular clumping (or thrombosis) of the sludged platelets. Peripheral blood vessels, particularly in the fingers and toes, are affected.

Thrombocytosis remains asymptomatic until the platelet count exceeds 1 million/mm^3. Other symptoms may include splenomegaly and easy bruising.

CLINICAL SIGNS AND SYMPTOMS

Thrombocytosis

- Thrombosis
- Splenomegaly
- Easy bruising

Thrombocytopenia

Thrombocytopenia, a decrease in the number of platelets (less than 150,000/mm^3) in circulating blood, can result from decreased or defective platelet production or from accelerated platelet destruction.

There are many causes of thrombocytopenia (Box 5-1). In a physical therapy practice the most common causes seen are bone marrow failure from radiation treatment, leukemia, or

BOX 5-1 CAUSES OF THROMBOCYTOPENIA

- Bone marrow failure
- Radiation
- Aplastic anemia
- Leukemia
- Metastatic carcinoma
- Cytotoxic agents (chemotherapy)
- Medications
  - Nonsteroidal antiinflammatory drugs (NSAIDs), including aspirin
  - Methotrexate
  - Gold
  - Coumadin/warfarin

metastatic cancer; cytotoxic agents used in chemotherapy; and drug-induced platelet reduction, especially among adults with rheumatoid arthritis treated with gold or inflammatory conditions treated with aspirin or other NSAIDs.

Primary bleeding sites include bone marrow or spleen; secondary bleeding occurs from small blood vessels in the skin, mucosa (e.g., nose, uterus, GI tract, urinary tract, and respiratory tract), and brain (intracranial hemorrhage).

Clinical Signs and Symptoms. Severe thrombocytopenia results in the appearance of multiple petechiae (small, purple, pinpoint hemorrhages into the skin), most often observed on the lower legs. GI bleeding and bleeding into the central nervous system (CNS) associated with severe thrombocytopenia may be life-threatening manifestations of thrombocytopenic bleeding.

The physical therapist must be alert for obvious skin, joint, or mucous membrane symptoms of thrombocytopenia, which include severe bruising, external hematomas, joint swelling, and the presence of multiple petechiae observed on the skin or gums. These symptoms usually indicate a platelet count well below $100,000/mm^3$. Strenuous exercise or any exercise that involves straining or bearing down could precipitate a hemorrhage, particularly of the eyes or brain. Blood pressure cuffs must be used with caution and any mechanical compression, visceral manipulation, or soft tissue mobilization is contraindicated without a physician's approval.

People with undiagnosed thrombocytopenia need immediate physician referral. Exercise guidelines for thrombocytopenia can be found in Table 39-7 in Goodman's *Pathology: Implications for the Physical Therapist,* second edition.

CLINICAL SIGNS AND SYMPTOMS

Thrombocytopenia

- Bleeding after minor trauma
- Spontaneous bleeding
  - Petechiae (small red dots)
  - Ecchymoses (bruises)
  - Purpura spots (bleeding under the skin)
  - Epistaxis (nosebleed)
- Menorrhagia (excessive menstruation)
- Gingival bleeding
- Melena (black, tarry stools)

Coagulation Disorders

Hemophilia

Hemophilia is a hereditary blood-clotting disorder caused by an abnormality of functional plasma-clotting proteins known as factors VIII and IX. In most cases, the person with hemophilia has normal amounts of the deficient factor circulating, but it is in a functionally inadequate state. Persons with hemophilia bleed longer than those with normal levels of functioning factors VIII or IX, but the bleeding is not any faster than would occur in a normal person with the same injury.

Clinical Signs and Symptoms. Bleeding into the joint spaces (hemarthrosis) is one of the most common clinical manifestations of hemophilia. It may result from an identifiable trauma or stress or may be spontaneous, most often affecting the knee, elbow, ankle, hip, and shoulder (in order of most common appearance).

Recurrent hemarthrosis results in hemophiliac arthropathy (joint disease) with progressive loss of motion, muscle atrophy, and flexion contractures. Bleeding episodes must be treated early with factor replacement and joint immobilization during the period of pain. This type of affected joint is particularly susceptible to being injured again, setting up a cycle of vulnerability to trauma and repeated hemorrhages.[20]

Hemarthroses are not common in the first year of life but increase in frequency as the child begins to walk. The severity of the hemarthrosis may vary (depending on the degree of injury) from mild pain and swelling, which resolves without treatment within 1 to 3 days, to severe pain with an excruciatingly painful, swollen joint that persists for several weeks and resolves slowly with treatment.

Bleeding into the muscles is the second most common site of bleeding in persons with hemophilia. Muscle hemorrhages can be more insidious and massive than joint hemorrhages. They may occur anywhere but are common in the flexor muscle groups, predominantly the iliopsoas, gastrocnemius, and flexor surface of the forearm, and they result in deformities such as hip flexion contractures, equinus position of the foot, or Volkmann's deformity of the forearm.[21] For a more in-depth discussion of hemophilia and the clinical signs and symptoms associated with it, see Table 14-8 in Goodman and Fuller's *Pathology: Implications for the Physical Therapist,* third edition.

When bleeding into the psoas or iliacus muscle puts pressure on the branch of the femoral nerve supplying the skin over the anterior thigh, loss of sensation occurs. Distention of the muscles with blood causes pain that can be felt in the lower abdomen, possibly even mimicking appendicitis when bleeding occurs on the right side. In an attempt to relieve the distention and reduce the pain, a position with hip flexion is preferred.

CLINICAL SIGNS AND SYMPTOMS

Acute Hemarthrosis

- Aura, tingling, or prickling sensation
- Stiffening into the position of comfort
- Decreased range of motion
- Pain
- Swelling
- Tenderness
- Heat

CLINICAL SIGNS AND SYMPTOMS

Muscle Hemorrhage

- Gradually intensifying pain
- Protective spasm of the muscle
- Limitation of movement at the surrounding joints
- Muscle assumes the position of comfort (usually shortened)
- Loss of sensation

CLINICAL SIGNS AND SYMPTOMS

CNS Involvement

- Intraspinal hemorrhage (rare)
- Intracranial hemorrhage
  - Irritability, lethargy
  - Seizures
  - Feeding difficulties (children)
  - Unequal pupils
  - Apnea
  - Vomiting
  - Paralysis
  - Tense, bulging fontanelles
  - Death

CLINICAL SIGNS AND SYMPTOMS

GI Involvement

- Abdominal pain and distention
- Melena (blood in stool)
- Hematemesis (vomiting blood)
- Fever
- Low abdominal/groin pain due to bleeding into wall of large intestine or iliopsoas muscle
- Flexion contracture of the hip due to spasm of the iliopsoas muscle secondary to retroperitoneal hemorrhage[22]

Two tests are used to distinguish an iliopsoas bleed from a hip bleed[3]:

1. When the client flexes the trunk, severe pain is produced in the presence of *iliopsoas bleeding,* whereas only mild pain is found with a hip hemorrhage.
2. When the hip is gently rotated in either direction, severe pain is experienced with a *hip hemorrhage* but is absent or mild with iliopsoas bleeding.

Over time, the following complications may occur:

- Vascular compression causing localized ischemia and necrosis
- Replacement of muscle fibers by nonelastic fibrotic tissue causing shortened muscles and thus producing joint contractures
- Peripheral nerve lesions from compression of a nerve that travels in the same compartment as the hematoma, most commonly affecting the femoral, ulnar, and median nerves
- Pseudotumor formation with bone erosion

PHYSICIAN REFERRAL

Understanding the components of a client's past medical history that can affect hematopoiesis (production of blood cells) can provide the physical therapist with valuable insight into the client's present symptoms, which are usually already well known to the attending physician.

For example, the effects of certain drugs, exposure to radiation, or recent cytotoxic cancer chemotherapy can affect bone marrow. Whenever uncertain, the physical therapist is encouraged to contact the physician by telephone for discussion and clarification of the client's medical symptoms.

A history of excessive menses, a folate-poor diet, alcohol abuse, drug ingestion, family history of anemia, and family roots in geographic areas where RBC enzyme or hemoglobin abnormalities are prevalent represent some important findings. The presence of any one or more of these factors should alert the physical therapist to the need for medical referral when the client is not already under the care of a physician or when new signs or symptoms develop.

In addition, exercise for *anemic* clients must be instituted with extreme caution and should first be approved by the client's physician. Clients with undiagnosed thrombocytopenia need immediate medical referral. The physical therapist must be alert for obvious skin or mucous membrane symptoms of *thrombocytopenia.* The presence of severe bruising, hematomas, and multiple petechiae usually indicates a platelet count well below normal. With clients who have been diagnosed with *hemophilia,* medical referral should be made when any painful episode develops in the muscle(s) or joint(s). Pain usually occurs before any other evidence of bleeding. Any unexplained symptom may be a signal of bleeding.

Guidelines for Immediate Medical Attention

- Signs and symptoms of thrombocytopenia (decreased platelets, e.g., excessive or spontaneous bleeding, petechiae, severe bruising) previously unseen or unreported to the physician

Guidelines for Physician Referral

- Consultation with the physician may be necessary when establishing or progressing an exercise program for a client with known anemia
- New episodes of muscle or joint pain in a client with hemophilia; pain usually occurs before any other evidence of bleeding. Any unexplained symptom(s) may be a signal of bleeding; coughing up blood in this population group must be reported to the physician

Clues to Screening for Hematologic Disease

- These clues will help the therapist in the decision-making process:
- Previous history (delayed effects) or current administration of chemotherapy or radiation therapy

- Chronic or long-term use of aspirin or other NSAIDs (drug-induced platelet reduction)
- Spontaneous bleeding of any kind (e.g., nosebleed, vaginal/menstrual bleeding, blood in the urine or stool, bleeding gums, easy bruising, hemarthrosis), especially with a previous history of hemophilia
- Recent major surgery or previous transplantation
- Rapid onset of dyspnea, chest pain, weakness, and fatigue with palpitations associated with recent significant change in altitude
- Observed changes in the hands and fingernail beds (see Table 5-1 and Fig. 4-30)

Key Points to Remember

- Anemia may have no symptoms until hemoglobin concentration and hematocrit fall below one half of normal.
- Weakness, fatigue, and dyspnea are early signs of anemia.
- Exercise for anyone who is anemic must be instituted gradually per tolerance and/or perceived exertion levels with physician approval.
- Platelet level below 10,000 (thrombocytopenia) can be life threatening. Platelet transfusions are usually given for platelet counts below this level in adults and children who have chemotherapy-induced thrombocytopenia. Multiple bruises and petechiae may be the only sign.
- For clients with known thrombocytopenia, exercise programs must avoid the Valsalva (or bearing down) movement, and caution must be used to avoid further injury by bumping against objects.
- During the inspection/observation portion of the objective examination, screen both hands for skin or nail bed changes indicative of hematologic involvement.
- For the client with hemophilia, bleeding episodes must be treated early with factor replacement and joint immobilization during the period of pain. Never apply heat to a bleeding or suspected bleeding area.
- Pain may be the only symptom of a joint or muscle bleed for the client with hemophilia. Any painful or unexplained symptom in this population must be screened medically. Coughing up blood is not a normal finding with hemophilia and should be reported to the physician immediately.
- The National Hemophilia Foundation (NHF) publishes additional materials for physical therapists. These can be ordered by calling the NHF at (212) 328-3700.

SUBJECTIVE EXAMINATION

Special Questions to Ask

Past Medical History

- Have you recently been told you are anemic?
- Have you recently had a serious blood loss (possibly requiring transfusion)? (**Anemia;** also consider **jaundice/hepatitis posttransfusion**)
- Have you ever been told that you have a congenital heart defect (also chronic lung/heart disorders)? (**Polycythemia; also possible with history of heavy tobacco use**)
- Do you have a history of bruising easily, nose bleeds, or excessive blood loss?* (**Polycythemia, hemophilia, thrombocytopenia**)

For example, do you bleed or bruise easily after minor trauma, surgery, or dental procedures?

Has any previous bleeding been severe enough to require a blood transfusion?

- Have you been exposed to occupational or industrial gases, such as chlorine gas, mustard gas, Agent Orange, napalm?

Associated Signs and Symptoms

- Do you experience shortness of breath, heart palpitations, or chest pain with slight exertion (e.g., climbing stairs) or even just at rest? (**Anemia**)
- Alternate or additional questions: Do you ever have trouble catching your breath?
- Are there any activities you have had to stop doing because you don't have enough energy or breath?

(Therapist: Be aware of the clients who stop doing certain activities because they become short of breath. For example, they no longer go up and down stairs in their homes and choose to avoid this activity . . . or the client who can't

*Symptoms beginning in infancy or childhood suggest a congenital hemostatic defect, whereas symptoms beginning later in life indicate an acquired disorder, such as secondary to drug-induced defect of platelet function, a common cause of easy bruising and excessive bleeding. This bruising or bleeding occurs usually in association with trauma, menstruation, dental work, or surgical procedures. Drug-induced bruising or bleeding may also occur with use of aspirin and aspirin-containing compounds; NSAIDs such as ibuprofen (Motrin) and naproxen (Naprosyn) (see Table 8-3); and penicillins because these drugs inhibit platelet function to some extent.

Continued

SUBJECTIVE EXAMINATION—*cont'd*

complete all of his or her shopping at one time. They may not report being short of breath because they have decreased their activity level to accommodate for the change in their pulmonary capacity.)

For persons at elevations above 3500 feet: Have you recently moved from one geographic location to another? **(Polycythemia)**

Do you ever have episodes of dizziness, blurred vision, headaches, fainting, or a feeling of fullness in your head? **(Polycythemia)**

Do you have recurrent infections and low-grade fever such as colds, influenza-like symptoms, or other upper respiratory infections? **(Abnormal leukocytes)**

- Do you have black, tarry stools **(bleeding into the gastrointestinal tract)** or blood in urine **(genitourinary tract)**?

For women (anemia, thrombocytopenia): Do you frequently have prolonged or excessive bleeding in association with your menstrual flow? (Excessive may be considered to be measured by the use of more than four tampons each day; prolonged menstruation usually refers to more than 5 days—both of these measures are subjective and must be considered along with other factors, such as the presence of other symptoms, personal menstrual history, placement in the life cycle [i.e., in relation to menopause].)

CASE STUDY

REFERRAL

You are working in a hospital setting and you have received a physician's referral to ambulate and exercise a patient who was involved in a serious automobile accident 10 days ago. The patient had internal injuries that required immediate abdominal surgery and 600 mL of blood transfused within 24 hours postoperatively. His condition is considered to be medically "stable."

CHART REVIEW

What specific medical information should you look for in the medical record before beginning your evaluation?

Name, age, and occupation:

Past medical history:

Previous myocardial infarcts, history of heart disease, diabetes (type)

Surgical report:

Type of surgery, locations of scar, any current contraindications

Were there any other injuries?

If yes, what were these and what is the current status of each injury?

Body weight:

Pulmonary status:

Is the patient a cigarette or pipe smoker (or other tobacco user)?

Is the patient currently receiving oxygen or respiratory therapy? Is there a recommendation for how many liters (L) of oxygen per minute can be used during exercise?

What was the patient's pulmonary status after the accident and postoperatively?

Laboratory report:

Hematocrit/hemoglobin levels. Anemia?

Current status:

Nursing reports of the patient's complaints of any kind (e.g., symptoms of dyspnea or heart palpitations from rapid loss of blood).

Has the patient been out of bed at all yet?

If yes, when? How far did he walk? How much assistance was required? Did he have symptoms of orthostatic hypotension?

Does the patient have any gastrointestinal symptoms?

Is the patient oriented to time, place, and person?

Are there any dietary or fluid restrictions to be observed while the patient is in the physical therapy department? Is he on an intravenous line?

Vital signs:

Blood pressure

Presence of fever

Resting pulse rate

Pulse oximetry (if available)

Pain assessment

Current medications:

Be aware of the purpose for each medication and its potential side effects.

Are there any known discharge plans at this time?

PRACTICE QUESTIONS

1. If rapid onset of anemia occurs after major surgery, which of the following symptom patterns might develop?
   a. Continuous oozing of blood from the surgical site
   b. Exertional dyspnea and fatigue with increased heart rate
   c. Decreased heart rate
   d. No obvious symptoms would be seen
2. Chronic GI blood loss sometimes associated with use of NSAIDs can result in which of the following problems?
   a. Increased incidence of joint inflammation
   b. Iron deficiency
   c. Decreased heart rate and bleeding
   d. Weight loss, fever, and loss of appetite
3. Under what circumstances would you consider asking a client about a recent change in altitude or elevation?
4. Preoperatively, clients cannot take aspirin or antiinflammatory medications because these:
   a. Decrease leukocytes
   b. Increase leukocytes
   c. Decrease platelets
   d. Increase platelets
   e. None of the above
5. Skin color and nail bed changes may be observed in the client with:
   a. Thrombocytopenia resulting from chemotherapy
   b. Pernicious anemia resulting from Vitamin B12 deficiency
   c. Leukocytosis resulting from AIDS
   d. All of the above
6. In the case of a client with hemarthrosis associated with hemophilia, what physical therapy intervention would be contraindicated?
7. Bleeding under the skin, nosebleeds, bleeding gums, and black stools require medical evaluation as these may be indications of:
   a. Leukopenia
   b. Thrombocytopenia
   c. Polycythemia
   d. Sickle cell anemia
8. Describe the two tests used to distinguish an iliopsoas bleed from a joint bleed.
9. What is the significance of *nadir?*
10. When exercising a client with known anemia, what two measures can be used as guidelines for frequency, intensity, and duration of the program?

REFERENCES

1. Bushnell BD: Perioperative medical comorbidities in the orthopaedic patient. *J Am Acad Orthop Surg* 16:216–227, 2008.
2. Hillman R: *Hematology in clinical practice*, ed 5, New York, 2011, McGraw-Hill.
3. Hoffman R: *Hematology: basic principles and practice*, ed 5, Philadelphia, 2008, Churchill-Livingstone.
4. Wehbi M: Acute gastritis. *eMedicine Specialties—Gastroenterology*. Updated Jan. 12, 2011. Available on-line at http://emedicine.medscape.com/article/175909-overview. Accessed January 26, 2011.
5. Abeloff M: *Abeloff's clinical oncology*, ed 4, Philadelphia, 2008, Churchill Livingstone.
6. Holcomb S: Anemia: pointing the way to a deeper problem. *Nursing 2001* 31(7):36–42, 2001.
7. Goodman CC, Fuller K: *Pathology: implications for the physical therapist*, ed 3, St Louis, 2009, WB Saunders.
8. Merriman NA: Hip fracture risk in patients with a diagnosis of pernicious anemia. *Gastroenterology* 138(4):1330–1337, 2010.
9. Sproule BJ: Cardiopulmonary physiological responses to heavy exercise in patients with anemia. *J Clin Invest* 39(2):378–388, 1960.
10. Callahan L, Woods K, et al: Cardiopulmonary responses to exercise in women with sickle cell anemia. *Am J Respir Crit Care Med* 165(9):1309–1316, 2002.
11. Goodman C, Helgeson K: *Exercise prescription for medical conditions: handbook for physical therapists*, Philadelphia, 2011, FA Davis.
12. Lombardi G, Rizzi E, et al: Epidemiology of anemia in older patients with hip fracture. *J Am Geriatr Soc* 44(6):740–741, 1996.
13. Mackenzie C: Hip fracture in the elderly, *Best practice of medicine*, Merck Medicus, Thomson Micromedex, March 2002.
14. Parker MJ: Iron supplementation for anemia after hip fracture surgery. *J Bone Joint Surg Am* 92:265–269, 2010.
15. Saini KS: Polycythemia vera-associated pruritus and its management. *Eur J Clin Invest* 40(9):828–834, 2010.
16. Rees DC: Sickle-cell disease. *Lancet* 376(9757):2018–2031, 2010.
17. Abbas MBBS, Lichtman AH: *Basic immunology updated edition: functions and disorders of the immune system*, ed 3, Philadelphia, 2011, WB Saunders.
18. Krishnan K: Thrombocytosis, Secondary. *eMedicine Specialities—Hematology*. Updated Oct. 4, 2009. Available on-line at http://emedicine.medscape.com/article/206811-overview. Accessed January 26, 2010.
19. Inoue S: Thrombocytosis. *eMedicine Specialties—Pediatrics/Hematology*. Updated April 19, 2010. Available on-line at http://emedicine.medscape.com/article/959378-overview. Accessed January 26, 2010.
20. Agaliotis DP: Hemophilia. *eMedicine Specialties—Coagulation, Hemostasis, and Disorders*. Updated November 22, 2010. http://emedicine.medscape.com/article/210104-overview.
21. Kumar V: *Robbins and Cotran pathologic basis of disease: professional edition*, ed 8, Philadelphia, 2009, WB Saunders.
22. Kliegman RM: Nelson textbook of pediatrics, ed 18, Philadelphia, 2007, WB Saunders.

CHAPTER

6

Screening for Cardiovascular Disease

The cardiovascular system consists of the heart, capillaries, veins, and lymphatics and functions in coordination with the pulmonary system to circulate oxygenated blood through the arterial system to all cells. This system then collects deoxygenated blood from the venous system and delivers it to the lungs for reoxygenation (Fig. 6-1).

Heart disease remains the leading cause of death in industrialized nations. In the United States alone, cardiovascular disease (CVD) is responsible for approximately one million deaths each year. One in three Americans has some form of cardiovascular disease. The American Heart Association (AHA) reports that about half of all deaths from heart disease are sudden and unexpected.[1]

Known risk factors include advancing age, hypertension, obesity, sedentary lifestyle, excessive alcohol consumption, oral contraceptive use (over age 35, combined with smoking), first-generation family history, tobacco use (including exposure to second-hand smoke), abnormal cholesterol levels, and race (e.g., African Americans, Mexican Americans, Native Americans, and Pacific Islanders are at greater risk).

Fortunately, during the last two decades cardiovascular research has greatly increased our understanding of the structure and function of the cardiovascular system in health and disease. Despite the formidable statistics regarding the prevalence of CVD, during the last 15 years a steady decline in mortality from cardiovascular disorders has been witnessed. Effective application of the increased knowledge regarding CVD and its risk factors will assist health care professionals to educate clients in achieving and maintaining cardiovascular health.

Information about heart disease is changing rapidly. Part of the therapist's intervention includes patient/client education. The therapist can access up-to-date information at many useful websites (Box 6-1).

SIGNS AND SYMPTOMS OF CARDIOVASCULAR DISEASE

Cardinal symptoms of cardiac disease usually include chest, neck and/or arm pain or discomfort, palpitation, dyspnea, syncope (fainting), fatigue, cough, diaphoresis, and cyanosis. Edema and leg pain (claudication) are the most common symptoms of the vascular component of a cardiovascular pathologic condition. Symptoms of cardiovascular involvement should also be reviewed by system (Table 6-1).

Chest Pain or Discomfort

Chest pain or discomfort is a common presenting symptom of cardiovascular disease and must be evaluated carefully. Chest pain may be cardiac or noncardiac in origin and may radiate to the neck, jaw, upper trapezius muscle, upper back, shoulder, or arms (most commonly the left arm).

Radiating pain down the arm follows the pattern of ulnar nerve distribution. Pain of cardiac origin can be experienced in the somatic areas because the heart is supplied by the C3 to T4 spinal segments, referring visceral pain to the corresponding somatic area (see Fig. 3-3). For example, the heart and the diaphragm, supplied by the C5-6 spinal segment, can refer pain to the shoulder (see Figs. 3-4 and 3-5).

Cardiac-related chest pain may arise secondary to angina, myocardial infarction (MI), pericarditis, endocarditis, mitral valve prolapse, or dissecting aortic aneurysm. Location and description (frequency, intensity, and duration) vary according to the underlying pathologic condition (see each individual condition).

Cardiac chest pain is often accompanied by associated signs and symptoms such as nausea, vomiting, diaphoresis, dyspnea, fatigue, pallor, or syncope. These associated signs and symptoms provide the therapist with red flags to identify musculoskeletal symptoms of a systemic origin.

Cardiac chest pain or discomfort can also occur when the coronary circulation is normal, as in the case of clients with anemia, causing lack of oxygenation of the myocardium (heart muscle) during physical exertion.

Noncardiac chest pain can be caused by an extensive list of disorders requiring screening for medical disease. For example, cervical disk disease and arthritic changes can mimic atypical chest pain. Chest pain that is attributed to anxiety, trigger points, cocaine use, and other noncardiac causes is discussed in Chapter 17.

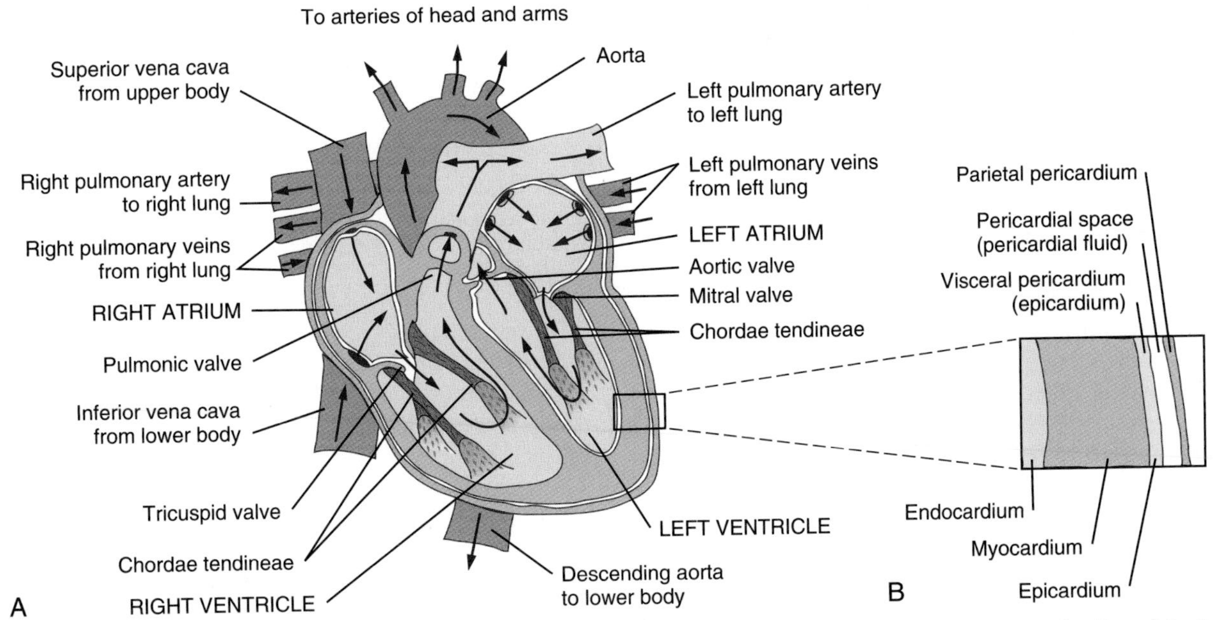

Fig. 6-1 Structure and circulation of the heart. Blood entering the left atrium from the right and left pulmonary veins flows into the left ventricle. The left ventricle pumps blood into the systemic circulation through the aorta. From the systemic circulation, blood returns to the heart through the superior and inferior venae cavae. From there the right ventricle pumps blood into the lungs through the right and left pulmonary arteries. A thick layer of connective tissue called the *septum* separates the left and right chambers of the heart. The top of the heart *(atria)* is also separated from the bottom of the heart *(ventricles)* by connective tissue, which does not conduct electrical activity and serves as an electrical barrier or insulator. (Redrawn from Black JM, Hawks JH: *Luckmann and Sorenson's medical-surgical nursing,* ed 8, Philadelphia, 2009, WB Saunders.)

TABLE 6-1 Cardiovascular Signs and Symptoms by System

System	Symptoms
General	Weakness Fatigue Weight change Poor exercise tolerance Peripheral edema
Integumentary	Pressure ulcers Loss of body hair Cyanosis (lips and nail beds)
Central nervous system	Headaches Impaired vision Dizziness or syncope
Pulmonary	Labored breathing, dyspnea Productive cough
Genitourinary	Urinary frequency Nocturia Concentrated urine Decreased urinary output
Musculoskeletal	Chest, shoulder, back, neck, jaw, or arm pain Myalgias Muscular fatigue Muscle atrophy Edema Claudication
Gastrointestinal	Nausea and vomiting Ascites (abdominal distention)

Modified from Goodman CC, Fuller K: *Pathology: implications for the physical therapist,* ed 3, Philadelphia, 2009, WB Saunders.

Palpitation

Palpitation, the presence of an irregular heartbeat, may also be referred to as arrhythmia or dysrhythmia, which may be caused by a relatively benign condition (e.g., mitral valve prolapse, "athlete's heart," caffeine, anxiety, exercise) or a severe condition (e.g., coronary artery disease, cardiomyopathy, complete heart block, ventricular aneurysm, atrioventricular valve disease, mitral or aortic stenosis).

The sensation of palpitations has been described as a bump, pound, jump, flop, flutter, or racing sensation of the heart. Associated symptoms may include lightheadedness or syncope. Palpated pulse may feel rapid or irregular, as if the heart "skipped" a beat.

Occasionally, a client will report "fluttering" sensations in the neck. Generally, unless accompanied by other symptoms, these sensations in the neck are caused by anxiety, random muscle fasciculation, or minor muscle strain or overuse.

Palpitations can be considered physiologic (i.e., when less than six occur per minute, this may be considered within normal function of the heart). However, palpitation lasting for hours or occurring in association with pain, shortness of breath, fainting, or severe lightheadedness requires medical evaluation. Palpitation in any person with a history of unexplained sudden death in the family requires medical referral.

Clients describing "palpitations" or similar phenomena may not be experiencing symptoms of heart disease.

BOX 6-1 INFORMATIONAL WEBSITES

American Heart Association (AHA)
http://www.americanheart.org
The AHA has also developed a validated tool to assess risk of heart attack, stroke, and diabetes. Available at www.americanheart.org/riskassessment.
The AHA also has a website just for health care professionals. They offer comprehensive information on cardiovascular and cerebrovascular medicine. You can access clinical summaries of new papers and journal articles from well-known publications. Available at http://www.my.americanheart.org/portal/professional

American Stroke Association
http://www.strokeassociation.org
The American Stroke Association is a division of the AHA with updated information for consumers on strokes, as well as a special link just for health care professionals.

National Cholesterol Education Program (NCEP)
http://www.nhlbi.nih.gov/about/ncep/
This website offers a risk assessment tool for estimating the 10-year risk of developing coronary vascular disease (heart attack and coronary death) based on recent data from the Framingham Heart Study. Available at http://hin.nhlbi.nih.gov/atpiii/calculator.asp?usertype=prof

NIHSeniorHealth
http://nihseniorhealth.gov
This website, designed especially for older adults, is a joint effort of the National Institute of Aging (NIA) and the National Library of Medicine (NLM). It contains answers to questions about preventing, detecting, and treating a heart attack. The information is provided in a senior-friendly format. Short, easy-to-read segments of information are featured that can be accessed in large-print type, open-captioned videos, and even an audio version.

Centers for Disease Control and Prevention (CDC)
Division for Heart Disease and Stroke Prevention (DHDSP)
http://www.cdc.gov/dhdsp/
The mission of this group is to provide public health leadership to improve cardiovascular health for all, reduce the burden, and eliminate disparities associated with heart disease and stroke. Its resources include information on heart disease and stroke topics and prevention programs, and publications and statistical information.

American College of Cardiology
http://www.acc.org
The American College of Cardiology offers the latest professional information on heart disease, research, and treatment. A special feature is the availability of clinical statements and guidelines that can be printed or downloaded.

Elsevier Science
http://www.cardiosource.com
This site is offered by collaboration between the American College of Cardiology Foundation and Elsevier Science. It includes a drug database, case studies for self-study, and a library with access to journal abstracts and reference texts.

National Heart, Lung, and Blood Institute (NHLBI)
http://www.nhlbi.nih.gov/index.htm
The NHLBI at the National Institutes of Health (NIH) offers information for health care professionals and consumers. Research results, clinical guidelines, and information for women and heart disease are available.

Heart Center Online
http://www.heartcenteronline.com
Physicians provide patient education on cardiac conditions, medical devices, procedures, and tests. There is also a prevention center with a focus on lifestyle issues and nutrition, video library, and an entire section on transtelephonic monitoring.

American College of Sports Medicine (ACSM)
www.acsm.org
The ACSM provides guidelines for exercise testing and prescription, including screening tests; pre-exercise evaluations; general principles for exercise prescription; and specifics for mode, frequency, intensity, and duration of exercise for individual diseases and conditions.

Framingham Risk Score
To calculate your Framingham risk score for heart attack in the next 10 years, go to www.health.harvard.edu/heartrisk. A score of 5% to 20% suggests the need for the hsCRP test (C-reactive protein). Using the results of the CRP test, the Reynolds model will provide a more accurate assessment of risk (www.reynoldsriskscore.org).

Palpitations may occur as a result of an overactive thyroid, secondary to caffeine sensitivity, as a side effect of some medications, and with the use of drugs such as cocaine. Encourage the client to report any such symptoms to the physician if this information has not already been brought to the physician's attention.

Dyspnea

Dyspnea, also referred to as breathlessness or shortness of breath, can be cardiovascular in origin, but it may also occur secondary to a pulmonary pathologic condition (see also Chapter 7), fever, certain medications, allergies, poor physical conditioning, or obesity. Early onset of dyspnea may be described as having to breathe too much or as an uncomfortable feeling during breathing after exercise or exertion.

Shortness of breath with mild exertion (dyspnea on exertion [DOE]), when caused by an impaired left ventricle that is unable to contract completely, results in the lung's inability to empty itself of blood. Pulmonary congestion and shortness of breath then occur. With severe compromise of the cardiovascular or pulmonary systems, dyspnea may occur at rest.

The severity of dyspnea is determined by the extent of disease. Thus the more severe the heart disease is, the easier it is to bring on dyspnea. Extreme dyspnea includes paroxysmal nocturnal dyspnea (PND) and orthopnea (breathlessness that is relieved by sitting upright with pillows used to prop the trunk and head).

PND and sudden, unexplained episodes of shortness of breath frequently accompany congestive heart failure (CHF). During the day the effects of gravity in the upright position and the shunting of excessive fluid to the lower extremities permit more effective ventilation and perfusion of the lungs, keeping the lungs relatively fluid free, depending on the degree of CHF. PND awakens the person sleeping in the recumbent position because the amount of blood returning to the heart and lungs from the lower extremities increases in this position.

Anyone who cannot climb a single flight of stairs without feeling moderately to severely winded or who awakens at night or experiences shortness of breath when lying down should be evaluated by a physician. Anyone with known cardiac involvement who develops progressively worse dyspnea must also notify the physician of these changes.

Dyspnea relieved by specific breathing patterns (e.g., pursed-lip breathing) or by specific body position (e.g., leaning forward on the arms to lock the shoulder girdle) is more likely to be pulmonary than cardiac in origin. Because breathlessness can be a terrifying experience for many persons, any activity that provokes the sensation is avoided, thus quickly reducing functional activities.

Cardiac Syncope

Cardiac syncope (fainting) or more mild lightheadedness can be caused by reduced oxygen delivery to the brain. Cardiac conditions resulting in syncope include arrhythmias, orthostatic hypotension, poor ventricular function, coronary artery disease, and vertebral artery insufficiency.

Lightheadedness that results from orthostatic hypotension (sudden drop in blood pressure [BP]) may occur with any quick change in a prolonged position (e.g., going from a supine position to an upright posture or standing up from a sitting position) or physical exertion involving increased abdominal pressure (e.g., straining with a bowel movement, lifting). Any client with aortic stenosis is likely to experience lightheadedness as a result of these activities.

Noncardiac conditions, such as anxiety and emotional stress, can cause hyperventilation and subsequent lightheadedness (vasovagal syncope). Side effects, such as orthostatic hypotension, may also occur during the period of initiation and regulation of cardiac medications (e.g., vasodilators).

Syncope that occurs without any warning period of lightheadedness, dizziness, or nausea may be a sign of heart valve or arrhythmia problems. Since sudden death can thus occur, medical referral is recommended for any unexplained syncope, especially in the presence of heart or circulatory problems or if the client has any risk factors for heart attack or stroke.

Examination of the cervical spine may include vertebral artery tests for compression of the vertebral arteries.[2-5] If signs of eye nystagmus, changes in pupil size, or visual disturbances and symptoms of dizziness or lightheadedness occur, care must be taken concerning any treatment that follows. It has been suggested, however, that other factors, such as individual sensitivity to extreme head positions, age, and vestibular responsiveness, could affect the results of these tests.[6]

The test may be contraindicated in individuals with cervical spine fusion, Down syndrome (due to cervical hypermobility and/or instability), or other cervical spine instabilities. Although there is controversy and uncertainty about the safety and accuracy of vertebral artery tests, long-term complications have not been reported as a result of administering these tests.[7,8] For a video of how to perform this test, go to www.youtube.com and type in "Gans video vertebral artery test."

Fatigue

Fatigue provoked by minimal exertion indicates a lack of energy, which may be cardiac in origin (e.g., coronary artery disease, aortic valve dysfunction, cardiomyopathy, or myocarditis) or may occur secondary to a neurologic, muscular, metabolic, or pulmonary pathologic condition. Often, fatigue of a cardiac nature is accompanied by associated symptoms such as dyspnea, chest pain, palpitations, or headache.

Fatigue that goes beyond expectations during or after exercise, especially in a client with a known cardiac condition, must be closely monitored. It should be remembered that beta-blockers prescribed for cardiac problems can also cause unusual fatigue symptoms.

For the client experiencing fatigue without a prior diagnosis of heart disease, monitoring vital signs may indicate a failure of the BP to rise with increasing workloads. Such a situation may indicate cardiac output that is inadequate in meeting the demands of exercise. However, poor exercise tolerance is often the result of deconditioning, especially in the older adult population. Further testing (e.g., exercise treadmill test) may be helpful in determining whether fatigue is cardiac-induced.

Cough

Cough (see also Chapter 7) is usually associated with pulmonary conditions, but it may occur as a pulmonary complication of a cardiovascular pathologic complex. Left ventricular dysfunction, including mitral valve dysfunction resulting from pulmonary edema or left ventricular CHF, may result in a cough when aggravated by exercise, metabolic stress, supine position, or PND. The cough is often hacking and may produce large amounts of frothy, blood-tinged sputum. In the case of CHF, cough develops because a large amount of fluid is trapped in the pulmonary tree, irritating the lung mucosa.

Cyanosis

Cyanosis is a bluish discoloration of the lips and nail beds of the fingers and toes that accompanies inadequate blood oxygen levels (reduced amounts of hemoglobin). Although cyanosis can accompany hematologic or central nervous system disorders, most often, visible cyanosis accompanies cardiac and pulmonary problems.

Edema

Edema in the form of a 3-pound or greater weight gain or a gradual, continuous gain over several days that results in swelling of the ankles, abdomen, and hands combined with shortness of breath, fatigue, and dizziness may be red-flag symptoms of CHF.

Other accompanying symptoms may include jugular vein distention (JVD; see Fig. 4-44) and cyanosis (of lips and appendages). Right upper quadrant pain described as a constant aching or sharp pain may occur secondary to an enlarged liver in this condition.

Right heart failure and subsequent edema can also occur secondary to cardiac surgery, venous valve incompetence or obstruction, cardiac valve stenosis, coronary artery disease, or mitral valve dysfunction.

Noncardiac causes of edema may include pulmonary hypertension, kidney dysfunction, cirrhosis, burns, infection, lymphatic obstruction, use of nonsteroidal antiinflammatory drugs (NSAIDs), or allergic reaction.

When edema and other accompanying symptoms persist despite rest, medical referral is required. Edema of a cardiac origin may require electrocardiogram (ECG) monitoring during exercise or activity (the physician may not want the client stressed when extensive ECG changes are present), whereas edema of peripheral origin requires treatment of the underlying etiologic complex.

Claudication

Claudication or leg pain occurs with peripheral vascular disease (PVD; arterial or venous), often occurring simultaneously with coronary artery disease. Claudication can be more functionally debilitating than other associated symptoms, such as angina or dyspnea, and may occur in addition to these other symptoms. The presence of pitting edema along with leg pain is usually associated with vascular disease.

Other noncardiac causes of leg pain (e.g., sciatica, pseudoclaudication, anterior compartment syndrome, gout, peripheral neuropathy) must be differentiated from pain associated with peripheral vascular disease. Low back pain associated with pseudoclaudication often indicates spinal stenosis. The discomfort associated with pseudoclaudication is frequently bilateral and improves with rest or flexion of the lumbar spine (see also Chapter 14).

Vascular claudication may occur in the absence of physical findings but is usually accompanied by skin discoloration and trophic changes (e.g., thin, dry, hairless skin) in the presence of vascular disease. Core temperature, peripheral pulses, and skin temperature should be assessed. Cool skin is more indicative of vascular obstruction; warm to hot skin may indicate inflammation or infection. Abrupt onset of ischemic rest pain or sudden worsening of intermittent claudication may be due to thromboembolism and must be reported to the physician immediately.

If people with intermittent claudication have normal-appearing skin at rest, exercising the extremity to the point of claudication usually produces marked pallor in the skin over the distal third of the extremity. This postexercise cutaneous ischemia occurs in both upper and lower extremities and is due to selective shunting of the available blood to the exercised muscle and away from the more distal parts of the extremity.

Vital Signs

The therapist may see signs of cardiac dysfunction as abnormal responses of heart rate and BP during exercise. The therapist must remain alert to a heart rate that is either too high or too low during exercise, an irregular pulse rate, a systolic BP that does not rise progressively as the work level increases, a systolic BP that falls during exercise, or a change in diastolic pressure greater than 15 to 20 mm Hg (Case Example 6-1).

Monitor vital signs in anyone with known heart disease. Some BP lowering medications can keep a client's heart rate from exceeding 90 bpm. For these individuals the therapist can monitor heart rate, but use perceived rate of exertion (PRE) as a gauge of exercise intensity. See Chapter 4 for more specific information about vital sign assessment.

CARDIAC PATHOPHYSIOLOGY

Three components of cardiac disease are discussed, including diseases affecting the heart muscle, diseases affecting heart valves, and defects of the cardiac nervous system (Table 6-2).

Conditions Affecting the Heart Muscle

In most cases, a cardiopulmonary pathologic condition can be traced to at least one of three processes:

1. Obstruction or restriction
2. Inflammation
3. Dilation or distention

Any combination of these can cause chest, neck, back, and/or shoulder pain. Frequently, these conditions occur sequentially. For example, an underlying *obstruction,* such as

CASE EXAMPLE 6-1

Cardiac Impairment Affecting Balance

Chief Complaint: An 84-year-old woman was referred to outpatient physical therapy for gait training with a diagnosis of ataxia and "at risk" status for falls. The client's goal is to walk without staggering.

Social History: Retired, lives alone in a two-story house.

Past Medical History: Atrial fibrillation, hypertension, arthritis in both hands, visual impairment (right eye), allergies to Novocain and antibiotics, transient ischemic attacks (TIAs), recurrent pneumonia.

Current Medications: Diltiazem (calcium channel blocker), Toprol (antihypertensive, beta blocker), aspirin (nonsteroidal antiinflammatory), Os-Cal (antacid, calcium supplement)

Clinical Presentation: Client is independent with activities of daily living but slow to complete tasks. She goes up and down one flight of steep stairs 3 to 4 times a day, using a handrail. She is an independent community ambulator but admits to frequently "staggering" (ataxia) with several falls.

Baseline testing:

1. Get-up-and-go test: 17 seconds (within normal limits [WNL])
2. Functional reach test: 8 inches (WNL)

Presence of other symptoms includes numbness along both feet that does not change with position. Shortness of breath with activity (short distance walking, ascending and descending 6 steps); recovers after a short rest. Thoracic kyphosis. Mild strength losses in hips, knees, and ankles. Reports of being "dizzy" and "lightheaded," feeling like she is going to "fall out."

Vital signs:

Blood pressure (before activity):	136/79 mm Hg (arm and position not recorded)
Blood pressure (after 10 minutes of activity):	132/69
Pulse (before activity):	82 bpm
Pulse (after activity):	73 bpm

Assessment: This 84-year-old client with a diagnosis of ataxia and risk of falls presented with fairly good bilateral lower extremity strength but had deficits in sensation and vision, increasing her risk for falling. She was quite resistant to using a cane for increased safety.

The client experienced an abnormal response to exercise in which the pulse rate and BP decreased. She required several rest breaks during the exercise session due to "fading."

The therapist referred the client to her cardiologist for evaluation to determine medical stability before further intervention.

Result: Nuclear stress testing revealed blockage of two major coronary arteries requiring cardiac catheterization with balloon angioplasty and placement of a stent. The cardiologist confirmed the therapist's suspicions that the client's balance deficits from neuropathy and decreased vision were made worse by shortness of breath and lightheadedness from cardiac impairment.

Monitoring vital signs before and during exercise was a simple way to screen for underlying cardiovascular impairment.

From Goff T: Case report presented in partial fulfillment of DPT 910. *Principles of differential diagnosis,* Institute for Physical Therapy Education, Chester, Pennsylvania, 2005, Widener University. Used with permission.

TABLE 6-2 Cardiac Diseases

Heart Muscle	Heart Valves	Cardiac Nervous System
Coronary artery disease Myocardial infarct Pericarditis Congestive heart failure Aneurysms	Rheumatic fever Endocarditis Mitral valve prolapse Congenital deformities	Arrhythmias Tachycardia Bradycardia

pulmonary embolus, leads to *congestion,* and subsequent *dilation* of the vessels blocked by the embolus.

The most common cardiovascular conditions to mimic musculoskeletal dysfunction are angina, MI, pericarditis, and dissecting aortic aneurysm. Other cardiovascular diseases are not included in this text because they are rare or because they do not mimic musculoskeletal symptoms.

Degenerative heart disease refers to the changes in the heart and blood supply to the heart and major blood vessels that occur with aging. As the population ages, degenerative heart disease becomes the most prevalent form of cardiovascular disease. Degenerative heart disease is also called atherosclerotic cardiovascular disease, arteriosclerotic cardiovascular disease, coronary heart disease (CHD), and coronary artery disease (CAD).

Hyperlipidemia

Hyperlipidemia refers to a group of metabolic abnormalities resulting in combinations of elevated serum total cholesterol (hypercholesterolemia), elevated low-density lipoproteins, elevated triglycerides (hypertriglyceridemia) and decreased high-density lipoproteins. These abnormalities are the primary risk factors for atherosclerosis and coronary artery disease.[9-11]

Statin medications (e.g., Zocor, Lipitor, Crestor, Lescol, Mevacor, Pravachol) are used to reduce low-density lipoprotein (LDL) cholesterol. While statins are generally well tolerated, there is a wide body of medical literature that associates the adverse reaction of myalgia and the more serious reaction of rhabdomyolysis with statin medications.[11,12] If detected early, statin-related symptoms can be reversible with reduction of dose, selection of another statin, or cessation of statin use.[13-15]

Screening for Side Effects of Statins. Myalgia and myopathy (including respiratory myopathy) are the most common myotoxic event associated with statins; joint pain is also reported.[16,17] The incidence of myotoxic events appears to be dose-dependent. Rates of adverse events from statins vary in the literature from 5% up to 18%[18] but with an increasing number of people taking these medications, physical therapists can expect to see a rise in the prevalence of this condition among the older age group.[19,20]

Monitoring for elevated serum liver enzymes and creatine kinase are significant laboratory indicators of muscle and liver impairment.[21] Symptoms of mild myalgia (muscle ache or weakness without increased creatine kinase [CK] levels), myositis (muscle symptoms with increased CK levels), or frank rhabdomyolysis (muscle symptoms with marked CK elevation; more than 10 times the normal upper limit) range from 1% to 7%.[11]

Muscle soreness after exercise that is caused by statin use may go undetected even by the therapist. Awareness of potential risk factors and monitoring in anyone taking statins and any of these additional risk factors is advised. Muscular symptoms are more common in older individuals (Case Example 6-2).[11,22] Other risk factors include[23]:

- Age over 80 (women more than men)
- Small body frame or frail
- Kidney or liver disease
- Drinks excessive grapefruit juice daily (more than 1 quart/day)
- Use of other medications (e.g., cyclosporine, some antibiotics, verapamil, human immunodeficiency virus [HIV] protease inhibitors, some antidepressants)[23]
- Alcohol abuse (independently predisposes to myopathy)

Muscle aches and pain, unexplained fever, nausea, vomiting and dark urine can potentially be signs of myositis and should be referred to a physician immediately. Risk for statin-induced myositis is highest in people with liver disease, acute infection, and hypothyroidism. Severe statin-induced myopathy can lead to rhabdomyolysis (enzyme leakage, muscle cell destruction, and elevated CK levels). Rhabdomyolysis is associated with impaired renal and liver function. Screening for liver impairment (see Chapter 9) in people taking statins is an important part of assessing for rhabdomyolysis (see further discussion of rhabdomyolysis in Chapter 9).[24]

CLINICAL SIGNS AND SYMPTOMS

Statin-Induced Side Effects

- Symptomatic myopathy (muscle soreness, pain, weakness, dyspnea); myositis (elevated CK level); look for weakness in several muscle groups
- Unexplained fever
- Nausea, vomiting
- Signs and symptoms of liver impairment:
  - Dark urine
  - Asterixis (liver flap)
  - Bilateral carpal tunnel syndrome
  - Palmar erythema (liver palms)
  - Spider angiomas
  - Nail bed changes, skin color changes
  - Ascites

The therapist will need to perform clinical tests and measures to differentiate exercise-related muscle fatigue and soreness from statin-induced symptoms. For example, exercise-induced muscle fatigue and soreness should be limited to the muscles exercised and resolve within 24 to 48 hours. Statin-related weakness may involve muscles not recently exercised and may progress or fail to show signs of improvement even after several days of rest.[25] Dynamometer testing can be used as a valid and reliable indicator of change in muscle strength.[26,27]

Strength testing, combined with client history, risk factors, and performance on the Stair-Climbing Test and Six-Minute Walk test, may prove to be an adequate means of assessment to detect meaningful declines in functional status; baseline measures are important.[25]

Coronary Artery Disease

The heart muscle must have an adequate blood supply to contract properly. As mentioned, the coronary arteries carry oxygen and blood to the myocardium. When a coronary artery becomes narrowed or blocked, the area of the heart muscle supplied by that artery becomes ischemic and injured, and infarction may result.

The major disorders caused by insufficient blood supply to the myocardium are angina pectoris and MI. These disorders are collectively known as coronary artery disease (CAD), also called coronary heart disease or ischemic heart disease. CAD includes atherosclerosis (fatty buildup), thrombus (blood clot), and spasm (intermittent constriction).

CAD results from a person's complex genetic makeup and interactions with the environment, including nutrition, activity levels, and history of smoking. Susceptibility to CVD may be explained by genetic factors, and it is likely that an "atherosclerosis gene" or "heart attack gene" will be identified.[28] The therapeutic use of drugs that act by modifying gene transcription is a well-established practice in the treatment of CAD and essential hypertension.[29,30]

Atherosclerosis. Atherosclerosis is the disease process often called arteriosclerosis or hardening of the arteries. It is a progressive process that begins in childhood. It can occur in any artery in the body, but it is most common in medium-sized arteries such as those of the heart, brain, kidneys, and legs. Starting in childhood, the arteries begin to fill with a fatty substance, or lipids such as triglycerides and cholesterol, which then calcify or harden (Fig. 6-2).

This filler, called *plaque,* is made up of fats, calcium, and fibrous scar tissue, and lines the usually supple arterial walls, progressively narrowing the arteries. These arteries carry blood rich in oxygen to the myocardium (middle layer of the heart consisting of the heart muscle), but the atherosclerotic process leads to ischemia and to necrosis of the heart muscle. Necrotic tissue gradually forms a scar, but before scar formation, the weakened area is susceptible to aneurysm development.

When fully developed, plaque can cause bleeding, clot formation, and distortion or rupture of a blood vessel (Fig. 6-3). Heart attacks and strokes are the most sudden and often fatal signs of the disease.

Thrombus. When plaque builds up on the artery walls, the blood flow is slowed and a clot (thrombus) may form on the plaque. When a vessel becomes blocked with a clot, it is called *thrombosis.* Coronary thrombosis refers to the formation of a clot in one of the coronary arteries, usually causing a heart attack.

CASE EXAMPLE 6-2

Statins and Myalgia

Referral: The client was a 53-year-old woman who complained of right-sided knee pain and stiffness and constant, bilateral thigh pain. She was referred by her orthopedic surgeon for physical therapy with a musculoskeletal diagnosis of osteoarthritis (OA) of both knees.

Medications: Lipitor (antilipemic for high cholesterol), Lopressor (antihypertensive, beta blocker), Ambien (sedative for insomnia), Naprosyn (antiinflammatory). Dosage of the Lipitor was increased from 20 mg to 40 mg at the last physician visit.

Past Medical History: Hypertension, hypercholesterolemia, insomnia.

Clinical Presentation: Pain pattern—Client reported constant but variable pain in the right knee, ranging from 2/10 at rest and 5/10 during and after weightbearing activities. Morning stiffness was prominent, and the client described difficulty transitioning from prolonged sitting to standing. The client reported increased pain in the right knee after weightbearing for approximately 5 minutes.

In addition to the right-sided knee pain, the client also complained of a constant anterior thigh pain. This pain was described as a "flushing" sensation and was unchanged by position or motion. The intensity of the bilateral constant thigh pain was rated 3-4/10.

Complete physical examination of integument and gait inspection, muscle strength, and joint range of motion (ROM) was conducted, and results were recorded (on file). Findings from the lower quarter screen (LQS) were unremarkable. Lumbar ROM was within normal limits (WNL).

Neurologic screening exam—WNL

Review of Systems (ROS): Unremarkable; no other associated signs or symptoms reported.

What are the red flags in this scenario?

Red Flags

- Age
- Constant, bilateral myalgic pain unchanged by position or motion
- Recent dosage change in medication

Is it safe to treat this client?

Physical therapy intervention can be implemented despite complaints of pain from an unknown origin. In the past 10 weeks prior to the initial physical therapy examination, this client had been evaluated by a physician four times. The most recent evaluation was by an orthopedic surgeon one week prior to her initial evaluation.

Additionally, the client appeared to be otherwise in good health, her ROS was unremarkable. The presence of three red flags warrants careful observation of response to intervention, progression of current symptoms, or onset of any new symptoms.

Symptoms related to OA were expected to improve, while symptoms from a non-musculoskeletal origin were not expected to improve. The plan of care was explained to the client, and it was mutually agreed that if the constant thigh pain did not significantly improve within 4 weeks, the client would be referred back to her physician.

Result: Four weeks after the initial physical therapy visit, symptoms of constant thigh pain had not resolved. The client had a follow-up appointment with her orthopedic surgeon at which time she presented documentation of the physical therapy intervention to date and the therapist's concerns about the thigh pain.

The orthopedic surgeon reiterated his belief that the origin of her symptoms was due to OA. The client was instructed to continue physical therapy and to take Naprosyn as prescribed.

The client was discharged after receiving 8 weeks of physical therapy intervention. At this juncture, the therapist believed that the client's right-sided knee symptoms were sufficiently improved and that the client could maintain or improve upon her current status by following her discharge instructions.

Furthermore, the myalgic thigh symptoms had not improved in the last 8 weeks and the therapist did not feel the bilateral thigh myalgia would be improved by further physical therapy interventions.

The client was instructed to contact her primary care physician regarding her myalgic symptoms. Three days later, the client called the physical therapy clinic. She had called her orthopedic surgeon rather than the primary care physician.

According to the client, the orthopedic surgeon dismissed the association between the thigh myalgia and the increased dosage of atorvastatin calcium (Lipitor). The client was again instructed to contact her primary care physician regarding the possibility of an adverse myalgic reaction to atorvastatin calcium (Lipitor). The client was also asked to contact the physical therapy clinic after being evaluated by her primary care physician.

Two weeks later, the client called and indicated her primary care physician evaluated her two days following our telephone conversation. The primary care physician discontinued the atorvastatin calcium (Lipitor). The client reported approximately a 50% reduction in the constant bilateral thigh myalgia after discontinuing the atorvastatin calcium (Lipitor).

The client was asked to contact the therapist in 2 or 3 weeks to provide an update of her status. Three weeks passed without hearing from the client. The therapist contacted the client by telephone. The client indicated that approximately 4 weeks following the discontinuation of the atorvastatin calcium (Lipitor) her myalgic thigh complaints were fully resolved.

She stated that she would remain off the atorvastatin calcium (Lipitor) for a total of 12 weeks and then would receive clinical laboratory testing to evaluate serum cholesterol levels. The primary care physician would consider prescribing a different statin to control her hypercholesterolemia based on future cholesterol levels.

Summary: Medication used in many situations may have a significant effect on the health of the client and may alter the clinical presentation or course of the individual's symptoms. It is important to ask if the client is taking any new medication (over the counter or by prescription), nutraceutical supplements, and if there have been any recent changes in the dosages of current medications.

There is a wide body of medical literature that associates the adverse reaction of myalgia and the more serious reaction of rhabdomyolysis with statin medications. Therapists must perform good pharmacovigilance.

From Trumbore DJ: Case report presented in partial fulfillment of DPT 910. *Principles of differential diagnosis,* Institute for Physical Therapy Education, Chester, Pennsylvania, 2005, Widener University. Used with permission.

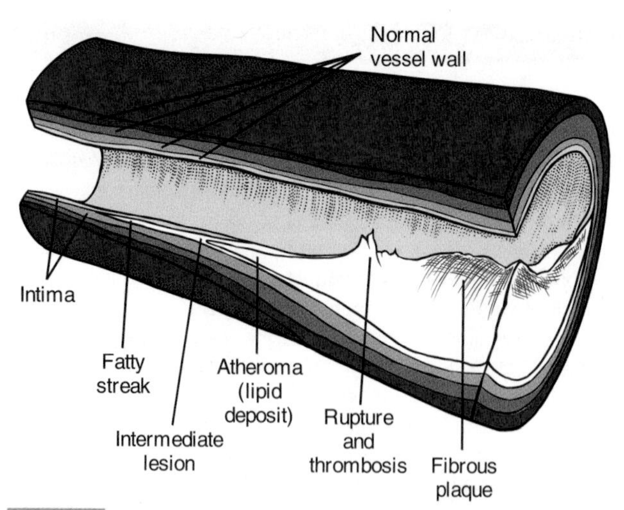

Fig. 6-2 Hardening of the arteries. Atherosclerosis begins with an injury to the endothelial lining of the artery (intimal layer) that makes the vessel permeable to circulating lipoproteins. Penetration of lipoproteins into the smooth muscle cells of the intima produces "fatty streaks." A fibrous plaque large enough to decrease blood flow through the artery develops. Calcification with rupture or hemorrhage of the fibrous plaque is the final advanced stage. Thrombosis (stationary blood clot) may occur, further occluding the lumen of the blood vessel.

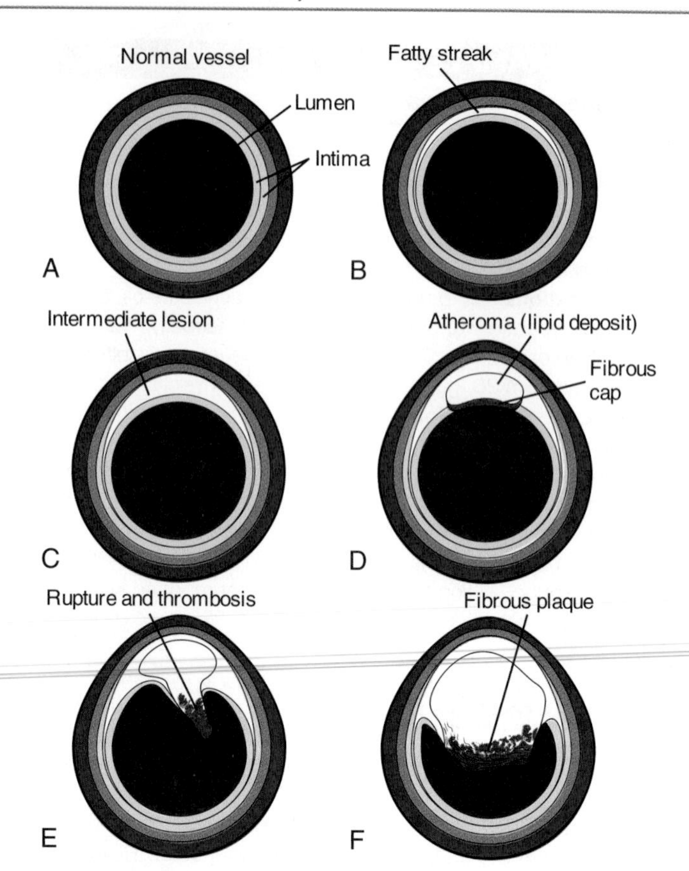

Fig. 6-3 Updated model of atherosclerosis. New technology using intravascular ultrasound shows the whole atherosclerotic plaque and has changed the way we view things. The traditional model held that an atherosclerotic plaque in the blood vessel, particularly a coronary blood vessel, kept growing inward and obstructing flow until it closed off and caused a heart attack. This is not entirely correct. It is more accurate to say that in the normal vessel **(A)**, penetration of lipoproteins into the smooth muscle cells of the intima produces fatty streaks **(B)** and the start of a coronary lesion forms. **C** and **D,** The coronary lesion grows outward first in a compensatory manner to maintain the open lumen. This is called *positive remodeling,* as the blood vessel tries to maintain an open lumen until it can do so no more. **E,** Only then does the plaque (atheroma) begin to build up, gradually pressing inward into the lumen with obstruction of blood flow and possible rupture and thrombosis, potentially leading to myocardial infarction (MI) or stroke. **F,** Vascular disease today is considered a disease of the wall. Some researchers like to say the disease is in the donut, not the hole of a donut, and that is a new concept.[102]

Spasm. Sudden constriction of a coronary artery is called a *spasm;* blood flow to that part of the heart is cut off or decreased. A brief spasm may cause mild symptoms that never return. A prolonged spasm may cause heart damage such as an infarct. This process can occur in healthy persons who have no cardiac history, as well as in those who have known atherosclerosis. Chemicals like nicotine and cocaine may lead to coronary artery spasm; other possible factors include anxiety and cold air.

Risk Factors. In 1948 the United States government decided to investigate the etiology, incidence, and pathology of CAD by studying the residents of Framingham, Massachusetts, a typical small town in the United States. Over the next multiple decades, various aspects of lifestyle, health, and disease were studied.

The research revealed important modifiable and nonmodifiable risk factors associated with death caused by CAD. Since that time, an additional category, contributing factors, has been added (Table 6-3). The AHA has also developed a validated health-risk appraisal instrument to assess individual risk of heart attack, stroke, and diabetes (see Box 6-1).

More recent research has identified other possible risk factors for and predictors of cardiac events, especially for those persons who have already had a heart attack. These additional risk factors include:

1. Exposure to bacteria such as *Chlamydia pneumoniae, Porphyromonas gingivalis,* and *Cytomegalovirus* organisms[31-33]
2. Excess levels of homocysteine, an amino acid by-product of food rich in proteins
3. High levels of α-lipoprotein, a close cousin of LDL that transports fat throughout the body
4. High levels of fibrinogen, a protein that binds together platelet cells in blood clots[34]
5. Large amounts of C-reactive protein (CRP), a specialized protein necessary for repair of tissue injury.[35,36]
6. The presence of troponin T, a regulatory protein that helps heart muscle contract[37]
7. The presence of diagonal earlobe creases (under continued investigation)[38-42]
8. Past history of cancer treatment (cardiotoxic chemotherapeutic agents, trunk radiation)[43]

TABLE 6-3 Risk Factors for Coronary Artery Disease

Modifiable Risk Factors	Nonmodifiable Risk Factors	Contributing Factors
Physical inactivity	Advancing age (65 or older)	Response to stress
Tobacco smoking	Male gender	Personality
Elevated serum Cholesterol	Family history	Peripheral vascular disease
High BP	Race	Hormonal status
Diabetes	Postmenopausal (female)	Alcohol consumption
Obesity		Obesity

AHA Scientific Statements on Prevention of Coronary Heart Disease and Stroke, 2010. Available on-line at www.heart.org (Risk Factors and Coronary Heart Disease). Accessed Sept. 1, 2010.

Therapists can assist clients in assessing their 10-year risk for heart attack using a risk assessment tool from the National Cholesterol Education program available at http://hin.nhlbi.nih.gov/atpiii/calculator.asp?usertype=prof.

Women and Heart Disease. Many women know about the risk of breast cancer, but in truth, they are 10 times more likely to die of cardiovascular disease. While 1 in 30 deaths is from breast cancer, 1 in 2.5 deaths are from heart disease.[44]

Women do not seem to do as well as men after taking medications to dissolve blood clots or after undergoing heart-related medical procedures. Of the women who survive a heart attack, 46% will be disabled by heart failure within 6 years.[45]

In general, the rate of CAD is rising among women and falling among men. Men develop CAD at a younger age than women, but women make up for it after menopause. African-American women have a 70% higher death rate from CAD than white women. So whenever screening chest pain, keep in mind the demographics: older men and women, menopausal women, and black women are at greatest risk.

Diabetes alone poses a greater risk than any other factor in predicting cardiovascular problems in women. Women with diabetes are seven times more likely to have cardiovascular complications and about half of them will die of CAD.[46]

Studies have shown that women and men actually differ in the symptoms of CAD and in the manner in which acute MI can present. Women experience symptoms of CAD, which are more subtle and are "atypical" compared to the traditional symptoms such as angina and chest pain.

One of the most important primary signs of CAD in women is unexplained, severe episodic fatigue and weakness associated with decreased ability to carry out normal activities of daily living (ADLs). Because fatigue, weakness, and trouble sleeping are general types of symptoms, they are not as easily associated with cardiovascular events and are many times missed by health care providers in screening for heart disease.[47]

Symptoms of weakness, fatigue and sleeping difficulty, and nausea have been reported as a common occurrence as much as a month prior to the development of acute MI in women (see Table 6-4). The classic pain of CAD is usually substernal chest pain characterized by a crushing, heavy, squeezing sensation commonly occurring during emotion or exertion. The pain of CAD in women, however, may vary greatly from that in men (see further discussion on MI in this chapter).

Risk reduction in women focuses on lifestyle changes such as smoking cessation; low fat, low cholesterol diet; increased intake of omega-3 fatty acids; increased fruit, vegetable, whole grain intake; salt and alcohol limitation; and increased exercise and weight loss. If the woman has diabetes, strict glucose control is extremely important.[48]

Clinical Signs and Symptoms. Atherosclerosis, by itself, does not necessarily produce symptoms. For manifestations to develop, there must be a critical deficit in blood supply to the heart in proportion to the demands of the myocardium for oxygen and nutrients (supply and demand imbalance). When atherosclerosis develops slowly, collateral circulation develops to meet the heart's demands. Often, symptoms of CAD do not appear until the lumen of the coronary artery narrows by 75% (see also Hypertension).

Although the arteries are rarely completely blocked, the deposits of plaque are often extensive enough to restrict blood flow to the heart, especially during exercise in a clinical practice when there is a need to deliver more oxygen-carrying blood to the heart. Like other muscles, the heart, when deprived of oxygen, may ache, causing chest pain or discomfort referred to as *angina.*

CAD is a progressive disorder, especially if left untreated. If the blood flow is entirely disrupted, usually by a clot that has formed in the obstructed region, some of the tissue that is supplied by the vessel can die, and a heart attack or even sudden cardiac death results.

When tissue loss is extensive enough to disrupt the electrical impulses that stimulate the heart's contractions, heart failure, chronic arrhythmias, and conduction disturbances may develop.

Angina

Acute pain in the chest, called *angina pectoris,* results from the imbalance between cardiac workload and oxygen supply to myocardial tissue. Angina is a symptom of obstructed or decreased blood supply to the heart muscle primarily from a condition called atherosclerosis.

Atherosclerosis is now recognized as an inflammatory condition affecting the coronary arteries as well as the peripheral vessels. It is often accompanied by hypertension and signs of PVD. Although the primary cause of angina is CAD, angina can occur in individuals with normal coronary arteries and with other conditions affecting the supply/demand balance.

As vessels become lined with atherosclerotic plaque, symptoms of inadequate blood supply develop in the tissues supplied by these vessels. A growing mass of plaque in the vessel collects platelets, fibrin, and cellular debris. Platelet aggregations are known to release prostaglandin capable of causing vessel spasm. This in turn promotes platelet aggregation, and a vicious spasm/pain cycle begins.

The present theory of heart pain suggests that pain occurs as a result of an accumulation of metabolites within an ischemic segment of the myocardium. The transient ischemia of angina or the prolonged, necrotic ischemia of an MI sets off pain impulses secondary to rapid accumulation of these metabolites in the heart muscle.

The imbalance between cardiac workload and oxygen supply can develop as a result of disorders of the coronary vessels, disorders of circulation, increased demands on output of the heart, or damaged myocardium unable to utilize oxygen properly.

Types of Anginal Pain. There are a number of types of anginal pain, including chronic stable angina (also referred to as walk-through angina), resting angina (angina decubitus), unstable angina, nocturnal angina, atypical angina, new-onset angina, and Prinzmetal's or "variant" angina.

Chronic stable angina occurs at a predictable level of physical or emotional stress and responds promptly to rest or to nitroglycerin. No pain occurs at rest, and the location, duration, intensity, and frequency of chest pain are consistent over time.

Resting angina, or *angina decubitus,* is chest pain that occurs at rest in the supine position and frequently at the same time every day. The pain is neither brought on by exercise nor relieved by rest.

Unstable angina, also known as crescendo angina, preinfarction angina, or progressive angina, is an abrupt change in the intensity and frequency of symptoms or decreased threshold of stimulus, such as the onset of chest pain while at rest. The most common trigger of unstable angina is the bursting of a cholesterol-filled plaque in the lining of a coronary artery. A blood clot forms at that site, partially blocking blood flow. The duration of these attacks is longer than the usual 1 to 5 minutes; they may last for up to 20 to 30 minutes and can progress into a full-blown heart attack. Pain or discomfort unrelieved by rest or nitroglycerin signals a higher risk for MI. Such changes in the pattern of angina require immediate medical follow up by the client's physician.

Nocturnal angina may awaken a person from sleep with the same sensation experienced during exertion. During sleep this exertion is usually caused by dreams. This type of angina may be associated with underlying CHF.

Atypical angina refers to unusual symptoms (e.g., toothache or earache) related to physical or emotional exertion. These symptoms subside with rest or nitroglycerin. New-onset angina describes angina that has developed for the first time within the last 60 days.

Prinzmetal's (variant) angina produces symptoms similar to those of typical angina but is caused by abnormal or involuntary coronary artery spasm rather than directly by a build-up of plaque from atherosclerosis. These spasms periodically squeeze arteries shut and keep the blood from reaching the heart. About two thirds of people with Prinzmetal's have severe coronary atherosclerosis in at least one major vessel. The spasm usually occurs very close to the blockage.

This form of angina typically occurs at rest, especially in the early hours of the morning, and can be difficult to induce by exercise. It is cyclic and frequently occurs at the same time each day. In postmenopausal women who are not undergoing hormone replacement therapy, the reduction in estrogen may cause coronary arteries to spasm, resulting in vasospastic (Prinzmetal's) angina.

Clinical Signs and Symptoms. The client may indicate the location of the symptoms by placing a clenched fist against the sternum. Angina radiates most commonly to the left shoulder and down the inside of the arm to the fingers; but it can also refer pain to the neck, jaw, teeth, upper back, possibly down the right arm, and occasionally to the abdomen (see Fig. 6-8).

Recognizing heart pain in women is more difficult because the symptoms are less reliable and often do not follow the classic pattern described earlier. Many women describe the pain in ways consistent with unstable angina, suggesting that they first become aware of their chest discomfort or have it diagnosed only after it reaches more advanced stages.

Some experience a sensation similar to inhaling cold air, rather than the more typical shortness of breath. Other women complain only of weakness and lethargy, and some have noted isolated pain in the midthoracic spine or throbbing and aching in the right biceps muscle (Fig. 6-4).

Pain associated with the angina and MI occurring along the inner aspect of the arm and corresponding to the ulnar nerve distribution results from common connections between the cardiac and brachial plexuses.

Cardiac pain referred to the jaw occurs through internuncial (neurons connecting other neurons) fibers from cervical spinal cord posterior horns to the spinal nucleus of the trigeminal nerve. Abdominal pain produced by referred cardiac pain is more difficult to explain and may be due to the overflow of segmental levels to which visceral afferent nerve

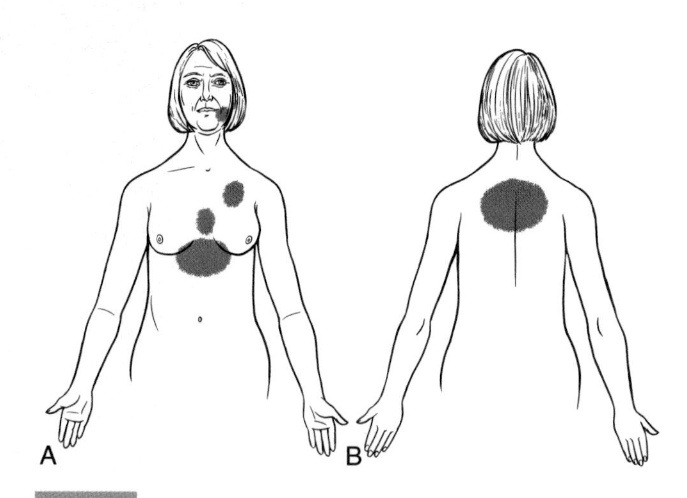

Fig. 6-4 Pain patterns associated with angina in women may differ from patterns in men. Many presenting symptoms are subjective such as extreme fatigue, lethargy, breathlessness, or weakness. Unusual patterns (e.g., temporomandibular joint [TMJ] pain) and failure to seek medical diagnosis may delay treatment with less optimal results. More classic pain patterns as shown in Fig. 6-8 are also possible.

pathways flow (see Fig. 3-3). This overflow increases the chances that final common pain pathways between the chest and the abdomen may occur.

The *sensation* of angina is described as squeezing, burning, pressing, choking, aching, or bursting. Chest pain can be brought on by a wide variety of noncardiac causes (see discussion of chest pain in Chapter 17).

In particular, angina is often confused with heartburn or indigestion, hiatal hernia, esophageal spasm, or gallbladder disease, but the pain of these other conditions is not described as sharp or knifelike.

The client often says the pain feels like "gas" or "heartburn" or "indigestion." Referred pain from a trigger point in the external oblique abdominal muscle can cause a sensation of heartburn in the anterior chest wall (see Fig. 17-7, *D*). A physician must make the differentiation between angina and heartburn, hiatal hernia, and gallbladder disease. The therapist can assess for trigger points; relief of symptoms with elimination of trigger points is an important diagnostic finding.

CLINICAL SIGNS AND SYMPTOMS

Heartburn

- Frequent "heartburn" attacks
- Frequent use of antacids to relieve symptoms
- Heartburn wakes client up at night
- Acid or bitter taste in the mouth
- Burning sensation in the chest
- Discomfort after eating spicy foods
- Abdominal bloating and gas
- Difficulty in swallowing

Severity is usually mild or moderate. Rarely is the pain described as severe. The five-grade angina scale ranks angina as:

Grade 0	No angina
Grade 1	Light, barely noticeable
Grade 2	Moderately bothersome
Grade 3	Severe, very uncomfortable
Grade 4	Most pain ever experienced

As to *location*, 80% to 90% of clients experience the pain as retrosternal or slightly to the left of the sternum. The *duration* of angina as a direct result of myocardial ischemia is typically 1 to 3 minutes and no longer than 3 to 5 minutes. However, attacks precipitated by a heavy meal or extreme anger may last 15 to 20 minutes. Angina is relieved by rest or nitroglycerin (a coronary artery vasodilator).

People who have had coronary artery stents placed can experience angina if an occlusion occurs above, below, or within the stent. Anyone with a stent who has chest pain should be immediately sent for referral to a physician.

Severity of pain is not a good prognostic indicator; some persons with severe discomfort live for many years, whereas others with mild symptoms may die suddenly. If the pain is not relieved by rest or up to 3 nitroglycerin tablets (taken one at a time at 5-minute intervals) in 10 to 15 minutes, the physician should be notified and the client taken to a cardiac care unit.

The client should take his or her own nitroglycerin. The therapist should not dispense medication but may assist the client in taking this medication. Nitroglycerin dilates the coronary arteries and improves collateral cardiac circulation, thus providing an increase in oxygen to the heart muscle and a decrease in symptoms of angina.

When screening for chest pain, a lack of objective musculoskeletal findings is always a red flag:

- Active range of motion (AROM), such as trunk rotation, side bending, or shoulder motions, does not reproduce symptoms.
- Resisted motion (horizontal shoulder abduction/adduction) does not reproduce symptoms.
- Heat and stretching do not reduce or eliminate symptoms.

CLINICAL SIGNS AND SYMPTOMS

Angina Pectoris

- Gripping, viselike feeling of pain or pressure behind the breast bone
- Pain that may radiate to the neck, jaw, back, shoulder, or arms (most often the left arm in men)
- Toothache
- Burning indigestion
- Dyspnea (shortness of breath); exercise intolerance
- Nausea
- Belching

Myocardial Infarction

Myocardial infarction (MI), also known as a heart attack, coronary occlusion, or a "coronary," is the development of ischemia and necrosis of myocardial tissue. It results from a sudden decrease in coronary perfusion or an increase in myocardial oxygen demand without adequate blood supply. If the requirements for blood are not eased (e.g., by decreased activity), the heart attempts to continue meeting the increased demands for oxygen with an inadequate blood supply, which leads to an MI. Myocardial tissue death is usually preceded by a sudden occlusion of one or more of the major coronary arteries.

The myocardium receives its blood supply from the two large coronary arteries and their branches. Occlusion of one or more of these blood vessels (coronary occlusion) is one of the major causes of MI. The occlusion may result from the formation of a clot that develops suddenly when an atheromatous plaque ruptures through the sublayers of a blood vessel, or when the narrow, roughened inner lining of a sclerosed artery leads to complete thrombosis.

Although coronary thrombosis is the most common cause of infarction, many interrelated factors may be responsible, including coronary artery spasm, platelet aggregation and embolism, thrombus secondary to rheumatic heart disease, endocarditis, aortic stenosis, a thrombus on a prosthetic mitral or aortic valve, or a dislodged calcium plaque from a calcified aortic or mitral valve.

Coronary blood flow is affected by the tonus (tone) of the coronary arteries. Arteries "clogged" by plaque formation become rigid, and resultant spasm may be provoked by cold and by exercise, which explains the adverse effect of both factors on clients with angina.

Clinical Signs and Symptoms. There are some well-known pain patterns specific to the heart and cardiac system. Sudden death can be the first sign of heart disease. In fact, according to the AHA, 63% of women who died suddenly of cardiovascular disease had no previous symptoms. Sudden death is the first symptom for half of all men who have a heart attack.

The onset of an infarct may be characterized by severe fatigue for several days before the infarct. The likelihood of having a heart attack in the morning hours is 40% higher than during the rest of the day.[49] The morning is when the body's clotting system is more active, BP surges, heart rate increases, and there may be reduced blood flow to the heart.

Additionally, the levels and activity of stress hormones (e.g., catecholamines), which can induce vasoconstriction, increase in the morning. Combined with these factors are the increased mental and physical stresses that typically occur after waking. The shift worker would experience this same phenomenon in the evening or on arising.

Persons who have MIs may not experience any pain and may be unaware that damage is occurring to the heart muscle as a result of prolonged ischemia. The presence of silent infarction (SI) increases with advancing age, especially SI without a history of CAD.

Cardiac Arrest. Researchers expect the number of Americans living with angina to grow as new treatments improve survival after heart attacks.[44] Failure to recognize prodromal symptoms in men or women may account for many cases of sudden cardiac death.[50]

Cardiac arrest strikes immediately and without warning. Signs of sudden cardiac arrest include[44]:

- Sudden loss of responsiveness. No response to gentle shaking.
- No normal breathing. The client does not take a normal breath when you check for several seconds.
- No signs of circulation. No movement or coughing.

If cardiac arrest occurs, call for emergency help and begin cardiopulmonary resuscitation (CPR) immediately, unless the client has a do not resuscitate (DNR) on file. Use an automated external defibrillator (AED) if available and appropriate.

Classic Warning Signs of Myocardial Infarction. Those who do have warning signs of MI may have severe unrelenting chest pain described as "crushing pain" lasting 30 or more minutes that is not alleviated by rest or by nitroglycerin. This chest pain may radiate to the arms, throat, and back, persisting for hours (see Fig. 6-9).

Other symptoms include pallor, profuse perspiration, and possibly nausea and vomiting. The pain of an MI may be misinterpreted as indigestion because of the nausea and vomiting. Nausea may be the only prodromal symptom. A medical evaluation may be difficult because many clients have coexisting hiatal hernia, peptic ulcer, or gallbladder disease.

Cardiac pain patterns may differ for men and women. For many men, the most common report is a feeling of pressure or discomfort under the sternum (substernal), in the mid-chest region, or across the entire upper chest. It can feel like uncomfortable pressure, squeezing, fullness, or pain.

Pain may occur just in the jaw, upper neck, mid-back, or down the arm without chest pain or discomfort. Pain may also radiate from the chest to the neck, jaw, mid-back or down the arm(s). Pain down the arm(s) affects the left arm most often in the pattern of the ulnar nerve distribution. Radiating pain down both arms is also possible.

An MI may occur during exertion, exercise, or exposure to extremes of temperature, or it may occur while the person is at rest. A subtle variation on ischemia during exertion is an important one for the therapist.

The onset of angina and a subsequent MI is known to be precipitated when working with the arms extended over the head. Oxygen requirements of the heart are greater during arm work compared to leg work at the same workload level. If the person becomes weak or short of breath while in this position, ischemia or infarction may be the cause of the pain and associated symptoms; workload should be cut in half to prevent cardiac ischemia.

Because the infarction process may take up to 6 hours to complete, restoration of adequate myocardial perfusion is important if significant necrosis is to be limited. Deaths generally result from severe arrhythmias, cardiogenic shock, CHF, rupture of the heart, and recurrent MI.

CLINICAL SIGNS AND SYMPTOMS

Myocardial Infarction

- May be silent (smokers, diabetics: reduced sensitivity to pain)
- Sudden cardiac death
- Prolonged or severe substernal chest pain or squeezing pressure
- Pain possibly radiating down one or both arms and/or up to the throat, neck, back, jaw, shoulders, or arms
- Feeling of nausea or indigestion
- Angina lasting for 30 minutes or more
- Angina unrelieved by rest, nitroglycerin, or antacids
- Pain of infarct unrelieved by rest or a change in position
- Nausea
- Sudden dimness or loss of vision or loss of speech
- Pallor
- Diaphoresis (heavy perspiration)
- Shortness of breath
- Weakness, numbness, and feelings of faintness

Warning Signs of Myocardial Infarction in Women. For women, symptoms can be more subtle or "atypical." Chest pain or discomfort is less common in women but still a key feature for some. Women describe heaviness, squeezing, or pain in the left side of the chest, abdomen, mid-back (thoracic), shoulder, or arm with no mid-chest symptoms.[51]

They often have prodromal symptoms up to 1 month before having a heart attack (Table 6-4).[50,52]

Fatigue, nausea, and lower abdominal pain may signal a heart attack. Many women pass these off as the flu or food poisoning. Other symptoms for women include a feeling of intense anxiety, isolated right biceps pain, mid-thoracic pain, or heartburn; sudden shortness of breath, or the inability to talk, move, or breathe; shoulder or arm pain; or ankle swelling or rapid weight gain.

In addition, she may describe palpitations or pain that is sharp and fleeting. Antacids may relieve it rather than rest or nitroglycerin. Women having an acute MI have also described a pain in the jaw, neck, shoulder, back, or ear and a feeling of intense anxiety, nausea, or shortness of breath. Many women do not associate these symptoms with having a heart attack, and they may do nothing to seek help.[51]

TABLE 6-4 Heart Attack Symptoms in Women

Timing	Symptoms
One month before a heart attack	Unusual fatigue (71%) Sleep disturbance (48%) Dyspnea (42%) Indigestion or GERD (39%) Anxiety (36%) Heart racing (27%) Arms weak/heavy (25%)
During a heart attack	Dyspnea (58%) Weakness (55%) Unusual fatigue (43%) Cold sweat (39%) Dizziness (39%) Nausea (36%) Arms weak/heavy (35%)

From McSweeney JC: Women's early warning symptoms of acute myocardial infarction, *Circulation* 108(21):2619-2623, 2003.
GERD, Gastroesophageal reflux disease.

CLINICAL SIGNS AND SYMPTOMS

Myocardial Ischemia in Women

- Heart pain in women does not always follow classic patterns.
- Many women do experience classic chest discomfort.
- Older female: mental status change or confusion may be common.
- Dyspnea (at rest or with exertion).
- Weakness and lethargy (unusual fatigue; fatigue that interferes with ability to perform ADLs).
- Indigestion, heartburn, or stomach pain; mistakenly diagnosed or assumed to have gastroesophageal reflux disease (GERD).
- Anxiety or depression.
- Sleep disturbance (woman awakens with any of the symptoms listed here).
- Sensation similar to inhaling cold air; unable to talk or breathe.
- Isolated, continuous midthoracic or interscapular back pain.
- Aching, heaviness, or weakness in one or both arms.
- Symptoms may be relieved by antacids (sometimes antacids work better than nitroglycerin).

Pericarditis

Pericarditis is an inflammation of the pericardium, the saclike covering of the heart. Specifically, it affects the parietal pericardium (fluidlike membrane between the fibrous pericardium and the epicardium) and the visceral (epicardium) pericardium (Fig. 6-5).

This inflammatory process may develop either as a primary condition or secondary to a number of diseases and conditions (e.g., influenza, HIV infection, tuberculosis, cancer, kidney failure, hypothyroidism, autoimmune disorders). Myocardial injury or trauma such as a heart attack, chest injury, chest radiation, or cardiac surgery can cause pericarditis. Very often the cause is unknown, resulting in a diagnosis of idiopathic pericarditis.

Pericarditis may be acute or chronic (recurring); it is not known why pericarditis may be a single illness in some persons and recurrent in others. Chronic or recurring pericarditis is accompanied by a pericardium that is rigid, thickened, and scarred.

Previous infection may be mild or asymptomatic with postinfectious onset of pain occurring 1 to 3 weeks later. Because this condition can occur in any age group, a history of recent pericarditis in the presence of new onset of chest, neck, or left shoulder pain is important.

Clinical Signs and Symptoms. At first, pericarditis may have no external signs or symptoms. The symptoms of acute pericarditis vary with the cause but usually include chest pain and dyspnea, an increase in the pulse rate, and a rise in temperature. Malaise and myalgia may occur.

Over time, the inflammatory process may result in an accumulation of fluid in the pericardial sac, preventing the heart from expanding fully. The inflamed pericardium may cause pain when it rubs against the heart. Chest pain from pericarditis (see Fig. 6-10) closely mimics that of an MI since it is substernal, is associated with cough, and may radiate to the left shoulder or supraclavicular area. It can be differentiated from MI by the pattern of relieving and aggravating factors (Table 6-5).

For example, the pain of an MI is unaffected by position, breathing, or movement, whereas the pain associated with pericarditis may be relieved by kneeling on all fours, leaning forward, or sitting upright. Pericardial chest pain is often worse with breathing, swallowing, belching, or neck or trunk movements, especially sidebending or rotation. The pain tends to be sharp or cutting and may recur in intermittent bursts that are usually precipitated by a change in body position. Pericarditis pain may diminish if the breath is held.

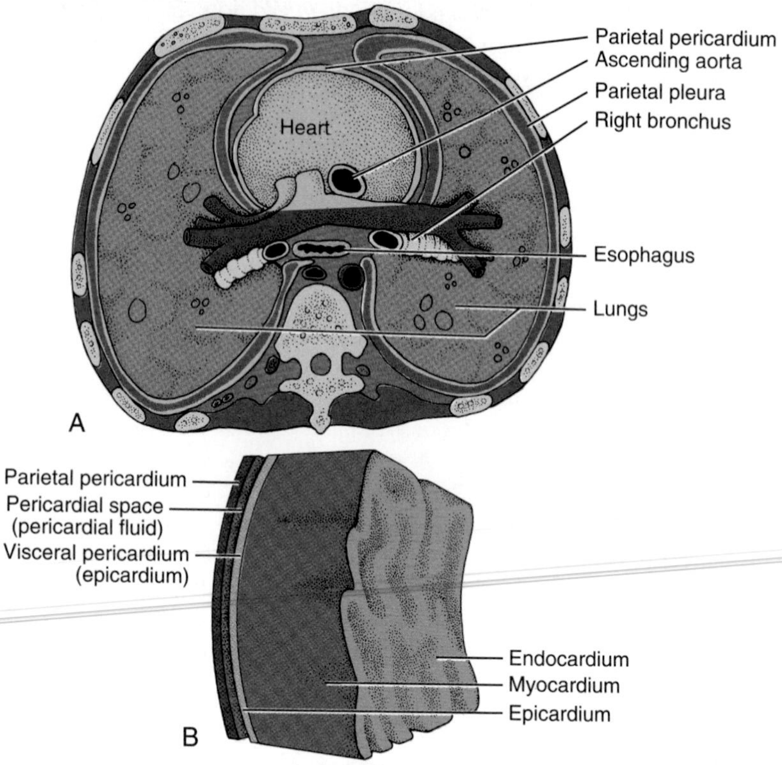

Fig. 6-5 The heart and associated layers of membranes. **A,** Cross-section through the thorax just above the heart, emphasizing the lining of the cavity that contains the lungs (parietal pleura) and the lining of the cavity that contains the heart (parietal pericardium). **B,** Sagittal view of the layers of the heart.

CLINICAL SIGNS AND SYMPTOMS

Pericarditis

- Substernal pain that may radiate to the neck, upper back, upper trapezius muscle, left supraclavicular area, down the left arm to the costal margins
- Difficulty in swallowing
- Pain relieved by leaning forward or sitting upright
- Pain relieved or reduced by holding the breath
- Pain aggravated by movement associated with deep breathing (laughing, coughing, deep inspiration)
- Pain aggravated by trunk movements (sidebending or rotation) and by lying down
- History of fever, chills, weakness, or heart disease (a recent MI accompanying the pattern of symptoms may alert the therapist to the need for medical referral to rule out cardiac involvement)
- Cough
- Lower extremity edema (feet, ankles, legs)

Congestive Heart Failure or Heart Failure

Heart failure, also called *cardiac decompensation* and *cardiac insufficiency,* can be defined as a physiologic state in which the heart is unable to pump enough blood to meet the metabolic needs of the body (determined as oxygen consumption) at rest or during exercise, even though filling pressures are adequate.

The heart fails when, because of intrinsic disease or structural defects, it cannot handle a normal blood volume, or in the absence of disease cannot tolerate a sudden expansion in blood volume (e.g., exercise). Heart failure is not a disease itself; instead, the term denotes a group of manifestations related to inadequate pump performance from either the cardiac valves or the myocardium.

Whatever the cause, when the heart fails to propel blood forward normally, congestion occurs in the pulmonary circulation as blood accumulates in the lungs. The right ventricle, which is not yet affected by congestive heart disease, continues to pump more blood into the lungs. The immediate result is shortness of breath and, if the process continues, actual flooding of the air spaces of the lungs with fluid seeping from the distended blood vessels. This last phenomenon is called pulmonary congestion or pulmonary edema.

Because a properly functioning heart depends on both ventricles, failure of one ventricle almost always leads to failure of the other ventricle. This is called *ventricular interdependence.* Right-sided ventricular failure (right-sided heart failure) causes congestion of the peripheral tissues and viscera. The liver may enlarge, the ankles may swell, and the client develops ascites (fluid accumulates in the abdomen).

Some clients have preexisting mild-to-moderate heart disease with no evidence of CHF. However, when the heart undergoes undue stress or deterioration from risk factors, compensatory mechanisms may be inadequate and the heart fails.

Conditions (risk factors) that precipitate or exacerbate heart failure include hypertension, CAD, cardiomyopathy, heart valve abnormalities, arrhythmia, fever, infection, anemia, thyroid disorders, pregnancy, Paget's disease, nutritional deficiency (e.g., thiamine deficiency secondary to alcoholism), pulmonary disease, spinal cord injury, and hypervolemia from poor renal function.

TABLE 6-5 Characteristics of Cardiac Chest Pain

Angina	Myocardial Infarction	Mitral Valve Prolapse	Pericarditis
Begins 3 to 5 minutes after exertion or activity ("lagtime"); lasts 1-5 minutes	30 minutes to 1 hour	Minutes to hours	Hours to days
Moderate intensity Tightness; chest discomfort Can occur at rest or during sleep Usually occurs with exertion, emotion, cold, or large meal	Severe (can be painless) Crushing pain; intolerable (can be painless) Exertion	Rarely severe May be asymptomatic; "sticking" sensation, not substernal Often occurs at rest	Varies; mild to severe Asymptomatic; sharp or cutting; can mimic MI Worse with breathing, swallowing, belching, neck or trunk movement
Subsides with rest or nitroglycerin; worse when lying down	Unrelieved by rest or nitroglycerin	Unrelieved by rest or nitroglycerin; may be relieved by lying down	Relieved by kneeling on all fours, leaning forward, sitting upright, or breathholding
Pain related to tone of arteries (spasm)	Pain related to ischemia	Mechanism of pain unknown	Pain related to inflammatory process

MI, Myocardial infarction.

Medications are frequently implicated in the development of CHF. Examples include cardiovascular drugs, antibiotics, central nervous system drugs (e.g., sedatives, hypnotics, antidepressants, narcotic analgesics), and antiinflammatory drugs (both nonsteroidal and steroidal).

Chemotherapy used with a variety of different types of cancer (including childhood cancers) has also been linked with increased risk of cardiovascular disease and congestive heart failure.[43] There may be a significant delay between treatment and the development of left ventricular dysfunction. Signs and symptoms of heart failure in young adults are unexpected. Consider cancer treatment in children who were treated successfully for cancer years ago a warning flag.

Clinical Signs and Symptoms. The incidence of CHF increases with advancing age. Because of the increasing age of the U.S. population and newer medications and technologies that have increased survival at the expense of increased cardiovascular morbidity, the population affected by CHF is markedly increasing. In view of this increase, many individuals with a wide variety of heart and lung diseases will very likely develop CHF at some time during their lives, manifesting itself as pulmonary congestion or edema.[53]

Left Ventricular Failure. Failure of the left ventricle causes either pulmonary congestion or a disturbance in the respiratory control mechanisms. These problems in turn precipitate respiratory distress. The degree of distress varies with the client's position, activity, and level of stress.

However, many persons with severely impaired ventricular performance may have few or no symptoms, particularly if heart failure has developed gradually. Breathlessness, exhaustion, and lower extremity edema are the most common signs and symptoms of CHF.

Dyspnea is subjective and does not always correlate with the extent of heart failure. To some degree, exertional dyspnea occurs in all clients. The increased fluid in the tissue spaces causes dyspnea, at first on effort and then at rest, by stimulation of stretch receptors in the lung and chest wall and by the increased work of breathing with stiff lungs.

Paroxysmal nocturnal dyspnea (PND) resembles the frightening sensation of suffocation. The client suddenly awakens with the feeling of severe suffocation. Once the client is in the upright position, relief from the attack may not occur for 30 minutes or longer.

Orthopnea is a more advanced stage of dyspnea. The client often assumes a "three-point position," sitting up with both hands on the knees and leaning forward. Orthopnea develops because the supine position increases the amount of blood returning from the lower extremities to the heart and lungs. This gravitational redistribution of blood increases pulmonary congestion and dyspnea. The client learns to avoid respiratory distress at night by supporting the head and thorax on pillows. In severe heart failure, the client may resort to sleeping upright in a chair.

Cough is a common symptom of left ventricular failure and is often hacking, producing large amounts of frothy, blood-tinged sputum. The client coughs because a large amount of fluid is trapped in the pulmonary tree, irritating the lung mucosa.

Pulmonary edema may develop when rapidly rising pulmonary capillary pressure causes fluid to move into the alveoli, resulting in extreme breathlessness, anxiety, frothy sputum, nasal flaring, use of accessory breathing muscles, tachypnea, noisy and wet breathing, and diaphoresis.

Cerebral hypoxia may occur as a result of a decrease in cardiac output, causing inadequate brain perfusion. Depressed cerebral function can cause anxiety, irritability, restlessness, confusion, impaired memory, bad dreams, and insomnia.

Fatigue and muscular cramping or weakness is often associated with left ventricular failure (Case Example 6-3). Inadequate cardiac output leads to hypoxic tissue and slowed removal of metabolic wastes, which in turn causes the client to tire easily. A common report is feeling tired after an activity

CASE EXAMPLE 6-3

Congestive Heart Failure—Muscle Cramping and Headache

A 74-year-old retired homemaker had a total hip replacement (THR) 2 days ago and remains an inpatient with complications related to congestive heart failure (CHF). She has a previous medical history of gallbladder removal 20 years ago, total hysterectomy 30 years ago, and surgically induced menopause with subsequent onset of hypertension. Her medications include intravenous furosemide (Lasix), digoxin, and potassium replacement.

During the initial physical therapy session, the client complained of muscle cramping and headache but was able to complete the entire exercise protocol. Blood pressure was 100/76 mm Hg. Systolic measurement dropped to 90 mm Hg when the client moved from supine to standing. Pulse rate was 56 bpm with a pattern of irregular beats. Pulse rate did not change with postural change. Platelet count was 98,000 cells/mm^3.

Result: With CHF, the heart will try to compensate by increasing the heart rate. However, the digoxin is designed to increase cardiac output and lower heart rate. In normal circumstances, postural changes result in an increase in heart rate, but when digoxin is used, this increase cannot occur so the person becomes symptomatic. Most of the clients like this one are also taking beta-blockers, which also prevent the heart rate from increasing when the BP drops.

In a clinical situation such as this one, the response of vital signs to exercise must be monitored carefully and charted. Any unusual symptoms, such as muscle cramping and headaches, and any irregular pulse patterns must also be reported and documented.

or type of exertion that was easily accomplished previously. Disturbances in sleep and rest patterns may aggravate fatigue.

The therapist must view this symptom in relation to the bigger picture. Is this someone who is taking diuretics? Is the diuretic a potassium-sparing drug? When does the muscle cramping occur? For example, muscle cramping and fatigue after working out in the garden under a hot sun may be related to fluid loss, dehydration, and exertion, whereas cramping that wakes the person up at night unrelated to exertion (including disturbing dreams) may indicate a different type of electrolyte imbalance.

Nocturia (urination at night) develops as a result of renal changes that can occur in both right- and left-sided heart failure (but more evident in left-sided failure). During the day the affected individual is upright and blood flow is away from the kidneys with reduced formation of urine. At night, urine formation increases as blood flow to the kidneys improves.

Nocturia may interfere with effective sleep patterns, contributing to the fatigue associated with CHF. As cardiac output falls, decreased renal blood flow may result in oliguria (reduced urine output), which is a late sign of heart failure.

CLINICAL SIGNS AND SYMPTOMS

Left-Sided Heart Failure

- Fatigue and dyspnea after mild physical exertion or exercise
- Persistent spasmodic cough, especially when lying down, while fluid moves from the extremities to the lungs
- Paroxysmal nocturnal dyspnea (occurring suddenly at night)
- Orthopnea (person must be in the upright position to breathe)
- Tachycardia
- Fatigue and muscle weakness
- Edema (especially of the legs and ankles) and weight gain
- Irritability/restlessness
- Decreased renal function or frequent urination at night

Right Ventricular Failure. Failure of the right ventricle may occur in response to left-sided CHF or as a result of pulmonary embolism (see cor pulmonale in Chapter 7). Right ventricular failure results in peripheral edema and venous congestion of the organs.

For example, as the liver becomes congested with venous blood, it becomes enlarged and abdominal pain occurs. If this occurs rapidly, stretching of the capsule surrounding the liver causes severe discomfort. The client may notice either a constant aching or a sharp pain in the right upper quadrant.

Dependent edema is one of the early signs of right ventricular failure. Edema is usually symmetric and occurs in the dependent parts of the body, where venous pressure is the highest. In ambulatory individuals, edema begins in the feet and ankles and ascends the lower legs. It is most noticeable at the end of a day and often decreases after a night's rest.

Many people experiencing this type of edema assume that it is a normal sign of aging and fail to report it to their physician. In the recumbent person, pitting edema may develop in the presacral area and, as it worsens, progress to the genital area and medial thighs (Case Example 6-4).

Cyanosis of the nail beds appears as venous congestion reduces peripheral blood flow. Clients with CHF often feel anxious, frightened, and depressed. Fears may be expressed as frightening nightmares, insomnia, acute anxiety states, depression, or withdrawal from reality.

CLINICAL SIGNS AND SYMPTOMS

Right-Sided Heart Failure

- Increased fatigue
- Dependent edema (usually beginning in the ankles)
- Pitting edema (after 5 to 10 pounds of edema accumulate)
- Edema in the sacral area or the back of the thighs
- Right upper quadrant pain
- Cyanosis of nail beds

CASE EXAMPLE 6-4

Congestive Heart Failure—Bilateral Pitting Edema

A 65-year-old man came to the clinic with a referral from his family doctor for "Hip pain—evaluate and treat." Past medical history included three total hip replacements of the right hip, open heart surgery 6 years ago, and persistent hypertension currently being treated with beta-blockers.

During the interview, it was discovered that the client had experienced many bouts of hip pain, leg weakness, and loss of hip motion. He was not actually examined by his doctor, but had contacted the physician's office by phone, requesting a new physical therapy referral.

On examination, large adhesed scars were noted along the anterior, lateral, and posterior aspects of the right hip, with significant bilateral hip flexion contractures. Pitting edema was noted in the right ankle, with mild swelling also observed around the left ankle. The client was unaware of this swelling. Further questions were negative for shortness of breath, difficulty in sleeping, cough, or other symptoms of cardiopulmonary involvement.

The bilateral edema could have been from compromise of the lymphatic drainage system following the multiple surgeries and adhesive scarring. However, with the positive history for cardiovascular involvement, bilateral edema, and telephone-derived referral, the physician was contacted by phone to notify him of the edema, and the client was directed by the physician to make an appointment.

The client was diagnosed in the early stages of congestive heart failure (CHF). Physical therapy to address the appropriate hip musculoskeletal problems was continued.

Diastolic Heart Failure. Diastolic heart failure describes a condition in which the left ventricle stiffens and hypertrophies. Open space inside the ventricle can become restricted by the thickened ventricle walls. The stiff heart muscle loses some of its flexibility. During diastole (when the muscle fibers relax and stretch), the chambers of the heart expand and fill with blood. Restrictions from a bulky heart muscle due to overwork or other causes make it more difficult for the muscle to relax between beats and thus unable to fill completely. This is different from systolic heart failure in which the left ventricle becomes weak and flabby. Although diastolic heart failure appears to be a "new" type of heart failure, it has been a factor in many people's health. Newer diagnostic technology has made it possible to differentiate diastolic from systolic heart failure. Both types of heart failure have the same end result: loss of blood supply (and oxygen) to the organs and tissues.

Risk factors for diastolic heart failure include advancing age, high BP, coronary artery disease (atherosclerosis), cardiac muscle damage from a previous heart attack, and valve dysfunction. Other medical conditions, such as diabetes, anemia, and thyroid disease, can also increase the risk of developing diastolic heart failure.

Clinical presentation is similar to systolic heart failure. A low ejection fraction (less than 35%) with symptoms suggests systolic heart failure, whereas a normal ejection fraction with symptoms is more typical with diastolic heart failure. Ejection fraction is the percentage of blood in the filled left ventricle that is pumped out during a contraction.

CLINICAL SIGNS AND SYMPTOMS

Diastolic Heart Failure

- Fatigue and dyspnea after mild physical exertion or exercise
- Orthopnea (dyspnea when lying down; person must be in the upright position to breathe)
- Edema (especially of the legs and ankles) and weight gain
- Jugular vein distention (see Fig. 4-44)

Aneurysm[54]

An aneurysm is an abnormal dilatation (commonly a saclike formation) in the wall of an artery, a vein, or the heart. Aneurysms occur when the vessel or heart wall becomes weakened from trauma, congenital vascular disease, infection, or atherosclerosis. This section could also be discussed under peripheral vascular diseases because aneurysms of arterial blood vessels can result in some form of PVD.

Aneurysms really fall under the broader category of *thoracic-aortic disease* (TAD), including aortic aneurysm and aortic dissection. Aneurysms can also be designated either venous or arterial and described according to the specific vessel in which they develop. *Thoracic aneurysms* usually involve the ascending, transverse, or descending portion of the aorta from the heart to the top of the diaphragm; *abdominal aneurysms* generally involve the aorta below the diaphragm between the renal arteries and the iliac branches; *peripheral arterial aneurysms* affect the femoral and popliteal arteries.[54]

Thoracic and Peripheral Arterial Aneurysms. A dissecting aneurysm (most often a thoracic aneurysm) occurs when a tear develops in the inner lining of the aortic wall. The inner and outer layers peel apart, creating an extra channel or "false vessel." Small tears may do no harm but aneurysms divert blood from organs and tissues and can result in heart attack, stroke, kidney damage, or other problems. Thoracic aneurysms occur most frequently in hypertensive men between the ages of 40 and 70 years. Marked elevation of BP may facilitate rapid disruption and final rupture (a break in all three layers of the aortic wall) when a small tear in the intima has occurred. Following a rupture, massive internal hemorrhage occurs as blood flows from the aorta into the chest.

The most common site for peripheral arterial aneurysms is the popliteal space in the lower extremities. Popliteal aneurysms cause ischemic symptoms in the lower limbs and an easily palpable pulse of larger amplitude. An enlarged area behind the knee may be present, seldom with discomfort.

Abdominal Aortic Aneurysms. An aneurysm is an abnormal dilation in a weak or diseased arterial wall causing a saclike protrusion. Aneurysms can occur anywhere in any blood vessel, but the two most common places are the aorta and cerebral vascular system. The aneurysm may be dissecting, which means a tear has occurred between two layers of the intima and blood is flowing between these two layers rather than through the lumen.

Abdominal aortic aneurysms (AAAs) occur about four times more often than thoracic aneurysms. The natural course of an untreated AAA is expansion and rupture in one of several places, including the peritoneal cavity, the mesentery, behind the peritoneum, into the inferior vena cava, or into the duodenum or rectum.

The most common site for an AAA is just below the kidney (immediately below the takeoff of the renal arteries), with referred pain to the thoracolumbar junction (see Fig. 6-11). Aneurysms can be caused by:

- Trauma/weight lifting (aging athletes)
- Congenital vascular disease
- Infection
- Atherosclerosis

Risk Factors. The therapist should look for a history of smoking,[55-57] known congenital heart disease (e.g., bicuspid aortic valve), surgery to replace or repair an aortic valve before age 70, recent infection, diagnosis of CAD (atherosclerosis), and some genetic conditions such as Marfan syndrome, Loeys-Dietz syndrome, Turner syndrome, or vascular Ehlers-Danlos syndrome. Many seniors are keeping active and fit by participating in activities at the gym, at home, or elsewhere that involve lifting weights. There is an increased risk of aneurysm in older adults, especially for these active clients. AAAs can be exacerbated by anticoagulant therapy (another risk factor).

The therapist may be prescribing progressive resistive exercises that can have an adverse effect in an older adult with any of these etiologies (Case Example 6-5). Monitoring vital signs is important among exercising senior adults. Teaching proper breathing and abdominal support without using a Valsalva maneuver is important in any exercise program, but especially for those clients at increased risk for aortic aneurysm.

The U.S. Preventive Services Task Force now recommends one-time ultrasonographic screening for abdominal aortic aneurysm for men ages 65 to 75 who presently smoke or who have smoked in the past. No recommendation is made for or against men who have never smoked. Routine screening for women is not advised.[58]

Clients who have had orthopedic surgery involving anterior spinal procedures of any kind (e.g., spinal fusion, spinal fusion with cages, artificial disk replacement) are at risk for trauma to the aorta (rather than aortic aneurysm) from damage to blood vessels moved out of the way during surgery. Internal bleeding can result in a distended abdomen, changes in BP, changes in stool (e.g., melena, bloody diarrhea), and possible back and/or shoulder pain.

Clinical Signs and Symptoms. Most AAAs are asymptomatic[59]; discovery occurs on physical or x-ray examination of the abdomen or lower spine for some other reason.[60]

The most common symptom is awareness of a pulsating mass in the abdomen, with or without pain, followed by abdominal pain and back pain. The therapist is most likely to observe rapid onset of severe neck or back pain (Case Example 6-6).

CASE EXAMPLE 6-5

Abdominal Aortic Aneurysm—Weight Lifting

A 72-year-old retired farmer has come to the physical therapist for recommendations about weight lifting. He had been following a regular program of weight lifting for almost 30 years, using a set of free weights purchased at a garage sale.

One year ago he experienced an abdominal aortic aneurysm that ruptured and required surgery. Symptoms at the time of the diagnosis were back pain at the thoracolumbar junction radiating outward toward the flanks bilaterally. The client is symptom-free and in apparent good health, taking no medications, and receiving no medical treatment at this time.

What are your recommendations for resuming his weight lifting program?

The hemodynamic stresses of weight lifting involve a rapid increase in systemic arterial BP without a decrease in total peripheral vascular resistance. This principle combined with aortic degeneration may have contributed to the aortic dissection.[103]

Weight lifting in anyone with a history of aortic aneurysm is considered a contraindication. The therapist suggested a conditioning program, combining a walking/biking program with resistive exercises using a lightweight elastic band alternating with an aquatic program for cardiac clients. Given this client's history, a medical evaluation before initiating an exercise program is necessary.

CASE EXAMPLE 6-6

Abdominal Aortic Aneurysm—Hip Replacement

A therapist was ambulating with a 76-year-old woman post-hip arthroplasty in an acute care (hospital) setting. The patient complained of back pain with every step and asked to sit down. The pain went away when she sat down. Once she started walking again, the pain started again.

She asked the therapist to walk her to the bathroom. After helping her onto the toilet, the therapist waited as a standby assist outside the bathroom. When several minutes went by with no call or sound, the therapist knocked on the door. There was no response. The therapist repeated knocking and calling the patient's name.

Upon opening the bathroom door, the therapist found the patient slumped over on the toilet. Emergency help was summoned immediately. She was later diagnosed with an abdominal aortic aneurysm.

What are the red flags in this scenario?

Age

Pain with activity that is relieved with rest

Vital signs should be taken in a case of this type after the first complaint of pain with activity.

The client may report feeling a heartbeat in the abdomen or stomach when lying down. Back pain may be the only presenting feature. Groin (scrotal), buttock, and/or flank pain may be experienced because of increasing pressure on other structures.

The pain is usually described as sharp, intense, severe, or knifelike in the abdomen, chest or anywhere in the back (including the sacrum). Pain may radiate to the chest, neck, between the scapulae, or to the posterior thighs.

The location of the symptoms is determined by the location of the aneurysm. Most aortic aneurysms (95%) occur just below the renal arteries. Extreme pain described as "tearing" or "ripping" may be felt at the base of the neck along the back, particularly in the interscapular area, while dissection proceeds over the aortic arch and into the descending aorta. Symptoms are not relieved by a change in position.

The physical therapist can palpate the width of the arterial pulses; these pulses (e.g., aortic, femoral) should be uniform in width from the midline outward on either side (see Fig. 4-55). The normal aortic pulse width is between 2.5 and 4.0 cm (some sources say the width must be no more than 3.0 cm; others list 4.0). Average pulse width is 2.5 cm or about 1.2 inches wide.[61,62]

According to the longitudinal community-based Framingham Heart Study, aortic root diameter increases with age in both men and women but is larger in men at any given age. Each 10-year increase in age is associated with a larger aortic root (by 0.89 mm in men and 0.68 mm in women) after adjustment for body size and BP. A 5-kg/m^2 increase in body mass index (BMI) was associated with a larger aortic root (by 0.78 mm in men and 0.51 mm in women) after adjustment for age and BP. Each 10-mm Hg increase in pulse pressure is related to age-related increase in stiffness and resultant increase in aortic pressure[63,64] and a smaller aortic root (by 0.19 mm in men and 0.08 mm in women) after adjustment for age and body size.[65] Cadaver studies also show significant reduction in tensile strength and stretch after age 30.[66]

As the aorta increases in diameter from an expanding aneurysm, the pulse width expands as well. The sensitivity of detection with abdominal palpation increases with the increasing diameter and has been reported as high as 82% with a diameter of 5 cm or more.[67] Sensitivity and specificity for detecting an abdominal aortic aneurysm with palpation have also been reported as 28% and 97%, respectively, for a definite pulsatile mass.[68] The risk of rupture approaches 25% for AAAs that are 6.0 cm (2.4 inches).[61]

Palpation is followed by auscultation for bruits (abnormal blowing or swishing sounds heard on auscultation of the arteries). Sensitivity for femoral bruit as a screening tool for detecting an abdominal aortic aneurysm has been reported as 17% with specificity of 87%. Abdominal bruit has an 11% sensitivity and 95% specificity.[68] Without a careful assessment, smaller diameter aneurysms may escape clinical detection.[69]

The abdominal aorta passes posterior to the diaphragm (aortic hiatus) at the level of the T12 vertebral body and bifurcates at the level of the L4 vertebral body to form the right and left common iliac arteries. Watch for a widening of the pulse width before reaching the umbilicus. The pulse width expands normally at the aortic bifurcation, usually observed just below the umbilicus. Ninety-five percent of all AAAs occur just below the renal arteries.

Systolic BP below 100 mm Hg and pulse rate over 100 beats per minute may indicate signs of shock. Other symptoms may include ecchymoses in the flank and perianal area; severe and sudden pain in the abdomen, paravertebral area, or flank; and lightheadedness and nausea with sudden hypotension.

The therapist may observe cold, pulseless lower extremities and/or BP differences (more than 10 mm Hg) between the arms. Consistent with the model for a screening examination the therapist must look for screening clues in the history, pain patterns, and associated signs and symptoms. Knowledge of the clinical signs and symptoms of impending rupture or actual rupture of the aortic aneurysm is important.

If a client (usually a postoperative inpatient) has internal bleeding (rather than an aneurysm) from complications of anterior spinal surgery the therapist may note

- Distended abdomen
- Changes in BP
- Changes in stool
- Possible back and/or shoulder pain

The client's recent history of anterior spinal surgery accompanied by any of these symptoms is enough to notify nursing or medical staff of these observations. Monitoring postoperative vital signs in these clients is essential.

CLINICAL SIGNS AND SYMPTOMS

Aneurysm

- Chest pain with any of the following:
- Palpable, pulsating mass (abdomen, popliteal space)
- Abdominal "heartbeat" felt by the client when lying down
- Dull ache in the midabdominal left flank or low back
- Hip, groin, scrotal (men), buttock, and/or leg pain (posterior thigh)
- Weakness or transient paralysis of legs

Ruptured Aneurysm

- Sudden, severe chest pain with a tearing sensation (see Fig. 6-10)
- Pain may extend to the neck, shoulders, between the scapulae, lower back, or abdomen; pain radiating to the posterior thighs helps distinguish it from a MI
- Pain is not relieved by change in position
- Pain may be described as "tearing" or "ripping"
- Pulsating abdominal mass
- Other signs: cold, pulseless lower extremities, BP changes (more than 10 mm Hg difference in diastolic BP between arms; systolic BP less than 100 mm Hg)
- Pulse rate more than 100 bpm
- Ecchymoses in the flank and perianal area
- Lightheadedness and nausea

Conditions Affecting the Heart Valves

The second category of heart problems includes those that occur secondary to impairment of the valves caused by disease (e.g., rheumatic fever or coronary thrombosis), congenital deformity, or infection such as endocarditis. Three types of valve deformities may affect aortic, mitral, tricuspid, or pulmonic valves: *stenosis, insufficiency,* or *prolapse.*

Stenosis is a narrowing or constriction that prevents the valve from opening fully, and may be caused by growths, scars, or abnormal deposits on the leaflets. *Insufficiency* (also referred to as *regurgitation*) occurs when the valve does not close properly and causes blood to flow back into the heart chamber. *Prolapse* affects only the mitral valve and occurs when enlarged valve leaflets bulge backward into the left atrium.

These valve conditions increase the workload of the heart and require the heart to pump harder to force blood through a stenosed valve or to maintain adequate flow if blood is seeping back. Further complications for individuals with a malfunctioning valve may occur secondary to a bacterial infection of the valves (endocarditis).

Persons affected by diseases of the heart valves may be asymptomatic, and extensive auscultation with a stethoscope and diagnostic study may be required to differentiate one condition from another. In its early symptomatic stages cardiac valvular disease causes the person to become fatigued easily. As stenosis or insufficiency progresses, the main symptom of heart failure (breathlessness or dyspnea) appears.

CLINICAL SIGNS AND SYMPTOMS

Cardiac Valvular Disease

- Easy fatigue
- Dyspnea
- Palpitation (subjective sensation of throbbing, skipping, rapid or forcible pulsation of the heart)
- Chest pain
- Pitting edema
- Orthopnea or paroxysmal dyspnea
- Dizziness and syncope (episodes of fainting or loss of consciousness)

Rheumatic Fever

Rheumatic fever is an infection caused by streptococcal bacteria that can be fatal or may lead to rheumatic heart disease, a chronic condition caused by scarring and deformity of the heart valves. It is called rheumatic fever because two of the most common symptoms are fever and joint pain.

The infection generally starts with strep throat in children between the ages of 5 and 15 years and damages the heart in approximately 50% of cases. Rheumatic fever produces a diffuse, proliferative, and exudative inflammatory process.

The aggressive use of specific antibiotics in the United States had effectively removed rheumatic fever as the primary cause of valvular damage. However, in 1985 a series of epidemics of rheumatic fever occurred in several widely diverse geographic regions of the continental United States. Currently, the prevalence and incidence of cases have not approximated the 1985 record, but they have remained above baseline levels.[70]

Clinical Signs and Symptoms. The most typical clinical profile of a child or young adult with acute rheumatic fever is an initial cold or sore throat followed 2 or 3 weeks later by sudden or gradual onset of painful migratory joint symptoms in the knees, shoulders, feet, ankles, elbows, fingers, or neck. Fever of 37.2° C to 39.4° C (99° F to 103° F) and palpitations and fatigue are also present. Malaise, weakness, weight loss, and anorexia may accompany the fever.

The migratory arthralgias may last only 24 hours, or they may persist for several weeks. Joints that are sore and hot and contain fluid completely resolve, followed by acute synovitis, heat, synovial space tenderness, swelling, and effusion present in a different area the next day. The persistence of swelling, heat, and synovitis in a single joint or joints for more than 2 to 3 weeks is extremely unusual in acute rheumatic fever.

In the acute full-blown sequelae, shortness of breath and increasing nocturnal cough will also occur. A rash on the skin of the limbs or trunk is present in fewer than 2% of clients with acute rheumatic fever. Subcutaneous nodules over the extensor surfaces of the arms, heels, knees, or back of the head may occur.

All layers of the heart (epicardium, endocardium, myocardium, and pericardium) may be involved, and the heart valves are affected by this inflammatory reaction. The most characteristic and potentially dangerous anatomic lesion of rheumatic inflammation is the gross effect on cardiac valves, most commonly the mitral and aortic valves. If untreated, as many as 25% of clients will have mitral valvular disease 25 to 30 years later.

Rheumatic chorea (also called chorea or St. Vitus' dance) may occur 1 to 3 months after the strep infection and always is noted after polyarthritis. Chorea in a child, teenager, or young adult is almost always a manifestation of acute rheumatic fever. Other uncommon causes of chorea are systemic lupus erythematosus, thyrotoxicosis, and cerebrovascular accident (CVA), but these are unlikely in a child.

The client develops rapid, purposeless, nonrepetitive movements that may involve all muscles except the eyes. This chorea may last for 1 week or several months or may persist for several years without permanent impairment of the central nervous system.

Initial episodes of rheumatic fever last months in children and weeks in adults. Twenty percent of children have recurrences within 5 years. Recurrences are uncommon after 5 years of good health and are rare after 21 years of age.

CLINICAL SIGNS AND SYMPTOMS

Rheumatic Fever

- Migratory arthralgias
- Subcutaneous nodules on extensor surfaces
- Fever and sore throat
- Flat, painless skin rash (short duration)
- Carditis
- Chorea
- Weakness, malaise, weight loss, and anorexia
- Acquired valvular disease

Endocarditis

Bacterial endocarditis, another common heart infection, causes inflammation of the cardiac endothelium (layer of cells lining the cavities of the heart) and damages the tricuspid, aortic, or mitral valve.

This infection may be caused by bacteria entering the bloodstream from a remote part of the body (e.g., skin infection, oral cavity), or it may occur as a result of abnormal growths on the closure lines of previously damaged valves or artificial valves. These growths called *vegetations* consist of collagen fibers and may separate from the valve, embolize, and cause infarction in the myocardium, kidney, brain, spleen, abdomen, or extremities.

Risk Factors. In addition to clients with previous valvular damage, injection drug users and postcardiac surgical clients are at high risk for developing endocarditis. Congenital heart disease and degenerative heart disease, such as calcific aortic stenosis, may also cause bacterial endocarditis. The prosthetic cardiac valve (valve replacement) has become more important as a predisposing factor for endocarditis because cardiac surgery is performed on a much larger scale than in the past.

This infection is often the consequence of invasive diagnostic procedures, such as renal shunts and urinary catheters, long-term indwelling catheters, or dental treatment (because of the increased opportunities for normal oral microorganisms to gain entrance to the circulatory system by way of highly vascularized oral structures). Individuals who are susceptible may take antibiotics as a precaution before undergoing any of these procedures.

Clinical Signs and Symptoms. A significant number of clients (up to 45%) with bacterial endocarditis initially have musculoskeletal symptoms, including arthralgia, arthritis, low back pain, and myalgias. Half these clients will have only musculoskeletal symptoms without other signs of endocarditis.

The early onset of joint pain and myalgia is more likely if the client is older and has had a previously diagnosed heart murmur. Musculoskeletal problems make up a significant part of the clinical picture of infective endocarditis diagnosed in an injection drug user.

The most common musculoskeletal symptom in clients with bacterial endocarditis is *arthralgia,* generally in the proximal joints. The shoulder is the most commonly affected site, followed (in declining incidence) by the knee, hip, wrist, ankle, metatarsophalangeal, and metacarpophalangeal joints, and acromioclavicular joints.

Most endocarditis clients with arthralgias have only one or two painful joints, although some may have pain in several joints. Painful symptoms begin suddenly in one or two joints, accompanied by warmth, tenderness, and redness. Symmetric arthralgia in the knees or ankles may lead to a diagnosis of rheumatoid arthritis. One helpful clue: as a rule, morning stiffness is not as prevalent in clients with endocarditis as in those with rheumatoid arthritis or polymyalgia rheumatica.

Osteoarticular infections are diagnosed infrequently and most commonly in association with injection drug use. Most commonly affected sites include the vertebrae, the wrist, the sternoclavicular joints, and the sacroiliac joints. Often, multiple joint involvement occurs.[71,72]

Endocarditis may produce destructive changes in the *sacroiliac joint,* probably as a result of seeding the joint by septic emboli. The pain will be localized over the sacroiliac joint, and the physician will use x-rays and bone scans to verify this diagnosis.

Almost one third of clients with bacterial endocarditis have *low back pain;* in many clients it is the principal musculoskeletal symptom reported. Back pain is accompanied by decreased ROM and spinal tenderness. Pain may affect only one side, and it may be limited to the paraspinal muscles.

Endocarditis-induced low back pain may be very similar to that associated with a herniated lumbar disk; it radiates to the leg and may be accentuated by raising the leg or by sneezing. The key difference is that neurologic deficits are usually absent in clients with bacterial endocarditis.

Widespread diffuse *myalgias* may occur during periods of fever, but these are not appreciably different from the general myalgia seen in clients with other febrile illnesses. More commonly, myalgia will be restricted to the calf or thigh. Bilateral or unilateral leg myalgias occur in approximately 10% to 15% of all clients with bacterial endocarditis.

The cause of back pain and leg myalgia associated with bacterial endocarditis has not been determined. Some suggest that concurrent aseptic meningitis may contribute to both leg and back pain. Others suggest that leg pain is related to emboli that break off from the infected cardiac valves. The latter theory is supported by biopsy evidence of muscle necrosis or vasculitis in clients with bacterial endocarditis.

Rarely, other musculoskeletal symptoms, such as osteomyelitis, nail clubbing, tendinitis, hypertrophic osteoarthropathy, bone infarcts, and ischemic bone necrosis, may occur.

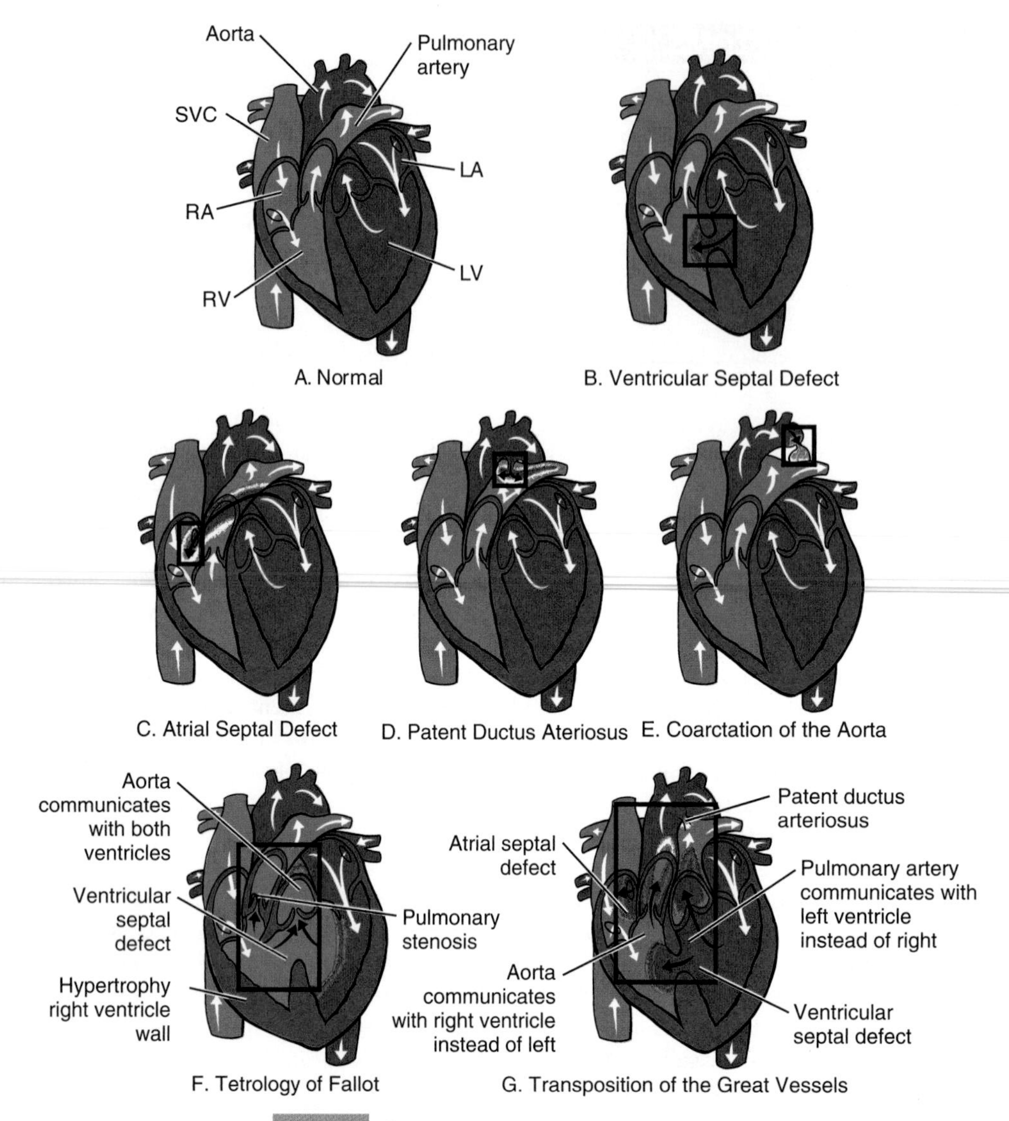

Fig. 6-6 Congenital malformations of the heart.

CLINICAL SIGNS AND SYMPTOMS

Endocarditis

- Arthralgias
- Arthritis
- Musculoskeletal symptoms
- Low back/sacroiliac pain
- Myalgias
- Petechiae/splinter hemorrhages
- Constitutional symptoms
- Dyspnea, chest pain
- Cold and painful extremities

Lupus Carditis

Systemic lupus erythematosus (SLE) is a multisystem clinical illness associated with the release of a broad spectrum of autoantibodies into the circulation (see Chapter 12). The inflammatory process mediated by the immune response can target the heart and vasculature of the client with SLE.

Except for pericarditis, clinically significant cardiac disease directly associated with SLE is relatively infrequent, but because of the musculoskeletal involvement, it may be of major importance for the therapist. Primary lupus cardiac involvement may include pericarditis, myocarditis, endocarditis, or a combination of the three.

Pericarditis is the most common cardiac lesion associated with SLE, appearing with the characteristic substernal chest pain that varies with posture, becoming worse in recumbency and improving with sitting or bending forward. *Myocarditis* may occur and is strongly associated with skeletal myositis in SLE.

Congenital Valvular Defects

Congenital malformations of the heart occur in approximately 1 of every 100 infants (1%) born in the United States.[73] The most common defects include (Fig. 6-6):

- Ventricular or atrial septal defect (hole between the ventricles or atria)
- Tetralogy of Fallot (combination of four defects)
- Patent ductus arteriosus (shunt caused by an opening between the aorta and the pulmonary artery)

- Congenital stenosis of the pulmonary, aortic, and tricuspid valves

These congenital defects require surgical correction and may be part of the client's past medical history. They are not conditions that are likely to mimic musculoskeletal lesions and are therefore not covered in detail in this text.

Congenital cardiovascular abnormalities, which are usually asymptomatic and often undiagnosed during life, are the main cause of sudden death in athletes. Aortic stenosis, hypertrophic cardiomyopathy, Marfan syndrome, congenital coronary artery anomalies, and ruptured aorta are the most commonly reported causes of sudden death during the practice of a sports activity.[74,75]

Family history of any of these conditions, premature sudden unexpected syncope, or family member death is an indication for a thorough cardiovascular evaluation of the athlete before participation in sports.[76]

Mitral Valve Prolapse. Echocardiographic studies have advanced our knowledge of mitral valve prolapse (MVP) in the last 2 decades. A more precise definition of MVP has resulted in a more accurate estimate of prevalence rate (2% to 3%).[77] This is equally distributed between men and women,[78,79] although men seem to have a higher incidence of complications.

MVP is characterized by mitral leaflet thickness with increased extensibility, decreased stiffness, and decreased strength compared to normal valves. This structural variation has many other names, including floppy valve syndrome, Barlow's syndrome, and click-murmur syndrome.

MVP appears to be due to connective tissue abnormalities in the valve leaflets or in response to abnormalities in left ventricular cavity geometry.[80] Normally, when the lower part of the heart contracts, the mitral valve remains firm and allows no blood to leak back into the upper chambers. In MVP the structural changes in the mitral valve allows one part of the valve, the leaflet, to billow back into the upper chamber during contraction of the ventricle.

One or both of the valve leaflets may bulge into the left atrium during ventricular systole. This protrusion can often be heard through a stethoscope as a sound known as a "click." Leaking of blood backward through the mitral valve can also be heard and is referred to as a heart murmur.

Risk Factors. MVP is a benign condition in isolation; however, it can be associated with a number of other conditions, especially the heritable connective tissue disorders such as Ehlers-Danlos syndrome, Marfan syndrome, and osteogenesis imperfecta. Other risk factors include endocarditis, myocarditis, atherosclerosis, SLE, muscular dystrophy, acromegaly, and cardiac sarcoidosis.

Clinical Signs and Symptoms. Two thirds of the individuals with MVP experience no symptoms. Approximately one third experience occasional symptoms that are mildly to moderately uncomfortable—enough to interfere with the person's ability to enjoy an unrestricted life. Only about 1% suffer severe symptoms and lifestyle restrictions.

Almost all the symptoms of MVP syndrome are due to an imbalance in the autonomic nervous system, called

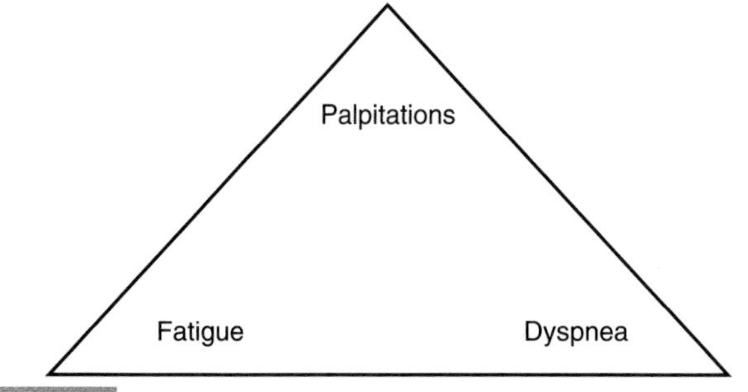

Fig. 6-7 The triad of symptoms associated with mitral valve prolapse.

dysautonomia. Frequently, when there is a slight variation in structure of the heart valve, there is also a slight variation in the function or balance of the autonomic nervous system (ANS).[80] This description in the autonomic innervation of the heart may account for the high incidence of MVP in fibromyalgia, a condition known to be associated with dysregulation or dysautonomia of the ANS.

Symptoms include profound fatigue that cannot be correlated with exercise or stress, cold hands and feet, shortness of breath, chest pain, and heart palpitations. The most common triad of symptoms associated with MVP is fatigue, palpitations, and dyspnea (Fig. 6-7). Frequently occurring musculoskeletal findings in clients with MVP include joint hypermobility, temporomandibular joint (TMJ) syndrome, and myalgias.

Although the fatigue that accompanies MVP is not related to exertion, deconditioning from prolonged inactivity may develop, further complicating the picture. Chest pain associated with MVP can be severe, but it differs from pain associated with MI (see Table 6-5). When there is an imbalance in the ANS, which controls contraction and relaxation of the chest wall muscles (the muscles of breathing), there may be inadequate relaxation between respirations. Over time these chest wall muscles go into spasm, resulting in chest pain.

It is important that the therapist evaluate the client with chest pain for trigger points. If palpation of the chest reproduces symptoms, especially radiating pain, deactivation of trigger points must be carried out followed by a reevaluation as part of the screening process for pain of a cardiac origin.

MVP is not life-threatening but may be lifestyle threatening for the small number of persons (rare) who have more severe structural problems that may progress to the point at which surgical replacement of the valve is required. Sudden death is a recognized risk for cases of severe mitral regurgitation. To prevent infective endocarditis, the client may be given antibiotics prophylactically before any invasive procedures.

MVP is included in this section because of its increasing prevalence in the physical therapy client population. At presentation, usually the client with MVP has some other unrelated primary (musculoskeletal) diagnosis. During physical therapy intervention, the symptomatic MVP client may experience symptoms associated with MVP and require assurance or education regarding exercise and MVP.

Most individuals with MVP do not have to restrict their activity level or lifestyle; regular exercise is encouraged.

Clients with mitral regurgitation (backward flow of blood into the left atrium) during exercise, but not at rest, have a higher rate of complications, such as heart failure, syncope, and progressive mitral regurgitation, requiring further medical treatment.[80]

Monitoring vital signs and observing or asking about additional signs and symptoms is important. Caution is advised in the use of weight training for the MVP client. Gradual buildup using lightweights and increased repetitions is recommended.

CLINICAL SIGNS AND SYMPTOMS

Mitral Valve Prolapse

- Profound fatigue; low exercise tolerance
- Chest pain; arm, back, or shoulder discomfort
- Palpitations or irregular heartbeat
- Tachycardia
- Migraine headache
- Anxiety, depression, panic attacks
- Dyspnea

Conditions Affecting the Cardiac Nervous System

The third component of cardiac disease is caused by failure of the heart's nervous system to conduct normal electrical impulses. The heart has its own intrinsic conduction system that allows the orderly depolarization of cardiac muscle tissue. Arrhythmias, also called *dysrhythmias,* are disorders of the heart rate and rhythm caused by disturbances in the conduction system.

Arrhythmias may cause the heart to beat too quickly (tachycardia), too slowly (bradycardia), or with extra beats and fibrillations. Arrhythmias can lead to dramatic changes in circulatory dynamics such as hypotension, heart failure, and shock.

Clients who are neurologically unstable owing to recent CVA, head trauma, spinal cord injury, or other central nervous system insult often exhibit new arrhythmias during the period of instability (Case Example 6-7). These may be due to elevation of intracranial pressure, and once this has been controlled and returned to normal range, arrhythmias usually disappear.

Arrhythmias also may be triggered by environmental factors, abnormal thyroid function, and some medications. Clients who have preexisting arrhythmias, CAD, or CHF, may progress from "transient arrhythmias" to arrhythmias that do not disappear, with the potential for serious complications. Dizziness and loss of consciousness may occur when the arrhythmia results in a serious reduction in cardiac output, owing to loss of brain perfusion (not caused by transient ischemic attacks, as is often suspected).

The therapist should monitor pulse carefully before, during, and after exercise when working with any client who has had a stroke. Any pulse irregularities not already documented should be reported to the physician immediately and an ECG should be done to determine the nature of the irregularity.

The therapist may be the first health care professional to identify an arrhythmia that appears during exercise. In the early recovery period, the therapist should monitor for these arrhythmias by taking the client's pulse. Arrhythmias should be reported to the physician (Case Example 6-8).

Fibrillation

The sinoatrial (SA) node (or cardiac pacemaker) initiates and paces the heartbeat. During an MI, damaged heart muscle cells, deprived of oxygen, can release small electrical impulses that may disrupt the heart's normal conduction pathway. These fibrillation impulses can occur in the atria or the ventricles.

If the heart attack develops suddenly into *ventricular* fibrillation, a potentially lethal arrhythmia, it can result in sudden death. Similarly, a heart damaged by CAD (with or without previous infarcts) can go into ventricular fibrillation. Ventricular fibrillation usually requires resuscitation and emergency electrical countershock (defibrillation) as life saving measures.

Atrial fibrillation (AF) is the most common cardiac dysrhythmia and although not immediately lethal, it can increase the risk of heart failure and stroke. It is characterized by a total disorganization of atrial activity without effective atrial contraction. The upper chambers of the heart contract in an unsynchronized pattern, causing the atrium to quiver rather than to contract and often causing blood to pool, which allows for clots to form. These clots can break loose and travel to the brain, causing a stroke.

Individuals with transient episodes (in and out of AF) or developing new onset of AF are at greater risk of clot migration than those who have a long-term history, especially if they are already on blood thinners. Medical consult is more imperative for anyone who has not previously been identified as having AF and/or who is not receiving anticoagulants.

The therapist can easily and quickly screen individuals at risk and teach them to screen themselves by checking the pulse for the telltale signs of an irregular heartbeat. A regular heartbeat is characterized by a series of even and continuous pulsations, whereas an irregular heartbeat often feels like an extra or missed beat. To help determine the steadiness of the heartbeat, the therapist or individual keeps time by tapping the foot.

Risk Factors. Persons at risk for fibrillation who require screening include those who have had a previous heart attack or a history that includes high BP, CHF, digitalis toxicity, pericarditis, or rheumatic mitral stenosis.

Other factors that can overstimulate the sinus node include emotional stress, excessive production of thyroid hormone (hyperthyroidism), alcohol and caffeine consumption, and high fevers. In many instances, particularly in younger persons, there is no apparent cause.

Recent studies have shown that presence of the organism *Helicobacter pylori* in the stomach is associated with persistent AF. In addition, high concentrations of CRP, which

CASE EXAMPLE 6-7

Cardiac Arrhythmia Cause of Transient Muscle Weakness

Referral: An 87-year-old man was admitted to a skilled nursing facility for short-term rehabilitation.

Goal: Safely and independently navigate home environment on level surfaces and stairs with discharge to home within 2 to 3 months.

Medical Diagnosis: Mild left hemiparesis secondary to recent cerebrovascular accident (CVA).

Past Medical History: Urosepsis, seizure disorder, deep venous thrombosis (DVT), hypertension (HTN), congestive heart failure (CHF), myocardial infarction (MI), frequent falls, subdural hematoma with evacuation.

Medications: Heparin (anticoagulant, antithrombotic), Lasix (diuretic for hypertension), Os-Cal (antacid, calcium supplement).

Systems Review

Integument: Well-healed scar on head, postsurgical evacuation of subdural hematoma.

Musculoskeletal: Head position in sitting and standing is held in approximately 30 degrees of lateral and forward flexion; full active head and neck range of motion (ROM) present; overall flexed/stooped posture (left more than right); muscle strength in right trunk and right upper extremity: 4–/5; muscle strength in left trunk and left lower extremity: 3–/5.

Neuromuscular: Neurologic screening results consistent with upper motor neuron lesion; nonambulatory and requires moderate assistance for bed mobility and transfers; able to sit, stand, and walk with assistance of two and the use of a walker; patellar deep tendon reflexes (DTRs): right: within normal limits; left: +3; positive Babinski response on the left

Cardiovascular/Pulmonary: Blood pressure: 142/80 mm Hg, measured in right upper extremity, sitting; pulse rate: 72 bpm; respiratory rate: 18 breaths per minute; pulse oximeter (oxygen saturation): 97%

First Progress Report: Client made many functional gains during the first 4 weeks of rehab and was planning a trip home for 2 days but experienced multiple falls with several incidences of lacerations to the head requiring stitches.

Client also presented with multiple episodes of transient generalized weakness, increased postural instability, and increased bradykinesia. Episodes of weakness and falling were without warning and unrelated to activity but interfering with progress toward functional independence and discharge plans.

Vital signs: Radial and apical heart rate 72 bpm with periodic and variable drops in heart rate to below 60 bpm with a low measurement of 37 bpm. Episodes of bradycardia lasted from 1 minute to 1 hour. Blood pressure fluctuated from 150/83 mm Hg to 92/40 mm Hg. Respiratory rate remained stable. Pulse oximetry levels varied from 91% to 97% with an occasional drop to 88% during transient episodes of weakness.

There was no complaint of nor apparent shortness of breath, no complaint of dizziness, and no complaint of syncope.

Referral: Client was referred to his primary physician with report of increasingly frequent periods of generalized muscle weakness with poor postural stability (cause unknown). Vital signs were reported.

Physician ruled out dehydration, renal insufficiency, anemia, and active bleeding as possible pathologies. Electrocardiogram (ECG) was obtained. He was eventually hospitalized and diagnosed with episodic bradycardia.

The client returned to the skilled nursing facility and continued to have multiple intermittent episodes of transient weakness and postural instability lasting from several minutes to several hours, affecting his progress toward independent functional mobility.

Second Progress Report: With each episode, vital signs were obtained. A significant drop in radial pulse was noted, along with decreased apical heart rate. This appeared to be the most significant finding. Blood pressure dropped during these incidents; however, it did not drop dangerously low except during several occurrences (82/40 mm Hg); respiratory rate appeared stable with minimal alteration.

The nursing staff and physician were notified at the time of each occurrence. A 24-hour Holter monitor was ordered. The results were inconclusive because no sustained arrhythmias were noted.

A second cardiac evaluation identified a tachy-brady syndrome. The client received a pacemaker and appropriate medications. The client was discharged to home 6 weeks later, independent in all functional mobility and activities of daily living (ADLs).

Summary: Cardiac arrhythmia may be an underlying concern in patients with diagnosis of CVA. With the increasing population of over 65 years old and the increasing need for rehabilitative services in the aging adult population, understanding the normal aging process versus disease states is imperative. Signs of pathology can be either overlooked or attributed to the age of the client or to the normal aging process.

Cardiac arrhythmia is an elusive state that may or may not produce overt symptoms. This case example suggests that although not documented in the literature yet, cardiac arrhythmia may cause transient weakness.

From Petherbridge A: Case report presented in partial fulfillment of DPT 910. *Principles of differential diagnosis,* Institute for Physical Therapy Education, Chester, Pennsylvania, 2005, Widener University. Used with permission.

confirm the presence of systemic inflammation, are present in people with AF. A potential noncardiovascular disease that predisposes to AF may be chronic gastritis caused by chronic *H. pylori* infection.[81]

Clinical Signs and Symptoms. Symptoms of fibrillation vary, depending on the functional state of the heart and the location of the fibrillation. Fibrillation may exist without symptoms. The affected individual is usually aware of the irregular heart action and reports feeling "palpitations." Careful questioning may be required to pinpoint the exact description of sensations reported by the client.

Some individuals experience the symptoms of inadequate blood flow and low oxygen levels, such as dizziness, chest pain, and fainting. Chronic AF may cause CHF, which is often experienced as shortness of breath during exercise and fluid accumulation in the feet and legs.

CASE EXAMPLE 6-8

Unstable Cardiac Arrhythmia

Chief Complaint: A 62-year-old woman with a diagnosis of left lower extremity weakness was referred by her cardiologist to physical therapy. She also reported weakness and tingling in both legs from the knees down.

Past Medical History: Recent hospitalization for lung infection. Past history of hypertension, dizziness, and cardiac arrhythmia. Previous surgical history included thoracic outlet release 14 years ago.

Clinical Presentation: The client experienced worsening of symptoms with walking and climbing stairs; rest relieved her symptoms. Manual muscle test revealed weakness in the left hip flexor and external rotator muscles.

Client became diaphoretic with functional strength and exercise tolerance testing (repeated sit to stand and back to sitting). She reported the same response when doing housework.

Review of systems (ROS): Significant for the cardiovascular system. Vital signs were as follows:

Blood pressure:	125/95 mm Hg (left arm, standing)/ 150/82 (supine) Lower standing BP is a risk factor for falls[104]
Apical pulse:	62 beats per minute; every third or fourth beat skipped; heart rate speeds up and then slows down after each skipped beat (cardiologist was aware of this pattern of irregularity; client was under medical treatment for cardiac arrhythmia)

What are the red flags here?

Age

History of cardiac arrhythmia

Diaphoresis is an abnormal response to exercise stress

Irregular heartbeat

Is it safe to proceed with physical therapy intervention since the client was referred by the cardiologist who is treating her for cardiac arrhythmia?

It is not always the case that the referring physician is aware of a client's current cardiovascular status. Referral may not always mean the client is appropriate for participation in an exercise program. Since the therapist noted the irregular heart rate, he was alert to the possibility of other signs of inappropriate exercise responses.

The main red flag here is the abnormal response to exercise stress in someone who is already being treated for a cardiac anomaly (arrhythmia). The variable BP and irregular heartbeat suggest an unstable situation.

Result: In this case, it just happened that the client was going from her appointment with the therapist to a visit with her family physician. A note summarizing the therapist's findings, including the vital signs and response to exercise, was sent to the physician.

A copy of the letter was also faxed to the physician's office and to the cardiologist. The therapist followed up with a phone call to the client for an update on her status. The client was given an electrocardiogram (ECG), which was found to be abnormal, and was admitted to the hospital for further testing.

The therapist was very instrumental in referring this client based on medical screening (i.e., systems review, exercise test). Information about the client's cardiovascular status gleaned by the therapist during the assessment was new and important for medical management.

From Vernier DA: The meaning of screening. The application of vital signs should be used in physical therapy practice, *ADVANCE for Physical Therapists & PT Assistants* 15(18):47-49, 2004.

More than six palpitations occurring in a minute or prolonged, repeated palpitations, especially if accompanied by chest pain, dyspnea, fainting, or other associated signs and symptoms, should be reported to the physician.

CLINICAL SIGNS AND SYMPTOMS

Fibrillation

- Subjective report of palpitations
- Sensations of fluttering, skipping, irregular beating or pounding, heaving action
- Dyspnea
- Chest pain
- Anxiety
- Pallor, fatigue
- Dizziness, lightheadedness, fainting
- Nervousness
- Cyanosis

Sinus Tachycardia

Sinus tachycardia, defined as an abnormally rapid heart rate, usually taken to be more than 100 bpm, is the normal physiologic response to such stressors as fever, hypotension, thyrotoxicosis, anemia, anxiety, exertion, hypovolemia, pulmonary emboli, myocardial ischemia, CHF, and shock.

Sinus tachycardia is usually of no physiologic significance; however, in clients with organic myocardial disease, the result may be reduced cardiac output, CHF, or arrhythmias. Because heart rate is a major determinant of oxygen requirements, angina or perhaps an increase in the size of an infarction may accompany persistent tachycardia in clients with CAD.

Clinical Signs and Symptoms. The symptoms of tachycardia vary from one person to another and may range from an increased pulse to a group of symptoms that would restrict normal activity of the client. Anxiety and apprehension may occur, depending on the pain threshold and emotional reaction of the client.

CLINICAL SIGNS AND SYMPTOMS

Sinus Tachycardia

- Palpitation (most common symptom)
- Restlessness
- Chest discomfort or pain
- Agitation
- Anxiety and apprehension

Sinus Bradycardia

In sinus bradycardia, impulses travel down the same pathway as in sinus rhythm, but the sinus node discharges at a rate less than 60 bpm. Bradycardia may be normal in athletes or young adults and is therefore asymptomatic.

In most cases, sinus bradycardia is a benign arrhythmia and may actually be beneficial by producing a longer period of diastole and increased ventricular filling. In some clients who have acute MI, it reduces oxygen demands and may help minimize the size of the infarction.

Eye surgery, meningitis, intracranial tumors, cervical and mediastinal tumors, and certain disease states (e.g., MI, myxedema, obstructive jaundice, and cardiac fibrosis) may produce sinus bradycardia.

Clinical Signs and Symptoms. Syncope may be preceded by sudden onset of weakness, sweating, nausea, pallor, vomiting, and distortion or dimming of vision. Signs and symptoms remit promptly when the client is placed in the horizontal position.

Physician referral for sinus bradycardia is needed only when symptoms, such as chest pain, dyspnea, lightheadedness, or hypotension, occur.

CLINICAL SIGNS AND SYMPTOMS

Sinus Bradycardia

- Reduced pulse rate
- Syncope

CARDIOVASCULAR DISORDERS

Hypertension

Blood pressure is the force against the walls of the arteries and arterioles as these vessels carry blood away from the heart. When these muscular walls constrict, reducing the diameter of the vessel, BP rises; when they relax, increasing the vessel diameter, BP falls (see also section on Blood Pressure in Chapter 4).

A high BP reading is usually a sign that the vessels cannot relax fully and remain somewhat constricted, requiring the heart to work harder to pump blood through the vessels. Over time the extra effort can cause the heart muscle to become enlarged and eventually weakened. The force of blood pumped at high pressure can also produce small tears in the lining of the arteries, weakening the arterial vessels. The evidence of this effect is most pronounced in the vessels of the brain, the kidneys, and the small vessels of the eye.

Hypertension is a major cardiovascular risk factor, associated with elevated risks of cardiovascular diseases, especially MI, stroke, PVD, and cardiovascular death. Although diastolic changes were always evaluated closely, research now shows that the risks increase progressively as systolic pressure goes up, especially in adults over age 50.[82,83]

Hypertension is often considered in conjunction with peripheral vascular disorders for several reasons: both are disorders of the circulatory system, the course of both diseases are affected by similar factors, and hypertension is a major risk factor in atherosclerosis, the largest single cause of PVD.

Hypertension is defined by a pattern of consistently elevated diastolic pressure, systolic pressure, or both measured over a period of time, usually several months. Medical researchers have developed classifications for BP based on risk (see Table 4-5).

The guidelines were updated in 2003 by the Joint National Committee on Prevention, Detection, Evaluation, and Treatment of Hypertension (Case Example 6-9).[84] Preliminary recommendations for a "new definition" of hypertension were issued in 2005 by the American Society of Hypertension.[85]

The new proposed definition/classification is based on the idea that hypertension is a complex cardiovascular disorder,

CASE EXAMPLE 6-9

Hypertension

Chief Complaint: A 70-year-old woman came to physical therapy with a diagnosis of left supraspinatus tendon strain.

Past Medical History: Cortisone injection to shoulder; recent normal electrocardiogram (ECG)

Clinical Presentation: The client reported posterior left shoulder pain and lateral arm pain with occasional radiating pain down the arm to her hand. No other symptoms were reported. Vital signs were assessed:

Blood pressure:	170/95 (left arm, sitting)
Pulse:	82 bpm, regular

Should the client be referred for high BP based on this information?

Hypertension is a medical diagnosis. However, one high reading is not sufficient to render this diagnosis. Many other factors can influence BP and should be evaluated (see Table 4-4).

The client was asked if she was ever diagnosed with high BP or hypertension. She denied any personal history of known elevated BP.

The client decided to have her BP checked at the local health department once a week for the next 2 weeks. Each time the readings were above normal. She made a self-referral to her medical doctor. After a medical evaluation, she was placed on appropriate medication.

Data from Vernier DA: The meaning of screening. The application of vital signs should be used in physical therapy practice, *ADVANCE for Physical Therapists & PT Assistants* 15(18):47-49, 2004.

not a scale of BP values. The new definition takes into account risk factors, early disease markers, and attempts to reflect the effects of hypertension on other organ systems. The goal of this risk-based approach is to identify individuals at any level of BP who have a reasonable likelihood of future cardiovascular events.[85]

Simply stated the new guidelines emphasize the continuous relationship between BP level and cardiovascular risk. The BP classification scale used for diagnosis and treatment is based on total cardiovascular risk, not just BP values. So for example, an individual with BP values of 140/90 mm Hg (defined as Stage I or borderline hypertension) may not begin medical therapy in the absence of other risk factors. On the other hand, someone with much lower BP values (e.g., 120/75 mm Hg) might be treated immediately if other risk factors are present such as overweight or tobacco use.[85]

Pulse Pressure

See Chapter 4 for more information on this topic.

Blood Pressure Classification

Hypertension can also be classified according to type (systolic or diastolic), cause, and degree of severity. *Primary (or essential) hypertension* is also known as *idiopathic hypertension* and accounts for 90% to 95% of all hypertensive clients.

Secondary hypertension results from an identifiable cause, including a variety of specific diseases or problems such as renal artery stenosis, oral contraceptive use, hyperthyroidism, adrenal tumors, and medication use.

Originally, birth control pills contained higher levels of estrogen, which was associated with hypertension, but today the estrogen and progestin contents of the pill are greatly reduced. The risk of high BP with oral contraceptive use is now considered quite low, but using oral contraceptives does still increase the risk of heart attack, stroke, and blood clots in certain women (e.g., age over 35, tobacco use, diabetes).

The risk may be increased for older women who smoke, but the risk for all women returns to normal after they discontinue the pill. The risk of venous thromboembolism associated with newer oral contraceptives remains under investigation.

Drugs that constrict blood vessels can contribute to hypertension. Among the most common are phenylpropanolamine in over-the-counter (OTC) appetite suppressants, including herbal ephedra, pseudoephedrine in cold and allergy remedies, and prescription drugs such as monoamine oxidase (MAO) inhibitors (a class of antidepressant) and corticosteroids when used over a long period.

Intermittent elevation of BP interspersed with normal readings is called *labile hypertension* or *borderline hypertension*. More and more adults over age 50 and many older adults have a type of high BP called *isolated systolic hypertension* (ISH) characterized by marked elevation of the systolic pressure (140 mm Hg or higher) but normal diastolic pressure (less than 90 mm Hg).[86]

ISH is a risk factor for stroke and death from cardiovascular causes. Elevated systolic pressure also raises the risk of heart attack, CHF, dementia, and end-stage kidney disease.[87]

Two other types of hypertension include *masked hypertension* (normal in the clinic but periodically high at home) and *white-coat hypertension*, a clinical condition in which the client has elevated BP levels when measured in a clinic setting by a health care professional. In such cases, BP measurements are consistently normal outside of a clinical setting.

Masked hypertension may effect up to 10% of adults; white-coat hypertension occurs in 15% to 20% of adults with Stage I hypertension.[88] These types of hypertension are more common in older adults. Antihypertensive treatment for white coat hypertension may reduce office BP but may not affect ambulatory BP. The number of adults who develop sustained high BP is much higher among those who have masked or white-coat hypertension.[88]

At-home BP measurements can help identify adults with masked hypertension, white-coat hypertension, ambulatory hypertension, and individuals who do not experience the usual nocturnal drop in BP (decrease of 15 mm Hg), which is a risk factor for cardiovascular events.[89] Excessive morning BP surge is a predictor of stroke in older adults with known hypertension and is also a red flag sign.[89] Medical referral is indicated in any of these situations.

Risk Factors

Modifiable risk factors for hypertension are primarily lifestyle factors such as stress, obesity, and poor diet or insufficient intake of nutrients (Table 6-6). Stress has been shown to cause increased peripheral vascular resistance and cardiac output and to stimulate sympathetic nervous system activity. Potassium deficiency can also contribute to hypertension.

The Seventh Report of the Joint National Committee on Prevention, Detection, Evaluation, and Treatment of High Blood Pressure (JNC-7; see Table 4-5) created a new BP category called "prehypertension" to identify adults considered to be at risk for developing hypertension and to alert both individuals and health care providers of the importance of adopting lifestyle changes. Screening for prehypertension provides important opportunities to prevent hypertension and cardiovascular disease.[90]

Nonmodifiable risk factors include family history, age, gender, postmenopausal status, and race. The risk of hypertension increases with age as arteries lose elasticity and

TABLE 6-6 Risk Factors for Hypertension

Modifiable	Nonmodifiable
Smoking or tobacco	African-American ethnicity
Type 2 diabetes	Age (60 or older)
High cholesterol	Postmenopausal status (including surgically-induced menopause)
Chronic alcohol use/abuse	Family history of cardiovascular disease (women younger than age 65; men younger than age 55)
Obesity	
Sedentary lifestyle	
Stress	
Diet, nutritional status; potassium deficiency	

become less able to relax. There is a poorer prognosis associated with early onset of hypertension.

A sex-specific gene for hypertension may exist[91] because men experience hypertension at higher rates and at an earlier age than women do until after menopause. Hypertension is the most serious health problem for African-Americans (both men and women and at earlier ages than for whites) in the United States (see further discussion of hypertension in African Americans in Chapter 4).

Clinical Signs and Symptoms

Clients with hypertension are usually asymptomatic in the early stages, but when symptoms do occur, they include occipital headache (usually present in the early morning), vertigo, flushed face, nocturnal urinary frequency, spontaneous nosebleeds, and blurred vision.

> **CLINICAL SIGNS AND SYMPTOMS**
>
> ***Hypertension***
>
> - Occipital headache
> - Vertigo (dizziness)
> - Flushed face
> - Spontaneous epistaxis
> - Vision changes
> - Nocturnal urinary frequency

Transient Ischemic Attack

Hypertension is a major cause of heart failure, stroke, and kidney failure. Aneurysm formation and CHF are also associated with hypertension. Persistent elevated diastolic pressure damages the intimal layer of the small vessels, which causes an accumulation of fibrin, local edema, and possibly intravascular clotting.

Eventually, these damaging changes diminish blood flow to vital organs, such as the heart, kidneys, and brain, resulting in complications such as heart failure, renal failure, and cerebrovascular accidents or stroke.

Many persons have brief episodes of transient ischemic attacks (TIAs). The attacks occur when the blood supply to part of the brain has been temporarily disrupted. These ischemic episodes last from 5 to 20 minutes, although they may last for as long as 24 hours. TIAs are considered by some as a progression of cerebrovascular disease and may be referred to as "mini-strokes."

TIAs are important warning signals that an obstruction exists in an artery leading to the brain. Without treatment, 10% to 20% of people will go on to have a major stroke within 3 months, many within 48 hours.[92] Immediate medical referral is advised for anyone with signs and symptoms of TIAs, especially anyone with a history of heart disease, hypertension, or tobacco use. Other risk factors for TIAs include age (over 65), diabetes, and being overweight.

> **CLINICAL SIGNS AND SYMPTOMS**
>
> ***Transient Ischemic Attack***
>
> - Slurred speech, sudden difficulty with speech, or difficulty understanding others
> - Sudden confusion, loss of memory, even loss of consciousness
> - Temporary blindness or other dramatic visual changes
> - Dizziness
> - Sudden, severe headache
> - Paralysis or extreme weakness, usually affecting one side of the body
> - Difficulty walking, loss of balance or coordination
> - Symptoms are usually brief, lasting only a few minutes but can persist up to 24 hours

Orthostatic Hypotension (See also discussion on Hypotension in Chapter 4)

Orthostatic hypotension is an excessive fall in BP of 20 mm Hg or more in systolic BP or a drop of 10 mm Hg or more of both systolic and diastolic arterial BP on assumption of the erect position with a 10% to 20% increase in pulse rate (Case Example 6-10). It is not a disease but a manifestation of abnormalities in normal BP regulation.

This condition may occur as a normal part of aging or secondary to the effects of drugs such as hypertensives, diuretics, and antidepressants; as a result of venous pooling (e.g., pregnancy, prolonged bed rest, or standing); or in association with neurogenic origins. The last category includes diseases affecting the autonomic nervous system, such as Guillain-Barré syndrome, diabetes mellitus, or multiple sclerosis.

Orthostatic intolerance is the most common cause of lightheadedness in clients, especially those who have been on prolonged bed rest or those who have had prolonged anesthesia for surgery. When such a client is getting up out of bed for the first time, BP, and heart rate should be monitored with the person in the supine position and repeated after the person is upright. If the legs are dangled off the bed, a significant drop in BP may occur with or without compensatory tachycardia. This drop may provoke lightheadedness, and standing may even produce loss of consciousness.

These postural symptoms are often accentuated in the morning and are aggravated by heat, humidity, heavy meals, and exercise.

> **CLINICAL SIGNS AND SYMPTOMS**
>
> ***Orthostatic Hypotension***
>
> - Change in BP (decrease) and pulse (increase)
> - Lightheadedness, dizziness
> - Pallor, diaphoresis
> - Syncope or fainting
> - Mental or visual blurring
> - Sense of weakness or "rubbery" legs

CASE EXAMPLE 6-10

Monitoring Vital Signs

Referral: An 83-year-old woman was referred to physical therapy for mobility training.

Goal: Improve balance to prevent nursing home placement.

Chief Complaints: Forgetfulness, two falls during the past three months, and inability to complete independent activities of daily living (ADLs).

Past/Current Medical History: Pernicious anemia, chronic venous insufficiency, non–insulin dependent diabetes, hypercholesterolemia, osteoporosis, progressive dementia, gastroesophageal reflux (GERD).

Medications/Supplements: Lasix for chronic edema in lower extremities secondary to venous insufficiency, Lipitor for elevated cholesterol, calcium for osteoporosis, Prilosec for GERD.

Systems Review

Integument: Integument intact with good turgor.

Musculoskeletal: Strength 4/5 upper and lower extremities. Range of motion (ROM) within functional limits (WNLs). Thoracic kyphosis present.

Neuromuscular: Intact cranial nerves, independent transfers, impaired gait and balance.

Cardiovascular/Pulmonary: Heart rate: 65 bpm at rest; respiratory rate: 16 breaths per minute; BP: 100/70 mm Hg (measured in sitting, right upper extremity).

Vital Signs: Oral temperature: 100° F with a regular pulse rate; finger pulse oximeter (oxygen saturation)—88% (rest) and 85% (walking with wheeled walker).

Further BP assessment was conducted comparing measures in supine (110/70 mm Hg), sitting (100/70 mm Hg), and standing (90/65 mm Hg) with only slight increase in pulse rate (from 16 to 20 bpm) suggesting postural orthostatic hypotension (for discussion on postural orthostatic hypotension, see Chapter 4). Client became diaphoretic during BP testing.

Decreased breath sounds heard in right lower lobe during auscultation. Client reported productive cough and unusual fatigue.

What are the red flags in this case?

- Age
- Constitutional symptoms (low-grade fever, fatigue, unexplained diaphoresis)
- Abnormal vital signs (BP changes with change in position accompanied by increase in pulse rate, low oxygen saturation levels)
- Productive cough, decreased breath sounds, right lower lobe

Result: The primary physician was contacted to discuss the therapist's concerns regarding the vital signs, constitutional symptoms, and signs and symptoms associated with the pulmonary system. The client was treated for unstable BP and pneumonia.

Physical therapists can take a leadership role in the management of their clients. Every physical therapy examination should include, at a minimum, a baseline measurement of vital signs even without red flags or specific symptoms to suggest it.

From Heins P: Case report presented in partial fulfillment of DPT 910. *Principles of differential diagnosis,* Institute for Physical Therapy Education, Chester, Pennsylvania, 2005, Widener University. Used with permission.

Peripheral arterial occlusive diseases also can be caused by embolism, thrombosis, trauma, vasospasm, inflammation, or autoimmunity. The cause of some disorders is unknown.

Arterial (Occlusive) Disease

Arterial diseases include acute and chronic arterial occlusion (Table 6-7). Acute arterial occlusion may be caused by

1. Thrombus, embolism, or trauma to an artery
2. Arteriosclerosis obliterans
3. Thromboangiitis obliterans or Buerger's disease
4. Raynaud's disease

Clinical manifestations of chronic arterial occlusion caused by peripheral vascular disease may not appear for 20 to 40 years. The lower limbs are far more susceptible to arterial occlusive disorders and atherosclerosis than are the upper limbs.

Risk Factors. Diabetes mellitus increases the susceptibility to coronary heart disease. People with diabetes have abnormalities that affect a number of steps in the development of atherosclerosis. Only the combination of factors, such as hypertension, abnormal platelet activation, and metabolic disturbances affecting fat and serum cholesterol, accounts for the increased risk.

Other risk factors include smoking, hypertension, hyperlipidemia (elevated levels of fats in the blood), and older age. Peripheral artery disease most often afflicts men older than 50, although women are at significant risk because of their increased smoking habits.

Clinical Signs and Symptoms. The first sign of vascular occlusive disease may be the loss of hair on the toes. The most important symptoms of chronic arterial occlusive disease are intermittent claudication (limping resulting from pain, ache, or cramp in the muscles of the lower extremities caused by ischemia or insufficient blood flow) and ischemic rest pain.

The pain associated with arterial disease is generally felt as a dull, aching tightness deep in the muscle, but it may be described as a boring, stabbing, squeezing, pulling, or even burning sensation. Although the pain is sometimes referred to as a cramp, there is no actual spasm in the painful muscles.

The location of the pain is determined by the site of the major arterial occlusion (see Table 14-7). Aortoiliac occlusive disease induces pain in the gluteal and quadriceps muscles. The most frequent lesion, which is present in about two thirds of clients, is occlusion of the superficial femoral artery between the groin and the knee, producing pain in the calf that sometimes radiates upward to the popliteal region and to the lower thigh. Occlusion of the popliteal or more distal arteries causes pain in the foot.

In the typical case of superficial femoral artery occlusion, there is a good femoral pulse at the groin but arterial pulses are absent at the knee and foot, although resting circulation appears to be good in the foot.

After exercise, the client may have numbness in the foot, as well as pain in the calf. The foot may be cold, pale, and chalky white, which is an indication that the circulation has been diverted to the arteriolar bed of the leg muscles. Blood

TABLE 6-7 Comparison of Acute and Chronic Arterial Symptoms

Symptom Analysis	Acute Arterial Symptoms	Chronic Arterial Symptoms
Location	Varies; distal to occlusion; may involve entire leg	Deep muscle pain, usually in calf, may be in lower leg or dorsum of foot
Character	Throbbing	Intermittent claudication; feels like cramp, numbness, and tingling; feeling of cold
Onset and duration	Sudden onset (within 1 hour)	Chronic pain, onset gradual following exertion
Aggravating factors	Activity such as walking or stairs; elevation	Same as acute arterial symptoms
Relieving factors	Rest (usually within 2 minutes); dangling (severe involvement)	Same as acute arterial symptoms
Associated symptoms	6 P's: Pain, pallor, pulselessness, paresthesia, poikilothermia (coldness), paralysis (severe)	Cool, pale skin
At risk	History of vascular surgery, arterial invasive procedure, abdominal aneurysm, trauma (including injured arteries), chronic atrial fibrillation	Older adults; more males than females; inherited predisposition; history of hypertension, smoking, diabetes, hypercholesterolemia, obesity, vascular disease

From American Heart Association (AHA): *Heart and stroke encyclopedia*. Available at http://www.americanheart.org. Accessed November 11, 2010.

in regions of sluggish flow becomes deoxygenated, inducing a red-purple mottling of the skin.

Painful cramping symptoms occur during walking and disappear quickly with rest. Ischemic rest pain is relieved by placing the limb in a dependent position, using gravity to enhance blood flow. In most clients the symptoms are constant and reproducible (i.e., the client who cannot walk the length of the house because of leg pain one day but is able to walk indefinitely the next does not have intermittent claudication).

Intermittent claudication is influenced by the speed, incline, and surface of the walk. Exercise tolerance decreases over time, so that episodes of claudication occur more frequently with less exertion. The differentiation between vascular claudication and neurogenic claudication is presented in Chapter 16 (see Table 16-5).

Ulceration and gangrene are common complications and may occur early in the course of some arterial diseases (e.g., Buerger's disease). Gangrene usually occurs in one extremity at a time. In advanced cases the extremities may be abnormally red or cyanotic, particularly when dependent. Edema of the legs is fairly common. Color or temperature changes and changes in nail bed and skin may also appear.

CLINICAL SIGNS AND SYMPTOMS

Arterial Disease

- Intermittent claudication
- Burning, ischemic pain at rest
- Rest pain aggravated by elevating the extremity; relieved by hanging the foot over the side of the bed or chair
- Color, temperature, skin, nail bed changes
- Decreased skin temperature
- Dry, scaly, or shiny skin
- Poor nail and hair growth
- Possible ulcerations and gangrene on weight bearing surfaces (e.g., toes, heel)
- Vision changes (diabetic atherosclerosis)
- Fatigue on exertion (diabetic atherosclerosis)

Raynaud's Phenomenon and Disease

The term *Raynaud's phenomenon* refers to intermittent episodes during which small arteries or arterioles in extremities constrict, causing temporary pallor and cyanosis of the digits and changes in skin temperature.

These episodes occur in response to cold temperature or strong emotion (anxiety, excitement). As the episode passes, the changes in color are replaced by redness. If the disorder is secondary to another disease or underlying cause, the term *secondary Raynaud's phenomenon* is used.

Secondary Raynaud's phenomenon is often associated with connective tissue or collagen vascular disease, such as scleroderma, polymyositis/dermatomyositis, SLE, or rheumatoid arthritis. Raynaud's may occur as a long-term complication of cancer treatment. Unilateral Raynaud's phenomenon may be a sign of hidden neoplasm.

Raynaud's phenomenon may occur after trauma or use of vibrating equipment such as jackhammers, or it may be related to various neurogenic lesions (e.g., thoracic outlet syndrome) and occlusive arterial diseases.

Raynaud's disease is a primary vasospastic or vasomotor disorder, although it is included in this section under occlusive arterial because of the arterial involvement. It appears to be caused by

1. Hypersensitivity of digital arteries to cold
2. Release of serotonin
3. Congenital predisposition to vasospasm

Eighty percent of clients with Raynaud's disease are women between the ages of 20 and 49 years. Primary Raynaud's disease rarely leads to tissue necrosis.

Idiopathic Raynaud's disease is differentiated from secondary Raynaud's phenomenon by a history of symptoms for at least 2 years with no progression of the symptoms and no evidence of underlying cause.

Clinical Signs and Symptoms. The typical progression of Raynaud's phenomenon is pallor in the digits, followed by cyanosis accompanied by feelings of cold, numbness, and occasionally pain, and finally, intense redness with tingling or throbbing.

The pallor is caused by vasoconstriction of the arterioles in the extremity, which leads to decreased capillary blood flow. Blood flow becomes sluggish and cyanosis appears; the digits turn blue. The intense redness (rubor) results from the end of vasospasm and a period of hyperemia as oxygenated blood rushes through the capillaries.

CLINICAL SIGNS AND SYMPTOMS

Raynaud's Phenomenon and Disease

- Pallor in the digits
- Cyanotic, blue digits
- Cold, numbness, pain of digits
- Intense redness of digits

Venous Disorders

Venous disorders can be separated into acute and chronic conditions. Acute venous disorders include thromboembolism. Chronic venous disorders can be separated further into varicose vein formation and chronic venous insufficiency.

Acute Venous Disorders. Acute venous disorders are due to formation of thrombi (clots), which obstruct venous flow. Blockage may occur in both superficial and deep veins. Superficial thrombophlebitis is often iatrogenic, resulting from insertion of intravenous catheters or as a complication of intravenous sites.

Pulmonary emboli (see Chapter 7), most of which start as thrombi in the large deep veins of the legs, are an acute and potentially lethal complication of deep venous thrombosis.

Thrombus formation results from an intravascular collection of platelets, erythrocytes, leukocytes, and fibrin in the blood vessels, often the deep veins of the lower extremities. When thrombus formation occurs in the deep veins, the production of clots can cause significant morbidity and mortality resulting in a floating mass (embolus) that can occlude blood vessels of the lungs and other critical structures.[93]

Risk Factors. Deep venous thrombosis (DVT) defined as blood clots in the pelvic, leg, or major upper extremity veins is a common disorder, affecting women more than men and adults more than children. Approximately one third of clients older than 40 who have had either major surgery or an acute MI develop DVT. The most significant clinical risk factors are age over 70 and previous thromboembolism[94,95] (Box 6-2).

Thrombus formation is usually attributed to (1) venous stasis, (2) hypercoagulability, or (3) injury to the venous wall. *Venous stasis* is caused by prolonged immobilization or absence of the calf muscle pump (e.g., because of illness, paralysis, or inactivity). Other risk factors include traumatic spinal cord injury, multiple trauma, CHF, obesity, pregnancy, and major surgery (orthopedic, oncologic, gynecologic, abdominal, cardiac, renal, or splenic)[96] (Case Example 6-11).

Hypercoagulability often accompanies malignant neoplasms, especially visceral and ovarian tumors. Oral contraceptives, selective estrogen receptor modulators (SERMs; e.g., raloxifene) often used for osteoporosis related to menopause, and hematologic (clotting) disorders also may increase the coagulability of the blood. In addition, previous spontaneous thromboembolism and increased levels of homocysteine are risk factors for venous as well as arterial thrombosis.[97]

The observed relationship of higher venous thrombosis risk with the use of third-generation oral contraceptives is an important consideration.[98,99] Third-generation contraceptives refer to the newest formulation of oral contraceptives with much lower levels of estrogen than those first administered.

The risk of having a blood clot depends on a number of factors. It increases with age and it also depends on what kind of oral contraceptive is being taken. Women using progestogen-only pills are at little or no increased risk of blood clots. The venous clots associated with the newest oral contraceptives typically develop in superficial leg veins and rarely result in pulmonary emboli.

Injury or trauma to the venous wall may occur as a result of intravenous injections, Buerger's disease, fractures and dislocations, sclerosing agents, and opaque mediator radiography.

Clinical Signs and Symptoms. Superficial thrombophlebitis appears as a local, raised, red, slightly indurated (hard), warm, tender cord along the course of the involved vein.

In contrast, symptoms of DVT are less distinctive; about one half of clients are asymptomatic. The most common symptoms are pain in the region of the thrombus and unilateral swelling distal to the site (Case Example 6-12).

Other symptoms include redness or warmth of the arm or leg, dilated veins, or low-grade fever, possibly accompanied

BOX 6-2 RISK FACTORS FOR PULMONARY EMBOLISM (PE) AND DEEP VENOUS THROMBOSIS (DVT)

Previous personal/family history of thromboembolism
Congestive heart failure
Age (over 50 years)
Oral contraceptive use
Immobilization or inactivity (blood stasis)
Obstetric/gynecologic conditions
Obesity
Neoplasm
Pacemaker
Trauma
- Recent surgical procedures
- Indwelling central venous catheter (upper extremity DVT)
- Fracture
- Burns
- Spinal cord injury
- Endothelial injury, stroke

Blood disorders (e.g., hypercoagulable state, clotting abnormalities)
History of infection, diabetes mellitus

CASE EXAMPLE 6-11

Deep Venous Thrombosis in a Spinal Cord–Injured Patient

Referral: An 18-year-old male with Down syndrome fell and sustained a fracture at C2 with resultant spinal cord injury and flaccid quadriparesis. After medical treatment and stabilization, he was transferred to a rehabilitation facility.

Medications: Lovenox (anticoagulant, antithrombotic for DVT prevention)

Summary: This client had a long and extensive recovery and rehabilitation due to his diagnosis of Down syndrome, English as a second language, high-level spinal cord injury, chronic pressure ulcers, and cardiovascular complications.

Eight months after the start of rehabilitation, unilateral swelling, pain, and warmth developed in the left lower extremity. Elevating the leg did not relieve symptoms.

Circumference measurements around the left thigh, calf, and foot were 1 inch greater than around the right leg.

A deep venous thrombosis (DVT) was highly suspected because the leg symptoms were unilateral. If the swelling were simply due to his legs being in the dependent position, one would expect swelling in both of his legs. The client was in a high-risk category for developing a DVT.

He was also taken off Lovenox injections just 1 month prior to developing his symptoms. Weaning a client off percutaneous anticoagulation therapy by 6 months is a generally accepted practice, continuing only with oral anticoagulation therapy such as Coumadin.[105] The client had been on Lovenox for 9 months.

Result: The client was diagnosed and treated for DVT and returned to the rehabilitation hospital. Repeat ultrasound the following month showed interval improvement without resolution of the DVT. Further medical treatment was instituted.

Although the risk of DVT is greater in the acute care phase, the therapist must remain alert to symptoms of DVT in at-risk clients with multiple comorbidities and complications. Use of the Autar Scale (see Table 4-11) or Wells' Clinical Decision Rule (see Table 4-12) is advised.

Even with medical treatment, it should not be assumed that the condition has resolved until confirmed by medical testing. All precautions must remain in effect until released by the physician.

From Rosenzweig K: Case report presented in partial fulfillment of DPT 910. *Principles of differential diagnosis,* Institute for Physical Therapy Education, Chester, Pennsylvania, 2005, Widener University. Used with permission.

by chills and malaise. Unfortunately, the first clinical manifestation may be pulmonary embolism. Frequently, clients have thrombi in both legs even though the symptoms are unilateral. For further discussion of DVT of the upper extremity, see Chapter 18.

Homans' sign (discomfort in the upper calf during gentle, forced dorsiflexion of the foot) is still commonly assessed during physical examination. Unfortunately, it is insensitive and nonspecific. It is present in less than one third of clients with documented DVT. In addition, more than 50% of clients with a positive finding of Homans' sign do not have evidence of venous thrombosis.

Other more specific risk assessment and physical assessment tools are available for assessment of DVT and PVD (see ankle-brachial index [ABI], Autar DVT Risk Assessment Scale, and Wells' Clinical Decision Rule [CDR] in Chapter 4). A simple model to predict upper extremity DVT has also been proposed and remains under investigation (see Table 18-3).[100]

The CDR may be used more widely in the clinic, but the Autar scale is more comprehensive and incorporates BMI, postpartum status, and the use of oral contraceptives as potential risk factors.

The Society of Interventional Radiology (SIR) now recommends that anyone being evaluated for PVD should have an ABI measurement done, since this is a significantly more accurate screening measure for PVD.[101] Symptoms of superficial thrombophlebitis are relieved by bed rest with elevation of the legs and the application of heat for 7 to 15 days. When local signs of inflammation subside, the client is usually allowed to ambulate wearing elastic stockings.

Sometimes, antiinflammatory medications are required. Anticoagulants, such as heparin and warfarin, are used to prevent clot extension.

CLINICAL SIGNS AND SYMPTOMS

Superficial Venous Thrombosis

- Subcutaneous venous distention
- Palpable cord
- Warmth, redness
- Indurated (hard)

CLINICAL SIGNS AND SYMPTOMS

Deep Venous Thrombosis

- Unilateral tenderness or leg pain
- Unilateral swelling (difference in leg circumference)
- Warmth
- Discoloration
- Pain with placement of BP cuff around calf inflated to 160 mm Hg to 180 mm Hg

Chronic Venous Disorders. Chronic venous insufficiency, also known as postphlebitic syndrome, is identified by chronic swollen limbs; thick, coarse, brownish skin around the ankles; and venous stasis ulceration. Chronic venous insufficiency is the result of dysfunctional valves that reduce venous return, which thus increases venous pressure and causes venous stasis and skin ulcerations.

Chronic venous insufficiency follows most severe cases of deep venous thrombosis but may take as long as 5 to 10 years to develop. Education and prevention are essential, and clients with a history of deep venous thrombosis must be monitored periodically for life.

CASE EXAMPLE 6-12

Deep Venous Thrombosis (DVT)

Referral: A 96-year-old woman was discharged from a hospital to a subacute center for rehabilitation.

Goal: Return to previous level of function if possible.

Chief Complaint: Fall with fracture of left pelvis; diffuse pain around left pelvic area rated 5/10 on the numeric rating scale; pain increases with weightbearing and movement. Conservative nonsurgical treatment was employed.

Past Medical History: Dementia, colon cancer

Current Medications: Acetaminophen 500 mg prn (for pain), warfarin daily (anticoagulant), Risperidone (for dementia), calcium carbonate/vitamin D (for osteoporosis)

Systems Review

Integument: Skin integrity within normal limits for client's age

Musculoskeletal: Muscle strength 3+/5 in both lower extremities; range of motion within functional limits for all but left hip; left hip flexion limited by pain at end of range; functional transfers, bed mobility, ambulation, and stairs with assistance; antalgic gait with decreased base of support, decreased stride/step length, minimal weight bearing on left leg

Neuromuscular: Standing balance fair, dynamic standing balance fair minus (F –) with a wheeled walker

Cardiovascular/pulmonary: Lower extremity pulses and circulation within normal limits for client's age (baseline). Two weeks later, affected foot and calf were edematous but reportedly pain free. Calf was tender to touch; leg was warm with discoloration of the affected limb.

Further Screening: Wells' Clinical Decision Rule for DVT

Result: Client was referred for medical evaluation of sudden change in the involved lower extremity. Recent pelvic fracture is a major risk factor for deep venous thrombosis, a potentially life-threatening condition. Duplex ultrasonography confirmed provisional medical diagnosis of DVT.

Clinical presentation	Possible score	Client's score
Active cancer (within 6 months of diagnosis or receiving palliative care)	1	0
Paralysis, paresis, or recent immobilization of lower extremity	1	1
Bedridden for more than 3 days or major surgery in the last 4 weeks	1	0
Localized tenderness in the center of the posterior calf, the popliteal space, or along the femoral vein in the anterior thigh/groin	1	1
Lower extremity swelling	1	0
Unilateral calf swelling (more than 3 mm larger than uninvolved side)	1	1
Unilateral pitting edema	1	0
Collateral superficial veins (nonvaricose)	1	0
An alternative diagnosis is as likely (or more likely) than DVT (e.g., cellulitis, postoperative swelling, calf strain)	–2	0
Total Points		3
Key:		
–2 to 0		Low probability of DVT (3%)
1 to 2		Moderate probability of DVT (17%)
3 or more		High probability of DVT (75%)

Medical consultation is advised in the presence of low probability; medical referral is required with moderate or high score.
From Wells PS, Anderson DR, Bormanis J, et al: Value of assessment of pretest probability of deep-vein thrombosis in clinical management, *Lancet* 350:1795-1798, 1997. Used with permission.

From Kehinde JA: Case report presented in partial fulfillment of DPT 910. *Principles of differential diagnosis,* Institute for Physical Therapy Education, Chester, Pennsylvania, 2005, Widener University. Used with permission.

Lymphedema

The final type of peripheral vascular disorder, lymphedema, is defined as an excessive accumulation of fluid in tissue spaces. Lymphedema typically occurs secondary to an obstruction of the lymphatic system from trauma, infection, radiation, or surgery.

Postsurgical lymphedema is usually seen after surgical excision of axillary, inguinal, or iliac nodes, usually performed as a prophylactic or therapeutic measure for metastatic tumor. Lymphedema secondary to primary or metastatic neoplasms in the lymph nodes is common.

CLINICAL SIGNS AND SYMPTOMS

Lymphedema

- Edema of the dorsum of the foot or hand
- Decreased range of motion, flexibility, and function
- Usually unilateral
- Worse after prolonged dependency
- No discomfort or a dull, heavy sensation; sense of fullness

LABORATORY VALUES

The results of diagnostic tests can provide the therapist with information to assist in client education. The client often reports test results to the therapist and asks for information regarding the significance of those results. The information presented in this text discusses potential reasons for abnormal laboratory values relevant to clients with cardiovascular problems.

A basic understanding of laboratory tests used specifically in the diagnosis and monitoring of cardiovascular problems can provide the therapist with additional information regarding the client's status.

Some of the tests commonly used in the management and diagnosis of cardiovascular problems include lipid screening (cholesterol levels, LDL levels, high-density lipoprotein [HDL] levels, and triglyceride levels), serum electrolytes, and arterial blood gases (see Chapter 7).

Other laboratory measurements of importance in the overall evaluation of the client with cardiovascular disease include red blood cell values (e.g., red blood cell count, hemoglobin, and hematocrit). Those values (see Chapter 5) provide valuable information regarding the oxygen-carrying capability of the blood and the subsequent oxygenation of body tissues such as the heart muscle.

Serum Electrolytes

Measurement of serum electrolyte values is particularly important in diagnosis, management, and monitoring of the client with cardiovascular disease because electrolyte levels have a direct influence on the function of cardiac muscle (in a manner similar to that of skeletal muscle). Abnormalities in serum electrolytes, even in noncardiac clients, can result in significant cardiac arrhythmias and even cardiac arrest.

In addition, certain medications prescribed for cardiac clients can alter serum electrolytes in such a way that rhythm problems can occur as a result of the medication. The electrolyte levels most important to monitor include potassium, sodium, calcium, and magnesium (see inside back cover).

Potassium

Serum potassium levels can be lowered significantly as a result of diuretic therapy (particularly with loop diuretics such as Lasix [furosemide]), vomiting, diarrhea, sweating, and alkalosis. Low potassium levels cause increased electrical instability of the myocardium, life-threatening ventricular arrhythmias, and increased risk of digitalis toxicity.

Serum potassium levels must be measured frequently by the physician in any client taking a digitalis preparation (e.g., digoxin), because most of these clients are also undergoing diuretic therapy. Low potassium levels in clients taking digitalis can cause digitalis toxicity and precipitate life-threatening arrhythmias.

Increased potassium levels most commonly occur because of renal and endocrine problems or as a result of potassium replacement overdose. Cardiac effects of increased potassium levels include ventricular arrhythmias and asystole/flat line (complete cessation of electrical activity of the heart).

Sodium

Serum sodium levels indicate the client's state of water/fluid balance, which is particularly important in CHF and other pathologic states related to fluid imbalances. A low serum sodium level can indicate water overload or extensive loss of sodium through diuretic use, vomiting, diarrhea, or diaphoresis.

A high serum sodium level can indicate a water deficit state such as dehydration or water loss (e.g., lack of antidiuretic hormone [ADH]).

Calcium

Serum calcium levels can be decreased as a result of multiple transfusions of citrated blood, renal failure, alkalosis, laxative or antacid abuse, and parathyroid damage or removal. A decreased calcium level provokes serious and often life-threatening ventricular arrhythmias and cardiac arrest.

Increased calcium levels are less common but can be caused by a variety of situations, including thiazide diuretic use (e.g., Diuril [chlorothiazide]), acidosis, adrenal insufficiency, immobility, and vitamin D excess. Calcium excess causes atrioventricular conduction blocks or tachycardia and ultimately can result in cardiac arrest.

Magnesium

Serum magnesium levels are rarely changed in healthy individuals because magnesium is abundant in foods and water. However, magnesium deficits are often seen in alcoholic clients or clients with critical illnesses that involve shifting of a variety of electrolytes.

Magnesium deficits often accompany potassium and calcium deficits. A decrease in serum magnesium results in myocardial irritability and cardiac arrhythmias, such as atrial or ventricular fibrillation or premature ventricular beats (PVCs).

SCREENING FOR THE EFFECTS OF CARDIOVASCULAR MEDICATIONS

When a client is physically challenged, as often occurs in physical therapy, signs and symptoms develop from side effects of various classes of cardiovascular medications (Table 6-8).

For example, medications that cause peripheral vasodilation can produce hypotension, dizziness, and syncope when combined with physical therapy interventions that also produce peripheral vasodilation (e.g., hydrotherapy, aquatics, aerobic exercise).

On the other hand, cardiovascular responses to exercise can be limited in clients who are taking beta-blockers because these drugs limit the increase in heart rate that can occur as exercise increases the workload of the heart. The available

TABLE 6-8 Cardiovascular Medications

Condition	Drug Class
Angina pectoris	Organic nitrates Beta-blockers Calcium channel (Ca^{2+}) blockers
Arrhythmias	Sodium channel blockers Beta-blockers Calcium channel (Ca^{2+}) blockers Agents prolonging depolarization
Congestive heart failure	Cardiac glycosides (digitalis) Diuretics ACE inhibitors Vasodilators
Hypertension	Diuretics Beta-blockers ACE inhibitors Vasodilators Calcium (Ca^{2+}) channel blockers Alpha (α-$_1$)-blockers

Courtesy Susan Queen, Ph.D., P.T., University of New Mexico School of Medicine, Physical Therapy Program, Albuquerque, New Mexico.
ACE, Angiotensin-converting enzyme.

BOX 6-3 POTENTIAL SIDE EFFECTS OF CARDIOVASCULAR MEDICATIONS

Abdominal pain*
Asthmatic attacks*
Bradycardia*
Cough
Dehydration
Difficulty swallowing*
Dizziness or fainting*
Drowsiness
Dyspnea* (shortness of breath or difficulty breathing)
Easy bruising
Fatigue
Headache
Insomnia†
Joint pain*
Loss of taste
Muscle cramps†
Nausea
Nightmares†
Orthostatic hypotension†
Palpitations†
Paralysis*
Sexual dysfunction†
Skin rash†
Stomach irritation†
Swelling of feet or abdomen†
Symptoms of congestive heart failure*
Shortness of breath
Swollen ankles
Coughing up blood
Tachycardia†
Unexplained swelling, unusual or uncontrolled bleeding†
Vomiting
Weakness

*Immediate physician referral.
†Notify physician.

pharmaceuticals used in the treatment of the conditions listed in Table 6-8 are extensive. Understanding of drug interactions and implications requires a more specific text.

The therapist must especially keep in mind that nonsteroidal antiinflammatory drugs (NSAIDs), often used in the treatment of inflammatory conditions, have the ability to negate the antihypertensive effects of angiotensin-converting enzyme (ACE) inhibitors. Anyone being treated with both NSAIDs and ACE inhibitors must be monitored closely during exercise for elevated BP.

Likewise, NSAIDs have the ability to decrease the excretion of digitalis glycosides (e.g., digoxin [Lanoxin] and digitoxin [Crystodigin]). Therefore levels of these glycosides can increase, thus producing digitalis toxicity (e.g., fatigue, confusion, gastrointestinal problems, arrhythmias).

Digitalis and diuretics in combination with NSAIDs exacerbate the side effects of NSAIDs. Anyone receiving any of these combinations must be monitored for lower-extremity (especially ankle) and abdominal swelling.

Diuretics

Diuretics, usually referred to by clients as "water pills," lower BP by eliminating sodium and water and thus reducing the blood volume. Thiazide diuretics may also be used to prevent osteoporosis by increasing calcium reabsorption by the kidneys. Some diuretics remove potassium from the body, causing potentially life-threatening arrhythmias.

The primary adverse effects associated with diuretics are fluid and electrolyte imbalances such as muscle weakness and spasms, dizziness, headache, incoordination, and nausea (Box 6-3).

Beta-Blockers

Beta-blockers relax the blood vessels and the heart muscle by blocking the beta receptors on the sinoatrial node and myocardial cells, producing a decline in the force of contraction and a reduction in heart rate. This effect eases the strain on the heart by reducing its workload and reducing oxygen consumption.

The therapist must monitor the client's perceived exertion and watch for excessive slowing of the heart rate (bradycardia) and contractility, resulting in depressed cardiac function. Other potential side effects include depression, worsening of asthma symptoms, sexual dysfunction, and fatigue. The generic names of beta-blockers end in "olol" (e.g., propranolol, metoprolol, atenolol). Trade names include Inderal, Lopressor, and Tenormin.

Alpha-1 Blockers

Alpha-1 blockers lower the BP by dilating blood vessels. The therapist must be observant for signs of hypotension and reflex tachycardia (i.e., the heart rate increases to compensate for the hypotension). The generic names of alpha-1 blockers end in "zosin" (e.g., prazosin, terazosin, doxazosin; trade names include Minipress, Hytrin, Cardura).

ACE Inhibitors

Angiotensin-converting enzyme (ACE) inhibitors are highly selective drugs that interrupt a chain of molecular messengers that constrict blood vessels. They can improve cardiac function in individuals with heart failure and are used for persons with diabetes or early kidney damage. Rash and a persistent dry cough are common side effects. The generic names of ACE inhibitors end in "pril" (e.g., benazepril, captopril, enalapril, lisinopril). Trade names include Lotensin, Capoten, Vasotec, Prinivil, and Zestril. Newest on the market are ACE II inhibitors, such as Cozaar (losartan potassium) and Hyzaar (losartan potassium-hydrochlorothiazide).

Calcium Channel Blockers

Calcium channel blockers inhibit calcium from entering the blood vessel walls, where calcium works to constrict blood vessels. Side effects may include swelling in the feet and ankles, orthostatic hypotension, headache, and nausea.

There are several groups of calcium channel blockers. Those in the group that primarily interact with calcium channels on the smooth muscle of the peripheral arterioles all end with "pine" (e.g., amlodipine, felodipine, nisoldipine, nifedipine). Trade names include Norvasc, Plendil, Sular, and Adalat or Procardia.

A second group of calcium channel blockers works to dilate coronary arteries to lower BP and suppress some arrhythmias. This group includes verapamil (Verelan, Calan, Isoptin) and diltiazem (Cardizem, Dilacor).

Nitrates

Nitrates, such as nitroglycerin (e.g., nitroglycerin [Nitrostat, Nitro-Bid], isosorbide dinitrate [Iso-Bid, Isordil]), dilate the coronary arteries and are used to prevent or relieve the symptoms of angina. Headache, dizziness, tachycardia, and orthostatic hypotension may occur as a result of the vasodilating properties of these drugs.

There are other classes of drugs to treat various aspects of cardiovascular diseases separate from those listed in Table 6-8. Hyperlipidemia is often treated with medications to inhibit cholesterol synthesis. Platelet aggregation and clot formation are prevented with anticoagulant drugs, such as heparin, warfarin (Coumadin), and aspirin, whereas thrombolytic drugs, such as streptokinase, urokinase, and tissue-type plasminogen activator (t-PA), are used to break down and dissolve clots already formed in the coronary arteries.

Anyone receiving cardiovascular medications, especially in combination with other medications or OTC drugs, must be monitored during physical therapy for red flag signs and symptoms and any unusual vital signs.

The therapist should be familiar with the signs or symptoms that require immediate physician referral and those that must be reported to the physician. Special Questions to Ask: Medications are available at the end of this chapter.

PHYSICIAN REFERRAL

Referral by the therapist to the physician is recommended when the client has any combination of systemic signs or symptoms discussed throughout this chapter at presentation. These signs and symptoms should always be correlated with the client's history to rule out systemic involvement or to identify musculoskeletal or neurologic disorders that would be appropriate for physical therapy intervention.

Clients often confide in their therapists and describe symptoms of a more serious nature. Cardiac symptoms unknown to the physician may be mentioned to the therapist during the opening interview or in subsequent visits.

The description and location of chest pain associated with pericarditis, MI, angina, breast pain, gastrointestinal disorders, and anxiety are often similar. The physician is able to distinguish among these conditions through a careful history, medical examination, and medical testing.

For example, compared with angina, the pain of true musculoskeletal disorders may last for seconds or for hours, is not relieved by nitroglycerin, and may be aggravated by local palpation or by exertion of just the upper body.

It is not the therapist's responsibility to differentiate diagnostically among the various causes of chest pain, but rather to recognize the systemic origin of signs and symptoms that may mimic musculoskeletal disorders.

The physical therapy interview presented in Chapter 2 is the primary mechanism used to begin exploring a client's reported symptoms; this is accomplished by carefully questioning the client to determine the location, duration, intensity, frequency, associated symptoms, and relieving or aggravating factors related to pain or symptoms.

Guidelines for Immediate Medical Attention

Sudden worsening of intermittent claudication may be due to thromboembolism and must be reported to the physician immediately. Symptoms of TIAs in any individual, especially those with a history of heart disease, hypertension, or tobacco use, warrant immediate medical attention.

In the clinic setting, the onset of an anginal attack requires immediate cessation of exercise. Symptoms associated with angina may be reduced immediately but should subside within 3 to 5 minutes of cessation of activity.

If the client is currently taking nitroglycerin, self-administration of medication is recommended. Relief from anginal pain should occur within 1 to 2 minutes of nitroglycerin administration; some women may obtain similar results with an antacid. The nitroglycerin may be repeated according to the prescribed directions. If anginal pain is not relieved in 20 minutes or if the client has nausea, vomiting, or profuse sweating, immediate medical intervention may be indicated.

Changes in the pattern of angina, such as increased intensity, decreased threshold of stimulus, or longer duration of pain, require immediate intervention by the physician. Pain associated with an MI is not relieved by rest, change of position, or administration of nitroglycerin or antacids.

Clients in treatment under these circumstances should either be returned to the care of the nursing staff or, in the case of an outpatient, should be encouraged to contact their physicians by telephone for further instructions before leaving the physical therapy department. The client should be advised not to leave unaccompanied.

Guidelines for Physician Referral

- When a client has any combination of systemic signs or symptoms at presentation, refer him or her to a physician.
- Women with chest or breast pain who have a positive family history of breast cancer or heart disease should always be referred to a physician for a follow up examination.
- Palpitation in any person with a history of unexplained sudden death in the family requires medical evaluation. More than 6 episodes of palpitations in 1 minute or palpitations lasting for hours or occurring in association with pain, shortness of breath, fainting, or severe lightheadedness require medical evaluation.
- Anyone who cannot climb a single flight of stairs without feeling moderately to severely winded or who awakens at night or experiences shortness of breath when lying down should be evaluated by a physician.
- Fainting (syncope) without any warning period of lightheadedness, dizziness, or nausea may be a sign of heart valve or arrhythmia problems. Unexplained syncope in the presence of heart or circulatory problems (or risk factors for heart attack or stroke) should be evaluated by a physician.
- Clients who are neurologically unstable as a result of a recent CVA, head trauma, spinal cord injury, or other central nervous system insult often exhibit new arrhythmias during the period of instability. When the client's pulse is monitored, any new arrhythmias noted should be reported to the nursing staff or the physician.
- Cardiac clients should be sent back to their physician under the following conditions:
- Nitroglycerin tablets do not relieve anginal pain.
- Pattern of angina changes is noted.
- Client has abnormally severe chest pain with nausea and vomiting.
- Anginal pain radiates to the jaw or to the left arm.
- Anginal pain is not relieved by rest.
- Upper back feels abnormally cool, sweaty, or moist to touch.
- Client develops progressively worse dyspnea.
- Individual with coronary artery stent experiencing chest pain.
- Client demonstrates a difference of more than 40 mm Hg in pulse pressure (systolic BP minus diastolic BP = pulse pressure).
- Client has any doubt about his or her present condition.

Clues to Screening for Cardiovascular Signs and Symptoms

Whenever assessing chest, breast, neck, jaw, back, or shoulder pain for cardiac origins, look for the following clues:

- Personal or family history of heart disease including hypertension
- Age (postmenopausal woman; anyone over 65)
- Ethnicity (Black women)
- Other signs and symptoms such as pallor, unexplained profuse perspiration, inability to talk, nausea, vomiting, sense of impending doom or extreme anxiety
- Watch for the three Ps.
    1. Pleuritic pain (exacerbated by respiratory movement involving the diaphragm, such as sighing, deep breathing, coughing, sneezing, laughing, or the hiccups; this may be cardiac if pericarditis or it may be pulmonary); have the client hold his or her breath and reassess symptoms—any reduction or elimination of symptoms with breath holding or the Valsalva maneuver suggests pulmonary or cardiac source of symptoms.
    2. Pain on palpation (musculoskeletal origin).
    3. Pain with changes in position (musculoskeletal or pulmonary origin; pain that is worse when lying down and improves when sitting up or leaning forward is often pleuritic in origin).
- If two of the three P's are present, an MI is very unlikely. An MI or anginal pain occurs in approximately 5% to 7% of clients whose pain is reproducible by palpation. If the symptoms are altered by a change in positioning, this percentage drops to 2%, and if the chest pain is reproducible by respiratory movements, the likelihood of a coronary event is only 1%.[72]
- Chest pain may occur from intercostal muscle or periosteal trauma with protracted or vigorous coughing. Palpation of local chest wall will reproduce tenderness. However, a client can have both a pulmonary/cardiac condition with subsequent musculoskeletal trauma from coughing. Look for associated signs and symptoms (e.g., fever, sweats, blood in sputum).
- Angina is activated by physical exertion, emotional reactions, a large meal, or exposure to cold and has a lag time of 5 to 10 minutes. Angina does not occur immediately

after physical activity. Immediate pain with activity is more likely musculoskeletal, thoracic outlet syndrome, or psychologic (e.g., "I do not want to shovel today").

- Chest pain, shoulder pain, neck pain, or TMJ pain occurring in the presence of coronary artery disease or previous history of MI, especially if accompanied by associated signs and symptoms, may be cardiac.
- Upper quadrant pain that can be induced or reproduced by lower quadrant activity, such as biking, stair climbing, or walking without using the arms, is usually cardiac in origin.
- Recent history of pericarditis in the presence of new onset of chest, neck, or left shoulder pain; observe for additional symptoms of dyspnea, increased pulse rate, elevated body temperature, malaise, and myalgia(s).
- If an individual with known risk factors for congestive heart disease, especially a history of angina, becomes weak or short of breath while working with the arms extended over the head, ischemia or infarction is a likely cause of the pain and associated symptoms.
- Insidious onset of joint or muscle pain in the older client who has had a previously diagnosed heart murmur may be caused by bacterial endocarditis. Usually there is no morning stiffness to differentiate it from rheumatoid arthritis.
- Back pain similar to that associated with a herniated lumbar disk but without neurologic deficits especially in the presence of a diagnosed heart murmur, may be caused by bacterial endocarditis.
- Watch for arrhythmias in neurologically unstable clients (e.g., spinal cord, new CVAs, or new traumatic brain injuries [TBIs]); check pulse and ask about/observe for dizziness.
- Anyone with chest pain must be evaluated for trigger points. If palpation of the chest reproduces symptoms, especially symptoms of radiating pain, deactivation of the trigger points must be carried out and followed by a reevaluation as part of the screening process for pain of a cardiac origin (see Fig. 17-7 and Table 17-4).
- Symptoms of vascular occlusive disease include exertional calf pain that is relieved by rest (intermittent claudication), nocturnal aching of the foot and forefoot (rest pain), and classic skin changes, especially hair loss of the ankle and foot. Ischemic rest pain is relieved by placing the limb in a dependent position.
- Throbbing pain at the base of the neck and/or along the back into the interscapular areas that increases with exertion requires monitoring of vital signs and palpation of peripheral pulses to screen for aneurysm. Check for a palpable abdominal heartbeat that increases in the supine position.
- See also section on clues to differentiating chest pain in Chapter 17.

CARDIAC CHEST PAIN PATTERNS

ANGINA (FIG. 6-8)

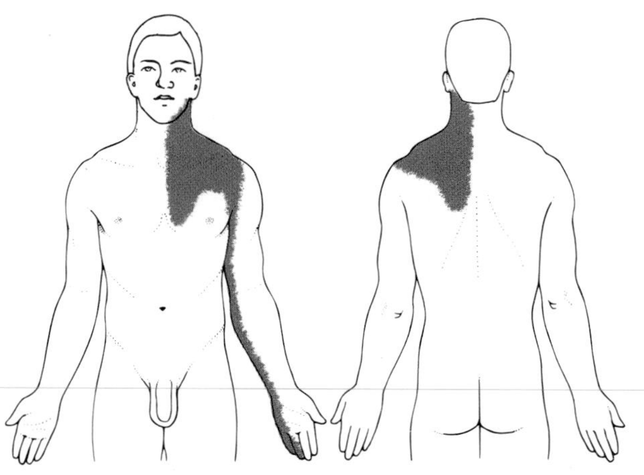

Fig. 6-8 Pain patterns associated with angina. *Left,* Area of substernal discomfort projected to the left shoulder and arm over the distribution of the ulnar nerve. Referred pain may be present only in the left shoulder or in the shoulder and along the arm only to the elbow. *Right,* Occasionally, anginal pain may be referred to the back in the area of the left scapula or the interscapular region. Women can have the same patterns as shown for men in this figure or they may present as shown in Fig. 6-4. There may be no pain but rather a presenting symptom of extreme fatigue, weakness, or breathlessness.

Continued

CARDIAC CHEST PAIN PATTERNS—*cont'd*

Location:	Substernal/retrosternal (beneath the sternum)
	Left chest pain in the absence of substernal chest pain (women)
	Isolated midthoracic back pain (women)
	Aching in one or both upper arm (biceps)
Referral:	Neck, jaw, back, shoulder, or arms (most commonly the left arm)
	May have only a toothache
	Occasionally to the abdomen
Description:	Viselike pressure, squeezing, heaviness, burning indigestion
Intensity*:	Mild to moderate
	Builds up gradually or may be sudden
Duration:	Usually less than 10 minutes
	Never more than 30 minutes
	Average: 3-5 minutes
Associated signs and symptoms:	Extreme fatigue, lethargy, weakness (women)
	Shortness of breath (dyspnea)
	Nausea
	Diaphoresis (heavy perspiration)
	Anxiety or apprehension
	Belching (eructation)
	"Heartburn" (unrelieved by antacids) (women)
	Sensation similar to inhaling cold air (women)
	Prolonged and repeated palpitations without chest pain (women)
Relieving factors:	Rest or nitroglycerin
	Antacids (women)
Aggravating factors:	Exercise or physical exertion
	Cold weather or wind
	Heavy meals
	Emotional stress

*For each pattern reviewed, intensity is related directly to the degree of noxious stimuli.

CARDIAC CHEST PAIN PATTERNS—*cont'd*

MYOCARDIAL INFARCTION (FIG. 6-9)

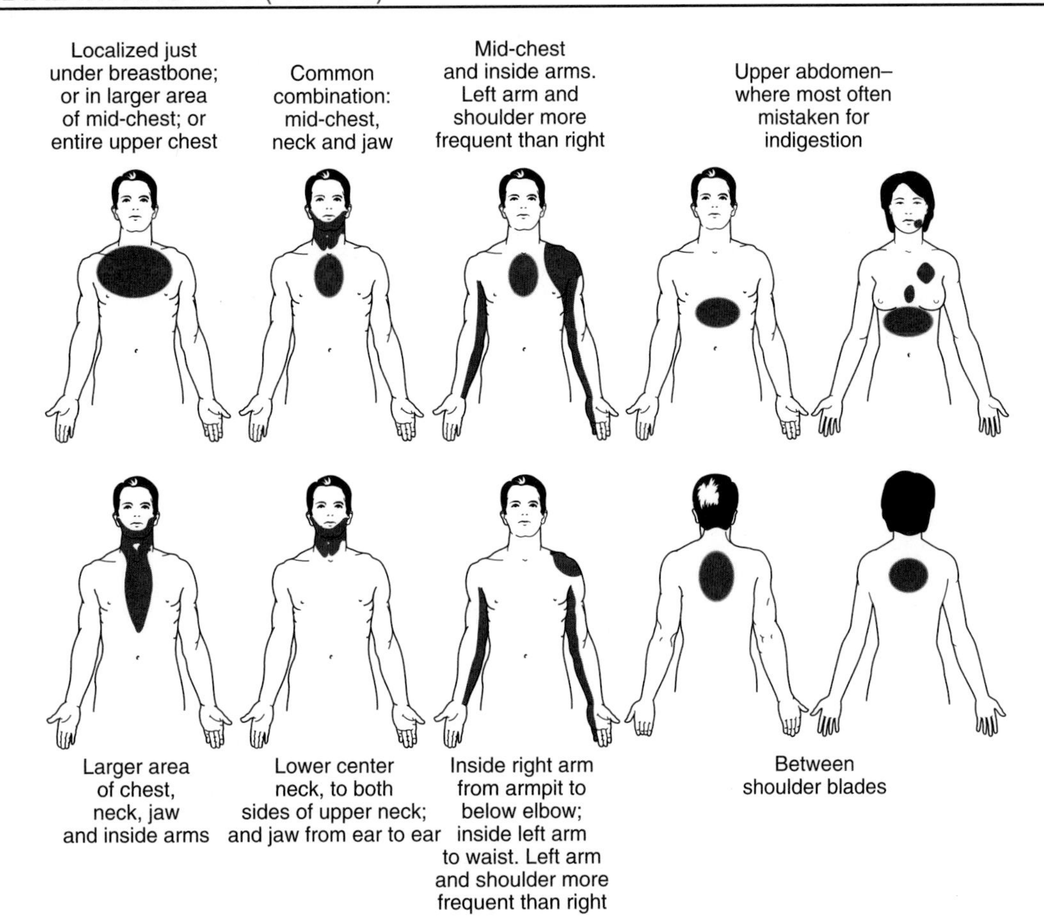

Most common warning signs of heart attack

- Uncomfortable pressure, fullness, squeezing or pain in the center of the chest (prolonged)
- Pain that spreads to the throat, neck, back, jaw, shoulders, or arms
- Chest discomfort with lightheadedness, dizziness, sweating, pallor, nausea, or shortness of breath
- Prolonged symptoms unrelieved by antacids, nitroglycerin, or rest

Atypical, less common warning signs (especially women)

- Unusual chest pain (quality, location, e.g., burning, heaviness; left chest), stomach or abdominal pain
- Continuous midthoracic or interscapular pain
- Continuous neck or shoulder pain (not shown in Fig. 6-9)
- Pain relieved by antacids; pain unrelieved by rest or nitroglycerin
- Nausea and vomiting; flu-like manifestation without chest pain/discomfort
- Unexplained intense anxiety, weakness, or fatigue
- Breathlessness, dizziness

Fig. 6-9 Early warning signs of a heart attack. Multiple segmental nerve innervations shown in Fig. 4-3 account for varied pain patterns possible. A woman can experience any of the various patterns described but is just as likely to develop atypical symptoms of pain as depicted here. (From Goodman CC, Fuller K: *Pathology: implications for the physical therapist,* ed 3, Philadelphia, 2009, WB Saunders.)

Location: Substernal, anterior chest
Referral: May radiate like angina, frequently down both arms
Description: Burning, stabbing, viselike pressure, squeezing, heaviness
Intensity: Severe
Duration: Usually at least 30 minutes; may last 1 to 2 hours
Residual soreness 1 to 3 days

Continued

CARDIAC CHEST PAIN PATTERNS—*cont'd*

Associated signs and symptoms: None with a silent MI
Dizziness, feeling faint
Nausea, vomiting
Pallor
Diaphoresis (heavy perspiration)
Apprehension, severe anxiety
Fatigue, sudden weakness
Dyspnea
May be followed by painful shoulder-hand syndrome (see text)

Relieving factors: None; unrelieved by rest or nitroglycerin taken every 5 minutes for 20 minutes

Aggravating factors: Not necessarily anything; may occur at rest or may follow emotional stress or physical exertion

PERICARDITIS (FIG. 6-10)

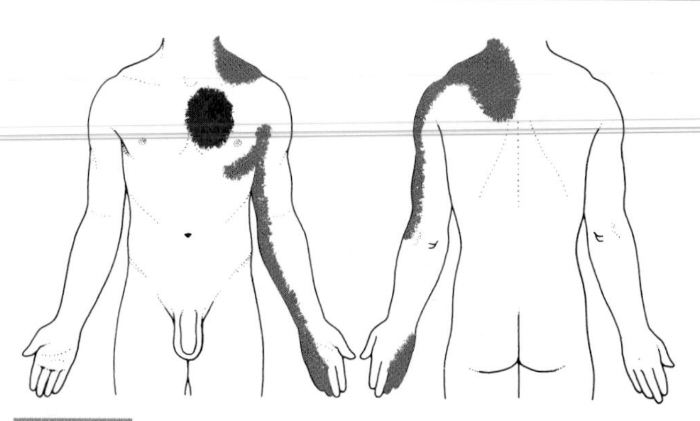

Fig. 6-10 Substernal pain associated with pericarditis *(dark red)* may radiate anteriorly *(light red)* to the costal margins, neck, upper back, upper trapezius muscle, and left supraclavicular area or down the left arm.

Location: Substernal or over the sternum, sometimes to the left of midline toward the cardiac apex

Referral: Neck, upper back, upper trapezius muscle, left supraclavicular area, down the left arm, costal margins

Description: More localized than pain of MI
Sharp, stabbing, knifelike

Intensity: Moderate-to-severe

Duration: Continuous; may last hours or days with residual soreness following

Associated signs and symptoms: Usually medically determined associated symptoms (e.g., by chest auscultation using a stethoscope); cough

Relieving factors: Sitting upright or leaning forward

Aggravating factors: Muscle movement associated with deep breathing (e.g., laughter, inspiration, coughing)
Left lateral (side) bending of the upper trunk
Trunk rotation (either to the right or to the left)
Supine position

GASTROINTESTINAL PAIN PATTERNS—*cont'd*

DISSECTING AORTIC ANEURYSM (FIG. 6-11)

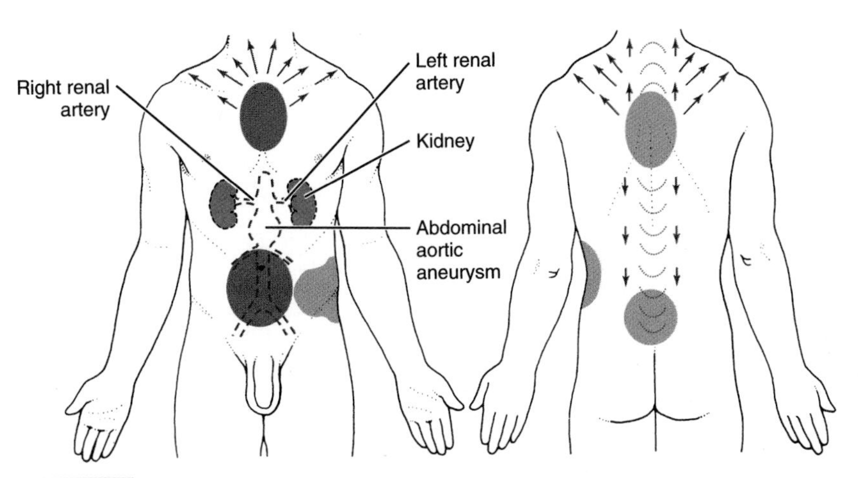

Fig. 6-11 Most aortic aneurysms (more than 95%) are located just below the renal arteries and extend to the umbilicus, causing low back pain. Chest pain *(dark red)* associated with thoracic aneurysms may radiate *(see arrows)* to the neck, interscapular area, shoulders, lower back, or abdomen. Early warning signs of an impending rupture may include an abdominal heartbeat when lying down (not shown) or a dull ache in the midabdominal left flank or lower back *(light red).*

Location:	Anterior chest (thoracic aneurysm) Abdomen (abdominal aneurysm) Thoracic area of back
Referral:	Pain may move in the chest as dissection progresses Pain may extend to the neck, shoulders, interscapular area, or lower back
Description:	Knifelike, tearing (thoracic aneurysm) Dull ache in the lower back or midabdominal left flank (abdominal aneurysm)
Intensity:	Severe, excruciating
Duration:	Hours
Associated signs and symptoms:	Pulses absent Person senses "heartbeat" when lying down Palpable, pulsating abdominal mass Lower BP in one arm Other medically determined symptoms
Relieving factors:	None
Aggravating factors:	Supine position accentuates symptoms

NONCARDIAC CHEST PAIN PATTERNS

- Musculoskeletal disorders, see Chapters 14-18.
- Neurologic disorders, see Chapters 14-18.
- Pleuropulmonary disorders, see Chapter 7.
- Gastrointestinal disorders, see Chapter 8.
- Breast diseases, see Chapter 17.
- Anxiety states, see Chapter 3.

Key Points to Remember

- Fatigue beyond expectations during or after exercise is a red-flag symptom.
- Be on the alert for cardiac risk factors in older adults, especially women, and begin a conditioning program before an exercise program.
- The client with stable angina typically has a normal BP; it may be low, depending on medications. Blood pressure may be elevated when anxiety accompanies chest pain or during acute coronary insufficiency; systolic BP may be low if there is heart failure.
- Cervical disk disease and arthritic changes can mimic atypical chest pain of angina pectoris, requiring screening through questions and musculoskeletal evaluation.
- If a client uses nitroglycerin, make sure that he or she has a fresh supply, and check that the physical therapy department has a fresh supply in a readily accessible location.
- Anyone being treated with both NSAIDs and ACE inhibitors must be monitored closely during exercise for elevated BP.
- A person taking medications, such as beta-blockers or calcium channel blockers, may not be able to achieve a target heart rate (THR) above 90 bpm. To determine a safe rate of exercise, the heart rate should return to the resting level 2 minutes after stopping exercise.
- Make sure that a client with cardiac compromise has not smoked a cigarette or eaten a large meal just before exercise.
- A 3-pound or greater weight gain or gradual, continuous gain over several days, resulting in swelling of the ankles, abdomen, and hands, combined with shortness of breath, fatigue, and dizziness that persist despite rest, may be red-flag symptoms of CHF.
- The pericardium (sac around the entire heart) is adjacent to the diaphragm. Pain of cardiac and diaphragmatic origin is often experienced in the shoulder because the heart and the diaphragm are supplied by the C5-6 spinal segment. The visceral pain is referred to the corresponding somatic area.
- Watch for muscle pain, cramps, stiffness, spasms, and weakness that cannot be explained by arthritis, recent strenuous exercise, a fever, a recent fall, or other common causes in clients taking statins to lower cholesterol.

SUBJECTIVE EXAMINATION

Special Questions to Ask

Past Medical History

- Has a doctor ever said that you have heart trouble? High BP?
- Have you ever had a heart attack?
  If yes, when? Please describe.
- Do you associate your current symptoms with your heart problems?
- Have you ever had rheumatic fever, twitching of the limbs called St. Vitus' dance, or rheumatic heart disease?
- Have you ever had an abnormal electrocardiogram (ECG)?
- Have you ever had an ECG taken while you were exercising (e.g., climbing up and down steps or walking on a treadmill) that was not normal?
- Do you have a pacemaker, artificial heart, or any other device to assist your heart?
- For the therapist: Remember to review smoking, diet, lifestyle, exercise, and stress history (see family/personal history, Chapter 2).

Angina/Myocardial Infarct

- Do you have angina (pectoris) or chest pain or tightness?
  If yes, please describe the symptoms and tell me when it occurs.
  If no, have you ever had chest pain, dizziness, or shortness of breath during or after activity, exercise, or sport?
- Can you point to the area of pain with one finger? (**Anginal pain is characteristically demonstrated with the hand or fist on the chest.**)
- Is the pain close to the surface or deep inside? (**Pleuritic pain is close to the surface; anginal pain can be close to the surface but always also has a "deep inside" sensation.**)
  If yes, what makes it better?
  If no, pursue further with the following questions:
- Do you ever have discomfort or tightness in your chest?
- Have you ever had a crushing sensation in your chest with or without pain down your left arm?
- Do you have pain in your jaw either alone or in combination with chest pain?
- If you climb a few flights of stairs fairly rapidly, do you have tightness or pressing pain in your chest?
- Do you get pressure or pain or tightness in the chest as if you were walking in the cold wind or facing a cold blast of air?
- Have you ever had pain or pressure or a squeezing feeling in the chest that occurred during exercise, walking, or any other physical or sexual activity?
- Have you been unusually tired lately (**possible new onset of angina in women**)?

SUBJECTIVE EXAMINATION—*cont'd*

- Do you get tired faster than others doing the same things?
- Has anyone in your family ever had or died from heart problems?
  If yes, did he/she die suddenly before age 50?

Associated Symptoms

- Do you ever have bouts of rapid heart action, irregular heartbeats, or palpitations of your heart?
- Have you ever felt a "heartbeat" in your abdomen when you lie down? *If yes,* is this associated with low back pain or left flank pain? **(Abdominal aneurysm)**
- Do you ever notice sweating, nausea, or chest pain when your current symptoms (e.g., back pain, shoulder pain) occur?
- Do you have frequent attacks of heartburn, or do you take antacids to relieve heartburn or acid indigestion? **(Non-cardiac cause of chest pain [men], abdominal muscle trigger point, gastrointestinal disorder)**
- Do you get very short of breath during activities that do not make other people short of breath? **(Dyspnea)**
- Do you ever wake up at night gasping for air or have short breaths? **(Paroxysmal nocturnal dyspnea)**
- Do you ever need to sleep on more than one pillow to breathe comfortably? **(Orthopnea)**
- Do you ever get cramps in your legs if you walk for several blocks? **(Intermittent claudication)**
- Do you ever have swollen feet or ankles? *If yes,* are they swollen when you get up in the morning? **(Edema/CHF; NSAIDs)**
- Have you gained unexpected weight during a fairly short period of time (i.e., less than 1 week)? **(Edema, CHF)**
- Do you ever feel dizzy or have fainting spells? **(Valvular insufficiency, bradycardia, pulmonary hypertension, orthostatic hypotension)**
- Have you had any significant changes in your urine (e.g., increased amount, concentrated urine, frequency at night, or decreased amount)? **(CHF, diabetes, hypertension)**
- Do you ever have sudden difficulty with speech, temporary blindness, or other changes in your vision? **(TIAs)**
- Have you ever had sudden weakness or paralysis down one side of your body or just in an arm or a leg? **(TIAs)**

Medications

- Have you ever taken digitalis, nitroglycerin, or any other drug for your heart?
- Have you been on a diet or taken medications to lower your blood cholesterol?
- For the therapist: Any clients taking anticlotting drugs should be examined for hematoma, nosebleed, or other sites of bleeding. Protect client from trauma.
- Anyone taking cardiovascular medications (especially ACE inhibitors or digitalis glycosides) in combination with NSAIDs must be monitored closely (see text explanation).
- Any woman older than 35 and taking oral contraceptives who has a history of smoking should be monitored for increases in BP.
- Any woman taking third-generation oral contraceptives should be monitored for venous thrombosis.

For clients taking nitroglycerin

- Do you ever have headaches, dizziness, or a flushed sensation after taking nitroglycerin? (Most common side effects)
- How quickly does your nitroglycerin reduce or eliminate your chest pain? (Use as a guideline in the clinic when the client has angina during exercise; refer to a physician if angina is consistently unrelieved with nitroglycerin or rest after the usual period of time.)

For clients with breast pain (see questions, Chapter 17)

For clients with joint pain

- Have you had any recent skin rashes or dot-like hemorrhages under the skin? **(Rheumatic fever, endocarditis)**
  If yes, did this occur after a visit to the dentist? **(Endocarditis)**
- Do you notice any increase in your joint pain or symptoms 1 to 2 hours after you take your medication? **(Allergic response)**
- For new onset of left upper trapezius muscle/left shoulder pain: Have you been treated for any infection in the last 3 weeks?

CASE STUDY

REFERRAL

A 30-year-old woman with 5 children comes to you for an evaluation on the recommendation of her friend, who received physical therapy from you last year. She has not been to a physician since her last child was delivered by her obstetrician 4 years ago.

Her chief complaint is pain in the left shoulder and left upper trapezius muscle with pain radiating into the chest and referred pain down the medial aspect of the arm to the thumb and first two fingers.

When the medical history is being taken, the client mentions that she was told 5 years ago that she had mitral valve prolapse (MVP) secondary to rheumatic fever, which she had when she was 12 years old. She is not taking any medication, denies any palpitations, but she complains of fatigue and has dyspnea after playing ball with her son for 10 or 15 minutes.

There is no reported injury or trauma to the neck or shoulder, and the symptoms subside with rest. Physical exertion, such as carrying groceries up the stairs or laundry outside, aggravates the symptoms, but she is uncertain whether just using her upper body has the same effect.

Despite the client's denial of injury or trauma, the neck and shoulder should be screened for any possible musculoskeletal or neurologic origin of symptoms. Your observation of the woman indicates that she is 30 to 40 pounds overweight. She confides that she is under physical and emotional stress by the daily demands made by seven people in her house.

She is not involved in any kind of exercise program outside of her play activities with the children. These two factors (obesity and stress) could account for her chronic fatigue and dyspnea, but that determination must be made by a physician. Even if you can identify a musculoskeletal basis for this woman's symptoms, the past medical history of rheumatic heart disease and absence of medical follow up would support your recommendation that the client should go to the physician for a medical checkup.

How do you rule out the possibility that this pain is not associated with MVP and is caused instead by true cervical spine or shoulder pain?

It should be pointed out here that the therapist is not equipped with the skills, knowledge, or expertise to determine that the MVP is the cause of the client's symptoms.

However, a thorough subjective and objective evaluation can assist the therapist both in making a determination regarding the client's musculoskeletal condition and in providing clear and thorough feedback for the physician on referral.

SCREENING FOR MITRAL VALVE PROLAPSE

- Pain of MVP must be diagnosed by a physician.
- MVP may be asymptomatic.
- Positive history for rheumatic fever.
- Carefully ask the client about a history of possible neck or shoulder pain, which the person may not mention otherwise.
- Musculoskeletal pain associated with the neck or shoulder is more superficial than cardiac pain.
- Total body exertion causing shoulder pain may be secondary to angina or myocardial ischemia and subsequent infarction, whereas movements of just the upper extremity causing shoulder pain are more indicative of a primary musculoskeletal lesion.
- Ask the client: Does your shoulder pain occur during exercise, such as walking, climbing stairs, mowing the lawn, or during any other physical or sexual activity that does not require the use of your arm or shoulder?
- Presence of associated signs and symptoms such as dyspnea, fatigue, or heart palpitations.
- X-ray findings, if available, may confirm osteophyte formation with decreased intraforaminal spaces, which may contribute to cervical spine pain.
- History of neck injury or overuse.
- History of shoulder injury or overuse.
- Results of objective tests to clear or rule out the cervical spine and shoulder as the cause of symptoms.
- Presence of other neurologic signs to implicate the cervical spine or thoracic outlet type of symptoms (e.g., abnormal deep tendon reflexes, subjective report of numbness and tingling, objective sensory changes, muscle wasting or atrophy).
- Pattern of symptoms; a change in position may relieve symptoms associated with a cervical disorder.

PRACTICE QUESTIONS

1. Pursed-lip breathing in the sitting position while leaning forward on the arms relieves symptoms of dyspnea for the client with:
   a. Orthopnea
   b. Emphysema
   c. CHF
   d. a and c
2. Briefly describe the difference between myocardial ischemia, angina pectoris, and MI.
3. What should you do if a client complains of throbbing pain at the base of the neck that radiates into the interscapular areas and increases with exertion?
4. What are the 3Ps? What is the significance of each one?
5. When are palpitations clinically significant?
6. A 48-year-old woman with TMJ syndrome has been referred to you by her dentist. How do you screen for the possibility of medical (specifically cardiac) disease?
7. A 55-year-old male grocery store manager reports that he becomes extremely weak and breathless when he is stocking groceries on overhead shelves. What is the possible significance of this complaint?
8. You are seeing an 83-year-old woman for a home health evaluation after a motor vehicle accident (MVA) that required a long hospitalization followed by transition care in an intermediate care nursing facility and now home health care. She is ambulating short distances with a wheeled walker, but she becomes short of breath quickly and requires lengthy rest periods. At each visit the client is wearing her slippers and housecoat, so you suggest that she start dressing each day as if she intended to go out. She replies that she can no longer fit into her loosest slacks and she cannot tie her shoes. Is there any significance to this client's comments, or is this consistent with her age and obvious deconditioning? Briefly explain your answer.
9. Peripheral vascular diseases include:
   a. Arterial and occlusive diseases
   b. Arterial and venous disorders
   c. Acute and chronic arterial diseases
   d. All of the above
   e. None of the above
10. Which statement is the most accurate?
    a. Arterial disease is characterized by intermittent claudication, pain relieved by elevating the extremity, and history of smoking.
    b. Arterial disease is characterized by loss of hair on the lower extremities and throbbing pain in the calf muscles that goes away by using heat and elevation.
    c. Arterial disease is characterized by painful throbbing of the feet at night that goes away by dangling the feet over the bed.
    d. Arterial disease is characterized by loss of hair on the toes, intermittent claudication, and redness or warmth of the legs that is accompanied by a burning sensation.
11. What are the primary signs and symptoms of CHF?
    a. Fatigue, dyspnea, edema, nocturia
    b. Fatigue, dyspnea, varicose veins
    c. Fatigue, dyspnea, tinnitus, nocturia
    d. Fatigue, dyspnea, headache, night sweats
12. When would you advise a client in physical therapy to take his/her nitroglycerin?
    a. 45 minutes before exercise
    b. When symptoms of chest pain do not subside with 10 to 15 minutes of rest
    c. As soon as chest pain begins
    d. None of the above
    e. All of the above

REFERENCES

1. Heart disease & stroke statistics—2009: a report from the American Heart Association Statistics Committee and Stroke Statistics Subcommittee. *Circulation* 119:e21–e181, 2009.
2. Aspinall W: Clinical testing for the craniovertebral hypermobility syndrome. *J Orthop Sports Phys Ther* 12:180–181, 1989.
3. Magee DJ: *Orthopedic physical assessment*, ed 5, Philadelphia, 2008, WB Saunders.
4. Childs JD, Flynn TW, Fritz JM, et al: Screening for vertebrobasilar insufficiency in patients with neck pain: manual therapy decision-making in the presence of uncertainty. *J Orthop Sports Phys Ther* 35(5):300–306, 2005.
5. Rivett DA: The premanipulative vertebral artery testing protocol: a brief review. *Physiotherapy* 23:9–12, 1995.
6. Thiel H, Wallace K, Donut J, et al: Effect of various head and neck positions on vertebral artery blood flow. *Clin Biomech* 9:105–110, 1994.
7. Childs JD: Screening for vertebrobasilar insufficiency in patients with neck pain: manual therapy decision-making in the presence of uncertainty. *J Orthop Sports Phys Ther* 35(5): 300–306, 2005.
8. Richter RR, Reinking MF: How does evidence on the diagnostic accuracy of the vertebral artery test influence teaching of the test in a professional physical therapist education program? *Phys Ther* 85(6):589–599, 2005.
9. Trumbore DJ: *Statins and myalgia: a case report of pharmacovigilance with implications for physical therapy case report presented in partial fulfillment of DPT 910, Principles of Differential Diagnosis, Institute for Physical Therapy Education*, Chester, Pennsylvania, 2005, Widener University.
10. Zhao H, Thomas G, Leung Y, et al: Statins in lipid-lowering therapy. *Acta Cardiol Sin* 19:1–11, 2003.
11. Rosenson RS: Current overview of statin-induced myopathy. *Am J Med* 116:408–416, 2004.
12. Roten L, Schoenenberger RA, Krahenbuhl S, et al: Rhabdomyolysis in association with simvastatin and amiodarone. *Ann Pharmacother* 38:978–981, 2004.
13. Tomlinson S, Mangione K: Potential adverse effects of statins on muscle: update. *Phys Ther* 85(5):459–465, 2005.
14. Pasternak RC, Smith SC, Bairey-Merz CN, et al: ACC/AHA/NHLBI clinical advisory on the use and safety of statins. *Circulation* 106:1024, 2002.
15. Mills EJ: Efficacy and safety of statin treatment for cardiovascular disease: a network meta-analysis of 170,255 patients from 76 randomized trials, *QJM* Oct. 7, 2010. Epub ahead of print.
16. Chatham K: Suspected statin-induced respiratory muscle myopathy during long-term inspiratory muscle training in a

patient with diaphragmatic paralysis. *Phys Ther* 89:257–266, 2009.
17. Tomlinson SS: Potential adverse effects of statins on muscle. *Phys Ther* 85:459–465, 2005.
18. Bruckert E: Mild to moderate muscular symptoms with high-dosage statin therapy in hyperlipidemic patients: the PRIMO study. *Cardiovasc Drugs Ther* 19:403–414, 2005.
19. Mann D: Trends in statin use and low-density lipoprotein cholesterol levels among US adults. *Ann Pharmacother* 42: 1208–1215, 2008.
20. Buettner C: Prevalence of musculoskeletal pain and statin use. *J Gen Intl Medicine* 23:1182–1186, 2008.
21. Baxter R, Moore J: Diagnosis and treatment of acute exertional rhabdomyolysis. *J Orthop Sports Phys Ther* 33(3):104–108, 2003.
22. Evans M, Rees A: Effects of HMG-CoA reductase inhibitors on skeletal muscle: are all statins the same? *Drug Saf* 25:649–663, 2002.
23. Using Crestor—and all statins—safely, *Harvard Heart Letter*, September 2005; p 3. More information available at www.health.harvard.edu/heartextra. Accessed: October 17, 2010.
24. Cholesterol drugs: very safe and highly beneficial. *Johns Hopkins Medical Letter: Health After 50* 13(12):3, 2002.
25. DiStasi SL: Effects of statins on skeletal muscle: a perspective for physical therapists. *Phys Ther* 90(10):1530–1542, 2010.
26. Dobkin BH: Underappreciated statin-induced myopathic weakness causes disability. *Neurorehabil Neural Repair* 19:259–263, 2005.
27. Kelln BM: Hand-held dynamometry: reliability of lower extremity muscle testing in healthy, physically active, young adults. *J Sport Rehab* 17:160–170, 2008.
28. Prager GW, Binder BR: Genetic determinants: is there an "atherosclerosis gene"? *Acta Med Austriaca* 31(1):1–7, 2004.
29. Kurtz TW, Gardner DG: Transcription-modulating drugs: a new frontier in the treatment of essential hypertension. *Hypertension* 32(3):380–386, 1998.
30. Benson SC, Pershadsingh HA, Ho CI: Identification of telmisartan as a unique angiotensin II receptor antagonist with selective PPAR gamma-modulating activity. *Hypertension* 43(5):993–1002, 2004.
31. Davidson M: Confirmed previous infection with Chlamydia pneumoniae (TWAR) and its presence in early coronary atherosclerosis. *Circulation* 98(7):628–633, 1998.
32. Muhlestein JB: Bacterial infections and atherosclerosis. *J Invest Med* 46(8):396–402, 1998.
33. Grayston JT, Kronmal RA, Jackson LA, et al: Azithromycin for the secondary prevention of coronary events. *N Engl J Med* 352(16):1637–1645, 2005.
34. Toss H, Gnarpe J, Gnarpe H: Increased fibrinogen levels are associated with persistent *Chlamydia pneumoniae* infection in unstable coronary artery disease. *Eur Heart J* 19(4):570–577, 1998.
35. Anderson JL, Carlquist JF, Muhlestein JB, et al: Evaluation of C-reactive protein, an inflammatory marker, and infectious serology as risk factors of coronary artery disease and myocardial infarction. *J Am Coll Cardiol* 32(1):35–41, 1998.
36. Toth PP: C-reactive protein as a potential therapeutic target in patients with coronary heart disease. *Curr Atheroscler Rep* 7(5):333–334, 2005.
37. Morrow DA, Rifai N, Antman EM, et al: C-reactive protein is a potent predictor of mortality independently of and in combination with troponin T in acute coronary syndromes: a TIMI 11A substudy-thrombolysis in myocardial infarction. *J Am Coll Cardiol* 31(7):1460–1465, 1998.
38. Elliot WJ, Powel LH: Diagonal earlobe creases and prognosis in patients with suspected coronary artery disease. *Am J Med* 100(2):205–211, 1996.
39. Bahcelioglu M, Isik AF, Demirel D, et al: The diagonal ear lobe crease as sign of some diseases. *Saudi Med* 26(6):947–951, 2005.
40. Shrestha I: Diagonal ear-lobe crease is correlated with atherosclerotic changes in carotid arteries. *Circ J* 73(10):1945–1949, 2009.
41. Friedlander AH: Diagonal ear lobe crease and atherosclerosis: a review of the medical literature and oral and maxillofacial implications. *J Oral Maxillofac Surg* Oct 22, 2010. Epub ahead of print.
42. Koracevic G: Point of disagreement in evidence-based medicine. *Am J Forensic Med Pathol* 30(1):89, 2009.
43. Yeh ET, Bickford CL: Cardiovascular complications of cancer therapy: incidence, pathogenesis, diagnosis, and management. *J Am Coll Cardiol* 16;53(24):2231–2247, 2009.
44. American Heart Association (AHA): Heart and stroke encyclopedia. Available at: http://www.americanheart.org. Accessed November 11, 2010.
45. Gender matters: Heart disease risk in women. *Harvard Women's Health Watch* 11(9):1–3, 2004.
46. Cheek D: What's different about heart disease in women? *Nursing2003* 33(8):36–42, 2003.
47. LaGrossa J: *Heart attack in women.* Advance Online Editions for Physical Therapists, February 2, 2004. Available at: www.advanceforpt.com. Accessed October 17, 2010.
48. Barclay L, Vega C: AHA Updates Guidelines for cardiovascular disease prevention in women, CME 2004. Available at: www.medscape.com. Accessed October 17, 2010.
49. Cohen MC, Rohtla KM, Mittleman MA et al: Meta-analysis of the morning excess of acute myocardial infarction and sudden cardiac death. *Am J Cardiol* 79(11):1512–1516, 1997.
50. McSweeney JC: Women's early warning symptoms of acute myocardial infarction. *Circulation* 108(21):2619–2623, 2003.
51. Is it a heart attack? If you're a woman, will you know? *Berkeley Wellness Letter* 17(2):10, 2000.
52. Marrugat J: Mortality differences between men and women following first myocardial infarction. *JAMA* 280:1405–1409, 1998.
53. Cahalin LP: Heart failure. *Phys Ther* 76(5):517–533, 1996.
54. Hiratzka LF, Bakris GL, Beckman JA, et al: Guidelines for the diagnosis and management of patients with thoracic aortic disease: executive summary. A report of the American College of Cardiology Foundation/American Heart Association Task Force on Practice Guidelines, American Association for Thoracic Surgery, American College of Radiology, American Stroke Association, Society of Cardiovascular Anesthesiologists, Society for Cardiovascular Angiography and Interventions, Society of Interventional Radiology, Society of Thoracic Surgeons, and Society for Vascular Medicine. *Catheter Cardiovasc Interv* 76(2):E43–86, 2010.
55. Lederle FA: Smokers' relative risk for aortic aneurysm compared with other smoking-related diseases: a systematic review. *J Vasc Surg* 38:329–334, 2003.
56. Dua MM, Dalman RL: Identifying aortic aneurysm risk factors in postmenopausal women. *Womens Health* 5(1):33–37, 2009.
57. Lederle FA: Abdominal aortic aneurysm events in the women's health initiative: cohort study. *BMJ* 337:1724–1734, 2008.
58. US Preventive Services Task Force (USPSTF): One-time screening in select subsets of men. *Ann Intern Med* 142:198–202, 2005.
59. Edwards JZ: Chronic back pain caused by an abdominal aortic aneurysm: case report and review of the literature. *Orthopedics* 26:191–192, 2003.
60. Chervu A: Role of physical examination in detection of abdominal aortic aneurysms. *Surgery* 117(4):454–457, 1995.
61. O'Gara PT: Aortic aneurysm. *Circulation* 107:e43, 2003. Available on-line at http://circ.ahajournals.org/cgi/content/full/107/6/e43. Accessed January 6, 2010.

62. Moore KL: *Clinically oriented anatomy*, ed 6, Philadelphia, 2010, Wolters Kluwer/Lippincott Williams & Wilkins.
63. O'Rourke MF: The cardiovascular continuum extended: aging effects on the aorta and microvasculature. *Vasc Med* 15(6):461–468, 2010.
64. Hickson SS: The relationship of age with regional aortic stiffness and diameter. *J Am Coll Cardiol: Cardiovascular Imaging* 3(12):1247–1255, 2010.
65. Lam CS: Aortic root remodeling over the adult life course: longitudinal data from the Framingham Heart Study. *Circulation* 122(9):884–890, 2010.
66. Guinea GV: Factors influencing the mechanical behavior of healthy human descending thoracic aorta. *Physiol Meas* 31:1553–1565, 2010.
67. Fink HA: The accuracy of physical examination to detect abdominal aortic aneurysm. *Arch Intern Med* 160:833–836, 2000.
68. Lederle FA: Selective screening for abdominal aortic aneurysms with physical examination and ultrasound. *Arch Intern Med* 148:1753, 1988.
69. Mechelli F: Differential diagnosis of a patient referred to physical therapy with low back pain: abdominal aortic aneurysm. *J Orthop Sports Phys Ther* 38(9):551–557, 2008.
70. Hillman ND, Tani LY, Veasy LG, et al: Current status of surgery for rheumatic carditis in children. *Ann Thoracic Surg* 78(4): 1403–1408, 2004.
71. Sapico FL, Liquette JA, Sarma RJ: Bone and joint infections in patients with infective endocarditis: review of a 4-year experience. *Clin Infect Dis* 22:783–787, 1996.
72. Vlahakis NE, Temesgen Z, Berbari EF, et al: Osteoarticular infection complicating enterococcal endocarditis. *Mayo Clin Proc* 78(5):623–628, 2003.
73. Petrini JR: Racial differences by gestational age in neonatal deaths attributable to congenital heart defects in the United States. *MMWR* 59(37):1208–1211, 2010.
74. Cava JR, Danduran MJ, Fedderly RT, et al: Exercise recommendations and risk factors for sudden cardiac death. *Pediatr Clin North Am* 51(5):1401–1420, 2004.
75. Berger S, Kugler JD, Thomas JA, et al: Sudden cardiac death in children and adolescents: introduction and overview. *Pediatr Clin North Am* 51(5):1201–1209, 2004.
76. Bader RS, Goldberg L, Sahn DJ: Risk of sudden cardiac death in young athletes: which screening strategies are appropriate? *Pediatr Clin North Am* 51(5):1421–1441, 2004.
77. Hayek E: Mitral valve prolapse. *Lancet* 365(9458):507–518, 2005.
78. Freed LA, Benjamin EJ, Levy D, et al: Mitral valve prolapse in the general population: the benign nature of echocardiographic features in the Framingham Heart Study. *J Am Coll Cardiol* 40(7):1298–1304, 2002.
79. Freed LA, Levy D, Levine RA, et al: Prevalence and clinical outcome of mitral valve prolapse. *N Engl J Med* 341(1):1–7, 1999.
80. Hayek E, Gring CN, Griffin BP: Mitral valve prolapse. *Lancet* 365(9458):507–518, 2005.
81. Montenero A, Mollichelli N, Zumbo F, et al: *Helicobacter pylori* and atrial fibrillation: a possible pathogenic link. *Heart* 91(7):960–961, 2005. Available at: http://www.heart.bmjjournals.com. Accessed October 17, 2010.
82. Strandberg TE: Isolated systolic blood pressure measurement. *Lancet* 372(9643):1033–1034, 2008.
83. Ntatsaki E: Isolated systolic blood pressure measurement. *Lancet* 372(9643):1033, 2008.
84. National Heart, Lung, and Blood Institute (NHLBI): The Seventh Report of the Joint National Committee on Prevention, Detection, Evaluation, and Treatment of High Blood Pressure, NIH Publication No. 03-5233, May 2003. Available at www.nhlbi.nih.gov/. Accessed October 17, 2010.
85. Brookes L: The definition and consequences of hypertension are evolving. *Medscape Cardiology* 9(1):2005. Available at: http://www.medscape.com/viewarticle/506463. Accessed October 20, 2010.
86. Chaudhry SI, Krumholz HM, Foody JM: Systolic hypertension in older persons. *JAMA* 292(9):1074–1080, 2004.
87. Your blood pressure: check that top number. *Johns Hopkins Medical Letter: Health After 50* 16(11):6–7, 2005.
88. Mancia G: Long-term risk of sustained hypertension in white-coat or masked hypertension. *Hypertension* 54(2):226–232, 2009.
89. Furie KL: Guidelines for the prevention of stroke in patients with stroke and transient ischemic attack. A guideline for healthcare professionals from the American Heart Association/American Stroke Association, *Stroke* epub ahead of print; October 21, 2010.
90. Miller ER, Jehn ML: New high blood pressure guidelines create new at-risk classification: changes in blood pressure classification by JNC 7. *J Cardiovasc Nurs* 19(6):367–371, 2004.
91. O'Donnell CJ, Lindpaintner K, Larson MG, et al: Evidence for association and genetic linkage with hypertension and blood pressure in men but not women in the Framingham heart study. *Circulation* 97(18):1766–1772, 1998.
92. Treating a "mini stroke" to prevent a "major" stroke. *Johns Hopkins Medical Letter: Health After 50* 17(8):6–7, 2005.
93. Tepper S, McKeough M: Deep venous thrombosis: risks, diagnosis, treatment interventions, and prevention. *Acute Care Perspectives* 9(1):1–7, 2000.
94. Rosenzweig K: *Differential diagnosis of deep vein thrombosis in a spinal cord injured client, Case report presented in partial fulfillment of DPT 910, Principles of Differential Diagnosis, Institute for Physical Therapy Education*, Chester, Pennsylvania, 2005, Widener University.
95. Powell M: Duplex ultrasound screening for deep vein thrombosis in spinal cord injured patients at rehabilitation admission. *Arch Phys Med Rehab* 80:1044–1046, 1999.
96. Agnelli G, Sonaglia F: Prevention of venous thromboembolism. *Thromb Res* 97(1):V49–V62, 2000.
97. Bauer K: Hypercoagulable states. *Hematology* 10(Suppl 1):39, 2005.
98. Wu O, Robertson L, Langhorne P, et al: Oral contraceptives, hormone replacement therapy, thrombophilias, and risk of venous thromboembolism: a systematic review. The Thrombosis: Risk and Economic Assessment of Thrombophilia Screening (TREATS) Study. *Thromb Haemost* 94(1):17–25, 2005.
99. Gomes MP, Deitcher SR: Risk of venous thromboembolic disease associated with hormonal contraceptives and hormone replacement therapy: a clinical review. *Arch Intern Med* 164(18):1965–1976, 2004.
100. Constans J: A clinical prediction score for upper extremity deep venous thrombosis. *Thromb Haemost* 99:202–207, 2008.
101. Sacks D, Bakal C, Beatty P, et al: Position statement on the use of the ankle brachial index in the evaluation of patients with peripheral vascular disease. *J Vasc Interv Radiol* 14(9 Pt 2):S389, 2003.
102. Horn HR: The impact of cardiovascular disease. On-line: http://www.medscape.com/viewarticle/466799_2, April 2004. Accessed Nov. 05, 2011.
103. de Virgilio C: Ascending aortic dissection in weight lifters with cystic medial degeneration. *Ann Thorac Surg* 49(4):638–642, 1990.
104. Kario K, Tobin J, Wolfson L, et al: Lower standing systolic blood pressure as a predictor of falls in the elderly: a community-based prospective study. *J Am Coll Cardiol* 38(1): 246–252, 2001.
105. Ageno W: Treatment of venous thromboembolism. *Thromb Res* 97(1):V63–V72, 2000.

CHAPTER

7

Screening for Pulmonary Disease

For the client with neck, shoulder, or back pain at presentation, it may be necessary to consider the possibility of a pulmonary cause requiring medical referral. The most common pulmonary conditions to mimic those of the musculoskeletal system include pneumonia, pulmonary embolism, pleurisy, pneumothorax, and pulmonary arterial hypertension.

As always, using the past medical history, risk factor assessment, and clinical presentation, as well as asking about the presence of any associated signs and symptoms guide the screening process. In the case of pleuropulmonary disorders, the client's recent personal medical history may include a previous or recurrent upper respiratory infection or pneumonia.

Pneumothorax may be preceded by trauma, overexertion, or recent scuba diving. Each pulmonary condition will have its own unique risk factors that can predispose clients to a specific respiratory disease or illness.

A previous history of cancer, especially primary lung cancer or cancers that metastasize to the lungs (e.g., breast, bone), is a red flag and a risk factor for cancer recurrence. Risk factor assessment also helps identify increased risk for other respiratory conditions or illnesses that can present as a primary musculoskeletal problem.

The material in this chapter will assist the therapist in treating both the client with a known pulmonary problem and the client with musculoskeletal signs and symptoms that may have an underlying systemic basis (Case Example 7-1).

SIGNS AND SYMPTOMS OF PULMONARY DISORDERS

Signs and symptoms of pulmonary disorders can be many and varied; the most common symptoms associated with pulmonary disorders are cough and dyspnea. Other manifestations include chest pain, abnormal sputum, hemoptysis, cyanosis, digital clubbing, altered breathing patterns, and chest pain.

Cough

As a physiologic response, cough occurs frequently in healthy people, but a persistent dry cough may be caused by a tumor, congestion, or hypersensitive airways (allergies). A productive cough with purulent sputum (yellow or green in color) may indicate infection, whereas a productive cough with nonpurulent sputum (clear or white) is nonspecific and indicates airway irritation. Rust colored sputum may be a sign of pneumonia and should also be investigated. Hemoptysis (coughing and spitting blood) indicates a pathologic condition—infection, inflammation, abscess, tumor, or infarction.

Dyspnea

Shortness of breath (SOB), or dyspnea, usually indicates hypoxemia but can be associated with emotional states, particularly fear and anxiety. Dyspnea is usually caused by diffuse and extensive rather than focal pulmonary disease; pulmonary embolism is the exception. Factors contributing to the sensation of dyspnea include increased work of breathing (WOB), respiratory muscle fatigue, increased systemic metabolic demands, and decreased respiratory reserve capacity. Dyspnea when the person is lying down is called *orthopnea* and is caused by redistribution of body water. Fluid shift leads to increased fluid in the lung, which interferes with gas exchange and leads to orthopnea. In supine and prone, the abdominal contents also exert pressure on the diaphragm, increasing the WOB and often limiting vital capacity.

The therapist must be careful when screening for dyspnea or SOB, either with exertion or while at rest. If a client denies compromised breathing, look for functional changes as the client accommodates for difficulty breathing by reducing activity or exertion.

Cyanosis

The presence of cyanosis, a bluish color of the skin and mucous membranes, depends on the oxygen saturation of arterial blood and the total amount of circulating hemoglobin. It may be observed as a bluish discoloration in the oral mucous membranes, lips, and conjunctivae and pale (white) or blue nail beds and nose.

Clubbing (see Chapter 4)

Thickening and widening of the terminal phalanges of the fingers and toes result in a painless clublike appearance recognized by the loss of the angle between the nail and the nail

CASE EXAMPLE 7-1

Bronchopulmonary Pain

A 67-year-old woman with a known diagnosis of rheumatoid arthritis has been treated as needed in a physical therapy clinic for the last 8 years. She has reported occasional chest pain described as "coming on suddenly, like a knife pushing from the inside out—it takes my breath away."

She missed 2 days of treatment because of illness, and when she returned to the clinic, the physical therapist noticed that she had a newly developed cough and that her rheumatoid arthritis was much worse. She says that she missed her appointments because she had the "flu."

Further questioning to elicit the potential development of chest pain on inspiration, the presence of ongoing fever and chills, and the changes in breathing pattern is recommended.

Positive findings beyond the reasonable duration of influenza (7 to 10 days) or an increase in pulmonary symptoms (shortness of breath [SOB], hacking cough, hemoptysis, wheezing or other changes in breathing pattern) raise a red flag, indicating the need for medical referral.

This clinical case points out that clients currently undergoing physical therapy for a known musculoskeletal problem may be describing signs and symptoms of systemic disease.

bed (see Figs. 4-36 and 4-37). Conditions that chronically interfere with tissue perfusion and nutrition may cause clubbing, including cystic fibrosis (CF), chronic obstructive pulmonary disease (COPD), lung cancer, bronchiectasis, pulmonary fibrosis, congenital heart disease, and lung abscess. Most of the time, clubbing is due to pulmonary disease and resultant hypoxia (diminished availability of blood to the body tissues), but clubbing can be a sign of heart disease, peripheral vascular disease, and disorders of the liver and gastrointestinal tract.

Altered Breathing Patterns

Changes in the rate, depth, regularity, and effort of breathing occur in response to any condition affecting the pulmonary system. Breathing patterns can vary, depending on the neuromuscular or neurologic disease or trauma (Box 7-1). Breathing pattern abnormalities seen with head trauma, brain abscess, diaphragmatic paralysis of chest wall muscles and thorax (e.g., generalized myopathy or neuropathy), heat stroke, spinal meningitis, and encephalitis can include apneustic breathing, ataxic breathing, or Cheyne-Stokes respiration (CSR).

Apneustic breathing (gasping inspiration with short expiration) localizes damage to the midpons and is most commonly a result of a basilar artery infarct. *Ataxic*, or *Biot's*, breathing (irregular pattern of deep and shallow breaths with abrupt pauses) is caused by disruption of the respiratory rhythm generator in the medulla.

CSR may be evident in the well older adult, as well as in compromised clients. The most common cause of CSR is severe congestive heart failure (CHF), but it can also occur

BOX 7-1 BREATHING PATTERNS AND ASSOCIATED CONDITIONS

Hyperventilation
- Anxiety
- Acute head injury
- Hypoxemia
- Fever

Kussmaul's
- Strenuous exercise
- Metabolic acidosis

Cheyne-Stokes
- Congestive heart failure
- Renal failure
- Meningitis
- Drug overdose
- Increased intracranial pressure
- Infants (normal)
- Older people during sleep (normal)

Hypoventilation
- Fibromyalgia syndrome
- Chronic fatigue syndrome
- Sleep disorder
- Muscle fatigue
- Muscle weakness
- Malnutrition
- Neuromuscular disease
  - Guillain-Barré
  - Myasthenia gravis
  - Poliomyelitis
  - Amyotrophic lateral sclerosis (ALS)
- Pickwickian or obesity hypoventilation syndrome
- Severe kyphoscoliosis

Apneustic
- Midpons lesion
- Basilar artery infarct

Biot's Respiration (Ataxia)
- Exercise
- Shock
- Cerebral hypoxia
- Heat stroke
- Spinal meningitis
- Head injury
- Brain abscess
- Encephalitis

From Ikeda B, Goodman CC, Fuller K: The respiratory system. In Goodman CC: *Pathology: implications for the physical therapist,* ed 3, St Louis, 2009, Elsevier.

with renal failure, meningitis, drug overdose, and increased intracranial pressure. It may be a normal breathing pattern in infants and older persons during sleep.

Exercise may induce pleural pain, coughing, hemoptysis, SOB, and/or other abnormal changes in breathing patterns. When asked if the client is ever short of breath, the individual may say "no" because he or she has reduced activity levels to avoid dyspnea (see Appendix B-12).

Pulmonary Pain Patterns

The most common sites for referred pain from the pulmonary system are the chest, ribs, upper trapezius, shoulder, and thoracic spine. The first symptoms may not appear until the client's respiratory system is stressed by the addition of exercise during physical therapy. On the other hand, the client may present with what appears to be primary musculoskeletal pain in any one of those areas. Auscultation may reveal the first signs of pulmonary distress (see Chapter 4 for screening examination by auscultation).

Pulmonary pain patterns are usually localized in the substernal or chest region over involved lung fields that may include the anterior chest, side, or back (Fig. 7-1). However, pulmonary pain can radiate to the neck, upper trapezius muscle, costal margins, thoracic back, scapulae, or shoulder. Shoulder pain may radiate along the medial aspect of the arm, mimicking other neuromuscular causes of neck or shoulder pain (see Fig. 7-10).

Pulmonary pain usually increases with inspiratory movements, such as laughing, coughing, sneezing, or deep breathing, and the client notes the presence of associated symptoms, such as dyspnea (exertional or at rest), persistent cough, fever, and chills. Palpation and resisted movements will not reproduce the symptoms, which may get worse with recumbency, especially at night or while sleeping.

The thoracic cavity is lined with pleura, or serous membrane. One surface of the pleura lines the inside of the rib cage (parietal) and the other surface covers the lungs (visceral). The parietal pleura is sensitive to painful stimulation, but the visceral pleura is insensitive to pain. Extensive disease may occur in the lung without occurrence of pain until the process extends to the parietal pleura. This explains why pathology of the lungs may be painless until obstruction or inflammation is enough to press on the parietal pleura.

Pleural irritation then results in sharp, localized pain that is aggravated by any respiratory movement. Clients usually note that the pain is alleviated by autosplinting, that is, lying on the affected side, which diminishes the movement of that side of the chest (see further discussion of pleural pain in this chapter).[1]

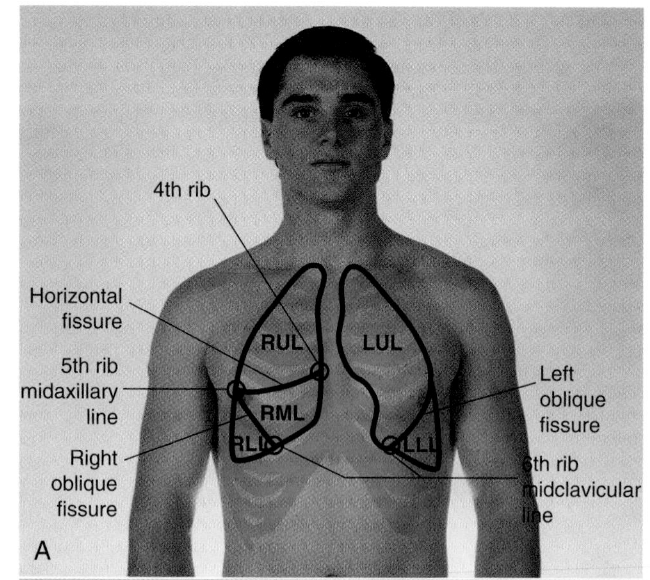

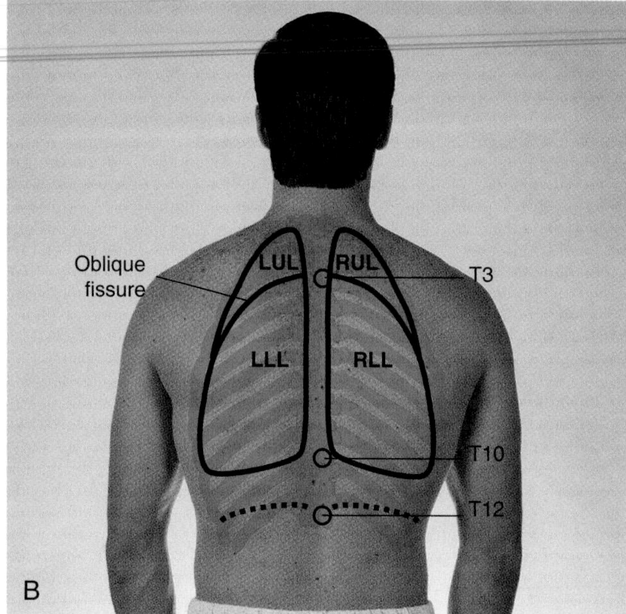

Fig. 7-1 Pulmonary pain patterns are localized over involved lung fields affecting the anterior chest, side, or back. Radiating pain can also cause neck, shoulder, upper trapezius, rib, and/or scapular pain. **A,** Anterior chest. **B,** Posterior chest. The posterior chest is comprised primarily of lower lung lobes. The upper lobes occupy a small area from T1 to T3 or T4. (From Jarvis C: *Physical examination and assessment,* ed 5, Philadelphia, 2007, WB Saunders.)

Tracheobronchial Pain

Within the pulmonary system, the trachea and large bronchi are innervated by the vagus trunks, whereas the finer bronchi and lung parenchyma appear to be free of pain innervation. Tracheobronchial pain is referred to sites in the neck or anterior chest at the same levels as the points of irritation in the air passages (Fig. 7-2). This irritation may be caused by inflammatory lesions, irritating foreign materials, or cancerous tumors.

Pleural Pain

When the disease progresses enough to extend to the parietal pleura, pleural irritation occurs and results in sharp, localized pain that is aggravated by any respiratory movement. Clients usually note that the pain is alleviated by lying on the affected side, which diminishes the movement of that side of the chest called *autosplinting.*

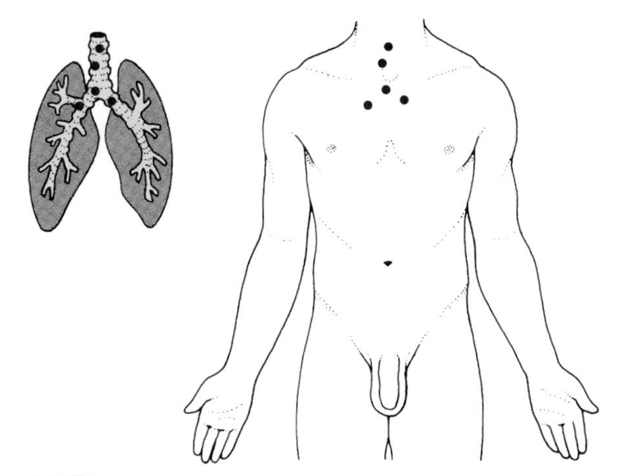

Fig. 7-2 Tracheobronchial pain is referred to sites in the neck or anterior chest at the same levels as the points of irritation in the air passages. The points of pain are on the same side as the areas of irritation.

Debate continues concerning the mechanism by which pain occurs in the parietal membrane. It has been long thought that friction between the two pleural surfaces (when the membranes are irritated and covered with fibrinous exudate) causes sharp pain. Other theories suggest that intercostal muscle spasm resulting from pleurisy or stretching of the parietal pleura causes this pain.

Pleural pain is present in pulmonary diseases such as pleurisy, pneumonia, pulmonary infarct (when it extends to the pleural surface, thus causing pleurisy), tumor (when it invades the parietal pleura), and pneumothorax. Tumor, especially bronchogenic carcinoma, may be accompanied by severe, continuous pain when the tumor tissue, extending to the parietal pleura through the lung, constantly irritates the pain nerve endings in the pleura.

Diaphragmatic Pleural Pain

The *diaphragmatic pleura* receives dual pain innervation through the phrenic and intercostal nerves. Damage to the phrenic nerve produces paralysis of the corresponding half of the diaphragm. The phrenic nerves are sensory and motor from both surfaces of the diaphragm.

Stimulation of the peripheral portions of the diaphragmatic pleura results in sharp pain felt along the costal margins, which can be referred to the lumbar region by the lower thoracic somatic nerves. Stimulation of the central portion of the diaphragmatic pleura results in sharp pain referred to the upper trapezius muscle and shoulder on the ipsilateral side of the stimulation (see Figs. 3-4 and 3-5).

Pain of cardiac and diaphragmatic origin is often experienced in the shoulder because the heart and diaphragm are supplied by the C5-6 spinal segment, and the visceral pain is referred to the corresponding somatic area.

Diaphragmatic pleurisy secondary to pneumonia is common and refers sharp pain along the costal margins or upper trapezius, which is aggravated by any diaphragmatic motion, such as coughing, laughing, or deep breathing.

There may be tenderness to palpation along the costal margins, and sharp pain occurs when the client is asked to take a deep breath. A change in position (sidebending or rotation of the trunk) does not reproduce the symptoms, which would be the case with a true intercostal lesion or tear.

Forceful, repeated coughing can result in an intercostal lesion in the presence of referred intercostal pain from diaphragmatic pleurisy, which can make differentiation between these two entities impossible without a medical referral and further diagnostic testing.

Pulmonary Physiology

The primary function of the respiratory system is to provide oxygen to and to remove carbon dioxide from cells in the body. The act of breathing, in which the oxygen and carbon dioxide exchange occurs, involves the two interrelated processes of ventilation and respiration.

Ventilation is the movement of air from outside the body to the alveoli of the lungs. Respiration is the process of oxygen uptake and carbon dioxide elimination between the body and the outside environment.

Breathing is an automatic process by which sensors detect changes in the levels of carbon dioxide and continuously direct data to the medulla. The medulla then directs respiratory muscles that adjust ventilation. Breathing patterns can be altered voluntarily when this automatic response is overridden by conscious thought.

The major sensors mentioned here are the central chemoreceptors (located near the medulla) and the peripheral sensors (located in the carotid body and aortic arch). The central chemoreceptors respond to increases in carbon dioxide and decreases in pH in cerebrospinal fluid.

As carbon dioxide increases, the medulla signals a response to increase respiration. The peripheral chemoreceptor system responds to low arterial blood oxygen and is believed to function only in pathologic situations such as when there are chronically elevated carbon dioxide levels (e.g., COPD).

Acid-Base Regulation

The proper balance of acids and bases in the body is essential to life. This balance is very complex and must be kept within the narrow parameters of a pH of 7.35 to 7.45 in the extracellular fluid. This number (or pH value) represents the hydrogen ion concentration in body fluid.

A reading of less than 7.35 is considered *acidosis*, and a reading greater than 7.45 is called *alkalosis*. Life cannot be sustained if the pH values are less than 7 or greater than 7.8.

Living human cells are extremely sensitive to alterations in body fluid pH (hydrogen ion concentration); thus various mechanisms are in operation to keep the pH at a relatively constant level.

Acid-base regulatory mechanisms include chemical buffer systems, the respiratory system, and the renal system. These systems interact very closely to maintain a normal acid-base ratio of 20 parts of bicarbonate to 1 part of carbonic acid and thus to maintain normal body fluid pH.

The blood test used most often to measure the effectiveness of ventilation and oxygen transport is the arterial blood gas (ABG) test (Table 7-1). The measurement of arterial blood gases is important in the diagnosis and treatment of ventilation, oxygen transport, and acid-base problems.

The ABG test measures the amount of dissolved oxygen and carbon dioxide in arterial blood and indicates acid-base status by measurement of the arterial blood pH. In simple terms a low pH reflects increased acid buildup, and a high pH reflects an increased base buildup.

Acid buildup occurs when there is an ineffective removal of carbon dioxide from the lungs or when there is excess acid production from the tissues of the body. These problems are corrected by adjusting the ventilation or buffering the acid with bicarbonate.

TABLE 7-1 Arterial Blood Gas Values*

pH	7.35-7.45
pCO_2 (partial pressure of carbon dioxide)	35-45 mm Hg
HCO_3 (bicarbonate ion)	22-31 mEq/L
pO_2 (partial pressure of oxygen)	75-100 mm Hg
O_2 saturation (oxygen saturation)	96%-100%
PANIC VALUES	
pH	≤7.20 or >7.6
pCO_2	<20 or >70 mm Hg
HCO_3	<10 or >40 mEq/L
pO_2	<40 mm Hg
O_2 saturation	≤60%

Normal pH level: The pH is inversely proportional to the hydrogen ion concentration in the blood. As the hydrogen ion concentration increases (acidosis), the pH decreases; as the hydrogen ion concentration decreases (alkalosis), the pH increases.

Normal pCO_2: The pCO_2 is a measure of the partial pressure of carbon dioxide (CO_2) in the blood. As the CO_2 increases, the pH decreases (respiratory acidosis); as the CO_2 level decreases, the pH increases (respiratory alkalosis). pCO_2 measures the effectiveness of the body's ventilation system as CO_2 is removed.

Bicarbonate ion: HCO_3 is a measure of the metabolic portion of the acid-base function. As the bicarbonate value increases, the pH increases (metabolic alkalosis); as the bicarbonate value decreases, the pH decreases (metabolic acidosis).

Partial pressure of oxygen: The pO_2 is a measure of the partial pressure of oxygen (O_2) in the blood and represents the status of alveolar gas exchange.

Oxygen saturation: O_2 saturation is an indication of the percentage of hemoglobin saturated with oxygen. When 95% to 100% of the hemoglobin binds and carries oxygen, the tissues are adequately perfused with oxygen. As the pO_2 decreases, the percentage of hemoglobin saturation also decreases. At oxygen saturation levels of less than 70%, the tissues are unable to carry out vital functions.

*Modified from Chernecky C, Berger B: *Laboratory tests and diagnostic procedures,* ed 5, Philadelphia, 2008, WB Saunders.

Pulmonary Pathophysiology

Respiratory Acidosis

Any condition that decreases pulmonary ventilation increases the retention and concentration of carbon dioxide (CO_2), hydrogen, and carbonic acid; this results in an increase in the amount of circulating hydrogen and is called respiratory acidosis.

If ventilation is severely compromised, CO_2 levels become extremely high and respiration is depressed even further, causing hypoxia as well.

During respiratory acidosis, potassium moves out of cells into the extracellular fluid to exchange with circulating hydrogen. This results in hyperkalemia (abnormally high potassium concentration in the blood) and cardiac changes that can cause cardiac arrest.

Respiratory acidosis can result from pathologic conditions that decrease the efficiency of the respiratory system. These pathologies can include damage to the medulla, which controls respiration, obstruction of airways (e.g., neoplasm, foreign bodies, pulmonary disease such as COPD, pneumonia), loss of lung surface ventilation (e.g., pneumothorax, pulmonary fibrosis), weakness of respiratory muscles (e.g., poliomyelitis, spinal cord injury, Guillain-Barré syndrome), or overdose of respiratory depressant drugs.

As hypoxia becomes more severe, diaphoresis, shallow rapid breathing, restlessness, and cyanosis may appear. Cardiac arrhythmias may also be present as the potassium level in the blood serum rises.

Treatment is directed at restoration of efficient ventilation. If the respiratory depression and acidosis are severe, injection of intravenous sodium bicarbonate and use of a mechanical ventilator may be necessary. Any client with symptoms of inadequate ventilation or CO_2 retention needs immediate medical referral.

CLINICAL SIGNS AND SYMPTOMS

Respiratory Acidosis

- Decreased ventilation
- Confusion
- Sleepiness and unconsciousness
- Diaphoresis
- Shallow, rapid breathing
- Restlessness
- Cyanosis

Respiratory Alkalosis

Increased respiratory rate and depth decrease the amount of available CO_2 and hydrogen and create a condition of increased pH, or alkalosis. When pulmonary ventilation is increased, CO_2 and hydrogen are eliminated from the body too quickly and are not available to buffer the increasingly alkaline environment.

Respiratory alkalosis is usually due to *hyperventilation.* Rapid, deep respirations are often caused by neurogenic or psychogenic problems, including anxiety, pain, and cerebral trauma or lesions. Other causes can be related to conditions that greatly increase metabolism (e.g., hyperthyroidism) or overventilation of clients who are using a mechanical ventilator.

If the alkalosis becomes more severe, muscular tetany and convulsions can occur. Cardiac arrhythmias caused by serum potassium loss through the kidneys may also occur. The kidneys keep hydrogen in exchange for potassium.

Treatment of respiratory alkalosis includes reassurance, assistance in slowing breathing and facilitating relaxation, sedation, pain control, CO_2 administration, and use of a rebreathing device such as a rebreathing mask or paper bag. A rebreathing device allows the client to inhale and "rebreathe" the exhaled CO_2.

Respiratory alkalosis related to hyperventilation is a relatively common condition and might be present more often in the physical therapy setting than respiratory acidosis. Pain and anxiety are common causes of hyperventilation, and treatment needs to be focused toward reduction of both of these interrelated elements. If hyperventilation continues in the absence of pain or anxiety, serious systemic problems may be the cause, and immediate physician referral is necessary.

If either respiratory acidosis or alkalosis persists for hours to days in a chronic and not life-threatening manner, the kidneys then begin to assist in the restoration of normal body fluid pH by selective excretion or retention of hydrogen ions or bicarbonate. This process is called *renal compensation.* When the kidneys compensate effectively, blood pH values are within normal limits (7.35 to 7.45) even though the underlying problem may still cause the respiratory imbalance.

CLINICAL SIGNS AND SYMPTOMS

Respiratory Alkalosis

- Hyperventilation
- Lightheadedness
- Dizziness
- Numbness and tingling of the face, fingers, and toes
- Syncope (fainting)

Chronic Obstructive Pulmonary Disease

COPD, also called chronic obstructive lung disease (COLD), refers to a number of disorders that have in common abnormal airway structures resulting in obstruction of air in and out of the lungs. The most important of these disorders are obstructive bronchitis, emphysema, and asthma.

Although bronchitis, emphysema, and asthma may occur in a "pure form," they most commonly coexist. For example, adult subjects with active asthma are as much as 12 times more likely to acquire COPD over time than subjects with no active asthma.[2-5]

COPD is a leading cause of morbidity and mortality among cigarette smokers. Other factors predisposing to COPD include air pollution; occupational exposure to aerosol pesticides, irritating dusts or gases, or art materials (e.g., paint, glass, ceramics, sculpture); hereditary factors; infection; allergies; aging; and potentially harmful drugs and chemicals.[6]

COPD rarely occurs in nonsmokers; however, only a minority of cigarette smokers develop symptomatic disease, suggesting that genetic factors or other underlying predisposition may contribute to the development of COPD.[7]

In all forms of COPD, narrowing of the airways obstructs airflow to and from the lungs (Table 7-2). This narrowing impairs ventilation by trapping air in the bronchioles and alveoli. The obstruction increases the resistance to airflow. The severity of symptoms depends on how much of the lungs have been damaged or destroyed.

Trapped air hinders normal gas exchange and causes distention of the alveoli. Other mechanisms of COPD vary with each form of the disease. In the healthy adult the bottom margin of the respiratory diaphragm sits at T9 when the lungs are at rest. Taking a deep breath expands the diaphragm (and lungs) inferiorly to T11. For the client with COPD the lower lung lobes are already at T11 when the lungs are at rest from overexpansion as a result of alveoli distention and hyperinflation.

TABLE 7-2 Respiratory Diseases: Summary of Differences

Disease	Primary Area Affected	Results
Bronchitis	Membrane lining bronchial tubes	Inflammation of lining
Bronchiectasis	Bronchial tubes (bronchi or air passages)	Bronchial dilation with inflammation
Pneumonia	Alveoli (air sacs)	Causative agent invades alveoli with resultant outpouring from lung capillaries into air spaces and continued healing process
Emphysema	Air spaces beyond terminal bronchioles (small airways)	Breakdown of alveolar walls; air spaces enlarged
Asthma	Bronchioles (small airways)	Bronchioles obstructed by muscle spasm, swelling of mucosa, thick secretions
Cystic fibrosis	Bronchioles	Bronchioles become obstructed and obliterated. Later, larger airways become involved. Plugs of mucus cling to airway walls, leading to bronchitis, bronchiectasis, atelectasis, pneumonia, or pulmonary abscess

COPD develops earlier in life than is usually recognized, making it the most underdiagnosed and undertreated pulmonary disease. Smoking cessation is the only intervention shown to slow decline in lung function. Identifying risk factors and recognizing early signs and symptoms of COPD increases the affected individual's chances of reduced morbidity through early intervention.[6]

Bronchitis

Acute. Acute bronchitis is an inflammation of the trachea and bronchi (tracheobronchial tree) that is self-limiting and of short duration with few pulmonary signs. This condition may result from chemical irritation (e.g., smoke, fumes, gas) or may occur with viral infections such as influenza, measles, chickenpox, or whooping cough.

These predisposing conditions may become apparent during the subjective examination (i.e., Personal/Family History form or the Physical Therapy Interview). Although bronchitis is usually mild, it can become complicated in older clients and clients with chronic lung or heart disease. Pneumonia is a critical complication.

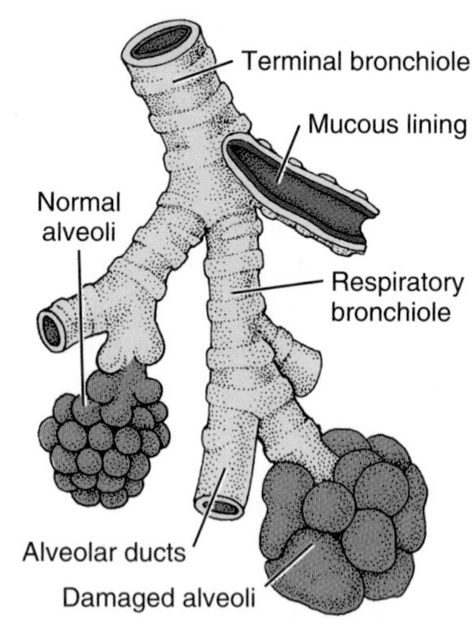

Fig. 7-3 Chronic bronchitis may lead to the formation of misshapen or large alveolar sacs with reduced space for oxygen and carbon dioxide exchange. The client may develop cyanosis and pulmonary edema.

CLINICAL SIGNS AND SYMPTOMS
Acute Bronchitis

- Mild fever from 1 to 3 days
- Malaise
- Back and muscle pain
- Sore throat
- Cough with sputum production, followed by wheezing
- Possibly laryngitis

Chronic. Chronic bronchitis is a condition associated with prolonged exposure to nonspecific bronchial irritants and is accompanied by mucus hypersecretion and structural changes in the bronchi (large air passages leading into the lungs). This irritation of the tissue usually results from exposure to cigarette smoke or long-term inhalation of dust or air pollution and causes hypertrophy of mucus-producing cells in the bronchi.

In bronchitis, partial or complete blockage of the airways from mucus secretions causes insufficient oxygenation in the alveoli (Fig. 7-3). The swollen mucous membrane and thick sputum obstruct the airways, causing wheezing, and the client develops a cough to clear the airways. The clinical definition of a person with chronic bronchitis is anyone who coughs for at least 3 months per year for 2 consecutive years without having had a precipitating disease.

To confirm that the condition is chronic bronchitis, tests are performed to determine whether the airways are obstructed and to exclude other diseases that may cause similar symptoms such as silicosis, tuberculosis, or a tumor in the upper airway. Sputum samples will be analyzed, and lung function tests may be performed.

Treatment is aimed at keeping the airways as clear as possible. Smokers are encouraged and helped to stop smoking. A combination of drugs may be prescribed to relieve the symptoms, including bronchodilators to open the obstructed airways and to thin the obstructive mucus so that it can be coughed up more easily.

Chronic bronchitis may develop slowly over a period of years, but it will not go away if untreated. Eventually, the bronchial walls thicken, and the number of mucous glands increases. The client is increasingly susceptible to respiratory infections, during which the bronchial tissue becomes inflamed and the mucus becomes even thicker and more profuse.

Chronic bronchitis can be incapacitating and lead to more serious and potentially fatal lung disease. Influenza and pneumococcal vaccines are recommended for these clients.

CLINICAL SIGNS AND SYMPTOMS
Chronic Bronchitis

- Persistent cough with production of sputum (worse in the morning and evening than at midday)
- Reduced chest expansion
- Wheezing
- Fever
- Dyspnea (SOB)
- Cyanosis (blue discoloration of skin and mucous membranes)
- Decreased exercise tolerance

Bronchiectasis. Bronchiectasis is a form of obstructive lung disease that is actually a type of bronchitis. It is a progressive and chronic pulmonary condition that occurs after infections such as childhood pneumonia or CF.

Although bronchiectasis was once a common disease because of measles, pertussis, tuberculosis, and poorly treated bacterial pneumonias, the prevalence of bronchiectasis has diminished greatly since the introduction of antibiotics. It is characterized by abnormal and permanent dilatation of the large air passages leading into the lungs (bronchi) and by destruction of bronchial walls.

Bronchiectasis is caused by repeated damage to bronchial walls. The resultant destruction and bronchial dilatation reduce bronchial wall movement so that secretions cannot be removed effectively from the lungs, and the person is predisposed to frequent respiratory infections.

This vicious cycle of bacterial infection and inflammation of the bronchial wall leads to loss of ventilation and irreversible lung damage. Advanced bronchiectasis may cause pneumonia, cor pulmonale, or right-sided ventricular failure.

All pulmonary irritants, especially cigarette smoke, should be avoided. Postural drainage, adequate hydration, good nutrition, and bronchodilator therapy in bronchospasm are important components in treatment. Antibiotics are used during disease exacerbations (e.g., increased cough, purulent sputum, hemoptysis, malaise, and weight loss). The use of immunomodulatory therapy to alter the host response directly and thereby reduce tissue damage is under investigation.[8,9]

CLINICAL SIGNS AND SYMPTOMS

Bronchiectasis

Clinical signs and symptoms of bronchiectasis vary widely, depending on the extent of the disease and on the presence of complicating infection, but may include:

- Chronic "wet" cough with copious foul-smelling secretions; generally worse in the morning after the individual has been recumbent for a length of time
- Hemoptysis (bloody sputum)
- Occasional wheezing sounds
- Dyspnea
- Sinusitis (inflammation of one or more paranasal sinuses)
- Weight loss
- Anemia
- Malaise
- Recurrent fever and chills
- Fatigue

Emphysema. Emphysema may develop in a person after a long history of chronic bronchitis in which the alveolar walls are destroyed, leading to permanent overdistention of the air spaces and loss of normal elastic tension in the lung tissue.

Air passages are obstructed as a result of these changes (rather than as a result of mucus production, as in chronic bronchitis). Difficult expiration in emphysema is due to the destruction of the walls (septa) between the alveoli, partial airway collapse, and loss of elastic recoil.

As the alveoli and septa collapse, pockets of air form between the alveolar spaces (called *blebs*) and within the lung parenchyma (called *bullae*). This process leads to increased ventilatory "dead space," or areas that do not participate in gas or blood exchange. The WOB is increased because there is less functional lung tissue to exchange oxygen and CO_2. Emphysema also causes destruction of the pulmonary capillaries, further decreasing oxygen perfusion and ventilation.

In advanced emphysema, oxygen therapy is usually necessary to treat the progressive hypoxemia that occurs as the disease worsens. Oxygen therapy is carefully titrated and monitored to maintain venous oxygen saturation levels at or slightly above 90%. Too much oxygen can depress the respiratory drive of a person with emphysema.

The drive to breathe in a healthy person results from an increase in the arterial carbon dioxide level (pCO_2). In the normal adult, increased CO_2 levels stimulate chemoreceptors in the brainstem to increase the respiratory rate. With some chronic lung disorders these central chemoreceptors may become desensitized to pCO_2 changes resulting in a dependence on the peripheral chemoreceptors to detect a fall in arterial oxygen levels (pO_2) to stimulate the respiratory drive. Therefore too much oxygen delivered as a treatment can depress the respiratory drive in those individuals with COPD who have a dampening of the CO_2 drive.

Monitoring respiratory rate, level of oxygen administered by nasal canula, and oxygen saturation levels is very important in this client population. Some pulmonologists agree that supplemental oxygen levels can be increased during activity without compromising the individual because they will "blow it (carbon dioxide) off" anyway. To our knowledge, there is no evidence yet to support this clinical practice.

Types of Emphysema. There are three types of emphysema. *Centrilobular emphysema* (Fig. 7-4), the most common type, destroys the bronchioles, usually in the upper lung regions. Inflammation develops in the bronchioles, but usually the alveolar sac remains intact.

Panlobular emphysema destroys the more distal alveolar walls, most commonly involving the lower lung. This destruction of alveolar walls may occur secondary to infection or to irritants (most commonly, cigarette smoke). These two forms of emphysema, collectively called centriacinar emphysema, occur most often in smokers.

Paraseptal (or *panacinar*) *emphysema* destroys the alveoli in the lower lobes of the lungs, resulting in isolated blebs along the lung periphery. Paraseptal emphysema is believed to be the likely cause of spontaneous pneumothorax.

Clinical Signs and Symptoms. The irreversible destruction reduces elasticity of the lung and increases the effort to exhale trapped air, causing marked dyspnea on exertion later progressing to dyspnea at rest. Cough is uncommon.

The client is often thin, has tachypnea with prolonged expiration, and uses the accessory muscles for respiration. The client often leans forward with the arms braced on the knees to support the shoulders and chest for breathing. The combined effects of trapped air and alveolar distention change the size and shape of the client's chest, causing a barrel chest and increased expiratory effort.

As the disease progresses, there is a loss of surface area available for gas exchange. In the final stages of emphysema, cardiac complications, especially enlargement and dilatation of the right ventricle, may develop. The overloaded heart

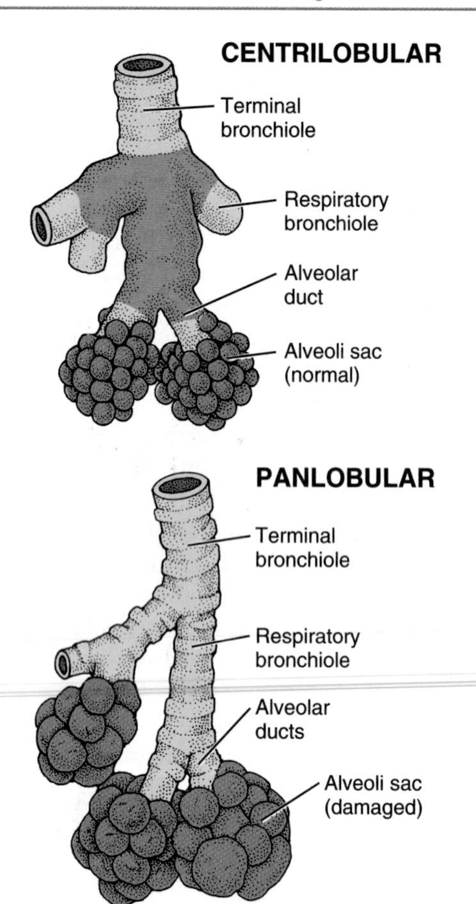

Fig. 7-4 Emphysema traps air in the lungs so that expelling air becomes increasingly difficult. *Centrilobular* emphysema affects the upper airways and produces destructive changes in the bronchioles. *Panlobular* emphysema affects the lower airways and is more diffusely scattered throughout the alveoli.

reaches its limit of muscular compensation and begins to fail (cor pulmonale).

The most important factor in the treatment of emphysema is smoking cessation. The main goals for the client with emphysema are to improve oxygenation and decrease CO_2 retention.

Pursed-lip breathing causes resistance to outflow at the lips, which in turn maintains intrabronchial pressure and improves the mixing of gases in the lungs. This type of breathing should be encouraged to help the client get rid of the stale air trapped in the lungs.

Exercise has not been shown to improve pulmonary function but is used to enhance cardiovascular fitness and train skeletal muscles to function more effectively. Routine progressive walking is the most common form of exercise.

Lung volume reduction surgery is available for some individuals and improves not only lung function and exercise performance, but also activities of daily function and quality of life.[10,11]

CLINICAL SIGNS AND SYMPTOMS

Emphysema

- SOB
- Dyspnea on exertion
- Orthopnea (only able to breathe in the upright position) immediately after assuming the supine position
- Chronic cough
- Barrel chest
- Weight loss
- Malaise
- Use of accessory muscles of respiration
- Prolonged expiratory period (with grunting)
- Wheezing
- Pursed-lip breathing
- Increased respiratory rate
- Peripheral cyanosis

INFLAMMATORY/INFECTIOUS DISEASE

Asthma

Asthma is a reversible obstructive lung disease caused by increased reaction of the airways to various stimuli. It is a chronic inflammatory condition with acute exacerbations that can be life-threatening if not properly managed. Our understanding of asthma has changed dramatically over the last decade.

Asthma was once viewed as a bronchoconstrictive disorder in which the airways narrowed, causing wheezing and breathing difficulties. Treatment with bronchodilators to open airways was the primary focus. Scientific evidence now supports the idea that asthma is primarily an inflammatory disorder in which the constriction of airways is a symptom of the underlying inflammation.

Asthma and other atopic disorders are the result of complex interactions between genetic predisposition and multiple environmental influences. The marked increase in asthma prevalence in the last 3 decades suggests environmental factors as a key contributor in the process of allergic sensitization.[12]

Fifteen million persons of all ages are affected by asthma in the United States. This represents a 61% increase over the last 15 years with a 45% increase in mortality during the last decade. Women are affected more than men, accounting for about 60% of the nearly 18 million cases of adult asthma. Hormones are thought to be a possible cause for this increase in incidence in women.[13]

Immune Sensitization and Inflammation

There are two major components to asthma. When the immune system becomes sensitized to an allergen, usually through heavy exposure in early life, an inflammatory cascade occurs, extending beyond the upper airways into the lungs.

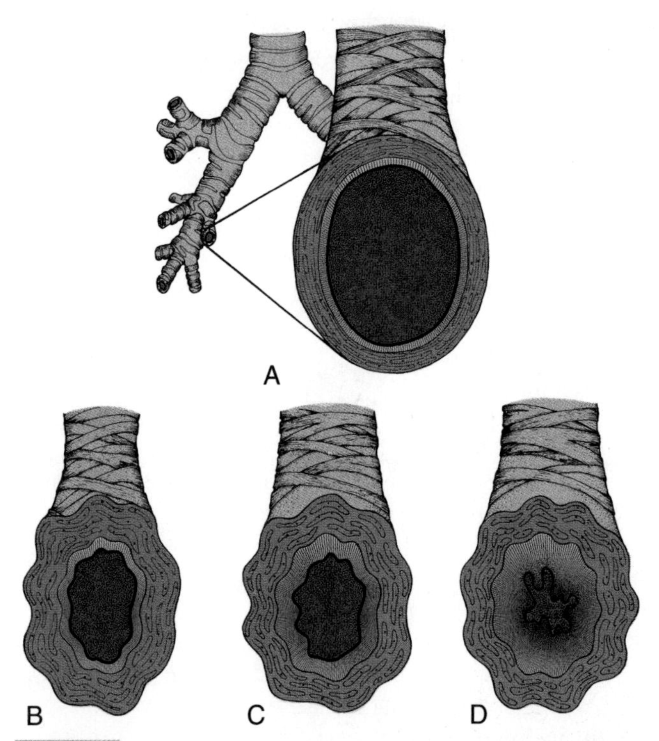

Fig. 7-5 Airway changes with asthma. **A,** Normal bronchus: cross-section of a normal bronchus (mucous membrane in color). Healthy bronchioles accommodate a constant flow of air when open and relaxed. **B,** Asthma: airway inflammation begins. The smooth muscle surrounding the bronchus contracts and causes narrowing of the airway, called *bronchospasm*. **C,** The airway tissue swells; this edema of the mucous membrane further narrows airways. **D,** Mucus is produced, further compromising airflow.

The lungs become hyperreactive, responding to allergens and other irritants in an exaggerated way. This hyperresponsiveness causes the muscles of the airways to constrict, making breathing more difficult (Fig. 7-5). The second component is inflammation, which causes the air passages to swell and the cells lining the passages to produce excess mucus, further impairing breathing.

Asthma may be categorized as conventional asthma, occupational asthma, or exercise-induced asthma (EIA), but the underlying pathophysiologic complex remains the same. Since the triggers or allergens vary, each person reacts differently. SOB, wheezing, tightness in the chest, and cough are the most commonly reported symptoms, but other symptoms may also occur.

Clinical Signs and Symptoms

Anytime a client experiences SOB, wheezing, and cough and comments, "I'm more out of shape than I thought," the therapist should ask about a past medical history of asthma and review the list of symptoms with the client. Therapists working with known asthmatic clients should encourage them to maintain hydration by drinking fluids to prevent mucous plugs from hardening and to take prescribed medications.

EIA or hyperventilation-induced asthma potentially can be prevented by exercising in a moist, humid environment and by grading exercise according to client tolerance using diaphragmatic breathing. Any type of sustained running or cycling or activity in the cold is more likely to precipitate EIA (Box 7-2).

Complications. Status asthmaticus is a severe, life-threatening complication of asthma. With severe bronchospasm the workload of breathing increases five to ten times, which can lead to acute cor pulmonale. When air is trapped, a severe paradoxic pulse develops as venous return is obstructed. This condition is seen as a blood pressure drop of more than 10 mm Hg during inspiration.

Pneumothorax can develop. If status asthmaticus continues, hypoxemia worsens and acidosis begins. If the condition is untreated or not reversed, respiratory or cardiac arrest will occur. An acute asthma episode may constitute a medical emergency.

Medical treatment for the underlying inflammation and resulting airway obstruction is with antiinflammatory agents and bronchodilators to prevent, interrupt, or terminate ongoing inflammatory reactions in the airways. A new class of antiinflammatory agents known as leukotriene modifiers works by blocking the activity of chemicals called leukotrienes, which are involved in airway inflammation. Reducing, eliminating, and avoiding allergens or triggers are important in self-care (see Box 7-2).

CLINICAL SIGNS AND SYMPTOMS

Asthma

Listen for
- Wheezing, however light
- Irregular breathing with prolonged expiration
- Noisy, difficult breathing
- Episodes of dyspnea
- Clearing the throat (tickle at the back of the throat or neck)
- Cough with or without sputum production, especially in the absence of a cold and/or occurring 5 to 10 minutes after exercise

Look for
- Skin retraction (clavicles, ribs, sternum)
- Hunched-over body posture; inability to stand, sit straight, or relax
- Pursed-lip breathing
- Nostrils flaring
- Unusual pallor or unexplained sweating

Ask about
- Restlessness during sleep
- Vomiting
- Fatigue unrelated to working or playing

BOX 7-2 FACTORS THAT MAY TRIGGER ASTHMA

- Respiratory infections, colds
- Cigarette smoke
- Allergic reactions to pollen, mold, animal dander, feather, dust, food, insects
- Indoor and outdoor air pollutants, including ozone
- Physical exertion or vigorous exercise
- Exposure to cold air or sudden temperature change
- Excitement or strong emotion, psychologic or emotional stress

Pneumonia

Pneumonia is an inflammation of the lungs and can be caused by (1) aspiration of food, fluids, or vomitus; (2) inhalation of toxic or caustic chemicals, smoke, dust, or gases; or (3) a bacterial, viral, or mycoplasmal infection. It may be primary or secondary (a complication of another disease); it often follows influenza.

The common feature of all types of pneumonia is an inflammatory pulmonary response to the offending organism or agent. This response may involve one or both lungs at the level of the lobe (lobar pneumonia) or more distally beginning in the terminal bronchioles and alveoli (bronchopneumonia). Bronchopneumonia is seen more frequently than lobar pneumonia and is common in clients postoperatively and in clients with chronic bronchitis, particularly when these two situations coexist.

There are three main types of pneumonia: hospital-acquired pneumonia (HAP; also known as nosocomial pneumonia), ventilator-associated pneumonia (VAP), and community-acquired pneumonia (CAP). By definition, HAP occurs 48 hours or more after hospital admission; when HAP develops in a mechanically ventilated patient after endotracheal intubation, it becomes VAP.

Risk Factors

Infectious agents responsible for pneumonia are typically present in the upper respiratory tract and cause no harm unless resistance is lowered severely by some other factor such as smoking, a severe cold, disease, alcoholism, or generally poor health (e.g., poorly controlled diabetes, chronic renal problems, compromised immune function).

Older or bedridden clients are particularly at risk because of physical inactivity and immobility. Limited mobility causes normal secretions to pool in the airways and facilitates bacterial growth. Other risk factors predisposing a client to pneumonia are listed in Box 7-3.

Pneumocystis carinii is a protozoan organism that rarely causes pneumonia in healthy individuals. *Pneumocystis carinii* pneumonia (PCP) has been the most common life-threatening opportunistic infection in persons with acquired immunodeficiency syndrome (AIDS). PCP also has been shown to be the first indicator of conversion from human immunodeficiency virus (HIV) infection to the designation of AIDS.

BOX 7-3 RISK FACTORS FOR PNEUMONIA

- Age: very young, very old
- Have not received a pneumococcal vaccination
- Smoking
- Air pollution
- Upper respiratory infection (URI)
- Altered consciousness: alcoholism, head injury, seizure disorder, drug overdose, general anesthesia
- Endotracheal intubation, nasogastric tube
- Recent chest surgery
- Prolonged immobility
- Immunosuppressive therapy: corticosteroids, cancer chemotherapy
- Nonfunctional immune system: acquired immunodeficiency syndrome (AIDS)
- Severe periodontal disease
- Prolonged exposure to virulent organisms
- Malnutrition, dehydration
- Chronic diseases: diabetes mellitus, heart disease, chronic lung disease, renal disease, cancer
- Prolonged debilitating disease
- Inhalation of noxious substances
- Aspiration of oral/gastric material (food or fluid), foreign materials (e.g., Petroleum products)
- Chronically ill, older clients who have poor immune systems, often residing in group-living situations; transfer from one health care facility to another (hospital-acquired or nosocomial pneumonia); hospitalization in the fall or winter

Clinical Signs and Symptoms

The onset of all pneumonias is generally marked by any of the following: fever, chills, sweats, pleuritic chest pain, cough, sputum production, hemoptysis, dyspnea, headache, or fatigue. PCP causes a dry, hacking cough without sputum production.

The older client can have full-blown pneumonia and may appear with altered mental status (especially confusion) rather than fever or respiratory symptoms because of the changes in temperature regulation as we age. Anytime an older person has shoulder pain and confusion at presentation, consider the possibility of diaphragmatic impingement by an underlying lung pathologic condition (Case Example 7-2).

The clinical manifestations of PCP are slow to develop; they include fever, tachypnea, tachycardia, dyspnea, nonproductive cough, and hypoxemia. A diffuse, bilateral pattern of alveolar infiltration is apparent on chest radiograph.

Hospitalization may be required for the immunocompromised client. Otherwise, if the client has an intact defense system and good general health, recuperation can take place at home with rest and supportive treatment. In the hospital, rigorous handwashing by medical personnel is essential for reducing the transmission of infectious agents.

CLINICAL SIGNS AND SYMPTOMS

Pneumonia

- Sudden and sharp pleuritic chest pain that is aggravated by chest movement
- Shoulder pain
- Hacking, productive cough (rust-colored or green, purulent sputum)
- Dyspnea
- Tachypnea (rapid respirations associated with fever or pneumonia) accompanied by decreased chest excursion on the affected side
- Cyanosis
- Headache
- Fever and chills
- Generalized aches and myalgia that may extend to the thighs and calves
- Knees may be painful and swollen
- Fatigue
- Confusion in older adult or increased confusion in client with dementia or Alzheimer's disease

CASE EXAMPLE 7-2

Pneumonia

A 42-year-old man came to an outpatient physical therapy clinic with complaints of painful, swollen knees. Symptoms were first observed 10 days ago, and the left knee was reportedly worse than the right. Stiffness was reported in the morning on rising, with pain increasing as the day progressed. As a nonsmoker, the man reported his general health as "good" and noted that he had sprained his left ankle 2 months before the onset of knee pain.

The knee joints were not tender, warm, or red. Observable and palpable "boggy" fluid could be demonstrated in the popliteal spaces bilaterally. There was no sign of effusion when viewed anteriorly, and the test for a wave of fluid was negative.

All special tests for the hip and knee were negative, with full active and passive range of motion present. There was no known history of Baker's cysts (herniation of synovial tissue through a weakening in the posterior capsule wall) reported. There were no palpable myalgias of the lower leg musculature and no trigger points present.

The left ankle demonstrated some residual stiffness with a mild loss of plantar flexion. Joint accessory motions were consistent with a grade 1 lateral ankle sprain. Standing posture was unremarkable for possible contributing alignment problems.

The clinical presentation of this client was puzzling to the evaluating therapist. A brief screening for possible systemic origin of symptoms elicited no red-flag symptoms or history. Ongoing evaluation continued as treatment for both knees and the left ankle was initiated. After 3 weeks, there were no changes in the clinical presentation of the knees. At that time the client developed a noticeable productive cough with greenish/yellow sputum but no other reported symptoms. Vital signs (including temperature) were unremarkable.

Given the unusual clinical presentation, lack of progress with treatment, and development of a productive cough, this client was referred to his family physician for a medical evaluation. A one-page letter outlining the therapist's findings and treatment protocol was sent with the client. A medical diagnosis of pneumonia was established. The physician noted that although the clinical presentation was unusual, knee involvement can occur with pneumonia. The pathophysiologic mechanism for this is unknown.

Tuberculosis

Tuberculosis (TB) is a bacterial infectious disease transmitted by the gram-positive, acid-fast bacillus *Mycobacterium tuberculosis.* Despite improved methods of detection and treatment, TB remains a worldwide health problem with increasing spread of a highly drug-resistant strain of TB present in almost every state in the United States.

Before the development of anti-TB drugs in the late 1940s, TB was the leading cause of death in the United States. Drug therapy, along with improvements in public health and general living standards, resulted in a marked decline in incidence. However, recent influxes of immigrants from developing Third World nations, rising homeless populations, and the emergence of HIV led to an increase in reported cases in the mid1980s, reversing a 40-year period of decline.

Risk Factors

Although TB can affect anyone, certain segments of the population have an increased risk of contracting the disease (Box 7-4). The mycobacterium is usually spread by airborne droplet nuclei, which are produced when actively infected persons sneeze, speak, sing, or cough.

Once released into the atmosphere, the organisms are dispersed and can be inhaled by a susceptible host. Brief exposure to a few bacilli rarely causes an infection. More commonly, it is spread with repeated close contact with an infected person.

Drug-resistant strains of TB have developed when the full course of treatment, lasting 6 to 9 months, is not completed. Once the infected person feels better and stops taking the prescribed medication, a new drug-resistant strain is passed along. Incomplete treatment among inner-city residents and the homeless presents a major factor in the failure to eradicate TB.

Drug-resistant strains are also developing globally. Areas of the world with increased rates of drug-resistant disease include countries of the former Soviet Union (e.g., Estonia, Kazakhstan, Latvia, Lithuania, Uzbekistan) and in Central Asia.

Families who adopt internationally should be aware of potential TB infection in children from some of the high-risk areas of the world. The vaccine BCG (bacille Calmette-Guerin) has been used in many foreign countries to attempt to prevent serious dissemination of TB infection in those countries.

BOX 7-4 RISK FACTORS FOR TUBERCULOSIS

- Health care workers, especially those working in older hospitals (centralized ventilation), homeless shelters, or extended care facilities. Health care workers, including physical therapists must be alert to the need to use a special mask (particulate respirator) when cough-inducing procedures are being performed with any client with a risk for, or active tuberculosis (TB).
- Older adults, who constitute nearly half of the newly diagnosed cases of TB in the United States.
- Overcrowded housing, most common among the economically disadvantaged; homeless, especially those people in crowded homeless shelters.
- People who are incarcerated.
- U.S.-born non-Hispanic Blacks.[47]
- Immigrants (including adopted children) from Southeast and Central Asia, Ethiopia, Mexico, Latin America, Eastern Europe.
- Clients who are dependent on alcohol or other chemicals with resultant malnutrition, debilitation, and poor health.
- Infants and children under the age of 5 years.
- Clients with reduced immunity or malnutrition (e.g., anyone undergoing cancer therapy or steroid therapy) and those with human immunodeficiency virus (HIV)-positive lung cancer or head and neck cancer.
- Persons with diagnosed rheumatoid arthritis. Data suggests increases in incidence could be due to new immunosuppressive treatments.[48]
- Persons with diabetes mellitus and/or end-stage renal disease.
- People with a history of gastrointestinal disease (e.g., chronic malabsorption syndrome, upper gastrointestinal carcinomas, gastrectomy, intestinal bypass).

The value of BCG is controversial, since the protection it confers is short term. The Centers for Disease Control and Prevention (CDC) and the American Academy of Pediatrics strongly recommend that history of BCG vaccination in a child from a high-risk part of the world should usually be ignored and all children adopted internationally should be skin tested for TB and treated if the disease is latent or active.[14]

TB most often involves the lungs, but extrapulmonary TB (XPTB) can also occur in the kidneys, bone growth plates, lymph nodes, and meninges and can be disseminated throughout the body.

Widespread dissemination throughout the body is termed *miliary tuberculosis* and is more common in people 50 years or older and very young children with unstable or underdeveloped immune systems.

On rare occasions, TB will affect the hip joints and vertebrae, resulting in severe, arthritis-like damage, possibly even avascular necrosis of the hip. Tuberculosis of the spine, referred to as Pott's disease, is rare but can result in compression fracture of the vertebrae.

Pyogenic vertebral osteomyelitis can be caused by atypical organisms such as tuberculosis. As with other pyogenic infection, back pain is the most common symptom, but it is less severe than in other infections. Individuals from high-risk areas of the world, high-risk living conditions, and the immunocompromised and malnourished should be considered suspect for this condition.[15]

Clinical Signs and Symptoms

Clinical signs and symptoms are absent in the early stages of TB. Many cases are found incidentally when routine chest radiographs are made for other reasons. When systemic manifestations of active disease initially appear, the clinical signs and symptoms listed here may appear.

Tuberculin skin testing is done to determine whether the body's immune response has been activated by the presence of the bacillus. A positive reaction develops 3 to 10 weeks after the initial infection. A positive skin test reaction indicates the presence of a tuberculous infection but does not show whether the infection is dormant or is causing a clinical illness.

Chest x-ray films and sputum cultures are done as a follow-up to positive skin tests. All cases of active disease are treated, and certain cases of inactive disease are treated prophylactically.

CLINICAL SIGNS AND SYMPTOMS

Tuberculosis

- Fatigue
- Malaise
- Anorexia
- Weight loss
- Low-grade fevers (especially in late afternoon)
- Night sweats
- Frequent productive cough
- Dull chest pain, tightness, or discomfort
- Dyspnea

Systemic Sclerosis Lung Disease

Systemic sclerosis (SS), or scleroderma, is a restrictive lung disease of unknown etiologic origin characterized by inflammation and fibrosis of many organs (see Chapter 12). Fibrosis affecting the skin and the visceral organs is the hallmark of SS.

The lungs, highly vascularized and composed of abundant connective tissue, are a frequent target organ, ranking second only to the esophagus in visceral involvement.

The most common pulmonary manifestation of SS is interstitial fibrosis, which is clinically apparent in more than 50% of cases. Autopsy results suggest a prevalence of 75%, indicating the insensitivity of traditional tests such as pulmonary function tests and chest radiographs.

Clinical Signs and Symptoms

As discussed in Chapter 12, skin changes associated with SS generally precede visceral alterations. Dyspnea on exertion and nonproductive cough are the most common clinical findings associated with SS. Rarely, these symptoms precede the occurrence of cutaneous changes of scleroderma.

Clubbing of the nails rarely occurs in SS because of the nearly universal presence of sclerodactyly (hardening and shrinking of connective tissues of fingers and toes). Peripheral edema may develop secondary to cor pulmonale, which occurs as the pulmonary fibrosis becomes advanced.

Pulmonary manifestations in systemic sclerosis include:

- *Common:* Interstitial pneumonitis and fibrosis and pulmonary vascular disease
- *Less common:* Pleural disease, aspiration pneumonia, pneumothorax, neoplasm, pneumoconiosis, pulmonary hemorrhage, and drug-induced pneumonitis

Pleural effusions may appear with orthopnea, edema, and paroxysmal nocturnal dyspnea if CHF occurs. Cystic changes in the parenchyma may progress to form pneumatoceles (thin-walled air-containing cysts) that may rupture spontaneously and produce a pneumothorax. Clients with SS have an increased incidence of lung cancer. Hemoptysis is often the first signal of pulmonary malignancy in individuals with SS.

The course of SS is unpredictable, from a mild, protracted course to rapid respiratory failure and death. Treatment of pulmonary complications, pulmonary hypertension, and interstitial lung disease remains difficult.

CLINICAL SIGNS AND SYMPTOMS

Systemic Sclerosis Lung Disease

- Dyspnea on exertion
- Nonproductive cough
- Peripheral edema (secondary to cor pulmonale)
- Orthopnea
- Paroxysmal nocturnal dyspnea (CHF)
- Hemoptysis

Neoplastic Disease

Lung Cancer (Bronchogenic Carcinoma)

Lung cancer is malignancy in the epithelium of the respiratory tract. At least a dozen different cell types of tumors are included under the classification of lung cancer.

Clinically, lung cancers are grouped into two divisions: small cell lung cancer (15% of all lung cancers) and non–small cell lung cancer. The four major types of lung cancer include small cell carcinoma (oat cell carcinoma), and the subtypes of non–small cell lung cancer (e.g., squamous cell carcinoma [25% to 30%], adenocarcinoma [40%], and large cell carcinoma [10% to 15%]).

Since the mid1950s, lung cancer has been the most common cause of death from cancer in men. In 1987, lung cancer surpassed breast cancer to become the leading cause of cancer death in women in the United States. It is now the second most commonly diagnosed cancer in both men and women and remains the number one cause of death in both groups.[15a] Lung cancer affects some races more than others; blacks have higher incidence and mortality rates than do whites.[16]

The rate of lung cancer among women in the United States may be declining for some age groups. This decline is expected to continue through at least the year 2025. Sustaining that trend will require continued reductions in the number of female children who start to smoke and continued smoking cessation among addicted female smokers.[17]

Risk Factors. Smoking is the major risk factor for lung cancer, accounting for 82% of deaths caused by lung cancer.[18] Other risk factors are listed in Box 7-5. Compared with nonsmokers, heavy smokers (i.e., those who smoke more than 25 cigarettes a day) have a twentyfold greater risk of developing cancer.

States with strong anti-tobacco programs (e.g., Arizona, California) have the fewest current smokers, the most people who have quit smoking in some age groups, and the greatest drop in the death rate from lung cancer.[19] Quitting smoking lowers the risk, but the decrease is gradual and does not approach that of a nonsmoker.

The risk of lung cancer is increased in the smoker who is exposed to other carcinogenic agents such as radon, asbestos, and chemical carcinogens. Internationally, the incidence of lung cancer is growing with industrialized nations having the highest rates.[20]

The increase of lung cancer mortality in the last decades can be entirely attributed to the trend of tobacco consumption. However, there is a lag time of many years between beginning smoking and the clinical manifestation of cancer. The therapist can have a key role in the prevention of lung cancer through risk-factor assessment and client education (see Chapter 2). Personal risk of lung cancer can be assessed at www.yourdiseaserisk.wustl.edu.

Metastases. Metastatic spread of pulmonary tumors is usually to the long bones, vertebral column (especially the

BOX 7-5 RISK FACTORS FOR LUNG CANCER

- Age greater than 50 years
- Smoking or other tobacco use
- Previous tobacco-related cancer
- Passive (environmental) smoke
- Low consumption of fruit and vegetables
- Genetic predisposition
- Exposure to air pollution, toxic chemicals (e.g., asbestos, uranium), fumes, radon gas
- Previous lung disease (e.g., chronic obstructive pulmonary disease [COPD], tuberculosis, pulmonary fibrosis, sarcoidosis)

thoracic vertebrae), liver, and adrenal glands. Brain metastasis is also common, occurring in as many as 50% of cases.

Local metastases by direct extension may involve the chest wall, pleura, pulmonary parenchyma, or bronchi. Further local tumor growth may erode the first and second ribs and associated vertebrae, causing bone pain and paravertebral pain associated with involvement of sympathetic nerve ganglia.

The respiratory system is a common site for complications associated with cancer and cancer therapy. Several factors can lead to pulmonary complications. Immunosuppression caused by the underlying disease or the cancer therapy can lead to infectious disease.

In addition, the lungs contain an enormous capillary bed through which flows the entire venous circulation, making it a common site of metastasis from other primary cancers and pulmonary emboli. Carcinomas of the kidney, breast, pancreas, colon, and uterus are especially likely to metastasize to the lungs.

Clinical Signs and Symptoms. Clinical signs and symptoms of lung cancer often remain silent until the disease process is at an advanced stage. In many instances, lung cancer may mimic other pulmonary conditions or may initially appear as chest, shoulder, or arm pain (Case Example 7-3).

Chest pain is a vague aching, and depending on the type of cancer, the client may have pleuritic pain on inspiration that limits lung expansion. Anorexia and weight loss occur in many clients with lung cancer and can be a symptom of advanced disease.[21]

Hemoptysis (coughing or spitting up blood) may occur secondary to ulceration of blood vessels. Wheezing occurs when the tumor obstructs the bronchus. Dyspnea, either unexplained or out of proportion, is a red flag indicating the need for medical screening, as is unexplainable weight loss accompanied by dyspnea.

Centrally located tumors cause increased cough, dyspnea, and diffuse chest pain that can be referred to the shoulder,

CASE EXAMPLE 7-3

Neurologic Deficits in a Smoker

A 66-year-old man was referred by his primary care physician to physical therapy for weakness in the lower extremities. He also reported dysesthesia (pain with touch) in both legs from the knees down. The symptoms had been present for about 1 month before he saw his doctor. At the time of his physical therapy evaluation, symptoms had been present for almost 2 months (client was delayed getting in to see a therapist due to his scheduling conflicts).

The client was a social worker who had never been married but had two children out of wedlock, had a history of chronic alcohol use, and reported a 60-pack year history of tobacco use (pack years = number of packs/day × number of years; in this case the client had smoked 2 packs a day for the last 30 years).

Past medical history was negative for any previous significant injuries, illnesses, or hospitalizations. Both parents were killed in a car accident when the client was a child. Any other family history was unknown for the parents and unremarkable for the siblings.

Clinical examination revealed mild weakness in the distal muscle groups of the lower extremities (left weaker than right). Altered sensation was circumferential and included both lower legs equally. Tests for clonus and Babinski were negative. Deep tendon reflexes were equal bilaterally and within normal limits (WNL). Other neurologic screening tests were negative. There were no constitutional signs or symptoms reported or observed.

A program of strengthening and conditioning was started based on the physician's referral requesting strength training and clinical findings of muscular weakness. In the first 2 weeks of treatment, the client's weakness increased and he developed bilateral foot drop.

He started reporting episodes of dropping anything he lifted over 2 pounds. A quick screening examination showed bilateral weakness developing in the hands and wrists, as well as the feet and ankles.

What are all the red flags in this case?

Age (over 50 years)
Significant smoking history
Alcohol use
Bilateral symptoms (hands and feet)
Progressive neurologic symptoms

Are there any other screening tests that can/should be done?

A screening physical examination should be conducted (see Chapter 4) and any significant findings noted and reported to the physician.

A general survey, vital signs, and chest auscultation would be a good place to start. It is possible that the client presentation (and certainly the new onset of symptoms) is unknown to the physician who saw him almost a month ago.

In fact, the therapist observed signs of digital clubbing (hands only), reported oxygen saturation levels (SaO_2) consistently at 90%, and noted bilateral basilar crackles on lung auscultation.

The client was advised to make a follow-up appointment with the physician, and the therapist faxed a letter of request for follow-up based on these new findings. The client also hand carried a copy of the therapist's letter to the medical appointment.

Result: The client was diagnosed with lung cancer with accompanying paraneoplastic syndrome (see discussion of paraneoplastic syndromes, Chapter 13). His condition worsened rapidly and he died 6 weeks later.

The family later came back to the therapist and expressed their appreciation for finding the problem early enough to make end-of-life decisions. Both children and three of the four siblings were able to visit with the gentleman before he died suddenly in his sleep.

scapulae, and upper back. This pain is the result of peribronchial or perivascular nerve involvement.

Other symptoms may include postobstructive pneumonia with fever, chills, malaise, anorexia, hemoptysis, and fecal breath odor (secondary to infection within a necrotic tumor mass). If these tumors extend to the pericardium, the client may develop a sudden onset of arrhythmia (tachycardia or atrial fibrillation), weakness, anxiety, and dyspnea.

Peripheral tumors are most often asymptomatic until the tumor extends through visceral and parietal pleura to the chest wall. Irritation of the nerves causes localized sharp, pleuritic pain that is aggravated by inspiration.

Metastases to the mediastinum (tissue and organs between the sternum and the vertebrae, including the heart and its large vessels; trachea; esophagus; thymus; lymph nodes) may cause hoarseness or dysphagia secondary to vocal cord paralysis as a result of entrapment or local compression of the laryngeal nerve.

Apical (Pancoast's) tumors of the lung apex do not usually cause symptoms while confined to the pulmonary parenchyma. They can extend into surrounding structures and frequently involve the eighth cervical and first thoracic nerves within the brachial plexus.

A constellation of symptoms referred to as Pancoast's syndrome present in the distribution of the C8, T1, and T2 dermatomes, mimicking thoracic outlet syndrome.[22] Tumor invasion of any anatomic structures of the lower trunks of the brachial plexus and/or the C8 and T1 nerve roots can result in significant disability and loss of hand function.[22] Extension of the tumor into the paravertebral sympathetic nerves results in Horner's syndrome, which consists of enophthalmos (backward displacement of the eye), ptosis (drooping eyelid), and miosis (pupil constriction).

The most common initial symptom is sharp (often posterior) shoulder pain produced by invasion of the brachial plexus and/or extension of the tumor into the parietal pleura, endothoracic fascia, first and second ribs, or vertebral bodies. There may be pain in the axilla and subscapular areas on the affected side (Case Example 7-4).

CASE EXAMPLE 7-4

Pancoast's Tumor

A 55-year-old man presented with shoulder pain radiating down the arm present for the last 3 months. His job as a mechanic required many hours with his arms raised overhead, which is what he thought was causing the problem.

He was diagnosed with cervical radiculopathy after cervical x-rays showed moderate osteoarthritic changes at the C678 levels. Electromyographic (EMG) studies confirmed the diagnosis, and he was sent to physical therapy.

The client gave a history as a nonsmoker but mentioned his parents were chain-smokers and his wife of 35 years also smokes heavily. There was no other significant social or personal history. The client was adopted and did not know his family history.

The therapist conducted a physical screening examination and noted a slight drooping of the left eyelid, which the client attributed to fatigue and changes with middle age. Vital signs were unremarkable. Muscle atrophy and weakness were present in the left hand consistent with a C78 neurologic impairment. There were changes in the thumbs and index fingers on both sides with what looked like early signs of digital clubbing. The nail beds were spongy with a definite change in the shape of the distal phalanx.

When asked if there were any other symptoms of any kind anywhere else in the body, the client mentioned a change in the way he perspires. He noticed his left face and armpit do not perspire like the right side. He could not remember when this change began but knew it was not something he had his whole life.

What are the red flag signs in this case?

Age
Exposure to passive tobacco smoke
Nail bed changes
Anhydrosis (lack of sweating)
Questionable changes in eyelid

What other screening tests might be appropriate?

A more careful neurologic examination is in order. Any findings to suggest impairment outside the parameters of the C678 nerve function might raise a yellow flag. Upper limb neurodynamic and neural tension tests are important; trigger point assessment is often helpful. A cranial nerve assessment also is advised due to the possible eye drooping observed.

Although the vital signs were unremarkable, the nailbed changes should prompt lung auscultation and a closer look at skin color, capillary refill time, and peripheral vascular assessment.

Result: No other neurologic findings were observed and the cranial nerve assessment was within normal limits with the exception of the eyelid drooping. The therapist also noticed the pupil in the left eye seemed smaller than the pupil on the right. Repeated attempts to use a penlight or darkness to change pupil size were unsuccessful.

Based on objective findings, the therapist started an intervention of neural mobilizations and gave the client a home program of postural exercises and self-neural mobilizations. A plan was outlined to integrate a strengthening program as soon as time would allow.

All findings were documented and sent to the physician. The client was not sent back to the physician, but the physician, upon reading the therapist's notes, called the therapist and asked for further explanation of the therapist's findings. The physician was concerned and asked to have the client make a follow-up appointment.

Further medical testing brought about a diagnosis of Pancoast's tumor in the left upper lung lobe involving the brachial plexus and the first and second ribs. Horner's syndrome was also recognized as the cause of his drooping eye and anhydrosis.

Symptoms from the Horner's syndrome resolved after radiotherapy; pain and weakness also improved after medical therapy. Physical therapy for rehabilitation was initiated after medical treatment, but the client developed progressive regional disease with distant metastases and died within 2 months.

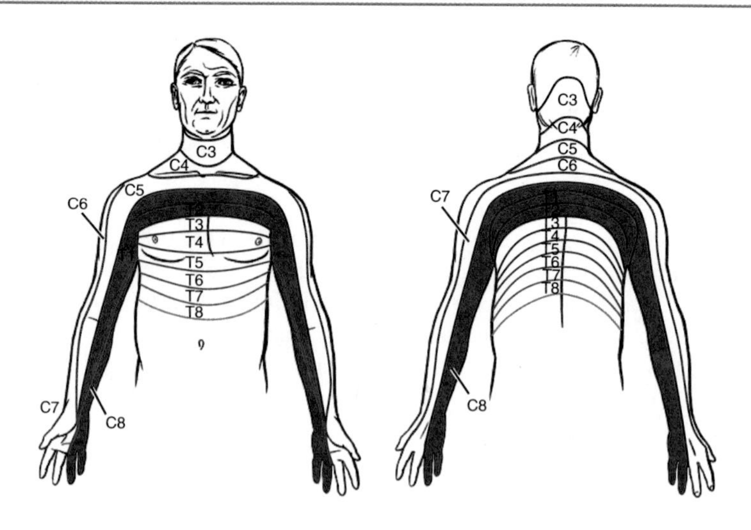

Fig. 7-6 Pancoast's tumors can present with changes in cutaneous dermatomal innervation. The shaded areas show the dermatomes affected when the superior (apical) sulcus tumors associated with Pancoast's syndrome invade the brachial plexus. Direct extension to the brachial plexus involving C8 and T1 result in symptoms affecting the C8, T1, and T2 dermatomes.

Pain may radiate up to the head and neck, across the chest, and/or down the medial aspect of the arm and hand (ulnar nerve distribution) (Fig. 7-6). There may be subsequent atrophy of the upper extremity muscles with weakness of the muscles of the hand.

Pulmonary symptoms, such as cough, hemoptysis, and dyspnea, are uncommon until late in the disease. Affected individuals are often treated for presumed cervical osteoarthritis or shoulder bursitis resulting in delay of diagnosis. The pain eventually progresses to become severe and constant, eventually resulting in more thorough testing and accurate diagnosis.[23] The onset of pulmonary symptoms in any client with neck, shoulder, and/or arm pain should be a red flag symptom for the therapist.

Trigger points of the serratus anterior muscle (see Fig. 17-7) also mimic the distribution of pain caused by the eighth cervical nerve root compression. Trigger points can be ruled out by palpation and lack of neurologic deficits and may be confirmed by elimination with appropriate physical therapy intervention.

Paraneoplastic syndromes (remote effects of a malignancy; see explanation in Chapter 13) occur in 10% to 20% of lung cancer clients and represent a feature of advanced disease. These usually result from the secretion of hormones by the tumor acting on target organs, producing a variety of symptoms. Occasionally, symptoms of paraneoplastic syndrome occur before detection of the primary lung tumor.

As mentioned earlier, brain metastasis is common, occurring in as much as 50% of cases. About 10% of all individuals with lung cancer have central nervous system (CNS) involvement at the time of diagnosis. CNS symptoms, such as muscle weakness, muscle atrophy, loss of lower extremity sensation, and localized or radicular back pain, may be associated with lung cancer and must be investigated by a physician to establish a medical diagnosis.

Other clinical symptoms of brain metastasis resulting from increased intracranial pressure may include headache, nausea, vomiting, malaise, anorexia, weakness, and alterations in mental processes. Localized motor or sensory deficits occur, depending on the location of lesions (see Chapter 13).

Metastasis to the spinal cord produces signs and symptoms of cord compression (see Table 13-5 and Appendix A-2), including back pain (localized or radicular), muscle weakness, loss of lower extremity sensation, bowel and bladder incontinence, and diminished or absent lower extremity reflexes (unilateral or bilateral).

CLINICAL SIGNS AND SYMPTOMS

Lung Cancer

- Any change in respiratory patterns
- Recurrent pneumonia or bronchitis
- Hemoptysis
- Persistent cough
- Change in cough or development of hemoptysis in a chronic smoker
- Hoarseness or dysphagia
- Sputum streaked with blood
- Dyspnea (SOB)
- Wheezing
- Sharp chest, upper back, shoulder, scapular, rib, or arm pain aggravated by inspiration or accompanied by respiratory signs and symptoms
- Sudden, unexplained weight loss; anorexia; fatigue
- Chest, shoulder, or arm pain; bone aching, joint pain
- Atrophy and weakness of the arm and hand muscles
- Fecal breath odor
- See also Clinical Signs and Symptoms of Paraneoplastic Syndrome, Brain Metastasis, and Metastasis to the Spinal Cord in Chapter 13; Table 13-5; and Appendix A-2

GENETIC DISEASE OF THE LUNG

Cystic Fibrosis

Cystic fibrosis (CF) is an inherited disease of the exocrine ("outward-secreting") glands primarily affecting the digestive and respiratory systems.

This disease is the most common genetic disease in the United States, inherited as a recessive trait: both parents must be carriers, each having a defective copy of the CF gene. Each time two carriers conceive a child, there is a 25% chance that the child will have CF, a 50% chance that the child will be a carrier, and a 25% chance that the child will be a noncarrier. In the United States, 5% of the population, or 12 million people, carry a single copy of the CF gene.

Because cysts and scar tissue on the pancreas were observed during autopsy when the disease was first being differentiated from other conditions, it was given the name *cystic fibrosis of the pancreas*. Although this term describes a secondary rather than primary characteristic, it has been retained.

In 1989, scientists isolated the CF gene located on chromosome 7. In healthy people a protein called CF transmembrane conductance regulator (CFTR) provides a channel by which chloride (a component of salt) can pass in and out of cells.

Persons with CF have a defective copy of the gene that normally enables cells to construct that channel. As a result, salt accumulates in the cells lining the lungs and digestive tissues, making the surrounding mucus abnormally thick and sticky. These secretions, which obstruct ducts in the pancreas, liver, and lungs, and abnormal secretion of sweat and saliva are the two main features of CF.

Usually, CF manifests itself in early childhood, but there are some individuals who have a variant form of the disease in which symptoms can appear during adolescence or adulthood. Symptoms tend to be milder and sweat chloride concentration may be normal.[24]

Obstruction of the bronchioles by mucous plugs and trapped air predisposes the client to infection, which starts a destructive cycle of increased mucus production with increased bronchial obstruction, infection, and inflammation with eventual destruction of lung tissue.

Clinical Signs and Symptoms

Pulmonary involvement is the most common and severe manifestation of CF. Obstruction of the airways leads to a state of hyperinflation and bronchiectasis. In time, fibrosis develops, and restrictive lung disease is superimposed on the obstructive disease.

Over time, pulmonary obstruction leads to chronic hypoxia, hypercapnia, and acidosis. Pneumothorax, pulmonary hypertension, and eventually cor pulmonale may develop. These are very poor prognostic indicators in adults. The course of CF varies from one client to another depending on the degree of pulmonary involvement.

Advances in treatment, including aerosolized antibiotics, mucus thinning agents, antiinflammatory agents, chest physical therapy, enzyme supplements, and nutrition programs, have extended the average life expectancy for CF sufferers into their early 20s, with maximal survival estimated at 30 to 40 years.

Because the genetic abnormality has been identified, considerable progress has been made in the development of gene therapy and preventive gene transfer for this disease.[25-27] Lung transplantation in older childhood and adolescence is a possible treatment option based on rapidly declining lung function.[28]

CLINICAL SIGNS AND SYMPTOMS
Cystic Fibrosis

In Early or Undiagnosed Stages
- Persistent coughing and wheezing
- Recurrent pneumonia
- Excessive appetite but poor weight gain
- Salty skin/sweat
- Bulky, foul-smelling stools (undigested fats caused by a lack of amylase and tryptase enzymes)

In Older Child and Young Adult
- Infertility
- Nasal polyps
- Periostitis
- Glucose intolerance

CLINICAL SIGNS AND SYMPTOMS
Pulmonary Involvement in Cystic Fibrosis

- Tachypnea (very rapid breathing)
- Sustained chronic cough with mucus production and vomiting
- Barrel chest (caused by trapped air)
- Use of accessory muscles for respiration and intercostal retraction
- Cyanosis and digital clubbing
- Exertional dyspnea with decreased exercise tolerance

Further complications include:
- Pneumothorax
- Hemoptysis
- Right-sided heart failure secondary to pulmonary hypertension

OCCUPATIONAL LUNG DISEASES

Lung diseases are among the most common occupational health problems. They are caused by the inhalation of various chemicals, dusts, and other particulate matter present in certain work settings. Not everyone exposed to occupational inhalants will develop lung disease. Prolonged exposure combined with smoking increases the risk of developing occupational lung disease and increases the severity of these diseases.[29]

During the interview process, the therapist will ask questions about occupational and smoking history to identify the possibility of an underlying pulmonary pathologic condition (see Chapter 2 and Appendix B-14).

The most commonly encountered occupational lung diseases are occupational lung cancer, occupational asthma (also known as work-related asthma), asbestosis, mesothelioma, and byssinosis (brown lung disease). Other less common occupational lung diseases include hypersensitivity pneumonitis, acute respiratory irritation, and pneumoconiosis (black lung disease, silicosis).

The greatest number of occupational agents causing *asthma* are those with known or suspected allergic properties such as plant and animal proteins (e.g., wheat, flour, cotton, flax, and grain mites). Exposures within the workplace can aggravate preexisting asthma.[30]

Asbestosis and *mesothelioma* occur as a result of asbestos exposure. Asbestos is the name of a group of naturally occurring minerals that separate into strong, very fine fibers. The fibers are heat-resistant and extremely durable, which are qualities that made asbestos useful in construction and industry.

Scarring of the lung tissue occurs in asbestosis as a result of exposure to the microscopic fibers of asbestos. Under certain circumstances, fibers can be released and pose a health risk such as lung cancer from inhaling the fibers. Mesothelioma is an otherwise rare cancer of the chest lining caused by asbestos exposure.

Byssinosis (brown lung disease) caused by dusts from hemp, flax, and cotton processing, results in chronic obstruction of the small airways impairing lung function. Textile workers are at greatest risk of disability from byssinosis.

Hypersensitivity pneumonitis, or allergic alveolitis, is most commonly due to the inhalation of organic antigens of fungal, bacterial, or animal origin. *Acute respiratory irritation* results from the inhalation of chemicals such as ammonia, chlorine, and nitrogen oxides in the form of gases, aerosols, or particulate matter. If such irritants reach the lower airways, alveolar damage and pulmonary edema can result. Although the effects of these acute irritants are usually short-lived, some may cause chronic alveolar damage or airway obstruction.

Pneumoconioses, or "the dust diseases," result from inhalation of minerals, notably silica, coal dust, or asbestos. These diseases are most commonly seen in miners, construction workers, sandblasters, potters, and foundry and quarry workers. Occupational exposure to dust, fumes, or gases (including diesel) increases mortality due to COPD, even among workers who have never smoked.[31]

Pneumoconioses usually develop gradually over a period of years, eventually leading to diffuse pulmonary fibrosis, which diminishes lung capacity and produces restrictive lung disease.[31]

Home Remodeling

Home remodeling projects in the United States have increased dramatically in the last decade. Whether it is a do-it-yourself project or the occupants remain in the home during remodeling, problems can occur from dust inhalation and exposure to hazardous materials such as lead, asbestos, and creosote. Creosote is toxic (inhaled as fumes) and is a skin and eye irritant.

Lead poisoning is a serious problem in home remodeling projects throughout the United States. Special precautions to avoid lead poisoning must be followed if the home was built prior to 1978.

Lead poisoning can occur from inhaling paint dust (the result of sanding or scraping painted surfaces) and lead can be found in soil (children come into contact during play). Both sources of poisoning are common problems associated with remodeling projects.

Anyone presenting with a constellation of integumentary, musculoskeletal, and/or neurologic symptoms accompanied by pulmonary involvement should be asked about the possibility of recent home remodeling projects and exposure to any of these materials.

Clinical Signs and Symptoms

Early symptoms of occupational-related lung disease depend on the specific exposure but may include noninflammatory joint pain, myalgia, cough, and dyspnea on exertion.

Chest pain, productive cough, and dyspnea at rest develop as the condition progresses. The therapist needs to be alert for the combination of significant arthralgias and myalgias with associated respiratory symptoms, accompanied by a past occupational and smoking history (see Appendix B-14).

CLINICAL SIGNS AND SYMPTOMS

Occupational Lung Diseases

- Arthralgia
- Myalgia
- Chest pain
- Cough
- Dyspnea on exertion (progresses to dyspnea at rest)
- See also signs and symptoms of lung cancer in this chapter

PLEUROPULMONARY DISORDERS

Pulmonary Embolism and Deep Venous Thrombosis

Pulmonary embolism (PE) involves pulmonary vascular obstruction by a displaced thrombus (blood clot), an air bubble, a fat globule, a clump of bacteria, amniotic fluid, vegetations on heart valves that develop with endocarditis, or other particulate matter. Once dislodged, the obstruction travels to the blood vessels supplying the lungs, causing SOB, tachypnea (very rapid breathing), tachycardia, and chest pain.

The most common cause of PE is deep venous thrombosis (DVT) originating in the proximal deep venous system of the lower legs. The embolism causes an area of blockage, which then results in a localized area of ischemia known as a *pulmonary infarct.* The infarct may be caused by small emboli that extend to the lung surface (pleura) and result in acute pleuritic chest pain.

Risk Factors

Three major risk factors linked with DVT are blood stasis (e.g., immobilization because of bed rest, such as with burn clients, obstetric and gynecologic clients, and older or obese populations), endothelial injury (secondary to neoplasm, surgical procedures, trauma, or fractures of the legs or pelvis), and hypercoagulable states (see Box 6-2).

Other people at increased risk for DVT and PE include those with CHF, trauma, surgery (especially hip, knee, and prostate surgery), age over 50 years, previous history of thromboembolism, malignant disease, infection, diabetes mellitus, inactivity or obesity, pregnancy, clotting abnormalities, and oral contraceptive use (see Chapter 6).

Prevention

Given the mortality of PE and the difficulties involved in its clinical diagnosis, prevention of DVT and PE is critical. A careful review of the Personal/Family History form (outpatient) or hospital medical chart (inpatient) may alert the therapist to the presence of factors that predispose a client to have a PE. Risk factor assessment is an important part of screening and prevention.

Although frequent changing of position, exercise, and early ambulation are necessary to prevent thrombosis and embolism, sudden and extreme movements should be avoided. Under no circumstances should the legs be massaged to relieve "muscle cramps," especially when the pain is located in the calf and the client has not been up and about.

Restrictive clothing and prolonged sitting or standing should be avoided. Elevating the legs should be accomplished with caution to avoid severe flexion of the hips, which will slow blood flow and increase the risk of new thrombi.

Deep Venous Thrombosis (see also Chapter 6)

Signs and symptoms of DVT include tenderness, leg pain, swelling (a difference in leg circumference of 1.4 cm in men and 1.2 cm in women is significant), and warmth. One may also see subcutaneous venous distention, discoloration, a palpable cord, and/or pain upon placement of a blood pressure cuff around the calf (considerable pain with the cuff inflated to 160 to 180 mm Hg).

Homans' sign is an unreliable test to diagnose DVT. The therapist should be aware that using the Homans' test to assess for DVT is no longer recommended or supported by evidence.[32-35] Homans' sign is elicited by gentle squeezing of the affected calf or slow dorsiflexion of the foot on the affected side to elicit deep calf pain. In theory, the inflamed nerves in the veins within the muscle are compressed or stretched causing deep calf pain. Only about half of all clients with DVT experience pain and Homans' sign.

Unfortunately, at least half the cases of DVT are asymptomatic,[34] and in up to one-third of all clients with apparent clinical appearance of DVT, there is no DVT demonstrable. A more sensitive and specific tool for predicting DVT is the Autar DVT Scale (see Table 4-11).

CLINICAL SIGNS AND SYMPTOMS

Deep Venous Thrombosis

- Unilateral tenderness or leg pain; dull ache or sensation of "tightness" in the area where the DVT is located
- Unilateral swelling (difference in leg circumference)
- Warmth
- Subcutaneous venous distention (superficial thrombus)
- Discoloration
- Palpable cord (superficial thrombus)
- Pain with placement around calf of blood pressure cuff inflated to 160 to 180 mm Hg

Pulmonary Embolism

Signs and symptoms of PE are nonspecific and vary greatly, depending on the extent to which the lung is involved, the size of the clot, and the general condition of the client.

Clinical presentation does not differ between younger and older persons. Dyspnea, pleuritic chest pain, and cough are the most common symptoms reported. Pleuritic pain is caused by an inflammatory reaction of the lung parenchyma or by pulmonary infarction or ischemia caused by obstruction of small pulmonary arterial branches.

Typical pleuritic chest pain is sudden in onset and aggravated by breathing. The client may also report hemoptysis, apprehension, tachypnea, and fever (temperature as high as 39.5° C, or 103.5° F). The presence of hemoptysis indicates that the infarction or areas of atelectasis have produced alveolar damage.

The therapist can use the *simplified Wells criteria* for clinical assessment of pulmonary embolism when assessing the need for referral for possible PE (Table 7-3). Previously, the *modified Wells rule* for assessing clinical probability for exclusion of pulmonary embolism was used with 1 to 3 points assigned for the variables listed.[36] Since that time, studies assigning only 1 point to the original variables have shown the simplified Wells rule to be a simpler, yet valid assessment tool[37-39]; prospective validation remains necessary.

The Wells criteria also outline factors that constitute the PE rule-out criteria (PERC). Combining information from the list below with the PE score can help the therapist recognize low versus high priority for medical consultation.[36] The following parameters reduce the likelihood of a pulmonary embolism:

- Age less than 50 years
- Heart rate less than 100 bpm
- Oxyhemoglobin saturation equal to or greater than 95%
- No hemoptysis
- No estrogen use
- No prior DVT or PE
- No unilateral leg swelling
- No surgery or trauma requiring hospitalization within the past 4 weeks

TABLE 7-3 Simplified Wells Criteria for the Clinical Assessment of Pulmonary Embolism

Criteria	Score
Clinical symptoms of DVT (leg swelling, pain with palpation)	1.0
Other diagnosis less likely than pulmonary embolism	1.0
Heart rate greater than 100 bpm	1.0
Immobilization for 3 or more days or surgery in the past four weeks	1.0
Previous history of DVT/PE	1.0
Hemoptysis	1.0
Malignancy	1.0
SCREENING CLINICAL PROBABILITY ASSESSMENT	
PE likely; medical consult advised	Total score: 2 or more
PE unlikely; review all other factors then document findings	Total score: 0 or 1

Data from Douma RA: Validity and clinical utility of the simplified Wells rule for assessing clinical probability for the exclusion of pulmonary embolism, *Thromb Haemost* 101(1):197-200, 2009, and Gibson NS: Further validation and simplification of the Wells clinical decision rule in pulmonary embolism, *Thromb Haemost* 99:229-234, 2008.
DVT, Deep venous thrombosis; *PE,* pulmonary embolism.

The PERC approach has a high negative predictive value and sensitivity when combined with a low probability of PE using the Wells criteria, but a low positive predictive value and specificity. In other words, low-risk patients who have all of the PERC are highly unlikely to have PE, but the absence of one or more of the PERC does not mean that a PE exists.[40]

CLINICAL SIGNS AND SYMPTOMS

Pulmonary Embolism

- Dyspnea
- Pleuritic (sharp, localized) chest pain
- Diffuse chest discomfort
- Persistent cough
- Hemoptysis (bloody sputum)
- Apprehension, anxiety, restlessness
- Tachypnea (increased respiratory rate)
- Tachycardia
- Fever

Cor Pulmonale

When a PE has been sufficiently massive to obstruct 60% to 75% of the pulmonary circulation, the client may have central chest pain, and acute cor pulmonale occurs. Cor pulmonale is a serious cardiac condition and an emergency situation arising from a sudden dilatation of the right ventricle as a result of PE.

As cor pulmonale progresses, edema, and other signs of right-sided heart failure develop. Symptoms are similar to those of CHF from other causes: dyspnea, edema of the lower extremities, distention of the veins of the neck, and liver distention. The hematocrit is increased as the body attempts to compensate for impaired circulation by producing more erythrocytes.

CLINICAL SIGNS AND SYMPTOMS

Cor Pulmonale

- Peripheral edema (bilateral legs)
- Chronic cough
- Central chest pain
- Exertional dyspnea or dyspnea at rest
- Distention of neck veins
- Fatigue
- Wheezing
- Weakness

Pulmonary Arterial Hypertension

Pulmonary arterial hypertension (PAH) is a condition of vasoconstriction of the pulmonary arterial vascular bed. PAH is medically defined as a mean pulmonary artery pressure of 25 mm Hg or more with a pulmonary capillary wedge pressure of 15 mm Hg or less (measured by cardiac catheterization).[41]

It can be either *primary* (rare), occurring three times more often in women in their 30s and 40s compared with men, or *secondary,* occurring as a result of other clinical conditions such as PE, chronic lung disease, sickle-cell disease, Graves' disease, polycythemia, collagen vascular disease, portal hypertension, heart abnormalities, and sleep apnea. Along with thromboemboli, tumors can also obstruct pulmonary circulation. Either type is probably a combination of genetic and environmental factors.[41,42]

Normally, the pulmonary circulation has a low resistance and can accommodate large increases in blood flow during exertion. When pulmonary arterial vasoconstriction occurs and pulmonary arterial pressures rise above normal, the condition becomes self-perpetuating inducing further vasoconstriction in the pulmonary vasculature, structural abnormalities, and eventual right-sided heart failure (cor pulmonale).

Clinical Signs and Symptoms

There may not be any symptoms in the early stages of PAH. Onset of symptoms can be very subtle and difficult to recognize initially, especially in secondary PAH since underlying lung disease is usually present. PAH may present as progressive dyspnea (present on exertion first and later developing at rest). Dull retrosternal chest pain, fatigue, and dizziness on exertion are common and often mimic angina pectoris.[41,42]

The right ventricle must pump very hard against a narrowed, resistant pulmonary vascular bed, thus resulting in pump failure. The right ventricle enlarges in its effort to overcome abnormally high PA pressure. Ascites (increased abdominal girth) is a common visible sign. With auscultation there may be an accentuated pulmonic component of S2 caused by the increased force of pulmonary valve closure in the presence of PAH. There may be a pulmonary regurgitation murmur as well.[41]

Pleurisy

Pleurisy is an inflammation of the pleura (serous membrane enveloping the lungs) and is caused by infection, injury, or tumor. The membranous pleura that encases each lung consists of two close-fitting layers: the visceral layer encasing the lungs and the parietal layer lining the inner chest wall. A lubricating fluid lies between these two layers.

If the fluid content remains unchanged by the disease, the pleurisy is said to be dry. If the fluid increases abnormally, it is a wet pleurisy or pleurisy with effusion (pleural effusion). If the wet pleurisy becomes infected with formation of pus, the condition is known as purulent pleurisy or empyema.

Pleurisy may occur as a result of many factors, including pneumonia, tuberculosis, lung abscess, influenza, systemic lupus erythematosus (SLE), rheumatoid arthritis, and pulmonary infarction. Any of these conditions is actually a risk factor for the development of pleurisy, especially in the aging adult population.

Pleurisy, with or without effusion associated with SLE, may be accompanied by acute pleuritic pain and dysfunction of the diaphragm.

Clinical Signs and Symptoms

The chest pain is sudden and may vary from vague discomfort to an intense stabbing or knifelike sensation in the chest. The pain is aggravated by breathing, coughing, laughing, or other similar movements associated with deep inspiration.

The visceral pleura is insensitive; pain results from inflammation of the parietal pleura. Because the latter is innervated by the intercostal nerves, chest pain is usually felt over the site of the pleuritis, but pain may be referred to the lower chest wall, abdomen, neck, upper trapezius muscle, and shoulder because of irritation of the central diaphragmatic pleura (Fig. 7-7).

CLINICAL SIGNS AND SYMPTOMS

Pleurisy

- Chest pain
- Cough
- Dyspnea
- Fever, chills
- Tachypnea (rapid, shallow breathing)

Pneumothorax

Pneumothorax, or free air in the pleural cavity between the visceral and parietal pleurae, may occur secondary to pulmonary disease (e.g., when an emphysematous bulla or other weakened area on the lung ruptures) or as a result of trauma and subsequent perforation of the chest wall. Other risk factors include scuba diving and overexertion.

Pneumothorax is not uncommon after surgery or after an invasive medical procedure involving the chest or thorax. Air may enter the pleural space directly through a hole in the chest wall (open pneumothorax) or diaphragm. Pneumothorax associated with surgical management of patent ductus arteriosus (PDA) in neonates has been reported.[43]

Air may escape into the pleural space from a puncture or tear in an internal respiratory structure (e.g., bronchus, bronchioles, or alveoli). This form of pneumothorax is called closed or spontaneous pneumothorax.

Pneumothorax associated with scuba diving occurs as a result of arterial gas embolism (AGE). AGE is caused by pulmonary overinflation if the breathing gas cannot be exhaled adequately during the ascent. Inert gas bubbles cause impairment of pulmonary functions due to hypoxia.[44]

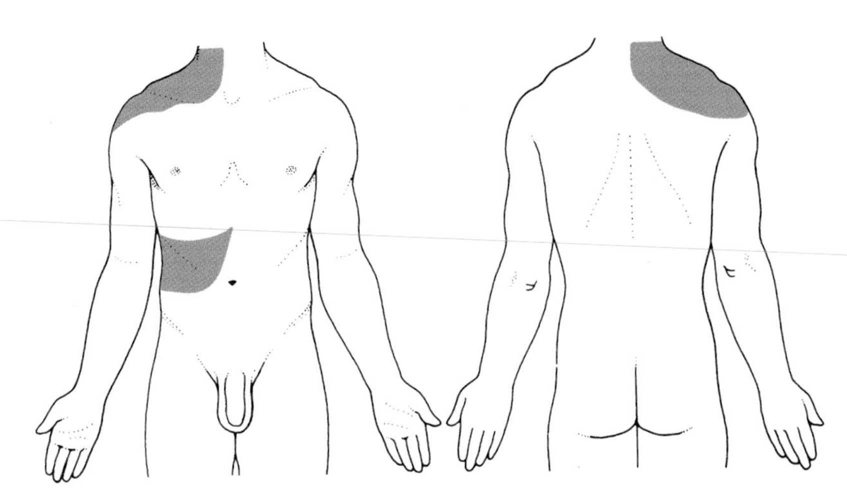

Fig. 7-7 Chest pain over the site of pleuritis is usually perceived by the client. Referred pain *(light red)* associated with pleuritis may occur on the same side as the pleuritic lesion affecting the shoulder, upper trapezius muscle, neck, lower chest wall, or abdomen.

Extraalveolar air (pulmonary barotrauma) from scuba diving can be overlooked, resulting in serious neurologic sequelae. Scuba diving is contraindicated in anyone with asthma, hypertension, coronary heart disease, diabetes, or a history of pneumothorax.

Spontaneous pneumothorax occasionally affects the exercising individual and occurs without preceding trauma or infection. In a healthy individual, abrupt onset of dyspnea raises the suspicion of a spontaneous pneumothorax. Peak incidence for this type of pneumothorax is in adults between 20 and 40 years. Spontaneous pneumothorax in term newborn infants is significantly more likely in males with higher birth weights and with vacuum delivery.[45]

Idiopathic spontaneous pneumothorax (SP) is the result of leakage of air from the lung parenchyma through a ruptured visceral pleura into the pleural cavity. This rupture may be caused by an increased pressure difference between parenchymal airspace and pleural cavity. Another theory is that peripheral airway inflammation leads to obstruction with airtrapping in the lung parenchyma, which precedes spontaneous pneumothorax.[46]

Clinical Signs and Symptoms

Symptoms of pneumothorax, whether occurring spontaneously or as a result of injury or trauma, vary, depending on the size and location of the pneumothorax and on the extent of lung disease. When air enters the pleural cavity, the lung collapses, producing dyspnea and a shift of tissues and organs to the unaffected side.

The client may have severe pain in the upper and lateral thoracic wall, which is aggravated by any movement and by the cough and dyspnea that accompany it. The pain may be referred to the ipsilateral shoulder (corresponding shoulder on the same side as the pneumothorax), across the chest, or over the abdomen (Fig. 7-8). The client may be most comfortable when sitting in an upright position.

Other symptoms may include a fall in blood pressure, a weak and rapid pulse, and cessation of normal respiratory movements on the affected side of the chest (Case Example 7-5).

CLINICAL SIGNS AND SYMPTOMS

Pneumothorax

- Dyspnea
- Change in respiratory movements (affected side)
- Sudden, sharp chest pain
- Increased neck vein distention
- Weak and rapid pulse
- Fall in blood pressure
- Dry, hacking cough
- Shoulder pain
- Sitting upright is the most comfortable position

CASE EXAMPLE 7-5

Tension Pneumothorax

An 18-year-old male, who was injured in a motor vehicle accident (MVA), has come into the hospital physical therapy department with orders to begin ambulation. He had a long leg cast on his left leg and has brought a pair of crutches with him. This is the first time he has been out of bed in the upright position; he has not ambulated in his room yet.

Blood pressure measurement taken while the client was sitting in the wheelchair was 110/78 mm Hg. Pulse was easily palpated and measured at 72 bpm. The therapist gave the necessary instructions and assisted the client to the standing position in the parallel bars. Immediately on standing, this young man began to experience the onset of sharp midthoracic back pain and shortness of breath (SOB). He became pale and shaky, breaking out in a cold sweat.

The therapist assisted him to a seated position and asked the client if he was experiencing pain anywhere else (e.g., chest, shoulder, abdomen) while reassessing blood pressure. His blood pressure had fallen to 90/56 mm Hg, and he was unable to respond verbally to the questions asked. A clinic staff person was asked to telephone for immediate emergency help. While waiting for a medical team, the therapist noted a weak and rapid pulse, distention of the client's neck veins, and diminished respiratory movements.

This young man was diagnosed with tension pneumothorax caused by a displaced fractured rib. Untreated, tension pneumothorax can quickly produce life-threatening shock and bradycardia. Monitoring of the client's vital signs by the therapist resulted in fast action to save this young man's life.

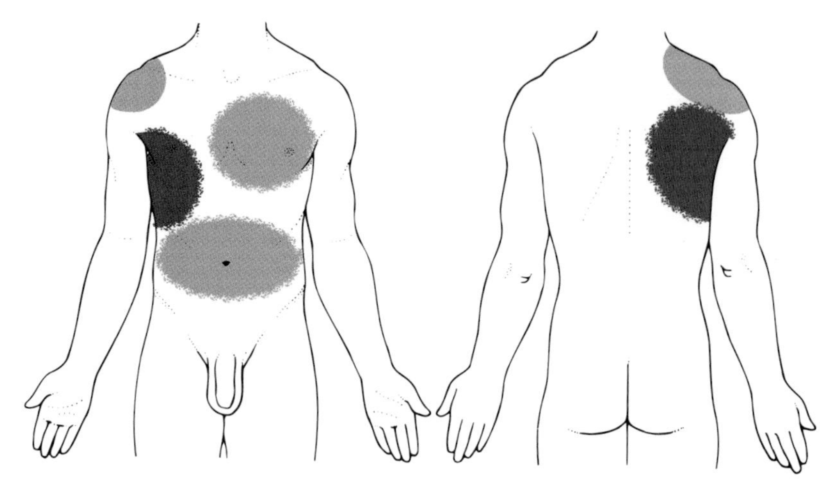

Fig. 7-8 Possible pain patterns associated with pneumothorax: upper and lateral thoracic wall with referral to the ipsilateral shoulder, across the chest, or over the abdomen.

PHYSICIAN REFERRAL

It is more common for a therapist to be treating a client with a previously diagnosed musculoskeletal problem who now has chronic, recurrent pulmonary symptoms than to be the primary evaluator and health care provider of a client with pulmonary symptoms.

In either case, the therapist needs to know what further questions to ask and which of the client's responses represent serious symptoms that require medical follow-up.

Shoulder or back pain can be referred from diseases of the diaphragmatic or parietal pleura or secondary to metastatic lung cancer. When clients have chest pain, they usually fall into two categories: those who demonstrate chest pain associated with pulmonary symptoms and those who have true musculoskeletal problems, such as intercostal strains and tears, myofascial trigger points, fractured ribs, or myalgias secondary to overuse.

Clients with chronic, persistent cough, whether that cough is productive or dry and hacking, may develop sharp, localized intercostal pain similar to pleuritic pain. Both intercostal and pleuritic pain are aggravated by respiratory movements, such as laughing, coughing, deep breathing, or sneezing. Clients who have intercostal pain secondary to insidious trauma or repetitive movements, such as coughing, can benefit from physical therapy.

For the client with asthma, it is important to maintain contact with the physician if the client develops signs of asthma or any bronchial activity during exercise. The physician must be informed to help alter the dosage or the medications to maintain optimal physical performance.

The therapist will want to screen for medical disease through a series of questions to elicit the presence of associated systemic (pulmonary) signs and symptoms. Aggravating and relieving factors may provide further clues that can assist in making treatment or referral decision.

In all these situations, the referral of a client to a physician is based on the family/personal history of pulmonary disease, the presence of pulmonary symptoms of a systemic nature, or the absence of substantiating objective findings indicating a musculoskeletal lesion.

Guidelines for Immediate Medical Attention

- Abrupt onset of dyspnea accompanied by weak and rapid pulse and fall in blood pressure (pneumothorax), especially following motor vehicle accident, chest injury, or other traumatic event
- Chest, rib, or shoulder pain with neurologic symptoms following recent recreational or competitive scuba diving
- Client with symptoms of inadequate ventilation or CO_2 retention (see Respiratory Acidosis)
- Any red flag signs and symptoms in a client with a previous history of cancer, especially lung cancer

Guidelines for Physician Referral

- Shoulder pain aggravated by respiratory movements; have the client hold his or her breath and reassess symptoms; any reduction or elimination of symptoms with breath holding or the Valsalva maneuver suggests pulmonary or cardiac source of symptoms.
- Shoulder pain that is aggravated by supine positioning; pain that is worse when lying down and improves when sitting up or leaning forward is often pleuritic in origin; abdominal contents push up against diaphragm and, in turn, against the parietal pleura.
- Shoulder or chest (thorax) pain that subsides with autosplinting (lying on the painful side).
- For the client with asthma: Signs of asthma or bronchial activity during exercise.
- Weak and rapid pulse accompanied by fall in blood pressure (pneumothorax).
- Presence of associated signs and symptoms such as persistent cough, dyspnea (rest or exertional), or constitutional symptoms (see Box 1-3).

Clues to Screening for Pulmonary Disease

These clues will help the therapist in the decision-making process:

- Age over 40 years.
- History of cigarette smoking for many years.
- Past medical history of breast, prostate, kidney, pancreas, colon, or uterine cancer.
- Recent history of upper respiratory infection, especially when followed by noninflammatory joint pain of unknown cause.
- Musculoskeletal pain exacerbated by respiratory movements (e.g., deep breathing, coughing, laughing).
- Respiratory movements are diminished or absent on one side (pneumothorax).
- Dyspnea (unexplained or out of proportion), especially when accompanied by unexplained weight loss.
- Unable to localize pain by palpation.
- Pain does not change with spinal motions (e.g., no change in symptoms with side bending, rotation, flexion, or extension).
- Pain does not change with alterations in position (possible exceptions: sitting upright is preferred with pneumothorax; symptoms may be worse at night with recumbency, sitting upright eases or relieves symptoms).
- Symptoms are increased with recumbency (lying supine shifts the contents of the abdominal cavity in an upward direction, thereby placing pressure on the diaphragm and in turn, the lungs, referring pain from a lower lung pathologic condition).
- Presence of associated signs and symptoms, especially persistent cough, hemoptysis, dyspnea, and constitutional symptoms, most commonly sore throat, fever, and chills.
- Autosplinting decreases pain.

- Elimination of trigger points resolves symptoms, confirming a musculoskeletal problem (or conversely, trigger point therapy does NOT resolve symptoms, raising a red flag for further examination and evaluation).
- Range of motion does not reproduce symptoms* (e.g., trunk rotation, trunk side bending, shoulder motions).
- Anytime an older person has shoulder pain and confusion at presentation, consider the possibility of diaphragmatic impingement by an underlying lung pathologic condition, especially pneumonia.

*There are two possible exceptions to this guideline. Painful symptoms from an intercostal tear (secondary to forceful coughing caused by diaphragmatic pleurisy) will be reproduced by trunk side bending to the opposite side and trunk rotation to one or both sides. In such a case there is an underlying pulmonary pathologic condition, and a musculoskeletal component. Pleuritic pain can also be reproduced by trunk movements, but the therapist will be unable to localize the pain on palpation.

PULMONARY PAIN PATTERNS

PLEUROPULMONARY DISORDERS (FIG. 7-9)

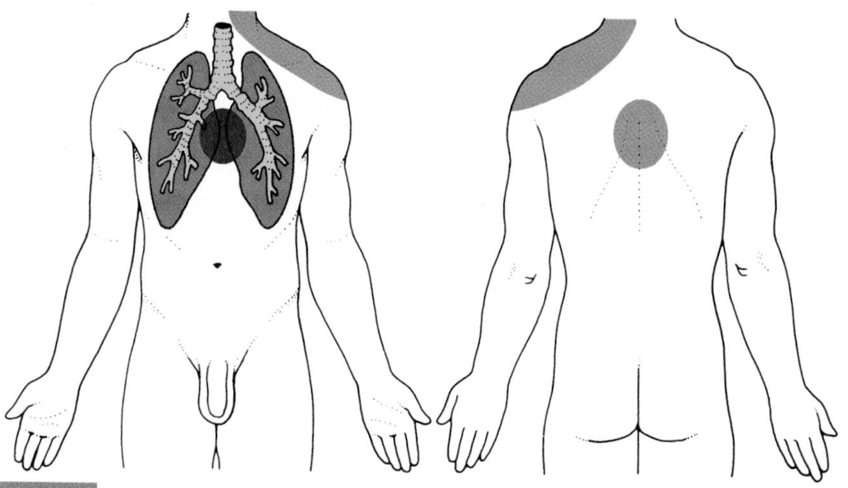

Fig. 7-9 Primary pain patterns *(dark red)* associated with pleuropulmonary disorders, such as pulmonary embolus, cor pulmonale, pleurisy, or spontaneous pneumothorax, may vary, but they usually include substernal or chest pain. Pain over the involved lung fields (anterior, lateral, or posterior) may occur (not shown). Pain may radiate *(light red)* to the neck, upper trapezius muscle, ipsilateral shoulder, thoracic back, costal margins, or upper abdomen (the latter two areas are not shown).

Location: Substernal or chest over involved lung fields—anterior, side, and back

Referral: Often well localized (client can point right to the exact site of pain) without referral
May radiate to neck, upper trapezius muscle, shoulder, costal margins, or upper abdomen
Thoracic back pain occurs with irritation of the posterior parietal pleura

Description: Sharp ache, stabbing, angina-like pressure, or crushing pain with pulmonary embolism
Angina-like chest pain with severe pulmonary hypertension

Intensity: Moderate

Duration: Hours to days

PULMONARY PAIN PATTERNS—*cont'd*

Associated signs and symptoms:
Preceded by pneumonia or upper respiratory infection
Wheezing
Dyspnea (exertional or at rest)
Hyperventilation
Tachypnea (increased respirations)
Fatigue, weakness
Tachycardia (increased heart rate)
Fever, chills
Edema
Apprehension or anxiety, restlessness
Persistent cough or cough with blood (hemoptysis)
Dry hacking cough (occurs with the onset of pneumothorax)
Medically determined signs and symptoms (e.g., by chest auscultation and chest radiograph)

Relieving factors:
Sitting
Some relief when at rest, but most comfortable position varies (pneumonia)
Pleuritic pain may be relieved by lying on the affected side

Aggravating factors:
Breathing at rest
Increased inspiratory movement (e.g., laughter, coughing, sneezing)
Symptoms accentuated with each breath

LUNG CANCER

Location: Anterior chest

Referral:
Scapulae, upper back, ipsilateral shoulder radiating along the medial aspect of the arm
First and second ribs and associated vertebrae and paravertebral muscles (apical or Pancoast's tumors)

Description:
Localized, sharp pleuritic pain (peripheral tumors)
Dull, vague aching in the chest
Neuritic pain of shoulders/arm (apical or Pancoast's tumors)
Bone pain caused by metastases to adjacent bone or to the vertebrae

Intensity: Moderate-to-severe

Duration: Constant

Associated signs and symptoms:
Dyspnea or wheezing
Hemoptysis (coughing up or spitting up blood)
Fever, chills, malaise, anorexia, and weight loss
Fecal breath odor
Tachycardia or atrial fibrillation (palpitations)
Muscle weakness or atrophy (e.g., Pancoast's tumor may involve the shoulder and the arm on the affected side)
Associated CNS symptoms:
- Headache
- Nausea
- Vomiting
- Malaise

Signs of cord compression:
- Localized or radicular back pain
- Weakness
- Loss of lower extremity sensation
- Bowel/bladder incontinence

Hoarseness, dysphagia (peripheral tumors)

Relieving factors: None without medical intervention

Aggravating factors: Inspiration: Deep breathing, laughing, coughing

Key Points to Remember

- Pulmonary pain patterns are usually localized in the substernal or chest region over involved lung fields, which may include the anterior chest, side, or back (Fig. 7-10; see also Fig. 7-9).
- Pulmonary pain can radiate to the neck, upper trapezius muscle, costal margins, thoracic back, scapulae, or shoulder.
- Shoulder pain caused by pulmonary involvement may radiate along the medial aspect of the arm, mimicking other neuromuscular causes of neck or shoulder pain.
- Pulmonary pain usually increases with inspiratory movements such as laughing, coughing, sneezing, or deep breathing.
- Shoulder pain that is relieved by lying on the involved side may be "autosplinting," a sign of a pulmonary cause of symptoms.
- Shoulder pain that is aggravated when lying supine (arm/elbow supported) may be an indication of a pulmonary cause of symptoms.
- For anyone with pain patterns pictured here as presenting symptoms, especially in the absence of trauma or injury, check the client's personal medical history for previous or recurrent upper respiratory infection or pneumonia.
- Any client with symptoms of inadequate ventilation, pneumothorax, or CO_2 retention needs immediate medical referral.
- Clients with COPD who tend to retain CO_2 must be monitored carefully. Since clients with CO_2 retention have a decreased ventilatory drive unless oxygen levels are low, oxygen delivered by nasal cannula cannot get too high or the client will become apneic. There is a standard practice to increase oxygen levels administered by cannula during exercise with some clients who are compromised. Maintaining these levels around 1 to 2 L/min may be required with some clients who have COPD. Consult with respiratory therapy or nursing staff for optimal levels for this particular group of clients.
- CNS symptoms, such as muscle weakness, muscle atrophy, headache, loss of lower extremity sensation, and localized or radicular back pain, may be associated with lung cancer.
- Any CNS symptom may be the silent presentation of a lung tumor.
- Posterior leg or calf pain postoperatively may be caused by a thrombus and must be reported to the physician before physical therapy begins or continues.
- Hemoptysis or exertional/at rest dyspnea, either unexplained or out of proportion to the situation or person, is a red-flag symptom requiring medical referral.
- Change in cough or change in sputum requires further assessment.
- Any client with chest pain should be evaluated for trigger points and intercostal tears.

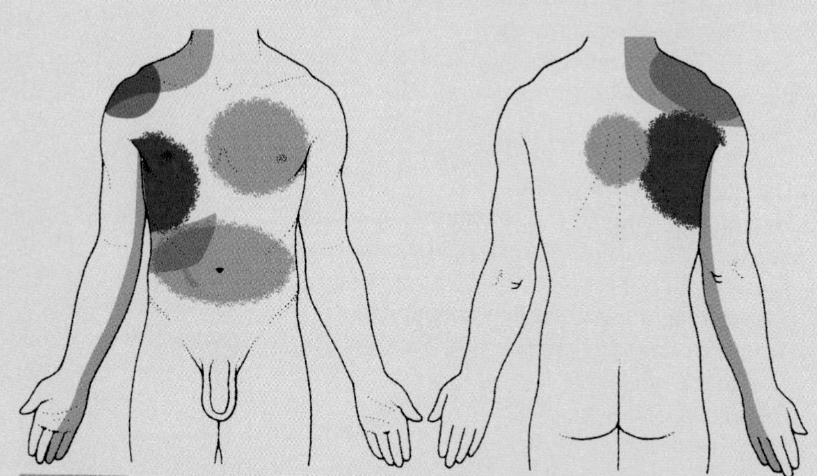

Fig. 7-10 A composite picture of the pain patterns associated with many different impairments of the pulmonary parenchyma including pleuritis, pneumothorax, pulmonary embolism, cor pulmonale, and pleurisy. No single individual will present with all of these patterns at the same time. A composite illustration gives an idea of the wide range of referred pain patterns possible with pulmonary diseases or conditions. Remember that viscerogenic pain patterns do not usually present as discrete circles or ovals of pain as depicted here. This figure is an approximation of what the therapist might expect to hear the client describe associated with a pulmonary problem.

SUBJECTIVE EXAMINATION

Special Questions to Ask

Past Medical History

- Have you ever had trouble with breathing or lung disease such as bronchitis, emphysema, asthma, pneumonia, or blood clots?
  - *If yes:* Describe what this problem was, when it occurred, and how it was treated.
  - If the person indicated yes to asthma, either on the Personal/Family History form or to this question, ask:
    - How can you tell when you are having an asthma episode?
    - What triggers an asthma episode for you?
    - Do you use medications during an episode?
    - Do you have trouble with asthma during exercise?
    - Do you time your medications with your exercise to prevent an asthma episode during exercise?
- Have you ever had tuberculosis?
  - *If yes:* When did it occur, and how was it treated? What is your current status?
  - When was your last test for tuberculosis? What was the test result?
- Have you had a chest x-ray film taken in the last 5 years?
  - *If yes:* What were the results?
- Have you ever broken your nose, been told that you have a deviated septum (nasal passageway), nasal polyps, or sleep apnea? **(Hypoxia)**
- Have you ever had lung or heart surgery?
  - *If yes:* What and when? **(Decreased vital capacity)**

Associated Signs and Symptoms

- Are you having difficulty breathing now?
- Do you ever have SOB or breathlessness or cannot quite catch your breath?
  - *If yes:* When does this happen? When you rest? When you lie flat, walk on level ground, or walk up stairs?
  - How far can you walk before you feel breathless?
  - What symptoms stop your walking (e.g., SOB, heart pounding, chest tightness, or weak legs)?
  - Are these episodes associated with night sweats, cough, chest pain, or bluish color around your lips or fingernails?
  - Does your breathlessness seem to be related to food, pollen, dust, animals, season, stress, or strong emotion? **(Asthma)**
  - Do you have any breathing aids (e.g., oxygen, continuous positive airway pressure (CPAP), nebulizer, inhaler, humidifier, air cleaner, or other aid)?
- Do you have a cough? (Note whether the client smokes, for how long, and how much.)
  - *If yes* to cough, separate this cough from a smoker's cough by asking: When did it start? Is it related to smoking?
  - Do you cough anything up? *If yes:* Describe the color, amount, and frequency.
  - Are you taking anything to prevent this cough? *If yes,* does it seem to help?
  - Are there occasions when you cannot seem to stop coughing?
  - Do you ever cough up blood or anything that looks like coffee grounds? (Bright red fresh blood; brown or black older blood)
  - Have you strained a muscle or your lower back from coughing?
  - Does it hurt to touch your chest or take a deep breath, cough, sneeze, or laugh?
- Have you unexpectedly lost or gained 10 or more pounds recently?

  Gained: **Pulmonary edema, CHF, fat deposits under the diaphragm in the obese client reduces ventilation**

  Lost: **Emphysema, cancer**
- Do your ankles swell? **(CHF)**
- Have you been unusually tired lately? **(CHF, emphysema)**
- Have you noticed a change in your voice? **(Pathology of left hilum or trachea)**

Environmental and Work History

Quick Survey (For full survey, see Appendix B-14):

- What kind of work do you do?
- Do you think your health problems are related to your work?
- Do you wear a mask at work?
- Are your symptoms better or worse when you're at home or at work?

  Follow-up if worse at work: Do others at work have similar problems?

  Follow-up if worse at home: Have you done any remodeling at home in the last 6 months?
- Have you been exposed to dusts, asbestos, fumes, chemicals, radiation, or loud noise?
- Have you ever served in any branch of the military?
  - *If yes,* were you ever exposed to dusts, fumes, chemicals, radiation, or other substances?

Follow-up: It may be necessary to ask additional questions based on past history, symptoms, and risk factors present.

CASE STUDY

REFERRAL

A 65-year-old man has come to you for an evaluation of low back pain, which he attributes to lifting a heavy box 2 weeks ago. During the course of the medical history, you notice that the client has a persistent cough and that he sounds hoarse.

After reviewing the Personal/Family History form, you note that the client smokes two packs of cigarettes each day and that he has smoked at least this amount for at least 50 years. (One pack per day for 1 year is considered "one pack year.") This person has smoked an estimated 100 pack years; anyone who has smoked for 20 pack years or more is considered to be at risk for the development of serious lung disease.

What questions will you ask to decide for yourself whether this back pain is systemic?

PHYSICAL THERAPY INTERVIEW

Introduction to Client

It is important for me to make certain that your back pain is not caused by other health problems, such as prostate problems or respiratory infection, so I will ask a series of questions that may not seem to be related to your back pain, but I want to be very thorough and cover every possibility to obtain the best and most effective treatment for you.

Pain

From your history form, I see that you associate your back pain with lifting a heavy box 2 weeks ago. When did you first notice your back pain (sudden or gradual onset)?

Have you ever hurt your back before or have you ever had pain similar to this episode in the past? (**Systemic disease: recurrent and gradually increases over time**)

Please describe your pain (supply descriptive terms if necessary)

How often do you have this pain?

Follow-ups (FUPs): How long does it last when you have it?

What aggravates your pain/symptoms?

What relieves your pain/symptoms?

How does rest affect your pain?

Have you noticed any changes in your pain/symptoms since they first started to the present time?

Do you have any numbness in the groin or inside your legs? (Saddle anesthesia: **cauda equina**)

Pulmonary

I notice you have quite a cough and you sound hoarse to me. How long have you had this cough and hoarseness (when did it first begin)?

Do you have any back pain associated with this cough? Any other pain associated with your cough?

If yes: Have the person describe where, when, intensity, aggravating and relieving factors.

How does it feel when you take a deep breath? Does your lower back hurt when you laugh or take a deep breath?

When you cough, do you produce phlegm or mucus?

If yes: Have you ever noticed any red streaks or blood in it?

Does your coughing or back pain keep you awake at night?

Have you been examined by a physician for either your cough or your back pain?

Have you had any recent chest or spine x-rays taken?

If yes: When and where? What were the results?

General Systemic

Have you had any night sweats, daytime fevers, or chills?

Do you have difficulty in swallowing (**Esophageal cancer, anxiety, cervical disc protrusion**)?

Have you had laryngitis over and over? (**Oral cancer**)

Urologic

Have you ever been told that you have a prostate problem or prostatitis?

If yes: Determine when this occurred, how it was treated, and whether the person had the same symptoms at that time that he is now describing to you.

Have you noticed any change in your bladder habits?

FUPS: Have you had any difficulty in starting or continuing to urinate?

Is there any burning or discomfort on urination?

Have you noticed any blood in your urine?

Have you recently had any difficulty with kidney stones or bladder or kidney infections?

Gastrointestinal

Have you noticed any change in your bowel pattern?

Have you had difficulty having a bowel movement?

Do you find that you have soiled yourself without even realizing it? (**Cauda equina lesion**—this would require immediate referral to a physician)

Does your back pain begin or increase when you are having a bowel movement?

Is your back pain relieved after having a bowel movement?

Have you noticed any association between when you eat and when your pain/symptoms increase or decrease?

Final Question

Is there anything about your current back pain or your general health that we have not discussed that you think is important for me to know?

(Refer to Special Questions to Ask in this chapter for other questions that may be pertinent to this client, depending on the answers to these questions.)

CASE STUDY—*cont'd*

PHYSICIAN REFERRAL

As always, correlation of findings is important in making a decision regarding medical referral. If the client has a positive family history for respiratory problems (especially lung cancer) and if clinical findings indicate pulmonary involvement, the client should be strongly encouraged to see a physician for a medical check-up.

If there are positive systemic findings, such as difficulty in swallowing, persistent hoarseness, SOB at rest, night sweats, fevers, bloody sputum, recurrent laryngitis, or upper respiratory infections *either in addition to or in association with* the low back pain, the client should be advised to see a physician, and the physician should receive a copy of your findings.

This guideline covers the client who has a true musculoskeletal problem but also has other health problems, as well as the client who may have back pain of systemic origin that is unrelated to the lifting injury 2 weeks ago.

PRACTICE QUESTIONS

1. If a client reports that the shoulder/upper trapezius muscle pain increases with deep breathing, how can you assess whether this results from a pulmonary or musculoskeletal cause?
2. Neurologic symptoms such as muscle weakness or muscle atrophy may be the first indication of:
   a. Cystic fibrosis
   b. Bronchiectasis
   c. Neoplasm
   d. Deep vein thrombosis
3. Back pain with radiating numbness and tingling down the leg past the knee does not occur as a result of:
   a. Postoperative thrombus
   b. Bronchogenic carcinoma
   c. Pott's disease
   d. Trigger points
4. Pain associated with pleuropulmonary disorders can radiate to the:
   a. Anterior neck
   b. Upper trapezius muscle
   c. Ipsilateral shoulder
   d. Thoracic spine
   e. a and c
   f. All of the above
5. The presence of a persistent dry cough (no sputum or phlegm produced) has no clinical significance to the therapist. True or false?
6. Dyspnea associated with emphysema is the result of:
   a. Destruction of the alveoli
   b. Reduced elasticity of the lungs
   c. Increased effort to exhale trapped air
   d. a and b
   e. All of the above
7. What is the significance of autosplinting?
8. Which symptom has greater significance: dyspnea at rest or exertional dyspnea?
9. The presence of pain and anxiety in a client can often lead to hyperventilation. When a client hyperventilates, the arterial concentration of carbon dioxide will do which of the following?
   a. Increase
   b. Decrease
   c. Remain unchanged
   d. Vary depending on potassium concentration
10. Common symptoms of respiratory acidosis would be most closely represented by which of the following descriptions?
    a. Presence of numbness and tingling in face, hands, and feet
    b. Presence of dizziness and lightheadedness
    c. Hyperventilation with changes in level of consciousness
    d. Onset of sleepiness, confusion, and decreased ventilation

REFERENCES

1. Scharf SM: History and physical examination. In Baum GL, Wolinsky E, editors: *Textbook of pulmonary diseases*, ed 5, Boston, 1989, Little, Brown.
2. Guerra S: Overlap of asthma and chronic obstructive pulmonary disease. *Curr Opin Pulm Med* 11(1):7–13, 2005.
3. Zuskin E: Respiratory function in pesticide workers. *J Occup Environ Med* 50(11):1299–1305, 2008.
4. Zuskin E: Occupational health hazards of artists. *Acta Dermatovenerol Croat* 15(3):167–177, 2007.
5. Schachter EN: Gender and respiratory findings in workers occupationally exposed to organic aerosols: a meta analysis of 12 cross-sectional studies. *Environ Health* 12(8):1–9, 2009.
6. Pauwels RA, Rabe KF: Burden and clinical features of chronic obstructive pulmonary disease (COPD). *Lancet* 364(9434):613–620, 2004.
7. Meyers DA, Larj MJ, Lange L: Genetics of asthma and COPD. Similar results for different phenotypes. *Chest* 126(2 Suppl): 105S-110S, 2004.
8. Amsden GW: Anti-inflammatory effects of macrolides: an underappreciated benefit in the treatment of community-acquired respiratory tract infections and chronic inflammatory pulmonary conditions? *J Antimicrob Chemother* 55(1):10–21, 2005.

9. Rubin BK, Henke MO: Immunomodulatory activity and effectiveness of macrolides in chronic airway disease. *Chest* 125(2 Suppl):70S-78S, 2004.
10. Goto Y, Kurosawa H, Mori N, et al: Improved activities of daily living, psychological state and health-related quality of life for 12 months following lung volume reduction surgery in patients with severe emphysema. *Respirology* 9(3):337–344, 2004.
11. Trow TK: Lung-volume reduction surgery for severe emphysema: appraisal of its current status. *Curr Opin Pulm Med* 10(2):128–132, 2004.
12. Upham JW, Holt PG: Environment and development of atopy. *Curr Opin Allergy Clin Immunol* 5(2):167–172, 2005.
13. Asthma in older women. *Harvard Women's Health Watch* 11(3):5–6, 2003.
14. Wisniewski A: Chronic bronchitis and emphysema: clearing the air. *Nursing 2003* 33(5):44–49, 2003.
15. Tay B, Deckey J, Hu S: Spinal infections. *J Am Acad Orthop Surg* 10(3):188–197, 2002.
15a. Jemal A: Cancer statistics, 2010. *CA Cancer J Clin* 60(5):277–300, 2010.
16. Fairley TL: Racial/ethnic disparities and geographic differences in lung cancer incidence. *MMWR* 59(44):1434–1438, 2010.
17. Jemal A, Ward E, Thun MJ: Contemporary lung cancer trends among U.S. women. *Cancer Epidemiol Biomarkers Prev* 14(3):582–585, 2005.
18. American Lung Association: Trends in tobacco use. Feb 2010. Available at http://www.lungusa.org/finding-cures/our-research/trend-reports/Tobacco-Trend-Report.pdf. Posted on October 27, 2010. Accessed January 28, 2011.
19. Jemal A: Lung cancer trends in young adults: an early indicator of progress in tobacco control (United States). *Cancer Causes Control* 14(6):579–585, 2003.
20. Porello P: Neoplasms, Lung. eMedicine. Available at www.emedicine.com/emerg/topic335.htm. Accessed January 28, 2011.
21. Kreamer K: Getting the lowdown on lung cancer. *Nursing 2003* 33(11):36–42, 2003.
22. Davis GA: Pancoast tumors. *Neurosurg Clin N Am* 19:545–557, 2008.
23. Arcasoy SM, Jett JR, Schild SE: Pancoast's syndrome and superior (pulmonary) sulcus tumors. UpToDate Patient Information. Sponsored by the American Society of General Internal Medicine and the American College of Rheumatology. Available at: http://patients.uptodate.com/topic.asp?file=lung_ca/12055 (version 13.2), Updated September 2010. Accessed January 28, 2011.
24. Donaldson SH: Update on pathogenesis of cystic fibrosis lung disease. *Curr Opin Pulm Med* 9(6):486–491, 2003.
25. Mitomo K: Toward gene therapy for cystic fibrosis using lentivirus pseudotyped with Sendai virus envelopes. *Mol Ther* 18(6):1173–1182, 2010.
26. Ostedgaard LS, Rokhlina T, Karp PH, et al: A shortened adeno-associated virus expression cassette for CFTR gene transfer to cystic fibrosis airway epithelia. *Proc Natl Acad Sci U S A* 102(8):2952–2957, 2005.
27. Sloane PA: Cystic fibrosis transmembrane conductance regulator protein repair as a therapeutic strategy in cystic fibrosis. *Curr Opin Pulm Med* 16(6):591–597, 2010.
28. Rosenbluth DB, Wilson K, Ferkol T, et al: Lung function decline in cystic fibrosis patients and timing for lung transplantation referral. *Chest* 126(2):412–419, 2004.
29. American Lung Association: Occupational lung disease fact sheet, January 2004. Available at: http://www.lungusa.org/. Accessed January 28, 2011.
30. National Institute for Occupational Safety and Health: *Chapter 2: Respiratory diseases. worker health chartbook,* Pub No. 2004–146, 2004.
31. Bergdahl IA, Toren K, Eriksson K, et al: Increased mortality in COPD among construction workers exposed to inorganic dust. *Eur Respir J* 23(3):402–406, 2004.
32. O'Donnell T, Abbott W, Athanasoulis C, et al: Diagnosis of deep venous thrombosis in the outpatient by venography. *Surg Gynecol Obstet* 150:69–74, 1980.
33. Molloy W, English J, O'Dwyer R, et al: Clinical findings in the diagnosis of proximal deep venous thrombosis. *Ir Med J* 75:119–120, 1982.
34. Delis KT: Incidence, natural history, and risk factors of deep vein thrombosis in elective knee arthroscopy. *Thromb Haemost* 86(3):817–821, 2001.
35. Shah NB: 82-year-old man with bilateral leg swelling. *Mayo Clinic Proc* 85(9):859–862, 2010.
36. van Belle A, Buller HR, Huisman MV, et al: Effectiveness of managing suspected pulmonary embolism using an algorithm combining clinical probability, D-dimer testing, and computed tomography. *JAMA* 295:172, 2006.
37. Wells PS: Derivation of a simple clinical model to categorize patients probability of pulmonary embolism: increasing the models utility with the simplified D-dimer. *Thromb Haemost* 83:416–430, 2000.
38. Douma RA: Validity and clinical utility of the simplified Wells rule for assessing clinical probability for the exclusion of pulmonary embolism. *Thromb Haemost* 101(1):197–200, 2009.
39. Gibson NS: Further validation and simplification of the Wells clinical decision rule in pulmonary embolism. *Thromb Haemost* 99(6):1134–1136, 2008.
40. Wolf SJ: Assessment of the pulmonary embolism rule-out criteria rule for evaluation of suspected pulmonary embolism in the emergency department. *Am J Emerg Med* 26:181, 2008.
41. Badesch DB: Medical therapy for pulmonary arterial hypertension: updated ACCP evidence-based clinical practice guidelines. *Chest* 131(6):1917–1928, 2007.
42. Pulmonary Hypertension Association: What is pulmonary hypertension? Available at http://www.phassociation.org. Accessed December 21, 2010.
43. Jog SM, Patole SK: Diaphragmatic paralysis in extremely low birthweight neonates: is waiting for spontaneous recovery justified? *J Paediatr Child Health* 38(1):101–103, 2002.
44. Taylor DM, O'Toole KS, Ryan CM: Experienced, recreational scuba divers in Australia continue to dive despite medical contraindications. *Wilderness Environ Med* 13(3):187–193, 2002.
45. Al Tawil K, Abu-Ekteish FM, Tamimi O, et al: Symptomatic spontaneous pneumothorax in term newborn infants. *Pediatr Pulmonol* 37(5):443–446, 2004.
46. Smit HJ, Golding RP, Schramel FM, et al: Lung density measurements in spontaneous pneumothorax demonstrate airtrapping. *Chest* 125(6):2083–2090, 2004.
47. Racial disparities in tuberculosis, selected Southeastern States, 1991-2002. *MMWR* 53(25):556–559, 2004.
48. Carmona L, Hernández-Garcia C, Vadillo C, et al: Increased risk of tuberculosis in patients with rheumatoid arthritis. *J Rheumatol* 30:1436–1439, 2003.

CHAPTER

8

Screening for Gastrointestinal Disease

A great deal of new understanding of the enteric system and its relationship to other systems has been discovered over the last decade. For example, it is now known that the lining of the digestive tract from the esophagus through the large intestine (Fig. 8-1) is lined with cells that contain neuropeptides and their receptors. These substances, produced by nerve cells, are a key to the mind-body connection that contributes to the physical manifestation of emotions.[1,2]

In addition to the classic hormonal and neural negative feedback loops, there are direct actions of gut hormones on the dorsal vagal complex. The person experiencing a "gut reaction" or "gut feeling" may indeed be experiencing the direct effects of gut peptides on brain function.[3]

The association between the enteric system, the immune system, and the brain (now a part of the research referred to as psychoneuroimmunology (PNI) has been clearly established and forms an integral part of gastrointestinal (GI) symptoms associated with immune disorders such as fibromyalgia, systemic lupus erythematosus, rheumatoid arthritis, chronic fatigue syndrome, and others.

Researchers estimate that more than two thirds of all immune activity occurs in the gut. There are more T cells in the intestinal epithelium than in all other body tissues combined. The gamma delta T cells form the forefront of the immune defense mechanism. They act as an early warning system in the cells lining the intestines, which are heavily exposed to microorganisms and toxins.[4,5] In some people, the wall of the gut seems to have been breached, either because the network of intestinal cells develops increased permeability (a syndrome referred to as "leaky gut") or perhaps because bacteria and yeast overwhelm it and migrate into the bloodstream.

Allowing undigested food or bacteria into the bloodstream sets in motion a chain of events as the immune system reacts. The body responds as if to an illness and expresses it in a number of ways such as a rash, diarrhea, GI upset, joint pain, migraines, and headache. The exact cause for these microscopic breaches remains unknown, but food allergies, too much aspirin or ibuprofen, certain antibiotics, excessive alcohol consumption, smoking, or parasitic infections may be implicated.

All of these associations and new findings support the need for the therapist to assess carefully the possibility of GI symptoms present but unreported. This is especially important when considering the fact that GI tract problems can sometimes imitate musculoskeletal dysfunction.

GI disorders can refer pain to the sternal region, shoulder and neck, scapular region, mid-back, lower back, hip, pelvis, and sacrum. This pain can mimic primary musculoskeletal or neuromuscular dysfunction, causing confusion for the physical therapist or for the physician assessing the client's chief complaint.

Although these neuromusculoskeletal symptoms can occur alone and far from the actual site of the disorder, the client usually has other systemic signs and symptoms associated with GI disorders that should give the therapist who does a thorough investigation grounds for suspicion.

A careful interview to screen for systemic illness should include a few important questions concerning the client's history, prescribed medications, and the presence of any associated signs or symptoms that would immediately alert the therapist about the need for medical follow-up. The most common intraabdominal diseases that refer pain to the musculoskeletal system are those that involve ulceration or infection of the mucosal lining. Drug-induced GI symptoms can also occur with delayed reactions as much as 6 or 8 weeks after exposure to the medication. The most common occurrences are antibiotic colitis; nausea, vomiting, and anorexia from digitalis toxicity; and nonsteroidal antiinflammatory drug (NSAID)–induced ulcers.

SIGNS AND SYMPTOMS OF GASTROINTESTINAL DISORDERS

Any disruption of the digestive system can create symptoms such as nausea, vomiting, pain, diarrhea, and constipation. The bowel is susceptible to altered patterns of normal motility caused by food, alcohol, caffeine, drugs, physical and

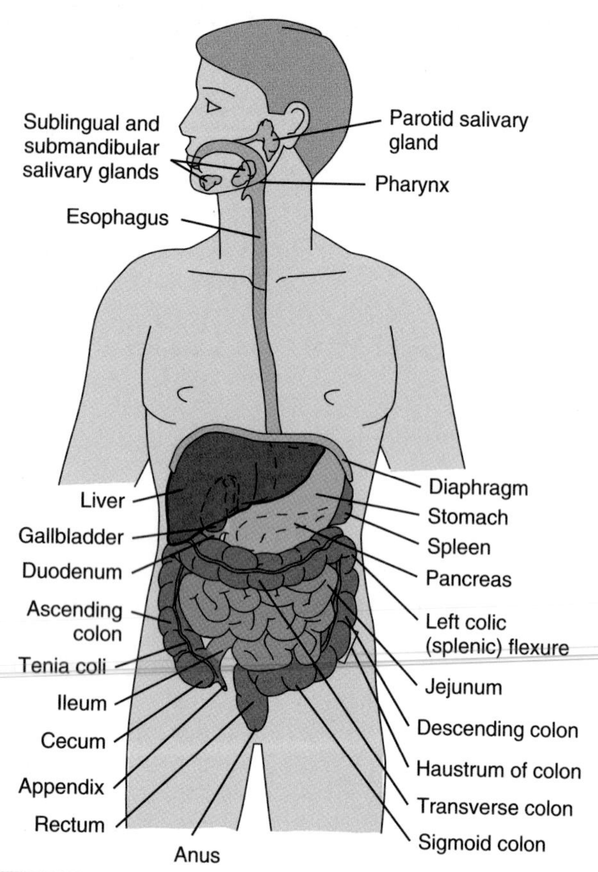

Fig. 8-1 Organs of the digestive system; see also Fig. 9-1. (From Hall JE: *Guyton and Hall textbook of medical physiology,* ed 12, Philadelphia, 2010, WB Saunders.)

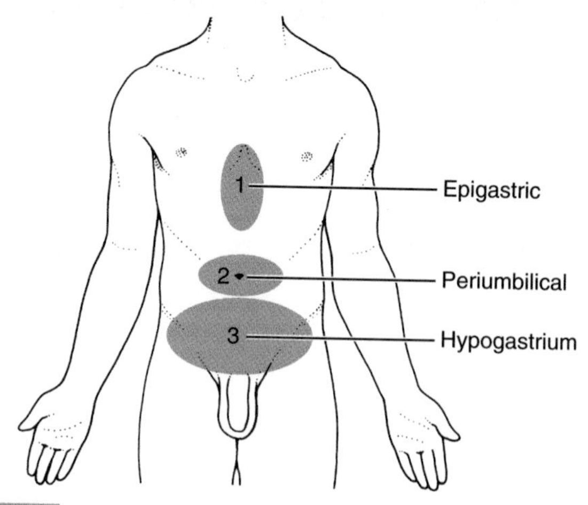

Fig. 8-2 Visceral pain. *1,* The epigastric region from the heart, esophagus, stomach, duodenum, gallbladder, liver, or pancreas and corresponding to T3 to T5 sympathetic nerve distribution; *2,* the periumbilical region from the pancreas, small intestine, appendix, or proximal colon (T9 to T11 sympathetic nerve distribution; the umbilicus is level with the disk located between the L3 and L4 vertebral bodies in the adult who is not overweight); and *3,* the lower midabdominal or hypogastrium region from the large intestine, colon, bladder, or uterus (T10 to L2 sympathetic nerve distribution).

emotional stress, and lifestyle (e.g., lack of regular exercise, tobacco use). GI effects of chemotherapy include nausea and vomiting, anorexia, taste alteration, weight loss, oral mucositis, diarrhea, and constipation.

Symptoms, including pain, can be related to various GI organ disturbances and differ in character, depending on the affected organ. The most clinically meaningful GI symptoms reported in a physical therapy practice include

- Abdominal pain
- Dysphagia
- Odynophagia
- GI bleeding (emesis, melena, red blood)
- Epigastric pain with radiation to the back
- Symptoms affected by food
- Early satiety with weight loss
- Constipation
- Diarrhea
- Fecal incontinence
- Arthralgia
- Referred shoulder pain
- Psoas abscess
- Tenderness over McBurney's point (see Appendicitis this chapter)
- Neuropathy

Abdominal Pain

As we enter into this next discussion on primary and referred abdominal pain patterns, be aware that each pain pattern has listed with it both the sympathetic nerve distribution to the viscera (i.e., autonomic nervous system innervation of the structure) and the anatomic location of radiating or referred pain from the viscera or GI segment involved in the primary pain patterns.

Whenever possible, labels are used to differentiate between sympathetic nerve innervations of the viscera and anatomic locations of the pain. For example, the small intestine (viscera) is innervated by T9 to T11 but refers (somatic) pain to the L3 to L4 (anatomic) lumbar spine.

Primary Gastrointestinal Visceral Pain Patterns

Visceral pain (internal organs) occurs in the midline because the digestive organs arise embryologically in the midline and receive sensory afferents from both sides of the spinal cord. The site of pain generally corresponds to dermatomes from which the visceral organs receive their innervation (see Fig. 3-3). Pain is not well localized because innervation of the viscera is multisegmental over up to eight segments of the spinal cord with fewer nerve endings than other sensitive organs.

The most common primary pain patterns associated with organs of the GI tract are depicted in Fig. 8-2. Reasons for abdominal pain fall into three broad categories: inflammation, organ distention (tension pain), and necrosis (ischemic pain). The underlying cause can be life-threatening, requiring a quick assessment and fast referral.

Pain in the *epigastric region* occurs anywhere from the midsternum to the xiphoid process from the heart, esophagus, stomach, duodenum, gallbladder, liver, and other mediastinal organs corresponding to the T3 to T5 sympathetic nerve distribution. The client may report the pain radiates around the ribs or straight through the chest to the thoracic spine at the T3 to T6 or T7 anatomic levels.

Pain in the *periumbilical region* (T9 to T11 nerve distribution) occurs with impairment of the small intestine (see Fig. 8-15), pancreas, and appendix. Primary pain in the periumbilical region usually sends the client to a physician. However, pain around the umbilicus may be accompanied by low back pain. In the healthy adult who is not obese and does not have a protruding abdomen, the umbilicus is level with the disk located anatomically between the L3 and L4 vertebral bodies.

The physical therapist is more likely to see a client with anterior abdominal and low back pain at the same level but with alternating presentation. In other words, first the client experiences periumbilical pain with or without associated GI signs and symptoms, then the painful episode resolves. Later, the client develops low back pain with or without GI symptoms but does not realize there is a link between these painful episodes. It is at this point the client presents in a physical therapy practice.

Pain in the *lower abdominal region* (hypogastrium) from the large intestine and/or colon may be mistaken for bladder or uterine pain (and vice versa) by its suprapubic location. Referred pain at the same anatomic level posteriorly corresponds to the sacrum (see Fig. 8-16). The large intestine and colon are innervated by T10 to L2, depending on the location (e.g., ascending, transverse, descending colon).

The abdominal viscera are ordinarily insensitive to many stimuli, such as cutting, tearing, or crushing, that when applied to the skin evokes severe pain. Visceral pain fibers are sensitive only to stretching or tension in the wall of the gut from neoplasm, distention, or forceful muscular contractions secondary to bowel obstruction or spasm.

Tension pain can occur as a result of bowel obstruction; constipation; and pus, fluid, or blood accumulation from infection or other causes. The rate that tension develops must be rapid enough to produce pain; gradual distention, such as with malignant obstruction, may be painless unless ulceration occurs. Rapid, peristalsis forces of the bowel trying to eliminate irritating substances can cause tension pain described as "colicky" pain. Individuals with tension pain have trouble finding a comfortable position. They will constantly shift positions to try and find a comfortable position.

Visceral organs of the GI tract (particularly hollow organs such as the intestines) respond to stretching and distention as pain, more so than typical tissue injury caused by cutting or crushing. Because of similar innervation, it is often difficult to distinguish pain associated with the heart from pain caused by an esophageal disorder.

One difference between visceral organ pain and pain from the parietal peritoneum is that the parietal peritoneum is innervated by nerves that travel with the somatic nerves, providing a more precise location of pain. This is noted with acute appendicitis, when early, vague pain (from inflammation of the appendix) is replaced by more localized pain at McBurney's point once the inflammation involves the parietal peritoneum.

Inflammatory pain arising from the visceral or parietal peritoneum (e.g., acute appendicitis) is described as steady, deep, and boring. It can be poorly localized as when the visceral peritoneum is involved or more localized with parietal peritoneum involvement. Individuals with inflammatory pain seek a quiet position (often with the knees bent or in a curled up/fetal position) without movement.

Ischemia (deficiency of blood) may produce visceral pain by increasing the concentration of tissue metabolites in the region of the sensory nerve. Pain associated with ischemia is steady pain, whether this ischemia is secondary to vascular disease or due to obstruction causing strangulation of bowel tissue. The pain is sudden in onset and extremely intense. It progresses in severity and is not relieved by analgesics.

Additionally, although the viscera experience pain, the visceral peritoneum (membrane enveloping organs) is not sensitive to cutting. Except in the presence of widespread inflammation or ischemia, it is possible to have extensive disease without pain until the disease progresses enough to involve the parietal peritoneum.

Visceral pain is usually described as deep aching, boring, gnawing, vague burning, or deep grinding as opposed to the sharp, pricking, and knifelike qualities of cutaneous pain. When referred to the somatic regions of the low back, hip, or shoulder, the sensation is vague and poorly localized because visceral afferents provide input over multiple segments of the spinal cord. As mentioned, afferents from different abdominal locations converge on the same dorsal nerve roots, which may be shared with the more precisely developed somatic sensory pathways.

Referred Gastrointestinal Pain Patterns

Sometimes visceral pain from a digestive organ is felt in a location remote from the usual anterior midline presentation. The referred pain site still lies within the dermatomes of the dorsal nerve roots serving the painful viscera. Referred pain is often more intense and localized than typical visceral pain. Afferent nerve impulses transmit pain from the esophagus to the spinal cord by sympathetic nerves from T5 to T10. Integration of the autonomic and somatic systems occurs through the vagus and the phrenic nerves. There can be referred pain from the esophagus to the mid-back and referred pain from the mid-back to the esophagus. For example, esophageal dysfunction can present as anterior neck pain or mid-thoracic spine pain and disk disease of the mid-thoracic spine can masquerade as esophageal pain.

Client history and the presence or absence of associated signs and symptoms will help guide the therapist. For example, a client with mid-back pain from esophageal dysfunction will not likely report numbness and tingling in the upper extremities or bowel and bladder changes such as you might see with disk disease. Likewise, disk involvement with referred pain to the esophagus will not cause melena or symptoms associated with meals.

Visceral afferent nerves from the liver, respiratory diaphragm, and pericardium are derived from C3 to C5 sympathetics and reach the central nervous system (CNS) via the phrenic nerve (see Fig. 3-3). The visceral pain associated with these structures is referred to the corresponding somatic area (i.e., the shoulder).

Afferent nerves from the gallbladder, stomach, pancreas, and small intestine travel through the celiac plexus (network of ganglia and nerves supplying the abdominal viscera) and the greater splanchnic nerves and enter the spinal cord from T6 to T9. Referred visceral pain from these visceral structures may be perceived in the mid-back and scapular regions.

Afferent stimuli from the colon, appendix, and pelvic viscera enter the 10th and 11th thoracic segments through the mesenteric plexus and lesser splanchnic nerves. Finally, the sigmoid colon, rectum, ureters, and testes are innervated by fibers that reach T11 to L1 segments through the lower splanchnic nerve and through the pelvic splanchnic nerves from S2 to S4. Referred pain may be perceived in the pelvis, flank, low back, or sacrum (Case Example 8-1).

Hyperesthesia (excessive sensibility to sensory stimuli) of skin and hyperalgesia (excessive sensibility to painful stimuli) of muscle may develop in the referred pain distribution. As mentioned in Chapter 3, in the early stage of visceral disease, sympathetic reflexes arising from afferent impulses of the internal viscera can be expressed first as sensory, motor, and/or trophic changes in the skin, subcutaneous tissues, and/or muscles. The client may present with itching, dysesthesia, skin temperature changes, perspiration, or dry skin.

The viscera do not perceive pain, but the sensory side is trying to get the message out that something is wrong by creating sympathetic sudomotor changes. When the afferent visceral pain stimuli are intense enough, discharges at synapses within the spinal cord cause this reflex phenomenon, usually transmitted by peripheral nerves of the same spinal segment(s). Thus the sudomotor changes occur as an automatic reflex along the distribution of the somatic nerve.

Remember from our discussion of viscerogenic pain patterns in Chapter 3 that any structure touching the respiratory diaphragm can refer pain to the shoulder, usually to the ipsilateral shoulder, depending on where the direct pressure occurs. Anyone with upper back or shoulder pain and symptoms should be asked a few general screening questions about the presence of GI symptoms.

Referred pain to the musculoskeletal system can occur alone, without accompanying visceral pain, but usually visceral pain (or other symptoms) precedes the development of referred pain. The therapist will find that the client does not connect the two sets of symptoms or fails to report abdominal pain and GI symptoms when experiencing a painful shoulder or low back, thinking these are two separate problems. For a more complete discussion of the mechanisms behind viscerogenic referred pain patterns, see Chapter 3.

Most of what has been presented here has dealt with the sensory side of the clinical presentation. There can be motor effects of GI dysfunction, too. For example, contraction, guarding, and splinting of the rectus abdominis and muscles above the umbilicus can occur with dysfunction of the stomach, gallbladder, liver, pylorus, or respiratory diaphragm. Impairment of the ileum, jejunum, appendix, cecum, colon, and rectum are more likely to result in muscle spasm of the rectus abdominis below the umbilicus.[6]

At the same time, impairment of these GI structures can cause muscle dysfunction in the back (thoracic and lumbar spine) with loss of motion of the involved spinal segments. The clinical picture is one that is easily confused with primary pathology of the spinal segment.[6] Once again, the history and associated signs and symptoms help the therapist sort through the clinical presentation to reach a differential diagnosis. A thorough screening process is essential in such cases.

CASE EXAMPLE 8-1

Colon Cancer

A 66-year-old university professor consulted with a physical therapist after twisting his back while taking the garbage out. He reported experiencing ongoing, painful, low back symptoms 3 weeks after the incident. The objective assessment was consistent with a strain of the right paraspinal muscles with overall diminished lumbar spinal motion consistent with this gentleman's age. Given the reported mechanism of injury and the results of the examination consistent with a musculoskeletal problem, a medical screening examination was not included in the interview. A home exercise program was initiated, including stretching and conditioning components.

When the client did not return for his follow-up appointment, telephone contact was made with his family. The client had been hospitalized after collapsing at work. A medical diagnosis of colon cancer was determined. The family reported he had been experiencing digestive difficulties "off and on" and low back pain for the past 3 years, always alternately and never simultaneously. The client died 6 weeks later.

In this case, the only red flag suggesting the need for medical screening was the client's age. However, the therapist did not ask about any associated signs and symptoms. We must always remember that even with a known and plausible reason for the injury, the client may wrongfully attribute symptoms to a logical event or occurrence. This man had been experiencing both abdominal symptoms and referred back pain, but since these episodes did not occur at the same time, he did not see a connection between them.

Always finish every interview with this question: "Are you having any other symptoms of any kind anywhere else in your body?"

Dysphagia

Dysphagia (difficulty swallowing) is the sensation of food catching or sticking in the esophagus. This sensation may occur (initially) just with coarse, dry foods and may eventually progress to include anything swallowed, even thin liquids and saliva. Dysphagia may be caused by achalasia, a process by which the circular and longitudinal muscular fibers of the

CASE EXAMPLE 8-2

Esophageal Cancer

An obese 88-year-old woman with a total knee replacement (TKR) was referred for rehabilitation because of loss of motion, joint swelling, and persistent knee pain. She was accompanied to the clinic for each session by one of her three daughters. Over a period of 2 or 3 weeks, each daughter commented on how much weight the mother had lost. When questioned, the client complained of a loss of appetite and difficulty in swallowing, but she had been evaluated and treated only for her knee pain by the orthopedist. She was encouraged to contact her family doctor for evaluation of these red flag symptoms and was subsequently diagnosed with esophageal cancer.

lower esophageal sphincter fail to relax, producing an esophageal obstruction.

Other possible GI causes of dysphagia include peptic esophagitis (inflammation of the esophagus) with stricture (narrowing), gastroesophageal reflux disease (GERD), and neoplasm (Case Example 8-2). Dysphagia may be a symptom of many other disorders unrelated to GI disease (e.g., stroke, Alzheimer's disease, Parkinson's disease). Certain types of drugs, including antidepressants, antihypertensives, and asthma drugs, can make swallowing difficult.

The presence of dysphagia requires prompt attention by the physician. Medical intervention is based on a subsequent endoscopic examination.

Odynophagia

Odynophagia, or pain during swallowing, can be caused by esophagitis or esophageal spasm. Esophagitis may occur secondary to GERD, the herpes simplex virus, or fungus caused by the prolonged use of strong antibiotics. Pain after eating may occur with esophagitis or may be associated with coronary ischemia.

To differentiate esophagitis from coronary ischemia: *upright positioning relieves esophagitis pain,* whereas *cardiac pain* is relieved by nitroglycerin or by supine positioning. Both conditions require medical attention.

Gastrointestinal Bleeding

Occult (hidden) GI bleeding can appear as mid-thoracic back pain with radiation to the right upper quadrant. Bleeding may not be obvious; serial Hemoccult tests and laboratory tests (checking for anemia and iron deficiency) are needed. A medical doctor should evaluate any type of bleeding. Ask about the presence of other signs such as blood in the vomit or stools (Box 8-1). *Coffee ground emesis* (vomit) may indicate a perforated peptic or duodenal ulcer.

Bloody diarrhea may accompany other signs of ulcerative colitis. Diarrhea and ulcerative colitis are discussed in greater depth separately in this chapter. *Bright red blood* usually represents pathology close to the rectum or anus and may be an indication of rectal fissures (e.g., history of anal intercourse) or hemorrhoids but can also occur as a result of colorectal cancer.

BOX 8-1 SIGNS OF GASTROINTESTINAL BLEEDING

Coffee ground emesis (vomit)
Bloody diarrhea
Bright red blood
Melena (dark, tarry stools)
Reddish or mahogany-colored stools

Melena, or black, tarry stool, occurs as a result of large quantities of blood in the stool. When asked about changes in bowel function, clients may describe black, tarry stools that have an unusual, noxious odor. The odor is caused by the presence of blood, and the black color arises as the digestive acids in the bowel oxidize red blood cells (e.g., bleeding esophageal varices, stomach or duodenal ulceration). Melena is very sticky and does not clean well.

It may be necessary to ask about bowel smears on the undergarments or difficulty getting wiped clean after a bowel movement. The following series may guide the therapist in this area:

FOLLOW-UP QUESTIONS

- I would like to ask a few questions that may not seem related to your shoulder (back, hip, pelvic) pain, but these are very important in finding out what is causing your symptoms.
- Have you noticed any blood in your stools or change in the color or consistency of your bowel movements?
- Do you have any trouble wiping yourself clean after a bowel movement?
- Have you noticed any bowel smears on your underwear later after a bowel movement?

After going through the questions, it may be helpful to leave the back door open. Perhaps leave the client with this thought:

- If you do not know the answer right now or if you just have not noticed, please feel free to let me or your physician know if you notice any changes.

Esophageal varices are dilated blood vessels, usually secondary to alcoholic cirrhosis of the liver. Blood that would normally be pumped back to the heart must bypass the damaged liver. The blood then "backs up" through the esophagus. Ruptured esophageal varices are an emergent, life-threatening condition. Vascular abnormalities of the stomach causing bleeding may include ulcers.

The client should be asked about the presence of any blood in the stool to determine whether it is melenic (from the upper GI tract; ask about a history of NSAID use) or bright red (from the distal colon or rectum). Bleeding from internal or external hemorrhoids (enlarged veins inside or outside the rectum), rectal fissures, or colorectal carcinoma can cause bright red blood in the stools. Rectal bleeding from anal lesions or fissures can occur in the homosexual population who are sexually active. Women engaging in anal intercourse can also be affected. A brief sexual history may be indicated in some cases.

Reddish or mahogany-colored stools can occur from eating certain foods, such as beets, or significant amounts of red food coloring but can also represent bleeding in the lower GI/colon. Medications that contain bismuth (e.g., Kaopectate, Pepto-Bismol, Bismatrol, Pink Bismuth) can cause darkened or black stools and the client's tongue may also appear black.

Clients who have received pelvic radiation for gynecologic, rectal, or prostate cancers have an increased risk for radiation proctitis, which can cause subsequent (delayed) rectal bleeding episodes. Be sure and ask about a past history of cancer and radiation treatment.

Epigastric Pain with Radiation

Epigastric pain perceived as intense or sharp pain behind the breastbone with radiation to the back may occur secondary to long-standing ulcers. For example, the client may be aware of an ulcer but does not relate the back pain to the ulcer. Close questioning related to GI symptoms can provide the therapist with knowledge of underlying systemic disease processes.

Anyone with epigastric pain accompanied by a burning sensation that begins at the xiphoid process and radiates up toward the neck and throat may be experiencing heartburn. Other common symptoms may include a bitter or sour taste in the back of the throat, abdominal bloating, gas, and general abdominal discomfort. Heartburn is often associated with GERD. It can be confused with angina or heart attack when accompanied by chest pain, cough, and shortness of breath (SOB). A physician must evaluate and diagnose the cause of epigastric pain or heartburn.

A screening interview and evaluation is especially helpful when clients have neglected medical treatment for so long that epigastric back pain may in turn have created biomechanical changes in muscular contractions and spinal movement. These changes eventually create pain of a biomechanical nature.[7] The client then presents with enough true musculoskeletal findings such that a diagnosis of back dysfunction can be supported. However, the symptoms may be associated with a systemic problem. A good medical history can be a valuable tool in revealing the actual cause of the back pain.

Symptoms Affected by Food

Clients may or may not be able to relate pain to meals. Pain associated with gastric ulcers (located more proximally in the GI tract) may begin within 30 to 90 minutes after eating, whereas pain associated with duodenal or pyloric ulcers (located distally beyond the stomach) may occur 2 to 4 hours after meals (i.e., between meals). Alternatively stated, food is not likely to relieve the pain of a gastric ulcer, but it may relieve the symptoms of a duodenal ulcer.

The client with a duodenal ulcer or cancer-related pain may report pain during the night between midnight and 3:00 AM. Ulcer pain may be differentiated from the nocturnal pain associated with cancer by its intensity (7 or higher on a scale from 0 to 10) and duration (constant). More specifically, the gnawing pain of an ulcer may be relieved by eating, but the intense, boring pain associated with cancer is not relieved by any measures.

Ask the client with nighttime shoulder, neck, or back pain to eat something and assess the effect of food on these symptoms. Anyone whose musculoskeletal pain is altered (increased or decreased) or eliminated by food should be screened more thoroughly and referred for further medical evaluation when appropriate. Anyone with a previous history of cancer and nighttime pain must also be evaluated more closely. This is true even if eating has no effect on the client's symptoms.

Early Satiety

Early satiety occurs when the client feels hungry, takes one or two bites of food, and feels full. The sensation of being full is out of proportion with the time of the previous meal and the initial degree of hunger experienced. This can be a symptom of obstruction, stomach cancer, gastroparesis (slowing down of stomach emptying), peptic ulcer disease, and other tumors. Vertebral compression fractures can occur from a variety of disorders including osteoporosis and can result in severe spinal deformity. This deformity, along with severe back pain, can cause early satiety resulting in malnutrition.[8]

Constipation

Constipation is defined clinically as being a condition of prolonged retention of fecal content in the GI tract resulting from decreased motility of the colon or difficulty in expelling stool.

The Rome III Diagnostic criteria for functional constipation defines this condition as hard, lumpy stools; stools that are difficult to expel; infrequent stools (less than three per week); or a feeling of incomplete evacuation after defecation and general discomfort.[9,10] Constipated clients with tender psoas trigger points (TrPs) may report anterior hip, groin, or thigh pain when the fecal bolus presses against the TrPs.[11]

Intractable constipation is called *obstipation* and can result in a fecal impaction that must be removed. Back pain may be the overriding symptom of obstipation, especially in older adults who do not have regular bowel movements or who cannot remember the last bowel movement was several weeks ago (Case Example 8-3).

Keep in mind the individual who has low back pain with constipation could also be manifesting symptoms of pelvic floor muscle overactivity or spasm. In such cases, pelvic floor assessment should be a part of the screening exam. Consultation with a physical therapist skilled in this area should be considered if the primary care therapist is unable to perform this examination.

CASE EXAMPLE 8-3

Obstipation

A 75-year-old Caucasian male was transported from his home to a hospital emergency department with acute onset of shortness of breath (SOB). He was intubated en route by ambulance personnel, secondary to hypoxemia and acute respiratory distress. Family members state that the patient has severe chronic obstructive pulmonary disorder (COPD) and uses continuous supplemental oxygen at home (usually 3 L per minute). The client had no complaints of chest pain leading up to or during the episode.

While in the hospital, the client was hypotensive and started on dopamine. Chest x-ray revealed acute pulmonary edema consistent with congestive heart failure (CHF). He was treated with intravenous Lasix. Following removal of the nasogastric (NG) tube, the client began to complain of severe low back pain and was started on Vicodin. Magnetic resonance imaging (MRI) of the lumbar/sacral spine showed multiple levels of lumbar stenosis and facet sclerosis.

Four days post hospital admission, the client's oxygen saturation was 90% on 4 L per minute of supplemental oxygen. The decision was made to transfer the client to a skilled nursing facility (SNF) with orders for activity as tolerated and physical therapy (evaluate and treat accordingly).

Medical Diagnoses

- Acute respiratory failure
- Hypoxemia
- Congestive heart failure (CHF)
- Pulmonary edema
- COPD
- Chronic low back pain
- Degenerative joint disease (DJD)
- Spinal stenosis

Past Medical History

- Pulmonary asbestosis
- Hypotension
- Benign prostatic hypertrophy (BPH)
- Non-Q myocardial infarction (MI)

Medications

- Bactrim (antiinfective)
- Ketoconazole (antifungal)
- Plavix (coronary artery disease prophylaxis; platelet aggregation inhibitor)
- Aspirin (ASA; coronary artery disease prophylaxis)
- Magnesium oxide (supplement)
- K-Dur (supplement)
- Vasotec (antihypertensive; angiotensin-converting enzyme [ACE] inhibitor)
- Lasix (loop diuretic; CHF)
- Percocet (opiate analgesic; back pain)
- Colace (laxative)

Current Complaints

Client reports increased SOB with minor exertion and severe lumbar/sacral pain that has been constant over the last 3 days and appears to be getting worse.

Pain is described as "a dull ache" and is aggravated by movement. Minor relief is obtained through rest and use of pain medication. The client also reported recent lower abdominal discomfort, which he attributed to something he "ate for breakfast."

When asked about elimination patterns, he states that his bowel movements are not regular, but he "must have had one in the hospital." He "urinates frequently," has trouble starting a flow of urine, and does not void completely due to an enlarged prostate.

He reports a long history of progressive back pain without traumatic onset, starting in his 40s. His immediate goal is relief of back pain. His "normal" back pain is described as a 4 to 6 on a 0 to 10 scale. His current level of intensity is described as an 8/10 on pain medication.

Review of Systems

General

- Oxygen saturation (pulse oximeter) while on 4 L oxygen per minute: 92%
- Blood pressure (BP): 110/65 (seated, left arm)
- Respiratory rate (RR): 16/minute
- Heart rate (HR): 86 bpm (regular taken for 1 full minute)
- Body temperature: 99.4° F

Neurologic

History of radicular pain symptoms in both lower extremities (LEs) above the knee but not present at this time

- No observable atrophy of LE musculature
- Deep tendon reflexes (DTRs): +1 in bilateral patellar tendon; 0 for bilateral Achilles
- Manual muscle test (MMT): Unable to perform due to back pain; decreased functional strength observed
- Proprioception: Decreased in feet and ankles, bilaterally

Cardiovascular

- Mild pitting edema (pedal: feet and ankles bilaterally) with 15-20 second rebound
- History of claudication with prolonged standing and ambulation

Pulmonary

- Diminished breath sounds, especially at bases (auscultation)
- Early exertional dyspnea
- 2-3 pillow orthopnea
- Digital clubbing, bilaterally

Gastrointestinal

- Low abdominal pain/discomfort
- Constipation

Genitourinary

- Frequent urge to urinate
- Difficulty starting flow
- Decreased urine output

Musculoskeletal

- Flexed postural stance (unable to straighten up due to pain)
- Mild age-related range of motion (ROM) limitations noted
- Muscle guarding and spasm in paravertebral musculature from T4 to L2
- Balance, mobility, ambulation assessed and recorded

Evaluation: Although the client's back pain was made worse by movement, the presence of intense pain and constitutional symptom (low-grade fever) alerted the therapist to a possible systemic or viscerogenic cause of pain. The fact that the client could not remember his last bowel movement combined with abdominal pain was of concern. Change in bladder function was also of concern.

Prior to initiation of physical therapy services, the client was referred back to the attending physician. A brief summary of the client's neuromusculoskeletal impairments was presented, along with a description of the proposed intervention. A simple statement at the end was highlighted:

Continued

CASE EXAMPLE 8-3—cont'd

Obstipation

Intense back pain accompanied by low-grade fever
Abdominal pain and no recall of last bowel movement
Difficulty initiating and maintaining a flow of urine
Urinary frequency without a sense of void completion

Doctor

These symptoms are outside the scope of physical therapy intervention. Would you please evaluate before we begin rehab? Please advise if there are any recommended changes in the proposed program of intervention. Thank you.

Outcome

Physician ordered a urine culture, but attempts to obtain a sample were unsuccessful. The physician was unable to insert a straight catheter, so the resident was sent to the hospital and a suprapubic catheter was inserted. He returned to the SNF 4 days later with the following diagnoses:

Bladder outlet obstruction
Urinary tract infection with *Escherichia coli*
Prostate cancer, probably metastatic
Obstipation

When the resident returned to the SNF and was seen by physical therapy, there were no complaints of low back pain (beyond his lifetime baseline) and no lower abdominal discomfort. The neurologic deficits previously identified in both lower extremities were absent.

Summary: This is a good case to point out that medical personnel occasionally miss things that a physical therapist can find when conducting a screening exam and a review of systems. Recognizing red flags sent this client back to the physician sooner rather than later and ended needless painful suffering on his part.

From Joseph R. Clemente, DPT (submitted as part of a t-DPT requirement), New York, 2003.

TABLE 8-1 Causes of Constipation

Neurogenic	Muscular	Mechanical	Rectal Lesions	Drugs/Diet
Cortical, voluntary, or involuntary evacuation	Atony (loss of tone)	Bowel obstruction	Thrombosed hemorrhoids	Anesthetic agents (recent general surgery)
Central nervous system lesions	Severe malnutrition	Neoplasm	Perirectal abscess	Antacids (containing aluminum or calcium)
Multiple sclerosis	Metabolic defects	Volvulus (intestinal twisting)		Anticholinergics
Cord tumors	Hypothyroidism	Diverticulitis		Anticonvulsants
Tabes dorsalis	Hypercalcemia	Extraalimentary tumors		Antidepressants
Spinal cord lesions or tumors	Potassium depletion	Pregnancy		Antihistamines
Parkinson's	Hyperparathyroidism	Colostomy		Antipsychotics
Irritable bowel syndrome	Inactivity; chronic back pain			Barium sulfate
Dementia				Cancer chemotherapy (e.g., Oncovin)
				Iron compounds
				Diuretics
				Narcotics
				Lack of dietary bulk
				Renal failure (due to fluid restriction, phosphate binders)
				Myocardial infarction (narcotics for pain control)

Changes in bowel habit may be a response to many other factors such as diet (decreased fluid and bulk intake), smoking, side effects of medication (especially constipation associated with opioids), acute or chronic diseases of the digestive system, extraabdominal diseases, personality, mood (depression), emotional stress, inactivity, prolonged bed rest, and lack of exercise (Table 8-1). Commonly implicated medications include narcotics, aluminum- or calcium-containing antacids (e.g., Alu-Tab, Basaljel, Tums, Rolaids), anticholinergics, tricyclic antidepressants, phenothiazines, calcium channel blockers, and iron salts.

Diets that are high in refined sugars and low in fiber discourage bowel activity. Transit time of the alimentary bolus from the mouth to the anus is influenced mainly by dietary fiber and is decreased with increased fiber intake. Additionally, motility can be decreased by emotional stress that has been correlated with personality. Constipation associated with severe depression can be improved by exercise.

People with low back pain may develop constipation as a result of muscle guarding and splinting that causes reduced bowel motility. Pressure on sacral nerves from stored fecal content may cause an *aching discomfort in the sacrum, buttocks, or thighs* (Case Example 8-4).

Because there are many specific organic causes of constipation, it is a symptom that may require further medical evaluation. It is considered a red flag symptom when clients

CASE EXAMPLE 8-4

Constipated Biker with Leg Pain

A 29-year-old male presented in the physical therapy clinic with inner thigh pain of the left leg of unknown cause over the last 3 weeks. The pain occurred most often when he had a bowel movement. He was training for an iron man competition (swimming, biking, running) but did not have any known injury or accident to attribute the symptom to.

When asked if there were any other symptoms anywhere else in his body, the client reported an inability to get an erection and a tendency toward constipation with hard stools. The therapist could find no clinical signs of muscle weakness, atrophy, or dysfunction. Postural alignment was symmetrical and without apparent problems. All provocation tests for hip, spine, sacrum, sacroiliac (SI), and pelvis were negative. The client could complete a full squat without difficulty. Hop test and heel strike were both negative.

The client was screened for signs and symptoms associated with other possible causes of erectile dysfunction such as diabetes, past history of testicular or prostate problems, past history of cancer, and possible sexual abuse. There was no red flag history or red flag signs and symptoms. Visual inspection of the lower half of the body revealed no signs of vascular compromise. The client denied any bladder problems or urinary incontinence.

Knowing that the pudendal nerve is responsible for penile erection, the therapist asked to see the client on his bicycle. Pressure on the nerve from a poorly constructed and minimally padded seat was a possible cause. The client was advised to change bike seats, change the seat height and tilt, and reassess symptoms in 2 weeks.

The client was also encouraged to stand up intermittently to relieve perineal pressure.

Result: The client reported complete cessation of all symptoms with the purchase of a bicycle seat with a cut-away middle. Since the obturator nerve passes below the symphysis pubis, it is likely bicycle seat compression on the nerve contributed to the inner thigh pain as well.

Bicycle seat neuropathy is not uncommon among long-distance bikers due to the cyclist supporting the body weight on a narrow seat. Vascular and/or neurologic compromise of the pudendal nerve is the most likely explanation for these symptoms.[80,81]

with unexplained constipation have sudden and unaccountable changes in bowel habits or blood in the stools.

Diarrhea

Diarrhea, by definition, is an abnormal increase in stool frequency and liquidity. This may be accompanied by urgency, perianal discomfort, and fecal incontinence. The causes of diarrhea vary widely from one person to another, but food, alcohol, use of laxatives and other drugs, medication side effects, and travel may contribute to the development of diarrhea (Table 8-2).

Acute diarrhea, especially when associated with fever, cramps, and blood or pus in the stool, can accompany invasive enteric infection. Chronic diarrhea associated with weight loss is more likely to indicate neoplastic or inflammatory bowel disease. Extraintestinal manifestations such as arthritis or skin or eye lesions are often present in inflammatory bowel disease. Any of these combinations of symptoms must be reported to the physician.

Drug-induced diarrhea is associated most commonly with antibiotics. Diarrhea may occur as a direct result of antibiotic use and the GI symptom resolves when the drug is discontinued. Symptoms may also develop 6 to 8 weeks after first ingestion of an antibiotic. A more serious, less frequent antibiotic-induced colitis with severe diarrhea is caused by *Clostridium difficile.*

This anaerobic bacterium colonizes the colon of 5% of healthy adults and over 20% of hospitalized patients. Clients receiving enteral (tube) feedings are at higher risk for acquisition of *C. difficile* and associated severe diarrhea. *C. difficile* is the major cause of diarrhea in patients hospitalized for more than 3 days. It is spread in an oral-fecal manner and is readily transmitted from patient to patient by hospital personnel. Fastidious handwashing, use of gloves, and extremely careful cleaning of bathroom, bed linen, and associated items are helpful in decreasing transmission.[12]

Athletes using creatine supplements to enhance power and strength in performance may experience minor GI symptoms. Muscle cramps, diarrhea, loss of appetite, weight gain, and dizziness occur in about 8% of the individuals taking

TABLE 8-2 Causes of Diarrhea

Malabsorption	Neuromuscular	Mechanical	Infectious	Nonspecific
Pancreatitis Pancreatic carcinoma Crohn's disease	Irritable bowel syndrome Diabetic enteropathy Hyperthyroidism Caffeine	Incomplete obstruction Neoplasm Adhesions Stenosis Fecal impaction Muscular incompetency Postsurgical effect (ileal bypass)	Viral Bacterial Parasitic Protozoal *(Giardia)*	Ulcerative colitis Diverticulitis Diet Laxative abuse Food allergy Antibiotics *(Clostridium difficile)* Creatine use Cancer chemotherapy (e.g., Fluorouracil) Lactose (milk) intolerance Psychogenic (nervous tension)

these supplements. Therapists working with athletes should keep this in mind when hearing reports of GI distress. Many sports players do not even know how much creatine they are taking or are taking more than the recommended dose. Players as young as 13 years old have reported using creatine supplements.[12a,13] The use of creatine for individuals under the age of 18 is not recommended; safety and efficacy of creatine has not been established in adolescents.[14]

For the client describing chronic diarrhea, it may be necessary to probe further about the use of laxatives as a possible contributor to this condition. Laxative abuse contributes to the production of diarrhea and begins a vicious cycle as chronic laxative users experience excessive secretion of aldosterone and resultant edema when they attempt to stop using laxatives. This edema and increased weight forces the person to continue to rely on laxatives. The abuse of laxatives is common in the eating disorder populations (e.g., anorexia, bulimia); affected persons may ingest up to 100 laxatives at a time.

Questions about laxative use can be asked tactfully during the Core Interview (see Chapter 2) when asking about medications, including over-the-counter (OTC) drugs such as laxatives. Encourage the client to discuss bowel management without drugs at the next appointment with the physician.

Fecal Incontinence

Fecal incontinence may be described as an inability to control evacuation of stool and is associated with a sense of urgency, diarrhea, and abdominal cramping. Causes include partial obstruction of the rectum (cancer), colitis, and radiation therapy, especially in the case of women treated for cervical or uterine cancer. The radiation may cause trauma to the rectum that results in incontinence and diarrhea. Anal distortion secondary to traumatic childbirth, hemorrhoids, and hemorrhoidal surgery may also cause fecal incontinence.

Arthralgia

The relationship between "gut" inflammation and joint inflammation is well known but not fully understood. Many inflammatory GI conditions have an arthritic component affecting the joints. For example, inflammatory bowel disease (ulcerative colitis and Crohn's disease) is often accompanied by rheumatic manifestations; peripheral joint arthritis and spondylitis with sacroiliitis are the most common of these manifestations.[15,16] Sacroiliac (SI) disease without inflammation has been documented as a primary cause of lower abdominal or inguinal pain.[17]

There may be a genetic component between inflammatory bowel disease and ankylosing spondylitis.[18] The relationship between intestinal problems and joint involvement may also be explained by some type of "interface" between the bowel and the articular surface of joints.[19,20] It is hypothesized that an antigen crosses the gut mucosa and enters the joint, which sets up an immunologic response. Arthralgia with synovitis and immune-mediated joint disease may occur as a result of this immunologic response.[19] It is likely that an impaired antibacterial host defense and an uncontrolled proinflammatory response of the innate immune system are at fault.[21]

Joint arthralgia associated with GI infection is usually asymmetric, migratory, and oligoarticular (affecting only one or two joints). This type of joint involvement is termed *reactive arthritis* when triggered by microbial infection such as *C. difficile* from the GI (and sometimes genitourinary or respiratory) tract. Other accompanying symptoms may include fever, malaise, skin rash or other skin lesions, nail bed changes (nails separate from the nail beds and become thin and discolored), iritis, or conjunctivitis.

The bowel and joint symptoms may or may not occur at the same time. Usually, this type of arthralgia is preceded 1 to 3 weeks by diarrhea, urethritis, regional enteritis (Crohn's disease), or other bacterial infection. The knees, ankles, shoulders, wrists, elbows, and small joints of the hands and feet (listed in order of decreasing frequency) are the peripheral joints affected most often.[22]

A large knee effusion is a common presentation, but some clients have joint pain with minimal or no signs of inflammation. Muscle atrophy occurs when a chronic condition is present; in which case, there will be a history of previous GI and joint involvement. Stiffness, pain, tenderness, and reduced range of motion may be present, but with proper medical intervention, there is no permanent deformity.

Spondylitis with sacroiliitis may present as low back pain and morning stiffness that improves with activity and restriction of chest and spinal movement. Radiographic findings are consistent with those of classic ankylosing spondylitis with bilateral SI joint involvement and bony erosion and sclerosis of the symphysis pubis, ischial tuberosities, and iliac crests. Ultimately, "bamboo spine" (see Fig. 12-4) will result.

Inflammation involving the sites of bony insertion of tendons and ligaments termed *enthesitis* is a classic sign of reactive arthritis. Tendon sheaths and bursae may also become inflamed. Ligaments along the spine and SI joints and around the ankle and midfoot may also show evidence of inflammation.

Heel pain is a frequent complaint, with swelling and tenderness located either posteriorly at the Achilles tendon insertion site, or inferiorly where the plantar fascia attaches to the calcaneus. Plantar fasciitis is common. Enthesopathy can also occur around the knee, ischial tuberosities, greater femoral trochanter, and costovertebral and manubriosternal joints.[23]

For a more complete discussion of joint pain and how to evaluate joint pain, see Chapter 3. A list of screening questions for joint pain is also reproduced in the Appendix as a quick reference in clinical practice.

Shoulder Pain

Pain in the left shoulder (Kehr's sign: pain with pressure placed on the upper abdomen; Danforth sign: shoulder pain with inspiration) can occur as a result of free air following laparoscopic surgery or blood in the abdominal cavity, usually from a ruptured spleen or retroperitoneal bleeding causing distention. Retroperitoneum refers to a position external or posterior to the peritoneum, the serous membrane lining the abdominopelvic walls. Retroperitoneal organs refer to viscera that lie against the posterior body wall and are covered by peritoneum on the anterior surface only (e.g., thoracic portion of the esophagus, pancreas, duodenal cap, ascending and descending colon, rectum).

The screening interview may help the client recall any precipitating trauma or injury such as a sharp blow during an athletic event, a fall, or perhaps even a minor automobile accident causing pressure from the steering wheel. The client may not connect these seemingly unrelated events with the present shoulder pain.

Perforated duodenal or gastric ulcers can leak gastric juices on the posterior wall of the stomach that irritate the diaphragm referring pain to the shoulder; although the stomach is on the left side of the body, the referral pattern is usually to the right shoulder.

A ruptured ectopic pregnancy with retroperitoneal bleeding into the abdominal cavity can also present as low abdominal and/or shoulder pain. Usually there is a history of sexual activity and missed menses in a woman of reproductive age.

Pancreatic cancer can refer pain to the shoulder and is often missed as the cause. Fluid in the pleural space as a result of pancreatitis can present as shoulder pain. When the head of the pancreas is involved, the client could have right shoulder pain, but more often it manifests as mid-back or midthoracic pain sometimes lateralized from the spine on either side. When the tail of the pancreas is diseased, pain can be referred to the left shoulder (see Fig. 3-4). Pain may also occur in the right shoulder when blood is present in the abdominal cavity due to liver trauma (Case Example 8-5). Accumulation of blood in this area from a slow bleed of the spleen, liver, or stomach can produce bilateral shoulder pain.

CASE EXAMPLE 8-5

Ruptured Spleen

A 23-year-old soccer player sustained a blow from the side as he was moving down the soccer field. He fell on his left side with the full force of his own body weight and the weight of the other player on top of him. He reported having "the wind knocked out of me" and sat out on the sidelines for 20 minutes. He resumed playing and completed the game. The next morning, he woke up with severe left shoulder pain and stopped by the office of a physical therapist located in the same building as his office. The objective examination was unremarkable for shoulder movement dysfunction, which was inconsistent with the client's complaint of "constant pain." The client was treated symptomatically and instructed in pendulum exercises to maintain the joint motion.

He made a follow-up appointment with the therapist for the next day, but before noon, he collapsed at work and was taken to a hospital emergency department. A diagnosis of ruptured spleen was made during emergency surgery. A ruptured spleen would have sent the typical adult for medical care much sooner, but this client was in excellent physical condition with a high tolerance for pain. Physical therapy intervention was not appropriate in this situation; an immediate medical referral was indicated given the history of trauma, sudden onset of symptoms, left shoulder pain (Kehr's sign), and constancy of pain.

Obturator or Psoas Abscess

Abscess of the obturator or psoas muscle is a possible cause of lower abdominal pain, usually the consequence of spread of inflammation or infection from an adjacent structure. Since these muscles lie behind abdominal structures with no protective barrier, any infectious or inflammatory process affecting the abdominal or pelvic cavity can cause an obturator or psoas abscess (Figs. 8-3 and 8-4).

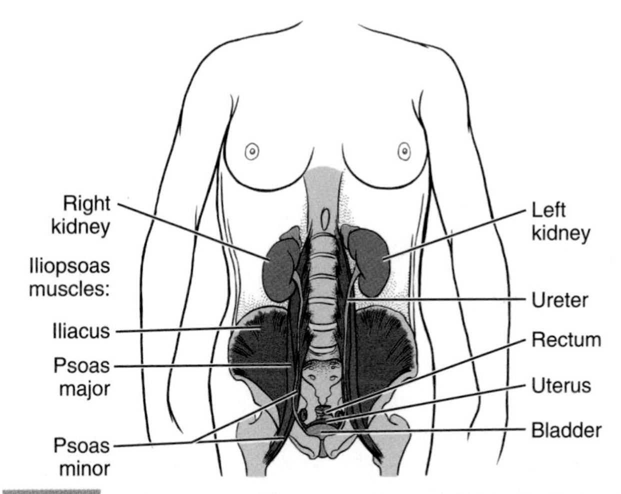

Fig. 8-3 The iliopsoas muscle is not separated from the abdominal or pelvic cavity. As this illustration shows, most of the viscera in the abdominal and pelvic cavities can come in contact with the iliopsoas muscle. Any infectious or inflammatory process present in either of these cavities can seed itself to the psoas muscle by direct extension.

Psoas abscesses most commonly result from direct extension of intraabdominal infections such as diverticulitis, Crohn's disease, pelvic inflammatory disease (PID), and appendicitis (see also the discussion on McBurney's point later in this chapter).[24] Kidney infection or abscess can also cause psoas abscess. *Staphylococcus aureus* (staph infection) is the most common cause of psoas abscess secondary to vertebral osteomyelitis.

Peritonitis as a result of any infectious or inflammatory process can result in psoas abscess. Besides the diseases and conditions mentioned here, peritonitis can occur as a surgical complication. Look for a history of abdominal surgery of any kind, especially the anterior approach to spinal surgery for disk removal, spinal fusion, and insertion of a cage or artificial disk implant.[25] In adult women, hematogenous psoas abscesses have been observed as a complication of spontaneous vaginal delivery.[26,27]

Regardless of the etiology, the abscess is usually confined to the psoas fascia but can spread to the hip, upper thigh, or buttock. The iliacus muscle in the iliac fossa joins with the lower portion of the psoas muscle. Osteomyelitis of the ilium or septic arthritis of the SI joint can penetrate the muscle

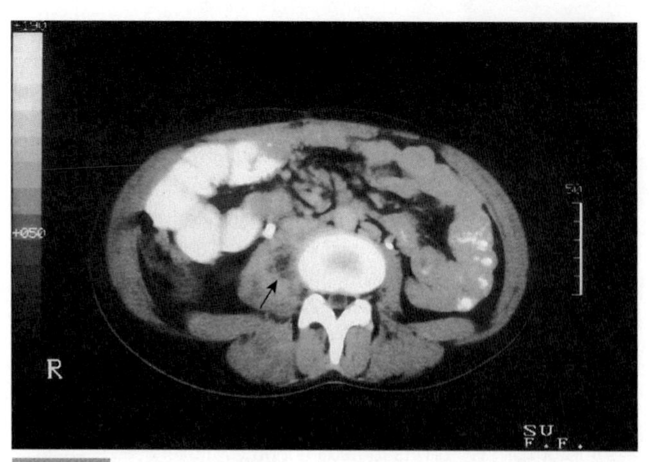

Fig. 8-4 Right psoas abscess in a 17-year-old girl *(arrow)*. (From Mandell GL, Bennett JE, Dolin R: *Mandell, Douglas, and Bennett's principles and practice of infectious diseases,* Philadelphia, 2009, Churchill Livingstone.)

sheath of either muscle, producing an abscess of either the iliacus or psoas portion of the muscle.[28]

In addition, abscesses of the pelvis, retroperitoneal area, and abdomen can spread bacteria or fungi to local vertebral areas, causing spinal infections such as pyogenic vertebral osteomyelitis. From the lumbar spine, abscess formation may track along the psoas muscle and into the buttock (piriformis fossa), the perianal region, the groin, and even the popliteal fossa.[29]

Clinical manifestations of a psoas or iliacus abscess include fever; night sweats; lower abdominal, pelvic, or back pain; or pain referred to the hip, medial thigh or groin (femoral triangle area), or knee. The right side is affected most often when associated with appendicitis. Both sides can be involved with generalized peritonitis but usually that person has a clear systemic presentation and seeks medical evaluation. It is the unusual cases that a therapist will see, making it necessary to know both the typical pain patterns associated with systemic disease, as well as the atypical presentations.

Antalgic gait may develop with a psoas abscess secondary to a reflex spasm pulling the leg into internal rotation and causing a functional hip flexion contracture. The affected individual may have pain with hip extension. Often a tender mass can be palpated in the groin. The therapist must assess for TrPs of the iliopsoas muscle. A psoas minor syndrome can be mistaken for appendicitis so be sure and assess for TrPs.[11]

Four tests can be performed to assess the possibility of systemic origin of painful hip or thigh symptoms (Box 8-2). Gently pick up the client's leg on the involved side and tap the heel. A painful expression and report of right lower quadrant pain may accompany peritoneal inflammation. If the client is willing and able, have him or her hop on one leg. The person with an inflamed peritoneum will clutch that side and be unable to complete the movement. The *iliopsoas muscle test* (Fig. 8-5) is performed when acute abdominal pain is a possible cause of hip or thigh pain. When an abscess forms on the iliopsoas muscle from an inflamed or perforated appendix or inflamed peritoneum, the iliopsoas muscle test

BOX 8-2 SCREENING TESTS FOR PSOAS ABSCESS

- Heel tap
- Hop test
- Iliopsoas muscle test
- Palpate iliopsoas muscle

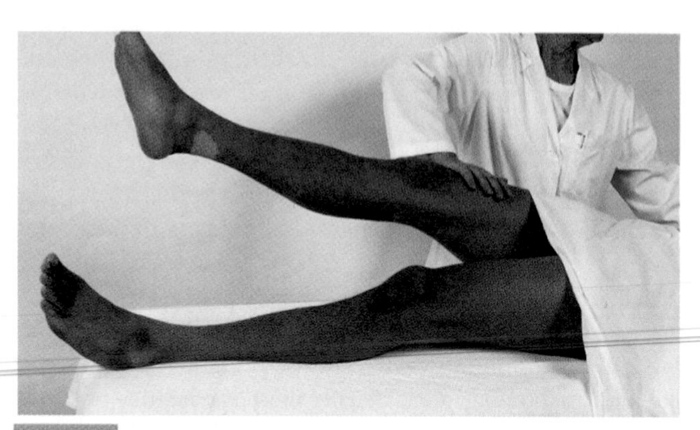

Fig. 8-5 Iliopsoas muscle test. In the supine position, have the client actively perform a straight leg raise; apply resistance to the distal thigh as the client tries to hold the leg up. Alternately, ask the client to turn onto his or her side. Extend the person's uppermost leg at the hip. Increased abdominal, flank, or pelvic pain on either maneuver constitutes a positive sign, suggesting irritation of the psoas muscle by an inflamed appendix or peritoneum. Only a handful of studies have been done to validate the accuracy of this test for appendicitis/peritonitis. One systematic review reported sensitivity value at 0.16 and specificity as 0.95 with a positive likelihood ratio (LR+) of 2.38 (reported range: 1.21-4.67) and negative likelihood range (LR–) of 0.90 (range: 0.83-0.98).[57] (From Jarvis C: *Physical examination and health assessment,* Philadelphia, 2000, WB Saunders.)

causes pain felt in the right lower abdominal quadrant. (Pain and tenderness in the lower left side of the abdomen and pelvis may be caused by bowel perforation associated with diverticulitis, constipation, or obstipation [impaction] of the sigmoid, or appendicitis when the appendix is located on the left side of the midline.)

Alternately, the client lies on the pain-free side, and the therapist gently hyperextends the involved leg to stretch the psoas major muscle. Additionally, palpate the iliopsoas muscle by placing the client in a supine position with hips and knees flexed and fully supported in a 90-degree position (Fig. 8-6). Palpate one third of the distance between the anterior superior iliac spine (ASIS) and the umbilicus. The client is asked to flex the hip gently to assist in isolating the iliopsoas muscle. Muscular tightness in the iliopsoas may result in radiating pain to the low back region during palpation, whereas inflammation or abscess will bring on painful symptoms in the right (or left depending on the underlying pathology) lower abdominal quadrant.

The *obturator muscle test* (Fig. 8-7) is also performed when the appendix could be the cause of referred pain to the hip. A perforated appendix or inflamed peritoneum can irritate

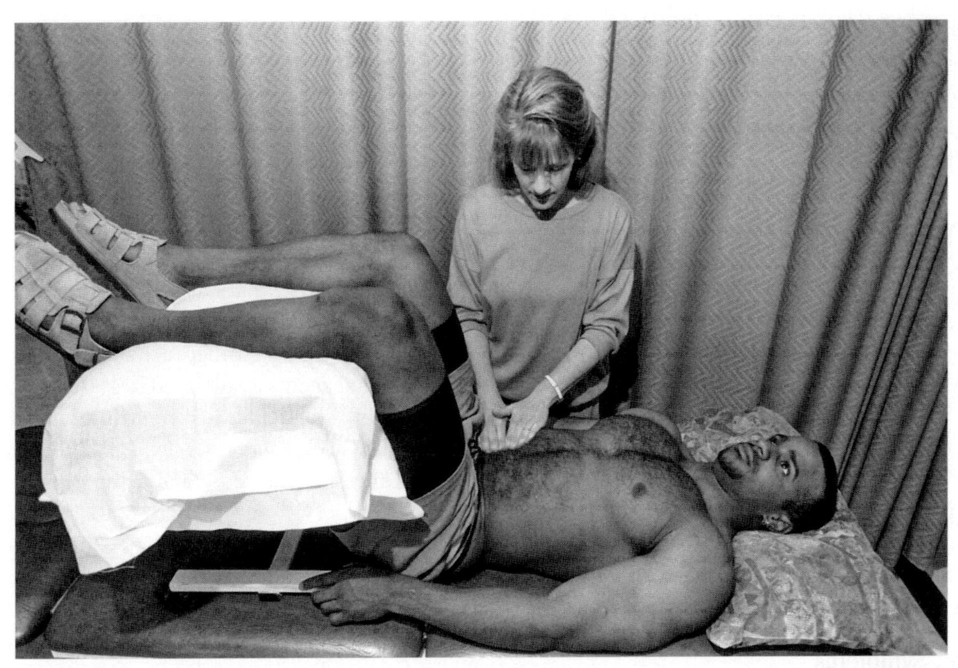

Fig. 8-6 Palpating the iliopsoas muscle. Place the client in a supine position with the hips and knees both flexed and supported at a 90-degree angle. Slowly press fingers into abdomen approximately one third the distance from the anterior superior iliac spine (ASIS) toward the umbilicus. It may be necessary to ask the client to initiate slight hip flexion to help isolate the muscle and avoid palpating the bowel. Reproducing or causing lower quadrant, pelvic, or abdominal pain is considered a positive sign for iliopsoas abscess. Palpation may produce back pain or local muscular pain from shortened or contracted muscle. (From Goodman CC, Fuller KS: *Pathology: special implications for the physical therapist,* ed 3, Philadelphia, 2009, WB Saunders.)

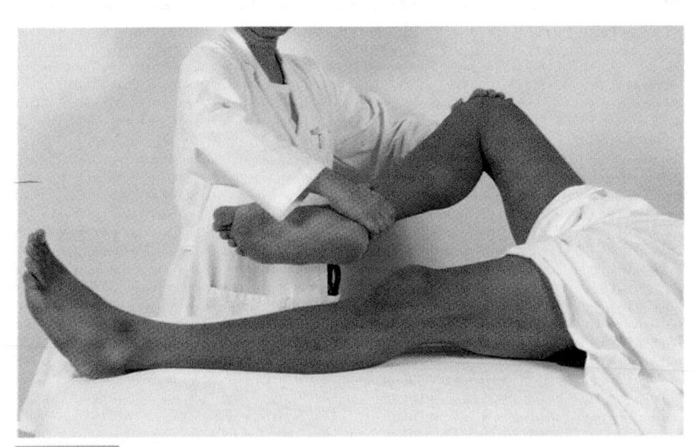

Fig. 8-7 Obturator muscle test. In the supine position, perform active assisted motion, flexing at the hip and 90 degrees at the knee. Hold the ankle and rotate the leg internally to stretch the obturator muscle. A negative or normal response is no pain. A positive test for muscle affected by peritoneal infection or inflammation from a perforated appendix reproduces right lower quadrant abdominal or pelvic pain with irritation of the muscle. Although the obturator test has not been studied independently of the psoas test, this sign is assumed to have a sensitivity and specificity similar to the psoas sign. Further research is needed to verify this as a valid and reliable evidence-based test for appendicitis/peritonitis. (From Jarvis C: *Physical examination and health assessment,* ed 5, Philadelphia, 2007, WB Saunders.)

the obturator muscle, producing right lower quadrant abdominal pain during the obturator test.

Although uncommon, psoas abscess still can be confused with a hernia. The therapist may perform evaluative tests to screen for a psoas abscess, but the physician must differentiate between an abscess and a hernia. Psoas abscess is often softer than a femoral hernia and has ill-defined borders, in contrast to the more sharply defined margins of the hernia. The major differentiating feature is the fact that a psoas abscess lies lateral to the femoral artery, whereas the femoral hernia is located medial to the femoral artery.[30]

CLINICAL SIGNS AND SYMPTOMS

Psoas Abscess

- Fever ("hectic" fever pattern: up and down)
- Night sweats
- Abdominal pain
- Loss of appetite or other GI upset
- Back, pelvic, abdominal, hip, and/or knee pain
- Antalgic gait
- Palpable, tender mass

Neuropathy

Numbness and weakness of the lower extremities have been reported as a result of vitamin B_{12} deficiency in the aging adult population or from thiamine deficiency after gastric bypass. Such events are infrequent but expected to increase as the number of gastric bypasses increases.[31] Symmetric paresthesias and ataxia associated with loss of vibration and position sense in the lower extremities may occur with vitamin B_{12} deficiency.

Other symptoms can include irritability, memory loss, and dementia. Thiamine deficiency after gastric bypass may present as early as 2 months or as late as 10 years after the procedure.[32,33] Besides neuropathy, presentation may include confusion, nystagmus, seizures, unsteady gait and ataxia, hearing loss, and lower limb hypotonia.[32] Symptoms may resolve with medical treatment.

GASTROINTESTINAL DISORDERS

Gastroesophageal Reflux Disease

GERD is an array of problems related to the backward movement of stomach acids and other stomach contents, such as pepsin and bile, into the esophagus, a phenomenon called *acid reflux.* Normally, some gastric contents move or reflux from the stomach into the esophagus, but in GERD, the process becomes pathologic, producing symptoms that point to tissue injury in the esophagus and sometimes the respiratory tract.[34,35] In adults, GERD is usually caused by intermittent relaxation of the lower esophageal sphincter (LES).

Clinical Signs and Symptoms

Symptoms can include heartburn, chest pain, dysphagia, and a sense of a lump in the throat. Symptoms are sometimes mistaken for a heart attack. Less frequent symptoms can include wheezing, hoarseness, coughing, earache, sore throat, and difficulty swallowing. Sleep disturbance from nighttime coughing and heartburn can lead to fatigue and decreased daytime functioning. Complications of GERD may range from discomfort to severe strictures of the esophagus, esophagitis, aspiration pneumonia, and asthma.

Other serious consequences can be related to weight loss, GI blood loss, and Barrett's esophagus, a precancerous condition. The relationship between GERD and asthma is poorly understood but is thought to be a consequence of aspiration of gastric acid contents into the lung, causing bronchospasm. Most adults with asthma also have GERD.[35,36]

GERD can occur in infants, but most "outgrow" it. Watch for frequent, forceful spitting up or vomiting, accompanied by irritability. Other alarm symptoms include respiratory distress, apnea, dysphagia, or failure to thrive. Watch for change in color, change in muscle tone, or choking and gagging.

Children may experience GERD in the same way adults do with abdominal or epigastric pain. Nighttime coughing, vomiting, and/or nausea are also possible. Neurologically impaired children and adults are at increased risk for reflux with aspiration. Fluid enters the upper airways from the esophagus, causing chronic respiratory problems, including recurrent pneumonia.

GERD should be treated in order to prevent a chronic condition from occurring with more serious consequences. Symptoms may be mild at first, but have a cumulative effect with increasing symptoms after the age of 40. Chronic GERD is a major risk factor for adenocarcinoma, an increasingly common cancer in white males in the United States.

Medical referral is advised for anyone who reports signs and symptoms of GERD. Some clients may need surgical treatment, now available with less invasive endoscopic techniques, but most can be treated with some simple changes in eating patterns, positioning, and medications. Drug treatment includes antacids, H_2-receptor blockers, and proton pump inhibitors (PPIs). Antacids, such as Mylanta, Maalox, Tums, and Rolaids, are available OTC and do not reduce the acid, but merely neutralize it. H_2-receptor blockers, such as Tagamet (cimetidine), Zantac (ranitidine), and Pepcid (famotidine), reduce the amount of stomach acid produced by the stomach and are available over the counter.

PPIs, such as Prilosec (omeprazole), Prevacid (lansoprazole), or Nexium (esomeprazole), are the most potent acid-suppressing agents available. These drugs actually inhibit acid formation rather than just neutralize it. The first PPI is now available OTC; others are expected to become available as well. Caution is needed when using PPIs to self-treat without medical supervision. They can mask symptoms of serious GI disorders such as esophageal or stomach cancer. Diagnosis at an early, treatable stage may be delayed with serious implications.

Therapists must listen for client reports of headache, constipation or diarrhea, abdominal pain, or dizziness in anyone taking these medications. The client should be advised to notify his or her medical doctor with a report of these side effects.

CLINICAL SIGNS AND SYMPTOMS

Gastroesophageal Reflux Disease

Typical Symptoms
- Heartburn
- Regurgitation with bitter taste in mouth
- Belching

Atypical Symptoms
- Chest pain unrelated to activity
- Sensation of a lump in the throat
- Difficulty swallowing (dysphagia)
- Painful swallowing (odynophagia)
- Wheezing, coughing, hoarseness
- Asthma
- Sore throat, laryngitis
- Weight loss
- Anemia

Peptic Ulcer

Peptic ulcer is a loss of tissue lining the lower esophagus, stomach, and duodenum. Gastric and duodenal ulcers are considered together in this section. Acute lesions that do not extend through the mucosa are called erosions. Chronic ulcers involve the muscular coat, destroying musculature, and replacing it with permanent scar tissue at the site of healing.

Originally, all ulcers in the upper GI tract were believed to be caused by the aggressive action of hydrochloric acid and pepsin on the mucosa. They thus became known as "peptic ulcers," which is actually a misnomer.

It is now known that many of the gastric and duodenal ulcers are caused by infection with *Helicobacter pylori,* a

TABLE 8-3 Nonsteroidal Antiinflammatory Drugs

Generic	Common Brand Names
OVER-THE-COUNTER	
Aspirin	Anacin, Ascriptin,* Bayer,* Bufferin,* Ecotrin,* Excedrin*
Ibuprofen	Advil, Motrin, Ibuprofen, various generic store brands
Ketoprofen	Nexcede
Naproxen sodium	Aleve, various generic store brands
PRESCRIPTION NONSELECTIVE (STANDARD) COX INHIBITORS	
Diclofenac sodium	Voltaren
Diflunisal	None
Etodolac	None
Fenoprofen calcium	Nalfon
Flurbiprofen	Ansaid
Ibuprofen	None
Indomethacin	Indocin
Ketorolac	None
Ketoprofen	None
Meclofenamate sodium	None
Mefenamic acid	Ponstel
Meloxicam	Mobic
Nabumetone	None
Naproxen	Naprosyn, Naprelan
Naproxen sodium	Anaprox, Anaprox DS, EC-Naprelan, others
Oxaprozin	Daypro
Piroxicam	Feldene
Salsalate	Salsitab, Salflex
Sulindac	Clinoril
Tolmetin sodium	None
PRESCRIPTION COX-2 SELECTIVE INHIBITORS	
Celecoxib	Celebrex

NSAIDs, Nonsteroidal antiinflammatory drugs; *COX,* cyclooxygenase.
Information in this table was reviewed and updated by the University of Montana College of Health Professional Biomedical Sciences Drug Information Services (Nicole Marcellus, PharmD candidate), 2010.
*These all have additives to minimize gastrointestinal (GI) side effects but are known as aspirin products. Many nonselective (standard) NSAIDs are available over-the-counter (OTC) at a lower dosage (e.g., 200 mg) and by prescription at a higher dosage (e.g., 500 mg). See discussion of peak effect for NSAIDs and time to impact underlying tissue impairment in Chapter 2.
Data from Drug Facts and Comparisons eAnswers (online), 2010, available from Wolters Kluwer Health, Inc (accessed June 23, 2010); Micromedex Healthcare Series (Internet database), Greenwood Village, Colorado, Thomson Reuters (Healthcare) Inc, updated periodically; and Drugs@FDA: FDA-approved drug products website, available at http://www.accessdata.fda.gov/scripts/cder/drugsatfda/ (accessed June 23, 2010).

corkscrew-shaped bacterium that bores through the layer of mucus that protects the stomach cavity from stomach acid. Ten percent of ulcers are induced by chronic use of NSAIDs, such as aspirin, ibuprofen, and naproxen, commonly taken by people with arthritis (Table 8-3).[37]

H. pylori ulcers are primarily located in the lining of the duodenum (upper portion of the small intestine that connects to the stomach) (Fig. 8-8). NSAID-induced ulcers occur primarily in the lining of the stomach, most frequently on the posterior wall, which can account for shoulder (usually right shoulder; depends on the extent of retroperitoneal bleeding) or back pain as an associated symptom.

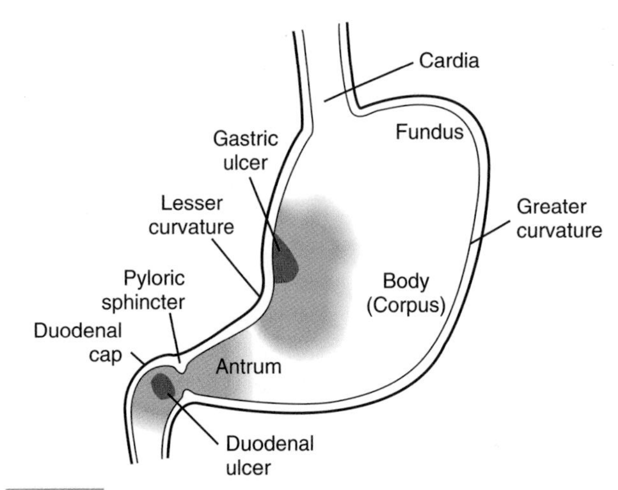

Fig. 8-8 Most common sites for peptic ulcers. Gastric ulcers are found along the distribution on the eighth thoracic nerve with a corresponding pain pattern as described in Fig. 8-13. Pain patterns associated with duodenal ulcers correspond to the tenth thoracic nerve. (Adapted from Ignatavicius DD, Bayne MV: *Medical-surgical nursing,* Philadelphia, 1991, WB Saunders.)

Ulcers can be dangerous if left untreated, eroding into the stomach arteries and causing life-threatening bleeding or perforating the stomach and spreading infection. *H. pylori*–induced ulcers can recur after treatment. The recurrence rate is higher in clients with gastric ulcers and in people who smoke, consume alcohol, and use NSAIDs.[38] A past medical history of peptic ulcers in anyone with new onset of back or shoulder pain is a red flag requiring further screening and possible medical referral.

Clinical Signs and Symptoms

The cardinal symptom of peptic ulcer is epigastric pain that may be described as "heartburn" or as burning, gnawing, cramping, or aching located over a small area near the midline in the epigastrium near the xiphoid. Gastric ulcers are found along the distribution of the eighth thoracic nerve which causes pain in the upper epigastrium about one to two inches to the right of a spot halfway between the xiphoid and the umbilicus (see Fig. 8-14). Duodenal pain tends to present more in the right epigastrium, specifically a localized spot one to two inches above and to the right of the umbilicus because of its innervation by the tenth thoracic nerve.

The pain comes in waves that last several minutes (rather than hours) and may radiate below the costal margins into the back or to the right shoulder. The daily pattern of pain is related to the secretion of acid and the presence of food in the stomach to act as a buffer.

Pain associated with duodenal ulcers is prominent when the stomach is empty such as between meals and in the early morning. The pain may last from minutes to hours and may be relieved by antacids. Gastric ulcers are more likely to cause pain associated with the presence of food. Symptoms often appear for 3 or 4 days or weeks and then subside, reappearing weeks or months later.

Other symptoms of uncomplicated peptic ulcer include nausea, vomiting, loss of appetite, sometimes weight loss, and occasionally back pain. In duodenal ulcers, steady pain near the midline of the back (see Fig. 8-14) between T6 and T10 with radiation to the right upper quadrant may indicate perforation of the posterior duodenal wall.

Back pain may be the first and only symptom. Complications of hemorrhage, perforation, and obstruction may lead to additional symptoms that the client does not relate to the back pain. Bleeding may occur when the ulcer erodes through a blood vessel. It may present as vomited bright red blood or coffee-ground vomitus and by dark tarry stools (melena). The bleeding may vary from massive hemorrhage to occult (hidden) bleeding that occurs over a long period of time.

Symptoms associated with *H. pylori* include halitosis (bad breath) and a form of facial acne called *rosacea.* Rosacea is characterized by a rosy appearance of the cheeks, nose, and chin. Facial flushing, red lines, and bumps over the nose may accompany rosacea.

CLINICAL SIGNS AND SYMPTOMS

Peptic Ulcer

- "Heartburn" or epigastric pain aggravated by food (gastric ulcer); relieved by food, milk, antacids, or vomiting (duodenal ulcer)
- Night pain (12 midnight to 3:00 AM)—same relief as for epigastric pain (duodenal ulcer)
- Radiating back pain
- Stomach pain
- Right shoulder pain
- Lightheadedness or fainting
- Nausea
- Vomiting
- Anorexia
- Weight loss
- Bloody stools
- Black, tarry stools

Gastrointestinal Complications of Nonsteroidal Antiinflammatory Drugs

NSAIDs (see Table 8-3) have become increasingly popular by virtue of their analgesic, antiinflammatory, antipyretic, and antithrombotic (platelet-inhibitory) actions. More than 70 million prescriptions are written each year in the United States. If you add OTC use, then 30 billion doses of NSAIDs are consumed annually in the United States.[39]

Most commonly taken NSAIDs really have few toxic effects. On the other hand, NSAIDs can have deleterious effects on the entire GI tract from the esophagus to the colon, although the most obvious clinical effect is on the gastroduodenal mucosa. GI impairment can be seen as subclinical erosions of the mucosa or more seriously, as ulceration with life-threatening bleeding and perforation.

The incidence of NSAID-related ulcer complications remains high despite the availability of newer gastroprotective NSAIDs such as cyclooxygenase-2 (COX-2) inhibitors (e.g., celecoxib, rofecoxib).[40] COX inhibitors, a group of enzymes that facilitate the production of prostaglandins, exist in two forms: COX-1 and COX-2. COX-1 promotes proper GI function and blood clotting. COX-2 has a role in preventing or reducing the inflammatory response.[41]

Standard NSAIDs nonselectively inhibit the actions of both types of COX inhibitors so the client gets the antiinflammatory effect but at the expense of the GI system. COX-2 inhibitors suppress COX-2, providing some benefit in reducing ulcer formation and GI bleeding, although they do not completely reduce the risk and have been shown to increase the incidence of myocardial events. If the newer agents are combined with even low-dose aspirin, the safety of the COX-2 agent is partially negated.[42,43] Low-dose aspirin is defined as 325 mg taken every other day or one "baby aspirin" containing 81 mg used for cardioprotection by people with or at risk for heart disease.

Infection with *H. pylori* bacteria increases the risk of ulcer disease threefold or more in people taking standard NSAIDs or low-dose aspirin.[44] COX-2 inhibitors are widely promoted as easier on the stomach than older NSAIDs, but not all clients are taking this newer generation of NSAIDs. For those who are taking COX-2 agents, preliminary studies show clients with a history of bleeding ulcer are at increased risk of recurrence so these clients must be monitored closely as well.[40] With a 6-month period of treatment with NSAIDs, dyspepsia (digestive discomfort potentially leading to ulceration) occurs in 15% to 30% (some reports put this figure closer to 50%) of adults using NSAIDs,[45,46] but we must remember that physical and occupational therapists are seeing a majority of these people.

People with NSAID-induced GI impairment can be asymptomatic until the condition is advanced. GI effects of NSAIDs are responsible for approximately 40% of hospital admissions among clients with arthritis. NSAID-induced GI bleeding is a major cause of morbidity and mortality among the aging adult population.[47]

For those who are symptomatic, the most common side effects of NSAIDs are stomach upset and pain, possibly leading to ulceration. GI complications of NSAID use include ulcerations, hemorrhage, perforation, stricture formation, and exacerbation of inflammatory bowel disease. Each NSAID has its own pharmacodynamic characteristics, and clients' responses to each drug may vary greatly.

Other possible adverse side effects of NSAIDs may include suppression of cartilage repair and synthesis, fluid retention and kidney damage, liver damage, skin reactions (e.g., itching, rashes, acne), and impairment of the nervous system such as headache, depression, confusion or memory loss, mood changes, and ringing in the ears.[39]

Many people diagnosed with painful musculoskeletal conditions, especially arthritis, rely on NSAIDs to relieve pain and improve function. Anyone with a current history of NSAID use presenting with back or shoulder pain, especially when accompanied by any of the associated signs and symptoms listed for peptic ulcer, must be evaluated by a physician. The therapist should remain alert for the client taking multiple NSAIDs and simultaneously combining prescription and OTC NSAIDs or other drugs.

These drugs are potent renal vasoconstrictors, so look for increased blood pressure and ankle/foot edema. Take vital signs and visually inspect clients at risk for NSAID-induced impairments. Ask about muscle weakness, unusual fatigue, restless legs syndrome, polyuria, nocturia, or pruritus (signs of renal failure). In the aging adult, NSAID use may be associated with confusion and memory loss or increased confusion in the client with dementia or Alzheimer's disease. Teach clients to recognize signs and symptoms of adverse effects from NSAIDs and report any associated signs and symptoms to the physician. Changing the dosage or switching to a different NSAID at the first sign of side effects can help clients avoid serious complications that can occur with prolonged use of an inappropriate dose or poorly tolerated NSAID.

CLINICAL SIGNS AND SYMPTOMS

NSAID-Induced Disease

- Asymptomatic
- Stomach upset (nausea) and stomach pain
- Indigestion, heartburn
- Skin reactions (itching, rash, acne)
- Increased blood pressure
- New-onset back (thoracic) or shoulder pain
- Melena
- Tinnitus (ringing in the ears)
- CNS changes
  - Headache
  - Depression
  - Confusion (older adult)
  - Memory loss (older adult)
  - Mood changes
- Renal involvement
  - Muscle weakness
  - Unusual fatigue
  - Restless legs syndrome
  - Polyuria
  - Nocturia
  - Pruritus (skin itching)

Risk factor assessment is especially important in the primary care setting. Any identified risk factors should serve as red flags in any setting. The most predictive risk factors of serious GI events include age, disability, NSAID use, previous GI hospitalization, prior GI symptoms with NSAIDs, and use of prednisone (Box 8-3 and Case Example 8-6). See Chapter 2 for information on screening for the use of NSAIDs.

Is Your Client At Risk for NSAID-Induced Gastropathy? Therapists also can estimate the risk of GI complications in clients with rheumatoid arthritis. The tool in Table 8-4 can be used with clients who are taking NSAIDs of any kind for rheumatoid arthritis. This tool may prove valuable

BOX 8-3 RISK FACTORS FOR NSAID-INDUCED GASTROPATHY

- Age older than 65 years
- History of peptic ulcer disease or GI disease
- Smoking, alcohol use
- Oral corticosteroid use
- Anticoagulation or use of other anticoagulants (even when used for heart patients at a lower dose [e.g., 81 to 325 mg aspirin/day])
- Renal complications in clients with hypertension or CHF or who use diuretics or ACE inhibitors
- Use of acid suppressants (e.g., H_2-receptor antagonists, antacids); these agents can mask the warning symptoms of more serious GI complications, leaving the client unaware of ongoing damage
- NSAIDs combined with selective serotonin reuptake inhibitors (SSRIs; antidepressants such as Prozac, Zoloft, Celexa, Paxil)

The newer COX-2 (cyclooxygenase) inhibitors have reduced the incidence of GI disturbances, but this does not mean a client taking a COX-2 inhibitor cannot have NSAID-induced GI complaints. The risk of complications with COX-2 inhibitors is increased in the presence of any of the risk factors listed above.

GI, Gastrointestinal; *CHF,* congestive heart failure; *ACE,* angiotensin-converting enzymes; *NSAIDs,* nonsteroidal antiinflammatory drugs; *SSRIs,* serotonin selective reuptake inhibitors; *COX-2,* cyclooxygenase-2.

TABLE 8-4 Calculating Your Client's Risk of NSAID-Induced Gastropathy

Risk Is Equal to the Sum of:	Calculation	Points
Age in years	Multiply × 2 =	
History of NSAID symptoms (e.g., upper abdominal pain, bloating, nausea, heartburn, loss of appetite, vomiting)	If yes, add 50 points	
ACR Class (see Table 8-5)	Add 0, 10, 20, or 30 based on Class 1-4	
NSAID dose (fraction of maximum recommended; see text explanation)	NSAID dose × 15	
If currently using prednisone	Add 40 points	
TOTAL SCORE		
*Risk/year = [TOTAL SCORE − 100] ÷ 40		

NSAID, Nonsteroidal antiinflammatory drug; *ACR,* College of Rheumatology.

*Higher total scores yield a greater predictive risk. The risk ranges from 0.0 (low risk) to 5.0 (high risk).

From Fries JF, et al: Nonsteroidal antiinflammatory drug-associated gastropathy: incidence and risk factor models, *Am J Med* 91(3):213-222, 1991.

CASE EXAMPLE 8-6

Nonsteroidal Antiinflammatory Drugs

Outpatient Orthopedic Client: A 72-year-old client is status-post (s/p) left total knee replacement (TKR) ×4 weeks. She did not attain 90-degrees knee flexion and continues to walk with a stiff leg. Her orthopedic surgeon has sent her to physical therapy for rehab.

Past Medical History: Client reports generalized osteoarthritis. Previous left shoulder replacement 18 months ago. Very slow recovery and still does not have full shoulder range of motion (ROM). Long-standing hearing impairment for 60 years. Lost her left eye to macular degeneration 2 years ago.

Medications: Client reports the following drug use—Darvocet for pain 3×/day. Vioxx daily for arthritis. Also takes Feldene when her shoulder bothers her and daily ibuprofen.

Walks with a Trendelenburg gait and drags left leg using wheeled walker.

Current symptoms include left knee and shoulder pain, intermittent dizziness, sleep disturbance, finger/hand swelling in the afternoons, early morning nausea.

How do you assess for nonsteroidal antiinflammatory drug (NSAID) complications?

Review risk factors:

- >65 years old
- Shoulder pain
- Ask about tobacco and alcohol use
- Nausea ... ask about other gastrointestinal (GI) symptoms and previous history of peptic ulcer disease
- Take blood pressure
- Observe for peripheral edema (sacral and pedal)

How do you carry out a Review of Systems from a screening perspective and a Systems Review in accordance with the *Guide to Physical Therapist Practice* (the *Guide*)?

After gathering all of the subjective and objective data, make a list of all the signs and symptoms. Are there any clusters or groups of signs and symptoms that fall into any particular category? These may or may not be associated with the primary neuromusculoskeletal problem as many clients have one or more other diseases, illnesses, or conditions (referred to as comorbidities) with additional clinical manifestations.

Start with general health. She reports:

- Hearing and vision loss
- Intermittent dizziness
- Early morning nausea
- Finger/hand swelling
- Sleep disturbance

There is not much in the report about her general health. Make a note to consider asking a few more questions about her past and current general health. Ask how she would describe her overall health in one or two words.

Review her medications. She reports:

- Darvocet 3/day for pain
- Vioxx daily (cyclooxygenase-2 [COX-2] NSAID)
- Feldene prn (standard or nonselective NSAID)
- Ibuprofen daily (standard or nonselective NSAID)

Given how many forms of NSAIDs she is taking, ask yourself: Did I ask if there were any other symptoms or problems of any kind anywhere else in the body?

The remaining symptoms noted (positive Trendelenburg gait and antalgic gait, left shoulder and knee pain) fall into the musculoskeletal category. No other symptoms are noted.

Think now about the Systems Review as outlined by the *Guide*. Are there isolated groupings or clusters of signs and symptoms that fall into any of the other three diagnostic categories?

- Neuromuscular
- Cardiovascular/Pulmonary
- Integumentary

Knowing what we do about the potential for GI and renal complications in some clients taking NSAIDs, make a mental note to do two things: (1) Assess risk factors for NSAID-induced gastropathy (see Box 8-3) and (2) Ask about the presence of previously unreported GI or renal signs and symptoms (see discussion of Clinical Signs and Symptoms of NSAID-Induced Impairment). If appropriate you can go through this list and ask

- Do you have any nausea? Stomach pain? Indigestion or heartburn?
- Have you had any skin changes? You may want to prompt with: itching? Rash anywhere on your body?
- Any ringing in the ears? Headaches? Depression or mood changes? Memory loss or confusion?
- Have you had any trouble getting up out of a chair or bed? Difficulty with stairs? (muscle weakness) Shortness of breath? Unusual fatigue?
- Are you urinating more often during the day? Getting up at night to empty your bladder? Do you have any trouble wiping yourself clean after a bowel movement? Any change in the color or smell of your stools?

Documentation, communication, and medical referral will be based on the results of your evaluation using a review mechanism like the one we just completed.

in assessing other patient populations as well.[48] This calculation can be used in one of several ways. First, clinical research is needed to substantiate the number of clients in a physical therapy practice who are at risk for serious NSAID-related gastropathy.

Second, charting a client's risk can help in the early identification of problems. Because prednisone use and NSAID dose are modifiable risk factors, early identification and referral to the physician can minimize the detrimental effects of NSAID-induced gastropathy. Clients with one or more risk factors for NSAID-associated GI ulcer should be prescribed preventive strategies, such as acid-suppressive drugs and/or COX-2 inhibitors, rather than standard NSAIDs.[49,50]

Third, from a fiscal point of view, every GI complication prevented lowers the cost of medical care in this country. Clients over 50 with comorbidities, such as heart disease, renal disease, a history of ulcers, or taking prednisone or warfarin, must be watched carefully.

The scoring system in Table 8-4 was designed to allow clinicians to estimate the risk of GI problems in clients with

TABLE 8-5 ACR Criteria for Classification of Functional Status in Rheumatoid Arthritis

Class 1	Completely able to perform usual ADLs (self-care, vocational, avocational)	0 points	Normal
Class 2	Able to perform usual self-care and vocational activities, but limited in avocational activities	10 points	Adequate
Class 3	Able to perform usual self-care activities, but limited in vocational and avocational activities	20 points	Limited
Class 4	Limited in ability to perform usual self-care, vocational, and avocational activities	30 points	Unable

ACR, American College of Rheumatology; *ADLs,* activities of daily living.

rheumatoid arthritis who are also taking NSAIDs.[51] The formula is based on age, history of NSAID symptoms, NSAID dose, and the American College of Rheumatology's (ACR) Functional Classes (Table 8-5).

NSAID dose used in this formulation is the fraction of the manufacturer's highest recommended dose. The manufacturer's highest recommended dose on the package insert is given a value of 1.00. The dose of each client is then normalized to this dose.

For example, the value 1.03 indicates the client is taking 103% of the manufacturer's highest recommended dose. Most often, clients are taking the highest dose recommended. They receive a 1.0. Anyone taking less will have a fraction percentage less than 1.0. Anyone taking more than the highest dose recommended will have a fraction percentage greater than 1.0. See formulation in Case Example 8-7.

To determine the risk (%) of hospitalization or death caused by GI complications over the next 12 months, subtract 100 from the total score obtained in Table 8-4 and divide the result by 40. Higher total scores yield a greater predictive risk. The risk ranges from 0.0 (low risk) to 5.0 (high risk) (see Case Example 8-7).

Note that although further studies validating this tool have not been published, additional efforts to predict the risk of GI bleeding due to NSAID use have shown a significant increase in the GI event rate associated with age, sex, prior GI bleeds, use of GI medications, and prednisone use. The use of disease-modifying antirheumatic drugs (DMARDs) was not linked with gastropathy.[52]

Diverticular Disease

The terms *diverticulosis* and *diverticulitis* are used interchangeably although they have distinct meanings. *Diverticulosis* is a benign condition in which the mucosa (lining) of the colon balloons out through weakened areas in the wall.

CASE EXAMPLE 8-7

Is Your Client At Risk for Nonsteroidal Antiinflammatory Drug–Induced Gastropathy?

A 66-year-old woman with a history of rheumatoid arthritis (class 3) has been referred to physical therapy after three metacarpophalangeal (MCP) joint replacements.

Although her doctor has recommended maximum dosage of ibuprofen (800 mg tid; 2400 mg), she is really only taking 1600 mg/day. She says this is all she needs to control her symptoms. She was taking prednisone before the surgery but tapered herself off and has not resumed its use.

She has been hospitalized 3 times in the past 6 years for gastrointestinal (GI) problems related to nonsteroidal antiinflammatory drug (NSAID) use but does not have any apparent GI symptoms at this time.

Calculating her risk for serious problems with NSAID use, we have

Age in years	66 × 2 =	132
History of NSAID symptoms (e.g., abdominal pain, bloating, nausea)	+50 points	50
ARA Class (see Table 8-5)	Add 0, 10, 20, or 30 based on Class 1-4	20
Daily NSAID dose (fraction of maximum recommended)	1600 mg/2400 mg × 15 (0.67 × 15)	10
If currently using prednisone	Add 40 points	0
TOTAL SCORE		212

Risk/year = [TOTAL SCORE − 100] ÷ 40
Risk/year = [212 − 100] ÷ 40
Risk/year = 112 ÷ 40 = 2.80

The scores range from 0.0 (very low risk) to 5.0 (very high risk). A predictive risk of 2.8 is moderately high. This client should be reminded to report GI distress to her doctor immediately. Periodic screening for GI gastropathy is indicated with early referral if warranted.

Up to 60% of people over age 65 have these saclike protrusions. Someone with diverticulosis is typically asymptomatic; the diverticula are diagnosed when screening for colon cancer or other problems.[53]

Diverticulitis describes the infection and inflammation that accompany a microperforation of one of the diverticula. Diverticulosis is very common, whereas complications resulting in diverticulitis occur in only 10% to 25% of people with diverticulosis. The most common cause of major lower intestinal tract bleeding is diverticulosis. A significant number of cases of diverticular bleeding are associated with the use of NSAIDs in combination with diverticulosis.[54]

There is some controversy regarding whether diverticulosis is symptomatic, but perforation and subsequent infection causes symptoms of left lower abdominal or pelvic pain and tenderness in diverticulitis. For the therapist performing the iliopsoas and obturator tests, abdominal pain in the left lower quadrant may be caused by diverticular disease and should be reported to the physician. The diagnosis of diverticulitis is confirmed by accompanying fever, bloody stools, elevated white blood cell count, and imaging studies.

CLINICAL SIGNS AND SYMPTOMS

Diverticulitis

- Generalized abdominal pain often with loss of appetite, nausea, abdominal bloating
- Left lower abdominal/pelvic (cramping) pain and tenderness (present in 70% of people with diverticulitis)[55]; possible positive pinch-an-inch test (see Fig. 8-11)
- Right lower abdominal pain (1.5% of cases)[55]
- Decreased or absent bowel sounds; palpable abdominal mass
- Flatulence (passing gas)
- Bloody stools
- Constipation or irregular bowel movements
- Urinary urgency and frequency if the colon near the bladder is affected
- Low grade fever (not always present)

Appendicitis

Appendicitis is an inflammation of the vermiform appendix that occurs most commonly in adolescents and young adults. It is a serious disease usually requiring surgery. When the appendix becomes obstructed, inflamed, and infected, rupture may occur, leading to peritonitis.

Diseases that can be mistaken for appendicitis include Crohn's disease (regional enteritis), perforated duodenal ulcer, gallbladder attacks, and kidney infection on the right side, and for women, ruptured ectopic pregnancy, twisted ovarian cyst, or a hemorrhaging ovarian follicle at the middle of the menstrual cycle. Right lower lobe pneumonia sometimes is associated with prominent right lower quadrant pain.

Clinical Signs and Symptoms

The classic symptoms of appendicitis are pain preceding nausea and vomiting and low-grade fever in adults. Children tend to have higher fevers. Other symptoms may include coated tongue and bad breath.

The pain usually begins in the umbilical region and eventually localizes in the right lower quadrant of the abdomen over the site of the appendix. In retrocecal appendicitis, the pain may be referred to the thigh or right testicle (see Fig. 8-10). Groin and/or testicular pain may be the only symptoms of appendicitis, especially in young, healthy, male athletes. The pain comes in waves, becomes steady, and is aggravated by movement, causing the client to bend over and tense the abdominal muscles or to lie down and draw the legs up to relieve abdominal muscle tension (Case Example 8-8).

Generalized peritonitis, whether caused by appendicitis or some other abdominal or pelvic inflammatory condition, can result in a "boardlike" abdomen due to the spasm of the rectus abdominis muscles. Lean muscle mass deteriorates with aging, especially evident in the abdominal muscles of the aging population. The very old person may not present with this classic sign of generalized peritonitis because of the lack of toned abdominal muscles.

CASE EXAMPLE 8-8

Appendicitis

Remember the 32-year-old female university student featured in Fig. 1-6? She had been referred to physical therapy with the provisional diagnosis: *Possible right oblique abdominis muscle tear/possible right iliopsoas muscle tear.* Her history included the sudden onset of "severe pain" in the right lower quadrant with accompanying nausea and abdominal distention. Aggravating factors included hip flexion, sit-ups, fast walking, and movements such as reaching, turning, and bending. Painful symptoms could be reproduced by resisted hip or trunk flexion, and tenderness/tightness was elicited on palpation of the right iliopsoas muscle compared with the left. A neurologic screen was negative. Screening questions for general health revealed constitutional symptoms, including fatigue, night sweats, nausea, and repeated episodes of severe, progressive pain in the right lower abdominal quadrant.

Although she presented with a musculoskeletal pattern of symptoms at the time of her initial evaluation with the physician, by the time she entered the physical therapy clinic her symptoms had taken on a definite systemic pattern. She was returned for further medical follow-up, and a diagnosis of appendicitis complicated by peritonitis was established. This client recovered fully from all her symptoms following emergency appendectomy surgery.

For this reason, the nursing home, skilled care facility, or home health therapist must evaluate the aging client who presents with hip or thigh pain for possible systemic origin (assess for signs of peritonitis and/or appendicitis as appropriate; see also McBurney's point, and specific tests for iliopsoas or obturator abscess).

CLINICAL SIGNS AND SYMPTOMS

Appendicitis

- Periumbilical and/or epigastric pain
- Right lower quadrant or flank pain
- Right thigh, groin, or testicular pain
- Abdominal involuntary muscular guarding and rigidity
- Positive McBurney's point and/or positive pinch-an-inch test
- Rebound tenderness (peritonitis)
- Positive hop test (hopping on one leg or jumping on both feet reproduces painful symptoms)
- Nausea and vomiting
- Anorexia
- Dysuria (painful/difficult urination)
- Low-grade fever
- Coated tongue and bad breath

McBurney's Point

Parietal pain caused by inflammation of the peritoneum in acute appendicitis or peritonitis (from appendicitis or other inflammatory/infectious causes) may be located at

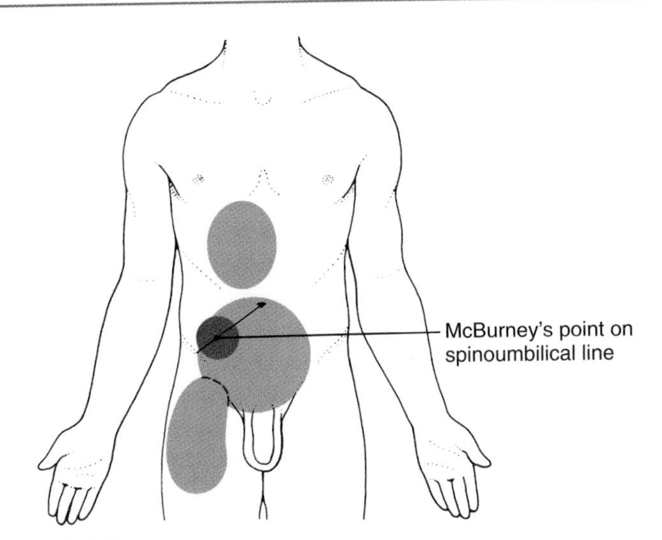

Fig. 8-9 The vermiform appendix and colon can refer pain to the area of sensory distribution for the eleventh thoracic nerve (T11). Primary *(dark red)* and referred *(light red)* pain patterns associated with the vermiform appendix are shown here with McBurney's point halfway between the ASIS and the umbilicus, usually on the right side. Gentle palpation of McBurney's point produces pain or exquisite tenderness. Pinch-an-inch test should also be assessed (see Fig. 8-11).

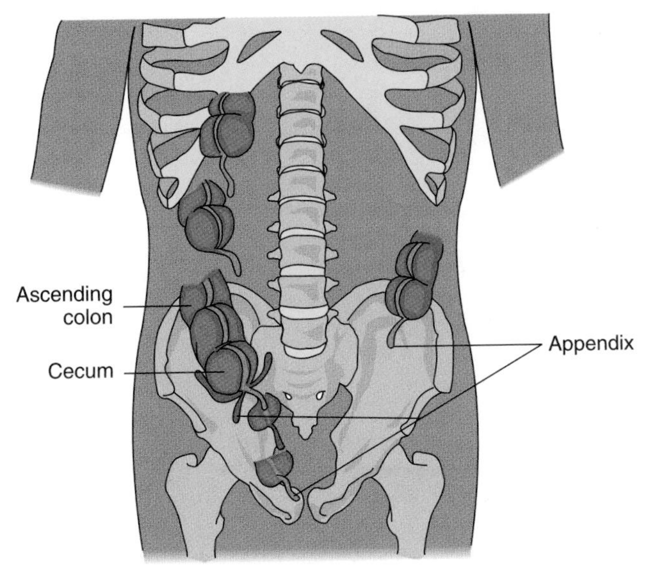

Fig. 8-10 Variations in the location of the vermiform appendix. Negative tests for appendicitis using McBurney's point may occur when the appendix is located somewhere other than at the end of the cecum. In 50% of cases the appendix is retrocecal (behind the cecum) or retrocolic (behind the colon). See Fig. 8-11 for an alternate test.

McBurney's point (Fig. 8-9). The vermiform appendix receives its sympathetic supply from the 11th thoracic segment. In some people, a branch of the 11th thoracic nerve pierces the rectus abdominis muscle and innervates the skin over McBurney's point. This may explain the hyperalgesia seen at this point in appendicitis.[6]

McBurney's point is located by palpation with the client in a fully supine position. Isolate the ASIS and the umbilicus, then palpate for tenderness halfway between these two surface anatomic points. This method differs from palpation of the iliopsoas muscle because the position used to locate the iliopsoas muscle is the client in a supine position, with hips and knees flexed in a 90-degree position, whereas McBurney's point is palpated with the client in the fully supine position.

The palpation point for the iliopsoas muscle is one-third the distance between the ASIS and the umbilicus, whereas McBurney's point is halfway between these two points. Be aware that the location of the vermiform appendix can vary from individual to individual making the predictive value of this test less accurate (Fig. 8-10). Since the appendix develops during the descent of the colon, its final position can be posterior to the cecum or colon. These positions of the appendix are called *retrocecal* or retrocolic, respectively. In about 50% of cases, the appendix is retrocecal or retrocolic.[56] Atypical locations of the appendix can lead to unusual clinical findings with poorly localized abdominal or pelvic pain, unusual symptoms of urinary and defecation urgency (due to irritation of the ureter and rectum), painful urination, and diarrhea.[57]

Both McBurney's point and the iliopsoas muscle are palpated for reproduction of symptoms to rule out appendicitis or iliopsoas abscess associated with appendicitis or peritonitis. Alternately, instead of palpating for McBurney's point (these tests can be very painful when positive), perform a pinch-an-inch test (Fig. 8-11). This test is a new technique for detecting peritonitis/appendicitis that is more comfortable and statistically equivalent to the traditional rebound tenderness technique.[58,59] Like a rebound tenderness test, a positive pinch-an-inch test is a classic sign of peritonitis and represents aggravation by stretching or moving the parietal layer of the peritoneum. A positive pinch-an-inch test, or alternately, rebound tenderness, may occur with any disease or condition affecting the peritoneum (including appendicitis when it has progressed to include peritonitis). If the pinch-an-inch test is negative, then proceed with the rebound tenderness test (Fig. 8-12) and/or palpation of McBurney's point.

Pancreatitis

Pancreatitis is an inflammation of the pancreas that may result in autodigestion of the pancreas by its own enzymes. Pancreatitis can be acute or chronic, but the therapist is most likely to see individuals with referred pain patterns associated with acute pancreatitis. The pancreas is both an exocrine gland and an endocrine gland. Its function in digestion is primarily an exocrine activity. This chapter focuses on digestive disorders associated with the pancreas. See Chapter 11 for pancreatic disorders associated with endocrine function.

Acute pancreatitis can arise from a variety of etiologic (risk) factors, but in most instances, the specific cause is unknown. Chronic alcoholism or toxicity from some other agent, such as glucocorticoids, thiazide diuretics, or acetaminophen, can bring on an acute attack of pancreatitis.

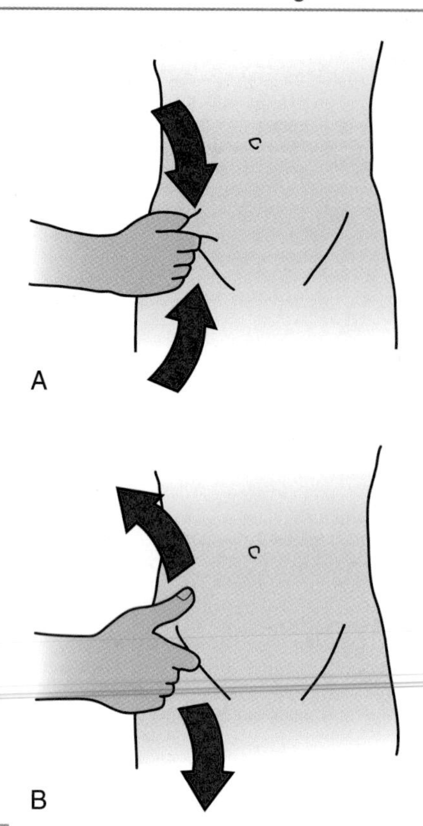

Fig. 8-11 Pinch-an-inch test. **A,** To avoid the discomfort of the classic rebound tenderness (Blumberg's) test, the pinch-an-inch test is recommended to assess for appendicitis or generalized peritonitis. To perform the test, a fold of abdominal skin over McBurney's point is gently grasped and elevated away from the peritoneum. **B,** The skin is then allowed to recoil back against the peritoneum quickly. If the individual has increased pain when the skin fold strikes the peritoneum (upon release of the skin), the test is positive for possible peritonitis. If the person being tested reacts to the pinch in an excessive fashion, he or she may have a very low pain threshold, a factor that should be taken into consideration when assessing the results.[58,59]

Chronic pancreatitis is caused by long-standing alcohol abuse in more than 90% of adult cases. In these cases, chronic pancreatitis is characterized by the progressive destruction of the pancreas with accompanying irregular fibrosis and chronic inflammation.[60]

A mechanical obstruction of the biliary tract may be present, usually because of gallstones in the bile ducts. Viral infections (e.g., mumps, herpesviruses, hepatitis) also may cause an acute inflammation of the pancreas.

Clinical Signs and Symptoms

The clinical course of most clients with *acute pancreatitis* follows a self-limited pattern. Symptoms can vary from mild, nonspecific abdominal pain to profound shock with coma and possible death. Abdominal pain begins abruptly in the midepigastrium, increases in intensity for several hours, and can last from days to more than a week.

The pain has a penetrating quality and radiates to the back. Pain is made worse by walking and lying supine and is relieved by sitting and leaning forward. The client may have a bluish discoloration of the periumbilical area (Cullen sign) as a physical manifestation of acute pancreatitis. This occurs in cases of severe hemorrhagic pancreatitis. Turner's sign is a reddish-brown discoloration of the flanks, also present in hemorrhagic pancreatitis.

Symptoms associated with *chronic pancreatitis* include persistent or recurrent episodes of epigastric and left upper quadrant pain with referral to the upper left lumbar region. Pathology of the head of the pancreas is more likely to cause epigastric and mid-thoracic pain from T5 to T9. Impairment of the tail of the pancreas (located to the left of midline; see Fig. 3-4) can refer pain to the left shoulder.

Anorexia, nausea, vomiting, constipation, flatulence, and weight loss are common. Attacks may last only a few hours or as long as 2 weeks; pain may be constant. In clients with

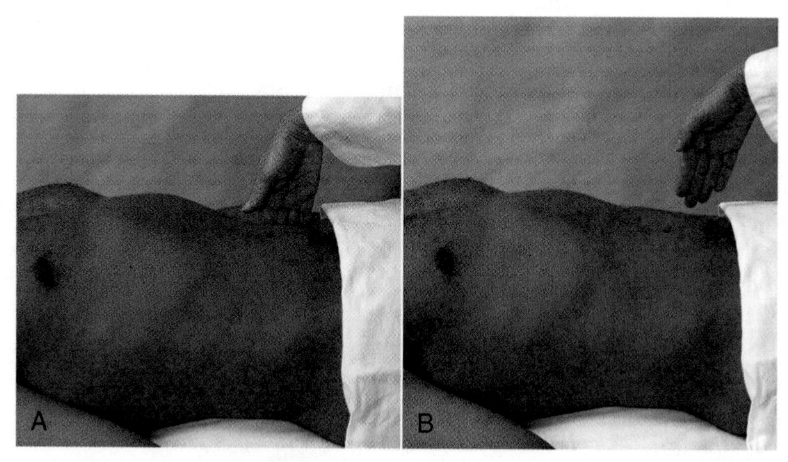

Fig. 8-12 Rebound tenderness or Blumberg's sign. **A,** To assess for appendicitis or generalized peritonitis, press your fingers gently but deeply over the right lower quadrant for 15-30 seconds. **B,** The palpating hand is then quickly removed. Pain induced or increased by quick withdrawal results from rapid movement of inflamed peritoneum and is called rebound tenderness. When rebound tenderness is present, the client will have pain or increased pain on the side of the inflammation when the palpatory pressure is released. Ask the client if it hurts as you are palpating or during the release. Since abdominal pain is increased uncomfortably with this test, save it for last when assessing abdominal pain during the physical examination. (From Jarvis C: *Physical examination and health assessment,* ed 5, Philadelphia, 2007, WB Saunders.)

alcohol-associated pancreatitis, the pain often begins 12 to 48 hours after an episode of inebriation. Clients with gallstone-associated pancreatitis typically experience pain after a large meal. Nausea and vomiting accompany the pain. Other symptoms include fever, tachycardia, jaundice, and malaise.

CLINICAL SIGNS AND SYMPTOMS

Acute Pancreatitis

- Epigastric pain radiating to the back
- Nausea, vomiting, diarrhea; anorexia
- Abdominal distention and pain
- Fever and sweating
- Tachycardia
- Malaise
- Weakness
- Bluish discoloration of abdomen or flanks (severe hemorrhagic acute pancreatitis)
- Jaundice

CLINICAL SIGNS AND SYMPTOMS

Chronic Pancreatitis

- Epigastric pain radiating to the back
- Upper left lumbar region pain
- Nausea and vomiting
- Constipation
- Flatulence
- Weight loss

Pancreatic Carcinoma

Pancreatic carcinoma is the fifth most common cause of death from cancer for women and fourth most common for men. The majority of pancreatic cancers (70%) arise in the head of the gland and only 20% to 30% occur in the body and tail (see Fig. 9-1). The latter usually have grown to a large size by the time the diagnosis is made due to the absence of symptoms.

Clinical Signs and Symptoms

The clinical features of pancreatic cancer initially are nonspecific and vague, contributing to a delay in diagnosis and high mortality. Symptoms do not usually appear until the tumor obstructs nearby bile ducts or grows large enough to cause abdominal pressure or pain.

The most common symptoms of pancreatic cancer are anorexia and weight loss, epigastric/upper abdominal pain with radiation to the back, and jaundice secondary to obstruction of the bile duct. Jaundice is characterized by fatigue and yellowing of the skin and sclera of the eye. The urine may become dark like the color of a cola soft drink.

As with any pancreatic impairment, involvement of the head of the pancreas is more likely to cause epigastric and mid-thoracic pain (T5-T9), whereas impairment of the tail of the pancreas (located to the left of midline; see Fig. 3-4) can refer pain to the left shoulder. Epigastric pain is often vague and diffuse. Radiation of pain into the lumbar region is common and sometimes the only symptom.

The pain may become worse after the person eats or lies down. Sitting up and leaning forward may provide some relief, and this usually indicates that the lesion has spread beyond the pancreas and is inoperable. Other signs and symptoms include light-colored stools, constipation, nausea, vomiting, loss of appetite, weight loss, and weakness.

CLINICAL SIGNS AND SYMPTOMS

Pancreatic Carcinoma

- Epigastric/upper abdominal pain radiating to the back
- Low back pain may be the only symptom
- Jaundice
- Anorexia and weight loss
- Light-colored stools
- Constipation
- Nausea and vomiting
- Weakness

Inflammatory Bowel Disease

Inflammatory bowel disease (IBD; *not* the same as irritable bowel syndrome [IBS]) refers to two inflammatory conditions discussed separately:

- Ulcerative colitis
- Crohn's disease (also referred to as regional enteritis or ileitis)

Crohn's disease (CD) and ulcerative colitis (UC) are disorders of unknown etiology involving genetic and immunologic influences on the GI tract. UC affects the large intestine (colon). CD can affect any portion of the intestine from the mouth to the anus. Both diseases not only cause inflammation inside the intestine but can also cause significant problems in other parts of the body.[61] These two diseases share many epidemiologic, clinical, and therapeutic features. Both are chronic, medically incurable conditions.

Extraintestinal manifestations occur frequently in clients with IBD and complicate its management. The client may not know these signs and symptoms are associated with CD. Manifestations involve the joints most commonly (see previous discussion of Arthralgia). The client with new onset of joint pain should be asked about a previous history of CD.

Skin lesions may occur as either erythema nodosum (red bumps/purple knots over the ankles and shins) or pyoderma (deep ulcers or canker sores) of the shins, ankles, and calves. Ask about a recent history (last 6 weeks) of skin lesions anywhere on the body. Uveitis may cause red and painful eyes that are sensitive to light, but this condition does not affect the person's vision.

Nutritional deficiencies are the most common complications of IBD. Evidence to suggest increased intestinal permeability allowing increased exposure to foreign antigens has been discovered.[62-64] Inflammation alone and the decrease in functioning surface area of the small intestine, increases food requirements, causing poor absorption.

Nutritional problems associated with the medical treatment of IBD may occur. The use of prednisone decreases vitamin D metabolism, impairs calcium absorption, decreases potassium supplies, and increases the nutritional requirement for protein and calories. Decreased vitamin D metabolism and impaired calcium absorption subsequently result in bone demineralization and osteoporosis.

Crohn's Disease

CD is an inflammatory disease that most commonly attacks the terminal end (or distal portion) of the small intestine (ileum) and the colon. However, it can occur anywhere along the alimentary canal from the mouth to the anus. It occurs more commonly in young adults and adolescents but can appear at any age.

Clinical Signs and Symptoms

CD may have acute manifestations, but the condition is usually slow and nonaggressive. The client may present with mild intermittent symptoms months before the diagnosis is made. Fever may occur, with acute inflammation, abscesses, or rheumatoid manifestations.

Terminal ileum involvement produces pain in the periumbilical region with possible referred pain to the corresponding segment of the low back. Pain of the ileum is intermittent and felt in the lower right quadrant with possible associated iliopsoas abscess causing hip pain (see previous discussion of Psoas Abscess). The client may experience relief of discomfort after passing stool or flatus. For this reason, it is important to ask whether low back pain is relieved after passing stool or gas.

Twenty-five percent of people with CD may present with arthritis or migratory arthralgias (joint pain). The person may present with monoarthritis (i.e., asymmetric pattern affecting one joint at a time), usually involving an ankle or knee, although elbows and wrists can be included.

Polyarthritis (involving more than one joint) or sacroiliitis (arthritis of the lower spine and pelvis) is common and may lead to ankylosing spondylitis in rare cases. Whether monoarthritic or polyarthritic, this condition comes and goes with the disease process and may precede repeat episodes of bowel symptoms by 1 to 2 weeks. With proper medical intervention, there is no permanent joint deformity.

Ulcerative Colitis

By definition, UC is an inflammation and ulceration of the inner lining of the large intestine (colon) and rectum. When inflammation is confined to the rectum only, the condition is known as ulcerative proctitis. UC is not the same as irritable bowel syndrome (IBS) or spastic colitis (another term for IBS).

Cancer of the colon is more common among clients with UC than among the general population. The incidence is greatly increased among those who develop UC before the age of 16 years and those who have had the condition for more than 30 years.

Clinical Signs and Symptoms

The predominant symptom of UC is rectal bleeding. Mainly the left colon is involved; the small intestine is never involved. Clients often experience diarrhea, possibly 20 or more stools per day. Nausea, vomiting, anorexia, weight loss, and decreased serum potassium may occur with severe disease. Fever is present during acute disease. Nocturnal diarrhea is usually present when daytime diarrhea is prominent.

The development of anemia depends on the degree of blood loss, severity of the illness, and dietary iron intake. Ankylosing spondylitis, anemia, and clubbing of the fingers are occasional findings. Clubbing (see Figs. 4-36 and 4-37) develops quickly within 7 to 10 days.

Medical testing and diagnosis are required to differentiate between these inflammatory conditions. Most often, the therapist is faced with clients presenting complaints of pain located in the shoulder, back, or groin that may have a GI origin and not be true musculoskeletal dysfunction at all.

CLINICAL SIGNS AND SYMPTOMS

Ulcerative Colitis and Crohn's Disease

- Diarrhea
- Constipation
- Fever
- Abdominal pain
- Rectal bleeding
- Night sweats
- Decreased appetite, nausea, weight loss
- Skin lesions
- Uveitis (inflammation of the eye)
- Arthritis
- Migratory arthralgias
- Hip pain (iliopsoas abscess)

Irritable Bowel Syndrome

Irritable bowel syndrome (IBS) has been called the "common cold of the stomach." It is a functional disorder of motility in the small and large intestines diagnosed according to specific bowel symptom clusters.

IBS is classified as a "functional" disorder because the abnormal muscle contraction identified in people with IBS cannot be attributed to any identifiable abnormality of the bowel. A lowered visceral pain threshold is commonly found with complaints of bloating and distention at lower volumes of colonic insufflation than normal controls.[65] In other words, affected individuals perceive unpleasant or inappropriate sensory experiences in the absence of any physiologic or pathophysiologic event. There is some evidence that a dysregulation in central pain processing similar to that seen in other chronic pain disorders may explain the symptoms.[66] IBS rarely progresses and is never fatal.

Other descriptive names for this condition are spastic colon, irritable colon, nervous indigestion, functional dyspepsia, pylorospasm, spastic colitis, intestinal neuroses, and laxative or cathartic colitis.

IBS is the most common GI disorder in Western society and accounts for 50% of subspecialty referrals. It is often linked with psychosocial factors. There may be an association with disturbances in circadian rhythm observed first in nurses working rotating shifts.[67] In cases in which symptoms are severe and refractory to treatment, a history of mental, physical, or sexual abuse is suspected.[68,69] IBS is most common in women in early adulthood, and there is a well-documented association between IBS and dysmenorrhea.[70,71] It is unclear whether this correlation represents diagnostic confusion or whether dysmenorrhea and IBS have a common physiologic basis.

As mentioned earlier in this chapter, emotional or psychologic responses to stress have a profound effect on brain chemistry, which in turn influences the enteric nervous system. Conversely, messages from the CNS are processed in the intestines by an elaborate neural network. Research is ongoing to find the biochemical links between psychosocial factors, physical disease, and somatic illness.

Clinical Signs and Symptoms

There is a highly variable complex of intermittent GI symptoms, including nausea and vomiting, anorexia, foul breath, sour stomach, flatulence, cramps, abdominal bloating, and constipation and/or diarrhea. The client may report white mucus in the stools.

Pain may be steady or intermittent, and there may be a dull deep discomfort with sharp cramps in the morning or after eating. The typical pain pattern consists of lower left quadrant abdominal pain, constipation, and diarrhea. Symptoms seem to come and go with no apparent cause and effect that can be identified by the affected individual. Abdominal pain or discomfort is relieved by defecation.

These primary symptoms occur when the natural motility of the bowel (rhythmic peristalsis) is disrupted by stress, smoking, eating, and drinking alcohol. Rapid alterations in the speed of bowel movement create an obstruction to the natural flow of stool and gas. The resultant pressure build-up in the bowel produces pain and spasm.

The therapist should also be alert for the client with a known history of IBS now experiencing unexplained weight loss or persistent, severe diarrhea, possibly signaling disorders such as malignancy, IBD, or celiac disease. Symptoms of IBS tend to disappear at night when the client is asleep. Nocturnal diarrhea, awakening the client from a sound sleep, is more often a result of organic disease of the bowel and is less likely to occur in IBS. Sudden return of symptoms after age 50 following prolonged remission must be evaluated medically, especially if there is blood in the stool.[72,73]

CLINICAL SIGNS AND SYMPTOMS
Irritable Bowel Syndrome

- Painful abdominal cramps
- Constipation
- Diarrhea
- Nausea and vomiting
- Anorexia
- Flatulence
- Foul breath

Colorectal Cancer

Colorectal cancer is the third leading cause of cancer deaths in the United States and the leading cause of cancer deaths among nonsmokers.[74] If current trends in health behaviors, screening, and treatment continue, U.S. residents can expect to see a 36% decrease in the colorectal mortality rate by 2020, compared with 2000.[75] Incidence increases with age, beginning around 40 years of age, and is higher in men than women. More African-American than Caucasian men and women are affected.[76]

Mortality can be significantly reduced by population screening by means of a simple fecal occult blood test (FOBT). Screening is particularly applicable to individuals belonging to high-risk groups, particularly those with a previous history of chronic IBD (e.g., CD, UC), adenomatous polyps, and hereditary nonpolyposis colon cancer.[77] High-quality colonoscopy screening is used to identify adenomatous polyps that raise the risk of colorectal cancer if not removed and flat and depressed nonpolypoid growths on the colon wall that blend in with surrounding tissue and may be premalignant lesions.[78]

Clinical Signs and Symptoms

The presentation of colorectal carcinoma is related to the location of the neoplasm within the colon. Individuals are asymptomatic in the early stages, then develop minor changes in their bowel patterns (e.g., increased frequency of morning evacuation, sense of incomplete evacuation), and experience occasional rectal bleeding. When vague cramping pain or an aching pressure sensation occurs, it is usually associated with a palpable abdominal mass, although these symptoms are experienced before the identification of the mass. Acute pain is often indistinguishable from that of cholecystitis or acute appendicitis.

Fatigue and shortness of breath may occur secondary to the iron deficiency anemia that develops with chronic blood loss. Mahogany-colored stools may be present when there is blood mixed with the stool. The reddish-mahogany color associated with bleeding in the lower GI/colon differs from the melena or dark, tarry stools that occur when blood loss in the upper GI tract is oxidized before being excreted. Bleeding with bright red blood is more common with a carcinoma of the left side of the colon. Pencil-thin stool may be described with cancer of the rectum.

When rectal tumors enlarge and invade the perirectal tissue, a sensation of rectal fullness develops and may progress to a dull, aching perineal or sacral pain that can radiate down the legs when peripheral nerves are involved.

CLINICAL SIGNS AND SYMPTOMS

Colorectal Cancer

Early Stages
- Rectal bleeding, hemorrhoids
- Abdominal, pelvic, back, or sacral pain
- Back pain that radiates down the legs
- Changes in bowel patterns

Advanced Stages
- Constipation progressing to obstipation
- Diarrhea with copious amounts of mucus
- Nausea, vomiting
- Abdominal distention
- Weight loss
- Fatigue and dyspnea
- Fever (less common)

Acute Colonic Pseudo-Obstruction

Acute colonic pseudo-obstruction (Ogilvie's syndrome) is a massive dilation of the cecum and proximal colon in the absence of actual mechanical causes such as colonic obstruction.[79] This severe dilation of the colon may lead to spontaneous perforation of the colon, which is a life-threatening problem.

Ogilvie's syndrome is most commonly detected in surgical patients after trauma, burns, and GI tract surgery or in medical patients who have severe metabolic, respiratory, and electrolyte disturbances. However, this complication has also been seen after hip arthroplasty. Possible explanations include acetabular trauma and heat generation from bone cement leading to damage to tissues close to the point of contact of the heated cement.

Other reported risks for development of this syndrome can be related to increased age, immobility, and use of client-controlled narcotic analgesia.[74] Symptoms include abdominal distention, nausea, vomiting, abdominal pain, and absent bowel movements. Bowel sounds may be absent or decreased, and rebound tenderness is not usually present unless colon perforation has occurred and peritonitis is present.

PHYSICIAN REFERRAL

A 67-year-old man is seeing you through home health care for a home program after discharge from the hospital 2 weeks ago for a total hip replacement. His recovery has been slowed by chronic diarrhea. A 25-year-old woman who is diagnosed as having SI pain and joint dysfunction asks you what exercises she can do for constipation. A 44-year-old man with biceps tendinitis reports several episodes of fever and chills, diarrhea, and abdominal pain, which he attributes to "the stress of meeting deadlines on the job."

These are common examples of symptoms of a GI nature that are described by clients and are unrelated to current physical therapy treatment. These people may be seeking the therapist's advice as the only medical person with whom they have contact. Knowing the pain patterns associated with GI involvement and which follow-up questions to ask can assist the therapist in deciding when to suggest that the client return to a physician for a medical examination and treatment.

The client may not associate GI symptoms or already diagnosed GI disease with his or her musculoskeletal pain, which makes it necessary for the therapist to initiate questions to determine the presence of such GI involvement.

Taking the client's temperature and vital signs during the initial evaluation is recommended for any person who has musculoskeletal pain of unknown origin. Fever, low-grade fever over a long period (even if cyclic), or sweats are indicative of systemic disease.

When appendicitis or peritonitis from any cause is suspected because of the client's symptoms, a physician should be notified immediately. The client should lie down and remain as quiet as possible. It is best to give her or him nothing by mouth because of the danger of aggravating the condition, possibly causing rupture of the appendix, or in case surgery is needed. Applications of heat are contraindicated for the same reason.

On the other hand, the therapist may be evaluating a client who presents with shoulder, back, or groin pain and limitations that are not caused by true musculoskeletal lesions but rather the result of GI involvement. The presence of associated GI symptoms in the absence of conclusive musculoskeletal findings will alert the therapist to the possible need for medical referral. Correlate the *history* with *pain patterns* and any *unusual findings* that may indicate systemic disease.

Guidelines for Immediate Medical Attention

- Anytime appendicitis or iliopsoas/obturator abscess is suspected (positive McBurney's test, positive iliopsoas/obturator test, positive pinch-an-inch test, positive test for rebound tenderness).
- Anytime the therapist suspects retroperitoneal bleeding from an injured, damaged, or ruptured spleen or ectopic pregnancy; or there is a history of trauma; missed menses; positive Kehr's sign.

Guidelines for Physician Referral

- Clients who chronically rely on laxatives should be encouraged to discuss bowel management without drugs with their physician.
- Joint involvement accompanied by skin or eye lesions may be reflective of inflammatory bowel disease and should be reported to the physician if the physician is unaware of these extraintestinal manifestations.
- Anyone with a history of NSAID use presenting with back or shoulder pain, especially when accompanied by any of

the associated signs and symptoms listed for peptic ulcer, must be evaluated by a physician.

- Back pain associated with meals or relieved by a bowel movement (especially if accompanied by rectal bleeding) or with back pain and abdominal pain at the same level requires medical evaluation.
- Back pain of unknown cause that does not fit a musculoskeletal pattern, especially in a person with a previous history of cancer.

Clues to Screening for Gastrointestinal Disease

These clues will help the therapist in the decision-making process:

- Age over 45.
- Previous history of NSAID-induced GI bleeding; NSAID use, especially chronic or multiple prescriptions and OTC NSAIDs taken simultaneously.
- Symptoms increase within 2 hours after taking NSAIDs or other medication.
- Symptoms are affected (increased or decreased) by food anywhere from immediately up to 2 to 4 hours later.
- Presence of abdominal or GI symptoms occurring within 4 to 6 weeks of musculoskeletal symptoms, especially recurring or cyclical symptoms (systemic pattern).
- Back pain and abdominal pain at the same level, simultaneously or alternately, especially when accompanied by constitutional symptoms.
- Shoulder, back, pelvic, or sacral pain:
  - Of unknown origin, especially with a past history of cancer.
  - Affected by food, milk, antacids, or vomiting.
  - Accompanied by constitutional symptoms.
- Back, pelvic, or sacral pain that is relieved or reduced by a bowel movement or accompanied by rectal bleeding.
- Low back pain accompanied by constipation may be a manifestation of pelvic floor overactivity or spasm; this requires a pelvic floor screening examination.
- Shoulder pain within 24 to 48 hours of laparoscopy, ruptured ectopic pregnancy, or traumatic blow or injury to the left side (Kehr's sign; see Chapter 18).
- Positive iliopsoas or obturator sign; positive McBurney's point; right (or left) lower quadrant abdominal or pelvic pain produced when palpating the iliopsoas muscle or tapping the heel of the involved side.
- Joint pain or arthralgias preceded by skin rash, especially in the presence of a history of CD.
- When evaluated during early onset of referred pain, there is usually full and painless range of motion, but as time goes on, muscle splinting and guarding secondary to pain can produce altered movements as well.

GASTROINTESTINAL PAIN PATTERNS

ESOPHAGEAL PAIN (FIG. 8-13)

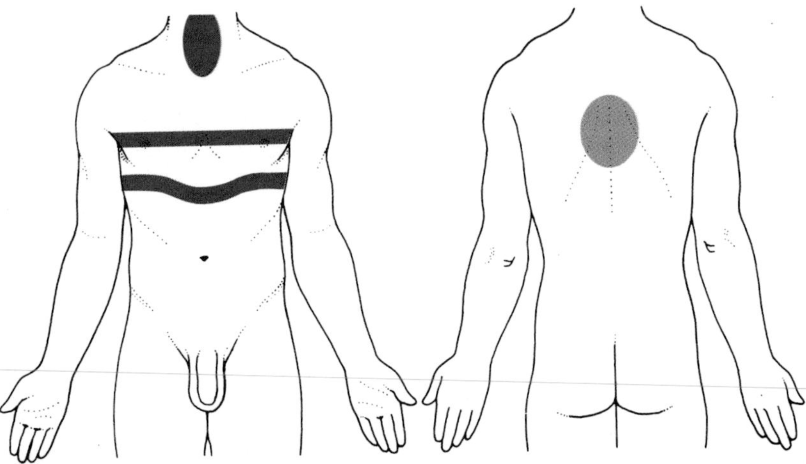

Fig. 8-13 Nerve distribution of the esophagus is through T5 to T6 with primary pain around the xiphoid. Esophageal pain may be projected around the chest at any level corresponding to the esophageal lesion. Only two of the possible bands of pain around the chest are shown here. Similar symptoms can occur anywhere a lesion appears along the length of the esophagus.

Continued

GASTROINTESTINAL PAIN PATTERNS—*cont'd*

Location: Substernal discomfort at the level of the lesion
Lesion of upper esophagus: pain in the (anterior) neck
Lesion of lower esophagus: pain originating from the xiphoid process, radiating around the thorax

Referral: Severe esophageal pain: pain referred to the middle of the back
Back pain may be the only symptom or may be the earliest symptom of esophageal cancer

Description: Sharp, sticking, knifelike, stabbing
Strong burning pain (esophagitis)

Intensity: Varies from mild discomfort to severe pain

Duration: May be constant; associated with meals

Associated signs and symptoms: Dysphagia, odynophagia, melena

Possible etiology: Obstruction of the esophagus (neoplasm)
Esophageal stricture secondary to acid reflux (peptic esophagitis)
Esophageal stricture of unknown cause
Achalasia
Esophagitis or esophageal spasm
Esophageal varices (usually asymptomatic except bleeding)

STOMACH AND DUODENAL PAIN (FIG. 8-14)

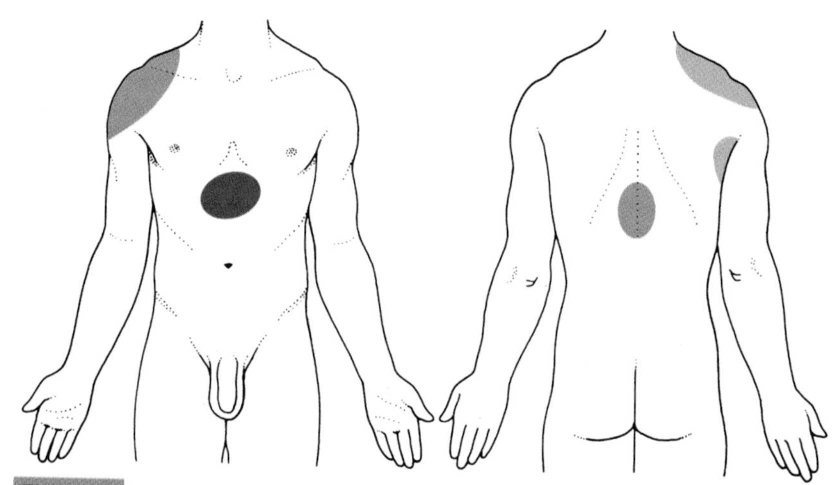

Fig. 8-14 Stomach or duodenal pain *(dark red)* may occur anteriorly in the midline of the epigastrium or upper abdomen just below the xiphoid process. There is a tendency for the stomach and duodenum to refer pain posteriorly. Referred pain *(light red)* to the back occurs at the anatomic level of the abdominal lesion (T6 to T10). Other patterns of referred pain (*light red*) may include the right shoulder and upper trapezius or the lateral border of the right scapula.

Location: Pain in the midline of the epigastrium
Upper abdomen just below the xiphoid process
One to two inches above and to the right of the umbilicus

Referral: Common referral pattern to the back at the level of the lesion (T6 to T10)
Right shoulder/upper trapezius
Lateral border of the right scapula

Description: Aching, burning ("heartburn"), gnawing, cramplike pain (true visceral pain)

Intensity: Can be mild or severe

Duration: Comes in waves

Associated signs and symptoms: Early satiety
Melena
Symptoms may be associated with meals

GASTROINTESTINAL PAIN PATTERNS—*cont'd*

Possible etiology: Peptic ulcers: gastric, pyloric, duodenal (history of NSAIDs)
Stomach carcinoma
Kaposi's sarcoma (most common malignancy associated with acquired immunodeficiency syndrome [AIDS])

SMALL INTESTINE PAIN (FIG. 8-15)

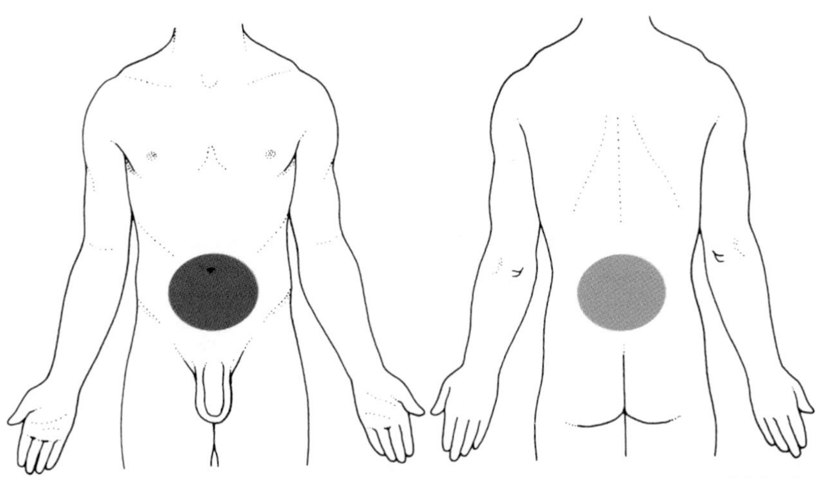

Fig. 8-15 Midabdominal pain *(dark red)* caused by disturbances of the small intestine is centered around the umbilicus (T9 to T11 nerve distribution) and may be referred *(light red)* to the low back area at the same anatomic level. Keep in mind the umbilicus is at the same level as the L3-L4 disk space in the average adult who is not obese or who has a protruding abdomen.

Location: Midabdominal pain (about the umbilicus)
Referral: Pain referred to the back if the stimulus is sufficiently intense or if the individual's pain threshold is low
Description: Cramping pain
Intensity: Moderate to severe
Duration: Intermittent (pain comes and goes)
Associated signs and symptoms: Nausea, fever, diarrhea
Pain relief may not occur after passing stool or gas
Possible etiology: Obstruction (neoplasm)
Increased bowel motility
Crohn's disease (regional enteritis)

Continued

GASTROINTESTINAL PAIN PATTERNS—*cont'd*

LARGE INTESTINE AND COLON PAIN (FIG. 8-16)

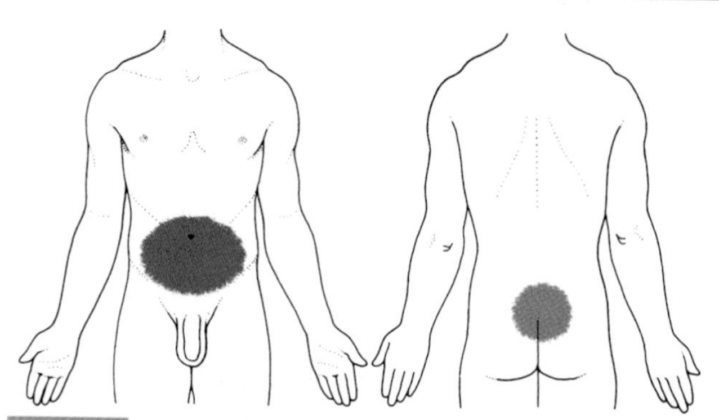

Fig. 8-16 Pain associated with the large intestine and colon *(dark red)* may occur in the lower abdomen across either or both abdominal quadrants. Pain may be referred to the sacrum *(light red)* when the rectum is stimulated. The pattern of nerve supply varies depending on the segment: vermiform appendix, cecum, and ascending colon are supplied by the T10 to T12 sympathetic fibers. Nerve distribution to the transverse colon is T12 to L1 and the descending colon is supplied by L1 to L2.

Location:	Lower midabdomen (across either or both quadrants) Poorly localized
Referral:	Pain may be referred to the sacrum when the rectum is stimulated
Description:	Cramping
Intensity:	Dull
Duration:	Steady
Associated signs and symptoms:	Bloody diarrhea, urgency Constipation Rectal pain; pain during defecation Pain relief may occur after defecation or passing gas
Possible etiology:	Ulcerative colitis Crohn's disease (regional enteritis) Carcinoma of the colon Long-term use of antibiotics Irritable bowel syndrome

GASTROINTESTINAL PAIN PATTERNS—*cont'd*

PANCREATIC PAIN (FIG. 8-17)

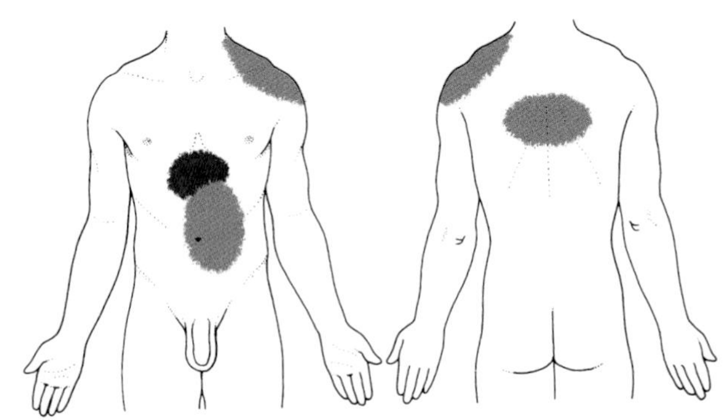

Fig. 8-17 Pancreatic pain *(dark red)* occurs in the midline or left of the epigastrium, just below the xiphoid process, but may be referred *(light red)* to the left shoulder or to the mid-thoracic spine. Posterior pain may radiate or lateralize from the spine away from the midline. Sensory nerve distribution is from T5 to T9.

Location: Midline or to the left of the epigastrium, just below the xiphoid process

Referral: Referred pain in the middle or lower back is typical with pancreatic disease; more rarely, pain may be referred to the upper back, midscapular region.

Somatic pain felt in the left shoulder may result from activation of pain fibers in the left diaphragm by an adjacent inflammatory process in the tail of the pancreas. Less often, pain is perceived in the right shoulder if/when the head of the pancreas is involved.

Description: Burning or gnawing abdominal pain

Intensity: Severe

Duration: Constant pain, sudden onset

Associated signs and symptoms:
Sudden weight loss
Jaundice
Nausea and vomiting
Constipation
Flatulence
Tachycardia
Light-colored stools (carcinoma)
Symptoms may be unrelated to digestive activities (carcinoma)
Weakness
Symptoms may be related to digestive activities (pancreatitis)
Fever
Malaise

Aggravating factors:
Walking and lying supine (pancreatitis)
Alcohol, large meals

Relieving factors: Sitting and leaning forward (pancreatitis, pancreatic carcinoma)

Possible etiology:
Pancreatitis
Pancreatic carcinoma (primarily disease of men, occurs during the 6th and 7th decade)

Continued

GASTROINTESTINAL PAIN PATTERNS—*cont'd*

APPENDICEAL PAIN (SEE FIG. 8-9)

Location: Right lower quadrant pain

Referral: Well localized; first referred to epigastric or periumbilical area
Referred pain pattern to the right hip and/or right testicle

Description: Aching, comes in waves

Intensity: Moderate to severe

Duration: Steadily progresses over time (usually 12 hours with acute appendicitis)

Associated signs and symptoms:
- Positive McBurney's point for tenderness
- Iliopsoas abscess may occur; positive iliopsoas muscle test or positive obturator test
- Anorexia, nausea, vomiting, low-grade fever
- Coated tongue and bad breath
- Dysuria (painful/difficult urination)

Figs. 8-18 and 8-19 provide a summary of all the GI pain patterns described that can mimic the pain and dysfunction usually associated with musculoskeletal lesions.

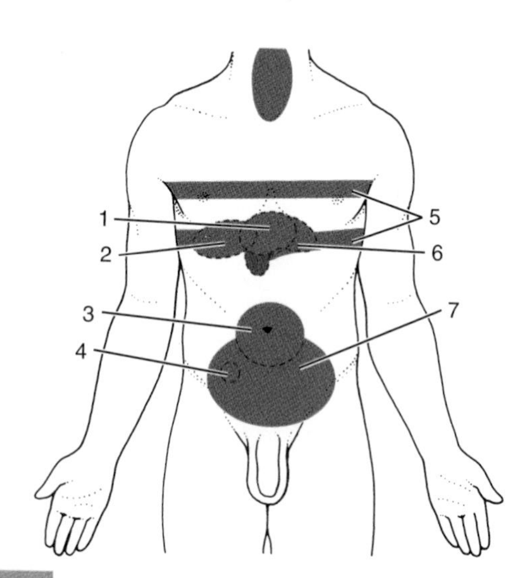

Fig. 8-18 Full-figure **primary** pain pattern. *1,* Stomach/duodenum; *2,* liver/gallbladder/common bile duct; *3,* small intestine; *4,* appendix; *5,* esophagus; *6,* pancreas; and *7,* large intestine/colon.

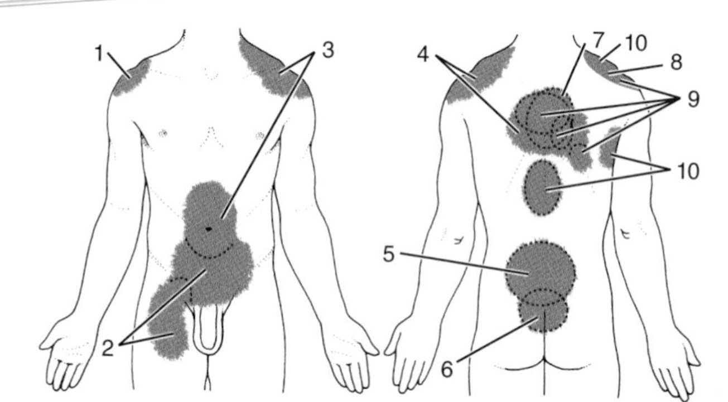

Fig. 8-19 Full-figure **referred** pain patterns. *1,* Liver/gallbladder/common bile duct; *2,* appendix; *3,* pancreas; *4,* pancreas; *5,* small intestine; *6,* colon; *7,* esophagus; *8,* stomach/duodenum; *9,* liver/gallbladder/common bile duct; and *10,* stomach/duodenum.

Key Points to Remember

- GI disorders can refer pain to the sternum, neck, shoulder, scapula, low back, sacrum, groin, and hip.
- When evaluated during early onset of referred pain, there is usually full and painless range of motion, but as time goes on, muscle splinting and guarding secondary to pain or as a component of motor nerve involvement will produce altered movements as well.
- The membrane that envelops organs (visceral peritoneum) is insensitive to pain so that, except in the presence of inflammation/ischemia, it is possible to have extensive disease without pain.
- Clients may not relate known GI disorders to current (or new) musculoskeletal symptoms.
- Sudden and unaccountable changes in bowel habits, blood in the stool, or vomiting red blood or coffee-ground vomitus are red flag symptoms requiring medical follow-up.
- Antibiotics and NSAIDs are the drugs that most commonly induce GI symptoms.
- Kehr's sign (left shoulder pain) occurs as a result of free air or blood in the abdominal cavity causing distention (e.g., trauma, ruptured spleen, laparoscopy, ectopic pregnancy).
- Epigastric pain radiating to the upper back or upper back pain alone can be the primary symptom of peptic ulcer, pancreatitis, or pancreatic carcinoma.
- Appendicitis and diseases of the intestines, such as Crohn's disease and ulcerative colitis can cause abscess of the iliopsoas muscle, resulting in hip, thigh, or groin pain.
- Arthritis and migratory arthralgias occur in 25% of Crohn's disease cases.

SUBJECTIVE EXAMINATION

Special Questions to Ask

After completing the initial intake interview, if there is cause to suspect GI involvement, include any of the following additional questions that seem pertinent. It may be helpful to let the client know you will be asking some questions about overall health issues that may seem unrelated to their current symptoms but that are nevertheless important.

When asking questions about medications, look for long-term use of antibiotics, corticosteroids, or other hepatotoxic drugs. See Table 8-1 for a list of medications that can cause constipation.

Past Medical History

- For the client with left shoulder pain: Have you sustained any injuries in the last week during a sports activity, fall, or automobile accident? Were you pushed down or pushed against something hard (assault)? **(Ruptured spleen: positive Kehr's sign)**
- Have you experienced any abdominal or intestinal problems, nausea, vomiting, episodes of night sweats, or fever?
  - *If yes,* have you seen a physician about these problems or reported them to your physician?
  - For further follow-up questions related to this area, see Associated Signs and Symptoms.
- Have you ever had an upset stomach or heartburn while taking your (NSAID) pain relievers like ibuprofen, naproxen (name the specific drug)?
- Have you ever been treated for an ulcer or internal bleeding while taking these (NSAID) pain relievers?
  - *If so,* when?
  - Do you still have any pain from your ulcer? Please describe.
- Have you ever had a colonoscopy, proctoscopy, or endoscopy?
  - *If yes,* why and how long ago?
- Have you ever been diagnosed with cancer of any kind?
  - *If yes,* what, when, and has there been any follow-up?
- Have you ever had radiation treatment? **(Rectal bleeding is a sign of radiation proctitis.)**
- Have you ever had abdominal or spine (anterior retroperitoneal approach) surgery?
  - *If yes,* when and what type was it?
- Do you have hemorrhoids?
  - *If yes,* have you had surgery for your hemorrhoids? **(Most common cause of bright red blood coating stools)**

Associated Signs and Symptoms: Effects of Eating/Drinking

- Do you have any problems chewing or swallowing food? Do you have any pain when swallowing food or liquids? **(Dysphagia, odynophagia)**
- Have you been vomiting? **(Esophageal varices, ulcers)**
  - *If so,* how often?
  - Is your vomitus ever dark brown or black or look like it has coffee grounds in it? **(Blood)**
  - Have you ever vomited, coughed up, or spit up blood?
- Have you experienced any loss of appetite or sudden weight loss in the last few weeks? (i.e., 10 to 15 pounds in 2 weeks without trying)
- Does eating relieve your symptoms? **(Duodenal or pyloric ulcer)**
  - *If yes,* how soon after eating?

Continued

SUBJECTIVE EXAMINATION—*cont'd*

- Does eating aggravate your symptoms? (**Gastric ulcer, gallbladder inflammation**)
- Does your pain occur 1 to 3 hours after eating or between meals? (**Duodenal or pyloric ulcers, gallstones, pancreatitis**)
- Have you ever had gallstones?
- Have you noticed any change in your symptoms after drinking alcohol? (**Alcohol-associated pancreatitis**)
- Have you ever awakened at night with pain? (**Duodenal ulcer, cancer**)
  - Approximately what time does this occur? (**12 midnight to 3:00 AM: ulcer**)
  - Can you relieve the pain in any way and get back to sleep. *If yes,* how? (**Ulcer: eating and antacids relieve/ Cancer: nothing relieves**)
- Do you have a feeling of fullness after only one or two bites of food? (**Early satiety: esophagus, stomach and duodenum, or gallbladder**)

Associated Signs and Symptoms: Change in bowel habits

- Have you had any changes in your bowel movements (Normal frequency varies from three times a day to once every 3 or more days)? (**Constipation/bowel obstruction**)
  - *If yes* to constipation (see Table 8-1), do you use laxative or stool softeners? How often?
- Do you have diarrhea? (**Ulcerative colitis, Crohn's disease, long-term use of antibiotics, colonic obstruction, amebic colitis, angiodysplasia, creatine supplementation**)
  - Do you have more than two loose stools a day? *If so,* do you take medication for this problem? What kind of medication do you use?
  - Have you traveled outside of the United States within the last 6 months to 1 year? (**Amebic colitis associated with bloody diarrhea**)
- Do you have a sense of urgency so that you have to find a bathroom immediately without waiting?
- Do you ever have any blood in your stool, reddish mahogany-colored stools, or dark, tarry stools that are hard to wipe clean? (**Bleeding ulcer, esophageal varices, colon or rectal cancer, hemorrhoids or rectal fissures; rectal lesions with bleeding can be caused by homosexual activity [men] or anal intercourse [women]**)
  - *If yes,* how often?
  - For the therapist: *If yes,* assess NSAID use and risk factors for NSAID-induced gastropathy.
  - Is the blood mixed in with the stool or does it coat the surface? (**Distal colon or rectum versus melena**)
- Do you ever have white mucus around or in your stools? (**Irritable bowel syndrome**)
- Do you ever have gray-colored stools? (**Lack of bile or caused by biliary obstruction such as hepatitis, gallstones, cirrhosis, pancreatic carcinoma, hepatotoxic drugs**)
- Are your stools ever pencil thin? (**Indicates bowel obstruction such as tumor or rectocele [prolapsed rectum] in women after childbirth**)
- Is your pain relieved after passing stool or gas? (**Yes: large intestine and colon; No: small intestine**)

CASE STUDY

Crohn's Disease

REFERRAL

A 21-year-old woman comes to you with complaints of pain on hip flexion when she lifts her right foot off the brake in the car. There are no other aggravating factors, and she is unaware of any way to relieve the pain when she is driving her car. Before the onset of symptoms, she jogged 5 to 6 miles/ day but could not recall any injury or trauma that might contribute to this pain. The Family/Personal History form indicates no personal illness but shows a complex, positive family history for heart disease, diabetes, ulcerative colitis, stomach ulcers, stomach cancer, and alcoholism.

PHYSICAL THERAPY INTERVIEW

It is suggested that the therapist use the physical therapy interview to assess the client's complaints today and follow-up with appropriate additional questions such as those noted here.

Introduction to Client

From your family history form, I notice that a number of your family members have reportedly been diagnosed with various diseases.

CASE STUDY—*cont'd*

Crohn's Disease

- Do you have any other medical or health-related problems?
- Have you sustained any injuries to the lower back, side, or abdomen in the last week—for example, during a sports activity, fall, or automobile accident? Were you pushed, kicked, or shoved against something?

Although the symptoms that you have described appear to be a musculoskeletal problem, I would like to check out the possibility of a urologic, abdominal, or gynecologic source of this irritation. I will ask you some additional questions that may seem to be unrelated to the problem with your hip, but which will help me put together the whole picture of the history, symptoms, and actual physical results from my examination today.

General Systemic

What other symptoms have you had with this problem? (After allowing the client to answer, you may prompt her by asking: For example, have you had any …)

- Numbness
- Fatigue
- Legs giving out from under you
- Burning, tingling sensation
- Weakness

Gastrointestinal

- Nausea
- Diarrhea
- Loss of appetite
- Feeling of fullness after only one or two bites of a meal
- Unexpected weight gain or loss (10 to 15 pounds without trying)
- Vomiting
- Constipation
- Blood in your stool

(If yes to any of these, follow-up with *Special Questions to Ask* in the Subjective Examination box.)

Have you noticed any association between when you eat and your symptoms? (After allowing the client to respond, you may want to prompt her by asking whether eating relieves the pain or aggravates the pain.)

Is your pain relieved or aggravated during or after you have a bowel movement?

Gynecologic

Since your hip/groin/thigh symptoms started, have you been examined by a gynecologist to rule out any gynecologic causes of this problem?

If no:

- Have you ever been told that you have ovarian cysts, uterine fibroids, retroverted uterus, endometriosis, an ectopic pregnancy, or any other gynecologic problem?
- Are you pregnant or have you recently terminated a pregnancy either by miscarriage or abortion?
- Are you using an intrauterine contraceptive device (IUD)?
- Are you having any unusual vaginal discharge?

(If yes to any of these questions, see the follow-up questions for women in Appendix B-37.)

Urologic

- Have you had any problems with your kidneys or bladder?
  - If yes, please describe.
- Have you noticed any changes in your ability to urinate since your pain or symptoms started? (If no, it may be necessary to provide examples of what changes you are referring to; for example, difficulty in starting or continuing the flow of urine, numbness or tingling in the groin or pelvis, painful urination, urinary incontinence, blood in the urine.)
- Have you had burning with urination during the last 1 to 3 weeks?

Objective Examination

Your objective examination reveals tenderness or palpation over the right anterior upper thigh muscles into the groin, with reproduction of the pain on resisted trunk flexion only. This woman attends daily ballet classes, stretches daily, and seems to be very active physically. All tests for flexibility were negative for tightness, including the Thomas' test for tight hip flexors.

Other special tests for hip and a neurologic screen had negative results. The client's temperature was normal when it was taken today during the intake screen of vital signs, but during the physical therapy interview, when specifically asked about fevers and night sweats, she indicated several recurrent episodes of night sweats during the last 3 months.

RESULTS

Although the client's complaints are primarily musculoskeletal, the absence of trauma, positive family history for systemic disease, limited musculoskeletal findings, and the client's remark concerning the presence of night sweats will alert the physical therapist to the need for a medical referral to rule out the possibility of a systemic origin of symptoms.

The client's condition gradually worsened during a 3-week period and reexamination by the physician led to an eventual diagnosis of Crohn's disease (regional gastroenteritis). The client was treated with medications that reduce abdominal inflammation and eliminated subjective reports of pain on active hip flexion. Performing the special tests for iliopsoas abscess may have provided valuable information and earlier medical referral if assessed during the initial evaluation.

PRACTICE QUESTIONS

1. Bleeding in the gastrointestinal (GI) tract can be manifested as:
   a. Dysphagia
   b. Melena
   c. Psoas abscess
   d. Tenderness over McBurney's point
2. What is the significance of Kehr's sign?
   a. Gas, air, or blood in the abdominal cavity
   b. Infection of the peritoneum (peritonitis, appendicitis)
   c. Esophageal cancer
   d. Thoracic disk herniation masquerading as chest or anterior neck pain
3. What areas of the body can GI disorders refer pain to?
   a. Sternum, shoulder, scapula
   b. Anterior neck, mid-back, lower back
   c. Hip, pelvis, sacrum
   d. All of the above
4. A 56-year-old client was referred to PT for pelvic floor rehab. His primary symptoms are obstructed defecation and puborectalis muscle spasm. He wakes nightly with left flank pain. The pattern is low thoracic, laterally, but superior to iliac crest. Sometimes he has buttock pain on the same side. He doesn't have any daytime pain but is up for several hours at night. Advil and light activity do not help much. The pain is relieved or decreased with passing gas. He has very tight hamstrings and rectus femoris. Change in symptoms with gas or defecation is possible with:
   a. Thoracic disk disease
   b. Obturator nerve compression
   c. Small intestine disease
   d. Large intestine and colon dysfunction
5. Name two of the most common medications taken by clients seen in a physical therapy practice likely to induce GI bleeding.
   a. Corticosteroids
   b. Antibiotics and antiinflammatories
   c. Statins
   d. None of the above
6. What is the significance of the psoas sign?
7. Which of the following are clues to the possible involvement of the GI system?
   a. Abdominal pain alternating with TMJ pain within a 2-week period
   b. Abdominal pain at the same level as back pain, occurring either simultaneously or alternately
   c. Shoulder pain alleviated by a bowel movement
   d. All of the above
8. A 65-year-old client is taking OxyContin for a "sore shoulder." She also reports aching pain of the sacrum that radiates. The sacral pain can be caused by:
   a. Psoas abscess caused by vertebral osteomyelitis
   b. GI bleeding causing hemorrhoids and rectal fissures
   c. Crohn's disease manifested as sacroiliitis
   d. Pressure on sacral nerves from stored fecal content in the constipated client taking narcotics
9. A 64-year-old woman with chronic rheumatoid arthritis fell and broke her hip. Six months after her total hip replacement, she is still using a walker and complains of continued loss of strength and function. Her family practice physician has referred her to physical therapy for a home program to "improve gait and increase strength."

   The client reports frequent episodes of lightheadedness when her legs feel rubbery and weak. She is taking a prescription NSAID along with an OTC NSAID 3 times each day and has been taking NSAIDs 3 years continuously. There are no reported GI complaints or associated signs and symptoms, but after completing the intake interview and objective examination, you think there may be weakness associated with blood loss and anemia secondary to chronic NSAID use. How would you handle a case like this?
10. Body temperature should be taken as part of vital sign assessment:
    a. For every client evaluated
    b. For any client who has musculoskeletal pain of unknown origin
    c. For any client reporting the presence of constitutional symptoms, especially fever or night sweats
    d. b and c

REFERENCES

1. Pert CB, Dreher HE, Ruff MR: The psychosomatic network: foundations of mind-body medicine. *Altern Ther Health Med* 4(4):30–41, 1998.
2. Pert C: Paradigms from neuroscience: when shift happens. *Mol Interv* 3(7):361–366, 2003.
3. Mayer EA: Gut feelings: what turns them on? *Gastroenterology* 108(3):927–931, 1995.
4. Groh V, Spies T: Recognition of stress-induced MHC molecules by intestinal epithelial gamma delta T cells. *Science* 279:1737–1740, 1998.
5. Wu J, et al: T-cell antigen receptor engagement and specificity in the recognition of stress-inducible MHC class I-related chains by human epithelial gamma delta T cells. *J Immunol* 169(3):1236–1240, 2002.
6. Rex L: *Evaluation and treatment of somatovisceral dysfunction of the gastrointestinal system*, Edmonds, WA, 2004, URSA Foundation.
7. Rose SJ, Rothstein JM: Muscle mutability: general concepts and adaptations to altered patterns of use. *Phys Ther* 62:1773, 1982.
8. Ledlie J, Renfro M: Balloon kyphoplasty: one-year outcomes in vertebral body height restoration, chronic pain, and activity levels. *J Neurosurg (Spine I)* 98:36–42, 2003.
9. Longstreth GF: Functional bowel disorders: revised Rome II diagnostic criteria for functional bowel disorders. *Gastroenterology* 130(5):1480–1491, 2006.
10. Spiller R: Do the symptom-based, Rome criteria of irritable bowel syndrome lead to better diagnosis and treatment outcomes? *Clin Gastroenterol Hepatol* 8(2):125–129, 2010.
11. Travell JG, Simons DG: *Myofascial pain and dysfunction: the trigger point manual*, vol 2, Baltimore, 1992, Williams and Wilkins.
12. CDC: *Preventing transmission of infectious agents in healthcare settings*, Centers for Disease Control and Prevention. Available at www.cdc.gov/. Accessed Sept. 24, 2010.

12a. Smith J, Dahm DL: Creatine use among select population of high school athletes. *Mayo Clin Proc* 75(12):1257–1263, 2000.

13. Graham AS, Hatton RC: Creatine: a review of efficacy and safety. *J Am Pharm Assoc* 39(6):803–810, 1999.
14. Metzl JD: Creatine use among young athletes. *Pediatrics* 108(2):421–425, 2001.
15. Inman RD: Arthritis and enteritis—an interface of protean manifestations. *J Rheumatol* 14:406–410, 1987.
16. Gran JT, Husby G: Joint manifestations in gastrointestinal diseases. *Dig Dis* 10:295–312, 1992.
17. Norman GF: Sacroiliac disease and its relationship to lower abdominal pain. *Am J Surg* 116:54–56, 1968.
18. Thjodleifsson B: A common genetic background for inflammatory bowel disease and ankylosing spondylitis. *Arthritis Rheum* 56:2633–2639, 2007.
19. Baeten D, et al: Influence of the gut and cytokine patterns in spondyloarthropathy. *Clin Exp Rheumatol* 20(6 Suppl 28):S38-S42, 2002.
20. Sieper J, et al: Diagnosing reactive arthritis: role of clinical setting in the value of serologic and microbiologic assays. *Arthritis Rheum* 46:319, 2002.
21. Mustafa K, Khan MA: Recognizing and managing reactive arthritis. *J Musculoskeletal Med* 13(6):28–41, 1996.
22. Burger EL: Lumbar disk replacement: restoring mobility. *Orthopedics* 27(4):386–388, 2004.
23. Tay B, et al: Spinal infections. *J Amer Acad Orthop Surg* 10(3):188–197, 2002.
24. Mallick IH: Iliopsoas abscesses. *Postgrad Med J* 80:459–462, 2004.
25. Goodman CC: The gastrointestinal system. In Goodman CC, Fuller KS, et al, editors: *Pathology: implications for the physical therapist*, ed 3, Philadelphia, 2009, WB Saunders.
26. Sokolov KM, Kreye E, Miller LG, et al: Postpartum iliopsoas pyomyositis due to community-acquired methicillin-resistant *Staphylococcus aureus*. *Obstet Gynecol* 2007; 110:535–538.
27. Shahabi S: Primary psoas abscess complicating a normal vaginal delivery. *Obstet Gynecol* 99:906–909, 2002.
28. Mandell GL: *Mandell, Douglas, and Bennett's principles and practice of infectious diseases*, ed 7, Philadelphia, 2009, Churchill Livingstone.
29. Rayhorn N, Argel N, Demchak K: Understanding gastroesophageal reflux disease. *Nursing2003* 33(10):37–41, 2003.
30. Sabesin SM, Fass R, Fisher R: Not all heartburn patients are equal: strategies for coping with gastroesophageal reflux disease (GERD), *Medscape Continuing Medical Education*.
31. Aluka KJ: Guillain-Barré syndrome and postbariatric surgery polyneuropathies. *JSLS* 13(2):250–253, 2009.
32. Teitleman M: Polyneuropathy after gastric bypass surgery. *Medscape General Medicine* 7(2):2005, 2010. Available on-line at www.medscape.com/viewarticle/499484. Accessed on Sept. 21.
33. Juhasz-Pocsine KN: Neurologic complications of gastric bypass surgery for morbid obesity. *Neurology* 68(21):1843–1850, 2007.
34. Asthma in older women. *Harvard Women's Health Watch* 11(3):5, 2003.
35. Margolis S: Getting the right cure for ulcers. *Johns Hopkins Medical Letter* 10(1):1–2, 1998.
36. Miwa, H, Sakaki N, Sugano K, et al: Recurrent peptic ulcers in patients following successful *Helicobacter pylori* eradication: a multicenter study of 4940 patients. *Helicobacter* 9(1):9–16, 2004.
37. Chan FKL, Graham DY: Prevention of non-steroidal anti-inflammatory drug gastrointestinal complications—review and recommendations based on risk assessment, *Medscape Continuing Medical Education*.
38. Lanas A: Gastrointestinal bleeding associated with low-dose aspirin use: relevance and management in clinical practice. *Expert Opin Drug Safety* July 20, 2010. Epub ahead of print.
39. Wiegand T: Toxicity, Nonsteroidal anti-inflammatory agents, *eMedicine Specialties: Emergency Medicine—Toxicity*. Updated May 20, 2010. Available on-line at http://emedicine.medscape.com/article/816117-overview. Accessed January 31, 2011.
40. McPhee S, Papadakis M, editors: *Current medical diagnosis and treatment*, ed 50, New York, 2011, Lange.
41. Bronstein AC, Spyker DA, Cantilena LR Jr, et al: 2007 Annual Report of the American Association of Poison Control Centers' National Poison Data System (NPDS): 25th Annual Report. *Clin Toxicol (Phila)* 46(10):927–1057, 2008.
42. Chan FK: Celecoxib versus diclofenac and omeprazole in reducing the risk of recurrent ulcer bleeding in patients with arthritis. *N Engl J Med* 347(26):2104–2110, 2002.
43. Sostres C: Adverse effects of non-steroidal anti-inflammatory drugs (NSAIDs, aspirin and coxibs) on upper gastrointestinal tract. *Best Pract Res Clin Gastroenterol* 24(2):121–132, 2010.
44. National Institute of Diabetes and Digestive and Kidney Diseases (NIDDK): *H. pylori* and peptic ulcers. Available on-line at http://digestive.niddk.nih.gov/ddiseases/pubs/hpylori/#2. Accessed January 31, 2011.
45. Peloso PM: NSAIDs: a Faustian bargain. *Am J Nurs* 100(6):34–43, 2000.
46. Wolfe MM: Gastrointestinal toxicity of nonsteroidal anti-inflammatory drugs. *N Engl J Med* 340(24):1888–1899, 1999.
47. Sturkenboom MC, Burke TA, Dieleman JP, et al: Underutilization of preventive strategies in patients receiving NSAIDs. *Rheumatology (Oxford)* 42(Suppl 3):iii23–31, 2003.
48. Goldstein JL: Challenges in managing NSAID-associated gastrointestinal tract injury. *Digestion* 69(Suppl 1):25–33, 2004.
49. Fries JF, et al: Nonsteroidal antiinflammatory drug-associated gastropathy: incidence and risk factor models. *Am J Med* 91(3):213–222, 1991.
50. Enns R: Acute lower gastrointestinal bleeding, Parts 1 and 2. *Can J Gastroenterol* 15:509–517, 2001.
51. Sadler TW: *Langman's medical embryology*, ed 11, Philadelphia, 2009, Lippincott, Williams & Wilkins.
52. Cheetham TC: Predicting the risk of gastrointestinal bleeding due to nonsteroidal antiinflammatory drugs: NSAID electronic assessment of risk. *J Rheumatol* 30(10):2241–2244, 2003.
53. National Institute of Diabetes and Digestive and Kidney Diseases: *Diverticulosis and diverticulitis*, NIH Publication 07–1163, October 2006. Available on-line at www.digestive.niddk.nih.gov. Accessed September 22, 2010.
54. Rayhorn N: Inflammatory bowel disease (IBD). *Nursing2003* 33(11):54–55, 2003.
55. Young-Fadok T: *Clinical manifestations and diagnosis of colonic diverticular disease*. Up-to-date November 19, 2008. Available (by subscription) on-line at www.uptodate.com. Accessed Sept. 24, 2010.
56. Ma TY: Intestinal epithelial barrier dysfunction in Crohn's disease. *Proc Soc Exp Biol Med* 214(4):318–327, 1997.
57. Wagner JM: Does this patient have appendicitis? *JAMA* 276(19):1589–1594, 1996.
58. Adams BD: Pinch-an-inch test for appendicitis. *South Med J* 98(12):1207–1209, 2005.
59. Adams BD: The Pinch-an-Inch test is more comfortable than rebound tenderness. *Internet J Surg* 12(2), 2007.
60. Banks PA: Practice Parameters Committee of the American College of Gastroenterology. Practice Guidelines in acute pancreatitis. *Am J Gastroenterol* 101(10):2379–2400, 2006.
61. Ma TY, Iwamoto GK, Hoa NT, et al: TNF-alpha-induced increase in intestinal epithelial tight junction permeability requires NF-kappa B activation. *Am J Physiol Gastrointest Liver Physiol* 286(3):G367–G376, 2004.
62. Salim SY: Importance of disrupted intestinal barrier in inflammatory bowel diseases. *Inflamm Bowel Dis* Aug 19, 2010. Epub ahead of print.
63. Han X: Intestinal permeability as a clinical surrogate endpoint in the development of future Crohn's disease therapies. *Recent Pat Inflamm Allergy Drug Discov* 4(2):159–176, 2010.

64. Shen L: Mechanisms and functional implications of intestinal barrier defects. *Dig Dis* 27(4):443–449, 2009.
65. Crowell MD, Dubin NH, Robinson JC, et al: Functional bowel disorders in women with dysmenorrhea. *Am J Gastroenterol* 89:1973, 1994.
66. Heymen S: Central processing of noxious somatic stimuli in patients with irritable bowel syndrome compared with healthy controls. *Clin J Pain* 26(2):104–109, 2010.
67. Nokjov B: The impact of rotating shift work on the prevalence of irritable bowel syndrome in nurses. *Am J Gastroenterol* 105(4):842–847, 2010.
68. National Institutes of Health (NIH): *Irritable bowel syndrome,* NIH Publication No. 03–693, April, 2003.
69. Jemal A: Cancer statistics 2010. *CA Cancer J Clin* 60(5):277–300, 2010.
70. Sargent C, Murphy D: What you need to know about colorectal cancer. *Nursing2003* 33(2):37–41, 2003.
71. Smith R, et al: American Cancer Society guidelines for early detection of cancer. *CA Cancer J Clin* 52(1):8–22, 2002.
72. Lucak S: Diagnosing irritable bowel syndrome: what's too much, what's enough? *Medscape Medical Continuing Education* on line, posted 3/12/04 on www.medscape.com/viewarticle 465760.
73. Schermer CR, et al: Ogilvie's syndrome in the surgical patient: a new therapeutic modality. *J. Gastroenterol Surg* 3(2):173, 1999.
74. el Maraghy A, et al: Ogilvie's syndrome after lower extremity arthroplasty. *Can J Surg* 42(2):133, 1999.
75. Richardson LC: Vital signs: colorectal cancer screening among adults ages 50–75 years in the United States. *MMWR* 59:1–5, 2010.
76. Sargent C, Murphy D: What you need to know about colorectal cancer. *Nursing2003* 33(2):37–41, 2003.
77. Smith R, et al: American Cancer Society guidelines for early detection of cancer. *CA Cancer J Clin* 52(1):8–22, 2002.
78. Soetikno RM: Prevalence of nonpolypoid (flat and depressed) colorectal neoplasms in asymptomatic and symptomatic adults. *JAMA* 299(9):1027–1035, 2008.
79. CR Schermer, et al: Ogilvie's syndrome in the surgical patient, a new therapeutic modality. *J Gastroenterol Surg* 3(2):173, 1999.
80. Oberpenning F, Roth S, Leusmann DB, et al: The Alcock syndrome: temporary penile insensitivity due to compression of the pudendal nerve within the Alcock canal. *J Urol* 151(2):423–425, 1994.
81. Weiss BD: Clinical syndromes associated with bicycle seats. *Clin Sports Med* 13(1):175–186, 1994.

CHAPTER

9

Screening for Hepatic and Biliary Disease

As with many of the organ systems in the human body, the hepatic and biliary organs (liver, gallbladder, and common bile duct) can develop diseases that mimic primary musculoskeletal lesions (Fig. 9-1). The musculoskeletal symptoms associated with hepatic and biliary pathologic conditions are generally confined to the mid-back, scapular, and right shoulder regions. These musculoskeletal symptoms can occur alone (as the only presenting symptom) or in combination with other systemic signs and symptoms discussed in this chapter.

HEPATIC AND BILIARY SIGNS AND SYMPTOMS

The major causes of acute hepatocellular injury include hepatitis, drug-induced hepatitis, and ingestion of hepatotoxins. The physical therapist is most likely to encounter liver or gallbladder diseases manifested by a variety of signs and symptoms outlined in this section.

Taking a careful history and making close observations of the client's physical condition and appearance can detect telltale signs of hepatic disease. Most of the liver is contained underneath the rib cage and is largely inaccessible (Fig. 9-2). An enlarged liver that is palpable is always a red flag (see Fig. 4-51). Medical diagnosis of liver or gallbladder disease is made by x-ray examination or ultrasonic scanning of the gallbladder and computed tomography (CT) scanning of the abdomen, including the liver.

Other tests, such as a cholescintigraphy, may be used to track the flow of radioactivity into and out of the gallbladder to confirm gallstones. Blood tests may be used to look for signs of infection, obstruction, or jaundice. Laboratory tests useful in the diagnosis and treatment of liver and biliary tract disease are listed inside the back cover.

Skin and Nail Bed Changes

Skin changes associated with impairment of the hepatic system include jaundice, pallor, and orange or green skin in a Caucasian individual. Change in skin tones may be visible in Black or Asian people, but these may only be observable to the affected individual or to those who know him or her well. In some situations jaundice may be the first and only manifestation of disease. It is first noticeable in people of all skin colors in the sclera of the eye as a yellow hue when bilirubin reaches levels of 2 to 3 mg/dL. When the bilirubin level reaches 5 to 6 mg/dL, changes in skin color occur.

Other skin changes may include pruritus (itching), bruising, spider angiomas (Fig. 9-3), and palmar erythema (see Fig. 9-5). *Spider angiomas* (arterial spider, spider telangiectasis, vascular spider), branched dilations of the superficial capillaries resembling a spider in appearance (Fig. 9-4), may be vascular manifestations of increased estrogen levels (hyperestrogenism). Spider angiomas and palmar erythema both occur in the presence of liver impairment as a result of increased estrogen levels normally detoxified by the liver.

Palmar erythema (warm redness of the skin over the palms, also called *liver palms*), caused by an extensive collection of arteriovenous anastomoses, is especially evident on the hypothenar and thenar eminences and pulps of the finger (Fig. 9-5). The person may complain of throbbing, tingling palms. The soles of the feet may be similarly affected. Throbbing and tingling may be associated with these anastomoses.

Various forms of nail disease have been described in cases of liver impairment such as the white nails of Terry (Fig. 9-6). Other nail bed changes, such as white bands across the nail plate (leukonychia), clubbed nails (see Fig. 4-36), or koilonychia (see Fig. 4-32), can occur, but these are not specific to liver impairment and can develop in the presence of other diseases as well.

Musculoskeletal Pain

Musculoskeletal pain associated with the hepatic and biliary systems includes thoracic pain between the scapulae, right shoulder, right upper trapezius, right interscapular, or right subscapular areas (see Fig. 9-10 and Table 9-1).

Referred shoulder pain may be the only presenting symptom of hepatic or biliary disease. Afferent pain signals from the superior ligaments of the liver and the superior portion of the liver capsule are transmitted by the phrenic

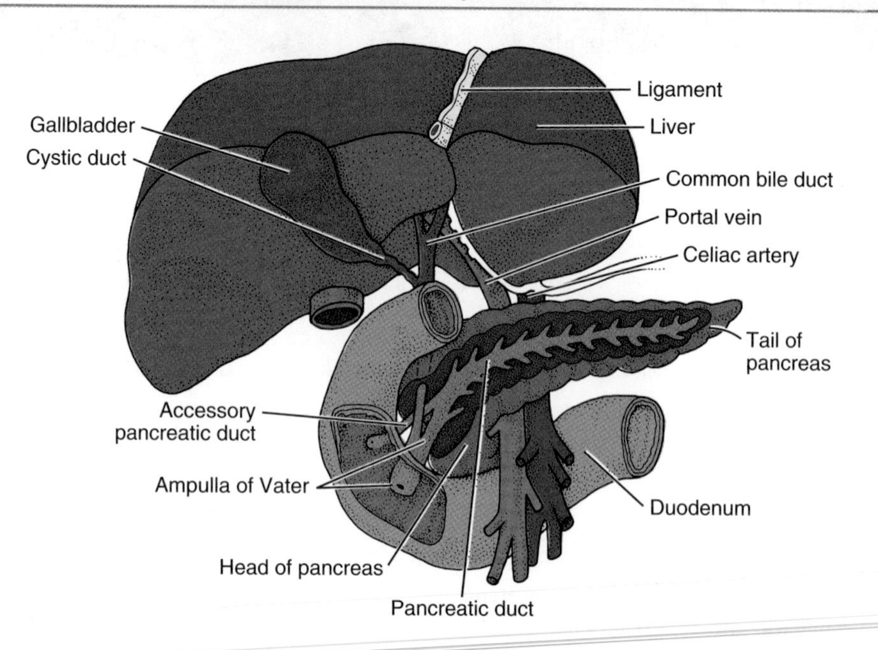

Fig. 9-1 Anatomy of the liver, gallbladder, common bile duct, and pancreas. The pear-shaped *gallbladder* is tucked up under the right side of the liver. The *pancreas* is located behind the stomach anterior to the L1 to L3 vertebral bodies. It is about 6 inches long, wide at one end (the head), then tapered through the body to the narrow end called the tail.

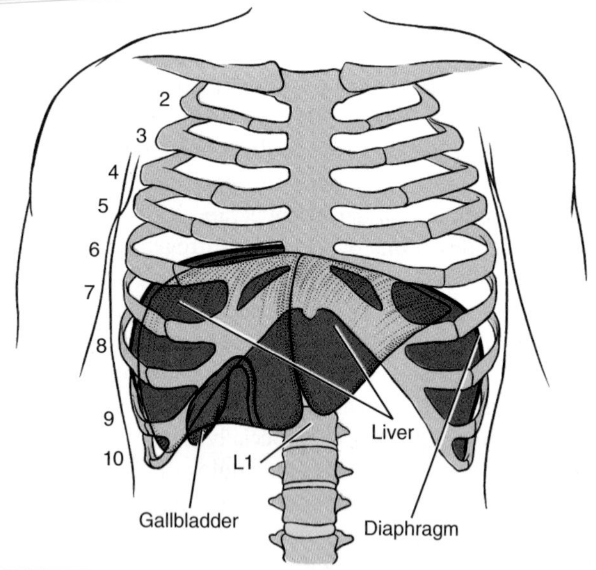

Fig. 9-2 Location of the liver and gallbladder. The *liver* is located just below the respiratory diaphragm, predominately on the right side, but with a portion crossing the midline to the left side. It is a large organ and spans many vertebral levels. The most superior part is the dome of the right lobe. The "peak" of the dome lies at about T8 or T9 during expiration. The inferior border of the left lobe is located just below the level of the left nipple and inclines downwards to the right at the tip of the 8th costal margin. The right lobe angles downward to the 9th and 10th costal margins. Posteriorly, the liver is located from approximately T9 to L1 at the midline. This varies from person to person and with inhalation (moves up a level or two) and exhalation (moves down). The fundus (base) of the *gallbladder* usually appears below the edge of the liver in contact with the anterior abdominal wall at the tip of the 9th right costal cartilage.

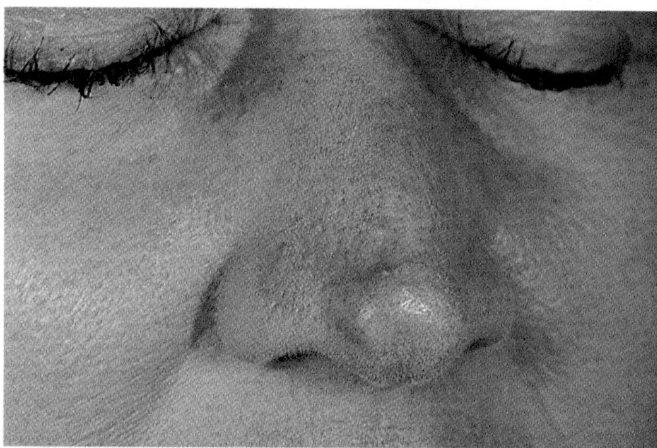

Fig. 9-3 Spider Angioma. Permanently enlarged and dilated capillaries visible on the surface of the skin caused by vascular dilation are called *spider angiomas*. These capillary radiations can be flat or raised in the center. They present on the upper half of the body, primarily on the face, neck, chest, or abdomen and occur as a normal development or in association with pregnancy, chronic liver disease, or estrogen therapy. They do not go away when the underlying condition is treated; laser therapy is available to remove them for cosmetic reasons. (From Swartz, MH: *Textbook of physical diagnosis: history and examination,* ed 6, Philadelphia, 2009, Saunders.)

nerves. Sympathetic fibers from the biliary system are connected through the celiac (abdominal) and splanchnic (visceral) plexuses to the hepatic fibers in the region of the dorsal spine (see Fig. 3-3).

The celiac and splanchnic connections account for the intercostal and radiating interscapular pain that accompanies gallbladder disease. Although the innervation is bilateral, most of the biliary fibers reach the cord through the right splanchnic nerves, synapsing with adjacent phrenic nerve fibers innervating the diaphragm and producing pain in the right shoulder (see Fig. 3-4).

Fig. 9-4 Arterial spider. Schematic diagram of an arterial spider formed by a coiled arteriole that spirals up to a central point and then branches out into thin-walled vessels that merge with normal capillaries resembling a spider in appearance. Exposure to heat (e.g., hot tubs, warm shower) will cause temporary vasodilation. The skin lesion will appear larger until vasoconstriction occurs.

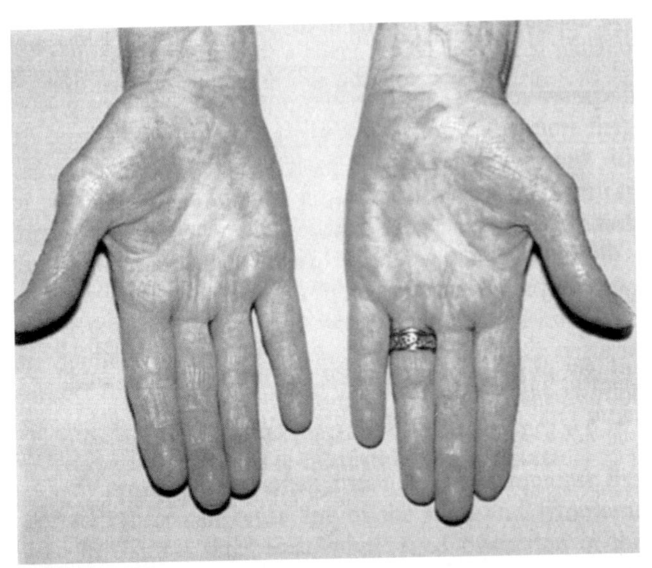

Fig. 9-5 Palmar erythema caused by liver impairment presents as a warm redness of the skin over the palms and soles of the feet in the Caucasian population. Darker skin tones may change from a tan color to a gray appearance. Look for other signs of liver disease such as nail bed changes, spider angiomas, liver flap, and bilateral carpal or tarsal tunnel syndrome. Palmar erythema can occur in healthy individuals and in association with nonhepatic diseases. (From Barrison I, ed: *Gastroenterology in practice,* St. Louis, 1992, Mosby.)

Hepatic osteodystrophy, abnormal development of bone, can occur in all forms of cholestasis (bile flow suppression) and hepatocellular disease, especially in the alcoholic person. Either osteomalacia or more often, osteoporosis frequently accompanies bone pain from this condition. Vertebral

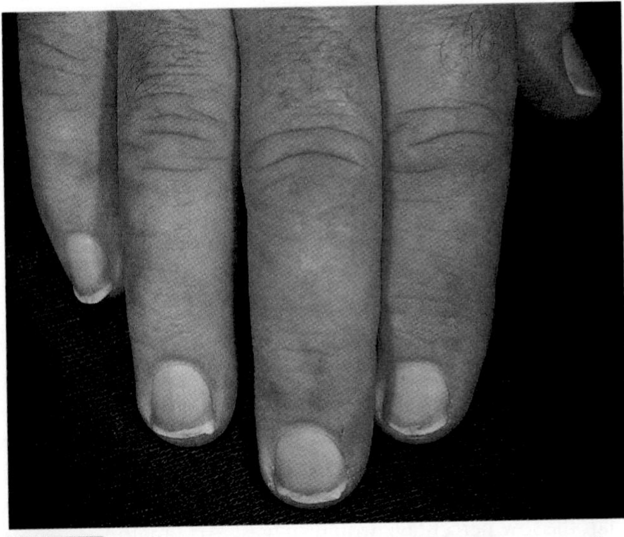

Fig. 9-6 Nails of Terry. Opaque white nails of Terry in a patient with cirrhosis. Various forms of nail disease have been described in patients with cirrhosis. This is an example of the classic white nails of Terry characterized by an opaque nail plate with a narrow line of pink at the distal end instead of the more normal pink nail plate in the Caucasian. Nails of Terry can also present as a result of malnutrition, diabetes mellitus, hyperthyroidism, trauma, and sometimes for unknown reasons (idiopathic). (From Callen JP, Jorizzo JL, editors: *Dermatological signs of internal disease,* ed 4, Philadelphia, 2009, WB Saunders.)

TABLE 9-1 Referred Pain Patterns: Liver, Gallbladder, Common Bile Duct

Systemic Causes	Location (see Fig. 9-10)
Liver disease (abscess, cirrhosis, tumors, hepatitis)	Thoracic spine (T7-T10; midline to the right) Right upper trapezius and shoulder
Gallbladder	Right upper trapezius and shoulder Right interscapular area (T4 or T5-T8) Right subscapular area

wedging, vertebral crush fractures, and kyphosis can be severe; decalcification of the ribcage and pseudofractures[1] occur frequently.

Pseudofractures, or Looser's zones, are narrow lines of radiolucency (areas of darkness on x-ray film), usually oriented perpendicular to the bone surface. This may represent a stress fracture that is repaired by laying down inadequately mineralized osteoid, or these sites may occur as a result of mechanical erosion caused by arterial pulsations, since arteries frequently overlie sites of pseudofractures.

Osteoporosis associated with primary biliary cirrhosis and primary sclerosing cholangitis parallels the severity of liver disease rather than its duration. Painful osteoarthropathy may develop in the wrists and ankles as a nonspecific complication of chronic liver disease. *Rhabdomyolysis* is a potentially fatal condition is which myoglobin and other muscle tissue contents are released into the bloodstream as a result of muscle tissue disintegration. This could occur with acute trauma (e.g., crush injuries, significant blunt trauma, high-voltage electrical burns, surgery[2]), severe burns, overexertion,[3] or in the case of liver impairment, from alcohol abuse or alcohol poisoning or the use of cholesterol-lowering drugs called *statins* (e.g., Zocor, Lipitor, Crestor, Mevacor, Pravachol).[4]

Rhabdomyolysis is characterized by muscle aches, cramps, soreness, and weakness. It may be accompanied by other symptoms of respiratory muscle myopathy (impaired diaphragmatic function)[5] or liver or renal involvement. Laboratory testing will show a creatine kinase (CK) level more than 10 times the upper limit of normal.

Although the literature reports the incidence of this severe myopathy with statin use as about 0.1% to 2.0% in clinical trials,[6] therapists (and others) report seeing cases more often than the low percentage would suggest. Any anatomic region can be affected, but the back and extremity musculoskeletal pain are the two areas of involvement reported most often.[7,8]

Statin-associated myopathy appears to occur more often in people with complex medical problems and/or those taking illegal drugs (e.g., cocaine, heroin, LSD) or abusing prescription drugs (e.g., barbiturates and amphetamines).[9] Other risk factors that increase the chances of this condition include excessive alcohol use, advancing age (over 80 years), recent history of surgery, and small physical stature.[4] For additional discussion, see Screening for Side Effects of Statins in Chapter 6.

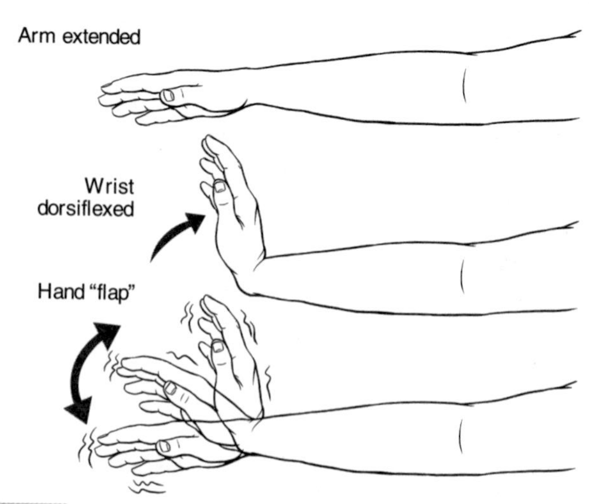

Fig. 9-7 To test for asterixis or liver flap, have the client extend the arms, spread the fingers, extend the wrist, and observe for the abnormal "flapping" tremor at the wrist. If a tremor is not readily apparent, ask the client to keep the arms straight while gently hyperextending the client's wrist. There is an alternate method of testing for this phenomenon: have the client relax the legs in the supine position with the knees bent. The feet are flat on the table. As the legs fall to the sides, watch for a flapping or tremoring of the legs at the hip. The knees appear to come back toward the midline repeatedly.[10]

Neurologic Symptoms

Neurologic symptoms, such as confusion, sleep disturbances, muscle tremors, hyperreactive reflexes, and asterixis, may occur. When liver dysfunction results in increased serum ammonia and urea levels, peripheral nerve function can be impaired.

Ammonia from the intestine (produced by protein breakdown) is normally transformed by the liver to urea, glutamine, and asparagine, which are then excreted by the renal system. When the liver does not detoxify ammonia, ammonia is transported to the brain, where it reacts with glutamate (excitatory neurotransmitter), producing glutamine.

The reduction of brain glutamate impairs neurotransmission, leading to altered central nervous system (CNS) metabolism and function. Symptoms of poor concentration, fatigue, and other symptoms of encephalopathy can result.

Another outward sign of liver disease producing CNS dysfunction is *asterixis.* Impaired inflow of joint and other afferent information to the brainstem reticular formation produces this movement dysfunction. Also called *flapping tremors* or *liver flap,* asterixis is described as the inability to maintain wrist extension with forward flexion of the upper extremities. It is tested by asking the client to actively hyperextend the wrist and hand with the rest of the arm supported on a firm surface or with the arms held out in front of the body (Fig. 9-7).[10] The test is positive if quick, irregular extensions and flexions of the wrist and fingers occur. Asterixis may also be observed when releasing the pressure in the arm cuff during blood pressure readings.

Asterixis and numbness/tingling (the latter are misinterpreted as carpal tunnel syndrome) can occur as a result of this ammonia abnormality, causing an intrinsic nerve pathologic condition (Case Example 9-1). There are many potential causes of carpal tunnel syndrome, both musculoskeletal and systemic (see Table 11-2). Careful evaluation is required (Box 9-1).

CASE EXAMPLE 9-1

Carpal Tunnel Syndrome from Liver Impairment

A 45-year-old truck driver was diagnosed by a hand surgeon as having bilateral carpal tunnel syndrome (CTS) and referred to physical therapy. A screening examination was not performed during the evaluation. During the course of treatment, the client commented that he was seeing an acupuncturist, who told him that liver disease was the cause of his bilateral CTS.

The therapist suspected a history of alcohol abuse, which is a risk factor for liver disease. Further questioning at that time indicated the lack of any other associated symptoms to suggest liver or hepatic involvement. However, because his symptoms were bilateral and there is a known correlation between liver disease and CTS, the referring physician was notified of these findings.

The client was referred for evaluation, and a diagnosis of liver cancer was confirmed. Physical therapy for CTS was appropriately discontinued.

A careful history and close observation of the client are important in determining whether a person may need a medical referral for possible liver disease. Jaundice in the postoperative client is not uncommon, but it can be a potentially serious complication of liver damage that occurs following surgery and anesthesia. Clues to screening for hepatic disease (see Clues to Screening for Hepatic Disease at the end of this chapter) should be taken into consideration when evaluating the clinical history and observations.

Gastrointestinal System

Normally, bilirubin, excreted in bile and carried to the small intestines, is reduced to a form that causes the stool to assume a brown color. Light-colored (almost white) stools and urine the color of tea or cola indicate an inability of the liver or biliary system to excrete bilirubin properly. Gallbladder disease, hepatotoxic medications, or pancreatic cancer blocking the bile duct may cause light stools.

CLINICAL SIGNS AND SYMPTOMS

Liver Disease

Gastrointestinal
- Sense of fullness of the abdomen
- Anorexia, nausea, and vomiting

Integumentary
- Skin color changes and nail bed changes
- Pallor (often linked to cirrhosis or carcinoma)
- Jaundice
- Bruising
- Spider angioma
- Palmar erythema
- White nails of Terry, other nail bed changes may be present

Hepatic
- Dark urine and light-colored or clay-colored stools
- Ascites (Fig. 9-8)
- Edema and oliguria (reduced urine secretion in relation to fluid intake)
- Right upper quadrant (RUQ) abdominal pain

Musculoskeletal
- Musculoskeletal pain, especially right shoulder pain
- Myopathy (rhabdomyolysis in severe cases)

Neurologic
- Confusion
- Sleep disturbances
- Muscle tremors
- Hyperactive reflexes
- Asterixis (motor disturbance resembling body or extremity flapping)
- Bilateral carpal tunnel syndrome (numbness, tingling, burning pain in thumb, index and middle fingers)
- Bilateral tarsal tunnel (tarsal tunnel characterized by pain around the ankle that extends to the palmar surfaces of the toes, made worse by walking)

Other
- Gynecomastia (enlargement of breast tissue in men)

CLINICAL SIGNS AND SYMPTOMS

Gallbladder Disease

Gastrointestinal
- Right upper abdominal pain
- Indigestion, nausea, feeling of fullness
- Excessive belching, flatulence (intestinal gas)
- Intolerance of fatty foods

Integumentary
- Jaundice (result of blockage of the common bile duct)
- Persistent pruritus (skin itching)

Musculoskeletal
- Sudden, excruciating pain in the midepigastrium with referral to the back and right shoulder (acute cholecystitis)
- Anterior rib pain (tip of 10th rib; can also affect ribs 11 and 12)

Constitutional
- Low-grade fever, chills

BOX 9-1 EVALUATING CARPAL TUNNEL SYNDROME ASSOCIATED WITH LIVER IMPAIRMENT

For any client presenting with bilateral carpal tunnel syndrome
- Ask about the presence of similar symptoms in the feet
- Ask about a personal history of liver or hepatic disease (e.g., cirrhosis, cancer, hepatitis)
- Look for a history of hepatotoxic drugs (see Box 9-3)
- Look for a history of alcoholism
- Ask about current or previous use of statins (cholesterol-lowering drugs such as Crestor, Lipitor, Mevacor, or Zocor)
- Look for other signs and symptoms associated with liver impairment (see Clinical Signs and Symptoms of Liver Disease)
- Test for signs of liver disease
  - Skin color changes
  - Spider angiomas
  - Palmar erythema (liver palms)
  - Nail bed changes (e.g., white nails of Terry, white bands, clubbing)
  - Asterixis (liver flap)

HEPATIC AND BILIARY PATHOPHYSIOLOGY

Liver Diseases

Hepatitis

Hepatitis is an acute or chronic inflammation of the liver. It can be caused by a virus, a chemical, a drug reaction, or alcohol abuse. In addition, hepatitis can be secondary to

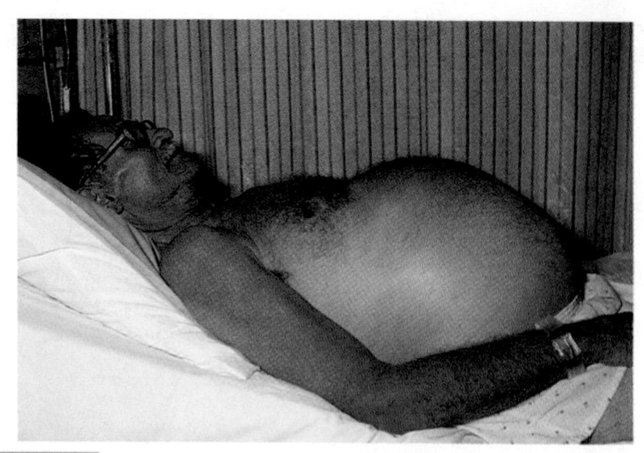

Fig. 9-8 Ascites is an abnormal accumulation of serous (edematous) fluid in the peritoneal cavity associated with liver impairment, especially the portal and hepatic venous hypertension that accompanies cirrhosis of the liver. This condition also may be associated with other disorders such as advanced congestive heart failure, constrictive pericarditis, cancer, chronic hepatitis, and hyperaldosteronism. Any condition affecting the peritoneum by producing increased permeability of the peritoneal capillaries and electrolyte disturbances can result in ascites. (From Swartz M: *Textbook of physical diagnosis: health and examination,* Philadelphia, 1989, WB Saunders.)

disease conditions, such as an infection with other viruses (e.g., Epstein-Barr virus or cytomegalovirus).

Viral Hepatitis. Viral hepatitis is an acute infectious inflammation of the liver caused by one of the following identified viruses: A, B, C, D, E, and G (Table 9-2).

Hepatitis is a major uncontrolled public health problem for several reasons: not all the causative agents have been identified, there are limited specific drugs for its treatment, its incidence has increased in relation to illicit drug use, and it can be communicated before the appearance of observable clinical symptoms.

Viral hepatitis is spread easily to others and usually results in an extended period of convalescence with loss of time from school or work. It is estimated that 60% to 90% of viral hepatitis cases are unreported because many cases are subclinical or involve mild symptoms.[11,12]

Hepatitis A and E are transmitted primarily by the fecal-oral route. Common source outbreaks result from contaminated food or water. Hepatitis A virus (HAV) must also be considered a potential problem in situations where fecal-oral communication along with food handling and/or unsanitary conditions occur. Some examples of potential sources of contact with HAV might include restaurants, day care centers, correctional institutions, sewage plants, and countries where these viruses are endemic.[13,14]

Hepatitis viruses B, C, D, and G are primarily bloodborne pathogens that can be transmitted from percutaneous or mucosal exposures to blood or other body fluids from an infected person.

Hepatitis B virus (HBV) is usually transmitted by inoculation of infected blood or blood products or by sexual contact and is also found in body fluids (e.g., spinal, peritoneal, pleural), saliva, semen, and vaginal secretions. Hepatitis D virus (HDV) must have HBV present to coinfect. Groups at risk include homosexuals and intravenous (IV) drug users; health care workers in any area in which contact with blood, blood products, or body fluids is likely; and residents and workers in correctional settings.[14,15]

Hepatitis C virus (HCV) is transmitted similarly to HBV and HDV. Risk factors are also very similar with the addition of people who have received blood transfusions or organ transplants, including anterior cruciate ligament (ACL) reconstruction allograft.[16,17] There has been growing concern worldwide about the risk of occupational transmission of HCV.

Occupational exposure to HCV accounts for approximately 4% of new infections. On average, the chance of acquiring HCV after a needle-stick injury involving an infected patient is 1.8% (range, 0% to 7%). Reports of HCV transmission from health care workers to patients are extremely uncommon.[14]

Hepatitis G virus (HGV) designation has been applied to a virus that is percutaneously transmitted and associated with bloodborne viral presence lasting approximately 10 years. HGV has been detected primarily in IV drug users, clients on hemodialysis, clients with hemophilia, and in a small percentage of blood donors. It does not appear to cause important liver disease or affect the response rate of those with chronic HBV or HCV to antiviral therapy.[18]

Hepatitis affects people in three stages: the initial or preicteric stage, the icteric or jaundiced stage, and the recovery period (Table 9-3). During the *initial* or *preicteric stage,* which lasts for 1 to 3 weeks, the person experiences vague gastrointestinal (GI) and general body symptoms. Fatigue, malaise, lassitude, weight loss, and anorexia are common.

Many people develop an aversion to food, alcohol, and cigarette smoke. Nausea, vomiting, diarrhea, arthralgias, and influenza-like symptoms may occur. There is a strong association between hepatitis-induced arthralgia and age with increasing incidence of joint involvement with increased age; arthralgia in children is much less common.[19,20] The liver becomes enlarged and tender (see Fig. 4-51), and intermittent itching (pruritus) may develop. From 1 to 14 days before the icteric stage, the urine darkens and the stool lightens as less bilirubin is conjugated and excreted.

The *icteric (acute) stage* is characterized by the appearance of jaundice, which peaks in 1 to 2 weeks and can persist for 6 to 8 weeks. During this stage, the acuteness of the inflammation subsides. The GI symptoms begin to disappear, and after 1 to 2 weeks of jaundice the liver decreases in size and becomes less tender. During the icteric stage the post-cervical lymph nodes and spleen are enlarged (see Fig. 4-53). Persons who have been treated with human immune serum globulin (ISG) may not develop jaundice.

The *recovery stage* lasts for 3 to 4 months, during which time the person generally feels well but fatigues easily.

People with mild-to-moderate acute hepatitis rarely require hospitalization. The emphasis is on preventing the spread of infectious agents and avoiding further liver damage

TABLE 9-2 Comparison of Major Types of Viral Hepatitis

Factor	Hepatitis A	Hepatitis B	Hepatitis C	Hepatitis D (Delta Agent)	Hepatitis E
Incidence	Endemic in areas of poor sanitation; reduced incidence with vaccine; occurs most often in fall and early winter	Reduced incidence with vaccine but present worldwide, especially in injection drug users, men who have sex with men, people exposed to blood products; occurs all year	Transfusion-related cases decreasing with blood screening; increased risk in those working around blood and blood products; occurs all year	Causes hepatitis only in association with hepatitis B and only in presence of HBsAg; endemic in Mediterranean area	Parts of Asia, Africa, and Mexico, where sanitation is poor
Incubation period	2-6 weeks	6 weeks-6 months	6-7 weeks	Same as hepatitis B	2-9 weeks
Risk factors	Close personal contact or by handling feces-contaminated food or water	Health care workers in contact with body secretions, blood, and blood products; hemodialysis and posttransfusion clients; men who have sex with men and injection drug users; morticians; those receiving tattoos; workers, residents of correctional settings	Similar to hepatitis B; healthcare workers in contact with blood and body fluids; blood transfusion recipients	Same as hepatitis B	Traveling or living in areas where incidence is high
Transmission	Infected feces, fecal-oral route*; shellfish from contaminated water; also rarely parenteral; no carrier state	Parenteral, sexual contact, and fecal-oral route; carrier state	Contact with blood and body fluids; source of infection uncertain in many clients; carrier state	Coinfects with hepatitis B, close personal contact; carrier state	Fecal-oral route, food-borne or waterborne; no carrier state
Severity	Mortality low; rarely causes fulminating hepatic failure	More serious; may be fatal; mortality rate is up to 60%	Can lead to chronic hepatitis	Similar to hepatitis B; more severe if occurs with chronic active hepatitis B	Illness self-limiting; mortality rate in pregnant women is 10% to 20%
Prophylaxis and active or passive immunity	Hygiene; vaccine available; immune globulin	Hygiene; avoidance of risk factors; immune globulin (passive); hepatitis B vaccine (active); treatment with antiviral agents	Hygiene; immune globulin (passive); treatment with interferon alfacon-1 (Infergen) or pegylated interferons (peginterferon alpha-2a) and Ribavirin (viral inhibitor)	Hygiene; hepatitis B vaccine (active) Interferon alpha-2b can inhibit HDV but effect ends when therapy ends	Hygiene; sanitation; recombinant vaccine

From Centers for Disease Control and Prevention (CDC): Division of Viral Hepatitis. Viral Hepatitis for Health Professionals. Available online at www.cdc.gov/hepatitis. Accessed Oct. 18, 2010.

HBsAg, Hepatitis B surface antigen; *HDV*, hepatitis D virus.

*The oral-fecal route of transmission is primarily from poor or improper handwashing and personal hygiene, particularly after using the bathroom and then handling food for public consumption. This route of transmission may also occur through shared use of razors and oral utensils such as straws, silverware, and toothbrushes.

when the underlying cause is drug-induced or toxic hepatitis. People with fulminant (sudden and severe or prolonged) hepatitis require special management because of the rapid progression of their disease and the potential need for urgent liver transplantation.

An entire spectrum of rheumatic diseases can occur concomitantly with HBV and HBC, including transient arthralgias, vasculitis, polyarteritis nodosa, rheumatoid arthritis (RA), fibromyalgia, lymphoma, Sjögren's syndrome, and persistent synovitis. Some conditions, such as RA and fibromyalgia, occur only in association with HCV, whereas others, such as polyarteritis nodosa, are observed in association with both forms of hepatitis.[20-22]

Rheumatic manifestations of hepatitis are varied early in the course of disease and can be indistinguishable from mild RA. The therapist should be suspicious of anyone with risk factors for hepatitis, including injection drug use; previous blood transfusion, especially before 1991; hemodialysis; or

TABLE 9-3 Four Phases or Stages of Hepatitis

Phase	Symptoms
Incubation/preclinical period (10-50 days)	Asymptomatic
Prodromal/preicteric (1-3 weeks)	Dark urine Light stools Vague GI symptoms Constitutional symptoms • Fatigue • Malaise • Weight loss • Anorexia • Nausea/vomiting • Diarrhea Aversion to food, alcohol, cigarette smoke Enlarged and tender liver Intermittent pruritus (itching) Arthralgias
Icteric (2-4 weeks)	Jaundice GI symptoms subside Liver decreases in size and tenderness Enlarged spleen Enlarged post cervical lymph nodes
Recovery/convalescence (3-4 months)	Easily fatigued

Modified from Goodman CC, Fuller KS: *Pathology: implications for the physical therapist,* ed 3, Philadelphia, 2009, WB Saunders.
Source: World Health Organization. Global alert and response to hepatitis. Available online at www.who.int. Accessed February 1, 2011.

BOX 9-2 RISK FACTORS FOR HEPATITIS

- Injection drug use
- Acupuncture
- Tattoo inscription or removal
- Ear or body piercing
- Recent operative procedure
- Liver transplant recipient
- Blood or plasma transfusion before 1991
- Hemodialysis
- Health care worker exposed to blood products or body fluids
- Exposure to certain chemicals or medications
- Unprotected homosexual/bisexual activity
- Severe alcoholism
- Travel to high risk areas
- Consumption of raw shellfish

other exposure to blood products/body fluids, such as a health care worker (Box 9-2), or a past history of hepatitis that currently appears with arthralgias (Case Example 9-2).

Other red flag symptoms include joint or muscle pain that is disproportionate to the physical findings and the presence of palmar tendinitis in someone with RA and positive risk factors for hepatitis.

CLINICAL SIGNS AND SYMPTOMS

Hepatitis A

Hepatitis A is often acquired in childhood as a mild infection with symptoms similar to the "flu" and may be misdiagnosed or ignored. It does not usually cause lasting damage to the liver, although the following symptoms may persist for weeks:

- Extreme fatigue
- Anorexia
- Fever
- Arthralgias and myalgias (generalized aching)
- Right upper abdominal pain
- Clay-colored stools
- Dark urine
- Icterus (jaundice)
- Headache
- Pharyngitis
- Alterations in the senses of taste and smell
- Loss of desire to smoke cigarettes or drink alcohol
- Low-grade fever
- Indigestion (varying degrees of nausea, heartburn, flatulence)

CLINICAL SIGNS AND SYMPTOMS

Hepatitis B

Hepatitis B may be asymptomatic but can include

- Jaundice (changes in skin and eye color)
- Arthralgias
- Rash (over entire body)
- Dark urine
- Anorexia, nausea
- Painful abdominal bloating
- Fever

Chronic Hepatitis. *Chronic hepatitis* is the term used to describe an illness associated with prolonged inflammation of the liver after unresolved viral hepatitis or associated with *chronic active hepatitis* (CAH) of unknown cause. *Chronic* is defined as inflammation of the liver for 6 months or more. The symptoms and biochemical abnormalities may continue for months or years. It is divided into CAH and *chronic persistent hepatitis* (CPH) by findings on liver biopsy.

Chronic Active Hepatitis. This type of hepatitis refers to seriously destructive liver disease that can result in cirrhosis. CAH is often a result of viral infection (HBV, HCV, and HDV), but it can also be secondary to drug sensitivity (e.g., methyldopa [Aldomet], an antihypertensive medication, and isoniazid [INH], an antitubercular drug).

Steroid therapy is sometimes recommended for clients with evidence of aggressive liver inflammation and necrosis (identified by liver biopsy) as a result of these drugs. If CAH

CASE EXAMPLE 9-2

Hepatitis C

A 43-year-old man, 1 year following traumatic injury to the right forearm, underwent surgery to transplant his great toe to function as a thumb. The surgery took place in another state, and the man, who had been a client in our facility before surgery, returned for postoperative rehabilitation.

Complaints of hives of the involved forearm, fatigue, depression, and increased perspiration were documented but attributed by his physician to recovery from the traumatic injury and the multiple operations. Medical records from the hospital consisted of therapy notes only.

Eventually, the client developed a yellowing of the sclerae (white outer coat of the eyeballs). Medical referral was requested, and the client was evaluated by an internal medicine specialist.

Hepatitis C was diagnosed, and full medical records then obtained revealed that although the man had donated his own blood in advance for the surgery, he was short by one unit of blood, which he received through a blood bank. The blood donation was attributed as the probable source of contamination.

Continued physical therapy intervention was modified to accommodate liver impairment with particular attention paid to activity level. The therapist also observed the client carefully for signs of fluid shift such as weight gain and orthostasis; dehydration; pneumonia; and vascular problems.

CLINICAL SIGNS AND SYMPTOMS

Chronic Active Hepatitis

The clinical signs and symptoms of chronic active hepatitis may range from asymptomatic to the person who is bedridden with cirrhosis and advanced hepatocellular failure. In the latter the prominent signs and symptoms may reflect multisystem involvement, including

- Fatigue
- Jaundice
- Abdominal pain
- Anorexia
- Arthralgia
- Fever
- Splenomegaly and hepatomegaly
- Weakness
- Ascites (see Fig. 9-8)
- Hepatic encephalopathy

CLINICAL SIGNS AND SYMPTOMS

Chronic Persistent Hepatitis

- Right upper quadrant (RUQ) pain
- Anorexia
- Mild fatigue
- Malaise

is left untreated, its course is unpredictable and may range from progressive deterioration of liver function to spontaneous remissions and exacerbations.

Steroids may be used to treat CAH. They are usually prescribed for a period of 3 to 5 years. In addition, recombinant interferon-alpha-2b injections in low doses over a 6-month period have been shown to improve hepatic function in persons with CAH. Treatment of HCV is relatively new and consists of the use of interferons (IFNs), a protein naturally occurring in the healthy body in response to infection such as the hepatitis virus.

Conventional IFN has been used for many years in the treatment of chronic HCV in clients who persistently maintain HCV/RNA blood levels. Combining IFNs with the drug ribavirin has resulted in better control of chronic HCV in some individuals, but the treatment is not well tolerated because of side effects from the ribavirin.[23,24]

Pegylated IFNs, such as Pegasys (peginterferon alpha-2a), are improved forms of IFNs that allow a decrease in dosage and offer improved efficacy. Peginterferons (PEGs) in combination with ribavirin are now considered the standard treatment for chronic HCV infection. These new PEG interferons do not eliminate the known side effects associated with classical interferon treatment (e.g., fatigue, headache, myalgia, fever, anxiety, irritability, GI upset).[19,25]

Metabolic Disease. The most common metabolic diseases that can cause chronic hepatitis and are of interest to a physical therapist are Wilson's disease and hematochromatosis, also termed *hemochromatosis.* Both of these diseases are dealt with in greater detail as metabolic disorders in Chapter 11.

Wilson's disease is an autosomal recessive disorder in which biliary excretion of copper is impaired, and as a consequence, total body copper is progressively increased. There may be mild-to-severe neurologic dysfunction, depending on the rate of hepatocyte injury.

Hemochromatosis is the most common genetic disorder (autosomal recessive defect in iron absorption) causing liver failure. Excessive iron is stored in various parenchymal organs with subsequent development of fibrosis. Arthralgias and arthropathy may develop and are often confused with RA or osteoarthritis. The second and third metacarpophalangeal joints are usually involved first. Knees, hips, shoulders, and lower back may be affected. Acute synovitis with pseudogout of the knees has been observed.

Nonviral Hepatitis. Nonviral hepatitis is considered to be a toxic or drug-induced form of liver inflammation. This type of hepatitis occurs secondary to exposure to alcohol, certain chemicals, or drugs such as antiinflammatories, anticonvulsants, antibiotics, cytotoxic drugs for the treatment of cancer, antituberculars, radiographic contrast agents for diagnostic testing, antipsychotics, and antidepressants (Box 9-3).

BOX 9-3 COMMON HEPATOTOXIC AGENTS

Analgesics
Acetaminophen
Aspirin
Diclofenac

Anesthetics
Halothane
Enflurane
Methoxyflurane
Chloroform

Anticonvulsants
Valproic acid
Phenytoin
Carbamazepine
Lamotrigine

Antidepressants/antipsychotics
Monamine oxidase (MAO) inhibitors
Chlorpromazine and other phenothiazines

Antineoplastics
Methotrexate (related to cumulative dose)
Mercaptopurine
L-asparaginase
Carmustine, lomustine
Streptozocin

Antimicrobials
Chloramphenicol
Isoniazid (antitubercular)
Oxacillin
Erythromycin estolate
Novobiocin
Ketoconazole (antifungal)
Nitrofurantoin
Sulfonamides (class)
Minocycline
Tetracyclines (class)
Efavirenz (antiviral)
Nevirapine (antiviral)
Ritonavir (antiviral)

Cardiovascular
Quinidine sulfate
Amiodarone
Methyldopa

Hormonal
Oral contraceptives
Anabolic steroids
Oral hypoglycemics

Recreational Drugs
Alcohol
Cocaine
Ecstasy

Vitamins
Vitamin A (large doses)
Niacin (large doses)

Other
Carbon tetrachloride
Poisonous mushrooms
Heavy metals
Phosphorus
Tannic acid
Propylthiouracil
Diagnostic contrast agents

Acetaminophen, the popular over-the-counter (OTC) pain reliever, has been found to be the leading cause of sudden liver failure in adults in the United States.[26] The drug is safe when taken properly, but even a small overdose in some people can trigger sudden liver failure. The use of this drug becomes even more dangerous with taken by individuals with an already impaired liver.[27]

The mechanism by which these agents induce overt injury may be dose-related and predictable or idiosyncratic and unpredictable, with the latter caused by an unusual susceptibility of the individual. Some drugs (e.g., oral contraceptives) may impair liver function and produce jaundice without causing necrosis, fatty infiltration of liver cells, or a hypersensitivity reaction.

CLINICAL SIGNS AND SYMPTOMS

Toxic and Drug-Induced Hepatitis

These vary with the severity of liver damage and the causative agent. In most individuals symptoms resemble those of acute viral hepatitis:

- Anorexia, nausea, vomiting
- Fatigue and malaise
- Jaundice
- Dark urine
- Clay-colored stools
- Headache, dizziness, drowsiness (carbon tetrachloride poisoning)
- Fever, rash, arthralgias, epigastric or right upper quadrant (RUQ) pain (halothane anesthetic)

Cirrhosis

Cirrhosis is a chronic hepatic disease characterized by the destruction of liver cells and by the replacement of connective tissue by fibrous bands. As the liver becomes more and more scarred (fibrosed), blood and lymph flow become impaired, causing hepatic insufficiency and increased clinical manifestations. The causes of cirrhosis can be varied, although alcohol abuse is the most common cause of liver disease in the United States.

In addition, about 25% of Americans have a problem called nonalcoholic fatty liver disease (NAFLD), defined as fatty infiltration of the liver exceeding 5% to 10% by weight. NAFLD is an illness closely associated with diabetes, obesity, and insulin resistance or metabolic syndrome.[28,29] It is an independent risk factor for cardiovascular disease[29] and may make liver damage caused by other agents (e.g., alcohol, industrial toxins, hepatotrophic viruses) worse.[30,31]

Ten percent to 20% of people with NAFLD will develop liver inflammation leading to liver scarring and cirrhosis.[32] Prevention and treatment of diabetes, obesity, and insulin resistance and protection of the liver from medications that cause fatty infiltration and toxins can help to limit the course of this disease.[31]

The activity level of the client with damage from chronic liver impairment is determined by the symptoms. Because hepatic blood flow diminishes with moderate exercise, rest periods are advised and are adjusted according to the level of fatigue experienced by the client both during the exercise and afterward at home.

The person may return to work with medical approval but is advised to avoid straining, such as lifting heavy objects if portal hypertension and esophageal varices are a problem. Because stress decreases hepatic blood flow, any reduction of stress at home, at work, or during treatment is therapeutic.

CLINICAL SIGNS AND SYMPTOMS

Cirrhosis

- Mild right upper quadrant (RUQ) pain (progressive)
- GI symptoms
- Anorexia
- Indigestion
- Weight loss
- Nausea and vomiting
- Diarrhea or constipation
- Dull abdominal ache
- Ease of fatigue (with mild exertion)
- Weakness
- Fever

Progression of Cirrhosis. As the cirrhosis progresses and hepatic insufficiency develops, a series of conditions emerges, including portal hypertension, ascites, and esophageal varices. Late symptoms affecting the entire body develop (Table 9-4).

Portal hypertension is elevated pressure in the portal vein (through which blood passes from the GI tract and spleen to the liver), occurring as portal blood meets increased resistance to flow in the fibrotic liver. The blood then backs up into esophageal, stomach, and splenic structures and bypasses the liver through collateral vessels.

CLINICAL SIGNS AND SYMPTOMS

Portal Hypertension

- Ascites (see Fig. 9-8)
- Dilated collateral veins
- Esophageal varices (upper GI)
- Hemorrhoids (lower GI)
- Splenomegaly (enlargement of the spleen)
- Thrombocytopenia (decreased number of blood platelets for clotting)

Ascites is an abnormal accumulation of fluid containing large amounts of protein and electrolytes in the peritoneal cavity as a result of portal backup and loss of proteins that presents as a distended abdomen, bulging flanks, and a protruding, displaced umbilicus (see Fig. 9-8). It is the result of free fluid in the peritoneal cavity. For the physical therapist, abdominal hernias and lumbar lordosis observed in clients with ascites may present symptoms that mimic musculoskeletal involvement, such as groin or low-back pain (Case Example 9-3).

Esophageal varices are dilated veins of the lower esophagus that occur as a result of portal vein blood backup. These varices are thin-walled and can rupture, causing severe hemorrhage and sometimes death.

TABLE 9-4 Clinical Manifestations of Cirrhosis

Body System	Clinical Manifestations
Respiratory	Limited thoracic expansion (caused by ascites) Hypoxia • Dyspnea • Cyanosis • Clubbing
Central nervous system (progressive to hepatic coma)	Subtle changes in mental acuity (progressive) Mild memory loss Poor reasoning ability Irritability Paranoia and hallucinations Slurred speech Asterixis (tremor of outstretched hands) Peripheral neuritis Peripheral muscle atrophy
Hematologic	Impaired coagulation/bleeding tendencies • Nosebleeds • Easy bruising • Bleeding gums Anemia (usually caused by GI blood loss from esophageal varices)
Endocrine (caused by liver's inability to metabolize hormones)	Testicular atrophy Menstrual irregularities Gynecomastia (excessive development of breasts in men) Loss of chest and axillary hair
Integument (cutaneous and skin)	Severe pruritus (itching) Extreme dryness Poor tissue turgor Abnormal pigmentation Prominent spider angiomas Palmar erythema
Hepatic	Hepatomegaly (enlargement of the liver) Ascites Edema of the legs Hepatic encephalopathy (see Table 9-5)
Gastrointestinal (GI)	Anorexia Nausea Vomiting Diarrhea

CLINICAL SIGNS AND SYMPTOMS

Hemorrhage Associated with Esophageal Varices

- Restlessness
- Pallor
- Tachycardia
- Cooling of the skin
- Hypotension

Hepatic Encephalopathy (Hepatic Coma)

Hepatic coma is a neurologic disorder resulting from the inability of the liver to detoxify ammonia (produced from protein breakdown) in the intestine. Increased serum levels of ammonia are directly toxic to central and peripheral

CASE EXAMPLE 9-3

Ascites

A 69-year-old man was seen at the Veteran's Administration (VA) Hospital outpatient physical therapy department following a left total hip replacement (THR) 2 weeks ago. The surgery was performed at a civilian hospital, but all his follow-up care is through the VA. He had a long history of alcohol and tobacco use and medical intervention for heart disease, hypertension, and peripheral vascular disease.

The medical problem list (established by the physician) included
Liver cirrhosis secondary to alcoholism
Ascites secondary to portal hypertension
Coronary artery disease with hypertension
Peripheral vascular disease (arterial)
Mild vision loss secondary to macular degeneration

The client was referred to physical therapy for rehabilitation following his THR. During the examination, the client reported various other musculoskeletal aches and pains, including chronic low back pain present off and on for the last 6 months and new onset of groin pain on the left side (just since the THR).

Ascites can be a cause of low back and/or groin pain. How do you screen this client for a medical (vascular, liver) cause of the groin pain?

Past Medical History

Past history of cancer of any kind
Past history of abdominal or inguinal hernia

Clinical Presentation

Ask additional questions about pain pattern as discussed in Chapter 3.
What do you think is causing your groin pain?
Watch for red flag for possible vascular involvement: Client describes pain as "throbbing."
Pain is worse 5 to 10 minutes after the start of activity involving the lower extremities and relieved by rest (intermittent claudication).
Visual inspection and palpation, including observing for postural components (e.g., lumbar lordosis associated with ascites) as a contributing factor; abdominal or inguinal hernia; liver palpation; and lymph node palpation.
Perform stretching and resistive movements to eliminate, reproduce, or aggravate symptoms; you may be limited in this assessment area because of THR precautions.
Red flag: pain is not altered by stretching or resistive movement; pain cannot be reproduced with palpation.
Assess for trigger points (e.g., adductor magnus), keeping in mind that common systemic perpetuating factors with myofascial pain include anemia and hypothyroidism, as well as vitamin deficiency common with chronic alcohol use. Further screening may require assessing for risk factors and associated signs and symptoms for each of these conditions.

Associated Signs and Symptoms

Ask the client about any other symptoms of any kind that may have developed just before or around the time of the onset of groin pain; offer some suggestions from the Overview section that appears later in this chapter.
As mentioned above, the therapist may have to ask about the presence of signs and symptoms associated with anemia and endocrine disease.

Should you send this client back to the doctor before continuing with physical therapy intervention?

It is very likely that this client will require referral to his physician. Your referral decision will be dependent on your findings, of course. For example, the presence of trigger points may warrant treatment first and reassessment for change in clinical presentation before making a final decision. Given the movement precautions for THR, positional release or stretch positions for trigger points may be contraindicated. You may have to use alternate methods of trigger point release.

Remember true hip pain is often felt in the groin or deep buttock. There could be a problem with the hip implant (e.g., fracture, infection, loosening) causing the groin pain. There will be pain with active or passive motion of the hip joint. The pain increases with weight bearing.[45] If the physician does not know about this new groin pain, medical referral to reevaluate the implant is needed before continuing with a THR rehab protocol.

By continuing the screening process, the therapist can provide the physician with additional information to describe the problem. Communication is an important key element in the referral process. Provide the physician with a *brief* summary of your findings including a list of any unusual findings (see further discussion regarding physician in Chapter 1).

nervous system function, causing an array of neurologic symptoms. Flapping tremors (asterixis) and numbness/tingling (misinterpreted as carpal/tarsal tunnel syndrome) are common symptoms of this ammonia abnormality.

Clinical Signs and Symptoms. Clinical manifestations of hepatic encephalopathy vary, depending on the severity of neurologic involvement, and develop in four stages as the ammonia level increases in the serum. The accompanying clinical features are presented in Table 9-5.

For the physical therapist, the inpatient with impending hepatic coma has difficulty in ambulating and is unsteady. Protection from falling and seizure precautions must be taken. Skin breakdown in a client who is malnourished because of liver disease, immobile, jaundiced, and edematous can occur in less than 24 hours. Careful attention to skin care, passive exercise, and frequent changes in position are required.

Newborn Jaundice

Jaundice affects approximately 60% of newborn infants because liver function is somewhat slow to develop in the first few days of life.[33] In a small percentage of infants, extreme jaundice can occur and if left untreated for too long can result in brain damage from toxic levels of bilirubin in the blood. It is critically important for all newborns to be screened for the development of this condition. Development of any color

TABLE 9-5 Stages of Hepatic Encephalopathy

Stage	Symptoms
Stage I (prodromal stage)	Subtle symptoms may be overlooked Slight personality changes: • Disorientation • Confusion • Euphoria or depression • Forgetfulness • Slurred speech
Stage II (impending stage)	Tremor progresses to asterixis (liver flap) Resistance to passive movement (increased muscle tone) Lethargy Aberrant behavior Apraxia* Ataxia Facial grimacing and blinking
Stage III (stuporous stage)	Client can still be aroused Hyperventilation Marked confusion Abusive and violent Noisy, incoherent speech Asterixis (liver flap) Muscle rigidity Positive Babinski† reflex Hyperactive deep tendon reflexes
Stage IV (comatose stage)	Client cannot be aroused; responds only to painful stimuli No asterixis Positive Babinski reflex Hepatic fetor (musty, sweet odor to the breath caused by the liver's inability to metabolize the amino acid methionine)

*This type of motor apraxia can be best observed by keeping a record of the client's handwriting and drawings of simple shapes such as a circle, square, triangle, rectangle. Check for progressive deterioration.
†A reflex action of the toes that is normal during infancy but abnormal after 12 to 18 months. It is elicited by a firm stimulus (usually scraping with the handle of a reflex hammer) on the sole of the foot from the heel along the lateral border of the sole to the little toe, across the ball of the foot to the big toe. Normally, such a stimulus causes all the toes to flex downward. A positive Babinski reflex occurs when the great toe flexes upward and the smaller toes fan outward.

change in newborns needs immediate referral and testing for abnormal bilirubin levels.[34,35]

Liver Abscess

Liver abscess is relatively rare but occurs when bacteria (e.g., *Escherichia coli*), fungi (e.g., *Candida*), or protozoa (ameba) destroy hepatic tissue and produce a cavity that fills with infectious organisms, liquefied liver cells, and leukocytes. Necrotic tissue then isolates the cavity from the rest of the liver. Biliary tract disease (obstruction of bile flow allows for bacterial invasion) is the most frequent cause of liver abscess. Less common causes include appendicitis, immune disorders, and infections in other organs.[36]

The development of new radiologic techniques, the improvement in microbiologic identification, and the advancement of medical treatment, as well as improved supportive care, have decreased mortality rates from 30% to 50% down to 5% to 30%; yet the prevalence of liver abscess has remained relatively unchanged. Untreated, this infection remains uniformly fatal.[36,37]

CLINICAL SIGNS AND SYMPTOMS

Liver Abscess

Clinical signs and symptoms of liver abscess depend on the degree of involvement; some people are acutely ill, others are asymptomatic. Depending on the type of abscess, the onset may be sudden or insidious. The most common signs include

- Right abdominal pain
- Right shoulder pain
- Weight loss
- Fever, chills, malaise
- Diaphoresis
- Nausea and vomiting, anorexia
- Anemia
- Tender hepatomegaly (with or without a palpable mass)
- Jaundice

Liver Cancer

Metastatic tumors to the liver occur 20 times more often than primary liver tumors. The liver filters blood coming from the GI tract, making it a primary metastatic site for tumors of the stomach, colorectum, and pancreas. It is also a common site for metastases from other primary cancers such as esophagus, lung, and breast.

Primary liver tumors (hepatocellular carcinoma [HCC]) are often associated with cirrhosis but can be linked to other predisposing factors such as fungal infection (common in moldy foods of Africa), viral hepatitis, excessive use of anabolic steroids, trauma, nutritional deficiencies, and exposure to hepatotoxins.

Cholangiocarcinoma (CCC), a serious and often fatal form of liver cancer, is the second most common form of hepatic malignancy. CCC originates from the epithelium of the bile ducts and has many of the same risk factors as HCC, but preexisting biliary disease is the primary risk factor.[38]

Several types of benign and malignant hepatic neoplasms can result from the administration of chemical agents. For example, adenoma (a benign tumor) can occur in recipients of oral contraceptives. Regression of the tumor occurs after withdrawal of the drug.

In most instances, interference with liver function does not occur until approximately 80% to 90% of the liver is replaced by metastatic carcinoma or primary carcinoma. Signs of liver impairment are often late in the presentation, making early detection and successful treatment less likely. The alert physical therapist may be the first to identify liver involvement when the neuromuscular or musculoskeletal systems are affected.

CLINICAL SIGNS AND SYMPTOMS

Liver Neoplasm

If clinical signs and symptoms of liver neoplasm do occur (whether of primary or metastatic origin), these may include

- Jaundice (icterus)
- Progressive failure of health
- Anorexia and weight loss
- Overall muscular weakness
- Epigastric fullness and pain or discomfort
- Constant ache in the epigastrium or mid-back
- Early satiety (cystic tumors)

BOX 9-4 RISK FACTORS FOR GALLSTONES

- Age: Incidence increases with age
- Sex: Women are affected more than men before age 60
- Elevated estrogen levels
  - Pregnancy
  - Oral contraceptives
  - Hormone therapy
  - Multiparity (woman who has had two or more pregnancies resulting in viable offspring)
- Obesity
- Diet: High cholesterol, low fiber
- Diabetes mellitus
- Liver disease
- Rapid weight loss or fasting
- Taking cholesterol-lowering drugs (statins)
- Ethnicity (stronger genetic predisposition in Native Americans, Mexican Americans)
- Genetics (family history of gallstones)

GALLBLADDER AND DUCT DISEASES

Cholelithiasis

Gallstones are stonelike masses called *calculi* (singular: calculus) that form in the gallbladder, possibly as a result of changes in the normal components of bile. Although there are two types of stones, pigment and cholesterol stones, most types of gallstone disease in the United States, Europe, and Africa are associated with cholesterol stones.

Cholelithiasis, the presence or formation of gallstones, can be asymptomatic, detected incidentally during medical imaging. Problems arise if a stone leaves the gallbladder and causes obstruction somewhere else in the biliary system, presenting as biliary colic, cholecystitis, or cholangitis.

Cholelithiasis is the fifth leading cause of hospitalization among adults and accounts for 90% of all gallbladder and duct diseases. The incidence of gallstones increases with age, occurring in more than 40% of people older than 70. See Box 9-4 for risk factors to watch for in a client's history that correlate with the incidence of gallstones.

Clients with gallstones may be asymptomatic or may have symptoms of a gallbladder attack described in the next section. The prognosis is usually good with medical treatment, depending on the severity of disease, presence of infection, and response to antibiotics.

Biliary Colic

With biliary colic, the stone gets lodged in the neck of the gallbladder (cystic duct). Pain results as the gallbladder contracts and tries to push the stone through. The classic symptom of this problem is right upper abdominal pain that comes and goes in waves. The pain builds to a peak and then fades away.

Obstructions of the gallbladder can result in biliary stasis, delayed gallbladder emptying, and subsequent mixed stone formation. Stasis and delayed gallbladder emptying can occur with any pathologic conditions of the liver, hormonal influences, and pregnancy (usually third trimester when the developing fetus compresses the mother's gallbladder up against the liver).

Cholecystitis

Cholecystitis, the blockage or impaction of gallstones in the cystic duct (Fig. 9-9), leads to infection or inflammation of the gallbladder. This condition may be acute or chronic, causing painful distention of the gallbladder. The affected individual may feel steady, severe pain that increases rapidly, lasting several minutes to several hours. Nausea, vomiting, and fever may be present.

Other causes of acute cholecystitis may be typhoid fever or a malignant tumor obstructing the biliary tract. Whatever the cause of the obstruction, the normal flow of bile is interrupted and the gallbladder becomes distended and ischemic.

Gallstones may also cause chronic cholecystitis (persistent gallbladder inflammation), in which the gallbladder atrophies and becomes fibrotic, adhering to adjacent organs. It is not unusual for affected clients to have repeated episodes before seeking medical attention.

Cholangitis

Gallstones lodged further down in the system in the common bile duct can cause cholangitis. Blocking the flow of bile at this point in the biliary tree can lead to jaundice. Infection can develop here and travel up to the liver, becoming a potentially life-threatening situation.

Clinical Signs and Symptoms

The typical pain of gallbladder disease has been described as colicky pain that occurs in the right upper quadrant (RUQ) of the abdomen after the person has eaten a meal that is high in fat (although food that provokes an attack of pain does not need to be "fatty"). However, the pain is not necessarily

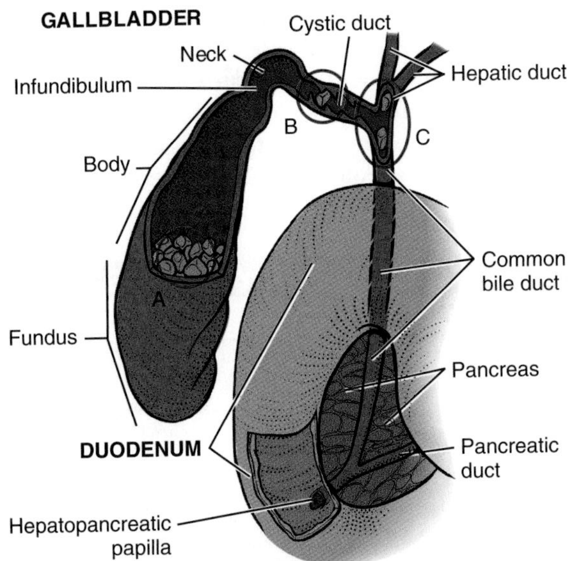

Fig. 9-9 The gallbladder and its divisions: fundus, body, infundibulum, and neck. **A,** Cholelithiasis, the presence or formation of gallstones, can be asymptomatic, detected incidentally during medical imaging. Problems arise if a stone leaves the gallbladder and causes obstruction somewhere else in the biliary system. **B,** If a gallstone enters the cystic duct and becomes lodged there, it can lead to cholecystitis (inflammation of the gallbladder). **C,** Obstruction of either the hepatic or common bile duct by stone or spasm blocks the exit of bile from the liver where it is formed. Jaundice is often the first symptom. If an infection develops and backs up into the liver, a condition called *cholangitis* can occur, a potentially life-threatening problem.

CASE EXAMPLE 9-4

Gallbladder Pain

A 48-year-old schoolteacher was admitted to the hospital following an episode of intense, sharp pain that started in the epigastric region and radiated around her thorax to the interscapular area. Her gallbladder had been removed 2 years ago, but she remarked that her current symptoms were "exactly like a gallbladder attack." The client was referred to physical therapy for "back care/education" on the day of discharge.

On examination, the client was in acute distress, unable to tolerate a full examination. She had not been able to transfer or ambulate independently. She was instructed in relaxation and breathing techniques to reduce her extreme level of anxiety associated with pain and given supportive reassurance. Instruction and assistance were provided in all transfers to minimize pain and maximize independent function. Given her discharge status, outpatient physical therapy was recommended for follow-up intervention.

She returned to physical therapy as planned and was provided with a back care program. She was also treated locally for scar tissue adhesion at the site of the gallbladder removal. Symptomatic relief was obtained in the first two sessions without recurrence of symptoms.

This case example is included to demonstrate how scar tissue associated with organ removal can reproduce visceral symptoms that are actually of musculoskeletal origin—the opposite concept of what is presented in this text. This may be more of an example of cellular memories sustaining a viscero-somatic reflex via the action of neuropeptides at the cellular level (see discussion of Psychoneuroimmunology in Chapter 3).

limited to the RUQ, and more likely than not, it is constant, not colicky.

Like the stomach, pylorus, and duodenum, the liver and gallbladder can cause spasm of the rectus abdominis muscles above the umbilicus. This occurs when disturbances within the hepatic and biliary systems as part of the overall GI system affect motor reflexes.

These disturbances can be reflected in muscular contractions of the spinal, abdominal, and other muscles supplied by the motor nerves from the anterior horn of the segment innervating the affected viscera.[39]

It looks just like a musculoskeletal problem, but the pain pattern is the result of viscero-somatic reflexes as discussed in Chapter 3 (Case Example 9-4). Ask about the timing of symptoms in relation to eating or drinking. Watch for symptoms that are worse immediately after eating (gallbladder inflammation) or pain and nausea 1 to 3 hours after eating (gallstones).

Muscle guarding and tenderness of the spinal musculature in the presence of constitutional symptoms (e.g., fever, sweats, chills, nausea) is another red flag. Ask about a previous history of GI, liver, or gallbladder problems, and review the client's risk factors for hepatic involvement.

In the case of gallbladder disease, it is also possible to get tender points in the soma corresponding to visceral innervation. A gallbladder problem can result in a sore 10th rib tip (right side anteriorly) when messages from the viscera entering the spinal cord at the same level as the innervation of the rib are misinterpreted as a somatic problem. The gallbladder has most of its innervation from the right side of the cervical ganglia to the splanchnic nerves, which explains the predominance of right-sided somatic symptoms.

When visceral and cutaneous fibers enter the spinal cord at the same level, the nervous system may respond with sudomotor changes, such as pruritus (itching of the skin) or a sore rib, instead of gallbladder symptoms. The clinical presentation appears as a biomechanical problem, such as a rib dysfunction, instead of nausea and food intolerances normally associated with gallbladder dysfunction.

Likewise, from our understanding of viscerogenic pain patterns based on embryologic development, we know that the visceral pericardium of the heart (see Fig. 6-5) is derived from the same embryologic tissue as the gallbladder. A gallbladder problem can also cause referred pain to the heart and must be ruled out by the physician as a possible cause of chest pain.

CLINICAL SIGNS AND SYMPTOMS

Acute Cholecystitis

- Chills, low-grade fever
- Jaundice
- GI symptoms
- Nausea
- Anorexia
- Vomiting
- Tenderness over the gallbladder
- Tenderness on the tip of the 10th rib (right side anteriorly); called a "hot rib"; can also affect 11th and 12th ribs (right anterior)
- Severe pain in the right upper quadrant (RUQ) and epigastrium (increases on inspiration and movement)
- Pain radiating into the right shoulder and between the scapulae

CLINICAL SIGNS AND SYMPTOMS

Chronic Cholecystitis

These may be vague or a sense of indigestion and abdominal discomfort after eating, unless a stone leaves the gallbladder and causes obstruction of the common duct (called *choledocholithiasis*), causing

- Biliary colic: severe, steady pain for 3 to 4 hours in the right upper quadrant (RUQ)
- Pain: may radiate to the mid-back between the scapulae (caused by splanchnic fibers synapsing with phrenic nerve fibers)
- Nausea (intolerance of fatty foods; decreased bile production results in decreased fat digestion)
- Abdominal fullness
- Heartburn
- Excessive belching
- Constipation and diarrhea

Primary Biliary Cirrhosis

Primary biliary cirrhosis (PBC) is a chronic, progressive, autoimmune disease of the liver that involves primarily the intrahepatic bile ducts and results in impairment of bile secretion. The disease, which often affects middle-aged women (in particular Native American women), begins with pruritus or biochemical evidence of cholestasis and progresses at a variable rate to jaundice, portal hypertension, and liver failure.[40]

The cause of PBC is unknown, although various factors are being investigated. An autoimmune pathogenesis is widely accepted, based on the presence of autoantibodies and autoreactive T cells.[41] Many clients have associated autoimmune features, particularly Sjögren's syndrome, autoimmune thyroiditis, and renal tubular acidosis.[42] In more rare cases, clients may exhibit sensory peripheral neuropathies of the hands and feet.

Fatigue and pruritus are the most common symptoms of PBC, but the majority of people are asymptomatic at first presentation.[43] The most significant clinical problem for clients with PBC is bone disease characterized by impaired osteoblastic activity and accelerated osteoclastic activity. Calcium and vitamin D should be carefully monitored and appropriate replacement instituted. Physical activity following an osteoporosis protocol should be encouraged.

CLINICAL SIGNS AND SYMPTOMS

Primary Biliary Cirrhosis

- Pruritus
- Jaundice
- GI bleeding
- Ascites (see Fig. 9-8)
- Fatigue
- Right upper quadrant (RUQ) pain (posterior)
- Sensory neuropathy of hands/feet (rare)
- Osteoporosis (decreased bone mass)
- Osteomalacia (softening of the bones)
- Burning, pins and needles, prickling of the eyes
- Muscle cramping

Gallbladder Cancer

Gallbladder cancer, closely associated with gallstone disease, often has a poor outcome due to delay in diagnosis. The primary associated risk factors include cholelithiasis (especially symptomatic, untreated), obesity, reproductive abnormalities, chronic gallbladder infections, and exposure to radon and certain industrial exposures including cellulose acetate fiber manufacturing. Testing and treatment of symptomatic gallstones is the only preventive measure identified at this time for gallbladder cancer.[44]

PHYSICIAN REFERRAL

A careful history and close observation of the client are important in determining whether a person may need a medical referral for possible hepatic or biliary involvement. Any client with mid-back, scapular, or right shoulder pain (see Table 9-1) without a history of trauma (e.g., forceful movement of the spine, repetitive movements of the shoulder or back, or easy lifting) should be screened for possible systemic origin of symptoms.

For the physical therapist treating the inpatient population, jaundice in the postoperative individual is not uncommon but can be a potentially serious complication of surgery and anesthesia.

Clinical management of jaundice is complicated by anything capable of damaging the liver, including physical stress associated with physical therapy intervention. Hypoxemia, blood loss, infection, and administration of multiple drugs can add additional physical stress.

When making the referral, it is important to report to the physician the results of your objective findings, especially when there is a lack of physical evidence to support a musculoskeletal lesion. The Special Questions to Ask may assist in assessing the client's overall health status.

Guidelines to Immediate Physician Referral

- New onset of myopathy in any client, but especially the older adult, with a history of statin use (cholesterol-lowering drugs); look for other risk factors and other signs and symptoms of liver or renal impairment.

Guidelines to Physician Referral

- Obvious signs of hepatic disease, especially with a history of previous cancer or risk factors for hepatitis (see Box 9-2)
- Development of arthralgias of unknown cause in anyone with a previous history of hepatitis or risk factors for hepatitis
- Presence of bilateral carpal tunnel syndrome accompanied by bilateral tarsal tunnel syndrome unknown to the physician, asterixis, or other associated hepatic signs and symptoms
- Presence of sensory neuropathy of unknown cause accompanied by signs and symptoms associated with hepatic system impairment

Clues to Screening for Hepatic Disease

- Right shoulder/scapular and/or upper mid-back pain of unknown cause (see also Clues to Screening Shoulder Pain in Chapter 18).
- Shoulder motion is not limited by painful symptoms; client is unable to localize or pinpoint pain or tenderness.
- Presence of GI symptoms, especially if there is any correlation between eating and painful symptoms.
- Bilateral carpal/tarsal tunnel syndrome, especially of unknown origin; check for other signs of liver impairment such as liver flap, liver palms, and skin or nail bed changes (see Box 9-1).
- Personal history of cancer, liver, or gallbladder disease.
- Personal history of hepatitis, especially with joint pain associated with rheumatoid arthritis or fibromyalgia accompanied by palmar tendinitis.
- Recent history of statin use (cholesterol-lowering drugs such as Zocor, Lipitor, Crestor) or other hepatotoxic drugs.
- Recent operative procedure (possible postoperative jaundice).
- Recent (within last 6 months) injection drug use, tattoo (receiving or removal), acupuncture, ear or body piercing, dialysis, blood or plasma transfusion, active homosexual activity, heterosexual sexual activity with homosexuals, consumption of raw shellfish (hepatitis).
- Changes in skin (yellow hue, spider angiomas, palmar erythema) or eye color (jaundice).
- Employment or lifestyle involving alcohol consumption (jaundice).
- Contact with jaundiced persons (health care worker handling blood or body fluids, dialysis clients, injection drug users, active homosexual sexual activity, heterosexual sexual activity with homosexuals).

LIVER/BILIARY PAIN PATTERNS

LIVER PAIN (FIG. 9-10)

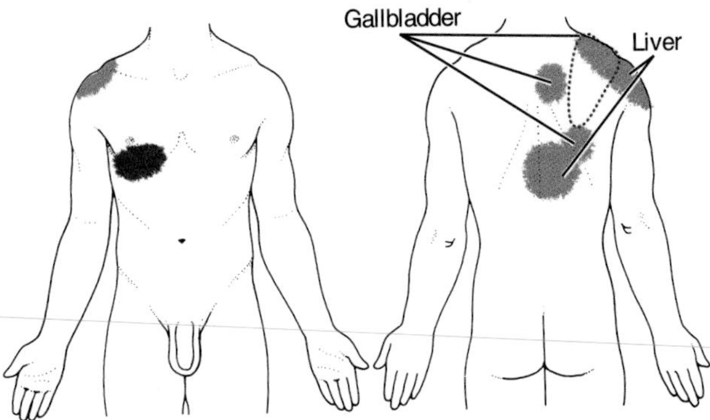

Fig. 9-10 The primary pain pattern from the liver, gallbladder, and common bile duct *(dark red)* presents typically in the midepigastrium or right upper quadrant of the abdomen. Innervation of the liver and biliary system is through the autonomic nervous system from T5 to T11 (see Fig. 3-3). Liver impairment is primarily reflected through the 9th thoracic distribution. Referred pain *(light red)* from the liver occurs in the thoracic spine from approximately T7 to T10 and/or to the right of midline, possibly affecting the right shoulder (right phrenic nerve). Referred pain from the gallbladder can affect the right shoulder by the same mechanism. The gallbladder can also refer pain to the right interscapular (T4 or T5 to T8) or right subscapular area.

Continued

LIVER/BILIARY PAIN PATTERNS—*cont'd*

Location: Pain in the midepigastrium or right upper quadrant (RUQ) of abdomen

Referral: Pain over the liver, especially after exercise (hepatitis)
RUQ pain may be associated with right shoulder pain
Both RUQ and epigastrium pain may be associated with back pain between the scapulae
Pain may be referred to the right side of the midline in the interscapular or subscapular area (T7-T10)

Description: Dull abdominal aching
Sense of fullness of the abdomen or epigastrium

Intensity: Mild at first, then increases steadily

Duration: Constant

Associated signs and symptoms: Nausea, anorexia (viral hepatitis)
Early satiety (cystic tumors)
Aversion to smoking for smokers (viral hepatitis)
Aversion to alcohol (hepatitis)
Arthralgias and myalgias (hepatitis A, B, or C)
Headaches (hepatitis A, drug-induced hepatitis)
Dizziness/drowsiness (drug-induced hepatitis)
Low-grade fever (hepatitis A)
Pharyngitis (hepatitis A)
Extreme fatigue (hepatitis A, cirrhosis)
Alterations in the sense of taste and smell (hepatitis A)
Rash (hepatitis B)
Dark urine, light- or clay-colored stools
Ascites (see Fig. 9-8)
Edema and oliguria
Neurologic symptoms (hepatic encephalopathy)

- Confusion, forgetfulness
- Muscle tremors
- Asterixis (liver flap)
- Slurred speech
- Impaired handwriting

Skin and nail bed changes
Skin pallor (often linked with cirrhosis or carcinoma)
Jaundice (skin and sclerae changes)
Spider angiomas
Palmar erythema (liver palms)
Nail beds of Terry; leukonychia; digital clubbing; koilonychia
Bleeding disorders

- Purpura
- Ecchymosis

Diaphoresis (liver abscess)
Overall muscular weakness (cirrhosis, liver carcinoma)

Possible etiology: Peripheral neuropathy (chronic liver disease)
Any liver disease

- Hepatitis
- Cirrhosis
- Metastatic tumors

Pancreatic carcinoma
Liver abscess
Medications: Use of hepatotoxic drugs

LIVER/BILIARY PAIN PATTERNS—*cont'd*

GALLBLADDER PAIN (SEE FIG. 9-10)

Location: Pain in the midepigastrium (may be perceived as heartburn)

Referral:
RUQ of abdomen
RUQ pain may be associated with right shoulder pain
Both may be associated with back pain between the scapulae; back pain can occur alone as the primary symptom
Pain may be referred to the right side of the midline in the interscapular or subscapular area
Anterior rib pain (soreness or tender) at the tip of the 10th rib (less often, can also affect ribs 11 and 12)

Description:
Dull aching
Deep visceral pain (gallbladder suddenly distends)
Biliary carcinoma is more persistent and boring

Intensity: Mild at first, then increases steadily to become severe

Duration: 2 to 3 hours

Aggravating factors:
Respiratory inspiration
Eating
Upper body movement
Lying down

Associated signs and symptoms:
Dark urine, light stools
Jaundice
Skin: Green hue (prolonged biliary obstruction)
Persistent pruritus (cholestatic jaundice)
Pain and nausea occur 1 to 3 hours after eating (gallstones)
Pain immediately after eating (gallbladder inflammation)
Intolerance of fatty foods or heavy meals
Indigestion, nausea
Excessive belching
Flatulence (excessive intestinal gas)
Anorexia
Weight loss (gallbladder cancer)
Bleeding from skin and mucous membranes (late sign of gallbladder cancer)
Vomiting
Feeling of fullness
Low-grade fever, chills

Possible etiology:
Gallstones (cholelithiasis)
Gallbladder inflammation (cholecystitis)
Neoplasm
Medications: Use of hepatotoxic drugs

COMMON BILE DUCT PAIN (SEE FIG. 9-10)

Location: Pain in midepigastrium or RUQ of abdomen

Referral:
Epigastrium: Heartburn (choledocholithiasis)
RUQ pain may be associated with right shoulder pain
Both may be associated with back pain between the scapulae
Pain may be referred to the right side of the midline in the interscapular or subscapular area

Description:
Dull aching
Vague discomfort (pressure within common bile duct increasing)
Severe, steady pain in RUQ (choledocholithiasis)
Biliary carcinoma is more persistent and boring

Intensity: Mild at first, increases steadily

Continued

LIVER/BILIARY PAIN PATTERNS—*cont'd*

Duration:
Constant
3 to 4 hours (choledocholithiasis)

Associated signs and symptoms:
Dark urine, light stools
Jaundice
Nausea after eating
Intolerance of fatty foods or heavy meals
Feeling of abdominal fullness
Skin: Green hue (prolonged biliary obstruction); pruritus (skin itching)
Low-grade fever, chills
Excessive belching (choledocholithiasis)
Constipation and diarrhea (choledocholithiasis)
Sensory neuropathy (primary biliary cirrhosis)
Osteomalacia (primary biliary cirrhosis)
Osteoporosis (primary biliary cirrhosis)

Possible etiology:
Common duct stones
Common duct stricture (previous gallbladder surgery)
Pancreatic carcinoma (blocking the bile duct)
Medications: Use of hepatotoxic drugs
Neoplasm
Primary biliary cirrhosis
Choledocholithiasis (obstruction of common duct)

Key Points to Remember

- Primary signs and symptoms of liver diseases vary and can include GI symptoms, edema/ascites, dark urine, light-colored or clay-colored feces, and right upper abdominal pain.
- Neurologic symptoms, such as confusion, muscle tremors, and asterixis, may occur.
- Skin changes associated with the hepatic system include pruritus, jaundice, pallor, orange or green skin, bruising, spider angiomas, and palmar erythema.
- Active, intense exercise should be avoided when the liver is compromised (jaundice or other active disease).
- Antiinflammatory and minor analgesic agents can cause drug-induced hepatitis. Nonviral hepatitis may occur postoperatively.
- When liver dysfunction results in increased serum ammonia and urea levels, peripheral nerve function is impaired. Flapping tremors (asterixis) and numbness/tingling (misinterpreted as carpal/tarsal tunnel syndrome) can occur.
- Musculoskeletal locations of pain associated with the hepatic and biliary systems include thoracic spine between scapulae, right shoulder, right upper trapezius, right interscapular, or right subscapular areas.
- Referred shoulder pain may be the only presenting symptom of hepatic or biliary disease.
- Gallbladder impairment can present as a rib dysfunction with tenderness anteriorly over the tip of the 10th rib (occasionally ribs 11 and 12 are also involved).

SUBJECTIVE EXAMINATION

Special Questions to Ask

Past Medical History

- Have you ever had an ulcer, gallbladder disease, your spleen removed, or hepatitis/jaundice?
  - *If yes* to hepatitis or jaundice: When was this diagnosed? How did you get this?
- Has anyone in your family ever been diagnosed with Wilson's disease **(excessive copper retention)** or hemochromatosis **(excessive iron absorption)**? **(Hereditary)**
- Do you work in a clinical laboratory, operating room, or with dialysis clients? **(Hepatitis)**
- Have you been out of the United States in the last 6 to 12 months? **(parasitic infection, country where hepatitis is endemic)**
- Have you worked in any setting that might be high risk for disease transmission such as a day care, correctional setting, or institutional setting? **(Hepatitis)**
- Have you had any recent contact with hepatitis or with a jaundiced person?
- Have you eaten any raw shellfish recently? **(Viral hepatitis)**
- Have you had any recent blood or plasma transfusion, blood tests, acupuncture, ear or body piercing, tattoos (including removal), or dental work done? **(Viral hepatitis)**
- Have you had a recent ACL reconstruction with an allograft? **(Hepatitis)**
- Have you had any kind of injury or trauma to your abdomen? **(Possible liver damage)**

For women: Are you currently using oral contraceptives? (Hepatitis, adenoma)

For the therapist:

- When asking about drug history, keep in mind that oral contraceptives may cause cholestasis (suppression of bile flow) or liver tumors. Some common over-the-counter drugs (e.g., acetaminophen) and some antibiotics, antitubercular drugs, anticonvulsants, cytotoxic drugs for cancer, antipsychotics, and antidepressants may have hepatotoxic effects. Ask about the use of cholesterol-lowering statins.
- Use questions from Chapter 2 to determine possible consumption of alcohol as a hepatotoxin.

Associated Signs and Symptoms

- Have you noticed a recent tendency to bruise or bleed easily? **(Liver disease)**
- Have you noticed any change in the color of your stools or urine? **(Dark urine, the color of cola and light- or clay-colored stools associated with jaundice)**
- Has your weight fluctuated 10 or 15 pounds or more recently without a change in diet? **(Cancer, cirrhosis, ascites, but also congestive heart failure)**
- *If no,* have you noticed your clothes fitting tighter around the waist from abdominal swelling or bloating? **(Ascites)**
- Do you have a feeling of fullness after only one or two bites of food? **(Early satiety: stomach and duodenum, cystic tumors, or gallbladder)**
- Does your stomach feel swollen or bloated after eating? **(Abdominal fullness)**
- Do you have any abdominal pain? (Abdominal pain may be *visceral* from an internal organ [dull, general, poorly localized], *parietal* from inflammation of overlying peritoneum [sharp, precisely localized, aggravated by movement], or *referred* from a disorder in another site.)
- How does eating affect your pain? **(When eating aggravates symptoms: gastric ulcer, gallbladder inflammation)**
  - Are there any particular foods you have noticed that aggravate your symptoms?
  - *If yes,* which ones? **(Gallbladder: intolerance to fatty foods)**
- Have you noticed any unusual aversion to odors, food, alcohol, or (for people who smoke) smoking? (Jaundice)
- *For clients with only shoulder or back pain:* Have you noticed any association between when you eat and when your symptoms increase or decrease?

CASE STUDY

Hepatitis

REFERRAL

A 29-year-old male law student has come to you (self-referral) with headaches that developed after a motor vehicle accident 12 weeks ago. He was evaluated and treated in the emergency department of the local hospital and is not under the care of a primary care physician.

The headaches occur two to three times each week, starting at the base of the occiput and progressing up the back of his head to localize in the forehead bilaterally. The client has a sedentary lifestyle with no regular exercise, and he describes his stress level as being 6 on a scale from 0 to 10.

The Family/Personal History form (see Fig. 2-2) indicates that he was diagnosed with hepatitis at the time of the accident.

PHYSICAL THERAPY INTERVIEW

What follow-up questions will you ask this client related to the hepatitis?

- I see from your History form that you have hepatitis.
- What type of hepatitis do you have?

Give the client a chance to respond, but you may need to prompt with "type A," "type B," or "types C or D." Remember that hepatitis A is communicable before the appearance of any observable clinical symptoms. If he has been diagnosed, he is probably past this stage.

- Do you know how you initially came in contact with hepatitis? (Depending on the answer to the previous question, you may not need to ask this question.)

Considerations requiring further questioning may include

- Illicit or recreational drug use
- Inadequate hygiene and poor handwashing in close quarters with travel companion
- Ingestion of contaminated food, water, milk, or seafood
- Recent blood transfusion or contact with blood/blood products
- For type B: Modes of sexual transmission

Remember the three stages when trying to determine whether this person may still be contagious. Hepatitis B can persist in body fluids indefinitely, requiring necessary precautions by you.

Hepatitis caused by medications or toxins is noninfectious hepatitis and is not communicable.

Transmissible hepatitis requires handwashing and hygiene precautions, including avoidance of any body fluids on your part through the use of protective gloves. This is especially true when treating a person with diabetes requiring fingerstick blood testing, when performing needle electromyograms, or providing open wound care, especially with debridement.

MEDICAL TREATMENT

- Did you receive any medical treatment? (**immune globulin**)

Immune serum globulin (ISG) is considered most effective in producing passive immunity for 3 to 4 months when administered as soon as possible after exposure to the hepatitis virus, but within 2 weeks after the onset of jaundice. Persons who have been treated with ISG may not develop jaundice, but those who have not received the gamma globulin usually develop jaundice.

- Are you currently receiving follow-up care for your hepatitis through a local physician?

This information will assist you in determining the appropriate medical source for further information if you need it and in a case like this, assist you in choosing further follow-up questions that may help you determine whether this person requires additional medical follow up.

Keep in mind that headaches can be persistent symptoms of hepatitis A. If the client is receiving no further medical follow-up (especially if no serum globulin was administered initially), consider these follow-up questions:

ASSOCIATED SYMPTOMS

- What symptoms did you have with hepatitis?
- Do you have any of those symptoms now?
- Are you experiencing any unusual fatigue or muscle or joint aches and pains?
- Have you noticed any unusual aversion to foods, alcohol, or cigarettes/smoke that you did not have before?
- Have you had any problems with diarrhea, vomiting, or nausea?
- Have you noticed any change in the color of your stools or urine? (1 to 4 days before the icteric stage, the urine darkens and the stool lightens)
- Have you noticed any unusual skin rash developing recently?
- When did you notice the headaches developing?

PRACTICE QUESTIONS

1. Referred pain patterns associated with hepatic and biliary pathologic conditions produce musculoskeletal symptoms in the:
   a. Left shoulder
   b. Right shoulder
   c. Mid-back or upper back, scapular, and right shoulder areas
   d. Thorax, scapulae, right or left shoulder
2. What is the mechanism for referred right shoulder pain from hepatic or biliary disease?
3. Why does someone with liver dysfunction develop numbness and tingling that is sometimes labeled carpal tunnel syndrome?
4. When a client with bilateral carpal tunnel syndrome is being evaluated, how do you screen for the possibility of a pathologic condition of the liver?
5. What is the first most common sign associated with liver disease?
6. You are treating a 53-year-old woman who has had an extensive medical history that includes bilateral kidney disease with kidney removal on one side and transplantation on the other. The client is 10 years posttransplant and has now developed multiple problems as a result of the long-term use of immunosuppressants (cyclosporine to prevent organ rejection) and corticosteroids (prednisone). For example, she is extremely osteoporotic and has been diagnosed with cytomegalovirus and corticosteroid-induced myopathy. The client has fallen and broken her vertebra, ankle, and wrist on separate occasions. You are seeing her at home to implement a strengthening program and to instruct her in a falling prevention program, including home modifications. You notice the sclerae of her eyes are yellow-tinged. How do you tactfully ask her about this?
7. Clients with significant elevations in serum bilirubin levels caused by biliary obstruction will have which of the following associated signs?
   a. Dark urine, clay-colored stools, jaundice
   b. Yellow-tinged sclera
   c. Decreased serum ammonia levels
   d. a and b only
8. Preventing falls and trauma to soft tissues would be of utmost importance in the client with liver failure. Which of the following laboratory parameters would give you the most information about potential tissue injury?
   a. Decrease in serum albumin levels
   b. Elevated liver enzyme levels
   c. Prolonged coagulation times
   d. Elevated serum bilirubin levels
9. Decreased level of consciousness, impaired function of peripheral nerves, and asterixis (flapping tremor) would probably indicate an increase in the level of:
   a. AST (aspartate aminotransferase)
   b. Alkaline phosphatase
   c. Serum bilirubin
   d. Serum ammonia
10. An inpatient who has had a total hip replacement with a significant history of alcohol use/abuse has a positive test for asterixis. This may signify:
    a. Renal failure
    b. Hepatic encephalopathy
    c. Diabetes
    d. Gallstones obstructing the common bile duct
11. A decrease in serum albumin is common with a pathologic condition of the liver because albumin is produced in the liver. The reduction in serum albumin results in some easily identifiable signs. Which of the following signs might alert the therapist to the condition of decreased albumin?
    a. Increased blood pressure
    b. Peripheral edema and ascites
    c. Decreased level of consciousness
    d. Exertional dyspnea

REFERENCES

1. Key L, Bell NH: Osteomalacia and disorders of vitamin D metabolism. In Stein JH, editor: *Internal medicine*, ed 5, St Louis, 1998, Mosby.
2. Baxter RE, Moore JH: Diagnosis and treatment of acute exertional rhabdomyolysis. *J Orthop Sports Phys Ther* 33(3):104–108, 2003.
3. Kumbhare D: Validity of serum creatine kinase as a measure of muscle injury produced by lumbar surgery. *J Spinal Disord Tech* 21(1):49–54, 2008.
4. Lenfant C: ACC/AHA/NHLBI Clinical advisory on the use and safety of statins. *Cardiol Rev* 20(4Suppl):9–11, 2003.
5. Chatham K: Suspected statin-induced respiratory muscle myopathy during long-term inspiratory muscle training in a patient with diaphragmatic paralysis. *Phys Ther* 89(3):1–10, 2009.
6. Newman CB, Palmer G, Silbershatz H, et al: Safety of atorvastatin derived from analysis of 44 completed trials in 9,416 patients. *Am J Cardiol* 92(6):670–676, 2003.
7. Tomlinson SS, Mangione KK: Potential adverse effects of statins on muscle. *Phys Ther* 85:459–465, 2005.
8. Buettner C: Prevalence of musculoskeletal pain and statin use. *J Gen Intern Med* 23:1182–1186, 2008.
9. Ballantyne CM, Corsini A, Davidson MH, et al: Risk for myopathy with statin therapy in high-risk patients. *Arch Intern Med* 163(5):553–564, 2003.
10. Parnes A: Asterixis. *Trinity Stud Med J* 1:58, 2000. Available at: http://www.tcd.ie/tsmj. Accessed Oct. 18, 2010.
11. Melnick JL: History and epidemiology of Hepatitis A virus. *J Infect Dis* 171(Suppl):S2-S8, 1995.
12. McPhee S, Papadakis M, editors: *Current medical diagnosis and treatment*, ed 50, New York, 2011, Lange.
13. Grande P, Cronquist A: Public health dispatch, multistate outbreak of hepatitis A among young adult concert attendees, United States, 2003. *MMWR* 52(35):844–845, 2003.
14. Wolf DC: Hepatitis, Viral. *eMedicine Specialties: Gastroenterology*. Updated June 25, 2010. Available online at http://emedicine.medscape.com/article/185463-overview. Accessed February 1, 2011.

15. Weinbaum C, Lyerla C: Prevention and control of infections with hepatitis viruses in correctional settings. *MMWR* 5201 1–33, Jan 24, 2003.
16. Parini S: Hepatitis C. *Nursing2003* 33(4):57, 2003.
17. Spencer KY, Chang MD: Anterior cruciate ligament reconstruction: allograft vs. autograft. *J Arthrosc Rel Surg* 19(5):453, 2003.
18. Friedman McPhee S, Papadakis M, editors: *Current medical diagnosis and treatment*, ed 50, New York, 2011, Lange.
19. Sanzone AM: Hepatitis and arthritis: an update. *Infect Dis Clin N Am* 20:877–889, 2006.
20. Khouqeer RA: Viral arthritis. *eMedicine Specialties: Rheumatology—Infectious arthritis*. Updated August 1, 2008. Available online at http://emedicine.medscape.com/article/335692-overview. Accessed February 1, 2011.
21. Lovy MR, Wener MH: Rheumatic disease: when is hepatitis C the culprit? *J Musculoskel Med* 13(4):27–35, 1996.
22. Rull M, Zonay L, Schumacher HR: Hepatitis C and rheumatic diseases. *J Musculoskel Med* 15(11):38–44, 1998.
23. Foster GR: Past, present, and future hepatitis C treatments. *Semin Liver Disease* Suppl 2:97–104, 2004.
24. Pearlman BL: Hepatitis C treatment update. *Am J Med* 117(5):344–352, 2004.
25. Pullen L: Hep-Hazard. *HemAware* 9(2):54–56, 2004.
26. Farrell SE: Toxicity, Acetaminophen, *eMedicine Specialties: Emergency medicine—Toxicology*. Updated Nov. 19, 2010. Available online at http://emedicine.medscape.com/article/820200-overview. Accessed February 1, 2011.
27. Liver function: two new threats. *John Hopkins Med Lett: Health after 50* 15(3):2–7, 2003. www.johnshopkinshealthalerts.com/health_after_50. Accessed Oct. 18, 2010.
28. Adams LA, Feldstein AE: Nonalcoholic steatohepatitis: risk factors and diagnosis. *Expert Rev Gastroenterol Hepatol* 4(5):623–635, 2010.
29. Feldstein AE: Novel insights into the pathophysiology of nonalcoholic fatty liver disease. *Semin Liver Dis* 30(4):391–401, 2010.
30. Salt WB: Nonalcoholic fatty liver disease (NAFLD): a comprehensive review. *J Insur Med* 36(1):27–41, 2004.
31. Pasumarthy L, Srour J: Nonalcoholic steatohepatitis: a review of the literature and updates in management. *South Med J* 103(6):547–550, 2010.
32. Younossi ZM, McCullough AJ, Ong JP, et al: Obesity and nonalcoholic fatty liver disease in chronic hepatitis C. *J Clin Gastroenterol* 38(8):705–709, 2004.
33. Cohen RS: Understanding neonatal jaundice: a perspective on causation. *Pediatr Neonatol* 51(3):143–148, 2010.
34. Neonatal jaundice, unwelcome return for kernicterus. *Nursing2003* 33(11):35, 2003.
35. Lease M, Whalen B: Assessing jaundice in infants of 35-week gestation and greater. *Curr Opin Pediatr* 22(3):352–365, 2010.
36. Peralta R: Liver abscess. *eMedicine Specialties: Abdomen*. Updated Sept. 15, 2009. Available online at http://emedicine.medscape.com/article/188802-overview. Accessed February 1, 2011.
37. Chen SC: Severity of disease as main predictor for mortality in patients with pyogenic liver abscess. *Am J Surg* 198(2):164–172, 2009.
38. Bisceglie A: Prevention of hepatocellular carcinoma complicating chronic hepatitis C. *J Gastroenterol Hepatol* 24(4):531–536, 2009.
39. Rex L: *Evaluation and treatment of somatovisceral dysfunction of the gastrointestinal system*, Edmonds, WA, 2004, URSA Foundation.
40. Hu CJ: Primary biliary cirrhosis: what do autoantibodies tell us? *World J Gastroenterol* 16(29):3616–3629, 2010.
41. Selmi C: The etiology mystery in primary biliary cirrhosis. *Dig Dis* 28(1):105–115, 2010.
42. Pyrsopoulos NT: Primary biliary cirrhosis, *eMedicine Specialties: Liver*. Updated Dec. 23, 2009. Available online at http://emedicine.medscape.com/article/171117-overview. Accessed February 1, 2011.
43. Abbas G: Fatigue in primary biliary cirrhosis. *Nat Rev Gastroenterol Hepatol* 7(6):313–319, 2010.
44. Mehrotra B: Gallbladder cancer: epidemiology, risk factors, clinical features, and diagnosis. *Up-To-Date* August 19, 2010. Available online (subscription) at http://www.uptodate.com/contents/gallbladder-cancer-epidemiology-risk-factors-clinical-features-and-diagnosis. Accessed February 1, 2011.
45. Kimbel DL: Hip pain in a 50-year-old woman with RA. *J Musculoskel Med* 16(11):651–652, 1999.

CHAPTER 10

Screening for Urogenital Disease

A 40-year-old athletic man comes to your clinic for an evaluation of back pain that he attributes to a very hard fall on his back while he was alpine skiing 3 days ago. His chief complaint is a dull, aching costovertebral pain on the left side, which is unrelieved by a change in position or by treatment with ice, heat, or aspirin. He stated that "even the skin on my back hurts." He has no previous history of any medical problems.

After further questioning, the client reveals that inspiratory movements do not aggravate the pain, and he has not noticed any change in color, odor, or volume of urine output. However, percussion of the costovertebral angle (see Fig. 4-54) results in the reproduction of the symptoms. This type of symptom complex may suggest renal involvement even without obvious changes in urine.

Whether secondary to trauma or of insidious onset, a client's complaints of flank pain, low back pain, or pelvic pain may be of renal or urologic origin and should be screened carefully through the subjective and objective examinations. Medical referral may be necessary.

SIGNS AND SYMPTOMS OF RENAL AND UROLOGIC DISORDERS

This chapter is intended to guide the physical therapist in understanding the origins and relationships of renal, ureteral, bladder, and urethral symptoms. The urinary tract, consisting of kidneys, ureters, bladder, and urethra (Fig. 10-1), is an integral component of human functioning that disposes of the body's toxic waste products and unnecessary fluid and expertly regulates extremely complicated metabolic processes. The ureters, bladder, and urethra function primarily as transport vehicles for urine formed in the kidneys. The lower urinary tract is the last area through which urine is passed in its final form for excretion.

Formation and excretion of urine is the primary function of the renal nephron (the functional unit of the kidney) (Fig. 10-2). Through this process the kidney is able to maintain a homeostatic environment in the body. Besides the excretory function of the kidney, which includes the removal of wastes and excessive fluid, the kidney plays an integral role in the balance of various essential body functions, including the following:

- Acid base balance
- Electrolyte balance
- Control of blood pressure with renin
- Formation of red blood cells (RBCs)
- Activation of vitamin D and calcium balance

The failure of the kidney to perform any of these functions results in severe alteration and disruption in homeostasis and signs and symptoms resulting from these dysfunctions (Box 10-1).[1]

THE URINARY TRACT

The upper urinary tract consists of the kidneys and ureters. The kidneys are located in the posterior upper abdominal cavity in a space behind the peritoneum (retroperitoneal space) (see Fig. 4-50). Their anatomic position is in front of and on both sides of the vertebral column at the level of T11 to L3. The right kidney is usually lower than the left to accommodate the liver.[2]

The upper portion of the kidney is in contact with the diaphragm and moves with respiration. The kidneys are protected anteriorly by the ribcage and abdominal organs (see Fig. 4-49) and posteriorly by the large back muscles and ribs. The lower portions of the kidneys and the ureters extend below the ribs and are separated from the abdominal cavity by the peritoneal membrane.

The lower urinary tract consists of the bladder and urethra. From the renal pelvis, urine is moved by peristalsis to the ureters and into the bladder. The bladder, which is a muscular, membranous sac, is located directly behind the symphysis pubis and is used for storage and excretion of urine. The urethra is connected to the bladder and serves as a channel through which urine is passed from the bladder to the outside of the body.

Voluntary control of urinary excretion is based on learned inhibition of reflex pathways from the walls of the bladder. Release of urine from the bladder occurs under voluntary control of the urethral sphincter.

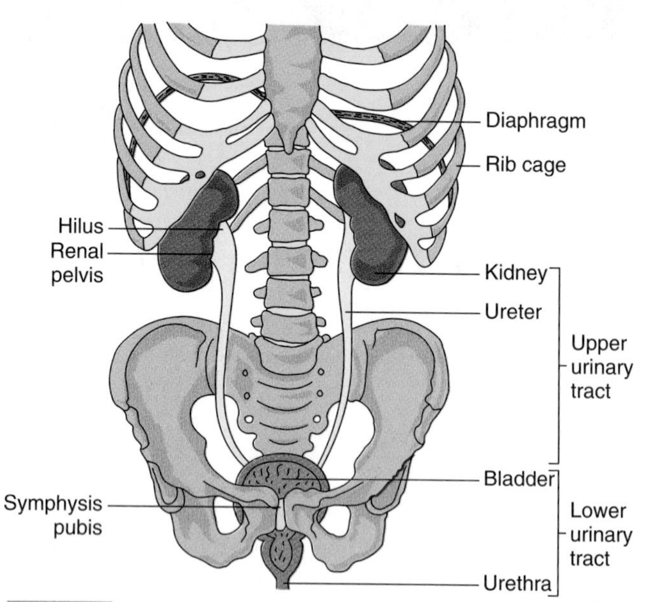

Fig. 10-1 Urinary tract structures. The upper urinary tract is composed of the kidneys and ureters while the lower urinary tract is made up of the bladder and urethra. The upper portion of each kidney is protected by the ribcage, and the bladder is partially protected by the symphysis pubis.

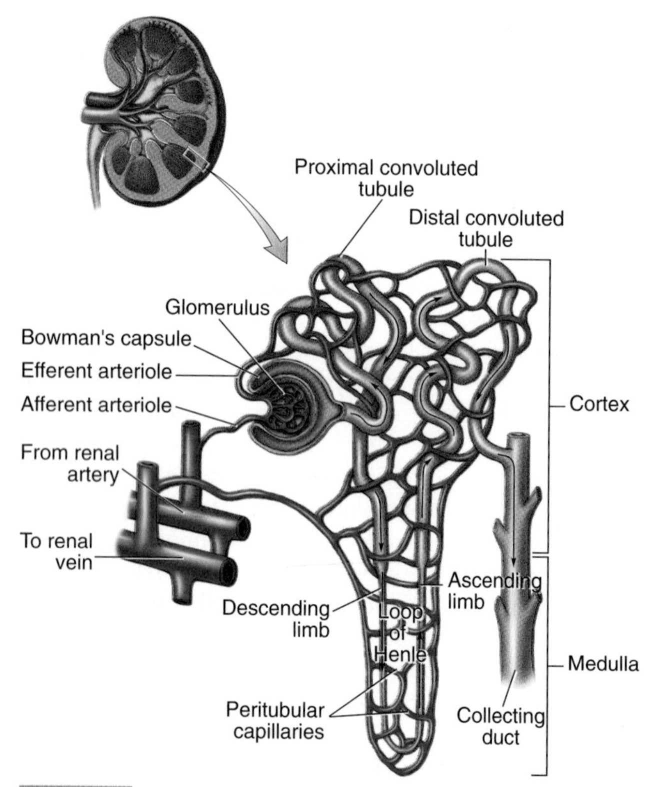

Fig. 10-2 Components of the nephron. The afferent arteriole carries blood to the glomerulus for filtration through Bowman's capsule and the renal tubular system. (From Herlihy B: *The human body in health and illness*, ed 4, Philadelphia, 2011, Saunders.)

The male genital or reproductive system is made up of the testes, epididymis, vas deferens, seminal vesicles, prostate gland, and penis (Fig. 10-3). These structures are susceptible to inflammatory disorders, neoplasms, and structural defects.

BOX 10-1 SIGNS AND SYMPTOMS OF GENITOURINARY DISEASE

Constitutional Symptoms
- Fever, chills
- Fatigue, malaise
- Anorexia, weight loss

Musculoskeletal
- Unilateral costovertebral tenderness
- Low back, flank, inner thigh, or leg pain
- Ipsilateral shoulder pain

Urinary Problems
- Dysuria (painful burning or discomfort with urination)
- Nocturia (getting up more than once at night to urinate)
- Feeling that bladder has not emptied completely but unable to urinate more; straining to start a stream of urine or to empty bladder completely
- Hematuria (blood in urine; pink or red-tinged urine)
- Dribbling at the end of urination
- Frequency (need to urinate or empty bladder more than every 2 hours)
- Hesitancy (weak or interrupted urine stream)
- Proteinuria (protein in urine; urine is foamy)

Other
- Skin hypersensitivity (T10-L1)
- Infertility

Women
- Abnormal vaginal bleeding
- Painful menstruation (dysmenorrhea)
- Changes in menstrual pattern
- Pelvic masses or lesions
- Vaginal itching or discharge
- Pain during intercourse (dyspareunia)

Men
- Difficulty starting or continuing a stream of urine
- Discharge from penis
- Penile lesions
- Testicular or penis pain
- Enlargement of scrotal contents
- Swelling or mass in groin
- Sexual dysfunction

In males the posterior portion of the urethra is surrounded by the prostate gland, a gland approximately 3.5 cm long by 3 cm wide (about the size of two almonds). Located just below the bladder, this gland can cause severe urethral obstruction when enlarged from a growth or inflammation resulting in difficulty starting a flow of urine, continuing a flow of urine, frequency, and/or nocturia.

The prostate gland is commonly divided into five lobes and three zones. Prostate carcinoma usually affects the

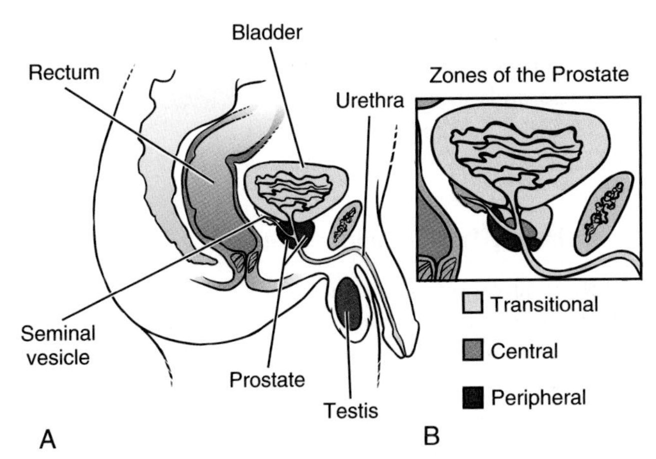

Fig. 10-3 **A**, The prostate is located at the base of the bladder, surrounding a part of the urethra. It is innervated by T11-L1 and S2-S4 and can refer pain to the sacrum, low back, and testes (see Fig. 10-10). As the prostate enlarges, the urethra can become obstructed, interfering with the normal flow of urine. **B**, The prostate is composed of three zones. The transitional zone surrounds the urethra as it passes through the prostate. This is a common site for benign prostatic hyperplasia (BPH). The central zone is a cone-shaped section that sits behind the transitional zone. The peripheral zone is the largest portion of the gland and borders the other two zones. This is the most common site for cancer development. Most early tumors do not produce any symptoms because the urethra is not in the peripheral zone. It is not until the tumor grows large enough to obstruct the bladder outlet that symptoms develop. Tumors in the transitional zone, which houses the urethra, may cause symptoms sooner than tumors in other zones.

posterior lobe of the gland; the middle and lateral lobes typically are associated with the nonmalignant process called *benign prostatic hyperplasia* (BPH).

RENAL AND UROLOGIC PAIN

Upper Urinary Tract (Renal/Ureteral)

The kidneys and ureters are innervated by both sympathetic and parasympathetic fibers. The kidneys receive sympathetic innervation from the lesser splanchnic nerves through the renal plexus, which is located next to the renal arteries. Renal vasoconstriction and increased renin release are associated with sympathetic stimulation. Parasympathetic innervation is derived from the vagus nerve, and the function of this innervation is not known.

Renal sensory innervation is not completely understood, even though the capsule (covering of the kidney) and the lower portions of the collecting system seem to cause pain with stretching (distention) or puncture. Information transmitted by renal and ureteral pain receptors is relayed by sympathetic nerves that enter the spinal cord at T10 to L1 (see Fig. 3-3).

Because visceral and cutaneous sensory fibers enter the spinal cord in close proximity and actually converge on some of the same neurons, when visceral pain fibers are stimulated, concurrent stimulation of cutaneous fibers also occurs. The visceral pain is then felt as though it is skin pain (hyperesthesia), similar to the condition of the alpine skier who stated that "even the skin on my back hurts." Renal and ureteral pain can be felt throughout the T10 to L1 dermatomes.

Renal pain (see Fig. 10-7) is typically felt in the posterior subcostal and costovertebral regions. To assess the kidney, the test for costovertebral angle tenderness can be included in the objective examination (see Fig. 4-54).

Ureteral pain is felt in the groin and genital area (see Fig. 10-8). With either renal pain or ureteral pain, radiation forward around the flank into the lower abdominal quadrant and abdominal muscle spasm with rebound tenderness can occur on the same side as the source of pain.

The pain can also be generalized throughout the abdomen. Nausea, vomiting, and impaired intestinal motility (progressing to intestinal paralysis) can occur with severe, acute pain. Nerve fibers from the renal plexus are also in direct communication with the spermatic plexus, and because of this close relationship, testicular pain may also accompany renal pain. Neither renal nor urethral pain is altered by a change in body position.

The typical renal pain sensation is aching and dull in nature but can occasionally be a severe, boring type of pain. The constant dull and aching pain usually accompanies distention or stretching of the renal capsule, pelvis, or collecting system. This stretching can result from intrarenal fluid accumulation such as inflammatory edema, inflamed or bleeding cysts, and bleeding or neoplastic growths. Whenever the renal capsule is punctured, a dull pain can also be felt by the client. Ischemia of renal tissue caused by blockage of blood flow to the kidneys results in a *constant* dull or a *constant* sharp pain.

Ureteral obstruction (e.g., from a urinary calculus or "stone" consisting of mineral salts) results in distention of the ureter and causes spasm that produces intermittent or constant severe colicky pain until the stone is passed. Pain of this origin usually starts in the costovertebral angle (CVA) and radiates to the ipsilateral lower abdomen, upper thigh, testis, or labium (see Fig. 10-8). Movement of a stone down a ureter can cause *renal colic,* an excruciating pain that radiates to the region just described and usually increases in intensity in waves of colic or spasm.

Chronic ureteral pain and renal pain tend to be vague, poorly localized, and easily confused with many other problems of abdominal or pelvic origin. There are also areas of *referred pain* related to renal or ureteral lesions. For example, if the diaphragm becomes irritated because of pressure from a renal lesion, shoulder pain may be felt (see Figs. 3-4 and 3-5). If a lesion of the ureter occurs *outside* the ureter, pain may occur on movement of the adjacent iliopsoas muscle (see Fig. 8-3).

Abdominal rebound tenderness results when the adjacent peritoneum becomes inflamed. Active trigger points along the upper rim of the pubis and the lateral half of the inguinal ligament may lie in the lower internal oblique muscle and possibly in the lower rectus abdominis. These trigger points can cause increased irritation and spasm of the detrusor and urinary sphincter muscles, producing urinary frequency, retention of urine, and groin pain.[3]

Pseudorenal Pain

Pseudorenal pain may occur secondary to radiculitis or irritation of the costal nerves caused by mechanical derangements of the costovertebral or costotransverse joints. Disorders of this sort are common in the cervical and thoracic areas, but the most common sites are T10 and T12.[4] Irritation of these nerves causes costovertebral pain that can radiate into the ipsilateral lower abdominal quadrant.

The onset is usually acute with some type of traumatic history such as lifting a heavy object, sustaining a blow to the costovertebral area, or falling from a height onto the buttocks. The pain is affected by body position, and although the client may be awakened at night when assuming a certain position (e.g., sidelying on the affected side), the pain is usually absent on awakening and increases gradually during the day. It is also aggravated by prolonged periods of sitting, especially when driving on rough roads in the car. It may be relieved by changing to another position (Table 10-1).

Radiculitis may mimic ureteral colic or renal pain, but true renal pain is seldom affected by movements of the shoulder or spine. Exerting pressure over the CVA with the thumb may elicit local tenderness of the involved peripheral nerve at its point of emergence, whereas gentle percussion over the angle may be necessary to elicit renal pain, indicating a deeper, more visceral sensation usually associated with an infectious or inflammatory process such as pyelonephritis, perinephric abscess, or other kidney problems.

Fig. 4-54 illustrates percussion over the CVA (Murphy's percussion or punch test). Although this test is commonly performed, its diagnostic value has never been validated. Results of at least one Finnish study[5] suggested that in acute renal colic loin tenderness and hematuria (blood in the urine) are more significant signs than renal tenderness.[6]

A diagnostic score incorporating independent variables, including results of urinalysis; presence of CVA and renal tenderness; and duration of pain, appetite level, and sex (male versus female), reached a sensitivity of 0.89 in detecting acute renal colic, with a specificity of 0.99 and an efficiency of 0.99.[5]

Lower Urinary Tract (Bladder/Urethra)

Bladder innervation occurs through sympathetic, parasympathetic, and sensory nerve pathways. Sympathetic bladder innervation assists in the closure of the bladder neck during seminal emission. Afferent sympathetic fibers also assist in providing awareness of bladder distention, pain, and abdominal distention caused by bladder distention. This input reaches the spinal cord at T9 or higher. Parasympathetic bladder innervation is at S2, S3, and S4 and provides motor coordination for the act of voiding. Afferent parasympathetic fibers assist in sensation of the desire to void, proprioception (position sensation), and perception of pain.

Sensory receptors are present in the mucosa of the bladder and in the muscular bladder walls. These fibers are more plentiful near the bladder neck and the junctional area between the ureters and bladder.

Urethral innervation, also at the S2, S3, and S4 level, occurs through the pudendal nerve. This is a mixed innervation of both sensory and motor nerve fibers. This innervation controls the opening of the external urethral sphincter (motor) and an awareness of the imminence of voiding and heat (thermal) sensation in the urethra.

Bladder or urethral pain is felt above the pubis (suprapubic) or low in the abdomen (see Fig. 10-9). The sensation is usually characterized as one of urinary urgency, a sensation to void, and dysuria (painful urination). Irritation of the neck of the bladder or the urethra can result in a burning sensation localized to these areas, probably caused by the urethral thermal receptors. See Box 10-2 for causes of pain outside the urogenital system that present like upper or lower urinary tract pain of either an acute or chronic nature.

TABLE 10-1 Assessment for Pseudorenal Pain

History	Trauma (fall, assault, blow, lifting) History of straining, lifting, accident or other mechanical injury to thoracic spine

Pain Pattern	• Back and/or flank pain occur at the same level as the kidney
	• Affected by change in position
	  • Lying on the involved side increases pain
	  • Prolonged sitting increases pain
	  • Symptoms are reproduced with movements of the spine
	  • Costovertebral angle tenderness present on palpation

Associated Signs and Symptoms	None Murphy's percussion (punch) test is negative Report of bowel and bladder changes unlikely

BOX 10-2 EXTRAUROLOGIC CONDITIONS CAUSING URINARY TRACT SYMPTOMS

Acute or chronic conditions affecting other viscera outside the urologic system can refer pain and symptoms to the upper or lower urinary tract. These can include

- Perforated viscus (any large internal organ)
- Intestinal obstruction
- Cholecystitis (inflammation of the gallbladder)
- Pelvic inflammatory disease
- Tubo-ovarian abscess
- Ruptured ectopic pregnancy
- Twisted ovarian cyst
- Tumor (benign or malignant)

RENAL AND URINARY TRACT PROBLEMS

Pathologic conditions of the upper and lower urinary tracts can be categorized according to primary causative factors. Inflammatory/infectious and obstructive disorders are presented in this section along with renal failure and cancers of the urinary tract.

When screening for any conditions affecting the kidney and urinary tract system, keep in mind factors that put people at increased risk for these problems (Case Example 10-1). Early screening and detection is recommended based on the presence of these risk factors.[7]

- Age over 60
- Personal or family history of diabetes or hypertension
- Personal or family history of kidney disease, heart attack, or stroke
- Personal history of kidney stones, urinary tract infections, lower urinary tract obstruction, or autoimmune disease
- African, Hispanic, Pacific Island, or Native American descent
- Exposure to chemicals (e.g., paint, glue, degreasing solvents, cleaning solvents), drugs, or environmental conditions
- Low birth weight

Inflammatory/Infectious Disorders

Inflammatory disorders of the kidney and urinary tract can be caused by bacterial infection, by changes in immune response, and by toxic agents such as drugs and radiation. Common infections of the urinary tract develop in either the upper or lower urinary tract (Table 10-2).

Upper urinary tract infections (UTIs) include kidney or ureteral infections. Lower UTIs include cystitis (bladder infection) or urethritis (urethral infection). Symptoms of UTI depend on the location of the infection in either the upper or lower urinary tract (although, rarely, infection could occur in both simultaneously).

Inflammatory/Infectious Disorders of the Upper Urinary Tract

Inflammations or infections of the upper urinary tract (kidney and ureters) are considered to be more serious because these lesions can be a direct threat to renal tissue itself.

TABLE 10-2 Urinary Tract Infections

Upper Urinary Tract Infection	Lower Urinary Tract Infection
Renal infections, such as pyelonephritis (renal parenchyma, i.e., kidney tissue) Acute or chronic glomerulonephritis (glomeruli) Renal papillary necrosis Renal tuberculosis	Cystitis (bladder infection) Urethritis (urethra infection)

The more common conditions include pyelonephritis (inflammation of the renal parenchyma) and acute and chronic glomerulonephritis (inflammation of the glomeruli of both kidneys). Less common conditions include renal papillary necrosis and renal tuberculosis.

CASE EXAMPLE 10-1

Screening in the Presence of Risk Factors for Kidney Disease

A 66-year-old African-American woman with a personal history of systemic lupus erythematosus (SLE) lost her balance and fell off the deck at her home. She sustained vertebral and rib fractures at T10 and T11. She is a retired paint factory worker. She reported daily exposure to paint and paint solvents during her 15 years of employment.

She was seen as a walk-in at the local medical clinic where she is a regular patient. She did not see the rheumatologist who was managing her SLE. The attending physician told her the injuries were "probably from the long-term use of prednisone for her lupus." She was referred to physical therapy by the attending physician for postural exercises.

During the interview, when asked, "Are you having any symptoms of any kind anywhere else in your body?" the client admitted to a pink color to her urine and some burning on urination. These symptoms have been present since the day after the fall 3 weeks ago.

There were no other signs or symptoms reported. Blood pressure measured 175/95 on three separate occasions. The client reported her blood pressure was elevated at the time of her visit to the doctor, but she thought it was caused by the stress of the fall.

Question: As you step back and conduct a Review of Systems, what are the red flags to suggest medical referral is needed? To whom do you refer this client?

Red flags:

- Age over 40 (age over 60 is a risk factor for kidney disease)
- African-American descent (at risk for diabetes, kidney disease)
- Long-term use of nonsteroidal antiinflammatory drugs (NSAIDs) (synergistic nephrotoxin in combination with certain chemicals such as paint and paint solvents)
- Elevated blood pressure
- Change in color and pattern of urination

The therapist may not recognize specific factors present that put the client at increased risk for kidney disease, but the obvious changes in urine color and pattern along with changes in blood pressure require medical referral.

Without the medical records, it is impossible to know what (if any) testing was done related to kidney function (e.g., urinalysis, blood test) at the time of the initial injury. A phone call to the referring physician is probably the best place to start. Documentation of the recent events and current red flag symptoms should be sent to the referring physician, the primary care physician, and the rheumatologist (if different from the primary care doctor).

Physical therapy intervention is still appropriate given her musculoskeletal injuries. Further medical assessment is warranted based on the development of symptoms unknown to the referring physician.

TABLE 10-3 Clinical Symptoms of Infectious/Inflammatory Urinary Tract Problems

Upper Urinary Tract (Kidney or Ureteral Infection)	Lower Urinary Tract (Cystitis or Urethritis)
Unilateral costovertebral tenderness	Urinary frequency
Flank pain	Urinary urgency
Ipsilateral shoulder pain	Low back pain
Fever and chills	Pelvic/lower abdominal pain
Skin hypersensitivity (hyperesthesia of dermatomes)	Dysuria (discomfort, such as pain or burning during urination)
Hematuria (blood [RBCs] in urine)	Hematuria
Pyuria (pus or white blood cells in urine)	Pyuria
Bacteriuria (bacteria in urine)	Bacteriuria
Nocturia (unusual or increased nighttime need to urinate)	Dyspareunia (painful intercourse)

Symptoms of upper urinary tract inflammations and infections are shown in Table 10-3. If the diaphragm is irritated, ipsilateral shoulder pain may occur. Signs and symptoms of renal impairment are also shown in Table 10-4 and if present, are significant symptoms of impending kidney failure.

Inflammatory/Infectious Disorders of the Lower Urinary Tract

Both the bladder and urine have a number of defenses against bacterial invasion. These defenses are mechanisms such as voiding, urine acidity, osmolality, and the bladder mucosa itself, which is thought to have antibacterial properties.

Urine in the bladder and kidney is normally sterile, but urine itself is a good medium for bacterial growth. Interferences in the defense mechanisms of the bladder such as the presence of residual or stagnant urine, changes in urinary pH or concentration, or obstruction of urinary excretion can promote bacterial growth.

Routes of entry of bacteria into the urinary tract can be *ascending* (most commonly up the urethra into the bladder and then into the ureters and kidney), *bloodborne* (bacterial invasion through the bloodstream), or *lymphatic* (bacterial invasion through the lymph system, the least common route).

A lower UTI occurs most commonly in women because of the short female urethra and the proximity of the urethra to the vagina and rectum. The rate of occurrence increases with age and sexual activity since intercourse can spread bacteria from the genital area to the urethra. Chronic health problems, such as diabetes mellitus, gout, hypertension, obstructive urinary tract problems, and medical procedures requiring urinary catheterization, are also predisposing risk factors for the development of these infections.[8]

TABLE 10-4 Systemic Manifestations of Chronic Kidney Disease

System	Manifestation
General	Fatigue, malaise
Skin and nail beds	Pallor, ecchymosis, pruritus, dry skin and mucous membranes, thin/brittle nail beds, urine odor on skin, uremic frost (white urea crystals) on the face and upper trunk, poor wound healing
Skeletal	Osteomalacia, osteoporosis,* bone pain, myopathy, tendon rupture, fracture, joint pain, dependent edema
Neurologic	*CNS*: Recent memory loss, decreased alertness, difficulty concentrating, irritability, lethargy/sleep disturbance, coma, impaired judgment *PNS*: Muscle weakness, tremors, and cramping; neuropathies with restless leg syndrome, cramps, carpal tunnel syndrome, paresthesias, burning feet syndrome, pruritus (itching)
Eye, ear, nose, throat	Metallic taste in mouth, nosebleeds, uremic (urine-smelling) breath, pale conjunctiva, visual blurring
Cardiovascular	Hypertension, friction rub, congestive heart failure, pericarditis, cardiomyopathy, arrhythmia, Raynaud's phenomenon
Pulmonary	Dyspnea, pulmonary edema, crackles (rales), pleural effusion
Gastrointestinal	Anorexia, nausea, vomiting, hiccups, gastrointestinal bleeding
Genitourinary	Decreased urine output and other changes in pattern of urination (e.g., nocturia)
Metabolic/endocrine	Dehydration, hyperkalemia, metabolic acidosis, hypocalcemia, hyperphosphatemia, fertility and sexual dysfunction (e.g., impotence, loss of libido, amenorrhea), hyperparathyroidism
Hematologic	Anemia Thrombocytopenia

*Bone demineralization leads to a condition called *renal osteodystrophy*.
CNS, Central nervous system; *PNS*, peripheral nervous system.
From Goodman CC, Fuller KS: *Pathology: implications for the physical therapist*, ed 3, Philadelphia, 2009, WB Saunders.

Individuals with diabetes are prone to complications associated with UTIs. Staphylococcus infection of the urinary tract may be a source of osteomyelitis, an infection of a vertebral body resulting from hematogenous spread or local spread from an abscess into the vertebra. The infected vertebral body may gradually undergo degeneration and destruction, with collapse and formation of a segmental scoliosis.[9]

This condition is suspected from the onset of nonspecific low back pain, unrelated to any specific motion. Local tenderness can be elicited, but the initial x-ray finding is negative. Usually, a low-grade fever is present but undetected, or it develops as the infection progresses. This is why anyone with low back pain of unknown origin should have his or her temperature taken, even in a physical therapy setting.

Older adults (both men and women) are at increased risk for UTI. They may present with nonspecific symptoms, such as loss of appetite, nausea, and vomiting; abdominal pain; or change in mental status (e.g., onset of confusion, increased confusion). Watch for predisposing conditions that can put the older client at risk for UTI. These may include diabetes mellitus or other chronic diseases (e.g., Alzheimer's disease, Parkinson's disease), immobility, reduced fluid intake, use of incontinence management products (e.g., pads, briefs, external catheters), indwelling catheterization, and previous history of UTI or kidney stones.

Cystitis

Cystitis (inflammation with infection of the bladder), *interstitial cystitis* (inflammation without infection), and *urethritis* (inflammation and infection of the urethra) appear with a similar symptom progression (Case Example 10-2).

According to the Interstitial Cystitis Association (ICA), interstitial cystitis (IC), also known as painful bladder syndrome, is a condition that consists of recurring pelvic pain, pressure, or discomfort in the bladder and pelvic region and affects more than 4 million people in the United States.[10] IC is often associated with urinary frequency and urgency. Men can be affected by this condition, but the majority of people living with IC are women. Several other disorders are associated with IC including allergies, inflammatory bowel syndrome, fibromyalgia, and vulvitis.[11]

Bladder pain associated with IC can vary from person to person and even within the same individual and may be dull, achy, or acute and stabbing. Discomfort while urinating also varies from mild stinging to intense burning. Sexual intercourse may ignite pain that lasts for days.[11]

Clients with any of the symptoms listed for the lower urinary tract in Table 10-3 at presentation should be referred promptly to a physician for further diagnostic workup and possible treatment. Infections of the lower urinary tract are potentially very dangerous because of the possibility of upward spread and resultant damage to renal tissue. Some individuals, however, are asymptomatic, and routine urine culture and microscopic examination are the most reliable methods of detection and diagnosis.

Obstructive Disorders

Urinary tract obstruction can occur at any point in the urinary tract and can be the result of *primary* urinary tract obstructions (obstructions occurring within the urinary tract) or *secondary* urinary tract obstructions (obstructions resulting from disease processes outside the urinary tract).

A primary obstruction might include problems such as acquired or congenital malformations, strictures, renal or ureteral calculi (stones), polycystic kidney disease, or neoplasms of the urinary tract (e.g., bladder, kidney).

CASE EXAMPLE 10-2

Bladder Infection

A 55-year-old woman came to the clinic with back pain associated with paraspinal muscle spasms. Pain was of unknown cause (insidious onset), and the client reported that she was "just getting out of bed" when the pain started. The pain was described as a dull aching that was aggravated by movement and relieved by rest (musculoskeletal pattern).

No numbness, tingling, or saddle anesthesia was reported, and the neurologic screening examination was negative. Sacroiliac (SI) testing was negative. Spinal movements were slow and guarded, with muscle spasms noted throughout movement and at rest. Because of her age and the insidious onset of symptoms, further questions were initiated to screen for medical disease.

This client was midmenopausal and was not taking any hormone replacement therapy (HRT). She had a bladder infection a month ago that was treated with antibiotics; tests for this were negative when she was evaluated and referred by her physician for back pain. Two weeks ago she had an upper respiratory infection (a "cold") and had been "coughing a lot." There was no previous history of cancer.

Local treatment to reduce paraspinal muscle spasms was initiated, but the client did not respond as expected over the course of five treatment sessions. Because of her recent history of upper respiratory and bladder infections, questions were repeated related to the presence of constitutional symptoms and changes in bladder function/urine color, force of stream, burning on urination, and so on. Occasional "sweats" (present sometimes during the day, sometimes at night) was the only red flag present. The combination of recent infection, failure to respond to treatment, and the presence of sweats suggested referral to the physician for early reevaluation.

The client did not return to the clinic for further treatment, and a follow-up telephone call indicated that she did indeed have a recurrent bladder infection that was treated successfully with a different antibiotic. Her back pain and muscle spasm were eliminated after only 24 hours of taking this new antibiotic.

Secondary obstructions produce pressure on the urinary tract from outside and might be related to conditions such as prostatic enlargement (benign or malignant); abdominal aortic aneurysm; gynecologic conditions such as pregnancy, pelvic inflammatory disease, and endometriosis; or neoplasms of the pelvic or abdominal structures.

Obstruction of any portion of the urinary tract results in a backup or collection of urine behind the obstruction. The result is dilation or stretching of the urinary tract structures that are positioned behind the point of blockage.

Muscles near the affected area contract in an attempt to push urine around the obstruction. Pressure accumulates above the point of obstruction and can eventually result in severe dilation of the renal collecting system (hydronephrosis) and renal failure. The greater the intensity and duration of the pressure, the greater is the destruction of renal tissue.

Because urine flow is decreased with obstruction, urinary stagnation and infection or stone formation can result. Stones are formed because urine stasis permits clumping or precipitation of organic matter and minerals.

Lower urinary tract obstruction can also result in constant bladder distention, hypertrophy of bladder muscle fibers, and formation of herniated sacs of bladder mucosa. These herniated sacs result in a large, flaccid bladder that cannot empty completely. In addition, these sacs retain stagnant urine, which causes infection and stone formation.

Obstructive Disorders of the Upper Urinary Tract

Obstruction of the upper urinary tract may be sudden (acute) or slow in development. Tumors of the kidney or ureters may develop slowly enough that symptoms are totally absent or very mild initially, with eventual progression to pain and signs of impairment. *Acute* ureteral or renal blockage by a stone (calculus consisting of mineral salts), for example, may result in excruciating, spasmodic, and radiating pain accompanied by severe nausea and vomiting.

Calculi form primarily in the kidney. This process is called *nephrolithiasis.* The stones can remain in the kidney (renal pelvis) or travel down the urinary tract and lodge at any point in the tract. Strictly speaking, the term *kidney stone* refers to stones that are in the kidney. Once they move into the ureter, they become *ureteral stones.*

Ureteral stones are the ones that cause the most pain. If a stone becomes wedged in the ureter, urine backs up, distending the ureter and causing severe pain. If a stone blocks the flow of urine, urine pressure may build up in the ureter and kidney, causing the kidney to swell (hydronephrosis). Unrecognized hydronephrosis can sometimes cause permanent kidney damage.[12]

The most characteristic symptom of renal or ureteral stones is sudden, sharp, severe pain. If the pain originates deep in the lumbar area and radiates around the side and down toward the testicle in the male and the bladder in the female, it is termed *renal colic. Ureteral colic* occurs if the stone becomes trapped in the ureter. Ureteral colic is characterized by radiation of painful symptoms toward the genitalia and thighs (see Fig. 10-8).

Since the testicles and ovaries form in utero in the location of the kidneys and then migrate at full term following the pathways of the ureters, kidney stones moving down the pathway of the ureters cause pain in the flank. This pain radiates to the scrotum in males and the labia in females. For the same reason, ovarian or testicular cancer can refer pain to the back at the level of the kidneys.

Renal tumors may also be detected as a flank mass combined with unexplained weight loss, fever, pain, and hematuria. The presence of any amount of blood in the urine always requires referral to a physician for further diagnostic evaluation because this is a primary symptom of urinary tract neoplasm.

CLINICAL SIGNS AND SYMPTOMS

Obstruction of the Upper Urinary Tract

- Pain (depends on the rapidity of onset and on the location)
  - Acute, spasmodic, radiating
  - Mild and dull flank pain
  - Lumbar discomfort with some renal diseases or renal back pain with ureteral obstruction
- Hyperesthesia of dermatomes (T10 through L1)
- Nausea and vomiting
- Palpable flank mass
- Hematuria
- Fever and chills
- Urge to urinate frequently
- Abdominal muscle spasms
- Renal impairment indicators (see inside front cover: Renal Blood Studies; see also Table 10-4)

Obstructive Disorders of the Lower Urinary Tract

Common conditions of (mechanical) obstruction of the lower urinary tract are bladder tumors (bladder cancer is the most common site of urinary tract cancer) and prostatic enlargement, either benign (BPH) or malignant (cancer of the prostate). An enlarged prostate gland can occlude the urethra partially or completely.

Mechanical problems of the urinary tract result in difficulty emptying urine from the bladder. Improper emptying of the bladder results in urinary retention and impairment of voluntary bladder control (incontinence). Several possible causes of mechanical bladder dysfunction include pelvic floor dysfunction, UTIs, partial urethral obstruction, trauma, and removal of the prostate gland.

The nerves that carry pain sensation from the prostate do not localize the source of pain very precisely, and therefore it may be difficult for the man to describe exactly where the pain is coming from. Discomfort can be localized in the suprapubic region or in the penis and testicles, or it can be centered in the perineum or rectum (see Fig. 10-10).

Prostatitis. Prostatitis is a relatively common inflammation of the prostate causing prostate enlargement. This condition affects up to 10% of the adult male population, accounting for the 2 million or more men who seek treatment annually in the United States.[13,14] It is often disabling, affecting men at any age, but typically found in men ages 40 to 70 years. Acute bacterial prostatitis occurs most often in men under age 35.

The National Institutes of Health (NIH) Consensus Classification of Prostatitis[15,16] includes four distinct categories:

Type I	Acute bacterial prostatitis
Type II	Chronic bacterial prostatitis
Type III	Chronic prostatitis/chronic pelvic pain syndrome (CP/CPPS) A. Inflammatory B. Noninflammatory
Type IV	Asymptomatic inflammatory prostatitis

Type I is an acute prostatic infection with a uropathogen, often with systemic symptoms of fever, chills, and hypotension. The prostate is inflamed and may block urinary flow without treatment. Type II is characterized by recurrent episodes of documented UTIs with the same uropathogen repeatedly and causes pelvic pain, urinary symptoms, and ejaculatory pain. The source of recurrent infections in the lower urinary tract must be identified and treated.

Chronic (type III, nonbacterial) prostatitis is characterized by pelvic pain for more than 3 of the previous 6 months, urinary symptoms, and painful ejaculation without documented urinary tract infections from uropathogens.

The symptoms of CP/CPPS appear to occur as a result of interplay between psychologic factors and dysfunction in the immune, neurologic, and endocrine systems.[17] Studies show a major impact on quality of life, urinary function, and sexual function along with chronic pain and discomfort (Fig. 10-4).[18,19]

The pain of prostatitis can be exacerbated by sexual activity, and some men describe pain upon ejaculation. A digital rectal examination by the physician will reproduce painful symptoms when the prostate is inflamed or infected (Fig. 10-5).

In men with chronic prostatitis, voiding complaints similar to those caused by BPH are the predominant symptoms. These complaints include urgency, frequency, and nocturia (getting up at nighttime more than once); less frequently, men may complain of difficulty starting the urinary stream or a slow stream.

These symptoms typically differ from symptoms of BPH in that they are associated with some degree of discomfort before, during, or after voiding. Physical or emotional stress and/or irritative components of the diet (e.g., caffeine in coffee, soft drinks) commonly exacerbate chronic prostatitis symptoms.

The causes of prostatitis are unclear. Although it can be the result of a bacterial infection, many men have nonbacterial prostatitis of unknown cause. Risk factors for bacterial prostatitis include some sexually transmitted diseases (e.g., gonorrhea) from unprotected anal and vaginal intercourse, which can allow bacteria to enter the urethra and travel to the prostate.

Other risk factors include bladder outlet obstruction (e.g., stone, tumor, BPH), diabetes mellitus, immunosuppression, and urethral catheterization. Neither prostatitis nor prostate enlargement is known to cause cancer, but men with prostatitis or BPH can develop prostate cancer as well.

The NIH Chronic Prostatitis Symptom Index (NIH-CPSI) provides a valid outcome measure for men with chronic (nonbacterial) prostatitis. The index may be useful in clinical practice, as well as research protocols.[20]

Anyone with significant symptoms assessed by the NIH-CPSI associated with constitutional symptoms should be rechecked by a physician. Individuals with significant symptoms but no constitutional symptoms and individuals nonresponsive to antibiotics should be assessed by a pelvic floor specialist. The index is available for clinical practice and may be useful for research protocols. It is available online at www.prostatitis.org/symptomindex.html.

A less complete list of questions for screening purposes are most appropriate for men with low back pain and any of the risk factors or symptoms listed for **prostatitis** and may include the following.

FOLLOW-UP QUESTIONS

- Do you ever have burning pain or discomfort during or right after urination?
- Does it feel like your bladder is not empty when you finish urinating?
- Do you have to go to the bathroom every 2 hours (or more often)?
- Do you ever have pain or discomfort in your testicles, penis, or the area between your rectum and your testicles (perineum)?
- Do you ever have pain in your pubic or bladder area?
- Do you have any discomfort during or after sexual climax (ejaculation)?

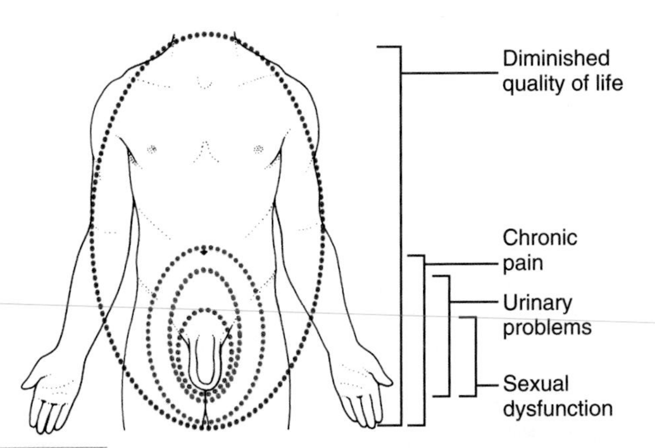

Fig. 10-4 Chronic prostatitis/chronic pelvic pain syndrome (CP/CPPS) can have a serious impact on a man's quality of life as a result of voiding problems, chronic pelvic pain and discomfort, and sexual dysfunction with painful ejaculation, cramping, or discomfort after ejaculation and infertility.

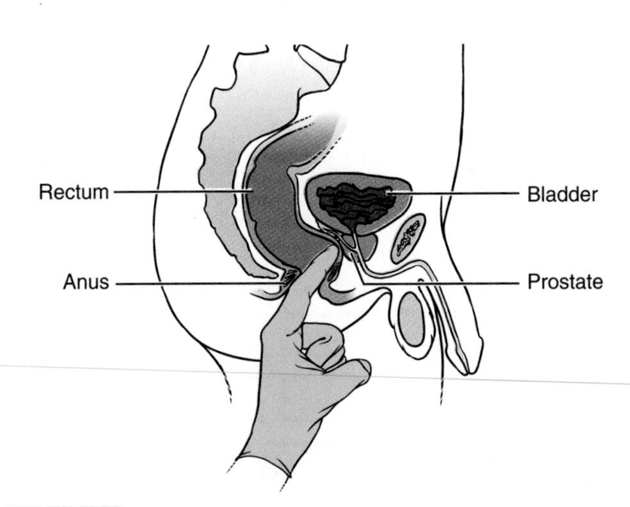

Fig. 10-5 Digital rectal examination performed by a medical doctor or trained health care professional, such as a nurse practitioner or physician's assistant, puts pressure on the inflamed prostate reproducing painful symptoms associated with prostatitis.

The therapist is reminded in asking these questions to offer clients a clear explanation for any questions asked concerning sexual activity, sexual function, or sexual history. There is no way to know when someone will be offended or claim sexual harassment. It is in your own interest to conduct the interview in the most professional manner possible.

There should be no hint of sexual innuendo or humor injected into any of your conversations with clients at any time. The line of sexual impropriety lies where the complainant draws it and includes appearances of misbehavior. This perception differs broadly from client to client.[21]

Prostatitis cannot always be cured but can be managed. Correct diagnosis is the key to the management of prostatitis. Screening men with red flag symptoms, history, and risk factors can result in early detection and medical referral.

Physical therapy has been shown to have some potential in helping men with chronic prostatitis. Physical therapy for this problem is more common in the European countries but is gaining support in the United States.[22] Other minimally invasive intervention strategies directed toward reducing pelvic floor muscle tone and improving urinary function include electrostimulation, transrectal or transurethral microwave hyperthermia, needle ablation hyperthermia, BOTOX injection, biofeedback, myofascial release, and transrectal mobilization of the pelvic ligaments.[23-27]

CLINICAL SIGNS AND SYMPTOMS

Prostatitis

- Sudden moderate-to-high fever
- Chills
- Low back, inner thigh, and perineal pain
- Testicular or penis pain
- Urinary frequency and urgency
- Nocturia (unusual voiding during the night)/sleep disturbance
- Dysuria (painful or difficult urination)
- Weak or interrupted urine stream (hesitancy)
- Unable to completely empty bladder
- Sexual dysfunction (e.g., painful ejaculation, cramping/discomfort after ejaculation, infertility)
- General malaise
- Arthralgia
- Myalgia

Benign Prostatic Hyperplasia. BPH (enlarged prostate) is a common occurrence in men older than 50. Like all cells in the body, cells in the prostate constantly die and are replaced by new cells. As men age, the ratio of new prostate cells to old prostate cells shifts in favor of lower cell death. With a lower cell turnover, there are more "old" cells than "new" ones and the prostate enlarges, squeezing the urethra and interfering with urination and sexual function. It is unclear why cell replacement is diminished, but it may be related to hormone changes associated with aging.

Prostate enlargement affects about half of all men between ages 60 and 69 and close to 80% of men between ages 70 and 90. Severity of signs and symptoms varies and only about half of men with prostate enlargement have problems noticeable enough to seek treatment.[28]

Because of the prostate's position around the urethra (see Fig. 10-3), enlargement of the prostate quickly interferes with the normal passage of urine from the bladder. Sexual function is not usually affected unless prostate surgery is required and sexual dysfunction occurs as a complication. If the prostate is greatly enlarged, chronic constipation may result.

Urination becomes increasingly difficult, and the bladder never feels completely empty. Straining to empty the bladder can stretch the bladder, making it less elastic. The detrusor becomes less efficient, and urine collecting in the bladder can foster urinary tract infections.

If left untreated, loss of bladder tone and damage to the detrusor may not be reversible. Continued enlargement of the prostate eventually obstructs the bladder completely, and emergency measures become necessary to empty the bladder.

Like prostatitis, BPH cannot be cured, but symptoms can be managed with medical treatment. Anyone with undiagnosed symptoms of BPH should seek medical evaluation as soon as possible. Screening questions for an **enlarged prostate** can include the following.

FOLLOW-UP QUESTIONS

- Does it feel like your bladder is not empty when you finish urinating?
- Do you have to urinate again less than 2 hours after the last time you emptied your bladder?
- Do you have a weak stream of urine or find you have to start and stop urinating several times when you go to the bathroom?
- Do you have to push or strain to start urinating or to keep the urine flowing?
- Do you have any leaking or dribbling of urine from the penis?
- Do you get up more than once at night to urinate?

CLINICAL SIGNS AND SYMPTOMS

Obstruction of the Lower Urinary Tract (Benign Prostatic Hyperplasia/Prostate Cancer)

Lower urinary tract symptoms of blockage are most commonly related to bladder or urethral pressure (e.g., prostate enlargement). This pressure results in bladder distention and subsequent pain. Common symptoms of lower urinary tract obstruction include

- Bladder palpable above the symphysis pubis
- Urinary problems
  - Hesitancy: difficulty in initiating urination or an interrupted flow of urine
  - Small amounts of urine with voiding (weak urine stream)
  - Dribbling at the end of urination
  - Frequency: need to urinate often (more than every 2 hours)
  - Nocturia (unusual voiding during the night)/sleep disturbance
- Lower abdominal discomfort with a feeling of the need to void
- Low back and/or hip, upper thigh pain or stiffness
- Suprapubic or pelvic pain
- Difficulty having an erection
- Blood in urine or semen

Prostate Cancer. Prostate cancer is a slow growing form of cancer causing microscopic changes in the prostate in one third of all men by age 50. Carcinoma in situ is present in 50% to 75% of American men by age 75. Most of these changes are latent, meaning they produce no signs or symptoms or they are so slow growing (indolent) that they never cause a health threat.[29]

Even so, prostate cancer is the most common type of cancer and second leading cause of death among men in this country. Of all the men who are diagnosed with cancer each year, about one third have prostate cancer.[30]

The number of new diagnosed cases of prostate cancer has dramatically increased over the last two decades (peaking in 1992), probably due to mass screening using a blood test to measure the prostate-specific antigen (PSA). PSA rises in men who have any changes in the prostate (e.g., tumor, infection, enlargement).[30] Despite the many controversies over "normal" levels of PSA, this test has shifted the detection of the majority of prostate cancer cases from late-stage to early-stage disease when prostate cancers are more likely to be curable.[31]

Because more men are living longer and the incidence of prostate cancer increases with age, prostate cancer is becoming a significant health issue. Risk factors include advancing age, family history, ethnicity, and diet. Most men with prostate cancer are older than 65; the disease is rare in men younger than 45.

A man's risk of prostate cancer is higher than average if his brother or father had the disease. In fact, the more first-degree family members affected, the greater the person's risk of prostate cancer. It is more common in African-American men compared to white or Hispanic men. It is less common in Asian and Native American men.[29]

Some studies suggest a diet high in animal fat or meat may be a risk factor. Other risk factors may include low levels of vitamins or selenium; multiple sex partners; viruses; and occupational exposure to chemicals (including farmers exposed to herbicides and pesticides), cadmium, and other metals.[29]

Early prostate cancer often does not cause symptoms. But prostate cancer can cause any of the signs and symptoms listed in Clinical Signs and Symptoms: Obstruction of the Lower Urinary Tract.

It is often diagnosed when the man seeks medical assistance because of symptoms of lower urinary tract obstruction or low back, hip, or leg pain or stiffness (Case Example 10-3). There are four stages of prostate cancer[29]:

- Stage I or Stage A: The cancer cannot be felt during a rectal exam. It may be found when surgery is done for another reason, usually for BPH. There is no evidence that the cancer has spread outside the prostate.
- Stage II or Stage B: The tumor is large enough that it can be palpated during a rectal exam or found with a biopsy. There is no evidence that the cancer has spread outside the prostate.
- Stage III or Stage C: The cancer has spread outside the prostate to nearby tissues.
- Stage IV or Stage D: The cancer has spread to lymph nodes or to other parts of the body

Back pain and sciatica can be caused by cancer metastasis via the bloodstream or the lymphatic system to the bones of the pelvis, spine, or femur. Lumbar pain is predominant, but the thoracolumbar pain can be painful as well, depending on the location of the metastases. Prostate cancer is unique in that bone is often the only clinically detectable site of metastases. The resulting tumors tend to be osteoblastic (bone forming, causing sclerosis), rather than osteolytic (bone lysing) (Fig. 10-6; see also Fig. 13-7).[32]

Symptoms of metastatic disease include bone pain, anemia, weight loss, lymphedema of the lower extremities and scrotum, and neurologic changes associated with spinal cord compression when spinal involvement occurs.

CASE EXAMPLE 10-3

Prostate Cancer

A 66-year-old man with low back pain was evaluated by a female physical therapist but treated by a male physical therapy aide. By the end of the third session, the client reported some improvement in his painful symptoms. During the second week there was no improvement and even a possible slight setback. During the treatment session he commented to the aide that he is impotent.

Given this man's age, inconsistent response to therapy, and report of impotency, a medical referral was necessary. A brief note was sent to the physician relating this information and requesting medical follow-up. (The therapist was careful to use the word *follow-up* rather than *medical reevaluation* since the impotency was present at the time of the initial medical evaluation.)

Result: A medical diagnosis of testicular cancer was established, and appropriate treatment was initiated. Physical therapy was discontinued until medical treatment was completed and systemic origin of the back pain could be ruled out.

Incontinence

Urinary incontinence (UI) is the involuntary leakage of urine. According to the U.S. Department of Health and Human Services, incontinence is a vastly underdiagnosed and underreported problem affecting millions of Americans each year. The incidence of incontinence is expected to grow dramatically as the U.S. population continues to age.[33]

UI is not a disease but rather a symptom of other underlying health conditions, including trauma (e.g., childbirth, incest), diabetes, multiple sclerosis, Parkinson's disease, spinal injury, spina bifida, surgery, hormonal changes, medications, stroke dysfunction, UTIs, neuromuscular conditions, constipation, or even dietary issues, including caffeine intake.

Incontinent people may restrict their activities for fear of urine loss and concerns about odors in public. This reduction in social activity and impact on lifestyle can have profound

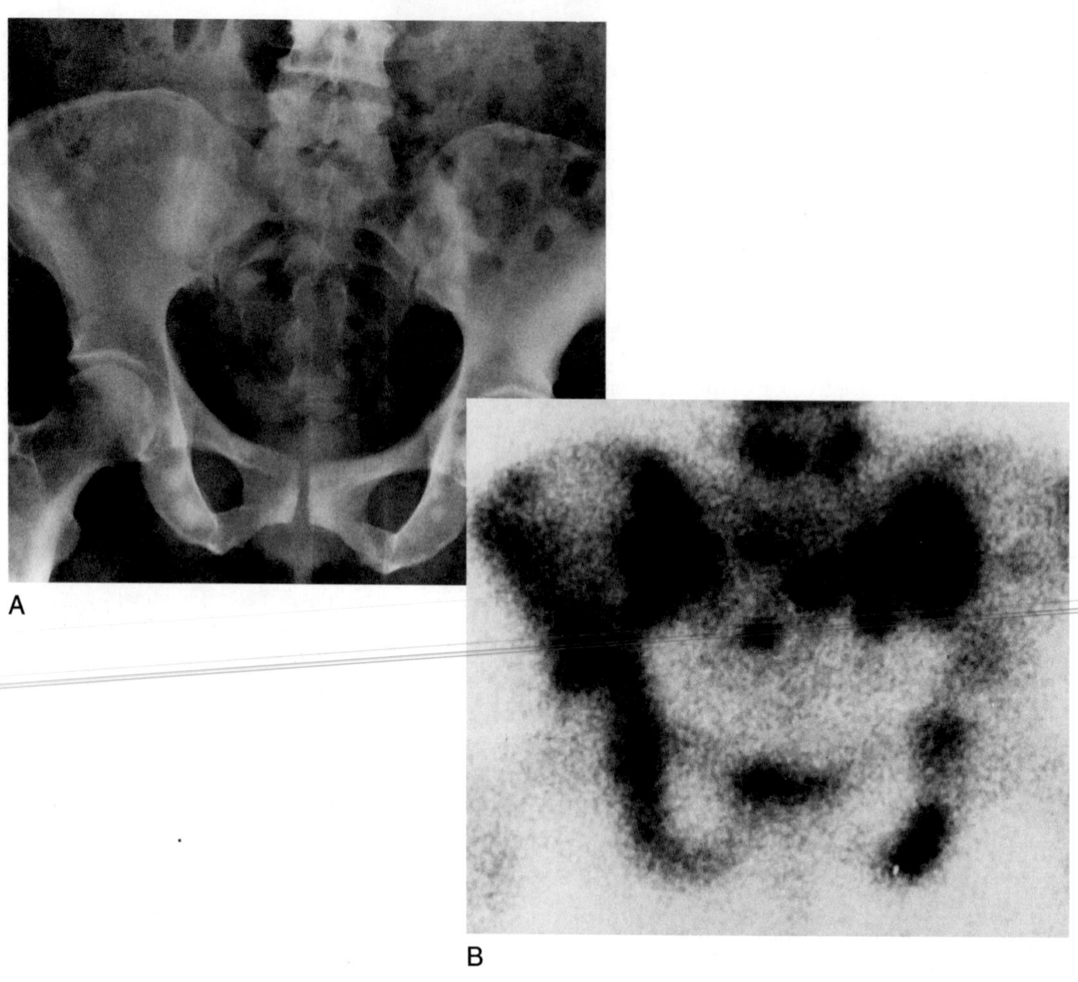

Fig. 10-6 Widespread osteoblastic skeletal metastases in prostate adenocarcinoma. **A**, Anteroposterior radiograph of pelvis shows multiple sclerotic foci. **B**, Radioisotopic bone scan shows multiple foci of increased uptake in pelvis from the same patient. (From Dorfman HD, Czerniak B: *Bone tumors*, St. Louis, 1998, Mosby.)

effects on psychologic well being and health, including depression, skin breakdown, UTIs, and urosepsis. The therapist can have an important role in the successful treatment of incontinence; therefore screening for this symptom is vital and should be a routine part of the health assessment for all adult clients, especially in a primary care setting.

There are four primary types of UI recognized in adults. These are based on the underlying anatomic or physiologic impairment and include stress, urge, mixed (combination of urge and stress), and overflow.

Stress incontinence occurs when the support for the bladder or urethra is weak or damaged, but the bladder itself is normal. With stress incontinence, pressure applied to the bladder from coughing, sneezing, laughing, lifting, exercising, or other physical exertion increases abdominal pressure, and the pelvic floor musculature cannot counteract the urethral/bladder pressure. This type of incontinence causes 75% of all cases of UI in women and is primarily related to urethral sphincter weakness, pelvic floor weakness, and ligamentous and fascial laxity.

Urge incontinence, now more commonly called *overactive bladder,* is the involuntary contraction of the detrusor muscle (smooth muscle of the bladder wall) with a strong desire to void (urgency) and loss of urine as soon as the urge is felt. The bladder involuntarily contracts or is unstable, or there may be involuntary sphincter relaxation.[34] Urge incontinence is often idiopathic but can be caused by medications, alcohol, bladder infections, bladder tumor, neurogenic bladder, or bladder outlet obstruction.

Overflow incontinence is overdistention of the bladder and the bladder cannot empty completely. Urine leaks or dribbles out so the client does not have any sensation of fullness or emptying.

It may be caused by an acontractile or deficient detrusor muscle, a hypotonic or underactive detrusor muscle secondary to drugs, fecal impaction, diabetes, lower spinal cord injury, or disruption of the motor innervation of the detrusor muscle (e.g., multiple sclerosis).

In men, overflow incontinence is most often secondary to obstruction caused by prostatic hyperplasia, prostatic carcinoma, or urethral stricture. In women, this type of incontinence occurs as a result of obstruction caused by severe genital prolapse or surgical overcorrection of urethral detachment.

The client with incontinence from overflow will report a feeling that the bladder does not empty completely with an urge to void frequently, including at night. Small amounts of urine are lost involuntarily throughout the day and night. There may be a weak stream or flow sometimes described as "dribbling."

The term *functional incontinence* describes another type of UI that occurs when the bladder is normal but the mind and body are not working together. Functional incontinence occurs from mobility and access deficits such as being confined to a wheelchair or needing a walker to ambulate.[35]

Deficits in dexterity, such as weakness from a stroke or neuropathy and loss of motion from arthritis, may keep the individual from getting pants unfastened or panties pulled down in time to avoid an accident. Altered mentation from dementia or Alzheimer's disease can also contribute to untimely urination without a urologic structural problem.

Causes of incontinence can range from urologic/gynecologic to neurologic, psychologic, pharmaceutical, or environmental. Anything that can interfere with neurologic function or produce obstruction can contribute to UI. There is a high prevalence of stress and urge incontinence in female elite athletes. The frequency of UI is significantly higher in athletes with eating disorders.[36]

Risk factors for developing UI are listed in Box 10-3. Chronic constipation at any time, but especially during pregnancy, can lead to increased abdominal pressure, which can cause UI. Any condition leading to an enlarged abdomen (e.g., ascites, weight gain, pregnancy) with increased pressure on the bladder can contribute to incontinence.

Chemotherapy, radiation, surgery, and medications can cause disruptions in the cycle of micturition (urination) for many different physiologic reasons. For example, chemotherapy can increase fat deposits and decrease muscle mass, which increase the risk of bowel and bladder dysfunction.

External radiation alters tissue viability in the surrounding area, which can affect circulation to the organs and support from muscle, fascia, ligaments, and tendons.[37,38] Radiation can cause fibrotic contracted bladder tissue and damaged sphincter, contributing to UI. Acute radiation prostatocystitis due to external beam radiation can cause frequency, nocturia, urgency, or urge incontinence, as well as hematuria or transient urine retention.[39,40]

Surgery to remove tumors, lymph nodes, or the prostate can affect bladder control through alterations of blood and lymphatic circulation, innervation, and fascial support. Edema secondary to lymphatic system compromise can increase bladder (and bowel) dysfunction. Brain, spinal cord, or pelvic surgery can affect nervous control of the bowel and bladder.[37] Urge incontinence can occur as a result of bladder denervation from surgical injury.[39] Postprostatectomy UI (when incontinence is defined as any leak) occurs in up to 70% of all cases, but the rate of urine leak decreases as a result of time, medical treatment, and physical therapy intervention. UI is two times more common after prostatectomy than after radiation; surgical clients are three times more likely to

BOX 10-3 RISK FACTORS FOR URINARY INCONTINENCE

Advancing age
Alzheimer's disease or dementia
Arthritis or other musculoskeletal problems
Overweight/obese
Chronic cough
Chronic constipation
History of recurrent urinary tract infections
History of sexually transmitted diseases
Enlarged abdomen (e.g., ascites, pregnancy, obesity, tumor)
Diabetes mellitus
Neurologic disorders
Medications
- Sedatives
- Diuretics
- Estrogens
- Anticholinergics
- Antibiotics
- Alpha-adrenergic blockers (antihistamines, decongestants)
- Calcium channel blockers
- Antipsychotics
- Antidepressants
- Antiparkinsonian drugs
- Laxatives
- Opioids
- Vincristine
- Angiotensin-converting enzyme (ACE) inhibitors

Caffeine, alcohol
Female gender (see below)

Specific to Women

Pregnancy (multiparity)
Vaginal or cesarean* birth
Previous bladder or pelvic surgery
Pelvic trauma or radiation
Bladder or bowel prolapse
Menopause (natural or surgically induced; estrogen deficiency)†
Tobacco use

Specific to Men

Enlarged prostate gland
Prostate or pelvic surgery
Radiation (acute and late complications), especially when combined with brachytherapy[39]

*Although the abdominal muscles are disrupted with a cesarean section and limit how much the woman can bear down on the bladder, abdominal tone and function are essential for pelvic muscle function.

†Urinary incontinence in middle-aged women may be more closely associated with mechanical factors, such as childbearing, history of urinary tract infections, gynecologic surgery, chronic constipation, obesity, and exertion, than with menopausal transition.[64]

use pads. Recovery occurs in most cases between 6 and 12 months after surgery.[39]

Incontinence is not a normal part of the aging process. When confronted with UI in an older adult, consider some of the following causes of this disorder: infection, endocrine disorders, atrophic urethritis or vaginitis, restricted mobility, stool impaction (especially in smokers), alcohol or caffeine intake, and medications.

Smoking contributes to constipation and is often accompanied by chronic cough, which stresses the bladder. Some medications can lead to UI or aggravate already existing UI. Medications commonly involved with alterations in urinary continence include anticholinergic agents, calcium channel blockers, diuretics, sedatives, beta-antagonists, and beta-agonists.[41]

With any kind of incontinence, the onset of cervical spine pain at the same time that UI develops is a red flag. These two findings would suggest there is a protrusion pressing on the spinal cord.

If a medical diagnosis for cervical disk protrusion has been established, referral would not be necessary. However, if incontinence is a new development from the time of the medical evaluation, the physician should be made aware of this information. Cervical spinal manipulation is contraindicated.

Many people are embarrassed about having an incontinence problem. It may help to introduce the subject by making a general statement such as "Many men and women have problems with bladder control. This is an area physical therapists can often help clients with so we routinely ask a few questions about bladder function."

FOLLOW-UP QUESTIONS

Screening questions for incontinence can include

General

Do you have any problems holding urine or emptying your bladder?

Do you ever leak urine or have accidents?

Do you wear pads to protect against urine leaking? Follow-up: How many do you use in a 24-hour period and how wet are they?

Are your activities limited because of urine leaking?

If the client answers "yes" to any of these questions, you may want to screen further with the following questions.

For Stress Incontinence

Do you ever lose urine or wet your pants when you cough, sneeze, or laugh?

Do you lose urine or wet your pants when getting out of a chair, lifting, or exercising?

For overactive bladder (urge incontinence)

Do you have frequent, strong, or sudden urges to urinate and cannot get to the bathroom in time? For example:

When arriving home and getting out of the car?

When using a key to open the door?

When you hear water running?

Or when you run water over your hands?

When you go out into cold weather or put your hands in the freezer?

Do you get to the toilet and lose urine as you are pulling down your panties/shorts?

Do you urinate more than eight times a day?

Do you get up to go more than twice a night?

For Overflow Incontinence

Do you dribble urine during the day and/or at night?

Can you urinate with a strong stream or does the urine dribble out slowly?

Does it feel like your bladder is empty when you are done urinating?

For Functional Incontinence

Can you get to the toilet easily?

Do you have trouble getting to the bathroom on time?

Do you have trouble finding the bathroom or toilet?

Do you have accidents in the bathroom because you cannot get your pants unfastened or pulled down?

Chronic Kidney Disease

Symptoms of renal failure generally cannot be mistaken for musculoskeletal disorders that are treated by physical or occupational therapists. However, patients/clients with chronic kidney disease leading to kidney failure may receive treatment in both inpatient and outpatient clinics for primary musculoskeletal lesions. Understanding symptoms associated with kidney disease and recognizing complications associated with dialysis shunt are imperative for the therapist.

Kidney failure exists when the kidneys can no longer maintain the homeostatic balances within the body that are necessary for life. Renal failure is classified as acute or chronic in origin and progression. *Acute renal failure* refers to the abrupt cessation of kidney activity, usually occurring over a period of hours to a few days. Acute renal failure is often reversible, with return of kidney function in 3 to 12 months.

Chronic renal failure, or irreversible renal failure (also known as end-stage renal disease [ESRD]), is defined as a state of progressive decrease in the ability of the kidney to filter fluids, metabolites, and electrolytes from the body, resulting in eventual permanent loss of kidney function. ESRD is the final stage (stage 5) of chronic kidney disease; it can develop slowly over a period of years or can result from an episode of acute renal failure that does not resolve.

ESRD is a complex condition with multiple systemic complications. Diabetic nephropathy is the primary cause of ESRD, accounting for approximately 40% of newly diagnosed cases of ESRD.[42] Individuals with diabetes and ESRD have higher morbidity and mortality rates than individuals with ESRD only.[43]

Risk factors for ESRD include advancing age, diabetes mellitus, hypertension, chronic urinary tract obstruction and infection (especially glomerulonephritis), and kidney transplantation. Hereditary defects of the kidneys, polycystic kidneys, and glomerular disorders such as glomerulonephritis can also lead to renal failure.

Chronic intake of certain medications and over-the-counter (OTC) drugs is also a factor in the development of renal disease. The increasing availability of OTC drugs has led to consumers treating themselves when they may lack the knowledge to do so safely. Age-related decline in renal function combined with multiple medication use in the aging adult population increases the risk of hepatotoxicity.[44] Excessive consumption of acetaminophen and nonsteroidal anti-inflammatory drugs, especially when combined with caffeine and/or codeine, are toxic to the kidneys.[45,46]

Clinical Signs and Symptoms

Failure of the filtering and regulating mechanisms of the kidney can be either acute (sudden in onset and potentially reversible) or chronic (called *uremia,* which develops gradually and is usually irreversible).

Individuals with diabetes and ESRD often have autonomic dysfunction and sensorimotor peripheral (uremic) neuropathies affecting the distal extremities. Symptoms tend to be symmetric and more subjective than objective such as restless legs syndrome, cramps, paresthesias, impaired vibration sense, burning feet syndrome, abnormal Achilles reflex, pruritus (itching of the skin), constipation or diarrhea, abdominal bloating, and decreased sweating. When present, a fall in blood pressure is one measurable sign of autonomic nervous system dysfunction.[47]

Individuals with either type of renal failure develop signs and symptoms characteristic of impaired fluid and waste excretion and altered renal regulation of other body metabolic processes such as pH regulation, RBC production, and calcium-phosphorus balance.

Signs of renal impairment are shown in Table 10-4. The signs of actual renal failure are the same but more pronounced. In most cases of renal failure, urine volume is significantly decreased or absent. Edema becomes severe and can result in heart failure. Renal anemia is usually associated with extreme fatigue and intolerance to normal daily activities, as well as a marked decrease in exercise capacity.[48]

In addition, the continuous presence of toxic waste products in the bloodstream (urea, creatinine, uric acid) results in damage to many other body systems, including the central nervous system (CNS), peripheral nervous system (PNS), eyes, gastrointestinal (GI) tract, integumentary system, endocrine system, and cardiopulmonary system.

Treatment of renal failure involves several elements designed to replace the lost excretory and metabolic functions of this organ. Treatment options include dialysis, dietary changes, and medications to regulate blood pressure and assist in replacement of lost metabolic functions, such as calcium balance and RBC production.

The choice of treatment options, such as dialysis, transplantation, or no treatment, depends on many factors, including the person's age, underlying physical problems, and availability of compatible organs for transplantation.[49] Untreated or chronic renal failure eventually results in death.

From a screening perspective, the therapist must be alert to the many complications associated with chronic renal failure and dialysis. Watch for signs and symptoms of fluid and electrolyte imbalances (see Chapter 11), dehydration (see Chapter 11), cardiac arrhythmias (see Chapter 6), and depression (see Chapter 3).

CLINICAL SIGNS AND SYMPTOMS

Renal Impairment

Symptoms of upper urinary tract infection, particularly renal infection, can be categorized according to urinary tract manifestations or systemic manifestations caused by renal impairment (see Table 10-4). Clinical signs and symptoms of urinary tract involvement can include

- Unilateral costovertebral tenderness
- Flank pain
- Ipsilateral shoulder pain
- Fever and chills
- Skin hypersensitivity
- Hematuria (blood in urine)
- Pyuria (pus in urine)
- Bacteriuria (presence of bacteria in urine)
- Hypertension
- Decreased urinary output
- Dependent edema
- Weakness
- Anorexia (loss of appetite)
- Dyspnea
- Mild headache
- Proteinuria (protein in urine, urine may be foamy)
- Abnormal blood serum level, such as elevated blood urea nitrogen (BUN) and creatinine
- Anemia

Cancers of the Urinary Tract

Bladder Cancer

Bladder cancer is a common, major public health concern that is strongly linked to cigarette smoking.[50] It is the fourth most common cancer in men and the tenth most common in women. Bladder cancer is nearly three times more common in men than in women, thus it is typically diagnosed later in women and often at a more advanced stage.[51]

The exact cause of bladder cancer is not known, but certain risk factors have been identified which increase chances of developing this type of cancer.[52]

- Age (over 40)
- Tobacco use (cigarette, pipe, and cigar smokers)
- Occupation (exposure to work place carcinogens such as paper, rubber, chemical, leather industries; hairdressers, machinists, metal workers, dental workers, printers, painters, auto workers, textile workers, truck drivers)
- Infections (parasitic, usually in tropical areas of the world)
- Treatment with cyclophosphamide or arsenic (for other cancers)
- Race (whites highest; Asians lowest)

- Gender (men two to three times more likely than women)
- Previous personal history of bladder cancer
- Family history (some association but not clearly defined)[53,54]

Common symptoms of bladder cancer include blood in the urine, pain during urination, and urinary urgency or the feeling of urinary urgency without resulting urination. Overactive bladder with or without hematuria (blood in the urine) may be a presenting symptom. This symptom is 10 times more common in women than men despite the fact that bladder cancer is more common in men than women.[55] These symptoms are not sure signs of bladder cancer, but anyone with these symptoms should be referred to a physician for further follow-up studies.

Measures that have been shown to reduce the risk of developing bladder cancer include cessation of smoking, adequate intake of fluids, intake of cruciferous vegetables, limiting exposure to workplace chemicals, and prompt treatment of bladder infections.

Renal Cancer

Cancer of the kidney (renal cancer) develops most often in people over the age of 40 and has some associated risk factors. Risk factors for renal cancer include

- Smoking (two times the risk as nonsmokers)
- Obesity
- Hypertension
- Long-term dialysis
- Von Hippel-Lindau (VHL) syndrome (genetic, familial syndrome)
- Occupation (coke oven workers in the iron and steel industry; asbestos and cadmium exposure)
- Gender (men twice more likely than women)

Common symptoms of renal cancer are very similar to those of bladder cancer and require immediate referral for follow-up. These symptoms can include blood in the urine, pain in the side that does not go away, a lump or mass in the side or abdomen, weight loss, fever, and general fatigue or feeling of poor health.[56]

CLINICAL SIGNS AND SYMPTOMS

Bladder and Renal Cancer

Bladder Cancer
- Blood in the urine
- Pain during urination
- Urinary urgency

Renal Cancer
- Blood in the urine
- Pain during urination
- Urinary urgency
- Flank or side pain
- Lump or mass in the side or abdomen
- Weight loss
- Fever
- General fatigue; feeling of poor health

Testicular Cancer[57]

The testicles (also called *testes* or *gonads*) are the male sex glands. They are located behind the penis in a pouch of skin called the *scrotum* (see Fig. 10-3). The testicles produce and store sperm and serve as the body's main source of male hormones. These hormones control the development of the reproductive organs and other male characteristics such as body and facial hair, low voice, wide shoulders, and sexual function.

Testicular cancer is relatively rare and occurs most often in young men between the ages of 15 and 35 years old, although any male can be affected at any time (including infants). According to the National Cancer Institute's Surveillance, about 8500 men are diagnosed with testicular cancer each year (350 deaths annually).[30] The incidence of testicular cancer around the world has doubled in the past 30 to 40 years.

The cause of testicular cancer and even the risk factors remain unknown. Risk is higher than average for boys born with an undescended testicle (cryptorchidism). The cancer risk for boys with this condition is increased even if surgery is done to move the testicle into the scrotum. In the case of unilateral cryptorchidism, the risk of testicular cancer is increased in the normal testicle as well. This fact suggests testicular cancer is due to whatever caused the undescended testicle.[58]

Having a brother or father with testicular cancer also increases an individual's risk. Other risk factors may include occupation (e.g., miners, oil and gas workers, leather workers, food and beverage processing workers, janitors, firefighters, utility workers) and human immunodeficiency virus (HIV) infection.

The risk of testicular cancer among white American men is about 4 times that of African-American men and more than twice that of Asian-American men.[59] The risk for Hispanics is between that of Asians and non-Hispanic whites. The reason for this difference is unknown.

The testicular cancer rate has more than doubled among white Americans in the past 40 years but has not changed for African Americans. However, African Americans present with more advanced disease at the time of diagnosis and African-American men with testicular cancer have a higher mortality rate compared with Caucasians.[59,60] Worldwide, the risk of developing this disease is highest among men living in the United States and Europe and lowest among African and Asian men.

Clinical Signs and Symptoms

Testicular cancer can be completely asymptomatic. The most common sign is a hard, painless lump in the testicle about the size of a pea. There may be a dull ache in the scrotum and the man may be aware of tender, larger breasts. Other symptoms are listed in the box Clinical Signs and Symptoms: Testicular Cancer.

There are three stages of testicular cancer:

- Stage I: The cancer is confined to the testicle.
- Stage II: The cancer has spread to the retroperitoneal lymph nodes, located in the posterior abdominal cavity below the diaphragm and between the kidneys.
- Stage III: The cancer has spread beyond the lymph nodes to remote sites in the body, including the lungs, brain, liver, and bones.

If found early, testicular cancer is almost always curable.[30] The American Cancer Society recommends monthly self-exam of the testicles for adolescents and men, starting at age 15. Testicular self-examination is an effective way of getting to know this area of the body and thus detecting testicular cancer at a very early, curable stage. The self-exam is best performed once each month during or after a warm bath or shower when the heat has relaxed the scrotum (see Appendix D-8).

Men who have been treated for cancer in one testicle have about a 3% to 4% chance of developing cancer in the remaining testicle. If cancer does arise in the second testicle, it is nearly always a new disease rather than metastasis from the first tumor.

Metastases occur via the blood or lymph system. The most common place for the disease spread is to the lymph nodes in the posterior part of the abdomen. Therefore lower back pain is a frequent symptom of later stage testicular cancer (Case Example 10-4). If the cancer has spread to the lungs, persistent cough, chest pain, and/or shortness of breath can occur. Hemoptysis (sputum with blood) may also develop.

Survivors of testicular cancer should be checked regularly by their doctors and should continue to perform monthly testicular self-examinations. Any unusual symptoms should be reported to the doctor immediately. Outcome even after a secondary testicular cancer is still excellent with early detection and treatment.

CLINICAL SIGNS AND SYMPTOMS

Testicular Cancer

- A lump in either testicle
- Any enlargement, swelling, or hardness of a testicle
- Significant loss of size in one of the testicles
- Feeling of heaviness in the scrotum and/or lower abdomen
- Dull ache in the lower abdomen or in the groin
- Sudden collection of fluid in the scrotum
- Pain or discomfort in a testicle or in the scrotum
- Enlargement or tenderness of the breasts
- Unexplained fatigue or malaise
- Infertility
- Low back pain (metastases to retroperitoneal lymph nodes)

CASE EXAMPLE 10-4

Testicular Cancer

A 20-year-old track star and college football player developed back, buttock, and posterior thigh pain after a football injury. He was sent to physical therapy by the team physician with a diagnosis of "Sciatica; L4-5 radiculopathy. Please treat using McKenzie exercise program."

During the physical therapy interview, the client reported left low back pain and left buttock pain present for the last 2 weeks after being tackled from the right side in a football game. Symptoms developed approximately 12 hours after the injury. Pain was always present but was worse after sitting and better after standing.

On examination the client presented with major losses of lumbar spine range of motion in all planes. There was no observable lateral shift and lumbar lordosis was not excessive or reduced. Overall postural assessment was unremarkable.

He was able to lie flat in the prone position and perform a small prone press up without increasing any of his symptoms, but he described feeling a "hard knot in my stomach" while in this position. When asked if he had any symptoms of any kind anywhere else in his body, the client replied that right after the injury, his left testicle swelled up but seemed better now. He denied any blood in the urine or difficulty urinating. Vital signs were within normal limits.

Even though the therapist thought the clinical findings supported a diagnosis of a derangement syndrome according to the McKenzie classification, there were enough red flags to warrant further investigation.

The client was given an appropriate self-treatment program to perform throughout the day with instructions for self-assessment of his condition. In the meantime, the therapist contacted the physician with the following concerns:

- Palpable (nonpulsatile) abdominal mass in the left upper abdominal quadrant (anterior)
- Reported left testicular swelling
- Age
- No imaging studies were done to confirm a disk lesion as the underlying cause of the symptoms

Result: Physician referral was made after a telephone discussion outlining the additional findings listed above. An abdominal CT scan showed a 20-cm (5-inch) abdominal mass pressing on the spinal nerves as the cause of the back pain. Further diagnostic testing revealed testicular cancer as the primary diagnosis, with metastases to the abdomen causing the abdominal mass.

Surgery was performed to remove the testicle. The back pain was relieved within 3 days of starting chemotherapy. Physical therapy was discontinued for back pain, but a new plan of care was established for exercise during cancer treatment.

PHYSICIAN REFERRAL

The proximity of the kidneys, ureters, bladder, and urethra to the ribs, vertebrae, diaphragm, and accompanying muscles and tendinous insertions often can make it difficult to identify the client's problems accurately.

Pain related to a urinary tract problem can be similar to pain felt from an injury to the back, flank, abdomen, or upper thigh. The physical therapist is advised to question the client further whenever any of the signs and symptoms listed in Table 10-3 are reported or observed. Further diagnostic testing and medical examination must be performed by the physician to differentiate urinary tract conditions from musculoskeletal problems.

The physical therapist must be able to recognize the systemic origin of urinary tract symptoms that mimic musculoskeletal pain. Many conditions that produce urinary tract pain also include an elevation in temperature, abnormal urinary constituents, and changes in color, odor, or amount of urine.

These types of changes would not be observed or reported with a musculoskeletal condition, and the client may not mention them, thinking these symptoms do not have anything to do with the back, flank, or thigh pain present. The therapist must ask a few screening questions to bring this kind of information to the forefront.

When the physical therapist conducts a Review of Systems, any signs and symptoms associated with renal or urologic impairment should be correlated with the findings of the objective examination and combined with the medical history to provide a comprehensive report at the time of referral to the physician or other health care provider.

Diagnostic Testing

Screening of the composition of the urine is called *urinalysis* (UA), and UA is the commonly used method of determining various properties of urine. This analysis is actually a series of several tests of urinary components and is a valuable aid in the diagnosis of urinary tract or metabolic disorders.

Normal urinary constituents are shown (see inside front cover: Urine Analysis). Urine cultures are also very important studies in the diagnosis of UTIs. Anyone at risk for chronic kidney disease should be tested for markers of kidney damage. This is done by urinalysis for albumin (protein in the urine) and by blood serum for creatinine (waste product of muscle metabolism).

Various *blood studies* can be done to assess renal function (see inside front cover: Renal Blood Studies). These studies examine both the serum and cellular components of the blood for specific changes characteristic of renal performance. Substances that must be examined in the serum are those that are a *direct* reflection of renal function, such as creatinine, and others that are more *indirect* in renal evaluation, such as blood urea nitrogen (BUN), pH-related substances, uric acid, various ions, electrolytes, and cellular components (RBCs). (For a more in-depth discussion of laboratory values the reader is referred to a more specific source of information.)[61-63]

Guidelines for Immediate Medical Attention

- The presence of any amount of blood in the urine always requires a referral to a physician. However, the presence of abnormalities in the urine may not be obvious, and a thorough diagnostic analysis of the urine may be needed. Careful questioning of the client regarding urinary tract history, urinary patterns, urinary characteristics, and pain patterns may elicit valuable information relating to potential urinary tract symptoms.
- Presence of cervical spine pain at the same time that urinary incontinence develops. If a diagnosis of cervical disk prolapse has been made, the physician should be notified of these findings; referral may not be necessary, but communication with the physician to confirm this is necessary.
- Client with bowel/bladder incontinence and/or saddle anesthesia secondary to cauda equina lesion.

Guidelines for Physician Referral

Although immediate (emergency) medical attention is not required, medical referral is needed under the following circumstances:

- When the client has any combination of systemic signs and symptoms presented in this chapter. Damage to the urinary tract structures can occur with accident, injury, assault, or other trauma to the musculoskeletal structures surrounding the kidney and urinary tract and may require medical evaluation if the clinical presentation or response to physical therapy treatment suggests it.

  For example, the alpine skier discussed at the beginning of the chapter had a dull, aching costovertebral pain on the left side that was unrelieved by a change of position or by ice, heat, or aspirin. His pain is related directly to a traumatic episode, and musculoskeletal injury is a definite possibility in his case. He has no medical history of urinary tract problems and denies any changes in urine or pattern of urination. Because the pain is constant and unrelieved by usual measures and the location of the pain is approximate to renal structures, a medical follow up and urinalysis would be recommended.
- Back or shoulder pain accompanied by abnormal urinary constituents (e.g., change in color, odor, amount, flow of urine).
- Positive Murphy's percussion (punch) test, especially with a recent history of renal or urologic infection.

Clues Suggesting Pain of Renal/Urologic Origin

- Men 45 years old or older
- In men, back pain accompanied by burning on urination, difficulty in urination, or fever may be associated with prostatitis; usually in such a case, there is no limitation of back motion and no muscle spasm (until symptoms progress, causing muscle guarding and splinting)
- Blood in urine
- Change in urinary pattern such as increased or decreased frequency, change in flow of urine stream (weak or dribbling), and increased nocturia
- Presence of constitutional symptoms, especially fever and chills; pain is constant (may be dull or sharp, depending on the cause)
- Pain is unchanged by altering body position; side bending to the involved side and pressure at that level is "more comfortable" (may reduce pain but does not eliminate it)

- Neither renal nor urethral pain is altered by a change in body position; pseudorenal pain from a mechanical cause can be relieved by a change in position
- True renal pain is seldom affected by movements of the spine
- Straight leg–raising test is negative with renal colic appearing as back pain
- Back pain at the level of the kidneys in a woman with previous breast or uterine cancer (ovarian cancer)
- Assessment for pseudorenal pain is negative (see Table 10-1)

RENAL AND UROLOGIC PAIN PATTERNS

KIDNEY (FIG. 10-7)

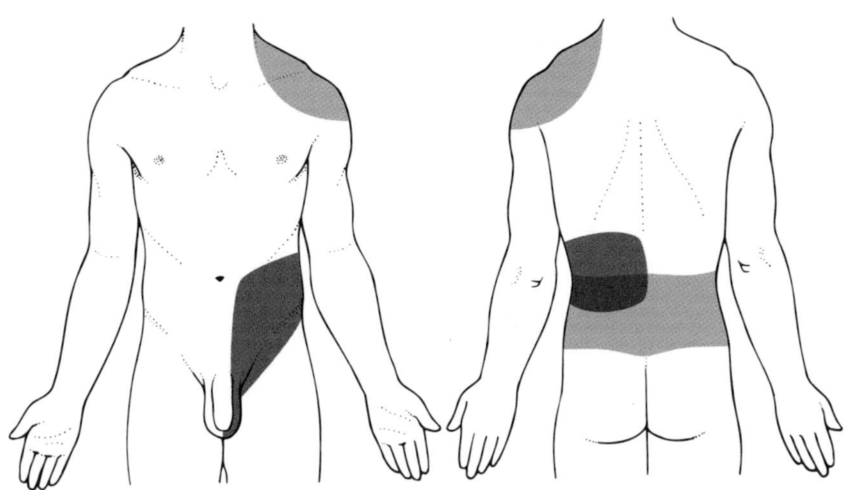

Fig. 10-7 Renal pain is typically felt in the posterior subcostal and costovertebral region *(dark red)*. It can radiate across the low back *(light red)* and/or forward around the flank into the lower abdominal quadrant. Ipsilateral groin and testicular pain may also accompany renal pain. Pressure from the kidney on the diaphragm may cause ipsilateral shoulder pain.

Location: Posterior subcostal and costovertebral region
Usually unilateral

Referral: Radiates forward, around the flank or the side into the lower abdominal quadrant (T11-12), along the pelvic crest and into the groin
Pressure from the kidney on the diaphragm may cause ipsilateral shoulder pain

Description: Dull, aching, boring

Intensity: Acute: Severe, intense
Chronic: Vague and poorly localized

Duration: Constant

Associated Signs and Symptoms: Fever, chills
Increased urinary frequency
Blood in urine
Hyperesthesia of associated dermatomes (T9 and T10)
Ipsilateral or generalized abdominal pain
Spasm of abdominal muscles
Nausea and vomiting when severely acute
Testicular pain may occur in men
Unrelieved by a change in position

Continued

RENAL AND UROLOGIC PAIN PATTERNS—*cont'd*

URETER (FIG. 10-8)

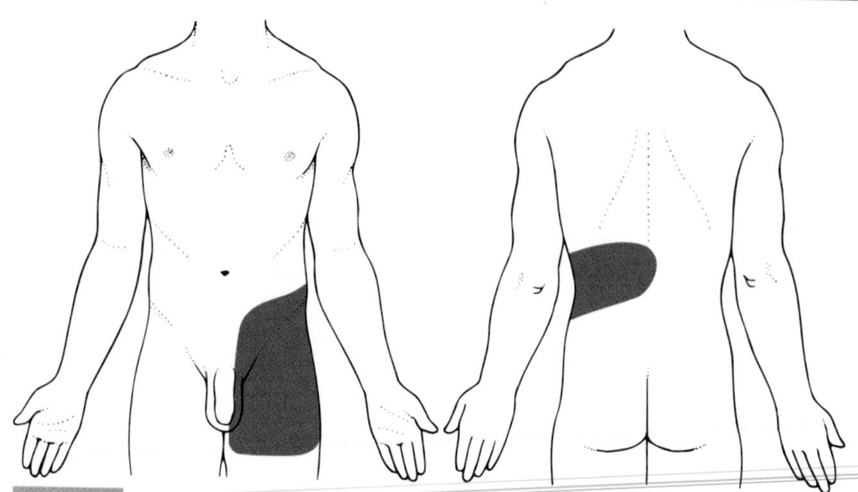

Fig. 10-8 Ureteral pain may begin posteriorly in the costovertebral angle. It may then radiate anteriorly to the ipsilateral lower abdomen, upper thigh, testes, or labium.

Location: Costovertebral angle
Unilateral or bilateral

Referral: Radiates to the lower abdomen, upper thigh, testis, or labium on the same side (groin and genital area)

Description: Described as crescendo waves of colic

Intensity: Excruciating, severe (Ureteral pain is commonly acute and caused by a kidney stone. Lesions outside the ureter are usually painless until advanced progression of the disease occurs.)

Duration: Ureteral pain caused by calculus is intermittent or constant without relief until treated or until the stone is passed

Associated Signs and Symptoms: Rectal tenesmus (painful spasm of anal sphincter with urgent desire to evacuate the bowel/bladder; involuntary straining with little passage of urine or feces)
Nausea, abdominal distention, vomiting
Hyperesthesia of associated dermatomes (T10-L1)
Tenderness over the kidney or ureter
Unrelieved by a change in position
Movement of iliopsoas may aggravate symptoms associated with a lesion outside the ureter (see Fig. 8-3)

RENAL AND UROLOGIC PAIN PATTERNS—*cont'd*

BLADDER/URETHRA (FIG. 10-9)

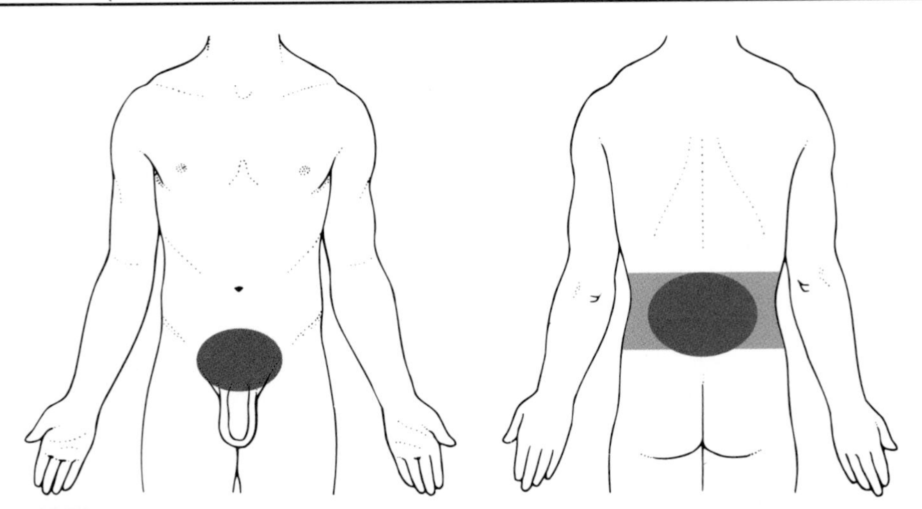

Fig. 10-9 *Left*, Bladder or urethral pain is usually felt suprapubically or ipsilaterally in the lower abdomen. This is the same pattern for gas pain from the lower gastrointestinal (GI) tract for some people. *Right*, Bladder or urethral pain may also be perceived in the low back area (*dark red:* primary pain center; *light red:* referred pain). Low back pain may occur as the first and only symptom associated with bladder/urethral pain, or it may occur along with suprapubic or abdominal pain or both.

Location: Suprapubic or low abdomen, low back
Referral: Pelvis
Can be confused with gas
Description: Sharp, localized
Intensity: Moderate-to-severe
Duration: Intermittent; may be relieved by emptying the bladder
Associated Signs and Symptoms: Great urinary urgency
Tenesmus
Dysuria
Hot or burning sensation during urination

PROSTATE (FIG. 10-10)

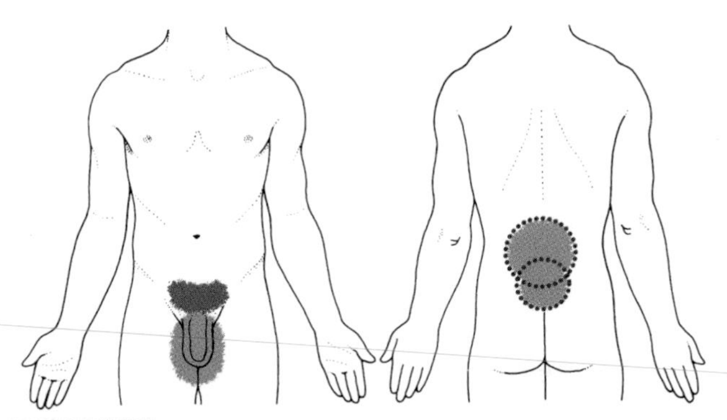

Fig. 10-10 The prostate is segmentally innervated from T11-L1, S2-S4. Prostate problems can be painless. When pain occurs, the primary pain pattern is in the lower abdomen, suprapubic region *(dark red),* and perineum (between the rectum and testes; not pictured). Pain can be referred to the low back, sacrum, testes, and inner thighs *(light red).*

Continued

RENAL AND UROLOGIC PAIN PATTERNS—*cont'd*

Symptoms of prostate involvement vary depending on the underlying cause (e.g., prostatitis versus BPH versus prostate cancer).

Location: May be pain free; lower abdomen, suprapubic region

Referral: Low back, pelvis, sacrum, perineum, inner thighs, testes; thoracolumbar spine with metastases (the latter is not pictured)

Description: Persistent aching pain; pain is reproduced with digital rectal exam

Intensity: Mild-to-severe; varies from person to person and can fluctuate for each individual on any given day

Duration: Varies according to underlying cause

Associated Signs and Symptoms:
Chills and fever (prostatitis)
Frequent and/or painful urination
Urgency, hesitancy
Nocturia
Incomplete emptying of bladder
Painful ejaculation
Hematuria
Arthralgia, myalgia

Key Points to Remember

- Renal and urologic pain can be referred to the shoulder or low back.
- Lesions outside the ureter can cause pain on movement of the adjacent iliopsoas muscle.
- Radiculitis can mimic ureteral colic or renal pain, but true renal pain is seldom affected by movements of the spine.
- Inflammatory pain may be relieved by a change in position. Renal colic remains unchanged by a change in position.
- Change in color, consistency, smell, or reduced volume or flow of urine requires further assessment and change in urgency and frequency, and pain with urination requires further evaluation.
- Low back, pelvic, or femur pain can be the first symptom of prostate cancer.
- Change in size, shape, or appearance of testicles or penis with or without urethral discharge requires further assessment.
- Change in the smell, volume, or consistency of ejaculate with or without pain during intercourse requires further assessment.
- Urinary incontinence is not a normal part of aging and should be evaluated carefully.
- With any kind of incontinence, the onset of cervical spine pain at the same time that urinary incontinence develops is a red flag and contraindicates the use of cervical spinal manipulation.
- Lower thoracic disk herniation can cause groin pain and/or leg pain, mimicking renal pain. The presence of neurologic changes, such as bladder dysfunction, can cause confusion when trying to differentiate a systemic from neuromusculoskeletal cause of symptoms. True renal pain is seldom affected by movements of the spine. Compare results of palpation and percussion tests.
- Testicular cancer with metastasis to the lymph system or bone can cause low back pain from pressure on the spinal nerves. Always watch for red flags even when an injury occurs; this is especially true in the young adult or athlete.
- Anyone with hypertension and/or diabetes (and/or other significant risk factors for renal disease) should be monitored carefully and consistently for any systemic signs and symptoms of renal impairment.
- People with diabetes are prone to complications associated with urinary tract infections.

The sudden onset of nonspecific low back pain, unrelated to any specific motion may be an indication of osteomyelitis from spread of infection to the spine. Take the client's body temperature and ask him/her to monitor temperature for a few days to uncover the possibility of a low-grade fever associated with osteomyelitis.

All the possible pain patterns discussed in this chapter are presented in the figure at the top of p. 405.

■ Key Points to Remember—cont'd

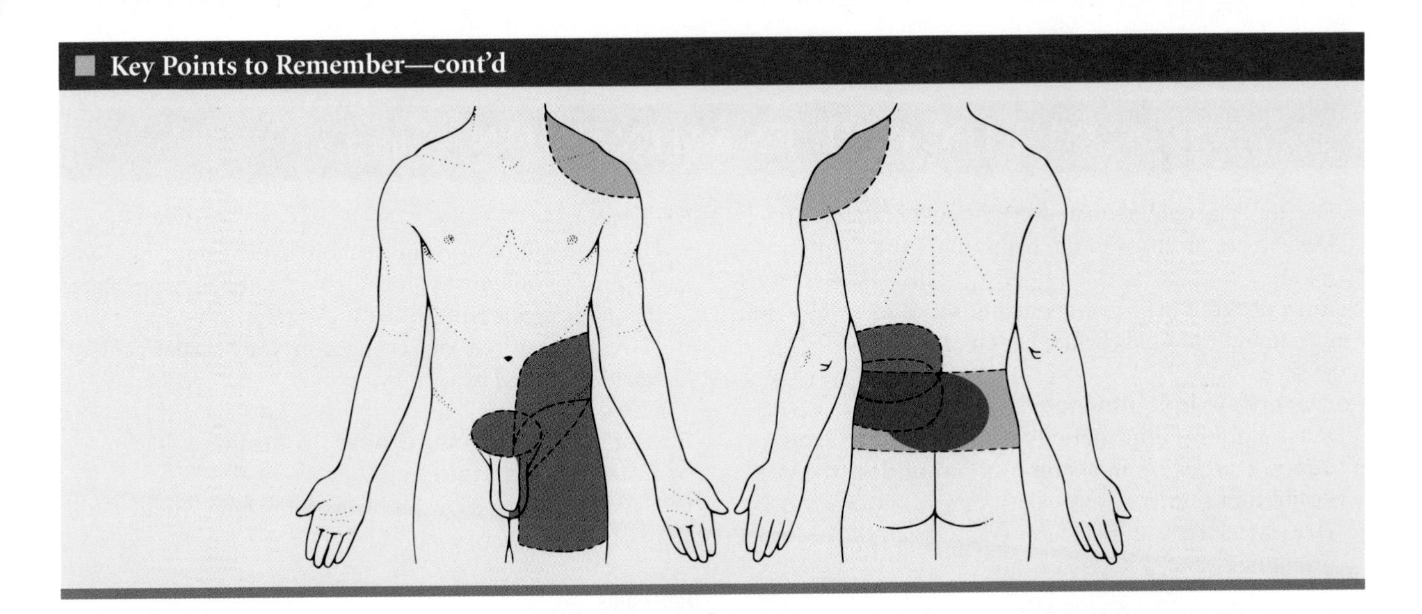

SUBJECTIVE EXAMINATION

Special Questions to Ask

Clients may be reluctant to answer the physical therapist's questions concerning bladder and urinary function. The physical therapist is advised to explain the need to rule out possible causes of pain related to the kidneys and bladder and to give the client time to respond if answers seem to be uncertain. For example, the physical therapist may ask the client to observe urinary function over the next 2 days. These questions should be reviewed again at the next appointment.

Past Medical History

- Have you had any problems with your prostate (for men), kidneys, or bladder? *If so,* describe.
- Have you ever had kidney or bladder stones? If so, when? How were these stones treated?
- Have you had an injury to your bladder or kidneys? *If so,* when? How was this treated? (**Be aware of unreported domestic abuse/assault.**)
- Have you had any kidney or bladder infections in the past 6 months? How were these infections treated? Were they related to any specific circumstances (**e.g., pregnancy, intercourse, after strep throat or strep skin infections**)?
- Have you ever had surgery on your bladder or kidneys? *If so,* when and what?
- Have you had any hernias? *If yes,* when and how was this treated?
- Have you ever had cancer of any kind?
- Have you ever had testicular, kidney, bladder, or prostate cancer?
- Have you ever been treated with radiation or chemotherapy?

Special Questions to Ask: Bladder Control/Incontinence

Begin with a lead-in introduction to these questions such as:

Many people are embarrassed about having an incontinence problem. It may help to introduce the subject by making a general statement such as: "Many men and women have problems with bladder control. This is an area physical therapists can often help clients with so we routinely ask a few questions about bladder function."

General

- Do you have any problems holding urine or emptying your bladder?
- Do you ever leak urine or have accidents?
- Do you wear pads to protect against urine leaking? Follow-up: How many do you use in a 24-hour period? How wet are they?
- Are your activities limited because of urine leaking?

If the male client answers "yes" to any of these questions, you may want to screen further with the following questions. See also Appendix B-30.

For Stress Incontinence

- Do you ever lose urine or wet your pants when you cough, sneeze, or laugh?
- Do you lose urine or wet your pants when getting out of a chair, lifting, or exercising?

For Overactive Bladder (Urge Incontinence)

- Do you have frequent, strong, or sudden urges to urinate and cannot get to the bathroom in time? For example:
    - When arriving home and getting out of the car?
    - When using a key to open the door?
    - When you hear water running?
    - Or when you run water over your hands?
    - When you go out into cold weather or put your hands in the freezer?
- Do you get to the toilet and lose urine as you are pulling down your panties/shorts?

Continued

SUBJECTIVE EXAMINATION—*cont'd*

- Do you urinate more than every 2 hours in the daytime?
- Do you get up to go to the bathroom more than once a night?

If yes, does this happen every night? Is it because you drink a large amount of fluids before bedtime?

For Overflow Incontinence

- Do you dribble urine during the day and/or at night?
- Can you urinate with a strong stream or does the urine dribble out slowly?
- Does it feel like your bladder is empty when you are done urinating?

For Functional Incontinence

- Can you get to the toilet easily?
- Do you have trouble getting to the bathroom on time?
- Do you have trouble finding the bathroom or toilet?
- Do you have accidents in the bathroom because you cannot get your pants unfastened or pulled down?

Special Questions to Ask: Urinary Tract Infection

- Have you had any side (flank) pain (**kidney or ureter**) or pain just above the pubic area (**suprapubic: bladder or urethra, prostate**)?
  - *If so,* what relieves this pain? Does a change in position affect it? (**Inflammatory pain** may be relieved by a change in position. **Renal colic** remains unchanged by a change in position.)
- During the last 2 to 3 weeks, have you noticed a change in the amount or number of times that you urinate? (**Infection**)
- Do you ever have pain or a burning sensation when you urinate? (**Lower urinary tract irritation; prostatitis; venereal disease**)
- Does your urine look brown, red, or black? (Changes in urine color may be normal with some medications and foods such as beets or rhubarb.)
- Is your urine clear or cloudy? If not clear, describe. How often does this happen? (Could indicate **upper or lower UTI.**)
- Have you noticed an unusual or foul odor coming from your urine? (**Infection, secondary to medication**; may be normal after eating asparagus.)

For Women

- When you urinate, do you have trouble starting or continuing the flow of urine? (**Urethral obstruction**)
- Have you noticed any unusual vaginal discharge during the time that you had pain (pubic, flank, thigh, back, labia)? (**Infection**)
- Have you noticed any changes in your sexual activity/function caused by your symptoms?

For Men

- Have you noticed any unusual discharge from your penis during the time that you had pain (especially pain above the pubic area)? (**Infection**)
- Have you noticed any changes in your sexual activity/function caused by your symptoms?

Screening Questions to Ask: Prostatitis or Enlarged Prostate

- Have you ever had any problems with your prostate in the past?

Prostatitis

- Do you ever have burning pain or discomfort during urination?
- Does it feel like your bladder is not empty when you finish urinating?
- Do you have to go to the bathroom every 2 hours (or more often)?
- Do you ever have pain or discomfort in your testicles, penis, or the area between your rectum and your testicles (perineum)?
- Do you ever have pain in your pubic or bladder area?
- Do you have any discomfort during or after sexual climax (ejaculation)?

Enlarged Prostate

- Does it feel like your bladder is not empty when you finish urinating?
- Do you have to urinate again less than 2 hours after you finished going to the bathroom last?
- Do you have a weak stream of urine or find you have to start and stop urinating several times when you go to the bathroom?
- Do you have an urge to go to the bathroom but very little urine comes out?
- Do you have to push or strain to start urinating or to keep the urine flowing?
- Do you have any leaking or dribbling of urine from the penis?
- How often do you get up to urinate at night?

The American Urologic Association recommends using the following scale when asking most of these screening questions. Some questions such as "How often do you get up at night?" require a single number response. A total score of seven or more suggests the need for medical evaluation:

0	1	2	3	4	5
Not at all	Less than one time in five	Less than half the time	About half the time	More than half the time	Almost always

CASE STUDY

REFERRAL

The client is self-referred and states that he has been to your hospital-based outpatient clinic in the past. He has a very extensive chart containing his entire medical history for the last 20 years.

BACKGROUND INFORMATION

He is a 44-year-old man who describes his current occupation as "errand boy/gopher," which requires minimal lifting, bending, or strenuous physical activity. His chief complaint today is pain in the lower back, which comes and goes and seems to be aggravated by sitting. The pain is poorly described, and the client is unable to specify any kind of descriptive words for the type of pain, intensity, or duration.

SPECIAL QUESTIONS TO ASK

See Chapter 14 for Special Questions to Ask about the back. The client's answer to any questions related to bowel and bladder functions is either "I don't know" or "Well, you know," which makes a complete interview impossible.

SUBJECTIVE/OBJECTIVE FINDINGS

There are radiating symptoms of numbness down the left leg to the foot. The client denies any saddle anesthesia. Deep tendon reflexes are intact bilaterally, and the client stands with an obvious scoliotic list to one side. He is unable to tell you whether his symptoms are relieved or alleviated on performing a lateral shift to correct the curve. There are no other positive neuromuscular findings or associated systemic symptoms.

RESULT

After 3 days of treatment over the course of 1 week, the client has had no subjective improvement in symptoms. Objectively, the scoliotic shift has not changed. A second opinion is sought from two other staff members, and the consensus is to refer the client to his physician. The physician performs a rectal examination and confirms a positive diagnosis of prostatitis based on the results of laboratory tests. These tests were consistent with the client's physical findings and previous history of prostate problems 1 year ago. The client was reluctant to discuss bowel or bladder function with the female therapist but readily suggested to his physician that his current symptoms mimicked an earlier episode of prostatitis.

It is not always possible to elicit thorough responses from clients concerning matters of genitourinary function. If the client hesitates or is unable to answer questions satisfactorily, it may be necessary to present the questions again at a later time (e.g., next treatment session), to ask a colleague of the client's sex to confer with the client, or to refer the client to his or her physician for further evaluation. Occasionally, the client will answer negatively to any questions regarding observed changes in urinary function and will then report back at the next session that there was some pathologic condition that was not noted earlier.

In this case, a close review of the extensive medical records may have alerted the physical therapist to the client's previous treatment for the same problem, which he was reluctant to discuss.

PRACTICE QUESTIONS

1. Percussion of the costovertebral angle that results in the reproduction of symptoms:
   a. Signifies radiculitis
   b. Signifies pseudorenal pain
   c. Has no significance
   d. Requires medical referral
2. Renal pain is aggravated by:
   a. Spinal movement
   b. Palpatory pressure over the costovertebral angle
   c. Lying on the involved side
   d. All of the above
   e. None of the above
3. Important functions of the kidney include all the following *except:*
   a. Formation and excretion of urine
   b. Acid-base and electrolyte balance
   c. Stimulation of red blood cell production
   d. Production of glucose
4. Who should be screened for possible renal/urologic involvement?
5. What do the following terms mean?
   - Dyspareunia
   - Dysuria
   - Hematuria
   - Urgency

Continued

PRACTICE QUESTIONS—*cont'd*

6. What is the difference between urge incontinence and stress incontinence?
7. What is the significance of "skin pain" over the T9/T10 dermatomes?
8. How do you screen for possible prostate involvement in a man with pelvic/low-back pain of unknown cause?
9. Explain why renal/urologic pain can be felt in such a wide range of dermatomes (i.e., from the T9 to L1 dermatomes).
10. What is the mechanism of referral for urologic pain to the shoulder?

REFERENCES

1. Cannon J: Recognizing chronic renal failure, the sooner, the better. *Nursing 2004* 34(1):50–53, 2004.
2. Netter FH: *Atlas of human anatomy*, ed 5, Philadelphia, 2010, WB Saunders.
3. Simons DG, Travell JG, Simons LS: *Travell & Simons' myofascial pain and dysfunction: the trigger point manual*, ed 2, vol 1, Baltimore, 1999, Williams & Wilkins.
4. Smith DR, Raney FL, Jr: Radiculitis distress as a mimic of renal pain. *J Urol* 116:269, 1976.
5. Eskelinen M: Usefulness of history-taking, physical examination and diagnostic scoring in acute renal colic. *Eur Urol* 34(6):467–473, 1998.
6. Houppermans RP, Brueren MM: Physical diagnosis—pain elicited by percussion in the kidney area. *Ned Tijdschr Geneeskd* 145(5):208–210, 2001.
7. National Kidney Foundation: *K/DOQI clinical practice guidelines for chronic kidney disease: evaluation, classification, and stratification*. Available online at http://www.kidney.org/professionals. Accessed January 1, 2011.
8. Banishing urinary tract infections. *Harvard Women's Health Watch* 10(4):4–5, 2002.
9. Cailliet R: *Low back pain syndrome*, ed 5, Philadelphia, 1995, FA Davis.
10. Interstitial Cystitis Association (ICA): *About interstitial cystitis*. Available online at http://www.ichelp.org/. Accessed Feb. 2, 2011.
11. Diagnosing and treating interstitial cystitis. *Harvard Women's Health Watch* 10(12):3, 2003.
12. Medical conditions, coping with kidney stones. *Harvard Women's Health Watch* 9(4):4–5, 2001.
13. Gurunadha Rao Tunuguntla HS, Evans CP: Management of prostatitis. *Prostate Cancer Prostatic Dis* 5(3):172–179, 2002.
14. Alexander RB: Treatment of chronic prostatitis. *Nat Clin Pract Urol* 1(1):2–3, 2004. Available online http://www.medscape.com/viewarticle/494378. Accessed January 3, 2011.
15. Krieger JN: NIH consensus definition and classification of prostatitis. *JAMA* 282(3):236–237, 1999.
16. Schaeffer AJ: Classification (Traditional and National Institutes of Health) and demographics of prostatitis. *Urology* 60:5–7, 2002.
17. Pontari MA, Ruggieri MR: Mechanisms in prostatitis/chronic pelvic pain syndrome. *J Urol* 172(3):839–845, 2004.
18. Tripp DA, Curtis NJ, Landis JR, et al: Predictors of quality of life and pain in chronic prostatitis/chronic pelvic pain syndrome: findings from the National Institutes of Health Chronic Prostatitis Cohort Study. *BJU* 94(9):1279–1282, 2004.
19. Schultz PL, Donnell RF: Prostatitis: the cost of disease and therapies to patients and society. *Curr Urol Rep* 5(4):317–319, 2004.
20. Litwin MS, McNaughton-Collins M, Fowler FL: *Prostatitis: The National Institutes of Health Chronic Prostatitis Symptoms Index (NIH-CPSI)*, Smithshire, IL, 2002, The Prostatitis Foundation. Available online at http://www.prostatitis.org/symptomindex.html. Accessed January 03, 2011.
21. Rex L: *Evaluation and treatment of somatovisceral dysfunction of the gastrointestinal system*, Edmonds, WA, 2004, URSA Foundation.
22. Cornel EB, van Haarst EP: *Chronic pelvic pain syndrome type 3 successfully treated with biofeedback physical therapy* (Abstract). Presented at the American Urological Association 2004 Annual Meeting, May 8–13, 2004, San Francisco, CA. Available online at http://www.prostatitis.org/AmericanUrologicalMeeting04.html. Accessed January 3, 2011.
23. Zvara P, Folsom JB, Plante MK: Minimally invasive therapies for prostatitis. *Curr Urol Rep* 5(4):320–326, 2004.
24. Sokolov AV: Transrectal microwave hyperthermia in the treatment of chronic prostatitis. *Urologiia* 5:20–26, 2003.
25. Wehbe SA: Minimally invasive therapies for chronic pelvic pain syndrome. *Curr Urol Rep* 11(4):276–285, 2010.
26. Murphy AB: Chronic prostatitis: management strategies. *Drugs* 69(1):71–84, 2009.
27. Kastner C: Update on minimally invasive therapy for chronic prostatitis/chronic pelvic pain syndrome. *Curr Urol Rep* 9(4):333–338, 2008.
28. Sheeler R: *Enlarged prostate. Know when to seek treatment.* Available online at http://www.mayoclinic.com/health/prostate-cancer/DS00043. Accessed January 3, 2011.
29. National Cancer Institute: *Prostate cancer*. Available online at http://www.nci.nih.gov/cancertopics/types/prostate. Accessed January 03, 2011.
30. Jemal A: Cancer statistics, 2010. *CA Cancer J Clin* 60(5):277–300, 2010.
31. Carroll PR, Nelson WG: Report to the nation on prostate cancer: introduction. *Medscape Hematology-Oncology* 7(2), 2004. Available online at http://www.medscape.com/viewarticle/489635. Accessed January 3, 2011.
32. Logothetis CJ, Lin SH: Osteoblasts in prostate cancer metastasis to bone. *Nat Rev Cancer* 5(1):21–28, 2005.
33. U.S. Department of Health & Human Services: *Diseases and conditions*. Accessed January 7, from http://www.hhs.gov/.
34. Shafik A, Shafik IA: Overactive bladder inhibition in response to pelvic floor muscle exercises. *World J Urol* 20(6):374–377, 2003.
35. Schultz JM: Urinary incontinence. Solving a secret problem. *Nursing 2003* (Suppl) 33(11):5–10, 2003.
36. Bo K, Borgen JS: Prevalence of stress and urge urinary incontinence in elite athletes and controls. *Med Sci Sports Exerc* 33(11):1797–1802, 2001.
37. Hulme J: *Regaining bowel and bladder control after cancer*, Missoula, MT, 2003, Phoenix Publishers.
38. D'Amico AV: Surrogate end point for prostate cancer-specific mortality after radical prostatectomy or radiation therapy. *J Natl Cancer Inst* 95(18):1376–1383, 2003.
39. Grise P, Thurman S: Urinary incontinence following treatment of localized prostate cancer. *Cancer Control* 8(6):532–539, 2002.

Available online at http://www.medscape.com/viewarticle/423513. Accessed January 3, 2011.
40. Wakamatsu MM: *Better bladder and bowel control,* Boston, 2009, Harvard Medical School.
41. Yim PS, Peterson AS: Urinary incontinence. *Postgrad Med* 99(5):137–150, 1996.
42. Burrows NR: Incidence of end-stage renal disease attributed to diabetes among persons with diagnosed diabetes in the United States and Puerto Rico. *MMWR* 59(42):1361–1366, 2010.
43. Evans N, Forsyth E: End-stage renal disease in people with type 2 diabetes: systemic manifestations and exercise implications. *Phys Ther* 84(5):454–463, 2004.
44. Peterson GM: Selecting nonprescription analgesics. *Am J Ther* 12(1):67–79, 2005.
45. Elseviers MM, De Broe ME: Analgesic abuse in the elderly. Renal sequelae and management. *Drugs Aging* 12(5):391–400, 1998.
46. National Kidney Foundation (NKF): *Can analgesics hurt kidneys?* Available online at http://www.kidney.org/atoz/atozPrint.cfm?id=23. Accessed January 3, 2011.
47. Malik J: Understanding the dialysis access steal syndrome: a review of the etiologies, diagnosis, prevention, and treatment strategies. *J Vasc Access* 9(3):155–166, 2008.
48. Holub C, Lamont M: The reliability of the six-minute walk test in patients with end stage renal disease. *Acute Care Perspect* 11(4):8–11, 2002.
49. Paton M: Continuous renal replacement therapy. *Nursing2003,* 33(6):40–50, 2003.
50. Best treatments for beating bladder cancer. *Johns Hopkins Med Lett* 15(1):6–7, 2004.
51. Bladder cancer in women: no time to wait. *Harvard Women's Health Watch* 11(7):3–5, 2004.
52. Jacobs BL: Bladder cancer in 2010: How far have we come? *CA Cancer J Clin* 60(4):244–272, 2010.
53. National Cancer Institute: *What you need to know about bladder cancer,* Available online at www.cancer.gov/cancertopics/wyntk/bladder. Accessed January 3, 2011.
54. Ongoing care of patients after primary treatment for their cancer: genitourinary cancers, bladder and kidney. *CA Cancer J Clin* 53(3):190–191, 2003.
55. Weiss J: Refractory overactive bladder without hematuria: a presenting symptom of bladder cancer. Presentation at the joint annual meeting of the International Continence Society (ICS) and the International Urogynecological Association (IUA). August 23-27, 2010. Available online at https://www.icsoffice.org/Abstracts/Publish/105/000348.pdf. Accessed December 2, 2010.
56. National Cancer Institute: *What you need to know about kidney cancer,* Available online at www.cancer.gov/cancertopics/wyntk/kidneys. Accessed January 3, 2011.
57. American Cancer Society: *Detailed guide: testicular cancer. What are the risk factors for testicular cancer?* Available online at http://www.cancer.org. Accessed January 3, 2011.
58. American Cancer Society: *Testicular cancer.* Available online at http://www.cancer.org/Cancer/TesticularCancer/DetailedGuide/testicular-cancer-risk-factors. Accessed Feb 2, 2011.
59. Gajendran VK: Testicular cancer patterns in African-American men. *Urology* 66(3):602–605, 2005.
60. Powe BD: Testicular cancer among African American college men. *Am J Mens Health* 1(1):73–80, 2007.
61. Goodman CC, Fuller K: *Pathology: implications for the physical therapist,* ed 3, Philadelphia, 2009, WB Saunders.
62. Lab Values Interpretation Resources: *Acute Care Section—APTA Task Force on Lab Values, 2008.* Available online at www.acutept.org [members only]. Accessed January 3, 2011.
63. Irion GL: Lab values update. *Acute Care Perspect* 13(1):1, 3–5, 2004.
64. Sherburn M, Guthrie JR, Dudley EC, et al: Is incontinence associated with menopause? *Obstet Gynecol* 98(4):628–633, 2001.

CHAPTER 11

Screening for Endocrine and Metabolic Disease

Endocrinology is the study of ductless (endocrine) glands that produce hormones. A hormone acts as a chemical agent that is transported by the bloodstream to target tissues, where it regulates or modifies the activity of the target cell.

The endocrine system cannot be understood fully without consideration of the effects of the nervous system on the endocrine system. The endocrine system works with the nervous system to regulate metabolism, water and salt balance, blood pressure, response to stress, and sexual reproduction.

The endocrine system is slower in response and takes longer to act than the nervous system in transferring biochemical information. The pituitary (hypophysis), thyroid, parathyroids, adrenals, and pineal are glands of the endocrine system whose functions are solely endocrine related and have no other metabolic functions (Fig. 11-1). The hypothalamus controls pituitary function and thus has an important indirect influence on the other glands of the endocrine system. Feedback mechanisms exist to keep hormones at normal levels.

The endocrine system meets the nervous system in a complex series of interactions that link behavioral-neural-endocrine-immunologic responses. The hypothalamus and the pituitary form an integrated axis that maintains control over much of the endocrine system. The discovery and study of this complex interface axis is called psychoneuroimmunology (PNI) and has provided a new understanding of interactive biologic signaling.

The hypothalamus exerts direct control over both the anterior and posterior portions of the pituitary gland and can synthesize and release hormones from its axon terminals directly into the blood circulation. These neurosecretory cells are so-called because the neurons have a hormone-secreting function. Although neurons can have a hormone-secreting function, the opposite pathway is also present. Hormones that can stimulate the neural mechanism (e.g., acetylcholine) are called *neurohormones.* Acetylcholine is a neurotransmitter and a neurohormone. It is released at synapses to allow messages to pass along a nerve network, resulting in the release of both hormones and chemicals.

ASSOCIATED NEUROMUSCULAR AND MUSCULOSKELETAL SIGNS AND SYMPTOMS

The musculoskeletal system is composed of a variety of connective tissue structures in which normal growth and development are influenced strongly and sometimes controlled by various hormones and metabolic processes. Alterations in these control systems can result in structural changes and altered function of various connective tissues, producing systemic and musculoskeletal signs and symptoms (Table 11-1).

Muscle Weakness, Myalgia, and Fatigue

Muscle weakness, myalgia, and fatigue may be early manifestations of thyroid or parathyroid disease, acromegaly, diabetes, Cushing's syndrome, and osteomalacia. Proximal muscle weakness associated with endocrine disease is usually painless and unrelated to either the severity or the duration of the underlying disease. The muscular system is sometimes but not always restored with effective treatment of the underlying condition.

Bilateral Carpal Tunnel Syndrome

Bilateral carpal tunnel syndrome (CTS), resulting from median nerve compression at the wrist, is a common finding in a variety of systemic and neuromusculoskeletal conditions[1-3] but especially with certain endocrine and metabolic disorders (Table 11-2).[4] The fact that the majority of persons with CTS are women at or near menopause suggests that the soft tissues about the wrist could be affected in some way by hormones.[5-8]

Thickening of the transverse carpal ligament in certain systemic disorders (e.g., acromegaly, myxedema) may be sufficient to compress the median nerve. Any condition that increases the volume of the contents of the carpal tunnel (e.g., neoplasm, calcium, and gouty tophi deposits) can compress the median nerve.

The signs and symptoms often associated with CTS include paresthesia, tingling, and numbness and/or pain (or burning pain) with cutaneous distribution of the median

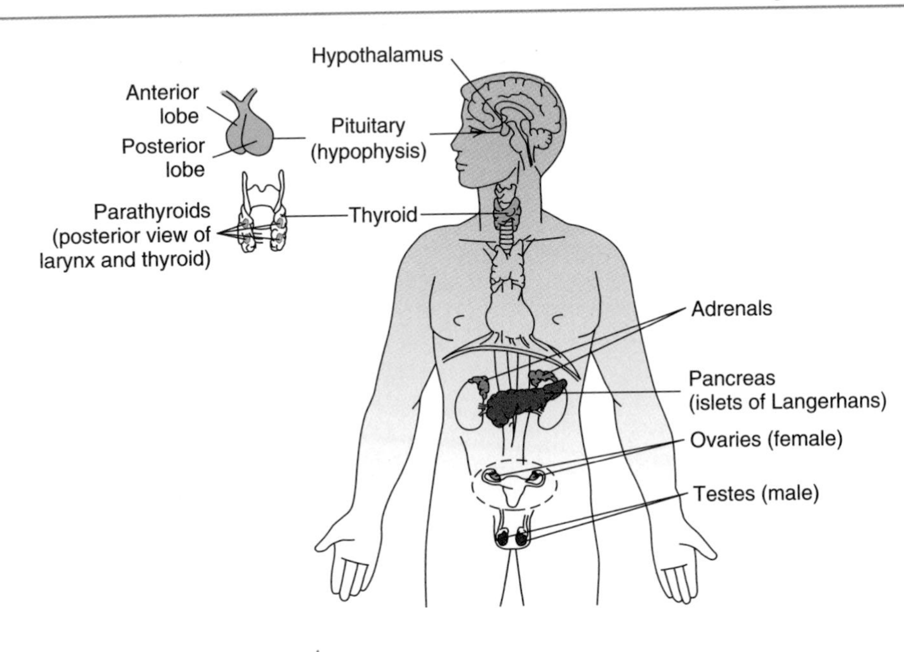

Fig. 11-1 Location of the nine endocrine glands. Not shown: adipose tissue (now classified as the largest endocrine gland in the body).

TABLE 11-1 Signs and Symptoms of Endocrine Dysfunction

Neuromusculoskeletal	Systemic
Signs and symptoms associated with rheumatoid arthritis	Excessive or delayed growth
Muscle weakness	Polydipsia
Muscle atrophy	Polyuria
Myalgia	Mental changes (nervousness, confusion, depression)
Fatigue	Changes in hair (quality and distribution)
Carpal tunnel syndrome	Changes in skin pigmentation
Synovial fluid changes	Changes in vital signs (elevated body temperature, pulse rate, increased blood pressure)
Periarthritis	Heart palpitations
Adhesive capsulitis (diabetes)	Increased perspiration
Chondrocalcinosis	Kussmaul's respirations (deep, rapid breathing)
Spondyloarthropathy	Dehydration or excessive retention of body water
Osteoarthritis	
Hand stiffness	
Arthralgia	

From Goodman CC, Fuller KS: *Pathology: implications for the physical therapist,* ed 3, Philadelphia, 2009, WB Saunders.

nerve to the thumb, index, middle, and radial half of the ring finger. Nocturnal paresthesia is a common complaint, and this discomfort causes sleep disruption. It can be partially relieved by shaking of the hand or changing wrist and hand position. Pain may radiate into the palm and up the forearm and arm.[9]

It should be noted that bilateral tarsal syndrome affecting the feet also can occur either alone or in conjunction with CTS, although the incidence of tarsal tunnel syndrome is not high (see further discussion of tarsal tunnel syndrome in relation to carpal tunnel syndrome in Chapter 9). Bilateral median nerve neuritis can be characteristic of many systemic diseases, including rheumatoid arthritis, myxedema, localized amyloidosis, sarcoidosis, and infiltrative leukemia.[10,11]

Whenever a client presents with bilateral symptoms, it represents a red flag. With bilateral CTS the therapist can screen for medical disease by using the Special Questions to Ask: Bilateral Carpal Tunnel Syndrome section (see Appendix B-4).

Periarthritis and Calcific Tendinitis

Periarthritis (inflammation of periarticular structures, including the tendons, ligaments, and joint capsule) and calcific tendinitis occur most often in the shoulders of people who have endocrine disease. Treatment of the underlying endocrine impairment often improves the clinical picture; physical therapy intervention may have a temporary palliative effect.

Chondrocalcinosis

Chondrocalcinosis is the deposition of calcium salts in the cartilage of joints. When accompanied by attacks of gout-like symptoms, it is called *pseudogout.* Chondrocalcinosis is commonly seen on x-ray films as calcified hyaline or fibrous cartilage. There is an associated underlying endocrine or metabolic disease in approximately 5% to 10% of individuals with chondrocalcinosis (Table 11-3).

Spondyloarthropathy and Osteoarthritis

Spondyloarthropathy (disease of joints of the spine) and osteoarthritis occur in individuals with various metabolic or endocrine diseases, including hemochromatosis (disorder of iron metabolism with excess deposition of iron in the tissues; also known as *bronze diabetes* and *iron storage disease*), ochronosis (metabolic disorder resulting in discoloration of body tissues caused by deposits of alkapton bodies), acromegaly, and diabetes mellitus.

TABLE 11-2 Causes of Carpal Tunnel Syndrome

Neuromusculoskeletal	Systemic
IDIOPATHIC	Chronic kidney disease (fluid imbalance)
Cause unknown	Congestive heart failure (fluid imbalance)
ANATOMIC (COMPRESSION)	Hemochromatosis
Small carpal canal, anomalous muscles/tendons	Leukemia (tissue infiltration)
Basal joint (thumb) arthritis	Liver disease
Cervical disk lesions	Medications
Cervical spondylosis	• Nonsteroidal antiinflammatory drugs (NSAIDs)
Congenital anatomic differences or anatomic change in nerve or carpal tunnel (e.g., shape, size, volume of structures; presence of palmaris longus)	• Oral contraceptives
History of wrist surgery, especially previous carpal tunnel surgery	• Statins
Injection: High pressure	• Alendronate[1] (Fosamax)
Peripheral neuropathy	• Lithium
Poor posture (may also be associated with TOS)	• Beta blocker
Tendinitis	Obesity
Trigger points	Pregnancy (fluid retention)
Tenosynovitis	Tumors (lipoma, hemangioma, ganglia, synovial sarcoma, fibroma, neuroma, neurofibroma)
Thoracic outlet syndrome (TOS)	Use of oral contraceptives
TRAUMA/EXERTIONAL	Vitamin deficiency
Swelling, hemorrhage, scar, wrist fracture, carpal dislocation	**ENDOCRINE**
Cumulative trauma disorders (CTD)*	Acromegaly
Repetitive strain injuries (RSI)*	Diabetes mellitus
Vibrational exposure (jackhammer or other manual labor equipment)	Gout (deposits of tophi and calcium)
	Hormonal imbalance (menopause; posthysterectomy)
	Hyperparathyroidism
	Hyperthyroidism (Graves' disease)
	Hypocalcemia
	Hypothyroidism (myxedema)
	INFECTIOUS DISEASE
	Atypical mycobacterium
	Histoplasmosis
	Rubella
	Sporotrichosis
	INFLAMMATORY
	Amyloidosis
	Arthritis (rheumatoid, gout, polymyalgia rheumatica)
	Dermatomyositis
	Gout/pseudogout
	Scleroderma
	Systemic lupus erythematosus
	NEUROPATHIC
	Alcohol abuse
	Chemotherapy (delayed, long-term effect)
	Diabetes
	Multiple myeloma (amyloidosis deposits)
	Thyroid disease
	Vitamin/nutritional deficiency (especially vitamin B_6; folic acid)
	Vitamin toxicity

*The role of repetitive activities and occupational factors (e.g., hand use of any type and keyboard or computer work in particular) has been questioned as a direct cause of carpal tunnel syndrome (CTS) and remains under investigation; sufficient evidence to implicate hand use of any type linked with CTS remains unproven.[3,4]

Modified from Goodman CC, Fuller K: *Pathology: implications for the physical therapist,* ed 3, Philadelphia, 2009, Saunders.

Hand Stiffness and Hand Pain

Hand stiffness and hand pain, as well as arthralgias of the small joints of the hand, can occur with endocrine and metabolic diseases. Hypothyroidism is often accompanied by CTS; flexor tenosynovitis with stiffness is another common finding.

ENDOCRINE PATHOPHYSIOLOGY

Disorders of the endocrine glands can be classified as primary (dysfunction of the gland itself) or secondary (dysfunction of an outside stimulus to the gland) and are a result of either an excess or an insufficiency of hormonal secretions.

TABLE 11-3 Endocrine and Metabolic Disorders Associated with Chondrocalcinosis

Endocrine	Metabolic
Hypothyroidism Hyperparathyroidism Acromegaly	Hemochromatosis Hypomagnesemia Hypophosphatasias Ochronosis Oxalosis Wilson's disease

Modified from Louthrenoo W, Schumacher HR: Musculoskeletal clues to endocrine or metabolic disease, *J Musculoskel Med* 7(9):41, 1990.

Secondary dysfunction may also occur (iatrogenically) as a result of chemotherapy, surgical removal of the glands, therapy for a nonendocrine disorder (e.g., the use of large doses of corticosteroids resulting in Cushing's syndrome), or excessive therapy for an endocrine disorder.

Pituitary Gland

Diabetes Insipidus

Diabetes insipidus (DI) is caused by a lack of secretion or action of vasopressin (antidiuretic hormone [ADH]). This hormone normally stimulates the distal tubules of the kidneys to reabsorb water. Without ADH, water moving through the kidney is not reabsorbed but is lost in the urine, resulting in severe water loss and dehydration through diuresis.

There are two main types of DI: central DI and nephrogenic DI. Central DI, which is the most common type, can be idiopathic (primary) or related to other causes (secondary), such as pituitary trauma, head injury (including neurosurgery), infections such as meningitis or encephalitis, pituitary neoplasm, anorexia, and vascular lesions such as aneurysms. Nephrogenic DI occurs as a result of some medications (e.g., lithium, phenytoin, corticosteroids, anticholinergics), alcohol, electrolyte imbalances such as hypercalcemia and hypokalemia, and diseases affecting the renal system (e.g., sarcoidosis, multiple myeloma, pyelonephritis, systemic lupus erythematosus).

If the person with DI is unconscious or confused and is unable to take in necessary fluids to replace those fluids lost, rapid dehydration, shock, and death can occur. Because sleep is interrupted by the persistent need to void (nocturia), fatigue and irritability result.

CLINICAL SIGNS AND SYMPTOMS

Diabetes Insipidus

- Polyuria (increased urination more than 3 L/day in adults)
- Polydipsia (increased thirst, which occurs subsequent to polyuria in response to the loss of fluid)
- Dehydration (dry, cracked lips/skin; fever; orthostatic hypotension; weakness; dizziness; fatigue)
- Decreased urine specific gravity (1.001 to 1.005)
- Nocturia, fatigue, irritability
- Increased serum sodium (more than 145 mEq/dL; resulting from concentration of serum from water loss)

Syndrome of Inappropriate Secretion of Antidiuretic Hormone

Syndrome of inappropriate secretion of ADH (SIADH) is an excess or inappropriate secretion of vasopressin that results in marked retention of water in excess of sodium in the body. Urine output decreases dramatically as the body retains large amounts of water. Almost all the excess water is distributed within body cells, causing intracellular water gain and cellular swelling (water intoxication).

Risk Factors. Risk factors for the development of SIADH include pituitary damage caused by infection, trauma, or neoplasm; secretion of vasopressin-like substances from some types of malignant tumors (particularly pulmonary malignancies); and thoracic pressure changes from compression of pulmonary or cardiac pressure receptors, or both.

Clinical Presentation. Symptoms of SIADH are the clinical opposite of symptoms of DI. They are the result of water retention and the subsequent dilution of sodium in the blood serum and body cells. Neurologic and neuromuscular signs and symptoms predominate and are directly related to the swelling of brain tissue and sodium changes within neuromuscular tissues.

CLINICAL SIGNS AND SYMPTOMS

Syndrome of Inappropriate Secretion of Antidiuretic Hormone

- Headache, confusion, lethargy (most significant early indicators)
- Decreased urine output
- Weight gain without visible edema
- Seizures
- Muscle cramps
- Vomiting, diarrhea
- Increased urine specific gravity (greater than 1.03)
- Decreased serum sodium (less than 135 mEq/dL; caused by dilution of serum from water)

Acromegaly

Acromegaly is an abnormal enlargement of the extremities of the skeleton resulting from hypersecretion of growth hormone (GH) from the pituitary gland. This condition is relatively rare and occurs in adults, most often owing to a tumor of the pituitary gland. In children, overproduction of GH stimulates growth of long bones and results in gigantism, in which the child grows to exaggerated heights. With adults, growth of the long bones has already stopped, so the bones most affected are those of the face, jaw, hands, and feet. Other signs and symptoms include amenorrhea (in women), diabetes mellitus, profuse sweating, and hypertension.

Clinical Presentation. Degenerative arthropathy may be seen in the peripheral joints of a client with acromegaly, most frequently attacking the large joints. On x-ray studies, osteophyte formation may be seen, along with widening of the joint space because of increased cartilage thickness. In late-stage disease, joint spaces become narrowed, and occasionally chondrocalcinosis may be present.

Stiffness of the hand, typically of both hands, is associated with a broad enlargement of the fingers from bony overgrowth and with thickening of the soft tissue. Thickening and widening of the phalangeal tufts are typical x-ray findings in soft tissue. In clients with these x-ray findings, much of the pain and stiffness is believed to be due to premature osteoarthritis.

CTS is seen in up to 50% of people with acromegaly. The CTS that occurs with this growth disorder is thought to be caused by compression of the median nerve at the wrist from soft tissue hypertrophy or bony overgrowth or by hypertrophy of the median nerve itself.

Myopathy in people with acromegaly is commonly reported but poorly understood. Changes in muscle size and strength are associated with acromegaly and are probably multifactorial in origin. Screening individuals with acromegaly for muscle weakness and poor exercise tolerance is now recommended.[12]

About half the individuals with acromegaly have back pain. X-ray studies demonstrate increased intervertebral disk spaces and large osteophytes along the anterior longitudinal ligament (ALL), mimicking diffuse idiopathic skeletal hyperostosis (DISH).

DISH (also known as *Forestier's disease*) is characterized by abnormal ossification of the ALL, resulting in an x-ray image of large osteophytes seemingly "flowing" along the anterior border of the spine. DISH is particularly common in the thoracic spine and has been reported to be more prevalent among persons with diabetes than among the nondiabetic population. DISH appears to be an age-related predisposition to ossification of tendon, joint capsule, and ligamentous attachments. Identification of the presence of DISH syndrome prior to surgery is important in the prevention of heterotropic bone formation.[13]

CLINICAL SIGNS AND SYMPTOMS

Acromegaly

- Bony enlargement (face, jaw, hands, feet)
- Amenorrhea
- Diabetes mellitus
- Profuse sweating (diaphoresis)
- Hypertension
- Carpal tunnel syndrome (CTS)
- Hand pain and stiffness
- Back pain (thoracic and/or lumbar)
- Myopathy and poor exercise tolerance

Adrenal Glands

The adrenals are two small glands located on the upper part of each kidney. Each adrenal gland consists of two relatively discrete parts: an outer cortex and an inner medulla. The outer cortex is responsible for the secretion of mineralocorticoids (steroid hormones that regulate fluid and mineral balance), glucocorticoids (steroid hormones responsible for controlling the metabolism of glucose), and androgens (sex hormones). The centrally located adrenal medulla is derived from neural tissue and secretes epinephrine and norepinephrine. Together, the adrenal cortex and medulla are major factors in the body's response to stress.

Adrenal Insufficiency

Primary Adrenal Insufficiency. Chronic adrenocortical insufficiency (hyposecretion by the adrenal glands) may be primary or secondary. Primary adrenal insufficiency is also referred to as *Addison's disease* (hypofunction), named after the physician who first studied and described the associated symptoms. It can be treated by the administration of exogenous cortisol (one of the adrenocortical hormones).

Primary adrenal insufficiency occurs when a disorder exists within the adrenal gland itself. This adrenal gland disorder results in decreased production of cortisol and aldosterone, two of the primary adrenocortical hormones. The most common cause of primary adrenal insufficiency is an autoimmune process that causes destruction of the adrenal cortex.

The most striking physical finding in the person with primary adrenal insufficiency is the increased pigmentation of the skin and mucous membranes. This discoloration may vary in the white population from a slight tan or a few black freckles to an intense generalized pigmentation, which has resulted in persons being mistakenly considered to be of a darker-skinned race. Members of darker-skinned races may develop a slate-gray color that may be obvious only to family members.

Melanin, the major product of the melanocyte, is largely responsible for the coloring of skin. In primary adrenal insufficiency, the increase in pigmentation is initiated by the excessive secretion of melanocyte-stimulating hormone (MSH) that occurs in association with increased secretion of adrenocorticotropic hormone (ACTH). ACTH is increased in an attempt to stimulate the diseased adrenal glands to produce and release more cortisol.

Most commonly, pigmentation is visible over extensor surfaces such as the backs of the hands; elbows; knees; and creases of the hands, lips, and mouth. Increased pigmentation of scars formed after the onset of the disease is common. However, it is possible for a person with primary adrenal insufficiency to demonstrate no significant increase in pigmentation.

Secondary Adrenal Insufficiency. Secondary adrenal insufficiency refers to a dysfunction of the gland because of insufficient stimulation of the cortex owing to a lack of pituitary ACTH. Causes of secondary disease include tumors of the hypothalamus or pituitary, removal of the pituitary, or rapid withdrawal of corticosteroid drugs. Clinical manifestations of secondary disease do not occur until the adrenals are almost completely nonfunctional and are primarily related to cortisol deficiency only.

CASE EXAMPLE 11-1

Cushing's Syndrome

A 53-year-old woman with Cushing's syndrome resulting from long-term use of cortisol for systemic lupus erythematosus reports the following problems:

- Hair and nail thinning and breaking easily
- Temperature intolerance (always cold)
- Muscle cramps
- Generalized weakness and fatigue

Her primary complaint and reason for referral to physical therapy is for sacroiliac (SI) joint pain as a result of stepping down off an uneven curb.

You realize the signs and symptoms are of an endocrine origin, but you do not know whether they are part of the Cushing's syndrome or a separate endocrine problem.

Should you send this client to a physician (or back to the referring physician)?

Not necessarily. This is more a case of need for additional information. Requesting a copy of the client's most recent physician's notes may answer all of your questions. Reading the physician's systems review portion of the exam may reveal a record of these signs and symptoms with a corresponding medical problem list and plan.

If there is no mention of any of these associated signs and symptoms, a phone call to the physician's office may be the next step. If you speak with the physician directly, identify yourself and your connection with the client by name. Briefly mention why you are seeing this client and make the following observation:

"Mrs. Jones reports muscle cramps and generalized weakness that do not seem consistent with her SI problem. She complains of temperature intolerance and hair and nail bed changes. These symptoms are outside the scope of my practice.

Can you help me understand this? Are they part of her lupus, Cushing's syndrome, or something else?"

CLINICAL SIGNS AND SYMPTOMS

Adrenal Insufficiency

- Dark pigmentation of the skin, especially mouth and scars (occurs only with primary disease; Addison's disease)
- Hypotension (low blood pressure causing orthostatic symptoms)
- Progressive fatigue (improves with rest)
- Hyperkalemia (generalized weakness and muscle flaccidity)
- Gastrointestinal (GI) disturbances
- Anorexia and weight loss
- Nausea and vomiting
- Arthralgias, myalgias (secondary only)
- Tendon calcification
- Hypoglycemia

Cushing's Syndrome

Cushing's syndrome (hyperfunction of the adrenal gland) is a general term for increased secretion of cortisol by the adrenal cortex. When corticosteroids are administered externally, a condition of hypercortisolism called *iatrogenic Cushing's syndrome* occurs, producing a group of associated signs and symptoms. Hypercortisolism caused by excess secretion of ACTH (e.g., from pituitary stimulation) is called *ACTH-dependent Cushing's syndrome.*[14]

Therapists often treat people who have developed Cushing's syndrome after these clients have received large doses of cortisol (also known as *hydrocortisone*) or cortisol derivatives (e.g., dexamethasone) for a number of inflammatory disorders (Case Example 11-1).

It is important to remember that whenever corticosteroids are administered externally, the increase in serum cortisol levels triggers a negative feedback signal to the anterior pituitary gland to stop adrenal stimulation. Adrenal atrophy occurs during this time, and adrenal insufficiency will result if external corticosteroids are abruptly withdrawn. Corticosteroid medications must be reduced gradually so that normal adrenal function can return.

Because cortisol suppresses the inflammatory response of the body, it can mask early signs of infection. *Any unexplained fever without other symptoms should be a warning to the therapist of the need for medical follow-up.*

CLINICAL SIGNS AND SYMPTOMS

Cushing's Syndrome

- "Moonface" appearance (very round face; Fig. 11-2)
- Buffalo hump at the neck (fatty deposits)
- Protuberant abdomen with accumulation of fatty tissue and stretch marks
- Muscle wasting and weakness
- Decreased density of bones (especially spine)
- Hypertension
- Kyphosis and back pain (secondary to bone loss)
- Easy bruising
- Psychiatric or emotional disturbances
- Impaired reproductive function (e.g., decreased libido and changes in menstrual cycle)
- Diabetes mellitus
- Slow wound healing
- *For women:* Masculinizing effects (e.g., hair growth, breast atrophy, voice changes)

Effects of Cortisol on Connective Tissue. Overproduction of cortisol or closely related glucocorticoids by abnormal adrenocortical tissue leads to a protein catabolic state. This overproduction causes liberation of amino acids from muscle tissue. The resultant weakened protein structures (muscle and elastic tissue) cause a protuberant abdomen, poor wound healing, generalized muscle weakness, and marked osteoporosis (demineralization of bone causing reduced bone mass), which is made worse by an excessive loss of calcium in the urine.

Excessive glucose resulting from this protein catabolic state is transformed mainly into fat and appears in characteristic sites, such as the abdomen, supraclavicular fat pads, and facial cheeks. The change in facial appearance may not be

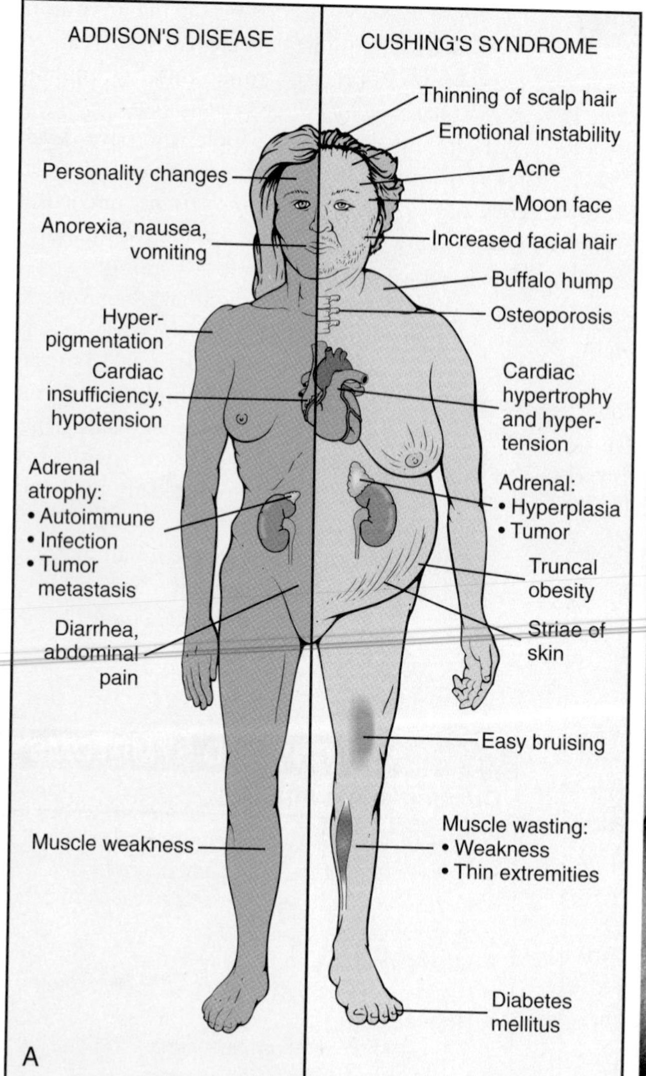

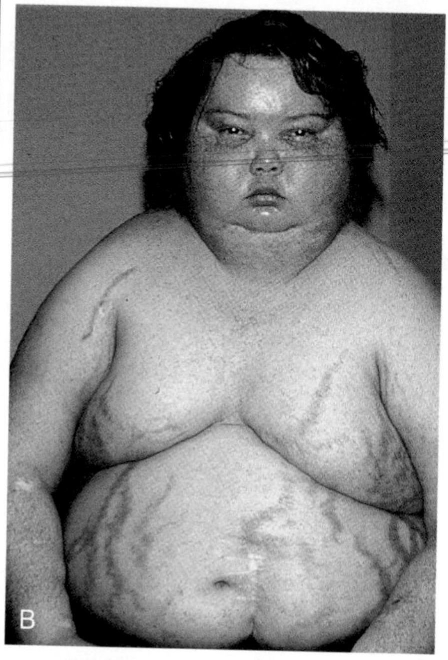

Fig. 11-2 **A,** Comparison of hyperfunction of the adrenal cortex (Addison's disease) and hypofunction (Cushing's syndrome). **B,** Individuals treated with corticosteroids can develop clinical features of Cushing's syndrome called *cushingoid features* including "moonface," obesity, and cutaneous striae as shown here. (From Damjanov I: *Pathology for the health-related profession,* ed 3, Philadelphia, 2006, WB Saunders. Used with permission.)

readily apparent to the client or to the therapist, but pictures of the client taken over a period of years may provide a visual record of those changes.

The effect of increased circulating levels of cortisol on the muscles of clients varies from slight to very marked. There may be so much muscle wasting that the condition simulates muscular dystrophy. Marked weakness of the quadriceps muscle often prevents affected clients from rising out of a chair unassisted. Those with Cushing's syndrome of long duration almost always demonstrate demineralization of bone. In severe cases, this condition may lead to pathologic fractures, but it results more commonly in wedging of the vertebrae, kyphosis, bone pain, and back pain

Obesity, diabetes, polycystic ovarian syndrome, and other metabolic/endocrine problems can resemble Cushing's syndrome. It is important to recognize critical indicators of this particular disorder, such as excessive hair growth, moonface, mood disorders, and increased muscle weakness, as indicators for further endocrine diagnostic testing.[15]

The poor wound healing that is characteristic of this syndrome becomes a problem when any surgical procedures are required. Inhibition of collagen formation with corticosteroid therapy is responsible for the frequency of wound breakdown in postsurgical clients.

Thyroid Gland

The thyroid gland is located in the anterior portion of the lower neck below the larynx, on both sides of and anterior to the trachea. The chief hormones produced by the thyroid are thyroxine (T_4), triiodothyronine (T_3), and calcitonin. Both T_3 and T_4 regulate the metabolic rate of the body and increase protein synthesis. Calcitonin has a weak physiologic effect on calcium and phosphorus balance in the body.

Genetics plays a role in thyroid disease. A family history of thyroid disease is a risk factor. Age and gender are also factors; most cases occur after age 50. Women are more likely than men to develop thyroid dysfunction.[15] Data gathered on the medical history of the orthopedic physical therapy outpatient population indicate a 7% incidence of thyroid disease in the female population.[16]

Thyroid function is regulated by the hypothalamus and pituitary feedback controls, as well as by an intrinsic regulator mechanism within the gland itself. Basic thyroid disorders of significance to physical therapy practice include goiter, hyperthyroidism, hypothyroidism, and cancer. Alterations in thyroid function produce changes in hair, nails, skin, eyes, gastrointestinal (GI) tract, respiratory tract, heart and blood vessels, nervous tissue, bone, and muscle.

The risk of having thyroid diseases increases with age, but in people older than 60 years of age, it becomes more difficult to detect because it masquerades as other problems such as heart disease, depression, or dementia. Fatigue and weakness may be the first symptoms among older adults, often mistaken or attributed to normal aging. New-onset depression in the older adult population and anxiety syndromes are also symptoms that can indicate thyroid dysfunction.[17]

On the other hand, thyroid dysfunction can mimic signs and symptoms of aging such as hair loss, fatigue, and depression. The therapist may recognize problems early and make a medical referral, minimizing the client's symptoms. A simple and inexpensive blood test called a *thyroid-stimulating hormone* (TSH) *test* is usually recommended to show whether the thyroid gland is hyperfunctioning or hypofunctioning.

Goiter

Goiter, an enlargement of the thyroid gland, occurs in areas of the world where iodine (necessary for the production of thyroid hormone) is deficient in the diet. It is believed that when factors (e.g., a lack of iodine) inhibit normal thyroid hormone production, hypersecretion of TSH occurs because of a lack of a negative feedback loop. The TSH increase results in an increase in thyroid mass.

Pressure on the trachea and esophagus causes difficulty in breathing, dysphagia, and hoarseness. With the use of iodized salt, this problem has almost been eliminated in the United States. Although the younger population in the United States may be goiter free, older adults may have developed goiter during their childhood or adolescent years and may still have clinical manifestations of this disorder.

CLINICAL SIGNS AND SYMPTOMS

Goiter

- Increased neck size
- Pressure on adjacent tissue (e.g., trachea and esophagus)
- Difficulty in breathing
- Dysphagia
- Hoarseness

Thyroiditis

Thyroiditis is an inflammation of the thyroid gland. Causes can include infection and autoimmune processes. The most common form of this problem is a chronic thyroiditis called *Hashimoto's thyroiditis.* This condition affects women more frequently than men and is most often seen in the 30- to 50-year-old age group. Destruction of the thyroid gland from this condition can cause eventual hypothyroidism (Case Example 11-2).

Usually, both sides of the gland are enlarged, although one side may be larger than the other. Other symptoms are related to the functional state of the gland itself. Early involvement may cause mild symptoms of hyperthyroidism, whereas later symptoms cause hypothyroidism.

CLINICAL SIGNS AND SYMPTOMS

Thyroiditis

- Painless thyroid enlargement
- Dysphagia, "tight" sensation when swallowing, or choking
- Anterior neck, shoulder, or rib cage pain without biomechanical changes
- Gland sometimes easily palpable over anterior neck (warm, tender, swollen)
- Fatigue, weight gain, dry hair and skin, constipation (these are later symptoms associated with hypothyroidism)

Hyperthyroidism

Hyperthyroidism (hyperfunction), or *thyrotoxicosis,* refers to those disorders in which the thyroid gland secretes excessive amounts of thyroid hormone. Graves' disease is a common type of excessive thyroid activity characterized by a generalized enlargement of the gland (or goiter leading to a swollen neck) and often, protruding eyes caused by retraction of the eyelids and inflammation of the ocular muscles.

Clinical Presentation. Excessive thyroid hormone creates a generalized elevation in body metabolism. The effects of thyrotoxicosis occur gradually and are manifested in almost every system (Fig. 11-3 and Table 11-4).

In more than 50% of adults older than 70, three common signs are tachycardia, fatigue, and weight loss. In clients younger than 50, clinical signs and symptoms found most often include tachycardia, hyperactive reflexes, increased sweating, heat intolerance, fatigue, tremor, nervousness, polydipsia, weakness, increased appetite, dyspnea, and weight loss.[18]

Chronic periarthritis is also associated with hyperthyroidism. Inflammation that involves the periarticular structures, including the tendons, ligaments, and joint capsule, is termed *periarthritis.* The syndrome is associated with pain and reduced range of motion. Calcification, whether periarticular or tendinous, may be seen on x-ray studies. Both periarthritis and calcific tendinitis occur most often in the shoulder, and both are common findings in clients who have endocrine disease (Case Example 11-3).

CASE EXAMPLE 11-2

Hashimoto's Thyroiditis

Referral: A 38-year-old woman with right-sided groin pain was referred to physical therapy by her physician. She says that the pain came on suddenly without injury. The pain is worse in the morning and hurts at night, waking her up when she changes position. The woman's symptoms are especially acute when she tries to stand up after sitting, with weight bearing impossible for the first 5 to 10 minutes.

The woman, who looks athletic, reports that before the onset of this problem she was running 5 miles every other day without difficulty. The x-ray finding is reportedly within normal limits for structural abnormalities. Erythrocyte sedimentation rate (ESR) was 16 mm/hour.* The client has chronic sinusitis and has had two surgeries for that condition in the last 3 years. She is not a smoker and drinks only occasionally on a social basis.

This client was seen 6 weeks ago by another physical therapist, who tried ultrasound and stretching without improvement in symptoms or function.

Clinical Presentation: The physical therapy evaluation today revealed a positive Thomas test for right hip flexion contracture. However, it was difficult to assess whether there was a true muscle contracture or only loss of motion as a result of muscle splinting and guarding. Patrick's test (FABER) for hip pathology and the iliopsoas test for intraabdominal infection were both negative. Joint accessory motions appeared to be within normal limits, given that the movements were tested in the presence of some residual muscle tension from protective splinting. A neurologic screen failed to demonstrate the presence of any neurologic involvement. Symptoms could be reproduced with deep palpation of the right groin area. There were no active or passive movements that could alter, provoke, change, or eliminate the pain. There were no trigger points in the abdomen or right lower quadrant that could account for the symptomatic presentation.

There was no apparent cause for her movement system impairment. Physical therapy intervention with soft tissue mobilization and proprioceptive neuromuscular facilitation techniques were initiated and used as a diagnostic tool. There was no change in the client's symptoms or clinical presentation as the therapist continued trying a series of physical therapy techniques.

Result: In a young and otherwise healthy adult, a lack of measurable, reportable, or observable progress becomes a red flag for further medical follow-up. The results of the physical therapy examination and lack of response to treatment constitute a valuable medical diagnostic tool.

Further laboratory results revealed a medical diagnosis of Hashimoto's thyroiditis. Treatment with thyroxine (T_4) resulted in resolution of the musculoskeletal symptoms. The correlation between groin pain and loss of hip extension with Hashimoto's remains unclear. Even so, response to the red flag (no change or improvement with intervention) resulted in a correct medical diagnosis.

*The ESR (an indication of possible infection or inflammation) was within normal limits for an adult woman.

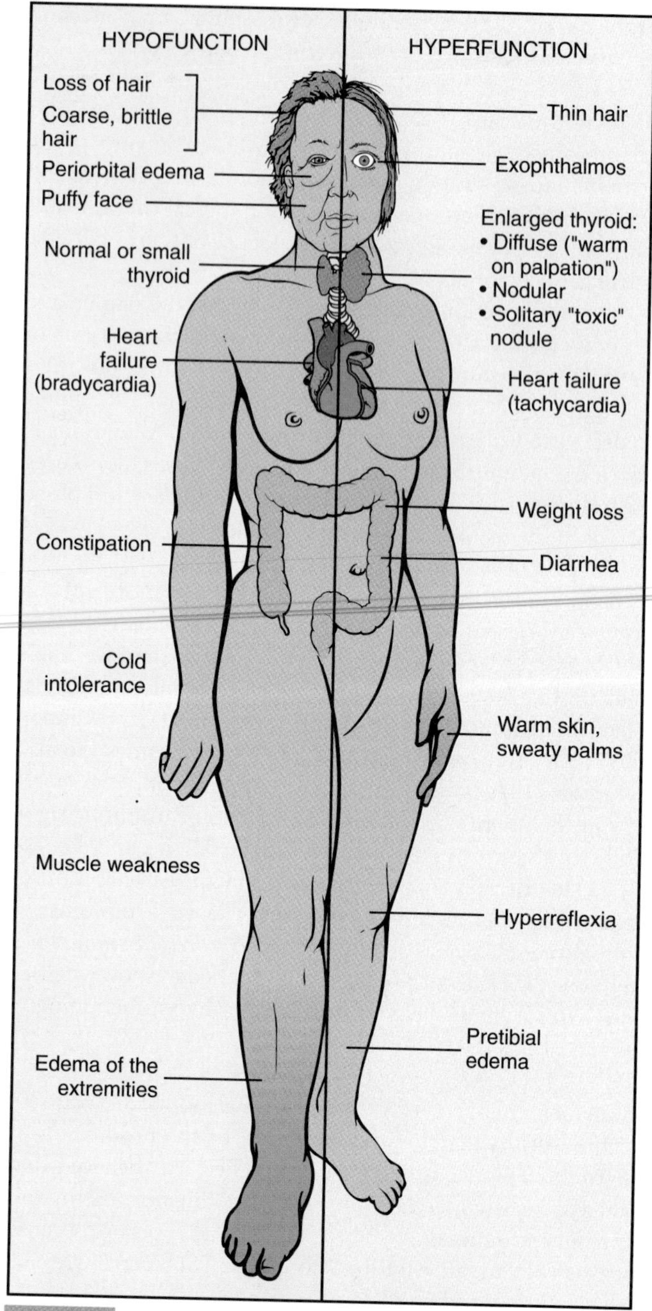

Fig. 11-3 Comparison of hyperthyroidism and hypothyroidism. (From Damjanov I: *Pathology for the health-related profession*, ed 3, Philadelphia, 2006, WB Saunders. Used with permission.)

Painful restriction of shoulder motion associated with periarthritis has been widely described among clients of all ages with hyperthyroidism. The involvement can be unilateral or bilateral and can worsen progressively to become adhesive capsulitis (frozen shoulder). Acute calcific tendinitis of the wrist also has been described in such clients. Although antiinflammatory agents may be needed for the acute symptoms, chronic periarthritis usually responds to treatment of the underlying hyperthyroidism.

Proximal muscle weakness (most marked in the pelvic girdle and thigh muscles), accompanied by muscle atrophy known as *myopathy,* occurs in up to 70% of people with

TABLE 11-4 Systemic Manifestations of Hyperthyroidism

CNS Effects	Cardiovascular and Pulmonary Effects	Joint and Integumentary Effects	Ocular Effects	GI Effects	GU Effects
Tremors Hyperkinesis (abnormally increased motor function or activity) Nervousness, irritability Emotional lability Weakness and muscle atrophy Increased deep tendon reflexes Fatigue	Increased pulse rate/ tachycardia/ palpitations Arrhythmias (palpitations) Weakness of respiratory muscles (breathlessness, hypoventilation) Increased respiratory rate Low blood pressure Heart failure	Chronic periarthritis Capillary dilation (warm flushed, moist skin) Heat intolerance Onycholysis (separation of the fingernail from the nail bed) Easily broken hair and increased hair loss Hyperpigmentation Hard, purple area over the anterior surface of the tibia with itching, erythema, and occasionally pain	Weakness of the extraocular muscles (poor convergence, poor upward gaze) Sensitivity to light Visual loss Spasm and retraction of the upper eyelids (bulging eyes), lid tremor	Hypermetabolism (increased appetite with weight loss) Increased peristalsis Diarrhea, nausea, and vomiting Dysphagia	Polyuria (frequent urination) Amenorrhea (absence of menses) Female infertility First-trimester miscarriage Increased frequency of bowel movements

CNS, Central nervous system; *GI,* gastrointestinal; *GU,* genitourinary.

CASE EXAMPLE 11-3

Graves' Disease (Hyperthyroidism)

A 73-year-old woman who has rheumatoid arthritis has just joined the Physical Therapy Aquatic Program. Despite the climate-controlled facility, she becomes flushed, demonstrates an increased respiratory rate that is inconsistent with her level of exercise, and begins to perspire profusely. She reports muscle cramping of the arms and legs and sudden onset of a headache.

Questions

- How would you handle this situation?
- Can this client resume the aquatic program when her symptoms have resolved?

Result: The client was quickly escorted from the pool. Her vital signs were taken and recorded for future reference. Later, the therapist reviewed the client's health history and noted that the "thyroid medication" she reported taking was actually an antithyroid medication for Graves' disease.

The heat intolerance associated with the Graves' disease (hyperthermia secondary to accelerated metabolic rate) presents a potential contraindication for aquatic or pool therapy. Heat intolerance contributes to exercise intolerance, and the client was exhibiting signs and symptoms of heat stroke, even when exercising in a climate-controlled facility. The physician was notified of the symptoms and how quickly the onset occurred (after only 5 minutes of warm-up exercises). Strenuous exercise or a conditioning program should be delayed until symptoms of heat intolerance, tachycardia, or arrhythmias are under medical control.

hyperthyroidism. Muscle strength returns to normal in about 2 months after medical treatment, whereas muscle wasting resolves more slowly. In severe cases normal strength may not be restored for months.

The incidence of myasthenia gravis is increased in clients with hyperthyroidism, which in turn can aggravate muscle weakness. If the hyperthyroidism is corrected, improvement of myasthenia gravis follows in about two-thirds of clients.

Thyroid Storm. Life-threatening complications with hyperthyroidism are rare but still important for the therapist to recognize. Unrecognized disease, untreated disease, or incorrect treatment can result in a condition called *thyroid storm.* In addition, precipitating factors, such as trauma, infection, or surgery, can turn well-controlled hyperthyroidism into a thyroid storm.

Thyroid storm is characterized by signs and symptoms of hypermetabolism including severe tachycardia with heart failure, shock, and hyperthermia (up to 105.3° F [40.7° C]). Restlessness, agitation, chest pain, abdominal pain, nausea and vomiting, and coma can occur. Immediate medical referral is required to return the client to a normal thyroid state and prevent cardiovascular or hyperthermic collapse. Look for a recent history of the precipitating factors mentioned.

Hypothyroidism

Hypothyroidism (hypofunction) is more common than hyperthyroidism, results from insufficient thyroid hormone, and creates a generalized depression of body metabolism. Hypothyroidism in fetal development and infants is usually a result of absent thyroid tissue and hereditary defects in thyroid hormone synthesis. Untreated congenital hypothyroidism is referred to as *cretinism.*

The condition may be classified as either primary or secondary. *Primary hypothyroidism* results from reduced functional thyroid tissue mass or impaired hormonal synthesis or release (e.g., iodine deficiency, loss of thyroid tissue, autoimmune thyroiditis). *Secondary hypothyroidism* (which accounts for a small percentage of all cases of hypothyroidism) occurs as a result of inadequate stimulation of the gland because of anterior pituitary gland dysfunction.

Risk Factors. Women are 10 times more likely than men to have hypothyroidism. More than 10% of women over age 65 and 15% over age 70 are diagnosed with this disorder. Risk factors include surgical removal of the thyroid gland, external irradiation, and some medications (e.g., lithium, amiodarone).

Clinical Presentation. As with all disorders affecting the thyroid and parathyroid glands, clinical signs and symptoms affect many systems of the body (Table 11-5). Because the thyroid hormones play such an important role in the body's metabolism, lack of these hormones seriously upsets the balance of body processes.

Among the primary symptoms associated with hypothyroidism are intolerance to cold, excessive fatigue and drowsiness, headaches, and weight gain. In women, menstrual bleeding may become irregular, and premenstrual syndrome (PMS) may worsen. Physical assessment often reveals dryness of the skin and increasing thinness and brittleness of the hair and nails. There may be nodules or other irregularities of the thyroid palpable during anterior neck examination.

Ichthyosis, or dry scaly skin (resembling fish scales; the word *ichthyosis* is derived from the Latin word *ichthus*, which means "fish"), may be an inherited dermatologic condition (Fig. 11-4). It may also be the result of a thyroid condition. It must not be assumed that clients who present with this condition are merely in need of better hydration or regular use of skin lotion. A medical referral is needed to rule out underlying pathology.

Myxedema. A characteristic sign of hypothyroidism and more rarely associated with hyperthyroidism (Graves' disease) is *myxedema* (often used synonymously with *hypothyroidism*). Myxedema is a result of an alteration in the composition of the dermis and other tissues, causing connective tissues to be separated by increased amounts of mucopolysaccharides and proteins.

This mucopolysaccharide-protein complex binds with water, causing a nonpitting, boggy edema, especially around the eyes, hands, and feet and in the supraclavicular fossae (Case Example 11-4). The binding of this protein-mucopolysaccharide complex causes thickening of the tongue and the laryngeal and pharyngeal mucous membranes. This results in hoarseness and thick, slurred speech, which are also characteristic of untreated hypothyroidism.

Clients who have myxedematous hypothyroidism may demonstrate synovial fluid that is highly distinctive. The fluid's high viscosity results in a slow fluid wave that creates a sluggish "bulge" sign visible at the knee joint. Often, the fluid contains calcium pyrophosphate dihydrate (CPPD)

TABLE 11-5 Systemic Manifestations of Hypothyroidism

CNS Effects	Musculoskeletal Effects	Pulmonary Effects	Cardiovascular Effects	Hematologic Effects	Integumentary Effects	GI Effects	GU Effects
Slowed speech and hoarseness Anxiety, depression Slow mental function (loss of interest in daily activities, poor short-term memory) Hearing impairment Fatigue and increased sleep Headache Cerebellar ataxia	Proximal muscle weakness Myalgias Trigger points Stiffness, cramps Carpal tunnel syndrome Prolonged deep tendon reflexes (especially Achilles) Subjective report of paresthesias without supportive objective findings Muscular and joint edema Back pain Increased bone density Decreased bone formation and resorption	Dyspnea Respiratory muscle weakness Pleural effusion	Bradycardia Congestive heart failure Poor peripheral circulation (pallor, cold skin, intolerance to cold, hypertension) Severe atherosclerosis; hyperlipidemias Angina Elevated blood pressure Increased cholesterol, triglycerides, LDL Cardiomyopathy	Anemia Easy bruising	Myxedema (periorbital and peripheral) Thickened, cool, and dry skin Scaly skin (especially elbows and knees) Carotenosis (yellowing of the skin) Coarse, thinning hair Intolerance to cold Nonpitting edema of hands and feet Poor wound healing Thin, brittle nails	Anorexia Constipation Weight gain disproportionate to caloric intake Decreased absorption of nutrients Decreased protein metabolism (retarded skeletal and soft tissue growth) Delayed glucose uptake Decreased glucose absorption	Infertility Menstrual irregularity Heavy menstrual bleeding

CNS, Central nervous system; *LDL,* low-density lipoprotein; *GI,* gastrointestinal; *GU,* genitourinary.

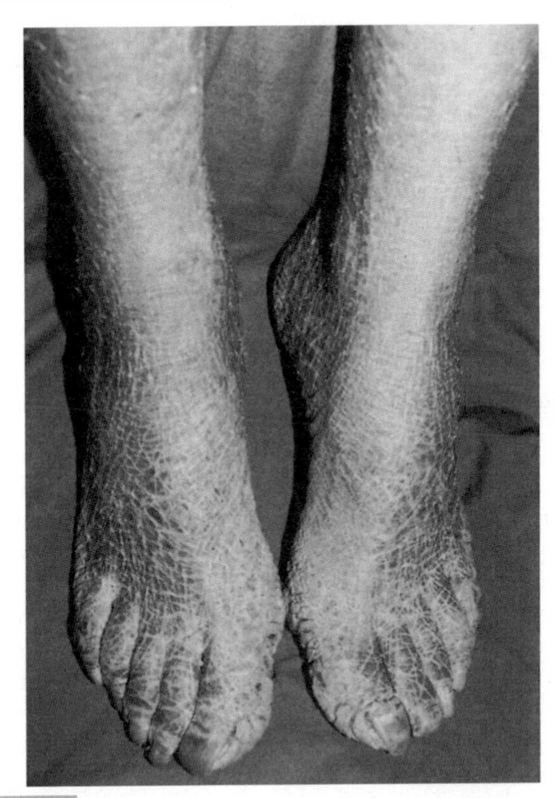

Fig. 11-4 Ichthyosis of the legs in a woman with severe hypothyroidism. (From Callen JP, Jorizzo J, Greer KE, et al: *Dermatological signs of internal disease,* Philadelphia, 1988, WB Saunders. Used with permission.)

crystal deposits that may be associated with chondrocalcinosis (deposit of calcium salts in joint cartilage). Thus a finding of a highly viscous, "noninflammatory" joint effusion containing CPPD crystals may suggest to the physician possible underlying hypothyroidism.

When such clients with hypothyroidism have been treated with thyroid replacement, some have experienced attacks of acute pseudogout caused by CPPD crystals remaining in the synovial fluid.

Neuromuscular Symptoms. Neuromuscular symptoms are among the most common manifestations of hypothyroidism. Flexor tenosynovitis with stiffness often accompanies CTS in people with hypothyroidism. CTS can develop before other signs of hypothyroidism become evident. It is thought that this CTS arises from deposition of myxedematous tissue in the carpal tunnel area. Acroparesthesias may occur as a result of median nerve compression at the wrist. The paresthesias are almost always located bilaterally in the hands. Most clients do not require surgical treatment because the symptoms respond to thyroid replacement.

Proximal muscle weakness sometimes accompanied by pain is common in clients who have hypothyroidism. As mentioned earlier, muscle weakness is not always related to either the severity or the duration of hypothyroidism and can be present several months before the diagnosis of hypothyroidism is made. Muscle bulk is usually normal; muscle hypertrophy is rare. Deep tendon reflexes are characterized by slowed muscle contraction and relaxation (prolonged reflex).

Characteristically, the muscular complaints of the client with hypothyroidism are aches and pains and cramps or stiffness. Involved muscles are particularly likely to develop persistent myofascial trigger points (TrPs). Of particular interest to the therapist is the concept that clinically any compromise of the energy metabolism of muscle aggravates and perpetuates TrPs. Treatment of the underlying hypothyroidism is essential in eliminating the TrPs,[19] but new research also supports the need for soft tissue treatment to achieve full recovery.[20]

CASE EXAMPLE 11-4

Myxedema

Referral: A 36-year-old African-American woman with a history of Graves' disease came to an outpatient hand clinic as a self-referral with painless swelling in both hands and feet. She had seen her doctor 6 weeks ago and was told that she did not have rheumatoid arthritis and should see a physical therapist.

Past Medical History: The woman had a 3-year history of Graves' disease, which was treated with thyroid supplementation. She had a family history of thyroid problems, maternal history of diabetes, and history of early death from heart attack (father). Aside from symptoms of hyperthyroidism, she did not have any health problems.

Clinical Presentation: There was a mild swelling apparent in the soft tissues of the fingers and toes. Presentation was painless and bilateral, although asymmetric (second and third digits of the right hand were affected; third and fourth digits of the left hand were symptomatic).

The therapist was alerted to the unusual clinical presentation by the following signs:

- Thickening of the skin over the affected digits in the hands and feet
- Clubbing of all digits (fingers and toes)
- Nonpitting edema and thickening of the skin over the front of the lower legs down to the feet

The client did not think these additional symptoms were present at the time she saw her physician 6 weeks ago, but she could not remember exactly.

Result: The therapist was unsure if the symptoms present were normal manifestations of Graves' disease or an indication that the client's thyroid levels were abnormal. The physician was contacted with information about the additional signs and questions about this client's clinical presentation.

The physician requested a return visit from the client, at which time further testing was done. The skin changes and edema of the lower legs are called *pretibial myxedema.* Myxedema is more commonly associated with hypothyroidism. When accompanied by digital clubbing and new bone formation, the condition is called *thyroid acropachy.* This condition is seen most often in individuals who have been treated for hyperthyroidism.

Drug therapy for the thyroid function does not change the acropachy; treatment is palliative for relief of symptoms. Physical therapy intervention can be prescribed but has not been studied to prove effectiveness for this condition.

There appears to be an association between hypothyroidism and fibromyalgia syndrome (FMS). Individuals with FMS and clients with undiagnosed myofascial symptoms may benefit from a medical referral for evaluation of thyroid function.[21-24]

Neoplasms

Cancer of the thyroid is a relatively uncommon, slow-growing neoplasm that rarely metastasizes. It is often the incidental finding in persons being treated for other disorders (e.g., musculoskeletal disorders involving the head and neck). Primary cancers of other endocrine organs are rare and are not encountered by the clinical therapist very often.

Risk factors for thyroid cancer include female gender, age over 40 years, Caucasian race, iodine deficiency, family history of thyroid cancer, and being exposed to radioactive iodine (I-131), especially as children. In addition, nuclear power plant fallout could expose large numbers of people to I-131 and subsequent thyroid cancer. The use of potassium iodide (KI) can protect the thyroid from the adverse effects of I-131 and is recommended to be made available in areas of the country near nuclear power plants in case of nuclear fallout.[25] The initial manifestation in adults and especially in children is a palpable lymph node or nodule in the neck lateral to the sternocleidomastoid muscle in the lower portion of the posterior triangle overlying the scalene muscles[26] (Fig. 11-5).

A physician must evaluate any client with a palpable nodule because a palpable nodule is often clinically indistinguishable from a mass associated with a benign condition. The presence of new-onset hoarseness, hemoptysis, or elevated blood pressure is a red-flag symptom for systemic disease.

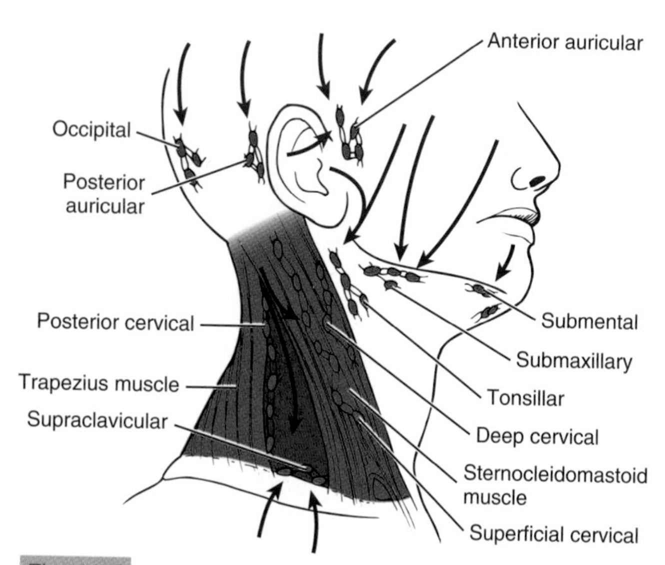

Fig. 11-5 Lymph node regions of the head and neck. Palpable nodal disease associated with thyroid carcinoma is commonly located lateral to the sternocleidomastoid muscle in the lower portion of the posterior triangle overlying the scalene muscles *(dark red triangle)*. (Modified from Swartz MH: *Textbook of physical diagnosis*, Philadelphia, 1989, WB Saunders.)

CLINICAL SIGNS AND SYMPTOMS

Thyroid Carcinoma

- Presence of asymptomatic nodule or mass in thyroid tissue
- Nodule is firm, irregular, painless
- Hoarseness
- Hemoptysis
- Dyspnea
- Elevated blood pressure

Parathyroid Glands

Two parathyroid glands are located on the posterior surface of each lobe of the thyroid gland. These glands secrete parathyroid hormone (PTH), which regulates calcium and phosphorus metabolism. Parathyroid disorders include hyperparathyroidism and hypoparathyroidism.

The therapist may see clients with parathyroid disorders in acute care settings and postoperatively because these disorders can result from diseases and surgical procedures. If damage or removal of these glands occurs, the resulting hypoparathyroidism (temporary or permanent) causes hypocalcemia, which can result in cardiac arrhythmias and neuromuscular irritability (tetany).

Disorders of the parathyroid glands may produce periarthritis and tendinitis. Both types of inflammation may be crystal induced and can be associated with periarticular or tendinous calcification.

Hyperparathyroidism

Hyperparathyroidism (hyperfunction), or the excessive secretion of PTH, disrupts calcium, phosphate, and bone metabolism. The primary function of PTH is to maintain a normal serum calcium level. Elevated PTH causes release of calcium by the bone and accumulation of calcium in the bloodstream.

Symptoms of hyperparathyroidism are related to this release of bone calcium into the bloodstream. This causes demineralization of bone and subsequent loss of bone strength and density. At the same time, the increase of calcium in the bloodstream can cause many other problems within the body, such as renal stones. The incidence of hyperparathyroidism is highest in postmenopausal women.[27]

The major cause of primary hyperparathyroidism is a tumor of a parathyroid gland, which results in the autonomous secretion of PTH. Renal failure, another common cause of hyperparathyroidism, causes hypocalcemia and stimulates PTH production. Hyperplasia of the gland occurs as it attempts to raise the blood serum calcium levels. Thiazide diuretics (used for hypertension) and lithium carbonate (used for some psychiatric problems) can exacerbate or even cause hyperparathyroid disorders.[28]

Clinical Presentation. Many systems of the body are affected by hyperparathyroidism (Table 11-6). Proximal muscle weakness and fatigability are common findings and may be secondary to a peripheral neuropathic process.

TABLE 11-6 Systemic Manifestations of Hyperparathyroidism

Early CNS Symptoms	Musculoskeletal Effects	GI Effects	GU Effects
Lethargy, drowsiness, paresthesia Slow mentation, poor memory Depression, personality changes Easily fatigued Hyperactive deep tendon reflexes Occasionally glove-and-stocking distribution of sensory loss	Mild-to-severe proximal muscle weakness of the extremities Muscle atrophy Bone decalcification (bone pain, especially spine; pathologic fractures; bone cysts) Gout and pseudogout Arthralgias involving the hands Myalgia and sensation of heaviness in the lower extremities Joint hypermobility	Peptic ulcers Pancreatitis Nausea, vomiting, anorexia Constipation	Renal colic associated with kidney stones Hypercalcemia (polyuria, polydipsia, constipation) Kidney infections

CNS, Central nervous system; *GI,* gastrointestinal; *GU,* genitourinary.

Myopathy of respiratory muscles with associated respiratory involvement often goes unnoticed. Striking reversal of muscle weakness and atrophy occur with successful treatment of the underlying hyperparathyroidism.

Other symptoms associated with hyperparathyroidism are muscle weakness, loss of appetite, weight loss, nausea and vomiting, depression, and increased thirst and urination (Case Example 11-5). Hyperparathyroidism can also cause GI problems, pancreatitis, bone decalcification, and psychotic paranoia (Fig. 11-6).

Bone erosion, bone resorption, and subsequent bone destruction from hypercalcemia associated with hyperparathyroidism occurs rarely today. In most cases, hypercalcemia is mild and detected before any significant skeletal disease develops. The classic bone disease *osteitis fibrosa cystica* affects persons with primary or renal hyperparathyroidism. Bone lesions called *Brown tumors* appear at the end stages of the cystic osteitis fibrosa. There are increasing reports of this condition in hyperparathyroidism secondary to renal failure because of the increasing survival rates of clients on hemodialysis.

Currently, skeletal manifestations of primary hyperparathyroidism are more likely to include bone pain secondary to osteopenia, especially diffuse osteopenia of the spine with possible vertebral fractures. In addition, a number of articular and periarticular disorders have been recognized in association with primary hyperparathyroidism. The therapist may encounter cases of ruptured tendons caused by bone resorption in clients with hyperparathyroidism.

Inflammatory erosive polyarthritis may be associated with chondrocalcinosis and CPPD deposits in the synovial fluid. This erosion is called *osteogenic synovitis.* Concurrent illness and surgery (most often parathyroidectomy) are recognized inducers of acute arthritic episodes.

Hypoparathyroidism

Hypoparathyroidism (hypofunction), or insufficient secretion of PTH, most commonly results from accidental removal or injury of the parathyroid gland during thyroid or anterior

CASE EXAMPLE 11-5

Rheumatoid Arthritis and Hyperparathyroidism

Referral: A 58-year-old man was referred to physical therapy by his primary care physician with a diagnosis of new-onset rheumatoid arthritis. Chief complaint was bilateral sacroiliac (SI) joint pain and pain on palpation of the hands and wrists.

When asked if he had any symptoms of any kind anywhere else in the body, he mentioned constipation, nausea, and loss of appetite. The family took the therapist aside and expressed concerns about personality changes, including apathy, depression, and episodes of paranoia. These additional symptoms were first observed shortly after the hand pain developed.

Past Medical History: The client had a motorcycle accident 2 years ago but reported no major injuries and no apparent residual problems. He had a family history of heart disease and hypertension but was not hypertensive at the time of the physical therapy interview. There was no other contributory personal or family past medical history.

Clinical Presentation: The therapist was unable to account for the sacroiliac joint pain. There were no particular movements that made it better or worse, and no objective findings to suggest an underlying movement system impairment.

Other red flags included age, bilateral hand and SI symptoms, gastrointestinal (GI) distress, and psychologic/behavioral changes observed by the family.

Result: The therapist contacted the referring physician with the results of her evaluation. During the telephone conversation, the therapist mentioned the family's concerns about the client's personality change and the fact that the client had bilateral symptoms that could not be provoked or relieved. Additional GI symptoms were also discussed.

At the physician's request, the client completed a short course of physical therapy intervention with an emphasis on posture, core training, and soft tissue mobilization. The client returned to the physician for a follow-up examination 4 weeks later. His symptoms were unchanged.

After additional testing, the client was eventually diagnosed with hyperparathyroidism and treated accordingly. Both his hand and SI pain went away, as well as most of the GI problems.

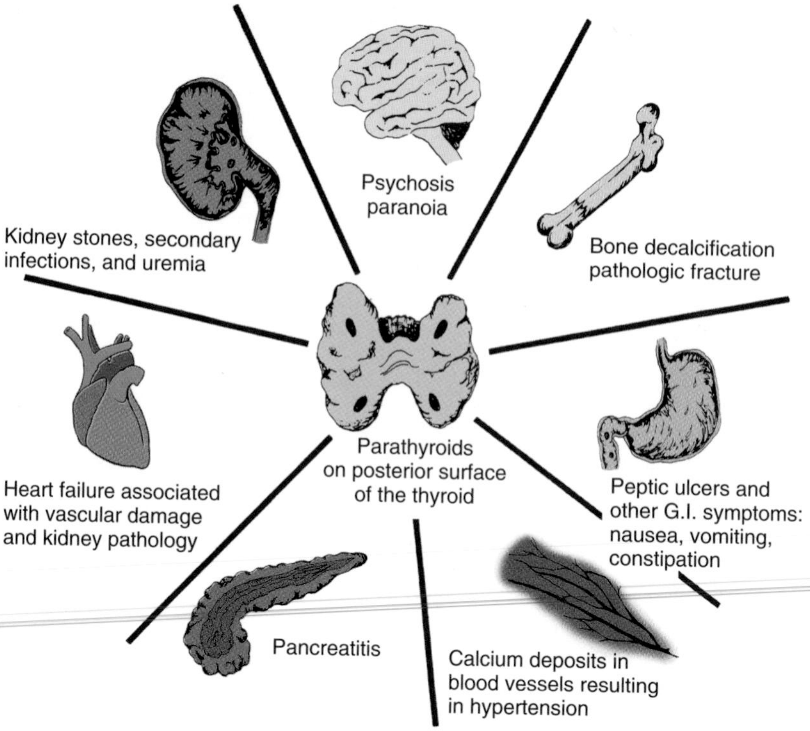

Kidney stones, secondary infections, and uremia
Psychosis paranoia
Bone decalcification pathologic fracture
Heart failure associated with vascular damage and kidney pathology
Parathyroids on posterior surface of the thyroid
Peptic ulcers and other G.I. symptoms: nausea, vomiting, constipation
Pancreatitis
Calcium deposits in blood vessels resulting in hypertension

Fig. 11-6 The pathologic processes of body structures as a result of excess parathyroid hormone. (From Muthe NC: *Endocrinology: a nursing approach,* Boston, 1981, Little, Brown.)

TABLE 11-7 Systemic Manifestations of Hypoparathyroidism

CNS Effects	Musculoskeletal Effects*	Cardiovascular Effects*	Integumentary Effects	GI Effects
Personality changes (irritability, agitation, anxiety, depression)	Hypocalcemia (neuromuscular excitability and muscular tetany, especially involving flexion of the upper extremity) Spasm of intercostal muscles and diaphragm compromising breathing Positive Chvostek's sign (twitching of facial muscles with tapping of the facial nerve in front of the ear)	Cardiac arrhythmias Eventual heart failure	Dry, scaly, coarse, pigmented skin Tendency to have skin infections Thinning of hair, including eyebrows and eyelashes Fingernails and toenails become brittle and form ridges	Nausea and vomiting Constipation or diarrhea Neuromuscular stimulation of the intestine (abdominal pain)

CNS, Central nervous system; *GI,* gastrointestinal.
*The most common and important effects for the therapist to be aware of are the musculoskeletal and cardiovascular effects.

neck surgery. A less common form of the disease can occur from a genetic autoimmune destruction of the gland. Hypofunction of the parathyroid gland results in insufficient secretion of PTH and subsequent hypocalcemia, hyperphosphatemia, and pronounced neuromuscular and cardiac irritability.

Clinical Presentation. Hypocalcemia occurs when the parathyroids become inactive. The resultant deficiency of calcium in the blood alters the function of many tissues in the body. These altered functions are described by the systemic manifestations of signs and symptoms associated with hypoparathyroidism (Table 11-7).

The most significant clinical consequence of hypocalcemia is neuromuscular irritability. This irritability results in muscle spasms, paresthesias, tetany, and life-threatening cardiac arrhythmias. Muscle weakness and pain have been reported along with hypocalcemia in clients with hypoparathyroidism.

Hypoparathyroidism is primarily treated through pharmacologic management with intravenous calcium gluconate, oral calcium salts, and vitamin D. Acute hypoparathyroidism is a life-threatening emergency and is treated rapidly with calcium replacement, anticonvulsants, and prevention of airway obstruction.

Pancreas

The pancreas is a fish-shaped organ that lies behind the stomach. Its head and neck are located in the curve of the duodenum, and its body extends horizontally across the posterior abdominal wall.

The pancreas has dual functions. It acts as both an *endocrine gland,* secreting the hormones insulin and glucagon, and an *exocrine gland,* producing digestive enzymes. Disorders of endocrine function are included in this chapter, whereas disorders of exocrine function affecting digestion are included in Chapter 8.

Diabetes Mellitus

Diabetes mellitus (DM) is a chronic disorder caused by deficient insulin or defective insulin action in the body. It is characterized by hyperglycemia (excess glucose in the blood) and disruption of the metabolism of carbohydrates, fats, and proteins. Over time, it results in serious small vessel and large vessel vascular complications and neuropathies.

Diabetes is the leading cause of end-stage renal disease (ESRD) [kidney failure requiring dialysis or transplantation], nontraumatic lower extremity amputations, and new cases of blindness among adults in the United States, and a major cause of heart disease and stroke.[29,30]

Type 1 DM is a condition in which little or no insulin is produced. It occurs in about 10% of all cases and usually occurs in children or young adults. Type 2 DM commonly occurs after age 40 and is a condition of defective insulin and/or impaired cell receptor binding of insulin. Table 11-8 depicts the major differences between type 1 and type 2 in presentation and treatment. There has been some discussion as to whether Alzheimer's disease is type 3 diabetes ("brain diabetes"), unique to the brain or if diabetes is just a risk factor for Alzheimer's disease.[31] A relationship between DM and dementia is undeniable, with numerous studies concluding that DM increases the risk of cognitive decline and dementia, including Alzheimer's disease.[32]

Native Americans, Latino Americans, Native Hawaiians, and some Asian Americans and Pacific Islanders have been identified at particularly high risk for type 2 DM and its complications.[33] Lack of exercise and obesity are two major risk factors for type 2 DM. As a result of these lifestyle factors (sedentary lifestyle, obesity), the overall number of persons in the United States with diabetes has increased from 10 million in 1977 to 24 million in 2007 and is projected to double or triple by 2050 if current trends in diabetes prevalence continue.[34,35]

Clinical Presentation. Specific physiologic changes occur when insulin is lacking or ineffective. Normally, the blood glucose level rises after a meal. A large amount of this glucose is taken up by the liver for storage or for use by other tissues such as skeletal muscle and fat. When insulin function is impaired, the glucose in the general circulation is not taken up or removed by these tissues; thus it continues to accumulate in the blood. Because new glucose has not been "deposited" into the liver, the liver synthesizes more glucose and releases it into the general circulation, which increases the already elevated blood glucose level.

TABLE 11-8 Primary Differences Between Type 1 and Type 2 Diabetes

Factors	Type 1	Type 2
Age of onset	Usually younger than 30	Usually older than 35 (Can be younger if history of childhood obesity)
Type of onset	Abrupt	Gradual
Endogenous (own) insulin production	Little or none	Below normal or above normal
Incidence	5%-10%	90%-95%
Ketoacidosis	May occur	Unlikely
Insulin injections	Required	Needed in 20% to 30% of clients
Body weight at onset	Normal or thin	80% are obese
Management	Diet, exercise, insulin	Diet, exercise, oral hypoglycemic agents or insulin
Etiology	Possible viral/autoimmune, resulting in destruction of islet cells	Obesity-associated insulin receptor resistance
Hereditary	Yes	Yes
Risk factors	May be autoimmune, environmental, genetic	Insulin resistance syndrome/metabolic syndrome Ethnicity • Native American • Hispanic/Latin • Native Hawaiian, Pacific Islanders

Protein synthesis is also impaired because amino acid transport into cells requires insulin. The metabolism of fats and fatty acids is altered, and instead of fat formation, fat breakdown begins in an attempt to liberate more glucose. The oxidation of these fats causes the formation of ketone bodies. Because the formation of these ketones can be rapid, they can build quickly and reach very high levels in the bloodstream. When the renal threshold for ketones is exceeded, the ketones appear in the urine as acetone (ketonuria).

The accumulation of high levels of glucose in the blood creates a hyperosmotic condition in the blood serum. This highly concentrated blood serum then "pulls" fluid from the interstitial areas, and fluid is lost through the kidneys (osmotic diuresis). Because large quantities of urine are excreted (polyuria), serious fluid losses occur, and the conscious individual becomes extremely thirsty and drinks large amounts of water (polydipsia). In addition, the kidney is unable to resorb all the glucose, so glucose begins to be excreted in the urine (glycosuria).

Certain medications can cause or contribute to hyperglycemia. Corticosteroids taken orally have the greatest glucogenic effect. Any person with diabetes taking corticosteroid medications must be monitored for changes in blood glucose levels.

Other hormones produced by the body also affect blood glucose levels and can have a direct influence on the severity of diabetic symptoms. Epinephrine, glucocorticoids, and growth hormone can cause significant elevations in blood glucose levels by mobilizing stored glucose to blood glucose during times of physical or psychologic stress.

When persons with DM are under stress, such as during surgery, trauma, pregnancy, puberty, or infectious states, blood glucose levels can rise and result in the need for increased amounts of insulin. If these insulin needs cannot be met, a hyperglycemic emergency such as diabetic ketoacidosis can result.

It is essential to remember that clients with DM who are under stress will have increased insulin requirements and may become symptomatic even though their disease is usually well controlled in normal circumstances.

CLINICAL SIGNS AND SYMPTOMS

Untreated or Uncontrolled Diabetes Mellitus

The classic clinical signs and symptoms of untreated or uncontrolled diabetes mellitus usually include one or more of the following:

- Polyuria: increased urination caused by osmotic diuresis
- Polydipsia: increased thirst in response to polyuria
- Polyphagia: increased appetite and ingestion of food (usually only in type 1)
- Weight loss in the presence of polyphagia: weight loss caused by improper fat metabolism and breakdown of fat stores (usually only in type 1)
- Hyperglycemia: increased blood glucose level (fasting level greater than 126 mg/dL)
- Glycosuria: presence of glucose in the urine
- Ketonuria: presence of ketone bodies in the urine (by-product of fat catabolism)
- Fatigue and weakness
- Blurred vision
- Irritability
- Recurring skin, gum, bladder, vaginal, or other infections
- Numbness/tingling in hands and feet
- Cuts/bruises that are difficult and slow to heal

Diagnosis. To be diagnosed with diabetes, a person must have fasting plasma glucose (FPG) readings of 126 mg/dL or higher on two different days. The previous cutoff, set in 1979, was 140 mg/dL. This change occurred as a result of research showing that individuals with readings as low as the mid-120s have already started developing tissue damage from diabetes. A value greater than 100 mg/dL is a risk factor for future diabetes and cardiovascular disease. It has been suggested that the term "prediabetes" is no longer used: the person either has diabetes or does not.[36]

The American Diabetes Association offers consumers a risk test for diabetes (http://www.diabetes.org/risk-test.jsp). All adults should take this risk test; anyone 45 or older should be tested for diabetes every 3 years. Individuals with elevated FPG values as described should be tested every 1 to 2 years. The therapist can offer at-risk clients information on increased activity and exercise as a means of lowering their risk of developing diabetes.[37]

Physical Complications. At presentation, the client with DM may have a variety of serious physical problems. Infection and atherosclerosis are the two primary long-term complications of this disease and are the usual causes of severe illness and death in the person with diabetes.

Blood vessels and nerves sustain major pathologic changes in the person affected by DM. Atherosclerosis in both large vessels (macrovascular changes) and small vessels (microvascular changes) develops at a much earlier age and progresses much faster in the individual with DM. The blood vessel changes result in decreased blood vessel lumen size, compromised blood flow, and resultant tissue ischemia. The pathologic end-products are cerebrovascular disease (CVD), coronary artery disease (CAD), renal artery stenosis, and peripheral vascular disease (PVD).

Microvascular changes, characterized by the thickening of capillaries and damage to the basement membrane, result in diabetic nephropathy (kidney disease) and diabetic retinopathy (disease of the retina). Diabetes is the leading cause of kidney failure and new cases of blindness in the United States as of 2007.[30,38]

Poorly controlled DM can lead to various tissue changes that result in impaired wound healing. Decreased circulation to the skin can further delay or diminish healing. Skin eruptions called *xanthomas* (Fig. 11-7) may appear when high lipid levels (e.g., cholesterol and triglycerides) in the blood cause fat deposits in the skin over extensor surfaces such as the elbows, knees, back of the head and neck, and heels. Yellow patches on the eyelids are another sign of hyperlipidemia. Medical referral is required to normalize lipid levels.

Physical Complications of Diabetes Mellitus.

- Atherosclerosis
  - Macrovascular disease
    - CVD
    - CAD
    - Renal artery stenosis
    - PVD
  - Microvascular disease
    - Nephropathy
    - Retinopathy
    - Decreased microcirculation to skin/body organs
- Infection/impaired wound healing
- Neuropathy
  - Autonomic (gastroparesis, diarrhea, incontinence, postural hypotension, decreased heart rate)
  - Peripheral (polyneuropathy, diabetic foot)
  - Diabetic amyotrophy
  - CTS (mononeuropathy; ischemia of median nerve)
  - Charcot's joint (diabetic arthropathy)
- Periarthritis
- Hand stiffness
  - Limited joint mobility (LJM) syndrome
  - Flexor tenosynovitis

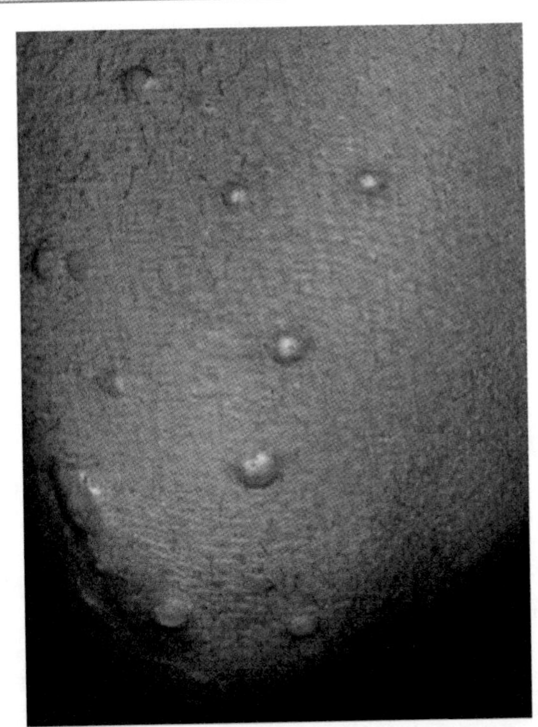

Fig. 11-7 Multiple eruptive xanthomas over the extensor surface of the elbow in a client with poorly controlled diabetes. These lipid-filled nodules characterized by an intracellular accumulation of cholesterol develop in the skin, often around the extensor tendons. Medical referral is required; xanthomas in this population are a sign that the health-care team, including the therapist, must work with the client to provide further education about diabetes, gain better control of glucose levels, and prevent avoidable complications. These skin lesions will go away when the diabetes is under control. Xanthomas can occur in any condition with disturbances of lipoprotein metabolism (not just diabetes). (From Callen JP, Jorizzo J, Greer KE, et al: *Dermatological signs of internal disease,* Philadelphia, 1988, WB Saunders. Used with permission.)

- Dupuytren's contracture
- Complex regional pain syndrome (CRPS)

Depression. Depression is common in individuals with type 2 diabetes (see Box 3-10) and is linked with increased mortality in this population.[39] Adults with diabetes and depression are less likely to follow recommendations for nutrition and exercise. They are less likely to check their blood glucose levels routinely and more likely to take drug "holidays" from their other medications (e.g., for hyperlipidemia or hypertension). Clients with diabetes who are depressed are more likely to miss health care appointments for prevention and intervention.[40,41]

Diabetic Neuropathy. Neuropathy is the most common chronic complication of long-term DM. Neuropathy in the client with DM is thought to be related to the accumulation in the nerve cells of sorbitol, a by-product of improper glucose metabolism. This accumulation then results in abnormal fluid and electrolyte shifts and nerve cell dysfunction. The combination of this metabolic derangement and the diminished vascular perfusion to nerve tissues contributes to the severe problem of diabetic neuropathy.

Risk factors. Other than glycemic control, there is no curative intervention for diabetic neuropathy. Identifying potentially modifiable risk factors for neuropathy is crucial; the therapist can have a key role in providing risk factor assessment for clients with diabetes.

Risk factors for the development of diabetic neuropathy include the duration and severity of diabetes, elevated triglycerides, higher body mass index (BMI), and a history of smoking or hypertension.[42,43]

Clinical presentation. Neuropathy may affect the central nervous system, peripheral nervous system, or autonomic nervous system. Peripheral neuropathy usually develops first as a sensory impairment of the extremities. Autonomic involvement is more common with long-standing disease.

Most common among the peripheral neuropathies are chronic sensorimotor distal symmetric polyneuropathy (DPN).[44] Polyneuropathy affects peripheral nerves in distal lower extremities, causing burning and numbness in the feet. It can result in muscle weakness, atrophy, and foot drop. Diabetic neuropathy can produce a syndrome of bilateral but asymmetric proximal muscle weakness called *diabetic amyotrophy.* Although the muscle enzyme levels are usually normal, muscle biopsy reveals atrophy of type II muscle fibers.

CTS (mononeuropathy) is also a common finding in persons with DM; it represents one form of diabetic neuropathy. As many as 5% to 16% of people with CTS have underlying diabetes. The mechanism is thought to be ischemia of the median nerve resulting from diabetes-related microvascular damage. This ischemia then causes increased sensitivity to even minor pressure exerted in the carpal tunnel area.

Autonomic involvement affects the pace of the heart beat, blood pressure, sweating, and bladder function and can cause symptoms such as erectile dysfunction and gastroparesis (delayed stomach emptying).[42]

CLINICAL SIGNS AND SYMPTOMS

Diabetic Neuropathy (at least two or more are present)

Peripheral (Motor and Sensory)

- Sensory, vibratory impairment of the extremities
- Burning, stabbing, pain, or numbness in distal lower extremities
- Extreme sensitivity to touch
- Muscle weakness and atrophy (diabetic amyotrophy)
- Absence of distal deep tendon reflexes (knee, ankle)
- Loss of balance
- Carpal tunnel syndrome

Autonomic

- Gastroparesis (delayed emptying of the stomach)
- Constipation or diarrhea
- Erectile dysfunction (sex drive unaffected; sexual function decreased)
- Urinary tract infections; urinary incontinence
- Profuse sweating
- Lack of oil production resulting in dry, cracked skin susceptible to bacteria and infection
- Pupillary adjustment restricted (difficulty seeing at night)
- Orthostatic hypotension
- Loss of heart rate variability

Charcot's joint, or neuropathic arthropathy, is a well-known complication of DM. This condition is due at least in part to the loss of proprioceptive sensation that marks diabetic neuropathy. Severe degenerative arthritis similar to Charcot's joint has been noted in clients with CPPD crystal deposition disease. Shoulder, hand, and foot disorders are very common, and evaluation of clients with DM should include examination of these areas (Case Example 11-6).[45,46]

CLINICAL SIGNS AND SYMPTOMS

Charcot's Joints

- Severe unilateral swelling (bilateral in 20% of cases but not bilateral at the same time)
- Increased skin warmth
- Redness
- Deep pressure sensation but significantly less pain than anticipated
- Normal x-rays initially but changes over time
- Joint deformity

The large- and small-vessel changes that occur with DM contribute to the changes in the feet of individuals with diabetes. Sensory neuropathy, which may lead to painless trauma and ulceration, can progress to infection. Neuropathy can result in drying and cracking of the skin, which creates more openings for bacteria to enter. The combination of all these factors can ultimately lead to gangrene and eventually require amputation. Prevention of these problems by meticulous care of the diabetic foot can reduce the need for amputation by 50% to 75%.

An annual foot screen by a health care provider is currently recommended for anyone with diabetes. This screen includes examination of toenails for length, thickness, and ingrown position. All calluses should be examined because ulceration can occur underneath them. General skin integrity, color, circulation, and structure should also be assessed.[47]

Whether a poorly controlled blood glucose level is a causative factor in the development of the long-term physical complications of diabetes is still controversial, but it does seem clear that these complications increase with the duration of the disease. Stable glycemic control (between 80 and 110 mg/dL), which prevents the fluctuation of blood glucose

CASE EXAMPLE 11-6

Charcot Shoulder (Neuroarthropathy)

Referral: A 44-year-old wheelchair-dependent man with type 2 diabetes who was well-known to the physical therapy clinic came in with new symptoms of right shoulder pain. There was no known trauma or injury to account for the changes in his shoulder. He had previously been evaluated for an exercise program as part of his diabetes management.

Past Medical History: The client was involved in a rock-climbing accident 15 years ago. He has had multiple reconstructive surgeries for broken bones and frostbite of the lower extremities associated with the accident. He was diagnosed with type 2 diabetes 3 years ago and uses an insulin pump but does not have consistent control of his blood glucose levels.

The man remains active and has resumed rock climbing along with many other outdoor activities. This new onset of shoulder pain has limited his activities and impaired his ability to propel his wheelchair.

There is no other significant history to report. The client is a nonsmoker, drinks only occasionally, and then drinks only socially (one or two glasses of wine). He has not had any other symptoms; there have been no constitutional symptoms, loss of appetite, or other gastrointestinal problems.

Clinical Presentation: Cervical spine and elbow were cleared for any loss of motion, weakness, or other problems that might contribute to shoulder pain. Gross examination of motion and strength of the left shoulder revealed no problems. The skin was normal on both sides, no cervical or supraclavicular lymph node changes were observed or palpated, and no other observable changes in the upper quadrant were evident.

Range of motion of the right shoulder:

- Active and passive abduction were equal and limited to 60 degrees and painful.
- Active and passive flexion were equal and limited to 65 degrees and painful.
- Biceps and deltoid strength were both 4/5; upper trapezius and triceps strength was normal (5/5).
- Grip strength appeared normal.

Further neurologic screening exam revealed severely decreased proprioception of the entire right upper extremity; no other neurologic changes were observed or reported.

Radial pulses intact and equal bilaterally.

Referral Decision: The therapist decided an x-ray might be helpful before initiating a program of physical therapy intervention. The client was very active and athletic and may have injured the joint or fractured the bone. Given the severity of his diabetic course over the last 3 years, an x-ray might be helpful in revealing any related arthritis that may be present.

The physician agreed with the therapist's assessment, and a radiographic examination was ordered.

Result: X-ray studies revealed destruction of two thirds of the right humeral head with microfractures and fragmentation throughout. The diagnosis of Charcot shoulder or neuroarthropathy was made. In this case, the therapist's knowledge of the client's past medical history and awareness of the physical complications possible with diabetes led to the referral decision before further damage was done to the bone and joint.

It is unusual for someone with this severe a condition to present with only mild symptoms. His extreme athleticism and stoic attitude may have masked the intensity of his symptoms.

levels, has been shown to be helpful in decreasing neuropathic pain (and of course other complications).[48]

Periarthritis. Musculoskeletal disorders of the hand and shoulder, including periarthritis of the shoulder, is five times as common in this group as it is in individuals who do not have diabetes. The condition most often affects insulin-dependent people, and involvement is typically bilateral.

The mechanism of this association is unclear, but it is believed to be related to fibroblast proliferation in the connective tissue structures around joints or to microangiopathy (disorder involving small blood vessels) involving the tendon sheaths. This periarthritic condition can behave unpredictably: It may regress spontaneously, remain stable, or progress to adhesive capsulitis or frozen shoulder.[49]

Hand Stiffness. Diabetic stiff hand, LJM syndrome, cheirarthritis (inflammation of the hand and finger joints), and diabetic contractures are common in both types of DM in direct relation to the presence and duration of microvascular complications.

Flexor tenosynovitis, caused by accumulation of excessive dermal collagen in the fingers, results in thickening and induration of the skin around the joints. This condition can lead to sclerodactyly (hardening and shrinking of fingers and toes), which in turn can mimic scleroderma.

Dupuytren's contracture has a strong association with DM. This syndrome is characterized by nodular thickening of the palmar fascia and flexion contracture of the digits. Clients usually have pain in the palm and digits, with decreased mobility and contracture of the fingers. In clients with diabetes, Dupuytren's contracture must be differentiated from LJM, which may involve the entire hand and is frequently bilateral, and from flexor tenosynovitis, which is marked by trigger finger.

Individuals with DM may develop CRPS (formerly called *reflex sympathetic dystrophy* [RSD] *syndrome*), which is characterized by pain, hyperesthesia, vasomotor and dystrophic skin changes, and tenderness and swelling around the hands and feet.

Intervention. Medical management of the client with diabetes is directed primarily toward maintenance of blood glucose values within the range of 80 to 120 mg/dL. The three primary treatment modalities used in the management of DM are diet, exercise, and medication (insulin and oral hypoglycemic agents; Table 11-9).

Recommended preventive care services, such as yearly eye and foot examinations, as well as measurements of glycosylated hemoglobin (A1C) two or more times per year are critical in the prevention of diabetic complications such as blindness, amputation, and cardiovascular disease.[50] A1C (also known as *glycosylated hemoglobin, glycated hemoglobin,* or *glycohemoglobin*) is an accurate, objective measurement of chronic glycemia in diabetes.

Most laboratories list the normal reference range as 4% to 6%. A1C equal or greater than 6.5% on two consecutive occasions is diagnostic of diabetes but must be confirmed by fasting glucose levels.[51] The goal is to maintain consistent

TABLE 11-9 Types of Insulin and Insulin Action

Type	Name	Onset	Peak	Duration
Rapid-acting	Insulin lispro (Humalog)	Begins to work 5 minutes after injection	Peaks in about 1 hour (range of 30-90 minutes)	Continues to work for 3-4 hours
	Insulin aspart (NovoLog)	15 minutes	1-3 hours (range of 45-90 minutes)	3-5 hours
Regular or short-acting	Humulin-R (human) Novolin-R (human)	Reaches bloodstream in first 30 minutes after injection	2-3 hours	Effective for about 3-8 hours
	Insulin glulisine (Apidra)	15 minutes	3-5 hours	Range: 3-8 hours
Intermediate-acting	NPH (Humulin N, Novolin N)	Reaches bloodstream as soon as 1.5 hours after injection	4-12	Effective for up to 24 hours
Long-acting	Ultralente (Humulin U)	Reaches bloodstream 6-10 hours after injection	No peak (maintains consistent level)	Effective for 20-24 hours
	Glargine (Lantus)	30-60 minutes	No peak	11-24 hours (usually 24 hours)
	Detemir (Levemir)	30 minutes	No peak	6-24 hours
Premixed insulins (combination of two types of insulin)	70/30 (%) NPH/regular 50/50 (%) NPH/regular 75/25 (%) (Humalog mix) 70/30 (%) (NovoLog mix)	10-15 minutes	Depends on mixture*	Effective for up to 24 hours

*When all mixtures are considered, the range is 1 to 12 hours.

Onset is how long it takes before the insulin reaches the bloodstream and starts to lower glucose levels.

Peak is the time when insulin reaches its maximum strength.

Duration defines how long the insulin continues to lower blood glucose.

Data from Micromedex Healthcare Series [Internet database]. Greenwood Village, CO, Thomson Reuters Healthcare Inc. Updated periodically. Compiled by Tanner Higginbotham, PharmD. University of Montana Drug Information Service, 2010.

A1C levels below 7% (American Diabetes Association recommendation),[36] which correlates to an average daily blood glucose below 150 to 154 mg/dL. This recommendation (and the plasma glucose levels used to diagnose diabetes) were determined based on the presence of retinopathy at these thresholds.

The guideline for A1C levels applies to the general population; individuals with a history of severe hypoglycemia, limited life expectancy, advanced diabetes-related complications, and extensive comorbid complications may be advised by their medical doctors to follow levels at or above 7%.[36] The Association of Endocrinologists recommends A1C levels of 6.5% or lower (average blood sugar reading of 135 mg/dL over a 2- to 3-month period).The A1C measurement gives the client and the therapist an indication of how successful diet, exercise, and medication are in controlling glucose levels over time. It can be used as a baseline from which to evaluate results of intervention. An A1C value greater than 10% warrants immediate medical attention (usually insulin treatment).[51]

For individuals with type 2 diabetes and the following factors, an A1C goal of less than 8% may be more appropriate than an A1C goal of less than 7%.[51]

- Known cardiovascular disease or high risk cardiovascular risk
- Inability to recognize and treat hypoglycemia, history of severe hypoglycemia requiring assistance
- Inability to comply with standard goals such as polypharmacy issues
- Limited life expectancy or estimated survival of less than 10 years
- Cognitive impairment
- Extensive comorbid conditions such as renal failure, liver failure, and end-stage disease complications

The therapist can conduct a careful screening examination (Box 11-1). All individuals with type 2 diabetes should be screened at the time of diagnosis and annually thereafter for diabetic peripheral neuropathies. Individuals with type 1 diabetes should be screened 5 years after diagnosis and annually thereafter. Screening should include checking knee and ankle reflexes, examining sensory function in the feet, asking about neuropathic symptoms, and examining the distal extremities for ulcers, calluses, and deformities.[44]

Exercise-Related Complications. Any exercise can improve the body's ability to use insulin. Exercise causes a decrease in the amount of insulin the pancreas releases because muscle contractions are increasing blood glucose uptake. For the person taking insulin, exercise adds to the effects of the insulin, dropping blood sugars to dangerously low levels. Exercise for the person with DM must be planned and instituted cautiously and monitored carefully because significant complications can result from exercise of higher intensity or longer duration.

Exercise-related complications can be prevented by careful monitoring of the client's blood glucose level before, during, and after strenuous exercise sessions. (Safe levels are individually determined but usually fall between 100 and 250 mg/dL; between 250 and 300 mg/dL is considered the "caution zone.") The following recommendations are general guidelines. These are not necessarily "fasting levels" (unless the person has not eaten for the last 12 hours for some reason). Exceptions are common, depending on the type of exercise, training level of the participant, expected glycemic pattern, and whether the individual is using an insulin pump.

If the blood glucose level is between 250 and 300 mg/dL at the start of the exercise, the client may be experiencing a state of insulin deficiency and should test urine for ketones, an indication that the body does not have enough insulin to control the blood sugar and is breaking down fat for energy. Exercise is likely to raise the blood sugars more; the exercise session should be postponed until the blood glucose level is under better control. Blood glucose levels of 300 mg/dL or higher indicate the blood sugar level is too high to exercise safely, putting the client at risk for ketoacidosis. Exercise should be postponed until the blood glucose level drops to a safe pre-exercise range (between 100 to 250 mg/dL, possibly up to 300 mg/dL as described).

If the blood glucose level is less than 100 mg/dL, a 10- to 15-g carbohydrate snack should be given and the glucose retested in 15 minutes to ensure an appropriate level.

Clients with active retinopathy and nephropathy should avoid high-intensity exercise that causes significant increases

BOX 11-1 ROLE OF THE PHYSICAL THERAPIST IN DIABETES SCREENING

The therapist can provide education and prevention through the screening process, including:

- Conduct periodic screening for neuropathy
- Assess for early signs of neuropathy (e.g., deep tendon reflexes, vibratory and position sense, touch)
- Education in avoiding late complications of neuropathy (e.g., annual foot and hand screening, preventive foot care; periodic footwear evaluation)
- Assess for signs of neuropathic arthropathy (Charcot's joint)
- Monitor blood glucose levels in association with exercise
- Screen for neuromusculoskeletal disorders (e.g., adhesive capsulitis, Dupuytren disease, flexor tenosynovitis, carpal tunnel syndrome, complex regional pain syndrome)
- Monitor vital signs (especially blood pressure)
- Conduct periodic lower extremity vascular examination (see Box 4-16; Table 4-10)
- Screen for depression; monitor depression (see Appendices B-9 and B-10; see Table 3-11)
- Encourage/remind client to get periodic A1C levels tested
- Reminder about annual eye examination

in blood pressure because such increases can cause further damage to the retinas and kidneys. Any exercise that places the head below the waist causing increased intrathoracic and intracranial pressures can also aggravate retinal problems. Screening for neuropathies by testing deep tendon reflexes and vibratory and position sense are also very important in the prevention of exercise-related complications such as ulcerations or fractures.

It is very important to have the client avoid insulin injection to active extremities within 1 hour of exercise because insulin is absorbed much more quickly in an active extremity. It is important to know the type, dose, and time of the client's insulin injections so that exercise is not planned for the peak activity times of the insulin.

Clients with type 1 diabetes may need to reduce the insulin dose or increase food intake when initiating an exercise program. During prolonged activities, a 10- to 15-g carbohydrate snack is recommended for each 30 minutes of activity. Activities should be promptly stopped with the development of any symptoms of hypoglycemia, and blood glucose should be tested. In addition, individuals with diabetes should not exercise alone. Partners, teammates, and coaches must be educated regarding the possibility of hypoglycemia and the way to manage it.

Insulin Pump During Exercise. People with type 1 diabetes (and some individuals with insulin-requiring type 2 diabetes) may be using an insulin pump. Continuous subcutaneous insulin infusion (CSII) therapy, known as *insulin pump therapy,* can bring the hormonal and metabolic responses to exercise close to normal for the individual with diabetes.

Although there are many benefits of pump use for active individuals with diabetes, there are a few drawbacks as well.[52] Exercise can speed the development of diabetic ketoacidosis (DKA) when there is an interruption of insulin delivery, which can quickly become a life-threatening condition.

Other considerations include the effect of excessive perspiration or water on the infusion set (needle into the skin at the infusion site gets displaced), ambient temperature (insulin degrades under extreme conditions of heat or cold), and the effect of movement or contact at the infusion site (this causes skin irritation).

Insulin pump users who have pre-exercise blood glucose levels less than 100 mg/dL may not need a carbohydrate snack because they can reduce or suspend base insulin levels during an activity. The insulin reductions and required level of carbohydrate intake needed depends on the intensity and duration of the activity.[52]

The therapist should become familiar with the features of each pump in use by clients. Knowledge of basic guiding principles for exercise with diabetes and general recommendations for insulin regimen changes is also helpful.

Severe Hyperglycemic States

The two primary life-threatening metabolic conditions that can develop if uncontrolled or untreated DM progresses to a state of severe hyperglycemia (more than 400 mg/dL) are DKA and hyperglycemic, hyperosmolar, nonketotic coma (HHNC; Table 11-10).

TABLE 11-10 Clinical Symptoms of Life-Threatening Glycemic States

Diabetic Ketoacidosis (DKA)	Hyperosmolar, Hyperglycemic State (HHS)	Hypoglycemia Insulin Shock
GRADUAL ONSET	**GRADUAL ONSET**	**SUDDEN ONSET**
Thirst	Thirst	**Sympathetic activity**
Hyperventilation	Polyuria leading quickly to decreased urine output	Pallor
Fruity odor to breath	Volume loss from polyuria leading quickly to renal insufficiency	Perspiration
Lethargy/confusion	Severe dehydration	Irritability/nervousness
Coma	Lethargy/confusion	Weakness
Muscle and abdominal cramps (electrolyte loss)	Seizures	Hunger
Polyuria, dehydration	Coma	Shakiness
Flushed face, hot/dry skin	Abdominal pain and distention	**CNS activity**
Elevated temperature	Blood glucose level >300 mg/dL	Headache
Blood glucose level >300 mg/dL		Double/blurred vision
Serum pH <7.3		Slurred speech
		Fatigue
		Numbness of lips/tongue
		Confusion
		Convulsion/coma
		Blood glucose level <70 mg/dL

DKA occurs with severe insulin deficiency caused by either undiagnosed DM or a situation in which the insulin needs of the person become greater than usual (e.g., infection, trauma, surgery, emotional stress). It is most often seen in the client with type 1 diabetes but can in rare situations occur in the client with type 2 diabetes. Medical treatment is necessary.

HHNC occurs most commonly in the older adult with type 2 diabetes. This complication is extremely serious and, in many cases, fatal. Factors that can precipitate this crisis are infections (e.g., pneumonia); medications that elevate the blood glucose level (e.g., corticosteroids); and procedures such as dialysis, surgery, or total parenteral nutrition (TPN).

There are specific clinical features that identify HHNC. Some of these are similar to those of DKA, such as severe hyperglycemia (1000 to 2000 mg/dL) and dehydration. The major differentiating feature between DKA and HHNC, however, is the absence of ketosis in HHNC.

Because it is likely that the therapist will work with clients who have diabetes, it is imperative that the clinical symptoms of DM and its potentially life-threatening metabolic states are understood. *If anyone with diabetes arrives for a clinical appointment in a confused or lethargic state or is exhibiting changes in mental function, fingerstick glucose testing should be performed. Immediate physician referral is necessary.*

Hypoglycemia

Hypoglycemia (blood glucose of less than 70 mg/dL) is a major complication of the use of insulin or oral hypoglycemic agents. Hypoglycemia is usually the result of a decrease in food intake or an increase in physical activity in relation to insulin administration. It is a potentially lethal problem. The hypoglycemic state interrupts the oxygen consumption of nervous system tissue. Repeated or prolonged attacks can result in irreversible brain damage and death.

Hypoglycemia Associated With Diabetes Mellitus. Hypoglycemia during or after exercise can be a problem for anyone with diabetes. This condition results as glucose is used by the working muscles, if the circulating level of injected insulin is too high, or both. The degree of hypoglycemia depends on such factors as pre-exercise blood glucose levels, duration and intensity of exercise, and blood insulin concentration.

Clinical Presentation. The severity and number of signs and symptoms depend on the individual client and the rapidity of the drop in blood glucose. It is important to note that clients can exhibit signs and symptoms of hypoglycemia when their elevated blood glucose level drops rapidly to a level that is still elevated (e.g., 400 to 200 mg/dL). The *rapidity* of the drop is the stimulus for sympathetic activity; even though a blood glucose level appears elevated, clients may still have hypoglycemia.

Clients receiving beta-adrenergic blockers (e.g., propranolol) can be at special risk for hypoglycemia by the actions of this medication. These beta-blockers inhibit the normal physiologic response of the body to the hypoglycemic state or block the appearance of the sympathetic manifestations of hypoglycemia. Clients may also have hypoglycemia during nighttime sleep (most often related to the use of intermediate- and long-acting insulins given more than once a day), with the only symptoms being nightmares, sweating, or headache.

Intervention. Hypoglycemia can be treated in the conscious client by immediate administration of sugar. It is always safer to give the sugar, even when there is doubt concerning the origin of symptoms (DKA and HHNC can also have similar central nervous system symptoms at presentation). Most often, 10 to 15 g of carbohydrate are sufficient to reverse the episode of hypoglycemia. Immediate-acting glucose sources should be kept in every physical therapy department (e.g., ½ cup of fruit juice or sugared cola, 8 ounces of milk, two packets of sugar, 2-ounce tube of honey or cake-decorating gel).

Most people with diabetes carry a rapid-acting source of carbohydrate, such as readily absorbable glucose tablets, so that it is available for use if a hypoglycemic episode occurs. Some individuals use intramuscular glucagon. If the client loses consciousness, emergency personnel must be notified, and glucose will be administered intravenously.

Any episode or suspected episode of hypoglycemia must be treated promptly and must be reported to the client's physician. It is important to question each client who has diabetes regarding his or her individual response to hypoglycemia. Information regarding individual symptoms, frequency of episodes, and precipitating factors may be invaluable to the therapist in preventing or minimizing a hypoglycemic attack.

Other Hypoglycemic States

Other conditions that can cause hypoglycemic states are usually related to hormonal deficiencies (e.g., cortisol, glucagon, ACTH) or overproduction of insulin or insulin-like material from tumors.

Reactive hypoglycemia, also known as *functional hypoglycemia,* occurs after the intake of a meal and usually results from stomach or duodenal surgery. This condition involves rapid stomach emptying with rapid rises of glucose levels. Glucose then rapidly falls to below normal levels as an exaggerated response of insulin secretion develops. The cause of reactive hypoglycemia is unknown.

Clinical Presentation. Clinical signs and symptoms of non–diabetes-related hypoglycemic states are the same as those described earlier for hypoglycemia related to DM. The client is warned to avoid fasting and simple sugars.

INTRODUCTION TO METABOLISM

As noted earlier, the endocrine system works with the nervous system to regulate and integrate the body's metabolic activities. The rate of metabolism can be increased by exercise, elevated body temperature (e.g., high fever), hormonal activity (e.g., thyroxine, insulin, epinephrine), and specific dynamic action that occurs after ingestion of a meal. All metabolic functions require proper fluid and acid-base balance. Although acid-base metabolism is not in itself a sign or a symptom, the consequences of an acid-base metabolism disorder can result in many clinical signs and symptoms.

Therapists are unlikely to evaluate someone with a primary musculoskeletal lesion that reflects an underlying metabolic disorder. However, many inpatients in hospitals and some outpatients may be affected by disturbances in acid-base metabolism and other specific metabolic disorders. Only those conditions that are likely to be encountered by a therapist are included in this text.

Fluid Imbalances

Fluid Deficit/Dehydration

Fluid deficit can occur as a result of two primary types of imbalance. There is either a loss of water without loss of solutes or a loss of both water and solutes.

The loss of body water without solutes results in the excess concentration of body solutes within the interstitial and intravascular compartments. To preserve equilibrium, water will then be forced to shift by osmosis from inside cells to these outside compartments.

If this state persists, large amounts of body water will be shifted and excreted (osmotic diuresis), and severe cellular dehydration will result. This type of imbalance can occur as a result of several conditions:

- Decreased water intake (e.g., unavailability, unconsciousness)
- Water loss without proportionate solute loss (e.g., prolonged hyperventilation, diabetes insipidus)
- Increased solute intake without proportionate water intake (tube feeding)
- Excess accumulation of solutes (e.g., high glucose levels such as in DM)

The second type of fluid imbalance results from a loss of *both* water and solutes. Causes of the loss of both water and solutes include hemorrhage, profuse perspiration (e.g., marathon runners), and loss of gastrointestinal tract secretions (e.g., vomiting, diarrhea, draining fistulas, ileostomy). Postsurgical patients who have had joint replacements, hip fractures, multiple trauma, or neurosurgery often lose blood and become hypovolemic despite efforts to maintain their homeostasis through blood transfusion and fluid replacement.

Severe losses of water or solutes (or both) can lead to dehydration and hypovolemic shock. It is important for the therapist to be aware of possible fluid losses or water shifts in any client who is already compromised by advanced age or by a situation, such as an ileostomy or tracheostomy, that results in a continuous loss of fluid. Because the response to fluid loss is highly individual, it is important to recognize the early clinical symptoms of fluid loss and to carefully monitor vital signs and clinical symptoms in clients who are at risk, especially the elderly, the very young, or the chronically ill.[53]

Athletes and normal adults may experience orthostatic hypotension when slightly dehydrated, especially when intense exercise increases the core body temperature. The normal vascular system can accommodate this effectively.

CLINICAL SIGNS AND SYMPTOMS

Dehydration or Fluid Loss

Early clinical signs and symptoms:

- Thirst
- Weight loss

As the condition worsens, other symptoms may include the following:

- Poor skin turgor
- Dryness of the mouth, throat, and face
- Absence of sweat
- Increased body temperature
- Low urine output
- Postural hypotension (increased heart rate by 10 beats/minute and decreased systolic or diastolic blood pressure by 20 mm Hg when moving from a supine to a sitting position)
- Dizziness when standing
- Confusion
- Increased hematocrit

Fluid Excess

Fluid excess can occur in two major forms: water intoxication (excess of water without an excess of solutes) or edema (excess of both solutes and water).

Because the etiologic complex, symptoms, and outcomes related to these problems are substantially different, these fluid imbalances are discussed separately.

Water Intoxication. Water intoxication (resulting in hyponatremia) is an excess of extracellular water in relationship to solutes. The extracellular fluid (ECF) becomes diluted, and water must then move into cells to equalize solute concentration on both sides of the cell membrane. High water consumption without solute replacement can result in hyponatremia, a potentially lethal situation.

Water excess can be caused by an accumulation of solute-free fluid. An increase in solute-free fluid usually occurs because of excess ADH (tumors, endocrine disorders) or intake of large amounts of only tap water without balanced solute ingestion. The latter situation occurs most often in older adults who drink additional water after having the flu, with its associated vomiting and diarrhea, or in athletes who have lost large amounts of body fluids during exercise that have been replaced with only water.

Symptoms of water intoxication are largely neurologic because of the shifting of water into brain tissues and resultant dilution of sodium in the vascular space.

CLINICAL SIGNS AND SYMPTOMS

Water Intoxication

- Decreased mental alertness

Other accompanying symptoms:

- Sleepiness
- Anorexia
- Poor motor coordination
- Confusion

In a severe imbalance, other symptoms may include the following:

- Convulsions
- Sudden weight gain
- Hyperventilation
- Warm, moist skin
- Signs of increased intracerebral pressure
    - Slow pulse
    - Increased systolic blood pressure (more than 10 mm Hg)
    - Decreased diastolic blood pressure (more than 10 mm Hg)
- Mild peripheral edema
- Low serum sodium
- Low hematocrit

Edema. An excess of solutes and water is called *isotonic volume excess.* The excess fluid is retained in the extracellular compartment and results in fluid accumulation in the interstitial spaces *(edema).* Edema can be produced by many different situations, most commonly including vein obstruction, decreased cardiac output, endocrine imbalances, and loss of serum proteins (e.g., burns, liver disease, allergic reactions).

CLINICAL SIGNS AND SYMPTOMS

Edema

- Weight gain (primary symptom)
- Excess fluid (several liters may accumulate before edema is evident)
- Dependent edema (collection of fluid in lower parts of the body)
- Pitting edema (finger pressed into edematous area leaves a persistent indentation in tissues)
- Increased blood pressure
- Neck vein engorgement (see Fig. 4-44)
- Effusions (pulmonary, pericardial, peritoneal)
- Congestive heart failure

Diuretic medications are used frequently to treat volume excess. Various diuretic medications may be used depending on the underlying cause of the problem and the desired effect of the drug. The most commonly used are the thiazide diuretics (e.g., chlorothiazide, hydrochlorothiazide). It is important to assess clients who take diuretic therapy for potential fluid loss and dehydration by observing for clinical symptoms of both.

These medications inhibit sodium and water resorption by the kidneys. Potassium is usually also lost with the sodium and water, so continuous replacement of potassium is a major concern for anyone receiving non–potassium-sparing diuretics. It is essential to monitor clients who take diuretics for signs and symptoms of potassium depletion.

It is also very important to check laboratory data for the potassium level in any client taking diuretics, particularly before exercise. Any value below the normal range (less than 3.5 mEq/L) is potentially dangerous and could result in a lethal cardiac arrhythmia even with moderate cardiovascular exercise.

For clients on diuretics, the therapist must observe for the appearance of symptoms consistent with dehydration or potassium depletion. Any concerns should be discussed with a physician before physical therapy intervention.

CLINICAL SIGNS AND SYMPTOMS

Potassium Depletion

- Muscle weakness
- Fatigue
- Cardiac arrhythmias
- Abdominal distention
- Nausea and vomiting

Metabolic Disorders

Metabolic Syndrome

Metabolic syndrome (sometimes referred to as *prediabetes type 2* but again, the term "prediabetes" is being questioned) is a group of signs and symptoms that are actually risk factors strongly linked to type 2 diabetes, cardiovascular disease, and stroke. This condition is characterized by insulin resistance and seems to be on the rise in Americans because of lifestyle and metabolic risk factors.

Insulin resistance is a generalized metabolic disorder in which the body cannot use insulin efficiently. Not only do the cells become resistant to insulin, but the cells themselves lose receptor sites (outside and inside the cell membrane). Insulin, which acts like a key to let glucose into the cells, cannot find a keyhole (receptor site) to open the door and let the glucose in.[54] This loss of receptor site and decreased receptor site receptivity is why the metabolic syndrome is also called the *insulin resistance syndrome.*

Some people are genetically predisposed to insulin resistance. Acquired factors, such as excess body fat and physical inactivity, can elicit insulin resistance and the metabolic syndrome in these people. Most people with insulin resistance have abdominal obesity. The biologic mechanisms at the molecular level between insulin resistance and metabolic risk factors are complex and not fully understood.[55]

Risk Factors and Red Flags. Serious health complications can be reduced by identifying risk factors early through screening. The dominant underlying risk factors for this syndrome appear to be sedentary lifestyle with little to no physical activity or exercise, abdominal obesity, and insulin resistance. Other risk factors include family history of metabolic syndrome, type 2 diabetes, hypertension, elevated fasting glucose (100 mg/dL or more), elevated triglyceride levels (150 mg/dL or more), and low high-density lipoprotein (HDL) (men: less than 40 mg/dL; women: less than 50 mg/dL).[56,57]

During any person's examination, the therapist should watch for red flags of metabolic syndrome, including BMI above 30, waist circumference (see Clinical Signs and Symptoms on the next page), elevated blood pressure, and signs of insulin resistance (e.g., acanthosis nigricans [Fig. 11-8], fatigue and decreased energy, depression, drowsiness after meals).

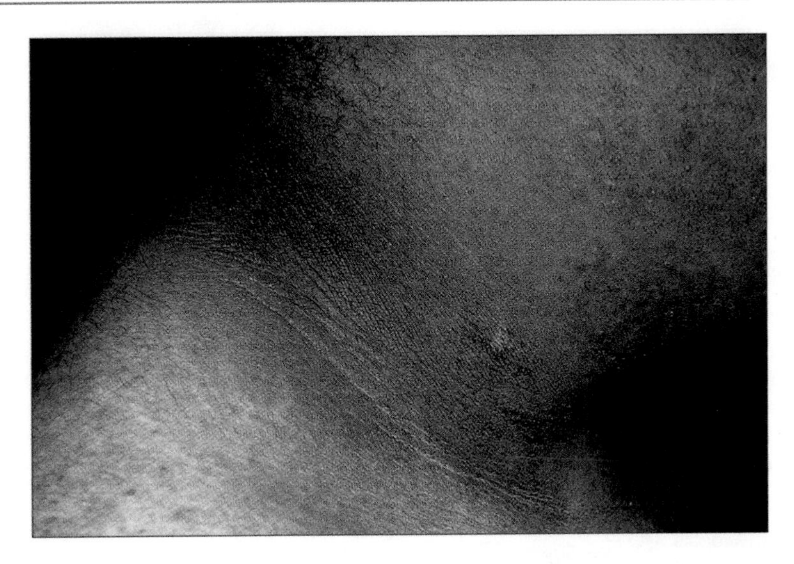

Fig. 11-8 Acanthosis nigricans is a brown to black hyperpigmentation of the skin with a soft to velvet-like feel to it. It is usually found in body folds, such as the posterior and lateral folds of the neck, the axilla, groin, umbilicus, forehead, and other areas. Acanthosis nigricans may be genetically inherited but is also associated with obesity or endocrine disorders such as hypothyroidism or hyperthyroidism, acromegaly, polycystic ovary disease, insulin-resistant diabetes, or Cushing's disease. (From Bolognia JL, Jorizzo JL, Rapini RP: *Dermatology,* ed 2, St. Louis, 2007, Mosby.)

CLINICAL SIGNS AND SYMPTOMS

Metabolic Syndrome

The metabolic syndrome is characterized by a group of metabolic risk factors in one person. The risk of serious illness increases in anyone with three or more of the following factors[58]:

- Abdominal obesity (excessive fat tissue in and around the abdomen; increased waist size; non-Asian men: equal or greater than 40 inches (102 cm); non-Asian women: equal or greater than 35 inches (88 cm). Predictive values for Asian men: 90 cm or more; Asian women: 80 cm or more.
- Atherogenic dyslipidemia (blood fat disorders—triglycerides more than 150 mg/dL, HDL cholesterol less than 50 mg/dL for a woman and less than 40 mg/dL for a man, and high low-density lipoprotein [LDL] cholesterol—that foster plaque buildups in artery walls) or on medication for elevated triglycerides
- Elevated blood pressure (130/85 mm Hg or more or on antihypertensive medication)
- Insulin resistance or glucose intolerance (the body cannot properly use insulin or blood sugar; fasting blood sugar is over 100 mg/dL)
- Prothrombotic state (e.g., high fibrinogen or plasminogen activator inhibitor–1 in the blood)
- Proinflammatory state (e.g., elevated C-reactive protein in the blood)

Metabolic Alkalosis

Metabolic alkalosis results from metabolic disturbances that cause either an increase in available bases or a loss of nonrespiratory body acids. Blood pH (hydrogen ion concentration in the ECF, a measure of metabolic process function and homeostasis) rises to a level greater than 7.45 (Table 11-11).

Common causes of metabolic alkalosis include excessive vomiting or upper gastrointestinal suctioning, diuretic therapy, or ingestion of large quantities of base substances such as antacids.

Decreased respirations may occur as the respiratory system attempts to compensate by buffering the basic environment. The lungs attempt to retain carbon dioxide (CO_2) and thus hydrogen ions (H).

It is important for the therapist to ask clients about the use of magnesium-containing antacids because symptoms of alkalosis can affect muscular function by causing muscle fasciculation and cramping. Prevention of problems related to alkalosis may be accomplished by education of the client regarding antacid use.

CLINICAL SIGNS AND SYMPTOMS

Metabolic Alkalosis

- Nausea
- Prolonged vomiting
- Diarrhea
- Confusion
- Irritability
- Agitation, restlessness
- Muscle twitching and muscle cramping
- Muscle weakness
- Paresthesias
- Convulsions
- Eventual coma
- Slow, shallow breathing

Metabolic Acidosis

Metabolic or nonrespiratory acidosis is an accumulation of fixed (nonvolatile) acids or a deficit of bases. Blood pH decreases to a level below 7.35 (see Table 11-11). Common causes of metabolic acidosis include DKA, lactic acidosis from a wide range of conditions (e.g., smoke inhalation, sepsis, cardiopulmonary failure, adverse reaction to drugs, alcohol abuse, liver disease, cancer), renal failure, severe diarrhea, and drug (including alcohol) or chemical toxicity.

Ketoacidosis occurs because insufficiency of insulin for the proper use of glucose (or increased insulin requirements) results in increased breakdown of fat. This accelerated fat breakdown produces ketones and other acids. These acids accumulate to high levels. While the body attempts to

TABLE 11-11 Laboratory Values: Metabolic Acidosis and Alkalosis

	Metabolic Acidosis	Metabolic Alkalosis
Lab values	• pH < 7.35 • Bicarbonate (HCO_3) decreased <22 mEq/L • $PaCO_2$ decreased <35 mm Hg (compensated)	• pH > 7.45 • Bicarbonate (HCO_3) increased > 26 mEq/L • $PaCO_2$ increased > 45 mm Hg (compensated)
GI effects*	• Loss of appetite (anorexia) • Nausea, vomiting • Abdominal pain, discomfort	• Nausea, vomiting • Diarrhea
CNS effects*	• Weakness • Lethargy • Confusion • Convulsions, coma	• Hyperactive deep tendon reflexes • Muscle weakness, cramping, twitching, tetany • Confusion, seizures • Irritability • Paresthesias
Integumentary effects*	• Warm and flushed	—
Cardiovascular effects*	• Peripheral vasodilation • Decreased heart rate • Cardiac dysrhythmias	• Hypotension • Cardiac dysrhythmias —
Skeletal system effects*	• Bone disease with chronic acidosis	—
Signs of compensation	• Increased respiratory rate and depth • Hyperkalemia • Increased ammonia in urine	• Decreased respiratory rate and depth

CNS, Central nervous system; *GI,* gastrointestinal.
*Signs and symptoms.
Normal values for pH = 7.35-7.45, $PaCO_2$ = 35-45 mm Hg; and HCO_3 = 22-36 mEq/L

neutralize these increased acids, the plasma bicarbonate (HCO_3) is used up.

Chronic kidney disease with renal failure results in acidosis because the failing kidney not only is unable to rid the body of excess acids but also cannot produce necessary bicarbonate.

Lactic acidosis occurs as excess lactic acid is produced during strenuous exercise or when oxygen is insufficient for proper use of carbohydrate (CHO), glucose, and water (H_2O).

Intestinal and pancreatic secretions are highly alkaline so that *severe diarrhea* depletes the body of these necessary bases. Metabolic acidosis can result from ingestion of large quantities of acetylsalicylic acid (salicylates); symptoms of possible metabolic acidosis should be carefully assessed in clients undergoing high-dose aspirin therapy.

Hyperventilation may occur as the respiratory system attempts to rid the body of excess acid by increasing the rate and depth of respiration. The result is an increase in the amount of carbon dioxide and hydrogen excreted through the respiratory system.

CLINICAL SIGNS AND SYMPTOMS

Metabolic Acidosis

- Headache
- Fatigue
- Drowsiness, lethargy
- Nausea, vomiting
- Diarrhea
- Muscular twitching
- Convulsions
- Coma (severe)
- Rapid, deep breathing (hyperventilation)

Gout

Primary gout is the manifestation of an inherited inborn error of purine metabolism characterized by an elevated serum uric acid (hyperuricemia). Excess uric acid in the blood can result in the formation of tiny uric acid crystals that collect in the joints, triggering a painful inflammatory response.

Gout affects men predominantly, and the usual form of primary gout is uncommon before the third decade, with its peak incidence in the 40s and 50s. The frequency of gout in women approaches that in men after menopause when estrogen, which helps clear uric acid from the kidneys, declines dramatically.[59] Gout may occur as a result of another disorder or of its therapy. This is referred to as *secondary gout.* Secondary gout may be associated with neoplasm, renal disease, or other metabolic disorders such as diabetes and hyperlipidemia (excess serum lipids).[60]

Risk Factors. Increased serum uric acid levels are associated with middle age, menopause, obesity, white race, stress (including surgery and medical illness), and high dietary intake of purine-rich foods. A variety of medications (e.g., penicillin, insulin, or thiazide diuretics) may increase the serum uric acid level or decrease uric acid excretion, as may a number of acute or chronic disorders other than gout (Table 11-12).

High intake of meat (beef, pork, lamb) and seafood consumption has been associated with increased risk of gout, whereas high intake of low-fat dairy products and vitamin C supplementation can lower urate levels and reduce the risk of gout. Purine-rich vegetables, protein intake, low-to-moderate alcohol consumption, and BMI are not associated with an increase in gout.[61-63]

TABLE 11-12 Causes of Secondary Hyperuricemia

Hematopoietic	Hemolytic anemia Myeloproliferative disorders Polycythemia vera Myeloma
Neoplastic	Leukemia Lymphoma Multiple myeloma
Endocrine	Hypoparathyroidism Hyperparathyroidism Hypothyroidism Diabetes mellitus
Renal	Hemodialysis Renal insufficiency Polycystic kidney disease
Drugs	Low-dose aspirin Diuretics Antineoplastic (cytotoxic) agents Alcohol Vitamin B_{12}
Other	Chondrocalcinosis Psoriasis Sarcoidosis Obesity Hyperlipidemia Starvation, dehydration Toxemia of pregnancy

From Wade JP, Liang MH: Avoiding common pitfalls in the diagnosis of gout, *J Musculoskel Med* 5(8):16–27, 1988.

Many diseases have a presentation similar to that of acute gouty arthritis. Gout and septic arthritis occasionally occur together. The diagnosis of gout must be based on the demonstration of monosodium urate crystals by synovial fluid analysis rather than on the clinical presentation alone.

Clinical Presentation. Uric acid is usually dissolved in the blood until it is passed through the kidneys into the urine and then excreted. In individuals with gout, the uric acid changes into crystals (urate) that deposit in joints (causing gouty arthritis) and other tissues such as the kidneys, causing renal disease.

Most renal disease in clients with gout is the result of coexisting conditions such as hypertension or atherosclerosis. Renal dysfunction can occur as a result of urate-related parenchymal damage without the existence of other comorbidities. Urate nephropathy is relatively rare; the therapist is more likely to see a client receiving cyclosporine after a heart or kidney transplant who develops gout during the first year after the transplant.[64]

The most usual symptom of gout is acute monarticular arthritis. The individual may be awakened from sleep with exquisite pain in the affected joint; any pressure (even the touch of clothes or bed sheets) on the joint is intolerable. Redness and swelling occur within a few hours, sometimes accompanied by low-grade fever and chills. Untreated, the attack lasts from 10 days to 2 weeks. Later, episodes may develop more gradually, affecting more than one joint as the disease progresses.

The peripheral joints of the hands and feet are involved, with 90% of gouty clients having attacks in the metatarsophalangeal joint of the great toe. Other typical sites of initial involvement (in order of frequency) are the instep, ankle, heel, knee, and wrist, although any joint in the body may be involved.

In chronic gouty arthritis, periarticular and subcutaneous deposits of sodium urate (or urate salts) form; these are referred to as *tophus (tophi)*. These deposits produce an acute inflammatory response that leads to acute arthritis and later to chronic arthritis. Enlarged tophi on the joints of the hands and feet may erupt and discharge chalky masses of urate crystals.

The formation of tophi is directly related to the elevation of serum urate; the higher the client's serum urate concentration, the higher the rate of urate deposition in soft tissue. Before urate-lowering agents became available, 30% to 50% of people with acute gouty arthritis developed tophi. Today, chronic tophaceous gout is rarely seen.

The therapist should refer anyone taking urate-lowering drugs for gout who is having recurrent symptoms. It may be necessary to adjust medication levels. The therapist can reinforce the need for compliance with the management program and provide more education about controlling hypertension and obesity through diet and exercise. Avoidance of alcohol (especially beer), dehydration, and trauma to the extremities are other important components of effective management.[64]

CLINICAL SIGNS AND SYMPTOMS

Gout

- Tophi: Lumps under the skin or actual eruptions through the skin of chalky urate crystals
- Joint pain and swelling (especially first metatarsal joint)
- Fever and chills
- Malaise
- Redness

Pseudogout. Pseudogout is an arthritic condition caused by calcium pyrophosphate dihydrate (CPPD) crystals. It occurs about one-eighth as often as gout and may be hereditary or secondary to other disease processes (hyperparathyroidism is the most common one; Case Example 11-7).

Pseudogout is marked by attacks of gout-like symptoms, usually affecting a single joint (particularly the knee) and associated with chondrocalcinosis (deposition of calcium salts in joint cartilage). In anyone with pseudogout, routine x-ray studies of the knee and wrist frequently demonstrate cartilage calcification, or chondrocalcinosis. Because these changes are found in up to 10% of older adults, diagnosis must be made through aspiration of synovial fluid to identify the CPPD crystals.

Hemochromatosis

Hemochromatosis, also termed *hematochromatosis,* is an inborn error of iron metabolism. Mutations of the hemochromatosis gene (HFE) have been identified to help in

CASE EXAMPLE 11-7

Pseudogout

A 69-year-old man in previously good health complained of steadily increasing pain that had developed in his hands over the past several months. There was no history of occupational or accidental trauma.

Although the pain was present bilaterally, the pain in the left hand was more severe than in the right. The gentleman was right-hand dominant. There was a pattern of symptoms of increasing pain (described as deep aching) from morning to evening with a corresponding decrease in function.

Objective findings included reduced wrist range of motion in all directions bilaterally. There was no observed edema, warmth, or redness of the forearms, wrists, or hands. Although there were no reported symptoms at the elbow, left elbow extension and left forearm supination were also decreased by 25% as compared with the right side. Grip strength was reduced by 50% bilaterally for age and sex.

Neurologic screening was without significant findings. There were no trigger points corresponding to the pain pattern present. No constitutional symptoms were reported, and vital signs were unremarkable.

Result: This man was treated by a hand therapist without significant changes in symptoms or function. In fact, he reported an increased inability to write with his right hand. The therapist suggested a medical evaluation with possible inclusion of x-ray examination. Physician assessment resulted in a diagnosis of calcium pyrophosphate dihydrate (CPPD) arthropathy (pseudogout) of unknown cause. Medical treatment included a prescription nonsteroidal antiinflammatory drug (NSAID) and return to physical therapy for continued symptomatic treatment addressing the loss of function.

Although a medical condition existed, physical therapy treatment was still warranted. In this case, a medical differential diagnosis provided the client with necessary medical treatment and the physical therapist with information necessary to treat the client more specifically.

better defining the type of disease present in an individual.[65]

The cardinal defect in hemochromatosis is the lack of regulation of iron absorption, but the exact mechanism is unknown. The intestinal tract absorbs more iron than is required, thus producing an excess with progressive tissue damage in parenchymal organs from iron retention.

Hemochromatosis is found five to ten times more often in men than in women because women lose blood through menstruation and pregnancy. Men seldom have symptoms until after 50 years of age and are rarely symptomatic before 30 years of age. Because of menstrual blood loss, women display symptoms 10 years later than men (median age: 60 years).

Ascorbic acid (vitamin C) and alcohol seem to accelerate the absorption of dietary iron. The high incidence of alcoholism among clients with hemochromatosis (40%) supports this concept.

Clinical Presentation. For many years, hemochromatosis was identified by a classic clinical triad of enlarged liver, skin hyperpigmentation, and diabetes. The term *bronze diabetes* was used to describe this presentation. Hyperpigmentation is caused by an increased number of melanocytes and a thinning of the epidermis. However, hemochromatosis may have many different signs and symptoms, confusing early diagnosis (Case Example 11-8).

In its early stages, hemochromatosis produces no symptoms because it takes many years of iron accumulation to produce warning signs or symptoms. Unfortunately, when the disease becomes evident, it is often too late because iron accumulation has caused irreversible tissue or end-organ damage in the heart, liver, endocrine glands, skin, joints, bone, and pancreas. About half the clients with hemochromatosis will develop arthritis.

Hemochromatosis has a well-known association with chondrocalcinosis (deposition of calcium salts in the cartilage of joints). Acute attacks of synovitis can occur, which may resemble a rheumatoid flare. A biopsy of synovial tissue reveals iron deposition in the cells of the synovial lining that is noninflammatory.

Arthritis may be the presenting symptom of hemochromatosis, but it usually occurs after diagnosis and is more severe in adults older than 50. Arthritic manifestations are diverse, and joint damage occurs not from iron but from deposition of CPPD crystals.

The distribution of joint involvement may resemble rheumatoid arthritis, affecting the metacarpophalangeal (MCP) joints, in particular the second and third MCP joints. However, reduced MCP flexion is not accompanied by ulnar deviation. The arthritis can progress, and large joints may become involved, particularly the hips, knees, and shoulders.

CLINICAL SIGNS AND SYMPTOMS

Hemochromatosis

- Arthropathy (joint disease)*
- Arthralgias
- Myalgias
- Progressive weakness
- Bilateral pitting edema (lower extremities)
- Vague abdominal pain
- Hypogonadism (lack of menstrual periods, impotence)*
- Congestive heart failure (CHF)
- Hyperpigmentation of the skin (gray/blue to yellow)
- Loss of body hair
- Diabetes mellitus

*Unfortunately, even with treatment (removal of accumulated iron), arthritis, impotency, and sterility are not reversed.

Metabolic Bone Disease

Of the numerous metabolic disorders involving connective tissue, only the most commonly occurring diseases that would appear in a physical therapy setting are discussed in

CASE EXAMPLE 11-8

Hemochromatosis

A 68-year-old man was admitted to the hospital after sustaining multiple fractures of unknown origin. He was referred to physical therapy for functional mobility, transfers, and active range of motion with prescribed limitations. The admitting physician was a third-year resident on an emergency department rotation.

When the client was seen by the physical therapist, there was obvious swelling and limited range of motion in the right shoulder. The skin was warm and tender over the shoulder joint.

The therapist also observed the following:

- Bony prominences involving the second and third metacarpophalangeal (MCP) joints
- Bony prominences over the wrists, elbows, knees, and ankles
- Palpable and audible crepitus of these same joints
- Gray discoloration to skin throughout the body and axial skeleton
- Very sparse axial hair and an unusual leathery texture to the skin

The client was a poor historian but mentioned a "liver problem" that he experienced years ago. When asked about any other problems anywhere else in the body, the client mentioned difficulty with sexual arousal, erection, and ejaculation over the last 6 months.

The therapist developed a plan of care based on the current medical problem list and physician's orders. She also made it a point to seek out the referring resident to review some of the more unusual findings and ask about the possible cause of these symptoms.

Result: Further testing revealed that this client had a hereditary disease called *hemochromatosis*. The condition is characterized by excessive iron absorption by the small intestine. Individuals with hemochromatosis lack an effective way to remove excess iron, and the iron begins to accumulate in the liver, pancreas, skin, heart, and other organs.

Excess iron accumulation in the body promotes oxidation and causes tissue injury, fatigue, arthralgia or arthritis, and skin changes. Complications can include hepatomegaly, diabetes, impotence (males), pulmonary involvement, and cardiac myopathy.

Medical treatment for this condition was required to prevent the condition from worsening. Treatment does not improve the associated arthritis in a case like this, but it does keep it from getting worse.

The therapist's careful observations and follow-up made a significant difference in this man's medical outcome.

Data from Sokolova Y: Acute shoulder pain and swelling in a 68-year-old man, *J Musculoskel Med* 17(11):699–700, November 2000.

this text. These include osteoporosis, osteomalacia, and Paget's disease.

Osteoporosis. Osteoporosis, meaning "porous bone" and defined as a decreased mass per unit volume of normally mineralized bone compared with age- and sex-matched controls, is the most prevalent bone disease in the world.

TABLE 11-13 Causes of Osteoporosis

Endocrine and Metabolic	Other
Diabetes mellitus (Type 1) Glucocorticoid excess (hyperadrenocorticism) • Iatrogenic Cushing's syndrome • Hyperadrenalism Hyperthyroidism (thyrotoxicosis) Hyperparathyroidism Hemochromatosis Acromegaly Testicular insufficiency	**MEDICATIONS** • Immunosuppressants (cyclosporine) • Excess thyroid hormone • Glucocorticoids • Methotrexate • Anticonvulsants/seizure medications (e.g., Dilantin, phenobarbital) **NUTRITIONAL** • Anorexia nervosa; any eating disorder • Chronic alcohol use • Calcium/vitamin D deficiency • Chronic liver disease • Gastric bypass • Malabsorption syndromes (e.g., celiac sprue) **COLLAGEN/GENETIC DISORDERS** • Ehlers-Danlos syndrome • Marfan syndrome • Osteogenesis imperfecta

Osteoporosis is classified as *primary* or *secondary.* Primary osteoporosis is the deterioration of bone mass unassociated with other chronic illnesses or diseases. It is usually related to the aging process, including decreased gonadal function. Idiopathic, postmenopausal, and senile osteoporosis are included in the primary osteoporosis classification. *Postmenopausal osteoporosis* is associated with accelerated bone loss in the perimenopausal and postmenopausal period, accompanied by high fracture rates, particularly involving the vertebrae. *Senile osteoporosis,* or age-related osteoporosis, increases with advancing age; it is caused by the bone loss that normally accompanies aging.

Secondary osteoporosis may accompany various endocrine and metabolic disorders (e.g., hyperthyroidism, hyperparathyroidism, hypogonadism, Cushing's disease, and diabetes mellitus) that can produce associated osteopenia conditions (Table 11-13). *Endocrine-mediated bone loss* can produce osteoporosis because numerous endocrine hormones affect skeletal remodeling and hence skeletal mass.

Secondary osteoporosis is associated with other disorders that contribute to accelerated bone loss, such as chronic renal failure, rheumatoid arthritis, malabsorption syndromes related to gastrointestinal and hepatic disease, chronic respiratory disease, malignancies, and chronic chemical dependency (e.g., alcoholism).

Transient osteoporosis of the hip (TOH), a rare but temporary presentation of spontaneous osteoporosis of the femoral head (and sometimes femoral neck and acetabulum), is discussed in Chapter 16.

Risk Factors. There is consensus that risk factors associated with osteoporosis and osteoporotic fractures have been identified but which target groups should be screened (e.g., all adults, all postmenopausal women, adults 50 years old and older) varies among groups and organizations.[66]

United States Preventive Services Task Force (USPSTF) now recommends that all women 65 years of age and older should be screened for osteoporosis. Despite the facts that approximately half of postmenopausal women will sustain an osteoporosis-related fracture and 15% will sustain a hip fracture in their lifetime, 75% of American women between the ages of 45 and 75 years have never discussed osteoporosis with their physician.[67,68] Plus, a gap exists in the number of people who present with osteoporotic fractures and those who are treated for the underlying cause of their fracture.[69]

Comparison of bone mineral density (BMD) to peak bone mass of women who are at peak bone mass is the designated T-score. T-scores are used as a method of describing severity of risk (Table 11-14). Therapists can be involved in primary prevention and education, encouraging and instructing consumers and clients in risk assessment and risk factor reduction (Fig. 11-9 and Box 11-2).

TABLE 11-14 Bone Mineral Density T-Scores

Status	T-Scores	Interpretation
Normal	−1.0 or above	T-score (BMD) is within (or above) 1 SD of the young adult reference mean.
Osteopenia (low bone mass)	−1.0 to −2.5	T-score is 1.0 to 2.5 SDs below young adult mean for age.
Osteoporosis	−2.5 or less	T-score is 2.5 or more SDs below mean for age.
Severe osteoporosis	−2.5 or less with one or more fragility fractures	BMD is 2.5 or more SDs below mean for age.

The T-score compares one person's BMD in standard deviations with the average peak BMD in healthy young persons. Sometimes, Z-scores are used, which compare one individual's BMD with the mean BMD of persons in the same age group, rather than with the normal values listed for young adults.

The World Health Organization (WHO) has proposed the clinical definition of osteoporosis in the table based on epidemiologic data that link low bone mass with increased fracture risk.

In study populations of Caucasian postmenopausal women, a BMD that was lower than 2.5 SD of normal peak bone mass was associated with a fracture prevalence; 50% of women with bone mass at this level had at least one bone fracture.[93]

On the basis of these data, the WHO defined osteoporosis as BMD 2.5 or more SD below peak bone mass, osteopenia as bone mass between 1.0 and 2.5 SD below peak, and normal as 1.0 SD below normal peak bone mass or higher.

The WHO criteria apply only to Caucasian, postmenopausal women, and not men, premenopausal women, or women of ethnicity other than Caucasian. T-score conversion based on the NHANES database is available for clinically significant low bone mass in other population groups at http://courses.washington.edu/bonephys/opbmdtz.html.

SD, Standard deviation; *BMD,* bone mineral density.

Data from World Health Organization: *Criteria for defining bone density,* WHO, 1994. Assessment of fracture risk and its application to screening for postmenopausal osteoporosis, Geneva, 1994, World Health Organization. Technical Report Series, No. 843.

The World Health Organization (WHO) has developed a computerized tool called the *Fracture Risk Assessment* (FRAX) to calculate the 10-year risk of sustaining a fracture (www.shef.ac.uk/FRAX). It is based on individual patient models that integrate the risks associated with clinical risk factors, as well as BMD at the femoral neck. Medication treatment is recommended for postmenopausal women and men age 50 or older who have a 3% to 10% risk of a hip fracture or 20% risk of major fracture elsewhere in the body (e.g., wrist, spine, upper arm) according to the test.

Screening should begin at 60 years of age in women with key risk factors[70] (Box 11-3). Screening begins earlier for anyone who has a fracture after age 50 or a maternal history of fracture after age 50.[71] Key risk factors include women who have body weight less than or equal to 125 pounds (57 kg)[71] and who do not currently use estrogen.

There is little known evidence to support the use of other individual risk factors, such as smoking, caffeine or alcohol use, low calcium, or low vitamin D intake, because those factors have not been shown to be independent predictors of low bone density.[72] Medications used in high doses or for long-term use, such as thyroid supplements, corticosteroids, anticoagulants, lithium, and anticonvulsants, can contribute to the development of secondary osteoporosis.[73]

Primary prevention and education begins in childhood and adolescence, which are recognized as critical time periods for the development of normal peak bone mass. Diet and bone-building exercises during this critical period are essential in the development of adequate bone mass.

Men can be affected, especially those who smoke, drink alcohol moderately, fail to maintain a calcium-rich diet, have a sedentary lifestyle, or have a family history of fractures or those undergoing dialysis or long-term steroid administration. Some data suggest that men do not receive treatment for osteoporosis or counseling for osteoporosis prevention as aggressively as women.[74]

Additionally, researchers are beginning to examine the environmental influences associated with industrialized countries such as the United States. For example, although menopause is universal and the resulting estrogen deficiency is presumably similar for all women, differences in the occurrence of osteoporosis among countries cannot be explained only on the basis of estrogen deficiency.[75] Countries with the highest incidence of osteoporosis also have a high incidence of heart disease and the highest consumption of carbohydrates, fat, protein, salt, and caffeine.

Clinical Presentation. Osteoporosis is a silent disease with no visible signs or symptoms until bone loss is sufficient to result in fracture. Osteoporosis associated with aging

Osteoporosis Screening Evaluation

Name ______________________ Date ________

	YES	NO
1. Are you 65 years old or older?	☐	☐
2. Is your weight below 57.6 kg (127 pounds)?	☐	☐
3. Are you Caucasian or Asian?	☐	☐
4. Have any of your blood-related family members had osteoporosis?	☐	☐
5. Are you a postmenopausal woman?	☐	☐
6. Do you drink 2 or more ounces of alcohol each day? (1 beer, 1 glass of wine, or 1 cocktail)	☐	☐
7. Do you smoke more than 10 cigarettes each day?	☐	☐
8. Are you physically inactive? (Walking or similar exercise at least three times per week is average.)	☐	☐
9. Have you had both ovaries (with or without a hysterectomy) removed before age 40 years without treatment (hormone replacement)?	☐	☐
10. Have you ever been treated for or told you have rheumatoid arthritis?	☐	☐
11. Have you been taking thyroid medication, antiinflammatories, or seizure medication for more than 6 months?	☐	☐
12. Have you ever broken your hip, spine, rib, or wrist?	☐	☐
13. Do you drink or eat four or more servings of caffeine (carbonated beverages, tea, coffee, chocolate) per day?	☐	☐
14. Is your diet low in dairy products and other sources of calcium? (Three servings of dairy products or two doses of a calcium supplement per day are average.)	☐	☐
15. Are you vegetarian or vegan?	☐	☐

Fig. 11-9 If you answer "yes" to three or more of these questions, you may be at greater risk for developing osteoporosis, or "brittle bone disease," and you should contact your physician for further information.

involves fractures of the proximal femur and vertebrae as well as the hip, pelvis, proximal humerus, distal radius, and tibia.

Postmenopausal osteoporosis is associated with accelerated bone loss in the perimenopausal period accompanied by high fracture rates, particularly involving the vertebrae. Vertebral compression fracture will be likely in 25% of women older than 65 and 50% of women 80 years and older (Case Example 11-9).[76]

Mild-to-severe back pain and loss of height may be the only early signs observed. Changes in bone density do not show up on x-ray films until there is a 30% loss. The cardinal features of established osteoporosis are bone fracture, pain, and deformity.

More than half the women in the United States who are 50 years of age or older are likely to have radiologically detectable evidence of abnormally decreased bone mass (osteopenia) in the spine. More than a third of these women develop major orthopedic problems related to osteoporosis. Most fractures sustained by women older than 50 are secondary to osteoporosis.

CLINICAL SIGNS AND SYMPTOMS

Osteoporosis

- Back pain: episodic, acute low thoracic/high lumbar pain
- Compression fracture of the spine (postmenopausal osteoporosis)
- Bone fractures (age-related osteoporosis)
- Decrease in height (more than 1 inch shorter than maximum adult height)
- Kyphosis
- Dowager's hump
- Decreased activity tolerance
- Early satiety

Osteomalacia. Osteomalacia is a softening of the bones caused by a vitamin D deficiency in adults and resulting from impaired mineralization in bone matrix. This failure in mineralization results in a reduced rate of bone formation. The deficiency may be due to lack of exposure to ultraviolet rays,

BOX 11-2 OSTEOPOROSIS RESOURCES

National Osteoporosis Foundation
www.nof.org
Offers information for consumers and professionals about osteoporosis and its prevention. Includes information on bone density testing and risk factor assessment, as well as a special web link, *Men and Osteoporosis.*

Harvard Center for Cancer Prevention
http://www.yourdiseaserisk.wustl.edu/
Offers consumers an opportunity to find out individual risk of developing five diseases, including cancer, diabetes, heart disease, osteoporosis, and stroke. Also includes tips on prevention for each of these diseases.

Medline Plus
http://www.nlm.nih.gov/medlineplus/ency/article/007197.htm
Information on bone mineral density testing, including normal values and how, when, and why the test is performed.

WebMD: Medical Tests
http://www.webmd.com/hw/osteoporosis/hw3738.asp
Offers the same information as provided by Medline Plus but also explains different techniques used to measure bone mineral density, how each one is done, and how to interpret the results.

Osteoporosis Education Project
www.betterbones.com/bonehealthprofile/default.aspx
In addition to the osteoporosis screening evaluation presented in this chapter (see Fig. 11-8), consumers can take an osteoporosis fracture risk questionnaire offered by the Osteoporosis Education Project, a nonprofit organization dedicated to education and research on the topic of osteoporosis. The quiz is designed for both men and women of all ages.

Johns Hopkins SCORE Screening Quiz
http://www.hopkins-arthritis.org/arthritis-info/osteoporosis/diagnosis.html
The Simple Calculated Osteoporosis Risk Estimation (SCORE) is a 6-question screening questionnaire for osteoporosis with 89% sensitivity and 50% specificity in an ambulatory population of postmenopausal women. A score of six or more is an indication that referral for bone density testing is advised.[81]

Fracture Index
An assessment tool for predicting fracture risk. This clinical assessment tool based on seven risk factors (age, T-score, personal or maternal fracture after age 50 years, weight, smoking status, use of arms to stand up from a chair) that can be used to assess a woman's risk of hip, vertebral, and nonvertebral fractures.

From Black DM, Steinbuch M, Palermo L, et al: An assessment tool for predicting fracture risk in postmenopausal women, *Osteoporos Int* 12(7):519–528, 2001.

inadequate intake of vitamin D in the diet, failure to absorb or use vitamin D, increased catabolism of vitamin D, a renal tubular defect, or a pathologically reduced number of vitamin D receptor sites in tissues.

The disease is characterized by decalcification of the bones, particularly those of the spine, pelvis, and lower extremities. X-ray examination reveals transverse, fracture-like lines in the affected bones and areas of demineralization in the matrix of the bone. These pseudofractures, known as *Looser's transformation zones,* are bilateral. The most common sites are the ribs, long bones, lateral scapular margin, upper femur, and pubic rami. As the bones soften, they become bent, flattened, or otherwise deformed. Looser's zones are believed to result from pressure on the softened bone by the nutrient arteries of its blood supply.

Severe bone pain, skeletal deformities, fractures, and severe muscle weakness and pain are common in people with osteomalacia. Clients typically complain of muscle weakness and pain that sometimes mimics polymyositis or muscular dystrophy.

A similar condition in children, occurring before epiphyseal plate closure, is called *rickets.* In children with rickets, x-ray findings include the well-known bowing of the long bones, in addition to widening, fraying, and clubbing of the areas of active bone growth. These areas especially include the metaphyseal ends of the long bones and the sternal ends of the ribs, the so-called rachitic rosary.

CLINICAL SIGNS AND SYMPTOMS

Osteomalacia

- Bone pain
- Skeletal deformities
- Fractures
- Severe muscle weakness
- Myalgia

Paget's Disease. Paget's disease (osteitis deformans), named after Sir James Paget from the mid-1880s, is a focal inflammatory condition of the skeleton that produces disordered bone remodeling. Bone is resorbed and formed at an increased rate and in a haphazard fashion. As a result, the new bone is larger, less compact, more vascular, and more susceptible to fracture than normal bone.

BOX 11-3 RISK FACTORS FOR OSTEOPOROSIS

Residents in a nursing home, extended care, or skilled nursing facility have a fivefold to tenfold increase in fracture risk compared with community dwellers.[82] Therapists in these work settings have the potential to improve the recognition and management of osteoporosis in these populations, including reducing the number of fractures and falls.

Women

- Caucasian and Asian women are more likely to develop osteoporosis; African-American and Hispanic women have a significant risk for developing osteoporosis.[83]
- Gender: more common in women than men
- Age: postmenopausal (older than 65)
- Early or surgically induced menopause; menstrual dysfunction (amenorrhea)
- Family history of osteoporosis
- Family history and/or personal history of fractures
- Lifestyle*: cigarette smoking, excessive alcohol intake, inadequate calcium, little or no weight-bearing exercise
- Prolonged exposure to certain medications (more than 6 months):

Thyroid medications, corticosteroids, antiinflammatories, anti-seizure medication, aluminum-containing antacids, lithium, methotrexate, anticoagulants (heparin, warfarin), benzodiazepines (e.g., lorazepam, diazepam), cyclosporine A (immunosuppressant), gonadotropic-releasing hormone agonists, Depo-Provera injections (contraceptives in adolescents)

- Some cancer treatments (oophorectomy, ovarian suppression, chemotherapy-induced ovarian failure, estrogen suppression, bone marrow transplantation)
- Thin, small-boned frame (weight less than or equal to 125 pounds or 57 kg)
- Chronic diseases that affect the kidneys, lungs, stomach, and intestines or alter hormones (especially if treated with corticosteroids); dialysis

Men

- Caucasian
- Gender: increasing incidence among men
- Advancing age
- Lifestyle: same as for women
- Prolonged exposure to medications (same as for women)
- Family history of osteoporosis
- History of prostate cancer with bilateral orchiectomy
- Undiagnosed low levels of testosterone
- Hypogonadism (long-term androgen deprivation therapy [ADT])
- Chronic diseases (as listed for women)

At least one study[84] reports that use of caffeine and antacids has no probable effect on bone mass in older women. The investigators also emphasized that it is weight, not body mass index (BMI), that is important. A 10 kg increase in weight reportedly implied a 6% increase in bone mineral density.

*Remains under investigation. There is little known evidence to support the use of other individual risk factors such as smoking, caffeine or alcohol use, low calcium, or low vitamin D intake because those factors have not been shown to be independent predictors of low bone density.[72]

CASE EXAMPLE 11-9

Osteoporosis

Referral: A 77-year-old Caucasian woman was referred to outpatient physical therapy 1 month ago by her primary physician because of her complaint of gradual onset of low back pain (LBP) over the last 2 months. The physician's diagnosis was LBP secondary to osteoarthritis and osteoporosis. A recent radiology report indicated moderate osteoarthritis at L1-5 and radiolucency of the spine suggesting severe osteoporosis. No fractures or abnormal curvatures were noted.

Past Medical and Social History

- Osteoarthritis
- LBP secondary to L4/5 herniated disk; status post-diskectomy
- Osteoporosis (2-year history)

The client denied diabetes, high blood pressure, other heart diseases, or other health concerns.

She is a retired teacher who lives alone in an adult complex and still drives a car. She lost her second son in a motor vehicle accident 3 months ago and appears emotionally stressed from her loss. She has declined any counseling or medication suggested by her physician.

The client is highly motivated to improve so that she can go back to walking about 1 mile every other day; currently her pain level prevents her from this activity.

Medications

- Relafen 500 mg twice daily
- 5% Lidoderm patch applied to the skin once a day for pain
- Norflex 100 mg twice daily to reduce muscle spasms

She has been on Fosamax 70 mg once a week for 2 years to improve her bone density loss caused by osteoporosis. The client reported little or no change in her pain level with the use of analgesics.

Clinical Presentation: During initial evaluation, LBP was graded as a 7/10 on the Visual Analogue Scale (VAS). Pain was localized to the low back without radiation; she described it as worse on getting up in the morning and after sitting or walking for a short period. Pain was progressively worse with walking, and the client stopped walking after 3 or 4 minutes. She denied any urinary or bowel incontinence.

On examination, the client presented with mild tenderness with palpation of L3-5 and mild paraspinal muscle spasms with slight

Continued

CASE EXAMPLE 11-9—cont'd

Osteoporosis

loss of lumbar lordosis. There was no sensory loss noted with either upper/lower extremities or trunk and no pedal edema.

Range of motion (ROM): ROM was within normal limits (WNL) in both upper and lower extremities. Trunk flexion 0-76 degrees, trunk extension 0-13 degrees; all other motion: WNL.

Manual muscle test (MMT): Muscle strength for all extremities was grossly 5/5. Trunk extensors and abdominals were graded 4/5.

Straight leg raise (SLR): Negative bilaterally; the client was unable to fully raise both legs because of hamstring tightness.

Normal deep tendon reflex (DTRs) for both quadriceps and Achilles tendons.

Intervention: Physical therapy intervention consisted of education on osteoporosis and its cause, prevention, treatment, and sequelae. Client was instructed in fall prevention and in making her apartment fall-proof.

Moist heat was applied to the low back for 15 minutes to reduce muscle spasm, increase muscle flexibility,[85] and reduce pain associated with osteoarthritis.[86]

Massage/soft tissue mobilization: This has been shown to be effective in reducing LBP when used in conjunction with other treatment modalities.[87]

Therapeutic exercise: Therapeutic exercise has been shown to be effective in the management of LBP.[88] In this case, single and double knee-to-chest exercises were done in the supine position, holding each one for 5 seconds. Single SLR supine and prone (double SLR was avoided because of its tendency to put great pressure on the spine, which may result in fracture in this client).

Walking on a treadmill: During the initial evaluation, the client was able to tolerate only 3½ minutes on the treadmill at 1.0 mph (zero grade) because of increasing pain. Treadmill walking was used to measure progress because one of the client's goals was to be able to walk up to 1 mile.

All exercises were progressed as client improved. A written handout was provided with drawings and instructions for each exercise. Precautions were given to stop the exercise if experiencing shortness of breath, palpitation, or increased pain and to report these symptoms to the doctor. Any exercise that increased the pain was to be discontinued until the client checked with the therapist.

Short-term and long-term goals were established; the prognosis was expected to be good.

Outcome: The client showed remarkable improvement with her treatment. She was very diligent in performing her home exercise program (HEP) and following the therapist's instructions. She was highly motivated, attended all scheduled sessions, and was dedicated to achieving her goals.

By the third week of treatment, her pain had reportedly decreased, reduced from 7 to 4/10 on the VAS, trunk flexion was 0-94 degrees, and she was independent in her home exercise program and she was able to verbalize her fall prevention plan.

At 3 to 4 weeks, the client reported sudden increase in her LBP while getting out of bed. Pain was rated 6/10 and reported as constant but not getting worse. On examination, there was tenderness over L3-5, but it was not worse than previously reported.

There were mild low back muscle spasms, but no neurologic signs were noted and no abnormal curvature observed. This appeared to be an exacerbation episode. The primary care physician was notified, and the therapist was advised to continue intervention as planned. Treatment was continued as planned for 1 week without much improvement.

Result: The client returned to the physician for reevaluation. X-rays at that time diagnosed a compression fracture at L1. Further physical therapy intervention was placed on hold pending orthopedic consult. She returned to physical therapy with a recommendation for lumbar corset, rest for 2 weeks, and continued physical therapy intervention.

Reflections: Compression fracture is a known complication of osteoporosis with or without neurologic deficit.[89] It is accepted that posterior midline tenderness is a red flag for spinal fracture; however, the absence of a posterior midline tenderness does not exclude significant spinal injury without trauma such as spinal compression fracture.[90] The therapist should remain alert to the possibility of a new vertebral fracture in anyone with osteoporosis who reports a substantial increase in low back pain.[91,92]

Signs and symptoms of compression facture may be difficult to recognize, especially in a client who is already being treated for chronic LBP 2 years after diskectomy for disk herniation.

It is not uncommon to see occasional flare-ups of pain in physical therapy clients who have been showing good improvement. A typical clinical scenario is the client who increases the frequency, intensity, or duration of activities, even adding activities he or she has been unable to enjoy previously because of back pain.

In some cases, clients overdo the home exercise program or add a new exercise suggested by a friend or seen on TV or at the gym. In some cases, there is no apparent reason for exacerbation of symptoms.

This case study demonstrates how any adverse change in pain level in an individual with osteoporosis undergoing physical therapy for back pain should not be dismissed as insignificant but should be thoroughly investigated, including medical referral when indicated.

From Nubi M: Case report presented in fulfillment of DPT 910, *Institute for Physical Therapy Education,* Widener University, 2005, Chester, PA. Used with permission.

Risk Factors. Paget's disease is the most common skeletal disorder after osteoporosis, affecting men more often than women by a 3:2 ratio. Although Paget's disease affects 2% to 5% of the population older than 40, it is most commonly seen in people older than 70, most of whom are asymptomatic.[77] It is more prevalent in Europe and Australia and in people of Anglo-Saxon descent.

Genetic factors are important in the pathogenesis of Paget's disease; in many families, the disease is inherited in an autosomal-dominant manner. Specific genetic mutations are being identified.[78] Although the cause of this condition remains unknown, available evidence points to a slow viral infection in genetically predisposed individuals.[79] Evidence for a major genetic component is supported by 40% of affected individuals having affected first-degree relatives.

There are no known ways to prevent Paget's disease. Eating a healthy diet with sufficient calcium and vitamin D and getting exercise are critical in maintaining skeletal and joint function.[79]

Clinical Presentation. The severity of involvement and associated clinical characteristics vary greatly. Although some people are asymptomatic, with very limited bone involvement, others manifest a disabling, painful form of Paget's disease that is characterized by skeletal pain and bones that are extremely deformed and easily fractured. Bones most commonly involved include (in decreasing order) the pelvis, lumbar spine, sacrum, femur, tibia, skull, shoulders, thoracic spine, cervical spine, and ribs.

Bone pain associated with Paget's disease is described as aching, deep and boring, worse at night, and diminishing but not disappearing with physical activity. Muscular pain may be referred from involved bony structures or as a result of mechanical changes caused by joint deformities.

Other complications include a variety of nerve compression syndromes, secondary osteoarthritis, and vertebral compression and collapse. Rarely, Paget's disease converts to a malignant neoplasm (osteogenic sarcoma of the femur or humerus) in older adults with extensive Paget's disease. Metastases are common at the time of diagnosis; survival rates are very poor.[80]

CLINICAL SIGNS AND SYMPTOMS

Paget's Disease

These depend on the location and severity of the bone lesions and may include the following:

- Pain and stiffness
- Fatigue
- Headaches and dizziness
- Bone fractures
- Vertebral compression and collapse
- Deformity
- Bowing of long bones
- Increased size and abnormal contour of clavicles
- Osteoarthritis of adjacent joints
- Acetabular protrusion
- Head enlargement
- Periosteal tenderness
- Increased skin temperature over long bones*
- Decreased auditory acuity (if skull is affected)
- Compression neuropathy
- Spinal stenosis
- Paresis
- Paraplegia
- Muscle weakness

*Increased skin temperature over affected long bones is a typical finding and is explained by soft tissue vascularity surrounding the bones.

PHYSICIAN REFERRAL

Disorders of the endocrine and metabolic systems may appear with recognizable clinical signs and symptoms but almost always require a combination of clinical and laboratory findings for accurate identification.

The therapist is encouraged to complete a thorough Family/Personal History form, augmented by the screening interview, and careful clinical observations, to provide the physician with pertinent screening information when making a referral. When appropriate, the Osteoporosis Screening Evaluation (see Fig. 11-9; see also Appendix C-6) may be helpful. In most cases, the client who has suffered from an endocrine disorder has already been diagnosed and may have been referred for physical therapy for some other musculoskeletal complaint. Such clients may have musculoskeletal problems that can be affected by symptoms associated with hormone imbalances (see Tables 11-4 through 11-7).

Diseases of the endocrine-metabolic system, such as diabetes, obesity, and thyroid abnormalities, account for some of the most common disorders encountered in a physical therapy practice. In recent years, new laboratory techniques have greatly enhanced the physician's ability to diagnose these diseases.

Nevertheless, in many cases, the disorder remains unrecognized until relatively late in its course; signs and symptoms may be attributed to some other disease process or musculoskeletal disorder (e.g., weakness may be the major complaint in Addison's disease). Thus any client who has any of the generalized signs and symptoms associated with the endocrine system (see Box 4-19) without an obvious or already known cause should be further evaluated by a physician.

Guidelines for Immediate Medical Attention

- Any person with diabetes who is confused, lethargic, exhibiting changes in mental function, profuse sweating (without exercise), or demonstrating signs of DKA should receive medical attention. (Perform a fingerstick glucose test to help evaluate the situation.)
- Likewise, any episode or suspected episode of hypoglycemia must be treated promptly and reported to the client's physician.
- Signs of potassium depletion (e.g., muscle weakness or cramping, fatigue, cardiac arrhythmias, abdominal distention, nausea and vomiting) or fluid dehydration in a client who is taking non–potassium-sparing diuretics requires medical attention. Consultation with the physician is advised before exercising the individual.
- Signs of thyroid storm (tachycardia, elevated core body temperature, restlessness, agitation, abdominal pain, nausea, vomiting); observe clients with known history of hyperthyroidism carefully postoperatively or following trauma or infection.

Guidelines for Physician Referral

- Any unexplained fever without other symptoms in a person taking corticosteroids may be an indication of infection and should be evaluated by a physician.
- Palpable nodules or a palpable mass in the supraclavicular area or the scalene triangle, or both (see Fig. 11-5), especially if accompanied by new-onset hoarseness, hemoptysis, or elevated blood pressure, must be evaluated by a physician.
- Any episode (especially a series of episodes) of hypoglycemia in the client with diabetes should be reported to the physician.
- The presence of multiple eruptive xanthomas on the extensor tendons of anyone with diabetes may signal uncontrolled glycemia and requires medical referral to normalize lipid levels; exercise remains a key to the management of this condition.
- Signs of fluid loss or dehydration in anyone taking diuretics should be reported to the physician.
- Recurrent arthritic symptoms in a client with gout who is already taking urate-lowering drugs require medical referral for review of medication.

Clues to Symptoms of Endocrine or Metabolic Origin

Past Medical History

- Endocrine or metabolic disease has been previously diagnosed. Bilateral CTS, proximal muscle weakness, and periarthritis of the shoulder(s) are common in persons with certain endocrine and metabolic diseases. Look for other associated signs and symptoms of endocrine or metabolic disease (see Box 4-19).
- Long-term use of corticosteroids can result in classic symptoms referred to as *Cushing's syndrome.*

Clinical Presentation

- Identified trigger points are not eliminated or relieved by trigger point therapy. Observe for signs and symptoms of hypothyroidism.
- Palpable lymph node(s) or nodule(s) in the scalene triangle (see Fig. 11-5), especially when accompanied by new-onset hoarseness, hemoptysis, or elevated blood pressure.
- Anyone with muscle weakness and fatigue who is taking diuretics may be experiencing symptoms of potassium depletion. Assess for cardiac arrhythmias, and ask about nausea and vomiting.
- Muscle fasciculation and cramping may be associated with antacid use (metabolic alkalosis).

Associated Signs and Symptoms

- Watch for anyone with arthralgias, hand pain and stiffness, or muscle weakness with an accompanying cluster of signs and symptoms of endocrine or metabolic disorders (see Box 4-19).

Clues to Recognizing Osteoporosis

- Pain is usually severe and localized to the site of fracture (usually mid-thoracic, lower thoracic, and lumbar spine vertebrae).
- Pain may radiate to the abdomen or flanks.
- Aggravating factors: prolonged sitting, standing, bending, or performing Valsalva's maneuver.
- Alleviating factors: sidelying with hips and knees flexed.
- Sitting up from supine requires rolling to the side first.
- Not usually accompanied by sciatica or chronic pain from nerve root impingement
- Tenderness to palpation over the fracture site
- Rib or spinal deformity, dowager's hump (cervical kyphosis)
- Loss of height

Key Points to Remember

- Clients with a variety of endocrine and metabolic disorders commonly complain of fatigue, muscle weakness, and occasionally muscle or bone pain.
- Muscle weakness associated with endocrine and metabolic disorders usually involves proximal muscle groups.
- Periarthritis and calcific tendinitis of the shoulder is common in clients with endocrine issues. Symptoms usually respond to treatment of the underlying endocrine pathologic condition and are not likely to respond to physical therapy treatment.
- CTS, hand stiffness, and hand pain can occur with endocrine and metabolic diseases.
- There is a correlation between hypothyroidism and FMS, which is being investigated. Any compromise of muscle energy metabolism aggravates and perpetuates TrPs. Treatment of the underlying endocrine disorder is necessary to eliminate the TrPs, but myofascial treatment must be part of the recovery process to restore full function.
- Anyone with diabetes taking corticosteroid medications must be monitored for changes in blood glucose levels because these medications can cause or contribute to hyperglycemia.
- Exercise for the client with diabetes must be carefully planned because significant complications can result from strenuous exercise.
- Clients with DM who are under physical, emotional, or psychologic stress (e.g., hospitalization, pregnancy, personal problems) have increased insulin requirements; symptoms may develop in the person who usually has the disease under control.
- Exercise for the client with insulin-dependent diabetes should be coordinated to avoid peak insulin dosage whenever possible. Any client with known diabetes who appears confused or lethargic must be tested immediately by fingerstick for glucose level. Immediate medical attention may be necessary. Other precautions regarding diabetes mellitus for the therapist are covered in the text.
- When it is impossible to differentiate between ketoacidosis and hyperglycemia, administration of some source of sugar (glucose) is the immediate action to take.
- Early osteoporosis has no visible signs and symptoms. History and risk factors are important clues.
- Cortisol suppresses the body's inflammatory response, masking early signs of infection. Any unexplained fever without other symptoms should be a warning to the therapist of the need for medical follow-up.
- Excessive use of antacids can result in muscle fasciculation and cramping (see section on Alkalosis).

SUBJECTIVE EXAMINATION

Special Questions to Ask

Endocrine and metabolic disorders may produce subtle symptoms that progress so gradually that the person may be unaware of the significance of such findings. This requires careful interviewing to screen for potential physical and psychologic changes associated with hormone imbalances or other endocrine or metabolic disorders.

As always, it is important to be aware of client medications (whether over-the-counter or prescribed), the intended purpose of these drugs, and any potential side effects.

Past Medical History/Risk Factors

- Have you ever had head/neck radiation or cranial surgery? **(thyroid cancer, pituitary dysfunction)**
- Have you ever had a head injury? **(pituitary dysfunction)**
- Have you ever been told you have diabetes or that you have "sugar" in your blood?
- Have you ever been told that you have osteoporosis or brittle bones, fractures, or back problems? **(wasting of bone matrix in Cushing's syndrome, osteoporosis)**
- Have you ever been told that you have Cushing's syndrome?

Clinical Presentation

- Have you noticed any decrease in your muscle strength recently? **(growth hormone imbalance, ACTH imbalance, Addison's disease, hyperthyroidism, hypothyroidism)**
- Have you had any muscle cramping or twitching? **(metabolic alkalosis)**
    - *If yes*, do you take antacids with magnesium on a daily basis? How much and how often?
- Do you have any difficulty in going up stairs or getting out of chairs? **(muscle wasting secondary to large doses of cortisol)**

Associated Signs and Symptoms

- Have you noticed any changes in your vision, such as blurred vision, double vision, loss of peripheral vision, or sensitivity to light? **(thyrotoxicosis, hypoglycemia, diabetes mellitus)**
- Have you had an increase in your thirst or the number of times you need to urinate? **(adrenal insufficiency, diabetes mellitus, diabetes insipidus)**
- Have you had an increase in your appetite? **(diabetes mellitus, hyperthyroidism)**

Continued

SUBJECTIVE EXAMINATION—*cont'd*

- Do you bruise easily? **(Cushing's syndrome, excessive secretion of cortisol causes capillary fragility; small bumps/injuries produce bruising)**
- When you injure yourself, do your wounds heal slowly? **(growth hormone excess, ACTH excess, Cushing's syndrome)**
- Do you frequently have unexplained fatigue? **(hyperparathyroidism, hypothyroidism, growth hormone deficiency, ACTH imbalance, Addison's disease)**
  - *If yes,* what activities seem to be too difficult or tiring? **(muscle weakness caused by cortisol and aldosterone hypersecretion and adrenocortical insufficiency, hypothyroidism)**
- Have you noticed any increase in your collar size (goiter growth), difficulty in breathing or swallowing? **(goiter, Graves' disease, hyperthyroidism)**
  - *To the therapist:* Observe also for hoarseness.
- Have you noticed any changes in skin color? **(Addison's disease, hemochromatosis)** (e.g., overall skin color has become a darker shade of brown or bronze; occurrence of black freckles; darkening of palmar creases, tongue, mucous membranes)

For the Client with Diagnosed Diabetes Mellitus

- What type of insulin do you take? (see Table 11-9)
- What is your schedule for taking your insulin?
  - *To the therapist:* Coordinate exercise programs according to the time of peak insulin action. Do not schedule exercise during peak times.
- Do you ever have episodes of hypoglycemia or insulin reaction?
  - *If yes,* describe the symptoms that you experience.
- Do you carry a source of sugar with you in case of an emergency?
  - *If yes,* what is it, and where do you keep it in case I need to retrieve it?
- Have you ever had diabetic ketoacidosis (diabetic coma)?
  - *If yes,* describe any symptoms you may have had that I can recognize if this occurs during therapy.
- Do you use the fingerstick method for testing your own blood glucose levels?
  - *To the therapist:* You may want to ask the client to bring the test kit for use before or during exercise.
- Do you have difficulty in maintaining your blood glucose levels within acceptable ranges (70 to 100 mg/dL)?
  - *If yes, to the therapist:* You may want to take a baseline of blood glucose levels before initiating an exercise program.
- Do you ever have burning, numbness, or a loss of sensation in your hands or feet? **(diabetic neuropathy)**

CASE STUDY

REFERRAL

Paul Martin, a 45-year-old client with type 1 diabetes mellitus, has been receiving wound care for a foot ulcer during the last 2 weeks. Today when he came to the clinic, he appeared slightly lethargic and confused. He indicated to you that he has had a "case of the flu" since early yesterday and that he had vomited once or twice the day before and once that morning before coming to the clinic. His wife, who had driven him to the clinic, said that he seemed to be "breathing fast" and urinating more frequently than usual. He has been thirsty, so he has been drinking "7-Up" and water, and those fluids "have stayed down okay."

PHYSICAL THERAPY INTERVIEW

- When did you last take your insulin? (Client may have forgotten because of his illness, forgetfulness, confusion, or just being afraid to take it while feeling sick with the "flu.")
- What type of insulin did you take?
- Do you have a source of sugar with you? If *yes,* where do you keep it? (This question should be asked during the initial physical therapy interview.)
- Have you contacted your physician about your condition?
- Have you done a recent blood glucose level (fingerstick)? If *yes,* when was the last time that this test was done?

WHAT WERE THE RESULTS?

To his wife: Your husband seems to be confused and is not himself. How long has he been like this? Have you observed any strong breath odor since this "flu" started? (Make your own observations regarding breath odor at this time.)

If possible, have the client perform a fingerstick blood glucose test on himself. This type of client should be sent immediately to his physician without physical therapy intervention. If he is hypoglycemic (unlikely under these circumstances), this condition should be treated immediately. It is more likely that this client is hyperglycemic and may have diabetic ketoacidosis. In either situation, he should not be driving, and arrangements should be made for transport to the physician's office.

PRACTICE QUESTIONS

1. What are the most common musculoskeletal symptoms associated with endocrine disorders?
2. What systemic conditions can cause carpal tunnel syndrome?
3. What are the mechanisms by which carpal tunnel syndrome occurs?
4. Disorders of the endocrine glands can be caused by:
   a. Dysfunction of the gland
   b. External stimulus
   c. Excess or insufficiency of hormonal secretions
   d. a and b
   e. b and c
   f. All the above
5. List three of the most common symptoms of diabetes mellitus.
   1. ______________________
   2. ______________________
   3. ______________________
6. What is the primary difference between the two hyperglycemic states: diabetic ketoacidosis (DKA) and hyperglycemic, hyperosmolar, nonketotic coma (HHNC)?
7. Is it safe to administer a source of sugar to a lethargic or unconscious person with diabetes?
8. Clients with diabetes insipidus (DI) would most likely come to the therapist with which of the following clinical symptoms?
   a. Severe dehydration, polydipsia
   b. Headache, confusion, lethargy
   c. Weight gain
   d. Decreased urine output
9. Clients who are taking corticosteroid medications should be monitored for the onset of Cushing's syndrome. You will need to monitor your client for which of the following problems?
   a. Low blood pressure, hypoglycemia
   b. Decreased bone density, muscle wasting
   c. Slow wound healing
   d. b and c
10. Signs and symptoms of Cushing's syndrome in an adult taking oral steroids may include:
    a. Increased thirst, decreased urination, and decreased appetite
    b. Low white blood cell count and reduced platelet count
    c. High blood pressure, tachycardia, and palpitations
    d. Hypertension, slow wound healing, easy bruising
11. Parathyroid hormone (PTH) secretion is particularly important in the metabolism of bone. The client with an oversecreting parathyroid gland would most likely have:
    a. Increased blood pressure
    b. Pathologic fractures
    c. Decreased blood pressure
    d. Increased thirst and urination
12. Which glycosylated hemoglobin (A1C) value is within the recommended range?
    a. 6%
    b. 8%
    c. 10%
    d. 12%
13. A 38-year-old man comes to the clinic for low back pain. He has a new diagnosis of Graves' disease. When asked if there are any other symptoms of any kind, he replies "increased appetite and excessive sweating." When you perform a neurologic screening examination, what might be present that would be associated with the Graves' disease?
    a. Hyporeflexia but no change in strength
    b. Hyporeflexia with decreased muscle strength
    c. Hyperreflexia with no change in strength
    d. Hyperreflexia with decreased muscle strength
14. All of the following are common signs or symptoms of insulin resistance *except:*
    a. Acanthosis nigricans
    b. Drowsiness after meals
    c. Fatigue
    d. Oliguria

REFERENCES

1. Schnetzler KA: Acute carpal tunnel syndrome. *J Am Acad Orthop Surg* 16:276–282, 2008.
2. Bickel KD: Carpal tunnel syndrome. *J Hand Surg* 35A:147–152, 2010.
3. Palmer KT: Carpal tunnel syndrome and its relation to occupation: a systematic literature review. *Occup Med* 57:57–66, 2007.
4. Lozano-Calderon S: The quality and strength of evidence for etiology: example of carpal tunnel syndrome. *J Hand Surg* 33A(4):525–538, 2008.
5. Phalen GS: The carpal tunnel syndrome: seventeen years' experience in diagnosis and treatment of six hundred and fifty-four hands. *J Bone Joint Surg* 48A(2):211–228, 1966.
6. Grossman LA, Kaplan HJ, Ownby FD: Carpal tunnel syndrome: initial manifestation of systemic disease. *JAMA* 176:259–261, 1961.
7. Wluka AE, Cicuttini FM, Spector TD: Menopause, oestrogens, and arthritis. *Maturitas* 35(3):183–189, 2000.
8. Ferry S, Hannaford P, Warskyj M, et al: Carpal tunnel syndrome: a nested case-control study of risk factors in women. *Am J Epidemiol* 151(6):566–574, 2000.
9. Michlovitz S: Conservative interventions for carpal tunnel syndrome. *J Orthop Sports Phys Ther* 34(10):591–598, 2004.
10. Grokoest AW, Demartini FE: Systemic disease and carpal tunnel syndrome. *JAMA* 155:635–637, 1954.
11. Katz JN: Clinical practice: carpal tunnel syndrome. *N Engl J Med* 346:1807, 2002.
12. McNab T, Khandwala H: Acromegaly as an endocrine form of myopathy: a case report and review of the literature. *Endocr Prac* 11(1):18–22, 2005. Available online at www.medscape.com/viewarticle/501408. Accessed January 21, 2011.
13. Rothschild B: Hyperostosis associated with hip surgery? *J Musculoskel Med* 21(5), 2004.
14. Holcomb S: Confronting Cushing's syndrome. *Nursing 2005*, 35(9):32–36, 2005.
15. Holcomb SS: Detecting thyroid disease. *Nursing 2005* 35(10):S4-S9, 2005.

16. Boissonnault WG, Koopmeiners MB: Medical history profile: orthopaedic physical therapy outpatients. *J Orthop Sports Phys Ther* 20(1):2–10, 1994.
17. President and Fellows of Harvard College, Harvard Medical School: Thyroid diseases—a special health report. *Harvard Health Newsletter*, 2004.
18. Trivalle C, Doucet J, Chassagne P, et al: Differences in the signs and symptoms of hypothyroidism in older and younger patients. *J Am Geriatr Soc* 1(44):50–53, 1996.
19. Lowe JC, Honeyman-Lowe G: *The metabolic treatment of fibromyalgia*, Lafayette, CO, 2000, McDowell Health Science Books.
20. Lowe JC, Lowe G: Facilitating the decrease in fibromyalgia pain during metabolic rehabilitation: an essential role for soft tissue therapies. *J Bodywork Movement Ther* 2(4):208–217, 1998.
21. Lowe JC, Reichman AJ, Honeyman GS, et al: Thyroid status of fibromyalgia patients. *Clin Bull Myofascial Ther* 3(1):47–53, 1998a.
22. Lowe JC, Reichman AJ, Yellin BA: A case-control study of metabolic therapy for fibromyalgia: long term follow-up comparison of treated and untreated patients. *Clin Bull Myofascial Ther* 3(1):65–79, 1998b.
23. Geenen R, Jacobs JW, Bijlsma JW: Evaluation and management of endocrine dysfunction in fibromyalgia. *Rheum Dis Clin North Am* 28(2):389–404, 2002.
24. Garrison RL, Breeding PC: A metabolic basis for fibromyalgia and its related disorders: the possible role of resistance to thyroid hormone. *Med Hypotheses* 61(2):182–189, 2003.
25. An anti-nuclear shield for your thyroid. *Johns Hopkins Med Lett* 14(8):5–7, 2002.
26. Gagel RF, Goepfert H, Callender DL: Changing concepts in the pathogenesis and management of thyroid carcinoma. *CA Cancer J Clin* 46(5):261–283, 1996.
27. Strewler G: Primary hyperparathyroidism does not progress in most patients. *JAMA* (293):1772–1779, 2005.
28. Utiger R: The physician's perspective. *NEJM Health News* 9(1):5, 2003.
29. Morbidity and Mortality Weekly Report: November is American Diabetes Month. *MMWR* 59(42):1361, 2010.
30. Burrows NR: Incidence of end-stage renal disease attributed to diabetes among persons with diagnosed diabetes in the United States and Puerto Rico. *MMWR* 59(42):1361–1366, 2010.
31. New York Academy of Sciences: *Is Alzheimer's disease type 3 diabetes?* Webinar presented by the Biochemical Pharmacology Discussion Group and the American Chemical Society. Oct. 27, 2009. Available online at http://www.nyas.org/events/WebinarDetail.aspx?cid=25aebb2b-0529-4cc1-92b0-c38c38b3d34f. Accessed March 3, 2011.
32. Wood L, Setter SM: Type 3 Diabetes: brain diabetes? *US Pharm* 35(5):36–41, 2010. Available online at http://www.uspharmacist.com/content/d/feature/c/20754/. Accessed March 3, 2011.
33. Centers for Disease Control and Prevention: Diabetes: National Diabetes Fact Sheet for the United States, 2010. Available online at http://www.cdc.gov/diabetes/statistics/index.htm. Accessed January 21, 2011.
34. Boyle JP: Projection of the year 2050 burden of diabetes in the US adult population: dynamic modeling of incidence, mortality, and prediabetes prevalence. *Popul Health Metr* 8:29, 2010.
35. Centers for Disease Control and Prevention (CDC): *Number of Americans with diabetes projected to double or triple by 2050. Older, more diverse populations and longer lifespans contribute to increase.* October 22, 2010. Available online at http://www.cdc.gov/media/pressrel/2010/r101022.html. Accessed March 3, 2011.
36. American Diabetes Association: Standards of medical care in diabetes—2010. *Diabetes Care* 33(suppl 1): S11-S61, 2010.
37. Norris SL, Zhang X, Avenell A, et al: Long-term non-pharmacological weight loss interventions for adults with prediabetes. *Cochrane Database Syst Rev* 2:CD005270, April 18, 2005.
38. US Renal Data System: *USRDS 2009 annual data report: Atlas of chronic kidney disease and end-stage renal disease in the United States.* Bethesda, MD, 2009, National Institutes of Health, National Institute of Diabetes and Digestive and Kidney Diseases.
39. Katon WJ, Rutter C, Simon G, et al: The association of comorbid depression with mortality in patients with type 2 diabetes. *Diabetes Care* 28(11):2668–2672, 2005.
40. Ciechanowski P, Russo J, Katon W, et al: Where is the patient? The association of psychosocial factors and missed primary care appointments in patients with diabetes. *Gen Hosp Psychiatry* 28(1):9–17, 2006.
41. Katon W, Cantrell CR, Sokol MC, et al: Impact of antidepressant drug adherence on comorbid medication use and resource utilization. *Arch Intern Med* 165(21):2497–2503, 2005.
42. Barclay L: Modifiable risk factors may be linked to risk of developing diabetic neuropathy. *N Engl J Med* 352:341–350, 408–409, 2005. Available online at http://www.medscape.com/viewarticle/498185. Accessed January 11, 2011.
43. Tesfaye S, Chaturvedi N, Eaton SE, et al: Vascular risk factors and diabetic neuropathy. *N Engl J Med* 352(4):341–350, 2005.
44. Said G: Diabetic neuropathy—a review. *Nat Clin Pract Neurol* 3(6):331–340, 2007. Available online at: http://www.medscape.com/viewarticle/558574. Accessed Dec. 13, 2010.
45. Cagliero E, Apruzzese W, Perlmutter GS: Watch for hand, shoulder disorders in patients with diabetes. *Am J Med* 112:487–490, 2002.
46. Cullen A, Oflouglu O, Donthineni R: Neuropathic arthropathy of the shoulder, Charcot shoulder. *Medscape* 7(1), 2005. Available online at www.medscape.com/viewarticle/496650). Accessed on January 21, 2011.
47. Scarborough P: Diabetes care: tests and measures for the foot and lower extremity. *Acute Care Perspectives* 11(4):1–6, 2002.
48. Boulton A: Treatment of symptomatic diabetic neuropathy. *Diabetes Metab Res Rev* 19(Suppl 1): S16–S21, 2003.
49. Kelly J: Manipulation for frozen shoulder. *J Musculoskel Med*, 20(2):58, 2003.
50. Mukhtar Q, Pan L, Jack L et al: Prevalence of receiving multiple preventive care services among adults with diabetes—United States, 2002–2004. *MMWR* 54(44):1130, 2005.
51. Agency for Healthcare Research and Quality (AHRQ): *Diagnosis and management of type 2 diabetes mellitus in adults.* Updated by ECRI Institute, 2010. Available online at http://guideline.gov/content.aspx?f=rss&id=24137. Accessed March 2, 2011.
52. Colberg SR, Walsh J: Pumping insulin during exercise. *Phys Sports Med* 30(4), April 2002. Available online at http://www.physsportsmed.com/index.php?article=255. Accessed January 20, 2011.
53. Auber G: Taking the heat off: how to manage heat injuries. *Nursing 2004* 34(7):50–52, 2004.
54. Appel SJ: Sizing up patients for metabolic syndrome. *Nursing 2005* 35(12):20–21, 2005.
55. Lorenzo C: The National Cholesterol Education Program—Adult Treatment Panel III, International Diabetes Federation and World Health Organization definitions of the metabolic syndrome as predictors of incident cardiovascular disease and diabetes. *Diabetes Care* 30(1):8–13, 2007.
56. Alberti KG, Eckel RH, Grundy SM, et al: Harmonizing the metabolic syndrome: A joint interim statement of the International Diabetes Federation Task Force on Epidemiology and Prevention; National Heart, Lung, and Blood Institute; American Heart Association; World Heart Federation; International Atherosclerosis Society; and International Association for the Study of Obesity. *Circulation* 120:1640–1645, 2009.

57. Olatunbosun ST: Insulin resistance, *eMedicine*. Updated 5/16/10. Available online at http://emedicine.medscape.com/article/122501-print. Accessed Sept. 15, 2010.
58. Rosenzweig JL: Primary prevention of cardiovascular disease and type 2 diabetes in patients at metabolic risk: an Endocrine Society clinical practice guideline. *J Clin Endocrinol Metab* 93(10):3671–3689, 2008.
59. Chen L, Schumacher R: Gout and gout mimickers: 20 clinical pearls. *J Musculoskel Med* 20(5):254–258, 2003.
60. Kurakula PC, Keenan RT: Diagnosis and management of gout: an update. *J Musculoskel Med* 27(suppl): S13-S19, 2010.
61. Barclay L, Lie D: Dietary risk factors for gout clarified. *Medscape Medical News March* 10, 2004. Available online at http://www.medscape.com/viewarticle/471444. Accessed January 21, 2011.
62. Choi HK: A prescription for lifestyle change in patients with hyperuricemia and gout. *Curr Opin Rheumatol* 22(2):165–172, 2010.
63. Choi HK: Alcohol intake and risk of incident gout in men: a prospective study. *Lancet* 363:1277–1281, 2004.
64. Wiese W, Sanders LS, Wortmann RL: Gout: effective strategies for acute and long-term control. *J Musculoskel Med* 21(10):510–519, 2004.
65. Stevens S, Edwards C: Recognizing and managing hemochromatosis and hemiarthropathy. *J Musculoskel Med* 21(4):212–225, 2004.
66. Agency for Healthcare Research & Quality (AHRQ): Screening and risk assessment for osteoporosis in postmenopausal women. Revised 2011. Available online at http://guideline.gov/syntheses/synthesis.aspx?f=rss&id=25304. Alternate website: http://www.guideline.gov. Accessed March 2, 2011.
67. Edwards BJ, Brooks ER, Langman CB: Osteoporosis screening of postmenopausal women in the primary care setting; a case-based approach. *Gend Med* 1(2):70–85, 2004.
68. Feldstein AC, Nichols GA, Elmer JP, et al: Older women with fractures: patients falling through the cracks of guideline-recommended osteoporosis screening and treatment. *J Bone Joint Surg* 85A(12):2294–2302, 2003.
69. Dipaola CP: Survey of spine surgeons on attitudes regarding osteoporosis and osteomalacia screening and treatment for fractures, fusion surgery, and pseudoarthrosis. *Spine J* 9(7):537–544, 2009.
70. Focus on Healthy Aging: Expert panel urges mass screening for osteoporosis, older women. *Mt Sinai School Med* 6(1):2, 2003.
71. Black DM, Steinbuch M, Palermo L, et al: An assessment tool for predicting fracture risk in postmenopausal women. *Osteoporosis Int* 12(7):519–528, 2001.
72. Maricic M, Gluck O: Osteoporosis: 20 clinical pearls. *J Musculoskel Med* 20(11):508–512, 2003.
73. Bones are big news. *Johns Hopkins Med Lett Health After 50* 17(1):4–5, 2005.
74. Kiebzak G, Beinart G, Perser K, et al: Undertreatment of osteoporosis in men with hip fracture. *Arch Intern Med* 162(19):2217–2222, 2002.
75. Simmons G: *Far Eastern osteoporosis study*, Hong Kong, 1996, The Gordon Simmons Research Group Ltd.
76. DePalma M, Slipman C: Managing osteoporotic vertebral compression fractures. *J Musculoskel Med* 22(9):445–454, 2005.
77. Papapoulos SE: Paget's disease of bone: clinical, pathogenetic, and therapeutic aspects. *Baillieres Clin Endocrinol Metab* 11(1):117–143, 1997.
78. Daroszewska A, Ralston SH: Genetics of Paget's disease of bone. *Clin Sci (London)* 109(3):257–263, 2005.
79. Roodman GD, Windle JJ: Paget disease of bone. *J Clin Invest* 115(2):200–208, 2005.
80. American Academy of Orthopedic Surgeons: Paget's disease of bone, Oct. 2005. Available online at: http://www.orthoinfo.aaos.org, (See tumors). Accessed January 21, 2011.
81. Lydick E, Cook K, Turpin J, et al: Development and validation of a simple questionnaire to facilitate identification of women likely to have low bone density. *Am J Man Care* 4:37–48, 1998.
82. Elliott ME, Drinka PJ, Krause P, et al: Osteoporosis assessment strategies for male nursing home residents. *Maturitas* 48(3):225–233, July 2004.
83. National Osteoporosis Foundation (NOF): Standing tall for you. Prevention: who's at risk? Available online at http://www.nof.org. Accessed January 10, 2011.
84. Orwoll ES, Bauer DC, Vogt TM, et al: Axial bone mass in older women. *Ann Intern Med* 124:187–196, 1996.
85. Funk D, Swank AM, Adam KJ, et al: Efficacy of moist heat pack application over static stretching on hamstring flexibility. *J Strength Cond Res* 15(1):123–126, 2001.
86. Daly MP, Berman BM: Rehabilitation of the elderly patient with osteoarthritis. *Clin Geriatr Med* 9:783–801, 1993.
87. Cottingham JT, Maitland J: A three-paradigm treatment model using soft tissue mobilization and guided movement-awareness techniques for a patient with chronic low back pain: a case study. *J Orthop Sports Phys Ther* 26(3):155–167, 1997.
88. Hayden JA, Van Tulder MW, Malmivaara AV, Koes BW: Meta-analysis: Exercise therapy for non-specific low back pain. *Ann Intern Med* 142(9):765–775, 2005.
89. Heggeness MH: Spinal fracture with neurological deficit in osteoporosis. *Osteoporos Int* 3(4):215–221, 1993.
90. D'Costa H, George G, Parry M, et al: Pitfalls in the clinical diagnosis of vertebral fractures: a case series in which posterior midline tenderness was absent. *Emerg Med J* 22:330–332, 2005.
91. Nevitt MC, Ettinger B, Black DM, et al: The association of radiographically detected vertebral fractures with back pain and function: a prospective study. *Ann Intern Med* 128(10):793–800, 1998.
92. Fink HA, Milaetz DL, Palermo L, et al: What proportion of incident radiographic vertebral deformities is clinically diagnosed and vice versa? *J Bone Miner Res* 20(7):1216–1222, 2005.
93. The WHO Study Group: *Assessment of fracture risk and its application to screening for postmenopausal osteoporosis.* WHO Technical Report Series 843, Geneva, 1994, World Health Organization.

CHAPTER

12

Screening for Immunologic Disease

Immunology, one of the few disciplines with a full range of involvement in all aspects of health and disease, is one of the most rapidly expanding fields in medicine today. Staying current is difficult at best, considering the volume of new immunologic information generated by clinical researchers each year. The information presented here is a simplistic representation of the immune system, with the main focus on screening for immune-induced signs and symptoms mimicking neuromuscular or musculoskeletal dysfunction.

Immunity denotes protection against infectious organisms. The immune system is a complex network of specialized organs and cells that has evolved to defend the body against attacks by "foreign" invaders. Immunity is provided by lymphoid cells residing in the immune system. This system consists of central and peripheral lymphoid organs (Fig. 12-1).

By circulating its component cells and substances, the immune system maintains an early warning system against both exogenous microorganisms (infections produced by bacteria, viruses, parasites, and fungi) and endogenous cells that have become neoplastic.

Immunologic responses in humans can be divided into two broad categories: humoral immunity, which takes place in the body fluids (extracellular) and is concerned with antibody and complement activities, and cell-mediated or cellular immunity, primarily intracellular, which involves a variety of activities designed to destroy or at least contain cells that are recognized by the body as being alien and harmful. Both types of responses are initiated by lymphocytes and are discussed in the context of lymphocytic function.

USING THE SCREENING MODEL

As always in the screening evaluation of any client, the medical history is the most important variable, followed by any red flags in the clinical presentation and an assessment of associated signs and symptoms. Many immune system disorders have a unique chronology or sequence of events that define them. When the immune system may be involved, some important questions to ask include the following:

- How long have you had this problem? (acute versus chronic)
- Has the problem gone away and then recurred?
- Have additional symptoms developed or have other areas become symptomatic over time?

Past Medical History

As mentioned, the family history is important when assessing the role of the immune system in presenting signs and symptoms. Persons with fibromyalgia or chronic pain often have a family history of alcoholism, depression, migraine headaches, gastrointestinal (GI) disorders, or panic attacks.

Clients with systemic inflammatory disorders may have a family history of an identical or related disorder such as rheumatoid arthritis (RA), systemic lupus erythematosus (SLE), autoimmune thyroid disease, multiple sclerosis (MS), or myasthenia gravis (MG). Other rheumatic diseases that are often genetically linked include seronegative spondyloarthropathy.

The seronegative spondyloarthropathies include a wide range of diseases linked by common characteristics such as inflammatory spine involvement (e.g., sacroiliitis, spondylitis), asymmetric peripheral arthritis, enthesopathy, inflammatory eye disease, and musculoskeletal and cutaneous features. All of these changes occur in the absence of serum rheumatoid factor (RF), which is present in about 85% of people with RA.[1]

This group of diseases includes ankylosing spondylitis (AS), reactive arthritis (ReA; such as Reiter's syndrome), psoriatic arthritis (PsA), and arthritis associated with inflammatory bowel disease (IBD; such as Crohn's disease or ulcerative colitis).

A recent history of surgery may be indicative of bacterial or reactive arthritis, which requires immediate medical evaluation.

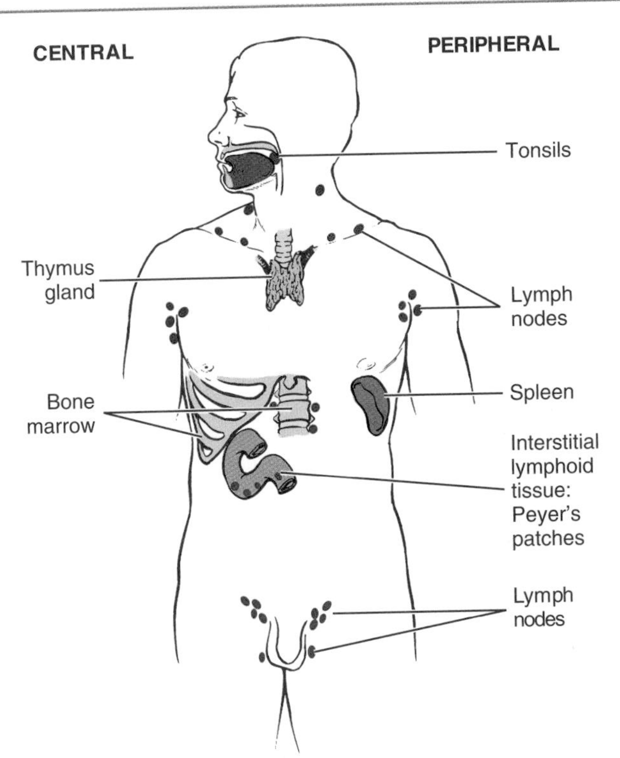

Fig. 12-1 Major organs of the immune system. Two-thirds of the immune system resides in the intestines (intestinal lymphoid tissue), emphasizing the importance of diet and nutrition on immune system function.

Risk Factor Assessment

The cause and risk factors for many conditions related to immune system dysfunction remain unknown. Past medical history with a positive family history for systemic inflammatory or related disorders may be the only available red flag in this area.

Clinical Presentation

Symptoms of rheumatic disorders often include soft tissue and/or joint pain, stiffness, swelling, weakness, constitutional symptoms, Raynaud's phenomenon, and sleep disturbances. Inflammatory disorders, such as RA and polymyalgia rheumatica (PMR), are marked by prolonged *stiffness* in the morning lasting more than 1 hour. This stiffness is relieved with activity, but it recurs after the person sits down and subsequently attempts to resume activity. This is referred to as the *gel phenomenon.*

Specific arthropathies have a predilection for involving specific joint areas. For example, involvement of the wrists and proximal small joints of the hands and feet is a typical feature of RA. RA tends to involve joint groups symmetrically, whereas the seronegative spondyloarthropathies tend to be asymmetric. PsA often involves the distal joints of the hands and feet.[1]

In anyone with *swelling,* especially single-joint swelling, it is necessary to distinguish whether this swelling is articular (as in arthritis), is periarticular (as in tenosynovitis), involves an entire limb (as with lymphedema), or occurs in another area (such as with lipoma or palpable tumors). The therapist will need to assess whether the swelling is intermittent, persistent, symmetric, or asymmetric and whether the swelling is minimal in the morning but worse during the day (as with dependent edema).

Generalized *weakness* is a common symptom of individuals with immune system disorders in the absence of muscle disease. If the weakness involves one limb without evidence of weakness elsewhere, a neurologic disorder may be present. Anyone having trouble performing tasks with the arms raised above the head, difficulty climbing stairs, or problems arising from a low chair may have muscle disease.

Nail bed changes are especially indicative of underlying inflammatory disease. For example, small infarctions or splinter hemorrhages (see Fig. 4-34) occur in endocarditis and systemic vasculitis. Characteristics of systemic sclerosis and limited scleroderma include atrophy of the fingertips, calcific nodules, digital cyanosis, and sclerodactyly (tightening of the skin). Dystrophic nail changes are characteristic of psoriasis. Spongy synovial thickening or bony hypertrophic changes (Bouchard's nodes) are present with RA and other hand deformities.

Associated Signs and Symptoms

With few risk factors and only the family history to rely upon, the clinical presentation is very important. Most of the immune system conditions and diseases are accompanied by a variety of associated signs and symptoms. Disease progression is common with different clinical signs and symptoms during the early phase of illness compared with the advanced phase.

Review of Systems

With many problems affecting the immune system, taking a step back and reviewing each part of the screening model (history, risk factors, clinical presentation, associated signs and symptoms) may be the only way to identify the source of the underlying problem. Remember to review Box 4-19 during this process.

For anyone with new onset of joint pain, a review of systems should include questions about symptoms or diagnoses involving other organ systems. In particular, the presence of dry, red, or irritated and itching eyes; chest pain with dyspnea; urethral or vaginal discharge; skin rash or photosensitivity; hair loss; diarrhea; or dysphagia should be assessed.

IMMUNE SYSTEM PATHOPHYSIOLOGY

Immune disorders involve dysfunction of the immune response mechanism, causing overresponsiveness or blocked, misdirected, or limited responsiveness to antigens. These disorders may result from an unknown cause, developmental defect, infection, malignancy, trauma, metabolic disorder, or drug use. Immunologic disorders may be classified as one of the following:

- Immunodeficiency disorder
- Hypersensitivity disorder
- Autoimmune disorder
- Immunoproliferative disorder

Immunodeficiency Disorders

When the immune system is underactive or hypoactive, it is referred to as being *immunodeficient* or *immunocompromised* such as occurs in the case of anyone undergoing chemotherapy for cancer or taking immunosuppressive drugs after organ transplantation.

Acquired Immunodeficiency Syndrome

Human immunodeficiency virus (HIV) is a cytopathogenic virus that causes acquired immunodeficiency syndrome (AIDS). HIV has been identified as the causative agent, its genes have been mapped and analyzed, drugs that act against it have been found and tested, and vaccines against the HIV infection have been under development.

Acquired refers to the fact that the disease is not inherited or genetic but develops as a result of a virus. *Immuno* refers to the body's immunologic system, and *deficiency* indicates that the immune system is underfunctioning, resulting in a group of signs or symptoms that occur together called a *syndrome.*

People who are HIV-infected are vulnerable to serious illnesses called *opportunistic* infections or diseases, so named because they use the opportunity of lowered resistance to infect and destroy. These infections and diseases would not be a threat to most individuals whose immune systems functioned normally. *Pneumocystis carinii* pneumonia (PCP) continues to be a major cause of morbidity and mortality in the AIDS population.

HIV infection is the fifth leading cause of death for people who are between 25 and 44 years old in the United States. Each year, about 2 million people worldwide die of AIDS. African Americans represent about 12% of the total US population but makeup over half of all AIDS cases reported. AIDS is the leading cause of death for African-American men between the ages of 35 and 44. Overall estimates are that 850,000 to 950,000 US residents are living with HIV infection, one-quarter of whom are unaware of their infection. Approximately 56,000 new HIV infections occur each year in the United States, and approximately 2.7 million new HIV cases occur each year worldwide (Box 12-1).[2]

Risk Factors. Population groups at greatest risk include commercial sex workers (prostitutes) and their clients, men having sex with men, injection drug users (IDUs), blood recipients, dialysis recipients, organ transplant recipients, fetuses of HIV-infected mothers or babies being breast-fed by an HIV-infected mother, and people with sexually transmitted diseases (STDs). The latter group is estimated to have a 3 to 5 times higher risk for HIV infection compared with those having no STDs.

BOX 12-1 OVERVIEW OF AIDS IN THE UNITED STATES

What it is: AIDS (acquired immunodeficiency syndrome) is a contagious disease that destroys the T cells, a key component of the body's immune system.

What causes it: AIDS is caused by the human immunodeficiency virus (HIV), spread through sexual contact, needles, or syringes shared by injection drug users (IDUs); transfusion of infected blood or blood products; or perinatal transmission (from infected birthing or breastfeeding mother to her infant).

Who gets it: Primary persons infected with HIV have been homosexual men (men who have sex with men [MSM] and MSMs who have sex with women) and IDUs. The Centers for Disease Control and Prevention (CDC) estimate that heterosexual contact is responsible for 3% of male cases and 34% of female cases. Although 1 million Americans are infected, one-fourth do not know they have it; as many as one-third of adults tested never come back for the results.[105]

Diagnosis: Screening for AIDS is conducted by testing a fingerstick sample of blood for the presence of antibodies to HIV-1. The test indicates only if a person has been exposed to the virus. A new "quick" test called OraQuick Rapid HIV Antibody Test is almost 100% accurate, and results are available within 20 minutes. A positive test requires additional confirmation testing.

Prognosis: At present, there is no cure, but many people in the United States remain healthy and active with combination antiretroviral medications designed to attack HIV in various stages of its life cycle; when death occurs, it is usually as a result of "opportunistic" infections or cancers that the immunosuppressed body cannot resist. IDUs are four times more likely to die of AIDS than individuals infected through sexual contact.[106]

The rate of new cases of HIV among bisexual men of all races has started to rise again after a period of relative stability. Experts suggest the increase is due to erosion of safe sex practices referred to as *prevention fatigue.* African Americans (both men and women) are still 8 times as likely as whites to contract HIV, although the rate of newly diagnosed HIV infections among African Americans is slowly declining.[3]

Transmission. Transmission of HIV occurs according to the following descending hierarchy[4]:

- Male-to-male sexual contact
- Injection drug use
- Both male-to-male sexual contact and injection drug use
- High-risk heterosexual contact (with someone of the opposite sex with HIV/AIDS or a risk factor for HIV)

Transmission occurs through *horizontal* transmission (from either sexual contact or parenteral exposure to blood and blood products) or through *vertical* transmission (from

HIV-infected mother to infant). HIV is not transmitted through casual contact, such as the shared use of food, towels, cups, razors, or toothbrushes, or even by kissing. Despite substantial advances in the treatment of HIV, the number of new infections has not decreased in the past 10 years. Prevention of infection transmission by reduction of behaviors that might transmit HIV to others is critical.[5]

Transmission always involves exposure to some body fluid from an infected client. The greatest concentrations of virus have been found in blood, semen, cerebrospinal fluid, and cervical/vaginal secretions. HIV has been found in low concentrations in tears, saliva, and urine, but no cases have been transmitted by these routes. Breast-feeding is a route of HIV transmission from an HIV-infected mother to her infant. The reduction of HIV transmission through breast milk remains a challenge in many resource-poor settings.[6,7]

Any injectable drug, legal or illegal, can be associated with HIV transmission. It is not injection drug use that spreads HIV but the sharing of HIV-infected intravenous (IV) drug needles among individuals. Despite the perception that only IV injection is dangerous, HIV also can be transmitted through subcutaneous and intramuscular injection. Use of needles contaminated with blood for tattooing or body piercing are included in this category.

Public health organizations have changed their terminology, substituting the abbreviation *IDU* (injection drug user) for the earlier term *IVDU* (IV drug user). IDUs who sterilize their drug paraphernalia with a 1:10 solution of bleach to water before passing the needles are less likely to spread HIV. For further information regarding this or other HIV/AIDS-related questions, contact the CDC-INFO (formerly the CDC National AIDS Hotline) 24 hours/day, 7 days a week, at 1-800-CDC-INFO (1-800-232-4636).

Blood and Blood Products. Parenteral transmission occurs when there is direct blood-to-blood contact with a client infected with HIV. This can occur through sharing of contaminated needles and drug paraphernalia ("works"), through transfusion of blood or blood products, by accidental needlestick injury to a health care worker, or from blood exposure to nonintact skin or mucous membranes. Health care workers who have contact with clients with AIDS and who follow routine instructions for self-protection are a very low risk group.

Almost all persons with hemophilia born before 1985 have been infected with HIV. Heat-treated factor concentrates, involving a method of chemical and physical processes that completely inactivate HIV, became available in 1985, effectively eliminating the transmission of HIV to anyone with a clotting disorder who is receiving blood or blood products.

The risk for acquiring HIV infection through blood transfusion today is estimated conservatively to be one in 1.5 million, based on 2007-2008 data.[8] A blood center in Missouri discovered that blood components from a donation in November 2008 tested positive for HIV infection.[9] A subsequent investigation determined that the blood donor had last donated in June 2008, at which time he incorrectly reported no HIV risk factors and his donation tested negative for the presence of HIV. One of the two recipients of blood components from this donation, an individual undergoing kidney transplantation, was found to be HIV infected, and an investigation determined that the recipient's infection was acquired from the donor's blood products. The CDC advises that even though such transmissions are rare, health care providers should consider the possibility of transfusion-transmitted HIV in HIV-infected transfusion recipients with no other risk factors.[10]

Additionally, HIV has been transmitted heterosexually from infected men with hemophilia to spouses or sexual partners in what is termed *the second wave* of infection and on to children born to infected couples. HIV infection in the United States is currently on the increase among women exposed via sexual intercourse with HIV-infected men. Minority women and women over the age of 50 are being affected more frequently than in prior years.[11]

Clinical Signs and Symptoms. Many individuals with HIV infection remain asymptomatic for years, with a mean time of approximately 10 years between exposure and development of AIDS. Systemic complaints, such as weight loss, fevers, and night sweats, are common. Cough or shortness of breath may occur with HIV-related pulmonary disease. GI complaints include changes in bowel function, especially diarrhea.

Cutaneous complaints are common and include dry skin, new rashes, and nail bed changes. Because virtually all of these findings may be seen with other diseases, a combination of complaints is more suggestive of HIV infection than any one symptom.

Many persons with AIDS experience back pain, but the underlying causes may differ. Decrease in muscle mass with subsequent postural changes may occur as a result of the disease process or in response to medications. It is not uncommon for pain to develop in the back or another musculoskeletal location where there may have been a previous injury. This is more likely to occur when the T-cell count drops.

Bone disorders such as osteopenia, osteoporosis, and osteonecrosis have been reported in association with HIV, but the etiology and mechanism of these disorders are unknown. Prevalence reported varies from study to study; scientists are researching the influence of antiretroviral therapy and lipodystrophy (absence or presence), severity of HIV disease, and overlapping risk factors for bone loss (e.g., smoking and alcohol intake). The therapist should conduct a risk factor assessment for bone loss in anyone with known HIV and educate clients about prevention strategies.[12]

Any woman at risk for AIDS should be aware of the possibility that recurrent or stubborn cases of vaginal candidiasis may be an early sign of infection with HIV. Pregnancy, diabetes, oral contraceptives, and antibiotics are more commonly linked to these fungal infections.

Side Effects of Medication. The therapist should review the potential side effects from medication used in the treatment of AIDS. Delayed toxicity with long-term treatment for HIV-1 infection with antiretroviral therapy occurs in a substantial number of affected individuals.[13-15] The more commonly occurring symptoms include rash, nausea, headaches, dizziness, muscle pain, weakness, fatigue, and insomnia. Hepatotoxicity is a common complication; the therapist should be alert for carpal tunnel syndrome, liver palms, asterixis, and other signs of liver impairment (see Chapter 9).

Body fat redistribution to the abdomen, upper body, and breasts occurs as part of a condition called *lipodystrophy* associated with antiretroviral therapy. Other metabolic abnormalities, such as dysregulation of glucose metabolism (e.g., insulin resistance, diabetes), combined with lipodystrophy are labeled lipodystrophic syndrome (LDS). LDS contributes to problems with body image and increases the risk for cardiovascular complications.[16-19]

CLINICAL SIGNS AND SYMPTOMS

Early Symptomatic HIV Infection

- Fever
- Sweats
- Chronic diarrhea
- Fatigue
- Minor oral infections
- Headache
- Women: Vaginal candidiasis
- Cough
- Shortness of breath
- Cutaneous changes (rash, nail bed changes, dry skin, telangiectasias, psoriasis, dermatitis)

Advanced Symptomatic HIV Infection

- Kaposi's sarcoma
  - Multiple purple blotches and bumps on skin
- Hypertension (pulmonary and/or cardiac)
  - Dyspnea, syncope, fatigue, chest pain, nonproductive cough
- Opportunistic diseases (e.g., tuberculosis [TB], *Pneumocystis carinii* pneumonia, lymphoma, thrush; herpes simplex virus [HSV] I and II; toxoplasmosis; candidiasis)
  - Persistent dry cough
  - Fever, night sweats
  - Easy bruising
  - Thrush (thick, white coating on the tongue or throat accompanied by a sore throat)
  - Muscle atrophy and weakness
  - Back pain
  - Side effects of medication (see text)
- Poor wound healing
- HIV-related dementia (memory loss, confusion, behavioral change, impaired gait)
- Distal symmetric polyneuropathy (pain, numbness, tingling, burning, weakness, atrophy)

AIDS and Other Diseases

AIDS is a unique disease—no other known infectious disease causes its damage through a direct attack on the human immune system. Because the immune system is the final mediator of human host–infectious agent interactions, it was anticipated early that HIV infection would complicate the course of other serious human diseases.

This has proved to be the case, particularly for TB and certain sexually transmitted infections such as syphilis and the genital herpes virus. Cancer has been linked with AIDS since 1981; this link was discovered with the increased appearance of a highly unusual malignancy, Kaposi's sarcoma. Since then, HIV infection has been associated with other malignancies, including non-Hodgkin's lymphoma (NHL), AIDS-related primary central nervous system lymphoma, and hepatocellular carcinoma.[20-22]

Kaposi's Sarcoma. Classic Kaposi's sarcoma (KS) was first recognized as a malignant tumor of the inner walls of the heart, veins, and arteries in 1873 in Vienna, Austria. Before the AIDS epidemic, KS was a rare tumor that primarily affected older people of Mediterranean and Jewish origin.

Clinically, KS in HIV-infected immunodeficient persons occurs more often as purplish-red lesions of the feet, trunk, and head (Fig. 12-2). The lesion is not painful or contagious. It can be flat or raised and over time frequently progresses to a nodule. The mouth and many internal organs (especially those of the GI and respiratory tracts) may be involved either symptomatically or subclinically.

Prognosis depends on the status of the individual's immune system. People who die of AIDS usually succumb to opportunistic infections rather than to KS.

Non-Hodgkin's Lymphoma. Approximately 3% of AIDS diagnoses in all risk groups and in all areas originate through discovery of NHL. The incidence of NHL increases with age and as the immune system weakens.

These malignancies are difficult to treat because clients often cannot tolerate the further immunosuppression that treatment causes. As with KS, prognosis depends largely on the initial level of immunity. Clients with adequate immune reserves may tolerate therapy and respond reasonably well.

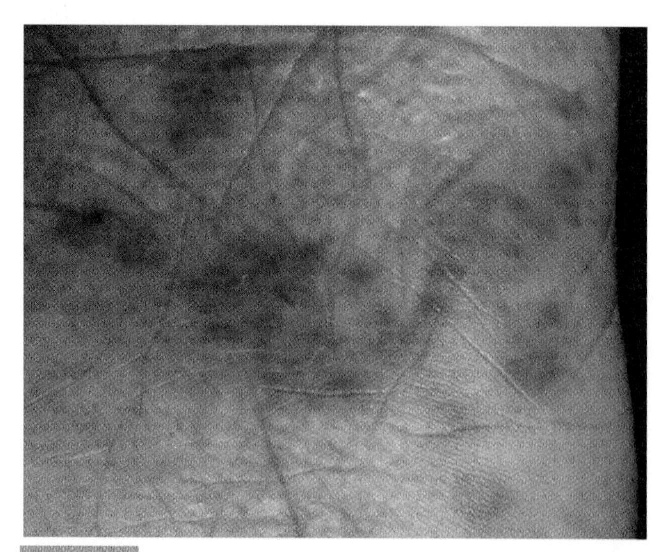

Fig. 12-2 AIDS-related Kaposi's sarcoma. The early lesions appear most commonly on the toes or soles as reddish or bluish-black macules and patches that spread and coalesce to form nodules or plaques. Lesions can appear anywhere on the body including the tongue and genitals. (From James WD: *Andrews' diseases of the skin: clinical dermatology,* ed 10, Philadelphia, 2006, WB Saunders.)

However, in people with severe immunodeficiency, survival is only 4 to 7 months on average. Clients diagnosed with HIV-related brain lymphomas have a very poor prognosis.

Tuberculosis. Tuberculosis (TB) was considered a stable, endemic health problem, but now, in association with the HIV/AIDS pandemic, TB is resurgent.[23] The recent emergence of multiple-drug–resistant TB, which has reached epidemic proportions in New York City, has created a serious and growing threat to the capacity of TB control programs (see Chapter 7).

In urban areas of the United States, the present upsurge in TB cases is occurring among young (aged 25 to 44 years) IDUs, ethnic minorities, prisoners and prison staff (because of poorly ventilated and overcrowded prison systems), homeless people, and immigrants from countries with a high prevalence of TB.

The first major interaction between HIV and TB occurs as a result of the weakening of the immune system in association with progressive HIV infection. The great majority of individuals exposed to TB are infected but not clinically ill. Their subclinical TB infection is kept in check by an active, healthy immune system. However, when a TB-infected person becomes infected with HIV, the immune system begins to decline, and at a certain level of immune damage from HIV, the TB bacteria become active, causing clinical pulmonary TB.

TB is the only opportunistic infection associated with AIDS/HIV that is directly transmissible to household and other contacts. Therefore each individual case of active TB is a threat to community health.

Clinical Signs and Symptoms. Pulmonary TB is the most common manifestation of TB disease in HIV-positive clients. When TB precedes the diagnosis of AIDS, disease is usually confined to the lung, whereas when TB is diagnosed after the onset of AIDS, the majority of clients also have extrapulmonary TB, most commonly involving the bone marrow or lymph nodes. Fever, night sweats, wasting, cough, and dyspnea occur in the majority of clients (see further discussion of TB in Chapter 7).

HIV Neurologic Disease. HIV neurologic disease may be the presenting symptom of HIV infection and can involve the central and peripheral nervous systems. HIV is a neurotropic virus and can affect neurologic tissues from the initial stages of infection. In the early course of the infection, the virus can cause demyelination of central and peripheral nervous system tissues.[24] Signs and symptoms range from mild sensory polyneuropathy to seizures, hemiparesis, paraplegia, and dementia.

Central Nervous System. Central nervous system (CNS) disease in HIV-infected clients can be divided into intracerebral space–occupying lesions, encephalopathy, meningitis, and spinal cord processes. Toxoplasmosis is the most common space–occupying lesion in HIV-infected clients. Presenting symptoms may include headache, focal neurologic deficits, seizures, or altered mental status.

AIDS dementia complex (HIV encephalopathy) is the most common neurologic complication and the most common cause of mental status changes in HIV-infected clients. It is characterized by cognitive, motor, and behavioral dysfunction. This disorder is similar to Alzheimer's dementia but has less impact on memory loss and a greater effect on time-related skills (i.e., psychomotor skills learned over time such as playing piano or reading).

Early symptoms of AIDS dementia involve difficulty with concentration and memory, personality changes, irritability, and apathy. Depression and withdrawal occur as the dementia progresses. Motor dysfunction may accompany cognitive changes and may result in poor balance, poor coordination, and frequent falls.

Progressive multifocal leukoencephalopathy (PML), which produces localized lesions within the brain, causes demyelination in the brain and leads to death within a few months.

In addition to the brain, neurologic disorders related to AIDS and HIV may affect the spinal cord, appearing as myelopathies. A vacuolar myelopathy often appears in the thoracic spine and causes gradual weakness, painless gait disturbance characterized by spasticity, and ataxia in the lower extremities that progresses to include weakness of the upper extremities.

Structural and inflammatory abnormalities in the muscles of people with HIV have been reported to impair the muscle's ability to extract or utilize oxygen during exercise. Clinical manifestations of HIV-associated myopathies include proximal weakness, myalgia, abnormal electromyogram (EMG) activity, elevated creatine kinase, and decreased functioning of the muscle.[25]

Peripheral Nervous System. Peripheral nerve disease is a common complication of the HIV infection. Peripheral nervous system syndromes include inflammatory polyneuropathies, sensory neuropathies, and mononeuropathies. An inflammatory demyelinating polyneuropathy similar to Guillain-Barré syndrome can occur in HIV-infected clients. Cytomegalovirus (CMV), a highly host-specific herpes virus that infects the nerve roots, may result in an ascending polyradiculopathy characterized by lower extremity weakness progressing to flaccid paralysis.

The most common neuropathy develops into painful sensory neuropathy with numbness and burning or tingling in the feet, legs, or hands. Immobility caused by painful neuropathies can result in deconditioning and eventual cardiopulmonary decline.

CLINICAL SIGNS AND SYMPTOMS

HIV Neurologic Disease

- Difficulty with concentration and memory
- Personality changes (depressions, withdrawal, apathy)
- Headaches
- Seizures
- Paralysis (hemiparesis, paraplegia)
- Motor dysfunction (balance and coordination)
- Gradual weakness of extremities
- Numbness and tingling (peripheral neuropathy)
- Radiculopathy

Hypersensitivity Disorders

Although the immune system protects the body from harmful invaders, an overactive or overzealous response is detrimental. When the immune system becomes overactive or hyperactive, a state of hypersensitivity exists, leading to immunologic diseases such as allergies.

Although the word *allergy* is widely used, the term *hypersensitivity* is more appropriate. Hypersensitivity designates an increased immune response to the presence of an antigen (referred to as an *allergen*) that results in tissue destruction.

The two general categories of hypersensitivity reaction are immediate and delayed. These designations are based on the rapidity of the immune response. In addition to these two categories, hypersensitivity reactions are divided into four main types (I to IV).

Type I Anaphylactic Hypersensitivity ("Allergies")

Allergy and Atopy. *Allergy* refers to the abnormal hypersensitivity that takes place when a foreign substance (allergen) is introduced into the body of a person likely to have allergies. The body fights these invaders by producing the special antibody immunoglobulin E (IgE). This antibody (now a vital diagnostic sign of many allergies), when released into the blood, breaks down mast cells, which contain chemical mediators, such as histamine, that cause dilation of blood vessels and the characteristic symptoms of allergy.

Atopy differs from allergy because it refers to a genetic predisposition to produce large quantities of IgE, causing this state of clinical hypersensitivity. The reaction between the allergen and the susceptible person (i.e., allergy-prone host) results in the development of a number of typical signs and symptoms usually involving the GI tract, respiratory tract, or skin.

Clinical Signs and Symptoms. Clinical signs and symptoms vary from one client to another according to the allergies present. With the Family/Personal History form used, each client should be asked what known allergies are present and what the specific reaction to the allergen would be for that particular person. The therapist can then be alert to any of these warning signs during treatment and can take necessary measures, whether that means grading exercise to the client's tolerance, controlling the room temperature, or appropriately using medications prescribed.

Anaphylaxis. Anaphylaxis, the most dramatic and devastating form of type I hypersensitivity, is the systemic manifestation of immediate hypersensitivity. The implicated antigen is often introduced parenterally such as by injection of penicillin or a bee sting. The activation and breakdown of mast cells systematically cause vasodilation and increased capillary permeability, which promote fluid loss into the interstitial space, resulting in the clinical picture of bronchospasms, urticaria (wheals or hives), and anaphylactic shock.

Initial manifestations of anaphylaxis may include local itching, edema, and sneezing. These seemingly innocuous problems are followed in minutes by wheezing, dyspnea, cyanosis, and circulatory shock. Clinical signs and symptoms of anaphylaxis are listed by system in Table 12-1.

Clients with previous anaphylactic reactions (and the specific signs and symptoms of that individual's reaction) should be identified by using the Family/Personal History form. Identification information should be worn at all times by individuals who have had previous anaphylactic reactions. For identified and unidentified clients, immediate action is required when the person has a severe reaction. In such situations, the therapist is advised to call for emergency assistance.

Type II Hypersensitivity (Cytolytic or Cytotoxic)

A type II hypersensitivity reaction is caused by the production of autoantibodies against self cells or tissues that have some form of foreign protein attached to them. The autoantibody binds to the altered self cell, and the complex is destroyed by the immune system. Typical examples of this type of hypersensitivity are hemolytic anemias, idiopathic thrombocytopenic purpura (ITP), hemolytic disease of the newborn, and transfusion of incompatible blood. Blood group incompatibility causes cell lysis, which results in a hemolytic transfusion reaction. The antigen responsible for

TABLE 12-1 Clinical Aspects of Anaphylaxis by System

System	Signs and Symptoms
General	Malaise, weakness
	Sense of illness
	Metallic taste
Integument	Hives, erythema
	Edema of the lips, tongue
Mucosal	Periorbital edema
	Nasal congestion and pruritus
	Flushing or pallor, cyanosis
Respiratory	Sneezing
	Wheezing
	Dyspnea
Upper airway	Hoarseness
	Tongue and pharyngeal edema
Lower airway	Dyspnea
	Acute emphysema
	Air trapping: asthma, bronchospasm
	Chest tightness; wheezing
Gastrointestinal	Increased peristalsis
	Vomiting
	Dysphagia
	Nausea
	Abdominal cramps
	Metallic taste in mouth
	Diarrhea (occasionally with blood)
Cardiovascular	Tachycardia
	Palpitations
	Hypotension
	Cardiac arrest
Central nervous system	Anxiety, seizures, headache

Modified from Adkinson NF: *Middleton's allergy: principles and practice,* ed 7, St. Louis, 2008, Mosby.

initiating the reaction is a part of the donor red blood cell (RBC) membrane.

Manifestations of a transfusion reaction result from intravascular hemolysis of RBCs.

CLINICAL SIGNS AND SYMPTOMS

Type II Hypersensitivity

- Headache
- Back (flank) pain
- Chest pain similar to angina
- Nausea and vomiting
- Tachycardia and hypotension
- Hematuria
- Urticaria (skin reaction)

Type III Hypersensitivity (Immune Complex)

Immune complex disease results from formation or deposition of antigen-antibody complexes in tissues. For example, the antigen-antibody complexes may form in the joint space, with resultant synovitis, as in RA. Antigen-antibody complexes are formed in the bloodstream and become trapped in capillaries or are deposited in vessel walls, affecting the skin (urticaria), the kidneys (nephritis), the pleura (pleuritis), and the pericardium (pericarditis).

Serum sickness is another type III hypersensitivity response that develops 6 to 14 days after injection with foreign serum (e.g., penicillin, sulfonamides, streptomycin, thiouracils, hydantoin compounds). Deposition of complexes on vessel walls causes complement activation with resultant edema, fever, inflammation of blood vessels and joints, and urticaria.

CLINICAL SIGNS AND SYMPTOMS

Type III Hypersensitivity

- Fever
- Arthralgias; synovitis
- Lymphadenopathy
- Urticaria
- Visceral inflammation (nephritis, pleuritis, pericarditis)

Type IV Hypersensitivity (Cell-Mediated or Delayed)

In cell-mediated hypersensitivity, a reaction occurs 24 to 72 hours after exposure to an allergen.

For example, type IV reactions occur after the intradermal injections of TB antigen. Graft-versus-host disease (GVHD) and transplant rejection are also type IV reactions. In GVHD, immunocompetent donor bone marrow cells (the graft) react against various antigens in the bone marrow recipient (the host), which results in a variety of clinical manifestations, including skin, GI, and hepatic lesions.

Contact dermatitis is another type IV reaction that occurs after sensitization to an allergen, commonly a cosmetic, adhesive, topical medication, drug additive (e.g., lanolin added to lotions, ultrasound gels, or other preparations used in massage or soft tissue mobilization), or plant toxin (e.g., poison ivy).

With the first exposure, no reaction occurs; however, antigens are formed. On subsequent exposures, hypersensitivity reactions are triggered, which leads to itching, erythema, and vesicular lesions. Anyone with known hypersensitivity (identified through the Family/Personal History form) should have a small area of skin tested before use of large amounts of topical agents in the physical therapy clinic. Careful observation throughout the episode of care is required.

CLINICAL SIGNS AND SYMPTOMS

Type IV Hypersensitivity

- Itching
- Erythema
- Vesicular skin lesions
- Graft-versus-host disease (GVHD): skin, GI, hepatic dysfunction

Autoimmune Disorders

Autoimmune disorders occur when the immune system fails to distinguish self from nonself and misdirects the immune response against the body's own tissues. The body begins to manufacture antibodies called *autoantibodies* directed against the body's own cellular components or specific organs. The resultant abnormal tissue reaction and tissue damage may cause systemic manifestations varying from minimal localized symptoms to systemic multiorgan involvement with severe impairment of function and life-threatening organ failure.

The exact cause of autoimmune diseases is not understood, but factors implicated in the development of autoimmune immunologic abnormalities may include genetics (familial tendency), sex hormones (women are affected more often than men by autoimmune diseases), viruses, stress, cross-reactive antibodies, altered antigens, or the environment.

Autoimmune disorders may be classified as organ-specific diseases or generalized (systemic) diseases. Organ-specific diseases involve autoimmune reactions limited to one organ. *Organ-specific autoimmune diseases* include thyroiditis, Addison's disease, Graves' disease, chronic active hepatitis, pernicious anemia, ulcerative colitis, and insulin-dependent diabetes. These diseases have been discussed in this text (see the chapter appropriate to the organ involved) and are not covered further in this chapter.

Generalized autoimmune diseases involve reactions in various body organs and tissues (e.g., fibromyalgia, RA, SLE, and scleroderma). Systemic autoimmune diseases lead to a sequence of abnormal tissue reaction and damage to tissue that may result in diffuse systemic manifestations.

Fibromyalgia Syndrome

Fibromyalgia syndrome (FMS) is a noninflammatory condition appearing with generalized musculoskeletal pain in conjunction with tenderness to touch in a large number of specific areas of the body and a wide array of associated symptoms. FMS is much more common in women than in men; it is 2 to 5 times more common than RA. It occurs in age groups from preadolescents to early postmenopausal women.[26] The condition is less common in older adults.

There is still much controversy over the exact nature of FMS and even debate over whether fibromyalgia is an organic disease with abnormal biochemical or immunologic pathologic aspects. Some theories suggest that it is a genetically predisposed condition with dysregulation of the neurohormonal and autonomic nervous systems.[27] It may be triggered by viral infection, a traumatic event, or stress. The role of inadequate thyroid hormone regulation as a main mechanism of fibromyalgia has been proposed and is under investigation.[28]

More recent research suggests that chronic pain of this type is *centrally mediated* since most of these individuals experience pain with input or stimuli that are not usually painful. The problem is with pain or sensory processing, rather than some disease, inflammation, or impairment of the area that actually hurts (e.g., the back, the hips, the wrists).

Functional brain imaging shows areas of the brain that light up when pressure is applied to painful areas of the body. All indications are that once the central pain mechanisms get turned on, they "wind up" until there is pain even when the stimulus (e.g., pressure, heat, cold, electrical impulses) is no longer there. This phenomenon is called *sensory augmentation*. There is some evidence that people with fibromyalgia have a decrease in their reactivity threshold. In other words, with a low threshold, it only takes a small amount of stimuli before the pain switch gets turned on. Exactly why this happens remains unknown.[27]

Controversy also existed regarding use of the American College of Rheumatism (ACR) criteria for tender point count in clinical diagnosis of FMS.[29] In fact, the original author of the ACR criteria suggested that counting the tender points was "perhaps a mistake" and advised against using it in clinical practice.[30] A Symptom Intensity Scale was subsequently developed and validated by Wolfe[31,32] to help differentiate FMS from other rheumatologic conditions (e.g., SLE or polymyalgia rheumatica) with similar widespread pain.

Since that time, Wolfe and associates proposed a new set of diagnostic criteria, which the ACR has adopted.[33] The new tool focuses on measuring symptom severity rather than relying on the tender point examination. The new criteria use a clinician-queried checklist of painful sites and a symptom severity scale that focuses on fatigue, cognitive dysfunction, and sleep disturbance. Tender point assessment still has value; people with fewer than 11 of the 18 tender points included in the ACR classification criteria may still be diagnosed with fibromyalgia if they have other clinical features consistent with fibromyalgia.[34]

Fibromyalgia has been differentiated from myofascial pain in that FMS is considered a systemic problem with multiple tender points as one of the key symptoms; there is usually a cluster of associated signs and symptoms. Myofascial pain is a localized condition specific to a muscle (trigger point [TrP]) and may involve as few as one or several areas without associated signs and symptoms.

The hallmark of myofascial pain syndrome is the TrP, as opposed to tender points in FMS. Both disorders cause myalgia with aching pain and tenderness and exhibit similar local histologic changes in the muscle. Painful symptoms in both conditions are increased with activity, although fibromyalgia involves more generalized aching, whereas myofascial pain is more direct and localized (Table 12-2).

There is some clinical evidence that myofascial pain syndrome is linked with the use of hormones containing synthetic progestin (e.g., birth control pills). The effects of progestin on women with fibromyalgia is unknown.[35]

FMS has striking similarities to chronic fatigue syndrome (CFS), with a mix of overlapping symptoms (about 70%) that have some common biologic denominator. Diagnostic criteria for CFS focus on fatigue, whereas the criteria for FMS focus on pain, the two most prominent symptoms of these syndromes. Studies have shown that CFS and FMS are characterized by greater similarities than differences and both involve the central and peripheral nervous systems as well as the body tissues themselves (Box 12-2).[36]

TABLE 12-2 Differentiating Myofascial Pain Syndrome from Fibromyalgia Syndrome

Myofascial Pain Syndrome	Fibromyalgia Syndrome
Trigger points (pain with deep pressure); often radiates locally	Tender points (pain with light touch); no radiation
Localized musculoskeletal condition	Systemic condition
Palpable taut band found in muscle; no associated signs and symptoms	No palpable or visible local abnormality; wide array of associated signs and symptoms
Etiology: Overuse, repetitive motions; reduced muscle activity (e.g., casting or prolonged splinting); hormones containing progestins (under investigation)[35]	Etiology: Neurohormonal imbalance; autonomic nervous system dysfunction
Risk factors: Immobilization, repetitive use	Risk factors: Trauma, psychosocial stress, mood (or other psychologic) disorders; other medical conditions
Pathophysiology: Unknown, possibly muscle spindle dysfunction	Pathophysiology: Sensitization of spinal neurons from excitatory nerve messenger substances
Prognosis: Excellent	Prognosis: Good with early diagnosis and intervention, variable with delayed diagnosis; often a chronic condition

Data from Lowe JC, Yellin JG: *The metabolic treatment of fibromyalgia*, Utica, Kentucky, 2000, McDowell Publications.

BOX 12-2 FIBROMYALGIA WEB LINKS

- **Fibromyalgia Network** provides educational materials on fibromyalgia syndrome (FMS) and chronic fatigue syndrome (CFS) and is a very reliable source of information.
  http://www.fmnetnews.com
- **American Fibromyalgia Syndrome Association,** a nonprofit organization, is dedicated to research, education and patient advocacy for FMS and CFS.
  http://www.afsafund.org
- **National Fibromyalgia Association,** a nonprofit organization, increases fibromyalgia awareness and improves treatment options.
  http://www.fmaware.org
- **Fibromyalgia Research Foundation** emphasizes the metabolic basis of fibromyalgia and its treatment.
  http://www.drlowe.com
- **Phoenix Publishing,** website of physical therapist Janet A. Hulme, MA, PT, includes personal care kits for the management of chronic pain/fibromyalgia. A self-care and treatment manual written by a physical therapist is available to help clients identify which one (or more) subtype they may be and offers specific treatment suggestions based on the subtypes, including modulating the autonomic nervous system through a process called Physiologic Quieting.
  http://www.phoenixpub.com

Risk Factors. Numerous studies have implicated a genetic predisposition related to brain and/or body chemistry, but it has also been shown that a history of childhood trauma, family issues, and/or physical/sexual abuse are significant risk factors.[37] Stress, illness, disease, or anything the body perceives as a threat are risk factors for those who develop FMS. But why one person develops this condition, whereas others with equal or worse situations do not, remains a mystery. Anxiety, depression, and posttraumatic stress disorder also seem to be linked with FMS.[38] Having a bipolar illness increases the risk of developing FMS dramatically.[39]

Clinical Signs and Symptoms. The core features of FMS include widespread pain lasting more than 3 months and widespread local tenderness in all clients (Fig. 12-3). Primary musculoskeletal symptoms most frequently reported are (1) aches and pains, (2) stiffness, (3) swelling in soft tissue, (4) tender points, and (5) muscle spasms or nodules. Fatigue, morning stiffness, and sleep disturbance with nonrefreshed awakening may be present but are not necessary for the diagnosis.[40]

Nontender control points (such as midforehead and anterior thigh) have been included in the examination by some clinicians. These control points may be useful in distinguishing FMS from a conversion reaction, referred to as *psychogenic rheumatism,* in which tenderness may be present everywhere. However, evidence suggests that individuals with FMS may have a generalized lowered threshold for pain

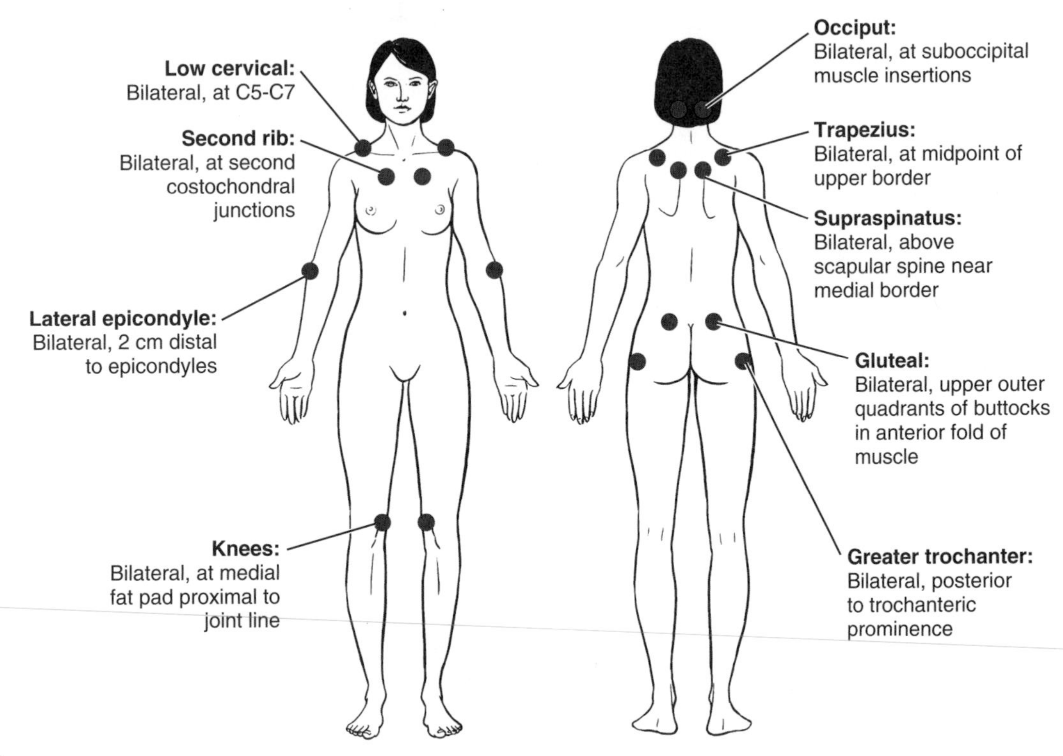

Fig. 12-3 Anatomic locations of tender points associated with fibromyalgia. According to the literature, digital palpation should be performed with an approximate force of 4 kg (enough pressure to indent a tennis ball), but clinical practice suggests much less pressure is required to elicit a painful response. For a tender point to be considered positive, the subject must state that the palpation was "painful." A reply of "tender" is not considered a positive response. As mentioned in the text, counting the number of points as part of the clinical diagnosis of fibromyalgia syndrome (FMS) has been discounted; however, the presence of multiple tender points is still a key feature of FMS.

on palpation and the control points may also be tender on occasion. There is also an increased sensitivity to sensory stimulation such as pressure stimuli, heat, noise, odors, and bright lights.[41]

Symptoms are aggravated by cold, stress, excessive or no exercise, and physical activity ("overdoing it"), including overstretching, and may be improved by warmth or heat, rest, and exercise, including gentle stretching. Smoking has been linked with increased pain intensity and more severe fibromyalgia symptoms, but not necessarily a higher number of tender points. Exposure to tobacco products may be a risk factor for the development of fibromyalgia, but this has not been investigated fully or proven.[42]

Sleep disturbances in stage 4 of nonrapid eye movement sleep (needed for healing of muscle tissues), sleep apnea, difficulty getting to sleep or staying asleep, nocturnal myoclonus (involuntary arm and leg jerks), and bruxism (teeth grinding) cause clients with FMS to wake up, feeling unrested or unrefreshed, as if they had never gone to sleep (Box 12-3).

BOX 12-3 FIBROMYALGIA SYNDROME (FMS) SCREEN

Do you have trouble sleeping through the night?	YES	NO
Do you feel rested in the morning?	YES	NO
Are you stiff and sore in the morning?	YES	NO
Do you have daytime fatigue/exhaustion?	YES	NO
Can you do the grocery shopping on your own?	YES	NO
Can you do your regular daily activities?	YES	NO
Do your muscle pain and soreness travel (move around the body)?	YES	NO
Do you have tension/migraine headaches?	YES	NO
Do you have irritable bowel symptoms (e.g., nausea, diarrhea, stomach cramping)?	YES	NO
Do you have swelling, numbness, or tingling in your arms or legs?	YES	NO
Are you sensitive to temperature and humidity or changes in the weather?	YES	NO
Can you read a book or watch a movie and follow what is happening?	YES	NO
Does "brain fog" interfere with your activities or work?	YES	NO

Key: Researchers have been unable to develop a reliable screening questionnaire for FMS because of the wide-ranging symptoms associated with this condition. This type of screening tool may help the therapist identify potential cases of FMS but should not be relied upon as the only evaluation instrument. See also the Symptom Intensity Scale.[31,32]

CLINICAL SIGNS AND SYMPTOMS

Fibromyalgia Syndrome

- Myalgia (generalized aching)
- Fatigue (mental and physical)
- Sleep disturbances, nocturnal myoclonus, nocturnal bruxism
- Restless legs syndrome
- Tender points of palpation
- Chest wall pain mimicking angina pectoris
- Tendinitis, bursitis
- Temperature dysregulation
  - Raynaud's phenomenon; cold-induced vasospasm (hypersensitivity to cold)
  - Hypothermia (mild decrease in core body temperature)
- Dyspnea, dizziness, syncope
- Headache (throbbing occipital pain)
- Morning stiffness (more than 15 minutes)
- Paresthesia (numbness and tingling)
- Mechanical low back pain with sciatica-like radiation of pain
- Subjective swelling
- Irritable bowel symptoms
- Urinary urgency; irritable bladder syndrome
- Dry eyes/mouth (Sicca syndrome)
- Depression/anxiety
- Cognitive difficulties (e.g., short-term memory loss, decreased attention span; sometimes referred to as *fibro fog*)
- Premenstrual syndrome (PMS)
- Weight gain from physical inactivity because of pain and fatigue

DIFFERENTIAL DIAGNOSIS

Fibromyalgia

Frequently misdiagnosed, fibromyalgia syndrome (FMS) is often confused with any of the following:

- Hypothyroidism, hyperparathyroidism
- Adult growth hormone deficiency (e.g., pituitary tumors, head trauma, AIDS)
- Polymyalgia rheumatica/giant cell arteritis
- Rheumatoid arthritis, seronegative
- Polymyositis/dermatomyositis
- Systemic lupus erythematosus
- Multiple sclerosis
- Myofascial pain syndrome
- Metabolic myopathy (e.g., alcohol)
- Lyme disease
- Neurosis (depression/anxiety)
- Metastatic cancer
- Chronic fatigue syndrome
- Temporomandibular joint dysfunction
- Disk disease
- Myalgic side effect of medication (e.g., statins)
- Parasitic infection
- Depression, anxiety

Researchers are beginning to identify various subtypes of fibromyalgia and recognize the need for specific intervention based on the underlying subtype. These classifications are based on impairment of the autonomic nervous system. They include the following[35,43]:

- Hypoglycemia
- Hypothyroid
- Neurally mediated
- Immune system
- Reproductive hormone imbalance

A multidisciplinary or interdisciplinary team approach to this condition requires medical evaluation and treatment as part of the intervention strategy for fibromyalgia. Therapists should refer clients suspected of having fibromyalgia for further medical follow-up.

Rheumatoid Arthritis

Rheumatoid arthritis (RA) is a chronic, systemic, inflammatory disorder of unknown cause that can affect various organs but predominantly involves the synovial tissues of the diarthrodial joints. There are more than 100 rheumatic diseases affecting joints, muscles, and extraarticular systems of the body.

Women are affected with RA 2 to 3 times more often than men; however, women who are taking or have taken oral contraceptives are less likely to develop RA. Although it may occur at any age, RA is most common in persons between the ages of 20 and 40 years.

Risk Factors. The etiologic factor or trigger for this process is as yet unknown. Support for a genetic predisposition comes from studies suggesting that RA clusters in families. One gene in particular (HLA-DRB1 on chromosome 6) has been identified in determining susceptibility. RA may be caused by a genetically susceptible person encountering an unidentified agent (e.g., virus, self-antigen), which then results in an immunopathologic response.[44] Researchers hypothesize that an infection could trigger an immune reaction that is mediated through multiple complex genetic mechanisms and continues clinically even if the organism is eradicated from the body.

Other nongenetic factors may also contribute to the development of RA. Because arthritis (and many related diseases) is more common in women, hormones have been implicated, but the relationship remains unclear. Environmental and occupational causes, such as chemicals (e.g., hair dyes, industrial pollutants), minerals, mineral oil, organic solvents, silica, toxins, medications, food allergies, cigarette smoking, and stress, remain under investigation as possible triggers for those individuals who are genetically susceptible to RA.[45] Many systemic disorders can express themselves through the musculoskeletal system often presenting first with rheumatic manifestations (Box 12-4).[46]

Clinical Signs and Symptoms. Clinical features of RA vary not only from person to person but also in an individual over the disease course. In most people, the symptoms begin gradually during a period of weeks or months. Frequently, malaise and fatigue prevail during this period, sometimes accompanied by diffuse musculoskeletal pain. The multidimensional aspects of rheumatoid arthritic pain can be assessed quantitatively using the Rheumatoid Arthritis Pain Scale (RAPS).[47] The complete RAPS is available in the appendix of this article.

BOX 12-4 SYSTEMIC CONDITIONS ASSOCIATED WITH RHEUMATIC MANIFESTATIONS

Malignant Disorders
Hypertrophic osteoarthropathy
Lymphoma
Leukemia
Carcinoma polyarthritis

Hematologic Disorders
Hemophilia
Sickle cell disease
Thalassemia
Multiple myeloma
Amyloidosis

Gastrointestinal Disorders
Spondyloarthropathies
Whipple disease
Hemochromatosis
Primary biliary cirrhosis

Endocrinopathies
Diabetes
Hypothyroidism
Hyperthyroidism
Hyperparathyroidism
Acromegaly

From Andreoli TE, Benjamin I, Griggs RC, et al: *Andreoli and Carpenter's Cecil essentials of medicine,* ed 8, Philadelphia, 2010, WB Saunders.

Symptoms of an inflammatory arthritis include the spontaneous onset of one or more swollen joints; morning stiffness lasting longer than 45 minutes; and diffuse joint pain and tenderness, particularly involving the metatarsophalangeal (MTP) or metacarpophalangeal (MCP) joints. Inactivity, such as sleep or prolonged sitting, is commonly followed by stiffness. "Morning" stiffness occurs when the person arises in the morning or after prolonged inactivity. The duration of this stiffness is an accepted measure of the severity of the condition.

Clients should be asked: "After you get up in the morning, how long does it take until you are feeling the best you will feel for the day?" Pain and stiffness increase gradually as RA progresses and may limit a person's ability to walk, climb stairs, open doors, or perform other activities of daily living (ADLs). Weight loss, depression, and low-grade fever can accompany this process.

The inflammatory process may be under way for some time before swelling, tissue reaction, and joint destruction are seen. Structural damage usually begins between the first and second year of the disease. Early medical referral, followed by expedited diagnosis and intervention, results in a much more favorable outcome for persons with RA.

Studies have shown that 70% to 90% of persons with RA have significant joint erosions on x-ray by only 2 years after disease onset and that halting or slowing erosions should be initiated very early on in the course of the disease.[48] Having awareness of the group of symptoms that suggest inflammatory arthritis is critical. It is recommended that the criteria for referral of a person with early inflammatory symptoms include significant discomfort on the compression of the metacarpal and metatarsal joints, the presence of three or more swollen joints, and more than 1 hour of morning stiffness.[49]

Shoulder. Chronic synovitis of the elbows, shoulders, hips, knees, and/or ankles creates special secondary disorders. When the shoulder is involved, limitation of shoulder mobility, dislocation, and spontaneous tears of the rotator cuff result in chronic pain and adhesive capsulitis.

Elbow. Destruction of the elbow articulations can lead to flexion contracture, loss of supination and pronation, and subluxation. Compressive ulnar nerve neuropathies may develop related to elbow synovitis. Symptoms include paresthesias of the fourth and fifth fingers and weakness in the flexor muscle of the little finger.

Wrists. The joints of the wrist are frequently affected in RA, with variable tenosynovitis of the dorsa of the wrists and, ultimately, interosseous muscle atrophy and diminished movement owing to articular destruction or bony ankylosis. Volar synovitis can lead to carpal tunnel syndrome.

Hands and Feet. Forefoot pain may be the only small-joint complaint and is often the first one. Subluxation of the heads of the MTP joints and shortening of the extensor tendons give rise to "hammer toe" or "cock up" deformities. A similar process in the hands results in volar subluxation of the MCP joints and ulnar deviation of the fingers. An exaggerated inflammatory response of an extensor tendon can result in a spontaneous, often asymptomatic rupture. Hyperextension of a proximal interphalangeal (PIP) joint and flexion of the distal interphalangeal (DIP) joint produce a swan neck deformity. The boutonnière deformity is a fixed flexion contracture of a PIP joint and extension of a DIP joint.

Cervical Spine. Involvement of the cervical spine by RA tends to occur late in more advanced disease. Clinical manifestations of early disease consist primarily of neck stiffness that is perceived through the entire arc of motion. Inflammation of the supporting ligaments of C1-C2 eventually produces laxity, sometimes giving rise to atlantoaxial subluxation. Spinal cord compression can result from anterior dislocation of C1 or from vertical subluxation of the odontoid process of C2 into the foramen magnum.

Extraarticular. Extraarticular features, such as rheumatoid nodules, arteritis, neuropathy, scleritis, pericarditis, lymphadenopathy, and splenomegaly, occur with considerable frequency (Table 12-3). Once thought to be complications of RA, they are now recognized as being integral parts of the disease and serve to emphasize its systemic nature.

The subcutaneous nodules, present in approximately 25% to 35% of clients with RA, occur most commonly in subcutaneous or deeper connective tissues in areas subjected to repeated mechanical pressure such as the olecranon bursae, the extensor surfaces of the forearms, the elbow, and the Achilles tendons.

TABLE 12-3 Extraarticular Manifestations of Rheumatoid Arthritis

Organ System	Extraarticular Manifestations
Skin	Cutaneous vasculitis Rheumatoid nodules Ecchymoses/petechiae (drug-induced)
Eye	Episcleritis Scleritis Scleromalacia perforans Corneal ulcers/perforation Uveitis Retinitis Glaucoma Cataract
Lung	Pleuritis Diffuse interstitial fibrosis Vasculitis Rheumatoid nodules Caplan's syndrome Pulmonary hypertension
Heart and blood vessels	Pericarditis Myocarditis Coronary arteritis Valvular insufficiency Conduction defects Vasculitis Felty's syndrome
Nervous system	Mononeuritis multiplex Distal sensory neuropathy Cervical spine instability (spinal cord compression)

Modified from Andreoli TE, Benjamin I, Griggs RC, et al: *Andreoli and Carpenter's Cecil essentials of medicine,* ed 8, Philadelphia, 2010, Saunders.

CLINICAL SIGNS AND SYMPTOMS

Rheumatoid Arthritis

- **S**welling in one or more joints
- **E**arly morning stiffness
- **R**ecurring pain or tenderness in any joint
- **I**nability to move a joint normally
- **O**bvious redness and warmth in a joint
- **U**nexplained weight loss, fever, or weakness combined with joint pain
- **S**ymptoms such as these that last for more than 2 weeks
- See also Table 12-3

Age-Related Differences. One-third of persons with RA acquire the disease after the age of 60 years. There are differences in presentation of the disease in older versus younger people. Onset in younger people is usually in 30- to 50-year-olds, with a 2:1 ratio of women to men; polyarticular, involving small joints; and gradual in onset with a positive RF. In the elderly population (age 60 years or older), joint involvement may be oligoarticular, involving large joints, and more abrupt in onset, with an equal ratio of women to men and a negative RF finding.[50]

Juvenile Idiopathic Arthritis. Juvenile idiopathic arthritis (JIA) replaces the term juvenile rheumatoid arthritis (JRA). JIA is a chronic inflammatory disorder that occurs during childhood and is made up of a heterogeneous group of diseases that share synovitis as a common feature. JIA has seven subcategories:

- Oligoarthritis JIA
- Polyarthritis JIA (positive RF)
- Polyarthritis JIA (negative RF)
- Systemic onset JIA
- Psoriatic JIA
- Enthesitis-related arthritis
- Other arthritis

Early recognition of JIA is key to timely initiation of treatment. For a diagnosis of JIA to be made, objective arthritis must be seen in one or more joints for at least 6 weeks in children younger than 16 years. Children should be screened for an array of symptoms, depending on the appropriate subcategory of disease. The number of joints involved, involvement of small joints, symmetry of joint involvement, uveitis risk, systemic features, and family history are important parts of this screening. It is important to educate parents regarding symptoms because parents are the first line of communication from their children to health care professionals.[51]

Diagnosis. The clinical diagnosis of RA is based on careful consideration of three factors: the clinical presentation of the client, which is elucidated through history taking and physical examination; the corroborating evidence gathered through laboratory tests and radiography; and the exclusion of other possible diagnoses.

The physical presence of rheumatoid nodules and the presence of RF measured by laboratory studies are two indicators of RA, although some persons with actual RFs are missed by commonly available methods.

Classification of RA (Table 12-4) is difficult in the early course of the disease, when articular symptoms are accompanied only by constitutional symptoms such as fatigue and loss of appetite, which are common to a number of chronic diseases. A full array of clinical signs and symptoms may not be manifest for 1 to 2 years. A diagnosis of RA is established on the presentation of 4 of the 7 listed criteria with duration of joint signs and symptoms for at least 6 weeks.

Additional laboratory tests of significance in the diagnosis and management of RA include white blood cell (WBC) count, erythrocyte sedimentation rate (ESR), hemoglobin and hematocrit, urinalysis, and RF assay. The elevation of C-reactive protein (CRP) has been discovered in significant levels within 2 years preceding a confirmed diagnosis of RA. The increased levels of CRP signal the kind of steady, low-grade inflammation that may be an early predictor of later symptomatic inflammation.[52] CRP as a predictive factor remains a topic of debate and under continued investigation.

The number of WBCs will increase in the presence of joint inflammation, as will the erythrocyte sedimentation rate. Anemia may be present, and the RF will be elevated in clients with active RA. If the client's urinalysis reveals any protein, blood cells, or casts, SLE should be suspected. This type of abnormal urinalysis would necessitate further diagnostic evaluation and immediate physician referral (Case Example 12-1).

Treatment. Aggressive treatment of early arthritis with new medications has shown marked improvement in outcome.[53] These medications have both immunosuppressive and biologic side effects that must be monitored. The emergence of therapeutic biologic agents, such as anti-tumor necrosis factor (TNF) and biologic response modifiers called disease-modifying antirheumatic drugs (DMARDs), has led to suppression of symptoms of RA and slowed disease progression. Side effects and safety concerns are critical in the

TABLE 12-4 American Rheumatism Association Criteria for Classification of Rheumatoid Arthritis*

Criteria	Definition
Morning stiffness	Morning stiffness in and around the joints lasting at least 1 hour
Arthritis of three or more joint areas	Simultaneous soft tissue swelling or fluid (not bony overgrowth alone) observed by a physician; the 14 possible joint areas are (right or left): PIP, MCP, wrist, elbow, knee, ankle, and MTP joints
Arthritis of hand joints	At least one joint area swollen as above in wrist, MCP, or PIP joint
Symmetric arthritis	Simultaneous bilateral involvement of the same joint areas as above (PIP, MCP, or MTP joints without absolute symmetry is acceptable)
Rheumatoid nodules	Subcutaneous nodules, over bony prominences or extensor surfaces
Serum rheumatoid factor	Abnormal amounts of serum rheumatoid factor
Radiographic changes	Radiographic changes typical of rheumatoid arthritis on posteroanterior hand and wrist radiographs, which must include erosions or bony decalcification localized to involved joints (osteoarthritis changes alone do not qualify)

Modified from Harris ED, Jr: Clinical features of rheumatoid arthritis. In Kelley WN, et al: *Textbook of rheumatology,* ed 4, Philadelphia, 1993, Saunders, p 874.

*For classification purposes, a client is said to have rheumatoid arthritis (RA) if he or she has satisfied at least four of the above seven criteria. Criteria 1 through 4 must be present for at least 6 weeks. Clients with two clinical diagnoses are not excluded. Designation as classic, definite, or probable RA is no longer made.

MCP, Metacarpophalangeal; *MTP,* metatarsophalangeal; *PIP,* proximal interphalangeal.

CASE EXAMPLE 12-1

Rheumatoid Arthritis

History: A 67-year-old woman with a 13-year history of rheumatoid arthritis (RA) requiring gold and methotrexate fell and fractured her right acetabulum, requiring a total hip replacement. She was referred to physical therapy through a home health agency for "aggressive rehabilitation." After 10 weeks, she was walking unassisted after having progressed from a walker to a cane and participating in a swimming program sponsored by the local arthritis organization. She was discharged with a home program to continue working on strength and balance activities.

About 6 weeks later, the therapist received a telephone call from the client's husband, who reported that there had been a gradual decline in her walking and asked the therapist for a reevaluation. The woman came into the outpatient clinic and was examined with the following findings.

Clinical Presentation: The client had resumed the use of a cane, and her gait was characterized by wide-based stance, shortened steps, and trunk instability. She frequently took a few steps forward before tottering backward without falling. The client was unable to stand from a sitting position without assistance. When moving from a standing to a sitting position, she consistently fell backward.

When asked about the new onset of any other symptoms, the client noted urinary urge incontinence, and her husband commented that she had just started having difficulty remembering the dates of their children's birthdays and the names of their grandchildren.

The therapist performed both an orthopedic and a neurologic screening examination and measured vital signs. The client's blood pressure was 135/78 mm Hg, resting pulse was 78 bpm, and body temperature was considered normal. The orthopedic examination was consistent with a total hip replacement 6 months ago, with mild hip flexor weakness and mild loss of hip motion on the left (compared with the right). However, the neurologic examination raised some red flags.

Muscle tone was increased in the lower extremities, with proprioception and deep tendon reflexes decreased in both feet (right more than left). Pinprick, light-touch, and two-point discrimination were normal. Romberg's sign was absent, but a test for dysmetria of the upper extremities revealed mild cogwheeling. There was an observable tremor when the client's arms were stretched out in front of her trunk.

Result: Given the history of a progressive gait disturbance, new onset of urge incontinence, and positive findings on a neurologic screening examination, this client was referred to her family physician for a medical evaluation. The therapist explained to the client and her husband that these findings were not typical of someone who has had a total hip replacement or someone with RA.

The client was examined by her family physician and referred to a neurologist. A magnetic resonance imaging (MRI) study was ordered, and a diagnosis of basilar impression was made. *Basilar impression* is the term used when the odontoid peg of C2 pushes up into the foramen magnum.

RA is a classic cause of this via atlantoaxial dislocation. The destructive inflammatory process of RA weakens ligaments that attach the odontoid to the atlas into the skull. The subsequent dislocation of the atlas on the axis can remain mobile, producing intermittent problems, or it can become fixed, producing persistent symptoms.

Modified from Williams ME, Richman J, Scatliff J: A 67-year-old woman with a progressive gait disturbance, *J Am Geriatr Soc* 44(7):843-846, 1996.

prevention of serious side effect–related problems, including the following:

- Congestive heart failure (CHF)
- Serious infections due to immunosuppression (e.g., TB, *Listeria monocytogenes,* coccidioidomycosis, histoplasmosis, viral hepatitis)
- Skin reactions (erythema, pruritus, rashes, urticaria, infection, eczema)[54]
- New onset symptoms of MS, optic neuritis, and transverse myelitis
- Hematologic abnormalities such as aplastic anemia, pancytopenia
- Increased risk of lymphomas

Screening should include a thorough epidemiologic history regarding previous exposure to TB.[55]

Polymyalgia Rheumatica

Polymyalgia rheumatica (PMR) is a systemic rheumatic inflammatory disorder with an unknown cause; there may be an autoimmune, viral, or stress-induced mechanism.

Risk Factors. PMR occurs almost exclusively in people over 55 years of age. The disease rarely occurs in persons under the age of 50, and it affects twice as many women as men. It is at least 10 times more prevalent in persons over 80 than in persons between the ages of 50 and 59, and it predominantly affects the Caucasian population.[56,57]

Clinical Presentation. PMR is characterized by severe aching and stiffness primarily in the muscles, as opposed to the joints. Onset is usually very sudden and insidious. Areas commonly affected include the neck, shoulder girdle, and pelvic girdle. Joint pain is possible, and headache, weakness, and fatigue are commonly reported. The pain is usually more severe when the person gets up in the morning.

PMR is closely linked with giant cell or temporal arteritis. Giant cell arteritis (GCA) primarily affects the medium-sized muscular arteries, such as the cranial and extracranial branches of the carotid artery that pass over the temples in the scalp.[58] The temporal arteries become inflamed, subjecting them to damage. In GCA, the most common symptom is a severe headache on one or both sides of the head.

Orofacial symptoms of pain, jaw claudication, hard end-feel limitation of jaw range of motion, and temporal headache can be misdiagnosed as a temporomandibular joint (TMJ) disorder.[59] A careful evaluation by the physical therapist can often pinpoint enough differences to recognize the need for referral when the clinical presentation and outcomes of treatment do not match expectations for TMJ impairment.

The ophthalmic arteries are affected in nearly half of all affected individuals, sometimes resulting in partial loss of vision or even sudden blindness. Early diagnosis of GCA is important to prevent blindness. The diagnosis of polymyalgia rheumatica is made on the basis of age, clinical presentation, and a very high ESR. The ESR measures the total inflammation in the body (Case Example 12-2).

CLINICAL SIGNS AND SYMPTOMS

Polymyalgia Rheumatica

- Muscle pain or aching (proximal muscle groups: neck; shoulder and pelvic girdle)
- Stiffness upon arising in the morning or after rest
- Weakness, fatigue, malaise
- Low-grade fever, sweats
- Headache (temporal arteritis)
- Weight loss
- Depression
- Vision changes

PMR is self-limiting, typically lasting 2 to 3 years. In some cases, it goes away for reasons unknown. However, some persons may have a longer course of disease, requiring low-dose steroids for much longer; a few have PMR for less than a year. Oral corticosteroids (especially prednisone) used to suppress the inflammation, treat the symptoms, and provide remission but do not cure the illness.[60,61] These drugs usually afford prompt relief of symptoms, providing further diagnostic confirmation that PMR is the underlying problem.[56] The therapist must remain alert to the possibility of steroid-induced osteopenia and diabetes.

With medical intervention, most people with PMR (with or without GCA) do not have lasting disability. However, in GCA, if one or both eyes develop blindness before treatment becomes effective, the blindness may be permanent.

Systemic Lupus Erythematosus

Systemic lupus erythematosus (SLE) belongs to the family of autoimmune rheumatic diseases. It is known to be a chronic, systemic, inflammatory disease characterized by injury to the skin, joints, kidneys, heart and blood-forming organs, nervous system, and mucous membranes.

Lupus comes from the Latin word for wolf, referring to the belief in the 1800s that the rash of this disease was caused by a wolf bite. The characteristic rash of lupus (especially a butterfly rash across the cheeks and nose) is red, leading to the term *erythematosus.*

There are two primary forms of lupus: discoid and systemic. *Discoid lupus* is a limited form of the disease confined to the skin presenting as coin-shaped lesions, which are raised and scaly (see Fig. 4-17). Discoid lupus rarely develops into systemic lupus. Individuals who develop the systemic form probably had systemic lupus at the outset, with the discoid lesions as the main symptom.

Systemic lupus is usually more severe than discoid lupus and can affect almost any organ or system of the body. For some people, only the skin and joints will be involved. In others, the joints, lungs, kidneys, blood, or other organs or tissues may be affected.[62]

Risk Factors. The exact cause of SLE is unknown, although it appears to result from an immunoregulatory disturbance brought about by the interplay of genetic, hormonal, chemical, and environmental factors.

Some of the environmental factors that may trigger the disease are infections (e.g., Epstein-Barr virus), antibiotics (especially those in the sulfa and penicillin groups) and other medications, exposure to ultraviolet (sun) light, and extreme physical and emotional stress, including pregnancy.

Although there is a known genetic predisposition, no known gene is associated with SLE. Lupus can occur at any age, but it is most common in persons between the ages of 15 and 40 years; it rarely occurs in older people. Women are affected 10 to 15 times more often than men, possibly because of hormones, but the exact relationship remains unknown.

SLE is more common in African-American, African-Caribbean, Hispanic-American, American-Indian, and Asian persons than in the Caucasian population.[63] Other risk factors among African-American women include early tobacco use (before age 19) but not alcohol intake.[64]

Clinical Signs and Symptoms. There is no single characteristic clinical pattern of symptoms. Clients may differ dramatically in the relative severity and pattern of organ involvement. SLE can appear in one organ or many. Common organ involvement includes cutaneous lupus, polyarthritis, nephritis, and hematologic lupus.[65] Although these symptoms may not be present at disease onset, most persons develop manifestations of multisystem disease.

Integumentary Changes. The classic butterfly rash associated with SLE often appears on the cheeks, bridge of the nose, forehead, chin, and V-area of the neck. Facial involvement is usually symmetric, and the nasolabial folds and upper eyelids are typically spared. Skin rash can also appear over the extensor surfaces of the arms, forearms, and hands/fingers. The hair may start to thin, causing an unkempt appearance referred to as *lupus hair.* Photosensitivity is common.[62]

Musculoskeletal Changes. Arthralgias and arthritis are the most common presenting manifestations of SLE. Acute migratory or persistent nonerosive arthritis may involve any joint, but typically the small joints of the hands, wrists, and knees are symmetrically involved. Lupus does not directly affect the spine, but syndromes, such as costochondritis and cervical myofascial syndrome associated with SLE, are commonly treated in a physical therapist practice.

One-fourth of all persons with lupus develop progressive musculoskeletal damage with deforming arthritis, osteoporosis with fracture and vertebral collapse, and osteomyelitis. Avascular necrosis has been detected in 6% of the adult SLE population (femoral head as the most common site of

CASE EXAMPLE 12-2

Polymyalgia Rheumatica

Current Complaint: The client was a 51-year-old female referred to physical therapy by her primary care physician for evaluation and treatment of persistent shoulder girdle pain and stiffness. The client reports that she had 1 or 2 days of fever and flu-like symptoms associated with neck stiffness. The problem did not go away after the fever passed, and the symptoms gradually worsened.

A month later, both shoulders were particularly worse in the morning when she was waking up. It usually took her 3 to 4 hours before she was able to function properly. Her best moment in the day was late afternoon. Symptoms were persisting and also waking her up at night when she was trying to change positions. She tried to exercise and take nonsteroidal antiinflammatory drugs (NSAIDs) in order to overcome the symptoms, but this did not help.

She saw her primary care physician for an annual physical but did not put much emphasis on these symptoms. She was given a prescription NSAID, which only minimally improved the symptoms. She again got in touch with her doctor, who referred her to physical therapy for evaluation of shoulder girdle pain.

Past Medical History: The client is postmenopausal and is no longer taking hormone replacement therapy (HRT). She has hypertension, which is controlled by medication. She had a cesarean section for delivery of twins in 1992 but no other surgeries. Family history reveals that her mother had Alzheimer's disease and passed away in December 2003. Her father had a stroke in 1991, which left him aphasic; he has hypertension and a history of coronary artery disease with coronary artery bypass graft surgery in 1975. There was no family history of cancer.

The client is married, has three children, and works full-time in health care. Her desired outcome is to decrease the pain and stiffness in the shoulder area and get back to her full activities.

Current medications included atenolol/chlorthalidone 50/25 daily for blood pressure management; potassium 10 mEq/L daily; aspirin 81 mg daily; calcium 1000 mg daily. She was started on 10 mg prednisone by her primary care physician. There are no known allergies.

Evaluation

Examination: Tests and Measures

Cardiovascular/pulmonary: On initial evaluation at 3 PM, the patient's blood pressure was 120/80 mm Hg and resting pulse was 68. The client routinely exercises using the treadmill and performs aerobic exercises three times per week. Lungs were clear to auscultation. There was no history of angina, dyspnea, or chronic cough. In the mornings, there is some pain when coughing and on deep breathing.

Pain: Using a Visual Analog Scale, from 0 (no pain) to 10 (worst pain imaginable), pain was rated as 8/10 in the shoulder girdle area.

Posture: Mild kyphosis and a mild forward head (possible sign of osteoporosis) but otherwise unremarkable.

Gait: Gait examination revealed an antalgic gait due to pain. She was not using any assistive devices or orthotics.

Range of motion (ROM) and strength: Full active and passive shoulder ROM with moderate stiffness/pain on movement. There were no gross deformities or evidence of impingement. Hands, wrists, and elbows present painless full ROM. Lower extremity evaluation revealed a prominent first left metatarsophalangeal (MTP) joint, but tibiotalar joints were fully mobile without pain. Knee and hip range of motion were normal on exam. There was no effusion or instability noted.

Neurologic exam: There was no focal or diffuse muscle strength deficit. Deep tendon reflexes were normal bilaterally. Cranial nerves and sensation were intact.

Work, community, and leisure: Client has reported she has had difficulty getting out of bed in the morning and difficulty with sleep because it is very painful when she changes positions at night. Lack of sleep has been affecting her work.

Red Flags

- Age: over 50
- Postmenopausal
- Persistent symptoms of shoulder girdle stiffness and pain for the last 3 months despite medication use
- The musculoskeletal examination of the upper quadrant is not consistent with rheumatoid arthritis (RA) or other musculoskeletal problems.
- No apparent red flags suggestive of cardiopulmonary involvement

Result: The therapist contacted the physician by phone to discuss findings and concerns. A written report was faxed to the physician's office prior to the phone call. The physician ordered blood tests that showed the following results:

	Client's value	Reference range (female)
WBC	7.7	4.5-11.0/mm^3
Hemoglobin	11.6	12-15 g/dL
Hematocrit	33.6	35%-47%
Platelets	410	150-400/mm^3
Sedimentation rate	71*	1-25 mm/hr
Rheumatoid factor	36.3	Less than 60 μ/mL (units per milliliter)

*Outside normal range.

Client was diagnosed with polymyalgia rheumatica (PMR) with a recommendation for continued physical therapy intervention. Because of her prednisone use and the fact that she is postmenopausal, attention to management of the potential for osteoporosis must be included in the treatment plan. Client was advised to have a baseline dual-energy x-ray absorptiometry (DEXA) scan to determine bone mineral density given her postmenopausal status and the fact that she is likely going to be on corticosteroids for some time.

Even though the client was seen by the physician before referral to physical therapy, there were enough red flags to warrant consultation before beginning physical therapy intervention. Early diagnosis can be important with PMR to prevent permanent disability, including visual loss when giant cell arteritis is present.

From Anita Bemis-Dougherty, PT, MAS. Case report presented in fulfillment of DPT 910, Institute for Physical Therapy Education, Widener University, Chester, PA, 2005. Used with permission.

involvement).[66] Often, these musculoskeletal complications occur as a result of the drugs necessary for treatment.

Approximately 30% of people with SLE have coexistent fibromyalgia, independent of race. Fibromyalgia is identified as a major contributor of pain and fatigue, but a medical differential diagnosis is required to rule out hypothyroidism, anemia, or pulmonary lupus (interstitial lung disease or pulmonary hypertension).

Peripheral Neuropathy. Peripheral neuropathy may be motor, sensory (stocking-glove distribution), or mixed motor and sensory polyneuropathy. These may develop subacutely in the lower extremities and progress to the upper extremities. Numbness on the tip of the tongue and inside the mouth is also a frequent complaint. Touch, vibration, and position sense are most prominently affected, and the distal limb reflexes are depressed (Case Example 12-3).

Neuropsychiatric Manifestations. Individuals with SLE are at increased risk of several neuropsychiatric manifestations sometimes referred to as *neurolupus.* Common (cumulative incidence >5%) manifestations include cerebrovascular disease (CVD) and seizures; relatively uncommon (1% to 5%) are severe cognitive dysfunction, major depression, acute confusional state (ACS), peripheral nervous disorders psychosis. Strong risk factors (at least fivefold increased risk) are previous or concurrent severe neuropsychiatric SLE (for cognitive dysfunction, seizures) and antiphospholipid antibodies (for CVD, seizures, chorea).[67,68]

CLINICAL SIGNS AND SYMPTOMS

Systemic Lupus Erythematosus

Although lupus can affect any part of the body, most people experience symptoms in only a few organs. The most common symptoms associated with lupus are listed here in order of declining prevalence.

- Constitutional symptoms (especially low-grade fever and fatigue)
- Achy joints (arthralgia)
- Arthritis (swollen joints)
- Arthralgia
- Skin rashes (malar)
- Pulmonary involvement (e.g., pleurisy, pleural effusion: chest pain, difficulty breathing, cough)
- Anemia
- Kidney involvement (e.g., lupus nephritis)
- Sun or light sensitivity (photosensitivity)
- Hair loss
- Raynaud's phenomenon (fingers turning white or blue in the cold)
- Nervous system involvement:
  - Seizures
  - Headache
  - Peripheral neuropathy
  - Cranial neuropathy
  - Cerebral vascular accidents
  - Organic brain syndrome
  - Psychosis
- Mouth, nose, or vaginal ulcers

CASE EXAMPLE 12-3

Systemic Lupus Erythematosus

Current Complaint: A 33-year-old woman with a known diagnosis of systemic lupus erythematosus (SLE) came to the physical therapy clinic with the following report: "About 3 weeks ago, I was carrying a heavy briefcase with a strap around my shoulder. I put weight on my right leg and felt my hip joint slip in the back with immediate pain, and I was unable to put any weight on that leg. I moved my hip around in the socket and was able to get immediate relief from the pain, but it felt like it could catch at any time." The client also reported that "it feels like my left hip is 2 inches higher than my right."

Past Medical History: The client reported prolonged (over 7 years) use of prednisone and a past medical history of proteinuria and compromised kidney function. Muscle weakness 2 years ago resulted in a muscle biopsy and a diagnosis of "abnormal" muscle tissue of unknown cause. The client developed a staph infection from the biopsy, which resolved very slowly.

Other past medical history included a motor vehicle accident 2 years ago, at which time her knees went through the dashboard, which left both knees "numb" for a year after the accident.

Clinical Presentation: Aggravating and relieving factors from this visit fit a musculoskeletal pattern of symptoms, and objective examination was consistent with lumbar/sacroiliac mechanical dysfunction with a multitude of other compounding factors, including bilateral posterior cruciate ligament laxity, poor posture, obesity, and emotional lability.

Physical therapy treatment was initiated, but a week later, when the client woke up at night to go to the bathroom, she swung her legs over the edge of her bed and experienced immediate hip and diffuse low back pain and lower extremity weakness.

She went to the emergency department by ambulance and later was admitted to the hospital. She was evaluated by a neurologist (results unknown), recovered from her symptoms within 24 hours, and was released after a 3-day hospitalization. She was directed by her primary care physician to continue outpatient physical therapy services.

Result: After consulting with this client's physician, conservative symptomatic treatment was planned. Within 2 weeks, she experienced another middle-of-the-night acute exacerbation of symptoms. A subsequent magnetic resonance image (MRI) resulted in a diagnosis of disk extrusion (annulus fibrosus perforated with diskal material in the epidural space) at two levels (L4-L5 and L5-S1).

This case example is included to point out the complexity of treating a musculoskeletal condition in a client with a long-term chronic inflammatory disease process requiring years of steroidal antiinflammatory medications. Before including any resistive exercises, muscle energy techniques, or joint or self-mobilization techniques, the therapist must be aware of any clinically significant changes in bone density and the presence of developing osteoporosis.

Scleroderma (Progressive Systemic Sclerosis)

Scleroderma, one of the lesser-known chronic multisystem diseases in the family of rheumatic diseases, is characterized by inflammation and fibrosis of many parts of the body, including the skin, blood vessels, synovium, skeletal muscle, and certain internal organs such as kidneys, lungs, heart, and GI tract.

There are two major subsets: limited cutaneous (previously known as the CREST syndrome) and diffuse cutaneous scleroderma. The major differences between these two types are the degree of clinically involved skin and the pace of disease.

Limited scleroderma (also known as *morphea*) is often characterized by a long history of Raynaud's phenomenon before the development of other symptoms. Skin thickening is limited to the hands, frequently with digital ulcers. Esophageal dysmotility is common. Although limited scleroderma is generally a milder form than diffuse scleroderma, life-threatening complications can occur from small intestine involvement and pulmonary hypertension.

Children affected by juvenile localized scleroderma develop multiple extracutaneous manifestations in 25% of all cases. These extracutaneous features can include joint, neurologic (e.g., epilepsy, peripheral neuropathy, headache), vascular, and ocular changes. These manifestations are often unrelated to the site of the skin lesions and can be associated with multiple organ involvement. Even so, the risk of developing systemic sclerosis (SSc) is very low.[69]

Diffuse scleroderma has a much more acute onset, with many constitutional symptoms, arthritis, carpal tunnel syndrome, and marked swelling of the hands and legs. Widespread skin thickening occurs, progressing from the fingers to the trunk. Internal organ problems, including GI effects and pulmonary fibrosis (see the section on systemic sclerosing lung disease in Chapter 7), are common, and severe life-threatening involvement of the heart and kidneys occurs.[70-72]

Risk Factors. Although the cause of scleroderma is unknown, researchers suspect a complex interaction of genetic and environmental factors. Scleroderma can occur in individuals of any age, race, or sex, but it occurs most commonly in young or middle-age women (ages 25 through 55).[73]

Clinical Signs and Symptoms.

Skin. Raynaud's phenomenon and tight skin are the hallmarks of SSc. Virtually all clients with SSc have Raynaud's phenomenon, which is defined as episodic pallor of the digits following exposure to cold or stress associated with cyanosis, followed by erythema, tingling, and pain. Raynaud's phenomenon primarily affects the hands and feet and less commonly the ears, nose, and tongue.

The appearance of the skin is the most distinctive feature of SSc. By definition, clients with diffuse SSc have taut skin in the more proximal parts of extremities, in addition to the thorax and abdomen. However, the skin tightening of SSc begins on the fingers and hands in nearly all cases. Therefore the distinction between limited and diffuse SSc may be difficult to make early in the illness.

Musculoskeletal. Articular complaints are very common in progressive systemic sclerosis (PSS) and may begin at any time during the course of the disease. The arthralgias, stiffness, and arthritis seen may be difficult to distinguish from those of RA, particularly in the early stages of the disease. Involved joints include the MCPs, PIPs, wrists, elbows, knees, ankles, and small joints of the feet.

Muscle involvement is usually mild, with weakness, tenderness, and pain of proximal muscles of the upper and lower extremities. Late scleroderma is characterized by muscle atrophy, muscle weakness, deconditioning, and flexion contractures.

Viscera. Skin changes, Raynaud's phenomenon, and involvement of the GI tract are the most common manifestation of SSc. Esophageal hypomotility occurs in more than 90% of clients with either diffuse or limited SSc. Similar changes occur in the small intestine, resulting in reduced motility and causing intermittent diarrhea, bloating, cramping, malabsorption, and weight loss. Inflammation and fibrosis can also affect the lungs, resulting in interstitial lung disease, a restrictive lung disease.[74,75]

The overall course of scleroderma is highly variable. Once remission occurs, relapse is uncommon. The diffuse form generally has a worse prognosis because of cardiac involvement such as cardiomyopathy, pericarditis, pericardial effusions, or arrhythmias.

CLINICAL SIGNS AND SYMPTOMS

Scleroderma

Limited Cutaneous Sclerosis (lSSc)

- CREST syndrome
  - **C**alcinosis (abnormal deposition of calcium salts in tissues; usually on the fingertips and over bony prominences)
  - **R**aynaud's phenomenon persisting for years
  - **E**sophageal dysmotility, dysphagia, heartburn
  - **S**clerodactyly (chronic hardening and shrinking of fingers and toes)
  - **T**elangiectasia (spiderlike hemangiomas formed by dilation of a group of small blood vessels; occurs most commonly on the face and hands)

Diffuse Cutaneous Sclerosis (dSSc)

- Raynaud's phenomenon (acute onset)
- Trunk and extremity skin changes (swelling, thickening, hardening)
- Ulcerations of the fingers secondary to constriction of small blood vessels
- Polyarthralgia (joint pain affecting large and small joints with inflammation, stiffness, swelling, warmth, and tenderness)
- Tendon friction rubs
- Flexion contractures of large and small joints
- Visceral involvement
  - Interstitial lung disease (dyspnea on exertion, chronic cough, pleurisy)
  - Esophageal involvement
  - Renal failure (headache, blurred vision, seizures, malaise)
  - GI disease (bloating, cramps, diarrhea or constipation)
  - Myocardial involvement (cardiomyopathy, pericarditis, pericardial effusions, arrhythmias)

Spondyloarthropathy

Spondyloarthropathy represents a group of noninfectious, inflammatory, erosive rheumatic diseases that target the sacroiliac joints, the bony insertions of the annulus fibrosi of the intervertebral disks, and the facet or apophyseal joints. This group of diseases is currently being reclassified with updated criteria for two major categories: inflammatory back pain (IBP) and axial spondyloarthritis (SpA)[76,77] and includes ankylosing spondylitis (AS; also known as Marie-Strümpell disease), Reiter's syndrome, PsA, and arthritis associated with chronic IBD (see discussion in Chapter 8).

Individuals with spondyloarthropathies are not seropositive for RF, and the progressive joint fibrosis present is associated with the genetic marker human leukocyte antigen (HLA-B27). Spondyloarthropathy is more common in men, who by gender have a familial tendency toward the development of this type of disease.

RISK FACTORS AND HISTORY

Spondyloarthropathy

- Insidious onset of each episode of backache
- First episode of backache occurs before 30 years of age
- Each episode lasts for months
- Pain intensifies after rest
- Pain lessens with movement
- Family history of a spondyloarthropathy

Ankylosing Spondylitis. Ankylosing spondylitis (AS) is a chronic, progressive inflammatory disorder of undetermined cause. It is actually more an inflammation of fibrous tissue affecting the entheses, or insertions of ligaments, tendons, and capsules into bone, than of synovium, as is common in other rheumatic disorders.

The sacroiliac joints, spine, and large peripheral joints are primarily affected, but this is a systemic disease with widespread effects. People with AS may experience arthritis in other joints, such as the hips, knees, and shoulders, along with fever, fatigue, loss of appetite, and redness and pain of the eyes.

Although AS has always been more common in men, some studies now suggest that the disease has a more uniform sex distribution, but it may be milder in women with more peripheral joint manifestations than spinal disease.[78,79]

Diagnosis may be delayed or inappropriate when reliance is only on x-rays because disease progression over time is required for a confirmed diagnosis. New diagnostic criteria using magnetic resonance imaging (MRI) have helped improve rates of early diagnosis.[80]

Clinical Signs and Symptoms. The classic presentation of AS is insidious onset of middle and low back pain and stiffness for more than 3 months in a person (usually male) under 40 years of age. It is usually worse in the morning, lasting more than 1 hour, and is characterized as achy or sharp ("jolting"), typically localized to the pelvis, buttocks, and hips; this pain can be confused with sciatica. A neurologic examination will be within normal limits.

Paravertebral muscle spasm, aching, and stiffness are common, but some clients may have slow progressive limitation of motion with no pain at all. Most clients have sacroiliitis as the earliest feature seen on x-ray films before clinical involvement extends to the lumbar spine. A MRI examination can demonstrate acute and chronic changes of sacroiliitis, osteitis, diskovertebral lesions, disk calcifications and ossification, arthropathic (joint) lesions, and complications such as fracture and cauda equina syndrome.[81]

On physical examination, decreased mobility in the anteroposterior and lateral planes will be symmetric. Reduction in lumbar flexion is an early sign of AS. The Schober test is used to confirm reduction in spinal motion associated with AS.[82] The sacroiliac joint is rarely tender by direct palpation. As the disease progresses, the inflamed ligaments and tendons around the vertebrae ossify (turn to bone), causing a rigid spine and the loss of lumbar lordosis. In the most severe cases, the spine becomes so completely fused that the person may be locked in a rigid upright position or in a stooped position, unable to move the neck or back in any direction.

Peripheral joint involvement usually (but not always) occurs after involvement of the spine. Typical extraspinal sites include the manubriosternal joint, symphysis pubis, shoulder, and hip joints. If the ligaments that attach the ribs to the spine become ossified, diminished chest expansion (<2 cm) occurs, making it difficult to take a deep breath. Chest wall stiffness seldom leads to respiratory disability as long as diaphragmatic movement is intact. This process of vertebral and costovertebral fusion results in the formation of syndesmophytes (Fig. 12-4). This reparative process also forms linear bone ossification along the outer fibers of the annulus fibrosus of the disk.

This bridging of the vertebrae is most prominent along the anterior longitudinal ligament and occurs earliest in the thoracolumbar region. Destructive changes of the upper and lower corners of the vertebrae (at the insertion of the annulus fibrosus of the disk) are responsible for the vertebral squaring. Late in the disease the vertebral column takes on an appearance that is referred to as *bamboo spine*.

Extraarticular features. Uveitis, conjunctivitis, colitis, psoriasis, enthesitis, or iritis occurs in nearly 25% of clients and follows a course that is unrelated to the severity of the joint disease.[76] Ocular symptoms may precede spinal symptoms by several weeks or even years. Pulmonary changes (chronic infiltrative or fibrotic bullous changes of the upper lobes) occur in 1% to 3% of persons with AS and may be confused with TB.

Cardiomegaly, conduction defects, and pericarditis are well-recognized cardiovascular complications of AS. Occasionally, renal manifestations precede other symptoms of AS.

Complications. The very stiff osteoporotic spine of clients with AS is prone to *fracture* from even minor trauma. It has been estimated that the incidence of thoracolumbar fractures in AS is four times higher than that in the general population.[83] The most common site of fracture is the lower cervical

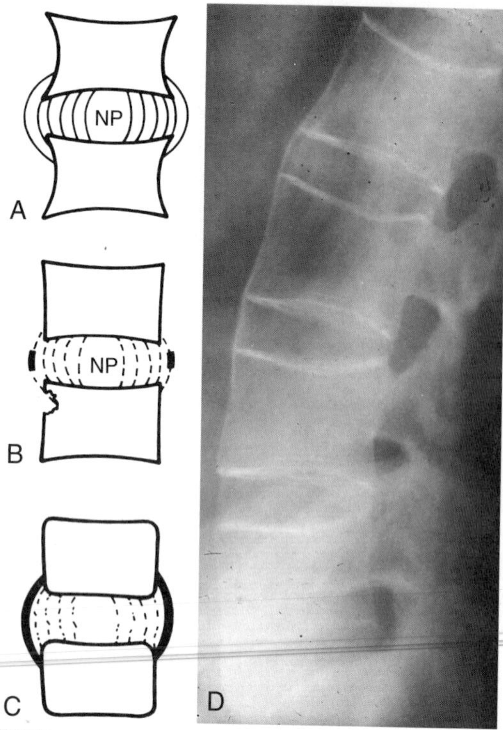

Fig. 12-4 Pathogenesis of the syndesmophyte. The syndesmophyte, along with destruction of the sacroiliac joint, is the hallmark of the inflammatory spondyloarthropathies such as ankylosing spondylitis (AS). It should be distinguished from the osteophyte, which is characteristic of degenerative spondylosis. **A,** Normal intervertebral disk. The inner fibers of the annulus fibrosus are next to the nucleus pulposus (*NP*). The outer fibers insert into the periosteum of the vertebral body at least one-third the distance toward the next endplate. **B,** With early inflammation, the corners of the bodies are reabsorbed and appear to be square or even eroded. Fine deposits of amorphous apatite (calcium phosphate, a mineral constituent of bone) first appear on radiographs as thin, delicate calcification in the outer fibers of the midannulus. **C** and **D,** The process progresses to bridging calcification, with the syndesmophyte extending from one midbody to the next. Thus the spine takes on its bamboo-like appearance on radiographs. (**A** to **C,** From Hadler NM: *Medical management of the regional musculoskeletal diseases,* Orlando, FL, 1984, Grune & Stratton, p 5; **D,** From Bullough PG: *Bullough and Vigorita's orthopaedic pathology,* ed 3, London, 1987, Mosby-Wolfe, p 68.)

spine. Risk of neurologic damage may be compounded by the development of epidural hematoma from lacerated vessels.

Severe neck or occipital pain possibly referring to the retroorbital or frontal area is the presenting symptom of *atlantoaxial subluxation.* This underappreciated entity may be either an early or a late manifestation, but it is frequently seen in clients with persistent peripheral arthritis.

Movement aggravates pain, and progressive myelopathy develops from cord compression, leading to motor/sensory disturbance in bladder and bowel control. The diagnosis of atlantoaxial subluxation is usually made from lateral x-ray views of the cervical spine in flexion and extension.

Spondylodiscitis (erosive and destructive lesions of vertebral bodies) is seen in clients with long-standing disease. Intervertebral disk lesions occur at multiple levels, especially in the thoracolumbar region.

Cauda equina syndrome is a late (rare) manifestation of the disease, with an average interval of 24 years between onset of AS and the syndrome.[84] The initial deficit is loss of sensation of the lower extremities, along with urinary and rectal sphincter disturbances and/or perineal pain and weakness or saddle anesthesia. Neurologic abnormalities in AS are usually related to nerve impingement or spinal cord trauma. Anyone with a known diagnosis of AS and a history of incontinence (bowel or bladder) or neurologic deficit should be evaluated for surgical intervention (e.g., laminectomy, lumboperitoneal shunting).[85]

Spinal stenosis occurs as a result of bony overgrowth of the spinal ligaments and facet joints. Symptoms of pain and numbness of the lower extremities are brought on by walking and relieved by rest.

CLINICAL SIGNS AND SYMPTOMS

Ankylosing Spondylitis

Early Stages

- Intermittent low back pain (nontraumatic, insidious onset; relieved by exercise or activity; persists beyond 3 months)
- Sacroiliitis (inflammation, pain, and tenderness in the sacroiliac joints)
- Spasm of the paravertebral muscles
- Loss of normal lumbar lordosis (positive Schober's test)
- Intermittent, low-grade fever
- Fatigue
- Anorexia, weight loss
- Anemia
- Painful limitation of cervical joint motion

Advanced Stages

- Constant low back pain
- Loss of normal lumbar lordosis
- Ankylosis (immobility and consolidation or fusion) of the sacroiliac joints and spine
- Muscle wasting in shoulder and pelvic girdles
- Marked dorsocervical kyphosis
- Decreased chest expansion
- Arthritis involving the peripheral joints (hips and knees)
- Hip flexion in standing

Extraskeletal

- Cauda equina syndrome
  - Low back pain with or without sciatica
  - Loss of sensation in the lower extremities
  - Bowel and/or bladder changes (decreased anal sphincter tone, urinary retention, overflow incontinence)
  - Perineal pain or loss of sensation (saddle anesthesia)
  - Muscle weakness and atrophy
- Iritis or iridocyclitis (inflammation of the iris; occurs in 25% of all cases)
- Conjunctivitis
- Enthesitis
- Carditis (10% occurrence)
- Pericarditis and pulmonary fibrosis (rare)
- Prostatitis

Reiter's Syndrome. Reiter's syndrome is characterized by a triad of arthritis, conjunctivitis, and nonspecific urethritis, although some clients develop only two of these three problems. Reiter's syndrome occurs mainly in young adult men between the ages of 20 and 40 years, although women and children can be affected.

Risk Factors. Reactive arthritis associated with Reiter's syndrome occurs in response to infection and typically begins acutely 2 to 4 weeks after venereal infections or bouts of gastroenteritis. Most cases occur in young men and are believed to result from venereal-acquired infections. Other infections, such as foodborne enteric infections, affect both men and women. The onset of Reiter's syndrome can be abrupt, occurring over several days or more gradually over several weeks.

HLA-B27, present in a high-frequency pattern, supports a genetic predisposition for the development of this syndrome after a person is exposed to certain bacterial infections or after sexual contact. Having HLA-B27 does not necessarily mean that the person will develop this syndrome but indicates that the person will have a greater chance of developing Reiter's syndrome than do persons without this marker.

Reiter's syndrome can be differentiated from AS by the presence of urethritis and conjunctivitis, the prominent involvement of distal joints, and the presence of asymmetric radiologic changes in the sacroiliac joints and spine.

Clinical Signs and Symptoms. Arthritis associated with Reiter's syndrome often occurs precipitously and frequently affects the knees and ankles, lasting weeks to months. The distribution of the arthritis begins in the weight-bearing joints, especially of the lower extremities.

The arthritis may vary in severity from absence to extreme joint destruction. Involvement of the feet and spine is most common and is associated with HLA-B27 positivity. Affected joints are usually warm, tender, and edematous, with pain on active and passive movement. A dusky-blue discoloration or frank erythema accompanied by exquisite tenderness is a sign of a septic joint. Although the joints usually begin to improve after 2 or 3 weeks, many people continue to have pain, especially in the heels and back.

Low back and buttock pain are common in reactive arthritis; such pain is caused by sacroiliac or other spinal joint involvement. Sacroiliac changes seen on x-ray films are usually asymmetric and similar to those of AS. Small joint involvement, especially in the feet, is more common in Reiter's syndrome than in AS and is often asymmetric.

In addition to arthritis, inflammation typically occurs at bony sites where tendons, ligaments, or fascia have their attachments or insertions (entheses). Enthesitis most commonly occurs at the insertions of the plantar aponeurosis and Achilles tendon, on the calcaneus, leading to heel pain—one of the most frequent, distinctive, and disabling manifestations of the disease. Other common sites for enthesitis include ischial tuberosities, iliac crests, tibial tuberosities, and ribs, with associated musculoskeletal pain at sites other than joints (Case Example 12-4).

CASE EXAMPLE 12-4

Reiter's Syndrome

Past Medical History: At presentation, a 22-year-old man had left heel pain that had developed 3 weeks before his appointment in physical therapy. He could not attribute any trauma to the foot and was not involved in any sports or athletic activities. Previous medical history was minimal except for an appendectomy when he was 18 years old.

Clinical Presentation: The client reported that his pain was worst when he first got out of bed in the morning but improved with stretching and taking aspirin. He did not wear any orthotics or special shoes. The therapist did not ask about the presence of associated signs or symptoms.

No obvious gait abnormalities were observed. On palpation of the foot, there was no warmth, bruising, or redness in the area of the plantar fascia or calcaneus. Tenderness was reported along the plantar fascia, with a painful response to palpation of the tendinous attachment to the calcaneus. Ankle range of motion and muscle strength of the left lower leg were within normal limits. There was no tenderness of the surrounding bones, tendons, or muscles. A neurologic screen was also considered normal.

Intervention: The therapist treated this client by using a treatment protocol for plantar fasciitis, including ultrasound, deep friction massage, and stretching exercises. Symptoms subsided, and the client was discharged. He returned 6 weeks later with recurrence of the original symptoms and new onset of low back pain.

The therapist reevaluated the client, including an in-depth evaluation of postural components and performance of a back screening examination, but again did not ask any questions related to associated signs and symptoms. Before his next appointment, the client called and canceled further physical therapy treatment.

Result: A follow-up call determined that this young man had developed other symptoms, such as fever, red and itching eyes, and frequent urination. He went to a walk-in clinic, was referred to an internist, and received a diagnosis of Reiter's syndrome.

Whenever a client has musculoskeletal pain or symptoms of unknown cause, a series of questions must be posed to screen for medical disease. This is especially important when symptoms do not respond to treatment, when symptoms recur, or when new musculoskeletal symptoms develop.

Although joint pain, heel pain, or back pain usually occurs after the development of conjunctivitis, enteritis, or urethritis in Reiter's syndrome, this young man developed musculoskeletal symptoms first. At the time of the initial physical therapy evaluation, he was experiencing fatigue, low-grade fever, and malaise that he did not report. Asking the question "Are there any other symptoms of any kind anywhere in your body?" might have elicited the early red flag–associated signs and symptoms for infection.

The *conjunctivitis* of Reiter's syndrome is mild and characterized by irritation with redness, tearing, and burning usually lasting a few days (or less commonly as long as several weeks). The process is ordinarily self-limiting.

Urethritis manifested by burning and urinary frequency is often the earliest symptom. A profuse and watery diarrhea can precede the onset of urethritis in Reiter's syndrome.

CLINICAL SIGNS AND SYMPTOMS

Reiter's Syndrome

Articular Manifestations

- Polyarthritis (occurs several days or weeks after symptoms of infection appear)
- Sacroiliac joint changes
- Low back and buttock pain
- Small joint involvement, especially the feet (heel pain)
- Plantar fasciitis
- Low-grade fever
- Urethritis (when present, precedes other symptoms by 1 to 2 weeks)
- Conjunctivitis and iritis, bilaterally

Extraarticular Manifestations

- Skin involvement: inflammatory hyperkeratotic lesions of the toes, nails, and soles resembling psoriasis
- May be preceded by bowel infection: diarrhea, nausea, vomiting
- Anorexia and weight loss

Psoriatic Arthritis. Psoriatic arthritis (PsA) is a chronic, recurrent, erosive, inflammatory arthritis associated with the skin disease psoriasis. It is not just a variant of RA but is a distinct disease that combines features of both RA (e.g., joint pain, erythema, swelling, stiffness) and the spondyloarthropathies (e.g., enthesopathy: inflammation at insertion points of tendon, ligament, capsule; iritis). Psoriasis is quite common, affecting 1% to 3% of the general population. This arthritis occurs in one third of clients with psoriasis.

In contrast to RA, there is no gender predilection in PsA. Both sexes are affected equally, although women tend to develop symmetric polyarthritis, and spinal involvement is more common in men. PsA can occur at any age, although it usually occurs between the ages of 20 and 30 years. The onset of the arthritis may be acute or insidious and is usually preceded by the skin disease.

Risk Factors. The cause of psoriasis and any risk factors for PsA are unknown. PsA is a complex, multifactorial disease; multiple genes are likely to influence disease susceptibility and severity.[86] The presence of the histocompatibility complex marker HLA-B27 and other HLA antigens is not uncommon, and they occur in clients with peripheral arthritis and spondylitis.

The presence of these genetic markers may be associated with an increased susceptibility to unknown infectious or environmental agents or to primary abnormal autoimmune phenomena. There is some evidence to support dysregulated angiogenesis as a primary pathogenic mechanism in PsA.[87]

Clinical Signs and Symptoms. Skin lesions that characterize psoriasis are readily recognized as piles of well-defined, dry, erythematous, often overlapping silver-scaled papules and plaques. These may appear in small, easily overlooked patches or may run together and cover wide areas. The scalp, extensor surfaces of the elbows and knees, back, and buttocks are common sites. The lesions, which do not usually itch, come and go and may be present for years (typically 5 to 10 years) before the onset of arthritis.

Nail lesions, including pitting, ridging (transverse grooves), cracking, onycholysis (loosening or separation of the nail; see Fig. 4-31), brown-yellow discoloration, and destruction of the nail, are the only clinical features that may identify clients with psoriasis in whom arthritis is likely to develop. The nail changes may be mistaken for those produced by a fungal infection.

Arthritis appears as an early and severe sign in a symmetric distal distribution (DIP joints of fingers and toes before involvement of MCP and MTP joints) in half of all clients with PsA, which distinguishes it from RA. Severe erosive disease may lead to marked deformity of the hands and feet, called arthritis mutilans. Wrists, ankles, knees, and elbows can also be involved.

Clients report pain and stiffness in the inflamed joints, with morning stiffness that lasts more than 30 minutes. Other evidence of inflammation includes pain on stressing the joint, tenderness at the joint line, and the presence of effusion. Painful symptoms are aggravated by prolonged immobility and are reduced by physical activity.

Marked vertebral involvement can result in *ankylosis of the spine.* This differs from AS in a number of respects, most notably in the tendency for many of the syndesmophytes to arise not at the margins of the vertebral bodies but from the lateral and anterior surfaces of the bodies. *Sacroiliac changes,* including erosions, sclerosis, and ankylosis similar to that in Reiter's syndrome, occur in 10% to 30% of clients with PsA.

Soft-tissue involvement, similar to clinical manifestations of spondyloarthropathy, occurs often in PsA. Enthesitis, or inflammation at the site of tendon insertion or muscle attachment to bone, is frequently observed at the Achilles tendon, plantar fascia, and pelvic bones. Also common is tenosynovitis of the flexor tendons of the hands, extensor carpi ulnaris, and other sites.

Dactylitis, which occurs in more than one-third of PsA clients, is marked by diffuse swelling of the whole finger. Inflammation in this typical "sausage finger" extends to the tendon sheaths and adjacent joints.

Extraarticular features similar to those seen in clients with other seronegative spondyloarthropathies are frequently seen. These extraarticular lesions include iritis, mouth ulcers, urethritis, and, less commonly, colitis and aortic valve disease.

CLINICAL SIGNS AND SYMPTOMS

Psoriatic Arthritis

- Fever
- Fatigue
- Dystrophic nail bed changes
- Polyarthritis
- Psoriasis
- Sore fingers (sometimes sausagelike swelling)

Unique Clinical Features of PsA

- DIP joint involvement
- Nail changes
- Dactylitis
- Spondylitis
- Iritis

Early recognition of this disorder is important because medical intervention with newer biologic agents can help prevent long-term complications such as permanent joint destruction and disability.[88]

Lyme Disease

In the early 1970s, a mysterious clustering of juvenile arthritis occurred among children in Lyme, Connecticut, and in surrounding towns. Medical researchers soon recognized the illness as a distinct disease, which they called Lyme disease. They were able to identify the deer tick infected with a spiral bacterium or spirochete (later named *Borrelia burgdorferi*) as the key to its spread.

The number of reported cases of Lyme disease, as well as the number of geographic areas in which it is found, has been increasing. Most cases are concentrated in the coastal northeast, the mid-Atlantic states, Wisconsin, Minnesota, Oregon, and northern California. Children may be more susceptible than are adults simply because they spend more time outdoors and are more likely to be exposed to ticks.

Clinical Signs and Symptoms. In most individuals, the first symptom of Lyme disease is a red rash, known as *erythema migrans,* that starts as a small red spot that expands over a period of days or weeks, forming a circular, triangular, or oval rash. Sometimes the rash resembles a bull's-eye because it appears as a red ring surrounding a central clear area. The rash can range in size from that of a dime to the entire width of a person's back, appearing within a few weeks of a tick bite and usually at the site of the bite, which is often the axilla or groin. As infection spreads, several rashes can appear at different sites on the body.

Erythema migrans is often accompanied by flu-like symptoms such as fever, headache, stiff neck, body aches, and fatigue. Although these symptoms resemble those of common viral infections, Lyme disease symptoms tend to persist or may occur intermittently over a period of several weeks to months.

Arthritis appears several months after infection with *Borrelia burgdorferi.* Slightly more than half of the people who are not treated with antibiotics develop recurrent attacks of painful and swollen joints that last a few days to a few months. About 10% to 20% of untreated clients will go on to develop chronic arthritis.[89]

In most clients, Lyme arthritis is monoarticular or oligoarticular (few joints), most commonly affecting the knee, but the arthritis can shift from one joint to another. Other large joints, such as the hip, shoulder, and elbow, are also commonly affected.[90] Involvement of the hands and feet is uncommon, and it is these features that help differentiate Lyme arthritis from RA.[91]

Neurologic symptoms (including cognitive dysfunction referred to as *neurocognitive symptoms*) may appear because Lyme disease can affect the nervous system. Symptoms include stiff neck; severe headache associated with meningitis; Bell's palsy; numbness, pain, or weakness in the limbs; or poor motor coordination. Memory loss, difficulty in concentrating, mood changes, and sleep disturbances have also been associated with Lyme disease. Nervous system involvement can develop several weeks, months, or even years following an untreated infection. These symptoms last for weeks or months and may recur.

Cardiac involvement occurs in less than 1% of the people affected by Lyme disease. Symptoms of irregular heartbeat, dizziness, and dyspnea occur several weeks after the infection and rarely last more than a few days or weeks. Recovery is usually complete.

Finally, although Lyme disease can be divided into early and later stages, each with a different set of complications, these stages may vary in duration, may overlap, or may even be absent. Clinical manifestations may first appear from 3 to 30 days after the tick bite but usually occur within 1 week.

Lyme disease is still mistaken for other ailments, including Guillain-Barré syndrome, MS, and fibromyalgia syndrome, and can be difficult to diagnose. The only distinctive hallmark unique to Lyme disease, the erythema migrans rash, is absent in at least one-fourth of those who become infected. Many people are unaware that they have been bitten by a tick (Case Example 12-5).

In general, the sooner treatment is initiated, the quicker and more complete the recovery, with less chance for the development of subsequent symptoms of arthritis and neurologic problems. Following treatment for Lyme disease, some persons still have persistent fatigue and achiness, which can take months to subside.

Unfortunately, having had Lyme disease once is no guarantee that the illness will be prevented in the future. The disease can strike more than once in the same individual if she or he is reinfected with the Lyme disease bacterium.

CASE EXAMPLE 12-5

Lyme Disease

A 54-year-old business executive developed searing neck and back pain and was diagnosed as having a cervical disk protrusion. He was sent to physical therapy but had a very busy travel schedule and was unable to make even half of his scheduled appointments.

He chose to discontinue physical therapy, but his symptoms worsened and the pain became so intense that he was unable to go to work some mornings. He also started experiencing numbness in his right arm along the ulnar nerve distribution. He returned to physical therapy, but there was no discernible improvement subjectively, by client report, or objectively, as measured by functional improvement.

Anterior cervical diskectomy was performed to remove the fifth cervical disk but with no change in symptoms postoperatively. There was significant right extremity paresis, with maximal functional loss of the right hand and continued neck and back pain.

This client was eventually discharged from further physical therapy services and underwent a second surgical procedure, with no improvement in his condition. A year later, he telephoned the therapist to report that he had been diagnosed with Lyme disease. This man spent his vacations in the woods of Connecticut and Long Island, but this important piece of information was never gleaned from his past medical history.

Despite the lengthy time before diagnosis, the client was almost entirely recovered and ready to return to work after completing a course of antibiotics.

CLINICAL SIGNS AND SYMPTOMS

Lyme Disease

Early Infection (one or more may be present at different times during infection)

- Red rash (erythema migrans)
- Flu-like symptoms (fever, headache, stiff neck, fatigue)
- Migratory musculoskeletal pain (joints, bursae, tendons, muscle, or bone)
- Neurologic symptoms:
  - Severe headache (meningitis)
  - Numbness, pain, weakness of extremities
  - Poor motor coordination
  - Cognitive dysfunction: Memory loss, difficulty in concentrating, mood changes, sleep disturbances

Less Common Symptoms

- Eye problems such as conjunctivitis
- Heart abnormalities and myocarditis

Late Infection (months to years after infection)

- Arthritis, intermittent or chronic
- Encephalopathy (mood and sleep disturbances)
- Neurocognitive dysfunction
- Peripheral neuropathy

Autoimmune-Mediated Neurologic Disorders

Some neurologic disorders encountered by the therapist display features that suggest an immunologic basis for the disorder. Such diseases include MS, Guillain-Barré syndrome, and MG. Other dysfunctions, such as amyotrophic lateral sclerosis (ALS) and acute disseminated encephalomyelitis, also associated with immunologic dysfunction but seen less often by the therapist, are not discussed.

Multiple Sclerosis

Multiple sclerosis (MS) is the most common inflammatory demyelinating disease of the CNS, affecting areas of the brain and spinal cord but sparing the peripheral nerves. Symptoms appear usually between 20 and 40 years of age, with a peak onset of age 30 years. Onset is rare in children and in adults older than 50.

Risk Factors. Women are affected twice as often as men, and a family history of MS increases the risk tenfold. MS is not considered a hereditary disease, but a person who has a first-degree relative affected by the disease has an above-average risk for it.[92]

It would appear that MS susceptibility and age at onset are to some extent under genetic control. Environmental factors may affect onset; MS is five times more prevalent in the temperate (colder) climates of North America and Europe than in tropical areas, even among people with similar genetic backgrounds. The reason may be lack of sunlight (less ultraviolet radiation needed for vitamin D).[93]

Evidence suggests that MS is an autoimmune disease, but the actual cause remains unknown. A virus or other infectious agent, toxins, vaccinations, stress,[94] and surgery are all thought to be possible triggers for the immune-mediated response, which is believed to destroy the CNS myelin.[95] According to the immune system hypothesis, T cells that have been called up against the virus, toxin, or stressor turn their focus to the myelin and continue to make intermittent attacks on it long after the initial infection or problem has resolved.

The disease is characterized by inflammatory demyelinating (destructive removal or loss) lesions that later form scars known as plaques, which are scattered throughout the CNS white matter, especially the optic nerves, cerebrum, and cervical spinal cord. When edema and inflammation subside, some remyelination occurs, but it is often incomplete. Axonal injury may cause permanent neurologic dysfunction.

The progression of MS is difficult to predict and depends on several factors, including the person's age and the intensity of onset, the neurologic status at 5 years after the onset, and the course of exacerbations and remissions. The survival rate after the onset of symptoms is usually good, and death typically results from either respiratory or urinary infection.

Clinical Signs and Symptoms. Clinically, MS is characterized by multiple and varying signs and symptoms and by unpredictable and fluctuating periods of remissions and exacerbations. Symptoms may vary considerably in character, intensity, and duration. Symptoms can develop rapidly over a course of minutes or hours; less frequently, the onset may be insidious, occurring during a period of weeks or months.

Symptoms depend on the location of the lesions, and early symptoms demonstrate involvement of the sensory, pyramidal, cerebellar, and visual pathways or disruption of cranial nerves and their linkage to the brainstem.

Motor Symptoms. Many persons with MS experience weakness in the extremities, leading to difficulty with ambulation, coordination, and balance, with ataxia or tremor present if lesions are in the cerebellum. Spasticity and hyperreflexia are common causes of disability with severe, uncontrollable spasms of the extremities. Profound fatigue or dysmetria (intention tremor) contribute to motor impairment.[96]

Difficulties with speech (slow, slurred) or chewing and swallowing can occur if the brainstem or cranial nerves are affected. Urinary frequency, urinary urgency, incontinence, urinary retention, or urinary hesitancy commonly characterizes motor and/or sensory bowel/bladder dysfunctions.

Sensory Symptoms. Unilateral visual impairment (e.g., double vision, visual loss, red-green color blindness) that comes and goes as a result of optic neuritis is often the first indication of a problem. Optic neuritis occurs in about 20% of persons initially presenting with MS, while 40% may present with optic neuritis during the course of their disease.[97]

Extreme sensitivity to temperature changes is evident in more than 60% of the people diagnosed with MS. Elevated temperatures shorten the duration of the nerve impulse and worsen symptoms, whereas cooler temperatures actually restore conduction in blocked nerves and improve symptoms.

Paresthesias (numbness and tingling) accompanied by burning in the extremities can result in injury to the hands or feet. Lhermitte's sign (electric shock–like sensation down the spine and radiating to the extremities, initiated by neck flexion) is very suggestive of MS but can also occur with disk protrusion.

CLINICAL SIGNS AND SYMPTOMS

Multiple Sclerosis

(Listed in declining order of frequency)

Symptoms
- Unilateral visual impairment
- Paresthesias
- Ataxia or unsteadiness
- Vertigo (sensation of rotation of self or surroundings)
- Fatigue
- Muscle weakness
- Bowel/bladder dysfunctions:
  - Frequency
  - Urgency
  - Incontinence
  - Retention
  - Hesitancy
- Speech impairment (slow, slurred speech)

Signs
- Optic neuritis
- Nystagmus
- Spasticity or hyperreflexia
- Babinski's sign
- Absent abdominal reflexes
- Dysmetria or intention tremor
- Labile or changed mood
- Lhermitte's sign

Guillain-Barré Syndrome (Acute Idiopathic Polyneuritis)

Guillain-Barré syndrome is an acute, acquired autoimmune disorder with demyelination of the peripheral nervous system (especially spinal nerves) and is characterized by an abrupt onset of paralysis.[98] The disease affects all age groups, and incidence is not related to race or sex.

Risk Factors. The exact cause of the disease is unknown, but it frequently occurs after an infectious illness. Upper respiratory infections, influenza, vaccinations, or viral infections such as measles, hepatitis, or mononucleosis commonly precede acute idiopathic polyneuritis by 1 to 3 weeks.[99] The immune system attacks its own myelin cells because they look similar to the molecules of the infecting virus. The immune system shifts into an accidental self-destructive overdrive.

Clinical Signs and Symptoms. The onset of acute idiopathic polyneuritis is generally characterized by a rapidly progressive weakness for a period of 3 to 7 days. It is usually symmetric, involving first the lower extremities, then the upper extremities, and then the respiratory musculature. Weakness and paralysis are frequently preceded by paresthesias and numbness of the limbs, but actual objective sensory loss is usually mild and transient.

Although muscular weakness is usually described as bilateral, progressing from the legs upward toward the arms, this syndrome may be missed when the client has unilateral symptoms that do not progress proximally.

Muscular weakness of the chest may appear early in this disease process as respiratory compromise. Respiratory involvement as such may be unnoticed until the person develops more severe symptoms associated with the Guillain-Barré syndrome.

The progression of paralysis varies from one client to another, often with full recovery from the paralysis. Usually symptoms develop over a period of 1 to 3 weeks, and the progression of paralysis may stop at any point. Once the weakness reaches a maximum (usually during the second week), the client's condition plateaus for days or even weeks before spontaneous improvement and eventual recovery begin, extending over a period of 6 to 9 months.

Cranial nerves, most commonly the facial nerve, can be involved. The tendon reflexes are decreased or lost early in the course of the illness. The incidence of residual neurologic deficits is higher than was previously recognized, and deficits may occur in as many as 50% of all cases.

Treatment. There is no immediate cure for this disease, but medical support is vital during the progression of symptoms, particularly in the acute phase when respiratory function may be compromised. Physical therapy is initiated at an early stage to maintain joint range of motion within the client's pain tolerance and to monitor muscle strength until active exercises can be initiated.

The usual precautions for clients immobilized in bed are required to prevent complications during the acute phase. A major precaution is to provide active exercise at a level consistent with the client's muscle strength. Overstretching and overuse of painful muscles may result in a prolonged recovery period or a lack of recovery (Case Example 12-6).

CLINICAL SIGNS AND SYMPTOMS

Guillain-Barré Syndrome (Acute Idiopathic Polyneuritis)

- Muscular weakness (bilateral, progressing from the legs to the arms to the chest and neck)
- Diminished deep tendon reflexes
- Paresthesias (without loss of sensation)
- Fever, malaise
- Nausea

CASE EXAMPLE 12-6

Guillain Barré Syndrome

A 67-year-old retired aeronautics engineer was referred to physical therapy by his physician for electrotherapy and therapeutic exercise. The physician's diagnosis was right-sided Bell's palsy. Past medical history was significant for an upper respiratory infection 2 weeks before the onset of his first symptoms.

The client reported difficulty in closing his eyes, chewing, and drinking, and he was unable to smile. There were no changes in sensation or hearing. During the neurologic examination, the client was unable to raise his eyebrows or close his eyes, and there was obvious facial drooping on both sides. A gross manual muscle test revealed full (5/5) muscle strength in all four extremities, but muscle stretch reflexes were absent in all four extremities.

Result: The therapist recognized three red-flag symptoms in this case: (1) recent upper respiratory infection followed by the development of neurologic symptoms; (2) progressive development of symptoms from right-sided to bilateral between the time the client was evaluated by the physician and went to the physical therapist; and (3) absent deep tendon reflexes, an inconsistent finding for Bell's palsy.

The therapist contacted the physician by telephone to relay this information and confirm the treatment plan given this new information. The physician requested that the client return for further medical testing, and a revised diagnosis of Guillain-Barré syndrome was made.

The client's clinical status stabilized, and he returned to the physical therapist. The treatment plan was modified accordingly. This case again demonstrates the importance of performing a careful examination, including screening for systemic disease and recognizing red-flag symptoms.

Myasthenia Gravis

Myasthenia gravis (MG) develops when, for unknown reasons, antibodies produced by the immune system block receptors in muscles that receive signals of acetylcholine (a chemical messenger generated by nerve impulses), thus impairing muscle function.

MG may begin at any time in life, including in the newborn infant, but there are two major peaks of onset. In early-onset MG, at age 20 to 30 years, women are more often affected than men. In late-onset MG, after age 50 years, men are more often affected. The incidence of MG in older patients is rising and not just because of improved disease recognition but the reason(s) for this remain unclear.[100]

Clinical Signs and Symptoms. Clinically, the disease is characterized by muscle weakness and fatigability, most commonly in muscles controlling eye movement, chewing, swallowing, and facial expressions. Antibodies to the postsynaptic acetylcholine receptor at the myoneural junction cause diminution of the force of muscle contractions leading to a feeling of fatigue. MG treatments increase the availability of acetylcholine and reduce antibody formation.[101]

Symptoms show fluctuations[102] in intensity and are more severe late in the day or after prolonged activity. Speech may become unintelligible after prolonged periods of talking. Fluctuations also occur with superimposed illness, menses, and air temperature (worse with warming, improved with cold). Common comorbid conditions include thymoma and hyperthyroidism.

Fatigable and rapidly fluctuating asymmetric ptosis is a hallmark of the problem, since ocular muscle dysfunction is usually one of the first symptoms. The ice pack test, rest test, sleep tests, and peek sign are all useful in confirming the presence of MG (Box 12-5).[103]

Proximal muscles are affected more than distal muscles, and difficulty in climbing stairs, rising from chairs, combing the hair, or even holding up the head occurs. Cranial muscles, neck muscles, respiratory muscles, and muscles of the proximal limbs are the primary areas of muscular involvement. Neurologic findings are normal except for muscle weakness. There is no muscular atrophy or loss of sensation. Muscular weakness ranges from mild to life threatening (when involving respiratory muscles).

BOX 12-5 ELICITING SIGNS AND SYMPTOMS OF MYASTHENIA GRAVIS

The characteristic finding in myasthenia gravis (MG) is decreased muscle strength that gets worse with repetition and improves with rest. Asymmetric drooping of the eyelids (ptosis) is one of the first signs and can be identified using the following tests.

The client sits and fixes his or her gaze on a distant object without blinking. The frontalis muscle should be relaxed although this may be difficult. The eye with the most noticeable ptosis is tested.

The Ice Pack Test

Place a latex-free glove filled with crushed ice over the eyelid for 2 minutes.

The Rest Test

Place a cotton-filled latex-free glove (rest) over the eyelid while holding the eyes closed for 2 minutes.

The Sleep Test

Client is placed in a dark room with eyes closed for 30 minutes.

Key: Evaluate response to these tests immediately following timed period. A positive response is complete or almost complete resolution of the ptosis. Improvement may be greater with the ice test compared to the rest test.[107] Medical referral is required.

Data from Scherer K, Bedlack RS, Simel L: Does this patient have myasthenia gravis? *JAMA* 293(15):1906-1914, 2005.

CLINICAL SIGNS AND SYMPTOMS

Myasthenia Gravis

- Muscle fatigability and proximal muscle weakness aggravated by exertion
- Respiratory failure from progressive involvement of respiratory muscles
- Ptosis (extraocular muscle weakness resulting in drooping of the upper eyelid)
- Diplopia (double vision)
- Dysarthria (slurred speech)
- Bulbar involvement
- Alteration in voice quality
- Dysphagia (difficulty swallowing)
- Nasal regurgitation
- Choking, difficulty in chewing

Immunoproliferative Disorders

Immunoproliferative disorders occur when abnormal reproduction or multiplication of the cells of the lymphoid system results in leukemia, lymphoma, and other related disorders. These have been covered in other parts of this text and are not discussed further in this chapter.

PHYSICIAN REFERRAL

In most immunologic disorders, physicians must rely on the client's history and clinical findings in association with supportive information from diagnostic tests to make a differential diagnosis. Often, there are no definitive diagnostic tests, such as in the case of MS. The physician instead relies on objectively measured CNS abnormalities, a history of episodic exacerbations, and remissions of symptoms with progressive worsening of symptoms over time.

In the early stages of treating disorders such as MS, Guillain-Barré syndrome, and myositis, factors, such as the effect of fatigue on the client's progress and fragile muscle fibers, necessitate that the therapist keep close contact with the physician, who will use a physical examination and laboratory tests to determine the most opportune time for an exercise program. While the physician is monitoring serum enzyme levels and the overall medical status of the client, the therapist will continue to provide the physician with essential feedback regarding objective findings such as muscle tenderness, muscle strength, and overall physical endurance.

A careful history and close clinical observations may elicit indications that the client is demonstrating signs and symptoms unrelated to a musculoskeletal disorder. Because the immune system can implicate many of the body systems, the therapist should not hesitate to relay to the physician any unusual findings reported or observed.

Guidelines for Immediate Medical Attention

- Anyone exhibiting signs and symptoms of anaphylactic shock, especially vocal hoarseness, difficulty breathing, and chest discomfort or tightness (see Table 12-1)
- New onset of joint pain with a recent history of surgery **(bacterial or reactive arthritis)**
- A dusky blue discoloration or erythema accompanied by exquisite tenderness is a sign of a septic (infected) joint; ask about a recent history of infection of any kind anywhere in the body; medical referral is advised.

Guidelines for Physician Referral

- New onset of joint pain within 6 weeks of surgery, especially when accompanied by constitutional symptoms, rash, or skin lesions
- Symmetric swelling and pain in peripheral joints may be an early sign of RA; early medical intervention is critical to prevent erosive joint disease and disability.[48]
- Development of progressive neurologic symptoms within 1 to 3 weeks of a previous infection or recent vaccination
- Evidence of spinal cord compression in anyone with cervical RA who has progressed from generalized stiffness to new onset of cervical laxity (C1-C2 subluxation or dislocation)

- Presence of incontinence (bowel or bladder) in anyone with AS requires medical referral; surgical treatment of the underlying dural ectasia may be helpful.[85]
- Positive ptosis tests for MG (ice pack, rest, sleep tests)

Clues to Immune System Dysfunction

- Client with a history of RA taking DMARDs who develops symptoms of drug toxicity (e.g., rash, petechiae/ecchymosis, photosensitivity, dyspnea, nausea/vomiting, lymph node swelling, edema, oral ulcers, diarrhea)
- Long-term use of nonsteroidal antiinflammatory drugs (NSAIDs) or other antiinflammatory drugs, especially with new onset of GI symptoms; back or shoulder pain of unknown cause
- Long-term use of immunosuppressives or corticosteroids with onset of constitutional symptoms, especially fever
- Insidious onset of episodic back pain in a person younger than 40 who has a family history of spondyloarthropathy
- Joint pain preceded or accompanied by burning and urinary frequency (urethritis) and/or accompanied by eye irritation, crusting, redness, tearing, or burning usually lasting only a few days **(conjunctivitis; Reiter's syndrome)**
- Joint pain preceded or accompanied by skin rash or lesions **(PsA; Lyme disease; rheumatic fever)**
- New onset of inflammatory joint pain (especially monoarticular joint involvement) postoperatively, especially accompanied by extraarticular signs or symptoms such as rash, diarrhea, urethritis **(reactive or bacterial arthritis)**; mouth ulcers **(Reiter's syndrome, SLE)**; raised skin patches **(PsA)**
- Development of neurologic symptoms 1 to 3 weeks after an infection **(Guillain-Barré syndrome)**

Key Points to Remember

- Pain in the knees, hands, wrists, or elbows may indicate an autoimmune disorder; aching in the bones can be caused by expanding bone marrow.
- True arthritis produces pain and limitation during both active and passive range of motion. Limitation from tendonitis is much worse during active range of motion.
- Any change in cough, pain, or fever and any change or new presentation of symptoms should be reported to the physician.
- Be alert to any warning signs of hypersensitivity response (allergic reaction) during therapy and be prepared to take necessary measures (e.g., graded exercise to client tolerance, control of room temperature, client use of medications).
- Immediate emergency procedures are required when a client has a severe allergic reaction (anaphylactic shock).
- For the client with Guillain-Barré syndrome, active exercise must be at a level consistent with the client's muscle strength. Overstretching and overuse of painful muscles may result in prolonged or lack of recovery.
- For the client with MS, treatment should take place in the coolest (temperature) setting possible.
- Increase in shoe size, marked fatigue, and onset of symptoms in the first year postpartum are important clues to the development of RA.
- For the client with RA or AS, the risk of fracture from the development of atlantoaxial subluxation necessitates the use of extreme caution in treatment procedures. The most common site of fracture is the lower cervical spine.

SUBJECTIVE EXAMINATION

Special Questions to Ask

Signs and symptoms of immune disorders can appear in any body system. A thorough review of the Family/Personal History form, subjective interview, and appropriate follow-up questions will help the therapist identify signs and symptoms that are not part of a musculoskeletal pattern. Special attention should be given to the question on the Family/Personal History form concerning general health. Clients with immune disorders or immunocompromised clients often have poor general health or recurrent infections.

When the immune system is involved, some important questions to ask include the following:

- How long have you had this problem? (acute versus chronic)
- Has the problem gone away and then recurred?
- Have additional symptoms developed or have other areas become symptomatic over time?

Past Medical History

- Have you ever been told that you had/have an immune disorder, autoimmune disease, or cancer? (Predisposes the person to other diseases)
- Have you ever had radiation treatment? (Diminishes blood cell production, predisposes to infection)
- Have you ever had an organ transplant (especially kidney) or removal of your thymus? **(MG)**

Associated Signs and Symptoms

- Do you have difficulty with combing your hair; raising your arms; getting out of a bathtub, bed, or chair; or climbing stairs? **(MG)**
- Do you have difficulty when raising your head from the pillow when you are lying down on your back? **(MG)**
- Do you have difficulty with swallowing, or have you noticed any changes in your voice? **(MG)**
- Have you noticed any changes in your skin texture or pigmentation? Do you have any skin rashes? **(Scleroderma, allergic reactions, SLE, RA, dermatomyositis, PsA, AIDS, Lyme disease)**
  - *If yes,* Have you noticed any association between the development of the skin rash and pain or swelling in any of your joints (or other symptoms)?
  - Do these other symptoms go away when the skin rash clears up?
  - Have you been exposed to ticks? For example, have you been out walking in the woods or in tall grass or in contact with pets? **(Lyme disease)**
- Have you had any recent vision problems? **(MS, SLE)**
- Have you had any body tattooing or ear/body piercing done in the last 6 weeks to 6 months? **(AIDS, hepatitis)**
- Have you had any difficulties with urination—for example, a change in appearance of urine, accidents, increased frequency? **(MS, MG, Reiter's syndrome)**

For the Person with Known Allergies (check the Family/Personal History form)

- What are the usual symptoms that you experience in association with your allergies?
- Describe a typical allergic reaction for you.
- Do the symptoms relate to physical changes (e.g., cold, heat, or dampness)?
- Do the symptoms occur in association with activities (e.g., exercise)?
- Do you take medication for your allergies?

For the Client Reporting Fatigue and Weakness

- Do you feel tired all the time or only after exertion?
- Do you get short of breath after mild exercise or at rest?
- How much sleep do you get at night?
- Do you take naps during the day?
- Have you ever been told by a physician that you are anemic?
- How long have you had this weakness?
- Does it come and go, or is it persistent (there all the time)?
- Are you able to perform your usual daily activities without stopping to rest or nap?

For the Client with Sudden Onset of Joint Pain (Reiter's syndrome; see also Appendix B-18)

- Have you recently noticed any crusting, redness, or burning of your eyes?
- Have you noticed any burning when you urinate?
- Have you noticed an increase in the number of times you urinate?
- Have you had any bouts of diarrhea over the last 1 to 3 weeks (before the onset of joint pain)?
  - *If yes* to any of these questions, have you ever been told you have a sexually transmitted infection such as herpes, genital warts, Reiter's disease, or other disease?

For the Client with Fever (fevers recurring every few days, fevers that rise and fall within 24 hours, and fevers that recur frequently should be documented and reported to the physician)

- When did you first notice this fever?
- Is it constant, or does it come and go?
- Does your temperature fluctuate?
  - *If yes,* over what period of time does this occur?

CASE STUDY

REFERRAL

A 28-year-old Hispanic man has come to physical therapy for an evaluation without a medical referral. He has seen no medical practitioner for his current symptoms, consisting of an unusual gait pattern and weakness of the lower extremities, which he noticed during the last 2 days. He speaks English with a heavy accent, making it difficult to obtain a clear medical history, but the Personal/Family History form (see example in Fig. 2-2) indicates no previous or current health or medical problems of any kind. He does note that he has had influenza in the last 3 weeks but that he is fully recovered now.

PHYSICAL THERAPY INTERVIEW

Using the format outlined in the chapter on Interviewing as a Screening Tool (Chapter 2), begin with an open-ended question and follow up with additional appropriate questions incorporating the following:

Current Symptoms

- Tell me why you are here (open-ended question). Or, you may prefer to say, "I notice from your intake form that you have had some weakness in your legs and a change in the way you walk. What can you tell me about this?"
- When did you first notice these changes?
- What did you notice that made you think that something was happening?
- Just before the development of these symptoms, did you injure yourself in any way that you can remember?
- Did you have a car accident, fall down, or twist your trunk or hips in any unusual way?
- Do you have any pain in your back, hips, or legs? *If yes*, use Fig. 3-6 to elicit a further description.

Associated Symptoms

- Have you had any numbness or tingling in your back, buttocks, or hips or down your legs?
- Have you had any other changes in sensation in these areas such as a burning or prickling feeling?
- Besides the flu, have you had any other infection recently (e.g., head cold, upper respiratory infection, urinary tract infection)?
- Have you had a fever or elevated temperature in the last 48 hours?
- Do you think that you have a temperature right now?
- Have you noticed any other symptoms that I should know about?

Give the client time to answer this last question. Prompt him or her if necessary with various suggestions (include any others that seem appropriate to the information and responses already given by the client; a similar checklist is provided in Fig. 3-6) such as the following:

- Nausea or dizziness
- Diarrhea or constipation
- Unusual fatigue
- Choking, difficulty with chewing
- Recent headaches
- Vomiting
- Cold sweats during the day or night
- Changes in vision or speech
- Skin rashes
- Joint pains
- Shortness of breath with mild exertion (e.g., walking to the car or even at rest)
- Have you noticed any other respiratory, lung, or breathing problems?

Final Question

Is there anything else you think that I should know about your current condition or general health that I have not asked yet?

PROCEDURES TO CARRY OUT DURING THE FIRST SESSION

Given the client's report of lower extremity weakness and antalgic gait of sudden onset without precipitating cause, the following possible problems should be assessed during the examination:

- Neurologic disease or disorder (immunologically based or otherwise), such as:
  - Discogenic lesion
  - Tumor
  - Myasthenia gravis (unlikely because of the man's age)
  - Guillain-Barré syndrome (recent history of the "flu")
  - Multiple sclerosis
- AIDS dementia (unlikely, given the way the history was presented)
- Psychogenic disorder (e.g., hysteria, anxiety, alcoholism, or drug addiction)

Observation/Inspection

- Take the client's vital signs.
- Note any obvious changes such as muscle atrophy, difficulty with breathing or swallowing, facial paralysis, intention tremor.
- Describe the gait pattern: Observe for ataxia, incoordination, positive Trendelenburg position, balance, patterns of muscular weakness or imbalance, other gait deviations.

Neurologic Screening Examination

- All deep tendon reflexes
- Manual muscle testing of proximal-to-distal large muscle groups, looking for a pattern of weakness
- Babinski sign and clonus
- Gross sensory screen, looking for any differences in perceived sensation, proprioception, or vibration from one side to the other
- Test for dysmetria, balance, and coordination

CASE STUDY—*cont'd*

Orthopedic Assessment

- Lower extremity range of motion (ROM): active and passive
- Back, lower quadrant evaluation protocol[104]

Testing Results

In the case of this client, the interview revealed very little additional information because he denied any other associated (systemic) signs or symptoms and denied bowel/bladder dysfunction, precipitating injury or trauma, and neurologic indications such as numbness, tingling, or paresthesias. Although he was difficult to understand, the therapist thought that the client had understood the questions and had answered them truthfully. Subjectively, he did not appear to be a malingerer or a hysterical/anxious individual.

The client's gait pattern could best be described as ataxic. His lower extremities would not support him fully, and he frequently lost his balance and fell down, although he denied any pain or warning that he was about to fall.

Objective findings revealed inconsistent results of muscle testing: The proximal muscles were more involved than the distal muscles (difference of one grade: proximal muscles = fair grade; distal muscles = good grade), but repeated tests elicited alternately strong, weak, or cogwheel responses, as if the muscles were moving in a ratcheting motion against resistance through the ROM.

The only other positive findings were slightly diminished deep tendon reflexes of the lower extremities compared with the upper extremities, but, again, these findings were inconsistent when tested over time.

Final Results

Because the subjective and objective examinations were so inconsistent and puzzling, the therapist asked another therapist to briefly examine this client. In turn, the second therapist decided to ask the client to return either at the end of the day or for the first appointment of the next day to reexamine him for any changes in the pattern of his symptoms. It was more convenient for him to return the next day, and he did.

At that time, it became clear that the therapist's difficulty in understanding the client had less to do with his use of English as a second language and more to do with an increasingly slurred speech pattern. His gait remained unchanged, but the muscle strength of the proximal pelvic muscles was consistently weak over several trials spread out during the therapy session, which lasted for 1 hour.

This time, the therapist checked the muscles of his upper extremities and found that the scapular muscles were also unable to move against any manual resistance. Deep tendon reflexes of the upper extremities were inconsistently diminished, and reflexes of the lower extremities were now consistently diminished.

The client was referred to a physician for further follow-up and was not treated at the physical therapy clinic that day. He was examined by his family physician, who referred him to a neurologist. A diagnosis of Guillain-Barré syndrome was confirmed when the client's symptoms progressed dramatically, requiring hospitalization.

PRACTICE QUESTIONS

1. Fibromyalgia syndrome is a:
   a. Musculoskeletal disorder
   b. Psychosomatic disorder
   c. Neurosomatic disorder
   d. Noninflammatory rheumatic disorder
2. Which of the following best describes the pattern of rheumatic joint disease?
   a. Pain and stiffness in the morning gradually improves with gentle activity and movement during the day.
   b. Pain and stiffness accelerate during the day and are worse in the evening.
   c. Night pain is frequently associated with advanced structural damage seen on x-ray.
   d. Pain is brought on by activity and resolves predictably with rest.
3. Match the following skin lesions with the associated underlying disorder:

a. Raised, scaly patches	______ Psoriatic arthritis
b. Flat or slightly raised malar on the face	______ Systemic lupus erythematosus
c. Petechiae	______ HIV infection
d. Tightening of the skin	______ Scleroderma
e. Kaposi's sarcoma	______ Rheumatoid arthritis
f. Erythema migrans	______ Allergic reaction
g. Hives	______ Lyme disease
h. Subcutaneous nodules	______ Thrombocytopenia

4. A new client has come to you with a primary report of new onset of knee pain and swelling. Name three clues that this client might give from his medical history that should alert you to the possibility of immunologic disease.

Continued

PRACTICE QUESTIONS—*cont'd*

5. A positive Schober's test is a sign of:
   a. Reiter's syndrome
   b. Infectious arthritis
   c. Ankylosing spondylitis
   d. a or b
   e. a or c
6. What is Lhermitte's sign, and what does it signify?
7. Proximal muscle weakness may be a sign of:
   a. Paraneoplastic syndrome
   b. Neurologic disorder
   c. Myasthenia gravis
   d. Scleroderma
   e. b, c, and d
   f. All of the above
8. Which of the following skin assessment findings in the HIV-infected client occurs with Kaposi's sarcoma?
   a. Darkening of the nail beds
   b. Purple-red blotches or bumps on the trunk and head
   c. Cyanosis of the lips and mucous membranes
   d. Painful blistered lesions of the face and neck
9. The most common cause of change in mental status of the HIV-infected client is related to:
   a. Meningitis
   b. Alzheimer's disease
   c. Space-occupying lesions
   d. AIDS dementia complex
10. Symptoms of anaphylaxis that would necessitate immediate medical treatment or referral are:
   a. Hives and itching
   b. Vocal hoarseness, sneezing, and chest tightness
   c. Periorbital edema
   d. Nausea and abdominal cramping

REFERENCES

1. Klippel JH, editor: *Primer on the rheumatic diseases*, ed 13, Atlanta, 2008, Arthritis Foundation.
2. Centers for Disease Control and Prevention (CDC): *HIV/AIDS & STDs: fact sheet.* Available at http://www.cdc.gov/std/hiv/STDFact-STD-HIV.htm. Accessed January 19, 2011.
3. Centers for Disease Control and Prevention National Prevention Information Network (NPIN): *African-Americans.* Available at http://www.cdcnpin.org/scripts/population/afram.asp. Accessed January 4, 2011.
4. West-Ojo T: Expanded HIV testing and trends in diagnoses of HIV infection, 2004-2008. *MMWR* 59(24):737–741, 2010.
5. Jaffe H, Janssen R: Incorporating HIV prevention into the medical care of persons living with HIV. *MMWR* 52(RR 12):1–24, 2003.
6. McIntyre J: Preventing mother-to-child transmission of HIV: successes and challenges. *BJOG* 112(9):1196–1203, 2005.
7. Coutsoudis A: Infant feeding dilemmas created by HIV: South African experiences. *J Nutr* 135(4):956–959, 2005.
8. Zou S: Prevalence, incidence, and residual risk of human immunodeficiency virus and hepatitis C virus infections among United States blood donors since the introduction of nucleic acid testing. *Transfusion* 50:1495–1504, 2010.
9. Stramer SL: Third reported US case of breakthrough HIV transmission from NAT screened blood. *Transmission* 43(suppl):40A, 2003.
10. Laffoon B: HIV transmission through transfusion. *MMWR* 59(41):1335–1339, 2010.
11. Workowski KA: Sexually transmitted diseases treatment guidelines, 2010. *MMWR* 59(RR12):1–110, 2010.
12. Qaqish RB, Sims KA: Bone disorders associated with the human immunodeficiency virus: pathogenesis and management. *Pharmacotherapy* 24(10):1331–1346, 2004.
13. Abrescia N, D'Abbraccio M, Figoni M, et al: Hepatotoxicity of antiretroviral drugs. *Curr Pharm Des* 11(28):3697–3710, 2005.
14. Baba M: Advances in antiviral chemotherapy. *Uirusu* 55(1):69–75, 2005.
15. Nunez M: Clinical syndromes and consequences of antiretroviral-related hepatotoxicity. *Hepatology* 52(3):1143–1155, 2010.
16. Malita FM, Karelis AD, Toma E, et al: Effects of different types of exercise on body composition and fat distribution in HIV-infected patients: a brief review. *Can J Appl Physiol* 30(2):233–245, 2005.
17. Sweet DE: Metabolic complications of antiretroviral therapy. *Top HIV Med* 13(2):70–74, 2005.
18. Magkos F: Body fat redistribution and metabolic abnormalities in HIV-infected patients on highly active antiretroviral therapy. *Metabolism* 60(6):749–753, 2011. Epub October 20, 2010.
19. Troll JG: Approach to dyslipidemia, lipodystrophy, and cardiovascular risk in patients with HIV infection. *Curr Atheroscler Rep* 13(1):51–56, 2011. Epub Dec. 23, 2010.
20. Carbone A, Gloghini A: AIDS-related lymphomas: from pathogenesis to pathology. *Br J Haematol* 130(5):662–670, 2005.
21. Kasamon YL, Ambinder RF: AIDS-related primary central nervous system lymphoma. *Hematol Oncol Clin North Am* 19(4):665–687, 2005.
22. Sherman M: Hepatocellular carcinoma: epidemiology, risk factors, and screening. *Semin Liver Dis* 25(2):143–154, 2005.
23. Trends in tuberculosis—United States, 2010. *MMWR* 60(11):333–337, 2011.
24. Deshpande A, Mrinal M, Patnaik M: Nonopportunistic neurologic manifestations of the human immunodeficiency virus: an Indian study, *eJIAS. Medscape Gen Med* 7(3):1–2, 2005. Available at http://www.medscape.com/viewarticle/511865. Accessed online January 26, 2010.
25. Cade W, Peralta L, Keyser R: Aerobic exercise dysfunction in HIV: A potential link to physical disability. *Phys Ther* 84(7):655–664, 2004.
26. Frey Law LA: Spreading of pain: women vs. men. *J Pain* 9(4 suppl 2):P9 Abs #134, 2008.
27. Williams DA, Clauw DJ: Understanding fibromyalgia: lessons from the broader pain research community. *J Pain* 10(8):777–791, 2009.

28. Lowe JC, Yellin J: Inadequate thyroid hormone regulation as the main mechanism of fibromyalgia: a review of the evidence. *Thyroid Science* 3(6):R1–14, 2008.
29. Khanis A: Diagnosing fibromyalgia: moving away from tender points. *J Musculoskel Med* 27(4):155–162, 2010.
30. Wolfe F: Stop using the American College of Rheumatology criteria in the clinic. *J Rheumatol* 30(8):1671–1672, 2003.
31. Wolfe F: Pain extent and diagnosis: development and validation of the regional pain scale in 12,799 patients with rheumatic disease. *J Rheumatol* 30:369–378, 2003.
32. Wolfe F, Rasker JJ: The Symptom Intensity Scale, fibromyalgia, and the meaning of fibromyalgia-like symptoms. *J Rheumatol* 22:2291–2299, 2006.
33. Wolfe F: The American College of Rheumatology preliminary diagnostic criteria for fibromyalgia and measurement of symptom severity. *Arthritis Care Res* 62:600–610, 2010.
34. Bernstein C, Marcus D: Fibromyalgia: Current concepts in diagnosis, pathogenesis, and treatment. *Pain Medicine News* 6(9):8–19, December 2008.
35. Lowe JR, Yellin JG: *The metabolic treatment of fibromyalgia*, Utica, KY, 2000, McDowell Publishing Company. (Available from McDowell Publishing Company; phone: (603) 769–9590; online at http://www.McDowellPublishing.com).
36. American Association for Chronic Fatigue Syndrome: CFS Conference Highlights: The merging of two syndromes. *Fibromyalgia Network* 61:4–70, 2003.
37. Weissbecker I: Childhood trauma and diurnal cortisol disruption in fibromyalgia syndrome. *Psychoneuroendocrinology* 31(3):312–327, 2006.
38. Bradley LA: Pathophysiologic mechanisms of fibromyalgia and its related disorders. *J Clin Psychiatr* 69(suppl 2):6–13, 2008.
39. Wallace DJ: Hypothesis: bipolar illness with complaints of chronic musculoskeletal pain is a form of pseudofibromyalgia. *Semin Arthritis Rheum* 37:256–259, 2008.
40. Leventhal L, Bouali H: Fibromyalgia: 20 clinical pearls. *J Musculoskel Med* 20(2):59–65, 2003.
41. Clauw D: Fibromyalgia: Correcting the misconceptions. *J Musculoskel Med* 20(10):467–472, 2003.
42. Weingarten TN: Impact of tobacco use in patients presenting to a multidisciplinary outpatient treatment program for fibromyalgia. *Clin J Pain* 25(1):39–43, 2009.
43. Hulme J: *Fibromyalgia: a handbook for self-care and treatment*, ed 3, Missoula, MT, 2001, Phoenix. www.phoenixpub.com; 1-800-549-8371.
44. Koch A: Targeting cytokines and growth factors in RA. *J Musculoskel Med* 22(3):130–136, 2005.
45. Yavari N: What role do occupational exposures play in RA? *J Musculoskel Med* 25(3):130–136, 2008.
46. Andreoli TE, Benjamin I, Griggs, RC, et al: *Andreoli and Carpenter's Cecil essentials of medicine*, ed 8, Philadelphia, 2010, WB Saunders.
47. Anderson D: Development of an instrument to measure pain in rheumatoid arthritis: Rheumatoid Arthritis Pain Scale (RAPS). *Arthritis Care Res* 45:317–323, 2001.
48. Freeston J, Keenan AM, Emery P: Spotting the early warning signs of aggressive RA. *J Musculoskel Med* 25(3):110–115, 2008.
49. Bykerk V, Keystone E: RA in primary care: 20 clinical pearls. *J Musculoskel Med* 21(3):133–146, 2004.
50. Yazici Y, Erkan D, Paget S: Inflammatory musculoskeletal diseases in the elderly, Part I. *J Musculoskel Med* 19(7):265–276, 2002.
51. Moorthy L, Onel K: Juvenile idiopathic arthritis: making the diagnosis. *J Musculoskel Med* 21(11):581–588, 2004.
52. Clinical Update: advances in rheumatology: new approaches to early diagnosis and treatment. *J Musculoskel Med* 21(10): 509–560, 2004.
53. Olson NY, Lindsley CB: Advances in pediatric rheumatology paving the way to better care. *J Musculoskel Med* 25:505–512, 2008.
54. Flendrie M, Vissers WH, Creemers MC, et al: Dermatological conditions during TNR-alpha-blocking therapy in patients with rheumatoid arthritis: a prospective study. *Arthritis Res Ther* 7(3):R666–R676, 2005.
55. Taylor M, Furst D: Biologic response modifiers: addressing the safety concerns. *J Musculoskel Med* 22(5):223–239, 2005.
56. Labbe P, Hardouin P: Epidemiology and optimal management of polymyalgia rheumatica. *Drugs Aging* 13(2):109–118, 1998.
57. Spiera R, Spiera H: Inflammatory diseases in older adults: polymyalgia rheumatica. *Geriatrics* 59(11):39–43, 2004.
58. Gonzalez-Gay MA: Giant cell arteritis: epidemiology, diagnosis, and management. *Curr Rheumatol Rep* 12(6):436–442, 2010.
59. Reiter S: Giant cell arteritis misdiagnosed as temporomandibular disorder: a case report and review of the literature. *J Orofac Pain* 23(4):360–365, 2009.
60. Salvarani C, Cantini F, Bioardi L, et al: Polymyalgia rheumatica. *Best Pract Res Clin Rheumatol* 18(5):705–722, 2004.
61. Hernandez-Rodriguez J: Medical management of polymyalgia rheumatica. *Expert Opin Pharmacother* 11(7):1077–1087, 2010.
62. Grau RH: Cutaneous cues to diagnosis of lupus. *J Musculoskel Med* 24(6):247–263, 2007.
63. Lupus Foundation of America (LFA): *Understanding lupus*. (One of five patient education booklets available at www.lupus.org.) Accessed March 3, 2011.
64. Formica MK, Palmer JR, Rosenberg L, et al: Smoking, alcohol consumption, and risk of systemic lupus erythematosus in the Black Women's Health Study. *J Rheumatol* 30:1222–1226, 2003.
65. Petri M: Systemic lupus erythematosus: new management strategies. *J Musculoskel Med* 22(3):108–116, 2005.
66. Sayarlioglu M: Risk factors for avascular bone necrosis in patients with systemic lupus erythematosus. *Rheumatol Int* Epub Aug 15, 2010.
67. Bertsias GK: EULAR recommendations for the management of systemic lupus erythematosus with neuropsychiatric manifestations. Report of a task force of the EULAR standing committee for clinical affairs. *Ann Rheum Dis* 69(12):2074–2082, 2010.
68. Muscal E, Brey RL: Neurologic manifestations of systemic lupus erythematosus in children and adults. *Neurol Clin* 28(1):61–73, 2010.
69. Zulian F, Vallongo C, Woo P, et al: Localized scleroderma in childhood is not just a skin disease. *Arthritis Rheum* 52(9): 2873–2881, 2005.
70. Steen VD: Clinical manifestations of systemic sclerosis. *Semin Cutan Med Surg* 17(1):48–54, 1998.
71. Steen VD: Autoantibodies in systemic sclerosis. *Semin Arthritis Rheum* 35(1):35–42, 2005.
72. Steen VD, Syzd A, Johnson JP, et al: Kidney disease other than renal crisis in patients with diffuse scleroderma. *J Rheumatol* 32(4):649–655, 2005.
73. Moxley G: *Scleroderma and related diseases, WebMD Scientific American® Medicine*. Posted 03/22/2004. Available at: www.medscape.com/viewarticle/472036. Accessed November 28, 2005.
74. Highland KB, Silver RM: New developments in scleroderma interstitial lung disease. *Curr Opin Rheumatol* 17(6):737–745, 2005.
75. Hassoun M: Lung involvement in systemic sclerosis. *Presse Med* 40(1 Pt 2):e3–e17, 2010.
76. Haroon N, Inman RD: Ankylosing spondylitis: new criteria, new treatments. *Bull NYU Hosp Joint Dis* 68(3):171–174, 2010.

77. Zeidler H, Amor B: The Assessment in Spondyloarthritis International Society (ASAS) classification criteria for peripheral spondyloarthritis and for spondyloarthritis in general: the spondyloarthritis concept in progress. *Ann Rheum Dis* 70:1–3, 2011.
78. Gomez KS, Raza K, Jones SD, et al: Juvenile onset ankylosing spondylitis: more girls than we thought? *J Rheumatol* 24(4):735–737, 1997.
79. Ostensen M, Ostensen H: Ankylosing spondylitis: the female aspect. *J Rheumatol* 25(1):120–124, 1998.
80. Toussirot E: Late-onset ankylosing spondylitis and spondylarthritis: an update on clinical manifestations, differential diagnosis, and pharmacological therapies. *Drugs Aging* 27(7): 523–531, 2010.
81. Levine DS, Forbat SM, Saifuddin A: MRI of the axial skeletal manifestations of ankylosing spondylitis. *Clin Radiol* 59(5): 400–413, 2004.
82. Lewis R, Creamer P: Ankylosing spondylitis: early diagnosis and management. *J Musculoskel Med* 20(4):184–198, 2003.
83. Hitchon P, From A, Brenton M, et al: Fractures of the thoracolumbar spine complicating ankylosing spondylitis. *J Neurosurg (Spine 2)* 97:218–222, 2002.
84. Tullous MW, Skerhut HEI, Story JL, et al: Cauda equina syndrome of long-standing ankylosing spondylitis: case report and review of the literature. *J Neurosurg* 73:441–447, 1990.
85. Ahn NU, Ahn UM, Nallamshetty L, et al: Cauda equina syndrome in ankylosing spondylitis (the CES-AS syndrome): meta-analysis of outcomes after medical and surgical treatments. *J Spinal Disord* 14(5):427–433, 2001.
86. Korendowych E, McHugh N: Genetic factors in psoriatic arthritis. *Curr Rheumatol Rep* 7(4):306–312, 2005.
87. Leong TT, Faron U, Veale DJ: Angiogenesis in psoriasis and psoriatic arthritis: clues to disease pathogenesis. *Curr Rheumatol Rep* 7(4):325–329, 2005.
88. Qureshi AA, Husni ME, Mody E: Psoriatic arthritis and psoriasis: need for a multidisciplinary approach. *Semin Cutan Med Surg* 24(1):46–51, 2005.
89. Milewski MD: Lyme arthritis in children presenting with joint effusions. *J Bone Joint Surg* 93A(2):252–260, 2011.
90. Smith BG: Lyme disease and the orthopaedic implications of Lyme arthritis. *J Am Acad Orthop Surg* 19(2):91–100, 2011.
91. Kalish R, Biggee B: Lyme disease: 20 clinical pearls. *J Musculoskel Med* 20(6):271–285, 2003.
92. Making sense of multiple sclerosis. *Harvard Women's Health Watch* 12(11):4–6, 2005.
93. Munger KL, Zhang SM, O'Reilly E, et al: Vitamin D intake and incidence of multiple sclerosis. *Neurology* 62(1):60–65, 2004.
94. Galea I, Newman TA, Gidron Y: Stress and exacerbations in multiple sclerosis. *BMJ* 328(7434):287, 2004.
95. Haslam C: Managing bladder symptoms in people with multiple sclerosis. *Nursing Times* 101(2):48–52, 2005.
96. Calabresi P: Diagnosis and management of multiple sclerosis. *Am Fam Physician* 70(10):1–14, 2004.
97. Lee AG, Berlie CL: Multiple sclerosis, *eMedicine* updated September 22, 2010. Available at: www.emedicine.com/oph/topic179.htm. Accessed March 3, 2011.
98. Ramchandren S: The immunopathogenesis of Guillain-Barré syndrome. *Clin Adv Hematol Oncol* 8(3):203–206, 2010.
99. Lehmann HC: Guillain-Barré syndrome after exposure to influenza virus. *Lancet Infect Dis* 10(9):643–651, 2010.
100. Farrugia ME, Vincent A: Autoimmune mediated neuromuscular junction defects. *Curr Opin Neurol* 23(5):489–495, 2010.
101. Cantor F: Central and peripheral fatigue: exemplified by multiple sclerosis and myasthenia gravis. *PM R* 2(5):399–405, 2010.
102. Angelini C: Diagnosis and management of autoimmune myasthenia gravis. *Clin Drug Investig* 31(1):1–14, 2011.
103. Scherer K, Bedlack RS, Simel L: Does this patient have myasthenia gravis? *JAMA* 293(15):1906–1914, 2005.
104. Magee DJ: *Orthopedic physical assessment*, ed 5, Philadelphia, 2008, WB Saunders.
105. Janssen RS, Onorato IM: Advancing HIV prevention: new strategies for a changing epidemic—United States, 2003. *MMWR* 52(15):329–332, 18, 2003.
106. Porter K, Babiker A, Bhaskaran K: Determinants of survival following HIV-1 seroconversion after the introduction of HAART. *Lancet* 362(9392):1267–1274, 2003.
107. Kubis KC, Danesh-Meyer HV, Savino PJ, et al: The ice test versus the rest test in myasthenia gravis. *Ophthalmology* 107(11):1995–1998, 2000.

CHAPTER

13

Screening for Cancer

A 56-year-old man has come to you for an evaluation without a referral. He has not seen any type of physician for at least 3 years. He is seeking an examination at the insistence of his wife, who has noticed that his collar size has increased two sizes in the last year and that his neck looks "puffy." He has no complaints of any kind (including pain or discomfort), and he denies any known trauma; however, his wife insists that he has limited ability in turning his head when backing the car out of the driveway.

- What questions would be appropriate for your first physical therapy interview with this client?
- What test procedures will you carry out during the first session?
- If you suggest to this man that he should see his physician, how would you make that recommendation? (See the Case Study at the end of the chapter.)

A large part of the screening process is identifying red flag histories and red flag signs and symptoms. Advancing age and previous history of any kind of cancer are two of the most important risk factors for cancer. Following the screening model presented in Chapters 1 and 2, the therapist will use past medical history, clinical presentation, and associated signs and symptoms as the basic tools to screen for cancer.

The client history with interview is the number one tool for cancer screening. Take the client's history, looking for the presence of any risk factors for cancer. Cancer in its early stages is often asymptomatic. Survival rates are increased with early detection and screening, making this element of client management extremely important.

Keep in mind that some cancers, such as malignant melanoma (skin cancer), do not have a highly effective treatment. Early detection and referral can make a life and death difference in the final outcome. Morbidity can be reduced and quality of life and function improved with early intervention.

Whether primary cancer, cancer that has recurred locally, or cancer that has metastasized, clinical manifestations can mimic neuromuscular or musculoskeletal dysfunction. The therapist's task is to identify abnormal tissue, not diagnose the lesion.

CANCER STATISTICS

Cancer accounts for more deaths than heart disease in the United States in persons under the age of 85 years. There are more than 1.5 million new cases of cancer in the United States each year; more than half a million will die from cancer this year. One in four deaths in the United States is attributed to cancer.[1]

Predicting lifetime risk of cancer is based on present rates of cancer. Using today's epidemiologic data, 44% of all men and 38% of all women will develop cancer at some time in their lifetime.[1] It is estimated that by 2030, 20% of the U.S. population will be 65 years old or older, accounting for 70% of all cancers and 85% of all cancer-related mortality.[2]

In the past, certain types of cancer were invariably fatal. Today, however, death rates continue to decline for most cancers, and there continues to be a reported reduced mortality from cancer. The percentage of people who have survived longer than 5 years after cancer diagnosis has increased over the past 2 decades.[3]

Fig. 13-1 summarizes current U.S. figures for cancer incidence and deaths by site and sex. Although prostate and breast cancers are the most common malignancies in men and women, respectively, the cancer that most commonly causes death is lung cancer.[1]

Carcinoma in situ is not included in the statistics related to invasive carcinoma or sarcoma as reported by the American Cancer Society (ACS) or the National Cancer Institute (NCI). Carcinoma in situ is considered a premalignant cancer that is localized to the organ of origin. As noted, it is reported separately and primarily relative to breast and skin cancer. Carcinoma in situ of the breast accounts for about 46,000 new cases every year, and in situ melanoma accounts for about 50,000 new cases annually.[1]

Cancer Cure and Recurrence

Cancer is considered cured or in remission when evidence of the disease cannot be found in the individual's body. Early diagnosis and aggressive intervention help people obtain a

Estimated New Cases*

Males			Females		
Prostate	240,890	29%	Breast	230,480	30%
Lung & bronchus	115,060	14%	Lung & bronchus	106,070	14%
Colon & rectum	71,850	9%	Colon & rectum	69,360	9%
Urinary bladder	52,020	6%	Uterine corpus	46,470	6%
Melanoma of the skin	40,010	5%	Thyroid	36,550	5%
Non-Hodgkin lymphoma	37,120	5%	Non-Hodgkin lymphoma	30,300	4%
Kidney & renal pelvis	36,060	4%	Melanoma of the skin	30,220	4%
Oral cavity & pharynx	27,710	3%	Kidney & renal pelvis	23,800	3%
Leukemia	25,320	3%	Ovary	21,990	3%
Pancreas	22,050	3%	Pancreas	21,980	3%
All Sites	**822,300**	**100%**	**All Sites**	**774,370**	**100%**

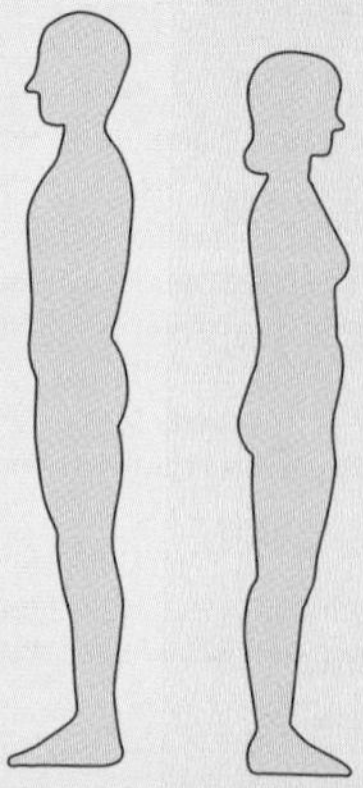

Estimated Deaths

Males			Females		
Lung & bronchus	85,600	28%	Lung & bronchus	71,340	26%
Prostate	33,720	11%	Breast	39,520	15%
Colon & rectum	25,250	8%	Colon & rectum	24,130	9%
Pancreas	19,360	6%	Pancreas	18,300	7%
Liver & intrahepatic bile duct	13,260	4%	Ovary	15,460	6%
Leukemia	12,740	4%	Non-Hodgkin lymphoma	9,570	4%
Esophagus	11,910	4%	Leukemia	9,040	3%
Non-Hodgkin lymphoma	10,670	4%	Uterine corpus	8,120	3%
Urinary bladder	9,750	3%	Liver & intrahepatic bile duct	6,330	2%
Kidney & renal pelvis	8,270	3%	Brain & other nervous system	5,670	2%
All Sites	**300,430**	**100%**	**All Sites**	**271,520**	**100%**

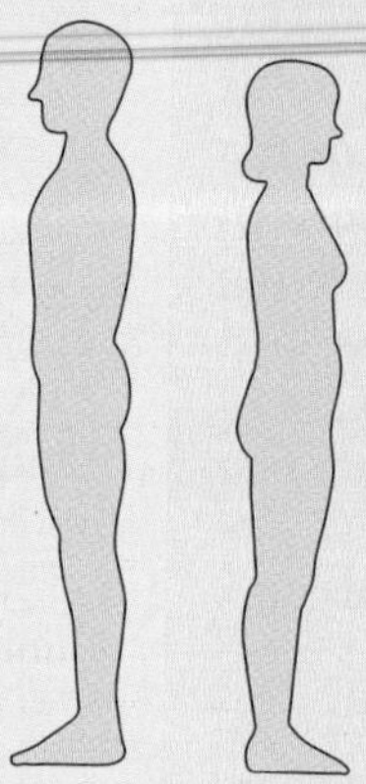

Fig. 13-1 Estimated new cases of cancer and cancer deaths by site for men and women. (From Siegel R, Ward E, Brawley O, et al: Cancer statistics, 2011: the impact of eliminating socioeconomic and racial disparities on premature cancer deaths, *CA Cancer J Clin* 61:212-236, 2011. © 2011 American Cancer Society.)
*Estimates are rounded to the nearest 10 and exclude basal and squamous cell skin cancers and in situ carcinoma except urinary bladder.

cure. In general, individuals with no evidence of cancer are considered to have the same life expectancy as those who never had cancer. However, late physical and psychosocial complications of disease and treatment are being recognized.

Cancer recurrence or a new cancer can occur in some individuals with a previous personal history of cancer. Causes of cancer recurrence can include inadequate surgical margin, skip metastases, tumor thrombus, and lymph node metastasis.

Additionally, many of the antineoplastic strategies (e.g., chemotherapy, hormone therapy, radiation therapy) mutate cells further and can initiate or stimulate new malignant tumors. Besides second malignancies, these treatments can come with many unintended long-term problems and adverse consequences (e.g., fibrosis, occlusive coronary artery disease and cardiotoxicity, impaired motion and strength). The therapist should consider it a red flag any time a client has a previous history of cancer or cancer treatment.

Childhood Cancers

Cancer is the second leading cause of death in children between the ages of 1 and 14, with accidents remaining the most frequent cause of death in this age group. The most frequently occurring cancers in children are leukemia (primarily, acute lymphocytic leukemia), brain and other nervous system cancers, soft tissue sarcomas, non-Hodgkin lymphoma, and renal (Wilms') tumor.[1]

Survival rates for childhood cancer have increased to 81% now because of improvements in treatment for many types of cancer over the past 2 decades. The result is an increasing population of long-term cancer survivors (or "overcomers" as some "survivors" prefer to call themselves). Currently, 1 in 900 young adults is a childhood cancer survivor. Long-term health problems related to the effects of cancer therapy are a major focus of this population group.[4]

It has been shown that, to varying degrees, long-term survivors of childhood cancer are at risk of developing second cancers and of experiencing organ dysfunction, such as cardiomyopathy, joint dysfunction, reduced growth and development, decreased fertility, and early death.[5]

The degree of risk of late effects may be influenced by various treatment-related factors such as the intensity, duration, and timing of therapy. Individual characteristics, such as the type of cancer diagnosis, the person's sex, age at time of intervention, and genetic factors as indicated by, for example, family history of cancer, may also play a role in cancer recurrence and late effects of treatment.[5,6]

RISK FACTOR ASSESSMENT

Risk factor assessment is a part of the cancer prevention model. Every health care professional has a role and a responsibility to help clients identify risk factors for disease. Knowing the various risk factors for different kinds of cancers is an important part of the medical screening process. Educating clients about their risk factors is a key element in risk factor reduction.

A new branch of medicine called *preventive oncology* has developed to address this important area. Preventive oncology or *chemoprevention* includes primary and secondary prevention. Chemoprevention is based on the hypothesis that certain nontoxic chemicals (e.g., retinoids, cyclooxygenase-2 [COX-2] inhibitors, hormonal agents) can be given preventatively to interrupt the biological processes involved in carcinogenesis and thus reduce its incidence. Currently, many clinical trials and studies have been devoted to the idea of chemical prevention of cancer development.[7]

Therapists can have an active role in both primary and secondary prevention through screening and education. Primary prevention involves stopping the processes that lead to the formation of cancer in the first place. According to the *Guide to Physical Therapist Practice* (the *Guide*),[8] physical therapists are involved in primary prevention by "preventing a target condition in a susceptible or potentially susceptible population through such specific measures as general health promotion efforts."[8] Risk factor assessment and risk reduction fall under this category.

Secondary prevention involves regular screening for early detection of cancer and the prevention of progression of known premalignant lesions such as skin and colon lesions. This does not prevent cancer but improves the outcome. The *Guide* outlines the physical therapist's role in secondary prevention as "decreasing duration of illness, severity of disease, and number of sequelae through early diagnosis and prompt intervention."[8]

Another way to look at this is through the use of screening and surveillance. *Screening* is a method for detecting disease or body dysfunction before an individual would normally seek medical care. Medical screening tests are usually administered to individuals who do not have current symptoms but who may be at high risk for certain adverse health outcomes.

Surveillance is the analysis of health information to look for problems that may be occurring in the workplace that require targeted prevention. Surveillance has often used screening results from groups of individuals to look for abnormal trends in health status.

Known Risk Factors for Cancer

Certain risk factors have been identified as linked to cancer in general (Table 13-1). More than half of all cancer deaths in the United States could be prevented if Americans adopted a healthier lifestyle and made better use of available screening tests.[9]

According to the NCI, environmental factors, whether linked to lifestyle issues such as smoking and diet or exposure to carcinogens in the air and water, are thought to be linked to an estimated 80% to 90% of cancer cases.[10]

Some of the most common risk factors for cancer include the following:

- Age over 50 (single most important risk factor)
- Ethnicity
- Family history (1st generation)
- Environment and lifestyle

TABLE 13-1 Risk Factors for Cancer

Nonmodifiable Risk Factors	Modifiable Risk Factors
Age	Smoking, use of smokeless tobacco
Previous history of cancer	Chemical or other exposure (e.g., paint, cadmium, dye, rubber, arsenic, asbestos, radon, benzene, ionizing radiation, Agent Orange, pesticides, herbicides, organic amines)
Ethnicity	Urban dwelling
Skin color	Alcohol consumption (more than 1-2 drinks per day)
Gender	Sedentary lifestyle; lack of exercise
Heredity (identified oncogenes)	Obesity; diet high in animal fat
Age of menarche, menopause	Insulin resistance (elevated serum insulin)
Adenomatous polyps	Radiation/chemotherapy treatment
Inflammatory bowel disease	Estrogen replacement therapy
Fat distribution patterns	Sexually transmitted diseases
Congenital immunodeficiencies	Ionizing radiation
Congenital diseases	HTLV-1 (virus, rare in United States)
Long-term *Helicobacter* infection	Previous lung scarring
	Organ transplantation (immunosuppression)
	HIV infection
	Chronic exposure to UV rays
	Geographic location
	Smoked foods, salted fish and meat (nitrates and nitrites)
	Tamoxifen use
	Nulliparity (never having children)
	Vitamin B_{12} deficiency
	Lack of access to or use of health care and screening tests

HTLV-1, Human T-lymphotropic virus type 1; *HIV,* human immunodeficiency virus; *UV,* ultraviolet.

Age

The majority of cancer incidence and mortality occurs in individuals aged 65 and older.[11] With estimates from the U.S. Census Bureau predicting a rapid rise in the number of people 65 years old and older, the therapist must pay close attention to the client's age, especially in correlation with a personal or family history of cancer. Many cancers, such as prostate, colon, ovarian, and some chronic leukemias, have increased incidence in older adults. The incidence of cancer doubles after 25 years of age and increases with every 5-year increase in age until the mid-80s, when cancer incidence and mortality reach a plateau and even decline slightly.

Other cancers occur within very narrow age ranges. Testicular cancer is found in men from about 20 to 40 years of age. Breast cancer shows a sharp increase after age 45. Ovarian cancer is more common in women older than 55. A number of cancers, such as Ewing sarcoma, acute leukemia, Wilms' tumor, and retinoblastoma, occur mainly in childhood.

Screening for age is discussed more completely in Chapter 2. Please refer to this section for information on screening for this red flag/risk factor.

Ethnicity

Racial/ethnic minorities account for a disproportionate number of newly diagnosed cancers. African Americans have a 10% higher incidence rate than whites and a 30% higher death rate from all cancers combined than whites.[1,12]

African Americans have the highest mortality and worst survival of any population, and diagnosis occurs at a later stage.[1] The statistics have gotten worse over the past 20 years, and studies have shown that equal treatment yields equal outcomes among individuals with equal disease.[13,14]

Compared with the general population, African Americans die of cancer at a 40% greater rate (they are 30% more likely to die from heart disease). The Institute of Medicine (IOM) document, *Unequal Treatment, Confronting Racial and Ethnic Disparities in Health Care,* suggests that care providers may be part of the problem.[14]

In terms of risk assessment, therapists must keep these figures in mind when examining and evaluating clients of African-American descent. For any ethnic group, the therapist is advised to be aware of cancer and disease demographics and epidemiology for that particular group.

Cancer statistics for Hispanic Americans compared with non-Hispanic Americans are becoming more available.[1,15,16] There are lower rates of incidence and mortality from the four major cancer killers (breast, prostate, lung, colorectal) but a higher incidence and mortality from cancer with an infectious etiology (stomach, liver, uterine cervix, gallbladder).

The use of cancer screening tests and early detection has been increasing among this group. Mammography among Hispanic women exceeds the national average, but screening for colorectal, cervical, and prostate cancers is below average.

Cancer statistics and epidemiology in this group are problematic. Hispanic people originate from 23 different countries and have enormous diversity among themselves. They are the poorest minority group and have the highest uninsured rate of all groups. The uninsured are less likely to get preventive care such as cancer screening.

The most common cancer among Hispanic women is breast cancer; second is lung cancer. Men are more likely to have prostate cancer but die more often of lung cancer. Hispanics have twice the incidence rate and a 70% higher death rate from liver cancer compared with non-Hispanics. This type of cancer is on the rise in Hispanic women. Cancer is typically diagnosed in Hispanics at a later stage than in non-Hispanic white Americans. Consequently, they have lower cure rates.

Therapists can offer health care education and cancer screening to this unique group of people. This will be increasingly common in our practice as our health care delivery system moves from an illness-based system to a health promotion–based system. For all groups, high-quality prevention and early detection and intervention can reduce cancer incidence and mortality.[3] Screening for ethnicity is discussed more completely in Chapter 2. Please refer to this section for information on screening for this important red flag/risk factor.

Family History and Genetics

Family history is often an important factor in the development of some cancers. This usually includes only first generation family members, including parents, siblings, and children.

Hereditary cancer syndromes account for approximately 5% of breast, ovarian, and colon cancers. Both clients and providers are becoming aware of the potential therapeutic advantages of early identification of hereditary cancer risk.[16a]

The hereditary syndromes most frequently identified are hereditary breast and ovarian cancer (HBOC) syndrome due to mutations in BRCA1 and BRCA2 genes; hereditary colon cancer (HCC), specifically, familial adenomatous polyposis (FAP); and hereditary nonpolyposis colorectal cancer.

The small percentage of people who may be suspected of a hereditary cancer syndrome can be screened regarding personal and family medical history. Critical details, such as the cancer site and age at diagnosis, are needed for risk assessment. The following are some basic hallmarks of families who could have a hereditary cancer syndrome[17]:

- Diagnosis of cancer in two or more relatives in a family
- Diagnosis of cancer in a family member under the age of 50
- Occurrence of the same type of cancer in several members of a family
- Occurrence of more than one type of cancer in one person
- Occurrence of a rare type of cancer in one or more members of a family[18]

BOX 13-1 CANCERS LINKED TO OBESITY, DIET, AND NUTRITION

Mouth, pharynx, esophagus	Colon, rectum
Larynx	Breast
Lung	Ovary
Stomach	Endometrium
Pancreas	Cervix
Gallbladder	Prostate
Liver	Kidney
Uterus	Bladder

Data from American Institute for Cancer Research (AICR): *Food, nutrition, and the prevention of cancer: A global perspective,* Washington, DC, 2007, AICR.

Environment and Lifestyle Factors

It is now apparent that, although genetic predisposition varies, the two key factors determining whether people develop cancer are environment and lifestyle. The most important way to reduce cancer risk is to avoid cancer-causing agents.

Obesity, diet, sedentary lifestyle, sexual practices, and the use of tobacco, alcohol, and/or other drugs make up the largest percentage of modifiable risk factors for cancer. Current data support the findings that obesity, inappropriate diet, and excess weight cause around one-third of all cancer deaths (Box 13-1).[19,20] Increased body weight and obesity (as measured by body mass index [BMI], an approximation of body adiposity) are associated with increased death rates for all cancers and for cancers at specific sites, especially when combined with a sedentary lifestyle.[21-25]

Overweight and obesity may account for 20% of all cancer deaths in U.S. women and 14% in U.S. men.[26] It is estimated that 90,000 cancer deaths could be prevented each year if Americans maintained a healthy body weight.[21]

Excess body weight increases amounts of circulating hormones, such as estrogens, androgens, and insulin,[27,28] all of which are associated with cellular and tumor growth. It has also been shown that physical activity reduces the risk of breast and colon cancers and may reduce the risk of several other types of cancer by decreasing excess body weight and by actually decreasing the circulation of some of the growth-related hormones.[29]

In 1999, the American Institute for Cancer Research (AICR) estimated that at least 20% of all cancers could be prevented if everyone ate at least 5 (½ cup) servings of fruits and vegetables each day.[29] Numerous resources on nutrition and its influence in preventing and treating cancer are available.[30-32]

Dietary guidelines were updated to 9 servings a day (equal to 4½ cups) for overall health and chemoprevention in January 2005 by the U.S. Department of Health and Human Services (HHS) in the publication *Dietary Guidelines for Americans 2005.* The *Guidelines* provide authoritative advice for people 2 years of age and older about how good dietary habits can promote health and reduce risks of major chronic diseases. The numbers of avoidable cancers through avoidance of excess weight are substantial.[33]

Specific factors associated with individual cancer types are known in some cases. For example, inadequate hydration is known to increase the risks of colon and bladder cancers. Alcohol consumption is linked with breast, head or neck, and gastrointestinal (GI) cancers. High dietary animal fat intake and tobacco use increase prostate cancer risk. Current smoking is an additive risk factor when combined with obesity for esophageal squamous cell carcinoma and lung and pancreatic cancers.[33] Adenomatous polyps in the colon are known precursors of colorectal cancer.

Sexually Transmitted Infections. Sexually transmitted diseases (STDs) or sexually transmitted infections (STIs) have been positively identified as a risk factor for cancer. Not all STIs are linked with cancer, but studies have confirmed that human papillomavirus (HPV) is the primary cause of cervical cancer (see Fig. 4-20). With current technology, high-risk HPV DNA can be detected in cervical specimens.[34]

HPV is the leading viral STI in the United States. More than 100 types of HPV have been identified; more than 30 types are transmitted sexually[35]: 23 infect the cervix and 13 types are associated with cancer (men and women). Infection with one of these viruses does not predict cancer, but the risk of cancer is increased.

In 1970, 1 of every 300 Americans had an STI. Today, 3 million teenagers contract STIs every year; STIs affect about 1 in 4 teens who are sexually active and 50% to 75% of sexually active adults will develop HPV sometime in their lives.[36] Nearly 50% of African-American teenagers have genital herpes.[37] For every unwed adolescent who gets pregnant this year, 10 teenagers will get an STI.[38,39]

Tobacco Use. Tobacco and tobacco products are known carcinogens, not just for lung cancer but also for leukemia and cancers of the cervix, kidney, pancreas, stomach, bladder, esophagus, and oropharyngeal and laryngeal structures. This includes second-hand smoke, pipes, cigars, cigarettes, and chewing (smokeless) tobacco. Combining tobacco with caffeine and/or alcohol brings on additional problems. More people die from tobacco use than from use of alcohol and all the other addictive agents combined.

In any physical therapy practice, clients should be screened for the use of tobacco products (see Fig. 2-2). Client education includes a review of the physiologic effects of tobacco (see Table 2-3). For a more complete discussion of screening for tobacco use, see Chapter 2.

If the client indicates a desire to quit smoking or using tobacco, the therapist must be prepared to help him or her explore options for smoking cessation. Pamphlets and other reading material should be available for any client interested in tobacco cessation. Referral to medical doctors who specialize in smoking cessation may be appropriate for some clients.

Occupation and Local Environment. Well-defined problems occur in people engaging in specific occupations, especially involving exposure to chemicals and gases. Exposure to carcinogens in the air and water and on our food sources may be linked to cancer. Isolated cases of excessive copy toner dust linked with lung cancer have been reported.[40,41]

Reactions can be delayed up to 30 years, making client history an extremely important tool in identifying potential risk factors. People may or may not even remember past exposures to chemicals or gases. Some may not be aware of childhood exposures. Taking a work or military history may be important (see Chapter 2 and Appendix B-14).

The industrial chemicals people are exposed to vary across the country and will depend on where the individual has lived or where the client lives now. Each state in the United States has its own unique environmental issues. For example, in Montana, there has been a significant chlorine spill, exposure to agricultural chemicals, vermiculite mining, and many other forms of mining.

In New York, Love Canal was the focus of concern in the 1980s and 1990s, when the effects of hazardous wastes dumped in the area were discovered. Alaskan oil spills, air pollution in Los Angeles, and hazardous and radioactive nuclear waste in Washington state burial grounds are a few more examples.

In Utah, Nevada, and Arizona, the Radiation Exposure Compensation Act (RECA) was passed by Congress in 1990 after studies showed a possible link between hundreds of above-ground nuclear tests in the late 1950s and early 1960s and various cancers and primary organ diseases.[42-45] Groundwater wells at old open-pit copper mines in various states have tested positive for uranium up to 40 times higher than legal limits. Hundreds of active wells tap into groundwater within 5 miles of these sites.

Wherever the therapist practices, it is important to be aware of local environmental issues and the impact these may have on people in the vicinity.

Ionizing Radiation. Exposure to ionizing radiation is potentially harmful. Ionizing radiation is the result of electromagnetic waves entering the body and acting on neutral atoms or molecules with sufficient force to remove electrons, creating an ion. The most common sources of ionizing radiation exposure in humans are accidental environmental exposure and medical, therapeutic, or diagnostic irradiation.

Nonionizing radiation is electromagnetic radiation that includes radio waves, microwaves, infrared light, and visible light. Nonionizing radiation does not have enough energy to ionize (i.e., break up) atoms. Electronic devices, such as laser scanners, high-intensity lamps, and electronic antitheft surveillance devices, expose the human to nonionizing radiation. There is not a proven link between exposure to nonionizing radiation and cancer, but there is considerable speculation that long-term exposure to electromagnetic fields may be correlated with the development of various illnesses and diseases.

Some studies have reported the possibility of increased cancer risks, especially of leukemia and brain cancer, for electrical workers and others whose jobs require them to be around electrical equipment. Additional risk factors, however, such as exposure to cancer-initiating agents, may also be involved.[46]

Some researchers have looked at possible associations between electromagnetic exposure and breast cancer, miscarriages, depression, suicides, Alzheimer's disease, and amyotrophic lateral sclerosis (ALS, or Lou Gehrig disease), but the general scientific consensus is that the evidence is not yet conclusive.[46]

Ultraviolet radiation (UVR), sometimes also called UV light, is invisible electromagnetic radiation of the same nature as visible light but having shorter wavelengths and higher energies. The main source of natural UVR is the sun. UVR is conventionally divided into three bands in order of increasing energy: UVA, UVB, and UVC.

In the electromagnetic spectrum, UVR extends between the blue end of the visible spectrum and low-energy x-rays, straddling the boundary between ionizing and nonionizing radiation (which is conventionally set at 100 nm). Because of the different wavelengths and energies, each of the three bands has distinct effects on biologic tissue.

The highest-energy band, UVC, can damage DNA and other molecules and is used in hospitals for sterilization. UVC is rapidly attenuated in air, and therefore it is not found in ground-level solar radiation. Exposure to UVC, however, can take place close to sources such as welding arcs or germicidal lamps.

UVB is the most effective UV band in causing tanning and sunburn (erythema), and it can affect the immune system. UVA penetrates deeper in the skin because of its longer wavelength and plays a role in skin photoaging. UVA can also affect the immune system. Exposure to UVA and UVB has been implicated in the development of skin cancer.

Tanning lamps emit mostly UVA radiation with a few percentage content of UVB. Use of tanning lamps and beds can lead to significant exposure to UVA radiation. The risk of melanoma reportedly increases 75% when the use of tanning devices starts before age 30.[47,48]

The greater the frequency and intensity of exposure, the greater the risk. The risk is even higher for individuals using high-intensity or high-pressure devices.[48] Despite known negative health effects from the use of indoor tanning devices, this practice is still very popular in the United States and Europe. Therapists have a role in client education, especially concerning reducing modifiable risk factors such as outdoor exposure to the sun without protection, exposure to sun lamps, and indoor tanning.[49]

Military Workers. Survivors of recent wars who have been exposed to chemical agents may be at risk for the development of soft tissue sarcoma, non-Hodgkin's lymphoma, Hodgkin's disease, respiratory and prostate cancers, skin diseases, and many more problems in themselves and their offspring.

Three million Americans served in the armed forces in Vietnam during the 1960s and early 1970s. Large quantities of defoliant agents, such as Agent Orange, were used to remove forest cover, destroy crops, and clear vegetation from around U.S. military bases.

At least half of the 3 million Americans in Vietnam were there during the heaviest spraying. Many of our military personnel were exposed to this toxic substance. Exposure could occur through inhalation, ingestion, and skin or eye absorption.

In early 2003, the military acknowledged that exposure to Agent Orange is associated with chronic lymphocytic leukemia among surviving veterans. There is also sufficient evidence of an association between Agent Orange and soft tissue sarcoma and non-Hodgkin's lymphoma.[50-52]

Taking an environmental, occupational, or military history may be appropriate when a client has a history of asthma, allergies, or autoimmune disease, along with puzzling, nonspecific symptoms such as myalgias, arthralgias, headaches, back pain, sleep disturbance, loss of appetite, loss of sexual interest, and recurrent upper respiratory symptoms.

The affected individual often presents with an unusual combination of multiorgan signs and symptoms. A medical diagnosis of chronic fatigue syndrome, fibromyalgia, or another more nonspecific disorder is a yellow flag. When and how to take the history and how to interpret the findings are discussed in Chapter 2. The mnemonic CH2OPD2 (Community, Home, Hobbies, Occupation, Personal habits, Diet, and Drugs) can be used as a tool to identify a client's history of exposure to potentially toxic environmental contaminants.[53]

Risk Factors for Cancer Recurrence

As cancer survivors live longer, the chance of recurring cancer increases. Positive lymph nodes, tumor size greater than 2 cm, and a high-grade histopathologic designation increase a client's risk of cancer recurrence. Recurrence can occur at the original location of the first cancer, in local or distant lymph nodes, or in metastatic sites such as the bone or lung tissues.

Each type of cancer has its own risk factors for cancer recurrence. For example, increased numbers of positive lymph nodes and negative estrogen/progesterone receptor (ER/PGR) status for breast cancer survivors are risk factors (Case Example 13-1). A positive ER/PGR status lowers the risk of breast cancer recurrence because it allows the woman to receive treatment for prevention of recurrence according to age and stage of cancer.

CASE EXAMPLE 13-1

Risk Factors for Cancer Recurrence

A 46-year-old woman presented with mid-thoracic back pain present for the past 2 weeks. She described the pain as sharp and rated it as a 7 on the numeric rating scale. The pain was increased when she raised her arms overhead and relieved when she put her arms down. There were no other aggravating or relieving factors.

The client also noted occasional shoulder pain, sometimes with back pain and sometimes by itself. There were no other reported symptoms of any kind.

Past medical history included breast cancer diagnosed and treated 8 years ago with no cancer recurrence. The client had 17 nodes removed (12 were positive) and a mastectomy, followed by chemotherapy (short-term) and tamoxifen (long-term). The client was estrogen negative.

The clinical presentation was consistent with a posterior rib dysfunction, but there was no identified trauma or cause attributed to the onset of the back pain. The therapist's judgment was that there were enough risk factors in the history for cancer recurrence combined with additional red flags (e.g., age over 40, pain level, insidious onset) to warrant medical evaluation before a plan of care was established.

Results: Client was diagnosed with cancer metastases to the thoracic vertebrae at T4-6. The physician called the therapist to ask what tipped her off to the need for medical referral. Knowing the risk factors for cancer AND for cancer recurrence made a difference in this case.

CANCER PREVENTION

Cancer prevention begins with risk factor assessment and risk reduction. The key to cancer prevention lies in minimizing as many of the individual modifiable risk factors as possible. The AICR estimates that recommended diets, together with maintenance of physical activity and appropriate body mass, can in time reduce cancer incidence by 30% to 40%. At current rates, on a global basis, this represents 3 to 4 million cases of cancer per year that could be prevented by dietary and associated means.[20]

There are some simple steps to take in starting this process. The first is to assess personal/family health history. Note any cancers present in first-generation family members. Some helpful tools are available for assessing cancer risk. The Harvard School of Public Health offers an interactive tool to estimate an individual's risk of cancer and offers cancer prevention strategies. Anyone can benefit from it, but accuracy is greatest for adults over age 40 who have never had any type of cancer. It is available at the Harvard Center for Cancer Prevention website at www.yourcancerrisk.harvard.edu.

Cancer screening is available and widely recommended for the following types of cancer: colorectal, breast, cervix (women), and prostate (men). Early detection at a localized stage is linked with less morbidity and lower mortality.

For example, 90% of colon cancer cases and deaths can be prevented. The ACS provides a summary of risk factors and early detection screening tests for many types of cancer, including colon cancer. (This information is available at www.cancer.org/Healthy/InformationforHealthCare Professionals/ColonMDClinicansInformationSource/index, on the ACS website.) The Gail Model Risk Assessment Tool can also be used to assess personal risk for breast cancer. It is available at www.halls.md/breast/riskcom.htm (Breast Cancer Risk [Gail Model] Calculation Methods).

As health care educators, therapists can make use of this information to promote cancer prevention for themselves, their families, and their clients.

Genomics and Cancer Prevention

With the advent of the Genome Project, the sequencing of all genes in humans is nearly completed. Along with this discovery has come the development of a new biology of genetics called genomics. Understanding gene–environment interaction will be a major focus of genomics-based public health.[54]

There are many known or suspected carcinogens that increase an individual's risk of cancer. Different people respond to carcinogens differently. It is still not clear why one person develops cancer and another does not when both have the same risk factors.

Toxicogenomics, the development of molecular signatures for the effects of specific hazardous chemical agents, will bring to our understanding ways to track multiple sources of the same agent, multiple media and pathways of exposures, multiple effects or risks from the same agent, and multiple agents that cause similar effects.[54]

Defects may occur in one or more genes. Damage may occur in genes that involve the metabolism of carcinogens or in genes that deal with the DNA-repair process.[55] An important discovery in the area of gene identification related to cancer suppression is discovery of the p53 tumor-suppressor gene.[56] The p53 gene encodes a protein with cancer-inhibiting properties. Loss of p53 activity predisposes cells to become unstable and more likely to take on mutations. Mutation of the p53 gene is the most common genetic alteration in human cancers.[57]

It is possible that genetic defects combined with lifestyle or environmental factors may contribute to the development of cancer. For example, there is a known increase in risk of breast cancer in American women born after 1940 who have BRCA1 or BRCA2 mutation. This suggests that changes in the environment or lifestyle increased the risk already conferred by these genes.[58]

Air pollution is moderately associated with increased lung cancer, but when combined with exposure to tobacco smoke, the risk increases dramatically. About 50% of all people lack the GSTM1 metabolic gene that can detoxify tobacco smoke and air pollution. People who have this genetic defect and who have heavy exposure to pollution may have a higher risk of developing lung cancer.[55]

Once it is understood how genes and the environment work to contribute to cancer development, this knowledge can be applied to intervention. Anyone with genes that lead to a higher risk of cancer may benefit from chemoprevention.

MAJOR TYPES OF CANCER

There are three major types of cancer: carcinoma, sarcoma, and bloodborne cancers such as lymphoma and the leukemias.

Carcinoma is a malignant tumor that comprises epithelial tissue and accounts for 85% of all cancers. Carcinomas affect structures such as the skin, large intestine, stomach, breast, and lungs. These can be fast-growing tumors because they are derived from the epithelial lining of the organ, which grows rapidly and replaces itself frequently.

Carcinomas spread by invading local tissues and by metastasis. Generally, carcinomas tend to metastasize via the lymphatics, whereas sarcomas are more likely to metastasize hematogenously.

Sarcoma is a fleshy growth and refers to a large variety of tumors arising in the connective tissues that are grouped together because of similarities in pathologic appearance and clinical presentation.

Tissues affected include connective tissue such as bone and cartilage (discussed subsequently under Bone Tumors), muscle, fibrous tissue, fat, and synovium. The different types of sarcomas are named for the specific tissues affected (e.g., fibrosarcomas are tumors of the fibrous connective tissue; osteosarcomas are tumors of the bone; and chondrosarcomas are tumors arising in cartilage) (Table 13-2).

TABLE 13-2 Classification of Soft Tissue and Bone Tumors

Tissue of Origin	Benign Tumor	Malignant Tumor
CONNECTIVE TISSUE		
Fibrous	Fibroma	Fibrosarcoma
Cartilage	Chondroma Enchondroma Chondroblastoma	Chondrosarcoma
Bone	Osteoma	Osteosarcoma
Bone marrow		Leukemia Multiple myeloma Ewing family of tumors (EFT)
Adipose (fat)	Lipoma	Liposarcoma
Synovial	Ganglion, giant cell of tendon sheath	Synovial sarcoma
MUSCLE		
Smooth muscle	Leiomyoma	Leiomyosarcoma
Striated muscle	Rhabdomyoma	Rhabdomyosarcoma
ENDOTHELIUM (VASCULAR/LYMPHATIC)		
Lymph vessels	Lymphangioma	Lymphangiosarcoma Kaposi's sarcoma
Lymphoid tissue		Lymphosarcoma (lymphoma) Lymphatic leukemia
Blood vessels	Hemangioma	Hemangiosarcoma
NEURAL TISSUE		
Nerve fibers and sheaths	Neurofibroma Neuroma Neurinoma (neurilemmoma/schwannoma)	Neurofibrosarcoma Neurogenic sarcoma
Glial tissue	Gliosis	Glioma
EPITHELIUM		
Skin and mucous membrane	Papilloma	Squamous cell carcinoma
	Polyp	Basal cell carcinoma
Glandular epithelium	Adenoma	Adenocarcinoma

Data from Purtilo DT, Purtilo RB: *A survey of human disease,* ed 2, Boston, 1989, Little, Brown; Phipps W, et al: *Medical-surgical nursing: concepts and clinical practice,* ed 4, St. Louis, 1990, Mosby.

TABLE 13-3 Subcategories of Malignancy by Cell Type of Origin

Subcategory	Cell Type of Origin
Carcinomas	Arise from epithelial cells: • Breast • Colon • Pancreas • Skin • Large intestine • Lungs • Stomach Metastasize via lymphatics
Sarcomas	Develop from connective tissues: • Fat • Muscle • Bone • Cartilage • Synovium • Fibrous tissue Metastasize hematogenously Local invasion
Lymphomas	Originate in lymphoid tissues: • Lymph nodes • Spleen • Intestinal lining Spread by infiltration
Leukemias	Cancers of the hematologic system: • Bone marrow Invasion and infiltration

As a general category, sarcoma differs from carcinoma in the origin of cells composing the tumor (Table 13-3). As mentioned, sarcomas arise in connective tissue (embryologic mesoderm), whereas carcinomas arise in epithelial tissue (embryologic ectoderm) (i.e., cellular structures covering or lining surfaces of body cavities, small vessels, or visceral organs).

Cancers of the blood and lymph system arise from the bone marrow and include leukemia, multiple myeloma, and lymphoma. These cancers are characterized by the uncontrolled growth of blood cells. Metastasis is hematogenous.

RESOURCES

Although there are many websites related to cancer, we recommend what the physicians use: the well-known and respected National Comprehensive Cancer Network (NCCN). The NCCN Clinical Practice Guidelines in Oncology are recognized by clinicians around the world as the standard for oncology care. The NCCN now has consumer versions of its clinical practice guidelines (www.nccn.com).

Other reliable sites include Abramson Cancer Center of the University of Pennsylvania at www.oncolink.org or www.oncolink.com and the National Cancer Institute (www.cancer.gov/). For self-assessment of cancer (and other diseases), go to www.yourdiseaserisk.wustl.edu.

METASTASES

Neoplasms are divided into three categories: benign, invasive, and metastatic. Benign neoplasms are noncancerous tumors that are localized, encapsulated, slow growing, and unable to move or metastasize to other sites.

Invasive carcinoma is a malignant cancer that has invaded surrounding tissue. The spread of cancer cells from the primary site to secondary sites is called *metastasis.* A regional metastasis is the local arrest, growth, and development of a malignant lesion to regional lymph nodes.

A distal or distant metastasis is the distant arrest, growth, and development of a malignant lesion to another organ (e.g., lung, liver, brain). Within the categories of invasive and metastatic tumors, four large subcategories of malignancy have been identified and classified according to the cell type of origin (see Table 13-3).

For the therapist, primary cancers arising from specific body structures are not as likely to present with musculoskeletal signs and symptoms. It is more likely that recurrence of a previously treated cancer will have metastasized from another part of the body (secondary neoplasm) with subsequent bone, joint, or muscular presentation.

Metastatic spread can occur as late as 15 to 20 years after initial diagnosis and medical intervention. As many as 70% of people who die of cancer have been shown on autopsy to have spinal metastases; up to 14% exhibit clinical symptomatic disease before death.[59] For these reasons, the therapist must take care to conduct a screening interview during the examination, including past medical history of cancer or cancer treatment (e.g., chemotherapy, radiation).

Use the personal/family history form (see Fig. 2-2) in Chapter 2 to assess for a personal or first-degree family history of cancer. When asked about a past medical history of cancer, clients may say "No," even in the presence of a personal history of cancer. This is especially common in those clients who have reached and/or passed the 5-year survival mark.

Always link these two questions together:

FOLLOW-UP QUESTIONS

- Have you ever had any kind of cancer?
- If no, have you ever had chemotherapy, radiation therapy, or immunotherapy of any kind?

Mechanisms and Modes of Metastases

Cancer cells can spread throughout the body through the bloodstream (hematogenous or vascular dissemination), via the lymphatic system, or by direct extension into neighboring tissue or body cavities (Fig. 13-2). Once a primary tumor is initiated and starts to move by local invasion, tumor angiogenesis occurs (blood vessels from surrounding tissue grow into the solid tumor). Tumor cells then invade host blood vessels and are discharged into the venous drainage.

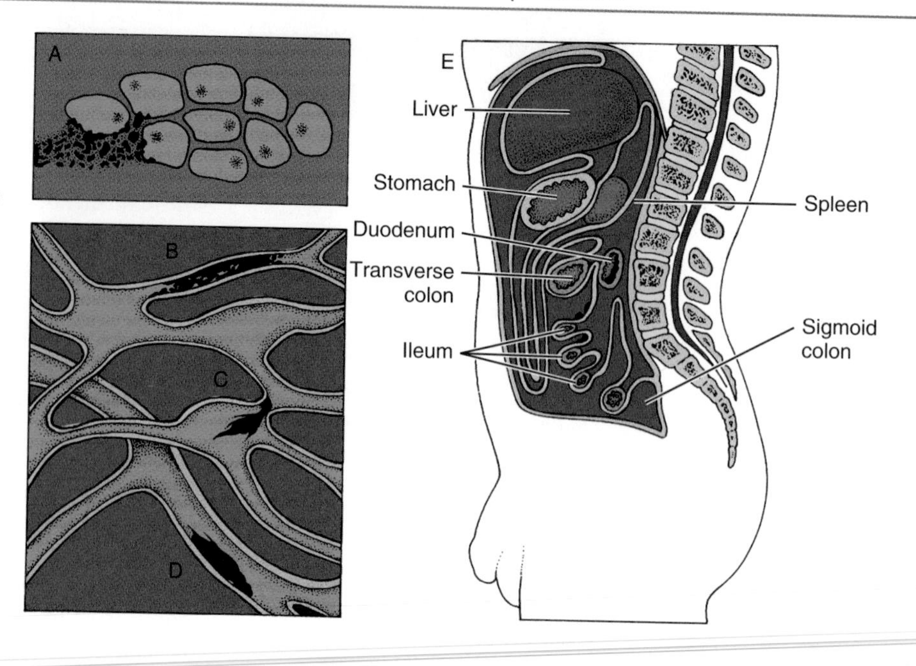

Fig. 13-2 Some modes of dissemination of cancer. **A,** Direct extension into neighboring tissue. **B,** Permeation along lymphatic vessels. **C,** Embolism via lymphatic vessels to the lymph nodes. **D,** Embolism via blood vessels (hematogenous spread). **E,** Invasion of a body cavity by diffusion. (Modified from Monahan et al, editors: *Phipps' medical-surgical nursing: health and illness perspective,* ed 8, St. Louis, 2007, Mosby.)

Many individuals develop multiple sites of metastatic disease because of the potential of cancers to spread. A metastatic colony is the end result of a complicated series of tumor–host interactions called the *metastatic cascade.*

Metastasis requires a good deal of coordination between cancer cells and the body. Fortunately, many early metastases die in transit for a number of reasons such as blood vessel turbulence and genes that normally suppress growth of micrometastases in new environments. Even so, some metastatic cells do survive and move on to other sites. At secondary sites, the malignant cells continue to reproduce, and new tumors or lesions develop.

Some clients with newly diagnosed cancers have clinically detectable metastases; remaining clients who are clinically free of metastases may harbor occult metastases.

The usual mode of spread and eventual location of metastases vary with the type of cancer and the tissue from which the cancer arises. Early clinical observations led to the idea that carcinomas spread by the lymphatic route and mesenchymal tumors, such as melanoma, spread through the bloodstream. We now know that both the lymphatic and vascular systems have many interconnections that allow tumor cells to pass from one system to the other.

During invasion, tumor cells can easily penetrate small lymphatic vessels and are then transported via the lymph. Tumor emboli may be trapped in the first draining lymph node, or they may bypass these regional lymph nodes (RLNs) to form noncontinuous and distant nodal metastases called "skip metastasis."

The relatively high incidence of anatomic skip metastasis can be attributed to aberrant distribution of lymph nodes.[60] Multiple interconnections between the lymphatic and hematogenous systems may also allow transport of tumor cells via the arterial or venous blood supply, bypassing some lymph nodes while reaching other more distant nodes.[61]

Patterns of blood flow, regional venous drainage, and lymphatic channels determine the distribution pattern of most metastases. For example, breast cancer spreads via the *lymphatics* and via the vertebral venous system to bones in the shoulder, hip, ribs, vertebrae, lungs, and liver.

Primary bone cancer, such as osteogenic sarcoma, initially metastasizes via *blood* to the lungs. Prostate cancer spreads via *lymphatics* to pelvic and vertebral bones, sometimes appearing as low back and/or pelvic pain radiating down the leg. The more common cancers and their metastatic pathways are provided in Table 13-4.

The high proportion of bone metastases in breast, prostate, and lung cancers is an example of selective movement of tumor cells to a specific organ. For example, in breast cancer, it is thought that the continuous remodeling of bone by osteoclasts and osteoblasts predisposes bone to metastatic lesions.[62,63] For some cancers, such as malignant melanoma, no typical pattern exists, and metastases may occur anywhere.

Increased tumor contact with the circulatory system provides tumors with a mechanism to enter the general circulation and colonize at distant sites. Both vascular endothelial growth factor (VEGF) and fibroblast growth factor stimulate proliferation of vascular cells and even allow the newly formed blood vessels to be easily invaded by the cancer cells that are closely adjacent to them.[64] Resection of tumors without clear margins has the potential to provide remaining tumor cells with a means of metastasizing as new blood vessels form during the healing process.

Benign Mechanical Transport

Mechanical transport rather than metastasis may be another mechanism of cancer spread. Two potential modes of benign mechanical transport (BMT) have been detected: lymphatic transport of epithelial cells displaced by biopsy of the primary

TABLE 13-4 Pathways of Cancer Metastases

Primary Cancer	Mode of Dissemination	Location of Primary Metastases
Breast	Lymphatics	Bone (shoulder, hips, sacrum, ribs, vertebrae); CNS (brain, spinal fluid, brachial plexus)
	Blood (vascular or hematogenous)	Lung, pleural cavity, liver, bone
Bone	Blood	Lungs, liver, bone, then CNS
Cervical (cervix)	Local extension and lymphatics	Retroperitoneal lymph nodes, bladder, rectum; paracervical, parametrial lymphatics
	Blood	CNS (brain), lungs, bones, liver
Chordoma	Direct extension	Neighboring soft tissues, spine
	Blood	Liver, lungs, heart, brain, spine
	Lymphatics	Lymph nodes, peritoneum
Colorectal	Direct extension	Bone (vertebrae, hip, sacrum)
	Peritoneal seeding	Peritoneum
	Blood	Liver, lung
Ewing sarcoma	Blood	Lung, bone, bone marrow
Giant cell tumor of bone	Blood	Lung
Kidney	Lymph	Pelvis, groin
	Blood	Lungs, pleural cavity, bone, liver, brain
Leukemia		Does not really "metastasize" as it is present throughout the body and therefore causes symptoms throughout body
Liver	Blood	CNS (brain)
Lung (bronchogenic sarcoma)	Blood	CNS (brain, spinal cord)
	Blood	Bone (ribs, sacrum, vertebrae)
	Direct extension, lymphatics	Mediastinum (tissue and organs between the sternum and vertebrae such as the heart, blood vessels, trachea, esophagus, thymus, lymph nodes)
Lung (apical or Pancoast's tumors)	Direct extension	8th cervical and 1st and 2nd thoracic nerves within the brachial plexus
	Blood	CNS (brain, spinal cord), bone
Lung (small cell)	Blood	CNS (brain, spinal fluid)
Lymphomas	Blood	CNS (spinal cord, spinal fluid), bone
	Lymphatics	Can occur anywhere, including skin, visceral organs, especially liver
Malignant melanoma	No typical pattern	Mets can occur anywhere; skin and subcutaneous tissue; lungs; CNS (brain, spinal fluid); liver; gastrointestinal tract; bone
Multiple myeloma	Blood	Bone (sacrum)
Nonmelanoma skin cancer	Usually remain local without metastases; local invasion	Bones underlying involved skin; brain
Osteogenic sarcoma (osteosarcoma)	Blood	Lungs, CNS (brain)
	Lymphatics	Lymph nodes, lungs, bone, kidneys
Ovarian	Direct extension into abdominal cavity	Nearby organs (bladder, colon, rectum, uterus, fallopian tubes); spread beyond abdomen is rare
	Lymphatics, peritoneal fluid through the abdomen	Liver, lungs; regional and distant
Pancreatic	Blood	Liver
Prostate	Lymphatics	Pelvic, sacrum, and vertebral bones, sacral plexus
		Bladder, rectum
		Distant organs (lung, liver, brain)
Soft Tissue Sarcoma	Blood; lymphatics (rare)	Lung (first) but also bone, brain, liver, soft tissue (distant)
Spinal cord	Local invasion; dissemination through the intervertebral foramina	CNS (brain, spinal cord)
Stomach, gastric	Blood	Liver, vertebrae, abdominal cavity (intraperitoneum)
	Local invasion	
Testes	Local invasion	Bone (pelvis, lumbar spine, hip)
	Blood, lymphatics	Lung
Thyroid	Direct extension	Bone; nearby tissues of neck
	Lymphatics	Regional lymph nodes (neck, upper chest, mediastinum)
	Blood	Distant (lung, bone, brain)

CNS, Central nervous system; *Mets,* metastases.
Use this table to identify the most likely location of symptoms associated with cancer recurrence/metastases. If the new symptoms match areas of common metastases for the client's primary cancer, further screening and/or consult or referral are indicated.

tumor and breast massage–assisted sentinel lymph node (SLN) localization.[65]

Samples of malignant tissues must be very carefully excised by surgeons who are expert in the biopsy of malignant tissues.[66,67] The risk of local recurrence is increased when intralesional curettage alone is performed.[68] Recurrence along the surgical pathway has been reported for some tumors following needle biopsy.[69,70] It is hypothesized that this recurrence is the result of intraoperative seeding. Poorly planned biopsies or incomplete tumor resection increases the risk of local recurrence and metastasis. The biopsy tract should be excised when complete tumor removal occurs.[71]

A second mode of BMT may be the pre-SLN breast massage used to facilitate the localization of SLNs during breast cancer staging. Mechanical transport of epithelial cells to SLNs has been verified. The significance of small epithelial clusters in SLN is unknown; further research is needed before changes are recommended for biopsy and SLN-localizing practices.[65]

The bottom line for the therapist is this: Anyone who has had a recent biopsy (within the past 6 months) must be followed carefully for any signs of local cancer recurrence.

CLINICAL MANIFESTATIONS OF MALIGNANCY

The therapist may be the first to see clinical manifestations of primary cancer but is more likely to see signs and symptoms of cancer recurrence or cancer metastasis. In general, the five most common sites of cancer metastasis are bone, lymph nodes, lung, liver, and brain. However, the therapist is most likely to observe signs and symptoms affecting one of the following systems:

- Integumentary
- Pulmonary
- Neurologic
- Musculoskeletal
- Hepatic

Each of these systems has a core group of most commonly observed signs and symptoms that will be discussed throughout this section (Table 13-5).

Early Warning Signs

For many years, the ACS has publicized the seven warning signs of cancer, the appearance of which could indicate the presence of cancer and the need for a medical evaluation. The mnemonic in Box 13-2 is often used as a helpful reminder of these warning signs.

Other early warning signs can include rapid, unintentional weight loss in a short period of time (e.g., 10% of the person's body weight in 2 weeks), unusual changes in vital signs, frequent infections (e.g., respiratory or urinary), and night pain. Bleeding is an important sign of cancer, but a cancer is generally well established by the time bleeding occurs. Bleeding develops secondary to ulcerations in the central areas of the tumor or by pressure on or rupture of local blood vessels. As the tumor continues to grow, it may enlarge beyond its capacity to obtain necessary nutrients, resulting in revitalization of portions of the tumor.

TABLE 13-5 Signs and Symptoms of Metastases*

Integumentary	Musculoskeletal	Neurologic (CNS)	Pulmonary	Hepatic
Any skin lesion or observable/palpable skin changes Any observable or palpable change in nailbeds (fingers or toes) Unusual mole (use ABCDE method of assessment; see Fig. 4-6) Cluster mole formation Bleeding or discharge from mole, skin lesion, scar, or nipple Tenderness and soreness around a mole; sore that does not heal	May present as an asymptomatic soft tissue mass Bone pain • Deep or localized • Increased with activity • Decreased tolerance to weight bearing; antalgic gait • Does not respond to physical agents Soft tissue swelling Pathologic fractures Hypercalcemia (see Table 13-7) • CNS • Musculoskeletal • Cardiovascular • Gastrointestinal Back or rib pain	Drowsiness, lethargy Headaches Nausea, vomiting Depression Increased sleeping Irritability, personality change Confusion, increased confusion Change in mental status, memory loss, difficulty concentrating Vision changes (blurring, blind spots, double vision) Numbness, tingling Balance/coordination problems Changes in deep tendon reflexes Change in muscle tone for individual with previously diagnosed neurologic condition Positive Babinski reflex Clonus (ankle or wrist) Changes in bowel and bladder function Myotomal weakness pattern Paraneoplastic syndrome (see text)	Pleural pain Dyspnea New onset of wheezing Productive cough with rust, green, or yellow-tinged sputum	Abdominal pain and tenderness Jaundice Ascites (see Fig. 9-8) Distended abdomen Dilated upper abdominal veins Peripheral edema General malaise and fatigue Bilateral carpal/tarsal tunnel syndrome Asterixis (liver flap) (see Fig. 9-7) Palmar erythema (liver palms) (see Fig. 9-5) Spider angiomas (over the abdomen) (see Fig. 9-3) Nail beds of Terry (see Fig. 9-6) Right shoulder pain

*Seen most often in a physical therapy practice.
CNS, Central nervous system; *ABCDE,* asymmetry, border, color, diameter, evolving.

BOX 13-2 EARLY WARNING SIGNS OF CANCER

Changes in bowel or bladder habits
A sore that does not heal in 6 weeks
Unusual bleeding or discharge
Thickening or lump in breast or elsewhere
Indigestion or difficulty in swallowing
Obvious change in a wart or mole
Nagging cough or hoarseness
Supplemental signs and symptoms (rapid unintentional weight loss, changes in vital signs, frequent infections, night pain, pathologic fracture, proximal muscle weakness, change in deep tendon reflexes)

For the physical therapist:
Changes in vital signs
Proximal muscle weakness
Change in deep tendon reflexes

This process of invading and compressing local tissue, shutting off blood supply to normal cells, is called *necrosis.* Tissue necrosis leads ultimately to secondary infection, severe hemorrhage, and the development of pain when regional sensory nerves become involved. Other symptoms can include pathologic fractures, anemia, and thrombus formation.

Awareness of these signals is useful, but it is generally agreed that these symptoms do not always reflect early curable cancer nor does this list include all possible signs for the different types of cancer.

Lumps, Lesions, and Lymph Nodes

The therapist should take special note of "T," which is *thickening or lump in breast or elsewhere.* Clients often point out a subcutaneous lesion (often a benign lipoma) and ask us to identify what it is. Baseline examination of a lump or lesion is important. Palpation of skin lesions and lymph nodes is presented in Chapter 4.

Whenever examining a lump or lesion, one should use the mnemonic in Box 4-11 to document and report findings on location, size, shape, consistency, mobility or fixation, and signs of tenderness (see Fig. 4-48). A clinically detectable tumor the size of a small pea already contains billions of cells. Most therapists will be able to palpate a lesion below the skin when it is half that size.[61]

Review Appendix B-21 for appropriate follow-up questions. Keep in mind that the therapist cannot know what the underlying pathology may be when lymph nodes are palpable and questionable. Performing a baseline assessment and reporting the findings are important outcomes of the assessment.

In previous editions of this text, it was noted that any changes in lymph nodes present for longer than 1 month in more than one location was a red flag. This recommendation is no longer appropriate. The recommendation today, based on an increased understanding of cancer metastases via the lymphatic system and the potential for cancer recurrence, is that all suspicious lymph nodes should be evaluated by a physician (Case Example 13-2).

Supraclavicular lymph nodes that are easily palpable on examination may indicate possible metastatic disease. Any lymph nodes that are hard, immovable, and nontender raise the suspicion of cancer, especially in the presence of a previous history of cancer.

Keep in mind that lymph nodes can fluctuate over the course of 10 to 14 days. When making the medical referral, look for a cluster of signs and symptoms, recent trauma (including recent biopsy), or a past history of chronic fatigue syndrome, mononucleosis, and allergies. Record and report all findings.

CASE EXAMPLE 13-2

Palpable and Observable Lymph Nodes

A 73-year-old woman was referred to a physical therapy clinic by her oncologist with a diagnosis of cervical radiculopathy. She had a history of uterine cancer 20 years ago and a history of breast cancer 10 years ago.

Treatment included a hysterectomy, left radical mastectomy, radiation therapy, and chemotherapy. She has been cancer-free for almost 10 years. Her family physician, oncologist, and neurologist all evaluated her before she was referred to the physical therapist.

Examination by a physical therapist revealed obvious lymphadenopathy of the left cervical and axillary lymph nodes. When asked if the referring physician (or other physicians) saw the "swelling," she told the therapist that she had not disrobed during her medical evaluation and consultation.

The question for us as physical therapists in a situation like this one is how to proceed?

Several steps must be taken. First, the therapist must document all findings. If possible, photographs of the chest, neck, and axilla should be obtained.

Second, the therapist must ascertain whether or not the physician is already aware of the problem and has requested physical therapy as a palliative measure. Requesting the physician's dictation or notes from the examination is essential.

Contact with the physician will be important soon after the records are obtained, either to confirm the request as palliative therapy or to report your findings and confirm the need for medical reevaluation.

If it turns out that the physician is, indeed, unaware of these physical findings, it is best to send a problem list identified as "outside the scope of a physical therapist" when returning the client to the physician. Be careful to avoid making any statements that could be misconstrued as a medical diagnosis.

We recommend writing a brief letter with the pertinent findings and ending with one of two one-liners:

What do you think?
Please advise.

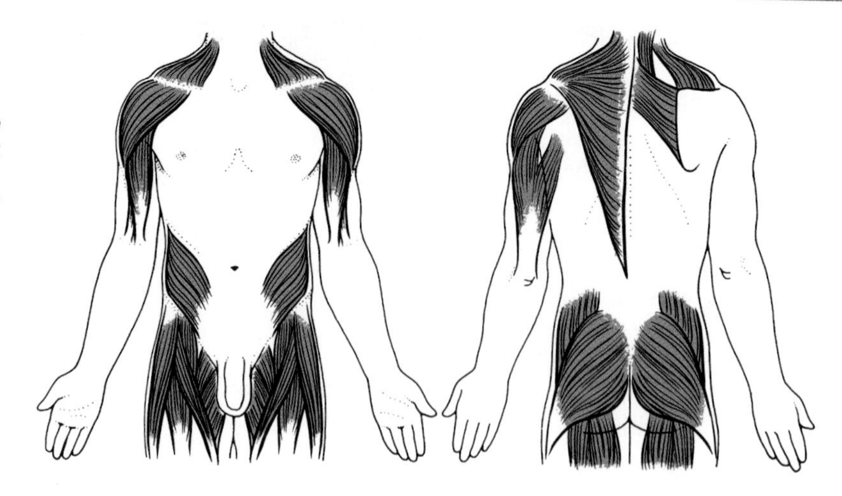

Fig. 13-3 Proximal muscle weakness can be observed clinically as a positive Trendelenburg test (usually present bilaterally) and abnormal manual muscle testing. It can also be observed functionally when the client has difficulty getting up from sitting or climbing stairs. As the weakness progresses, the client may have trouble getting into and out of a vehicle and/or the bathtub. Respiratory muscle weakness may be seen as shortness of breath or reported as altered activity to avoid dyspnea.

Proximal Muscle Weakness

For the therapist, idiopathic proximal muscle weakness may be an early sign of cancer (Fig. 13-3). This syndrome of proximal muscle weakness is referred to as *carcinomatous neuromyopathy.* It is accompanied by changes in two or more deep tendon reflexes (ankle jerk usually remains intact). Muscle weakness may occur secondary to hypercalcemia, which occurs as an indirect humoral effect on bone (see the later discussion on Paraneoplastic Syndrome in this chapter). Clients with advanced cancer, multiple myeloma, or breast or lung cancer are affected most often by hypercalcemia.

Screening for muscle weakness is not always a straightforward process. Sometimes, questions must be directed toward function to find out this information. If a client is asked whether he or she has any muscle weakness, difficulty getting up from sitting, trouble climbing stairs, or shortness of breath, the answer may very well be "No" on all accounts. Consider using the following flow of questions:

FOLLOW-UP QUESTIONS

- Do you have any muscle weakness in your arms, legs, back, or chest?
- Do you have any trouble getting into and out of a chair?
- Are there any activities you would like to be able to do that you currently can't do?
- Are there any activities you used to be able to do that you can't do now?
- Are there activities you can do now that used to be much easier?
- Do you have any trouble going up and down stairs without stopping?
- Can you do all your grocery shopping without sitting down or stopping?
- Are you able to complete your household chores (e.g., make a meal, wash and dry clothes) without stopping?

Pain

Pain is rarely an early warning sign of cancer, even in the presence of unexplained bleeding. Night pain that is constant and intense (often rated 7 or higher on the Numeric Rating Scale; see Fig. 3-6) is a red flag symptom of primary or recurring cancer. But not all people with musculoskeletal cancers experience night pain.[72]

Pain is usually the result of destruction of tissue or pressure on tissue due to the presence of a tumor or lesion. The lesion or lesions must be of significant size or location to create pressure and/or occlusion of normal structures; pain will be dependent on the area of the body affected.

Acute and chronic cancer-related pain syndromes can occur in association with diagnostic and therapeutic interventions such as bone marrow biopsy, lumbar puncture, colonoscopy, percutaneous biopsy, and thoracentesis. Chemotherapy and radiation toxicities can result in painful peripheral neuropathies.

Likewise, many different chronic pain syndromes (e.g., tumor-related radiculopathy, phantom breast pain, postsurgical pelvic or abdominal pain, burning perineum syndrome, postradiation pain syndrome) can occur as a result of tumors or cancer therapy. See further discussion under Oncologic Pain in this chapter.

Change in One or More Deep Tendon Reflexes

When a neurologic screening examination is performed, testing of deep tendon reflexes (DTRs) is usually included. Some individuals have very brisk reflexes under normal circumstances; others are much more hyporeflexive.

Tumors (whether benign or malignant) can also press on the spinal nerve root, mimicking a disk problem. A lesion that is small enough can put just enough pressure to irritate the nerve root, resulting in a hyperreflexive DTR. A large tumor can obliterate the reflex arc, resulting in diminished or absent reflexes.

Either way, changes in DTRs must be considered a red flag sign (possibly of cancer) that should be documented and further investigated. For example, a hyporesponsive patellar tendon reflex that is unchanged with distraction or repeated testing and is accompanied by back, hip, or thigh pain, along with a past history of prostate cancer, presents a different clinical picture altogether. Guidelines for assessing reflexes are discussed in Chapter 4.

Integumentary Manifestations

Internal cancers can invade the skin through vascular dissemination or direct extension. Metastases to the skin may be the first sign of malignancy, especially for breast or upper respiratory tract cancer. Integumentary carcinomatous metastases often present as asymmetrical, firm, skin-colored, red, purple, or blue nodules near the site of the primary tumor (see Fig. 4-26).

Distant cutaneous metastasis can result from lymphoma (see Fig. 4-29), multiple myeloma (see Fig. 4-27), and stomach/colon, ovarian, pancreatic, kidney, and breast cancer (Case Example 13-3). The scalp is a common site for such lesions (see Fig. 4-28), which are sometimes accompanied by hair loss called *alopecia neoplastica.*

The integumentary screening examination, including assessment of common skin lesions and nailbed assessment, is presented in Chapter 4. Cancer-related skin lesions (e.g., pinch purpura, renal nodule, local cancer recurrence, Kaposi's sarcoma, xanthomas) are also included in Chapter 4.

During observation and inspection, the therapist should be alert to any potential signs of primary skin cancer or integumentary metastases. When a suspicious skin lesion is noted, the therapist should conduct a risk factor assessment and ask three questions:

- How long have you had this area of skin discoloration/mole/spot (use whatever brief description seems most appropriate)?
- Has it changed in the past 6 weeks to 6 months?
- Has your physician examined this area?

No matter what the therapist's own cultural background, as a health care professional, his or her responsibility to screen skin lesions is clear. *How* questions are posed is just as important as *what* is said.

The therapist may want to introduce the subject by saying that as health care professionals, we are trained to observe many body parts (skin, joints, posture, and so on). You notice that the client has an unusual mole (or rash or whatever has been observed), and you wonder whether this is something that has been there for years. Has it changed in the past 6 weeks to 6 months? Has the client ever shown it to the doctor?

A client with a past medical history of cancer now presenting with a suspicious skin lesion that has not been evaluated

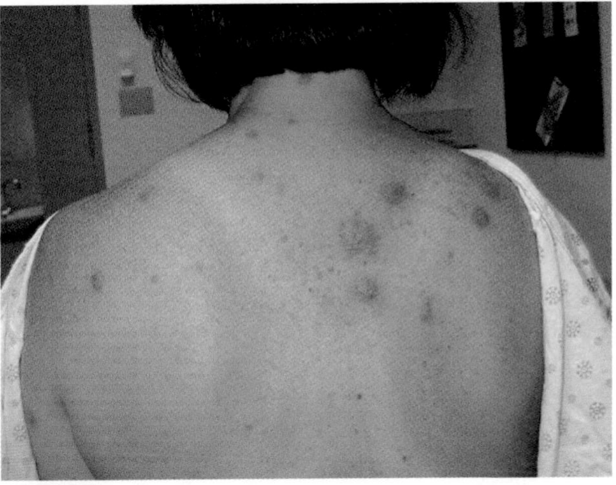

Fig. 13-4 Day 1. Skin rash on upper back associated with Sweet's syndrome (acute febrile neutrophilic dermatosis), a disorder usually associated with significant constitutional symptoms and involvement of the lungs and joints. Red, tender papules, plaques, or nodules appear on the face, extremities, and upper trunk. The surface appears vesicular and may produce pustules. Most cases are idiopathic, associated with inflammatory bowel disease, or preceding upper respiratory tract infection, but some have been associated with malignancies. (Courtesy Insley Puma Flaig, MD, University of Minnesota Medical Center, Fairview, MN, 2003. Used with permission.)

CASE EXAMPLE 13-3

Cancer-Related Skin Rash

A 42-year-old woman with a previous history of breast cancer and breast lumpectomy asked a fellow clinician to examine her scar for any sign of cancer recurrence. She had just had her 6-month cancer check-up and was not scheduled to see her oncologist for another 6 months.

In the meantime, she had developed a skin rash over the upper chest wall and axilla of the involved side (upper back and left thigh) (Figs. 13-4 and 13-5). When asked if there were any other symptoms present, she reported feeling feverish and a bit nauseous, and noted slight muscle aching. On examination, the client's vital signs were taken. (See Chapter 2 about the importance of vital signs in the screening process; see also discussion of constitutional symptoms.)

All vital signs were within normal limits for the client's age, except body temperature, which was 102.2° F. The client reported that her normal body temperature was usually 98° F. She was not aware of an elevated body temperature, although she stated she had awakened in the night feeling feverish and took some Tylenol.

Upper quadrant examination was unremarkable, except for skin rash and the presence of bilateral anterior cervical adenopathy. There was a fullness of lymph node tissue without firmness or distinct nodes palpated in the axilla on the involved side. The clinician was unable to palpate as far into the Zone II space (see Figs. 4-42 and 4-43) as would be expected.

Results: This client had three red flags: recent history of cancer, skin rash, and a constitutional symptom (fever). Even though there was no external sign of local cancer recurrence and even though she was just seen by her oncologist, these new findings warranted a return visit to her physician.

The skin rash turned out to be Sweet's syndrome, a disorder usually associated with significant constitutional symptoms and involvement of the lungs and joints. Most cases are idiopathic, but some have been associated with malignancies.[159-161]

In this case, no further findings were made despite laboratory and medical tests performed. The use of systemic corticosteroids is usually recommended for Sweet's syndrome, but the client declined and opted to use vitamin supplements, as her symptoms were resolving by that time. She was followed more closely for any cancer recurrence with more frequent testing thereafter.

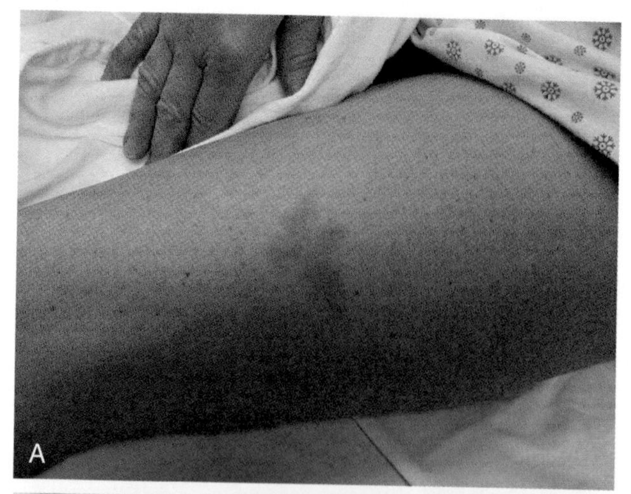

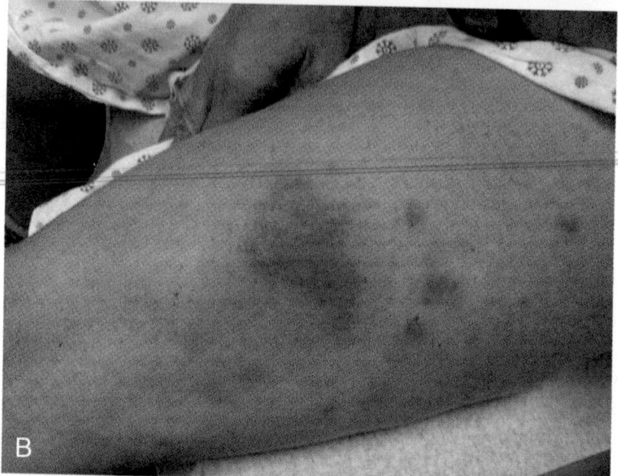

Fig. 13-5 **A,** Day 1 (same client as in Fig. 13-4). Skin rash on upper thigh associated with Sweet's syndrome. **B,** Day 5. Rash progressed quickly to cover large areas of the upper chest wall, axilla, upper back, and left thigh. (Courtesy Insley Puma Flaig, MD, University of Minnesota Medical Center, Fairview, MN, 2003. Used with permission.)

by the physician must be advised to have this evaluated as soon as possible.

For any client with a previous history of cancer with surgical removal, it is always a good idea to look at the surgical site(s) for any sign of local cancer recurrence. Start by asking the client if he or she has noticed any changes in the scar. Continue by asking the following:

FOLLOW-UP QUESTIONS

- Would you have any objections if I looked at (or examined) the scar tissue?

Any suspicious scab or tissue granulation, redness, or discoloration must be noted (photographed, if possible) (see Figs. 4-9 and 17-5). Again, three screening questions apply in this situation. The therapist has a responsibility to report these findings to the appropriate health care professional and to make every effort to ensure client compliance with follow-up.

Skin Cancers

Skin cancers are the most common of all types of cancer and are usually classified as nonmelanoma skin cancer (NMSC) or melanoma. Most skin cancers are classified as nonmelanoma and are slow growing, easy to recognize, and responsive to intervention, if found early. Nonmelanoma skin cancers are further classified as basal cell or squamous cell, depending on the tissue affected. They rarely metastasize to other parts of the body and have a nearly 100% rate of cure.

Melanoma, the most serious of the skin cancers, has a 96% 5-year survival rate if localized, but only a 13% 5-year survival if it is invasive or has spread to other parts of the body. More than 77% of cancer deaths result from invasive melanoma.

The primary warning sign for melanoma is a flat, colored, irregularly shaped lesion that can be mottled with light brown to black colors. It may turn various shades of red, blue, or white or crust on the surface and bleed. A changing mole, the appearance of a new mole, or a mole that is different or growing requires prompt medical attention.[73] The Skin Cancer Foundation advocates use of the ABCD (asymmetry, border irregularity, color variegation, and a diameter of 6 mm or greater) method of early detection of melanoma and dysplastic (abnormal in size or shape) moles (see discussion, Chapter 4 and Fig. 4-6).

Actinic keratosis is a premalignant form of skin cancer (see Fig. 4-7). With actinic keratosis, overexposure to sunlight results in abnormal cell growth, causing a well-defined, crusty patch or bump on sun-exposed parts of the body.

Clients often point out skin lesions or ask the therapist about various lumps and bumps. In addition, the therapist may observe changes in skin, skin lesions, or aberrant tissue during the visual inspection and palpation portion of the examination (see Chapter 4) that need further medical investigation. Mortality is reduced when lesions are found early and treated promptly. Therapists can and should be a part of the screening process for skin cancer.

The cause of skin cancer is well known. Prolonged or intermittent exposure to UVR from the sun, especially when it results in sunburn and blistering, damages DNA. The majority of all NMSCs occur on parts of the body unprotected by clothing (i.e., face, neck, forearms, and backs of hands) and in persons who have received considerable exposure to sunlight.

Risk Factor Assessment. All adults, regardless of skin tone and hair color, are at risk for skin cancer; however, some people are at much greater risk than others (Box 13-3). In general, individuals with red, blonde, or light brown hair with light complexion and maybe freckles, many of Celtic or Scandinavian origin, are most susceptible; persons of African or Asian origin are least susceptible.

The most severely affected people usually have a history of long-term occupational or recreational sun exposure. Australia and New Zealand have the highest incidence of

BOX 13-3 RISK FACTOR ASSESSMENT FOR SKIN CANCER

- Advancing age
- Personal or family history of skin cancer (particularly melanoma)
- Moles with any of the ABCDE features, or moles that are changing in any way
- Complexion that is fair or light with green, blue, or gray eyes
- Skin that sunburns easily; skin that never tans
- History of painful sunburns with blistering during childhood or the adolescent years
- Use of tanning beds or lamps
- Short, intense episodes of sun exposure: the indoor worker who spends the weekend out in the sun without skin protection (or any sporadic exposure to strong sunshine of normally covered skin)
- Transplant recipient

Data from Skin Cancer Foundation, 2011.

melanoma in the world. New Zealand has nearly 5 times the amount of skin cancer that occurs in the United States.

Melanoma occurs in every part of the North American continent. In the United States, the five states with the highest predicted incidence of new cases are California, Florida, Texas, New York, and Pennsylvania. Men are more likely than women to develop nonmelanoma and melanoma skin cancers. The rate of melanoma is 10 times higher for whites than blacks because blacks have the protective effects of skin pigment.[74]

Many older adults assume that skin changes are a "normal" sign of aging and do not see a physician when lesions first appear. Early detection and referral is always the key to a better prognosis. In asking the three important questions, the therapist plays an instrumental part in the cancer screening process.

FOLLOW-UP QUESTIONS

- How long have you had this?
- Has it changed in the past 6 weeks to 6 months?
- Has your physician seen it?

An increased incidence of skin cancers has been noted after solid organ transplantation, especially liver, heart, and kidney transplants. Squamous and basal cell carcinomas are 250 and 10 times more frequent, respectively, in transplant recipients compared with the general population.[75] Skin cancers developing in transplant recipients are more aggressive, making early detection and intervention imperative. Renal transplant recipients have a cumulative increase that corresponds with the number of years posttransplantation (e.g., 7% after 1 year of immunosuppression, 45% after 11 years, 70% after 20 years).[75]

Basal Cell Carcinoma. Basal cell carcinoma involves the bottom layer of the epidermis and occurs mainly on any hair-bearing area exposed to the sun (e.g., face, neck, head, ears, hands). Occasionally, basal cell carcinoma may appear on the trunk, especially the upper back and chest. These lesions grow slowly, attaining a size of 1 to 2 cm in diameter, often after years of growth. Metastases almost never occur, but neglected lesions may ulcerate and produce great destruction, ultimately invading vital structures.

There are a number of common forms of basal cell carcinoma:

- Pearly papule, 2 to 3 mm in diameter and covered by tightly stretched epidermis laced with small, delicate, branching vessels (telangiectasia)
- Pearly papule with a small crater in the center
- Scaly, red, sharply outlined plaque
- Ill-defined pale, tough, scar-like tumor

Squamous Cell Carcinoma. Squamous cell carcinoma arises from the top of the epidermis and is found on areas often exposed to the sun, which are typically the rim of the ear, the face, the lips and mouth, and the dorsa of the hands. These lesions appear as small, red, hard nodules with a smooth or warty surface. The central portion may be scaly, ulcerated, or crusted. Premalignant lesions include sun-damaged skin or dysplasias (whitish-discolored areas), scars, radiation-induced keratosis, actinic keratosis (rough, scaly spots), and chronic ulcers.

Metastases are uncommon but are much more likely to occur in lesions arising in chronic leg ulcers, burn scars, and areas of prior x-ray exposure. Although these tumors do not usually metastasize, they are potentially dangerous. They may infiltrate surrounding structures and metastasize to lymph nodes and eventually to distant sites, including bone, brain, and lungs, to become fatal. Invasive tumors are firm and increase in elevation and diameter. The surface may be granular and may bleed easily.

Malignant Melanoma. Malignant melanoma (MM) is the most serious form of skin cancer. It arises from pigmented cells in the skin called *melanocytes.* In contrast to basal and squamous cell carcinomas, the majority of MMs appear to be associated with the intensity rather than the duration of sunlight exposure.

The average lifetime risk of developing invasive melanoma is 1 in 58 (a 2000% increase from 1930).[1] An individual's risk is much greater if any of the risk factors listed in Box 13-3 are present.

Melanoma can appear anywhere on the body, not just on sun-exposed areas. The clinical characteristics of early malignant melanoma are similar, regardless of anatomic site. Unlike benign pigmented lesions, which are generally round and symmetric, the shape of an early MM is often asymmetric.

Whereas benign pigmented lesions tend to have regular margins, the borders of early MM are often irregular (see Fig. 4-6). Round, symmetric skin lesions such as common moles, freckles, and birthmarks are considered "normal." If an existing mole or other skin lesion starts to change and a line drawn down the middle shows two different halves, medical evaluation is needed.

Compared with benign pigmented lesions, which are more uniform in color, MMs are usually variegated, ranging from various hues of tan and brown to black, sometimes intermingled with red and white. The diameters of MM are often 6 mm or larger when first identified.

The most common sites of distant metastasis associated with MM are the skin and subcutaneous tissue, lungs, and surrounding visceral pleura, although any anatomic site may be involved. In-transit metastases (unique malignancies that have spread from the primary tumor but may not have reached the regional lymph nodes) typically develop multiple bulky tumors on an arm or leg. Often, these tumors cause pain, swelling, bleeding, ulceration, and decreased mobility.[76]

Other signs that may be important include irritation and itching; tenderness, soreness, or new moles developing around the mole in question; or a sore that keeps crusting and does not heal within 6 weeks. Benign moles tend to be flat, hairless, round or oval, and less than 6 mm in diameter. Pigmentation is generally even. Although there may be color variations, especially in shades of brown, benign moles, freckles, "liver spots," and other benign skin changes are usually of a single color (most often, a single shade of brown or tan). A single lesion with more than one shade of black, brown, or blue may be a sign of malignant melanoma.

Adolescents frequently have nevi with irregular borders, multiple shades of pigment, or both. Most are normal variations of benign nevi, but a physician should examine any lesion that arouses clinical suspicion or is of concern to the client.

If any of these signs and symptoms is present in a client whose skin lesion has not been examined by a physician, a medical referral is recommended. If the client is planning a follow-up visit with the physician within the next 2 to 4 weeks, the client is advised to point out the mole or skin changes at that time. If no appointment is pending, the client is encouraged to make a specific visit either to the family/personal physician or to a dermatologist.

CLINICAL SIGNS AND SYMPTOMS

Early Melanoma

A = Asymmetry: Uneven edges, lopsided in shape, one-half unlike the other half

B = Border: Irregularity, irregular edges, scalloped or poorly defined edges

C = Color: Black, shades of brown, red, white, occasionally blue

D = Diameter: Larger than a pencil eraser

E = Evolving: Mole or skin lesion that looks different from the rest or is changing in size, shape, or color

The ABCD (now including E) criteria have been verified in multiple studies, documenting the effectiveness and diagnostic accuracy of this screening technique. Their efficacy has been confirmed with digital image analysis; sensitivity ranges from 57% to 90% and specificity from 59% to 90%.[77]

Resources. The Skin Cancer Foundation (www.skincancer.org) has many public education materials available to help the therapist identify suspicious skin lesions. In addition to its website, the Skin Cancer Foundation has posters, brochures, videos, and other materials available for use in the clinic. It is highly recommended that these types of education materials be available in waiting rooms as part of a nationwide primary prevention program.

Other websites, such as the Melanoma Education Foundation at www.skincheck.com and the Melanoma Foundation of the University of Sydney, Australia at http://melanomafoundation.com.au/, provide additional photos of suspicious lesions with additional screening guidelines. For information on ratings of sunscreen sold in the United States, see the Environmental Awareness Group's Sunscreen Guide at www.ewg.org/2010sunscreen.

The therapist must become as familiar as possible with what suspicious skin aberrations may look like in order to refer as early as possible. Prognosis in melanoma is directly related to the depth of the neoplasm. Melanoma typically start growing horizontally within the epidermis (in situ) but then become invasive as tumor cells penetrate into the dermis. The vertical depth of the melanoma correlates with prognosis.[77] That is why early detection and referral is so important. See also the discussion on Examining a Mass or Skin Lesion in Chapter 4.

Pulmonary Manifestations

Pulmonary metastases are the most common of all metastatic tumors because venous drainage of most areas of the body passes through the superior and inferior venae cavae into the heart, making the lungs the first organ to filter malignant cells. *Primary* bone tumors (e.g., osteogenic sarcoma) metastasize first to the lungs.

Pleural pain and dyspnea may be the first two symptoms experienced by the person (Case Example 13-4). When either or both of these pulmonary symptoms occur, look for increased symptoms with deep breathing and activity. Ask about a productive cough with bloody or rust-colored sputum. Ask about new onset of wheezing at any time or difficulty breathing at night. Symptoms that are relieved by sitting up are indicative of pulmonary impairment and must be reported to the physician.

Symptoms may not occur until tumor cells have expanded and become large enough or invasive enough to reach the parietal pleura, where pain fibers are stimulated. The lining surrounding the lungs allows no pain perception, so it is not until the tumor is large enough to press on other nearby structures or against the chest wall that symptoms may first appear.

Lung cancer is the most common primary tumor to metastasize to the brain. Tumor cells from the lung embolizing via the pulmonary veins and carotid artery can result in metastases to the central nervous system (CNS). Anyone with a history of lung cancer should be screened for neurologic

CASE EXAMPLE 13-4

Lung Cancer

A 69-year-old man with a recent total hip replacement (THR) was referred to home health for physical therapy. He did not have a good postoperative recovery and has been slow to regain range of motion, strength, and function.

He experienced dyspnea and chest pain within the first 10 ft of ambulation. He has a past medical history of cancer.

What are the red flags here? How should you proceed?

Red Flags

Age (>40 years old)

Past medical history of cancer

Cardiopulmonary symptoms: shortness of breath and chest pain

How to Proceed: Ask the client how long he has had these symptoms. Take all vital signs as discussed in Chapter 2.

Your next steps may depend in part on any procedural instructions you have received from your home health agency. If there is a case manager, contact him or her with your concerns. Ask for a copy of the medical file. Contact the physician's office with your findings.

Difficulty in referral arises when a client has been seen by an orthopedic surgeon but is demonstrating signs and symptoms of possible systemic disease. Diplomacy and communication are the keys to success here.

Document your findings, and make sure these are sent to the primary care physician AND the orthopedic surgeon. The medical record may already indicate awareness of these red flags, and no further follow-up is needed.

If not, then a brief cover letter with your full report should be sent to the physician. The letter should contain the usual "thanks for this referral" kind of introduction with a paragraph about physical therapy intervention.

Then, include a medical problem list such as:

Patient reports shortness of breath and chest pain within the first 3 minutes of ambulation. This has just started in the last few days. His vital signs are: [list these].

Given the patient's age, past medical history of cancer, and new onset of cardiopulmonary symptoms, we would like medical clearance before progressing his exercise and rehab program. Please advise if there are any contraindications for exercise at this time. Thank you.

Make sure you call the physician's office and alert the staff of your concerns and that this letter/fax is on its way. Make telephone contact again within 3 days (sooner if the information is faxed or emailed to the doctor's office).

Results: The orthopedic surgeon advised the client to see his primary care physician for follow-up of this problem. After medical examination and testing, the final diagnosis was lung cancer. The medical doctor surmised that the stress of the surgery was enough to advance the cancer from subclinical to clinical status with new onset of symptoms that were not present before the orthopedic surgery.

involvement. In any individual, any neurologic sign may be the presentation of a silent lung tumor.[78]

Neurologic Manifestations

As just mentioned, cancer metastasis to the CNS is a common problem. Secondary metastases to the brain are 10 times more common than primary brain tumors. In all, 20% to 25% of individuals with primary sites outside of the CNS will develop brain metastases.[79] The most common primary cancers with metastases to the brain are lung, colon, kidney, skin (melanoma), and breast cancer (Case Example 13-5).

Tumor cells can easily embolize via the pulmonary veins and carotid artery to the brain. The blood–brain barrier does not prevent invasion of the brain parenchyma by circulating metastatic cells. Metastatic brain tumors can increase intracranial pressure, obstruct the normal flow of cerebrospinal fluid, change mentation, and reduce sensory and motor function.

Whether the pressure-causing lesion is a primary cancer of the brain or spinal cord or whether it is a cancer that has metastasized to the CNS, clinical signs and symptoms of pressure will be the same because in both cases, the same system is affected.

Primary tumors can also cause peripheral nervous system (PNS) problems when tumors compress, impinge, or infiltrate any of the nerve plexuses. No matter where neural compression occurs, the primary sign is unrelenting pain (worse at night) followed by development of weakness. Watch for focal sensory disturbances or weakness in the distribution of the affected plexus or spinal cord segment involved. Brachial plexopathy most commonly occurs in carcinoma of the breast and lung; lumbosacral plexopathy is most common with colorectal and gynecologic tumors, sarcomas, and lymphomas.[80,81]

Clinical Signs and Symptoms

Brain tumors can be asymptomatic. When symptoms do occur, they are usually general or focal symptoms, depending on the size and location of the lesion. For example, if a tumor is growing in the motor cortex, the client may develop isolated extremity weakness or hemiparesis. If the tumor is developing in the cerebellum, coordination may be affected with ataxia as an observable sign.

Two of the most common clinical manifestations of brain tumor are headache and personality change, but personality change is often attributed to depression, delaying the diagnosis of brain tumor. Tumors that affect the frontal lobes are most likely to produce personality changes. Seizures occur in approximately one-third of persons with metastatic brain tumors.

CASE EXAMPLE 13-5

Bone Metastases and Wrist Sprain

A 75-year-old woman fell and sprained her wrist. Her family doctor sent her to physical therapy. After the interview, her daughter took the therapist aside and commented that her mother seems confused. Other family members are wondering if her fall had anything to do with mental deterioration.

There is a positive personal history for breast cancer. Past medical history included breast cancer, diverticulosis, gallbladder removal, and hysterectomy. There were no current health concerns expressed by the client or her family. She is not taking any medication (prescription or over-the-counter).

Since the wrist was obviously not broken, no x-rays were taken.

What are the red flags in this case? Since she just came to physical therapy from a medical doctor, is follow-up medical attention needed?

Red Flags: Age, confusion, past medical history of cancer, recent loss of balance and fall, lack of diagnostics.

This client actually presents with a cluster of four significant red flags in the screening process. The therapist should carry out a balance and vestibular function screening examination and neurologic screening examination (see Chapter 4). Additional key information may be obtained from this testing.

The next step is to inquire of the client or family member if the doctor is aware of the past history of cancer. Older adults moving closer to family members may give up their lifelong family provider. The new physician may not have all the history compiled. This is especially true when patients visit a "Doc-in-a-Box" at the local mall or convenience care facility.

Likewise, check with the family to see whether the physician has been notified of the client's new onset of confusion. This is the number one sign of nervous system impairment in older adults.

The therapist is advised to document these findings and report them to the physician. As always, a letter of appreciation for the referral is a good idea. State the physical therapy diagnosis in terms of the human movement system (see the *Guide to Physical Therapist Practice* and discussion of physical therapy diagnosis in Chapter 1 of this text).

Include a follow-up paragraph with this information:

> I am concerned that the combination of the patient's age, new onset of confusion as described by her family, and recent history of falls resulting in this episode of care may be an indication of significant underlying pathology.
>
> What do you think? I will treat the musculoskeletal impairment, but please advise if any further follow-up is needed.

Results: Given the client's past history of cancer, and knowing that confusion is not a "normal" sign of aging and that any neurologic sign can be an indicator of cancer, the physical therapist suggested that the family should also talk with the referring physician about these observations.

The client progressed well with the wrist rehabilitation program. The family reported that the physician did not seem concerned about the developing confusion or recent falls. No further medical testing was recommended. Six weeks later, the client fell and broke her hip. At that time, she was given a diagnosis of metastases to the bone and brain (central nervous system [CNS]).

Headaches occur in 30% to 50% of persons with brain tumors and are usually bioccipital or bifrontal. They are usually intermittent and of increasing duration and may be intensified by a change in posture or by straining.

The headache is characteristically worse on awakening because of differences in CNS drainage in the supine and prone positions; it usually disappears soon after the person arises. It may be intensified or precipitated by any activity that increases intracranial pressure, such as straining during a bowel movement, stooping, lifting heavy objects, or coughing.

Often, the pain can be relieved by taking aspirin, acetaminophen, or other moderate painkillers. Vomiting with or without nausea (unrelated to food) occurs in about 25% to 30% of people with brain tumors and often accompanies headaches when there is an increase in intracranial pressure. If the tumor invades the meninges, the headaches will be more severe.

Focal manifestations of a space-occupying brain lesion are caused by the local compression or destruction of the brain tissue, as well as by compression secondary to edema. Papilledema (edema and hyperemia of the optic disc) may be the first sign of intracranial tumors. Visual changes do not occur until prolonged papilledema causes optic atrophy.

Nerve and Cord Compression

Symptoms of nerve and/or cord compression may occur when tumors invade and impinge directly on the spinal cord, thecal sac, or nerve root.[82] Severe destructive osteolytic lesions of the vertebral bodies from metastases can lead to pathologic fracture, fragility, and subsequent deformity of one or more vertebral bodies. Bone collapse can occur spontaneously or following trivial injury, sometimes with bone fragments adding to the compression.[83]

Compressive pathologies affecting the spinal cord and nerve roots affect 5% to 10% of all people with cancer.[82] The thoracic spine is affected most often (70% of all cases), usually secondary to metastatic lung and breast cancer. Twenty per cent develop in the lumbosacral region as a result of metastases from prostate and gastrointestinal cancers or melanoma. A small number of cases (10%) arise in the cervical region of the vertebral column.[84,85]

Other (more rare) cancer-related causes of spinal cord compression include radiation myelopathy, malignant plexopathy, and paraneoplastic disorders. Chronic progressive radiation myelopathy can occur in anyone who has received irradiation to the spine or nearby structures. Localized spinal cord dysfunction within the area of the radiation port occurs with numbness and upper motor neuron findings.[86]

Whether from a primary cord tumor or a metastasis, compression of the cord can be the first symptom of cancer. Prostate, lung, and breast cancers are the most common tumors to metastasize to the spine, leading to epidural spinal cord compression, but lymphoma, multiple myeloma, and carcinomas of the colon or kidney and sarcomas can also result in spinal cord and nerve root compression.[87]

Individuals with lymphoma or retroperitoneal tumors may suffer cord compression from tumors that grow through the intervertebral foramen and compress the cord without involving the vertebra.[87] Cord compression is becoming increasingly common, as individuals affected by cancer survive longer with medical treatment.

Signs and Symptoms of Cord Compression. Spinal cord compression with resultant quadriplegia, paraplegia, and possible death is the most common pathologic feature of all tumors within the spinal column. Pain and sensory symptoms usually occur in the body below the level of the tumor but not necessarily at predictable levels. For example, 54% of individuals with T1-T6 compression have lumbosacral pain, and a similar number with lumbosacral compression have thoracic pain.[88]

The location of the metastasis is proportionate to the volume or mass of bone in each region: 60% of metastases occur in the thoracic spine, 30% in the lumbosacral spine, and 10% in the cervical spine.[89,90] Compression at the level of the cauda equina is relatively rare (0.7%).[91] The therapist must observe carefully for subtle objective neurologic deficit (e.g., decreased sensory function, decreased but useful motor function, change in reflexes) that might otherwise be interpreted as side effects of medication.[59]

Breast and lung cancers typically cause thoracic lesions, whereas colon and pelvic carcinomas are more likely to affect the lumbosacral spine. In up to one-third of affected individuals, spinal cord compression occurs at multiple sites.[87]

Early characteristics of spinal cord compression include pain, sensory loss, muscle weakness, and muscle atrophy. Back pain at the level of the spinal cord lesion occurs in up to 95% of cases, presenting hours to months before the compression is diagnosed.

Pain is caused by the expanding tumor in the bone, bone collapse, and/or nerve damage. Pain is usually described as sharp, shooting, deep, or burning and may be aggravated by lying down, weight bearing, bending, sneezing, or coughing.[87]

Discomfort may occur as thoracolumbar back pain in a beltlike distribution; the pain may extend to the groin or the legs. The pain may be constant or intermittent and occurs most often at rest; pain occurring at night can awaken an individual from sleep; the person reports that it is impossible to go back to sleep.

Symptoms of severe pain preceding the onset of motor weakness generally correlate with epidural compression, whereas muscle weakness and bowel/bladder sphincter dysfunction with very little pain indicates intramedullary metastasis.[92]

Weakness in an individual with cancer may be incorrectly attributed to fatigue, anemia, pain medication, or metabolic derangement. The therapist must remain alert to any subtle signs and symptoms of spinal cord compression as the underlying etiology and report these to the physician immediately.[84,86]

Over half of individuals present with sensory changes, either starting in the toes and moving caudally in a stocking-like pattern to the level of the lesion or starting 1 to 5 levels below the level of the actual cord compression.[89]

Less commonly, chest or abdominal pain may occur, caused by nerve root compression from epidural tumor(s). Progressive cord compression is manifested by spastic weakness below the level of the lesion, decreased sensation, and increased weakness. Bowel and bladder dysfunction are late findings.

Cauda Equina Syndrome. Cauda equina syndrome is defined as a constellation of symptoms that result from damage to the cauda equina, the portion of the nervous system below the conus medullaris (i.e., lumbar and sacral spinal nerves descending from the conus medullaris). Although tumors are the focus here, other causes of cauda equina syndrome include acute lumbar disk herniation, spinal stenosis, spinal infection, epidural hematoma, and spinal fracture or dislocation. This syndrome involves peripheral nerves (sensory and motor) within the spinal canal and thecal sac.[93]

Individuals with cauda equina syndrome present differently from those with spinal cord compression. The three most common symptoms of cauda equina syndrome include saddle anesthesia, bowel or bladder dysfunction, and lower extremity weakness.[94]

Diminished sensation over the buttocks and posterior-superior thighs is also common.

Decreased anal (rectal) sphincter tone, urinary retention, and overflow incontinence occur in 60% to 80% of patients at the time of diagnosis. About half of clients need urinary catheters.[89]

CLINICAL SIGNS AND SYMPTOMS

Cauda Equina Syndrome

- Low back pain
- Sciatica
- Saddle and/or perianal hypesthesia or anesthesia
- Bowel and/or bladder changes or dysfunction (e.g., difficulty initiating flow of urine, urine retention, urinary or fecal incontinence, constipation, decreased rectal tone and sensation)
- Lower extremity weakness (variable); gait disturbance
- Sexual dysfunction:
  - Men: Erectile dysfunction (inability to attain or sustain an erection)
  - Women: Dyspareunia (painful intercourse)
- Decreased rectal tone
- Decreased perineal reflexes
- Diminished or absent lower extremity reflexes (patellar, Achilles)

Individuals with cauda equina syndrome caused by neoplasm may present with a long history of back pain and paresthesias; urinary difficulties are very common.[95] The presentation may mimic a diskogenic source, causing a delay in diagnosis, especially in the young adult with a primary tumor. Individuals with metastatic tumors are older with a previous history of cancer.[93]

Associated signs and symptoms of primary or metastatic tumors causing cauda equina syndrome may include abnormal weight loss, hematuria, hemoptysis, melena, and/or constipation. The medical diagnosis of cauda equina syndrome is not always straightforward. Abnormal rectal tone may be delayed in individuals presenting with cauda equina syndrome. This is because sensory nerves are smaller and more sensitive than motor nerves; even so, some people present with abnormal rectal tone (motor) without saddle anesthesia.[96]

Peripheral Neuropathy. Peripheral neuropathy with loss of vibratory sense, proprioception, and deep tendon reflexes is most often chemotherapy-related (e.g., cisplatin, Taxol, vincristine). Numbness, tingling, and burning pain in the hands and feet and loss of balance and difficulties with mobility are common with this problem.[97] It is important to differentiate the type and etiology of peripheral neuropathy before planning treatment intervention.

For example, chemotherapy-induced peripheral neuropathy (CIPN) may not be as likely to respond to lymph drainage and compression bandaging, whereas good results have been seen when this treatment intervention is used for weakness and paresthesias from lymphedema-induced nerve compression (e.g., breast cancer, ovarian cancer, testicular cancer).

In other words, resolution of neuropathy symptoms utilizing principles of manual lymphatic drainage may confirm subclinical lymphedema as the major etiologic factor in some clients.

Paraneoplastic Syndromes

Other neurologic problems occur frequently in individuals with cancer. These may be nonmetastatic and associated with cancer-related opportunistic infections, metabolic disturbances, vascular complications, treatment neurotoxicity, and paraneoplastic syndromes.

When tumors produce signs and symptoms at a distance from the tumor or its metastasized sites, these "remote effects" of malignancy are collectively referred to as *paraneoplastic syndromes.* This can be the first sign of malignancy and may show up months (even years) before the cancer is detected. They are usually caused by one of three phenomena:

- Tumor metastases to the brain
- Endocrine, fluid, and electrolyte abnormalities
- Remote effects of tumors on the CNS

The causes of these syndromes are not well understood. In contrast to the hormone syndromes in which the cancer directly produces a substance that circulates within blood to produce symptoms, the neurologic syndromes are a group of syndromes mediated by the immune response.

Tumors involved in this type of syndrome stimulate the production of immunologically active nervous system proteins. These immune responses are frequently associated with antineuronal antibodies that can be used as diagnostic markers of paraneoplastic disorders. As a result of these immune responses, discrete or multifocal areas of nervous system degeneration can occur, causing diverse symptoms and deficits.[98]

These are not direct effects of either the tumor or its metastases. Cancer cells can acquire new cellular functions uncharacteristic of the originating tissue. Many of these syndromes involve ectopic hormone production by tumor cells. These hormones are distributed by the circulation and act on target organs at a site other than the location of the tumor. Some tumor cells secrete biochemically active substances that can also cause metabolic abnormalities.

The reported frequency of paraneoplastic syndromes ranges from 10% to 15% to 2% to 20% of malignancies. However, these could be underestimates. Neurologic paraneoplastic syndromes are estimated to occur in fewer than 1% of patients with cancer.[99]

The neuromusculoskeletal system is often affected, and the clinical presentation is unusual. The clinical manifestation of paraneoplastic syndrome depends on the tumor effects. The therapist is often the first health care professional to see and/or recognize the incongruence of the signs and symptoms.

In fact, the presentation may confound the medical staff. When the client fails to respond to palliative treatment, physical therapy is recommended. The alert therapist will recognize the unusual presentation and will follow up with a screening examination.

The paraneoplastic syndromes are of considerable importance because they may accompany relatively limited neoplastic growth and provide an early clue to the presence of certain types of cancer (e.g., osteoarthropathy caused by bronchogenic carcinoma, hypercalcemia from osteolytic skeletal metastases). The most common cancer associated with paraneoplastic syndromes is small cell cancer of the lungs (produces adrenocorticotrophic hormone [ACTH] and causes Cushing's syndrome).

Clinical Signs and Symptoms of Paraneoplastic Syndromes. Clinical findings of paraneoplastic syndromes may resemble those of primary endocrine, metabolic, hematologic, or neuromuscular disorders. Depending on which system is compromised, symptoms can include rheumatologic, renal, GI, vascular, hematologic, cutaneous, metabolic, endocrine, neurologic, and/or neuromuscular physical findings.[100-102]

For example, the Lambert-Eaton myasthenic syndrome (LEMS), often secondary to small cell lung carcinoma, results in muscle weakness when autoantibodies directed against the presynaptic calcium channels at the neuromuscular junction cause impaired release of acetylcholine from presynaptic nerve terminals. The clinical presentation is distinct from myasthenia gravis (MG), with lower limb muscle fatigability and autonomic symptoms appearing first in the paraneoplastic form.[103]

Gradual, progressive muscle weakness during a period of weeks to months (especially of the pelvic girdle muscles) may occur. Proximal muscles are most likely to be involved (see Fig. 13-3). The weakness does stabilize. Reflexes of the involved extremities are present but diminished. The weakness often improves, and deep tendon reflexes may return with exercise.

In clients who develop myopathies, such as dermatomyositis (DM) or polymyositis (PM), the myositis may precede, follow, or arise concurrently with the malignancy. No particular type of cancer has been found to predominate in such cases, but the clients affected are generally older and respond poorly to medical treatment for the myositis.

The course of the paraneoplastic syndrome usually parallels that of the tumor. Therefore effective medical intervention (rather than physical therapy) should result in resolution of the syndrome. A paraneoplastic syndrome may be the first sign of a malignancy or recurrence of cancer that may be cured if detected early. Paraneoplastic syndromes with musculoskeletal manifestations are listed in Table 13-6.

Even such nonspecific symptoms as anorexia, malaise, weight loss, and fever are truly neoplastic and are probably due to the production of specific factors by the tumor itself. For example, anorexia is a common symptom in clients with cancer that is attributed to tumor production of the protein tumor necrosis factor (TNF), also called *cachectin*. Fever may be seen in clients with cancer in the absence of infection when it is produced by tumor induction of pyrogen formation by host white cells or by direct tumor production of a pyrogen.

TABLE 13-6 Paraneoplastic Syndromes Having Musculoskeletal Manifestations

Malignancy	Rheumatic Disease	Clinical Features
Lymphoproliferative disease (leukemia, lymphomas)	Vasculitis	Necrotizing vasculitis
	Polyarthritis	Polymyalgia, swelling
Plasma cell dyscrasia	Cryoglobulinemia	
	Immune complex disease	Vasculitis; Raynaud's phenomenon; arthralgia; neurologic symptoms
		Nephrotic syndrome
Hodgkin's disease	Reflex sympathetic dystrophy	Palmar fasciitis and polyarthritis
Ovarian cancer		
Carcinoid syndrome (breast, uterus, lung cancers)	Scleroderma	Scleroderma-like changes; anterior tibia
	Pyogenic arthritis	Enteric bacteria cultured from joint
Colon cancer	Osteogenic osteomalacia	Bone pain; stress fractures
Mesenchymal tumors	Severe Raynaud's phenomenon	Digital necrosis
Renal cell cancer (and other tumors)	Panniculitis	Subcutaneous nodules, especially in males
Pancreatic cancer	Hypertrophic osteoarthropathy	Digital clubbing, excess bone formation
Lung cancers		

Data from Santacroce L: Paraneoplastic syndromes, *eMedicine Specialties*, 2010. Available online at www.emedicine.com/med/TOPIC1747.HTM. Accessed March 7, 2011.

CLINICAL SIGNS AND SYMPTOMS

Paraneoplastic Syndromes

Constitutional Symptoms
- Fever
- Fatigue
- Anorexia, malaise, weight loss, cachexia
- Confusion (also a neurologic symptom)

Cardiovascular
- Hypertension
- Thrombophlebitis
- Endocarditis

Integument
- Skin rash, skin flushing, pigmentation changes (see Figs. 4-10, 4-11, 4-26)
- Clubbing of the fingers or toes
- Itching, ichthyosis (dry, flaking skin; see Fig. 11-4)
- Alopecia (hair loss)
- Herpes zoster
- Acanthosis nigricans (see Fig. 11-8)

Rheumatic
- Arthralgias, polyarthritis
- Palmar fasciitis
- Bone pain

Neurologic
- Proximal muscle weakness
- Change in deep tendon reflexes (most often hyporeflexia)
- Sensory neuropathy (progressive sensory loss of hands and feet; may or may not be symmetric)
- CNS (cerebellar degeneration): gait difficulties, dizziness, nausea, diplopia (double vision), ataxia, dysphagia

Hematologic
- Anemia
- Polycythemia
- Signs and symptoms of hypercalcemia (see Table 13-7)
- Thrombocytosis (platelet level greater than 500,000/dL)

GI
- Diarrhea (malabsorption, electrolyte imbalance)

Renal/Urologic
- Nephropathy

Rheumatologic Manifestations. Cancer can be associated with arthritis and can present as a paraneoplastic syndrome called *carcinoma polyarthritis*. Paraneoplastic rheumatic disorders of this type are induced by the malignancy through hormones, peptides, autocrine and paracrine mediators, antibodies, and cytotoxic lymphocytes.[104] Polyarthritis has been reported in adults ages 43 to 80 years of age when associated with solid tumors and in individuals from 12 to 65 years old with hematologic malignancies.[105]

Cancer-associated rheumatic syndromes are characterized by a relatively short interval between the appearance of the rheumatic disorder and diagnosis of its associated neoplasm (usually less than 2 years).[104]

Rheumatic disorders caused by cancer have been associated most often with breast and lung cancers. Palmar fasciitis and polyarthritis have been reported in association with metastatic ovarian carcinoma (Case Example 13-6).[106] Even though cancer polyarthritis is a fairly uncommon occurrence, the therapist is more likely than most other health care professionals to see this. Timely recognition can reduce morbidity and mortality.

The medical diagnosis can be missed or delayed without careful evaluation. Sometimes, the diagnosis of polymyalgia rheumatica is made in error. Anyone with a sudden onset of rheumatic disease that is seronegative and monarticular and occurs in the presence of a past history of cancer may be demonstrating signs of metastatic cancer or an occult malignancy.

Rheumatologic complaints have a sudden onset and may spare the small joints of the hands and wrists. Clinical features of carcinoma polyarthritis primarily affect asymmetric joints of the lower extremities, often the result of metastasis to the joint or periarticular bone.

Other rheumatologic conditions and muscular disorders can be associated with malignancy (Box 13-4). These conditions often disappear after successful treatment of the underlying malignancy.[105,107]

There is also a link between longstanding rheumatic disorders and cancer. The risk of malignant transformation during the course of chronic rheumatic disorders, including rheumatoid arthritis, Sjögren's syndrome, and systemic sclerosis is well-known. The underlying mechanism is likely the result of immune dysregulation.[108]

Digital Clubbing. Digital clubbing is another possible sign of paraneoplastic syndrome, especially when associated with pulmonary malignancy (see Fig. 4-36). Clubbing of the fingers and toes is seen most often with chronic conditions such as congenital heart disease with cyanosis, cystic fibrosis, or chronic obstructive pulmonary disease (COPD). It can also develop with paraneoplastic syndromes and within 10 days of acute systemic illness such as acute pulmonary abscess, heart disease, and ulcerative colitis.

BOX 13-4 MUSCULAR DISORDERS ASSOCIATED WITH MALIGNANCY

Dermatomyositis and polymyositis
Type II muscle atrophy
Myasthenia gravis (MG)
Lambert-Eaton myasthenic syndrome (LEMS)
Metabolic myopathies
Primary neuropathic diseases
  Amyotrophic lateral sclerosis (ALS)
  Amyloidosis

From Gilkeson GS, Caldwell DS: Rheumatologic associations with malignancy, *J Musculoskel Med* 7:70, 1990.

CASE EXAMPLE 13-6

Arthritis Associated with Ovarian Carcinoma

A 56-year-old woman was sent to physical therapy by a hand surgeon with a provisional diagnosis of rheumatoid arthritis, pending results of laboratory studies. She described a 3-month history of bilateral finger stiffness with swelling and pain. Most recently, she developed nodules at the proximal interphalangeal joints (PIPs) and thickening of the palms with erythema, both bilaterally.

At the time of her first physical therapy visit, she also reported new onset of right shoulder pain and loss of motion. When asked if she had noticed any symptoms or changes of any kind anywhere else in the body, she mentioned pain and a sense of "fullness" in the left lower abdominal quadrant. She denied having any hip pain on that side or any gastrointestinal (GI) or genitourinary (GU) signs and symptoms.

There was no previous history of any significance. When asked about birth histories and deliveries, she reported never being married and never being pregnant. She had her last menstrual period 3 years ago. Her last Papanicolaou (PAP) smear and clinical breast examination were performed 2 years ago, and results were reportedly within normal limits.

During a screening physical examination, the therapist noted visible asymmetry of the lower abdominal quadrant with distention observed on the left compared with the right. There was no warmth or tenderness to abdominal palpation, but an unidentified mass could be felt just to the midline of the left anterior superior iliac spine (ASIS). Because the client was postmenopausal, there was no need to screen for possible pregnancy.

How would you proceed in a situation like this? Do you suggest the client call and report new onset of shoulder pain and "fullness" to the referring physician? Or should you suggest she go to her gynecologist for a pelvic examination and updated PAP smear?

The new onset of shoulder pain is important information, given the physician's "provisional diagnosis" while waiting for lab results. Although the apparent pelvic mass is not usually of interest to a hand surgeon, it will be up to the referring physician to decide what further medical testing is needed.

The therapist should provide the physician with the new and additional information obtained, present a plan for physical therapy intervention, and request approval before proceeding, given the new signs and symptoms present. Until a final medical diagnosis is made and cancer ruled out, ultrasound should not be used.

Results: Laboratory tests revealed a normal complete blood count (CBC) and erythrocyte sedimentation rate (ESR) and routine chemistry results. Special tests for markers to indicate rheumatoid arthritis (rheumatoid factor, antinuclear antibody) were normal.

When this additional information was presented, further tests were ordered. A diagnosis of ovarian cancer (Stage IV) was made, indicating distant metastases. Physical therapy intervention was put on hold until medical treatment (i.e., surgery and chemotherapy) could be completed. She received occupational therapy as an inpatient for home adaptive aids and stretching exercises. Her hand symptoms resolved with medical treatment of the carcinoma.

Digital clubbing occurs in the distal phalanx and causes the ends of the digits to become round and wide, like "little clubs." The thumb and index finger are affected first and can be assessed by the Schamroth method (see Fig. 4-37).

Look for recent onset of other signs and symptoms (e.g., pulmonary, hepatic, cardiac, gastrointestinal). For example, digital clubbing accompanied by a recent, unexplained weight loss, hemoptysis, and a significant smoking history may be a red flag sign associated with lung cancer.

Skeletal Manifestations

Primary bone cancer is uncommon; primary cancers of the musculoskeletal system are discussed later in this chapter. The skeleton is, however, the most common organ affected by metastatic cancer. Tumors arising from the breast, prostate, thyroid, lung, and kidney possess a special propensity to spread to bone.

Tumor cells commonly metastasize to the most heavily vascularized parts of the skeleton, particularly the red bone marrow of the axial skeleton and the proximal ends of the long bones (humerus and femur), and the vertebral column, pelvis, and ribs (Box 13-5).

Occasionally, a growing bone mass is the first sign of disease. Diagnosis is made by x-ray study and surgical biopsy, requiring immediate attention to suspicious symptoms by referral to the client's physician.

Local swelling can be detected when the lesion protrudes beyond the normal confines of the bone. The swelling of a benign lesion is usually firm and nontender. In the presence of a rapidly growing malignant neoplasm, however, the swelling is more diffuse and frequently tender (Fig. 13-6).

The overlying skin may be warm because of the highly vascularized nature of neoplasms. If the lesion is close to a joint, function in that joint may be disturbed, with painful and restricted range of motion.

Bone Pain

Bone pain, resulting from structural damage, rate of bone resorption, periosteal irritation, and nerve entrapment, is the most common complication of metastatic disease to the skeletal system. A history of sudden onset of severe pain usually indicates the complication of a pathologic fracture (a break in an already weakened bone). Pathologic fractures are the result of metastatic disease of primary cancers most often affecting the lung, prostate, and breast.

BOX 13-5 MOST COMMON SITES OF BONE METASTASES (IN ORDER OF FREQUENCY)

Vertebrae (thoracic 60%/lumbosacral 30%)
Pelvis
Ribs (posterior)
Skull
Femur (proximal)
Others: Sternum, cervical spine

Data from Smuckler A, Govindan R: Management of bone metastasis, *Contemp Oncol* 1(13):1-10, 2002.

Pathologic fractures tend to affect the vertebral body at both the thoracic and lumbar levels. Kyphotic deformity can occur with compression of the cord or cauda equina (see further discussion on Cauda Equina Syndrome in this chapter).

Bone pain is usually deep, intractable, and poorly localized, sometimes described as burning or aching and accompanied by episodes of stabbing discomfort (Case Example 13-7). The pain may be cyclic and progressive until it becomes constant. The pain is made worse by activity, especially weight bearing. The pain is often associated with trauma during a game or exercise and may be dismissed in children as "growing pains."

It is often worse at night, awakening the person; neither sleep nor lying down provides relief. Pain at night that is unrelieved by rest or change in position is a red flag. Assessing night pain is discussed in detail in Chapter 3 in Night Pain and Cancer (see Box 3-7).

Beware of the client who reports disproportionate (excessive) pain relief with aspirin, as this may be a sign of a particular bone cancer called *osteoid osteoma.* Pain subsiding with aspirin (contains salicylates) is the hallmark of this entity. Salicylates inhibit the prostaglandins that are produced by osteoid osteomas.

Bone pain associated with skeletal metastases can often be reproduced with a heel strike when an undiagnosed fracture

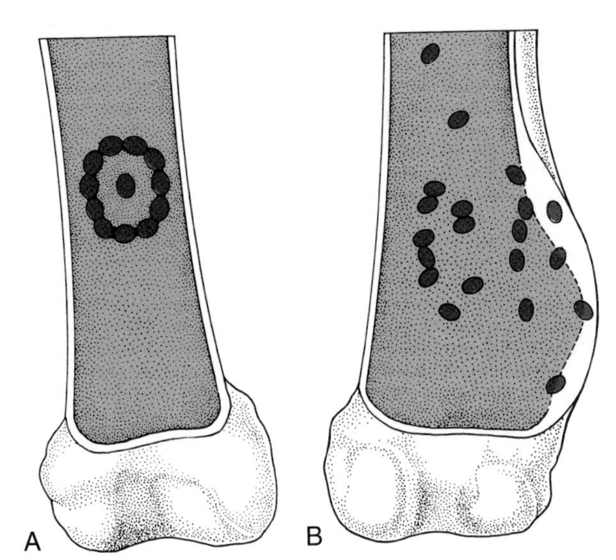

Fig. 13-6 **A,** Benign bone tumors have a characteristic sclerotic rim around the periphery of the lesion. The lesion is usually well defined, and there is no evidence of erosion of the cortex or a soft tissue mass. **B,** Malignant bone tumors can have lytic or sclerotic components. It is frequently difficult to know the extent of the lesion within the bone because there is no well-defined sclerotic rim around the tumor. The destructive process is diffuse within the medullary cavity of the bone, and the tumor may break through the cortex of the bone, producing Codman's triangle. Frequently, an associated soft tissue mass is present. Medical differential diagnosis of this lesion is between an osteogenic sarcoma and a chondrosarcoma.

CASE EXAMPLE 13-7

Uterine Cancer With Bone Metastasis

A 44-year-old slender, athletic woman with isolated left knee pain of unknown cause was referred to physical therapy by her physician for a "strengthening program." She was actively involved in a variety of physical activities, including a co-ed baseball team, a hiking club, and church basketball intramurals but could not recall any specific injury, fall, or other impact to her leg. She had a pair of shoe orthotics prescribed by a podiatrist 5 years ago "to compensate for my excessive Q-angle."

The physical therapy examination was unremarkable for any joint swelling, redness, or palpable warmth. There was point tenderness along the medial joint line and a palpable though asymptomatic plica. Joint integrity was intact, and all special tests were negative.

A neurologic screening examination was also considered within normal limits, although muscle strength for the quadriceps and hamstrings was diminished by pain. Pain was present on weightbearing activities but did not prevent the woman from participating in all activities. There was no reported night pain, fever, or other associated signs and symptoms.

Without a definitive physical therapy diagnosis, a treatment plan was outlined to include modalities for pain and a stretching and strengthening program. Within a week's time, this client's pain level escalated on the numeric rating scale (NRS) from 3 to 10 (on a scale from 1 to 10) with constant pain that kept her awake at night for hours. When she returned to the physical therapy clinic, she was using crutches and was not bearing weight on the left leg.

Results: Therapists should be careful about assuming that physical therapy treatment has exacerbated a client's symptoms and instituting a change in program. If the treating therapist decided to continue physical therapy, with the use of some other approach, the physician should have been notified of the change in status.

Given the insidious onset of this joint pain and the rapidly progressive nature of the symptoms, this client was immediately sent back to her physician. A diagnosis of bone metastasis was made, with early stage endometrial carcinoma appearing as an unusual, isolated skeletal lesion.

She was treated with aggressive multidisciplinary therapy, including limb salvage and physical therapy as part of her rehabilitation program. The early referral most likely contributed to her favorable prognosis and cancer-free status 2 years later.

is present in the lower extremities. Watch for pain on weight bearing with a positive heel strike test or reproduced symptoms when hopping on one leg (in the younger client; this is not a likely test to use in the older adult). Perform translational/rotational tests for stress fracture.

The pain does not respond to physical agents or physical therapy intervention. Sometimes, the client has some relief after the first few sessions of physical therapy, but pain returns and may even be worse than before. The therapist may think the chosen intervention has been unsuccessful and is at fault. Consider it a red flag whenever a client fails to improve or improves and then gets worse. Further investigation and screening is advised under these circumstances.

Pain may occur around joints because of mechanical, chemical, or bony change; pain and the rate of bone resorption appear to be linked. There is often disturbance of the highly innervated periosteum, giving bone pain its neurogenic-like qualities, especially its unrelenting, intractable quality.

Fracture

Pathologic fractures (e.g., vertebrae, long bones) occur in half of all people with osteolytic metastases. In fact, this may be the presenting sign of bone cancer. An injury with subsequent medical evaluation reveals the fracture and the cancer simultaneously.

Back Pain

Neoplastic disease can cause backache, particularly in older adults, or shoulder pain in the presence of breast cancer. Although primary neoplasms of the spine are rare, myeloma and metastatic disease are more common. Malignancy as a cause of low back pain in primary care clients accounts for only 1% of the affected population.[109]

In anyone with a known cancer, the onset of back pain could suggest spinal metastases. An insidious onset of waist-level or midback pain that becomes progressively more severe and more persistent often occurs. The pain is usually unrelieved by lying down and frequently becomes worse at night. Unexplained weight loss with severe back pain aggravated by rest may point to metastatic carcinoma of the spine. Other bone-related cancers, such as multiple myeloma, can cause severe, unremitting backaches that are present at rest and become worse when lying down.

Cancer causing low back pain can be ruled out with 100% sensitivity if the client is less than 50 years old, has not experienced unexplained or unintentional weight loss, has never had cancer before, and has responded to the physical therapist's intervention.[110,111] Referral to a spine specialist may not be needed if radiographs have not been taken yet and/or laboratory tests have not been ordered to complete a simple screening strategy.[110,112] If the therapist is not in a setting that allows these next steps to be generated from within the department, then referral may be advised.

Hypercalcemia from Skeletal Metastases

Hypercalcemia (greater than normal amounts of calcium in the blood) occurs frequently in clients with metastatic bone disease who have osteolytic lesions (Fig. 13-7). Normal serum calcium levels range between 8.2 and 10.2 mg/dL. Mild hypercalcemia occurs when this level drops to around 12 mg/dL; severe hypercalcemia is defined by serum calcium at 14 mg/dL or more.

Hypercalcemia is very common in cases of breast cancer and myeloma, primarily because of an increase in bone resorption, which is caused in turn by tumor cell production of parathyroid hormone–related protein that stimulates osteoclastic bone resorption.[90,113]

Other tumors associated with hypercalcemia may include carcinomas of the lung (most commonly, small cell lung cancers), squamous cell carcinoma of the head and neck,

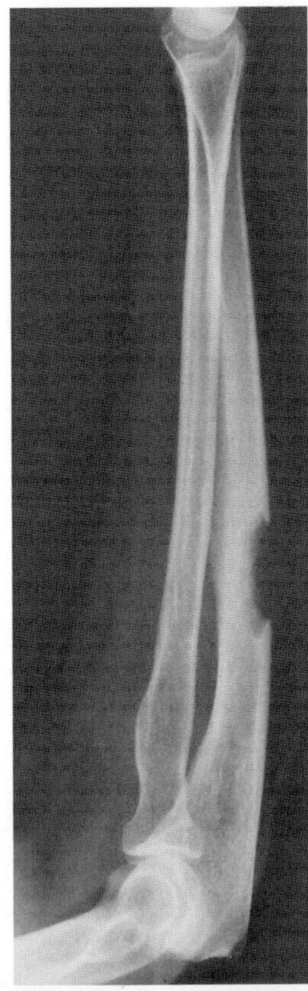

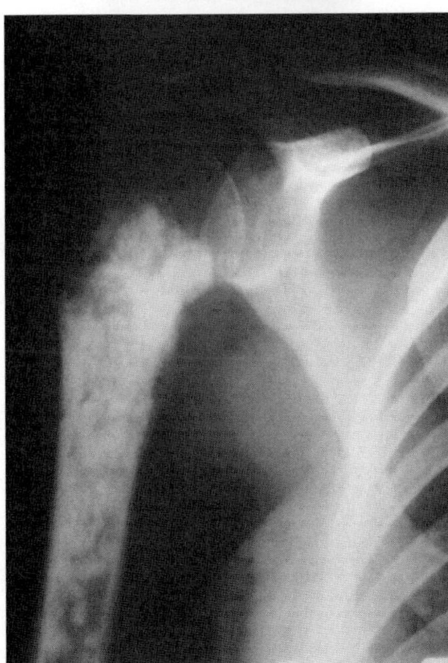

Fig. 13-7 Lytic versus blastic bone. In the top x-ray, you will see a lytic bone lesion from breast cancer. Notice how the bone has distinct punched out segments. This is characteristic of a lytic lesion. On the bottom, you will see blastic bone lesions from osteosarcoma. The blastic form of bone cancer is a more diffuse pattern of degeneration. (From Dorfman HD, Czerniak B: *Bone tumors,* St. Louis, 1998, Mosby.)

renal cell cancer, prostate cancer, lymphoma and leukemia, thyroid cancer, and parathyroid carcinoma (rare). In most cases, hypercalcemia is an indication of progression of disease. Hypercalcemia associated with metastatic breast cancer involving bone may occur with hormone therapy.

Hypercalcemia is characterized by musculoskeletal, nervous system, cardiovascular/pulmonary, and GI symptoms (Table 13-7). The therapist may see the first signs and symptoms of hypercalcemia in the musculoskeletal system but should watch for others as well.

Signs and symptoms of CNS-related hypercalcemia are similar to other causes of CNS problems and include confusion, drowsiness, lethargy, headache, depression, and irritability. Hypercalcemia can also affect the GI system. The most common hypercalcemia-induced GI signs and symptoms are anorexia, nausea, vomiting, constipation, dehydration, and thirst.

Finally, in the clinical practice of a therapist, hypercalcemia secondary to bone cancer or metastases to the bone may affect the cardiac system. These clients are usually inpatients or are known to have cancer. Hypertension may be the only outward sign of hypercalcemia-induced cardiovascular changes. Vital sign assessment may help identify early signs of cardiac involvement. However, cardiac arrest may present as the first sign of a problem.

Bisphosphonates (bone resorption inhibitors, such as Fosamax [alendronate], Actonel [risedronate], Evista

TABLE 13-7 Hypercalcemia

System	Symptom
Central nervous system (CNS)	Drowsiness, lethargy, coma Irritability, personality change Confusion, increased confusion Headaches Depression, memory loss, difficulty concentrating Visual disturbances Balance/coordination problems Changes in deep tendon reflexes (hyporeflexive or hyperreflexive) Change in muscle tone for individual with neurologic condition Positive Babinski and/or clonus reflex Changes in bowel/bladder function
Musculoskeletal	Muscle pain or tenderness and weakness Muscle spasms Bone pain (worse at night and on weight bearing) Pathologic fracture
Cardiovascular	Hypertension Arrhythmia Cardiac arrest
Gastrointestinal	Anorexia (loss of appetite) Nausea Vomiting Constipation Dehydration Thirst

[raloxifene], and Miacalcin [calcitonin-salmon]) are drugs used to control hypercalcemia and limit or prevent bone loss. In emergent or predictable situations, intravenous use of bisphosphonates (e.g., pamidronate, zoledronic acid) can be used to stabilize and/or prevent hypercalcemia. With their use, health care professionals expect to see fewer cases of hypercalcemia than in the past. These drugs also reduce bone pain, delay skeletally related events (SREs), reduce the number of pathologic fractures, and in some cases, prolong survival.

Hepatic Manifestations

Liver metastases are among the most ominous signs of advanced cancer. The liver filters blood coming in from the GI tract, making it a primary metastatic site for tumors of the stomach, colorectum, and pancreas.

Symptoms observed in a physical therapy practice include bilateral carpal/tarsal tunnel syndrome, possibly accompanied by abdominal pain and tenderness with general malaise and fatigue. Right upper quadrant pain with possible referral to the right shoulder may also occur with or without carpal tunnel syndrome (see Table 13-5).

Carpal Tunnel Syndrome

Carpal tunnel syndrome (CTS) can be caused by a wide range of both neuromusculoskeletal and systemic conditions and illnesses (see Table 11-2). Whenever anyone presents with *bilateral* symptoms of any kind, it is considered a "red flag" symptom. In Chapter 2 of this text, we discussed the various bilateral symptoms the therapist might encounter in a clinical practice.

A common systemic cause of CTS involves the hepatic system (see Chapter 9 for an explanation). Briefly, liver dysfunction results in increased serum ammonia and urea levels. When these toxins are no longer absorbed into the portal vein and removed from the body, they pass directly to the brain.

Ammonia transported to the brain reacts with glutamate (an excitatory neurotransmitter), producing glutamine. The reduction of brain glutamate impairs neurotransmission. This leads to altered CNS metabolism and function. As blood ammonia levels rise, many unusual compounds (e.g., octopamine) form and serve as false neurotransmitters in the CNS. Asterixis (also known as liver flap) and numbness/tingling occur as a result of this ammonia abnormality, causing intrinsic nerve pathology. This can be misinterpreted as CTS (or tarsal tunnel syndrome) (Case Example 13-8).

When screening for bilateral CTS as a result of liver impairment, always ask about the presence of similar symptoms in the feet. Look for a history of alcoholism, cirrhosis, previous cancer, other liver disease, and the use of statins (cholesterol-lowering drugs such as simvastatin [Zocor] and atorvastatin calcium [Lipitor]; liver damage occurs in some people taking these medications).

Ask about the presence of other GI signs and symptoms. A client presenting with shoulder or upper back pain may not think nausea and abdominal bloating are related in any way to the symptoms in the wrists and hands. Perform a quick

CASE EXAMPLE 13-8

Carpal Tunnel Syndrome Associated With Liver Cancer

A 52-year-old male who was employed as an over-the-road (OTR) trucker was referred by a hand surgeon for bilateral carpal tunnel syndrome (CTS). The client did not want surgery and opted for a more conservative, nonoperative approach.

He was hostile and verbally abusive, refusing to even sit down for his treatment. His wife reported a history of alcohol use/abuse. He was not screened for medical disease, just treated with the CTS protocol in a hand clinic.

During a treatment session, he commented that he had just seen an acupuncturist who told him he has liver disease. Because symptoms of bilateral numbness and tingling in the hands and feet can be a sign of liver impairment, a screening examination was performed.

The client was tested for liver flap and was observed for palmar erythema and nailbed and skin changes. Liver flap was not present, but tremoring of the hands was observed along with palmar erythema. No obvious ascites or angiomas were present.

The client was later given a medical diagnosis of liver cancer.

What are the red flags in this case? How do you return the client to the referring physician for further follow-up?

Red Flags

Age over 50
Reported history of alcohol use/abuse
Bilateral symptoms
Liver impairment diagnosis by acupuncturist
Palmar erythema, motor tremor

Physician Referral: This may depend upon the therapist's relationship with the physician. It may be possible to telephone the physician with exactly what happened and what the therapist sees as "red flags."

If that is not feasible, then a letter (brief and to the point) with a quick summary of the findings and an open-ended question should be faxed or sent. For example,

Date (very important for documentation)

Dear Dr. Lowell,

Thank you for your recent referral of Mr. Smith for hand therapy. We are following our usual protocol for carpal tunnel syndrome. Something has come up that concerns me. Mr. Smith saw Dr. Jyn, the local acupuncturist, who mentioned liver impairment.

Given his age, drinking history, and bilateral CTS, I'm wondering if there isn't something else going on. I noticed a fine tremor in both hands (present at rest and with activity) and color changes in his hands suggestive of palmar erythema.

We will continue treating him, but perhaps an appointment with you sooner than his scheduled 4-week follow-up is in order. What do you think? (Alternate: Please advise.)

Signature, etc.

liver screen and look for signs of liver disease (see Box 9-1 or Appendix B-4).

ONCOLOGIC PAIN

As mentioned earlier, pain is rarely an early warning sign of cancer and is uncommon in some cancers such as leukemia. However, pain occurs in 60% to 80% of clients with solid tumors.

This pain syndrome has multiple causes, and the therapist must always keep in mind common patterns of referred pain (see Chapter 3; see also Table 3-8). Some pain is caused by pressure on peripheral nerves or displacement of these nerves. Pain may also result from interference with blood supply or from blockage within hollow organs.

A common cause of cancer pain is metastasis of cancer to the bone. This type of pain can occur as a result of pathologic fracture with resultant muscle spasms; if the spine is involved, nerves may be affected. Pain may also result from iatrogenic causes such as surgery, radiation therapy, and chemotherapy. Immobility and inflammation also can lead to pain.

Signs and Symptoms Associated with Levels of Pain

The severity of pain varies from one client to another, but certain signs and symptoms are characteristic of particular levels of pain. For example, in *mild-to-moderate superficial pain,* a sympathetic nervous system response is usually elicited with hypertension, tachycardia, and tachypnea (rapid, shallow breathing).

In *severe* or *visceral pain,* a parasympathetic nervous system response is more characteristic, with hypotension, bradycardia, nausea, vomiting, tachypnea, weakness, or fainting. Depression and anxiety may increase the client's perception of pain, requiring additional psychologic and emotional support.

Biologic Mechanisms

Five biologic mechanisms have been implicated in the development of chronic cancer pain. The characteristics of the pain depend on tissue structure, as well as on the mechanisms involved.

Bone Destruction

Bone destruction secondary to infiltration by malignant cells or resulting from metastatic lesions is the first and most common of the biologic mechanisms causing chronic cancer pain. Bone metastases cause increased release of prostaglandins and subsequent bone breakdown and resorption.

The client's pain threshold is reduced through sensitization of free nerve endings. Bone pain may be mild to intense. Maladaptive outcomes of bone destruction may include sharp, continuous pain that increases on movement or ambulation. The rich supply of nerves and tension or pressure on the sensitive periosteum or endosteum may cause bone pain.

Other factors contributing to the intense discomfort reported by clients include limited space for relief of pressure, altered local metabolism, weakening of the bone structure, and pathologic fractures ranging in size from microscopic to large.

Visceral Obstruction

Obstruction of a hollow visceral organ and ducts, such as the bowel, stomach, or ureters, is a second physiologic factor in the development of chronic cancer pain.

Viscus obstruction is most often due to the obstruction of an organ lumen by tumor growth. In the GI or genitourinary (GU) tract, obstruction results in either a severe, colicky, crampy pain or true visceral pain that is dull, diffuse, boring, and poorly localized.

If a vein, artery, or lymphatic channel is obstructed, venous engorgement, arterial ischemia, or edema, respectively, will result. In these cases, pain is described as dull, diffuse, burning, and aching. Obstruction of the ducts leading from the gallbladder and pancreas is common in cancer of these organs, although jaundice is more frequently an earlier symptom than pain. Cancer of the throat or esophagus can obstruct these organs, leading to difficulties in eating or speaking.

Nerve Compression

Infiltration or compression of peripheral nerves is the third physiologic factor that produces chronic cancer pain and discomfort. Pressure on nerves from adjacent tumor masses and microscopic infiltration of nerves by tumor cells result in continuous, sharp, stabbing pain that generally follows the pattern of nerve distribution. The invading cells affect the conduction of impulses by the nervous system and sometimes result in constant, dull, poorly localized pain and altered sensation.

Blockage of the blood in arteries and veins, again both by pressure from tumor masses nearby and by infiltration, can decrease oxygen and nutrient supply to tissues. This deficiency can be perceived as pain that is similar in origin and character to cardiac pain or angina pectoris, which is chest pain from an insufficient supply of oxygen to the heart. Hyperesthesia or paresthesia may result.

Skin or Tissue Distention

Infiltration or distention of the integument (skin) or tissue is the fourth physiologic phenomenon resulting in chronic, severe cancer pain. This type of pain is secondary to the painful stretching of skin or tissue because of underlying tumor growth. This stretching produces severe, dull, aching, and localized pain, with severity of pain increasing concurrently with increase in tumor size.

Pain associated with headache secondary to brain tumor is thought to be due to traction on pain-sensitive intracranial structures.

Tissue Inflammation, Infection, and Necrosis

Inflammation, infection, and necrosis of tissue may be the fifth and final cause of cancer pain. Inflammation, with its accompanying symptoms of redness, edema, pain, heat, and loss of function, may progress to infection, necrosis, and sloughing of tissue.

If the inflammatory process alone is present, the pain is characterized by a sensitive tenderness. If, however, necrosis and tissue sloughing have occurred, pain may be excruciating.

SIDE EFFECTS OF CANCER TREATMENT

Conventional cancer treatment has many side effects because the goal of treatment is to remove or to kill certain tissues. In any situation, healthy tissue also is usually sacrificed. It is not always possible to differentiate between cancer recurrence and the acute or long-term effects of cancer treatment. For this reason, knowledge of both immediate and delayed side effects of cancer treatment is helpful.

For many years, three basic modalities of cancer treatment have been used, either alone or in combination: surgery, radiation therapy, and chemotherapy. In recent years, immunotherapies involving the use of cells of the immune system to prompt a tumor-killing response have been developed. Immunotherapy may be most effective when combined with conventional treatments, such as chemotherapy and radiation, to improve the success of treatment and decrease the side effects of conventional modalities.

The pharmaceuticals used in chemotherapy are cytotoxic (destructive) and are designed to kill dividing cells selectively by blocking the ability of DNA and RNA to reproduce and by lysing cell membranes. All types of rapidly dividing cells, not just the cancer cells, are affected. Damage to otherwise healthy tissue, such as bone marrow, hair follicles, mucosal cells in the mouth, digestive tract, and reproductive system, not just to cancer cells, is the cause of most side effects.

In addition, a combination of drugs (each causing cell death through different pharmacologic mechanisms) is traditionally used for greater efficacy in the systemic treatment of some cancers (e.g., breast cancer). Hence, an overlap of toxicities may result in greater side effects.

Common Physical Effects

The effects of treatment for cancer can be debilitating physiologically, physically, and psychologically. Common physical side effects include bone marrow suppression, severe mucositis, mouth sores, nausea and vomiting, fluid retention, pulmonary edema, cough, headache, CNS effects, peripheral neuropathies, malaise, fatigue, dyspnea, and loss of hair. Emotional and psychologic side effects are present but less evident (Table 13-8).

Bone marrow suppression (myelosuppression) is a common and serious side effect of many chemotherapeutic agents and can be a side effect of radiation therapy in some instances. This condition may lead to significant decreases in production of white blood cells (leukopenia), red blood cells (anemia), and, in some cases, platelets (thrombocytopenia).

Leukopenia (neutropenia) and resultant opportunistic infections have been shown to result in dose reductions, treatment delays, and hospitalizations. People at risk for leukopenia are taught infection prevention techniques and are often supportively or emergently treated with injections of

TABLE 13-8 Side Effects of Cancer Treatment

The health care professional must remember that some of the delayed effects of radiation, such as cerebral injury, pericarditis, pulmonary fibrosis, hepatitis, intestinal stenosis, other GI disturbances, and nephritis, may also be signs of recurring cancer. The physician must be notified by the affected individual of any new symptoms, change in symptoms, or increase in symptoms.

Surgery	Radiation	Chemotherapy	Biotherapy	Hormonal Therapy	Transplant (Bone Marrow, Stem Cell)
Fatigue	Fatigue	Fatigue	Fatigue	Nausea	Severe bone marrow suppression
Disfigurement	Radiation sickness	GI effects	Fever	Vomiting	Mucositis
Loss of function	Immunosuppression	Anorexia	Chills	Hypertension	Nausea and vomiting
Infection	Decreased platelets	Nausea	Nausea	Steroid-induced diabetes	Graft versus host disease (allogenic only)
Increased pain	Decreased WBCs	Vomiting	Vomiting	Myopathy (steroid-induced)	Delayed wound healing
Deformity	Infection	Diarrhea	Anorexia	Bone loss, fractures	Veno-occlusive disease
Scar tissue	Fibrosis	Ulcers	Fluid retention	Weight gain	Infertility
Fibrosis	Mucositis	Constipation	CNS effects	Altered mental status	Cataract formation
Hemorrhage, bleeding	Diarrhea	Hemorrhage	Anemia	Hot flashes	Thyroid dysfunction
	Edema	Bone marrow suppression	Leukopenia	Sweating	Growth hormone deficiency
	Hair loss	Anemia	Altered taste/ sensation	Decreased libido, sexual dysfunction	Osteoporosis
	Delayed wound healing	Leukopenia		Morning stiffness	Secondary malignancy
	PNS/CNS effects	Neutropenia		Arthralgias, myalgias	
	Malignancy	Thrombocytopenia		Vaginal dryness	
	Osteonecrosis (mandible, clavicle, humerus, femur)	Skin rashes			
	Radiation recall	Neuropathies			
		Hair loss			
		Infertility, sexual dysfunction			
		Phlebitis			
		Anxiety, depression			
		Weight gain/loss			

Adapted from Goodman CC, Fuller KS: *Pathology: implications for the physical therapist*, ed 3, Philadelphia, 2009, WB Saunders.
GI, Gastrointestinal; *WBC*, white blood cell; *PNS*, peripheral nervous system; *CNS*, central nervous system.

colony-stimulating factors, such as granulocyte colony-stimulating factor (GCSF) or a newer version colony-stimulating factor, pegfilgrastim (Neulasta), to stimulate increased production of needed white blood cells.[114]

Another relatively common bone marrow treatment toxicity is *anemia*. A drop in the production of red blood cells and in associated hemoglobin levels causes a loss of oxygenation to many body tissues and results in the many associated symptoms of anemia such as severe fatigue, muscle weakness, dizziness, dyspnea, pallor, and tachycardia.

Red blood cell transfusions and/or the use of injectable epoetin alfa (Epogen), the recombinant form of human erythropoietin, or darbepoetin (Aranesp), a newer version of Epogen, is very useful in the treatment of anemia.

Closely related to anemia is *fatigue*. Cancer-related fatigue is a frequent, difficult, and often debilitating problem. It differs from fatigue of healthy people because it happens independently of rest and activity patterns.[115] Factors contributing to fatigue can include many physical and emotional components of cancer such as anemia, poor nutrition, infection, low thyroid output, tumor breakdown by products, depression, pain, and medications.

Fatigue has been identified as a major determinant of perceived quality of life; it may be temporary, may persist throughout the episode of care, and may even continue many months after treatment has concluded. Adequate hydration, exercise, dietary measures, and treatment of anemia and depression, and addressing barriers to exercise are all measures used to help in the treatment of patients with cancer-related fatigue.[116-120]

Aggressive chemotherapeutic agents and chest irradiation can cause *cardiopulmonary dysfunction*, especially in the treatment of Hodgkin's disease and breast and lung cancers. High-dose radiation can result in pericardial fibrosis (scarring of the pericardium) and constrictive pericarditis (inflammation of the pericardium). These conditions are usually asymptomatic until the client starts to exercise, and then, exertional *dyspnea* is the first symptom.

Other causes of dyspnea include deconditioning, anemia, peripheral arterial disease, and increased physiologic demand for oxygen because of fever or infection. During radiation therapy, the client may be more tired than usual. Resting throughout exercise is important, as are adequate nutrition and hydration.

The skin in the irradiated area may become red or dry and should be exposed to the air but protected from the sun and from tight clothing. Gels, lotions, oils, or other topical agents should not be used over the irradiated skin without a physician's approval. Clients may have other side effects, depending on the areas treated. For example, radiation to the lower back may cause nausea, vomiting, or diarrhea because the lower digestive tract is exposed to the radiation.

Radiation recall is a severe skin reaction that can occur when certain chemotherapy drugs (e.g., Actinomycin, doxorubicin, methotrexate, fluorouracil, hydroxyurea, paclitaxel, liposomal doxorubicin) are given during or soon after radiation treatment.

The skin reaction appears like a severe sunburn or rash on the area of skin where the radiation was previously administered. It can appear weeks to months after the last dose of radiation. It is very important that this reaction be immediately reported because symptoms may be severe enough that chemotherapy must be delayed until the skin has healed.

Bone necrosis and demineralization *(radiation osteonecrosis)* can also result from radiation therapy and are usually not reversible. Individuals with this problem have an increased likelihood of pathologic fractures and need to be carefully handled by the therapist. Any activities, including weight-bearing activities and range of motion, should be addressed prior to the initiation of therapeutic exercise.[121]

Monitoring Laboratory Values

It is very important to review hematologic values in clients receiving these treatment modalities before any type of vigorous physical therapy is initiated. A guideline still used by some physical therapy exercise programs is the Winningham Contraindications for Aerobic Exercise. According to these guidelines, aerobic exercise is contraindicated in chemotherapy clients when laboratory values are as follows[122]:

Platelet count	$<50,000/mm^3$
Hemoglobin	<10g/dL
White blood cell count	$<3000/mm^3$
Absolute granulocytes	$<2500/mm^3$

These guideline values have not been tested under today's patient or treatment parameters and may not be considered valid or reliable. More specific exercise guidelines for all diseases are available.[123] All treatment facilities (and even individual physicians within the center) establish their own parameters and protocols. Many centers use hemoglobin levels exclusively because hematocrit is linked with hydration and may not provide the information needed. It does not appear that there is an "industry standard" for these measures.

In an outpatient setting without the benefit of laboratory values for guidance, the therapist is advised to use vital signs as discussed in Chapter 3, along with rate of perceived exertion (RPE) during exercise. Observe for clinical signs and symptoms of infection and fever, thrombocytopenia, deep vein thrombosis, dehydration, and electrolyte imbalance.

Late and Long-Term Physical Effects

Today, more than 10 million individuals treated for various types of cancers are surviving disease-free (but not "free of their disease") for many years following surgery, chemotherapy, immunotherapy, stem cell transplantation, and/or radiation. Many treatments are not highly specific and put normal cells, organs, and systems (especially the nervous system) at risk. Adjunctive medications, such as corticosteroids, antiepileptic medications, immunosuppressive agents, opioids, hypnotics, and antiemetics, contribute to impaired cognition.

Late effects from treatment, recurrence, secondary malignancies, and issues related to body image and quality of life are often part of the survivorship experience.[124,125] Any body system can be affected with long-term sequelae determined by the specific treatment received (Box 13-6). Therapists are increasingly becoming a part of the rehabilitation of cancer survivors and will need to be aware of the potential issues faced by cancer survivors.[126]

These may include fatigue, lymphedema, endocrine and fertility effects in male survivors, reproductive and hormonal changes in women, sleep disturbances, impaired cognitive function, osteoporosis, spiritual issues, financial problems, poor quality of life, pain, disfigurement, neuropathy, sexuality and body image, and dental changes. There can be functional decline, poor quality of life, and psychosocial stress and difficulty coping.[127]

Primary muscle shortening and secondary loss of muscle activity may produce movement disorders. Radiation-induced changes in vascular networks resulting in ischemia may affect muscle contractility. Magnetic resonance imaging (MRI) studies have shown radiation-induced muscle morbidity in cervical, prostate, and breast cancers.[128]

The controversial term *chemobrain* is often used to describe mental fogginess experienced by some people during the course of chemotherapy. But this type of cognitive function is reported to persist well after treatment has ended.[129] Memory lapses, difficulty concentrating or staying focused on a task, trouble remembering details (e.g., names, dates, phone numbers), and difficulty with word retrieval are just a few examples of the specific experiences represented by chemobrain. Depression, insomnia, and difficulty doing two things at once (e.g., talking on the phone and cleaning or cooking while carrying on a conversation) are also typical of people with chemobrain.[129,130]

BOX 13-6 LATE AND LONG-TERM EFFECTS OF CANCER TREATMENT

Cardiovascular
Cardiomyopathy
Pericarditis
Coronary artery disease
Congestive heart failure
Valvular heart disease
Sinus node dysfunction

Lymphatic
Lymphedema

Pulmonary
Fibrosis

Neurologic/sensory
Hearing loss
Visual impairment
Vestibular impairment (balance loss)
Neuropathy*/plexopathy
Postsurgical pain syndrome
Thoracic outlet syndrome

Musculoskeletal
Muscle and joint pain
Raynaud's phenomenon
Avascular necrosis
Altered muscle activity/movement disorder syndromes/mobility decline
Bone loss, pathologic fractures

Integument
Radiation recall

Gastrointestinal
Malabsorption
Bowel dysfunction, chronic pain (scarring, fibrosis, ischemia)
Increased risk of second malignancy (e.g., radiation for prostate cancer)
Fecal incontinence

Rheumatologic (various arthritic conditions)

Renal (chronic kidney disease)

Endocrine/Metabolic
Thyroid impairment
Diabetes mellitus
Diabetes insipidus
Adrenal insufficiency
Osteoporosis

Data from Miller KD, Triano LR: Medical issues in cancer survivors—a review, *Cancer J* 14(6):375–387, 2008.
*Referred to as chemotherapy-induced peripheral neuropathy (CIPN).

CANCERS OF THE MUSCULOSKELETAL SYSTEM

In addition to increasing age as a red flag and risk factor for cancer, we now add "young age" as a possible red flag factor. Primary bone cancer is more likely to occur in the population under age 25. Both primary bone cancer and cancer that has metastasized to the bone present with the same subset of clinical signs and symptoms because in both cases, the same system (skeletal) is affected.

Sarcoma

Malignant neoplasms or new growths that develop as *primary* lesions in the musculoskeletal tissues are relatively rare, representing less than 1% of malignant disease in all age groups and 15% of annual pediatric malignancies.[131]

Secondary neoplasms that develop in the connective tissues as metastases from a primary neoplasm elsewhere (especially metastatic carcinoma) are common. Fibrosarcomas occurring after radiotherapy can occur usually after a significant latent period (4 years or longer).[132]

High grade (higher grade represents greater likelihood of metastasis based on measures of cell differentiation and growth) and evidence of metastasis are associated with a poor prognosis for all neoplasms of bone or soft tissue. The prognosis for clients with soft tissue sarcoma depends on several factors. Factors associated with a poorer prognosis include age older than 60 years, tumors larger than 5 cm, and histology of high grade.[133]

Soft Tissue Tumors

Soft tissue sarcomas make up a group of relatively rare malignancies. Little is known about important epidemiologic or etiologic factors in clients with soft tissue sarcomas. There is no proven genetic predisposition to the development of soft tissue sarcomas, but studies do indicate that workers exposed to phenoxyacetic acid in herbicides and chlorophenols in wood preservatives and persons who had radiation to the tonsils, adenoids, and thymus may have an increased risk of developing soft tissue sarcoma. There are two peaks of incidence in human sarcoma development: early adolescence and the middle decades.

Soft tissue sarcomas can arise anywhere in the body. In adults, most soft tissue sarcomas arise in the extremities (usually the lower extremity, at or below the knee), followed by the trunk and the retroperitoneum. In their early stages, these sarcomas do not usually cause symptoms because soft tissue is relatively elastic, allowing tumors to grow rather large before they are felt or seen.

In contrast, the overwhelming majority of childhood soft tissue sarcomas are rhabdomyosarcomas, and the anatomic distribution of these lesions is entirely different (18% in the extremities, 35% in the head and neck region, and 20% in GU sites).[134] Many of the primary sites in children (e.g., orbit of the eye, paratesticular region, prostate) are never primary sites for soft tissue sarcoma in adults.

Risk Factors. Soft tissue sarcomas occur more frequently in persons who have one of the following conditions:

- von Recklinghausen's disease
- Gardner's syndrome
- Werner's syndrome
- Tuberous sclerosis
- Basal cell nevus syndrome
- Li-Fraumeni syndrome (p53 suppressor-gene mutations)
- Exposure to radiation, herbicides, wood preservatives, vinyl chloride
- AIDS (Kaposi's sarcoma)

Metastases. In children, tumors of the extremities tend to behave relatively aggressively, with a high incidence of nodal spread and distant metastases. In adults, soft tissue sarcomas rarely spread to regional lymph nodes, instead invading aggressively into surrounding tissues with early hematogenous dissemination, usually to the lungs and to the liver.

Even with pulmonary metastases, the survival rate has greatly improved over the past decade with a multidisciplinary approach that includes multiagent chemotherapy and limb-sparing surgery.[134] However, as more people survive for increasingly longer periods, serious and potentially life-threatening complications of such therapy can develop months to years later.

Clinical Signs and Symptoms. Soft tissue sarcomas most often appear as asymptomatic soft tissue masses. Because these lesions arise in compressible tissues and are often far from vital organs, symptoms are few unless they are located close to a major nerve or in a confined anatomic space.

The most common manifestations of these neoplasms are swelling and pain. Pelvic sarcomas may appear with swelling of the leg or pain in the distribution of the femoral or sciatic nerve. Some people attribute swelling to a minor injury, reporting a misleading cause of onset to the therapist. The therapist must always keep this in mind when evaluating a client of any age.

More often, the neoplasm goes unnoticed until some trauma or injury requires medical attention and an x-ray study reveals the lesion. When pain is the most significant symptom, it is usually mild and intermittent, progressively becoming more severe and more constant with rapidly growing neoplasms.

No reliable physical signs are present to distinguish between benign and malignant soft tissue lesions. Consequently, all soft tissue lumps that persist or grow should be reported immediately to the physician.

CLINICAL SIGNS AND SYMPTOMS

Soft Tissue Sarcoma

- Persistent swelling or lump in a muscle (most common finding)
- Pain
- Pathologic fracture
- Local swelling
- Warmth of overlying skin

Bone Tumors

Malignant (primary) bone tumors are relatively rare, accounting for 1% of total deaths from cancer. Excluding multiple myeloma, the ratio of benign to malignant bone tumors is approximately 7:1.

Primary bone cancer affects children and young adults most commonly, whereas secondary bone tumors or metastatic neoplasms occur in adults with primary cancer (e.g., cancer of the prostate, breast, lungs, kidneys, thyroid).

Symptoms are not necessarily different between primary and secondary bone cancer, but the *history* is very different. Medical screening with possible referral is essential for anyone with clinical manifestations discussed in this chapter (see Clinical Manifestations of Malignancy: Skeletal) who also has a past medical history of any kind of cancer.

This text is limited to the most common forms of bone tumors. The two most common childhood sarcomas of the bone are osteosarcoma (osteogenic sarcoma) and Ewing family of tumors (EFT).

Osteosarcoma. Osteosarcoma (also known as osteogenic sarcoma) is the most common type of bone cancer, occurring between the ages of 10 and 25 years and also in adulthood. It is slightly more common in boys.

Although it can involve any bones in the body, because it arises from osteoblasts, the usual site is the epiphyses of the long bones, where active growth takes place (e.g., lower end of the femur, upper end of the tibia or fibula, upper end of the humerus).

In general, 80% to 90% of osteosarcomas occur in the long bones; the axial skeleton is rarely affected. The growth spurt of adolescence is a peak time for the development of osteosarcoma. Half of all osteosarcomas are located in the upper leg above the knee, where the most active epiphyseal growth occurs.

Risk Factors. There appears to be an association between rapid bone growth and risk of tumor formation. Young people previously treated with radiation for an earlier cancer have an increased risk of developing osteosarcoma later.

With chemotherapy given before and after surgical removal, many people can now be cured of osteosarcoma. Survival lessens if metastases are present. Limb-sparing surgery rather than amputation is effective in 50% to 80% of people.[135]

Metastases. Bone tumors, unlike carcinomas, disseminate almost exclusively through the blood; bones lack a lymphatic system. Metastases to the lungs, pleura, lymph nodes, kidneys, and brain and to other bones are common and occur early in the disease process.

Hematogenous spread occurs to the lungs first and to other bones second. In some cases, surgery can be attempted to remove pulmonary metastases, but survival decreases if metastatic sites are present.

Clinical Signs and Symptoms. Osteosarcoma usually appears with pain in a lesional area, usually around the knee in clients with femur or tibia involvement. The pain is initially mild and intermittent but becomes progressive and more severe and more constant over time.

Most lesions produce pain as the tumor starts to expand the bony cortex and stretch the periosteum. A tender lump may develop, and a bone weakened by erosion of the metaphyseal cortex may break with little or no stress. This pathologic fracture often brings the person into the medical system, at which time a diagnosis is established by x-ray study and surgical biopsy. This neoplasm is highly vascularized, so that the overlying skin is usually warm.

CLINICAL SIGNS AND SYMPTOMS

Osteosarcoma

- Pain and swelling of the involved body part
- Loss of motion and functional movement of adjacent joints
- Tender lump
- Pathologic fracture
- Occasional weight loss
- Malaise
- Fatigue

Ewing Sarcoma. Four percent of all childhood tumors are in the Ewing family of tumors (EFT). In the United States, approximately 300 to 400 children and adolescents are diagnosed with a Ewing tumor each year. Although rare, it is the third most frequent primary sarcoma of bone after osteosarcoma and chondrosarcoma. Almost any bone can be involved, but typically, the pelvis, femur, tibia, ulna, and metatarsus are the most common sites for Ewing sarcoma. Soft tissue involvement is rare.[136]

Risk Factors. Ewing sarcoma is most common between the ages of 5 and 16 years, with a slightly greater incidence in boys than in girls. Anyone of any age can develop Ewing sarcoma but most people who have Ewing tumors are in the second decade of life and Caucasian, either Hispanic or non-Hispanic. This tumor is rare in other racial groups.[137] A spontaneous (rather than environmentally or trauma-induced) gene translocation in chromosome 22 has been found with EFT, and there is study in this area. No other risk factors have been identified in the development of EFTs.

Metastases. Metastases are predominantly hematogenous (to lungs and bone), although lymph node involvement may occur. Metastases usually occur late in the disease process, but aggressive chemotherapy has increased 5-year survival rates from 10% to 70%.

Clinical Signs and Symptoms. Ewing sarcoma is a rapidly growing tumor that often outgrows its blood supply and quickly erodes the bone cortex, producing a painful, soft, tender, palpable mass. Intramedullary tumors that erode into the periosteum often result in an "onion skin" appearance as the periosteum is elevated and replaced by a new periosteal bone.

The most common symptom of EFT, bone pain, appears in about 85% of people with bone tumors. The pain may be caused by periosteal erosion from a break or fracture of a bone weakened by the tumor.

The pain may be intermittent and may not be accompanied by swelling, resulting in a physical therapy referral. Systemic symptoms, such as fatigue, weight loss, and intermittent fever, may be present, especially in clients with metastatic disease. Fever may occur when products of bone degeneration enter the bloodstream. In addition, the blood supply to local areas of bone may be compromised, with resultant avascular necrosis of bone.

Ewing sarcoma occurs most frequently in the long bones and the pelvis, the most common site being the distal metaphysis and the diaphysis of the femur. The next most common sites are the pelvis, tibia, fibula, and humerus (Case Example 13-9). About 30% of the time, the bone tumor may be soft and warm to the touch and the child may have a fever.

Less common presentations of Ewing sarcoma include primary rib tumor associated with a pleural effusion and respiratory symptoms, mandibular lesions presenting with chin and lip paresthesias, primary vertebral (cervical, lumbar) tumor with symptoms of nerve root or spinal cord compression, primary sacral tumor with neurogenic bladder, and pelvic tumor with pain and bowel/bladder disturbances. Neurologic symptoms may occur secondary to nerve

CASE EXAMPLE 13-9

Back Pain Associated With Ewing Sarcoma

A 17-year-old male high school athlete noted low back pain 6 weeks ago. He was unable to identify a specific traumatic event or injury but noted that he had been "training pretty hard" the last 2 weeks. Spinal motions were all within normal limits with no apparent step suggestive of spondylolisthesis. There were no obvious postural changes such as a scoliotic shift or unusual kyphosis/lordosis of the spinal curves.

The only positive evaluation findings included a mild left foot drop and an absent left ankle jerk. Pain was intensified on weight bearing and movement of any kind. Pain was not relieved by rest or aspirin.

There was no fever or recent history of sore throat, upper respiratory or ear infection, and so on. After 1 week in physical therapy, pain began radiating into the posterior aspect of the left thigh. The client had also noted for the first time paresthesias along the lateral side of the left leg. The pain had increased rather markedly over the past 2 weeks.

The therapist might assume that the physical therapy intervention aggravated this client's condition, causing increased symptoms. However, given the unknown cause of pain, symptoms inconsistent with musculoskeletal conditions (e.g., unrelieved by rest), combined with the recent change in symptoms and the presence of positive neurologic symptoms, this client was returned to the physician before continuing further therapy.

Results: A blood test performed at that time indicated that the WBC count was 10,000/mm^3 (normal range = 4300 to 10,800/mm^3). Further testing, including an x-ray, resulted in a diagnosis of Ewing sarcoma.

entrapment by the tumor, and misdiagnosis as lumbar disk disease can occur.[138] If the tumor has spread, the child may feel very tired or may lose weight.

CLINICAL SIGNS AND SYMPTOMS

Ewing Family of Tumors

- Increasing and persistent pain
- Increasing and persistent swelling over a bone (localized over the area of tumor)
- Decrease in movement if a limb bone is involved
- Fever
- Fatigue
- Weight loss
- Bowel and/or bladder disturbances

Chondrosarcoma. Chondrosarcoma, the most common malignant cartilage tumor (and second most common sarcoma of bone after osteosarcoma), occurs most often in adults older than 40. However, when it does occur in a younger age group, it tends to be a higher grade of malignancy and capable of metastases.

It occurs most commonly in some part of the pelvic or shoulder girdles or long bones such as the femurs (Fig. 13-8). Chondrosarcomas are the most common malignant tumors of the sternum and scapula.

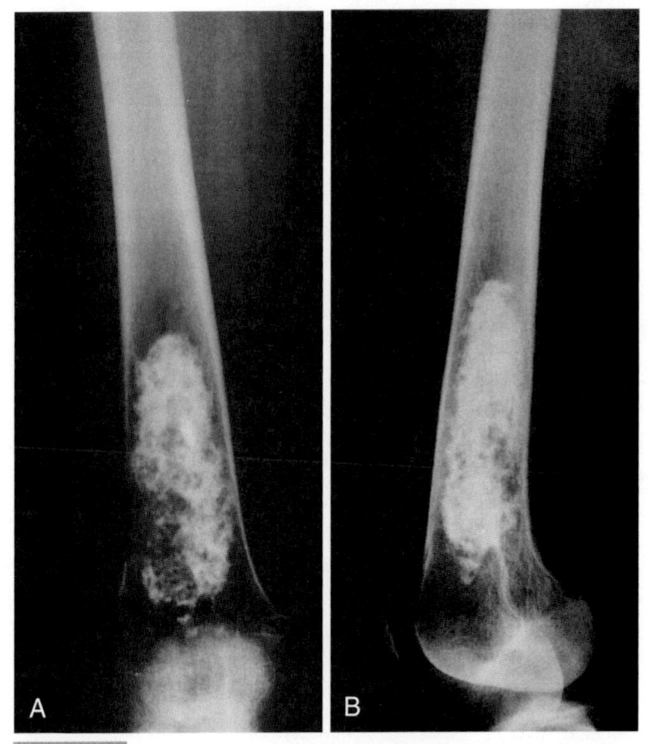

Fig. 13-8 Large intramedullary calcified lesion diagnosed as a low-grade chondrosarcoma of the femur. **A,** Anteroposterior radiograph. **B,** Lateral view. (From Dorfman HD, Czerniak B: *Bone tumors,* St. Louis, 1998, Mosby.)

Chondrosarcoma is usually a relatively slow-growing malignant neoplasm that arises either spontaneously in previously normal bone or as a result of malignant change in a preexisting benign bone tumor (osteochondromas and enchondromas or chondromas). The latter are referred to as *secondary* chondrosarcomas, which are composed entirely of cartilaginous tissue and usually represent a low-grade malignancy.[139]

Risk Factors. See information related to soft tissue sarcomas.

Metastases. Although slow growing, chondrosarcoma has a high tendency for thrombus formation in the tumor blood vessels, with an increased risk for pulmonary embolism and metastatic spread to the lungs. Metastases develop late, so the prognosis of chondrosarcoma is considerably better than that of osteosarcoma.

Clinical Signs and Symptoms. Clinical presentation of chondrosarcoma varies. *Peripheral chondrosarcomas* (arising from bone surface) grow slowly and may be undetected and quite large. Local symptoms develop only because of mechanical irritation; otherwise, pain is not a prominent symptom.

Pelvic chondrosarcomas are often large and appear with pain referred to the back or thigh, sciatica caused by sacral plexus irritation, urinary symptoms from bladder neck involvement, or unilateral edema caused by iliac vein obstruction.

Conversely, *central chondrosarcomas* (arising within bone) appear with dull pain, and a mass is rare. Pain, which indicates active growth, is an ominous sign of a central cartilage lesion.

CLINICAL SIGNS AND SYMPTOMS

Chondrosarcoma

- Palpable mass
- Back, pelvis, or thigh pain
- Sciatica
- Bladder symptoms
- Unilateral edema

Osteoid Osteoma. Osteoid osteoma is a noncancerous osteoblastic tumor that accounts for approximately 10% of benign bone tumors. It occurs predominantly in children and young adults between the ages of 7 and 25, affecting males 2 to 3 times more often than females.[140]

Osteoid osteoma is a noncancerous lesion with distinct histologic features, consisting of a central core of vascular osteoid tissue and a peripheral zone of sclerotic bone (Fig. 13-9). This type of tumor commonly occurs in the diaphysis of long bones such as the proximal femur, accounting for more than half of all cases; less often, the hands and feet and posterior elements of the spine are involved.

On x-ray, the lesion is seen as a translucent area representing the nidus, usually measuring less than 1 cm, that is surrounded by bone sclerosis. The nidus may be uniformly radiolucent or may contain variable amounts of calcification. Computed tomography (CT) shows a well-circumscribed small area of low attenuation, representing the nidus, surrounded by a larger area of higher attenuation, representing the reactive bone formation.

Clinical Signs and Symptoms. The clinical presentation typically consists of pain, which is often worse at night, increased skin temperature, sweating, and tenderness in the affected region. In many cases, pain is completely relieved by aspirin (a salicylate compound), which is a hallmark finding for this particular type of bone cancer.[141]

The pathogenesis of this pain may be related to production of prostaglandins by the tumor cells. Prostaglandins can cause changes in vascular pressure, which result in local stimulation of sensory nerve endings. Salicylates in the aspirin inhibit the prostaglandins, reducing painful symptoms.[140]

CLINICAL SIGNS AND SYMPTOMS

Osteoid Osteoma

- Bone pain (femur, 50% of cases), worse at night; relieved by aspirin
- Warmth and tenderness over the involved site

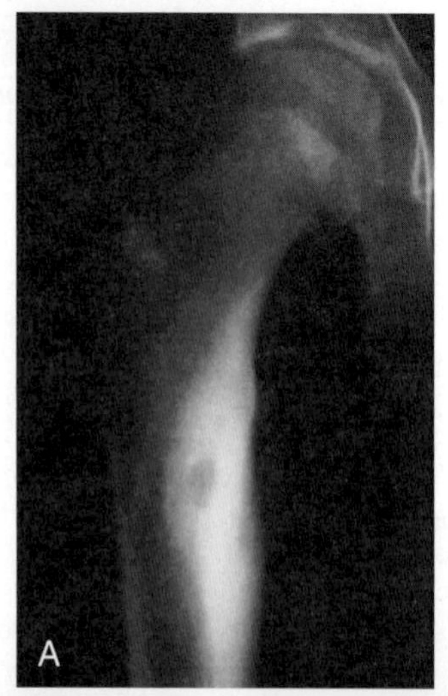

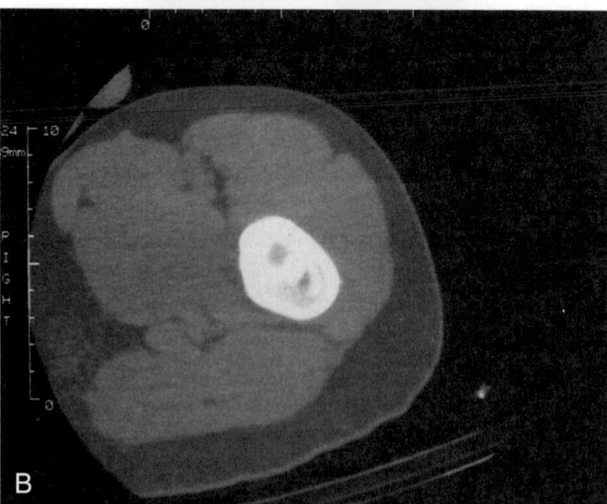

Fig. 13-9 Osteoid osteoma. **A,** A 15-year-old boy whose pain was worse at night and relieved by aspirin was found to have a well-defined lytic intracortical lesion in the proximal femoral shaft. Note the thickening around and extending above and below the nidus. **B,** Computed tomography (CT) of the same osteoid osteoma shows marked sclerosis of the bone around the nidus. (From Dorfman HD, Czerniak B: *Bone tumors,* St. Louis, 1998, Mosby.)

PRIMARY CENTRAL NERVOUS SYSTEM TUMORS

Primary tumors of the CNS once thought to be rare are now recognized as a significant problem in the United States. The incidence among persons over age 70 has increased sevenfold since 1970, with the incidence in people of all ages rising by 9% in the past 3 decades.[142] The majority of people who develop primary brain tumors are over the age of 40, but these tumors can cause morbidity and mortality in younger individuals as well.

CNS neoplasms include tumors that lie within the spinal cord (intramedullary), within the dura mater (extramedullary), or outside the dura mater (extradurally). About 80% of CNS tumors occur intracranially, and 20% affect the spinal cord and peripheral nerves. Of the intracranial lesions, about 60% are primary, and the remaining 40% are metastatic lesions, often multiple and most commonly from the lung, breast, kidney, and GI tract.

More common benign tumors are meningiomas (from meninges) and schwannomas (from nerve sheaths), both arising from tissues of origin. These tumors are considered noncancerous histologically, but sometimes malignant by

location, meaning that they can be located in an area where tumor removal is difficult and deficit and death can result.

Any CNS tumor, even if well differentiated and histologically benign, is potentially dangerous because of the lethal effects of increased intracranial pressure and tumor location near critical structures. For example, a small, well-differentiated lesion in the pons or medulla may be more rapidly fatal than a massive liver cancer.

Primary CNS tumors rarely metastasize outside the CNS; there is no lymphatic drainage available, and hematogenous spread is also unlikely. In most cases, CNS spread is contained within the cerebrospinal axis, involving local invasion or CNS seeding through the subarachnoid space and the ventricles.

Whether from primary CNS tumors or cancer that has metastasized to the CNS from some other source, the effects are the same. Clinical manifestations of CNS involvement were discussed earlier in this chapter.

Risk Factors

The incidence of primary CNS lymphoma is increasing among older adults who are immunocompetent and even more so among aging adults who are immunodeficient. The etiology of most primary brain tumors is unknown, although ionizing radiation has been implicated most often.

Children with a history of cranial radiation therapy, such as low-dose radiation of the scalp for fungal infection, are at significantly greater risk of developing primary brain tumors. However, most radiation-induced brain tumors are caused by radiation to the head used for the treatment of other types of cancer.[143]

Occupational exposure to gases and chemicals has been proposed as a risk factor. Several congenital genetic disorders are directly linked with an increased risk of primary brain tumors (e.g., neurofibromatosis, Turcot syndrome, tuberous sclerosis). Other potential environmental risk factors, such as exposure to vinyl chloride and petroleum products, have shown inconclusive results. Current preliminary evidence has shown altered brain activity with exposure to electromagnetic fields from cell phones or other wireless devices.[144] Whether this effect on the brain is neutral, positive, or negative remains under investigation.

Brain Tumors

The incidence of primary brain tumor is increasing in persons of all ages; in children, it is second only to leukemia as a cause of death. Although the causes for this overall increase remain unknown, it is clear that it is not simply a matter of better diagnostic techniques. Adding the number of people who survive other primary cancers but later develop metastatic brain tumors increases the overall incidence dramatically.

Individuals with mental disorders are more likely to be diagnosed with brain tumors and at younger ages than people without mental illness. This increased risk for brain tumors may reflect the early presence of mental symptoms, or there may be a true association between the two conditions. The exact relationship remains unknown at this time.[145]

Primary Malignant Brain Tumors

The most common primary malignant brain tumors are astrocytomas. Low-grade astrocytomas (Grade I), such as juvenile pilocytic astrocytomas, have an excellent prognosis after surgical excision.

At the other extreme is the Grade IV glioma such as glioblastoma multiforme, which is an aggressive, high-grade tumor with a very poor prognosis (usually less than 12 months). There are intermediate histologic grades (II, III) with intermediate survival statistics. Low-grade tumors are more common in children than in adults.[143]

Metastatic Brain Tumors

Cancer can spread through bloodborne metastasis or via cerebrospinal fluid pathways. Therefore the brain is a common site of metastasis. In fact, metastatic brain tumors are probably the most common form of malignant brain tumors.

The cerebrum is the most common site of metastasis. Cerebellar metastases are less frequent; brainstem metastases are the least common. Approximately two-thirds of people with brain metastases will present with multiple metastases.

Up to 50% of individuals affected with cancers develop neurologic symptoms resulting from brain metastases. Headache, seizures, loss of motor function, and cerebellar signs are the most common early symptoms.

As mentioned previously, any neurologic sign can be the silent presentation of cancer metastasized to the CNS. The most common sources of brain metastases are cancers of the lung and breast, followed by metastases from melanomas and cancers of the colon and kidney.

CLINICAL SIGNS AND SYMPTOMS

Brain Tumor

- Increased intracranial pressure
- Headache, especially retroorbital; sometimes worse upon awakening, improves during the day
- Vomiting (with or without nausea)
- Visual changes (blurring, blind spots, diplopia, abnormal eye movements)
- Changes in mentation (impaired thinking, difficulty concentrating or reading, memory, or speech)
- Personality change, irritability
- Unusual drowsiness, increased sleeping
- Seizures (without previous history)
- Sensory changes
- Muscle weakness or hemiparesis
- Bladder dysfunction
- Increased lower extremity reflexes compared with upper extremity reflexes
- Decreased coordination, gait changes, ataxia
- Positive Babinski reflex
- Clonus (ankle or wrist)
- Vertigo, head tilt

Spinal Cord Tumors

Spinal tumors are similar in nature and origin to intracranial tumors but occur much less often. They are most common in young and middle-aged adults, and they occur most often in the thoracic spine because of its length, proximity to the mediastinum, and likelihood of direct metastatic extension from lymph nodes involved with lymphoma, breast cancer, or lung cancer.

Metastases

Most metastasis is disseminated by local invasion. Spinal cord tumors account for less than 15% of brain tumors. As mentioned, 10% of spinal tumors are themselves metastasized neoplasms from the brain.

One other means of dissemination is through the intervertebral foramina. The extradural space communicates through the intervertebral foramina with adjacent extraspinal compartments such as the mediastinum and retroperitoneal space.

In most cases, extradural tumors are metastatic, reaching the extradural space and then adjacent extraspinal spaces through this foraminal connection. Tumors within the spinal cord (intramedullary) or outside the spinal cord (extramedullary) may metastasize to the dural tube to become intradural tumors.

Clinical manifestations of spinal tumors vary according to their location. See previous discussion in this chapter on Clinical Manifestations of Malignancy: Neurologic. Pain associated with *extramedullary tumors* can be located primarily at the site of the lesion or may refer down the ipsilateral extremity, with radicular involvement from nerve root compression, irritation, or occlusion of blood vessels supplying the cord. Progressive cord compression is manifested by spastic weakness below the level of the lesion, decreased sensation, and increased weakness.

Intramedullary tumors produce more variable signs and symptoms. High cervical cord involvement causes spastic quadriplegia and sensory changes. Tumors in descending areas of the spinal cord produce motor and sensory changes appropriate to functions of that level.

CLINICAL SIGNS AND SYMPTOMS

Spinal Cord Tumors

- Pain
- Decreased sensation
- Spastic muscle weakness
- Progressive muscle weakness
- Muscle atrophy
- Paraplegia or quadriplegia
- Thoracolumbar pain
- Unilateral groin or leg pain
- Pain at rest and/or night pain
- Bowel/bladder dysfunction (late finding)

CANCERS OF THE BLOOD AND LYMPH SYSTEM

Cancers arising from the bone marrow include acute leukemias, chronic leukemias, multiple myelomas, and some lymphomas. These cancers are characterized by the uncontrolled growth of blood cells.

The major lymphoid organs of the body are the lymph nodes and the spleen (see Figs. 12-1 and 4-41). Cancers arising from these organs are called *malignant lymphomas* and are categorized as either Hodgkin's disease or non-Hodgkin's lymphoma.

Leukemia

Leukemia, a malignant disease of the blood-forming organs, is the most common malignancy in children and young adults. One-half of all leukemias are classified as *acute,* with rapid onset and progression of disease resulting in 100% mortality within days to months without appropriate therapy.

Acute leukemias are most common in children from 2 to 4 years of age, with a peak incidence again at age 65 years and older. The remaining leukemias are classified as *chronic,* which have a slower course and occur in persons between the ages of 25 and 60 years. From these two broad categories, leukemias are further classified according to specific malignant cell line (Table 13-9).

Leukemia develops in the bone marrow and is characterized by abnormal multiplication and release of white blood cell (WBC) precursors. The disease process originates during WBC development in the bone marrow or lymphoid tissue. In effect, leukemic cells become arrested in "infancy," with most of the clinical manifestations of the disease being related to the absence of functional "adult" cells, which are products of normal differentiation.

With rapid proliferation of leukemic cells, the bone marrow becomes overcrowded with immature WBCs, which then spill over into the peripheral circulation. Crowding of the bone marrow by leukemic cells inhibits normal blood cell production.

Decreased red blood cell (RBC; erythrocyte) production results in anemia and reduced tissue oxygenation. Decreased platelet production results in thrombocytopenia and risk of hemorrhage. Decreased production of normal WBCs results in increased vulnerability to infection, especially because leukemic cells are functionally unable to defend the body against pathogens.

Leukemic cells may invade and infiltrate vital organs such as the liver, kidneys, lung, heart, or brain.

Risk Factors

Several predisposing factors for the development of leukemia have been identified. Exposure to ionizing radiation remains the most conclusively identified causative factor in humans. Prior drug therapies, such as chloramphenicol, phenylbutazone, chemotherapy alkylating agents, and benzene, have been implicated in the development of acute leukemia.

TABLE 13-9 Overview of Leukemia

	Acute Lymphoblastic Leukemia (ALL)	Acute Myeloid Leukemia (AML)	Chronic Lymphocytic Leukemia (CLL)	Chronic Myelogenous Leukemia (CML)
INCIDENCE				
Percentage of all leukemias	20%	20%	25%-40%	15%-20%
Adults	20%	85%	100%	95%-100%
Children	80%-85%	10%-20%	—	3%
AGE	Peak: 3-7 years 65+ (older adults)	15-40 years Incidence increases with age from 40-80+	50+	25-60 years
ETIOLOGY	Unknown Chromosomal abnormality Environmental factors Down syndrome (high incidence)	Benzene Alkylating agents Radiation Myeloproliferative disorders Aplastic anemia	Chromosomal abnormalities Slow accumulation of CLL lymphocytes	Philadelphia chromosome; BCR-ABL gene Radiation exposure
PROGNOSIS	Adult: Poor Child: 66% with aggressive treatment; 90.8% under age 5	Poor even with treatment 10%-15% survival	2-10 years survival Median survival: 6 years	Poor; 2-8 years Median survival: 3-4 years

From Goodman CC, Fuller KS: *Pathology: implications for the physical therapist*, ed 3, Philadelphia, 2009, WB Saunders.
There are many alternate names for these four main types of leukemia. For further details, see the Leukemia and Lymphoma Society (www.lls.org).

Hereditary syndromes associated with development of leukemia include Bloom syndrome, Down syndrome, Klinefelter's syndrome, neurofibromatosis, and others. In addition, viruses and immunodeficiency disorders have been associated as causative factors.[146]

Clinical Signs and Symptoms

Most of the clinical findings in acute leukemia are due to bone marrow failure, which results from replacement of normal bone marrow elements by malignant cells. Infections are due to a depletion of competent WBCs needed to fight infection. Abnormal bleeding is caused by a lack of blood platelets required for clotting, and severe fatigue is due to a lack of red blood cells. The most common symptoms of leukemia include infection, fever, pallor, fatigue, anorexia, bleeding, anemia, neutropenia, and thrombocytopenia.

For women, the abnormal bleeding may be prolonged menstruation leading to anemia. The Special Questions for Women (see Appendix B-32) may elicit this kind of valuable information, which would then require medical referral. Less common manifestations include direct organ infiltration; clients may experience easy bruising of the skin or abnormal bleeding from the nose, urinary tract, or rectum.

The appearance of a painless, enlarged lymph node or skin lesion (see Fig. 4-29) is followed by weakness, fever, and weight loss. A history of chronic immunosuppression (e.g., antirejection drugs for organ transplants, chronic use of immunosuppressant drugs for inflammatory or autoimmune diseases, cancer treatment) in the presence of this clinical presentation is a major red flag.

Clients on immunosuppressants do not usually have an elevated body temperature, so the presence of a fever is a red flag in this group. Significant weight loss with inactivity secondary to pain is another red flag; most inactive individuals experience weight gain not weight loss. A good rule of thumb to use in recognizing significant weight loss is 10% of the individual's total body weight in 10 to 14 days without trying.

Lymphoproliferative malignancies, such as leukemia and lymphoma, may also involve extramedullary areas and can present with localized or generalized symptoms such as enlarged liver and spleen, bone and joint pain, bone fracture, and parotid gland and testicular infiltration.[146]

Involvement of the synovium may lead to symptoms suggestive of rheumatic disease. A possible presentation in a child with acute lymphoblastic leukemia can be joint pain and swelling that mimics juvenile idiopathic arthritis (JIA) (formerly JRA). Leukemic arthritis is present in about 5% of leukemia cases. Acute leukemia can cause joint pain that is severe and episodic and disproportionately severe in comparison to the minimal heat and swelling that are present.[147]

Arthritic symptoms in such a child may be a consequence of leukemic synovial infiltration, hemorrhage into the joint, synovial reaction to an adjacent tumor mass, or crystal-induced synovitis.

CLINICAL SIGNS AND SYMPTOMS

Acute and Chronic Leukemias

- Infections, fever
- Abnormal bleeding
- Easy bruising of the skin
- Petechiae
- Epistaxis (nosebleeds) and/or bleeding gums
- Hematuria (blood in the urine)
- Rectal bleeding
- Weakness
- Easy fatigability
- Enlarged lymph nodes
- Bone and joint pain
- Weight loss
- Loss of appetite
- Pain or enlargement in the left upper abdomen (enlarged spleen)

Multiple Myeloma

Multiple myeloma is a cancer caused by uncontrolled growth of plasma cells in the bone marrow. Excessive growth of plasma cells originating in the bone marrow destroys bone tissue and is associated with widespread osteolytic lesions (decreased areas of bone density). Plasma cells are part of the immune system, and in multiple myeloma, they grow uncontrolled, forming tumors in the bone marrow.

Bone lesions and hypercalcemia can occur in some hematologic malignancies such as multiple myeloma. Multiple myeloma to date is an incurable disease, but the prognosis has improved, with lengthy remissions possible with treatment.

Risk Factors

There are no clear predisposing factors, other than age or exposure to ionizing radiation. This disease can develop at any age from young adulthood to advanced age but peaks among persons between the ages of 50 and 70 years. It is more common in men and African Americans. Molecular genetic abnormalities have been identified in the etiologic complex of this disease. Recent data suggest that a virus (Kaposi-associated herpes virus, HHV-8) may be implicated in acquired immunodeficiency syndrome (AIDS)–related cases.[148]

Clinical Signs and Symptoms

Multiple myeloma causes symptoms in many areas of the body. It normally originates in the bone marrow and then causes difficulties in other organs. The onset of multiple myeloma is usually gradual and insidious. Most clients pass through a long presymptomatic period that lasts 5 to 20 years.

Early symptoms involve the skeletal system, particularly the pelvis, spine, and ribs. Some clients have backache or bone pain that worsens with movement (Case Example 13-10). Bone disease is the most common complication and results in severe bone pain, pathologic fractures, hypercalcemia, and spinal cord compression. Renal failure, anemia, cardiac failure, and infection are serious and often fatal complications of this process.[149]

CLINICAL SIGNS AND SYMPTOMS

Multiple Myeloma

- Recurrent bacterial infections (especially pneumococcal pneumonias)
- Anemia with weakness and fatigue
- Bleeding tendencies

Bone Destruction

- Skeletal/bone pain (especially pelvis, spine, and ribs)
- Spontaneous fracture
- Osteoporosis
- Hypercalcemia (confusion, increased urination, loss of appetite, abdominal pain, vomiting, and constipation)

Renal Involvement

- Kidney stones
- Renal insufficiency

Neurologic Abnormalities

- CTS
- Back pain with radicular symptoms
- Spinal cord compression (motor or sensory loss, bowel/bladder dysfunction, paraplegia)

Bone Destruction. Bone pain is the most common symptom of myeloma. It is caused by infiltration of the plasma cells into the marrow with subsequent destruction of bone. Initially, the skeletal pain may be mild and intermittent, or it may develop suddenly as severe pain in the back, rib, leg, or arm, often the result of an abrupt movement or effort that has caused a spontaneous (pathologic) bone fracture.

The pain is often radicular and sharply cutting to one or both sides and is aggravated by movement. As the disease progresses, more and more areas of bone destruction and hypercalcemia develop. Symptoms associated with bone pain usually subside within days to weeks after initiation of antiresorptive agents. If left untreated, this disease will result in skeletal deformities, particularly of the ribs, sternum, and spine.

Hypercalcemia. Bone fractures are a result of osteoclast activity and bone destruction. This process results in calcium release from the bone and hypercalcemia. Hypercalcemia is considered an oncologic emergency. To rid the body of excess calcium (hypercalcemia), the kidneys increase the output of urine, which can lead to serious dehydration if there is an inadequate intake of fluids. Vomiting may compound this dehydration. Clients who have symptoms of hypercalcemia (see Table 13-7) should seek immediate medical care because this condition can be life-threatening.

Renal Effects. Drainage of calcium and phosphorus from damaged bones eventually leads to the development of renal stones, particularly in immobilized clients. Renal

CASE EXAMPLE 13-10

Rib Metastases Associated With Multiple Myeloma

At presentation, a 57-year-old man had rib pain that began 1 month ago. He could not think of any possible cause and denied any repetitive motions, recent trauma, forceful coughing, or history of tobacco use.

Past medical history was significant for hepatitis A (10 years ago) and benign prostatic hyperplasia (BPH). The BPH is reportedly well controlled with medication, but he noticed a decreased need to urinate and mentioned that he has been meaning to have a recheck of this problem.

On examination, there was bilateral point tenderness over the posterior seventh and eighth ribs. Symptoms were not increased by respiratory movements, trunk movements, position, or palpation of the intercostal spaces. Trunk and extremity movements were considered within normal limits, and a neurologic screening examination was unremarkable.

The therapist could not account for the client's symptoms, given the history and clinical presentation. Further screening revealed that the client had noticed progressive fatigue and generalized aching over the previous 2 weeks that he attributed to the flu. He had lost about 10 pounds during the week of vomiting and diarrhea. Vital signs were taken and were within normal limits. There were no other red flag symptoms to suggest a systemic origin or symptoms.

Results: Without a proper physical therapy diagnosis on which to base treatment, the therapist did not have a clear plan of care. There was no well-defined musculoskeletal problem, but a variety of systemic variables were present (e.g., sudden weight loss and constitutional symptoms accounted for by the flu, oliguria attributed to BPH, insidious onset of rib pain).

The therapist decided to treat the client for 7 to 10 days and reassess at that time. Intervention consisted of stretching, manual therapy, postural exercises, and Feldenkrais techniques. At the end of the prescribed time, there was no change in clinical presentation.

The therapist asked a colleague in the same clinic for a second opinion at no charge to the client. An ultrasound test for possible rib fracture was performed and was considered negative. No new findings were uncovered, and it was agreed collaboratively (including the client) to request a medical evaluation of the problem from the client's family physician. The client was subsequently given a diagnosis of multiple myeloma and was treated medically.

NOTE: The use of ultrasound over the painful rib was contraindicated in the presence of bone metastases.

insufficiency is the second most common cause of death, after infection, in clients with multiple myeloma.

In addition to bone destruction, multiple myeloma is characterized by disruption of RBC, leukocyte, and platelet production, which results from plasma cells crowding the bone marrow. Impaired production of these cell forms causes anemia, increased vulnerability to infection, and bleeding tendencies.

Neurologic Complications. Approximately 10% of persons with myeloma have amyloidosis, deposits of insoluble fragments of a monoclonal protein resembling starch. These deposits cause tissues to become waxy and immobile and may affect nerves, muscles, tendons, and ligaments, especially the carpal tunnel area of the wrist. CTS with pain, numbness, or tingling of the hands and fingers may develop. Excess immunoglobulins, caused by multiple myeloma, can cause a hyperviscosity syndrome characterized by changes in mental status, vision, fatigue, angina, and bleeding disorders.[150]

More serious neurologic complications may occur in 10% to 15% of clients with multiple myeloma. Spinal cord compression is usually observed early or in the late relapse phase of disease. Back pain is usually present as the initial symptom, with radicular pain that is aggravated by coughing or sneezing. Motor or sensory loss and bowel/bladder dysfunction are signs of more extensive compression. Paraplegia is a later, irreversible event.

Hodgkin's Disease

Hodgkin's disease is a chronic, progressive, neoplastic disorder of lymphatic tissue characterized by the painless enlargement of lymph nodes with progression to extralymphatic sites such as the spleen and liver.

In Caucasians, Hodgkin's disease demonstrates constant incidence rates, with a first peak occurring in adolescents and young adults and a second peak occurring in older adults. Men are affected more often than women, and boys are affected more often than girls.[151]

Epidemiologic and clinical-pathologic features of Hodgkin's disease suggest that an infectious agent may be involved in this disorder. Recently, accumulated data provide direct evidence supporting a causal role of Epstein-Barr virus (EBV) in a significant portion of cases, and a greater incidence of Hodgkin's disease has been observed in young individuals who previously had infectious mononucleosis.

Risk Factors

Infection with EBV and infectious mononucleosis has been associated with Hodgkin's lymphoma. No other risk factors have been identified.

Metastases

The exact mechanism of growth and spread of Hodgkin's disease remains unknown. The disease may progress by extension to adjacent structures or via the lymphatics because lymphoreticular cells inhabit all tissues of the body except the CNS. Hematologic spread may also occur, possibly by means of direct infiltration of blood vessels.

Clinical Signs and Symptoms

Hodgkin's disease usually appears as a painless, enlarged lymph node, often in the neck, underarm, or groin. The therapist may palpate these nodes during a cervical spine, shoulder, or hip examination (see Figs. 4-41 through 4-43).

Lymph nodes are evaluated on the basis of size, consistency, mobility, and tenderness. Lymph nodes up to 1 cm in diameter of soft-to-firm consistency that move freely and easily without tenderness are considered within normal limits.

Lymph nodes more than 1 cm in diameter that are firm and rubbery in consistency or tender are considered suspicious.

Enlarged lymph nodes associated with infection are more likely to be tender, soft, and movable than slow-growing nodes associated with cancer. Lymph nodes enlarged in response to infections throughout the body require referral to a physician, especially in someone with a current or previous history of cancer. The physician should be notified of these findings, and the client should be advised to have the lymph nodes checked at the next follow-up visit with the physician if not sooner, depending on the client's particular circumstances.

As always, *changes* in size, shape, tenderness, and consistency raise a red flag. Supraclavicular nodes are common metastatic sites for occult lung and breast cancers, whereas inguinal nodes implicate tumors arising in the legs, perineum, prostate, or gonads.

Other early symptoms may include unexplained fevers, night sweats, weight loss, and pruritus (itching). The itching occurs more intensely at night and may result in severe scratches because the client is unaware of scratching during the sleep state. The fever typically peaks in the late afternoon, and night sweats occur when the fever breaks during sleep. Fatigue, malaise, and anorexia may accompany progressive anemia. Some clients with Hodgkin's disease experience pain over the involved nodes after ingesting alcohol.

Symptoms may arise when enlarged lymph nodes obstruct or compress adjacent structures, causing edema of the face, neck, or right arm secondary to superior vena cava compression or causing renal failure secondary to urethral obstruction.

Obstruction of bile ducts as a result of liver damage causes bilirubin to accumulate in the blood and discolor the skin. Mediastinal lymph node enlargement with involvement of lung parenchyma and invasion of the pulmonary pleura progressing to the parietal pleura may result in pulmonary symptoms, including nonproductive cough, dyspnea, chest pain, and cyanosis.

Dissemination of disease from lymph nodes to bones may cause compression of the spinal cord, leading to paraplegia. Compression of nerve roots of the brachial, lumbar, or sacral plexus can cause nerve root pain.

CLINICAL SIGNS AND SYMPTOMS

Hodgkin's Disease

- Painless, progressive enlargement of unilateral lymph nodes, often in the neck
- Pruritus (itching) over entire body
- Unexplained fevers, night sweats
- Anorexia and weight loss
- Anemia, fatigue, malaise
- Jaundice
- Edema
- Nonproductive cough, dyspnea, chest pain, cyanosis
- Nerve root pain
- Paraplegia

Non-Hodgkin's Lymphoma

Non-Hodgkin's lymphoma (NHL) is a group of lymphomas affecting lymphoid tissue and occurring in persons of all ages. It is more common in adults in their middle and older years (40 to 60 years).

Risk Factors

Males are affected more often than females and individuals with congenital or acquired immunodeficiencies (e.g., those undergoing organ transplantation and anyone with autoimmune diseases are all at increased risk for development of NHL).[151,152] In addition, some people who have been exposed to large levels of radiation (e.g., nuclear reactor accidents) or extensive radiation and chemotherapy for a different cancer site may be at increased risk for lymphoma.

Individuals infected with the human immunodeficiency virus (HIV) are at increased risk for developing NHL and to a lesser extent, Hodgkin's disease as well. AIDS-related lymphoma (ARL) is now the second most common cancer associated with HIV after Kaposi's sarcoma. The relative risk of developing lymphoma within 3 years of an AIDS diagnosis is increased by 165-fold when compared with people without AIDS.[152]

Several possible etiologic mechanisms are hypothesized for NHL. Immunosuppression, possibly in combination with viruses or exposure to certain infectious agents, could be the primary cause. Chemicals, UV light, blood transfusion, acquired and congenital immune deficiency, and autoimmune disorders increase the risk for NHL.[153]

Other studies link the disease to widespread environmental contaminants such as benzene found in cigarette smoke, gasoline, automobile emissions, and industrial pollution.

Clinical Signs and Symptoms

NHL presents a clinical picture broadly similar to that of Hodgkin's disease, except that the disease is usually initially more widespread and less predictable. The disease starts in the lymph nodes, although early involvement of the oropharyngeal lymphoid tissue or the bone marrow is common, as is abdominal mass or GI involvement with complaints of vague back or abdominal discomfort.[151]

The most common manifestation is painless enlargement of one or more peripheral lymph nodes. Systemic symptoms are not as commonly associated with NHL as with Hodgkin's disease. Clients with NHLs often have remarkably few symptoms, even though many node areas or extranodal sites are involved.

Most NHLs fall into two broad categories related to their clinical activity: indolent and aggressive lymphomas. Indolent disease may be minimally active and treatable for many years. However, the disease is frequently disseminated at the time of diagnosis. Surgery is usually used only for staging or debulking purposes. Combination chemotherapy, biotherapy (targeted monoclonal antibodies), and radiation therapies are now used as treatment for NHL. Radioactive isotope combinations with monoclonal antibodies are also in use for some types of NHL.[151]

CLINICAL SIGNS AND SYMPTOMS

Non-Hodgkin's Lymphoma

- Enlarged lymph nodes
- Fever
- Night sweats
- Weight loss
- Bleeding
- Infection
- Red skin and generalized itching of unknown origin

Acquired Immunodeficiency Syndrome–Non-Hodgkin's Lymphoma

Only recently has AIDS-NHL emerged as a major sequela of HIV infection. It now occurs frequently in clients who survive other consequences of AIDS.

The etiologic basis of AIDS-NHL is still under investigation; profound cellular immunodeficiency plays a central role in lymphoma genesis. The molecular pathogenesis is a complex process involving both host factors and genetic alterations.[154]

Nearly 95% of all HIV-associated malignancies are either NHL or Kaposi's sarcoma. People with CNS lymphoma usually have advanced AIDS, are severely debilitated, and are usually thought to be at terminal stages of the disease.[151]

EBV often accompanies NHL. It is generally accepted that EBV acts in the pathogenesis of lymphoma owing to the alteration in balance between host and latent EBV infection in immunodeficiency states, with increased activity of the virus.

Risk Factors

Infection with HIV and related immunodeficiencies resulting from HIV are the primary risk factors for this disease. NHL is more likely to develop among clients who have Kaposi's sarcoma, a history of herpes simplex infection, and a lower neutrophil count.

Clinical Signs and Symptoms

The most common presentations of HIV-related NHL are systemic B symptoms (which may suggest an infectious process), a rapidly enlarging mass lesion, or both. At the time of diagnosis, approximately 75% of clients will have advanced disease. Extranodal disease frequently involves any part of the body, with the most common locations being the CNS, bone marrow, GI tract, and liver.

Diagnosis of NHL in areas of the body other than the CNS is complicated by a history of fevers, night sweats, and weight loss and loss of appetite, which are also common symptoms related to HIV infection and AIDS.

Although musculoskeletal lesions are not reported as commonly as pulmonary or CNS abnormalities in HIV-positive individuals, a wide variety of osseous and soft tissue changes are seen in this group. Diffuse adenopathy, lower extremity pain and swelling, subcutaneous nodules, and lytic lesions of the extremities are common.[155]

CLINICAL SIGNS AND SYMPTOMS

AIDS-NHL

- Painless, enlarged mass
- Subcutaneous nodules
- Constitutional symptoms (fever, night sweats, weight loss)
- Musculoskeletal lesions (lytic bone, pain, swelling)

PHYSICIAN REFERRAL

Early detection of cancer can save a person's life. Any suspicious sign or symptom discussed in this chapter should be investigated immediately by a physician. This is true especially in the presence of a positive family history of cancer, a previous personal history of cancer, and environmental risk factors, and/or in the absence of medical or dental (oral) evaluation during the previous year.

The therapist is not responsible for diagnosing cancer. The primary goal in screening for cancer is to make sure the client's problem is within the scope of a physical therapist's practice. In this regard, documentation of key findings and communication with the physician are both very important.

When trying to sort out neurologic findings, remember to look for changes in DTRs, a myotomal weakness pattern, and changes in bowel/bladder function. These findings will not give you a definitive diagnosis but will provide you with valuable information to offer the physician if further medical testing is advised.

Pain on weight bearing that is unrelieved by rest or change in position and does not respond to treatment, unremitting pain at night, and a history of cancer are all red flags indicating that medical evaluation is needed.

Any recently discovered lumps or nodules must be examined by a physician. Any suspicious finding by report, on observation, or by palpation should be checked by a physician.

If any signs of skin lesions are described by the client or if they are observed by the therapist and the client has not been examined by a physician, a medical referral is recommended.

If the client is planning a follow-up visit with the physician within the next 2 to 4 weeks, that client is advised to indicate the mole or skin changes at that time. If no appointment is pending, the client is encouraged to make a specific visit either to the family/personal physician or to a dermatologist.

Guidelines for Immediate Physician Referral

- Presence of recently discovered lumps or nodules or changes in previously present lumps, nodules, or moles, especially in the presence of a previous history of cancer or when accompanied by carpal tunnel or other neurologic symptoms.

- Detection of palpable, fixed, irregular mass in the breast, axilla, or elsewhere requires medical referral or a recommendation to the client to contact a physician for evaluation of the mass. Suspicious lymph node enlargement or lymph node changes; generalized lymphadenopathy.
- Recurrent cancer can appear as a single lump, a pale or red nodule just below the skin surface, a swelling, a dimpling of the skin, or a red rash. Report any of these changes to a physician immediately.
- Notify physician of any suspicious changes in lymph nodes; note the presence of lymphadenopathy and describe the location and any observed or palpable characteristics.
- Presence of any of the early warning signs of cancer, including idiopathic muscle weakness accompanied by decreased DTRs.
- Any unexplained bleeding from any area (e.g., rectum, blood in urine or stool, unusual or unexpected vaginal bleeding, breast, penis, nose, ears, mouth, mole, skin, or scar).
- Any sign or symptom of metastasis in someone with a previous history of cancer (see individual cancer types for specific clinical signs and symptoms; see also Clues to Screening for Cancer).
- Any man with pelvic, groin, sacroiliac, or low back pain accompanied by sciatica and a past history of prostate cancer.

Clues to Screening for Cancer

- Age older than 50 years
- Previous personal history of any cancer, especially in the presence of bilateral carpal tunnel symptoms, back pain, shoulder pain, or joint pain of unknown or rheumatic cause at presentation
- Previous history of cancer treatment (late physical complications and psychosocial complications of disease and treatment can present in a somatic presentation)
- Any woman with chest, breast, axillary, or shoulder pain of unknown cause, especially with a previous history of cancer and/or over the age of 40
- Anyone with back, pelvic, groin, or hip pain accompanied by abdominal complaints, palpable mass
- *For women:* Prolonged or excessive menstrual bleeding (or in the case of the postmenopausal woman who is not taking hormone replacement, breakthrough bleeding)
- *For men:* Additional presence of sciatica and past history of prostate cancer
- Recent weight loss of 10% of total body weight (or more) within a 2-week to 1-month period of time without trying; weight gain is more typical with true musculoskeletal dysfunction because pain has limited physical activities
- Musculoskeletal symptoms are made better or worse by eating or drinking (GI involvement)
- Shoulder, back, hip, pelvic, or sacral pain accompanied by changes in bowel and/or bladder function or changes in stool or urine
- Hip or groin pain is reproduced by heel strike/hopping test or translational/rotational stress (bone fracture from metastases)
- When a back "injury" is not improving as expected or if symptoms are increasing
- Early warning signs, including proximal muscle weakness and changes in DTRs
- Constant pain (unrelieved by rest or change in position); remember to assess constancy by asking, "Do you have that pain right now?"
- Intense pain present at night (rated 7 or higher on a numeric scale from 0 for "no pain" to 10 for "worst pain")
- Signs of nerve root compression must be screened for cancer as a possible cause
- Development of new neurologic deficits (e.g., weakness, sensory loss, reflex change, bowel or bladder dysfunction)
- Changes in size, shape, tenderness, and consistency of lymph nodes, especially painless, hard, rubbery lymph nodes present in more than one location
- A growing mass, whether painless or painful, is assumed to be a tumor unless diagnosed otherwise by a physician. A hematoma should decrease in size over time, not increase
- Disproportionate pain relieved with aspirin may be a sign of bone cancer (osteoid osteoma)
- Signs or symptoms seem out of proportion to the injury and persist longer than expected for physiologic healing of that type of injury; no position is comfortable (remember to conduct a screening examination for emotional overlay)
- Change in the status of a client currently being treated for cancer

CANCER PRESENCE AND PAIN

METASTASES (MOST COMMONLY SEEN IN A PHYSICAL THERAPY SETTING)

Location: Integumentary system
Pulmonary system
Neurologic system
Musculoskeletal
Hepatic

Referral: See Table 13-5

SKIN (MELANOMA ONLY)

Location: Anywhere on the body
Women: Arms, legs, back, face
Men: Head, trunk
African Americans: Palms, soles, under the nails

Referral: None

Description: Usually painless; see ABCDE method of detection (see text)
Sore that does not heal
Irritation and itching
Cluster mole formation
Tenderness and soreness around a mole

Intensity: Mild

Duration: Constant

Associated signs and symptoms: None

PARANEOPLASTIC SYNDROMES

Location: Remote sites from primary neoplasm

Referral: Organ dependent

Description: Asymmetric joint involvement
Lower extremities primarily
Concurrent arthritis and malignancy
Explosive onset at late age
See Tables 13-6 and Box 13-4

Intensity: Symptom dependent

Duration: Symptom dependent

Associated signs and symptoms: Fever
Skin rash
Clubbing of the fingers
Pigmentation disorders
Arthralgias
Paresthesias
Thrombophlebitis
Proximal muscle weakness
Anorexia, malaise, weight loss
Rheumatologic complaints

Continued

CANCER PRESENCE AND PAIN—*cont'd*

ONCOLOGIC (CANCER) PAIN

Location: Localized bone pain; referred pain
Referral: May follow nerve distribution
Description: Bone pain: Sharp, intense, constant
Viscera: Colicky, cramping, dull, diffuse, boring, poorly localized
Vein, artery, lymphatic channel: Dull, diffuse, burning, aching
Nerve compression: Sharp, stabbing; follows nerve distribution or dull, poorly localized
Inflammation: Sensitive tenderness
Intensity: Varies from mild to severe or excruciating
Bone pain: Increases on movement or weight bearing
Duration: Usually constant; may be worse at night
Associated signs and symptoms: With mild-to-moderate superficial pain: Sympathetic nervous system response (e.g., hypertension, tachycardia, tachypnea)
With severe or visceral pain: Parasympathetic nervous system response (e.g., hypotension, tachypnea, weakness, fainting)
Organ dependent (e.g., esophagus: difficulty eating or speaking; gallbladder: jaundice, nausea; nerve involvement: altered sensation, paresthesia; see individual visceral cancers)

SOFT TISSUE TUMORS

Location: Any connective tissue (e.g., tendon muscle, cartilage, fat, synovium, fibrous tissue)
Referral: According to the tissue involved
Description: Persistent swelling or lump, especially in the muscle
Intensity: Mild, increases progressively to severe
Duration: Intermittent, increases progressively to constant
Associated signs and symptoms: Local swelling with tenderness and skin warmth
Pathologic fracture

BONE TUMORS

Location: Can affect any bone in the body, depending on the specific type of bone cancer
Referral: According to pattern and location of metastases
Description: Sharp, knifelike, aching bone pain
Occurs on movement and weight bearing, with pathologic fractures
Pain at night, preventing sleep
Intensity: Initially mild, progressing to severe
Duration: Usually intermittent, progressing to constant
Associated signs and symptoms: Fatigue and malaise
Significant unintentional weight loss
Swelling and warmth over localized areas of tumor
Soft, tender palpable mass over bone
Loss of range of motion and joint function if limb bone is involved
Fever
Sciatica
Unilateral edema

CANCER PRESENCE AND PAIN—*cont'd*

PRIMARY CENTRAL NERVOUS SYSTEM: BRAIN TUMORS

Location: Intracranial

Referral: Specific symptoms depend on tumor location
Headaches

Description: Bioccipital or bifrontal headache

Intensity: Mild to severe

Duration: Worse in morning on awakening
Diminishes or disappears soon after rising

Aggravating factors: Activity that increases intracranial pressure (e.g., straining during bowel movements, stooping, lifting heavy objects, coughing, bending over)
Prone/supine position at night during sleep

Relieving factors: Pain medications, including aspirin, acetaminophen

Associated signs and symptoms: Papilledema
Altered mentation:
- Increased sleeping
- Difficulty in concentrating
- Memory loss
- Increased irritability
- Poor judgment
- Vomiting unrelated to food accompanies headaches
- Seizures

Neurologic findings:
- Positive Babinski reflex
- Clonus (ankle or wrist)
- Sensory changes
- Decreased coordination
- Ataxia
- Muscle weakness
- Increased lower extremity deep tendon reflexes
- Transient paralysis

PRIMARY CENTRAL NERVOUS SYSTEM: SPINAL CORD TUMORS

Location: Intramedullary (within the spinal cord)
Extramedullary (within the dura mater)
Extradural (outside the dura mater)

Referral: Back pain at the level of the spinal cord lesion
Pain may extend to the groin or legs

Description: Dull ache; sharp, knifelike sensation

Intensity: Mild to severe, progressive; night pain

Duration: Intermittent, progressing to constant, or constant

Aggravating factors: (Back pain) Lying down/rest
Weight bearing
Sneezing or coughing

Associated signs and symptoms: Muscle weakness
Muscle atrophy
Sensory loss
Paraplegia/quadriplegia
Chest or abdominal pain
Bowel/bladder dysfunction (late findings)

Continued

CANCER PRESENCE AND PAIN—*cont'd*

LEUKEMIA

Location: Usually painless; may have pain in the left abdomen; bone and joint pain possible
Referral: None
Description: Dull pain in the abdomen; may occur only on palpation
Intensity: Mild to moderate
Duration: Intermittent (with applied pressure)
Associated signs and symptoms:
Enlarged lymph nodes
Unusual bleeding from the nose or rectum, or blood in urine
Prolonged menstruation
Easy bruising of the skin
Fatigue
Dyspnea
Weight loss, loss of appetite
Fevers and sweats

MULTIPLE MYELOMA

Location: Skeletal pain, especially in the spine, sternum, rib, leg, or arm
Referral: According to the location of the tumor
Description: Sharp, knifelike
Intensity: Moderate to severe
Duration: Intermittent, progressing to constant
Associated signs and symptoms:
Hypercalcemia: Dehydration (vomiting), polyuria, confusion, loss of appetite, constipation
Bone destruction with spontaneous bone fracture
Neurologic: CTS; back pain with radicular symptoms; spinal cord compression (motor or sensory loss, bowel/bladder dysfunction, paraplegia)

HODGKIN'S DISEASE

Location: Lymph glands, usually unilateral neck or groin
Referral: According to the location of the metastases
Description: Usually painless, progressive enlargement of lymph nodes
Intensity: Not applicable
Duration: Not applicable
Associated signs and symptoms:
Fever peaks in the late afternoon, night sweats
Anorexia and weight loss
Severe itching over the entire body
Anemia, fatigue, malaise
Jaundice
Edema
Nonproductive cough, dyspnea, chest pain, cyanosis

NON-HODGKIN'S LYMPHOMA (INCLUDING AIDS-NHL)

Location: Peripheral lymph nodes
Referral: Not applicable
Description: Usually painless enlargement
Intensity: Not applicable
Duration: Not applicable
Associated signs and symptoms:
Constitutional symptoms (fever, night sweats, weight loss)
Bleeding
Generalized itching and reddened skin
AIDS-NHL: Musculoskeletal lesions, subcutaneous nodules

Key Points to Remember

- When put to the task of screening for cancer, always remember our three basic clues:
  - Past Medical History
  - Clinical Presentation
  - Associated Signs and Symptoms
- Any suspicious lesions or red flag symptoms, especially in the presence of a past medical history of cancer or risk factors for cancer, should be investigated further. With the increasing number of people diagnosed with cancer, recognizing hallmark findings of cancer is important.
- Knowing the systems most often affected by cancer metastasis and the corresponding clinical manifestations is a good starting point. Any time a client reports a past medical history of cancer, we must be alert for signs or indications of cancer recurrence (locally or via metastasis).
- Knowing the most common risk factors for cancer in general and risk factors for specific cancers is the next step. Risk factor assessment and cancer prevention are a part of every health care professional's role as educator and in primary prevention.
- Whether you are working in an oncology setting or in a general practice with an occasional client, good resource information is available. Thorough, reliable, and up-to-date information about specific types of cancer, cancer treatments, and recent breakthroughs in cancer research is available from The Abramson Cancer Center of the University of Pennsylvania (Philadelphia) at http://oncolink.upenn.edu
- Spinal malignancy involves the lumbar spine more often than the cervical spine and is usually metastatic rather than primary.
- Spinal cord compression from metastases may appear as back pain, leg weakness, and bowel/bladder symptoms.
- Fifty percent of clients with back pain from a malignancy have an identifiable preceding trauma or injury to account for the pain or symptoms. Always remember that clients may erroneously attribute symptoms to an event.
- Back pain may precede the development of neurologic signs and symptoms in any person with cancer.
- The presence of jaundice in association with any atypical presentation of back pain may indicate liver metastasis.
- Signs of nerve root compression may be the first indication of cancer, in particular, lymphoma, multiple myeloma, or cancer of the lung, breast, prostate, or kidney.
- The five most common sites of metastasis are the lymph nodes, liver, lung, bone, and brain.
- Lung, breast, prostate, thyroid, and the lymphatics are the primary sites responsible for most metastatic bone disease.
- Monitoring physiologic responses (vital signs) to exercise is important in the immunosuppressed population. Watch closely for early signs (dyspnea, pallor, sweating, and fatigue) of cardiopulmonary complications of cancer treatment.
- To determine appropriate exercise levels for clients who are immunosuppressed, review blood test results (WBCs, RBCs, hematocrit, platelets). When these are not available, monitor vital signs and use RPE as a guideline.
- Besides the seven early warning signs of cancer, the therapist should watch for idiopathic muscle weakness accompanied by decreased DTRs.
- Changes in size, shape, tenderness, and consistency of lymph nodes raise a red flag. Supraclavicular nodes and inguinal nodes are common metastatic sites for cancer.
- No reliable physical signs distinguish between benign and malignant soft tissue lesions. All soft tissue lumps that persist or grow should be reported immediately to the physician.
- Malignancy is always a possibility in children with musculoskeletal symptoms.

SUBJECTIVE EXAMINATION

Special Questions to Ask

Special questions to ask will vary with each client and the clinical signs and symptoms presented at the time of evaluation. The therapist should refer to the specific chapter representing the client's current complaints. The case study provided here is one example of how to follow up with necessary questions to rule out a systemic origin of musculoskeletal findings.

A previous history of drug therapy and current drug use may be important information to obtain because prolonged use of drugs such as phenytoin (Dilantin) or immunosuppressive drugs such as azathioprine (Imuran) and cyclosporine may lead to cancer. Postmenopausal use of estrogens has been linked with breast cancer.[156-158]

Past Medical History

A previous personal/family history of cancer may be significant, especially any history of breast, colorectal, or lung cancer that demonstrates genetic susceptibility.

- Have you ever had cancer or do you have cancer now?

If no, have you ever received chemotherapy, hormone therapy, or radiation therapy?

If yes, what was the treatment for?

If yes to previous history of cancer, ask about type of cancer, date of diagnosis, stage (if known), treatment, and date of most recent follow-up visit with oncologist or other cancer specialist.

- Has your physician said that you are cancer-free?
- Have you ever been exposed to chemical agents or irritants, such as asbestos, asphalt, aniline dyes, benzene, herbicides, fertilizers, wood dust, or others? **(Environmental causes of cancer;** see complete environmental/occupational screening survey in Chapter 2 and Appendix B-14)

Clinical Presentation: Early Warning Signs

When using the seven early warning signs of cancer as a basis for screening (see Box 13-2), one or all of the following questions may be appropriate:

- Have you noticed any changes in your bowel movement or in the flow of urination?
  - *If yes,* ask pertinent follow-up questions as suggested in Chapter 10; see also Appendix B-5.
  - *If the client answers no,* it may be necessary to provide prompts or examples of what changes you are referring to (e.g., difficulty in starting or continuing the flow of urine, numbness or tingling in the groin or pelvis).
- Have you noticed any sores that have not healed properly?
  - *If yes,* where are they located? How long has the sore been present? Has your physician examined this area?
- Have you noticed any unusual bleeding (*for women:* including prolonged menstruation or *any* bleeding for the postmenopausal woman who is not taking hormone replacement) or prolonged discharge from any part of your body?
  - *If yes,* where? How long has this been present? Has your physician examined this area?
- Have you noticed any thickening or lump of any muscle, tendon, bone, breast, or anywhere else?
  - *If yes,* where? How long has this been present? Has your physician examined this area?*
  - *If no (for women):* Do you examine your own breasts? How often do you examine yourself?
- When was the last time you did a breast self-examination (see Appendix D-6)?
- Do you have any pain, swelling, or unusual tenderness in the breasts? **(Pain can be a symptom of cancer; cyclic pain is common with normal breasts, use of oral contraceptives, and fibrocystic disease.)**
  - *If yes,* is this pain brought on by strenuous activity? **(Spontaneous/systemic or related to specific musculoskeletal cause** [e.g., use of one arm])
- Have you noticed any rash on the breast or discharge from the nipple? **(Medications such as oral contraceptives, phenothiazines, diuretics, digitalis, tricyclic tranquilizers, reserpine, methyldopa, and steroids can cause clear discharge from the nipple; blood-tinged discharge is always significant.)**
- Have you noticed any difficulty in eating or swallowing? Have you had a chronic cough, recurrent laryngitis, hoarseness, or any difficulty with speaking?
  - *If yes,* how long has this been happening? Have you discussed this with your physician?
- Have you had any change in digestive patterns? Have you had increasing indigestion or unusual constipation?
  - *If yes,* how long has this been happening? Have you discussed this with your physician?
- Have you had a recent, sudden weight loss without dieting? **(10% of client's total body weight in 10 days to 2 weeks is significant.)**
- Have you noticed any obvious change in color, shape, or size of a wart or mole?
  - *If yes,* what have you noticed? How long has this wart or mole been present? Have you discussed this problem with your physician?
- Have you had any unusual headaches or changes in your vision?

*An asymptomatic mass that has been present for years and causes only cosmetic concern is usually benign, whereas a painful mass of short duration that has caused a decrease in function may be malignant.

SUBJECTIVE EXAMINATION—*cont'd*

- *If yes,* please describe. **(Brain tumors: bioccipital or bifrontal)**
- Can you attribute these to anything in particular?
- Do you vomit (unrelated to food) when your headaches occur? **(Brain tumors)**

- Have you been more tired than usual or experienced persistent fatigue during the last month?
- Can you think of any time during the past week when you may have bumped yourself, fallen, or injured yourself in any way? (Ask when in the presence of local swelling and tenderness.) **(Bone tumors)**
- Have you noticed any bone pain or problems with any of your bones? Is the pain affected by movement? **(Fractures cause sharp pain that increases with movement. Bone pain from systemic causes usually feels dull and deep and is unrelated to movement.)**

Associated Signs and Symptoms

- Are you having any symptoms of any kind anywhere else in your body?

CASE STUDY

REFERRAL

A 56-year-old man has come to you for an evaluation without referral. He has not been examined by a physician of any kind for at least 3 years. He is seeking an evaluation on the insistence of his wife, who has noticed that his collar size has increased two sizes in the last year and that his neck looks "puffy." He has no complaints of any kind (including pain or discomfort), and he denies any known trauma, but his wife insists that he has limited ability in turning his head when backing the car out of the driveway.

PHYSICAL THERAPY SCREENING INTERVIEW

First, read the client's Family/Personal History form with particular interest in his personal or family history of cancer, presence of allergies or asthma, use of medications or over-the-counter drugs, previous surgeries, available x-ray studies of the neck or spine, and/or history of cigarette smoking (or other tobacco use).

An appropriate lead-in to the following series of questions may be: "Because you have not seen a physician before your appointment with me, I will ask you a series of questions to find out if your symptoms require examination by a physician rather than treatment in this office."

CURRENT SYMPTOMS

- What have you noticed different about your neck that brings you here today?
- When did you first notice that your neck was changing (in size or shape)?
- Can you remember having any accidents, falls, twists, or any other kind of potential trauma at that time?
- Do you ever notice any pain, stiffness, soreness, or discomfort in your neck or shoulders?
- *If yes,* please describe (as per the outline in the Core Interview, Chapter 2).
- Does this or any pain ever awaken you at night or keep you awake? **(Night pain associated with cancer)**
    - *If yes,* follow-up with appropriate questions (see the Core Interview, Chapter 2).

ASSOCIATED SYMPTOMS

- Have you noticed any numbness or tingling in your arms or hands?
- Have you noticed any swollen glands, lumps, or thickened areas of skin or muscle in your neck, armpits, or groin? **(Cancer screen)**
- Do you have any difficulty in swallowing? Do you have recurrent hoarseness, flulike symptoms, or a persistent cough or cold that never seems to go away? **(Cancer screen)**
- Have you noticed any low-grade fevers or night sweats? **(Systemic disease)**
- Have you had any recent unexplained weight gain or loss? (You may need to explain that you mean a gain or loss of 10 to 15 pounds in as many days without dieting.) Have you had a loss of appetite? **(Cancer screen or other systemic disease)**
- Do you ever have any difficulty with breathing or find yourself short of breath at rest or after minimal exercise? **(Dyspnea)**
- Do you have frequent headaches, or do you experience any dizziness, nausea, or vomiting? **(Systemic disease, carotid artery affected)**

FUNCTIONAL CAPACITY

- What kind of work do you do?
- Do you have any limitations caused by this condition that affects you in any way at work or at home? **(Occupational disease, limitations of activities of daily living [ADL] skills)**

Continued

CASE STUDY—*cont'd*

FINAL QUESTIONS

- How would you describe your general health?
- Have you ever been diagnosed with cancer of any kind?
- Is there anything that you would like to tell me that you think is important about your neck or your health in general?

FIRST VISIT: ASSESSING THE MUSCULOSKELETAL SYSTEM

- Observation/Inspection
- Observe for the presence of swelling anywhere, tender or swollen lymph nodes (cervical, supraclavicular, and axillary), changes in skin temperature, and unusual moles or warts. Perform a brief posture screen (general postural observations may be made while you are interviewing the client). Palpate for carotid artery and upper extremity pulses. Check vital signs and **Take the Client's Temperature!**
- Cervical active range of motion (AROM)/passive range of motion (PROM)
- Assess for muscle tightness, loss of joint motion (including accessory movements, if indicated by a loss of passive motion). Assess for compromise of the vertebral artery, and if negative, clear the cervical spine by using a quadrant test with overpressure (e.g., Spurling's test) and assess accessory movements of the cervical spine. Perform tests for thoracic outlet syndrome. Palpate the anterior cervical spine for pathologic protrusion while the client swallows.
- Temporomandibular joint (TMJ) screen
  - Clear the joint above (i.e., TMJ) using AROM, observation, and palpation specific to the TMJ.
- Shoulder screen
  - Clear the joint below (i.e., shoulder) by using a screening examination (e.g., AROM/PROM and quadrant testing).
- Neurologic screen (see Chapter 4)

Deep tendon reflexes (DTRs), sensory screen (e.g., gross sensory testing for light touch), manual muscle test (MMT) screening using break testing of the upper quadrant, grip strength. If test(s) is abnormal, consider further neurologic testing (e.g., balance, coordination, stereognosis, in-depth sensory examination, dysmetria). Ask about the presence of recent visual changes, headaches, numbness, or tingling into the jaw or down the arm(s).

It is always recommended that the therapist give the client ongoing verbal feedback during the examination regarding evaluation results, such as: "I notice you can't turn your head to the right as much as you can to the left—from checking your muscles and joints, it looks like muscle tightness, not any loss of joint movement." or "I notice your reflexes on each side aren't the same (your right arm reacts more strongly than the left)—let's see if we can find out why."

RECOMMENDATION FOR PHYSICIAN VISIT

- I noticed on your intake form that you haven't listed the name of a personal or family physician. Do you have a physician?
- If *yes,* when was the last time you saw your physician? Have you seen your physician for this current problem?

Give the client a brief summary of your findings while making your recommendations, for example, "Mr. X., I notice today that although you don't have any ongoing neck pain, the lymph nodes in your neck and armpit are enlarged but not particularly tender. Otherwise, all of my findings are negative. Your loss of motion on turning your head is not unusual for a person your age and certainly would not cause your neck to increase in size or shape.

"Given the fact that you have not seen a physician for almost 3 years, I strongly recommend that you see a physician of your choice, or I can give you the names of several to choose from. In either case, I think some medical tests are necessary to rule out any underlying medical problem. For instance, a neck x-ray exam would be recommended before physical therapy treatment is started."

If the client has indicated a positive family history of cancer, it might be appropriate to suggest, "Given your positive family history of previous medical illnesses, the 3 years since you have seen a physician, and the lack of musculoskeletal findings, I strongly recommend ..." It is important to provide the client with all the information available to you but without causing undue alarm and emotional stress, which could actually prevent the client from seeking further testing.

If the client does give the name of a physician, you may ask for written permission (disclosure release) to send a copy of your results to the physician. If the client does not have a physician and requests recommendations from you, you may offer to send a copy of your results to the physician with whom the client makes an appointment.

If you think that a problem may be potentially serious and you want this person to receive adequate follow-up without causing alarm, you may offer to let him make the appointment from your office, suggest that your secretary or receptionist make the appointment for him, or even offer to make the initial telephone contact yourself.

RESULTS

This client did comply with the therapist's suggestion to see a physician and was diagnosed with Hodgkin's disease (a cancer of the lymph system) without constitutional symptoms (i.e., without evidence of weight loss, fever, or night sweats). Medical intervention was initiated, and physical therapy treatment was not warranted.

PRACTICE QUESTIONS

1. Name three predisposing factors to cancer that the therapist must watch for during the interview process as red flags.
2. How do you monitor exercise levels in the oncology patient without laboratory values?
3. In a physical therapy practice, clients are most likely to present with signs and symptoms of metastases to:
   a. Skeletal system, hepatic system, pulmonary system, central nervous system
   b. Cardiovascular system, peripheral vascular system, enteric system
   c. Hematologic and lymphatic systems
   d. None of the above
4. What is the significance of nerve root compression in relation to cancer?
5. Complete the following mnemonic:
   C
   A
   U
   T
   I
   O
   N
   S
6. Whenever a therapist observes, palpates, or receives a client report of a lump or nodule, what three questions must be asked?
7. How can the therapist determine whether a client's symptoms are caused by the delayed effects of radiation as opposed to being signs of recurring cancer?
8. Give a general *description* and *explanation* of the changes seen in deep tendon reflexes associated with cancer.
9. Why is weight loss a significant red flag sign in a physical therapy practice?
10. When tumors produce signs and symptoms at a site distant from the tumor or its metastasized sites, these "remote effects" of malignancy are called:
    a. Bone metastases
    b. Vitiligo
    c. Paraneoplastic syndrome
    d. Ichthyosis
11. A client who has recently completed chemotherapy requires immediate medical referral if he has which of the following symptoms?
    a. Decreased appetite
    b. Increased urinary output
    c. Mild fatigue but moderate dyspnea with exercise
    d. Fever, chills, sweating
12. A suspicious skin lesion requiring medical evaluation has:
    a. Round, symmetric borders
    b. Notched edges
    c. Matching halves when a line is drawn down the middle
    d. A single color of brown or tan
13. What is the significance of Beau's lines in a client treated with chemotherapy for leukemia?
    a. Impaired nail formation from death of cells
    b. Temporary longitudinal groove or ridge through the nail
    c. Increased production of the nail by the matrix as a sign of healing
    d. A sign of local trauma
14. A 16-year-old boy was hurt in a soccer game. He presents with exquisite right ankle pain on weight bearing but reports no pain at night. Upon further questioning, you find he is taking Ibuprofen at night before bed, which may be masking his pain. What other screening examination procedures are warranted?
    a. Perform a heel strike test.
    b. Review response to treatment.
    c. Assess for signs of fracture (edema, exquisite tenderness to palpation, warmth over the painful site).
    d. All of the above
15. When is it advised to take a work or military history?
    a. Anyone with head and/or neck pain who uses a cell phone more than 8 hours/day
    b. Anyone over age 50
    c. Anyone presenting with joint pain of unknown cause accompanied by multiple other signs and symptoms
    d. This is outside the scope of a physical therapist's practice
16. A 70-year-old man came to outpatient physical therapy with a complaint of pain and weakness of his fingers and morning stiffness lasting about an hour. He presented with bilateral swelling of the metacarpophalangeal (MCP) joints of the index and ring fingers. He saw his family doctor 4 weeks ago and was given diclofenac, which has not changed his symptoms. Now he wants to try physical therapy. Since he last saw his physician, he has developed additional joint pain in the left knee and right shoulder. How can you tell if this is cancer, polyarthritis, or a paraneoplastic disorder?
    a. Ask about a previous history of cancer and recent onset of skin rash.
    b. You can't. This requires a medical evaluation.
    c. Look for signs of digital clubbing, cellulitis, or proximal muscle weakness.
    d. Assess vital signs.
17. A 49-year-old man was treated by you for bilateral synovitis of the proximal interphalangeal (PIP) joints in the second, third, and fourth fingers. His symptoms went away with treatment, and he was discharged. Six weeks later, he returned with the same symptoms. There was obvious soft tissue swelling with morning stiffness worse than before. He also reports problems with his bowels but isn't able to tell you exactly what's wrong. There are no other changes in his health. He is not taking any medications or over-the-counter drugs and does not want to see a doctor. Are there enough red flags to warrant medical evaluation before resumption of physical therapy intervention?
    a. Yes; age, bilateral symptoms, progression of symptoms, report of GI distress
    b. No; treatment was effective before—it's likely that he has done something to exacerbate his symptoms and needs further education about joint protection.

Continued

PRACTICE QUESTIONS—*cont'd*

18. A client with a past medical history of kidney transplantation (10 years ago) has been referred to you for a diagnosis of rheumatoid arthritis. His medications include tacrolimus, methotrexate, Fosamax, and Wellbutrin. During the examination, you notice a painless lump under the skin in the right upper anterior chest. There is a loss of hair over the area. What other symptoms should you look for as red flag signs and symptoms in a client with this history?
    a. Fever, muscle weakness, weight loss
    b. Change in deep tendon reflexes, bone pain
    c. Productive cough, pain on inspiration
    d. Nose bleeds or other signs of excessive bleeding
19. A 55-year-old man with a left shoulder impingement also has palpable axillary lymph nodes on both sides. They are firm but movable, about the size of an almond. What steps should you take?
    a. Examine other areas where lymph nodes can be palpated.
    b. Ask about history of cancer, allergies, or infections.
    c. Document your findings and contact the physician with your concerns.
    d. All of the above

REFERENCES

1. Siegel R, Ward E, Brawley O, et al: Cancer statistics 2011. *CA Cancer J Clin* 61:212–236, 2011.
2. Brem SN, Nabors LB, Raizer JJ: Central nervous system cancers, National Comprehensive Cancer Guidelines (2009). Available at http://www.nccn.org/professionals/physician_gls/f_guidelines.asp. Accessed November 01, 2010.
3. National Cancer Institute (NCI): Annual report: Continued declines in annual cancer rates, 2009. Available at www.nci.nih.gov/. Accessed Oct 29, 2010.
4. Oeffinger KC, Mertens AC, Hudson MM, et al: Health care of young adult survivors of childhood cancer: a report from the Childhood Cancer Survivor Study. *Ann Fam Med* 2:61–70, 2004.
5. University of Minnesota Cancer Center: Childhood Cancer Survivor Study (CCSS). Available at http://www.cancer.umn.edu/ltfu. Accessed Oct. 11, 2010.
6. Hawkins M: Long-term survivors of childhood cancers: what knowledge have we gained? Available at www.medscape.com/viewarticle/492506. Accessed Oct. 11, 2010.
7. Tsao A, Kim E, Hong W: Chemoprevention of cancer. *CA Cancer J Clin* 54:150–180, 2004.
8. *Guide to Physical Therapist Practice*, ed 2, Alexandria, VA, 2003, American Physical Therapy Association.
9. American Cancer Society (ACS): *Cancer prediction and early detection (CPED)*, Atlanta, 2010, ACS. Available at www.cancer.org. Accessed Oct. 11, 2010.
10. National Cancer Institute: Cancer Statistics. Available at http://www.nci.nih.gov/statistics/. Accessed Oct. 11, 2010.
11. Pal SK, Katheria V, Hurria A: Evaluating the older patient with cancer: understanding frailty and the geriatric assessment. *CA Cancer J Clin* 60(2):120–132, 2010.
12. Ries LAG, Eisner MP, Kosary CL: *SEER Cancer Statistics Review*, Bethesda, MD, 1973-1999, NCI. Available at http://seer.cancer.gov/csr/1973_1999/. Accessed September 2002.
13. Bach PB, Schrag D, Brawley OW: Survival of blacks and whites after a cancer diagnosis. *JAMA* 287:2106–2113, 2002.
14. Institute of Medicine (IOM): *Unequal treatment, confronting racial and ethnic disparities in health care*, Washington DC, 2003, IOM. Available at http://www.iom.edu/ [type in search window: Unequal Treatment…]. Accessed Oct. 11, 2010.
15. O'Brien K: Cancer statistics for Hispanics, 2003. *CA Cancer J Clin* 53:208–226, 2003.
16. Huerta EE: Cancer statistics for Hispanics, 2003: good news, bad news, and the need for a health system paradigm change. *CA Cancer J Clin* 53:205–207, 2003.

16a. Ziogas A: Clinically relevant changes in family history of cancer over time. *JAMA* 306:172–178, 208–210, 2011.

17. Sifri R, Gangadharappa S, Acheson L: Identifying and testing for hereditary susceptibility to common cancers. *CA Cancer J Clin* 54:309–326, 2004.
18. National Society of Genetic Counselors (NSGC): Your family history: your future, Chicago, IL, NSGC. Available at www.nsgc.org. Accessed Oct. 11, 2010.
19. American Institute for Cancer Research (AICR): *Food, nutrition, physical activity and the prevention of cancer: a global perspective*, Washington, DC, 2007, AICR.
20. American Institute for Cancer Research (AICR): *Expert Panel Report: Summary—food nutrition and the prevention of cancer: a global perspective*, Washington, DC, 2005, AICR. Available at http://www.aicr.org/. Accessed Oct. 10, 2010.
21. Calle EE, Rodriquez C, Walker-Thurmond K, et al: Overweight, obesity, and mortality from cancer in a prospectively studied cohort of U.S. adults. *N Engl J Med* 348:1625–1638, 2003.
22. Patel AV, Rodriquez C, Bernstein L, et al: Obesity, recreational physical activity, and risk of pancreatic cancer in a large U.S. cohort. *Cancer Epidemiol Biomarkers Prev* 14:459–466, 2005.
23. Key TJ, Schatzkin A, Willett WC, et al: Diet, nutrition, and the prevention of cancer. *Public Health Nutr* 7:187–200, 2004.
24. Roberts DL: Biological mechanisms linking obesity and cancer risk: new perspectives. *Ann Rev Med* 61:301–316, 2010.
25. Harriss DJ: Lifestyle factors and colorectal cancer risk: systematic review and meta-analysis of associations with mass index. *Colorectal Dis* 11(6):547–563, 2009.
26. Mai V, Kant AK, Flood A, et al: Diet quality and subsequent cancer incidence and mortality in a prospective cohort of women. *Int J Epidemiol* 34:54–60, 2005.
27. Kabat GC: Repeated measures of serum glucose and insulin in relationship to postmenopausal breast cancer. *Int J Cancer* 125(11):2704–2710, 2009.
28. Duggan C: Associations of insulin resistance and adiponectin with mortality in women with breast cancer. *J Clin Oncol* 29(1):32–39, 2011. Epub Nov 29, 2010.
29. Eyre H, Kahn R, Robertson RM: Preventing cancer, cardiovascular disease and diabetes: a common agenda for the ACS, American Diabetes Association, American Heart Association. *CA Cancer J Clin* 54:190–207, 2004.
30. National Cancer Institute (NCI): *Nutrition in cancer care 2010*. Available on-line at http://www.cancer.gov/cancertopics/pdq/supportivecare/nutrition/Patient/page1. Accessed March 7, 2011.

31. American Cancer Society (ACS): *Nutrition for people with cancer.* Available on-line at www.cancer.org. Accessed March 7, 2011.
32. American Institute for Cancer Research: Diet and Cancer Materials. Food, Nutrition, Physical Activity and the Prevention of Cancer. Available on-line at www.dietandcancerreport.org. Accessed March 7, 2011.
33. Renehan AG: Interpreting the epidemiological evidence linking obesity and cancer. *Eur J Cancer* 46(145):2581–2592, 2010.
34. Chen YC, Hunter D: Molecular epidemiology of cancer. *CA Cancer J Clin* 55:45–54, 2005.
35. National Cancer Institute: *Fact sheet: human papillomaviruses and cancer: questions and answers.* Available on-line at http://www.cancer.gov/cancertopics/factsheet/Risk/HPV. Accessed October 11, 2010.
36. CDC: *Sexually transmitted diseases. Genital HPV infection—CDC fact sheet, 2010.* Available online at http://www.cdc.gov/std/hpv/stdfact-hpv.htm. Accessed October 11, 2010.
37. Fleming DT: Herpes simplex virus type 2 in the United States, 1976 to 1994. *N Engl J Med* 337:1105–1160, 1997.
38. Centers for Disease Control and Prevention, National Center for Health Statistics: *National Vital Statistics Report: Sexually transmitted infections*, Hyattsville, MD, 2009, CDC.
39. Centers for Disease Control and Prevention (CDC): *Tracking the hidden epidemics 2000: Trends in STDs in the United States,* Hyattsville, MD, 2000, CDC. Available at http://www.cdc.gov/. Accessed Oct. 10, 2010.
40. Armbruster C, Dekan G, Hovorka A: Granulomatous pneumonitis and mediastinal lymphadenopathy due to photocopier toner dust. *Lancet* 348:1518–1519, 1996.
41. Gallardo M, Romero P, Sanchez-Quevedo MC, et al: Siderosilicosis due to photocopier toner dust. *Lancet* 412–413, 1994.
42. US Justice Department: Radiation Exposure Compensation Program. Available at www.usdoj.gov/civil/torts/const/reca/. Accessed Oct. 10, 2010.
43. National Cancer Institute (NCI): *About radiation fallout. report on exposure to iodine 131 from atomic bombs detonated above ground at the nevada test site: fact sheets, a dose calculator, and state and county exposures.* Available at rex.nci.nih.gov/INTRFCE_GIFS/radiation_fallout/radiation_131.html. Accessed Oct. 10, 2010.
44. Centers for Disease Control and Prevention (CDC): *A Feasibility study of the health consequences to the American population from nuclear weapons tests conducted by the United States and other nations.* Available at www.cdc.gov/nceh/radiation/fallout/default.htm. Accessed Oct. 11, 2010.
45. United States Department of Energy: *DOE Nevada: Detailed, historical reports on nuclear testing at the Nevada test site.* Available at www.nv.doe.gov/news&pubs/publications/historyreports/default.htm. Accessed Oct. 10, 2010.
46. National Safety Council (NSC): *Understanding radiation. Sources of nonionizing radiation,* Posted Dec. 2, 2005. Available at http://www.nsc.org/. Accessed Oct 10, 2010.
47. El Ghissassi F, Baan R, Straif K, et al: A review of human carcinogens. Part D: radiation. *Lancet Oncol* 10(8):751–752, 2009.
48. Lazovich D: Indoor tanning and risk of melanoma: a case-control study in a highly exposed population. *Cancer Epidemiol Biomarkers Prev* 19(6):1557–1568, 2010.
49. Woo DK: Tanning beds, skin cancer, and vitamin D: An examination of the scientific evidence and public health implications. *Dermatol Ther* 23(1):61–71, 2010.
50. Frumkin H: Agent orange and cancer: an overview for clinicians. *CA Cancer J Clin* 53:245–255, 2003.
51. Air Force Research Laboratory: *Air Force health study.* Available at http://www.wpafb.af.mil/AFRL/. Accessed Oct. 10, 2010.
52. Institute of Medicine (IOM): *Reports: veterans and agent orange: update 2008.* Available at http://www.iom.edu. Accessed Oct 30, 2010.
53. Marshall L: Identifying and managing adverse environmental health effects: taking an exposure history. *Can Med Assoc J* 166:1049–1054, 2002. Available at www.cmaj.ca/cgi/reprint/166/8/1049.pdf. Accessed Oct. 10, 2010.
54. Omenn GS: Genomics and prevention: a vision for the future. *Medscape Public Health & Prevention* 3(1). Available at http://www.medscape.com/viewarticle/501299. Accessed Oct. 10, 2010.
55. Lindsey H: Environmental factors & cancer: research roundup. *Oncology Times* 27(4):8–10, 2005.
56. Joerger AC, Fersht AR: The tumor suppressor p53: from structures to drug discovery. *Cold Spring Harb Perspect Biol* 2(6):a000919, 2010.
57. Maslon MM, Hupp TR: Drug discovery and mutant p53. *Trends Cell Biol* 20(9):542–555, 2010.
58. Couzin J: Choices—and uncertainties—for women with BRCA mutations. *Science* 302:592, 2003.
59. Rose PS, Buchowski JM: Metastatic disease in the thoracic and lumbar spine: evaluation and management. *J Am Acad Orthop Surg* 19(1):37–48, 2011.
60. Kitagawa Y, Fujii H, Mukai M, et al: Intraoperative lymphatic mapping and sentinel lymph node sampling in esophageal and gastric cancer. *Surg Oncol Clin N Am* 11:293–304, 2002.
61. McGarvey CL: *Principles of oncology for the physical therapist,* Long Island, NY, 2003, Stony Brook University.
62. Fagan A: Bone metastases in breast cancer. *Rehabil Oncol* 22:23–26, 2004.
63. Lipton A: Bone continuum of cancer. *Am J Clin Oncol* 33(3):Suppl. S1–S7, 2010.
64. Eatock AM, Scatzlein A, Kayes L: Tumour vasculature as a target for anticancer therapy. *Cancer Treat Rev* 26:191–204, 2000.
65. Diaz NM, Vrcel V, Centeno BA, et al: Modes of benign mechanical transport of breast epithelial cells to axillary lymph nodes. *Adv Anat Pathol* 12:7–9, 2005.
66. Mankin HJ, Mankin CJ, Simon MA: The hazards of the biopsy revisited. *J Bone Joint Surg* 78A:656–663, 1996.
67. Springfield DS, Rosenberg A: Biopsy: complicated, risky (editorial). *J Bone Joint Surg* 78A:639–643, 1996.
68. Abdu WA, Provencher M: Primary bone and metastatic tumors of the cervical spine. *Spine* 23:2767–2777, 1998.
69. Austin JP: Probable causes of recurrence in patients with chordoma and chondrosarcoma of the base of skull and cervical spine. *Int J Radiat Oncol Biol Phys* 25:439–444, 1993.
70. Fischbein NJ: Recurrence of clival chordoma along the surgical pathway. *Am J Neuroradiol* 21:578–583, 2000.
71. Bergh P: Prognostic factors in chordoma of the sacrum and mobile spine: a study of 39 patients. *Cancer* 88:2122–2134, 2000.
72. Slipman CW: Epidemiology of spine tumors presenting to musculoskeletal physiatrists. *Arch Phys Med Rehab* 84:492–495, 2003.
73. American Academy of Dermatology (ADD): *Melanoma fact sheet 2010.* Available at www.aad.org. Accessed Oct. 11, 2010.
74. American Cancer Society: *Skin cancer 2010.* Available at www.cancer.org. Accessed Oct. 30, 2010.
75. Chen K, Craig JC, Shumack S: Oral retinoids for the prevention of skin cancers in solid organ transplant recipients: a systematic review of randomized controlled trials. *Br J Dermatol* 152:518–523, 2005.
76. Ross MI: Aid for patients with limb metastases. *World Melanoma Update* 1:15, 1997.
77. Rigel DS: The evolution of melanoma diagnosis: 25 years beyond the ABCDs. *CA Cancer J Clin* 60(5):301–316, 2010.

78. Gudas S: The physical therapy challenge in disseminated cancer. *Oncol Sect News APTA* 5:3, 1987.
79. Thomas SS, Dunbar EM: Modern multidisciplinary management of brain metastases. *Curr Oncol Rep* 12(1):34–40, 2010.
80. Jaeckle KA: Neurological manifestations of neoplastic and radiation-induced plexopathies. *Semin Neurol* 24:385–393, 2004.
81. Jaeckle KA: Neurologic manifestations of neoplastic and radiation-induced plexopathies. *Semin Neurol* 30(3):254–262, 2010.
82. Prasad D, Schiff D: Malignant spinal cord compression. *Lancet Oncol* 6:15–25, 2005.
83. Pigott KH, Baddeley H, Maher EJ: Pattern of disease in spinal cord compression on MRI scan and implications for treatment. *Clin Oncol (R Coll Radiol)* 6:7–10, 1994.
84. Morris GS: Oncologic emergencies. *Acute Care Perspectives* 16(4):1, 3–7, 2007.
85. Coleman RE: Management of bone metastases. *Oncologist* 5:463–470, 2000.
86. Schiff D: Peer viewpoint. *J Support Oncol* 2:398, 401, 2004.
87. Abrahm JL: Assessment and treatment of patients with malignant spinal cord compression. *J Support Oncol* 2:377–401, 2004.
88. Levack P, Graham J, Collie D, et al: Scottish Cord Compression Study Group. Don't wait for a sensory level—listen to the symptoms: a prospective audit of the delays in diagnosis of malignant cord compression. *Clin Oncol (R Coll Radiol)* 14:472–480, 2002.
89. Schiff D: Spinal cord compression. *Neurol Clin* 21:67–86, 2003.
90. Tatu B: Physical therapy intervention with oncological emergencies. *Rehabil Oncol* 23:4–6, 2005.
91. Ampil FL, Mills GM, Burton GV: A retrospective study of metastatic lung cancer compression of the cauda equina. *Chest* 120:1754–1755, 2001.
92. Orendacova J, Cizkova D, Kafka J, et al: Cauda equina syndrome. *Prog Neurobiol* 64:613–637, 2001.
93. Bagley CA, Gokaslan ZL: Cauda equina syndrome caused by primary and metastatic neoplasms. *Neurosurg Focus* 16(6), 2004. Available online at http://www.medscape.com/viewarticle/482042_print. Accessed Oct. 11, 2010.
94. Small SA, Perron AD, Brady WJ: Orthopedic pitfalls: cauda equina syndrome. *Am J Emerg Med* 23:159–163, 2005.
95. Uchiyama T, Sakakibara R, Hattori T, et al: Lower urinary tract dysfunctions in patients with spinal cord tumors. *Neurourol Urodyn* 23:68–75, 2002.
96. McCarthy MJH: Cauda equina syndrome. Factors affecting long-term functional and sphincteric outcome. *Spine* 32(2): 207–216, 2007.
97. Hile ES: Persistent mobility disability after neurotoxic chemotherapy. *Phys Ther* 90(11):1649–1657, 2010.
98. Bataller A, Dalman J: Paraneoplastic disorders of the central nervous system: update on diagnosis and treatment. *Semin Neurol* 24:461–471, 2004.
99. Santacroce L: *Paraneoplastic syndromes*. eMedicine Specialties, 2010. Available online at www.emedicine.com/med/TOPIC1747.HTM. Accessed March 7, 2011.
100. Velez A, Howard MS: Diagnosis and treatment of cutaneous paraneoplastic disorders. *Dermatol Ther* 23(6):662–675, 2010.
101. Briani C: Spectrum of paraneoplastic disease associated with lymphoma. *Neurology* 76(8):705–710, 2011.
102. Maverakis E: The etiology of paraneoplastic autoimmunity. *Clin Rev Allergy Immunol* Jan 19, 2011. Epub ahead of print.
103. Farrugia ME: Myasthenic syndromes. *J R Coll Physicians Edinb* 41(1):43–48, 2011.
104. Naschitz JE: Musculoskeletal syndromes associated with malignancy. *Curr Opin Rheumatol* 20(1):100–105, 2008.
105. Stummvoll GH, Aringer M, Machold KP, et al: Cancer polyarthritis resembling rheumatoid arthritis as a first sign of hidden neoplasms. *Scand J Rheumatol* 30:40–44, 2001.
106. Martorell EA, Murray PM, Peterson JJ, et al: Palmar fasciitis and arthritis syndrome associated with metastatic ovarian carcinoma: a report of four cases. *J Hand Surg* 29A:654–660, 2004.
107. Stummvoll GH, Graninger WB: Paraneoplastic rheumatism—musculoskeletal diseases as a first sign of hidden neoplasms. *Acta Med Austriaca* 29:36–40, 2002.
108. Klippel JH: *Primer on the rheumatic diseases*, ed 13, Arthritis Foundation, New York, 2008, Springer Science Media.
109. Joines JD, McNutt RA, Carey TS, et al: Finding cancer in primary care outpatients with low back pain: a comparison of diagnostic strategies. *J Gen Intern Med* 16:14–29, 2001.
110. Deyo RA, Diehl AK: Cancer as a cause of back pain: frequency, clinical presentation, and diagnostic strategies. *J Gen Intern Med* 3:230–238, 1988.
111. Jarvik J, Deyo R: Diagnostic evaluation of low back pain with emphasis on imaging. *Ann Intern Med* 137:586–597, 2002.
112. Ross MD, Boissonnault WG: Red flags: to screen or not to screen? *J Orthop Sports Phys Ther* 40(11):682–684, 2010.
113. Deftos LJ: Hypercalcemia in malignant and inflammatory diseases. *Endocrinol Metab Clin North Am* 31:141–158, 2002.
114. Nirenberg A: Managing hematologic toxicities. *Cancer Nursing* 26:32S–37S, 2003.
115. Barsevick A: Energy conservation and cancer-related fatigue. *Rehabil Oncol* 20:14–17, 2002.
116. National Comprehensive Cancer Network (NCCN): *Causes of cancer related fatigue, 2010.* Available at www.nccn.org. Accessed on Oct. 11, 2010.
117. Litterini AJ, Fieler VK: The change in fatigue, strength, and quality of life following a physical therapist prescribed exercise program for cancer survivors. *Rehabil Oncol* 26(3):11–17, 2008.
118. Schmitz KH, Courneya KS: American College of Sports Medicine roundtable on exercise guidelines for cancer survivors. *Med Sci Sports Exerc* 42(7):1409–1426, 2010.
119. Blaney J: The cancer rehabilitation journey: barriers to and facilitators of exercise among patients with cancer-related fatigue. *Phys Ther* 90(8):1135–1147, 2010.
120. Van Weert E: Cancer-related fatigue and rehabilitation. *Phys Ther* 90(10):1413–1425, 2010.
121. Volk K, Wruble E: Irradiation side effects and their impact on physical therapy. *Acute Care Perspect* 10:11–13, 2001.
122. Winningham ML, McVicar M, Burke C: Exercise for cancer patients: guidelines and precautions. *Phys Sportsmed* 14:121–134, 1986.
123. Goodman CC, Fuller K: *Pathology: Implications for the physical therapist*, ed 3, Philadelphia, 2009, WB Saunders.
124. Jacobs AL: Adult cancer survivorship: evolution, research, and planning care. *CA Cancer J Clin* 59:391–410, 2009.
125. Miller KD, Triano LR: Medical issues in cancer survivors—a review. *Cancer J* 14(6):375–387, 2008.
126. Hile ES: Persistent mobility disability after neurotoxic chemotherapy. *Phys Ther* 90(11):1649–1657, 2010.
127. Asher A: Cancer rehabilitation and survivorship: Cedars-Sinai Medical Center experience. *Oncol Nurse* 3(6):1, 18–21, 2010.
128. Shamley DR: Changes in shoulder muscle size and activity following treatment for breast cancer. *Breast Cancer Res Treat* 106(1):19–27, 2007.
129. Ahles TA: Candidate mechanisms for chemotherapy-induced cognitive changes. *Nat Rev Cancer* 7:192–201, 2007.
130. Ferguson RJ: Management of chemotherapy-related cognitive dysfunction. In Feuerstein M, editor: *Handbook of cancer survivorship*, New York, 2006, Springer.
131. Siegel HJ, Pressey JG: Current perspectives on the surgical and medical management of soft tissue sarcomas, *Medscape 2008:*

Available at http://www.medscape.com/viewarticle/579883. Accessed Oct. 13, 2010.
132. Borman H, Safak T, Ertoy D: Fibrosarcoma following radiotherapy for breast cancer: a case report and review of the literature. *Ann Plast Surg* 41:201–204, 1998.
133. National Cancer Institute (NCI): Treatment statement for health professionals: adult soft tissue sarcoma, *Med News* 2005. Available at www.meb.unibonn.de/cancer.gov/CDR0000062820.html. Accessed Oct. 13, 2010.
134. Roll L: Cancer in children and adolescents. In Varricchio C, editor: *ACS: A cancer source book for nurses*, ed 8, Boston, 2004, Jones and Bartlett, pp 229–242.
135. American Cancer Society (ACS): *Cancer reference information. overview: osteosarcoma.* Available at http://www.cancer.org/docroot/CRI/CRI_2_1x.asp?dt=52. Accessed Oct. 13, 2010.
136. Maheshwari AV, Cheng EY: Ewing sarcoma family of tumors. *J Am Acad Orthop Surg* 18(2):94–107, 2010.
137. American Cancer Society (ACS): *Cancer Reference information. Detailed guide: Ewing family of tumors.* Available at http://www.cancer.org/Cancer/EwingFamilyofTumors/DetailedGuide/index. Accessed Oct. 12, 2010.
138. Grubb MR, Currier BL, Pritchard DJ, et al: Primary Ewing sarcoma of the spine. *Spine* 19:309–313, 1994.
139. Lin PP: Secondary chondrosarcoma. *J Am Acad Orthop Surg* 18(10):608–615, 2010.
140. Resnick D: *Diagnosis of bone and joint disorders*, ed 4, Philadelphia, 2002, WB Saunders.
141. Payne WT, Merrell G: Benign bone and soft tissue tumors of the hand. *J Hand Surg* 35A(11):1901–1910, 2010.
142. Better outlook for people with brain tumors. *Johns Hopkins Med Lett* 14:3, 2002.
143. Kasper D, longo D, Fauci A, et al: *Harrison's principles of internal medicine*, ed 18, New York, 2011, McGraw-Hill.
144. Volkow ND: Effects of cell phone radiofrequency signal exposure on brain glucose metabolism. *JAMA* 305(8):808–813, 2011.
145. Carney CP, Woolson RF, Jones L, et al: Occurrence of cancer among people with mental health claims in an insured population. *Psychosom Med* 66:735–743, 2004.
146. Hong WK: *Holland-Frei cancer medicine*, Shelton, CT, 2010, Peoples Medical Publishing House, 2010.
147. Abu-Shakra M, Buskila D, Ehrenfeld M, et al: Cancer and autoimmunity: autoimmune and rheumatic features in patients with malignancies. *Ann Rheum Dis* 60:433–441, 2001.
148. Zaidi A, Vesole H: Multiple myeloma: an old disease with new hope for the future. *CA Cancer J Clin* 51:273–285, 2001.
149. Paulson B, Gudas S: Multiple myeloma. *Rehabil Oncol* 21:8–10, 2003.
150. Volker D: Other cancers: multiple myeloma. In Varricchio C, editor: *ACS: Cancer source book for nurses*, ed 8, Boston, 2004, Jones and Bartlett, pp 324–336.
151. Jones A: Lymphomas. In Varricchio C, editor: *ACS: A cancer source book for nurses*, ed 8, Boston, 2004, Jones and Bartlett, pp 265–276.
152. Lim ST, Levine AM: Recent advances in acquired immunodeficiency syndrome (AIDS)-related lymphoma. *CA Cancer J Clin* 55:229–241, 2005.
153. Hardell L, Axelson O: Environmental and occupational aspects on the etiology of non-Hodgkin lymphoma. *Oncol Res* 10:1–5, 1998.
154. Gaidano G, Carbone A, Dalla-Favera R: Genetic basis of acquired immunodeficiency syndrome–related lymphomagenesis. *J Natl Cancer Inst Monogr* 23:95–100, 1998.
155. Aboulafia AJ, Khan F, Pankowsky D, et al: AIDS-associated secondary lymphoma of bone: a case report with review of the literature. *Am J Orthop* 27:128–134, 1998.
156. Gapstur SM, Morrow M, Sellers TA: Hormone replacement therapy and risk of breast cancer with a favorable histology: results of the Iowa Women's Health Study. *JAMA* 281:2091–2097, 1999.
157. Aubuchon M, Santoro N: Lessons learned from the WHI: HRT requires a cautious and individualized approach. *Geriatrics* 59:22–26, 2004.
158. Rossouw JE, Anderson GL, Prentice RL, et al: Risks and benefits of estrogen plus progestin in healthy postmenopausal women: principal results from the Women's Health Initiative randomized controlled trial. *JAMA* 288:321–333, 2002.
159. Holden AF: Sweet's syndrome in association with generalized granuloma annulare in a patient with previous breast carcinoma. *Clin Exp Dermatol* 26:668–670, 2001.
160. Bourke JF, Keohane S, Long CC, et al: Sweet's syndrome and malignancy in the UK. *Br J Dermatol* 137:609–613, 1997.
161. Cohen PR, Kurzrock R: Sweet's syndrome revisited: a review of disease concepts. *Int J Dermatol* 42:761–778, 2003.

SECTION III: Systemic Origins of Neuromuscular or Musculoskeletal Pain and Dysfunction

The potential for referral of pain from systemic diseases to specific muscles and joints is well documented in the medical literature. These referral patterns most often affect the back and shoulder but may also appear in the chest, thorax, hip, pelvis, groin, sacrum, or sacroiliac joint.

Up to this point the text has focused on each organ system and the pain or other signs and symptoms referred from organs to musculoskeletal sites. In this third section the focus is turned around so that the reader can quickly refer to the site of presenting pain or other symptoms and determine possible systemic involvement.

The therapist may then question the client, as suggested, and determine the possible need for referral to a physician or other appropriate resource. The reader is referred to the individual chapters within this text for an in-depth discussion of the specific visceral, medical, or systemic causes of musculoskeletal signs or symptoms.

DECISION-MAKING PROCESS

In Chapter 1, a model for decision making in the screening process was presented, including:

- Client history (client demographics, past medical history, personal and family history, psychosocial history)
- Risk-factor assessment
- Clinical presentation, including assessment of pain patterns and pain types
- Associated signs and symptoms of systemic diseases
- Review of Systems

The therapist uses this screening model during the screening interview to gather important information and then correlate the subjective findings with the objective findings to recognize presenting conditions that require medical follow-up.

Accordingly, the therapist will want to obtain the client's history, assess types of pain, pain patterns, and signs and symptoms that may suggest systemic origins of problems appearing in the musculoskeletal or neuromuscular system. Taking a step back and looking at the entire case presentation, called the Review of Systems (see Box 4-19), is often the final step in the screening process.

These guidelines for collecting and correlating subjective and objective information are suggested for any client who demonstrates one or more of the characteristics outlined in Chapter 1. In addition, with a specific focus on local regions, diagnostic imaging may be needed or perhaps already available for review. Hazle* offers some screening guidance regarding imaging with the following suggestions:

Consider whether the origin of pathology has been imaged. Keep in mind the symptomatic area may not be the origin of symptoms but rather a site of referred pain.

Check the side of involvement against the side imaged (it is possible that the uninvolved side was imaged by mistake).

Evaluate the type of imaging modality used to identify or rule out suspected pathology.

Ask yourself these questions: was the imaging used to detect a suspected problem sensitive/specific enough? Would some other type of imaging yield more reliable and accurate information?

Finally, when looking at images, consider whether the results observed are anatomic variations, developmental disorders, or typical of degenerative processes. Many people have significant anatomic deviations from the norm without symptoms. The corollary is also true: individuals with severe painful symptoms can have no visible pathologic explanation—at least as far as imaging can visualize the bony and soft tissue structures.

*Hazle CR: *Diagnostic imaging of the spine: Medical screening and integration into clinical decision-making. PT online PT10,* Alexandria, VA, 2010, American Physical Therapy Association.

CHAPTER

14

Screening the Head, Neck, and Back

It is estimated that 80% to 90% of the western population will experience an episode of acute back pain at least once during their lifetime,[1] making it one of the most common problems physical therapists evaluate and treat.[2-4]

It has been suggested that mechanical low back pain (LBP) and leg pain with spinal causes compose approximately 97% of all cases.[5] Nonmechanical spinal disease can be attributed to neoplasm, infection, or inflammation in 1% of all cases with another 2% accounted for by visceral disorders (pelvic organs, gastrointestinal [GI] dysfunction, renal involvement, abdominal aneurysms).[6]

Most cases of back pain in adults are associated with age-related degenerative processes, physical loading, and musculoligamentous injuries. Many mechanical causes of back pain resolve within 1 to 4 weeks without serious problems. It has been estimated that fewer than 2% of individuals presenting with LBP present with significant neurologic involvement or other signs that require referral or imaging.[7] Up to 10% of LBP patients have no identifiable cause.[8]

Sacroiliac (SI) joint dysfunction can mimic LBP and disk-ogenic disease with pain referred below the knee to the foot. Studies show SI joint dysfunction is the primary source of LBP in 18% to 30% of people with LBP.[8-13] As always, when conducting a physical examination the therapist must consider the possibility of a mechanical problem above or below the area of pain or symptom presentation.

A smaller number of people will develop chronic pain without organic pathology or they may have an underlying serious medical condition. The therapist must be aware that many different diseases can appear as neck pain, back pain, or both at the same time (Table 14-1). For example, rheumatoid arthritis affects the cervical spine early in the course of the disease but may go unrecognized at first.[14-16] Neck pain may be a feature of any disorder or disease that occurs above the shoulder blades; it is a rare symptom of neoplasm or infection.[17]

In this chapter, general information is offered about back pain with a focus on clinical presentation, while keeping in mind risk factors and associated signs and symptoms typical of each visceral system capable of referring pain to the head, neck, and back. Neck and back pain may arise in the spine from infection, fracture, or inflammatory, metabolic, or neoplastic disorders.

Additionally LBP can be referred from abdominal or pelvic disease. Nonsteroidal antiinflammatory drug (NSAID) use is a typical cause of intraperitoneal or retroperitoneal bleeding causing LBP. People most often taking NSAIDs have a history of inflammatory conditions such as osteoarthritis.

Although the incidence of back pain from NSAIDs is fairly low (i.e., number of people on NSAIDs who develop GI problems and referred pain), the prevalence (number seen in a physical therapist's practice) is much higher.[18-20] In other words physical therapists are seeing a majority of people with arthritis or other inflammatory conditions who are taking one or more prescription and/or over-the-counter (OTC) NSAID.[21]

Screening for medical disease is an important part of the evaluation process that may take place more than once during an episode of care (see Fig. 1-4). The clues about the quality of pain, the age of the client, and the presence of systemic complaints or associated signs and symptoms indicate the need to investigate further.

USING THE SCREENING MODEL TO EVALUATE THE HEAD, NECK, OR BACK

Past Medical History

A carefully taken, detailed medical history is the most important single element in the evaluation of a client who has musculoskeletal pain of unknown origin or cause. It is essential for the recognition of systemic disease or medical conditions that may be causing integumentary, muscle, nerve, or joint symptoms.

The history combined with the physical therapy examination provides essential clues in determining the need for referral to a physician or other appropriate health care provider. A history of cancer is most important, however long ago. If a client has had a low backache for years, progressive serious disease is unlikely, though the therapist should not be misled by a chronic history of back pain because the client may be presenting with a new episode of serious back pain. Six weeks to 6 months of increasing backache, often in an older client, may be a signal of lumbar metastases, especially in a person with a past history of cancer.

TABLE 14-1 Viscerogenic Causes of Neck and Back Pain

	Cervical	Thoracic/Scapular	Lumbar/Sacrum*
Cancer	Metastatic lesions (leukemia, Hodgkin's disease) Cervical bone tumors Cervical cord tumors Lung cancer; Pancoast's tumor Esophageal cancer Thyroid cancer	Mediastinal tumors Metastatic extension Pancreatic cancer Breast cancer Multiple myeloma	Primary bone tumors Neurogenic tumors (sacrum) Metastatic lesions Prostate cancer Testicular cancer Pancreatic cancer Colorectal cancer Multiple myeloma Lymphoma
Cardiovascular	Angina Myocardial infarction Aortic aneurysm Occipital migraine Cervical artery ischemia or dissection Arteritis	Angina Myocardial infarction Aortic aneurysm	Abdominal aortic aneurysm Endocarditis Myocarditis Peripheral vascular: • Postoperative bleeding from anterior spine surgery
Pulmonary	Lung cancer; Pancoast's tumor Tracheobronchial irritation Chronic bronchitis Pneumothorax Pleuritis involving the diaphragm	Respiratory or lung infection Empyema Chronic bronchitis Pleurisy Pneumothorax Pneumonia	
Renal/urologic		Acute pyelonephritis Kidney disease	Kidney disorders: • Acute pyelonephritis • Perinephritic abscess • Nephrolithiasis • Ureteral colic (kidney stones) • Urinary tract infection • Dialysis (first-use syndrome) • Renal tumors
Gastrointestinal	Esophagitis Esophageal cancer	Esophagitis (severe) Esophageal spasm Peptic ulcer Acute cholecystitis Biliary colic Pancreatic disease	Small intestine: • Obstruction (neoplasm) • Irritable bowel syndrome • Crohn's disease Colon: • Diverticular disease Pancreatic disease Appendicitis
Gynecologic			Gynecologic disorders: • Cancer • Retroversion of the uterus • Uterine fibroids • Ovarian cysts • Endometriosis • Pelvic inflammatory disease (PID) • Incest/sexual assault • Rectocele, cystocele • Uterine prolapse Normal pregnancy Multiparity
Infection	Vertebral osteomyelitis Meningitis Lyme disease Retropharyngeal abscess; epidural abscess (post-steroid injection)	Vertebral osteomyelitis Herpes zoster Human immunodeficiency virus (HIV) Epidural abscess	Vertebral osteomyelitis Herpes zoster Spinal tuberculosis Candidiasis (yeast) Psoas abscess HIV
Other	Osteoporosis Fibromyalgia Psychogenic (nonorganic causes; see Chapter 3) Fracture Rheumatoid: • Rheumatoid arthritis and atlantoaxial subluxation • Psoriatic arthritis • Polymyalgia rheumatica • Ankylosing spondylitis Viral myalgias Cervical lymphadenitis Thyroid disease	Osteoporosis Fibromyalgia Psychogenic (nonorganic) Acromegaly Cushing's syndrome Fracture	Osteoporosis Fibromyalgia Psychogenic (nonorganic) Fracture Cushing's syndrome Type III hypersensitivity disorder (back/flank pain) Postregional anesthesia Ankylosing spondylitis

*Sacral sources of low back pain (LBP) are discussed separately in Chapter 15.

Watch for history of diabetes, immunosuppression, rheumatologic disorders, tuberculosis, and any recent infection (Case Example 14-1). A history of fever and chills with or without previous infection anywhere in the body may indicate a low-grade infection.

Symptoms are likely to appear some time before striking physical signs of disease are evident and before laboratory tests are useful in detecting disordered physiology. Thus an accurate and sufficiently detailed history provides historical clues that can be significant in determining when the client should be referred to a physician or other appropriate health care provider.

The therapist must always ask about a history of motor vehicle accident, blunt impact, repetitive injury, sudden stress caused by lifting or pulling, or trauma of any kind. Even minor falls or lifting when osteoporosis is present can result in severe fracture in older adults (Case Example 14-2). Anyone who cannot bear weight through the legs and hips should be considered for an immediate medical evaluation.[21a]

Surgery of any kind can result in infection and abscess leading to hip, pelvic, abdominal, and/or LBP.[22] A recent history of spinal procedures (e.g., fusion, diskectomy, kyphoplasty, vertebroplasty) can be followed by back pain, motor impairment, and/or neurologic deficits when complicated by hematoma, infection, bone cement leakage, or subsidence (graft or instrumentation sinking into the bone).[23] Infection following spinal epidural injection is an infrequent but potentially serious complication.[24,25]

CASE EXAMPLE 14-1

Bilateral Facial Pain

Background: A 79-year-old woman was in a rehabilitation facility following a stroke with resultant left hemiplegia. She told the therapist she was starting to have some new symptoms in her face. She could not smile on her "good" side and was having trouble closing her eyes, which was not a problem after her stroke.

Clinical Presentation: There were no apparent changes in hearing, sensation, or motor control of the right arm. The therapist conducted a new neurologic screening examination and found the following results:

Cranial Nerve VII: Client was unable to raise and lower either eyebrow or close the eyes tightly; there was bilateral facial drooping; as reported, the client was unable to smile with the right side of her face.

There was no change in sensory or motor findings from the initial evaluation post-cerebral vascular accident (post-CVA). However, deep tendon reflexes were absent in both arms, leading the therapist to check deep tendon reflexes in the lower extremities, which were also absent. There were no other significant neurologic changes from the initial evaluation.

The therapist reviewed the Special Questions to Ask: Neck or Back (Pain Assessment and General Systemic) to look for any other screening questions and asked about a recent history of infection. The client reported a mild upper respiratory infection 2 weeks ago. There were no other obvious red flag findings.

Result: The therapist reported the new episode of signs and symptoms. Red flags observed included bilateral symptoms, absent muscle stretch reflexes, and recent history of infection. A medical evaluation was carried out, and a diagnosis of Guillain-Barré was made. The client continued to get worse with involvement of the respiratory muscles, foot drop, and numbness in the hands and feet.

A new episode of care was initiated to include physical therapy to strengthen facial musculature and prevent atrophy on the right side and to prevent pneumonia from respiratory muscle involvement.

CASE EXAMPLE 14-2

Minimal Trauma

Background: An inpatient acute care therapist was working with a 75-year-old woman who was 1-day status post (S/P) right total hip replacement (THR). The patient reported getting out of bed by herself early in the morning and falling against the night stand. She complained of low back pain (LBP) when the therapist arrived to help her sit up in bed and stand. The pain was in the left lumbar area without radiation.

Past Medical History: Past medical history included osteoporosis (treated with bisphosphonate medication, calcium, and vitamin D), breast cancer with mastectomy 30 years ago, and hypothyroidism treated with medication (Synthroid).

Clinical Presentation: No preoperative baseline information was available regarding the client's physical function, gait pattern, or range of motion (ROM) for the spine or hips. There was moderate tenderness to palpation and percussion of the sacrum on the left side. Mild tenderness was reported with percussion to the upper and lower lumbar spine. There were no apparent skin changes, bruising, warmth, or swelling.

The patient could ambulate slowly with a walker but reported pain in both hips with each step. She could only take small steps, moving approximately 2 to 4 inches forward with each step. Lumbar ROM was very limited in flexion, side bending, and extension. She was unable to straighten up to a fully upright standing position due to her low back/sacral pain.

Outcome: The therapist filed an incident report with the hospital unit clerk and spoke directly with the nursing supervisor requesting an ortho consult before continuing with the standard THR rehabilitation protocol.

The patient was diagnosed with a sacral insufficiency fracture on the left at S3. X-rays and magnetic resonance imaging (MRI) also revealed scoliosis of the lumbosacral spine, moderate degenerative arthritis, marked narrowing of the intervertebral disk spaces throughout the lumbar spine, and old compression fractures at T11 and T12. There was no evidence of bone lesions suggestive of breast cancer metastasis. Moderate foraminal stenosis was observed at the right L3 nerve root.

The client returned to physical therapy with an altered rehabilitation program consisting of weightbearing exercises on the left (to stimulate osteoblastic bone formation) as tolerated given the compromise on both sides. She had a minimally invasive hip procedure, so aquatic therapy was approved when there were no openings in the skin at the incision site (1 week later).

A few key questions to ask about the history might include:

FOLLOW-UP QUESTIONS

- What do you think caused this pain?
- When did the pain (numbness, weakness, stiffness) start?
- Have you ever had this type of problem before?
- Have you ever had back surgery, seen a chiropractor (physical therapist or other health care professional), or had injection therapy for this problem?

Risk Factor Assessment

Understanding who is at risk and what the risk factors are for various illnesses, diseases, and conditions will alert the therapist early on as to the need for screening, education, and prevention as part of the plan of care. Educating clients about their risk factors is a key element in risk factor reduction.

Risk factors vary, depending on family history, previous personal history, and disease, illness, or condition present. For example, risk factors for heart disease will be different from risk factors for osteoporosis or vestibular/balance problems. When it comes to the musculoskeletal system, risk factors, such as heavy nicotine use, injection drug use, alcohol abuse, diabetes, history of cancer, or corticosteroid use, may be important.

Always check medications for potential adverse side effects causing muscular, joint, neck, or back pain. Long-term use of corticosteroids can lead to vertebral compression fractures (Case Example 14-3). Fluoroquinolones (antibiotic) can cause neck, chest, or back pain. Headache is a common side effect of many medications.

Keep in mind that physical and sexual abuse are risk factors for chronic head, neck, and back pain for men, women, and children (see Appendix B-3).

Age is a risk factor for many systemic, medical, and viscerogenic problems. The risk of certain diseases associated with back pain increases with advancing age (e.g., osteoporosis, aneurysm, myocardial infarction, cancer). Under the age of 20 or over the age of 50 are both red flag ages for serious spinal pathology. The highest likelihood of vertebral fracture occurs in females aged 75 years or older.[26]

As with all decision-making variables, a single risk factor may or may not be significant and must be viewed in context of the whole patient/client presentation. See Appendix A-2 for a list of some possible health risk factors.

Routine screening for osteoporosis, hypertension, incontinence, cancer, vestibular or balance problems, and other potential problems can be a part of the physical therapist's practice. Therapists can advocate disease prevention, wellness, and promotion of healthy lifestyles by delivering health care services intended to prevent health problems or maintain health and by offering wellness screening as part of primary prevention.

CASE EXAMPLE 14-3

Corticosteroid Use

Referral: A 73-year-old man was referred to a physical therapist by his family practitioner for evaluation of middle-to-low back pain that started when he stepped down from a curb. He was not experiencing radiating pain or sciatica and appeared to be in good general health. His medical history included bronchial asthma treated with oral corticosteroids and an abdominal hernia repaired surgically 10 years ago. There were no diagnostic imaging tests ordered.

Clinical Presentation: Vital signs were measured and appeared within normal limits for the client's age. There were no constitutional symptoms, no fever present, and no other associated signs or symptoms reported.

There was a marked decrease in thoracic and lumbar range of motion from T10 to L1 and tenderness throughout this same area. No other objective findings were noted despite a careful screening examination.

The client was treated conservatively over a 2-week period but without change in his painful symptoms and without improvement in spinal movement. A second therapist in the same clinic was consulted for a reevaluation without significant differences in findings. Several suggestions were made for alternative treatment techniques. After 1 more week without change in client symptoms, the client was reevaluated.

What is the next step in the screening process?

Using Table 14-1, the therapist can scan down the Thoracic/Scapular and Lumbar columns for any screening clues. Prostate and testicular cancers are listed along with metastatic lesions. Given the client's age, questions should be asked about a past history of cancer and any associated urinary signs and symptoms.

Given his age, cardiovascular causes of back pain are also possible. Review past medical history, risk factors, and ask about signs and symptoms associated with angina, myocardial infarction, and aneurysm.

The therapist can continue to review Table 14-1 for potential pulmonary and gastrointestinal causes of this client's back pain and ask any further questions regarding possible risk factors and past history. Record all positive findings and conduct a final Review of Systems.

Use the Special Questions to Ask: Neck or Back at the end of this chapter to reassess the client's general health and clinical presentation. Not all questions must be asked; the therapist will use his or her judgment based on known history for this client and current clinical findings.

Result: In this case the client's age, lack of improvement with a variety of treatment techniques, lack of diagnostic imaging studies to rule out fracture or infection, and history of long-term corticosteroid use necessitated a return to the referring physician for further medical evaluation.

Long-term corticosteroid therapy and radiation therapy for cancer are risk factors for ischemic or avascular necrosis. Hip or back pain in the presence of these factors should be examined carefully.

Radiographic testing demonstrated ischemic vertebral collapse secondary to chronic corticosteroid administration. Diffuse osteopenia and a compression fracture of the tenth thoracic vertebral body were also mentioned in the medical report.

Clinical Presentation

During the examination the therapist will begin to get an idea of the client's overall clinical presentation. The client interview, systems review of the cardiopulmonary, musculoskeletal, neuromuscular, and integumentary systems, and assessment of pain patterns and pain types form the basis for the therapist's evaluation and eventual diagnosis.

Assessment of pain and symptoms is often a large part of the interview. In this final section of the text, pain and dysfunction associated with each anatomic part (e.g., back, chest, shoulder, pelvis, sacrum/SI, hip, and groin) are discussed and differentiated as systemic from musculoskeletal whenever possible.

Characteristics of pain, such as onset, description, duration, pattern, and aggravating and relieving factors, and associated signs and symptoms are presented in Chapter 3 (see Table 3-2; see also Appendix C-7). Reviewing the comparison in Table 3-2 will assist the therapist in recognizing systemic versus musculoskeletal presentation of signs and symptoms.

Effect of Position

When seen early in the course of symptoms, neck or back pain of a systemic, medical, or viscerogenic origin is usually accompanied by full and painless range of motion (ROM) without limitations. When the pain has been present long enough to cause muscle guarding and splinting, then subsequent biomechanical changes occur.

Typically, systemic back pain or back pain associated with other medical conditions is not relieved by recumbency. In fact, the bone pain of metastasis or myeloma tends to be more continuous, progressive, and prominent when the client is recumbent.

Beware of the client with acute backache who is unable to lie still. Almost all clients with regional or nonspecific backache seek the most comfortable position (usually recumbency) and stay in that position. In contrast, individuals with systemic backache tend to keep moving trying to find a comfortable position.

In particular, visceral diseases, such as pancreatic neoplasm, pancreatitis, and posterior penetrating ulcers, often have a systemic backache that causes the client to curl up, sleep in a chair, or pace the floor at night.

Back pain that is unrelieved by rest or change in position or pain that does not fit the expected mechanical or neuromusculoskeletal pattern should raise a red flag. When the symptoms cannot be reproduced, aggravated, or altered in any way during the examination, additional questions to screen for medical disease are indicated.

Night Pain

Pain at night can signal a serious problem such as tumor, infection, or inflammation. Long-standing night pain unaltered by positional change suggests a space-occupying lesion such as a tumor.

Systemic back pain may get worse at night, especially when caused by vertebral osteomyelitis, septic diskitis, Cushing's disease, osteomalacia, primary and metastatic cancer, Paget's disease, ankylosing spondylitis, or tuberculosis of the spine (see Chapter 3 and Appendix B-25).

Associated Signs and Symptoms

After reviewing the client history and identifying pain types or pain patterns, the therapist must ask the client about the presence of additional signs and symptoms. Signs and symptoms associated with systemic disease or other medical conditions are often present but go unidentified, either because the client does not volunteer the information or the therapist does not ask. To assess for associated signs and symptoms, the therapist can end the client interview with the following question:

FOLLOW-UP QUESTION

- Are there any other symptoms anywhere else in your body that you haven't told me about or we haven't discussed? They do not have to be related to your back pain or symptoms.

The client with back pain and bloody diarrhea or the person with mid-thoracic or scapular pain in the presence of nausea and vomiting may not think the two symptoms are related. If the therapist only focuses on the chief complaint of back, neck, shoulder, or other musculoskeletal pain and does not ask about the presence of symptoms anywhere else, an important diagnostic clue may be overlooked.

Other possible associated symptoms may include fatigue, dyspnea, sweating after only minor exertion, and GI symptoms (see also Appendix A-2 for a more complete list of possible associated signs and symptoms).

If the therapist fails to ask about associated signs and symptoms, the Review of Systems offers one final step in the screening process that may bring to light important clues.

Review of Systems

Clusters of these associated signs and symptoms usually accompany the pathologic state of each organ system (see Box 4-19). As part of the physical assessment, the therapist must conduct a Review of Systems. General questions about fevers, excessive weight gain or loss, and appetite loss should be followed by questions related to specific organ systems. Medications should be reviewed for possible adverse side effects.

Throughout the interview the therapist must remain alert to any yellow (caution) or red (warning) flags that may signal the need for further screening. Review of Systems is important even for clients who have been examined by a medical doctor. It has been reported that only 5% of physicians assess patients for "red flags."[27,28] In contrast, documentation of red flags by physical therapists (at least for patients with LBP) has been reported as high as 98%.[29]

During the Review of Systems a pattern of systemic, medical, or viscerogenic origin may be seen as the therapist combines information from the client history, risk factors present, associated signs and symptoms, and yellow or red flag findings.

Yellow Flag Findings[30]

Yellow flags are indicators that findings may be present requiring special attention but not necessarily immediate action. One of the primary yellow flag findings that is prognostically important in individuals with LBP is the presence of psychosocial risk factors (e.g., work, attitudes and beliefs, behaviors, affective presentation).[31-34] The presence of these yellow flags suggests a poor response to traditional intervention and the need to address the underlying psychosocial aspects of health and healing. A management approach using cognitive behavioral therapy and/or referral to a mental health professional may be warranted.[34a,34b]

Work. In particular, belief that pain is harmful resulting in fear-avoidance behavior and belief that all pain must be gone before going back to work or normal, daily activities contribute to yellow (psychosocial) warning flags. Poor work history, unsupportive work environment, and belief that work is harmful all fall under the category of yellow work flags.[30]

Beliefs. People with chronic LBP who demonstrate yellow flag beliefs also have an increased risk for poor prognosis. This category includes catastrophizing, thinking the worst, a belief that pain is uncontrollable, poor compliance with exercise, low educational background, and the expectation of a quick fix for pain.

Behaviors. Beliefs extend into behaviors such as passive attitude toward rehabilitation, use of extended rest, reduced activity, increased intake of alcohol and other drugs to "manage" the pain, and avoidance or withdrawal from daily and/or social activities.

Affective. Depressed mood, irritability, and heightened awareness of bodily sensations along with anxiety represent affective psychosocial yellow flags (also prognostic of poor outcome for chronic LBP). Other affective yellow flags include feeling useless and not needed, disinterest in outside activities, and lack of family or personal support systems.

The assessment of psychosocial yellow flags should be part of any ongoing management of LBP at any time in the course of the problem. The New Zealand Guidelines[34] recommend the administration of a screening questionnaire at 2 to 4 weeks after onset of pain (see Appendix C-4 for a checklist of red/yellow flag indicators).

There is no evidence that this is the optimal time. This is early in the natural history of complaints of LBP, and other interventions may take this long to achieve their effects. In fact, over this time frame, practitioners may still be concerned about red flag conditions, and their time with the client may still be consumed with ensuring compliance with home rehabilitation and analgesics.[34]

On the other hand, waiting until someone develops chronic pain (3 months) may be too late; the window of opportunity to prevent chronicity will have passed, by definition. Therefore, in anyone with persisting pain, formal exploration of yellow flags should occur no later than 2 months after onset of pain, and possibly by the end of the first month. A practical clinical approach would be to begin screening for yellow flag issues at the 1-month follow-up appointment.[35]

FOLLOW-UP QUESTIONS

- What do you understand is the cause of your back pain?
- What are you doing to cope with your symptoms?
- Do you expect to get back to work after treatment?
- (Alternate): Do you expect to fully return to work?

Red Flag Signs and Symptoms

Watch for the most common red flags associated with back pain of a systemic origin or other medical condition (Box 14-1) but be aware that some recommended red flags have

BOX 14-1 MOST COMMON RED FLAGS ASSOCIATED WITH BACK PAIN OF SYSTEMIC ORIGIN

- Age less than 20 or over 50 (malignancy)/over 70 (fracture)*
- Previous history of cancer*
- Constitutional symptoms (e.g., fever, chills, unexplained weight loss*)
- Failure to improve with conservative care (usually over 4 to 6 weeks)*
- Recent urinary tract infection, blood in urine (or stools), difficulty with urination
- History of injection drug use
- Immunocompromised condition (e.g., prolonged use of corticosteroids, transplant recipient, autoimmune diseases)
- Pain is not relieved by rest or recumbency.
- Severe, constant nighttime pain
- Progressive neurologic deficit; saddle anesthesia; urinary or fecal incontinence
- Back pain accompanied by abdominal, pelvic, or hip pain
- History of falls or trauma (screen for fracture, osteoporosis, domestic violence, alcohol use)
- Significant morning stiffness with limitation in all spinal movements (ankylosing spondylitis or other inflammatory disorder)
- Skin rash (inflammatory disorder [e.g., Crohn's disease, ankylosing spondylitis])

*Recommendations made on the use of red flags to screen for serious pathology have come under scrutiny in recent years. Therapists are advised to make special note of the first four red flags listed above. The presence of any of these signals the need for additional screening (e.g., radiographs and simple blood tests) to detect cancer.[111]

high false-positive rates when used in isolation.[36] Each condition (e.g., infection, malignancy, fracture) will likely have its own predictive risk factors. A recent systematic review in the (medical) primary care setting reported that only three red flags are associated with fracture (prolonged use of corticosteroids, age older than 70 years, and significant trauma).[36] Individuals with serious spinal pathology almost always have at least one red flag that can be missed when the clinician (physician or therapist) assumes the client's symptoms are the result of mechanical-induced back pain. (See also Appendix A-2.)

Key findings are advancing age, significant recent weight loss, previous malignancy, and constant pain that is not relieved by positional change or rest and is present at night, disturbing the person's sleep. Poor response to conservative care or poor success with comparable care is an additional red flag in the diagnosis and management of musculoskeletal spine pain.[37] According to one source, cancer as a cause of LBP can be ruled out with 100% sensitivity when the affected individual is younger than 50 years old, has no prior history of cancer, no unexplained or unintended weight loss, and responds to conservative care.[6]

According to a recent systematic review, five red flags have been identified to screen for vertebral fractures in clients presenting with acute LBP including age over 70, female sex, major trauma, pain and tenderness, and a distracting painful injury.[38] Older females, especially older adults who have used corticosteroids, are predisposed to osteoporosis and increased risk of fracture from even minor trauma.[39,40]

More recent evidence to suggest that backache is a frequent finding in children and adolescents and is seldom associated with serious pathology has been published.[41-43] But back pain in children is still considered a red flag, especially in young children[43a] and/or if it has been present for more than 6 weeks because of the concern for infection or neoplasm.[43b,43c] Children are less likely to report associated signs and symptoms and must be interviewed carefully. Ask about any other joint involvement, swelling anywhere, changes in ROM, and the presence of any constitutional and GI symptoms. A recent history of viral illnesses may be linked to myalgias and diskitis. Most common causes of back pain in children are listed in Table 14-2.

Red flags requiring medical evaluation or reevaluation include back pain or symptoms that are not improving as expected, steady pain irrespective of activity, symptoms that are increasing, or the development of new or progressive neurologic deficits such as weakness, sensory loss, reflex changes, bowel or bladder dysfunction, or myelopathy.[38]

Indications for the use of plain films of the lumbar spine include any of the following features[44]:

- History of trauma
- History of cancer
- Older adults with minimal trauma
- Failure to respond to treatment

Use the Quick Screen Checklist (see Appendix A-1) to conduct a consistent and complete screening examination.

A few key screening questions might include:

FOLLOW-UP QUESTIONS

- Have you had an injury or trauma to your head, face, neck, or back?
- Do you have (or have you recently had) a fever? Headache? Sore throat? Skin rash?
- Have you ever had cancer of any kind? Ever been treated with chemotherapy or radiation therapy?
- Are you taking any medications?
- Have you had any problems with your bowels or bladder?

For the Therapist[45]

- Is it possible/probable there is a serious systemic disease or medical condition causing the pain?
- Is there neurologic compromise that might require surgical intervention?
- Is there social or psychologic distress that may amplify or prolong pain?

TABLE 14-2 Causes of Back Pain in Children

Inflammatory Conditions	Developmental Conditions	Trauma	Neoplastic Disease	Other
Diskitis (most common before age 6) Vertebral osteomyelitis Spinal abscess Nonspinal infections (e.g., pancreatitis, pyelonephritis) Rheumatoid arthritis (cervical spine involved most often) Reiter's syndrome Psoriatic arthritis Ankylosing spondylitis (presents during adolescence) Inflammatory bowel disease	Spondylolysis Spondylolisthesis Scheuermann's syndrome Scoliosis (especially left thoracic)	Muscle strain Vertebral stress or compression fracture Overuse syndrome Physical abuse	Leukemia Hodgkin's disease Non-Hodgkin's lymphoma Ewing's sarcoma (primary) Osteogenic sarcoma (osteosarcoma) [primary] Rhabdomyosarcoma (rare; skeletal metastasis)	Mechanical (hip and pelvic anomalies, upper cervical spine instability) Herniated disk Psychosomatic (conversion reaction) Benign tumors (osteoid osteoma) After lumbar puncture Juvenile osteoporosis

From Kliegman RM, editor: *Nelson essentials of pediatrics*, ed 5, Philadelphia, 2006, WB Saunders. Used with permission.

LOCATION OF PAIN AND SYMPTOMS

There are many ways to examine and classify head, neck, and back pain. Pain can be divided into anatomic location of symptoms (where is it located?): Cervical, thoracic, scapular, lumbar, and SI joint/sacral (as shown in Table 14-1). For example, intrathoracic disease refers more often to the neck, mid-thoracic spine, shoulder, and upper trapezius areas. Visceral disease of the abdomen and/or pelvis is more likely to refer pain to the low back region. Later in this section, spine pain is presented by the source of symptoms (what is causing the problem?).

Whenever faced with the need to screen for medical disease the therapist can review Table 14-1. First identify the location of the pain. Then scan the list for possible causes. Given the client's history, risk factors, clinical presentation, and associated signs and symptoms, are there any conditions on this list that could be the possible cause of the client's symptoms? Is age or sex a factor? Is there a positive family or personal history?

Sometimes reviewing the possible causes of pain based on location gives the therapist a direction for the next step in the screening process. What other questions should be asked? Are there any tests that will help differentiate symptoms of one anatomical area from another? Are there any tests that will help identify symptoms that point to one system versus another?

Head

The therapist may evaluate pain and symptoms of the face, scalp, or skull. Headaches are a frequent complaint given by adults and children. It may not be the primary reason for seeing a physical therapist but is often mentioned when asked if there are any other symptoms of any kind anywhere else in the body.

The brain itself does not feel pain because it has no pain receptors. Most often the headache is caused by an extracranial disorder and is considered "benign." Headache pain is related to pressure on other structures such as blood vessels, cranial nerves, sinuses, and the membrane surrounding the brain. Serious causes have been reported in 1% to 5% of the total cases, most often attributed to tumors and infections of the central nervous system (CNS).[1,46] In the past, headache was viewed as many disorders along a continuum. Better headache classifications have brought about the development of many discrete entities among these disorders.[47,48] The International Headache Society (HIS) has published commonly used *International Classification of Headache Disorders* (second edition, revised), which divides headaches into three parts: primary headache, secondary headache, and cranial neuralgias.[49,50]

Primary headache includes migraine, tension-type headache, and cluster headache. Secondary headaches, of which there is a large number, are attributed to some other causative disorder specified in the diagnostic criteria attached to them.

The therapist often provides treatment for secondary headache called *cervicogenic headache* (CGH). This type of headache is defined as referred pain in any part of the head (e.g., musculoskeletal tissues innervated by these nerve roots) caused by spondylitic, fibrotic, or vascular compression or compromise of cervical nerves (C1-C4).[51] CGHs are frequently associated with postural strain or chronic tension, acute whiplash injury, intervertebral disk disease, or progressive facet joint arthritis (e.g., cervical spondylosis, cervical arthrosis) (Table 14-3).

Causes of Headaches

Headache can be a symptom of neurologic impairment, hormonal imbalance, neoplasm, side effect of medication,[48] or other serious condition (Box 14-2). Headache may be the only symptom of hypertension, cerebral venous thrombosis, or impending stroke.[52,53] Sudden, severe headache is a classic symptom of temporal vasculitis (arteritis), a condition that can lead to blindness if not recognized and treated promptly.

Recognizing associated signs and symptoms and performing vital sign assessment, especially blood pressure monitoring, are important screening tools for vascular-induced headaches (see Chapter 4 for information on monitoring blood pressure).

Stress and inadequate coping are risk factors for persistent headache. Headache can be part of anxiety, depression, panic disorder, and substance abuse.[54,55] Headaches have been linked with excessive caffeine consumption or withdrawal in children, adolescents, and adults.[56]

Therapists often encounter headaches as a complaint in clients with posttraumatic brain injury, postwhiplash injury, or postconcussion injury. A constellation of other symptoms are often present such as dizziness, memory problems, difficulty concentrating, irritability, fatigue, sensitivity to noise, depression, anxiety, and problems with making judgments. Symptoms may resolve in the first 4 to 6 weeks following the injury but can persist for months to years causing permanent disability.[57,58]

Cancer. The greatest concern is always whether or not there is brain tumor causing the headaches. Only a minority of individuals who have headaches have brain tumors. Risk factors include occupational exposure to gases and chemicals and history of cranial radiation therapy for fungal infection of the scalp or for other types of cancer.

A previous history of cancer, even long past history, is a red flag for insidious onset of head and occipital neck pain. Metastatic lesions of the upper cervical spine are difficult to diagnose. Plain radiographs generally appear negative, which can delay diagnosis of clients with C1-C2 metastatic disease.[59]

The alert therapist may recognize the need for further imaging studies or medical evaluation. Persistent documentation of clinical findings and nonresponse to physical therapy intervention with repeated medical referral may be required.

TABLE 14-3 Clinical Signs and Symptoms of Major Headache Types

Migraine	Tension	Cervicogenic
Can be headache-free Migraines with headache are often described as throbbing or pulsating Often one-sided (unilateral); often around or behind one eye Associated with nausea, vomiting Light and/or sound sensitivity (photophobia and phonophobia) Common triggers: • Alcohol • Food • Hormonal changes • Hunger • Lack of sleep • Perfume • Stress • Medications • Environmental factors (e.g., pollutants, air pressure changes, temperature) May be preceded by prodromal symptoms: • Visual changes (aura): described as spots, balloons, lights, colors • Motor weakness • Dizziness • Paresthesias (numbness/tingling) • Confusion Facial pallor, cold hands and feet History of headaches in childhood; family history of migraines	Described as dull pressure Sensation of band or vise around the head; sometimes described as a painful, "tight" scalp Headache pain is bilateral or global (entire head) Muscular tenderness or soreness in soft tissues of the upper cervical spine Not usually accompanied by associated signs and symptoms May get worse with loud sounds or bright lights Current diagnosis or history of anxiety, depression, or panic disorder	Pain starts in the occipital region and spreads anteriorly toward the frontal area Usually bilateral Pain intensity fluctuates from mild to severe Often made worse by neck movements or sustained postures Decreased neck range of motion Forward head posture Trigger points or tender points in muscles Cervical muscle weakness or dysfunction Can resemble migraines with throbbing pain, nausea, phonophobia, photophobia History of trauma (e.g., whiplash), disk disease, or arthritis may be helpful

BOX 14-2 SYSTEMIC ORIGINS OF HEADACHE

Cancer
Primary neoplasm
Chemotherapy; brain radiation

Cardiovascular
Migraine
Ischemia (atherosclerosis; vertebrobasilar insufficiency; internal carotid artery dysfunction)
Cerebral vascular thrombosis
Arteriovenous malformation
Subarachnoid hemorrhage
Giant cell arteritis; vascular arteritis; temporal vasculitis
Hypertension
Febrile illnesses
Hypoxia
Systemic lupus erythematosus

Pulmonary
Obstructive sleep apnea
Hyperventilation (e.g., associated with anxiety or panic attacks)

Renal/Urologic
Kidney failure; renal insufficiency
Dialysis (first-use syndrome)

Gynecologic
Pregnancy
Dysmenorrhea

Neurologic
Postseizure
Disorder of cranium, cranial structures (e.g., nose, eyes, ears, teeth, neck)
Cranial neuralgia (e.g., trigeminal, Bell's palsy, occipital, Herpes zoster, optic neuritis)
Brain abscess
Hydrocephalus

Other
History of physical or sexual abuse
Side effect of medications
Allergens/toxins (environmental or food)
Overuse of medications (analgesic rebound effect)
Psychogenic/psychiatric disorder
Substance abuse/withdrawal (drugs and/or alcohol)
Caffeine use/withdrawal
Candidiasis (yeast)
Trauma (e.g., cervicogenic headache, fracture, eating disorders with forced vomiting)
Infection (e.g., meningitis, sinusitis, syphilis, tuberculosis, sarcoidosis, herpes)
Postdural puncture
Scuba diving
Hantavirus
Paget's disease (when skull is affected)
Hypoglycemia
Fibromyalgia
Temporomandibular joint dysfunction

Although primary head and neck cancers can cause headaches, neck pain, facial pain, and/or numbness in the face, ear, mouth, and lips are more likely. Other signs and symptoms can include sore throat, dysphagia, a chronic ulcer that does not heal, a lump in the neck, and persistent or unexplained bleeding. Color changes in the mouth known as leukoplakia (white patches) or erythroplakia (red patches) may develop in the oral cavity as a premalignant sign.[60]

Cancer recurrence is not uncommon within the first 3 years after treatment for cancers of the head and neck; often these cancers are not diagnosed until an advanced stage due to neglect on the part of the affected individual. Cervical spine metastasis is most common with distant metastases to the lungs, although any part of the body can be affected.[61] Anyone with a history of head and neck cancer should be screened for cancer recurrence when seen by a therapist for any problem.

As always, prevention and early detection improve survival rates. Education is important because most of the risk factors (tobacco and alcohol use, betel nut, syphilis, nickel exposure, woodworking, sun exposure, dental neglect) are modifiable.

Tension-type or migraine headaches can occur with tumors. Rapidly growing tumors are more likely to be associated with headache and will eventually present with other signs and symptoms such as visual disturbances, seizures, or personality changes.[62,63] Headaches associated with brain tumors occur in up to half of all cases and are usually bioccipital or bifrontal, intermittent, and of increasing duration. Presence of tumor headache varies, depending on size, location, and type of tumor.[64]

The headache is worse on awakening because of differences in CNS drainage in the supine and prone positions and usually disappears soon after the person arises. It may be intensified or precipitated by any activity that increases intracranial pressure such as straining during a bowel movement, stooping, lifting heavy objects, or coughing.

Often, the pain can be relieved by taking aspirin, acetaminophen, or other moderate painkillers. Vomiting with or without nausea (unrelated to food) occurs in about 25% to 30% of people with brain tumors and often accompanies headaches when there is an increase in intracranial pressure. If the tumor invades the meninges, the headaches will be more severe.

Recognizing the need for medical referral for the client with complaints of headaches can be difficult. Past medical history can be complex in adults and screening clues are often confusing. Careful review of the clinical presentation is required. For example, although pain associated with the CGH can be constant (a red flag symptom) the intensity often varies with activity and postures. Sustained posture consistently increases intensity of painful symptoms.

Migraines. Migraine headaches are often accompanied by nausea, vomiting, and visual disturbances, but the pain pattern is also often classic in description. Age is a yellow (caution) flag because migraines generally begin in childhood to early adulthood. Migraines can first occur in an individual beyond the age of 50 (especially in perimenopausal or menopausal women); advancing age makes other types of headaches more likely. A family history is usually present, suggesting a genetic predisposition in migraine sufferers. In addition to the typical clinical presentation, there are usually normal examination results.

Migraines can present with paralysis or weakness of one side of the body mimicking a stroke. A medical examination is required to diagnose migraine, especially in cases of hemiplegic migraines. Medical evaluation and treatment for migraines in general is recommended.

There is a role for the physical therapist because the beneficial effects of exercise on migraine headaches have been documented.[65,66] Physical therapy is most effective for the treatment of migraine when combined with other treatments such as biofeedback[67] and relaxation training.[68]

When present, associated signs and symptoms offer the best yellow or red flag warnings. For example, throbbing headache with unexplained diaphoresis and elevated blood pressure may signal a significant cardiovascular event. Daytime sleepiness, morning headache, and reports of snoring may point to obstructive sleep apnea. Headache-associated visual disturbances or facial numbness raises the suspicion of a neurologic origin of symptoms. Other red flags are listed in Box 14-3.

BOX 14-3 RED FLAG SIGNS AND SYMPTOMS ASSOCIATED WITH HEADACHE

The therapist should watch for any of the following red flags (listed in descending order of importance) and report them to a medical doctor. A complete screening interview and examination can establish a baseline of information and aid in the medical referral decision-making process.

- Headache that wakes the individual up or is present upon awakening (e.g., hypertension, tumor)
- Headache accompanied by documented elevated blood pressure changes
- Insidious or new onset of headache (less than 6 months)
- New onset of headache with associated neurologic signs and symptoms (e.g., confusion, dizziness, gait or motor disturbances, fatigue, irritability or mood changes)
- New onset of headache accompanied by constitutional symptoms (e.g., fever, chills, sweats) or stiff neck (infection, arteritis)
- Episodes of "blacking out" during headache (seizures, hemorrhage, tumor)
- Sudden severe headache accompanied by flu-like symptoms, aching muscles, jaw pain when eating, and visual disturbances (temporal arteritis)
- No previous personal or family history of migraine headaches

The therapist is advised to follow the same screening decision-making model introduced in Chapter 1 (see Box 1-7) and reviewed briefly at the beginning of this chapter. Physical examination should include measurement of vital signs, a general assessment of cardiac and vascular signs, and a thorough head and neck examination. A screening neurologic examination should address mental status (including pain behavior), cranial nerves, motor function, reflexes, sensory systems, coordination, and gait (see Chapter 4). Special Questions to Ask: Headache are listed at the end of this chapter and in Appendix B-17.

Cervical Spine

Neck pain is very common and has many mechanical and systemic causes. Neck and shoulder pain and neck and upper back pain often occur together making the differential diagnosis more difficult.

Traumatic and degenerative conditions of the cervical spine, such as whiplash syndrome and arthritis, are the major primary musculoskeletal causes of neck pain.[69] The therapist must always ask about a history of motor vehicle accident or trauma of any kind, including domestic violence.

Cervical or neck pain with or without radiating arm pain or symptoms may be caused by a local biomechanical dysfunction (e.g., shoulder impingement, disk degeneration, facet dysfunction) or a medical problem (e.g., infection, tumor, fracture). Referred pain presenting in these areas from a systemic source may occur from infectious disease, such as vertebral osteomyelitis, or from cancer, cardiac, pulmonary, or abdominal disorders (see Table 14-1).

Rheumatoid arthritis is often characterized by polyarthritic involvement of the peripheral joints, but the cervical spine is often affected early on (first 2 years) in the course of the disease. Deep aching pain in the occipital, retroorbital, or temporal areas may be present with pain referred to the face, ear, or suboccipit from irritation of the C2 nerve root. Some clients may have atlantoaxial (AA) subluxation and report a sensation of the head falling forward during neck flexion or a clunking sensation during neck extension as the AA joint is reduced spontaneously. Symptoms of cervical radiculopathy are common with AA joint involvement.[14]

Radicular symptoms accompanied by weakness, coordination impairment, gait disturbance, bowel or bladder retention or incontinence, and sexual dysfunction can occur whenever cervical myelopathy occurs, whether from a mechanical or medical cause. Cervical spondylotic myelopathy has been verified as a potential cause of LBP as well.[70] The Babinski test may be the most reliable screening test. There is no single reliable or valid clinical screening test or combination of tests that can be used to confirm spinal cord compression myelopathy.[71] An imaging study is usually needed to differentiate biomechanical from medical cause of radicular pain, especially when conservative care fails to bring about improvement.[72]

CLINICAL SIGNS AND SYMPTOMS

Cervical Myelopathy

- Neck pain and/or shoulder pain, stiffness
- Wide-based clumsy, incoordinated gait
- Loss of hand dexterity
- Paresthesias in one or both arms or hands
- Visible change in handwriting
- Difficulty manipulating buttons or handling coins
- Hyperreflexia
- Positive Babinski test
- Positive Hoffman sign
- Lhermitte's sign (electric shock sensation down spine/arms with neck flexion/extension)
- Urinary retention followed by overflow incontinence (severe myelopathy)
- Low back pain[70]

Torticollis of the sternocleidomastoid muscle may be a sign of underlying thyroid involvement. Anterior neck pain that is worse with swallowing and turning the head from side to side may be present with thyroiditis. Ask about associated signs and symptoms of endocrine disease (e.g., temperature intolerance; hair, nail, skin changes; joint or muscle pain; see Box 4-19) and a previous history of thyroid problems.[73]

Palpate the anterior spine and have the client swallow during palpation. Palpation of a soft tissue mass or lump should be noted. See guidelines for palpation in Chapter 4. Palpation of a firm, fixed, and immoveable mass raises a red flag of suspicion for neoplasm. Visually inspect and palpate the trachea for lateral deviation to either side.[74]

Anterior disk bulge into the esophagus or pharynx and/or anterior osteophyte of the vertebral body may give the sensation of difficulty swallowing or feeling a lump in the throat when swallowing. Anxiety can also cause a sensation of difficulty swallowing with a lump in the throat. Conduct a cranial nerve assessment for cranial nerves V and VII (see Table 4-9; see also Appendix B-21).

Vertebral artery syndrome caused by structural changes in the cervical spine is characterized by the client turning the whole body instead of turning the head and neck when attempting to look at something beyond his or her peripheral vision. Combined cervical motions, such as extension, rotation, and side bending, cause dizziness, visual disturbances, and nystagmus.

Headache/neck pain may be the early presentation of an underlying vascular pathology. Decreased blood flow to the brain, referred to as cerebral ischemia, may be caused by vertebrobasilar insufficiency (VBI)/cervical arterial dysfunction[75] secondary to atherosclerosis or other arterial dysfunction. Arterial compression can also occur when decreased vertebral height, osteophyte formation, postural changes, and ligamentous changes reduce the foraminal space and encroach on the vertebral artery. Premanipulative screening tests for vertebral artery patency and other tests to "clear" the upper cervical spine before using upper cervical manipulative techniques (e.g., cervical rotation, alar and transverse ligament stress tests, tectorial membrane stress test) may help identify the underlying cause of neck pain. Consensus on the need to conduct these tests has not been reached because the validity of tests for VBI has not been established.[76,77]

Caution is advised with older adults, anyone with a history of hypertension, rheumatoid arthritis, or long-term use of corticosteroids. A careful history, blood pressure measurements, observing for vascular pain patterns, and conducting a neurologic screening exam (possibly including cranial nerves) are advocated by some prior to upper cervical manipulation.

Thoracic Spine

As with the cervical spine and any musculoskeletal part of the body, the therapist must look for the cause of thoracic pain at the level above and below the area of pain and dysfunction. Possible musculoskeletal sources of thoracic pain include muscle strain, vertebral or rib fracture, zygapophyseal joint arthropathy,[78] active trigger points, spinal stenosis, costotransverse and costovertebral joint dysfunction, ankylosing spondylitis, intervertebral disk herniation, intercostal neuralgia, diffuse idiopathic skeletal hyperostosis (DISH), and T4 syndrome.[79] Shoulder impingement and mechanical problems in the cervical spine also can refer pain to the thoracic spine.

Systemic origins of musculoskeletal pain in the thoracic spine (Table 14-4) are usually accompanied by constitutional symptoms and other associated symptoms. Often, these additional symptoms develop after the initial onset of back pain, and the client may not relate them to the back pain and therefore may fail to mention them.

The close proximity of the thoracic spine to the chest and respiratory organs requires careful screening for pleuropulmonary symptoms in anyone with back pain of unknown cause or past medical history of cancer or pulmonary problems. Thoracic pain can also be referred from the kidney, biliary duct, esophagus, stomach, gallbladder, pancreas, and heart.

Thoracic aortic aneurysm, angina, and acute myocardial infarction are the most likely cardiac causes of thoracic back pain. Usually, there is a cardiac history and associated signs and symptoms such as weak or thready pulse, extremely high or extremely low blood pressure, or unexplained perspiration and pallor.

Tumors occur most often in the thoracic spine because of its length, the proximity to the mediastinum, and direct metastatic extension from lymph nodes with lymphoma, breast, or lung cancer. The client may report symptoms typical of cancer. Tumor involvement in the thoracic spine may produce ischemic damage to the spinal cord or early cord compression since the ratio of canal diameter to cord size is small, resulting in rapid deterioration of neurologic status (Case Example 14-4).

CASE EXAMPLE 14-4

Mid-Thoracic Back Pain

Background: A 55-year-old woman presents with sharp pain in the mid-back region around T5 to T6. The pain started after vacuuming her house last week. She has been taking Tylenol, but the pain is unrelieved. She reports being unable to find a comfortable position; the pain is keeping her awake at night.

History reveals a previous episode of pain in the same area 2 months ago. The pain started after she went grocery shopping and carried the heavy bags into her house. At that time, Tylenol quickly relieved her symptoms. The pain from the previous episode was described as "aching," not sharp like today.

Past Medical History: Past medical history includes breast cancer 15 years ago, surgical hysterectomy 10 years ago, and hypothyroidism. She does not remember what kind of breast cancer she had. She was treated with a lumpectomy and radiation. She has not had a mammography or clinical breast exam in the past 5 years. She does not perform self-breast examination on a regular or consistent basis.

She takes Synthroid for her thyroid problem but is not taking any other prescription medication. She takes a daily vitamin and 1200 mg of calcium but no other supplements. Tylenol is the only other over-the-counter product she takes.

She does not smoke or drink, even socially. She does not use any other substances of any kind. She reports there are no other symptoms of any kind anywhere else in her body.

Clinical Presentation: Vital signs are normal. There are no visible or palpable lesions in the upper quadrant on either side. Axillary and supraclavicular lymph nodes are not enlarged or palpable. Submandibular lymph nodes are palpable but not tender or hard.

Neurologic screening exam is normal, including bowel and bladder function, although the client reports a sensation of intermittent "weakness" in her left arm. There is exquisite pain on palpation of the thoracic spine from T4 to T6. There was no apparent movement dysfunction observed.

How can you differentiate between a disk problem and bony metastases?

A differential diagnosis of this type is outside the scope of the physical therapist's practice and requires a medical evaluation. The physician's differential diagnosis may include mammography, x-rays, and computed tomography (CT) scan or magnetic resonance imaging (MRI) to assist in the diagnosis.

Severe back pain that is unrelieved by rest or change of position and present at night in a woman with a past history of breast cancer requires immediate referral. Breast cancer has a predilection for axial skeletal bony metastases. Metastases can also occur hematogenously to the lungs (see Table 13-5). The therapist can perform a pulmonary system screening examination and ask about specific pulmonary signs and symptoms.

Reviewing Table 14-1 for possible viscerogenic causes of mid-thoracic back pain in a 55-year-old, the screening process can also include a brief cardiovascular examination and questions about GI function. Baseline information of this type can be extremely helpful later when documenting change in status or condition.

Rather than provide physical therapy intervention and assessing the results, immediate medical evaluation is in the best interest of this client. If the medical tests come back negative or if there is a disk problem, then the appropriate physical therapy intervention can be prescribed.

TABLE 14-4 Origin of Thoracic/Scapular Pain

Systemic Origin	Location	Neuromusculoskeletal
CARDIAC		
Myocardial infarct	Mid-thoracic spine	• Trauma (including motor vehicle accident, domestic violence, assault) • Muscle strain; overuse from repetitive motions • Degenerative disk disease; disk calcification or other disk lesions • Spinal stenosis • Rib syndromes or zygapophyseal joint disorders (e.g., costovertebral [rib] dysfunction, slipping rib syndrome, 12th rib syndrome, osteoarthritis, costovertebral or costotransverse joint hypomobility) • Thoracic outlet syndrome • Trigger points: Trapezius (middle), multifidi, rotators, rectus abdominis, latissimus dorsi, rhomboids, infraspinatus, serratus posterior • Vertebral or rib fracture or dislocation • Bone or soft tissue ossification (e.g., spinal ligaments, bone spurs) • Psychogenic (e.g., anxiety, depression, somatoform disorders) • Scoliosis; spinal deformity; Scheuermann's disease • Scapular dyskinesia
Aortic aneurysm	Thoracic spine; thoracolumbar spine	
Angina	Mid-thoracic spine; radiating down from shoulder (usually left side)	
PULMONARY		
Basilar pneumonia	Right upper back	
Empyema	Scapula	
Pleurisy	Scapula	
Pneumothorax	Ipsilateral scapula	
RENAL		
Acute pyelonephritis	Costovertebral angle (posterior)	
GASTROINTESTINAL		
Esophagitis	Mid-back between scapulae	
Peptic ulcer: Stomach/duodenal	6th through 10th thoracic vertebrae	
Gallbladder disease	Mid-back between scapulae; right upper scapula or subscapular area	
Biliary colic	Right upper back; mid-back between scapulae or subscapular areas	
Pancreatic carcinoma	Midthoracic or lumbar spine	
INFLAMMATORY/INFECTIOUS		
Rheumatoid arthritis Ankylosing spondylitis Osteomyelitis Pott's disease (tuberculosis of the spine) Spinal abscess or infection	Variable locations	
CANCER		
Primary: Osteoid osteoma, spinal canal, spinal nerve roots *Metastases*: Breast cancer, lung cancer, thyroid cancer, Hodgkin's disease, esophageal cancer, skin cancer	Variable locations	
OTHER		
Acromegaly	Mid-thoracic or lumbar spine	
Breast cancer	Mid-thoracic spine or upper back	
Osteoporosis	Variable locations	
Paget's disease of bone	Variable locations	
Pregnancy	Variable locations	
Blood disorders (sickle cell disease)	Variable locations	

Peptic ulcer can refer pain to the mid-thoracic spine between T6 and T10. The therapist should look for a history of NSAID use and ask about blood in the stools and the effect of eating food on pain and bowel function (see further discussion in Chapter 2).

Scapula

Most causes of scapular pain occur along the vertebral border and result from various primary musculoskeletal lesions. However, cardiac, pulmonary, renal, and GI disorders can cause scapular pain.

Specific questions to rule out potential systemic or medical origin of symptoms are listed in each individual chapter. For example, if the client reports any renal involvement, the therapist can use the questions at the end of Chapter 10 to screen further for urologic involvement. Appendix A contains a series of screening questions based on the presence of specific factors (e.g., sex, joint pain, night pain, shortness of breath).

Lumbar Spine

Low back pain (LBP) is very prevalent in the adult population, affecting up to 80% of all adults sometime in their lifetimes. In most cases, acute symptoms resolve within a few weeks to a few months. Individuals reporting persistent pain and activity limitation must be given a second screening examination.

As Table 14-1 shows, there is a wide range of potential systemic and medical causes of LBP. Older adults with more comorbidities are at increased risk for LBP. Bone and joint diseases (inflammatory and noninflammatory), lung and heart diseases, and enteric diseases top the list of conditions contributing to LBP in older adults.[11,80]

Pain referred to the lumbar spine and low back region from the pelvic and abdominal viscera may come directly from the organ structures, but some experts suspect the referred pain pattern is really produced by irritation of the posterior abdominal wall by pus, blood, or leaking enzymes. If that is the case, the pain is not referred but rather arises directly from the anterior aspect of the back.[30]

Sacrum/Sacroiliac

Sacral or SI pain in the absence of trauma and in the presence of a negative spring test (posterior-anterior glide of sacrum between the innominates) must be evaluated more closely. The most common etiology of serious pathology in this anatomic region comes from the spondyloarthropathies (disease of the joints of the spine) such as ankylosing spondylitis, Reiter's syndrome, psoriatic arthritis, and arthritis associated with chronic inflammatory bowel (enteropathic) disease.

Spondyloarthropathy is characterized by morning pain accompanied by prolonged stiffness that improves with activity. There is limitation of motion in all directions and tenderness over the spine and SI joints. The most significant finding in ankylosing spondylitis is that the client has night (back) pain and morning stiffness as the two major complaints, but asymmetric SI involvement with radiation into the buttock and thigh can occur.

In addition to back pain, these rheumatic diseases usually include a constellation of associated signs and symptoms, such as fever, skin lesions, anorexia, and weight loss, that alert the therapist to the presence of systemic disease or other medical conditions. Such symptoms present a red flag identifying clients who should be referred to a physician.

Age, sex, and risk factors are important in assessing for systemic origin of symptoms associated with any of these inflammatory conditions. Clients with these diseases have a genetic predisposition to these arthropathies, which are triggered by a number of environmental factors such as trauma and infection. Each of these clinical entities has been discussed in detail in Chapter 12.

Polymyalgia rheumatica and fibromyalgia syndrome are muscle syndromes associated with lumbosacral pain. Fibromyalgia syndrome refers to a syndrome of pain and stiffness that can occur in the low back and sacral areas with localized tender areas. Both these disorders are also discussed in Chapter 12.

Anyone under the age of 45 with low back, hip, buttock, and/or sacral pain lasting more than 3 months should be asked these four questions:

- Do you have morning back *stiffness* that lasts more than 30 minutes?
- Does the back pain wake you up during the second half of the night?
- Does the pain alternate from one buttock to the other (shift from side to side)?
- Does rest relieve the pain?

There is a 70% sensitivity and 81% specificity for inflammatory back pain if two of the four questions are positive. Sensitivity drops to 33% if three of the four questions are answered "yes" but the specificity increases to nearly 100%.[80a]

SOURCES OF PAIN AND SYMPTOMS

Pain can be evaluated by the source of symptoms (what is causing the problem?). It could be visceral, neurogenic, vasculogenic, spondylogenic, or psychogenic in origin. Specific symptoms and characteristics of pain (frequency, intensity, duration, description) help identify sources of back pain (Table 14-5).

The therapist must look at the history and risk factors, too. Any associated signs and symptoms that might reflect any one (or more) of these sources should be identified. Again, the therapist can use the tables in this chapter along with screening questions provided in Appendix A to help guide the screening process.

Viscerogenic

Visceral pain is not usually confused with pain originating in the head, neck, and back because sufficient specific symptoms and signs are often present to localize the problem correctly.

TABLE 14-5 Neck and Back Pain: Symptoms and Possible Causes

Symptom	Possible Cause
Night pain unrelieved by rest or change in position; made worse by recumbency; back pain, scoliosis, sensory and motor deficits in adolescents[163]	Tumor
Fever, chills, sweats	Infection
Unremitting, throbbing pain	Aortic aneurysm
Abdominal pain radiating to mid-back; symptoms associated with food; symptoms worse after taking NSAIDs	Pancreatitis, gastrointestinal disease, peptic ulcer
Morning stiffness that improves as day goes on	Inflammatory arthritis
Leg pain increased by walking and relieved by standing	Vascular claudication
Leg pain increased by walking, unaffected by standing but sometimes relieved by sitting or prolonged rest	Neurogenic claudication
"Stocking glove" numbness	Referred pain, nonorganic pain
Global pain	Nonorganic pain
Long-standing back pain aggravated by activity	Deconditioning
Pain increased by sitting	Discogenic disease
Sharp, narrow band of pain radiating below the knee	Herniated disk
Chronic spinal pain	Stress/psychosocial factors (unsatisfying job, fear-avoidance behavior, work or family issues, attitudes and beliefs)
Back pain dating to specific injury or trauma	Strain or sprain, fracture; failed back surgery
Back pain in athletic teenager[21a]	Developmental, trauma, epiphysitis, juvenile discogenic disease, hyperlordosis, spondylosis, spondylolysis, or spondylolisthesis
Exquisite tenderness over spinous process	Tumor, fracture, infection
Back pain preceded or accompanied by skin rash	Inflammatory bowel disease

Modified from Nelson BW: A rational approach to the treatment of low back pain, *J Musculoskel Med 10*(5):75, 1993.
NSAIDs, Nonsteroidal antiinflammatory drugs.

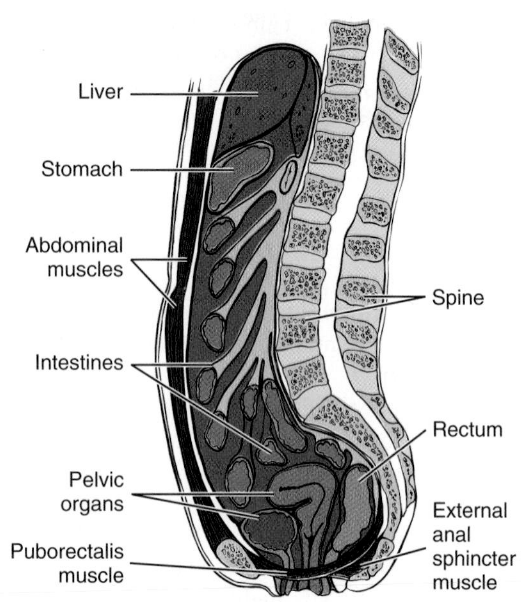

Fig. 14-1 Sagittal view of abdominal and pelvic cavities to show the proximity of viscera to the spine. The abdominal muscles and muscles of the pelvic floor provide anterior and inferior support, respectively. Any dysfunction of the musculature can alter the relationship of the viscera; likewise anything that impacts the viscera can affect the dynamic tension and ultimately the function of the muscles. Pathology of the organs can refer pain through shared pathways or by direct distention as a result of compression from inflammation and tumor.

It is the unusual presentation of systemic disease in the therapist's practice that will make it more difficult to recognize.

LBP is more likely to result from disease in the abdomen and pelvis than from intrathoracic disease, which usually refers pain to the neck, upper back, and shoulder. Disorders of the GI, pulmonary, urologic, and gynecologic systems can cause stimulation of sensory nerves supplied by the same segments of the spinal cord, resulting in referred back pain.[81]

As discussed in Chapter 3, the CNS may not be able to distinguish which part of the body is responsible for the input into common neurons.

Back pain can be associated with distention or perforation of organs, gynecologic conditions, or gastroenterologic disease. Pain can occur from compression, ischemia, inflammation, or infection affecting any of the organs (Fig. 14-1).

Referred pain can also originate in organs that share pain innervation with areas of the lumbosacral spine. Colicky pain is associated with spasm in a hollow viscus. Severe, tearing pain with sweating and dizziness may originate from an expanding abdominal aortic aneurysm. Burning pain may originate from a duodenal ulcer.

Muscle spasm and tenderness along the vertebrae may be elicited in the presence of visceral impairment. For example, spasm on the right side at the 9th and 10th costal cartilages can be a symptom of gallbladder problems. The spleen can cause tenderness and spasm at the level of T9 through T11 on the left side. The kidneys are more likely to cause tenderness, spasm, and possible cutaneous pain or sensitivity at the level of the 11th and 12th ribs.

Most often, past medical history, clinical presentation, and associated signs and symptoms will alert the therapist to an underlying systemic origin of musculoskeletal symptoms. Any client older than 50 with back pain, especially with insidious onset or unknown cause, must have vital signs taken, including body temperature.

Careful questioning can elicit important information that the client withheld, thinking it was irrelevant to the problem, such as LBP alternating with abdominal pain at the same level or back pain alternating with bouts of bloody diarrhea.

The therapist should look for clusters of signs and symptoms that may suggest involvement of a particular system. The Review of Systems chart in Chapter 4 (see Box 4-19) can be very helpful in identifying visceral sources of symptoms.

Neurogenic

Neurogenic pain is not easily differentiated. Radicular pain results from irritation of axons of a spinal nerve or neurons in the dorsal root ganglion, whereas referred pain results from activation of nociceptive free nerve endings (nociceptors) in somatic or visceral tissue.

Neurologic signs are produced by conduction block in motor or sensory nerves, but conduction block does not cause pain. Thus, even in a client with back pain and neurologic signs, whatever causes the neurologic signs is not causing the back pain by the same mechanism. Therefore finding the cause of the neurologic signs does not always identify the cause of the back pain.[35] The therapist must look further.

Conditions, such as radiculitis, may cause both pain and neurologic signs, but in that case the pain occurs in the lower limb, not in the back or in the upper extremity, not in the neck. If root inflammation also happens to involve the nerve root sleeve, neck or back pain might also arise. In such a case the individual will have three problems each with a different mechanism: neurologic signs due to conduction block, radicular pain due to nerve-root inflammation, and neck or back pain due to inflammation of the dura.[35]

Identifying a mechanical cause of pain does not always rule out serious spinal pathology. For example, neurogenic pain can be caused by a metastatic lesion applying pressure or traction on any of the neural components. Positive neural dynamic tests do not reveal the underlying cause of the problem (e.g., tumor versus scar tissue restriction). The therapist must rely on history, clinical presentation, and the presence of any neurologic or other associated signs and symptoms to make a determination about the need for medical referral.

Sciatica alone or sciatica accompanying back pain is an important but unreliable symptom. Although 90% of cases of sciatica are caused by a herniated disk,[82] there are those other 10% the therapist must also be aware of in the screening process. For example, diabetic neuropathy can cause nerve root irritation. Prostatic metastases to the lumbar and pelvic regions or other neoplasms of the spine can create a clinical picture that is indistinguishable from sciatica of musculoskeletal origin (see Tables 16-1 and 16-6). This similarity may lead to long and serious delays in diagnosis. Such a situation may require persistence on the part of the therapist and client in requesting further medical follow-up.

Spinal stenosis caused by a narrowing of the vertebral (spinal) canal, lateral recess, or intervertebral foramina may produce neurogenic claudication (Fig. 14-2). The canal tends to be narrow at the lumbosacral junction, and the nerve roots in the cauda equina are tightly packed. Pressure on the cauda equina from tumor, disk protrusion, spinal fracture or dislocation, infection, or inflammation can result in cauda equina syndrome, which is a neurologic medical emergency.[83,84] Cauda equina syndrome is defined as a constellation of symptoms that result from damage to the cauda equina, the portion of the nervous system below the conus medullaris (i.e., lumbar and sacral spinal nerves descending from the conus medullaris).

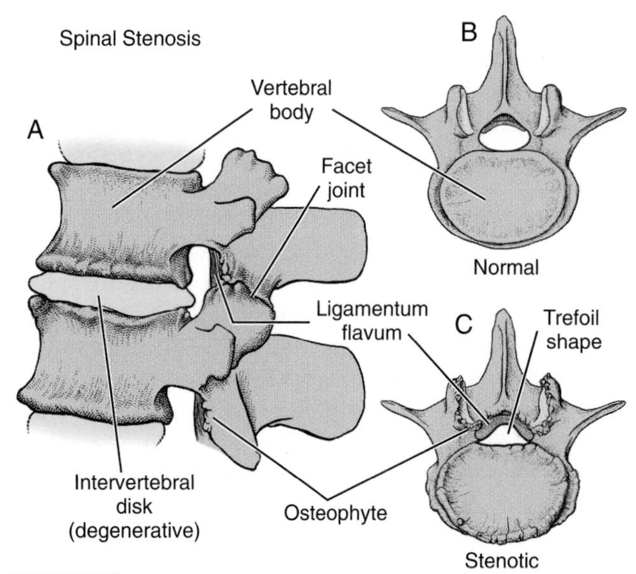

Fig. 14-2 Spinal stenosis. **A,** Aging causes a loss of disk height and compression of the vertebral body. The bone attempts to cushion itself by forming a lip or extra rim around the periphery of the endplates. This lipping can extend far enough to obstruct the opening to the vertebral canal. At the same time, the ligamentum flavum begins to hypertrophy or thicken and osteophytes (bone spurs) develop. Degenerative disease can cause the apophyseal (facet) joints to flatten out or become misshapen. Any or all of these variables can contribute to spinal stenosis. **B,** Normal, healthy vertebral body with a widely open vertebral canal. **C,** Stenotic spine from a variety of contributing factors. Many clients have all of these changes, but some do not. The presence of pathologic changes is not always accompanied by clinical symptoms.

CLINICAL SIGNS AND SYMPTOMS
Cauda Equina Syndrome

- LBP
- Unilateral or bilateral sciatica
- Saddle anesthesia; perineal hypoesthesia
- Change in bowel and/or bladder function (e.g., difficulty initiating flow of urine, urine retention, urinary or fecal incontinence, constipation, decreased rectal tone and sensation)
- Sexual dysfunction:
  - Men: Erectile dysfunction (inability to attain or sustain an erection)
  - Women: Dyspareunia (painful intercourse)
- Lower extremity motor weakness and sensory deficits; gait disturbance
- Diminished or absent lower extremity deep tendon reflexes (patellar, Achilles)

The medical diagnosis of cauda equina syndrome is not always straightforward. Abnormal rectal tone may be delayed in individuals presenting with cauda equina syndrome. This is because sensory nerves are smaller and more sensitive than motor nerves; even so, some people present with abnormal rectal tone (motor) without saddle anesthesia.[85]

The emerging nerve root exits through a shallow lateral recess and also may be compressed easily. Any combination of degenerative changes, such as disk protrusion, osteophyte formation, and ligamentous thickening, reduces the space needed for the spinal cord and its nerve roots (see Fig. 14-2).

Confusion with spinal stenosis syndromes may occur when atheromatous change in the internal iliac artery results in ischemia to the sciatic nerve. The subsequent sciatic pain with vascular claudication-like symptoms may go unrecognized as a vascular problem. The therapist may be able to recognize the need for medical intervention by combining a careful subjective and objective examination with knowledge of vascular and neurogenic pain patterns (Table 14-6). This is especially true in the treatment of unusual cases of sciatica or back pain with leg pain.

The client with a neurogenic source of back pain may develop a characteristic pattern of symptoms, with back pain, discomfort in the buttock, thigh, or leg and numbness and paresthesia in the leg developing after the person walks a few hundred yards (neurogenic claudication). The person may be forced to stop walking and obtains relief after long periods of rest. The pattern of symptoms is similar to that of intermittent claudication associated with vascular insufficiency, the major differences being immediate response to rest and position of the spine (see Fig. 14-4; see also Fig. 14-2).

The vertebral canal is wider when the spine is flexed, so relief from neurogenic pain may be obtained when the spine is flexed forward. Some individuals will bend over or squat as if to tie their shoelaces to assume a flexed spine position in public situations. Position of the spine (e.g., flexion, extension, side bending, or rotation) does not affect symptoms of a cardiac origin.

TABLE 14-6 Back Pain: Vascular or Neurogenic?

Vascular	Neurogenic
Throbbing	Burning
Diminished, absent pulses	No change in pulses
Trophic changes (skin color, texture, temperature)	No trophic changes; look for subtle strength deficits (e.g., partial foot drop, hip flexor or quadriceps weakness; calf muscle atrophy)
Pain present in all spinal positions	Pain increases with spinal extension, decreases with spinal flexion
Symptoms with standing: No	Symptoms with standing: Yes
Pain increases with activity; promptly relieved by rest or cessation of activity	Pain may respond to prolonged rest

Vasculogenic

Pain of a vascular origin may be mistaken for pain from a wide variety of musculoskeletal, neurologic, and arthritic disorders. Conversely, in a client with known vascular disease, a primary musculoskeletal disorder may go undiagnosed (e.g., diskogenic disease, spinal cord tumor, peripheral neuritis, arthritis of the hip) because all symptoms are attributed to cardiovascular insufficiency.

Vasculogenic pain can originate from both the heart (viscera) and the blood vessels (soma), primarily peripheral vascular disease. Back pain has been linked to atherosclerotic changes in the posterior wall of the abdominal aorta in older adults.[86] The therapist can rely on special clues regarding vasculogenic-induced pain in the screening process (Box 14-4).

Vascular injury to the great vessels, which are in proximity to the vertebral column, can occur during lumbar disk surgery or can present as a complication postoperatively. In rare cases, severe bleeding can result in back pain and hypotension in the acute care phase. Late complications of back pain from pseudoaneurysms are rare (less than 0.05%) but can occur years after spine surgery (diskectomy).[87,88]

Once the history has been reviewed, the therapist assesses the pain pattern present on clinical examination, asks about associated signs and symptoms, and conducts a Review of Systems.

Vascular back pain may be described as "throbbing" and almost always is increased with any activity that requires greater cardiac output and diminished or even relieved when

BOX 14-4 CLUES TO VASCULOGENIC PAIN

Pain of a vascular origin may be:

- Described as "throbbing"
- Accompanied by leg pain that is relieved by standing still or rest
- Accompanied by leg pain that is described as "aching, cramping or tired"
- Present in all spinal positions and increased by exertion
- Accompanied by a pulsing sensation in abdomen or palpable abdominal pulse
- Caused by a back injury (lifting) in someone with known heart disease or past history of aneurysm
- Accompanied by pelvic pain, leg pain, or buttock pain
- Presented as arm pain when working with the arms overhead
- Accompanied by temperature changes in the extremities
- An early or late complication of lumbar surgery; ask about a history of previous spine surgery

the workload or activity is stopped. A "throbbing" headache may be a vascular headache from a variety of causes.

Women in the perimenopausal and menopausal states may experience vascular headaches from fluctuating hormonal levels. Clients on cardiac medication, such as glyceryl trinitrate (which relaxes smooth muscle, especially blood vessels) and is used to prevent angina, may also report episodes of throbbing headaches. Vascular symptoms of this kind require medical evaluation.

Atherosclerosis and the resulting peripheral arterial disease are the underlying causes of most vascular back pain. Often, the client history will reveal significant cardiovascular risk factors such as smoking, hypertension, diabetes, advancing age, or elevated serum cholesterol (see Table 6-3 and discussion of peripheral vascular disease in Chapter 6).

Older age is an important red flag when assessing for pain of a vasculogenic origin. Most often, clients with back pain and any of the vascular clues listed are middle-aged and older. A personal or family history of heart disease is a second red flag. Continuous mid-thoracic pain can be a symptom of myocardial infarction, especially in a postmenopausal woman with a positive family history of heart disease.

Older clients with long-term nonspecific lower back pain may have occluded lumbar/middle sacral arteries associated with disk degeneration. Back pain and neurogenic symptoms in the presence of high serum low-density lipoprotein (LDL) cholesterol levels raises a red flag.[89]

Spondylogenic

Bone tenderness and pain on weight bearing usually characterize spondylogenic back pain (or the symptoms produced by bone lesions). Associated signs and symptoms may include weight loss, fever, deformity, and night pain. There are numerous conditions capable of producing bone pain, but the most common pathologic disorders are fracture from any cause, osteomalacia, osteoporosis, Paget's disease, infection, inflammation, and metastatic bone disease (Case Example 14-5).

The acute pain of a compression fracture superimposed on chronic discomfort, often in the absence of a history of trauma, may be the only presenting symptom. The client may recall a "snap" associated with mild pain, or there may have been no pain at all after the "snap." More intense pain may not develop for hours or until the next day.

Back pain over the thoracic or lumbar spine that is intensified by prolonged sitting, standing, and the Valsalva maneuver may resolve after 3 or 4 months as the fractures of the vertebral bodies heal. Clients who undergo kyphoplasty or vertebroplasty often have immediate pain relief.

The pain of untreated vertebral compression fractures may persist because of microfractures from the biomechanical effects of deformity. Other symptoms include pain on percussion over the fractured vertebral bodies, paraspinal muscle spasms, loss of height, and kyphoscoliosis.

When asking about the presence of any associated symptoms the therapist must keep in mind that older adults with vertebral compression fractures or kyphotic posture for any reason may report other pulmonary, digestive, and skeletal problems. These symptoms may not be indicative of back pain from a systemic or medical cause but rather organic dysfunction from a skeletal cause (i.e., somato-visceral response from the effects of a forward bent, kyphotic posture on the viscera).[90]

Sacral stress fractures should be considered in LBP of postmenopausal women with risk factors and athletes, particularly runners, volleyball players, and field hockey players (see further discussion on spondylogenic causes of sacral pain in Chapter 15).

CASE EXAMPLE 14-5

Osteoporosis

A 59-year-old man came to physical therapy for mid-thoracic back pain that seemed to come on gradually over the last few weeks and was starting to make his job as a janitor more difficult. There were no other symptoms to report: no neck, chest, or arm pain.

Past medical history was without incident. The client had never missed a day of work due to illness, never been hospitalized, and had no previous history of surgery. He has a 40-pack year history of smoking and "throws back a few beers" every night (6-pack daily for the last 15 years).

Clinical Presentation: Postural examination revealed a significant thoracic kyphosis with limited passive and active extension to neutral. Range of motion (ROM) in the lumbar spine was within normal limits. ROM in the hip and knee was also normal.

The client could take a deep breath without increasing his pain but not without setting off a long spell of coughing. There was local tenderness palpable in the mid-thoracic paraspinal and rhomboid muscles without evidence of erythema, swelling, or other skin changes.

Neurologic screening examination was normal.

What are the red flags? Is a medical referral needed before initiating treatment?

Red flags include age and a significant history of tobacco and alcohol abuse. All three are risk factors for reduced bone mass and fracture. Osteopenia and osteoporosis are often overlooked in men and occur more often than previously appreciated.[164-168] Thirty percent of osteoporotic fractures occur in men.[169]

An x-ray would be a good idea in this case before beginning a program of back extension exercises or applying any manual therapy.

Psychogenic

Psychogenic pain is observed in the client who has anxiety that amplifies or increases the person's perception of pain. Depression has been implicated in many painful conditions as the primary underlying problem. The prevalence of depression in physical therapy patients being treated for LBP has been reported as high as 26%.[91] There is a concern that depression may go unrecognized or inappropriately managed. Screening for depression can be done quickly and easily with the following two questions:

- During the past month, have you been bothered often by feeling down, depressed, or hopeless?
- During the past month, have you been bothered often by little interest or pleasure in doing things?

This two-question tool has been proved to have a 96% sensitivity with a negative likelihood ratio (LR–) of 0.07, and a negative predictive value of 98%. Specificity was reported at 57% with a LR+ of 2.2, and a positive predictive value of 33%.[92,93] A *yes* response to either or both of these questions warrants further evaluation.[91] (See additional questions in Appendices B-9 and B-10.)

Anxiety, depression, and panic disorder (see Chapter 3 for further discussion of anxiety, depression, and panic disorder) can lead to muscle tension, more anxiety, and then to muscle spasm. Signs and symptoms of these conditions are listed in Tables 3-9 and 3-10. Other signs of psychogenic-induced back pain may be:

- Paraplegia with only stocking glove anesthesia
- Reflexes inconsistent with the presenting problem or other symptoms present
- Cogwheel motion of muscles for weakness
- Straight leg raise (SLR) in the sitting versus the supine position (person is unable to complete SLR in supine but can easily perform an SLR in a sitting position)
- SLR supine with plantar flexion instead of dorsiflexion reproduces symptoms

The client may use words to describe painful symptoms characterized as "emotional." Recognizing these descriptors will help the therapist identify the possibility of an underlying psychologic or emotional etiology. An "exploding" or "vicious" headache, "agonizing" neck pain, or "punishing" backache are all red-flag descriptors of psychogenic origin (see Table 3-1).

The client who is unable to concentrate on anything except the symptoms and who reports that the symptoms interfere with every activity may need a psychologic/psychiatric referral. The therapist can screen for illness behavior as described in Chapter 3. Recognizing illness behavior helps the therapist clarify the physical assessment and alerts the therapist to the need for further psychologic assessment.[1]

Many studies have now shown a link between psychosocial distress and chronic neck or back pain.[94-99] Factors associated with chronic LBP may include job dissatisfaction, depression, fear-avoidance behavior, and compensation issues.[100,101] It may be necessary to conduct a social history to assess the client's recent life stressors and history of depression or drug or alcohol abuse.

The presence of psychosocial risk factors does not mean the pain is any less real nor does it reduce the need for symptom control. The therapist concentrates on pain management issues and improving function. Tools to screen for emotional overlay and fear-avoidance behavior are available in Chapter 3 of this text.

SCREENING FOR ONCOLOGIC CAUSES OF BACK PAIN

Cancer is a possible cause of referred pain. Tumors of the spine reportedly account for anywhere between 0.1% and 12% of back pain patients seen in a general medical practice.[102] Autopsy reports show up to 70% of adults who die of cancer have spinal metastases; up to 14% exhibit clinically symptomatic disease before death.[103]

Most reports place malignancy as a source of LBP in less than 1% of primary care patients.[104] Brain tumors (e.g., meningioma) in the motor cortex can present as LBP; abnormal symptom behavior and atypical responses to treatment may be observed and represent red flags for referral.[105] Head and neck pain from cancer is discussed earlier in this chapter (see Causes of Headaches).

Multiple myeloma is the most common primary malignancy involving the spine often resulting in diffuse osteoporosis and pain with movement that is not relieved while the person is recumbent. Back pain with radicular symptoms can develop with spinal cord compression. There can be a long period of development (5 to 20 years) with a chronic presentation of LBP.

For most oncologic causes of back pain, the thoracic and lumbosacral areas are affected. As a general rule, thoracic pain must be screened for metastatic carcinoma. Pain and dysfunction in the lumbosacral area may be caused by direct spread of cancer from the abdomen or pelvic areas. When the lumbar spine is affected by metastases, it is usually from breast, lung, prostate, or kidney neoplasm. GI cancer, myelomas, and lymphomas can also spread to the spine via the paravertebral venous plexus. This thin-walled and valveless venous system probably accounts for the higher incidence of metastases in the thoracic spine from breast carcinoma and in the lumbar region from prostatic carcinoma.

Past Medical History

Prompt identification of malignancy is important, starting with knowledge of previous cancers. Past history of cancer anywhere in the body is a red flag warning that careful screening is required. Always ask clients who deny a previous personal history of cancer about any previous chemotherapy or radiation therapy.

Early recognition and intervention does not always improve prognosis for survival from metastatic cancer, but it does reduce the risk of cord compression and paraplegia. It is important to remember that the history can be misleading. For example, almost 50% of clients with back pain from a malignancy have an identifiable (or attributable) antecedent injury or trauma[106] (Case Example 14-6).

It is unclear if this is a coincidence or merely reflective of weakness in the musculoskeletal system leading to loss of balance and strength and ultimately an injury. If the trauma results in significant injury (e.g., fracture), then the underlying cancer is usually identified right away. But if soft tissue injury does not necessitate an x-ray or other imaging study,

CASE EXAMPLE 14-6

Multiple Myeloma Presenting as Back Pain

Background: A 41-year-old woman presented with low back pain (LBP) after a skiing accident 6 months ago. She continued skiing but reinjured her back a month later while loading bicycles onto a car.

She did not seek help at that time, thinking the pain would resolve with healing and time. She took acetaminophen and over-the-counter (OTC) nonsteroidal antiinflammatory drugs (NSAIDs) but did not think these helped with her symptoms.

She reports her stress level as "high" due to family problems. She reports her fatigue level to be "high" also because of caring for four preschool-aged children and a sick husband. She has lost 6 pounds in the last month trying to keep up with work and home activities. She currently reports her height and weight as 5 feet 4 inches tall and 108 pounds.

She reports her LBP is "always there," but it gets worse with activity or movement. There is no numbness or tingling, but the pain does radiate into the buttocks on both sides. When asked if there were any symptoms anywhere else in her body, she mentioned a mild discomfort in the lower thorax/chest that gets worse when she coughs or takes a deep breath.

She has seen her family physician and been told that the LBP is postrepetitive trauma and that she needs to give it time to heal. She was advised to avoid activities that could strain her back. She decided to see a physical therapist for exercises.

Past Medical History

- Benign breast cyst reported as negative 5 months ago
- Cesarean section delivery of all four children without complications

Clinical Presentation

- Posture: Standing and sitting postures appeared natural; normal lumbar lordosis
- Thin and pale but in no acute distress
- Vital signs: All normal
- Alert and oriented to time, place, and person

Neurologic Screen

Cranial nerves	Within normal limits (WNL)
Manual muscle testing (MMT)	WNL (5/5 all extremities)
Sensory exam	WNL (light touch, pinprick)
Deep tendon reflexes (DTR)	Brisk 3+, equal in all 4 extremities
Straight-leg raise (SLR)	Limited to 25 degrees, bilaterally because of back pain and apprehension
Romberg	WNL

Unable to test physiologic (accessory, joint play) motions of the spine due to painful response

Unable to test for hip motion or overpressure of the sacroiliac (SI) joint because of pain

Positive tapping test (percussion over spinous processes) from L4 to S1

Walking pattern unremarkable; no antalgic gait

Able to walk in tandem and squat

Able to stand and walk on both heels and toes, bilaterally

Associated Signs and Symptoms

No report of fever, chills, night sweats, or night pain

No report of gastrointestinal (GI) or genitourinary (GU) dysfunction

Mild discomfort in the lower thorax/chest that gets worse when she coughs or takes a deep breath

What else do you need to know in the screening process?

Past history of infections of any kind? Cancer?

Recent or current medications besides OTC NSAIDs?

Tobacco use? Substance use (especially injection drugs with back pain)?

Did the physician examine your spine?

Were any x-rays or other imaging studies done?

Did you have a urinalysis or blood test done?

Recheck her vital signs on another day. Ask her to report any sweats, chills, or fever over the next 24 to 72 hours.

Any cough or shortness of breath? (Remember to ask about any functional limitations, not just ask if the client is having these symptoms.)

Any other respiratory signs and symptoms or red flags?

Take a more detailed birth/delivery history.

Type of birth control used (intrauterine contraceptive device?)

Date of last pap smear and mammogram.

Has she had a hysterectomy (consider surgical menopause and osteoporosis)? Ask about STIs or the possibility of physical or sexual assault.

Any pelvic symptoms? Vaginal discharge? Unusual bleeding? Missed menses?

What other steps can you take in the screening process?

Turn to Table 14-1. As you look this over, does anything else come to mind given the client's age, sex, and history? Vertebral osteomyelitis is one possibility. Review the risk factors for this condition. Making a diagnosis of vertebral osteomyelitis would be outside the scope of a physical therapist's practice, but identifying risk factors and associated signs and symptoms aids the therapist in making a referral decision.

Review Clues to Screening Head, Neck, or Back Pain at the end of this chapter. After looking this list over, the therapist may be prompted to ask if there are any other painful or symptomatic joints anywhere else in the body.

The therapist can scan the Special Questions to Ask: Back to see if there have been any questions left out or that now seem appropriate to ask based on the information gathered so far. Review Special Questions for Women.

Given the information you have, would you treat or refer this client?

Even though the vital signs are unremarkable and the neurologic screen appears negative, there are plenty of red flags here. Weight loss of 6 pounds even with emotional or psychologic stress in a thin person must be considered significant until proven otherwise.

Her age is borderline at 41, but there is an increased risk for diseases and illnesses with increasing age. Her pain appears to be constant but can be made worse with activity or movement. The fact that she injured her back 6 months ago but is still too acute to examine today is a red flag for possible orthopedic involvement that requires additional medical testing. This is not the expected clinical picture. The positive tapping test with percussion over the spine is another orthopedic red flag.

Radiating pain into the buttocks on both sides (bilateral) raises a red flag. It may be neurologic from a disk problem. There is

Continued

CASE EXAMPLE 14-6—cont'd

Multiple Myeloma Presenting as Back Pain

also a possibility of vascular cause of bilateral buttock pain. The client is not as old as one might expect with vascular claudication, but at age 41 it still must be considered. Palpate for abdominal pulse (possible aneurysm). Check the width of the aortic pulse.

Pain on inspiration should prompt auscultation of respiratory sounds.

Screening for psychogenic or emotional overlay may be appropriate. If the therapist decides to treat the client as part of the diagnostic process without the aid of imaging studies, caution is advised with any intervention. Obtaining the medical records is important, especially the physician's notes from the client's most recent visit.

Do not hesitate to contact the physician with your findings first and wait for agreement with your treatment plan. What the therapist observes during the examination may not be what the physician saw (e.g., acute presentation, positive tapping test, bilateral buttock pain).

If the client does not respond to physical therapy intervention, consider it the final red flag and refer immediately.

Result: The therapist made a judgment for immediate medical consultation by phone and by sending a faxed copy of the physical therapy evaluation. After conferring with the physician, a magnetic resonance imaging (MRI) scan was requested along with a complete blood cell count. The client had a compression fracture involving the central aspect of both the superior and inferior endplates of L5.

Blood cell counts were significantly decreased below normal (white blood cell [WBC], hemoglobin, hematocrit, and platelets). Erythrocyte sedimentation rate (ESR, or sed rate) and total protein levels were elevated.

Further diagnostic testing revealed a diagnosis of multiple myeloma. The diagnosis was confirmed by bone marrow biopsy, which showed infiltration of plasma cells. Further radiologic imaging revealed metastatic involvement of several ribs on both sides of the thoracic cage, right tibial head, and left ulna.

Physical therapy intervention was not appropriate in this case. A 41-year-old woman with LBP following repetitive injuries can be very deceiving. Multiple myeloma is unusual in people younger than 40 years and affects more men than women and more blacks than whites.

Exposure to radiation, wood dust, or pesticides can contribute to the development of multiple myeloma. The therapist did not ask any questions about occupational or environmental exposures because there was nothing in the history or clinical presentation to suggest it.

Data from Dajoyag-Mejia MA, Cocchiarella A: Multiple myeloma presenting as low back pain, *J Musculoskel Med* 21(4):229-232, 2004.

then the underlying oncologic cause may go undetected. Once again the therapist may be the first to recognize the cluster of clinical signs and symptoms and/or red flag findings to suggest a more serious underlying pathology.

Red Flags and Risk Factors

A combination of age (50 or older), previous history of cancer, unexplained weight loss, and failure to improve after 1 month of conservative care has a reported sensitivity of 100%.[104] Of these four red flags, a previous history of cancer is the most informative with a pooled LR+ of 23.7 compared with a LR+ of 3 for the other three.[99]

Until now, there has been an emphasis in this text on advancing age as a key red flag. Back pain at a young age (younger than 20 years old) may be considered a red flag as well. As a general rule, persistent backache due to extraspinal causes is rare in children. As mentioned, mechanical back pain in children is possibly linked with heavy backpacks,[107,108] sports, and sedentary lifestyle. However, primary bone cancer occurs most often in adolescents and young adults, hence the addition of this red flag: Age younger than 20 years. Bones of the appendicular skeleton (limbs) are affected more than the spine in this age group, but secondary metastases to the vertebrae can occur.

CLINICAL SIGNS AND SYMPTOMS

Oncologic Spine Pain

- Severe weakness without pain
- Weakness with full range
- Sciatica caused by metastases to bones of pelvis, lumbar spine, or femur
- Pain (nonmechanical) does not vary with activity or position (intense, constant); night pain
- Skin temperature differences from side to side
- Progressive neurologic deficits[109]
    - Sensory changes in myotome/dermatome pattern
    - Decreased motor function
    - Radiculopathy (rapid onset)
    - Myelopathy or cauda equina syndrome
- Positive percussive tap test to one or more spinous process
- Occipital headache, neck pain, palpable external mass in neck or upper torso
- Cervical pain or symptoms accompanied by urinary incontinence
- Look for signs and symptoms associated with other visceral systems (e.g., GI, genitourinary [GU], pulmonary, gynecologic)

Clinical Presentation

Back pain associated with cancer is usually constant, intense, and worse at night or with weight-bearing activities, although vague, diffuse back pain can be an early sign of non-Hodgkin's lymphoma and multiple myeloma. Pain with metastasis to the spine may become quite severe before any radiologic manifestations appear.[103]

Back pain associated with malignant retroperitoneal lymphadenopathy from lymphomas or testicular cancers is characterized as persistent, poorly localized LBP present at night but relieved by forward flexion. Pain may be so excruciating while lying down that the person can sleep only while sitting in a chair hunched forward over a table.

Palpate the midline of the spinous processes for any abnormality or tenderness. Perform a tap test (percussion over the involved spinous process).[110] Reproduction of pain or exquisite tenderness over the spinous process(es) is a red-flag sign requiring further investigation and possible medical referral.

Neoplasm (whether primary or secondary) may interfere with the sympathetic nerves; if so, the foot on the affected side is warmer than the foot on the unaffected side. Paresis in the absence of nerve root pain suggests a tumor. Severe weakness without pain is very suggestive of spinal metastases. Gross muscle weakness with a full range of SLR and without a history of recent acute sciatica at the upper two lumbar levels is also suggestive of spinal metastases.

A careful assessment of motor strength, sensory levels, proprioception, and reflexes is recommended. These findings can provide a baseline against which to compare future responses that might represent deterioration or undiagnosed lesions at other levels. Abnormal or new findings should be reported.[103]

A short period of increasing central backache in an older person is always a red-flag symptom, especially if there is a previous history of cancer. The pain spreads down both lower limbs in a distribution that does not correspond with any one nerve root level. Bilateral sciatica then develops, and the back pain becomes worse.

X-rays do not show bone destruction from metastatic lesions until the lytic process has destroyed 30% to 50% of the bone. The therapist cannot assume metastatic lesions do not exist in the client with a past medical history of cancer now presenting with back pain and "normal" x-rays.[111-113]

CASE EXAMPLE 14-7

Skin Lesions

A 52-year-old woman presented in physical therapy with low back pain (LBP) radiating down the right leg to the knee. She had recently completed chemotherapy for acute myelocytic leukemia and was referred to physical therapy by the oncology nurse. Bone marrow biopsy 1 month ago was negative for leukemic cells.

Clinical Presentation: The client presented with acute LBP described as "going across my low back area." She had a normal gait pattern but decreased lumbar motions in forward bending, right side bending, and left rotation.

Her pain was relieved by forward bending. Pain was too intense to conduct accessory motion testing because the client was unable to lie down for more than a minute before having to sit up.

Neurologic screen revealed a positive straight leg raise (SLR) on the right, intact sensation, and decreased ankle reflex on the right (patellar tendon reflex was assessed as normal). Manual muscle strength testing was deferred due to the client's extreme agitation during testing. There were no reported changes in bowel or bladder.

When asked if there were any other symptoms of any kind anywhere else, the client raised her shirt and showed the therapist several nodules on her skin. They were not tender or oozing any discharge. The client reported she first noticed them about a week before her back pain started. She had not remembered to tell the nurse or her doctor about them.

Outcome: This is a good case to point out that even though the client has a known condition, such as cancer, and the referral comes from a health care professional, screening for medical disease as the cause of the pain or symptoms is still very important.

The therapist made phone contact with the referring nurse and reported findings from the evaluation. Of particular concern were the skin lesions and neurologic changes. The nurse was unaware of these changes. The therapist requested a medical evaluation before starting a physical therapy program.

The client was diagnosed with cancer metastases to the spine and cauda equina syndrome. Cauda equina syndrome, caused by mechanical compression of the spinal nerve roots by tumor (or infection), requires immediate medical attention.

The client underwent urgent total spine irradiation, which did relieve her back pain. She declined further medical care (i.e., chemotherapy) and decided to continue with physical therapy to regain motion and strength.

Associated Signs and Symptoms

Clinical signs and symptoms accompanying back pain from an oncologic cause may be system related (e.g., GI, GU, gynecologic, spondylogenic), depending on where the primary neoplasm is located and the location of any metastases (Case Example 14-7).

The therapist must ask about the presence of constitutional symptoms, symptoms anywhere else in the body, and assess vital signs as part of the screening process. Unexplained weight loss is a common feature in anyone with tumors of the spine. Review the red flags in Box 14-1 and conduct a Review of Systems to identify any clusters of signs and symptoms.

SCREENING FOR CARDIAC CAUSES OF NECK AND BACK PAIN

Vascular pain patterns originate from two main sources: cardiac (heart viscera) and peripheral vascular (blood vessels). The most common referred cardiac pain patterns seen in a physical therapy practice are angina, myocardial infarction, and aneurysm.

Pain of a cardiac nature referred to the soma is based on multisegmental innervation. For example, the heart is innervated by the C3 through T4 spinal nerves. Pain of a cardiac source can affect any part of the soma (body) also innervated by these levels. This is why someone having a heart attack can experience jaw, neck, shoulder, arm, upper back, or chest pain. See Chapter 3 for an in-depth discussion of the origins of viscerogenic pain patterns affecting the musculoskeletal system.

On the other hand, pain and symptoms from a peripheral vascular problem are determined by the location of the underlying pathology (e.g., aortic aneurysm, arterial or venous obstruction). Peripheral vascular patterns will be reviewed later in this chapter.

Angina

Angina may cause chest pain radiating to the anterior neck and jaw, sometimes appearing only as neck and/or jaw pain and misdiagnosed as temporomandibular joint (TMJ) dysfunction. Postmenopausal women are the most likely candidates for this type of presentation. If the jaw pain is steady, lasts a long time, or is worst when first waking up in the morning, it could be that the individual is grinding the teeth while sleeping. But jaw pain that comes and goes with physical activity or stress may be a symptom of angina.

Angina and/or myocardial infarction can appear as isolated mid-thoracic back pain in men or women (see Figs. 6-4 and 6-8). There is usually a lag time of 3 to 5 minutes between increase in activity and onset of musculoskeletal symptoms caused by angina.

Myocardial Ischemia

Heart disease and myocardial infarction (MI), in particular, can be completely asymptomatic. In fact, sudden death occurs without any warning in 50% of all MIs. Back pain from the heart (cardiac pain pattern) can be referred to the anterior neck and/or mid-thoracic spine in both men and women.

When pain does present, it may look like one of the patterns shown in Fig. 6-9. There are usually some associated signs and symptoms such as unexplained perspiration (diaphoresis), nausea, vomiting, pallor, dizziness, or extreme anxiety. Age and past medical history are important when screening for angina or MI as possible causes of musculoskeletal symptoms. Vital sign assessment is a key clinical assessment (Case Example 14-8).

CASE EXAMPLE 14-8

Back Pain and Dizziness after Colonoscopy

An 87-year-old woman visiting her daughter from out of town fell and suffered a compression fracture of L1. She reported having "heart problems" during a colonoscopy several weeks before this fall. She has had extreme back pain and is being given Vicodin (opioid analgesic for mild pain).

She is nauseated and attributes this to the pain medication. Blood pressure is 200/90 mm Hg with pulse in the low 80s. There is no respiratory distress, no heart palpitations, and no fever. She reports being on many blood pressure and heart medications and thyroid meds.

The family reports she has dizzy spells and is weak. She frequently loses her balance but does not fall. She is extremely tired and the family reports she sleeps much during the day.

She has been referred to physical therapy through a home health agency. Since she is from out of town, she does not have a primary care physician. The daughter took her to a local walk-in clinic. The nurse practitioner then referred her to home health. Physical therapy was prescribed for the dizziness and falling.

You suspect the symptoms of dizziness, drowsiness, and weakness may be drug-induced. What do you do in a case like this?

Conduct an evaluation and gather as much information as you can from the client and family members. Use the Quick Screen Checklist and complete a Review of Systems. Organize the information you obtain from the evaluation so that the need for any other screening questions can be identified.

Look up potential side effects of Vicodin and ask the client about the presence of any other symptoms of any kind. See if any of the reported signs and symptoms point to side effects of medication. Conduct a cardiovascular screening examination (see Chapter 4).

Do not hesitate to contact the local clinic/nurse practitioner and ask if the client's symptoms could be cardiac or drug-induced. Report the abnormal vital signs. There may be a change in drug dosage, suggested drug administration (with or without food, time of day), or change in prescribed drug that can alleviate symptoms while still controlling pain. Vital signs may return to normal with better pain control unless there is an underlying cardiovascular reason for her symptoms.

Assess muscle weakness, vestibular function, and balance. Look for modifiable risk factors. Offer as much intervention as possible, given the temporary visiting situation and short-term episode of care.

Document findings, problem list, and plan of care and communicate these results with the referring agency. Medical referral may be advised given the client's age, vital signs, history of heart disease, and use of multiple medications.

Abdominal Aortic Aneurysm

On occasion, an abdominal aortic aneurysm (AAA) can cause severe back pain (see Fig. 6-11 and additional discussion in Chapter 6). An aneurysm is an abnormal dilation in a weak or diseased arterial wall causing a saclike protrusion. Prompt medical attention is imperative because rupture can result in death. Aneurysms can occur anywhere in any blood vessel, but the two most common places are the aorta and cerebral vascular system. AAA occurs most often in men in the sixth or seventh decade of life.

Risk Factors

The major risk factors for AAA include older age, male sex,[114] smoking, and family history.[115-118] Although the underlying cause is most often atherosclerosis, the therapist should be aware that aging athletes involved in weight lifting are at risk for tears in the arterial wall, resulting in an aneurysm. There is often a history of intermittent claudication and decreased or absent peripheral pulses. Other risk factors include congenital malformation and vasculitis. Often the presence of these risk factors remains unknown until an aneurysm becomes symptomatic.

Clinical Presentation

Pain presents as deep and boring in the midlumbar region. The pattern is usually described as sharp, intense, severe, or knifelike in the abdomen, chest, or anywhere in the back (including the sacrum). The location of the symptoms is determined by the location of the aneurysm (see Fig. 6-11).

Most aortic aneurysms (95%) occur just below the renal arteries. An objective examination may reveal a pulsing abdominal mass or abnormally widened aortic pulse width (see Fig. 4-55).

Obesity and abdominal ascites or distention makes this examination more difficult. The therapist can also listen for bruits. Bruits are abnormal blowing or swishing sounds heard on auscultation of the arteries.

Bruits with both systolic and diastolic components suggest the turbulent blood flow of partial arterial occlusion. The client will be hypertensive if the renal artery is occluded as well. Peripheral pulses may be diminished or absent. Other historical clues of coronary disease or intermittent claudication of the lower extremities may be present.

Monitoring vital signs is important, especially among exercising senior adults. Teaching proper breathing and abdominal support without using a Valsalva maneuver is important in any exercise program, but especially for those clients at increased risk for aortic aneurysm.

CLINICAL SIGNS AND SYMPTOMS

Impending Rupture or Actual Rupture of an Aortic Aneurysm

- Rapid onset of severe neck or back pain (buttock, hip, and/or flank pain possible)
- Pain may radiate to chest, between the scapulae, or to posterior thighs
- Pain is not relieved by change in position
- Pain is described as "tearing" or "ripping"
- Other signs: Cold, pulseless lower extremities, blood pressure differences between arms (more than 10 mm Hg diastolic)

The U.S. Preventive Services Task Force (USPSTF) updated its guidelines for medical screening for AAA in 2005. The new guidelines recommend ultrasound screening for men ages 65 to 75 years who are current or former smokers.[115] The therapist should advise men in this age group who have ever smoked to discuss their risk for AAA with a medical doctor. Any male with these two risk factors, especially presenting with any of these signs or symptoms, must be referred immediately.

The cost-effectiveness of screening women for AAA is under investigation. Preliminary studies suggest that despite a lower prevalence of this disease among women (1.1%), when aneurysms occur in women, they seem to enlarge faster than in men[119] and there is a higher rupture rate.[120]

The orthopedic or acute care therapist must be aware that aortic damage (not an aneurysm but sometimes referred to as a *pseudoaneurysm*) can occur with any anterior spine surgery (e.g., spinal fusion, spinal fusion with cages). Blood vessels are moved out of the way and can be injured during surgery. If the client (usually a postoperative inpatient) has internal bleeding from this complication, there may be:

- Distended abdomen
- Changes in blood pressure
- Changes in stool
- Possible back and/or shoulder pain

In such cases, the client's recent history of anterior spinal surgery accompanied by any of these symptoms is enough to notify nursing or medical staff of concerns. Monitoring postoperative vital signs in these clients is essential.

SCREENING FOR PERIPHERAL VASCULAR CAUSES OF BACK PAIN

Most physical therapists are very familiar with the signs and symptoms of peripheral vascular disease (PVD) affecting the extremities, including both arterial and venous disease (see discussion in Chapter 6).

When assessing back pain for the possibility of a vascular cause, remember peripheral vascular disease can cause back pain. The location of the pain or symptoms is determined by the location of the pathology (Fig. 14-3).

With obstruction of the aortic bifurcation, the client may report back pain alone, back pain with any of the following features, or any of these signs and symptoms alone (Table 14-7):

- Bilateral buttock and/or leg pain or discomfort
- Weakness and fatigue of the lower extremities
- Atrophy of the leg muscles
- Absent lower extremity pulses
- Color and/or temperature changes in the feet and lower legs

Symptoms are often (but not always) bilateral because the obstruction occurs before the aorta divides (i.e., before it becomes the common iliac artery and supplies each leg separately). Frequently, someone with symptomatic atherosclerotic disease in one blood vessel has similar pathology in other blood vessels as well. Over time, there may be a progression of symptoms as the disease worsens and blood vessels become more and more clogged with plaque and debris.

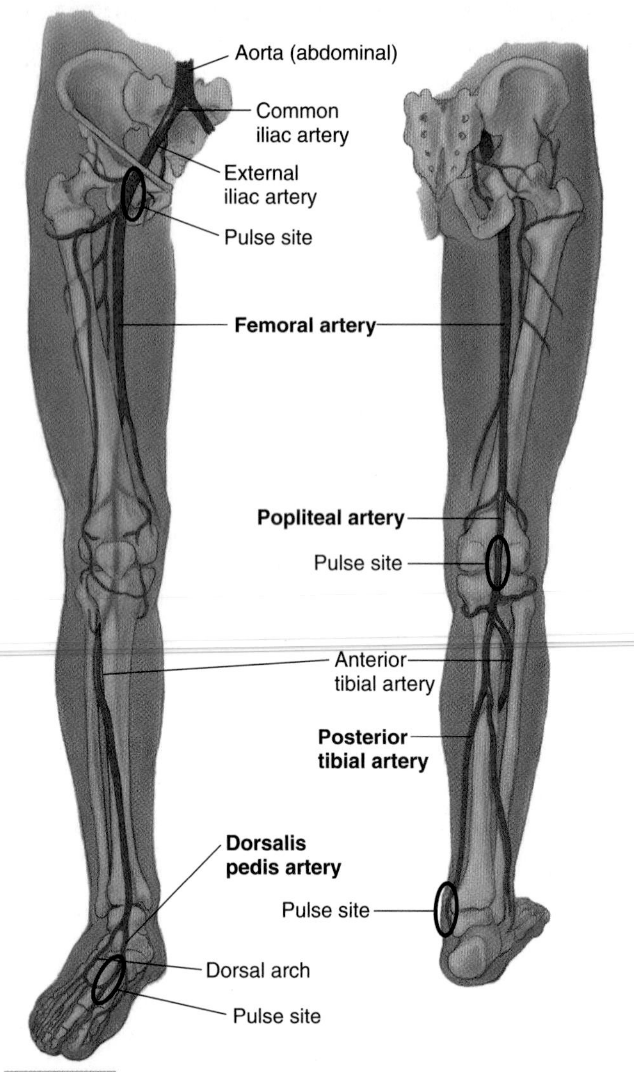

Fig. 14-3 Arteries in the lower extremities. As you look at this illustration, note the location of the arteries in the lower extremities starting with the aorta branching into the common iliac artery, which descends on both sides into the legs. Once the common iliac artery passes through the pelvis to the femur, it becomes the femoral artery and then the popliteal artery behind the knee before branching into the popliteal artery. The final split comes as the popliteal artery divides to form the anterior tibial artery down the front of the lower leg and the posterior tibial artery down the back of the lower leg. The anterior tibial artery also becomes the dorsalis pedis artery. Note the pulse points shown with bold, black ovals and remember that distal pulses disappear with aging and the presence of atherosclerosis causing peripheral vascular disease. (From Jarvis C: *Physical examination and health assessment*, ed 5, Philadelphia, 2008, WB Saunders.)

With obstruction of the iliac artery the client is more likely to present with pain in the low back, buttock, and/or leg of the affected side and/or numbness in the same area(s). Obstruction of the femoral artery can result in thigh and/or calf pain, again with distal pulses diminished or absent.

Ipsilateral calf/ankle pain or discomfort (intermittent claudication) occurs with obstruction of the popliteal artery and is a common first symptom of PVD.

Adults over the age of 50 presenting with back pain of unknown cause and mild-to-moderate elevation of blood pressure should be screened for the presence of PVD.

TABLE 14-7 Back and Leg Pain from Arterial Occlusive Disease

The location of discomfort, pain, or other symptoms is determined by the location of the pathology (arterial obstruction).

Site of Occlusion	Signs and Symptoms

Aortic bifurcation	• Sensory and motor deficits
	  • Muscle weakness and atrophy
	  • Numbness (loss of sensation)
	  • Paresthesias (burning, pricking)
	  • Paralysis
	• Intermittent claudication (pain or discomfort relieved by rest): bilateral buttock and/or leg, low back, gluteal, thigh, calf
	• Cold, pale legs with decreased or absent peripheral pulses

Iliac artery	• Intermittent claudication (pain or discomfort in the buttock, hip, thigh of the affected leg; can be unilateral or bilateral; relieved by rest) • Diminished or absent femoral or distal pulses • Impotence in males
Femoral and popliteal artery	• Intermittent claudication (pain or discomfort; calf and foot; may radiate) • Leg pallor and coolness • Dependent rubor • Blanching of feet on elevation • No palpable pulses in ankles and feet • Gangrene
Tibial and common peroneal artery	• Intermittent claudication (calf pain or discomfort; feet occasionally) • Pain at rest (severe disease); possibly relieved by dangling leg • Same skin and temperature changes in lower leg and foot as described above • Pedal pulses absent; popliteal pulses may be present

From Goodman CC, Fuller KS: *Pathology: implications for the physical therapist*, ed 3, Philadelphia, 2009, WB Saunders.

Back Pain: Vascular or Neurogenic?

The medical differential diagnosis is difficult to make between back pain of a vascular versus neurogenic origin. Frequently, vascular and neurogenic claudication occurs in the same age group (over 60 and even more often, after age 70). Sometimes clients are referred to physical therapy to help make the differentiation (Case Example 14-9).

Vascular and neurogenic disease often coexists in the same person with an overlap of symptoms of each. There are several major differences to look for but especially response to rest (i.e., activity pain), position of the spine, and the presence of any trophic (skin) changes (see Tables 14-6 and 16-5).

Vascular-induced back and/or leg pain or discomfort is alleviated by rest and usually within 1 to 3 minutes.

CASE EXAMPLE 14-9

Spinal Stenosis

Background: A 68-year-old woman with a long history of degenerative arthritis of the spine was referred to physical therapy for conservative treatment toward a goal of improving function despite her painful symptoms. She was a nonsmoker with no other significant previous medical history.

Her symptoms were diffuse bilateral lumbosacral back pain into the buttocks and thighs, which increased with walking or any activity and did not subside substantially with rest (except for prolonged rest and immobility).

Clinical Presentation: On examination, this client moved slowly and with effort, complaining of the painful symptoms described. There was no tenderness of the sacroiliac joint or sciatic notch but a subjective report of tenderness over L4 to L5 and L5 to S1. Tap test was negative; the client reported mild diffuse tenderness. There was no palpable step-off or dip of the spinous processes for spondylolisthesis and no paraspinal spasm, but a marked right lumbar scoliosis was noted. The client reported knowledge of scoliosis since she was a child.

A neurologic screening examination revealed normal straight leg raise (SLR) and normal sensation and reflexes in both lower extremities. Motor examination was unremarkable for an inactive 68-year-old woman. Dorsalis pedis and posterior tibialis pulses were palpable but weak bilaterally.

Despite physical therapy treatment and compliance on the part of the client with a home program, her symptoms persisted and progressively worsened.

What is the next step in the screening process?

Reevaluate the client's movement dysfunction and the selected intervention to date. Was the right treatment approach taken? Reassess red flag findings (age, lack of improvement with intervention) and conduct a review of systems (if this has not already been done).

In this case the client's age, negative neurologic screening examination, and diminished lower extremity pulses suggested a second look for vascular cause of symptoms.

Vital signs were assessed along with a peripheral vascular screening examination. The Bike Test was administered, but the results were unclear with increased pain reported in both extension and flexion.

Result: She returned to her physician with a report of these findings. Further testing showed that in addition to degenerative arthritis of the lumbosacral spine, there was secondary stenosis and marked aortic calcification, indicating a vascular component to her symptoms.

Surgery was scheduled: An L4 to L5 laminectomy with fusion, iliac crest bone graft, and decompression foraminotomies. Postoperatively, the client subjectively reported 80% improvement in her symptoms with an improvement in function, although she was still unable to return to work.

Conversely, activity (usually walking) brings the symptoms on within 1 to 3, sometimes 3 to 5 minutes. Neurogenic-induced symptoms often occur immediately with use of the affected body part and/or when adopting certain positions. The client may report the pain is relieved by prolonged rest or not at all.

What is the effect of changing the position of the spine on pain of a vascular nature? Are the vascular structures compromised in any way by forward bending, side bending, or backward bending (Fig. 14-4)? Are we asking the diseased heart or compromised blood vessels to supply more blood to this area?

It is not likely that movements of the spine will reproduce back pain of a vascular origin. What about back pain of a neurogenic cause? Forward bending opens the vertebral canal (vertebral foramen) giving the spinal cord (through L1) additional space. This is important in preventing painful symptoms when spinal stenosis is present as a cause of neurogenic claudication.

Unless there is a spinal neuroma, a true stenosis with spinal cord pressure does not occur in the lumbar region since the spinal cord ends at L1 in most people. Neural symptoms at L1 to L3 are rare and more likely indicate a spinal tumor rather than disk or facet pathology. Nerve pressure leading to radicular symptoms (e.g., pain, numbness, myotomal weakness) below L2 is not true stenosis of the vertebral canal but rather intervertebral foraminal stenosis with encroachment of the peripheral nerve as it leaves the spinal canal through the neural foramina.

The position of comfort for someone with back pain associated with spinal stenosis is usually lumbar flexion. The client may lean forward and rest the hands on the thighs or lean the upper body against a table or cupboard.

The Bicycle Test

The Bicycle test of van Gelderen[121] is one way to assess the cause of back pain (Fig. 14-5). Although sensitivity, specificity, and LRs have not been established for this test, from clinical experience it is believed to offer clues to the source (neurogenic or vascular). It is not a definitive test by itself. The Bicycle test is based on two of the three variables listed earlier: (1) response to rest and (2) position of the spine. Trophic (skin) changes are assessed separately.

In theory, if someone has back/buttock pain of a vascular origin, what is the effect of pedaling a stationary bicycle? Increased demand for oxygen can result in back/buttock pain when the cardiac workload/oxygen need is greater than the ability of the affected coronary arteries to supply the necessary oxygen.

Normally, the response would be angina (chest pain or discomfort or whatever pattern the client typically experiences). In the case of referred pain patterns, the client may experience mid-thoracic or even lumbar pain. How soon do these symptoms appear? With musculoskeletal pain of a cardiac origin, there is a 3- to 5-minute lag time before the onset of symptoms. Immediate reproduction of painful symptoms is more indicative of neuromusculoskeletal involvement.

After pedaling for 5 minutes and observing the client's response, ask him or her to lean forward and continue pedaling. What is the expected response if the back, buttock, or leg pain is vascular-induced? In other words, what is the response to a change in position when someone has back pain of a vascular origin?

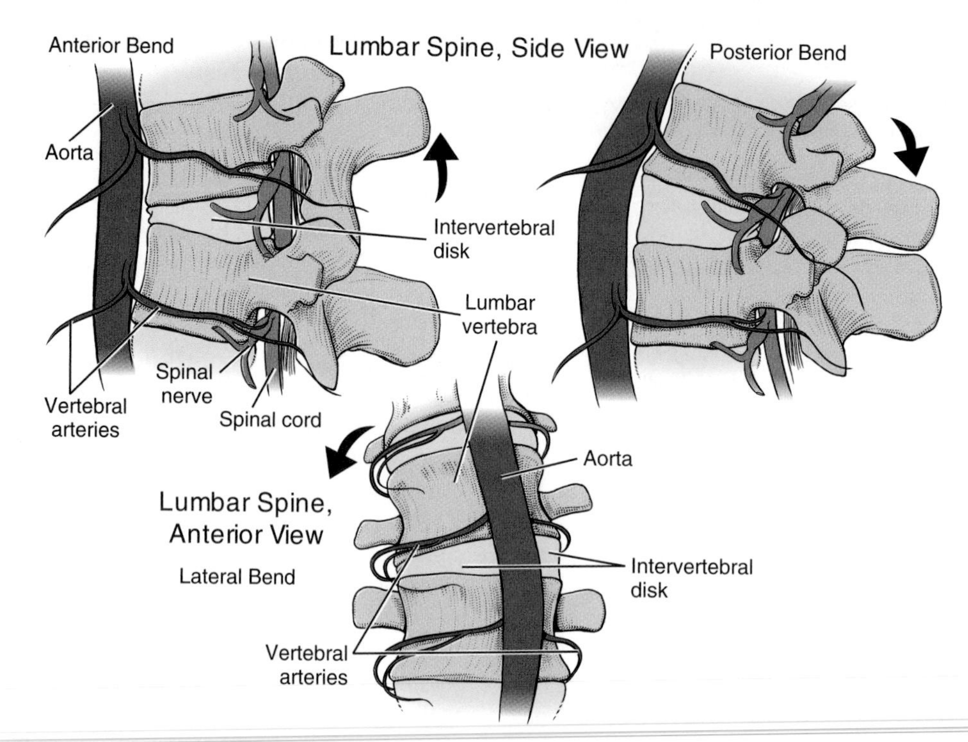

Fig. 14-4 Vascular supply is not compromised by position of the spine, so there is usually no change in back pain that is vascular-induced with change of position. Forward bend, extension, and side bending do not aggravate or relieve symptoms. Rather, increased activity requiring increased blood supply to the musculature is more likely to reproduce symptoms; likewise, rest may relieve the symptoms. Watch for a lag time of 3 to 5 minutes after the start of activity or exercise before symptoms appear or increase as a sign of a possible vascular component.

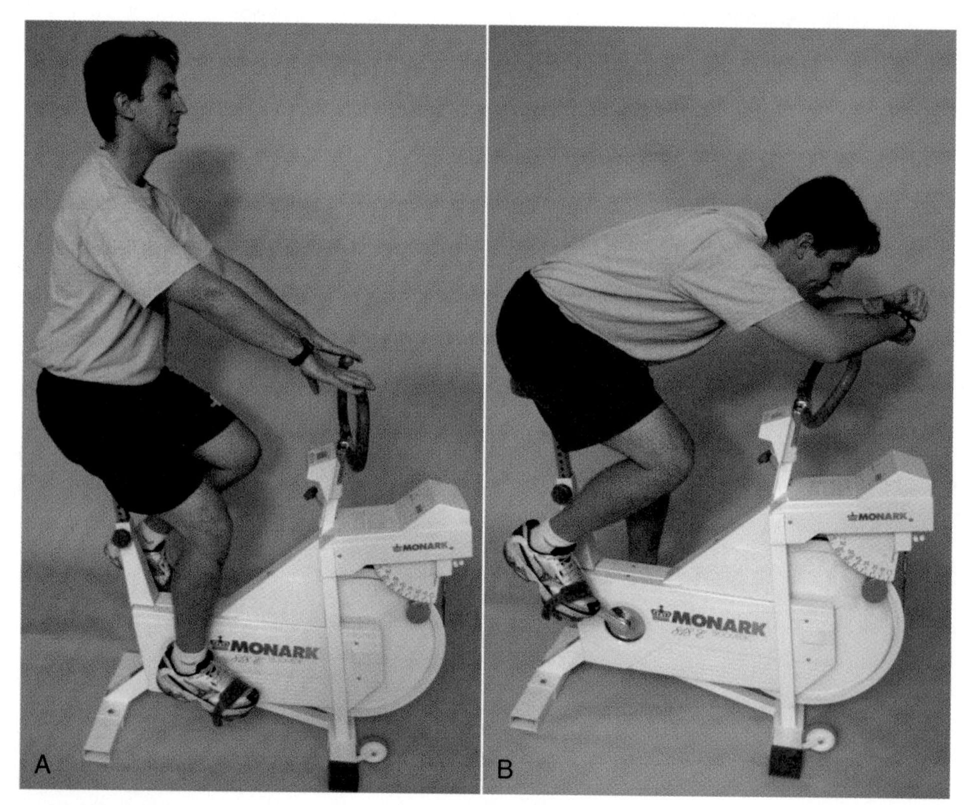

Fig. 14-5 Assessing the underlying cause of intermittent claudication: Vascular or neurogenic? The effect of stooping over while pedaling on vascular claudication is negligible, whereas a change in spine position can aggravate or relieve claudication of a neurogenic origin. **A,** The client is seated on an exercise bicycle and asked to pedal against resistance without using the upper extremities except for support. If pain into the buttock and posterior thigh occurs, followed by tingling in the affected lower extremity, the first part of the test is positive, but whether it is vascular or neurogenic remains undetermined. **B,** While pedaling, the client leans forward. If the pain subsides over a short time, the second part of the test is positive for neurogenic claudication but negative for vascular-induced symptoms. The test is confirmed for neurogenic cause of symptoms when the client sits upright again and the pain returns. (From Magee DJ: *Orthopedic physical assessment,* ed 5, Philadelphia, 2008, WB Saunders.)

Typically, there is no change because a change in position does not reproduce or alleviate vascular symptoms. The therapist can palpate pulses before and after the test to confirm the presence of vascular symptoms. What about neurogenic impairment? The client with neurogenic back pain may report a decrease in pain intensity or duration with forward flexion. Leaning forward (spinal flexion) can increase the diameter of the spinal canal, reducing pressure on the neural tissue.

When using the Bicycle test to look for neurogenic claudication, the client starts pedaling while leaning back slightly. This position puts the lumbar spine in a position of extension. If the pain is reproduced, the first part of the test is positive for a neurogenic source of symptoms. The client then leans forward while still pedaling. If the pain is less or goes away, the second part of the test is positive for neurogenic claudication. With neurogenic claudication, the pain returns when the individual sits upright again.

There is one major disadvantage to this test. Many clients in their sixth and seventh decades have both spinal stenosis and atherosclerosis contributing to painful back and/or leg symptoms. What if the client has back pain before even getting on the bicycle that is not relieved when bending forward? What diagnostic information does that provide?

The client could be experiencing neurogenic back pain that would normally feel better with flexion, but now while pedaling, vascular compromise occurs. In some cases, neurogenic pain lasts for hours or days despite change in position because once the neurologic structures are irritated, pain signals can persist.

The Bicycle test has its greatest use when only one source of back pain is present: Either vascular or neurogenic and even then, chronic neurogenic pain may not be modulated by change in position. An alternate test to distinguish neurogenic claudication (pseudoclaudication) from vascular claudication is the Stoop test.[122] The individual being tested walks quickly until symptoms develop. Relief of symptoms in response to sitting or bending forward is a positive test for a neurogenic source of pain. Straightening up and/or extending the spine reproduces the symptoms.

The therapist must rely on results of the screening interview and examination, taking time to perform a Review of Systems to identify clients who may need further medical evaluation. In some cases, medical referral is not required. Identifying the underlying pathologic mechanism directs the therapist in choosing the most appropriate intervention.

SCREENING FOR PULMONARY CAUSES OF NECK AND BACK PAIN

There are many potential pulmonary causes of back pain. The lungs occupy a large area of the upper trunk (see Fig. 7-1), with an equally large anterior and posterior thoracic area where pain can be referred. The most common conditions known to refer pulmonary pain to the somatic areas are pleuritis, pneumothorax, pulmonary embolus, cor pulmonale, and pleurisy.

Past Medical History

A recent history of one of these disorders in a client with neck, shoulder, chest, or back pain raises a red flag of suspicion. In keeping with the model for screening the therapist should review (1) past medical history, (2) risk factors, (3) clinical presentation, and (4) associated signs and symptoms (Box 14-5).

Clinical Presentation

Pulmonary pain patterns vary in their presentation based in part on the lobe(s) or segment(s) involved and on the underlying pathology. Several different pain patterns are presented in Chapter 7 (see Fig. 7-10).

Autosplinting is considered a valuable red flag of possible pulmonary involvement. Autosplinting occurs when the client prefers to lie on the involved side. Because pain of a

BOX 14-5 SCREENING FOR PULMONARY-INDUCED NECK OR BACK PAIN

History

Previous history of cancer (any kind, but especially lung, breast, bone, myeloma, lymphoma)
Previous history of recurrent upper respiratory infection (URI) or pneumonia
Recent scuba diving, accident, trauma, or overexertion (pneumothorax)

Risk Factors

Smoking
Trauma (e.g., rib fracture, vertebral compression fracture)
Prolonged immobility
Chronic immunosuppression (e.g., corticosteroids, cancer chemotherapy)
Malnutrition, dehydration
Chronic diseases: Diabetes mellitus, chronic lung disease, renal disease, cancer
Upper respiratory infection or pneumonia

Pain Pattern

Sharp, localized
Aggravated by respiratory movements
Prefer to sit upright
Autosplinting decreases the pain
ROM does not reproduce symptoms (e.g., shoulder and/or trunk movements)

Associated Signs and Symptoms

Dyspnea
Persistent cough
Constitutional symptoms: Fever, chills
Weak and rapid pulse with concomitant fall in blood pressure (e.g., pneumothorax)

pulmonary source is referred from the ipsilateral side, putting pressure on the involved lung field reduces respiratory movements and therefore reduces pain. It is uncommon for a person with a true musculoskeletal problem to find relief from symptoms by lying on the involved side.

The therapist should perform the following tests for clients with back pain who have a suspicious history or concomitant respiratory symptoms:

- Vital sign assessment
- Auscultation
- Assess the effect of reproducing respiratory movements on symptoms (e.g., does deep breathing, laughing, or coughing reproduce the painful symptoms?)
- ROM: Assess all active trunk movements (flexion, extension, side bending, and rotation)
- Can pain or symptoms be reproduced with palpation (e.g., palpate the intercostals)?

Although reproducing pain or increased pain on respiratory movements is considered a hallmark sign of pulmonary involvement, symptoms of pleural, intercostal, muscular, costal, and dural origin all increase with coughing or deep inspiration.

Only pain of a cardiac origin is ruled out when symptoms increase in association with respiratory movements. For this reason the therapist must always carefully correlate clinical presentation with client history and associated signs and symptoms when assessing for pulmonary disease.

Forceful coughing from an underlying pulmonary problem can cause an intercostal tear, which can be palpated. Even if some symptoms can be reproduced with palpation, the problem may still be pulmonary-induced, especially if the cause is repeated, forceful coughing from a pulmonary etiology.

Pancoast's tumors of the lung may invade the roots of the brachial plexus causing entrapment as they enlarge, appearing as pain in the C8 to T1 region, possibly mimicking thoracic outlet syndrome. Faulty data collection leads to inaccurate findings or incorrect diagnosis and treatment, resulting in less than optimal outcomes.[123] Other signs may include wasting of the muscles of the hand and/or Horner's syndrome with unilateral constricted pupil, ptosis, and loss of facial sweating (see the section on lung cancer in Chapter 7).

Tracheobronchial irritation can cause pain to be referred to sites in the neck or anterior chest at the same levels as the points of irritation in the air passages (see Fig. 7-2). This irritation may be caused by inflammatory lesions, irritating foreign materials, or cancerous tumors.

Associated Signs and Symptoms

Assessing for associated signs and symptoms will usually bring to light important red flags to assist the therapist in recognizing an underlying pulmonary problem. Neck or back pain that is reproduced, increased with inspiratory movements, or accompanied by dyspnea, persistent cough, cyanosis, or hemoptysis must be evaluated carefully. Clients with respiratory origins of pain usually also show signs of general malaise or constitutional symptoms.

SCREENING FOR RENAL AND UROLOGIC CAUSES OF BACK PAIN

When considering the possibility of a renal or urologic cause of back pain, the therapist can use the same step-by-step approach of looking at the history, risk factors, clinical presentation, and associated signs and symptoms.

For example, in anyone with back pain reported in the T9 to L1 area corresponding to pain patterns from the kidney or urinary tract (see Figs. 10-7 and 10-8), ask about a history of kidney stones, urinary tract infections (UTIs), and trauma (fall, blow, lift).

Origin of Pain Patterns

As discussed in Chapter 3, there can be at least three possible explanations for visceral pain patterns, including embryologic development, multisegmental innervation, and direct pressure on the diaphragm.

All three of these mechanisms are found in the urologic system. The *embryologic* origin of urologic pain patterns begins with the testicles and ovaries. These reproductive organs begin in utero where the kidneys are in the adult and then migrate during fetal development following the pathways of the ureters. A kidney stone down the pathway of the ureter causes pain in the flank radiating to the scrotum (male) or labia (female).

Evidence of the influence of *multisegmental innervation* is observed when skin pain over the kidneys is reported. Visceral and cutaneous sensory fibers enter the spinal cord close to each other and converge on the same neurons. When visceral pain fibers are stimulated, cutaneous fibers are stimulated, too. Thus visceral pain can be perceived as skin pain.

None of the components of the lower urinary tract comes in contact with the diaphragm, so the bladder and urethra are not likely to refer pain to the shoulder. Lower urinary tract impairment is more likely to refer pain to the low back, pelvic, or sacral areas. However, the upper urinary tract can impinge the diaphragm with resultant referred pain to the costovertebral area or shoulder.

Past Medical History

Kidney disorders such as acute pyelonephritis and perinephric abscess of the kidney may be confused with a back condition. Most renal and urologic conditions appear with a combination of systemic signs and symptoms accompanied by pelvic, flank, or LBP.

The client may have a history of recent trauma or a past medical history of UTIs to alert the clinician to a possible renal origin of symptoms.

Clinical Presentation

Acute pyelonephritis, perinephric abscess, and other kidney conditions appear with aching pain at one or several costovertebral areas, posteriorly, just lateral to the muscles at T12 to L1, from acute distention of the capsule of the kidney.

The pain is usually dull and constant, with possible radiation to the pelvic crest or groin. The client may describe febrile chills, frequent urination, hematuria, and shoulder pain (if the diaphragm is irritated). Percussion to the flank areas reveals tenderness; the therapist can perform Murphy's percussion (punch) test (see Fig. 4-54). Although this test is commonly performed, its diagnostic value has never been validated. Results of at least one Finnish study[124] suggested that in acute renal colic loin tenderness and hematuria (blood in the urine) are more significant signs than renal tenderness.[125] A diagnostic score incorporating independent variables, including results of urinalysis, presence of costovertebral angle tenderness and renal tenderness, duration of pain, appetite level, and sex (male versus female), reached a sensitivity of 0.89 in detecting acute renal colic, with a specificity of 0.99 and an efficiency of 0.99.[124]

Nephrolithiasis (kidney stones) may appear as back pain radiating to the flank or the iliac crest (see Fig. 10-7) (Case Example 14-10). Kidney stones may occur in the presence of diseases associated with hypercalcemia (excess calcium in the blood) such as hyperparathyroidism, metastatic carcinoma, multiple myeloma, senile osteoporosis, specific renal tubular disease, hyperthyroidism, and Cushing's disease. Other conditions associated with calculus formation are infection, urinary stasis, dehydration, and excessive ingestion or absorption of calcium.

Ureteral colic, caused by passage of a kidney stone (calculus), appears as excruciating pain that radiates down the course of the ureter into the urethra or groin area. The pain is unrelieved by rest or change in position. These attacks are intermittent and may be accompanied by nausea, vomiting, sweating, and tachycardia. Localized abdominal muscle spasm may be present. The urine usually contains erythrocytes or is grossly bloody.

Urinary tract infection affecting the lower urinary tract is related directly to irritation of the bladder and urethra. The intensity of symptoms depends on the severity of the infection. Although LBP may be the client's chief complaint, further questioning usually elicits additional urologic symptoms. The therapist should ask about

- Urinary frequency, urgency, dysuria (burning pain on urination), nocturia (frequency at night)
- Constitutional symptoms (fever, chills, nausea, vomiting)
- Blood in urine
- Testicular pain

Clients can be asymptomatic with regard to urologic symptoms, making the physical therapy diagnosis more difficult.

Screening Questions: Renal and Urologic System

It is important to ask questions about the presence of urologic symptoms (see Appendix B-5). Many people (therapists and clients alike) are uncomfortable discussing the details of bladder (or bowel) function. If presented in a professional manner with a brief explanation, both parties can be put at ease. For example, the interview may go something like this:

"I'm going to ask a few other questions that may not seem like they fit with the back pain (shoulder pain, pelvic pain) you're having. There are many possible causes of back pain and I want to make sure I don't leave anything out.

If I ask you anything you don't know, please pay attention over the next few days and see if you notice something. Don't hesitate to bring this information back to me. It could be very important."

To the therapist: The important thing to look for is CHANGE. Many people have problems with incontinence, nocturia, or frequency. If someone has always experienced a delay before starting a flow of urine, this may be normal for him (or her).

Many women have nocturia after childbirth, but most men do not get up at night to empty their bladders until after age 65. They may not even be aware that this has changed for them. Often, it is the wife or partner who answers the question about getting up at night as "yes!" Likewise, if a man has always had a delay in starting a flow of urine, he may not be aware that the delay is now twice as long as before. Or he may not recognize that being unable to continue a flow of urine is not "normal" and in fact, requires medical evaluation.

Pseudorenal Pain

Sometimes clients appear to have classic symptoms of a kidney problem but without any associated signs and symptoms. Such a situation can occur with someone who has a mechanical derangement of the costovertebral or costotransverse joint or irritation of the costal nerve (radiculitis, T10-T12).

What does this look like in the clinic? How does the therapist make the differentiation? Use the same guidelines for decision making in the screening process presented throughout this text (e.g., history, risk factors, associated signs, and symptoms).

History

Trauma is often the underlying etiology. The client may or may not report assault. The individual may not remember any specific trauma or accident. Pseudorenal pain can occur when floating ribs become locked with the ribs above, but this is a rare cause of these symptoms. Radiculitis or mechanical derangement of the T10 to T12 costovertebral or costotransverse joint(s) is more likely.

Risk Factors

Unknown or none for this condition.

Clinical Presentation

Pain pattern is affected by change in position:

- Lying on that side increases pain (remember clients with renal pain prefer pressure on the involved side; musculoskeletal symptoms are often made worse by lying on the affected side).
- Prolonged sitting increases pain; slumped sitting especially increases pain; the therapist can have the client try this position and see what effect it has on symptoms.

CASE EXAMPLE 14-10

Back and Flank Pain

Background and Description of Client: JH is a 57-year-old male with a history of mild mental retardation, seizure disorder, obesity, osteoarthritis, hypertension, and cervical disk disease (magnetic resonance imaging [MRI] reveals herniation at C7-T1 and spondylosis at C5-C6). He resides at a residential facility and is well known to physical therapy over the past 6 years because of five separate physical therapy examinations related to complaints of insidious onset of back pain.

These previous episodes of back pain resolved without physical therapy intervention. JH presented in physical therapy this time with complaints of low back and right hip pain that he and his primary physician attributed to a minor fall 2 months prior to the physical therapy examination. Physical therapy was not consulted during the initial period after the fall because x-rays were unremarkable and JH had not complained of any symptoms at that time.

When asked to point to the area of pain, JH indicated his right lower lumbar area and along the right hip and flank. He was unable to describe the pain due to some cognitive limitations, but he did report that it was unrelieved with rest and occurred intermittently.

JH works full-time in a sheltered workshop doing piecework. He reported that the pain kept him from performing his job fully, and he found that lifting boxes was particularly difficult due to the bending. He also reported that prolonged ambulation or exercise caused an increase in the flank pain. He was taking over-the-counter (OTC) ibuprofen for his pain; however, it was not effective.

JH is on the following medications: Colace (for constipation), Allegra (for allergy), Tegretol (for seizures), Zoloft (for obsessive-compulsive disorder), Risperdal (for psychosis), BuSpar (for anxiety), and ibuprofen (prn for pain).

Clinical Presentation: Vital signs were as follows: HR: 65 bpm; BP: 130/70 mm Hg; RR: 12; Temp: 99° F. These were not significantly different from JH's normal vital signs.

Gait analysis was significant for an antalgic gait, slight increase in base of support, decreased trunk and pelvic rotation, significant ankle pronation, and pes planus bilaterally (JH does not like to wear his orthotics). He is an independent ambulator on all surfaces without the use of an assistive device. He lives in a 2-story home and is able to ascend and descend stairs independently without complaints of pain.

Posture in standing was significant for decreased lumbar lordosis, rounded shoulders, and forward head, left shoulder mildly depressed.

Strength testing revealed strength of 4+/5 throughout upper extremities, trunk, and left lower extremity. JH was very hesitant with resisted strength testing on his right lower extremity for fear of pain; therefore no formal data was obtained. JH did report pain upon mildly resisted right hip flexion, abduction, and adduction.

Passive range of motion (PROM) was all within functional limits. There was no apparent evidence of inflammation in bilateral knees or hips. The physical therapist was unable to reproduce symptoms with palpation along spine and bilateral hips and knees.

Right knee extension active ROM (AROM) in sitting revealed pain in right flank. Right straight leg raise (SLR) test in supine also revealed similar pain in right flank. Right side bending produced right flank pain. Left side bending produced no symptoms.

Neurologic examination revealed intact sensation to light touch along dermatomal pattern. Deep tendon reflexes (DTRs) were 1+ throughout.

Evaluation: JH's symptoms appeared inconsistent and dependent upon level of physical activity. It seemed counterintuitive that a minor fall 2 months prior to this examination could cause the current symptoms. The location of the pain also raised some concerns because JH had never before complained of flank pain.

The physical therapist did not have access to the prior x-rays taken at the time of the fall. Therefore the therapist requested further x-rays of JH's hip and spine from the orthopedic surgeon serving as consultant to rule out a more serious orthopedic, systemic, or other medical issue. Physical therapy was deferred until the x-ray results were examined and reviewed by the orthopedic consultant and therapist.

Outcome and Discussion: AP pelvis and frog view x-rays of hips were reviewed by the orthopedic consultant and the physical therapist, and it was concluded that the x-rays were unremarkable. AP and lateral x-rays of JH's TLS spine at first glance also appeared to be unremarkable, and the x-ray report agreed with our initial assessment.

However, upon closer inspection, there was a circular 2-cm suspicious area that appeared on film at the level and location of JH's right kidney. The orthopedic surgeon ordered further imaging to confirm a diagnosis of kidney stone. An intravenous pyelogram (IVP) did confirm the diagnosis. After appropriate treatment for the kidney stones, JH reported that the pain on his right side had resolved.

JH was well known to the physical therapy department due to his previous examinations. JH was a challenging case because of the previous "false alarms" and because he does not always accurately communicate his symptoms due to his mild cognitive limits.

He also has comorbidities that warrant a more cautious approach in treating and assessing his complaints. These include hypertension and a seizure disorder and the multitude of medication that he takes.

It is up to the physical therapist to understand him and try to interpret his meanings as closely as she/he can. Fortunately in this case, JH's chief complaint of flank pain was different enough from previous complaints, and the films clearly showed a systemic cause of JH's symptoms.

Instructor's Comments: Some additional screening questions/information that might help with a case like this:

1. Did he have any symptoms of genitourinary distress (pain on urination? blood in the urine? difficulty starting or continuing a flow of urine? nocturia? frequency? or changes in bladder function)?
2. Did he have a past medical history of kidney stones?
3. Was Murphy's percussion (punch) test positive?
4. Was there a report of any constitutional symptoms (night sweats? spiked temps? flu-like symptoms)?

From Yee J, DPT: *Case report submitted as part of course requirements in fulfillment of DPT 910,* New York, 2002, Stony Brook. Used with permission.

- Symptoms are reproduced with movements of the spine (especially forward flexion and side bending).
- Presence of costovertebral angle tenderness: the therapist may be able to reproduce pain with palpation; Murphy's percussion (punch) test is negative (see Fig. 4-54).

A positive Murphy's test for renal involvement elicits kidney pain or reproduces the referred back pain and must be reported to the physician. A negative response occurs when there is no discomfort or pain or pain that can be reproduced by local palpation at the costovertebral angle. The therapist must ask about the presence of signs and symptoms associated with renal disease.

One final note about pseudorenal back pain: Thoracic disk disease can mimic kidney disease and presents with flank, buttock, and/or leg pain. Magnetic resonance imaging (MRI) scans are negative but may show only the lumbar spine.[126]

In the case of a possible thoracic disk mimicking renal involvement, the therapist can provide the physician with clinical findings and the reason for the referral. Look for a history of straining, lifting, accident, or other mechanical injury to the thoracic spine.

The therapist must look carefully for evidence of neurologic involvement. Perform a screening neurologic assessment as outlined in Chapter 4. There may be bladder changes, which can be confusing; are these urologic-induced or disk-related? Report any suspicious symptoms.

Associated Signs and Symptoms

Usually none when pseudorenal pain is present.

SCREENING FOR GASTROINTESTINAL CAUSES OF BACK PAIN

Back pain of a visceral origin occurs most often as a result of GI problems. Pain patterns associated with the GI system can present as sternal, shoulder, scapular, mid-back, low back, or hip pain and dysfunction. If the client had primary symptoms of GI impairment (abdominal pain, nausea, diarrhea, or constipation; see Fig. 8-18), he or she would see a medical doctor. Keep in mind the individual who has LBP with constipation could also be manifesting symptoms of pelvic floor muscle overactivity or spasm. In such cases, pelvic floor assessment should be a part of the screening exam. Consultation with a physical therapist skilled in this area should be considered if the primary care therapist is unable to perform this examination.

As it is, the referred pain patterns are quite convincing that the musculoskeletal region described is the problem. Referred pain patterns for the GI system are presented in Fig. 8-19 (anterior and posterior). These are the pain patterns the therapist is most likely to see.

Past Medical History and Risk Factors

Taking a closer look at past medical history, risk factors, and clinical presentation and asking about associated signs and symptoms may reveal important red flags and clues pointing to the GI system. The most significant and common history is one of long-term or chronic use of NSAIDs. Risk factors and assessment of risk for NSAID-induced gastropathy are discussed in detail in Chapter 8.

Other significant risk factors in the history include the long-term use of immunosuppressants, past history of cancer, history of Crohn's disease (also known as regional enteritis), or previous bowel obstruction.

Signs and Symptoms of Gastrointestinal Dysfunction

The most common signs and symptoms associated with the GI system are listed in Box 14-6 and discussed in greater detail in Chapter 8. Back pain (as well as hip, pelvic, sacral, and lower extremity pain) with any of these accompanying features should be considered a red flag for the possibility of GI impairment.

Anterior neck (esophageal) pain may occur, usually with a burning sensation ("heartburn") or other symptoms related to eating or swallowing (e.g., dysphagia, odynophagia). Esophageal varices associated with chronic alcoholism may appear as anterior neck pain but usually occur at the xiphoid process and are attributed to heartburn.

Anterior neck pain can also occur as a result of a diskogenic lesion requiring a careful history and neurologic screening to document findings. Clients with eating disorders who repeatedly binge and then purge by vomiting may report anterior neck pain without realizing the correlation between eating behaviors and symptoms.

When assessing neck pain, the therapist should look for other associated signs and symptoms, such as sore throat; pain that is relieved with antacids, the upright position, fluids, or avoidance of eating; and pain that is aggravated by eating, bending, or recumbency.

Dysphagia or difficulty swallowing, *odynophagia* (painful swallowing), and *epigastric pain* are indicative of esophageal involvement. Certain types of drugs (e.g., antidepressants,

BOX 14-6 SIGNS AND SYMPTOMS OF GASTROINTESTINAL DYSFUNCTION

Anterior neck pain or back pain accompanied by any of the following is a red flag:
- Esophageal pain
- Epigastric pain with radiation to the back
- Dysphagia (difficulty swallowing)
- Odynophagia (pain with swallowing)
- Early satiety; symptoms associated with meals
- Bloody diarrhea
- Fecal incontinence
- Melena (dark, tarry, sticky stools caused by oxidized blood)
- Hemorrhage (blood in the toilet)

antihypertensives, asthma medications) can make swallowing difficult, requiring a careful evaluation during the client interview.

Early satiety (the client takes one or two bites of food and is no longer hungry) is another red-flag symptom of the GI system (Case Example 14-11). In general, back pain made better, worse, or altered in any way by eating is a red-flag symptom. If the change in symptom(s) occurs immediately to within 30 minutes of eating, the upper GI tract or stomach/duodenum may be a possible cause. Change in symptoms 2 to 4 hours *after* eating is more indicative of the lower GI tract (intestines/colon).

Bloody diarrhea, fecal incontinence, and *melena* are three additional signs of lower GI involvement. It is important to ask the client about the presence of specific signs that may be too embarrassing to mention (or the client may not see the connection between back pain and bowel smears on the underwear). Asking someone with back pain about bowel function can be accomplished in a very professional manner. The therapist may tell the client:

"I am going to ask you a series of questions about your bowels. This may not seem like it is connected to your current problem, so just bear with me. These are important questions to make sure we have covered every possibility. If you do not know the answer to the question, pay attention over the next day or two to see how everything is working. If you notice anything unusual or different, please let me know when you come in next time."

CASE EXAMPLE 14-11

Early Satiety and Weight Loss

Background: A 78-year-old female was referred to physical therapy by her orthopedic surgeon 6 weeks status post (S/P) total knee replacement (TKR). Her active knee flexion was 70 degrees; passive knee flexion was only 86 degrees. There was a 15-degree extensor lag.

During the course of her rehabilitation program, her adult daughters took turns bringing her to the clinic. They all commented on how much weight she had lost, though the therapist thought she looked quite obese.

When asked about the weight loss, she replied, "Oh, I take a bite or two and then I'm not very hungry." This symptom (early satiety with weight loss) had been present for the last 2 months (starting prior to the TKR).

She did not have any other signs or symptoms associated with the gastrointestinal (GI) system. There were no reported changes in bowel function or the appearance of her stools, no blood in the stools, no back or sacral pain, no night pain that was not directly related to her knee, and no other changes in her health.

Her social history included the recent death of a spouse. She had taken care of her husband at home for the last 3 years after he had a severe stroke. She knew she needed a knee replacement, but put it off because of her husband's poor health. Within 6 weeks of his death, she scheduled the needed operation.

Could her weight loss be a delayed grieving reaction? Emotional overlay? How can you tell?

The screening process often begins with the recognition and categorization of red flags. It is not within the scope of a physical therapist's practice to diagnose psychologic or emotional problems. Clearly, many of the clients and patients in our clinics have significant psychologic needs and emotional responses to their illnesses, injuries, or conditions.

Identifying a cluster of signs and symptoms suggestive of a psychologic or behavioral component may help determine the need for behavioral counseling or a psych consult. However, the therapist's plan of care may include the use of specific client management skills based on observation of particular behavioral patterns.

What do you see in the history, clinical presentation, and associated signs and symptoms as they are presented here that raise a red flag?

- History: Age and positive social history for recent personal loss
- Clinical Presentation: Unremarkable; consistent with orthopedic diagnosis
- Associated Signs and Symptoms: Early satiety with weight loss

Viewing the whole client or patient and identifying the presence of emotional overlay to symptoms can be accomplished using the McGill Pain questionnaire, Waddell's nonorganic signs adapted for the knee, and listening to the client's response to her condition and the rehabilitation program (symptom magnification). These three assessment tools are discussed in Chapter 3.

There are really only two red flags here (age and early satiety with weight loss), but they are significant enough to warrant contact with her physician. The next question is: to whom do you send her? The referring orthopedist or her family doctor (if she has one)?

It may be best to communicate all findings with the referring physician or health care provider. The therapist can leave the door open by asking any one of the following questions:

- Do you want to see Mrs. So-and-So back in your office or shall I send her to her family physician?
- Do you want Mr. X/Mrs. Y to check with his/her family doctor or do you prefer to see him/her yourself?
- How do you want to handle this? or How do you want me to handle this?

Outcome: The orthopedic surgeon recommended referral to her primary care physician. Examination and diagnostic tests resulted in a diagnosis of esophageal cancer (early stage). The client was treated successfully for the cancer while completing her rehabilitation program.

FOLLOW-UP QUESTIONS

- When was your last bowel movement? (Look for a change of any kind in the client's normal elimination pattern. Additionally, failure to have a bowel movement over a much longer period of time than expected for that client may be a sign of impaction/obstruction/obstipation.)
- Are you having any diarrhea?
- Is there any blood in your stool?
- Have you ever been told you have hemorrhoids or do you know that you have hemorrhoids?
- Do you have difficulty wiping yourself clean?
- Do you find smears on your underwear later after a bowel movement?
- Do you have small amounts of stool leakage?

Again, when it comes to something like bowel smears on the underpants, it is important to distinguish between pathology and poor hygiene. The key to look for is change such as the new appearance of a problem that was not present before the onset of back pain or other symptoms. With blood in the stools, a medical doctor must differentiate between internal versus external bleeding.

Melena is a dark, tarry stool caused by oxidation of blood in the GI tract (usually the upper GI tract, but it can be the lower GI tract). The most common causes of abdominal bleeding are chronic use of NSAIDs, leading to ulceration, Crohn's disease, or ulcerative colitis, and diverticulitis or diverticulosis. Anyone with a history of these problems presenting with new onset of back pain must be screened for medical disease.

Hemorrhage or visible blood in the toilet may be a sign of anal fissures, hemorrhoids, or colon cancer. The etiology must be determined by a medical doctor. Be aware that there is an increased incidence of rectal bleeding from anal fissures and local tissue damage associated with anal intercourse. This occurs predominantly in the male homosexual or bisexual population but can be seen in heterosexual partners who engage in anal intercourse. There are also increasing reports of adolescents engaging in oral and anal intercourse as a form of birth control.

It may be necessary to take a sexual history. The therapist should offer the client a clear explanation for any questions concerning sexual activity, sexual function, or sexual history. There is no way to know when someone will be offended or claim sexual harassment. It is in the therapist's best interest to maintain the most professional manner possible.

There should be no hint of sexual innuendo or humor injected into any of the therapist's conversations with clients at any time. The line of sexual impropriety lies where the complainant draws it and includes appearances of misbehavior. This perception differs broadly from client to client.[110]

You may need to include the following questions (see also Appendix B-32). Always offer an explanation for taking a sexual history. For example, *"There are a few personal questions I'll need to ask that may help sort out where your symptoms are coming from. Please answer these as best you can."*

FOLLOW-UP QUESTIONS

- Are you sexually active?

  "Sexually active" does not necessarily mean engaging in sexual intercourse. Sexual touch is enough to transmit many STIs. The therapist may have to explain this to the client to clarify this question. Oral and anal intercourse are often not viewed as "sexual intercourse" and will result in the client answering the question with "No" when, in fact, for screening purposes, the answer is "Yes."
- Have you had more than one sexual partner (one at a time or during the same time period)?
- Have you ever been told you have a STI or STD such as herpes, chlamydia, gonorrhea, venereal disease, human immunodeficiency virus (HIV), or other disease?
- Is there any chance the bleeding you are having could be related to sexual activity?

For Women:

- What form of birth control are you using? (Risk factor: Intrauterine contraceptive device [IUCD])
- Is there any possibility you could be pregnant?
- Have you ever had an abortion?
  - *If* yes, follow up with careful (sensitive) questions about how many, when, where, and any immediate or delayed complications (physical or psychologic).

Back pain from any cause may impair sexual function. Many health care professionals do not address this issue; the therapist can offer much in the way of education, pain management, improved function, and proper positioning for work and recreation. Some publications are available to assist therapists in discussing sexual function and pain control for the client with back pain.[127,128]

Esophagus

Esophageal pain will occur at the level of the lesion and is usually accompanied by epigastric pain and heartburn. Severe esophagitis (see Fig. 8-13) may refer pain to the anterior cervical or more often, the mid-thoracic spine.

The pain pattern will most likely present in a band of pain starting anteriorly and spreading around the chest wall to the back. Rarely, pain will begin in the mid-back and radiate around to the front. Referred pain to the mid-thoracic spine occurs around T5-6.

As with cervical pain of GI origin, there may be a history of alcoholism with esophageal varices, cirrhosis, or an underlying eating disorder. If liver impairment is an underlying factor, there may be signs such as asterixis (liver flap or flapping tremor), palmar erythema, spider angiomas, and carpal (tarsal) tunnel syndrome (see discussion in Chapter 9).

Keep in mind that this same type of mid-thoracic back pain can occur with thoracic disk disease. Look for a history of trauma and neurologic changes typically associated with disk degeneration (e.g., bowel and bladder changes, numbness and tingling or paresthesias in the upper extremities); these are not usually present with esophageal impairment. Lower thoracic disk herniation can cause groin pain, leg pain, or mimic kidney pain.

Stomach and Duodenum

Long-term use of NSAIDs is the most common cause of back pain referred from the stomach or duodenum. Ulceration and bleeding into the retroperitoneal area can cause pain in the back or shoulder. The primary and referred pain patterns for pain of a stomach or duodenal source are shown in Fig. 8-14.

The referred pain to the back is at the level of the lesion, usually between T6 and T10. For the client with mid-thoracic spine pain of unknown cause or which does not fit the expected musculoskeletal presentation, ask about associated signs and symptoms such as

- Blood in the stools
- Symptoms associated with meals
- Relief of pain after eating (immediately or 2 hours later)
- Increased symptoms with or during a bowel movement
- Decreased symptoms after a bowel movement

The pain of peptic ulcer (see Figs. 8-8 and 8-14) occasionally occurs only in the mid-thoracic back between T6 and T10, either at the midline or immediately to one side or the other of the spine. Posterior penetration of the retroperitoneum with blood loss and resultant referred thoracic pain is most often caused by long-term use of NSAIDs. The therapist should look for a correlation between symptoms and the timing of meals, as well as the presence of blood in the feces or relief of symptoms with antacids.

Small Intestine

Diseases of the small intestine (e.g., Crohn's disease, irritable bowel syndrome, obstruction from neoplasm) usually produce midabdominal pain around the umbilicus (see Fig. 8-2), but the pain may be referred to the back if the stimulus is sufficiently intense or if the individual's pain threshold is low (see Fig. 8-15) (Case Example 14-12).

For the client with LBP of unknown cause or suspicious presentation, ask if there is ever any abdominal pain present. Alternating abdominal/LBP at the same level is a red flag that requires medical referral. Since both symptoms do not always occur together, the client may not recognize the relationship or report the symptoms. The therapist must be sure and ask appropriate screening questions (Case Example 14-13).

Look for a known history of Crohn's disease (regional enteritis), irritable bowel syndrome, bowel obstruction, or cancer. Low back, sacral, or hip pain may be a new symptom of an already established disease. The client may not be aware that 25% of the people with GI disease have concomitant back or joint pain.

CASE EXAMPLE 14-12

Crohn's Disease and Back Pain

A 23-year-old ballet dancer with "shin splints" comes to you from a sports medicine doctor. Besides anterior lower leg pain, she also reports low back pain (LBP) that seems to come and go with overuse. She has a history of Crohn's disease.

Can symptoms of anterior compartment syndrome be caused by Crohn's disease?

It is very unlikely. There are no reported cases to date. Crohn's disease is linked with low back, hip, and sometimes knee pain (knee pain is usually associated with hip pain and usually does not occur alone).

Anterior compartment syndrome is easily reproducible with tenderness on palpation of the anterior tibial region. The pain pattern and etiology is fairly typical and symptoms respond to treatment. If the soft tissues are acutely inflamed, surgical intervention may be required.

What questions can you ask to rule out a gastrointestinal (GI) cause for her back pain?

- Ask about the presence of GI signs and symptoms:
  Are you having any nausea, vomiting, diarrhea, or constipation?
  Any change in your bowel movements? Any trouble wiping yourself clean after a bowel movement?
  Any blood in the stools?
- Any other symptoms of any kind? (headaches, sweats, fever)
- Is there abdominal pain and is it at the same level as the back pain?
- Does the abdominal and/or back pain change with food intake (assess from 30 minutes to 2 hours after eating)?
- Is there relief of back pain with passing gas or having a bowel movement?
- Is there a recent (chronic) history of antibiotic and/or NSAID use?
- Has the client experienced any joint pain anywhere else in the body?
  Any skin rashes anywhere?

A "yes" answer to any of these questions is a significant red flag and must be evaluated in context of the overall clinical presentation and findings from the Review of Systems.

Enteric-induced arthritis can be accompanied by a skin rash that comes and goes. A flat red or purple rash or raised skin lesion(s) is possible, usually preceding the joint or back pain. The therapist must ask the client if he/she has had any skin rashes in the last few weeks.

The therapist may treat joint or back pain when there is an unknown or unrecognized enteric cause. Palliative intervention for musculoskeletal symptoms or apparent movement impairment can make a difference in the short-term but does not affect the final outcome.

Eventually the GI symptoms will progress; symptoms that are unrelieved by physical therapy intervention are red flags. Medical treatment of the underlying disease is essential to correcting the musculoskeletal component.

CASE EXAMPLE 14-13

Abdominal and Back Pain at the Same Level

Background: A 68-year-old accountant came to physical therapy as a self-referral for low back pain (LBP). He reported slipping on a patch of ice as the mechanism of injury.

Symptoms were mild but distressing to this gentleman. He reported pain as "sore" and "aching" with any spinal twisting or side bending to the right. The pain was present across the low back on both sides.

The client reported symptoms of stomach distress from time to time. He attributed this to his trips overseas, eating foods from Ireland, Scotland, Germany, and the Netherlands.

Lumbar range of motion (ROM) was fairly typical of a nearly 70-year-old man with most of his functional forward flexion from the hips and thoracic spine. True physiologic motion in the lumbar spine was negligible. Accessory spinal motions were also limited globally. Active rotation and side bending were stiff and limited to both sides, but only painful to the right.

Neurologic screening exam was negative. The therapist did not ask about the presence of any other symptoms of any kind anywhere else in his body. No questions were asked about changes in the pattern of his bowel movements or appearance of his stools.

Given the examination results as tested, a conditioning exercise program seemed most appropriate. The client began a stationary bicycling program alternating with walking when the weather permitted. He reported gradual relief from his symptoms and return of motion and function to his previous levels.

Four months later this same client reported another injury while walking with subsequent back pain.

What are the red flag findings? What is the next step in the screening process?

The client's age (over 50) is the first red flag. Back pain across both sides can be considered bilateral and therefore a red flag until further assessment is completed. The presence of back pain and abdominal pain or discomfort warrants some additional questions.

The therapist should conduct a more thorough pain assessment and ask about the location of the symptoms as well as the presence of any additional gastrointestinal (GI) symptoms. Back pain and abdominal pain at the same level is always a red flag.

Screening questions related to the back and GI dysfunction are available at the end of this chapter. Questions about changes in bowel function may reveal some important clues. A screening physical assessment of the abdomen including visual inspection, palpation, and auscultation as described in Chapter 4 may be helpful. Vital sign assessment is always recommended.

Result: The key red flag in this case was alternating back and abdominal pain at the same level. The client did not see a connection between these two episodes of pain. When his back hurt, he did not have any abdominal pain and vice versa.

The client was advised to see his regular physician for an evaluation. He was diagnosed with colon cancer in advanced stages and died 6 weeks later. Earlier detection may have made a difference in this case, but the cyclical nature of his presentation masked the true significance of his symptoms.

SCREENING FOR LIVER AND BILIARY CAUSES OF BACK PAIN

The primary pain pattern for liver disease is right over the liver. In primary liver pathology, palpation of the organ will reproduce the symptoms and the examiner can feel the liver distention. The normal, healthy liver is located up under the right side of the diaphragm and ribs. The gallbladder is tucked up under the liver (see also Fig. 9-2).

When a referred pain pattern occurs, there may be pain on palpation of the liver, but the primary complaint is of back pain. There is no report of anterior pain to alert the examiner to the need for liver palpation. In anyone with the referred pain patterns depicted and described in Fig. 9-10, liver palpation may be required as part of the physical assessment (see Figs. 4-51 and 4-52). In addition to a painful and distended liver, the client may report:

- Pain/nausea 1 to 3 hours after eating (gallstones)
- Pain immediately after eating (gallbladder inflammation)
- Muscle guarding/tenderness and fever/chills in the right upper quadrant (posterior)

Other signs and symptoms associated with liver impairment are discussed in detail in Chapter 9 and include:

- Liver flap (asterixis)
- Nail bed changes (nail of Terry)
- Palmar erythema (liver palms)
- Spider angiomas
- Ascites, jaundice

Gallbladder and biliary disease may also refer pain to the interscapular or right subscapular area. The therapist should be observant for any report of fever and chills, nausea and indigestion, changes in urine or stool, or signs of jaundice. The client may not associate GI symptoms with the scapular pain or discomfort. The therapist can use specific questions to rule out potential GI problems (see Special Questions to Ask in this chapter and in greater detail in Chapter 8).

The Pancreas

Acute pancreatitis may appear as epigastric pain radiating to the mid-thoracic spine (see Fig. 8-17). Pain from the head of the pancreas is felt to the right of the spine, whereas pain from the body and tail is perceived to the left of the spine. More rarely, pain may be referred to the upper back and midscapular areas.

There may be a history of alcohol and tobacco use. Associated symptoms, which are usually GI related, may include diarrhea, anorexia, pain after a meal, and unexplained weight loss. The pain is relieved initially by heat, which decreases muscular tension, and may be relieved by leaning forward, sitting up, or lying motionless.

The therapist should remain alert for the client with LBP who reports benefit from a heating pad or other heat modalities but then suddenly gets worse and does not improve with physical therapy intervention.

SCREENING FOR GYNECOLOGIC CAUSES OF BACK PAIN

Gynecologic disorders can cause midpelvic or LBP and discomfort. Gynecologic-induced back pain occurs most often in women of childbearing ages (commonly between ages 20 and 45). How can the therapist recognize when a woman may be experiencing back pain from a gynecologic cause?

As always, the model for screening includes history, presence of any risk factors, clinical presentation, and associated signs and symptoms. Obviously, female sex is a clear flag of possible gynecologic involvement in the case of back (or pelvic, groin, hip, sacral or SI) symptoms, especially with a history of insidious onset or previous history of reproductive cancer.

Whenever there is an absence of objective musculoskeletal findings, a history of gynecologic involvement, or associated signs and symptoms of gynecologic disorders, the therapist is encouraged to ask appropriate questions to determine the need for a gynecologic evaluation (Case Example 14-14).

The therapist must determine what phase the woman is in her reproductive life cycle (see previous discussions of Life Cycles and Menopause in Chapter 2). If the client is an adolescent, has she begun her menstrual cycle (menses)? If a young to middle-aged adult, is she menstruating, or has she had a hysterectomy and experienced surgically induced menopause?

Past Medical History

Gynecologic conditions causing back pain can include retroversion (tipping back) of the uterus, ovarian cysts, uterine fibroids, endometriosis, pelvic inflammatory disease, or normal pregnancy (Case Example 14-15).

Usually, there is a history of a chronic or long-standing gynecologic disorder, and the association between back pain and gynecologic disorder has been established. There may be a history of sexual assault, incest, sexually transmitted disease (STD), ectopic pregnancy, use of an IUCD, dysuria, or abortion.

Risk Factors

Often the history and risk factors for back or pelvic pain are synonymous, especially multiple pregnancies and births, with administration of an epidural during delivery, prolonged pushing, and/or use of forceps. Other risk factors include abnormal uterine position, endometriosis, ovarian cysts and uterine fibroids, ectopic pregnancy, and the use of an IUCD.

Back pain is common during pregnancy beginning most often during the second trimester between the fifth and seventh months of gestation; intensity varies throughout pregnancy.[129-132] Women who have had multiple pregnancies or births may have SI pain or LBP associated with poor abdominal or pelvic floor tone and ligamentous laxity. Studies repeated over time have not shown an increased risk of back pain in women who receive epidural anesthesia during delivery.[133,134]

Additionally, women who have had one or more abortions may seek health care months to years later with a variety of physical and psychologic symptoms referred to as postabortion syndrome or postabortion survivor's syndrome. This condition has not been classified in the *Diagnostic and Statistical Manual* and its existence remains controversial.

CASE EXAMPLE 14-14

Human Movement Impairment

A 28-year-old woman in the twentieth week of her first pregnancy reported low back pain (LBP) of approximately 2 weeks' duration. She could not recall any injury or cause for her pain and attributed it to her pregnancy. She did report a 6-year history of back pain caused by exercise (military press); prior to this episode her back pain could be relieved by rest, heat, and massage therapy.

The current back pain was located bilaterally in the thoracolumbar paraspinal region and described as a "nagging ache." The client rated her pain as a 7 to 9 on the Numeric Rating Scale (NRS; see Fig. 3-6), worse in the afternoon and evening. Pain was aggravated by sitting more than 20 minutes and bending forward. She reported episodes of night pain that could be relieved by a change in position.

There were no other symptoms anywhere in her body; she was not taking any medications except for prenatal vitamins. She reported her pregnancy was "normal" with appropriate weight gain. There has been no spotting or vaginal bleeding during the pregnancy. Vital signs were within normal limits (WNL).

Is a medical screening examination needed?

The client's age is not a red flag at this time. Although she reports an insidious onset for her symptoms, the pain is not constant and can be relieved with a change in position. The pain wakes her up at night, but she is able to get back to sleep by getting up and walking or by changing position. Vital signs were normal and there were no constitutional symptoms.

At this point the evaluation can proceed as usual. The therapist should include a screening neurologic assessment as part of the examination. Keep in mind that hormonal changes can unmask a preexisting but asymptomatic musculoskeletal condition, which is something the physical therapist can address. Movement testing further confirmed an extension syndrome with worse symptoms during trunk flexion and improved pain after repetitive trunk extension.

No further medical screening is required unless additional red flag symptoms develop. The client's improvement with physical therapy intervention confirmed the decision that medical referral was not necessary.

Data from Requejo SM, Barnes R, Kulig K, et al: The use of a modified classification system in the treatment of low back pain during pregnancy: A case report, *J Orthop Sports Phys Ther* 32(7):318-326, 2002.

Multiple Pregnancies and Births

Even though pregnancy and childbirth are natural physiologic processes, these events can be traumatic to the soft tissues of the pelvic floor. The risk of postpartum pain

CASE EXAMPLE 14-15

Back Pain During Pregnancy

A 32-year-old Native American woman in the third trimester of her second pregnancy presented with acute onset of mid- to right-sided lumbar pain. She reported pain radiating around to the right side. An abdominal sonogram was negative, and all lab values were within normal limits. The client declined any further imaging studies and requested a referral to physical therapy.

What will you need to do to make sure this client's problem is within the scope of a physical therapy practice?

Take a thorough history (including childbirth histories) and evaluate pain pattern(s) carefully.

Screen for domestic abuse sometime during the evaluation or early treatment intervention.

Ask about the presence of any other symptoms, even if they seem unrelated to her pregnancy or back pain.

See if you can reproduce the symptoms by palpation or through position or movement; assess for trigger points.

Take all vital signs and ask about the presence of constitutional symptoms.

Assess for rectus abdominis diastasis (separation of the rectus abdominal muscles) as a possible contributing factor.

Outcome: During palpation of the ribs, the therapist noted an outward flaring of the lower ribs. There were pain and tenderness at the interchondral junctions between the eighth and tenth ribs.

The history was significant for chronic cough from smoking. The woman reported feeling the child in a horizontal position pushing against the lower ribs.

Based on these findings, the therapist telephoned the physician and asked if there was any chance a rib fracture could be causing the painful symptoms. The client agreed to an x-ray and the radiograph showed a fracture of the right tenth rib.

referred to the low back (with or without pelvic girdle pain) increases as a result of multiple pregnancies and births. If the woman's history includes a recent birth or multiple previous births, she may not recognize the association between her current symptoms and her pregnancy/delivery history.

Abnormal Uterine Positions

Having an understanding of the normal female reproductive anatomy (see Fig. 15-3) can help the therapist better appreciate musculoskeletal pain and dysfunction that can occur with abnormal uterine positions (see Fig. 15-4).

Taking a careful history and correlating symptoms with a woman's monthly cycle can help the therapist determine when to refer a client for a possible gynecologic cause of back, pelvic, or sacral pain/symptoms. Many problems affecting the pelvic floor musculature can be treated successfully by a physical therapist and do not require medical referral.

Endometriosis

Endometriosis is an estrogen-dependent disorder defined by the presence of endometrial tissue (lining of the uterus) outside of the uterus. Each month as the woman's body prepares for a fertilized egg, the uterus becomes engorged with blood, providing a fertile place for the egg to attach and begin growing. If and when the unfertilized egg passes out of the body, the uterus sloughs off the lining of blood and the woman has a flow of menstrual blood for about 3 to 5 days.

One theory to explain endometriosis suggests there may be *retrograde menses*, which means the blood goes up into the body, rather than down and out through the vagina. Whatever the mechanism, implants of cells that line the uterine cavity become transplanted and form small cysts outside of the uterus. These cysts are found on other organs or structures within the pelvic cavity. The cysts respond each month the same way as the endometrium during the menstrual cycle. The misplaced tissue engorges with blood just as it would when lining the uterus. The blood cannot drain out of the body and the result is lesions filled with a thick chocolate-type material or "chocolate cysts" wherever the endometrial tissue is located, with subsequent swelling, bleeding, and scar tissue formation.[135,136]

These pockets of blood can be deposited anywhere in the body. Whereas it was once thought that the blood just reached the pelvic and abdominal cavities, coating the viscera contained within, it is clear now that endometrial tissue migrates throughout the body. It has been recovered from bone, lungs, and even the brain.[137,138]

Pain can occur anywhere, but often the woman experiences back, pelvic, hip, and/or sacral pain that can be mistaken for a musculoskeletal, musculoligamentous, or neuromuscular impairment of the lumbar spine (Case Example 14-16).

The key to recognizing this condition is that often it is cyclical. Symptoms come and go with the menstrual cycle. After menopause, pain can persist from scar tissue. There may be urinary tract and bowel involvement with associated symptoms ranging from urinary frequency, intermittent dysuria, and bloody stools to ureteral or bowel obstruction.

This condition is more common than previously thought. It is estimated that up to 50% of the female population who are infertile are affected by endometriosis.[137,139] It is not clear what, if any, risk factors increase a woman's risk of developing endometriosis. Endometriosis has been linked with other health problems such as chronic fatigue syndrome, hypothyroidism, fibromyalgia, rheumatoid arthritis, multiple sclerosis, and systemic lupus erythematosus.[140-142] Endometriosis is a risk factor for ovarian and breast cancer.[137,143]

A cure has not been found at the present time, but for many women, it can be managed with medications and/or surgery. The therapist can be helpful in providing pain management strategies that can reduce sick leave and improve daily function. See Box 15-5 for more information on this condition.

CASE EXAMPLE 14-16

Endometriosis

Case Description: A 25-year-old female was referred for physical therapy with a diagnosis of nonspecific low back pain (LBP). She presented with the sudden onset of pain in the left lumbosacral region, left lower abdominal quadrant, and left buttock and anterior thigh which was constant and severe.

Medical examination ruled out a renal source of pain and diagnosed the client with a low back sprain. X-rays and magnetic resonance imaging (MRI) scans were negative and ruled out a spondylogenic, oncologic, or discogenic lesion. She was given an injection of Demerol and prescription for nonsteroidal antiinflammatory drugs (NSAIDs) and antispasmodics and referred to physical therapy.

Past Medical History and Risk Factors: The client was a nonsmoker and consumed alcohol only on occasion. Personal family history was unremarkable; she reported that her mother had rheumatoid arthritis and hypothyroidism.

Clinical Presentation: The client was seen in physical therapy 3 weeks after the initial painful episode. She presented with a chief complaint of sharp, constant pain in the left lumbosacral region, which occasionally radiated into the left lower abdominal quadrant and into the left buttock and the anterior thigh as far distally as the knee.

The pain was worse when sitting or walking. She was only able to sleep 1 to 2 hours at a time because of the severity of the pain. There was no report of bowel or bladder changes. The hip and sacroiliac joint were ruled out as the sources of pain. A neurologic screening examination was negative.

Trunk motions were mildly restricted with increased pain during forward flexion. There was a positive left straight leg raise (SLR) test at 60 degrees. The client appeared to have a musculoskeletal-based movement impairment.

Physical examination determined the most significant clinical finding to be exquisite tenderness in the left lower abdominal quadrant. The client reported marked tenderness with palpation over the left lower abdominal quadrant just proximal to the ASIS. She also reported tenderness with palpation directly over the left lumbar paraspinal region just superior to the iliac crest.

Red Flags: The sudden onset, intensity, severity, and duration of the client's back pain raised a red flag. The left lower quadrant was the location of greatest tenderness and severe subjective pain, both experienced at rest and with activity. The client's sex and childbearing age raise yellow (caution) flags.

Should the therapist treat this client and reassess symptoms and clinical presentation in 2 weeks or refer immediately?

Once again, the decision to carry out a physical therapy plan of care with direct intervention versus making a medical referral is based on clinical judgment. Given the presentation of this case, either decision could be justified.

Since she was evaluated by a medical doctor who sent her to physical therapy, a telephone call would be more appropriate than suggesting the client go back to her doctor.

In this case the therapist made the decision not to treat the client given the fact that a delay in diagnosis with risk for increased morbidity and possible mortality is possible with LBP from serious pelvic pathology.

Outcomes: The therapist conferred with the referring orthopedic surgeon and a referral was made to a gynecologist. Further testing provided a diagnosis of endometriosis and ovarian cyst. The client underwent laparoscopy; the diagnosis of endometriosis was confirmed.

Following medical and surgical intervention, the lower quadrant pain was abolished, and the LBP and leg pain significantly diminished in frequency and intensity, enabling the client to return to her normal activities.

Discussion: Given the prevalence of endometriosis, physical therapists are likely to encounter clients with this disorder in orthopedic physical therapy practice. Proper differential diagnosis is necessary to identify the risk factors and physical findings that would provide early diagnosis of endometriosis and avoid the morbidity associated with this and other pelvic disorders.

From Troyer MR: *Differential diagnosis of endometriosis in a patient with nonspecific low back pain.* Case report presented in partial fulfillment of DPT 910, Principles of Differential Diagnosis, Institute for Physical Therapy Education, Chester, PA, 2005, Widener University. Used with permission.

CLINICAL SIGNS AND SYMPTOMS

Endometriosis

- Intermittent, cyclical, or constant pelvic and/or back pain (unilateral or bilateral)
- Pain during or after sexual intercourse
- Painful bowel movements or painful urination during menstrual period
- Small blood loss (spotting) before or between periods
- Heavy or irregular menstrual bleeding
- Bleeding anywhere else (nose bleeds, coughing up blood, blood in urine or stools)
- Fatigue
- History of ectopic pregnancy, miscarriage, infertility
- GI problems (abdominal bloating and cramping, nausea, diarrhea, constipation)

Ovarian Cysts and Uterine Fibroids

Ovarian cysts are often asymptomatic until they grow large enough to pull the ovary out of its normal position, sometimes cutting off the blood supply to the ovary. As the weight of the ovary causes a change in position, pressure is exerted against the uterus, bladder, intestines, or vagina, causing a variety of symptoms.

Lower abdominal or pelvic pain is most common, but back pain associated with ovarian cysts and uterine fibroids can occur, usually presenting in a cyclical pattern associated with the menstrual cycle similar to endometriosis. A physician must determine the underlying gynecologic cause of back (hip, pelvic, sacral) pain or symptoms.

In a screening context, we look for red-flag histories, clinical presentation, risk factors, and associated signs and symptoms. Obesity may be a risk factor because more than half of the women affected by this disorder are obese but other risk factors remain unknown. Ovarian cysts present as part of the polycystic ovarian syndrome put the woman at increased risk for insulin resistance and potentially at increased risk for cardiovascular disease as a result.[144-146]

If the Review of Systems points to a gynecologic source of pain/symptoms, further questions can be asked and a referral made if appropriate. Low back pain is a late finding for some women with ovarian cancer (see Chapter 15).

CLINICAL SIGNS AND SYMPTOMS

Ovarian Cysts

- Abdominal pressure, pain, or bloating
- Discomfort during urination, bowel movements, or sexual intercourse
- Irregular menses, infertility
- Dull aching low back, buttock, pelvic, or groin pain
- Sudden, sharp pain with rupture or hemorrhage

Ectopic Pregnancy

An ectopic pregnancy is a live pregnancy that takes place outside the uterus. As shown in Fig. 15-5, this may occur in a variety of places such as the ovary, the tube (tubal pregnancy), outside lining of the uterus, or along the peritoneal cavity. None of these locations can sustain a viable ovum, and the woman will have a spontaneous abortion (miscarriage).

Risk factors include STDs, prior tubal surgery, and current use of an IUCD. Depending on the location of the ectopic pregnancy, symptoms can include back, hip, sacral, abdominal, pelvic, and/or shoulder pain. Shoulder pain is more likely to occur if there is retroperitoneal bleeding when rupture of the developing embryo and hemorrhage occurs with pressure on the diaphragm.

It is usually unilateral on the same side as the bleeding but can cause bilateral shoulder pain if the hemorrhage is significant enough to impinge both sides of the diaphragm. The pain is usually of a sudden onset (when rupture and hemorrhage occur) with intense, constant pain. Situations of this type represent a medical emergency. Most likely the client did not come to the therapist for this problem but may develop emerging symptoms while being treated for some other orthopedic or neurologic problem.

Consider it a red flag when any woman of childbearing age who is sexually active has sudden, intense pain as described. Take her blood pressure and other vital signs while asking appropriate screening questions. Seek immediate medical assistance.

CLINICAL SIGNS AND SYMPTOMS

Ectopic Pregnancy

- Amenorrhea or irregular bleeding and spotting
- Diffuse, aching lower abdominal quadrant or LBP; can cause ipsilateral shoulder pain
- May progress to a sharper, intermittent type of pain

Intrauterine Contraceptive Device

The intrauterine contraceptive device (IUCD is the current medical term; known by most women as an IUD) has become popular once again, having gone out of favor in the 1970s when the copper T caused so many problems. Although this contraceptive device has been improved, there are still potential problems (Fig. 14-6). The body may recognize this as a foreign object and set up an immune response or try to wall it off. The IUCD can become embedded in the tissue of the uterus, causing inflammation, infection, and scarring.

For any woman with low back, pelvic, sacral, or hip pain who is in the reproductive age range, it may be necessary to ask about her method of birth control: Are you using an IUD for birth control?

Clinical Presentation

Back pain that is associated with the menstrual cycle occurs most often at or around the point of ovulation (between day 10 and day 14 for most women) and again just prior to or during menstrual flow (between days 23 and 28 for most women). Day 1 is counted as the first day the woman experiences bleeding with her menstrual cycle.

Back pain associated with the menstrual cycle may be a regular feature for a woman, it may occur intermittently, or it may be new onset, and the woman is unaware of the link between the two until she charts her monthly cycle and correlates it with her back pain.

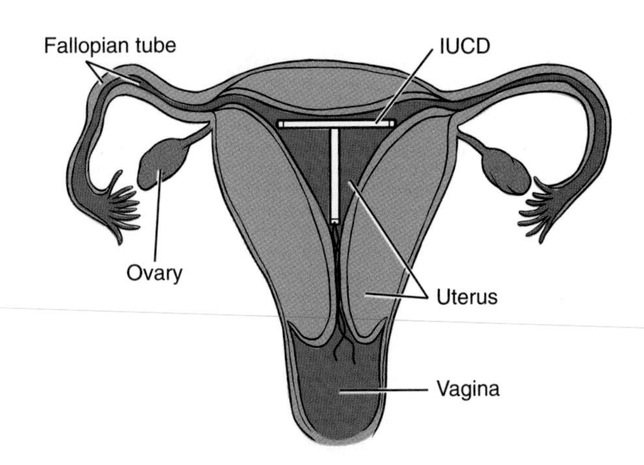

Fig. 14-6 Intrauterine contraceptive device (IUCD or IUD), a potential source of low back, pelvic, sacral, or even hip pain in any woman of reproductive age who is using this form of birth control.

A woman may have back pain accompanied by or alternating with sharp, bilateral, and cramping pain in the lower abdominal and/or pelvic quadrants. Menstrual pain can be referred to the rectum, lower sacrum, or coccyx. Tumors, masses, or even endometriosis may involve the sacral plexus or its branches, causing severe, burning pain.

Associated Signs and Symptoms

After gathering information during the examination, the therapist performs a Review of Systems looking for clusters of signs and symptoms suggesting a gynecologic cause of LBP. If appropriate, the next step is to ask a few final screening questions.

CLINICAL SIGNS AND SYMPTOMS

Gynecologic Disorders

- Missed menses, irregular menses, history of menstrual disturbances, painful menstruation
- Tender breasts
- Nausea, vomiting
- Chronic constipation (with laxative and enema dependency)
- Pain on defecation
- Fever, night sweats, chills
- Low blood pressure (hemorrhaging with ectopic pregnancy)
- Vaginal discharge
- Abnormal vaginal bleeding
  - Late menstrual periods with persistent bleeding
  - Spotting before period or between periods
  - Irregular, longer, heavier menstrual periods, no specific pattern
  - Any postmenopausal bleeding
- Urinary problems (intermittent dysuria, frequency, urgency, hematuria)

SCREENING FOR MALE REPRODUCTIVE CAUSES OF BACK PAIN

Men can experience back pain (as well as hip, groin, SI, and sacral pain) caused by referred pain from the male reproductive system. See Chapter 10 for a complete discussion of prostate impairments (e.g., prostatitis, benign prostatic hypertrophy, chronic pelvic pain syndrome, prostate and testicular cancer).

Prostate cancer is the second most common cancer in males over the age of 60 in the United States.[147] The incidence of prostate cancer has risen 60% to 75% in the Western world in the last 15 years[148] and with continued improvements in early detection is expected to continue to rise over the next 20 years,[149] making it very likely that the therapist will treat clients with prostate pathology.

Testicular cancer, though relatively rare, is the most common cancer in males ages 15 to 35 years and on the rise.[147] Details of both conditions are discussed in Chapter 10. Benign prostatic hyperplasia (BPH) is one of the most common disorders of the aging male population affecting 50% of men over age 50.

Risk Factors

Risk factors for prostate dysfunction include advancing age, family history, ethnicity (greater risk for African-American men), diet, and possibly exposure to chemicals. Not all disorders of this system occur with aging, so the therapist must remain alert for red flag symptoms in males of any age.

Clinical Presentation

Back pain, changes in bladder function, and sexual dysfunction are the most common symptoms associated with male reproductive disorders. Any obstruction, growth, or inflammation of the prostate can directly affect the urethra, resulting in difficulty starting a flow of urine, continuing a flow of urine, frequency, and/or nocturia.

Prostate cancer is often asymptomatic and only diagnosed when the man seeks medical assistance because of symptoms of urinary obstruction or sciatica. Sciatic pain affects the low back, hip, and leg and is caused by metastasis to the bones of the pelvis, lumbar spine, or femur.

Associated symptoms may include melena, sudden moderate to high fever, chills, and changes in bowel or bladder function. Men who have reached the fifth decade or more are most commonly affected.

Testicular cancer presents most often as a painless swelling nodule in one gonad, noted incidentally by the client or his sexual partner. This is described as a lump or hardness of the testis, with occasional heaviness or a dull, aching sensation in the lower abdomen or scrotum. Acute pain is the presenting symptom in about 10% of affected men.

Involvement of the epididymis or spermatic cord may lead to pelvic or inguinal lymph node metastases, although most tumors confined to the testis itself will spread primarily to the retroperitoneal lymph nodes. Subsequent cephalad drainage may be to the thoracic duct and supraclavicular nodes. Hematogenous spread to the lungs, bone, or liver may occur as a result of direct tumor invasion.

In about 10% of affected individuals, dissemination along these pathways results in thoracic, lumbar, supraclavicular, neck, or shoulder pain or mass as the first symptom. Other symptoms related to this pathway of dissemination may include respiratory symptoms or GI disturbance.

As discussed earlier, back pain caused by neoplasm is typically progressive, is more pronounced at night, and may not have a clear association with activity level (as is more characteristic of mechanical back pain). The usual progression of symptoms in clients with cord compression is back pain followed by radicular pain, lower extremity weakness, sensory loss, and finally, loss of sphincter (bowel and bladder) control.

Associated Signs and Symptoms

Besides changes in urinary patterns, the therapist must ask about discharge from the penis, constitutional symptoms, and pain in any of the nearby soft tissue areas (groin, rectum, scrotum). Is there any blood in the urine (or change in color from yellow to orange or red)? Recurrent UTI is common in prostatitis but does not lead to prostate cancer.

Because the therapist is not going to be treating any of these problems, any red flags should be reported to the physician. A rectal exam may be needed. Access to the prostate is easiest through this type of exam. By pressing on the inflamed or infected prostate, the physician can reproduce painful symptoms as part of the differential diagnosis (see Fig. 10-5).

Many men are reluctant to pursue diagnosis and treatment whenever the male reproductive system is involved. Early detection and treatment of these conditions can result in a good outcome. Screening questions for men are a good way to elicit red-flag history, risk factors, and signs or symptoms. The therapist must follow up with the client and make sure contact is made with the appropriate health care professional.

CLINICAL SIGNS AND SYMPTOMS

Prostate Pathology

- May be asymptomatic early on
- Urinary dysfunction (hesitancy, frequency, urgency, nocturia, dysuria)
- Low back, inner thigh, or perineal pain or stiffness
- Suprapubic or pelvic pain
- Testicular or penis pain
- Sciatica (prostate cancer metastases)
- Bone pain, lymphedema of the groin, and/or lower extremities (prostate cancer metastases)
- Neurologic changes from spinal cord compression (prostate cancer metastases to the vertebrae)
- Sexual dysfunction (difficulty having an erection, painful ejaculation, cramping/discomfort after ejaculation)
- Constitutional symptoms with prostatitis
- Blood in urine or semen
- See full discussion in Chapter 10; see also Appendices B-24 and B-30)

SCREENING FOR INFECTIOUS CAUSES OF BACK PAIN

Drug abuse, immune suppression, and human immunodeficiency virus (HIV) may predispose to infection. Fever in anyone taking immunosuppressants is a red flag symptom indicating a possible underlying infection. Many people with a spinal infection do not have a fever; they are more likely to have a red flag history or risk factors.[1]

Vertebral Osteomyelitis

Vertebral osteomyelitis is a bone infection most often affecting the first and second lumbar vertebrae, causing LBP. There are many causative factors. Osteomyelitis may occur in individuals with diabetes, injection drug users (IDUs), alcoholics, clients taking corticosteroid drugs, clients with spinal cord injury and neurogenic bladder, and otherwise debilitated or immune-suppressed clients. Older children can be affected, although the most common peak is after the third decade of life.

Vertebral osteomyelitis is increasingly being reported as a complication of nosocomial bacteremia. Methicillin-resistant *Staphylococcus aureus* (MRSA) is the most common causative organism. Osteomyelitis also can occur after surgery, open fractures, penetrating wounds, skin breakdown and ulcers, and systemic infections. It may result from a hematogenous spread through arterial and venous routes secondary to surgically implanted hardware for internal fixation of the spine, pelvic inflammatory disease, or genitourinary tract infection.

A physician should evaluate new onset of back pain in anyone who has been treated with vancomycin therapy for MRSA. Vancomycin therapy may give the appearance of being effective with resolution of fever and the return of white blood cell counts to normal ranges but in fact be insufficient to prevent or reverse the progression of hematogenous MRSA vertebral osteomyelitis.[150,151]

In the adult, usually two adjacent vertebrae and their intervening disk are involved, and the vertebral body(ies) may undergo destruction and collapse. Abscess formation may result, with possible neurologic involvement. The abscess can advance anteriorly to produce an abscess that can extend to the psoas muscle producing hip pain.

The most consistent clinical finding is marked local tenderness over the spinous process of the involved vertebrae with "nonspecific backache." The classic history describes pain that has been increasing in severity over a period of 1 to 3 weeks. Movement is painful, and there is marked muscular guarding and spasm of the paravertebral muscles and the hamstrings. The involved vertebrae are usually exquisitely sensitive to percussion, and pain is more severe at night.

There may be no rise in temperature or abnormality in white blood cell count because generalized sepsis is not present, but an elevated erythrocyte sedimentation rate (ESR) is likely. A low-grade fever is most common in adults when body temperature changes do occur.

Children are more likely to present with acute, severe complaints including high fever, intense pain, and localized manifestations such as edema, erythema, and tenderness. Acute hematogenous osteomyelitis seen in children usually originates in the metaphysis of a long bone. Precipitating trauma is often present in the history, and well-localized, acute bone pain of 1 day to several days' duration is the primary symptom. The pain is most commonly severe enough to limit or restrict the use of the involved extremity, and fever and malaise consistent with sepsis are usual.

CLINICAL SIGNS AND SYMPTOMS

Vertebral Osteomyelitis

- Pain and local tenderness over the involved spinous process(es); possible swelling, redness, and warmth in the affected area
- Night pain
- Stiff back with difficulty bearing weight, moving, walking
- Paravertebral muscle guarding or spasm
- Positive SLR
- Hip pain if infection spreads to the psoas muscle
- May be constitutional symptoms (fever, malaise)
- Recent history of bacterial infection (e.g., pharyngitis, otitis media in children)

Disk Space Infection

Disk space infection is a form of subacute osteomyelitis involving the vertebral end-plates and the disk in both children and adults. The lower thoracic and lumbar spines are the most common sites of infection.

Symptoms associated with postoperative disk space infection occur 2 to 8 weeks after diskectomy. Diskitis of an infectious type occurs following bacteremia secondary to UTI, with or without instrumentation (e.g., catheterization or cystoscopy). Low-grade viral or bacterial infection (e.g., gastroenteritis, upper respiratory infection, UTI) is most often implicated in young children with diskitis (4 years old and younger). Ask the parent, guardian, or caretaker of any young child with back pain if there has been a recent history of sore throat, cold, ear infection, or other upper respiratory illness.

Adults with disk space infection often complain of LBP localized around the disk area. The pain can range from mild to "excruciating" and sometimes is described as "knifelike." Such severe pain is accompanied by restricted movement and constant pain, present both day and night. The pain is usually made worse by activity, but unlike most other causes of back pain, it is *not* relieved by rest. If the condition becomes chronic, pain may radiate into the abdomen, pelvis, and lower extremities.

Children present with a history of increasingly severe localized back pain often accompanied by a limp or refusal to walk. There may be an increased lumbar lordosis. Pain may occur in the flank, abdomen, or hip. Symptoms may get worse with passive SLR testing or other hip motion. A neurologic screening examination is usually negative.

Physical examination may reveal localized tenderness over the involved disk space, paraspinal muscle spasm, and restricted lumbar motion. SLR may be positive, and fever is common (Case Example 14-17).

CASE EXAMPLE 14-17

Septic Diskitis

Background: A 72-year-old man with leg myalgia and stabbing back pain of 2 weeks' duration was referred to physical therapy for evaluation by a rural nurse practitioner. When questioned about past medical history, the client reported a prostatectomy 22 years ago with no further problems. He was not aware of any other associated signs and symptoms but reported a recurring dermatitis that was being treated by his nurse practitioner. There were no skin lesions associated with the dermatitis present at the time of the physical therapist evaluation.

Clinical Presentation: The examination revealed spasm of the thoracolumbosacral paraspinal muscles bilaterally. The client reported extreme sensitivity to palpation of the spinous processes at L3 and L4; tap test reproduced painful symptoms. Spinal accessory motions could not be tested because of the client's state of acute pain and immobility.

Hip flexion and extension reproduced the symptoms and produced additional radiating flank pain. A straight leg raise (SLR) caused severe back pain with each leg at 30 degrees on both sides. A neurologic examination was otherwise within normal limits. Vital signs were taken: blood pressure of 180/100 mm Hg; heart rate of 100 bpm; temperature of 101° F.

What are the red flags in this case?

- Age
- Recurring dermatitis
- Positive tap test
- Bilateral SLR
- Vital signs

Result: The therapist contacted the nurse practitioner by telephone to report the findings, especially the vital signs and results of the SLR. It was determined that the client needed a medical evaluation, and he was referred to a physician's center in the nearest available city. A summary of findings from the physical therapist was sent with the client along with a request for a copy of the physician's report.

The client returned to the physical therapist's clinic with a copy of the physician's report with the following diagnosis: *Clostridium perfringens* septic diskitis (made on the basis of blood culture). The prescribed treatment was intravenous antibiotic therapy for 6 weeks, progressive mobilization, and a spinal brace to be provided and fitted by the physical therapist. The client's back pain subsided gradually over the next 2 weeks, and he was followed up at intervals until he was weaned from the brace and resumed normal activities.

Septic diskitis may occur following various invasive procedures, or it may be related to occult infections, urinary tract infections, septicemia, and dermatitis. Contact dermatitis was the most likely underlying cause in this case.

Bacterial Endocarditis

Bacterial endocarditis often presents initially with musculoskeletal symptoms, including arthralgia, arthritis, LBP, and myalgias. Half of these clients will have only musculoskeletal symptoms, without other signs of endocarditis.

The early onset of joint pain and myalgia is more likely if the client is older and has had a previously diagnosed heart murmur or prosthetic valve (risk factors). Other risk factors include injection drug use, previous cardiac surgery, recent dental work, and recent history of invasive diagnostic procedures (e.g., shunts, catheters).

Almost one-third of clients with bacterial endocarditis have LBP. In many persons, LBP is the principal musculoskeletal symptom reported. Back pain is accompanied by decreased ROM and spinal tenderness. Pain may affect only one side, and it may be limited to the paraspinal muscles.

Endocarditis-induced LBP may be very similar to the pain pattern associated with a herniated lumbar disk; it radiates to the leg and may be accentuated by raising the leg, coughing, or sneezing. The key difference is that neurologic deficits are usually absent in clients with bacterial endocarditis. The therapist can review history and risk factors and conduct a Review of Systems to help in the screening process.

PHYSICIAN REFERRAL

Most adults with an episode of acute back pain experience recovery within 1 to 4 weeks. As many as 90% of affected individuals resume normal activity levels during this time.[152,153]

All clients who have not regained usual activity after 4 weeks should be formally reassessed, including a review of the history and examination, looking for yellow (caution) or red (warning) flags, testing for any neurologic deficit, and conducting a Review of Systems to identify any evidence of systemic disease or other medical condition requiring referral.

Reassessment of movement dysfunction is critical at this stage to look for alternate impairments not previously observed or identified. The therapist must consider whether the underlying primary problem is spinal or nonspinal, mechanical or medical, and what specific structures are involved.

Review concepts from the screening physical assessment in Chapter 4 to make sure the evaluation is complete. Inspection, palpation, and auscultation may reveal key findings previously missed. Assessment of fear-avoidance may be needed as discussed in Chapter 3. The therapist should not rely on his or her own perception of patient/client's fear-avoidance behaviors. Tools such as the Fear-Avoidance Beliefs Questionnaire (FABQ; see Table 3-7), Tampa Scale of Kinesophobia (TSK-11), and Pain Catastrophizing Scale (PSC) are available to identify fear-avoidance beliefs.[154]

Medical referral is made on the basis of a comparison of baseline data with findings upon reassessment. Providing the physician with concise but comprehensive information about findings and concerns is a helpful part of the medical differential diagnostic process.

Guidelines for Immediate Medical Attention

Immediate medical referral is not always required when a client presents with any one of the red flags listed in Box 14-1. When viewed as a whole, the history, risk factors, and any cluster of red flag findings will guide the therapist in making a final intervention versus referral decision.

- Neck pain with evidence of VBI (e.g., reproduction of symptoms with vertebral artery testing such as vertigo, visual changes, headaches, nausea) requires medical attention. VBI can develop into cerebral or brainstem ischemia, leading to severe morbidity or death.[155]
- Immediate medical attention is required when anyone with LBP presents with symptoms of cauda equina syndrome (e.g., saddle anesthesia, fecal incontinence, motor weakness of the legs, radiculopathy, unable to heel or toe walk, altered knee or ankle deep tendon reflexes). Acute mechanical compression of nerves in the lower extremities, bowel, and bladder as they pass through the caudal sac may be a surgical emergency.
- Massive midline rupture of a disk in the lower lumbar levels can lead to LBP, rapidly progressive bilateral motor weakness and sciatica, saddle anesthesia (buttock and medial and posterior thighs; the area that would come in contact with a saddle when sitting on a horse), and bowel and bladder incontinence or urinary retention.[156]
- Men between the ages of 65 and 75 who ever smoked should undergo medical screening for AAA. Any male with these two risk factors, especially presenting with signs or symptoms of AAA, must be referred immediately.
- Sudden, intense back and/or shoulder pain in a sexually active woman of childbearing age may signal the end of an ectopic pregnancy. Sudden change in blood pressure, pallor, pain, and dizziness will alert the therapist to the need for immediate medical attention.
- Inability to bear weight, especially with fever and/or a history of cancer, diabetes, immunosuppression, or trauma (even if radiographs have already been obtained and declared "negative") [infection, fracture].

Guidelines for Physician Referral

- Red flags requiring physician referral or reevaluation include back pain or symptoms that are not improving as expected, steady pain irrespective of activity, symptoms that are increasing, or the development of new or progressive neurologic deficits such as weakness, sensory loss, reflex changes, bowel or bladder dysfunction, or myelopathy.
- A positive Sharp-Purser test for AA subluxation in the client with rheumatoid arthritis (sensation of head falling forward during neck flexion and clunking during neck extension) must be evaluated by an orthopedic surgeon.[14]
- The ESR, serum calcium level, and alkaline phosphatase level are usually elevated if bone cancer is present.[157] Back pain in the presence of elevated alkaline phosphatase levels can also indicate hyperparathyroidism, osteomalacia, pregnancy, and/or rickets.
- Reproduction of pain or exquisite tenderness over the spinous process(es) is a red-flag sign requiring further investigation and possible medical referral.

Clues to Screening Head, Neck, or Back Pain

General

- Age younger than 20 and older than 50 with no history of a precipitating event
- Back pain in children is uncommon and constitutes a red-flag finding, especially back pain that lasts more than 6 weeks
- Nocturnal back pain that is constant, intense, and unrelieved by change in position
- Pain that causes constant movement or makes the client curl up in the sitting position
- Back pain with constitutional symptoms: Fatigue, nausea, vomiting, diarrhea, fever, sweats
- Back pain accompanied by unexplained weight loss
- Back pain accompanied by extreme weakness in the leg(s), numbness in the groin or rectum, or difficulty controlling bowel or bladder function (**cauda equina syndrome;** rare but requires immediate medical attention)
- Back pain that is insidious in onset and progression (remember to assess for unreported sexual assault or physical abuse)
- Back pain that is unrelieved by recumbency
- Back pain that does not vary with exertion or activity
- Back pain that is relieved by sitting up and leaning forward (**pancreas**)
- Back pain that is accompanied by multiple joint involvement (**GI, rheumatoid arthritis, fibromyalgia**) or by sustained morning stiffness (**spondyloarthropathy**)
- Severe, persistent back pain with full and painless movement of the spine
- Sudden, localized back pain that does not diminish in 10 days to 2 weeks in postmenopausal women or osteoporotic adults (**osteoporosis with compression fracture**)

Past Medical History

- Previous history of cancer, Crohn's disease, or bowel obstruction
- Long-term use of nonsteroidal antiinflammatory drugs (**GI bleeding**), steroids, or immunosuppressants (**infectious cause**)
- Recent history or previous history of recurrent upper respiratory infection or pneumonia
- Recent history of surgery, especially back pain 2 to 8 weeks after diskectomy (**infection**)
- History of osteoporosis and/or previous vertebral compression fracture(s) (**fracture**)
- History of heart murmur or prosthetic valve in an older client who currently has LBP of unknown cause (**bacterial endocarditis**)
- History of intermittent claudication and heart disease in a man with deep midlumbar back pain; assess for pulsing abdominal mass (**AAA**)
- History of diseases associated with hypercalcemia such as hyperparathyroidism, multiple myeloma, senile osteoporosis, hyperthyroidism, Cushing's disease, or specific renal tubular disease not appearing with back pain radiating to the flank or iliac crest (**kidney stone**)

Oncologic

- Back pain with severe lower extremity weakness without pain, with full ROM and recent history of sciatica in the absence of a positive SLR
- Bilateral leg pain with motor and reflex impairments
- Bone tenderness over the spinous processes (**infection or neoplasm**)
- Temperature differences: Involved side warmer when tumor interferes with sympathetic nerves
- Associated signs and symptoms: significant weight loss; night pain disturbing sleep; extreme fatigue; constitutional symptoms such as fever, sweats; other organ/system-dependent symptoms such as urinary changes (urologic), cough, and dyspnea (pulmonary); abdominal bloating or bloody diarrhea (GI)

Cardiovascular

- Back pain that is described as "throbbing"
- Back pain accompanied by leg pain that is relieved by standing still or rest
- Back pain that is present in all spinal positions and increased by exertion
- Back pain accompanied by a pulsating sensation or palpable abdominal pulse (possibly a palpable pulsating abdominal mass)
- Low back, pelvic, and/or leg pain with temperature changes from one leg to the other (involved side warmer: Venous occlusion or tumor; involved side colder: arterial occlusion)
- Back injury that occurred during weight lifting in someone with known heart disease or past history of aneurysm

Pulmonary

- Associated signs and symptoms (dyspnea, persistent cough, fever and chills)
- Back pain aggravated by respiratory movements (deep breathing, laughing, coughing)
- Back pain relieved by breath holding or Valsalva maneuver
- Autosplinting by lying on the involved side or holding firm pillow against the chest/abdomen decreases the pain
- Spinal/trunk movements (e.g., trunk rotation, trunk side bending) do not reproduce symptoms (exception: An intercostal tear caused by forceful coughing from underlying diaphragmatic pleurisy can result in painful movement but is also reproduced by local palpation)
- Weak and rapid pulse accompanied by fall in blood pressure (pneumothorax)

Renal/Urologic

- Renal and urethral pain is felt throughout T9 to L1 dermatomes; pain is constant but may crescendo (kidney stones)

- Kidney pain of an inflammatory nature can be relieved by a change in position. However, renal colic (e.g., infection) remains unchanged by a change in position. But there are usually constitutional symptoms associated with either inflammation or infection to tip off the alert therapist.
- Back pain at the level of the kidneys can be caused by ovarian or testicular cancer
- Back pain and shoulder pain, either simultaneously or alternately, may be renal/urologic in origin
- Side bending to the same side and pressure placed along the spine at that level is "more comfortable"; pain may be reduced, but it is not eliminated when the kidney is involved. The client with kidney disease/disorder may prefer this position because it moves the kidney out away from the spine and away from any compressive forces causing painful symptoms.
- Associated signs and symptoms (blood in urine, fever, chills, increased urinary frequency, difficulty starting or continuing stream of urine, testicular pain in men, painful erection and/or ejaculation)
- Assess for costovertebral angle tenderness; pain is affected by change of position **(pseudorenal pain)**
- History of traumatic fall, blow, lift **(musculoskeletal)**
- Persistent back (pelvic, groin, or testicular) pain in a male with history of chronic prostatitis that does not respond to medical treatment such as antibiotics suggests the need to assess (or reassess) for pelvic floor impairment; watch for a history of improved symptoms without complete resolution with an orthopedic treatment approach

Gastrointestinal

- Back and abdominal pain at the same level (may occur simultaneously or alternately); check for GI history or associated signs and symptoms
- Back pain with abdominal pain at a lower level than the back pain; look for its source in the back
- Back pain associated with food or meals (increase or decrease in symptoms)
- Back pain accompanied by heartburn or relieved by antacids
- Associated signs and symptoms (dysphagia, odynophagia, melena, unexplained or unintended weight loss especially if accompanied by early satiety, abdominal distention, tenderness over McBurney's point, positive iliopsoas or obturator sign, bloody diarrhea, nausea, vomiting)
- LBP accompanied by constipation may be a manifestation of pelvic floor overactivity or spasm; this requires a pelvic floor screening examination.
- Sacral pain occurs when the rectum is stimulated, such as during a bowel movement or when passing gas, and relieved after each of these events

Gynecologic

- History or current gynecologic disorder (e.g., uterine retroversion, ovarian cysts, uterine fibroids, endometriosis, pelvic inflammatory disease, sexual assault/incest, IUCD, multiple births with prolonged labor or forceps use)
- Associated signs and symptoms (missed or irregular menses, tender breasts, cyclic nausea and vomiting, chronic constipation, vaginal discharge, abnormal uterine bleeding or bleeding in a postmenopausal woman)
- Low back and/or pelvic pain developing soon after a missed menstrual cycle; blood pressure may be significantly low, and there may be concomitant shoulder pain when hemorrhaging occurs **(ectopic pregnancy)**
- Low back and/or pelvic pain occurring intermittently but with regularity in response to menstrual cycle (e.g., ovulation around days 10 to 14 and onset of menses around days 23 to 28)

Nonorganic (Psychogenic) (see discussion in Chapter 3)

- Widespread, nonanatomic low back tenderness with overreaction to superficial palpation
- Assess for nonorganic signs such as axial loading (downward pressure on the top of the head) or shoulder-hip rotation (client rotates shoulder and hips with feet planted); Waddell's nonorganic signs (see Table 3-12)
- Regional (whole leg) pain, numbness, weakness, sensory disturbances
- Chronic use of (or demand for) narcotics

Infectious

- Infection, such as osteomyelitis or inflammatory arthritis (e.g., ankylosing spondylitis, psoriatic or reactive arthritis), may present as back pain with intermittent reports of fever, chills, sweats, fatigue, and/or adenopathy. Watch for other associated signs and symptoms specific to each condition (e.g., urethritis, psoriasis, severe morning stiffness).

Pediatrics

- Children presenting with back pain are very different from adults with the same problem; children are less likely than adults to report symptoms when there is no organic cause for the complaint.[158]
- Eighty-five percent of children with back pain lasting more than 2 months have a diagnosable lesion.[159]
- Children with persistent reports of LBP must be evaluated and reevaluated until a diagnosis is reached; x-rays and laboratory values are needed.

REFERRED BACK PAIN PATTERNS (FIG. 14-7)

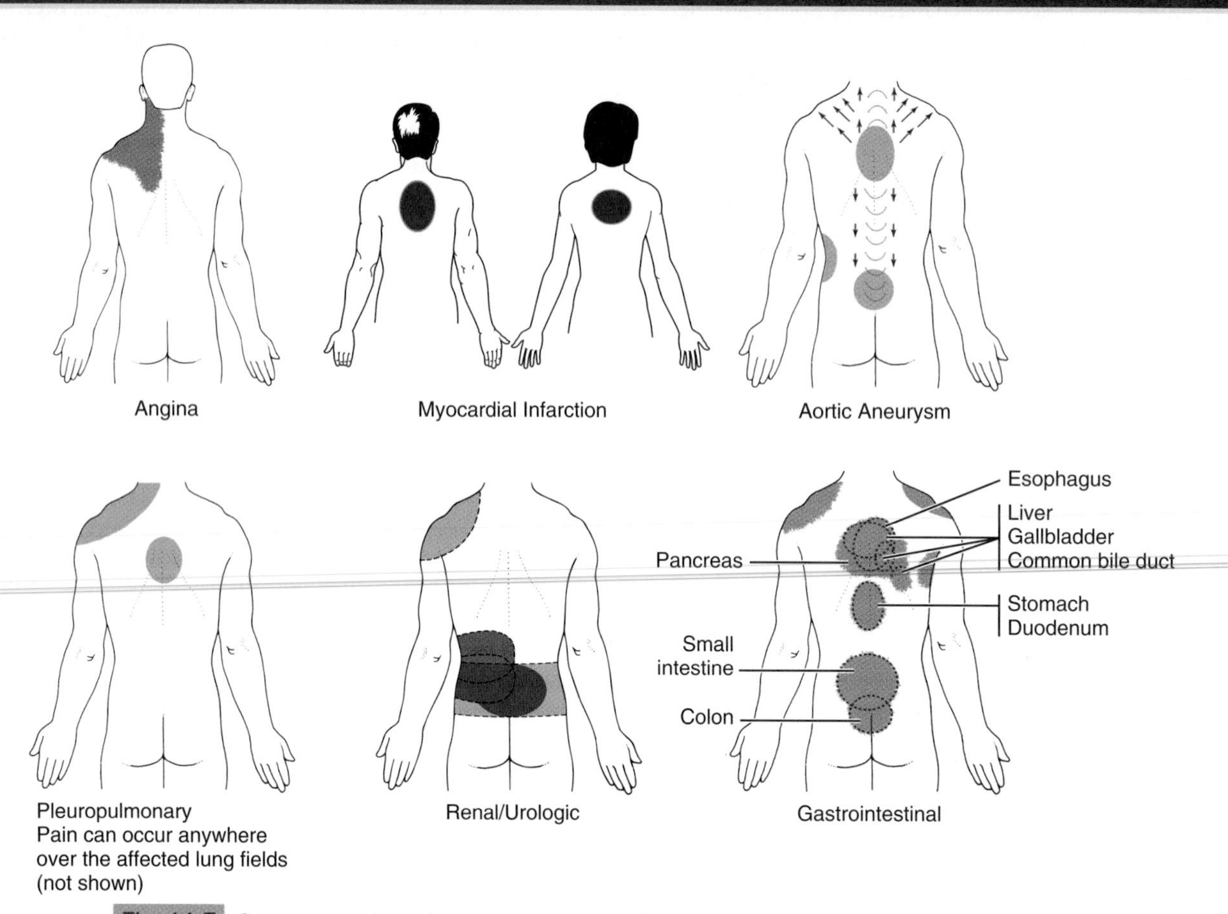

Fig. 14-7 Composite picture of referred back pain patterns. Not pictured: gynecologic pain patterns.

Key Points to Remember

- Clients may inaccurately attribute symptoms to a particular incident or activity, or they may fail to recognize causative factors.
- At presentation, any person with musculoskeletal pain of unknown cause and/or a past medical history of cancer should be screened for medical disease. Special Questions for Men and Women may be helpful in this screening process.
- Consider visceral origin of back pain in the absence of muscular spasm, tenderness, and impaired movement.[110]
- Perform a breast exam on any woman with upper back pain, especially women with a history of cancer.
- Backache may be the earliest and only manifestation of visceral disease.[110]
- Neck or back pain in the presence of normal ROM and strength is a yellow (caution) flag symptom.
- Persistent backache due to extraspinal pathology is rare in children but common in adults.[43a,110]
- Children, adolescents, and especially athletes reporting back pain of more than 3 weeks duration may need medical referral, depending on the history and clinical presentation.[43a,160]
- Lumbar spasm may occur in the presence of severe pain from retroperitoneal diseases (e.g., renal tumors, abscesses, appendicitis, kidney stones, lymphoma)[110]
- Back pain accompanied by recent history of infection (especially UTI) or in the presence of constitutional symptoms (e.g., fever, chills, nausea; see Box 1-3) must be screened more carefully.
- Nonpainful paresthesias can be the result of neural compression but also occur from ischemia (atherosclerosis, tumor, prodromal sign of a migraine headache); painful paresthesias are more likely indicative of an inflammatory or mechanical process.
- When symptoms seem out of proportion to the injury, or if they persist beyond the expected time for the nature of the injury, medical referral may be indicated.
- Pain that is unrelieved by rest or change in position or pain/symptoms that do not fit the expected mechanical or neuromusculoskeletal pattern should serve as red-flag warnings.
- When symptoms cannot be reproduced, aggravated, or altered in any way during the examination, additional questions to screen for medical disease are indicated.
- Always rule out trigger points as a possible cause of musculoskeletal symptoms before referring the client elsewhere.
- Postoperative infection of any kind may not appear with any clinical signs/symptoms for weeks or months.
- Muscle weakness without pain, without history of sciatica, and without a positive straight leg raising is suggestive of spinal metastases.
- Sciatica may be the first symptom of prostate cancer metastasized to the bones of the pelvis, lumbar spine, or femur.
- Back pain may be a symptom of depression.
- Urinary incontinence concomitant with cervical spine pain requires a neurologic screening examination and possible medical referral.
- Anterior neck pain, movement dysfunction, and torticollis of the sternocleidomastoid muscle may be a sign of underlying thyroid involvement.
- The therapist may need to screen for illness behavior and the need for psychologic evaluation. Many clients with chronic back pain have both a physical problem and varying degrees of illness behavior. A single behavioral sign or symptom may be normal; multiple findings of several different kinds are much more significant.[1]
- Remember to screen for fear-avoidance behaviors; this is considered a yellow flag pointing to psychosocial factors that can direct treatment.

SUBJECTIVE EXAMINATION

Special Questions to Ask: Headache

See Appendix C-7 for a complete pain assessment.

History

- Do other family members have similar headaches?
- What major life changes or stressors have you had in the last 6 months?
- Have you ever had a head injury? Cancer of any kind? A hysterectomy? High blood pressure? A stroke? Seizures?
- Have you been hit or kicked in the head, neck, or face? Pushed against a wall or other object? Pulled or thrown by the hair?
- For women of childbearing age: Is it possible you are pregnant?

Site

- Where do you feel the headache? Can you point to it with one finger (localized versus diffuse)? Does it move?

Onset

- Do you recall your first headache of this type?
- Was it caused by a fall or trauma? (Therapist may have to screen for trauma associated with domestic violence as a potential cause.)

Frequency

- How often do you have this type of headache?

Intensity

- On a scale from 0 (no pain) to 10 (worst pain), how would you rate your headache now? Worst it has been?
- Does the pain keep you from your daily activities? From exercise or recreation? From work?

Duration

- How long do your headaches last?

Description

- What do your headaches feel like? (The client may have more than one type of headache.)
- Alternate question: What words would you use to describe the pain?

Pattern

- Is there a pattern to your headaches (e.g., weekly? Monthly? Morning to evening?)
- Do you wake up in the early morning hours with a headache? **(occipital pain: hypertension)**
- For women who are perimenopausal or menopausal (natural or surgically induced): Are the headaches cyclical? (Monthly? Right before or right after the menstrual flow?)

Aggravating Factors

- What makes the headache worse?
- Are you aware of any triggers that can bring the headache on? (Alcohol, noise, lights, food, coughing or sneezing, fatigue or lack of sleep, stress, caffeine withdrawal; for women: menstrual cycle)
- Do you grind your teeth during the day or at night?
  - *If yes,* assessment of the cervical spine and temporomandibular joints is indicated. Referral to a dentist may be required.
- Are you taking any medications? (Headache can be a side effect of many different medications, but especially NSAIDs, muscle relaxants, antianxiety and antidepressant agents, and food and drugs containing nitrates, calcium, and beta blockers.)

Relieving Factors

- Is there anything you can do to make the headache better?
  - *If yes,* how? (caffeine, medications, sleep, avoid certain foods, alcohol, cigarettes) [Ask follow-up questions about use of OTC or prescription drugs and/or herbs or pharmaceuticals.]
  - How does rest affect your symptoms?

Associated Symptoms

- Do you have any symptoms of any kind anywhere else in your head or body? (Follow-up with questions about vision changes, dizziness, ringing in the ears, mood changes, nausea, vomiting, nasal congestion, nose bleeds, light or sound sensitivity, paresthesias such as numbness and tingling of the face or fingers, difficulty swallowing, hoarseness, fever, chills.)

For the Therapist

- Take the client's blood pressure and pulse and assess for cardiovascular risk factors.
- Auscultate for bruits in the temporal and carotid arteries **(temporal arteritis, carotid stenosis).**
- Headaches that cannot be linked to a neuromuscular or musculoskeletal cause (e.g., dysfunction of the cervical spine, thoracic spine, or temporomandibular joints; muscle tension, poor posture, nerve impingement) may need further medical referral and evaluation.

Special Questions to Ask: Neck or Back

To the Therapist: If a more complete screening interview is required, see *Special Questions to Ask* at the end of each chapter for further questions related to the individual organ systems.

SUBJECTIVE EXAMINATION—*cont'd*

Pain Assessment (see also Appendix C-7)

- When did the pain or symptoms start?
- Did it (they) start gradually or suddenly? **(Vascular versus trauma problem)**
- Was there an illness or injury before the onset of pain?
- Have you noticed any changes in your symptoms since they first started to the present time?
- Is the pain aggravated or relieved by coughing or sneezing? **(Nerve root involvement, muscular)**
- Is the pain aggravated or relieved by activity?
- Are there any particular positions (sitting, lying, standing) that make your back pain feel better or worse?
- Does the pain go down the leg? *If so,* how far does it go?
- Have you noticed any muscular weakness?
- Have you been treated previously for back disorders?
- How has your general health been both before the beginning of your back problem and today?
- How does rest affect the pain or symptoms?
- Do you feel worse in the morning or evening ... **OR** ... What difference do you notice in your symptoms from the morning when you first wake up until the evening when you go to bed?

General Systemic

Most of these questions may be asked of clients who have pain or symptoms anywhere in the musculoskeletal system.

- Have you ever been told that you have osteoporosis or brittle bones?
- Have you ever fractured your spine?
- Have you ever been diagnosed or treated for cancer in any part of your body?
  - *If no,* have you ever had chemotherapy or radiation therapy for anything? **(Rectal bleeding is a sign of radiation proctitis.)**
- Do you ever notice sweating, nausea, or chest pains when your current symptoms occur?
- What other symptoms have you had with this problem? Wait for an answer but consider offering, for example, have you had:
  - Numbness
  - Burning, tingling
  - Nausea, vomiting
  - Loss of appetite
  - Unexpected or significant weight gain or loss
  - Diarrhea, constipation, blood in your stool or urine
  - Difficulty in starting or continuing the flow of urine or incontinence (inability to hold your urine)
  - Hoarseness or difficulty in swallowing
  - Heart palpitations or fluttering
  - Difficulty in breathing while just sitting or resting or with mild effort (e.g., when walking from the car to the house)
  - Unexplained sweating or perspiration
  - Night sweats, fever, chills
  - Changes in vision: blurred vision, black spots, double vision, temporary blindness
  - Fatigue, weakness, sudden paralysis of one side of your body, arm, or leg **(Transient ischemic attack)**
  - Headaches
  - Dizziness or fainting spells
- Have you had a recent cold, sore throat, upper respiratory infection, or the flu? Have you ever been diagnosed with HIV?

Cardiovascular

- Have you ever been told you have high blood pressure or heart trouble?
- Do you ever have chest pain or discomfort when your back hurts or just before your back starts hurting?
- Do you ever have swollen feet or ankles? *If yes,* are they swollen when you get up in the morning? **(Edema/congestive heart failure)**
- Do you ever get cramps in your legs if you walk for several blocks or in your arms if you work with your arms overhead? **(Intermittent claudication)**
- Do you ever have bouts of rapid heart action, irregular heartbeats, or palpitations of your heart?
- Have you ever felt a "heartbeat" in your abdomen when you lie down?
  - *If yes,* is this associated with LBP or left flank pain? **(Abdominal aneurysm)**
- Do you ever notice sweating, nausea, or chest pain when your current symptoms (e.g., head, neck, jaw, back pain) occur?

Pulmonary

- Are you able to take a deep breath?
- Do you ever have shortness of breath or breathlessness with your back pain?
  - How far can you walk before you feel breathless?
  - What symptoms stop your walking/activities (e.g., shortness of breath, heart pounding, chest tightness, or weak arms/legs)?
- Have you had any trouble with coughing lately?
  - *If yes,* have you strained your back from coughing?

Renal/Urologic

- Have you noticed any changes in the flow of urine since your back/groin pain started?
  - If *no,* it may be necessary to provide prompts or examples of what changes you are referring to (e.g., difficulty in starting or continuing the flow of urine, numbness or tingling in the groin or pelvis, increased frequency, getting up at night)
- Have you had burning with urination during the last 3 to 4 weeks? Fever and/or chills?

Continued

SUBJECTIVE EXAMINATION—*cont'd*

- Do you ever have blood in your urine or notice blood in the toilet going to the bathroom?
- Have you noticed any changes in color or blood in your urine?
- Do you have any problems with your kidneys or bladder? *If so,* describe.
- Have you ever had kidney or bladder stones? *If so,* how were these stones treated?
- Have you ever had an injury to your bladder or to your kidneys? *If so,* how was this treated?
- Have you had any infections of the bladder, and how were these infections treated?
  - Were they related to any specific circumstances (e.g., pregnancy, intercourse)?
- Have you had any kidney infections, and how were these treated?
  - Were they related to any specific circumstances (e.g., pregnancy, after bladder infections, a strep throat, or strep skin infections)?
- Do you ever have pain, discomfort, or a burning sensation when you urinate? **(Lower urinary tract irritation)**

Gastrointestinal

- Are you having any stomach or abdominal pain either at the same time as the back pain or at other times?
  - *If yes,* assess the location and the presence of any GI symptoms.
- Have you noticed any association between when you eat and when your symptoms increase or decrease?
  - Do you notice any change in your symptoms 1 to 3 hours after you eat?
  - Do you notice any pain beneath the breastbone (epigastric) or just beneath the shoulder blade (subscapular) 1 to 2 hours after eating?
- Do you have a feeling of fullness after only one or two bites of food? **(Early satiety)**
- Is your back pain relieved after having a bowel movement? **(GI obstruction)**
- Do you have rectal, low back, or SI pain when passing stool or having a bowel movement?
- Do you have any blood in your stools or change in the normal color of your bowel movements (e.g., black, red, mahogany color, gray color)? **(Hemorrhoids, prostate problems, cancer)**
- Are you having any diarrhea, constipation, or other changes in your bowel function?
- Do you have frequent heartburn or take antacids to relieve heartburn or acid indigestion?
- Have you had any skin rashes or skin lesions in the last 6 weeks **(Regional enteritis or Crohn's disease)**

To the therapist: It may be necessary to conduct a risk factor assessment for NSAID-induced back pain (see discussion in Chapter 8) or screen for eating disorders (see Appendix B-13, *A*).

Special Questions to Ask: Sexual History

There are a wide range of reasons why it may be necessary to ask questions about sexual function, birth control, and STDs. For example, joint pain can be caused by STIs. Low back, sacral, and pelvic pain can be caused by sexual trauma or sexual violence. Sciatica accompanied by unreported impotence can be caused by prostate cancer metastasized to the skeletal system.

Whenever taking a sexual history seems appropriate, remember to offer your clients a clear explanation for any questions asked concerning sexual activity, sexual function, or sexual history. The therapist may want to introduce the series of questions by saying, "When evaluating LBP sometimes it's necessary to ask some more personal questions. Please answer as accurately as you can."

The personal nature of some questions sometimes leads clients to feel embarrassed. It is important to assure them they will not be judged and that providing accurate information is crucial to providing good care. Investing in good history taking can lead to early detection and early treatment with less morbidity and better outcomes.[161]

Try to avoid medical terminology and jargon—a common pitfall among healthcare providers when they feel embarrassed. Listen to the words the clients use to describe sexual activities and practices and then use their preferred words when appropriate.

Men who have sex with men may identify themselves as homosexual, bisexual, or heterosexual. No matter what label is used, these men are at increased risk for STDs, as well as psychologic and behavioral disorders, drug abuse, and eating disorders. Avoid terms such as "gay," "queer," and "straight" when talking about sexual practices or sexual identity.[162]

There is no way to know when someone will be offended or claim sexual harassment. It is in your own interest to behave in the most professional manner possible. There should be no hint of sexual innuendo or humor injected into any of your conversations with clients at any time. The line of sexual impropriety lies where the complainant draws it and includes appearances of misbehavior. This perception differs broadly from client to client.[110]

It is also true that clients sometimes behave inappropriately; there may be times when the therapist must remind clients of appropriate personal boundaries. At the same time, the therapist must be prepared to hear just about anything if and when it is necessary to ask questions about sexual history or sexual practices. Be aware of your facial expressions, body language, and verbal remarks in response to a client's answers to these questions.

SUBJECTIVE EXAMINATION—*cont'd*

What if a man or woman with pelvic or sacral pain tells you he or she has been the victim of repeated violent sexual acts? What if a client admits to being the victim of physical or emotional assault? The therapist must be prepared to respond in a professional and responsible way. Additional training in this area may be helpful. Many local organizations, such as Planned Parenthood, Lambda Alliance, and AIDS Council, may offer helpful information and/or training.

The therapist may want to introduce the series of questions by saying, *"When evaluating low back pain sometimes it's necessary to ask some more personal questions. Please answer as accurately as you can."*

- Are you sexually active?
    - Follow-up question: How does sexual activity affect your symptoms?

"Sexually active" does not necessarily mean engaging in sexual intercourse. Sexual touch is enough to transmit many STIs. You may have to explain this to your client to clarify this question. Oral and anal intercourse are often not viewed as "sexual intercourse" and will result in the client answering the question with "No" when, in fact, for screening purposes, the answer is "Yes."

- Do you have pain with certain positions? (e.g., for the therapist: a position with the woman on top can be more difficult with **prolapsed uterus; the penis or other object touching inflamed cervix also can cause pain**)
- Have you had more than one sexual partner? **(Increases risk of STDs)**
- Have you ever been told you have a STI or STD such as herpes, genital warts, Reiter's disease, syphilis, "the clap," chlamydia, gonorrhea, venereal, HIV, or other disease?
- Have you ever had sexual intercourse without wanting to? Alternate question: Have you ever been raped?
- Do you have any blood in your urine or your stools? Alternate question: Do you have any bleeding when you go to the bathroom?
    - What do you think could be causing this?
    - *If yes,* do you have a history of hemorrhoids?

Special Questions to Ask: Women Experiencing Back, Hip, Pelvic, Groin, Sacroiliac, or Sacral Pain

Not all these questions will need to be asked. Use your professional judgment to decide what to ask based on what the woman has told you and what you have observed during the examination.

Past Medical History

Have you ever been told that you have:

- Retroversion of the uterus (tipped back)
- Ovarian cysts
- Fibroids or tumors
- Endometriosis
- Cystocele (sagging bladder)
- Rectocele (sagging rectum)
- Pelvic inflammatory disease (PID)?
- Have you had vaginal surgery or a hysterectomy? **(Hysterectomy: Joint pain and myalgias may occur; vaginal surgery: Incontinence)**
- Have you had a recent history of bladder or kidney infections? **(Referred back pain)**
- Have you ever been told you have "brittle bones" or osteoporosis?
- Have you ever had a compression fracture of your back?

Menstrual History

A menstrual history may be helpful when evaluating back or shoulder pain of unknown cause in a woman of reproductive age. Not all these questions will need to be asked. Use your professional judgment to decide what to ask based on what the woman has told you and what you have observed during the examination.

- Is there any connection between your (back, hip, SI) pain/symptoms and your menstrual cycle (related to either ovulation, midcycle, or menses)?
- Since your back/SI (or other) pain/symptoms started, have you seen a gynecologist to rule out any gynecologic cause of this problem?
- Where were you in your menstrual cycle when your injury or illness occurred?
- Where are you in your menstrual cycle today (premenstrual/midmenstrual/postmenstrual)? **(Appropriate question for shoulder or back pain of unknown cause)**
- Please describe any other menstrual irregularity or problems not already discussed.

For the Young Female Adolescent/Athlete

- Have you ever had a menstrual period?
    - *If yes,* do you have a menstrual period every month? (Amenorrhea or irregular cycles can be a natural part of development but also the result of an eating disorder.)
- Have you ever gone 3 months without having a period?
- Do your periods change with your training regimen?
    - *If yes,* please describe.
- Are you taking birth control pills or using a patch or injection?
    - *If yes,* are you using them for birth control, to regulate your menstrual cycle, or both?
    - Assess risk factors and monitor blood pressure.
    - How long have you been on birth control?
    - When was the last time you saw the doctor who prescribed birth control for you?
- Please describe any other menstrual irregularity or problems not already discussed.

Continued

SUBJECTIVE EXAMINATION—*cont'd*

Reproductive History

- Is there any possibility you could be pregnant?
- Was your last period normal for you?
- What form of birth control are you using? (If the client is using birth control pills, patches, or injections check her blood pressure.)
- Do you have an intrauterine coil or loop contraceptive device (IUD or IUCD)? **(PID and ectopic pregnancy can occur.)**
- **For the pregnant woman:** Are you under the care of a physician? Have you had any spotting or bleeding during your pregnancy?
- Have you recently had a baby? **(Birth trauma)**
  - *If yes,* did you have an epidural (anesthesia)? **(Postpartum back pain)**
  - *If yes,* did you have any significant medical problem during your pregnancy or delivery?
- Have you ever had a tubal or ectopic pregnancy? Is it possible that you may be pregnant now?
- How many pregnancies have you had?
- How many live births have you had?
- Have you ever had an abortion or miscarriage?
  - *If yes,* follow-up with careful (sensitive) questions about how many, when, where, and any immediate or delayed complications (physical or psychologic). (Weakness secondary to blood loss, infection, scarring; blood in **peritoneum irritating diaphragm causing lumbar and/or shoulder pain**); ask about the onset of symptoms in relation to the incident.
- Do you ever experience a "falling out" feeling or pelvic heaviness after standing for a long time? **(Uterine prolapse; pelvic floor weakness; incontinence)**
- Do you ever leak urine with coughing, laughing, lifting, exercising, or sneezing? **(Stress incontinence; tension myalgia of pelvic floor)**
  - *If yes to incontinence:* Ask several additional questions to determine the frequency, the amount of protection needed (as measured by the number and type of pads used daily), and how much this problem interferes with daily activities and lifestyle. See also Appendix B-5.
- Do you have an unusual amount of vaginal discharge or vaginal discharge with an obvious odor? **(Referred back pain)**
  - *If yes,* do you know what is causing this discharge? Is there any connection between when the discharge started and when you first noticed your back/SI (or other) symptoms?
- For the postmenopausal woman: Are you taking hormone replacement therapy (HRT) or any natural hormone products?

Special Questions to Ask: Men Experiencing Back, Hip, Pelvic, Groin, or Sacroiliac Pain

- Have you ever had prostate problems or been told you have prostate problems?
- Have you ever been told you have a hernia? Do you think you have one now?
  - *If yes,* follow-up with medical referral. Strangulation of the bowel can lead to serious complications. If the client has been evaluated by a physician and has declined treatment (usually surgery), encourage him to follow up on this recommendation.
- Have you recently had kidney stones or bladder or kidney infections?
- Have you had any changes in urination recently?
- Do you ever have blood in your urine?
- Do you ever have pain, burning, or discomfort on urination?
- Do you urinate often, especially during the night?
- Can you easily start a flow of urine?
- Can you keep a steady stream without stopping and starting?
- When you are done urinating, does it feel like your bladder is empty or do you feel like you still have to go, but you can't get any more out?
- Do you ever dribble or leak urine?
- Do you have trouble getting an erection?
- Do you have trouble keeping an erection?
- Do you have trouble ejaculating? (Therapists beware: This term may not be understood by all clients.)
- Any unusual discharge from your penis?

To the therapist: If the client is having difficulty with sexual function, it may be necessary to conduct a screening examination for bladder or prostate involvement. See Appendices B-5 and B-30.

CASE STUDY

Steps in the Screening Process

REFERRAL

A 47-year-old man with LBP of unknown cause has come to you for exercises. After gathering information from the client's history and conducting the interview, you ask him:

FOLLOW-UP QUESTIONS

- Are there any other symptoms of any kind anywhere else in your body?

The client tells you he does break out into an unexpected sweat from time to time but does not think he has a temperature when this happens. He has increased back pain when he passes gas or a bowel movement, but then the pain goes back to the "regular" pain level [reported as 5 on a scale from 0 to 10]. Other reported symptoms include:

- Heartburn and indigestion
- Abdominal bloating after meals
- Chronic bronchitis from smoking [3 packs/day]
- Alternating diarrhea and constipation

Use the list of signs and symptoms in Box 4-19 to review this case.

Do these symptoms fall into any one category?

It appears that many of the symptoms may be GI in nature.

What is the next step in the screening process?

Since the client has mentioned unexplained sweating but no known fevers, take the time to measure all vital signs, especially body temperature. Turn to Special Questions to Ask at the end of Chapter 8 and scan the list of questions for any that might be appropriate for this client.

For example, find out about the use of NSAIDs (prescription and OTC; be sure to include aspirin). Follow-up with:

FOLLOW-UP QUESTIONS

- Have you ever been treated for an ulcer or internal bleeding while taking any of these pain relievers?
- Have you experienced any unexpected weight loss in the last few weeks?
- Have you traveled outside the United States in the last year?
- What is the effect of eating or drinking on your abdominal pain? Back pain?
- Have the client pay attention to his symptoms over the next 24 to 48 hours:
  - Immediately after eating
  - Within 30 minutes of eating
  - 1 to 2 hours later
- Do you have a sense of urgency so that you have to find a bathroom for a bowel movement or diarrhea right away without waiting?

Ask any further questions that may be appropriate as listed in this chapter or from the more complete Special Questions to Ask section of Chapter 8 (see the subsection: Associated Signs and Symptoms: Change in bowel habits).

You will make your decision to refer this client to a physician depending on your findings from the clinical examination and the client's responses to these questions. Use the Quick Screen Checklist in Appendix A-1 to see if you have left anything out that might be important.

This does not appear to be an emergency since the client is not in acute distress. An elevated temperature or other unusual vital signs might speed the referral process along.

PRACTICE QUESTIONS

1. The most common sites of referred pain from systemic diseases are:
   a. Neck and back
   b. Shoulder and back
   c. Chest and back
   d. None of the above
2. To screen for back pain caused by systemic disease:
   a. Perform special tests (e.g., Murphy's percussion, Bicycle test)
   b. Correlate client history with clinical presentation and ask about associated signs and symptoms
   c. Perform a Review of Systems
   d. All of the above
3. What are two ways of classifying back pain (as presented in the text)?
4. Which statement is the most accurate?
   a. Arterial disease is characterized by intermittent claudication, pain relieved by elevating the extremity, and history of smoking.
   b. Arterial disease is characterized by loss of hair on the lower extremities, throbbing pain in the calf muscles that goes away by using heat and elevation.
   c. Arterial disease is characterized by painful throbbing of the feet at night that goes away by dangling the feet over the bed.
   d. Arterial disease is characterized by loss of hair on the toes, intermittent claudication, and redness or warmth of the legs that is accompanied by a burning sensation.

Continued

PRACTICE QUESTIONS—*cont'd*

5. Pain associated with pleuropulmonary disorders can radiate to:
   a. Anterior neck
   b. Upper trapezius muscle
   c. Ipsilateral shoulder
   d. Thoracic spine
   e. All of the above
6. Which of the following are clues to the possible involvement of the GI system?
   a. Abdominal pain alternating with TMJ pain within a 2-week period of time
   b. Abdominal pain at the same level as back pain occurring either simultaneously or alternately
   c. Shoulder pain alleviated by a bowel movement
   d. All of the above
7. Percussion of the costovertebral angle resulting in the reproduction of symptoms signifies:
   a. Radiculitis
   b. Pseudorenal pain
   c. Has no significance
   d. Medical referral is advised
8. A 53-year-old woman comes to physical therapy with a report of leg pain that begins in her buttocks and goes all the way down to her toes. If this pain is of a vascular origin she will most likely describe it as:
   a. Sore, hurting
   b. Hot or burning
   c. Shooting or stabbing
   d. Throbbing, "tired"
9. Twenty-five percent of the people with GI disease, such as Crohn's disease (regional enteritis), irritable bowel syndrome, or bowel obstruction, have concomitant back or joint pain.
   a. True
   b. False
10. Skin pain over T9 to T12 can occur with kidney disease as a result of multisegmental innervation. Visceral and cutaneous sensory fibers enter the spinal cord close to each other and converge on the same neurons. When visceral pain fibers are stimulated, cutaneous fibers are stimulated, too. Thus visceral pain can be perceived as skin pain.
    a. True
    b. False
11. Autosplinting is the preferred mechanism of pain relief for back pain caused by kidney stones.
    a. True
    b. False
12. Back pain from pancreatic disease occurs when the body of the pancreas is enlarged, inflamed, obstructed, or otherwise impinging on the diaphragm.
    a. True
    b. False
13. A 53-year-old postmenopausal woman with a history of breast cancer 5 years ago with mastectomy presents with a report of sharp pain in her mid-back. The pain started after she lifted her 2-year-old granddaughter 3 days ago. Tylenol seems to help, but the pain is keeping her awake at night. Once she wakes up, she cannot find a comfortable position to go back to sleep. What are the red flags? What will you do to screen for a medical cause of her symptoms?

REFERENCES

1. Waddell G: *The back pain revolution*, ed 2, Edinburgh, 2004, Churchill Livingstone.
2. Jette AM, Davis KD: A comparison of hospital-based and private outpatient physical therapy practices. *Phys Ther* 71(5): 366–381, 1991.
3. Jette AM, Smith K, Haley SM, et al: Physical therapy episodes of care for patients with low back pain. *Phys Ther* 74(2):101–115, 1994.
4. Freburger JK, Carey TS, Holmes GM: Management of back and neck pain: who seeks care from physical therapists? *Phys Ther* 85(9):872–886, 2005.
5. Deyo RA, Weinstein JN: Low back pain. *N Engl J Med* 344(5):363–370, 2001.
6. Jarvik J, Deyo R: Diagnostic evaluation of low back pain with emphasis on imaging. *Ann Intern Med* 137:586–597, 2002.
7. Spitzer WO: Quebec Task Force on Spinal Disorders: Scientific approach to the assessment and management of activity-related spinal disorders: a monograph for clinicians. *Spine* 12(suppl 1):51–59, 1987.
8. Sembrano JN, Polly DW: How often is low back pain not coming from the back? *Spine* 34:E27–E32, 2009.
9. Bernard TN, Jr, Kirkaldy-Willis WH: Recognizing specific characteristics of nonspecific low back pain. *Clin Orthop* 217:266–280, 1987.
10. Shaw JA: The role of the sacroiliac joint as a cause of low back pain and dysfunction. In Vleeming A, Mooney V, Snijders C, et al, editors: *The First Interdisciplinary World Congress on low back pain and its relation to the sacroiliac joint*, Rotterdam, Netherlands, 1992, ECO, pp 67–80.
11. Depalma MJ: What is the source of chronic low back pain and does age play a role? *Pain Med* 12(2):224–233, 2011.
12. Vora AJ: Functional anatomy and pathophysiology of axial low back pain: discs, posterior elements, sacroiliac joint, and associated pain generators. *Phys Med Rehabil Clin N Am* 21(4):679–709, 2010.
13. Vanelderen P: Sacroiliac joint pain. *Pain Pract* 10(5):470–478, 2010.
14. Kim DH, Hilibrand AS: Rheumatoid arthritis in the cervical spine. *J Am Acad Orthop Surg* 13(7):463–474, 2005.
15. Magarelli N: MR imaging of atlantoaxial joint in early rheumatoid arthritis. *Radiol Med* 115(7):1111–1120, 2010.
16. Krauss WE: Rheumatoid arthritis of the craniovertebral junction. *Neurosurgery* 66(3 Suppl):83–95, 2010.
17. Guzman J: A new conceptual model of neck pain. *Spine* 33(4S):S14–S23, 2008.

18. Boissonnault WG, Koopmeiners MB: Medical history profile: orthopaedic physical therapy outpatients. *J Orthop Sports Phys Ther* 20:2–10, 1994.
19. Boissonault WG: Prevalence of comorbid conditions, surgeries, and medication use in a physical therapy outpatient population: a multi-centered study. *J Orthop Sports Phys Ther* 29:506–519; discussion 520–525, 1999.
20. Boissonnault WG, Meek PD: Risk factors for antiinflammatory drug or aspirin induced gastrointestinal complications in individuals receiving outpatient physical therapy services. *J Orthop Sports Phys Ther* 32:510–517, 2002.
21. Biederman RE: Pharmacology in rehabilitation: non-steroidal antiinflammatory agents. *J Orthop Sports Phys Ther* 35:356–367, 2005.

21a. Daniels JM: Evaluation of low back pain in athletes. *Sports Health* 3(4):336–345, 2011.

22. Witkin LR: Abscess after a laparoscopic appendectomy presenting as low back pain in a professional athlete. *Sports Health* 3(1):41–45, 2011.
23. Cosar M: The major complications of transpedicular vertebroplasty. *J Neurosurg: Spine* 11(5):607–613, 2009.
24. Johnson BA: Epidurography and therapeutic epidural injections: technical considerations and experience with 5334 cases. *Am J Neuroradiol* 20:697–705, 1999.
25. Davenport TE: Subcutaneous abscess in a patient referred to physical therapy following spinal epidural injection for lumbar radiculopathy. *J Orthop Sports Phys Ther* 38(5):287, 2008.
26. van den Bosch MAAJ: Evidence against the use of lumbar spine radiography for low back pain. *Clin Radiol* 59:69–76, 2004.
27. Bishop PB, Wing PC: Compliance with clinical practice guidelines in family physicians managing worker's compensation board patients with acute lower back pain. *Spine J* 3(6):442–450, 2003.
28. Bishop PB, Wing PC: Knowledge transfer in family physicians managing patients with acute low back pain: a prospective randomized control trial. *Spine J* 6(3):282–288, 2006.
29. Leerar PJ, Boissonnault W, Domholdt E, et al: Documentation of red flags by physical therapists for patients with low back pain. *J Man Manip Ther* 15(1):42–49, 2007.
30. Bogduk N, McQuirk B: *Medical management of acute and chronic low back pain: an evidence based approach*, Amsterdam, 2002, Elsevier.
31. Burton AK: Psychosocial predictors of outcome in acute and subchronic low back trouble. *Spine* 20:722–728, 1995.
32. McCarthy CJ: The reliability of the clinical tests and questions recommended in International Guidelines for Low Back Pain. *Spine* 32(6):921–926, 2007.
33. Moore JE: Chronic low back pain and psychosocial issues. *Phys Med Rehabil Clin N Am* 21(4):801–815, 2010.
34. Kendall NAS, Linton SJ, Main CJ: *Guide to assessing psychosocial yellow flags in acute low back pain: risk factors for long-term disability and work loss*, Wellington, New Zealand, 1998, Accident Rehabilitation and Compensation Insurance Corporation of New Zealand and the National Health Committee.

34a. Burns SA: A treatment-based classification approach to examination and intervention of lumbar disorders. *Sports Health* 3(4):363–372, 2011.

34b. George SZ, Fritz JM, Childs JD: Investigation of elevated fear-avoidance beliefs for patients with low back pain: a secondary analysis involving patients enrolled in physical therapy clinical trials. *J Orthop Sports Phys Ther* 38(2):50–58, 2008.

35. Bogduk N: *Evidence-based clinical guidelines for the management of acute low back pain: The national musculoskeletal medicine initiative*, Australia, 2002, Australian Association of Musculoskeletal Medicine.
36. Henschke N: Prevalence of and screening for serious spinal pathology in patients presenting to primary care settings with acute low back pain. *Arthritis Rheum* 60(10):3072–3080, 2009.
37. Sizer PS, Brismee JM, Cook C: Medical screening for red flags in the diagnosis and management of musculoskeletal spine pain. *Pain Pract* 7(1):53–71, 2007.
38. Henschke N: A systematic review identifies five "red flags" to screen for vertebral fracture in patients with low back pain. *J Clin Epidemiol* 61:110–118, 2008.
39. Papaionnou A: Diagnosis and management of vertebral fractures in elderly adults. *Am J Med* 113:220–228, 2002.
40. Grigoryan M: Recognizing and reporting osteoporotic vertebral fractures. *Eur Spine J* 12(Suppl 2):S104–S112, 2003.
41. Sanpera I: Bone scan as a screening tool in children and adolescents with back pain. *J Pediatr Orthop* 26(2):221–225, 2006.
42. Feldman DS: The use of bone scan to investigate back pain in children and adolescents. *J Pediatr Orthop* 20:790–795, 2000.
43. Taimela S: The prevalence of low back pain among children and adolescents: a nationwide, cohort-based questionnaire survey in Finland. *Spine* 22:1132–1136, 1997.

43a. Micheli LJ: Back pain in young athletes: significant differences from adults in causes and patterns. *Arch Pediatr Adolesc Med* 149(1):15–18, 1995.

43b. Bhatia N: Diagnostic modalities for the evaluation of pediatric back pain: a prospective study. *J Pediatr Orthop* 28(2):230–233, 2008.

43c. Nigrovic PA: Evaluation of the child with back pain. http:www.uptodate.com/index. Published January 25, 2010. Accessed July 15, 2011.

44. Deyo RA, Diehl AK: Lumbar spine films in primary care: current use and effects of selective ordering criteria. *J Gen Intern Med* 1:20–25, 1986.
45. Royal College of General Practitioners: *Clinical guidelines of the management of acute low back pain*, National Low Back Pain Clinical Guidelines, 1997.
46. Leon-Diaz A, Gonzalez-Rabelino G, Alonso-Cervino M: Analysis of the etiologies of headaches in a pediatric emergency service. *Rev Neurol* 39(3):217–221, 1–15, 2004.
47. Olesen J, Steiner TJ: The international classification of headache disorders, ed 2 (ICHD-II). *J Neurol Neurosurg Psychiatry* 75(6):808–811, 2004.
48. Silberstein SD, Olesen J, Bousser MG, et al: The international classification of headache disorders, ed 2 (ICHD-II)—revision of criteria for 8.2 medication overuse headache. *Cephalalgia* 25(6):460–465, 2005.
49. Headache Classification Committee of the International Headache Society: Classification and diagnostic criteria for headache disorders, cranial neuralgias, and facial pain. *Cephalalgia* 8(Suppl 7):1–96, 1988.
50. Headache Classification Committee of the International Headache Society: Classification and diagnostic criteria for headache disorders, cranial neuralgias and facial pain, ed 2 (revised). *Cephalalgia* 25(12):460–465, 2004.
51. Petersen SM: Articular and muscular impairments in cervicogenic headache. *J Orthop Sports Phys Ther* 33(1):21–30, 2003.
52. Agostoni E: Headache in cerebral venous thrombosis. *Neurol Sci* 25(suppl 3):S206–S210, 2004.
53. Agostoni E, Aliprandi A: Alterations in the cerebral venous circulation as a cause of headache. *Neurol Sci* 30(suppl 1):S7–S10, 2009.
54. Jacobson SA, Folstein MF: Psychiatric perspectives on headache and facial pain. *Otolaryngol Clin North Am* 36(6):1187–1200, 2003.
55. Farmer K: Psychologic factors in childhood headaches. *Semin Pediatr Neurol* 17(2):93–99, 2010.
56. Hering-Hanit R, Gadoth N: Caffeine-induced headache in children and adolescents. *Cephalalgia* 23(5):332–335, 2003.
57. Ryan LM, Warden DL: Post concussion syndrome. *Int Rev Psychiatry* 15(4):310–316, 2003.

58. Seiffert TD, Evans RW: Posttraumatic headache: a review. *Curr Pain Headache Rep* 14(4):292–298, 2010.
59. Phillips E, Levine AM: Metastatic lesions of the upper cervical spine. *Spine* 14(10):1071–1077, 1989.
60. O'Reilly MB: Nonresectable head and neck cancer. *Rehab Oncology* 22(2):14–16, 2004.
61. Neville BW, Day TA: Oral cancer and precancerous lesions. *CA Cancer J Clin* 52(4):195–215, 2002.
62. Purdy RA, Kirby S: Headaches and brain tumors. *Neurol Clin* 22(1):39–53, 2004.
63. Kirby S: Headache and brain tumours. *Cephalalgia* 30(4):387–388, 2010.
64. Valentinis L: Headache attributed to intracranial tumours: a prospective cohort study. *Cephalalgia* 30(4):389–398, 2010.
65. Narin SO, Pinar L, Erbas D, et al: The effects of exercise and exercise-related changes in blood nitric oxide level on migraine headache. *Clin Rehabil* 17(6):624–630, 2003.
66. Sandor PS, Afra J: Nonpharmacologic treatment of migraine. *Curr Pain Headache Rep* 9(3):202–205, 2005.
67. Andrasik F: Biofeedback in headache: an overview of approaches and evidence. *Cleve Clin J Med* 77(Suppl 3):S72–S76, 2010.
68. Biondi DM: Physical treatments for headache: a structured review. *Headache* 45(6):738–746, 2005.
69. Gorski JM, Schwartz LH: Shoulder impingement presenting as neck pain. *J Bone Joint Surg* 85A(4):635–638, 2003.
70. Lee L, Elliott R: Cervical spondylotic myelopathy in a patient presenting with low back pain. *J Orthop Sports Phys Ther* 38(12):798, 2008.
71. Cook C: Reliability and diagnostic accuracy of clinical tests for myelopathy in patients seen for cervical dysfunction. *J Orthop Sports Phys Ther* 39(3):172–178, 2009.
72. Slipman CW, Issac Z, Patel R, et al: Chronic neck pain: the specific syndromes. *J Musculoskel Med* 20(1):24–33, 2003.
73. Koopmeiners MB: *Personal communication*, 2003.
74. Boissonnault WG: *Personal communication*, 2003.
75. Kerry R: Manual therapy and cervical arterial dysfunction, directions for the future: a clinical perspective. *J Man Manip Ther* 16(1):39–48, 2008.
76. Kerry R, Taylor AJ: Cervical arterial dysfunction assessment and manual therapy. *Man Ther* 11:243–253, 2006.
77. Kerry R, Taylor AJ: Cervical arterial dysfunction: knowledge and reasoning for manual physical therapists. *J Orthop Sports Phys Ther* 39(5):378–387, 2009.
78. Mooney V, Robertson J: The facet syndrome. *Clin Orthop* 115:149–156, 1976.
79. Fruth SJ: Differential diagnosis and treatment in a patient with posterior upper thoracic pain. *Phys Ther* 86(2):154–268, 2006.
80. Hartvigsen J, Christensen K, Frederiksen H: Back pain remains a common symptom in old age. A population-based study of 4,486 Danish twins aged 70-102. *Eur Spine J* 12(5):528–534, 2003.
80a. Braun J, Inman R: Clinical significance of inflammatory back pain for diagnosis and screening of patients with axial spondyloarthritis. *Ann Rheum Dis* 69(7):1264–1268, 2010.
81. O'Neill CW, Kurgansky ME, Derby R, et al: Disc stimulation and patterns of referred pain. *Spine* 27(24):2776–2781, 2002.
82. Koes BW: Diagnosis and treatment of sciatica. *BMJ* 334:1313–1317, 2007.
83. Crowell MS, Gill NW: Medical screening and evacuation: cauda equina syndrome in a combat zone. *J Orthop Sports Phys Ther* 39(7):541–549, 2009.
84. O'Laughlin SJ, Kokosinski E: Cauda equina syndrome in a pregnant woman referred to physical therapy for low back pain. *J Orthop Sports Phys Ther* 38(11):721, 2008.
85. McCarthy MJH: Cauda equina syndrome: Factors affecting long-term functional and sphincteric outcome. *Spine* 32(2):207–216, 2007.
86. Kauppila LI: Atherosclerosis and disc degeneration/low-back pain—a systematic review. *Eur J Endovasc Surg* 37(6):661–670, 2009.
87. Bingol H, Cingoz F, Yilmaz AT, et al: Vascular complications related to lumbar disc surgery. *J Neurosurg: Spine* 100(3):249–253, 2004.
88. Lacombe M: Vascular complications of lumbar disk surgery. *Ann Chir* 131(10):583–589, 2006.
89. Kauppila LI, Mikkonen R, Mankinen P, et al: MR aortography and serum cholesterol levels in patients with long-term nonspecific lower back pain. *Spine* 29(19):2147–2152, 2004.
90. Silverman SL: The clinical consequences of vertebral compression fracture. *Bone* 13:S27–S31, 1992.
91. Badke MB, Boissonnault WG: Changes in disability following physical therapy intervention for patients with low back pain: dependence on symptom duration. *Arch Phys Med Rehab* 87(6):749–756, 2006.
92. Whooley MA: Case-finding instruments for depression. Two questions are as good as many. *J Gen Intern Med* 12:439–445, 1997.
93. Haggman S: Screening for symptoms of depression by physical therapists managing low back pain. *Phys Ther* 84:1157–1165, 2004.
94. Hoogendoorn WE, van Poppel MN, Bongers PM, et al: Systematic review of psychosocial factors at work and private life as risk factors for back pain. *Spine* 25:2114–2125, 2000.
95. Marras WS, Davis KG, Heaney CA, et al: The influence of psychosocial stress, gender, and personality on mechanical loading of the lumbar spine. *Spine* 25(23):3045–3054, 2000.
96. Thorbjornsson CO, Alfredsson L, Fredriksson K, et al: Physical and psychosocial factors related to low back pain during a 24-year period. *Occup Environ Med* 55(2):84–90, 1998.
97. McCarthy CJ: The reliability of the clinical tests and questions recommended in International Guidelines for Low Back Pain. *Spine* 32(6):921–926, 2007.
98. Ramond A: Psychosocial risk factors for chronic low back pain in primary care—a systematic review. *Fam Prac* 28(1):12–21, 2011.
99. van Tulder M: European guidelines for the management of acute nonspecific low back pain in primary care. European Commission, Geneva. *European Spine Journal* 15(Suppl 2): S169–S191, 2006.6Available on-line at http://www.backpaineurope.org/. Accessed December 1, 2010.
100. Kendall NAS, Linton SJ, Main CJ: *Guide to assessing psychological yellow flags in acute low back pain: risk factors for long-term disability and work loss*, Wellington, New Zealand, 1997, Accident Rehabilitation and Compensation Insurance Corporation of New Zealand and the National Health Committee.
101. Borkan J, Van Tulder M, Reis S, et al: Advances in the field of low back pain in primary care. A report from the Fourth International Forum. *Spine* 27(5):E128–E132, 2002.
102. Slipman C: Epidemiology of spine tumors presenting to musculoskeletal physiatrists. *Arch Phys Med Rehabil* 84:492–495, 2003.
103. Rose PS, Buchowski JM: Metastatic disease in the thoracic and lumbar spine: evaluation and management. *J Am Acad Orthop Surg* 19(1):37–48, 2011.
104. Henschke N: Screening for malignancy in low back pain patients: a systematic review. *Eur Spine J* 16(10):1673–1679, 2007.
105. Briggs HK: The physical therapist's management of a patient with low back pain following an atypical response to treatment: a case report. *J Orthop Sports Phys Ther* 41(1):A16, 2011.
106. Mazanec DJ, Segal AM, Sinks PB: Identification of malignancy in patients with back pain: red flags. *Arthritis Rheum* 36(suppl):S251–S258, 1993.

107. Skoffer B: Low back pain in 15- to 16-year-old children in relation to school furniture and carrying of the school bag. *Spine* 32(24):E713–E717, 2007.
108. Neuschwander TB: The effect of backpacks on the lumbar spine in children. *Spine* 35(1):83–88, 2009.
109. Patchell RA: Direct decompressive surgical resection in the treatment of spinal cord compression caused by metastatic cancer: a randomized trial. *Lancet* 366(9486):643–648, 2005.
110. Rex L: *Evaluation and treatment of somatovisceral dysfunction of the gastrointestinal system*, Edmonds WA, 2004, URSA Foundation.
111. Deyo RA, Diehl AK: Cancer as a cause of back pain: frequency, clinical presentation, and diagnostic strategies. *J Gen Intern Med* 3(3):230–238, 1988.
112. Wong DA, Fornasier VL, MacNab I: Spinal metastases: the obvious, the occult, and the imposters. *Spine* 15(1):1–4, 1990.
113. Ross MD, Bayer E: Cancer as a cause of low back pain in a patient seen in a direct access physical therapy setting. *J Orthop Sports Phys Ther* 35(10):651–658, 2005.
114. Cosford PA, Leng GC: Screening for abdominal aortic aneurysm. *Cochrane Database Syst Rev* 18(2):CD002945, 2007.
115. Fleming C, Whitlock EP, Beil TL, et al: Screening for abdominal aortic aneurysm: a best evidence systematic review for the U.S. Preventive Services Task Force. *Ann Intern Med* 142(3):203–211, 2005.
116. Lederle FA: Smokers' relative risk for aortic aneurysm compared with other smoking-related diseases: a systematic review. *J Vasc Surg* 38:329–334, 2003.
117. Dua MM, Dalman RL: Identifying aortic aneurysm risk factors in postmenopausal women. *Womens Health* 5(1):33–37, 2009.
118. Lederle FA: Abdominal aortic aneurysm events in the women's health initiative: cohort study. *BMJ* 337:1724–1734, 2008.
119. Mofidi R: Influence of sex on expansion rate of abdominal aortic aneurysms. *Brit J Surg* 94:310–314, 2007.
120. Wanhainen A, Lundkvist J, Bergqvist D, et al: Cost-effectiveness of screening women for abdominal aortic aneurysm. *J Vasc Surg* 43:908–914, 2006.
121. Dyck P, Doyle JB: "Bicycle test" of van Gelderen in diagnosis of intermittent cauda equina compression syndrome. *J Neurosurg* 46:667–670, 1977.
122. Dyck P: The stoop-test in lumbar entrapment radiculopathy. *Spine* 4:89–92, 1979.
123. Yung E: Screening for head, neck, and shoulder pathology in patients with upper extremity signs and symptoms. *J Hand Ther* 23(2):173–186, 2010.
124. Eskelinen M: Usefulness of history-taking, physical examination and diagnostic scoring in acute renal colic. *Eur Urol* 34(6):467–473, 1998.
125. Houppermans RP, Brueren MM: Physical diagnosis—pain elicited by percussion in the kidney area. *Ned Tijdschr Geneeskd* 2001;145(5):208–210.
126. Herkowitz HN, editor: *The spine*, Philadelphia, 1999, WB Saunders.
127. Sex and back pain video, Dixfield, Maine, IMPACC, Inc. Available at www.impaccusa.com. Accessed March 9, 2011.
128. Sex and back pain patient manual, Dixfield, ME, IMPACC, Inc. Available at www.impaccusa.com. Accessed March 9, 2011.
129. Padua L, Caliandro P, Aprile I, et al: Back pain in pregnancy. *Eur Spine J* 14(2):151–154, 2005.
130. Borg-Stein J, Dugan SA, Gruber J: Musculoskeletal aspects of pregnancy. *Am J Phys Med Rehabil* 84(3):180–192, 2005.
131. Quaresma C: Back pain during pregnancy: a longitudinal study. *Acta Rheumatol Port* 35(3):346–351, 2010.
132. Han IH: Pregnancy and spinal problems. *Curr Opin Obstet Gynecol* 22(6):477–481, 2010.
133. Macarthur AJ: Is epidural anesthesia in labor associated with chronic low back pain? A prospective cohort study. *Anesth Analg* 85(5):1066–1070, 1997.
134. Leighton BL, Halpern SH: The effects of epidural analgesia on labor, maternal, and neonatal outcomes: a systematic review. *Am J Obstet Gynecol* 186(5 Suppl Nature):S69–S77, 2002.
135. Deevey S: Endometriosis: Internet resources. *Medical Ref Serv Quart* 24(1):67–77, 2005.
136. Jones KD, Sutton CJ: Recurrence of chocolate cysts after laparoscopic ablation. *J Am Assoc Gynecol Laparosc* 9(3):315–320, 2002.
137. Giudice LC, Kao LC: Endometriosis. *Lancet* 364(9447):1789–1799, 2004.
138. Sarma D: Cerebellar endometriosis. *Am Roentgen Ray Society* 182:1543–1546, 2004.
139. Carrell DT, Peterson CM, editors: *Reproductive endocrinology and infertility*, New York, 2010, Springer Science.
140. Sinaii N, Cleary SD, Ballweg ML, et al: High rates of autoimmune and endocrine disorders, fibromyalgia, chronic fatigue syndrome, and atopic diseases among women with endometriosis: a survey analysis. *Hum Reprod* 17(10):2715–2724, 2002.
141. Sundqvist J: Endometriosis and autoimmune disease. *Fertil Steril* 95(1):437–440, 2011.
142. Barrier BF: Immunology of endometriosis. *Clin Obstet Gynecol* 53(2):397–402, 2010.
143. Nissenblatt M: Endometriosis-associated ovarian carcinomas. *N Engl J Med* 364(5):482–485, 2011.
144. Svendsen PF, Nilas L, Norgaard K, et al: Polycystic ovary syndrome. New pathophysiological discoveries. *Ugeskr Laeger* 167(34):3147–3151, 2005.
145. Dokras A, Bochner M, Hollinrake E, et al: Screening women with polycystic ovary syndrome for metabolic syndrome. *Obstet Gynecol* 106(1):131–137, 2005.
146. Wild RA: Assessment of cardiovascular risk and prevention of cardiovascular disease in women with the polycystic ovary syndrome: a consensus statement by the Androgen Excess and Polycystic Ovary Syndrome Society. *J Clin Endocrinol Metab* 95(5):2038–2049, 2010.
147. Siegel R, et al: Cancer statistics 2011. *CA Cancer J Clin* 61:212–236, 2011.
148. Swan J: Data and trends in cancer screening in the United States. *Cancer* 116(20):4872–4881, 2010.
149. Centers for Disease Control and Prevention: Prostate cancer: statistics. Available online at http://www.cdc.gov/cancer/prostate/. Accessed March 9, 2011.
150. Gelfand MS, Cleveland KO: Vancomycin therapy and the progression of methicillin-resistant *Staphylococcus aureus* vertebral osteomyelitis. *South Med J* 97(6):593–597, 2004.
151. Van Hal SJ: Emergence of daptomycin resistance following vancomycin-unresponsive *Staphylococcus aureus* bacteraemia in a daptomycin-naive patient—a review of the literature. *Eur J Clin Microbiol Infect Dis* 30(5):603–610, 2011.
152. Patel RK, Everett CR: Low back pain: 20 clinical pearls. *J Musculoskel Med* 20(10):452–460, 2003.
153. Walton DM: Recovery from acute injury: clinical, methodological and philosophical considerations. *Disabil Rehabil* 32(10):864–874, 2010.
154. Calley D: Identifying patient fear-avoidance beliefs by physical therapists managing patients with low back pain. *J Orthop Sports Phys Ther* 40(12):774–783, 2010.
155. Asavasopon S, Jankoski J, Godges JJ: Clinical diagnosis of vertebrobasilar insufficiency: resident's case problem. *J Orthop Sports Phys Ther* 35(10):645–650, 2005.
156. Wiesel BB, Wiesel SW: Radiographic evaluation of low back pain: a cost-effective approach. *J Musculoskel Med* 21(10):528–538, 2004.
157. Mazanec DJ: Recognizing malignancy in patients with low back pain. *J Musculoskel Med* 13(1):24–31, 1996.
158. King H: Evaluating the child with back pain. *Pediatr Clin North Am* 33(6):1489–1493, 1986.

159. Behrman R, Kliegman RM, Arvin AM, editors: *Nelson's textbook of pediatrics*, ed 17, Philadelphia, 2004, WB Saunders.
160. McTimoney CA, Micheli LJ: Managing back pain in young athletes. *J Musculoskel Med* 21(2):63–69, 2004.
161. Goode B: *Personal communication*, Raleigh, NC, 2006, Centers for Disease Control and Prevention.
162. Knight D: Health care screening for men who have sex with men. *Amer Fam Phys* 69(9):2149–2156, 2004.
163. Ozgen S: Lumbar disc herniation in adolescence. *Pediatr Neurosurg* 43:77–81, 2007.
164. Seeman E: The dilemma of osteoporosis in men. *Am J Med* 98(2A):765S–788S, 1995.
165. Orwoll ES, Klein RF: Osteoporosis in men. *Endocr Rev* 16:87–116, 1995.
166. Kiebzak G, Beinart G, Perser K, et al: Undertreatment of osteoporosis in men with hip fracture. *Arch Intern Med* 162(19): 2217–2222, 2002.
167. Ebeling PR: Osteoporosis in men. *N Engl J Med* 358(14):1474–1482, 2008.
168. Ebeling PR: Androgens and osteoporosis. *Curr Opin Endocrinol Diabetes Obes* 17(3):284–292, 2010.
169. Blain H: Osteoporosis in men: epidemiology, physiopathology, diagnosis, prevention, and treatment. *Rev Med Interne* 25(suppl 5):S552–S559, 2005.

CHAPTER 15

Screening the Sacrum, Sacroiliac, and Pelvis

Following the model for decision making in the screening process outlined in Chapter 1 (see Box 1-7), we now turn our attention to pain from medical conditions, illnesses, and diseases referred to the sacrum, sacroiliac (SI), and pelvic regions.

The basic premise is that physical therapists must be able to identify signs and symptoms of systemic origin or associated with medical conditions that can mimic neuromuscular or musculoskeletal (neuromusculoskeletal [NMS]) impairment in these areas.

In the screening process, therapists will watch for yellow (caution) or red (warning) flags to direct them. Clinicians rely on special questions to ask men and women with significant risk factors, significant past medical history, suspicious clinical presentation, or associated signs and symptoms.

With a careful interview and the right screening questions, the therapist can identify clues suggestive of a problem outside the scope of a physical therapist's practice that may require medical referral. Specific tests to screen for an underlying infectious or inflammatory source of pelvic or abdominal pain are also presented with a suggested order of testing.

When dealing with painful symptoms of the sacral and pelvic areas, the therapist may need to ask questions about sexual history or sexual practices. The therapist must remain aware of facial expressions, body language, and verbal remarks in response to a client's answers to these questions.

The therapist must be prepared to respond in a professional and responsible way if a man or woman with pelvic or sacral pain reports that he or she has been the victim of repeated violent sexual acts, or if a client admits to physical or emotional assault. More about the client interview, the screening interview, and screening for assault and domestic (intimate partner) violence is included in Chapter 2 (see also Appendices B-3 and B-32).

THE SACRUM AND SACROILIAC JOINT

Evaluating the SI joint can be difficult in that no single physical examination finding can predict a disorder of the SI joint. Pain originating from the SI joint can mimic pain referred from lumbar disk herniation, spinal stenosis, facet joint impairment, or even a disorder of the hip.[1-3]

The most common clinical presentation of sacroiliac pain is associated with a memorable *physical event* that initiated the pain such as a misstep off a curb, a fall on the hip or buttocks, lifting of a heavy object in a twisted position, or childbirth (Case Example 15-1). A history of previous spine surgery is very common in clients with SI intraarticular pain.[1]

The most typical *medical conditions* that refer pain to the sacrum and SI joint include endocarditis, prostate cancer or other neoplasm,[4] gynecologic disorders, rheumatic diseases that target the SI area (e.g., spondyloarthropathies such as ankylosing spondylitis, Reiter's syndrome, or psoriatic arthritis), and Paget's disease (Table 15-1).[5]

Disorders of the large intestine and colon, such as ulcerative colitis, Crohn's disease (regional enteritis), carcinoma of the colon, and irritable bowel syndrome (IBS), can refer pain to the sacrum when abscess develops or when the rectum is stimulated.[6] Likewise, primary SI problems can refer pain to the lower abdomen.[7]

A medical differential diagnosis may be needed to exclude fracture, infection, or tumor. Insufficiency fractures of the sacrum can occur after pelvic radiotherapy for cancer[8] and in osteoporotic bone with minimal or unremembered trauma.[9] (See further discussion in this chapter on spondylogenic causes of sacral pain.)

Using the Screening Model to Evaluate Sacral/Sacroiliac Symptoms

The principles guiding evaluation of SI joint or sacral pain are consistent with the information presented throughout this text and, in particular, in the chapter on back pain (see Chapter 14).

Each of the disorders listed in Table 15-1 usually has its own unique *clinical presentation* with clues available in the *past medical history.* The presence of *associated signs and symptoms* is always a red flag. Most of these conditions have clear red flag clues that come to light if the client is interviewed carefully.

Clinical Presentation

Insidious onset or unknown cause is always a red flag. Without a clear cause, the therapist looks for something else in the history or accompanying signs and symptoms. Even with a known or assigned cause, it is important to keep other possibilities in mind and to watch for red flags (Box 15-1). Sacral pain in the absence of a history of trauma or overuse is a clue to the presentation of systemic backache.

CASE EXAMPLE 15-1

Sacroiliac Pain Caused by Pelvic Floor Impairment

Background: A 33-year-old woman referred by her orthopedic surgeon presented with low back pain centered over the sacroiliac (SI) region. She described it as "sharp" and "knifelike." It comes and goes with no warning. Sometimes, it is so severe she cannot catch her breath and falls to her knees. After that, she cannot stand up straight for several hours and walks "hunched over."

The pain presented on both sides intermittently, but the primary pain pattern was localized in the left SI area. Heat seems to help for a short time, but nothing brings complete relief all the time.

She has a previous history of disk herniation with diskectomy and laminectomy and complete resolution of symptoms. No cause is known for this new onset of SI symptoms. No radiating symptoms are apparent, and recent magnetic resonance imaging (MRI) shows no sign of disk protrusion at this time. (She tried doing her previous program of McKenzie exercises, but no change in symptoms occurred.)

Clinical Presentation: Physical therapy examination reveals the following:

Antalgic gait secondary to pain. Trendelenburg sign: Negative. Slight left lumbar lateral shift; posture within normal limits otherwise. Active lumbar motions are full, with a normal capsular end feel and no reproduction of symptoms. Repeated trunk and lumbar motions do not elicit painful symptoms.

Neurologic screen: Negative for abnormal reflexes, abnormal sensation, decreased strength, or altered neural tension. Hamstrings are tight bilaterally, but a straight leg raise does not increase symptoms. In fact, it is the only time in the assessment when the client reports a slight decrease in pain.

Examination of the SI area revealed an upslip on the left (anterior superior iliac spine [ASIS] and posterior superior iliac spine [PSIS] on the left are higher than ASIS and PSIS on the right, indicating an upward movement of the ilium on the sacrum on the high side; leg length discrepancy or muscle spasm from a disk lesion can also cause an upslip). Given her past history of diskogenic lesion, it is possible that altered muscle activation is the cause. This will have to be examined further.

Is a screening examination for systemic origin of symptoms warranted? Why, or why not?

Using our screening model, review the past medical history. Are there any red flags here? No, but the history is very incomplete. We know she had a previous diskogenic lesion treated operatively. Nothing of her personal or family history is included.

Even in a musculoskeletal assessment, we will want to know about pregnancy and birth histories; use of medications, over-the-counter drugs, and illicit drugs; smoking and drinking history or current use; levels of activity before the onset of symptoms; correlation of symptoms with menses or births; occupation and work-related activities; and history of cancer.

A general screening interview will ask about recent history of infection, the presence of joint pain or skin rash anywhere else, and the presence of any constitutional or other symptoms.

Next, review the clinical presentation. Are there any red flags here? Not really. There is no night pain. There is the fact that nothing seems to make it better or worse, but one red flag by itself usually is not highly significant. We will tuck that bit of information in the back of our minds as we continue the evaluation process.

Hamstring stretching brings some mild, temporary relief. This suggests a muscular component, but that has to be further evaluated. The SI upslip could be the cause of the symptoms, but this will not be determined until the alignment and cause of the upslip are corrected.

A trigger point assessment may be needed as well.

Step three involves a review of associated signs and symptoms. We do not know about constitutional symptoms, relationship of SI pain to menses, or the presence of any other symptoms associated with the viscera (e.g., gastrointestinal, urologic). It is always recommended to take the client's temperature in the presence of pain of unknown cause.

What to Do: Several strategies are presented here. Intervention for the upslip may be the first step with reassessment of symptoms. If a lack of progress occurs, the therapist can go back and ask more specific questions. Or, the therapist can treat the upslip while continuing to interview the client each day, obtaining additional pertinent information before making a final decision.

Result: In the end, it was discovered that the client had significant pelvic floor impairment with overactivity of the pelvic floor muscles (levator ani) and detrusor imbalance with urinary incontinence. She reported a complicated birth history with her first child, which was repeated with less severity during the births of her second and third children.

Intercourse was extremely painful, but the client was too embarrassed to bring this up until the therapist asked directly about sexual activity. The client finally described a sensation as if "trying to deliver a baby through my rectum" during intercourse (a sign of levator ani impairment).

Once all the additional information had been brought out and organized, the client shared the signs and symptoms with her gynecologist. An internal vaginal examination reproduced her symptoms exactly. The evaluating therapist was not trained in pelvic floor assessment and did not make this finding directly.

In looking back, it is likely that development of the diskogenic lesion was linked to birth/delivery problems (or perhaps, vice versa; it was never known for sure). Closer examination revealed a loss of lumbar stabilization because of multifidus impairment. Muscle impairment at the time of the disk lesion and births probably contributed to the gradual development of pelvic floor impairment.

Changes were also noted in the abdominal muscles with a loss of co-contraction between the multifidus and the transversus abdominis. The levator ani and pelvic floor muscles were in a contract-hold pattern, contributing to the painful symptoms described.

Heat relaxed the muscles but only for a short time. Hamstring stretching may have brought about an inhibition to the pelvic floor muscles, reducing pain.

A program directed at restoring normal muscle tone and function in the lumbar spine, abdominal muscles, and pelvic floor resulted in immediate reduction and eventual elimination of painful symptoms and return of comfortable coitus. Symptoms of urinary incontinence also were resolved.

Although the SI upslip could be corrected, the client could not maintain the correction. Because she was pain free, she did not return to physical therapy for further evaluation of the underlying biomechanics around the SI upslip.

TABLE 15-1 Causes of Sacral and Sacroiliac Pain

Systemic	Neuromuscular/Musculoskeletal[17]
INFECTIOUS/INFLAMMATORY Spondyloarthropathy: • Ankylosing spondylitis • Reiter's syndrome • Psoriatic arthritis • IBD (arthritis associated with IBD) Vertebral osteomyelitis Endocarditis Tuberculosis (uncommon) SPONDYLOGENIC Fracture (traumatic, insufficiency, pathologic), metabolic bone disease • Osteoporosis (insufficiency fractures) • Paget's disease • Osteodystrophy • Osteoarthritis GYNECOLOGIC Reproductive cancers Retroversion of the uterus Uterine fibroids Ovarian cysts Endometriosis PID Incest/sexual assault Rectocele, cystocele Uterine prolapse Normal pregnancy; multiparity (more than one pregnancy) GASTROINTESTINAL Ulcerative colitis Colon cancer IBS Crohn's disease (regional enteritis) CANCER Primary tumors* (rare: giant cell, chondrosarcoma, chondroma, synovial villoadenomas, schwannoma, neurofibroma, ependymomas, ganglioneuromas) Metastatic lesions (history of cancer): • Prostate • Lung • Thyroid • Colorectal • Breast • Gastrointestinal • Multiple myeloma • Kidney OTHER Fibromyalgia	Idiopathic (unknown) Trauma Myofascial or kinetic chain imbalance Enthesis (tendon insertion)/ligamentous sprain Degenerative joint disease Bone harvesting for grafts (may cause secondary instability) Lumbar spine fusion or hip arthrodesis Myofascial syndromes (mimics SI joint pain) Diskogenic disease (mimics SI joint pain) Nerve root compression (mimics SI joint pain) Zygapophyseal joint pain (mimics SI joint pain)

IBD, Inflammatory bowel disease; *SI,* sacroiliac; *PID,* pelvic inflammatory disease; *IBS,* irritable bowel syndrome.
*Includes benign and malignant osseous and neurogenic tumors affecting the sacrum.

The amount and direction of pain radiation can offer helpful clues. Low back or sacral pain radiating around the flank suggests the renal or urologic system. In such cases, the therapist should ask questions about bladder or urologic function.

Low back or sacral pain radiating to the buttock or legs may be vascular. Questions about the effects of activity on symptoms and history of cardiovascular or peripheral vascular diseases are important (see discussion in Chapter 14). Sorting out pain of a vascular versus neurogenic cause is also discussed in Chapter 14.

Most commonly, unless pain causes muscle spasm, splinting, and subsequent biomechanical changes, clients affected by systemic, medical, or viscerogenic causes of sacral or SI pain demonstrate a remarkable lack of objective findings to implicate the SI joint or sacrum as the causative factor for the presenting symptoms. Pain elicited by pressing on the sacrum with the client in a prone position suggests

BOX 15-1 RED FLAGS ASSOCIATED WITH SACROILIAC/SACRAL PAIN OR SYMPTOMS

History
- Sacroiliac/sacral pain without a history of trauma or overuse (rule out assault, anal intercourse)
- Previous history of cancer
- Previous history of gastrointestinal (GI) disease (ulcerative colitis, Crohn's disease, irritable bowel syndrome)
- Previous history of infection (skin, genitourinary tract, heart, septic sacroiliitis)

Risk Factors
- Osteoporosis
- Sexually transmitted infection or other organ system infection
- Intravenous drug abuse
- Long-term use of antibiotics (colitis)

Clinical Presentation
- Insidious onset/unknown cause
- Lack of objective findings
- Anterior pelvic, suprapubic, or low abdominal pain at the same level as the sacrum

Associated Signs and Symptoms
- Pain relieved by passing gas or having a bowel movement
- Presence of GI, gynecologic, or urologic signs and symptoms
- Fever (septic sacroiliitis); other constitutional symptoms

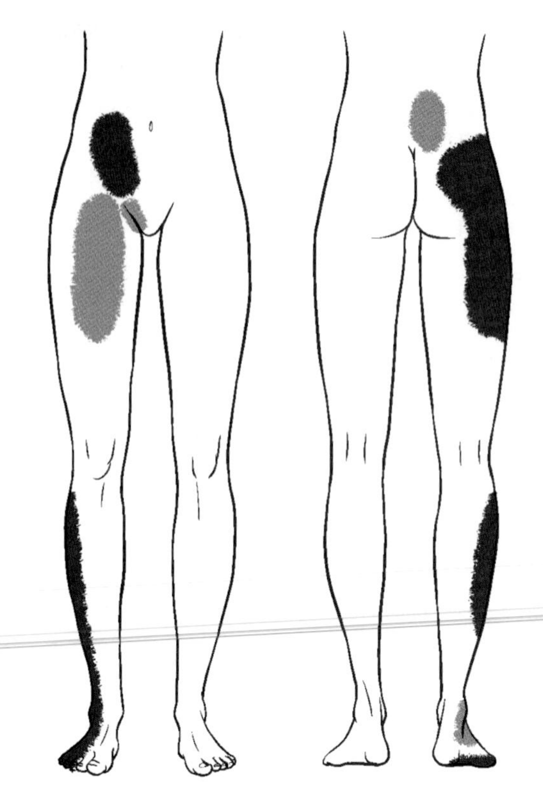

Fig. 15-1 Unilateral sacroiliac (SI) pain pattern. Pain coming from the sacroiliac joint is usually centered over the area of the posterior superior iliac spine (PSIS), with tenderness directly over the PSIS. Lower lumbar pain occurs in 72% of cases; it rarely presents as upper lumbar pain above L5 (6%). It may radiate over the buttocks (94%), down the posterior–lateral thigh (50%), and even past the knee to the ankle (14%) and lateral foot (8%). Paresthesias in the leg are not a typical feature of SI joint pain. The affected individual may report abdominal (2%), groin or pubic (14%), or anterior thigh pain (10%). Anterior symptoms may occur alone or in combination with posterior symptoms. Occasionally, a client will report bilateral pain. (Data from Slipman CW, Jackson HB, Lipetz JS, et al: Sacroiliac joint pain referral zones, *Arch Phys Med Rehab* 81:334-338, 2000.)

sacroiliitis (inflammation of the SI joint) or mechanical derangement.

Sacroiliac Joint Pain Pattern. Whether from a mechanical or a systemic origin, the patient usually experiences pain over the posterior SI joint and buttock, with or without lower extremity pain. Pain may be unilateral or bilateral (Fig. 15-1)[10] and can be referred to a wide referral zone, including the lumbar spine, abdomen, groin, thigh, foot, and ankle.[1,11]

Clients with SI joint pain rarely have pain at or above the level of the L5 spinous process, although it is possible. The presence of midline lumbar pain tends to exclude the SI joint as a potential pain generator.[12,13]

A wide range of SI joint–referred pain patterns occur because innervation is highly variable and complex or because pain may be somatically referred, as discussed in Chapter 3. Adjacent structures, such as the piriformis muscle, sciatic nerve, and L5 nerve root, may be affected by intrinsic joint disease and can become active nociceptors. Pain referral patterns also may be dependent on the distinct location of injury within the SI joint.[13,14]

SI pain can mimic diskogenic disease with radicular pain down the leg to the foot.[15] People who report midline lumbar pain when they rise from a sitting position are likely to have diskogenic pain. Clients with unilateral pain below the level of the L5 spinous process and pain when they rise from sitting are likely to have a painful SI joint.[12,13]

Pain from SI joint syndrome may be aggravated by sitting or lying on the affected side. Pain gets worse with prolonged driving or riding in a car, weight bearing on the affected side, the Valsalva maneuver, and trunk flexion with the legs straight.[14]

SI pain can also mimic the pain pattern of kidney disease with anterior thigh pain, but with SI impairment, no signs and symptoms (e.g., constitutional symptoms, bladder dysfunction) are associated, as would be the case with thigh pain referred from the renal system.

Screening for Infectious/Inflammatory Causes of Sacroiliac Pain

Joint infections spread hematogenously through the body and can affect the SI joint. Usually, the infection is unilateral and is caused by *Pseudomonas aeruginosa, Staphylococcus aureus*, *Cryptococcus* organisms, or *Mycobacterium tuberculosis*.

Risk factors for joint infection include trauma, endocarditis, intravenous drug use, and immunosuppression. Postoperative infection of any kind may not appear with any clinical signs or symptoms for weeks or months. Infections causing bacterial sacroiliitis as a complication of dilatation and curettage (D and C) after incomplete abortions have been reported.[16]

Infection can cause distention of the anterior joint capsule, irritating the lumbosacral nerve roots.[17] Inflammation of the SI joint may result from metabolic, traumatic, or rheumatic causes. Sacroiliitis is present in all individuals with ankylosing spondylitis.[18]

Rheumatic Diseases as a Cause of Sacral or Sacroiliac Pain

The most common systemic causes of sacral pain are noninfected, inflammatory erosive rheumatic diseases that target the SI, including ankylosing spondylitis, Reiter's syndrome, psoriatic arthritis, and arthritis associated with inflammatory bowel disease (IBD) such as regional enteritis (Crohn's disease).

Reiter's syndrome (see Chapter 12) occurs most often in young men with venereal disease. Reiter's syndrome often presents as a triad of symptoms, including arthritis, conjunctivitis, and urethritis. These three symptoms in the presence of sacral pain raise a red flag. The therapist must ask about pain in other joints, urologic symptoms, and a recent (or current) history of conjunctivitis (red, painful inflammation of the eye).

A positive sexual history or known diagnosis of venereal disease is helpful information. With sacral or SI pain, the therapist should always consider taking a sexual history (see Special Questions to Ask in Chapter 14 or Appendix B-32).

Crohn's disease (see Chapter 8) may be accompanied by skin rash and joint pain. This enteric condition is well known for its arthritic component, which is present in up to 25% of all cases. The client may have had Crohn's disease for years and may not recognize the onset of these new symptoms as part of that condition. Skin rash may precede joint pain by days or weeks. The hips, thighs, and legs are affected most often; the rash may be raised or flat, purple or red. Knowing the history and association between skin lesions and joint pain can help the therapist direct screening questions and make a reasonable decision about referral.

Screening for Spondylogenic Causes of Sacral/Sacroiliac Pain

Metabolic bone disease (MBD) such as osteoporosis, Paget's disease, and osteodystrophy can result in loss of bone mineral density and deformity or fracture of the sacrum. The therapist should review cases of sacral pain for the presence of risk factors for any of these metabolic bone diseases (see the discussion on metabolic bone disease in Chapter 11). Neoplasm and fracture are two other possible bony causes of sacral pain. Neoplasm is discussed separately in this chapter.

Metabolic Bone Disease

Mild-to-moderate MBD may occur with no visible signs. Advanced cases of MBD include constipation, anorexia, fractured bones, and deformity.

Osteoporosis. Osteoporosis can cause insufficiency fractures of the sacrum. The therapist must assess for risk factors (see Boxes 15-2 and 11-3) in anyone with sacral pain, especially those in whom pain has an unknown cause, postmenopausal women, older men (over 65), and anyone with a known history of osteoporosis or Paget's disease. See further discussion on osteoporosis in Chapter 11 and discussion on fractures at the end of this section.

Paget's Disease. Paget's disease as a cause of lumbar, sacral, SI, or pelvic pain occurs most commonly in men over 70 years of age (although it can occur earlier and in women). It is the second most common metabolic bone disease after osteoporosis.

Characterized by slowly progressive enlargement and deformity of multiple bones, it is associated with unexplained acceleration of bone deposition and resorption. The bones become weak, spongy, and deformed. Redness and warmth may be noted over involved areas, and the most common symptom is bone pain (see further discussion on Paget's disease in Chapter 11 and an excellent online article as referenced here).[19]

Fracture

Three types of fractures affect the sacrum: Traumatic, insufficiency, and pathologic. Trauma resulting in fracture occurs most often with lateral compression injuries seen in motor vehicle accidents or vertical shear injuries resulting from a fall from height onto the lower limbs. Less commonly, direct stress to the sacrum from a fall landing on the buttocks or athletic injury can cause traumatic sacral fracture.[20,21] Other risk factors for sacral fracture are listed in Box 15-2.

Trauma-related fatigue or stress fracture of the sacrum occurs most often in young active persons and older adults with osteoporosis. Fatigue or stress fractures can develop as a result of submaximal repetitive forces over time such as occur with overuse or overtraining in military personnel and

BOX 15-2 RISK FACTORS FOR SACRAL FRACTURES

- Osteoporosis (see also Box 11-3)
- Paget's disease
- Gender (female)
- Athletes, military personnel (overuse, overtraining, improper footwear or training surface)
- Athletic pregnant or postpartum women
- Pelvic radiation
- Lumbosacral fusion (early postoperative)
- Osteomyelitis
- Multiple myeloma
- Trauma (motor vehicle accident, fall, assault)
- Prolonged use of corticosteroids

athletes (e.g., runners, volleyball and field hockey players). Less often, pregnant or postpartum women experience sacral stress fractures, especially if they are participating in athletic training activities or running.[22-24]

Insufficiency fractures of the sacrum result from a normal stress acting on bone with deficient elastic resistance. Reduced bone integrity is most often associated with postmenopausal or corticosteroid-induced osteoporosis and radiation therapy.[20] Insufficiency fractures occur insidiously or as a result of minor trauma, possibly even from weight bearing transmitted through the spine.[25]

Pathologic fracture describes fractures that occur as a result of bone weakened by neoplasm or other disease conditions (e.g., osteomyelitis, giant cell tumor, chordoma, Ewing sarcoma, multiple myeloma). Insufficiency fractures are actually a subset of pathologic fractures confined to bones with structural alterations due to MBD.[20]

Clinical manifestations of sacral fractures can present with a wide range of signs and symptoms, many of which are present inconsistently and are considered nonspecific.[26] Bilateral or multiple stress fractures of the sacrum or pelvis have been reported.[27]

The client may report or demonstrate localized pain, tenderness with palpation, antalgic gait, and leg length discrepancy. With all sacral fractures, hip, low back, sacral, groin, or buttock pain may occur, especially with multiple stress fractures of the pelvic and sacral bones. Symptoms may mimic other conditions such as disk disease, recurrence of a local tumor, or metastatic disease.[20]

Diagnostic imaging may be needed to make the final medical diagnosis. Radiographic studies (x-rays) are often negative in the early phases of stress reactions or fractures. More advanced diagnostic bone imaging may show changes when the client becomes symptomatic.

New onset of sacral or buttock pain 1 to 2 weeks after multilevel lumbosacral fusion with instrumentation should be evaluated for sacral insufficiency fractures, especially if the patient has a recent history of osteoporosis, prolonged sitting, and kyphosis.[28-30]

Screening for Gynecologic Causes of Sacral Pain

See later discussion on gynecologic causes of pelvic pain in this chapter.

Screening for Gastrointestinal Causes of Sacral/Sacroiliac Pain

The *primary* pain pattern for gastrointestinal (GI) disease involves the midabdominal region around the umbilicus. It is not likely that the therapist will see clients with this chief complaint; they are more likely to see a doctor or go to the emergency department.

However, the therapist may be evaluating or treating a client for an orthopedic or neurologic problem who reports GI symptoms. When a client relates symptoms associated with the viscera or abdomen, the therapist must think in terms of screening questions to discern whether these symptoms require immediate medical assessment and intervention.

The therapist is more likely to see clients with *referred* low back or sacral pain from the small or large intestine as it presents in the low back or sacral area (see Figs. 8-15 and 8-16). Although these illustrations depict the pain in small, very round areas, actual pain patterns can vary quite a bit. The location will be approximately the same, but individual variation does occur.

The therapist must ask about the presence of abdominal pain or GI symptoms, occurring either simultaneously or alternating but at the same anatomic level as back or sacral pain. See Case Example 14-13 to review the importance of looking for this particular red flag.

Sacral pain from a GI source may be reduced or relieved after the person passes gas or completes a bowel movement. It may be appropriate to ask a client the following:

FOLLOW-UP QUESTIONS

- Is your pain relieved by passing gas or having a bowel movement?
- The patient may have a history of GI disease and medication used to treat such a condition. Keep the following conditions or past histories in mind when asking questions of anyone with lumbar spine or sacral pain patterns:
  - Ulcerative colitis
  - Crohn's disease
  - Irritable bowel syndrome
  - Colon cancer
  - Long-term use of antibiotics (colitis)

Screening for Tumors as a Cause of Sacral/Sacroiliac Pain

Primary sacral tumors include benign and malignant growths. Benign neoplasms include osteochondroma, giant cell tumor, and osteoid osteoma. The more common primary malignant lesions directly affecting the sacrum include chordoma, chondrosarcoma, osteosarcoma, and myeloma.

MBD to the sacrum from primary breast, lung, colon, and prostate is far more common. Sacral insufficiency fractures after pelvic radiation for rectal, prostate, or reproductive cancers can occur, although these are rare.[8,31]

Although rare, sacral neoplasms usually are not diagnosed early in the disease course because of mild symptoms resembling low back, buttock, or leg pain (sciatica).[32,33] Sacral tumors are not easy to see on x-rays and are easily overlooked due to the curvature of the sacrum, location deep within the pelvis, and frequent presence of overlying bowel gas. It is not uncommon for diagnostic delays as the person is treated for a presumed lumbar pathology before the sacrum is finally identified as the source of pathology.[33] Referral to a physical therapist before a correct medical diagnosis is made is not unusual.

Giant cell tumor is a highly aggressive local tumor of the bone. The sacrum is the third most common site of involvement. Clients present with localized pain in the lower back and sacrum that may radiate to one or both legs. Swelling may be noted in the involved area. When asked about the presence of other symptoms anywhere in the body, the client may report abdominal complaints and neurologic signs and symptoms (e.g., bowel and bladder or sexual dysfunction, numbness and weakness of the lower extremity).[4,34,35]

Colorectal or anorectal cancer as a cause of sacral pain is possible as the result of local invasion. Severe sacral pain in the presence of a previous history of uterine, abdominal, prostate, rectal, or anal cancer requires immediate medical referral.

Prostatic (males) *or reproductive cancers* in men and women can result in sacral pain. See further discussions on testicular cancer in Chapters 10 and 14, prostate cancer in Chapter 10, and gynecologic conditions in this chapter.

THE COCCYX

The coccyx or tailbone is a small triangular bone that articulates with the bottom of the sacrum at the sacrococcygeal joint. Injury or trauma to this area can cause coccygeal pain called *coccygodynia.*

Coccygodynia

Most cases of coccygodynia or coccydynia (pain in the region of the coccyx) seen by the physical therapist occur as a result of trauma, such as a fall directly on the tailbone, or events associated with childbirth.

Symptoms include localized pain in the tailbone that is usually aggravated by direct pressure such as that caused by sitting, passing gas, or having a bowel movement. Moving from sitting to standing may also reproduce or aggravate painful symptoms.

In the case of *persistent* coccygodynia with a history of trauma, the therapist must keep in mind the possibility of rectal or bladder lesions (Box 15-3). When asked about the presence of other symptoms, clients with coccygodynia after a traumatic fall may also report bladder, bowel, or sexual symptoms. The therapist must ask whether bladder, bowel, or rectal symptoms were present before the fall. Because 50% of all clients with back or sacral pain from a malignancy have preceding trauma or injury, the apparent trauma (especially if the client reports associated symptoms that were present before the trauma) may be something more serious.

For possible clues to treating a client with coccygodynia, the therapist should review Box 15-3, keeping in mind the risk factors for each of these conditions. The therapist should also conduct a neurologic screening examination to identify any signs or symptoms of disk disease. Past history of any of the problems listed is a yellow (warning) flag. Blood in the toilet after a bowel movement may be a sign of anal fissures, hemorrhoids, or colorectal cancer and requires medical evaluation.

BOX 15-3 CAUSES OF COCCYGEAL PAIN

- Diskogenic disease (herniation)
- Degenerative spondylolysis or spondylolisthesis
- Lumbar spinal stenosis
- Sacroiliac joint impairment
- Anal fissures
- Inflammatory cysts
- Prostatitis
- Thrombosed hemorrhoids
- Chordoma (neoplasm)
- Pilonidal cysts
- Trauma (fall, childbirth, anal intercourse)
- Nonunion fracture (sacrum, coccyx)
- Coccygeal disk injury (rare)

Data from Wood KB, Mehbod AA: Operative treatment for coccygodynia, *J Spinal Disord Tech* 17(6):511-515, 2004.

THE PELVIS

Once again, the principles used in screening for systemic, medical, or viscerogenic causes of back, sacral, and SI pain also apply to pelvic pain. The history and associated signs and symptoms may vary somewhat according to the cause, but many of the causes are the same (e.g., cancer, GI, vascular, urogenital) (Table 15-2).

The most common primary causes of pelvic pain are musculoskeletal, neuromuscular, gynecologic, infectious, vascular, cancer, and GI (in descending order). For example, chronic pelvic pain is most commonly associated with endometriosis, adhesions, IBS, and interstitial cystitis. Infectious disease is the most common systemic cause of pelvic pain.[36,37]

The goal in screening is to identify individuals with infectious, vascular, or neoplastic causes of pelvic pain and refer those people appropriately while at the same time making sure that those individuals we treat have a problem within the scope of our practice. Therapists must keep in mind that pelvic pain and symptoms can be referred to the pelvis from the hip, sacrum, SI area, or lumbar spine. At the same time, pelvic diseases can refer pain or symptoms to the abdomen, low back, buttocks, groin, and thigh. This means that anytime a client presents with pain or impairment in any of these areas, pelvic disease must be considered as a possible cause. At the same time, keep in mind that pelvic floor muscle spasm can be associated with disorders such as IBS or adhesions secondary to interstitial cystitis; such problems can be aided by the therapist who is skilled in management of pelvic floor muscle impairments.

The anterior pelvic wall is part of the musculature of the abdominal cavity. The lateral walls are covered by the iliopsoas and obturator muscles, and inferiorly, the outlet is guarded by the levator ani and pubococcygeus (pelvic floor) muscles, with which the corresponding muscles of the opposite side form the pelvic diaphragm.

These two anatomic regions are separated only by walls of muscle. Because the pelvic cavity is in direct communication with the abdominal cavity (see Fig. 14-1), any organ disease or systemic condition of the pelvic or abdominal cavity can

TABLE 15-2 Causes of Pelvic Pain

Systemic	Neuromuscular/Musculoskeletal/Soft Tissue
GYNECOLOGIC Pregnancy (including ectopic, ruptured or unruptured) Uterovaginal, urethral, and/or rectal prolapse Vulvodynia Dysmenorrhea Endometriosis Premenstrual tension Uterine or cervical tumors, fibroids, adhesions, polyps, stenosis Ovarian cysts, varicosities, torsion, ovulation; any ovarian anomaly Gynecalgia Intrauterine contraceptive device (IUCD) Adnexal torsion (ovaries, fallopian tubes twisted) (rare) **INFECTION/INFLAMMATION** Spontaneous, therapeutic, or incomplete abortion; postabortion syndrome Septic arthritis; reactive arthritis Ankylosing spondylitis Ileal Crohn's disease Acute or chronic appendicitis Herpes zoster Osteomyelitis PID Sexually transmitted infection Postpartum infection **VASCULAR DISORDERS** Arterial occlusion; ischemia Abdominal angina Abdominal aneurysm Pelvic congestion Varicosities or pelvic thrombophlebitis **CANCER** Reproductive Urologic **GASTROINTESTINAL DISORDERS** IBD • Crohn's disease • Ulcerative colitis IBS Diverticular disease Constipation (common in older adults) Neoplasm Hernia Bowel obstruction	Hip, sacroiliac joint, low back, sacral, or coccyx impairment* Muscle impairment (hamstrings, abdominals, rectus femoris, piriformis, adductor muscles, pelvic floor muscles, pelvic girdle)† Psoas abscess (abdominal or pelvic infectious process) Total hip arthroplasty (polyethylene wear debris) Stress reactions/fractures Vertebral compression fractures (lumbar spine) Spondylolysis Pubic strain/sprain/separation (pubis symphysis) Osteitis pubis Sexual, birth, or activity-related trauma or injury: Levator ani syndrome Tension myalgia Coccygodynia Neurologic disorders: Nerve entrapment (surgical scar in lower abdomen) Incomplete spinal cord lesion; neoplasia of spinal cord or sacral nerve Multiple sclerosis Pudendal neuralgia Shingles (herpes zoster) Diskogenic lesions (herniation) Complex regional pain syndrome Abdominal migraine Scoliosis Osteoporosis Somatization disorders Adhesions Abdominal wall myofascial pain, trigger points Postural alignment issues Hernia (obturator, sciatic, inguinal, femoral, umbilical)

*The combined medical and physical therapy differential diagnosis includes many origins of pathokinesiologic conditions, including joint laxity; subluxations or displacements; thoracolumbar hypermobility; bursitis; osteoarthritis; spondyloarthropathy; fracture; and postural, ligamentous, or osteoporosis/osteomalacia. (This list is not exhaustive.)

†As with joint impairment, the differential diagnosis of muscle pathokinesiologic conditions can include many origins (e.g., trigger points, tendinous avulsion, strain/sprain/tear, weakness, loss of flexibility, pelvic floor overactivity [pain and spasm] or underactivity [laxity, weakness, and leaking], diastasis recti).

TABLE 15-2 Causes of Pelvic Pain—cont'd

Systemic	Neuromuscular/Musculoskeletal/Soft Tissue
UROGENITAL	
Chronic urinary tract infection	
Detrusor-sphincter dyssynergia (bladder spasm)	
Interstitial cystitis/radiation cystitis	
Acute pyelonephritis	
Kidney stones (ureteric calculus, urolithiasis)	
Chronic nonbacterial prostatitis, benign prostatic hypertrophy, prostatodynia, prostate cancer	
Chronic orchalgia	
Urethral strictures	
Vulvovaginitis	
OTHER	
Psychogenic; somatization disorder; depression	
Sleep disorder	
Trauma/sexual assault; physical abuse	
Surgery (abdominal/laparoscopic, tubal, pelvic)	
Fibromyalgia	
Autonomic nervous system impairment	
Paget's disease	
Lead or mercury toxicity	
Substance abuse (cocaine)	
Sickle cell anemia	
Chronic visceral pain syndrome	

IUCD, Intrauterine contraceptive device; *PID,* pelvic inflammatory disease; *IBD,* inflammatory bowel disease, *IBS,* irritable bowel syndrome.

cause primary pelvic pain or referred musculoskeletal pain, as is described in this section.

The therapist should keep in mind that pelvic pain, pelvic girdle pain, and low back pain often occur together or alternately. Whenever discussing pelvic pain, the therapist should ask about the presence of unreported low back pain (including pelvic girdle pain). Pelvic girdle pain can occur separately or combined with low back pain and is defined as generally present between the posterior iliac crest (posterior superior iliac spine [PSIS]) and the gluteal fold in the vicinity of the SI joint.

Using the Screening Model to Evaluate the Pelvis

When our screening model is followed, the same steps are always taken. A personal or family history is obtained, and risk factor assessment is performed. Once the history has been established, the pelvic pain pattern is reviewed. The therapist looks for red flags that may suggest systemic, medical, or viscerogenic causes. Additional questions may be needed to complete the screening process. These questions are presented for all causes of pelvic pain at the end of this chapter.

History Associated With Pelvic Pain

With so many possible causes of pelvic pain, many different factors in the past medical history can raise a red flag. Pelvic pain is a very complex problem. Many medical texts are written about just this one anatomic area.

This text does not attempt to explain or discuss all the possible causes of pelvic floor or pelvic girdle pain. Rather, the intent is for the reader to learn how to screen for the possibility of systemic or viscerogenic sources of pelvic pain or symptoms. With a good understanding of what is important in the history and a list of possible follow-up questions, the therapist assesses each client, keeping in mind that medical referral may be needed.

Some of the more common red flag histories associated with pelvic pain are listed in Box 15-4. With the use of categories from the screening model, risk factors, clinical presentation, and associated signs and symptoms also are listed.

Most conditions that affect the pelvic structures are found in women, but men may also experience pelvic floor impairment and pain. Sexual assault, anal intercourse, prostate or colon cancer, and sexually transmitted disease (STD) are the most common causes for men. Prostate problems such as benign prostatic hyperplasia (BPH) or prostatitis can cause lower abdominal, back, thigh, or pelvic pain. These conditions are discussed in Chapter 10.

Clinical Presentation

In the screening process, clinical presentation and especially pain patterns are very important. Mechanisms of viscerogenic pain (i.e., how these patterns develop) are discussed in Chapter 3.

Pelvic pain may be visceral pain, caused by stimulation of autonomic nerves (T11 to S3); somatic pain, caused by

BOX 15-4 RED FLAGS ASSOCIATED WITH PELVIC PAIN OR SYMPTOMS

History*
- History of reproductive, colon, or breast cancer
- History of dysmenorrhea, ovarian cysts, pelvic inflammatory disease, sexually transmitted disease
- Endometriosis
- Chronic bladder or urinary tract infections
- Chronic irritable bowel syndrome
- Previous history of pelvic/bladder surgeries, especially hysterectomy
- Recent abortion or miscarriage
- History of assault, incest, trauma
- History of prolonged labor; use of forceps or vacuum extraction during delivery
- History of multiple births
- Chronic yeast/vaginal infections
- History of varicose veins in the lower extremities (risk factor for pelvic congestion syndrome)

Risk Factors
- Recent intrauterine contraceptive device (rejection) or long-term use, especially without medical follow-up (scar tissue)
- Perimenopause, menopause (vaginitis, vaginal atrophy)
- Sexual activity without use of a condom (sexually transmitted diseases, pelvic inflammatory disease [PID])
- Multiple sexual partners (sexually transmitted diseases, PID)
- Pregnancy, childbirth, recent abortion, multiple abortions

Clinical Presentation
- Insidious onset; unknown cause
- Poorly localized, diffuse; client unable to point to one spot
- Aggravated by increased intraabdominal pressure (e.g., standing, walking, sexual intercourse, coughing, constipation, Valsalva's maneuver)
- Pelvic pain is not affected by specific movements but gets worse toward the end of the day or after standing for a long time
- May be temporarily relieved by position change (e.g., getting off feet, resting or elevating the legs, putting the feet up)
- Pelvic pain is not reduced or eliminated by scar or soft tissue mobilization or by trigger point release of myofascial structures in the pelvic cavity
- Positive McBurney's, Pinch an Inch, or iliopsoas/obturator sign (see Chapter 8)
- Presence of vulvar varicosities (seen most often in women with pelvic floor congestion syndrome and pregnancy)

Associated Signs and Symptoms
- Discharge from vagina or penis
- Urologic signs or symptoms
- Unreported abdominal pain
- Dyspareunia (painful or difficult intercourse)
- Constitutional symptoms
- Missed menses or unexplained/unexpected spotting (light staining of blood) (e.g., ectopic pregnancy) ask about shoulder pain
- Headache, fatigue, irritability

*Many of the histories listed here are also *risk factors* for pelvic pain. Regarding pelvic girdle pain, according to the European Guidelines for the Diagnosis and Treatment of Pelvic Girdle Pain, red and yellow flags are the same for low back pain and pelvic girdle pain with the possible exception of age (pelvic girdle pain affects younger individuals less than 30 years old and is less likely to be caused by malignancy).[38]

stimulation of sensory nerve endings in the pudendal nerves (S2, S3); or peritoneal pain, caused by pressure from inflammation, infection, or obstruction of the lining of the pelvic cavity.

Peritoneal pain may be caused by disruption of the autonomic nerve supply of the visceral pelvic peritoneum, which covers the upper third of the bladder, the body of the uterus, and the upper-third of the rectum and the rectosigmoid junction. It is not sensitive to touch but responds with pain on traction, distention, spasm, or ischemia of the viscus.

Peritoneal pain may also occur in relation to the parietal pelvic peritoneum, which covers the upper half of the lateral wall of the pelvis and the upper two thirds of the sacral hollow—all supplied by somatic nerves. These somatic nerves also supply corresponding segmental areas of skin and muscles of the trunk and the anterior abdominal wall. Painful stimulation of the parietal pelvic peritoneum may cause referred segmental pain and spasm of the iliopsoas muscle and muscles of the anterior abdominal wall.

Knowing the characteristics of pain patterns typical of each system is essential. When the client describes these patterns, it is possible for the therapist to recognize them for what they are and to see how the clinical presentation differs from neuromuscular or musculoskeletal impairment and dysfunction.

Pelvic disease may cause primary pelvic pain and may also refer pain to the low back, thigh, groin, and rectum. Usually, pelvic disease appears as acute illness with sudden onset of severe pain accompanied by nausea and vomiting, fever, and abdominal pain. Mild-to-moderate back or pelvic pain that gets worse as the day progresses may be associated with gynecologic disorders. The therapist is more likely to see the atypical presentation of systemically related central lumbar and sacral pain, which is easily mistaken for mechanical pain.

Associated Signs and Symptoms

While collecting pertinent personal and family history, conducting a risk factor assessment, and evaluating the client's pain pattern, the therapist listens and looks for any yellow or red flags. From there, the therapist formulates any additional questions that may be appropriate on the basis of data collected so far. Before leaving the screening task, the therapist asks a few final questions. The first is about the presence of any associated signs and symptoms.

For example, perhaps the client has pelvic pain and unreported shoulder pain. She may not think her previously unreported shoulder pain has any connection with the current pelvic pain, or she may not see that the presence of a vaginal discharge is linked in any way to her low back and pelvic pain. Discharge from the vagina or penis (yellow or green, with or without an odor) in the presence of low back, pelvic, or sacral pain may be a red flag.

To bring this information out and make any of these connections, the therapist must ask about the presence of any associated signs and symptoms. Ask the client the following:

FOLLOW-UP QUESTIONS

- Do you have any symptoms anywhere else in your body? Tell me even if you don't think they are related to your pelvic pain.

If the client says "No," then ask about the presence of urologic symptoms and constitutional symptoms, and look for a connection between the menstrual cycle and symptoms. If it appears that there may be a gynecologic basis for the client's symptoms, the therapist may want to ask some additional questions about missed menses, shoulder pain, and spotting or bleeding.

The therapist should assess for the presence of dysmenorrhea, defined as painful cramping during menstruation. Dysmenorrhea may be primary (of unknown cause) or secondary as a result of a pelvic pathologic condition related to endometriosis, intrauterine tumors or polyps (myomas), uterine prolapse, pelvic inflammatory disease (PID), cervical stenosis, and adenomyosis (benign invasive growths of the endometrium into the muscular layers of the uterus).

Dysmenorrhea is characterized by spasmodic, cramp-like pain that comes and goes in waves and radiates over the lower abdomen and pelvis, thighs, and low back, sometimes accompanied by headache, irritability, mental depression, fatigue, and GI symptoms.

Screening for Neuromuscular and Musculoskeletal Causes of Pelvic and Pelvic Floor and Pelvic Girdle Pain

The therapist is most likely to see pelvic floor pain and/or pelvic girdle pain that are caused by neuromuscular or musculoskeletal problems. Pelvic pain or symptoms may be referred from systemic or neuromusculoskeletal origins from the hip, SI joint, sacrum, or low back.

Pelvic floor pain can present suprapubically, perineally, and/or in the low buttock/anal areas. Pelvic girdle pain can occur separately or combined with low back pain and is defined as generally present between the posterior iliac crest (PSIS) and the gluteal fold in the vicinity of the SI joint. Pain may radiate to the posterior thigh; endurance for standing, walking, and sitting is decreased.[38] Pelvic girdle pain occurs most often during pregnancy or continues many years postpartum. Many women have both pelvic floor and pelvic girdle pain, requiring each to be addressed externally and internally.

Likewise, pelvic diseases can refer pain and symptoms to the low back, groin, and thigh. When evaluating low back or pelvic pain, the therapist must assess for pelvic floor laxity or tension, psoas abscess, trigger points, history of birth or sexual trauma, and the presence of any associated signs and symptoms.

Neurologic disorders (e.g., nerve entrapment, incomplete spinal cord lesion, multiple sclerosis, Parkinson's, stroke, pudendal neuralgia) can cause pelvic pain and dysfunction. Pudendal nerve entrapment is characterized by pain relief when one is sitting on a toilet seat or standing; elimination of symptoms after a pudendal nerve block is diagnostic.

Pregnancy-related and postpartum low back pain, pelvic floor pain, and pelvic girdle pain are also common and have an impact on daily life for many women. Prevention and treatment of symptoms is an important issue for therapists who work in the area of women's health.[39,40]

Musculoskeletal impairment of the pelvic floor and low back may manifest as dyspareunia (pain before, during, or after intercourse). Overactivity (pain and spasm; muscles contract when they should relax or do not relax completely)[41] of the pelvic floor and pelvic floor trigger points can contribute to entrance (superficial or deep) dyspareunia. Deep thrust dyspareunia may also be related to SI or low back impairment. Dyspareunia symptoms that are reduced in alternate positions may indicate a musculoskeletal component, especially when other signs and symptoms characteristic of musculoskeletal impairment are also present.[42,43]

One of the most common musculoskeletal sources of pelvic floor pain in men and women is the trigger point. Muscles most likely to cause or refer pain to the pelvic area include the levator ani, abdominals, quadratus lumborum, and iliopsoas.[44-46]

Typical aggravating and relieving factors for pain from a neuromuscular or musculoskeletal source include the following:

- Aggravated by exercise, weight bearing
- Aggravated by trunk/lumbar rotation
- Relieved by rest or stretching
- Pain or altered movement pattern produced by trunk and lumbar rotation
- Eliminated by trigger point therapy

The therapist looks for a contributing history such as a fall on the buttocks, pregnancy, or trauma. Avulsion of hamstrings from a sports injury may be reported. Trauma from physical or sexual assault may remain unreported. Screening for assault is an important part of many evaluations (see Chapter 2).

The therapist also looks for muscle impairment. For therapists trained in pelvic floor muscle examination, external and internal palpation of the pelvic floor musculature is helpful.[47,48] Examination also includes observation for varicosities and assessment of muscle tone (muscle overactivity [pain and spasm] or underactivity [laxity with weakness and leaking] and the presence of trigger points.[44,49,50] Transabdominal ultrasound and vaginal dynamometer are two tools used by some physical therapists to assess pelvic floor muscle contraction. Many clients who experience low back, pelvic, SI, sacral, or groin pain have unrecognized pelvic floor impairment.[51]

Pain provocation tests for the symphysis pubis and SI joint (e.g., Patrick's/Faber's, modified Trendelenburg, Gaenslen's, shear, P4, gapping, and compression tests), palpation, and mobility testing help point to pelvic girdle impairment, but this could be associated with primary pelvic floor impairment so that both problems coexist.[52]

The treatment strategy may be to address the pelvic girdle pain first; if it does not resolve, then a pelvic floor muscle examination may be needed to confirm pelvic floor impairment. P4 is a pelvic floor muscle examination that refers to the "provocation of posterior pelvic pain" screening test for pregnancy-related pelvic girdle pain.[52] Reliability and validity of the provocation tests mentioned here have been evaluated and reported; for details see Vleeming[38] and Olsen.[52] Imaging tests, such as x-rays, computed tomography (CT), and magnetic resonance imaging (MRI), help rule out problems such as fractures, ankylosing spondylitis, and reactive arthritis.[38]

Fig. 15-2 gives a simple representation of how the puborectalis muscle acts as a sling around various structures of the pelvis. The condition and position of the pelvic sling are very important in the maintenance of normal pelvic floor health.

Fig. 14-1 provides a visual reminder that the muscles of the pelvic floor support the reproductive organs and the viscera in the peritoneum. Any impairment of these organs may cause impairment of the pelvic floor and vice versa. Any weakness or impairment of the pelvic floor can lead to problems with the viscera located in the abdominal or pelvic cavities.

Anterior Pelvic Pain

Anterior pelvic pain occurs most often as a result of any disorder that affects the hip joint, including inflammatory arthritis; upper lumbar vertebrae disk disease (rare at these segments); pregnancy with separation of the symphysis pubis; local injury to the insertion of the rectus abdominis, rectus femoris, or adductor muscle; femoral neuralgia; and psoas abscess.

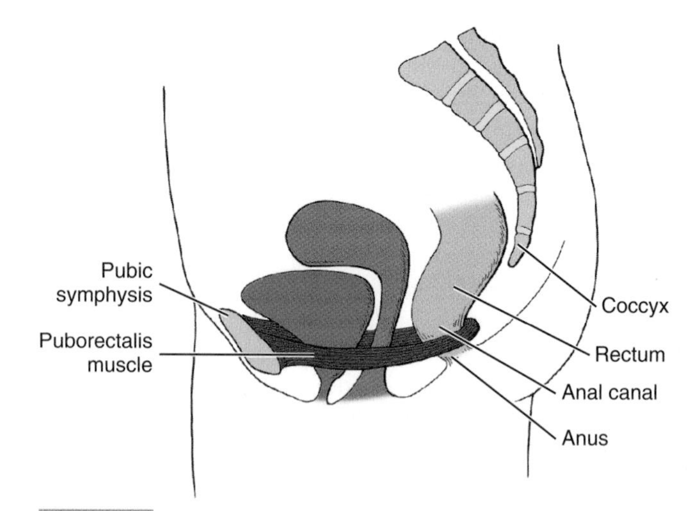

Fig. 15-2 Pelvic sling. Puborectalis muscle forms a U-shaped sling encircling the posterior aspect of the rectum and returns along the opposite side of the levator hiatus to the posterior surface of the pubis. This shows how the condition and position of the pelvic sling contribute to the function of the pelvic floor and the encircled viscera. Obesity, multiparity, and prolonged pushing during labor and delivery are just a few of life's events that can disrupt the integrity of the pelvic sling and the pelvic floor. (From Myers RS: *Saunders manual of physical therapy practice,* Philadelphia, 1995, WB Saunders.)

Stress reactions of the pubis or ilium, sometimes called *stress fractures* (disruption of the bone at the tendon-bone interface without displacement from repetitive contraction), can occur during traumatic labor and delivery, but they are more common in osteomalacia and Paget's disease and produce anterior pelvic pain. Traumatic stress reactions may also occur in joggers, military personnel, and athletes.

Although the underlying pathology differs, symptoms are similar to separation of the symphysis pubis and pelvic ring disruption and may include pain in the involved areas that is aggravated by active motion of the limb or deep pressure and weight bearing during ambulation. Symptoms from pelvic instability as a result of pelvic ring injury or disruption (whether from birth trauma in women of childbearing age or pelvic stress fracture in an older adult with osteoporosis) may be aggravated by single-leg-stance.[53,54]

Femoral hernia, which accounts for 20% of hernias in women, may cause lateral wall pelvic pain when the hernia strangulates. The referred pain pattern is located down the medial side of the thigh to the knee; inguinal hernias are likely to cause groin pain. Immediate surgical repair is indicated.

Posterior Pelvic Pain

Posterior pelvic pain originating in the lumbosacral, sacroiliac, coccygeal, and sacrococcygeal regions usually appears as localized pain in the lower lumbar spine, pelvic girdle, and over the sacrum, often radiating over the sacroiliac ligaments. Pain radiating from the SI joint can commonly be felt in both the buttock and the posterior thigh and is often aggravated by rotation of the lumbar spine on the pelvis. A proximal hamstring injury, including avulsion of the ischial epiphysis in the adolescent, may also cause posterior pelvic and buttock pain.

Coccygodynia and sacrococcygeal pain are common presentations in women and are often associated with a fall on the buttocks or traumatic childbirth. They manifest with the person having difficulty sitting on firm surfaces and having pain in the coccygeal region on defecation or straining.

Levator ani syndrome and tension myalgia may produce symptoms of pain, pressure, and discomfort in the rectum, vagina, perirectal area, or low back and can mimic a diskogenic problem. Overactivity (pain and spasm) and tenderness in the levator ani may occur in men and women and may be caused by chronic prostatitis that does not resolve with antibiotics (men), birthing trauma (women), neurologic abnormalities in the lumbosacral spine, sexual assault or trauma, or anal fissures from anal intercourse. Pain or rectal pressure may occur during sexual intercourse, as may throbbing pain during bowel movement with accompanying constipation and impaired bowel and bladder function.

Screening for Gynecologic Causes of Pelvic Pain

Pregnancy, multiparity, and prolonged labor and delivery (especially combined with obesity) are risk factors for gynecologic conditions that can alter the normal position of the bladder, uterus, and rectum in relation to one another (Fig. 15-3), resulting in pelvic organ prolapse such as rectocele, cystocele, and prolapsed uterus with concomitant pelvic floor pain and impairment.

Gynecologic causes of pelvic floor pain are most often produced by congenital anomaly, inflammatory processes (including infection), neoplasia, or trauma. In addition, pelvic girdle pain may be associated with pregnancy, endometriosis, and altered uterine position (Fig. 15-4). Variations in the angle and position of the uterus occur from woman to woman. Many women are unaware of their uterine position. Only if the physician tells her, "You have a tipped uterus," or "You have a retroverted uterus," will she know whether any change from the normal position of the uterus has occurred. Other women experience extreme pain associated with the menstrual cycle, which may be linked with uterine position.

Children younger than 14 years rarely experience pelvic pain of gynecologic origin. Infection is the most likely cause and is limited to the vulva and vagina. Theoretically, infection can ascend to involve the peritoneal cavity, causing iliopsoas abscess and pelvic, hip, or groin pain, but this rarely happens in this age group.

Pregnancy

Pelvic pain associated with normal pregnancy is similar to low back pain, as was discussed earlier in Chapter 14. About 1% of all pregnancies take place outside the endometrium (or ectopic), with most ectopic implantations occurring in the fallopian tube (Fig. 15-5). Risk factors include tubal ligation; STD; pelvic inflammatory disease; infertility or infertility treatment; previous tubal, pelvic, or abdominal surgery; or the use of intrauterine contraceptive devices (IUCDs) such as rings, loops, coils, or Ts (see Fig. 14-6).

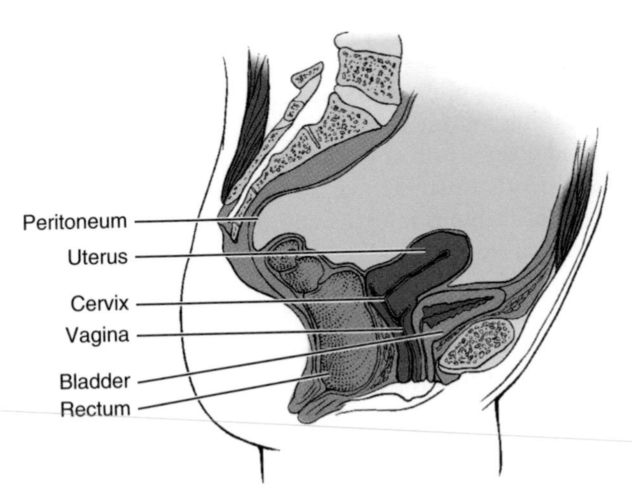

Fig. 15-3 Normal female reproductive anatomy (sagittal view). Locate the rectum, uterus, bladder, vagina, and cervix in this illustration. Note the size, shape, and orientation of each of these structures. The rectum turns away from the viewer in this sagittal section, giving it the appearance of ending with no connection to the intestines. Understanding the normal orientation of these structures will help when each of the diseases that can cause low back pain is considered.

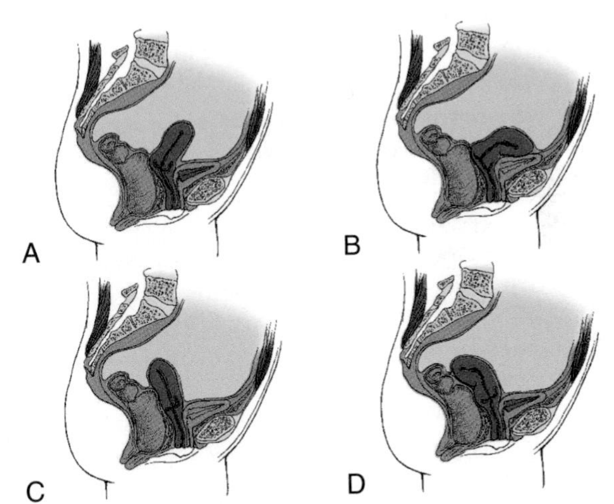

Fig. 15-4 Abnormal positions of the uterus. Variations in the angle and position of the uterus occur from woman to woman. Each illustration depicts a slightly different anatomic position of the uterus. **A,** Midline position. Usually, the uterus is above and parallel to the bladder. In the midline position, the uterus is more vertical. **B,** Anteflexed uterus. The uterus is in its proper position above the bladder, but the upper one-third to one-half of the body is flexed forward. **C,** Retroverted uterus. About 20% of American women have a tilted, or retroverted, uterus. The top of the uterus naturally slants toward the spine rather than toward the umbilicus. **D,** Retroflexed uterus. An extremely tilted uterus called *retroflexion* may even bend down toward the tailbone. A woman with a retroflexed uterus may be unable to use a tampon or a diaphragm. Back pain is more likely to occur with pregnancy and labor for the woman with a retroverted or retroflexed uterus.

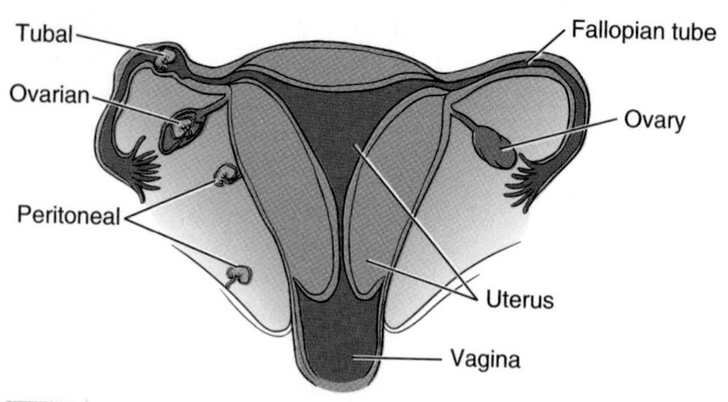

Fig. 15-5 Ectopic pregnancy. An ectopic pregnancy can occur when the egg is fertilized and implanted outside the uterus. The ovum can be embedded inside the ovary (ovarian pregnancy), inside the fallopian tube (tubal pregnancy), or anywhere between the ovary and the uterus, including along the outside lining of the uterus (extra-uterine) or inside the abdominal cavity along the peritoneum as shown. Rupture of the ovum and hemorrhage is the usual result. If this occurs early in the menstrual cycle, the woman may experience heavier bleeding than usual but remain unaware of the failed pregnancy.

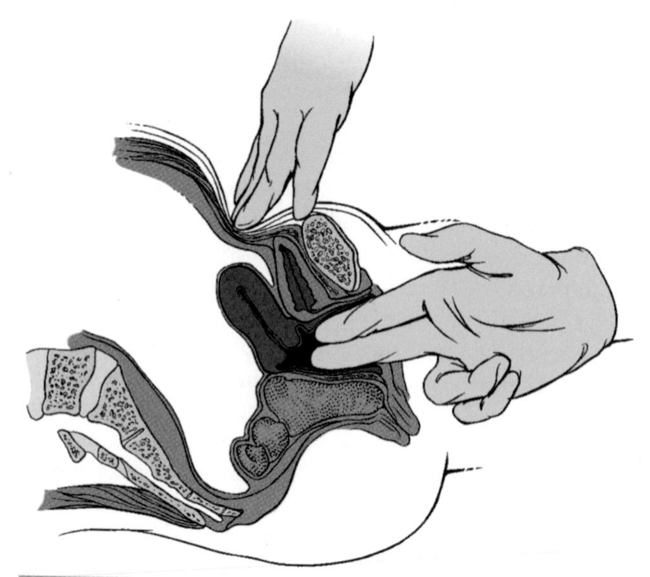

Fig. 15-6 Pelvic examination. With the woman in the lithotomy position (supine with hips and knees flexed and feet in stirrups), the examiner inserts one or two gloved fingers into the vaginal canal up to the point of the cervix or soft tissue obstruction. The examiner applies firm pressure in the lower abdomen above the bladder while the woman bears down slightly as if performing a Valsalva maneuver. The examiner evaluates the tone of the pelvic floor and the position of the uterus during this test. Integrity of the pelvic floor (e.g., muscle tone, laxity, trigger points) can also be tested.

Symptoms of ectopic pregnancy most often include unexplained vaginal spotting, bursts of bleeding, and sudden lower abdominal and pelvic cramping shortly after the first missed menstrual period. At first, the pain may be a vague "twinge" or soreness on the affected side; later it can be sharp and severe.

Gradual hemorrhage causes pelvic (and sometimes low back or shoulder) pain and pressure, but rapid hemorrhage results in hypotension or shock. Tubal rupture is common and requires medical attention and diagnosis.

CLINICAL SIGNS AND SYMPTOMS

Ectopic Pregnancy

- Unexplained vaginal bleeding (spotting), missed menses
- Sudden, unexplained lower abdominal and pelvic cramping (especially after first missed menstrual period); usually unilateral
- Pain may be mild, progressing to severe over a matter of hours to days
- Low back (unilateral or bilateral) or shoulder pain (unilateral)
- Hypotension (low blood pressure and pulse rate), shock (tubal rupture)

Prolapsed Conditions

Prolapse is the collapse, falling down, or downward displacement of structures such as the uterus, bladder, or rectum. A pelvic examination is performed by a physician or other trained professional, such as a physical therapist, to identify prolapse (Fig. 15-6).

Uterovaginal prolapse can cause low-grade and persistent pelvic pain. Prolapse may result from a combination of basic anatomic structure, effects of pregnancy and labor, postmenopausal hormone deficiency, and poor general muscular fitness. Pelvic floor tension myalgia and prolapse often occur together. Obesity combined with chronic cough, constipation, and multiparity is a common contributing factor to pelvic floor problems.

Uterine Prolapse. Uterine prolapse occurs most often after childbirth and is graded as first-, second-, or third-degree prolapse (Fig. 15-7). Secondary prolapse may occur with prolonged pushing during labor and delivery, large intrapelvic tumors, or sacral nerve disorders, or it may follow pelvic or abdominal surgery.

The pain of prolapse is central, suprapubic, and dragging in the groin, and a sensation of a lump at the vaginal opening is noted. Pain is primarily due to stretching of the ligamentous supports (uterosacral ligament attaches to the sacrum; loss of ligamentous integrity contributes to significant biomechanical changes) and secondarily to excoriation (scratch or abrasion) of the prolapsed cervical or vaginal tissue, which may occur.

Third-degree prolapse is often accompanied by low back pain with or without pelvic, sacral, or abdominal cramping or heaviness. Symptoms are relieved by rest and lying down and are often aggravated by prolonged standing, walking, coughing, sexual intercourse, or straining. Urinary incontinence is commonly associated with uterine prolapse.

Sexual intercourse is possible because the soft tissues of the uterus and vagina can be pushed or pressed out of the way. However, excoriation (scratching or abrasion) of the tissue may occur, accompanied by bleeding and local pain. Care must be taken when anything is inserted into the vagina. Excessive, repetitive force should be avoided.

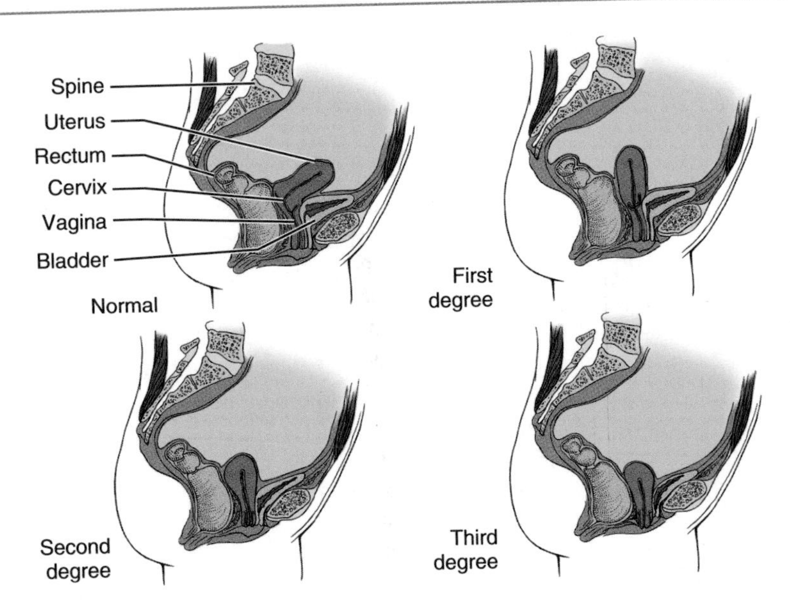

Fig. 15-7 Uterine prolapse. *First-degree prolapse*: The uterus has dropped up to one-third of the way into the vaginal canal. *Second-degree prolapse*: The uterus has descended fully into the vaginal canal, right down to the vaginal opening. *Third-degree prolapse*: the uterus is displaced downward even further and bulges outside the vaginal opening.

Some women use a removable device called a *pessary* for a prolapsed uterus, bladder, or rectum. It is placed in the vagina to support the prolapsed structure. These devices are usually considered temporary and should be used in conjunction with a program to rehabilitate the pelvic floor impairment. Long-term use of such devices may be required when surgical repair is not possible or the woman is not a good surgical candidate.

Identifying the presence of uterine prolapse does not necessarily require medical referral. Conservative care such as a program of pelvic floor recovery and management of sexual intercourse can be very helpful for the woman and may be the first step in treatment. Client education about positions in which gravity is used to assist the uterus in resuming its normal position can be very helpful. For example, supine with a pillow or wedge support under the pelvis is a helpful rest position and can be used while the patient is doing pelvic floor exercises. It is also a more comfortable position for sexual intercourse for some women.

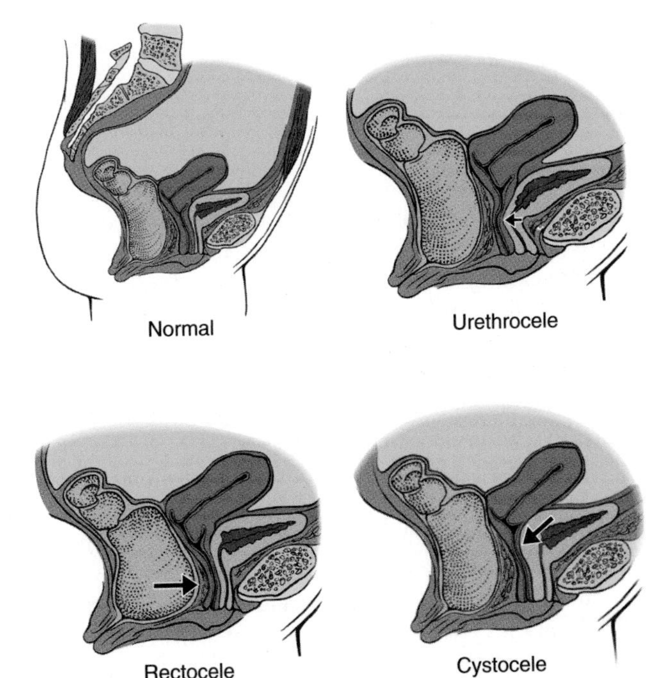

Fig. 15-8 Pelvic organ prolapse. **A,** Normal alignment of uterus, bladder, and rectum. **B,** *Urethrocele*. Urethra pressing against the vaginal canal. **C,** *Rectocele*. The uterus and bladder are in their proper anatomic place, but the rectum has prolapsed and is compressing against the vaginal canal. Many women have more than one of these conditions at the same time as a result of pregnancy and childbirth. **D,** *Cystocele*. The arrow shows displacement of the bladder against the vaginal canal.

CLINICAL SIGNS AND SYMPTOMS

Uterine Prolapse

- Lump in vaginal opening
- Pelvic discomfort, backache
- Abdominal cramping
- Symptoms relieved by lying down
- Symptoms made worse by prolonged standing, walking, coughing, or straining
- Urinary incontinence

Cystocele and Rectocele. Cystocele is the protrusion or herniation of the urinary bladder against the wall of the vagina. Rectocele is a protrusion or herniation of the rectum and posterior wall of the vagina into the vagina (Fig. 15-8).

Similar to the prolapsed uterus, these two pelvic floor disorders occur most often after pregnancy and childbirth but may also be associated with surgery and obesity (especially obesity combined with multiple pregnancies and births). These conditions are the result of pelvic floor relaxation or structural overstretching of the pelvic musculature or ligamentous structures. Patient history may include prolonged labor, bearing down before full dilation, forceful delivery of the placenta, instrument delivery (e.g., forceps, vacuum suction), chronic cough, or lifting of heavy objects.

Trauma to the pudendal or sacral nerves during birth and delivery is an additional risk factor. Decreased muscle tone due to aging, complications of pelvic surgery, or excessive straining during bowel movements may also result in prolapse. Pelvic tumors and neurologic conditions, such as spina bifida and diabetic neuropathy, which interrupt the innervation of pelvic muscles, can also increase the risk of prolapse.

CLINICAL SIGNS AND SYMPTOMS

Cystocele

- Urinary frequency and urgency
- Difficulty emptying the bladder
- Cystitis (bladder infection)
- Painful lump or bearing down sensation in the perineal area
- Urinary stress incontinence

Rectocele

- Pelvic, perineal pain and difficulty with defecation
- Feeling of incomplete rectal emptying
- Constipation
- Painful intercourse
- Aching or pressure after a bowel movement

Endometriosis

Endometriosis (see Chapter 14) is a pathologic condition of retrograde menstruation. Tissue resembling the mucous membrane lining the uterus occurs outside the normal location in the uterus but within the pelvic cavity, including the ovaries, pelvic peritoneum, bowel, and diaphragm. It occurs most often during the reproductive years and in up to 50% of women with infertility.[55-57] Severity of pain is related more to the site than to the extent of disease.

Pelvic pain associated with endometriosis can be referred to the low back, rectum, and lower sacral or coccygeal region, starting before or after the onset of menstruation and improving after cessation of menstrual flow, with *cyclic recurrence* (a key finding). As the condition progresses, pain continues throughout the cycle, with exacerbation at menstruation and, finally, constant severity.

Other symptoms may include rectal discomfort during bowel movements, diarrhea, constipation, recurrent miscarriage, and infertility. Box 15-5 has more information on this condition.

Gynecalgia

Although a pathologic cause can be identified for most cases of chronic pelvic pain, a small percentage remains for which no physical cause can be determined, and the term gynecalgia is used. Women with gynecalgia syndrome are usually 25 to 40 years of age and have at least one child. The symptoms are of at least 2 years' duration (and often many more), with acute exacerbation from time to time.

Pain associated with gynecalgia is vague and poorly localized, although it is usually confined to the lower abdomen and pelvis, radiating to the groin and upper and inner thighs. Other symptoms include dyspareunia, menstrual changes, low back pain, urinary and bowel changes, fatigue, and obvious anxiety and depression.

Screening for Infectious Causes of Pelvic Girdle Pain

Infection is the most common cause of systemically induced pelvic pain. Infection or inflammation within the pelvis from acute appendicitis, diverticulitis, Crohn's disease, osteomyelitis, septic arthritis of the SI joint, urologic disorders, sexually transmitted infection (STI; e.g., *Chlamydia trachomatis*), and salpingitis (inflammation of the fallopian tube) can produce visceral and somatic pelvic pain because of the involvement of the parietal peritoneum.

Secondary pelvic infection may follow surgery, septic abortion, pregnancy, or recent birth as a result of the entry of endogenous bacteria into the damaged pelvic tissues. PID and STI are the most common causes of infection in women.

All these disorders have similar signs and symptoms during the acute phase. The client may not have any pain but will report low back or pelvic "discomfort," or there may be a report of acute, sharp, severe aching on both sides of the pelvis. Accompanying groin discomfort may radiate to the inner aspects of the thigh.

Keep in mind that in the older adult, the first sign of any infection might not be an elevated temperature, but rather, confusion, increased confusion, or some other change in mental status.

Right-sided abdominal or pelvic inflammatory pain is often associated with appendicitis, whereas left-sided pain is more likely associated with diverticulitis, constipation, or obstipation (left sigmoid impaction). Left-sided appendicitis is possible and not that uncommon.[58] Bilateral pain may indicate infection. The pain may be aggravated by increased abdominal pressure (e.g., coughing, walking). Knowing these pain patterns helps the therapist quickly decide what questions to ask and which associated signs and symptoms to look for. The therapist should test for iliopsoas or obturator abscesses (see Chapter 8).

Other red flag symptoms may be reported in response to specific questions about disturbances in urination, odorous vaginal discharge, tachycardia, dyspareunia (painful or difficult intercourse), or constitutional symptoms such as fever, general malaise, and nausea and vomiting.

Pelvic Inflammatory Disease

PID consists of a variety of conditions (i.e., it is not a single entity), including endometritis, salpingitis, tubo-ovarian abscess, and pelvic peritonitis. Any inflammatory condition that affects the female reproductive organs (uterus, fallopian tubes, ovaries, cervix) may come under the diagnostic label of PID.[59]

PID is a bacterial infection that occurs whenever the uterus is traumatized; it is often associated with STI/STD and may occur after birth or after an abortion. Infection can be introduced from the skin, vagina, or GI tract. It can be an acute, one-time episode or may be chronic with multiple recurrences.

BOX 15-5 RESOURCES

Endometriosis*

- Endometriosis Zone, a service of The Universe of Women's Health—a commercial organization directed by a board of obstetricians and gynecologists. Information is directed at medical professionals, the medical industry, and women. The Endometriosis Zone is found at both of the following websites:
  http://www.endozone.org
  http://www.endometriosiszone.org
- Endometriosis Research Center (ERC) was started as a lobbying organization. The goal of the ERC is to bring science and support together through education.
  http://www.endocenter.org or call (800) 239-7280.
- The International Endometriosis Association (IEA) was established by Mary Lou Ballweg, RN, PhD, as an advocacy organization for endometriosis; offers online support for women diagnosed with endometriosis.
  http://www.endometriosisassn.org
- The National Library of Medicine offers an interactive tutorial about endometriosis in both English and Spanish at:
  http://www.nlm.nih.gov/medlineplus/tutorials/endometriosis

Pelvic Inflammatory Disease (PID)

- The National Women's Health Information Center (NWHIC) offers information on all aspects of women's health, including PID; (1-800-994-9662) or online at:
  http://www.4woman.gov/
- Centers for Disease Control and Prevention (CDC) provide a PID Fact Sheet:
  http://www.cdc.gov/std/PID/STDFact-PID.htm
- Mount Auburn Obstetrics & Gynecologic Associates, Cincinnati, OH, a group of obstetric and gynecologic (OBGYN) professionals offers online education about endometriosis and other OBGYN topics:
  http://www.mtauburnobgyn.com/pid.html

Pelvic Pain

- Fall M, Baranowski AP, Elneil S: *Guidelines on chronic pelvic pain,* Arnhem, The Netherlands, 2008, European Association of Urology (EAU).
- Vleeming A: European guidelines for the diagnosis and treatment of pelvic girdle pain, *Eur Spine J* 17(6):794-819, 2008.
- The American College of Obstetricians and Gynecologists (ACOG) offers information, education, and publications related to a wide variety of women's health issues:
  http://www.acog.org/

The ACOG has recently issued a new practice bulletin on chronic pelvic pain in women. The guidelines were published as: ACOG Practice Bulletin No. 51. Chronic Pelvic Pain, *Obstet Gynecol* 103(3):589-605, 2004. To read more about these guidelines, go to: http://www.medscape.com/viewarticle/471545

- The International Pelvic Pain Society is a professional organization with the goal to enhance and improve the treatment of diseases that cause pelvic pain in men and women. Education for health care professionals is a major focus of this organization, which can be reached at: http://www.pelvicpain.org/

*Data from Deevey S: Endometriosis: Internet resources. *Med Ref Serv Q* 24(1):67-77, Spring 2005; contact information updated 2011.

It is estimated that two-thirds of all cases are caused by STIs such as chlamydia and gonorrhea.[60] Chlamydia is a bacterial STI that is acquired through vaginal, oral, or anal intercourse. It is often asymptomatic but can present with vaginal bleeding and discharge and burning during urination. Pelvic pain does not occur until chlamydia leads to PID. When detected and treated early, chlamydia is relatively easy to cure.

A direct relationship has been observed between early age of first sexual intercourse, the number of sexual partners a woman has, recent new partner (within previous 3 months), and a past history of STD (in the client or her partner) or risk of STD (especially human papillomavirus, or HPV, a risk factor for cervical cancer).[61-63] PID may occur if chlamydia is not treated; even if it is treated, damage to the pelvic cavity cannot be reversed. The more partners a woman has, the greater is the risk of PID.[59]

Other risk factors include interruption of the cervical barrier by means of pregnancy termination, insertion of intrauterine device within the past 6 weeks, in vitro fertilization/intrauterine insemination, or any other instrumentation of the uterus.[59]

PID associated with scarring in the pelvic organs, including the ovaries, fallopian tubes, bowel, and bladder, may cause chronic pain. Women can be left infertile because of damage and scarring to the fallopian tubes. After a single episode of PID, a woman's risk of ectopic pregnancy increases sevenfold compared with the risk for women who have no history of PID.[64,65]

CLINICAL SIGNS AND SYMPTOMS

Pelvic Inflammatory Disease

- Often asymptomatic
- Abnormal vaginal discharge or bleeding
- Burning on urination (dysuria)
- Moderate (dull aching) to severe lower abdominal and/or pelvic pain; back pain is possible
- Painful intercourse (dyspareunia)
- Painful menstruation
- Constitutional symptoms (fever, chills, nausea, vomiting)

STIs such as chlamydia and syphilis are on the rise among America's sexually active young adult population (ages 18 to 25). In fact, chlamydia was the most commonly reported infectious disease in the United States in 2004. According to the Centers for Disease Control and Prevention (CDC) annual report, the highest rates of chlamydia occur in sexually active women ages 15 to 19. Syphilis predominates in men who engage in risky sexual behavior (e.g., unprotected vaginal, anal, oral sex) with men or with men and women.[65-69]

It does not happen often, but there may be times when the therapist must ask about the possibility of an STI. Sexually active women with vague symptoms are the most likely group to be interviewed about STIs/STDs. See specific screening questions in Chapter 14 and Appendix B-32.

Any of the red flags listed in Box 15-4 in the presence of pelvic pain raises the suspicion of a medical problem. Medical referral must be made as quickly as possible. Early medical intervention can prevent the spread of infection and septicemia, and can preserve fertility. Damage to the pelvic floor from any of these conditions can result in pelvic floor impairment that is within the scope of a physical therapist's practice. See Box 15-5 for resources that can provide more information on this and other conditions.

Screening for Vascular Causes of Pelvic Girdle Pain

Vascular problems that affect the pelvic cavity and pelvic floor musculature have two primary causes. The first is the general condition of peripheral vascular disease (PVD); the second is a specific example of PVD called *pelvic congestion syndrome* from ovarian and/or vulvar varicosities (abnormal enlargement of veins). Other conditions, such as abdominal angina and abdominal aneurysm, are less common vascular causes of pelvic pain; these conditions are discussed in greater detail in Chapter 6.

Peripheral Vascular Disease

The iliac arteries may become gradually occluded by atherosclerosis or may be obstructed by an embolus. The resultant ischemia produces pain in the affected limb but may also give rise to pelvic pain. Whether the occlusion is thrombotic or embolic, the client may report pain in the pelvis, affected limb, and possibly the buttocks.

The pain is characteristically aggravated by exercise (claudication). Typically, symptoms develop 5 or 10 minutes after the client has started the activity. This lag time is characteristic of a vascular pain pattern associated with atherosclerosis or blood vessel occlusion.

Musculoskeletal causes of pelvic pain are also made worse by activity and exercise, especially weight-bearing exercise, but the timing is not as predictable as with pain from vascular causes. Musculoskeletal conditions may cause pain immediately (e.g., with muscle strain or trigger points) or, more likely, after prolonged activity or exercise. The affected limb becomes colder and paler. In sudden occlusion, diminished sensation to pinprick may be observed on examination. Femoral and distal arteries should be palpated for pulsation.

Thrombosis of the large iliac veins may occur spontaneously after injury to the lower limb and pelvis, or it may appear after pelvic surgical procedures. An estimated 30% of clients have asymptomatic deep vein thrombosis after major surgery. Thrombosis that occludes the iliac vein produces an enlarged, warm, and painful leg; occasionally, discomfort in the pelvis is noted.

Anyone with PVD can demonstrate the same kind of symptoms in the pelvic floor structures. The most likely age group to be affected by vascular disease is adults over 60, especially women who are postmenopausal.

Watch for a history of heart disease with a clinical presentation of pelvic, buttock, and leg pain that is aggravated by activity or exercise (claudication). Look for changes in skin and temperature on the affected side (arterial occlusion or venous thrombosis), especially in the presence of known heart disease or recent pelvic surgery (see Box 4-13; Case Example 15-2).

Pelvic Congestion Syndrome

Varicose veins of the ovaries (varicosities) cause the blood in the veins to flow downward rather than up toward the heart. They are a manifestation of PVD and a potential cause of chronic pelvic pain. The condition has been called *pelvic congestion syndrome* (PCS) or ovarian varicocele.

The specific impairment associated with PCS is an incompetent and dilated ovarian vein with retrograde blood flow (Fig. 15-9). Ovarian venous reflux and stasis produce venous dilatation, congestion, and pelvic pain. Imaging studies have verified the fact that very few venous valves are found in the blood vessels of the pelvic area.[70-73]

Any compromise of the valves (or blood vessels) in the area can lead to this condition. It can also occur as the result of kidney removal or donation because the ovarian vein is cut

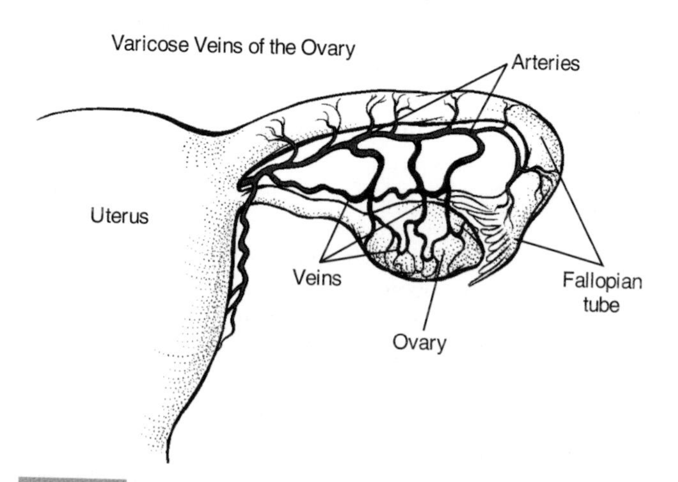

Fig. 15-9 Ovarian varicosities associated with pelvic congestion syndrome are the cause of chronic pelvic pain for women. This form of venous insufficiency is often accompanied by prominent varicose veins elsewhere in the lower quadrant (buttocks, thighs, calves). Men may have similar varicosities of the scrotum (not shown).

CASE EXAMPLE 15-2

Pelvic and Buttock Pain

A 34-year-old man with leukemia had a routine bone marrow biopsy near the left posterior superior iliac crest. No problems were noted at the time of biopsy, but 2 days later, the man came into physical therapy complaining of pelvic girdle pain.

He said his platelet count was 50,000/uL and international normalized ratio (INR), which is a measure of clotting time, was "normal." Laboratory values were recorded on the day of the biopsy.

The only clinical findings were a positive Faber's (Patrick's) test on the left and tenderness to palpation over the left sciatic notch, about an inch below the biopsy site. No abnormal neurologic signs were observed.

What are the red flags in this scenario?

Use the screening model to find the red flags and decide what to do.

History: Current history of cancer; recent history of biopsy

Clinical Presentation: Reduced platelet count (normal is >100,000); new onset of painful symptoms within 48 hours of biopsy; tenderness to palpation in left buttock

Associated Signs and Symptoms: None. Client had no other signs and symptoms.

The therapist has to make a clinical judgment in a case like this. The platelet level is low, putting the client at risk for poor clotting and spontaneous bleeding, but the INR suggests that the body is able to initiate the coagulation cascade.

Given the timing between the biopsy and the symptoms, it is likely that the procedure caused an intramuscular hematoma. The diagnosis can be made with a computed tomography (CT) scan. The location of biopsy needle entry indicates that the gluteus medius was punctured. No major blood vessel is located in this area, so the problem is rare.

Pain after bone marrow biopsy is usually mild to moderate and gradually gets better. The use of ice, massage, and later, moist heat is safe when properly applied. Worsening buttock pain over the next 24 to 48 hours would necessitate a medical referral.

It is always a good idea to contact the primary care physician and report your findings and intended intervention. This gives the doctor the option of following up with the client immediately if he or she thinks it is warranted.

when the kidney is removed. Varicosity of the gonadal venous plexus can occur in men and is more readily diagnosed by the presentation of observable varicosities of the scrotum.

Many women are unaware that they have this problem and remain asymptomatic. Women of childbearing age are affected most often. Many have had three or four (or more) pregnancies and are 40 years old or older.

Symptoms of ovarian varicosities reflect the vascular incompetence associated with venous insufficiency. These symptoms include pelvic pain that worsens toward the end of the day or after standing for a long time, pain after intercourse, sensation of heaviness in the pelvis, and prominent varicose veins elsewhere on the body, especially the buttocks and thighs.[70,73-75]

CLINICAL SIGNS AND SYMPTOMS

Pelvic Congestion Syndrome (Ovarian Varicosities)

- Lower abdominal/pelvic pain (intermittent or continuous, described as "dull aching" but can be sharp and severe)
- Tenderness on deep palpation of the ovarian point (located on the imaginary line drawn from the anterior superior iliac spine (ASIS) to the umbilicus where the upper one-third meets the lower two-thirds)[76]
- Unilateral or bilateral
- Pain that worsens with prolonged standing or at the end of the day
- Pain that is worse before or during menses
- Pain or "aching" that occurs after intercourse (dyspareunia)
- Presence of varicose veins in the buttocks, thighs, or lower extremities
- Low backache is a common feature, made worse by standing
- Other associated signs and symptoms (these vary; see text below)

Other associated symptoms may vary and include vaginal discharge, headache, emotional distress, GI distress, constipation, and urinary frequency and urgency. An undetermined number of women also have endometriosis, but the relationship is unknown.[77,78] Varicosities may be large enough to compress the ureter, leading to these urologic symptoms. Fatigue (loss of energy) and insomnia are common in women who experience headache with PCS (Case Example 15-3).

Screening for Cancer as a Cause of Pelvic Pain

The female pelvis is a depository for malignant tissue after incomplete removal of a primary carcinoma within the pelvis, for recurrence of cancer after surgical resection or radiotherapy of a pelvic neoplasm, or for metastatic deposits from a primary lesion elsewhere in the abdominal cavity.

Metastatic spread can occur from any primary tumor in the abdominal or pelvic cavity (see Fig. 13-2). For example, colon cancer can metastasize to the pelvic cavity by direct extension through the bowel wall to the musculoskeletal walls of the pelvic cavity or to surrounding organs. This may produce fistulas into the small intestine, bladder, or vagina. Advanced rectal tumors can become "fixed" to the sacral hollow. Deep pain within the pelvis may indicate spread of neoplasm into the sacral nerve plexuses.

Cancer recurrence can also occur after radiotherapy or surgery to the abdominal or pelvic cavity. This happens most often when incomplete removal of the primary carcinoma has occurred.

Using the Screening Model for Cancer

In the case of cancer as a cause of pelvic pain, a *past history* of cancer is usually present, most commonly, cancer within the pelvic or abdominal cavity (e.g., GI, renal, reproductive). A history of cancer with recent surgical removal of tumor

CASE EXAMPLE 15-3

Pelvic Congestion Syndrome

If a woman presents with chronic pelvic girdle pain, how would you assess for a vascular problem? Or ovarian varicosities as a possible cause? Remember, we are not trying to make a medical diagnosis, but rather to look for clues to suggest when medical referral is required. Use the overall clues in our screening model.

What kind of *past medical history* and *risk factors* would you expect to see? With a vascular cause of pelvic girdle pain, a history of heart disease is often reported in a postmenopausal woman. With ovarian varicosities, multiparity is usually present (i.e., a woman who has had several full-term pregnancies and deliveries).

What would you expect to see in the *clinical presentation?* With a vascular cause of pelvic girdle pain, the client often reports pelvic, buttock, and leg pain or "discomfort" that is aggravated by activity or exercise. With varicosities, the client usually has a generalized dull ache in the lower abdominal/low back area that is worse after standing, after intercourse, or just at the end of the day.

When you ask the client what other symptoms are present, she may not have any other symptoms, but if she does, look for a cluster of vascular signs and symptoms. These can be found in Box 4-19.

With ovarian varicocele, visually observe for varicose veins in the legs. These are a prominent feature in the clinical presentation of most women with ovarian varicosities. Ask about the presence of associated signs and symptoms such as vaginal discharge, headache, gastrointestinal distress, insomnia, and urologic symptoms.

tissue followed by back, hip, sacral, pelvic, or pelvic girdle pain within the next 6 months is a major red flag. Even if it appears to be a clear neuromuscular or musculoskeletal problem, referral is warranted for medical evaluation.

A common *clinical presentation* of pelvic or abdominal cancer referred to the soma is one of back, sacral, or pelvic pain described as one or more of the following: Deep aching, colicky, constant with crescendo waves of pain that come and go, or diffuse pain. Usually, the client cannot point to it with one finger (i.e., pain does not localize).

The therapist must remember to ask whether the client is having any symptoms of any kind anywhere else in the body. This is vitally important! Signs and symptoms associated with pelvic pain can range from constitutional symptoms to symptoms more common with the GI, genitourinary (GU), or reproductive system.

The therapist must ask about blood in the urine or stools. Once the physical therapy examination has been completed, including the history, risk factor assessment, pain patterns, and any associated signs and symptoms, it is time to step back and conduct a Review of Systems (see Chapters 1 and 4).

The *Review of Systems* is part of the evaluation described in the *Guide's* Elements of Patient/Client Management that leads to optimal outcomes (see Fig. 1-4). It is part of the dynamic process in which the therapist makes clinical judgments on the basis of data gathered during the examination.

In the screening process, the therapist reviews the following:

- Do any red flags in the history or clinical presentation suggest a systemic origin of symptoms?
- Are any red flags associated signs and symptoms?
- What additional screening tests or questions are needed (if any)?
- Is referral to another health care provider needed, or is the therapist clear to proceed with a planned intervention (Case Example 15-4)?

Keep in mind the Clues Suggesting Systemic Pelvic Pain, which are listed at the end of this chapter. If hip or groin pain is an accompanying feature with pelvic pain, review Clues to Screening Lower Quadrant Pain (see Chapter 16); likewise for anyone with pelvic and back pain see Clues to Screening Head, Neck, or Back Pain (see Chapter 14).

The therapist can use the Special Questions to Ask at the end of Chapter 14. It may not be necessary to ask all these questions. The therapist can use the overall clues gathered from the *history, risk factor assessment, clinical presentation,* and *associated signs and symptoms,* while reviewing the list of special questions to see whether there is anything appropriate to ask the individual client.

Gynecologic Cancers

Cancers of the female genital tract account for about 12% of all new cancers diagnosed in women. Although gynecologic cancers are the fourth leading cause of death from cancer in women in the United States, most of these cancers are highly curable when detected early. The most common cancers of the female genital tract are uterine endometrial cancer, ovarian cancer, and cervical cancer.[79]

Endometrial (Uterine) Cancer. Cancer of the uterine endometrium, or lining of the uterus, is the most common gynecologic cancer, usually occurring in postmenopausal women between the ages of 50 and 70 years. Its occurrence is associated with obesity, endometrial hyperplasia, prolonged unopposed estrogen therapy (hormone replacement therapy without progesterone), and more recently, tamoxifen used in the treatment of breast cancer.[80,81]

Clinical Signs and Symptoms. Seventy-five percent of all cases of endometrial cancer occur in postmenopausal women. The most common symptom is abnormal vaginal bleeding or discharge at presentation. However, 25% of these cancers occur in premenopausal women, and 5% occur in women younger than 40 years.

In a physical therapy practice, the most common presenting complaint is pelvic pain without abnormal vaginal bleeding. Abdominal pain, weight loss, and fatigue may occur but remain unreported. Unexpected or unexplained vaginal bleeding in a woman taking tamoxifen (chemoprevention for breast cancer) is a red flag sign. Tamoxifen as a risk factor for endometrial carcinoma has come under question.[82]

CASE EXAMPLE 15-4

Peripheral Neuropathy of the Pelvic Floor

A 57-year-old woman presented with an unusual triad of symptoms. She reported numbness and tingling of the feet, urinary incontinence, and migrating arthralgias and myalgias of the lower body (e.g., low back or hip, sometimes hip adductor spasm or aching, a "heavy" sensation in the pelvic region).

Past Medical History: Significant previous medical history included a hysterectomy 10 years ago for uncontrolled bleeding, and oophorectomy 2 years ago followed by pelvic radiation for ovarian cancer.

She is a nonsmoker and a nondrinker and is in apparent good health after cancer treatment. She is not taking any medications or using any drugs or supplements. All follow-up checks have detected no signs of cancer recurrence. She is active in a women's cancer support group and exercises 4 or 5 times a week.

She has kept a journal of activities, foods, and symptoms but cannot find a pattern to explain any of her symptoms. Urinary incontinence is present continually with constant dripping and leaking. It is not made worse by exercise, the sound or feel of running water, putting the key in the door, or other triggers of urge or stress incontinence.

Bowel function is reportedly "normal." The client is a widow and is not currently sexually active.

Where do you go from here? What are the red flags? What questions do you ask? What tests do you perform? Is medical referral needed?

Red Flags

- Age
- Previous history of cancer
- Bilateral symptoms (numbness and tingling in both feet)

Screening Questions

Menstrual history, including pregnancies, miscarriages or abortions, births; current menstrual status (perimenopausal, postmenopausal, hormone replacement therapy)

Any symptoms or other problems anywhere else in the body?

Screening Tests

Can you reproduce any of the muscle or joint pain?

Neurologic Screen: Besides the usual manual muscle testing, deep tendon reflexes, and sensation, the therapist should test for lower extremity proprioception and assess feet more closely to identify the level of peripheral nerve impairment.

- Ask about the presence of other neurologic symptoms such as headache, muscle weakness, confusion, depression, irritability, blurred vision, balance/coordination problems, memory changes, and sleepiness.
- Some of these are more likely when the central nervous system is impaired; for now it looks as though we are looking at a problem in the peripheral nervous system, but paraneoplastic syndrome or metastases to the central nervous system can occur.

Assess for signs of skin or soft tissues, including the presence of lymphedema

Palpate the lymph nodes

Assess vital signs

Medical Referral

Immediate medical referral is warranted if the patient has not been evaluated recently. It is impossible to tell whether her symptoms are radiation induced or are signs of cancer recurrence. A phone conversation between therapist and the oncologist may be all that is needed. Information gathered during the interview and examination should be summarized for the physician.

Result: The client had peripheral neuropathies that affected the bladder, pelvic floor muscles, and feet because the same nerves innervate these two areas. Physical therapy intervention remained appropriate, and cancer recurrence was ruled out.

Radiation therapy is well known to cause significant delayed, chronic effects on connective tissue and the nervous system. Fibrosis of connective tissue can result in impairment of the soft tissues, such as pelvic adhesions, with subsequent functional limitations.

The incidence of plexopathy after radiation therapy has been reduced significantly with improved treatment, but it still occurs in a small number of cases. Younger women seem more vulnerable to radiation-induced peripheral neuropathy.

CLINICAL SIGNS AND SYMPTOMS

Endometrial (Uterine) Cancer

- Unexpected or unexplained vaginal bleeding or vaginal discharge after menopause (extremely significant sign)
- Persistent irregular or heavy bleeding between menstrual periods, especially in obese women
- Watery pink, white, brown, or bloody discharge from the vagina
- Abdominal or pelvic pain (more advanced disease)
- Weight loss, fatigue

Ovarian Cancer. Ovarian cancer is the second most common reproductive cancer in women and the leading cause of death from gynecologic malignancies, accounting for more than half of all gynecologic cancer deaths in the Western world.[79] It affects women of all races and ethnic groups.

Risk Factors. Risk increases with advancing age, and the incidence of ovarian cancer peaks between the ages of 40 and 70 years. Other factors that may influence the development of ovarian cancer include the following:

- Nulliparity (never being pregnant), giving birth to fewer than two children, giving birth for the first time when over age 35
- Personal or family history of breast, endometrial, or colorectal cancer
- Family history of ovarian cancer (mother, sister, daughter; especially at a young age); carrying the *BRCA1* or *BRCA2* gene
- Infertility
- Early menarche, late menopause, prolonged postmenopausal hormone (estrogen) therapy (long-term, additive exposure to estrogen)
- Obesity (body mass index [BMI] of 30 or more)

- Exposure to cosmetic talc or asbestos (conflicting evidence)[83-87]

Identification of the *BRCA1* or *BRCA2* gene and subsequent evidence for a family of genes that may play a role in the breast–ovarian syndrome and familial ovarian cancer offer the possibility of identifying women truly at risk for this disease.[88,89]

No reliable screening test can detect ovarian cancer in its early, most curable stages. Two diagnostic tests are used, but both lack sensitivity and specificity. The CA-125 blood test (carcinoembryonic antigen, a biologic marker) shows elevation in about half of women with early-stage disease and about 80% of those with advanced disease. Transvaginal ultrasonography helps determine whether an existing ovarian growth is benign or cancerous. Because early-stage symptoms are nonspecific, most women do not seek medical attention until the disease is advanced.

The ovaries begin in utero, where the kidneys are located in the fully developed human, and then migrate along the pathways of the ureters. Following the viscerosomatic referral patterns discussed in Chapter 3, ovarian cancer can cause back pain at the level of the kidneys. Murphy's percussion test (see Chapter 10) would be negative; other symptoms of ovarian cancer might be present but remain unreported if the woman does not recognize their significance.

CLINICAL SIGNS AND SYMPTOMS

Ovarian or Primary Peritoneal Cancer

Retrospective studies indicate that more than 70% of women with ovarian cancer have symptoms for 3 months or longer before diagnosis.[90] Over two-thirds of women have late-stage cancer with metastatic disease at the time of diagnosis. Early symptoms are often vague, nonspecific, and easily overlooked.

In one study, when symptoms listed below were present for less than 1 year and occurred more than 12 days per month, there was a cancer symptom index with a sensitivity of 56.7 for early-stage disease and 79.5% for advanced-stage disease. Specificity was 90% for women age older than 50 years and 86.7% for women age younger than 50 years.[91] Even early-stage ovarian cancer can produce these signs and symptoms[92]:

- Persistent vague GI complaints
- Pelvic/abdominal discomfort, bloating, increase in abdominal or waist size (ascites)
- Indigestion, belching
- Early satiety
- Mild anorexia in a woman age 40 or older
- Vaginal bleeding
- Changes in bowel or bladder habits, especially constipation, urinary frequency, or severe urinary urgency
- Pelvic discomfort or pressure; back pain
- Ascites, pain, and pelvic mass (advanced disease)

An ovarian symptom index for advanced ovarian cancer has also been developed.[93]

Rarely, reproductive carcinomas, including ovarian carcinoma, will present first with a paraneoplastic syndrome such as polyarthritis syndrome, carpal tunnel syndrome, myopathy, plantar fasciitis, or palmar fasciitis (swelling, digital stiffness or contractures, palmar erythema). The condition may be misdiagnosed as chronic regional pain syndrome (formerly reflex sympathetic dystrophy), Dupuytren's contracture, or a rheumatologic disorder.[94,95]

Hand and upper extremity manifestations often appear before the tumor is clinically evident. Treatment of the symptoms will have little effect on these conditions. Only successful treatment of the underlying neoplasm will affect symptoms favorably.[94]

The therapist should consider it a red flag whenever someone does not improve with physical therapy intervention. Failure to respond or worsening of symptoms requires a second screening examination. Progression of disease is often accompanied by a cluster of new signs and symptoms.

Extraovarian Primary Peritoneal Carcinoma. Extraovarian primary peritoneal carcinoma (EOPPC) is an abdominal cancer (peritoneal carcinomatosis) without ovarian involvement. It arises in the peritoneum and mimics the symptoms, microscopic appearance, and pattern of spread of endothelial ovarian cancer with no identifiable disease of the ovaries.[96]

EOPPC develops only in women and accounts for most extraovarian causes of symptoms with a presumed but inaccurate diagnosis of ovarian cancer.[97] EOPPC has been reported after bilateral oophorectomy performed for benign disease or prophylaxis.[98] The occurrence of EOPPC with the same histology as neoplasms arising within the ovary may be explained by the common origin of the peritoneum and the ovaries from the coelomic epithelium.[99]

Cervical Cancer. Cancer of the cervix is the third most common gynecologic malignancy in the United States. It is the most common cause of death from gynecologic cancer in the world. Since the widespread introduction of the Papanicolaou (Pap) smear as a standard screening tool, the diagnosis of cervical cancer at the invasive stage has decreased significantly. Even so, nearly half of all women diagnosed with cervical cancer are diagnosed at a late stage, with locally or regionally advanced disease and a poor prognosis.[79]

At the same time that rates of invasive cervical carcinoma have been on the decline, the highly curable preinvasive carcinoma in situ (CIS) has increased. CIS is more common in women 30 to 40 years of age, and invasive carcinoma is more frequent in women over age 40 years.

Risk Factors. Risk factors associated with the development of cervical cancer are many and varied and include the following:

- Early age at first sexual intercourse
- Early age at first pregnancy
- Tobacco use, including exposure to passive smoke[100]
- Low socioeconomic status (lack of screening)
- History of any STD, especially human papillomavirus (HPV) and human immunodeficiency virus (HIV)
- History of multiple sex partners
- History of childhood sexual abuse
- Intimate partner abuse

- Women whose mothers used the drug diethylstilbestrol (DES) during pregnancy

Research into the health effects of intimate partner abuse points to a higher risk of STD and prevention of women from seeking health care; both contribute to an increased risk of cervical cancer.[101] Women with a past history of childhood sexual abuse may avoid regular gynecologic care because being examined triggers painful memories. A history of childhood sexual abuse also increases a woman's risk of exposure to STIs that may contribute to the development of cervical cancer.[101]

The American Cancer Society (ACS) has issued updated recommendations for the early detection of cervical cancer.[102,103] The ACS advises all women to start cervical cancer screening 3 years after beginning to have vaginal intercourse, but no later than age 21. Pap smears should be done regularly, usually every year. After a total hysterectomy (including removal of the cervix) or after age 70, the Pap smear is discontinued.[104]

In the normal healthy adult female age 30 years or older, after three negative annual examinations, the Pap may be performed less frequently at the advice of the physician. Women with certain risk factors for cervical cancer (e.g., HIV infection, long-term steroid use, immunocompromised status, DES exposure before birth) should be advised to have an annual Pap smear.[102,104]

Clinical Signs and Symptoms. Early cervical cancer has no symptoms. Clinical symptoms related to advanced disease include painful intercourse; postcoital, coital, or intermenstrual bleeding; and a watery, foul-smelling vaginal discharge.

Disease usually spreads by local extension and through the lymphatics to the retroperitoneal lymph nodes (see Table 13-4). Metastases to the central nervous system can occur hematogenously late in the course of the disease and are generally rare. Clinical presentation of brain metastases depends on the site of the metastasized lesion; hemiparesis and headache are the most commonly reported signs and symptoms.[105]

CLINICAL SIGNS AND SYMPTOMS

Cervical Cancer

- May be asymptomatic (early stages)
- Painful intercourse or pain after intercourse
- Unexplained or unexpected bleeding
- Watery, foul-smelling vaginal discharge
- Hemiparesis, headache (cancer recurrence with brain metastases)

Screening for Gastrointestinal Causes of Pelvic Pain

GI conditions can cause pelvic pain. The most common causes of pelvic pain referred from the GI system are the following:

- Acute appendicitis
- IBD, Crohn's disease or regional enteritis, ulcerative colitis)
- Diverticulitis
- IBS

The small bowel, sigmoid, and rectum can be affected by gynecologic disease; abdominal, low back, and/or pelvic pain may result from pressure or displacement of these organs. Swelling, reaction to an adjacent infection, or reaction to the spilling of blood, menstrual fluid, or infected material into the abdominal cavity can cause pressure or displacement.

Bowel function is usually altered, but sometimes, the client experiences periods of normal bowel function alternating with intermittent bowel symptoms, and the client does not see a pattern or relationship until asked about current (or recent) changes in bowel function.

For all of these conditions, the symptoms as seen or reported in a physical therapy practice are usually the same. The client may present with one or more of the following:

- GI symptoms (see Box 4-19)
- Symptoms aggravated by increased abdominal pressure (coughing, straining, lifting, bending)
- Iliopsoas abscess (see Figs. 8-3 through 8-7; a positive test is indicative of an inflammatory/infectious process)
- Positive McBurney's point (see Figs. 8-9 and 8-10; appendicitis)
- Rebound tenderness or Blumberg's sign (see Fig. 8-12; appendicitis or peritonitis)

Appendicitis can cause peritoneal inflammation with psoas abscess, resulting in referred pain to the low back, hip, pelvis, or groin area (Case Example 15-5). The position of the vermiform appendix in the abdominal cavity is variable (see Fig. 8-10). Negative tests for appendicitis that use McBurney's point may occur when the appendix is located somewhere other than at the end of the cecum. See Fig. 8-12 for an alternate test (Blumberg's sign). Clinical signs and symptoms of appendicitis are listed in Chapter 8.

Blumberg's sign, a test for rebound tenderness, is usually positive in the presence of peritonitis, appendicitis, PID, or any other infection or inflammation associated with abdominal or pelvic conditions. Acute appendicitis is rare in older adults, but half of all those who die from a ruptured appendix are over 65.[106]

The test for rebound tenderness can be very painful for the client. The therapist is advised to do this test last. Some clinicians prefer to start out with this test (the only screening test used for abdominal or pelvic inflammation or infection) and to make a medical referral immediately when it is positive.

This is really a matter of professional preference based on experience and clinical judgment. In our experience, the iliopsoas and obturator tests are useful tools. If back pain (rather than abdominal quadrant pain) is the response, the therapist is alerted to the need to assess these muscles further and to consider their role in low back pain.

If the iliopsoas test is negative for lower quadrant pain, the therapist can palpate the integrity of the iliopsoas muscle and assess for trigger points (see Fig. 8-6). If the tests are negative (i.e., they do not cause abdominal pain), then the therapist can palpate McBurney's point for the appendix. If

CASE EXAMPLE 15-5

Appendicitis

Background: A 23-year-old woman who was training for a marathon developed groin and pelvic girdle pain—first just on the right side, but then on both sides. She reported that the symptoms came on gradually over a period of 2 weeks. She could not point to a particular spot as the source of the pain, but rather indicated a generalized lower abdominal, pelvic, and inner thigh area.

She denied ever being sexually active and had never been diagnosed with a sexually transmitted disease. She was on a rigorous training schedule for the marathon, did not appear anorexic, and seemed in overall good health.

No signs of swelling, inflammation, or temperature change were noted in the area. Running made the pain worse, rest made it better.

Range of motion of the hip and back was full and painless. A neurologic screening examination was normal. Resisted hip abduction was "uncomfortable" but did not exactly reproduce the symptoms.

What are the red flags here? What do you ask about or do next?

Not very many red flags are present: The bilateral presentation and overall size and location of the symptoms are the first two to be considered. Aggravating and relieving factors seem consistent with a musculoskeletal problem, but objective findings to support an impairment of the movement system are significantly lacking.

What do you ask about or do next?

Take the client's vital signs, including body temperature, blood pressure, respiratory rate, and heart rate. If you are pressed for time, at least take the body temperature and blood pressure.

Perform one or more of the tests for abdominal or pelvic infection/inflammation. You can go right to the rebound (Blumberg's) test, or you can assess the soft tissues one at a time as discussed in the text. If this is negative, consider trigger points as a possible source of painful symptoms.

Ask the client about constitutional symptoms or other symptoms anywhere else in the body.

Your next step or steps in interviewing or assessing the client will depend on the results of your evaluation so far. Once you have compiled the clinical presentation, step back and conduct a Review of Systems. If a cluster of signs and symptoms is associated with a particular visceral system, look over the Special Questions to Ask at the end of the chapter that address that system.

Check the Special Questions for Men and Women (Appendices A-24 and B-37). Have you left out or missed any that might be appropriate to this case?

Results: The client had normal vital signs but reported "night and day sweats" from time to time. The iliopsoas and obturator tests caused some general discomfort but were considered negative.

McBurney's point was positive, eliciting extreme pain. Blumberg's test for rebound was not performed. The client was referred to the emergency department immediately because she did not have a primary care physician.

It turned out that this client had peritonitis from a ruptured appendix. The doctors think she was in such good shape with a high pain threshold that she presented with minimal symptoms (and survived). Her white blood count was almost 100,000 at the time that laboratory work was finally ordered.

McBurney's is negative but an infectious cause of symptoms is suspected, the test for rebound tenderness can be conducted last.

Clients with symptoms of a possible inflammatory or infectious origin usually have a history of the conditions mentioned earlier (e.g., appendicitis, IBD or IBS, other GI disease). PID is another common cause of pelvic pain that can cause psoas abscess and a subsequent positive iliopsoas or obturator test. In this case, it is most likely a young woman with multiple sexual partners who has a known or unknown case of untreated chlamydia.

Crohn's disease, chronic inflammation of all layers of the bowel wall (see Chapter 8), may affect the terminal ileum and cecum or the rectum and sigmoid colon in the pelvis. In addition to pelvic and low back pain, systemic manifestations of Crohn's disease may include intermittent fever with sweats, malaise, anemia, arthralgias, and bowel symptoms.

Diverticular disease of the colon (diverticulosis), an acquired condition most common in the fifth to seventh decades, appears with intermittent symptoms. Moderate-to-severe pain in the left lower abdomen and the left side of the pelvis may be accompanied by a feeling of bowel distention and bowel symptoms such as hard stools, alternating diarrhea and constipation, mucus in the stools, and rectal bleeding.

IBS produces persistent, colicky lower abdominal and pelvic pain associated with anorexia, belching, abdominal distention, and bowel changes. Symptoms are produced by excessive colonic motility and spasm of the bowel (spastic colon).

See Chapter 8 for additional details about the referred pain patterns and most common associated signs and symptoms for each of these diseases.

Screening for Urogenital Causes of Pelvic Pain

Infection of the bladder or kidney, kidney stones, renal failure (chronic kidney disease), spasm of the urethral smooth muscle, and tumors in any of the urogenital organs can refer pain to the lower lumbar and pelvic regions, mimicking musculoskeletal impairment. Pelvic floor tension myalgia can develop in response to these conditions and create pelvic pain. The primary pain pattern may radiate around the flanks to the lower abdominal region, the genitalia, and the anterior/medial thighs (see Figs. 10-7 to 10-10).

Usually, the most common diseases of this system appear as obvious medical problems. In the physical therapy setting, past medical history, risk factors, and associated signs and symptoms provide important red-flag clues. The therapist needs to ask the client about the presence of painful urination or changes in urination and constitutional symptoms such as fever, chills, sweats, and nausea or vomiting.

Deep, aching pelvic pain that is worse on weight bearing or is accompanied by sciatica or numbness and tingling in the groin or lower extremity may be associated with cancer recurrence or cancer metastases.

Screening for Other Conditions as a Cause of Pelvic Girdle Pain

Psychogenic pain is often ill defined, and its anatomic distribution depends more on the person's concepts than on clinical disease processes. Pelvic pain may co-evolve with psychologic impairments.[107]

Such pain does not usually radiate; commonly, the client has multiple unrelated symptoms, and fluctuations in the course of symptoms are determined more by crises in the person's psychosocial life than by physical changes. (See also Screening for Emotional and Psychologic Overlay in Chapter 3.)

A history of sexual abuse in childhood or adulthood (men and women) may contribute to chronic pelvic pain or symptoms of a vague and diffuse nature.[108] In some cases, the link between abuse and pelvic pain may be psychologic or neurologic, or may result from biophysical changes that heighten a person's physical sensitivity to pain. Taking a history of sexual abuse may be warranted.[108,109]

Occasionally, a woman has been told there is no organic cause for her distressing pelvic pain. Chronic vascular pelvic congestion, enhanced by physical or emotional stress, may be the underlying problem. The therapist may be instrumental in assessing for this condition and providing some additional clues to the medical community that can lead to a medical diagnosis.

Surgery, in particular hysterectomy, is associated with varying amounts of pain from problems such as nerve damage, scar formation, or hematoma formation with infection, which can cause backache and pelvic pain. Lower abdominal discomfort, vaginal discharge, and fatigue may accompany pelvic pain or discomfort months after gynecologic surgery.

Other types of abdominal, pelvic, or tubal surgery, such as laparotomy, tubal ligation, or laminectomy, can also be followed by pelvic pain, usually associated with low back pain. The use of "candy cane" stirrups that wrap around the foot, ankle, and calf and place the hip in extreme hip abduction may contribute to hip labral tears and subsequent hip and/or pelvic pain. There is evidence that these stirrups are linked with lumbosacral nerve plexus injuries and transient postoperative neuropathy.[110] The association between the stirrups and labral tears is based on clinical observation and anecdotal data. Research is needed to verify this relationship. During the client interview, the therapist must include questions about recent surgical procedures.

PHYSICIAN REFERRAL

Guidelines for Immediate Medical Attention

- Immediate medical attention is required anytime the therapist identifies signs and symptoms that point to fracture, infection, or neoplasm. For example, a positive rebound test for appendicitis or peritonitis requires immediate medical referral. Likewise, severe sacral pain in the presence of a previous history of uterine, abdominal, prostate, rectal, or anal cancer requires immediate medical referral.
- Suspicion of any infection (e.g., STD, PID) requires immediate medical referral. Early medical intervention can prevent the spread of infection and septicemia and preserve fertility.
- A sexually active female with shoulder or back pain of unknown cause may need to be screened for ectopic pregnancy. Onset of symptoms after a missed menstrual cycle or in association with unexplained or unexpected vaginal bleeding requires immediate medical attention. Hemorrhage from ectopic pregnancy can be a life-threatening condition.

Guidelines for Physician Referral

- Blood in the toilet after a bowel movement may be a sign of anal fissures or hemorrhoids but can also signal colorectal cancer. A medical differential diagnosis is needed to make the distinction. History of an unrepaired hernia or suspected undiagnosed hernia requires medical referral. Lateral wall pelvic pain referred down the anteromedial side of the thigh to the knee can occur with femoral hernias; inguinal hernias are more likely to cause groin pain.
- A history of cancer with recent surgical removal of tumor tissue followed by back, hip, sacral, or pelvic pain within 6 months of surgery is a red flag for possible cancer recurrence. Even in the presence of apparent movement system impairment, referral for medical evaluation is warranted.
- All adolescent females and adult women who are sexually active or over the age of 21 should be asked when their last Pap smear was done and what the results were. The therapist can play an important part in client education and disease prevention by teaching women about the importance of an annual Pap smear and encouraging them to schedule one, if appropriate.
- Women with conditions such as endometriosis, pelvic congestion syndrome, STI, and PID can be helped with medical treatment. Medical referral is advised anytime a therapist identifies signs and symptoms that suggest any of these conditions.
- Failure to respond to physical therapy intervention is usually followed by reevaluation that includes a second

screening and a Review of Systems. Any red flags or cluster of suspicious signs and symptoms must be reported. Depending on the therapist's findings, medical evaluation may be the next step.

Clues to Screening the Sacrum/Sacroiliac

Past Medical History

- Previous history of Crohn's disease; presence of skin rash with new onset of sacral, hip, or leg pain
- Previous history of other GI disease
- Previous history of rheumatic disease
- Previous history of conjunctivitis or venereal disease (Reiter's syndrome)
- History of heart disease or PVD; the therapist should ask about the effect of activity on symptoms
- Remember to consider unreported assaults or anal intercourse (partnered rape, teens, homosexual men with men). Please note that many of today's teens are resorting to anal intercourse and oral sex in an effort to prevent pregnancy. These forms of sexual contact do not prevent STD. In addition, they can result in sacral pain and other lesions (e.g., rectal fissures) caused by trauma.

Clinical Presentation

- Constant (usually intense) pain; pain with a "catch" or "click" (sacral fracture)
- Sacral pain occurs when the rectum is stimulated (pain occurs when passing gas or having a bowel movement).
- Pain relief occurs after passing gas or having a bowel movement.
- Sacral or SI pain in the absence of a (remembered) history of trauma or overuse
- Assess for trigger points, a common musculoskeletal (not systemic) cause of sacral pain. If trigger point therapy relieves, reduces, or eliminates the pain, further screening may not be necessary.
- Lack of objective findings; special tests (e.g., Patrick's test, Gaenslen's maneuver, Yeoman's test, central posterior–anterior overpressure or spring test on the sacrum) are negative. Soft tissue and contractile tissue can usually be provoked during a physical examination by palpation, resistance, overpressure, compression, distraction, or motion.
- Look for other pelvic floor impairment.

Associated Signs and Symptoms

- Presence of urologic or GI symptoms along with sacral pain (the therapist must ask to find out)

Clues to Screening the Pelvis

Frequently, pelvic and low back pain occur together or alternately. Whenever pelvic pain is listed, the reader should consider this as pelvic pain with or without pelvic girdle pain, low back, or sacral pain.

Past Medical History/Risk Factors

- History of dysmenorrhea, ovarian cysts, inflammatory disease, STD, fibromyalgia, sexual assault/incest/trauma, chronic yeast/vaginal infection, chronic bladder or urinary tract infection, chronic IBS
- History of abdominal, pelvic, or bladder surgery
- History of pelvic or abdominal radiation
- Recent therapeutic or spontaneous abortion
- Recent IUCD in the presence of PID or in women with a history of PID
- History of previous gynecologic, colon, or breast cancer
- History of prolonged labor, use of forceps or vacuum extraction, and/or multiple births
- Obesity, chronic cough

Clinical Presentation

- Pelvic pain that is described as "achy" or "comes and goes in waves" and is poorly localized (person cannot point to one spot)
- Pelvic pain that is aggravated by walking, sexual intercourse, coughing, or straining
- Pain that is not clearly affected by position changes or specific movements, especially when accompanied by night pain unrelieved by change in position
- Pelvic pain that is not reduced or eliminated by scar tissue mobilization, soft tissue mobilization, or release of trigger points of the myofascial structures in the pelvic cavity

Associated Signs and Symptoms

- Pelvic pain in the presence of yellow, odorous vaginal discharge
- Positive McBurney's or iliopsoas/obturator tests (see Chapter 8)
- Pelvic pain with constitutional symptoms, especially nausea and vomiting, GI symptoms (possible enteropathic origin)
- Presence of painful urination; urinary incontinence, urgency, or frequency; nocturia; blood in the urine; or other urologic changes

Gynecologic

- Pelvic pain that is relieved by rest, placing a pillow or support under the hips and buttocks in the supine position, or "getting off your feet"
- Pelvic pain that is correlated with menses or sexual intercourse
- Pelvic pain that occurs after the first menstrual cycle is missed, especially if the woman is using an IUCD or has had a tubal ligation (see text for other risk factors), with shoulder pain also present (ruptured ectopic pregnancy); assess for low blood pressure
- Presence of unexplained or unexpected vaginal bleeding, especially after menopause
- Presence of pregnancy

Vascular

- History of heart disease with a clinical presentation of pelvic, buttock, and leg pain that is aggravated by activity or exercise (claudication)
- Pelvic pain accompanied by buttock and leg pain with changes in skin and temperature on the affected side (arterial occlusion or venous thrombosis), especially in the presence of known heart disease or recent pelvic surgery
- Pain that worsens toward the end of the day, accompanied by pain after intercourse and in the presence of varicose veins elsewhere in the body (ovarian varicosities)

Key Points to Remember

Many of the Key Points to Remember in Chapter 14 also apply to the sacrum and sacroiliac joints. These will not be repeated here.

Sacrum/Sacroiliac Joint

- Sacral pain, in the absence of a history of trauma or overuse, that is not reproduced with anterior–posterior overpressure (spring test) on the sacrum is a red flag presentation that indicates a possible systemic cause of symptoms.
- Pain above the L5 spinous process is not likely from the sacrum or SI joint.
- Midline lumbar pain, especially if present when rising from sitting, more often comes from a diskogenic source; clients with unilateral pain below L5 when rising from sitting are more likely to have a painful SI joint.
- Insufficiency fractures of the spine are not uncommon with individuals who have osteoporosis or who are taking corticosteroids; apparent insidious onset or minor trauma is common.
- The most common cause of noninfected, inflammatory sacral/SI pain is ankylosing spondylitis; other causes may include Reiter's syndrome, psoriatic arthritis, and arthritis associated with IBD.
- Infection can seed itself to the joints, including the SI joint. Watch for a history of recent dental surgery (endocarditis), intravenous drug use, trauma (including surgery), and chronic immunosuppression.
- Anyone with joint pain of unknown cause should be asked about a recent history of skin rash (delayed allergic reaction, Crohn's disease).

Pelvis

- Pelvic and low back pain often occur together; either may be accompanied by unreported abdominal pain, discomfort, or other symptoms. The therapist must ask about the presence of any unreported pain or symptoms.
- Yellow or green discharge from the vagina or penis (with or without an odor) in the presence of low back, pelvic, or sacral pain may be a red flag. The therapist must ask additional questions to determine the need for medical evaluation.
- The first sign of pelvic infection in the older adult might not be an elevated temperature, but rather, confusion, increased confusion, or some other change in mental status.
- Bilateral anterior pelvic pain may be a symptom of inflammation; the therapist can test for iliopsoas or obturator abscess, appendicitis, or peritonitis (see discussion in Chapter 8).
- Pelvic girdle pain that is aggravated by exercise and starts 5 or more minutes after exercise begins may be vascularly induced.
- A history of sexual abuse at any time in the person's past may contribute to chronic pelvic pain or nonspecific symptoms. Taking a history of sexual abuse may be needed.

SUBJECTIVE EXAMINATION

Special Questions to Ask: Sacrum, Sacroiliac, and Pelvis

See Special Questions to Ask: Neck or Back, and Special Questions to Ask Men/Women Experiencing Back, Hip, Pelvic, Groin, Sacroiliac, or Sacral Pain in Chapter 14.

Not all the special questions listed in Chapter 14 will have to be asked. Use your professional judgment to decide what to ask based on what the client has told you and what you've observed during the examination.

Sacral/Sacroiliac Pain

- Have you ever been diagnosed with ulcerative colitis, Crohn's disease, IBS, or colon cancer?
- Are you taking any antibiotics? (Long-term use of antibiotics can result in colitis.)
- Have you ever been diagnosed or treated for cancer of any kind? **(Metastases to the bone, especially common with breast, lung, or prostate cancer, but also with pelvic or abdominal cancer)**
- Do you have any abdominal pain or GI symptoms? (assess for lower abdominal or suprapubic pain at the same level as the sacral pain; if the client denies GI symptoms, follow-up with a quick list: Any nausea? Vomiting? Diarrhea? Change in stool color or shape? Ever have blood in the toilet?)
- If sacral pain occurs when the rectum is stimulated:
  - Is your pain relieved by passing gas or by having a bowel movement?
- Sacral or SI pain without a history of trauma or overuse

Remember to consider unreported assault, anal intercourse (partnered rape; adolescents may use anal intercourse to prevent pregnancy, homosexual men with men). Please note that many of today's teens are resorting to anal intercourse and oral sex in an effort to prevent pregnancy.

These forms of sexual contact do not prevent STD. In addition, they can result in sacral pain and other lesions (e.g., rectal fissures) resulting from trauma.

Pelvic Pain

- Have you ever been diagnosed or treated for cancer of any kind?
- Have you had recent abdominal or pelvic surgery (including hysterectomy, bladder reconstruction, prostatectomy)?
- Have you ever been told that you have (or do you have) varicose veins? **(Pelvic congestion syndrome)**
- Do you ever have blood in the toilet?

For women with low back, sacral, or pelvic pain: See Special Questions to Ask: Women Experiencing Back, Hip, Pelvic, Groin, Sacroiliac (SI), or Sacral Pain in Chapter 14.

For anyone with low back, sacral, or pelvic pain of unknown cause: It may be necessary to conduct a sexual history as part of the screening process (see Chapter 14 or Appendix B-32).

For men with sciatica, pelvic, sacral, or low back pain: See Special Questions to Ask: Men Experiencing Back, Hip, Pelvic, Groin, or Sacroiliac Pain (Chapter 14 or Appendix B-24).

CASE STUDY

Steps in the Screening Process

When a client presents with pelvic pain, how do you get started with the screening process?

First, review the possible causes of pelvic pain (see Table 15-2).

- Was there anything in the history or presentation to suggest one of the categories in this table?
- From looking at the table, do additional questions come to mind?
- Review Special Questions to Ask: Sacrum, SI, or Pelvis presented in this chapter. Are any of these questions appropriate or needed?
- Did you ask about associated signs and symptoms?
- Remember to ask the client the following:
  - Is there anything else about your general health that concerns you?
  - What other symptoms are you having that may or may not be connected to your current problem?

Next, review Clues to Screening the Pelvis. Is there anything here to raise your suspicion of a systemic disorder?

- If necessary, conduct a general health screening examination:
  - Have you had any recent infections or illnesses?
  - Have you had any fevers, sweats, or chills?
  - Any unusual discharge from the vagina or penis?
  - Any unusual skin rashes or muscle/joint pain?
  - Any unusual fatigue, irritability, or difficulty sleeping?
- Is there anything to suggest a pelvic floor impairment as the source of symptoms? Look for the following:

CASE STUDY—*cont'd*

Steps in the Screening Process

- Pain that comes and goes and changes location
- Pain that is not predictably reproducible
- Pain that is alleviated by heat to the lower abdomen, groin, or front of the upper thighs
- Pain made worse by hip flexion or straight leg raise (muscle impairment associated with disk pathology)
- Rectal pain or discomfort that is worse during intercourse or penetration
- Pain or discomfort (better or worse) before, during, or after menstrual cycle

- Mentally conduct a Review of Systems
  - Did the past medical history, age, medications, or associated signs and symptoms point to anything?
  - Use your Review of Systems table (see Box 4-19) to look for possible clusters of symptoms, or to remind you what to look for
    - If you identify a specific system in question, ask additional questions for that system:
    - For example, if a significant past medical history or current signs and symptoms of GI involvement are reported, review the Special Questions to Ask in Chapter 8. Would any questions listed be appropriate to ask, given your client's clinical presentation? Or, if you suspected a renal/urologic cause of symptoms, look at the questions posed in Chapter 10.

Sometimes, the initial screening process does not raise any suspicious history or red-flag symptoms. As discussed in Chapter 1, screening can take place anywhere in the Guide's patient/client management model (see Box 1-5).

The therapist may begin to carry out the intervention without seeing any red flags that suggest a systemic disorder and may then find that the client does not improve with physical therapy. This in itself is a red flag.

If someone is not improving with physical therapy intervention, the therapist reviews the findings (i.e., what you are doing and why you are doing it), while evaluating the need to repeat some steps in the screening process. Because systemic disease progresses over time, new signs and symptoms may have developed since the time of the first interview and client history taking.

The therapist may want to have someone else review the case. Often, this can provide some clarity and add insight to the evaluation process. Asking a few screening questions may bring to light some new information to be included in the ongoing evaluation. Now may be the time to repeat (or perform for the first time) specific and appropriate screening tests and measures.

PRACTICE QUESTIONS

1. Pelvic pain that is made worse after 5 to 10 minutes of physical activity or exertion but goes away with rest or cessation of the activity describes:
   a. Constitutional symptom
   b. Infectious process
   c. Symptom of osteoporosis
   d. Vascular pattern of ischemia
2. Pain that is relieved by placing a pillow or support under the hips and buttocks describes:
   a. Constitutional symptom
   b. Infectious process
   c. Response to vascular congestion
   d. Trigger point pattern
3. A positive Blumberg's sign indicates:
   a. Pelvic infection
   b. Ovarian varicosities
   c. Arthritis associated with IBD
   d. Sacral neoplasm
4. A 33-year-old pharmaceutical sales representative reports pain over the mid-sacrum radiating to the right PSIS. Overpressure on the sacrum does not reproduce symptoms. This signifies:
   a. Neoplasm is present
   b. Red flag sign of sacral insufficiency fracture
   c. A lack of objective findings
   d. Coccygodynia
5. A 67-year-old man was seen by a physical therapist for low back pain rated 7 out of 10 on the visual analogue scale. He was evaluated, and a diagnosis was made by the physical therapist. The client attained immediate relief of symptoms, but after 3 weeks of therapy, the symptoms returned. What is the next step from a screening perspective?
   a. The client can be discharged. Maximum benefit from physical therapy has been achieved.
   b. The client should be screened for systemic disease even if you have already included screening during the initial evaluation.
   c. The client should be sent back to the physician for further medical follow-up.
   d. The client should receive an additional modality to help break the pain–spasm cycle.

Continued

PRACTICE QUESTIONS—*cont'd*

6. McBurney's point for appendicitis is located:
   a. Approximately one-third the distance from the ASIS toward the umbilicus, usually on the left side
   b. Approximately one-half the distance from the ASIS toward the umbilicus, usually on the left side
   c. Approximately one-third the distance from the ASIS toward the umbilicus, usually on the right side
   d. Approximately one-half the distance from the ASIS toward the umbilicus, usually on the right side
   e. Impossible to tell because the appendix can be located anywhere in the abdomen
7. Which one of the following is a yellow (caution) flag?
   a. Sacral pain occurs when the examiner performs a sacral spring test (posterior-anterior glide of the sacrum).
   b. Sacral pain is relieved when the client passes gas or has a bowel movement.
   c. Sacral pain occurs following a history of overuse.
   d. Sacral pain is reduced or relieved by release of trigger points.
8. Cancer as a cause of sacral or pelvic pain is usually characterized by:
   a. A previous history of reproductive cancer
   b. Constant pain
   c. Blood in the urine or stools
   d. Constitutional symptoms
   e. All of the above
9. Reproduced or increased abdominal or pelvic pain when the iliopsoas muscle test is performed suggests:
   a. Iliopsoas trigger point
   b. Inflammation or abscess of the muscle from an inflamed appendix or peritoneum
   c. Abdominal aortic aneurysm
   d. Neoplasm
10. A 75-year-old woman with a known history of osteoporosis has pain over the sacrum radiating to the right PSIS and right buttock. How do you rule out an insufficiency fracture?
    a. Perform Blumberg's test.
    b. Conduct a sacral spring test (posterior–anterior overpressure of the sacrum).
    c. Perform Murphy's percussion test.
    d. Diagnostic imaging is the only way to know for sure.
11. What is the importance of the pelvic floor musculature in relation to the abdominal and pelvic viscera?

REFERENCES

1. Buchowski JC, Kebaish KM, Sinkov V, et al: Functional and radiographic outcome of sacroiliac arthrodesis for the disorders of the sacroiliac joint. *Spine J* 5(5):520–528, 2005.
2. Vora AJ: Functional anatomy and pathophysiology of axial low back pain: discs, posterior elements, sacroiliac joint, and associated pain generators. *Phys Med Rehabil Clin N Am* 21(4):679–709, 2010.
3. Vanelderen P: Sacroiliac joint pain. *Pain Pract* 10(5):470–478, 2010.
4. Martin C, McCarthy EF: Giant cell tumor of the sacrum and spine: series of 23 cases and a review of the literature. *Iowa Orthop* 30:69–75, 2010.
5. Haanpaa M, Paavonen J: Transient urinary retention and chronic neuropathic pain associated with genital herpes simplex virus infection. *Acta Obstet Gynecol Scand* 83(10):946–949, 2004.
6. Smith C, Kavar B: Extensive spinal epidural abscess as a complication of Crohn disease. *J Clin Neurosci* 17(1):144–166, 2010.
7. Norman GF: Sacroiliac disease and its relationship to lower abdominal pain. *Am J Surg* 116:54–56, 1968.
8. Igdem S: Insufficiency fractures after pelvic radiotherapy in patients with prostate cancer. *Int J Radiat Oncol Biol Phys* 77(3):818–823, 2010.
9. Blake SP, Connors AM: Sacral insufficiency fracture. *Br J Radiol* 77(922):891–896, 2004.
10. Fortin J, Aprill CN, Dwyer A, et al: Sacroiliac joint: Pain referral maps upon applying a new injection/arthrography technique. I. Asymptomatic volunteers. *Spine* 19(13):1475–1482, 1994.
11. Fortin JD, Aprill CN, Ponthieux B, et al: Sacroiliac joint: Pain referral maps upon applying a new injection/arthrography technique. II. Clinical evaluation. *Spine* 19(13):1483–1489, 1994.
12. Young S, Aprill C, Laslett M: Correlation of clinical examination characteristics with three sources of low back pain. *Spine J* 3(6):460–465, 2003.
13. Depalma MJ: Does location of low back pain predict its source? *PM R* 3(1):3–39, 2011.
14. Slipman CW, Patel RK, Whyte WS, et al: Diagnosing and managing sacroiliac pain. *J Musculoskel Med* 18(6):325–332, 2001.
15. Cheng DS: Sacral insufficiency fracture: a masquerader of diskogenic low back pain. *PM R* 2(2):162–164, 2010.
16. Yansouni CP: Bacterial sacroiliitis and gluteal abscess after dilation and curettage for incomplete abortion. *Obstet Gynecol* 114(2 Pt 2):440–443, 2009.
17. Dreyfuss P, Dreyer SJ, Cole A, et al: Sacroiliac joint pain. *J Am Acad Orthop Surg* 13(4):255–265, July/August 2004.
18. Schumacher HR Jr, Klippel JH, Koopman WJ, editors: *Primer on the rheumatic diseases*, ed 12, Atlanta, 2001, Arthritis Foundation.
19. Betancourt-Albrecht M, Roman F, Marcelli M: Grand rounds in endocrinology, diabetes, and metabolism from Baylor College of Medicine: A man with pain in his bones. *Medscape Diabetes Endocrinol* 5(1):2003. Available online (free service but requires login and password): http://www.medscape.com/viewarticle/445158. Accessed March 10, 2011.
20. White JH, Hague C, Nicolaou S, et al: Imaging of sacral fractures. *Clin Radiol* 58:914–921, 2003.
21. Southam JD: Sacral stress fracture in a professional hockey player. *Orthopedics* 33(11):846, 2010.

22. Lin JT, Lane JM: Sacral stress fractures. *J Womens Health (Larchmt)* 12(9):879–888, November 2003.
23. Beltran LS, Bencardino JT: Lower back pain after recently giving birth: postpartum sacral stress fractures. *Skeletal Radiol* 40(4):481–482, 2011.
24. Rodrigues LM: Sacral stress fracture in a runner. *Clinics (Sao Paulo)* 64(11):1127–1129, 2009.
25. Leroux JL, Denat B, Thomas E, et al: Sacral insufficiency fractures presenting as acute low-back-pain-biomechanical aspects. *Spine* 18(16):2502–2506, 1993.
26. Boissonnault WG, Thein-Nissenbaum JM: Differential diagnosis of a sacral stress fracture. *J Orthop Sports Phys Ther* 12(32):613–621, December 2002.
27. Ahovuo JA, Kiuru MJ, Visuri T: Fatigue stress fractures of the sacrum: diagnosis with MR imaging. *Eur Radiol* 14(3):500–505, March 2004.
28. Khan MH, Smith PN, Kang JD: Sacral insufficiency fractures following multilevel instrumented spinal fusion. *Spine* 30(16):E484–E488, August 15, 2005.
29. Klineberg E: Sacral insufficiency fractures caudal to instrumented posterior lumbosacral arthrodesis. *Spine* 33(16):1806–1811, 2008.
30. Vavken P: Sacral fractures after multi-segmental lumbosacral fusion: a series of four cases and systematic review of literature. *Eur Spine J* Suppl 2: S285–S290, 2008.
31. Parikh VA, Edlund JW: Sacral insufficiency fractures—rare complication of pelvic radiation for rectal carcinoma. *Dis Colon Rectum* 41(2):254–257, February 1998.
32. Zileli M, Hoscoskun C, Brastiano P, et al: Surgical treatment of primary sacral tumors: complications with sacrectomy. *Neurosurg Focus* 15(5):E9, November 15, 2003.
33. Scuibba DM: Diagnosis and management of sacral tumors. *J Neurosurg Spine* 10(3):244–256, 2009.
34. Randall RL: Giant cell tumor of the sacrum. *Neurosurg Focus* 15(2):E13, August 15, 2003.
35. Payer M: Neurological manifestation of sacral tumors. *Neurosurg Focus* 15(2):E1, August 15, 2003.
36. Howard FM, El-Minawi AM, Sanchez RA: Conscious pain mapping by laparoscopy in women with chronic pelvic pain. *Obstet Gynecol* 96(6):934–939, December 2000.
37. Howard FM: Chronic pelvic pain. *Obstet Gynecol* 101(3):594–611, March 2003.
38. Vleeming A: European guidelines for the diagnosis and treatment of pelvic girdle pain. *Eur Spine J* 17(6):794–819, 2008.
39. Vollestad NK, Stuge B: Prognostic factors for recovery from postpartum pelvic girdle pain. *Eur Spine J* 18(5):718–726, 2009.
40. Stuge B, Hilde G, Vollestad N: Physical therapy for pregnancy-related low back and pelvic pain: a systematic review. *Acta Obstet Gynecol Scand* 82(11):983–990, 2003.
41. Messelink B: Standardization of terminology of pelvic floor muscle function and dysfunction: report from the pelvic floor clinical assessment group of the International Continence Society. *Neurourol Urodyn* 24:374–380, 2005.
42. Baker PK: Musculoskeletal problems. In Steege JF, Metzger DA, Levy BS: *Chronic pelvic pain: an integrated approach*, Philadelphia, 1998, WB Saunders, pp 215–240.
43. Steege JF, Zolnoun DA: Evaluation and treatment of dyspareunia. *Obstet Gynecol* 113(5):1124–1136, 2009.
44. Simons DG, Travell JG: *Travell & Simons' myofascial pain and dysfunction: the trigger point manual*, vol 2, Baltimore, 1993, Williams & Wilkins, 1993. (Myopain Seminars: http://www.myopainseminars.com/).
45. Bassaly R: Myofascial pain and pelvic floor dysfunction in patients with interstitial cystitis. *Int Urogynecol J Pelvic Floor Dysfunc* 22(4):413–418, 2011.
46. Anderson RU: Painful myofascial trigger points and pain sites in men with chronic prostatitis/chronic pelvic pain syndrome. *J Urol* 182(6):2753–2758, 2009.
47. Bø K, Sherburn M: Evaluation of female pelvic-floor muscle function and strength. *Phys Ther* 85(3):269–282, March 2005.
48. Riesco ML: Perineal muscle strength during pregnancy and postpartum: the correlation between perineometry and digital vaginal palpation. *Rev Lat Am Enfermagem* 18(6):1138–1144, 2010.
49. Headley B: *When movement hurts: a self-help manual for treating trigger points*, Minneapolis, MN, 1997, Orthopedic Physical Therapy Products.
50. Kostopoulos D, Rizopoulos K: *The manual of trigger point and myofascial therapy*, Thorofare, NJ, 2001, Slack, Inc.
51. Arab AM: Correlation of digital palpation and transabdominal ultrasound for assessment of pelvic floor muscle contraction. *J Man Manip Ther* 17(3):e75–e79, 2009.
52. Olsen MF: Self-administered screening test for pregnancy-related pain. *Europ Spine J* 18(8): 2009.
53. MacAvoy MC: Stability of open-book pelvic fractures using a new biomechanical model of single-limb stance. *J Orthop Trauma* 11:590–593, 1997.
54. Siegel J: Single-leg-stance radiographs in the diagnosis of pelvic instability. *J Bone Joint Surg* 90A:2119–2125, 2008.
55. Giudice LC: Status of current research on endometriosis. *J Reprod Med* 43(3 suppl):252–262, 1998.
56. Giudice LC, Kao LC: Endometriosis. *Lancet* 364(9447):1789–1799, November 2004.
57. Carrell DT, Peterson CM, editors: *Reproductive endocrinology and infertility*, New York, 2010, Springer Science.
58. Akbulut S: Left-sided appendicitis: review of 95 published cases and a case report. *World J Gastroenterol* 16(44):5598–5602, 2010.
59. Ross J: European guideline for the management of pelvic inflammatory disease. 2008. Available at http://www.iusti.org/regions/europe/PID_v5.pdf. Accessed March 10, 2011.
60. Soper DE: Pelvic inflammatory disease. *Obstet Gynecol* 116(2 Pt 1):419–428, 2010.
61. Kahn JA, Kaplowitz RA, Goodman E, et al: The association between impulsiveness and sexual risk behaviors in adolescent and young adult women. *J Adolesc Health* 30(4):229–232, April 2002.
62. Kahn JA, Rosenthal SL, Succop PA, et al: Mediators of the association between age of first sexual intercourse and subsequent papillomavirus infection. *Pediatrics* 109(1):E5, January 2002.
63. Shikary T: Epidemiology and risk factors for human papillomavirus infection in a diverse sample of low-income young women. *J Clin Virol* 46(2):107–111, 2009.
64. Centers for Disease Control and Prevention (CDC): *Policy guidelines for prevention and management of pelvic inflammatory disease (PID)*, Washington, DC, 1991, U.S. Department of Health and Human Services. Available at http://www.cdc.gov/mmwr/preview/mmwrhtml/00031002.htm. Accessed March 10, 2011.
65. Centers for Disease Control and Prevention: *Sexually transmitted disease surveillance: 2009*, Washington, DC, 2011, U.S. Department of Health and Human Services. Available at http://www.cdc.gov/std/stats09/default.htm. Accessed March 14, 2011.
66. Anderton JP, Valdiserri RO: Combating syphilis and HIV among users of internet chat rooms. *J Health Commun* 10(7):665–771, October-November 2005.
67. Douglas JM, Jr, Peterman TA, Fenton KA: Syphilis among men who have sex with men: challenges to syphilis elimination in the United States. *Sex Transm Dis* 32(10 suppl):S80–S83, October 2005.

68. Centers for Disease Control and Prevention (CDC): Syphilis Profiles 2009. Updated February 28, 2011. Available at http://www.cdc.gov/std/syphilis/. Accessed March 14, 2011.
69. Centers for Disease Control and Prevention (CDC): Syphilis & MSM (Men who have sex with men)—CDC fact sheet. Updated April 28, 2010. Available at http://www.cdc.gov/std/syphilis/STDFact-MSM-Syphilis.htm. Accessed March 14, 2011.
70. Tarazov PG, Prozorovskij KV, Ryzhkov VK: Pelvic pain syndrome caused by ovarian varices. *Acta Radiol* 38(6):1023–1025, 1997.
71. El-Minawi AM: Pelvic varicosities and pelvic congestion syndrome. In Howard FM, et al, editors: *Pelvic pain diagnosis and management*, Philadelphia, 2000, Lippincott, Williams & Wilkins, pp 171–183.
72. Hobbs JT: Varicose veins arising from the pelvis due to ovarian vein incompetence. *Int J Clin Pract* 59(10):1195–1203, October 2005.
73. Freedman J: Pelvic congestion syndrome: the role of interventional radiology in the treatment of chronic pelvic pain. *Postgrad Med* 86(1022):704–710, 2010.
74. Gasparini D, Geatti O, Orsolon PG, et al: Female "varicocele." *Clin Nucl Med* 23(7):420–422, 1998.
75. Tu FF: Pelvic congestion syndrome-associated pelvic pain: a systematic review of diagnosis and management. *Obstet Gynecol Surv* 65(5):332–340, 2010.
76. Hantes J: History and physical of the chronic pelvic pain patient. *VISION: The International Pelvic Pain Society* 9(1):1–4, 2004.
77. Hartung O, Grisoli D, Boufi M, et al: Endovascular stenting in the treatment of pelvic vein congestion caused by nutcracker syndrome: lessons learned from the first five cases. *J Vasc Surg* 42(2):275–280, 2005.
78. Singh MK: Chronic pelvic pain. *Emedicine Specialties*. Updated Sep 13, 2010. Available at http://emedicine.medscape.com/article/258334-overview. Accessed March 14, 2011.
79. Jemal A, Murray T, Ward E, et al: Cancer statistics, 2010. *CA Cancer J Clin* 60(5):277–300, September/October 2010.
80. Ferguson SE, Soslow RA, Amsterdam A, et al: Comparison of uterine malignancies that develop during and following tamoxifen therapy. *Gynecol Oncol* 101(2):322–326, 2006.
81. Carter J, Pather S: An overview of uterine cancer and its management. *Expert Rev Anticancer Ther* 6(1):33–42, January 2006.
82. Creasman WT: Endometrial carcinoma. *Emedicine Specialties*. 2010. Available at http://emedicine.medscape.com/article/254083-overview. Accessed March 14, 2011.
83. Huncharek M, Geschwind JF, Kupelnick B: Perineal application of cosmetic talc and risk of invasive epithelial ovarian cancer: a meta-analysis of 11,933 subjects from sixteen observational studies. *Anticancer Res* 23(2C):1955–1960, March-April 2003.
84. Langseth H, Kjaerheim K: Ovarian cancer and occupational exposure among pulp and paper employees in Norway. *Scand J Work Environ Health* 30(5):356–361, 2004.
85. Mills PK, Riordan DG, Cress RD, et al: Perineal talc exposure and epithelial ovarian cancer risk in the Central Valley of California. *Int J Cancer* 112(3):458–464, November 2004.
86. Wu AH: Markers of inflammation and risk of ovarian cancer in Los Angeles county. *Int J Cancer* 124(6):1409–1415, 2009.
87. Muscat JE, Huncharek MS: Perineal talc use and ovarian cancer: a critical review. *Eur J Cancer Prev* 17(2):139–146, 2008.
88. Study questions ovary removal during hysterectomy: what factors affect ovarian cancer risk? *Harvard Women's Health Watch* 13(2):7, October 2005.
89. Daly MB: Genetic/familial high risk assessment: breast and ovarian. *J Natl Comp Canc Netw* 8(5):562–594, 2010.
90. Goff BA, Mandel LS, Melancon CH, et al: Frequency of symptoms of ovarian cancer in women presenting to primary care clinics. *JAMA* 291(22):2705–2712, June 2004.
91. Goff BA: Development of an ovarian cancer symptom index: possibilities for earlier detection. *Cancer* 109(2):221–227, 2007.
92. Liu S: Patterns of symptoms in women after gynecologic surgery. *Oncol Nurs Forum* 37(2):156, 2010.
93. Jensen SE: A new index of priority symptoms in advanced ovarian cancer. *Gynecol Oncol* 120(2):214–219, 2011.
94. Martorell EA, Murray PM, Peterson JJ, et al: Palmar fasciitis and arthritis syndrome associated with metastatic ovarian carcinoma: a report of four cases. *J Hand Surg* 29A(4):654–660, 2004.
95. Krishna K: Palmar fasciitis with polyarthritis syndrome in a patient with breast cancer. *Clin Rheumatol* Oct 16, 2010. Epub ahead of print.
96. Ozat M: Extraovarian conditions mimicking ovarian cancer. *Arch Gynecol Obstet* Oct. 15, 2010. Epub ahead of print.
97. Roffers SD, Wu XC, Johnson CH, et al: Incidence of extra-ovarian primary cancers in the United States, 1992–1997. *Cancer* 97(10 suppl):2643–2647, May 15, 2003.
98. Eltabbakh GH, Piver MS: Extraovarian primary peritoneal carcinoma. *Oncology (Williston Park)* 12(6):813–819, June 1998.
99. Kunz J, Rondez R: Correlation between serous ovarian tumors and extra-ovarian peritoneal tumors of the same histology. *Schweiz Rundsch Med Prax* 87(6):191–198, February 1998.
100. Trimble CL, Genkinger JM, Burke AE, et al: Active and passive cigarette smoking and the risk of cervical neoplasia. *Obstet Gynecol* 105(1):174–181, January 2005.
101. Coker AL: Violence against women raises risk of cervical cancer. *J Womens Health (Larchmt)* 18(8):1179–1185, 2009.
102. Saslow D, Runowicz CD, Solomon D, et al: American Cancer Society guideline for the early detection of cervical neoplasia and cancer. *CA Cancer J Clin* 52(6):342–362, November/December 2002.
103. Schiffman M, Solomon D: Screening and prevention methods for cervical cancer. *JAMA* 302(16):1809–1810, 2009.
104. American Cancer Society (ACS): ACS cancer detection guidelines. Atlanta, Georgia. Available at http://http://www.cancer.org/Healthy/FindCancerEarly/CancerScreeningGuidelines/american-cancer-society-guidelines-for-the-early-detection-of-cancer. Accessed March 10, 2011.
105. Amita M, Sudeep G, Rekha W, et al: Brain metastasis from cervical carcinoma. *Medscape Gen Med* 7(1): 2005. Available at http://www.medscape.com/viewarticle/496603_3. Accessed March 15, 2011.
106. Storm-Dickerson TL, Horattas MC: What have we learned over the past 20 years about appendicitis in the elderly? *Am J Surg* 185(3):198–201, March 2003.
107. Mathias SD, Kuppermann M, Liberman RF, et al: Chronic pelvic pain: prevalence, health-related quality of life, and economic correlates. *Obstet Gynecol* 87(3):321–327, 1996.
108. Paras ML: Sexual abuse and lifetime diagnosis of somatic disorders: a systematic review and meta-analysis. *JAMA* 302(5):550–561, 2009.
109. Hilden M, Schei B, Swahnberg K, et al: A history of sexual abuse and health: a Nordic multicentre study. *BJOG* 111(10): 1121–1127, October 2004.
110. Bohrer JC: Pelvic nerve injury following gynecologic surgery: a prospective cohort study. *Am J Obstet Gynecol* 201(531): E1–E7, 2009. Available at www.AJOG.org. Accessed April 19, 2010.

CHAPTER

16

Screening the Lower Quadrant: Buttock, Hip, Groin, Thigh, and Leg

The causes of lower quadrant pain or dysfunction vary widely; presentation of symptoms is equally wide ranging. Vascular conditions (e.g., arterial insufficiency, abdominal aneurysm), infectious or inflammatory conditions, gastrointestinal (GI) disease, and gynecologic and male reproductive systems may cause symptoms in the lower quadrant and lower extremity,[1] including the pelvis, buttock, hip, groin, thigh, and knee. Some overlap may occur, but unique differences exist.

Cancer may present as primary hip, groin, or leg pain or symptoms. Primary cancer can metastasize to the low back, pelvis, and sacrum, thus referring pain to the hip and groin. Primary cancer may also metastasize to the hip, causing hip or groin pain and symptoms.

Pain may be referred from other locations such as the scrotum, kidneys, abdominal wall, abdomen, peritoneum, or retroperitoneal region. Lower quadrant pain may be referred through conditions that affect nearby anatomic structures, such as the spine, spinal nerve roots, or peripheral nerves, and overlying soft tissue structures (e.g., hernia, bursitis, fasciitis).[1a]

One of the keys to accurate and quick screening is knowledge of the types of conditions, illnesses, and systemic disorders that can refer pain to the lower quadrant, especially the hip and groin. Much of the information related to screening of the back (see Chapter 14), sacrum, sacroiliac (SI), and pelvis (see Chapter 15) also applies to the hip and groin.

USING THE SCREENING MODEL TO EVALUATE THE LOWER QUADRANT

When screening is called for, the therapist looks at the client's personal and family history, clinical presentation, and associated signs and symptoms. Knowledge of problems that can affect the lower quadrant, along with the likely history, pain patterns, and associated signs and symptoms, shows us the steps to follow in screening.

Most often, the screening process takes place through a series of special questions. A few special tests may be used as well. Recognition of red flag signs and symptoms of systemic or viscerogenic problems can direct the client toward the necessary medical attention early in the disease process. In many cases, early detection and treatment may result in improved outcomes.

Past Medical History

Some of the more common histories associated with lower extremity, hip, or groin pain of a visceral nature are listed in Box 16-1. A previous history of cancer, such as prostate cancer (men), any reproductive cancers (women), or breast cancer, is a red flag as these cancers may be associated with metastases to the hip.

Past history of joint replacement (especially hip arthroplasty) combined with recent infection of any kind and new onset of hip, groin, or knee pain is suspicious. Postoperatively, orthopedic pins may migrate, referring pain from the hip to the back, tibia, or ankle. Loose components, improper implant size, muscular imbalance, and infection that occur any time after joint arthroplasty may cause lower quadrant pain or symptoms (Case Example 16-1).

There have been reports of hip, groin, and/or pelvic pain and/or mass associated with wear debris from hip arthroplasty. Polyethylene wear debris can also cause deep vein thrombosis, lower extremity edema, ureteral or bladder compression, or sciatic neuropathy.[2]

Risk Factors

Each condition, illness, or disease that can cause referred pain to the buttock, hip, thigh, groin, or lower extremity has its own unique risk factors. Many of the items listed as past medical history are risk factors. For example, femoral artery catheterization used to monitor ongoing hemodynamic status (arterial line; status post burn injuries, and/or individuals in the intensive care unit [ICU]) or used for individuals with poor upper extremity intravenous access can cause retroperitoneal hematoma formation or septic arthritis and subsequent hip pain.

Most known risk factors for systemically induced problems have been discussed in the individual chapters on each

BOX 16-1 RED FLAG HISTORIES ASSOCIATED WITH THE LOWER EXTREMITY

- Previous history of cancer
- Previous history of renal or urologic disease such as kidney stones and urinary tract infections (UTIs)
- Trauma/assault (fall, blow, lifting)
- Femoral artery catheterization
- History of infectious or inflammatory condition
  - Crohn's disease (regional enteritis) or ulcerative colitis
  - Diverticulitis
  - Pelvic inflammatory disease (PID)
  - Reiter's syndrome
  - Appendicitis
- History of gynecologic condition(s):
  - Recent pregnancy, childbirth, or abortion
  - Multiple births (multiparity)
  - Other gynecologic conditions
- History of alcoholism (e.g., hip osteonecrosis)
- Long-term use of immunosuppressants (e.g., Crohn's disease, sarcoidosis, cancer treatment, organ transplant, autoimmune disorders)
- History of heart disease (e.g., arterial insufficiency, peripheral vascular disease)
- Receiving anticoagulation therapy (risk factor for hemarthrosis)
- History of acquired immunodeficiency syndrome (AIDS)-related tuberculosis
- History of hematologic disease such as sickle cell anemia or hemophilia

specific condition. For example, arterial insufficiency as a cause of low back, hip, buttock, or leg pain is presented as part of the discussion of peripheral vascular disease in Chapter 6 and again in Chapter 14 because it relates just to low back pain. Likewise, known risk factors for bone cancer or metastases as a cause of hip, groin, or lower extremity pain are presented in Chapter 13.

Many conditions with overlap symptoms (e.g., back and hip pain, pelvic and groin pain) are presented throughout this third text section (Systemic Origins of Neuromusculoskeletal Pain and Dysfunction) as part of the discussion of back pain (see Chapter 14) or pelvic pain (see Chapter 15).

Awareness of risk factors for various problems can help alert the therapist early to the need for medical intervention, as well as for direct education and prevention efforts. Many risk factors for disease are modifiable. Exercise often plays a key role in prevention and treatment of pathologic conditions. Recognizing red flags in the history and clinical presentation and knowing when to refer versus when to treat are topics of focus in this chapter.

TABLE 16-1 Causes of Buttock Pain

Systemic/Medical Conditions	Neuromusculoskeletal
Sciatica from tumor, infection, endometriosis (see Table 16-6)	Sciatica (nerve compression from surrounding soft tissues; piriformis syndrome; see Table 16-6)
Sacroiliitis (pyogenic infection, Reiter's syndrome, spondyloarthritis)	Posterior facet joint dysfunction
Neoplasm (primary or regional metastases via lymph nodes)	Hip joint disease
Osteoid osteoma of the upper femur	Disk disease (thoracic or lumbar)
Osteomyelitis of the upper femur	SI joint dysfunction
Fracture (sacrum, ilium, pubic ramus)	Bursitis (psoas, gluteal, ischial, trochanteric)
Septic arthritis (hip, SI)	TrPs
Abscess from aseptic necrosis, Crohn's disease, or other retroperitoneal infection; anorectal abscess	Hamstring strain
Sacral neuralgia from genital herpes	Myofascial syndromes
Ischemia (e.g., claudication from PVD, peripheral arterial aneurysm)	Pelvic upslip
	Spondylolisthesis
	Direct trauma

SI, Sacroiliac; *PVD*, peripheral vascular disease; *TrPs*, trigger points.

Clinical Presentation

If no neuromuscular or musculoskeletal cause of the client's symptoms can be identified, then the therapist must consider the following:

FOLLOW-UP QUESTIONS

- Are red flags suggestive of a viscerogenic cause of pain or symptoms? (See Box 14-1; the lack of diagnostic testing or imaging studies may be an additional red flag[3].)
- What kind of pain patterns do we expect to see with each of the viscerogenic causes?
- Are any associated signs and symptoms suggestive of a particular organ system?

Hip and Buttock

The physical therapist is well acquainted with hip or buttock pain (Table 16-1) as a result of regional neuromuscular or musculoskeletal disorders. The therapist must be aware that disorders affecting the organs within the pelvic and abdominal cavities can also refer pain to the hip region, mimicking a primary musculoskeletal lesion. A careful history and physical examination usually differentiate these entities from true hip disease.[4]

Pain Pattern. True hip pain, whether from a neuromusculoskeletal or systemic cause (Table 16-2), is usually felt posteriorly deep within the buttock or anteriorly in the groin, sometimes with radiating pain down the anterior thigh. Pain

CASE EXAMPLE 16-1

Screening After Total Hip Replacement

A 74-year-old retired homemaker had a total hip replacement (THR) 2 days ago. She remains an inpatient with complications related to congestive heart failure. She has a previous medical history of gallbladder removal 20 years ago, total hysterectomy 30 years ago, and surgically induced menopause with subsequent onset of hypertension.

Her medications include intravenous furosemide (Lasix), digoxin, and potassium replacement.

During the initial physical therapy intervention, the client reported muscle cramping and headache but was able to complete the entire exercise protocol. Blood pressure was 100/76 mm Hg (measured in the right arm while lying in bed). Systolic measurement dropped to 90 mm Hg when the client moved from supine to standing. Pulse rate was 56 bpm with a pattern of irregular beats. Pulse rate did not change with postural change. Platelet count was 98,000 cells/mm^3 when it was measured yesterday.

How would you screen a client with this history and current comorbidities?

Neuromusculoskeletal	Systemic
Assess orthopedic complications such as signs of infection, increased skin temperature, localized swelling, pain.	Monitor all vital signs.
Observe patient's adherence to hip precautions; note surgical technique and approach used, type of implant, and location of incision.	Monitor platelet levels, international normalized ratio: If low, observe for bruising, joint bleeds, deep venous thrombosis; follow precautions and exercise guidelines.[120]
Be aware that orthostatic hypotension can cause dizziness, loss of balance, falls—a very dangerous situation with a recent THR. This can be compounded by osteoporosis, if present as a result of surgical menopause.	Watch for signs and symptoms of cardiovascular/pulmonary impairments such as: • Fatigue and muscle weakness • Tachycardia • Fluid migration from the legs to the lungs during the supine position • Dyspnea, orthopnea,* spasmodic cough (check sputum) • Peripheral edema; check jugular distention (see Fig. 4-44) • Check nail beds for signs of decreased perfusion
	Observe for side effects of medications or drug interactions: • Diuresis from Lasix (loop diuretic) can result in potassium depletion and lead to increased sensitivity of myocardium to digoxin (digitalis); monitor serum electrolytes, and observe for signs/symptoms of potassium imbalance; observe for urinary frequency and headache. • Common adverse effects of Lasix include dehydration, muscle cramping, fatigue, weakness, headache, paresthesias, nausea, confusion, orthostatic hypotension, blurred vision, rash • Digoxin: Headache, drowsiness, other central nervous system disturbance, bradycardia, arrhythmia, gastrointestinal upset, blurred vision, halos

*Ask if the patient must use pillows and sit up or have the head of the bed elevated; often described as "1-pillow orthopnea" or "2-pillow orthopnea."

What signs and symptoms should be reported to the medical staff?

Nurses will be closely monitoring the patient's signs and symptoms. Read the medical record to stay up with what everyone else knows or has observed about the patient. Read the physician's notes to see whether medical intervention has been ordered.

Report anything observed but not already recorded in the chart such as muscle cramping, headache, irregular heartbeat with bradycardia, low pulse, and orthostatic hypotension.

Bradycardia is one of the first signs of digitalis toxicity. In some hospitals, a pulse less than 60 bpm in an adult would indicate that the next dose of digoxin should be withheld and the physician contacted. The protocol may be different from institution to institution.

The therapist is advised to report the following:

- Irregular heartbeat with bradycardia (a possible sign of digoxin/digitalis toxicity)
- Muscle cramping (possible adverse effect of Lasix) and headache (possible adverse effect of digoxin)
- Charting of vital signs; her blood pressure was not too unusual and pulse rate did not change with position change (probably because of medications), so she does not have medically defined orthostatic hypotension.
- Monitor vital signs throughout intervention; record the time it takes for vital signs to return to normal after exercise or treatment for your own documentation of measurable outcomes.

TABLE 16-2 Causes of Hip Pain

Systemic/Medical Conditions	Neuromusculoskeletal†
Cancer • Metastasis • Bone tumors* Osteoid osteoma Chondrosarcoma Giant cell tumor Ewing's sarcoma Vascular • Arterial insufficiency • Abdominal aortic aneurysm • Avascular necrosis Urogenital • Kidney (renal) impairment; kidney stones • Urinary tract infection • Testicular cancer Infectious/inflammatory conditions • Abdominal or peritoneal inflammation (psoas abscess; see Box 16-3) • Ankylosing spondylitis • Appendicitis • Ischial rectal abscess* • Crohn's disease; ulcerative colitis • Diverticulitis • Osteomyelitis (upper femur)* • PID • Reiter's syndrome • Inflammatory arthritis (RA, SLE, seronegative arthropathies, gout) • Septic hip or SI arthritis* • Septic hip bursitis* • Tuberculosis Metabolic disease • Osteomalacia, osteoporosis • Gaucher's disease • Paget's disease • Ochronosis • Hemochromatosis • Diabetes mellitus (associated with neuropathy) Other • Sickle cell crisis • Hemophilia • Ectopic pregnancy • Femoral artery catheterization	Lumbar spine (especially spondylosis, stenosis, disk), SI joint, sacral, or knee pathology Osteoarthritis Synovitis Femoral, inguinal, or sports hernia/athletic pubalgia FAI Bursitis (trochanteric, iliopectineal, iliopsoas, ischial) Fasciitis, myofascial pain Muscle impairment (weakness, loss of flexibility, hypertonus, hypotonus sprain/strain/tear/avulsion); snapping hip syndrome Tendinopathy (tendinitis, tendinosis) Piriformis syndrome Stress reactions/fractures Occult fracture of the femoral neck Peripheral nerve injury or entrapment; meralgia paresthetica Total hip arthroplasty • Infection • Implant loosening • Intraoperative blood vessel injury • Bone loss; subsidence Acetabular labral or cartilage lesions Developmental hip dysplasia; hip dislocation Legg-Calvé-Perthes disease SCFE Osteitis pubis (pubic pain radiates to anterior hip)

PID, Pelvic inflammatory disease; *RA,* rheumatoid arthritis; *SLE,* systemic lupus erythematosus; *SI,* sacroiliac; *FAI,* femoroacetabular impingement; *SCFE,* slipped capital femoral epiphysis.
*Most common causes of the "Sign of the Buttock."
†This is not an exhaustive, all-inclusive list, but rather, it includes the most commonly encountered adult neuromuscular or musculoskeletal causes of hip pain.

BOX 16-2 SIGN OF THE BUTTOCK

James Cyriax, M.D., was the first to write about the "Sign of the Buttock," which is actually made up of seven signs that indicate serious disease posterior to the axis of flexion and extension of the hip. These signs of neural tension deficit suggest severe central nervous system compromise, requiring medical referral. When positive, this test may help the therapist to identify serious extracapsular hip or pelvic disease.

- Primary sign of the buttock: Passive hip flexion more limited and more painful than the straight leg raise
- Limited (and painful) straight leg raise
- Trunk flexion limited to the same extent as hip flexion
- Painful weakness of hip extension
- Noncapsular pattern of restriction (hip); the capsular pattern is marked limitation of hip medial rotation first, then hip flexion with some limitation of abduction and little or no limitation of adduction and lateral rotation.
- Swelling (and tenderness) in the buttocks region
- Empty end feel with hip flexion

Data from Cyriax J: *Textbook of orthopaedic medicine. Diagnosis of soft tissue lesions,* ed 8, Philadelphia, 1983, WB Saunders.

perceived on the outer (lateral) side or posterior aspect of the hip is usually not caused by an intraarticular problem but more likely results from a trigger point, bursitis, knee, SI, or back problem.

With true hip joint disease, pain will occur with active or passive motion of the hip joint; this pain increases with weight bearing.[5] Often, an antalgic gait pattern is observed as the individual leans away from the affected hip and shortens the swing phase to avoid weight bearing.

When the underlying problem is related to soft tissue (e.g., abductor weakness) rather than to the joint as the source of symptoms, the client may lean toward the affected side to compensate for the downward rotation of the pelvis.[6] With soft tissue involvement of the bursa or tendons (e.g., gluteus medius, gluteus minimus) pain may radiate from the buttock, greater trochanter, and/or lateral thigh down the leg to the level of insertion of the iliotibial tract on the proximal tibia.[7-9]

Pain with medial rotation and decreased hip medial range of motion is associated with hip osteoarthritis.[10] Cyriax's "Sign of the Buttock" (Box 16-2) can help differentiate between hip and lumbar spine disease.[11-13] The presence of any of these signs may be an indication of osteomyelitis, neoplasm (upper femur, ilium), fracture (sacrum), abscess, or other infection.[12]

Neuromusculoskeletal Presentation. Identifying the hip as the source of a client's symptoms may be difficult because pain originating in the hip may not localize to the hip but rather may present as low back, buttock, groin, SI, anterior thigh, or even knee or ankle pain (Fig. 16-1).

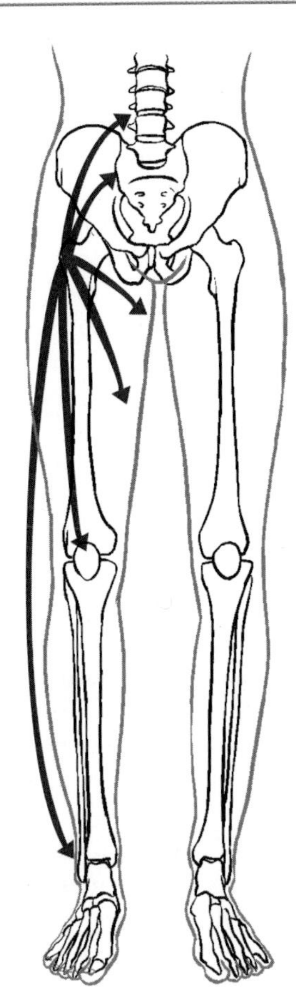

Fig. 16-1 Pain referred *from* the hip *to* other structures and anatomic locations. Pain from a pathologic condition of the hip can be referred to the low back, sacroiliac or sacral area, groin, anterior thigh, knee, or ankle.

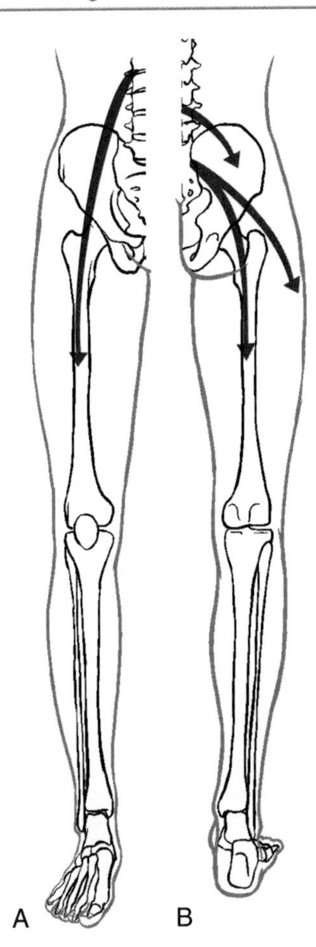

Fig. 16-2 Pain referred *to* the hip *from* other structures and anatomic locations. **A,** Hip pain referred from the upper lumbar vertebrae can radiate into the anterior aspect of the thigh. **B,** Hip pain from the lower lumbar vertebrae and sacrum is usually felt in the gluteal region, with radiation down the back or outer aspect of the thigh.

On the other hand, regional pain from the low back, SI, sacrum, or knee can be referred to the hip. SI pain that localizes to the base of the spine may be accompanied by radicular pain extending across the buttock and down the leg. It can also cross the lateral hip area. Additionally, SI joint dysfunction can cause groin pain and, with referred pain to the hip, may be accompanied by an ipsilateral decrease in hip joint internal rotation of 15 degrees or more, thereby confusing the clinical picture even further.[14,15]

Overlying soft tissue structure disorders such as femoral hernia, bursitis, or fasciitis; muscle impairments such as weakness, loss of flexibility, hypertonus or hypotonus, strain, sprain, or tears; and peripheral nerve injury or entrapment, including meralgia paresthetica, can also cause localized hip (and/or groin) pain.

Hip pain referred from the upper lumbar vertebrae can radiate into the anterior aspect of the thigh, whereas hip pain from the lower lumbar vertebrae and sacrum is usually felt in the gluteal region, with radiation down the back or outer aspect of the thigh (Fig. 16-2).

The client with pain caused by component instability following total hip arthroplasty may report hip or groin pain with activity, pain at rest, or both. Clinically, a history of "start up" pain may indicate a loose component. After 5 or 10 steps, the groin pain subsides. Pain may increase again after a moderate amount of walking. Groin or thigh pain is most common with micromotion at the bone–prosthesis interface or other loose component, periosteal irritation, or an undersized femoral stem.[16-18]

The client reports a dull aching pain in the thigh with no history of systemic illness or recent trauma. Often, the pain is localized to the site of the prosthetic stem tip. The client points to a specific spot along the anterolateral thigh. Pain on initiation of activity that resolves with continued activity should raise suspicion of a loose prosthesis. Persistent pain that is not relieved with rest and continues through the night suggests infection, requiring medical referral.[16,19]

Systemic Presentation. A *noncapsular* pattern of restricted hip motion (e.g., limited hip extension, adduction, lateral rotation) may be a sign of pathology other than a joint problem associated with osteoarthritis, potentially a serious underlying disease (Case Example 16-2). The pattern of movement restriction most common with a *capsular* pattern for the hip is limitation of hip medial rotation, flexion,

CASE EXAMPLE 16-2

Noncapsular Hip Pattern

A 46-year-old male long-distance runner developed sudden onset of right hip pain. He was given a diagnosis of trochanteric bursitis (now called *greater trochanteric pain syndrome* [GTPS]) by an orthopedic physician and was referred to physical therapy.

Objective Findings

– For tenderness on palpation over the greater trochanter
– Trigger points (TrPs) of the hip and low back region
+ Noncapsular pattern of restriction of the hip (capsular pattern in the hip is flexion, abduction, and medial rotation); client was limited in extension and lateral rotation
+ Heel strike test

The major criteria for a medical diagnosis of trochanteric bursitis (GTPS) consist of marked tenderness to deep palpation of the greater trochanter and relief of pain after peritrochanteric injection with a local anesthetic and corticosteroid.

The absence of greater trochanter tenderness and the presence of a noncapsular pattern of restriction of the hip were not consistent with the given diagnosis. Local injection was not administered. If an injection had been given, trochanteric bursitis/GTPS may have been eliminated from the list of possible diagnoses.

Objective findings are not consistent with trochanteric bursitis/GTPS. What do you do now?

More tests, of course, and more questions! Is there any history of cancer or prostate problems? Take his vital signs. Can he squat? Clear the hip. Conduct a Review of Systems to look for a pattern in the past medical history, clinical presentation, and any associated signs and symptoms.

Look for a pattern of symptoms that suggests a particular visceral system. Hip pain can be caused by gastrointestinal (GI), vascular, infectious, or cancerous causes. Ask a few screening questions directed at each of these systems. For example:

GI: Are you having any nausea? Vomiting? Abdominal pain? Changes in bowel function? Blood in the stool? Test for psoas abscess.

Vascular: Any throbbing pain? Presence of varicose veins? Trophic changes? History of heart disease?

Infectious: Any history of inflammatory bowel conditions such as Crohn's disease, ulcerative colitis, or diverticulitis? Ever have appendicitis? Any recent skin rashes in the legs?

Cancerous: Previous history of cancer? Bone pain at night? Night sweats? Palpate the lymph nodes in the inguinal and popliteal regions.

Result: Red flags included:

- Age
- Past history of prostate cancer at age 44
- Positive heel strike test
- Noncapsular hip pattern
- Inconsistent symptoms with diagnosis

The results of the physical therapy examination warranted further medical evaluation, and the client was returned to the physician with a recommendation for imaging studies. Magnetic resonance imaging (MRI) results indicated a nondisplaced, complete fracture of the femoral neck from prostate cancer that had metastasized to the bone.

Data from Jones DL, Erhard RE: Differential diagnosis with serious pathology: A case report, *Phys Ther* 76:S89-S90, 1996.

abduction, and, sometimes, slight limitation of hip extension. Empty end feel can be an indicator of potentially serious disease such as infection or neoplasm. Empty end feel is described as limiting pain before the end range of motion is reached but with no resistance perceived by the examiner.[12]

Whenever assessing hip joint pain for a systemic or viscerogenic cause, the therapist should look at hip rotation in the neutral position and perform the log-rolling test. With the client in the supine position, the examiner supports the client's heels in the examiner's hands and passively rolls the feet in and out. Decreased range of motion (usually accompanied by pain) is positive for an intraarticular source of symptoms. If normal hip rotation is present in this position but the motion reproduces hip pain, then an extraarticular cause should be considered.

Log-rolling of the hip back and forth, though not sensitive, is generally considered to be the most specific examination maneuver for intraarticular hip pathology because it rotates the femoral head back and forth in relation to the acetabulum and capsule, not stressing any of the surrounding extraarticular structures.[20] The test does not identify the specific disease present but identifies the source of the symptoms as intraarticular.

Keep in mind that if normal rotations are present but painful, the problem may still be musculoskeletal in origin (e.g., SI, early sign of arthritic changes in the hip joint). Full motion is also possible in the early stages of avascular necrosis and sickle cell anemia. The log-rolling test should be combined with Patrick's or Faber's (flexion, abduction, and external rotation) test, long-axis distraction, compressive hip loading, and the scour (quadrant) test to determine whether the hip is a possible source of symptoms.

The presence of GI symptoms (e.g., nausea, vomiting, diarrhea, constipation, abdominal bloating or cramping) or urologic symptoms (e.g., urinary frequency, nocturia, dysuria, or flank pain) along with hip pain is cause to take a closer look. Palpable reproduction of painful symptoms is generally considered extraarticular.[21]

Negative radiographs of the hip may not rule out bone lesions. When intervention by the physical therapist does not yield relief of symptoms (or only temporary relief), further imaging studies may be needed. A careful review of risk factors and clinical presentation will guide this decision.[22]

Groin

The physical therapist may see a client with an isolated groin problem, especially in the sports or military populations (Case Example 16-3), but more often, the individual has low back, pelvic, hip, knee, or SI problems with a secondary complaint of groin pain. Possible systemic and/or visceral causes of groin pain are wide ranging, whether appearing as an isolated symptom or in combination with pelvic, hip, low back, or thigh pain (Table 16-3 and Case Example 16-4).

Palpating the groin area is usually necessary in making a differential diagnosis. This can be a sensitive issue, and the therapist is advised to have a third person in the examination area. This person should be the same gender as the client.

CASE EXAMPLE 16-3

Groin Pain in a 13-Year-Old Skateboarder

Referral: A 13-year-old boy presented with a 2-week history of left groin pain. He reported a skateboarding accident as the cause of the symptoms. He was coming down a flight of stairs, hit the last step by mistake, and caught his foot on the stair railing. His leg was forced into wide abduction and external rotation. No (heard or felt) pop or snap was perceived at the time of injury.

The client continued skateboarding but experienced increasing pain 2 hours later. At that time, he could "hardly walk" and has had trouble walking without limping ever since. He tried getting back to skateboarding but was stopped by sharp pain in the groin. No other symptoms were reported (no saddle anesthesia, no numbness and tingling, no bladder changes, no constitutional symptoms).

Clinical Presentation: An antalgic gait was observed as the boy avoided putting full weight through the hip during the stance phase. Trendelenburg gait or Trendelenburg test was not positive. He could not do a squat test because of pain. He could not put enough weight on the left leg to try heel walking or toe walking.

Generalized pain occurred along the inner thigh and was described as "tenderness." The child cannot internally rotate the hip past midline. Abduction was limited to 30 degrees with painful empty end feel. During active hip flexion, the hip automatically flexes, abducts, and externally rotates. Pain increases with active assisted or passive hip flexion when one is trying to keep the hip in neutral alignment.

Associated Signs and Symptoms: When asked about symptoms of any kind anywhere else in his body, the boy replied, "No." When offered a list of possible symptoms, these were all negative. He did admit to being slightly constipated because of the pain. Vital signs were all within normal limits.

Is referral indicated in the absence of any signs or symptoms of viscerogenic or systemic disease?

Some red flags are identified here, even though they do not point to a viscerogenic or systemic origin. Trauma, young age, and failure to complete a squat screening test for orthopedic clearance of the hip, knee, and ankle all suggest the need for medical referral before physical therapy intervention is initiated.

Turn to Table 16-3. As you look at the left column of Systemic Causes, what clinical presentation and signs and symptoms might be expected with each of these conditions? Does the current clinical presentation fit any of these?

Now look at the musculoskeletal causes of groin pain (right column, Table 16-3). Are past medical history, risk factors, or clinical presentation consistent for any of these problems? For example, pain in the hip or groin area in anyone who is not skeletally mature raises the suspicion of an orthopedic injury. Abduction and external rotation forces on the hip can produce a slipped capital femoral epiphysis (SCFE).

This is the case here, which required imaging studies for diagnosis. Anteroposterior x-rays were negative, but a lateral view showed slippage to confirm SCFE.

Data from Learch T, Resnick D: Groin pain in a 13-year-old skateboarder, *J Musculoskel Med* 20:513-515, 2003.

TABLE 16-3 Causes of Groin Pain

Systemic/Medical Conditions	Neuromusculoskeletal
Cancer • Spinal cord tumors • Osteoid osteoma • Hodgkin's disease/ lymphoma • Leukemia • Testicular • Prostate • Soft tissue masses Osteoporosis Fluid in peritoneal cavity • Ascites (cirrhosis) • Congestive heart failure • Cancer • Hyperaldosteronism Hemophilia • GI bleeding Abdominal aortic aneurysm, peripheral arterial aneurysm Gynecologic conditions • Cancer (uterine/ovarian masses) • Uterine fibroids • Ovarian cyst • Endometriosis (causing pubalgia) • Ectopic pregnancy (not common) • Sexually transmitted infection • PID Infection, usually intraabdominal or intraperitoneal infection (see Box 16-3) Urologic • Prostate impairment (prostatitis, BPH, prostate cancer) • Epididymitis; testicular torsion • Urethritis/urinary tract infection • Upper urinary tract problems affecting the kidneys or ureters (inflammation, infection, obstruction) • Hydrocele/varicocele GI • Diverticulitis • Inflammatory bowel disease Seronegative spondyloarthropathy	Musculotendinous strain (adductors, hamstrings, iliopsoas, abdominals, tensor fascia lata, gluteus medius)[35] Internal oblique avulsion Nerve compression or entrapment (ilioinguinal, obturator, lateral femoral cutaneous, sciatic nerves) Stress reaction, stress fracture, avulsion fracture, or complete bone fracture [femoral neck, pubic ramus] Bursitis (iliopectineal) Pubalgia* Osteitis pubis Apophysitis (young athletes) Trauma (physical, sexual, birth) Sports, inguinal or femoral hernia Hip joint impairment • Subluxation, dislocation, dysplasia • Avascular necrosis (osteonecrosis) • Total hip arthroplasty (loosening, infection, bone loss, subsidence) • SCFE • Legg-Calvé-Perthes disease • Labral tear with or without femoroacetabular impingement • Arthritis, arthrosis SI joint impairment Lumbar spine impairment (spinal stenosis, disk disease) TrPs Thoracic disk disease (lower thoracic spine)

GI, Gastrointestinal; *PID,* pelvic inflammatory disease; *BPH,* benign prostatic hyperplasia; *SCFE,* slipped capital femoral epiphysis; *SI,* sacroiliac; *TrPs,* trigger points.

*Pubalgia is really a description of painful symptoms of the groin that can be caused by a wide range of muscular, tendinous, osseous, and even visceral structures. This condition may be labeled osteitis pubis when there is articular involvement such as arthritis, articular instability, or other articular lesions involving the pubic symphysis.[35]

CASE EXAMPLE 16-4

Soft Tissue Sarcoma

A 38-year-old female patient was referred to physical therapy by a primary care clinic physician assistant with a diagnosis of "groin strain." The client denied any injury or trauma. Little to no pain was reported, but a feeling of "fullness" in the left proximal thigh was described. She was unable to cross her legs when sitting because of this fullness. No other constitutional symptoms or associated symptoms were noted.

When asked, "How long have you had this?" the client thought it had been present for the past 3 months. When asked, "Has it changed since you first noticed it?" she stated that she thought it was getting larger.

Examination: There was an obvious area of edema or tissue mass identified in the proximal medial left thigh. No tenderness, bruising, erythema, or skin temperature changes were reported. The area in question had a boggy feel on palpation. Lower extremity range of motion and manual muscle testing were within normal limits.

Screening and Differential Diagnosis: Look at Table 16-3. As you review the possible systemic and musculoskeletal causes of groin pain, what additional questions and tests or measures must be asked/carried out to complete your screening examination?

On the Systemic Side

- Spinal cord tumors—No temperature changes, dermatomal changes, or associated bowel and bladder changes; no further testing required at this time
- Hodgkin's disease/lymphoma/leukemia—Ask about previous history of cancer, family history of cancer; palpate lymph nodes (quick screen of lymph nodes above and below the groin and careful examination of inguinal lymph nodes)
- Urinary tract involvement—No history of recent fever, chills, difficulty urinating, or urinary tract infection; no blood in the urine; no further questions at this time
- Ascites—No apparent abdominal ascites, no history of alcoholism; check for asterixis, liver palms (palmar erythema); ask about symptoms of carpal tunnel syndrome, look for spider angiomas during inspection, and observe nail beds for any changes (nails of Terry)
- Hemophilia—It is a long shot, but ask about personal/family history
- Abdominal aortic aneurysm (AAA)—Ask about bounding pulse sensation in the abdomen; palpate aortic pulse width (see Fig. 4-55); ask about the presence of chest or back pain at any time, especially with exertion
- Gynecologic—Ask about a history of pelvic pain, pelvic inflammatory disease, or sexually transmitted infection
- Appendicitis—Perform McBurney's test, Blumberg's sign, and iliopsoas and obturator tests (see Chapter 8 for descriptions)

On the Musculoskeletal Side

- Muscle strain—As already tested, no loss of motion or strength; no pain with resisted movement; no history of trauma or overuse. Red flag: Clinical presentation is not consistent with the medical diagnosis.
- Internal oblique avulsion/stress reaction or fracture—As above
- Pubalgia—As above; no painful symptoms reported, no pain on palpation
- Sexual assault/domestic violence—Even though the client denies trauma, consider a screening interview for nonaccidental trauma (see Chapter 2 or Appendix B-3); absence of erythema, skin bruising, or other skin changes makes this type of trauma unlikely
- Total hip arthropathy—Negative history
- Avascular necrosis—Not likely, given the clinical presentation; ask about a history of long-term use of immunosuppressants (corticosteroids for Crohn's disease, sarcoidosis, autoimmune disorders)
- Trigger points (TrPs)—Atypical presentation for a trigger point; check for latent TrPs of the adductors, iliopsoas, vastus medialis, and sartorius

Special Questions to Ask: Take a final look at *Special Questions to Ask* in this chapter. Have you missed anything? Left anything out?

Result: On the basis of lack of objective findings and red flags of mass increasing in size and clinical presentation inconsistent with medical diagnosis, the therapist consulted with an orthopedic surgeon in the same health care facility. The orthopedic surgeon ordered x-rays, which were normal, and advised a short period of observation before ordering magnetic resonance imaging (MRI).

After 3 weeks, no changes were observed, and an MRI was ordered. The MRI showed a soft tissue tumor, later diagnosed on biopsy as a stage IIIB high-grade soft tissue sarcoma.

The client underwent multiple surgical procedures, including removal of the medial compartment musculature and limb salvage with an eventual hemiarthroplasty. Physical therapy included gait training, regaining safe hip active range of motion, an aquatic rehabilitation program, use of an underwater treadmill, and both open and closed kinetic chain strengthening.

Adapted from Baxter RE: Identification of neoplasm mimicking musculoskeletal pathology: A case report involving groin symptoms. Poster presented at Combined Sections Meeting, 2004, New Orleans, LA. Used with permission.

The therapist should explain the examination procedure and obtain the client's permission.

During examination of the groin, the physical therapist may palpate enlarged lymph nodes, or the client may indicate these nodes to the examiner. Painless, progressive enlargements of lymph nodes or lymph nodes that are aberrant or suspicious for any reason, especially if present in more than one area or in the presence of a past medical history of cancer, are an indication of the need for medical referral.

Changes in lymph nodes without a previous history of cancer continue to represent a yellow or red flag. Tender, movable inguinal lymph nodes may be a sign of food intolerance or allergies or an indication that the body is fighting off an infectious process. The therapist should use his or her best clinical judgment in deciding what to do but should always err on the side of caution. When doubt arises, one should contact the physician and communicate any concerns, observations, or questions.

CASE EXAMPLE 16-5

Groin Pain—Musculoskeletal Cause

A 44-year-old male patient came to physical therapy with a 7-year history of right groin pain. X-rays, bone scan, and arthrogram of the hip were negative. At the time of initial examination, the client was taking morphine for pain that was described as constant, severe, and sharp and that was rated 8 out of 10 on the Numeric Rating Scale (NRS; see Chapter 3). Sitting and driving made the symptoms worse, and he was unable to work as a mechanic because prolonged squatting was required. Lying supine relieved the pain.

Physical examination revealed extreme hip medial rotation associated with active hip flexion, abduction, and knee extension; each of these movements reproduced his symptoms. Passive range of motion of the right hip was painful and was limited to 95 degrees of flexion and 0 degrees of lateral rotation.

Visual inspection during movement and palpation of the greater trochanter indicated that the proximal femur had medially rotated and moved anteriorly during hip flexion. Through application of a posteroinferior glide over the proximal femur during hip flexion, groin pain was decreased and motion increased. The client was able to moderate his symptoms by avoiding hip medial rotation during hip and knee movements.

Consider: Are any red flags present? Is further screening indicated to rule out systemic origin of symptoms? If yes, what questions or tests might you consider carrying out?

Red Flags: Age (over 40); constant, intense pain

Further Screening Required: The length of time that symptoms have been present without accompanying signs and symptoms of a urologic or gastrointestinal (GI) nature (7 years) is not typical of systemic origin of musculoskeletal symptoms.

The fact that no aggravating and relieving factors are known further rules out a viscerogenic cause of pain. It would be appropriate to ask the *Special Questions for Men* at the end of Chapter 14 (see also Appendix B-24).

It is always a good idea to ask one final question: *Are any other symptoms of any kind anywhere else in your body?* Special tests might include the heel strike test (fracture), translational rotation tests for stress reaction (fracture), iliopsoas and obturator tests (abscess; see Chapter 8), and trigger point assessment.

Result: The client was treated for femoral anterior glide with medial rotation (movement impairment diagnosis).[23] Training to teach the client to modify hip medial rotation during sustained postures and functional activities was a key component of the intervention. Exercises were given to strengthen the right iliopsoas muscle, hip lateral rotator muscles, and posterior gluteus medius muscle.

The client was pain-free and off pain medications 2 months later after 6 treatment sessions. He was able to return to full-time work.

Comment: Knowledge of red-flag signs and symptoms, risk factors for various systemic conditions and illnesses, associated signs and symptoms of viscerogenic pain, and typical clinical presentations for neuromuscular and musculoskeletal problems can guide the therapist in quickly sizing up a situation and deciding whether or not further screening is warranted.

In this case, the therapist can see that only a few screening questions are in order. The application of any additional special tests depends on the client's answers to screening questions. The client's immediate response to intervention is another way to verify a correct physical therapy diagnosis. Failure to progress with intervention is a red flag that indicates the need for reevaluation.

Data from Bloom NJ, Sahrmann SA: Groin pain caused by movement system impairments: A case report. Poster presented at Combined Sections Meeting, 2004, New Orleans, LA.. Used with permission.

Neuromusculoskeletal Presentation. Neuromuscular or musculoskeletal causes of groin pain should also be considered (Case Example 16-5).[23,24] Keep in mind that intraarticular pathology of the hip can manifest as groin pain owing to the innervation of the hip capsule. Extraarticular hip conditions radiate to the lateral or posterior aspects of the hip.[25]

Groin pain is a common complaint in sports that involve kicking and rapid change of direction (e.g., soccer, hockey). The most common musculoskeletal cause of groin pain is strain of the adductor muscles, most often involving the adductor longus. The history includes a specific trauma, repetitive motion, or injury, which occurs primarily at the junction of the muscle fibers and the extended tendon of origin. Acutely, this injury causes unilateral or bilateral pain during or after activity, with local palpation of the adductor longus origin, and during passive stretching or active contraction; eccentric activation may be even more painful.[26,27] Acute injury may be followed in several days by ecchymosis.

Chronic groin or inguinal pain in the active athletic, sports, or military groups is often referred to as *athletic pubalgia.* Athletic pubalgia is sometimes used interchangeably to describe a *sports* or *athletic hernia,* which is a tear in the muscles of the inner thigh, lower abdomen, and/or the fascia.[28] The term *sports hernia* may be a bit misleading because experts in this area do not consider this condition the same as a true inguinal or femoral hernia.[29]

Symptoms associated with athletic pubalgia are often described as deep groin or lower abdominal pain with exertion (usually unilateral). There may be a localized sharp burning sensation in the lower abdomen and/or inguinal region. Symptoms are relieved with rest but aggravated by activity, especially sport-related activities. As the condition progresses, symptoms may radiate to the adductor region, testes (male), and labia (female).[30,31]

Labral tears of the acetabulum can also cause groin pain. There may be a history of trauma but acetabular labral tears can occur without trauma. The clinical presentation can vary and include night pain, activity-related pain, positive Trendelenburg sign, and positive impingement sign (pain reproduced with hip flexion, adduction, and internal rotation). In young, active individuals with a primary complaint of groin pain with or without a history of trauma, the diagnosis of a labral tear should be suspected and investigated further.[32]

Femoroacetabular impingement presents as groin pain in young adults. Onset is gradual and progressive with intermittent groin pain after prolonged walking, prolonged sitting, or athletic activities that stress the hip. The impingement test (internal hip rotation and adduction while the hip is flexed) is always positive. Referral for a medical orthopedic examination and imaging studies may be warranted.[33]

Another common problem in the young athlete or long distance runner is osteitis pubis. Repetitive stress of the adductor group can cause inflammation at the musculotendinous attachment on the pubic bone, contributing to sclerosis and bony changes.[34]

Osteitis pubis with inflammation and sclerosis of the pubic symphysis can cause both acute and chronic groin pain. Individuals affected most often include competitive sports athletes involved in running, leaping and landing with force, repetitive kicking motions, or training on concrete, uneven, or other hard surfaces. Osteitis pubis can also occur as a result of leg length differences, faulty foot and body mechanics, or muscular imbalances and during pregnancy. Tenderness on palpation of the pubic symphysis helps identify this condition.[26] Onset of midline pain that radiates to the groin is typical. Pain is reproduced by palpation of the pubis (anterior), passive hip abduction, and resisted hip adduction. Articular lesions involving the pubis symphysis can also lead to pubalgia.[35]

Insertional injuries of the upper attachment of the rectus abdominis muscle over the anteroinferior pubis (just lateral to the pubic symphysis) can lead to tendinopathy presenting as pubalgia. Without magnetic resonance imaging (MRI), insertional abdominis pathology cannot be differentiated from adductor pathology as the abdominis pubic attachment and the thigh adductor tendon blend to form one unit.[35]

Chronic, unresolved groin pain in the athletic population also has been linked with altered neuromotor control.[36] The therapist may need to evaluate groin pain from a motor control point of view. See further discussion of stress reaction/fractures in the section on Trauma as a Cause of Hip, Groin, or Lower Quadrant Pain in this chapter.

Older adults are more likely to experience hip, buttock, or groin pain associated with arthritis, lumbar stenosis, insufficiency fractures, or hip arthroplasty. Arthritis is characterized by radiating pain to the knee, but not below, with decreased hip range of motion. Gait disturbances may be seen as arthritis progresses.[17] Insufficiency fracture of the pubic rami can also cause hip/groin pain, resulting in a reluctance to bear weight on the affected side along with an antalgic gait.[37]

Hip and groin pain secondary to lumbar stenosis can manifest as low back pain that radiates to the lower extremities. The pain begins and gets worse with ambulation. Standing and walking may also increase symptoms when the lumbar spine assumes a more lordotic position and the ligamentum flavus folds in on itself, pinching the foramina closed. The client who has stenosis bends forward or sits to avoid painful symptoms. Clients who have a total hip arthroplasty for hip pain may have continued groin and buttock pain, secondary to sciatica or lumbar spinal stenosis.[17]

Systemic Presentation. The clinical presentation of groin pain from a systemic source does not vary from musculoskeletally induced groin pain. Once again, the key is to look at the client's age (e.g., atherosclerotically induced vascular problems in the older adult), past medical history (e.g., previous history of cancer, liver disease, hemophilia), and gender (e.g., ectopic pregnancy, prostate or testicular problems).

In addition, asking about the presence of other symptoms and conducting a Review of Systems may help the therapist identify any one of the systemic causes listed in Table 16-3.

Thigh

Once again, we cannot emphasize enough the importance of conducting a thorough physical examination to rule out systemic or viscerogenic disease as the source of thigh pain; client history and lower quadrant screening examination should be performed (see Box 4-16).

Anterior thigh pain is more common (Table 16-4), but posterior thigh pain may occur, with ruptured abdominal aortic aneurysm. Local anterior or posterior thigh pain of systemic origin generally occurs as a deep aching generated by soft tissue irritation or bone involvement. Radicular pain is usually a sharp, stabbing pain that projects in dermatomal distributions caused by compression of the dorsal nerve roots.

Neuromusculoskeletal Presentation. The lower lumbar vertebrae and sacrum can refer pain to the gluteal and hip region, with pain radiating down the posterior or posterolateral thigh. Pain down the lateral aspect of the thigh to the knee may also be caused by inflammation of the tensor fascia lata with iliotibial band syndrome.[5] A similar pattern has been reported in association with irritability, injury, or disease of the thoracolumbar transitional segments,[38,39] and at least one case of synovial cell sarcoma presenting as iliotibial band syndrome has been reported.[40]

Anterior thigh pain is commonly disk related, resulting from L3-L4 disk herniation and occurring most often in older clients with a previous history of lumbar spine surgery. The clinical presentation varies among affected individuals, but thigh pain alone is most common (Case Example 16-6).

Use of the extreme lateral interbody fusion (XLIF) technique has been linked with thigh weakness and/or numbness postoperatively as a possible consequence of trauma to the psoas muscle or femoral nerve during the approach. Symptoms are temporary and appear to resolve with soft tissue healing following surgery.[41]

Back and thigh pain, a positive reverse straight leg raise (SLR) test, and depressed knee reflex are described more often in clients with disk herniation at the L3-L4 level than in clients with L4-L5 and L5-S1 levels.[42,43] A positive reverse SLR is defined as pain traveling down the ipsilateral leg when the person is prone and the leg is extended at the hip and the

TABLE 16-4 Causes of Thigh Pain

Systemic/Medical Conditions	Neuromusculoskeletal
Retroperitoneal or intraabdominal tumor or abscess (see Box 16-3) Kidney stones (nephrolithiasis, ureteral or renal colic) Peripheral neuropathy (bilateral, symmetric) • Diabetes mellitus • Neoplasm • Chronic alcohol use Thrombosis (femoral artery, great saphenous vein) Bone tumor (primary or metastases) Bone fracture associated with long-term bisphosphonates (rare; under investigation)*	Musculotendinous strains (e.g., adductor, abductor, quadriceps) Iliopectineal bursitis (anterior and medial thigh pain); trochanteric bursitis/greater trochanteric pain syndrome (lateral thigh) Peripheral neuropathy (unilateral, asymmetric) Contusions (collisions with balls, hockey pucks, the ground, other athletes) Nerve compression (e.g., meralgia paresthetica from compression of the LFCN) Myositis ossificans (injury with contusion and hematoma formation) Femoral shaft or subtrochanteric stress reaction or fracture; insufficiency fracture/stress reaction Hip disease (osteoarthritis, labral tear) Total hip arthroplasty (loose component, polyethylene wear debris, undersized/oversized femoral stem, periosteal irritation) SI joint dysfunction Upper lumbar spine dysfunction; spondylolisthesis, herniated disk, previous surgery TrPs Inguinal hernia

SI, Sacroiliac; *LFCN,* lateral femoral cutaneous nerve; *TrPs,* trigger points.

*Reports of thigh pain and weakness in affected thigh for weeks to months before a low-energy fracture occurs. See Update: Thighbone fractures in women taking bisphosphonate drugs, *Harvard Women's Health Watch* 17(7):6-7, 2010; and Abrahamsen B: Subtrochanteric and diaphyseal femur fractures in patients treated with alendronate: A register-based national cohort study, *J Bone Miner Res* 24(6):1095-1102, 2009.

knee. A positive test is caused by tension on the femoral nerve and its roots.[44]

Objective neurologic findings, such as hyperreflexia or hyporeflexia, decreased sensation to light touch or pinprick, and decreased motor strength, can occur with soft tissue problems such as bursitis. However, clients with true nerve root irritation experience pain extending into the lower leg and foot. Clients with bursitis exhibit a positive "jump" sign when pressure is applied over the greater trochanter; no jump sign is seen with nerve root irritation.[7]

A common neuromuscular cause of anterior or anterolateral thigh pain is lateral femoral cutaneous nerve (LFCN) neuralgia. Entrapment or compression of the LFCN causes pain or dysesthesia, or both, in the anterolateral thigh—a condition called *meralgia paresthetica.* Compression of the LFCN may occur at the level of the L2 and L3 roots through upper lumbar disk herniation or tumor in the second lumbar vertebra. LFCN neuropathy may occur after spine surgery to repair nerve damage that occurred during harvesting of the iliac bone graft or that resulted from pressure on the pelvis from prone positioning or with use of the Relton-Hall frame.[45]

Other causes of injury to the LFCN include positioning during hip arthroplasty (at risk: obese individuals)[46]; abnormal posture; chronic muscle spasm; tight-fitting braces, corsets, or pants; and thigh injury.[47] For clients with hip arthroplasty, implant loosening, fracture, or subsidence (sinking down into the bone) can cause thigh pain as the first symptom of instability.[19] Both passive and active range of motion should be evaluated to assess implant stability. X-rays are needed to look at component position, bone–prosthesis interface, and signs of fracture or infection.[16]

Systemic Presentation. The pain pattern for anterior thigh pain produced by systemic causes is often the same as that presented for pain resulting from neuromusculoskeletal causes. The therapist must rely on clues from the history and the presence of associated signs and symptoms to help guide the decision-making process.

For example, obstruction, infection, inflammation, or compression of the ureters may cause a pattern of low back and flank pain that radiates anteriorly to the ipsilateral lower abdomen and upper thigh. The client usually has a past history of similar problems or additional urologic symptoms such as pain with urination, urinary frequency, low-grade fever, sweats, or blood in the urine. Murphy's percussion test (see Fig. 4-54) may be positive when the kidney is involved.

The same pain pattern can occur with lower thoracic disk herniation. However, instead of urologic signs and symptoms, the therapist should look for a history of back pain and trauma and the presence of neurologic signs and symptoms accompanying diskogenic lesions.

Retroperitoneal or intraabdominal tumor or abscess may also cause anterior thigh pain. A past history of reproductive or abdominal cancer or the presence of any condition listed in Box 16-3 is a red flag.

Thigh pain has been reported as a prodromal symptom of unilateral low-energy subtrochanteric and femoral shaft (diaphyseal) stress reactions and fractures in a small number of people on long-term bisphosphonate therapy.[48]

BOX 16-3 CAUSES OF PSOAS ABSCESS

- Diverticulitis
- Crohn's disease
- Appendicitis
- Pelvic inflammatory disease (PID)
- Diabetes mellitus
- Any other source of infection, including dental[102]
  - Renal infection
  - Infective spondylitis (vertebra)
  - Osteomyelitis
  - Sacroiliac (SI) joint infection

CASE EXAMPLE 16-6

Buttock Pain Post Prostatectomy

A 62-year-old male patient was examined by a physical therapist for a chief complaint of severe left buttock and lateral thigh pain. No injury or trauma was reported; the client noticed low back pain 3 days ago. He lifted a couple of sand bags but did not think that was the cause of his pain. He has seen the chiropractor twice this week and felt that the electrical stimulation he had on one visit "usually does it" (helped relieve the pain). Pain relief was of a very short-term nature and had no lasting effects.

Past Medical History: Prostatectomy 4 years ago for cancer followed by 36 radiation treatments. The bowel was resected, and the patient received a stoma at that time.

Current Health Report: Prostate-specific antigen has increased from 0 to 0.4 in a stepwise fashion over the past year. The patient has not seen his oncologist for any follow-up "for quite some time." At this time, the client is not taking any medications except for over-the-counter pain relievers. Supplements include calcium and fish oil.

Clinical Presentation

Pain Pattern: Pain is reported as "constant," but it "has its highs and lows." The client prefers lying on his left (involved) side. He cannot sit for longer than 1 minute without onset of radicular symptoms.

Physical Examination: Visual inspection showed flattened lumbar spine. What appeared to be atrophy was seen in the right gluteal; this was confirmed with comparative palpation. Pelvic landmarks were slightly elevated (L higher than R). Lumbar range of motion was limited in all planes with remarkably minimal flexion, which the patient said was normal for him. No centralization of pain occurred with side glides or with repeated extension in standing.

Vascular Examination: No signs of peripheral vascular disease (PVD) were noted in the lower extremities. Blood pressure was not assessed.

Neurologic Screening Examination: Hyperreflexive patellar deep tendon reflexes (DTRs) on the right (L3); this was difficult to assess: He may have been notably hyporeflexive on the left. Achilles deep tendon reflexes (S1) appeared equal, with grading of 2/4 bilaterally. Clonus, Babinski's, and Oppenheim's were negative. Manual muscle testing (MMT) showed fatiguing weakness on the left at L2 (hip flexors), L3 (quadriceps), L5 (extensor hallucis longus and gluteus medius), and S1 (hamstring). No loss of light touch sensation was observed.

Associated Signs and Symptoms: No nausea or vomiting was reported. No recent significant weight loss or gain occurred. No changes in bowel or bladder function were described. The patient reported feeling chills of late, intermittently, which he says are caused by the bouts of severe pain. He showed no diaphoresis during the physical therapist's examination.

Red Flags

- Insidious onset of radicular pain in a 62-year-old with a previous history of cancer
- Constitutional symptom (chills)
- Constant, intense pain
- Notable proximal muscle weakness; multisegmental weakness on the left
- No improvement with chiropractic care or physical therapy

Result: The therapist applied some direct intervention for pain relief (positioning, Pain Reflex Release Technique (PRRT), trigger point release) with no immediate relief of painful symptoms. The therapist explained his concerns regarding the red-flag symptoms and advised the client to make an appointment with his oncologist for further evaluation. The client was instructed to call the therapist with the name and number for the oncologist, so his findings could be relayed to her.

The client left a message on the therapist's answering machine (received the next morning) that he was "going to the ER: I've got to do something about the pain."

The client followed up midday to state that he had gone to the emergency department. Diagnostic tests were ordered, and MRI revealed a herniated nucleus pulposus (HNP) of the L3/4 disk with effacement on the L3 nerve root. The L5/S1 disk was also reportedly herniated, although this did not affect the adjacent nerve root. The client is to see a neurosurgeon next week.

Knee and Lower Leg

Pain in the lower leg is most often caused by injury, inflammation, tumor (malignant or benign), altered peripheral circulation, deep venous thrombosis (DVT), or neurologic impairment (Table 16-5). Assessment of limb pain follows the series of pain-related questions presented in Fig. 3-6. The therapist can use the information in Boxes 4-13 and 4-16 to conduct a screening examination.

Neuromusculoskeletal Presentation. In addition to screening for medical problems, the therapist must remember to clear the joint above and below the area of symptoms or dysfunction. True knee pain or symptoms are often described as mechanical (local pain and tenderness with locking or giving way of the lower leg) or loading (poorly localized pain with weight bearing).

There are many musculoskeletal or neuromuscular conditions well known to the therapist as a potential cause of generalized knee pain, including muscle spasm, strain, or tear; patellofemoral pain syndrome; tendinitis; ligamentous disruption, meniscal tear, or osteochondral lesion; stress fracture[49]; and nerve entrapment.[50,51]

Degenerative joint disease of the hip[52] or other hip pathology can masquerade as knee pain in adults.[53] Neurologic problems, including spinal stenosis, complex regional pain syndrome (Type 1), neurogenic claudication, and lumbar radiculopathy are common disorders that can produce knee pain. Isolated knee pain involving SI dysfunction has also been reported.[54]

Pain and impaired function from a variety of intraarticular or extraarticular etiologies can also develop following a total knee arthroplasty.[55] Client history and clinical

TABLE 16-5 Symptoms and Differentiation of Leg Pain

	Vascular Claudication	Neurogenic Claudication	Peripheral Neuropathy	Restless Legs Syndrome
Description	Pain* is usually bilateral No burning or dysesthesia	Pain is usually bilateral but may be unilateral Burning and dysesthesia in the back, buttocks, and/or legs	Pain, aching, and numbness of feet (and hands) Motor, sensory, and autonomic changes: burning, prickling, or tingling may be present; extreme sensitivity to touch (or numbness); weakness, falling (foot drop), muscle atrophy; infection, ulcers, gangrene	Crawling, creeping sensation in legs; involuntary Involuntary contractions of calf muscles, occurring especially at night Pain† can be mild to severe, lasting seconds, minutes, or hours
Associated signs and symptoms	Decreased or absent pulses Color and skin changes in feet Normal DTRs; may be absent in people older than 60 Sciatica possible (ischemia)	Normal pulses Good skin nutrition Depressed or absent ankle jerks Positive SLR Sciatica Positive "shopping cart" sign (leaning forward on supportive surface; unable to straighten up due to painful symptoms)	Pulses may be affected, depending on underlying pathologic condition (e.g., diabetes) DTRs diminished or absent May have positive SLR May have sciatica	Sleep disturbance, paresthesias
Location	Usually calf first but may occur in the buttock, hip, thigh, or foot	Low back, buttock, thighs, calves, feet	Feet and hands in stocking-glove pattern	Feet, calves, legs
Aggravating factors	Pain is consistent in all spinal positions; brought on by physical exertion (e.g., walking, positive van Gelderen bicycle test); increased by climbing stairs or walking uphill (increased metabolic demand)	Increased in spinal extension Increased with walking; increased by walking downhill (increased lumbar lordosis); less painful when walking uphill	Depends on underlying cause (e.g., uncontrolled glucose levels with diabetes; progressive alcoholism)	Caffeine, pregnancy, iron deficiency
Relieving factors	Relieved promptly by standing still, sitting down, or resting (1-5 minutes)	Pain decreased by sitting, lying down, bending forward, or flexion exercises (may persist for hours)	Relieved by pain medications and relaxation techniques; treatment of underlying cause	Eliminate caffeine; increase iron intake, movement, walking, moderate exercise; medications; stretching; maintain hydration; heat or cold
Ages affected	40-60+	40-60+	Varies, depending on underlying cause	Variable
Cause	Atherosclerosis in peripheral arteries	Neoplasm or abscess Disk protrusion Osteophyte formation Ligamentous thickening	More than 100 causes: diabetes; medications; accidents; nerve compression; metal toxicity; nutritional deficiency; diseases such as RA, SLE, AIDS; cancer, hypothyroidism, alcoholism	Cause unknown; may be a sleep disorder, arterial disorder, or dysautonomic disorder of the autonomic nervous system; may occur with dehydration or as a side effect of many medications

DTRs, Deep tendon reflexes; *SLR,* straight leg raise; *RA,* rheumatoid arthritis; *SLE,* systemic lupus erythematosus; *AIDS,* acquired immunodeficiency syndrome.

*"Pain" associated with vascular claudication may also be described as an "aching," "cramping," or "tired" feeling.

†"Pain" associated with restless legs syndrome may not be painful but may be described as a "frantic," "unbearable," or "compelling" need to move the legs.

examination will help establish the diagnosis. Assessment of trigger points (TrPs) is also essential as pain referral to the knee from TrPs in the lower quadrant is well recognized but sometimes forgotten.[56,57]

Many therapists over the years have shared with us stories of clients treated for knee pain with a total knee replacement only to discover later (when the knee pain was unchanged) that the problem was really extraarticular (i.e., coming from the back or hip). On the flip side, it is not as likely but is still possible that hip pain can be caused by knee disease. Individual case reports of hip fracture presenting as isolated knee pain have been published[58] (Case Example 16-7).

Systemic Presentation. Systemic or pathologic conditions presenting as *generalized knee pain* can include fractures, Baker's cyst, tumors (benign or malignant), arthritis, infection, and/or DVT.[50] Other types of cancer can also cause knee pain such as lymphoma, leukemia, and myeloma. Watch for unusual bleeding, easy bruising, unintentional weight loss, fatigue, fevers, worsening pain (duration and intensity), sweats, dyspnea, and lymphadenopathy.[49]

A history of trauma accompanied by persistent or worsening symptoms despite restricted loading of the area are typical with bone or soft tissue tumors. *Night pain, localized swelling or warmth, locking,* and *palpable mass* with any of the other symptoms listed raise the suspicion of bone or soft tissue tumor.[59]

Burning and pain in the legs and feet at night are common in older adults; this is also a potential side-effect of some chemotherapy drugs. The exact mechanism is often unknown; many factors should be considered, including allergic response to the fabric in clothing and socks, poorly fitting shoes, long-term alcohol use, adverse effects of medications, diabetes, pernicious anemia, and restless legs syndrome.

Leg cramps, especially those occurring in the lower leg and calf, are common in the adult population.[60,61] Older adults, athletes, and pregnant women are at increased risk.[62] The history and physical examination are key elements in identifying the cause. The most common causes of leg cramps include dehydration, arterial occlusion from peripheral vascular disease, neurogenic claudication from lumbar spinal stenosis,[62] neuropathy, medications, metabolic disturbances, nutritional (vitamin, calcium) deficiency, and anterior compartment syndrome from trauma, hemophilia, sickle cell anemia, burns, casts, snakebites, or revascular perfusion injury.

Athletes often experience leg cramps preceded by muscle fatigue or twitching. Fractures and ligament tears can mimic a cramp. Cramping associated with severe dehydration may be a precursor to heat stroke.[63]

Heel pain is often a symptom of plantar fasciitis, heel spurs, nerve compression (e.g., tarsal tunnel syndrome), or stress fractures. Heel pain can also be a symptom of systemic conditions such as rheumatoid arthritis (RA), seronegative arthritides, primary bone tumors or metastatic disease, gout, sarcoidosis, Paget's disease of the bone, inflammatory bowel disease, osteomyelitis, infectious diseases, sickle cell disease, and hyperparathyroidism.[64]

CASE EXAMPLE 16-7

Total Knee Arthroplasty

A 78-year-old woman went to the emergency department over a weekend for knee pain. She reported a knee joint replacement 6 months ago because of arthritis. X-ray examination showed that the knee implant was intact with no complications (i.e., no infection, fracture, or loose components). She was advised to contact her orthopedic surgeon the following Monday for a follow-up visit. The woman decided instead to see the physical therapist who was involved with her postoperative rehabilitation.

The physical therapist's interview and examination revealed the following information. No pain was perceived or reported anywhere except in the knee. The pain pattern was constant (always present) but was made worse by weight-bearing activities. The knee was not warm, red, or swollen. No other associated signs and symptoms or constitutional symptoms were present, and vital signs were within normal limits for her age range.

Range of motion was better than at the time of previous discharge, but painful symptoms were elicited with a gross manual muscle screening examination. After a test of muscle strength, the woman was experiencing intense pain and was unable to put any weight on the painful leg.

The physical therapist insisted that the woman contact her physician immediately and arranged by phone for an emergency appointment that same day.

Result: Orthopedic examination and pelvic and hip x-ray films showed a hip fracture that required immediate total hip replacement the same day. The knee can be a site for referred pain from other areas of the musculoskeletal system, especially when symptoms are monoarticular. Systemic origin (or medical conditions causing) symptoms is more likely when multiple joints are involved or migrating arthralgias are present.

No history or accompanying signs and symptoms suggested a systemic origin of knee pain, but the pain on weight bearing made worse after muscle testing was a red-flag symptom for bone involvement. Hip fractures or other hip disease can masquerade as knee pain.

Prompt diagnosis of hip fracture is important in preventing complications. This therapist chose the conservative approach with medical referral rather than proceeding with physical therapy intervention. Sometimes, the "treat-and-see" approach to symptom assessment works well, but if any red flags are identified, a physician referral is advised.

The resolution of heel pain of a musculoskeletal origin is variable and can take weeks to months. Knowing when to request additional diagnostic assessment is not always clear cut. The therapist must keep each potential systemic cause in mind when looking for clues in the client's profile that might point to any one of these conditions.

For example, RA is more common between the third and fifth decades, affecting women 2 to 2½ times more often than men. Ankylosing spondylitis usually affects the spine and SI joints first. Heel pain as a secondary symptom would be suspicious. Men are affected more often than women.

A history of inflammatory bowel disease (e.g., Crohn's disease or ulcerative colitis) or cancer with new onset of ankle and/or heel pain must be evaluated medically. The calcaneus is the most common site of metastasis to the foot.[65] Subdiaphragmatic disease (especially genitourinary or colorectal neoplasm) tends to metastasize to the feet,[66] but cases of supradiaphragmatic disease, such as breast cancer metastasizing to the heel, have been reported.[67] X-rays and blood tests will be needed to look for an inflammatory, infectious, or metastatic cause of heel pain.[64]

No matter what area of the lower quadrant is affected, asking about the presence of other signs and symptoms, conducting a Review of Systems, and identifying red-flag symptoms will help the therapist in the clinical decision-making process. The therapist can use the red flags (see Appendix A-2) to guide screening questions. Always ask every client the following:

FOLLOW-UP QUESTION

- Are there any other symptoms of any kind anywhere else in your body?

If the client says, "No," the therapist may want to ask some general screening questions, including questions about constitutional symptoms.

Failure to improve with physical therapy intervention may be part of the medical differential diagnosis and should be reported within a reasonable length of time, given the particular circumstances of each client.

TRAUMA AS A CAUSE OF HIP, GROIN, OR LOWER QUADRANT PAIN

Trauma, including accidents, injuries, physical or sexual assault, or birth trauma, can be the underlying cause of buttock, hip, groin, or lower extremity pain.

Birth Trauma

Birth trauma is one possible cause of low back, pelvic, hip, or groin pain, with pain radiating down the leg in some cases. Multiple births, prolonged labor and delivery, forceps/vacuum delivery, and postepidural complications are just a few of the more common birth-related causes of hip, groin, and lower extremity pain. Gynecologic conditions are discussed more completely in Chapter 15.

Stress Reaction or Fracture

An undiagnosed stress reaction or stress fracture is a possible cause of hip, thigh, groin, knee, shin, heel, or foot pain. A stress reaction or fracture is a microscopic disruption, or break, in a bone that is not displaced; it is not seen initially on regular x-rays. Exercise-induced groin, tibial, or heel pain are the most common stress fractures.

There are two types of stress fractures. *Insufficiency* fractures are breaks in abnormal bone under normal force. *Fatigue* fractures are breaks in normal bone that has been put under extreme force. Fatigue fractures are usually caused by new, strenuous, very repetitive activities such as marching, jumping, or distance running.

Fatigue fractures are more likely in distance runners, sprinters,[68] military recruits, or other high-intensity athletes affecting the pubic ramus, calcaneus, femoral neck, anterior tibia most often.[69,70] Older adults are more likely to present with insufficiency hip fractures. Depending on the age of the client, the therapist should look for a history of high-energy trauma, prolonged activity, or abrupt increase in training intensity. Traction from attached muscles such as the adductor magnus on the inferior pubic ramus is a contributing factor to pubic ramus stress fractures.

Other risk factors include changes in running surface, use of inadequately cushioned footwear, and the presence of the female athlete triad of disordered eating, osteoporosis, amenorrhea, and menopause.[71-73] Anything that can lead to poor bone density should be considered a risk factor for insufficiency stress fractures including radiation and/or chemotherapy,[74] prolonged use of corticosteroids, renal failure, metabolic disorders affecting bone, Paget's disease, and coxa vara.[75,76] A smaller cross-sectional diameter of the long bones of the leg in male distance runners is a unique risk factor for tibial stress fractures.[77]

Femoral shaft stress fractures are rare in the general population but are not uncommon among distance runners and military recruits involved in repetitive loading activities such as running and marching. Pain presentation is not always predictable.[75] Vague anterior thigh pain that radiates to the hip or knee with activity or exercise is the most common clinical presentation. The affected individual usually has full but painful active hip motion.[78] The fulcrum test (Fig. 16-3) has high clinical correlation with femoral shaft stress injury.[79]

Likewise, heel pain from calcaneal fractures can occur in the athlete following significant increases in athletic activities or after a plantar fascia rupture. Posterior and plantar heel pain and swelling are often misdiagnosed as plantar fasciitis. A medial-lateral squeeze test may help identify the need for further imaging studies, especially when x-rays have been read as "normal."[80]

Osteopenia or osteoporosis, especially in the postmenopausal woman or older adult with arthritis, can result in injury and fracture or fracture and injury (Case Example 16-8). The client has a small mishap, perhaps losing her footing on a slippery surface or tripping over an object. As she tries to "catch herself," a torsional force occurs through the hip, causing a fracture and then a fall. This is a case of fracture then fall, rather than the other way around. Often, but not always, the client is unable to get up because of pain and instability of the fracture site.[81,82]

Pain on weight bearing is a red flag symptom for stress reaction or fracture in any individual. In the case of bone pain (deep pain, pain on weight bearing), the therapist can perform

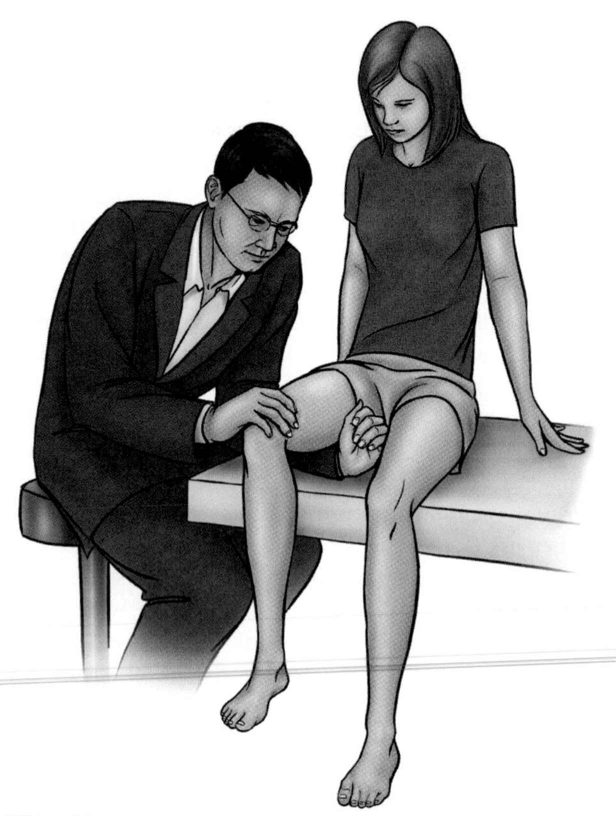

Fig. 16-3 Fulcrum test for femoral shaft stress reaction or fracture. With the client in a sitting position, the examiner places his or her forearm under the client's thigh and applies downward pressure over the anterior aspect of the distal femur. A positive test is characterized by reproduction of thigh pain often described as "sharp," with considerable apprehension on the part of the client.[79]

a heel strike test. This is done by applying a percussive force with the heel of the examiner's hand through the heel of the client's foot in a non–weight-bearing (supine) position. Reproduction of painful symptoms with axial loading is positive and highly suggestive of a bone fracture or stress reaction.[83]

The therapist can ask a physically capable client to hop on the uninvolved side and to do a full squat to clear the hip, knee, and ankle. These tests are used to screen for pubic ramus or hip stress fracture (reaction). Palpation over the injured bone may reproduce the painful symptoms, but when the stressed bone lies deep within the tissue, the therapist may be able to reproduce the pain by stressing the bone with translational (resisted active adduction) or rotational force (resisted active adduction combined with hip external rotation). Swelling is not usually evident early in the course of a stress reaction or fracture, but it does develop if the person continues athletic activity.

Look for the following clues suggestive of hip, groin, or thigh pain caused by a stress reaction or stress fracture.

CASE EXAMPLE 16-8

Insufficiency Fracture

A 50-year-old Caucasian woman was referred to physical therapy with a 4-year history of rheumatoid arthritis (RA). She had been taking prednisone (5 to 30 mg/day) and sulfasalazine (1 g twice a day).

She has a history of hypertension, smokes a pack of cigarettes a day, and drinks a six-pack of beer every night. She lives alone and no longer works outside the home. She admits to very poor nutrition and does not take a multivitamin or calcium.

Clinical Presentation: Symmetric arthritis with tenderness and swelling of bilateral metacarpophalangeal (MCP) joints, proximal interphalangeal (PIP) joints, wrists, elbows, and metatarsophalangeal (MTP) joints.

The patient reported "hip pain," which started unexpectedly 2 weeks ago in the right groin area. The pain went down her right leg to the knee but did not cross the knee. Any type of movement made it hurt more, especially on walking.

Hip range of motion was limited because of pain; formal range of motion (active, passive, accessory motions) and strength testing were not possible.

What are the red flags in this case?

- Age
- Insidious onset with no known or reported trauma
- Cigarette smoking
- Alcohol use
- Poor diet
- Corticosteroid therapy

Result: The client was showing multiple risk factors for osteoporosis. Further questioning revealed that surgical menopause took place 10 years ago; this is another risk factor.

The patient was unable to stand on the right leg unsupported. She could not squat because of her arthritic symptoms. Heel strike test was negative. Patrick's (Faber's) test could not be performed because of the acuteness of her symptoms.

The patient was referred to her rheumatologist with a request for a hip x-ray before any further physical therapy was provided. The therapist pointed out the risk factors present for osteoporosis and briefly summarized the client's current clinical presentation.

The client was given a diagnosis of insufficiency fracture of the right inferior and superior pubic rami. An insufficiency fracture differs from a stress fracture in that it occurs when a normal amount of stress is placed on abnormal bone. A stress fracture occurs when an unusual amount of stress is placed on normal bone.

Conservative treatment was recommended with physical therapy, pain medications, and treatment of the underlying osteoporosis. Weight bearing as tolerated, a general conditioning program, and an osteoporosis exercise program were prescribed by the physical therapist. Client education about managing active RA and synovitis was also included.

Data from Kimpel DL: Hip pain in a 50-year-old woman with RA, *J Musculoskel Med* 16:651-652, 1999.

CLINICAL SIGNS AND SYMPTOMS

Stress Reaction/Stress Fracture

- Pain described as aching or deep aching in hip and/or groin area; may radiate to the knee
- Pain increases with activity and improves with rest
- Muscle weakness (reduced grade on manual muscle testing; involved muscles vary depending on location of the fracture)
- Compensatory gluteus medius gait
- Pain localizing to a specific area of bone (localized tenderness)
- Positive Patrick's or Faber's test
- Pain reproduced by weight bearing, heel strike, or hopping test; positive medial-lateral squeeze test (calcaneal stress test)
- Pain reproduced by translational/rotational stress (exquisite pain in response to active resistance to hip adduction/hip adduction combined with external rotation)
- Thigh pain reproduced by the fulcrum test (femoral shaft fracture)
- Possible local swelling
- Increased tone of hip adductor muscles; limited hip abduction
- Night pain (femoral neck stress fracture)

Radiographs may not show the fracture, especially during its early stages.[35] The therapist should also keep in mind that some fractures of the intertrochanteric region do not show up on standard anteroposterior or lateral x-ray. An oblique view may be needed. If an x-ray has been ruled negative for hip fracture but the client cannot put weight on that side and a heel strike test is positive, communication with the physician may be warranted.

Fig. 16-4 Sciatica pain pattern. Perceived or reported pain associated with compression, stretch, injury, entrapment, or scarring of the sciatic nerve depends on the location of the lesion in relation to the nerve root. The sciatic nerve is innervated by L4, L5, S1, S2, and sometimes S3 with several divisions (e.g., common fibular [peroneal] nerve, sural nerve, tibial nerve).

Assault

The client may not report assault as the underlying cause, or he or she may not remember any specific trauma or accident. It may be necessary to take a sexual history (see Appendix B-32) that includes specific questions about sexual activity (e.g., incest, partner assault or rape) or the presence of sexually transmitted infection. Appropriate screening questions for assault or domestic violence are included in Chapter 2; see also Appendix B-3.

SCREENING FOR SYSTEMIC CAUSES OF SCIATICA

Sciatica, described as pain radiating down the leg below the knee along the distribution of the sciatic nerve, is usually related to mechanical pressure or inflammation of lumbosacral nerve roots (Fig. 16-4). *Sciatica* is the term commonly used to describe pain in a sciatic distribution without overt signs of radiculopathy.

Radiculopathy denotes objective signs of nerve (or nerve root) irritation or dysfunction, usually resulting from involvement of the spine. Symptoms of radiculopathy may include weakness, numbness, or reflex changes. Sciatic *neuropathy* suggests damage to the peripheral nerve beyond the effects of compression, often resulting from a lesion outside the spine that affects the sciatic nerve (e.g., ischemia, inflammation, infection, direct trauma to the nerve, compression by neoplasm or piriformis muscle).

The terms *radiculopathy, sciatica,* and *neuropathy* are often used interchangeably, although there is a pathologic difference.[84] Electrodiagnostic studies, including nerve conduction studies (NCS), electromyography (EMG), and somatosensory evoked potential studies (SSEPs), are used to make the differentiation.

Sciatica has many neuromuscular causes, both diskogenic and nondiskogenic; systemic or extraspinal conditions can produce or mimic sciatica (Table 16-6). Risk factors for a

TABLE 16-6 Causes of Sciatica

NEUROMUSCULAR CAUSES			SYSTEMIC/EXTRASPINAL CAUSES*
Disorder	**Symptoms**	**Physical Signs**	**Disorders**
DISKOGENIC			Vascular • Ischemia of sciatic nerve • PVD • Intrapelvic aneurysm (internal iliac artery) Neoplasm (primary or metastatic) Diabetes mellitus (diabetic neuropathy) Megacolon Pregnancy; vaginal delivery Endometriosis Infection • Bacterial endocarditis • Wound contamination[89,90] • Herpes zoster (shingles) • Psoas muscle abscess (see Box 16-3) • Reiter's syndrome Total hip arthroplasty DVT (blood clot)
Disk herniation	Low back pain with radiculopathy and paravertebral muscle spasm; Valsalva's maneuver and sciatic stretch reproduce symptoms	Restricted spinal movement; restricted spinal segment; positive Lasègue's sign or restricted SLR	
Lateral entrapment syndrome (spinal stenosis)	Buttock and leg pain with radiculopathy; pain often relieved by sitting, aggravated by extension of the spine	Similar to disk herniation	
NONDISKOGENIC			
Sacroiliitis	Low back and buttock pain	Tender SI joint; positive lateral compression test; positive Patrick's test	
Piriformis syndrome	Low back and buttock pain with referred pain down the leg to the ankle or midfoot	Pain and weakness on resisted abduction/external rotation of the thigh	
Iliolumbar syndrome	Pain in iliolumbar ligament area (posterior iliac crest); referred leg pain	Tender iliac crest and increased pain with lateral or side bending	
Trochanteric bursitis	Buttock and lateral thigh pain; worse at night and with activity	Tender greater trochanter; rule out associated leg-length discrepancy; positive "jump sign" when pressure is applied over the greater trochanter	
GTPS	Mimics lumbar nerve root compression	Low back, buttock, or lateral thigh pain; may radiate down the leg to the iliotibial tract insertion on the proximal tibia; inability to sleep on the involved side[7]	
Ischiogluteal bursitis	Buttock and posterior thigh pain; worse with sitting	Tender ischial tuberosity; positive SLR and Patrick's tests; rule out associated leg-length discrepancy	
Posterior facet syndrome	Low back pain	Lateral bending in spinal extension increases pain; side bending and rotation to the opposite side are restricted at the involved level	
Fibromyalgia	Back pain, difficulty sleeping, anxiety, depression	Multiple tender points (see Fig. 12-3)	

Data from Namey TC, An HC: Sorting out the causes of sciatica, *Mod Med* 52:132, 1984.
SLR, Straight leg raise; *PVD,* peripheral vascular disease; *SI,* sacroiliac; *DVT,* deep venous thrombosis; *GTPS,* greater trochanteric pain syndrome.
*Clinical symptoms of systemic/extraspinal sciatica can be very similar to those of sciatica associated with disk protrusion.

mechanical cause of sciatica include previous trauma to the low back, taller height, tobacco use, pregnancy, and work- and occupational-related posture or movement.[85]

Risk Factors

Risk factors for systemic or extraspinal causes vary with each condition (Table 16-7). For example, clients with arterial insufficiency are more likely to be heavy smokers and to have a history of atherosclerosis. Increasing age, past history of cancer, and comorbidities, such as diabetes mellitus, endometriosis, or intraperitoneal inflammatory disease (e.g., diverticulitis, Crohn's disease, pelvic inflammatory disease), are risk factors associated with sciatic-like symptoms (Case Example 16-9).

Total hip arthroplasty is a common cause of sciatica because of the proximity of the nerve to the hip joint. Possible mechanisms for nerve injury include stretching, direct trauma

TABLE 16-7 Risk Factors for Sciatica

Musculoskeletal or Neuromuscular Factors	Systemic/Medical Factors
Previous low back injury or trauma; direct fall on buttock(s); gunshot wound	Tobacco use
Total hip arthroplasty	History of diabetes mellitus
Pregnancy	Atherosclerosis
Work- or occupation-related postures or movements	Previous history of cancer (metastases)
Fibromyalgia	Presence of intraabdominal or peritoneal inflammatory disease (abscess): • Crohn's disease • Pelvic inflammatory disease (PID) • Diverticulitis
Leg-length discrepancy	Endometriosis of the sciatic nerve
Congenital hip dysplasia; hip dislocation	Radiation therapy (delayed effects; rare)
Degenerative disk disease	Recent spinal surgery, especially with instrumentation
Piriformis syndrome	
Spinal stenosis	

CLINICAL SIGNS AND SYMPTOMS

Sciatica/Sciatic Radiculopathy

Symptoms are variable and may include the following:

- Pain along the sciatic nerve anywhere from the spine to the foot (see Fig. 16-4)
- Numbness or tingling in the groin, rectum, leg, calf, foot, or toes
- Diminished or absent deep tendon reflexes
- Weakness in the L4, L5, S1, S2 (and sometimes S3) myotomes (distal motor deficits more prominent than proximal)
- Diminished or absent deep tendon reflexes (especially of the ankle)
- Ache in the calf

Sciatic Neuropathy

- Symptoms of sciatica as described above
- Dysesthetic* pain described as constant burning or sharp, jabbing pain
- Foot drop (tibialis anterior weakness) with gait disturbance
- Flail lower leg (severe motor neuropathy)

*Dysesthesia is the distortion of any sense, especially touch; it is an unpleasant sensation produced by normal stimuli.

from retractors, infarction, hemorrhage, hip dislocation, and compression.[86] Sciatica referred to as sciatic nerve "burn" has been reported as a complication of hip arthroplasty caused by cement extrusion. The incidence of this complication has decreased with its increased recognition and the increasing use of cementless implants,[19] but even small amounts of cement can cause heat production or direct irritation of the sciatic nerve.[87]

Propionibacterium acnes, a cause of spinal infection, has been linked to sciatica.[88] Bacterial wound contamination during spinal surgery has been traced to this pathogen on the patient's skin. Minor trauma to the disk with a breach to the mechanical integrity of the disk may also allow access by low virulent microorganisms, thereby initiating or stimulating a chronic inflammatory response. These microorganisms may cause prosthetic hip infection but also may be associated with the inflammation seen in sciatica; they may even be a primary cause of sciatica.[89,90]

Endometriosis at the sciatic notch and pelvic endometriosis affecting the lumbosacral plexus or proximal sciatic nerve can present as sciatica/buttock pain that extends down the posterior aspect of the thigh and calf to the ankle. The pain is cyclic and corresponds with the menstrual cycle.[91,92]

Anyone with pain radiating from the back down the leg as far as the ankle has a greater chance that disk herniation is the cause of low back pain. This is true with or without neurologic findings. Unremitting, severe pain and increasing neurologic deficit are red-flag findings. Sciatica caused by extraspinal bone and soft tissue tumors is rare but may occur when a mass is present in the pelvis, sacrum, thigh, popliteal fossa, and calf.[93,94]

The therapist can conduct an examination to look for signs and symptoms associated with systemically induced sciatica. Box 4-13 offers guidelines on conducting an assessment for peripheral vascular disease. Box 4-16 provides a checklist for the therapist to use when examining the extremities. These tools can help the therapist define the clinical presentation more accurately.

The SLR test and other neurodynamic tests are widely used but do not identify the underlying cause of sciatica. For example, a positive SLR test does not differentiate between diskogenic disease and neoplasm.

Without a combination of imaging and laboratory studies, the clinical picture of sciatica is difficult to distinguish from that of conditions such as neoplasm and infection. Erythrocyte sedimentation rate (ESR or sed rate) is the rate at which red blood cells settle out of unclotted blood plasma within 1 hour. A high ESR is an indication of infection or inflammation (see top table, Inside Front Cover). Elevated ESR and abnormal imaging are effective tools to use in screening for occult neoplasm and other systemic disease.[95]

Imaging studies are an essential part of the medical diagnosis, but even with these diagnostic tests, errors in conducting and interpreting imaging studies may occur. Symptoms can also result from involvement outside the area captured on computed tomography (CT) scan or MRI.

SCREENING FOR ONCOLOGIC CAUSES OF LOWER QUADRANT PAIN

Many clients with orthopedic or neurologic problems have a previous history of cancer. The therapist must recognize signs and symptoms of cancer recurrence and those associated

CASE EXAMPLE 16-9

Low Back Pain with Sciatica

A 52-year-old man with low back pain and sciatica on the left side has been referred to you by his family physician. He underwent diskectomy and laminectomy on two separate occasions about 5 to 7 years ago. No imaging studies have been done since that time.

What follow-up questions would you ask to screen for systemic disease?

1. The first question should always be "Did you actually see your doctor?" (Of course, communication with the physician is the key here in understanding the physician's intended goal with physical therapy and his or her thinking about the underlying cause of the sciatica.)
2. Assess for the presence of constitutional symptoms. For example, after paraphrasing what the client has told you, ask, "Are you having any other symptoms of any kind in your body that you haven't mentioned?" If no, ask more specifically about the presence of associated signs and symptoms; name constitutional symptoms one by one.
3. Follow-up with *Special Questions for Men* (see Appendix B-24). Include questions about past history of prostate health problems, cancer of any kind, and current bladder function.
4. Take a look at Table 16-6. By reviewing the possible systemic/extraspinal causes of sciatica, we can decide what additional questions might be appropriate for this man.

Vascular ischemia of the sciatic nerve can occur at any age as a result of biomechanical obstruction. It can also result from peripheral vascular disease. Check for skin changes associated with ischemia of the lower extremities. Ask about the presence of known heart disease or atherosclerosis.

Intrapelvic aneurysm: Palpate aortic pulse width and listen for femoral bruits.

Neoplasm (primary or metastatic): Consider this more strongly if the client has a previous history of cancer, especially cancer that might metastasize to the spine. We know from Chapter 13 that the three primary sites of cancer most likely to metastasize to the bone are lung, breast, and prostate. Other cancers that metastasize to the bone include thyroid, kidney, melanoma (skin), and lymphoma. A previous history of any of these cancers is a red-flag finding.

Primary bone cancer is not as likely in a middle-aged male as in a younger age group. Cancer metastasized to the bone is more likely and is most often characterized by pain on weight bearing that is deep and does not respond to treatment modalities.

Diabetes (diabetic neuropathy): Ask about a personal history of diabetes. If the client has diabetes, assess further for associated neuropathy. If not, assess for symptoms of possible new-onset but as yet undiagnosed diabetes.

Megacolon: An unlikely cause unless the client is much older or has recently undergone major surgery of some kind.

Pregnancy: Not a consideration in this case.

Infection: Ask about a recent history of infection (most likely bacterial endocarditis, urinary tract infection, or sexually transmitted infection, but any infection can seed itself to the joints or soft tissues). Ask about any other signs or symptoms of infection (e.g., flu-like symptoms, such as fever and chills or skin rash, in the last few weeks).

Remember from Chapter 3 to ask the following:

FOLLOW-UP QUESTIONS

- Are you having any pain anywhere else in your body?
- Are you having symptoms of any other kind that may or may not be related to your main problem?
- Have you recently (last 6 weeks) had any of the following:
- Fractures
  - Bites (human, animal)
  - Antibiotics or other medications
  - Infections (you may have to prompt with specific infections such as strep throat, mononucleosis, urinary tract, upper respiratory [cold or flu], gastrointestinal (GI), hepatitis, sexually transmitted diseases)

Total hip arthroplasty: Has the client had a recent (cemented) total hip replacement (e.g., cement extrusion, infection, implant fracture, loose component)?

Result: The client had testicular cancer that had already metastasized to the pelvis and femur. By asking additional questions, the physical therapist found out that the client was having swelling and hardness of the scrotum on the same side as the sciatica. He was unable to maintain an erection or to ejaculate. The physician was unaware of these symptoms because the client did not mention them during the medical examination.

Testicular carcinoma is relatively rare, especially in a man in his 50s. It is most common in the 15- to 39-year-old male group. Metastasis usually occurs via the lymphatics, with the possibility of abdominal mass, psoas invasion, lymphadenopathy, and back pain. Palpation revealed a dominant mass (hard and painless) in the ipsilateral groin area.

Sending a client back to the referring physician in a case like this may require tact and diplomacy. In this case, the therapist made telephone contact to express concerns about the reported sexual dysfunction and palpable groin lymphadenopathy.

By alerting the physician to these additional symptoms, further medical evaluation was scheduled, and the diagnosis was made quickly.

with cancer treatment such as radiation therapy or chemotherapy. The effects of these may be delayed by as long as 10 to 20 years or more (see Table 13-8; Case Example 16-10).

Until now, the emphasis has been on advancing age as a key red flag for cancer. Anyone older than 50 years of age may need to be screened for systemic origin of symptoms. With cancer and specifically, musculoskeletal pain caused by primary cancer or metastases to the bone, young age is a red flag as well. Primary bone cancer occurs most often in adolescents and young adults, hence the new red flag: age younger than 20 years, or bone pain in an adolescent or young adult.

CASE EXAMPLE 16-10

Evaluating a Client for Cancer Recurrence

Referral: A 54-year-old man is self-referred to physical therapy on the recommendation of his personal trainer who is a friend of yours. He is experiencing leg weakness (greater on the right), with occasional pain radiating into the groin area on both sides.

He reports a twisting back injury 5 years ago when he was shoveling snow. At that time, he saw a physical therapist but did not get any better until he started working out at the YMCA.

Leg weakness has been present about 2 weeks. Last weekend, he went to the emergency department because his leg was numb and he could not lift his ankle. He was told to rest. The leg was better the next day.

Past Medical History: Renal calculi, surgery for parathyroid and thyroid cancer 10 years ago, pneumonia 20 years ago. Currently seeing a counselor for emotional problems.

Objective Findings:

Neurologic Screen

- Alert; oriented to time, place, person
- Pupils equal and equally reactive to light; eye movements in all directions without difficulty
- No tremor, upper extremity weakness, or changes in deep tendon reflexes (DTRs)
- Straight leg raise (SLR) was mobile and pain free to 90 degrees bilaterally
- Iliopsoas, gluteal, hamstring manual muscle testing (MMT) = 3/5 on the right side. MMT within normal limits on the left side
- Tibialis anterior, plantar evertors and flexors: MMT = 2/5 (right); 3+ to 4 on the left
- No ankle clonus, no Babinski's, no changes in DTRs of lower extremities (LEs)
- Increased muscle tone in both LEs

No pain was reported with any movements performed during the examination.

Name 3 red-flag symptoms in this case.

Age is the first red flag: A man over 40 (and especially over 50 years of age) with a previous history of cancer (second red flag) and new onset of painless neurologic deficit (third red flag) is significant.

Now that we have identified three red flags, what is next? Does this signify an automatic referral to the physician? We do not think so: The need for physician referral may depend on the specific red flags that are present. For example, in the case just presented, the three red flags are pretty significant. Take a closer look, and gather as much information as possible. In this case, it appears likely that an immediate referral is warranted.

Can we tell whether this is a recurrence of his previous cancer now metastasized or the presence of prostate cancer? No, but we can ask some additional questions to look for clusters of associated signs and symptoms that might point to prostate involvement. First, ask about bladder function, urination, and finally, sexual function. Remember, you may have to explain the need to ask a few personal questions.

- Have you ever had prostate problems or been told you have prostate problems?
- Have you had any changes in urination recently?
- Can you easily start a flow of urine?
- Can you keep a steady stream without stopping and starting?
- When you are finished urinating, does it feel as though your bladder is empty? Or, do you feel like you still have to go, but you can't get any more out?
- Do you ever dribble urine?
- Do you have trouble getting an erection?
- Do you have trouble keeping an erection?
- Do you have trouble ejaculating?

Because the patient is seeing a counselor for emotional problems, you may wish to screen him for emotional overlay. You can use the three tools discussed in Chapter 3 (Symptom Magnification, McGill's Pain Questionnaire, Waddell's nonorganic tests).

After you have completed your examination, step back and put all the pieces together. Is there a cluster of signs and symptoms that point to any particular system? The answer to this question may lead you to ask some additional questions or to confirm the need for medical attention.

Special Note: Palpating the groin area is usually necessary when performing a thorough evaluation. This can be a sensitive issue. In today's litigious culture, you may want to have a third person in the examination area with you. This person should be the same gender as your client. You will certainly want to explain everything you are doing and obtain the client's permission.

For men, give the client time to make any necessary "adjustments" before beginning palpation. If the client has an erection during palpation, do not make any joking or unprofessional comments. This may seem self-evident but we have observed a wide range of responses when supervising others that supports the need to provide specific guidelines as stated here.

Cancer Recurrence

The therapist is far more likely to encounter clinical manifestations of metastases from cancer recurrence than from primary cancer. Breast cancer often affects the shoulder, thoracic vertebrae, and hip first, before other areas. Recurrence of colon (colorectal) cancer is possible with referred pain to the hip and/or groin area.

Beware of any client with a past history of colorectal cancer and recent (past 6 months) treatment by surgical removal. Reseeding the abdominal cavity is possible. Every effort is made to shrink the tumor with radiation or chemotherapy before attempts are made to remove the tumor. Even a small number of tumor cells left behind or introduced into a nearby (new) area can result in cancer recurrence.

Hodgkin's Disease

Hodgkin's disease arises in the lymph glands, most commonly on a single side of the neck or groin, but lymph nodes also enlarge in response to infection throughout the body.

Lymph nodes in the groin area can become enlarged specifically as a result of sexually transmitted disease.

The presence of painless, hard lymph nodes that are also similarly present at other sites (e.g., popliteal space) is always a red-flag symptom. As always, the therapist must question the client further regarding the onset of symptoms and the presence of any associated symptoms, such as fever, weight loss, bleeding, and skin lesions. The client must seek a medical diagnosis to be certain of the cause of enlarged lymph nodes.

Spinal Cord Tumors

Spinal cord tumors (primary or metastasized) present as dull, aching discomfort or sharp pain in the thoracolumbar area in a beltlike distribution, with pain extending to the groin or legs. Depending on the location of the lesion, symptoms may be unilateral or bilateral with or without radicular symptoms. The therapist should look for and ask about associated signs and symptoms (e.g., constitutional symptoms, bleeding or discharge, lymphadenopathy).

Symptoms of thoracic disk herniation can mimic spinal cord tumor. In isolated cases, thoracic disk extrusion has been reported to cause groin pain and lower extremity weakness that gets progressively worse over time. A tumor is suspected if the client has painless neurologic deficit, night pain, or pain that increases when supine.

Testing the cremasteric reflex may help the therapist identify neurologic impairment in any male with suspicious back, pelvic, groin (including testicular), or anterior thigh pain. The cremasteric reflex is elicited by stroking the thigh downward with a cotton-tipped applicator (or handle of the reflex hammer). A normal response in males is upward movement of the testicle (scrotum) on the same side. The absence of a cremasteric reflex is an indication of disruption at the T12-L1 level.

Additionally, groin pain associated with spinal cord tumor is disproportionate to that normally expected with disk disease. No change in symptoms occurs after successful surgery for herniated disk. Age is an important factor: teenagers with symptoms of disk herniation should be examined closely for tumor.[96,97]

Spinal metastases to the femur or lower pelvis may appear as hip pain. With the exception of myeloma and rare lymphoma, metastasis to the synovium is unusual. Therefore joint motion is not compromised by these bone lesions. Although any tumor of the bone may appear at the hip, some benign and malignant neoplasms have a propensity to occur at this location.

Bone Tumors

Osteoid osteoma, a small, benign but painful tumor, is relatively common, with 20% of lesions occurring in the proximal femur and 10% in the pelvis. The client is usually in the second decade of life and complains of chronic dull hip, thigh, or knee pain that is worse at night and is alleviated by activity and aspirin and nonsteroidal antiinflammatory drugs (NSAIDs). Usually, an antalgic gait is present, along with point tenderness over the lesion with restriction of hip motion.

A great many varieties of benign and malignant tumors may appear differently, depending on the age of the client and the site and duration of the lesion (Case Example 16-11).[98,99] Malignant lesions compressing the LFCN can cause symptoms of meralgia paresthetica, delaying diagnosis of the underlying neoplasm. Other bone tumors that cause hip pain, such as chondroblastoma, chondrosarcoma, giant cell tumor, and Ewing's sarcoma, are discussed in greater detail in Chapter 13.

CLINICAL SIGNS AND SYMPTOMS

Buttock, Hip, Groin, or Lower Extremity Pain Associated with Cancer

- Bone pain, especially on weight bearing; positive heel strike test
- Antalgic gait
- Local tenderness
- Night pain (constant, intense; unrelieved by change in position)
- Pain relieved disproportionately by aspirin
- Fever, weight loss, bleeding, skin lesions
- Vaginal/penile discharge
- Painless, progressive enlargement of inguinal and/or popliteal lymph nodes

SCREENING FOR UROLOGIC CAUSES OF BUTTOCK, HIP, GROIN, OR THIGH PAIN

Ureteral pain usually begins posteriorly in the costovertebral angle but may radiate anteriorly to the upper thigh and groin (see Fig. 16-6), or it may be felt just in the groin and genital area. These pain patterns represent the pathway that genitals take as they migrate during fetal development from their original position, where the kidneys are located in the adult, down the pathways of the ureters to their final location. Pain is referred to a site where the organ was located during fetal development. A kidney stone down the pathway of the ureters causes pain in the flank that radiates to the scrotum (male) or labia (female).

The lower thoracic and upper lumbar vertebrae and the SI joint can refer pain to the groin and anterior thigh in the same pain pattern as occurs with renal disease. Irritation of the T10-L1 sensory nerve roots (genitofemoral and ilioinguinal nerves) from any cause, especially from diskogenic disease, may cause labial (women), testicular (men), or buttock pain.[100] The therapist can evaluate these conditions by conducting a neurologic screening examination and using the screening model.

Referred symptoms from ureteral colic can be distinguished from musculoskeletal hip pain by the history, the presence of urologic symptoms, and the pattern of pain. Is

CASE EXAMPLE 16-11

Ischial Bursitis

Referral: A 30-year-old dentist was referred to physical therapy by an orthopedic surgeon for ischial bursitis, sometimes referred to as "Weaver's bottom." He reported left buttock pain and "soreness" that was intermittent and work related. As a dentist, he was often leaning to the left, putting pressure on the left ischium.

Background: Magnetic resonance imaging (MRI) showed local inflammation on the ischial tuberosity to confirm the medical diagnosis. He was given a steroid injection and was placed on an antiinflammatory (Celebrex) before he went to physical therapy.

The client reported a mild loss of hip motion, especially of hip flexion, but no other symptoms of any kind. The pain did not radiate down the leg. No significant past medical history and no history of tobacco use were reported; only an occasional beer in social situations was described. The client described himself as being "in good shape" and working out at the local gym 4 to 5 times/week.

Intervention/Follow-Up: Physical therapy intervention included deep friction massage, iontophoresis, and stretching. The client modified his dentist's chair with padding to take pressure off the buttock. Symptoms did not improve after 10 treatment sessions over the next 6 to 8 weeks; in fact, the pain became worse and was now described as "burning."

The client went back to the orthopedic surgeon for a follow-up visit. A second MRI was done with a diagnosis of "benign inflammatory mass." He was given a second steroid injection and was sent back to physical therapy. He was seen at a different clinic location by a second physical therapist.

The physical therapist palpated a lump over the ischial tuberosity, described as "swelling"; this was the only new physical finding since his previous visits with the first physical therapist.

Treatment concentrated deep friction massage in that area. The therapist thought the lump was getting better, but it did not resolve. The client reported increased painful symptoms, including pain at work and pain at night. No position was comfortable; even lying down without pressure on the buttocks was painful. He modified every seat he used, including the one in his car.

Result: The orthopedic surgeon did a bursectomy, and the pathology report came back with a diagnosis of epithelioid sarcoma. The diagnosis was made 2½ years after the initial painful symptoms. A second surgery was required because the first excision did not have clear margins.

It is often easier to see the red flags in hindsight. As this case is presented here with the final outcome, what are the red flags?

Red Flags

- No improvement with physical therapy
- Progression of symptoms (pain went from "sore" to "burning," and intermittent to constant)
- Young age

Clinical signs of all types of bursitis are similar and include local tenderness, warmth, and erythema. The latter two signs may not be obvious when the inflamed bursa is located deep beneath soft tissues or muscles, as in this case.[98]

The presence of a "lump" or swelling as presented in this case caused a delay in medical referral and diagnosis because MRI findings were consistent with a diagnosis of inflammatory mass. In this case, symptoms progressed and did not fit the typical pattern for bursitis (e.g., pain at night, no position comfortable).

Other Tests

When a client is sent back a second time, the therapist's reevaluation is essential for documenting any changes from the original baseline and discharge findings. Reevaluation should include the following:

- Recheck levels above and below for possible involvement, including lumbar spine, sacroiliac joint, hip, and knee; perform range of motion and special tests, and conduct a neurologic screening examination (see Chapter 4).
- Test for the sign of the buttock to look for serious disease posterior to the axis of flexion and extension of the hip (see Box 16-2). A positive sign may be an indication of abscess, fracture, neoplasm, septic bursitis, or osteomyelitis.[12]

A noncapsular pattern is typical with bursitis and by itself is not a red flag. A capsular pattern with a diagnosis of bursitis would be more suspicious. Limited straight leg raise with no further hip flexion after bending the knee is a typical positive buttock sign seen with ischial bursitis. The absence of this sign would raise clinical suspicion that the diagnosis of bursitis was not accurate.[12]

With an ischial bursitis, expect to see equal leg length, negative Trendelenburg test, and normal sensation, reflexes, and joint play movements.[99] Anything outside these parameters should be considered a yellow (caution) flag.

- Assess for trigger points (TrPs) that may cause buttock pain, especially quadratus lumborum, gluteus maximus, and hamstrings, but also gluteus medius and piriformis.
- Reassess for the presence of constitutional symptoms or any associated signs and symptoms of any kind anywhere in the body.

Case Report courtesy of Jason Taitch, DDS, Spokane, WA, 2005.

there any history of urinary tract impairment? Is there a recent history of other infection? Are any signs and symptoms noted that are associated with the renal system?

Active TrPs along the upper rim of the pubis and the lateral half of the inguinal ligament may lie in the lower internal oblique muscle and possibly in the lower rectus abdominis. These TrPs can cause increased irritability and spasm of the detrusor and urinary sphincter muscles, producing urinary frequency, retention of urine, and groin pain.[57]

The therapist can perform Murphy's percussion test to rule out kidney involvement (see Chapter 10; see also Fig. 4-54). A positive Murphy's percussion test (pain is reproduced with percussive vibration of the kidney) points to the possibility of kidney infection or inflammation. When this test is positive, ask about a recent history of fever, chills, unexplained perspiration ("sweats"), or other constitutional symptoms.

SCREENING FOR MALE REPRODUCTIVE CAUSES OF GROIN PAIN

Men can experience groin pain caused by disease of the male reproductive system such as prostate cancer, testicular cancer, benign prostatic hyperplasia (BPH), or prostatitis. Isolated groin pain is not as common as groin pain that is accompanied by low back, buttock, or pelvic pain. Risk factors, clinical presentation, and associated signs and symptoms for these conditions are discussed in Chapter 14.

SCREENING FOR INFECTIOUS AND INFLAMMATORY CAUSES OF LOWER QUADRANT PAIN

Anyone with joint pain of unknown cause who presents with current or recent (i.e., within the past 6 weeks) skin rash or recent history of infection (e.g., hepatitis, mononucleosis, urinary tract infection, upper respiratory infection, sexually transmitted infection, streptococcus, dental infection)[101,102] must be referred to a health care clinic or medical doctor for further evaluation.

Conditions affecting the entire peritoneal cavity such as pelvic inflammatory disease (PID) or appendicitis may cause hip or groin pain in the young, healthy adult. Widespread inflammation or infection may be well tolerated by athletes, sometimes for up to several weeks (Case Example 16-12).

Clinical Presentation

The clinical presentation can be deceptive in young people. The fever is not dramatic and may come and go. The athlete may dismiss excessive or unusual perspiration ("sweats") as part of a good workout. Loss of appetite associated with systemic disease is often welcomed by teenagers and young adults and is not recognized as a sign of physiologic distress.

With an infectious or inflammatory process, laboratory tests may reveal an elevated ESR. Questions about the presence of any other symptoms may reveal constitutional symptoms such as elevated nocturnal temperature, sweats, and chills, suggestive of an inflammatory process (Case Example 16-13).

Psoas Abscess

Any infectious or inflammatory process affecting the abdominal or pelvic region can lead to psoas abscess and irritation of the psoas muscle. For example, lesions outside the ureter, such as infection, abscess, or tumor, or abdominal or peritoneal inflammation, may cause pain on movement of the adjacent iliopsoas muscle that presents as hip or groin pain. (See discussion of Psoas Abscess in Chapter 10.)

PID is another common cause of pelvic, groin, or hip pain that can cause psoas abscess and a subsequent positive iliopsoas or obturator test. In this case, it is most likely a young woman with multiple sexual partners who has a known or unknown case of untreated *Chlamydia*.

The psoas muscle is not separated from the abdominal or pelvic cavity. Fig. 8-3 shows how most of the viscera in the abdominal and pelvic cavities can come into contact with the iliopsoas muscle. Any infectious or inflammatory process (see Box 16-3) can seed itself to the psoas muscle by direct extension, resulting in a psoas abscess—a localized collection of pus.

Hip pain associated with such an abscess may involve the medial aspect of the thigh and femoral triangle areas (Fig. 16-5). Soft tissue abscess may cause pain and tenderness to palpation without movement. Once the abscess has formed, muscular spasm may be provoked, producing hip flexion and even contracture. The leg also may be pulled into internal rotation. Pain that increases with passive and active motion can occur when infected tissue is irritated. Pain elicited by stretching the psoas muscle through extension of the hip, called the *positive psoas sign,* may be present.

CLINICAL SIGNS AND SYMPTOMS

Psoas Abscess

- Pain that is usually confined to the psoas fascia but that may extend to the buttock, hip, groin, upper thigh, or knee
- Pain located in the anterior hip in the area of the medial thigh or femoral triangle, often accompanied by or alternating with abdominal pain
- Psoas spasm causing functional hip flexion contracture
- Leg pulled into internal rotation
- Positive psoas sign (i.e., pain elicited by stretching the psoas muscle by extending the hip)
- Fever up and down (hectic fever pattern)
- Sweats
- Loss of appetite or other GI symptoms
- Palpable mass in the inguinal area (present with distal extension of the abscess)
- Positive iliopsoas or obturator test (see Figs. 8-5 through 8-7)

A positive response for any of these tests is indicative of an infectious or inflammatory process. Direct back, pelvic, or hip pain that results from these palpations is more likely to have a musculoskeletal cause. Besides the iliopsoas and obturator tests, another test for rebound tenderness used more often is the *pinch-an-inch* test (see Fig. 8-11). It may be appropriate to conduct these tests with a variety of clinical presentations involving the pelvic area, sacrum, hip, or groin.

Psoas abscess must be differentiated from TrPs of the psoas muscle, causing the psoas minor syndrome, which is easily mistaken for appendicitis. Hemorrhage within the psoas muscle, either spontaneous or associated with anticoagulation therapy for hemophilia, can cause a painful compression syndrome of the femoral nerve.

CASE EXAMPLE 16-12

Dancer with Appendicitis

A 21-year-old dance major was referred to the physical therapy clinic by the sports medicine clinic on campus with a medical diagnosis of "strained abdominal muscle."

She described her symptoms as pain with hip flexion when shifting the gears in her car. Some dance moves involving hip flexion also reproduced the pain, but this was not consistent. The pain was described as "deep," "aching," and "sometimes sharp, sometimes dull."

Past medical history was significant for Crohn's disease, but the client was having no gastrointestinal (GI) symptoms at this time. On examination, no evidence of abdominal trigger points (TrPs) or muscle involvement was found. The pain was not reproduced with superficial palpation of the abdominal muscles on the day of initial examination.

Intervention with stretching exercises did not change the clinical picture during the first week.

Result: The client was a no-show for her Monday afternoon appointment, and the physical therapy clinic receptionist received a phone call from the campus clinic with information that the client had been hospitalized over the weekend with acute appendicitis and peritonitis.

The surgeon's report noted massive peritonitis of several weeks' duration. The client had a burst appendix that was fairly asymptomatic until peritonitis developed with subsequent symptoms. Her white blood cells were in excess of 100,000 at the time of hospitalization.

In retrospect, the client did relate some "sweats" occurring off and on during the last 2 weeks and possibly a low-grade fever.

What additional screening could have been conducted with this client?

1. Ask the client whether she is having any symptoms of any kind anywhere in her body. If she answers, "No," be prepared to offer some suggestions such as:
   - Any headaches? Fatigue?
   - Any change in vision?
   - Any fevers or sweats, day or night?
   - Any blood in your urine or stools?
   - Burning with urination?
   - Any tingling or numbness in the groin area?
   - Any trouble sleeping at night?
2. Even though she has denied having any GI symptoms associated with her Crohn's disease, it is important to follow-up with questions to confirm this:
   - Any nausea? Vomiting?
   - Diarrhea or constipation?
   - Any change in your pattern of bowel movements?
   - Any blood in your stools? Change in color of your bowel movements?
   - Any foods or smells you can't tolerate?
   - Any change in your symptoms when you eat or don't eat?
   - Unexpected weight gain or loss?
   - Is your pain any better or worse during or after a bowel movement?
3. As part of the past medical history, it is important with hip pain of unknown cause to know whether the client has had any recent infections, sexually transmitted diseases, use of antibiotics or other medications, or skin rashes.
4. In a woman of reproductive years, it may be important to take a gynecologic history:
   - Have you been examined by a gynecologist since this problem started?
   - Is there any chance you could be pregnant?
   - Are you using an intrauterine contraceptive device (IUD or IUCD)?
   - Have you had an abortion or miscarriage in the last 6 weeks?
   - Are you having any unusual vaginal discharge?
5. Check vital signs. The presence of a fever (even low grade) is a red flag when the cause of symptoms is unknown. With a burst appendix, she may have had altered pulse and blood pressure that could alert the therapist of a systemic cause of symptoms.
6. Test for McBurney's point (Fig. 8-9), rebound tenderness using the pinch-an-inch test (Fig. 8-11), and the obturator or iliopsoas sign (Figs. 8-5 to 8-7). Check for Murphy's percussion (Fig. 4-54; kidney involvement).

Systemic causes of hip pain from psoas abscess are usually associated with loss of appetite or other GI symptoms, fever, and sweats. Symptoms from an iliopsoas trigger point are aggravated by weight-bearing activities and are relieved by recumbency or rest. Relief is greater when the hip is flexed.[57]

SCREENING FOR GASTROINTESTINAL CAUSES OF LOWER QUADRANT PAIN

The relationship of the gut to the joint is well known but poorly understood. Intestinal bypass syndrome, inflammatory bowel disease, ankylosing spondylitis, celiac disease, postdysenteric reactive arthritis, bowel bypass syndrome, and antibiotic-associated colitis all share the fact that some "interface" exists between the bowel and the hip articular surface. It is possible that the clinical expression of immune-mediated joint disease results from an immunologic response to an antigen that crosses the gut mucosa with an autoimmune response against self.[103-110]

For the client with hip pain of unknown cause or suspicious presentation, ask whether any back pain or abdominal pain is ever present. Alternating abdominal pain with low back pain at the same level, or alternating abdominal pain with hip pain is a red flag that requires medical referral.

The therapist may treat a patient with joint or back pain with an underlying enteric cause before he or she realizes what the underlying problem is. Palliative intervention can make a difference in the short term but does not affect the final outcome. Symptoms that are unrelieved by physical therapy intervention are always a red flag. Symptoms that improve after physical therapy but then get worse again are also a red flag, revealing the need for further screening.

CASE EXAMPLE 16-13

Limp After Total Hip Arthroplasty

A 70-year-old man was referred to physical therapy by his doctor 1 year after a right total hip replacement (THR) for osteoarthritis. The client reports that he is in good general health without pain. His primary problem is a persistent limp, despite completion of a THR rehabilitation protocol.

How can you tell whether this is an infectious versus biomechanical problem?

First of all, laboratory tests, such as erythrocyte sedimentation rate (ESR or "sed" rate) and C-reactive protein level, can be done to screen for infection. The therapist can request this information from the medical record.

The absence of pain usually rules out infection or implant loosening. An x-ray may be needed to rule out implant loosening. Again, check the record to see whether this was part of the medical diagnostic workup.

Besides infection, a limp after THR may have many possible causes. Loosening of the prosthesis, neurologic dysfunction, altered joint biomechanics, and muscle weakness or dysfunction (e.g., hip abductors) are a few potential causes. As always, in an orthopedic examination, check the joints above (low back, sacrum, sacroiliac) and below (knee) the level of impairment. In the case of joint replacement, evaluate the contralateral hip as well.

Test for abdominal muscle weakness. This can be confirmed with manual muscle testing or a Trendelenburg test. An anterolateral approach to THR is more likely to cause partial or complete abductor muscle disruption than is a posterior approach.

With either approach, the superior gluteal nerve can be damaged by stretching or by cutting one of its branches. The therapist may be able to get some clues to this by looking at the incision site. Disruption of the nerve is more likely when the gluteus medius is split more than 5 cm proximal to the tip of the greater trochanter. If nerve damage has occurred, the client may not regain full strength. Electromyography (EMG) testing may be needed to document muscle denervation.

Physical therapy may be a diagnostic step for the physician. If muscle strengthening does not recondition the remaining intact muscle, a revision operation to repair the muscle may be needed. It may be helpful to communicate with the physician to see what his or her thinking is on this client.

Data from Farrell CM, Berry DJ: Persistent limping after primary total hip replacement, *J Musculoskel Med* 19:484-486, 2002.

In the case of enterically induced joint pain, the client will get worse without medical intervention. Without early identification and referral, the client will eventually return to his or her gastroenterologist or primary care physician. Medical treatment for the underlying disease is essential in affecting the musculoskeletal component. Physical therapy intervention does not alter or improve the underlying enteric disease. It is better for the client if the therapist recognizes as soon as possible the need for medical intervention.

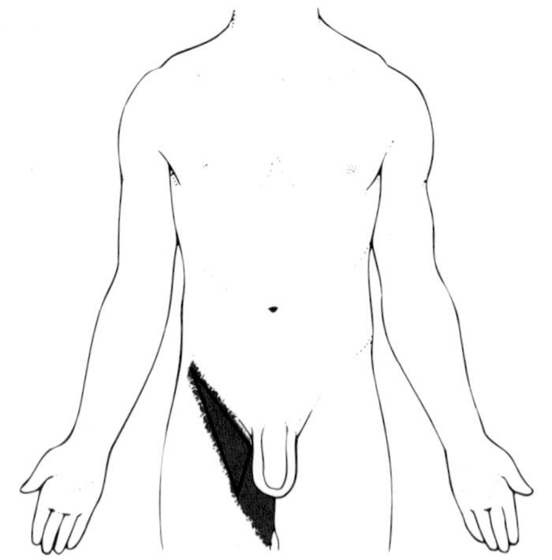

Fig. 16-5 Femoral triangle: Referred pain pattern from psoas abscess. Hip pain associated with such an abscess may involve the medial aspect of the thigh and femoral triangle areas. The femoral triangle is the name given to the anterior aspect of the thigh formed as different muscles and ligaments cross each other, producing an inverted triangular shape.

Crohn's Disease

In anyone with hip or groin pain of unknown cause, look for a known history of PID, Crohn's disease (regional enteritis), ulcerative colitis, irritable bowel syndrome, diverticulitis, or bowel obstruction.

It is possible that new onset of low back, sacral, buttock, or hip pain is merely a new symptom of an already established enteric (GI) disease. Twenty-five percent of those with inflammatory enteric disease (particularly Crohn's disease) have concomitant back or joint pain that are symptoms of spondyloarthritis/spondyloarthropathy.

A skin rash that comes and goes can accompany enterically induced arthritis. A flat rash or raised skin lesion of the lower extremities is possible; it usually precedes joint or back pain. Be sure to ask the client whether he or she has had skin rashes of any kind over the past few weeks.

Several tests can be done to assess for hip pain resulting from psoas abscess caused by abdominal or intraperitoneal infection or inflammation. These were discussed in the previous section.

A positive response for each of these tests is NOT a reproduction of the client's hip or groin pain, but rather, lower quadrant abdominal pain on the side of the test. This is a symptom of an infectious or inflammatory process. Hip or back pain in response to these tests is more likely musculoskeletal in origin such as a trigger point of the iliopsoas or muscular tightness.

Reactive Arthritis

In the case of reactive arthritis, joint symptoms occur 1 to 4 weeks after an infection, usually GI or genitourinary (GU).[110] The joint is not septic (infected), but rather, it is aseptic (without infection). Affected joints often occur at a site that is remote from the primary infection. Prosthetic joints are not immune to this type of infection and may become infected years after the joint is implanted.

Whether the infection occurs in the natural joint or in the prosthetic implant, the client is unable to bear weight on the joint. An acute arthritic presentation may occur, and the client often has a fever (commonly of low grade in older adults or in anyone who is immunosuppressed). Screening questions for clients with joint pain are listed in Box 3-5 and in Appendix B-18. These questions may be helpful for the client with joint pain of unknown cause or with an unusual presentation/history that does not fit the expected pattern for injury, overuse, or aging.

SCREENING FOR VASCULAR CAUSES OF LOWER QUADRANT PAIN

Vascular pain is often throbbing in nature and exacerbated by activity. With atherosclerosis, a lag time of 5 to 10 minutes occurs between when the body asks for increased oxygenated blood and when symptoms occur because of arterial occlusion. The client is older, often with a personal or family history of heart disease. Other risk factors include hyperlipidemia, tobacco use, and diabetes.

Peripheral Vascular Disease

Peripheral vascular disease (PVD), also known as peripheral arterial disease (PAD) or arterial insufficiency, in which the arteries are occluded by atherosclerosis, can cause unilateral or bilateral low back, hip, buttock, groin, or leg pain, along with intermittent claudication and trophic changes of the affected lower extremities.

Intermittent claudication of vascular origin may begin in the calf and may gradually make its way up the lower extremity. The client may report the pain or discomfort as "burning," "cramping," or "sharp." Pain or other symptoms begin several minutes after the start of physical activity and resolve almost immediately with rest. As discussed in Chapter 14, the site of symptoms is determined by the location of the pathology (see Fig. 14-3) (Case Example 16-14).

PVD is a rare cause of lower quadrant pain in anyone under the age of 65, but leg pain in recreational athletes caused by isolated areas of arterial stenosis has been reported.[111]

The therapist must include assessment of vital signs and must look for trophic skin changes so often present with chronic arterial insufficiency. Pulse oximetry may be helpful when thrombosis is not clinically obvious; for example, pulses can be present in both feet with oxygen saturation (SaO_2) levels at 90% or less.[112] When assessing for PVD as a possible cause of back, buttock, hip, groin, or leg pain, look for other signs of PVD. See further discussion of this topic in Chapters 4, 6, and 14.

DVT as a cause of lower leg pain may present as loss of knee or ankle motion, swelling of the knee, calf, or ankle, with calf tenderness and erythema. There can be increased local

CASE EXAMPLE 16-14

Intermittent Claudication with Sciatica

Referral: A 41-year-old woman who was referred by her primary care physician with a medical diagnosis of sciatica reported bilateral lower extremity weakness with pain in the left buttock and left sacroiliac (SI) area. She also noted that she had numbness in her left leg after walking more than half a block.

She said both her legs felt like they were going to "collapse" after she walked a short distance and that her left would go "hot and cold" during walking. She also experienced cramping in her right calf muscle after walking more than half a block.

Symptoms are made worse by walking and better after resting or by standing still. Symptoms have been present for the last 2 months and came on suddenly without trauma or injury of any kind. No night pain was reported.

No medical tests or imaging studies have been done at this time.

Past Medical History: Significant positive for family history of heart disease (both sides of the family); smoking history: 1 pack of cigarettes/day for the past 26 years.

Clinical Presentation

Neurologic Screening Examination: Negative/within normal limits (WNL)

Neural Tissue Mobility: Tests were all negative; tissue tension WNL

Complete Lumbar Spine Examination: Unremarkable; ruled out as a source of client's symptoms

Diminished dorsalis pedis pulse on the left side

Bike Test (reviewed in Chapter 14; this test can be used to stress the integrity of the vascular supply to the lower extremities): Cycling in a position of lumbar forward flexion reproduced leg weakness and eliminated dorsalis pedis pulse on the left; no change was noted on the right.

Associated Signs and Symptoms

None.

What are the red flags in this case?

- Lower extremity (LE) weakness without pain accompanied by "giving out" sensation
- Symptoms brought on by specific activity, relieved by rest or standing still
- Significant family history of heart disease
- No known cause; onset of symptoms without trauma or injury
- Temperature changes in LEs
- Positive smoking history

Result: Given the severity of her family history of heart disease (sudden death at a young age was very common), she was sent back to the doctor immediately. The therapist briefly outlined the red flags and asked the physician to reevaluate for a possible vascular cause of symptoms.

Medical testing revealed a high-grade circumferential stenosis (narrowing) of the distal aorta at the bifurcation. The client underwent surgery for placement of a stent in the occluded artery. After the operation, the client reported complete relief from all symptoms, including buttock and SI pain.

Data from Gray JC: Diagnosis of intermittent vascular claudication in a patient with a diagnosis of sciatica: Case report, *Phys Ther* 79:582-590, 1999.

skin temperature, local edema, and decreased distal pulses in the lower extremity.[51] Further discussion and information on assessment of DVT are presented in Chapters 4 and 6.

Abdominal Aortic Aneurysm

Abdominal aortic aneurysm (AAA) may be asymptomatic; discovery occurs on physical or x-ray examination of the abdomen or lower spine for some other reason. The most common symptom is awareness of a pulsating mass in the abdomen, with or without pain, followed by abdominal and back pain. Groin pain and flank pain may occur because of increasing pressure on other structures. (For more detailed information, see Chapter 6.)

Be aware of the client's age. The client with an AAA can be of any age because this may be a congenital condition, but usually, he or she is over age 50 and more likely, is 65 or older. The condition remains asymptomatic until the wall of the aorta grows large enough to rupture. If that happens, blood in the abdomen causes searing pain accompanied by a sudden drop in blood pressure. Other symptoms of impending rupture or actual rupture of the aortic aneurysm include the following:

- Rapid onset of severe groin pain (usually accompanied by abdominal or back pain)
- Radiation of pain to the abdomen or to posterior thighs
- Pain not relieved by change in position
- Pain described as "tearing" or "ripping"
- Other signs such as cold, pulseless lower extremities

An increasingly prevalent risk factor in the aging adult population is initiation of a weight-lifting program without prior medical evaluation or approval. The presence of atherosclerosis, elevated blood pressure, or an unknown aneurysm during weight training can precipitate rupture.

The therapist can palpate the aortic pulse to identify a widening pulse width, which is suggestive of an aneurysm (see Fig. 4-55). Place one hand or one finger on either side of the aorta as shown. Press firmly deep into the upper abdomen just to the left of midline. You should feel aortic pulsations. These pulsations are easier to appreciate in a thin person and are more difficult to feel in someone with a thick abdominal wall or a large anteroposterior diameter of the abdomen.

Obesity and abdominal ascites or distention make this more difficult. For therapists who are trained in auscultation, listen for bruits. Bruits are abnormal blowing or swishing sounds heard on auscultation of the arteries. Bruits with both systolic and diastolic components suggest the turbulent blood flow of partial arterial occlusion. If the renal artery is occluded as well, the client will be hypertensive.

Avascular Osteonecrosis

Avascular osteonecrosis (also known as *osteonecrosis* or *septic necrosis*) can occur without known cause but is often associated with trauma (e.g., hip dislocation or fracture), as well as various other nontraumatic risk factors.[113] Chronic use and abuse of alcohol is a common risk factor for this condition. Screening for alcohol or drug use and abuse is discussed in Chapter 2 (see also Appendices B-1 and B-2).

Osteonecrosis is also associated with many other conditions such as systemic lupus erythematosus, pancreatitis, kidney disease, blood disorders (e.g., sickle cell disease, coagulopathies, leukemia), diabetes mellitus, Cushing's disease, and gout. Long-term use of corticosteroids or immunosuppressants or use of medications for human immunodeficiency virus (HIV) or acquired immunodeficiency syndrome (AIDS), or any condition that causes immune deficiency, can also result in osteonecrosis.[113] Other individuals who are taking immunosuppressants include organ transplant recipients, clients with cancer, and those with RA or another chronic autoimmune disease.[114]

The femoral head is the most common site of this disorder. Bones with limited blood supply are at enhanced risk for this condition. Hip dislocation or fracture of the neck of the femur may compromise the already precarious vascular supply to the head of the femur. Ischemia leads to poor repair processes and delayed healing. Necrosis and deformation of the bone occur next.

The client may be asymptomatic during the early stages of osteonecrosis. Hip pain is the first symptom. At first, it may be mild, lasting for weeks. As the condition progresses, symptoms become more severe, with pain on weight bearing, antalgic gait, and limited motion (especially internal rotation, flexion, and abduction). The client may report a distinct click in the hip when moving from the sitting position and increased stiffness in the hip as time goes by.

CLINICAL SIGNS AND SYMPTOMS

Osteonecrosis

- May be asymptomatic at first
- Hip pain (mild at first, progressively worse over time)
- Groin or anteromedial thigh pain possible
- Pain worse on weight bearing
- Antalgic gait with a gluteus minimus limp
- Limited hip range of motion (internal rotation, flexion, abduction)
- Tenderness to palpation over the hip joint
- Hip joint stiffness
- Hip dislocation

SCREENING FOR OTHER CAUSES OF LOWER QUADRANT PAIN

Osteoporosis

Osteoporosis may result in hip fracture and accompanying hip pain, especially in postmenopausal women who are not taking hormone replacement. Osteoporosis accompanying the postmenopausal period—when combined with circulatory impairment, postural hypotension, or some medications—may increase a person's risk of falling and incurring hip fracture.

Transient osteoporosis of the hip can occur during third-trimester pregnancy, although the incidence is fairly low. There have been reports (rare) of transient osteoporosis in nonpregnant women, children, adolescents, and men as well.[115] Symptoms include spontaneous acute and progressive hip pain. In some cases, pain is referred to the lateral thigh and severe enough to result in an antalgic gait (limp). There is usually minimal night discomfort. Hip range-of-motion is usually spared though the individual may report pain at the end of internal rotation. Often, the pain subsides in 6 to 8 weeks; this corresponds with resolution of bone edema. During pregnancy, the pain develops shortly before or during the last trimester and is aggravated by weight bearing. There is a classic left-sided predominance seen in pregnant women that is not present in nonpregnant individuals. The pain subsides, and the x-ray appearance returns to normal within several months after delivery.[115,116]

The natural history in nonpregnant individuals is for spontaneous regression and recovery within 6 to 9 months with no permanent problems. X-rays are often normal at presentation but later show progressive osteoporosis of the femoral head (and sometimes the femoral neck and acetabulum).[115]

Any evaluation procedures that produce significant shear through the femoral head of a pregnant woman must be performed by the physical therapist with extreme caution. The transient osteoporosis of pregnancy is not limited to the hip, and vertebral compression may also occur.

Extrapulmonary Tuberculosis

Tubercular disease of the hip or spine is rare in developed countries, but it may occur as an opportunistic disease associated with AIDS that causes hip or back pain. Usually, the diagnosis of AIDS and tuberculosis is known, which alerts the therapist about the underlying systemic cause.

With hip involvement, the client usually appears with a chronic limp and describes pain in the hip that persists at rest. Approximately 60% of affected individuals do not have constitutional symptoms, although the tuberculin skin test is usually positive, and radiographs are similar to those for septic arthritis.

Sickle Cell Anemia and Hemophilia

Sickle cell anemia resulting in avascular necrosis (death of cells caused by lack of blood supply) of the hip and hemarthrosis (blood in the joint) associated with *hemophilia* are two of the most common hematologic diseases that cause pain in the hip, groin, knee, or leg.

Hemophilia may involve GI bleeding accompanied by low abdominal, hip, or groin pain caused by bleeding into the wall of the large intestine or the iliopsoas muscle. This retroperitoneal hemorrhage produces a muscle spasm of the iliopsoas muscle. The subsequent bleeding–spasm cycle produces increased hip pain and hip flexion spasm or contracture. Other symptoms may include melena, hematemesis, and fever.

CLINICAL SIGNS AND SYMPTOMS

Hip Hemarthrosis

- Pain in the groin and thigh
- Fullness in the hip joint, both anterior in the groin and over the greater trochanter
- Limited motion in hip flexion, abduction, and external rotation (allows most room for the blood in the joint capsule)

Liver (Hepatic) Disease

Tarsal tunnel syndrome characterized by pain around the ankle that extends to the plantar surfaces of the toes possibly made worse by walking may be the result of tibial nerve compression from any space-occupying lesion. Causes of compression include a history of trauma (nonunion or displaced fracture), varicosities, lipomas, ganglion cysts, or tumors.[117]

Additional symptoms can include burning pain and numbness on the plantar surface of the foot. Similar symptoms misinterpreted as tarsal tunnel syndrome can occur with neuropathy associated with diabetes mellitus and/or alcoholism. Tinel's sign (reproduction of characteristic pain or tingling with tapping or compression of the tibial nerve) may be positive but does not differentiate between a musculoskeletal cause versus systemic origin of symptoms.[118]

Ascites is an abnormal accumulation of serous (edematous) fluid in the peritoneal cavity; this fluid contains large quantities of protein and electrolytes as the result of portal backup and loss of proteins (see Fig. 9-8). This condition is associated with liver disease and alcoholism. For the physical therapist, the distended abdomen, abdominal hernias, and lumbar lordosis observed in clients with ascites may present musculoskeletal symptoms such as groin or low back pain.

The presence of ascites as it is linked with groin pain would be physically evident. If abdominal distention is present, then the therapist should ask about a past medical history of liver impairment, chronic alcohol use, and the presence of carpal or tarsal tunnel syndrome associated with liver impairment. The therapist can carry out the four screening tests for liver impairment discussed in Chapter 9, including the following:

- Liver flap (asterixis; see Fig. 9-7)
- Palmar erythema (liver palms; see Fig. 9-5)
- Scan for angiomas (upper body and abdomen; see Fig. 9-3)
- Assessment of nail beds for change in color (nail beds of Terry; see Fig. 9-6)
- Asking about the presence of tarsal tunnel (and carpal tunnel) symptoms

PHYSICIAN REFERRAL

Guidelines for Immediate Medical Attention

- Painless, progressive enlargement of lymph nodes, or lymph nodes that are suspicious for any reason and that persist or that involve more than one area (groin and popliteal areas); immediate medical referral is required for a client with a past medical history of cancer
- Hip or groin pain alternating or occurring simultaneously with abdominal pain at the same level **(Aneurysm, colorectal cancer)**
- Hip or leg pain on weight bearing with positive tests for stress reaction or fracture

Guidelines for Physician Referral

- Hip, thigh, or buttock pain in a client with a total hip arthroplasty that is brought on by activity but resolves with continued activity **(Loose prosthesis),** or who has persistent pain that is unrelieved by rest **(Implant infection)**
- Sciatica accompanied by extreme motor weakness, numbness in the groin or rectum, or difficulty controlling bowel or bladder function
- One or more of Cyriax's Signs of the Buttock (see Box 16-2)
- New onset of joint pain in a client with a known history of Crohn's disease, requiring careful screening and possible referral based on examination results

Clues to Screening Lower Quadrant Pain

- See also Clues to Screening Head, Neck, or Back Pain; general concepts from the back also apply to the hip and the groin (see especially the discussion on Cardiovascular)
- Client does not respond to physical therapy intervention or gets worse, especially in the presence of a past medical history of cancer or an unknown cause of symptoms

Past Medical History

- History of AIDS-related tuberculosis, sickle cell anemia, or hemophilia
- History of endometriosis in women **(Extrapelvic endometriosis)**
- Hip or groin pain in a client who has a long-term history of use of NSAIDs or corticosteroids **(Avascular necrosis)**
- History of alcohol abuse or injection drug abuse
- Femoral artery catheterization **(Septic hip arthritis, retroperitoneal hematoma formation)**

Clinical Presentation

- Symptoms are unchanged by rest, movement, or change in position
- Limited passive hip range of motion with empty end feel, especially in someone with a previous history of cancer, insidious onset, or an unknown cause of painful symptoms
- Palpable soft tissue mass in the anterior hip or groin **(Psoas abscess, hernia)**
- Presence of rebound tenderness, positive McBurney's, iliopsoas, or obturator test (see Chapter 8)
- Abnormal cremasteric response in male with groin or anterior thigh pain
- Hip pain in a young adult that is worse at night and is alleviated by activity and aspirin (osteoid osteoma)
- Sciatica in the presence of night pain and an atypical pattern of restricted hip range of motion[119]
- No change in symptoms of sciatica with trigger point release, neural gliding techniques, soft tissue stretching, or postural changes
- Painless neurologic deficit **(Spinal cord tumor)**
- Insidious onset of groin or anterior thigh pain with a recent history of increased activity (e.g., runners who increase their mileage)
- Symptoms are cyclical and related to menstrual cycle **(Endometriosis)**

Associated Signs and Symptoms

- Hip or groin pain accompanied by or alternating with signs and symptoms associated with the GI, urologic/renal, hematologic, or cardiovascular system, or with constitutional symptoms, especially fever and night sweats
- Groin pain in the presence of fever, sweats, weight loss, bleeding, skin lesions, or vaginal/penile discharge; night pain
- Hip or groin pain, with any clues suggestive of cancer (see Chapter 13), especially anyone with a previous history of cancer and men between the ages of 18 and 24 years who experience hip or groin pain of unknown cause **(Testicular cancer)**
- Buttock, hip, thigh, or groin pain accompanied by fever, weight loss, bleeding or other vaginal/penile discharge, skin lesions, or other discharge

REFERRED LOWER QUADRANT PAIN PATTERNS (FIG. 16-6)

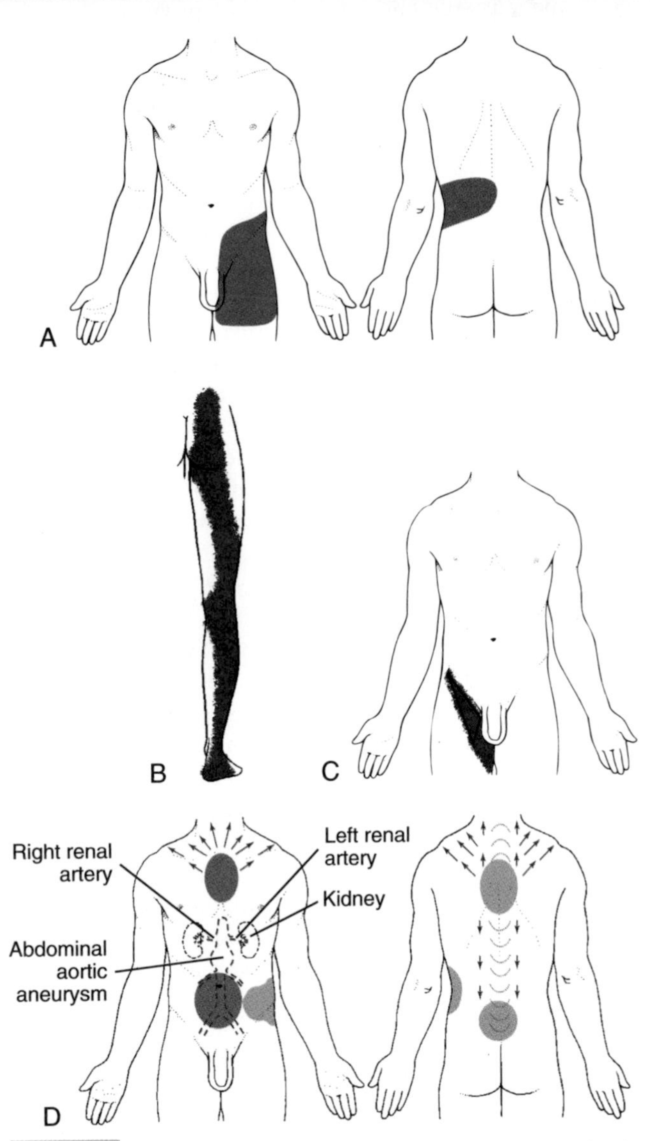

Fig. 16-6 Overview: Composite figure. **A,** Ureteral pain may begin posteriorly in the costovertebral angle, radiating anteriorly to the ipsilateral lower abdomen, upper thigh, or groin area. Isolated anterior thigh pain is possible, but uncommon. **B,** Pain pattern associated with sciatica from any cause. **C,** Pain pattern associated with psoas abscess from any cause. **D,** Abdominal aortic aneurysm can cause low back pain that radiates into the buttock unilaterally or bilaterally (not shown), depending on the underlying location and size of the aneurysm.

Key Points to Remember

- See also Key Points to Remember in Chapter 14.
- Identifying the hip as the source of a client's symptoms may be difficult in that pain originating in the hip may not localize to the hip, but rather may present as low back, buttock, groin, SI, anterior thigh, or even knee or ankle pain.
- Hip pain can be referred from other locations such as the scrotum, kidneys, abdominal wall, abdomen, peritoneum, or retroperitoneal region.
- In addition to screening for medical problems, the therapist must remember to clear the joint above and below the area of symptoms or dysfunction.
- True hip pain from any cause is usually felt in the groin or deep buttock, sometimes with pain radiating down the anterior thigh. Pain perceived on the outer (lateral) side of the hip is usually not caused by an intraarticular problem but likely results from a trigger point or from bursitis, SI, or back problems.
- Hip pain referred from the upper lumbar vertebrae can radiate into the anterior aspect of the thigh, whereas hip pain from the lower lumbar vertebrae and sacrum is usually felt in the gluteal region, with radiation down the back or outer aspect of the thigh.
- Systemic, medical, or viscerogenic causes of lower quadrant pain or symptoms mimic a neuromuscular or musculoskeletal cause, but usually, a red-flag history, risk factors, or associated signs and symptoms are identified during the screening process; this facilitates identification of the underlying problem.
- Cancer recurrence most likely to metastasize to the hip includes breast, bone, and prostate.
- Changes in lymph nodes with or without a previous history of cancer are a yellow or red flag.
- Normal but painful hip rotations (log-rolling test) present when the client is tested in the supine position with the hips in neutral extension (zero degrees of hip flexion) may be a yellow warning flag.
- Cyriax's "Sign of the Buttock" can help differentiate between hip and lumbar spine disease.
- Anyone with lower quadrant pain and a past history of hip or knee arthroplasty must be evaluated for component problems (e.g., infection, subsidence, looseness), regardless of the client's perceived cause of the problem. Watch for pain on initiation of activity that gets better with continued activity (loose prosthesis); also watch for signs of infection (recent history of infection anywhere else in the body, fever, chills, sweats, pain that is not relieved with rest, night pain, pain on weight bearing).
- A noncapsular pattern of restricted hip motion (e.g., limited hip extension, adduction, lateral rotation) may be a sign of serious underlying disease.
- Anyone with pain radiating from the back down the leg as far as the ankle has a greater chance for disk herniation to be the cause of low back pain; this is true with or without neurologic findings.
- The SLR and other neurodynamic tests are widely used but do not identify the underlying cause of sciatica. A positive SLR test does not differentiate between diskogenic disease and neoplasm; imaging studies may be needed.
- Tests for the presence of hip pain caused by psoas abscess are advised whenever an infectious or inflammatory process is suspected on the basis of past medical history, clinical presentation, and associated signs and symptoms.
- New onset of low back, buttock, sacral, or hip pain in a client with a previous history of Crohn's disease, especially in the presence of a recent history of skin rash, requires screening for GI signs and symptoms.
- Long-term use of corticosteroids or immunosuppressants or any condition that causes immune deficiency may also result in hip pain from osteonecrosis. As the condition progresses, symptoms become more severe with pain on weight bearing, antalgic gait, and limited motion.

SUBJECTIVE EXAMINATION

Special Questions to Ask: Lower Quadrant

It is not necessary to ask every client every question listed. Sometimes, we ask some general screening questions because of something the client has told us. At other times, we screen because of something we saw in the clinical presentation. We may need to ask some specific questions based on gender. Finally, sometimes, the Review of Systems has pinpointed a particular system (e.g., GI, GU, vascular, pulmonary, gynecologic), and we go right to the end of the chapter dealing with that system and look for any screening questions that may be pertinent to the client.

The more often the therapist conducts screening interviews, the faster the process will get, and the easier it will become to remember which questions make the most sense to ask. The beginner may ask more questions than are really needed, but with practice and experience, the screening process will smooth out. Generally, it takes about 3 to 5 minutes to conduct a screening interview and another 5 minutes to carry out any special tests.

Because hip pain may be caused by referred pain from disorders of the low back, abdomen, and reproductive and urologic structures, special questions should include consideration of the following:

- Special Questions for Women Experiencing Back, Hip, Pelvic, Groin, or Sacroiliac Pain (see Appendix B-37)
- Special Questions to Ask: Men Experiencing Back, Hip, Pelvic, Groin, or Sacroiliac Pain (see Appendix B-24)

SUBJECTIVE EXAMINATION—*cont'd*

- Special questions for clients (see Chapter 14: Special Questions to Ask: Neck or Back):
- General systemic questions
- Pain assessment
- GI questions
- Urologic questions
- For anyone with lower quadrant pain of unknown cause: It may be necessary to conduct a sexual history as part of the screening process (see Chapter 14 or Appendix B-32).
- A quick screening interview and additional questions may include the following:

Pain Assessment

See Appendix C-7 for a complete pain assessment.

- Have you had a recent injury?
  - *If yes,* tell me what happened.
  - Did you hear any popping, snapping, or cracking when the injury occurred?
- How is the pain affected by putting weight on it?
- Does your leg "give out" on you (or feel like it is going to give out)?
- Does your pain feel better, same, or worse after walking on it for awhile? **(With joint arthroplasty, pain may improve after walking in the presence of loose components.)**

Past Medical History

- Have you ever been told (or have you known) that you have a sexually transmitted infection or disease?
- Have you been treated with cortisone, prednisone, other corticosteroids, or any other drug of that type?
- Do you have a known history of Crohn's disease, diverticulitis, or PID?
- Have you ever had cancer of any kind?

  If no, Have you ever been treated with chemotherapy or radiation therapy?
- Have you ever had a bone tumor?

Associated Signs and Symptoms

- Do you have any other symptoms anywhere else in your body?
  - Any fatigue? Fever? Chills? Swollen joints?

CASE STUDY

Steps in the Screening Process

A 34-year-old woman was referred to physical therapy for pelvic pain from a nonrelaxing puborectalis muscle. She reported bilateral groin pain that was superficial and affected the skin area. She also said the area feels "warm." The pain was worse when sitting, better when standing, and had lasted longer than a month. The physician ruled out shingles and sent her to physical therapy for further evaluation.

WHAT ARE SOME STEPS YOU CAN TAKE TO START THE SCREENING PROCESS?

Have the client complete a past medical history form, and review it for any clues that might help direct the screening process. Ask the usual questions about bowel and bladder function (see Appendices B-5 and B-6).

Superficial skin changes are usually a sudomotor response; messages arrive via the spinal cord, but the system has no way to know the specific source of the problem (i.e., viscerogenic versus somatic), so it sends out a "distress" signal that something is wrong at the S2-S3 level. The therapist must consider what could be involved.

Using Table 16-3 as a guide, the therapist can assess the likelihood of each condition listed on the basis of age, gender, past medical history, and associated signs and symptoms. Screening tests may be conducted, as appropriate. For example, a neurologic screening examination may help identify diskogenic disease or possible spinal cord tumor.

The client is young to have developed an AAA from atherosclerosis, but a congenital aneurysm may be present. Palpating the abdomen and the aortic pulse and listening with a stethoscope for femoral bruits may be helpful.

A stress fracture would likely have a suspicious history such as prolonged activity requiring axial loading or trauma of some kind. It may be necessary to ask about physical or sexual assault. Conduct screening tests such as heel strike, rotational/translational stress test of the pubis, hop on one leg, and full squat. Assess for TrPs.

Ureteral problems are usually accompanied by bladder changes (e.g., dysuria, hematuria, frequency) and constitutional symptoms such as fever, sweats, or chills. Take vital signs.

Gynecologic causes of low back, pelvic, groin, hip, or SI pain are usually accompanied by a significant history of gynecologic conditions or traumatic or multiple birth/delivery history. Some additional questions along these lines may be needed if the past medical history form is not sufficient. Sexually transmitted infection or ectopic pregnancy is possible, although rare causes of groin pain may occur in sexually active women.

Appendicitis or another infectious process can cause a wide range of symptoms outside of the typical or expected right lower abdominal quadrant pain, including isolated groin pain or combined hip and groin pain. McBurney's test (see Fig. 8-9) or Blumberg's sign for rebound tenderness (see Figs. 8-11 and 8-12) can help the therapist to recognize when medical referral is required.

PRACTICE QUESTIONS

1. The screening model used to help identify viscerogenic or systemic origins of hip, groin, and lower extremity pain and symptoms is made up of:
   a. Past medical history, risk factors, clinical presentation, and associated signs and symptoms
   b. Risk factors, risk reduction, and primary prevention
   c. Enteric disease, systemic disease, and neuromusculoskeletal dysfunction
   d. Physical therapy diagnosis, Review of Systems, and physician referral
2. When would you use the iliopsoas, obturator, or Blumberg's test?
3. Hip and groin pain can be referred from:
   a. Low back
   b. Abdomen
   c. Retroperitoneum
   d. All of the above
4. Screening for cancer may be necessary in anyone with hip pain who:
   a. Is younger than 20 or older than 50
   b. Has a past medical history of diabetes mellitus
   c. Reports fever and chills
   d. Has a total hip arthroplasty (THA)
5. Pain on weight bearing may be a sign of hip fracture, even when x-rays are negative. Follow-up clinical tests may include:
   a. McBurney's, Blumberg's, Murphy's test
   b. Squat test, hop test, translational/rotational tests
   c. Psoas and obturator tests
   d. Patrick's or Faber's test
6. Abscess of the hip flexor muscles from intraabdominal infection or inflammation can cause hip and/or groin pain. Clinical tests to differentiate the cause of hip pain resulting from psoas abscess include:
   a. McBurney's, Blumberg's, or Murphy's test
   b. Squat test, hop test, translational/rotational tests
   c. Iliopsoas and obturator tests
   d. Patrick's or Faber's test
7. Anyone with hip pain of unknown cause must be asked about:
   a. Previous history of cancer or Crohn's disease
   b. Recent infection
   c. Presence of skin rash
   d. All of the above
8. Vascular diseases that may cause referred hip pain include:
   a. Coronary artery disease
   b. Intermittent claudication
   c. Aortic aneurysm
   d. All of the above
9. True hip pain is characterized by:
   a. Testicular (male) or labial (female) pain
   b. Groin or deep buttock pain with active or passive range of motion
   c. Positive McBurney's test
   d. All of the above
10. Hip pain associated with primary or metastasized cancer is characterized by:
    a. Bone pain on weight bearing; may not be able to stand on that leg
    b. Night pain that is relieved by aspirin
    c. Positive heel strike test with palpable local tenderness
    d. All of the above

REFERENCES

1. Hammond NA: Left lower-quadrant pain: guidelines from the American College of Radiology appropriateness criteria. *Am Fam Phys* 82(7):766–770, 2010.

1a. Lachiewicz PF: Abductor tendon tears of the hip: evaluation and management. *J Am Acad Orthop Surg* 19(7):385–391, 2011.

2. Lachiewicz PF: Thigh mass resulting from polyethylene wear of a revision total hip arthroplasty. *Clin Orthop Relat Res* 455:274–276, 2007.
3. Browder DA, Erhard RE: Decision making for a painful hip: A case requiring referral. *J Orthop Sports Phys Ther* 35:738–744, 2005.
4. Lesher JM: Hip joint pain referral patterns: a descriptive study. *Pain Med* 9:22–25, 2008.
5. Kimpel DL: Hip pain in a 50-year-old woman with RA. *J Musculoskel Med* 16:651–652, 1999.
6. Bertot AJ, Jarmain SJ, Cosgarea AJ: Hip pain in active adults: 20 clinical pearls. *J Musculoskel Med* 20:35–55, 2003.
7. Tortolani PJ, Carbone JJ, Quartararo LG: Greater trochanteric pain syndrome in patients referred to orthopedic spine specialists. *Spine J* 2:251–254, 2002.
8. Strauss EJ: Greater trochanteric pain syndrome. *Sports Med Arthroscop* 18(2):113–119, 2010.
9. Williams BS, Cohen SP: Greater trochanteric pain syndrome: a review of anatomy, diagnosis, and treatment. *Anesth Analg* 108(5):1662–1670, 2009.
10. Lyle MA, Manes S, McGuinness M, et al: Relationship of physical examination findings and self-reported symptom severity and physical function in patients with degenerative lumbar conditions. *Phys Ther* 85:120–133, 2005.
11. Greenwood MJ, Erhard RE, Jones DL: Differential diagnosis of the hip vs. lumbar spine: Five case reports. *J Orthop Sports Phys Ther* 27:308–315, 1998.
12. Cyriax J: *Textbook of orthopaedic medicine*, ed 8, London, 1982, Bailliere Tindall.
13. VanWye WR: Patient screening by a physical therapist for nonmusculoskeletal hip pain. *Phys Ther* 89(3):248–256, 2009.
14. Cibulka MT, Sinacore DR, Cromer GS, et al: Unilateral hip rotation range of motion asymmetry in patients with sacro-iliac joint regional. *Spine* 23:1009–1015, 1998.
15. Cibulka MT: Symmetrical and asymmetrical hip rotation and its relationship to hip rotator muscle strength. *Clin Biomech* 25(1):56–62, 2010.

16. Brown TE, Larson B, Shen F, et al: Thigh pain after cementless total hip arthroplasty: Evaluation and management. *J Am Acad Orthop Surg* 10:385–392, 2002.
17. Fogel GR, Esses SI: Hip spine syndrome: Management of coexisting radiculopathy and arthritis of the lower extremity. *Spine J* 3:238–241, 2003.
18. Kim YH, Oh SH, Kim JS, et al: Contemporary total hip arthroplasty with and without cement in patients with osteonecrosis of the femoral head. *J Bone Joint Surg* 85:675–681, 2003.
19. Khanuja HS: Cementless femoral fixation in total hip arthroplasty. *J Bone Joint Surg* 93(5):500–507, 2011.
20. Byrd JWT: Investigation of the symptomatic hip: Physical examination. In Byrd JWT, editor: *Operative hip arthroscopy*, ed 2, New York, 2005, Springer, pp 36–50.
21. Kelly BT: Hip arthroscopy: current indications, treatment options, and management issues. *Am J Sports Med* 31:1020–1037, 2003.
22. Hair LC. Deyle G: Eosinophilic granuloma in a patient with hip pain. *J Orthop Sports Phys Ther* 41(2):119, 2011.
23. Sahrmann SA: *Diagnosis and treatment of movement impairment syndromes*, St. Louis, 2002, Mosby.
24. Sahrmann S: *Movement system impairment syndromes of the extremities, cervical, and thoracic spines*. St. Louis, 2010, Mosby.
25. Grumet RC: Lateral hip pain in an athletic population: differential diagnosis and treatment options. *Sports Health* 2(3):191–196, 2010.
26. Schilders E: Adductor-related groin pain in recreational athletes. *J Bone Joint Surg* 91A(10):2455–2460, 2009.
27. Kluin J: Endoscopic evaluation and treatment of groin pain in the athlete. *Am J Sports Med* 32(4):944–949, 2004.
28. Swan KG, Wolcott M: The athletic hernia: a systematic review. *Clin Ortho Relat Res* 455:78–87, 2006.
29. Larson CM: Athletic Pubalgia: current concepts and evolving management. *Orthop Today* 31(2):46–52, 2011.
30. van Veen RN: Successful endoscopic treatment of chronic groin pain in athletes. *Surg Endosc* 21:189–193, 2007.
31. Kachingwe AF, Grech S: Proposed algorithm for the management of athletes with athletic pubalgia (sports hernia): a case series. *J Orthop Sports Phys Ther* 38(12):768–781, 2008.
32. Burnett SJ: Clinical presentation of patients with tears of the acetabular labrum. *J Bone Joint Surg* 88A(7):1448–1552, 2006.
33. Parvizi J: Femoroacetabular impingement. *J Am Acad Ortho Surg* 15(69):561–570, 2007.
34. Mcintyre J: Groin pain in athletes. *Curr Sports Med Report* 5(6):293–299, 2006.
35. Zajick D, Zoga A, Omar I, et al: Spectrum of MRI findings in clinical athletic pubalgia. *Sem Musculoskel Radiol* 12(1):3–12, 2008.
36. Cowan SM, Schache P, Brukner KL, et al: Onset of transversus abdominis in long-standing groin pain. *Med Sci Sports Exerc* 36:2040–2045, 2004.
37. Mabry LM: Insufficiency fracture of the pubic rami. *J Orthop Sports Phys Ther* 40(10):666, 2010.
38. Maigne J-Y: Upper thoracic dorsal rami: anatomic study of their medial cutaneous branches. *Surg Radiol Anat* 13:109–112, 1991.
39. Giles LGF, Singer KP: *The clinical anatomy and management of thoracic spine pain*. Oxford, 2000, Butterworth Heinemann.
40. Mesiha M: Synovial sarcoma presenting as iliotibial band friction syndrome: Case report. *J Knee Surg* 22(4):376–378, 2009.
41. Youssef JA: Minimally invasive surgery: lateral approach interbody fusion. *Spine* 35(26S):S302–S311, 2010.
42. Tamir E, Anekshtein Y, Melamed E, et al: Clinical presentation and anatomic position of L3-L4 disc herniation. *J Spinal Disord Tech* 17:467–469, 2004.
43. Reverse straight leg raise test. Available online at http://courses.washington.edu/hubio553/glossary/reverse.html. Accessed March 16, 2011.
44. Foster MR: Herniated nucleus pulposus. *eMedicine Specialties*. Updated Jan 8, 2010. Available online at http://emedicine.medscape.com/article/1263961-overview. Accessed March 16, 2011.
45. Cho KT: Prone position-related meralgia paresthetica after lumbar spinal surgery: a case report and review of the literature. *J Korean Neurosurg Soc* 44(6):392–395, 2008.
46. Weier CA: Meralgia paresthetica of the contralateral leg after total hip arthroplasty. *Orthopedics* 16:265–268, 2010.
47. Yang SH, Wu CC, Chen PQ: Postoperative meralgia paresthetica after posterior spine surgery. *Spine* 30:E547–E550, 2005.
48. Capeci CM, Tejwani NC: Bilateral low-energy simultaneous or sequential femoral fractures in patients on long-term alendronate therapy. *J Bone Joint Surg* 91A(11):2556–2561, 2009.
49. Rosenthal MD: Diagnosis of medial knee pain: atypical stress fracture about the knee joint. *J Orthop Sports Phys Ther* 36(7):526–534, 2006.
50. Constantinou M: Differential diagnosis of a soft tissue mass in the calf. *J Orthop Sports Phys Ther* 35:88–94, 2005.
51. Fink ML, Stoneman PD: Deep vein thrombosis in an athletic military cadet. *J Orthop Sports Phys Ther* 36(9):686–697, 2006.
52. Poppert E, Kulig K: Hip degenerative joint disease in a patient with medial knee pain. *J Orthop Sports Phys Ther* 41(1):33, 2011.
53. Emms NW: Hip pathology can masquerade as knee pain in adults. *Age Ageing* 31:67–69, 2002.
54. Vaughn DW: Isolated knee pain: a case report highlighting regional interdependence. *J Orthop Sports Phys Ther* 38(10):616–623, 2008.
55. Brown EC: The painful total knee arthroplasty: diagnosis and management. *Orthopedics* 29(2):129–138, 2006.
56. Cummings M: Referred knee pain treated with electroacupuncture to iliopsoas. *Acupunct Med* 21:32–35, 2003.
57. Travell JG, Simons DG: *Myofascial pain and dysfunction: the lower extremities*, vol 2, Baltimore, 1992, Williams and Wilkins.
58. Guss DA: Hip fracture presenting as isolated knee pain. *Ann Emerg Med* 29:418–420, 1997.
59. Muscolo DL: Tumors about the knee misdiagnosed as athletic injuries. *J Bone Joint Surg Am* 85A:1209–1214, 2003.
60. Abdulla AJ: Leg cramps in the elderly: prevalence, drug, and disease associations. *Int J Clin Pract* 53:494–496, 1999.
61. Butler JV: Nocturnal leg cramps in older people. *Postgrad Med J* 78:596–598, 2002.
62. Matsumoto M: Nocturnal leg cramps. *Spine* 34(5):E189–E194, 2009.
63. Steele MK: Relieving cramps in high school athletes. *J Musculoskel Med* 20:210, 2003.
64. Lui E: Systemic causes of heel pain. *Clin Podiatr Med Surg* 27:431–441, 2010.
65. Maheshwari AV: Metastatic skeletal disease of the foot: case reports and literature review. *Foot Ankle Int* 29:699–710, 2008.
66. Berlin SJ: Tumors of the heel. *Clin Podiatr Med Surg* 7:307–321, 1990.
67. Groves MJ: Metastatic breast cancer presenting as heel pain. *J Am Podiatr Med Assoc* 88:400–405, 1998.
68. Krause DA, Newcomer KL: Femoral neck stress fracture in a male runner. *J Orthop Sports Phys Ther* 38(8):517, 2008.
69. Thelen MD: Identification of a high-risk anterior tibial stress fracture. *J Orthop Sports Phys Ther* 40(12):833, 2010.
70. Duquette TL, Watson DJ: Femoral neck stress fracture in a military trainee. *J Orthop Sports Phys Ther* 40(12):834, 2010.
71. Brukner P, Bennell KM, Matheson G: *Stress fractures*, Australia, 1999, Blackwell Publishing.

72. Seidenberg PH, Childress MA: Managing hip pain in athletes. *J Musculoskel Med* 22:246–254, 2005.
73. Kelly AK, Hame SL: Managing stress fractures in athletes. *J Musculoskel Med* 27(12):480–486, 2010.
74. Cho CH: Sacral fractures and sacroplasty. *Neuroimaging Clin N Am* 20(2):179–186, 2010.
75. Gurney B, Boissonnault WG, Andrews R: Differential diagnosis of a femoral neck/head stress fracture. *J Orthop Sports Phys Ther* 36(2):80–88, 2006.
76. Carpintero P: Stress fractures of the femoral neck and coxa vara. *Arch Orthop Trauma Surg* 123(6):273–277, 2003.
77. Tommasini SM: Relationship between bone morphology and bone quality in male tibias: implications for stress fracture risk. *J Bone Min Res* 20(8):1372–1380, 2005.
78. Weishaar MD, McMillian DJ, Moore JH: Identification and management of 2 femoral shaft stress injuries. *J Orthop Sports Phys Ther* 35:665–673, 2005.
79. Johnson AW, Weiss CB, Jr, Wheeler DL: Stress fractures of the femoral shaft in athletes—more common than expected: A new clinical test. *Am J Sports Med* 22:248–256, 1994.
80. Hunt KJ, Anderson RB: Heel pain in the athlete. *Sports Health* 1(5):427–434, 2009.
81. Salter RB: *Textbook of disorders and injuries of the musculoskeletal system*, ed 3, Baltimore, 1999, Williams and Wilkins.
82. Hoppenfeld S, Murthy VL: *Treatment and rehabilitation of fractures*, Philadelphia, 2000, Lippincott Williams & Wilkins.
83. Ozburn MS, Nichols JW: Pubic ramus and adductor insertion stress fractures in female basic trainees. *Milit Med* 146:332–334, 1981.
84. Valat JP: Sciatica. *Best Pract Res Clin Rheum* 24(2):241–252, 2010.
85. Jewell DV, Riddle DL: Interventions that increase or decrease the likelihood of a meaningful improvement in physical health in patients with sciatica. *Phys Ther* 85(11):1139–1150, 2005.
86. Yuen EC, So YT: Sciatic neuropathy. *Neurol Clin* 17:617–631, 1999.
87. Martin WN, Dixon JH, Sandhu H: The incidence of cement extrusion from the acetabulum in total hip arthroplasty. *J Arthroplasty* 18:338–341, 2003.
88. Carricajo A: Propionibacterium acnes contamination in lumbar disc surgery. *J Hosp Infect* 66(3):275–277, 2007.
89. Stirling A, Worthington T, Rafiq M, et al: Association between sciatica and *Propionibacterium acnes*. *Lancet* 357:2024–2025, 2001.
90. McLorinn GC, Glenn JV, McMullan MG, et al: *Propionibacterium acnes* wound contamination at the time of spinal surgery. *Clin Orthop Relat Res* 437:67–73, 2005.
91. Hulbert A, Deyle GD: Differential diagnosis and conservative treatment for piriformis syndrome: a review of the literature. *Curr Orthop Pract* 20(3):313–319, 2009.
92. Floyd JR, 2nd: Cyclic sciatica from extrapelvic endometriosis affecting the sciatic nerve. *J Neurosurg Spine* 14(2):281–289, 2011.
93. Bickels J, Kahanvitz N, Rubert CK, et al: Extraspinal bone and soft-tissue tumors as a cause of sciatica: Clinical diagnosis and recommendations: analysis of 32 cases. *Spine* 24:1611, 1999.
94. Chin KR, Kim JM: A rare anterior sacral osteochondroma presenting as sciatica in an adult: a case report and review of the literature. *Spine J* 10(5):e1–e4, 2010.
95. Deyo RA, Diehl AK: Cancer as a cause of back pain: Frequency, clinical presentation, and diagnostic strategies. *J Gen Intern Med* 3:230–238, 1988.
96. Guyer RD, Collier RR, Ohnmeiss DD, et al: Extraosseous spinal lesions mimicking disc disease. *Spine* 13:328–331, 1988.
97. Bose B: Thoracic extruded disc mimicking spinal cord tumor. *Spine J* 3:82–86, 2003.
98. Arromdee E, Matteson EL: Bursitis: Common condition, uncommon challenge. *J Musculoskel Med* 18:213–224, 2001.
99. Magee DJ: *Orthopedic physical assessment*, ed 5, Philadelphia, 2008, WB Saunders.
100. Doubleday KL, Kulig K, Landel R: Treatment of testicular pain using conservative management of the thoracolumbar spine: A case report. *Arch Phys Med Rehabil* 84:1903–1905, 2003.
101. Keulers BJ, Roumen RH, Keulers MJ, et al: Bilateral groin pain from a rotten molar. *Lancet* 366:94, 2005.
102. Todkar M: Case report: Psoas abscess—Unusual etiology of groin pain. *Medscape Gen Med* 7. Available at: http://www.medscape.com/viewarticle/507610_print. Accessed online March 12, 2011.
103. Inman RD: Arthritis and enteritis—An interface of protean manifestations. *J Rheumatol* 14:406–410, 1987.
104. Inman RD: Antigens, the gastrointestinal tract, and arthritis. *Rheum Dis Clin North Am* 17:309–321, 1991.
105. Gran JT, Husby G: Joint manifestations in gastrointestinal diseases. 1. Pathophysiological aspects, ulcerative colitis and Crohn's disease. *Dig Dis* 10:274–294, 1992.
106. Gran JT, Husby G: Joint manifestations in gastrointestinal diseases. 2. Whipple's disease, enteric infections, intestinal bypass operations, gluten-sensitive enteropathy, pseudomembranous colitis and collagenous colitis. *Dig Dis* 10:295–312, 1992.
107. Keating RM, Vyas AS: Reactive arthritis following *Clostridium difficile* colitis. *West J Med* 162:61–63, 1995.
108. Tu J: Bowel bypass syndrome/bowel-associated dermatosis arthritis syndrome post laparoscopic gastric bypass surgery. *Australas J Dermatol* 52(1):e5–7, 2011.
109. Brakenhoff LK: The joint-gut axis in inflammatory bowel disease. *J Crohns Colitis* 4(3):257–268, 2010.
110. Prati C: Reactive arthritis due to Clostridium difficile. *Joint Bone Spine* 77(2):190–192, 2010.
111. Lundgren JM, Davis BA: End artery stenosis of the popliteal artery mimicking gastrocnemius strain. *Arch Phys Med Rehabil* 85:1548–1551, 2004.
112. Brau SA, Delamarter RB, Schiffman ML, et al: Vascular injury during anterior lumbar surgery. *Spine J* 4:409–441, 2004.
113. Babis GC: Osteonecrosis of the femoral head. *Orthopedics* 34(1):39–48, 2011.
114. Norton R: Hip pain in young adults: making a difficult diagnosis. *J Musculoskel Med* 23(12):857–872, 2006.
115. Holzer I: Transient osteoporosis of the hip: long-term outcomes in men and nonpregnant women. *Curr Orthop Pract* 20(2):161–163, 2009.
116. Boissonnault WB, Boissonnault JS: Transient osteoporosis of the hip associated with pregnancy. *J Orthop Sports Phys Ther* 31:359–367, 2001.
117. Kline AJ: Current concepts in managing chronic ankle pain. *J Musculoskel Med* 24(11):477–484, 2007.
118. Bilstrom E: Injection of the carpal and tarsal tunnels. *J Musculoskel Med* 24(11):472–474, 2007.
119. Ross MD, Bayer E: Cancer as a cause of low back pain in a patient seen in a direct access physical therapy setting. *J Orthop Sports Phys Ther* 35:651–658, 2005.
120. Goodman CC, Snyder TEK: Laboratory tests and values. In Goodman CC, Fuller K, editors: *Pathology: implications for the physical therapist*, ed 3, Philadelphia, 2009, WB Saunders.

CHAPTER

17

Screening the Chest, Breasts, and Ribs

Clients do not present in a physical therapy clinic with chest or breast pain as the primary symptom very often. The therapist is more likely to see the individual with an orthopedic or neurologic impairment who experiences chest or breast pain during exercise or during other intervention by the therapist.

In other situations, the client reports chest or breast pain as an additional symptom during the screening interview. The pain may occur along with (or alternating with) the presenting symptoms of jaw, neck, upper back, shoulder, breast, or arm pain. When chest pain is the primary complaint, it is often an atypical pain pattern (possibly in a young athlete) that has misled the client and/or the physician.[1]

On the other hand, it is also possible for clients to have primary chest pain from a human movement system impairment, particularly spinal referred pain.[2] Symptoms persist or recur, often with months in between when the client is free of any symptoms. Countless medical tests are performed and repeated with referral to numerous specialists before a physical therapist is consulted (see Case Example 1-7).

Finally, so many of today's aging adults with movement system impairments have multiple medical comorbidities that the therapists must be able to identify signs and symptoms of systemic disease that can mimic neuromuscular or musculoskeletal dysfunction. Systemic or viscerogenic pain or symptoms that can be referred to the chest or breast include the cardiovascular, pulmonary, and upper gastrointestinal (GI) systems, as well as other causes such as cancer, anxiety, steroid use, and cocaine use (Table 17-1).[3] Various neuromusculoskeletal (NMS) conditions, such as thoracic outlet syndrome, costochondritis, trigger points, and cervical spine disorders, can also affect the chest and breast.[4]

When faced with chest pain, the therapist must know how to assess the situation quickly and decide if medical referral is required and whether medical attention is needed immediately. As experts in understanding and assessing the human movement system, we are the most capable health care professional when it comes to differentiating NMS from systemic origins of symptoms.

The therapist must especially know how and what to look for to screen for cancer, cancer recurrence, and/or the delayed effects of cancer treatment. Cancer can present as primary chest pain with or without accompanying neck, shoulder, and/or upper back pain/symptoms. Basic principles of cancer screening are presented in Chapter 13; specific clues related to the chest, breast, and ribs will be discussed in this chapter. Breast cancer is always a consideration with upper quadrant pain or dysfunction.

USING THE SCREENING MODEL TO EVALUATE THE CHEST, BREASTS, OR RIBS

There are many causes of chest pain, both cardiac and noncardiac in origin (see Table 17-1). Two conditions may be present at the same time, each contributing to chest pain. For example, someone with cervicodorsal arthritis could also experience reflux esophagitis or coronary disease. Either or both of these conditions can contribute to chest pain.

Chest pain can be evaluated in one of two ways: cardiac versus noncardiac or systemic versus neuromusculoskeletal (NMS). Physicians and nurses assess chest pain from the first paradigm: cardiac versus noncardiac. The therapist must understand the basis for this screening method while also viewing each problem as potentially systemic versus NMS. Throughout the screening process, it is important to remember we are not medical cardiac specialists; we are just screening for systemic disease masquerading as NMS symptoms or dysfunction.

Paying attention to past medical history, recognizing unusual clinical presentation for a neuromuscular or musculoskeletal condition, and keeping in mind the clues to differentiating chest pain will help the therapist evaluate difficult cases.

Additionally, the woman with chest, breast, axillary, or shoulder pain of unknown origin at presentation must be questioned regarding breast self-examinations. Any recently discovered lumps or nodules must be examined by a physician. The client may need education regarding breast self-examination, and the physical therapist can provide this valuable information.[5,6] Techniques of breast self-examination are commonly available in written form for the physical

TABLE 17-1 Causes of Chest Pain

Systemic/Medical Conditions	Neuromusculoskeletal
Cancer • Mediastinal tumors Cardiac • Myocardial ischemia (unstable angina) • Myocardial infarct • Cardiomyopathy • Myocarditis • Pericarditis • Dissecting aortic aneurysm • Aortic aneurysm • Aortic stenosis or regurgitation • Mitral valve prolapse* • Tachycardia Pleuropulmonary • Asthma/COPD • Pulmonary embolism • Pneumothorax • Pulmonary hypertension* • Cor pulmonale • Pneumonia with pleurisy or pleuritis • Mediastinitis Epigastric/upper GI • Esophagitis* • Esophageal spasm* • Reflux or other motility disorders • Hiatal hernia • Upper GI ulcer • Cholecystitis • Pancreatitis Breast (see Table 17-2) Hematologic • Anemia • Polycythemia • Sickle cell crisis Other • Rheumatic diseases (sternoclavicular joint) • Infection (sepsis): sternoclavicular joint (injection drug use) • Anxiety, panic attack* • Vertebroplasty (possible pulmonary embolism) • Cocaine use • Anabolic steroids • Collagen vascular disorders with pleuritis or pericarditis • Fibromyalgia • Hyperthyroidism • Dialysis (first-use syndrome) • Type III hypersensitivity reaction • Herpes zoster (shingles) • Psychogenic	Tietze's syndrome Costochondritis, sternochondritis Sternoclavicular joint strain Hypersensitive xiphoid, xiphodynia Slipping rib syndrome TrPs (see Table 17-4) Myalgia Cervical spine disorders, arthritis Neurologic • Nerve root compression • Intercostal neuritis • Dorsal nerve root irritation • TOS • Thoracic disk disease Postoperative pain Breast • Mastodynia • TrPs • Trauma (including motor vehicle accident, assault) • Rib fracture, costochondral • Dislocations, chest contusion

COPD, Chronic obstructive pulmonary disease; *GI*, gastrointestinal; *TrPs*, trigger points; *TOS*, thoracic outlet syndrome.
*Relieved by nitroglycerin because it relaxes smooth muscle.

therapist or the client who is unfamiliar with these methods (see Appendix D-6).

Past Medical History

Although the past medical history (PMH) is important, it cannot be relied upon to confirm or rule out medical causes of chest pain. PMH does alert the therapist to an increased risk of systemic conditions that can masquerade as NMS disorders. Like risk factors, PMH varies according to each system affected or condition present and is reviewed individually in each section of this chapter.

Risk Factors

Any suspicious findings should be checked by a physician, especially in the case of the client with identified risk factors for cancer or heart disease. Identifying red-flag risk factors and PMH and then correlating this information with objective findings are important steps in the screening process.

Risk for cardiac-caused symptoms increases with advancing age, tobacco use, menopause (women), family history of hypertension or premature coronary artery disease, and high cholesterol. Risk factors associated with noncardiac conditions vary with each individual condition (e.g., infectious, rheumatologic, pulmonary, or other systemic causes).

Clinical Presentation

When the clinical presentation suggests further screening is needed, the therapist can follow the guide to pain assessment (see Chapter 3) and physical assessment for the upper quadrant as presented in Table 4-13. Assess vital signs and watch for trends in heart rate and blood pressure. Keep in mind that tachycardia may be a compensatory response to reduced cardiac output and bradycardia may be an indication of myocardial ischemia or (unreported) trauma.

The client's general appearance, along with vital sign assessment, will offer some idea of the severity of the condition. Watch for uneven pulses from side to side, diminished or absent pulses, elevated blood pressure, or extreme hypotension. Auscultation for breath or lung sounds and chest percussion may provide additional cardiopulmonary clues.

Check to see if the pain can be reproduced or made worse by palpation or with pressure on the chest; and, of course, ask about associated symptoms such as nausea and shortness of breath. The key features that point to spinal referred pain are chest pain reproduced on movement (especially with resistive movement), tenderness and tightness of musculoskeletal structures at a spinal level supplying the painful area, and an absence or lack of symptoms suggestive of a nonmusculoskeletal cause.[2]

Chest Pain Patterns

From the previous discussion in Chapter 3, we know that there are at least three possible mechanisms for referred pain patterns to the soma from the viscera (embryologic development, multisegmental innervations, direct pressure on the diaphragm). Pain in the chest may be derived from the chest wall (dermatomes T1-12), the pleura, the trachea and main airways, the mediastinum (including the heart and esophagus), and the abdominal viscera. From an embryologic point of view, the lungs are derived from the same tissue as the gut, so problems can occur in both areas (lung or gut), causing chest pain and other related symptoms.

Certain chest pain patterns are more likely to point to a medical rather than musculoskeletal cause. For example, pain that is positional or reproduced by palpation is not as suspicious as pain that radiates to one or both shoulders or arms or that is precipitated by exertion. Physicians agree that the chest pain history by itself is not enough to rule out cardiac or other systemic origin of symptoms. In most cases, some diagnostic testing is needed.[7]

Chest pain associated with increased activity is a red flag for possible cardiovascular involvement. In such cases, the onset of pain is not immediate but rather occurs 5 to 10 minutes after activity begins. This is referred to as the "lag time" and is a screening clue used by the physical therapist to assess when chest pain may be caused by musculoskeletal dysfunction (immediate chest pain occurs with movement of the arms and/or trunk) or by possible vascular compromise (chest pain occurs 5 to 10 minutes after activity begins).

Parietal pain may appear as unilateral chest pain (rather than midline only) because at any given point the parietal peritoneum obtains innervation from only one side of the nervous system. It is usually not reproduced by palpation. Thoracic disk disease can also present as unilateral chest pain, requiring careful screening.[8,9]

The four types of pain discussed in Chapter 3 (cutaneous, deep somatic or parietal, visceral, and referred) also apply to the chest. *Parietal (somatic) chest pain* is the most common systemic chest discomfort encountered in a physical therapy practice. Parietal pain refers to pain generating from the wall of any cavity, such as the chest or pelvic cavity (see Fig. 6-5). Although the visceral pleura are insensitive to pain, the parietal pleura are well supplied with pain nerve endings. It is usually associated with infectious diseases but is also seen in pneumothorax, rib fractures, pulmonary embolism with infarction, and other systemic conditions.

Pain fibers, originating in the parietal pleura, are conveyed through the chest wall as fine twigs of the intercostal nerves. Irritation of these nerve fibers results in pain in the chest wall that is usually described as knifelike and is sharply localized close to the chest wall, occurring cutaneously (in the skin).

Pain from the thoracic viscera and true chest wall pain are both felt in the chest wall, but *visceral pain* is referred to the area supplied by the upper four thoracic nerve roots. Report of pain in the lower chest usually indicates local disease, but upper chest pain may be caused by disease located deeper in the chest.

There are few nerve endings (if any) in the visceral pleurae (linings of the various organs), such as the heart or lungs. The exception to this statement is in the area of the pericardium (sac enclosed around the entire heart), which is adjacent to the diaphragm (see Fig. 6-5). Extensive disease may develop within the body cavities without the occurrence of pain until the process extends to the parietal pleura. Neuritis (constant irritation of nerve endings) in the parietal pleura then produces the pain described in this section.

Pleural pain may be aggravated by any respiratory movement involving the diaphragm, such as sighing, deep breathing, coughing, sneezing, laughing, or the hiccups. It may be referred along the costal margins or into the upper abdominal quadrants. Palpation usually does not reproduce pleural pain; change in position does not relieve or exacerbate the pain. In some cases of pleurisy, the individual can point to the painful spot but deep breathing (not palpation) reproduces it.

Associated Signs and Symptoms

If the client has an underlying infectious or inflammatory process causing chest or breast pain or symptoms, there may be changes in vital signs and/or constitutional symptoms such as chills, night sweats, fever, upper respiratory symptoms, or GI distress.

Signs and symptoms associated with noncardiac causes of chest pain vary according to the underlying system involved. For example, cough, sputum production, and a recent history of upper respiratory infection may point to a pleuropulmonary origin of chest or breast pain. Anyone with persistent coughing or asthma can experience chest pain related to the strain of the chest wall muscles.

Chest or breast pain associated with GI disease is often food related in the presence of a history of peptic ulcer, gastroesophageal reflux disease (GERD), or gallbladder problems. Blood in the stool or vomitus, along with a history of chronic nonsteroidal antiinflammatory drug (NSAID) use, may point to a GI problem and so on.

Many of the conditions affecting the breast are not accompanied by other systemic signs and symptoms. Risk factors, client history, and clinical presentation provide the major clues as to a viscerogenic, systemic, or cancerous origin of chest and/or breast pain or symptoms.

SCREENING FOR ONCOLOGIC CAUSES OF CHEST OR RIB PAIN

Cancer can present as primary chest, neck, shoulder, and/or upper back pain and symptoms. A previous history of cancer of any kind is a major red flag (Case Example 17-1). Primary cancer affecting the chest with referred pain to the breast is not as common as cancer metastasized to the pulmonary system with subsequent pulmonary and chest/breast symptoms.

Clinical Presentation

The most common symptoms associated with metastases to the pulmonary system are pleural pain, dyspnea, and persistent cough. As with any visceral system, symptoms may not occur until the neoplasm is quite large or invasive because the

CASE EXAMPLE 17-1

Rib Metastases Associated with Ovarian Cancer

Referral: A 53-year-old university professor came to the physical therapy clinic with complaints of severe left shoulder pain radiating across her chest and down her arm. She rated the pain a 10 on the numeric rating scale (NRS; see explanation in Chapter 3).

Past Medical History: She had a significant personal and social history, including ovarian cancer 10 years ago, death of a parent last year, filing for personal bankruptcy this year, and a divorce after 30 years of marriage.

Clinical Presentation (First Visit): During the screening examination for vital signs, the client's blood pressure was 220/125 mm Hg. Pulse was 88 bpm. Pulse oximeter measured 98%. Oral temperature: 98.0° F. She denied any previous history of cardiovascular problems or current feelings of stress.

Intervention: She was referred for medical attention immediately on the basis of her blood pressure readings but returned a week later with a medical diagnosis of "rib bruise." Electrocardiography (ECG) and heart catheterization ruled out a cardiac cause of symptoms. She was put on Prilosec for gastroesophageal reflux disease (GERD) and an antiinflammatory for her rib pain.

Clinical Presentation (Second Visit): The therapist was able to reproduce the symptoms described above with moderate palpation of the eighth rib on the left side and sidebending motion to the left side. The client described the symptoms as constant, sharp, burning, and intense. She had pain at night if she slept too long on either side.

Sidelying on the involved side and slump sitting did not reproduce the symptoms. There was no obvious mechanical cause for the painful symptoms (e.g., intercostal tear, costovertebral dysfunction, neuritis from nerve entrapment).

The therapist considered the possibility of a somato-visceral reflex response (e.g., a biomechanical dysfunction of the tenth rib can cause gallbladder changes), but there were no accompanying associated signs and symptoms and the tenth rib was not painful.

Result: The therapist decided to contact the referring physician to discuss the client's clinical presentation before initiating treatment, especially given the constancy and intensity of the pain in the presence of a past medical history of cancer.

The physician directed the therapist to have the client return for further testing. A bone scan revealed metastases to the ribs and thoracic spine. Physical therapy intervention was not appropriate at this time.

lining surrounding the lungs has no pain perception. Symptoms first appear when the tumor is large enough to press on other nearby structures or against the chest wall. The presence of any skin changes, lesions, or masses should be documented using the information presented in Box 4-11. Skeletal pain from metastases to the bone or primary cancers such as multiple myeloma affecting the sternum can present much like costochondritis.

Skin Changes

Ask the client about any recent or current skin changes. Metastatic carcinoma can present with a cellulitic appearance on the anterior chest wall as a result of carcinoma of the lung (see Fig. 4-26). The skin lesion may be flat or raised and any color from brown to red or purple.

Liver impairment from cancer or any liver disease can also cause other skin changes, such as angiomas over the chest wall. An angioma is a benign tumor with blood (or lymph, as in lymphangioma) vessels. Spider angioma (also called spider nevus) is a form of telangiectasis, a permanently dilated group of superficial capillaries (or venules; see Fig. 9-3).

In the presence of skin lesions, ask about a recent history of infection of any kind, use of prescription drugs within the last 6 weeks, and previous history of cancer of any kind. Look for lymph node changes. Report all of these findings to the physician.

Palpable Mass

Occasionally, the therapist may palpate a painless sternal or chest wall mass when evaluating the head and neck region. Most mediastinal tumors are the result of a metastatic focus from a distant primary tumor and remain asymptomatic unless they compress mediastinal structures or invade the chest wall.

The primary tumor is usually a lymphoma (Hodgkin's lymphoma in a young adult or non-Hodgkin's lymphoma in a child or older adult; see Fig. 4-29), multiple myeloma (primarily observed in people over 60 years of age), or carcinoma of the breast, kidney, or thyroid.

When involvement of the chest wall and nerve roots results in pain, the pattern is more diffuse, with radiation of pain to the affected nerve roots (Case Example 17-2). Irritation of an intercostal nerve from rib metastasis produces burning pain that is unilateral and segmental in distribution. Sensory loss or hyperesthesia over the affected dermatomes may be noted.

SCREENING FOR CARDIOVASCULAR CAUSES OF CHEST, BREAST, OR RIB PAIN

Cardiac-related chest pain may arise secondary to angina, myocardial infarction, pericarditis, endocarditis, mitral valve prolapse, or aortic aneurysm. Despite diagnostic advances, acute coronary syndromes and myocardial infarctions are missed in 2% to 10% of patients.[7] There is no single element of chest pain history powerful enough to predict who is or who is not having a coronary-related incident. Medical referral is advised whenever there is any doubt; medical diagnostic testing is almost always required.[7]

Cardiac-related chest pain also can occur when there is normal coronary circulation, as in the case of clients with pernicious anemia. Affected clients may have chest pain or angina on physical exertion because of the lack of nutrition to the myocardium.

CASE EXAMPLE 17-2

Lymphoma Masquerading as Nerve Entrapment

Referral: A 72-year-old woman was referred to physical therapy for a postural exercise program and home traction by her neurologist with a diagnosis of "nerve entrapment." She was experiencing symptoms of left shoulder pain with numbness and tingling in the ulnar nerve distribution. She had a moderate forward head posture with slumped shoulders and loss of height from known osteoporosis.

Past Medical History: The woman's past medical history was significant for right breast cancer treated with a radical mastectomy and chemotherapy 20 years ago. She had a second cancer (uterine) 10 years ago that was considered separate from her previous breast cancer.

Clinical Presentation: The physical therapy examination was consistent with the physician's diagnosis of nerve entrapment in a classic presentation. There were significant postural components to account for the development of symptoms. However, the therapist palpated several large masses in the axillary and supraclavicular fossa on both the right and left sides. There was no local warmth, redness, or tenderness associated with these lesions. The therapist requested permission to palpate the client's groin and popliteal spaces for any other suspicious lymph nodes. The rest of the examination findings were within normal limits.

Associated Signs and Symptoms: Further questioning about the presence of associated signs and symptoms revealed a significant disturbance in sleep pattern over the last 6 months with unrelenting shoulder and neck pain. There were no other reported constitutional symptoms, skin changes, or noted lumps anywhere. Vital signs were unremarkable at the time of the physical therapy evaluation.

Result: Returning this client to her referring physician was a difficult decision to make given that the therapist did not have the benefit of the medical records or results of neurologic examination and testing. With the significant past medical history for cancer, the woman's age, presence of progressive night pain, and palpable masses, no other reasonable choice remained. When asked if the physician had seen or felt the masses, the client responded with a definite "no."

There are several ways to approach handling a situation like this one, depending on the physical therapist's relationship with the physician. In this case, the therapist had never communicated with this physician before. It is possible that the physician was aware of the masses, knew from medical testing that there was extensive cancer, and chose to treat the client palliatively.

Because there was no indication of such, the therapist notified the physician's staff of the decision to return the client to the physician. A brief (one-page) written report summarizing the findings was given to the client to hand-carry to the physician's office.

Further medical testing was performed, and a medical diagnosis of lymphoma was made.

Risk Factors

Gender and age are nonmodifiable risk factors for chest pain caused by heart disease. The rate of coronary artery disease (CAD) is rising among women and falling among men. Men develop CAD at a younger age than women, but women make up for it after menopause. Many women know about the risk of breast cancer, but in truth, they are 10 times more likely to die of cardiovascular disease. While one in 30 women's deaths is from breast cancer, one in 2.5 deaths is from heart disease.[10]

Women do not seem to do as well as men after taking medications to dissolve blood clots or after undergoing heart-related medical procedures. Of the women who survive a heart attack, 46% will be disabled by heart failure within 6 years.[11] African-American women have a 70% higher death rate from CAD compared with Caucasian women.[10] Whenever screening individuals who have chest pain, keep in mind that older men and women, menopausal women, and African-American women are at greatest risk for cardiovascular causes.

A common treatment for CAD after heart attack is angioplasty with insertion of a stent. A stent is a wire mesh tube that props open narrowed coronary arteries. Sometimes, the stent malfunctions or gets scarred over. Cardiologists have realized that such treatments, while effective at alleviating chest pain, do not reduce the risk of heart attacks for most people with stable angina.

When the client presents with chest pain, he or she often does not think it can be from the heart because there is a stent in place, but this may not be true. Anyone with a history of stent insertion presenting with chest pain should be assessed carefully. Take vital signs, and ask about associated signs and symptoms. Evaluate the effect of exercise on symptoms. For example, does the chest, neck, shoulder, or jaw pain start 3 to 5 minutes after exercise or activity? What is the effect on pain in the upper body when the individual is using just the lower extremities, such as walking on a treadmill or up a flight of stairs?

Other risk factors for CAD are listed in Table 6-3. Efforts are being made to determine evidence-based risk factors for low- versus high-risk chest pain of unknown origin. Predictive values for ischemia resulting in myocardial infarction or death include two or more episodes of chest pain typical of a heart attack in an adult 55 years old or older who has a family history of heart disease and/or a personal history of diabetes.[12]

Clinical Presentation

There are some well-known pain patterns specific to the heart and cardiac system. Sudden death can be the first sign of heart disease. In fact, according to the American Heart Association (AHA), 63% of women who died suddenly of cardiovascular disease had no previous symptoms. Sudden death is the first symptom for half of all men who have a heart attack. Cardiac arrest strikes immediately and without warning.

CLINICAL SIGNS AND SYMPTOMS

Cardiac Arrest

- Sudden loss of responsiveness; no response to gentle shaking
- No normal breathing; client does not take a normal breath when you check for several seconds.
- No signs of circulation; no movement or coughing

Cardiac Pain Patterns

Doctors and nurses often use "the three Ps" when screening for chest pain of a cardiac nature. The presence of any or all of these Ps suggests the client's pain or symptoms are *not* caused by a myocardial infarction (MI):

- ***P***leuritic pain (exacerbation by deep breathing is more likely pulmonary in nature)
- Pain on ***p***alpation (musculoskeletal cause)
- Pain with changes in ***p***osition (musculoskeletal cause)

Cardiac pain patterns may differ for men and women. For many men, the most common report is a feeling of pressure or discomfort under the sternum (substernal), in the midchest region, or across the entire upper chest. It can feel like uncomfortable pressure, squeezing, fullness, or pain.

Pain may occur just in the jaw, upper neck, midback, or down the arm without chest pain or discomfort. Pain may also radiate from the chest to the neck, jaw, midback, or down the arm(s). Pain down the arm(s) affects the left arm most often in the pattern of the ulnar nerve distribution. Radiating pain down both arms is also possible.

For women, symptoms can be more subtle or atypical (Box 17-1). Chest pain or discomfort is less common in women but still a key feature for some. They often have prodromal symptoms (e.g., pain in the chest, pain in the shoulder or back, radiating pain or numbness in the arms, dyspnea, and fatigue) 12 months prior and up to 1 month before having a heart attack (see Table 6-4).[13-15] Black women younger than 50 years are more likely to report frequent and intense prodromal symptoms.[16]

Fatigue, nausea, and lower abdominal pain may signal a heart attack. Many women pass these off as the flu or food poisoning. Other symptoms for women include a feeling of intense anxiety, isolated right biceps pain, or midthoracic pain. Heartburn; sudden shortness of breath or the inability to talk, move, or breathe; shoulder or arm pain; or ankle swelling or rapid weight gain are also common symptoms with MI.

BOX 17-1 SIGNS AND SYMPTOMS OF MYOCARDIAL ISCHEMIA IN WOMEN

- Heart pain in women does not always follow classic patterns.
- Many women do experience classic chest discomfort.
- In older women, mental status change or confusion may be common.
- Dyspnea (at rest or with exertion)
- Weakness and lethargy (unusual fatigue; fatigue that interferes with ability to perform activities of daily living)
- Indigestion or heartburn; mistakenly diagnosed or assumed to have gastroesophageal reflux disease (GERD)
- Lower abdominal pain
- Anxiety or depression
- Sleep disturbance (woman awakens with any of the symptoms listed here)
- Sensation similar to inhaling cold air; unable to talk or breathe
- Isolated mid-thoracic back pain
- Symptoms may be relieved by antacids (sometimes antacids work better than nitroglycerin).

Chest Pain Associated with Angina

The therapist should keep in mind that coronary disease may go unnoticed because the client has no anginal or infarct pain associated with ischemia. This situation occurs when collateral circulation is established to counteract the obstruction of the blood flow to the heart muscle. Anastomoses (connecting channels) between the branches of the right and left coronary arteries eliminate the person's perception of pain until challenged by physical exertion or exercise in the physical therapy setting.

Chest pain caused by angina is often confused with heartburn or indigestion, hiatal hernia, esophageal spasm, or gallbladder disease, but the pain of these other conditions is not described as sharp or knifelike. The client often says the pain feels like "gas" or "heartburn" or "indigestion." Referred pain from a trigger point in the external oblique abdominal muscle can cause a sensation of heartburn in the anterior chest wall (see Fig. 17-7).

Episodes of stable angina usually develop slowly and last 2 to 5 minutes. Discomfort may radiate to the neck, shoulders, or back (Case Example 17-3). Shortness of breath is common. Symptoms of angina may be similar to the pattern associated with a heart attack. One primary difference is duration. Angina lasts a limited time (a few minutes up to a half hour) and can be relieved by rest or nitroglycerin. When screening for angina, a lack of objective musculoskeletal findings is always a red flag:

- Active range of motion (AROM), such as trunk rotation, side bending, or shoulder motions, does not reproduce symptoms.
- Resisted motion does not reproduce symptoms (horizontal shoulder abduction/adduction).
- Heat and stretching do not reduce or eliminate symptoms.

The therapist should also watch for unstable angina in a client with known angina. Unlike stable angina, rest or nitroglycerin does not relieve symptoms associated with an MI,

CASE EXAMPLE 17-3

Adhesive Capsulitis

Referral: A 56-year-old man returned to the same physical therapist with his third recurrence of left shoulder adhesive capsulitis of unknown cause.

Past Medical History: There was no reported injury, trauma, or repetitive motion as a precipitating factor in this case. The client was a car salesman with a fairly sedentary job. He reported a past history of prostatitis, peptic ulcers, and a broken collarbone as a teenager. He reported being a "social" drinker at work-related functions but did not smoke or use tobacco products. He was taking ibuprofen for his shoulder but no other over-the-counter or prescription medications or supplements.

The two previous episodes of shoulder problems resolved with physical therapy intervention. The client had a home program to follow to maintain range of motion and normal movement. At the time of his most recent discharge 6 months ago, he had attained 80% of motion available on the uninvolved side with some continued restricted glenohumeral movement and altered scapulohumeral rhythm. The client reported that he did not continue with his exercise routine at home and "that's why I got worse again."

Clinical Presentation

Shoulder flexion and abduction	Left: 105/100	Right: 170/165
Shoulder medial (internal) rotation	0-70	0-90
Shoulder lateral (external) rotation	0-45	0-80

Accessory motions: Reduced inferior and anterior glide on the left; within normal limits on the right. The client reports pain during glenohumeral flexion, abduction, and medial and lateral rotations.

Clinical impressions: Decreased physiologic motion with capsular pattern of restriction and compensatory movements of the shoulder girdle; humeral superior glide syndrome.

Associated Signs and Symptoms: When asked if there were any symptoms of any kind anywhere else in the body, the client reported "chest tightness" whenever he tried to use his arm for more than a few minutes. Previously, he was used to "working through the pain," but he can't seem to do that anymore.

He also reported "a few bouts of nausea and sweating" when his shoulder started aching. He denied any shortness of breath or constitutional symptoms such as fever or sweats. There were no other gastrointestinal-related symptoms.

What are the red flags in this case? How would you screen further?

- Age over 50
- Nausea and sweating concomitant with shoulder pain; chest tightness
- Insidious onset
- Recurring pattern of symptoms

Screening can begin with something as simple as vital sign assessment. The therapist can consult Box 4-19 for a list of other associated signs and symptoms and look for a cluster or pattern associated with a particular system.

Given his age, sedentary lifestyle, and particular clinical presentation, a cardiovascular screening examination seems most appropriate. The therapist can also consult the Special Questions to Ask box at the end of Chapter 6 for any additional pertinent questions based on the client's responses to questions and examination results. A short (3- to 5-minute) bike test also can be used to assess the effect of lower extremity exertion on the client's symptoms.

Result: The client's blood pressure was alarmingly high at 185/120 mm Hg. Although this is an isolated (one time) reading, he was under no apparent stress, and he revealed that he had a history of elevated blood pressure in the past. The bike test was administered while his heart rate and blood pressure were being monitored. Symptoms of chest and/or shoulder pain were not reproduced by the test, but the therapist was unwilling to stress the client without a medical evaluation first.

Referral was made to his primary care physician with a phone call, fax, and report of the therapist's findings and concerns. Although there is a known viscero-somatic effect between heart and chest and heart and shoulder, there is no reported direct cause and effect link between heart disease and adhesive capsulitis. Comorbid factors, such as diabetes or heart disease, have been shown to affect pain levels and function.[73]

Likewise, adhesive capsulitis is known to occur in some people following immobility associated with intensive care, coronary artery bypass graft, or pacemaker complications/revisions.

The physician considered this an emergency situation and admitted the client to the cardiology unit for immediate workup. The electrocardiogram results were abnormal during the exercise stress test. Further testing confirmed the need for a triple bypass procedure. Following the operation and phase 1 cardiac rehab in the cardiac rehab unit, the client returned to the original outpatient physical therapist for his phase 2 cardiac rehab program. Shoulder symptoms were gone, and range of motion was unimproved but regained rapidly as the rehab program progressed.

The therapist shared this information with the cardiologist, who agreed that there may have been a connection between the chest/shoulder symptoms before surgery, although he could not say for sure.

unless administered intravenously. Without intervention, symptoms of an MI may continue without stopping. A sudden change in the client's typical anginal pain pattern suggests unstable angina. Pain that occurs without exertion, lasts longer than 10 minutes, or is not relieved by rest or nitroglycerin signals a higher risk for a heart attack. Immediate medical referral is required under these circumstances.

SCREENING FOR PLEUROPULMONARY CAUSES OF CHEST, BREAST, OR RIB PAIN

Pulmonary chest pain usually results from obstruction, restriction, dilation, or distention of the large airways or large pulmonary artery walls. Specific diagnoses include pulmonary artery hypertension, pulmonary embolism, mediastinal emphysema, asthma, pleurisy, pneumonia, and pneumothorax. Pleuropulmonary disorders are discussed in detail in Chapter 7.

Past Medical History

A previous history of cancer of any kind, recent history of pulmonary infection, or recent accident or hospitalization may be significant. Look for other risk factors, such as age, smoking, prolonged immobility, immune system suppression (e.g., cancer chemotherapy, corticosteroids), and eating disorders (or malnutrition from some other cause).

Mechanical alterations related to the overload of respiratory muscles with chronic asthma can lead to chest pain from musculoskeletal dysfunction and alterations in muscle length and posture.[17]

Clinical Presentation

Pulmonary pain patterns differ slightly depending on the underlying pathology and the location of the disease. For example, tracheobronchial pain is referred to the anterior neck or chest at the same levels as the points of irritation in the air passages. Chest pain that tends to be sharply localized or that worsens with coughing, deep breathing, other respiratory movements or motion of the chest wall and that is relieved by maneuvers that limit the expansion of a particular part of the chest (e.g., autosplinting) is likely to be pleuritic in origin.

Symptoms that increase with deep breathing and activity or the presence of a productive cough with bloody or rust-colored sputum are red flags. The therapist should ask about new onset of wheezing at any time or difficulty breathing at night. Be careful when asking clients about changes in breathing patterns. It is not uncommon for the client to deny any shortness of breath.

Often, the reason for this is because the client has stopped doing anything that will bring on the symptoms. It may be necessary to ask what activities he or she can no longer do that were possible 6 weeks or 6 months ago. Symptoms that are relieved by sitting up are indicative of pulmonary impairment and must be evaluated more carefully.

SCREENING FOR GASTROINTESTINAL CAUSES OF CHEST, BREAST, OR RIB PAIN

GI causes of upper thorax pain are a result of epigastric or upper GI conditions. GERD ("heartburn" or esophagitis) accounts for a significant number of cases of noncardiac chest pain, in the young as well as older adults.[3,18-20] Stomach acid or gastric juices from the stomach enter the esophagus, causing irritation to the protective lining of the lower esophagus. Whether the client is experiencing GERD or some other cause of chest pain, there is usually a telltale history or associated signs and symptoms to red flag the case.

Past Medical History

Watch for a history of alcoholism, cirrhosis, esophageal varices, and esophageal cancer or peptic ulcers. Any risk factors associated with these conditions are also red flags such as long-term use of NSAIDs as a cause of peptic ulcers or chronic alcohol use associated with cirrhosis of the liver.

Clinical Presentation

The GI system has a broad range of referred pain patterns based on embryologic development and multisegmental innervations, as discussed in Chapter 3. Upper GI and pancreatic problems are more likely than lower GI disease to cause chest pain. Chest pain referred from the upper GI tract can radiate from the chest posteriorly to the upper back or interscapular or subscapular regions from T10 to L2 (Fig. 17-1).

Esophagus

Esophageal dysfunction will present with symptoms such as anterior neck and/or anterior chest pain, pain during swallowing (odynophagia), or difficulty swallowing (dysphagia) at the level of the lesion. Symptoms occur anywhere a lesion is present along the length of the esophagus. Early satiety, often with weight loss, is a common symptom with esophageal carcinoma.

Lesions of the upper esophagus may cause pain in the (anterior) neck, whereas lesions of the lower esophagus are

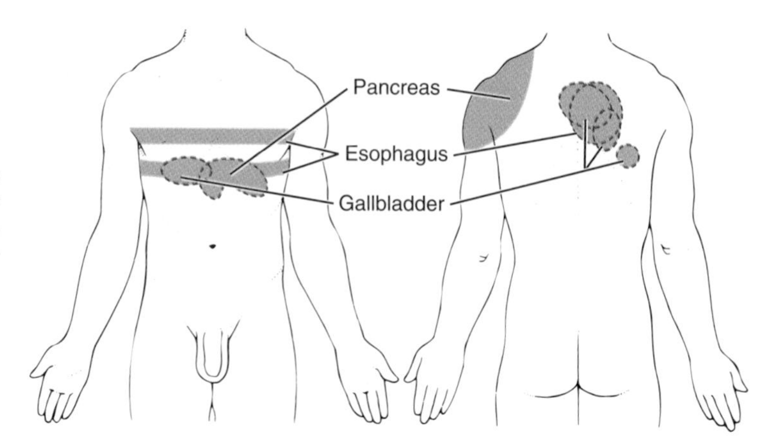

Fig. 17-1 Chest pain caused by gastrointestinal (GI) disease with referred pain to the shoulder and back. Upper GI problems can refer pain to the anterior chest with radiating pain to the thoracic spine at the same level. Look for accompanying GI symptoms and red-flag history.

more likely to be characterized by pain originating from the xiphoid process, radiating around the thorax to the middle of the back.

Chest pain with or without accompanying or alternating midthoracic back pain from an esophageal or other upper GI problem is usually red flagged by a suspicious history or cluster of associated signs and symptoms. The pain pattern associated with thoracic disk disease can be the same as for esophageal pathology. In the case of disk disease, there may be bowel and/or bladder changes and sometimes numbness and tingling in the upper extremities. The therapist should ask about a traumatic injury to the upper back region and conduct a neurologic screening examination to assess for this possibility as a cause of the symptoms.

Epigastric Pain

Epigastric pain is typically characterized by substernal or upper abdominal (just below the xiphoid process) discomfort (see Fig. 17-1). This may occur with radiation posteriorly to the back secondary to long-standing duodenal ulcers. Gastric duodenal peptic ulcer may occasionally cause pain in the lower chest rather than in the upper abdomen. Antacid and food often immediately relieve pain caused by an ulcer. Ulcer pain is not produced by effort and lasts longer than angina pectoris. The therapist will not be able to provoke or eliminate the client's symptoms. Likewise, physical therapy intervention will not have any long-lasting effects unless the symptoms were caused by trigger points (TrPs).

Pain in the lower substernal area may arise as a result of reflux esophagitis (regurgitation of gastroduodenal secretions), a condition known as *gastroesophageal reflux disease* (GERD). It may be gripping, squeezing, or burning, described as "heartburn" or "indigestion." Like that of angina pectoris, the discomfort of reflux esophagitis may be precipitated by recumbency or by meals; however, unlike angina, it is not precipitated by exercise and is relieved by antacids.

Hepatic and Pancreatic Systems

Epigastric pain or discomfort may occur in association with disorders of the liver, gallbladder, common bile duct, and pancreas, with referral of pain to the interscapular, subscapular, or middle/low back regions. This type of pain pattern can be mistaken for angina pectoris or myocardial infarction (e.g., hypotension occurring with pancreatitis produces a reduction of coronary blood flow with the production of angina pectoris).

Hepatic disorders may cause chest pain with radiation of pain to the shoulders and back. Cholecystitis (gallbladder inflammation) appears as discrete attacks of epigastric or right upper quadrant pain, usually associated with nausea, vomiting, and fever and chills. Dark urine and jaundice indicate that a stone has obstructed the common duct.

The pain has an abrupt onset, is either steady or intermittent, and is associated with tenderness to palpation in the right upper quadrant. The pain may be referred to the back and right scapular areas. A gallbladder problem can result in a sore tenth rib tip (right side anteriorly) as described in Chapter 9 (Case Example 17-4). Rarely, pain in the left upper quadrant and anterior chest can occur.

Acute pancreatitis causes pain in the upper part of the abdomen that radiates to the back (usually anywhere from T10 to L2) and may spread out over the lower chest. Fever and abdominal tenderness may develop.

CLINICAL SIGNS AND SYMPTOMS

Gastrointestinal Disorders

- Chest pain (may radiate to back)
- Nausea
- Vomiting
- Blood in stools
- Pain on swallowing or associated with meals
- Jaundice
- Heartburn or indigestion
- Dark urine

CASE EXAMPLE 17-4

Chest Pain During Pregnancy

Referral: A 33-year-old woman in her twenty-ninth week of gestation with her first pregnancy was referred to a physical therapist by her gynecologist. Her abdominal sonogram and lab tests were normal. A chest x-ray was read as negative.

Past Medical History: None. The client had the usual childhood illnesses but had never broken any bones and denied use of tobacco, alcohol, or substances of any kind. There was no recent history of infections, colds, viruses, coughs, trauma or accidents, and changes in gastrointestinal function and no history of cancer.

Clinical Presentation: Although there were no signs and symptoms associated with the respiratory system, the client's symptoms were reproduced when she was asked to take a deep breath. Palpation of the upper chest, thorax, and ribs revealed pain on palpation of the right tenth rib (anterior).

Thinking about the role of the gallbladder causing tenth rib pain, the therapist asked further questions about past history and current gastrointestinal symptoms. The client had no red flag symptoms or history in this regard.

Knowing that transient osteoporosis can be associated with pregnancy,[74-79] the therapist gave the client the Osteoporosis Screening Evaluation (see Appendix C-6). The client replied "yes" to three questions (Caucasian or Asian, mother diagnosed with osteoporosis, physically inactive), suggesting the possibility of rib fracture.[75,77]

Result: The therapist initiated a telephone consultation with the physician to review her findings. Although the original x-ray was read as negative, the physician ordered a different view (rib series) and identified a fracture of the tenth rib.

The physician explained that the mechanical forces of the enlarging uterus on the ribs pull the lower ribs into a more horizontal position. Any downward stress from above (e.g., forceful cough or pull from the external oblique muscles) or upward force from the serratus anterior and latissimus dorsi muscles can increase the bending stress on the lower ribs.[75]

An aquatics therapy program was initiated and continued throughout the remaining weeks of this client's pregnancy.

SCREENING FOR BREAST CONDITIONS THAT CAUSE CHEST OR BREAST PAIN

Occasionally, a client may present with breast pain as the primary complaint, but most often the description is of shoulder or arm or neck or upper back pain. When asked if any symptoms occur elsewhere in the body, the client may mention breast pain (Case Example 17-5).

During examination of the upper quadrant, the therapist may observe suspicious or aberrant changes in the integument, breast, or surrounding soft tissues. The client may report discharge from the nipple. Discharge from both nipples is more likely to be from a benign condition; discharge from one nipple can be a sign of a precancerous or malignant condition.

Asking the client about history, risk factors, and the presence of other signs and symptoms is the next step (see Box 4-18). Knowing possible causes of breast pain can help guide the therapist during the screening interview (see Table 17-2).

Past Medical History

A past history of breast cancer, heart disease, recent birth, recent upper respiratory infection (URI), overuse, or trauma (including assault) may be significant for the client presenting with breast pain or symptoms. Any component of heart disease, such as hypertension, angina, myocardial infarction, and/or any heart procedure such as angioplasty, stent, or coronary artery bypass, is considered a red flag.

Any woman experiencing chest or breast pain should be asked about a personal history of previous breast surgeries, including mastectomy, breast reconstruction, or breast implantation or augmentation. A past history of breast cancer is a red flag even if the client has completed all treatment and has been cancer free for 5 years or more.

On the flip side, a past history of breast cancer in a client who presents with musculoskeletal symptoms with or without a history of trauma does not always mean cancer metastases. A complete evaluation with advanced imaging may be needed to uncover the true underlying etiology as in the reported case of fibular pain in a patient with a history of breast cancer that turned out to be an incomplete nondisplaced distal fibular stress fracture with no evidence of tumor or mass (Case Example 17-6).[21]

Breast cancer and cysts develop more frequently in individuals who have a family history of breast disease. A previous history of cancer is always cause to question the client further regarding the onset and pattern of current symptoms. This is especially true when a woman with a previous history of breast cancer or cancer of the reproductive system appears with shoulder, chest, hip, or sacroiliac pain of unknown cause.

If a client denies a previous history of cancer, the therapist should still ask whether that person has ever received chemotherapy or radiation therapy. It is surprising how often the answer to the question about a previous history of cancer is

CASE EXAMPLE 17-5

Breast Pain and Trigger Points

Referral: A 67-year-old woman came to physical therapy after seeing her primary care physician with a report of decreased functional left shoulder motion. She was unable to reach the top shelf of her kitchen cabinets or closets. She felt that at 5 feet 7 inches this is something she should be able to do.

Past Medical History: During the Past Medical History portion of the interview, she mentioned that she had had a stroke 10 years ago. Her referring physician was unaware of this information. She had recently moved here to be closer to her daughter, and no medical records have been transferred. There was no other significant history.

At the end of the interview, when asked, "Is there anything else you think I should know about your health or current situation that we haven't discussed?" she replied, "Well, actually the reason I really went to see the doctor was for pain in my left breast."

She had not reported this information to the physician.

Clinical Presentation: Examination revealed mild loss of strength in the left upper extremity accompanied by mild sensory and proprioceptive losses. Palpation of the shoulder and pectoral muscles produced breast pain. The client had been aware of this pain, but she had attributed it to a separate medical problem. She was reluctant to report her breast pain to her physician. Objectively, there were positive trigger points of the left pectoral muscles and loss of accessory motions of the left shoulder (see Fig 17-7).

- Active trigger point of the left pectoralis major with pain centered in the left breast
- Decreased left shoulder accessory motions (caudal glide, posterior glide and lateral traction); no shoulder pain or discomfort reported
- Range of motion limited by 20% compared with the right shoulder in flexion, external rotation, and abduction
- Mild strength deficit
- Mild sensory and proprioceptive losses
- Vital signs:

Blood pressure (sitting, left arm)	142/108 mm Hg
Heart rate	72 bpm
Pulse oximeter	98%
Oral temperature	98.0° F

Intervention: Physical therapy treatment to eliminate trigger points and restore shoulder motion resolved the breast pain during the first week.

Should you make a medical referral for this client? If so, on what basis?

Despite this woman's positive response to physical therapy treatment, given the age of this client, her significant past medical history for cerebrovascular injury (reportedly unknown to the referring physician), current blood pressure (although an isolated measurement), report of breast pain (also unreported to her physician), and the residual paresis, medical referral was still indicated.

At the first follow-up visit, a letter was sent with the client that briefly summarized the initial objective findings, her progress to date, and the current concerns. She returned for an additional week of physical therapy to complete the home program for her shoulder. A medical evaluation ruled out breast disease, but medical treatment (medication) was indicated to address cardiovascular issues.

CASE EXAMPLE 17-6

Fibular Pain with History of Breast Cancer

In a published case report, physical therapists evaluated a 46-year-old woman with left ankle pain who also had a past medical history of breast cancer. The client gave a month-old history of an ankle sprain while running, symptoms were made worse by running, and there were no advanced imaging studies to rule out cancer metastases.

Standard radiographs of the ankle were read as normal by the radiologist, but the therapist was suspicious of an observed irregularity in the distal fibula. Local tenderness was palpated just above the distal tip of the fibula.

A bone scan was ordered to differentiate between an old injury and new pathology. There was increased metabolic activity around the area in question. A follow-up magnetic resonance imaging (MRI) showed an incomplete, nondisplaced stress fracture of the distal fibula. No metastatic lesions were present.

This is a case where bone fracture could not be ruled out with a standard x-ray. Age over 40, past medical history of breast cancer, and new onset of unresolving bone pain led to a differential diagnosis that confirmed a true musculoskeletal problem.

From Ryder M, Deyle GD: Differential diagnosis of fibular pain in a patient with a history of breast cancer, *J Orthop Sports Phys Ther* 39(3):230, 2009.

"no" but the answer to the question about prior treatment for cancer is "yes."

Clinical Presentation

For the most part, breast pain (mastalgia), tenderness, and swelling are the result of monthly hormone fluctuations. Cyclical pain may get worse during perimenopause when hormone levels change erratically. These same symptoms may continue after menopause, especially in women who use hormone replacement therapy (HRT). Noncyclical breast pain is not linked to menstruation or hormonal fluctuations. It is unpredictable and may be constant or intermittent, affecting one or both breasts in a small area or the entire breast.

The typical referral pattern for breast pain is around the chest into the axilla, to the back at the level of the breast, and occasionally into the neck and posterior aspect of the shoulder girdle (Fig. 17-2). The pain may continue along the medial aspect of the ipsilateral arm to the fourth and fifth digits, mimicking pain of the ulnar nerve distribution.

Jarring or movement of the breasts and movement of the arms may aggravate this pain pattern. Pain in the upper inner arm may arise from outer quadrant breast tumors, but pain in the local chest wall may point to any pathologic condition of the breast.

Nipple discharge in women is common, especially in pregnant or lactating women, and does not always signal a serious underlying condition. It may occur as a result of some

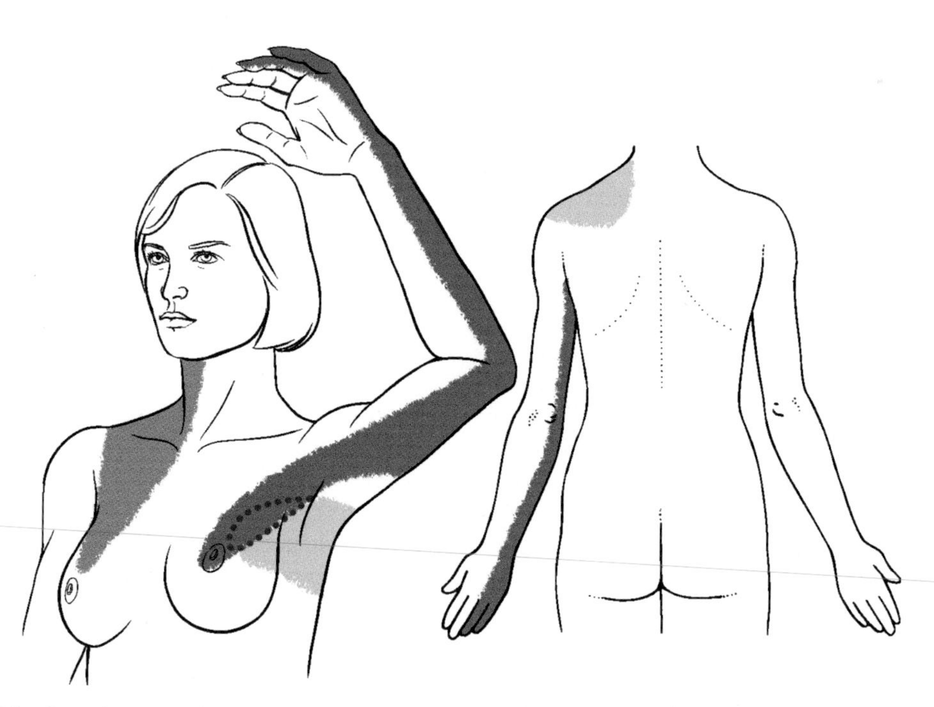

Fig. 17-2 Pain arising from the breast (mastalgia) can be referred into the axilla along the medial aspect of the arm. Referral pattern can also extend to the supraclavicular level and into the neck. Breast pain may be diffuse around the thorax through the intercostal nerves. Pain may be referred to the back and the posterior shoulder. Ask the client about the presence of lumps, nipple discharge, distended veins, or puckered or red skin (or any other skin changes).

medications (e.g., estrogen-based drugs, tricyclic antidepressants, benzodiazepines, and others).

The fluid may be thin to thick in consistency and various colors (e.g., milky white, green, yellow, brown, or bloody). Any unusual nipple discharge should be evaluated by a medical doctor. Injury, hormonal imbalance, underactive thyroid, infection or abscess, or tumors are just a few possible causes of nipple discharge.

CLINICAL SIGNS AND SYMPTOMS

Breast Pathology

- Family history of breast disease
- Palpable breast nodules or lumps and previous history of chronic mastitis
- May be painless
- Breast pain with possible radiation to inner aspect of arm(s)
- Skin surface over a tumor may be red, warm, edematous, firm, and painful.
- Firm, painful site under the skin surface
- Skin dimpling over the lesion with attachment of the mass to surrounding tissues, preventing normal mobilization of skin, fascia, and muscle
- Unusual nipple discharge or bleeding from the nipple(s)
- Pain aggravated by jarring or movement of the breasts
- Pain that is not aggravated by resistance to isometric movement of the upper extremities

Causes of Breast Pain

There is a wide range of possible causes of breast pain, including both systemic or viscerogenic and NMS etiologies (Table 17-2). Not all conditions are life threatening or even require medical attention.

Although it is more typical in women, both men and women can have chest, back, scapular, and shoulder pain referred by a pathologic condition of the breast. Only those conditions most likely to be seen in a physical therapist's practice are included in this discussion.

Mastodynia

Mastodynia (irritation of the upper dorsal intercostal nerve) that causes chest pain is almost always associated with ovulatory cycles, especially premenstrually. The association between symptoms and menses may be discovered during the physical therapist's interview when the client responds to Special Questions to Ask: Breast (see end of this chapter or Appendix B-7). The presentation is usually unilateral breast or chest pain and occurs initially at the premenstrual period and later more persistently throughout the menstrual cycle.

Mastitis

Mastitis is an inflammatory condition associated with lactation (breast feeding). Mammary duct obstruction causes the duct to become clogged. The breast becomes red, swollen, and painful. The involved breast area is often warm or even hot.

TABLE 17-2 Causes of Breast Pain

Systemic/Medical Conditions	Neuromusculoskeletal
Infection • Mastitis (lactating women) • Abscess Paget's disease Tumors, cysts, calcific changes Inflammatory carcinoma of the breast Acute fat necrosis (after trauma) Lymph disease PMS, menstrual or hormonal influences (including early pregnancy) Shingles (herpes zoster) Pleuritis GERD Other: • Medications (e.g., some hormone, cardiovascular, psychiatric drugs) • Anxiety	Pectoral myalgia or other conditions affecting the pectoralis muscles TrPs Mastodynia (mammary neuralgia) Breast implants, augmentation, reduction • Scar tissue Trauma or injury (e.g., assault, breast biopsy, or surgery) Cervical radiculopathy TOS Costochondritis Connective tissue disorders Heavy, pendulous breasts

PMS, Premenstrual syndrome; *GERD,* gastroesophageal reflux disease; *TrPs,* trigger points; *TOS,* thoracic outlet syndrome.

Constitutional symptoms such as fever, chills, and flulike symptoms are common. Acute mastitis can occur in males (e.g., nipple chafing from jogging); the presentation is the same as for females.

Risk factors include previous history of mastitis; cracked, bleeding, painful nipples; and stress or fatigue. Bacteria can enter the breast through cracks in the nipple during trauma or nursing. Subsequent infection may lead to abscess formation. Obstructive and infectious mastitis are considered as two conditions on a continuum. Mastitis is often treated symptomatically, but the client should be encouraged to let her doctor know about any breast signs and symptoms present. Antibiotics may be needed in the case of a developing infection.

Benign Tumors and Cysts

Benign tumors and cysts were once lumped together and called "fibrocystic breast disease." With additional research over the years, scientists have come to realize that a single label is not adequate for the variety of benign conditions possible, including fibroadenomas, cysts, and calcifications that can occur in the breast.

An unchanged lump of long duration (years) is more likely to be benign. Many lumps are hormonally induced cysts and resolve within two or three menstrual cycles. Cyclical breast cysts are less common after menopause.

Other conditions can include intraductal papillomas (wartlike growth inside the breast), fat necrosis (fat breaks down and clumps together), and mammary duct ectasia (ducts near the nipple become thin-walled and accumulate secretions). Some of these breast changes are a variation of the norm, and others are pathologic but nonmalignant. A medical diagnosis is needed to differentiate between these changes.

Paget's Disease

Paget's disease of the breast is a rare form of ductal carcinoma arising in the ducts near the nipple. The woman experiences itching, redness, and flaking of the nipple with occasional bleeding (Fig. 17-3). Paget's disease of the breast is not related to Paget's disease of the bone, except that the same physician (Dr. James Paget, a contemporary of Florence Nightingale, 1877) named both conditions after himself.

Breast Cancer

The breast is the second most common *site* of cancer in women (the skin is first). Cancer of the breast is second only to lung cancer as a *cause of death* from cancer among women. Male breast cancer is possible but rare, accounting for 1% of all new cases of breast cancer (2,140 cases in 2011 for men compared with 230,480 for women).[22]

Although the frequency of breast cancer in men is strikingly less than that in women, the disease in both sexes is remarkably similar in epidemiology, natural history, and response to treatment. Men with breast cancer are 5 to 10 years older than women at the time of diagnosis, with mean or median ages between 60 and 66 years. This apparent difference may occur because symptoms in men are ignored for a longer period and the disease is diagnosed at a more advanced state.

Risk Factors. Despite the discovery of a breast cancer gene (BRCA-1 and BRCA-2), researchers estimate that only 5% to 10% of breast cancers are a result of inherited genetic susceptibility. Normally, BRCA-1 and BRCA-2 help prevent cancer by making proteins that keep cells from growing abnormally. Inheriting either mutated gene from a parent does increase the risk of breast cancer.[23] But a much larger proportion of cases are attributed to other factors, such as advancing age, race, smoking, obesity, physical inactivity, excess alcohol intake, exposure to ionizing radiation, and exposure to estrogens (Table 17-3).[24]

Women who received multiple fluoroscopies for tuberculosis or radiation treatment for mastitis during their adolescent or childbearing years are at increased risk for breast cancer as a result of exposure to ionizing radiation. In the past, irradiation was used for a variety of other medical conditions, including gynecomastia, thymic enlargement, eczema of the chest, chest burns, pulmonary tuberculosis, mediastinal lymphoma, and other cancers. Most of these clients are in their 70s now and at risk for cancer because of advancing age as well.

As a general principle, the risk of breast cancer is linked to a woman's total lifelong exposure to estrogen. The increased incidence of estrogen-responsive tumors (tumors that are rich in estrogen receptors proliferate when exposed to estrogen) has been postulated to occur as a result of a variety of

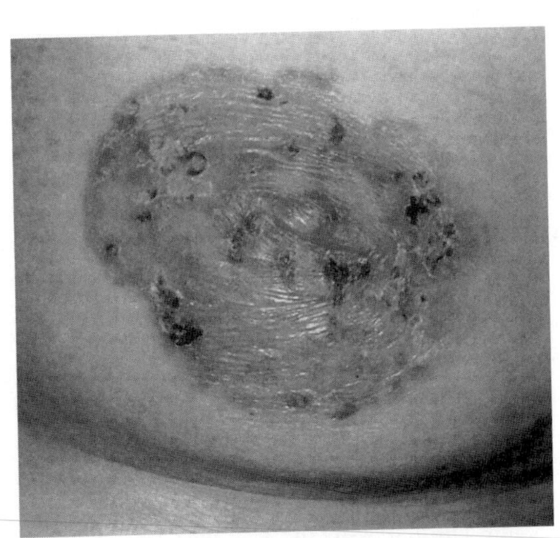

Fig. 17-3 Paget's disease of the breast is a rare form of breast cancer affecting the nipple. It is characterized by a red (sometimes scaly) rash on the breast that often surrounds the nipple and areola, as seen in this photograph. Other presentations are possible such as a red pimple or sore on the nipple that does not heal. Symptoms are unilateral, and the breast may be sore, itch, or burn. Diagnosis is often delayed because the symptoms seem harmless or the condition is misdiagnosed as dermatitis. (From Callen JP, Jorizzo J, Greer KE, et al: *Dermatological signs of internal disease,* Philadelphia, 1988, Saunders.)

TABLE 17-3 Factors Associated With Breast Cancer

Factor	
Gender	Women > men
Race	White
Age	Advancing age, >60 years; younger age at menarche, at first live birth, and at diagnosis is associated with inflammatory breast cancer[72] Peak incidence: 45-70 Mean and median age: 60-61 (women) 60-66 (men)
Genetic	BRCA1/BRCA2 gene mutations
Family history	First-degree relative with breast cancer Premenopausal Bilateral Mother, daughter, or sister
Previous medical history	Previous personal history of cancer Breast Uterine Ovarian Colon Number of previous breast biopsies (positive or negative)
Exposure to estrogen	Age at menarche <12 Age at menopause >55 Nulliparous (never pregnant) First live birth after age 35 Environmental estrogens (esters)

For a more detailed guide to risk factors for breast cancer, see the American Cancer Society's document, *What are the risk factors for breast cancer?* Available at http://www.cancer.org/docroot/CRI/content/CRI_2_4_2x_what_are_the_risk_factors_for_breast_cancer_5.asp.

factors, such as prenatal and lifelong exposure to synthetic chemicals and environmental toxins, earlier age of menarche (first menstruation), improved nutrition in the United States, delayed and decreased childbearing, and longer average lifespan.

At the same time, it should be remembered that many women diagnosed with breast cancer have no identified risk factors. More than 70% of breast cancer cases are not explained by established risk factors.[23,25] There is no history of breast cancer among female relatives in more than 90% of clients with breast cancer. However, first-degree relatives (mother, daughters, or sisters) of women with breast cancer have two to three times the risk of developing breast cancer than the general female population, and relatives of women with bilateral breast cancer have five times the normal risk.[23]

Risk factors for men are similar to those for women, but at least half of all cases do not have an identifiable risk factor. Risk factors for men include heredity, obesity, infertility, late onset of puberty, frequent chest x-ray examinations, history of testicular disorders (e.g., infection, injury, or undescended testes), and increasing age. Men who have several female relatives with breast cancer and those in families who have the BRCA-2 mutation have a greater risk potential.

The presence of any of these factors may become evident during the interview with the client and should alert the physical therapist to the potential for neuromusculoskeletal complaints from a systemic origin that would require a medical referral. There are several easy-to-use screening tools available. In addition to screening for current risk, clients should be given this information for future use (Box 17-2).

Clinical Presentation. Breast cancer may be asymptomatic in the early stages. The discovery of a breast lump with or without pain or tenderness is significant and must be investigated. Physical signs associated with advanced breast cancer have been summarized using the acronym BREAST: **B**reast mass, **R**etraction, **E**dema, **A**xillary mass, **S**caly nipple, and **T**ender breast.[26] Less common symptoms are breast pain; nipple discharge; nipple erosion, enlargement, itching, or redness; and generalized hardness, enlargement, or shrinking of the breast. Watery, serous, or bloody discharge from the nipple is an occasional early sign but is more often associated with benign disease.

BOX 17-2 RESOURCES FOR ASSESSING AND LOWERING BREAST CANCER RISK

Breastcancer.Org

- www.breastcancer.org

Breastcancer.Org is a nonprofit organization with a website dedicated to providing current and accurate information on every aspect of cancer. A professional advisory board that consists of more than 60 practicing medical professionals around the world review all information available on the website.

National Cancer Institute

- http://bcra.nci.nih.gov/brc/

The National Cancer Institute (NCI) offers a Breast Cancer Risk Assessment, which is an interactive tool to measure the risk of invasive breast cancer. This tool was designed to assist health care professionals in guiding individual clients to estimate the risk of invasive breast cancer. It is only part of a woman's options for assessing risk and screening for breast cancer. More information is available by calling the Cancer Information Service (CIS) at 1-800-4-CANCER.

Breast Cancer Risk Calculator

- http://www.halls.md/breast/risk.htm

This calculator uses the Gail model but with some added risk modifier questions. The author of the website (Steven B. Halls, MD) notes that the methods on the website have been gathered from peer-reviewed journals but have not been peer reviewed. Results provided are estimates.

Oncolink

- http://www.oncolink.com/

Abramson Cancer Center of the University of Pennsylvania offers a comprehensive website with information about various types of cancers, risk factors, cancer treatment, and cancer resources. Click on Cancer Types>Breast Cancer.

The Harvard Center for Cancer Prevention

- http://www.yourdiseaserisk.harvard.edu/

The Harvard Center for Cancer Prevention offers an easy-to-use tool to assess risk factors for a variety of diseases, including breast cancer.

Susan G. Komen for the Cure

- www.komen.org

This is a website devoted to detecting and understanding breast cancer. This website provides the consumer with interactive tools and videos in many different languages to teach women (and men) how to examine their own breasts, what their individual risk factors are, and tips on making healthy lifestyle choices as cancer prevention. Regarding the controversy over self-breast examination, Susan G. Komen for the Cure says: Women should be aware of how their breasts normally look and feel. Knowing what is normal for you may enable you to note changes in your breast in the time between your yearly mammogram and/or clinical breast exam. Breast self-exam (BSE) is a tool that may help you become familiar with the way your breasts normally look and feel. Women who practice BSE should also be sure to get mammograms and clinical breast exams at the appropriate age. BSE should not be substituted for these screening tests.

Breast cancer usually consists of a nontender, firm, or hard lump with poorly delineated margins that is caused by local infiltration. Breast cancer in women has a predilection for the outer upper quadrant of the breast and the areola (nipple) area (Fig. 17-4) involving the breast tissue overlying the pectoral muscle. During palpation, breast tissue lumps move easily over the pectoral muscle, compared with a lump within the muscle tissue itself. Later signs of malignancy include fixation of the tumor to the skin or underlying muscle fascia.

Male breast cancer begins as a painless induration, retraction of the nipple, and an attached mass progressing to include lymphadenopathy and skin and chest wall lesions. A tumor of any size in male breast tissue is associated with skin fixation and ulceration and deep pectoral fixation more often than a tumor of similar size in female breast tissue is because of the small size of male breasts.

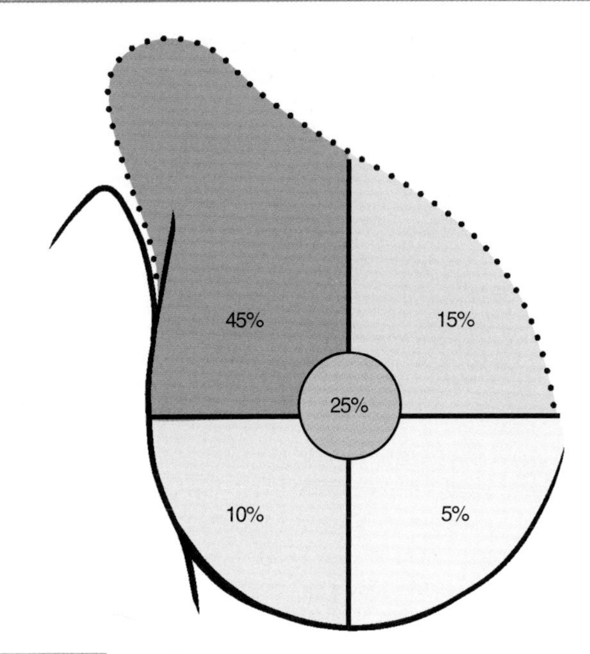

Fig. 17-4 Most breast cancer presents in the upper outer quadrant of the breast (45%) or around the nipple (25%). Metastases occur via the lymphatic system at the axillary lymph nodes to the bones (shoulder, hip, ribs, vertebrae) or central nervous system (brain, spinal cord). Breast cancer can also metastasize hematogenously to the lungs, pleural cavity, and liver.

CLINICAL SIGNS AND SYMPTOMS

Breast Cancer

- Nontender (painless), firm, or hard lump
- Unusual discharge from nipple
- Skin or nipple retraction dimpling; erosion, retraction, itching of nipple
- Redness or skin rash over the breast or nipple
- Generalized hardness, enlargement, shrinking, or distortion of the breast or nipple
- Unusual prominence of veins over the breast
- Enlarged rubbery lymph nodes
- Axillary mass
- Swelling of arm
- Bone or back pain
- Weight loss

Clinical Breast Examination. Breast cancer mortality is reduced when women are screened by both clinical breast examination (CBE) and mammography. CBE alone detects 3% to 45% of diagnosed breast cancers that screening mammography misses. Studies show the sensitivity of CBE is 54% (test's ability to determine a true positive) and specificity is 94% (test's ability to determine a true negative).[27] However, the rate of false-positive tests may be higher when CBE is performed without mammography.[28]

A previous edition of this text (*Differential Diagnosis in Physical Therapy*, edition 3) specifically stated, "breast examination is not within the scope of a physical therapist's practice." This practice is changing. As the number of cancer survivors increases in the United States, physical therapists treating postmastectomy women and clients of both genders with lymphedema are on the rise.

With direct and unrestricted access of consumers to physical therapists in many states, advanced skills have become necessary. For some clients, performing a CBE is an appropriate assessment tool in the screening process.[29] The American Cancer Society (ACS) and National Cancer Institute (NCI) support the provision of cancer screening procedures by qualified health specialists. With additional training, physical therapists can qualify.[29,30] Guidelines for CBE are provided in Appendix D-7.

Therapists who are trained to perform CBEs must make sure this examination is allowed according to the state practice act. In some states, it is allowed by exclusion, meaning it is not mentioned and therefore included. Discussion of the role of the physical therapist in primary care and cancer screening as it relates to integrating CBE into an upper quarter examination is available.[29] A form for recording findings from the CBE is provided in Fig. 4-48 and Appendix C-10.

The physical therapist does not diagnose any kind of cancer, including breast cancer; only the pathologist diagnoses cancer. The therapist can identify aberrant soft tissue and refer the client for further evaluation. Early detection and intervention can reduce morbidity and mortality.

For the therapist who is not trained in CBE, the client should be questioned about the presence of any changes in breast tissue (e.g., lumps, distended veins, skin rash, open sores or lesions, or other skin changes) and nipple (e.g., rash or other skin changes, discharge, distortion). Visual inspection is also possible and may be very important postmastectomy (Fig. 17-5). Ask the client if he or she has noticed any changes in the scar. Continue by asking:

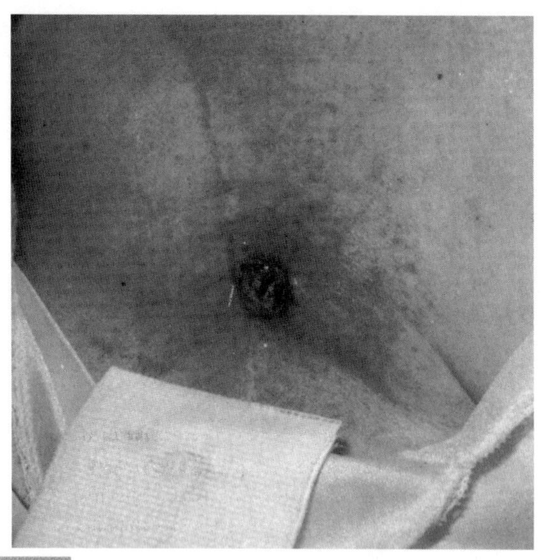

Fig. 17-5 This photo shows the chest of a woman who has had a right radical mastectomy. There is a metastatic nodule in the mastectomy scar as a result of local cancer recurrence. Breast cancer can occur (recur) if a mastectomy has been done. A closer look at the lesion suggests that the skin changes have been present for quite some time. Even in this black and white photo, the change in skin coloration is obvious in a large patch around the nodule. Anytime a woman with a past medical history of cancer develops neck, back, upper trapezius, or shoulder pain, or other symptoms, examining the site of the original cancer removal is a good idea. (From Callen JP, Jorizzo J, Greer KE, et al: *Dermatological signs of internal disease,* Philadelphia, 1988, Saunders.)

FOLLOW-UP QUESTIONS

- Would you have any objections if I looked at (or examined) the scar tissue?

If the client declines or refuses, the therapist should follow up with counsel to perform self-inspection, emphasizing the need for continued CBEs and the importance of reporting any changes to the physician immediately.

Therapists have an important role in primary prevention and client education. The ACS offers recommendations for breast cancer screening. The therapist can encourage women (and men) to follow these guidelines (Box 17-3).

Lymph Node Assessment. Palpation of the underlying soft tissues (chest wall, axilla) and lymph nodes in the supraclavicular and axillary regions should be part of a screening exam in any client with chest pain (see Chapter 4 for description of lymph node palpation). Any report of palpable breast nodules, lumps, or changes in the appearance of the breast requires medical follow-up, especially when there is a personal or family history of breast disease.[31]

"Normal" lymph nodes are not palpable or visible, but not all palpable or visible lymph nodes are a sign of cancer. Infections, viruses, bacteria, allergies, and food intolerances can cause changes in the lymph nodes. Lymph nodes that are hard, immovable, irregular, and nontender raise the suspicion of cancer, especially in the presence of a previous history of cancer. The skin surface over a tumor may be red, warm, edematous, firm, and painful. There may be skin dimpling over the lesion, with attachment of the mass to surrounding tissues preventing normal mobilization of skin, fascia, and muscle.

In the past, therapists were taught that any changes in lymph nodes present for more than 1 month in more than one region were a red flag. This has changed with the increased

BOX 17-3 SUMMARY OF GUIDELINES FOR BREAST CANCER SCREENING

The debate is not over about these revised recommendations. Some experts advise using these guidelines as a starting point for discussion but should not be used to justify delaying or avoiding screening. Recognizing that the majority of cancers that would be missed by postponing mammography would not be immediately life-threatening, many physicians advise women to make an individual decision about screening based on these guidelines and their own risk factors and to do so in consultation with their primary care physician.

From the American College of Obstetricians and Gynecologists (ACOG), the American Cancer Society, and the National Comprehensive Cancer Network:

- Routine (annual) screening mammography is still recommended for women beginning at age 40 and continuing for as long as a woman is in good health; however, evidence is lacking in determining guidelines for screening mammography after age 74.
- Mammography may begin before age 40 based on individual risk factors and personal preferences. Women younger than 40 should become aware of the benefits and harms of routine mammograms when making this decision.
- Clinical breast exam (CBE) about every 3 years for women in their 20s and 30s and every year for women 40 and over is recommended.
- Women should know how their breasts normally look and feel and report any breast change promptly to their health care provider. Breast self-exam (BSE) is an option for women starting in their 20s.
- Insufficient evidence makes it impossible to develop guidelines at this time regarding the effectiveness of digital mammography or breast MRI instead of film mammography. Some women (due to family history, genetic tendency, or other risk factors) may be encouraged by their physicians to be screened with MRI in addition to mammograms.

From Practice Bulletin no. 122; breast cancer screening. *Obstet Gynecol* 118(Pt. 1):372–382, 2011 and Smith RA: Cancer screening in the United States, 2010: A review of current American Cancer Society guidelines and issues in cancer screening, *CA Cancer J Clin* 60(2):99–119, 2010. Available on-line at www.cancer.org. Accessed August 4, 2011.

understanding of cancer metastases via the lymphatic system and the potential for cancer recurrence. A physician must evaluate all suspicious lymph nodes.

Metastases. Metastases have been known to occur up to 25 years after the initial diagnosis of breast cancer. On the other hand, breast cancer can be a rapidly progressing, terminal disease. Approximately 40% of clients with stage II tumors experience relapse.

Knowledge of the usual metastatic patterns of breast cancer and the common complications can aid in early recognition and effective treatment. Because bone is the most frequent site of metastases from breast cancer in men and women, a past medical history of breast cancer is a major red flag in anyone presenting with new onset or persistent findings of NMS pain or dysfunction.

All distant visceral sites are potential sites of metastases. Other primary sites of involvement are lymph nodes, remaining breast tissue, lung, brain, central nervous system (CNS), and liver. Women with metastases to the liver or CNS have a poorer prognosis.

Spinal cord compression, usually from extradural metastases, may appear as back pain, leg weakness, and bowel/bladder symptoms. Rarely, an axillary mass, swelling of the arm, or bone pain from metastases may be the first symptom. Back or bone pain, jaundice, or weight loss may be the result of systemic metastases, but these symptoms are rarely seen on initial presentation.

Medical referral is advised before initiating treatment for anyone with a past history of cancer presenting with symptoms of unknown cause, especially without an identifiable movement system impairment.

A medical evaluation is still needed in light of new findings even if the client has been rechecked by a medical oncologist recently. It is better to err on the side of caution. Failure to recognize the need for medical referral can result in possible severe and irreversible consequences of any delay in diagnosis and therapy.[32]

CLINICAL SIGNS AND SYMPTOMS

Metastasized Breast Cancer

- Palpable mass in supraclavicular, chest, or axillary regions
- Unilateral upper extremity numbness and tingling
- Back, hip, or shoulder pain
- Pain on weight bearing
- Leg weakness or paresis
- Bowel/bladder symptoms
- Jaundice

SCREENING FOR OTHER CONDITIONS AS A CAUSE OF CHEST, BREAST, OR RIB PAIN

Breast Implants

Scar tissue or fibrosis from a previous breast surgery, such as reconstruction following mastectomy for breast cancer or augmentation or reduction mammoplasty for cosmetic reasons, is an important history to consider when assessing chest, breast, neck, or shoulder symptoms. Likewise, the client should be asked about a history of radiation to the chest, breast, or thorax.

Women who have silicone or saline implants for reconstruction following mastectomy for breast cancer are more likely to have complications including late complications[33] (e.g., pain, capsular contracture, rupture, rippling, infection, hematoma, seroma) than those who receive implants for cosmetic reasons only.[34,35] The rate of fibrosis and capsular contracture is significantly higher for irradiated breasts than for nonirradiated breasts.[36,37]

Studies show that ruptures are rare (0.4 %) in women who have breast implants after mastectomy; thick, tight scarring, implant malposition, and infection are more common.[38,39]

Other complications of breast implantation may include gel bleed, implant leaking, calcifications around the implant, chronic breast pain, prolonged wound healing, and formation of granulation tissue.

Anxiety

An anxiety state, or in its extreme form, panic attack, can cause chest or breast pain typical of a heart attack. The client experiences shortness of breath, perspiration, and pallor. It is the most common noncardiac cause of chest pain, accounting for half of all emergency department admissions each year for chest or breast pain (just ahead of chest pain caused by cocaine use).

Risk Factors

The first panic attack often follows a period of extreme stress, sometimes associated with being the victim of a crime or the loss of a job, partner, or close family member. The presence of another mental health disorder, such as depression or substance abuse (drugs or alcohol), increases the risk of developing panic disorder. There may be a familial component, but it is not clear if this is hereditary or environmental (learned behavior).[40,41]

Drugs such as over-the-counter (OTC) decongestants and cold remedies can trigger panic attacks. Excessive use of caffeine and stimulants, such as amphetamines and cocaine combined with a lack of sleep, can also trigger an attack. Menopause, quitting smoking, or caffeine withdrawal can also bring on new onset of panic attacks in someone who has never experienced this problem before. See Chapter 3 for further discussion.

Clinical Presentation

There are several types of chest or breast discomfort caused by anxiety. The pain may be sharp, intermittent, or stabbing and located in the region of the left breast. The area of pain is usually no larger than the tip of the finger but may be as large as the client's hand. It is often associated with a local area of hyperesthesia of the chest wall. The client can point to it with one finger. It is not reproduced with palpation or activity. It is not changed or altered by a change in position.

Anxiety-related pain may be located precordially (region over the heart and lower part of the thorax) or retrosternally (behind the sternum). It may be of variable duration, lasting no longer than a second or for hours or days. This type of pain is unrelated to effort or exercise. Distinguishing this sensation from myocardial ischemia requires medical evaluation.

Discomfort in the upper portion of the chest, neck, and left arm, again unrelated to effort, may occur. There may be a sense of persistent weakness and unpleasant awareness of the heartbeat. In the past, radiation of chest discomfort to the neck or left arm was considered to be diagnostic of atherosclerotic coronary heart disease. More recently, stress testing and coronary arteriography have shown that chest discomfort of this type can occur in clients with normal coronary arteriograms.

Some individuals with anxiety-related chest pain may have a choking sensation in the throat caused by hysteria. There may be associated hyperventilation. Palpitation, claustrophobia, and occurrence of symptoms in crowded places are common.

Hyperventilation occurs in persons with and without heart disease and may be misleading. Such clients have numbness and tingling of the hands and lips and feel as if they are going to "pass out." For a more detailed explanation of anxiety and its accompanying symptoms (e.g., hyperventilation), see Chapter 3.

CLINICAL SIGNS AND SYMPTOMS

Chest Pain Caused by Anxiety

- Dull, aching discomfort in the substernal region and in the anterior chest
- Sinus tachycardia
- Fatigue
- Fear of closed-in places
- Diaphoresis
- Dyspnea
- Dizziness
- Choking sensation
- Hyperventilation: numbness and tingling of hands and lips

Cocaine

Cocaine (also methamphetamine, known as *crank,* and phencyclidine [PCP]) is a stimulant that has profound effects on all organs systems of the body, especially cardiotoxic effects, including cocaine-dilated cardiomyopathy, angina, and left ventricular dysfunction. Injection or inhalation can precipitate MI, cardiac arrhythmias, and even sudden cardiac death.[42-45]

Chronic use of cocaine or any of its derivatives is the number-one cause of stroke in young people today. The incidence of stroke associated with substance use and abuse is increasing. Use of these stimulants also has an effect on anyone with a congenital cerebral aneurysm and can lead to rupture.

The physiologic stress of cocaine use on the heart accounts for an increasing number of heart transplants. Acute effects of cocaine include increased heart rate, blood pressure, and vasomotor tone.[44,46] Cocaine remains the most common illicit drug–related cause of severe chest pain bringing the person to the emergency department.[42,45] In fact, chest pain is the most common cocaine-related medical complaint.

Many people with chest pain have used cocaine within the last week but deny its use. The use of these substances is not uncommon in middle-aged and older adults of all socioeconomic backgrounds. The therapist should not neglect to ask clients about their use of substances because of preconceived ideas that only teenagers and young adults use drugs. Careful questioning (see Chapter 2; see also Appendix B-36) may assist the physical therapist in identifying a possible correlation between chest pain and cocaine use.

Always end this portion of the interview by asking:

FOLLOW-UP QUESTIONS

- Are there any drugs or substances you take that you haven't mentioned?

Anabolic-Androgenic Steroids

Anabolic steroids are synthetic derivatives of testosterone used to enhance athletic performance or cosmetically shape the body. Used in supraphysiologic doses (more than the body produces naturally), these drugs have a potent effect on the musculoskeletal system, including the heart, potentially altering cardiac cellular and physiologic function.[47,48] Effects persist long after their use has been discontinued.[49]

The use of self-administered anabolic-androgenic steroids (AASs) is illegal but continues to increase dramatically among both athletes and nonathletes.[48] It is used among preteens who do not compete in sports for cosmetic reasons. The goal is to advance to a more mature body build and enhance their looks. AASs do have medical uses and were added to prescribed controlled substances in 1990 under the control of the Drug Enforcement Administration.

In spite of stricter control of the manufacture and distribution of AASs, illegal supplies come from unlicensed sources all over the world. When dispensed without a regulating agency, the purity and processing of chemicals is unknown. The quality of black market supplies is a major concern. There is no guarantee that the products obtained are correctly labeled. Contents and dosage may be inaccurate. Some athletes are using injectable anabolic steroids intended for veterinary use only. There is a trend for self-administration of higher doses and for combining AAS with other potentially harmful drugs.[50]

Clinical Presentation

Any young adult with chest pain of unknown cause, possibly accompanied by dyspnea and elevated blood pressure and without clinical evidence of NMS involvement, may have a history of anabolic steroid use. Consider anabolic steroid use as a possibility in men and women presenting with chest pain in their early 20s who have used this type of steroid since age 11 or 12.

In the pediatric population, there is a risk of decreased or delayed bone growth. Tendon and muscle strains are common and take longer than normal to heal. *Injuries that take longer than the expected physiologic time to heal* are an important red flag. Delayed healing occurs because the soft tissues are working under the added strain of extra body mass.

The alert therapist may recognize the associated signs and symptoms accompanying chronic use of these steroids. Changes in personality are the most dramatic signs of steroid use. The user may become more aggressive or experience mood swings (hypomanic or manic symptoms) and psychologic delusions (e.g., believe he or she is indestructible; sometimes referred to as "steroid psychosis"). "Roid rages," characterized by sudden outbursts of uncontrolled emotion, may be observed. Severe depression leading to suicide can occur with AAS withdrawal.[48]

CLINICAL SIGNS AND SYMPTOMS

Anabolic Steroid Use

- Chest pain
- Elevated blood pressure
- Ventricular tachycardia
- Weight gain (10 to 15 pounds in 2 to 3 weeks)
- Peripheral edema
- Acne on the face, upper back, chest
- Altered body composition with marked development of the upper torso
- Stretch marks around the back, upper arms, and chest
- Needle marks in large muscle groups (e.g., buttocks, thighs, deltoids)
- Development of male pattern baldness
- Gynecomastia (breast tissue development in males); breast tissue atrophy in females
- Frequent hematoma or bruising
- Personality changes called "steroid psychosis" (rapid mood swings, sudden increased aggressive or even violent tendencies)
- Females: Secondary male characteristics (deeper voice, breast atrophy, abnormal facial and body hair); menstrual irregularities
- Jaundice (chronic use)

The therapist who suspects a client may be using anabolic steroids should report findings to the physician or coach if one is involved. The therapist can begin by asking about the use of nutritional supplements or performance-enhancing agents. In the well-muscled male athlete, observe for common side effects of AAS such as acne, gynecomastia, and cutaneous striae in the deltopectoral region. Women who use AAS may exhibit muscular hypertrophy; male pattern baldness; excess hair growth on the face, breasts, and arms; and breast tissue atrophy.[47] Asking about the presence of common side effects of AAS and testing for elevated blood pressure may provide an opportunity to ask if the client is using these chemicals.

SCREENING FOR MUSCULOSKELETAL CAUSES OF CHEST, BREAST, OR RIB PAIN

It is estimated that half of all chest pain is of noncardiac nature and 20% to 25% of noncardiac chest pain has a musculoskeletal basis.[51,52] Musculoskeletal causes of chest (wall) pain must be differentiated from pain of cardiac, pulmonary, epigastric, and breast origin (see Table 17-1) before physical therapy treatment begins. Careful history taking to identify red flag conditions differentiates those who require further investigation.

Movement system impairment is most often characterized by pain during specific postures, motion, or physical activities. Reproducing the pain by movement or palpation often directs the therapist in understanding the underlying problem.

Chest pain can occur as a result of cervical spine disorders because nerves originating as high as C3 and C4 can extend as far down as the nipple line. Pectoral, suprascapular, dorsal scapular, and long thoracic nerves originate in the lower cervical spine, and impingement of these nerves can cause chest pain.

Musculoskeletal disorders such as myalgia associated with muscle exertion, myofascial TrPs, costochondritis, osteomyelitis, or xiphoiditis can produce pain in the chest and arms. Compared with angina pectoris, the pain associated with these conditions may last for seconds or hours, and prompt relief does not occur with the ingestion of nitroglycerin.

Tietze's syndrome, costochondritis, a hypersensitive xiphoid, and the slipping rib syndrome must be differentiated from problems involving the thoracic viscera, particularly those of the heart, great vessels, and mediastinum, as well as from illness originating in the head, neck, or abdomen.[53]

Rib pain (with or without neck, back, or chest pain or symptoms) must be evaluated for systemic versus musculoskeletal origins (Box 17-4). The same screening model used for all conditions can be applied.

Costochondritis

Costochondritis, also known as *anterior chest wall syndrome, costosternal syndrome,* and *parasternal chondrodynia* (pain in a cartilage), is used interchangeably with Tietze's syndrome, although these two conditions are not the same. Costochondritis is more common than Tietze's syndrome.

Although both disorders are characterized by inflammation of one or more costal cartilages (costochondral joints where the ribs join the sternum), costochondritis refers to pain in the costochondral articulations without swelling. This disorder can occur at almost any age but is observed most often in people older than 40. It tends to affect the second,

BOX 17-4 CAUSES OF RIB PAIN

Systemic/Medical Conditions
- Gallbladder disease (tenth rib)
- Shingles (herpes zoster)
- Pleurisy
- Osteoporosis
- Cancer (metastasized to the bone)

Musculoskeletal
- Trauma (e.g., bruise, fracture)
- Slipping rib syndrome
- Tietze's syndrome or costochondritis
- Trigger points
- Thoracic outlet syndrome (TOS)

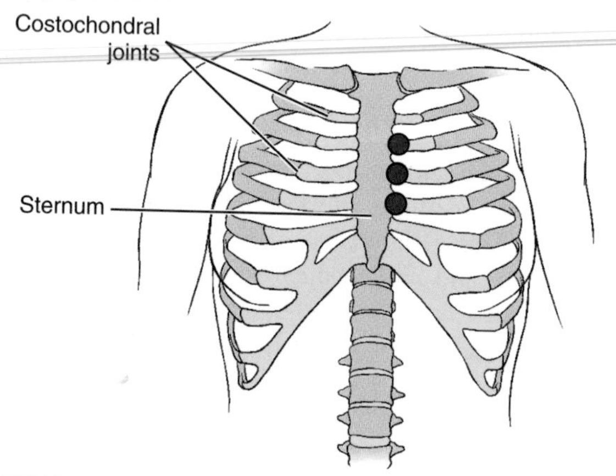

Fig. 17-6 Costochondritis is an inflammation of any of the costochondral joints (also called *costal cartilages*) where the rib joins the sternum. Sharp, stabbing, or aching pain can occur on either side of the sternum but tends to affect the left more often, even radiating down the left arm sometimes or to upper back. Many people mistake the symptoms for a heart attack. In most cases, symptoms occur at a single site involving the second or third costochondral joint, although any of the joints can be affected as shown.

third, fourth, and fifth costochondral joints; women are affected in 70% of all cases (Fig. 17-6). Other risk factors include trauma (e.g., driver striking steering wheel with chest during a motor vehicle accident, upper chest surgery, helmet tackle in sports, or other sports injury to the chest)[54] or repetitive motion (e.g., grocery-store clerk lifting and scanning items, competitive swimming).[55]

Costochondritis is characterized by a sharp pain along the front edges of the sternum, especially on the left side with possible radiation to the arms, back, or shoulders, often misinterpreted as a heart attack.

In fact, it is estimated that one-third of all people who present at the emergency department with chest pain have costochondritis.[56] The pain may radiate widely (to the arms, back, shoulders), stimulating intrathoracic or intraabdominal disease. It differs from a myocardial infarction because during a heart attack, the initial pain is usually in the center of the chest, under the sternum, not along the edges.

Costochondritis can be similar to muscular pain and (unlike cardiac-related pain) is elicited by palpatory pressure over the costochondral junctions. Occasionally, the affected individual will report a burning sensation in the breast(s) associated with this condition. Absence of associated signs and symptoms such as dyspnea, nausea, vomiting, and diaphoresis helps differentiate this condition from cardiac- or pulmonary-related chest pain.

Costochondritis may follow trauma or may be associated with systemic rheumatic disease. It can come and go (especially in conjunction with activities involving the upper extremities or rest) or persist for months. Inflammation of upper costal cartilages may cause chest pain, whereas inflammation of lower costal cartilages is more likely to cause abdominal or low back discomfort. In some cases, costochondritis is associated with an URI but may be a result of the stress of coughing rather than the body's response to the virus.

Tietze's Syndrome

Tietze's syndrome (inflammation of a rib and its cartilage; costal chondritis) may be one possible cause of anterior chest wall pain, manifested by painful swelling of one or more costochondral articulations.

In most cases, the cause of Tietze's syndrome is unknown. Other causes of sternal swelling may include an infectious process in the immunocompromised person resulting from tuberculosis, aspergillosis, brucellosis, staphylococcal infection, or pseudomonal disease producing sternal osteomyelitis. Onset is usually before 40 years of age, with a predilection for the second and third decades. However, it can occur in children.

Approximately 80% of clients have only single sites of involvement, most commonly the second or third costal cartilage (costochondral joint). Anterior chest pain may begin suddenly or gradually and may be associated with increased blood pressure, increased heart rate, and pain radiating down the left arm. Pain is aggravated by sneezing, coughing, deep inspirations, twisting motions of the trunk, horizontal shoulder abduction and adduction, or the "crowing rooster" movement of the upper extremities.

These symptoms may seem similar to those of a heart attack, but the raised blood pressure, reproduction of painful symptoms with palpation or pressure, and aggravating factors differentiate Tietze's syndrome from myocardial infarction (Case Example 17-7). In rare cases, the individual has been diagnosed with Tietze's syndrome only to find out later the precipitating cause was cancer (e.g., lymphoma, squamous cell carcinoma of the mediastinum).[57,58] Tietze's syndrome can also be confused with TrPs (pectoralis major, internal intercostalis), an often overlooked cause of the same symptoms.[59]

CASE EXAMPLE 17-7

Tietze's Syndrome

Referral: A 53-year-old woman was referred by her physician with a diagnosis of left anterior chest pain. The woman is employed at a sawmill and performs tasks that require repetitive shoulder flexion and extension in using a hydraulic apparatus on a sliding track. Lifting (including overhead lifting) is required occasionally but is limited to items less than 20 pounds.

Past Medical History: Her past medical history was significant for a hysterectomy 10 years ago for prolonged bleeding. She has been a 4- to 5-pack/day smoker for 30 years but has cut down to ½ pack/day for the last 2 months.

Clinical Presentation

Pain Pattern: The woman described the onset of her pain as sudden, crushing chest pain radiating down the left arm, occurring for the first time 6 weeks ago. She was transported to the emergency department, but tests were negative for cardiac incident. Blood pressure at the time of the emergency admittance was 195/110 mm Hg. She was released from the hospital with a diagnosis of "stress-induced chest pain."

The client experienced the same type of episode of chest pain 10 days ago but described radiating pain around the chest and under the armpit to the upper back. Today, her symptoms include extreme tenderness and pain in the left chest with deep pain described as penetrating straight through her chest to her back. There is no numbness or tingling and no pain down the arm but a residual soreness in the left arm.

The client believes that her symptoms may be "stress-induced" but expresses some doubts about this because her symptoms persist and no known cause has been found. She relates that because of divorce proceedings and child custody hearings, she is under extreme stress at this time.

Examination: The neurologic screen was negative. The deep tendon reflexes were within normal limits; strength testing was limited by pain but with a strong initial response elicited; and no changes in sensation, two-point discrimination, or proprioception were observed.

There was exquisite pain on palpation of the left pectoral muscle with tenderness and swelling noted at the second, third, and fourth costochondral joints. Painful and radiating symptoms were reproduced with resisted shoulder horizontal adduction. Active shoulder range of motion was full but with a positive painful arc on the left. There was also painful reproduction of the radiating symptoms down the arm with palpation of the left supraspinatus and biceps tendons.

The painful chest/arm/upper back symptoms were not altered by respiratory movements (deep breathing or coughing), but the client was unable to lie down without extreme pain.

Result: The physical therapy assessment was suggestive of Tietze's syndrome secondary to repetitive motion and exacerbated by emotional stress* with concomitant shoulder dysfunction. Physical therapy intervention resulted in initial rapid improvement of symptoms with full return to work 6 weeks later.

*It should be noted that although the physical therapist's assessment recognized emotional stress as a factor in the client's symptoms, it may not be in the client's best interests to include this information in the documentation. Although the medical community is increasingly aware of the research surrounding the mind-body connection, worker's compensation and other third-party payers may use this information to deny payment.

CLINICAL SIGNS AND SYMPTOMS

Tietze's Syndrome or Costochondritis

- Sudden or gradual onset of upper anterior chest pain
- Pain/tenderness of costochondral joint(s)
- Bulbous swelling of the involved costal cartilage (Tietze's syndrome)
- Mild-to-severe chest pain that may radiate to the left shoulder and arm
- Pain aggravated by deep breathing, sneezing, coughing, inspiration, bending, recumbency, or exertion (e.g., push-ups, lifting grocery items)

Hypersensitive Xiphoid

The hypersensitive xiphoid (xiphodynia) is tender to palpation, and local pressure may cause nausea and vomiting. This syndrome is manifested as epigastric pain, nausea, and vomiting.

Slipping Rib Syndrome

The slipping, or painful, rib syndrome (sometimes also referred to as the *clicking rib syndrome*) can present as chest pain and occurs most often when there is hypermobility of the lower ribs.[53] In this condition, inadequacy or rupture of the interchondral fibrous attachments of the anterior ribs allows the costal cartilage tips to sublux, impinging on the intercostal nerves. This condition can occur alone or can be associated with a broader phenomenon such as myofascial pain syndrome.[60]

Rib syndrome can occur at any age, including during childhood[61] but most commonly occurs during the middle-aged years. The physical therapist is usually able to identify readily a rib syndrome as the cause of chest pain after a careful musculoskeletal examination. In some cases, persistent upper abdominal and/or low thoracic pain occurs, leaving physicians, chiropractors, and therapists puzzled.[62,63] A sonogram may be needed to make the diagnosis. Pain is made worse by slump sitting or side bending to the affected side. Reduction or elimination of symptoms following rib mobilization helps confirm the differential diagnosis.

Gallbladder impairment can also cause tenderness or soreness of the tip of the tenth rib on the right side. The affected individual may or may not have gallbladder symptoms. Because visceral and cutaneous fibers enter the spinal cord at the same level for the ribs and gallbladder, the nervous system may respond to the afferent input with sudomotor changes such as pruritus (itching of the skin) or a sore rib instead of gallbladder symptoms.

The clinical presentation appears as a biomechanical problem, such as a rib dysfunction, instead of nausea and food intolerances normally associated with gallbladder dysfunction. Symptoms will not be alleviated by physical therapy intervention, eventually sending the client back to his or her physician.

Trigger Points

The most common musculoskeletal cause of chest pain is TrPs, sometimes referred to as *myofascial TrPs* (MTrPs). TrPs (hypersensitive spots in the skeletal musculature or fascia) involving a variety of muscles (Table 17-4) may produce precordial pain (Fig. 17-7). Abdominal muscles have multiple referred pain patterns that may reach up into the chest or midback and produce heartburn or deep epigastric pain. Although these patterns strongly mimic cardiac pain, myofascial TrP pain shows a much wider variation in its response to daily activity than does angina pectoris to activity.[59]

In addition to mimicking pain of a cardiac nature, TrPs can occur in response to cardiac disorders. A viscero-somatic response can occur when biochemical changes associated with visceral disease affect somatic structures innervated by the same spinal nerves. In such cases, the individual has a past history of visceral disease. TrPs accompanied by symptoms, such as vertigo, headache, visual changes, nausea, and syncope, are yellow flags warning of autonomic involvement not usually present with TrPs strictly from a somatic origin.

Chest pain that persists long after an acute MI may be due to myofascial TrPs. In acute MI, pain is commonly referred from the heart to the midregion of the pectoralis major and minor muscles (see discussion of viscero-somatic sources of pain in Chapter 3). The injury to the heart muscle initiates a viscero-somatic process that activates TrPs in the pectoral muscles.[59]

After recovery from the infarction, these self-perpetuating TrPs tend to persist in the chest wall. As with all myofascial syndromes, inactivation of the TrPs eliminates the client's symptoms of chest pain. If the client's symptoms are eliminated with TrP release, medical referral may not be required. However, communication with the physician is essential; the therapist is advised to document all findings and report them to the client's primary care physician.

TABLE 17-4 Trigger Point Pain Guide

Location	Potential Muscles Involved
Front of chest pain	Pectoralis major Pectoralis minor Scaleni Sternocleidomastoid (sternal) Sternalis Iliocostalis cervicis Subclavius External abdominal oblique
Side of chest pain	Serratus anterior Latissimus dorsi
Upper abdominal/lower chest pain	Rectus abdominis Abdominal obliques Transversus abdominis

Modified from Travell JG, Simons DG: *Myofascial pain and dysfunction: the trigger point manual,* Baltimore, 1983, Williams & Wilkins, p 574.

Past Medical History

There may be a history of URI with repeated forceful coughing. There is often a history of immobility (e.g., cast immobilization after fracture or injury). The therapist should also ask about muscle strain from lifting weights overhead, from pushups, and from prolonged, vigorous activity that requires forceful abdominal breathing, such as severe coughing, running a marathon, or repetitive bending and lifting.

Clinical Presentation

TrPs are reproduced with palpation or resisted motions. On examination, the physical therapist should palpate for tender points and taut bands of muscle tissue, squeeze the involved muscle, observe for increased pain with palpation, test for increased pain with resisted motion, and correlate symptoms with respiratory movements.

Chest pain from serratus anterior TrPs may occur at rest in severe cases. Clients with this myofascial syndrome may report that they are "short of breath" or that they are in pain when they take a deep breath. *Serratus anterior* TrPs on the left side of the chest can contribute to the pain associated with myocardial infarction. This pain is rarely aggravated by the usual tests for range of motion at the shoulder but may result from a strong effort to protract the scapula. Palpation reveals tender points that increase symptoms, and there is usually a palpable taut band present within the involved muscles.

One of the most extensive patterns of pain from irritable TrPs is the complex pattern from the *anterior scalene* muscle. This may produce ipsilateral sternal pain, anterior chest wall pain, breast pain, or pain along the vertebral border of the scapula, shoulder, and arm, radiating to the thumb and index finger.

Breast pain may be differentiated from the aching pain arising from the scalene or pectoral muscles by a history of upper extremity overuse usually associated with myalgia. Resistance to isometric movement of the upper extremities reproduces the symptoms of a myalgia but does not usually aggravate pain associated with breast tissue. Additionally, palpation of the underlying muscle reproduces the painful symptoms.

When active TrPs occur in the left *pectoralis major* muscle, the referred pain (anterior chest to the precordium and down the inner aspect of the arm) is easily confused with that of coronary insufficiency. Pacemakers placed superficially can

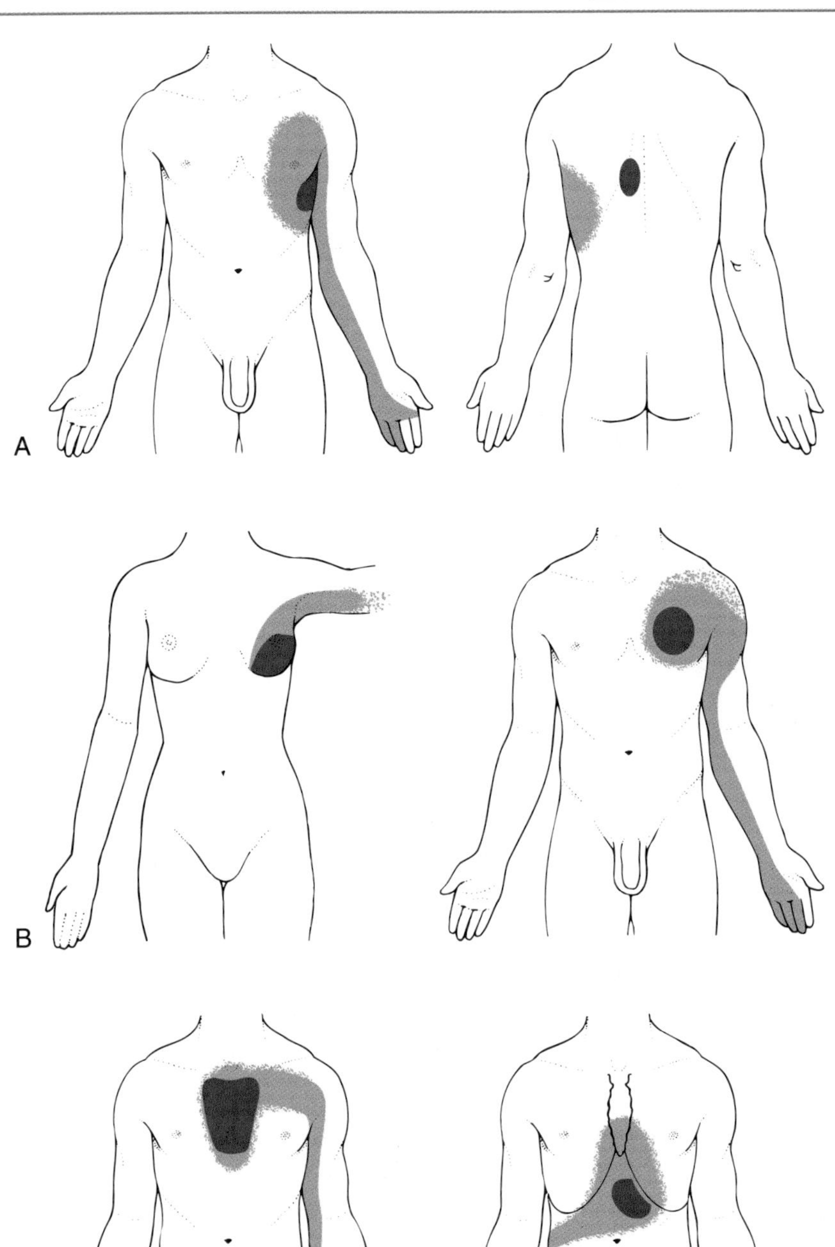

Fig. 17-7 **A,** Referred pain pattern from the left serratus anterior muscle. **B,** Left pectoralis major muscle: referred pain pattern in a woman and a man. **C,** Referred pain pattern from the left sternalis muscle. **D,** Referred pain from the external oblique abdominal muscle can cause "heartburn" in the anterior chest wall. Marathon runners may report chest pain mimicking a heart attack from this trigger point.

cause pectoral trigger points. In the case of pacemaker-induced TrPs, the physical therapist can teach the client TrP self-treatment to carry out at home.

Myalgia

Myalgia, or muscular pain, can cause chest pain separate from TrP pain but with a similar etiologic basis of prolonged or repeated movement. As mentioned earlier, the physical therapy interview must include questions about recent URI with repeated forceful coughing and recent activities of a repetitive nature that could cause sore muscles (e.g., painting or washing walls; calisthenics, including pushups; or lifting heavy objects or weights).

Three tests must be used to confirm or rule out muscle as the source of symptoms: (1) palpation, (2) stretch, and (3) contraction. If the muscle is not sore or tender on palpation, stretch, or contraction, the source of the problem most likely lies somewhere else.

With true myalgia, squeezing the muscle belly will reproduce painful chest symptoms. The discomfort of myalgia is almost always described as aching and may range from mild to intense. Diaphragmatic irritation may be referred to the ipsilateral neck and shoulder, lower thorax, lumbar region, or upper abdomen as a muscular aching pain. Myalgia in the respiratory muscles is well localized, reproducible by palpation, and exacerbated by movement of the chest wall.

Rib Fractures

Periosteal (bone) pain associated with fractured ribs can cause sharp, localized pain at the level of the fracture with an increase in symptoms associated with trunk motions and respiratory movements, such as deep inspiration, laughing, sneezing, or coughing. The pain may be accompanied by a grating sensation during breathing. This localized pain pattern differs from bone pain associated with chronic disease affecting bone marrow and endosteum, which may result in poorly localized pain of varying degrees of severity.

Occult (hidden) rib fractures may occur, especially in a client with a chronic cough or someone who has had an explosive sneeze. Fractures may occur as a result of trauma (e.g., motor vehicle accident, assault), but painful symptoms may not be perceived at first if other injuries are more significant.

A history of long-term steroid use in the presence of rib pain of unknown cause should raise a red flag. Rib fractures must be confirmed by x-ray diagnosis. Rib pain without fracture may indicate bone tumor or disease affecting bone, such as multiple myeloma.

Cervical Spine Disorders

Cervicodorsal arthritis may produce chest pain that is seldom similar to that of angina pectoris. It is usually sharp and piercing but may be described as a deep, boring, dull discomfort. There is usually unilateral or bilateral chest pain with flexion or hyperextension of the upper spine or neck. The chest pain may radiate to the shoulder girdle and down the arms and is not related to exertion or exercise. Rest may not alleviate the symptoms, and prolonged recumbency makes the pain worse.

Diskogenic disease can also cause referred pain to the chest, but there is usually evidence of disk involvement observed with diagnostic imaging and the presence of neurologic symptoms (Case Example 17-8).

SCREENING FOR NEUROMUSCULAR OR NEUROLOGIC CAUSES OF CHEST, BREAST, OR RIB PAIN

There are several possible neurologic disorders that can cause chest and/or breast pain, including nerve root impingement or inflammation, herpes zoster (shingles), thoracic disk disease, postoperative neuralgia, and thoracic outlet syndrome (TOS; see Table 17-1).

Neurologic disorders such as intercostal neuritis and dorsal nerve root radiculitis or a neurovascular disorder such as thoracic outlet syndrome also can cause chest pain. The two most commonly recognized noncardiac causes of chest pain seen in the physical therapy clinic are herpes zoster (shingles) and TOS.

Intercostal Neuritis

Intercostal neuritis, such as herpes zoster or shingles produced by a viral infection of a dorsal nerve root, can cause neuritic chest wall pain, which can be differentiated from coronary pain.

Risk Factors

Shingles may occur or recur at any age, but there has been a recent increase in the number of cases in two distinct age groups: college-aged young adults and older adults (over 70 years). Health care experts suggest that stress is the key

CASE EXAMPLE 17-8

Pediatric Occupational Therapist with Chest Pain

Referral: A 42-year-old woman presented with primary chest pain of unknown cause. She is employed as an independent pediatric occupational therapist. She has been seen by numerous medical doctors who have ruled out cardiac, pulmonary, esophageal, upper GI, and breast pathology as underlying etiologies.

Since her symptoms continue to persist, she was sent to physical therapy for an evaluation.

She reports symptoms of chest pain/discomfort across the upper chest rated as a 5 or 6 and sometimes as an 8 on a scale from 0 to 10. The pain does not radiate down her arms or up her neck. She cannot bring the symptoms on or make them go away. She cannot point to the pain, but reports it as being more diffuse than localized.

Past Medical History: She denies any shortness of breath, but admits to being "out of shape" and has not been able to exercise due to a failed bladder neck suspension surgery 2 years ago. She reports fatigue, but states that this is not unusual for her with her busy work schedule and home responsibilities.

She has not had any recent infections, no history of cancer or heart disease, and her mammogram and clinical breast exam are up to date and normal. She does not smoke or drink, but by her own admission has a "poor diet" due to time pressure, stress, and fatigue.

How do we proceed in the screening process with a client like this?

First review Table 17-1. According to the client, the doctors have ruled out most of the systemic causes in the left column. We know that medical specialization and progression of disease can account for missed cases of somatic symptoms of a viscerogenic or systemic origin.

In the case of medical specialization, the doctors can be looking so carefully for a problem within their own specialty area that they miss the obvious somewhere else. Progression of disease means that if enough time passes and the disease process goes unchecked, eventually the client will present with additional signs and symptoms to clear up the diagnostic mystery.

In the case of disease progression, it's possible that what the physician observed in his or her office is not the same as what you see weeks to months later. By carefully interviewing the client, you may be able to bring to light any new changes and report these to the physician.

CASE EXAMPLE 17-8—cont'd

Pediatric Occupational Therapist with Chest Pain

Systemic Causes

Anemia: She has complained of fatigue, a hallmark finding in anemia. You may think the doctors would have already found any evidence of anemia, but symptoms may not be recognized until hemoglobin concentration is reduced to half of normal. Since this is a Caucasian woman, we can skip looking at sickle cell anemia at this time.

As we look at the general information about anemia in Chapter 5, we are reminded to ask the following questions:

- Have you experienced any unusual or prolonged bleeding from any part of your body?
- Have you noticed any blood in your urine or stools? Have you noticed any changes in the color of your stools (dark, tarry, sticky stools may signal melena from blood loss in the GI tract)?
- Have you been taking any over-the-counter or prescribed antiinflammatory drugs (NSAIDs and peptic ulcer with GI bleeding)?
- Have you ever been told you have rheumatoid arthritis, lupus, HIV/AIDS, or anemia?

You notice from the text that there can be nail bed changes. Quickly inspect the nail beds, palms, and skin. Remember, observation of the hands should be done at the level of the client's heart, and the hands should be warm if possible.

Be sure to assess ALL vital signs. This includes pulse, respirations, oxygen saturation, skin temperature, core body temperature, and blood pressure. For a review of vital signs as a red flag, see Vital Signs in chapter 4.

Cancer: Palpate lymph nodes for generalized lymphadenopathy.

As we look at the items listed under "Other" in Table 17-1 are there any conditions here that might apply to this woman? What additional questions will you want to ask this client?

Rheumatic Diseases: First under "Other" is "Rheumatic diseases." Ask if the client has ever been diagnosed with arthritis of any kind or had any arthritic-like symptoms anywhere in her body. When you conduct your examination, keep in mind that RA is a systemic condition that can cause chest pain. Osteoarthritis of the cervical spine can also cause chest pain, so we will make a note to look at that more closely later.

Fibromyalgia: While you are asking about arthritis, go ahead and ask about fibromyalgia also in this list. You may want to ask about the presence of symptoms commonly associated with fibromyalgia.

Pain patterns typical of fibromyalgia can be found in chapter 12. You can find a list of clinical signs and symptoms of fibromyalgia in that section.

Anxiety, Cocaine, or Steroid Use: Finally, from the list in the left column of Table 17-1, we have anxiety and cocaine or anabolic steroid use. We have already discussed anxiety as a potential cause of chest pain, but cocaine and steroid use are also covered in this chapter.

Additional signs and symptoms present with anxiety, anxiety as a factor in pain assessment, and screening for anxiety are discussed in chapter 3. Table 3-9 lists the physical, behavioral, cognitive, and psychologic symptoms of anxiety. You may just want to take a look at this list and ask your client if she experiences any of these symptoms on a regular basis.

Anabolic steroid use isn't likely in this client, but cocaine use may be an issue. Do not neglect to ask about the use of any recreational drugs of any kind, including cocaine, crack, PCP, marijuana, hash, and so forth.

Neuromusculoskeletal Causes: Now use the neuromusculoskeletal causes of chest pain listed in the right column as a springboard for your examination. Perform appropriate orthopedic and other clinical tests and measures to identify any of the conditions listed.

Most of these conditions are associated with painful symptoms that can be reproduced if you know what movements to suggest and/or where and when to palpate. In the case of rib fractures or breast pain from trauma, don't forget to screen for domestic violence or assault as covered in Chapter 2.

Use Table 17-4 to identify trigger points in muscles that can cause chest pain. Sometimes, failure to respond to physical therapy intervention is considered a red flag for systemic origin of somatic symptoms.

Beware when your client fails to respond to trigger point therapy. This may not be a red flag suggesting screening for systemic or other causes of muscle pain. Muscle recovery from trigger points isn't always so simple.

According to Headley, muscles with active trigger points fatigue faster and recover more slowly. They show more abnormal neural circuit dysfunction. The pain and spasm of trigger points may not be relieved until the aberrant circuits are corrected.[59a]

Results: After completing the evaluation with appropriate questions, tests and measures, a Review of Systems pointed to the cervical spine as the most likely source of this client's symptoms. The jaw and shoulder joint were cleared, although there were signs of shoulder movement dysfunction from a possible impingement syndrome.

She had limited cervical spine range of motion with obvious limitations in passive intervertebral movements (joint play or accessory motions) at the C4-5 level. There were no neurologic findings except for global neck muscle weakness, which was consistent with a chronic pain pattern. There were latent trigger points of the pectoralis major and minor, but eliminating these did not change the primary chest pain pattern.

A trial course of manual therapy improved the client's symptoms temporarily. The client was discharged with a home program to maintain neck range of motion and gradually increase neck musculature strength. Symptoms returned intermittently over the next 6 weeks.

After relaying these findings to the client's primary care physician, radiographs of the cervical spine were ordered. Interestingly, despite the thousands of dollars spent on repeated diagnostic work-ups for this client, a simple x-ray had never been taken.

Results showed significant spurring and lipping throughout the cervical spine from early osteoarthritic changes of unknown cause. Cervical spine fusion was recommended and performed for instability in the midcervical region.

The client's chest pain was eliminated and did not return even up to 2 years after the cervical spine fusion. The physical therapist's contribution in pinpointing the location of referred symptoms brought this case to a successful conclusion and closure.

factor in the first group, and immune system failure is the key factor in the second group.

Anyone who is immunocompromised as a result of advancing age, underlying malignancy, organ transplantation, or acquired immunodeficiency syndrome (AIDS) is at risk for shingles. There is an increased incidence of herpes zoster in clients with lymphoma, tuberculosis, and leukemia, but it can be triggered by trauma or injection drugs or occur with no known cause.

Anyone in good health who had the chickenpox as a child is not at great risk for shingles. The risk of developing shingles increases for anyone who is immunocompromised for any reason or who has never had chickenpox.

Herpes zoster is a communicable disease and requires some type of isolation. Anyone in contact with the client before the outbreak of the skin lesions has already been exposed. Specific precautions depend on whether the disease is localized or disseminated and the condition of the client. Persons susceptible to chickenpox should avoid contact with the affected client and stay out of the client's room.

Clinical Presentation

Herpes zoster is characterized by raised fluid-filled clusters of grouped vesicles that appear unilaterally along cranial or spinal nerve dermatomes 3 to 5 days after transmission of the virus (see Figs. 4-23 and 4-24). The affected individual experiences 1 to 2 days of pain, itching, and hyperesthesia before the outbreak of skin lesions.

The skin changes are referred to as "shingles" and are easily recognizable as they follow a dermatome anywhere on the body. The lesions do not cross the body midline as they follow nerve pathways, although nerves of both sides may be involved. The skin eruptions evolve into crusts on the skin and clear in about 2 weeks, unless the period between the pain and the eruption is longer than 2 days. Postherpetic neuralgia, with its burning and paroxysmal stabbing pain, may persist for long periods.

Neuritic pain occurs unrelated to effort and lasts longer (weeks, months, or years) than angina. The pain may be constant or intermittent and can vary from light burning to a deep visceral sensation. It may be associated with chills, fever, headache, and malaise. Symptoms are confined to the somatic distribution of the involved spinal nerve(s).

CLINICAL SIGNS AND SYMPTOMS

Herpes Zoster (Shingles)

- Fever, chills
- Headache and malaise
- 1 to 2 days of pain, itching, and hyperesthesia before skin lesions develop
- Skin eruptions (vesicles) that appear along dermatomes 4 or 5 days after the other symptoms

Dorsal Nerve Root Irritation

Dorsal nerve root irritation of the thoracic spine is another neuritic condition that can refer pain to the chest wall. This condition can be caused by infectious processes (e.g., radiculitis or inflammation of the spinal nerve root dural sheath; shingles can also fit in this category). However, the pain is more likely to be the result of mechanical irritation caused by spinal disease or deformity (e.g., bone spurs secondary to osteoarthritis or the presence of cervical ribs placing pressure on the brachial plexus).

The pain of dorsal nerve root irritation can appear as lateral or anterior chest wall pain with referral to one or both arms through the brachial plexus. Although it mimics the pain pattern of coronary heart disease, such pain is more superficial than cardiac pain. Like cardiac pain, dorsal nerve root irritation can be aggravated by exertion of only the upper extremities. However, unlike cardiac pain, exertion of the lower extremities has no exacerbating effect. It is usually accompanied by other neurologic signs such as muscle atrophy and numbness or tingling.

CLINICAL SIGNS AND SYMPTOMS

Dorsal Nerve Root Irritation

- Lateral or anterior chest wall pain
- History of back pain
- Pain that is aggravated by exertion of only the upper body
- May be accompanied by neurologic signs
  - Numbness
  - Tingling
  - Muscle atrophy

Thoracic Outlet Syndrome

Thoracic outlet syndrome (TOS) refers to compression of the neural and/or vascular structures that leave or pass over the superior rim of the thoracic cage (see Fig. 17-10). Various names have been given to this condition according to the presumed site of major neurovascular compression: First thoracic rib, cervical rib, scalenus anticus, costoclavicular, and hyperabduction syndromes.

Past Medical History

History of associated back pain may be the only significant past medical history. The presence of anatomic anomalies, such as an extra rib or unusual sternoclavicular and/or acromioclavicular angle, may be the only known history linked to the development of TOS.

Risk Factors

Symptoms may be related to occupational activities (e.g., carrying heavy loads, working with arms overhead), poor posture, sleeping with arms elevated over the head, or acute injuries such as cervical flexion/extension (whiplash). Athletes such as swimmers, volleyball players, tennis players, and baseball pitchers are also at increased risk for compression of the neurovascular structures. Most people become symptomatic in the third or fourth decade, and women (especially during pregnancy) are affected three times more often than are men.

Clinical Presentation

Chest/breast pain can occur (and may be the only symptom of TOS) as a result of cervical spine disorders, an underlying etiology in TOS. This is because spinal nerves originating as high as C3-4 can extend down as low as the nipple line.

The compressive forces associated with this problem usually affect the upper extremities in the ulnar nerve distribution but can result in episodic chest pain mimicking coronary heart disease. Neurogenic pain associated with TOS may be described as stabbing, cutting, burning, or electric. The pain is often unrelated to effort and lasts hours to days.[64] Pectoralis minor syndrome has been identified as one cause of recurrent neurogenic TOS contributing to an estimated 75% or more of all cases.[65]

There may be radiating pain to the neck, shoulder, scapula, or axilla, but usually the superficial nature of the pain and associated changes in sensation and neurologic findings point to chest pain with an underlying neurologic cause (Table 17-5). Paresthesias (burning, pricking sensation) and hypoesthesia (abnormal decrease in sensitivity to stimulation) are common. Anesthesia and motor weakness are reported in about 10% of the cases.

When a vascular compressive component is involved, there may be more diffuse pain in the limb, with associated fatigue and weakness. With more severe arterial compromise, the client may describe coolness, pallor, cyanosis, or symptoms of Raynaud's phenomenon. Although vascular in origin, these symptoms are differentiated from CAD by the local or regional presentation, affecting only a single extremity or only the upper extremities.

Palpation of the supraclavicular space may elicit tenderness or may define a prominence indicative of a cervical rib. The effect on pulse of the Adson or Halstead maneuvers (Fig. 17-8), the hyperabduction or Wright test (Fig. 17-9), and the costoclavicular test (exaggerated military attention posture) should be compared in both arms.

Despite the widespread use of these tests, the reliability remains unknown. Specificity reported ranges from 18% to 87%, but sensitivity has been documented at 94%.[66] During assessment for vascular origin of symptoms, a change in pulse rate or rhythm is a positive test; however, because more than 50% of normal, asymptomatic individuals have pulse rate changes, it is better to reproduce the client's symptoms as a true indicator of TOS.[66,67]

Other clinical tests are described in orthopedic assessment texts.[66,68] With the use of special tests, patterns of positive objective findings may help characterize TOS as vascular, neural, or a combination of both (neurovascular).

Response to nerve blocks has not proved to be a predictable or reliable diagnostic or treatment approach to neurogenic TOS.[69] Although no specific testing for thoracic outlet has proven valid in detecting upper extremity pain of a neurogenic origin, Table 17-5 may help guide the therapist in assessing for this condition.

Knowing what the tests are for can be helpful in guiding intervention. For example, Fig. 17-10 gives a visual representation of the effect of the hyperabduction test. A positive hyperabduction test may point to the need to restore normal function and movement of the pectoralis minor muscle.

Likewise, if there is a neural component, assess for location (upper plexus versus lower plexus). Reproduction of pain or paresthesias with light pressure on the scalene, supraclavicular area, and pectoralis minor points to thoracic outlet rather than a cervical radiculopathy or more distal entrapment but several pathologic conditions can be present at the same time.[70]

TOS should be considered when persistent chest pain occurs in the presence of a normal coronary angiogram and normal esophageal function tests. Other conditions in the differential diagnosis include cervical degenerative disk disease or abnormality of the lung or chest wall.[71]

TABLE 17-5 Assessing Symptoms of Thoracic Outlet Syndrome*

Component	Symptoms
Vascular component	3-minute elevated test Adson's test Swelling (arm/hand) Discoloration of hand Costoclavicular test Hyperabduction test Upper extremity claudication Differences in blood pressure Skin temperature changes Cold intolerance
Neural	**Upper plexus** Point tenderness of C5-C6 Pressure over lateral neck elicits pain and/or numbness. Pain with head turned and/or tilted to opposite side Weak biceps Weak triceps Weak wrist Hypoesthesia in radial nerve distribution 3-minute abduction stress test **Lower plexus** Pressure above clavicle elicits pain Ulnar nerve tenderness when palpated under axilla or along inner arm Tinel's sign for ulnar nerve in axilla Hypoesthesia in ulnar nerve distribution Serratus anterior weakness Weak hand grip

*Although no specific testing for thoracic outlet has proven valid in detecting upper extremity pain of a neurogenic origin, the use of these special tests may help identify patterns of positive objective findings to help characterize thoracic outlet syndrome.

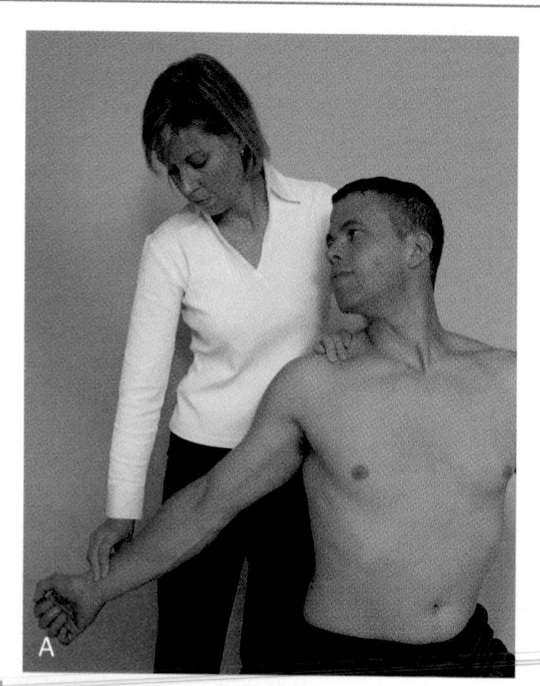

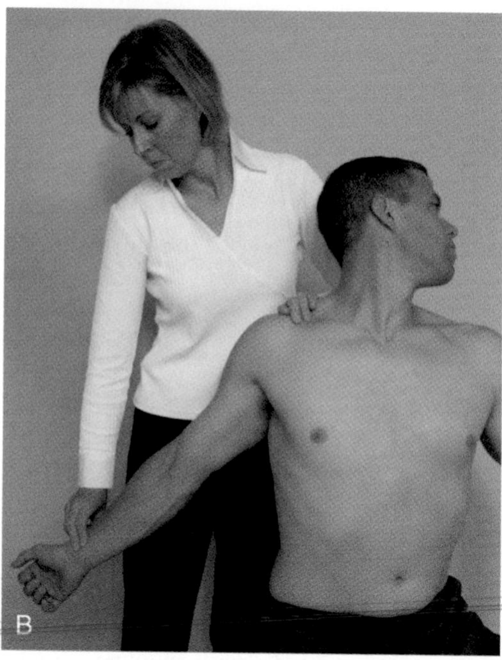

Fig. 17-8 Adson maneuver. The client begins in the sitting position with arms at his or her sides and face forward. The examiner takes a baseline, resting radial pulse rate for 1 minute. **A,** The client then turns his or her head toward the test arm. The head and neck are extended slightly while the examiner laterally rotates and extends the shoulder. The client is asked to take a deep breath and hold it. Reproduction of the symptoms is the best indication of thoracic outlet syndrome (TOS), but a disappearance of the pulse is considered a positive test. **B,** Halstead maneuver. Baseline radial pulse is obtained before the client hyperextends and rotates his head to the opposite side. The examiner applies a downward, traction force on the involved side. Once again, the test is considered positive for a vascular component of a TOS when there is a change in pulse rate or rhythm. (From Magee D: *Orthopedic physical assessment,* ed 5, Philadelphia, 2008, Saunders.)

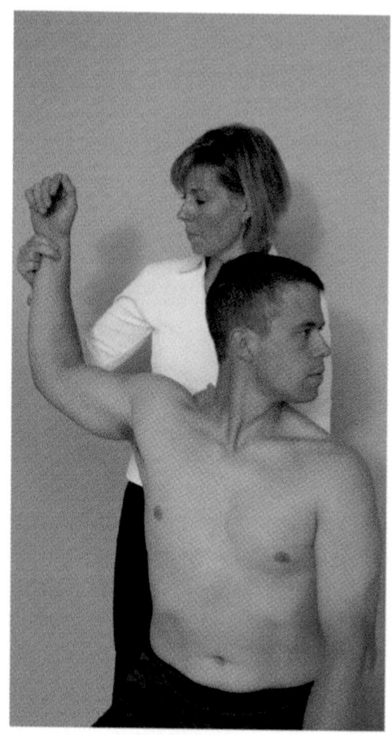

Fig. 17-9 Modified Wright test, also known as the Allen test or maneuver. The hyperabduction test can help screen for vascular compromise in thoracic outlet syndrome (TOS). Start with the client's arm resting at his or her side. Take the client's resting radial pulse for a full minute. Make note of any irregular or skipped beats. Raise the client's arm as shown with the client's face turned away, and recheck the pulse. This test is used to detect compression in the costoclavicular space. Diminished or thready pulse or absence of the pulse is a positive sign for (vascular) TOS. In the standard test, the examiner waits up to 3 minutes before palpating to give time for an accurate assessment. In our experience, clients with a positive hyperabduction test almost always demonstrate early changes in symptoms, skin color, and skin temperature. Having the client take a breath and hold it may have an additional effect. Tests for other aspects of neurologic or vascular compromise are available.[66,68] (From Magee D: *Orthopedic physical assessment,* ed 5, Philadelphia, 2008, Saunders.)

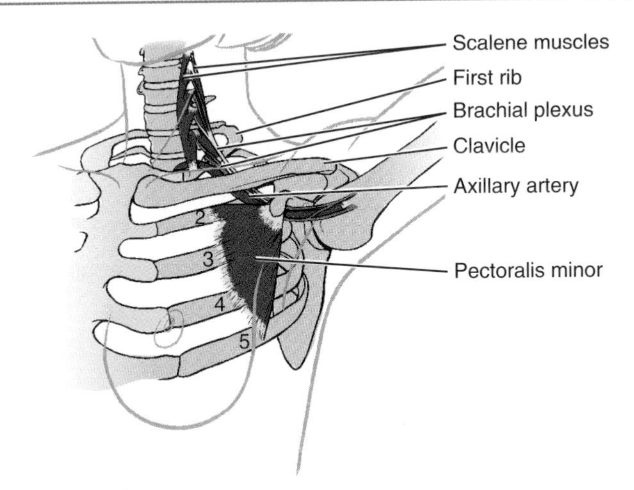

Fig. 17-10 The neurovascular bundle associated with TOS can become compressed by nearby soft tissue structures such as the pectoralis minor. This illustration shows why the hyperabduction test can alter the client's pulse or reproduce symptoms. Effecting a change in the pectoralis minor may result in a change in the client's symptoms and can be measured by a return of the normal pulse rate and rhythm in the hyperabducted position.

CLINICAL SIGNS AND SYMPTOMS

Thoracic Outlet Syndrome

Vascular

- Swelling, sometimes described as "puffiness," of the supraclavicular fossa, axilla, arm, and/or hand
- Cyanotic (blue or white) appearance of the hand; especially notable when the arm is elevated over head; sometimes referred to as the white hand sign
- Coldness or blanching of the hand during exercise
- Subjective report of "heaviness" in arm or hand
- Chest, neck, and/or arm pain described as "throbbing" or deep aching
- Upper extremity fatigue and weakness
- Difference in blood pressure from side to side (more than 10-mm Hg difference in diastolic)

Neurologic

- Numbness and/or tingling, usually ulnar nerve distribution
- Atrophy of the hand; difficulty with fine motor skills
- Pain in the upper extremity (proximal to distal); described as stabbing, cutting, burning, or electric
- Numbness and tingling down the inner aspect of the arm (ulnar nerve distribution)

A vascular component to TOS may present with significant differences in blood pressure from side to side (a change of 10 mm Hg or more in diastolic is most likely). This does not mean that a medical referral is required immediately. Assess client age, PMH, and presence of comorbidities (e.g., known hypertension), and ask about any associated signs and symptoms that might point to heart disease as a cause of the underlying symptoms.

Physical therapy intervention can bring about a change in the soft tissue structures, putting pressure on the blood vessels in this area. In fact, blood pressure can be used as an outcome measure to document the effectiveness of the intervention. If blood pressure does not normalize and equalize from side to side, then medical referral may be required.

If there is a cluster of cardiac symptoms, especially in the presence of a significant history of hypertension or heart disease, medical referral may be required before initiating treatment. If the Review of Systems does not provide cause for concern, documentation and communication with the physician are still important while initiating a plan of care.

Postoperative Pain

Postoperative chest pain following cardiac transplantation or other open heart procedures is usually due to the sternal incision and musculoskeletal manipulation during surgery. Coronary insufficiency does not appear as chest pain because of cardiac denervation.

PHYSICIAN REFERRAL

Never dismiss chest pain as insignificant. Chest pain that falls into any of the categories in Table 6-5 requires medical evaluation. This table offers some helpful clues in matching client's clinical presentation with the need for medical referral.

It may be impossible for a physician to differentiate anxiety from myocardial ischemia without further testing; such a differentiation is outside the scope of a physical therapist's practice. The therapist must confine himself or herself to a screening process before conducting a differential diagnosis of movement system impairments.

The therapist is not making the differential diagnosis between angina, MI, mitral valve prolapse, or pericarditis. The therapist is screening for systemic or viscerogenic causes of chest, breast, shoulder or arm, jaw, or neck or upper back symptoms.

Knowing the chest and breast pain patterns and associated signs and symptoms of conditions that masquerade as NMS dysfunction will help the therapist recognize a condition requiring medical attention. Likewise, quickly recognizing red-flag signs and symptoms is important in providing early medical referral and intervention, preferably with improved outcomes for the client.

Guidelines for Immediate Medical Attention

- Sudden onset of acute chest pain with sudden dyspnea could be a life-threatening condition (e.g., pulmonary embolism, MI, ruptured abdominal aneurysm), especially in the presence of red-flag risk factors, personal medical history, and vital signs.
- A sudden change in the client's typical anginal pain pattern suggests unstable angina. For the client with known angina, pain that occurs without exertion, lasts longer than 10 minutes, or is not relieved by rest or nitroglycerin signals a higher risk for a heart attack.
- The woman with chest, breast, axillary, or shoulder pain of unknown origin at presentation must be questioned regarding breast self-examinations. Any recently

discovered breast lumps or nodules or lymph node changes must be examined by a physician.

Guidelines for Physician Referral

- No change is noted in uneven blood pressure from one arm to the other after intervention for a vascular TOS component.
- The therapist who suspects a client may be using anabolic steroids should report findings to the physician or coach if one is involved.
- Symptoms are unrelieved or unchanged by physical therapy intervention.
- Medical referral is advised before initiating treatment for anyone with a past history of cancer presenting with symptoms of unknown cause, especially without an identifiable movement system impairment.

Clues to Screening Chest, Breast, or Rib Pain

Past Medical History

- History of repetitive motion; overuse; prolonged activity (e.g., marathon); long-term use of steroids, assault, or other trauma
- History of flu, trauma, URI, shingles **(herpes zoster),** recurrent pneumonia, chronic bronchitis, or emphysema
- History of breast cancer or any other cancer; history of chemotherapy or radiation therapy
- History of heart disease, hypertension, previous MI, heart transplantation, bypass surgery, or any other procedure affecting the chest/thorax (including breast reconstruction, implantation, or reduction)
- Prolonged use of cocaine or anabolic steroids
- Nocturnal pain, pain without precise movement aggravation, or pain that fails to respond to treatment
- Weight loss in the presence of immobility when weight gain would otherwise be expected
- Recent childbirth and/or lactation (breast feeding) **(Pectoral myalgia, mastitis)**

Risk Factors (see also Table 6-3)

- Age
- Tobacco use
- Obesity
- Sedentary lifestyle, prolonged immobilization

Clinical Presentation

- Range of motion (e.g., trunk rotation of side bending, shoulder motions) does not reproduce symptoms (exception: intercostal tear caused by forceful coughing associated with diaphragmatic pleurisy).
- There is a lack of musculoskeletal objective findings; squeezing the underlying pectoral muscles does not reproduce symptoms; resisted motion (e.g., horizontal shoulder abduction or adduction) does not reproduce symptoms; heat and stretching do not reduce or eliminate the symptoms; pain or symptoms are not altered or eliminated with TrP therapy or other physical therapy intervention.
- Chest pain relieved by antacid **(reflux esophagitis),** rest from exertion or taking nitroglycerin **(angina),** recumbency **(mitral valve prolapse),** squatting **(hypertrophic cardiomyopathy),** passing gas **(gas entrapment syndrome)**
- Presence of painless sternal or chest wall mass or painless, hard lymph nodes
- Unusual vital signs; changes in breathing

Cardiovascular

- Timing of symptoms in relation to physical or sexual activity (immediate, 5 to 10 minutes after engaging in activity, after activity ends). **(Lag time is associated with angina; symptoms occurring immediately or after an activity may be a sign of TOS, asthma, myalgias, or TrPs.)**
- Assess the effect of exertion; reproduction of chest, shoulder, or neck symptoms with exertion of only the lower extremities may be cardiovascular.
- Chest, neck, or shoulder pain that is aggravated by physical exertion, exposure to temperature changes, strong emotional reactions, or a large meal **(coronary artery disease)**
- Atypical chest pain associated with dyspnea, arrhythmias, and light-headedness or syncope
- Other signs and symptoms such as pallor, unexplained profuse perspiration, inability to talk, nausea, vomiting, sense of impending doom, or extreme anxiety
- Symptoms can be precipitated by working with arms overhead; the client becomes weak or short of breath 3 to 5 minutes after raising the arms above the heart.

Pleuropulmonary (see also Clues to Screening in Chapter 7)

- Autosplinting (lying on the involved side) quiets chest wall movements and reduces or eliminates chest or rib pain; symptoms are worse with recumbency (supine position).
- Pain is not reproduced by palpation.
- Assess for the three *ps*: Pleural pain, palpation, position (***p***leuritic pain exacerbated by respiratory movements, pain on ***p***alpation associated with musculoskeletal condition, pain with changes in neck, trunk, or shoulder ***p***osition, indicating musculoskeletal origin).
- Musculoskeletal: Symptoms do not increase with pulmonary movements (unless there is an intercostal tear or rib dysfunction associated with forceful coughing from a concomitant pulmonary problem) but can be reproduced with palpation.
- Pleuropulmonary: Symptoms increase with pulmonary movements and cannot be reproduced with palpation (unless there is an intercostal tear or rib dysfunction associated with forceful coughing).
- Increased symptoms occur with recumbency (abdominal contents push up against diaphragm and in turn push against the parietal pleura).
- Increased chest pain with exercise or increased movement can also be a sign of asthma; ask about a personal or family history of asthma or allergies.

- Presence of associated signs and symptoms such as persistent cough, dyspnea (rest or exertional), or constitutional symptoms
- Chest pain with sudden drop in blood pressure or symptoms such as dizziness, dyspnea, vomiting, or unexplained sweating while standing or ambulating for the first time after surgery, an invasive medical procedure, assault, or accident involving the chest or thorax **(Pneumothorax)**

Gastrointestinal (Upper GI/Epigastric; see also Clues to Screening in Chapter 8)

- Effect of food on symptoms (better or worse); presence of GI symptoms, simultaneously or alternately with somatic symptoms
- Pain on swallowing
- Symptoms are relieved by antacids, food, passing gas, or assuming the upright position.
- Supine position aggravates symptoms **(upper GI problem);** symptoms are relieved by assuming an upright position.
- Symptoms radiate from the chest posteriorly to the upper back, interscapular, subscapular, or T10 to L2 areas.
- Symptoms are not reproduced or aggravated by effort or exertion.
- Presence of associated signs and symptoms such as nausea, vomiting, dark urine, jaundice, flatulence, indigestion, abdominal fullness or bloating, blood in stool, pain on swallowing

Breast (Alone or in Combination with Chest, Neck, or Shoulder Symptoms)

- Appearance (or report) of lump, nodule, discharge, skin puckering, or distended veins
- Jarring or movement of the breast tissue increases or reproduces the pain.
- Pain is palpable within the breast tissue.
- Assess for TrPs (sternalis, serratus anterior, pectoralis major; see Fig. 17-7); breast pain in the absence of TrPs or failure to respond to TrP therapy must be investigated further.
- Resisted isometric shoulder horizontal adduction or abduction does *not* reproduce breast pain.
- Breast pain is reproduced by exertion of the lower extremities **(Cardiac).**
- Association between painful symptoms and menstrual cycle **(Ovulation or menses)**
- Presence of aberrant or suspicious axillary or supraclavicular lymph nodes (e.g., large, firm, hard, or fixed)
- Skin dimpling especially with adherence of underlying tissue; ask about or visually inspect for:
  - Lump or nodule
  - Red, warm, edematous, firm, and painful area over or under skin
  - Changes in size or shape or color of either breast or surrounding area
  - Unusual rash or other skin changes (e.g., puckering, dimpling, peau d'orange)
  - Distended veins
  - Unusual sensations in nipple or breast
  - Unusual nipple ulceration or discharge

Anxiety (see Table 3-9)

- Pain pattern:
  - Sharp, stabbing pain: left breast region
  - Dull aching: substernal
  - Discomfort: upper chest, neck, left arm
  - Fingertip size; does not radiate
  - Unable to palpate locally
  - Lasts seconds to hours to days
- Not aggravated by respiratory or other (shoulder, arm, back) movements
- Unchanged by rest or change in position
- Unrelated to effort or exertion
- Associated signs and symptoms
  - Local hyperesthesia of chest wall
  - Choking sensation (hysteria/panic)
  - Claustrophobia
  - Sense of persistent weakness
  - Unpleasant awareness of heartbeat
  - Hyperventilation (can also occur with heart attack; watch for sighing respirations and numbness/tingling of face and fingertips)

Neuromusculoskeletal

- Symptoms described using words typical of NMS origin (e.g., aching, burning, hot, scalding, searing, cutting, electric shock)
- Pain is superficial compared with pain of a cardiac or pleuropulmonary origin.
- Symptoms are confined to somatic or spinal nerve root distribution.
- History of associated back pain
- Positive hyperabduction test or other tests for TOS
- Presence of TrPs; elimination of TrPs reduces or eliminates symptoms (see Table 17-4 and Fig. 17-7)
- Symptoms are elicited easily by palpation (e.g., squeezing the pectoral muscle belly, palpating the chest wall, intercostal spaces, or costochondral junction).
- Symptoms are reproduced by resisted horizontal shoulder abduction, adduction, or other shoulder movements.
- Symptoms are relieved by heat and stretching.
- Soft tissues (tendon and muscle) take longer than the expected time to heal **(Anabolic steroids).**
- Costochondritis or Tietze's syndrome may be accompanied by an increase in blood pressure but is usually palpable and aggravated by trunk movements.
- Presence of neurologic involvement (e.g., numbness, tingling, muscle atrophy); consider age and history of trauma or injury **(Degenerative disk disease)**
- Pain referred along peripheral nerve pathway **(Dorsal nerve root irritation)**
- Pain is unrelated to effort and lasts hours or weeks to months.
- Associated signs and symptoms: Numbness and tingling, muscle atrophy **(Neurologic);** rash, fever, chills, headache, malaise **(Constitutional symptoms; neuritis or shingles)**

REFERRED CHEST, BREAST, RIB PAIN PATTERNS

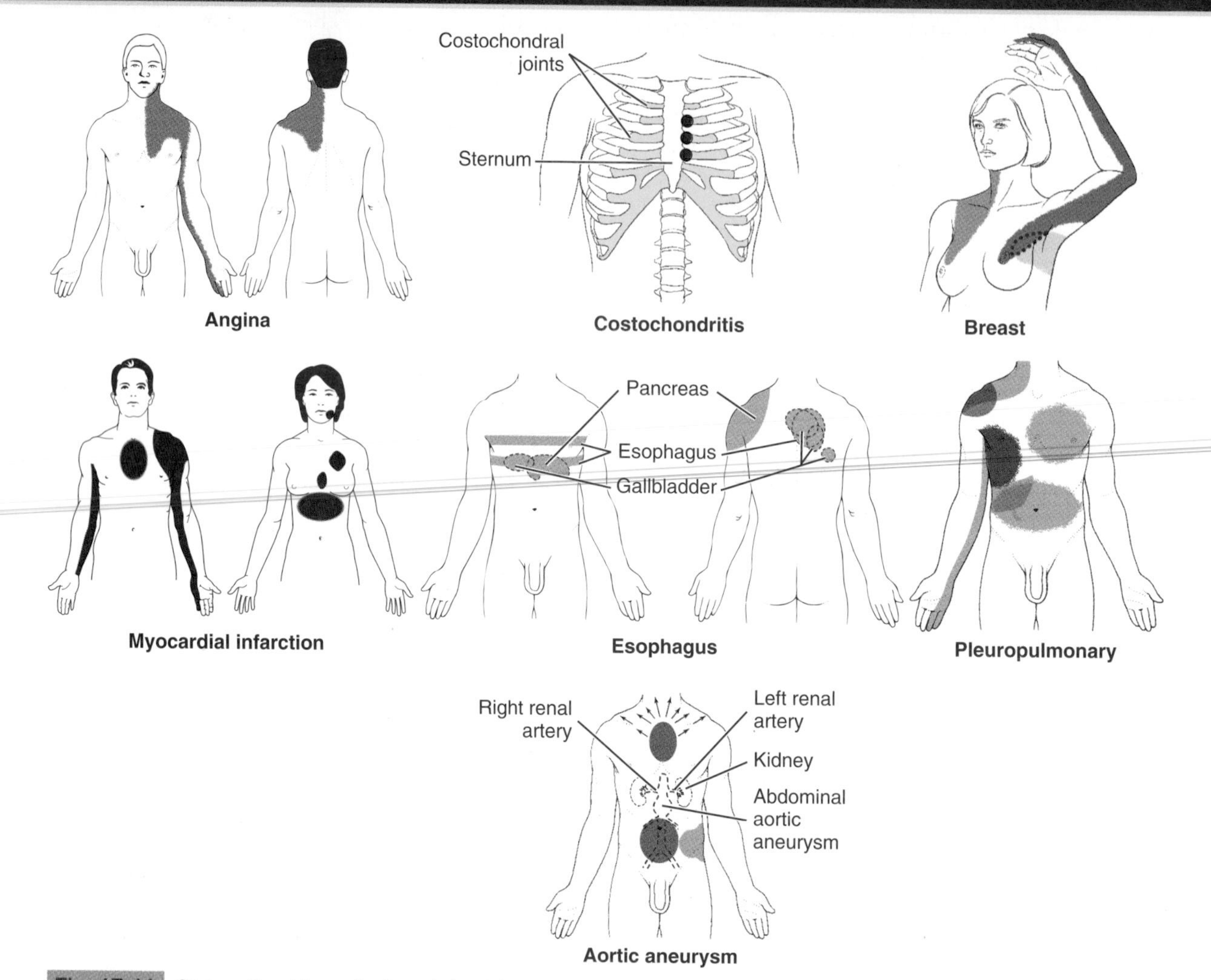

Fig. 17-11 Composite picture of referred chest, breast, and rib pain patterns. Not shown: trigger point patterns (see Fig. 17-7).

Key Points to Remember

- When faced with chest pain, the therapist must know how to assess the situation quickly and decide if medical referral is required and whether medical attention is needed immediately. Therapists must be able to differentiate neuromusculoskeletal from systemic origins of symptoms.
- Although the PMH is important, it cannot be relied upon to confirm or rule out medical causes of chest pain. PMH does alert the therapist to an increased risk of systemic conditions that can masquerade as neuromusculoskeletal disorders.
- Likewise, chest pain history by itself is not enough to rule out cardiac or other systemic origin of symptoms; in most cases, some diagnostic testing is needed. The physical therapist can offer valuable information from the screening process to aid in the medical differential diagnosis.
- Chest pain associated with increased activity is a red flag for possible cardiovascular involvement. The physical therapist can assess when chest pain may be caused by musculoskeletal dysfunction (immediate chest pain occurs with use) or by possible vascular compromise (chest pain occurs 5 to 10 minutes after activity begins).
- Anyone with a history of stent insertion presenting with chest pain should be screened carefully. The stent can get scarred over and/or malfunction. Stents are effective at alleviating chest pain but do not reduce the risk of heart attacks for most people with stable angina.
- Cardiac pain patterns may differ for men and women; the therapist should be familiar with known pain patterns for both genders.
- TrPs can cause chest, breast, or rib pain, even mimicking cardiac pain patterns; a viscero-somatic response can also occur following an MI, causing persisting symptoms of myocardial ischemia (angina); releasing the TrP relieves the symptoms.
- The therapist must especially know how and what to look for to screen for cancer, cancer recurrence, and/or the delayed effects of cancer treatment. Cancer can present as primary chest pain with or without accompanying neck, shoulder, and/or upper back pain/symptoms.
- When a woman with a PMH of cancer develops neck, back, upper trapezius or shoulder pain, or other symptoms, examining the site of the original cancer removal is a good idea.
- The ACS and the NCI support breast cancer screening by qualified health care specialists. With adequate training, the physical therapist can incorporate CBE as a screening tool in the upper quarter examination for appropriate clients (e.g., individuals with neck, shoulder, upper back, chest, and/or breast signs or symptoms of unknown cause or insidious onset).[6]
- A physical therapist conducting a CBE could miss a lump (false negative), but this will most certainly happen if the therapist does not conduct a CBE at all to assess skin integrity and surrounding soft tissues of the breast or axilla.[6]
- The physical therapist does not diagnose any kind of cancer, including breast cancer; only the pathologist diagnoses cancer. The therapist can identify aberrant soft tissue and refer the client for further evaluation. Early detection and intervention can reduce morbidity and mortality.
- Thoracic disk disease can also present as unilateral chest pain and requires careful screening.
- Chest pain of unknown cause in the adolescent or young adult athlete may be the result of anabolic steroid use. Watch for injuries that take longer than expected to heal, personality changes, and any of the physical signs listed in the text.
- A history of long-term steroid use in the presence of rib pain of unknown cause raises a red flag for rib fracture.
- Many people with chest pain have used cocaine within the last week but deny its use; the therapist should not neglect asking clients of all ages about their use of substances.

SUBJECTIVE EXAMINATION

Special Questions to Ask: Chest/Thorax

Musculoskeletal

- Have you strained a muscle from (repeated, forceful) coughing?
- Have you ever injured your chest?
- Does it hurt to touch your chest or to take a deep breath (e.g., coughing, sneezing, sighing, or laughing)? **(Myalgia, fractured rib, costochondritis, myofascial TrP)**
- Do you have frequent attacks of heartburn, or do you take antacids to relieve heartburn or acid indigestion? **(Noncardiac cause of chest pain, abdominal muscle TrP, GI disorder)**
- Does chest movement or body/arm position make the pain better or worse?

Neurologic

- Do you have any trouble taking a deep breath? **(Weak chest muscles secondary to polymyositis, dermatomyositis, myasthenia gravis)**
- Does your chest pain ever travel into your armpit, arm, neck, or wing bone (scapula)? **(TOS, TrPs)**
  - *If yes,* do you ever feel burning, prickling, numbness, or any other unusual sensation in any of these areas?

Pulmonary

- Have you ever been treated for a lung problem?
  - *If yes,* describe what this problem was, when it occurred, and how it was treated.
- Do you think your chest or thoracic (upper back) pain is caused by a lung problem?
- Have you ever had trouble with breathing?
- Are you having difficulty with breathing now?
- Do you ever have shortness of breath, breathlessness, or can't quite catch your breath?
  - *If yes,* does this happen when you rest, lie flat, walk on level ground, walk up stairs, or when you are under stress or tension?
  - How long does it last?
  - What do you do to get your breathing back to normal?
- How far can you walk before you feel breathless?
- What symptom stops your walking (e.g., shortness of breath, heart pounding, or weak legs)?
- Do you have any breathing aids (e.g., oxygen, nebulizer, humidifier, or ventilation devices)?
- Do you have a cough? (Note whether the person smokes, for how long, and how much.) Do you have a smoker's hack?
  - *If yes* to having a cough, distinguish it from a smoker's cough. Ask when it started.
  - Does coughing increase or bring on your symptoms?
  - Do you cough anything up? *If yes,* please describe the color, amount, and frequency.
  - Are you taking anything for this cough? *If yes,* does it seem to help?
- Do you have periods when you can't seem to stop coughing?
- Do you ever cough up blood?
  - *If yes,* what color is it? (Bright red: fresh; brown or black: older)
  - *If yes,* has this been treated?
- Have you ever had a blood clot in your lungs?
  - *If yes,* when and how was it treated?
- Have you had a chest x-ray film taken during the last 5 years?
  - *If yes,* when and where did it occur? What were the results?
- Do you work around asbestos, coal, dust, chemicals, or fumes? *If yes,* describe.
- Do you wear a mask at work? *If yes,* approximately how much of the time do you wear a mask?
- If the person is a farmer, ask what kind of farming (because some agricultural products may cause respiratory irritation).
- Have you ever had tuberculosis or a positive skin test for tuberculosis?
  - *If yes,* when did it occur and how was it treated? What is your current status?
- When was your last test for tuberculosis? Was the result normal?

Cardiac

- Has a physician ever told you that you have heart trouble?
- Have you recently (or ever) had a heart attack? *If yes,* when? Describe.
  - *If yes,* to either question: Do you think your current symptoms are related to your heart problems?
- Do you have angina (pectoris)?
  - *If yes,* describe the symptoms, and tell me when it occurs.
  - *If no,* pursue further with the following questions.
- Do you ever have discomfort or tightness in your chest? **(Angina)**
- Have you ever had a crushing sensation in your chest with or without pain down your left arm?
- Do you have pain in your jaw, either alone or in combination with chest pain?
- If you climb a few flights of stairs fairly rapidly, do you have tightness or pressing pain in your chest?
- Do you get pressure or pain or tightness in the chest if you walk in the cold wind or face a cold blast of air?
- Have you ever had pain or pressure or a squeezing feeling in the chest that occurred during exercise, walking, or any other physical or sexual activity?
- Do you ever have bouts of rapid heart action, irregular heartbeats, or palpitations of your heart?

SUBJECTIVE EXAMINATION—*cont'd*

- *If yes,* did this occur after a visit to the dentist? **(Endocarditis)**
- Have you noticed any skin rash or dots under the skin on your chest in the last 3 weeks? **(Rheumatic fever, endocarditis)**
- Have you noticed any other symptoms (e.g., shortness of breath, sudden and unexplained perspiration, nausea, vomiting, dizziness or fainting)?
- Have you used cocaine, crack, or any other recreational drug in the last 6 weeks?
- Does your pain wake you up at night? (Therapist: distinguish between awakening *from* pain and awakening *with* pain; awakening from pain is more likely with **cardiac ischemia,** whereas awakening with pain is characteristic of sleep disturbances and more common with **psychogenic or stress-induced** chest pain; this information will help in deciding whether referral is needed immediately or at the next follow-up appointment)

Epigastric

- Have you ever been told that you have an ulcer?
- Does the pain under your breast bone radiate (travel) around to your back, or do you ever have back pain at the same time that your chest hurts?
- Have you ever had heartburn or acid indigestion?
  - *If yes,* how is this pain different?
  - *If no,* have you noticed any association between when you eat and when this pain starts?

Special Questions to Ask: Breast

- Have you ever had any breast surgery (implants, lumpectomy, mastectomy, reconstructive surgery, or augmentation)?
  - *If yes,* has there been any change in the incision line, nipple, or breast tissue?
  - May I look at the incision during my exam?
- Do you have a history of cystic or lumpy breasts?
  - *If yes,* do the lumps come and go or change with your periods?
- Is there a family history of breast disease?
  - *If yes,* ask about type of disease, age of onset, treatment, and outcome.
- Have you ever had a mammogram or ultrasound?
  - *If yes,* when was your last test? What were the results?
- Have you ever had a lump or cyst drained or biopsied?
  - *If yes,* what was the diagnosis?
- Have you ever been treated for cancer of any kind? *If yes,* when? What?
- Have you examined yourself for any lumps or nodules and found any thickening or lump in the breast or armpit area?
  - *If yes,* has your physician examined/treated this?
  - *If no,* do you examine your own breasts? (Follow-up questions regarding last breast examination by self or other health care professional)
- Do you have any discharge from your breasts or nipples?
  - *If yes,* do you know what is causing this discharge? Have you received medical treatment for this problem?
- Are you nursing or breastfeeding an infant (lactating)?
  - *If yes,* are your nipples sore or cracked?
  - Is your breast painful or hot? Are there any areas of redness?
  - Have you had a fever? **(Mastitis)**
- Have you noticed any other changes in your breast(s)? For example, are there any noticeable bulging or distended veins, puckering, swelling, tenderness, rash, or any other skin changes?
- Do you have any pain in your breasts?
  - *If yes,* does the pain come and go with your period? **(Hormone-related)**
  - Does squeezing the breast tissue cause the pain?
  - Does using your arms in any way cause the pain?
- Have you been involved in any activities of a repetitive nature that could cause sore muscles (e.g., painting, washing walls, push-ups or other calisthenics, heavy lifting or pushing, overhead movements, prolonged running, or fast walking)?
- Have you recently been coughing excessively? **(Pectoral myalgia)**
- Have you ever had angina (chest pain) or a heart attack? **(Residual trigger points)**
- Have you been in a fight or hit, punched, or pushed against any object that injured your chest or breast? **(Assault)**

Special Questions to Ask: Lymph Nodes

Use the lymph node assessment form (Fig. 4-48) to record and report baseline findings.

- General screening question: Have you examined yourself for any lumps or nodules and found any thickening or lump?
  - *If yes,* has your physician examined/treated this?

If any suspicious or aberrant lymph nodes are observed during palpation, ask the following questions.

- Have you recently had any skin rashes anywhere on your face or body?
- Have you recently had a cold, URI, the flu, or other illness? (Enlarged lymph nodes)
- Have you ever had:
  - Cancer of any kind?

If no, have you ever been treated with radiation or chemotherapy for any reason?

  - Breast implants
  - Mastectomy or prostatectomy

Continued

SUBJECTIVE EXAMINATION—*cont'd*

- Mononucleosis
- Chronic fatigue syndrome
- Allergic rhinitis
- Food intolerances, food allergies, or celiac sprue
- Recent dental work
- Infection of any kind
- Recent cut, insect bite, or infection in the hand or arm
- A sexually transmitted disease of any kind
- Sores or lesions of any kind anywhere on the body (including genitals)

Special Questions to Ask: Soft Tissue Lumps or Skin Lesions

- How long have you had this?
- Has it changed in the last 6 weeks to 6 months?
- Has your doctor seen it?
- Does it itch, hurt, feel sore, or burn?
- Does anyone else in your household have anything like this?
- Have you taken any new medications (prescribed or over-the-counter) in the last 6 weeks?
- Have you traveled somewhere new in the last month?
- Have you been exposed to anything in the last month that could cause this? (consider exposure due to occupational, environmental, and hobby interests)
- Do you have any other skin changes anywhere else on your body?
- Have you had a fever or sweats in the last 2 weeks?
- Are you having any trouble breathing or swallowing?
- Have you had any other symptoms of any kind anywhere else in your body?

CASE STUDY

Steps in the Screening Process

If a client comes to you with chest pain, breast pain, or rib pain (either alone or in combination with neck, back, or shoulder pain), start by looking at Tables 17-1 and 17-2 and Box 17-4. As you look down these lists, does your client have any red-flag histories, unusual clinical presentation, or associated signs and symptoms to point to any particular category? Just by looking at these lists, you may be prompted to ask some additional questions that have not been asked yet.

COULD IT BE CANCER?

The therapist does not make a determination as to whether or not a client has cancer; only the pathologist makes this kind of determination. The therapist's assessment determines whether the client has a true neuromuscular or musculoskeletal problem that is within the scope of our practice.

However, knowing red flags for the possibility of cancer helps the therapist know what questions to ask and what red flags to look for. Early detection often means reduced morbidity and mortality for many people. Watch for the following:

- Previous history of cancer (any kind, but especially breast or lung cancer).
- Be sure to assess for trigger points (TrPs). Reassess after TrP therapy (e.g., Were the symptoms alleviated? Did the movement pattern change?).
- Conduct a neurologic screening exam.
- Look for skin changes or other trophic changes, and ask about recent rashes or lesions (see Box 4-11).

COULD IT BE VASCULAR?

- Consider the client's age, menopausal status (women), past medical history, and the presence of any cardiac risk factors. Do any of these components suggest the need to screen further for a vascular cause?
- Are there any reported associated signs and symptoms (e.g., unexplained perspiration without physical activity, nausea, pallor, unexplained fatigue, palpitations; see Box 4-19)?
- Is there a significant difference in blood pressure from one arm to the other? Have you checked? Do the symptoms suggest the need to conduct this assessment?
- Have you assessed for the 3 Ps? (*p*leuritic pain, *p*alpation, *p*osition)

COULD IT BE PULMONARY?

- Consider the age of the client and any recent history of pneumonia or other URIs. Again, consider the 3 Ps.
- Have you observed or heard any reports from the client to suggest changes in the breathing pattern? Are there other pulmonary symptoms present (e.g., dry or productive cough, symptoms aggravated by respiratory movements)?
- Are the symptoms made better by sitting up, worse by lying down, or better in sidelying on the affected side (autosplinting)? *If yes,* further screening may be warranted.

CASE STUDY—*cont'd*

Steps in the Screening Process

COULD IT BE UPPER GI?

- Follow the same line of thinking in terms of mentally reviewing the client's past medical history (e.g., chronic NSAID use, GERD, gallbladder, or liver problems) and the presence of any GI signs or symptoms. Is there anything here to suggest a potential GI cause of the current symptoms? If yes, then review the Special Questions to Ask box for any further screening questions.
- Have you asked the client about the effect of eating or drinking on the symptoms? It is a quick and simple screening question to help identify any GI component.
- Be sure and assess for trigger points as a potential cause of what might appear to be GI-induced symptoms.

COULD IT BE BREAST PATHOLOGY?

- Consider red flag histories, risk factors, and pain pattern for men and women when considering breast tissue as a possible cause of upper quadrant pain.
- Is there any cyclical aspect to the symptoms linked to menstruation or hormonal fluctuations?
- Ask if jarring or squeezing the breast reproduces the pain.
- Ask if there have been any obvious changes in the breast tissue or nipple.
- Have you palpated the axillary or supraclavicular lymph nodes? This is a quick and easy screening test that can easily be incorporated into your examination.

COULD IT BE TRAUMA OR OTHER CAUSES?

- Remember to consider trauma (including assault) as a possible cause of symptoms.
- Is there any reason to suspect drug use (e.g., cocaine, anabolic steroids)?
- Should you consider screening for emotional overlay or psychogenic source of symptoms (see Chapter 3; see Appendix B-31)?
- Consider anemia as a possible cause; without a laboratory test, this is impossible to know for certain. In the screening process, the therapist can ask some questions to help formulate a referral decision. For example, has the client complained of fatigue, a hallmark finding in anemia? Some additional questions may include the following:
- Have you experienced any unusual or prolonged bleeding from any part of your body?
- Have you noticed any blood in your urine or stools? Have you noticed any changes in the color of your stools? (Dark, tarry, sticky stools may signal melena from blood loss in the GI tract.)
- Have you been taking any over-the-counter or prescribed antiinflammatory drugs (NSAIDs and peptic ulcer with GI bleeding)?
- Have you ever been told you have rheumatoid arthritis (RA), lupus, human immunodeficiency virus/acquired immunodeficiency syndrome (HIV/AIDS), or anemia?
- RA is a systemic condition that can cause chest pain; osteoarthritis of the cervical spine, fibromyalgia, and anxiety can also cause chest pain. When completing the Review of Systems, look for a cluster of associated signs and symptoms that might suggest any of these conditions.
- Do not forget to consider screening for anabolic steroid use, cocaine or other substance use, and domestic violence or assault.

Finally, review the clues to differentiating chest, breast, or rib pain, and then scan the Special Questions to Ask: Chest/Thorax or Special Questions to Ask: Breast in this chapter (depending on the chief complaint and presenting symptoms). Have you left anything out?

PRACTICE QUESTIONS

1. Chest pain can be caused by trigger points of the:
   a. Sternocleidomastoid
   b. Rectus abdominis
   c. Upper trapezius
   d. Iliocostalis thoracis
2. During examination of a 42-year-old woman's right axilla, you palpate a lump. Which characteristics most suggest the lump may be malignant?
   a. Soft, mobile, tender
   b. Hard, immovable, nontender
3. A client complains of throbbing pain at the base of the anterior neck that radiates into the chest and interscapular areas and increases with exertion. What should you do first?
   a. Monitor vital signs, and palpate pulses
   b. Call the physician or 911 immediately
   c. Continue with the exam; find out what relieves the pain
   d. Ask about past medical history and associated signs and symptoms

Continued

PRACTICE QUESTIONS—*cont'd*

4. A 55-year-old grocery store manager reports becoming extremely weak and breathless whenever stocking groceries on overhead shelves. What is the possible significance of this complaint?
   a. TOS
   b. Myocardial ischemia
   c. TrP
   d. All of the above
5. Chest pain of a pleuritic nature can be distinguished by:
   a. Increases with autosplinting (lying on the involved side)
   b. Reproduced with palpation
   c. Exacerbated by deep breathing
   d. All of the above
6. A 66-year-old woman has come to you with a report of anterior neck pain radiating down the left arm. Her past medical history is significant for chronic diabetes mellitus (insulin dependent), coronary artery disease, and peripheral vascular disease. About 6 weeks ago, she had an angioplasty with stent placement. Which test will help you differentiate a musculoskeletal cause from a cardiac cause of neck and arm pain?
   a. Stair climbing or stationary bike test
   b. Using arms overhead for 3 to 5 minutes
   c. TrP assessment
   d. All of the above
7. You are evaluating a 30-year-old woman with left chest pain that starts just below the clavicle and extends down to the nipple line. The majority of test results point to thoracic outlet syndrome. Her blood pressure is 120/78 mm Hg on the right (sitting) and 125/100 on the left (sitting). She is in apparent good health with no history of surgeries or significant health problems. What plan of action would you recommend?
   a. Refer her to a physician before initiating treatment.
   b. Carry out a plan of care, and reassess after three sessions or 1 week, whichever comes first.
   c. Document your findings, and contact the physician by phone or by fax while initiating treatment.
   d. Eliminate trigger points, and then reassess symptoms.
8. A 60-year-old woman with a history of left breast cancer (10 years postmastectomy) presents with pain in her midback. The pain is described as "sharp" and radiates around her chest to the sternum. She gets some relief from her pain by lying down. Her vital signs are normal, and there are no palpable or aberrant lymph nodes. She denies any changes in breast tissue on the right or the scar and soft tissue on the left. You do not have adequate training to perform a clinical breast examination, but the client agrees to visual inspection, which reveals nothing unusual. All other findings are within normal limits; you are unable to provoke or aggravate her symptoms. Neurologic screening examination is within normal limits. The client denies any history of trauma. What plan of action would you recommend?
   a. Refer her to a physician before initiating treatment.
   b. Carry out a plan of care, and reassess after three sessions or 1 week, whichever comes first.
   c. Document your findings, and contact the physician by phone or by fax while initiating treatment.
   d. Eliminate TrPs, and then reassess symptoms.
9. You are working with a client in his home who had a total hip replacement 2 weeks ago. He describes chest pain with increased activity. Knowing what could cause this symptom will help guide you in asking appropriate screening questions. Can this be a symptom of:
   a. Asthma
   b. Angina
   c. Pleuritis or pleurisy
   d. All of the above
10. Cardiac pain in women does not always follow classic patterns. Watch for this group of symptoms in women at risk:
   a. Indigestion, food poisoning, jaw pain
   b. Nausea, tinnitus, night sweats
   c. Confusion, left biceps pain, dyspnea
   d. Unusual fatigue, shortness of breath, weakness, or sleep disturbance

REFERENCES

1. Sik EC: Atypical chest pain in athletes. *Curr Sports Med Rep* 8(2):52–58, 2009.
2. Harding G, Yelland M: Back, chest, and abdominal pain—is it spinal referred pain? *Aust Fam Phys* 36(6):422–429, 2007.
3. Lenfant C: Chest pain of cardiac and noncardiac origin. *Metabolism* 59(suppl 1):S41–S46, 2010.
4. Bono CM: An evidence-based clinical guideline for the diagnosis and treatment of cervical radiculopathy from degenerative disorders. *Spine J* 11(1):64–72, 2011.
5. Lovelace-Chandler V, Bassar M, Dow D, et al: The role of physical therapists assisting women in skill development in performing breast self-examination. Poster presentation. Combined Sections Meeting, New Orleans, February 2005.
6. Goodman CC, McGarvey CL: The role of the physical therapist in primary care and cancer screening: integrating clinical breast examination (CBE) in the upper quarter examination. *Rehabil Oncol* 21(2):4–11, 2003.
7. Swap C, Nagurney JT: Value and limitations of chest pain history in the evaluation of patients with suspected acute coronary symptoms. *JAMA* 294(20):2623–2629, 2005.
8. Bruckner FE, Greco A, Leung AW: Benign thoracic pain syndrome: role of magnetic resonance imaging in the detection and localization of thoracic disc disease. *J R Soc Med* 82:81–83, 1989.
9. Baranto A: Acute chest pain in a top soccer player due to thoracic disc herniation. *Spine* 34(10):E359–362, 2009.
10. American Heart Association: Heart News. Available on-line at http://www.americanheart.org. Accessed March 17, 2011.
11. Heart and stroke statistics: American Heart Association 2011. Available online at: http://www.heart.org/HEARTORG/General/Heart-and-Stroke-Association-Statistics_UCM_319064_SubHomePage.jsp#. Accessed March 17, 2011.
12. Sanchis J: Identification of very low risk chest pain using clinical data in the emergency department. *Int J Cardiol* 150(3):260–263, 2011. Epub ahead of print May 5, 2010.

13. Marrugat J, et al: Mortality differences between men and women following first myocardial infarction. *JAMA* 280:1405–1409, 1998.
14. McSweeney JC: Women's early warning symptoms of acute myocardial infarction. *Circulation* 108(21):2619–2623, 2003.
15. Lovlien M: Early warning signs of an acute myocardial infarction and their influence on symptoms during the acute phase, with comparisons by gender. *Gend Med* 6(3):444–453, 2009.
16. McSweeney JC: Cluster analysis of women's prodromal and acute myocardial infarction symptoms by race and other characteristics. *J Cardiovasc Nurs* 25(4):311–312, 2010.
17. Lunardi AC: Musculoskeletal dysfunction and pain in adults with asthma. *J Asthma* 48(1):105–110, 2011.
18. Rathod NR: Extra-oesophageal presentation of gastro-oesophageal reflux disease. *J Indian Med Assoc* 108(1):18–22, 2010.
19. Chait MM: Gastoesophageal reflux disease: important considerations for the older patients. *World J Gastrointest Endosc* 2(12):388–396, 2010.
20. Seo TH: Clinical distinct features of noncardiac chest pain in young patients. *J Neurogastroenterol Motil* 16(2):166–171, 2010.
21. Ryder M, Deyle GD: Differential diagnosis of fibular pain in a patient with a history of breast cancer. *J Orthop Sports Phys Ther* 39(3):230, 2009.
22. Siegel R, Ward E, Brawley O, et al: Cancer statistics 2011. *CA Cancer J Clin* 61(4):212–236, 2011.
23. American Cancer Society (ACS): What are the risk factors for breast cancer? Available at http://www.cancer.org/docroot/CRI/content/CRI_2_4_2X_What_are_the_risk_factors_for_breast_cancer_5.asp. Accessed March 18, 2011.
24. Travis RC: Gene-environment interactions in 7610 women with breast cancer: prospective evidence from the Million women study. *Lancet* 375(9732):2143–2151, 2010.
25. Garfinkel L: Current trends in breast cancer. *CA Cancer J Clin* 43(1):5–6, 1993.
26. Coleman EA, Heard JK: Clinical breast examination: an illustrated educational review and update. *Clin Excell Nurse Pract* 5:197–204, 2001.
27. Barton MB, Harris R, Fletcher SW: Does this patient have breast cancer? The screening clinical breast examination: should it be done? How? *JAMA* 282:1270, 1999.
28. Chiarelli AM: The contribution of clinical breast examination to the accuracy of breast screening. *J Natl Cancer Inst* 101(18):1236–1243, 2009.
29. Goodman CC, McGarvey CL: An introductory course to breast cancer and clinical breast examination for the physical therapist is available. (Charlie McGarvey, PT, MS and Catherine Goodman, MBA, PT present the course in various sites around the U.S. and upon request.)
30. A certified training program is also available through *MammaCare Specialist*. The program is offered to health care professionals at training centers in the United States. The course teaches proficient breast examination skills. For more information, contact: http://www.mammacare.com/index.php.
31. Schwartz GF: Proceedings of the international consensus conference on breast cancer risk, genetics, and risk management. *Breast J* 15(1):4–16, 2009.
32. Rubin RN: Woman with sharp back pain. *Consultant* 39(11):3065–3066, 1999.
33. Hall-Findlay EJ: Breast implant complication review: double capsules and late seromas. *Plast Reconstr Surg* 127(1):56–66, 2011.
34. Gabriel SE, Woods JE, O'Fallon WM, et al: Complications leading to surgery after breast implantation. *N Engl J Med* 336(10):718–719, 1997.
35. Codner MA: A 15-year experience with primary breast augmentation. *Plast Reconstr Surg* 127(3):1300–1310, 2011.
36. Benediktsson K, Perback L: Capsular contracture around saline-filled and textured subcutaneously-placed implants in irradiated and non-irradiated breast cancer patients: five years of monitoring of a prospective trial. *J Plast Reconstr Aesthet Surg* 59(1):27–34, 2006.
37. Lipa JE: Pathogenesis of radiation-induced capsular contracture in tissue expander and implant breast reconstruction. *Plast Reconstr Surg* 125(2):437–445, 2010.
38. Henriksen TF, Fryzek JP, Holmich LR, et al: Reconstructive breast implantation after mastectomy for breast cancer: clinical outcomes in a nationwide prospective cohort study. *Arch Surg* 140(12):1152–1159, 2005.
39. Scuderi N: Multicenter study on breast reconstruction outcome using Becker implants. *Aesthetic Plast Surg* 35(1):66–72, 2011.
40. National Institute of Mental Health: Health Information—Anxiety Disorders. Available at http://www.nimh.nih.gov/. Updated 3/16/11. Accessed March 18, 2011.
41. Rubio M: Psychopathology risk and protective factors research program. National Institute of Mental Health (NIMH), 2009. Available online at http://www.nimh.nih.gov/about/organization/datr/adult-psychopathology-and-psychosocial-intervention-research-branch/psychopathology-risk-and-protective-factors-research-program.shtml. Accessed March 18, 2011.
42. Velasquez EM, Anand RC, Newman WP, et al: Cardiovascular complications associated with cocaine use. *J La State Med Soc* 156(6):302–310, 2004.
43. Bamberg F: Presence and extent of coronary artery disease by cardiac computed tomography and risk for acute coronary syndrome in cocaine users among patients with chest pain. *Am J Cardiol* 103(5):620–625, 2009.
44. Milroy CM, Parai JL: The histopathology of drugs of abuse. *Histopathology* Jan 25, 2011. Epub ahead of print.
45. Schwartz BG: Cardiovascular effects of cocaine. *Circulation* 122(24):2558–2569, 2010.
46. Pletcher MJ, Kiefe CI, Sidney S, et al: Cocaine and coronary calcification in young adults: the Coronary Artery Risk Development in Young Adults (CARDIA) study. *Am Heart J* 150(5):921–926, 2005.
47. van Amsterdam J: Adverse health effects of anabolic-androgenic steroids. *Regul Toxicol Pharmacol* 57(1):117–123, 2010.
48. Kanayama G: Illicit anabolic-androgenic steroid use. *Horm Behav* 58(1):111–121, 2010.
49. Sullivan ML, Martinez CM, Gennis P, et al: The cardiac toxicity of anabolic steroids. *Prog Cardiovasc Dis* 41(1):1–15, 1998.
50. Sanchez-Orio M: Anabolic-androgenic steroids and liver injury. *Liver Int* 28(2):278–282, 2008.
51. Jensen S: Musculoskeletal causes of chest pain. *Aust Fam Phys* 30(9):834–839, September 2001.
52. Cohen SP: Noncardiac chest pain during war. *Clin J Pain* 27(1):19–26, 2011.
53. Stochkendahl MJ: Chest pain in focal musculoskeletal disorders. *Med Clin North Am* 94(2):259–273, 2010.
54. Peterson LL, Cavanaugh DG: Two years of debilitating pain in a football spearing victim: slipping rib syndrome. *Med Sci Sports Exerc* 35(10):1634–1637, 2003.
55. Cubos J: Chronic costochondritis in an adolescent competitive swimmer: a case report. *J Can Chiropr Assoc* 54(4):271–275, 2010.
56. Karnath B: Chest pain: differentiating cardiac from noncardiac causes. *Hospital Physician* 36(4):24–38, 2000.
57. Thongngarm T, Lemos LB, Lawhon N, et al: Malignant tumor with chest pain mimicking Tietze's syndrome. *Clin Rheumatol* 20(4):276–278, 2001.
58. Fioravanti A, Tofi C, Volterrani L, et al: Malignant lymphoma presenting as Tietze's syndrome. *Arthritis Rheum* 49(5):737, 2003.

59. Simons DG, Travell JG, Simons LS: *Travell & Simons' myofascial pain and dysfunction: the trigger point manual. Volume 1: Upper half of body*, ed 2, Baltimore, 1999, Williams & Wilkins.

59a. Headley BJ: *When movement hurts: a self-help manual for treating trigger points*, Minneapolis, 1997, Orthopedic Physical Therapy Products.

60. Hughes KH: Painful rib syndrome: a variant of myofascial pain syndrome. *AAOHN* 46(3):115–120, 1998.

61. Saltzman DA, Schmitz ML, Smith SD, et al: The slipping rib syndrome in children. *Paediatr Anaesth* 11(6):740–743, 2001.

62. Meuwly JY, Wicky S, Schnyder P, et al: Slipping rib syndromes: a place for sonography in the diagnosis of a frequently overlooked cause of abdominal or low thoracic pain. *J Ultrasound Med* 21(3):339–343, 2002.

63. Udermann BE, Cavanaugh DG, Gibson MH, et al: Slipping rib syndrome in a collegiate swimmer: a case report. *J Athl Train* 40(2):120–122, 2005.

64. Christo PJ, McGreevy K: Updated perspectives on neurogenic thoracic outlet syndrome. *Curr Pain Headache Rep* 15(1);14–21, 2011.

65. Sanders RJ: The forgotten pectoralis minor syndrome: 100 operations for pectoralis minor syndrome alone or accompanied by neurogenic thoracic outlet syndrome. *Ann Vasc Surg* 701–708, 2010.

66. Dutton M: *Orthopaedic examination, evaluation, and intervention*, ed 2, New York, 2008, McGraw-Hill.

67. Selke FW, Kelly TR: Thoracic outlet syndrome. *Am J Surg* 156:54–57, 1988.

68. Magee D: *Orthopedic physical assessment*, ed 5, Philadelphia, 2008, Saunders.

69. Sanders RJ: Recurrent neurogenic thoracic outlet syndrome stressing the importance of pectoralis minor syndrome. *Vasc Endovascular Surg* 45(1):33–38, 2011.

70. Yung E: Screening for head, neck, and shoulder pathology in patients with upper extremity signs and symptoms. *J Hand Ther* 23(2):173–186, 2010.

71. Braun RM: Thoracic outlet syndrome: a primer on objective methods of diagnosis. *J Hand Surg* 35A(9):1539–1541, 2010.

72. Robertson FM: Inflammatory breast cancer. The disease, the biology, the treatment. *CA Cancer J Clin* 60(6):351–375, 2010.

73. Wolf JM, Green A: Influence of comorbidity on self-assessment instrument scores of patients with idiopathic adhesive capsulitis. *J Bone Joint Surg* 84A(7):1167–1173, 2002.

74. Smith R, Athanasou NA, Ostlere SJ, et al: Pregnancy-associated osteoporosis. *QJM* 88:865–878, 1995.

75. Baitner AC, Bernstein AD, Jazrawi AJ: Spontaneous rib fracture during pregnancy: a case report and review of the literature. *Bull Hosp Jt Dis* 59(3):163–165, 2000.

76. Boissonnault WG, Boissonnault JS: Transient osteoporosis of the hip associated with pregnancy. *J Orthop Sports Phys Ther* 31(7):359–367, 2001.

77. Michalakis K: Pregnancy- and lactation-associated osteoporosis: a narrative minireview. *Endocr Regul* 45(1):43–47, 2011.

78. Kovacs CS: Calcium and bone metabolism during pregnancy and lactation. *J Mammary Gland Biol Neoplasia* 10(2):105–118, 2005.

79. Debnah UK, Kishore R, Black RJ: Isolated acetabular osteoporosis in TOH in pregnancy: a case report. *South Med J* 98(11):1146–1148, 2005.

CHAPTER

18

Screening the Shoulder and Upper Extremity

The therapist is well aware that many primary neuromuscular and musculoskeletal conditions in the neck, cervical spine, axilla, thorax, thoracic spine, and chest wall can refer pain to the shoulder and arm. For this reason, the physical therapist's examination usually includes assessment above and below the involved joint for referred musculoskeletal pain (Case Example 18-1).

In this chapter, we explore systemic and viscerogenic causes of shoulder and arm pain and take a look at each system that can refer pain or symptoms to the shoulder. This will include vascular, pulmonary, renal, gastrointestinal (GI), and gynecologic causes of shoulder and upper extremity pain and dysfunction. Primary or metastatic cancer as an underlying cause of shoulder pain also is included. The therapist must know how and what to look for to screen for cancer.

Systemic diseases and medical conditions affecting the neck, breast, and any organs in the chest or abdomen can present clinically as shoulder pain (Table 18-1).[1] Peptic ulcers, heart disease, ectopic pregnancy, and myocardial ischemia are only a few examples of systemic diseases that can cause shoulder pain and movement dysfunction. Each disorder listed can present clinically as a shoulder problem before ever demonstrating systemic signs and symptoms.

USING THE SCREENING MODEL TO EVALUATE SHOULDER AND UPPER EXTREMITY

Past Medical History

As you look over the various potential systemic causes of shoulder symptomatology listed in Table 18-1, think about the most common risk factors and red flag histories you might see with each of these conditions. For example, a history of any kind of cancer is always a red flag. Breast and lung cancer are the two most common types of cancer to metastasize to the shoulder.

Heart disease can cause shoulder pain, but it usually occurs in an age specific population.[2,3] Anyone over 50 years old, postmenopausal women, and anyone with a positive first generation family history is at increased risk for symptomatic heart disease. Younger individuals may be more likely to demonstrate atypical symptoms such as shoulder pain without chest pain.[4]

Alternately, although atherosclerosis has been demonstrated in the blood vessels of children, teens, and young adults, they are rarely symptomatic unless some other heart anomaly is present.[5,6]

Hypertension, diabetes, and hyperlipidemia are other red flag histories associated with cardiac-related shoulder pain. Of course, a history of angina,[7] heart attack, angiography, stent or pacemaker placement, coronary artery bypass graft (CABG), or other cardiac procedure is also a yellow (caution) flag to alert the therapist of the potential need for further screening.

Knowledge of risk factors associated with pathologic conditions, illnesses, and diseases helps the therapist navigate the screening process. For example, pulmonary tuberculosis (TB) is a possible cause of shoulder pain.[8-10] Who is most likely to develop TB? Risk factors include:

- Health care workers
- Homeless population
- Prison inmates
- Immunocompromised individuals (e.g., transplant recipients, long-term users of immunosuppressants, anyone treated for long-term rheumatoid arthritis [RA], anyone treated with chemotherapy for cancer)
- Older adult (over 65 years)
- Immigrants from areas where TB is endemic
- Injection drug users
- Malnourished (e.g., eating disorders, alcoholism, drug users, cachexia)

In a case like tuberculosis, there will usually be other associated signs and symptoms such as fever, sweats, and cough. When completing a screening examination for a client with shoulder pain of unknown origin or an unusual clinical presentation, the therapist might look at vital signs, auscultate the client, and see what effect increased respiratory movements have on shoulder symptoms (Case Example 18-2).

Clinical Presentation

Differential diagnosis of shoulder pain is sometimes especially difficult because any pain that is felt in the shoulder often affects the joint as though the pain were originating in the joint.[3] Shoulder pain with any of the components listed

CASE EXAMPLE 18-1

Evaluation of a Professional Golfer

Referral: A 38-year-old male, professional golfer presented to physical therapy with a diagnosis of shoulder impingement syndrome, with partial thickness tears of the supraspinatus tendon.

Prior to the physical therapy intervention, x-rays taken were reported as negative for fracture or tumor. Magnetic resonance imaging (MRI) was reported as positive for bursitis and supraspinatus tendinitis with some partial tears. The shoulder specialist also provided the client with one corticosteroid injection, which gave him some relief of his shoulder pain.

Past Medical History: Past medical history and Review of Systems were negative for any systemic issues. He was on no medication at the time of evaluation.

Clinical Presentation: Functional deficits were reported as pain with the take-away phase of the golf swing and with the adduction motion of the shoulder in follow-through. He also reported a loss of distance associated with his drive by 20 to 30 yards. He had trouble sleeping and reported pain would wake him up if his head were turned into left rotation. He also had pain when turning his head to the left (e.g., when driving a car).

Upper Quarter Screen

Shoulder Range of Motion (ROM)

Active ROM:

Left		Right
160 degrees	Flexion (flex)	170 degrees
165 degrees	Abduction (abd)	170 degrees
50 degrees	Internal rotation (IR)	55 degrees
55 degrees	External rotation (ER)	85 degrees

Passive ROM:

Left		Right
170 degrees	Flex	175 degrees
170 degrees	Abd	175 degrees
55 degrees	IR	60 degrees
60 degrees	ER	75 degrees

Isometric muscle testing of rotator cuff	
Abd	Painful/strong
Abd with IR	Painful/strong
IR	Painless/strong
ER	Painless/strong

Special tests

Hawkins/Kennedy +
Neer +
Speed +
ER lag test –
IR lag test –

Cervical ROM	
Flexion 40 degrees	
Extension (ext) 20 degrees	Report of left scapular pain
Left side bend 20 degrees	Report of left scapular pain
Right side bend 25 degrees	No report of pain
Left rotation 45 degrees	Report of left scapular pain
Right rotation 70 degrees	No report of pain
Quadrant position	Right and left: Reproduced left posterior scapular pain with radicular pain to the thumb and second finger area

Deep Tendon Reflexes (DTRs)

Left	DTRs	Right
2+	Biceps	2+
0	Triceps	2+
2+	Brachioradialis	2+

Strength

Left		Right
5/5	Shoulder flex	5/5
4/5	Shoulder abd	5/5
5/5	Elbow flex	5/5
2/5	Elbow ext	5/5
3/5	Wrist ext	5/5
5/5	Wrist flex	5/5
5/5	Thumb ext	5/5
5/5	Finger abd	5/5

He did have intact sensation to light touch and proprioceptive sense. Strength testing on the Cybex weight-lifting machines showed he was able to do 10 triceps extensions on the right with four plates while on the left, he was only able to do one repetition with one plate.

Result: With the data obtained in the examination, the conclusion was made that he did have an impingement syndrome as described by Neer, with involvement of the bursa and rotator cuff tendons.[72] Cyriax muscle testing revealed some musculotendon involvement with the strong/painful tests.[61]

The cervical findings required consultation with the referring physician. A provisional medical diagnosis was made of cervical radiculopathy with a C5-C6 herniated disk. The client was referred to a neurosurgeon for evaluation. An MRI confirmed the diagnosis and the client underwent an anterior cervical fusion with diskectomy.

Summary: This case example helps highlight the importance of a complete examination process, even if a physician specialist refers a client for physical therapy services. The therapist must "clear" or examine the joints above and below the region thought to be the cause of the dysfunction. The major reason for the symptoms or a secondary diagnosis may be missed if the screening step is left out because of a lack of time or assuming someone else checked out the entire client.

Voshell S: Case report presented in fulfillment of DPT 910, Institute for Physical Therapy Education, Widener University, Chester, PA, 2005. Used with permission.

TABLE 18-1 Systemic and Medical Conditions as Causes of Shoulder and Upper Extremity Symptoms

	Neck	Chest/Trunk/Back	Abdomen
Cancer	Metastases (leukemia, Hodgkin's lymphoma) Cervical cord tumors Bone tumors	Metastases to nodes in axilla or mediastinum Metastases to lungs from: Bone Breast Kidney Colorectal Pancreas Uterus Bone metastases to thoracic spine: Breast Lung Thyroid Breast cancer Lung cancer	Pancreatic cancer Spinal metastases Kidney Testicle Prostate
Cardiovascular/vascular	TOS	Angina/MI Acute coronary syndrome ICU/s/p CABG Pacemaker (complications) Bacterial endocarditis Pericarditis Thoracic aortic aneurysm Empyema and lung abscess Collagen vascular disease	Dissecting aortic aneurysm
Pulmonary	Pulmonary tuberculosis	Pulmonary embolism Pulmonary tuberculosis Spontaneous pneumothorax Pancoast tumor Pneumonia	
Renal/urologic			Kidney stones Obstruction, inflammation, or infection of upper urinary tract
Gastrointestinal/hepatic		Hiatal hernia	Peptic/duodenal ulcer (perforated) Ruptured spleen Liver disease Gallbladder disease Pancreatic disease Subphrenic abscess
Infection		Septic arthritis Necrotizing fasciitis Mononucleosis Osteomyelitis/transverse myelitis Syphilis/gonorrhea Herpes zoster (shingles) Pneumonia Cellulitis (skin anywhere in neck, chest, arm, hand)	
Gynecologic			Ectopic pregnancy (rupture) Endometriosis (cysts)
Other	Cervical central cord lesion Trauma: Cervical fractures or ligamentous instability; whiplash	Mastodynia (breast) Diabetes mellitus (adhesive capsulitis) Sickle cell disease Hemophilia	Diaphragmatic hernia Anterior spinal surgery (postoperative hemorrhage)

TOS, Thoracic outlet syndrome; *MI,* myocardial infarction; *ICU/s/p CABG,* intensive care unit status post coronary artery bypass graft.

in this chapter should be approached as a manifestation of systemic visceral illness, even if shoulder movements exacerbate the pain or if there are objective findings at the shoulder.

Many visceral diseases present as unilateral shoulder pain (Table 18-2). Esophageal, pericardial (or other myocardial diseases), aortic dissection, and diaphragmatic irritation from thoracic or abdominal diseases (e.g., upper GI, renal, hepatic/biliary) all can appear as unilateral pain.

Adhesive capsulitis, a condition in which both active and passive glenohumeral motions are restricted, can be associated with diabetes mellitus, hyperthyroidism,[11,12] ischemic

CASE EXAMPLE 18-2

Homeless Man with Tuberculosis

Referral: A 36-year-old man was referred to physical therapy as an inpatient for a short-term hospitalization. He was a homeless man brought to the hospital by the police and admitted with an extensive medical problem list including:

- Malnutrition
- Alcoholism
- Depression
- Hepatitis A
- Broken wrist
- Shoulder pain
- Dehydration

There was no past medical history of cancer. The client was a smoker when he could get cigarettes. He would like to support a one-pack/day habit.

Medical service requested an evaluation of the client's shoulder pain. X-rays were not taken because the man had full active ROM, no history of trauma, and no insurance to cover additional testing.

Clinical Presentation: The therapist was unable to reproduce the shoulder pain with palpation, position, or provocation testing. There was no sign of rotator cuff dysfunction, adhesive capsulitis, tendinitis, or trigger points in the upper quadrant. There was a noticeable stiffening of the neck with very limited cervical ROM in all planes and directions.

Vital signs were unremarkable, but the client was perspiring heavily despite being in threadbare clothing and at rest. He reported getting the "sweats" every day around this same time.

The therapist asked the client to take a deep breath and cough. He went into a paroxysm of coughing, which he said caused his shoulder to start aching. The cough was productive, but the client swallowed the sputum. Auscultation of lung sounds revealed rales (crackles) in the right upper lung lobe. Supraclavicular lymph nodes were palpable, tender, and moveable on both sides.

The therapist contacted the charge nurse and reported the following concerns:

- Constitutional symptoms of sweats and fatigue (although fatigue could be caused by his extreme malnutrition)
- Pulmonary impairment with reproduction of symptoms with respiratory movements
- Suspicious (aberrant) lymph nodes (bilateral)
- Cervical spine involvement with no apparent cause or recognizable musculoskeletal pattern

Result: Consult with the physician on-call resulted in a medical evaluation and x-ray. Client was diagnosed with pulmonary tuberculosis, which was confirmed by a skin test. Shoulder and neck pain and dysfunction were attributed to a pulmonary source and not considered appropriate for physical therapy intervention.

The client was sent to a halfway house where he could receive adequate nutrition and medical services to treat his tuberculosis.

TABLE 18-2 Location of Shoulder Pain

Systemic Origin	Right Shoulder Location	Systemic Origin	Left Shoulder Location
Peptic ulcer	Lateral border, right scapula	Internal bleeding: Spleen (trauma, rupture); Postoperative laparoscopy	Left shoulder (Kehr's sign)
Myocardial ischemia	Right shoulder, down arm	Myocardial ischemia	Left pectoral/left shoulder
		Thoracic aortic aneurysm	Left shoulder (or between shoulder blades)
Hepatic/biliary:		Pancreas	Left shoulder
Acute cholecystitis	Right shoulder; between scapulae; right subscapular area		
Gallbladder	Right upper trapezius, right shoulder	Infectious mononucleosis (hepatomegaly, splenomegaly)	Left shoulder/left upper trapezius
Liver disease (hepatitis, cirrhosis, metastatic tumors, abscess)	Right shoulder, right subscapular		
Pulmonary: Pleurisy; Pneumothorax; Pancoast's tumor; Pneumonia	Ipsilateral shoulder; upper trapezius	Pulmonary: Pleurisy; Pneumothorax; Pancoast's tumor; Pneumonia	Ipsilateral shoulder; upper trapezius
Kidney	Ipsilateral shoulder	Kidney	Ipsilateral shoulder
Gynecologic: Endometriosis	Reported in right shoulder[68]; possible in either shoulder, depending on location of cysts	Gynecologic: Ectopic pregnancy	Ipsilateral shoulder

heart disease, infection, and lung diseases (tuberculosis, emphysema, chronic bronchitis, Pancoast's tumors) (Case Example 18-3).[9,10,13-15]

Shoulder pain (unilateral or bilateral) progressing to adhesive capsulitis can occur 6 to 9 months after CABG. Similarly, anyone immobile in the intensive care unit (ICU) or coronary care unit (CCU) can experience loss of shoulder motion resulting in adhesive capsulitis (Case Example 18-4). Clients with pacemakers who have complications and revisions that result in prolonged shoulder immobilization can also develop complex regional pain syndrome (CRPS) and/or adhesive capsulitis.[16]

The Shoulder Is Unique

It has been stressed throughout this text that the basic clues and approach to screening are similar, if not the same, from system to system and anatomic part to anatomic part.

So, for example, much of what was said about screening the neck and back (Chapter 14) applied to the sacrum, sacroiliac (SI), and pelvis (Chapter 15); buttock, hip, and groin (Chapter 16); and chest, breast, and rib (Chapter 17). Presenting the shoulder last in this text is by design. These principles do apply to the shoulder but beyond that:

> *Shoulder pain is difficult to diagnose because any pain felt in the shoulder will affect the joint as though the pain was originating in the joint.*
>
> ***John Mennell***[17]
>
> *... even when there is a known cause, especially in the older adult.*
>
> ***Catherine Goodman***

It is not uncommon for the older adult to attribute "overdoing" it to the appearance of physical pain or neuromusculoskeletal (NMS) dysfunction. Any adult over age 65 presenting with shoulder pain and/or dysfunction must be screened for systemic or viscerogenic origin of symptoms, even when there is a known (or attributed) cause or injury.

In Chapter 2, it was stressed that clients who present with no known cause or insidious onset must be screened along with anyone who has a known or assumed cause of symptoms. Whether the client presents with an unknown etiology of injury or impairment or with an assigned cause, always ask yourself these questions:

FOLLOW-UP QUESTIONS

- Is it really insidious?
- Is it really caused by such and such (whatever the client told you)?

The client may wrongly attribute onset of symptoms to an activity. The alert therapist may recognize a true causative factor.

Shoulder Pain Patterns

In Chapter 3, we presented three possible mechanisms for referred pain patterns from the viscera to the soma (embryologic development, multisegmental innervations, and direct pressure on the diaphragm). Multisegmental innervations (see Fig. 3-3) and direct pressure on the diaphragm (see Figs. 3-4 and 3-5) are two key mechanisms for referred shoulder pain.

Multisegmental Innervations. Because the shoulder is innervated by the same spinal nerves that innervate the diaphragm (C3 to C5), any messages to the spinal cord from the diaphragm can result in referred shoulder pain. The nervous system can only tell what nerves delivered the message. It does not have any way to tell if the message sent along via spinal nerves C3 to C5 came from the shoulder or the diaphragm. So it takes a guess and sends a message back to one or the other.

This means that any organ in contact with the diaphragm that gets obstructed, inflamed, or infected can refer pain to the shoulder by putting pressure on the diaphragm, stimulating afferent nerve signals, and telling the nervous system that there is a problem.

Diaphragmatic Irritation. Irritation of the peritoneal (outside) or pleural (inside) surface of the central diaphragm refers sharp pain to the ipsilateral upper trapezius, neck and/or supraclavicular fossa (Fig. 18-1). Shoulder pain from diaphragmatic irritation usually does not cause anterior shoulder pain. Pain is confined to the suprascapular, upper trapezius, and posterior portions of the shoulder.

If the irritation crosses the midline of the diaphragm, then it is possible to have bilateral shoulder pain. This does not happen very often and is most common with cardiac ischemia or pulmonary pathology affecting the lower lobes of the lungs on both sides. Irritation of the peripheral portion of

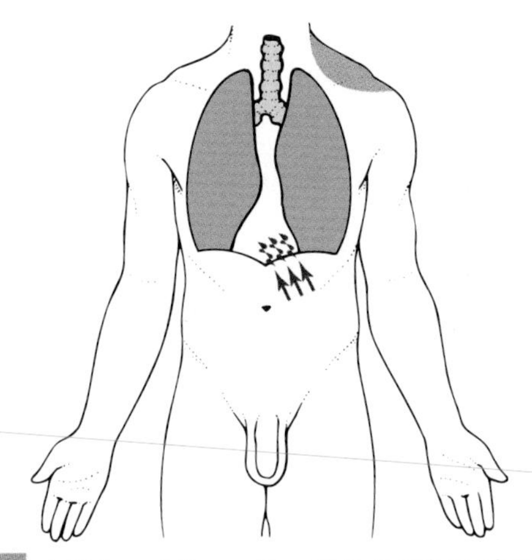

Fig. 18-1 Irritation of the peritoneal (outside) or pleural (inside) surface of the *central* area of the diaphragm can refer sharp pain to the upper trapezius muscle, neck, and supraclavicular fossa. The pain pattern is ipsilateral to the area of irritation. Irritation to the *peripheral* portion of the diaphragm can refer sharp pain to the ipsilateral costal margins and lumbar region (not shown).

CASE EXAMPLE 18-3

Cardiac Cause of Shoulder Pain

A 65-year-old retired railroad engineer has come to you with a left "frozen shoulder." During the course of the subjective examination, he tells you he is taking two cardiac medications.

What questions would you ask that might help you relate these two problems or rule out a cardiac condition as a possible cause? (shoulder/cardiac)

Try to organize your thoughts using these categories:

- Onset/history of shoulder involvement
- Medical testing
- Clinical presentation
- Past medical history

Physical Therapy Screening Interview

Onset/History

- What do you think is the cause of your shoulder problem?
- When did it occur, or how long have you had this problem (sudden or gradual onset)?
- Can you recall any specific incident when you injured your shoulder, for example, by falling, being hit by someone or something, automobile accident?
- Did you ever have a snapping or popping sensation just before your shoulder started to hurt? **(Ligamentous or cartilaginous lesion)**
- Did you injure your neck in any way before your shoulder developed these problems?
- Have you had a recent heart attack? Have you had nausea, fatigue, sweating, chest pain, or pressure? Any pain in your neck, jaw, left shoulder, or down your left arm?
- Has your left hand ever been stiff or swollen? **(CRPS after myocardial infarction [MI])**
- Do you think your shoulder pain is related to your heart problems?
- Shortly before you first noticed difficulty with your shoulder were you involved in any kind of activities that would require repetitive movements, such as painting, gardening, playing tennis or golf?

Medical Testing

- Have you had any recent x-rays taken of the shoulder or your neck?
- Have you received medical or physical therapy treatment for shoulder problems before?
  - *If yes*, where, when, why, who, and what (see Chapter 2 for specific questions)?
- Have you had any (extensive) medical testing during the past year?

Clinical Presentation

Pain/Symptoms

Follow the usual line of questioning regarding the pattern, frequency, intensity, and duration outlined in Fig. 3-6 to establish necessary information regarding pain.

- Is your shoulder painful?
  - *If yes,* how long has the shoulder been painful?

Aggravating/Relieving Activities

- How does rest affect your shoulder symptoms? **(True muscular lesions are relieved with prolonged rest [i.e., more than 1 hour], whereas angina is usually relieved more immediately by cessation of activity or rest [i.e., usually within 2 to 5 minutes, up to 15 minutes].)**
- Does your shoulder pain occur during exercise (e.g., walking, climbing stairs, mowing the lawn or any other physical or sexual activity? **(Evaluate the difference between total body exertion causing shoulder symptoms versus movements of the upper extremities only reproducing symptoms. Total body exertion causing shoulder pain may be secondary to angina or MI, whereas movements of just the upper extremities causing shoulder pain are indicative of a primary musculoskeletal lesion.)**

Past Medical History

- Have you had any surgery during the past year?
- How has your general health been? (**Shoulder pain is a frequent site of referred pain from other internal medical problems;** see Fig. 18-2.)
- Did you have rheumatic fever when you were a child?
- What is your typical pattern of chest pain or angina?
- Has this pattern changed in any way since your shoulder started to hurt? For example, does the chest pain last longer, come on with less exertion, and feel more intense?
- What medications are you taking?
- Do your heart medications relieve your shoulder symptoms, even briefly?
  - *If yes*, how long after you take the medications do you notice a difference?
  - Does this occur every time that you take your medications?

Evaluating subacute/acute/chronic musculoskeletal lesion versus systemic pain pattern (see Chapter 3 for specific meaning to the client's answers to these questions):

- Can you lie on that side?
- Does the shoulder pain awaken you at night?
  - *If yes*, is this because you have rolled onto that side?
- Do you notice any chest pain, night sweats, fever, or heart palpitations when you wake up at night?
- Have you ever noticed these symptoms (e.g., chest pain, heart palpitations) with your shoulder pain during the day?
- Do these symptoms wake you up separately from your shoulder pain, or does your shoulder pain wake you up and you have these additional symptoms? **(As always, when asking questions about sleep patterns, the person may be unsure of the answers to the questions. In such cases the physical therapist is advised to ask the client to pay attention to what happens related to sleep during the next few days up to 1 week and report back with more information.)**

Other Clinical Tests: In addition to an orthopedic screening examination, the therapist should review potential side effects and interactions of cardiac medications, take vital signs, auscultate (including femoral bruits), and palpate for the aortic pulse (see Fig. 4-55).

CASE EXAMPLE 18-4

Pleural Effusion with Fibrosis, Late Complication of Coronary Artery Bypass Graft

Referral: A 53-year-old man was referred to physical therapy by his primary care physician for left shoulder pain.

Past Medical History: The client had a recent (6 months ago) history of cardiac bypass surgery (also known as coronary artery bypass graft [CABG]) and had completed phase 1 and phase 2 cardiac rehab programs. He was continuing to follow an exercise program (phase 3 cardiac rehab) prescribed for him at the time of his physical therapy referral.

Clinical Presentation: The client looked in good health and demonstrated good posture and alignment. Shoulder range of motion (ROM) was equal and symmetric bilaterally, but the client reported pain when the left arm was raised over 90 degrees of flexion or abduction. His position of preference was left sidelying. The pain could be reduced in this position from a rated level of 6 to a 2 on a scale from 0 (no pain) to 10 (worst pain).

Scapulohumeral motion on the left was altered compared to the right. Medial and lateral rotations were within normal limits (WNL) with the upper arm against the chest. Lateral rotation reproduced painful symptoms when performed with the shoulder in 90 degrees of abduction. Physiologic motions were fully present in all directions on the left but seemed "sluggish" compared to the right.

Neurologic screen was negative.

Vital signs:

Blood pressure:	122/68 mm Hg
Resting pulse:	60 bpm
Body temperature:	98.6° F

Cardiopulmonary screening exam:

Diminished basilar (lower lobes) breath sounds on the left compared to the right

Decreased chest wall excursion on the left; increased shoulder pain with deep inspiration

Dyspnea was not observed at rest

When asked if there were any symptoms of any kind anywhere else in the body, the client reported ongoing but intermittent chest pain and shortness of breath for the last 3 months. The client had not reported these "new" symptoms to the physician.

What are the red flags (if any)? Is an immediate medical referral indicated?

Red Flags

- Age over 40
- Previous (recent) history of cardiac surgery
- Unequal basilar breath sounds
- Unreported symptoms of chest pain and dyspnea
- Autosplinting (lying on the affected side diminishes lung movement, reducing shoulder pain)

Medical Consultation: Shoulder problems are not uncommon following CABG, but the number and type of red flags present caught the therapist's attention. The client was not in any apparent physiologic distress and vital signs were WNL (although he was on antihypertensive medications). Since he was referred by his primary care physician, the therapist made telephone contact with the physician's office and faxed a summary of findings immediately.

A program of physical therapy intervention was determined, but the therapist insisted on speaking with the physician first before proceeding with the program. The physician approved the therapist's treatment plan but requested immediate follow-up with the client who was seen the next day.

Result: The client was diagnosed with pleural effusion causing pleural fibrosis, a rare long-term complication of cardiac bypass surgery. The physician noted that the left lower lobe was adhered to the chest wall.

Pleural effusion is a common complication of cardiac surgery and is associated with other postoperative complications. It occurs more often in women and individuals with associated cardiac or vascular comorbidities and medications used to treat those conditions.[73-76]

The client was treated medically but also continued in physical therapy to restore full and normal motion of the shoulder complex. The physician also asked the therapist to review the client's cardiac rehab program and modify it accordingly due to the pulmonary complications.

the diaphragm is more likely to refer pain to the costal margins and lumbar region on the same side.

As you review Fig. 3-4, note how the heart, spleen, kidneys, pancreas (both the body and the tail), and the lungs can put pressure on the diaphragm. This illustration is key to remembering which shoulder can be involved based on organ pathology. For example, the spleen is on the left side of the body so pain from spleen rupture or injury is referred to the left shoulder (called *Kehr's sign*) (Case Example 18-5).[18]

Either shoulder can be involved with renal colic or distention of the renal cap from any kidney disorder, but it is usually an ipsilateral referred pain pattern depending on which kidney is impaired (see Fig. 10-7; again, via pressure on the diaphragm). Bilateral shoulder pain from renal disease would only occur if and when both kidneys are compromised at the same time.

Look for history of a recent surgery as part of the past medical history and the presence of accompanying urologic symptoms.

The body of the pancreas lies along the midline of the diaphragm. When the body of the pancreas is enlarged, inflamed, obstructed, or otherwise impinging on the diaphragm, back pain is a possible referred pain pattern. Pain felt in the left shoulder may result from activation of pain fibers in the left diaphragm by an adjacent inflammatory process in the tail of the pancreas.

Postlaparoscopic shoulder pain (PLSP) frequently occurs after various laparoscopic surgical procedures. During the procedure air is introduced into the peritoneum to expand the area and move the abdominal contents out of the way. The mechanism of PLSP is commonly assumed to be overstretching of the diaphragmatic muscle fibers due to the

CASE EXAMPLE 18-5

Rugby Injury: Kehr's Sign

Referral: A 27-year-old male accountant who has an office in the same complex with a physical therapy practice stopped by early Monday morning complaining of left shoulder pain.

When asked about repetitive motions or recent trauma or injuries, he reported playing in a rugby tournament over the weekend. "I got banged up quite a few times, but I had so much beer in me, I didn't feel a thing."

Clinical Presentation: Pain was described as a deep, sharp aching over the upper trapezius and shoulder area on the left side. There were no visual bruises or signs of bleeding in the upper left quadrant.

Vital signs:

Pulse:	89 bpm
Respirations:	12 per minute
Blood pressure:	90/48 mm Hg (recorded sitting, left arm)
Temperature:	97° F (reported as the client's "normal" morning temperature)
Pain:	Rated as a 5 on a scale from 0 to 10

Range of motion was full in all planes and movements. No particular movement increased or decreased the pain. Gross manual muscle test of the upper extremities was normal (5/5 for flexion, abduction, extension, rotations).

Neurologic screen was negative. All special shoulder tests (e.g., impingement, anterior and posterior instability, quadrant position) were unremarkable.

What are the red flags here? What are your next questions, steps, or screening tests?

Red Flags

- Hypotension
- Left shoulder pain within 24 hours of possible trauma or injury
- Unable to alter, provoke, or palpate painful symptoms
- Clinical presentation is not consistent with expected picture for a shoulder problem; lack of objective findings.

What are your next questions, steps, or screening tests?

Repeat blood pressure measurements, bilaterally. Perform percussive tests for the spleen (see Fig. 4-53).

Depending on the results of these clinical tests, referral might be needed immediately. In this case, the percussive test for enlarged spleen was inconclusive, but there was an observable and palpable "fullness" in the left flank compared to the right.

Result: This client was told:

"Mr. Smith, your exam does not look like what I would expect from a typical shoulder injury. Since I cannot find any way to make your pain better or worse and I cannot palpate or feel any areas of tenderness, there may be some other cause for your symptoms.

Given your history of playing rugby over the weekend, it is possible you have some internal injuries. I am not comfortable treating you until a medical doctor examines you first. Bleeding from the spleen can cause left shoulder pain. When I tapped over the area of your spleen, it did not sound quite like I expected it to, and it seems like there is some fullness along your left side that I am not seeing or feeling on the right.

I do not want to alarm you, but it may be best to go over to the emergency department of the hospital and see what they have to say. You can also call your regular doctor and see if you can get in right away. You can do that right from our clinic phone."

Final Result: This accountant had clients already scheduled starting in 10 minutes. He did not feel he had the time to go check this out until his lunch hour. About 45 minutes later an ambulance was called to the building. Mr. Smith had collapsed, and his coworkers called 9-1-1.

He was rushed to the hospital and diagnosed with a torn and bleeding spleen, which the doctor called a "slow leak." It eventually ruptured, leaving him unconscious from blood loss.

pressure of a pneumoperitoneum (residual carbon dioxide [CO_2] gas after surgery).[19] Pressure from distention causes phrenic nerve–mediated referred pain to the shoulder.[20]

Keep in mind that shoulder pain also can occur from diaphragmatic dysfunction. For anyone with shoulder pain of an unknown origin or which does not improve with intervention, palpate the diaphragm and assess its excursion and timing during respiration. Reproduction of shoulder symptoms with direct palpation of the diaphragm and the presence of altered diaphragmatic movement with breathing offer clues to the possibility of diaphragmatic (muscular) involvement.

Fig. 18-2 reminds us that shoulder pain can be referred from the neck, back, chest, abdomen, and elbow. During orthopedic assessment, the therapist always checks "above and below" the impaired level for a possible source of referred pain. With this guideline in mind, we know to look for potential musculoskeletal or neuromuscular causes from the cervical and thoracic spine[21] and elbow.

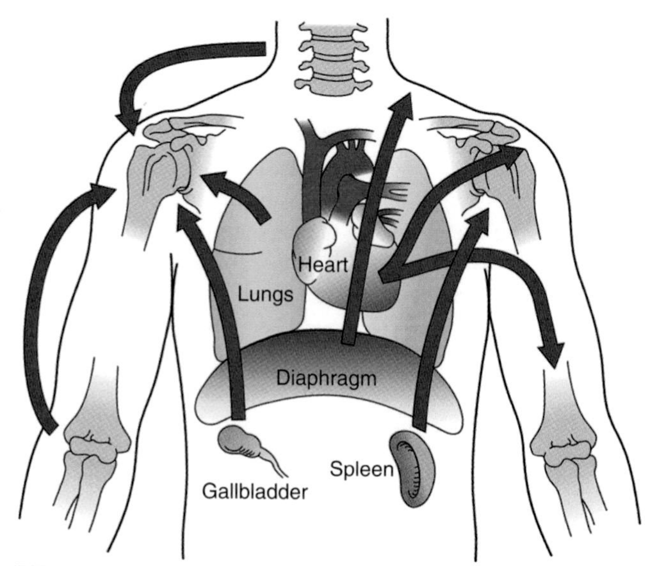

Fig. 18-2 Musculoskeletal and systemic structures referring pain to the shoulder.

Associated Signs and Symptoms

One of the most basic clues in screening for a viscerogenic or systemic cause of shoulder pain is to look for shoulder pain accompanied by any of the following features:

- Pleuritic component
- Exacerbation by recumbency
- Recent history of laparoscopic procedure (risk factor)[18,22,23]
- Coincident diaphoresis (cardiac)
- Associated GI signs and symptoms
- Exacerbation by exertion unrelated to shoulder movement (cardiac)
- Associated urologic signs and symptoms

Shoulder pain with any of these present should be approached as a manifestation of systemic visceral illness. This is true even if the pain is exacerbated by shoulder movement or if there are objective findings at the shoulder.[24]

Using the past medical history and assessing for the presence of associated signs and symptoms will alert the therapist to any red flags suggesting a systemic origin of shoulder symptoms. For example, a ruptured ectopic pregnancy with abdominal hemorrhage can produce left shoulder pain (with or without chest pain) in a woman of childbearing age.[25-27] The woman is sexually active, and there is usually a history of missed menses or recent unexplained/unexpected bleeding.

The client may not recognize the connection between painful urination and shoulder pain or the link between gallbladder removal by laparoscopy and subsequent shoulder pain. It is the therapist's responsibility to assess musculoskeletal symptoms, making a diagnosis that includes ruling out the possibility of systemic disease.

Review of Systems

Associated signs and symptoms feature heavily in the Review of Systems as we step back and look to see if a cluster of any particular organ-dependent signs and symptoms is present. Based on the results of this review, we formulate our final screening questions, tests, and measures. Always remember to end each client interview with the following (or similar) question:

FOLLOW-UP QUESTIONS

- Do you have any symptoms of any kind anywhere else in your body that we haven't talked about yet?

SCREENING FOR PULMONARY CAUSES OF SHOULDER PAIN

Extensive disease may occur in the periphery of the lung without pain until the process extends to the parietal pleura. Pleural irritation then results in sharp, localized pain that is aggravated by any respiratory movement.

Clients usually note that the pain is alleviated by lying on the affected side, which diminishes the movement of that side of the chest (called "autosplinting") whereas shoulder pain of musculoskeletal origin is usually aggravated by lying on the symptomatic shoulder.

Shoulder symptoms made worse by recumbence are a yellow flag for pulmonary involvement. Lying down increases the venous return from the lower extremities. A compromised cardiopulmonary system may not be able to accommodate the increase in fluid volume. Referred shoulder pain from the taxed and overworked pulmonary system may result.

At the same time, recumbency or the supine position causes a slight shift of the abdominal contents in the cephalic direction. This shift may put pressure on the diaphragm, which in turn presses up against the lower lung lobes. The combination of increased venous return and diaphragmatic pressure may be enough to reproduce the musculoskeletal symptoms.

Pneumonia in the older adult may appear as shoulder pain when the affected lung presses on the diaphragm; usually there are accompanying pulmonary symptoms, but in older adults, confusion (or increased confusion) may be the only other associated sign.

The therapist should look for the presence of a pleuritic component such as a persistent or productive cough and/or chest pain. Look for tachypnea, dyspnea, wheezing, hyperventilation, or other noticeable changes. Chest auscultation is a valuable tool when screening for pulmonary involvement.

SCREENING FOR CARDIOVASCULAR CAUSES OF SHOULDER PAIN

Pain of cardiac and diaphragmatic origin is often experienced in the shoulder because the heart and diaphragm are supplied by the C5 to C6 spinal segment, and the visceral pain is referred to the corresponding somatic area (see Fig. 3-3).

Exacerbation of the shoulder symptoms from a cardiac cause occurs when the client increases activity that does not necessarily involve the arm or shoulder. For example, walking up stairs or riding a stationary bicycle can bring on cardiac-induced shoulder pain.

In cases like this, the therapist should ask about the presence of nausea, unexplained sweating, jaw pain or toothache, back pain, or chest discomfort or pressure. For the client with known heart disease, ask about the effect of taking nitroglycerin (men) or antacids/acid-relieving drugs (women) on their shoulder symptoms.

Vital sign and physical assessment including chest auscultation are important screening tools. See Chapter 4 for details.

Angina or Myocardial Infarction

Angina and/or myocardial infarction (MI) can appear as arm and shoulder pain that can be misdiagnosed as arthritis or other musculoskeletal pathologic conditions (see complete discussion in Chapter 6 and see Figs. 6-8 and 6-9).

Look for shoulder pain that starts 3 to 5 minutes after the start of activity, including shoulder pain with isolated lower extremity motion (e.g., shoulder pain starts after the client climbs a flight of stairs or rides a stationary bicycle). If the client has known angina and takes nitroglycerin, ask about the influence of the nitroglycerin on shoulder pain.

Shoulder pain associated with MI is unaffected by position, breathing, or movement. Because of the well-known association between shoulder pain and angina, cardiac-related shoulder pain may be medically diagnosed without ruling out other causes, such as adhesive capsulitis or supraspinatus tendinitis, when, in fact, the client may have both a cardiac and a musculoskeletal problem (Case Example 18-6).

CASE EXAMPLE 18-6

Strange Case of the Flu

Referral: A 53-year-old butcher at the local grocery store stopped by the physical therapy clinic located in the same shopping complex with a complaint of unusual shoulder pain. He had been seen at this same clinic several years ago for shoulder bursitis and tendinitis from repetitive overuse (cutting and wrapping meat).

Clinical Presentation: His clinical presentation for this new episode of care was exactly as it had been during the last episode of shoulder impairment. The therapist reinstituted a program of soft tissue mobilization and stretching, joint mobilization, and postural alignment. Modalities were used during the first two sessions to help gain pain control.

At the third appointment, the client mentioned feeling "dizzy and sweaty" all day. His shoulder pain was described as a constant, deep ache that had increased in intensity from a 6 to a 10 on a scale from 0 to 10. He attributed these symptoms to having the flu.

It was not until this point that the therapist conducted a screening exam and found the following red flags:

- Age
- Recent history (past 3 weeks) of middle ear infection on the same side as the involved shoulder
- Constant, intense pain (escalating over time)
- Constitutional symptoms (dizziness, perspiration)
- Symptoms unrelieved by physical therapy treatment

Result: The therapist suggested the client get a medical checkup before continuing with physical therapy. Even though the clinical presentation supported shoulder impairment, there were enough red flags and soft signs of systemic distress to warrant further evaluation.

Taking vital signs would have been a good idea.

It turns out the client was having myocardial ischemia masquerading as shoulder pain, the flu, and an ear infection. He had an angioplasty with complete resolution of all his symptoms and even reported feeling energetic for the first time in years.

This is a good example of how shoulder pain and dysfunction can exactly mimic a true musculoskeletal problem—even to the extent of reproducing symptoms from a previous condition.

This case highlights the fact that we must be careful to fully assess our clients with each episode of care.

Using a review of symptoms approach and a specific musculoskeletal shoulder examination, the physical therapist can screen to differentiate between a medical pathologic condition and mechanical dysfunction[28] (Case Example 18-7).

Complex Regional Pain Syndrome

Complex regional pain syndrome (CRPS; types I and II) characterized by chronic extremity pain following trauma is sometimes still referred to by the outdated term *shoulder-hand syndrome* (see Case Example 1-5). CRPS-I was formerly known as *reflex sympathetic dystrophy* (RSD). CRPS-II was referred to as *causalgia*.

CRPS was first recognized in the 1800s as causalgia or burning pain in wounded soldiers. Similar presentations after lesser injuries were labeled as RSD.[29] Shoulder-hand syndrome was a condition that occurred after an MI (heart attack), usually after prolonged bedrest. This condition (as it was known then) has been significantly reduced in incidence by more up-to-date and aggressive cardiac rehabilitation programs.

Today CRPS-I, primarily affecting the limbs, develops after bone fracture or other injury (even slight or minor trauma, venipuncture, or an insect bite) or surgery to the upper extremity (including shoulder arthroplasty) or lower extremity. Type I is not associated with a nerve lesion, whereas Type II develops after trauma with a nerve lesion.[30,31]

CRPS-I is still associated with cerebrovascular accident (CVA), heart attack, or diseases of the thoracic or abdominal viscera that can refer pain to the shoulder and arm, which is why it is included here instead of in a section on neurologic conditions. CRPS secondary to deep venous thrombosis (DVT) has also been reported. Individuals developing limb pain and edema after DVT will need further diagnostic investigation to differentiate the cause of symptoms.[32]

Shoulder, arm, or hand pain and ischemia (usually acute) associated with CRPS that develop without a history of trauma may be attributed to cardiac embolism.[33] Structural cardiac causes of upper limb ischemia include a wide variety of conditions (e.g., atrial fibrillation, cardiomyopathy, prosthetic valve, endocarditis, atrial septal defects, aortic dissection).[34]

This syndrome occurs with equal frequency in either or both shoulders and except when caused by coronary occlusion, is most common in women. The shoulder is generally involved first, but the painful hand may precede the painful shoulder.

When this condition occurs after MI, the shoulder initially may demonstrate pericapsulitis. Tenderness around the shoulder is diffuse and not localized to a specific tendon or bursal area. The duration of the initial shoulder stage before the hand component begins is extremely variable. The shoulder may be "stiff" for several months before the hand becomes involved or both may become stiff simultaneously. Other accompanying signs and symptoms are usually present such as edema, skin (trophic) changes, and vasomotor (temperature, hidrosis) changes.

CASE EXAMPLE 18-7

Angina Versus Shoulder Pathology

Referral: A 54-year-old man was referred to physical therapy for pre-prosthetic training after a left transtibial (TT) amputation.

Past Medical History

A right transtibial amputation was done 4 years ago
Coronary artery disease (CAD) with coronary artery bypass graft (CABG), myocardial infarction (heart attack), and angina
Peripheral vascular disease (PVD)
Long-standing diabetes mellitus (insulin dependent ×47 years)
Gastroesophageal reflux disease (GERD)

Clinical Presentation: At the time of the initial evaluation for the left TT amputation, the client reported substernal chest pain and left upper extremity pain with activity. Typical anginal pain pattern was described as substernal chest pain. The pain occurs with exertion and is relieved by rest.

Arm pain has never been a part of his usual anginal pain pattern. He reports his arm pain began 10 months ago with intermittent pain starting in the left shoulder and radiating down the anterior-medial aspect of the arm, halfway between the shoulder and the elbow.

The pain is made worse by raising his left arm overhead, pushing his own wheelchair, and using a walker. He was not sure if the shoulder pain was caused by repetitive motions needed for mobility or by his angina. The shoulder pain is relieved by avoiding painful motions. He has not received any treatment for the shoulder problem.

Neurologic screen was negative.

Vital signs

Heart rate:	88 bpm
Blood pressure:	120/66 mm Hg (position and extremity, not recorded)
Respirations:	WNL

Vital signs (after transfer and pregait activities)

Heart rate:	92 bpm
Blood pressure:	152/76 mm Hg
Respirations:	"Minimal shortness of breath" recorded

Special tests

Yergason's sign:	Positive
Apprehension test:	Positive
Relocation test:	Positive
Speed's test:	Positive

Palpation of the biceps and supraspinatus tendons increased the client's shoulder pain.

Active range of motion (AROM): left shoulder

Flexion:	100 degrees
Abduction:	70 degrees
Internal/external rotation:	60 degrees

There is a capsular pattern in the left glenohumeral joint with limitations in rotation and adduction. Significant capsular tightness is demonstrated with passive or physiologic motions (joint play) of the humerus on the glenoid.

Manual muscle test (gross)

Bilateral upper extremity:	4/5 (throughout available active ROM)

Review of Systems: Dyspnea, fatigue, sweats with pain; when grouped together, these three symptoms fall under the Cardiovascular category; these do not occur at the same time as the shoulder pain.

- **How can you differentiate between medical pathology and mechanical dysfunction as the cause of this client's shoulder pain?**
- **Is a medical referral advised?**

1. Complete special tests for shoulder impingement, tendonitis, and capsulitis as demonstrated.
2. Assess for trigger points; eliminate trigger points and reassess symptoms.
3. Carry out a Review of Systems to identify clusters of systemic signs and symptoms. In this case, a small cluster of cardiovascular symptoms were identified.
4. Correlate symptoms from Review of Systems with shoulder pain (i.e., Do the associated signs and symptoms reported occur along with the shoulder pain or do these two sets of symptoms occur separately from each other?).
5. Assess the effect of using just the lower extremities on shoulder pain; this was difficult to assess given this client's status as a bilateral amputee without a prosthetic device on the left side.

Result: Test results point to an untreated biceps and supraspinatus tendinitis. This tendinitis combined with adhesive capsulitis most likely accounted for the left shoulder pain. This assessment was based on the decreased left glenohumeral AROM and decreased joint mobility.

With objective clinical findings to support a musculoskeletal dysfunction, medical referral was not required. There were no indications that the shoulder pain was a signal of a change in the client's anginal pattern.

Left shoulder impairments were limiting factors in his mobility and rehabilitation process. Shoulder intervention to alleviate pain and to improve upper extremity strength were included in the plan of care. The desired outcome was to improve transfer and gait activities.

Left shoulder pain resolved within the first week of physical therapy intervention. This gain made it possible to improve ambulation from 3 feet to 50 feet with a walker while wearing a right lower extremity prosthesis.

The client gained independence with bed mobility and supine-to-sit transfers. The client continued to make improvements in ambulation, range of motion, and functional mobility.

Physical therapy intervention for the shoulder impairments had a significant impact on the outcomes of this client's rehab program. By differentiating and treating the shoulder movement dysfunction, the intervention enabled the client to progress faster in the transfer and gait training program than he would have had his left shoulder pain been attributed to angina.[28]

Data from Smith ML: Differentiating angina and shoulder pathology pain, *Phys Ther Case Rep* 1(4):210–212, 1998.

CLINICAL SIGNS AND SYMPTOMS

Complex Regional Pain Syndrome (Type I)

Stage I (acute, lasting several weeks)
- Pain described as burning, aching, throbbing
- Sensitivity to touch
- Swelling
- Muscle spasm
- Stiffness, loss of motion and function
- Skin changes (warm, red, dry skin changes to cold [cyanotic], sweaty skin)
- Accelerated hair growth (usually dark hair in patches)

Stage II (subacute, lasting 3 to 6 months)
- Severity of pain increases
- Swelling may spread; tissue goes from soft to boggy to firm
- Muscle atrophy
- Skin becomes cool, pale, bluish, sweaty
- Nail bed changes (cracked, grooved, ridges)
- Bone demineralization (early onset of osteoporosis)

Stage III (chronic, lasting more than 6 months)
- Pain may stay same, improve, or get worse; variable
- Irreversible tissue damage
- Muscle atrophy and contractures
- Skin becomes thin and shiny
- Nails are brittle
- Osteoporosis

Thoracic Outlet Syndrome

Compression of the neurovascular bundle consisting of the brachial plexus and subclavian artery and vein (see Fig. 17-10) can cause a variety of symptoms affecting the arm, hand, shoulder girdle, neck, and chest (Case Example 18-8). Risk factors and clinical presentation are discussed more completely in Chapter 17.

Bacterial Endocarditis

The most common musculoskeletal symptom in clients with bacterial endocarditis is arthralgia, generally in the proximal joints. The shoulder is affected most often, followed (in declining incidence) by the knee, hip, wrist, ankle, metatarsophalangeal and metacarpophalangeal joints, and by acromioclavicular involvement.

Most clients with endocarditis-related arthralgias have only one or two painful joints, although some may have pain in several joints. Painful symptoms begin suddenly in one or two joints, accompanied by warmth, tenderness, and redness. One helpful clue: As a rule, morning stiffness is not as prevalent in clients with endocarditis as in those with rheumatoid arthritis or polymyalgia rheumatica.

Pericarditis

The inflammatory process accompanying pericarditis may result in an accumulation of fluid in the pericardial sac, preventing the heart from expanding fully. The subsequent chest pain of pericarditis (see Fig. 6-10) closely mimics that of a MI because it is substernal, is associated with cough, and may radiate to the shoulder.[35] It can be differentiated from MI by the pattern of relieving and aggravating factors.

For example, pericarditis pain is sharp and relieved by leaning over when seated. If there is irritation of the diaphragm, it can cause shoulder pain. The pain of MI is unaffected by position, breathing, or movement, whereas the chest and shoulder pain associated with pericarditis may be relieved by kneeling with hands on the floor, leaning forward, or sitting upright. Pericardial pain is often made worse by deep breathing, swallowing, or belching.

Aortic Aneurysm

Aortic aneurysm appears as sudden, severe chest pain with a tearing sensation (see Fig. 6-11), and the pain may extend to the neck, shoulders, lower back, or abdomen but rarely to the joints and arms, which distinguishes it from MI.

Isolated shoulder pain is not associated with aortic aneurysm; shoulder pain (usually left shoulder) occurs when the primary pain pattern radiates up and over the trapezius and upper arm(s) (see Fig. 6-11).[36] The client may report a bounding or throbbing pulse (heartbeat) in the abdomen. Risk factors and other associated signs and symptoms help distinguish this condition.

Deep Venous Thrombosis of the Upper Extremity

DVT of the upper extremity is not as common as in the lower extremity but incidence may be on the rise due to the increasing use of peripherally inserted central catheters (PICC lines) or central venous catheters (CVC).[37,38] Thrombosis affects the subclavian vein, axillary vein, or both most often with less common sites being the internal jugular and brachial veins.[39]

CVCs are frequently used in people with hematologic/oncologic disorders in order to administer drugs, stem cell infusions, blood products, parenteral alimentation, and blood sampling. Other risk factors include blood clotting disorders,[40] clavicle fracture,[41] insertion of pacemaker wires, and arthroscopy of the shoulder or reconstructive shoulder arthroplasty.[42,43] Thrombosis is the second-leading cause of death in cancer patients, and cancer is a major risk factor of venous thromboembolism (VTE), due to activation of coagulation, use of long-term CVC, and the thrombogenic effects of chemotherapy and anti-angiogenic drugs.[44]

Symptoms (when present) are similar as for the lower extremity (see discussion in Chapter 6). The therapist should be aware of the presence of any risk factors and watch for pain and pitting edema or swelling of the entire (usually upper) limb and/or an area of the limb that is 2 cm or more larger than the surrounding area indicating swelling requiring further investigation.

Other symptoms include redness or warmth of the arm, dilated veins, or low-grade fever possibly accompanied by chills and malaise. Bruising or discoloration of the area or

CASE EXAMPLE 18-8

House Painter

Referral: A 44-year-old female referred herself to physical therapy for a 2-month-long history of right upper trapezius and right shoulder pain. She works as a house painter and thinks the symptoms came on after a difficult job with high ceilings.

She reports new symptoms of dizziness when getting up too fast from bed or from a chair. She is seeing a chiropractor and a naturopathic physician for a previous back injury 2 years ago when she fell off a ladder.

She wants to try physical therapy since she has reached a "plateau" with her chiropractic care.

Past Medical History: Other significant past medical history includes a total hysterectomy 4 years ago for unexplained heavy menstrual bleeding. She does not smoke or use tobacco products but admits smoking marijuana occasionally and being a "social drinker" (wine coolers and beer on the weekends or at barbeques).

She is nulliparous (never pregnant). She is not on any medications except ibuprofen as needed for headaches. She takes a variety of nutritional supplements given to her by the naturopath. No recent history of infections or illness.

Clinical Presentation: There is no numbness or tingling anywhere in her body. No changes in vision, balance, or hearing. The client reports normal bowel and bladder function. Neurologic screen was within normal limits (WNL).

Postural screen:	Moderate forward head position, rounded shoulders, arms held in a position of shoulder internal rotation, minimal lumbar lordosis
Temporomandibular joint (TMJ) screen:	Negative
Vertebral artery tests:	Negative
Upper extremity (UE) range of motion (ROM):	Limited right shoulder internal rotation; all other motions in both UEs were full and pain free
Spurling's test:	Negative
Cervical spine mobility test:	Restriction of the left C4-5; no apparent cervical instabilities; tenderness along the entire right cervical spine with mild hypertonus
Trigger points (TrPs):	Positive for right sternocleidomastoid, right upper trapezius, and right levator scapula TrPs

Are there any red flags to suggest the need to screen for medical disease? What other tests (if any) would you like to do before making this decision?

- Age
- Unexplained dizziness
- Failure to progress with chiropractic care
- Surgical menopause and nulliparity (both increase her risk for breast cancer; early menopause puts her at risk for osteoporosis and accelerated atherosclerosis/heart disease)

Assessment: It is likely the client's symptoms are directly related to postural overuse. Long hours with her arms overhead may be contributing factors. A more complete exam for thoracic outlet syndrome (TOS) is warranted. Physical therapy intervention can be initiated, but must be reevaluated on an on-going basis. Eliminating the TrPs, improving her posture, and restoring full shoulder and neck motion will aid in the differential diagnosis.

The therapist should assess vital signs, including blood pressure measurements in both arms (looking for a vascular component of TOS) and from supine to sit to stand to assess for postural orthostatic hypotension. True postural hypotension must be accompanied by both blood pressure and pulse rate changes.

Depending on the results, medical evaluation may be warranted, especially if no underlying cause can be found for the dizziness. Although there is no reported visual change or loss of balance with the dizziness, a vestibular screening examination is warranted.

Given her age and risk factors, she should be asked when her last physical exam was done. If she has not been seen since her hysterectomy or within the last 12 months, she should be advised to see her personal physician for follow-up.

She should be encouraged to exercise on a regular basis (more education can be provided depending on her level of knowledge and the therapist's level of expertise in this area).

If baseline bone density studies have not been done, then she should pursue this now. Likewise, she should ask her doctor about baseline testing for thyroid, glucose, and lipid values if these are not already available.

In a primary care practice, risk factor assessment is a key factor in knowing when to carry out a screening evaluation. Patient education about personal health choices is also essential.

In any practice, we must know what impact medical conditions can have on the neuromuscular and musculoskeletal systems and watch for any links between the visceral and the somatic systems.

proximal to the thrombosis has been observed in some cases.[45] Swelling can contribute to decreased neck or shoulder motion. Severe thromboses can cause superior vena cava syndrome; symptoms include edema of the face and arm, vertigo, and dyspnea.[46]

Unfortunately, the first clinical manifestation of *deep* thrombosis may be pulmonary embolism (PE; see also Box 6-2 for overall risk factors for DVT and PE); superficial venous thrombosis is usually self-limiting and does not cause PE since the blood flow to deeper veins is through small perforating venous channels.[47]

PE as a consequence of upper extremity DVT can be fatal.[43] Chronic venous insufficiency or postthrombotic syndrome are possible sequelae to upper extremity DVT.[45,48]

To our knowledge, at this time, a validated screening tool, such as the Wells' Clinical Decision Rule for DVT, has not been investigated for the upper extremity. A simple model to predict upper extremity DVT has also been proposed and remains under investigation (Table 18-3).[49] The best available test for the diagnosis of upper extremity DVT is contrast venography; color Doppler ultrasonography may be preferred for some people because it is noninvasive.[50]

TABLE 18-3 Possible Predictors of Upper Extremity Deep Venous Thrombosis*

Independent Variable	Absent	Present
Venous material (catheter or access device in subclavian or jugular vein; pacemaker)	0	1.0
Localized pain	0	1.0
Unilateral pitting edema	0	1.0
Other diagnosis at least as plausible (negative association)	0	−1.0

*Concepts presented here are based on one preliminary study validated in a second sample but with a limited patient population; diagnosis was confirmed with ultrasound study.[49]

NOTE: As with lower extremity deep venous thrombosis (DVT), a low clinical probability does not exclude the diagnosis of upper extremity DVT. The scoring provides a tool to use in determining the need for additional testing (e.g., ultrasonography, venography).

Key: Total score of:

−1.0 or 0: Low probability of upper extremity DVT

1: Intermediate probability

2-3: High probability

CLINICAL SIGNS AND SYMPTOMS

Upper Extremity Deep Venous Thrombosis

- Numbness or heaviness of the extremity
- Itching, burning, coldness of the extremity
- Swelling, discoloration, warmth, or redness of the extremity; pitting edema
- Limited range of motion (ROM) of neck, shoulder
- Low-grade fever, chills, malaise
- For individuals with a PICC line (in addition to any of the signs and symptoms listed above):
  - Pain or tenderness at or above the insertion site

SCREENING FOR RENAL CAUSES OF UPPER QUADRANT/SHOULDER PAIN

The anatomic position of the kidneys (and ureters) is in front of and on both sides of the vertebral column at the level of T11 to L3. The right kidney is usually lower than the left.[51] The lower portions of the kidneys and the ureters extend below the ribs and are separated from the abdominal cavity by the peritoneal membrane. Because of its location in the posterior upper abdominal cavity in the retroperitoneal space, touching the diaphragm, the upper urinary tract can refer pain to the (ipsilateral) shoulder on the same side as the involved kidney.

Renal sensory innervation is not completely understood; the capsule (covering of the kidney) and the lower portions of the collecting system seem to cause pain with stretching (distention) or puncture. Information transmitted by renal and ureteral pain receptors is relayed by sympathetic nerves that enter the spinal cord at T10 to L1; therefore renal and ureteral pain is typically felt in the posterior subcostal and costovertebral regions (flank).[52-54]

Renal pain is aching and dull in nature but can occasionally be a severe, boring type of pain. The distention or stretching of the renal capsule, pelvis, or collecting system from intrarenal fluid accumulation (e.g., inflammatory edema, inflamed or bleeding cysts, and bleeding or neoplastic growths) accounts for the constant, dull, and aching quality of reported pain. Ischemia of renal tissue caused by blockage of blood flow to the kidneys can produce either a *constant dull* or *sharp* pain. True renal pain is seldom affected by change in position or movements of the shoulder or spine.

If the diaphragm becomes irritated because of pressure from a renal lesion, ipsilateral shoulder pain can be the only symptom or may occur in conjunction with other pain and associated signs and symptoms. For example, generalized abdominal pain may develop accompanied by nausea, vomiting, and impaired intestinal motility (progressing to intestinal paralysis) when pain is acute and severe. Nerve fibers from the renal plexus are also in direct communication with the spermatic plexus, and because of this close relationship, testicular pain may also accompany renal pain in males.[55]

Elevation in temperature or changes in color, odor, or amount of urine (flow, frequency, nocturia) presenting with shoulder pain should be reported to a physician. Shoulder pain that is not affected by movement or provocation tests requires a closer look.

The presence of constitutional symptoms, constant pain (even if dull), and failure to change the symptoms with a position change will also alert the therapist to the need for a more thorough screening examination. A past medical history of cancer is always an important risk factor requiring careful assessment. This is true even when patients/clients have a known or traumatic cause for their symptoms.

Flank pain combined with unexplained weight loss, fever, pain, and hematuria should be reported to the physician. The presence of any amount of blood in the urine always requires referral to a physician for further diagnostic evaluation because this is a primary symptom of urinary tract neoplasm.

Additionally, therapists need to be cognizant that those at high risk for chronic renal disease with associated neuropathies include anyone with diabetes and those with history of significant nonsteroidal antiinflammatory drug (NSAID) or acetaminophen use.[56]

SCREENING FOR GASTROINTESTINAL CAUSES OF SHOULDER PAIN

Upper abdominal or GI problems with diaphragmatic irritation can refer pain to the ipsilateral shoulder. Perforated gastric or duodenal ulcers, gallbladder disease, and hiatal hernia are the most likely GI causes of shoulder pain seen in the physical therapy clinic. Usually there are associated signs and symptoms, such as nausea, vomiting, anorexia, melena, or early satiety, but the client may not connect the shoulder pain with GI disorders. A few screening questions may be all that is needed to uncover any coincident GI symptoms.

The therapist should look for a history of previous ulcer, especially in association with the use of NSAIDs. Shoulder pain that is worse 2 to 4 hours after taking the NSAID can be suggestive of GI bleeding and is considered a yellow (caution) flag. With a true musculoskeletal problem, peak NSAID dosage (usually 2 to 4 hours after ingestion; variable with each drug) should reduce or alleviate painful shoulder symptoms. Any pain increase instead of decrease may be a symptom of GI bleeding.

The therapist must also ask about the effect of eating on shoulder pain. If eating makes shoulder pain better or worse (anywhere from 30 minutes to 2 hours after eating), there may be a GI problem. The client may not be aware of the link between these two events until the therapist asks. If the client is not sure, follow-up at a future appointment and ask again if the client has noticed any unusual symptoms or connection between eating and shoulder pain.

SCREENING FOR LIVER AND BILIARY CAUSES OF SHOULDER/UPPER QUADRANT SYMPTOMS

As with many of the organ systems in the human body, the hepatic and biliary organs (liver, gallbladder, and common bile duct) can develop diseases that mimic primary musculoskeletal lesions.

The musculoskeletal symptoms associated with hepatic and biliary pathologic conditions are generally confined to the midback, scapular, and right shoulder regions. These musculoskeletal symptoms can occur alone (as the only presenting symptom) or in combination with other systemic signs and symptoms. Fortunately, in most cases of shoulder pain referred from visceral processes, shoulder motion is not compromised and local tenderness is not a prominent feature.

Diagnostic interviewing is especially helpful when clients have avoided medical treatment for so long that shoulder pain caused by hepatic and biliary diseases may in turn create biomechanical changes in muscular contractions and shoulder movement. These changes eventually create pain of a biomechanical nature.[57]

Referred shoulder pain may be the only presenting symptom of hepatic or biliary disease. Sympathetic fibers from the biliary system are connected through the celiac and splanchnic plexuses to the hepatic fibers in the region of the dorsal spine. These connections account for the intercostal and radiating interscapular pain that accompanies gallbladder disease (see Fig. 9-10). Although the innervation is bilateral, most of the biliary fibers reach the cord through the right splanchnic nerves, producing pain in the right shoulder.

Carpal Tunnel Syndrome

There are many potential causes of carpal tunnel syndrome (CTS), both musculoskeletal and systemic (see Table 11-2). Careful evaluation is required (see Box 9-1). The presence of bilateral carpal tunnel syndrome warrants a closer look. For example, liver dysfunction resulting in increased serum ammonia and urea levels can result in impaired peripheral nerve function.

Ammonia from the intestine (produced by protein breakdown) is normally transformed by the liver to urea, glutamine, and asparagine, which are then excreted by the renal system. When the liver does not detoxify ammonia, ammonia is transported to the brain, where it reacts with glutamate (excitatory neurotransmitter), producing glutamine.

The reduction of brain glutamate impairs neurotransmission, leading to altered central nervous system metabolism and function. Asterixis and numbness/tingling (misinterpreted as carpal tunnel syndrome) can occur as a result of this ammonia abnormality, causing an intrinsic nerve pathologic condition (see Case Example 9-1).

For any client presenting with bilateral carpal tunnel syndrome:

- Ask about the presence of similar symptoms in the feet
- Ask about a personal history of liver or hepatic disease (e.g., cirrhosis, cancer, hepatitis)
- Look for a history of hepatotoxic drugs (see Box 9-3)
- Look for a history of alcoholism
- Ask about current or previous use of statins (cholesterol-lowering drugs such as Crestor, Lipitor, or Zocor)
- Look for other signs and symptoms associated with liver impairment (see Clinical Signs and Symptoms of Liver Disease in Chapter 9)
- Test for signs of liver disease
  - Skin color changes
  - Spider angiomas
  - Palmar erythema (liver palms)
  - Nail bed changes (e.g., white nails of Terry, white bands, clubbing)
  - Asterixis (liver flap)

SCREENING FOR RHEUMATIC CAUSES OF SHOULDER PAIN

A number of systemic rheumatic diseases can appear as shoulder pain, even as unilateral shoulder pain. The HLA-B27–associated spondyloarthropathies (diseases of the joints of the spine), such as ankylosing spondylitis, most frequently involve the SI joints and spine. Involvement of large central joints, such as the hip and shoulder, is common, however.

Rheumatoid arthritis (RA) and its variants likewise frequently involve the shoulder girdle. These systemic rheumatic diseases are suggested by the details of the shoulder examination, by coincident systemic complaints of malaise and easy fatigability, and by complaints of discomfort in other joints either coincidental with the presenting shoulder complaint or in the past.

Other systemic rheumatic diseases with major shoulder involvement include polymyalgia rheumatica and polymyositis (inflammatory disease of the muscles). Both may be somewhat asymmetric but almost always appear with bilateral involvement and impressive systemic symptoms.

SCREENING FOR INFECTIOUS CAUSES OF SHOULDER PAIN

The most likely infectious causes of shoulder pain in a physical therapy practice include infectious (septic) arthritis (see discussion in Chapter 3 and also Box 3-6), osteomyelitis, and infectious mononucleosis (mono). Immunosuppression for any reason puts people of all ages at risk for infection (Case Example 18-9).

Septic arthritis of the acromioclavicular joint (ACJ) or hand can present as insidious onset of shoulder pain. Likewise, septic arthritis of the sternoclavicular joint (SCJ) can present as chest pain. Usually, there is local tenderness at the affected joint. A possible history of intravenous drug use,

CASE EXAMPLE 18-9

Osteomyelitis

Referral: SC, an active 62-year-old cardiac nurse, was referred by her orthopedic surgeon for "PT [for] possible rotator cuff tear (RCT), 3 times a week for 4 weeks." SC reported an "open" magnetic resonance imaging (MRI) was negative for RCT and plain films were also negative. She noted that laboratory testing was not done.

Past Medical History

Medications: Current medications included Motrin 800 mg tid for pain; Decadron 0.75 mg qid for atypical dermatitis and asthma (45-year use of corticosteroids); Avapro 75 mg qid to control hypertension; HydroDIURIL 25 mg qid to counteract fluid retention from corticosteroids; and Chlor-Trimeton 12 mg qid to suppress the high level of blood histamine resulting from the long-term comorbid condition of atypical dermatitis and asthma.

Social History: The client consumes one glass of wine per day, quit smoking 20 years ago, and has never done illicit drugs.

Clinical Presentation

Pain Pattern: The client presented with primary complaints of severe and limiting pain of nearly four weeks duration with any active movement at her left shoulder and at rest. Her pain was rated on the visual analogue scale (VAS) as 7/10 at rest and 9/10 to 10/10 with motion at glenohumeral (GH) joint. Pain onset was gradual over a 3-day period; she was not aware of injury or trauma.

She reported an inability to (1) use her left upper extremity (UE); (2) lie on or bear weight on left side; (3) perform activities of daily living (ADLs); (4) sleep uninterrupted due to pain, awakening 4 or 5 times nightly; or (5) participate in regular weekly Yoga classes.

Vital Signs: Temperature: 37° C (98.6° F.); blood pressure: 120/98 mm Hg. SC reported that her medication combination of Decadron and Chlor-Trimeton had been implicated in the past by her physician as acting to suppress low-grade fevers.

Observation: Slight puffiness, minimal swelling, observed in the left supraclavicular area. SC holds left UE at her side with the elbow flexed to 90 degrees and the shoulder held in internal rotation.

Standing posture: Forward head position with increased cervical spine lordosis and thoracic spine kyphosis, with an inability to attain neutral or reverse either spinal curve.

Palpation revealed exquisite tenderness at distal clavicle and both anterior and posterior aspects of proximal humerus.

Cervical spine screen: Spurling's compression, distraction, and Cervical Quadrant testing were all negative; deep tendon reflexes (DTRs) at C5, C6, and C7 were symmetrically increased bilaterally; dermatomal testing was within normal limits (WNL); myotomes could not be reliably tested due to pain.

Special tests at the shoulder could not be performed or were unreliable due to pain limitation.

Range of motion (ROM): Left GH joint active ROM (AROM) and passive ROM (PROM) were severely limited. AROM: Unable to actively perform flexion or abduction at left shoulder. PROM left shoulder (measured in supine with arm at side and elbow flexed to 90 degrees):

Flexion:	35 degrees
Abduction:	35 degrees
Internal rotation:	50 degrees
External rotation:	−10 degrees

All ranges were pain limited with an "empty" end feel.

Evaluation/Assessment: SC's signs, symptoms, and examination findings were consistent with those of a severe, full-thickness RCT, including severity of pain and functional loss with empty end feel at GH joint ROM. However, the inability to perform special tests limited the certainty of the RCT diagnosis.

Red flags included age over 50, severe loss of motion with empty end feel, constancy and severity of pain, inability to relieve pain or obtain a comfortable position, bony tenderness, and insidious onset of the condition. Additional risk factors included long-term use of corticosteroids to treat atypical dermatitis with asthma.

Based on the objective examination findings, including swelling, bone tenderness, along with the severity and unrelenting nature of her pain, the presence of a more serious underlying systemic medical condition was considered (in addition to a possible unconfirmed RCT).

Associated Signs and Symptoms: SC denied a fever, chills, night sweats, pain in other joints or bones, weight loss, abdominal pain, nausea or vomiting.

Outcomes: The client made very little progress after the prescribed physical therapy intervention. The severity of pain and functional loss remained unchanged. Numerous attempts were made by the client and the therapist to discuss this case with the referring physician. The client eventually referred herself to a second physician.

Result: The client was diagnosed with osteomyelitis as a result of a repeat MRI and a triple-phase bone scan, and laboratory test results of elevated levels of erythrocyte sedimentation rate (ESR) and C-reactive protein (CRP) values. A surgical biopsy confirmed the diagnosis. She underwent three different surgical procedures culminating in a total shoulder arthroplasty (TSA) along with repair of the full-thickness RCT.

From West PR: Case report presented in fulfillment of DPT 910, Institute for Physical Therapy Education, Widener University, Chester, PA, 2005. Used with permission.

diabetes, trauma (puncture wound, surgery, human or animal bite), and infection is usually present. Punching someone in the mouth (hand coming in contact with teeth resulting in a puncture wound) has been reported as a potential cause of septic arthritis. With infection of this type, there may or may not be constitutional symptoms.[58,59]

Osteomyelitis (bone or bone marrow infection) is caused most commonly by *Staphylococcus aureus.* Children under 6 months of age are most likely to be affected by *Haemophilus influenzae* or *Streptococcus.* Hematogenous spread from a wound, abscess, or systemic infection (e.g., fracture, tuberculosis, urinary tract infection, upper respiratory infection, finger felons) occurs most often. Osteomyelitis of the spine is associated with injection drug use.

Onset of clinical signs and symptoms is usually gradual in adults but may be more sudden in children with high fever, chills, and inability to bear weight through the affected joint. In all ages there is marked tenderness over the site of the infection when the affected bone is superficial (e.g., spinous process, distal femur, proximal tibia). The most reliable way to recognize infection is the presence of both local and systemic symptoms.

Mononucleosis is a viral infection that affects the respiratory tract, liver, and spleen. Splenomegaly with subsequent rupture is a rare but serious cause of left shoulder pain (Kehr's sign).[60] There is usually left upper abdominal pain and, in many cases, trauma to the enlarged spleen (e.g., sports injury) is the precipitating cause in an athlete with an unknown or undiagnosed case of mono. Palpation of the upper left abdomen may reveal an enlarged and tender spleen (see Fig. 4-53).

The virus can be present 4 to 10 weeks before any symptoms develop so the person may not know mono is present. Acute symptoms can include sore throat, headache, fatigue, lymphadenopathy, fever, myalgias, and sometimes, skin rash. Enlarged tonsils can cause noisy or difficult breathing. When asking about the presence of other associated signs and symptoms (current or recent past), the therapist may hear a report of some or all of these signs and symptoms.

SCREENING FOR ONCOLOGIC CAUSES OF SHOULDER PAIN

A past medical history of cancer anywhere in the body with new onset of back or shoulder pain (or impairment) is a red-flag finding. Brachial plexus radiculopathy can occur in either or both arms with cancer metastasized to the lymphatics (Case Example 18-10).

Questions about visceral function are relevant when the pattern for malignant invasion at the shoulder emerges. Invasion of the upper humerus and glenoid area by secondary malignant deposits affects the joint and the adjacent muscles (Case Example 18-11).

Muscle wasting is greater than expected with arthritis and follows a bizarre pattern that does not conform to any one neurologic lesion or any one muscle. Localized warmth felt at any part of the scapular area may prove to be the first sign

CASE EXAMPLE 18-10

Upper Extremity Radiculopathy

Referral: A 72-year-old woman was referred to physical therapy by her neurologist with a diagnosis of "nerve entrapment" for a postural exercise program and home traction. She was experiencing symptoms of left shoulder pain with numbness and tingling in the ulnar nerve distribution. She had a moderate forward head posture with slumped shoulders and loss of height from known osteoporosis.

Past Medical History: The woman's past medical history was significant for right breast cancer treated with a radical mastectomy and chemotherapy 20 years ago. She had a second cancer (uterine) 10 years ago that was considered separate from her previous breast cancer.

Clinical Presentation: The physical therapy examination was consistent with the physician's diagnosis of nerve entrapment in a classic presentation. There were significant postural components to account for the development of symptoms. However, the therapist palpated several large masses in the axillary and supraclavicular fossa on both the right and left sides. There was no local warmth, redness, or tenderness associated with these lesions. The therapist requested permission to palpate the client's groin and popliteal spaces for any other suspicious lymph nodes. The rest of the examination findings were within normal limits.

Associated Signs and Symptoms: Further questioning about the presence of associated signs and symptoms revealed a significant disturbance in sleep pattern over the last 6 months with unrelenting shoulder and neck pain. There were no other reported constitutional symptoms, skin changes, or noted lumps anywhere. Vital signs were unremarkable at the time of the physical therapy evaluation.

Result: Returning this client to her referring physician was a difficult decision to make since the therapist did not have the benefit of the medical records or results of neurologic examination and testing. Given the significant past medical history for cancer, the woman's age, presence of progressive night pain, and palpable masses, no other reasonable choice remained. When asked if the physician had seen or felt the masses, the client responded with a definite "no."

There are several ways to approach handling a situation like this one, depending on the physical therapist's relationship with the physician. In this case the therapist had never communicated with this physician before. A telephone call was made to ask the clerical staff to check the physician's office notes (the client had provided written permission for disclosure of medical records to the therapist).

It is possible that the physician was aware of the masses, knew from medical testing that there was extensive cancer, and chose to treat the client palliatively. Since there was no indication of such, the therapist notified the physician's staff of the decision to return the client to the physician. A brief (one-page) written report summarizing the findings was given to the client to hand-carry to the physician's office.

Further medical testing was performed, and a medical diagnosis of lymphoma was made.

CASE EXAMPLE 18-11

Shoulder and Leg Pain

Referral: A 33-year-old woman came to a physical therapy clinic located inside a large health club. She reported right shoulder and right lower leg pain that is keeping her from exercising. She could walk but had an antalgic gait secondary to pain on weight bearing.

She linked these symptoms with heavy household chores. She could think of no other trauma or injury. She was screened for the possibility of domestic violence with negative results.

Past Medical History: There was no past history of disease, illness, trauma, or surgery. There were no other symptoms reported (e.g., no fever, nausea, fatigue, bowel or bladder changes, sleep disturbance).

Clinical Presentation: The right shoulder and right leg were visibly and palpably swollen. Any and all (global) motions of either the arm or the leg were painful. The skin was tender to light touch in a wide band of distribution around the painful sites. No redness or skin changes of any kind were noted.

Pain prevented strength testing or assessment of muscle weakness. There was no sign of scoliosis. Trendelenburg test was negative, bilaterally. Functionally, she was able to climb stairs and walk, but these and other activities (e.g., exercising, biking, household chores) were limited by pain.

How do you screen this client for systemic or medical disease?

You may have done as much screening as is possible. Pain is limiting any further testing. Assessing vital signs may provide some helpful information.

She has denied any past medical history to link with these symptoms. Her age may be a red flag in that she is young. Bone pain with these symptoms in a 33-year-old is a red flag for bone pathology and needs to be investigated medically.

Immediate medical referral is advised.

Result: X-rays of the right shoulder showed complete destruction of the right humeral head consistent with a diagnosis of metastatic disease. X-rays of the right leg showed two lytic lesions. There was no sign of fracture or dislocation. Computed tomography (CT) scans showed destructive lytic lesions in the ribs and ilium.

Additional testing was performed, including lab values, bone biopsy, mammography, and pelvic ultrasonography. The client was diagnosed with bone tumors secondary to hyperparathyroidism.

A large adenoma was found and removed from the left inferior parathyroid gland. Medical treatment resulted in decreased pain and increased motion and function over a period of 3 to 4 months. Physical therapy intervention was prescribed for residual muscle weakness.

Data from Insler H: Shoulder and leg pain in a 33-year-old woman, *J Musculoskel Med* 14(6)36–37, 1997.

of a malignant deposit eroding bone. Within 1 or 2 weeks after this observation, a palpable tumor will have appeared, and erosion of bone will be visible on x-ray films.[61]

Primary Bone Neoplasm

Bone cancer occurs chiefly in young people, in whom a causeless limitation of movement of the shoulder leads the physician to order x-rays. If the tumor originates from the shaft of the humerus, the first symptoms may be a feeling of "pins and needles" in the hand, associated with fixation of the biceps and triceps muscles and leading to limitation of movement at the elbow (Case Example 18-12).

Pulmonary (Secondary) Neoplasm

Occasionally, the client requires medical referral because shoulder pain is referred from metastatic lung cancer. When the shoulder is examined, the client is unable to lift the arm beyond the horizontal position. Muscles respond with spasm that limits joint movement.

If the neoplasm interferes with the diaphragm, diaphragmatic pain (C3 to C5) is often felt at the shoulder at each breath (at the fourth cervical dermatome [i.e., at the deltoid area]), in correspondence with the main embryologic derivation of the diaphragm.[62] Pain arising from the part of the pleura that is not in contact with the diaphragm is also brought on by respiration but is felt in the chest.

Although the lung is insensitive, large tumors invading the chest wall set up local pain and cause spasm of the pectoralis major muscle, with consequent shoulder pain and/or limitation of elevation of the arm.[63] If the neoplasm encroaches on the ribs, stretching the muscle attached to the ribs leads to sympathetic spasm of the pectoralis major. By contrast, the scapula is mobile, and a full range of passive movement is present at the shoulder joint.

Pancoast's Tumor

Pancoast's tumors of the lung apex usually do not cause symptoms while confined to the pulmonary parenchyma. Shoulder pain occurs if they extend into the surrounding structures, infiltrating the chest wall into the axilla. Occasionally, brachial plexus involvement (eighth cervical and first thoracic nerve) presents with radiculopathy.[64]

This nerve involvement produces sharp neuritic pain in the axilla, shoulder, and subscapular area on the affected side, with eventual atrophy of the upper extremity muscles. Bone pain is aching, exacerbated at night, and a cause of restlessness and musculoskeletal movement.[65]

Usually, general associated systemic signs and symptoms are present (e.g., sore throat, fever, hoarseness, unexplained weight loss, productive cough with blood in the sputum). These features are not found in any regional musculoskeletal disorder, including such disorders of the shoulder.

CASE EXAMPLE 18-12

Osteosarcoma

Referral: A 14-year-old boy presented to a physical therapist at a sports medicine clinic with a complaint of left shoulder pain that had been present off and on for the last 4 months. There was no reported history of injury or trauma despite active play on the regional soccer team.

Past Medical History: He has seen his pediatrician for this on several occasions. It was diagnosed as "tendinitis" with the suggestion to see a physical therapist of the family's choice. No x-rays or other diagnostic imaging was performed to date. The client could not remember if any laboratory work (blood or urinalysis) had been done.

The client reports that his arm feels "heavy." Movement has become more difficult just in the last week. The only other symptom present was intermittent tingling in the left hand. There is no other pertinent medical history.

Clinical Presentation: Physical examination of the shoulder revealed moderate loss of active motion in shoulder flexion, abduction, and external rotation with an empty end feel and pain during passive range of motion. There was no pain with palpation or isometric resistance of the rotator cuff tendons. Gross strength of the upper extremity was 4/5 for all motions.

There was a palpable firm, soft, but fixed mass along the lateral proximal humerus. The client reported it was "tender" when the therapist applied moderate palpatory pressure. The client was not previously aware of this lump.

Upper extremity pulses, deep tendon reflexes, and sensation were all intact. There were no observed skin changes or palpable temperature changes. Since this was an active athlete with left shoulder pain, screening for Kehr's sign was carried out but was apparently negative.

What are the red flags?

- Age
- Suspicious palpable lesion (likely not present at previous medical evaluation)
- Lack of medical diagnostics
- Unusual clinical presentation for tendinitis with loss of motion and empty end feel but intact rotator cuff

Result: The therapist telephoned the physician's office to report possible changes since the physician's last examination. The family was advised by the doctor's office staff to bring him to the clinic as a walk-in the same day. X-rays showed an irregular bony mass of the humeral head and surrounding soft tissues. The biopsy confirmed a diagnosis of osteogenic sarcoma. The cancer had already metastasized to the lungs and liver.

For example, a similar pain pattern caused by trigger points (TrPs) of the serratus anterior can be differentiated from neoplasm by the lack of true neurologic findings (indicating trigger point) or by lack of improvement after treatment to eliminate the trigger point (indicating neoplasm).

Breast Cancer

Breast cancer or breast cancer recurrence is always a consideration with upper quadrant pain or shoulder dysfunction (Case Example 18-13). The therapist must know what to look for as red flags for cancer recurrence versus delayed effects of cancer treatment. See Chapter 13 for a complete discussion of cancer screening and prevention. Breast cancer is discussed in Chapter 17.

CASE EXAMPLE 18-13

Breast Cancer

Referral: A 53-year-old woman with severe adhesive capsulitis was referred to a physical therapist by an orthopedic surgeon. A physical therapy program was initiated. When the client's shoulder flexion and abduction allowed for sufficient movement to place the client's hand under her head in the supine position, ultrasound to the area of capsular redundancy before joint mobilization was added to the treatment protocol.

During the treatment procedure, the client was dressed in a hospital gown wrapped under the axilla on the involved side. With the client in the supine position, the upper outer quadrant of breast tissue was visible and the physical therapist observed skin puckering (peau d'orange) accompanied by a reddened area.

Result: It is always necessary to approach situations like this one carefully to avoid embarrassing or alarming the client. In this case the therapist casually observed, "I noticed when we raised your arm up for the ultrasound that there is an area of your skin here that puckers a little. Have you noticed any changes in your armpit, chest, or breast areas?"

Depending on the client's response, follow-up questions should include asking about distended veins, discharge from the nipple, itching of the skin or nipple, and the approximate time of the client's last breast examination (self-examination and physician examination). Although not all therapists are trained to perform a clinical breast exam (CBE), palpation of lymph nodes and muscles such as the pectoral muscle groups can be performed.

There was no previous history of cancer, and further palpation did not elicit any other suspicious findings. The physical therapist recommended a physician evaluation, and a diagnosis of breast cancer was made.

Magnetic resonance imaging (MRI) studies have shown radiation-induced muscle morbidity in cervical, prostate, and breast cancer. Axillary radiation is a predictive factor for the development of shoulder morbidity.[66] Soft tissue changes from radiotherapy is dose dependent and may develop immediately or develop several years later.

Primary muscle shortening and secondary loss of muscle activity may produce movement disorders of the shoulder and/or upper quadrant. Radiation-induced changes in vascular networks resulting in ischemia may affect muscle contractility.[67]

SCREENING FOR GYNECOLOGIC CAUSES OF SHOULDER PAIN

Shoulder pain as a result of gynecologic conditions is uncommon, but still very possible. Occasionally a client may present with breast pain as the primary complaint, but most often the description is of shoulder or arm, neck, or upper back pain.

When asked if the client has any symptoms anywhere else in the body, breast pain may be mentioned.

Pain patterns associated with breast disease along with a discussion of various breast pathologies are included in Chapter 17. Many of the breast conditions discussed (e.g., tumors, infections, myalgias, implants, lymph disease, trauma) can refer pain to the shoulder either alone or in conjunction with chest and/or breast pain. Shoulder pain or dysfunction in the presence of any of these conditions as part of the client's current or past medical history raises a red flag.

Ectopic Pregnancy

The therapist must be aware of one other gynecologic condition commonly associated with shoulder pain: ectopic (extrauterine [i.e., outside the uterus]) pregnancy. This type of pregnancy occurs when the fertilized egg implants in some other part of the body besides inside the uterus. It may be inside the fallopian tube, inside the ovary, outside the uterus or even within the lining of the peritoneum (see Fig. 15-5).[25-27]

If the condition goes undetected, the embryo grows too large for the confined space. A tear or rupture of the tissue around the fertilized egg will occur. An ectopic pregnancy is not a viable pregnancy and cannot result in a live birth. This condition is life threatening and requires immediate medical referral.

The most common symptom of ectopic pregnancy is a sudden, sharp or constant one-sided pain in the lower abdomen or pelvis lasting more than a few hours. The pain may be accompanied by irregular bleeding or spotting after a light or late menstrual period.

Shoulder pain does not usually occur alone without preceding or accompanying abdominal pain, but shoulder pain can be the only presenting symptom with an ectopic pregnancy. When these two symptoms occur together (either alternating or simultaneously), the woman may not realize the abdominal and shoulder pain are connected. She may think there are two separate problems. She may not see the need to tell the therapist about the pelvic or abdominal pain, especially if she thinks it is menstrual cramps or gas. In addition, ask about the presence of lightheadedness, dizziness, or fainting.

The most likely candidate for an ectopic pregnancy is a woman in the childbearing years who is sexually active. Pregnancy can occur when using any form of birth control, so do not be swayed into thinking the woman cannot be pregnant because she is on the pill or some other form of contraception. Factors that put a woman at increased risk for an ectopic pregnancy include:

- History of endometriosis[68]
- Pelvic inflammatory disease (PID)
- Previous ectopic pregnancy
- Ruptured ovarian cysts or ruptured appendix
- Tubal surgery

Many of these conditions can also cause pelvic pain and are discussed in greater detail in Chapter 15. If the therapist suspects a gynecologic basis for the client's symptoms, some additional questions about past history, missed menses, shoulder pain, and spotting or bleeding may be helpful.

PHYSICIAN REFERRAL

Here in the last chapter of the text there are no new guidelines for physician referral that have not been discussed in the previous chapters. The therapist must remain alert to yellow (caution) or red (warning) flags in the history and clinical presentation, and ask about associated signs and symptoms.

When symptoms seem out of proportion to the injury or persist beyond the expected time of healing, medical referral may be needed.[69] Likewise, pain that is unrelieved by rest or change in position or pain/symptoms that do not fit the expected mechanical or NMS pattern should serve as red-flag warnings. A past medical history of cancer in the presence of any of these clinical presentation scenarios may warrant consultation with the client's physician.

Guidelines for Immediate Medical Attention

- Presence of suspicious or aberrant lymph nodes, especially hard, fixed nodes in a client with a previous history of cancer
- Clinical presentation and history suggestive of an ectopic pregnancy
- Trauma followed by failure of symptoms to resolve with treatment; pain out of proportion to the injury (**Fracture, acute compartment syndrome**)

Clues to Screening Shoulder/Upper Extremity Pain

- See also Clues to Screening Chest, Breast, or Rib Pain in Chapter 17
- Simultaneous or alternating pain in other joints, especially in the presence of associated signs and symptoms such as easy fatigue, malaise, fever
- Urologic signs and symptoms
- Presence of hepatic symptoms, especially when accompanied by risk factors for jaundice
- Lack of improvement after treatment, including trigger point therapy
- Shoulder pain in a woman of childbearing age of unknown cause associated with missed menses (**Rupture of ectopic pregnancy**)
- Left shoulder pain within 24 hours of abdominal surgery, injury, or trauma (**Kehr's sign, ruptured spleen**)

Past Medical History

- History of rheumatic disease
- History of diabetes mellitus (**Adhesive capsulitis**)
- "Frozen" shoulder of unknown cause in anyone with coronary artery disease, recent history of hospitalization in CCU or **ICU/s/p CABG**
- Recent history (past 1-3 months) of MI (**CRPS;** formerly RSD)

- History of cancer, especially breast or lung cancer (**Metastasis**)
- Recent history of pneumonia, recurrent upper respiratory infection, or influenza (**Diaphragmatic pleurisy**)
- History of endometriosis

Cancer

- Pectoralis major muscle spasm with no known cause; limited active shoulder flexion but with full passive shoulder motions and mobile scapula (**Neoplasm**)
- Presence of localized warmth felt over the scapular area (**Neoplasm**)
- Marked limitation of movement at the shoulder joint
- Severe muscular weakness and pain with resisted movements

Cardiac

- Exacerbation by exertion unrelated to shoulder movement (e.g., using only the lower extremities to climb stairs or ride a stationary bicycle)
- Excessive, unexplained coincident diaphoresis
- Shoulder pain relieved by leaning forward, kneeling with hands on the floor, sitting upright (**Pericarditis**)
- Shoulder pain accompanied by dyspnea, toothache, belching, nausea, or pressure behind the sternum (**Angina**)
- Shoulder pain relieved by nitroglycerin (men) or antacids/acid-relieving drugs (women) (**Angina**)
- Difference of 10 mm Hg or more in blood pressure in the affected arm compared to the uninvolved or a symptomatic arm (**Dissecting aortic aneurysm, vascular component of TOS**)

Pulmonary

- Presence of a pleuritic component such as a persistent, dry, hacking, or productive cough; blood-tinged sputum; chest pain; musculoskeletal symptoms are aggravated by respiratory movements
- Exacerbation by recumbency despite proper positioning of the arm in neutral alignment (**Diaphragmatic or pulmonary component**)
- Presence of associated signs and symptoms (e.g., tachypnea, dyspnea, wheezing, hyperventilation)
- Shoulder pain of unknown cause in older adults with accompanying signs of confusion or increased confusion (**Pneumonia**)
- Shoulder pain aggravated by the supine position may be an indication of mediastinal or pleural involvement. Shoulder or back pain alleviated by lying on the painful side may indicate autosplinting. (**Pleural**)

Renal

- Shoulder pain accompanied by elevation in temperature or changes in color, odor, or amount of urine (flow, frequency, nocturia); pain is not affected by movement or provocation tests.
- Shoulder pain accompanied by or alternating with flank pain, abdominal pain, or pelvic pain or, in men, testicular pain

Gastrointestinal

- Coincident nausea, vomiting, dysphagia; presence of other GI complaints such as anorexia, early satiety, epigastric pain or discomfort and fullness, melena
- Shoulder pain relieved by belching or antacids and made worse by eating
- History of previous ulcer, especially in association with the use of NSAIDs

Gynecologic

- Shoulder pain preceded or accompanied by one-sided lower abdominal or pelvic pain in a sexually active woman of reproductive age may be a symptom of **ectopic pregnancy;** there may be irregular bleeding or spotting after a light or late menstrual period.
- Shoulder pain with reports of lightheadedness, dizziness, or fainting in a sexually active woman of reproductive age (**Ectopic pregnancy**)
- Presence of endometrial cysts and/or scar tissue impinging diaphragm, nerve plexus, or the shoulder itself

REFERRED SHOULDER AND UPPER EXTREMITY PAIN PATTERNS

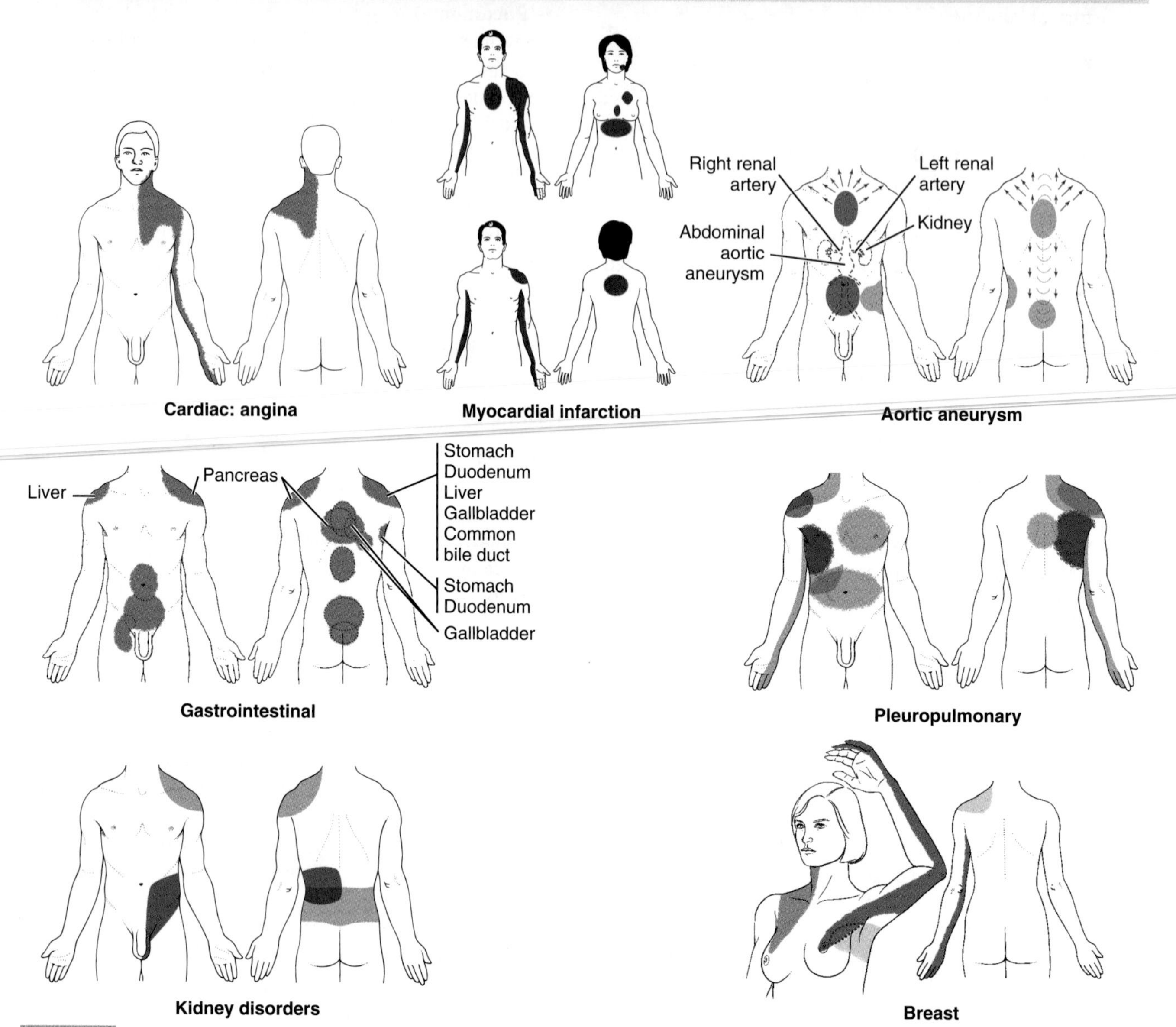

Fig. 18-3 Composite picture of referred shoulder and upper extremity pain patterns. Not pictured: trigger point referred pain (see Fig. 17-7).

Key Points to Remember

- Shoulder dysfunction can look like a true neuromuscular or musculoskeletal problem and still be viscerogenic or systemic in origin.
- Any adult over age 65 presenting with shoulder pain and/or dysfunction must be screened for systemic or viscerogenic origin of symptoms, even when there is a known (or attributed) cause or injury.
- Knowing the key red flags for cancer, vascular disease, pulmonary, GI, and gynecologic causes of shoulder pain and/or dysfunction will help the therapist screen quickly, efficiently, and accurately.
- Painless weakness of insidious onset is most likely a neurologic problem; painful, insidious weakness may be caused by cervical radiculopathy, chronic rotator cuff problems, tumors, or arthritis. A medical differential diagnosis is required.[70,71]
- As mentioned throughout this text, the therapist can collaborate with colleagues in asking questions and reviewing findings before making a medical referral. Perhaps someone else will see the answer or a solution to the client's unusual presentation, or perhaps another opinion will confirm the findings and give you the confidence you need to guide your professional decision making.
- Postoperative infection of any kind may not appear with any clinical signs/symptoms for weeks or months, especially in a client who is on corticosteroids or immunocompromised.
- Consider unreported trauma or assault as a possible etiologic cause of shoulder pain.
- Palpate the diaphragm and assess breathing patterns; shoulder pain reproduced by diaphragmatic palpation may point to a primary diaphragmatic (muscular) problem.

SUBJECTIVE EXAMINATION

Special Questions to Ask: Shoulder and Upper Extremity

General Systemic

- Does your pain ever wake you at night from a sound sleep? **(Cancer)**
  - Can you find any way to relieve the pain and get back to sleep?
  - If yes, how? (Cancer: Pain is usually intense and constant; nothing relieves it or if relief is obtained in any way, over time pain gets progressively worse)
- Since the beginning of your shoulder problem, have you had any unusual perspiration for no apparent reason, sweats, or fever?
- Have you had any unusual fatigue (more than usual with no change in lifestyle), joint pain in other joints, or general malaise? **(Rheumatic disease)**
- Have you sustained any injuries in the last week during a sports activity, car accident, etc?
  - (Ruptured spleen associated with pain in the left shoulder: Positive Kehr's sign)
- *For the therapist:* Has the client had a laparoscopy in the last 24 to 48 hours? **(Left shoulder pain: Positive Kehr's sign)**

Cardiac

- Have you recently (ever) had a heart attack? (**Referred pain via viscerosomatic zones,** see explanation Chapter 3)
- Do you ever notice sweating, nausea, or chest pain when the pain in your shoulder occurs?
- Have you noticed your shoulder pain increasing with exertion that does not necessarily cause you to use your shoulder (e.g., climbing stairs, stationary bicycle)?
- Do(es) your mouth, jaw, or teeth ever hurt when your shoulder is bothering you? **(Angina)**
- For the client with known angina: Does your shoulder pain go away when you take nitroglycerin? (Ask about the effect of taking antacids/acid-relieving drugs for women.)

Pulmonary

- Have you been treated recently for a lung problem (or think you have any lung or respiratory problems)?
- Do you currently have a cough?
  - *If yes,* is this a smoker's cough?
  - *If no,* how long has this been present?
  - Is this a productive cough (can you bring up sputum), and is the sputum yellow, green, black, or tinged with blood?
  - Does coughing bring on your shoulder pain (or make it worse)?
- Do you ever have shortness of breath, have trouble catching your breath, or feel breathless?
- Does your shoulder pain increase when you cough, laugh, or take a deep breath?
- Do you have any chest pain?
- What effect does lying down or resting have on your shoulder pain? (In the supine or recumbent position, a

Continued

SUBJECTIVE EXAMINATION—*cont'd*

pulmonary problem may be made worse, whereas a musculoskeletal problem may be relieved; on the other hand, pulmonary pain may be relieved when the client lies on the affected side, which diminishes the movement of that side of the chest.)

Gastrointestinal

- Have you ever had an ulcer?
  - *If yes*, when? Do you still have any pain from your ulcer?
  - Have you noticed any association between when you eat and when your symptoms increase or decrease?
- Does eating relieve your pain? (**Duodenal or pyloric ulcer**)
  - How soon is the pain relieved after eating?
- Does eating aggravate your pain? (**Gastric ulcer, gallbladder inflammation**)
- Does your pain occur 1 to 3 hours after eating or between meals? (Duodenal or pyloric ulcers, gallstones)
- For the client taking NSAIDs: Does your shoulder pain increase 2 to 4 hours after taking your NSAIDs? If the client does not know, ask him or her to pay attention for the next few days to the response of shoulder symptoms after taking the medication.
- Have you ever had gallstones?
- Do you have a feeling of fullness after only one or two bites of food? (**Early satiety: stomach and duodenum or gallbladder**)
- Have you had any nausea, vomiting, difficulty in swallowing, loss of appetite, or heartburn since the shoulder started bothering you?

Gynecologic

- Have you ever had a breast implant, mastectomy, or other breast surgery? (**Altered lymph drainage, scar tissue**)
- Have you ever had a tubal or ectopic pregnancy?
- Have you ever been diagnosed with endometriosis?
- Have you missed your last period? (**Ectopic pregnancy, endometriosis; blood in the peritoneum irritates diaphragm causing referred pain**)
- Are you having any spotting or irregular bleeding?
- Have you had any spontaneous or induced abortions recently? (**Blood in peritoneum irritating diaphragm**)
- Have you recently had a baby? (**Excessive muscle tension during birth**)
  - *If yes*: Are you breastfeeding with the infant supported on pillows?
  - Do you have a breast discharge, or have you had mastitis?

Urologic

- Have you had any recent kidney infections, tumors, or kidney stones? (**Pressure from kidney on diaphragm referred to shoulder**)

Trauma

- Have you been in a fight or been assaulted?
- Have you ever been pulled by the arm, pushed against the wall, or thrown by the arm?

If the answer is "Yes" and the history relates to the current episode of symptoms, then the therapist may need to conduct a more complete screening interview related to domestic violence and assault. Specific questions for this section have been discussed in Chapter 2; see also Appendix B-3.

CASE STUDY

Steps in the Screening Process

If a client comes to you with shoulder pain with any of the red-flag histories and/or red-flag clinical findings to suggest screening, start by asking yourself these questions:

- Which shoulder is it?
- Which organs could it be? (Use Fig. 3-4 showing the viscera in relation to the diaphragm and Tables 18-1 and 18-2 to help you.)
- What are the associated signs and symptoms of that organ? Are any of these signs or symptoms present?
- What is the history? Does anything in the history correlate with the particular shoulder involved and/or with the associated signs and symptoms? Conduct a Review of Systems as discussed in Chapter 4 (see Box 4-19).
- Can you palpate it, make it better or worse, or reproduce it in any way?

COULD IT BE CANCER?

Remember, the therapist does not make a determination as to whether a client has cancer. The therapist's assessment determines whether the client has a true neuromuscular or musculoskeletal problem that is within the scope of our practice. However, knowing red flags for the possibility of cancer helps the therapist know what questions to ask and what red flags to look for. Watch for:

- Previous history of cancer (any kind, but especially breast or lung cancer)
- Pectoralis major muscle spasm with no known cause, but full passive ROM and a mobile scapula. Be sure to assess for trigger points (TrPs). Reassess after TrP therapy.

CASE STUDY—*cont'd*

Steps in the Screening Process

- Were the symptoms alleviated? Did the movement pattern change?
- Conduct a neurologic screening exam.
- Shoulder flexion and abduction limited to 90 degrees with empty end feel.
- Presence of localized warmth over scapular area. Look for other trophic changes.

COULD IT BE VASCULAR?

Watch for

- Exacerbation by exertion unrelated to shoulder movements
- Does the shoulder pain and/or symptoms get worse when the client is just using the lower extremities? What is the effect of riding a stationary bike or climbing stairs without using the arms?
- Excessive, unexplained coincident diaphoresis (i.e., the client breaks out in a cold sweat just before or during an episode of shoulder pain; this may occur at rest but is more likely with mild physical activity).
- Shoulder pain relieved by leaning forward, kneeling with hands on the floor, sitting upright (pericarditis).
- Shoulder pain accompanied by dyspnea, temporomandibular joint (TMJ) pain, toothache, belching, nausea, or pressure behind the sternum.
- Bilateral shoulder pain that comes on after using the arms overhead for 3 to 5 minutes.
- Shoulder pain relieved by nitroglycerin (men) or antacids/acid-relieving drugs (women) [angina]
- Difference of 10 mm Hg or more (at rest) in diastolic blood pressure in the affected arm (aortic aneurysm; vascular component of thoracic outlet syndrome)

Remember to correlate any of these symptoms with:

- Client's past medical history (e.g., personal and/or family history of heart disease)
- Age (over 50, especially postmenopausal women)
- Characteristics of pain pattern (see Table 6-5; these characteristics of cardiac related chest pain can also apply to cardiac-related shoulder pain)

COULD IT BE PULMONARY?

- Ask about the presence of pleuritic component
  - Persistent cough (dry or productive)
  - Blood-tinged sputum; rust, green, or yellow exudate
  - Chest pain
  - Musculoskeletal symptoms are aggravated by respiratory movements; ask the client to take a deep breath. Does this reproduce or increase the pain/symptoms?
- Watch for the exacerbation of symptoms by recumbence even with proper positioning of the arm. Lying down in the supine position can put the shoulder in a position of slight extension.
- This can put pressure on soft tissue structures in and around the shoulder, causing pain in the presence of a true neuromuscular or musculoskeletal problem.
- For this reason, when assessing the effect of recumbence, make sure the shoulder is in a neutral position. You may have to support the upper arm with a towel roll under the elbow and/or put a pillow on the client's abdomen to give the forearms a place to rest.
- Pain is relieved or made better by sidelying on the involved side. This is called autosplinting.
- Pressure on the ribcage prevents respiratory movement on that side thereby reducing symptoms induced by respiratory movements. This is quite the opposite of a musculoskeletal or neuromuscular cause of shoulder pain; the client often cannot lie on the involved side without increased pain.
- Ask about the presence of associated signs and symptoms. Remember to ask our final question:
- Are there any symptoms of any kind anywhere else in your body?

In the older adult, listen for a self-report or family report of unknown cause of shoulder pain/dysfunction and/or any signs of confusion (confusion or increased confusion is a common first symptom of pneumonia in the older adult).

COULD IT BE GASTROINTESTINAL OR HEPATIC?

- Ask about a history of chronic (more than 6 months) NSAID use and history of previous ulcer, especially in association with NSAID use. This is the most common cause of medication-induced shoulder pain in all ages, but especially adults over 65.
- History of other GI disease that can refer pain to the shoulder such as:
  - Gallbladder
  - Acute pancreatitis
  - Reflex esophagitis
- Watch for coincident (or alternating) nausea, vomiting, dysphagia, anorexia, early satiety, or other GI symptoms. Clients often think they have two separate problems. The client may not think the therapist treating the shoulder needs or wants to know about their GI problems. The therapist who is not trained to screen for medical disease may not think to ask.
- Ask if shoulder pain is relieved by belching or antacids. This could signal an underlying GI problem or for women, cardiac ischemia.
- Look for shoulder pain that is changed by eating (better or worse within 30 minutes or worse 1 to 3 hours after eating).

Continued

CASE STUDY—*cont'd*

Steps in the Screening Process

The therapist does not have to identify the specific area of the GI tract that is involved or the specific pathology present. It is important to know that true NMS shoulder pain is not relieved or exacerbated by eating.

If there is a peptic ulcer in the upper GI tract causing referred pain to the shoulder, there is often a history of NSAID use. This client will have that red flag history along with shoulder pain that gets better after eating. There may also be other GI symptoms present such as nausea, loss of appetite, or melena from oxidized blood in the upper GI tract.

If there is liver impairment as well, there can be symptoms of carpal tunnel syndrome (CTS). For a list of possible NMS and systemic causes of CTS, see Table 11-2. Again, CTS in the presence of any of these systemic conditions should be assessed carefully. Likewise, CTS may be the first symptom of some of these pathologies.

The client with shoulder pain (GI bleed) and symptoms of CTS (liver impairment) may demonstrate other signs of liver impairment such as:

- Liver flap (asterixis)
- Liver palms (palmar erythema)
- Nail bed changes (white nails of Terry)
- Spider angiomas (over the abdomen)

These tests along with photos and illustrations are discussed in detail in Chapter 9.

COULD IT BE BREAST PATHOLOGY?

Remember that men can have breast diseases too, although not as often as women. Red-flag clinical presentation and associated signs and symptoms of breast disease referred to the shoulder may include:

- Jarring or squeezing the breast refers pain to the shoulder
- Resisted shoulder motions do not reproduce shoulder pain but do cause breast pain or discomfort
- Obvious change in breast tissue (e.g., lump[s], dimpling or peau d'orange, distended veins, nipple discharge or ulceration, erythema, change in size or shape of the breast)
- Suspicious or aberrant axillary or supraclavicular lymph nodes

PRACTICE QUESTIONS

1. A 66-year-old woman has been referred to you by her physiatrist for preprosthetic training after an above-knee amputation. Her past medical history is significant for chronic diabetes mellitus (insulin dependent), coronary artery disease with recent angioplasty and stent placement, and peripheral vascular disease. During the physical therapy evaluation, the client experienced anterior neck pain radiating down the left arm. Name (and/or describe) three tests you can do to differentiate a musculoskeletal cause from a cardiac cause of shoulder pain.
2. Which of the following would be useful information when evaluating a 57-year-old woman with shoulder pain?
   a. Influence of antacids on symptoms
   b. History of chronic NSAID use
   c. Effect of food on symptoms
   d. All of the above
3. Referred pain patterns associated with impairment of the spleen can produce musculoskeletal symptoms in:
   a. The left shoulder
   b. The right shoulder
   c. The mid- or upper back, scapular, and right shoulder areas
   d. The thorax, scapulae, right or left shoulder
4. Referred pain patterns associated with hepatic and biliary pathology can produce musculoskeletal symptoms in:
   a. The left shoulder
   b. The right shoulder
   c. The mid or upper back, scapular, and right shoulder areas
   d. The thorax, scapulae, right or left shoulder
5. The most common sites of referred pain from systemic diseases are:
   a. Neck and hip
   b. Shoulder and back
   c. Chest and back
   d. None of the above
6. A 28-year-old mechanic reports bilateral shoulder pain (right more than left) whenever he has to work on a car on a lift overhead. It goes away as soon as he puts his arms down. Sometimes, he has numbness and tingling in his right elbow going down the inside of his forearm to his thumb. The most likely explanation for this pattern of symptoms is:
   a. Angina
   b. Myocardial ischemia
   c. Thoracic outlet syndrome
   d. Peptic ulcer
7. A client reports shoulder and upper trapezius pain on the right that increases with deep breathing. How can you tell if this results from a pulmonary or a musculoskeletal cause?
   a. Symptoms get worse when lying supine but better when right sidelying when it is pulmonary
   b. Symptoms get worse when lying supine but better when right sidelying when it is musculoskeletal

PRACTICE QUESTIONS—*cont'd*

8. Organ systems that can cause simultaneous bilateral shoulder pain include:
   a. Spleen
   b. Heart
   c. Gallbladder
   d. None of the above
9. A 23-year-old woman was a walk-in to your clinic with sudden onset of left shoulder pain. She denies any history of trauma and has only a past history of a ruptured appendix three years ago. She is not having any abdominal pain or pain anywhere else in her body. How do you know if she is at risk for ectopic pregnancy?
   a. She is sexually active, and her period is late.
   b. She has a history of uterine cancer.
   c. She has a history of peptic ulcer.
   d. None of the above.
10. The most significant red flag for shoulder pain secondary to cancer is:
    a. Previous history of coronary artery disease
    b. Subscapularis trigger point alleviated with trigger point therapy
    c. Negative neurologic screening exam
    d. Previous history of breast or lung cancer

REFERENCES

1. Walsh RM, Sadowski GE: Systemic disease mimicking musculoskeletal dysfunction: A case report involving referred shoulder pain. *J Orthop Sports Phys Ther* 31(12):696–701, 2001.
2. Berg J: Symptoms of a first acute myocardial infarction in men and women. *Gend Med* 6(3):454–462, 2009.
3. Lovlien M: Early warning signs of an acute myocardial infarction and their influence on symptoms during the acute phase, with comparisons by gender. *Gend Med* 6(3):444–453, 2009.
4. Hwang SY: Comparison of factors associated with atypical symptoms in younger and older patients with acute coronary syndromes. *J Korean Med Sci* 24(5):789–794, 2009.
5. Rubba F: Vascular preventive measures: The progression from asymptomatic to symptomatic atherosclerosis management. Evidence on usefulness of early diagnosis in women and children. *Future Cardiol* 6(2):211–220, 2010.
6. Vercoza AM: Cardiovascular risk factors and carotid intima-media thickness in asymptomatic children. *Pediatr Cardiol* 30(8):1055–1060, 2009.
7. Smith ML: Differentiating angina and shoulder pathology pain. *Phys Ther Case Rep* 1(4):210–212, 1998.
8. Ogawa K: Advanced shoulder joint tuberculosis treated with debridement and closed continuous irrigation and suction. *Am J Orthop* 39(2):E15–18, 2010.
9. Ba-Fall K: Shoulder pain revealing tuberculosis of the humerus. *Rev Pneumonol* 65(1):13–15, 2009.
10. Nagaraj C: Tuberculosis of the shoulder joint with impingement syndrome as initial presentation. *J Microbiol Immunol Infect* 41(3):275–278, 2008.
11. Wohlgethan JR: Frozen shoulder in hyperthyroidism. *Arthritis Rheum* Aug 1987;30(8):936–939.
12. Roy A: Adhesive capsulitis in physical medicine and rehabilitation. *eMedicine Specialties.* Updated Oct. 15, 2009. Available online at http://emedicine.medscape.com/article/326828-overview. Accessed March 21, 2011.
13. Lebiedz-Odrobina D: Rheumatic manifestations of diabetes mellitus. *Rheum Dis Clin North Am* 36(4):681–699, 2010.
14. Garcilazo C: Shoulder manifestations of diabetes mellitus. *Curr Diabetes Rev* 6(5):334–340, 2010.
15. Saha NC: Painful shoulder in patients with chronic bronchitis and emphysema. *Am Rev Respir Dis* 94:455–456, 1966.
16. Okada M, Suzuki K, Hidaka T, et al: Complex regional pain syndrome type I induced by pacemaker implantation, with a good response to steroids and neurotrophin. *Intern Med* 41:498–501, 2002.
17. Mennell JM: *The musculoskeletal system: differential diagnosis from symptoms and physical signs*, Sudbury, MA, 1992, Jones and Bartlett.
18. Leff D: Ruptured spleen following laparoscopic cholecystectomy. *JSLS* 11(1):157–160, 2007.
19. Sharami SH: Randomised clinical trial of the influence of pulmonary recruitment manoeuvre on reducing shoulder pain after laparoscopy. *J Obstet Gynaecol* 30(5):505–510, 2010.
20. Shin HY: The effect of mechanical ventilation tidal volume during pneumoperitoneum on shoulder pain after a laparoscopic appendectomy. *Surg Endosc* 24(8):2002–2007, 2010.
21. Giles LGF, Singer KP: *The clinical anatomy and management of thoracic spine pain*, Oxford, 2000, Butterworth Heinemann.
22. Kandil TS: Shoulder pain following laparoscopic cholecystectomy: Factors affecting the incidence and severity. *J Laparoendosc Adv Surg Tech A* 20(8):677–682, 2010.
23. Chang SH: An evaluation of perioperative pregabalin for prevention and attenuation of postoperative shoulder pain after laparoscopic cholecystectomy. *Anesth Analg* 109(4):1284–1286, 2009.
24. Hadler NM: The patient with low back pain, *Hosp Pract* October 30, 1987:17–22.
25. Biolchini F: Emergency laparoscopic splenectomy for haemoperitoneum because of ruptured primary splenic pregnancy: a case report and review of literature. *ANZ J Surg* 80(102):55–57, 2010.
26. Dennert IM: Ectopic pregnancy. *J Minim Invasive Gynecol* 15(3):377–379, 2008.
27. Bildik F: Heterotopic pregnancy presenting with acute left chest pain. *Am J Emerg Med* 26(7):835.e1–2, 2008.
28. Smith ML: Differentiating angina and shoulder pathology pain. *Phys Ther Case Rep* 1(4):210–212, 1998.
29. Oaklander AL, Rissmiller JG, Gelman LB, et al: Evidence of focal small-fiber axonal degeneration in complex regional pain syndrome-I (reflex sympathetic dystrophy). *Pain* 120(3):235–243, 2006.
30. Jänig W, Baron R: Is CRPS I a neuropathic pain syndrome? *Pain* 120(3):227–229, 2006.

31. Jänig W: The fascination of complex regional pain syndrome. *Exp Neurol* 221(1):1–4, 2010.
32. Duman I: Reflex sympathetic dystrophy secondary to deep venous thrombosis mimicking post-thrombotic syndrome. *Rheumatol Int* 30(2):249–252, 2009.
33. Eyers P, Earnshaw JJ: Acute non-traumatic arm ischemia. *Br J Surg* 85:1340–1346, 1998.
34. Brinkley DM, Hepper CT: Heart in hand: structural cardiac abnormalities that manifest as acute dysvascularity of the hand. *J Hand Surg* 35A(12):2101–2103, 2010.
35. American Heart Association: Pericardium and pericarditis. Available online at http://www.americanheart.org/presenter.jhtml?identifier=4683. Accessed March 22, 2011.
36. Texas Heart Institute: Aneurysms and dissections. Available online at http://www.texasheartinstitute.org/hic/topics/cond/aneurysm.cfm. Accessed March 22, 2011.
37. Tran H: Deep venous thromboses in patients with hematological malignancies after peripherally inserted central venous catheters. *Leuk Lymphoma* 51(8):1473–1477, 2010.
38. Jones MA: Characterizing resolution of catheter-associated upper extremity deep venous thrombosis. *J Vasc Surg* 51(1):108–113, 2010.
39. Shah MK: Upper extremity deep vein thrombosis. *South Med J* 96(7):669–672, 2003.
40. Linneman B: Hereditary and acquired thrombophilia in patients with upper extremity deep-vein thrombosis. *Thromb Haemost* 100(3):440–446, 2008.
41. Jones RE: Upper limb deep vein thrombosis: a potentially fatal complication of clavicle fracture. *Ann R Coll Surg Engl* 92(5):W36–38, 2010.
42. Garofalo R: Deep vein thromboembolism after arthroscopy of the shoulder: two case reports and review of the literature. *BMC Musculoskel Disord* 11:65, 2010.
43. Willis AA: Deep vein thrombosis after reconstructive shoulder arthroplasty: a prospective observational study. *J Shoulder Elbow Surg* 18(1):100–106, 2009.
44. Farge D: Lessons from French National Guidelines on the treatment of venous thrombosis and central venous catheter thrombosis in cancer patients. *Thromb Res* 125(Suppl 2):S108–S116, 2010.
45. Lancaster SL: Upper-extremity deep vein thrombosis. *AJN* 110(5):48–52, 2010.
46. Gaitini D: Prevalence of upper extremity deep venous thrombosis diagnosed by color Doppler duplex sonography in cancer patients with central venous catheters. *J Ultrasound Med* 25(10):1297–1303, 2006.
47. Joffe HV: Upper extremity deep vein thrombosis: a prospective registry of 592 patients. *Circulation* 110:1605–1611, 2004.
48. Otten TR: Thromboembolic disease involving the superior vena cava and brachiocephalic veins. *Chest* 123(3):809–812, 2003.
49. Constans J: A clinical prediction score for upper extremity deep venous thrombosis. *Thromb Haemost* 99:202–207, 2008.
50. Di Nisio M: Accuracy of diagnostic tests for clinically suspected upper extremity deep vein thrombosis: a systematic review. *J Thromb Haemost* 8(4):684–692, 2010.
51. Netter FH: *Atlas of human anatomy*, ed 5, Philadelphia, 2010, WB Saunders.
52. Moore KL: *Clinically oriented anatomy*, ed 6, Baltimore, 2009, Lippincott, Williams and Wilkins.
53. Pedersen KV: Flank pain in renal and ureteral calculus. *Ugeskr Laeger* 173(7):503–505, 2011.
54. Pedersen KV: Visceral pain originating from the upper urinary tract. *Urol Res* 38(5):345–355, 2010.
55. Delavierre D: Symptomatic approach to referred chronic pelvic and perineal pain and posterior ramus syndrome. *Prog Urol* 20(12):990–994, 2010.
56. Myslinski MJ: NSAIDs: The good, the bad, and the ugly. Lecture presented at the APTA Combined Sections Meeting, Las Vegas, NV, February 11, 2009.
57. Rose SJ, Rothstein JM: Muscle mutability: general concepts and adaptations to altered patterns of use. *Phys Ther* 62:1773, 1982.
58. Yung E: Screening for head, neck, and shoulder pathology in patients with upper extremity signs and symptoms. *J Hand Ther* 23:173–186, 2010.
59. McKay P: Osteomyelitis and septic arthritis of the hand and wrist. *Curr Ortho Pract* 21(6):542–550, 2010.
60. Bonsignore A: Occult rupture of the spleen in a patient with infectious mononucleosis. *G Chir* 31(3):86–90, 2010.
61. Cyriax J: *Textbook of orthopaedic medicine*, ed 8, Baltimore, 1982, Williams and Wilkins.
62. Schumpelick V: Surgical embryology and anatomy of the diaphragm with surgical applications. *Surg Clin North Am* 80(1):213–239, 2000.
63. Tateishi U: Chest wall tumors: radiologic findings and pathologic correlation. *RadioGraphics* 23:1491–1508, 2003. Available online at http://radiographics.rsna.org/content/23/6/1491.full. Accessed March 22, 2011.
64. Bhimji S: Pancoast tumor. *eMedicine Specialties—Thoracic Surgery (Tumors)*. August 3, 2010. Available online at http://emedicine.medscape.com/article/428469-overview. Accessed March 2, 2011.
65. Cailliet R: *Shoulder pain*, ed 3, Philadelphia, 1991, FA Davis.
66. Reitman J: Late morbidity after treatment of breast cancer in relation to daily activities and quality of life: a systematic review. *Eur J Surg Oncol* 29:229–238, 2003.
67. Shamley DR: Changes in shoulder muscle size and activity following treatment for breast cancer. *Breast Cancer Res Treat* 106(1):19–27, 2007.
68. Seoud AA: Endometriosis: a possible cause of right shoulder pain. *Clin Exp Obstet Gynecol* 37(1):19–20, 2010.
69. Prasarn ML, Ouellette EA: Acute compartment syndrome of the upper extremity. *JAAOS* 19(1):49–58, 2011.
70. McFarland EG, Sanguanjit P, Tasaki A, et al: Shoulder examination: established and evolving concepts. *J Musculoskel Med* 23(1):57–64, 2006.
71. McFarland EG: Clinical and diagnostic tests for shoulder disorders: a critical review. *Br J Sports Med* 44(5):328–332, 2010.
72. Neer CS: Anterior acromioplasty for the chronic impingement syndrome in the shoulder: a preliminary report. *J Bone Joint Surg* 54(1):41–50, 1972.
73. Labidi M: Pleural effusions following cardiac surgery: prevalence, risk factors, and clinical features. *Chest* 136(6):1604–1611, 2009.
74. Ashikhmina EA: Pericardial effusion after cardiac surgery: risk factors, patient profiles, and contemporary management. *Ann Thorac Surg* 89(1):112–118, 2010.
75. Ahmed WA: Survival after isolated coronary artery bypass grafting in patients with severe left ventricular dysfunction. *Ann Thorac Surg* 87(4):1106–1112, 2009.
76. Jensen L: Risk factors for postoperative pulmonary complications in coronary artery bypass graft surgery patients. *Eur J Cardiovasc* 6(3):241–246, 2007.

APPENDICES

The following appendices contain examples of forms, questionnaires, and checklists that can be used in client evaluation. The more streamlined your paperwork process is, the more time you will have to treat your clients. These appendices will provide you with some of the tools needed to screen for systemic diseases and medical conditions that can mimic neuromuscular or musculoskeletal problems.

The following material can be found on EVOLVE, **www.DifferentialDiagnosisforPT.com** or by scanning the QR code with your mobile device. You may customize them to suit your needs or print them on your letterhead. For the clinic, it may be beneficial for you to print several copies at one time for ease of use.

APPENDIX A: SCREENING SUMMARY

APPENDIX B: SPECIAL QUESTIONS TO ASK (Screening for)

APPENDIX C: SPECIAL FORMS TO USE

APPENDIX D: SPECIAL TESTS TO PERFORM

INDEX

Page references followed by an "f" indicate figures, by "b" indicate boxes, and by "t" indicate tables.

C

D

F

G

I

P

T